16. October 19      Bransgore

Hermann Jacobsen

*Lexicon
of
Succulent
Plants*

BLANDFORD PRESS LTD.

With 1069 figures on 200 plates

Short descriptions, habitats and synonymy of succulent plants other than Cactaceae

# Lexicon of Succulent Plants

Hermann Jacobsen
F.C.S.S.A., F.A.S.P.S., F.L.S.

Blandford Press Ltd. London

FIRST PUBLISHED IN THIS EDITION 1974
BY BLANDFORD PRESS, LONDON WC1V 6 PH

ENGLISH EDITION, AFTER THE FIRST GERMAN EDITION, JENA 1970
REVISED AND ENLARGED
© BLANDFORD PRESS LTD.

---

Title of the first edition, published in the German language:

DAS SUKKULENTENLEXIKON

Kurze Beschreibung, Herkunftsangaben und Synonymie der sukkulenten Pflanzen mit Ausnahme der Cactaceae

von **HERMANN JACOBSEN**

© VEB GUSTAV FISCHER VERLAG · JENA 1970

All rights reserved

ISBN 07137 0652 X

Printed and bound
in the German Democratic Republic
by "Offizin Andersen Nexö" at Leipzig
Licence number 261 700/149/74

**Plate 1. Cyphostemma** trees about 500 years old; in cultivation in the Botanical Garden at Kiel. (Presse Photo MAGNUSSEN). – Left and right: **C. crameriana** (SCHINZ) B. DESC.; centre: **C. juttae** v. **ternata** JACOBS.

AUTHORISED TRANSLATION
AND PROOF CORRECTION
by
LOIS GLASS
HIGH WYCOMBE, BUCKS, ENGLAND

ENGLISH SUPERVISION OF TRANSLATION
AND CONTROL OF NOMENCLATURE
by
GORDON D. ROWLEY, B.Sc.
READING, BERKS, ENGLAND

Dr. rer. nat. h.c. HERMANN JACOBSEN, Kiel, Germany
Ex-Curator of the Botanical Garden of the University of Kiel, Honorary member of Deutsche Kakteengesellschaft (German Cactus Society), Vice-President and Fellow of The African Succulent Plant Society, England, Vice-President of National Cactus and Succulent Society, England, Fellow of the Cactus and Succulent Society of America, Fellow of The Linnéan Society of London, Founder-member of I.O.S. (International Organisation for Succulent Plant Study) and a member of other foreign societies for succulent plant study.

Figures without credits are produced by the author.
Many figures are from the archives of Professor Dr. G. SCHWANTES (these are marked with G. SCHWANTES).

Cover-Figure: Front: Stapelia gigantea N. E. BR. (Photo: H. STRAKA). Reverse: The author in the big American Succulent House of the Botanical Garden at Kiel with Chorisia speciosa A. SAINT HIL. (Photo: G. D. ROWLEY).
Vignette: Jacobsenia: open capsule.

In the present work are briefly described 366 genera and intergeneric hybrids distributed in 52 families, and including about 8,600 species and lower taxa, hybrids and cultivars. – In "Invalid Designations" some 7,500 synonyms etc. are given.

To my friends

GORDON D. ROWLEY,
   B.Sc., Lecturer in Horticultural
   Botany at the University of Reading,
   England,

LEONARD E. NEWTON,
   M.Sc., F.L.S., Lecturer in Botany
   at the University of Kumasi, Ghana,
   and

CYRIL A. E. PARR,
   F. C. S. S. A., F. L. S., Hon. Editor of The
   African Succulent Plant Society,
   Hitchin, Herts, England,

best informed connoisseurs of succulents,
this work is respectfully dedicated.

# PREFACE

The grateful thanks of all enthusiasts and students of succulent plants, especially in the English-speaking countries, will go to the publishers, BLANDFORD PRESS Ltd. of London, for presenting this 'Lexicon of Succulent Plants'. A German version appeared in 1970. As a result of many friendly and constructive criticisms the author was encouraged to continue his careful collection, study and classification of further material, while letters from all over the world drawing attention to errors and suggesting amendments have made it possible to include in the present volume additional genera, species and subordinate taxa.

The translation of this revised and expanded text was undertaken by Mrs. LOIS GLASS of High Wycombe, England. In little over six months Mrs. GLASS completed the text of the English version, and I extend to her my sincere thanks for the care and efforts she has brought to our joint task.

With his customary generosity Mr. G. D. ROWLEY B.Sc. of Reading, England, undertook responsibility for checking nomenclature and examining the accuracy of the English manuscript. His comprehensive knowledge of the subject and ready access to the relevant literature, together with his own studies and published works, have enabled Mr. ROWLEY to make valuable additions to the text and to include new material. He drew up keys or classifications for several of the larger genera (Anacampseros in collaboration with R. RAWÉ, Bulbine, Lewisia, Othonna, Senecio, Testudinaria) and provided names (in collaboration with D. BRAMWELL) for many hybrids. The present book is thus a permanent testimony to many months of tireless and disinterested effort on the part of my friend GORDON ROWLEY, serving perhaps to illustrate close co-operation between members of the I.O.S. – "personal satisfaction, not work", as Mrs. GLASS recently described to me the efforts of our team.

The Pedaliaceae have been revised in conformity with the work of H.-D. IHLENFELDT, Hamburg. The Crassulaceae and Mesembryanthemaceae of S.W. Africa have been revised according to H.-CH. FRIEDRICH (see Bibliography), while the more recent work of J. INGRAM and C. WEBER, published in "Baileya", has been taken into consideration. A key by B. FEARN has been included for the ever-popular Lithops. As the manuscript was nearing completion Dr. H. HARTMANN of Hamburg made available her revision of the genus Argyroderma (Mesembryanthemaceae) as well as a key to its species, and this work will also be found here.

The chapter "Invalid Designations" covers both Sections I and II of the text, the genera of the two parts being united in a **single** alphabetical sequence. A departure has been made from the German edition to simplify the finding of valid names so that instead of a page number the reader is now given the valid name in abbreviated form.

The list of "Specialist Organisations for the cultivation and study of succulent plants" and "Bibliography" was controlled and revised by Mr. LEONARD E. NEWTON of Kumasi, Ghana.

May I take this opportunity of expressing my warm appreciation to all who have helped with the present work, to my scientific colleagues as well as printers and publishers in Leipzig, Jena and London.

Further revisions of nomenclature and taxonomy must be expected, especially for the Mesembryanthemaceae. Some of the larger genera require fresh study which could result in many "species" being either included in an enlarged concept of what constitutes a species, or reduced to varietal status.

In the present Lexicon the attempt has been made to give due weight to all relevant studies, bringing the nomenclature into line with the most recent thinking. The author is well aware that objective and subjective views do not always coincide.

Dr. HERMANN JACOBSEN.

Kiel, 31st December 1973.

# PREFACE TO THE FIRST GERMAN EDITION, 1970

Since the appearance of my "Handbuch der sukkulenten Pflanzen", Jena 1955, and of the expanded English edition, "A Handbook of Succulent Plants", London 1959, there has been an ever-increasing enthusiasm for succulent plants. In addition, new species have been discovered, notably in South Africa, and these have been described by competent botanists. Moreover the taxonomists have become interested in completing existing descriptions, in examining the orders as they currently stand and where necessary even in revising them. Publications of these descriptions from South Africa, Europe, Mexico, etc., have stimulated the enlargement of existing plant-collections and the founding of new ones. It is particularly noteworthy that the abundance of material accumulating in the meantime has led to the formation of new specialist collections. The unusual succulents of Madagascar have aroused special attention and more recently, especially among English plant-lovers, the curious "caudiciform" plants have been a subject of interest, although it must be noted that in this field difficulties are encountered in defining a "succulent".

Moreover, during the intervening years, my existing connections with scientists all over the world have become closer, while new contacts have been made. As a result of my receiving in such generous measure almost all published works on succulent plants, I have been able to include and classify this information in the present work, to make some systematic amendments, and thus quite considerably to extend the scope of "A Handbook of Succulent Plants". Botanists specializing in this work have in some cases made a complete re-appraisal of certain genera, and they have placed their work at my disposal for inclusion in the text. The number of illustrations has also been increased, thanks to photographs most kindly lent to me by the owners, by the reproduction of other illustrations, and by photographs taken by me, both in the Botanical Gardens at Kiel, and in various private collections.

Let no-one assume that the subject of research into the classification of succulent plants has been exhausted. Nevertheless it is noteworthy that over the past few years there have been fewer new works in the international specialist literature on succulent plants. In view of this fact, the time seems ripe for presenting a lexicon of succulent plants (excluding the Cactaceae).[1] Mention must be made of the fact that many large genera of the family Mesembryanthemaceae, e.g. Delosperma, Lampranthus and Ruschia, are still in need of thorough systematic revision; while for many of the large genera, practical keys to the species still have to be worked out in order to simplify recognition of the various plants.

The present lexicon includes in alphabetical order all cultivated species of succulent plants known to me to date (30th June 1969). The brief descriptions have been simplified and standardized as far as possible. The information provided includes the source locality of the different species (precise place-names have had to be omitted owing to lack of space), and a synonymy. Where available, generic sub-divisions and keys have been included.

Even after assembling and evaluating the literature, I could never have completed my task without the generous support of many friends throughout the world. To all of these I therefore express my sincerest thanks for their help over the years.

Assistance has been provided by so many Botanical Institutes and Libraries in both Eastern and Western Europe that it is impossible to make individual acknowledgments here; I must content myself with a general word of thanks to their superintendents and librarians for their kind support.

My especial gratitude goes to a number of friends for their invaluable help in providing me with specialist literature and referring me to relevant texts, or for their constructive criticism of my earlier text, as well as the provision of photographs. I am above all indebted to those colleagues mentioned in my earlier books, as well as to the following: –

First and foremost my thanks go to Dr. LOUISA BOLUS of Claremont, Cape Province, South Africa. Right up to her 90th birthday on 31. 7. 1967, Mrs. Bolus gave me all her written work on Mesembryanthemums, often prior to publication, including new species descriptions as well as keys to the different species, and keys to the genera which she had checked, which are now published herein for the first time in their complete form;

Professor Dr. W. RAUH of Heidelberg, who has in recent times built up at the Botanical Gardens in Heidelberg what is probably the most outstanding collection of succulent plants of modern times; in word and deed he has given me most notable assistance;

Dr. A. TISCHER of Heidelberg allowed me to use his revised and extended "Classification of the Genus Conophytum N. E. BR." for inclusion in my text;

Professor Dr. H. MERXMUELLER and Dr. H.-CH. FRIEDRICH, of the Botanical Institute and Botanical Garden at Munich-Nymphenberg, helped me with problems concerning species from South-West Africa;

---

[1] See CURT BACKEBERG, Kakteenlexikon, Jena 1966.

Dr. H. W. DE BOER of Haren (Gron.), Holland, made available to me his work on the genus Lithops, and in particular his key to the different species as well as numerous photographs of Lithops;

Dr. H. PFENNIG of Herford, who has made a re-appraisal of the genus Sansevieria, was kind enough to make his text available to me, and it too has been printed in this lexicon;

Dr. HERBERT HUBER of Würzburg abstracted the succulent species of Ceropegia from his new work while still in manuscript. This extract has also found its place in the present work;

I have had generous support from Dr. R. V. MORAN of San Diego, Calif., USA, whose contributions include those on the genera Dudleya, Graptopetalum, Pachyphytum and Thompsonella while the genus Echeveria has been revised;

Kind assistance in obtaining from English libraries various rarer works in the literature has always been forthcoming from Mr. C. A. E. PARR of London; Mr. G. D. ROWLEY, B.Sc., Reading, Berks., England; Mr. L. E. NEWTON, M.Sc., F.L.S., Kumasi, Ghana; and Mr. C. W. PITCHER of Guiseley, Leeds, England;

Dr. G. W. REYNOLDS of Mbabane, Swaziland, sent me over the years all his new descriptions of the Aloes which were collected together in his books "The Aloes of South Africa" and "The Aloes of Tropical Africa and Madagascar";

Much helpful information has been received on numerous occasions from Professor Dr. H. STRAKA of Kiel, H. HERRE of Stellenbosch, C.P., South Africa, Dr. B. K. BOOM, Wageningen, Holland, JULIEN MARNIER-LAPOSTOLLE, Paris, France, Professor Dr. E. R. SVENTENIUS, Puerto de la Cruz, Tenerife, Canary Is., Spain, Professor Dr. J. A. HUBER, Dillingen/Donau, L. C. LEACH, Nelspruit, Transvaal, South Africa, Mr. J. J. LAVRAOS, Johannesburg, South Africa and – last but not least – P. R. O. BALLY of Nairobi, Kenya, who has not only provided me with much information but has also obtained printed material for me and checked some of the text.

Dr. H.-CH. FRIEDRICH-Munich had the kindness to revise the difficult chapter Crassula L.; many thanks to him for his generosity.

May I express my sincere thanks to all these friends as well as to those who have shared in the work in any way to a greater or lesser extent.

My warm thanks to the VEB GUSTAV FISCHER Verlag of Jena, for their decision to publish this extensive work, above all to their Reader, Frau JOHANNA SCHLUETER, for providing the stimulus to undertake the writing of this book, and for her unfailing advice. Herr RUDOLF HUBER merits gratitude and warmest praise for the magnificent reproduction of the illustrations as well as for supervising the printing of the entire Lexicon.

Thanks go to the typesetters, page regulators and printers for the high quality of the finished book, and to the blockmakers for their faithful reproduction of the plates. My gratitude goes in addition to Herr H. TÄUBERT of Gotha, who checked the innumerable geographical references and names of countries, and brought these completely up to date.

It is my hope that this book may be of assistance to all lovers of succulent plants by ensuring that their plants are correctly named; and that all interested scientists may find this Lexicon of service as a basic work and a summary of the entire complex as well as being of assistance with more specialized studies. Above all, I trust my endeavours will contribute to ensuring standardization of plant-names throughout our collections.

Dr. HERMANN JACOBSEN

Kiel,
Summer 1969.

# CONTENTS

| | |
|---|---|
| PREFACE | 9 |
| EXPLANATION OF SIGNS, ABBREVIATIONS AND GEOGRAPHICAL NAMES IN THE TEXT | 15 |
| EARLIER AND PRESENT POLITICAL DESIGNATIONS OF AFRICA | 17 |
| SPECIALIST ORGANISATIONS FOR THE CULTIVATION AND STUDY OF SUCCULENT PLANTS | 19 |
| INTRODUCTION | 21 |
| PLANT FAMILIES AND THEIR GENERA WITH SUCCULENT SPECIES IN ALPHABETICAL ORDER (WITHOUT CACTACEAE) | 23 |

SUCCULENT PLANTS

**PART I. FAMILIES AGAVACEAE TO ZYGOPHYLLACEAE** .......... 25

(EXCLUDING MESEMBRYANTHEMACEAE)
Genera, Intergeneric hybrids, Species and lower taxa, Hybrids, Cultivars, Origins and Descriptions

**PART II. FAMILY MESEMBRYANTHEMACEAE**

| | |
|---|---|
| 1. Systematic Classification of the Family | 395 |
| 2. Key to the Genera of the Family | 398 |
| 3. Genera, Intergeneric hybrids, Species and lower taxa, Hybrids, Cultivars, Origins and Descriptions | 404 |
| SUPPLEMENT | 584 |

**BIBLIOGRAPHY**

| | |
|---|---|
| A. Periodical Publications | 585 |
| B. Books and individual important papers | 587 |

**INVALID DESIGNATIONS** (Basionyms, Homonyms, Synonyms, nomina nuda, insufficiently known species) in alphabetical order of the Genera (collected Genera of Parts I and II) with references upon the valid names .......... 591

# EXPLANATIONS OF SIGNS, ABBREVIATIONS AND GEOGRAPHICAL NAMES IN THE TEXT

**Signs**

⊙ = annual
⊙ = biennial
♃ = perennial
♄ = shrub (s)
♄ = tree (s)
× = hybrid
× = hybridised with
⌀ = diameter
○ = circumference
± = more or less
♀ = female
♂ = male
⚥ = bisexual
§ = Section, also SG., series, row

**Abbreviations**

acc. = according to
Afr. = Africa
Am. = America
As. = Asia
Austr. = Australia
Bo. = body, bodies
Br. = branch(es)
c. = **circa,** about
Cal. = calyx
Cape = Cape Province
Carp. = carpel(s)
Cinc. = cincinnus
Cor. = corolla
cv. = cultivar
Cy. = cyathium
D. = District, Division
Dept. = Department
E. = East, eastern
emend. = **emendatus,** corrected
Eur. = Europe
exc. = except
f. = **forma,** form
Fi. = fissure(s)
Fl. = flower(s)
Fr. = fruit(s)
G. = genus
Gr. = group(s); Great (geogr. names)
hort. = **hortorum,** of garden origin
Infl. = inflorescence(s)
Int. = **internodium,** internode(s)
L. = leaf, leaves
l.c. = **locus citatus,** in the place cited
Lit. = Little (in name of countries)
M. = median, middle
Madag. = Madagascar
MS. = manuscript (-name)
mut. = **mutatio,** a change or "sport"
N. = North, northern

nm. = **nothomorph,** a segregate within a hybrid group
nom. = **nomen,** name, designation
nom.err. = **nomen erratum,** erroneous name
nom.ill. = **nomen illegetimum,** illegal name
nom.nud. = **nomen nudum,** name without a description
nom.prov. = **nomen provisorum,** provisional name
non = not
Ov. = ovary, ovaries
p. = page
p.part. = **pro parte,** partly
p.max.part. = **pro maximo parte,** for the most part
p.min.part. = **pro minimo parte,** for the smaller part
Pan. = panicle
Ped. = peduncle
Per. = perianth
Pet. = petal(s)
Pl. = plate
R. = row(s)
Rac. = raceme(s)
reg. = region
Repr. = reproduction
Ros. = rosette(s)
s. = see
S. = South, southern
Sc. = scape
Sect. = Section
sens.lat. = **sensu lato,** in the broad sense
Sep. = sepal(s)
Ser. = Series
SF. = Subfamily
SG. = Subgenus
SGr. = Subgroup
Sh. = sheath(s)
Shi. = shield(s)
Sp. = spine(s)
spec. = species
sphalm. = **sphalmate,** wrongly
SSect. = Subsection
SSer. = Subseries
ssp. = subspecies
St. = stem, trunc
Stip. = stipule(s)
STr. = Subtribe
T. = tooth, teeth
Tep. = tepal(s)
Th. = thorn(s)
Th.shi. = thorn-shield(s)
Tub. = tuber
v. = **varietas,** variety
veg. = vegetation
W. = West, western
Wi. = window(s)

The **valid name** (in bold type) is followed by the relevant *synonyms* (abbreviated, and in italics).

**Example: Conophytum bilobum** (Marl.) N. E. Br. (*Mesembryanthemum bilobum* Marl., *Derenbergia biloba* (Marl.) Schwant., *Conophytum exsertum* N. E. Br.).
**As abbreviated in the text:** –
**C. bilobum** (Marl.) N. E. Br. (*Mes. b.* Marl., *Derenbergia b.* (Marl.) Schwant., *C. exsertum* N. E. Br.).

Where a plant name cannot be found in the main text, either **Part I** or **Part II,** see under '**Invalid Designations**', where the valid name in abbreviated form appears after the invalid name. When identifying individual species, consult the characteristics set out in the classifications of the genera and in the keys.

Author's names: – The following abbreviations have been employed: –
    Bgr. instead of Berg. for A. Berger
    Dtr. instead of Dint. for K. Dinter
    Schltd. instead of Schlecht. for D. F. L. von Schlechtendahl
    Schltr. instead of Schlechter for R. Schlechter.
    Salm instead Salm-Dyck or S. D. for
        Prince von Salm-Reifferscheid-Dyck

# DEFINITIONS OF TECHNICAL TERMS

Of necessity, many technical terms are used in the Lexicon that will be unfamiliar to non-botanical readers. This is inevitable, because it is impossible to describe plants briefly without them. Rather than increase the bulk and cost of the volume by including a glossary, a list follows of readily available dictionaries to which the reader is referred:

Ivimey-Cooke, B., The Terminology of Succulent Plants. National Cactus & Succulent Society, 1974.

Stearn, W. T., Botanical Latin, 1956.

Stearn, W. T., & Smith, I. L. L., A Gardener's Dictionary of Plant Names (Revised from the original by A. W. Smith), 1972.

Parr, C. A. E., Glossary of Botanical Terms, in Bull. Afr. Succ. Pl. Soc. III: 203, Nov.-Dec. 1968 et seq.

# EARLIER AND PRESENT POLITICAL DESIGNATIONS OF AFRICA

put together by H. TÄUBERT

**Previously**

1. Abyssinia
2. Aden Protectorate (S. Arabia)
3. Algeria
4. Anglo-Egyptian Sudan
5. Angola
6. Basutoland
7. Belgian Congo
8. Bechuanaland Protectorate
9. British Bechuanaland
10. Cameroon; Cameroons
11. Chad Territory
12. Dahomey
13. Egypt
14. Erythrea
15. French Guinea
16. French Somaliland

17. French Sudan
18. Gabon
19. Gold Coast
20. Italian Somaliland (Somalia) and British Somaliland
21. Ivory Coast
22. Kenya
23. Liberia
24. Libya
25. Mauritania
26. Middle Congo
27. Morocco
28. Mozambique
29. Niger Colony
30. Nigeria
31. Northern Rhodesia
32. Nyassaland
33. Portuguese Guinea
34. Rio de Oro
35. Ruanda Urundi
36. Senegal
37. Sierra Leone
38. South Africa; Union of South Africa

39. Southern Rhodesia
40. South West Africa
41. Swaziland
42. Tanganyika; Tanganyika Territory (together with Zanzibar)
43. Togo
44. Tunisia
45. Ubangi Shari
46. Uganda

**Now**

Ethiopia
Southern Yemen (S. Arabia)
Algeria
Sudan
Angola
Lesotho
Zaire
Botswana
Bechuanaland
Cameroun; Cameroon; Cameroons
Chad
Dahomey
Egypt; Arab Republic of Egypt
Eritrea; Erythrea
Guinea
French Territory of the Afars and Issas; French Somali Coast
Mali, partly Upper Volta
Gabon
Ghana
Somalia

Ivory Coast
Kenya
Liberia
Libya
Mauritania
Congo
Morocco
Mozambique
Niger
Nigeria
Zambia
Malawi
Guinea-Bissau
Spanish Sahara
1. Rwanda; 2. Burundi
Senegal
Sierra Leone
Republic of South Africa[1] (Cape of Good Hope, Transvaal, Natal, Orange Free State); Republiek van Suid-Africa; South Africa
Rhodesia
Namibia; South West Africa
Swaziland
Tanzania

Togo
Tunisia
Central African Republic
Uganda

---

[1] Because of the exceptionally frequent occurrence of succulents in the Republic of South Africa, this political name is not repeated in full every time. Instead four names of provinces are entered: Cape of Good Hope, Transvaal, Natal, Orange Free State.

| Now | Previously |
|---|---|
| 1. Algeria | Algeria |
| 2. Angola | Angola |
| 3. Bechuanaland | British Bechuanaland |
| 4. Botswana | Bechuanaland Protectorate |
| 5. Cameroun; Cameroon; Cameroons | Cameroon; Cameroons |
| 6. Central African Republic | Ubangi Shari |
| 7. Chad | Chad Territory |
| 8. Congo | Middle Congo |
| 9. Dahomey | Dahomey |
| 10. Egypt; Arab Republic of Egypt | Egypt |
| 11. Eritrea; Erythrea | Erythrea |
| 12. Ethiopia | Abyssinia |
| 13. French Territory of the Afaras and Issas; French Somali Coast | French Somaliland |
| 14. Gabon | Gabon |
| 15. Ghana | Gold Coast |
| 16. Guinea | French Guinea |
| 17. Ivory Coast | Ivory Coast |
| 18. Kenya | Kenya |
| 19. Lesotho | Basutoland |
| 20. Liberia | Liberia |
| 21. Libya | Libya |
| 22. Malawi | Nyassaland |
| 23. Mali, partly Upper Volta | French Sudan |
| 24. Mauritania | Mauritania |
| 25. Mozambique | Mozambique |
| 26. Morocco | Morocco |
| 27. Namibia; South West Africa | South West Africa |
| 28. Niger | Niger Colony |
| 29. Nigeria | Nigeria |
| 30. Guinea-Bissau | Portugese Guinea |
| 31. Rhodesia | Southern Rhodesia |
| 32. 1. Rwanda; 2. Burundi | Ruanda Urundi |
| 33. Senegal | Senegal |
| 34. Sierra Leone | Sierra Leone |
| 35. Somalia | Italian Somaliland (Somalia) and British Somaliland |
| 36. South Africa; Republic of South Africa (Cape of Good Hope, Transvaal, Natal, Orange Free State); Republiek van Suid-Afrika | South Africa; Union of South Africa |
| 37. Southern Yemen (S. Arabia) | Aden Protectorate (S. Arabia) |
| 38. Spanish Sahara | Rio de Oro |
| 39. Sudan | Anglo-Egyptian Sudan |
| 40. Swaziland | Swaziland |
| 41. Tanzania | Tanganyika; Tanganyika Territory together with Zanzibar |
| 42. Togo | Togo |
| 43. Tunisia | Tunisia |
| 44. Uganda | Uganda |
| 45. Zaire | Belgian Congo |
| 46. Zambia | Northern Rhodesia |

# SPECIALIST ORGANISATIONS FOR THE CULTIVATION AND STUDY OF SUCCULENT PLANTS

**Australia.** The Cactus and Succulent Society of Australia, Newport West, Victoria. – The Cactus and Succulent Society of New South Wales, Kingsford, N.S.W. – The Cactus and Succulent Society of South Australia, Adelaide. – Also several local organisations.
**Austria.** Gesellschaft Österreichischer Kakteenfreunde, Vienna. – Steirische Kakteenfreunde, Knittelfeld.
**Belgium.** Belgische Vereniging voor Liefhebbers van Cactussen en Kamerplanten, Wilrijk. – 'Cactus', B 1200 Bruxelles.
**Canada.** Burnaby Cactus and Succulent Club, Burnaby, B.C.
**Ceylon.** Cactus and Succulent Society of Ceylon, Godakewela.
**ČSSR** (the Czechoslovak Socialist Republic). Svaz českých kaktusářů se sídlem v Brně (Czech Cactus Friends Association, P.O. Box 19, 602 00 Brno 2); Klub kaktusaru Praha (Prague Cactus Friends Club, the most important Branch Society of the Association).
**Denmark.** Nordisk Kaktus Selskab, Odense.
**Egypt.** Egyptian Society of Cactus Amateurs, Cairo.
**France.** Amis des Plantes de Serres et d'Acclimation, 45200 Montargis.
**Germany – Federal Republic of Germany.** Deutsche Kakteen-Gesellschaft e.V.Nürnberg, 2860 Osterholz-Scharmbeck). – Vereinigung der Kakteenfreunde Württembergs, Stuttgart.
**Germany – German Democratic Republic.** Fachgruppe für Kakteen und andere Sukkulenten im Kulturbund der DDR, Berlin.
**Great Britain.** The African Succulent Plant Society, Hitchin, Herts. – The Cactus and Succulent Society of Great Britain, Cheam, Surrey. – The National Cactus and Succulent Society, Oxford. – The Sempervivum Society, Burgess Hill, Sussex. – The Succulent Plant Institute, Morden, Surrey.
**Hawaii.** Cactus and Succulent Society of Hawaii, Honolulu.
**Hungary.** Kaktuszkedvelö Szakkör (Csili), Budapest 62. – Magyar Kaktuszgyüjtök Orszagos Egyesülete, Budapest.
**India.** The Cactus and Succulent Society of India, Bombay.
**International.** The International Organisation for Succulent Plant Study (I.O.S.), Natters, Austria. – International Succulent Institute (I.S.I.), Orinda, California, U.S.A.
**Japan.** The Cactus and Succulent Society of Japan, Kushimoto-cho. – Conophytum Society, Tokyo. – Shaboten-sha, Zushi. – Tokyo Cactus Club, Tokyo. – Also many local organisations.
**Malaya.** Penang Cactus Society.
**Malta G.C.** L-Chaqda Malija Tal-Kaktus u Sukkulenti Ohra, Zejtun. (The Cactus and Succulent Society of Malta, Zejtun.
**Mexico.** La Sociedad Mexicana de Cactologia, Mexico, D.F.
**Netherlands.** Nederlands-Belgische Vereniging van Liefhebbers van Cactussen en andere Vetplanten, Beekbergen.
**New Zealand.** Cactus and Succulent Society of New Zealand, Auckland. – Also several local organisations.
**Poland.** Polskie Towarzystwo Milosnikow Kaktusow, Katowice.
**Rhodesia.** The Aloe, Cactus and Succulent Society of Rhodesia, Salisbury.
**Romania.** Cactus and Succulent Society of Romania, Braila.
**South Africa.** The Natal Cactus and Succulent Club, Durban. – The South African Aloe and Succulent Society, Pretoria.
**Soviet Union.** Koskowskii kaktusowyi sojus, Moscow. – Society Cacti-Friends in Kasachstan, Alma-Ata.
**Spain.** Primera Asociación Cultural es España, para los Cactofilos-Amigos de los Cactos – y el Fomento de las Plantas Cactaceas, Valencia 2.
**Switzerland.** Schweizerische Kakteen-Gesellschaft, Zürich.
**U.S.A.** The Cactus and Succulent Society of America, Los Angeles, California. – Also many local organisations affiliated to The Cactus and Succulent Society of America.
**Yugoslavia.** Društwo prijareljev kaktej Slovenije, Ljubljana.

In 'Cactus and Succulent Journal of America' vol. 38, 1966, p. 18 DAN NEUMANN published a comprehensive catalogue of journals of succulent research.

# INTRODUCTION

Succulent, i.e. water-storing, plants are a living expression of particular conditions of climate and soil. They are denizens of the arid zones, deserts and semi-deserts. High temperatures and drought compel whatever plants grow in such areas to collect and conserve water so that they may survive long periods without water. Often the only source of water for years on end is the moisture from mist or fog. Whereas plants in the wetter regions of the tropics are mostly characterized by amply spreading foliage, succulent plants as a rule have more or less small leaves which, for the most part, are also fleshy, i.e. water is stored in the leaves. The stems can also be more or less thickened and equally capable of conserving water. This results in plant-forms which differ to a marked degree from those of either tropical forests or the temperate zones. In the deserts and semi-deserts of the American continent, the members of the Cactaceae are the outstanding examples of adaptation to the environment. Here we find shrubby species (Pereskia), columnar plants (Cerei) and globular types (Echinocactus, etc.). Wherever similar conditions prevail in the Old World, it will be noticed that plants from many different families have been compelled to make these modifications. But even in America, there are numerous examples of succulent plants in other families as well as the Cactaceae. All these succulents, i.e. those not within the family of the Cactaceae, are named in this Lexicon and described briefly.

Where very similar (convergent) external plant-forms arise in different families, confusion frequently occurs. Stem and leaf succulents from different families can often look extraordinarily similar (e.g. the leaf succulents Titanopsis calcarea (Mesembr.) and Crassula mesembryanthemopsis (Crass.), the stem succulents Didierca (Didiereac.) and Eulychnia (Cactac.). The surest means of establishing the true genus relies on the morphology of flowers and fruits, since these parts of the plant are unchanged in their essentials by external influences (climate, habitat). In the case of non-flowering plants, it is necessary to rely for identification on other morphological data, e.g. spination, leaf arrangement, or even on the sap – starch granules in milky sap (e.g. Euphorbia pendula, Sarcostemma viminalis), as well as on the presence or absence of any stored crystals of calcium oxalate. Readers are referred to the following detailed illustrated exposition of the subject: – "Succulent Plant-Form and Mode of Life" in the Cact. Succ. Journ. Am. **38**, 1966, 112–120.

Orchids (Orchidaceae), while they may have stems or even leaf-bases which are succulent, have not been included in the present work. For these plants, the reader is referred to the specialist literature (e.g. R. SCHLECHTER "Die Orchideen", Berlin/Hamburg, 1970).

Cultural instructions can only be indicated briefly in this Lexicon. For fuller information, see the section "The Cultivation of Succulents" in "A Handbook of Succulent Plants", Vol. I.

Throughout the last 200 years, scientists throughout the world, as well as numerous lovers of plant rarities, have been preoccupied with methods of cultivation. Moreover, many such enthusiasts have given serious attention to the questions of classification and nomenclature, and many valuable studies have been contributed by amateur botanists.

In retrospect, it is noticeable that there have been periods when a heightened interest has been taken in succulents. Towards the end of the 18th and the early 19th century, the outstanding names among botanists and collectors were those of ADRIAN HARDY HAWORTH and Count SALM-REIFFERSCHEIDT-DYCK. At the beginning of the 20th century, N. E. BROWN, ALWIN BERGER and Professor KURT DINTER spearheaded another vogue. At this period, the importation of numerous new plant-species resulted in the establishing of large collections. After the end of World War I came what was probably the most significant epoch of all. The vast amount of available material, as well as further imports, demanded systematic investigation. Extensive specialized works were the result. Only a few names can be mentioned here: – P. R. O. BALLY, Dr H. M. L. BOLUS, Dr. N. E. BROWN, Dr. N. Lord BRITTON, Prof. Dr. A. DYER, V. HIGGINS, C. P. HUTCHISON, Prof. Dr. R. MARLOTH, T. MARSHALL, Dr. R. V. MORAN, Prof. Dr. G. C. NEL, Dr. K. v. POELLNITZ, Prof. Dr. W. RAUH, Dr. G. W. REYNOLDS, G. D. ROWLEY, J. N. ROSE, Prof. Dr. G. SCHWANTES, B. L. SLOANE, Dr. E. WALTHER, Prof. Dr. E. WERDERMANN, A. WHITE, and many others. (For full information, see Kakt. u. a. Sukk. **15**, 1964, p. 74–78).

In all countries where there existed an interest in the cultivation of succulent plants, many societies have been founded to promote their study from the late 19th century onwards. The wealth of journals now being printed testifies to the enormous interest currently being taken in the succulents (see the previous page).

# PLANT FAMILIES AND THEIR GENERA WITH SUCCULENT SPECIES IN ALPHABETICAL ORDER (WITHOUT CACTACEAE)

Generic names not included here will be found in the chapter **Invalid Designations**.

The genera are assigned to families according to the International Code of Nomenclature (I.C.B.N.), Utrecht, 1972, Suppl. II, as undertaken by Dr. GÜNTHER BUCHHEIM (see Kakt. u. a. Sukk. **14**, 1963, p. 73 p.p.).

**Agavaceae** ENDL.: Agave, Beaucarnea, Beschorneria, Calibanus, Dasylirion, Furcraea, Hesperaloe, Nolina, Samuela, Sansevieria.
**Aizoaceae** RUDOLPHI[1]): Aizoanthemum, Sesuvium.
**Amaryllidaceae** JAUME ST.-HIL.: Ammocharis, Haemanthus.
**Anacardiaceae** LINDL.: Pachycormus.
**Apiaceae** LINDL. see **Umbelliferae** A. L. DE JUSS.
**Apocynaceae** A. L. DE JUSS: Adenium, Pachypodium, Plumeria.
**Araliaceae** A. RICH.: Cussonia.
**Asclepiadaceae** R. BR.: Brachystelma, Caralluma, Ceropegia, Cynanchum, Decabelone, Decanema, Diplocyatha, Dischidia, Duvalia, Echidnopsis, Edithcolea, Fockea, Folotsia, Hoodia, Hoodiopsis, Hoya, Huernia, Huerniopsis, Karimbolea, Kinepetalum, Luckhoffia, Pectinaria, Piaranthus, Pseudolithos, Pseudopectinaria, Raphionacme, Rhytidocaulon, Sarcostemma, Stapelia, Stapelianthus, Stapeliopsis, Stultitia, Trichocaulon, Whitesloanea.
**Asteraceae** DUMORTIER, see **Compositae** GISEKE.
**Balsaminaceae** A. RICH.: Impatiens.
**Basellaceae** MOQUIN-TANDON: Ullucus.
**Batidaceae** BENTH. et HOOK.: Batis.
**Begoniaceae** C. A. AGARDH: Begonia.
**Bombacaceae** KUNTH: Adansonia, Bombax, Cavanillesia, Chorisia.
**Brassicaceae** BURNETT, see **Cruciferae** JUSS.
**Bromeliaceae** A. L. DE JUSS.: Abromeitiella, Dyckia, Hechtia.
**Burseraceae** KUNTH: Bursera, Commiphora.
**Campanulaceae** A. L. DE JUSS.: Brighamia, Lobelia.
**Chenopodiaceae** VENT.: Allenrolfia, Arthrocnemum, Microcnemum, Pachycornia, Salicornia, Salsola, Suaeda.
**Commelinaceae** R. BR.: Cyanotis, Tradescantia.
**Compositae** GISEKE: Baeriopsis, Coreopsis, Coulterella, Espeletia, Gynura, Hertia, Othonna, Pteroneura, Senecio.
**Convolvulaceae** A. L. DE JUSS.: Ipomoea.
**Crassulaceae** A. P. DE CAND.: Aeonium, Adromischus, Afrovivella, Aichryson, Chiastophyllum, Cotyledon, Crassula, Diamorpha, Dinacria, Diopogon, Dudleya, × Dudleveria, Echeveria, Graptopetalum, × Graptoveria, × Greenonium, Greenovia, Hypagophytum, Kalanchoe, Lenophyllum, Meterostachys, Mucizonia, Orostachys, Pachyphytum, × Pachysedum, × Pachyveria, Pagella, Pistorinia, Pseudosedum, Rhodiola, Rochea, × Rocheassula, Rosularia, × Sedeveria, Sedum, Sempervivella, Sempervivum, Sinocrassula, Thompsonella, Umbilicus, Vauanthes, Villadia.
**Cruciferae** JUSS.: Caulanthes (not exactly a succulent plant).
**Cucurbitaceae** A. L. DE JUSS.: Acanthosicyos, Anisosperma, Apodanthera, Ceratosanthes, Corallocarpus, Cucurbita, Dendrosicyos, Echinocystis, Gerrardanthus, Ibervillea, Kedrostis, Momordica, Neoalsomitra, Pisosperma, Seyrigia, Telfaria, Tumamoca, Xerosicyos, Zehneria.
**Didiereaceae** DRAKE et CASTILLO: Alluaudia, Alluaudiopsis, Decaryia, Didierea.
**Dioscoreaceae** R. BR.: Testudinaria.
**Euphorbiaceae** A. L. DE JUSS.: Elaeophorbia, Euphorbia, Jatropha, Monadenium, Pedilanthus, Synadenium.
**Fabaceae** LINDL. see **Leguminosae**.
**Ficoidaceae** JUSS. see **Invalid Designations**.

---

[1]) In a narrow sense, that is without the separated families (formerly Sub-families) Mesembryanthemaceae, Molluginaceae, Tetragoniaceae etc. according to G. SCHWANTES, H.-D. IHLENFELDT and H. STRAKA. "Die höheren Taxa des Mesembryanthemaceae" in "Taxon", **XI**, No. 2, 1962.

**Fouquieriaceae** A. P. DE CAND.: Fouquieria, Idria.
**Geraniaceae** A. L. DE JUSS.: Pelargonium, Sarcocaulon.
**Gesneriaceae** BENTH. et HOOK.: Streptocarpus.
**Icacinaceae** MIERS: Pyrenacantha, Trematosperma.
**Labiatae** A. L. DE JUSS.: Aeolanthus, Coleus, Plectranthus.
**Lamiaceae** LINDL., see **Labiatae** A. L. DE JUSS.
**Leguminosae** A. L. DE JUSS.: Dolichos, Neorautenenia.
**Liliaceae** A. L. DE JUSS.: × Alchamaloe, × Aleptoe, × Allauminia, Aloe, × Aloella, × Alolirion, × Aloloba, × Alworthia, Astroloba, × Astroworthia, Bowiea, Bulbine, Bulbinopsis, Chamaealoe, × Chamaeleptaloe, Chortolirion, Drimia, × Gasterhaworthia, Gasteria, × Gastrolea, × Gastrolirion, Haworthia, × Leptaloinella, × Leptauminia, Litanthus, × Lomataloe, × Lomateria, Lomatophyllum, × Poellneria, Poellnitzia.
**Melastomataceae** BENTH. et HOOK.: Monolena.
**Menispermaceae** A. L. DE JUSS.: Stephania.
**Mesembryanthemaceae** FENZL: see Part II.
**Molluginaceae** HUTCHINS.: Hypertelis.
**Moraceae** LINK.: Dorstenia, Ficus.
**Moringaceae** DUM.: Moringa.
**Oxalidaceae** R. BR.: Oxalis.
**Passifloraceae** A. L. DE JUSS. ex KUNTH: Adenia, Modecca.
**Pedaliaceae** R. BR.: Harpagophytum, Holubia, Pedaliodiscus, Pedalium, Pterodiscus, Rogeria, Sesamothamnus, Uncarina.
**Piperaceae** G. A. AGARDH: Peperomia.
**Portulacaceae** A. L. DE JUSS.: Anacampseros, Calandrinia, Ceraria, Lewisia, Portulaca, Portulacaria, Talinum, Talinopsis.
**Rubiaceae** A. L. DE JUSS.: Hydnophytum, Myrmecodia.
**Scrophulariaceae** A. L. DE JUSS.: Castilleja, Chamaegigas, Dermatobotrys.
**Umbelliferae** A. L. KE JUSS.: Crithmum.
**Urticaceae** A. L. DE JUSS.: Pilea.
**Violaceae** BATSCH.: Hymenanthera.
**Vitaceae** A. L. DE JUSS.: Cissus, Cyphostemma.
**Welwitschiaceae:** Welwitschia (not exactly a succulent plant).
**Zygophyllaceae** R. BR.: Augea, Zygophyllum.

# SUCCULENT PLANTS

# PART I. FAMILIES AGAVACEAE TO ZYGOPHYLLACEAE (EXCLUDING MESEMBRYANTHEMACEAE)

Genera, Intergeneric hybrids, Species and Lower Taxa, Hybrids, Cultivars, Origins and Descriptions

**Abromeitiella** MEZ. Bromeliaceae. – Argentine. – ⚄, dwarf, mat-forming, consisting of small dense **Ros.** of projecting L.; **L.** linear-lanceolate or triangular-ovate, margins having small Sp., these sometimes absent, hard, fleshy; **Fl.** solitary or several, lateral, greenish. – Greenhouse, warm. Propagation: seed, division.

**A. abstrusa** CASTELL. – Argentine. – **L.** c. 5 cm long, 8 mm across, 4.5 mm thick, convex, grey-green, stiff, terminal Sp. 1 cm long; **Fl.** 1–3, c. 3.5 cm long.

**A. brevifolia** (GRISEB.) CASTELL. (Pl. 2/1) (*Navia b.* GRISEB., *Dyckia grisebachii* BAK.). – Argentine. – **L.** ovate-triangular, 2–3 cm long, 8–14 mm across, pointed, margins entire, thick, fleshy; **Fl.** c. 3 cm long, 6–7 mm wide, greenish.

**A. chlorantha** (HAUM.) MEZ. (*Lindmannia c.* HAUM., *A. pulvinata* MEZ.). – Argentine. – **Ros.** somewhat squarrose, 3 cm ⌀; **L.** long-triangular, 2 cm long, 5 mm across, Sp.-tipped, margins with small T. 1 mm long; **Fl.** 2 cm long, greenish.

**Acanthosicyos** WELW. Cucurbitaceae. – Greenhouse, warm. Propagation: seed.

**A. horrida** WELW. ('Naras'). (Pl. 2/2) – Angola: coastal zone; S.W.Afr.: Gr. Namaqualand. – ♄ with long, thick taproot, rigid, freely branching, covered with Th. 2–2.5 cm long, upright, the height of a man, 10 m ⌀; leafless; **Fl.** small, yellow-green; **Fr.** prickly, weighing up to ½ kg, flesh reddish-cream colour, edible.

**Adansonia** L. Bombacaceae. – Trop. W. and E.Afr.; Madag.; Comoro; Austr. – ♄, large; **St.** stout, conspicuous, water-storing, young plants have tuberous **St.**; **L.** initially simple, later digitate, sessile or stalked, shed during dry season, L.-lobes with margin entire, or less often saw-toothed; **Fl.** solitary, axillary, with pedicel, Cor. ovoid-clubshaped, lobes 5, triangular or elongated, white, yellow or red; **Fr.** variable in shape, ± tufted-hairy, edible. – Greenhouse, warm. Propagation: seed.

**A. alba** JUM. et PERR. – Madag. – 10–15 m tall; **St.** straight, narrowing to crown; **L.** later 5–7 lobed, ovate-conical, pointed; **Fl.** white.

**A. digitata** L. ('Baobab', 'Monkey-bread Tree'). – Trop. W. & E.Afr.; W.Madag.; Comoro. – To 18 m tall; **St.** circumference quite often 9–11 m; **L.** with stalk 7–12 cm long, 5–7 lobed, glabrous, L.-lobes almost sessile or short-stalked, blades narrowing to base, pointed at tip; **Fl.** pendulous, Pet. fan-shaped, 10–12 cm ⌀, white.

**A. fony** H. BAILL. v. **fony**. – S. & S.W.Madag. – 10–12 m tall; **St.** cigar-shaped; **L.** with stalk 3 cm long, L. lobes sessile, oval-lanceolate, 12–13 cm long, 1.2–2 cm across, margins saw-toothed; **Fl.** erect, digitate, pedicel short, Pet. 12–15 cm long, 1–1.5 cm across, pale yellow.

**A. — v. rubristipa** (JUM. et PERR.) PERR. (*A. rubristipa* JUM. et PERR.). – W.Madag. – With red bracts; stamens longer than in type.

**A. grandidieri** H. BAILL. (Pl. 2/3) – W.Madag. – **St.** cylindrical, fibrous; **L.** with stalk 10 cm long, stalk hairy, blade 5–7 lobed, L.-lobes with stalk 5 mm long, lanceolate, (3–)7(–9) cm long, 1.4–2 cm across, both surfaces hairy, pale bluish; **Fl.** small, yellow.

**A. madagascariensis** H. BAILL. – W.Madag. – 10–35 m tall; **St.** cylindrical; **L.** with long stalk, 5–7 lobes, these being sessile, ovate-spatulate, 9–10 cm long, 3–4 cm across, tip rounded and with a tiny point, margin entire; **Fl.** erect, 12–18 cm long, Pet. 13 cm long, 1.5 cm across, vivid red.

**A. suarezensis** PERR. – N.W.Madag. – 20–30 m tall; **St.** smooth; **L.** glabrous, with stalk 12–16 cm long, 3–9 lobes, these having stalk 4–8 mm long, elongate, 10.5–15.5 cm long, 4–6 cm across, tip blunt-pointed, margin entire; **Fl.** sessile, large, lovely vivid red.

**A. za** H. BAILL. v. **za**. – W.Madag. – 10–30 m tall; **St.** cylindrical, narrowing towards crown; **L.** with stalk, 6–12 cm long, 3–7 lobed, L.-lobes with slender stalks, oval-lanceolate, 5–10 cm long, 1.5–2.5 cm across, with a pointed tip; **Fl.** erect, Pet. 15–20 cm long, 1.2–1.5 cm across, vivid red.

**A. — v. boinensis** PERR. (*A. za* JUM. et PERR.). – **L.** 5–7 lobed, smaller, stiffer, L.-lobes first sessile, later with short stalk.

**A. — v. bozy** (JUM. et PERR.) PERR. (*A. bozy* JUM. et PERR.). – **L.** 7–11 lobed, L.-lobes 8–18 cm long, 3.5–7.5 cm across, with stalk 2–12 mm long.

**Adenia** FORSK. Passifloraceae. – From Somalia, Kenya and Tanzania to S.W.Afr.; Socotra; Madag.; Burma. – Highly succulent plant, often with thick, shapeless **caudex**; **Br.** ± long, thin, liana-like, thorny, some having tendrils; **L.** in part missing, otherwise L. ± large, margin entire, lobed or feathery; **Fl.** inconspicuous, greenish or yellowish; **Fr.** conical to egg-shaped. Plants sometimes dioecious. Greenhouse, warm. Propagation: seed, cuttings. (Lit.: WILDE, W. J. J. I. DE: A Monograph of the genus Adenia. Wageningen 1971).

**A. aculeata** (OLIV.) ENGL. – Somalia. – ♄, succulent; **Br.** long, 6–8 mm ⌀, 4-angled, angles projecting and forming conical Sp. 4–5 mm long, at intervals of 5 mm, tendrils single; leafless; ♀ **Fl.**, 4–5, almost sessile, ♂ **Fl.** on pedicel 10 cm long.

**A. apiculata** (MAST.) CHAKRAVARTY (*Modecca a.* MAST.). – Burma. – **St.** furrowed, cylindrical, bare; vines long; **L.** cordate, stalk 3–5 cm long, with deep and narrow indentations, lobes 5–8 cm long, 1 cm wide; **Infl.** few flowered, **Fl.** 5–6 mm ∅. – Scarcely differentiated from **A. pinnatisecta**, which has caudate filaments, while those of **A. apiculata** have a Th.-like tip.

**A. ballyi** VERDC. – Somalia. – **Caudex** ± globular, about 1 m ∅, fleshy, green; **Br.** thick, numerous, curved to somewhat climbing, with longitudinal stripes and minute lenticels, lateral Br. 1–2 cm long, Th.-tipped; **L.** not known; **Fl.** in tufts, Cal. greenish to brownish outside, lemon-yellow inside, Pet. 4 mm long; **Fr.** ovoid, c. 4 cm long, 3 cm ∅, blue-green, tinged red.

**A. digitata** (HARV.) ENGL. (*Modecca d.* HARV., *M. senensis* MAST., *A. multiflora* POTT., *A. angustisecta* BURTT-DAVY, *A. buchananii* HARMS ex ENGL.). – Transvaal: Pretoria D., Moçambique. – **Caudex** globular, with grey rind, tapering above to an upright, thin St.; **L.** in close tufts, digitate.

**A. firingalavensis** (DRAKE) HARMS (*Ophiocaulon f.* DRAKE). – Madag. – **Caudex** conical, c. 80 cm tall, 70 cm ∅ at base, rind with greenish, waxy coating, **Br.** climbing, 2–3 m long; **L.** 3-lobed.

**A. fruticosa** BURTT-DAVY. – Transvaal: Pietersburg D. – **St.** with lower part thickened, club-shaped, about 1 m tall, near soil-level 40–50 cm ∅, higher up dividing into a few thick **Br.**, then branching freely, St. and Br. with light grey rind, very fleshy; **L.** very numerous, rounded-ovate.

**A. globosa** ENGL. ssp. **globosa** (Pl. 3/1). – Kenya: Kilifi D.; Tanzania: Pare D. – **Caudex** tuberous, resembling a lump of stone, 1 m tall and thick, warty, green; **Br.** whip-shaped, erect, stiff with thick Th., striped lengthwise with minute lenticels, sometimes smooth, grey to yellowish-green, side-Br. up to 2.2 cm long, with Th. at tips; **L.** narrowly lanceolate, small, soon dropping; **Infl.** glomerate, Fl. star-shaped, brilliant red, scented.

**A. —** ssp. **curvata** (VERDC.) DE WILDE (*A. pseudoglobosa* ssp. *c.* VERDC.). – Tanzania: Mbulu D. – **Br.** curved or somewhat climbing.

**A. —** ssp. **pseudoglobosa** (VERDC.) DE WILDE (*A. p.* VERDC.). – Kenya. – **Caudex** irregular-globular to depressed-globular, 1.8–2.4 m ∅; **Br.** thick, numerous, up to 200, erect or curved, 1–2 m long, 3–9 mm thick; **L.**-stalk c. 2.5 mm long, blade round or 3-angled, 7 mm long and across, somewhat 3-lobed; **Fl.** numerous, in clusters, c. 8 mm long; **Fr.** ovoid-globular 2–2.8 cm long, 2–2.3 cm ∅, green.

**A. gummifera** (HARV.) HARMS (*Ophiocaulon g.* HARV., *Modecca g.* (HARV. et SOND.). – S.Afr.; Congo to the Seychelles. – Climbing to 30 m long 10 cm thick at base; **L.** entire to deeply 3(–5)-lobed, orbicular to ovate or rhombic; **Fl.** pale green to pale yellow, ♂ up to 35, campanulate, with Pet. 6–11 mm long, ♀ 2–6, smaller; **Fr.** subglobular to ovoid or ellipsoid, 25–45 mm long.

**A. huillensis** (WELW.) FERNANDES (*Machoda h.* WELW.). – Ethiopia: Huilla. – **Caudex** fleshy; St. c. 2 cm ∅, furrowed,. c. 20 cm long; **L.** long-linear, 10–15 cm long, 8–10 mm across, membranous, spine-tipped, almost leathery, margins wavy; **Fl.** campanulate, Pet. linear, veined; **Fr.** berry-like, yellow.

**A. keramanthus** HARMS (Pl. 3/2). – Kenya; Tanzania. – Stout **caudex**; **St.** erect, 50–80 cm tall, 6–8 cm ∅, with several upright **Br.**, hairy; **L.** at Br.-tips, with long stalks, cordate to almost circular, noticeably hairy, fairly large; **Fl.** elongated pitcher-shaped, yellowish; **Fr.** size of a hen's egg, brilliant coralered.

**A. lobata** (JACQ.) ENGL. (*Modecca l.* JACQ.). – Trop.Afr.: Sierra Leone. – **St.** thickened at the base, prostrate at first, later ascending by means of simple tendrils, ± gnarled, up to 2 m long, 10–15 cm ∅ below, tuberculate, spiny; **L.** with margins almost entire or with 3–7 tapering lobes, with 2 glandular, ovate, fleshy auricles at the base of the petiole, these being concave on the underside, with a white, glossy, secretory gland; **Fl.** small, yellowish-green, urn-shaped or campanulate, scented, 5-cleft, 3 lobes having wavy-fringed margins, 2 lobes always flat.

**A. pechuelii** (ENGL.) HARMS (Pl. 2/5) (*Echinothamnus p.* ENGL., *Paschanthus p.* ENGL.). – S.W.Afr.: Damaraland. – Very deeply penetrating **taproot**: caudex very large, hemispherical, tessellate, grey-green, up to 50 cm ∅ or more; lateral St. in clusters, ± branching, whitish-green, thorny; **L.** sessile, lanceolate, small; **Infl.** small, 3 Fl.; **Fr.** red.

**A. repanda** (BURCH.) ENGL. (Pl. 2/4) (*Paschanthus r.* BURCH., *Jäggidia r.* SCHINZ, *P. jäggii* SCHINZ). – Transvaal; S.W.Afr. – **Caudex** turnip-like, 20–30 cm long, 15 cm ∅, often even larger; **St.** thin, climbing by tendrils; **L.** lanceolate, blunt-tipped, 6 cm long, 8–10 mm across, blue-green, rather waxy-pruinose; **Infl.** from axils of L., Fl. 3–5, 1 cm long, tube-shaped, yellowish.

**A. spec.** W. RAUH in Sukk. Kd. Schweiz. Kakt. Ges. VII/VIII, 1963, 119 is identical with **A. globosa** ssp. **pseudoglobosa** VERDC.

**A. spinosa** BURTT-DAVY. – Transvaal. – **St.** tuber up to 2 m ∅, lying mostly above ground, rind smooth, green; new **Br.** continually growing, ultimately over the entire caudex, even the lower surfaces, thorny; **L.** simple, rather small, soon dropping.

**A. tisserantii** A. et R. FERNANDES. – Angola; Benguela. – **Caudex** thick, tuberous, 10 cm high, 13 cm thick; **St.** erect, c. 4 cm tall; **L.** sessile, linear-lanceolate, narrowing to base, shortly auriculate, 20 cm long, 4 cm across, margin entire; **Infl.** axillary, with only few yellowish Fl., Pet. linear, fringed.

**A. venenata** FORSK. – Ethiopia to E.Afr. – Almost tree-like; **St.** columnar, thick, fleshy, with green rind, 1–1.5 m tall, in upper part tapering, almost whiplike and climbing liana-like up the Br. of neighbouring trees; **L.** 5-lobed, from L.-axils, with tendrils.

**A. volkensii** HARMS. – Tanzania. Resembles

A. **keremanthus;** Caudex long, turnip-like, fleshy; **Br.** projecting; **L.** pinnate.

**Adenium** ROEM. et SCHULT. Apocynaceae. ('Desert Rose'). – From Arabia through Kenya to N.Tanzania; S.W.Afr.: Damaraland. – **St.-** succulents; **caudex** massive, often shaped like a sugarloaf; **Br.** thick in upper part; **L.** arranged spirally; **Fl.** in close-set cymes, large, funnel-shaped, red or pink. Very free-flowering. – Greenhouse, warm. Propagation: seed. – Milky sap very poisonous.

A. **boehmianum** SCHINZ. (Pl. 3/3). – S.W.Afr.: Damaraland. – ♄ 1.5–2 m tall; newer **Br.** thin with fine hairs; **L.** 1–3 cm long, almost leathery, obovate to broadly wedge-shaped, rounded or ± deeply indented, Sp.-tipped, with dense, soft hairs, especially on under-surface; **Infl.** with several Fl., Cor.-tube c. 4 cm long, 15 mm ⌀, pink.

A. **obesum** (FORSK.) ROEM. et SCHULT. v. **obesum** (Pl. 3/4) (*Nerium o.* FORSK., *A. arabicum* BALF. f., *A. micranthum* STAPF, *A. coetaneum* STAPF, *A. honghel* A.D.C., *A. speciosum* FENZL). – S.Arabia; Uganda to Moçambique; Kenya; Tanzania. – ♄ to almost 2 m tall; **caudex** thick, fleshy, twisted; **Br.** short, fleshy; **roots** frequently forming swellings above ground-level; **L.** arranged spirally in clusters at tips of Br., ovate or wedge-shaped, upper surface glossy green, underside dull, pale green, initially often with soft hairs, 3–10 cm long, 1.8–3 cm across;

Fl. in cymes of 2–10, Cor. cylindrical at base, lobes large and broad, pink. Very free-flowering (cf. J. J. LAVRANOS in Cact. Succ. J. Am. **38,** 1966, 19–23).

A. — ssp. **multiflorum** (KLOTZSCH) CODD. (*A. m.* KLOTZSCH). – Transvaal. – **Fl.** dark crimson, very numerous.

A. — ssp. **socotranum** (VIERH.) LAVR. (*A. s.* VIERH., *A. multiflorum* BALF. f.). – Socotra. – **Caudex** shapeless, several m tall, 1–2 m ⌀, columnar or globose, simple or divided into several thick, conical St.; **Br.** in part erect and very large, unbranched, upper groups of Br. directed ± upwards or curving back; **L.** at tips of the few short Br., in thick clusters, both surfaces glabrous; **Fl.** bright pink, unscented, woolly hairs inside.

A. **oleifolium** STAPF (*A. lugardii* N. E. BR.). – S.W.Afr.; Gr. Namaqualand; S.Botswana. – **Caudex** tuberous, head-sized, with brown rind, upper half throwing out roots growing vertically upwards (!); **Br.** 1–5, erect, 20 cm long, 1 to 2 cm ⌀, with brown rind; **L.** in close bunches of 10–15, linear, narrowing to short stalk, 5–15 cm long, 10–15 mm across, grooved, light grey-green, some hairs; **Fl.** 2–4, c. 4 cm long, with pedicel, pink. Very poisonous!

A. **somalense** BALF. f. – Somalia. – Resembles A. **obesum; caudex** initially regularly globular, later irregular in shape, 1 m ⌀ or more; with several thin **Br.** above up to 2 m long; **L.** linear-lanceolate, grey-blue, margins wavy; **Fl.** small, white to pink, lobes having a distinct red. rib.

**Adromischus** LEM. Crassulaceae. – S.W. Afr.; Cape. – Succulent ♃ or small ♄; **St.** often with aerial roots, sometimes quickly dying off, sometimes persistent; **L.** mostly alternate or decussate, sometimes arranged as a Ros., varying in shape, flat or roundish or even club-shaped, sometimes widening at the tip, sessile or narrowing to a short stalk, mostly glabrous, some species with fine hairs, green, in some species with dark purple spots; **Fl.** Ped. terminal, axis often short and with only one Fl., otherwise lengthened, mostly unbranched or forked, less frequently more branched, Infl. clustered, almost spicate, Fl. from the axils of the upper bracts, either upright or projecting, less frequently pendulous, often with very short Ped., Cal.-lobes very short, Cor.-tube narrow, slender, broadening out either little or not at all, lobes united[1]) or incised[2]), very short, initially projecting, later recurved, colour whitish or reddish, flowering period: Summer. – Greenhouse, warm, in summer also in garden-frames. Propagation: seed, cuttings, to some extent also L.-cuttings. (Lit.: KIMNACH, M.: The Genus Adromischus. In Cact. Succ. Jour. Amer. **25,** 41–48, 1953.)

## Classification of the Genus Adromischus in Sections by K. v. Poellnitz emend. Jacobsen

Type-species: A. hemisphaericus.
**Sect. I. Brevipedunculatae** v. POELLN. (*Cotyledon* L. Sect. *Paniculatae* SCHOENL. SSect. *Caryophyllaceae* SCHOENL. p. part.). – **Sc.** mostly short, less often to 15 cm long; **Fl.** 1–5, less often up to 10, with stalks erect; **St.** short, seldom long; plant small; **L.** small to medium. – Type-species: A. humilis. – Species: A. antidorcatum, herrei, humilis, nanus, schuldtianus.
**Setc. II. Adromischus** (Sect. *Longipedunculatae* v. POELLN.). – **Sc.** extended, infrequently short; **Fl.** numerous, ± sessile or with stalk, erect to projecting, in one species pendulous (A. phillipsiae), Infl. spicate, racemose or occasionally paniculate; **St.** often rather long; plant larger, **L.** more numerous and larger.
**SSect. 1. Cristati** (SCHOENL.) v. POELLN. (*Cotyledon* L. Sect. *Spicatae* HARV. SSect. *Cristati* SCHOENL.). – **L.** covered with soft hairs. – Type-species: A. cristatus. – Species: A. cristatus, poellnitzianus, schoenlandii, zeyheri.

---
[1]) Sect. *Connatilobatae* UITEW.   } This classification
[2]) Sect. *Incisilobatae* UITEW.      } is not adopted here.

**SSect. 2. Hemisphaerici** (SCHOENL.) v. POELLN. (*Cotyledon* L. Sect. *Spicatae* HARV. SSect. *Hemisphaericae* SCHOENL.; *Cotyledon* L. Sect. *Paniculatae* SCHOENL. SSect. *Caryophyllaceae* SCHOENL. p. part.). – **L.** glabrous, only having a few hairs in juvenile stage. – Type-species: A. hemisphaericus. – Species: A. alstonii, alveolatus, bicolor, blosianus, bolusii, caryophyllaceus, clavifolius, cooperi, festivus, fragilis, geyeri, grandiflorus, halesowensis, hallii, hemisphaericus, juttae, kesselringianus, kubusensis, leucophyllus, liebenbergii, maculatus, mamillaris, marianae, maximus, nussbaumerianus, pachylophus, procurvus, pulchellus, rhombifolius, roaneanus, rodinii, rotundifolius, saxicolus, subcompressus, subpetiolatus, subrubellum, tricolor, triebneri, triflorus, trigynus, umbraticolus.

**SSect. 3. Pendenti** JACOBS. – **L.** dimorphic; **Fl.** pendulous. – Type-species: A. phillipsiae. – No further species.

**A. alstonii** (SCHOENL. et BAK. f.) C. A. SMITH (§ II/2) (*Cotyledon a.* SCHOENL. et BAK. f., *C. trigyna* SCHOENL., *A. t.* v. POELLN. p.part.). – S.W.Afr.: Gr. Namaqualand. – ♄, dwarf; **L.** oblanceolate or narrowly obovate, upper part rounded or somewhat pointed, dull red, 4–7 cm long, 20–23 mm across; **Fl.** 3, lobes pink on upper surface (Acc. v. POELLNITZ, identical with **A. trigynus** (BURCH.) v. POELLN.).

**A. alveolatus** P. C. HUTCH. (§ II/2). – Cape: Lit. Namaqualand. – ⚁ : **Root** tuberous; **St.** erect, 1–2 cm long, to 5 mm ⌀, often reddish, rough; **L.** sometimes in Ros., usually symmetrical, short-stalked, round or ovate to lanceolate, to 3.5 cm long, 2 cm across, 12 mm thick, upper side broad with shallow grooves, whitish-green, epidermis rough, often alveolate, often finely tuberculate, tip somewhat pointed, margins rather thickened; **Infl.** simple, **Fl.** whitish-brown.

**A. antidorcatum** v. POELLN. (§ I) (incorrectly *A. anticordatum* v. POELLN.). – Cape: Lit. Namaqualand. – ♄, dwarf, 4 cm tall, juvenile **Br.** papillose, later rough; **L.** arranged spirally, 2–3.5 cm long, 5–6 mm across and thick, ovate-lanceolate, narrowing to petiole, almost circular, margins somewhat angular, green, reddish-brown markings, waxy, young L. with light papillae; **Fl.** unknown (but placed in § I).

**A. bicolor** P. C. HUTCH. (§ II/2). – Cape: Steytlerville D. – ⚁; **Root** tuberous, stemless; **L.** triangular-ovate, tip mostly rounded, margins rounded but pointed in upper part, underside convex and keeled in upper part, to 4 cm long and across, green, mottled silver or red, tip pink; **Infl.** simple, to 25 cm tall, tube 13 mm long, at mouth laciniate, green or yellow-green, lower part red, inside yellow.

**A. blosianus** P. C. HUTCH. (§ II/2). – Cape: Lit. Namaqualand. – ⚁; **Root** tuberous; **St.** 1–2 cm long, to 1 cm ⌀, mostly smooth; **L.** forming dense Ros., short-petiolate, symmetrical, obovate, pointed, usually longer than broad, to 3.5 cm long and 2.5 cm across, 15 mm thick, convex on both surfaces, grey-green and pale pink-purple in colour, bordered in upper part, lower part stalk-like, margins thickened, wavy, very dark red; **Infl.** simple, to 25 cm tall, tube 14 mm long, blue-green, purple at base, inside pale green, lobes pale reddish-purple, at base papillose.

**A. bolusii** (SCHOENL.) BGR. (§ II/2) (*Cotyledon b.* SCHOENL., *C. b.* v. *karrooensis* SCHOENL., *A. b.* v. *k.* (SCHOENL.) JACOBS.). – Cape: Riversdale, Graaff-Reinet D. – Perhaps identical with **A. caryophyllaceus,** only differentiated by unbranching Infl.

**A. caryophyllaceus** (BURM. f.) LEM. (§ II/2) (*Cotyledon c.* BURM. f., *A. jasminiflorus* (SALM) LEM., *C. c.* SALM). – Cape: Karroo. – ♄, dwarf; **St.** erect, later with grey rind; **L.** spatulate or obcuneate, oblong, or elongate, 2.5–4 cm long, 1–2 cm across, 3–8 mm thick, green, waxy, margins with horny edge, ± dark red, ± sinuate; **Infl.** branching, Fl. dark red inside.

**A. clavifolius** (HAW.) LEM. (§ II/2) (*Cotyledon c.* HAW., *Esula species* HORT. ex BGR., *A. vanderheydeni* HORT. ex BGR., *C. cristata* HARV. p. min. part.). – Cape: Uitenhage D. – ⚁, low-growing; **St.** short with aerial roots; **L.** 6–8, club-shaped, narrowing to stalk, upper part abruptly truncate, with or without a tiny blunt tip, 3–7 cm long, 9–15 mm thick, smooth, hairless, green, often with indistinct reddish markings; **Infl.** 10–30 cm tall, Fl. 1 cm long, greenish, lobes reddish.

**A. cooperi** (BAK.) BGR. (Pl. 4/1) (§ II/2) (*Cotyledon c.* BAK., *Echeveria c.* OTTO, *A. clavifolius* HORT. ex v. ROEDER). – Cape: Uitenhage D. – ⚁; **St.** short in age, grey-brown, branched; **L.** quite numerous, oblanceolate, spatulate or ovate-cuneate, 25–50 mm long, 12–20 mm across the upper 3/4, below almost rounded and terete, often rather wider than thick, underside rounded, surface broad-triangular, rounded or truncate, margins quite angular, ± sinuate or curled, glabrous, somewhat glossy, greenish, purple markings in upper part; **Infl.** c. 25 cm tall, tube cylindrical, 5-angled, 10 mm long, below purple-red, above greenish, lobes purple, margins whitish.

**A. — cv. Cristata.** – Monstrous cultivar; **L.** initially almost circular, later elongate-ovate, unmottled.

**A. cristatus** (HAW.) LEM. (Pl. 4/2) (§ II/1) (*Cotyledon c.* HAW., *Esula c.* HORT. ex v. ROEDER). – Cape: Graaff-Reinet to Bathurst D. – ⚁; **St.** short, later mostly branching, with stiff, red, curving, aerial roots; **L.** cuneate, narrowing to stem, 2.5–4 cm long, 4–16 mm across, hatchet-shaped, both sides convex, above bluntly truncate with wavy margin, light green, with dense soft hairs; **Infl.** 15–20 cm long, Fl. greenish, whitish-red.

**A. festivus** C. A. SMITH (Pl. 4/3) (§ II/2) (*A. clavifolius* v. ROED. non LEM., *C. speciosa* HORT. ex v. ROEDER, *A. cooperi* v. POELLN.). – Cape: Graaf-Reinet D. – ⚁; **St.** short; **L.** few, almost ovate, tip truncate, margin wavy or curly, 3–4 cm long, base 1 cm long, narrowed teretely and 5 mm ⌀, otherwise 1.5–2 cm across and thick, upper surface flattish-rounded, lower surface much rounded, grey-green to chalky-white, upper part having large purple markings.

**A. fragilis** P. C. HUTCH. (§ II/2). – Cape: Lit. Namaqualand. – ♃; **Roots** fibrous; **St.** eventually prostrate, occasionally with 1 small **Br.**, to 50 cm long, to 1 cm ⌀, purple-brown, later scabby grey; **L.** almost erect, round or oval in cross-section, usually rounded or almost pointed towards tip, infrequently with indistinct yellow margins, dull grey-green, irregular markings green to purple, often pink-purple, glandular-waxy; **Infl.** simple, tube greenish-white or pale green striped, lobes united.

**A. — v. fragilis. – L.** to 3 cm long, 1.5 cm across, 1 cm thick, pointed, with small markings; **Fl.** green, 13 mm long.

**A. — v. numeesensis** P. C. HUTCH. (Pl. 4/4). – L. 5.5 cm long, 1.5–2.5 cm across, to 2 cm thick, large markings; **Fl.** green to brown-red in colour, inside green, lobe-margins whitish.

**A. geyeri** P. C. HUTCH. (§ II/2). – S.W.Afr.: Southern Lüderitz Bay. – ♃; **Roots** tuberous; **St.** 1–1.5 cm long, to 1 cm ⌀, purple-brown, rough, white-waxy; **L.** symmetrical, short, with thick stalk, obovate, rather pointed or round-tipped, above broadly furrowed, to 3.5 cm long, 2.5 cm across, velvety white-grey, irregular brown markings; **Infl.** simple, to 40 cm tall, Fl. to 23, erect, Cal.-lobes pruinose, tube bluish, at base yellowish, above greenish, at tip purple-brown, lobes papillose, pale purple, divisions 3–4 mm deep.

**A. grandiflorus** UITEW. (§ II/2). – Cape: Bonnievale. – ♄, small, very much branched; **St.** ± erect, 15 cm tall or more, later grey-brown, about 1 cm thick; **L.** ovate-cuneate, broad-triangular, rounded at end and with a terminal Sp., 10–15 mm long, 10–20 mm across, 4–7 mm thick, green, often with reddish-brownish margins, glossy, waxy, teretely narrowing below; **Infl.** 8–20 cm tall, with 2–3 Fl., Cor. cylindrical, 15 mm long, tube purple, lobes white.

**A. halesowensis** UITEW. (§ II/2) (*A. cuneatus* v. POELLN.). – Cape: Halesowen near Cradock. – ♃; **St.** elongated, ascending; **L.** to 7 cm long, up to 2.5 cm across below tip-area, gradually widening from base to tip, below narrowing only slightly, ± terete at base, underside noticeably convex, somewhat pruinose, with waxy dots, towards tip with brownish-green markings, tip broadly triangular-rounded, margins angular, horny; **Infl.** simple, Fl. single or 2–4, tube brownish-green, lobes pink with purple, lower part papillose.

**A. hallii** P. C. HUTCH. (§ II/2). – Cape: Lit. Namaqualand. – ♃; **Roots** tuberous; **St.** 1 to 2 cm long, to 1 cm ⌀, wrinkled; **L.** arranged almost Ros.-like, mostly asymmetrical, circular to 2.5 cm across and long, to 13 mm thick, underside convex, upperside also with shallow, irregular furrows, grey-white, upper margin-area thickened, in juvenile stage often wavy, concolorous to bluish-brown; **Infl.** simple, tube blue-green outside, light green inside, lobes white, margins dark brown.

**A. hemisphaericus** (L.) LEM. (§ II/2) (*Cotyledon h.* L., *C. rhombifolius* ECKL. et ZEYH., *C. maculata* ECKL. et ZEYH., *C montium-klinghardtii* DTR., *A. m.-k.* (DTR.) BGR., *T. crassifolia* SALISB.). – S.Afr.: S.W.Cape; S.W.Afr. – **St.** woody, fleshy, branching, with rind grey or grey-brown, to 20 cm tall, erect or prostrate; **L.** borne in upper part, broad ovate-elongate to ± rounded, less frequently hemispherical, often apiculate, occasionally narrowing cylindrically to base, margins blunt below, angular above, often with horny edge, very convex, light green, waxy, often reddish, 1.5–2.5 cm long, 10–12 mm across; **Infl.** to 40 cm long, pruinose, tube cylindrical, reddish, lobes whitish or pale pink with purple.

**A. herrei** (BARK.) v. POELLN. (§ I) (*Cotyledon h.* BARK.). – Cape: Lit. Namaqualand. – ♄, c. 10 cm tall; **L.** up to 10, arranged spirally, round-ovate and pointed or elliptical-fusiform, small, with a slender point at tip, almost cylindrical, upper part flatter, narrowing below to a St. 5–9 mm long, greenish, greatly shrunken in resting period and ± black-brown, waxy, 3–3.5 cm long, 8–14 mm thick, rough with papillae; **Infl.** scape 4–5 cm long with 2–3 erect Fl., Cor. 10–13 mm long, green, red and grey tinged, lobes purple-red or pink, 2 mm long.

**A. humilis** (MARL.) C. A. SMITH (§ I) (*Cotyledon h.* MARL., *A. h.* (MARL.) v. POELLN., *C. nana* MARL., ? *C. parvula* BURCH.). – Cape: Beaufort West D. – ♃; almost or completely stemless; **Root** thickened-tuberous; **L.** c. 10 in a dense Ros., obovate-cuneate, blunt, upper side concave or grooved, green with red streaks, underside convex, red, to 1.5 cm long, 5–8 mm across, 2.5–4 mm thick, L. in cultivated plants are larger; **Infl.** a loose cyme, 10–15 cm tall, Cor. tube cylindrical, yellow-green, edge purple-red, lobes recurved-spreading.

**A. juttae** v. POELLN. (§ II/2). – S.W.Afr.; Bushmanland. – ♄, dwarf, to 7 cm tall; **St.** with grey rind; **L.** spatulate or obovate-cuneate or elongate, mostly narrowing teretely, triangular or ovate-triangular above, with a small point at tip, infrequently rounded, upper side somewhat convex, underside convex in lower half, margins below are rounded, angular above, young L. have horny margin, 2.5–4 cm long, 1–2 cm across, 3–8 mm thick, green, with numerous waxy dots; **Infl.** 20–30 cm tall, Fl. 6–12, green to red-brown.

**A. kesselringianus** v. POELLN. (§ II/2). – S.Afr.: origin unknown. – ♃; very short-stemmed, sprouting from soil-level, with red-brown aerial roots; **L.** 3–4 cm long, below tip 1–1.5 cm across, obovate-elongate, towards base almost cylindrical, underside convex in upper part or ± flat, at base often reddish, waxy, this waxy coating later ± splitting, mottled, tip-area triangular, margins angular, with wavy, whitish, horny edge, tiny terminal point; **Infl.** simple, 40 cm tall, Fl. numerous, greenish-reddish.

**A. kubusensis** UITEW. (§ II/2). – Cape: Lit. Namaqualand. – ♄, dwarf with few branches; **St.** to 9 cm tall, 6–7 mm ⌀, waxy; **L.** projecting or pendulous, elongate-clavate to almost spindle-shaped, cylindrical, pointed at tip, ± rounded, narrowing teretely at base, 5–8 cm long, 1–1.5 cm thick, green, later waxy; **Infl.** 13–30 cm tall with 6–12 Fl., Cor. 12 mm long, purple-red, lobes purple-red.

**A. leucophyllus** UITEW. (§ II/2). – Cape: Montagu D. – ♄, low-growing; **St.** 10–15 mm thick, branching, later brownish, ± densely foliate towards crown, rather knotted; **L.** waxy-pruinose, sessile, ± elongate-ovate to round-ovate, tip rounded, pointed, apiculate, underside more convex, margins almost angular, horny, in age brownish, c. 3 cm long, 15–20 mm across, occasionally with small purple markings in upper 2/3; **Infl.** simple or divided, c. 15 cm tall; Fl. single or to 3, Cor. light green, often pruinose, lobes pink.

**A. liebenbergii** P. C. HUTCH. (§ II/2) (*A. turgidus* nom. nud.). – Cape: Laingsburg D. – ♃; **Roots** fibrous; **St.** erect, to 6 cm tall; **L.** cuneate, with stalk to 1 cm long, pale green, waxy, 2 cm long and across, 14 mm thick, blunt-tipped, apiculate, margins angular; **Infl.** simple, to 30 cm tall, Fl. 15–25, solitary with united lobes, pale whitish-pink.

**A. maculatus** (SALM) LEM. (§ II/2) (*Cotyledon maculatus* SALM, *Crassula m.* HORT. ex v. ROEDER, *Cot. alternans* HAW.). – Cape: Worcester, Robertson, Oudtshoorn D. – ♄, dwarf; **St.** somewhat branching from base, brownish; **L.** rounded or ovate-spatulate, somewhat keeled at base and rather auriculate, blunt, apiculate or cordately notched, margins horny, often wavy, fresh green, glossy, mottled very dark red; Fl. solitary, in a loose spike, Cor.-tube green, lobes white or purple.

**A. mamillaris** (L. f.) LEM. v. **mamillaris** (§ II/2) (*Cotyledon m.* L. f.). – Cape: Gr. Karroo, Bushmanland, Lit. Namaqualand; S.W.Afr.: Gr. Namaqualand. – ♃; **St.** creeping or ascending, **Br.** thin, rooting; **L.** well-spaced, occasionally almost Ros.-like, almost ovate, becoming only a little wider at the tip and with a terminal point, narrowing teretely to base, bluish-green, pruinose, waxy, 1.5–7 cm long, c. 1–3 cm across; **Infl.** c. 30 cm tall, Fl. mostly 2–3, green-brown.

**A.** – v. **filicaulis** (ECKL. et ZEYH.) JACOBS. (*Cotyledon f.* ECKL. et ZEYH., *A. f.* (ECKL. et ZEYH.) C. A. SMITH). – S.W.Afr.: Gr. Namaqualand. – **L.** fusiform, narrowing to both ends.

**A.** – v. **fusiformis** (ROLFE) JACOBS. (*Cotyledon f.* ROLFE, *A. f.* (ROLFE) C. A. SMITH). – **L.** fusiform, mottled purple.

**A.** – v. **marlothii** (SCHOENL.) JACOBS. (*Cotyledon m.* SCHOENL., *A. m.* (SCHOENL.) BGR.). – Cape: Laingsburg D. – **St.** very short; **L.** elongate, narrowing to base, blunt.

**A.** – v. **ruber** v. POELLN. (*Cotyledon mamillaris* BAK., *A. kleinioides* C. A. SMITH). – S.W.Afr.: Gr. Namaqualand. – **L.** without mottling; Fl. intense red.

**A. marianae** (MARL.) BGR. v. **marianae** (Pl. 4/5) (§ II/2) (*Cotyledon m.* MARL.). – Cape: Clanwilliam, Van Rhynsdorp, Calvinia D., Bushmanland. – ♄, low-growing, clump-forming; **L.** dense, c. 10 cm long, 12–14 mm across,

c. 8 mm thick, upper side flat, often grooved and concave, margins thickened and rolled slightly inwards, upper part pointed, underside semi-cylindrical, dark grey-green, brown-grey mottled, white-grey margins; **Infl.** simple, Cor.-tube green, lobes red.

**A.** – v. **immaculata** UITEW. – Cape: Lit. Namaqualand. – **L.** without coloured markings.

**A. maximus** P. C. HUTCH. (§ II/2). – Cape: Van Rhynsdorp D. – ♃; **Roots** fibrous; **St.** 6 cm long and more, to 2.5 cm ⌀, brown, later grey; **L.** elongate-spatulate, to 12.5 cm long, 5.5 cm across, thickest above middle, to 1.5 cm thick, pale yellowish-green, margins rather angular; **Infl.** 1–3, c. 60 cm tall, tube greenish, lobes white with pink tip.

**A. nanus** (N. E. BR.) v. POELLN. (§ I) (*Cotyledon n.* N. E. BR., *A. casmithianus* v. POELLN. nom. ill., *A. pauciflorus* P. C. HUTCH.). – S.W.Afr.: S. of Aus; Cape: Lit. Namaqualand. – ♃; **Roots** with tuberous thickening; **St.** erect, much branched, c. 4 cm tall; **L.** dense, broad-ovate to almost circular, blunt, 6–14 mm long and across, light green or yellowish-green, with waxy spots, underside rounded; Fl. single, c. 2 cm long, reddish-green.

**A. nussbaumerianus** (v. POELLN.) v. POELLN. (§ II/2) (*Cotyledon n.* v. POELLN.). – S.Afr.: origin unknown. – ♃; **L.** 4–4.5 cm long, 17 to 21 mm across, 7–8 mm thick below centre, above that rather thinner, narrowing teretely at ends, upper side spatulate-concave, cartilaginous margin with few minute papillae; **Infl.** 40 cm tall, Fl. 1 cm long, 2 mm ⌀.

**A. pachylophus** C. A. SMITH (§ II/2) (*Cotyledon cooperi* v. *immaculata* SCHOENL. et BAK., *A. clavifolius* v. POELLN. p. min. part.). – Cape: Graaff Reinet, Queenstown D. – Similar to **A. cooperi**; **L.** without mottling, grey-green, tip broader.

**A. phillipsiae** (MARL.) v. POELLN. (§ II/3) (*Cotyledon p.* MARL.). – Cape: S. of Roggeveld. – ♄, clump-forming, stoloniferous; **L.** arranged almost Ros.-like, obovate or linear-lanceolate, 3–4 cm long, c. 8 mm across, c. 8 mm thick, concave above, dark green, underside reddish; Fl. pendulous, almost 3 cm long, orange-red. – This species is transitional to the genus **Cotyledon** L.

**A. poellnitzianus** WERDERM. (§ II/1). – Cape: near E.London. – ♃; **St.** to 5 cm long, somewhat branching, with numerous red aerial roots, 1 cm long; **L.** arranged almost Ros.-like, light green, with minute light hairs, widening gradually from base to tip, first cylindrical, flattened towards tip, (5–)6(–10) cm long, lower part of tip-area much widened, rounded-truncate above, wavy, 12–22 mm long and across; **Infl.** to 40 cm tall with some 40 Fl. c. 12 mm long, white, tips pink.

**A. procurvus** (N. E. BR.) C. A. SMITH (§ II/2) (*Cotyledon p.* N. E. BR., *A. p.* (N. E. BR.) v. POELLN.). – S.Afr.: origin unknown. – ♃, glabrous; **L.** narrow-cuneate, upper part almost truncate, narrowing to a long, round stalk, olive-green or somewhat purple-pruinose, darker markings above, 3–5 cm long, 15 to 22 mm across at tip, 7–8 mm thick; **Infl.** to 30 cm long, simple, **Fl.** 1–3 together, 10 mm long, dirty red, lobes reddish-pink.

**A. pulchellus** P. C. HUTCH. (Pl. 4/6) (§ II/2). – Cape: Lit. Namaqualand. – ♃; **Roots** fibrous; **St.** to 6 cm tall, c. 1 cm ⌀, purple-brown, scabby grey in age; **L.** elongate-spatulate, often obovate, rather pointed, upperside usually convex with upper part flattened, underside convex, glossy dark green, often purple, silvery-white waxy coating and large mottlings, to 5.5 cm long, 2.5 cm across, often shorter and 4 cm across and oval; **Infl.** simple, tube green, lobes united.

**A. rhombifolius** (HAW.) LEM. v. **rhombifolius** (§ II/2) (*Cotyledon r.* HAW., *C. hemisphaerica* HARV. p. min. part., *Echeveria r.* OTTO). – Cape. – ♄, small; **L.** obovate-spatulate, 5–9 cm long, 3.5–4.5 cm across, upperside flat, underside convex, margins somewhat cartilaginous, apiculate, grey-green, with minute markings above, rough to touch; **Fl.** pale red. Insufficiently clarified species.

**A.** — v. **bakeri** v. POELLN. (*Cotyledon maculata* BAK., *A. m.* BGR. ex ENGL. et PRANTL). – ♄, branching, 10–15 cm long to 1.6 cm thick, bark grey, grained; **L.** ovate-spatulate, 6 to 7.5 cm long, c. 4 cm across, deep grey-green, both surfaces with longitudinal red-brown markings.

**A.** — v. **sphenophyllus** (C. A. SMITH) JACOBS. (*A. s.* C. A. SMITH, *Cotyledon rhombifolia* BAK., *A. r.* BGR., *C. triflora* SALM). – Willowmore D. – **L.** very thickened in upper part, no mottling.

**A. roanianus** UITEW. (§ II/2). – Cape: Lit. Namaqualand. – ♃; **St.** often zigzagging at nodes, later elongated and branching, prostrate, rooting, red-brown, later grey-brown, 7 mm thick at base; **L.** broad-ovate to almost circular, below ± sharply tapering to a stalk, this part curved and 2–4 mm thick, tip apiculate, margins angular, blade 1.5–3 cm long and almost as wide, both surfaces rather swollen, to 1 cm thick, grey-green, waxy-mottled; **Infl.** c. 15 to 20 cm long, Fl. loosely racemose, Cor. green, lobes whitish and purple.

**A. rodinii** P. C. HUTCH. (§ II/2). – Cape: Lit. Namaqualand. – ♃; **St.** finally prostrate, rooting at nodes, thick, brownish; **L.** noticeably petiolate, lanceolate-spatulate, tip either rounded or somewhat pointed, upper side ± convex, underside very convex, to 6 cm long, 22 mm across, light green, with indistinct lighter mottling, upper 2 cm with keeled margins, purple-brown, stalk 4 mm thick, L. otherwise to 22 mm thick; **Infl.** with 6–12 Fl., Cor. bluish-green, lobes free, pale yellowish or pale greenish-white, bordered dark red.

**A. rotundifolius** (HAW.) v. POELLN. (§ II/2) (*Cotyledon r.* HAW., *A. hemisphaericus* LEM. ex JACOBSEN, *A. r.* (HAW.) SM. p. part.). – Cape: Clanwilliam, Somerset-West D. – ♃; **St.** sturdy, simple or branched at base, later prostrate, to 20 cm long, rind grey; **L.** numerous, rounded or rounded-obovate, rounded or almost truncate above, with minute point, less often narrowing teretely to base, green, with numerous waxy dots and deciduous waxy coating, underside slightly convex and often blunt-keeled below tip, upperside concave, margins horny towards tip, 1–2 cm long, upper L. 2.5–3.5 cm long, all rather longer than broad; **Infl.** simple, 25 cm long, Fl. numerous, light pink.

**A. saxicolus** C. A. SMITH (§ II/2?). – Cape: Pretoria D. – ♃; **St.** virtually absent; **L.** arranged almost Ros.-like, both surfaces flattened, without mottling, often reddish or light pink towards tip; **Fl.** purple-red to mauve-red.

**A. schoenlandii** (PHILLIPS) v. POELLN. (§ II/1) (*Cotyledon s.* PHILLIPS). – Cape: Origin unknown. – Similar to **A. clavifolius,** and classified by C. A. SMITH as synonymous; **St.** some 4 cm long, with aerial roots; **L.** narrowly elongate, narrowing teretely below, almost completely cylindrical up to tip-area, c. 3.5 cm long, 1 cm across, tip-area flattened, somewhat wavy and bent or having 1–2 shallow or deep notches, pointed, green, hairy; **Fl.** Cor. greenish, lobes white and pink.

**A. schuldtianus** (v. POELLN.) v. POELLN. (§ I) (*Cotyledon s.* v. POELLN.). – S.W.Afr.; Cape: Lit. Namaqualand, Lit. Karroo. – ♃; **L.** c. 25 mm long, 10–13 mm across, 3–4 mm thick, underside more convex than upper, middle to tip with very narrow, rather wavy, greenish-brownish margin, slightly glossy, with numerous tiny whitish dots, ± waxy; **Fl.** not known, but classified in § I.

**A. subcompressum** v. POELLN. (§ II/2). – Cape: Lit. Namaqualand, Bushmanland. – ♃; **L.** well spaced, elongate or ovate-elongate, broadest in centre, 4.5–6 cm long, about 2 cm across, narrowing teretely to base, similarly to tip, margins rounded, flat towards tip and with horny angle, dark green, waxy, often with dark red markings; **Infl.** 20–30 cm tall, Fl. numerous, brownish.

**A. subpetiolatus** v. POELLN. (§ II/2). – Cape: Uniondale D. – ♄, dwarf, dainty; **St.** elongated, 8–10 cm tall, with red-brown rind, 8–10 mm ⌀; **L.** almost round or obovate, terete towards base, both surfaces rather convex, 15–25 mm long, 12–25 mm across, 3–4 mm thick, grey-green or green, red-brown mottled, waxy, upper part rounded or triangular-round, usually with short pointed tip; **Fl.** not known, but placed in § II/2.

**A. subrubellus** v. POELLN. (§ II/2). – Cape: Van-Rhynsdorp D. – ♃; **St.** 10 cm long, with some **Br.** from near ground-level, lower part 1.5 cm thick, rind grey; **L.** obovate-cuneate or subcuneate, truncate above or truncate-rounded, often emarginate, rounded to a stalk at base and 7 mm across, both surfaces convex, in the middle 5–8 mm thick, upper part of margins slightly horny, green, tinged red; **Infl.** 30 cm long, racemose, with Fl. green-reddish, lobes white.

**A. tricolor** C. A. SMITH (§ II/2) (*Cotyledon cucullata* SCHOENL. ex v. POELLN., *A. marianae* v. POELLN. p. min. part.). – Cape: Clanwilliam D. – ♃; **L.** – base teretely narrowed, no flattening above, green, flecked darker green; acc. v. POELLNITZ, a variety of **A. marianae.**

**A. triebneri** v. POELLN. (§ II/2). – Cape: Lit. Namaqualand. – ♄, dwarf, about 10 cm tall, branching, glabrous; **St.** much thickened, to 2.5 cm ⌀, rind grey; **L.** obovate-cuneate, narrowing teretely or broadly rounded below, 3–6 cm long, almost as broad in upper part, upperside flat, underside rather convex, margins in lower part rounded, angular above, often with horny wavy margins, greenish-reddish, having numerous little waxy dots towards tip; **Infl.** simple, to 50 cm tall, Fl. reddish-green.

**A. triflorus** (THUNBG.) BGR. (§ II/2) (*Cotyledon t.* THUNBG., *A. robustus* LEM.). – Cape: Clanwilliam D. – ♄, dwarf, somewhat branching at base; **St.** fleshy, thick, glabrous 8–10 cm tall; **L.** sessile, thick, obovate, very blunt, almost truncate, at base subcylindrical 5 cm long, more than 2.5 cm across, light green; **Fl.** spike 20 cm long, Fl. usually 3, Cor. tube reddish-green, striped, lobes white on upperside, reddish underneath.

**A. trigynus** (BURCH.) V. POELLN. (§ II/2) (*Cotyledon t.* BURCH., *C. rhombifolia* v. *spathulata* N. E. BR., *A. rupicolus* C. A. SMITH). – From Transvaal to Great Karroo, Lit. Namaqualand and to S.W.Afr.: Gr. Namaqualand. – ♃; **St.** short; **L.** close-set, almost round to much longer than broad, dark green to brown-red, both surfaces mottled; **Infl.** 50–60 cm long, Fl. purple-brown.

**A. umbraticolus** C. A. SMITH (§ II/2) (*Cotyledon trigyna* BURTT-DAVY p. min. part., *A. trigynus* v. POELLN. p. min. part.). – Transvaal: near Pretoria. – ♃; **St.** present; **L.** flattened on both sides, elongate to elongate-cuneate or obovate-cuneate.

**A. zeyheri** (HARV.) v. POELLN. (§ II/1) (*Cotyledon z.* HARV., *A. cristatus* BGR., p. min. part.). – Cape: Swellendam, Riversdale D. – ♃; **St.** semi-erect, rooting at nodes, 7–10 cm long; **L.** hairy, 2–4 cm long, fanlike towards tip, narrowing below middle, short-stalked, both surfaces convex, rounded above, wavy; **Infl.** simple, Fl. sessile.

**Aeolanthus** MART. Labiatae. – Greenhouse, moist warm. Propagation: seed, cuttings.

**A. repens** OLIV. – Trop. Afr. – **St.** woody at base, creeping, ascending, with felt-like hairs, bluntly 4-angled, rooting; **L.** fleshy, elongate to ovate-elongate, 13–20 mm long, 4 mm across, margin entire or dentate, underside hairy; **Infl.** 2–3 forked, Fl. sessile, mauve.

**Aeonium** WEBB et BERTH. Crassulaceae. – Canary Is.; Cape Verde Is.; Madeira; N.Africa; Mediterranean. – Succulent ♄ or sub-♄ with woody and ± branched **St.**, with distinct **L.**-scars, usually with aerial roots; **L.** alternate, usually crowded into **Ros.** at the ends of the Br., some species have no distinct St. so that the Ros. lie close to the ground, flat or saucer-shaped; **L.** narrower towards the base, margins entire or ciliate; **Fl.** in pyramidal, dichotomously branched, urcved Rac., bright yellow, white or red, in spring. The Fl.-bearing Br. die in many species after the seeds have ripened, or new shoots may break below the dead Infl.-Summer: in the open. Winter: cold-house. Propagation: seeds, cuttings.

## Division of the Genus Aeonium into Sections by J. A. Huber

A. Large to medium-large plants.
  a) Short-stemmed species with large **Ros.** to 60 cm ⌀, simple or little branched at the base.
    1. **L.** large, soft; **St.** herbaceous, very short, with dense velvety hairs; **Fl.** yellow.

      **Sect. I. Canariensia** (CHRIST) HUBER (*Patinaria* LOWE). – Species: A. canariense, cuneatum, glandulosum, longithyrsum, palmense, subplanum, tabuliiforme, virgineum.

    2. **L.** very large, glabrous, 15–20 cm wide, 20–30 cm long; **St.** longer, up to 5 cm thick, woody at the base; **Fl.** intense red, monocarpic.

      **Sect. II. Megalonium** BGR. – Species: A. nobile.

    3. **L.** not as large as above, glabrous, glutinous, 3.5–5.5 cm wide, 8–10 cm long; **Fl.** golden.

      **Sect. III. Pittonium** BGR. – Species: A. glutinosum.

  a) ♄ or sub-♄, up to 60–120 cm tall, St. ± pencil-thick or even more.
    1. **L.** wide and flat at the base, usually thin, light green, glabrous, without red edges; **Fl.** golden-yellow, (8–)10(–12)-partite.

      **Sect. IV. Aeonium** (*Euaeonium* BGR., *Holochrysum* PRAEG.). – Species: A. arboreum, balsamiferum, gorgonum, holochrysum, leucoblepharum, manriqueorum, rubrolineatum, undulatum, vestitum, webbii.

    2. **L.** narrowed at the base, usually dark green or bluish, often with red margins; **Fl.** greenish, whitish or red, 6–9-partite.

      **Sect. V. Leuconium** BGR. (*Urbica* CHRIST). – Species: A. burchardii, castello-paivae, ciliatum, decorum, gomerense, haworthii, hierrense, lanzerottense, percarneum, urbicum, valverdense.

Plate 2. **1. Abromeitiella brevifolia** (GRISEB.) CASTELL.; **2. Acanthosicyos horrida** WELW. with fruits (Photo: W. TRIEBNER); **3. Adansonia grandidieri** H. BAILL. (Photo: W. RAUH); **4. Adenia repanda** (BURCH.) ENGL.; **5. Adenia pechuelii** (ENGL.) HARMS (Photo: G. C. NEL).

Plate 3. 1. Adenia globosa ENGL. ssp. globosa; 2. A. keramanthus HARMS (Photo: P. R. O. BALLY); 3. Adenium boehmianum SCHINZ; 4. A. obesum (FORSK.) ROEM. et SCHULT. ssp. obesum (Photo: as 2).

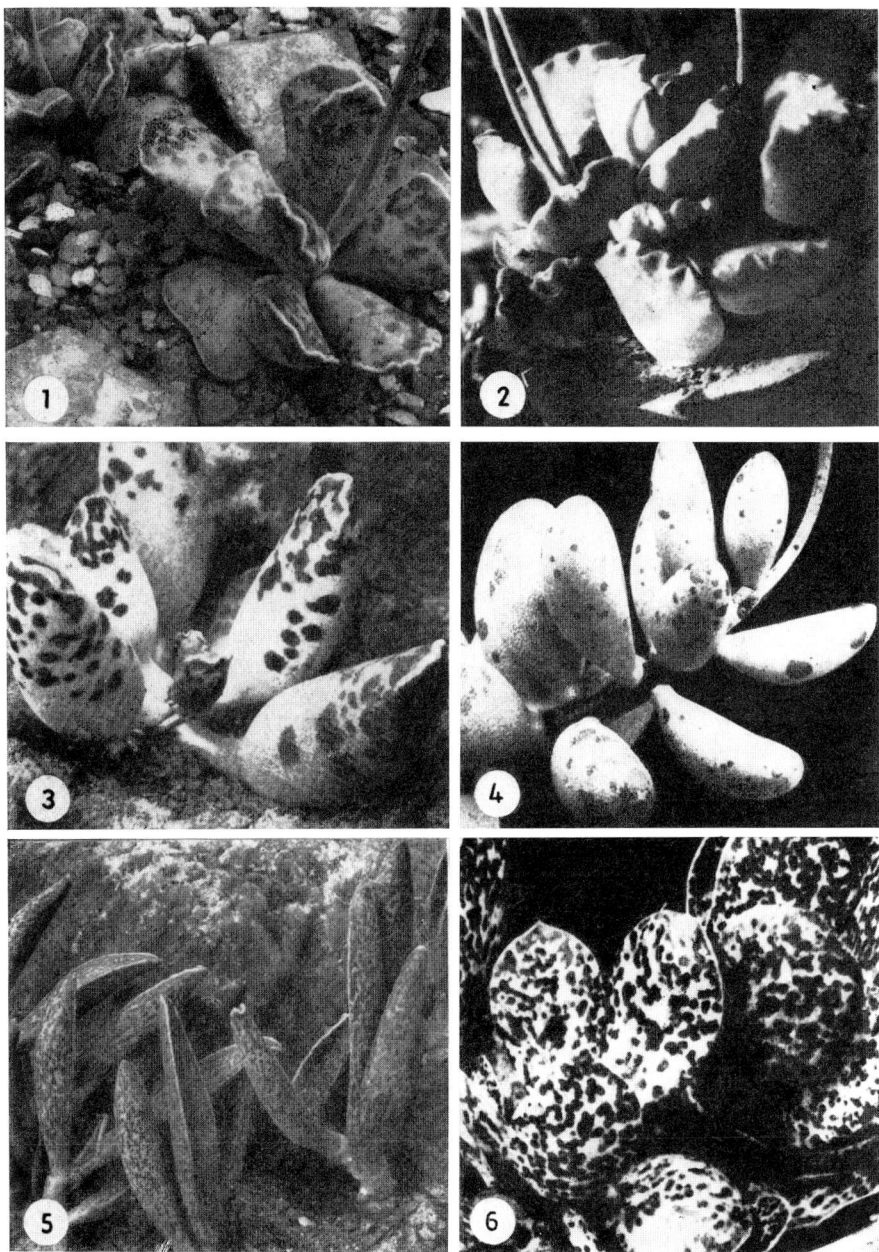

Plate 4. 1. **Adromischus cooperi** (BAK.) BGR.; 2. **A. cristatus** (HAW.) LEM. (Photo: G. KAISER); 3. **A. festivus** C. A. SMITH (Photo: A. J. A. UITEWAAL); 4. **A. fragilis** v. **numeesensis** P. C. HUTCHIS. (Repr.: Cact. Succ. Journ. Am. XXXI, 1959, 169. Photo: P. C. HUTCHISON); 5. **A. marianae** (MARL.) BGR. v. **marianae**; 6. **A. pulchellus** P. C. HUTCHIS. (Repr.: l.c. p. 120, Photo: as 4).

Plate 5. 1. Aeonium burchardii PRAEG. (Photo: O. BURCHARD); 2. A. nobile PRAEG. (Photo: as 1); 3. A. sedifolium (WEBB) PIT. et PROUST; 4. A. tabuliforme (HAW.) WEBB et BERTH.; 5. A. undulatum WEBB et BERTH. (Photo: as 1).

Plate 6. 1. Agave fourcroydes LEM. Infl. (Photo: F. RIVIERE DE CARALT); 2. A. americana cv. Marginata Aurea ; 3. A. attenuata SALM (Photo: J. VATRICAN).

Plate 7. 1. **Agave celsii** Hook. (Photo: J. Marnier-Lapostolle); **2. A. bracteosa** S. Wats. (Photo: F. Riviere de Caralt); **3. A. ferox** C. Koch (Photo: C. Faust); **4. A. ferdinandi-regis** Bgr.; **5. A. filifera** Salm v. **filifera**; **6. A. echinoides** Jacobi (Photo: as 1).

Plate 8. 1. Agave triangularis v. subintegra TREL.; 2. A. micrantha SALM; 3. A. parrasana BGR. (Photo: J. MARNIER-LAPOSTOLLE); 4. A. parviflora TORR.; 5. A. potatorum v. verschaffeltii (LEM.) BGR. (Photo: as 3); 6. A. striata ZUCC. v. striata.

Plate 9. 1. **Agave victoria-reginae** T. MOORE f. **victoria-reginae** (Photo: J. MARNIER-LAPOSTOLLE); 2. **Allenrolfia vaginata** (GRISEB.) O. KTZE.; 3. **Aichryson laxum** (HAW.) BRAMW.; 4. **A. bollei** WEBB (Photo: as 1).

A. Small, dense, to only 20–30 cm tall, with little **Ros.**; **L.** 1–3 cm long.

### Sect. VI. Goochiae CHRIST
a) **L.** without linear glands; margins not dentate or ciliate.

**SSect. Anodontium** (BGR.) HUBER (Sect. *Anodontium* BGR.). – Species: A. domesticum, goochiae, lindleyi, saundersii, sedifolium, tortuosum, viscatum.

a) **L.** with linear sunken glands; margins ciliate or with short cartilaginous T.

b) **St.** rough, only short-haired.

**SSect. 2. Aeonium** (BGR.) HUBER. – Species: A. simsii, spathulatum.

b) **St.** with long, white, spreading hairs.

**SSect. 3. Trichonium** (BGR.) HUBER (Sect. *Trichonium* BGR.). – Species: A. smithii.

**A. aizoides** (DC.) BGR. (*Sedum a.* DC., *Sempervivum a.* DC.). – Acc. PRAEGER, possibly a hybrid: **A. tortuosum** × **Aichryson divaricatum**, or **A. tortuosum** × **Aichryson punctatum**. – **St.** glabrous; **L.** ovate-spatulate, petiolate, pointed, fresh green, glabrous; **Infl.** with fine hairs, branching dichotomously.

**A.** × **anagense** BRAMW. et ROWL. – Canary Is.: Tenerife. – Acc. PRAEGER a spontaneous hybrid: **A. lindleyi** × **A. tabuliforme**. – Like a subshrubby **A. tabuliforme** with much smaller, less dense Ros.; **L.** with strongly compound ciliation.

**A. arboreum** (L.) WEBB et BERTH. (§ IV) (*Sempervivum a.* L.). – Circum-Mediterranean. – To 1 m tall; **St.** erect with few Br.; **Ros.** c. 20 cm ⌀; **L.** spatulate, blunt, with a tiny tip, light green, margins having white cilia; **Infl.** 25–30 cm tall, Fl. golden yellow.

**A.** — cv. Albovariegatum (*Sempervivum a.* v. a. WEST., *A. a. v. a.* (WEST.) BOOM, *S. a. v. variegatum* G. DON.). – White-margined cultivar.

**A.** — cv. Atropurpureum (*Sempervivum a.* v. a. NICH., *A. a. v. a.* (NICH.) BGR., *S. mutabile* SCHLTR. ex BREITER). – **L.** dark brownish purple, becoming green in winter.

**A.** — cv. Cristata. – Monstrous cultivar.

**A.** — cv. Luteovariegatum (*Sempervivum a.* v. l. WEST., *A. a. v. l.* (WEST.) BOOM). – Yellow-margined cultivar.

**A.** — cv. Zwartkop (cv. Blackhead, cv. Tete noire, cv. Schwarzkopf). – **L.** more pendent, glossy, deep dark red-brown, scarcely turning green even in winter.

**A. balsamiferum** WEBB et BERTH. (§ IV) (*Sempervivum b.* WEBB et BERTH.). – Canary Is.: Lanzarote, Fuerteventura. – Very tall; **St.** sturdy, branching; **Ros.** saucer-shaped, 20 cm ⌀; **L.** spatulate, narrowing to a slender tip, pointed, light green, sticky; **Fl.** light yellow.

**A.** × **bollei** KUNKEL. – Canary Is.: Gran Canaria. – Acc. PRAEGER a spontaneous hybrid: **A. percarneum** × **A. undulatum**. – **St.** erect, thick, unbranched, to 45 cm tall; **Ros.** dense; **L.** oblanceolate-spatulate, narrower above than in **A. percarneum**, broader below than in **A. undulatum**, 12–15 cm long, 5 cm wide; **Infl.** very large, pyramidal as in **A. undulatum**; **Fl.** paler yellow.

**A.** × **bramwellii** ROWL. – Canary Is.: Tenerife. – Acc. PRAEGER a spontaneous hybrid: **A. canariense** × **A. cuneatum**. – **Ros.** numerous, cup-shaped; **L.** spatulate, fresh green, glabrous or lightly pubescent, usually ciliate; **Infl.** densely glandular-pubescent; **Fl.** pale yellow, Pet. 7 mm long.

**A.** × **bravoanum** BRAMW. et ROWL. – Canary Is.: Gomera. – Acc. PRAEGER a spontaneous hybrid: **A. castello-paivae** × **A. viscatum**. – Like a small, greenish **A. castello-paivae**, but without cilia and with a fine glandular pubescence exactly like **A. viscatum**.

**A. burchardii** PRAEG. (Pl. 5/1) (§ V) (*Sempervivum b.* PRAEG.). – Canary Is.: Tenerife. – 20–30 cm tall, branching, with a glossy brown rind; **Ros.** 8–10 cm ⌀; **L.** thick, upperside keeled, dark green; **Fl.** soft ochreous colour.

**A. canariense** (L.) WEBB et BERTH. v. **canariense** (§ I) (*Sempervivum c.* L., *A. exsul* BORNM.). – Canary Is.: Tenerife. – **St.** very short; **Ros.** appears almost sessile, with numerous offsets, to 50 cm ⌀; **L.** broadly spatulate, rounded above, deep green, both surfaces with dense white hairs, rather sticky; **Infl.** to 80 cm tall, Fl. pale green.

**A.** — v. **latifolium** BURCH. – **L.** broader.

**A. castello-paivae** BOLLE (§ V) (*Sempervivum c.-p.* (BOLLE) CHRIST). – Canary Is.: Gomera. – **St.** ± freely branching; **Ros.** c. 10 cm ⌀; **L.** spatulate, 5 cm long, 2 cm across, fresh green, grey-pruinose, soft, older leaves with red border; **Infl.** glandular, sticky, Fl. greenish-white.

**A.** × **castelloplanum** BRAMW. et ROWL. – Canary Is.: Gomera. – Acc. PRAEGER a spontaneous hybrid: **A. castello-paivae** × **A. subplanum**. – **St.** erect, 7 mm thick; **Br.** few, patent; **Ros.** dense, flattish; **L.** 5–8 cm long, 3–4 cm broad, spatulate, bright green, margin ciliate and glandular-pubescent, otherwise glabrous; **Fl.** pale greenish-yellow.

**A. ciliatum** (WILLD.) WEBB et BERTH. (§ V) (*Sempervivum c.* WILLD.). – Canary Is.: La Palma, Tenerife. – **St.** numerous, branching from soil-level, Br. very bent; **Ros.** large; **L.** with rounded-spatulate tip, green to bluish-green, T. ciliate-serrate; **Infl.** broad, Fl. whitish-green.

**A. cuneatum** WEBB et BERTH. (§ I) (*Sempervivum c.* (WEBB et BERTH.) CHRIST). – Canary

Is.: Tenerife. – **St.** very short, with numerous offsets; **Ros.** large, roughly semi-ovoid; **L.** very long, narrowly spatulate, smooth, pointed, bluish-green; **Fl.** light yellow.

**A. decorum** WEBB et BOLLE (§ V) (*Sempervivum d.* (WEBB et BOLLE) CHRIST). – Canary Is.: Gomera. – Bush to 25 cm tall, hemispherical; **Ros.** 4–6 cm ∅; **L.** obovate-lanceolate, green, tinged vivid copper-red, border red and dentate; **Fl.** in slender Rac., pale pink.

**A. domesticum** (PRAEG.) BGR. (§ VI/1) (*Sempervivum d.* PRAEG., *Aichryson d.* PRAEG., *S. tortuosum* DC., *Aichryson t.* (DC.) WEBB et BERTH. – Origin unknown. – ♄, small; **Ros.** 2–5 cm ∅; **L.** almost circular, narrowing to base, with fine hairs; **Fl.** yellow.

**A. —** cv. Variegatum. – **L.** white-mottled.

**A. glandulosum** (AIT.) WEBB et BERTH. (§ I) (*Sempervivum g.* AIT., *A. meyerheimii* BOLLE). – Madeira. – **Ros.** sessile, flat plate-shaped; **L.** rhombic, spatulate, blunt or truncate, with shortly pointed tip, ± glandular, with soft hairs, sticky, green, somewhat ciliate; **Fl.** golden yellow.

**A. glutinosum** (AIT.) BGR. (*Sempervivum g.* AIT.). – Madeira. – Semi-♄; **Br.** spreading, glossy, sticky; **Ros.** loose, open; **L.** broad rounded-obovate, blunt above, with a shortly pointed tip, glossy light green, very sticky, 5–6 cm long, 3.8 cm across, margins with fine, cartilaginous cilia; **Fl.** light yellow.

**A. gomerense** PRAEG. (§ V). – Canary Is.: Gomera. – **St.** and **Br.** smooth, freely branching; **L.** obovate-spatulate, 6–10 cm long, 2–3 cm across, 5 mm thick in the middle, deep glossy green, less often somewhat bluish; **Infl.** flat, with fine glandular hairs, Fl. ochre-coloured.

**A. goochiae** WEBB et BERTH. (§ VI/1) (*Sempervivum g.* WEBB et BERTH.) CHRIST). – Canary Is.: La Palma. – ♄, **Br.** widespreading, slender, prostrate, often pendulous; **L.** at branch-ends often in cushions of Ros., petiolate, rhombic-ovate, small, somewhat sticky, deep olive-green to purple; **Fl.** pink.

**A. gorgonum** J. A. SCHMIDT (§ IV). – Cape Verde Is.: Santo Antão. – **St.** with numerous short side-Br.; **Ros.** broad; **L.** grey-green, margins reddened.

**A. × hawbicum** BRAMW. et ROWL. – Canary Is.: Tenerife. – Acc. PRAEGER a spontaneous hybrid: **A. haworthii** × **A. urbicum.** – **St.** thicker and **Br.** fewer than in **A. haworthii**; **Ros.** denser; **L.** longer; **Infl.** larger, as in **A. urbicum**; **Fl.** greenish or pinkish-white.

**A. haworthii** (S.D.) WEBB et BERTH. (§ V) (*Sempervivum h.* S.D.). – Canary Is.: Tenerife. – Bush; to 60 cm tall, freely branching, **St.** thin; **Ros.** numerous, dense, in cultivation usually 6–8 cm ∅, in habitat the plants are often simple, Ros. to 15 cm ∅; **L.** same width throughout, then abruptly pointed, bluish-green, margins red-brown with cartilaginous T., underside keeled; **Fl.** white.

**A. —** cv. Major. – **Ros.** larger.

**A. hierrense** (MURR.) PIT. et PROUST (§ V) (*Sempervivum h.* MURR.). – Canary Is.: Hierro. – **St.** short, silver-grey, much swollen, 50 cm tall;

Ros. about 50 cm ∅, open; L. narrow, smooth, curving upwards towards tip, bluish-green, tinged purple to violet, margins with stiff, fine cilia; **Infl.** c. 40 cm tall and wide, Pet. light pink with darker stripe.

**A. holochrysum** WEBB et BERTH. (§ IV) (*Sempervivum h.* (WEBB et BERTH.) CHRIST). – Canary Is.: Hierro, La Palma, Gomera, Tenerife. – **St.** tall, free-branching, initially erect, later somewhat pendent; **Ros.** dense, about 20 cm ∅; **L.** narrow-spatulate, smooth, yellowish-green, with red M. stripe, margins red and ciliate, thin, in resting-period L. may turn brown and fall, leaving only a tiny Ros.; **Fl.** golden-yellow.

**A. × hybridum** (HAW.) ROWL. (*Sempervivum ciliare* β *h.* HAW., *S. h.* SWEET, *A. barbatum* WEBB et BERTH., *A. × floribundum* BGR.). – Hybrid: **A. simsii** × **A. spathulatum.** – **L.** ribbon-like to spatulate, margins with roundish white T., ciliate; **Fl.** very numerous.

**A. × jacobsenii** BRAMW. et ROWL. – Canary Is.: Hierro; hybrid swarms abundant where the parents grow together. Acc. PRAEGER a spontaneous hybrid: **A. hierrense** × **A. palmense.** – Variable; monocarpic or more rarely branched and ♃, erect, to 60 cm tall; **St.** to 25 mm thick; **Ros.-L.** bright green, glabrous or minutely pubescent, ciliate at the base; **Infl.** c. 25 cm broad, intermediate between the parents; **Fl.** also intermediate, whitish, reddish or yellowish.

**A. × junoniae** BRAMW. et ROWL. – Canary Is.: La Palma. – Acc. PRAEGER a spontaneous hybrid: **A. ciliatum** × **A. palmense.** – Often monocarpic; **Ros.** large, 15–20 cm ∅; **L.** broadly spatulate, apiculate, soft and fleshy, often reddish above, usually glandular-pubescent, margin serrate-ciliolate and pubescent; **Infl.** much like **A. palmense**; **Fl.** pale yellow.

**A. × kunkelii** BRAMW. et ROWL. – Canary Is.: La Palma. – Acc. PRAEGER a spontaneous hybrid: **A. ciliatum** × **A. holochrysum.** – Similar to **A. holochrysum**, but L. 3–4 mm thick and dark green with a red edge, as in **A. ciliatum**.

**A. × lambii** BRAMW. et ROWL. – Canary Is.: Hierro. – Acc. PRAEGER a spontaneous hybrid: **A. palmense** × **A. valverdense.** – Intermediate between the parents; ♃, erect, branched, 60 cm tall; **St.** 2 cm thick; **L.** rough like **A. valverdense** but ± pubescent like **A. palmense**; **Fl.** yellowish pink.

**A. lanzerottense** PRAEG. (§ V). – Canary Is.: Lanzarote. – ♄ 30–45 cm tall, freely branching; **Ros.** dense; **L.** obovate-spatulate, pointed, 7–8 cm long, 3.5–4.5 cm across, 4–5 mm thick, underside convex, bare at base, margins otherwise ciliate, reddish; **Pet.** white with a red stripe.

**A. × lemsii** KUNKEL. – Canary Is.: Gran Canaria. – Acc. PRAEGER a spontaneous hybrid: **A. percarneum** × **A. virgineum.** – Shrubby; **St.** short, erect, branched; **L.** pale green like those of **A. virgineum**, sparsely and minutely pubescent, red-margined and streaked with purple as in **A. percarneum**.

**A. leucoblepharum** WEBB v. **leucoblepharum** (§ IV) (*Sempervivum l.* (WEBB) HUTCHINS. et

E. A. BRUCE, *S. chrysanthum* HOCHST., *A. c.* (HOCHST.) BGR.). – Ethiopia: Eritrea; Arabia; Kenya. – **St.** with numerous short **Br.**; **L.** light green, slightly reddened towards margins, to 9 cm long, 3.5 cm across; **Infl.** broad.

**A.** — v. **glandulosum** (CHIOV.) CUF. (*Sempervivum chrysanthum* v. *g.* CHIOV.). – **L.** glandular-hairy.

**A.** × **lidii** SUNDING et KUNKEL. – Canary Is.: Gran Canaria. – Acc. PRAEGER a spontaneous hybrid: **A. simsii** (*A. caespitosum*) × **A. percarneum.** – Intermediate between the parents. Erect, branched, compact, 25–30 cm high; **L.** linear-lanceolate, 10 cm long, 1–2 cm wide, margins ciliate; **Fl.** yellow.

**A. lindleyi** WEBB et BERTH. (§ VI/1) (*Sempervivum l.* WEBB et BERTH., *S. tortuosum* LINK., *A. t.* PIT. et PROUST, *S. viscosum* WEBB). – Canary Is.: La Palma, Tenerife. – Dense, hemispherical bush, 15–30 cm tall; **Br.** numerous, thin, gnarled; **Ros.** small; **L.** rounded-rhombic, thick, hairy, sticky, light green, often reddened; **Fl.** golden-yellow.

**A. longithyrsum** (BURCHARD) SVENT. (§ I) (*Sempervivum canariensis* ssp. *l.* BURCHARD nom. nud.). – Canary Is.: Ombrion Is. – Plant with white-felted glands, balsam-scented; **Ros.** single or several, lateral **St.** elongated to 50 cm, some offsets at ground-level; **L.** cuneate-spatulate, tip rounded or pointed, margins ± wavy or curly; **Infl.** long, bushy, densely glandular, Fl. about 15 mm ⌀, dirty yellow-greenish.

**A.** × **lowei** BRAMW. et ROWL. – Madeira. – Acc. PRAEGER a spontaneous hybrid: **A. glandulosum** × **A. glutinosum.** – **St.** short, 25 mm ⌀ at base, little branched; **L.** loosely rosulate, 8–10 cm long, 4 cm broad, spatulate or oblanceolate-spatulate to spatulate-rhomboidal, broadly cuneate above, linear below, glabrous except for the ciliate, pubescent margins; **Infl.** 20–30 cm wide, viscous-pubescent; **Fl.** 2 cm ⌀.

**A. manriqueorum** BOLLE (§ IV) (*Sempervivum m.* (BOLLE) CHRIST, *A. doramae* WEBB). – Canary Is.: Gran Canaria. – **St.** and Br. stout; **Br.** few, spreading or prostrate; **Ros.** dense, 18–25 cm ⌀; youngest L. of Ros. ovate-obcuneate, flat-concave, forming tight buds, older L. projecting, 8–9 cm long, 3 cm across immediately below tip, light green, with red lines along central nerve; **Infl.** with fine hairs.

*A. mauriqueorum* sphalm. = **A. manriqueorum** BOLLE (named after Spain. family MANRIQUE DE LARA).

**A. nobile** PRAEG. (Pl. 5/2) (§ II) (*Sempervivum n.* PRAEG.). – Canary Is.: La Palma. – Short-stemmed; **Ros.** to 50 cm ⌀; **L.** to 30 cm long, 12–20 cm across, very fleshy, with broad channel, margins curved up and over, olive to light green, sticky; **Infl.** a cyme 50 cm tall, with purple, succulent involucral L., **Fl.** copper-scarlet.

**A.** × **ombriosum** BRAMW. et ROWL. – Canary Is.: Hierro. – Acc. PRAEGER a spontaneous hybrid: **A. hierrense** × **A. valverdense.** – Intermediate in all characters between the parents.

**A. palmense** WEBB et CHRIST (§ I) (*Sempervivum p.* (WEBB et CHRIST) CHRIST, *S. christii* FRAEG., *S. canariense* v. *c.* BURCH.). – Canary Is.: Hierro, La Palma. – **St.** branching; **Ros.** flat, saucerlike, open; inner **L.** few, those above narrower and somewhat pointed, densely covered with short, coarse hairs, Infl. similarly; **Fl.** dark yellow.

**A. percarneum** (MURR.) PIT. et PROUST (§ V) (*Sempervivum p.* MURR.). – Canary Is.: Gran Canaria. – **St.** erect, stout, much-branching, rind silver-grey, producing thick adventitious roots; **Ros.** 8–12 cm ⌀; **L.** thick-fleshy, green, tinged-reddish; **Fl.** delicate pink or flesh-coloured.

**A.** × **praegeri** KUNKEL. – Acc. PRAEGER a spontaneous hybrid: **A. simsii** × **A. undulatum.** – Gran Canaria: Beo. – **St.** stout, brown, long; **L.** c. 12 cm long, 4 cm across, narrow obovate-spatulate, pointed, glabrous, underside with close-set, long, glandular hairs, margins with fine, membranous hairs, those at base being long, erect to recurved.

**A.** × **rowleyi** BRAMW. – Canary Is.: Tenerife. – Acc. PRAEGER a spontaneous hybrid: **A. smithii** × **A. spathulatum.** – Habit of **A. spathulatum** but more vigorous; **Ros.** larger and laxer; glands on **L.**-underside longer; lacks the shaggy hairs of **A. smithii** on the St.

**A. rubrolineatum** SVENT. (§ IV). – Canary Is.: Lesser Junonia, Gomera. – ♄ 50–80 cm tall, branching; **St.** stout, silvery red-grey rind with leaf-scars; **Br.** stiff, erect; **Ros.** dense, 15 to 20 cm ⌀, concave in middle; **L.** long-spatulate, short pointed at tip, apiculate, underside blunt-keeled in lower part, thick, 7–8 cm long, above about 2–2.5 cm across, lower part narrowing teretely, middle L. wavy and curling inwards, green, with dark purple flecks or stripes, margins lacking cartilaginous hairs; **Infl.** hemispherical-conical, Fl. sulphur-yellow.

**A.** × **sanctisebastianii** BRAMW. et ROWL. – Canary Is.: Gomera. – Acc. PRAEGER a spontaneous hybrid: **A. saundersii** × **A. subplanum,** *Sempervivum saundersii* × *S. canariensis* ssp. *latifolium*). – **St.** short, rather slender, with persistent dead L.; **Ros.** 5–8 cm ⌀; **L.** suberect, fresh green, obovate-spatulate, 3–4 cm long, 2–3 cm wide, glandular-pubescent, broadly cuneate above; **Infl.** and **Fl.** intermediate between the parents.

**A.** × **santosianum** BRAMW. et ROWL. – Canary Is.: La Palma. – Acc. PRAEGER a spontaneous hybrid: **A. goochiae** × **A. palmense.** – 30 cm tall, with a few bright brown **Br.** 4 mm thick; **L.** loosely rosulate, similar to **A. palmense** in shape and colour but less densely glandular-pubescent.

**A. saundersii** BOLLE (§ VI/1) (*Sempervivum s.* (BOLLE) CHRIST). – Canary Is.: Gomera. – ♄, freely branching, small; **Ros.** 3–5 cm ⌀; **L.** round-ovate, densely white-haired, very succulent, green, in resting period Ros. closes up to form a cherry-sized ball; **Fl.** yellow.

**A. sedifolium** (WEBB) PIT. et PROUST (Pl. 5/3) (§ VI/1) (*Aichryson s.* WEBB, *Sempervivum s.* CHRIST, *S. masferreri* HILLEBR.). – Canary Is.: La Palma, Tenerife. – Bushes, cushion-shaped,

10–15 cm tall; **St.** woody; **Br.** later pendent; **Ros.** small; **L.** set close together, short, rather thick, upper part convex, 8 mm long, mostly reddish-brown; **Fl.** large, yellow.

**A. simsii** (SWEET) STEARN (§ VI/2) (*Sempervivum s.* SWEET, *S. caespitosum* C. SM., *A. c.* (C. SM.) WEBB et BERTH., *S. ciliatum* C. SM., *S. ciliare* HAW.). – Canary Is.: Gran Canaria. – Bush forming cushions, with spreading **Br.** which are densely leafy; **Ros.** broad; **L.** linear-lanceolate, fresh green with reddish lines, margins with numerous, dense, white hairs 2–4 mm long; **Fl.** dark yellow. In resting period, Ros. closes to become onion-shaped.

**A. smithii** (SIMS) WEBB et BERTH. (§ VI/3) (*Sempervivum s.* SIMS). – Canary Is.: Tenerife. – ♄; **Br.** few, with dense hairs up to 1 cm long; **Ros.** c. 10 cm ⌀, open; **L.** fairly broad, softly hairy, light green or yellowish-green, often bronze-coloured, both surfaces with red markings running longitudinally, margins wavy; **Fl.** light yellow.

**A. spathulatum** (HORNEM.) PRAEG. v. **spathulatum** (§ VI/2) (*Sempervivum s.* HORNEM., *S. lineare* HAW., *S. barbatum* C. SMITH, *S. villosum* LINDL., *A. bentejui* WEBB, *S. strepsicladum* BORNM., *A. s.* WEBB et BERTH.). – Canary Is.: Hierro, La Palma, Tenerife, Gran Canaria. – ♄ pendet growth; **Br.** brown, to 90 cm long, bearing leaves; **Ros.** small; **L.** spatulate, 2 to 3 cm long, glabrous, sticky, with raised reddish lines running longitudinally, margins with blunt, cartilaginous T.; **Fl.** yellow.

**A.** — v. **cruentum** (WEBB et BERTH.) PRAEG. (*A. c.* WEBB et BERTH., *Aichryson pulchellum* C. A. MEY.). – Canary Is.: La Palma, Hierro. – **St.** more upright, red; **L.** longer and more cuneate at base, also blunter.

**A.** — cv. Minor. – **Ros.** smaller.

**A.** × **splendens** BRAMW. et ROWL. – Canary Is.: La Palma. – Acc. PRAEGER a spontaneous hybrid: **A. ciliatum** × **A. nobile.** – **St.** over 25 cm ⌀ and over 30 cm high, bearing a very large flat **Ros.**; **L.** dark green, 15 cm long, 7 cm wide, 10 mm thick at the centre and 3 mm thick at the margins, rounded-spatulate, ciliate and very finely pubescent, with red margins; **Infl.** more flat-topped than in **A. ciliatum**, with long Br., Pet. whitish above, red below with short crimson lines.

**A. subplanum** PRAEG. (§ I). – Canary Is.: Gomera. – **St.** rather lengthened; **Ros.** scarcely depressed, almost flat; **L.** elliptical in cross-section, narrowing to linear towards base.

**A.** × **sventenii** KUNKEL. – Canary Is.: Gran Canaria. – Acc. PRAEGER a spontaneous hybrid: **A. simsii** × **A. manriqueorum.** – **Ros.** like that of a glabrous **A. simsii** but otherwise like **A. manriqueorum**, with crowded cilia.

**A. tabuliforme** (HAW.) WEBB et BERTH. (Pl. 5/4) (§ I) (*Sempervivum t.* HAW., *A. bertoletianum* BOLLE, *A. macrolepum* WEBB, *S. complanatum* A. DC.). – Canary Is.: Tenerife. – Very dwarf semi-♄ or ♃; **Ros.** sessile, flat, plate-shaped, to 50 cm ⌀; **L.** narrow, spatulate, imbricate, margins with fine cilia, grass-green; Ros. later producing a much-branched Infl., 60 cm tall; **Fl.** sulphur-yellow.

**A.** — cv. Cristatum. – Monstrous cultivar.

**A.** × **tabulicum** BRAMW. et ROWL. – Canary Is.: Tenerife. – Acc. PRAEGER a spontaneous hybrid: **A. tabuliforme** × **A. urbicum.** – Ros. almost as flat and dense as in **A. tabuliforme**, but **L.** shining dark green with a trace of red in the margin, like **A. urbicum**; ciliation intermediate between the parents.

**A.** × **tenensis** BRAMW. et ROWL. – Canary Is.: Tenerife. – Acc. PRAEGER a spontaneous hybrid: **A. ciliatum** × **A. haworthii.** – Shrubby, to 15 cm or more high; **L.** intermediate between those of the parents, green, red-tipped, 5–8 cm long, 2.5 cm wide; **Infl.** as in **A. haworthii**; **Fl.** white with a greenish tinge.

**A.** × **teneriffae** BRAMW. et ROWL. – Canary Is.: Tenerife; hybrid swarms abundant where the parents grow together. – Acc. PRAEGER a spontaneous hybrid: **A. ciliatum** × **A. urbicum.** – Intermediate between the parents.

**A. tortuosum** (AIT.) BGR. (§ VI/1) (*Sempervivum t.* AIT., *Aichryson t.* PRAEG., *S. pygmaeum* WEBB et BERTH., *S. radicescens* LOWE, *Aichryson pulvinatum* BURCH., *S. villosum* HAW.). – Canary Is.: Lanzarote, Fuerteventura. – Small, cushion-shaped; **Br.** wide-spreading; **Ros.** small; **L.** 10–15 mm long, 5–7 mm across, 4 mm thick, light green, often brownish, densely haired; **Fl.** golden-yellow.

**A. undulatum** WEBB et BERTH. (Pl. 5/5) (§ IV) (*Sempervivum u.* WEBB, *S. youngianum* BOURG.). – Canary Is.: Gran Canaria. – ♄, sturdy, thick-stemmed, little branching, to 1 m tall, silver-grey, L.-scars red-brown; **Ros.** to 30 cm ⌀; **L.** very broad at tip, almost spoon-shaped, margin in upper part bordered and wavy, finely ciliate, glossy dark green; **Infl.** broad-pyramidal, **Fl.** very dark yellow.

**A. urbicum** (C. SM.) (WEBB et BERTH.) (§ V) (*Sempervivum u.* C. SM.). – Canary Is.: Tenerife. – **St.** simple, little branching, producing aerial roots; **Br.** mostly arising from ground-level, bent upwards; **Ros.** 20–25 cm ⌀; **L.** elongate-spatulate, ± sharply pointed, margins reddish to violet-purple; **Fl.** greenish-white or pink-white.

**A. valverdense** PRAEG. (§ V) (*Sempervivum v.* PRAEG.). – Canary Is.: Hierro. – (Perhaps nat. hybrid). – **St.** 60–90 cm tall, branched in whorls; **Ros.** 20–25 cm ⌀; **L.** obovate-spatulate, with a recurved point at tip, 10–12 cm long, 5–6 cm across, 8 mm thick, ashy-grey, tip and margins reddened, with short, fine hairs, cilia large and projecting; **Infl.** broad-conical, with fine white hairs, Fl. white.

**A.** × **vegamorai** BRAMW. et ROWL. – Canary Is.: Gomera. – Acc. PRAEGER a spontaneous hybrid: **A. subplanum** × **A. viscatum.** – Half the size of **A. subplanum**; Ros. dense, flattish, to 15 cm ⌀; L. spatulate, 7 cm long, 2.5 cm broad, soft, fleshy, sticky, glandular-pubescent, dark green, bluntly apiculate.

**A.** × **velutinum** PRAEG. ex HILL. (*Sempervivum v.* N. E. BR.). – Artificial hybrid: **A.**

caespitosum × **A. canariense.** – Habit as for **A. canariense**; **L.** downy, more numerous and closer spaced; **Fl.** light yellow.

**A. vestitum** SVENT. (§ IV). – Canary Is.: La Palma. – ♄, branching, erect, 40–60 cm tall; **Br.** standing out horizontally or pointing upwards, rind black; **Ros.** 8–15 cm ⌀, half-closed; **L.** closely spaced, lanceolate-spatulate, rounded at tip, apiculate, almost stalkless, 5–10 cm long, 2–3 cm across, glabrous, some stripes and purple dots towards tip, margins ciliate; **Infl.** a conical Pan., Fl. sulphur-yellow.

**A. virgineum** WEBB ex CHRIST (§ I) (*Sempervivum v.* (WEBB ex CHRIST) CHRIST, *S. canariense* v. v. BURCH.). – Canary Is.: Gran Canaria. – Matforming; **St.** elongated; **Ros.** ± depressed, somewhat funnel-shaped; **L.** obliquely erect, ± spatulate, reddish, with dense, soft, white hairs.

**A. viscatum** BOLLE (§ VI/1) (*Sempervivum v.* (BOLLE) CHRIST). – Canary Is.: Gomera. – ♄ 30 cm tall, freely branching; young **Br.** softly hairy; **L.** with glandular soft hairs, very sticky, lanceolate-spatulate, pointed, fairly long stalked, underside keeled, upper part thickened, convex; **Fl.** yellow.

**A. webbii** BOLLE (§ IV). – Cape Verde Is.: São Vicente. – **St.** short, thick; **Ros.** 30 cm ⌀; **L.** glossy green, thickened at base; **Infl.** tall, Fl. golden-yellow.

**Afrovivella** BGR. Crassulaceae. – Ethiopia. – Summer: planted out, Winter: unheated greenhouse. Propagation: seed, cuttings.

**A. simensis** (HOCHST.) BGR. (*Sempervivum s.* HOCHST., *Cotyledon s.* BR., *Umbilicus semiensis* J. GAY.). – Ethiopia: mountains near Ras Dashan. – ♄, small, succulent; **St.** woody-fleshy, branching, cushion-forming; **L.** in thick Ros., obovate-spatulate, with very short pointed tips, fleshy, glabrous, with a white apiculus, marginal cilia large, white, cartilaginous; **Infl.** laterally from Ros., leafy, Fl. 2–5, light pink.

**Agave** L. – Agavaceae. – Southern N. America; Mexico; Central America; northern S. America; the West Indies; widespread and wild in the Mediterranean. – ± large, succulent or xerophilous ♃, semi-shrubs or ♄; **L.** in Ros., of very firm, hard and ± fleshy texture, lanceolate or sword-shaped, triangular to dagger-shaped or even very narrow, sometimes with disjunctive threads, edge entire or usually armed with large and powerful teeth which are often connected by horny borders, mostly with a stout terminal spine at the tip, grey, green or bluish-green, often even with coloured markings, young L. in conical buds, often bearing the impress of the Sp. of the older L.; **Fl. Sc.** terminal, with leaf-like bracts which become smaller towards the end; **Fl.** numerous, large and fleshy, in spikes or Rac.; adventitious buds often in the axils of the panicle-branches, Per. ± funnel-shaped, usually with a short tube and almost equal Tep., the outer 3 covering the inner 3 at the edges, 6 stamens inserted into the tube, usually longer than the Tep., ovary epigynous, cylindrical to spindle-shaped, with 3 carpels, style elongated with a head-like stigma; **Fr.** a woody, roundish to elongate capsule, bluntly 3-angled, seed flat, blackish. – Greenhouse, warm or cold, in summer also in the open ground, in winter in a coldhouse. Propagation: seeds, offsets, adventitious buds. – (Lit.: A. BERGER, Die Agaven, Jena 1915; A. J. BREITUNG, Cult. a. Nat. Agaves in the U.S.A. in "Cact.Succ.Journ.Am." XXXI–XXXVII, 1959–1965; A. G. POMPA, El Gen. Agave, in "Cact. Succ.Mex." VIII, 1963.)

## Division of the Genus Agave into Subgenera, Sections and Series according to A. Berger

A. **Fl.** in long spikes or racemose Infl.; Tep. when blooming not at once withering.

  a) **Fl.** solitary along the Infl.; plants with a tuberous rootstock.

  **SG. I. Manfreda** (SALISB.) BAK. (*Manfreda* SALISB. as a genus). – Small ♃ with a tuberous or bulbous rootstock; **L.** fleshy, unarmed, sometimes annually deciduous; **Fl.** usually solitary in a loose spike, Tep. united into a ± long tube. – Species: A. alibertii, brachystachys, brunnea, debilis, gracillima, guttata, hauniensis, jaliscana, lata, maculata, maculosa, oliveriana, planifolia, pratensis, pubescens, revoluta, sessiliflora, singuliflora, undulata, variegata, virginica.

  a) **Fl.** usually in twos or up to several together.

  **SG. II. Littaea** (TAGLIAB.) BAK. – (*Littaea* TAGLIAB. as a genus.) – Plants with a woody **St.**; Fl. in an elongated, cylindrical Infl. (1–)2(–8) from the axil of a bract.

  **Sect. 1. Anacamptagave** BGR. – **Ros.** often branched after flowering, forming clumps of St. when old, the L. of the barren Ros. usually broader than the following, barren plants rarely branching, for instance **A. ousselghemiana**; **L.** in close Ros., from hard-fleshy and stiff to soft-fleshy and very flexible, the marginal teeth very variable as well; **Infl.** erect or ascending, with ± triangular, at first scarious, bracts; **Fl.** in spike-shaped Rac., usually two or more together, rarely solitary, to the tip of the Infl., Per. united into a ± long funnel-shaped tube. – Species: A. albicans, aloides, beguinii, botteri, bouchei, carbibaea, celsii, chiapensis, decaisneana, densiflora, ehrenbergii, flaccifolia, goeppertiana, haseloffii, horizontalis, lamprochlora, lindleyi, macrantha, maximowicziana, muilmannii, nizandensis, ousselghemiana, perlucida, polyacantha, sartorii, terracianoi, warelliana, xalapensis, yuccifolia.

**Sect. 2. Xysmagave** BGR. – **Ros.** usually with many L. and making offsets, rarely simple; **L.** ± strap- to rush-shaped, firm-fleshy but not hard, to soft and flexible, edge entire, with horny margin which splits into threads, and sharp, horny, terminal Sp.; **Infl.** slender and strong, with narrow-triangular, long tapering bracts; **Fl.** usually two together, on short, thick Ped., in loose or not very close Rac., Per. united into a ± long tube. – Species: A. angustissima, colimana, filifera, geminiflora, × leopoldii, parviflora, × romanii, schidigera, schottii, × taylori, toumeyana, villarum, wrightii.

**Sect. 3. Schoenoagave** BGR. – **Ros.** usually with numerous L., branching after flowering and forming short St. with many heads; **L.** narrow, rigid and stout, triangular or rhombic in transverse section, closely traversed by fine grey and green longitudinal nerves, edged with fine cartilaginous margins and rough with fine but sharp teeth, with a stout, pungent terminal Sp.; **Sc.** slender, green or brown-green, with numerous, pointed, white, soft and papery bracts; **Fl.** very short stalked in spike-shaped clusters, usually 2 together, solitary towards the tip, Per. united into a long funnel-shaped tube. – Species: A. echinoides, falcata, striata, stricta.

**Sect. 4. Chonanthagave** BGR. – **Ros.** with numerous, densely packed L., branching after flowering; **L.** strap-shaped or narrowly sword-shaped, stout, striped, similar to the Schoenoagaves, and also having fine cartilaginous T.; **Infl.** a pendent long spike, with numerous, dry-skinned, subulate bracts and similar floral bracteoles; **Per.** bell- to funnel-shaped. – Species: A. dasylirioides.

**Sect. 5. Pericamptagave** BGR. – **L.** in a ± stemless Ros., ± sword-shaped to elongate-spatulate or triangular, tough-leathery, usually with a continuing or even interrupted, woody or horny edge and mostly stout Sp., rarely entire, often with a pale or yellowish central band, especially in young plants, often even on the back surface with peculiar, dark green, fine stripes; **Infl.** slender, with narrow, empty bracts, **Fl.** in spike-shaped arrangements, 2 or more together, Per. only quite shortly united at the base, hardly forming a tube or only a very flat one. – Species: A. arizonica, aspera, expatriata, ferdinandi-regis, funkiana, glomeruliflora, hanburyi, haynaldii, horrida, kerchovei, lechuguilla, maigretiana, noli-tangere, obscura, pumila, purpusorum, roezliana, 'Simonii', splendens, triangularis, univittata, utahensis, victoria-reginae, vittata, washingtonensis, xylonacantha.

**Sect. 6. Brachysolenagave** BGR. – **Ros.** stemless; **L.** toughleathery-fleshy, with a continuous horny margin and ± large T. and a stout terminal Sp., younger L. often with a pale central ribbon; **Fl.** in a spike-shaped cluster, 2 or more together, greenish or brownish, Per. united into a tube 5–10 mm long, funnel-shaped at the base. – Species: A. ghiesbreghtii, leguayana, muelleriana, pavoliniana, peacockii, rohanii.

**Sect. 7. Anoplagave** BGR. – **Ros.** stemless or forming St. which are smooth, with a smooth rind, set with the scars of the cleanly falling L., sometimes shooting out of the old axils of the L. and making roots; **L.** soft-fleshy, fibrous, flexible with a soft tip (only rarely with a terminal Sp.), margins smooth or very finely cartilaginous-dentate, green or usually grey and pruinose; **Infl.** with numerous, narrow-triangular, subulate bracts, **Fl.** (2–)4(–8) together on thin, slender stalks, very numerous, in close, erect or pendent racemose spikes, Per. united only at the base and thus forming hardly any tube or only a short one, straight, spreading or broad bell-shaped or funnel-shaped, pale yellowish or greenish.

**SSect. a. Dracontagave** BGR. – Per. with a short tube. Forming **St.**; **L.** elliptical-lanceolate, marked along the nerves as if with stripes of grey. – Species: A. attenuata, cernua.

**SSect. b. Leptagave** BGR. – Per. without tube; **stemless** species; the grey colour of the L. not in stripes along the nerves. – Species: A. bracteosa, chrysoglossa, ellemeetiana, guiengola.

**Status unknown. c.** – L. with terminal Sp. – Species: A. eduardii, vilmoriniana.

**AA. Fl.** usually in paniculate, more rarely almost spike-shaped contracted Infl.; Tep. when blooming at once fading from the tip.

**SG. III. Agave** (*Euagave* BAK.). – **Ros.** of L. usually large, stemless or forming St., often making offsets, but always dying after flowering and not growing again by division of the Ros.; **L.** rough or soft-fleshy, usually thornily armed and with a stout terminal Sp.; **Infl.** usually much higher than the rosette, stout, with empty bracts; **Fl.** ± clustered to umbellate on short stalks and soondrying up, with small, triangular floral bracts, usually numerous together, erect, on ± large, spreading Br. in a spike-shaped, elongate or pyramidal Pan., the branchjets at the tip repeatedly ternately or umbellately branched, Tep. thick and fleshy, united into a ± long campanulate-globular tube at the base.

**Series 1. Salmianae** BGR. – **Ros.** very large, with large fleshy **L.**, marginal Sp. stout, terminal Sp. ± long decurrent; **Sc.** much longer than the Pan., very tall and extraordinarily thick and stout, with apressed bracts long remaining fresh, Pan. tall, pyramidal, with spreading or sinuately curved Br., **Fl.** very large, yellow, in close tufts or on repeatedly ternately branched twigs, Tep. very fleshy. – Species: A. asperrima, atrovirens, brevipetala, brevispina, cinerascens, cochlearis, crassispina, ekmanii, ferox, gracilis, gracilispina, masipaga, mitriformis, quiotifera, tecta.

Series 2. **Americanae** BAK. – Similar to the preceding order. **L.** usually numerous, not quite as thick-fleshy, the terminal Sp. less decurrent. **Ros.** with large and fleshy L., marginal Sp. stout; **Sc.** stout, as long as or shorter than the Pan., with numerous and smaller triangular bracts which have broad sheathing bases, suddenly and sharply tapering, soon drying up; **Pan.** large, elongate-pyramidal, with spreading and somewhat sinuately curved Br., the lower Br. shorter than the middle ones, all repeatedly ternately branched, **Per.** large, greenish-yellow or yellow, Tep. fleshy. – Species: A. americana, bourgaei, cyanophylla, deflexispina, expansa, franzosinii, ingens, massiliensis, mirabilis, mortolensis, palmeri, picta, schlechtendahlii, × winteriana.

Series 3. **Gloriosae** BGR. – **Ros.** large; **L.** lanceolate, spinous; Fl. **Sc.** stout, Pan. very large and freely branched, **Fl.** in umbels, short stalked, Per. large, with long tube, greenish. – Species: A. fuerstenbergii.

Series 4. **Campaniflorae** TREL. – **Ros.** stemless, rather large; **L.** fleshy, lanceolate, marginal Sp. rather large, terminal Sp. little decurrent; **Fl.-Sc.** stout, as long as the Pan. or shorter, with triangular, soon recurved bracts, similar to the Americanae, Pan. elongate-pyramidal, the Br. ± horizontally spreading, several times ternately branched, the lower twigs shorter, **Per.** rather large, with a broad-campanulate tube, golden yellow. – Species: A. aurea, promontorii.

Series 5. **Umbelliflorae** TREL. – **Ros.** spherical or elliptical, densely and many leaved; **L.** hard-fleshy, stiff, straight, usually short and broad, with a stout, flatly grooved, decurrent terminal Sp. and large, sometimes confluent marginal Sp.; Fl.-Sc. very stout, two to three times longer than the Pan., densely covered with triangular, adpressed bracts, Pan. short, with stiff, erect or almost horizontal Br., only at the tip short-branched. – Species: A. goldmaniana, sebastiana, shawii.

Series 6. **Applanatae** TREL. – **Ros.** simple, but often forming clumps, medium large or small, ± spherical; **L.** numerous, hard-fleshy, usually short, rarely linear-triangular, ± rigidly spreading, smooth, grey or white, with a stout, straight or little curved, subulate or flat terminal Sp., long decurrent at the edges, marginal Sp. rather large, distant; **Fl.-Sc.** stout, with numerous broad-triangular bracts which soon dry and spread, Pan. elongate, almost as long as the Sc. or even longer, with slightly curved twigs, spreading upwards or horizontal, ternately branched at the tip, **Per.** medium large or rather large, yellow or greenish-yellow. – Species: A. applanata, chihuahuana, demeesteriana, flexispina, gracilipes, havardiana, parrasana, parryi, patonii, scabra, vandervinnenii.

Series 7. **Scolymoides** BGR. – **Ros.** small to medium large, rarely large, with ± numerous ± spatulate-lanceolate **L.**, with a broadly furrowed terminal Sp., usually somewhat twisted and a little decurrent; **Fl.-Sc.** with triangular bracts which soon dry and fade, Pan. elongate to clustered with very shortened Br., **Fl.** of intermediate size.

SSeries 1. **Multiflorae** BGR. – Marginal teeth of the L. very fine; **Fl.** tufted on short stalks in almost spike-shaped or racemose Infl., Tep. as long as the tube or only a little longer. – –Species: A. engelmannii multiflora.

SSeries 2. **Euscolymoides** BGR. – **Ros.** small to intermediate; marginal teeth of the **L.** large; **Fl.** in Pan. or Infl. racemose and ± umbellate on the Br., Tep. as long as the tube or up to twice as long. – Species: A. baxteri, cucullata, erosa, galeottei, grandibracteata, littaeoides, megalacantha, pampaniniana, potatorum, saundersii, × simonii, viridissima, weissenburgensis.

SSeries 3. **Crenatae** BGR. – Plants large; marginal T. of the **L.** big; **Fl.** in Pan., Per. up to three times as long as the tube. – Species: A. colorata, conjuncta, crenata, cupreata, fenzliana, latissima, longisepala, mescal, wocomahii.

SSeries 4. **Guatemalenses** BGR. – Plants rather large, still insufficently known and uncertain in position. (From Guatemala.) – Species: A. carolischmidtii, eichlamii, guatemalensis, kellermannii, pachycentra, samala, seemanniana, tenuispina.

SSeries 5. **Costaricenses** BGR. – Plant rather large; **L.** grey; **Fl.** of medium size, yellow, Tep. twice as long as the tube. (From Costa Rica.) – Species: A. werklei.

Series 8. **Bahamanae** TREL. – **Ros.** stemless, simple, rather large; **L.** numerous, fleshy, dull grey, with a long, furrowed, parchmenty, decurrent terminal Sp. and small or larger, rather distant marginal Sp.; **Fl.-Sc.** rather tall, with triangular bracts, **Fl.** rather large, light yellow, rather loose in large ovate Pan. with subhorizontally spreading Br., sometimes with adventitious buds. – Species: A. acklinicola, bahamana, braceana, cacozela, indagatorum, millspaughii.

**Series 9. Antillanae** TREL. – **Ros.** large or very large, stemless, without offsets; **L.** numerous, fleshy, green, ± glossy, terminal Sp. long, furrowed, decurrent, marginal Sp. rather large, distant; **Fl.-Sc.** stout, panicles elongated, often bearing adventitious buds, Fl. often umbellately crowded, large, yellow or orange. – Species: A. acicularis, anomala, antillarum, grisea, harrisii, intermixta, legrelliana, longipes, missionum, morrisii, portoricensis, shaferi, sobolifera, underwoodii.

**Series 10. Caribaeae** TREL. – **Ros.** large, not making offsets, usually stemless; **L.** numerous, fleshy, green, terminal Sp. short, passing over into the subulate rolled and hardened tip, usually deeply projecting into the L.-tissue on the back side, marginal Sp. small to very small; **Infl.** stout, with triangular, rather crowded bracts, Pan. tall, elongate, with ascending branches, eventually with numerous adventitious plantlets; **Fl.** medium large to large, yellow or golden yellow. – Species: A. barbadensis, caribaeicola, dussiana, eggersiana, grenadina, karatto, kewensis, martiana, medioxima, montserratensis, nevidis, obducta, scheuermaniana, trankeera, unguiculata, vangrolae, ventum-versa.

**Series 11. Columbianae** BGR. – **Ros.** stemless, without offsets; **L.** ± lanceolate, large, marginal Sp. small; **Infl.** paniculate. – Not sufficiently known. – Species: A. cundinamarcensis, wallisii.

**Series 12. Viviparae** TREL. – **Ros.** small or medium large, somewhat caespitose; **L.** rather numerous, fleshy, smooth, green or pale grey-green, usually with a long, slender, furrowed or incurved terminal Sp. and medium large distant marginal Sp.; **Infl.** stout, Pan. medium large, elongate, **Fl.** medium large. – Species: A. boldinghiana, cocui, evadens, petiolata, vicina, vivipara.

**Series 13. Rigidae** BGR. – **Ros.** stemless or forming a St., medium large to large; **L.** ± linear-lanceolate or sword-shaped, with stout, conical or subulate, distinct or decurrent terminal Sp. and usually medium large to small or no marginal Sp.; **Fl.-Sc.** proportionately slender, with rather distant and small triangular bracts, Pan. elongate, rather loose, with ternately divided Br., usually with adventitious buds, **Fl.** medium large to large, usually greenish, sometimes with brownish markings.

**SSeries 1. Sisalanae** TREL. – Terminal **Sp.** sharply set off, not decurrent. – Species: A. ? amaniensis, angustifolia, brauniana, bromeliaefolia, cantala, decipiens, desmetiana, fourcroydes, humboldtiana, ixtli, neglecta, sisalana, smithiana, thomsoniana, vera-cruz, zapupe.

**SSeries 2. Tequilanae** TREL. – Terminal **Sp.** ± decurrent at the margins. – Species: A. aboriginum, bergeri, collina, deweyana, elongata, endlichiana, karwinskii, kirchneriana, lespinassei, macroacantha, murpheyi, pacifica, pseudotequilana, pugioniformis, rubescens, schneideriana, tequilana, weberi.

**Series 14. Datyliones** TREL. – **Ros.** stemless; **L.** fibrously fleshy, stiff, straight, dagger-like, terminal Sp. stout, decurrent at the base, marginal Sp. broad-triangular; **Fl.-Sc.** slender, panicle narrow-elongate, **Fl.** medium large to small, greenish. – Species: A. datylio.

**Series 15. Deserticolae** TREL. – **Ros.** stemless, often laterally shooting and forming mats; **L.** fibrously fleshy and straight or sickle-shaped, narrow- or broad-lanceolate, terminal Sp. slender, narrowly furrowed, decurrent, marginal Sp. medium large up to large, sometimes rather fragile; **Fl.-Sc.** slender, with rather distant, ± ascending, triangular bracts, Pan. short, with rather simple, slender, ascending Br., **Fl.** small or medium large, with a short tube. – Species: A. desertii, mckelvyana, margaritae, scobria.

**Series 16. Inaguenses** TREL. – **Ros.** small, stemless, making offsets, with a few L. only; **L.** erect, stiff, grey or very grey, ± lanceolate, with a long, narrowly furrowed, somewhat decurrent terminal Sp. and small marginal Sp. standing close together; **Fl.-Sc.** slender, Pan. loose, without adventitious buds, **Fl.** rather small, flat-umbellate, crowded, yellow. – Species: A. inaguensis, nashii.

**Series 17. Marmoratae** BGR. – **Ros.** large; **L.** lanceolate, very stout and coarsely dentate, extremely rough, with a short terminal Sp.; **Fl.-Sc.** rather slender, with distant, small, triangular bracts, Pan. elongate with ternate branches, **Fl.** small, yellow, Tep. united at the base into a very short and wide tube. – Species: A. marmorata.

**Series 18. Antillares** BGR. – **Ros.** medium large; **L.** lanceolate, soft-fleshy, shining grass-green to whitish-grey, with a small terminal Sp. and horny, pointed marginal teeth; **Pan.** with umbellate or capitate Fl.-clusters at the ends of the Br., **Fl.** small, yellow, with a short tube. – Species: A. albescens, brittoniana, papyrocarpa, tubulata, willdingii.

**A. aboriginum** TREL. (§ III/13/2). – Mexico. – Cultivated plant, used for its fibres. – Acauline, producing offsets; **Ros.** dense; **L.** long-pointed, 70–150 cm long, 5–11 cm across, rather fleshy, grey, very slightly striped, terminal Sp. 3.5 to 5 cm long, 4 mm thick, marginal Sp. bent forward hooklike, 5–8 mm long and broad, 2–2.5 cm apart, often with an intermediate Sp.; Fl. unknown.

**A. acicularis** TREL. (§ III/9). – Cuba. – **L.** lanceolate, c. 1 m long, 12 cm across, slightly grey, dull, terminal Sp. 25 mm long, 4 mm across, marginal Sp. brown, 10–15 mm apart, 2–3 mm long, bent slightly forwards, Sp. below middle recurved, 4 mm long; **Infl.** a panicle, Fl. 4 to 4.5 cm long, yellow.

**A. acklinicola** TREL. (§ III/8). – Southern Bahamas. – Habit as for **A. bahamana**; terminal Sp. red-brown, later grey, c. 25 mm long, 4–6 mm across, marginal Sp. 5–10 mm apart, 1–1.5 mm long.

**A. albescens** TREL. (§ III/18). – Cuba. – **Ros.** acauline, simple; **L.** white-grey, pruinose, some horizontal striping, rather rough, oblong-oblanceolate, slightly folded, c. 45 cm long, 15 cm across, terminal Sp. blackish-brown, 15 mm long, 4 mm across, marginal Sp. c. 1 cm apart, 2–3 mm long, broad-triangular; **Infl.** a Pan. 5 m tall, Fl. 3–3.5 cm long, goldenyellow.

**A. albicans** JACOBI (§ II/1) (*A. micrantha* v. *albidor* SALM., *A. mitis* v. *albicans* TERR.). – Mexico. – **St.** very short, with several **Ros.** of 20–30 L., broader than tall; **L.** c. 30 cm long or more, spatulate, 10 cm across above the middle, rather short-pointed, the inrolled edges uniting to form a terminal Sp., margins often sinuate, softly fleshy, flexible, easily broken off, pale greenish or white, terminal Sp. soft, bristle-shaped, margins with narrow cartilaginous edge and T. similar; **Infl.** c. 1 m tall, Fl. 4 cm long, green. (Acc. TRELEASE, synonymous with **A. micrantha.**)

**A. alibertii** BAK. (§ I) (*Alibertia intermedia* MARION). – Mexico. – ♃; **Root**stock with few leaflike scales; **L.** 10–12 in the Ros., 20–25 cm long, lanceolate, margins finely toothed; Fl. Sc. over 1 m tall, **Fl.** about 2.5 cm long.

**A. aloides** JACOBI (§ II/1). – Closely related to **A. mitis**; short-stemmed; **L.** c. 27 cm long, in middle 9 cm across, lower part 4.5 cm across, underside angular-keeled, soft-fleshy, pale grey, slightly pruinose, terminal Sp. strong, dark brown, marginal Sp. fairly widely spaced, short, brown, margins between Sp. sinuate.

**A. amaniensis** TREL. et NOWELL (? § III/13/1). – Origin unknown, in cultivation in Tanzania: Amani, Afr. Agr. Res. Sta. – Acauline; **L.** ensiform, rigid, leathery, yellow-green, c. 1.5 m long, 16 cm across at base, narrowing to 10 cm and then widening to 15 cm, margins towards tip curving in, entire or with some rudimentary Sp. towards base, with a translucent marginal line, terminal Sp. brown, twisted, 15 mm long; **Fl.** unknown. Plant used for its fibres.

**A. americana** L. (§ III/2) (*A. altissima* ZUMAGL., *A. vera-cruz* DRUM. et PRAIN.). – Mexico; Eur., coasts of S. and N.Afr., Atlantic islands and peninsular India, partly naturalized. – **Ros.** acauline, producing many offsets, in age 2–3 m ⌀; **L.** 20–30, stiffly projecting, pendent in upper part, long-lanceolate, to 1.75 m long, at base 30 cm across, gradually narrowing towards inrolled tip, terminal Sp. 3 cm long margins crenate-dentate, terminal Sp. 8–10 mm long, bent, black-brown; **Infl.** 5–8 m tall, 25–30 Br., Fl. 9 cm long.

**A.** — cv. Marginata (*A. a. v. m.* TREL.). – **L.** margins yellowish-white to deep golden-yellow.

**A.** — cv. Marginata Alba (*A. a. v. m. a.* TREL.). – **L.** margins cream-coloured, younger L. often pink-tinged.

**A.** — cv. Marginata Aurea (*A. a. v. m. a.* TREL.). – **L.** margins light yellow to greenish-yellow.

**A.** — cv. Marginata Pallida (*A. a. v. m. p.* BGR., *A. celsiana* HORT. ex BAK.). – **L.** margins pale green.

**A.** — cv. Medio-picta (*A. a. v. m. p.* TREL.). – **L.** with central yellowish stripe.

**A.** — cv. Medio-picta Alba. – **L.** with central white stripe.

**A.** — cv. Striata (*A. a. v. s.* TREL.). – **L.** with narrow central stripes, cream or yellow.

**A. ananassoides** JACOBI. – Inadequately described species; **L.** light pallid-green, underside white-green.

**A. angustifolia** HAW. v. **angustifolia** (§ III/13/1) (? *A. lurida* β AIT., *A. l.* JACQ., *A. jaquiniana* SCHULT. *A. l.* v. *j.* SALM, *A. veracruz* v. *j.* ASCHERS., *A. vivipara* WIGHT, ? *A. ixtlioides* HOOK., *A. flavovirens* JACOBI, *A. excelsa* JACOBI, *A. e.* BAK., *A. ixtli* v. *e.* TERR., *A. rigida* DE SPIN., *A. wightii* DRUMM. et PRAIN). – Origin unknown, cultivated plant of trop. zones. – **St.** 10–40 cm tall, offsetting from base; **Ros.** dense, ± globular, about 1 m ⌀; **L.** stiff, rigid, leathery, smooth, light green, ensiform, linear-lanceolate, 50–90 cm long, 7.5 cm across, coming gradually to a point, at base 4 cm across, terminal Sp. subulate, 18 mm long, marginal Sp. 15–20 mm apart, to 5 mm long, with curved tip; **Infl.** c. 2.65 m long, Fl. c. 5 cm long.

**A.** — cv. Marginata (*A. a. v. m.* TREL.). – **L.** margins white mottled.

**A.** — cv. Marginata Woodrowi (*A. vivipara* v. *w.* HORT. ex WATS., *A. a. v. w.* (WATS.) TREL.). – **L.** margins white.

**A.** — v. **sargentii** TREL. – Not offsetting; **L.** 25 cm long, 2.5 cm across, both surfaces convex; **Infl.** 1 m tall, Fl. c. 4 cm long.

**A. angustissima** ENGELM. (§ II/2). – W.Mexico – **Ros.** acauline; **L.** c. 15–20, linear, 25 to 45 cm long, 8–10 cm across, margins with a few fibres splitting off, terminal Sp. 3angled, brown; **Infl.** almost 2 m tall, Fl. yellow.

**A. anomala** TREL. (§ III/9). – E.Cuba. – **L.** lanceolate, elongated, gradually tapering, roundish-furrowed up to the middle; margins straight, with only a few, very short T. towards the base; **Infl.** a panicle, Fl. 5.5–7 cm long, yellow.

**A. antillarum** DESC. v. **antillarum** (§ III/9) (*A. vivipara* LAM. p.part., *A. sobolifera* SALM p.part., *A. americana* URBAN). – Haiti. – **Ros.** acauline, simple; **L.** lanceolate, gradually tapering, somewhat concave, about 1 m long, 8 cm across, light green, terminal Sp. brown, 1.5–2 cm long, 3 mm thick, marginal Sp. 2 to 3 mm long, 10–25 mm apart; **Infl.** to 5 m (?) tall, Fl. 4–5 cm long, orange-coloured.

**A.** — v. **grammontensis** TREL. – Haiti. – **L.** bluish, marginal Sp. 5 mm long, 5–10 mm apart; **Fl.** scarlet-red inside.

**A. applanata** C. KOCH (§ III/6) (*A. schnittspahnii* JACOBI). – S.E.Mexico. – **Ros.** very dense, making scarcely any offsets; **L.** numerous, rigid, to 1 m long, broadly linear-lanceolate, 10 cm across, tapering gradually, margins horny, terminal Sp. 6–7 cm long, marginal Sp. with tip bent at right-angles; **Infl.** 9–12 m tall, Fl. numerous, about 7 cm long, greenish-yellow.

**A. arizonica** GENTRY et WEBER (§ II/5). – N.Am.: Arizona, New River Mts. – **Ros.** small, acauline, simple or mat-forming, c. 30 cm tall, 40 cm ⌀; **L.** numerous, forming a tight dome, 17–24 cm long, in middle 2–4 cm across, linear-lanceolate, pointed, stiff, fleshy, dark green, fibrous, smooth, upperside concave, underside thick-convex; terminal Sp. 1–2.5 cm long, subulate, upperside grooved, darkbrown, later grey, merging into margins, these being 1–2 mm across, T. ± bent downwards, to 2.5 mm long, 1.5–2 cm apart; **Infl.** 3–4 m tall, racemose-paniculate, Fl. 25–32 mm long, pale yellow.

**A. aspera** TERR. (§ II/5). – Obviously within the **A. xylonacantha** complex.

**A. asperrima** JACOBI (§ III/1) (*A. caeciliana* BGR., *A. a.* J. MULF.). – USA: Texas; Mexico. – **Ros.** with few L., 1–1.5 m ⌀ or even larger, acauline, forming offsets; **L.** 60–75 cm long, (8–)10(–20) cm across, lanceolate, sharp-tipped or pointed, to deeply concave in middle, rolled inwards towards tip, ash-grey or grey-green, very rough, terminal Sp. thin, 3–3.5 cm long, slightly sinuate, marginal Sp. 7–12 mm long, 2–3 cm apart, on fleshy cushions; **Infl.** 4–5 m tall, Fl. 7.5 cm long, light yellow.

**A. atrovirens** KARW. v. **atrovirens** (§ III/1) (*A. jacobiana* SALM, *A. a.* v. *stigmatophylla* BGR., *A. lehmannii* JACOBI, *A. ottonis* JACOBI, *A. tehuacanensis* KUNTH). – Mexico: Oaxaca, Puebla. – **Ros.** (3–)6(–8) m ⌀, offsetting; **L.** 2.3–4.5 m long, 20–30 cm across, lanceolate to oblanceolate, upperside very concave, dark green, smooth, terminal Sp. 4–10 cm long, thin, marginal Sp. (5–)7(–10) cm long, up to 10 cm apart; **Infl.** 8–10 m tall, Fl. 10 cm long, greenish-yellow.

**A.** — v. **cochlearis** (JACOBI) TREL. (*A. c.* JACOBI, *A. a.* E. NEUB.). – **L.** broader.

**A.** — v. **salmiana** (OTTO ex SALM) TREL. (*A. s.* OTTO ex SALM, *A. s.* v. *recurvata* JACOBI, *A. s.* v. *angustifolia* BGR., *A. s.* v. *glauca* BECKER, *A. dyckii* HORT. ex BESAUC., *A. whitackeri* HORT.). – **L.** grey, rather rough.

**A. attenuata** SALM (Pl. 6/3) (§ II/7a) (*A. glaucescens* HOOK.). – Mexico: Hidalgo. – **St.** to 1 m tall, 8–10 cm ⌀, branching from base; **Ros.** of 6–15 L. 70 cm long, elliptical, 20–24 cm across, narrowing abruptly to base, tip soft, margins somewhat curved upwards, smooth, green, light grey or almost whitish-coated, margins entire; **Infl.** c. 1.50 m tall, with many adventitious plantlets. (J. A. BREITUNG considers this synonymous with **A. cernua**.)

**A. aurantiaca** HORT. is perhaps a smaller, spiny form of **A. fenzliana** JACOBI.

**A. aurea** BRANDEG. (§ III/4) (*A. campaniflora* TREL., *A. promontori* TREL.). – Mexico: Baja Calif. – Resembles **A. americana**; L. narrower, more deeply grooved, margins reddened; **Ros.** sessile, c. 1.5 m ⌀, of some 50 narrow-lanceolate, pointed **L.**, 70–75 cm long, 4 cm thick at base, 6 cm across, towards upper part up to 10 cm across, terminal Sp. about 3 cm long, sharp, subulate, marginal Sp. 2–4 mm long, 9–25 mm apart; **Infl.** 3.50 m tall, Fl. very numerous, 6 cm long, golden-yellow.

**A. bahamana** TREL. (§ III/9). – Bahamas. – **Ros.** acauline, simple; with very many dull green **L.**, narrow-lanceolate, sometimes rather folded, narrowing gradually to a point, terminal Sp. brownish, broad-conical, 10–15 mm long, marginal Sp. usually 5–10 mm apart, adpressed-recurved, triangular; **Infl.** c. 10 m tall, Fl. 3–3.5 cm long, golden-yellow.

**A. barbadensis** TREL. (§ III/10) (*A. americana* DILL., *A. vivipara* SALM). – Lesser Antilles: Barbados. – **Ros.** acauline, occasionally off-setting; **L.** broad-lanceolate, rather short-pointed, spatulate and plicate towards the tip, underside keeled, 1.50–2 m long, 25–30 cm across, dull dark green, terminal Sp. 10–15 mm long, 7 mm thick, marginal Sp. 10–12 mm apart, 2–3 mm long, margins initially reddish; **Infl.** 5–6 m tall, with numerous adventitious plantlets.

**A. baxteri** BAK. (§ III/7/2). – ? Mexico. – Acauline; **L.** c. 30, oblanceolate-spatulate, 7.5 cm across above middle, at base 5 cm across, rather grey-coloured, terminal Sp. c. 2.5 cm long, marginal Sp. 5–15 mm apart, with hooked tips, brown; **Infl.** c. 1.20 m tall, paniculate with Fl. in umbel-like clusters, 5.5 cm long.

**A.** × **beguinii** BGR. (§ II/1). – Hybrid from La Mortola. – **Ros.** initially simple, later mat-forming; **L.** long-lanceolate, 50–60 cm long, pointed, terminal Sp. 10–11 mm long, black-brown, margins with a line 1–1.5 mm across, finely saw-toothed; **Infl.** c. 2 m tall, Fl. 2.4 cm long, brown-red.

**A. bergeri** TREL. (§ III/13/2). – ? Mexico. – **Ros.** acauline, c. 1 m tall, 1.5 cm ⌀, **L.** 60–80, terminal bud bent somewhat to one side; **L.** 75–88 cm long, in middle 7–8 cm across, tapering gradually to a point, upperside grooved, at base up to 2.5 cm thick, leathery, grey-green, somewhat pruinose, somewhat dull grey striped, terminal Sp. 23 mm long, triangular, pointed, marginal Sp. 15–20 mm apart, slender, curved; **Infl.** 3.30 m tall, Fl. 7 cm long, green-yellow.

**A. bernhardii** JACOBI (§ II/1). – ? Mexico. – **Ros.** depressed-hemispherical; **L.** few 18 cm long, above soil 5.5 cm and in middle 7 cm across, almost elliptical-lanceolate, grooved towards tip, underside very keel-shaped convex, at base 2.5 cm thick, rigid, both surfaces glossy, light rich green, terminal Sp. 4 mm long, margins toothed, old L. spineless.

**A. boldinghiana** TREL. (§ III/12). – Caribbean Is. – Acauline or nearly so, offsetting; **L.** fairly numerous, sinuate, spreading, narrow-lanceolate, narrowing to a point above middle, underside rounded, somewhat folded towards tip, 1 m long, 15 cm across, grey-green, terminal Sp. subulate, grooved, 2–3 cm long, marginal Sp. 10–15 mm apart, 4–7 mm long, curved; **Infl.** c. 5 m tall, Fl. 4.5 cm long, golden-yellow.

**A. botteri** BAK. (§ II/1). – Mexico. – **Ros.** acauline, with some 50 long-spatulate **L.**, about 60 cm long, 15–20 cm across, fleshy, pallid green, terminal Sp. 1.5 cm long, marginal Sp. hooked, 3–4 mm long; **Infl.** 2.50 m long, Fl. 5 cm long, funnel-shaped, greenish-yellow.

**A. bouchei** Jacobi (§ II/1). – Mexico. – **St.** with several heads, to 70 cm tall; **L.** c. 20, lanceolate, pointed, 50–60 cm long, very fleshy at base, 2–2.5 cm thick and 6 cm across, narrowing in upper part, grooved, light green, soft, smooth, somewhat glossy, terminal Sp. 10 to 15 mm long, pointed, marginal Sp. very closely set, small, hooked, 1–1.5 mm long; **Infl.** 1.80 to 2 m tall, Fl. c. 5 cm long, light green.

**A. bourgaei** Trel. (§ III/2). – Mexico. – **Ros.** c. 3 m ⌀, acauline, freely offsetting; **L.** 1.5 m long, 10–15 cm across, fairly thin and flexible, oblanceolate, pointed, lightly tranversely banded, smooth or a little rough, terminal Sp. 3 cm long, marginal Sp. 2–3 mm long, usually 1 cm apart, triangular; **Infl.** 3–3.5 m tall, Fl. 7 to 7.5 cm long, yellowish.

**A. braceana** Trel. (§ III/8). – Bahamas. – **Ros.** acauline, simple; **L.** broad-lanceolate, almost flat, c. 65 cm long, 19 cm across, grey, terminal Sp. 10–15 mm long, marginal Sp. 5–10 mm apart, 2–3 mm long, triangular, curved; **Infl.** c. 7 m tall, Fl. 4–4.5 cm long, golden-yellow.

**A. brachystachys** Cav. (§ I) (*Manfreda b.* (Cav.) Rose, *A. spicata* DC., *A. polyacanthoides* Cham. et Schltr., *A. saponaria* Lindl., *A. humilis* Roem.). – Mexico to Guatemala. – ♃; thick rhizome; **Ros.** of 12–15 linear-lanceolate **L.** 25–35 cm long, 3–4 cm across at base, somewhat grooved, green to light grey-green, margins paler, indistinctly or not all toothed; **Infl.** 1 m tall, Fl. numerous. (Juice of roots used for soap.)

**A. bracteosa** S. Wats. (Pl. 7/2) (§ II 7/b). – N.E.Mexico. – **Ros.** acauline, about 60–80 cm ⌀, offsetting; **L.** recurved at tip, 35–50 cm long, to 4 cm across at base, narrowing gradually to a slender tip, no terminal Sp., upper side with raised longitudinal nerves, underside somewhat curled inwards, somewhat obliquely keeled, flexible, grey-green or dull green, margins very sharp, with fine T.; **Infl.** c. 2 m tall, unbranched, Fl. very numerous, stamens about 5 cm long.

**A. brauniana** Jacobi (§ III/13/1). – Mexico: San Luis Potosi. – **Ros.** short-stemmed, almost hemispherical; **L.** fairly numerous, recurved at tip, fleshy up to midway, then fibrous-leathery, dull ashy to light green, linear-oblanceolate, narrowing gradually to a point, terminal Sp. thin, marginal Sp. close to one another, triangular, tip finely pointed, curved upwards, cartilaginous. (Trelease includes this species in **A. celsii.**)

**A. brevipetala** Trel. (§ III/2). – Haiti. – Acauline; **L.** broad-lanceolate, 1 m long and more, dull green, terminal Sp. chestnut-brown, curved, 2–2.5 cm long, marginal Sp. 10–15 mm apart, 5–10 mm long, broad-triangular, curved; **Infl.** a Pan., Fl. c. 3.5 cm long.

**A. brevispina** Trel. (§ III/2). – Haiti. – Acauline. **L.** broad-lanceolate, 1 m long or more, 10 cm across or more, dark green, terminal Sp. somewhat granular, straight, marginal Sp. 5–15 mm apart, 1–2 mm long, lower Sp. curved; **Infl.** 4 m tall, Fl. 4 cm long, yellow.

**A. brittoniana** Trel. v. **brittoniana** (§ III/18). – Cuba. – **L.** broad-lanceolate, gradually or more abruptly narrowing to the tip, 1 m long, 20 cm across, grey-green, slightly pruinose at first, terminal Sp. brown, almost subulate, 10–15 mm long, marginal Sp. 10–15 mm apart, curved; **Infl.** to 9 m tall, Fl. 3–3.5 cm long, yellow, with pedicel 1 cm long.

**A. — v. brachypus** Trel. – Ped. shorter than in type.

**A. bromeliifolia** Salm (§ III/13/1). – Mexico. – Acauline or very short-stemmed; **Ros.** open, leafy; **L.** 75–90 cm long, 2.5 cm across and 6.5 cm midway, linear-lanceolate, gradually tapering, leathery-fibrous, almost parchment-like, rather glossy green, underside towards base with small, grey-green, longitudinal stripes, terminal Sp. scarcely developed, marginal Sp. triangular, straight or bent forwards.

**A. brunnea** S. Wats. (§ I) (*Manfreda b.* (S. Wats.) Rose). – Mexico. – ♃; **L.** few, 10 cm long, 12–30 mm across the middle, margins with curved T.; **Infl.** 60 cm tall, of some 60 Fl., these being 3–3.5 cm long.

**A. cacozela** Trel. (§ III/8). – Southern Bahamas. – **Ros.** acauline, large, simple; **L.** lanceolate, 1.5–2 m long, yellowish-green, rough, terminal Sp. 15–20 mm long, subulate, marginal Sp. 10–15 mm apart, 2–5 mm long, larger Sp. recurved; **Infl.** 5–7 m tall, Fl. 5–6 cm long, golden-yellow.

**A. canartiana** Jacobi (§ III/2). – ? Mexico. – **L.** at soil-level 4 cm and above middle 8 cm across, terminating in a long, thin Sp., somewhat subulate, underside angular-convex at base, upperside flat, convex at tip, rigid, fleshy, vivid green, both surfaces rough, margins toothed and finely serrate, cartilaginous T. having recurved tips.

**A. cantala** Roxb. (§ III/13/1) (*A. c.* Jacobi, *Fourcroya c.* Voigt, *A. cantula* Roxb., *A. vivipara* Dalz. et Gibs., *A. laxa* Karw. ex Otto, *A. rumphi* Jacobi, *A. candelabrum* Tod.). – Origin unknown, growing wild in trop. zones of Old World. – Acauline or short-stemmed, offsetting; **Ros.** large, L. arrangement rather open; **L.** 1–1.4 m long, 8–9 cm across the middle, at base to 5.5 cm across, long linear-lanceolate or ensiform, tapering gradually, light green or dark green, dull-glossy, smooth, underside rough, flexible, thus often pendent, margins recurved, terminal Sp. subulate, 17–20 mm long, marginal Sp. with hook directed upwards, 1.3 mm apart, 4–5 mm long; **Infl.** 4–6 m tall, with numerous adventitious plantlets, Fl. 6–6.5 cm long, yellow.

**A. caribaea** J. Versch. (§ II/1). – Related to **A. micracantha,** perhaps identical with this.

**A. caribaeicola** Trel. (§ III/10) (*A. caribaea* Bak., *A. americana* Urb.). – Caribbean Is. – Habit as for **A. karatto;** **L.** ensiform, gradually tapering, terminal Sp. 15–20 mm long, 3–4 mm thick, marginal Sp. arranged irregularly, small, triangular, scarcely 1 mm long, 5 mm apart; **Infl.** a panicle, Fl. 6–7 cm long, yellow.

**A. caroli-schmidtii** BGR. (§ III/7/4). – Guatemala. – **Ros.** acauline, simple; **L.** ovate-spatulate, narrowing abruptly to base, concave above the middle since margins are curved upwards, light grey, rough at base, underside almost white, terminal Sp. 2.5–3 cm long, subulate, twisted, marginal Sp. with horny tip, narrow-triangular, bent, 8–10 mm long, Sp. towards base being smaller.

**A. celsii** HOOK. (Pl. 7/1) (§ II/1) (*A. celsiana* JACOBI). – ? Mexico. – **Ros.** acauline; **L.** c. 25, soft-fleshy, apple-green, grey-pruinose, 45–60 cm long, spatulate, 10–15 cm across above middle, in upper part abruptly narrowing and terminating in a slender tip, at base 2 cm thick, margins at base curved-undulate, terminal Sp. 1–5 mm long, marginal Sp. weak, cartilaginous, white; **Infl.** c. 1.3 m tall, Fl. 5 cm long, yellow-green. (TRELEASE regards the following as synonymous: **A. brauniana** JACOBI, **A. humboldtiana** JACOBI, **A. lamprochlora** JACOBI, **A. perlucida** JACOBI, **A. rupicola** RGL., **A. smithiana** JACOBI, **A. thomsoniana** JACOBI).

**A. cernua** BGR. v. **cernua** (§ II/7/1) (*A. attenuata* v. *compacta* JACOBI, *A. spectabilis* HORT., *A. attenuata* AUCT., *A. virens* HORT. ex BESAUC., *A. elliptica* HORT. ex BESAUC., *A. compacta* HORT. ex BESAUC.). – Mexico. – Related to **A. attenuata** but **St.** freely offsetting; **Ros.** of up to 35 more elongated **L.**, less constricted above soil, 60 cm long, 16–17 cm across, dark green, light-grey pruinose; **Infl.** 3–4 m tall, curving gracefully over, Sc. very short, with many Fl., these being 5–6 cm long, greenish. (J. A. BREITUNG regards this species as synonymous with **A. attenuata**.)

**A.** — v. **serrulata** (TERR.) BGR. (*A. attenuata* v. *s.* TERR.). – **L.** with fine, cartilaginous T.

**A. chiapensis** JACOBI (§ II/1). – Mexico. – **Ros.** hemispherical, about 1 m ⌀; **L.** dense, obovate-spatulate, 50 cm long, 14 cm across above middle, sturdy, light green or grey-green, terminal Sp. 2.5 cm long, marginal Sp. triangular, 3 mm long, 10–15 mm apart; **Infl.** to 3 m tall, Fl. 6–7 cm long, yellowish-green.

**A. chihuahua** TREL. (§ III/6). – N.Mexico: Chihuahua. – Habit as for **A. scabra**; **L.** rigid, pointed, smooth, light grey-green, broad elongate-lanceolate or ovate-elongate, acuminate, upperside concave, with flat longitudinal depressions towards apex, terminal Sp. almost straight, 2.5–3.5 cm long, 7 mm across, marginal Sp. 15–25 mm apart, c. 6 mm long, recurved; **Infl.** a Pan., Fl. 6.7–7 cm long, yellowish.

**A. chrysoglossa** I. M. JOHNSTON (§ II/7/b). – Mexico: Sonora. – **Ros.** to 1.5 m ⌀; **L.** 50 to 150 cm long, 4–6 cm across, linear-lanceolate, terminal Sp. decurrent, merging into the toothless horny margin; **Infl.** to 3 m tall, often pendulous, Fl. light yellow.

**A. cinerascens** JACOBI (§ III/1). – Mexico. – **Ros.** compact; **L.** rigid, about 25 cm long, 7.5 cm across at base and below middle, at base 3.5 cm thick, long-ovate, tapering above, ashy light green, terminal Sp. 3 cm long, 4 to 6 mm across, marginal Sp. to 8 mm long, Sp.-cushions united, with horny margins, intermediate Sp. smaller, with twin points.

**A. cochlearis** JACOBI (§ III/1) (*A. atrovirens* W. NEUB.). – Mexico. – Resembles **A. salmiana**; **L.** lanceolate-spatulate, above middle about 36 cm across, underside very convex, olive-green, dull, terminal Sp. 6.5–7 cm long, marginal Sp. small, directed downwards, central Sp. larger; **Infl.** 8 m tall, Fl. yellow.

**A. cocui** TREL. (§ III/12) (*A. americana* HUMB.). – Venezuela. – **Ros.** large, of many **L.**, offsetting; **L.** curved S-shaped, spreading, broad-lanceolate, tapering above middle, upperside very concave, margins curved upwards, folded keelwise below apex, about 1.1 m long, 30 cm across, grey-green, later glossy green, terminal Sp. to 3 cm long, 3–4 mm across, marginal Sp. 10–20 mm apart, triangular, pointed, lower Sp. recurved; **Infl.** 9 m tall, Fl. 5–6 cm long, yellow.

**A. colimana** GENTRY (§ II/2). – Mexico: Colima. – **St.** short; **Ros.** with many **L.**, simple, to 1.20 m ⌀; **L.** straight, linear, thin, upperside flat, 40–70 cm long, 1–1.5 cm across, smooth, green, margins narrow, brown, splitting off long, brown threads, T. soft, 5–8 mm long, brown; **Infl.** 2–3 m tall, Fl. 4–5 cm long, pale yellow.

**A. collina** GREENM. (§ III/13/2). – Mexico: Morelos. – Acauline; **L.** 30–40, narrow-linear, gradually tapering, 60–80 cm long, underside convex, blue-green, terminal Sp. 2 cm long, margins with a fine, brown, horny line between the 8–12 T., these being directed upwards and hooked; **Infl.** 3–4 m tall, Fl. 7 cm long, greenish-yellow.

**A. colorata** GENTRY (§ III/7/3). – Mexico: Sonora. – **Ros.** simple or with few offsets, 50–80 cm ⌀; **L.** 30–40 cm long, 10–15 cm across, ovate, abruptly pointed, leathery, 1 to 2 cm thick, both surfaces smooth, ash-grey or with pink transverse hands, terminal Sp. 25 mm long, 6–8 mm across, purple-brown, marginal Sp. 5–10 mm long; **Infl.** c. 3 m tall, Fl. c. 3.5 cm long, yellow.

**A. conjuncta** BGR. (§ III/7/3). – ? Mexico. – **Ros.** simple, acauline, of 25–35 stout leathery-fleshy **L.**, both surfaces smooth, dull pale green, or slightly grey-green, oblanceolate-spatulate, short-tapering, about 55 cm long, 10 cm across base and 14–15 cm midway, underside convex, 4 cm thick at base, terminal Sp. 2.5–3 cm long, 5–6 mm across, marginal Sp. 5 mm long, intermediate Sp. smaller.

**A. crassispina** TREL. (§ III/1). – Mexico: San Luis Potosí and Durango. – **Ros.** 2 m ⌀, acauline, simple or with only few offsets; **L.** 90–120 cm long, 20–25 cm across, oblanceolate, more or less concave, apex pointed, narrowing below to very stout base, striped light bluish or grey and green, underside slightly rough, terminal Sp. 5–8 cm long, to 18 mm across, conical, marginal Sp. to 15 mm long, to 7 cm apart, curved forwards or recurved, on thick, fleshy cushions, intermediate Sp. small; **Infl.** 8–10 m tall, Fl. yellow.

**A. crenata** JACOBI (§ III/7/3) (*A. amoena* HORT. ex JACOBI, *A. heterodon* HORT., *A. mescal* C. KOCH). – Mexico. – **Ros.** depressed; **L.** recurved above middle and pendulous, subovate-elliptical, short-pointed, narrowing abruptly to base, soft, fleshy, vivid green, terminal Sp. very stout, bent, margins curved upwards, wavy, marginal Sp. 2–3 cm apart, with triangular straight or curved tip, intermediate Sp. smaller.

**A. cucullata** LEM. (§ III/7/2). – Mexico. – **Ros.** small, acauline, of 15–30 very short and broad, depressed-ovate, blunt. **L.** c. 10 cm long, 8 cm across above middle, narrowed above base to 5 cm, fleshy, dull, whitish grey-green, glabrous, terminal Sp. 2.5 cm long, marginal Sp. often set at an angle to base, straight or bent, usually very small.

**A. cundinamarcensis** BGR. (§ III/11). – Colombia. – **Ros.** acauline, without offsets; **L.** very thick above base, only 15 cm across, projecting horizontally, then curved upwards and becoming abruptly wider, above middle 45 cm across, then tapering, yellowish-green to blue-grey, terminal Sp. rather short, margins with very short, not very sharp T.; **Infl.** with very few adventitious buds.

**A. cupreata** TREL. et BGR. (§ III/7/3). – Mexico. – Closely related to **A. crenata**; **L.** 75 cm long, obovate, short-pointed, light green, terminal Sp. 5 cm long, bent, margins sinuate-dentate, T. on large, broad Sp.-cushions and with tip which is straight or curved, cushions 2–2.5 cm across, Sp. 1.5–2 cm long, intermediate Sp. smaller.

**A. cyanophylla** JACOBI (§ III/2). – Mexico. – Closely resembles **A. franzosinii**; **Ros.** somewhat caulescent, with some 15 L.; **L.** lanceolate, c. 53 cm long, 6.5 cm at base and 10 cm across middle, gradually tapering, underside thickened keel-like at base, fleshy as far as midway, then fleshy-leathery, vivid light seagreen, margins fleshy, terminal Sp. stout, 4 cm long, marginal Sp. well-spaced, broad-triangular, bent, brown.

**A. dasylirioides** JACOBI et BOUCHE v. **dasylirioides** (§ II/4). – Guatemala. – **Ros.** simple, dense, dividing after flowering; **L.** 80–100, linear, flat, very bluish, 40–60 cm long, 10 to 15 cm across at base, rather stiffly erect, flat, with very fine longitudinal stripes, margin entire, terminal Sp. 1 cm long, brown; **Infl.** 1.5–2 m tall, Fl. 3.5–4.5 cm long, yellow-green.

**A. —** v. **dealbata** (LEM. ex JACOBI) BAK. (*A. dealbata* LEM. ex JACOBI, *A. intrepida* GREENM.). – Mexico. – **L.** 50–100 cm long, broader, margins with very fine T.

**A. datylio** SIMON v. **datylio** (§ III/14). – Mexico: southern Baja Calif. – Acauline, mat-forming; **L.** elongate-lanceolate, upperside grooved, 30–75 cm long, 3–4 cm across, yellowish grey-green, fleshy, rigid, terminal Sp. almost triangular-conical, 2–3 cm long, marginal Sp. 2–3 cm apart, 3–5 mm long; **Infl.** 4–5 m tall, Fl. 5–10 cm long.

**A. —** v. **vexans** (TREL.) I. M. JOHNSTON (*A. v.* TREL.). – Mexico: E. coast of Baja Calif. – **L.** 20–45 cm long, 2 cm across; **Fl.** 4 cm long.

**A. debilis** (ROSE) (§ I) (*Manfreda d.* ROSE). – Mexico. – ♃; **L.** linear-lanceolate, 30 cm long, 1.5–2 cm across, green, brown mottled; **Infl.** to 1 m tall, Fl. in a compact raceme.

**A. decaisneana** JACOBI (§ II/1). – Mexico. – **Ros.** c. 25 cm ⌀; **L.** dense, broad-obovate, pointed, 10 cm long, at base 5 cm and in upper half 7.5 cm across, vivid glossy green, terminal Sp. black, marginal Sp. small, bent, black.

**A. decipiens** BAK. (§ II/13/1) (*A. spiralis* BRANDG., *A. laxifolia* BAK.). – Origin unknown, growing wild in Florida (USA). – **Ros.** with St. up to 3 m tall, covered in dead L., and with numerous basal offsets; **L.** (30–)70(–130) cm long, linear-lanceolate, thickened at base, above that somewhat constricted, below middle 6 to 10 cm across, gradually tapering, deeply grooved towards tip, underside very convex and at base almost triangular-keeled, glossy fresh green, very leathery-fleshy, terminal Sp. 10–15 cm long, sharply pointed, marginal Sp. 1.5–3 cm apart, 3–4 mm long, mostly recurved; **Infl.** 5–6 m tall, with numerous adventitious plantlets, Fl. 7.5 cm long, greenish-yellow.

**A. deflexispina** JACOBI (§ III/2). – ? Mexico. – **Ros.** with few L., about 70 cm ⌀, 45 cm tall; **L.** linear-lanceolate, 42 cm long, at base 7 cm across, then constricted to 5 cm, tapering gradually to a point, hard-fleshy, glabrous, dirty dark green, terminal Sp. 3 cm long, thin marginal Sp. inconspicuous, rather widely spaced, with recurved tips.

**A. demeesteriana** JACOBI (§ III/6). – Insufficiently known species.

**A. densiflora** HOOK. (§ II/1). (*A. polyacantha* v. *d.* TERR., *A. hookeri* HORT. ex BESAUC.). – Mexico. – Closely resembles **A. polyacantha** in age multi-headed, mat-forming; **L.**-margins slightly reddened, T. close together; **Fl.** very numerous, 2.5 cm long.

**A. —** cv. Angustifolia (*A. d. v. a.* BESAUC.). – L. narrower.

**A. —** cv. Striata Aureis (*A. d. v. s. a.* BESAUC.). – **L.** with yellowish stripes.

**A. desertii** ENGELM. (§ III/15) (*A. pringlei* ENGELM., *A. d.* ORC., *A. dentiens* TREL., *A. disjuncta* TREL., *A. consociata* TREL., *A. nelsoni* TREL.). – USA: southern Calif.; Arizona; Mexico: Sonora, island of S. Sebastian. – **Ros.** 40–80 cm ⌀, usually forming a mat or more open colonies by means of long stolons, acauline; **L.** 15–40 cm long, 3–6 cm across, long-triangular or elongate-lanceolate, gradually tapering, shallow-concave, later grooved, rigid, bluish, grey-green, often with horizontal bands, slightly granular-rough, terminal Sp. acicular-conic, 2.5–5 cm long, sulcate, margins either with T. or entire, any T. being 1 mm long, 5–15 mm apart; **Infl.** 2–7 m tall, Fl. 3.5–5 cm long, chromeyellow.

**A. desmetiana** JACOBI (§ III/13/1) (*A. regeliana* JACOBI, *A. miradorensis* JACOBI, *A. franceschiana* TREL., *A. elizae* BGR., *A. paupera* BGR.). – Mexico: Veracruz. – **Ros.** 1.5–1.8 m ⌀, acauline or short-stemmed, offsetting; **L.** 50 to 150 cm long, (3–)7(–11) cm across, lanceolate, bluish, with distinct green transverse stripes, almost straight, rather soft and flexible, terminal Sp. 2–2.5 cm long; 4–5 mm across, margin in upper part entire, below with minute marginal Sp., almost colourless, 3–5 cm apart; **Infl.** 3–5 m tall, with numerous adventitious plantlets, Fl. 5 cm long, green with brown dots.

**A. deweyana** TREL. (§ III/3/2). – Acauline or almost so, offsetting; **Ros.** dense; **L.** 1.5 m long, 5–10 cm across, tapering gradually, thin, yellowish-green, underside often zoned grey, terminal Sp. 1.5–4 cm long, marginal Sp. 2 to 3 mm long, tip hooked and recurved; **Infl.** 3–6 m tall.

**A. dussiana** TREL. (§ III/10) (*A. americana* GRISEB. p. part.). – Is. of Guadelupe. – Habit as for **A. karatto**; **L.** lanceolate, tapering gradually, 1–1.6 m long, 20–40 cm across, grey-green, terminal Sp. blackish, 5–7 mm long, 4 mm across, marginal Sp. 5–10 mm apart, 2 mm long, hooked and recurved, margin somewhat parchment-like; **Infl.** 5–9 m tall, Fl. 6.5 cm long.

**A. echinoides** JACOBI (Pl. 7/6) (§ II/3) (*A. striata* v. *echinoides* BAK.). – Mexico. – Resembles **A. striata**; **Ros.** to 30 cm ⌀, short-stemmed; **L.** very numerous, rigid, tapering from a broadly bottle-shaped base, 12–15 cm long, linear, lower part 8 mm across, terminating in a lanceolate tip and a 3–4-angled brown terminal Sp., upperside with M. keel, underside slightly convex, grey-green.

**A. eduardii** TREL. (§ II/7/c). – Mexico: Durango. – **Ros.** simple, 1 m tall; **L.** c. 15, 80–100 cm long, gradually tapering, upperside concave, underside convex, bluish, terminal Sp. rigid, thin, 1 cm long, 1 mm across, margins with narrow brown edge with minute T.; **Fl.** 4 cm long, yellow. (Possibly identical with **A. vilmoriniana**.)

**A. eggersiana** TREL. (§ III/10) (*A. americana* GRISEB. p. part.). – Caribbean Is. – **Ros.** acauline, simple; **L.** narrow-lanceolate, gradually tapering, 1.5 m long, 10–15 cm across, dull green, terminal Sp. 10–15 mm long, brown, marginal Sp. red-brown, 10 mm apart, 1 mm long; **Infl.** 5 m tall, Fl. 5–6 cm long, yellow.

**A. ehrenbergii** JACOBI (§ II/1). – Mexico. – Acauline; **L.** numerous, soft-fleshed, rich dark green, slightly pruinose, about 50 cm long, 7.5 cm across at base, rather narrower, above almost spatulate, terminal Sp. horny, brown, margins curved upwards, T. ciliate, small and close together; **Infl.** almost 2 m tall, Fl. c. 2 cm long.

**A. eichlamii** BGR. (§ III/7/4). – Guatemala. – **Ros.** acauline, simple; **L.** lanceolate-spatulate, tapering gradually, light grey, ± rough, terminal Sp. straight, subulate, about 3 cm long, marginal Sp. hooked and recurved, with tip 8–10 mm long, 5–8 mm apart.

**A. —** v. **injecta** BGR. – Marginal Sp. often (2–)3–4(–5), united.

**A. ekmanii** TREL. (§ III/2). – Cuba. – **L.** grey-green, less often bluish, blunt, marginal Sp. about 2 cm apart, slender; **Infl.** a panicle, Fl. 2.5–3 cm long, yellow.

**A. ellemeetiana** JACOBI (§ II/7/b) (*A. pruinosa* LEM.). – Mexico. – Acauline; **Ros.** c. 80 cm ⌀, of 30–35 oblanceolate **L.**, narrowing above base, 48–65 cm long, 10–11 cm across above middle, upper part short-pointed, tip hardened, underside much thickened, margin entire, with a cartilaginous edge; **Infl.** to 4.5 m tall, Fl. 15–16 mm long, light green, campanulate.

**A. elongata** JACOBI (§ III/13/2) (*A. spectabilis* TOD.). – ? Mexico. – **Ros.** acauline, in age rather elongated, producing few offsets, about 3 m ⌀; **L.** numerous, very rigid, projecting stiffly, 1.8–2.1 m long, 10 cm across base and 12 cm midway, gradually tapering, underside convex and keeled, light grey, white-grey pruinose, fibrous to leathery-fleshy, terminal Sp. stout, to 4 cm long, 2–8 mm across, margins with a fine cartilaginous line, marginal Sp. with broad base, usually hooked, black-brown, 5 mm long, 12 to 20 mm apart; **Infl.** 5 m tall, Fl. 7–7.5 cm long, green, pruinose.

**A. endlichiana** TREL. (§ III/13/b). – Mexico. – Acauline, offsetting; **Ros.** dense; **L.** 80–125 cm long, 5–9 cm across, light to dark green, terminal Sp. 15–30 mm long, red to brown, marginal Sp. stout, triangular, hooked, 3–4 mm long, 10–30 mm apart, margin translucent.

**A. engelmannii** TREL. (§ III/7/1). – ? Mexico. – **Ros.** simple, acauline, 1.5 m ⌀; **L.** 30–40, c. 65 cm long, 17–21 cm across, oblanceolate-spatulate, pointed, terminal Sp. subulate, 2.5 to 3 cm long, L. 9–10 cm across at base, margins curved upwards, underside keeled below tip, stout-leathery, fleshy, rigid, grey-green, light grey-pruinose, marginal Sp. 5 mm apart, 2 mm long, triangular; **Infl.** to 2.5 m tall.

**A. erosa** BGR. (§ III/7/2). – ? Mexico. – **Ros.** simple, densely leaved; **L.** stout, rigid, light green, slightly grey, underside distinctly grey, 36–38 cm long, 7 cm across above base and 14 cm midway, obovate to spatulate, underside very convex, terminal Sp. 3–4 cm long, somewhat twisted, marginal Sp. 2–4 cm apart, hooked, 12–14 mm long, intermediate T. small.

**A. evadens** TREL. (§ III/12) (? *A. polyacantha* BAK.). – Caribbean Is. – **Ros.** short-stemmed; **L.** not very numerous, narrow-oblanceolate, gradually tapering above the middle, upperside rather folded, margins curved upwards, terminal Sp. short and stout, marginal Sp. close together; **Infl.** slender, Fl. 4.5 cm long.

**A. expansa** JACOBI (§ III/2). – Mexico. – **Ros.** 3 m ⌀, offsetting; **L.** spreading, rather short-pointed, almost 2 m long, 40 cm across the base and above that 15 cm, the upper half being 20 cm across, considerably swollen below, upper part deeply angular-concave, margins folded upwards, both surfaces with longitudinal grooves, hard-fleshy, bluish grey-green, terminal Sp. 2.5–3 cm long, margins very sinuate, T. widely spaced, stout, tip bent in at a right-angle; **Infl.** 7–8 m tall.

**A. expatriata** ROSE (§ II/5). – Origin unknown. – **Ros.** acauline or very short-stemmed with some 40 L. 60–75 cm long, in middle 9 cm across, lower part 6–6.5 cm across and 2–2.5 cm thick, thinner towards tip, margins with horny edge throughout, terminal Sp. 2.5 cm long, T. numerous, 1 cm apart, hooked,

5–8 mm long, intermediate T. smaller; **Infl.** 4–5 m tall, Fl. pale green or cream-coloured.

**A. falcata** ENGELM. v. **falcata** (§ II/3) (? *A. californica* BAK., *A. paucifolia* TOD., *A. striata* v. *c.* TERR.). – Similar to **A. striata**; **L.** rigid, stout, 40–50 cm long, light-grey, often purple or red-brown, upperside flat, 12–15 mm across, underside thickened, straight, terminal Sp. 22–25 mm long, very sharp-pointed, margins with fine cartilaginous T.; **Infl.** open, 1.1 m tall, Fl. to 4 cm long.

**A.** — v. **espadina** BGR. – **L.** longer, narrower, terminal Sp. smaller. Yields fibres known as 'espadin'.

**A.** — v. **microcarpa** BGR. – **L.** shorter, terminal Sp. stout; **Fr.** 12–13 cm long.

**A. fenzliana** JACOBI (§ III/7/3) (*A. flaccida* JACOBI, *A. maximiliana* BAK., *A. calodonta* BGR., *A. scolymus* BGR., *A. gustaviana* HORT. ex BAK.). – Mexico. – **Ros.** 1–2 m ⌀, simple; **L.** 90–150 cm long, 15–20 cm across, lanceolate, thick and fleshy at base, then abruptly thin and leathery, terminal Sp. 15–55 mm long, 5 mm across, T. on conspicuous cushions, brown, tip curved forwards; **Infl.** 6–8 m tall, Fl. c. 8 cm long, yellow.

**A. ferdinandi-regis** BGR. (Pl. 7/4) (*A. victoriae-reginae* v. *laxior* BGR., ? *A. nickelsiae* HORT., *A. nickelsii* ROL.-GOSS., *A. v.-r.* f. *n.* (ROL.-GOSS.) TREL.). – Mexico. – **Ros.** small, simple; **L.** triangular or triangular-ovate, pointed, stiffly erect, hard and firm-fleshed, dull dark green, both surfaces with white convergent lines, upperside very concave, almost folded, underside obliquely and sharply keeled below middle, about 13 cm long, 5 cm across in lower part, margins horny, broad, black to grey, terminal Sp. stout, 12–15 mm long and across at base, sharply 3-angled, each angle having 1–2 stout, pointed, lateral T., glossy black.

**A. ferox** C. KOCH (Pl. 7/3) (§ III/1) (*A. coelum* HORT. ex BESAUC.). – Mexico. – **Ros.** large, acauline, with few offsets; **L.** 20–30, to rather over 1 m long, oblong-spatulate, 30–35 cm across, narrowing abruptly to a point, the tip grooved on the upperside, underside very convex-keeled, rigid, fleshy, dark green, margins sinuate-dentate, terminal Sp. 4–8 cm long, T. 1.5–2.5 cm long, hooked; **Infl.** 8–10 m tall, Fl. almost 9 cm long.

**A. filifera** SALM v. **filifera** (Pl. 7/5) (§ II/2). – Mexico: Pachuca, Hidalgo, San Luis Potosí. – **Ros.** acauline, globular, offsetting laterally, with a thick, conical L.-bud, to 65 cm ⌀; **L.** numerous, 20–25 cm long, 3 cm across, lanceolate, gradually tapering, rigid, somewhat glossy, green, with 2–3 white lines and pale horny margin with 5–6 long fibres splitting off, terminal Sp. 1.5–2 cm long; **Infl.** to 2.5 m tall, Fl. 5 cm long, green-yellow.

**A.** — v. **compacta** TREL. (*A. perplexana* TREL.). – **Ros.** smaller; **L.** shorter, broader, scarcely 10 cm long.

**A.** — v. **filamentosa** (SALM) BAK. (*A. filamentosa* SALM, ? *A. pseudofilifera* ROSE et LANZA, *A. filifera* v. *major* HORT.). – **L.** to 30 cm long, narrower, more spreading and arranged less regularly.

**A.** — SALM v. **filamentosa** (SALM) BAK. f. **ortgiesiana** (TOD.) JACOBS. (*A. o.* TOD., *A. o.* v. *brevifolia* HORT., *A. nigromarginata* HORT.). – **L.** very narrow, green, margins black.

**A. flaccifolia** (BAK.) BGR. (§ II/1) (*A. micrantha* BAK.). – ? Mexico. – Very freely branching, forming large mats; **L.** to 60 cm long, 9–9.5 cm across, lanceolate, 5.5 cm across at base, almost 2 cm thick, soft-fleshy, thin in the middle, upperside grooved, glabrous, pale dull-green with indistinct grey-green longitudinal stripes, upper part mottled, terminal Sp. pointed, 4–9 mm long, margins with a reddish or whitish cartilaginous line, T. tiny, hooked; **Infl.** to 1.8 m tall, Fl. 5 cm long, green-brown.

**A. flexispina** TREL. (§ III/6). – Mexico: Durango. – **Ros.** c. 30 cm ⌀, rather compressed-globular, simple or with few offsets; **L.** 12 to 20 cm long, 6–10 cm across, broad-oval, abruptly pointed, upperside very concave, bluish-green, glabrous, terminal Sp. 3 cm long, curved, T. 6–9 mm long, 7–15 mm apart, thin; **Fl.** 4.5–5 cm long.

**A. fourcroydes** LEM. (Pl. 6/1) (§ III/13/1) (*A. ixtlioides* LEM. ex JACOBI, *A. ixtli* v. *elongata* BAK., *A. rigida* v. *e.* BAK., *A. e.* BGR., *A. r.* v. *longifolia* ENGELM., *A. r.* HORT., *A. ixtli* HORT., *A. l.* HORT., *A. sullivani* TREL.). – Mexico: Yucatan. – **St.** 75–120 cm tall, about 25 cm ⌀, offsetting; **Ros.** 2–2.5 m ⌀; **L.** rigid, projecting stiffly, 1.4 m long, above middle 10.5 cm across, oblanceolate, ensiform, narrowing to base, pointed at tip, terminal Sp. almost 3 cm long, upperside grooved towards tip and margins curved upwards, underside almost keeled, grey-green, somewhat pruinose, T. 2–4 mm long, 22–30 mm apart, hooked; **Infl.** 6–7 m tall, with numerous adventitious plantlets, Fl. 6–7 cm long, yellow-green.

**A. franzosinii** BAK. (§ III/2) (*A. f.* NISSEN ex RICASOLI nom. nud., *A. beaulueriana* JACOBI, *A. frederici* BGR.). – Mexico. – **Ros.** to 3 m tall, offsetting; **L.** 40–50, usually gracefully curved, grooved, attractive white-grey to blue-grey, very rough, to 2.3 m long, at base 13–15 cm thick, constricting to 17cm across, then widening towards middle to 40 cm, upper part deeply grooved, margins wavy, curved, tip conical, terminal Sp. subulate, 6–7 cm long, T. triangular, hooked, 6 mm long, 15 mm apart, upper T. larger and further apart; **Infl.** to 11.5 m tall, Fl. 6–7 cm long, yellow.

**A. fuerstenbergii** JACOBI (§ III/3). – ? Central Am. – **Ros.** acauline, with many L., about 3.6 m ⌀; **L.** later recurved, oblong-lanceolate, light grey-green, 1.7–1.8 m long, 33 cm across at base, then much narrowed, 13–16 cm across midway, tapering gradually, upperside grooved-concave, terminal Sp. long, margins sinuate-dentate, T. triangular-pointed, tip directed downwards; **Infl.** almost 7 m tall, Fl. numerous, 5 cm long, green.

**A. funkiana** C. KOCH et BOUCHÉ (§ II/5). – Mexico. – **Ros.** acauline, offsetting, to 1 m ⌀;

L. 25–30, tapering gradually, 55 cm long, 3 cm across base, 27 mm across midway, upper part somewhat grooved, light green, light blue pruinose, pale M. band 1 cm across, underside very pruinose, with dark-green longitudinal lines, horny margin very narrow, somewhat reddish-brown, terminal Sp. 11 cm long, brown, T. small, hooked-recurved, 15–35 mm apart.

**A. galeottei** BAK. (§ III/7/2). – ? Mexico. – Acauline; **Ros.** 60–90 cm ⌀; **L.** 30–40, oblong-spatulate, 30–45 cm long, 10–15 cm across above the middle, 5–8 cm across at base and 2–2.5 cm thick, green to grey-green, terminal Sp. 2 to 2.5 cm long, T. close-set, sharp, 5–6 mm long, black-brown.

**A. geminiflora** (TAGL.) KER-GAWL. v. **geminiflora** (§ II/2) (*Littaea g.* TAGL., *A. g.* v. *filamentosa* HOOK., *Dracaena boscii* HORT. CELS., *Yucca b.* DESF., *Bonaparte juncea* WILLD., *B. flagelliformis* C. HENKEL v. DONNERSM.). – ? Mexico. – Short-stemmed or acauline, branching in age; **L.** 100–200, forming a dense Ros., linear, rush-like, 60–90 cm long, 3–6 mm across, both surfaces convex, dark-green, margins having a fine white, cartilaginous line and curled threads, terminal Sp. 4 mm long, triangular; **Infl.** 3–4 m tall, Fl. 4–5 cm long.

**A. — v. atricha** TREL. (*A. knightiana* DRUMM., *A. g.* v. *k.* DRUMM.). – **L.** 30–35 cm long, marginal threads not present.

**A. — v. filifera** TERR. – **L.** margins with many more threads splitting away.

**A. ghiesbreghtii** C. KOCH (§ II/6). – Mexico. – **Ros.** acauline, offsets numerous; **L.** many, up to 60, rigid, c. 45 cm long, obovate-lanceolate, to 10 cm across, gradually tapering, underside almost rounded-keeled, glossy dark green, younger L. having a pale M. band, horny margin 2 mm across, terminal Sp. 15 mm long, T. 3 to 4 mm long, in upper part virtually absent; **Infl.** to 3 m tall, Fl. 5 cm long, greenish-brown.

**A. glomeruliflora** (ENGELM.) BGR. (§ II/5) (*A. heteracantha* f. *g.* ENGELM., *A. lechuguilla* f. *g.* (ENGELM.) TREL. ,*A. chisosensis* MÜLLER). – USA: Texas; Mexico. – **Ros.** to 1 m ⌀; **L.** fleshy, narrow ovate-lanceolate, bluish-green, glabrous, both surfaces with darker lines, 30 to 60 cm long, 4–8 cm across, terminal Sp. 2.5 to 5 cm long, dark brown, T. 6–13 mm long, recurved, 3–5 cm apart; **Infl.** 5–6 m tall, Fl. in compact clusters. (Description acc. BREITUNG.)

**A. goeppertiana** JACOBI (§ II/1). – Mexico. – **St.** to 20 cm tall, branching in age; **L.** spreading, broad-lanceolate, 60–70 cm long, at base 10 cm and in middle 13 cm across, curled inwards below the 3–4 mm long terminal Sp., 2–4 cm thick at base, underside almost keeled-convex, upper part with several grooves, fleshy, very brittle, glabrous, light sap-green, T. close set, triangular, curved, 3–4 mm apart, 1–1.5 mm long; **Infl.** to 2 m tall, Fl. 3.5 cm long, smelling of turpentine.

**A. goldmaniana** TREL. (§ III/5). – Mexico: Baja Calif. – **St.** prostrate, leafy, forming a mat from the base; **L.** ovate-lanceolate or lanceolate, pointed, stiffly erect, upperside flat-concave, 50 cm long, 10 cm across, terminal Sp. almost straight, 4 cm long, 7 mm across, T. 1.5–3 cm apart, often 1 cm long, black-grey, mostly recurved, margin somewhat sinuate; **Infl.** 5 to 7 m tall.

**A. gracilipes** TREL. (§ III/6) (*A. applanata* TREL., *A. americana* v. *latifolia* TORR.). – USA: W.Texas. – **Ros.** acauline, mat-forming; **L.** oblong-lanceolate, tapering above the middle, very flat-concave, 20–30 cm long, (7–)8(–12) cm across, white-grey, terminal Sp. 4–5 cm long, T. 15–20 cm apart, 2–10 mm long, upper T. curved upwards, narrow-triangular, margin slightly sinuate; **Infl.** to 5 m tall, Fl. 3.5 cm long, yellow.

**A. gracilispina** ENGELM. (§ III/1) (*A. salmiana* v. *g.* ROL.-GOSS.). – Mexico: San Luis Potosí. – **Ros.** with numerous offsets, 2–3 m ⌀; **L.** to 3 m long, 19 cm across above the middle, to 15 cm thick at base, thin, arching, green, terminal Sp. 3.5–7 cm long, brown, T. thin, brittle 1.5–5 cm apart, intermediate Sp. smaller; **Infl.** slender, Fl. yellow.

**A. gracillima** (ROSE) BGR. (§ I) (*Manfreda elongata* ROSE). – Mexico. – **L.** very furrowed, 30 cm long or more, 5 cm across, green, margin entire; **Fl.** 12–14 mm long, brownish.

**A. gracilis** JACOBI (§ III/1). – Mexico. – **Ros.** acauline, c. 72 cm ⌀; **L.** numerous, straight, very rigid, oblong-lanceolate, 35 cm long, 75 cm across midway, long-pointed, terminal Sp. stout, upperside of L. flat-convex, grooved at tip, underside convex, vivid olive to grey-green, glabrous, margins straight, T. close-set, slender, with tip curved-triangular, intermediate Sp. smaller.

**A. grandibracteata** Ross. (§ III/7/2). – ? Mexico. – **Ros.** acauline, simple, to 1.2 m ⌀; **L.** c. 50, 38–60 cm long, obovate-lanceolate or spatulate, short-pointed, narrowing to the base, c. 6 cm across at the base, 12–16 cm in the middle, 2.5 cm thick at the base, grooved towards tip, fresh green or light grey-green, rather rough towards the tip, terminal Sp. 4.5–6 cm long, about 5 mm across, T. triangular, projecting, 25 mm apart, 6–7 mm long, margins sinuate, lowest T. smaller; **Infl.** 3–5 m tall, involucral L. large, Fl. 4.5 cm long, green to brown.

**A. grenadina** TREL. (§ III/10). – Caribbean Is.: Grenada (Windward Is.). – Habit as for **A. karatto**; **L.** broad-lanceolate, blunt or gradually tapering, 2 m long, 25–30 cm across, deep-green, terminal Sp. on stouter, incurved L. – tip, about 15 mm long, 5–8 mm thick, L. margin straight, initially reddish, T. c. 1 cm apart, 1–2 mm long, triangular, slightly bent; **Infl.** 8 m tall, Fl. 6 cm long.

**A. grisea** TREL. v. **grisea** (§ III/9). – Cuba. – **Ros.** acauline, simple; **L.** lanceolate, somewhat concave, 1.5–2 cm long, (10–)20(–25) cm across, green to grey-green, terminal Sp. reddish-brown, triangular-conical, 1–2 cm long, 4–5 mm across, T. (15–)25(–45) mm apart, (2–)3(–5) mm long, slightly bent, broad-triangular, margins somewhat sinuate; **Infl.** 6–8 m tall, Fl. 4–4.5 cm long, golden-yellow.

**A. — v. cienfuegosana** TREL. – **L.** margins sinuate and rather wavy, marginal Sp. 2 mm long.

**A. — v. obesispina** TREL. – T. more triangular, with hardened tip, terminal Sp. thick, with shallow furrows.

**A. guatemalensis** BGR. (§ III/7/4). – Guatemala. – **Ros.** simple, acauline; **L.** ± recurved above middle, firm-fleshy, ovate-spatulate, short-pointed, upper side concave, ± grey-green, glabrous, underside rough, white-grey, terminal Sp. about 3 cm long, T. triangular, with a crescent-shaped base and grey, horny tip, about 8 mm long, 1–2 cm apart, the top 5–6 cm of the margins being spineless.

**A. guiengola** GENTRY (§ II/7/1). – Mexico: Oaxaca. – **Ros.** simple or with a few lateral Ros., 70–130 cm ∅; **L.** 25–30, bluish, 30–50 cm long, 15–25 cm across, ovate-lanceolate, short-pointed, upper side flat, narrowing and grooved towards tip, terminal Sp. 1.5–2 cm long, 3–4 mm across, T. 3–5 mm apart, 2–3 mm long, brown; **Fl.** to 3.5 cm long, pale yellow.

**A. guttata** JACOBI et BOUCHÉ (§ I) (*Manfreda g.* (JACOBI et BOUCHÉ) ROSE, *A. protuberans* ENGELM., *Leichtlinia p.* ROSS.). – Mexico. – **Root**-tuber spherical, 6 cm ∅; **L.** 6–12, oblong-lanceolate, 40–50 cm long, 5 cm across in the middle, narrowing to the base, margins curved upwards, tip conical, underside with projecting midrib, somewhat glossy-green, both surfaces mottled greenish-brown, margins rough; **Infl.** 90 cm tall, Fl. 3.5 cm long, light yellowish-green.

**A. hanburyi** BAK. (§ II/5). – Origin unknown. – **Ros.** simple, acauline, c. 70 cm ∅; **L.** 35–40, stiffly spreading, 7 cm across at the base, almost 2.5 cm thick, tapering gradually, 36 cm long, grooved in upper part, underside convex, slightly grey-green, glabrous, terminal Sp. 3 cm long, margins with a narrow, horny edge, T. bent, to 8 mm long, intermediate Sp. 1–2, smaller.

**A. harrisii** TREL. (§ III/9). – Jamaica. – **Ros.** acauline; **L.** narrow-lanceolate, tapering gradually, 1–1.25 m long, 15 cm across, dark green, fairly glossy, terminal Sp. 10–15 mm long, 2 mm thick, T. 1–2 cm apart, scarcely 2 mm long, straight or bent; **Fl.** 4.5–5 cm long, yellow.

**A. haseloffii** JACOBI (§ II/1). – Mexico. – **St.** branching, Br. retaining old L., 30 cm tall; **L.** 65 cm long, lanceolate, 9 cm across midway, 5 cm across at the base, gradually tapering, margins curved upwards, underside strongly keel-like, convex, fibrous-fleshy, terminal Sp. short, thin, T. 3–4 mm apart, 1 mm long; **Infl.** 1.5 m tall, Fl. 4 cm long, brownish-violet.

**A. hauniensis** J. BOYCE PETERSEN (§ I). – Origin unknown, in cultivation in Botanical Garden, Copenhagen. – **Rootstock** tuberous; **L.** linear-lanceolate, gradually tapering, c. 1 m long, 10–11 cm across, margins horny and slightly dentate; **Infl.** 60 cm tall, a compressed spike, Fl. greenish-brown.

**A. havardiana** TREL. (§ III/6). – USA: Texas. – Habit as for **A. parryi**; **L.** oblong-lanceolate, pointed, flattened-concave, 25–40 cm long, 8 to 15 cm across, rigid, glabrous, grey, terminal Sp. 3–4.5 cm long, T. (15–)25(–35) mm apart, 3–8 mm long, hooked, margins sinuate; **Infl.** 4–5 m tall, Fl. 6.5 cm long, yellow.

**A. haynaldii** TOD. (§ II/5). – ? Mexico. – **Ros.** acauline, to 2 m ∅; **L.** c. 80, ensiform, 1 m long, 9–11 cm across, dark green, younger L. with pale M. band, margins with a horny line, terminal Sp. slender, 2–4 cm long, T. mostly hooked and bent forwards; **Infl.** 8 m tall.

**A. horizontalis** JACOBI (§ II/1). – ? Mexico. – **St.** to 30 cm tall; **L.** not very numerous, oblanceolate, 60 cm long, narrowed below to 6.5 cm across, 11.5 cm across midway, vivid light green, underside angular-keeled, upperside with broad M. groove, terminal Sp. thin, margins cartilaginous, T. close-set, small, with dark brown tip; **Infl.** c. 1 m tall, Fl. 4 cm long, greenish-brown.

**A. horrida** LEM. v. **horrida** (§ II/5) (*A. desmetiana* HORT., *A. regeliana* HORT., *A. gilbeyi* BGR., *A. killisckii* HORT., *A. regeli* HORT. ex BESAUC.). – Mexico. – **Ros.** without offsets, to 60 cm ∅; **L.** numerous, obovate-lanceolate, to 40 cm long, short-pointed, rigid, hard, fleshy, fairly wide, dark green, with a horny margin, terminal Sp. 2.5–3 cm long, T. variously curved, stout, 1–2 cm apart; **Infl.** 3–4 m tall, Fl. 5 cm long, pale yellowish-green.

**A. — v. gilbeyi** BAK. (*A. g.* HORT. ex HAAGE et SCHMIDT, *A. roezliana* v. *nana* (LAURENTIUS) TREL.). – **Ros.** 30 cm ∅; **L.** 15 cm long, terminal Sp. 2–3 cm long, T. 3–6 mm long, margin horny, 2–3 mm across.

**A. — v. latifrons** BESAUC. – **L.** broader.

**A. — v. macrodonta** BAK. – More luxuriant variety.

**A. — v. monstruosa** BESAUC. – Monstrous variety.

**A. — v. recurvispina** BESAUC. – Marginal Sp. very long and recurved.

**A. humboldtiana** JACOBI (§ III/13/1). – Mexico: San Luis Potosí. – Acauline; **L.** fairly numerous, lanceolate, c. 90 cm long, 13 cm across midway, tapering gradually, tip grooved, underside convex to keel-like, dull green, fleshy, terminal Sp. short, stout, T. close-set, very small, triangular. (Acc. TRELEASE, synonymous with **A. celsii**.)

**A. inaguensis** TREL. (§ II/16). – Southern Bahamas. – Habit as for **A. nashii**; **L.** oblong or oblanceolate, contracting abruptly to a blunt tip, often somewhat folded, 40–60 cm long, 6–9 cm across, white-grey, terminal Sp. short, T. uneven, 2–3 cm apart, narrow-triangular, recurved, usually with 1–2 smaller between 2 larger T., ± confluent with black margin; **Fl.** 5 cm long.

**A. indagatorum** TREL. (§ III/8). – Bahamas. – Acauline, without offsets; **L.** lanceolate, gradually tapering, 1.5–2 m long, 20–25 cm across, upperside concave, somewhat grey, terminal Sp. 12 mm long, 3 mm thick, T. 5–12 mm apart, narrow-triangular, about 1 mm long, margin cartilaginous; **Infl.** 9 m tall.

**A. ingens** (BGR. (§ III/2) (*A. americana* AUCT., *A. picta* BGR.). - Mexico. - **Ros.** 2-3 m ⌀; **L.** 30-40, curving gracefully upwards, 2 m long, long-lanceolate, gradually tapering, 35 cm across at base and 12 cm thick, then narrowing in a slender neck to 15 cm across, in the middle 20 cm across and grooved, dark green, terminal Sp. thin, 4-5 cm long, T. widely spaced, absent from tip-area; **Infl.** 10 m tall. Closely resembles **A. americana,** habit more graceful.

**A. intermixta** TREL. (§ III/9). - Haiti. - Similar to **A. antillarum;** insufficiently known species.

**A. ixtli** KARW. (§ III/13/1). - Mexico: Yucatan. - **Ros.** tending to form a St.; **L.** not very numerous, linear-lanceolate, rather blunt, narrowing towards base, 30-38 cm long, subteretely rounded near base, compressed laterally, curving above to a short tip, intense green, terminal Sp. stout, T. small, widely spaced, curved upwards, margins somewhat sinuate.

**A. jaliscana** (ROSE) BGR. (§ I) (*Manfreda j.* ROSE). - Mexico. - **L.** almost 60 cm long, 1 cm across; **Per.** 6-8 mm long, Tep. 12-14 mm long, reddish.

**A. karatto** MILL. (§ III/10) (*A. keratto* HAW.). - Caribbean Is. - **Ros.** acauline, simple; **L.** lanceolate, pointed, upperside concave, 1.5 m long, 20 cm across, green, terminal Sp. black-brown, 3-4 mm long, T. (5-)15(-20) mm apart, bent, margin straight; **Infl.** 5-6 m tall, Fl. 6-6.5 cm long, golden yellow.

**A. karwinskii** ZUCC. (§ III/13/2) (*A. corderoyi* BAK., *A. bakeri* ROSS.). - Mexico. - **St.** long, (1-)3(-4) m tall; **L.** narrow-oblong or lanceolate, 35-70 cm long, 4 cm across, 5-6 mm thick, upper surface grooved, underside convex, green, rigid, hard-fleshy, terminal Sp. 3-angled, 2.5 to 5 cm long, 3-6 mm thick, black, T. 2.5-4.5 cm apart, 3-5 mm long, triangular, tip bent; **Infl.** 4-6 m tall, Fl. 5-6 cm long, brownish-reddish.

**A. kellermannii** TREL. (§ III/7/4). - Guatemala. - **Ros.** 2-2.3 m ⌀, with few offsets; **L.** 1 m long and more, 8-10 cm across, long-lanceolate, slightly pointed, rather concave, very bluish, glabrous, tip of underside slightly granular, terminal Sp. 3-3.5 cm long, 4-5 mm across, brown, T. 3-5 mm long, 10-15 mm apart, triangular on fleshy base.

**A. keratto** SALM ex BONPL. (? § II/1) (*A. salm-dyckii* BAK.). - Insufficiently known species.

**A. kerchovei** LEM. v. **kerchovei** (§ II/5) (*A. poselgeri* v. *k.* TERR.). - Mexico: Puebla. - **Ros.** 60-70 cm ⌀, turning somewhat to one side; **L.** 34 cm long, 6.5 cm across base, gradually tapering, upperside flat-grooved, underside very rounded, light green-grey, younger L. with narrow M. band, terminal Sp. 4 cm long, 3 mm across, margins with horny edge 2-3 mm across, T. 4-7 mm long, thin, well spaced, bent unevenly in different directions; **Infl.** stout, Fl. 3.5 cm long.

**A.** — v. **brevifolia** BESAUC. (*A. beaucarnei* LEM.). - **L.** shorter.

**A.** — v. **diplacantha** LEM. - **L.** narrower, T. always twin-pointed, with 2-3 on each margin.

**A.** — v. **distans** LEM. - T. more widely spaced.

**A.** — v. **glauca** BESAUC. - **L.** blue-green.

**A.** — v. **macrodonta** LEM. - T. larger.

**A.** — v. **miniata** BESAUC. - Small variety.

**A.** — v. **pectinata** BAK. - **L.** broader, without pale M. band, T. 5-15 (usually 10) mm apart, intermediate T. 1-2, smaller.

**A.** — v. **variegata** BESAUC. - **L.** with lighter, very distinctive M. band.

**A. kewensis** JACOBI (§ III/10). - Origin unknown. - **Ros.** short-stemmed, depressed-hemispherical, large; **L.** fleshy at base, thinner in upper part, oblong-lanceolate, upper side concave, almost spoonlike near tip, with several plait-like furrows and keeled, vivid yellowish to light green, terminal Sp. slender, almost black, margins thin and angular, curved upwards, T. flat-triangular, cartilaginous, very small; **Infl.** 4.5 m tall, Fl. 5.5 cm long, golden-yellow.

**A. kirchneriana** BGR. (§ III/13/2). - Mexico. - **L.** ensiform, gradually tapering, grey-green, 1.25 m long, 7-8 cm across, terminal Sp. about 2.5 cm long, conical, 6-7 mm across, black-brown, T. 12-20 mm apart, 3-4 mm long, hooked and curved upwards, black-brown; **Infl.** 5 m tall, Fl. 6 cm long, green.

**A. laetevirens** HORT. is a vivid green form of **A. sobolifera** SALM.

**A. lamprochlora** JACOBI (§ II/1). - Origin unknown. - **Ros.** depressed-globular; **L.** lanceolate, gradually tapering, 53 cm long, 10 cm across, hard-fleshy, bright green, terminal Sp. strong, black-brown, T. close-set, curved backwards. - Insufficiently known species. (Acc. TRELEASE, synonymous with **A. celsii.**)

**A. lata** H. SHINNERS (§ I). - Similar to **A. virginica; L.** shorter and broader, 10-18 cm long, 2-7 cm across; **Infl.** dense, Fl. 2.6-3.5 cm long.

**A. latissima** JACOBI (§ III/7/3; acc. A. BERGER § III/1) (*A. macroculmis* TOD., *A. coccinea* HORT., *A. c.* ROEZL.). - ? Mexico. - **Ros.** c. 3 m ⌀, acauline, **L.** up to c. 50, leathery-fleshy, lanceolate, 1.5-1.6 m long, 20 cm across at the base and 28-30 cm in the middle, then tapering, upperside flat-concave, grooved towards tip, underside much thickened and keel-like, 10 cm thick in lower part, thinner above, margins straight, somewhat recurved, with several grey-green zones, terminal Sp. 4-5 mm long, T. variable in size, merging into the horny margin, ± triangular, brown-red; Fl. pale.

**A. lechuguilla** TORR. (§ II/5) (*A. poselgeri* S. D., *A. lophantha* v. *p.* (S. D.) BGR., *A. l.* v. *pallida* BGR.). - USA: Texas. - **Ros.** 75 cm ⌀, numerous offsets; **L.** sickle-shaped, green or bluish, 20-35 cm long, 2.5-3.5 cm across, upperside indistinctly striped, underside with close-spaced lines, terminal Sp. 1.5-2 cm long, brown, T. curved, 3-5 mm long, 1-2 cm apart; **Infl.** 4-5 m tall.

**A. legrelliana** JACOBI v. **legrelliana** (§ III/9) (*A. laurentiana* JACOBI). - N.Cuba. - **Ros.** acauline, simple; **L.** numerous, oblanceolate, 1-2 m long, narrowing to base, above the middle 20-30 cm across then gradually tapering, fleshy, rather soft, dull light to dark green, upperside folded towards tip, terminal Sp. 18–20 mm long, 4 mm across, T. rostrate, upper T. smaller, horny tip directed backwards; **Infl.** 8–9 m tall, Fl. 7–8 cm long, orange-coloured.

**A. — v. breviflora** TREL. – **Fl.** shorter.

**A. leguayana** BAK. (§ II/6). – Related to **A. ghiesbreghtii**, and possibly only a variety of this; L. only 15 cm long, 4 cm across.

**A. —** HORT. may be related to **A. univitta** HAW.

**A.** cv. Leopoldii (§ II/2) (*A. disceptata* J. R. DRUMM.). – Garden hybrid: **A. filifera** × **A. schidigera** var. – L. numerous, in a dense Ros., thickened at base, 30–40 cm long, 2.5 cm across, 10 mm across upper part, tip rather blunt, terminal Sp. 5 mm long, light green, both surfaces with some white stripes.

**A. lespinassei** TREL. (§ III/13/2). – Mexico. – Acauline or almost so, offsetting; **Ros.** dense; L. about 1.5 m long, 6–7 cm across, gradually tapering, rather thin, yellowish-green, slightly grey-pruinose, terminal Sp. 3–3.5 cm long, 5–6 mm across, T. 15–20 mm apart, towards tip smaller and almost absent, 1–2 mm long, tip slender, recurved, margin thin, translucent.

**A. lindleyi** JACOBI (§ III/1). – Origin unknown. – Short-stemmed; L. numerous, thick-fleshy, glabrous, yellowish-white to grey-green, oblong or almost elliptical, much narrowed towards base, terminal Sp. grooved, black, T. close-set. Related to A. chiapensis.

**A. littaeoides** PAMPANINI (§ III/7/2). – Origin unknown. – **Ros.** acauline, simple; L. c. 30, oblong-spatulate, 25–28 cm long, c. 9 cm across, both surfaces convex at base, upper part gradually tapering and grooved-concave, grey-green, very leathery-fleshy, terminal Sp. very stout, 3.5–4 cm long, T. 10–12 mm long, tip brown, curved back and then up, marginal Sp. rising from broad, cartilaginous base, intermediate T. smaller; **Infl.** 2.4 cm tall, Fl. c. 3 cm long, greenish-yellow.

**A. longipes** TREL. (§ III/9). – Jamaica. – Similar to **A. sobolifera**; L. broad; terminal Sp. subulate with narrow grooves, T. 10–18 mm apart, hooked and recurved, margin straight; Fl. 6–7 cm long, yellow.

**A. longisepala** TOD. (§ III/7/3). – ? Mexico. – Ros. acauline, large; L. oblong-obovate to almost spatulate, with a lanceolate tip, narrowing to 9 cm at the base, in the middle 20 cm across and flat, grooved towards the tip, terminal Sp. 2.5–3 cm long, black-brown, margins straight, T. small, about 5 cm apart, 5 to 7 mm long, tip bent, upper 10 cm of margin entire; Fl. 7 cm long, yellowish.

**A. macrantha** TOD. (§ II/1) (*A. macroacantha* v. *macrantha* TERR.). – ? Mexico. – **Ros.** acauline; L. about 46, oblong-spatulate c. 14 cm across above the middle, terminating quite abruptly in a sharp tip, upperside concave, margins with a continuous horny line, terminal Sp. 2 cm long, brown, T. rather small, hooked, 10–11 mm apart; **Infl.** 1.5 m tall, Fl. 7–8 cm long, Tep. very fleshy, brownish.

**A. macroacantha** ZUCC. v. **macroacantha** (§ III/13/2) (*A. macracantha* HERB., *A. flavescens* v. *macroacantha* JACOBI, *A. besseriana* HORT. ex BAK., *A. b.* JACOBI, *A. b. hystrix* HORT. ex HOOK. *A. candida* HORT., *A. b. longifolia glauca* et *viridis* JACOBI, *A. flavescens* SALM, *A. subfalcata* JACOBI, ? *A. linearis* JACOBI, *A. paucifolia* BAK., *A. oligophylla* BAK., *A. concinna* BAK., *A. sudburiensis* BAK.). – Mexico. – **Ros.** hemispherical to globular or even short-stemmed, offsetting; L. oblong-lanceolate to linear-lanceolate, very rigid, stiff, grey-green to whitish grey, (17–) 25(–55) cm long, 2–4 cm across middle, tapering, terminal Sp. 3 cm long, underside more distinctly convex, 1.5 cm thick at base, T. 6–8, 2–5 mm long, hooked and bent forwards, 3 to 3.5 cm apart, margins cartilaginuous, slightly rough; **Infl.** 3 m tall, Fl. 5.5 cm long, reddish.

**A. —** v. **integrifolia** TREL. – L. with margins entire.

**A. —** v. **latifolia** TREL. – L. broader, sinuate. (Acc. TRELEASE, possibly a hybrid: **A. macroacantha** × **A. verschaffeltii**.)

**A. maculata** RGL. (§ 1). – ? S.Mexico. – Related to **A. maculosa.** Insufficiently known species.

**A. maculosa** (ROSE) HOOK. v. **maculosa** (§ I) (*Manfreda m.* ROSE, *A. maculata* ENGELM.). – USA: S.Texas. – **Rootstock** tuberous, mat-forming; L. herbaceous, 15–30 cm long, 1–2 cm across, linear-lanceolate, grey-green with brown mottling, T. cartilaginous; **Infl.** 1 m tall, Fl. 4–5 cm long, greenish-white, agreeably scented.

**A. —** v. **brevituba** ENGELM. – Per. tube shorter.

**A. —** v. **minor** JACOBI. – L. smaller.

**A. maigretiana** JACOBI (§ II/5). – Mexico. – Ros. simple, about 50 cm ⌀; L. 23–25 cm long, 8 cm across base and 7 cm in the middle, oblong-lanceolate, very thick at base, both surfaces convex, black-green, terminal Sp. long-pointed, somewhat twisted this way and that, T. widely spaced, large, flattened, tip bent, margins woody.

**A. margaritae** BRANDEG. (§ III/15). – Mexico: Baja Calif., Is. of Sta. Margarita and Magdalena. – **Ros.** acauline, offsetting; L. 30–40, rather rounded or oblanceolate, 10–15 cm long and almost as wide, dull green, with grey transverse bands, upper part grooved, terminal Sp. stout, 3 cm long, T. 6–8 along each margin, 4–8 mm long, margin sinuate; **Infl.** 3–4 m tall, Fl. 4.5–5 cm long, pale yellow.

**A. marmorata** ROEZL. (§ III/17) (*A. todaroi* BAK., *A. troubetskoyana* HORT. ex BAK., ? *A. fourcroydes* JACOBI, ? *A. ixtlii* C. KOCH). – Mexico: Puebla. – **Ros.** 2–3 m ⌀, acauline, simple or with few offsets; L. 1–1.8 m long, 25–40 cm across, lanceolate to oblong-lanceolate, pointed, leathery, concave and plicate, tip rolled somewhat inwards, lower part much thickened and narrowed, grey, with green zones, very rough, terminal Sp. 10–20 mm long, 3 to 5 mm thick, dull red, T. 2–3 cm apart and on fleshy cushions with pointed indentations between Sp., the margins in the upper part being straight; **Infl.** 3–5 m tall, Fl. 4 cm long, yellow.

**A. martiana** C. KOCH (§ III/1). – Origin unknown. – **Ros.** acauline, broader than tall; **L.** c. 20–30, linear-lanceolate, about 60 cm long, 9 cm across the middle, gradually tapering, upperside concave, vivid green, soft-fleshy, terminal Sp. conical, T. horny, arranged irregularly.

**A. masipaga** TREL. (§ III/1). – Mexico. – **Ros.** 4–5 m ⌀, offsetting; **L.** narrow, 2–2.8 m long, 12–15 cm across, fairly straight, gradually tapering, upperside/concave, green, with slight transverse bands, terminal Sp. 3–6 cm long, 4–8 mm across, conical, T. 1.5–2 mm long, 2–3 cm apart, often with small intermediate T.; **Infl.** 8–9 m tall, Fl. 7 cm long, greenish-yellow.

**A. massiliensis** BGR. (§ III/2). – Cultivated plant. – **Ros.** 1.3 m ⌀, with many lanceolate-spatulate to oblanceolate L., 70 cm long, 17–18 cm across above the middle, 12 cm across at the base, fleshy, grey-green, with grey zonal stripes, terminal Sp. 18–20 mm long, 4–5 mm thick, T. 4–5 mm long, hooked.

**A. maximowicziana** RGL. (§ II/1). – Origin unknown. – Related to **A. polyacantha**; **Ros.** short-stemmed, densely leafy; **L.** obovate-lanceolate, 6–6.5 cm across at the base and about 9 cm above the middle, gradually tapering, thick, rigid, rich green, terminal Sp. short, T. arranged irregularly, margins sinuate; **Fl.** 1.5–2 cm long, green.

**A. mckelvyana** GENTRY (§ III/15) (*A. aquariensis* TREL. nom.nud.). – Arizona. – Related to **A. desertii** and **A. sobria**; **L.** 3–5 cm broad; **Fl.** 30–40 mm long, 3–4 mm broad midway, Ov. constricted below the tube.

**A. medioxima** TREL. (§ III/10) (*A. americana* GRISEB. p. part.). – Caribbean Is.: Dominica (Windward Is.). – Habit as for **A. karatto**; **L.** gradually tapering, grooved towards tip, 1.25 m long, 15 cm across, green, underside grey-green, terminal Sp. c. 15 mm long, 5–6 mm across, T. 4–5 mm apart, 0.5–1.5 mm long, intermediate T. smaller, margin with a parchment-like line; Fl. 6 cm long, golden yellow.

**A. megalacantha** HEMSL. (§ III/7/2) (*A. reginae* HORT. ex BGR., *A. guadalajara* TREL.). – Mexico: Jalisco. – **Ros.** simple, acauline, 35 to 45 cm ⌀, of 20–25 broad-obovate, short-pointed **L.** 18–20 cm long, 13–13.5 cm across, narrowing to the base where L. is 2.5–3 cm thick, both surfaces grey-green to almost white-grey, upperside slightly rough, terminal Sp. 16–17 mm long, slender, T. 5–6, on fleshy cushions, narrow-triangular, hooked, 7–8 mm long, often with a smaller intermediate T.; **Fl.** 4.5 cm long, yellow.

**A.** — LEM. – Probably a form of **A. legrelliana** JACOBI. – Marginal Sp. more bent.

**A. mescal** K. KOCH (§ III/7/3) (*A. inaequidens* C. KOCH, *A. hookeri* JACOBI, *A. katherinae* BGR., *A. brevicornuta* GENTRY, acc. BREITUNG.). – Mexico. – **Ros.** 2–3 m ⌀, simple; **L.** 1–1.5 m long, 20–30 cm across middle, narrowing to the base, deep green, terminal Sp. 4–5 cm long, 5 mm across, brown, T. on fleshy cushions, 10–18 mm long, 2–3 cm apart, intermediate T. 1–4, smaller, margins sinuate; **Infl.** 3–4 m tall, Fl. about 6 cm long, yellow.

**A. micracantha** SALM (Pl. 8/2) (§ II/1) (*A. chloracantha* SALM, *A. mitis* SALM, *A. oblongata* JACOBI, ? *A. rupicola* RGL.; TRELEASE considers **A. albicans** JACOBI to be yet another synonym). – Mexico: Hidalgo. – **Ros.** acauline or short-stemmed, much branching in age, mat-forming; L. 40–60 cm long, 8–12 cm across middle, ovate-oblong, gradually tapering, narrowing to 4 cm wide at the base, very thick up to the middle, soft-fleshy, sappy, very brittle, underside rather convex-keeled, light green, slightly grey-pruinose, terminal Sp. fine-bristly, margins somewhat cartilaginous, T. numerous, closely spaced, small, brown; **Infl.** 1–1.7 m tall, Fl. 5–5.5 cm long, green.

**A. milleri** HAW. – Insufficiently known species.

**A.** — SALM. – Insufficiently known, possibly a form of **A. americana** L.

**A. millspaughii** TREL. (§ III/8). – Central Bahamas. – **Ros.** acauline; **L.** narrow-lanceolate, 1.25 m long, 15 cm across, upperside concave, green, quite glossy, terminal Sp. 15–20 mm long, 3–4 mm thick, T. 15–20 mm apart, 3–5 mm long, tip curved upwards, margin very straight; **Infl.** large, Fl. 5 cm long, ? yellow.

**A. mirabilis** TREL. (§ III/2). – Mexico: Veracruz. – **Ros.** 4–4.5 m ⌀, acauline, tending to form mats; **L.** 2–2.5 m long, 40 cm across, oblanceolate, pointed or tapering, slightly concave, bright dark green with bluish transverse bands, glabrous, terminal Sp. thin, 6–8 cm long, T. 6–15 mm long, 3–6 cm apart, on fleshy cushions; **Infl.** 8–10 m tall, Fl. 7–8 cm long.

**A. missionum** TREL. (§ III/9) (*A. vivipara* OLDENDORP, *A. americana* AUCT., *A. morrisii* AUCT.). – St. Thomas and Virgin Is. – **Ros.** acauline, of many broad-lanceolate, gradually tapering, concave **L.**, 2.5–2.75 m long, 20 cm across, dark green or slightly grey-green, terminal Sp. 15–25 mm long, triangular-subulate, T. mostly 10–15 mm apart, 3–5 mm long, margin straight; **Infl.** 5–7 m tall, Fl. 5.5 cm long, yellow.

**A. mitriiformis** JACOBI (§ III/1) (*A. coarctata* JACOBI, *A. potatorum* HORT., ? *A. p.* C. KOCH, *A. salmiana* v. *m.* CELS). – Mexico: Pueblo, Veracruz. – **Ros.** 1.5–1.8 m ⌀, offsetting; **L.** 0.75–1.2 m long, broadly elliptical, abruptly pointed or tapering, narrowing from 30–35 cm across middle to the base which is 12 cm across and 12 mm thick, becoming much thinner towards the concave tip, terminal Sp. 4–9 cm long, 5–8 mm across, T. 8–15 mm long, 2–3 cm apart, upper T. curved, the lower ones downwards; **Infl.** 4–5 m tall, Fl. yellowish.

**A. montserratensis** TREL. (§ III/10). – Caribbean Is.: Montserrat (Leeward Is.). – **Ros.** acauline, simple; **L.** oblong-lanceolate, somewhat finely grooved in upper part., tapering gradually or more abruptly, 1.75 m long, 15 cm across, blue-green, terminal Sp. quite slender with finely-toothed margins, T. (3–)10(–15) mm apart, (2–)3(–5) mm long, hooked; **Fl.** 6–6.5 mm long, yellow.

**A. morrissii** BAK. (§ III/9). – Jamaica. – Habit as for **A. sobolifera; L.** variable in shape, dull green, terminal Sp. brown, T. narrow-triangular, very hooked and bent forward, 6–7 mm long, margin sinuate; **Infl.** with numerous adventitious buds, Fl. 3.5–6 cm long.

**A.** × **mortolensis** BGR. (§ III/2). – La Mortola Nursery, N.Italy. – Hybrid: probably **A. ingens** × **A. salmiana.** – **Ros.** very large, acauline, offsetting; **L.** 25–70, spreading gracefully, linear-lanceolate or narrowing gradually from the very thick base which is 20 cm across, L. 1.6 m long, underside convex, upperside with broad grooves, curled inwards towards the tip, grey-green, glabrous, terminal Sp. 4 cm long, T. numerous, broadly triangular, curved, 4 to 5 mm long, 15–30 mm apart, intermediate T. smaller.

**A. muelleriana** BGR. (§ II/6). – Origin unknown. – Hybrid? – **Ros.** acauline; **L.** 12–15, lanceolate, c. 35 cm long, 5 cm across at base and 7.5 cm in the middle, tapering, dull dark grey-green, somewhat rough, somewhat grooved towards tip, underside very convex, terminal Sp. 8–10 mm long, 4 mm thick, margins with or without horny edge, T. very irregular, 7–15 mm apart, intermediate T. (1–)2(–3), smaller.

**A. muilmannii** JACOBI (§ II/1). – ? Mexico. – St. freely branching, about 30 cm tall, retaining remains of old L.; **L.** numerous, soft-fleshy, brittle, glabrous, glossy light green, 35–42 cm long, oblong-lanceolate, 6.5 cm across in the middle, long-pointed, underside very convex, 2 cm thick at base, terminal Sp. 1 cm long, margins with a horny, cartilaginous line, T. small; **Fl.** 16 mm long, light-green.

**A. multiflora** TOD. (§ III/7/1). – ? Mexico. – **Ros.** acauline, simple, of c. 20 oblong-spatulate, long-pointed L., narrowing at base to a long neck, 30–40 cm long, 9 cm across middle and 5.5–6 cm at base, leathery-fleshy, underside convex, terminal Sp. stout, 2–4 cm long, T. numerous, about 9 mm apart, 2–3 mm long, black-brown; **Infl.** 2.5 m tall, Fl. greenish.

**A. murpheyi** GIBSON (§ III/13/2). – Central Arizona. – **Ros.** 1.3 m ⌀, acauline, offsetting; **L.** 60–65 cm long, 5–8 cm across, narrow-lanceolate, broadest in the upper half, concave from middle to tip, bluish-green, transverse-banded, terminal Sp. 15 mm long, 4 mm across, T. stout, curved outwards, 3–5 mm long, 2.5 to 4.5 mm apart; **Infl.** 4–6 m tall, Fl. 3.5–4 cm long, greenish-brown.

**A. nashii** TREL. (§ III/16). – Southern Bahamas. – **Ros.** acauline, offsetting, soon forming a mat; **L.** narrow-oblong, 30–50 cm long, 4–5 cm across, upperside concave, grey-green, often tinged red, terminal Sp. 15 mm long, 3 mm thick, T. usually 3–5 mm apart, scarcely 2 mm long, often confluent; **Fl.** 3.5 cm long, golden yellow.

**A. neglecta** SMALL (§ III/13/1). – USA: Florida. – **Ros.** acauline; **L.** numerous, 1.5 to 2.8 m long, thick at base, broadest in the middle, then gradually tapering, grey, terminal Sp. brown, T. closely set, numerous, small; **Infl.** c. 13 m tall, Fl. 5.5 cm long, yellowish-green.

**A. nevidis** TREL. (§ III/10). – Caribbean Is.: Nevis (Leeward Is.). – Habit as for **A. karatto;** terminal Sp. conical, about 10 mm long, 4 to 5 mm across, T. 10–15, about 20 mm apart, 2 mm long, triangular.

**A. nissonii** BAK. – Little known species, perhaps related to **A. lophantha** SCHIEDE.

**A. nizandensis** CUTAK (§ II/1?). – Mexico: Oaxaca. – **Ros.** acauline, offsetting, of 10–15 L. spreaed almost horizontally, linear-lanceolate, sappy, fleshy, 30–35 cm long, 2.5–3 cm across, 13 mm thick, dark green, with a lighter M. stripe, terminal Sp. red, not sharp, T. small, white; **Infl.** 90 cm tall, Fl. 2 cm long, yellowish-green.

**A. noli-tangere** BGR. (§ II/5) (*A. horrida* HORT.). – ? Mexico. – **Ros.** with as many as 60 L. 45–50 cm long, 5.5–7 cm across middle and 4.5 cm across base which is 2.5 cm thick and very convex, lanceolate-ensiform, gradually tapering, with broad grooves, dull dark green or grey-green, glabrous, terminal Sp. to 4 cm long, 5 mm across, T. ± directed downwards, stout and woody, triangular or sickle-shaped, margins sinuate with a broad, horny line; **Infl.** compact, Fl. 3.5 cm long, pale green.

**A. obducta** TREL. (§ III/10) (*A. americana* GRISEB. p. part.). – Caribbean Is.: Antigua (Leeward Is.). – Habit as for **A. karatto; L.** narrow, green, dull, rather grey, terminal Sp. 10–15 mm long, 5 mm across, T. 10–15 mm apart, 2–3 mm long, narrow-triangular, straight or variously bent, margin straight or slightly sinuate; **Fl.** 6–6.5 cm long.

**A. obscura** SCHIEDE (§ II/5) (*A. grandidentata* JACOBI, *A. horrida* v. *micrantha* BAK.). – Mexico. – **Ros.** simple, 1.3 m ⌀; **L.** 45–60 cm long, 10 cm across, rigid, dark green, terminal Sp. 2.5–3 cm long, margins horny, T. triangular, straight or variously bent, 1–4 cm apart, 4 to 7 mm long.

**A. offoyana** JACOBI. – Insufficiently known species, possibly identical with **A. legrelliana** JACOBI.

**A. oliverana** (ROSE) BGR. (§ I) (*Manfreda o.* ROSE). – Mexico: Jalisco. – **L.** c. 6, up to 50 cm long, lower part deeply grooved, upper part flat, 31 mm across, pointed, light green, reddened towards base; **Infl.** 1.8 m tall, Fl. greenish-yellow.

**A. opacidens** TREL., acc. BREITUNG a form of **A. tenuispina** TREL. – **L.** lower surface and margins rough, terminal Sp. truncate, triangular.

**A. ornata** JACOBI is a mottled form of **A. sobolifera** SALM.

**A. ousselghemiana** JACOBI (§ II/1). – Mexico. – **Ros.** acauline, c. 30 cm ⌀, mat-forming; **L.** 40–45 cm long, linear-lanceolate, 8 cm across at base and 11 cm in the middle, then narrowing, 2.5 cm thick at base, underside convex-keeled, sides thin and curved upwards, soft-fleshy, light grey-green, white-grey pruinose, terminal Sp. 15 mm long, T. 0.5–1 mm long, triangular, ± hooked; **Infl.** 1.6 m tall, Fl. 5 cm long, brownish.

**A. pachycentra** TREL. (§ III/7/4). – Guatemala: Sierra de Santa Cruz. – **Ros.** 2.5–3.5 m ⌀, simple, acauline; **L.** 35–60 cm long, 15–20 cm across, broad-oblong to lanceolate, pointed, upperside concave, underside minutely granular, bluish, terminal Sp. 4–6 cm long, somewhat wavy, T. 5–10 mm long, (2.5–)4(–5) cm apart, on a fleshy base, margin somewhat indented; **Infl.** 5 m tall, with numerous adventitious buds, Fl. in dense clusters.

**A. pacifica** TREL. (§ III/13/2) (*A. yaquiana* TREL., *A. owenii* I. M. JOHNSTON). – Mexico. – **Ros.** 1–1.5 m ⌀, acauline, offsetting; **L.** 75 cm long, 3.5–5 cm across, rigid, yellow-green, lightly glaucous and zoned, terminal Sp. 15 to 25 mm long, 5 mm across, red-brown, T. 3 to 6 mm long, 15–25 mm apart, curved upwards, triangular, margin horny, straight; **Infl.** 3 m tall, Fl. 5 cm long, greenish-yellow.

**A. pallida** SART. – Insufficiently known species from the complex of **A. miradorensis** JACOBI.

**A. palmeri** ENGELM. v. **palmeri** (§ III/2). – S.Arizona (USA) to S.W.Mexico. – **Ros.** acauline, dense; **L.** to 1.5 m long, 5–12 cm across, lanceolate, gradually tapering, upperside concave-grooved, terminating in a conical tip, dull dark green or ± grey-green, terminal Sp. 2–3 cm long, T. fairly close-set, variable in size, curved, margins slightly sinuate; **Infl.** 2.5–6.5 m tall, Fl. 4–5.5 cm long, greenish to yellowish-white.

**A.** — v. **chrysantha** (PEEBLES) LITTLE (*A. c.* PEEBLES). – Fl. orange-cloured.

**A. pampaniniana** BGR. (§ III/7/2). – ? Mexico. – **Ros.** offsetting, 80–90 cm ⌀, of 20–30 stiffly projecting; **L.** which are hard-fleshy to fibrous, elliptical or obovate-lanceolate, tapering above the middle, 45–50 cm long, 8–8.5 cm across at base and 15–16 cm above the middle, convex at the base, then concave and grooved, underside of tip with keel-like thickening, light grey-green to white-grey, terminal Sp. 3 cm long, 6 mm across, T. 2.5–3.5 cm apart, on fleshy cushions, narrow-triangular, 8–9 mm long.

**A. papyrocarpa** TREL. (§ III/18). – Gr. Antilles. – **Ros.** acauline; **L.** not very numerous, oblong to long-oblanceolate, tapering upperside somewhat concave and frequently rather folded, 75–125 cm long, 15 cm across, terminal Sp. 8–15 mm long, 3 mm across, marginal Sp. 10–15 mm apart, 1–4 mm long, mostly directed backwards, often with 1–2 intermediate Sp.; **Infl.** 4 m tall, Fl. 4 cm long, light yellow.

**A. parrasana** BGR. (Pl. 8/3) (§ III/6). – Mexico: Coahuila. – **Ros.** almost spherical, dense, 60 cm ⌀, seldom offsetting; **L.** very stiff, thick, obovate, to 30 cm long, 11–16 cm across, upper part coming abruptly to a short tip, glabrous, dull green, light blue-grey pruinose, underside very convex, terminal Sp. 2.5 cm long, brown, margins sinuate-dentate, upper part having curved Sp. 17–20 mm long.

**A. parryi** ENGELM. v. **parryi** (§ III/6) (*A. parryi* HGE. et SCHMIDT nom. nud., *A. americana* v. *latifolia* TORR., *A. neoamericana* WOOTON et STANDLEY, *A. p.* v. *couesii* (ENGELM. et TREL.) KEARNEY et PEEBLES, *A. couesii* ENGELM., *A. applanata* v. *p.* MULF., *A. marcusii* L. DE SMET, *A. marcusae* HORT. ex TREL., *A. marensii* HORT. ex TREL., *A. parreyi* HORT. ex TREL., *A. paryi* HORT. ex TREL., *A. parayi* HORT. ex TREL.). – N. Arizona (USA) to Mexico. – **Ros.** acauline, simple or possibly mat-forming; **L.** stiff, broad-oblong, pointed or tapering, flat-concave, 25 to 30 cm long, 6–10 cm across, glabrous, grey, terminal Sp. almost straight, 2–2.5 cm long, 5–6 mm across, marginal Sp. 1.5–2 cm apart, 3–5 mm long, narrow-triangular, straight or recurved; **Infl.** 3–5 m tall, Fl. 5.5–6 cm long, creamy yellow. Virtually frost-resistant.

**A.** — v. **huachucensis** (BAK.) LITTLE (*A. huachucensis* BAK., *A. applanata* v. *h.* (BAK.) MULFORD). – **L.** broad-oblong, to 65 cm long, to 35 cm across, terminal Sp. 5–6 mm long; Fl. to 7.5 cm long.

**A.** — v. **integrifolia** BREITUNG. – **L.** margin entire, marginal Sp. absent.

**A. parviflora** TORR. (Pl. 8/4) (§ II/2) (*A. hartmanii* S. WATS.). – Mexico: Sonora, Chihuahua; USA: Arizona: Pimera Alta. – **Ros.** simple, seldom offsetting, 15 cm ⌀, of many rigid, hard, narrow-lanceolate **L.**, 10 cm long, 12 mm across, dark green, upperside with white lines, terminal Sp. 5 mm long, margins in upper part with white threads and in lower part toothed; **Infl.** 1–1.5 m tall, Fl. 15 mm long.

**A. patonii** TREL. (§ III/6). – Mexico: Durango. – **Ros.** acauline; **L.** oblong, abruptly pointed, flat-concave, 30 cm long, 20 cm across, glabrous, slightly grey-green, terminal Sp. 3–3.5 cm long, 6 mm across, T. 2–2.5 cm apart, 6–7 mm long, very bent, red-brown, margin slightly furrowed; Fl. 6.5 cm long.

**A. pavoliniana** PAMPANINI (§ II/6). – ? Mexico. – **Ros.** acauline, of some 25 L. 33–35 cm long, dark green, lanceolate-spatulate, very convex at base, somewhat grooved at tip, terminal Sp. rather sinuous, 2.5–3 cm long, stout, margins with a grey horny edge, T. often hooked; Infl. 2.6 m tall, Fl. 4 cm long.

**A. peacockii** CROUCHER (§ II/6) (*A. roezliana* v. *p.* (CROUCHER) TREL., *A. ghiesbreghtii* v. *p.* TERR., *A. henriquesii* BAK., *A. rohanii* HORT.). – Mexico: Tehuacan. – **Ros.** simple, 4–5 m ⌀, of c. 20 **L.** 60–75 cm long, 7–10 cm across, somewhat plicate, 6 cm thick at base, rather thinner in upper part, boat-shaped, terminal Sp. 2 to 3.5 cm long, T. 1.5–4 cm apart, often twin-tipped, 3–5 mm long, intermediate Sp. 1–3, smaller; **Infl.** 2.5 m tall, Fl. 2 cm long, purple.

**A. perlucida** JACOBI (§ II/1). – Origin unknown. – **Ros.** depressed-spherical; **L.** lanceolate, 60 cm long, 10–13 cm across, extended in a long, thin, fleshy, terete point, very glossy, sap-green, upper part having a dark edge, terminal Sp. thin, brown, T. close-set, small, blunt-triangular, brown. (TRELEASE considers this species synonymous with **A. celsii** HOOK.)

**A. petiolata** TREL. (§ III/12). – Caribbean Is. – **St.** 1 m tall, offsetting (?); **L.** lanceolate, gradually tapering, narrowing abruptly at the base into a long neck, 1.1 m long, about 17 cm across, blue-grey, terminal Sp. 2.5–6 cm long, 3–4 mm across, subulate, T. (15–)30(–50) mm apart, with a base 3–5 mm deep, narrow-triangular, straight or hooked; **Fl.** 3.5–4 cm long.

**A. picta** SALM (§ III/2) (*A. ingens* v. *p.* (SALM) BGR., *A. americana* v. *p.* TREL., *A. longifolia* v. *p.* RGL., *A. mexicana* v. *p.* CELS., *A. milleri* v. *p.* VAN HOUTTE). – Mexico. – **Ros.** 2–3 m ⌀, freely offsetting; **L.** 2 m long, 17 to 20 cm across, lanceolate, pointed, thick at base, rather thin above the middle, leathery, upper part hanging over gracefully, glabrous, dark and pure green, margins having a yellow band, terminal Sp. 4–5 cm long, 3–5 mm across, T. 5–10 mm long, 3–4 cm apart, on fleshy cushions, tuberculate on the front or back, intermediate Sp. small; **Infl.** 9–10 m tall.

**A. planifolia** S. WATS. (§ I) (*Manfreda p.* (S. WATS.) ROSE). – Mexico: Chihuahua. – **Rootstock** persisting 3–4 years; **L.** 18–25 cm long, 2.5–6 cm across, lanceolate, finely-toothed; **Infl.** 1.2–1.5 m tall, Fl. 2 cm long.

**A. polyacantha** JACOBI (§ II/1) (*A. kerratto* BAK.). – ? Mexico. – **Ros.** initially simple, but after flowering is over, or even earlier, dividing and forming a mat; **L.** 55–70 cm long, 11–15 cm across middle, lanceolate, narrowing at the base to 7–9 cm, 3 cm thick, leathery-fleshy, tapering above and thinner, rather rigid and stiff, dark green or slightly grey-green, terminal Sp. 7 to 15 mm long, black-brown, T. bent, numerous, 1.5–4 mm long, merging in the upper part into a brown horny line; **Infl.** 2–2.5 m tall, Fl. 2.5 cm long, light-green. Variable species.

**A. portoricensis** TREL. (§ III/9). – Puerto Rico. – **Ros.** acauline; **L.** broad-lanceolate, tapering, upperside concave, somewhat folded at tip, 1–1.5 m long, 15–20 cm across, glossy dark green, terminal Sp. 1–2 cm long, subulate, brown, T. 1.5–3 cm apart, 2–5 mm long, broad-triangular, straight or bent; **Infl.** 5–6 m tall, Fl. 5.5 cm long, greenish-yellow.

**A. potatorum** ZUCC. v. **potatorum** (§ III/7/2) (*A. scolymus* KARW., *A. elegans* HORT. ex SALM, *A. latifolia* HORT. ex SALM, *A. pulchra* HORT. ex SALM). – Central Mexico. – **Ros.** 20–25 cm ⌀, usually simple; **L.** obovate-spatulate, 20–30 cm long, 9–11 cm across, 7.5 cm across base and 15 mm thick, grey-green, pruinose surface easily rubbed off, margins sinuate-dentate, terminal Sp. 20–28 mm long, T. triangular on a broad base, sharply pointed, 6 mm long, lower T. 1 cm across, 12–20 mm apart, black-brown; **Infl.** to 3.7 m tall, Fl. 5.5 cm long, yellow-green.

**A. —** v. **verschaffeltii** (LEM.) BGR. (Pl. 8/5) (*A. v.* LEM., *A. tehuacanensis* KARW.). – **L.** attractive white-grey, coming abruptly to a short point, T. red or yellow-brown. (BREITUNG maintains that **A. verschaffeltii** LEM. is a valid species.)

**A. pratensis** (ROSE) BGR. (§ I) (*Manfreda p.* ROSE). – W.Mexico. – **Root-**Tub. oblong, closely covered with brown fibres; **L.** linear-lanceolate, 20–25 cm long, 8–10 mm across, pointed, glabrous, green; **Fl.** dark-red.

**A. promontorii** TREL. (§ III/4) (*A. brandegeei* TREL.). – Mexico: Baja Calif. – Habit as for **A. aurea; L.** oblanceolate, rather short-pointed, upperside concave, grey-green, terminal Sp. rather conical, T. horny, sharp; **Infl.** 4–6 m tall, Fl. 5–5.5 cm long, golden-yellow.

**A. pseudotequilana** TREL. (§ III/13/1). – Mexico: Jalisco. – **Ros.** 1–1.2 m ⌀; **L.** 60–65 cm long, 2.3 cm across, rather thin, flat, green, with slightly bluish transverse bands, terminal Sp. 8 mm long, 2 mm across, T. triangular, curved upwards, 1–2 mm long, 4–5 mm apart, margin cartilaginous; **Fl.** unknown.

**A. pubescens** RGL. et ORTG. (§ I) (*A. brachystachys* v. *p.* TERR., *Manfreda maculata* ROSE, *Polianthes maculata* MART.). – Mexico. – **L.** 12–15 forming a Ros., 25–30 cm long, 4 cm across, dull dark green, with brown markings towards base, very papillose and softly-haired; **Infl.** to 1 m tall, Fl. 3 cm long.

**A. pugioniformis** ZUCC. (§ III/13/2). – Insufficiently known species. **L.** narrow-linear, gradually tapering, upperside slightly concave, bluish, terminal Sp. stout, T. small, widely spaced.

**A. pumila** J. G. BAK. (§ II/5) (*A. simonis* HORT.). – ? Mexico. – **Ros.** 3–4 cm ⌀, branching, of 5–8 very short, ovate-triangular, very thick **L.**, upperside concave, underside very convex and somewhat keeled, blunt, constricted abruptly to a small tip, terminal Sp. short, sharp, T. small, bent, L. upper surface grey-green, underside with dark-green stripes.

**A. purpusorum** BGR. (§ II/5) (*A. roezli* HORT.). – Mexico: Puebla. – **Ros.** acauline, offsetting, 35 cm ⌀; **L.** hard, firm, stiff, about 17 cm long, 5 cm across base, ± triangular, tapering, upperside with broad, deep grooves, light green, with broad, light-coloured M. bands, underside much rounded, terminal Sp. c. 2.5 cm long, T. narrow triangular.

**A. quiotifera** TREL. (§ III/1). – Mexico: Durango. – **Ros.** 2.5–2.8 m ⌀, offsetting; **L.** 1.2 m long, 30 cm across, fairly concave, ash-grey, terminal Sp. (3–)10(–15) cm long, T. curved downwards, with horny bases, margin straight; **Infl.** 6 m tall, Fl. 7–8 cm long, yellow.

**A. ragusae** A. TERR. (§ III/7/2). – Origin unknown. – **Ros.** acauline of many **L.** thickened at the base, widening in upper part and oblong to oblanceolate, upperside concave, green, underside convex and very rough, margins deeply sinuate, terminal Sp. a little decurrent, T. large with stout, forward-pointing tips; **Infl.** 2–3 m tall, Fl. 6–7 cm long, greenish-yellow.

**A. revoluta** (ROSE) KLOTZSCH (§ I) (*Manfreda r.* ROSE). – Mexico. – ♃; **L.** herbaceous, c. 30 cm long, 2 cm across, narrowing above base and terminating in a tip which is rolled inwards and terete, underside keeled, green to grey-green, both surfaces with a lighter M. stripe, margins with narrow, parchment-like borders; **Infl.** c. 1.2 m tall, Fl. 32–34 mm long, greenish.

**A. rigida** MILL. – Venezuela. – Species now lost, related to **A. fourcroydes** LEM.

**A. roezliana** BAK. v. **roezliana** (§ II/5). – Mexico: Puebla. – **Ros.** 1.6 m ⌀, offsetting, mat-forming; **L.** 30–65 cm long, 8–12 cm across, light green, upperside having a broad M. stripe, terminal Sp. 2–3 cm long, T. 8–40 mm apart,

5–10 mm long, usually curved upwards, often with 1–2 smaller intermediate Sp.

**A. — v. inghamii** BAK. – **L.** 4–5 cm across, narrowing towards the base to 2.5–3 cm.

**A.** cv. Romanii (§ III/2). – Hybrid: **A. filifera** × **A. spec.** – **Ros.** with numerous linear-lanceolate, dark green L., somewhat concave towards tip, terminal Sp. 1 cm long, margins with a horny edge which is almost entire but tends to split away.

**A. rubescens** SALM (§ III/13/2) (*A. flaccida* SALM, *A. punctata* SALM, ? *A. serrulata* KARW., *A. sobolifera* v. s. TERR., ? *A. erubescens* ELLEMEET). – Mexico: Oaxaca. – **Ros.** acauline; L. rather thin, flexible, long-lanceolate, 75 cm long, 5 cm across, grey-green, often reddened, terminal Sp. 2.5 cm long, 4 mm thick, T. (1–)2(–2.5) cm apart, with a tip 3–5 mm long, curved forwards and often doubly curved, margin straight; **Infl.** 3 m tall.

**A. samala** TREL. (§ III/7/4) (*A. weingartii* BGR.). – Guatemala. – **Ros.** 1–1.8 m ⌀, acauline, tending to form a mat; L. quite numerous, 60 cm long, 15 cm across, oblanceolate, somewhat tapering, upperside slightly concave, underside minutely rough, bluish, terminal Sp. 3 cm long, 3–4 mm across, almost aciculate, T. variable, (1–)3(–5) mm long, (8–)15(–20) mm apart, broad-triangular, on fleshy cushions, margin sinuate; **Infl.** with numerous adventitious buds.

**A. sartorii** C. KOCH (§ II/1) (*A. aloinea* C. KOCH, *A. pendula* SCHNITTSP., *A. rubrocincta* JACOBI, *A. noackii* JACOBI, *A. caespitosa* TOD., *A. s.* v. *caespitosa* TERR.). – Mexico; ? Guatemala. – **Ros.** short-stemmed at maturity, mat-forming; L. narrow-lanceolate, 60 cm long, 7 cm across middle, vivid light green with a pale M. band, gradually tapering, terminal Sp. 4 mm long, T. small, triangular, not sharp-tipped; **Infl.** 1.1 m long, Fl. 3.5 cm long, green.

**A. saundersii** HOOK. f. (§ III/7/2). – ? Mexico. – **Ros.** acauline; L. 15–20, oblong-oblanceolate, sharply tapering, 45–60 cm long, fairly thick, upperside rather concave, terminal Sp. rather long, T. pointed, from a triangular base, margins slightly sinuate; **Infl.** 4 m tall, Fl. about 5 cm long, yellowish.

**A. scabra** SALM (§ III/6) (*A. wislizenii* ENGELM., *A. noah* HORT. ex TREL.). – Mexico: Coahuila. – Resembles **A. applanata**; **Ros.** acauline, of about 30 stiffly erect, oblong L., 20–30 cm long, 10–15 cm across middle, very concave, rather thin and flexible, almost white, terminal Sp. 15–20 mm long, 3–4 mm thick, T. 6–8 mm long, (15–)20(–25) mm apart, hooked, lower Sp. smaller and closer together; **Infl.** 4 m tall, Fl. 5.5–6.5 cm long, yellowish.

**A. scheuermaniana** TREL. (§ III/10). – Caribbean Is. – Habit as for **A. karatto**; L. lanceolate, gradually tapering, 1.5–1.75 m long, 20 cm across, grey-green, terminal Sp. almost black, 1.5–2 cm long, 4–5 mm thick, T. 15–25 mm apart, about 2 mm long, narrow-triangular, tip recurved; **Infl.** with numerous adventitious buds.

**A. schidigera** LEM. v. **schidigera** (§ II/2) (*A. filifera* v. s. TERR., *A. f.* v. *pannosa* SCHEIDW., *A. f.* v. *adornata* SCHEIDW., *A. vestita* S. WATS.) – Mexico. – **Ros.** flat-globular without lateral shoots, with many L., and a central bud which is thick and conical and further thickened by the swollen L. bases, 90–100 cm ⌀; L. with a thickened, triangular-ovate base 7 cm across, abruptly constricted to narrow-linear, up to 50 cm long, 14 mm across the middle, tapering gradually above, the tip abruptly constricted, terminal Sp. 7–12 mm long, margins with fine cartilaginous lines and numerous, thin threads splitting away; **Infl.** 3.5 m tall, Fl. 5–5.5 cm long, brown-red.

**A. — v. taylori** (BESAUC.) JACOBS. (*A. t.* BESAUC.). – L. broader.

**A. schlechtendahlii** JACOBI (§ III/2). – Mexico: Sonora. – **Ros.** acauline with few offsets, about 1.1 m ⌀; L. oblanceolate, 65 cm long, narrowing above the base to 8 cm across, 12–13 cm across in the upper third, then tapering, swollen at base, flat-concave in the upper part, grey-green or whitish-green, the grey colour appearing as either longitudinal or transverse bands, minutely rough, terminal Sp. 3 cm long, T. 3–3.5 cm apart, 9–10 mm long, hook-tipped, black-brown, margins sinuate, intermediate Sp. smaller.

**A. schneideriana** BGR. (§ III/13/2). – Mexico. – St. loosely covered by short L.-Sh.; **Ros.** of 17–20 linear-lanceolate L., 4 cm across middle, then gradually tapering, narrowing at base to 2.5 cm, grooved towards tip, underside convex, leathery, dull dark green, rough, terminal Sp. 17–30 mm long, 5 mm thick, margins with a pale cartilaginous line, T. irregular, larger Sp. 3–4 cm apart on fleshy cushions, 6 mm long, hooked, intermediate Sp. smaller.

**A. schottii** ENGELM. v. **schottii** (§ II/2) (*A. geminiflora* v. (?) *sonorae* TORR., *A. mulfordiana* TREL.). – USA: S. Arizona. – **Ros.** often bent to one side, on a caudex 5 cm thick from which offsets arise, Ros. L. 20–27 cm long, with a broad-ovate base which is almost 4 cm long and across, abruptly narrowing to narrow-linear, 6–7 mm thick, upperside flat-convex, 5–8 mm across the middle, narrowing in the upper part, both surfaces green, minutely rough, margins with a fine cartilaginous line and fine threads splitting off, terminal Sp. 7–10 mm long, subulate; **Infl.** 1.5–1.75 m tall, Fl. 5 cm long, light yellow.

**A. — v. treleasei** (TOUMEY) KEARNEY et PEEBLES (*A. t.* TOUMEY). – L. 30–40 cm long, 15–25 mm across, upperside almost flat, dark green.

**A. sebastiana** GREENE (§ III/5) (*A. shawii* v. s. (GREENE) GENTRY, *A. avellanidens* TREL., ? *A. applanata* v. *parryi* PURP.). – Mexico; central Baja Calif. – **Ros.** 80–100 cm ⌀, short-stemmed, tending to be mat-forming; L. 15 to 60 cm long, 6–12 cm across, ovate to lanceolate, pointed or tapered, broadly concave, bluish with green transverse bands, glabrous or granular-rough, terminal Sp. 2–7 cm long, 5–6 mm across, T. (3–)10(–15) mm long, (10–)20(–50) mm apart, narrow to broadly triangular, thickened below in a lentiform base merging into the cartilaginous margin; **Infl.** 2–3 m tall, Fl. 6–8.5 cm long, yellow.

**A. seemanniana** JACOBI (§ III/7/4) (*A. s.* BESAUC.). – Guatemala; Nicaragua. – **Ros.** acauline, small, of c. 20 **L.**, 13–23 cm long, oblong-spatulate, 8–9 cm across, narrowing abruptly to the base, with a short, broad tip, upperside flat, somewhat hollowed towards tip, thickened base very convex on the underside which is keeled in upper part of L., soft-fleshy, slightly fibrous, very brittle, somewhat glossy green, with lighter and dark green longitudinal lines, both sides rough, margins deeply and narrowly sinuate, terminal Sp. 15 mm long, T. fine, very pointed, tip bent, intermediate T. smaller.

**A. sessiliflora** HEMSL. (§ I). – Central Mexico. – **L.** 30–45 cm long, 2.5 cm across middle, gradually tapering, margins with closely-spaced fine T.; **Infl.** 60 cm tall or more, Fl. 4–5 cm long.

**A. shaferi** TREL. (§ III/9). – E.Cuba. – **L.** long-lanceolate, gradually tapering, 75 cm long, 10 cm across, green, terminal Sp. 1 cm long, 3 mm across, T. 1–2 cm apart, about 1 mm long, somewhat bent, margins slightly sinuate; **Infl.** 6–7 m tall, Fl. 5 cm long, light yellow.

**A. shawii** ENGELM. (§ III/5) (*A. s.* BRANDEG., *A. orcuttiana* TREL., *A. pachycantha* TREL.). – Mexico: Baja Calif. – Mat-forming, **St.** often 3 m tall; **L.** 20–50 cm long, 6–12 cm across, ovate to ovate-lanceolate, somewhat concave, stiff, green, glossy, terminal Sp. 2–4 cm long, 4–9 mm across, T. 10–15 mm long, 10–25 mm apart, variously curved, connected by the horny margin; **Infl.** 3–3.5 m tall, involucral L. and buds purple-brown, Fl. 7–10 cm long, greenish-yellow.

**A.** × **simonii** ANDRÉ (§ III/7/2). – Cultivated hybrid: **A. vandervinnenii** × **A. verschaffeltii.** – Similar to **A. verschaffeltii** LEM.

**A.** cv. Simonii (§ II/5). – Cultivated hybrid: **A. xylonacantha** × **A. univittata.** – L. c. 45 cm long, lanceolate-ensiform, 5.5 cm across above the base and 7.5 cm above the middle, slightly concave, terminal Sp. 2 cm long, margins straight, T. as for **A. xylonacantha** but considerably smaller, upper ones being larger, 25 mm apart.

**A. singuliflora** (S. WATS.) BGR. (§ I) (*Bravoa s.* S. WATS., *Manfreda s.* (S. WATS.) ROSE). – Mexico: Chihuahua. – **Root-**Tub. about 4 cm ⌀; L. 8–10, about 30 cm long, 4–6 cm across, with a dry, wavy margin; **Infl.** 60–90 cm tall, Fl. 3 cm long, reddish.

**A. sisalana** PERR. v. sisalana (§ III/13/1) (*A. rigida* v. *s.* ENGELM., ? *A. houlettii* JACOBI, ? *A. houlettiana* CELS. ex JACOBI, *A. laevis* HORT. ex BAK.). – Mexico: from Yucatan area, cultivated plant, grown to supply fibre. – **St.** to 1 m tall, 20 cm ⌀, offsetting; **Ros.** dense; **L.** stiff, narrowly lanceolate-ensiform, 1.1 to 1.8 m long, 3.5–4.5 cm thick at the base and 7–7.5 cm across, slightly grooved, grey-green, with a cartilaginous margin, terminal Sp. small, black-brown; **Infl.** 6–7 m tall, Fl. 6.5 cm long, green.

**A.** — v. **armata** TREL. – **L.** margins with horny Sp. 2–3 mm long.

**A. smithiana** JACOBI (§ III/13/1). – Mexico: San Luis Potosí. – **Ros.** acauline, of only few fleshy L., upper part fleshy-leathery, rich dark sap-green, oblong-lanceolate, narrowing to base, upper part with a fairly short point, the tip being rolled inwards, conical and grooved, underside very convex-keeled, terminal Sp. short, stout, brown, T. close-set, very small, blunt-triangular. (TRELEASE regards this species as being **A. celsii** HOOK.)

**A. sobolifera** SALM (§ III/9) (*A. americana sobolifera* HERMANN, *A. secunda* SLOANE, *A. americana* LAM., *A. vivipara* LAM. p. part.). – Jamaica. – **Ros.** large, acauline, with many S-shaped, broadly or narrowly-lanceolate L., tapering shortly or gradually, 1.25 m long, 15–25 cm across, upperside concave or folded, light green, terminal Sp. 1.5–2.5 cm long, 2–4 mm across, marginal Sp. about 1 cm apart, 2–4 mm long, straight or somewhat bent, margin between the T. notched or sinuate; **Infl.** 3–6 m tall, bearing many adventitious buds, Fl. 5 cm long, yellow.

**A. sobria** BRANDEG. v. **sobria** (§ III/15) (*A. cerulata* TREL., *A. carminis* TREL., *A. affinis* TREL., *A. subsimplex* TREL. acc. BREITUNG). – Mexico: Baja Calif., Is. in Gulf of Calif., Sonora. – **Ros.** 30–110 cm ⌀, simple or forming dense clumps, acauline; **L.** linear or oblong-triangular to broad-oblong or elliptical-lanceolate, short-pointed or tapering, very concave, grooved towards tip, thick, (15–)50(–100) cm long, 2–5 cm across, rigid, bluish grey-green, usually somewhat transverse-banded, glabrous to granular-rough, terminal Sp. 2–5 cm long, 2–4 cm across, often wavy, T. 3–10 mm long, (1–)3(–4) cm apart, narrow-triangular, straight or variously curved, margin sinuate between T.; **Infl.** 2–3.5 m tall, Fl. 3.6–5 cm long, yellow.

**A.** — v. **roseana** (TREL.) I. M. JOHNSTON (*A. roseana* TREL., *A. connochaeton* TREL.). – Marginal T. stout, 10–15 mm long, 10 mm across, mostly broad-triangular, variously bent or twisted, with the terminal Sp. twisted, 5–7 cm long.

**A. splendens** JACOBI (§ II/5) (*A. heteracantha* v. *s.* TORR.). – Origin unknown. – **Ros.** mound-shaped, with offsets from the axils; **L.** 15–20 cm long, 5 cm across, green, with a pale M. stripe, terminal Sp. 1.5 cm long, 0.5–1 mm across, T. 1–2 cm apart, 3–5 mm across, mostly curved downwards or bent upwards.

**A. stenophylla** JACOBI is probably **Furcraea.**

**A. striata** ZUCC. v. **striata** (Pl. 8/6) (§ II/3). – Mexico: Hidalgo. – In age with short St.; **Ros.** with many stiffly projecting **L.** 40–45 cm long, thickened at base, then abruptly narrowing and linear-tapering, 7 mm across, initially triangular, upper part rhombic in cross-section, grey-green, with darker stripes, margin entire, terminal Sp. 6–8 mm long, brown; **Infl.** very tall, Fl. 2.5 cm long, green.

**A. — v. mesae** BGR. — **L.** very long and very convex, but not angular, green, terminal Sp. small, triangular.

**A. — v. recurva** BAK. (*A. striata* HOOK., *A. r.* ZUCC.). — **L.** 60 cm long, gracefully recurved; **Fl.** c. 3 cm long.

**A. stricta** SALM (§ II/3) (*A. striata* v. *stricta* BAK., *A. hystrix* HORT. CELS.). — Mexico: Puebla. — In age developing many heads; **St.** thick, dividing several times; **Ros.** spherical, very dense with many upright **L.**, about 35 cm long, very broad at base, then abruptly tapering, linear, tapering sharply at tip, 8 mm across, both surfaces slightly keeled, green, with a fine cartilaginous margin, terminal Sp. 2–2.5 cm long, 3–4-angled, very pointed; **Infl.** 2.25 m long, Fl. c. 2 cm long.

**A. cv. Taylori** (§ II/2). — Probably a hybrid: **A. geminiflora** × **A. filifera** v. **filamentosa**. — Hort. WILLIAMS.

**A. tecta** TREL. (§ III/1). — Guatemala. — **Ros.** 3–4 m $\varnothing$, acauline, offsetting; **L.** 3.8 m long, 50 cm across, grey-green, glabrous, broad-oblanceolate to obovate, the abruptly-pointed tip being curved upwards, upperside concave, upper part grooved, terminal Sp. curved slightly S-shaped, 4.5–6.5 cm long, 5–7 mm across, T. about 8 mm long, 4–7 cm apart, recurved, broadening to the base, 15 mm across, 3-angled; **Infl.** 5–6 m tall, Fl. 7 cm long, yellow.

**A. tenuispina** TREL. (§ III/7/4). — Guatemala. — **Ros.** 1.3–1.5 m $\varnothing$, acauline, tending to form clumps; **L.** not numerous, 50–75 cm long, 8–12 cm across, oblanceolate, pointed, very concave, somewhat plicate, rather thin and flexible, upperside green with bluish transverse bands, underside rough and bluish, terminal Sp. 3–4 cm long, 2–4 mm across, aciculate, somewhat wavy, T. 2–3 mm long, 2–4 cm apart, curved either up or down, narrow-triangular, on fleshy basal cushions, margin straight or sinuate between T.

**A. tequilana** WEB. (§ III/13/2). — Mexico: Jalisco. — **Ros.** acauline, offsetting; **L.** straight, stiffly projecting, over 1 m long, 8 cm across, very gradually tapering, flat-grooved, rather thin, leathery, grey-green, margin straight, terminal Sp. short, stout, 10–12 mm long, T. about 1 cm apart, 2–3 mm long, triangular, bent forwards; **Infl.** c. 6 m tall, Fl. 6 cm long.

**A. terracina** PAX (§ II/1). — Mexico or ? USA: Texas. — **Ros.** short-stemmed, of c. 20–25 lanceolate, gradually tapering **L.** 25–50 cm long, 6 cm across the middle, upperside flat-grooved, with several shallow grooves towards tip, underside very convex up to the middle, soft-fleshy, dull light green with dark red mottling, terminal Sp. 5 mm long, slender, pointed, margins with a fine cartilaginous line, T. numerous, triangular, brown; **Infl.** 1.5 m tall, Fl. 2 cm long.

**A. thomsoniana** JACOBI (§ III/13/1). — Mexico: San Luis Potosí. — **Ros.** acauline, with fairly numerous, fairly straight **L.** tapering gradually towards base and above the middle, upper part flat-concave, underside very convex, at base almost semi-terete, fleshy, fairly soft, sap-green, terminal Sp. 10–12 mm long, stout, T. close-set, small, flat-triangular, base cartilaginous, intermediate T. (1–)2(–3), smaller. (TRELEASE considers this species to be synonymous with **A. celsii** HOOK.)

**A. toeniata** HORT. BESAUC. — Possibly related to **A. univittata** HAW.

**A. toneliana** HORT. BESAUC. — Possibly related to **A. univittata** HAW.

**A. toumeyana** TREL. v. **toumeyana** (§ II/2). — USA: southern central Arizona. — **Ros.** matforming, 30–80 cm $\varnothing$; **L.** 15–25 cm long, 10–25 mm across, upperside concave, underside rounded and slightly keeled, light green with white markings, margins with fine threads in upper 2/3, finely toothed at the base, terminal Sp. 5–8 mm long; **Infl.** 2–3 m tall, Fl. 18–25 mm long, pale yellow.

**A. — v. bella** BREITUNG. — **L.** shorter, more numerous, forming a dense Ros.

**A. trankeera** TREL. (§ III/10). — Lesser Antilles: Curaçao. — **Ros.** acauline or in age very short-stemmed; **L.** numerous, lanceolate, gradually tapering, upperside concave, almost troughlike towards the tip and with several shallow grooves, 1.5 m long, 18–20 cm across, dark green, dull, terminal Sp. 1 cm long, 4–7 mm thick, T. about 1 cm apart, straight or somewhat bent, uppermost T. much smaller; **Infl.** 6–7 m tall, Fl. 6 cm long, orange-coloured.

**A. triangularis** JACOBI v. **triangularis** (§ II/5) (*A. horrida* v. *t.* BAK., *A. regeliana* HORT. ex JACOBI, *A. kerchovei* HORT. ex JACOBI). — Mexico. — **Ros.** c. 21 cm $\varnothing$, offsetting; **L.** hard-fleshy, very rigid, almost triangular-lanceolate, upperside concave towards tip, underside convex and almost flat-keeled, glabrous, dull olive-grey-green, margins with a grey-brown horny band, terminal Sp. very stout, 2 cm long, T. 5–7 mm long, pointed, long-triangular, tip often bent, intermediate T. smaller; **Infl.** 3–5 m tall, Fl. cream-coloured.

**A. — v. rigidissima** (JACOBI) TREL. (*A. r.* JACOBI). — **Ros.** to 60 cm $\varnothing$; **L.** 35 cm long, 13 mm across, 3 mm thick.

**A. — v. subintegra** TREL. (Pl. 8/1) (*A. kerchovei* v. *inermis* BAK., *A. difformis* BGR.). — **L.** 65–70 cm long, thinner, T. few, smaller or absent.

**A. tubulata** TREL. (§ III/18). — N.W.Cuba. — **L.** broad-lanceolate, often folded, tapering gradually or more sharply, 60–75 cm long, 15–20 cm across, glossy green, terminal Sp. 15 mm long, 2 mm across, T. 15–20 mm apart, 1–3 mm long, upper T. bent forwards and the lower ones backwards, slender-tipped, margin sinuate; **Infl.** 2–5 m tall, Fl. 3–3.5 cm long, yellow.

**A. uncinata** JACOBI (§ II/1) (*A. myriacantha* HORT. ex BESAUC.). — ? Mexico. — Related to **A. polyacantha**; **Ros.** c. 90 cm $\varnothing$; **L.** c. 30, lanceolate, 7 cm across at the base and 10–11 cm in the middle, tapering, lower part 2 cm thick, thinner above, rather stiff, dull light green, terminal Sp. 2 cm long, stout, margins with a continuous horny band, T. 3–4 mm long, narrow-triangular, curved; **Infl.** c. 2 m tall, Fl. 2.5 cm long, violet-brownish.

**A. underwoodii** TREL. (§ III/9). – S.E.Cuba. – **Ros.** acauline; **L.** ± narrow-lanceolate, tapering, upperside concave, 1–2 m long, 20–25 cm across, terminal Sp. triangular, 1.5–2.5 cm long, 4–6 mm across, T. (10–)20(–30) mm apart, 2–5 mm long, straight or bent, occasionally hooked; **Infl.** 4–8 m tall, Fl. 5–5.5 cm long, golden-yellow.

**A. undulata** KLOTZSCH (§ I) (*Manfreda u.* (KLOTZSCH) ROSE, *A. drimiaefolia* HORT.). – Mexico. – ♃; **Ros.** of 12–15 **L.** 25–45 cm long, 3 cm across, gradually tapering, tip conical, curled inwards, green, both surfaces striped, underside keeled, margins very wavy, with a narrow, somewhat reddish, cartilaginous border; **Infl.** c. 1 m tall, Fl. 4 cm long, brownish.

**A. unguiculata** TREL. (§ III/10). – Caribbean Is. – Habit as for **A. karatto**; **L.** broad-lanceolate, upperside concave, about 2 m long, 28 cm across, green, terminal Sp. stout, 10–15 mm long, 5–6 mm across. T. 5–10 mm apart, 1–2 mm long; **Fl.** 7.5–8 cm long, golden-yellow.

**A. univittata** HAW. v. **univittata** (§ II/5) (*A. lophantha* SCHIEDE ex KUNTH). – Mexico: Coahuila, Veracruz. – **Ros.** to 85 cm ⌀; **L.** 30–45 cm long, 3–4 cm across, mid-green, often with a pale M. stripe, underside with darker lines, terminal Sp. 1–1.5 cm long, margins horny, 2 mm wide, T. triangular, variously curved, irregular, 3–7 mm long, 2–3 cm apart; **Infl.** 4 m tall, stiff. (BREITUNG considers **A. lophantha** to be synonymous, consequently also its varieties.)

**A. —** v. **angustifolia** (BGR.) JACOBS. (*A. lophantha* v. *a.* BGR.). – **L.** not numerous, 75 cm long, 2.5 cm across, dark green, underside more grey-green, with numerous dark transverse lines, T. small.

**A. —** v. **brevifolia** (JACOBI) JACOBS. (*A. lophantha* v. *b.* JACOBI). – **L.** 32–33 cm long, 5 cm across at base and 3.5 cm above, tapering gradually to tip, blunt, without green longitudinal lines.

**A. —** v. **carchariodonta** (PAMPANINI) BREITUNG (*A. c.* PAMPANINI). – Distinguished from v. **latifolia** by the discontinuous horny edge; **L.** to 60 cm long, 10 cm across the middle.

**A. —** v. **coerulescens** (SALM) JACOBS. (*A. c.* SALM, *A. lophantha* v. *c.* (SALM) JACOBI). – **L.** more numerous, somewhat shorter, ± rough, light blue-grey, no paler stripe, T. fairly stout, directed downwards.

**A. —** v. **gracilior** (JACOBI) JACOBS. (*A. lophantha* v. *g.* JACOBI). – **Ros.** very regular; **L.** numerous, narrower, margins more light-grey, T. much closer together, with brown tips.

**A. —** v. **heteracantha** (BGR.) BREITUNG (*A. h.* BGR., *A. ensifera* JACOBI). – **L.** grey-green, ± rough, T. often in pairs, variously twisted and curved.

**A. —** v. **latifolia** (BGR.) BREITUNG (*A. lophantha* v. *l.* BGR., *A. heteracantha* BAK.). – **L.** broader, underside without darker line, T. variously curved and twisted.

**A. —** v. **subcanescens** (JACOBI) JACOBS. (*A. lophantha* v. *s.* JACOBI). – **L.** margins and T. almost white, L. colour grey-green.

**A. —** v. **taumalipasana** [BGR.) JACOBS. (*A. lophantha* v. *t.* BGR.). – **L.** 30–40 cm long, 2.5–3.5 cm across, terminal Sp. conical, margins having a horny border which is liable to split off, T. 1.5–2 cm apart, 3–7 mm long.

**A. —** × **A. ghiesbreghtii** C. KOCH. – Hybrid. – **L.** lanceolate, 35–40 cm long, 5 cm across at the base and 7 cm in the middle, gradually tapering, dull green, margins with a stout cartilaginous line, terminal Sp. 3 cm long, T. very stout, very hooked, usually bent upwards, intermediate T. 1–2, smaller.

**A. utahensis** ENGELM. (§ II/5). – USA: Utah, Arizona, Calif. – **Ros.** 20–100 cm ⌀, simple or mat-forming; **L.** rigid, 12–35 cm long, blue to light green, terminal Sp. 1–8.5 cm long, margins straight or sinuate, T. 1.5–7 mm long, 1.5–3 cm apart; **Infl.** 1.5–4.5 m tall, Fl. yellow.

**A. —** v. **discreta** M. E. JONES (*A. newberryi* ENGELM.). – USA: Arizona. – **L.** light green, T. brown at base, margins straight.

**A. —** v. **eborispina** (HESTER) BREITUNG (*A. e.* HESTER). – USA: Nevada. – **L.** light green, terminal Sp. ivory-coloured, 1–2 cm long, T. 5–7 mm long.

**A. —** v. **kaibabensis** (MCKELVEY) BREITUNG (*A. k.* MCKELVEY). – USA: Arizona. – **Ros.** 70–100 cm ⌀, simple, occasionally having 1–3 offsets; **L.** 30–35 cm long, 5–6.5 cm across, light green, terminal Sp. 1–3 cm long, T. 1–2 cm apart, 1.5–3.5 mm long, brown, usually bent backwards rather than forwards; **Infl.** 4.5 m tall.

**A. —** v. **nevadensis** ENGELM. (*A. n.* ENGELM. ex (GREENM. et ROUSH) HESTER). – USA: Calif. – **L.** bluish-green as for v. **utahensis**, terminal Sp. 3–8.5 cm long.

**A. —** v. **utahensis** (*A. scaphoidea* GREENM. et ROUSH, *A. u. v. s.* (GREENM. et ROUSH) M. E. JONES, *A. haynaldii* v. *u.* TERR.). – USA: Utah. – Offsets numerous, mat-forming, **Ros.** 20 to 30 cm ⌀; **L.** 12–20 cm long, 2–3 cm across, bluish, terminal Sp. 1–2 cm long, T. slightly curved, 1.5–2 mm long, 2–3 cm apart, edges sinuate and soft; **Infl.** 1.5–2.5 m tall, Fl. 22 to 30 mm long, yellow.

**A. vandervinnenii** LEM. (§ III/6). – Insufficiently known species.

**A. vangrolae** TREL. (§ III/10) (*Furcraea gigantea* BOLDINGH). – Caribbean Is. – Habit as for **A. karatto**; **L.** lanceolate, gradually tapering, up to 2 m long, grey or grey-green, often somewhat mottled, terminal Sp. with base incurved and with finely dentate edges, 10–15 mm long, 5–7 mm across, the 2 mm tip being recurved, T. 5–10 mm apart, 2 mm long, triangular or bent, merging at base into a parchment-like line; **Infl.** 5 m tall, Fl. 4.5–5 cm long, golden yellow.

**A. variegata** JACOBI (§ I) (*Manfreda v.* (JACOBI) ROSE). – USA: S.E.Texas. – **Rootstock** tuberous; freely offsetting, mat-forming; **L.** 20–45 cm long, 2–4 cm across, indistinctly toothed, with large brown markings; **Infl.** 1–1.3 m tall, Fl. black-brown.

**A. ventum-versa** TREL. (§ III/10). – Caribbean Is. – Habit as for **A. karatto; L.** lanceolate, 1 m long or more, about 10 cm across, green, terminal Sp. red-brown, 2–2.5 cm long, 4–6 mm across, T. 5–10 mm apart, 2–3 mm long, broad-triangular, margins parchment-like, with isolated intermediate T.; **Fl.** 5 cm long, yellow.

**A. vera-cruz** MILL. (§ III/13/1) (*A. lurida* AIT., ? *A. vera-crucis* HAW., ? *A. mexicana* LAM. p. part., *A. vernae* BGR., *A. prainiana* BGR. acc. BREITUNG). – Mexico; in cultivation in Eur. – **Ros.** c. 2 m ∅, **St.** 40–50 cm tall, offsetting; **L.** linear-lanceolate, tapering, 1 m long, 15–16 cm across midway, narrowing towards base to 11 cm, upperside concave or flat, rolled inwards towards the tip, ± bluish or with grey transverse bands, glabrous, terminal Sp. 1.5–3 cm long, 3–4 mm across, rough, T. 2–3 mm long, 1 cm apart, triangular, straight or having tip curving forwards or back; **Infl.** 4–6 m tall, Fl. 3–3.5 cm long, yellow-green.

**A. vicina** TREL. (§ III/12). – Caribbean Is. – **Ros.** acauline; **L.** broad-oblanceolate, sharply tapering, upperside concave, c. 50 cm long, 15 cm across, dull-green, terminal Sp. 1.5–2.5 cm long, 3–4 mm across, T. 1.5–2 cm apart, broad-triangular, about 5 mm long, margins deeply sinuate, upper T. curved upwards and lower ones downwards; **Infl.** 4 m tall.

**A. victoria-reginae** T. MOORE (Pl. 9/1) (§ II/5). – Mexico. – **Ros.** usually simple, otherwise with only a few offsets, 50–70 cm ∅; **L.** numerous, 10–30 cm long, 3–7 cm across, tapering, narrowing abruptly to a rounded tip, very rigid, straight or curved inwards, dark green, with distinct irregular white lines, these occasionally absent, upperside concave, underside convex and sharply keeled in the upper part, terminal Sp. 1.5–5 mm long, usually having 1–2 small lateral Sp., edges entire, occasionally with small, white T.; **Infl.** 3–4 m tall, Fl. cream-coloured.

**A. — f. dentata** BREITUNG. – **L.** with several short white T., pointed and directed downwards, situated in the middle between the margins.

**A. — f. latifolia** BREITUNG. – **L.** 4–6 cm across, dark green, the white lines being particularly conspicuous.

**A. — f. longifolia** BREITUNG. – **L.** gradually tapering, 20–30 cm long.

**A. — f. longispina** BREITUNG. – Terminal Sp. thin, 2.5–3.5 cm long, black, somewhat twisted, 1.5 mm across the base.

**A. — f. ornata** BREITUNG. – **L.** with ± stronger, broader, white lines.

**A. — f. stolonifera** JACOBS. – **Ros.** forming underground offsets.

**A. — f. victoria-reginae** (*A. consideranti* CARR.). – **L.** 10–15 cm long, 3–4 cm across, tip short-rounded, terminal Sp. 2–5 mm long.

**A. — f. viridis** BREITUNG. – **L.** without white lines.

**A.** cv. Villarum (§ II/2) (*A. villaepirottii* HORT.). – Hybrid: **A. filifera** × **A. xylonacantha; Ros.** acauline, lax, 60 cm ∅; **L.** 20–40, 3.5 cm across base, gradually tapering, upperside flat-concave, underside convex, both surfaces slightly rough, underside rather darker green, terminal Sp. 3 cm long, margins with a horny border 1.5 mm across, occasionally splitting off; **Infl.** spicate, Fl. 17 mm long.

**A. vilmoriniana** BGR. (§ II/7/3) (*A. mayoensis* GENTRY). – Mexico. – **Ros.** with c. 25 soft-fleshy, green-blue, narrow-linear **L.** c. 80 cm long, tapering, upperside grooved, margins entire, sharp and horny, terminal Sp. 3–4 cm long, subulate, light-brown.

**A. virginica** L. v. **virginica** (§ I) (*Manfreda v.* (L.) SALISB.). – E.USA. – ⚇; Rhizome thick; **L.** 6–15, lanceolate or oblong-spatulate, sharply tapering, 15–45 cm long, 2.5 cm across, concave, rather grooved, limp, dark green, margins sinuate, indistinctly serrate; **Infl.** 90–100 cm tall, Fl. 2.5–3.5 cm long, greenish or brownish-yellow, agreeably scented.

**A. — v. tigrina** ENGELM. – USA: S.Carolina. – L. with large, brown spots.

**A. viridissima** BAK. (§ III/7/2). – Origin unknown. – **Ros.** acauline of 30 oblanceolate **L.** c. 30 cm long, 7.5 cm across above the middle and 6 cm across at the base, very rigid, light-green, terminal Sp. about 2.5 cm long, stout, T. close-set, irregular, 3–6 mm long, upper Sp. bent upwards.

**A. vittata** RGL. (§ II/5) (? *A. haynaldii* TOD., *A. toneliana* BAK.; synonyms acc. BREITUNG; BERGER regarded **A. toneliana** as a distinct species). – Mexico: Nuevo León. – **Ros.** simple or with a few offsets, 1.3 m ∅; **L.** 50 cm long, 3.5 cm across, dark green, terminal Sp. purple-brown, 1 cm long, margins horny, T. 1.5–2 cm apart, c. 1 mm long; **Infl.** 4 m tall.

**A. vivipara** L. (§ III/12) (*A. americana* COMMEL.). – Caribbean Is. – **Ros.** acauline or almost so, offsetting; **L.** broad-lanceolate, tapering, upperside flat-concave, 40–60 cm long, 12–20 cm across, grey-green, terminal Sp. somewhat triangular, 2.5–3 cm long, 3–4 mm across, T. 10–15 mm apart, 3–4 mm long, triangular, on raised cushions, lower Sp. recurved; **Infl.** 3 m tall, Fl. 4–4.5 cm long, yellow.

**A. wallisii** JACOBI (§ III/11). – Colombia. – **Ros.** acauline; **L.** thick-fleshy at base, upper part thinner and fibrous, glossy, bright green, slightly pruinose, broad-lanceolate, broadest in the middle, above that tapering, concave and having several grooves, underside thickened and keel-like, terminal Sp. thin, c. 23 mm long, T. somewhat sinuous and recurved, triangular, directed forwards, on fleshy bases.

**A. warelliana** BAK. (§ II/1). – ? Mexico. – Very closely related to **A. macrantha; Ros.** acauline, of very many L., almost spherical, 80–100 cm ∅, with only few offsets; **L.** lanceolate-spatulate, gradually tapering, 13–14 cm aross above middle, much thickened at base and 9–10 cm across, convex up to midway with almost riblike thickening, upper part flat-concave, fairly rigid and hard, light green, ± grey-pruinose, with a distinctly dark horny margin, terminal Sp. decurrent, T. 1 mm long, triangular, curved; **Infl.** 2 m tall, Fl. 5 cm long, yellowish.

**A. washingtoniensis** ROSE (§ II/5). – Origin unknown. – **Ros.** of 20–25 **L.** 75 cm long,

7–10 cm across middle, oblong-spatulate, 5 cm across base, flat, margins with a brown cartilaginous strip, terminal Sp. short, sharp, T. 1 cm apart, 2–3 mm long, almost triangular, hooked; **Infl.** c. 3 m tall.

**A. weberi** CELS. (§ III/13/2); acc. BREITUNG § III/1). – Mexico. – **Ros.** acauline, offsetting, about 1.6 m $\emptyset$; **L.** not very fleshy, flexible, 1.2–1.4 m long, 10 cm across base and 20 to 25 cm across the middle, gradually tapering, upperside flat to rather concave, folded several times towards tip into keels and grooves, margins spineless, terminal Sp. 2 cm long, black-brown.

**A. weissenburgensis** WITTM. (§ III/7/2). – Origin unknown. – Related to **A. potatorum**; **Ros.** acauline, of 30–40 **L**. 22–25 cm long, 5–6 cm across upper part, stiff, green, terminal Sp. 12 mm long, T. 3–4 mm apart, triangular, hooked; **Infl.** 2–2.5 m tall, Fl. more than 5 cm long.

**A. werklei** WEBER (§ III/7/5). – Costa Rica. – **Ros.** fairly large; **L.** very numerous, narrow and thick at base, short-tapering, intense white-grey, terminal Sp. 22 mm long, 3 mm across, T. black, somewhat recurved; **Infl.** 8 m tall, Fl. c. 6 cm long, golden yellow.

**A. willdingii** TOD. (§ III/18) (*A. w.* TOD. ex BAK.). – ? W.Cuba. – **Ros.** acauline, of about 15–30 **L**. 50–80 cm long, oblanceolate or oblong-spatulate, 12–16 cm across, narrowing at base to 5–9 cm across, thick in lower part, becoming thinner above, margins curved up, fresh green, terminal Sp. 10–15 mm long, slender, T. 3 to 4 mm long, straight, central Sp. 15 mm apart, becoming smaller in upper part and finally absent; **Infl.** 3–4 m tall, Fl. c. 4 cm long, orange-yellow.

**A.** × **winteriana** BGR. (§ III/2). – Cultivated hybrid ex Nursery of L. WINTER, Bordighera: **A. franzosinii** × ? **A. americana.** – **Ros.** imposing, offsetting, of c. 35 **L.**, these being oblanceolate, c. 1 m long, 11 cm across at the base and 18–20 cm above the middle, thereafter tapering, upperside slightly convex, then very concave and in upper part grooved, underside convex and keel-like at base, c. 6 cm thick, both surfaces light green to bluish, underside initially rough, margins sinuate, terminal Sp. 5 cm long, T. on fleshy bases, triangular-hooked, 10–11 mm long, 2–3 cm apart; **Infl.** almost 10 m tall, Fl. 5 cm long, greenish.

**A. wocamahii** GENTRY (§ III/7/3). – Mexico: Chihuahua. – **Ros.** 1.5–2 m $\emptyset$, simple or occasionally with 1–2 offsets; **L.** 75–150 cm long, 15–20 cm across, narrow-lanceolate, light green, terminal Sp. 5 cm long, 8–12 mm across, T. 6–10 mm long, 1–2 cm apart, straight or curved up or down, widening abruptly below into a horny base, intermediate T. small, margins slightly sinuate; **Infl.** 4–5 m tall, Fl. 4–4.5 cm long, yellow.

**A. wrightii** J. R. DRUMM. (§ II/2). – ? Central Mexico. – **St.** very short, concealed by the thickened **L.** bases; **L.** numerous, leathery-fleshy, 40–45 cm long, narrowing to a slender point, ovate at base, 3.5 cm across, 2.5 cm thick, upperside flat-convex, somewhat obliquely keeled, underside almost triangular and thickened, light green with small white dots, margins with a narrow, horny border, threads tearing away, with terminal Sp. 1 cm long; **Infl.** c. 3 m tall, Fl. 4 cm long, greenish.

**A. xalapensis** ROEZL. (§ II/1). – Mexico: Jalapa Enríquez. – Resembles **A. polyacantha**; **L.** smaller, grooved, 55–60 cm long, terminal Sp. 7–10 mm long, T. stouter, less regular, hooked forwards; **Infl.** 2.5 m tall, Fl. green-brown.

**A. xylonacantha** SALM v. **xylonacantha** (§ II/5) (*A. cornuta* HORT., *A. amurensis* JACOBI, *A. kochii* JACOBI). – Mexico: Hidalgo. – **Ros.** simple, spreading and laxly leafy; **L.** tapering gradually from the base, upperside grooved, 60–90 cm long, 6–8 cm across, robust and hard, underside convex and laterally keeled towards tip, dull grey-green, with a broad horny margin, terminal Sp. 4–5 cm long, T. 3–5 cm apart, on broad cushions, 11 mm long, 15 mm across; **Infl.** more than 3.3 m tall, Fl. 3.5–4 cm long, greenish.

**A.** — v. **latifolia** JACOBI (*A. maximiliana* HORT ex BESAUC.). – **L.** broader.

**A.** — v. **macracantha** JACOBI. – Marginal T. especially large.

**A.** — v. **torta** JACOBI. – **L.** irregularly curved.

**A.** — v. **vittata** JACOBI (*A. x. v. medio-picta* TREL., *A. perbella* HORT., *A. hybrida* HORT.). – **L.** upperside with a broad M. band, either simple or divided.

**A. yucciifolia** DC. (§ II/1). – Mexico: Río-del-Monte D. – **Ros.** short-stemmed or acauline, offsetting, of 12–15 **L.** 50 cm long, strap-shaped, gradually tapering, upperside grooved, light grey-green, with a pale M. band and small red dots, margin with a narrow, red border, dentate, terminal Sp. sharp, 6–7 mm long; **Infl.** 2–3 m tall, Fl. 4 cm long, green.

**A. zapupe** TREL. (§ III/13/1). – Mexico: Veracruz, cultivated for its fibres; origin unknown. – **Ros.** acauline or virtually so, densely leaved; **L.** thin, gradually tapering, 1.5–2 m long, 8–10 cm across, green, with grey-pruinose stripes, terminal Sp. 1.5–2.5 cm long, 4 mm thick, T. 1.5–3 cm apart, 2–3 mm long, margins straight, palely translucent; **Infl.** 3–6 m tall, Fl. 2.5 cm long, greenish.

**Aichryson** WEBB et BERTH., Crassulaceae. – Canary Is.: Madeira; Azores. – ♄, ○ or ⊙ plants, dying after flowering, 20–40 cm tall, succulent; **St.** erect, forked, loosely leaved; **L.** only towards ends of shoots, arranged rather Ros.-like, alternate, petiolate, softly hairy; **Fl.** in paniculate cymes, yellow or possibly reddish. – Greenhouse, moderately warm. Propagation: from seed. (Lit.: BRAMWELL, D.: Notes on the Taxonomy and Nomenclature of the genus Aichryson. In Inst. Nac. Invest. Agron. 28: 203–213, 1968.)

**A. bethencourtianum** WEBB (*Aeonium b.* (WEBB) WEBB. – Canary Is.: Fuerteventura. – Small, dense ♄ 20–30 cm tall with small Ros.; **Br.** hairy; **L.** 20–30 mm long, 10 mm wide, up to 3 mm thick, broad-spatulate to nearly circular, ± attenuated stalk-like at the base, hairy; **Infl.** short stalked, with 5–6 Br., Fl. yellow.

**A. bollei** WEBB (Pl. 9/4) (*Sempervivum b.* CHRIST). – Canary Is.: La Palma. – **St.** with adpressed long white hairs; **L.** trapezoid-spatulate, 4.5 cm long, 2 cm across, narrowed teretely at base, somewhat fleshy, slightly grooved, margins sinuate and toothed, T. black, surfaces covered with thick white hairs; **Infl.** 8–10 cm tall, Fl. pale-yellow.

**A.** × **bramwellii** ("bramwelli") KUNKEL. – Canary Is.: Gran Canaria. – Acc. PRAEGER a spontaneous hybrid: **A. porphyrogenetos** × **A. punctatum.** – Plant c. 20 cm high and 30 cm broad; **St.** thick, with widely divaricate Br.; **L.** and **Fl.** intermediate between those of the parents.

**A. brevipetalum** PRAEG. – Canary Is.: La Palma. – **St.** softly hairy; **L.** rhomboid-spatulate, finely toothed towards tip, softly hairy **Infl.** with 6–12 reddish Fl.

**A. divaricatum** (AIT.) PRAEG. v. **divaricatum** (*Sedum d.* AIT., *Sempervivum d.* v. *polita* LOWE). – Madeira. – **St.** stiffly erect; **L.** ovate, blunt, narrowing abruptly to a long petiole; **Infl.** with many Fl.

**A. —** v. **pubescens** LOWE. – **Infl.** and Cal. with glandular hairs.

**A. dumosum** (LOWE) PRAEG. (*Sempervivum d.* LOWE). – Madeira. – **St.** glandular and softly hairy above, irregularly forked; **L.** narrow-spatulate or obovate-lanceolate, narrowing into a long petiole, pale grey-green, not Ros.-forming; **Fl.** fairly large.

**A. gattefossei** (BATT. et JAHAND.) BRAMWELL (*Sedum g.* BATT. et JAHAND.). – Morocco. – ⊙; ovate-oblong, blunt; **Fl.** 6-merous, small, yellow. Species of little horticultural importance.

**A.** × **intermedium** BRAMW. et ROWL. – Canary Is.: Tenerife, La Palma. – Acc. PRAEGER a spontaneous hybrid: **A. laxum** × **A. punctatum** (*A. dichotomum* × **A. punctatum**). – Habit intermediate between the parents; **L.** more hairy than in **A. punctatum**; **Infl.** more lax than in **A. laxum**, very hairy.

**A. laxum** (HAW.) BRAMW. (Pl. 9/3) (*Sempervivum l.* HAW., *S. dichotomum* DC., *A. d.* (DC.) WEBB et BERTH., *S. annuum* CHR. SM.). – Canary Is. – **St.** dichotomously branching; **L.** broad-ovate to ovate-rhomboid, densely tomentose, purple to bronze-coloured like St., otherwise vivid green; **Fl.** pale-yellow.

**A. palmense** WEBB. – Canary Is.: La Palma. – **St.** with brown felt; **L.** broadly ovate-spatulate, rounded, very fleshy, petiole very hairy; **Infl.** viscous, glandular and coarsely hairy, **Fl.** golden-yellow.

**A. parviflorum** BOLLE (*Sempervivum p.* CHRIST). – Canary Is.: La Palma. – Resembles **A. divaricatum**; **Infl.** glabrous, Fl. small.

**A. porphyrogennetos** BOLLE (*Sempervivum p.* CHRIST). – Canary Is.: Tenerife, Gran Canaria. – **St.** white-felted to short-haired, Br. spreading and vivid red, like the rest of the plant; **L.** trapezoid-spatulate, blunt; **Fl.** golden-yellow.

**A.** × **praegeri** KUNKEL. – Canary Is.: Gran Canaria, Tenerife. – Acc. PRAEGER a spontaneous hybrid: **A. laxum** × **A. porphyrogenetos.** – Intermediate between the parents.

**A. punctatum** (C. SM.) WEBB et BERTH. v. **punctatum** (*Sempervivum p.* C. SM., *A. parlatorei* BOLLE). – Canary Is. – **St.** stiff, erect to spreading; **L.** rhomboid-obovate, indistinctly dentate; **Infl.** and upper part of St. both pubescent.

**A. —** v. **pachycaulon** (BOLLE) PRAEG. (*A. p.* BOLLE, *Sempervivum p.* (BOLLE) CHRIST, *A. immaculatum* WEBB). – Canary Is.: La Palma, Tenerife. – Ecotype from a moist habitat.

**A. —** v. **subvillosum** (LOWE) PIT et PROUST. – Canary Is. – **L.** finely hairy, margin entire. – Ecotype from a dry habitat.

**A. villosum** (AIT.) WEBB et BERTH. (*Sempervivum* v. AIT., *S. barreti* MENEZ. ex PRAEG.). – Madeira; Azores. – **St.** rough, sticky, with projecting white glandular hairs, ascending, 10–20 cm tall, branching above; **L.** broad-rhomboid, with dense, long hair; cyme flat.

**Aizoanthemum** DTR. ex FRIEDR. Aizoaceae. – S.W.Afr.; Angola. – ⊙; densely papillose, creeping, bushy or small tree-like plants, **St.** sometimes lengthened and prostrate, Br. dichotomously forked; **L.** alternate, somewhat fleshy, short-petiolate or sessile, margin entire; Fl. terminal, solitary, Cal. united for 1/3 or more cup-like, lobes 5; capsule almost globular, 5–10-celled, dehiscing starlike, as in **Mesembryanthemum**, seeds numerous. Summer: in the open. Propagation: from seed. – Acc. K. DINTER transitional between **Aizoon** and **Mesembryanthemum.**

**A. dinteri** (SCHINZ) FRIEDR. (*Aizoon d.* SCHINZ, *Aizoanthemum bossii* DTR. in sched., *A. sphingis* DTR. in sched., *A. stellatum* DTR. in sched.). – S.W.Afr. – Small tree-like to a sub-shrub, up to 40 cm tall; **St.** dichotomously branching, terete or furrowed, papillae small; **L.** rhomboid to round, narrowing towards the petiole which is 2–7 mm long, 1–6 cm long, 7–35 mm across, somewhat fleshy, finely papillose; **Fl.** usually sessile, c. 7 mm long, lobe-margins white-membranous. (Acc. H. CH. FRIEDRICH **A. stellatum, A. bossii** and **A. sphingis** from herbarium specimens are developmental stages of the above species.)

**A. galenioides** (FENZL ex SOND.) FRIEDR. (*Aizoon g.* FENZL ex SOND.). – S.W.Afr. – Some 30 cm tall, branching from base, with large papillae; **L.** lanceolate to oblong-elliptical, blunt to pointed, petiolate, 7–20 mm long, 3–7 mm across; **Fl.** sessile, 3–4 mm long.

**A. membrum-connectens** DTR. ex FRIEDR. – S.W.Afr. – Creeping; **St.** 5–12 cm long, with large papillae, young twigs papillose-hairy; **L.** broad-lanceolate to oblong-elliptical, blunt to somewhat pointed, almost sessile, rather fleshy, papillose, 1–3 cm long, 0.5–1 cm across; **Fl.** sessile, 1 cm long.

**A. mossamedense** (WELW.) FRIEDR. (*Aizoon m.* WELW.). – Angola: Moçâmedes. – **St.** branching dichotomously, papillose-hairy, 10–50 cm long;

L. oval to broad-elliptical, blunt, with a distinct petiole, fleshy, shining-papillose, 1.5–4.5 cm long, 2–12 mm across; **Fl.** sessile, c. 15 mm ⌀, light yellow.

× **Alchamaloe** ROWL. Liliaceae. – Intergeneric hybrid: **Aloe** × **Chamaealoe**. – Cultivation as for **Aloe**.

× **A.** cv. Marianne North. – **Aloe ballii** REYN. × **Chamaealoe africana** (HAW.) BGR. – Intermediate between the parents. Hort. Kew.

× **A. 1.** – **Aloe humilis** (L.) MILL. × **Chamaealoe africana** (HAW.) BGR. Ros. dense, of slender, soft-fleshy, erect to spreading, glaucous green **L.** rounded below, flat above, covered in long, soft, creamy white Sp. HORT. Hummel.

× **Aleptoe** ROWL. Liliaceae. – Intergeneric hybrid: **Aloe** × **Leptaloe**. Acc. REYNOLDS, **Leptaloe** STAPF is returned to the Genus **Aloe**, so this would be an intersectional hybrid listed under **Aloe**. – Cultivation as for **Aloe**.

× **A.** cv. Atrovirens (*Leptaloe* cv. A. POIND.). – **Aloe striatula** HAW. × **Leptaloe** sp. – Densely tufted rosettes of c. 18 linear, dark green, mostly unspotted **L.** 45–50 cm long, 20 mm wide and 4 mm thick at base; **Infl.** up to 165 cm tall, Fl. 24 mm long, yellowish-green tinged with dull red. HORT. Pointdexter.

× **Allauminia** ROWL. Liliaceae. – Intergeneric hybrids: **Aloe** × **Guillauminia**. Acc. REYNOLDS, **Guillauminia** BERTR. is returned to the genus **Aloe**, so that these would be intersectional hybrids listed under **Aloe**. – Cultivation as for **Aloe**.

× **A.** cv. Bountiful. – **Aloe thompsoniae** GROENEW. × **Guillauminia albiflora** BERTR. – Vigorous, offsetting freely to form large clumps; **L.** like those of **Guillauminia** but larger, to 25 cm long; **Infl.** to 50 cm tall, very freely produced; **Fl.** intermediate, pinkish orange. Hort. Rowley.

× **A.** cv. Frank Reinelt. – **Aloe brevifolia** MILL. × **Guillauminia albiflora** BERTR. – Intermediate between the parents. HORT. Kimnach.

× **A. 1.** – **Aloe bellatula** REYN. × **Guillauminia albiflora** BERTR. Intermediate between the parents. HORT. Rowley.

× **A. 2.** – **Aloe rauhii** REYN. × **Guillauminia albiflora** BERTR. Intermediate between the parents. HORT. Rowley.

**Allenrolfia** O. KTZE. Chenopodiaceae. – N. & S.Am. – **St.** erect, jointed, apparently leafless, the joints broadened, almost 2-lipped; **Fl.** 3–5 in the axils of shield-like scale-L., membranous, 4–5-lobed. Halophytes. – Cultivation as for **Salicornia**.

**A. occidentalis** (S. WATS.) O. KTZE. (*Spirostachys o.* S. WATS., *Arthrocnemum fruticosum* TORR., *A. macrostachys* TORR., *Halostachys o.* S. WATS., *Salicornia o.* GREENE). – USA: from Oregon to W.Texas and Baja Calif. Characteristic plant of strongly alkaline, moist soils of the plains of the Great Basin.

**A. patagonica** (MOQ.) O. KTZE. (*Halostachys p.* MOQ., *Spirostachys p.* GRISEB., *Halopeplis p.* (MOQ.) UNG.-STERNB.). – Patagonia & S.Argentine, salt-marshes. – ♄, 50 cm tall.

**A. vaginata** (GRISEB.) O. KTZE. (Pl. 9/2) (*Spirostachys v.* GRISEB.). – Argentine: Santiago del Estero and La Riojo, Patagonia, salt-marshes. ♄ to 2 m tall, freely branching.

**Alluaudia** DRAKE. Didiereaceae. – Madag. – ♄ or ♄, 10–15 m high; **Trunk** succulent; **Br.** ascending or somewhat spreading, branched, twigs with solitary Th. or these arranged in pairs 2–2.5 cm long, rarely without Th.; leafless or **L.** two together below the Th., ± roundish, sessile, somewhat fleshy; **Fl.** unisexual in ± pseudo-umbels, Cor. constricted, 7–8 mm long, the Cal. surrounding the Fr. – Greenhouse, warm. Propagation: seeds, cuttings.

## Key to the Species of the Genus Alluaudia by W. Rauh

| | | |
|---|---|---|
| 1. | Plants always leafless, only the young shoots develop small L. which soon fall off; Th. very short, often absent. Large ♄ or ♄. Br. erect or ascending | **dumosa** |
| 1a. | Plants with L. only in the growing season, subtended by groups of 2 long and strong Th. | 2. |
| 2. | Fl. in umbellate Infl. below the Th. directly developed from the shoot; Sep. of female Fl. winged at the back. Small strongly branched ♄, Br. all equally tall so that the top of the ♄ looks as if shorn | **comosa** |
| 2a. | Fl. in many-branched Infl. or clusters | 3. |
| 3. | Infl. large, to 30 cm ⌀, in a subterminal thyrse | 4. |
| 3a. | Infl. much reduced in circumference, to 15 cm long and 15 cm wide, not forming a subterminal thyrse, but arranged in steps on the upper lateral Br. | 5. |
| 4. | Young plants strongly branched from the base; mature plants with a thin trunk and numerous lateral Br.; L. of the short-shoots obovate, rounded at the tip, in the juvenile stage finely puberulous | **procera** |
| 4a. | Young plants not branched from the base; mature plants with a thick trunk, lateral Br. less numerous; L. of the short-shoots circular, incised at the tip, glabrous | **montagnacii** |

5. ♄ 5–12 m high, trunk thick and with less erect lateral Br.; L. of the short shoots circular, incised at the tip; Sep. of the female Fl. forming a helm ....................... **ascendens**
5 a. ♄ or small ♄ to 5 m high, the trunk thin, elongated, ± supporting itself; L. of the short shoots obovate, rounded at the tip, obtuse and faintly crenate ....................... **humbertii**

**A. ascendens** (DRAKE) DRAKE (*Didierea a.* DRAKE). – S.Madag. – ♄ 5–12 m tall; **St.** thick, with few erect Br., Th. 1.5–2 cm long, conical, arranged irregularly; **L.** circular to oblique-obcordate, swollen at tip, thick, 13.5–25 mm long and across, dark green; **Fl.** small.

**A. comosa** DRAKE (Pl. 10/4). – S. & S.W.Madag. – ♄ or ♄, 1–10 m tall, dividing into 4–5 erect Br., with many twigs, these having numerous, thin Th. which are 1.5–3.5 cm long; **L.** usually standing singly, obovate or almost circular, often having a slight indentation in the upper part, (1–)1.5(–2.2) cm long, (9–)14(–22) mm across; **Fl.** always several in groups, small.

**A. dumosa** DRAKE (Pl. 10/1). – S.Madag. – ♄ 2 m tall, in age ♄ up to 8 m tall; **Br.** erect, succulent, dividing; twigs with isolated Th. 2 mm long and 1 mm thick; **L.** short, cylindrical, soon falling; ♂ **Fl.** globular, ♀ Fl. oblong, only few mm long.

**A. humbertii** CHOUX. – Madag. – ♄ 6–7 m tall, forking, with thin twisted Br., Th. isolated, projecting, slender (5–)15(–23) mm long; **L.** in pairs below the Th., ovate-obcordate, 5–16 mm long, 5.5–10 mm across, fleshy; **Fl.** small.

**A. montagnacii** RAUH (Pl. 10/12). – S.W.Madag. – ♄ in age, with either no Br. or only a few, St. 6–8 m tall, columnar, with annular constrictions, the St. apex in age curving strongly towards the S., Th. on new growth up to 2.5 cm long, silver-grey pruinose, black-tipped, broadening below; **L.** on shorter St. short-petiolate, broad-oval to almost round, 1.5 cm long, 1.2 cm across, the tip ± deeply indented, L. on longer St. oblong-oval, 1.6–1.7 cm long, narrowing teretely towards the base, initially slightly hairy; **Infl.** on newest growth.

**A. procera** (DRAKE) DRAKE (Pl. 10/3) (*Didierea p.* DRAKE). – S. & S.W.Madag. – ♄ to 20 m tall; **Br.** ascending, curved slightly outwards, Th. dense, arranged spirally, 2–2.5 cm long, much broader at the base; **L.** in pairs, obovate or obovate-oblong, 7–25 mm long, 4–12 mm across, somewhat fleshy; **Fl.** 2–4 mm long.

**Alluaudiopsis** H. HUMB. et P. CHOUX. Didiereaceae. – Madagascar – ♄ 2–4 m tall, much branching from the base, with stout Th. which are borne singly or in pairs; **L.** oblong, fleshy; **Fl.** quite large, many in an Infl., or in groups of (2–)3(–5) on some short shoots, Pet. white-yellowish or vivid crimson. Plants dioecious. – Greenhouse, warm. Propagation: from seeds or cuttings.

## Key to the Species of the Genus Alluaudiopsis, by W. Rauh

1. Th. solitary; L. on short shoots above the Th. oblong-linear, to 4 cm long; Infl. of many Fl., Pet. white-yellowish ......................................... **fiherensis**.
1 a. Th. in pairs; L. on short shoots, oblong to oval, 1–1.5 cm long; Fl. in groups of (2–)3(–5) at the apex of short shoots, Pet. vivid crimson ............................ **marnierana**.

**A. fiherensis** H. HUMB. et P. CHOUX. (Pl. 10/5) – S.W.Madag. – ♄ to 3 m tall, ± branching; **Br.** long, slender, Th. 7–15 mm long, 1.5–2 cm apart, solitary; **L.** fleshy, 9–40 mm long, 2.5 mm across, about 5 mm thick; **Fl.** 10–15 mm long, white-yellowish.

**A. marnierana** RAUH. – Madag.; N.Tuléar. – ♄ with spreading Br., 1–2 m tall, Th in pairs, 1–1.5 cm long, thin; **L.** 1–1.5 cm long, ♀ Fl. crimson, ♂ Fl. not known.

**Aloe** L. Liliaceae. – S., trop. and N.Afr.; Ethiopia; Arabia; Madag.; wide spread and partly wild in the Mediteranean Region, Atlantic islands, in S.Am., Is. Reunion and in E.India. – Stemless plants or with short or even tall **St.** and ♄-like, sometimes ♄-like, branched and occasionally with thin, almost climbing shoots; **L.** arranged in Ros., often distichous or scattered, mostly fleshy, more rarely thinner and leathery, usually lanceolate, entire, often cartilaginous-edged or sinuately toothed, often even with the surface spinous, green, grey-green or bluish, sometimes with markings or stripes; **Infl.** from the leaf-axils, often appearing as if terminal, simple or branched, usually with bracts, Rac. nodding or erect, ± densely flowered, Fl. ± long stalked, mostly cylindrical to 3-angled, sometimes very much inflated at the base and constricted, sometimes curved, numerous, red, yellow or orange, sometimes with green-yellow tips, rarely white. – Greenhouse, warm, in summer partly in the open ground. Propagation: seeds, cuttings. – Important Literature see: A. BERGER, A. H. HAWORTH, H. HUMBERT, G. W. REYNOLDS et alia.

Plate 10. 1. **Alluaudia dumosa** DRAKE. Shoot with male Fl. (Photo: W. RAUH); 2. **A. montagnacii** RAUH (Photo: as 1); 3. **A. procera** (DRAKE) DRAKE (Photo: as 1); 4. **A. comosa** DRAKE (Photo: J. BOGNER); 5. **Alluaudiopsis fiherensis** HUMB. et CHOUX (Photo: L. SCHATTAT).

Plate 11. 1. **Aloe antandroi** (R. Decary) Perr.; 2. **A. africana** Mill. (Photo: J. Marnier-Lapostolle); 3. **A. albiflora** (Guill.) Bertr. Infl.; 4. **A. arborescens** Mill.; 5. **A. albiflora** (Guill.) Bertr.

Plate 12. 1. Aloe arenicola REYN.; 2. A. aristata HAW. v. aristata; 3. A. bainesii TH. DYER (Photo: W. TRIEBNER); 4. A. camperi SCHWEINF.

Plate 13. 1. Aloe ciliaris HAW. v. ciliaris; 2. A. capitata BAK. v. capitata (Repr.: EDMOND FRANCOIS, Plant. d. Madag.); 3. A. bakeri SCOTT ELLIOT; 4. A. contigua (PERR.) REYN. (Photo: as 2); 5. A. × delaetii RADL.

Plate 14. 1. Aloe deltoideodonta v. candicans f. latifolia PERR. (Photo: J. MARNIER-LAPOSTOLLE); 2. A. distans HAW.; 3. A. dichotoma L. v. dichotoma; 4. A. greenii BAK. (Photo: as 1).

Plate 15. 1. Aloe ferox MILL. (Photo: J. MARNIER-LAPOSTOLLE); 2. A. hereroensis ENGL. v. hereroensis; 3. A. humilis v. echinata (WILLD.) BAK. (Photo: as 1); 4. A. isaloensis PERR. (Photo: W. RAUH); 5. A. haemanthifolia MARL. et BGR. (Photo: A. BERG); 6. A. haworthioides BAK. v. haworthioides (Photo: as 1).

Plate 16. 1. **Aloe melanacantha** BGR. (Photo: J. MARNIER-LAPOSTOLLE); 2. **A. longistyla** BAK. (Photo: as 1); 3. **A. karasbergensis** PILL. (Photo: W. TRIEBNER); 4. **A. mitriformis** MILL. v. **mitriformis**.

Plate 17. 1. **Aloe pearsonii** SCHOENL. (Photo: H. HERRE); 2. **A. plicatilis** MILL.; 3. **A. pillansii** GUTHRIE (Photo: W. TRIEBNER); 4. **A. parvula** BGR.; 5. **A. polyphylla** SCHOENL. ex PILL. (Repr.: Nat. Cact. Succ. J. **21,** 1966, 17, Photo: L. KOFLER).

## Synopsis of the Genus Aloe acc. A. Berger and G. W. Reynolds
(Additional Madagascan species are also collected in XI).

**Sect. I. Aloinella** BGR. (incl. *Aloinella* LEMÉE & *Guillauminia* BERTR. as genera). – **Plants** small, stemless; **Roots** fusiform; **L.** in close Ros., triangular-lanceolate, papillose, aristate; **Infl.** slender, Rac. dense, Fl. erect, small, subsessile, with broad bracts. – Type-species: A. haworthioides. – Madag.-Species: A. albiflora, andringitrensis, bellatula, haworthioides, parvula, perrieri.

**Sect. II. Graminialoe** REYN. (*Leptaloe* STAPF as a genus). – **Roots** spindle-shaped; plants small, stemless; **L.** 4–10, narrowly linear, mostly rosulate-multifarious, sometimes distichous; **Infl.** simple, Ped. slender, with small bracts in the upper part, Rac. capitate or conical-capitate, Fl. stalked 10–20 mm long, Per. 10–20 mn long, narrowed teretely at the base, mouth to 3-angled or with two lobes, usually curved upwards. – Type-species: A. myriacantha. – S.Afr. and trop. Afr.–Species: A. albida, buchananii, caricina, graminifolia, minima, myriacantha, parviflora, saundersiae.

**Sect. III. Leptoaloe** BGR. (*Microcantha* BGR.). – Usually stemless, sometimes with short **St.**; **L.** distichous to rosulate, narrowly linear, lorate, or large, rigid or flabby, sometimes keeled, green or blue, mostly without spots on the upper surface, reverse often with white markings basally; margins rarely entire, usually minutely to prominently dentate; **Infl.** simple, with bracts, Rac. capitate to elongate, Per. ± narrowed into the pedicel 12–15 mm long, cylindrical to 3-angled, cylindrical-ventricose to broad at the base, pedicels 10–40 mm long. – Type-species: A. ecklonis. – S.Afr., trop. Afr., Madag.

   Series 1. Small plants with about 10 multifariously arranged **L.** 3–10 mm wide, margins finely toothed. – Species: A. bullockii, cannellii, chortolirioides, dominella, kniphofioides, plowesii, richardsiae, wildii, wolleyana.

   Series 2. Larger plants with more fleshy **L.**; **Fl.** reddish to scarlet and orange, 25–35 mm long. – Species: A. hazeliana, hlangapies, howmannii, musapanana, nubigena, soutpansbergensis, thompsoniae, torrei, verecunda, vossii.

   Series 3. Plants with yellow Fl. 12–22 mm long. – Species: A. ecklonis, integra, kraussii, linearifolia, modesta.

   Series 4. Fl. large, red, 36–40 mm long. – Species: A. boylei, cooperi, inyangensis, microcantha, nuttii, rhodesiana.

**Sect. IV. Aloe** (*Eualoe* BGR.). – Stemless plants or with ± tall **St.**, this often elongated; **Roots** usually cylindrical, rarely fusiform; **L.** ± fleshy and dentate, mostly long-lanceolate, distichous or arranged in spiral Ros. or multifarious; **Infl.** simple or branched.

  SSect. A. **Parvae** BGR. – Plants small.

   Series 1. **Haemanthifolia** BGR. – Stemless; **L.** distichously arranged, lorate, leathery, fleshy, margins entire; **Infl.** simple, Rac. short, dense, bracts acute, Per. narrowed into the long stalk. – Type-species: A. haemanthifolia. – S.Afr.–Species: A. haemanthifolia.

   Series 2. **Longistyla** BGR. – Stemless, small to large plants; **L.** multifariously arranged, lanceolate, stiff, sometimes with acute tubercles; **Infl.** simple, stout, short or long, with bracts, Rac. short or elongate, densely flowered, Fl. short-stalked, Per. straight or slightly curved, style long. – Type-species: A. longistyla. – S.Afr.–Species: A. broomii, chlorantha, hemmingii, longistyla, peglerae.

   Series 3. **Aristatae** BGR. – Stemless; **Ros.** spherical, later forming mats; **L.** very numerous, triangular-lanceolate, tapering, back and margins with white, horny, fleshy T.; **Infl.** simple, rarely branched, Rac. loose, bracts small, triangular-subulate, Fl. long-stalked, spreading, Per. clavate-cylindrically curved. – Type-species: A. aristata. – S.Afr.-Species: A. aristata.

  SSect. B. **Humiles** BGR. – **L.** less numerous, aristate, arranged in Ros. which branch to form mats.

   Series 4. **Virentes** BGR. – Stemless; **Ros.** making offsets; **L.** sublanceolate, somewhat fleshy, marginal T. without sharp tips; **Infl.** simple, bare, Cor. thickened below the centre. – Type-species: A. virens. – S.Afr.-Species: A. virens.

   Series 5. **Echinatae** S. D. – Stemless or with short, procumbent **St.**, solitary or forming groups; **L.** lanceolate or narrow-lanceolate, fleshy or leathery, upper surface or both sides tuberculate-spiny, the Sp. with soft to hard sharp points; **Infl.** simple, bracts tapering, Rac. elongated, Per. slightly inflated or cylindrical at the middle. – Type-species: A. humilis. – S.Afr.–Species: A. gloveri, humilis, krapohliana, melanacantha, schelpei.

   Series 6. **Proliferae** S. D. – Stemless, forming offsets; **L.** triangular or triangular-lanceolate, fleshy, blue, with firm horny T.; **Infl.** simple, with bracts, Rac. elongated. – Type-species: A. brevifolia. – S.Afr.-Species: A. brevifolia, jucunda.

**Series 7. Madagascarienses** BGR. see **XI. Madagascan species.**

**Series 8. Rhodacanthae** S. D. – Stemless or with a **St.**; **L.** triangular-lanceolate to lanceolate, blue, lineate, with pungent horny brown **T.**; **Infl.** simple, with tapering bracts, Fl. at first capitately congested. – Type-species: A. glauca. – S.Afr.-Species: A. glauca, lineata, polyphylla, pratensis.

**Series 9. Serrulatae** S. D. – Stemless or very short-stemmed; **L.** ± imbricate, trifarious, subspirally twisted, rigid, fleshy, cuticle leathery, reverse keeled, margins and keel cartilaginous crenate-dentate, surface glossy, with white spots irregularly arranged in transverse bands; **Infl.** simple or branched, Rac. loose, bracts small, Fl. short-stalked, Per. cylindrical, somewhat bellied at the base, swollen at the centre. – Type-species: A. variegata. – S.Afr., SW.Afr.–Species: A. ausana, dinteri, duckeri, serrulata, sladeniana, variegata.

**Series 10. Saponariae** BGR. – Plants mostly medium size, stemless or short-stemmed, simple or forming mats; **L.** in dense Ros., fleshy, with a smooth, usually hard surface, with longtudinal nerves, rarely without spots, frequently with numerous white markings arranged in ± irregular transverse bands, margins frequently sinuate-dentate, with pungent T.; **Infl.** rarely simple, mostly dichotomously branched, sometimes rebranched, Rac. short-headed to cylindrically tapering, densely or loosely flowered, bracts acute, Per. rounded at the base, not narrowed into the stalk, basally somewhat swollen and constricted above the ovary, thence decurved and enlarging towards the throat, or Per. clavate, not constricted above the ovary. – Type-species: A. saponaria. – S.- and E.Afr.; Ethiopia; Somalia.

A. **Per.** with a definite swelling at the base, constricted above the ovary, then elongated.

  **Group 1. Rac.** capitate, dense.
  a) **Rac.** flatly headed. – Species: A. branddraaiensis, latifolia, leptophylla, saponaria, swynnertonii.
  b) **Rac.** round-headed. – Species: A. umfoloziensis.

  **Group 2. Rac.** sub-capitate or more conical, rather dense.
  a) **Rac.** rounded above. – Species: A. affinis, boehmii, graminicola, greatheadii, immaculata, kilifensis, lateritia, macracantha, mudenensis, scobinifolia, wollastonii.
  b) **Rac.** more conical. – Species: A. davyana, graciliflora, verdoorniae.

  **Group 3. Rac.** cylindrically tapering.
  a) **Rac.** more narrow, rather dense. – Species: A. amudatensis, cremnophila, decurvidens, dyeri, fosteri, greenii, komatiensis, macrocarpa, pruinosa.
  b) **Rac.** wider, 8–9 cm ⌀, loose. – Species: A. lettyae, ? morgoroensis, vogtsii.

  **Group 4. Rac.** more narrow, long-cylindrical, tapering, not very densely flowered; **L.** with markings on the upper surface, reverse not spotted. – Species: A. barbertoniae, dewetii, longibracteata, mutans.

  **Group 5. Rac.** loose.
  a) Plants small, forming offsets. – Species: A. ammophila, neckii, vandermerwei.
  b) Plants larger; **Infl.** 100 to 175 cm high. – Species: A. burgersfortensis, constricta, keithii, parvibracteata, simii, transvaalensis, zebrina.
  c) Plants of median size; **Infl.** about 80 cm high, branched. – Species: A. monotropa.

B. **Per.** clavate, not constricted above the ovary; **L.** spotted on both sides, markings in transverse bands. – Species: A. grandidentata, greenwayi, prinslooi.

The following species are difficult to place within **Series 10**: A. commutata, comosibracteata, deflexidens, ellenbeckii, gasterioides, grahamii, heterocantha, labiaflava, leptosiphon, menyharthii, obscura, runcinata, spuria.

**Series 11. Paniculata** S. D. (*Striatae* REYN.). – Plants stemless or with a short procumbent **St.**; **L.** in a dense Ros., ovate-lanceolate, tapering, margins entire or minutely dentate and crenate; **Infl.** an umbellate or subcorymbose panicle, Per. up to 30 mm long with a subspherical swelling at the base, slightly constricted above the ovary, thence elongated. – Type-species: A. striata. – S. and SW.Afr.–Species: A. buhrii, karasbergensis, reynoldsii, striata.

**Series 12. Superpositae** POLE EVANS. – Stemless to short-stemmed; **L.** in a dense Ros., broad to narrow-lanceolate, without markings, obscurely or not linear; **Infl.** simple or branched, Rac. narrow to broad conical, pedicels 10–25 mm long, Fl. long cylindrical. – Type-species: A. suprafoliata. – S.Afr.
  a) **Infl.** up to 1 m high, simple, Fl. 20 mm long. – Species: A. suprafoliata, thorncroftii.
  b) **Infl.** 2–3 m high with 5–10 Br., Fl. 45 mm long. – Species: A. christianii, pretoriensis.

**Series 13. Asperifoliae** BGR. – Stemless or with creeping, rooting, procumbent **St.**; **Ros.** obliquely erect, older L. almost sword-shaped; **L.** thick, with rough surface, with isolated short horny marginal T.; **Infl.** rather loose, bracts ovate-cuspidate, longer than the pedicels, Per. narrowed into a stalk at the base, ± united up to the middle into a tube, incurved at apex. – Type-species: A. asperifolia. – S.Afr.–Species: A. asperifolia, claviflora, falcata, namibensis, pachygaster, viridiflora. –

**Series 14. Hereroensis** REYN. – Stemless or short-stemmed; **L.** lanceolate-triangular, densely rosulate, sometimes with elongated H-shaped spots, margins sinuate-dentate; **Infl.** a branched Pan., bracts narrowly lanceolate, Per. cylindrical-trigonous, slightly inflated at the middle, suddenly narrowed into the stalk, not at all constricted above the ovary, free above the centre, tips curved. – Type-species: A. hereroensis. – SW.Afr.–Species: A. hereroensis.

## SSect. C. Grandes BGR.

**Series 15. Percrassae** BGR. – Plants up to tall-stemmed; **Ros.** large; **L.** large, fleshy, blue-green, often spotted; **Infl.** branched, Rac. lax, bracts large, often reflexed, Per. cylindrical, not narrowed into the stalk. – Type-species: A. percrassa. – S.Afr.; Ethiopia; Angola; Kenya. – Species: A. babatiensis, berhana, classenii, esculenta, littoralis, macleayi, percrassa.

**Series 16. Verae** BGR. – Stemless or forming **St.**, often freely shooting from the base; **L.** in a ± dense Ros., sword-shaped, long-tapering, fleshy, greenish, often spotted in young plants, margins often with cartilaginous edge and with ± stout T.; **Infl.** stout, simple or with several erect Br., Rac. elongated, loose, with imbricate bracts tufted at the apex, floral bracts white, ± long, ovate-acute, often recurved, ± longer than the pedicels, Per. basally narrowed into the stalk, cylindrical, red or yellow, often with recurved tip, united tube-like up to the middle, the whole Infl. glabrous or ± densely covered with soft tangled hairs. – Type-species: A. vera. – Cape Verde Islands; E.Afr.; Somalia; S.Arabia; E.India. – Species: A. andhalica, barbadensis, breviscapa, desertii, dhalensis, dhufarensis, doei, eremophila, fulleri, harmsii, lavranosii, massawana, menachensis, metallica, mitis, niebuhriana, officinalis, otallensis, puberula, pubescens, rigens, serriyensis, tomentosa, trichosantha, turkanensis, vacillans.

**Series 17. Latebracteatae** BGR. – Stemless or forming a **St.**; **L.** sword-shaped; **Bracts** wide, obtuse, broadly ovate and basally amplexicaul, suddenly tapering, shorter than the pedicels, Per. not all narrowed into the stalk, cylindrical, constricted above the ovary. – Type-species: A. cryptopoda. – S.Afr.; Moçambique; Zanzibar. – Species: A. brachystachys, cryptopoda, lastii, lutescens, wickensii.

**Series 18. Tropicales** BGR. – Stemless or forming **St.**; **L.** fleshy, often large; **Infl.** dichotomously branched; **Bracts** lanceolate-acute, half as long as the Fl.-stalks, erect, pedicels about the length of the Per., which is yellow or red, basally narrowed into the stalk. – Type-species: A. abyssinica. – Ethiopia; trop. Afr.; Somalia.
a) Stemless; **L.** with markings. – Species: A. agaviifolia, somaliensis, wilsonii.
b) Stemless; **L.** with large T. – Species: A. venenosa.
c) Stemless; **L.** large. – Species: A. ? abyssinica, harlana, monticola, schliebenii, ukambensis, wrefordii.
d) Stemless; **L.** long. – Species: A. angolensis, buettneri, compacta, congolensis, crassipes, luapulana, trigonantha, trothai.
e) Forming stout **St.** – Species: A. andongensis.
f) Forming slender **St.** – Species: A. ballyi, gracilicaulis, medishiana, penduliflora.

**Series 19. Aethiopicae** BGR. – Stemless or forming **St.**; **L.** sword-shaped, often spotted; **Infl.** dichotomously branched, Rac. often cylindrical, bracts small, triangular, much shorter than the Fl. stalks, pedicels often half as long as the Per. or almost the same length, **Fl.** yellow-red or yellow. – Type-species: A. elegans. – Ethiopia; Nubia; Uganda; S.Afr.; Is. Rodriguez, Comoro; Socotra; Tanzania.
a) Stemless with large **L.** – Species: A. bukobana, chabaudii, elegans, grata, mcloughlinii, milne-redheadii, mzimbana.
b) Stemless with narrow **L.** – Species: A. lomatophylloides, mayottensis, rabaiensis.
c) Forming **St.**; **L.** serrate, with small marginal T. – Species: A. catengiana, gossweileri, hendrickxii, jacksonii, keayi, mubendiensis, palmiformis, perryi, retrospiciens, rupicola, schweinfurthii, seretii, splendens, stuhlmannii, suffulta, volkensii.
d) Forming **St.**; **L.** broad, T. large. – Species: A. adigratana, calidophila, camperi, dawei, marsabitensis, megalacantha, pungens, rivae, sinana, sinkatana, tororoana.

**Series 20. Cernuae** BGR. – Stemless or forming **St.**; **L.** narrow-lanceolate, rounded towards the tips, with curved brown T. in the middle; **Infl.** somewhat branched, Rac. short, truncate, Fl. densely and capitately arranged, nodding, long-stalked, Per. cylindrical-campanulate, Pet. free, slightly spreading, recurved at apex. – Type-species: A. capitata. – Madag.– Species: A. capitata, trachyticola.

## SSect. D. Prolongatae BGR.

**Series 21. Macrifoliae** HAW. (*Striatulae* BGR.). – **St.** thin, elongated, sometimes climbing, loose-leaved; **L.** linear-lanceolate, thin and slightly fleshy, soft, with minute, remote marginal T., amplexicaul, the Sh. striped; **Infl.** lateral, simple, glabrous, Rac. often rather lax, bracts small, Per. red or yellow, ± swollen. – Type-species: A. ciliaris. – S.Afr.; Madag.–Species: A. acutissima, antandroi, bakeri, ciliaris, commutata, decaryi, gigas, gracilis, kedongensis, laeta, millotii, ngobitensis, pearsonii, striatula, tenuior, yavellana.

**Series 22. Monostachyae** BGR. – Forming rather weak, loose-leaved **St.**; **L.** narrow, long tapering, fleshy, sinuate-dentate, T. ± stout; **Infl.** simple, Rac. rather lax, bracts small, triangular, Fl. short-stalked, Per. club-shaped to cylindrical, slightly constricted above the ovary and then recurved. – Type-species: A. cameronii. – Somalia; Uganda; Malawi; Tanzania; Zanzibar; Socotra. – Species: A. boscawenii, cameronii, dorotheae, elgonica, flexilifolia, monteiri, squarrosa, vituensis.

**Series 23. Pleurostachyae** BGR. – Plants similar to the **Ser. Monostachyae.** Forming rather weak, often elongated, laxly leaved **St.**; **L.** lanceolate or sword-shaped; **Infl.** paniculate, Br. slender, spreading, bracts small, Per. rather inflated, ± constricted above the ovary. – Type-species: A. secundiflora. – Somalia; Tanzania; Ethiopia; Madag.-Species: A. christianii, divaricata, erensii, guerrai, hildebrandtii, inermis, kirkii, leachii, leucantha, pirottae, ruspoliana, secundiflora, tweediae.

**Series 24. Fruticosae** BGR. – Shrubby; **St.** thin, erect or often curved, forming offsets from the base, laxly-leaved and **L.** distichous for the most part, fleshy, lanceolate or sword-shaped; **Infl.** branched, Br. and Rac. ± elongated, bracts rather long, triangular or lanceolate tapering, Per. united up to the apex. – Type-species: A. consobrina. – S.Afr.; Transvaal, S.Arabia. – Species: A. bussei, cinnabarina, confusa, consobrina, enotata, inamara, macrosiphon, mendesii, nyerensis, pendens, venusta, veseyi.

**Series 25. Mitriformes** SALM. – **St.** procumbent or erect, firm; **L.** laxly arranged in Ros., fleshy, rigid, ovate to sword-shaped, margins and keel with ± stout T.; **Infl.** simple, often branched, Rac. short, mitre-shaped or often longer and conical-cylindrical, bracts broadly triangular, shorter than the Fl.-stalks, pedicels spreading, ± as long as the Per., these red, cylindrical, Pet. free or ± united. – Type-species: A. mitriformis. – S.Afr.–Species: A. arenicola, brownii, comptonii, mitriformis, nobilis, sororia, stans.

**SSect. E. Magnae** BGR.

**Series 26. Comosae** BGR. – Plants arborescent with stout **St.** 1–2 m high; **L.** in dense Ros., lanceolate to sword-shaped; **Infl.** simple, Rac. much elongated, lax, Fl. long-stalked, bracts large, tapering, crowded in imbricate tufts at the ends, Per. cylindrical, Pet. free. – Type-species: A. comosa. – S.Afr.–Species: A. comosa.

**Series 27. Purpurascentes** SALM. – Stemless or shrubby with stout **St.**, low to high, often dichotomously branched, growing in dense groups; **L.** in dense Ros., ± sword-shaped, ± spotted; **Infl.** simple or branched, Sc. rather stout with bracts, Rac. elongated, denser at apex, bracts imbricate, large, Fl.-stalks long, nutant, Per. 25–40 mm long, red, yellow or orange. – Type-species: A. succotrina. – S.Afr.; Ethiopia.
a) **Infl.** simple. – Species: A. gariepensis, microstigma, purpurascens, succotrina.
b) **Infl.** branched. – Species: A. framesii, khamiesensis, steudneri.

**Series 28. Arborescentes** S. D. – Tall ♄ to ♄-like; **L.** in dense Ros., sword-shaped, fleshy, with cartilaginous T. marginal, hard and horny; **Infl.** simple or branched, bracted, Rac. densely flowered, bracts imbricate, Per. with long nodding pedicels, straight, cylindrical, red, Pet. free. – Type-species: A. arborescens. – S.Afr.; Madag.
a) Plants small, stemless or with creeping **St.** – Species: A. bulbillifera, mutabilis, vanbalenii.
b) Plants tall, stout, ♄-like. – Species: A. arborescens, pluridens.

**Series 29. Principales** BGR. – Tall ♄ to ♄-like plants; **L.** large, sword-shaped, fleshy, soft, not leathery, margins with cartilaginous, triangular, firm T.; **Infl.** large and broad, simple or branched, with bracts, Rac. elongated, with numerous mostly stalked Fl. which are pendulous, bracts large, imbricate, covering the buds, Per. straight, cylindrical, Pet. free or ± united when the Fl.are sessile. – Type-species: A. speciosa. – S.Afr.; Ethiopia; Madag. – Species: A. ? drepanophylla, longiflora, macroclada, platylepis, rubroviolacea, schoelleri, speciosa, ? spicata.

**Sect. V. Anguialoe** REYN. – **St.** often short, procumbent or usually erect, robust; **L.** in a dense Ros., fleshy, sword-shaped, without markings, margins toothed; **Infl.** simple, 2–5 from a Ros., Rac. cylindrical, elongated, very densely and many-flowered, Fl. sessile, rarely short-stalked, Per. campanulate, 9–20 mm long, Pet. free. – Type-species: A. sessiliflora. – S.Afr.–Species: A. alooides, castanea, dolomitica, recurviflora, sessiliflora, tauri, vryheidensis.

**Sect. VI. Pachydendron** HAW. – Stemless or short-stemmed, simple or branched; **L.** in a dense Ros., ± lanceolate or sword-shaped, rigid, leathery, margins and sometimes the surface armed with stout pungent T.; **Infl.** a branched Pan., frequently broadly to compactly candelabra-shaped, with erect, spreading or subhorizontal laterals, mostly furrowed, Rac. cylindrical,

± erect, dense or exceedingly densely flowered, or subhorizontal with secund Fl., Per. cylindrical-bellied or cylindrical to club-shaped, red, orange or yellow, Pet. ± free up to the middle. – Type-species: A. ferox. – S.Afr.; Rhodesia.

**Group 1.** (SSect. *Ortholophae* CHRISTIAN). – Stemless, forming dense groups; Rac. oblique, 30–40 cm long; Fl. turning to one side, Per. 26 mm long. – Species: A. decurva, globuligemma, mawii, ortholopha.

**Group 2.** St. 2–4 m high; L. concave on the upperside, recurved; Rac. erect, capitate, 8–10 cm long, pedicels 25 mm long, Per. 25 mm long. – Species: A. angelicae.

**Group 3.** St. 1–8 m high; L. 70–150 cm long; Rac. erect, broad-cylindrical, somewhat truncate, Fl. yellowish. – Species: A. excelsa, rupestris.

**Group 4.** Rac. erect, cylindrical, densely many-flowered, 30–60 cm long.
  a) Stemless to somewhat stemmed; L. stiff, up to 60 cm long, 10–12 cm wide. – Species: A. aculeata, gerstneri, petricola.
  b) St. 2–4 m high; L. 65–100 cm long; Rac. cylindrical to cylindrical-tapering. – Species: A. africana, candelabrum, ferox.

**Group 5.** St. 2–4 m high; Rac. suberect, densely flowered, 25 cm long. – Species: A. spectabilis.

**Group 6.** St. simple, 2–4 m high; L. up to 150 cm long; Rac. horizontally to suberect; Fl. turned to one side; Per. 30–35 mm long, orange. – Species: A. marlothii.

**Sect. VII. Dracoaloe** BGR. – Plants large ♄ to tall ♄, dichotomously branched; L. rosulate, margins minutely dentate; Infl. branched, Rac. broadly cylindrical, somewhat large, Fl. short-stalked, Per. cylindrical-ventricose, fleshy, yellow, Pet. united to beyond the middle. – Type-species: A. dichotoma. – S.Afr.–Species: A. dichotoma, pillansii, ramosissima.

**Sect. VIII. Aloidendron** BGR. – Dichotomously branched ♄; L. long, large, concave on the upper surface, dentate; Infl. mostly 3-branched, Rac. usually in one vertical plane, densely flowered, broadly cylindrical; Fl. short-stalked, Per. cylindrical, almost three-angled, Pet. free almost to base. – Type-species: A. bainesii. – S.Afr.–Species: A. bainesii.

**Sect. IX. Sabaealoe** BGR. – Tall, dichotomously branched ♄; L. long, sword-shaped, fleshy; Infl. branched, Rac. densely flowered, bracts triangular-ovate, 15 mm long. – Type-species: A. sabaea. – Arabia. – Species: A. eminens, gillilandii, sabaea. –

**Sect. X. Kumara** MED. – ♄-like, freely dichotomously branched; L. distichous, strap-shaped, rounded above, margins exceedingly minutely dentate; Infl. simple, Rac. lax, Fl. cylindrical, red. – Type-species: A. plicatilis. – S.Afr. Species: A. plicatilis.

**Complex XI. Madagascan species.** – Acc. G. W. REYNOLDS the species occurring in Madagascar have no close relation to the African species. – W. RAUH divided the Madagascan species according to growth habit into 4 Groups. –
  a) **Tree-like.** – St. mostly simple, erect, at least 2 m high, seldom branched, with large L.-Ros. at the tip. – S. and SW. Madag., in dry regions. – Species: A. helenae, suzannae, vaombe, vaotsanda.
  b) **Shrubby.** – St. seldom simple, mostly to several or many, branching from the base; L. lax, amplexicaul. – W. and SW. Central-Madag.–Species: A. acutissima, antandroi, decaryi, divaricata, isaloensis, millotii, subacutissima.
  c) **Cushion- and mat-forming.** – St. to 40 cm high, branching from near the base and forming groups, or St. thin, many-branched, often St.-less, mats formed by branching at the base; L. in extensive Ros. – Southern part of the Central Highland; SW. and SE.Madag.–Species: A. albiflora, bakeri, bellatula, buchlohii, calcairophila, deltoideodonta, descoingsii, parallelifolia, perrieri, rauhii, versicolor, viguieri.
  d) **Forming Ros.** – Stemless, with a solitary Ros.
    α) **Infl.** branched and paniculate. – S., SW. and central-Madag.–Species: A. andringitrensis, betsileensis, bulbillifera, capitata, contigua, ericetorum, erythrophylla, fievetii, ibitiensis, itremensis, madecassa, silicola, suarezensis, trachyticola.
    β) **Infl.** not branched. – Central Madag.
      1. **Infl.** spicate. – Species: A. conifera, cryptoflora, haworthioides, humbertii, macroclada.
      2. **Infl.** racemose. – Species: A. boiteaui, compressa, decorsei, laeta, leandrii, parvula, schomeri.

**A. abyssinica** LAM. (§ ? 4/C/18c). – ? Ethiopia. – Habitat not known; L. in fairly loose Ros., 75 cm long, 10 cm across lower part, thick, pure green, upperside grooved, margins with red T.; Infl. 1 m tall, Fl. 32 mm long, yellowish-green. Insufficiently known species.

**A. aculeata** POLE EVANS (§ VI/4a). – N.Transvaal. – Acauline or with short prostrate St.; L. 30 in an open Ros., broad-lanceolate, 50 to 60 cm long, 8–12 cm across base, upperside concave, underside distinctly keeled above, dark green to bluish, marginal T. pointed, triangular, reddish-brown, 5–6 mm long, 1–2 cm apart, upper and lower surfaces dentate; **Infl.** over 1 m tall, Fl. 2.5–4 cm long, lemon yellow.

**A. acutissima** PERR. (§ IV/D/21; XI/b). – Central Madag. – ♄ to 1 m ⌀, **St.** erect or obliquely ascending or prostrate, to 3 cm ⌀, with 10–20 L. almost down to the base; **L.** 1 cm apart above, gradually tapering, those at tip pointed and sharp, those at base 4 cm across and 1 cm thick, margins with well-spaced T., these being stout and red-brown, sap yellow; **Infl.** to 50 cm long, Fl. pendulous, scarlet.

**A. —** v. **antanimorensis** REYN. – SW.Madag. – Daintier in habit; **L.** shorter and narrower.

**A. adigratana** REYN. (§ IV/C/19d) (*A. abyssinica* HOOK. f., *A. eru* v. *hookeri* BGR.). – Ethiopia: Tigre. – ♄, St. to 1 m tall and erect, or 1–2 m and prostrate, 12 mm ⌀, often forming small bushes as a result of new St.; **L.** 16–20 arranged in a Ros., ensiform, 60–80 cm long, 15 cm across base and 1.5–2 cm thick, upperside grooved above, dull green, with numerous pale green, lenticular markings in the lower third, underside mottled, margins with T. which are up to 10 mm long, 2.5–3.5 cm apart, and reddish-brown; **Infl.** to 90 cm tall, Fl. 18–33 mm long, orange-yellow.

**A. affinis** BGR. (§ IV/B/10/A/2a). – Transvaal. – Acauline; **L.** 5–7 cm across base, then tapering, 20–25 cm long, fleshy, without mottling, margins sinuate-dentate, T. triangular, pointed, 5–8 mm long, about 10 mm apart; **Fl.** 3.5–4 cm long, brick-red.

**A. africana** MILL. (Pl. 11/2) (§ VI/4b) (*A. a.* v. *latifolia* HAW., *Pachydendron a.* v. *luteum* HAW., *A. a.* v. *angustior* HAW., *A. pseudoafricana* S. D., *P. a.* HAW., *P. angustifolium* HAW., *A. ang.* HAW., *A. perfoliata africana* AIT., *A. bolusii* BAK.). – Transvaal; Cape. – **St.** simple, 2–4 m tall, sometimes branching, thickly covered with old, dried L.; **L.** c. 30, arranged in a close Ros., to 65 cm long, 12 cm across base, dark green to bluish, upperside with scattered Sp. in the lower part, underside with Sp. on the M. line, reddish and ± spiny in the upper part; **Infl.** 60–80 cm tall, Fl. 3.5 cm long, yellow to orange-yellow.

**A. agavifolia** TOD. (§ IV/C/18a). – Trop. c. Afr. – Acauline, with few offsets; **L.** 20–25 in a close Ros., broadly linear to lanceolate, tapering, upperside grooved and concave, lower part 1 cm thick, 45–55 cm long, 9–15 cm across, green to purple with distinct lines and spots, margins with strong, horny, white T.; **Infl.** to 1 m tall, Fl. to 3 cm long, red.

**A. albida** (STAPF) REYN. (§ II) (*Leptaloe a.* STAPF, *A. krausii* v. *minor* BAK., *A. k.* SCHOENL., *A. myriacantha* v. *minor* BGR.). – E.Transvaal. – Acauline, small; **root** fusiform; **L.** 6–12, rosulate, almost linear, 10–15 cm long, 4–5 mm across, 10 mm across at base, and clasping the stem, upperside slightly grooved, dark green, underside slightly keeled, margins with T. 0.5 mm long, white, 1 mm apart; **Infl.** 9–15 cm tall, simple, Fl. 18 mm long, creamy-white, spotted with green.

**A. albiflora** GUILL. (Pl. 11/3, 5) (§ I; XI/c) (*Guillauminia a.* (GUILL.) A. BERTR.). – S.Madag. – Acauline. **L.** c. about 7, arranged spirally, broad linear-tapering, 12–15 cm long, 9 mm across at base, upperside slightly grooved, green, white-spotted, both surfaces very rough with raised, white tubercles, margins with sharp, white, cartilaginous T.; **Infl.** 25 cm tall, Fl. short and broadly campanulate, Tep. pure white, with brownish-green M. nerve, anthers orange-coloured.

**A. alooides** (BOLUS) VAN DRUTEN (§ V) (*Urginea a.* BOLUS, *Notoceptrum a.* BENTH., *A. recurvifolia* GROEN.). – E.Transvaal. – **St.** unbranching, 1.2 m tall; **L.** 40, densely rosulate, 1 m long, 17 cm across, upperside very concave, green with red margin, marginal T. 2 mm long, 1 mm apart; **Infl.** 1 m tall or more, Fl. 1 cm long, yellow-green.

**A. ammophila** REYN. (§ IV/B/10/A/5a). – Transvaal. – Forming dense clumps, acauline; **L.** 10–14, densely rosulate, lanceolate, tapering, c. 22 cm long, 5–6 cm across, upperside slightly concave, green, both sides with white ± translucent spots in wavy transverse bands, margins sinuate-dentate, T. 4–5 mm long, 8–12 mm apart, sharp; **Infl.** c. 60 cm tall, Fl. 2–2.5 cm long, coral-red.

**A. amudatensis** REYN. (§ IV/B/10/A/3a). – Uganda. – Acauline, mat-forming; **L.** c. 12 in a dense Ros., c. 22 cm long, 5 cm across, 8 mm thick at base, upperside dull green, often reddish-brown, slightly concave, with numerous, whitish dots, underside with pale green spots ± running into one another, margins sinuate-dentate, margin cartilaginous, T. stout, 5–8 mm apart, red-brown; **Infl.** 50–65 cm tall, Fl. 23 mm long, pink-red to coral-red.

**A. andongensis** BAK. (§ IV/C/18e). – Angola. – **St.** 30–60 cm tall, with 2–3 Br.; **L.** forming dense Ros., lanceolate, tapering, bluish, thick, 20 to 23 cm long, 30–36 mm across, margins thin, horny, dentate, T. 2–3 mm long, brown; **Infl.** branching, Fl. 18–20 mm long, yellow.

**A. cv. Andrea.** – Cultivated hybrid: ? **A. eru** v. **cornuta** × **A. striata**. Hort. DEL. – Acauline or short-stemmed, branching dichotomously; **Ros.** to 2 m tall; **L.** numerous, 70–100 cm long, 25–30 cm across base, very concave, fleshy, blue, with a red horny edge, T. 1 cm apart, triangular; **Infl.** short, of c. 30 Br. arranged candelabra-fashion, raceme dense, 20–30 cm long, Fl. coral-red.

**A. andringitrensis** PERR. (§ I; XI/d/a). – Central Madag. – Usually acauline, **root** fusiform; **Ros.** of 10–12 L., 40–50 cm long, 6–7 mm across, with a rounded tip, 1–2 cm thick, rough with tiny papillae, marginal T. 1.5 mm long, red; **Infl.** 2–5-branched, Fl. in a group of 25–30, yellow.

**A. angolensis** BAK. (§ IV/C/18d). – Angola. – Almost acauline; **L.** in a dense Ros., 4–5 cm across base, ensiform, blue, margins with T. 2 mm long, 12–20 mm apart and triangular; **Infl.** 80 cm tall, Fl. 2–2.4 cm long, sulphur-yellow.

**A. antandroi** (R. Decary) Perr. (Pl. 11/1) (§ IV/D/21; XI/b) (*Gasteria a.* R. Decary, *A. leptocaulon* Boj.). – S. & S.W.Madag. – **Root** very woody; **St.** 4–5, simple or branching, 6–8 mm $\varnothing$, also liana-like and climbing, (0.6–)1(–3) m long; **L.** 15–25 or more, 8–15 cm apart, so curved as to be hook-like, Sh. with marked vertical stripes, L. 6–12 cm long, 6–8 mm across base, marginal T. small, arranged irregularly; **Infl.** 15–20 cm long, Fl. 22 mm long, scarlet.

**A.** × **antonii** Bgr. (*A. hanburyi* A. Borzi). – Cultivated hybrid: ? **A. ferox** × **Aloe** spec. – **St.** 1–1.5 m tall; **L.** in a dense Ros., lanceolate-ensiform, 50 cm long or more, rather grooved towards the tip, green to bluish, margins with triangular, horny T.; **Infl.** 60 cm tall or more, Fl. red.

**A. arborescens** Mill. (Pl. 11/4) (§ IV/E/28b) (*A. perfoliata* v. *a.* Sol., *A. arborea* Med., *A. fruticosa* Lam.). – Natal, Cape; Malawi. – **St.** simple, 1–4 m tall, about 5 cm $\varnothing$; **L.** in a dense Ros., up to 60 cm long, 3.5–5.5 cm across, ensiform, grey-green or dark green, fleshy, upperside somewhat concave towards base, margins sinuate-dentate, T. up to 5 mm long, pointed; **Infl.** simple, Fl. 4–4.5 cm long, red.

**A.** — v. **arborescens** (*A. a.* v. *milleri* Bgr.). – L. 5 cm across, sea-green, very fleshy.

**A.** — v. **frutescens** (Salm) Link. (*A. f.* Salm). – Producing more offsets to form dense bushes; **Fl.** somewhat smaller.

**A.** — v. **natalensis** (Wood et Ev.) Bgr. (*A. n.* Wood et Ev.). – Natal, Cape. – Many-stemmed, often several m $\varnothing$, densely covered with dead L.; **L.** 20–50, ensiform, gradually tapering, 45–50 cm long, 5 cm across, bluish, margins sharp, horny, T. 4 mm long, 1 cm apart; **Fl.** to 4.5 cm long, purple-red.

**A.** — v. **pachythyrsa** Bgr. – Resembles v. **natalensis**; Rac. denser, Fl. longer, dark red.

**A.** — v. **ucriae** (Terr. f.) Bgr. (*A. ucriae* Terr. f.). – L. very numerous, 60 cm long, 3.5–4.5 cm across, very long-tipped, T. 3–4 mm long, 8 mm apart.

**A.** — v. **viridifolia** Bgr. – Roundish and low-growing bushes; **L.** 50–52 cm long, 5–5.5 cm across, green, T. close together; **Fl.** paler red.

**A. arenicola** Reyn. (Pl. 12/1) (§ IV/D/25). – S.W.Afr.: coastal zone. – **St.** branching or simple, 30–40 cm long; **Ros.** somewhat open; L. 18 cm long, 5.5 cm across, underside convex, blue-green, with scattered white marks, margins horny, Sp. 0.5 mm long, 5–8 mm apart; **Infl.** 50 cm long, Fl. 4 cm long, brilliant red.

**A. aristata** Haw. v. **aristata** (Pl. 12/2) (§ IV/A/3) (*A. longiaristata* Roem. et Schult., *A. ellenbergeri* Guill.). – Cape: Graaff-Reinet D; Orange Free State; Natal. – Acauline, usually in compact groups of up to 12 Ros. 10–15 cm $\varnothing$; L. 100–150, arranged in many rows, (4–)8–10 cm long, 10–11 mm across base, lanceolate, terminating in a dry, translucent awn, underside convex, green, the underside especially with short, white, soft, tuberculate Sp., these often arranged in transverse bands and forming 1–2 longitudinal rows, margins white, cartilaginous, with horny T. 1–2 mm long and 1–2 mm apart; **Infl.** to 50 cm tall, Fl. 4 cm long, orange-red. (For crosses with **Gasteria** see × **Gastrolea**.)

**A.** — v. **leiophylla** Bak. – **L.** smaller and thinner, upperside virtually without tuberculate Sp., T. smaller; **Infl.** more slender.

**A.** — v. **parvifolia** Bak. – L. 4–5 cm long, 18 mm across, bluish-green, tuberculate Sp. on underside arranged in double rows, shorter and translucent.

**A. asperifolia** Bgr. (§ IV/B/13). – S.W.Afr.: Gr. Namaqualand. – **Ros.** in dense clumps of 20–40, **St.** short, creeping; **L.** lanceolate, bluish to almost whitish, fleshy, 15–25 cm long, 4–7 cm across, both sides rough, warty, underside convex and keeled above, keel with several horny T., margins with T. 2–3 mm long 5–15mm apart, horny and sharp; **Infl.** 50–75 cm tall, Fl. 28 mm long, scarlet.

**A. audhalica** Lavr. et Hardy (§ IV/C/16). – S.W.Arabia. – Mostly simple, occasionally forming groups of several Ros.; **L.** up to 30, dense, triangular, pointed, to 45 cm long, 15 cm across base, upperside somewhat grooved in upper part, somewhat rough, bluish-green to reddish-brown, underside convex, margins pink, horny, T. sharp, dark-brown, 4 mm long, 4–6 mm apart; **Infl.** 80–100 cm tall, Fl. 3 cm long, scarlet, lobes with creamy margins and brown tip.

**A. ausana** Dtr. (§ IV/B/9) (*A. variegata* Dtr.). – S.W.Afr.: Gr. Namaqualand. – Resembles **A. variegata** L.; **L.** arranged in 5–7 Ros., 12 cm long, 4–4.5 cm across, deeply grooved, dots in indistinct rows. (Reynolds regarded this species as being **A. variegata** L.)

**A. babatiensis** Christian et Verdoorn (§ IV/C/15). – Tanzania. – **St.** simple, c. 50 cm tall, with dense Ros. above, lower part covered with dry, old L.; **L.** ovate-lanceolate to lanceolate, pointed, $\pm$ 27 cm long, 6 cm across, upperside glossy, dark green to coppery, broad-concave, often with dark lines or with some mottling towards the base, underside pale green, lined, convex, margins dentate, T. triangular, 4 mm long, 8–12 mm apart, red-brown tipped; **Infl.** 40–80 cm tall, Fl. about 37 mm long, orange to salmon-pink, lobes yellow.

**A. bainesii** Th. Dyer (Pl. 12/3) (§ VIII) (*A. bainesii* v. *barberae* (Dyer) Bak., *A. b.* Dyer). – Natal, Cape: E.London D.: E.Transvaal; Swaziland; Moçambique. – ♄ to 20 m tall, up to 5 m thick at the base, branching dichotomously; **L.** in a dense Ros., sheathing at the base, ensiform, leathery, broadly grooved, green, 60–100 cm long, margins with T. 3–5 mm long; **Infl.** sturdy, Fl. 3–4 cm long, red-yellow.

**A. bakeri** Scott Elliot (Pl. 13/3) (§ IV/D/21; XI/c). – Madag. – Forming mats up to 1 m $\varnothing$, with 100 **Ros.** or more; **St.** 10–15 cm long, thin, freely branching; **L.** 12, narrow-linear, up to 7 cm long, 8 mm across base, brown-red, white-mottled, margins with white, cartilaginous T.; **Infl.** 20–30 cm tall, Fl. to 23 mm long, orange-red, yellow-orange to yellow towards the tip.

**A. ballii** REYN. (§ I). – Rhodesia: Melsetter D. – Mostly pendulous; **St.** 1–1.5 m long, spirally twisted, glabrous, the top 20–30 cm bearing L., 9 mm thick; **L.** 8–10 arranged in 2 series, clasping the stem at their bases, 20–30 cm long, 1 cm across base, narrowing towards the tip, upperside concave, green, with several white, longish spots in the lower part, underside convex, green, with many white spots at base only, margins with tiny T. set 2–4 mm apart; **Infl.** 50–60 cm long, Fl. 12–16 mm long, flame-coloured to pale reddish-orange.

**A. ballyi** REYN. (§ IV/C/18f). – Tanzania. – **St.** simple, slender, 6–8 m tall, 10–15 cm thick; **L.** c. 25 in a dense Ros., c. 90 cm long, 14 cm across base, gradually tapering, upperside grey-green, marginal T. white, 4–5 mm long, 10 to 15 mm apart; **Infl.** c. 60 cm tall, Fl. 5.5 mm long, orange-red.

**A. barbadensis** MILL. (§ IV/C/16) (*A. perfoliata* v. *vera* L., *A. p.* v. *b* AUCT., & v. *littoralis* KOENIG ex BAK., & v. *lanzae* BGR. v. *chinensis* BGR., *A. lanzae* TOD., *A. chinensis* BAK., *A. b.* v. c. HAW., *A. arabica* LAM., *A. indica* ROYLE, *A. elongata* MURR., *A. vulgaris* LAM., *A. rubescens* DC., *A. flava* PERSOON, *A. vera* v. *wratislaviensis* HORT.). – Origin unknown, possibly S.Arabia; widely distributed and growing wild in Cape Verde Is., Santo Antão, Canary Is., Madeira, S.Mediterr. region, Is. of Reunion, S.Am.: at foot of Andes, peninsular India. – **Roots** fibrous, fleshy; acauline or with short **St.**, offsetting, forming dense clumps; **L.** c. 16, densely rosulate, 40–50 cm long, 6–7 cm across base, narrowing to tip, rather thick and fleshy, upperside grey-green, initially mottled, slightly grooved above, margins slightly pink with stout T. 2 mm long and 15–20 mm apart; **Infl.** 60–70 cm tall, Fl. c. 3 cm long, yellow.

**A. barbertoniae** POLE EVANS (§ IV/B/10/A/4). – Transvaal. – **Ros.** usually simple, acauline or short-stemmed, **St.** covered with old L.; **L.** 20–30, dense, lanceolate, pointed, 30–40 cm long, 10–11 cm across, extended by a dry tip 10 cm long, upperside slightly concave, green to reddish, with numerous oblong, white marks merging into wavy transverse bands, underside pale greenish, unmottled, margins sinuate-dentate, brown, horny, T. 5–6 mm long, 10–15 mm apart, triangular; **Infl.** to 1 m tall, Fl. 3.5–4 cm long, deep pink-red.

**A. bellatula** REYN. (§ I; XI/c). – Central Madag. – Acauline; **root** cylindrical, offsetting from base to form dense clumps; **L.** c. 15, densely rosulate, 10–13 cm long, 9–10 mm across base, 3 mm thick, gradually widening towards the tip, upperside slightly grooved, underside convex, both surfaces finely warty or rough-papillose, with numerous white, somewhat lenticular spots 1 mm long, 0.5 mm across, margins with soft, horny, triangular T. 1 mm long, 1 mm apart; **Infl.** to 60 cm tall, Fl. 13 mm long, coral-red.

**A. berhana** REYN. (§ IV/C/15). – Ethiopia: Shewa. – **Ros.** solitary or in small groups, acauline or with short creeping stems; **L.** 24 or more, 50–60 cm long, 15 cm across, gradually tapering to the tip, upperside green, without mottling, slightly grooved in upper part, underside convex, margins horny, red-brown, T. 3 to 4 mm long, about 15 mm apart; **Infl.** 1 m tall, or more, Fl. 3–3.5 cm long, scarlet.

**A. betsileensis** PERR. (§ XId/α). – Central Madag.: Tuléar. – **Ros.** acauline, large, of 12–16 **L.** 28–40 cm long, 7–9 cm across, gradually tapering, somewhat rounded at tip, 1.5–2 cm thick, marginal T. robust, 3 mm long, 1 cm apart, red; **Infl.** 1–4-branched, Fl. very numerous, to 1.5 cm long, orange-coloured.

**A. boehmii** ENGL. (§ IV/B/10/1/2a). – Tanzania. – **Ros.** acauline; **L.** elongated-triangular, gradually tapering, 25–30 cm long, 5.5 cm across base, margins white, cartilaginous and sinuate-dentate, T. 4–5 mm long, 9–12 mm apart, horny, brown; **Infl.** a corymb, Fl. c. 3 cm long, golden-yellow.

**A. boiteaui** GUILL. (§ XId/β2). – Tanzania. – Acauline; **L.** approximately 9, c. 15 cm long, 1 cm across, c. 2 mm thick, margins parallel, with a pointed tip, olive-green, margins and tip pink, marginal T. small, triangular, pale red; **Infl.** short, Fl. 2.5 cm long, red.

**A. boranensis** CUFOD. – S.Ethiopia. – Acc. REYNOLDS a hybrid: **A. secundiflora** × **A. otallensis** v. **elongata**.

**A.** × **bortiana** TERR. – Cultivated hybrid: parents unknown, possibly related to **A. striata**. Acauline or short-stemmed, simple; **L.** in a dense Ros., long-lanceolate, 25 cm long, 7–8 cm across, blue, stiff, fleshy, mottled, margins cartilaginous, dentate; **Infl.** as for **A. striata**.

**A. boscawenii** CHRISTIAN (§ IV/D/22). – Tanzania. – ♄, Br. from base, St. to 1 m tall, 5–7 cm ⌀, laxly leafy; **L.** ovate-lanceolate, 44–50 cm long, 8 cm across base, upperside grooved, light green, striped, underside rounded, margins narrow, T. 2–5 mm long, 7–18 mm apart, pointed, triangular; **Infl.** 90 cm tall, Fl. 3 cm long, yellow, the tips being brownish.

**A. boylei** BAK. (§ III) (*A. agrophila* REYN.). – Cape: Transvaal. – **Ros.** acauline, of 6–8 **L.** c. 55 cm long, 6–7 cm across, with white spots in the lower part, margins deeply sinuate and rather stiff, T. 1 mm long; **Infl.** 55 cm tall, Fl. 3.3 cm long, orange-red.

**A. brachystachys** BAK. (§ IV/C/17). – Zanzibar. – **St.** simple, slender; **L.** in an open Ros., ensiform, gradually tapering, 50–60 cm long, 5 cm across, green, margins horny, T. triangular, white; **Infl.** simple, Fl. 3–3.5 cm long, pale red. (Acc. REYNOLDS an insufficiently known species.)

**A.branddraaiensis** GROENEW.(§IV/B/10/1/1a). – E.Transvaal. – Acauline, often offsetting from base; **L.** 20–25, rosulate or frequently also in 2 series or spirally arranged, lanceolate, tapering, c. 35 cm long, 8–10 cm across base, tip often dying back, green, brownish towards the tip and with numerous, white longitudinal stripes and numerous, somewhat H-shaped marks, underside paler, spots more numerous, margins sinuate-dentate, T. pointed, 2–3 mm long, 10–15 mm apart; **Infl.** 1–1.5 m tall, Fl. 27 mm long, vivid scarlet.

**A. brevifolia** MILL. v. **brevifolia** (§ IV/B/6) (*A. prolifera* HAW.). – Cape: Riversdale D. – Freely offsetting, forming round groups; **Ros.** 10–12 cm ⌀, short-stemmed; **L.** triangular-oblong, 7–18 cm long, 2–3 cm across base, up to 1 cm thick, upperside concave in the upper part, underside convex with a few Sp. above and keeled, marginal T. 2–3 mm long; **Infl.** 50 cm tall, Fl. 1.5 cm long, red.

**A.** — v. **depressa** (HAW.) BAK. (*A. d.* HAW., *A. serra* DC., *A. b.* v. *s.* BGR.). – **L.** 60, very variable, 12–15 cm long, 6 cm across, often glabrous and unmottled to ± warty-spinous, margins white, horny, T. 2–4 mm long, 8 mm apart; **Fl.** 4 cm long, fiery scarlet.

**A.** — v. **postgenita** (ROEM. et SCHULT.) BAK. (*A. p.* ROEM. et SCHULT., *A. prolifera* v. *major* (S. D.). – **L.** 10–13 cm long, 4 cm across.

**A. breviscapa** REYN. et BALLY (§ IV/C/16). – Somalia. – Forming ± large groups; **St.** 50 to 100 cm long, prostrate; **L.** c. 24 in a dense Ros., lanceolate-tapering, mostly 30–35 cm long, 8–10 cm across base, narrowing to the pointed tip, upperside bluish-green with a reddish shimmer, somewhat concave above, underside rounded, margins in lower 1/4 with very short, blunt T. 1–2 mm long and 1 cm apart; **Infl.** 50 cm tall, Fl. to 3 cm long, scarlet.

**A. broomii** SCHOENL. (§ IV/A/2). – Cape: Lesotho. – **Ros.** acauline; **L.** dense, ovate-lanceolate, 25 cm long, 8 cm across, green, indistinctly lined, underside keeled below the tip and with 4–5 Sp., marginal T. 3 mm long, 7 mm apart, pointed; **Infl.** simple, Fl. 1.5 cm long, greenish yellow.

**A.** — v. **tarkaensis** REYN. – Growth more luxuriant; **L.** 2–3 times broader at base, reddish-brown, margins more sinuate-dentate; **Fl.** to 3 cm long.

**A. brownii** BAK. (§ IV/D/25) (*A. nobilis* v. *densifolia* BAK., *A. flavescens* BOUCHÉ). – Cape. – **St.** simple, 5–8 cm ⌀; **L.** lanceolate, tapering, 30–45 cm long, 7–10 cm across, thick, green, marginal T. triangular, 3–4 mm long, horny; **Infl.** 30–45 cm tall, Fl. 3 cm long, yellow-red.

**A. buchananii** BAK. (§ II). – Malawi. – Acauline; **L.** 8–12, distichous, narrow-linear, 30 to 45 cm long, 1 cm across, upperside concave, underside rounded with several small whitish markings at the base, margins with small T.; **Fl.** 3.4 cm long, pale red.

**A. buchlohii** RAUH (§ XI/c). – S.E.Madag. – **Ros.** densely leaved, sprouting from base; **L.** stiffly erect, 30–60 cm long, 2.5 cm across base, margins with horny, broadly triangular T.; **Infl.** 60–90 cm tall, Per. 2–2.5 cm long, pale yellow, Tep. with almost white margins and green tips, anthers orange-red.

**A. buettneri** BGR. (§ IV/C/18/d) (*A. paedogona* BGR., *A. bulbicaulis* CHRISTIAN, *A. barteri* BAK. p. part., *A. barteri* v. *sudanica* A. CHEV., *A. b.* v. *dahomensis* A. CHEV., *A. paludicola* A. CHEV.). – Togo; Mali; Ghana; Dahomey; W. and N.Nigeria; Central Afr. Rep.; Congo (Brazzaville); Angola; Malawi; Zambia. – Acauline, simple, the L.-bases broadened below soil-level to form a bulb-like swelling, this "bulb" being 6 to 10 cm ⌀, 6–8 cm long, sometimes with a rather woody St. below the "bulb", this St. being 10 cm long and 3–4 cm thick; **L.** about 16, rosulate, deciduous, somewhat leathery, 35 to 55 cm long, 8 cm across base, 9–10 cm across the upper third, then narrowing to a point, upperside green, margins sinuate-dentate with very narrow, white to pale pink horny edge and hard T., these being 1.5–3 mm long, 3–5 mm apart, or 10 mm apart above; **Infl.** 70–90 cm tall, Fl. c. 38 mm long, greenish-yellow or olive-green.

**A. buhrii** LAVR. (§ IV/B/11). – Cape: E.Bokkeveld plateau. – Acauline, several-headed, **Ros.** usually erect; **L.** c. 16, lanceolate-triangular, curved-erect or ascending, 40 cm long, 9 cm across base, fleshy, fairly soft, upperside slightly concave towards the tip, bluish with reddish tinge, distinctly striped and in part with pale, oblong or H-shaped marks, margins with pale red, border which is dentate, horny or almost entire and 1.5–2 mm across, T. blunt, less than 1 mm long, 3.5 mm apart; **Infl.** laxly paniculate, to 60 cm tall, with 7–15 side-Br., racemes subcapitate, Fl. lax, orange-red.

**A. bukobana** REYN. (§ IV/C/19a). – Tanzania. – Acauline, offsetting to form small, dense groups; **L.** c. 16, densely rosulate, lanceolate-tapering, to 30 cm long, 8 cm across base, about 14 mm thick, slightly grooved on upperside above, green, underside convex, grey-green, margins sinuate-dentate, T. 4 mm long, 1 cm apart and tipped with a small Sp.; **Infl.** 70 to 90 cm tall, Fl. 3–3.5 cm long, scarlet with paler tips.

**A. bulbillifera** PERR. v. **bulbillifera** (§ IV/E/28a; XI/d/a). – W.Madag.: Majunga. – **Ros.** large, usually acauline; **L.** 20–30, some 40–60 cm long, 6–10 cm across base, gradually tapering, margins with hard, sharp T.; **Infl.** to 2.5 m tall, bearing numerous adventitious buds, Fl. numerous, 22–25 mm long, scarlet with yellow tips.

**A.** — v. **pauliana** REYN. – Smaller in all parts; adventitious buds only on Ped. and lower parts of Br.

**A. bullockii** REYN. (§ III/1). – Tanzania: Kahama D. – **Root** thick, fusiform, with subterranean bulbs, 3 cm tall and 3–4 cm ⌀; **L.** 8–10, rosulate, linear-lanceolate, 10 cm long and 2 cm across during a dry season, or 20 cm long and 3 cm across during a rainy season, grooved on upperside, green, lined, margins narrow, horny, T. close together, soft, pale pink, 0.5 to 1 mm long, 0.5–1 mm apart; **Infl.** 35–40 cm tall, involucral L. membranous, Fl. 3 cm long, pale scarlet or coral-red.

**A. burgersfortensis** REYN. (§ IV/10/A/5b). – Transvaal: Lydenburg D. – **Ros.** simple, acauline, offsetting to form large groups; **L.** 10–20, c. 30 cm long extended by a dry tip, 5–10 cm long, 7–8 cm across base, gradually tapering, brownish-green above, with oblong, white spots in ± wavy transverse bands, underside convex, paler, somewhat lined, usually without spots, margins sinuate-dentate, T. horny, curved, triangular, brown, sharp, 3–4 mm long, 10 to

14 mm apart; **Infl.** 1–1.3 m tall, Fl. 3 cm long, deep red.

**A. bussei** BGR. (§ IV/D/24). – Zanzibar. – ♄ to 40 cm tall; **L.** to 20 cm long, 4.5 cm across midway, tapering, red-brown to green-purple, margins sinuate-dentate, cartilaginous, T. 3 to 5 mm long, 7–8 mm apart; **Infl.** with 1 or 2 Br. Insufficiently known species.

**A.** × **caesia** SALM. – Hybrid: **A. arborescens** × **A. ferox.** – Cape. – **St.** 1–2.5 m tall, branching from base; **L.** numerous, lanceolate-ensiform, tapering, fleshy, soft, 45–50 cm long, 12 cm across, marginal T. red; **Infl.** 80–100 cm tall, Fl. pale red, later whitish.

**A. calcairophila** REYN. (§ XI/c). – Central Madag. – **Root** cylindrical, acauline, forming small groups by means of offsets; **Ros.** to 5 cm ⌀; **L.** c. 10, distichous, 5–6 cm long, 14 mm across base, narrowing to a point, grey-green on upperside, grooved, underside rounded, margins with thin, pointed, horny, rather soft, white T. 2–3 mm long and 2–3 mm apart; **Infl.** 20–25 cm tall, Fl. 1 cm long, white.

**A. calidophila** REYN. (§ IV/C/19d). – Ethiopia: Sidamo. – **St.** short or 1–2 m long and prostrate in older plants, covered with remains of dead L., often offsetting freely from base and thus forming dense groups; **L.** c. 20, densely rosulate, 16 cm across base, gradually tapering, 80 cm long, deeply grooved on upperside, marginal T. white, 4–5 mm long, 2–2.5 cm apart; **Infl.** 1.3 m tall, Fl. 22 mm long, scarlet, yellow above.

**A. cameronii** HEMSL. v. **cameronii** (§ IV/D/22). – Malawi to Rhodesia. – ♄, branching from base; **St.** 1.6 m tall, slender, elongated, laxly leafy; **L.** lanceolate-ensiform, 22–30 cm long, about 3 cm across, underside convex, about 6–8 mm thick, green, often reddish, margins with triangular, incurved T. 4 mm long; **Infl.** c. 30 cm tall, Fl. 4–5 cm long, red with green tips.

**A.** — v. **bondana** REYN. – Rhodesia: Inyanga D. – **St.** usually simple, 60 cm tall, 4 cm thick; **L.** c. 35 cm long, 7 cm across, dark green to coppery; **Fl.** tube more club-shaped.

**A.** — v. **dedzana** REYN. – Moçambique: Dedza D., Central Prov. – **St.** 50–80 cm tall; **L.** 50 cm long, 6 cm across, coppery-red; **Infl.** 60 cm tall, Fl. 4 cm long.

**A. camperi** SCHWEINF. (Pl. 12/4) (§ IV/C/19d) (*A. abyssinica* S. D., *A. eru* BGR., *A. spicata* BAK., *A. albopicta* HORT. LIG., *A. e.* v. *cornuta* BGR., and cultivars: *A. eru* cv. Erecta, cv. Glauca, cv. Maculata, and cv. Parvipuncta). – Ethiopia, incl. Eritrea. – ♄, branching from base to form groups 1–2 m ⌀; **St.** usually 50 cm tall, 5 cm ⌀, erect to prostrate, the upper 10–20 cm being densely leafy; **L.** 12–16, rosulate, 50–60 cm long, 8–12 cm across base, sheathing, narrowing towards the tip, dark green on upperside, grooved, unmottled or with a few whitish, lentiform marks, underside convex, dark green, with or without markings, margins reddish with triangular, sharp, brownish-red T. 3–5 mm long and 1–2 cm apart; **Infl.** 70–100 cm tall, Fl. 20–22 mm long, orange or yellow. (Usually distributed under the name *A. eru*, but the name **A. camperi** has priority.)

**A. candelabrum** BGR. (§ VI/4b). – Natal. – Tree-like, simple, 2–4 m tall, **St.** bearing many dry old L.; **L.** densely rosulate, to 1 m long, 15 cm across base, gradually tapering, deeply concave on upperside, underside convex with Sp. along the M. line, often with Sp. scattered over the surface, deep green to blue, margins reddish, horny, T. reddish, triangular, sharp, 3 mm long, 1.5–2 cm apart; **Fl.** 32 mm long, scarlet or pink to orange.

**A. cannellii** LEACH (§ III/1). – Moçambique: Mania and Sofala D. – Group-forming, offsetting vigorously on all sides; **caudex** almost tuberous, roots fleshy, almost fusiform; **L.** green, normally 4–5, to 26 cm long, 4–8 cm across, narrowing to a point, outspreading or limply recurved, border narrow, membranous, upperside flat-grooved, lower part mottled with white, underside convex and more strongly mottled, T. tiny, translucent, pointed, 0.25 mm long, about 1 mm apart, sometimes absent; **Infl.** simple, 20–30 cm tall, Rac. lax, Fl. almost pendulous and with pedicels, to 25 mm long, orange-scarlet with green tips.

**A. capitata** BAK. v. **capitata** (Pl. 13/2) (§ IV/C/20; XI/d/α) (*A. cernua* TOD.). – Central Madag. – Acauline, robust; **Ros.** 60–80 cm tall, of 12–15 L., surrounded by dead L.; **L.** 50 to 70 cm long, 5–6 cm across, thick, rigid, with parallel sides narrowing to the tip, this being rounded, vivid red in full sunshine, marginal T. red, 2 mm long, 6–12 mm apart; **Infl.** simple, Rac. capitate, Fl. numerous, pendent, yellow-orange.

**A.** — v. **cipolinicola** PERR. – **St.** 1–3 m tall, up to 20 cm ⌀, less frequently branching; **L.** 60–100, c. 60 cm long, 5–6 cm across, dark brown-red; **Fl.** yellow-orange.

**A.** — v. **gneissicola** PERR. – Acauline; **L.** less numerous, marginal T. larger, white.

**A.** — v. **quartziticola** PERR. – Acauline; **L.** 30–40 cm long, 9–12 cm across, grey-blue or reddish-bluish, marginal T. red, 5 mm long, 7–20 mm apart.

**A.** — v. **silvicola** PERR. – Epiphytic; **L.** 50 to 60 cm long, 3–4 cm across, marginal T. smaller, often absent; **Fl.** smaller.

**A. caricina** BGR. (§ II). – Similar to **A. graminifolia**; **L.** shorter, narrower, grooved on upperside, with distinct longitudinal lines, margin entire; **Infl.** dainty. – Insufficiently known species.

**A. castanea** SCHOENL. (§ V). – Transvaal. – **St.** 1–5 m tall, branching; **L.** 43 cm long, 8 cm across, deeply concave on upperside, underside convex, marginal T. 1.5 mm long, 8–11 mm apart, somewhat curved; **Infl.** 1–3, to 2 m tall, Fl. 19 mm long, red-brown.

**A. catengiana** REYN. (§ IV/C/19c). – Angola: Benguela. – ♄ 1–2 m ⌀; **St.** thin, simple or branching from lower part, (1.5–)2(–3) m long, the upper 30 cm being somewhat laxly leafy, L.-sheaths with linear markings; **L.** narrow lanceolate to tapering, to 30 cm long, 3.5 cm across, pale yellowish grey-green on upperside, somewhat grooved towards tip, with numerous pale green, lentiform marks on the lower half, these marks being more scattered in upper part, underside paler, with marks over the entire surface, margins with stout, pale, triangular T. 3 mm long and 8–10 mm apart; **Infl.** slender, Fl. 28 mm long, scarlet.

**A. chabaudii** SCHOENL. v. **chabaudii** (§ IV/C/19a). – Rhodesia to N.Transvaal; Malawi. – Acauline; **L.** in open Ros., lanceolate-tapering above a broad base, 40–50 cm long, about 12 cm across base, concave on upperside, underside much rounded, pale blue-green, margins armed with stout T.; **Infl.** almost 1 m tall, Fl. red.

**A.** — v. **mlanjeana** CHRIST. – Malawi. – Smaller variety, offsetting from base; **L.** light green to brick-red, margins with a white, horny edge; **Infl.** small.

**A.** — v. **verekeri** CHRIST. – E.Transvaal. – **L.** olive-green to reddish; **Fl.** yellowish.

**A. chlorantha** LAVR. (§ IV/A/2). – Cape: Fraserburg D. – Stemless or often **St.** procumbent, up to 1.50 m long; **Ros.** solitary or up to 10; **L.** rosulate, deltoid, acute, slightly falcate, to 40 cm long, to 8 cm broad at base, leathery, upper surface almost resinous, flat or slightly convex, ± striate, bright green shading into purplish, lower surface similar to the upper but often bearing white spots, margins cartilaginous, dark brown-red, armed with rigid, deltoid, dark brown-red T. which are 2 mm long and 10–30 mm distant; **Infl.** up to 3 from each Ros., to 1.60 m high, Rac. to 35 cm long, bracts 12–20 mm long, 5–8 mm wide, fleshy, yellow green, many nerved, Fl. on stalks 15–22 mm long, 10 mm long, 4 mm ⌀, yellow green.

**A. chloroleuca** BAK. – ? S.Afr. – Insufficiently known, possibly a hybrid. – **St.** simple, 1 m tall; **L.** 60–75 cm long, 5 cm across base; **Fl.** 25 mm long, pale yellowish-white. Related to **A. drepanophylla.**

**A. chortolirioides** BGR. v. **chortolirioides** (§ III). – Transvaal. – **St.** c.5 cm long, branching, forming bushes; **Ros.** of 7–14 L. 6–15 cm long, 12 mm across base, triangular-ovate, then narrow-linear, 2–3 mm across tip, concave on upperside, marginal T. small; **Infl.** 16 cm tall, Fl. 3 cm long, pink-red with yellow tips and greenish stripes.

**A.** — v. **boastii** (LETTY) REYN. (*A. b.* LETTY). – Swaziland. – **St.** short; **L.** linear, 10–20 cm long, 8 mm across, narrowing to 4 mm across tip, marginal T. small; **Fl.** 2 cm long, orange-coloured with green spots.

**A. christianii** REYN. (§ IV/B/12b). – Rhodesia; Tanzania; Moçambique; Malawi. – **St.** simple, 1–1.5 m tall, erect or prostrate, 10 to 15 cm ⌀, covered with dry, old L., often forming groups; **L.** 30–40, densely rosulate, lanceolate-tapering to 75 cm long, with a desiccated tip, 10–15 cm long, 10–12 cm across lower part, somewhat concave on upperside, dark green, underside rounded, dark bluish-green, very faintly banded, margins sinuate-dentate, rather cartilaginous, T. 3–5 mm long, 1–2 cm apart, triangular, sharp, stiff, often directed forwards; **Infl.** 2–3 m tall, Fl. 4.5 cm long, red.

**A. ciliaris** HAW. v. **ciliaris** (Pl. 13/1) (§ IV/D/21). – E.Cape. – **St.** long, climbing or scrambling, to 5 m long or more, 10–15 mm ⌀, nodose, branching from the nodes, Br. with L. on the final 30–60 cm; **L.** linear-lanceolate, gradually tapering, Sh. 5–15 mm long, lined with green, with white T. 3 mm long, L. being 10–15 cm long, 10–25 mm across, green, underside convex, margins with white, horny T. about 1 mm long and 3 mm apart; **Infl.** 20–30 cm long, Fl. 3 cm long, scarlet, mouth green-yellow.

**A.** — f. **flanaganii** (SCHOENL.) RES. (*A. c.* v. *f.* SCHOENL.). – **L.** ovate-lanceolate.

**A.** — f. **haworthii** RES. – **L.** 8–15 cm long, 2–3 cm across, T. on L.-Sh. and margins 4–5 mm long.

**A.** — f. **tidmarshii** (SCHOENL.) RES. (*A. c.* v. *t.* SCHOENL., *A. t.* (SCHOENL.) MÜLLER). – T. on L. margins smaller, and those on L.Sh. less numerous.

**A. cinnabarina** DIELS et BGR. (§ IV/D/24). – Transvaal. – **L.** ensiform, tapering, c. 30 cm long, 3 cm across, margins horny, T. blue, triangular, 9 mm long, 12 mm apart; **Infl.** robust, Fl. 2.5 cm long. (Acc. REYNOLDS not known in S.Afr.)

**A. classenii** REYN. (§ IV/C/15). – Kenya. – Acauline or with **St.** 50 cm long, offsetting to form clumps several m ⌀; **L.** c. 24, densely rosulate, lanceolate, 35–40 cm long, 7–8 cm across base, deep olive-green to reddish-bronze on upperside, underside convex, with several small, oblong, pale spots towards base, margins sinuate-dentate, T. triangular, sharp, pale brown, up to 5 mm long, 10–15 mm apart, with terminal Sp.; **Infl.** to 60 cm tall, pruinose, Fl. 2–2.5 cm long, brownish-green.

**A. claviflora** BURCH. (§ IV/B/13) (*A. schlechteri* SCHOENL., *A. decora* SCHOENL.). – Cape: Bushmanland. – Acauline, usually forming groups of 1–2 m ⌀; **L.** 30–40 in dense Ros., ovate-lanceolate, about 30 cm long, 6–8 cm across, bluish on upperside, underside convex, with 1–2 keels in the upper third, these keels having 4–6 brownish Sp. 2–4 mm long, margins with T. 2–4 mm long, about 1 cm apart; **Infl.** to 50 cm tall, Fl. 3–4 cm long, yellowish orange-red, clubshaped-cylindrical.

**A. commixta** BGR. (§ IV/D/21) (*A. gracilis* BAK.). – Cape: Cape D. – Bushy; **St.** in dense groups, to 1 m long, 2–2.5 cm ⌀, some L. on upper parts; **L.** lanceolate-tapering, c. 20 cm long, 3 cm across base, Sh. with green stripes, upperside dark green, underside convex, indistinctly lined, margins with white T. 1–2 mm long and 2–4 mm apart; **Infl.** 30–35 cm tall, Fl. c. 4 cm long. yellowish to orange.

**A. commutata** TOD. v. **commutata** (§ IV/B/10x) (*A. grandidentata* HORT. PAN.). – Scarcely differentiated from **A. saponaria**; **Infl.** 80–90 cm tall, Fl. 3 cm long, red. (Acc. REYNOLDS, this is a cultivar, possibly a hybrid: **A. grandidentata** × **A. saponaria.**)

**A.** — v. **tricolor** (BAK.) BGR. (*A. t.* BAK.). – **L.** 3–5 cm across, mottled.

**A. comosa** MARL. et BGR. (§ IV/E/26). – S.W.Cape. – Tree-like, **St.** simple, 1–2 m tall, very robust, densely covered with dead L.; L. in dense Ros., lanceolate-ensiform, 35–50 cm long, 5 cm across middle, concave on upperside, fleshy, blue, margins with small, horny T. situated 4–8 mm apart; **Infl.** to 2 m tall, Fl. 2.5 cm long, greenish-white.

**A. comosibracteata** REYN. (§ IV/B/10x). – Transvaal. – Dubious species, acc. REYNOLDS related to **A. longibracteata** and **A. barbertoniae.**

**A. compacta** REYN. (§ IV/C/18d). – Tanzania: Tabora. – Acauline or short-stemmed; **St.** to 50 cm long, prostrate; **L.** c. 20, densely rosulate, lanceolate-tapering, slightly grooved above on upperside, dull green with reddish sheen, with several scattered, small, white markings towards the base, underside mottled over entire surface, margins pink and with T. 2–3 mm long and 1 cm apart; **Infl.** 1 m tall, Fl. 3.5 cm long, pale scarlet.

**A. compressa** PERR. v. **compressa** (§ XI/d/β 2). – Central Madag.: Tananarive. – Acauline or very short-stemmed; **Ros.** dense, fan-shaped, of 15–20 L. arranged distichously, 12–15 cm long, 5 cm across, narrowing towards the base, 5–6 mm thick, pruinose, marginal T. compressed 2 mm long, red-tipped, 4–5 mm apart; **Infl.** 60–70 cm tall, Fl. 3.5 cm long, white.

**A.** — v. **rugosquamata** PERR. – Central Madag. – **L.** to 23 cm long, 3–3.5 cm across, covered with small roundish protuberances, marginal T. smaller.

**A.** — v. **schistophila** PERR. – **Ros.** more compressed; **L.** more numerous, 12 cm long, 2 to 2.5 cm across, marginal T. longer.

**A. comptonii** REYN. (§ IV/D/25). – Cape. – Acauline, short-stemmed, or St. 1 m tall, normally forming dense groups; **Ros.** compact, 60 cm ⌀; **L.** c. 20, lanceolate-tapering, to 30 cm long, 9 cm across base, concave on upperside towards tip, bluish-green, often reddish, keeled on underside in upper half, keel with about 6 Sp., margins with broad, pale brown T. 2–3 mm long and 1–1.5 cm apart, set on a whitish base; **Infl.** 80–100 cm tall, Fl. 3.5–4 cm long, scarlet.

**A. confusa** ENGL. (§ IV/D/24). – Tanzania. – **St.** slender, branched, prostrate, L. arranged in a loose spiral, Sh. c. 10 cm long; **L.** linear-lanceolate, gradually tapering, 25–27 cm long, 2.5 cm across, margins with triangular T. 1 to 2 mm long; **Fl.** 2 cm long, red-yellow.

**A. congolensis** DE WILD. (§ IV/C/18d). – Congo (Kinshasa). – Acauline; **L.** 32–48 cm long, 5 cm across at base, ensiform, tapering, longitudinally banded on upperside, margins with triangular curved T. 4 mm long; **Infl.** dichotomously divided, Fl. about 3.5 cm long, red-yellow. (Acc. REYNOLDS an insufficiently known species.)

**A. conifera** PERR. (XI/d/β 1). – Central Madag. – Acauline; **L.** when immature distichous, bright reddish to blue-violet, densely spined, L. later 12–15 in a Ros., (16–)18(–22) cm long, 3–6 cm across, 18–20 mm thick, tapering, tip somewhat rounded, blue-grey to reddish-violet, margins shining red, marginal T. 2–3 mm tall, red, 6–10 mm apart; **Infl.** 50–70 cm tall, Fl. 14 mm long, lemon-yellow to orange, almost completely covered by the large bracts.

**A. consobrina** SALM (§ IV/D/24). – Cape: Cape D. (Acc. REYNOLDS not known in S.Afr.). – **St.** slender, elongated, ± simple; **L.** initially almost distichous, finally ± rosulate, 20 cm long, 2–2.5 cm across, 7–10 mm thick, somewhat convex on underside, bluish with oblong-roundish spots, marginal T. 3 mm long; **Infl.** to 50 cm tall, Fl. about 2.5–3 cm long, red-yellow.

**A. constricta** BAK. (§ IV/B/10/A/5b). – Moçambique. – **L.** ensiform, 45 cm long, 5 cm across base, with irregular, oblong marks, about 15 mm in length, margins dentate, T. 5–6 mm long, 12–15 mm apart, hard, pointed; **Fl.** 3.5 to 3.8 cm long, red, swollen spherically at base.

**A. contigua** (PERR.) REYN. (Pl. 13/4) (§ XI/d/x) (*A. deltoideodonta* v. c. PERR., *A. d.* v. *d. f. latifolia* PERR., *A. d.* v. c. f. l. sf. *variegata* BOIT., *A. d.* v. c. f. *longifolia* PERR., *A. imalotensis* REYN.). – S.W.Madag. – Simple or forming small groups; **St.** short, creeping, 10–20 cm long, 3 cm ⌀; **L.** 20–24, densely rosulate, broad ovate-pointed, to 30 cm long, 12–15 mm across, very fleshy, both surfaces bluish-green with reddish sheen, indistinctly lined, margins pink, horny, 1 mm across, and with triangular or blunt pink T.; **Infl.** 1–1.5 mm long and 1–4 mm apart, coralred. 50–65 cm tall, Fl. 30–34 mm long. (REYNOLDS gives specific status to **A. imalotensis**, with f. **latifolia** and f. **longifolia**.)

**A. cooperi** BAK. (§ III/4) (*A. schmidtiana* RGL.). – Cape; Transvaal; Natal. – Acauline; **L.** 10–14 in a distichous Ros., 30–50 cm long, 2–2.5 cm across base, with small, white spots, margins with white T.; **Infl.** 30–40 cm tall, Fl. orange-coloured.

**A. corifolia** PILLANS (? § IV/C/18). – Cape: Willowmore D. – Insufficiently known species which, acc. REYNOLDS, may be a hybrid; **St.** about 12 cm tall; **L.** rosulate, 30–40 cm long, 7–9 cm across, lanceolate, tapering, convex on underside, both surfaces grey-green, with sinuate-dentate margins and triangular, horny, sharp, red-brown T. 3 mm long, directed forwards and 1–1.5 cm apart; **Infl.** 50–60 cm tall, Fl. 26–30 mm long, red.

**A. crassipes** BAK. (§ IV/C/18d) (*A. c.* BAK. ex BGR.). – Sudan; Zambia. – Acauline or very short-stemmed; **L.** c. 20–24 in a dense Ros., about 40 cm long, 7 cm across base, gradually narrowing to the tip, bluish-green on upperside, very slightly grooved in upper part, slightly convex on underside, margins with stout, brown-tipped, triangular T. 5 mm long and 15 mm apart; **Infl.** 50–60 cm tall, Fl. 38 mm long, yellow-green.

**A. cremnophila** REYN. et BALLY (§ IV/B/10/A/3a). – Somalia. – ♄, pendulous; **St.** 10–20 cm long, 8–10 mm ⌀, freely offsetting from base; **L.** 6–8, rosulate, with Sh. at base, grey-green on upperside, slightly grooved in upper part, underside rounded, margins with triangular, fairly sharp, pale brown T. 2 mm long and 3–5 mm apart; **Infl.** 25–30 cm tall, at first pendulous, afterwards curved upwards, Fl.

2.5 cm long, scarlet, with yellowish-green mouth.

**A. cryptoflora** REYN. (§ XI/D/β 1). – Madag.: Fianarantsoa. – **Ros.** simple, acauline or with short St.; **L.** 15–20, slightly spirally twisted, lanceolate-tapering, 20–25 cm long, 6.5 cm across, deep green with a slight reddish tint on upperside, convex on underside, margins with continuous, brownish-red border and triangular T. of the same colour which are 2–3 mm apart; **Infl.** 40 cm tall, buds initially concealed by the densely imbricate involucral L., Fl. 1 cm long, greenish-yellow in lower part, and orange-yellow towards the mouth.

**A. cryptopoda** BAK. (§ IV/C/17) (*A. pienaari* POLE EVANS). – Moçambique; Malawi; Rhodesia; N.Transvaal; Botswana. – **Ros.** acauline or very short-stemmed, mostly simple, but frequently in small groups; **L.** 40–50, lanceolate-ensiform or gradually tapering, about 90 cm long, 12–15 cm across base, deep green or reddish to bluish-green on upperside, convex on underside, margins with red-brown, sharp, triangular T. which are 2 mm long and 5–7 mm apart, or else red-brown, horny; **Infl.** 1.25 to 1.75 m tall, Fl. 3.5–4 cm long, scarlet, later yellowish in the upper half.

**A.** cv. Cyanea. – Hybrid: **A. glauca** × **A. humilis** v. **incurva**. – **Ros.** acauline or forming clumps, densely leafy; **L.** 20–25 cm long, 3–4 cm across, tapering, bluish, somewhat grooved towards tip and with a few Sp.-tubercles, convex on underside, Sp.-tubercles more numerous and arranged in lengthwise rows, margins with cartilaginous, triangular, incurved T.; **Infl.** 25–30 cm tall, Fl. 1.5 cm long, scarlet.

**A. davyana** SCHOENL. v. **davyiana** (§ IV/B/10/A/2b). – Transvaal. – **Ros.** acauline, of some 12 **L.**, these being triangular-lanceolate, fleshy, convex on underside and keeled towards tip, with 1–2 Sp., upperside dark green to reddish with light spots in transverse bands, marginal T. sharp, brown, 3–4 mm long, 1–2 cm apart, set on a brown, horny border; **Infl.** 60–70 cm tall, Fl. 32 mm long, pale flesh-coloured.

**A.** — v. **sobolifera** GROENEW. – Very freely offsetting; **L.** unmottled.

**A. dawei** BGR. (§ IV/C/19d). – Uganda. – Caulescent, branching from base, forming clumps up to 1.8 m ⌀; **St.** simple, 4–5 cm ⌀; **L.** 40–50 cm long, ensiform, 6–7 cm across base, fleshy, blue-green to reddish, margins sinuate-dentate, T. triangular with horny tips, 4 mm long, 12–15 mm apart; **Infl.** robust, Fl. 3 to 3.5 cm long, red.

**A. decaryi** GUILL. (§ XI/b). – Madag.: Ambovombe. – Bushy; **St.** slender, 6–9 mm ⌀, somewhat climbing; **L.** 10–13, about 2–2.5 cm apart, Sh. with prominent nerve-stripes, blade 15–19 cm long, 8–9 mm across base, semiterete, grey-green, with widely spaced white marginal T.; **Infl.** with some 7 Fl., pale pink. (Acc. REYNOLDS an insufficiently known species; acc. W. RAUH only a variety of **A. antandroi**.)

**A. decorsei** PERR. (§ XI/d/β 2). – Central Madag. – **Ros.** acauline; **L.** 12–15, 60–70 cm long, 5–6 cm across, narrowing to base, tip pointed, green, marginal T. 1–1.5 mm long; **Infl.** 80–120 cm tall, stiff.

**A. decurva** REYN. (§ VI/1). – Moçambique. – Acauline or very short-stemmed, simple or forming 2 Ros.; **L.** 20–24, 9 cm across base, 1.5–2 cm thick, gradually narrowing towards the tip, to 55 cm long, dull green with a reddish tinge on upperside, grooved in upper part, rounded on underside, margins sinuate-dentate, T. sharp, triangular, 3 mm long, 8–15 mm apart; **Infl.** to 90 cm tall, Fl. about 38 mm long, light red.

**A. decurvidens** GROENEW. (§ IV/B/10/A/3a). – E.Transvaal. – Related to **A. transvaalensis**. Acc. REYNOLDS perhaps a smaller form of **A. komatiensis**.

**A. deflexidens** PILL. (§ IV/B/10x). – Natal: Zululand. – Resembles **A. saponaria**. – Offsetting; **L.** mottled on both surfaces, marginal T. distinctly recurved, 5–6 mm long; **Infl.** with 3–4 Br., Fl. 40–43 mm long. (Acc. REYNOLDS probably a hybrid.)

**A.** × **delaetii** RADL. (Pl. 13/5). – Hybrid: **A. ciliaris** × **A. succotrina**. – Branching from the base; **St.** and Br. 50–100 cm tall, 2–5 cm ⌀; **L.** arranged spirally, oblong-lanceolate, 30 cm long, 5 cm across base, gradually tapering, dark green, convex on upperside, rounded on underside and keeled at tip, margins and keel with horny, white, somewhat curved T. 3–4 mm long and 10–12 mm apart; **Infl.** 60 cm across, Fl. 2–2.5 cm long, orange-red.

**A.** cv. Deleuilii. – Hybrid: **A. abyssinica** × **A. ferox**. – Acauline or very short-stemmed; **Ros.** (2–)2.5(–3) m ⌀; **L.** numerous, 1–1.25 m long, 25–30 cm across base, 5–7 cm thick, gradually tapering, concave on upperside, tip rounded, bluish, margins purple with a red horny border, T. triangular; **Infl.** robust, Fl. red.

**A. deltoideodonta** BAK. (§ IV/B/7; XI/c) (*A. rossii* TOD.). – Central Madag. – Plant small, acauline or short-stemmed, probably offsetting; **L.** 12–16, densely rosulate, lanceolate-triangular, 7.5–30 cm long, 2.5–7 cm across, with a continuous horny, narrow, straw-coloured border which has yellow, triangular T. 2 mm long and 3–5 mm apart; **Infl.** 40–60 cm tall, Fl. 2.5–3 cm long, scarlet.

**A.** — v. **brevifolia** PERR. – S.W.Madag. – Acc. REYNOLDS only differentiated from v. **deltoideodonta** by the smaller **L.**; **Infl.** 30 cm tall, Rac. shorter.

**A.** — v. **candicans** PERR. – **L.** 15–30 cm long, 5–7 cm across, with dark lines; **Fl.** 2.5–3 cm long, pale scarlet.

**A.** — v. — f. **latifolia** PERR. (Pl. 14/1). – **L.** 16–18 cm long, 5–7 cm across.

**A.** — v. — f. **longifolia** PERR. – **L.** 30 cm long, 5 cm across.

**A.** — v. **deltoideodonta** (*A. d.* v. *typica* PERR.). – **L.** 7.5–10 cm long, 2.5 cm or more across, marginal T. triangular.

**A. descoingsii** REYN. (§ XI/c). – S.W.Madag.: Tuléar. – Acauline or very short-stemmed, forming cushions of more than 100 Ros. and up to 30 cm ⌀; **Ros.** 3–5 cm ⌀, of 8–10 **L.** which are ovate-tapering, 3–4 cm long, 1.5 cm across base, somewhat plicate on upperside in upper part, sharply mucronate, rough, dull green, both surfaces with numerous, dull white, warty protuberances, rounded on underside, with colour and structure as above, margins somewhat incurving, with stout, horny, white, triangular T. 1.5 mm long and 10–15 mm apart, these T. being absent in the upper part; **Infl.** 12–15 cm tall, Fl. campanulate-cylindrical, 7–8 mm long, scarlet to vermilion with mouth more yellow-orange.

**A. desertii** BGR. (§ IV/C/16). – Kenya. – **St.** 2 m tall; **L.** 5 cm across base, margins sinuate-dentate, T. triangular, 5–10 mm long, 15 mm apart; **Infl.** to 1.5 m tall, Fl. 2.5–2.7 cm long, light flesh-coloured. Insufficiently known species.

**A.** cv. Desmetiana. – Hybrid: **A. variegata** × **A. humilis** v. **echinata**. – Acauline; **L.** 40 in a dense Ros., lanceolate, 10–13 cm long, 3 cm across base, gradually tapering, rounded on underside with oblong, white-green warts, margins with small white T.; **Infl.** 30 cm tall, Fl. 3 cm long.

**A. dewetii** REYN. (§ IV/B/10/A/4). – Zululand. – Acauline, simple, without offsets; **L.** about 20, densely rosulate, lanceolate-tapering, about 48 cm long, 13 cm across base, slightly concave on upperside, dark green, with numerous oblong marks arranged either irregularly or ± in wavy transverse bands, convex on underside which is unmottled and lined, margins sinuate-dentate, brown, cartilaginous, T. triangular, 1 cm long, 1–1.5 cm apart; **Infl.** to 2 m tall, Fl. 35–42 mm long, dark scarlet.

**A. dhalensis** LAVR. (§ IV/C/16). – S.W.Arabia. – Simple or offsetting, acauline or short-stemmed; **L.** c. 18 in a dense Ros., narrowing to the tip, pointed, about 40 cm long, 8 cm across base, 5–8 mm thick, bluish-green on upperside, slightly tinged with brown, flat to slightly grooved, convex on underside, margins with sharp, brown, triangular T. 2–3 mm long and 5 mm apart; **Infl.** to 70 cm tall, Fl. 3 cm long, lemon-yellow.

**A. dhufarensis** LAVR. (§ IV/C/16). – S.Arabia. – Plant single, acauline; **L.** 14–20, triangular-tapering, ± erect, inflexed, soft, 45 cm long, 14 cm across base, 12–15 mm thick, bluish-grey, flat to somewhat grooved on upperside, convex on underside, margin entire apart from a few T. on immature L.; **Infl.** with 1–2 Br., to 90 cm tall, Rac. cylindrical-tapering, Fl. lax, 28–30 mm long and coral-red, with pedicel 12–15 mm long.

**A. dichotoma** L.f. v. **dichotoma** ('Kokerboom') (Pl. 14/3) (§ VII). – Cape: Lit. Namaqualand; S.W.Afr.: Gr. Namaqualand. – ♄ branching dichotomously in age, 8–10 m tall, up to 1 m thick at base; **L.** in dense Ros., arranged ± spirally, 15–25 cm long, 5–7 cm across base, linear-lanceolate, rounded above, fleshy, slightly convex on upperside, blue-green, margins with fine, horny, brown T.; **Infl.** dense, 3-branched, Fl. 3–3.5 cm long, yellow.

**A. —** v. **montana** (SCHINZ) BGR. (*A. m.* SCHINZ). – S.W.Afr.: Gr. Namaqualand. – Fl. and Fr. larger than in the type.

**A. dinteri** BGR. (§ IV/B/9). – S.W.Afr.: Gr. Namaqualand. – Acauline; **Ros.** 26 cm ⌀; **L.** in 3 series, 20–26 cm long, 8 cm across below the middle, very pointed, narrowing towards base. sharply falcate-recurved, deeply grooved, almost plicately keeled, keel and margins fine, cartilaginous, almost transparent, having fine, projecting T. below midway, both surfaces black-green to brown-green, with 6–10 narrow, white marks arranged in transverse bands; **Infl.** 50–75 cm tall, Fl. 28–30 mm long, pale red.

**A. dispar** BGR. (? §). – S.W.Afr. – Plant known only from a Fl.-description by A. BERGER.

**A. distans** HAW. (Pl. 14/2) (§ IV/D/25) (*A. brevifolia* HAW., *A. mitraeformis* v. *angustior* LAM., *A. perfoliata* v. *b*. AIT., *A. mitriformis* v. *b*. SIMS, *A. m.* v. *humilior* WILLD.). – W.Cape. – Branching from base; **St.** elongated, creeping, about 3 cm thick, often 2–3 m long, laxly leafy; **L.** rather closely arranged at the top, broad-ovate, tapering, sharp-pointed, 8–9 cm long, 5–6 cm across, fleshy, blue-green, convex on underside, with several warty marks, keel with warty T., margins with horny T. 2–3 mm long, a few similar T. on the upper part of the underside; **Infl.** 40–50 cm tall, Fl. 2.5–4.5 cm long, reddish.

**A. divaricata** BGR. v. **divaricata** (§ IV/D/23; XI/b) (*A. sahundra* BOJ. nom. nud., *A. vahontsohy* PERR. nom. nud., *A. vaotsohy* DECORSE et POISS.). – W. and S.W.Madag. – Many-stemmed, forming bushes; main **St.** erect, 0.6–2 m tall, often branching from below; **L.** in a spiral, 55–65 cm long, 5.5–7.5 cm across base, grey-green, margins with triangular, robust, red-brown T. set 5–10 mm apart, dry old L. remaining on plant; **Infl.** freely branching, Fl. 20–30, scarlet.

**A. —** v. **rosea** (R. DECARY) REYN. (*A. vaotsohy* v. *rosea* R. DECARY). – Fl. pale pink.

**A. doei** LAVR. v. **doei** (§ IV/C/16). – S.Yemen. – **Ros.** usually single, sometimes forming groups; **L.** 12–14, 30–35 cm long, 8–10 cm across base, dull green on upperside, sometimes with several scattered, round, white marks, slightly grooved towards tip, margins horny and with hard, sharp, red-brown, triangular T. up to 3 mm long and 8–12 mm apart; **Infl.** to 70 cm tall, Fl. yellow, with white hairs externally.

**A. doei** v. **lavranosii** J. MARN.-LAP. – Fl. yellowish-white, Cal. and Pet. thicker, with longer white hairs.

**A. dolomitica** GROENEW. (§ V). – N.Transvaal. – **St.** 2 m tall, simple, 20 cm ⌀; **Ros.** dense; **L.** c. 50, c. 30 cm long, 6–8 cm across, 1.5 cm thick, rigid, convex on underside, dark green with red-brown markings and tip, margins stiff, with strong, recurved T. 2 mm long and 1 cm apart; **Infl.** 48 cm long, Fl. 12 mm long, light red.

**A. dominella** REYN. (§ III). – Natal. – **St.** 15 cm tall, forming groups of up to 50; **Ros.** of 20 L. 35 cm long, 1 cm across, clasping the St., somewhat concave on upperside, convex on underside, dark green with white spots towards base, marginal T. 0.5–1 mm long, 2–5 mm apart; **Infl.** 35 cm long, Fl. 2 cm long, yellow with green stripes.

**A. dorotheae** BGR. (§ IV/D/22). – Tanzania. – **Ros.** dense; **L.** narrow-lanceolate, tapering, convex on underside, fleshy, c. 14 cm long, 2.5 to 3 cm across, margins cartilaginous with triangular, whitish-tipped T. 4 mm long and ± 9 mm apart, L. mottled on upperside, underside with oblong marks in longitudinal rows; **Infl.** c. 14 cm long, Fl. 3.5 cm long, red.

**A. drepanophylla** BAK. (§ IV/E/19). – Cape: Somerset East, Ladismith and Calitzdorp D. – Probably a hybrid: **A. arborescens** × **A. speciosa.** – **St.** erect, 3–4 m tall; **Ros.** dense; **L.** narrow, falcate, 60–90 cm long, 3–4 cm across, convex on underside, marginal T. 1–2 mm long; **Infl.** 10 cm tall, Fl. 22–23 mm long, red with green stripes.

**A. duckeri** CHRISTIAN (§ IV/B/9). – Malawi. – Acauline or **St.** 20 cm tall, 9 cm ∅; **L.** numerous, densely rosulate, ovate-lanceolate, tapering, to 54 cm long, 14 cm across, 1 cm thick in the middle, deep green on the upperside, underside pale milky-green, both surfaces with dark green longitudinal lines and stripes, deeply grooved above on upperside, margins sharp, narrow, white, translucent, horny, T. pale brown, triangular, 2–3 mm long, 12–25 mm apart; **Infl.** 1.5 m tall, Fl. c. 3.5 cm long, orangecoloured.

**A. dyeri** SCHOENL. (§ IV/B/10/A/3a). – Transvaal. – **St.** short, simple; **L.** 15 in a dense Ros., tapering, concave on upperside, light green, usually with some spots in distinct, broken lines, whitish-green on underside, with light transverse bands, marginal T. triangular, curved, 4–6 mm long, 1.5–2 mm apart on the horny border; **Infl.** c. 90 cm tall, Fl. 34 mm long, red.

**A. ecklonis** SALM (§ III/3). – Lesotho: N.Swaziland; S.E.Cape; E.Orange Free State; E.Transvaal. – **Ros.** single or in groups, acauline or very short-stemmed; **L.** 14–20, c. 40 cm long, 9 cm across, slightly concave on upperside, dark green, often rather reddish, slightly convex on underside, with several spots near base, margins horny with 1–3 triangular T.; **Infl.** to 50 cm tall, Fl. 2–2.4 cm long, yellow to red.

**A. elegans** TOD. (§ IV/C/19a) (*A. schweinfurthii* HOOK., *A. s.* BAK., *A. aethiopica* (SCHWEINF.) BGR., *A. vera* v. *a.* SCHWEINF., *A. peacockii* BGR., *A. abyssinica* BGR., *A. a.* v. *p.* BAK., *A. percrassa* v. *saganeitiana* BGR.). – W.Ethiopia; W.Afr. – **Ros.** normally single, less often 2–3, 60 cm ∅, of 16–20 L.; **St.** 20–30 cm long in mature plants, prostrate; **L.** 60 cm long, 15–18 cm across base, 2.5 cm thick, narrowing to the shortly-dentate tip, grey-green on upperside, slightly grooved in upper part, convex on underside, margins with a red border and brown, sharp, triangular T. 3–4 mm long and 1.5 to 2.5 mm apart; **Infl.** 1 m tall, Fl. 2.5–3 cm long, orange or scarlet.

**A. elgonica** BULLOCK (§ IV/D/22). – Kenya. – ♄ up to several m ∅; **St.** short or to 1 m tall, usually densely leafy; **L.** 22–24 in a dense Ros., to 40 cm long, 9 cm across base, flat to slightly grooved on the upperside, often tinged with red, margins sinuate-dentate, T. sharp, triangular, 8–9 mm long, 10–15 mm apart; **Infl.** 50–70 cm tall, Fl. c. 4 cm long, orange-scarlet.

**A. ellenbeckii** BGR. (§ IV/B/10x). – S.Somalia. – Insufficiently known species. – **L.** ensiform, gradually tapering, 25–30 cm long, 2.2–5 cm across, thin, striped, margins thin, cartilaginous, finely toothed; **Infl.** branching, Fl. 2.8 cm long, vermilion.

**A. eminens** REYN. et BALLY (§ IX). – Somalia. – ♄ 10–15 m tall; **St.** to 1.5 m ∅ at base, thinner above, irregularly branching, Br. thin; L. 16–20 at Br. apex, imbricate with Sh. at base, this part having a white, horny border 5 mm across, and horny, white T. which, in lower part, are 2–3 mm long and 3–5 mm apart, the upper T. being smaller and more widely-spaced or missing; **Infl.** 50–60 cm tall, Fl. 4 cm long, red.

**A. enotata** LEACH. (§ IV/D/24). – Zambia: Abercorn D. – Plant hanging; **St.** simple, up to 60 cm long, 3 cm ∅, sometimes forming clumps from offsets from the base, with a Ros. of ± 12 **L.** 45–60 cm long, 5–6 cm wide towards the base, widely spreading but all becoming strongly falcate and pointing downwards, upper surface flat or slightly concave, pale green and suffused with pink, lower surface convex, somewhat paler in colour, margins very narrowly cartilaginous, sharp edged, whitish, T. widely spaced, small, usually forwardly hooked, pungent, 2–4 cm apart; **Infl.** ± 85 cm long, branched, Rac. laxly flowered, Fl. 25–28 mm long, dull red, with obscurely purplish stripes.

**A. eremophila** LAVR. (§ IV/C/16). – S.Yemen. – **Ros.** usually single, but also in groups of 2–6, acauline or with short St., prostrate; **L.** 10–22, triangular, pointed, very rigid, 35 cm long, 10 cm across base, 1.5–2.5 cm thick, grey-green on upperside, with a brownish sheen, broadly grooved in upper part, margins with dark brown, sharp, triangular T. to 4 mm long and 1–1.5 mm apart; **Infl.** 40–75 cm tall, Fl. 3 cm long, scarlet.

**A. erensii** CHRISTIAN (§ IV/D/23). – Kenya. – **Ros.** acauline; **L.** 16–19, ovate-pointed, tapering above middle, 21 cm long, 8 cm across base, 13 mm thick, green to milky-green on upperside with dark lines and scattered white spots, slightly concave midway, markings on the underside in irregular transverse bands, not running into one another, margins translucent, T. flat, small, whitish, triangular; **Infl.** erect, Fl. 29 mm long, fleshy-pink.

**A. ericetorum** J. BOSSER (§XI/d/α). – Central Madag. – **Ros.** acauline, single; **L.** 15–20, fleshy, bluish, narrow-lanceolate, 18–19 cm long, 3.5 to 4.5 cm across base, concave on upperside, blunt-tipped, margins horny, thin, dentate with pale yellow T. 1.5–2 mm long, 7–10 mm apart; **Infl.** erect, 50–70 cm tall, stiff, simple or with 1–2 Br., Ped. with 5–12 sterile involucral L., Rac. c. 15 cm long, pedicel 1.5–3 cm long, Fl. pendulous, numerous, 3.5–3.7 cm long, yellow.

**A. erythrophylla** J. Bosser (XI d/a). – Central Madag. – Ros. acauline or short-stemmed, simple; L. 6–8, brownish-red, inner L. green and red-tinged; thick, lanceolate to linear, tapering, recurved, 10–17 cm long, 2–4 cm across, slightly concave on upperside, margins horny, narrow, dentate, T. sharp, red, 1–1.5 mm long, 4–6 mm apart; **Infl.** erect, 15–40 cm tall, Rac. simple or branched, lax, pyramidal, 6–20 cm long, of 20 or more Fl. on pedicels 7–12 mm long, pendulous, 2–2.5 cm long, red.

**A. esculenta** Leach (§ IV/C/15). – Angola: Huila D.; Zambia; S.W.Afr.; Botswana. – Acauline or with **St.** 40 cm tall, crowded, prostrate, forming clumps by means of offsets; **L.** c. 20, rosulate, (40–)50(–60) cm long, (7–) 8(–10.5) cm across, narrowing to a pointed tip, often recurved, flat in the lower part of upperside, much grooved towards upper part, grey or grey-green to pink-brown with numerous white lentiform dots arranged irregularly in ± wavy transverse bands, underside convex, like upperside in colour and marking, and having a M. row of stout Sp., often sharply keeled, margins slightly sinuate-dentate, T. triangular, 3–5 mm long, 10–20 mm apart; **Infl.** 1.5–2.2 m tall, branching, Rac. narrow-cylindrical, 30 to 40 cm long, pedicels 5–6 mm long, Fl. c. 3 cm long, scarlet, later yellow. In some localities the L. sap is not bitter.

**A. excelsa** Bgr. (§ VI/3). – Rhodesia. – **St.** simple, 5–6 m tall, very thick; **L.** in dense Ros., long-lanceolate, tapering, soft, 60–70 cm long, 6–7 cm across base, concave on upperside, convex on underside, margins and underside with T.; **Infl.** several in number.

**A. falcata** Bak. (§ IV/B/13). – S.W.Afr.; Cape: Lit. Namaqualand. – **L.** falcate, 22 cm long, 3.5–4.5 cm across, clasping St., lanceolate, tapering, surface rough-granular, blue, somewhat concave on upperside, keeled on underside towards tip, margins and keel with brown T. 7–10 mm apart, 3–4 mm long, curved; **Infl.** 40–45 cm tall, Fl. to 3 cm long.

**A. ferox** Mill. (Pl. 15/1) (§ VI/4b) (*A. subferox* Spreng., *A. f. v. s.* (Spreng.) Bak., *A. f. v. incurva* Bak., *A. galpinii* Bak., *A. f. v. g.* (Bak.) Reyn., *A. supralaevis* Haw., *A. f. v. hanburyi* Bak., *A. f. v. erythrocarpa* Bgr., *Pachydendron s.* Haw., *P. f.* Haw., *P. pseudoferox* Haw., *P. p.* S. D., *A. socotrina* Masson, *? A. horrida* Haw., *A. perfoliata* v. *f.* Ait., *A. p.* Thunbg., *A. muricata* Haw.). – Cape: widespread, Orange Free State, Natal; Lesotho. – **St.** simple, (2–)3(–5) m tall, densely covered with dry old L.; **L.** 50–60, densely rosulate, lanceolate-ensiform, to 1 m long, 15 cm across base, fleshy, concave on upperside, glabrous or irregularly spined, underside similarly spined, margins sinuate-dentate, T. triangular, reddish to brown-red, 6 mm long, 1–2 cm apart; **Infl.** erect, Fl. 33 mm long, scarlet-orange.

**A. fievetii** Reyn. (§ XI/d/α). – Madag.: Fianarantsoa. – Ros. acauline or short-stemmed, usually simple; **L.** 12–16, narrowly tapering-lanceolate about 35 cm long, 5–6 cm across base, green on upperside with a slight reddish tinge, grooved towards tip, margins reddish-pink with triangular T. 2–3 mm long and 7–10 mm apart, these becoming smaller towards the left sinistorse tip; **Infl.** to 50 cm tall, Fl. 27 to 30 mm long, orange yellow.

**A. flexilifolia** Christian (§ IV/D/22). – Tanzania. – ♄ branching freely from the base and forming groups about 2 m ⌀; **St.** 60–100 cm tall, 6–7 cm ⌀, often pendent; **L.** rosulate, arranged rather laxly in upper part, ensiform, gradually tapering, c. 50 cm long, 6–7 cm across, c. 1 cm thick, somewhat concave on upperside above, bluish-green with slight bluish sheen, initially somewhat mottled, convex on underside, sometimes with a few transverse folds or grooves near the base, margins narrow, horny with small brownish triangular T. 1–2 mm long and 1–2 cm apart; **Infl.** 50–60 cm tall, somewhat oblique, Fl. 33–35 mm long, scarlet to brownish-red.

**A. fosteri** Pillans (§ IV/B/10/A/33a). – Transvaal. – Ros. single, acauline, occasionally short-stemmed; **L.** 16–24, 40–50 cm long, the uppermost 10 cm usually desiccated, 6–8 cm across base, slightly concave on upperside, mostly dark grey-green, with spots arranged mostly in wavy transverse bands and elongated longitudinally, pale green and H-shaped; underside convex, grey-green, with light longitudinal veins, margins sinuate-dentate, horny, T. sharp, pale brown, 5 mm long, 1 cm apart; **Infl.** 1–1.5 m tall, Fl. 30–38 mm long, orange-red, golden yellow in the upper part.

**A. framesii** L. Bol. (§ IV/E/27b) (*A. amoena* Pillans). – Cape: Lit. Namaqualand. – **St.** short, thin, prostrate, branching, forming dense groups of up to 20 Ros. 50 cm ⌀, in groups of 2–3 m ⌀; **L.** lanceolate, tapering, 30–35 cm long, 7–8 cm across base, slightly concave on upperside towards tip, convex on underside, both surfaces dark grey-green to bluish-green, either with or without white, irregularly scattered spots, margins sinuate-dentate, T. triangular, sharp, reddish-brown, 3 mm long, 1 cm apart; **Infl.** c. 70 cm tall, Fl. 3.5 cm long, deep scarlet.

**A. fulleri** Lavr. (§ IV/C/16). – S.Arabia. – Plants solitary; **St.** short; **L.** c. 12, ensiform, stiff, thick, 45 cm long, 12 cm across base, more than 3 cm thick, bluish-green with a yellowish sheen, flat or slightly grooved on upperside, underside convex, margins straight, edge reddish-brown, horny, T. short, brown, sharp, 1–2 mm long, 12–25 mm apart; **Infl.** 40–70 cm tall, Rac. cylindrical to tapering, 30–35 cm long, pedicels 5–6 mm long, Fl. lax, pendulous, 35 mm long, coral-red.

**A. gariepensis** Pillans (§ IV/E/27a) (*A. gariusana* Dtr.). – Cape: Lit. Namaqualand; S.W.Afr. – Closely related to **A. microstigma**; L. longer, also the Fl., Ros. mostly simple, the short **St.** occasionally branching and thus forming small groups to 1 m tall, densely covered with dry old L.; **L.** densely rosulate, 30–40 cm long, 5–8 cm across base, lanceolate-tapering, flat on upperside, somewhat lined, dark green to reddish-brown, either with or without white marks, convex on underside, margins horny, T. red-brown, sharp, 2–3 mm long, about 1 cm apart; **Infl.** 0.8–1.2 m tall, Fl. 23–27 mm long, yellow or greenish-yellow.

**A. gasterioides** BAK. (§ IV/B/10x). – S.Afr. – **Ros.** acauline or short-stemmed, ? simple; **L.** lanceolate, 10–13 cm long, 4–5 cm across, thick, green, with oblong, whitish spots in irregular transverse bands, marginal T. 3 mm long; **Infl.** 30 cm tall, Fl. 2.5 cm long, red.

**A. gerstneri** REYN. (§ VI/4a). – Natal. – **Ros.** acauline or short-stemmed, usually unbranched; **L.** 20–30, 60 cm long, 9 cm across, somewhat concave on upperside, initially spinous, convex on underside with persistent Sp. down the M. line, marginal T. 4–5 mm long, 1–1.5 cm apart, brown; **Infl.** 1–2, Fl. 3 cm long, orange-red.

**A.** cv. Gigantea. – Cultivated hybrid: **A. vera** × **A. ferox.** – **Ros.** 1.25–1.5 m ⌀, densely leafy; **L.** ensiform, 75 cm long, 12–15 cm across, concave on upperside, fleshy, bluish, margins red, T. pointed; **Infl.** sturdy, Fl. 30–32 mm long, yellow.

**A. gigas** RES. (§ IV/E/21) (*A. ciliaris* f. *gigas* RES.). – Bot. Gard., Coimbra, Portugal. – Habit as for **A. ciliaris**; **L.** 15–30 cm long, about 4 cm across, 4.5 mm thick, T. on L. Sh. and L. margins not more than 1mm long.

**A. gillilandii** REYN. (§ IX). – S.Yemen: Hadramaut. – **St.** quite thin, 2 m or more tall; **L.** densely rosulate, tapering gradually to the tip, c. 65 cm long, grey-green on upperside, grooved towards the tip, rounded on underside, grey-green, margins continuous, pale pink, horny, T. fairly soft, pale pink, horny, 1–1.5 mm long, spaced irregularly 5–10 mm apart; **Infl.** 90 cm tall, Fl. 3 cm long, scarlet, with paler mouth.

**A. glauca** MILL. v. glauca (§ IV/B/8) (*A. g.* v. *elatior* S. D., *A. g.* v. *major* HAW., *A. rhodacantha* DC., *A. g.* v. *humilior* S. D., *A. g.* v. *minor* HAW.). – Cape: Caledon, Swellendam and Laingsburg D., Lit. Namaqualand. – **Ros.** almost acauline; **L.** 30–40, lanceolate, 30–40 cm long, 10–15 cm across base, slightly concave on upperside above, bright blue, convex on underside, warty-pointed at tip, margins with red-brown, sharp T. 4–5 mm long; **Infl.** 60–80 cm tall, Fl. 4 cm long, pale red.

**A.** — v. **muricata** (SCHULT.) BAK. (*A. m.* SCHULT., *A. g.* v. *spinosior* HAW.). – S.Afr.: W.Cape. – **L.** to 40 cm long, broad at base, bluish-green, lined, margins with dark brown, horny edge, T. 5 mm long and 1 cm apart; **Fl.** 3.5 cm long, salmon-orange.

**A. globuligemma** POLE EVANS (§ VI/1). – Transvaal; Rhodesia. – **St.** creeping and rooting, to 50 cm long, offsetting to form dense groups; **L.** in a dense Ros., lanceolate-tapering, 45–60 cm long, 8–9 cm across, somewhat concave on upperside above, bluish, convex on underside, margins white to pale pink, horny, T. about 2 mm long, about 1 cm apart, white, brown-tipped, curved; **Infl.** 1 m tall, buds spherical, Fl. 26 mm long, sulphur-yellow, somewhat reddish-tinged.

**A. gloveri** REYN. et BALLY (§ IV/B/5). – Somalia. – ♄ with Br. from base; **St.** spreading or prostrate, 3–4 cm ⌀, 50–100 cm long; **L.** lanceolate-tapering, 20–30 cm long, 4–6 cm across base, dull green on upperside, with or without white, lentiform marks, slightly concave towards tip, margins with sharp, triangular, reddish-brown T., 2–3 mm long and 8–10 mm apart; **Infl.** c. 50 cm tall, Fl. 26–30 mm long, yellow, orange or dull scarlet.

**A. gossweileri** REYN. (§ IV/C/19c). – Angola. – ♄ forming thickets; **St.** 1–1.5 m tall, 3–5 cm ⌀, offsetting from base, the top 10–20 cm quite densely leafy; **L.** c. 16 in a Ros., c. 30 cm long, 5 cm across base, narrowing gradually to the tip, grooved on upperside, green, underside also green, rounded, margins with triangular T. 3 to 4 mm long and 15 mm apart; **Infl.** to 40 cm tall, Fl. 3 cm long, scarlet.

**A. gracilicaulis** REYN. (§ IV/C/18f). – Somalia. – **St.** simple or branching from base, then becoming bushy, 2–4 m tall, 8–10 cm ⌀ at base, the top 30–50 cm bearing persistent, dry L.; **L.** about 20 in a Ros., ensiform, with Sh. at base where L. is 8 cm across, narrowing to pointed tip and sharply recurved, grey-green on upperside, very concave in upper part, rounded on underside, margins white, horny, 1 mm wide with white, horny T. 1 mm long, irregularly spaced, 2–10 mm apart; **Infl.** 60 cm tall, Fl. 18 mm long, yellow.

**A. graciliflora** GROENEW. (§ IV/B/10/A/2b). – E.Transvaal. – **Ros.** acauline or with short **St.**, usually forming small groups; **L.** 15–20, lanceolate-triangular, c. 25 cm long, 10 cm across base, slightly concave on upperside, light green, with oblong, white spots in ± distinct transverse bands, convex on underside, paler, unmottled to somewhat lined or spotted; **Infl.** c. 60–80 cm tall, Fl. up to 52 mm long, scarlet.

**A. gracilis** HAW. v. gracilis (§ IV/D/21) (*A. laxiflora* N. E. BR.). – Cape: Port Elizabeth, Uitenhage D. – ♄ branching from base; **St.** to 2 m long, 2 cm ⌀, laxly leafy in upper 30 to 60 cm; **L.** bases clasping the St., Sh. 10–15 cm long, L. with pale green stripes, almost lanceolate, 25 cm long, 2.5 cm across base, gradually tapering over upper third, dark green, margins thin, horny, with white T. which are curved in lower part, 1 mm long and 2–5 mm apart; **Infl.** simple or with 2–3 Br., Fl. 4–4.5 cm long, scarlet, with yellow mouth.

**A.** — v. **decumbens** REYN. – Having several St. to 75 cm long, prostrate; **L.** lanceolate-triangular, about 15 cm long, 1.5 cm across base; **Fl.** scarlet.

**A. grahamii** SCHOENL. (§ IV/B/10x). – S.E.-Cape. – Probably a form of **A. saponaria**; **St.** 45 cm long; **L.** 50 in a dense Ros., 60 cm long, 10 cm across, light green, striped, margins with brown T. 3 mm long; **Infl.** 60–90 cm tall, Fl. 37–45 mm long, red.

**A. graminicola** REYN. (§ IV/B/10/A/2a). – Kenya. – Very similar to **A. lateritia**; the capitate Ped. somewhat smaller, denser, the Fl. more curved.

**A. graminifolia** BGR. (§ II). – Tanzania. – Acauline; **L.** numerous in a dense Ros., with a translucent, striped Sh. at base, linear, 10 to 20 cm long, 3 mm across, margins in lower part with small, fine, cartilaginous, ciliate Sp., entire above; **Infl.** 40 cm tall, Fl. 15–17 mm long, greenish.

**A. grandidentata** SALM (§ IV/B/10/B). – Cape: S.W.Kalahari. – Short-stemmed; **L.** 30 to 50 cm long, 6–8 cm across, lanceolate, green, indistinctly striped, with whitish spots merging in irregular rows, margins having closely set, red-brown T. 4 mm long; **Infl.** 80–100 cm tall, Fl. 3 cm long, yellow.

**A. grata** REYN. (§ IV/C/19a). – Angola: Bié. – **Ros.** acauline or short-stemmed, forming dense groups, fairly compact; **L.** 16–20, lanceolate-tapering, 20–25 cm long, 7–8 cm across, somewhat grooved in upper part, green, with a reddish-brown sheen, convex on underside, paler bluish-green, with numerous, close-spaced spots below, these spots being pale green to white, round, c. 1 mm $\varnothing$, margins sinuate-dentate, often even serrate, T. 2–3 mm long, 5–8 mm apart; **Infl.** 70–90 cm tall, Fl. c. 3 cm long, scarlet.

**A. greatheadii** SCHOENL. (§ IV/B/10/A/2a) (*A. pallidiflora* BGR., *A. termetophila* DE WILD.). – Cape: N.Kalahari, Transvaal; Lesotho; Malawi. – **Ros.** short-stemmed; **St.** 10–12 cm $\varnothing$; **L.** linear-lanceolate, gradually tapering, about 35 cm long, 6–8 cm across, 13–15 mm thick, concave on upperside, glossy dark green, with white spots and pale lines, pale green and unmottled on the underside, margins whitish-striped, T. triangular, horny, brown, sharp, 3–6 mm long; **Infl.** 80–130 cm tall, Fl. 3 cm long, pale pink.

**A. greenii** BAK. (Pl. 14/4) (§ IV/B/10/A/3a). – Natal. – Acauline or with short **St.**; **L.** 12–15, 35–40 cm long, 6–7 cm across base, long-lanceolate, flat or slightly concave on upperside, underside rounded, light green, with dark lines and broad transverse bands of merging, oblong, white spots which, on the underside, are very protuberant, margins with brown, horny T.; **Infl.** to 1.5 m tall, Fl. 3 cm long, pale red.

**A. greenwayi** REYN. (§ IV/B/10/B). – Tanzania: Tanga. – **Ros.** single or forming small groups, acauline or with St. to 30 cm long or longer where the plant hangs down among rocks; **L.** c. 16, dense, to 30 cm long, 5 cm across, flat to slightly grooved on upperside, dull green, tinged reddish-bronze, usually having several scattered, very pale green spots, the underside paler green and the entire surface covered with pale green spots, margins with small, triangular, sharp T. 2 mm long and 5–8 mm apart; **Infl.** c. 60 cm tall, Fl. 3 cm long, red-orange to yellow.

**A.** × **grusonii** HENZE. – Hybrid: **A. humilis** × **A. schimperi.** – Short-stemmed; **L.** densely rosulate, gradually tapering, margins irregularly dentate; Fl. red.

**A. guerrae** REYN. (§ IV/D/23). – Angola: Bié. – **Ros.** single, acauline or very short-stemmed, usually slightly inclined; **L.** c. 24, narrowed-lanceolate, gradually tapering, about 40 cm long, 6–7 cm across base, slightly grooved above on upperside, dull green, indistinctly lined, margins sinuate-dentate, T. sharp, pale brown or reddish-brown, 4–5 mm long, 10 to 15 mm apart; **Infl.** 90–100 cm tall, Fl. 4 cm long, scarlet.

**A. haemanthifolia** MARL. et BGR. (Pl. 15/5) (§ IV/A/1). – Cape: Cape D. – Acauline; **L.** distichous, 10–20 cm long, 5–6 cm across, linear, rounded above, rather thick, very concave on upperside, convex on underside, grey-green to bluish-green, with a red margin; **Infl.** 30 cm tall, Fl. in compact clusters, red.

**A. harlana** REYN. (§ IV/C/18c). – Ethiopia: Harrar. – Acauline or with short **St.**, usually single or with 2–4 Ros., to 60 cm tall; **L.** c. 24, dense, lanceolate-tapering, 50 cm long, 12 to 15 cm across, pale to dark olive-green on upperside, slightly grooved in upper part, convex on underside, often with several pale green spots towards the base, margins sinuate-dentate, often with a horny, red-brown edge, with triangular, sharp T. 3–4 mm long and 1–1.5 cm apart; **Infl.** 70–90 cm tall, Fl. 33 mm long, deep red.

**A. harmsii** BGR. (§ IV/C/16). – S.Somalia; Tanzania. – Habit not known; **L.** c. 13 cm long, 4.5 cm across base, gradually tapering, grooved on upperside, leathery-fleshy, reddish to brown, glossy, underside with several whitish, oblong spots arranged in longitudinal groups, ± translucently mottled, margins cartilaginous, T. triangular, whitish, pointed, 2 mm long, 1 cm apart; **Infl.** stout, Fl. c. 3 cm long, yellow. (Acc. REYNOLDS, an insufficiently known species.)

**A. haworthioides** BAK. v. **haworthioides** (Pl. 15/6) (§ I; XI/d/β 1) (*Aloinella h.* (BAK.) LEMÉE). – Central Madag. – Acauline; **Root** fleshy, fusiform; **Ros.** dense, (3–)5(–10) cm $\varnothing$; **L.** (30–)35(–100), 3–6 cm long, 6–8 mm across base where L. is white and thin, tapering to a Sp. at the tip, grey-green, marginal T. white, cilia white, L. surface having white protuberances; **Infl.** 20–30 cm tall, Fl. pale orange-coloured, filaments projecting, ribbon-like, brilliant orange-coloured.

**A. —** v. **aurantiaca** PERR. – Fl. orange-red, filaments yellow.

**A. hazeliana** REYN. (§ III/2). – Rhodesia: Melsetter D. – **St.** several, erect, thin, to 50 cm long, 15 mm $\varnothing$, with c. 12 L., laxly arranged on upper 20–30 cm distichous, linear, to 20 cm long, 1–1.5 cm across, with the basal Sh. 10–15 mm long, flat to slightly grooved on the upperside, unmottled or with several pale green elliptical marks, green and rounded on the underside, with numerous spots, margins very narrow, often translucent, with stout, white, triangular, sharp T. about 5 mm long and about 5 mm apart, the T. towards the tip either smaller or absent, L. tip blunt, with a finely dentate apiculus; **Infl.** 30–40 cm tall, Fl. 2.5 cm long, scarlet with green tips.

**A. helenae** P. DANGUY (§ XI/a). – S.Madag.: Fort Dauphin D. – **St.** unbranched, to 4 m tall, 20 cm $\varnothing$, densely covered with dried L.; **L.** about 40 in a terminal Ros., spreading or ultimately drooping, to 1.4 m long, to 15 cm across base, very concave, marginal T. triangular, compressed; **Infl.** 1–8, to 60 cm tall, Fl. very numerous, greenish-red.

**A. hemmingii** REYN. et BALLY (§ IV/A/2). – N.Somalia. – **Ros.** single or in small groups, acauline or very short-stemmed; **L.** c. 10, ovate or tapering-lanceolate, 10–12 cm long, 3–3.5 cm across base, slightly grooved above on upperside, with margins somewhat incurved, brownish green, covered all over with numerous, dull white, oblong lines, these being smaller and more numerous on the underside, margins with sharp, triangular T., whitish at base and brown-tipped, those midway being 2 mm long and 4–6 mm apart, those towards the tip being smaller, the underside sometimes also having T. along the M.line towards the tip; **Infl.** 30 to 35 cm long, Fl. 24 mm long, flamingo-red or pale rose-pink.

**A. hendrickxii** REYN. (§ IV/C/19c). – Congo (Kinshasa), N. and S.Kivu. – ♄, acauline or very short-stemmed, with numerous offsets, forming dense groups; **L.** c. 16, rosulate, tapering-lanceolate, at the base clasping the St., 28 cm long, 3.4 cm across base, rich green on upperside, somewhat grey, slightly concave below, somewhat grooved above, underside often with several indistinct marks, margins with stout, green-white T. 3 mm long and 5–8 mm apart; **Infl.** 25–30 cm tall, Fl. 3 cm long, scarlet.

**A.** cv. Henzei. – Hybrid: **A.** × **grusonii** × **A. variegata.** – Intermediate hybrid from the Magdeburg Municipal Gardens.

**A. hereoensis** ENGL. v. **hereoensis** (Pl. 15/2) (§ IV/B/14). – S.W.Afr.: Damaraland, Gr. Namaqualand; Angola. – Acauline or with creeping **St.** to 1 m long; **L.** in compact Ros., 25 to 30 cm long, 4.5–8 cm across base, triangular-lanceolate, grooved above on upperside, convex on underside, blue, indistinctly lined, with several spots on underside, margins with a few horny T. 3 mm long; **Infl.** 1 m tall, Fl. 25–27 mm long, yellowish to scarlet.

**A.** — v. **lutea** BGR. – Fl. yellow.

**A.** — v. **orpeniae** (SCHOENL.) BGR. (*A. o.* SCHOENL.). – Less robust variety; **St.** short; **Ros.** dense; **L.** somewhat striate, with several light marks.

**A.** × **hertrichii** E. WALTH. – Hybrid: **A. vera** × **A. lineata.** – Ros. acauline, freely offsetting; **L.** rather triangular-lanceolate, 30 to 40 cm long, 7 cm across base, 1–1.5 cm thick, convex on both surfaces, greenish-bluish, margins with horny, brown T. 3–5 mm long and 3–15 mm apart; **Infl.** 40–50 cm tall, Fl. pendulous, 3 cm long.

**A. heteracantha** BAK. (§ IV/B/10x) (*A. paradoxa* HORT.). – ? S.Afr.: Cape. – Probably a hybrid: **A. striata** × **A. mitriformis** or × **A. distans.** – Almost acauline, forming dense cushions; **L.** triangular-lanceolate, tapering, tip shortly carinate, dark green, mottled, 20–25 cm long, 6.5–7.5 cm across, somewhat concave on upperside, marginal T. curved-triangular, 3 mm long; **Infl.** 60–90 cm tall, Fl. 3–3.5 cm long, vivid red.

**A. hexapetala** S. D. (§ ?). – Dubious species, possibly identical with **A. speciosa.**

**A. hildebrandtii** BAK. (§ IV/D/23). – Rhodesia. – ♄; **St.** c. 1 m tall; **L.** in an open Ros., long-lanceolate, tapering, c. 60 cm long, 10 cm across base, flat or slightly convex on upperside, often coppery-green, often with a few white marks towards base, margins with stout T.; **Infl.** often several.

**A. hlangapies** GROENEW. (§ III/2). – E.Transvaal. – **St.** short, often several, with 10–15 L. arranged almost distichously; **L.** 20 cm long, 5 cm across, linear, pointed at tip, margins with fine T.; **Infl.** 35 cm tall, Fl. 3 cm long, yellowish with green spots.

**A. howmanii** REYN. (§ III/2). – Rhodesia: Melsetter D. – **St.** several, hanging down from cliffs, simple or branching from base, 20–30 cm long, 12 mm ⌀; **L.** 6–12, distichous, overlapping sheath-like at bases, linear, 15–20 cm long, 12–15 mm across, green on upperside, underside sometimes having several small, white marks at base, margins translucent, membranous, 1.5 mm across, with small T. 2–4 mm apart, sometimes these absent; **Infl.** 20–25 cm long, Fl. 24 mm long, fiery-scarlet.

**A. humbertii** PERR. (§ XI/d/β 1). – Central Madag. – **Ros.** acauline or short-stemmed; **L.** 7–12, 25–30 cm long, 5–6 cm across, narrowing at both ends, marginal T. 3–6 mm apart, yellow; **Infl.** 35–40 cm tall, Fl. yellow-reddish.

**A. humilis** (L.) MILL. v. **humilis** (§ IV/B/5) (*A. perfoliata* v. *h.* L., *A. h.* v. *candollei* BAK.). – E.Cape. – **Ros.** 6–7 cm ⌀, forming clumps by offsets; **L.** 30–40, 5–10 cm long, 1–1.5 cm across, linear-lanceolate, light green, ± curved on upperside, warty, whitish, with almost translucent T. 2–3 mm long, the underside with thick, white-green warts arranged ± distinctly in transverse rows, these warts being dentate, especially above, margins dentate; **Infl.** 25 to 40 cm tall, Fl. 3 cm long, coral-red.

**A.** — v. **acuminata** (HAW.) BAK. (*A. a.* HAW., *A. suberecta* v. *a.* HAW., *A. humilis* KER.). – L. 10–12 cm long, bluish, marginal T. larger.

**A.** — v. **echinata** (WILLD.) BAK. (Pl. 15/3) (*A. e.* WILLD., *A. h.* v. *e.* sv. *minor* S. D., *A. tuberculata* HAW.). – L. smaller, with fleshy Sp. on the upperside.

**A.** — v. **incurva** HAW. – L. fleshier, larger, 2 cm across base, blue-green, incurved at tip.

**A.** — v. **macilenta** (HAW.) BAK. – Cultivated plant from Gt. Britain. **L.** very concave, purple.

**A.** — v. **suberecta** (HAW.) BAK. (*A. s.* HAW., *A. h.* v. *s.* sv. *semiguttata* HAW., *A. acuminata* v. *major* S. D.). – **L.** broader, 15–17 cm long, bluish, glabrous and flat on upperside, warty on underside.

**A.** — v. **subtuberculata** (HAW.) BAK. (*A. s.* HAW.). – L. bluish, the lower warts on the underside somewhat raised, marginal T. smaller and more curved.

**A. ibitiensis** PERR. (§ XI/d/œ). – Central Madag. – **Ros.** single, less often forming small groups, acauline; **L.** 8–12, 18–40 cm long,

3–6 cm across, narrowing to both ends, yellow to olive-green, distinctly nerved, margins with small, cartilaginous, yellowish T. 1.5–2.5 mm long and 4–6 mm apart; **Infl.** 40–80 cm tall, Fl. scarlet.

**A.** × **imerinensis** J. BOSSER. – Spontaneous hybrid: **A. capitata** × **A. macroclada**. – Central Madag. – **Ros.** acauline, single; **L.** c. 25, green, lanceolate, 60–70 cm long, 8–10 cm across, margins narrow, horny, almost red, dentate, T. 2 to 2.5 mm long, reddish to fiery-red, 7–20 mm apart; **Infl.** erect c. 1.5 m tall, branching pedicels short, 2–3.5 cm long, Rac. with many Fl. which are cylindrical, 25 cm long and 7 to 8 cm ⌀, yellow.

**A. immaculata** PILLANS (§ IV/B/10/A/2a). – Transvaal. – **Ros.** dense, St. to 10 cm tall; **L.** 15–20 cm long, 6–8 cm across, ovate-lanceolate, tapering, usually flat on upperside, pale green, lined with green, very distinctly lined on underside, margins sinuate-dentate, red-brown, T. 4 to 5 mm long, triangular, sharp, directed forwards, 7–13 mm apart; **Infl.** 50–60 cm tall, Fl. c. 2.5 cm long, fleshy-pink.

**A. inamara** LEACH (§ IV/D/24). – Angola: Cuanza-Sul D. – Plant pendulous, offsetting from base; **St.** to 2 m long, with a few Br. above, closely covered with dried L. remains, Int. about 1 cm long; **L.** c. 9, rosulate, (45–)60(–90) cm long, 4–5 cm across base, widely spreading, the tip curved-ensiform and directed downwards, about 1 cm thick at base, flat to slightly concave on upperside, more swollen towards tip, pale greenish-yellow, the tip turning brown, indistinctly and finely lined with ± numerous, ± H-shaped, small, whitish marks (or even unmottled), convex on underside, paler, markings as on upperside but arranged ± in wavy transverse bands, margins somewhat horny-translucent, with small, pointed T. which are triangular or hooked, to 1 mm long, 4–20 mm apart; **Infl.** pendulous, paniculate, 40–55 cm long, Rac. cylindrical-capitate, Fl. 3 cm long, dull red, on pedicels almost 3 cm long.

**A. inermis** FORSK. (§ IV/D/23) (*A. luntii* BAK.). – S.Yemen. – **St.** short, erect; **L.** 7–10 in a loose Ros., ensiform, about 30 cm long, 5 cm across, fleshy, concave on upperside, pale green to reddish, immature L. somewhat mottled, margins horny; **Infl.** 4–5-branched, Fl. c. 2.5 cm long, red.

**A.** × **insignis** N. E. BR. – Hybrid: **A. humilis** × **A. drepanophylla**. – **Ros.** short-stemmed, densely leafy; **L.** 20–28 cm long, 2.5–4 cm across base, gradually tapering, concave on upperside, blue-green, with white warty Sp., margins with triangular, incurved, unevenly arranged T.; **Infl.** simple, Fl. 32 mm long, white-green.

**A. integra** REYN. (§ III/3). – Transvaal. – **Root** thick, 15 mm ⌀; **St.** simple, to 20 cm long, 4–6 cm ⌀; **L.** 15–20, dense, 10–12 cm long, glabrous, usually unarmed, light green, often somewhat striate; **Infl.** 40 cm tall, Fl. 15–18 mm long, lemon-yellow with green.

**A. inyangensis** CHRISTIAN (§ III/4). – Rhodesia. – ♄; **St.** numerous, dense, 20–40 cm tall; **L.** 8–10, distichous, to 25 cm long, 1 cm across base, linear, usually ± intertwined, broadly concave on upperside, flat towards tip, dark green, with scattered, lentiform, greenish-white marks towards the base, underside furrowed towards the tip where the markings are more numerous, with small, whitish, round warts among them, margins hirsute-dentate, borders white, cartilaginous, with horny T. 0.5 mm long in the lower half, but often entire towards the tip; **Infl.** c. 15 cm tall, Fl. 4 cm long, scarlet with a greenish mouth.

**A. isaloensis** PERR. (Pl. 15/4) (XI/b). – Central Madag. – Small ♄; **St.** (12–)15(–50) cm tall, 1 cm ⌀, branching from base; **L.** 10–14, loosely scattered, with a long Sh. clasping the St., 14–18 cm long, 12–15 mm across, narrowing to the base, somewhat rounded at the tip, blue to grey waxy-pruinose, margins with T. 1 mm long and 1–5 mm apart; **Infl.** simple, to 50 cm long, buds scarlet, Fl. vermilion with Pet. tips pale yellow.

**A. itremensis** REYN. (§ XI/d/α). – Madag. – **Ros.** simple; **St.** to 20 cm long, to 3 cm ⌀; **L.** 12–16, to 30 cm long, 4.5 cm across base, gradually tapering, slightly grooved towards tip, deep green with a reddish sheen, margins with broad compressed, deltoid, brownish T. 1 to 1.5 mm long and 5–8 mm apart, these being larger towards the base and smaller towards the tip, finally missing; **Infl.** 1–1.2 m tall, Fl. pendulous, 2.5 cm long, scarlet.

**A. jacksonii** REYN. (§ IV/C/19c). – Ethiopia. – **St.** erect, 10–20 cm long, 8–10 mm ⌀, or branching from the base to form groups 30 cm ⌀; **L.** on Br.-tips almost subulate, tapering, 10 to 15 cm long, 10–14 mm across base where L. are Sh.like and striate, L.tips blunt, with 1–2 stout white Sp. 1–2 mm long, L. grooved on upperside, green, with several pale green marks on the lower half, rounded on the underside, marks numerous and often in transverse bands, margins straight, with isolated, stout, triangular, white, red-tipped T. 1 mm long and 3–6 mm apart; **Infl.** 22–25 cm tall, Fl. pendulous, 27 mm long, scarlet.

**A.** × **jacobseniana** MARN.-LAP. – Hybrid: **A. perrieri** × **A. deltoideodonta**. – Similar to **A. perrieri**, stouter; L. to 21 cm long, 2 cm across, reddish, both surfaces with white-pink marks 2–4 mm long, marginal T. 1–5 mm long, 4–5 mm apart; **Infl.** c. 42 cm tall, Fl. 2 cm long, pink-red.

**A. jucunda** REYN. (§ IV/B/6). – Somalia. – **Ros.** acauline or very short-stemmed, offsetting and forming dense groups of about 50 cm ⌀, individual Ros. being 8–9 cm ⌀; **L.** c. 12, broad to tapering-ovate, 4 cm long, 2.5 cm across base, fleshy, dark green on upperside, grooved above, brownish, with numerous pale green to white spots which are often translucent, deep green on underside, with smaller spots, margins with horny, red-brown, sharp, triangular T. 2 mm long and 3–4 mm apart, with rather small Sp. near the tip along the M. line; **Infl.** 33 cm tall, Fl. 2 cm long, pale pink.

**A. karasbergensis** PILLANS (Pl. 16/3) (§ IV/B/11). – S.W.Afr.: Gr. Namaqualand; Cape: Lit. Namaqualand. – Related to **A. striata**; L. more curved and striate, 90 cm long, 11 cm across, 1.5 cm thick; **Infl.** 50 cm tall, Fl. 1 cm long, red with green-white stripes.

**A. keayi** REYN. (§ IV/C/19c). – Ghana. – **Ros.** acauline or short-stemmed, often with 1–2 offsets; L. 20–30, dense, to 60 cm long, 10–12 cm across base, gradually narrowing to the tip, green on the upperside, with several pale green spots, slightly grooved above, rounded on the underside, usually without markings, margins with triangular, sharp T. 5–6 mm long and 10–15 mm apart; **Infl.** 1 m tall, Fl. 3.5 cm long, scarlet, green-tipped. – Acc. NEWTON probably a hybrid of **A. schweinfurtii**.

**A. kedongensis** REYN. (§ IV/D/21). – Kenya. – ♄; **St.** thin, somewhat twisted, about 4 cm thick, to 4 m tall, usually branching from the base, often also elsewhere, the final 30–60 cm with somewhat lax L., Br. covered with dry, old L.; L. with basal striate Sh., 30 cm long, 3.5 cm across, gradually tapering, grooved on upperside, young L. initially mottled, margins with pale, triangular T. with red-brown tips and forward-curved hooks, 2–3 mm long and 1 cm apart; **Infl.** 50 cm tall, Fl. 3.5 cm long, scarlet.

**A. keithii** REYN. (§ IV/B/10/A/5b). – E.Swaziland. – **St.** 30 cm tall, single or in small groups; L. c. 20, lanceolate-tapering, to 60 cm long, 9–11 cm across base, slightly concave on upperside, green, often brown towards the tip, often indistinctly lined, with numerous scattered white marks in irregular transverse bands, underside paler, usually not mottled, otherwise as for upperside, margins sinuate-dentate, T. deltoid, sharp, brownish, 6–8 mm long, 1.5 to 2.5 cm apart, straight or somewhat incurving; **Infl.** to 1.75 m tall, Fl. 36 mm long, coral-red.

**A. khamiesensis** PILLANS (§ IV/E/27b). – Cape: Lit. Namaqualand. – **St.** about 1.2 m tall, simple, 10 cm ⌀ at base; L. crowded on Br.-tip, 60–80 cm long, 5–7 cm across, 1 cm thick, lanceolate, tapering, rich green, both surfaces convex with roundish, white marks, margins sinuate-dentate, T. triangular, sharp, horny, brown, directed forwards, 2–4 mm long and 2.8 mm apart; **Infl.** 80–90 cm tall, Fl. c. 28 mm long, orange-red.

**A. kilifiensis** CHRISTIAN (§ IV/B/10/A/2a). – Kenya: Kilifi. – **Ros.** acauline; L. c. 15, tapering gradually from the base or tapering–ovate, 27 cm long, 7 cm across base, upperside green, in the dry season coppery, with many scattered spots, these being oval or H-shaped, the upperside grooved above, the underside more spotted, these spots merging, margins brown, sinuate-dentate, T. triangular, brown, horny, 3 mm long, 4 mm apart; **Infl.** to 57 cm tall, Fl. 3 cm long, deep wine-red.

**A. kirkii** BAK. (§ IV/D/23). – Zanzibar. – **Ros.** short-stemmed, dense; L. 30–40, 26–30 cm long, lanceolate, tapering, 5–6 cm across, 5–6 mm thick, somewhat concave towards the tip, glossy green, often reddish, margins with triangular T. 5–6 mm long and 1–1.5 cm apart, set on a horny border; **Infl.** 60 cm tall, Fl. 3 cm long, red-yellow. (Acc. REYNOLDS, an insufficiently known species.)

**A. kniphofioides** BAK. (§ III/1) (*A. marshallii* WOOD. et EVANS). – Pondoland, Natal; N.Swaziland. – Acauline, simple, with an underground, ovoid, bulb-like swelling and fusiform roots; L. c. 20, arranged in many rows, almost linear, 20–30 cm long, 6–7 mm across, somewhat grooved on upperside, convex on underside, both surfaces green, margin entire or with small, white T.; **Infl.** c. 35 cm tall, Fl. 3.5–5 cm long, scarlet with green tips.

**A. komatiensis** REYN. (§ IV/B/10/A/3a) (*A. lusitanica* GROENEW.). – Moçambique; E.Transvaal. – **Ros.** acauline or with St. 20 cm long, often solitary but usually in small groups; L. c. 20, lanceolate, tapering, about 40 cm long, 8–10 cm across, pale green, lined and with translucent spots in wavy transverse bands, convex on underside, dark green, lined, unspotted, margins sinuate-dentate, T. triangular, sharp, 3–4 mm long, 1–1.5 cm apart; **Infl.** to 2 m tall, Fl. 3 cm long, pale brick-red.

**A. krapohliana** MARL. v. **krapohliana** (§ IV/B/5). – Cape: Lit. Namaqualand. – **Ros.** acauline, 15–20 cm ⌀; L. 20–30, 6–10 cm long, 12–15 mm across, almost lanceolate, tapering, both surfaces convex, blue-pruinose, margins reddish, with numerous small curved T.; **Fl.** scarlet, mouth greenish.

**A. —** v. **dumoulinii** LAVR. – Near Alexander Bay. – **Ros.** smaller; L. much shorter, three-angled and strongly curved inwards; **Sc.** c. 15 cm high, Rac. 12–18-flowered.

**A. —** × **A. glauca** MARL. – Natural hybrid. – Cape: Lit. Namaqualand. – In size and L. colour resembling **A. krapohliana**, L.shape more like **A. glauca**; Fl. 2.5–3 cm long, red and greenish-white.

**A. krausii** BAK. (§ III/3). – Pondoland, Natal, E.Transvaal. – Plants solitary or in small groups; **root** fleshy, fusiform; L. 8–10, distichous or virtually so, in mature plants rosulate, broad-linear, tapering, slightly concave on upperside, dark green, somewhat convex on underside, with several spots near the base, margins white, horny, with small T.; **Infl.** c. 40 cm tall, Fl. 12–18 mm long, lemon-yellow to yellow.

**A. labiaeflava** GROENEW. (§ IV/B/10x). – Transvaal: Pretoria D. – Possibly a hybrid: **A. davyana** × **A. longibracteata** (acc. REYNOLDS). – Small, acauline, offsetting only occasionally; L. short, not much longer than broad, spotted on upperside, underside with only indistinct spots; **Fl.** scarlet, inside tip of Pet. being yellow.

**A. laeta** BGR. v. **laeta** (§ IV/21; XI/d/β 2). – Central Madag. – **Ros.** acauline, compact, to 30 cm ⌀; L. deltoid, flat or spreading, 15–23 cm long, 7–9 cm across base, bluish grey to darkest red, both surfaces with small protuberances, marginal T. regular and close-set, horny, 2 mm tall; **Infl.** 30–45 cm tall, Fl. carmine.

**A. — v. maniaensis** PERR. - **L.** 8 cm long, 2 cm across.

**A. × laetococcinea** BGR. - Cultivated hybrid from 'La Mortola', Italian Riviera. - Parents unknown. - **Ros.** acauline, densely leafy; **L.** 20 cm long, 5 cm across base, gradually tapering, flat on upperside, blue, underside convex, with irregularly arranged oblong, white marks, margins with subulate-triangular, pink, horny T. 5 mm long; **Infl.** large, Fl. 3.5 cm long, scarlet, tipped with green.

**A. lastii** BAK. (§ IV/C/17). - ? Zanzibar. - **St.** simple (? always), thin, c. 45 cm long; **L.** densely rosulate, 35 cm long, 7.5 cm across, upperside indistinctly lined, margins sinuate-dentate, T. 3–4 mm long, 8–10 mm apart; **Infl.** 50 cm tall, Fl. c. 3 cm long, pale yellow.

**A. lateritia** ENGL. (§ IV/B/10/A/2a) (*A. comphylosiphon* BGR., *A. angiensis* and v. *kitaliensis* REYN., *A. solaiana* CHRISTIAN, *A. amanensis* BGR). - Tanzania. - **Ros.** acauline, dichotomously branching and mat-forming; **L.** 15–20, lanceolate, c. 30 cm long, 5–7 cm across the base, tip recurved and somewhat carinate, deep dull green, broadly concave, indistinctly striate, with oblong dots in broad transverse bands, the marking on the underside being more distinct, whiter and larger, margins narrow, cartilaginous, sinuate-dentate, T. triangular, whitish to brown, pointed, 2–4 mm long, (5–)7(–13) mm apart; **Infl.** branched, Fl. 27 mm long, red.

**A. latifolia** HAW. (§ IV/B/10/A/1a) (*A. saponaria* v. *l.* HAW.). - E.Cape. - Resembles **A. saponaria;** more robust in all parts; **L.** larger, greener, spots oblong and less numerous.

**A. lavranosii** REYN. (§ IV/C/16). - S.Arabia. - Acauline or St. prostrate, to 15 cm long, usually in groups of 5–8 **Ros.; L.** 10–14, deltoid-pointed, 28 cm long, 7.5 cm across base, 1.5 cm thick, very rigid, olive-green on upperside with brownish sheen, broadly concave above, margins horny, continuous, reddish-brown, T. sharp, 3 mm long, 6–12 mm apart, L. tip pointed, with 2–4 sharp T.; **Infl.** 60 cm tall, Fl. 3 cm long, light yellow.

**A. leachii** REYN. (§ IV/D/23). - Tanzania: E. provinces. - Acauline or short-stemmed, solitary or forming groups by means of offsets; **L.** about 20, rosulate, to 35 cm long, 6 cm across base, narrowing regularly towards the tip and slightly concave, dark green with a reddish tinge, convex on underside, marginal T. sharp, reddish, triangular, c. 5 mm long, 1–2 cm apart; **Infl.** to 1 m tall, Fl. 3 cm long, scarlet, slightly striate.

**A. leandrii** J. BOSSER (§ XI/d/β 2). - Central Madag. - **Ros.** acauline to short-stemmed, solitary or offsetting from the base; **L.** 8–15, green, narrow-lanceolate to linear, 14–25 cm long, 1–2 cm across base, tip pointed to blunt, dentate, concave on upperside, rounded on underside, margins horny, narrow, dentate, T. triangular, 1–1.5 mm long, 3.5–8 mm apart; **Infl.** racemose, solitary, simple, Ped. 70–85 cm long, with 5–9 involucral L. above, their axils with 1 or numerous adventitious buds, Rac. short, open, Fl. 2.5–2.7 cm long and yellow, on pedicels 1.5–2 cm long.

**A. leptophylla** N. E. BR. ex BAK. v. **leptophylla** (§ IV/B/10/A/1b) (*A. cooperi* HORT.). - S.W.Cape. - **Ros.** almost acauline, solitary, densely leafy; **L.** linear-lanceolate, tapering, 50 cm long, 5 cm across, sinuous, vivid green to purple, with numerous lines, margins sinuate-dentate, T. 5–7 mm long, 8–10 mm apart; **Infl.** c. 60 cm tall, Fl. 3 cm long, pale yellow.

**A. — v. stenophylla** BAK. - **L.** only 4 cm across.

**A. leptosiphon** BGR. (§ IV/B/10x). - Tanzania: W.Usambara. - **L.** lanceolate-ensiform, tapering gradually from soil-level, 20–30 cm long, 3.5–4.5 cm across base, somewhat grooved, fleshy, with numerous, whitish, oblong marks, margins horny, T. 2–3 mm long; **Infl.** 35–45 cm tall, Fl. c. 26 mm long.

**A. lettyae** REYN. (§ IV/B/10/A/3b). - W.-Transvaal. - **Ros.** acauline, solitary; **L.** 20, lanceolate-tapering, to 45 cm long, 9 cm across base, slightly concave on upperside, dark green with numerous, white, oblong marks in wavy transverse bands, underside convex with larger marks, margins sinuate-dentate, T. triangular, brownish, 3–4 mm long, 1–1.5 cm apart; **Infl.** 1.75 m tall, Fl. c. 4 cm long, pink-red.

**A. leucantha** BGR. (§ IV/D/23). - Ethiopia. - **St.** about 1 m tall; **L.** lanceolate, fleshy; **Infl.** branching, Fl. 17 mm long, white. (Acc. REYNOLDS an insufficiently known species.)

**A. linearifolia** BGR. (§ III/3). - Natal. - **Root** thick-fleshy, cylindrical; acauline or short-stemmed, with 1–2 Br.; **Ros.** of 6–8 L. arranged distichously, 25 cm long, 5–8 mm across, clasping the St., somewhat twisted, convex on underside and often carinate, with white and brown spots, margins with fine T. towards base; **Infl.** 20–25 cm tall, Fl. 12 mm long, yellow-green.

**A. lineata** (AIT.) HAW. v. **lineata** (§ IV/B/8 acc. REYNOLDS; § IV/E/27 acc. BERGER) (*A. perfoliata* v. *l.* AIT., *A. dorsalla* HAW.). - E.Cape. - **St.** to 70 cm tall, the upper part covered with dead L.; **L.** in a dense Ros., broadly ovate-lanceolate, tapering, 40–50 cm long, 8–10 cm across base, convex on the underside, light green or light blue, both surfaces lined, margins with brownish T.; **Infl.** solitary, Fl. pendulous, red.

**A. — v. muirii** (MARL.) REYN. (*A. m.* MARL.). - **L.** light yellow-green or somewhat orange-green, with distinct longitudinal lines, T. larger.

**A. littoralis** BAK. (§ IV/C/15) (*A. rubrolutea* SCHINZ, *A. schinzii* BAK.). - Angola to S.W.Afr. - ♄ 2–3 m tall, principal St. stout, erect; **L.** densely rosulate, ensiform, tapering, 60–90 cm long, 6–7 cm across, margins with triangular, horny T. 3–4 mm long; **Infl.** 1.2–1.5 m tall, Fl. 2.5 cm long, coral-red.

**A. lomatophylloides** BALF. f. (§ IV/C/19b). - Rodriguez Is. - **Ros.** acauline with only few L. which are ensiform, 45–60 cm long, 7–8 cm across base, marginal T. triangular, pale, 1–2 mm long, 1–2 cm apart; **Infl.** c. 60 cm tall, Fl. 16 to 18 mm long, red.

**A. longibracteata** POLE EVANS (§ IV/B/10/A/4). - Transvaal. - **Ros.** solitary; **L.** 20–26, triangular-lanceolate, 10–15 cm long, 9–10 cm across, very thick and fleshy, extended by a dried, red tip 10 cm long, upperside flat-concave, dark green, slightly reddish, with numerous oblong, pale green to white, distinct or very indistinct spots in ± wavy transverse bands, margins sinuate-dentate, red-brown, horny, T. sharp, triangular, 5–7 mm long, 9–12 mm apart, brown to red-brown; **Infl.** 80–100 cm tall, Fl. 4–4.5 cm long, a delicate red.

**A. longiflora** BAK. (§ IV/E/29). – S.Afr. – **St.** simple or with some Br. from the base, c. 1.5 m tall; **L.** recurved, ensiform, 40–45 cm long, 4–5 cm across, green, flat on upperside, underside convex, marginal T. triangular, rather sharp, 3–4 mm long, 7–8 mm apart; **Infl.** c. 50 cm tall, Fl. 3.5–4 cm long, yellow.

**A. longistyla** BAK. (Pl. 16/2) (§ IV/A/2). – S. & Central Cape. – **Ros.** acauline, 18–23 cm ⌀; **L.** sublanceolate, 10–15 cm long, 2–2.5 cm across, light green with distinct longitudinal stripes, rounded on underside, with irregularly placed tubercles which are often dentate, upperside flat, with only few tubercles; **Infl.** 20 cm long, Fl. 4–5 cm long, salmon-red.

**A. luapulana** LEACH. (§ IV/C/18d). – Zambia: Luapula D. – Plants acauline, solitary; **L.** about 16, rosulate, spreading, ovate-attenuate, ± 30 cm long, 6–7.5 cm wide low down, usually with the apical two-thirds or more dry and twisted, upper surface pale green, obscurely striate, slightly concave low down, lower surface grey–green, more clearly striate, both surfaces tinged purplish, margins narrowly cartilaginous, whitish, T. ± deltate, pungent, 1–3.5 mm long, 2–5 mm apart, brownish–orange; **Infl.** 110 cm high, branched, Rac. 16–26 cm long, Fl. c. 2 cm long, coral-red.

**A.** × **luteobrunnea** BGR. (*A. thraskii* DE WILD.). – Hybrid: **A. ferox** × ? **A. principis.** – ♄, **St.** simple, stout; **L.** fleshy, gradually tapering, 60 cm long, 8 cm across, blue, broadly concave on upperside, margins with a thin, red line and curved, brown T. 2 mm long and 1.5–2 cm apart; **Infl.** branching, Fl. 3.5 cm long, golden yellow.

**A. lutescens** GROENEW. (§ IV/C/17). – Transvaal. – **St.** 5–30 cm long, merging into the root and there offsetting; **Ros.** dense; **L.** about 36, rigid, yellowish–green, 40–45 cm long, 8 cm across, convex on underside, margins with T. 2 mm long and 3–5 mm apart; **Infl.** 1–4, Fl. 3 cm long, dark yellow.

**A. macleayi** REYN. (§ VI/C/15). – Sudan. – **Ros.** simple, acauline; **L.** c. 24, tapering-lanceolate, up to 50 cm long, 12 cm across base, deep green to olive-green on underside, indistinctly lined, deeply grooved towards tip, rounded on underside, margins yellowish-white with stout, white, hooked, triangular T. 4 mm long and 8–15 mm apart; **Infl.** 90 cm tall, Fl. 26 mm long, scarlet, with orange-yellowish tips.

**A. macracantha** BAK. (§ IV/B/10/A/2a). – Natal, Cape. – **Ros.** acauline, with underground offsets; **L.** c. 20, broad-lanceolate and tapering, 35–40 cm long, 8 cm across, convex on underside, green, brownish, upperside with whitish spots, underside lighter, with indistinct longitudinal lines, margins with sharp, long T.; **Fl.** yellow.

**A. macrocarpa** TOD. v. **macrocarpa** (§ IV/B/10/A/3a) (*A. edulis* A. CHEV., *A. barteri* SCHNELL). – Ethiopia. – **Ros.** acauline or very short-stemmed, single or forming groups by means of offsets; **L.** 16–20, lanceolate to tapering-lanceolate (20–)30(–40) cm long, 6 to 7 cm across, flat to slightly grooved on upperside, green, all surfaces with numerous, dirty white or pale green oval spots and marks scattered irregularly or in wavy transverse rows, underside convex, with markings as for upperside, margins sinuate-dentate with a horny edge, T. pale brown, sharp, triangular, 3 mm long, 8–10 mm apart; **Infl.** 80–100 cm tall, Fl. 2.5 cm long, reddish-scarlet.

**A.** — v. **major** BGR. (*A. commutata* ENGL.). – Ethiopia; W. & E.Afr. – **Fl.** 2.5–3.5 cm long.

**A. macroclada** BAK. – (§ IV/E/29; XI/d/β 1).– Central Madag. – **Ros.** acauline; **L.** 20–50, 60–100 cm long, 17–22 cm across the base, 1–2 cm thick, with a short Sp. at the tip, marginal T. triangular, 3 mm long, 1 cm apart, reddish or yellowish; **Infl.** 1.2–1.8 m tall, Fl. 2 cm long, yellowish-red.

**A. macrosiphon** BAK. (§ IV/D/24) (*A. mwanzana* CHRISTIAN). – Tanzania; Uganda; Ruanda. – **Ros.** acauline, offsetting to form large, dense groups; **L.** c. 10, to 60 cm long, 8 cm across the base, narrowing to the pointed tip, usually almost erect, upperside slightly concave in the upper part, green to reddish-brown, with numerous dull white, lentiform markings 10 to 15 mm long and 2 mm across, underside convex with pale greenish markings, margins with sharp T. 4–5 mm long and 1–1.5 cm apart; **Infl.** 1–1.5 m tall, Fl. to 33 mm long, light pink with pale yellow mouth.

**A. madecassa** PERR. v. **madecassa** (§ XI/d/x). – Central Madag.; Tananarive. – **Ros.** acauline or very short-stemmed, dense; **L.** 15–20, 18 to 30 cm long, 5.5–7.5 cm across base, tapering, margins usually with triangular, pale pink to greenish T. 2 mm lang and 3–8 mm apart; **Infl.** up to 1 m tall, Fl. scarlet.

**A.** — v. **lutea** GUILL. – **L.** with pink T.; perianth tube lemon–yellow. Dubious variety.

**A. marlothii** BGR. v. **marlothii** (§ VI/6) (*A. supralaevis* v. *β hanburyi* BAK.). – Botswana; Transvaal, Natal. – **St.** simple, 2–4 m tall, thickly covered with dry, old L.; **L.** 40–50 in a dense Ros., broadly lanceolate, gradually tapering, usually concave on the upperside, rounded and distinctly carinate on the underside, 50–60 cm long, 8–10 cm across at the base, pale blue or greenish, margins and both surfaces densely and stoutly spinous, Sp. 3–4 mm long and sharp; **Infl.** much branched, Fl. orange to orange-yellow.

**A.** — v. **bicolor** REYN. – **L.** less strongly spined.

**A. marsabitensis** VERDOORN et CHRISTIAN (§ IV/C/19d). – Kenya: N.E.Prov. – **Ros.** single, or forming small groups; **St.** short or prostrate,

50–100 cm long, covered with dry, old L.; **L.** c. 20, broadly lanceolate-tapering, to 80 cm long, 16–18 cm across, 2 cm thick at the base, deeply grooved on the upperside, grey-green with a reddish sheen, underside rounded, grey-green, margins with pink-brown, triangular, sharp T. 2–3 mm long and 1.5–2 cm apart, the T. in the lower half connected by a pink-brown, horny border; **Infl.** 1–1.2 m tall, Fl. c. 28 mm long, 10–12 mm ⌀, coral-red.

**A. massawana** REYN. (§ IV/C/16). – Ethiopia. – **Ros.** acauline or short-stemmed, with numerous side-shoots, forming groups 2–3 m ⌀; **L.** about 16, c. 50 cm long, 10 cm across at the base and 2 cm thick, gradually narrowing towards the tip, slightly grooved on the upperside towards the tip, dull grey-green, with or without several dull white spots towards the tip, convex on the underside, with colour and markings as for the upperside, margins with soft or stiff T. which are usually white although those at the tip are reddish-brown, T. small and 1 cm apart; **Infl.** 1.2–1.5 m tall, Fl. about 3 cm long, pale scarlet.

**A. mawii** CHRISTIAN (§ VI/1). – Malawi: Zomba Plateau. – Tree-like, 1.3–2.6 m tall; **St.** robust, branching above, 8–12 cm ⌀; **L.** densely rosulate, ensiform, broadly concave or grooved, the underside striate and convex, margins sharp with a narrow, reddish border, T. green-pink, pointed, 3 mm long, 16 mm apart; **Infl.** up to 1 m tall, Fl. 3.5–4 cm long, red.

**A. mayottensis** BGR. (§ IV/C/19b). – Comoro: Mayotte. – Habit not known; **L.** linear-lanceolate, tapering, c. 25 cm long, 3.5 cm across, margins with triangular, cartilaginous, sharp T. 1–2 mm long and 5 mm apart, on a narrow cartilaginous border; **Infl.** 40 cm tall, Fl. 27 mm long.

**A. mcloughlinii** CHRISTIAN (§ IV/C/19a). – Ethiopia: Diredawa. – **Ros.** acauline; **L.** 14–20, ovate-tapering, narrowing gradually to the tip, to 45 cm long, 7–8 cm across at the base, green on the upperside with narrow cream-coloured lentiform markings, margins thin, sharp, narrow, horny, sinuate-dentate, T. brown-tipped, 3–4 mm long, 1 cm apart; **Infl.** 60–100 cm tall, Fl. 2 cm long, strawberry-coloured.

**A. medishiana** REYN. et BALLY (§ IV/C/18f). – Somalia. – **St.** simple, often becoming bushy by branching from the base, smooth, to 2 m tall, 3–3.5 cm ⌀; **L.** 24, crowded on the final 20 cm of the St., ensiform, to 30 cm long, 5.5 cm across at the base, gradually narrowing to the tip, grey-green on the upperside, flat to slightly convex, rounded on the underside, margins continuous, narrow, white, horny, T. stout, white, 1 mm long, c. 5 mm apart, more widely spaced or missing towards the L. tip; **Infl.** c. 50 cm tall, Fl. 19 mm long, dull scarlet.

**A. megalacantha** BAK. (§ IV/C/19d (*A. magnidentata* VERDOORN et CHRISTIAN). – Ethiopia: Harrar. – Habit not known; **L.** ensiform, 45 (?) cm long, 8–10 cm across at the base, gradually tapering, fleshy, margins with T. 1 cm long and 2–3 cm apart; **Fl.** 28 mm long.

**A. melanacantha** BGR. (Pl. 16/1) (§ IV/B/4 acc. REYNOLDS; § IV/B/13 acc. BERGER). – Cape: Lit. Namaqualand. – Short-stemmed or with prostrate **St.** in mature plants, to 50 cm and more in length, densely covered with dry old L., occasionally single, usually grouping; **Ros.** spherical, to 30 cm ⌀; **L.** triangular-lanceolate, to 20 cm long, 4 cm across at the base, terminating in a long black point, concave on upperside towards the tip, deep green to brown-green, underside slightly keeled above, this keel having 5 black Sp., margins with sharp, horny, triangular, well-spaced, small white and large black T. 1–1.5 cm apart; **Infl.** to 1 m tall, Fl. c. 4.5 cm long, scarlet to yellow.

**A. menachensis** (SCHWEINF.) BLATTER (§ IV/C/16) (*A. percrassa* v. *m.* SCHWEINF., *A. trichosantha* v. *m.* SCHWEINF.). – S.Arabia: W.Yemen. – **Ros.** with St. 50 cm tall; **L.** 40 cm long, 16 cm across, triangular-lanceolate, green-purple, margins red; **Fl.** c. 3 cm long, yellow to scarlet, with very short felty hairs.

**A. mendesii** REYN. (§ IV/D/24). – Angola: Huila D. – Plant pendulous; **St.** to 1 m long, 4 cm ⌀; **L.** about 10, pendulous, ensiform, 50 cm long, 7–8 cm across, green on the upperside, distinctly lined, flat to slightly grooved, margins horny, T. stout, horny, blunt, 1–2 mm long, irregularly 1–1.5 cm apart; **Infl.** pendulous, Br. ascending, c. 60 cm long, Fl. 2.5 cm long, scarlet.

**A. menyharthii** BAK. (§ IV/B/10x). – Moçambique: Zambezi reg. – **L.** short (?), lanceolate, tapering, 5 cm (?) across, margins horny with triangular, brown T. 3 mm long; **Fl.** 25–28 mm long, pale red. Insufficiently known species. (Acc. REYNOLDS may be identical with **A. swynnertonii**.)

**A. metallica** ENGL. et GILG (§ IV/C/16). – Angola: Bié D. – Plant c. 40 cm tall; **L.** linear-lanceolate, 37 cm long, 6.5 cm across, glossy, metallic green, marginal T. curved, narrowly triangular, 2–3 mm long, 12–15 mm apart, brown-tipped; **Infl.** 1.2 m tall, Fl. 3 cm long, pale red.

**A. microcantha** HAW. (§ III/4) (*A. micracantha* LK. et OTTO). – Transvaal; Cape: Port Elizabeth D. – Plant usually solitary; **Root** thick, fusiform; acauline or very short-stemmed, simple or with 1–2 Br.; **L.** 12–18, multifarious, rigid, to 50 cm long, 3 cm across at the base, gradually tapering, concave on the upperside, almost plicate, both surfaces deep green to yellow-green, with white, somewhat tuberculate-pointed spots, margins slightly involute, white, horny, with T. 2 mm long and 1–3 mm apart; **Infl.** 45–50 cm tall, Fl. 38 mm long, salmon-pink.

**A. microdonta** CHIOV. (§ IV/C/15). – Central & S.Somalia. – **Ros.** large; **L.** deeply grooved, 50–70 cm long, 9–11 cm across; **Infl.** 1.3 m tall, Fl. 23 mm long, scarlet.

**A. microstigma** SALM (§ IV/E/27a) (*A. arabica* SALM, *A. juttae* DTR., *A. brunnthaleri* BGR. ex CAMERLOHER). – E.Cape; S.W.Afr. – **Ros.** acauline or short-stemmed; **St.** simple or branched, usually prostrate, covered with remains of old L.; **L.** broadly lanceolate to gradually tapering, 30 cm long, 6.5 cm across at the base, light green with a pink sheen, both surfaces with whitish, rounded to H-shaped markings, margins horny with small brown T. 2–4 mm long and 5–10 mm apart; **Infl.** 60 to 80 cm tall, Fl. 2.5–3 cm long, initially orange, becoming greenish-yellow.

**A. millotii** REYN. (§ IV/D/21; XI/b). – Madag.: Cape Ste. Marie. – Small ♁ freely branching from the base and forming lax clumps; **St.** prostrate, 20–25 cm long, 7–9 mm ∅ somewhat thicker above and with 10–20 Br., the top 5–7 cm being laxly leafy; **L.** 8–10, distichous on young St., later more rosulate, sheathlike at the base, 8–10 cm long, 7–9 mm across at the base, gradually tapering, tip blunt-rounded with some 5 white, horny Sp. 0.5–1 mm long, grey-green on the upperside, often reddish, grooved, with or without whitish marks towards the base, the marks on the underside being more numerous and often prickly, margins with T. 0.5–1 mm long and 5–10 mm apart; **Infl.** 12–15 cm long, Fl. 22 mm long, scarlet.

**A. milne-redheadii** CHRISTIAN (§ IV/C/19). – Angola: Moxico. – **Ros.** acauline, offsetting, dense; **L.** 16–18, ovate-lanceolate, tapering, 22 cm long, 7 cm across, 1 cm thick midway, concave on the underside, biconvex towards the margins, milky-green on the upperside with dark lines and scattered pale green, elliptical marks in irregular transverse bands, convex on the underside, paler green, the markings more distinct, margins sharp, horny, dentate, T. triangular, white, brown-tipped, somewhat hooked, 3 mm long, up to 12 mm apart; **Fl.** 3.5 cm long, light red.

**A. minima** BAK. v. **minima** (§ II) (*Leptaloe m.* (BAK.) STAPF). – Natal. – Acauline; **Root** fusiform; **L.** 6–10, rosulate, linear, somewhat stiff, 25–30 cm long, 4–6 mm across, wider at the base and clasping the St., grooved on the upperside, convex on the underside, with a slight, rounded keel, indistinctly spotted below, margins with hairlike, whitish T. below; **Infl.** 20–50 cm tall, Fl. 11 mm long, deep pink.

**A. — v. blyderivierensis** (GROENEW.) REYN. (*Leptaloe b.* GROENEW.). – Transvaal. – **L.** usually 3–4, distichous, margins and underside spinous; **Fl.** 12 mm long, initially light green, becoming dark pink.

**A. mitis** BGR. (§ IV/C/16). – ? Somalia. – **Ros.** large, acauline or very short-stemmed, forming a mat; **L.** 50 cm long, 5 cm across at the base, gradually tapering, upperside compressed-grooved, underside convex, green, with oblong spots arranged longitudinally, margins sinuate-dentate, T. 3–4 mm long, cartilaginous; **Infl.** c. 90 cm tall, Fl. 22 mm long.

**A. mitriformis** MILL. v. **mitriformis** (Pl. 16/4) (§ IV/D/25) (*A. m.* v. *elatior* HAW., *A. perfoliata* v. *mitraeformis* AIT., *A. parvispina* SCHOENL. p. part., *A. xanthacantha* SALM). – Cape: Cape, Ladismith & Montagu D. – **St.** to 1–2 m tall, 4–6 cm ∅, creeping, the upper part bearing L. and ascending, offsetting from the base to form dense groups of Ros.; **L.** ovate-lanceolate, very fleshy, slightly concave on the upperside towards the tip, blue-green to green, the upper part of the underside slightly carinate, the keel being dentate, margins with triangular, pale to whitish or yellow, sharp T. 4–6 mm long, 1–1.5 cm apart; **Infl.** 40–60 cm tall, Fl. 4–4.5 cm long, scarlet. – The under-named varieties are known only from the literature.

**A. — v. albispina** (HAW.) BGR. (*A. a.* HAW.). – Marginal T. 8–10 mm long, white.

**A. — v. commelinii** (WILLD.) BAK. (*A. c.* WILLD., *A. m.* v. *humilior* HAW.). – Plant smaller; **L.** bluish, T. smaller.

**A. — v. flavispina** (HAW.) BAK. (*A. f.* HAW.). – Marginal T. paler.

**A. — v. pachyphylla** BAK. – **L.** thick, spreading, very reddened, flat on the upperside, the underside with 2 reduced Sp. towards the tip, marginal T. small.

**A. — v. spinulosa** (S. D.) BAK. (*A. s.* S. D.). – L. more spinous on the upperside.

**A. — v. xanthacantha** (WILLD.) BAK. (*A. x.* WILLD., *A. mitraeformis* S. D.). – Marginal T. smaller.

**A. modesta** REYN. (§ III/3). – Transvaal. – Small, acauline, unbranched; **Root** thick, fleshy; **L.** forming a basal bulb 15 mm long and 20 mm ∅, L. in a Ros. of 4–6, 15–20 cm long, 8–9 mm across at the base, linear in the lower half, then narrowing to a point, slightly concave on the upperside, dull deep green, rounded on the underside, with some white markings towards the base, margins narrow, horny; **Infl.** 25–30 cm tall, Fl. 13 mm long, yellow-green.

**A. monotropa** VERDOORN (§ IV/B/10/A/5c). – Transvaal. – **Ros.** short-stemmed; **St.** creeping, c. 30 cm long, often elongated; **L.** c. 20, ovate-lanceolate, gradually tapering, c. 33 cm long, 6 cm across and 5 mm thick near the base, upperside concave towards the tip, fleshy, lined and with irregularly arranged, oblong or H-shaped marks, convex on the underside, distinctly lined and marked, these marks sometimes merging to form transverse bands or pale green zones, marginal T. 2 mm long and brown-tipped; **Infl.** c. 80 cm tall, Fl. c. 3 cm long, red-brown.

**A. monteiri** BAK. (§ IV/D/22). – S.Moçambique: Nova Sofala. – **Ros.** short-stemmed; **L.** c. 12, ensiform, gradually tapering, 30 cm long, 2.5 cm across, upperside concave, indistinctly spotted, marginal T. triangular-tapering; **Infl.** 60 cm tall, Fl. 2.5 cm long, red. Insufficiently known species.

**A. monticola** REYN. (§ IV/C/18c). – Ethiopia: Tigre. – **Ros.** solitary, acauline or with short St., dense; **L.** c. 24, lanceolate-tapering, 60–70 cm long, 14–16 mm across at the base, olive-green on the upperside, slightly grooved above, convex on the underside, margins with very prominent, brown border, T. pale brown, triangular, sharp, 6 mm long, smaller towards the tip, 1–1.5 cm apart; **Infl.** 1 m tall, Fl. c. 38 mm long, usually yellow, often light scarlet.

**A. morogoroensis** CHRISTIAN (? § IV/B/10/A/3b). – Tanzania. – **Ros.** acauline, offsetting; **L.** c. 20, lanceolate-tapering, 29 cm long, 6 cm across, flat to slightly convex on the upperside, grooved towards the tip, green, underside with darker stripes, margins narrow, white, horny, T. triangular, 4 mm long, 12–14 mm apart; **Infl.** c. 55 cm tall, Fl. 3.5 cm long, brown-pink.

**A. mubendiensis** CHRISTIAN (§ IV/C/19c). – Uganda: Buganda. – Offsetting very freely to form quite large, dense groups; **Ros.** acauline or short-stemmed, dense; **L.** 16, 30–35 cm long, 6.5 cm across, narrow-lanceolate, upperside grooved towards the tip, dull grey-green, usually with scattered whitish spots, and indistinctly lined, the underside usually unspotted, margins sinuate-dentate, somewhat pink, horny, T. red-brown on a whitish base, 3–4 mm long, 1–1.5 cm apart; **Infl.** 70–90 cm tall, Fl. c. 3 cm long, dark brick-red.

**A. mudenensis** REYN. (§ IV/B/10/A/2a). – Natal. – **Ros.** solitary or in small groups; **St.** to 80 cm tall, 10 cm ⌀, erect or prostrate; **L.** about 20, dense, 25–30 cm long, 8–9 cm across, flat or somewhat concave on the upperside, blue-green with numerous white, oblong spots irregularly arranged, often lined, underside paler, lined, unmottled or only indistinctly spotted, with these spots in transverse bands, margins sinuate-dentate, T. 7 mm long, 1–2 cm apart, triangular, brown, often curved; **Infl.** to 1 m tall, Fl. 3.5 cm long, salmon-orange.

**A. munchii** H. B. CHRISTIAN (§ IV/E/28 or 29). – Rhodesia. – Treelike, to 6 m tall; **St.** simple, 4.5–7 cm ⌀; **L.** in a terminal Ros., 45–60 cm long, 6–10 cm across at the base, thin and flexible, narrowing regularly towards the tip, the upperside very grooved towards the tip, pale grey-green, rounded on the underside, margins coral-red, horny, T. red, almost triangular, 1 mm long, 11 mm apart midway; **Infl.** to 60 cm tall, Fl. 4.5 cm long, scarlet to golden-yellow.

**A. musapana** REYN. (§ III/2). – Rhodesia: Melsetter D. – Usually pendulous; **St.** several, 20–30 cm long, 1 cm thick, branching freely either from the base or higher to form dense groups; **L.** c. 10, distichous, clasping the St. basally, linear, narrowing evenly towards the tip, 30–40 cm long, 1.5 cm across at the base, dark green on the upperside, concave, with or without spots, convex on the underside, with several dirty white spots below, margins with whitish, horny, minute T. towards the base; **Infl.** 30–40 cm tall, Fl. usually pendulous, about 3 cm long, scarlet with a pale green mouth.

**A. mutabilis** PILLANS (§ IV/E/18a). – Transvaal. – Usually hanging down from cliffs; **St.** short to more than 1 m long, 10–15 cm ⌀, simple or branched; **Ros.** dense, 60–70 cm ⌀; **L.** 8–9 cm across at the base, gradually tapering, very fleshy, bluish-green on the upperside, indistinctly lined, almost flat, convex on the underside, margins thin, brownish-yellow, pale, T. pale yellow to orange-yellow, 2 mm long, 1.5–2.5 cm apart; **Infl.** 2–3, Fl. 3–3.5 cm long, greenish-yellow to yellow.

**A. mutans** REYN. (§ IV/B/10/A/4). – Transvaal. – **Ros.** acauline or very short-stemmed, grouping; **L.** 10–16, about 10 cm long, with a dried tip 5 cm long, 6 cm across at the base, broadly and shortly lanceolate, the upperside grooved towards the tip, with numerous white, oblong spots in ± irregular transverse bands, the underside convex, indistinctly spotted, margins horny, sinuate-dentate, T. brown, 3 to 4 mm long, 8–10 mm apart, stiff, sharp, curved; **Infl.** 60–80 cm tall, Fl. c. 3 cm long, pink to orange-yellow.

**A. myriacantha** (HAW.) ROEM. et SCHULT. (§ II) (*Bowiea m.* HAW., *Leptaloe* (HAW.) STAPF, *A. johnstonii* BAK.). – Cape: Albany D., Zululand; Tanzania; Kenya. – Acauline; **root** fusiform; **L.** 8–10, rosulate, almost linear, 25 cm long, 8–10 mm across, clasping the St., grooved on the upper surface, dark green, with several white spots below, convex on the underside, indistinctly spotted at the base, these spots being tuberculate-pointed, margins with small white T.; **Infl.** 20–25 cm tall, Fl. 2 cm long, red.

**A. mzimbana** CHRISTIAN (§ IV/C/19a). – Malawi. – **Ros.** acauline; **L.** elliptical, short-tapering, 24 cm long, 7 cm across, flat to concave on the upperside, grooved towards the tip, striate, paler on the underside, margins pink, horny, T. triangular, brown, 2 mm long, 8 to 10 mm apart; **Infl.** to 55 cm tall, Fl. 3.5 cm long, coral-red.

**A. namibensis** GIESS (§ IV/B/13). – S.W.Afr.: Namib, S. and N. of Swakopmund. – Usually solitary or with divided heads forming up to 2 Ros., stemless or with a prostrate St., with dried L. persistent; **L.** lanceolate, up to 40, up to 50 cm long and 7 cm broad at the base and slightly rough to touch, upper and lower surfaces glaucous, margins with small, evenly spaced, deltoid, brown T.; **Infl.** up to 95 cm high with 2–4 arcuate-ascending Br., Rac. densely flowered, Fl. cylindric-ventricose, up to 35 mm long, glossy scarlet-red to pale coral-red, stamens yellow.

**A. ngobitensis** REYN. (§ IV/D/21). – Kenya. – ♄ to 2 m tall and across; **St.** 2 m tall, 4–5 cm thick, branching from the base, laxly leafy over the uppermost 30–50 cm, old L. dry and persistent; **L.** Sh.-like and 5 cm across at the base, gradually tapering, 35 cm long, flat to slightly concave on the upperside, rounded on the underside, both surfaces grey-green, margins and T. grey-green, T. pale brown at the tip, 3–4 mm long, 8–12 mm apart, margins sinuate between the T.; **Infl.** 80 cm tall, Fl. 36 mm long, scarlet.

**A. niebuhriana** LAVR. (§ IV/C/16). – S.W. Arabia. – **Ros.** single or grouping, acauline or with short St., prostrate; **L.** 15–25, tapering-lanceolate, narrowing to an apiculate tip, to 45 cm long, 10 cm across at the base, 5–8 mm thick, upperside grey-green with a purple flush, broadly grooved, convex on the underside, margins with dark brown, blunt or pointed T. 1.5–2 mm long, 12–15 mm apart; **Infl.** 50 to 100 cm tall, Fl. pendulous, 28–30 mm long, scarlet, less often greenish-yellow, the lobes usually being papillose-felted.

**A. nobilis** HAW. (§ IV/D/25) (*A. mitriformis* v. *spinosior* HAW.). – Cape. – **St.** short, with scaly L., eventually creeping; **Ros.** 25 cm ⌀;

L. lanceolate-triangular, 5–5.5 cm across at the base, 12–15 cm long, fleshy, 1.5 cm thick, flat on the upperside, convex on the underside and keeled above, with several distinct lines and small spots, margins with horny white twisted T. 5 mm long and c. 8 mm apart; **Infl.** 80 cm tall, Fl. 4 cm long, reddish.

**A. nubigena** GROENEW. (§ III/2). – E.Transvaal. – Grouping; **St.** 15 cm and more in length, 1 cm ∅; **Ros.** of 10 L., arranged distichously; **L.** 30 cm long, 1.5 cm across, 1 mm thick, the underside boat-shaped, light grey, marginal T. 0.5 mm long where present, 1–4 mm apart; **Infl.** 20–25 cm tall, Fl. 2.5 cm long, yellow-orange.

**A. nuttii** BAK. (§ III/4) (*A. brunneo-punctata* ENGL. et GILG., *A. corbisieri* DE WILD., *A. mketiensis* CHRISTIAN). – Tanzania; Malawi; Zambia; Congo (Kinshasa); Angola. – **St.** (1–)2–3(–12), very short or 10–20 cm long, 4–5 cm thick; **L.** 16–20, rosulate, 3–4 cm across at the base, narrowing abruptly just above the base to 15–20 mm, otherwise linear and forming a slender point, 40–50 cm long, concave on the upperside, sometimes indistinctly lined, with several greenish lentiform spots below, rounded on the underside, not always carinate, spotted towards the base, these spots often with a spiny tubercle, margins horny, up to 1 mm across, T. soft, 1 mm long, horny; **Infl.** 60–80 cm tall, Fl. 38–42 mm long, peach or strawberry-coloured or salmon-pink.

**A. nyeriensis** CHRISTIAN (§ IV/D/24). – Kenya. – Caulescent, offsetting; **St.** 2.3 m tall, 4 to 5 cm ∅, leafy; **L.** lanceolate, gradually tapering, pointed at the tip, green, immature L. white-spotted, 55 cm long, 9 cm across, convex at the base, margins sinuate-dentate, T. 2.5 mm long, 1 cm apart; **Infl.** c. 60 cm tall, Fl. 4 cm long, coral-red, lobes yellow with green markings.

**A. obscura** MILL. (§ IV/B/10/x) (*A. perfoliata* v. o. AIT., *A. saponaria* v. o. HAW., *A. picta* THUNBG., *A. p.* v. *major* WILLD., *A. maculosa* LAM.). – Cape; S.W.Afr. – **Ros.** short-stemmed, dense; **L.** lanceolate-triangular, 20 cm long, 7–8 cm across at the base, 1.4 cm thick, flat or somewhat concave on the upperside, green, striate, with a few oblong spots, margins sinuate-dentate, horny, T. triangular, reddish, 3 mm long and 4–6 mm apart; **Infl.** 20 cm tall, Fl. 3–3.5 cm long, red.

**A. officinalis** FORSK. v. **officinalis** (§ IV/16) (*A. vera* v. o. (FORSK.) BAK.). – S.Arabia; Yemen. – **Ros.** with short, prostrate St., normally with offsets and forming dense clumps; **L.** gradually tapering, 70 cm long, 12 cm across, 10–18 mm thick, rigid, very fleshy, upperside grooved towards the tip, grey-green, often with white spots, underside convex, sometimes with 8–9 T. along the mid-line, marginal T. compressed, close-set, somewhat hooked, almost triangular; **Infl.** 90–100 cm tall, Fl. about 3 cm long, red or yellow to orange-coloured.

**A.** — v. **angustifolia** (SCHWEINF.)LAVR. (*A. vera* v. *a.* SCHWEINF.). – **L.** 35 cm long, 2.4 cm across, 1.7 cm thick.

**A. ortholopha** CHRISTIAN et MILNE-READHEAD (§ VI/1). – Rhodesia. – **Ros.** acauline; **L.** 30 or more, lanceolate to ovate-lanceolate, to 55 cm long, 14 cm across the middle and 14 mm thick, tip pointed, upperside striate towards the tip, margins dentate, horny, T. triangular, 4 mm long, 7–15 mm apart, red-brown; **Infl.** 1 m tall, Fl. 42 mm long, pale yellow-red.

**A. otallensis** BAK. v. **otallensis** (§ IV/C/16). – Somalia; S.Ethiopia; Kenya. – **Ros.** ? acauline; **L.** triangular-lanceolate, 4–5 cm across at the base, somewhat tapering, about 30 cm long, rigid, fleshy, blue-green with oblong white spots arranged longitudinally in groups, margins with triangular brown-tipped T. about 2 mm long, 8–10 mm apart; Fl. 27 mm long, ? yellow-red.

**A.** — v. **elongata** BGR. – Somalia. – **L.** 6 cm across the base, gradually tapering, rough, margins sinuate-dentate.

**A. pachygaster** DTR. (§ IV/B/13). – S.W.Afr.: Gr. Namaqualand. – **Ros.** to 20 cm ∅, often 5 close together; **L.** 25–32, 12–16 cm long, about 2.5 cm across, 12–14 mm thick, both surfaces very convex, light grey-green, rough, marginal T. initially yellow, finally almost black, 2 mm long, 5 mm apart, underside with some T. below the tip; **Fl.** 3–3.5 cm long, red.

**A. × pallancae** GUILL. – Acc. A. GUILLAUMIN a hybrid raised by Pallanca of Bordighera, Italy in 1932: **A. ferox × A. spectabilis.** – Caulescent; **L.** lanceolate, 20 cm long, 5 cm across, both margins white-striped, the underside with several, regularly spaced brown Sp., marginal T. brown; **Infl.** tall, Fl. 2 cm long, yellow-orange.

**A. palmiformis** BAK. (§ IV/C/19c). – Angola. – ♄; **St.** 90–120 cm tall, simple or forked; **L.** in a dense Ros., thick, stiff, brittle, to 30 cm long, 42–48 mm across at the base, gradually tapering, margins sinuous, dentate-serrate, T. 1 to 1.5 mm apart, 4 mm long; **Fl.** 2.5 cm long, coral-red.

**A. × panormitana** GUILL. (§ IV/B). – Natural hybrid: ? **A. brevifolia** v. **depressa × A. spec.** – **St.** 10 cm tall; **L.** c. 15, rosulate, triangular-lanceolate, 20–25 cm long, 6–8 cm across, flat on the upperside with a few Sp., the underside convex with a green, slightly mottled surface, marginal T. pointed, white, 2 mm long, c. 8 mm apart; **Infl.** 65 cm tall, Fl. 2.5 cm long, pink, green-tipped.

**A. parallelifolia** PERR. (§ XI/c). – Central Madag. – **St.** thin, to 40 cm tall, branching basally to form groups; **L.** 2–7, in a lax Ros., with a long Sh. clasping the St., narrowly linear, very fleshy, convex on the upperside, often almost terete, to 15 cm long, to 1 cm across, olive-green, margins irregularly dentate, T. yellow, 1–1.5 mm long, 4–5 mm apart; **Infl.** to 40 cm tall, Fl. pendulous, pink-red at base and whitish towards the tips.

**A. parvibracteata** SCHOENL. v. **parvibracteata** (§ IV/B/10/A/5b) (*A. pongolensis* REYN., ? *A. affinis* POLE EVANS). – Moçambique; E.Transvaal, N.Zululand; Swaziland. – **Ros.** acauline or short-stemmed, offsetting to form clumps; **L.** 10–15, lanceolate, gradually tapering, 30 to

40 cm long, 6–8 cm across the base, somewhat concave on the upperside, green to brownish-green, with numerous oblong, white, scattered spots, often arranged in wavy transverse bands, the underside usually without spots, margins sinuate-dentate, T. triangular, sharp, brown, 3–5 mm long, 1–1.5 cm apart, curved; **Infl.** 1 to 1.5 m tall, Fl. to 33 mm long, dark glossy red.

**A. — v. zuluensis** (REYN.) REYN. (*A. pongolensis* v. *zuluensis* REYN.). – Cape, Natal; Zululand. – **St.** 30–40 cm tall, not grouping; **L.** spreading; **Fl.** 37–40 mm long, coral-red.

**A. parviflora** BAK. (§ II) (*Leptaloe p.* (BAK.) STAPF). – Natal. – **Roots** fleshy, thick, 8–10 cm long; **L.** c. 4, distichous, linear, drooping, 20–25 cm long, 6–8 mm across mid-way, narrowing to the base, with rounded tip, the underside distinctly nerved, tuberculate-spiny, margins ciliate; **Infl.** 40 cm tall, Fl. 8 mm long, pink.

**A. parvula** BGR. (Pl. 17/4) (§ I; XI/d/β 2) (*A. sempervivoides* PERR.). – Central Madag. – **Ros.** acauline, solitary, to 10 cm $\varnothing$; **L.** c. 24, dense, 4–6 cm long, 8 mm across at the base, narrowing evenly towards the apiculate tip, flat on the upperside, warty, all surfaces rather prickly with rough, bristle-like hairs 0.5–1 mm long, surfaces grey-green to vivid bluish-violet, margins with soft to stout narrowly triangular, white, horny T. 1–2 mm long and 1–2 mm apart, the central vein protruding like a keel on the L. underside, with only a single row of bristle-hairs; **Infl.** c. 35 cm tall, Fl. 24 mm long, light coral-red.

**A. pearsonii** SCHOENL. (Pl. 17/1) (§ IV/D/21). – Cape: Lit. Namaqualand; S.W.Afr. – **St.** 60–120 cm tall, 1.5 cm $\varnothing$; **Ros.** dense; **L.** 9 cm long, 3–5 cm across, 1 cm thick, clasping the stem, L.Sh. striped red, L. curved, striped red-green, marginal T. 1–5 mm long, 3 mm apart; **Infl.** 30–35 cm tall, Fl. 2 cm long, greenish-yellow or even red.

**A. peckii** BALLY et VERDOORN (§ IV/B/10/A/5a). – Somalia. – **Ros.** acauline, offsetting; **L.** 14–16, fleshy, 10–16 cm long, 5.2 cm across, 2.2 cm thick, tapering, both surfaces convex, the upperside somewhat grooved towards the tip, initially glossy green with pale markings, these ultimately becoming greener, margins with triangular, brownish T. 2 mm long and 3–7 mm apart; **Infl.** to 86 cm tall, Fl. 3 cm long, straw-coloured.

**A. peglerae** SCHOENL. (§ IV/A/2). – Transvaal. – **Ros.** acauline, dense; **L.** broad ovate-lanceolate, gradually tapering, c. 35 cm long, 5–6 cm across at the base, underside rounded and carinate, pale green-blue, the upper third of the underside having a few Sp., margins dentate; **Fl.** at first pink, then becoming pale lemon-yellow.

**A. pendens** FORSK. (§ IV/D/24) (*A. dependens* STEUD.). – S.Arabia: Yemen. – ♄, offsetting; **St.** usually pendulous, 30–40 cm long, 1–1.5 cm thick; **L.** with a basal Sh. 1–2 cm long, initially distichous, later arranged in a dense Ros., striped with white, gradually tapering, upperside swollen, underside convex, 1–2 cm thick, 30–40 cm long, 4–8 mm across, pale green, margins narrow, red, horny, T. 1 mm long, 7–8 mm apart; **Infl.** 80–90 cm tall, Fl. 22 mm long, red-yellow.

**A. penduliflora** BAK. (§ IV/C/18d). – Zanzibar. – **St.** with few Br., 2–3 m long, 3–4 cm $\varnothing$; **L.** ensiform, 30–40 cm long, concave on the upperside, pale green, convex on the underside, marginal T. c. 2 mm long, 2 cm apart; **Infl.** slender and pendulous, Fl. 27 mm long, pendulous, pale green-yellow.

**A. percrassa** TOD. (§ IV/C/15) (*A. abyssinica* v. *p.* BAK., *A. schimperi* SCHWEINF., *A. oligospila* BAK., *A. debrana* CHRISTIAN). – Ethiopia. **Ros.** acauline, stout, usually simple, dense; **L.** 30–40, oblong-triangular, gradually tapering, 60 cm long, 15–18 cm across, upper surface flat at the base, later concave, underside convex, blue-green with several spots, margins cartilaginous, with T 3.5 mm long; **Infl.** robust, Fl. 1.5–3 cm long, pale red.

**A. perrieri** REYN. (§ I; XI/c) (*A. parvula* PERR.). – Madag.: Fianarantsoa. – **Ros.** acauline, with cylindrical roots, offsetting basally, forming dense groups; **L.** c. 16, 10–13 cm long, 9–10 mm across at the base, 3 mm thick, narrowing gradually to a pointed tip, slightly grooved on the upperside, underside convex, both surfaces finely rough-tuberculate with papillose spots, dark green, both surfaces with many spots which are pale green, lentiform, c. 1 mm long and 0.5 mm across, margins with soft, horny, triangular T. 1 mm long, 1 mm apart, T. becoming smaller above and finally missing; **Infl.** to 60 cm tall, Fl. 13 mm long, light coral-red.

**A. perryi** BAK. (§ IV/C/19c) (*A. forbesii* BALF.). – Socotra. – **St.** simple, c. 30 cm tall, 5 cm $\varnothing$; **L.** 15–20, densely rosulate, lanceolate, 35 cm long, 5–7 cm across, gradually tapering, grooved on the upperside, 4–5 mm thick, blue-green, indistinctly lined, often reddish, margins narrow, horny, T. triangular, brown-tipped, 4 mm long, 6 mm apart; **Infl.** 50–60 cm tall, Fl. 2.5 cm long, red with green tips.

**A. petricola** POLE EVANS (§ VI/4a). – E.Transvaal. – **Ros.** acauline, solitary or in large groups; **L.** 20–30, tapering-lanceolate, c. 60 cm long, 10 cm across at the base, concave on the upperside above, bluish, glabrous or with a few scattered T., underside convex, also with several T. on the keel towards the tip; **Infl.** to 1 m tall, Fl. 28–30 mm long, greenish-white to pale orange-coloured.

**A. petrophila** PILLANS (§ IV/B/10/A/1a). – N.Transvaal. – **Ros.** acauline or with St. 5–6 cm long, simple or offsetting basally and thus forming small groups; **L.** 10–20, oblong-lanceolate, 20–25 cm long, 5–6 cm across, tapering, slightly concave on the upper part of the upperside, light green, with scattered whitish, oblong to H-shaped spots and greenish white lines, convex on the underside, bright green with dull white spots, margins sinuate-dentate, T. 3 to 5 mm long, triangular, sharp, dark brown, 8–12 mm apart; **Infl.** c. 50 cm tall, Fl. 28 mm long, dull scarlet.

**A. pillansii** L. GUTHRIE (Pl. 17/3) (§ VI). – Cape: Lit. Namaqualand. – Resembles **A. dichotoma, St.** thicker, 10 m and more in height, 1–2 m ⌀ at base, branching dichotomously in the upper half, Br. in turn dividing dichotomously; **L.** in a dense Ros. at the Br. apex, tapering-lanceolate, 50–60 cm long, 10–20 cm across at the base and c. 1.5 cm thick, glabrous, grey-green to brownish-green, upperside concave above, underside convex, margins white, dentate, T. curved, 1 mm long below and 2 mm long above, 5–8 mm apart; **Infl.** much branched, Fl. 3 cm long, thick, fleshy, lemon-yellow to yellow.

**A. pirottae** BGR. (§ IV/D/23). – Somalia. – Habit unknown; **L.** 50 cm long, 8 cm across, ± ensiform, gradually tapering, fleshy, with numerous spots in bands 2–3 cm long, margins with curved, triangular, brown T. 4 mm long and 1.5 cm apart; **Infl.** corymbose, Fl. 20–23 mm long.

**A. platylepis** BAK. (? § IV/E/29). – Related to **A. speciosa** and **A. ferox,** possibly a hybrid. – S.Afr. – **St.** 3–4 m tall, simple or forked, with a grey rind; **L.** in a dense Ros., broadly ensiform, 40–60 cm long, blue-green, marginal T. 4–6 mm long, 16 mm apart; **Infl.** stout, Fl. yellow to golden-yellow with green tips.

**A. plicatilis** (L.) MILL. (Pl. 17/2) (§ X) (*A. p.* v. *major* S. D., *A. tripetala* MEDIC., *A. lingua* THUNBG., *A. linguaeformis* L. f., *A. disticha* v. *p.* L., *A. d.* v. *p.* N. L. BURM.). – Cape: Cape, Wellington D. – ♄ or small ♄ 3–5 m tall; **St.** short, divided dichotomously; **L.** 12–16 on Br. ends, distichous, basally sheathed, broad-linear to strap-shaped, c. 30 cm long, 4 cm across, tip rounded, both surfaces dark green to bluish-green, flat or slightly convex, margins fine, horny, very finely dentate in the upper third; **Infl.** c. 50 cm long, Fl. 5.5 cm long, scarlet.

**A. plowesii** REYN. (§ III/1). – Rhodesia: Melsetter D. – Acauline; **Roots** fusiform; **L.** c. 10, rosulate, broader at the base, amplexicaul, narrow-linear, 20–30 cm long, 6–10 mm across, narrowing evenly towards the tip, the upperside dull green and grooved-concave, having several scattered spots in the lower part, the underside with numerous small, white marks, margins with white T. 0.5 mm long, these T. missing towards the tip; **Infl.** 30–45 cm tall, Fl. 3–3.5 cm long, scarlet, the tips being paler or green.

**A. pluridens** HAW. (§ IV/E/28 a) (*A. p.* v. *beckeri* SCHOENL., *A. atherstonei* BAK.). – Cape: Grahamstown D. – Almost treelike; **St.** simple or dichotomously branching, 2–3 m tall, 10 cm thick at the base; **L.** numerous, densely rosulate, ensiform, gradually tapering to an acuminate tip, gracefully recurved, 70–80 cm long, 6–7 cm across at the base, flat to convex on the upperside, the underside convex, both surfaces pale green, distinctly longitudinally striate, margins grooved, horny, T. numerous, 2–3 mm long, 5–10 mm apart; **Infl.** 80–90 cm tall, Fl. 3.5 cm long, salmon-pink to scarlet.

**A. pole-evansii** CHRISTIAN (§ IV/D/24). – Kenya. – ♄ to 1.5 m tall, offsetting basally; **L.** lanceolate, tapering, 26–34 cm long, clasping the St. basally to form a distinct cup, 5 cm across, green, occasionally with a few spots, slightly convex on the underside, the cup-like base distinctly nerved, margins dentate, T. brown, 2 mm long, 11–13 mm apart; **Infl.** to 60 cm tall, Fl. 32 mm long, dark red.

**A. polyphylla** SCHOENL. ex PILL. (Pl. 17/5) (§ IV/B/8). – Lesotho. – **St.** scarcely 10 cm tall; **Ros.** solitary, in pairs, or in dense groups of 12 or more, 50–60 cm ⌀, consisting of 75–100 **L.,** arranged in 5 spirally ascending rows, either left or right-handed, 20–30 cm long, 6–10 cm across, 15–18 mm thick at the base, ovate-lanceolate, tapering, slightly falcate-sinuous, with a strong red-brown tip, margins paler, cartilaginous, with up to 12 triangular, curved, pale T. 5–8 mm long in the upper half; **Infl.** 50–60 cm tall, Fl. 3–4 cm long, green with purple tips.

**A. pratensis** BAK. (§ IV/B/8). – Cape (Pondoland), Natal; Lesotho. – Related to **A. variegata; Ros.** acauline, seldom offsetting; **L.** 30–40, 10–15 cm long, 30–33 mm across, 6 mm thick, ovate-lanceolate, upperside flat, underside convex, bluish, striped, fleshy, stiff, margins with rough sharp hard T. 4–6 mm long; **Infl.** c. 60 cm tall, Fl. 3.5–4 cm long, deep pink.

**A. pretoriensis** POLE EVANS (§ IV/B/12). – Transvaal. – **St.** short or up to 1 m tall, 8 to 12 cm ⌀; **Ros.** of 30–60 light green, hard **L.,** the tip usually dried, red, with marginal T. 3–4 mm long, 10–17 mm apart; **Infl.** 2–2.5 m tall, Fl. 4–4.5 cm long, red with yellow-green tips.

**A.** × **principis** (HAW.) STEARN (*Pachydendron p.* HAW., *A. fulgens* TOD., *A. africana* SALM, *A. salmdyckiana* ROEM. et SCHULT.). – Cape. – Acc. REYNOLDS a hybrid: **A. arborescens** × **A. ferox.** – Freely branching from base; **St.** to 2 m tall, 15 cm ⌀, thickly covered with dry old L.; **L.** up to 60 in a dense Ros., gracefully curved, long tapering-ensiform, very fleshy, about 75 cm long, 8–9 cm across midway, blue to blue-green, underside carinate below the tip, margins with widely-spaced T. 4 mm long; **Infl.** 80–100 cm tall, Fl. 3.5–4 cm long.

**A. prinslooi** VERDOORN et HARDY (§ IV/B/10/B). – Natal: Estcourt D. – Acauline, simple or with several Ros. growing from one rootstock; **L.** 16–30, rosulate, fleshy, 14–20 cm long, 4–8.5 cm across, c. 1.7 cm thick, both surfaces convex or upperside flat to slightly concave, with irregularly placed white spots, either closely or widely spaced, usually in transverse bands; **Infl.** c. 60 cm tall, Fl. c. 17 mm long, pale to deep pink.

**A. pruinosa** REYN. (§ IV/B/10/A/3a). – Natal. – **St.** simple, 30 cm tall; **L.** 16–24 in a dense Ros., tapering-lanceolate, c. 70 cm long, 8–10 cm across at the base, concave on the upperside, green with numerous white, H-shaped spots often running together into indistinct transverse bands, underside convex, with more elliptical markings, margins sinuate-dentate, T. triangular, sharp, pale pink-brown, 4 mm long,

14–20 mm apart; **Infl.** to 2 m tall, Fl. 30–33 mm long, dark brown-red.

**A.** × **pseudomacroclada** REYN. – Hybrid: **A. macroclada** × **A. capitata** v. **quartziticola**. – **Infl.** branching, Fl. with very short pedicels.

**A.** × **pseudopicta** BGR. – Natural hybrid related to **A. obscura**. – Acauline; **L.** 30 in a dense Ros., lanceolate-triangular, tapering above the middle, 15–30 cm long, 6 cm across, 8–10 mm thick, fleshy, flat on the upperside, underside convex and keeled towards the tip, pale green, infrequently bluish or reddish, with dark lines, upperside with oblong white spots 3–4 mm long, underside more strongly marked, margins thin, horny, T. 3 mm long, 6–8 mm apart; **Infl.** 60 cm long, Fl. 4 cm long, red.

**A.** × **puberula** (SCHWEINF.) BGR. (§ IV/C/16) (*A. vera* v. *p.* SCHWEINF.). – Ethiopia. – Natural hybrid: **A. camperi** × **A. trichosantha** (acc. REYNOLDS). – L. c. 50 cm long, 10 cm across, 2 cm thick, narrowing to the base, green to reddish, upperside flat, margins with horny, triangular, incurved T. 4–5 mm long, 10 to 15 mm apart; **Infl.** 25–35 cm long, Fl. 3 cm long, yellow.

**A. pubescens** REYN. (§ IV/C/16). – Ethiopia: Arusi. – **Ros.** solitary or in groups, acauline or with **St.** 20–30 cm tall; **L.** c. 16, tapering-lanceolate to 45 cm long, slightly concave in the upper part, convex on the underside, grass-green, margins with triangular, sharp T., red-brown above, white below, 2–3 mm long and close-set; **Infl.** 70–100 cm tall, 42 mm long, coral-red, with short hairs.

**A. pungens** BGR. (§ IV/C/19 d). – ? Tanzania. – **St.** short, with few offsets; **L.** 20, triangular-lanceolate, gradually tapering, 50 cm long, 12 cm across at the base and 2.5 cm thick, green, glossy, margins sinuate-dentate, T. brown, sharp, 5–7 mm long, triangular, 10–12 mm apart, with curved tips; **Infl.** 1.2 m tall, Fl. 4 cm long, red.

**A. purpurascens** (AIT.) HAW. (§ IV/E/27 a) (*A. perfoliata* v. *p.* AIT., *A. sinuata* THUNBG., *A. ramosa* HAW., *A. succotrina* v. *p.* KER.). – Cape. – **St.** 50–80 cm tall; **Ros.** dense; **L.** numerous, 45–50 cm long, ensiform, 6–7 cm across at the base and 1.5 cm thick, marginal T. 4 mm long; **Infl.** 80–90 cm tall, Fl. 4 cm long, light red with green spots.

**A. rabaiensis** RENDLE (§ IV/C/19 b) (*A. ngongensis* CHRISTIAN). – Kenya; Tanzania; S.Somalia. – Forming small or large bushes, often several m $\varnothing$; **St.** c. 2 m long, 6–8 cm $\varnothing$, the upper half rather laxly leafy, the top 30 cm densely leafy, dry old L. persisting; **L.** 16–20, rosulate, sheathing basally, 40–60 cm long, 8–9 cm across at the base, narrowing to the tip, convex on the upperside above, olive-green to tinged with red, underside convex, margins sinuate-dentate, T. triangular, sharp, reddish-brown, (3–)4(–5) mm long, 1–1.5 cm apart; **Infl.** 60–90 cm tall, Fl. 33–35 mm long, dull scarlet to light scarlet.

**A. ramosissima** PILL. (Pl. 18/2) (§ VII). – Cape: Lit. Namaqualand; S.W.Afr.: N. of the Orange R. – Bushy, 2–3 m tall and across; **St.** freely and dichotomously branching from ground-level, about 8 cm thick, with a waxy powdery-grey surface, rebranching; **L.** 10–14 in a dense capitate Ros., lanceolate-linear, 15 to 20 cm long, 22 mm across base, 7 mm thick, flat to slightly concave on the upperside, bluish-green, convex on the underside, margins pale yellow, somewhat horny, T. 1 mm long, 1–4 mm apart, pale brown; **Infl.** 15–20 cm tall, Fl. green-yellow to canary yellow.

**A. rauhii** REYN. (§ XI/c). – Madag.: Tuléar. – **Ros.** acauline or very short-stemmed, c. 10 cm $\varnothing$, forming groups; **L.** up to 20, lanceolate-triangular, 7–10 cm long, 1.5–2 cm across base, gradually narrowing to the tip, concave on the upperside, grey-green, often with a brownish tinge, with numerous H-shaped spots, underside rounded but otherwise as for the upperside, margins white, horny, T. triangular, horny, white, about 5 mm long, 1–2 mm apart; **Infl.** to 30 cm tall, Fl. 2.5 cm long, pink-scarlet.

**A. reitzii** REYN. (§ VI/4a). – E.Transvaal. – Acauline, or **St.** 60 cm tall; **Ros.** dense; **L.** stiff, 65 cm long, 12 cm across, slightly concave on the upperside, convex on the underside with Sp. 2 mm long and 1–2 cm apart near the tip which is also spiny; **Infl.** 1–1.3 m tall, Fl. 5 cm long, scarlet.

**A. retrospiciens** REYN. et BALLY (§ IV/C/19 c). – Somalia. – ♄, often with a single St., but usually with 2–6 Br. from the base; **St.** 1 to 1.25 m tall, c. 3 cm $\varnothing$; **L.** c. 12 in a dense, terminal Ros., to 25 cm long, 5–6 cm across at the base, gradually narrowing to the blunt tip which has several stout white T., upperside of the L. somewhat concave above, bluish-grey with a reddish sheen, with several scattered, small, white spots below, margins continuous, white, horny, T. stout, white, 1 mm long, 5 mm apart, becoming smaller and finally missing towards the tip; **Infl.** 45 cm tall, Fl. 2 cm long, yellow with green tips.

**A. reynoldsii** LETTY (§ IV/B/11). – S.Afr.: Pondoland. – Related to **A. striata**; acauline or **St.** 15–20 cm tall; **Ros.** of 14–20 **L.** 25 cm long, 8–10 cm across, oval, coming to a point, flat on the upperside, convex on the underside, light green with dark stripes and scattered whitish spots; **Infl.** 50 cm tall, Fl. 2.5 cm across, yellow, somewhat orange.

**A. rhodesiana** RENDLE (§ III/4) (*A. eylesii* CHRISTIAN). – Rhodesia; Melsetter D.; Moçambique. – **Roots** fusiform; acauline or with a **St.** 10 cm tall, 3–4 cm thick, simple or with 2–3 Br. from ground-level; **L.** 8–10, densely rosulate, broader at the base, imbricate and 4–5 cm across, narrowing towards the tip, 25–30 cm long, slightly grooved on the upperside, dull green, underside convex, sometimes with several elliptical spots below, margins very narrow, horny, T. stout, 0.5–1 mm long, 1–4 mm apart, becoming smaller above and finally missing; **Infl.** often 2–3 from one Ros., 40–45 cm tall, Fl. to 3.5 cm long, salmon-red.

**A.** cv. Riccobono. – Hybrid from the Bot. Gard. Palermo, Italy, 1912: **A. arborescens** × **A. capitata**.

**A. richardsiae** REYN. (§ III/1). – Tanzania: Iringa. – **Root** thick, fleshy, narrowed to gradually tapered, from an underground bulb 35 mm long and 30 mm thick; **L.** 8–10, rosulate, rather fleshy, 20–25 cm long, 15 mm across, grooved on the upperside, green, distinctly lined, underside convex, lined, margins with white, horny triangular T. 1.5 mm long, 1–2 mm apart; **Infl.** 35–45 cm tall, Fl. almost 5 cm long, pale orange to scarlet.

**A. rigens** REYN. et BALLY v. **rigens** (§ IV/C/16). – Somalia. – Acauline or short-stemmed, usually solitary, sometimes forming small groups; **L.** c. 24 arranged in a dense Ros., very rigid, 60–80 cm long, 10–15 cm across at the base, gradually narrowing to the tip, 12–15 mm thick below, upperside pale to darker greygreen, slightly concave, underside convex, margins with triangular T. reddish-brown above, 4–6 mm long and 2–3.5 mm apart; **Infl.** 1.25 to 1.75 m tall, Fl. 30–34 mm long, rosy red to dull scarlet.

**A.** — v. **glabrescens** REYN. et BALLY. – **L.** 40–50 cm long, 10–12 cm across; **Infl.** 0,75–1 m tall, Fl. 32 mm long, glabrous or minutely hairy, yellow.

**A.** — v. **mortimeriana** LAVR. – S.Arabia: Hadramaut. – **L.** less rigid, quite soft, 40–60 cm long, 6–10 cm across; **Fl.** with minute hairs, brick-red.

**A. rivae** BAK. (§ IV/C/19d). – Somalia. – Caulescent; **L.** 7–8 cm or more across at the base, ensiform, fleshy, c. 1.2 cm thick, rigid, bluish-green, marginal T. horny, triangular, 4–5 mm long, 12–15 mm apart; **Infl.** c. 40 cm tall, Fl. 32 mm long.

**A.** cv. Robertii. – Hybrid. HORT. DELEUIL. – **Ros.** acauline or short-stemmed, c. 2 m $\varnothing$; **L.** numerous, 70–80 cm long, 15–20 cm across at the base, green-bluish, marginal T. brown, 2 cm apart; **Fl.** numerous, yellow.

**A. rubroviolacea** SCHWEINF. (§ IV/E/19). – Arabia: Yemen. – **St.** thick, 1 m tall, somewhat curved, single-headed; **L.** in a dense Ros., broadly lanceolate-ensiform, 60 cm long, 10 to 11 cm across, deep bluish-violet, flat on the upperside, convex on the underside, with horny, red, dentate border, T. 2–3 mm long, 2–2.5 cm apart; **Fl.** pendulous, red.

**A. runcinata** BGR. (§ IV/B/10x) (*A. obscura* BGR. ex SCHOENL.). – Cape. – **St.** short; **Ros.** of 15–30 **L.** 35–40 cm long, 6–9 cm across, 15–18 mm thick, green, somewhat striate and spotted, marginal T. curved, 3–8 mm long, 8–20 mm apart; **Infl.** 90–120 cm tall, Fl. c. 4 cm long, light red.

**A. rupestris** BAK. (§ VI/3) (*A. pycnantha* MACOWAN, *A. nitens* BAK.). – Natal, Zululand; E.Swaziland; Moçambique. – Treelike; **St.** usually simple, 6–8 m tall, 20 cm $\varnothing$, the upper part covered with dry old L.; **L.** 30–40 in a dense Ros., c. 70 cm long, 7–10 cm across at the base, gradually tapering, deep green, concave on the upperside above, underside convex, margins sinuate-dentate, pink, T. triangular, sharp, red-brown, 4 mm long, 8–12 mm apart, those nearer the tip being larger and curved;

**Infl.** 1–1.25 m tall, Fl. 2 cm long, lemon-yellow to orange-yellow.

**A. rupicola** REYN. (§ IV/C/19c). – Angola: Bié. – ♄; **St.** (2–)3(–5) m tall, 10–12 cm $\varnothing$, usually branching basally; **L.** c. 40, densely rosulate, 30–35 cm long, 6 cm across, with a $\pm$ dry, brittle tip, usually flat on the upperside, green, indistinctly lined, underside slightly convex, with more distinct lines above, margins sinuate-dentate, T. sharp, reddish-brown, 4 to 5 mm long, 1 cm apart, the lower T. smaller and closer together; **Infl.** 70–90 cm tall, Fl. c. 4.2 cm long, orange-scarlet.

**A. ruspoliana** BAK. v. **ruspoliana** (§ IV/D/23) (*A. jex-blakeae* CHRISTIAN, *A. stephaninii* CHIOV.). – Ethiopia: Harrar; S.Somalia. – Acauline, or with **St.** c. 50 cm long, prostrate or ascending, forming groups of several m $\varnothing$; **L.** c. 16 in a dense Ros., tapering-lanceolate, 50–60 cm long, 12 cm across, yellowish-green on the upperside, sometimes with a few narrow, lentiform spots below, somewhat grooved above, underside convex, margins narrow, horny, T. white, 0.5 mm or less in length, 5–8 mm apart; **Infl.** 1.5 m tall, Fl. 16–20 mm long, yellow.

**A.** — v. **dracaeniformis** BGR. – Somalia. – Treelike, resembling a **Dracaena**; **L.** 5.5 cm across.

**A. sabaea** SCHWEINF. (§ IX). – S.Arabia. – Treelike; **St.** dividing dichotomously, to 9 m tall; **L.** densely rosulate, ensiform, gradually tapering, oblique-falcate, 60–100 cm long, 5 to 12 cm across, 15–16 mm thick, upperside deeply grooved, underside convex with several Sp. below the tip, margins serrate-dentate, T. triangular, pallid, curved, not spine-tipped, compressed and oblique-falcate; **Infl.** robust, Fl. 25–33 mm long, pink to flesh-coloured, with red longitudinal lines.

**A. saponaria** HAW. v. **saponaria** (Pl. 18/1) (§ IV/B/10/A/1a) (*A. disticha* MILL., *A. umbellata* S. D., *A. perfoliata* v. *s.* AIT., *A. s.* v. *minor* HAW.). – Natal, Cape: Pondoland. – **Ros.** short-stemmed, offsetting, densely leafy; **L.** 15 to 20 cm long, 4–6 cm across, almost 1 cm thick, lanceolate, narrowing towards the tip, convex on the underside, with oblong white marks in well-spaced rows, margins with triangular, brownish T.; **Infl.** 50–70 cm tall, Fl. 4–4.5 cm long, red-yellow.

**A.** — v. **brachyphylla** BAK. – **L.** 7–10 cm long, ovate, tapering, T. smaller.

**A.** — v. **ficksburgensis** REYN. – **L.** smaller than in the type; **Fl.** 3.5 cm long.

**A. saundersiae** (REYN.) REYN. (§ II) (*Leptaloe s.* REYN.). – Natal. – Acauline; **Ros.** dense; **L.** 12–15, 5–6 cm long, 3 mm across at the base, both surfaces green, the underside with several dark spots below, margins with white T. 0.5 mm long, 1 mm apart; **Infl.** 15–20 cm long, Fl. 8 mm long, pink to red below and brown-tipped.

**A. schelpei** REYN. (§ IV/B/5). – Ethiopia: Shewa. – ♄; **St.** prostrate, up to 50 cm long, 5–6 cm thick, offsetting either from the base or higher to form dense clumps; **L.** 16–20 in a Ros., lanceolate-tapering, to 45 cm long, 10–12 cm across base, upperside green with a bluish sheen, sometimes with several lentiform, pale green spots near the base, deeply grooved above, convex on the underside, somewhat deeper green, with some spots in the lower part, margins raised, reddish-pink, T. stout to somewhat sharp, 2–3 mm long, c. 1.5 cm apart, reddish; **Infl.** 50 cm tall, Fl. orange-red.

**A.** × **schimperi** TOD. – Garden hybrid: **A. saponaria** × **A. striata.** – **Ros.** acauline, simple; **L.** broadly linear-lanceolate, short-tapering, flat on the upperside, fleshy, blue, indistinctly striped, margins with small, irregular T.; **Infl.** 90–100 cm tall, Fl. 4 cm long, red.

**A. schliebenii** LAVR. (§ IV/C/19c). – Tanzania: Morogoro D. – Plant solitary, acauline or very short-stemmed; **L.** strap-shaped, very straight, narrowing to the pointed tip, c. 50 cm long, 5–5.5 cm across at the base, rather thin, margins sinuate-dentate, T. horny, curved, with a hard, brown tip, about 3 mm long, 7–13 mm apart; **Infl.** c. 1 m tall, Rac. with Fl. fairly close together, bracts ovate-round, Fl. c. 3 cm long, orange-red.

**A. schoelleri** SCHWEINF. (§ IV/E/29). – Ethiopia. – **St.** robust; **L.** densely rosulate, rather thick, lanceolate, 35 cm long, 15 cm across at the base, blue-green, margins red, horny, T. triangular, irregular; **Infl.** 0.6–1 m tall.

**A.** × **schoenlandii** BAK. – Hybrid: **A. striata** × **A. spec.** – Acauline; **L.** 18 in a dense Ros., ovate-lanceolate, blue, carinate below the tip, old L. ovate, reddish, c. 30 cm long, 12–15 cm across, c. 12 mm thick, upperside indistinctly striate, underside convex with spots in longitudinal rows, margins reddish, cartilaginous and dentate; **Infl.** 0.9–1.2 m tall, Fl. 2.5–3 cm long, red.

**A. schomeri** RAUH (§ XI/d/β 2). – S.E.Madag. – **Ros.** simple, acauline or very short-stemmed, to 60 cm ⌀; **L.** c. 30, 3.5 cm across at the base, 20–30 cm long, gradually tapering, rich dark green, margins with conspicuously yellow, horny, very hard T., often pointed-triangular, often double-tipped, 2 mm across and 5–8 mm apart; **Infl.** c. 65 cm tall, Fl. ± 2 cm long, pale yellow to almost white, Pet. with green nerves.

**A. schweinfurthii** BAK. v. **schweinfurthii** (§ IV/C/19c) (*A. barteri* BAK. p. part., *A. b.* v. *lutea* A. CHEV., *A. trivialis* A. CHEV.). – Sudan: Equatoria Prov.; Congo (Kinshasa): N. & S.Kivu; Ghana; N. & W.Nigeria; W.Cameroon; Central Afr. Rep.; Ubangi. – **Ros.** acauline or short-stemmed, forming small or larger groups by means of offsets; **L.** 16–20, (45–)50(–60) cm long, 6–7 cm across at the base, narrowing evenly towards the tip, grey-green on the upperside with a bluish sheen, grooved above, having several elongated, dull white spots in the upper third, duller green on the underside with spots near the base, margins sinuate-dentate, T. triangular with red-brown tips, about 4 mm long, 10–12 mm apart, bent slightly forwards; **Infl.** c. 90 cm tall, Fl. c. 28 mm long, orange.

**A.** – v. **labworana** REYN. – Uganda: Northern Prov. – **L.** longer, T. larger; **Fl.** yellow.

**A. scobinifolia** REYN. et BALLY (§ IV/B/10/. A/2a). – Somalia. – **Ros.** often solitary, but usually forming small groups by means of offsets, acauline or short-stemmed; **L.** 16–20, narrowly lanceolate-tapering, usually 30 cm long, 7 cm across at the base, 12–15 mm thick, narrowing to a spiny point, upperside flat to somewhat concave, both surfaces dull green, noticeably rough, margins very narrow, pale pink, horny, without T.; **Infl.** 60–70 cm tall, Fl. 22 mm long, yellow, orange or scarlet.

**A. secundiflora** ENGL. (§ IV/D/23) (*A. engleri* BGR., *A. floramaculata* CHRISTIAN). – Tanzania; Kenya. – **Ros.** acauline, dense; **L.** 30, to 45 cm long, 12–14 cm across base, green, margins sinuate-dentate, horny, T. triangular, 3–4 mm long, 1.5–2 cm apart, brown, projecting, sharp; **Infl.** 75 cm tall, Fl. 23–25 mm long, red.

**A. seretii** DE WILD. (§ IV/C/19c). – Congo (Kinshasa): Eastern Prov., on cliffs. – **Ros.** acauline or short-stemmed, offsetting basally to form small or larger groups, densely leafy; **L.** tapering-lanceolate, the top third being twisted, the upperside grooved above, grey to bluish-green with a reddish sheen, often with indistinct dull white markings, the underside convex, usually unspotted, margins sinuate-dentate, pink, T. sharp, usually 3–4 mm long, 8–10 mm apart, more widely spaced above; **Infl.** 60–70 cm tall, Fl. 28–33 mm long, dull to bright scarlet.

**A. serrulata** (AIT.) BAK. (§ IV/B/9) (*A. perfoliata* v. *s.* AIT., ? *A. pallescens* HAW.). – Cape. – Acauline or with short **St.**; **L.** arranged almost spirally, ovate-lanceolate, tapering, c. 20 cm long, 5 cm across, underside convex and obliquely keeled under the tip, irregularly spotted, margins serrate-dentate; **Infl.** c. 45 cm tall, Fl. 4 cm long, red.

**A. serriyensis** LAVR. (§ IV/C/16). – S.Arabia; S.Yemen: S.W.Hadramaut. – **Ros.** offsetting to form small groups; **L.** 30–35 cm long, 6–7 cm across at the base, narrowing evenly to a pointed tip, 5–8 mm thick, rather soft, upperside green, often with a brownish sheen, inconspicuously striate, broad-grooved, margins with horny, brown, blunt T. 2–4 cm apart; **Infl.** 40 cm tall, Fl. 27 mm long, powdery scarlet-pink, the lobes having 3 red nerves.

**A. sessiliflora** POLE EVANS (§ V). – Transvaal; Cape: Natal (Zululand). – **St.** 1.5 m tall; **Ros.** densely leafy; **L.** 45–70 cm long, 6–8 cm across, flat on the upperside, underside convex, marginal T. triangular, short, fairly widely spaced; **Infl.** 60–75 cm tall, Fl. 14 mm long, campanulate, red with yellow-green stripes.

**A.** cv. Sherman Hoyt. – **A. dawei** × **A. rauhii.** – Vigorous triploid hybrid, intermediate between the parents. Hort. Kew.

**A. signoidea** BAK. (§ VI). – Natal: Tongaland. – Possibly a variety of **A. ferox**, acc. GROENEWALD; acc. REYNOLDS, an insufficiently known species, possibly a hybrid with **A. arborescens** as one parent.

**A. silicola** PERR. (XI/α/α). – Central Madag.: Tananarive. – **Ros.** of 40–50 L.; **St.** up to 2 m tall, 5 cm ⌀, simple; **L.** thin, glossy, 45–50 cm long, 7–8 cm across, narrowing towards the

Plate 18. 1. **Aloe saponaria** HAW. v. **saponaria** (Photo: J. MARNIER-LAPOSTOLLE); 2. **A. ramosissima** PILL.; 3. **A. spectabilis** REYN. (Photo: as 1); 4. **A.** cv. Spinosissima.

Plate 19. 1. Aloe succotrina LAM.; 2. A. striata HAW.; 3. A. variegata L. v. variegata; 4. A. viguieri PERR. (Photo: W. RAUH).

Plate 20. 1. **Aloe squarrosa** BAK. et BALF. (Photo: J. MARNIER-LAPOSTOLLE); 2. **A. zebrina** BAK.; 3. **Ammocharis coranica** (KER-GAWL.) HERB. (Photo: W. KABEL).

Plate 21. 1. Anacampseros alstonii SCHOENL. (Photo: C. BACKEBERG); 2. A. recurvata SCHOENL. enlarged (Photo: J. MARNIER-LAPOSTOLLE); 3. A. bremekampii v. POELLN. (Photo: H. LANG); 4. A. filamentosa (HAW.) SM. (Photo: as 3); 5. A. rufescens (HAW.) SWEET; 6. A. tomentosa BGR. v. tomentosa.

Plate 22. **1. Astroloba** Infl. (Photo: J. A. Uitewaal); **2.** × **Astroworthia bicarinata** (Haw.) Rowl. (Photo: de Laet); **3. Astroloba foliolosa** (Willd.) Uitew.; **4. A. deltoidea** (Hook. f.) Uitew. v. **deltoidea** (Photo: J. Marnier-Lapostolle); **5.** × **Astroworthia bicarinata** nm. **skinneri** (Bgr.) Rowl.; **6. Astroloba herrei** Uitew. (Repr.: Shaboten, No. 57, 1966, 12); **7. A. pentagona** (Haw.) Uitew.

Plate 23. 1. **Augea capensis** THUNBG.; 2. **Baeriopsis guadalupensis** J. T. HOWELL; 3. **Batis maritima** L. Left: male, right: female plant. (Photo: R. MORAN); 4. **Beaucarnea stricta** LEM. (Repr.: Cact. y Suc. Mex. **V,** 1960, 93, Photo: T. MACDOUGALL).

Plate 24. 1. Begonia natalensis HOOK. (Photo: M. MASON); 2. The same (Repr.: Bot. Mag. Pl. 4841); 3. B. venosa SKAN; 4. B. incana LINDL.

Plate 25. 1. **Bombax ellipticum** H. B. et K.; 2. **Bowiea volubilis** Harv. et Hook. f. (Photo: J. Marnier-Lapostolle); 3. **Brachystelma pulchellum** (Harv.) Schltr. (Repr.: Fl. Pl. Afr. **XXIX**, Pl. 1121); 4. **B. floribundum** R. A. Dyer (Repr.: l.c. Pl. 1224. Drawing: R. Brown).

base, the tip pointed and with a terminal Sp., marginal T. 1–1.5 mm tall, somewhat translucent, about 7 mm apart; **Infl.** 50–60 cm tall, Fl. yellowish-orange.

**A. simii** POLE EVANS (§ IV/B/10/A/5b). – Transvaal. – **Ros.** solitary, less frequently 2–3 in a cluster; **L.** 15–20, dense, 40–60 cm long, 9–12 cm across at the base, very concave on the upperside, light milky-green, lined, often with a few spots, underside dull dark green, indistinctly lined, margins sinuate-dentate, T. ± horny, 3–4 mm long, 1–1.5 cm apart; **Infl.** 1–1.5 m tall, Fl. 3.5–4 cm long, pink.

**A. sinana** REYN. (§ IV/C/19d). – Ethiopia: Shewa. – **St.** to 1 m tall, 8–10 cm ∅, erect or divergent, with L. on the top 20 cm; **L.** 12–16, rosulate, sheathed basally and 10–13 cm across, gradually narrowing to the tip, 60–70 cm long, grooved on the upperside above, dull grey-green, the upper half usually with several scattered, pale green, elongated, lentiform spots, the underside convex, grey-green, with numerous spots in the upper half, margins horny, T. sharp, triangular, reddish-brown, 3–4 mm long, (10–)15(–20) mm apart; **Infl.** c. 1 m tall, Fl. c. 28 mm long, orange-yellow below, paler above.

**A. sinkatana** REYN. (§ IV/C/19d). – Sudan. – **Ros.** acauline, single or in groups; **L.** 16–20, somewhat leathery, 50–60 cm long, 6–8 cm across at the base, gradually narrowing to a rounded tip with 3–5 very tiny reddish T., upperside of L. grooved, dull grey-green, with or without spots, underside rounded, spotted only in the lower part, margins often twisted inwards, reddish, T. 2–3 mm long, 1.5–2 cm apart, stout, fairly sharp; **Infl.** 75–90 cm tall, Fl. c. 22 mm long, orange to yellow.

**A. sladeniana** POLE EVANS (§ IV/B/9) (*A. carowii* REYN.). – S.W.Afr.: Nauchas. – **Ros.** acauline, c. 9 cm ∅, offsetting basally; **L.** 6–8, in 3 irregular rows, 7–9 cm long, 3–4 cm across, convex on the upperside, margins curved upwards to form a U-shape, underside carinate, the keel having small hard Sp., L. light pale green, with numerous white oblong ± translucent spots which ± merge into one another, and are arranged in distinct transverse bands, margins thin, white, T. white; **Fl.** 3 cm long, deep pink.

**A. somaliensis** WATSON v. **somaliensis** (§ IV/C/18a). – Somalia. – **Ros.** simple or forming small groups by means of offsets, acauline or short-stemmed; **L.** 12–16, usually lanceolate-tapering, 20 cm long, 7 cm across at the base, 8–10 mm thick, slightly concave on the upperside above, fairly glossy brownish-green, with numerous pale green spots, those on the underside often being translucent, margins sinuate-dentate, T. reddish-brown, sharp, triangular, 4 mm long, 8–10 mm apart; **Infl.** 60–80 cm tall, Fl. 28–30 mm long, pink-scarlet.

**A. —** cv. Kew Green. – Selected cultivar with paler **L.** – Hort. Kew.

**A. —** v. **marmorata** REYN. et BALLY. – Larger than the type; **L.** darker green with darker lines and broader, dark green longitudinal markings which give the L. its marbled appearance; **Fl.** deeper red, lobes turning ivory-white.

**A. sororia** BGR. (§ IV/D/25). – Cape. – Possibly a variety of **A. mitriformis;** caulescent; **L.** 23 cm long, 7 cm across, marginal T. 5 mm long, 12–15 mm apart; **Fl.** 4 cm long, light red.

**A. soutpansbergensis** VERD. (§ III/2). – Transvaal: Soutpansberg. – Plant simple or offsetting below soil-level; **roots** fleshy, cylindrical to fusiform; **St.** c. 5 cm long, 8 mm ∅; **L.** c. 7, somewhat fleshy, initially distichous, soon becoming rosulate, 2.5 cm across at the base, and clasping the St., then abruptly narrowing to the linear blade, 1 cm across and 25 cm long, 3.5 mm thick at the base, with a rounded tip, convex on the upperside, with very few indistinct white marks, rounded on the underside with many white spots, either smooth or slightly raised, these being more numerous below, margins incurved, horny, with translucent T. 0.5 mm long and 3–4 mm apart, these T. being more frequent below; **Infl.** subcapitate, Fl. c. 27 mm long, scarlet-orange.

**A. speciosa** BAK. (§ IV/E/29). – E.Cape. – Treelike, to 8 m tall; **L.** in a dense Ros., ensiform, gradually tapering, blue-green, margins narrow, horny, T. pink, small, irregular; **Fl.** 3–3.5 cm long, whitish-green striate.

**A. cv. Speciosa.** – Garden hybrid: parents not known, possibly related to **A. heteracantha.** – **Ros.** elongated, short-stemmed; **L.** triangular-tapering, 18–20 cm long, 8 cm across at the base, green, flat on the upperside with several white spots, convex on the underside with more numerous spots, margins with horny, triangular, incurved T. 2 mm long and 8–10 mm apart; **Infl.** robust, 55 cm tall, Fl. 47 mm long, red.

**A. spectabilis** REYN. (Pl. 18/3) (§ VI/5) (*A. ferox* v. *xanthostachys* BGR.). – S.Afr.: Natal. – Tree-like; **St.** 2–4 m tall; **Ros.** dense; **L.** 50, 1 m long, 12–15 cm across, dark green with a reddish sheen, somewhat concave on the upperside, convex on the underside, both surfaces spiny, margins dentate, T. 5–7 mm long, red-brown, 1–2 cm apart; **Infl.** 1–3, Fl. 5 cm long, reddish.

**A. spicata** L. f. (§ IV/E/29?). – Cape. – Dubious species, probably identical with **A. speciosa.**

**A. cv. Spinosissima** (Pl. 18/4). – Acc. A. BERGER a hybrid of **A. humilis** × **A. arborescens** v. **pachythyrsa** raised at 'La Mortola', Ital. Riviera. – Short-stemmed, in age developing many **St.** 1 m or more tall; **L.** in a spiral Ros., lanceolate, gradually tapering, 25–30 cm long, 3–4 cm across at the base, very concave above, underside rounded, margins with broadly triangular, horny, curved T. 8 mm long, with numerous small Sp. on the underside and a few on the upperside; **Infl.** 60 cm tall, Fl. brilliant orange colour.

**A. splendens** LAVR. (§ IV/C/19d). – S.W.-Arabia. – **Ros.** solitary or in groups of 8, usually acauline; **St.** often 40 cm long, prostrate; **L.** 20–24, tapering-triangular, to 60 cm long, 15 cm across at the base, 1 cm thick, dull green on the upperside, often with a purple sheen, grooved towards the tip, underside convex, margins dark brown, horny, T. dark brown, hard, fairly blunt, triangular, 2 mm long, 1.5–3 cm apart; **Infl.** 1.4 m tall, Fl. 30–33 mm long, light scarlet, lobes with 3 dark brown nerves.

**A.** cv. Spuria (§ IV/B/10x). – Cultivated hybrid from 'La Mortola', Ital. Riviera. – **Ros.** acauline, forming groups; **L.** 10–12, broadly lanceolate, gradually tapering, upperside concave in the upper part, 30–45 cm long, 5–7 cm across, green to reddish, underside pale pink, indistinctly lined and with longitudinal marks in bands, marginal T. 3–4 mm long, 1–2 cm apart; **Infl.** 70–120 cm tall, Fl. c. 4 cm long, pale red.

**A. squarrosa** BAK. ex BALF. (Pl. 20/1) (§ IV/D/22) (*A. concinna* BAK., *A. zanzibarica* MILNE-REDHEAD). – W.Socotra; Zanzibar. – **St.** 10–20 cm long, 7–8 mm ⌀, L.Sh. clasping the St.; **L.** in a Ros., triangular-lanceolate, tapering, c. 8 cm long, 2 cm across at the base, green with spots and transverse bands, margins cartilaginous, T. 3–4 mm long, 4 mm apart; **Infl.** slender, Fl. 2 cm long, red.

**A. stans** BGR. (§ IV/D/25) (*A. nobilis* BAK.). – E.Cape. – **St.** 2 m tall, 4–5 cm ⌀; **Ros.** 60 cm ⌀; **L.** lanceolate-ensiform, c. 30 cm long, 8 cm across, tapering, green, margins with horny, brown T. 4 mm long and 9–10 mm apart; **Fl.** 3 cm long, light red.

**A. steudneri** SCHWEINF. (§ IV/E/27b). – Ethiopia. – **Ros.** short-stemmed, offsetting basally; **L.** c. 25, densely rosulate, c. 60 cm long, 12–15 cm across at the base, narrowing to the pointed tip, with a narrow, membraneous, pinkish border, T. 2 mm long, those below being 7–10 mm apart, those above 3–4 cm apart, the upperside of the L. slightly grooved, the underside carinate; **Infl.** 70–90 cm tall, Fl. 4–4.5 cm long, deep red.

**A. striata** HAW. (Pl. 19/2) (§ IV/B/11) (*A. paniculata* JACQ., *A. albo-cincta* HAW., *A. hanburyana* NAUD., *A. rhodocincta* HORT.). – S.W.Afr.: Gr. Namaqualand; Cape: Albany, Graaff-Reinet D. – **Ros.** dense, almost acauline; **L.** 45 to 50 cm long, 10–15 cm across, c. 1 cm thick, lanceolate, often reddened, pruinose, indistinctly spotted and striped, margin entire, white, cartilaginous, 2 mm across; **Fl.** orange-red.

**A. striatula** HAW. v. **striatula** (§ IV/D/21) (*A. aurantiaca* BAK., *A. macowanii* BAK., *A. cascadensis* KTZE.). – S.E.Cape. – **St.** numerous, 1–2 m tall; **L.** sheathed, these Sh. being striate, linear-lanceolate, tapering, 20–23 cm long, 2.5 to 3.5 cm across, flat or slightly concave on the upperside, underside convex, dark green, striate, margins with T. 1–10 mm long and 2–7 mm apart; **Infl.** 30–40 cm tall, Fl. 4 cm long, yellow.

**A. —** v. **caesia** REYN. f. **caesia** (*A. s.* v. *c.* f. *typica* RES. and f. *haworthii* RES.). – **L.** rosulate, milky-green, Sh. indistinctly striped, margins horny, T. 1 mm long; **Fl.** 3 cm long.

**A. —** v. — f. **conimbricensis** RES. (*A. s.* f. mut. *c.* RES.). – **L.** margins horny, T. 2 mm long.

**A. stuhlmannii** BAK. (§ IV/C/19c). – Zanzibar. – Caulescent ?; **L.** fleshy, 5–5.5 cm across at the base, ensiform, c. 45 cm long, margins horny, T. 4–5 mm long, 1.5 cm apart; **Fl.** c. 2.5 cm long.

**A. suarezensis** PERR. (§ XI/d/α). – W.Madag. – Acauline or **St.** 20 cm to 1 m tall, with 10–20 spreading **L.** 40–60 cm long, 9–10 cm across at the base, gradually tapering, marginal T. small, sharp, pale red, 1 cm apart; **Infl.** rarely simple, Fl. shortly hairy, vivid red.

**A. subacutissima** ROWL. (§ XI/b) (*A. deltoideodonta* v. *intermedia* PERR., *A. i.* REYN. non HAW.). – Madag. – Bushy; **St.** erect, divergent or prostrate, creeping, rooting, to 1 m long, 3 cm ⌀, the last 20 cm or more being almost densely leafy, otherwise covered with the remains of old L.; **L.** 20–26, rosulate, Sh.-like at the base, tapering-lanceolate, 25–30 cm long, 5–6 cm across, dull green on upperside, flat to somewhat concave, margins with reddish-brown, triangular T. 3–4 mm long, 1 cm apart midway but smaller and farther apart towards the tip, and more closely spaced towards the base; **Infl.** c. 60 cm tall, Fl. c. 3 cm long, scarlet.

**A. succotrina** LAM. (Pl. 19/1) (§ IV/E/17a) (*A. vera* MILL., *A. perfoliata* v. *s.* AIT., *A. s.* v. *saxigena* BGR., *A. s.* GARSAULT, *A. s.* I. A. et J. A. SCHULTES, *A. purpurea* HAW.). – Cape: Cape D. – Acauline or **St.** to 1 m tall, 10 to 15 cm ⌀, covered with dry old L., branching dichotomously; **Ros.** to 80 cm ⌀; **L.** numerous, 50 cm long, 5 cm across, ensiform to gradually tapering, blue-green to grey-green, somewhat lined, upperside somewhat spotted basally, margins white, horny, T 2–4 mm long, close-set, pale; **Infl.** 60 cm tall, Fl. 3.5 cm long, red with green stripes.

**A. suffulta** REYN. (§ IV/C/19c). – Moçambique. – **St.** simple, 10–20 cm long, 1.5–2 cm ⌀; **L.** c. 16, sheathing basally and 4 cm across, gradually tapering, 40–50 cm long, very concave on the upperside, green, entire surface with spots 5 mm long and 2 mm across, arranged in wavy transverse bands, underside convex, darker green, spots larger and translucent, margins with whitish T. 1–2 mm long, 5–10 mm apart; **Infl.** 1.75 m tall, Fl. 3–3.5 cm long, pale reddish turning whitish at the mouth.

**A. suprafoliata** POLE EVANS (§ IV/B/12a). – Swaziland. – Acauline or with short **St.**; **L.** distichous in immature plants, later rosulate, 32 cm long, 7–9 cm across, 7 mm thick, terminating in a point, upperside concave, underside convex, grey-green, margins angular, T. 4 to 6 mm long, 10–12 mm apart, brown; **Infl.** 36 cm tall, Fl. 4 cm long, coral-red with green tips.

**A. suzannae** R. DECARY (XIa). – S.W.Madag. – **St.** 3–4 m tall, 20–30 cm ⌀, rigid, only occasionally branching; **Ros.** with flaccid spreading L. 80–100 cm long, 7–8 cm across at the base, 3.5–4 cm thick, with rounded tips, marginal T. yellowish, small, irregularly spaced in groups of 2–3; **Infl.** 2–3.5 m tall, Fl. numerous, yellow.

**A. swynnertonii** RENDLE (§ IV/B/10/A/1a) (*A. chimanimaniensis* CHRISTIAN, *A. melsetterensis* CHRISTIAN). – Rhodesia; Moçambique: Gaza. – **Ros.** acauline, or with scarcely recognizable St., solitary or in small groups of 3–4 Ros.; **L.** c. 20, tapering-lanceolate, 25–30 cm long, with a further dry and twisted tip, 10 cm long, L. 8 to 10 cm across at the base ± grooved on the upperside towards the tip, dark green, the entire surface with H-shaped spots arranged ± in wavy transverse bands, convex on the underside which is paler green, indistinctly lined, usually unspotted, margins sinuate-dentate, T. sharp, red-brown, 4 mm long, 1–1.5 cm apart; **Infl.** 1.5–1.75 m tall, Fl. 3 cm long, fleshy-pink to dull coral-red.

**A. taurii** LEACH (§ V). – Rhodesia: Belingwe D. – Acauline or short-stemmed, offsetting to form clumps; **St.** 5–7 cm ⌀, usually creeping, less often up to 30 cm tall; **Ros.** dense; **L.** c. 25, narrowly tapering-ovate, 34–35 cm long, 6–8 cm across at the base, spreading-recurved, dark yellow-green, becoming coppery-red, marginal T. sharp, hooked, 2–3 mm long, interspaces rounded below but straight in the upper part; **Infl.** to 1 m tall, often with one Br., Fl. yellow-orange.

**A. tenuior** HAW. v. **tenuior** (§ IV/D/21) (*A. t.* v. *glaucescens* A. ZAHLBR.). – E.Cape. – **Rootstock** 36–60 cm in width or more, tuberous in the lower part; **St.** thin, clambering, branching, 1–3 m long, 1–1.5 cm ⌀; **L.** at Br.ends in lax Ros. of 8–10 cm ⌀, linear-lanceolate, thin or somewhat fleshy, 10–15 cm long, 10–15 mm across, sheathing basally, somewhat concave on the upperside above, underside convex, margins horny, white, T. 0.5 mm long, 1–2 mm apart; **Infl.** 10–20 cm tall, Fl. 11–14 mm long, yellow.

**A.** — v. **decidua** REYN. – **St.** erect, not clambering, 30–60 cm tall; **Ros.** lax; **L.** to 18 cm long, 22 mm across, rather bluish; **Infl.** 50 cm tall.

**A.** — v. **densiflora** REYN. – **Infl.** crowded.

**A.** — v. **rubriflora** REYN. – **Fl.** 15 mm long, red with yellow spots.

**A. thompsonii** GROENEW. (§ III/2). – Transvaal. – **St.** 4 cm or more tall, 8 mm ⌀, branching from base; **Ros.** of 10 L. c. 15 cm long, 2 cm across at the base, 8 mm across midway, 4 mm thick, flat on the upperside, underside boat-shaped, light green, margins narrow, white, T. pale, whitish, 1 mm long, the upperside having a few Sp. below; **Infl.** 8 cm tall, Fl. 2.5 cm long.

**A. thorncroftii** POLE EVANS (§ IV/B/12a). – Transvaal. – **Ros.** acauline or short-stemmed; **L.** 25–30, lanceolate, c. 25 cm long, 7–9 cm across at the base, underside rounded, bluish, margins with short T.; **Infl.** almost 1 m tall, Fl. 3.5–4 cm long, pink.

**A. thraskii** BAK. (§ VI/3) (*A. candelabrum* ENGL. et DRUDE). – Natal: Zululand. – **St.** simple, (1–)2(–4) m tall, covered with dry old L.; **Ros.** dense; **L.** to 1.6 m long, gracefully recurved, 22 cm across at the base, gradually tapering, very concave on the upperside, margins incurving and U-shaped, green to bluish, underside convex, often with several Sp. on the M.line, margins thin, reddish or brownish-red, T. triangular, 2 mm long, 1–2 cm apart; **Infl.** with up to 6 Br., Fl. 2.5 cm long, lemonyellow to orange.

**A.** × **todari** BORZI. – Hybrid: **A. humilis** × **A. spec.** – **St.** short or acauline; **L.** 30–40 in a dense Ros., lanceolate, tapering, flat on the upperside, 20–30 cm long, 6–8 cm across at the base, green to bluish, with dark lines and several oblong pale irregularly scattered spots, margins with red, triangular T. 8–12 mm apart; **Fl.** 4 cm long, red.

**A.** × — nm. **praecox** BORZI. – Variety which flowers earlier.

**A. tomentosa** DEFLERS (§ IV/C/16). – S.Arabia: Yemen; Somalia. – **St.** short; **Ros.** twisted, forming large, dense groups; **L.** 15–20, lanceolate-triangular, to 35 cm long, 8 cm across at the base, narrowing to a pointed, spiny tip, upperside grey-green with a reddish sheen, slightly grooved towards the tip, underside convex, margins narrow, pink-brown, horny, T. blunt, 0.5–1 mm long, 2–4 cm apart, often missing; **Infl.** c. 70 cm tall, Fl. 24–28 mm long, pink, distinctly tufted-hairy.

**A.** × **tomlinsonii** MARL. – Cape: Swellendam D. – Natural hybrid: **A. ferox** × **A. speciosa**. (REYNOLDS does not recognize this as a hybrid since the parent-plants grow too far apart.)

**A. tororoana** REYN. (§ IV/C/19d). – Uganda: Tororo. – **St.** never more than 20 cm long, slender, 15 mm ⌀; **Ros.** dense, 25 cm ⌀; **L.** c. 12, lanceolate, tapering, 15 cm long, 3–5 cm across at the base, 1 cm thick, flat on the upperside, dark green with small oblong whitish marks, convex on the underside where the marks are more numerous, margins sinuate-dentate, T. white, 4–5 mm long, 1–1.5 cm apart; **Infl.** c. 40 cm tall, Fl. 2 cm long, coral-red with green tips.

**A. torrei** VERDOORN et CHRISTIAN (§ III/2). – Moçambique: Quelimane D. – **St.** 10 cm tall; **L.** 10–12 in a Ros., 45 cm long, narrow, limp, deflexed, ± hanging downwards; **Infl.** 50 cm tall, Fl. about 20.

**A. trachyticola** (PERR.) REYN. (§ IV/C/20; XI/d/α) (*A. capitata* v. *t.* PERR.). – Madag.: Tananarive D., Fianarantsoa Prov. – Plant solitary, usually acauline, often with a prostrate **St.**; **L.** initially 6–10, distichous, later up to 14, arranged spirally to almost rosulate, 10–15 cm long, 3–4 cm across, the tip somewhat obliquely rounded and shortly dentate, upperside bluish-grey with a reddish sheen, flat to slightly concave, underside convex, margins with reddish-brown, triangular, sharp T. 1 to 1.5 mm long and 3–5 mm apart; **Infl.** 65–90 cm tall, Fl. 3.5 cm long, red.

**A. transvaalensis** KTZE. v. **transvaalensis** (§ IV/B/10/A/5b) (*A. laxissima* REYN.). – Transvaal. – Acc. BERGER, identical with **A. davyana**; acc. REYNOLDS **A. davyana** p. part. – **Ros.** acauline, offsetting to form small groups, dense; **L.** 12–16, lanceolate-tapering, 20–25 cm long, 6–7 cm across at the base, slightly concave on the upperside above, dark milky green, with numerous white oval spots ± in wavy transverse bands, the underside convex, paler green, lined, spots indistinct, margins sinuate-dentate, with triangular, sharp, light brown, horny T. 3 to 4 mm apart; **Infl.** 1–1.5 m tall, Fl. 3.6 cm long, fleshy-pink to coral-red.

**A. — v. stenacantha** GROENEW. – **L.** 10–15, to 20 cm long, 3–4 cm across, 6 mm thick basally, tapering evenly, upperside with distinct white spots; **Fl.** 33–38 mm long, light pink.

**A. trichosantha** BGR. v. **trichosantha** (§ IV/C/16) (*A. percrassa* SCHWEINF.). – Ethiopia. – Acauline; **L.** densely rosulate, wider basally, gradually tapering, 75 cm long, 17.5 cm across, 2 cm thick at the base, upperside concave at the tip, fleshy, keeled below the tip with 4–5 Sp., green, slightly striate with intermediate white spots, margins with triangular, pointed, brown, incurved T.; **Infl.** to 3 m tall, Fl. 27 to 30 mm long, white, Infl. and Fl. finely hairy.

**A. — v. albo-picta** SCHWEINF. – **L.** with large, white, roundish to oblong spots.

**A. trigonantha** LEACH (§ IV/C/18d). – Ethiopia: Begemdir Prov. – Acauline, forming dense groups from basal offsets; **L.** c. 24 in a Ros., spreading-recurved, somewhat flaccid towards the dry tip, 30–45 cm long, 5–10 cm across, upperside pale green, with only a few oblong white spots, the upperside somewhat grooved above, the underside rounded with stronger white markings, border horny, with triangular or hooked T. ± 3 mm long; **Infl.** 60–90 cm tall, a lax, cylindrical-tapering Rac. 7 cm long, Fl. with pedicels 15 mm long, Fl. tube thick, fleshy, triangular, 3.5 cm long, orange-scarlet.

**A. trothai** BGR. (§ IV/C/18d). – Tanzania: Tabora. – Acauline; **L.** to 60 cm long, 3 cm across, linear-ensiform; **Infl.** 1 m tall, Fl. 4 to 4.5 cm long. Insufficiently known species.

**A. turkanensis** CHRISTIAN (§ IV/C/16). – Kenya: Turcana Desert (W. of L. Rudolph). – St. c. 45 cm tall, leafy, becoming prostrate; **L.** 14–18, ovate-oblong, tapering, 40 cm long, 6–7 cm across, 1.5 cm thick, both sides without many spots, spots being oblong and pale green, the upperside broadly concave, the underside convex, margins narrow, horny, T. whitish, triangular, spreading, 2 mm long and 12 mm apart; **Infl.** to 92 cm tall, Fl. 2.5 cm long, red with fine spots, the lobe-tips having a pink border.

**A. tweediae** CHRISTIAN (§ IV/D/23). – Kenya: on the Uganda border. – **Ros.** acauline, solitary; **L.** 14–16, ovate-tapering, pointed, 45–60 cm long, 14 cm across the lower part, 17 mm thick, green on the upperside with scattered, pale green, elliptical spots, or else unmarked, broadly concave, the underside more strongly spotted, convex, margins very narrow, sharp, yellowish-white, horny T. brown-tipped, triangular, curved, 4 mm long, 14–18 mm apart; **Infl.** to 1.45 m tall, Fl. 2.5 cm long, red with yellow margin.

**A. ukambensis** REYN. (§ IV/C/18c). – Kenya. – **Ros.** acauline or very short-stemmed, dense; **L.** 30–40, to 50 cm long, 10–12 cm across at the base, upperside slightly grooved above, grey-green with a reddish tinge, distinctly longitudinally striped, with or without scattered, oval-lentiform, white spots in the lower part, the underside convex, with spots as for the upperside but H-shaped, margins sinuate-dentate, T. compressed, deltoid, red-brown, 3–4 mm long, 6–10 mm apart; **Infl.** 50 cm tall, Fl. 4 cm long, light red.

**A. umfoloziensis** REYN. (§ IV/B/10/A/b). – Natal. – **Ros.** acauline or with St. 30 cm tall, either solitary or forming groups; **L.** c. 20, lanceolate-tapering, 20–25 cm long, the tip often drying up, 8–9 cm across at the base, upperside flat or slightly concave, green or brown-green, with numerous white, oblong marks irregularly arranged in ± indistinct transverse bands, the underside convex, paler green, ± spotted, somewhat lined, margins sinuate-dentate, T. horny, sharp, triangular, 3–5 mm long, 10–15 mm apart, straight or curved; **Infl.** 1–1.5 m tall, Fl. 33–38 mm long, coral-red.

**A. vacillans** FORSK. (§ IV/C/15) (*A. vaccillans* err. est). – S.Arabia: Yemen. – **St.** usually prostrate; **L.** ensiform, gradually tapering, dull deep green, almost leathery, upperside concave towards the tip, 45–65 cm long, 10–13 cm across, 15–18 mm thick, margins red-brown, T. 2–3 mm long, 1–1.5 cm apart; **Infl.** 1.5–2 m tall, Fl. 3 cm long, red.

**A. vanbalenii** PILLANS (§ IV/E/28a). – Natal: Zululand. – **St.** 20–30 cm tall, simple; **L.** rosulate, 70–80 cm long, 12–14 cm across, 12–14 mm thick at the base, lanceolate to oblong-tapering, often curved sideways, much grooved, underside convex, light green with numerous dark green nerves, margins sinuate-dentate, horny, red-brown, T. 3–4 mm long, triangular, sharp, red-brown; **Infl.** c. 80 cm tall, Fl. 3.5 cm long, reddish.

**A. vandermerwei** REYN. (§ IV/B/10/A/5a) (*A. angustifolia* GROENEW.). – N.Transvaal. – **Ros.** acauline, offsetting to form large groups; **L.** 15–20, c. 60 cm long, 3.5 cm across, tapering to the tip, grooved on the upperside above, dark green with large, white, oblong marks arranged ± in transverse bands, the underside convex, paler green, less distinctly marked, margins sinuate-dentate, T. 3–4 mm long, 1–1.5 cm apart, green, curved; **Infl.** 1 m tall, Fl. 3 cm long, pale pink-red.

**A. vaombe** DECORSE et POISSON v. **vaombe** (§ XI/α). – S.W.Madag. – **St.** simple, to 6 m tall, covered with dry L. remains, c. 20 cm ⌀; **Ros.** dense; **L.** 30–40, to 1 m long, 15–20 cm across at the base, margins much curved upwards, T. triangular, 6 mm long and across, 1–1.5 cm apart, roundish-tapered; **Infl.** 2–4, Fl. to 28 mm long, vivid carmine.

**A. — v. poissonii** R. DECARY. – **St.** 4–5 m tall, weaker and thinner; flowering St. regularly curved upwards so that they project somewhat like a coronet above the L. Ros.

**A. vaotsanda** R. DECARY (§ XI/a). – S.W.-Madag. – **St.** to 4 m tall, covered throughout its length with dry L. remains, simple; **L.** to 1.4 m long, in mature plants much recurved, otherwise similar to **A. vaombe**; **Infl.** much branched, Fl. to 22 mm long, yellow-orange.

**A. variegata** L. v. **variegata** (Pl. 19/3) (§ IV/B/9) (*A. punctata* HAW.). – Cape: Lit. Namaqualand, Karroo; Botswana. – **St.** not normally discernible, L. dense, in 3 series or rosulate, with underground offsets, St. to 30 cm tall; **L.** erect-inclined, 12 cm or more in length, to 3.5 cm across, lanceolate, concave on the upperside, keeled on the underside, dark green with oblong white marks in irregular transverse bands, margins cartilaginous with white T.; **Infl.** c. 30 cm tall, Fl. 3.5–4.5 cm long, flesh-pink to scarlet.

**A.** — v. **haworthii** BGR. – L. ovate-triangular, spots less numerous, arranged on the upperside in narrow bands.

**A. venenosa** ENGL. (§ IV/C/18b). – Angola: Lunda D. – Acauline; **L.** lanceolate, 7 cm across at the base, gradually tapering, c. 35 cm long, thick, margins sinuate-dentate, T. deltoid, 6 to 7 mm long with curved tips, Fl. 27–30 mm long.

**A. venusta** REYN. (§ IV/D/24). – Tanzania: Tabora. – **Ros.** acauline, single or in small groups; **L.** c. 20, c. 50 cm long, 9 cm across at the base, gradually narrowing to the incurved tip, the L. tips often intertwined, the upper surface somewhat grooved above, dull grey-green, with many spots over the entire surface, underside rounded, with spots more numerous and smaller, margins sinuate-dentate, pink, T. triangular, curved forwards, 3 mm long, 8–10 mm apart, pink; **Infl.** much branched, Fl. 32 mm long, finely hairy, pale scarlet.

**A. verdoorniae** REYN. (§ IV/B/10/A/2b). – Transvaal. – **Ros.** acauline, single; **L.** 20 cm long, lanceolate, tapering, to 30 cm long, the 10 cm tip being desiccated, 8–9 cm across, upperside somewhat concave, grooved towards the tip, bluish-green, either with or without whitish oval marks, underside convex, margins sinuate-dentate, T. triangular, sharp, reddish-brown, 4–5 mm long, c. 1 cm apart, connected by a thick, red-brown, horny margin; **Infl.** c. 1 m tall, Fl. 36 mm long, coral-red.

**A. verecunda** POLE EVANS (§ III/2). – Transvaal: Middelburg D. – **St.** short; **L.** 8–10, sometimes distichous, 25–35 cm long, 8–10 mm across, linear, underside convex, with white spots at the base, marginal T. white, small, 6–7 mm apart; **Infl.** 25 cm tall, Fl. to 3 cm long, peach-coloured to scarlet.

**A. versicolor** GUILL. (§ XI/c). – S.E.Madag. – **Ros.** almost acauline, forming mats by means of offsets; **L.** 10–12, linear, to 15 cm long, c. 2 cm across, grey-green, grey-waxy-pruinose, often with transverse stripes, the dentate tip rounded, margins white, cartilaginous, serrate, T. whitish, 1 mm long, 3–8 mm apart; **Infl.** 40 cm tall, Fl. pendulous, coral-red.

**A. veseyi** REYN. (§ IV/D/24). – Zambia. – Plant pendulous; **St.** 30–40 cm long, 2 cm ⌀, offsetting basally to form 5 or more **Ros.**; **L.** c. 12, c. 40–50 cm long, 3 cm across at the base, gradually narrowing to a long, pointed tip, upperside dull grey-green with a reddish sheen, slightly grooved, the entire surface having numerous, dull white, lentiform spots, the underside rounded but otherwise as for the upperside, margins with stout, triangular, horny T. 1 to 2 mm long, to 2 cm apart, the T. in the upper part being smaller and finally missing; **Infl.** c. 60 cm long, pendulous, Fl. 2.5 cm long, pale yellow.

**A. viguieri** PERR. (Pl. 19/4) (§ XI/c). – S.W. Madag.: Tuléar. – Acauline, offsetting basally to form dense mats, the offsets frequently hanging down, up to 1 m long, the Ros. often appearing asymmetrical since most of the L. are directed downwards; **Ros.** of 12–15 L., usually very falcate, 25–60 cm long, 7–9 cm across at the base, narrowing evenly towards the pointed tip, whitish-green because of the very waxy coating, with a narrow cartilaginous border, T. 0.5–0.7 mm long, closely set; **Infl.** c. 1 m long, Fl. pendulous, scarlet.

**A. virens** HAW. v. **virens** (§ IV/B/4). – S.Afr. – Acauline, freely offsetting, forming groups; **L.** numerous, 20 cm long, 2–3 cm across, narrow-lanceolate, gradually tapering, light green with somewhat darker stripes and with scattered, light, round spots, marginal T. widely spaced, light-coloured; **Infl.** 50–60 cm tall, Fl. 4 cm long, red.

**A.** — v. **macilenta** BAK. – Smaller variety; L. thinner, upperside somewhat grooved, reddish.

**A. viridiflora** REYN. (§ IV/B/13). – S.W.Afr. – **Ros.** solitary, dense; **L.** 50–60, c. 40 cm long, 8 cm across, light green, somewhat striate, the upperside somewhat concave towards the tip, underside convex with several light spots, marginal T. red-brown, 2 mm long, 2–5 mm apart; **Infl.** to 1.5 m tall, Fl. 53 mm long, green.

**A. vituensis** BAK. emend. LEACH (§ IV/D/22). – Kenya: Tana R. area. – ♄ with a few Br. from the base, erect to prostrate; **St.** thin, to 40 cm long; **L.** very narrowly ovate-tapering, to 27 cm long, 3 cm across, 12 mm thick, arranged in a spiral, tip compressed, with a basal Sh., upperside slightly grooved towards the tip, yellow-green, bronze-coloured to brown towards the tip and the margins, indistinctly striate, with numerous lentiform to H-shaped white spots, margins sharp, thin, whitish, sinuate-dentate, T. sharp, straight or hooked, 3–4 mm long, orange-brown; **Infl.** simple, Rac. 7–8 cm long, Fl. 27–29 mm long.

**A. vogtsii** REYN. (§ IV/B/10/A/3b). – Transvaal. – **St.** 20 cm tall; **L.** 16–20, rosulate, 20 to 25 cm long, 5–6 cm across, tapering-lanceolate, terminating in a short, hard and sharp point, upperside somewhat concave, grey-green, indistinctly lined, with numerous white spots in ± wavy transverse bands, the elliptical spots often merging to an H-shape, underside convex, darker green, spots numerous, with a pale brown M.line, margins sinuate-dentate, T. deltoid, horny, pale brown, 3 mm long, 10–15 mm apart; **Infl.** c. 60 cm tall, Fl. 34 mm long, scarlet.

**A. volkensii** ENGL. (§ IV/C/19c). – Tanzania: Kilimanjaro area. – **St.** 4–5 m tall; **L.** in a dense Ros., ensiform, 7 cm across at the base, c. 60 cm long, green, margins with triangular, curved T.

2–3 mm long and 10 mm apart; **Infl.** corymbose, Fl. 3 cm long, red.

**A. vossii** REYN. (§ III/2). – N.Transvaal. – Acauline or very short-stemmed; **L.** 14–22 in 2 series or normally rosulate, 50 cm long, 3 cm across at the base, the upperside somewhat concave, the underside convex and boat-shaped, the upperside having scattered white marks and several Sp. on the upper edge, the spots on the underside being round, white, also with a few Sp.; **Infl.** 50 cm tall, Fl. 28 mm long, scarlet.

**A. vryheidensis** GROENEW. (§ V). – Natal: Vryheid D. – Acauline or short-stemmed, with few Br.; **Ros.** dense; **L.** 20–30, 60 cm long, 12 cm across, 1.5–2.5 cm thick, light green, the tip recurved, marginal T. 2–3 mm long, 1 to 1.5 cm apart, reddish; **Infl.** 1.5 m tall, Fl. 18 mm long, pink-red.

**A.** × **weingartii** BGR. – Hybrid: **A. variegata** × **A. humilis**. – **Ros.** acauline, 10 cm ⌀; **L.** arranged spirally or in 3 series, erect, triangular, narrowing to the base, upperside very concave, underside obliquely keeled, 8–10 cm long, dull green, underside towards the tip with white, sharp tubercles, margins dentate, cartilaginous; **Infl.** 20 cm tall, Fl. 3 cm long, purple-red.

**A. wickensii** POLE EVANS v. **wickensii** (§ IV/C/17). – N.Transvaal: Pietersburg D. – **Ros.** acauline or very short-stemmed, often grouping; **L.** c. 45, lanceolate-ensiform or gradually tapering, 68–80 cm long, 11–12 cm across at the base, somewhat concave on the upperside, underside rounded, both surfaces grey-green to bluish-green, leathery, margins with dark brown to black, triangular, sharp T. 2 mm long, 2–10 mm apart; **Infl.** 1–1.5 m tall, Fl. 3.5 cm long, chrome-yellow.

**A.** — v. **lutea** REYN. – E.Transvaal. – **Fl.** lemon-yellow to yellow.

**A. wildii** (REYN.) REYN. (§ III/1) (*A. torrei* v. *w.* REYN.). – Rhodesia: Melsetter D. – **Ros.** acauline; **L.** 4–6, distichous, 15–20 cm long, margins with horny, white T. 0.5 mm long, 1–2 mm apart; **Infl.** 25 cm tall, Fl. orange.

**A. wilsonii** REYN. (§ IV/C/18a). – Uganda: N.Reg. – **Ros.** single or in small groups, St. short or often 50–80 cm tall, 6 cm ⌀, prostrate; **L.** c. 24, tapering-lanceolate, to 25 cm long, 9 cm across at the base, 2 cm thick below, upperside flat to slightly concave, concolorous deep green with a yellowish sheen, underside sometimes with white spots in the bottom 1/4, margins sinuate-dentate, pale brown, horny, T. flat-triangular, sharp, reddish, 3–4 mm long, 1 cm apart; **Infl.** 50–60 cm tall, Fl. 28 mm long, pale scarlet.

**A. wollastonii** RENDLE (§ IV/B/10/A/2a). – Uganda. – Insufficiently known species, acc. REYNOLDS possibly a form of **A. lateritia**; **L.** ensiform, 15 cm long, 2.5–3 cm across at the base, narrowing to the tip; **Fl.** 3 cm long.

**A. woolleyana** POLE EVANS (§ III/1). – Transvaal: Barberton. – Related to **A. chortolirioides**; **St.** 6–12 cm tall; **L.** 5–10 in even rows, 6–10 cm long, 4–5 mm across, glabrous, linear, marginal T. small; **Infl.** 18–30 cm tall, Fl. 32 mm long, red, Tep. with dark nerves.

**A. wrefordii** REYN. (§ IV/C/18c). – Uganda: N.Reg. – **Ros.** solitary, acauline or very short-stemmed; **L.** 24, tapering-lanceolate, c. 60 cm long, 15 cm across at the base, 1.5 cm thick, terminating in a red-brown Sp., upperside dull greenish-grey, often with a reddish-brown tinge, very indistinctly lined, grooved in the upper part, underside convex, unspotted, margins sinuate-dentate, T. triangular, sharp, red-brown, 4 mm long, 1–1.5 cm apart, hooked forwards; **Infl.** to 1.2 m tall, Fl. 22 mm long, scarlet or orange-coloured.

**A. yavellana** REYN. (§ IV/D/21). – Ethiopia: Sidamo. – ♄; **St.** 1 m tall, or prostrate and 1–3 m long, with L. on the top 20 cm; **L.** 16–20, Sh. striate, blade c. 40 cm long, 6–8 cm across, tapering, coppery-brown, upperside flat and somewhat grooved towards the tip, underside convex, marginal T. triangular, sharp, 2–3 mm long, 1–1.5 cm apart; **Infl.** 60–90 cm tall, Fl. 17 mm long, scarlet.

**A. zebrina** BAK. (Pl. 20/2) (§ IV/B/10/A/5b) (*A. platyphylla* BAK., *A. lugardiana* BAK., *A. baumii* ENGL., *A. bamangwatensis* SCHOENL.). – Transvaal; Botswana; Angola; S.W.Afr. – **Ros.** acauline or short-stemmed, offsetting; **L.** 15–25, linear-lanceolate, tapering above the middle, 15–30 cm long, 6–7 cm across, flat on the upperside, grooved towards the tip, underside convex, fleshy, green, often reddish, with ± numerous, oblong spots merging in irregular transverse bands, marginal T. brown, 6–7 mm long, 10–16 mm apart; **Infl.** 1–1.5 m tall, Fl. 3–3.5 cm long, deep red.

× **Aloella** ROWL. Liliaceae. – Intergeneric hybrids: **Aloe** × **Aloinella**. Acc. ROWLEY **Aloinella haworthioides** (BAK.) LEMÉE is at least as worthy of generic rank as the other segregates from **Aloe**. If returned to **Aloe**, the following would be listed as intersectional hybrids under **Aloe**. – Cultivation as for **Aloe**.

× **A. 1.** – **Aloe bellatula** REYN. × **Aloinella haworthioides** (BAK.) LEMÉE. Similar to **A. haworthioides** but **L.** dull purplish and spotted, broader, more coarsely toothed with the T. mainly confined to the margins. Hort. Makin.

× **A. 2.** – **Aloe descoingsii** REYN. × **Aloinella haworthioides** (BAK.) LEMÉE. Intermediate between the parents. – Hort. Rowley.

× **Alolirion** ROWL. Liliaceae. – Intergeneric hybrid: **Aloe** × **Chortolirion**. – Cultivation as for **Aloe**.

× **A. 1.** – **Aloe striatula** HAW. × **Chortolirion tenuifolium** (ENGL.) BGR. Vigorous, erect growing, with a lax rosette of linear-lanceolate, soft, grass-green, not very succulent **L.** which are U-shaped in cross-section and finely serrulate at the margin. Hort. Kimnach.

× **Aloloba** ROWL. Liliaceae. – Intergeneric hybrid: **Aloe** × **Astroloba**. – Cultivation as for **Aloe**. – Several such hybrids are known to exist, but so far none has been named or described. Hort. Makin.

× **Alworthia** ROWL. Liliaceae. – Intergeneric hybrid: **Aloe** × **Haworthia.** – Only one cross is on record.

× **A. 'Black Gem'.** – **Haworthia cymbiformis** (HAW.) DUV. × **Aloe spec.** – **Ros.** compact, of few, thick, acute, very fleshy **L.** of a uniform blackish green colour. – Needs sharp drainage and warmth in winter.

**Ammocharis** HERB. Amaryllidaceae. – Africa. – Bulbous plant. – Greenhouse, warm. Propagation: seed.

**A. coranica** (KER-GAWL.) HERB. (Pl. 20/3) *Amaryllis c.* KER-GAWL., *Brunswigia c.* (KER-GAWL) HERB., *A. c. v. pallida* LINDL., *A. coccinea* PAX, *Crinum c.* (PAX) FRITSCH, *A. paveliana* SCHINZ, *C. p.* (SCHINZ) FRITSCH p. part., *C. falcatum* JACQ., *A. f.* (JACQ.) HERB., *Haemanthus f.* (JACQ.) THUNBG. p. part., *Brunswigia uitenhagensis* ECKL.). – Angola; Rhodesia; S.W.Afr.; Botswana; Lesotho; Cape; Natal; Orange Free State; widely distributed in all these areas. – Bulb ovoid, 16–20 cm ⌀, 25 cm long including the neck which is 5 cm long, bulb covered with stout, papery, dry, brown, glossy layers of skin which are striate, hairless and smooth; **L.** c. 15 or fewer, strap-shaped, varying considerably in length and breadth, 0.6–7.5 cm across, 2.3–115 cm long, striate, smooth, glabrous, green or bluish, margins rough; **Fl.-Sc.** 6–35 cm long, umbel of 3–56 Fl., bracts filiform, Fl. with pedicel up to 6 cm long, tube cylindrical, up to 2.5 cm long, lobes lanceolate, to 5.5 cm long, 5 mm across, twisted spirally towards the tip, copper-brown or carmine, the Pet. often with a white M.line both inside and outside, Fl. sweetly scented.

**Anacampseros** L. Portulacaceae. South West Africa; Central and East Africa; South Australia. – Low, succulent ♃ of different forms; Sep. 2, Pet. 5, soon withering; Fr. an elongate-conical papery capsule; seeds numerous. Greenhouse, warm. Propagation: seeds, cuttings.

## Division of the Genus Anacampseros into Sections by K. v. Poellnitz

**Sect. I. Avonia** E. MEY. – **Roots** tuberous; **shoots** simple or branched; **L.** very small, ± roundish, completely covered by the silver-white parchment-like stipules, between which are often ± long hairs; **Fl.** mostly solitary, terminal, often cleistogamous. – Species: A. albissima, alstonii, bremekampii, buderiana, decipiens, dinteri, fissa, herreana, meyeri, neglecta, ombonensis, omaruruensis, papyracea, quinaria, recurvata, rhodesica, ruschii, schmidtii, somaliensis, ustulata, variabilis, wischkonii.

**Sect. II. Anacampseros** (Sect. *Telephiastrum* DILL.). – Perennials; **roots** often somewhat tuberously thickened; **Shoots** ± numerous, often dichotomously branched; **L.** alternate, often spirally arranged, subovate-acute, thick, very fleshy, often with bristly hairs or bristles from the leaf axils, leaves often cobwebby; **Infl.** terminal, ± scorpioid, Fl. 2–4, with whitish or red petals, often opening for only a few hours, or not opening at all, nevertheless producing seed by self-fertilization. – Species: A. affinis, albidiflora, alta, angustifolia, arachnoides, baeseckii, comptonii, crinita, densifolia, depauperata, filamentosa, gracilis, karasmontana, lanceolata, lanigera, marlothii, namaquensis, nebrownii, nitida, paradoxa, parviflora, poellnitziana, retusa, rubroviridis, rufescens, schoenlandii, starkiana, subnuda, telephiastrum, tomentosa, truncata.

**Sect. III. Tuberosae** v. POELLN. – Subterranean **tuber** large; **L.** rosulate crowded, with hairs from the leaf-axils; **Infl.** scorpioid, Fl. hardly opening and cleistogamous. – Species: A. australiana.

**Sect. IV. Rosulatae** DTR. – Shoots much shortened; **L.** rosulate, fleshy, large, with hairs from the L.-axils; **Infl.** scorpioid. – Species: A. dielsiana.

## Key to the Species of Anacampseros by K. v. Poellnitz[1]), completed by R. Rawé & G. D. Rowley

I. L. small, mostly hidden beneath scale-like, parchmenty Stip.; Fl. mostly solitary.

Sect. I. Avonia E. MEY.
1. Stip. with bristles at the base ................................................ 2.
1a. Stip. without bristles at the base ............................................. 8.
2. Br. 6–10 mm ⌀ ............................................................... 3.
2a. Br. 4 mm ⌀ ................................................................. 6.
3. Stip. hooked-recurved ................................................... ruschii
3a. Stip. not recurved ........................................................... 4.
4. Stip. loosely adpressed .................................................. meyeri
4a. Stip. tightly adpressed, tongue-shaped to broad ovate ....................... 5.
5. Margins entire ........................................................ papyracea
5a. Margins serrate to dentate ............................................ variabilis

---

[1]) Botan. Jahrb. **LXV**, part 4/5.

6. Stip. tightly adpressed .................................................. 7.
6a. Stip. not adpressed, often a little recurved .......................... neglecta
7. Stip. tongue-shaped, plants densely ramose ........................... buderana
7a. Stip. often with a brown spot, plants sparsely ramose ............... albissima
7b. Stip. acute, with a brown median nerve ............................. herreana
8. Stip. divergent from the branches, recurved ........................... 9.
8a. Stip. adpressed, not always completely, but never recurved ........... 12.
9. Stip. split at the tip ................................................. fissa
9a. Stip. not split at the tip ............................................ 10.
10. Br. ramose, stipules with a yellow-green median nerve ............... recurvata
10a. Br. not ramose ....................................................... 11.
11. Stip. round-ovate, acute, median nerve prominent ................... wischkonii
11a. Stip. narrow-ovate ................................................ omaruruensis
11b. Stip. round-ovate, acute, without a median nerve ................... rhodesiaca
12. Br. 4–5 mm ∅ ........................................................ 13.
12a. Br. 2–3 mm ∅ ....................................................... 16.
13. Br. creeping, Stip. very long pointed ............................... somaliensis
13a. Br. not creeping, Stip. blunt to acute ................................ 14.
14. L. 3 mm ∅, Stip. 3–4 mm long, margin entire ......................... dinteri
14a. L. and Stip. distinctly smaller, margin entire or serrated ............ 15.
15. Stip. tongue-shaped ................................................. buderiana
15a. Stip. broad-ovate, tip acuminate ................................... schmidtii
16. Br. leaved in straight rows, Stip. deltoid ............................ alstonii
16a. Br. spirally-leaved ................................................... 17.
17. Br. ramose; Stip. broad-ovate, tipped brown, acuminate ............. ustulata
17a. Br. not ramose ....................................................... 18.
18. Stip. acute, white .................................................. ombonensis
18a. Stip. bluntish, often with one or two spots ....................... bremekampii
18b. Stip. truncated at the tip, white .................................. decipiens
18c. Stip. broad-ovate to ovate-triangular, white or with a brown spot ... quinaria

II. L. large, fleshy, mostly rosetted; Stip. when present resolved into bristly hairs; Fl. in cymose Infl.

 i. L. almost terete to clavate, covered when young with matted hairs so that the Ros. appear as grey, woolly tufts.

Sect. IV. **Rosulatae** Dtr. ... (1 spec. only) .............................. dielsiana

 ii. L. otherwise, flattened, not densely hairy

  A. Evergreen; roots slender or at most fusiform

Sect. II. **Anacampseros** (*Telephiastrum* Dill.).
1. L. naked, never tomentose ............................................. 2.
1a. L. tomentose .......................................................... 9.
2. St. caudiciform, dwarf plant, leaves up to 5 mm long ................. comptonii
2a. St. not caudiciform ................................................... 3.
3. L. truncated at right angles, plant 2 cm high ........................ truncata
3a. L. and plant larger .................................................... 4.
4. L. diverging, narrow linear-lanceolate ............................ angustifolia
4a. L. ascending, longish, elongate to ovate .......................... marlothii
4b. L. spreading, shorter ................................................ 5.
5. Stip. hairs shorter than the L. ....................................... 6.
5a. Stip. hairs longer than the leaves .................................... 8.
6. L. up to 20 mm long, very thick, rounded-obovate ...... arachnoides v. hispidula
6a. L. 20–25 mm long ..................................................... 7.
6b. L. larger, round-obovate to lanceolate, smooth or rough,
 rarely reddened, Stip. hairs few ................................... telephiastrum
7. Seeds broadly winged, leaves obovate-lanceolate to lanceolate ........ lanceolata
7a. Seeds not broadly winged, leaves ovate to oblanceolate, smooth,
 lower side reddened, Stip. hairs numerous .......................... rufescens
8. L. ovate or wedge-shaped ............................................. affinis
8a. L. rounded or obovate .............................................. nebrownii
8b. L. ovate, tip truncated, Stip. hairs almost absent .................... nitida
9. Only the younger L. tomentose, older L. naked ........................ 10.
9a. L. very tomentose, even the old L. rarely naked ...................... 16.
10. L. tip truncated, plants to 6 cm high .............................. retusa
10a. L. tip not truncated ................................................. 11.
11. Young L. sparsely tomentose ......................................... 12.
11a. Young L. very tomentose ............................................. 15.

| | | |
|---|---|---|
| 12. | Stip. hairs varied in length | 13. |
| 12a. | Stip.-hairs shorter than the leaves | 14. |
| 13. | L. tipped with an acute, 2–4 mm long point | schoenlandii |
| 13a. | L. point very small, often absent | rubroviridis |
| 14. | L. obovate to round-obovate, very thick | arachnoides |
| 14a. | L. obovate-lanceolate, not very thick | gracilis |
| 15. | L. 7–12 mm long | starkiana |
| 15a. | L. 4 mm long | subnuda |
| 16. | Pet. shorter than sepals | namaquensis |
| 16a. | Pet. as long as or longer than the Sep. | 17. |
| 17. | Stip. hairs shorter than the L., Pet. bicoloured | baeseckei |
| 17a. | Stip. hairs as long as the truncated leaves | 18. |
| 17b. | Stip. hairs longer than the L. | 19. |
| 18. | Plants up to 35 cm high, densely tomentose | alta |
| 18a. | Plants up to 9 cm high, less tomentose | alta v. humilis |
| 19. | L. red-brown | tomentosa v. margaretae |
| 19a. | L. green | 20. |
| 20. | L. truncated at right angles | tomentosa |
| 20a. | L. truncated obliquely | filamentosa |
| 20b. | L. rounded, ovate or wedge-shaped | 21. |
| 21. | Stip. hairs numerous | 22. |
| 21a. | Stip. hairs sparse | 23. |
| 22. | L. ends warty | paradoxa |
| 22a. | L. ends not warty | crinita |
| 23. | L. long-haired | lanigera |
| 23a. | L. short-haired | 24. |
| 24. | Pet. small, red | parviflora |
| 24a. | Pet. large | 25. |
| 25. | Fl. white | albidiflora |
| 25a. | Fl. pink | 26. |
| 26. | Bristle hairs almost absent | depauperata |
| 26a. | Bristle hairs numerous | 27. |
| 27. | L. c. 8 mm long | densifolia |
| 27a. | L. 10–12 mm long | 28. |
| 28. | L. 5 mm wide; roots slender | karasmontana |
| 28a. | L. 8 mm wide; roots thickened | poellnitziana |

A. Shoots deciduous, from large subterranean tubers.
**Sect. III. Tuberosae** v. POELLN. .... (1 spec. only) .......................... australiana

**A. affinis** PEARS. et STEPH. (§ II). – Cape: Lit. Namaqualand. – **St.** short; **L.** crowded at Br.tips, obovate to conical, pointed to tapering, underside slightly convex, 8–12 mm long, 4–6 mm across, with numerous bristle-hairs, white, glossy, 3 cm long; **Infl.** 15 cm tall, Fl. 1–4. Similar to **A. lanceolata** and **A. rufescens**.

**A. albidiflora** v. POELLN. (§ II). – Cape: origin unknown. – Similar to **A. lanigera**; **L.** oblong-cuneate, underside rounded, reddish-green, felted, c. 7 mm long, c. 5 mm across, bristle-hairs numerous, curling, glossy, white, 2 cm long; Fl. white.

**A. albissima** MARL. v. **albissima** (§ I) (*A. avasmontana* DTR.). – Cape: Lit. Namaqualand; S.W.Afr.: Gr. Namaqualand. – **St.** numerous, branching, ± curved upwards, 4 cm long, 3 to 4 mm ⌀; **L.** arranged in a spiral, very small, hidden by the firmly appressed, ovate-blunt, often brownish-flecked stipules with entire or laciniate margins, some hairs present between the Stip.; **Fl.** 1–3, white.

**A. —** v. **caespitosa** v. POELLN. (*A. avasmontana* v. c. v. POELLN.). – S.W.Afr. – **St.** with numerous terminal Br., only the latter projecting from the soil and forming mats; **Br.** simple, about 8 cm long, erect to almost creeping.

**A. —** v. **laciniata** v. POELLN. – S.W.Afr. – Stip. lanceolate, laciniate.

**A. alstonii** SCHOENL. (Pl. 21/1) (§ I). – S.W.-Afr.: Gr. Namaqualand; Cape: Lit. Namaqualand. – **Caudex** napiform, flattened above and up to 6 cm or more in ⌀; **St.** simple, very numerous, up to 300, 2–3 cm or less in length, 2 mm thick; **L.** in 5 straight rows, concealed by the silvery, pointed, triangular, appressed Stip. 2 mm long; **Fl.** about 3 cm ⌀, white.

**A. alta** v POELLN. v. **alta** (§ II). – Cape: Lit. Namaqualand. – Branching, c. 35 cm tall, Br. grey; **L.** crowded, brownish, ± obovate, c. 5 mm long, c. 3 mm thick, the upperside slightly convex, the underside more so, tuberculate towards the tip, bristle-hairs solitary, curling, white; **Fl.** deep pink.

**A. —** v. **humilis** v. POELLN. – Only 8 cm tall; L. 6 mm long and thick, very finely felted.

**A. angustifolia** (HAW.) SWEET (§ II) (*Rülingia a.* HAW.). – Origin unknown. – **St.** with short Br.; **L.** crowded, narrowly linear-lanceolate, spreading. Insufficiently known species.

**A. arachnoides** (HAW.) SIMS v. **arachnoides** (§ II) (*Portulaca a.* HAW., *Rülingia a.* HAW., *A. a.* v. *rubens* SOND., *P. r.* HAW., *R. r.* HAW., *A. r.* SWEET., *Talinum a.* AIT., *T. retusum* WILLD., *T. rubens* HORT. ex STEUD.). – Transvaal, and widely distributed in E.Cape. – **St.** c. 5 cm long, densely leafy; **L.** ± ovate-round, tapering, thick, green, c. 2 cm long, c. 1.5 cm across, tuberculate at the tip, covered with a few white threads like a spider's web, bristle hairs few in number, not longer than the L., white; **Fl.** c. 3 cm ⌀, whitish-pink.

**A.** — v. **hispidula** (BGR.) v. POELLN. (*A. h.* BGR., *A. asperula* BGR.). – Cape: Uitenhage D. – **L.** always white-felty; **Fl.** white.

**A. australiana** J. M. BLACK (§ III). – S.W. Austral. – **Root** tuberous, 6 cm long, 4 cm thick; **St.** thin, subterranean; **L.** in a tight Ros., lanceolate, 15–25 mm long, 5–10 mm across, fleshy, glabrous, bristle-hairs short, whitish; **Fl.** pink.

**A. baeseckei** DTR. (§ II). – S.W.Afr.: Gr. Namaqualand; Cape: Lit. Namaqualand. – Branching, 5 cm tall, **Br.** c. 1 cm ⌀; **L.** crowded, upperside roundish, underside ± tuberculate, 3–4 mm long and across, white-felted, with isolated whitish brown hairs on Br. tips, these being as long as the L.; **Fl.** c. 6 mm ⌀, carmine, Pet. with white border.

**A. bremekampii** v. POELLN. (Pl. 21/3) (§ I). – Cape: Langeberg. – **Root**stock short, c. 1 cm ⌀; **St.** numerous, simple, c. 1 cm long, c. 2 mm ⌀; **L.** arranged somewhat spirally, small, half-hidden by the Stip. which are 2 mm long and across, L. broadly ovate to triangular, silvery with somewhat brownish spots.

**A. buderana** v. POELLN. v. **buderana** (§ I). – Cape: Lit. Namaqualand. – Similar to **A. papyracea,** but smaller in all parts; **St.** small, forming circular mounds 15 mm ⌀, erect, c. 15 mm long, 4–5 mm ⌀, simple; **L.** arranged in a somewhat ascending spiral, very small, concealed by the closely appressed silvery-white Stip. 2.5–4 mm long, often with several basal hairs.

**A.** — v. **multiramosa** v. POELLN. – **St.** very freely branching.

**A. comptonii** N. S. PILLANS (§ II). – Cape: Calvinia D.; S.W.Afr.: near Lüderitz Bay. – Dwarf; **caudex** short, 10–15 mm tall, 1–2 cm ⌀, glabrous; **St.** short; **L.** 2–4, sessile, semi-ovate to roundish, with short, tapering tip, flat or somewhat grooved on the upperside, flesh-coloured to green, 3–5 cm long, 2–2.8 cm across, bristle-hairs 4–5 mm long; **Fl.** solitary, 5 to 6 mm ⌀, red.

**A. crinita** DTR. (§ II) (*A. baeseckei* v. c. DTR.). – S.W.Afr.: Gr. Namaqualand; Cape: Lit. Namaqualand. – **St.** 8 cm tall; **L.** ± crowded, c. 4 mm long, ± ovate-round, light green, immature L. white-felted, bristle-hairs about 15 mm long, numerous, wavy, whitish, reddish or reddish-brown; **Fl.** 2–4, reddish, 2 cm ⌀, Pet. with a white border.

**A. decipiens** v. POELLN. (§ I). – Transvaal: near Pretoria. – **Caudex** short, c. 2 cm ⌀; **St.** numerous, simple, c. 6 mm long, 2–3 mm ⌀ at the base; **L.** ± spirally arranged, partially concealed by the Stip., small, green, fleshy, Stip. broad-cordate, truncate, 2 mm across, c. 1.5 mm long, pure white, silvery.

**A. densifolia** DTR. (§ II). – S.W.Afr.: Gr. Namaqualand. – **St.** branching; **L.** crowded, ± obovate, both surfaces convex, slightly tuberculate towards the tip, glossy, red-green, white-felted, 8 mm long, c. 5 mm across, bristle-hairs numerous, 2 cm long, glossy, glabrous, white with brown tips; **Fl.** 2 cm ⌀, deep pink.

**A. depauperata** (BGR.) v. POELLN. (§ II) (*A. filamentosa* v. *d.* BGR.). – S.Afr.: Cape: Ceres Karroo. – As for **A. filamentosa; L.** somewhat smaller, bristle-hairs virtually absent; **Fl.** less numerous, pale pink.

**A. dielsiana** DTR. (§ IV). – S.W.Afr.: Gr. Namaqualand. – **Ros.** acauline, 2–2.5 cm ⌀, dense; **L.** 18–20, almost terete, 12–15 mm long, somewhat club-shaped at the tip, immature L. covered with floccose hairs, so that the Ros. is usually covered with grey wool; **Infl.** 6–10 cm tall, with 2–3 Fl. 1.5 cm ⌀, pink.

**A. dinteri** SCHINZ (§ I). – S.W.Afr. – **St.** numerous, creeping, c. 10 cm long, 4 mm ⌀; **L.** arranged spirally beneath the appressed pointed silvery stipules which are 3–5 mm long and c. 3 mm across; **Fl.** red.

**A. filamentosa** (HAW.) SIMS (Pl. 21/4) (§ II) (*A. intermedia* NICH., *Portulaca f.* HAW., *Rülingia f.* HAW., *Talinum f.* AIT.). – Cape: widely distributed. – Very variable species; **root** tuberous; **St.** 5 cm long, densely leafy; **L.** ovate to spherical, thick, more or less truncate-tipped, short-tapering, 6–10 mm long, tip rough, covered with fine white threads, bristle-hairs long, numerous, curly, whitish; **Infl.** 8 cm tall, Fl. 3–5, 3 cm ⌀, pink.

**A. fissa** v. POELLN. (§ I). – Transvaal: near Pretoria. – **Caudex** 2 cm long, 4 cm ⌀, flat above; **St.** numerous, ± erect, simple, 3 mm ⌀ at the base, thinner towards the tip; **L.** arranged ± spirally, half-concealed by the silvery Stip., with basal hairs, long-oval to truncate, c. 2 mm long and across, the tip being slit.

**A. gracilis** v. POELLN. (§ II). – Cape. – **St.** branching, Br. ± erect, 5–7 cm long; **L.** ovate-lanceolate, c. 8 mm long, c. 5 mm across, with a distinct apiculus, red-green, flat on the upperside, convex on the underside, immature L. somewhat felted on the upperside, bristle-hairs bushy, white, sometimes also solitary; **Fl.** 2.5 cm ⌀, white, Pet. striped with red.

**A. herreana** v. POELLN. (§ I). – Cape: Lit. Namaqualand. – **Root** thickened above; **St.** branching dichotomously, ± prostrate, to 1 cm long, 3 mm thick; **L.** arranged spirally, round, very small, concealed by the pointed, firmly adpressed Stip. which have a brown M.nerve.

**A. karasmontana** DTR. (§ II). – S.W.Afr.: Gr. Namaqualand. – **L.** crowded, ± cuneate, flat to somewhat convex on the upperside, underside convex, green to reddish, 10–12 mm long, c. 5 mm across, 4 mm thick, covered with white felt, bristle-hairs numerous, glossy, c. 2 cm long, somewhat curly; **Fl.** 2 cm ⌀, pink.

**A. lanceolata** (HAW.) SWEET v. **lanceolata** (§ II) (*Portulaca l.* HAW., *Rülingia l.* HAW., *Talinum l.* LINK.). – Cape: Karroo. – **L.** numerous on a short St., narrow-lanceolate to lanceolate, 25 mm long, 6–8 mm across, coming to a short, oblique point with a tiny Sp., underside roundish, bristle-hairs numerous, 5 mm long, white, curly; **Infl.** 8 cm tall, Fl. 1–4, 3 cm ⌀, red.

**A.** — v. **albiflora** v. POELLN. – Fl. white.

**A. lanigera** BURCH. (§ II). – Cape: Van-Rhynsdorp D., Karroo. – **St.** 2 cm ⌀; **L.** crowded, c. 4 mm long, across and thick, ± roundish to ± obovate, white-felted, tuberculate at the tip, bristle-hairs numerous, curly, white, as long as the L.; **Fl.** 2 cm ⌀, pink.

**A. marlothii** v. POELLN. (§ II). – Cape. – **Caudex** c. 2 cm long; **St.** c. 15 mm long; **L.** crowded at the St.tips, ± erect, narrowly oblong cuneate, rounded above, c. 26 mm long, c. 5 mm across, green, reddish at the base, conspicuously spotted, upperside somewhat convex, underside transversely grooved, bristle-hairs few, 5 mm long; **Infl.** c. 3 cm long, Fl. 2–3, 16 mm ⌀, carmine.

**A. meyeri** v. POELLN. v. **meyeri** (§ I). – Cape: Lit. Namaqualand. – Similar to **A. papyracea**; **St.** tuberously thickened, Br. fairly erect, 7 cm long, 1 cm ⌀; Stip. loosely appressed, whitish, usually with yellowish M.nerve.

**A.** — v. **minor** v. POELLN. – Smaller than the type.

**A. namaquensis** PEARS. et STEPHENS (§ II). – Cape: Lit. Namaqualand; S.W.Afr.: Gr. Karasberg. – Branching, c. 12 cm tall; **L.** 4–5 mm long and across, obovoid-spherical to cuneate, underside tuberculate, covered with white hairy felt, bristle-hairs yellowish, c. 6 mm long; **Fl.** solitary, 8–10 mm ⌀.

**A. nebrownii** v. POELLN. (§ II). – Cape. – Branching, 3 cm tall; **L.** crowded on Br.tips, roundish to elliptical-ovate, pointed to tapering, convex on both surfaces, 6–10 mm long, 4–7 mm across, bristle-hairs numerous, 25 mm long, ± curved, white; **Fl.** solitary, white.

**A. neglecta** v. POELLN. (§ I). – S.W.Afr.: Gr. Namaqualand. – Forming small mats; **St.** c. 4 mm ⌀; **L.** arranged spirally, reniform, 0.5 to 0.75 mm long, 0.5 mm across, covered by the silvery Stip. which are curved outwards at the tips and broadly ovate, pointed, c. 2 mm long, 1.5 mm across with numerous basal hairs.

**A. nitida** v. POELLN. (§ II). – Cape: Lit. Namaqualand. – **Root** thickened in its upper part; **L.** crowded, ± obovate, dark green, glossy, somewhat grooved at the tip, the underside somewhat convex, finely tuberculate, c. 12 mm long, c. 7 mm across, c. 5 mm thick, bristle-hairs few, somewhat longer than the L., rather curly.

**A. omaruruensis** DTR. (§ I). – S.W.Afr.: Omaruru. – Insufficiently known species which has not been found again.

**A. ombonensis** DTR. et v. POELLN. (§ I) (*A. omburensis* DTR.). – S.W.Afr. – **Caudex** tuberous, c. 7 mm ⌀, hemispherical above; **St.** numerous, simple, c. 1 cm long, 3 mm ⌀; **L.** short-reniform, arranged spirally, half-concealed by the silvery Stip., with a raised yellowish M.nerve, ovate-lanceolate, c. 1.5 mm across, 2 mm long; **Fl.** solitary, c. 12 mm ⌀, carmine.

**A. papyracea** E. MEY. (§ I). – Cape: Gr. Karroo; S.W.Afr. – **Root** tuberously thickened;

**St.** short with many Br., these usually prostrate, 5–6 cm long, 8–10 mm thick; **L.** arranged spirally, small, hidden by the firmly appressed, broad-ovate, blunt, white Stip.; **Fl.** greenish-white.

**A. paradoxa** v. POELLN. (§ II). – S.W.Afr. – Dwarf, branching ♄ c. 3 cm tall; **L.** arranged spirally, green to brownish green, white-felted, ± obovate, underside slightly convex, the tip blunt and tuberculate, 9 mm long, 5 mm across, 3 mm thick, bristle-hairs numerous, 2 cm long, curly, yellowish-tipped.

**A. parviflora** v. POELLN. (§ II). – Cape: Lit. Namaqualand. – Similar to **A. lanigera**; **L.** felted, even when mature; **Fl.** small, pink.

**A. poellnitziana** DTR. (§ II). – S.W.Afr. – **Root** thickened, St. to 25 mm tall; **L.** ± cuneate, c. 12 mm long, 8 mm across at the tip, 5 mm thick, initially somewhat convex, becoming flat to ± concave, the tip abruptly truncate with isolated tubercles, these sometimes merging, concolorous, glossy brown and ± felted, bristle-hairs numerous, short, glossy, white; **Fl.** 15 to 20 mm ⌀, pink.

**A. polyphylla** (HAW.) SWEET (*Rülingia p.* HAW., *Talinum p.* LINK.). – Insufficiently known species, probably misclassified.

**A. quinaria** E. MEY. (§ I). – S.W.Afr.: Gr. Namaqualand; Cape: Bushmanland. – **Caudex** fleshy, 8–25 mm long; **St.** to 15 mm long, Br. numerous, simple, 2.5–5 cm long, 2 mm ⌀; **L.** arranged spirally, hemispherical, covered by the broadly ovate to triangular silvery Stip. which have brownish spots; **Fl.** 12–15 mm ⌀, purple.

**A. recurvata** SCHOENL. (Pl. 21/2) (§ I). – Cape: Lit. Namaqualand. – **Root** with tuberous thickening; **St.** branching, Br. 4 mm ⌀, with **L.** arranged in a spiral, hemispherical, 1 mm long, 2 mm across, Stip. lanceolate, c. 3 mm long, curved outwards, with a light green M.nerve.

**A. retusa** v. POELLN. (§ II). – Cape: Clanwilliam D. – **Caudex** 15–20 mm long, thick, grey; **Br.** 4 cm long; **L.** crowded, initially only slightly hairy, the underside somewhat carinate towards the tip, both surfaces of the tip-area rough, the tip very abruptly truncate and grooved, bristle-hairs numerous, 15 mm long, white to yellowish; **Fl.** 2.5 cm ⌀, pink.

**A. rhodesica** N. E. BR. (§ I). – Rhodesia. – **Br.** numerous, simple, 4–10 mm long, 3 mm ⌀; **L.** arranged spirally and half-concealed by the silvery, roundish-ovate, short-pointed Stip. which are 2.5 mm long, 2 mm across and curved outwards at the tip; **Fl.** solitary, white to pale flesh-coloured.

**A. rubroviridis** v. POELLN. (§ II). – Cape: Lit. Namaqualand. – Dwarf ♄ c. 4 cm tall; **Br.** short, erect; **L.** c. 7 mm long, roundish, somewhat constricted midway, rather blunt and ± apiculate above, conspicuously white-cobwebby, with isolated white bristle-hairs; **Fl.** c. 15 mm ⌀, pink.

**A. rufescens** (HAW.) SWEET (Pl. 21/5) (§ II) (*A. arachnoides* HORT., *A. a.* v. *grandiflora* SOND., *A. filamentosa* DE WILD., *A. rubens* HORT. *Rülingia rufescens* HAW.). – Cape: Karroo. – Mat-forming; **caudex** thick; **St.** 5–8 cm tall, erect or creeping, branching dichotomously; **L.** in a dense spiral, obovate-lanceolate, tapering, c. 2 cm long, 1 cm across, thick, green, reddish on the underside, bristle-hairs numerous, c. 2 cm long, often wavy, white to yellow; **Infl.** 10 cm tall, Fl. 2–4, 3–4 cm ∅, pink.

**A. ruschii** DTR. et v. POELLN. (§ I). – S.W.-Afr.: Gr. Namaqualand; Cape: Lit. Namaqualand. – Resembles **A. papyracea**; **Br.** c. 4 cm long, c. 7 mm ∅; **L.** arranged spirally, Stip. somewhat yellowish, hooked and curved outwards at the tip, with numerous hairs at the base.

**A. schmidtii** (BGR.) v. POELLN. (§ I) (*A. quinaria* v. *s.* BGR., *A. qu.* KRAUS). – S.W.Afr. – **Caudex** fleshy, 2 cm ∅; **St.** numerous, c. 5 cm long, 4 cm ∅, creeping, simple, ± whitish; **L.** half-concealed by the Stip., lunate, c. 1.5 mm across, 0.5 mm long, arranged spirally, Stip. silvery, broad-ovate, tapering, c. 2 mm long; **Fl.** reddish.

**A. schoenlandii** v. POELLN. (§ II). – Cape. – Dwarf ♄ 8–10 cm tall; **Br.** short, erect or creeping, leafy above, with persistent bristle-hairs below; **L.** variable, obovate to obcordate to almost roundish, underside hemispherical, coming to a distinct point 2–4 mm long, lower L. conical to cylindrical-pointed and 2 mm long, upper ones 10 mm long, 7 mm across and thick, all L. covered with reddish felt, bristle-hairs few, whitish to yellow-grey; **Fl.** 18 mm ∅, pink.

**A. somaliensis** v. POELLN. (§ I). – Ethiopia: Ogaden. – **St.** creeping, rhizome-like, and rooting on the underside, c. 3 mm ∅; **Br.** to 25 mm long, c. 5 mm ∅; **L.** arranged spirally, half-covered by the silvery Stip., ovate-lanceolate, c. 2 mm long, 1 mm across, Stip. ovate-lanceolate to lanceolate, gradually coming to a point, c. 5 mm long, 1–2 mm across.

**A. starkiana** v. POELLN. (§ II). – Transvaal. – Dwarf ♄ c. 7 cm tall; **Br.** c. 2 cm ∅; **L.** green-brown, ± obovate to cuneate, underside somewhat convex, 12 mm long, c. 4 mm across and thick, the upperside of the tip tuberculate, with white hairs, mature L. hairless, bristle-hairs few, curly, as long as the L.; **Infl.** 4 cm tall, Fl. 2.5 cm ∅, white.

**A. subnuda** v. POELLN. (§ II). – Cape: Karroo, Bushmanland. – Resembles **A. lanigera**; c. 5 cm tall, branching; immature **L.** slightly felted, mature L. ± glabrous; **Fl.** white.

**A. telephiastrum** DC. (§ II) (*A. intermedia* HAW., *A. rotundifolia* SWEET, *A. varians* SWEET, *Ruelingia anacampseros* L., *Portulaca a.* L., *Talinum a.* WILLD. also MNCH., *A. a.* (L.) ASHERS. et GRAEBN.). – Cape: Karroo. – Older plants forming mats; **L.** compressed-rosulate, ovate or roundish, short-pointed, c. 18 mm long and across, thick, glabrous, green or brownish, bristle-hairs short, few; **Infl.** 15 cm tall, Fl. 1–4, 3–3.5 cm ∅, pink-carmine.

**A. tomentosa** BGR. v. **tomentosa** (Pl. 21/6) (§ II). – S.W.Afr.: Gr. Namaqualand. – **Root** thick; **St.** several, c. 5 cm tall; **L.** closely imbricate, obovate, the short, roughly tuberculate tip being abruptly truncate, 10 mm long, 8 mm across, 5 mm thick, upperside flat, underside much thickened, brownish-green, thickly white-felted, bristle-hairs few, as long as the L., curved; **Infl.** 6 cm tall, Fl. few, 3 cm ∅, pink.

**A.** — v. **crinita** DTR. – Bristle-hairs more numerous and longer.

**A.** — v. **margaretae** (DTR.) v. POELLN. (*A. margaretae* DTR.). – **L.** smaller, reddish-brown, bristle-hairs more numerous and longer; **Fl.** darker.

**A. trigona** DC. (§ I) (*Portulaca t.* THUNBG.). – Insufficiently known species.

**A. truncata** v. POELLN. (§ II). – Cape: Lit. Namaqualand. – **Root** somewhat thickened; **St.** short; **L.** crowded, c. 5 mm long and across, 3 mm thick, greenish-brown, upperside flat, underside slightly convex, abruptly truncate at the tip, tuberculate, young L. with fine white felt, bristle-hairs as long as the L.; **Fl.** c. 18 mm ∅, brownish.

**A. ustulata** E. MEY. (§ I). – Cape: Prince Albert D.; Orange Free State; Botswana. – **Br.** numerous, 2 cm long, 2 mm ∅; **L.** arranged spirally, short-reniform, small, concealed by the brownish-grey stipules, these being 1.5–2 mm long, 1–1.25 mm across, adpressed, silvery, broad-ovate, tapering, margins only slightly laciniate, with a short brown apiculus, often curved outwards; **Fl.** solitary, conspicuous.

**A. variabilis** v. POELLN. (§ I). – Cape: Lit. Namaqualand. – **Root** short, thick; **St.** simple, c. 3 cm long, 7 mm ∅; **L.** arranged in a spiral, variable in shape, usually roundish at the tip and pointed, but also with indentations or incisions up to midway, or even 2 incisions with a large point between them, hidden by the appressed white Stip. which are 3 mm long and broad-ovate.

**A. wischkonii** DTR. et v. POELLN. v. **wischkonii** (§ I). – S.W.Afr. – **Caudex** short, c. 1 cm ∅; **St.** erect to creeping, simple, c. 1 cm long, 3–5 mm ∅; **L.** arranged spirally, short-reniform, c. 1 mm long, c. 2 mm across, Stip. silvery, curved outwards at the tip, broad-ovate or broad-truncate, tapering, 2–2.5 mm long, 2 mm across, with distinct raised yellowish M.nerve; **Fl.** solitary, carmine.

**A.** — v. **levis** v. POELLN. – Stip. glabrous, M.nerve not raised.

**Anisosperma** MANSO. Cucurbitaceae. – S.America. – Greenhouse, warm. Propagation: seed.

**A. passiflora** (VELL.) MANSO (*Fevillea p.* VELL.). – E.Brazil: trop. reg. – **Caudex** fleshy, c. 4–5 cm ∅; **St.** long with 7 furrows, young St. densely puberulous, becoming glabrous; **L.** with slender petioles 1–2 cm long and glandular, lamina membranous, ovate-oblong, short-tapering, basally rounded, 3-nerved, green, initially puberulous, later glabrous, 10–15 cm long, 5 to 8 cm across; tendrils threadlike, short; ♂ **Fl.** in Infl., 3–4 mm ∅, ♀ Fl. 2–4 on a short stalk; **Fr.** ovoid to oblong, somewhat triangular, smooth to irregularly tuberculate, 8–15 mm long, 5 to 11 mm ∅.

**Apodanthera** ARN. Cucurbitaceae. – N.America. – Greenhouse, warm. Propagation: seed.
**A. undulata** A. GRAY. – USA: Arizona, W.Texas; Mexico. – ♃; **caudex** large, thickened; **St.** clambering or trailing, tendrils short, little coiled, herbaceous parts rough with adpressed hairs; **L.** reniform, broader than long, very shallow-lobed or with sinuate-dentate margins; **Fl.** few, ♂ Fl. in umbels or corymbs, ♀ Fl. solitary, Cor. yellow; **Fr.** resembling a pumpkin, about 4 cm ⌀, with longitudinal ribs.

**Arthrocnemum** MOQ. Chenopodiaceae.
**A. indicum** MOQ. – Coasts of trop. Afr. and peninsular India. – Creeping, succulent herb throwing out many fleshy, fertile spikes which completely enclose the Fl. – A halophyte, of little interest in cultivation.

**Astroloba** UITEW. (*Apicra* HAW.). Liliaceae. – South-east Cape. – Related to the genus **Haworthia**; the free segments of the Cor. (Pl. 22/1) are very short and curve open like a star (the Fl. is actinomorphic, whereas the Fl. of **Haworthia** is zygomorphic). (C.A.E. PARR considers these species to be **Haworthia** § **Quinquefariae**). – L.-Ros. usually extended stemlike; L. always in 5 straight or ± contorted rows, not dentate, usually with a sharp tip. – Greenhouse, warm. Propagation: seed, cuttings.
**A. aspera** (WILLD.) UITEW. v. **aspera** (*Aloe a.* WILLD., *Apicra a.* (WILLD.) HAW., *Haworthia a.* HAW.). – S.E.Cape. – **St.Ros.** erect, 10–15 cm tall, with few basal offsets; **L.** in 3–4 conspicuously spiral rows, triangular, short-tapering, older L. projecting horizontally, 12–14 mm long and across, green, upperside slightly convex, the underside very convex, often boat-shaped, with a pointed keel and large tubercles.
**A.** — v. **major** (HAW.) UITEW. (*Apicra a.* v. *m.* HAW., *Haworthia a.* v. *m.* (HAW.) PARR). – **L.** 18–24 mm long and across.
**A. bullulata** (JACQ.) UITEW. (*Aloe b.* JACQ., *Apicra pentagona* v. *b.* BAK., *Apicra b.* (JACQ.) WILLD., *Haworthia b.* (JACQ.) PARR). – S.Cape. – **St.Ros.** ± 6 cm tall with **L.** arranged ± in 5 spiral rows, 3.5–4.5 cm long, 13–15 mm across, somewhat inflated at the base, fresh green, underside very convex with numerous tubercles, singly or doubly keeled, tubercles greenish and often in transverse rows, keel and margins rough-tuberculate.
**A. congesta** (SALM) UITEW. (*Aloe c.* SALM, *Apicra c.* (SALM) BAK., *Haworthia c.* (SALM) PARR). – S.Cape. – **St.Ros.** 7.5 cm long, with L. in a dense spiral, ovate-deltoid, ± 3.5 cm long, 2.5 cm across, projecting almost horizontally, glabrous, dull-glossy, underside convex and carinate, keel finely crenate, margins double-crenate.
**A. deltoidea** (HOOK f.) UITEW. v. **deltoidea** (Pl. 22/4) (*Aloe d.* HOOK. f., *Apicra d.* (HOOK. f.) BAK., *Haworthia d.* (HOOK. f.) PARR). – S.Cape. – **St.Ros.** 20–25 cm long, ± prostrate; **L.** in 5 little contorted rows, broad-triangular, pointed, c. 2.5 cm long and across, rigid, glabrous, underside roundish and carinate, edges and keel cartilaginous and rough.
**A.** — v. **intermedia** (BGR.) UITEW. (*Apicra d.* v. *i.* BGR., *Haworthia d.* v. *i.* (BGR.) PARR). – Intermediate between the type and v. **turgida.**
**A.** — v. **turgida** (BAK.) UITEW. (*Apicra d.* v. *t.* (BAK.) BGR., *Haworthia d.* v. *t.* (BAK.) PARR). – **St.Ros.** 6.5 cm ⌀; **L.** more convex and more conspicuously spiral in their arrangement.
**A. dodsoniana** UITEW. (*Haworthia d.* (UITEW.) PARR). – **St.Ros.** erect, becoming prostrate in age, 20 cm or more in length, 3.5–6 cm ⌀, off-setting basally; **L.** in 5 spiral rows, crowded or somewhat more widely spaced, ovate-lanceolate, gradually tapering to a distinct point, 2.5 to 3.5 cm long, c. 1.5 cm across, 6–8 mm thick, blue-green, young L. with waxy coating, upperside ± concave, underside convex, keeled above, with numerous dark green longitudinal lines, margins and keel finely crenate.
**A. egregia** (v. POELLN.) UITEW. v. **egregia** (*Apicra e.* v. POELLN., *Haworthia e.* (v. POELLN.) PARR). – Cape: Oudtshoorn D. – **St.Ros.** erect, 10–15 cm tall, 3 cm ⌀, offsetting basally; **L.** in 5 straight or occasionally contorted rows, ovate-triangular, 17 mm long, 15 mm across the base, with a short apiculus, underside distinctly obliquely keeled, the border here having several green longitudinal stripes, borders and keel cartilaginous and rough.
**A.** — v. **fardeniana** UITEW. (*Haworthia e.* v. *f.* (UITEW.) PARR). – **L.** distinctly spiral in their arrangement, c. 2.5 cm long, c. 1.5 cm across, ovate-triangular, tuberculate, tubercles 2–10, irregularly arranged, margins and keel with small roundish glistening T.
**A. foliolosa** (WILLD.) UITEW. (Pl. 22/3) (*Haworthia f.* WILLD., *Apicra f.* (WILLD.) HAW.). – Cape. – **St.Ros.** slender, to 30 cm tall; **L.** in 5 spiral rows, 12 mm long and across, roundish-triangular, sharp-tipped, underside rounded-keeled, edges somewhat cartilaginous, glossy green.
**A. herrei** UITEW. (Pl. 22/6) (*Haworthia harlandiana* PARR). – Cape: Uniondale D. – **St.Ros.** densely leafy, 20 cm or more tall, c. 4 cm ⌀, somewhat branching, often dichotomously, either from the base or higher up; **L.** somewhat imbricate, in 5 ± spiral rows, ovate or round-ovate, terminating abruptly in a long, rigid and rather sharp point, 20–25 mm long, 12–14 mm across, 4–5 mm thick, deep green, young L. with a waxy coating and deep blue-green, underside rounded, sharply keeled towards the tip, with numerous dark green longitudinal lines, margins and keel slightly thickened, glossy, finely tuberculate.
**A. pentagona** (HAW.) UITEW. v. **pentagona** (Pl. 22/7) (*Aloe p.* HAW., *Haworthia p.* HAW., *H. spiralis* HAW., *Apicra p.* (HAW.) WILLD.). – S.Cape. – **St.Ros.** 25 cm tall, 6.5–8 cm ⌀; **L.** in 5 straight or only slightly spiral rows, triangular-lanceolate, pointed, 4 cm long, 13–19 mm across, underside rounded, with 1–2 blunt keels, slightly glossy, light green, somewhat rough.
**A.** — v. **spiralis** (HAW.) UITEW. (*A. s.* HAW., *Apicra p.* v. *willdenowii* BAK., *Aloe p.* v. *s.* SALM,

*Apicra s.* WILLD., *Haworthia p. v. s.* (SALM) PARR). – **L.** 5 cm long, arranged in 5 distinct spiral rows.
**A. — v. spirella** (HAW.) UITEW. (*Aloe s.* HAW., *Apicra p. v. s.* (HAW.) BAK., *Haworthia p. v. s.* (HAW.) PARR). – **L.** 2.5 cm long, narrower, arranged in 5 less regular rows.
**A. — v. torulosa** (HAW.) UITEW. (*A. p. v. t.* HAW., *Aloe spirella v. quinquangularis* SALM, *Haworthia p. v. t.* (HAW.) PARR). – **L.** 2.5 cm long; **Fl.** inflated.
**A. spiralis** (L.) UITEW. (*Aloe s.* L., *Apicra s.* (L.) BAK., *Aloe imbricata* HAW., *Haworthia i.* HAW., *H. gweneana* PARR). – S.Cape. – **St.Ros.** 10–20 cm tall, 2.5–3.5 cm ⌀, densely leafy; **L.** spirally imbricate, grey to blue-green, ovate-lanceolate to triangular-lanceolate, tip short and sharp, L. 3 cm long, c. 1.5 cm across and thick, upperside concave or slightly convex, obliquely keeled below the tip, margins and keel very finely crenate.
**A. turgida** (BAK.) JACOBS. (*Apicra t.* BAK., *Haworthia shieldsiana* PARR). – Cape: Albany D. – **St.Ros.** 15–20 cm tall, 5–6 cm ⌀; **L.** in 5 spirally-contorted rows, triangular, c. 25 mm long, 18 mm across, margins rough, lower L. much swollen on the upperside, 6–8 mm thick midway, green, with several darker green transverse ribs.

× **Astroworthia** ROWL. Liliaceae. – Intergeneric hybrids: **Astroloba** × **Haworthia.** – Intermediate between the parents. – Cultivation as for **Haworthia.**
× **A. bicarinata** (HAW.) ROWL. (Pl. 22/2) (*Apicra b.* HAW., *Astroloba b.* (HAW.) UITEW., *Haworthia b.* (HAW.) PARR). – **Haworthia margaritifera** × **Astroloba aspera.** – Spontaneous hybrid acc. H. HALL & B. BAYER. – Cape: Graaff-Reinet D. – **L.** in only slightly contorted rows, long-triangular, 2–2.5 cm long, underside with 2 distinct keels, light green, rough, borders and keel and often the upperside having a prominent line with small whitish warts, the underside with small whitish warts in ± distinct transverse lines.
× **A. —** nm. **skinneri** (BGR.) ROWL. (Pl. 22/5) (*Apicra s.* BGR., *Astroloba s.* (BGR.) UITEW., *Haworthia olivetteana* PARR). – Spontaneous hybrid of same parentage as the preceding, acc. H. HALL & B. BAYER. – S.Cape. – **St.Ros.** c. 15 cm tall; **L.** in a dense spiral 8 cm ⌀, dark green, ovate-lanceolate, gradually tapering, 5 to 6 cm long, 4 cm across, apiculate, upperside boat-shaped, concave, underside very rounded and keeled, underside with thick, whitish-green tubercles in ± distinct transverse rows, keel and margins tuberculate.
× **A.** 'Fardeniana' (× *Apworthia fardeniana* v. POELLN. nom. nud.). – × **Astroworthia bicarinata** nm. **skinneri** × **Haworthia caespitosa.** – Perhaps lost to cultivation.

**Augea** THUNBG. Zygophyllaceae. – S.Africa. – Greenhouse, warm. Propagation: seed.
**A. capensis** THUNBG. (Pl. 23/1). – Cape: Laingsburg D., W. Karroo; Botswana; S.W.Afr.: coastal deserts. – Shrubby ☉, succulent, with a long taproot; **St.** simple or divided, with numerous erect Br.; **L.** shaped like a cucumber, or digitate, upperside flat, tapering, 3–4 cm long, 10–12 mm thick, extremely watery (certainly 99 %); **Fl.** solitary or 2–3, from L. axils, whitish-green, succulent; **Fr.** a berry, about 1 cm thick, amber-yellow. Halophytic plant.

**Baeriopsis** J. T. HOWELL. Compositae. – Mexico. – Insignificant dwarf ♄; **L.** somewhat succulent. – Summer: planted out in the open, winter: unheated greenhouse. Propagation: seed, cuttings.
**B. guadalupensis** J. T. HOWELL (Pl. 23/2). – Mexico: Guadalupe Is. – Glabrous ♄ to 30 cm tall, 70 cm across; **St.** woody, to 1.5 cm ⌀; **Br.** 1–2 mm thick, densely leafy in the upper part; **L.** linear-oblanceolate, rounded to blunt at the tip, semi-terete, light green, 2–7 cm long, 2.5 mm across; **Infl.** terminal, solitary, 1–3 cm across, with stalk 1–5 cm long, yellow.

**Batis** L. Batidaceae. – East and West coasts of America; Jamaica; New Guinea. – Of the three known species of these shrubs of the seashore, only one is briefly described here. Of little significance in cultivation.
**B. maritima** L. (Pl. 23/3). – E. coastal reg.: from Florida (USA) to Brazil; USA: Calif.; Mexico: Baja Calif.; Hawaii. – ♄ 1–1.25 m tall; branching; **Br.** opposite, bluntly 4-angled; **L.** almost sessile, fleshy, linear-oblong, tapering, 1.5–3 cm long; **Fl.** dioecious, ♂ Fl. in 4 series in catkins, inconspicuous, ♀ Fl. without Cal.; **Fr.** ovoid-oblong, fleshy.

**Beaucarnea** LAM. Agavaceae. – Mexico. – Succulent trees, resembling **Nolina; caudex** short, succulent; **L.** crowded in clusters at the stem-tips, thin, linear; **Infl.** paniculate, with the flowers on short pedicels. – Greenhouse, warm. Propagation: seed.
**B. gracilis** LEM. (*B. oedipus* ROSE, *Nolina hystrix* HORT.). – S. central Mexico. – Habit as for **Nolina recurvata; St.** branching above; **L.** straight, about 50 cm long, 6 mm across, very grey, margins rough.
**B. stricta** LEM. (Pl. 23/4) (*Dasylirion s.* MACBR., *B. recurvata* v. *s.* BAK., *B. glauca* ROEZL, *B. purpusii* ROSE). – S. central Mexico: Puebla, Oaxaca. – ♄ 6–8 m tall; **caudex** thickened conically at the base, with a corky rind, branching above; **L.** crowded in clusters, erect, old L. recurved or deciduous, 50–60 cm long, 10–12 mm across, bluish, margins entire, rough, yellowish.

**Begonia** L. Begoniaceae. – Subtropical and tropical North America; Africa; Asia. – More than 1000 species are known, but only a minority of these are ± succulent, and only three can be included here. – Greenhouse, warm, ± moist. Propagation: seed, cuttings.
**B. incana** LINDL. (Pl. 24/4). – Mexico: trop. reg. – ♄; **St.** fleshy, erect, with L.-scars; **L.** thick-fleshy, with petiole 10 cm long, scutate, ovate-elongate, rather blunt, c. 12 cm long,

9 cm across, green, underside light green, densely white-scaly, the margins with well-spaced T.; **Fl.** whitish-pink.

**B. natalensis** HOOK. (Pl. 24/1, 2). – Natal. – **Caudex** large, compressed-tuberous, grey-brown, 15–20 cm or more in length, half being subterranean, roots fibrous; **St.** nodular and branching, very succulent, dirty green to coppery, articulate, 30–45 cm tall; **L.** semi-cordate, tapering, margins coarsely lobed-serrate, dull deep green, upperside white-spotted; **Infl.** branching dichotomously, Fl. yellowish-white.

**B. venosa** SKAN. (Pl. 24/3). – Brazil. – Under-♄; **St.** erect, stout, 60 cm or more tall; **L.** reniform or auriculate, with a thick petiole arising in the middle, 6–7 cm long, 10–12 cm across, 1 mm thick, dark green, both surfaces grey-felted, stipules much inflated, 4–5 cm long, 3–4 cm across, almost translucent, veined; **Fl.** numerous, white.

**Beschorneria** KUNTH. Agavaceae. – Mexico. – Succulent ♃, branching eventually to form clumps; **L.** rosulate, linear to lanceolate, leathery to fleshy, flaccid, with a fleshy median keel, light grey striate, margins very finely dentate; **Sc.** straight or curved obliquely; **Infl.** racemose or paniculate, with membranous, large and ± vivid red bracts, flowers pedicellate, pendulous, almost cylindrical, lobes little spreading, greenish or reddish. – Summer: planted out, winter: unheated greenhouse. Propagation: seed, division.

**B. bracteata** JACOBI. – Mexico. – **L.** large, grey-green; **Infl.** 1.5–2 m tall, bracts light red.

**B. dekosteriana** C. KOCH (*B. argyrophylla* HORT.). – Mexico. – Resembles **B. yuccoides**; **L.** light grey; **Infl.** with a light-brown axis, bracts large, vivid red.

**B. pubescens** BGR. – Mexico. – Resembles **B. tubiflora**; **L.** larger and broader, underside rough beneath the tip, margins rather more dentate; **Infl.** with a light red axis, bracts ovate, Fl. green to yellow, softly hairy.

**B. tonellii** JACOBI. – Resembles **B. dekosteriana**; **Fl.** axis purple-red.

**B. tubiflora** KUNTH. – Mexico. – **L.** 30 cm long, to 2.5 cm across, both surfaces rough; **Infl.** almost 1 m tall, brown-green, bracts violet-red, Fl. reddish-green.

**B. wrightii** HOOK. f. – Mexico. – **L.** large; **Infl.** tall, freely branching, Fl. greenish, softly hairy.

**B. yuccoides** HOOK. f. – Mexico. – **L.** c. 20, rough and grey on the underside, 50 cm long, 5 cm across; axis of **Infl.** curved over laterally, over 1 m tall, vivid red.

**Bombax** L. Bombacaceae. – Greenhouse, warm. Propagation: seed, cuttings.

**B. ellipticum** H. B. et K. (Pl. 25/1). – Mexico. – ♄; **caudex** succulent, club-shaped or almost spherical in the lower part, with a grey-brown, much torn rind, tapering above, with L. scars, often with several short St. sprouting from the thickened base; **L.** large, 3–5-lobed, dropping during the dry season.

**Bowiea** HARV. ex HOOK. f. Liliaceae. – South and tropical Africa. – Bulbous plants with twining **St.** and small **L.** – Greenhouse, warm. Propagation: seed.

**B. kilimandscharica** MILDBR. (*Schizobasopsis k.* (MILDBR.) BARSCHUS). – Tanzania: Kilimanjaro area. – Closely resembles **B. volubilis**; clambering over ♄ and ♃ and forming large groups; **Fl.** yellow, hyaline; **Fr.** pointed, to 3 cm long.

**B. volubilis** HARV. et HOOK. f. (Pl. 25/2) (*Schizobasopsis v.* (HARV. et HOOK. f.) FR. MACBR.). – S.Afr. – Bulb light green, growing above soil-level, spherical, to 20 cm ⌀; **St.** long, twining, thin; **L.** short-linear, soon dropping; **Fl.** greenish-white, about 8 mm ⌀; **Fr.** obtuse, not over 14 mm long.

**Brachystelma** R. BR. Asclepiadaceae. – Tropical Africa; Ethiopia; South Africa; South-West Africa. – ♃ with a tuberous caudex or numerous fleshy roots; **St.** simple or freely branching, erect, creeping or less often clambering, leafy throughout; **L.** opposite; **Fl.** small to medium, solitary, situated laterally at the nodes, or 2 or many in lateral or terminal umbels or umbel-like Infl.; Cal. 5-merous; Cor. round, disc-like or with a short campanulate tube, 5-lobed, lobes free or united at the tips, valvate or reduplicate in the bud-stage, corona double. – The species given below are the only ones where the available description mentions either a caudex or thickened roots. (Lit.: N. E. BROWN, Fl. Cap. IV, i. 1909, XLI, **Brachystelma** R. BR.). – Greenhouse, warm. Propagation: seed. The tubers can be baked and eaten.

**B. arnotii** BAK. (*B. grossarthii* DTR.). – Cape: Colesberg D.; S.W.Afr.: Waterberg, between Windhoek and Gobabis. – **Caudex** flattened, about 6.5 cm ⌀; **St.** solitary or several, simple or branching, c. 7.5 cm tall, with minute hairs; **L.** lanceolate to ovate-lanceolate, narrowing to the stalk, 10–17 mm long, 2–3 mm across, longitudinally plicate, margins wavy, underside densely grey hispid; **Fl.** 2–4, Cor. small, lobes spreading, tube beaker-shaped, underside puberulous, innerside glabrous, dark purple-brown, with a green warty thickening at the lobe-tip.

**B. barberae** HARV. ex HOOK. f. (*B. barbertonensis* HORT.). – Cape: Transkei. – **Caudex** compressed, turnip-size; **St.** very short; **L.** 7–10 cm long, linear-oblong, pointed; **Fl.** in a spherical, capitate Infl. 10–12 cm ⌀, dirty purple, yellow in the centre, spotted, Cor. margin 2–5 mm across, lobes 3-angled, drawn out into thin tails, 2.5 cm long and united by their tips.

**B. bingeri** A. CHEV. – Sudan. – **Caudex** nearly spherical or oblong, 4–8 cm ⌀; **St.** branching basally, 5–10 cm tall; **L.** in 3–5 pairs, short-petiolate, 2.5–6 cm long, 1.25–2.5 cm across, broadly elliptical to oblanceolate, pointed or blunt-tipped, upperside puberulous, underside with a few hairs; **Fl.** 3–4 in an Infl., or solitary, Cor. about 1.2 cm ⌀, reddish-white, puberulous on the outside, tube 2–3 mm long, lobes 7 mm long, triangular-ovate, rather pointed.

**B. blepharanthera** H. HUBER (*Blepharanthera dinteri* SCHLTR., *B. edulis* SCHLTR.). – S.W.Afr.; Damaraland. – **Caudex** 4 cm tall, up to 6 cm ⌀ - **St.** branching, with L. arranged laxly; **L.** oblong, spatulate, blunt, margins slightly wavy; **Fl.** 1–2, Cor. 9 mm long.

**B. bolusii** N. E. BR. – Cape: Graaff-Reinet D. – **Caudex** flattened; **St.** 10–15 cm tall, branching basally, puberulous; **L.** 5–10 mm long, 2–3 mm across, spatulate or cuneate-elliptical, margins very undulating, underside minutely hairy; **Fl.** 1–2, Cor. cage-like, lobed almost to the base, lobes united at the tips, 12–16 mm long, narrow-linear, hairy on the underside, spotted below, green in the upper part.

**B. buchananii** N. E. BR. – Malawi. – **Caudex** large, fleshy; **St.** puberulous; **L.** almost sessile, 4–12 cm long, 2–2.5 cm across, elliptical-ovate, blunt-rounded, with or without a short triangular tip, both surfaces short-puberulous, ciliate; **Fl.** 20–30 in an umbel, Cor. 2–2.5 cm ⌀, plate-shaped, blackish-purple, 5-lobed to midway, lobes triangular-pointed.

**B. caffrum** N. E. BR. – Cape: King William D. – **Caudex** compressed-spherical with a long neck; **St.** numerous, prostrate, branching, minutely rough; **L.** ovate, or the upper L. lanceolate, fairly thick, with minute rigid cilia; **Fl.** 1–2, Cor. round, the upper 2/3 consisting of lobes, 10–13 mm ⌀, fleshy, light pure yellow, lobes free, 2–4 mm long, 1–2 mm across, ovate, pointed, margins minutely ciliate.

**B. campanulatum** N. E. BR. – Cape: Bathurst D. – **Caudex** flattened or napiform, c. 5 cm ⌀; **St.** simple or with only few Br., 4–5 cm tall; **Br.** prostrate, c. 2.5 cm long, puberulous; **L.** 2–4 cm long, 6–10 mm across, elliptical, ovate to obovate, flat or longitudinally plicate and ± wavy, both surfaces puberulous; **Fl.** with Cor. campanulate, shortly 5-lobed, densely woolly, outside green, inside greenish-yellow, the entire upper part with purple-brown spots and longitudinal veins, with stiff hairs, lobes 3–5 mm long, triangular-ovate, pointed.

**B. cathcartense** R. A. DYER. – Cape: Cathcart D. – **Caudex** 4–5 cm ⌀, 2 cm tall; **St.** 5 cm tall, with few Br., minutely rough; lower **L.** almost sessile, lanceolate, less than 1 cm long, upper L. linear-lanceolate, to 2.5 cm long, both surfaces thinly and roughly hispid, margins folded slightly upwards; **Fl.** 1–2, axillary, nodal, foul-smelling, on a pedicel 8 mm long, with bracts, Sep. about 5 mm long, Cor. 2–2.2 cm long, 10-ribbed, with minute recurved T. in the indentations, glabrous or with a few scattered hairs outside, tube almost campanulate, 9 to 9.5 mm long, rather abruptly expanding above, 1.8–2 cm ⌀, lobes 11–13 mm long, 9–10 mm across at the base, triangular, ciliate towards the indentations, with chestnut-brown spots on the outside of the tube and lobe-bases, the lower half of the tube yellow, with chestnut-brown stripes and spots, the tips green with short hairs, the margins somewhat recurved, the tip curved inwards.

**B. circinatum** E. MEY. (*Dichaelia cinerea* SCHLTR., *B. cinereum* (SCHLTR.) N. E. BR., *D.*

*filiformis* SCHLTR., *B. f.* HARV., *D. forcipata* SCHLTR., *D. galpinii* SCHLTR., *B. galpinii* (SCHLTR.) N. E. BR., *D. macra* SCHLTR., *D. microphylla* S. MOORE, *B. ovatum* OLIV., *D. pallida* SCHLTR., *B. p.* (SCHLTR.) N. E. BR., *D. undulata* SCHLTR., *B. u.* (SCHLTR.) N. E. BR., *D. zeyheri* SCHLTR., *B. z.* (SCHLTR.) N. E. BR.). – E.Cape, N.W. into Transvaal, W. to northern S.W.Afr. – **Caudex** 5–13 cm ⌀, flattened; **St.** 1 or several, branching basally, to 25 cm tall, puberulous; **L.** linear or elliptical-lanceolate, with a short petiole, 5–18 cm long, 2–5 mm across, the underside being minutely puberulous; **Fl.** in clusters of up to 8 at the nodes, also solitary, on pedicels up to 3 mm long, Sep. pointed, to 2 mm long, Cor. cage-like, almost spherical, lobes free almost to the base, rather threadlike, much curved, united at the tips, outside glabrous, inside puberulous, yellowish or purple-brown or whitish in the lower part, tube plate-like.

**B. coddii** R. A. DYER. – Swaziland. – **Caudex** 2–5 cm ⌀, 1.5–2.5 cm tall; **St.** simple or branching basally; **Br.** creeping, 5–15 cm long; **L.** on a short petiole, ovate to broad-ovate, 2.5 cm long, 2 cm across; **Fl.** solitary, 17–20 mm ⌀, tube 3 mm long, 7 mm ⌀, parchment-coloured, lobes alate, wine-coloured, spotted with red, 6–7 mm long, 5 mm across the base.

**B. comptum** N. E. BR. – Cape: Uitenhage D. – **Caudex** long, thick, fleshy, narrowly fusiform; **St.** 2.5–7.5 cm tall, simple or branching, puberulous; **L.** 4–6 pairs on very short petioles, 7–12 mm long, 5–7 mm across, oblong to rounded-ovate, blunt, both surfaces thinly puberulous; **Fl.** solitary, tube rather amorphous to flat or beaker-shaped, c. 1 mm deep, glabrous, lobes free, 7–8 mm long, 2 mm across, margins slightly incurving, with a tuft of long, clavate, purple hairs.

**B. constrictum** J. B. HALL. – Ghana. – **Caudex** disc-like, c. 6 cm ⌀, 2 cm thick; **St.** branching from base; **Br.** prostrate, to 10 cm long; **L.** sessile, narrow-linear, up to 6 cm long, 1 mm across, margins recurved; **Fl.** 1–2, Cor. purple, tube 5 mm ⌀, 4.5 mm long, constricted below the lobes, these being 9–12 mm long, caudate, with a leather-coloured longitudinal band on the inside, the outside of the Cor. being minutely warty, inside with white hairs.

**B. crispum** GRAH. (*B. caudatum* N. E. BR., *B. spathulatum* LINDL., *B. s.* DECNE, *Stapelia tuberosa* MEERBURG, *S. caudata* THUNBG.). – Cape: Malmesbury, Cape D. – **Caudex** flat, 7.5–10 cm ⌀; **St.** often several, 10–15 cm tall, simple or branching basally, ± puberulous; **L.** 12–40 mm long, 2–5 mm across, linear-lanceolate to obovate-spatulate, pointed or blunt, margins normally wavy, both surfaces short-puberulous; **Fl.** 2–8, Cor. green outside, puberulous, the tube and the base of the lobes being spotted purple-brown, the inside of the tube and the base of the lobes glabrous, yellow or whitish (?), with tuberculate, purple-brown spots, lobes 2–3 cm long, broadly ovate, narrowing to long-linear.

Plate 26. 1. Bulbine Infl.; 2. B. alooides WILLD.; 3. B. frutescens (L.) WILLD. v. frutescens; 4. Bulbinopsis semibarbata (R. BR.) BORZI; 5. Bulbine mesembryanthoides HAW. (Photo: H. HERRE).

Plate 27. 1. Bursera microphylla (Rose) A. Gray (Photo: C. S. Perkins); 2. Calibanus hookeri Trel. (Photo: H. Baum); 3. Calandrinia spectabilis Otto; 4. Caralluma dummeri (N. E. Br.) White et Sloane (Photo: W. Rauh).

Plate 28. 1. **Caralluma aperta** (MASS.) N. E. BR. (Photo: H. LANG); 2. **C. caudata** N. E. BR. v. **caudata** (Photo: as 1); 3. **C. frerei** ROWL. (Photo: G. REESE); 4. **C. knobelii** (PHILL.) PHILL. (Photo: as 1).

7/B*

Plate 29. 1. Caralluma lugardii N. E. Br. (Photo: K. Dinter); 2. C. lutea N.E. Br. ssp. lutea (Photo: H. Lang); 3. C. maculata N. E. Br. v. maculata (Photo: as 1); 4. C. lutea ssp. vaga (N. E. Br.) Leach (Photo: as 2).

Plate 30. **1. Caulanthes inflatus** (GREENE) S. WATS. (Repr.: Fiori, Ed. Art. Maest.); **2. Ceraria namaquensis** (SOND.) PEARS. et STEPH. (Photo: H. CHR. FRIEDRICH); **3. Cavanillesia arborea** K. SCH. (Photo: E. WERDERMANN); **4. Ceraria pygmaea** (PILL.) PILL. (Repr.: Nat. Cact. Succ. J. **15,** 1960, 43, Photo: E. LAMB).

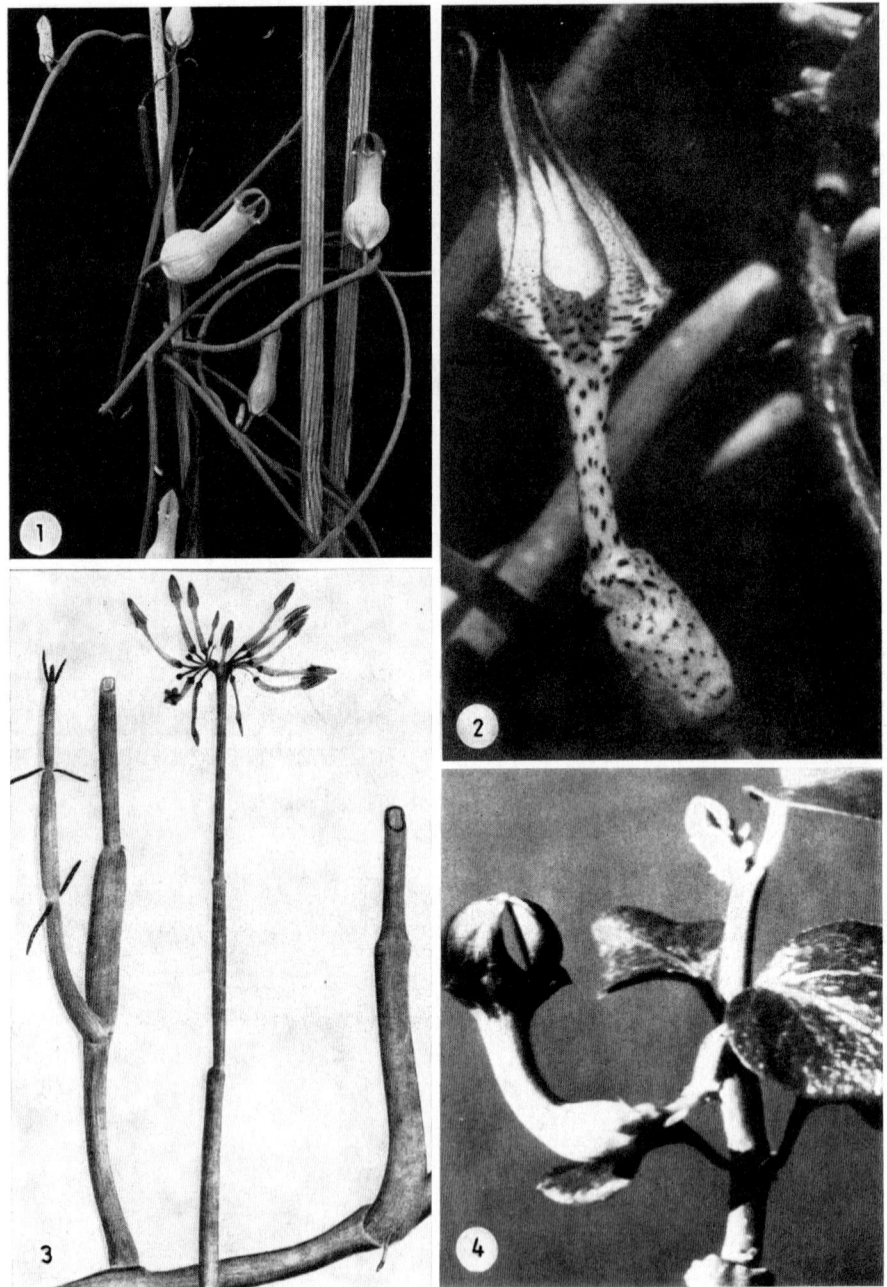

Plate 31. 1. Ceropegia ampliata E. MEY. v. ampliata; 2. C. ballyana BULLOCK (Photo: J. BOGNER); 3. C. chrysantha SVENT. (Drawing: E. R. SVENTENIUS); 4. C. aristolochioides DECNE. ssp. aristolochioides (Photo: as 2).

Plate 32. 1. Ceropegia cimiciodora OBERM. (Photo: H. LANG); 2. C. hians SVENT. (Drawing: R. E. SVENTENIUS); 3. C. distincta ssp. haygarthii (SCHLTR.) H. HUBER; 4. C. dichotoma HAW.

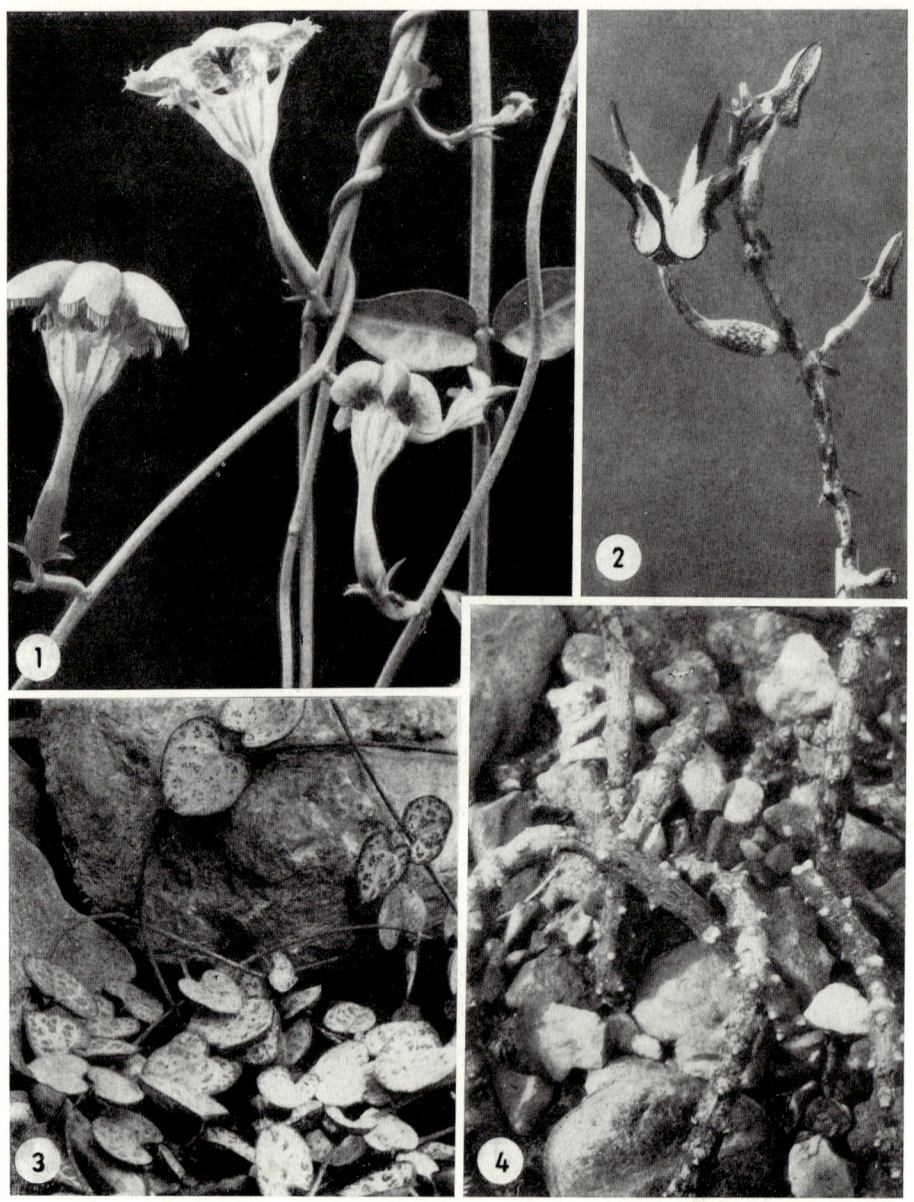

Plate 33. 1. Ceropegia sandersonii HOOK. f.; 2. C. stapeliiformis HAW. v. stapeliiformis; 3. C. woodii SCHLTR. ssp. woodii; 4. C. stapeliiformis HAW. v. stapeliiformis.

**B. cupulatum** R. A. DYER. – S.W.Afr.: Grootfontein. – **Caudex** about 10 cm ∅, compressed; **St.** simple or only somewhat branching, to 15 cm tall, with only few hairs; **L.** ovate, lanceolate, elliptical or linear, 3–8 cm long, 4 to 15 mm across, constricted basally into a short stalk, both surfaces shortly puberulous; **Fl.** several, extra-axillary, with pedicels 2–3 mm long, Sep. ovate to lanceolate, 1.5–2 mm long, Cor. green, 6–9 mm long, with some hairs outside, inside glabrous, tube 1.5–2.5 mm long, lobes ± oblong, narrowing at the tip, margins slightly recurved, outer corona beaker-shaped.

**B. dinteri** SCHLTR. (*B. brevipedicillatum* TURRILL, *B. ringens* E. A. BRUCE). – S.W.Afr.: Gr. Namaqualand. – **Caudex** 3–5 cm ∅, flat above; **St.** seldom branching, 15–20 cm long, 3 mm thick, hairy; **L.** broadly to narrowly lanceolate, pointed, 4 cm long, 16 mm across, both surfaces with fine velvety hairs, light green above, underside grey-green; **Fl.** 9–12 in a spherical umbel, Cor. flat, 10 mm ∅, upperside green with brown spots.

**B. distinctum** N. E. BR. – Cape: Albany D. (Grahamstown), Stockenstrom D. (Seymour). – **Caudex** flattened, about 7.5 cm (? or more) ∅; **St.** 5–15 cm tall, branching from the base, puberulous; **L.** 6–12 pairs, with very short petioles, 2–4 cm long, 2–6 mm across, linear to linear-lanceolate, blunt to almost pointed, longitudinally plicate, underside shortly puberulous; **Fl.** solitary or 2–6, Cor. cage-like, tube 3 mm long, campanulate, with whitish-purple spots, outside glabrous or with short, spreading hairs, with clavate white hairs down to the base of the lobes which are 12–20 mm long, almost linear, united at the tips.

**B. ellipticum** A. RICH. – Ethiopia. – **Caudex** compressed; **St.** 7.5–10 cm long, branching, slightly mealy- puberulous; **L.** 7–12 mm long, 2–3 mm across, oblong-lanceolate, pointed; **Fl.** not known. Insufficiently known species, possibly referrable to the G. **Raphionacme.**

**B. exile** BULLOCK. – Nigeria; Cameroon. – **Caudex** disc-like, about 3–5 cm ∅, 1–2 cm thick; **St.** 30–40 cm tall; **L.** sessile, lanceolate, c. 7 cm long, 9–12 mm across, pointed; **Fl.** solitary, tube cup-shaped, c. 2 mm long, lobes narrowly subulate-linear, 2.5–4 cm long, 2 mm across, with purple, versatile hairs 4 mm long.

**B. flavidum** SCHLTR. – Natal: Alexander County. – **Caudex** compressed, up to 4.5 cm ∅; **St.** several (?), branching, c. 5 cm tall, puberulous; **L.** almost sessile, up to 12 mm long, 2–3 mm across, lanceolate-elliptical, pointed; **Fl.** several, Cor. 4 mm long, yellowish, the upper 2/3 lobed, lobes free, gradually tapering from an ovate base.

**B. floribundum** R. A. DYER. (Pl. 25/4). – Orange Free State: Luckhoff D. – **Caudex** up to 5 cm tall, 4.5 cm ∅, with short perennial **St.** 1–2 cm long, 5–6 mm thick, up to 15 cm tall, thin, hairy; **L.** very shortly petiolate, oblong-lanceolate, up to 1 cm long, 3 mm across, with very fine papillose hairs, blue-green, very wavy; **Fl.** 2–3, Cor. 2.2–2.4 cm ∅, tube campanulate, 2–3 mm deep, with several stiff hairs inside, lobes 1–1.2 cm long, terete by reason of the recurved margins, light green, with 3 mauve veins at the base.

**B. foetidum** SCHLTR. – Transvaal. – **Caudex** compressed; **St.** branching basally; **Br.** with velvety hairs, densely leafy; **L.** ovate to ovate-spatulate, short-pointed, with velvet-like hairs, 1–2.5 cm long, 4 mm across, margins ± wavy; **Fl.** 1–2, Cor. hairy outside, inside dark purple-brown, with the lobes united tube-like in the lower part, the upper part free, lanceolate-triangular, tapering, 1.3 cm long.

**B. johnstonii** N. E. BR. – Uganda. – **St.** branching, 15–20 cm tall, puberulous; **L.** 1.25 to 2.5 cm long, 3–4 mm across, linear, pointed, longitudinally plicate, margins curved inwards, ± wavy, underside sparsely puberulous; **Fl.** 4–5, Cor. glabrous, purple or greenish-brown outside, inside of lobes white-hairy, tube purple-brown with whitish or yellowish concentric zones at the mouth, lobes 7.5 mm long, 5 mm across at the base, with a threadlike extension.

**B. lineare** A. RICH. – Ethiopia. – **Caudex** compressed-spherical, about 5 cm ∅; **St.** 7.5–10 cm tall, branching, puberulous; **L.** 3–4 cm long, 2–4 mm across, linear, pointed; **Fl.** 1–2, tube campanulate, about 4 mm long, lobes about 1 cm long, tapering-linear.

**B. macrorhizum** E. MEY. – Cape: Graaf-Reinet D. – **Caudex** the size of a human head, almost spherical; **L.** 10–15 mm long, on short petioles, ovate, undulate, glossy, whitish grey. – Insufficiently known species.

**B. mafekingense** N. E. BR. (*B. ramosissima* SCHLTR.). – Botswana. – **Caudex** compressed, about 7.5 cm ∅, 5 cm thick; **St.** 4–7 cm long, branching freely from the base, puberulous; **L.** 1.5 to 2 cm long, 4–7 mm across, lanceolate, pointed, longitudinally plicate, underside puberulous; **Fl.** 10–12 in an umbel, tube 4–5 mm long, c. 4 mm ∅, bluntly 5-angled, lobes 5–6 mm long, free, rolled tightly back, 2 mm across, blackish.

**B. magicum** N. E. BR. – Tanzania. – Related to **B. buchananii**; **St.** compressed (?); **L.** c. 9 cm long, 4 cm across, oblanceolate-oblong, rather blunt, both surfaces puberulous; **Cor.** glabrous, 2.5 cm ∅, round or broadly disc-shaped, with yellowish and blackish concentric zones, lobes reduced to short, triangular T., c. 2–3 mm long.

**B. micranthum** E. MEY. – Cape: Queenstown D. – **Caudex** flattened; **St.** several, branching, 5–7.5 cm tall, puberulous; **L.** 8–10 mm long, 1.5–5 mm across, linear-lanceolate to elliptical, blunt; **Fl.** 1–2, Cor. minute, lobes ovate-lanceolate, spreading. – Insufficiently known species.

**B. modestum** R. A. DYER. – Natal. – **Caudex** 2–3 cm ∅, 1–2 cm tall; **St.** simple or branching basally, 2–3 cm tall, glabrous to slightly hairy; **L.** ± oblong-elliptical, 1–3 cm long, 5–10 mm across, glabrous or with the upperside finely hairy; **Fl.** 1–2, tube 5 mm long, 10 mm ∅, campanulate, white, inside with dark red stripes, lobes triangular, alate, dark red.

**B. montanum** R. A. DYER. – Cape: Wodehouse D. – **Caudex** 2–5 cm across, c. 2 cm thick, hemispherical, with 1–2 underground St. up to

5 cm in length; **St.** 1–4, to 6 cm tall, branching above; **L.** oblong to linear-oblong, lower L. 1.5–2.3 cm long, 2–4 mm across, those above being smaller; **Fl.** usually 2, Cor. up to 1.5 cm long, tubular at the base, tube campanulate, to 4.5 mm long and ⌀, lobes linear from a broad base, 1 cm long, scarcely 1 mm across above, hairy in lower half.

**B. natalense** N. E. BR. – Natal: N. of Durban. – **Caudex** thick, fleshy, fusiform; **St.** solitary, simple or with 1 Br., 15–30 cm tall, puberulous; **L.** 2.5–6 cm long, 17–45 mm across, upper L. smaller, ovate to round-ovate, pointed or blunt and Sp.-tipped, both surfaces hairy, cilia short and dense; **Fl.** 2–4, Cor. nodding, c. 3–4 mm ⌀, the upper 2/3 lobed, flat cup-shaped below, lobes 2 mm long, 1.5 mm across, ovate-lanceolate, underside with or without a few hairs, dull green.

**B. occidentalis** SCHLTR. (*Brachystelmaria o.* SCHLTR.). – Cape D. – **Caudex** almost spherical; **St.** c. 7.5 cm tall, branching from the base, thin, hairy; **L.** 3–12 mm long, c. 1 mm across, linear, pointed, often somewhat fleshy, margins minutely ciliate; **Fl.** usually 2 together, tube c. 2 mm long, cup-shaped, lobes free, campanulate, spreading, c. 2 mm long, oblong-lanceolate, pointed, margins incurved midway, the tips much thickened and incurved-hooked, yellowish white, spotted white-red inside.

**B. oianthum** SCHLTR. – Transvaal. – Caudiciform; **St.** 6–8 cm tall, hairy; **L.** dense, linear-lanceolate, blunt to pointed, hairy, margins somewhat wavy, 1.5–5 cm long, 4–12 mm across; **Cor.** ovate, constricted throat-like, hairy on the inside, 2.2 cm long, 1.7 cm ⌀, dark purple-brown or yellowish with many dark purple-brown spots, lobe-tips free, ovate-triangular, tapering, margins with ciliate hairs.

**B. omissum** BULLOCK. – W.Cameroon. – **St.** stout, sparingly branched, up to 30 cm tall, from a fleshy disciform **caudex**; **L.** oblong-oblanceolate, up to 6 cm long and 1.5 cm across, ± hispid on both surfaces; **Fl.** several, in an umbel at apex of the leafy Br., pedicels c. 4 cm long, Cor. purple, densely tomentose inside, with a very short tube and broad flat disc, nearly 2 cm ⌀, the lobes very short, deltoid.

**B. pachypodium** R. A. DYER. – N.E.Transvaal: Letaba D. – **Caudex** of irregular shape, grey, 10 cm tall, 13 cm across, with stump-like, persistent St. above; **St.** several, branching; **Br.** 7–20 cm long, with 4–10 L.pairs; **L.** with petioles 5–10 mm long, broad-ovate or broadly elliptical-ovate, 2.5–5 cm long, 2–4 cm across, green; **Fl.** 1–2, Cor. cage-like, 20–23 mm long, tube white inside, lobes linear-lanceolate, c. 17 mm long, greenish-yellow inside, margins curved back, tips united.

**B. phyteumoides** K. SCHUM. Ethiopia. – **Rhizome** thick (or tuberous?); **St.** several, thin, slightly branching, c. 10 cm long; **L.** 4–6 cm long, 2–5 mm across, linear, tapering, with margins curled up; **Fl.** many in an umbel, tube short, campanulate, 2.5 mm long, 3 mm ⌀, lobes 2 cm long, spreading, filiform from a deltoid base, with recurved margins, glabrous inside, outside slightly hispid.

**B. pilosum** R. A. DYER. – Transvaal: Potgietersrust D. – **St.** 4–5 mm ⌀, branching from near soil-level, 25 cm tall, Br. rough-hairy; **L.** sessile, ± oblong, 1–1.5 cm long, 3–5 mm across, margins somewhat folded upwards and slightly wavy; **Fl.** usually 2 together, Cor. 16–18 mm long, the base saucer-shaped, 1–1.5 mm long, the lobes lanceolate below, united above, the underside hairy in the lower half.

**B. plocamoides** OLIV. – Tanzania. – **Caudex** large, fleshy, napiform, tasting somewhat of liquorice; **St.** repeatedly dividing into 2–3 from ground level, 20–25 cm tall; **L.** 5–8 cm long, 1–2 mm across, linear, terminating abruptly in a short hooked Sp.; **Fl.** solitary, Cor. stellate, round, the divisions between the 5 lobes reaching almost to the base, lobes about 15 mm long, 2–4 mm across at the base, pointed, dark purple.

**B. pulchellum** (HARV.) SCHLTR. (Pl. 25/3) (*Micraster p.* HARV.). – Transvaal: Nelspruit D. – **Caudex** spherical to conical, pale brown, c. 7 cm long, 5 cm across, 3 cm thick, above that a small trunk of 5 cm ⌀, 3 cm thick, protracted above into a neck; **St.** creeping, branching, 6–10 cm long, purple-red, with curly hairs; **L.** 4–8 pairs, broad-ovate, somewhat fleshy, 10–15 mm long, 6–10 mm across, upperside dark purple-green, margins minutely curly-haired; **Fl.** solitary, Cor. purple to chestnut-brown, 11–15 mm ⌀, the lower part campanulate to flat, inside glabrous, outside minutely spotted, lobes ± short-pointed, 4–6 mm long, 2–3 mm across at the base.

**B. pygmaeum** (SCHLTR.) N. E. BR. v. **pygmaeum** (*Dichaelia p.* SCHLTR.). – Cape: Bedford D. – **Caudex** napiform; **St.** several, simple or somewhat branching, 3–7.5 cm long, hispid; **L.** up to 1 cm long, up to 4 mm across, linear-spatulate; **Fl.** 1–3, Cor. cage-like, lobes either attached or free at the tips and outspread, tube about 1 mm deep, flat cup-shaped, outside glabrous, inside puberulous, yellow or greenish, lobes 4–5 mm long, linear from a triangular base, olive-green or purple-brown.

**B.** — v. **breviflorum** N. E. BR. – Botswana: Kalahari. – **L.** only developing after the flowering period; **Fl.** tube and the base of the 4–5 mm long lobes glabrous inside.

**B. ramosissimum** N. E. BR. – Botswana: Kalahari reg. – **Roots** 7.5–12.5 cm long, 7 mm or more thick, fleshy, terete, bunched; **St.** up to 15 cm tall, usually much branching basally, Br. angular, sparsely puberulous; **L.** almost sessile, 13–20 mm long, 3–5 mm across, linear-oblong to lanceolate, pointed, minutely ciliate; **Fl.** 2–4 in clusters, Cor. c. 6 mm ⌀, pale yellow or white, the united part usually flat, lobes free, 3–4 mm long, margins involute, with a few incurving hairs at the base.

**B. sandersonii** (OLIV.) N. E. BR. (*Lasiostelma s.* OLIV., *Dichaelia natalensis* SCHLTR., *Brachystelmaria n.* SCHLTR.). – Natal: Tupela. – **Roots** clustered, long, fleshy, cylindrical or narrow-fusiform; **St.** 20–45 cm tall, simple or branching from the upper half; **L.** very short-petiolate, almost leathery, 12–30 mm long, 3 to 12 mm across, linear, oblong to elliptical-oblong, pointed or blunt, margins rough-ciliate, also the underside along the M.rib; **Fl.** in clusters of 2–6, tube 1–2 mm deep, cup-shaped, lobes free, 3–5 mm long, 2 mm across the base, the sides folded back, ciliate, white with stripes of pink or light purple.

**B. schinzii** (K. Schum.) N. E. Br. (*Craterostemma s.* K. Schum.). – S.W.Afr.: Ovamboland. – **Caudex** c. 2.5 cm ⌀, fleshy, conical-ovate; **St.** single, with a few Br. from the base; **Br.** creeping, 2–6 cm long, shortly hairy; **L.** 2 to 5 cm long, 1–2 mm across, sessile, linear or almost terete, margins recurved, ± hairy; **Fl.** usually in pairs, Cor. round, almost 1 cm ⌀, divided halfway into 5 lobes, purple-brown with a white centre, lobes 3 mm long and across, pointed, the inside covered with long purple hairs.

**B. schizoglossoides** N. E. Br. – Cape: Albany D. (Grahamstown). – **Roots** clustered, thick, fleshy, long and narrow-fusiform; **St.** solitary, simple, 10–20 cm tall, with L. on the upper part; **L.** 1–4 cm long, 1–3 mm across, linear, pointed; **Fl.** 1–3 in a cluster, tube cup-shaped, lobes free, campanulate to spreading with incurved tips, 1.5–2 mm long and across, short-pointed, the tip bordered on the inside with minute white hairs.

**B. stellatum** Bruce et Dyer. – Transvaal: Lydenburg D. – **Caudex** ± flat-spherical, 2 to 5 cm ⌀; **St.** 1–4, branching basally; **Br.** almost prostrate, covered with minute hairs, 2–5 cm long; **L.** ovate to round, about 1 cm long, 5–10 mm across, rather fleshy, shortly petiolate, only slightly hairy, the tip pointed and the margins ciliate; **Fl.** 1–2, 9–10 mm ⌀, tube 2–2.5 mm deep, inside cream-coloured with purple markings and covered with purple hairs, lobes projecting, white at the base, the upper 2/3 yellow-green, covered with long white hairs, 2 mm long, 1.5 mm across at the base.

**B. stenophyllum** (Schltr.) R. A. Dyer (*Siphonostelma s.* Schltr.). – S.W.Afr.: Gr. Namaqualand. – **Caudex** flat, 5–6 cm ⌀; **St.** 1.3, 8–15 cm tall, Int. 10–15 mm long, 2 mm thick, minutely hairy; **L.** narrowly linear-lanceolate, 7–9 cm long, 4 mm across, or spatulate, 3.5 cm long, 15 mm across, upperside glabrous, underside slightly hairy; **Infl.** sessile, umbel-like, Fl. 2–4, on pedicels 6–10 mm long, Cor. campanulate, 15 mm long, lobes very long and united at the tips, green to brownish, inside light yellow-green.

**B. tavalla** K. Schum. – Tanzania: Iringa. – **Caudex** spherical; **St.** 20–25 cm tall, mostly simple, somewhat woolly above; **L.** sessile, 1.25–3 cm long, 3–4 mm across, pointed, both surfaces rather woolly; **Fl.** solitary or in pairs, Cor. round, violet, green inside, puberulous outside near the base, inside with dense wool-like purple hairs, tube 3 mm long, lobes 12 mm long, linear, somewhat broader at the base.

**B. tenue** R. A. Dyer. – Natal. – **Caudex** up to c. 4 cm ⌀, slightly compressed; **St.** somewhat branching; **Br.** erect, 5–10 cm long, covered with translucent hairs up to 1 mm long; **L.** short-petiolate or almost sessile, ovate to oblong-elliptical, 1–2 cm long, 1.5–5 mm across, underside and margins sparsely-haired, upperside almost glabrous; **Fl.** 1–2 at the nodes, bracts filiform, pedicels 8–10 mm long, Sep. linear, 3 mm long, Cor. yellowish-brown, about 1.5 cm long, cage-like, outside sparsely hairy, tube ± cup-shaped, 1 mm deep, slightly recurved in the indentations, lobes linear, tips united.

**B. togoense** Schltr. – Ghana; Togo; Dahomey; Nigeria. – **St.** stout with only few Br. up to 30 cm tall, from a fleshy disciform **caudex**; **L.** oblong-oblanceolate, up to 11 cm long and 4 cm across, ± pubescent on both surfaces; **Fl.** several, in umbels at the apex of the leafy Br., pedicels to 14 mm long, Cor. glabrous or pubescent inside, yellow, green or purple, the tube broadly funnel-shaped, 5 mm long, lobes triangular, 5–10 mm long and c. 4 mm across the base.

**B. tuberosum** R. Br. – S.Afr. – **Caudex** flat, 3–5 cm ⌀; **St.** 1 or several, with only few Br., 7.5–10 cm tall, leafy; **L.** 2–4 cm long, 3–9 mm across, mostly linear or linear-lanceolate, lower L. sometimes obovate, pointed or rather blunt, ± longitudinally plicate, the underside puberulous, minutely ciliate; **Fl.** 2–4, Cor. c. 2 cm ⌀, the outside green with purple-brown spots, the campanulate tube inside yellow with purple-brown transverse lines, lobes dark purple-brown inside, free, 6 mm long, margins recurved with quite long hairs at the base.

**B. vahrmeijeri** R. A. Dyer. – Natal. – **Caudex** c. 5 cm ⌀, red, slightly compressed; **St.** several, to 10 cm tall, almost unbranched, puberulous or glabrous; **L.** ± elliptical-lanceolate, up to 3 cm long, narrowing to the short petiole or sessile; **Fl.** 2–3, extra-axillary, with pedicels 5–10 mm long, bracts short, linear-lanceolate, Sep. 2.5 mm long, glabrous, Cor. yellowish-green or cream-coloured, less frequently white or somewhat chestnut-brown, 4 mm ⌀ at the mouth, lobes ovate-triangular, 5 mm long, 2.5 mm across at the base, somewhat spreading, fleshy.

**Brighamia** A. Gray. Campanulaceae. – Hawaii. – Greenhouse, warm. Propagation: seeds.

**B. insignis** A. Gray. – N. cliffs on Kuauo and Molokai. – **Caudex** thick and succulent, up to 1 m tall, tapered to a broad, almost spherical base; **L.** fleshy, entire, in a terminal lettuce-like Ros.; **Fl.** with a long tube, white.

× **Bryokalanchoe** Res. Crassulaceae. Intersectional hybrids made by F. Resende: *Bryophyllum* (**Kalanchoe** Sect. **Bryophyllum**) × **Kalanchoe**. (Lit.: s.F. Resende.)

× **B.** cv. Cancroidea. – Hybrid: **Kalanchoe blossfeldiana** cv. Feuerblüte × **K. pinnata** (*Bryophyllum calycinum*).

× **B.** cv. Lisbonensis. – Hybrid: **Kalanchoe blossfeldiana** cv. Feuerblüte × **K. daigremontianum** (*Bryophyllum d.*).

**Bulbine** L. Liliaceae – S. and E.Afr. – Stemless or caulescent herbs, some with a tuberous **caudex**, one ☉ (**B. annua**); L. radical, in Ros. or alternate along the Br., soft, green, ± fleshy, lanceolate, or subglobose; Fl. (Pl. 26/1) small, numerous, in a terminal Rac., yellow or rarely orange or white; stamens with all the filaments bearded. – Greenhouse, warm. Propagation: seed, sometimes also by cuttings or offsets.

## Synopsis of the Genus Bulbine by G. D. Rowley

after v. **Poellnitz** in Fedde Repert. LIII: 51–52, 1944

A. Plants caulescent, with visible Int.
    Sect. I. Bulbine (*Caulescentes* v. POELLN.)
        Type spec. **B. frutescens** (L.) WILLD.
  1. L. stout, ± straight; Rac. many-flowered .................................................. **2.**
  1a. L. slender, variously curved; Rac. few-flowered .................................... **curvifolia**
  2. L. glaucous; St. 10 mm or more thick ........................................ **frutescens** v. **rostrata**
  2a. L. green; St. 6–8 mm thick .......................................................... **3.**
  3. L. 4–5 cm long; Rac. lax, nodding ............................................. **frutescens** v. **incurva**
  3a. L. longer; Rac. dense, erect. .................................................. **frutescens** v. **frutescens**
A. Plants stemless
    Sect. II. Acaules v. POELLN.

B. Plant ☉
    SSect. 1. Annuae v. POELLN.
        One spec. only ................................................................ **annua**
B. Plant ♃, rhizomatous or tuberous
C. L. ± St.-clasping, with membranous marginal wings at the base, persistent.
    SSect. 2. Alatae v. POELLN.
        Type spec. **B. asphodeloides** (L.) SPRENG.
  4. L. in Aloe-like Ros., ± flat and tapering gradually ................................... **5.**
  4a. L. reed-like, linear, abruptly tapered from a broad base .......................... **7.**
  5. L. lanceolate, dark green; Fl. not crowded, Rac. lax .............................. **6.**
  5a. L. oblong-lanceolate, pale green; Fl. crowded in a dense Rac. ............. **latifolia**
  6. L. margins not ciliate ................................................................ **alooides**
  6a. L. margins ciliate .................................................................. **fallax**
  7. Fl. white or whitish ................................................................ **8.**
  7a. Fl. orange ......................................................................... **crocea**
  7b. Fl. yellow ......................................................................... **10.**
  8. L. up to 10 cm long ................................................................. **parviflora**
  8a. L. 30 or more cm long ............................................................ **9.**
  9. Infl. 30–50 cm tall ................................................................ **decurvata**
  9a. Infl. 60 cm or more tall ......................................................... **triebneri**
  10. Fl. crowded into a short terminal head or corymb not more than $1/4$ the length of the Ped. **11.**
  10a. Fl. in a lax or dense Rac. half as long as the Ped. or longer ............ **asphodeloides**..
    Here also belong the following: B. **altissima, angustifolia, coetzeei, filifolia, inflata, praemorsa, pugioniformis**.
    Many other names belong here, although mostly the plants are hardly succulent. The following names by v. POELLNITZ are probably merely synonyms of **asphodeloides**: *hamata, huilensis, latibracteata, latitepala, lydenburgensis*.
  11. L. up to 15 cm long, 1 mm broad ................................................ **capitata**
  11a. L. 35–40 cm long, 5–8 mm broad ............................................. **breviracemosa**
C. L. not St.-clasping, without marginal wings, ± deciduous in the dry season; dwarf geophytes
    SSect. 3. Integrae v. POELLN.
        Type spec. **B. mesembryanthoides** HAW.
  12. Caudex crowned by a tuft of spiny or membranous scales .................... **13.**
  12a. Caudex without Sp. or scales .................................................. **15.**
  13. Caudex with slender roots below ............................................... **succulenta**
  13a. Caudex divided below into several thick tapering roots ................... **14.**
  14. L. 10 cm or more long; Infl. 20–30 cm tall .................................. **tetraphylla**
  14a. L. 2–3 cm long; Infl. 3–7 cm tall ........................................... **rhopalophylla**
  15. L. reduced to 1 only .............................................................. **monophylla**
  15a. L. reduced to 2, on the Infl. axis ........................................... **diphylla**
  15b. L. usually more than 2, all radical .......................................... **16.**
  16. L. boat-shaped, reticulate, in a spiral Haworthia-like Ros. ................. **haworthioides**
  16a. L. terete, erect, not reticulate .............................................. **17.**

| | | |
|---|---|---|
| 17. | L. fusiform, slender, to 4 cm long | 18. |
| 17a. | L. very squat, truncate, up to 3 cm long | **mesembryanthoides** |
| 18. | L. straight | **minima** |
| 18a. | L. tortuous | **torta** |

**B. alooides** WILLD. (§ II/2) (Pl. 26/2). – S. Afr. –Offsetting to form clumps; **L.** in compact Aloe-like Ros., lanceolate, linguiform, very soft and fleshy, light green, flat above, convex below, up to 25 mm wide at the base, 15–22 cm long, gradually tapering; **Infl.** often several, 20–30 cm tall; **Fl.** numerous, yellow.

**B. altissima** (MILL.) FOURC. (§ II/2) (*Anthericum a.* MILL., *A. longiscapum* JACQ., *B. l.* (JACQ.) WILLD. – S.Afr. – **L.** 12–20, radical from a many-headed **caudex**, subulate, acute, amplexicaul, fleshy, glaucous, 20–30 cm long, 6 mm ⌀, flat above, rounded below; Ped. at least 30 cm tall, with a dense Rac. as long again and 4 cm ⌀; **Fl.** yellow.

**B. angustifolia** v. POELLN. (§ II/2). – S.W.-Afr.: Hereroland: Lichtenstein. – **Caudex** 1 to 1.5 cm thick; **L.** linear, 10–15, 30–40 cm long, 2 mm wide broadening to 4–5 mm at the base which has membranous margins, old L. persisting; **Infl.** 20–40 cm tall including the cylindrical Rac. 15–25 cm long; bracts lanceolate, longer than pedicels; **Fl.** numerous, yellow.

**B. annua** WILLD. (§ II/1). – Cape: Camp's Bay. – ☉; **roots** fibrous only; **L.** 10–20, erect, subulate, 20–25 cm long, 2–5 mm thick, green, slightly grooved above; Ped. subterete, 15 to 22 cm long; Rac. very lax, 10–15-flowered; **Fl.** bright yellow.

**B. asphodeloides** (L.) SPRENG. (§ II/2) (*Anthericum a.* L.). – **Caudex** short, thick, tuberous; **L.** 10–20, subterete from a sheathing base 13 mm wide, 2–3 mm wide above, 15–30 cm long, nearly flat above, tapering, glaucous; Ped. 15–30 cm long, Rac. dense, c. 15 cm long, 4–5 cm ⌀, pedicels 12–18 mm long, erecto-patent, bracts lanceolate-deltoid with a long cusp; **Fl.** bright yellow.

**B. breviracemosa** v. POELLN. (§ II/2). – Eritrea. – Habit of **B. asphodeloides**; differs in having broader **L.,** Rac. very short and narrow, and bracts equalling the pedicels and finally decurved.

**B. capitata** v. POELLN. (§ II/2). – Transvaal: Lydenburg Distr. – **L.** numerous, radical, 12 to 15 cm long, c. 1 mm broad, linear, surrounded by the old dead remains; Infl. 20 cm tall, very compact, capitate; **Fl.** many.

**B. coetzeei** OBERM. (§ II/2). – Transvaal; Swaziland; Natal. – **Caudex** bulb-shaped, c. 2 cm long and 1.5 cm across with swollen spreading roots; **L.** c. 7 in a radical Ros., filiform, scarcely succulent, c. 50 cm long and 1 mm thick, sheathing below, apple green; Rac. c. 50 cm long, dense at first; **Fl.** yellow.

**B. crocea** L. GUTHRIE (§ II/2) (*B. dielsii* v. POELLN.?). – Cape Prov.: Upper reg., near Queenstown. – **Caudex** small, 8 mm ⌀; **L.** 7 or more, subterete, light green, fleshy, to 17 cm long, 5 mm broad; Ped. to 30 cm long, bearing a dense Rac. 16 cm long, 3.5 cm ⌀; **Fl.** bright orange with a green keel to each tepal.

**B. curvifolia** v. POELLN. (§ I). – Western Cape: Mierkraal. – Differs from **B. frutescens** in having narrower, variously curved **L.,** lax few-flowered Rac. and glabrous bracts.

**B. decurvata** PETER (§ II/2). – E.Afr.: Unyamwezi. – Very similar to **B. asphodeloides,** but differs in having white Fl., filaments bearded almost to the base and pedicels recurved after flowering.

**B. diphylla** SCHLTR. ex v. POELLN. (§ II/3). – Cape: Van Rhynsdorp D. – **Caudex** divided below into 3–7 conical fleshy roots 5–10 cm long, producing a main axis 10–25 cm tall on which is borne the single **L.** pair 1–2 cm above ground; **L.** 2 only, very succulent and soft, light green, the larger channelled above and 3–5 cm long, up to 1.5 cm wide, with an obtuse apex, the smaller leaf slightly higher on the axis, terete; **Fl.** 5–10, yellow.

**B. fallax** v. POELLN. (§ II/2). – Western Cape: Van Rhynsdorp D. – Close to **B. alooides** and probably only a variety. **L.** and bracts minutely ciliate at the margins; bracts amplexicaul.

**B. filifolia** BAKER (§ II/2). – Graaff Reinet D. – **Caudex** globose, 2–2.5 cm ⌀, crowned by a dense ring of erect bristles 2.5–4 cm long; **L.** 6–8, subterete, 15–30 cm long, 1 mm ⌀; Ped. slender, 30–45 cm tall, bearing a lax Rac. 4–5 cm long; **Fl.** pale yellow.

**B. frutescens** (L.) WILLD. v. **frutescens** (§ I) (Pl. 26/3) (*B. caulescens* L. ex STEUD.). – Cape: Sutherland, Willowmore, E. London D. – Branching ♄ to 60 cm tall, with fibrous **roots** only; St. brown, 6–8 mm thick, with Int. 12–25 mm long; **L.** distichous, subterete, bright green, (4–)15(–22) cm long, 4–8 mm ⌀; Ped. c. 30 cm long, bearing a usually dense Rac. 15–30 cm long and 2.5 cm ⌀; bracts fimbriate along their enlarged bases; **Fl.** many, bright yellow, white or rich orange. A very variable species in size and Fl. colour.

**B. —** v. **incurva** (THUNBG.) ROWL. (*Anthericum i.* THUNBG., *B. i.* (THUNBG.) SPRENG., *B. caulescens* v. *i.* (THUNBG.) ROEM. et SCHULT. – Cape. – **L.** much shorter, 4–5 cm long, subulate to subtrigonous, incurved; **Rac.** more lax, nodding.

**B. —** v. **rostrata** (JACQ.) ROWL. (*Anthericum r.* JACQ., *B. r.* (JACQ.) WILLD., *B. caulescens* v. *r.* (JACQ.) v. POELLN. – Cape: Albert D. near Bushman River. – **St.** stouter and shorter; **L.** glaucous; **Fl.** yellow.

**B. haworthioides** B. NORD. (§ II/3). – Cape: Van Rhynsdorp D. – **Caudex** broadly oblong or rounded, c. 15 mm ⌀, fleshy, shortly 5–7-lobed below; **L.** rosulate, c. 14 in a flat spiral resembling **Haworthia tessellata**, very fleshy, boat-shaped, c. 1 cm long, 5 mm across, blunt, underside convex, upperside more or less flat with white reticulations, margins finely and densely ciliate; **Infl.** c. 15 cm tall, with c. 10 yellow Fl. in a lax Rac. 3–4 cm long.

**B. inflata** OBERM. (§ II/2). – Transvaal; Swaziland; Natal. – **Rhizome** small, compact, with radical Ros. of 10–15 erect, terete, light green **L.** c. 50 cm long, 1 cm ⌀, grooved, expanded below to broad bases; **Infl.** 60–125 cm tall with over 100 scented yellow **Fl.**; **Fr.** inflated, globose, c. 13 mm ⌀.

**B. latifolia** (L. f.) HAW. (§ II/2) (*Anthericum l.* L. f. non SCHULT. nec KNUTH, *B. l.* (L. f.) SPRENG. – Cape: Somerset S. – **Rootstock** stout, sparingly branched above, with few Aloe-like dense Ros. of 12–20 oblong-lanceolate, tapering, flat, soft-fleshy, pale green **L.** 20 to 35 cm long, 5–7 cm across the broad amplexicaul base, 5–6 mm thick, margins curled up; **Infl.** 60 or more cm tall with a dense Rac. of many yellow **Fl.**

**B. mesembryanthoides** HAW. (§ II/3) (Pl. 26/5) – Cape: Little Namaqualand. – **Caudex** subglobose, forking below into conical fleshy roots, 15–20 mm ⌀; **L.** 1–2, lemon-shaped, up to 16 mm long, 5–10 mm thick, very soft and watery, pale green with a translucent tip, broadly striate, slightly grey-pruinose; **Infl.** very slender, stiffly erect, 5–15 cm tall, with 3–6 widely spaced golden yellow **Fl.** Eaten as a thirst-quencher by native children. Haworth consistently used the above spelling, and the alteration to "*mesembryanthemoides*" is unwarranted.

**B. minima** BAKER (§ II/3) (*B. concinna* BAKER, ? *B. inops* N. E. BR.). – Cape: Caledon & Tulbagh D. – Dwarf geophyte with **caudex** 12–18 mm ⌀; **L.** c. 6, very slender, subulate, 2.5–4 cm long, without an amplexicaul base; Ped. very slender, 5–7.5 cm long, bearing a lax Rac. 2.5–7.5 cm long of few yellow **Fl.**

**B. monophylla** v. POELLN. (§ II/3). – Western Cape: Piquetberg D. – Dwarf geophyte with a globose or elongated **caudex** 1–2 cm wide bearing a single, short, fusiform **L.** 2–3 mm broad and enclosed in 1–2 membranous sheathing scales; **Infl.** 11–15 cm tall, bearing a Rac. of c. 12 **Fl.** Probably no more than a dwarf ecotype of **B. minima** or ally.

**B. parviflora** BAKER (§ II/2). – Cape: Worcester D. – **L.** linear, subulate, up to 10 cm long, 2 mm thick, surrounded by several large membranous "sheath – L."; Ped. 30 cm long or more, with a dense cylindrical Rac. 15 cm long; **Fl.** whitish with a distinct brown keel to each tepal.

**B. praemorsa** (JACQ.) SPRENG. (§ II/2) (*Anthericum p.* JACQ., *B. p.* (JACQ.) ROEM. et SCHULT. comb. superfl.). – S.Afr. – **Caudex** tuberous with fleshy protuberances below; **L.** 8–12, distichous, 30–45 cm long, 6–10 mm ⌀ midway, deeply channelled above, obtuse below, soft fleshy, with sheathing bases; Ped. 30 to 45 cm long with a dense Rac. 30 cm long and 4 cm ⌀; **Fl.** bright yellow.

**B. pugioniformis** (JACQ.) LINK (§ II/2) (*Anthericum p.* JACQ., *B. p.* (JACQ.) ROEM. et SCHULT. comb. superfl.). – Cape: Riversdale D. – Similar to the preceding but **caudex** globose; probably not specifically distinct.

**B. rhopalophylla** DTR. (§ II/3). – S.W.Afr. – **Caudex** almost 2 cm ⌀, forking below into several thickened tapering roots c. 8 cm long; **L.** erect, 4–5, clavate, very fleshy, 2–3 cm long, 3–6 mm thick, apiculate, light sap-green, flat above, somewhat channelled towards the tip, surrounded by 2–3 membranous scales; **Infl.** 3–7 cm tall with 2–7 brownish yellow **Fl.**

**B. succulenta** COMPT. (§ II/3). – S.Afr. – **Caudex** with slender roots and a tuft of spiny scales at the apex; **L.** 2–4, radical, erect, stout, 6–7 cm long, pale green, very fleshy, drying up completely in the summer; **Fl.** yellow.

**B. tetraphylla** DTR. (§ II/3). – S.W.Afr.: Gr. Namaqualand. – **Caudex** 3–4 cm broad, divided below into c. 6 thick conical roots; **L.** always 4, flat, slender, soft-fleshy, 10–14 cm long, gradually tapered from a base 1–1.5 cm wide, surrounded by brownish-white striate scale L.; **Infl.** solitary, 20–30 cm long including the 8 to 10 cm Rac., 1.5–2 cm wide; **Fl.** 15–30, brownish yellow.

**B. torta** N. E. BR. (§ II/3). – Cape: Worcester D. – Similar to **B. minima** but **L.** thicker and strongly tortuous.

**B. triebneri** DTR. (§ II/2). – S.W.Afr.: Gr. Namaqualand. – Similar to **B. asphodeloides,** but **Infl.** taller and **Fl.** white.

**Bulbinopsis** BORZI. Liliaceae. – E.Austr. – Very close to **Bulbine,** but distinguished by the ovules being only 2 per locule, as compared with 4 or more in **Bulbine.** Chromosome evidence upholds the separation, as does geographical (**Bulbine** is exclusively African).

**B. bulbosa** (R. BR.) BORZI (*Anthericum b.* R.BR., *Bulbine b.* (R.BR.) HAW.). – **Caudex** bulb-shaped; **L.** all radical, fleshy, linear, subulate, grooved in front, 15–30 cm long, 3–4 mm thick, with very short sheathing bases; **Infl.** 30–60 cm tall; **Fl.** yellow, with all 6 filaments bearded.

**B. semibarbata** (R. BR.) BORZI (Pl. 26/4) (*Anthericum s.* R. BR., *Bulbine s.* (R. BR.) HAW.). – ⊙ with fibrous roots; **L.** linear, subulate, forming a dense radical Ros.; **Infl.** 30 cm high, taller in cultivation, with lax Rac. of yellow **Fl.**; 3 filaments bearded, 3 not.

**Bursera** L. Burseraceae. – Tropical America. – Some 40 tree-like species are known, but only a few of these show any characteristics of St.-succulence. **L.** thin, mostly fern-like, imparipinnate or trifoliate with petiolate leaflets or entire; margins entire, or notched or serrate; **Fl.** small, grouped, often in a racemose Pan. – Greenhouse, cool. Propagation: seed, cuttings.

**B. hindsiana** (BENTH.) ENGL. (*Elaphrium hindsianum* BENTH.). – Mexico: Baja Calif. – **St.** and **Br.** not as swollen as in **B. microphylla;** bark reddish-brown, only the St. spherically swollen; **L.** simple or 3-lobed, blunt or sharply crenate.

**B. microphylla** (ROSE) A. GRAY (Pl. 27/1) (*Elaphrium m.* ROSE, *B. morelensis* RAMFREZ, *Terebinthus multifolius* ROSE). – Western N.Am.; USA (Colorado Desert, Calif., Arizona) to Mexico (Sonora). – 'Elephant Tree'. – ♄ to 10 m tall; **caudex** very thick, about 1 m tall; **St.** thick and freely branching, wood soft and spongy, with a white milky sap, bark thin, papery, peeling, reddish-brown; **L.** soon falling, small, pinnate, with 10–25 linear-oblong leaflets, 3–8 mm long, smelling strongly of cedar; **Fl.** yellow.

**Calandrinia** H. B. et K. Portulacaceae. – Western North America to Tierra del Fuego (Chile); North and West Australia. – Glabrous or hairy annual or ⚁ or subshrubs; **L.** ovate, linear, spatulate or rhombic; **Fl.** axillary on long pedicels, or terminal, arranged in a lax or compact-capitate Infl., Pet. usually 5, free or united, ± large, usually red; **Fr.** a capsule, 3-valved; seeds very numerous. – Cultivated in the open, or the bushy species can be over-wintered in the unheated greenhouse. Propagation: seed, cuttings. – Not all species of **Calandrinia** are succulent.

**C.** cv. Burridgei. – S.Am. – ☉; **St.** prostrate or ascending, up to 35 cm tall; **L.** linear-lanceolate, glabrous; **Fl.** in a leafy Rac., numerous, small, coppery-pink or brick-red.

**C. ciliata** DC. (*C. caulescens* H. B. et K.). – Peru; Ecuador. – ☉; **St.** 30–40 cm tall; **L.** linear-spatulate, to 30 cm long, margins glabrous or hairy; **Fl.** solitary, axillary, c. 1 cm ⌀, purple or white.

**C. discolor** SCHRAD. (*C. elegans* HORT., *C. speciosa* LEHM.). – Chile. – ☉ to subshrub up to 80 cm tall; **St.** somewhat fleshy; radical **L.** rosulate, spatulate to spatulate-lanceolate, thick, bluntish, margin entire, upperside grey-green, underside purple; **Fl.** in a terminal Rac., c. 7 mm ⌀, violet or light purple.

**C. grandiflora** LINDL. – Chile. – ☉ to subshrub up to 1 m tall; **L.** round-ovate, pointed, narrowing to the petiole, almost rhombic in shape, flat, thick, green, margins smooth, 10–20 cm long, almost 2 cm across; **Fl.** 1.5 cm ⌀, light purple.

**C. menziesii** HOOK. (*C. speciosa* HORT.). – Chile. – ☉; **St.** simple, 10–15 cm tall, green, the lowest L. almost 5 cm long; **Fl.** almost 2.5 cm ⌀, vivid purple.

**C. polyandra** (HOOK) BENTH. (*Talinum p.* HOOK., *Claytonia p.* F. MUELL.). – S. and W.Austr. – ☉; **St.** 12–50 cm tall; **L.** spatulate or broad-linear; **Infl.** with few, pink Fl.

**C. speciosa** LINDL. (*C. lindleyana* HORT.). – USA: Calif.; Mexico: Baja Calif. – ☉; **St.** prostrate, 20–40 cm long, glabrous, flaccid-branched; **L.** simple, lanceolate, pointed, 8 cm long, narrowing to the base, green, lowest L. spatulate-lanceolate; **Fl.** with 2-winged pedicels, opposite the L., 2.5 cm ⌀, glossy purple-red.

**C. spectabilis** OTTO et DIETR. (Pl. 27/3). – Chile. – ☉ to subshrubs up to 60 cm tall; **St.** fleshy, as are also all Br. and L.; **L.** crowded, 3–4 cm long, almost rhombic or spatulate-lanceolate, rather pointed, pruinose, margin entire; **Fl.** c. 5 cm ⌀, vivid purple-red; rarely producing seed.

**C. umbellata** DC. (*Talinum u.* RUIZ. et PAV., *Portulaca prostrata* DOMB.). – Peru; Chile. – ☉ to ⚁ herb or subshrub; **St.** prostrate, up to 15 cm long, reddish, branching; **L.** apical, narrow-linear, pointed, pubescent, 1.5–2 cm long; **Fl.** numerous, corymbose, 2 cm ⌀, brilliant dark violet or purple-violet.

**Calibanus** ROSE. Agavaceae. – Greenhouse, warm. Propagation: seed.

**C. hookeri** TREL. (Pl. 27/2) (*Dasylirion hookeri* LEM., *D. hartwegianum* HOOK., *C. caespitosum* ROSE). – E. and central Mexico: Tamaulipas Prov. – Caudex hemispherical, 30–40 cm ⌀, 25 cm tall, with thick, fissured and corky rind, often forming further lateral hemispherical caudices, and with a few Br. above; **L.** only few, filiform, fairly tough; **Infl.** 50–60 cm tall, branching, Fl. inconspicuous.

**Caralluma** R. BR. Asclepiadaceae. – South and North Africa; Ethiopia; Somalia; Sudan; Socotra; Arabia; South West Mediterranean Region; East India. – Small succulent ⚁; 4–6 – angled, sometimes creeping in the ground like a stolon, later erect, edges ± dentate, partly tapering into the elongated Infl. or uniformly thick; **L.** in one species (**C. frerei**) with a stalked ovate lamina, otherwise rudimentary, scale-like, tiny, green or grey-green; **Fl.** small to large, with a ± pungent odour, solitary or several together or many in large umbels, basal, lateral or terminal on young St., Cor. bell-shaped or rotate, deeply lobed, annulus ring or cup-shaped, united or wanting, corona variously coloured and dotted. – Greenhouse, warm. Propagation: Seeds, cuttings.

## Division of the Genus Caralluma into Sections according to G. D. Rowley adapted from P. R. O. Bally, A. White and B. L. Sloane

A. L. flat and fleshy.
    Sect. I. **Frerea** (DALZIEL) ROWL. (*Frerea* DALZIEL as a genus). – Only species: C. frerei.
A. L. rudimentary, scale-like or undeveloped.
B. St. narrowed to the tip, tapering into an elongated Infl.
    Sect. II. **Caralluma** (*Eucaralluma* K. SCHUM. emend. WHITE et SLOANE).
        SSect. 1. **Priogonium** BALLY – St. sturdy, their vegetative portion saw-edged with closely set teeth along their angles; corona sessile. Follicles flattened, lozenge-shaped in cross-section. – Type species: C. priogonium. – Species: C. congestiflora, mogadoxensis, priogonium.
        SSect. 2. **Lalacruma** (K. SCHUM.) BALLY (Sect. *Lalacruma* K. SCHUM.). – St. comparatively slender, teeth distant; Corona subsessile to stipitate. Follicles circular in cross – section. Type species: C. gracilipes.

a) **Series Lalacruma** – Lobes of the mature Cor. rigidly expanded, sublinear. – Species: C. adscendens, dalzielii, edulis, (?) furta, gracilipes, longidens, mireillae, mogadoxensis, moniliformis, mouretii, sinaica, stalagmifera, subulata, vittata.
b) **Series Flaccidiflorae** BALLY. – Lobes of the mature Cor. flaccidly pendulous, sublinear to elliptic-ovate, longitudinally replicate. – Species: C. dicapuae, C. peckii, turneri.

B. St. not tapering to the tip.
  **Sect. III. Boucerosia** (WIGHT et ARN.) K. SCHUM. (*Boucerosia* WIGHT et ARN. as a genus). –
C. Fl. in umbels at the tip or near the tip of the St.
  1. **Umbellata-europaea Group.** – Species: C. aaronensis, adenensis, arabica, aucherana, awdeliana, burchardii, crenulata, diffusa, dodsoniana, edithae, europaea, flava, foetida, hexagona, indica, joannis, kalmbacherana, munbyana, oxydonta, penicillata, plicatiloba, procumbens, retrospiciens, socotrana, solenophora, somalica, speciosa, subterranea, truncato-coronata, tuberculata, umbellata, vibratilis.
C. Fl. solitary or few together at the tip or other part of the St. or in fascicles at the sides of the St.
D. Species from East India, South Arabia and Socotra.
  a) Cor. lobes glabrous or only at the margins with vibratile clavate hairs. Corona small.
    2. **Pauciflora-quadrangula Group.** – Species: C. anemoniflora, chrysostephana, cicatricosa, foulcheri-delboscii, luntii, pauciflora, quadrangula, torta.
  a) Cor.-lobes with a tuft of vibratile clavate hairs at the tip. Corona large, hairy.
    3. **Lavranii Group.** – Species: C. lavrani.
D. Species from Africa.
  b) Species from North of the equator.
    4. **Ango Group.** – Species: C. ango, baldratii, commutata, decaisneana, deflersiana, hesperidum, sacculata, sprengeri, venenosa, wilsonii.
  b) Species from South of the equator.
  c) St. principally 5–6 – angled.
    5. **Mammillaris Group.** – Species: C. armata, chlorantha, framesii, longipes, mammillaris, maughanii, winkleri.
  c) St. principally 4-angled.
  d) Cor 2.5 cm or less in ∅.
    e) Cor. with a distinct tube.
      6. **Carnosa-incarnata Group.** – Species: C. acutiloba, aurea, carnosa, cincta, distincta, ericata, gerstneri, hottentotorum, huernioides, incarnata, inversa, keithii, kochii, linearis, parviflora, pillansii, ramosa, schweinfurthii, tubiformis, umdausensis, villetii, wilfriedii.
    e) Cor. without a distinct tube.
      7. **Marlothii Group.** – Species: C. arida, dependens, intermedia, marlothii, ortholoba, peschii, pruinosa, reflexa, simulans, swanepoelii, ubomboensis, virescens.
  d) Cor. more than 2.5 cm in ∅.
    8. **Maculata-lutea Group.** – Species: C. albo-castanea, aperta, arenicola, bredae, caudata, dummeri, gossweileri, huillensis, knobelii, longicuspis, lugardii, lutea, maculata, melanantha, praegracilis, rogersii, tsumebensis, valida.

**C. aaronis** (HART.) N. E. BR. (§ III/1) (*Boucerosia a.* HART.). – Jordan.; Israel. – **St.** 4-angled, thick, angles blunt, finely dentate; **Fl.** several, 1 cm ∅, tube flat, Cor. lobes triangular-ovate, pointed, reddish with transverse callosities.
**C. acutiloba** N. E. BR. (§ III/6). – Cape: Lit. Namaqualand. – Branching, bushy; **St.** 15 to 20 cm tall, 4-angled, 12–18 mm thick, T. 4 to 5 mm long; **Fl.** 1–2, small, glabrous, light green, spotted with dark red-brown, Cor.-lobes projecting, triangular-ovate, pointed, 4–4,5 mm long, 3 mm across.
**C. adenensis** (DEFL.) BGR. (§ III/1) (*Boucerosia a.* DEFL.). – S.Arabia: Aden peninsula. – ♄, up to 60 cm tall, branching dichotomously, repeatedly and irregularly; **St.** 4-angled, sides concave, 3.5 cm across, green, glabrous, angles somewhat sinuate-dentate; **Fl.** 25–40 in a spherical umbel, evil-smelling, campanulate, tube 1 cm long, Cor.-lobes triangular, pointed, 8 to 9 mm long, glabrous, greenish outside, dark purple and warty inside.

**C. adscendens** (ROXB.) R. BR. v. **adscendens** (§ II/2a) (*Stapelia a.* ROXB.). – Peninsular India. – **St.** at first creeping, becoming erect, 30–60 cm tall, 1.5 cm across the base, 4 mm ∅ above, 4-angled, slightly sinuate-dentate; **Fl.** few in a Rac., 23 mm ∅, Cor.-lobes projecting, lanceolate, bluntly tapering, green with small spots, brownish at the tips.
**C.** — v. **attenuata** (WIGHT) GRAVELY et MAYURANATHAN (*C. a.* WIGHT, *C. fimbriata* HOOK. f.). – **St.** initially growing very fast underground, very freely branching, with red lines, angles rounded; **Fl.** pubescent, 15 mm ∅, with distinct purple markings.
**C.** — v. **carinata** GRAVELY et MAYURANATHAN. – **St.** unbranched, angles acute; **Fl.** pubescent, purple, usually pendulous.
**C.** — v. **fimbriata** (WALLICH) GRAVELY et MAYURANATHAN (*C. f.* WALLICH). – **St.** angles rounded; **Fl.** rather small, pendulous, hirsute.
**C.** — v. **geniculata** GRAVELY et MAYURANATHAN. – **St.** with thin secondary shoots, these

sometimes numerous, angles acute; **Fl.** 15 mm ⌀, less hairy, rotate, not pendulous, markings chestnut-brown.

**C.** — v. **gracilis** GRAVELY et MAYURANATHAN. - **St.** very thin, branching; **Fl.** pubescent, rotate, 1 cm ⌀, the dark markings chestnut-brown, the pale spreading part of the Cor.-lobes larger than the dark plicate part.

**C. albo-castanea** (MARL.) LEACH (Pl. 134/1) (§ III/8) (*Stapelia a.-c.* MARL., *St. carolischmidtii* DTR. et BGR.). – S.W.Afr.: Gr. Namaqualand. – **St.** 4–8 cm long, with new St. from the base so that the plant extends outwards, 4-angled, red-brown with pallid spots; T. large, projecting, 8–9 mm long; **Cor.** 3 cm ⌀, lobes 10–12 mm long, the outside rough, papillose, green with red spots, inside yellowish-white, with large, roundish spots, margins with very dark red, club-shaped hairs 3–4 mm long.

**C. anemoniflora** (DEFL.) BGR. (§ III/2) (*Stapelia a.* DEFL.). – S.Arabia. – Mat-forming; **St.** 2–6 cm long, thick as a finger, club-shaped, somewhat 4-angled; **Fl.** 1–2, campanulate-rotate, deeply 5-cleft, Cor.-lobes ovate-oblong, blunt-tapering, almost 3 cm long, the outside glabrous, pale violet, with 10 raised longitudinal lines, inside with purple-red hairs.

**C. ango** (A. RICH.) N. E. BR. (§ III/4) (*Stapelia a.* A. RICH.). – Ethiopia. – **St.** c. 20 cm tall, branching, 3–4-angled, coarsely dentate; **Cor.** 22 mm ⌀, deeply 5-cleft, lobes ovate-lanceolate, underside green, upperside black-purple.

**C. aperta** (MASS.) N. E. BR. (Pl. 28/1) (§ III/8) (*Stapelia a.* MASS., *Orbea a.* SWEET, *Caruncularia a.* SWEET). – Cape: Lit. Namaqualand. – **St.** 5–7 cm long, erect or prostrate, bluntly 4-angled, grey-green, scarcely dentate; **Fl.** 2.5–3.5 cm ⌀, the base being tubular, the Cor.-lobes oblong, rather blunt, the margins rolled back, inside finely papillose and dark brown at the base, above whitish or yellowish, with red-brown furrows and spots.

**C. arabica** N. E. BR. (§ III/1). – S.Arabia. – **St.** 4-angled, 12–15 mm ⌀; **Fl.** 20–25 in a terminal umbel, campanulate, Cor.-lobes triangular, ovate, pointed, 4 mm long, 3 mm across, glabrous, very dark red.

**C. arenicola** N. E. BR. (§ III/8). – Cape: Laingsburg, Prince Albert, Oudtshoorn D. – Mat-forming; **St.** numerous, 4-angled, 7–10 cm tall, 2–2.5 cm thick, angles with 3–5 mm long, stout, hard T.; **Fl.** in small clusters, tube campanulate, 4 mm ⌀ and deep, Cor.-lobes 10 to 14 mm long, 3–4 mm across at the base, tapering, margins revolute, glabrous outside, inside chestnut red-brown, finely tuberculate-papillose with a white spot at the base.

**C. arida** (MASS.) N. E. BR. (*Stapelia a.* MASS., *Orbea a.* SWEET, *Obesia a.* SWEET, *Piaranthus a.* G. DON.). – Cape: Swellendam D. – Mat-forming; **St.** 5–7 cm long, 1.5–2 cm thick, with 4-angles with pointed T.; **Fl.** solitary, rotate, 11 mm ⌀, yellowish, Cor.-lobes ovate-lanceolate, spotted basally and with fine bristly hairs towards the tip.

**C. armata** N. E. BR. (§ III/5). – Cape: Lit. Namaqualand; S.W.Afr.: Gr. Namaqualand. –
St. c. 2 cm ⌀, 5–6-angled, sinuate-dentate, T. 12 mm long; **Fl.** 1–2, campanulate, tube short, Cor.-lobes linear-lanceolate, curved hook-like at the tip, 6 mm long, tomentose inside.

**C. aucherana** (DECNE.) N. E. BR. (§ III/1) (*Boucerosia a.* DECNE.). – S.E.Arabia. – Robust; **St.** c. 15 cm tall, finger-thick, with 4 acute angles, serrate-dentate, T. hard; **Fl.** in clusters, Cor. 11 mm ⌀, yellow. Insufficiently known.

**C. aurea** LUCKH. (§ III/6). – Cape: Clanwilliam D. – Very robust species related to **C. incarnata;** forming clumps 50 cm tall and 45 cm ⌀; **St.** erect, compact, branching basally, 1.5–2 cm thick, angles with low tubercles bearing sharp, conical T., grey-green; **Fl.** 1–6, from the base of young shoots, on a 4 mm long pedicel, Sep. 2 mm long and pointed, Cor. 2.2 cm ⌀, tube flat, 2 mm deep, outside whitish, mottled purple-brown, inside canary-yellow, becoming white, uniformly covered with tiny bristles, lobes spreading, 9 mm long, 2.7 mm across, recurved and usually revolute.

**C. awdeliana** (DEFL.) BGR. (§ III/1) (*Boucerosia a.* DEFL.). – S.Arabia. – Freely branching; **St.** 20–30 cm tall, 4-angled, 1–3 cm across, green with dark red spots, angles sinuate-dentate, T. short; **Fl.** 5–15, Cor. with a campanulate tube, lobes erect, 8–11 mm long, outside greenish-red with tiny red spots and pubescent, inside glabrous, yellow, mottled with black-purple spots and dots, lobes triangular, tapering, margins recurved.

**C. baldratii** WHITE et SLOANE (§ III/4) (*C. meintjesiana* LAVR.). – Ethiopia; S.W.Arabia. – **St.** 4-angled, sides somewhat sulcate, with pallid reddish spots, T. curved upwards; **Fl.** solitary or several, 28 mm ⌀, light mahogany-coloured with reddish spots, minutely white-haired, lobes long lanceolate-pointed, 2 mm across.

**C. bredae** R. A. DYER (§ III/8). – Cape.

**C.** — v. **bredae.** – Beaufort West D. – **Rhizome** 6–7 cm long, 6–8 mm thick; **St.** 4-angled, to 5 cm tall, 1 cm thick below, tapering above, minutely warty along the angles; **Infl.** of 1–4 Fl., Cor. 1.5–1.7 cm ⌀, red to wine-coloured, the middle cup-shaped, with or without a slightly raised annulus, lobes triangular-ovate, 5 mm long, 3.5 mm across the base, rough, margins slightly recurved.

**C.** — v. **thomallae** R. A. DYER. – Cradock D. – Cor. centre less depressed, lobes lanceolate, 7 mm long, margins much recurved.

**C. burchardii** N. E. BR. (§ III/1). – Canary Is.; Morocco.

**C.** — v. **burchardii.** – Canary Is.: Fuerteventura Is. – Mat-forming; **St.** 15–20 cm tall, 4-angled, the T. on the angles being directed downwards, olive-green or grey-green; **Fl.** in clusters, 13 mm ⌀, Cor. glabrous, olive-green, densely white-haired.

**C.** — v. **maura** MAIRE. – S.Morocco. – **Fl.** 7–9 mm ⌀, campanulate.

**C.** — v. **sventenii** LAMB. – Canary Is.: Fuerteventura Is. – **Fl.** with glabrous red-brown Cor., the lobe-margins having snow-white hairs.

**C. carnosa** STENT. (§ III/6). – Transvaal: Pretoria, Rustenberg, Potgietersrust D. – **St.** branching, 6–15 cm tall, c. 4.5 cm thick, grey-green, with reddish spots and 4 angles with pointed, hard T. 12 mm long; **Fl.** 1–3, 9 to 10 mm $\varnothing$, campanulate, fleshy, greenish-mauve, outside with reddish spots, inside deep cream-coloured, with minute and dense tuberculate hairs, spotted dark red, Cor. lobes triangular, pointed, 5 mm long, with a 5-angled, red-spotted ring round the tube-mouth.

**C. caudata** N. E. BR. v. **caudata** (Pl. 28/2) (§ III/8). – Malawi; Rhodesia; S.W.Afr. – **St.** up to 10 cm tall, 5–6 mm thick, with 4 angles, rounded and dentate, T. up to 15 mm long, horny; **Fl.** solitary to several in a cluster, up to almost 10 cm $\varnothing$, deeply 5-lobed, Cor.-lobes up to 4 cm long, 7 mm across at the base, tapering, both surfaces glabrous, yellow with red spots, pubescent at the base, margins with dispersed red hairs.

**C. —** v. **chibensis** (LUCKH.) LUCKH. (*C. c.* LUCKH.). – Rhodesia: Chibi. – **Fl.** solitary, 5.5 cm $\varnothing$, Cor.-lobes 22 mm long, tapering, inside canary-yellow, with pink spots and fine white hairs, with red hairs below.

**C. —** v. **fusca** LUCKH. – **Fl.** entirely dark purple.

**C. —** v. **milleri** NEL. – S.W.Afr. – **St.** 8 cm tall, 5 mm $\varnothing$, with 2–3 Br., bluish, indistinctly 4-angled, T. pointed, 15 mm long; **Fl.** 4–5, Cor. 5 cm $\varnothing$, outside green-yellow with red-purple spots, tube 8 mm $\varnothing$ and deep, appearing purple because of the dense hairs, lobes 2.1 cm long, 8 mm across at the base, linear-lanceolate, deep green-yellow with purple spots, pubescent, somewhat ciliate below.

**C. —** v. **stevensonii** OBERM. – Rhodesia. – **St.** 9 cm tall, T. 2–2.2 mm long; **Fl.** 1–2, Cor. 9 cm $\varnothing$, tube 5-angled, 4 mm deep, 12 mm $\varnothing$, whitish yellow with purple spots, papillose, lobes 4 cm long, tapering, whitish-yellow with purple spots and a few ciliate hairs below.

**C. chlorantha** SCHLTR. (§ III/5). – Cape: George D. – **St.** prostrate, c. 4 cm tall, 1 cm thick, indistinctly 6-angled, often with little tubercles arranged in 6 rows; **Fl.** 1–2, campanulate, tube short, Cor.-lobes linear-lanceolate, hooked, sinuous at the tip, 6 mm long, glabrous outside, inside densely hairy, greenish.

**C. chrysostephana** (DEFL.) BGR. (§ III/2) (*Stapelia c.* DEFL.). – S.Arabia. – Mat-forming; **St.** 8–10 cm tall, 4-angled, sides $\pm$ furrowed, about 1 cm across, dark green with pallid mottling, T. thick, conical, 1 cm long; **Fl.** 1–2, 20–25 mm $\varnothing$, rather fleshy, outside grey-green with fine red stripes, inside purple-brown, lobes recurved-sinuous, triangular-ovate, margins recurved, inside with white hairs.

**C. cicatricosa** (DEFL.) N. E. BR. (§ III/2) (*Boucerosia c.* DEFL., *B. forskahlii* DECNE., p. part., *Desmidorchis f.* DEFL.). – S.Arabia: Yemen. – Irregularly branched, up to 60 cm tall; **St.** 4-angled, then becoming round, with a grey waxy coating, finely dentate, scarred; **Fl.** solitary, tube campanulate, 4 mm long, 5 mm $\varnothing$, Cor.-lobes triangular-ovate, 8 mm long, tapering, red-brown inside, tuberculate, lighter at the base and with lamellar excrescences forming a ring round the tube.

**C. cincta** LUCKH. (§ III/6)). – Cape: Van Rhynsdorp D. – **St.** 5 cm tall, 12 mm thick, branching midway, with 4 sharp angles resolving into T. 3–5 mm long, green with brown spots; **Fl.** solitary, 21 mm $\varnothing$, tube short, outside light green, inside deep dark purple with several yellow concentric lines around the mouth of the tube, Cor.-lobes 8 mm long, 5 mm across at the base, lanceolate, margins recurved, with red hairs at the base and towards the tip, underside red-striate.

**C. commutata** BGR. (§ III/4). – ? S.Arabia. – ♄ 10–11 cm tall; **St.** about 2 cm thick, branching, 4-angled, light green with reddish spots and stripes, angles very blunt, T. conical-subulate, 5–7 mm long; **Fl.** 1–2, c. 23 mm $\varnothing$, fleshy, glabrous, tube short, 4-cleft through half its length, Cor.-lobes triangular-ovate, short pointed, inside brown, minutely tuberculate, outside glabrous.

**C. congestiflora** BALLY (§ II/1) (*C. plurifasciculata* nom. nud.). – Somalia. – Closely related to **C. priogonium**; much-branched, glabrous, up to 28 cm tall, with spreading stolons; **St.** 10–25, up to 23 cm long, 20 to 22 mm $\varnothing$, with 4 acute, dentate angles, the sides $\pm$ deeply grooved, T. 1–2 mm long, upper part of the St. much thinner, whiplike; **Infl.** an umbel of c. 24 Fl., Cor. round, deeply 5-cleft through half its length, lobes green, inside yellow with green stripes, lanceolate, margins recurved, with 2–5 motile hairs at the tip, these having a tuberous swelling below the tips.

**C. crenulata** WALL. (§ III/1) (*Desmidorchis c.* DECNE., *Boucerosia c.* WIGHT et ARN.). – Burma. – **St.** produced from underground rhizomes, slender, erect, with spreading Br. 7–15 cm long, 7–8 mm across, grey-green, the 4 angles acute, T. recurved, small; **Fl.** 8–9 in a terminal umbel, campanulate-rotate, 23 mm $\varnothing$, lobes triangular-ovate, brown-red, with yellow spots and stripes and brown hairs.

**C. dalzielii** N. E. BR. (§ II/2a). – N.Nigeria: Katagum D.; Sudan; Ghana. – **St.** branching basally, 15–28 cm tall, 8 mm thick at the base, narrowing to the tip, the 4 angles dentate; **Fl.** in pendulous clusters of 2–3, Cor. 1 cm $\varnothing$, tube cup-like, lobes spreading, 4.5–5 mm long, 1.5 mm across, lanceolate-subulate, tapering, margins rolled back, inside with a few fine hairs, with long, fusiform, dark purple hairs in the upper half, whitish at the base with purple spots, above dark purple.

**C. decaisneana** (LEM.) N. E. BR. (§ III/4) (*Boucerosia d.* LEM., *Stapelia d.* A. CHEV.). – Senegal; Sudan. – **St.** erect, cylindrical to 4-angled, angles much projecting, glossy, whitish-green and faded brown, T. pointed; **Fl.** few, 2.5 cm $\varnothing$, tube campanulate, Cor.-lobes ovate, pointed, 10–12 mm long, chestnut-red, with fine white papillae. (BULLOCK, in 'Kew Bull.' **17**, 1, 1963, 195–196, gives the following synonyms in addition to the above: *Stapelia dentata* FORSK. nom., *Boucerosia dentata* (FORSK.)

DECNE., nom., *C. d.* (FORSK.) BGR. nom., *S. ango* A. RICH., **C. ango** (A. RICH.) N. E. BR., **C. sprengeri** (SCHWEINF.) N. E. BR., *Huernia s.* SCHWEINF., *H. multangula* SCHWEINF., nom., **C. venenosa** MAIRE, **C. commutata** BGR., *C. c.* ssp. *eu-commutata* JEHAND. et MAIRE, **C. hesperidum** MAIRE, *C. h.* ssp. *h.* JEHAND. et MAIRE).
**C. deflersiana** LAVR. (§ III/4). – S.W.Arabia. – **St.** densely crowded, 5–7 cm long, 12–15 mm thick, grey-green, with numerous oblong brownish spots, T. extending horizontally, 8–12 mm long; **Fl.** solitary, Cor. campanulate, grey-green outside with numerous brownish spots, tube cylindrical, 13 mm long, 10 mm ⌀, inside pink and glabrous, purple-brown above, densely and minutely tuberculate, Cor.-lobes ovate-lanceolate, 18 mm long, 9 mm across the base, margins somewhat curved back, inside dark purple-brown, bristly-papillose.
**C. dependens** N. E. BR. (§ III/7) (*C. parviflora* SCHLTR.). – Cape: Clanwilliam, Van Rhynsdorp D. – Branching and bushy; **St.** 20 cm tall, 15–20 mm thick, with 4 rounded angles, T. pointed; **Fl.** 2–3, Cor. rotate, deeply 5-cleft, 9–10 mm ⌀, Cor.-lobes narrow-oblong, blunt, 1.5 mm across, margins ciliate with long, soft, purple or white hairs, at the base yellowish with dark brown transverse stripes, towards the tip dark brown.
**C. dicapuae** (CHIOV.) CHIOV. (§ II/2b) (*Spathulopetalum d.* CHIOV., *C. quadrangula* DI CAPUA). – N.E.Afr. – **St.** with 3–4 acute angles, T. conical, 5–8 mm long, sides 7–10 mm across, ashen grey, the St. prolonged above up to 60 cm, 2–3 mm thick, the stemlike portion breaking off after the flowering period; **Fl.** in clusters, rarely solitary, Infl. a lax Rac. 10–20 cm long, Cor. spreading, tube c. 0.5 mm long, lobes becoming recurved-pendulous, linear-spatulate, gradually tapering below, short-pointed above, 11–14 mm long, 1 mm across below and 3 mm below the tip, outside glabrous, dark purple, inside dark purple and densely covered with clavate hairs.
**C.** — ssp. **dicapuae**. – Ethiopia; northern Somalia; Kenya. – Outer corona-lobes short, erect, rather blunt, 2-cleft; inner corona-lobes almost linear.
**C.** — ssp. **seticorona** BALLY. – Northern Somalia. – Corona-lobes covered with stiff bristle-hairs; outer corona-lobes erect-spreading, rather bluntly 2-cleft, inner corona-lobes broadly strap-shaped, blunt.
**C.** — ssp. **turneri** (E. A. BRUCE) BALLY (*C. t.* E. A. BRUCE). – Kenya; Uganda. – Cor.-lobes dark chestnut-brown, with large green-yellow spots; outer corona-lobes deeply 2-cleft, with spreading-erect or recurved subulate divisions.
**C.** — ssp. **ukambensis** BALLY. – Kenya. – Outer corona-lobes shortly 2-cleft with recurved and almost pointed tips; inner corona-lobes linear, extending beyond the staminal column, and curved over it.
**C. diffusa** (WIGHT) N. E. BR. (§ III/1) (*Boucerosia d.* WIGHT). – S. peninsular India. – Mat-forming; **St.** 4-angled, 1–2 cm thick, furrows flat, T. small; **Fl.** numerous, in umbels, campanulate, lobes projecting, triangular-ovate, tapered, dark purple inside, with minute white transverse callosities, margins ciliate.
**C. dioscoridis** LAVR. (Sect. not yet determined). – Socotra. – Forming dense groups; **St.** 5–10 cm long, 10–15 mm ⌀, ascending or creeping, rooting, glabrous, pale green or brown-green, angles compressed, sinuate-dentate; **Fl.** solitary, less often 2, pedicel 15–35 mm long, Cor. round, 4–5 cm ⌀, very fleshy, outside pale green and minutely warty, inside pale cream-coloured, tube flat, 8–11 mm ⌀, entire surface having some long purple-brown clavate hairs, lobes ovate-triangular, 15–17 mm long, corona simple, light pink, 7.5 mm ⌀.
**C. distincta** E. A. BRUCE (§ III/6). – Tanzania; Kenya. – **St.** creeping, cylindrical, angles dentate; **Cor.** campanulate, lobes spotted.
**C. dodsoniana** LAVR. (§ III/1). – Somalia: N. reg. – Branching; **St.** erect or ascending, 4-angled, irregularly tuberculate-tessellate or rough, 4–6 cm long; **L.** minute, fleshy, deciduous; **Fl.** 2–3, terminal, pedicel 2 cm long, Cor. round, deep reddish-brown, glabrous, 6 mm ⌀, tube c. 1 mm long, lobes broad-triangular and somewhat revolute; corona double, 3.5 mm ⌀.
**C. dummeri** (N. E. BR.) WHITE et SLOANE (Pl. 27/4) (§ III/8) (*Stapelia d.* N. E. BR.). – Tanzania; Kenya. – **St.** prostrate, 6–9 cm long, c. 15 mm ⌀, indistinctly 4-angled or rounded, light grey-green, reddish striate, T. c. 15 mm long, with a slender apiculus; **Fl.** 4–6, almost 4 cm ⌀, olive-green, cup-shaped, lobes gradually tapering, outside glabrous, inside hairy.
**C. edithae** N. E. BR. (§ III/1). – Somalia. – **St.** robust, usually with 4 infrequently 5 angles, 2.5 cm and more thick, angles with broadly triangular T.; **Fl.** 60–70 in a terminal umbel of 6–7 cm ⌀, Cor. 12 mm ⌀, outside glabrous, inside warty-callused, dark red-brown, lobes triangular-ovate, pointed, with a few tiny hairs at the tip.
**C. edulis** (EDGEW.) BENTH. et HOOK. (§ II/2a) (*Boucerosia e.* EDGEW., *B. stocksiana* BOISS.). – Peninsular India. – **St.** scarcely branching, 15–45 cm tall, round to 4-angled, in the juvenile stage with ovate-lanceolate deciduous **L.**; **Fl.** 8 mm ⌀, purple inside, lobes ovate-lanceolate, glabrous.
**C. ericata** NEL (§ III/6). – Cape: Oudtshoorn D. – **St.** erect, 4 cm tall, 7 mm ⌀, 4-angled, T. with a short subulate apiculus; **Fl.** 2 together, yellow outside and below, then purple, upper part light green-yellow.
**C. europaea** (GUSS.) N. E. BR. (§ III/1) (*Stapelia e.* GUSS., *Boucerosia e.* (GUSS.) HOOK. f.). – E., W. and S. Mediterranean reg., Lampedusa Is., S.Spain. Comprising the following: –
**C.** — v. **affinis** (DE WILD.) BGR. (*C. a.* DE WILD., *C. maroccana* BGR.). – Morocco. – **St.** stouter than in v. **europaea**, 2.5 cm thick; **Fl.** 16–18 mm ⌀, almost hairless, stripes broader.
**C.** — v. — f. **parviflora** MAIRE. – **Fl.** smaller.
**C.** — v. **albotigrina** MAIRE. – Morocco. – **Fl.** with white stripes.
**C.** — v. **confusa** (FONT Y QUER) FONT Y QUER (*C. confusa* FONT Y QUER, *Stapelia e.* WEBB, *Apteranthes gussoneana* BOISS., *Boucerosia mun-*

*byana* v. *hispanica* JIM. et IBAN.). – Spain: Prov. Murcia. – Geographical race distinguished by the deeply 2-cleft lobes of the corona.

**C. — v. europaea.** – S.Spain. – **St.** branching, Br. often curved downwards and rooting, variable in height, 1–1.5 cm thick, 4-angled, grey-green, with rather pale reddish spots, angles blunt, slightly sinuate, dentate; **Fl.** 10–13 in sessile umbels, Cor. 13–16 mm ⌀, 5-cleft to midway, lobes ovate, pointed, margins slightly ciliate, greenish-yellow with brown transverse stripes, the tips brown-red.

**C. — v. marmariensis** BGR. – Libya: coast. – **St.** somewhat thinner; **Fl.** pale yellow, lobe-tips greenish, with red-brown transverse bands and fine white hairs inside.

**C. — v. maroccana** (HOOK. f.) BGR. (*Boucerosia m.* HOOK f., *C. m.* N. E. BR., *C. m.* v. *confusa* SEEMEN et MAIRE). – Morocco. – **Fl.** somewhat larger, lobes red-brown below with yellowish transverse bands, tips purple-brown.

**C. — v. micrantha** MAIRE. – Morocco. – **Fl.** 8 mm ⌀, with a few hairs on the margins of the lobe-tips.

**C. — v. muelleri** CH. SAUV. et J. VINDT. – **Fl.** olive-green, somewhat red below.

**C. — v. simonis** BGR. (*Boucerosia s.* HORT. ex BGR.). – Algeria; Tunisia; Morocco. – **Fl.** 13 mm ⌀, with white or brown-red hairs, lobes broader than long, red-brown, with white hairs below.

**C. flava** N. E. BR. (§ III/1). – S.Arabia. – **St.** 15–20 cm tall, 12–20 mm thick, branching, angles acute, T. short and broadly triangular; **Fl.** numerous in sessile umbels, short-campanulate, 2 cm ⌀, lobes oblong-pointed, margins recurved, yellow, spotted.

**C. foetida** E. A. BRUCE (§ III/1). – Tanzania. – **St.** 10–15 cm tall, c. 2 cm thick; **Fl.** in an umbel of 30–40, 14 mm ⌀, densely papillose and blackish to red-brown inside; otherwise as for **C. speciosa** but smaller in habit.

**C. foulcheri-delboscii** LAVR. v. **foulcheri-delboscii** (§ III/2). – S.Arabia: Hadramaut. – **St.** branching, forming groups, 10–40 cm tall, 12–15 mm thick, blue-green, the 4 angles sinuate-dentate, T. blunt; **Fl.** 3–8 in clusters, Cor. round-campanulate, to 12 mm ⌀, pale green with a few red-brown spots outside, inside creamy-white with spots closer together and merging into one another at the lobe-tips, the base of the tube unspotted, glabrous, with dark red-brown motile hairs only at the lobe-tips, these hairs being 2 mm long, tube 2 mm long and 4–5 mm ⌀, lobes 4 mm long, ovate, pointed.

**C. — v. greenbergiana** LAVR. – S.W.Arabia. – Forming dense clumps by branching; **Infl.** spherical-capitate with 40–50 densely crowded Fl.

**C. framesii** PILL. (§ III/5). – Cape: VanRhynsdorp D. – **St.** 20 cm tall, branching, 4–6-angled, 1.5–2 cm thick, green with reddish spots, angles ± distinctly spiraled, T. 3 mm long and hard-tipped; **Fl.** several, 16–18 mm ⌀, tube 3 mm ⌀, short, campanulate, glabrous outside, inside pale yellow with isolated yellow hairs, Cor.-lobes 7–8 mm long, 1.5–2 mm across, oblong, pointed, lemon-yellow outside, inside canary-yellow with isolated grey hairs.

**C. frerei** ROWL. (Pl. 28/3) (§ I) (*Frerea indica* DALZIEL). – Peninsular India. – **St.** whitish, round, fleshy, little branching, up to 10 cm tall, 2 cm thick; **L.** short-petiolate, oblong-oval, fleshy, 2–3 cm long, 1 cm across; **Fl.** at the Br.-apices, solitary, with a short pedicel, c. 1 cm ⌀, lobes broad-triangular, pubescent, red-brown with yellow or white markings resembling an oriental pattern.

**C. furta** BALLY (? § II/2a). – N.Somalia. – **St.** bluntly 4-angled, c. 7 cm tall, 1 cm thick at the base, freely branching basally, Br. 6–8 mm thick at the base, 3–6 cm long, angles dentate, T. 2 mm long, more closely spaced towards the apex, pale green-grey with a purple band along the angles; **Fl.** 2 (or more?), Cor. glabrous, 6 mm long, very deeply 5-lobed, tube 1 mm long, 2.5 mm across, flat cup-shaped, pale green below, lobes longitudinally plicate, 5 mm long, 2 mm across at the base, narrowing abruptly to a point, white, pale green in the upper part, inside with purple spots.

**C. gerstneri** LETTY ssp. **gerstneri** (§ III/6). – Natal: Zululand. – Clump-forming; **St.** arising from underground stolons, 6–7 cm tall, 2.5 to 3 cm ⌀, 4-angled, deeply furrowed, T. c. 13 mm long, with 2 tiny lateral T. near the tip; **Fl.** 2–6 in clusters, Cor. 3.5 cm ⌀, tube cup-shaped, pale cobalt-green outside, inside purple red with fine velvety hairs, lobes 1.5 cm long, 8 mm across the base, inside parchment-coloured and covered with dark red spots, velvety-rough, quite glossy, margins with purple cilia 1.5 mm long.

**C. — ssp. elongata** DYER. – Transvaal. – Cor.-lobes 2.5–3 cm long.

**C. gossweileri** S. MOORE (§ III/8). – Angola; Zambia; Rhodesia. – **St.** stout, rather prostrate, bluish-green with slight purple mottling, 4-angled, up to 10 cm long, c. 1.5 cm ⌀, sides somewhat furrowed, T. stout and conical-pointed, up to 2 cm long, 7.5 mm thick, with 2 small lateral T. below the tip; **Fl.** 6–15 in clusters, evil-smelling, Cor. deeply 5-lobed, tube short-campanulate, glabrous outside, inside minutely pubescent, transversely rough at the base of the lobes, and granular towards the tips with longitudinal clefts, c. 7 mm long, 8–12 mm ⌀, the Cor.-lobes up to 4.5 cm long, 12 mm across at the base, tapering, margins slightly curved back, with a short, blunt tooth in the cleft.

**C. gracilipes** K. SCHUM. (§ II/2a). – N.E.Afr. – **St.** 30–40 cm tall, 5–6 mm thick at the base, narrowing towards the tip, the 4 angles finely dentate; **Fl.** c. 15, on the thin St.-tip, Cor. 6–7 mm long, the tube widening.

**C. — ssp. arachnoidea** BALLY (*C. a.* BALLY nom. nud.). – Uganda; Tanzania; Kenya; Sudan. – Outer corona-lobes deeply 2-cleft, each section subulate; staminal column almost sessile or on a short stalk.

**C. — ssp. breviloba** BALLY. – Kenya. – Inner corona-lobes much reduced, of the same height as the staminal column.

**C. — ssp. gracilipes.** – Kenya. – Inner corona-lobes erect and projecting beyond the staminal column; outer corona-lobes much reduced, bluntly 2-cleft, staminal column on a distinct stalk.

**C. hesperidum** MAIRE (§ III/4). – Morocco. – St. simple or branching from a whitish **caudex**, greenish-white with reddish spots, the 4 angles much rounded, T. very fleshy, pointed, spine-like; **Fl.** 2–10, 2 cm $\varnothing$, rotate, 5-lobed, cleft to midway, velvety-papillose and dark brown-purple inside, the tube glabrous inside, yellowish, Cor.-lobes ovate-pointed, margins and tips somewhat recurved.

**C. hexagona** LAVR. (§ III/1). – S.W.Arabia. – Densely mat-forming; **St.** (3–)5(–8) cm long, 1.5–3 cm thick, green or grey-green, with 6 or less often 4–5 compressed, sinuate-dentate angles; **Fl.** in clusters, Cor. round, up to 22 mm $\varnothing$, greenish-white outside, with many tiny red tuberculate spots, the lobes similar, inside cream-coloured with a greenish tinge, with a narrow, dark purple border, the entire surface covered with dark purple hairs, tube up to 3 mm long, 6 mm $\varnothing$, lobes triangular, pointed, 8 mm long, 8 mm across the base.

**C. hottentotorum** (N. E. BR.) N. E. BR. v. **hottentotorum** (§ III/6) (*Quaqua h.* N. E. BR.). – Cape: Lit. Namaqualand. – **St.** 10–15 cm tall, up to 2.5 cm thick, grey-green, 4-angled, sharply furrowed, angles rounded, T. pointed; **Fl.** 6–10, about 6 mm $\varnothing$, campanulate, cleft to midway into 5 lobes, pale yellow-green, the inside of the throat having a few minute hairs, lobes triangular-ovate, upperside slightly keeled.

**C. — v. major** N. E. BR. (*C. ausana* DTR.). – Cape: Lit. Namaqualand; S.W.Afr.: near Aus. – **Fl.** 1 cm $\varnothing$.

**C. — v. minor** LUCKH. – Cape: Lit. Namaqualand. – **Fl.** 5 mm $\varnothing$, yellow.

**C. huernioides** BALLY (§ III/6). – Somalia. – St. c. 12 cm long, 1.4 cm thick, creeping to ascending, bluntly 4-angled, T. to 1 cm long, fleshy, 4 mm thick at the base, pale green, spotted dark red to completely purple-chestnut colour; **Fl.** 4–6, Cor. campanulate, curved downwards, 15 mm long, 16 mm $\varnothing$ at the mouth, outside pale green with a few green or red spots, inside yellow, reddish-brown papillose, these papillae being very compressed at the base and forming a star-like pattern, the lobes broadly 3-angled, 8.5–9 mm long, 8 mm across at the base, in colour and texture as for the tube, subsidiary lobes 1 mm long, recurved.

**C. huillensis** HIERN (§ III/8). – Angola: Huila. – Forming broad cushions; **St.** 6–7 cm tall, thick, clavate, 4-angled, soft, velvety, angles with compressed T.; **Fl.** 6–10 in clusters, Cor. deeply 5-lobed, dark red, finely pruinose outside, inside almost glabrous and rough, tube about 7 mm long, hemispherical, lobes 3–4 cm long, linear-lanceolate.

**C. incarnata** (L. f.) N. E. BR. v. **incarnata** (§ III/6) (*Stapelia i.* L. f., *Podanthes i.* SWEET, *Piaranthus i.* G. DON., *Boucerosia i.* N. E. BR.). – Cape: Ceres, Piquetberg, Malmesbury D., Lit. Namaqualand. – Much branching, bushy, about 30 cm tall; **Br.** erect, 12–15 mm thick, 4-angled, T. conical, 3–4 mm long, with a hard tip, grey-green; **Fl.** in small clusters, c. 8 mm $\varnothing$, tube campanulate, outside glabrous, Cor.-lobes 3 to 4 mm long, 2 mm across the base, linear-lanceolate, pointed, margins recurved, pink, hairy below.

**C. — v. alba** (G. DON.) N. E. BR. (*Piaranthus i.* v. *a.* G. DON.). – **St.** with grey coating; **Fl.** white inside.

**C. indica** (WIGHT et ARN.) N. E. BR. (§ III/1) (*Hutchinia i.* WIGHT et ARN., *Boucerosia i.* DECNE., *B. hutchinia* DECNE.). – Peninsular India. – **St.** slender, branching, Br. spreading, 10–15 cm long, 5–7 mm thick, 4-angled, somewhat sinuate-dentate, T. directed backwards, pointed; **Fl.** 5 or more in a terminal umbel, campanulate-rotate, lobes triangular-ovate, tapering, greenish, finely tuberculate, purple-spotted, the margins ciliate.

**C. intermedia** (N. E. BR.) SCHLTR. (§ III/7) (*Stapelia i.* N. E. BR.). – Cape: Clanwilliam D. – **St.** 10–15 cm tall, 4-angled, T. projecting from a sinuate base, large; **Fl.** solitary, 2.5 cm $\varnothing$, rotate, green, finely callused-warty, with tiny purple spots, lobes triangular-ovate, tapering, ? ciliate.

**C. inversa** N. E. BR. (§ III/6). – Cape: Clanwilliam D. – Bushy and branching, 10–20 cm tall; **Br.** with 4 rounded angles, T. projecting, sharp; **Fl.** 2–3, Cor. 6–10 mm $\varnothing$, rotate, lobes narrow-oblong, rather blunt, c. 5 mm long, 3–4 mm across, short-pointed, in the lower part purple-brown, grass-green above, both surfaces glabrous, tube with red hairs.

**C. joannis** MAIRE (§ III/1). – S.W.Morocco. – **St.** branching, 4-angled, smooth, green, 6 to 10 cm tall, 13–15 mm $\varnothing$, pendulous where growing on rocks and over 1 m long, angles rounded, sinuate-dentate, T. small; **Fl.** 2–10 in clusters, 15–25 mm $\varnothing$, tube campanulate, 5 mm long, olive-yellow, reddish spotted, Cor.-lobes 6–7 mm long, 5 mm across at the base, velvety inside, purple midway, tuberculate around the tube-mouth, margin-tips with ciliate hairs 1.8 mm long.

**C. kalmbacherana** LAVR. (§ II/1). – S.Arabia: Hadramaut. – Extremely succulent, very freely branching, 25 cm tall, c. 35 cm $\varnothing$; **St.** ascending, branching and re-branching, pale yellowish green to brownish green, 4-angled, principal St. up to 2 cm $\varnothing$, Br. 10–12 mm $\varnothing$, angles sinuate-dentate, rounded; **Fl.** 35–50 in a terminal spherical Infl., bracts linear, 2–3 mm long, pedicels 7 mm long, Cor. 2 cm $\varnothing$, completely glabrous, outside yellow with light orange-red spots, inside yellow, densely covered with small, rounded warty and often merging spots, tube 6 mm $\varnothing$, slightly urn-shaped, inside of mouth 4 mm $\varnothing$, Cor.-lobes deltoid, curved slightly upwards, 5–6 mm long, 6–9 mm across at the base.

**C. keithii** R. A. DYER (§ III/6) (*C. fosteri* PILL., *C. schweickerdtii* OBERM., *C. carnosa* SCHWEICK.). – Swaziland; Natal; Transvaal; Moçambique; Rhodesia. – Clumping; **St.** branching basally, 7–9 cm tall, 1.5–3 cm across, blue-green, red-spotted, 4-angled, T. 1–1.5 cm long, sharply pointed; **Fl.** 1–3, 1.2 cm ∅, tube campanulate, finely tuberculate, with a 5-angled annulus, purple, Cor.-lobes triangular-ovate, pointed, 5 mm long, 6 mm across, warty, hairy, dark purple, white-spotted, margins hairy.

**C. knobelii** (PHILL.) PHILL. (Pl. 28/4) (§ III/8) (*Stapelia k.* PHILL., *C. langii* WHITE et SLOANE, *C. k.* v. *l.* (WHITE et SLOANE) WHITE et SLOANE, *C. kalaharica* NEL). – Botswana; Cape. – **St.** c. 10 cm tall, 1 cm ∅, acutely 4-angled, T. 6 mm long, conical; **Fl.** 10 in clusters, 3.5 cm ∅, tube 1 cm ∅, Cor.-lobes ovate-pointed, 13 mm long, 1 cm across at the base, white, the upper part greenish, with black-purple spots, margins with dark hairs.

**C. kochii** (§ III/6). – Somalia: Mijertein Prov. – Grouping by means of offsets, both below and above ground; **St.** erect, 4-angled, leafless, glabrous, with prominent, upward-pointing T., bluish-green with brown spots, 4 to 10 cm long, 15–20 mm ∅; **Fl.** 2–3 at the St.-apex, pedicel 3–4 mm long, Sep. 3 mm long, fleshy, Cor. fleshy, campanulate, 15 mm long, tube 7 mm long, outside glabrous, greenish, purple-spotted, yellowish inside at the base and blood-red towards the mouth with whitish papillae, these having a short bristly hair at the tip, lobes ovate-deltoid, erect, 8 mm long, dark scarlet with white papillae.

**C. lavranii** RAUH et WERTEL (§ III/3). – S.E.Arabia. – ♄ 30–40 cm tall and up to 30 cm ∅; **St.** 1–3 cm thick, grey waxy-pruinose, branching coral-like, the 4 angles bluntly crenate because of the prominent L.cushions; **Fl.** solitary, terminal, tube short, flat, plate-shaped, Cor.-lobes spreading like a star, c. 1 cm long, 5 mm across at the base, margins curled back, glabrous, the lobe-tips having a tuft of motile wine-red clavate hairs 2 mm long, the underside smooth, glabrous, grey-green, the upperside roughly papillose, ochreous-yellow with numerous wine-red spots.

**C. linearis** N. E. BR. (§ III/6). – Cape: Prince Albert D. – **St.** 4-angled, glabrous, T. short, sharp, hard; **Fl.** several, c. 18 mm ∅, short-campanulate, Cor.-lobes linear, tapering, margins revolute, black-purple inside, tube whitish.

**C. longicuspis** N. E. BR. (§ III/8). – Cape: Griqualand West, ? Prieska D. – **St.** 5–8 cm tall, prostrate or erect and then again prostrate, 3–15 cm long, 6–20 mm ∅, grey-green, brown-spotted, the 4 angles rounded, T. 2–3 mm long, pointed, soon falling; **Fl.** 3–10 in an umbel, tube short funnel-shaped, 8 mm across, 4 mm deep, Cor.-lobes 2–2.5 cm long, 5–6 mm across, linear-lanceolate, pointed, margins recurved, glabrous outside, inside pale purple, densely and finely papillose, the indentations between the lobes finely ciliate.

**C. longidens** N. E. BR. (§ II/2a). – Sudan. – **St.** erect, branching, 12–18 cm tall, narrowing above, the 4 angles somewhat dentate; **Fl.** 1–2, Cor. dark purple-brown, glabrous, tube 5–6 mm long, 3–4 mm ∅, campanulate, lobes ovate-lanceolate, pointed, 4–5 mm long.

**C. longipes** N. E. BR. (§ III/5). – Cape: Sutherland D. – **St.** ± prostrate, 2–3 cm long, 6–15 mm thick, dark green, T. 2–3 mm long; **Fl.** 1–2, 12–13 mm ∅, flat, Cor.-lobes ovate, pointed, 4–5 mm long, pale dull yellow, the tips green.

**C. lugardii** N. E. BR. (Pl. 29/1) (§ III/8). – Cape: Prieska D.; Botswana; S.W.Afr. – **St.** prostrate or ascending, often spreading stolon-like below the soil-surface, not much branched, 4-angled, 10–15 cm long, 1 cm thick, grey-green, with darker green or brown spots or stripes, angles blunt and separated by angular furrows, T. projecting, pointed; **Fl.** 5–6, Cor. 5–6 cm ∅, broad-campanulate at the base, deeply cleft, lobes 3 cm long, 7 mm across at the base, narrow-linear, gradually tapering, brown inside to above the base of the lobes, finely hairy, green-yellow towards the tips, the margins shortly hairy.

**C. luntii** N. E. BR. (§ III/2). – S.Arabia. – **St.** 10–20 cm tall, 2 cm thick, grey-green with reddish spots, glabrous, 4-angled, T. 8–10, large, projecting; **Fl.** 1–3, Cor. deeply 5-cleft, lobes 16–18 mm long, narrow deltoid-lanceolate, 2 mm across, margins revolute, glabrous outside, inside finely and shortly hairy, greenish-yellow at the base, often red-brown.

**C. lutea** N. E. BR. (§ III/8). – Widely distributed: E. to S.Afr., S.W.Afr. – **St.** branching basally, 5–10 cm tall, up to 2 cm ∅, the 4 angles coarsely sinuate-dentate; **Fl.** numerous in large umbels, 5–7 cm ∅, rotate, Cor.-lobes narrowly or sometimes broadly lanceolate, gradually tapering, finely callused, golden-yellow to almost black or chestnut-brown, concolorous or spotted, margins with red ciliate hairs, not free-flowering, very evil-smelling.

**C.** — ssp. **lutea** (Pl. 29/2) (*C. lateritia* N. E. BR., *C. lut.* v. *lat.* (N. E. BR.) NEL, *C. l.* v. *stevensonii* WHITE et SLOANE, *C. vansonii* BREMEK. et OBERM., *C. lut.* v. *v.* (BREMEK. et OBERM.) LUCKH., *Stapelia vaga* sensu HUBER p. part.). – Botswana; Rhodesia; Transvaal; Natal; Cape. – Cor.-lobes narrowed-tapering, 3–5 times as long as broad but very variable, from 2.8 to 7.4 : 1; number and shape of the Fl. as well as the length of the pedicel also very variable; Fl. colour from yellow to darkest red to brown, concolorous to spotted in the North, but in the South more yellowish, brownish, almost green and less often orange-coloured, sometimes also striped.

**C.** — ssp. **vaga** (N. E. BR.) LEACH (Pl. 29/4) (*Stapelia v.* N. E. BR., *C. v.* (N. E. BR.) WHITE et SLOANE, *C. nebrownii* BGR., *C. pseudo-nebrownii* DTR., *C. p.* v. *p.* (DTR.) WHITE et SLOANE, *C. n.* v. *discolor* NEL, *C. brownii* DTR. et BGR., *C. hahnii* NEL). – S.W.Afr.; Angola; Cape. – Fl. Cor. with much broader lobes, 1.5 to 2.5 times as long as broad.

**C. maculata** N. E. BR. v. **maculata** (Pl. 29/3) (§ III/8) (*C. grandidens* VERD.). – Botswana: N.Kalahari. – **St.** numerous, simple or branching basally, c. 7 cm tall, 4-angled, green with red-brownish markings, arising from long underground stolons, T. c. 15 mm long, deltoid, pointed with 2 lateral T.; **Fl.** 3, c. 4 cm ⌀, both surfaces glabrous, smooth, deeply 5-lobed, yellow-green with wine-red spots, Cor.-lobes c. 18 mm long, 10 mm across, oblong to oblong-spatulate, pointed, margins revolute, ciliate over 3/4 of the length.

**C. — v. brevidens** H. HUBER (*C. rangeana* DTR. et BGR., *C. rangei* DTR. et BGR., *Piaranthus streyianus* NEL). – S.W.Afr.: Gr. Namaqualand. – T. 2–6 mm long, broadly triangular.

**C. mammillaris** (L.) N. E. BR. (§ III/5) (*Stapelia m.* L., *S. pulla* AIT., *Piaranthus p.* R. BR., *P. m.* G. DON., *Pectinaria m.* SWEET, *Boucerosia m.* N. E. BR.). – Cape, widely distributed. – **St.** robust, branching, 15 cm or more in height, 3 cm thick, fresh green, irregularly spirally 5–6-angled, T. projecting, 15 mm long; **Fl.** 3–20, Cor. deeply 5-cleft, tube short, lobes narrow-lanceolate, margins revolute, outside whitish, spotted, glabrous, inside finely bristly-tuberculate, black-purple, tube whitish with dark spots.

**C. marlothii** N. E. BR. (III/7). – Cape.

**C. — v. marlothii.** – Cape: Calvinia D., Karroo. – **St.** branching basally, 8–10 cm tall, 12–20 mm thick, 4-angled, dull green-violet, T. 2–4 mm long, pointed, with a white, hard tip; **Fl.** 2–4, Cor. yellowish to pale green, pale brown towards the lobe-tips, with dark spots, covered with filiform blackish-purple hairs.

**C. — v. viridis** LAMB. – Cape: Calvinia D. – Fl. Cor. apple-green, pale in the centre, lobe-tips deep pink, spots larger, hairs both filiform and clavate.

**C. maughanii** R. A. DYER (§ III/5). – Cape: Van-Rhynsdorp D. – **St.** numerous, branching basally, c. 7 cm tall, 1 cm ⌀, 6-angled, T. spreading, 2–3 mm long, with a hard tip; **Fl.** 1–2, 14–16 mm ⌀, Cor. lobes unequal, 5–7 mm long, margins recurved, yellow, reddish below.

**C. melanantha** (SCHLTR.) N. E. BR. v. **melanantha** (§ III/8) (*Stapelia m.* SCHLTR., *S. furcata* N. E. BR., *C. leendertziae* N. E. BR., *C. rubiginosa* WERD., *C. australis* NEL). – Transvaal: Pietersburg D. – Mat-forming; **St.** prostrate, 5–7 cm long, 2–2.5 cm thick, 4-angled, T. projecting, pointed, 6–10 mm long, triangular; **Fl.** 5 cm ⌀, Cor. rotate, 5-cleft to midway, lobes triangular-ovate, tapering, 17 mm long, 12 mm across, glabrous outside, inside black-brown, with ± concentric transverse wrinkles, margins ciliate.

**C. — v. sousae** GOMES et SOUSA. – Moçambique. – **Fl.** deep strong purple-red, lobes wrinkled inside and covered with fine black hairs.

**C. mireillae** LAVR. (§ II/2a). – Somalia: W. of Djibouti. – Succulent dwarf ♄ branching from the trunk which is 4-angled, **St.** narrowing towards the tip, angles rounded, not dentate, 3–10 cm tall, 5–10 mm ⌀, blue-green with brown spots; L. round-ovate, pointed, 2–3 mm long, soon falling; **Fl.** 3 mm long, with pedicel, in Infl. of only a few Fl. on young St., Cor. campanulate, c. 10 mm ⌀, tube 2 mm long, pale yellow, lobes 3.5 mm long, 2 mm across at the base, pale yellow with dark red marks near the base.

**C. mogadoxensis** CHIOV. (§ II/1a). – Somalia: E. of the Juba R. – Mat-forming; **St.** from a 5–10 cm long perennial trunk, 20–55 cm long, with 4 acute angles, T. subulate-conical, 2–3 mm long, the flowering portion long-subulate, 17 to 36 cm long, narrowing teretely to the tip; **Fl.** in a lax Rac., Cor. spreading, tube 1–2 mm deep, lobes linear, 25 mm long, 1.5 mm across, very pointed, purple outside, inside marked with transverse bands, or with white and deep purple spots alternating, margins ciliate, the hairs being 1.5 mm long.

**C. moniliformis** BALLY (§ II/2a). – Somalia. – Resembles **C. subulata,** Fl.St. cylindrical to moniliform; **St.** only slightly branched, bluntly 4-angled, dentate, 10–14 cm tall, 2 cm thick, sterile Br. somewhat shorter, T. 2–4 mm long, with a blunt, upcurved apiculus, T. and angles dark purple, sides bluish-green; **Fl.St.** 22 cm or more in length, 4-angled and thick-fleshy in the lower part, cylindrical above, c. 4 mm ⌀, with regular constrictions 6–8 mm long, densely and minutely purple-spotted; **Infl.** almost racemose, with 1–2 groups of 2–3 Fl., Cor.-tube cup-shaped, 1.3 mm deep, 2 mm ⌀, dark purple-red inside, outside pale green with purple markings, lobes narrow-triangular, deep purple-red inside, outside green with red spots, 6 mm long, c. 2 mm across, longitudinally plicate, margins purple-haired.

**C. mouretii** CHEV. (§ II/2a) (*C. edulis* CHEV.). – Mauretania. – **St.** rooting at the base, much branching, 20–30 cm tall, sub-cylindrical, almost 4-angled, 2–3 mm ⌀, not dentate; **Fl.** along the St. apices, scattered, 1–2, Cor. campanulate to cup-shaped, 10–12 mm long, tube 7–8 mm long and across, yellow-green, with some 20 longitudinal ribs on the inside, lobes 5 mm long, 2.5 mm across the base, long-triangular, glabrous inside, rough outside.

**C. munbyana** (DECNE.) N. E. BR. v. **munbyana** (§ III/1) (*Boucerosia m.* DECNE.). – S.Spain. – Mat-forming; **St.** 5–15 cm tall, c. 15 mm thick, soft, fresh green, 4-angled with depressed sides, angles sinuate-dentate, T. directed upwards; **Fl.** 4–10, Cor. deeply 5-cleft, lobes narrow-linear, velvety, brown, c. 7 mm long, margins very recurved; Fl. has a disagreeable odour.

**C. — v. hispanica** (DE COINCY) MAIRE (*Boucerosia h.* DE COINCY, *B. m.* v. *h.* DE COINCY). – Prov. Murcia; Morocco. – **St.** often with underground stolons, 10–20 cm long, with 4 sinuate angles, distinctly dentate; **Fl.** in small clusters, Cor. 6–8 mm ⌀, fleshy, somewhat campanulate, lobes linear, narrowing to the tip which is spine-like, margins recurved, red-brown inside, much paler outside, whitish at the base.

**C. ortholoba** LAVR. (§ III/6). – Cape: Calvinia and Van-Rhynsdorp D. – Related to **C. hottentotorum**; **St.** densely congested, 4–10 cm long, 2 cm thick, 4–5-angled, T. brown-tipped, pungent; **Fl.** 2–5, arising from a cushion-like Ped.; Sep. 2 mm long, deltoid, acute, Cor. glabrous, campanulate, yellowish with well-spaced minute irregular purple blotches, 9 mm ⌀, tube c. 2.3 mm long, 4 mm across, lobes erect, broadly deltoid, acute, 3 mm long and across, corona double, dark purple, 3.5 mm ⌀, outer corona cupular.

**C. oxydonta** CHIOV. (§ III/1). – Somalia. – **St.** growing horizontally from the base, 10–15 cm tall, 4-angled, branching above, Br. very furrowed, somewhat spirally twisted, 6–7 mm tall, angles slightly winged, white, horny, T. conical, whitish-horny; **Fl.** c. 12 in spherical umbels, Cor. campanulate, 15 mm ⌀, 2 mm long, yellowish below, black-purple above, lobes ovate-oblong, tapering, 10 mm long, 5–6 mm across, margins hairy.

**C. parviflora** (MASS.) N. E. BR. (§ III/6) (*Stapelia p.* MASS., *Piaranthus p.* SWEET). – Cape: Lit. Namaqualand. – **St.** 20 cm tall, 1 cm thick, branching, with 4 acute angles, T. recurved, pointed; **Fl.** 1–3, c. 8 mm ⌀, Cor.-lobes lanceolate, pointed, 4 mm long, 1.5 mm across, flat, rough, yellow-green with red spots, margins hairy.

**C. pauciflora** (WIGHT) N. E. BR. (§ III/2) (*Boucerosia p.* WIGHT, *Desmidorchis p.* DECNE.). – S.E.India. – **St.** slender, branching, with 4 acute angles, T. directed downwards; **Fl.** usually solitary, 2.5 cm ⌀, broad-campanulate, lobes ovate-tapering with ciliate tips, pale yellow with brown transverse stripes, wrinkled, minutely hairy.

**C. peckii** BALLY (§ II/2b). – Kenya: N. Frontier Prov. – Resembles **C. dicapuae**, **St.** shorter and more slender, branching only a little from the base, up to 20 cm tall, bluntly 4-angled, bluish with brown marks, much tapering above, pointed, angles dentate, T. 10–24 mm apart; **Fl.** on the upper part, Cor. c. 10 mm long, 5 mm ⌀, lobes drooping open, elliptical, somewhat pointed at the tip, narrowing and plicate at the base, green outside, pale yellow with red spots inside, with a few motile purple clavate hairs at the base.

**C. penicillata** (DEFL.) N. E. BR. v. **penicillata** (§ III/1) (*Boucerosia p.* DEFL., *Echidnopsis golathii* SCHWEINF.). – S.Arabia: Yemen. – **St.** branching basally, 2–3 cm thick, branching irregularly above, 40–100 cm tall, light green, 4-angled, the angles compressed and alate, the crest horny, 1–2 mm wide, T. directed downwards, deltoid, 2 cm long; **Fl.** in dense umbels of 30–50, Cor. 11–12 mm ⌀, 5-cleft to more than midway, lobes ovate-lanceolate, greenish with brownish spots outside, inside light brown with pale yellow spots, with clavate red hairs towards the lobe-tips.

**C.** — v. **robusta** (N. E. BR.) WHITE et SLOANE (*C. r.* N. E. BR.). – Ethiopia; Sudan. – Sturdier variety; **St.** bluish-green, T. rather shorter, spreading horizontally; **Fl.**-lobes tuberculate inside.

**C. peschii** NEL (§ III/7). – S.W.Afr.: Damaraland. – **St.** branching and bushy, 5–7 cm tall, grey-green, the 4 angles separated by shallow furrows; **Fl.** 6–8 in clusters, 9 mm ⌀, Cor.-lobes 4–5 mm long, 2–3 mm across the base, ovate, pointed, glabrous outside, inside yellow-green and hairy.

**C. pillansii** N. E. BR. (§ III/6). – Cape: Robertson D. – **St.** robust, 20–30 cm tall, very freely branching, Br. 2–2.5 cm thick, 4-angled, dark green with pale red blotches and spots, T. much compressed laterally, 8–12 mm long and across, with Sp.-like tips; **Fl.** several in dense clusters, tube campanulate, 5–6 mm deep, Cor.-lobes 8 to 10 mm long, 6 mm across at the base, ovate-oblong, abruptly tapering, margins replicate, whitish-green outside with tiny red spots, inside minutely rough-tuberculate, reddish-green, red-spotted and minutely red-hairy.

**C. plicatiloba** LAVR. (§ III/1). – S.W.Arabia. – **St.** branching and rebranching, lateral shoots rooting on touching the soil, St. erect, 2–10 cm long, (7–)15(–25) mm ⌀, bluish-green, initially slightly pruinose, bluntly 4- (or less frequently 5-)angled, T. roundish; **Fl.** 5–12 in a Rac., Cor. campanulate, c. 10 mm long, glabrous outside, rather rough, with a prominent M.nerve on the reverse of the lobes, yellowish-green, with close-spaced brownish red markings, inside vivid yellow, minutely hairy, purple-spotted, purple towards the lobe-tips, margins with motile, white, purple-tipped hairs 1–1.5 mm long, lobes broadly ovate-triangular, narrowing to a bifid, completely reduplicate tip.

**C. praegracilis** OBERM. (§ III/8). – Zululand; Rhodesia. – **St.** c. 10 cm tall, 4-angled, brownish-green to olive-green, with red spots and stripes, T. 9 mm long, roundish, curved upwards; **Fl.** 3–6, Cor. flat, star-shaped, whitish outside, red-spotted, the centre a vivid coppery-red, velvety with a few hairs, lobes c. 2.5 cm long, ovate at the base, 7 mm across, gradually tapering, light yellow with numerous red blotches.

**C. priogonium** K. SCHUM. (§ II/1) (*C. elata* CHIOV.). – Tanzania. – Mat-forming; **St.** tapering to a slender tip, 30–40 cm tall, with 4 projecting angles, T. sharp; **Fl.** in clusters at the St.tips, with short pedicels, Cor. purple, the lobe-bases having white bands and blotches.

**C. procumbens** GREV. et MAYUR. (§ III/1). – Peninsular India. – Mat-forming; **St.** arising from underground rhizomes, 15 mm across, acutely-angled, with few Br., T. scarcely distinguishable; **Fl.** in small Rac., Cor. campanulate, 8 mm ⌀, lobes broad-deltoid, whitish with red, often with reddish marks outside.

**C. pruinosa** (MASS.) N. E. BR. v. **pruinosa** (§ III/7) (*Stapelia p.* MASS., *Tromotriche p.* HAW.). – S.Afr.: Cape: Lit. Namaqualand. – **St.** erect, irregularly branching, 30 cm or more in height, grey-green, often reddish, with 4 rounded angles, T. small; **Fl.** 1–3, Cor. rotate, 14 mm ⌀, lobes ovate, tapering, dark brown, minutely white-haired, brownish and smooth outside.

**C.** — v. **nigra** LUCKH. – Cape: Lit. Namaqualand. – **Fl.** 9 mm ⌀, blackish inside, not so densely hairy, lobes very pointed.

**C. quadrangula** (FORSK.) N. E. BR. (§ III/2) (*Stapelia q.* FORSK., *Desmidorchis q.* DECNE., *D. forskahlii* DECNE. p. part., *Boucerosia q.* DECNE., *B. f.* DECNE. p. part., *Echidnopsis q.* DEFL.). – S.Arabia. – ♄, 25–40 cm tall, to 80 cm ⌀; **St.** branching irregularly to form a mound, green, with 4 rather blunt angles, T.

small, blunt, sides almost flat, 1–3 cm across; **Fl.** solitary, tube 2.5 mm long, then campanulate with ovate, tapering lobes 8–10 mm long, inside pale yellow, glabrous, smooth.

**C. ramosa** (MASS.) N. E. BR. (§ III/6) (*Stapelia r.* MASS., *Piaranthus r.* SWEET). – Cape: Karroo. – **St.** c. 20 cm tall, much branched, Br. 3 cm thick, with 4 blunt angles, scarcely dentate, grey-green; **Fl.** in small clusters, Cor. short-campanulate, lobes projecting, ovate-lanceolate, pointed, margins recurved, 6 mm long, black-red inside.

**C. reflexa** LUCKH. (§ III/7). – Cape: Van-Rhynsdorp D. – **St.** up to 15 cm long, glabrous, grey-green, 4-angled, angles with conical, compressed tubercles 4 mm long; **Fl.** 1–3 from midway up the St., with pedicels 2–3 mm long, Sep. 2–3 mm long, pointed, Cor. 9–10 mm ⌀, round, the united portion very small, flat or slightly cup-shaped, lobes curved outwards and recurved, 4.5 mm long, 1.5 mm across, glabrous outside, greenish-brown, the upper half dark green-brown inside, the basal portion and the united part of the Cor. whitish with narrow, concentric, green-brown lines, densely ciliate with long, wavy, light purple hairs.

**C. retrospiciens** (EHRENBG.) N. E. BR. v. **retrospiciens** (§ III/1) (*Desmidorchis r.* EHRENBG., *C. respiciens* K. SCHUM., *Boucerosia russeliana* COURBON, *C. r.* (COURBON) CUF.). – Red Sea: Dahlak Is.; Sudan; Ethiopia; Somalia. – Forming large ♄; **St.** branching freely and irregularly, 8 cm across, with 4 compressed angles, alate, sinuate-dentate, T. directed downwards and often recurved, hamate, pointed; **Fl.** numerous, in spherical umbels 12 cm ⌀, Cor. 2 cm ⌀, tube campanulate, lobes deltoid-ovate, acuminate, reddish outside, glabrous, deep dark red-brown or almost black-brown inside, with violet-red hairs and cilia; Fl. with a strong and disagreeable smell.

**C. — v. acutangula** (DECNE.) WHITE et SLOANE (*Desmidorchis a.* DECNE., *Boucerosia a.* N. E. BR.). – Senegal. – **St.** 3-angled, angles very acute; **Cor.** densely warty inside, and hairy over the greater part of the surface.

**C.—v. glabra** N. E. BR. – Kenya. – **Fl.** not hairy.

**C. — v. hirtiflora** (N. E. BR.) BGR. (*C. h.* N. E. BR.). – Red Sea: Nanish Is.; Ethiopia. – Fl. 2.5–3 cm ⌀, densely hairy inside.

**C. — v. tombuctuensis** (CHEV.) WHITE et SLOANE (*Boucerosia t.* CHEV., *C. t.* N. E. BR.). – W.Afr.: Mali; Timbuktu; Mauretania; Sudan; Nigeria. – Up to 1.5 m tall, forming clumps 75 cm ⌀; **St.** pale green, often with a waxy coating; **Fl.** in umbels 10 cm ⌀, 12–15 mm ⌀, black-purple, lobes wrinkled-papillose inside, margins with red ciliate hairs.

**C. rogersii** (L. BOL.) E. A. BRUCE et R. A. DYER (§ III/8) (*Stapelia r.* L. BOL.). – Botswana: Mahalapye. – **St.** c. 10 cm tall, 8 mm ⌀, 4-angled with furrows between the angles, T. 10 to 18 mm long, 5 mm thick at the base, pointed; **Fl.** in clusters, 3–3.5 cm ⌀, Cor. lobes linear-pointed, margins recurved, glabrous outside, inside minutely papillose with a few hairs below, yellow.

**C. sacculata** N. E. BR. (§ III/4). – S.Ethiopia. – **St.** erect or prostrate, 5–8 cm long, over 1 cm thick, 4-angled, angles with very pointed T. 8–11 mm long; **Fl.** in clusters of 3–4, Cor. tubular-campanulate, tube 8–10 mm deep, 6 to 8 mm ⌀ at the base, both surfaces glabrous, greenish, lobes 5 mm long, 3 mm across the base, deltoid-ovate, margins revolute, glabrous outside, inside pubescent, dark purple.

**C. schweinfurthii** BGR. (§ III/6) (*Huernia sprengeri* HORT., *C. piaranthoides* OBERM.). – Congolese Rep.; Rwanda; Rhodesia; Tanzania; Zambia. – **St.** freely branching, Br. 6–10 cm long, 15 mm thick, with 4 blunt angles, T. 15 mm long, with a small cartilaginous tip, light green, glabrous, with minute pale reddish spots or stripes; **Fl.** 4–5, Cor. deeply 5-cleft, 23 mm ⌀, rotate, light green outside, with light brown marks, lobes deltoid-ovate, tapering from the base, tube flat and almost annular, densely hairy, brownish or yellowish, sparsely hairy like the lobes.

**C. simulans** N. E. BR. (§ III/7). – Cape: Prince Albert D. – **St.** branching basally, 5 to 8 cm tall, 10–14 mm thick, 4-angled, angles acute, T. projecting horizontally, with cartilaginous tips 2–3 mm long, green or soft grey-green, often reddish; **Fl.** solitary, Cor. flat, rotate, 7 mm ⌀, deeply partite, lobes ovate or ovate-lanceolate, somewhat thickened towards the tip, 3 mm long, light yellow, both surfaces glabrous, with a few hairs inside at the base and also at the tip.

**C. sinaica** (DECNE.) BENTH. et HOOK. (§ II/2a) (*Boucerosia s.* DECNE.). – Sinai Peninsula. – **St.** extended, thin, old St. angular and furrowed, young growth more terete; **Fl.** solitary, Cor. c. 4 mm ⌀, lobes almost triangular, densely hairy inside.

**C. socotrana** (BALF. f.) N. E. BR. (§ III/1) (*Boucerosia s.* BALF. f., *C. corrugata* N. E. BR., *C. rivae* CHIOV., *C. rosengreenii* VIERHAPPER). – Socotra; Somalia; Israel. – **St.** dividing above into 2–3 Br., grey, often reddish, angles compressed, sinuate-dentate, T. deltoid, sharp, thorny, hard, directed downwards; **Fl.** in tufts, tube 2 cm long, lobes pointed, darkest red.

**C. solenophora** LAVR. (§ III/1). – S.W.Arabia. – **St.** densely branching, up to 20 cm tall, Br. under 8 cm in length, 2 cm ⌀, 4-angled, glabrous, green or brownish-green, angles compressed, T. rather blunt with the tip directed downwards; **Fl.** grouped in Infl., Cor. long-tubular, up to 22 mm long, glabrous outside, yellowish-brown with a suggestion of lengthwise lines or furrows, inside with a narrow dark purple border, the base of the tube creamy-yellow with dark brown markings, then yellow and densely covered with narrow, dark purple, transverse bands, these merging towards the lobe-tip, tube c. 19 mm long, spherical, 9 mm ⌀ at the base, then constricted to 6 mm ⌀, thereafter abruptly expanding to 8 mm ⌀, and then straight and cylindrical for 1 mm in length, Cor.-lobes deltoid, pointed, 5–6 mm long, 7 mm across at the base, and tipped with a cluster of dark purple, motile hairs 1–2 mm long.

**C. somalica** N. E. Br. (§ III/1). – Somalia: near Mogadishu. – **St.** with 4 acute and slightly dentate angles; **Fl.** very numerous, in spherical umbels of 5 cm ⌀, Cor. 12–14 mm ⌀, tube campanulate, glabrous outside, velvety inside, lobes triangular-ovate, pointed, 5 mm long, 4 mm across at the base.

**C. speciosa** (N. E. Br.) N. E. Br. (§ III/1) (*Sarcocodon s.* N. E. Br., *C. codonoides* N. E. Br.). – From Somalia, through Kenya to Tanzania. – Habit very similar to **C. retrospiciens** v. **tombuctuensis; Infl.** almost 13 cm ⌀, Fl. 4.5 cm ⌀, Cor. deep cup-shaped, lobes ovate-pointed, 12–14 mm long and across, of an even, deep brown, tube c. 18 mm long and across, orange-coloured to yellow. – A local form from Magadi has a blotchy zone between the lobes and the yellow tube.

**C. sprengeri** (Schweinf.) N. E. Br. (§ III/4) (*Huernia s.* Schweinf.). – Ethiopia. – **St.** freely branching, Br. 8–10 cm long, 15 mm across, 4-angled, light green, minutely reddish spotted or striped, angles with large and very pointed T. up to 15 mm long; **Fl.** 4–5, Cor. 23 mm ⌀, deeply 5-cleft, rotate, light green outside, with small blotches and stripes, lobes deltoid-ovate, tapering almost from the base, tube short, flat, almost annular, densely hairy, brownish or yellowish and, like the lobes, pubescent.

**C. stalagmifera** C. E. C. Fisch. (§ II/2a). – Peninsular India. – **St.** erect, 5–8 mm thick, somewhat attenuated, 4-angled, with thin and distinctly angular Br.; **Fl.** solitary, borne on the tubercles, Cor. campanulate, 1 cm ⌀, lobes ovate-oblong, green outside, inside darkest red with a whitish apiculus and minute red hairs.

**C. subterranea** Bruce et Bally v. **subterranea** (§ III/1). – Kenya; Tanzania. – Semi-subterranean stolons persistent, the aerial St. dying off; **St.** 5–8 cm tall; **Fl.** in terminal umbels, Cor. c. 18 mm ⌀, lobes dark brown or even lemon-yellow, covered with minute silvery hairs.

**C. —** v. **minutiflora** Bally. – Kenya. – Fl. 12 mm ⌀.

**C. subulata** (Forsk.) Decne. (§ II/2a) (*Stapelia s.* Forsk.). – S. Arabia: Yemen. – **St.** slender, 20–30 cm tall, branching, 4-angled at the base, tapering above, c. 1 cm ⌀, T. small, projecting; **Fl.** 1–2, pendulous, Cor. 15 mm ⌀, lobes oblong, tapering to the base, tapered and awned above, hairy inside, margins ciliate.

**C. swanepoelii** Lavr. (§ III/2). – W. Cape: Calvinia D. – Resembles **C. dependens**; **St.** densely congested, 2–6 cm long, 10–12 mm thick, ascending, 4-angled, the angles rounded, tuberculate-tessellate, each tubercle armed with an acute hard brown T.; **Fl.** produced from the apical part of the St., not pendulous, pedicels straight and 1.75 mm long, Sep. 2 mm long, deltoid and acute, Cor. 12 mm ⌀, rotate with a rather fleshy disc 4.5 mm ⌀, whitish with violet-brown spots, lobes 3.75 mm long, 3 mm across, ovate-triangular, whitish with violet-brown spots at the base, thereafter entirely violet-brown, glabrous except for violet-reddish hairs on the slightly reflexed margins, corona elevated 1.75 mm above the Cor.

**C. torta** N. E. Br. (§ III/2). – S. Arabia or? Socotra. – St. obtuse 4-angled, 8–10 mm thick, finely wrinkled and hairy, angles faintly dentate: **Fl.** solitary, Cor. finely hairy, outside, 16 to 27 mm ⌀, lobes linear, globosely b nt together at the base and let 5 window-like openings, twisted together into a long, slender terminal colurn, glabrous inside, red-brown mottled at the base.

**C. truncato-coronata** (Sedgwick) Grav. et May. (§ III/1) (*Boucerosia t.-c.* Sedgwick). – Peninsular India. – **St.** numerous, up to 15 cm tall, 6 mm thick, not tapering above, 4-angled, sides depressed, angles tuberculate; **Fl.** c. 13 in umbels, Cor. campanulate, tube 6 mm long and 6 mm across, green outside, lightly purple-spotted at the mouth, purple or almost black inside, with concentric yellow furrows, lobes with clavate hairs in the indentations.

**C. tsumebensis** Oberm. (§ III/8). – S.W.Afr.: Ovamboland. – **St.** branching, c. 25 cm tall, 3–6 cm ⌀ including the T., usually 5-angled, pale green with red blotches, T. up to 3 cm long and across, compressed, with 2 lateral T. towards the tip; **Fl.** numerous in umbel-like clusters, Cor. c. 6 cm ⌀, tube campanulate, 7 mm long and 12 mm ⌀, lobes ovate-lanceolate, pointed, 25 mm long, 7 mm across at the base, indistinctly nerved outside, deep chocolate-brown inside, glabrous, sometimes with transverse folds.

**C. tuberculata** N. E. Br. (§ III/1) (*Boucerosia aucheri* Decne., *B. aucheriana* Hook. f.). – Peninsular India; Afghanistan. – **St.** 5–15 cm tall, 6–12 mm thick, with short Br., the 4 angles dentate; **Fl.** in clusters, c. 8 mm ⌀, Cor.-lobes narrow-lanceolate, dark red with blister-like tubercles on the upperside.

**C. tubiformis** Bruce et Bally (§ III/6). – Related to **C. distincta**; **St.** erect, the 4 angles dentate; **Fl.** campanulate, dark purple, green outside with red blotches, the Cor.-lobes and the upper part of the tube having long white hairs.

**C. turneri** E. A. Bruce (§ II/2b). – Kenya. – **St.** 4-angled, angles acutely dentate, deep green with purple blotches; **Fl.** resembling a little tassel, Cor.-lobes pendulous, the width being only 1/3 of the length, sharply revolute, dark chestnut-brown with green-yellow marks. Varieties or regional variants show colour-differences.

**C. ubomboensis** Verd. (§ III/7). – Natal: Ubombo; Swaziland. – **St.** 4 cm tall, branching, 4-angled, 1 cm ⌀, angles dentate, T. conical; **Fl.** 2–3 together, Cor. 9 mm ⌀, dark purple, lobes 3–4 mm long, ovate.

**C. umbellata** Haw. (§ III/1) (*Stapelia u.* Roxb., *Boucerosia u.* Wight et Arn., *Desmidorchis u.* Decne., *Boucerosia lasiantha* Wight, *C. l.* N. E. Br., *B. campanulata* Wight, *C. c.* N. E. Br.). – S. India. – Mat-forming; **St.** 30 to 60 cm tall, 2–5 cm ⌀, freely branching, angles compressed, alate and sinuate-dentate; **Fl.** numerous in terminal umbels, 5–10 cm across, Cor.-lobes broad-ovate, abruptly tapering, yellowish or red-brown with darker transverse stripes, wrinkled, ± hairy.

**C. umdausensis** NEL (§ III/6). – Cape: Lit. Namaqualand. – **St.** c. 9 cm tall, c. 1 cm thick, the angles separated by distinct furrows, roughly tuberculate towards the apex; **Fl.** solitary, 22 mm ⌀, deeply 5-cleft, yellowish-green outside, with dark lobe-margins and red M.nerve, Cor. 5 mm deep, densely hairy inside, light red to purple, with an annulus around the tube-base, thereafter 20 whitish-green stripes, lobes 11 mm long, 5 mm across, ovate-lanceolate, pointed, with irregular yellowish-green spots, glabrous, rough above.

**C. valida** N. E. BR. (§ III/8). – ? Botswana; Transvaal; Rhodesia; Zambia. – **St.** prostrate, with an ascending tip, rooting, stout, 4-angled, ± 10 cm (occasionally up to 50 cm) long, c. 1.5 cm thick, green with purple mottling, T. stout, triangular, pointed, up to 1.5 cm long, with 2 minute lateral T. below the tip; **Fl.** in dense clusters of 20–40, Cor. deeply 5-cleft, tube short-campanulate, glabrous outside, inside minutely hairy, sometimes with a few stiff hairs up to 0.4 mm long, 7 mm deep with an abruptly spreading margin, 10–12 mm ⌀ at the mouth, lobes 15–25 mm long, 5–9 mm across the base, narrowing towards the tip, very rough inside, densely covered with motile, clavate hairs 3 mm long, with short fleshy T. in the indentations, Fl. deep chocolate or blood-red to almost black.

**C. venenosa** MAIRE (§ III/4). – Afr.: Central Sahara. – **St.** freely branching, Br. 6–10 cm tall, 15 mm ⌀, with 4 dentate angles, T. up to 15 mm long, very pointed and with a minute cartilaginous tip; **Fl.** 2–4, up to 2 cm ⌀, tube campanulate, 5–6 mm deep, 7–8 mm ⌀, minutely papillose, reddish-brown, grey-green outside, with very dark red spots, Cor.-lobes ovate-pointed, 7 mm long, 5–6 mm across, hairy.

**C. vibratilis** E. A. BRUCE et BALLY (§ III/1). – Kenya; Uganda; Tanzania. – In habit similar to **C. subterranea**; **Fl.** with campanulate Cor., purple-red with small yellow-green blotches, greenish outside, lobes with numerous clavate, motile hairs.

**C. villetii** LUCKH. (§ III/6). – Cape: Van-Rhynsdorp D. – **St.** c. 25 cm tall, 12 mm thick, branching basally, with 4 acute, dentate angles, T. compressed, pointed, grey-green with brown marks; **Fl.** solitary, 21 mm ⌀, light-green outside, tube short, light yellow inside, Cor.-lobes 7 mm long, 5 mm across at the base, lanceolate, margins recurved, with purple longitudinal lines outside, wine-red inside at the base, light yellow above, margins with red hairs.

**C. virescens** LUCKH. (§ III/7). – Cape: Van-Rhynsdorp D. – **St.** as for **C. reflexa**; **Fl.** 1–3, with pedicels up to 6 mm long, Sep. 2–3 mm long, Cor. 10–11 mm ⌀, the united portion short, flat or slightly cupped, lobes spreading, 5.5 mm long, 1.5 mm across, green in the upper half, the basal portion whitish with a few deep purple-brown transverse lines, with long, wavy, light purple ciliate hairs.

**C. vittata** N. E. BR. (§ II/2a). – Sudan. – **St.** branching, 15–20 cm tall, Br. 4-angled, 6 to 7 mm ⌀, much tapering above, angles somewhat dentate; **Fl.** 1–4 on the slender Br.-apex, tube campanulate, whitish with dark purple longitudinal stripes, both surfaces glabrous, Cor.-lobes ± spreading, 3.2 mm long, 1.6 mm across at the base, oblong-triangular, pointed, dark purple, glabrous, margins recurved.

**C. wilfriedii** DTR. (§ III/6). – S.W.Afr. – Dwarf plant with small dark Fl. – Insufficiently known species.

**C. winkleri** (DTR.) WHITE et SLOANE (§ III/5) (*Sarcophagophilus w.* DTR., *S. winklerianus* DTR., ? *C. namaquana* WELW. ex JACOBS.). – S.W.Afr.: Gr. Namaqualand. – With many St., and forming clumps; **St.** grey-green, 8–10 cm tall, c. 4 cm thick, with numerous T. arranged in 6 close spirals, these T. being hard, broadly compressed, 12–14 mm long, projecting horizontally and thorny; **Fl.** 18 mm ⌀, campanulate, Cor.-lobes lanceolate, darkest red, with a few tiny hairs at the base.

**C. wilsonii** BALLY (§ III/4). – Uganda. – **St.** prostrate to semi-erect, 4-angled, 5–6 cm long, bluish-green with brown blotches, angles dentate, thin, 1–2 cm long, acute; **Fl.** few, arranged almost umbel-like, tube flat, 6 mm ⌀, 2–2.5 mm deep, Cor.-lobes triangular, pointed, 3.5 mm long, 3.2 mm across at the base, glabrous outside, greenish-yellow, upperside lemon-yellow with small purple blotches and stiff purple bristles.

**Castilleja** MUTIS ex L.f. Scrophulariaceae. – N.Am., Mexico. – In part semi-parasitic plants. Mention is made here of only one semi-shrubby species with fleshy leaves. – Unheated greenhouse.

**C. fruticosa** MORAN. – Mexico: Guadelupe Is. – Woody sub-shrub (20–)50(–90) cm tall, 50 cm ⌀, herbaceous parts with fine glandular hairs, in part with tufted hairs; **St.** at first straight, erect, with several horizontal to ascending Br., after flowering becoming irregularly bent and densely concrescent, 2(–4) cm thick, the woody part hard and brittle; **Br.** with tufted hairs; L. light green, somewhat fleshy, smaller L. almost terete, otherwise linear to linear-oblanceolate, rounded above or even 3-lobed, 0.5–2.5 cm long, 1.5–3 mm across, 1–2 mm thick, constricted below to 0.5–1.5 mm across, upper L. often up to 6 cm long, 5 mm across, M.rib translucent; **Infl.** terminal, spicate, up to 7 cm long, Fl. 5–20, often tufted-hairy, involucral L. and Sep. red-tipped and glandular-hairy, Cor. 17–26 mm long, with 19 veins, hairy towards the centre, yellow-green.

**Caulanthes** S. WATS. Cruciferae. – North America. – Showy succulent plants with thickened but hollow stems. – Conspicuous annual, biennial to perennial plants, erect, with long Br. from the base; **St.** and Br. round and leafy; lower L. slender, tapering teretely, upper L. petiolate or amplexicaul; **Fl.** in a lax Rac., nodding, medium-large, whitish, with 4 Pet.; Fr.: a pod. – Propagation: seed.

**C. crassicaulis** (TORREY) S. WATS. (*Streptanthes c.* TORREY). – 'Indian cabbage'. – N.Am.: USA, from Calif. to Utah and Wyoming. – ♃;

**St.** simple or with only a few **Br.**, thickened in the middle, hollow; lower **L.** almost rosulate, long-petiolate, almost lyrate, upper **L.** sparse, narrow; **Infl.** with its short pedicels almost spicate.

**C. inflatus** (GREENE) S. WATS. (Pl. 30/1) (*Strepthanthus i.* GREENE). 'Candle plant'. – USA: California. – ☉; **St.** much inflated above and up to Fr.-cluster, tubular, golden-yellow even in the flowering period; **L.** forming a basal Ros. and amplexicaul.

**Cavanillesia** RUIZ et PAV. Bombacaceae. – Greenhouse, warm. Propagation: seed.

**C. arborea** K. SCHUM. (Pl. 30/3). – 'Barrel tree'. – E.Brazil. – ♄; **St.** swollen to a barrel-shape, fleshy, up to 20 m tall, up to 5 m ⌀, tapering both above and below the middle; **Fl.** appear before the L. come into growth. Probably not in cultivation.

**Ceraria** PEARS. et STEPH. Portulacaceae. – S. and S.W.Afr. – Dwarf or taller ♄, often with short, spreading opposite or alternate **Br.** which are ± fleshy, ± scarred; **L.** small, oval-cuneate, rather thick, green; **Fl.** in Infl. of 2–6, pedicellate, small, Pet. 5, pink. Plants dioecious. – Greenhouse, warm. Propagation: seed, cuttings.

**C. fruticulosa** PEARS. et STEPH. – Bushmanland. – Much-branched virgate ♄ 30–60 cm high with false-dichotomous or monopodial Br.; **St.** slender, only 1 mm ⌀ at tips, glabrous, wrinkled, reddish brown, whitish when young, with small, raised leaf cushions in 4 ranks; **L.** sessile, paired, fleshy, flattened, obovate to oblong-obovate, minutely apiculate, 4–6 mm long; **Fl.** 1–6 together, on pedicels 2–4 mm long.

**C. gariepina** PEARS. et STEPH. – Bushmanland. – Close to **C. namaquensis**, but **St.** pale yellow instead of dull grey, and **L.** smaller, less than 2 mm long.

**C. longipedunculata** MERZM. et PODL. – S.W. Afr.: Kaokoveld, Outjo. – Freely branching ♄; **Br.** red-brown, up to 3 mm ⌀; **L.** fleshy, cuneate, rounded at the tip, 6–8 mm long, 1–2 mm across; **Infl.** 1–4 cm long, Fl. about 5 mm ⌀, pink.

**C. namaquensis** (SOND.) PEARS. et STEPH. (Pl. 30/2) (*Portulacaria n.* SOND.). – Cape: Lit. Namaqualand; S.W.Afr. – ♄, up to almost 2 m; **Br.** ascending, rather thick, cortex smooth and papery, L. scars arranged in 4–16 vertical rows; **L.** 2–5 on short shoots, grey-green, fleshy, 7 to 8 mm long; **Fl.** white and pink.

**C. pygmaea** (PILL.) PILL. (Pl. 30/4) (*Portulacaria p.* PILL.). – Cape: Lit. Namaqualand. – Dwarf, branching, compact ♄ about 20 cm tall, 30 cm ⌀; **St.** branching almost dichotomously, rigid, fleshy; lower **Br.** spreading, upper ones projecting to curving downwards; **L.** 10–14 mm long, 7–9 mm across, 3–4 mm thick, sessile, ovate-cuneate, tip rounded and apiculate, rough with roundish papillae, blue-green, becoming yellowish-green; **Infl.** of 2–5 pale pink Fl.

**C. schaeferi** ENGL. et SCHLTR. – S.W.Afr. – Small ♄ with widespread Br. and a grey cortex; **Br.** gnarled, 3–4 mm thick; **L.** spatulate-cordate, about 15 mm long, 6 mm across above, c. 1 mm thick, bluish-green, deciduous; **Fl.** small, pink.

**Ceratosanthes** BURM. Cucurbitaceae. – Caribbean Is.; S.Am. – Greenhouse, warm. Propagation: seed.

**C. tuberosa** J. F. GMEL. (*Trichosanthes corniculata* LAM., *C.c.*(LAM.) COGN.). – Caribbean Is.; Venezuela. – **Caudex** tuberous, spherical to napiform, fleshy, 15–20 cm thick; **St.** rough, thick as a finger; **Br.** slender, long, angular-furrowed, glabrous; **L.** with petiole 2–3 cm long, glabrous to pubescent, lamina vivid green on the upperside, with scattered tufts of hairs, underside pale green, becoming glabrous, slightly emarginate at the base, 8–15 cm long and across, 3-lobed from the base; tendrils hair-like, short; ♂ **Fl.** 5–10 on a stalk 10–16 cm long, Cor. whitish-yellow, ♀ **Fl.** solitary; **Fr.** narrowly ovoid, longitudinally striate, with pale or dark green mottling, 3–4 cm long, 1.5–2 cm ⌀.

**Ceropegia** L. Asclepiadaceae. – Canary Is.; trop. and S.E.Afr.; Madag.; Comoro; trop. Arabia; the Himalayas; W. and S.China (N.W. as far as Hupeh and Szechuan); peninsular India; Ceylon; Indonesia; Philippines; New Guinea; Austr.: Queensland. – The text is a shortened version of a contribution by **H. Huber**. – ⚄ or half-♄, climbing or erect, quite frequently with a **caudex** or fleshy, thickened **roots**; **L.** opposite, occasionally in whorls of 3, in the case of the non-succulent species always well developed, usually petiolate, cordate (especially in the climbing species) to lanceolate, linear to almost filiform (more usual in the erect species); in the case of the succulent species the L. are well developed and smaller, or even reduced to scales or soon falling; **Fl.** in a cyme or quite frequently solitary, either pedicellate or sessile, arising laterally between two petioles, Cal. and Cor. 5-merous, Cor. united to form a tube which is usually longer than it is broad, swollen below and cylindrical or funnel-like in the central and upper parts, straight or curved, the central and upper parts of the tube sometimes completely undeveloped; the lobes seldom free, usually shortly united at the tips; the corona is double; the outer corona being cup-shaped, with margin either entire or 5–10-lobed, or with 5 lobes which are bifid or bipartite and ± fused with the inner corona, less often missing or reduced to minute pockets; the inner corona consisting of 5 linear lobes, incumbent on the anthers and projecting beyond these; **Fr.** a follicle, lanceolate to cylindrical, pointed, glabrous, seeds flat with a deciduous tuft of white hairs. – Type-species: **C. candelabrum.** There are some 200 known species, but only some 70 succulent species can be briefly described here. – Greenhouse: in summer moderately moist and well-ventilated; in winter: keep rather dry at approximately 13 °C. – Propagation: seeds, cuttings.

## Summary of the Sections and Series of the Genus Ceropegia

**Sect. I. Ceropegia.** – **Caudex** spherical; **St.** slightly succulent, twining, glabrous; **L.** usually petiolate; **cymes** pedunculate; corona glabrous outside, the Cor. tube considerably swollen below, glabrous inside or with only some scattered hairs midway, the mouth little widened or not at all, intermediate lobes absent, lobes with tips united, sparsely hairy on the inside. – Peninsular India; Ceylon. – Type species: C. candelabrum. – Only succulent species: C. bulbosa.

**Sect. II. Janthina** H. Huber. – **Caudex** either irregularly shaped or absent; **St.** ± succulent, sometimes dimorphous, usually twining, less often pendulous or erect, glabrous; **L.** petiolate or reduced to scales; **cymes** mostly pedunculate, the **Cor.** glabrous outside, the Cor.-tube much swollen either at the base or in the lower half, with a narrow ring of hairs inside, these separating the swollen venter from the slender central section, otherwise glabrous, the mouth of the tube widened to a funnel-shape, the central and upper part of the tube being absent in some species, which then have only the urn-shaped lower venter with a narrow ring of hair at the mouth, otherwise glabrous inside; intermediate lobes absent; the lobes united by their tips, hairy inside at the base. – S. India; Ceylon; Madag.; Comoro; trop. Afr. – Type species: C. elegans.

**Ser. 1. Elegantes** H. Huber. – Not tuberous; the **St.** not recognizably succulent, not dimorphous, even the lower part lacking projecting leaf-cushions; **L.** well developed or reduced to scales; the **Infl.** distinctly pedunculate, the Fl. tube much swollen either basally or in the lower half, the middle and upper parts of the tube gradually widening to the funnel-shaped mouth. – Type species: C. elegans. Succulent species: C. albisepta, ballyana, elegans, robynsiana, sankuruensis, succulenta, viridis.

**Ser. 2. Dimorpha** H. Huber. – **Caudex** irregular in shape or absent; **St.** dimorphous, contracted before flowering age is reached, very succulent, 4-angled and alate because of the projecting leaf-cushions, rather like the St. of Stapelia, prostrate or erect, not twining; Fl.-bearing sections slender, extended, terete, erect or twining; L. falling at the start of the dry season; **Infl.** usually short-pedunculate.
– Type species: C. dimorpha.

a) Cor. tube much widened in the lower third, middle and upper parts funnel-form (C. bosseri) or slender-cylindrical up to the mouth (C. leroyi). – Species: C. bosseri, leroyi.

b) Cor. tube urn-shaped, the middle and upper tube parts being undeveloped. – Species: C. armandii, dimorpha.

**Sect. III. Sarcodactylus** H. Huber. – **Caudex** absent; **St.** strongly succulent, up to finger-thickness, terete, erect or weakly scrambling over ♄, glabrous, mature plants normally densely white-pruinose; **L.** often soon falling; **cymes** sessile, only C. rupicola being short-pedunculate; **Fl.** Cor. glabrous outside, the Cor.-tube usually much swollen below, the central part having some scattered hairs inside, the mouth moderately widened, or very little or not at all; intermediate lobes absent; lobes united by their tips, often later separating, glabrous. – Arabia (1 species); Canary Is. – Type species: C. dichotoma.

**Ser. 1. Arabicae** H. Huber. – **L.** broadly ovate; **cymes** short-pedunculate; outer subsidiary Cor. shortly ciliate. – Species: C. rupicola.

**Ser. 2. Canarienses** H. Huber. – **L.** narrow-lanceolate to linear; **cymes** sessile; outer subsidiary Cor. glabrous. – Type species: C. dichotoma. – Species: C. ceratophora, chrysantha, dichotoma, fusca, hians, krainzii.

**Sect. IV. Phalaena** H. Huber. – **Caudex** usually absent (only C. crassifolia sometimes forming a Tub.); **St.** only moderately succulent or not at all, twining, glabrous; **L.** well developed, usually petiolate, only in C. juncea reduced and insignificant; **cymes** pedunculate; **Fl.**-Cor. glabrous outside or minutely downy; Cor.-tube swollen below, with scattered hairs inside, in C. juncea and sometimes in C. volubilis with a narrow ring of hairs as found in Section II, or completely glabrous, the mouth being much, moderately or only slightly widened to a funnel-shape; intermediate lobes absent; lobes united at their tips, with hair inside or margins ciliate, less often completely glabrous. – Peninsular India (C. juncea); Madag. (C. racemosa ssp. glabra); trop. and S.E.Afr. – Type species: C. aristolochioides.

**Ser. 1. Junceae** H. Huber. – **L.** reduced to sessile scales, often completely absent; tube as in Section II, having a narrow ring of hairs inside or completely glabrous, rarely extensively hairy in the swollen basal part. – Species: C. juncea.

**Ser. 2. Aristolochioides** H. Huber. – **L.** well developed and normally petiolate; tube inside ± extensively hairy (but cf. C. volubilis). – Type species: C. aristolochioides. – Succulent species: C. affinis, aristolochioides, carnosa, copleyae, crassifolia, distincta, rhynchantha, seticorona, somaliensis, volubilis, yorubana.

**Sect. V. Callopegia** H. Huber. – **Caudex** usually absent, but roots often a fleshy cluster; **St.** slightly or very succulent, twining or creeping, glabrous; **L.** usually well developed, less often reduced; **Infl.** pedunculate or sessile, corona glabrous outside, Cor.-tube basally much swollen, hairy inside below and usually also in the central section, mouth widened to a funnel-shape; intermediate lobes absent; lobes united by their tips, in C. sandersonii forming an umbrella-shaped canopy. – Trop. and S.E.Afr. – Type species: C. nilotica.

- **Ser. 1. Niloticae** H. Huber. – The venter (whether or not obliquely constricted) distinctly hairy in the lower half, glabrous above, the middle tube-section with scattered hairs inside, Cor.-lobes attached only at their tips; outer subsidiary corona glabrous. – Type species: C. nilotica. – Species: C. denticulata, gemmifera, nilotica, radicans.

- **Ser. 2. Sandersoniae** H. Huber. – The venter always hairy inside, central tube-section glabrous inside or almost so, Cor.-lobes united to form a broad umbrella-shape, outer subsidiary corona finely hairy. Only species: C. sandersonii.

- **Ser. 3. Hybrids of C. sandersonii:** Presence of hairs and lobe-attachment as for Series 1, but the outer subsidiary corona shortly hairy. – Hybrids: C. elegans × C. sandersonii (C. × hybrida), C. radicans × C. sandersonii (C. × rothii).

**Sect. VI. Coreosma** H. Huber. – **Caudex** usually absent; **St.** very succulent, terete, without projecting leaf-cushions but often tuberculate and rough with tiny projections, glabrous, not pruinose; **L.** reduced to sessile scales, soon falling; **cymes** short-pedunculate to almost sessile, Fl. corona glabrous outside, Cor.-tube much swollen basally, with hairs inside only in the swollen section, usually also in the funnel-shaped mouth, the central part glabrous, intermediate lobes present in C. cimiciodora, lobes free or only attached irregularly by their tips, hairy inside, at least towards the base. – S.E. Afr. – Type species: C. stapeliiformis. – Species: C. cimiciodora, stapeliiformis.

**Sect. VII. Ceropegiella** H. Huber. – **Caudex** usually present, either spherical or flattened; **St.** usually slightly succulent, twining, prostrate or less often erect, glabrous or sometimes hairy; **L.** always present, sometimes rather small; **Infl.** either pedunculate or sessile, corona glabrous outside, finely downy only in C. pachystelma, Cor.-tube swollen basally, with scattered hairs inside the middle section, rarely completely glabrous, the mouth little widened or not all, intermediate lobes missing or else minute, Cor.-lobes with tips united, umbrella-like, in C. rendallii, or free (C. multiflora ssp. tentaculata), variously hairy or completely glabrous. – Arabia; trop. and S.E.Afr. – Type species: C. africana.

- **Ser. 1. Africanae** H. Huber. – **Infl.** usually short-pedunculate; segments of the inner subsidiary corona usually with tips spreading, often curved hook-like outwards (if straight and erect, then the Infl. is always pedunculate). – Type species: C. africana.

  a) Segments of the outer subsidiary corona united into a cup-shape, equal in length to, or more often exceeding, the length of the staminal tube; this cup is slightly 5-lobed adaxially, very short-ciliate or glabrous. – Species: C. africana, barklyi, euryacme, rendallii, stentiae.

  b) Segments of the outer subsidiary corona free, almost rectangular, glabrous, slightly longer than the staminal tube, the two upper angles having an incurved T. – Species: C. linophyllum.

  c) Segments of the outer subsidiary corona truncate, usually shorter than the staminal tube, rarely of equal length, glabrous. – Species: C. ? boerhaaviifolia, cancellata, linearis, maiuscula, pachystelma, senegalensis, vignaldiana, woodii.

- **Ser. 2. Multiflorae** H. Huber. – **Infl.** usually sessile. Segments of the inner subsidiary corona always with decumbent tips. – Type species: C. floribunda. – Species: C. floribunda, occulta.

**Sect. VIII. Psilopegia** H. Huber. – **Caudex** present or absent; **St.** moderately or very succulent, twining, glabrous; **L.** mostly reduced to scales and dropping, except in C. filiformis where they are narrow-linear to almost filiform and more persistent; **cymes** of a few Fl., short-pedunculate or sessile, Cor. with fine downy hairs outside or glabrous, the venter much swollen, the mouth narrow or widened to a funnel-shape, completely glabrous inside or with hairs only at the mouth (which is then widened to a funnel-shape, only C. subaphylla having some hairs towards the base of the narrowed tube-section), intermediate lobes absent, the Cor.-lobes united only by their tips or else forming an umbrella-shaped canopy, glabrous or variously hirsute. – Trop. Arabia; Somalia; Kenya; S.E.Afr. – Type species: C. zeyheri.

- **Ser. 1. Zeyherae** H. Huber. – Any **L.** present are scale-like, up to 1 mm long; **Cor.**-lobes united only by their tips, hirsute towards the base. Outer subsidiary corona usually glabrous. – Type species: C. zeyheri. – Species: C. arabica, subaphylla, zeyheri.

- **Ser. 2. Galeatae** H. Huber. – **L.** as in Series 1, **Cor.**-lobes united in the upper third to form an umbrella-shape, the margins with long motile ciliate hairs, the outer subsidiary corona glabrous. – Type species: C. galeata. – Species: C. fimbriata, galeata.

**Ser. 3. Filiformis** H. Huber. – **L.** normally present, narrow-linear to almost filiform, 1–5 cm long; **Cor.**-lobes united only by the tips, completely glabrous; outer subsidiary corona sparsely hairy. – Species: C. filiformis.

**Sect. IX. Loligo** Chiov. – **Caudex** sometimes present; **St.** very succulent, twining or creeping, glabrous; **L.** transformed into small pointed conical T.; **cymes** usually pedunculate, Cor. glabrous outside, tube much swollen below, this venter constricted, with a little annular cap inside at the level of the constriction, the tube hirsute inside midway or in the upper third, otherwise glabrous, the mouth widened to a funnel-shape, intermediate lobes conspicuously prominent, either sack- or spur-shaped, opening by a longitudinal cleft on the upperside, Cor.-lobes united by the tips, hirsute towards the base on the inside. – Arabia; E.Afr. – Type species: C. variegata. – Species: C. de vecchi, variegata.

**Sect. X. Amphorina** H. Huber. – With or without a small **caudex**; **St.** not succulent or only slightly so, twining, hirsute or (in the case of the only succulent species) glabrous; **L.** well developed or reduced and soon falling; **cymes** sessile or pedunculate, in C. ampliata often reduced to a single Fl., Cor. usually glabrous outside, tube much swollen below, above that broadly tubular, with a little widening at the mouth, rarely slender-tubular, hirsute or glabrous inside, intermediate lobes absent or rudimentary, Cor.-lobes united by the tips, glabrous or hirsute. – S.E.Afr. – Type species: C. sobolifera. – Only succulent species: C. ampliata.

**Sect. XI. Laguncula** H. Huber. – With or without a **caudex**; **St.** not succulent or scarcely so, twining or erect, hairy at least on new growth; **L.** always well developed, sometimes sessile; **cymes** sessile or pedunculate, Fl.-tube swollen basally, usually not widening towards the mouth, the central and upper parts often reduced more or less to an urn-shape, glabrous inside, intermediate lobes often present, small, Cor.-lobes free or united by the tips, variously hirsute or glabrous. – Trop. and S.E.Afr. – Type species: C. abyssinica. – Only somewhat succulent species: C. patriciae.

# Key to the succulent Species of the Genus Ceropegia by H. Huber

1. Plants with conical or flattened caudex. Roots slender, finely branched. St. faintly succulent, never dimorphic, always with well-developed L. ............................................. **2.**
1a. Plants without caudex, rarely tuberous, then the roots fleshy, sometimes nearly spindle-shaped, thickened and little or not branched or the St. strongly succulent or dimorphic. L. well developed, reduced to scales or absent ................................. **25.**
2. Infl. sessile ............................................................................ **3.**
2a. Infl. stalked ............................................................................ **8.**
3. Cor.-tube widened in the lower 2/3 or 3/4, St. and L. (these at least on the lower surface) downy and other non-succulent species of the Sect. Laguncula ................. **patriciae**
3a. Cor.-tube only in the lower 1/3 or 1/4 ventricosely widened, above that slender cylindrical. St. and L. glabrous or hairy ............................................................. **4.**
4. L. absent or reduced to sessile scales at most 1 cm long. Cor. at least at first hairy outside ..................................................................................... **subaphylla**
4a. L. well developed, stalked. Corolla glabrous outside ................................. **5.**
5. Cor.-lobes 2–3 cm long .............................................................. **stentiae**
5a. Cor.-lobes 5–12 mm long ........................................................... **6.**
6. Cor.-lobes obovate ................................................................. **occulta**
6a. Cor.-lobes linear or filiform ......................................................... **7.**
7. Cor.-lobes linear, almost equally wide throughout ............................. **floribunda**
7a. Cor.-lobes hairlike from a triangular base, 5–7 mm long and with the tips united (= ssp. **multiflora**) or 7–15 mm long and the tips mostly permanently separating (= ssp. **tentaculata**) ........................................................................... **multiflora**
8. Cor.-lobes with the upper half united umbrella-like ........................... **rendallii**
8a. Cor.-lobes only united wholly at the tip ............................................. **9.**
9. Cor. finely downy outside, at least on the outside of the lobes (examine buds!) . **pachystelma 10.**
9a. Cor. completely glabrous outside ................................................... **11.**
10. Cor.-lobes 5–7 mm long, not or only faintly widened at the tip ........ — ssp. **pachystelma**
10a. Cor.-lobes 8–12 mm long, widened spatulately at the tip .................. — ssp. **undulata**
11. Cor.-lobes narrow linear, as long as the tube or somewhat longer ................. **barklyi**
11a. Cor.-lobes shorter than the tube ................................................... **12.**
12. Cor.-lobes widened spatulately towards the tip .................................... **13.**
12a. Cor.-lobes linear, not widened spatulately towards the tip ....................... **19.**
13. Outer corona united into a 5-lobed cup, as high as the staminal-tube or higher ......... **14.**
13a. Outer corona truncate, not covering the staminal-tube ............................. **15.**
14. Cor.-lobes 8–25 mm long, faintly spatulate ....................................... **barklyi**
14a. Cor.-lobes 4–6 mm long, broad spatulate ...................................... **euryacme**

**Ceropegia** 136

15. The 5 lobes of the inner corona narrow linear to nearly subulate, overtopping the staminal-tube .................................................................. **vignaldiana**
15a. The 5 lobes of the inner corona lanceolate to spatulate, distinctly widened in the middle or upper part ........................................................................... **16.**
16. St. downy-haired at first. Stalk of the Infl. unifariously hairy ........... **? boerhaaviifolia**
16a. St. and stalk of the Infl. glabrous ................................................. **17.**
17. Cor.-lobes ciliate at the margin, otherwise glabrous .......................... **linearis**
17a. Cor.-lobes hairy at the margin and inside at the keel ....................... **woodii 18.**
18. L. lanceolate, ovate, triangular or reniform. Frequently with globose tubers at the nodes of the stems ..................................................................... — ssp. **woodii**
18a. L. linear, semiterete. Without tubers at the nodes of the St. ............... — ssp. **debilis**
19. Cor.-lobes glabrous on both sides, not ciliate .................................. **cancellata**
19a. Cor.-lobes ciliate inside or at the margin ........................................ **20.**
20. Outer corona well developed, with free or cup-shaped united lobes, as high as the staminal-tube or higher ........................................................................ **21.**
20a. Outer corona truncate, not covering the staminal-tube ....................... **23.**
21. The 5 lobes of the outer corona free, nearly rectangular, with a T. at each corner which is turned inwards. Lobes of the inner corona narrow lanceolate to linear ....... **linophyllum**
21a. Outer corona united into a shallow 5-lobed cup hiding the staminal-tube. Lobes of the inner corona broad lanceolate to dolabriform ................................... **africana 22.**
22. Cor.-tube glabrous inside in the cylindrical middlepart ................. — ssp. **africana**
22a. Cor.-tube hairy inside in the cylindrical middlepart .................... — ssp. **fortuita**
23. The 5 lobes of the inner corona hiding the staminal-tube, straight, erect or bending together with the tips. Cor.-lobes 6–8 mm long ............................................. **bulbosa**
23a. The 5 lobes of the inner corona with the tips recurved or hooked. Cor.-lobes 8–15 mm long ............................................................................ **24.**
24. Cor.-tube widened ventricosely in the lower 1/3 or 1/4. Cor.-lobes very narrow, nearly filiform ......................................................................... **senegalensis**
24a. Cor.-tube widened ventricosely in the lower 1/3 or 1/2. Cor.-lobes linear ........ **maiuscula**
25. Cor.-lobes united in the upper half or higher into a 1.5–4 cm broad canopy ........... **26.**
25a. Cor.-lobes not united with each other into a canopy ............................... **28.**
26. Venter hairy inside. Outer corona also hairy .................................. **sandersonii**
26a. Cor.-tube quite glabrous inside. Outer corona also glabrous ....................... **27.**
27. Cor.-tube widened ventricosely only in the lower 1/3 or 1/4 ..................... **fimbriata**
27a. Cor.-tube widened ventricosely in the lower half ............................... **galeata**
28. Cor.-tube glabrous on both sides and not ciliate ................................ **29.**
28a. Cor.-lobes hairy or ciliate at least inside or at the base or margin ............... **39.**
29. St. twining ................................................................. **30.**
29a. St. never twining, mostly erect, always strong, usually 5 mm thick or more, thick as a finger in age, white or light grey puberulous ........................................... **34.**
30. L. narrow cylindrical to nearly filiform, 1–5 cm long. Cor.-lobes hair-thin, 12–20 mm long ......................................................................... **filiformis**
30a. L. ovate or reduced to scales. Cor.-lobes broad linear to ovate, 4–14 mm long ....... **31.**
31. Cor.-tube broad cylindrical, bright green outside. Cor.-lobes dark green. L. only rarely well developed ................................................................ **ampliata 32.**
31a. Cor.-tube narrow in the middle. Cor. dark purplish brown. L. well developed
.................................................................. **aristolochioides 33.**
32. Cor.-tube 24–50 mm long, the lobes 8–14 mm long, nearly obtuse ... **ampliata** ssp. **ampliata**
32a. Cor.-tube 14–24 mm long, the lobes 4–6 mm long, acute ................. — ssp. **oxyloba**
33. Cor.-lobes broad ovate, 4–8 mm long ................. **aristolochioides** ssp. **aristolochioides**
33a. Cor.-lobes linear, 7–14 mm long ........................................... — ssp. **albertina**
34. Cor. red-brown ............................................................. **35.**
34a. Cor. yellow ................................................................ **36.**
35. L. broad ovate. Cor.-lobes broad linear, to 20 mm long ...................... **rupicola**
35a. L. linear, early deciduous. Cor.-lobes narrow linear from a triangular base, to 14 mm long
.......................................................................... **fusca**
36. At least the lower Infl. extremely profuse, 20–70-flowered ..................... **krainzii**
36a. Infl. throughout fewer-flowered .............................................. **37.**
37. L. mostly more than 2mm wide. Cor.-lobes with the tips easily separating ...... **dichotoma**
Remark **C. dichotoma** in the narrow sense is restricted to Tenerife. On Isle La Palma it is represented by **C. hians** which is hardly more than a ssp.

37a. L. hardly 2 mm wide. Cor.-lobes remaining united by the tips ...................... **38.**
38. The 5 sections of the outer corona broadly rounded, only shallowly incised at the tip
.......................................................................... **chrysantha**

| | | |
|---|---|---|
| 38 a. | The 5 sections of the outer corona drawn out into 2 erect teeth | **ceratophora** |
| 39. | St. dimorphic; the barren St. or St.-sections compact, with tooth-shaped projecting podaria, faintly four-angled and recalling Stapelia St.; not twining. Fertile St.-sections slender, terete erect or twining | **40.** |
| 39 a. | St. or St.-sections not dimorphic, not like Stapelia, with projecting toothed podaria | **42.** |
| 40. | Cor.-tube 25–40 mm long, ventricosely widened in the lower third, thereafter narrowed midway Cor.-lobes less than half as long as the tube | **leroyi** |
| 40 a. | Cor.-tube 5–10 mm long, inflated lengthwise, narrowed like a pitcher at the mouth. Cor.-lobes as long as the tube or longer | **41** |
| 41. | Fertile St.-sections erect. L. linear-lanceolate. Cor.-lobes about as long as the tube | **dimorpha** |
| 41 a. | Fertile St.-sections twining. L. ovate or elliptical. Cor.-lobes about twice as long as the tube | **armandii** |
| 42. | Venter hairy inside all over or only in the lower half | **43.** |
| 42 a. | Venter glabrous inside, but not rarely bordered above by a hair-wreath from the cylindrical tube | **63.** |
| 43. | Outer corona glabrous, truncate, not covering the staminal-tube. Venter sometimes transversely constricted | **44.** |
| 43 a. | Outer corona hairy, mostly well developed, with the lobes bifid or drawn out into 2 teeth. Venter never transversely constricted | **48.** |
| 44. | St. creeping, procumbent and rooting at the nodes | **radicans 45.** |
| 44 a. | St. not rooting at the nodes, twining | **46.** |
| 45. | Cor.-lobes linear from a triangular base | **radicans ssp. radicans** |
| 45 a. | Cor.-lobes obovate | — **ssp. smithii** |
| 46. | Plants with thickened, easily detached side shoots consisting of 1–3 St.-joints. Venter not constricted | **gemmifera** |
| 46 a. | Plants without thickened side shoots. Venter often transversely constricted | **47.** |
| 47. | Cor.-lobes elongated from a triangular base into a linear beak 10–20 mm long, which is occasionally faintly spatulate at the tip and has the margin ciliate with long vibratile clavate hairs. Venter not constricted (v. **denticulata**) or often transversely constricted (v. **brownii**) | **denticulata** |
| 47 a. | Cor.-lobes broad to elongate, ovate or triangular, only rarely pulled out into a linear beak 6–12 mm long, finely and short-hairy inside, without clavate hairs. Venter always transversely constricted | **nilotica** |
| 48. | Cor.-tube 10–20 mm long, not or slightly widened at the mouth. Cor.-lobes elongate-ovate to triangular-ovate, 3–12 mm long, never pulled out into linear or spatulate tails | **49.** |
| 48 a. | Cor.-lobes pulled out into linear or spatulate tails, rarely elongate-ovate or broad-triangular, but then the Cor.-tube 20–50 mm long | **55.** |
| 49. | Cor.-tube glabrous inside in the middle part | **50.** |
| 49 a. | Cor.-tube hairy inside in the middle part. Infl. mostly with extended internodes | **51.** |
| 50. | Cor. densely hairy outside | **somalensis** |
| 50 a. | Cor. glabrous outside | **carnosa** |
| 51. | Cor.-lobes triangular-ovate, at most as long as the mouth of the Cor.-tube is wide | **affinis** |
| 51 a. | Cor.-lobes elongate-ovate, longer than the ⌀ of the Cor.-tube at its mouth | **racemosa 52.** |
| 52. | Cor. finely hairy outside | — **ssp. secamoides** |
| 52 a. | Cor. glabrous outside | **53.** |
| 53. | L. glabrous on both sides, elliptical or ovate-lanceolate | — **ssp. glabra** |
| 53 a. | L. hairy at least on the underside on the midrib | **54.** |
| 54. | The 5 lobes of the outer corona split up into 2 erect or nearly erect teeth | — **ssp. racemosa** |
| 54 a. | The 5 lobes of the outer corona split up into 2 horizontally spreading teeth | — **ssp. setifera** |
| 55. | Cor.-lobes broad-triangular, much broader than long. Cor.-tube elongated at the mouth into 5 star-like spreading intermediate lobes. L. absent or reduced to scales | **cimiciodora** |
| 55 a. | Cor.-lobes longer than wide | **56.** |
| 56. | Cor.-lobes ovate or elongate-ovate, not beaked at the tip | **57.** |
| 56 a. | Cor.-lobes elongated from an ovate or triangular base into a narrow linear, sometimes spatulate beak | **58.** |
| 57. | Cor.-lobes 20–25 mm long | **hybrida** |
| 57 a. | Cor.-lobes 12–15 mm long | **rothii** |
| 58. | Cor.-lobes spatulately widened towards the tip | **59.** |
| 58 a. | Cor.-lobes spatulately widened at the tip | **60.** |
| 59. | Plants faintly succulent. L. well developed, stalked. Cor.-lobes 10–15 mm long, with the tips permanently united | **yorubana** |
| 59 a. | Plants strongly succulent. L. absent or reduced to scales. Cor.-lobes 20–30 mm long, with the tips loosely united (= v. **stapeliiformis**) or tongue-shaped and permanently united (= v. **serpentina**) | **stapeliiformis** |
| 60. | Cor. densely downy-haired outside. Cor.-lobes nearly as long as the tube | **somalensis** |

**60 a.** Cor. glabrous outside, rarely thinly downy. Cor.-lobes distinctly shorter than the tube
**distincta 61.**
**61.** Cal.-lobes 8–14 mm long ......................................... — ssp. **distincta**
**61 a.** Cal.-lobes 3–6 mm long ........................................... **62.**
**62.** Cor.-lobes gradually elongated from a triangular or ovate base into a linear, spatulate beak
— ssp. **lugardae**
**62 a.** Cor.-lobes elongated from a broad-ovate or truncate base into a hair-thin, spatulate beak
— ssp. **haygarthii**
**63.** Venter transversely constricted. Cor.-tube elongated at the mouth in the sinuses into 5 spur-like intermediate lobes ........................................................ **64.**
**63 a.** Venter not transversely constricted. No intermediate lobes in the sinuses between the Cor.-lobes ..................................................................... **65.**
**64.** With short conical, oblique to backwardly directed spurs in the sinuses between the Cor.-lobes ............................................................... **variegata**
**64 a.** With long, cylindrical spurs in the sinuses between the Cor.-lobes ........... **devecchii**
**65.** Cor.-tube quite glabrous inside. Lobes of the inner corona straight, erect, with the tips never hooked ........................................................................ **66.**
**65 a.** Cor.-tube hairy inside at the border between the venter and the cylindrical middle part, or the whole middle part hairy, very rarely fully glabrous inside, but then the lobes of the inner corona hooked at the tip ...................................................... **67.**
**66.** Cor. glabrous outside. Cor.-lobes narrowed to the tip ..................... **arabica**
**66 a.** Cor. at least initially finely downy outside. Cor.-lobes faintly spatulate at the tip .... **zeyheri**
**67.** Lobes of the inner corona hooked at the tip ............................. **68.**
**67 a.** Lobes of the inner corona straight, erect ................................ **70.**
**68.** Cor.-lobes abruptly elongated from an ovate base into a hair-thin beak ...... **rhynchantha**
**68 a.** Cor.-lobes not beaked, or wholly narrowed into a linear or spatulate tip ............. **69.**
**69.** L. mostly well developed. Cor.-lobes 6–12 mm long, ovate or gradually elongated into a short spatulate beak ............................................................ **volubilis**
**69 a.** L. absent or reduced to scales. Cor.-lobes 15–25 mm long, elongated from a triangular base into linear tails .................................................................. **juncea**
**70.** Cor.-tube extensively hairy inside in the cylindrical middle part. Plants with thick fleshy roots (except **C. subaphylla**) ................................................ **71.**
**70 a.** Cor.-tube with a narrow hair-wreath inside at the border of the venter and the cylindrical middle-part. Roots (as far as known) not fleshy ............................. **73.**
**71.** L. absent or at most reduced to 1 cm long scales. Cor. at least initially finely downy outside. Cor.-lobes from a triangular base, narrow linear, 10–25 mm long. Roots thin ... **subaphylla**
**71 a.** L. well developed. Cor. always glabrous. Cor.-lobes broad ovate, 3–12 mm long. Roots fleshy ........................................................................ **72.**
**72.** L. mostly flat. Cor.-tube 15–32 mm long. Cor.-lobes 5–12 mm long ............. **crassifolia**
**72 a.** L. semiterete. Cor.-tube c. 10 mm long. Cor.-lobes 3–4 mm long ............. **copleyae**
**73.** Infl. mostly helicoidal ................................................ **albisepta**
**73 a.** Infl. cymose, false-umbellate ............................................. **74.**
**74.** Cor.-lobes broad-ovate, broad-triangular or truncate, as broad as long or broader than long. Faintly succulent plants ................................................. **elegans 75.**
**74 a.** Cor.-lobes elongate-ovate to linear, distinctly longer than broad ..................... **76.**
**75.** Cor.-tube 1–2 cm in ∅ at the mouth, not projecting in the sinuses between the lobes
— v. **elegans**
**75 a.** Cor.-tube 2.5–3.5 cm in ∅, the sinuses between the lobes arching outwards to form rounded pockets ................................................................ — v. **gardneri**
**76.** Cor.-lobes 6–20 mm long. Outer corona ciliate ............................. **77.**
**76 a.** Cor.-lobes 20–70 mm long. Outer corona glabrous ......................... **78.**
**77.** Plants not or faintly succulent. Cor.-lobes 6–10 mm long. Outer corona with a few long, stiff, hairs ................................................................ **sankuruensis**
**77 a.** Plants perceptibly succulent. Cor.-lobes 10–20 mm long .................... **succulenta**
**78.** Cal.-lobes 14–30 mm long. Cor.-lobes 60–70 mm long ....................... **halliana**
**78 a.** Cal.-lobes 5–8 mm long. Cor.-lobes at most 35 mm long .................... **robynsiana**

**C. affinis** VATKE (§ IV/2). – N.E.Afr.; Ethiopia; Somalia. – **Roots** thickened, tufted; **St.** slender, trailing; **L.** ovate-lanceolate to almost linear; **Infl.** often an extended helix, Cor.-tube 8–15 mm long, evenly hirsute inside, lobes broadly ovate-triangular, 3–6 mm long, sparsely hairy on the inside.

**C. africana** R. BR. v. africana (§ VII/1a) (*C. wightii* GRAH. et WIGHT.). – E.Cape. – **St.** slender, trailing; **L.** ovate, elliptical or lanceolate, fleshy; **Cor.** glabrous outside, tube 1–2.5 cm long, greenish outside, brown-violet striate towards the mouth, glabrous inside, lobes linear, with dark purple hairs inside along the keel.

**C.** — ssp. **fortuita** (R. A. DYER) H. HUBER (*C. f.* R. A. DYER). – Natal. – Cor.-tube hirsute inside in the central part.

**C. albisepta** JUM. et PERR. (§ II/1). – Central and S.Madag. – **St.** trailing, slender to moderately robust; **L.** ovate, elliptical or obovate, somewhat fleshy, petiole scarcely 1 cm long, blade 3–7 cm long; **Infl.** extended, Cor.-tube 2–3 cm long, widening at the mouth to 1 to 1.5 cm, lobes 1–2 cm long, extended from a triangular-ovate base to a narrow-linear pointed beak, margins with violet-coloured hairs.

**C. ampliata** E. MEY. v. **ampliata** (Pl. 31/1) (§ X). – Transvaal; Natal; E.Cape. – **St.** moderately succulent, leafless or with scale-like **L.**, in cultivated plants the L. often ovate or elliptical; **Fl.**tube 2.5–5 cm long, continuing as a broad cylinder, light green outside, with a purple transverse band inside, lobes oblong-lanceolate on a triangular base, dark green, glabrous.

**C. —** v. **oxyloba** H. HUBER. – Tanzania. – **Fl.** smaller, much swollen in the lower half, tube 14–24 mm long, lobes 4–6 mm long, narrowly triangular or lanceolate, pointed, ciliate.

**C. arabica** H. HUBER (§ VIII/1). – Arabia. – **Roots** tufted, fleshy; **St.** glabrous, twining, usually leafless; **Cor.** glabrous outside, tube 2.5–3.5 cm long, with purple markings outside, glabrous inside, hirsute only in the widened, funnel-shaped mouth, lobes narrow-linear from a triangular base, 1.5–2.5 cm long, white-haired on the inside, almost glabrous towards the tip.

**C. aristolochioides** DECNE. ssp. **aristolochioides** (Pl. 31/4) (§ IV/2). – Senegal to Ethiopia. – **Caudex** usually absent; **St.** slender, glabrous, slightly succulent; **L.** ovate, usually cordate at the base; **Cor.** dark purple outside, glabrous, tube 12–15 mm long, hairy inside only in the central part, the mouth usually widened funnel-like, lobes broad-ovate, 4–8 mm long, glabrous, not ciliate.

**C. —** ssp. **albertina** (S. MOORE) H. HUBER (*C. a.* S. MOORE). – Lobes oblong-linear, 7 to 16 mm long, tube less widened at the mouth.

**C. armandii** RAUH (§ II/2). – S.W.Madag. – **St.** stapelia-like, up to 2 cm thick, prostrate, 10–15 cm long, verruculose, twining and over a meter in length; **L.** ovate-lanceolate, about 7 mm long; **Fl.** Cor.-tube urn-shaped, 5 mm long and almost as wide, opening 1–2 mm wide, lobes linear, 10–12 mm long.

**C. ballyana** BULLOCK (Pl. 31/2) (§ II/1) (*C. helicoides* BRUCE et BALLY). – Kenya. – **St.** robust, twining; **L.** ovate or elliptical, fleshy, with a petiole 1–2 cm long; **Fl.** greenish-white with red-brown spots, tube 3.5–5 cm long, the lobes linear from a triangular base, the upper part twisted helically together, 6–7 cm long.

**C. barklyi** HOOK. f. (§ VII/1a) (*C. b.* v. *tugelensis* N. E. BR.). – Natal. – **St.** slender, glabrous or finely downy; **L.** ovate, lanceolate or linear, fleshy; **Cor.** glabrous outside, tube 1–2 cm long, hirsute on the inside of the slender cylindrical central part, especially towards the base, lobes narrow-linear, as long as the tube, hairy inside, particularly towards the base.

**C. —** × **C. rendallii** H. HUBER (§ VII/1a). – Garden hybrid. – **St.** slender, twining; **L.** ovate-lanceolate, fleshy; **Cor.** glabrous outside, tube about 18 mm long, hairy on the inside of the central part, lobes narrow-linear for 3/4 of their length, then abruptly cordate-spatulate, 10 to 12 mm long, united only by the tips.

**C. —** × **C. woodii** H. HUBER (§ VII/1a × c). – Garden-hybrid. Closely resembles the previous hybrid; Cor.-tube some 18 mm long, lobes widening above only slightly and gradually to the somewhat spatulate tip.

**C. boerhaaviifolia** DEFL. (§ VII/1c). – Arabia. – **St.** slender, twining, at first downily-hirsute; **L.** pointed-ovate; **Cor.** glabrous outside, tube 1 cm long, hairy inside towards the base of the narrow cylindrical part, lobes narrow-linear, widening gradually towards the slightly spatulate tip, 6 mm long, borders with red cilia. Dubious species.

**C. bosseri** RAUH et BUCHLOH (§ II/2a). – Central Madag. – Lower **St.** sections very succulent, prostrate, with short Int., much flattened, up to 2 cm across at the nodes, finely tuberculate; **L.** roundish-elliptical, c. 1 cm long, L.bases initially projecting horizontally, later curved-hooked, flowering St.sections slender, twining, with scale-L.; **Fl.**-tube up to 3 cm long, inflated-obovate in the bottom 1/3, above that narrow-tubular, widening towards the mouth to a broad funnel-shape, Cor.-lobes almost linear from a triangular-ovoid base, 13–15 mm long, with thin ciliate hairs, white and violet, along the borders and in the lower half.

**C. bulbosa** ROXB. (§ I) (*C. acuminata* ROXB., *C. lushii* GRAH.). – Peninsular India. – **St.** slender, twining; **L.** very variable, round, ovate, lanceolate or linear; **Fl.**-tube 12–18 mm long, lobes linear, 6–8 mm long. Edible.

**C. cancellata** REICHENB. (§ VII/1c) (*C. assimilis* N. E. BR.). – E.Cape. – **St.** slender, twining; **L.** ovate-oblong to linear, fleshy; **Cor.** glabrous outside, tube 12–20 mm long, greenish outside, hairy inside in the middle part, lobes linear, 6–12 mm long, purple-brown, glabrous.

**C. carnosa** E. MEY. (§ IV/2). – Natal; E.Cape. – **Roots** fleshy, thickened, tufted; **St.** slender, glabrous; **L.** small, fleshy, ovate, ovate-lanceolate, or triangular; **Infl.** often little extended, Cor. red-brown, glabrous outside, tube 12 to 19 mm long, hirsute inside towards the base and in the slightly widened mouth, glabrous in the central part, lobes ovate with sparse white hairs inside, 4–6 mm long.

**C. ceratophora** SVENT. (§ III/2). – Canary Is.: Gomera. – **St.** stiffly erect, 1–1.5 m tall; **L.** ephemeral, up to 2 cm long, less than 2 mm across, sessile; **Infl.** perennial, with 10–20 pale yellow Fl., Cor.-tube scarcely bent, c. 2 cm long, lobes remaining united by their tips, 1 cm long.

**C. chrysantha** SVENT. (Pl. 31/3) (§ III/2). – Canary Is.: Tenerife. – **St.** ascending, 0.5–1 m tall; **L.** ephemeral, linear, 2.3 cm long, less than 2 mm across, sessile; **Infl.** perennial, of 5–10 Fl., Cor.-tube yellowish-white, slightly bent, 2 to 2.5 cm long, lobes yellow, usually remaining united by the tips, c. 1 cm long.

**C. cimiciodora** OBERM. (Pl. 32/1) (§ VI). – Transvaal: Soutpansberg D. – Resembles **C. stapeliiformis**; **Fl.** solitary, the tube-mouth extended into 5 intermediate lobes, spreading star-like; Cor.-lobes much shortened, broadly triangular, broader than long.

**C. copleyae** E. A. BRUCE et BALLY (§ IV/2). – Kenya. – Closely related to **C. crassifolia,** possibly only a variety of this; erect; **L.** narrow, semi-terete; **Fl.** smaller, tube 1 cm long, lobes 3–4 mm long.

**C. crassifolia** SCHLTR. (§ IV/2) (*C. crispata* N. E. BR., *C. thorncroftii* N. E. BR., *C. tuberculata* DTR.). – From S.W.Afr. to Natal. – With or without an irregularly shaped **caudex**, roots fleshily thickened, tufted; **St.** very succulent, glabrous, twining; **L.** circular, ovate, lanceolate to linear; **Cor.** greenish-white, speckled redbrown, glabrous outside, tube 1.5–3 cm long, glabrous inside the venter, above that hairy, the funnel-like opening moderately widened, the lobes broad-ovate, 5–12 mm long, usually sparsely hirsute on the inside.

**C. denticulata** K. SCHUM. v. **denticulata** (§ V/1) (*C. nilotica* v. *simplex* H. HUBER p. part.). – Kenya; Uganda; Tanzania. – Closely resembles **C. nilotica**; the much swollen venter not constricted, the lobes prolonged from a triangular base into a linear beak 1–2 cm long, and sometimes slightly spatulate-widened at the tips, the margins ciliate with long motile purple clavate hairs. An Angolan form, with conspicuously long (2–2.5 cm) and extremely narrow lobes, has been described as **C. gossweileri.**

**C. —** v. **brownii** (LEDGER) BALLY (*C. b.* LEDGER). – Trop. Afr., widely distributed. – Cor.tube with the venter obliquely constricted, lobes with long ciliate clavate hairs.

**C. devecchii** CHIOV. v. **devecchii** (§ IX) (*C. variegata* v. *cornigera* H. HUBER). – Ethiopia; Somalia?; Yemen. – Closely related to **C. variegata**; tube 2–3 cm long, intermediate lobes 1–2 cm long and open above when the Fl. is fully out, otherwise closed, forming slender spurs narrowing conically towards the tip, Cor.-lobes distinctly longer than broad, the subulate, spirally twisted tips often being missing.

**C. —** v. **adelaidae** BALLY. – Kenya; N.Tanzania. – Intermediate lobes to 27 mm long, Cor.-lobes broad-triangular, c. 0.5 cm long and across, with a short apical appendage.

**C. dichotoma** HAW. (Pl. 32/4) (§ III/2). – Canary Is.: Tenerife. – **St.** erect or ascending, 33–100 cm tall; **L.** linear, 2.5–8 cm long, 2 to 9 mm across, with a petiole 2 mm long; **Infl.** of 1–9 Fl., tube pale yellow, slightly curved, (2–)2.5(–3) cm long, lobes bright yellow, separating as the flower dies, 1–1.5 cm long.

**C. dimorpha** H. HUMB. (§ II/2b). – S.W.-Madag. – Non-flowering, Stapelia-like **St.** up to 1.5 cm thick, erect, up to 15 cm tall; flowering St.segments slender, terete, erect; **L.** linear, lanceolate, up to 3.5 cm long; **Fl.**-tube urn-shaped, 1 cm long, 8 mm $\varnothing$, lobes linear.

**C. distincta** N. E. BR. (§ IV/2). – Zanzibar; S. and S.W.Afr.; Angola.

**C. —** ssp. **distincta.** – Zanzibar. – **St.** stout, slightly succulent, twining; **L.** ovate or elliptical; **Cor.** glabrous outside, tube 1.5–3 cm long, hirsute inside the venter and usually also in the much-widened funnel-shaped mouth, the middle part glabrous, the lobes with an ovate base gradually extended to a distinctly broad-spatulate tip, about 1 cm long with white cilia.

**C. —** ssp. **haygarthii** (SCHLTR.) H. HUBER (Pl. 32/3) (*C. h.* SCHLTR.). – Cape: Stockenstroom D.; Natal; Transvaal: Lydenburg D. – **Cor.**-lobes with a broad-rounded, almost hemispherical base, narrowing to a filiform stalk 5–14 mm long, then widening into an ovate tip.

**C. —** ssp. **lugardae** (N. E. BR.) H. HUBER (*C. l.* N. E. BR., *C. apiculata* SCHLTR., *C. cyrtoides* WERDERM., *C. tristis* HUTCH.). – Angola: northern S.W.Afr.: Botswana; trop. E.Africa, south of the equator. – **Cor.** glabrous outside or sometimes, in plants from Angola, Zambia and Rhodesia, with a fine down (= f. **pubescens** H. HUBER), lobes on a triangular-ovate base, gradually extending to a linear or ± spatulate tip 1.5–2 cm long.

**C. elegans** WALL. (§ II) (*C. ledgeri* N. E. BR., *C. similis* N. E. BR.). – Peninsular India; Ceylon. – **St.** thin, twining; **L.** with petiole 1–2 cm long, slightly fleshy, ovate or lanceolate; **Fl.** white with purple spots, tube 2–4 cm long, 1–3.5 cm $\varnothing$ at the mouth, Cor.-lobes broadly ovate to triangular, their length equalling the radius of the funnel-opening.

**C. —** v. **elegans.** – Mouth of the **Cor.**-tube 1–2 cm $\varnothing$, the indentations between the lobes not visibly projecting; the bud, seen from above, is pentagonal.

**C. —** v. **gardneri** (THWAITES ex HOOK.) H. HUBER (*C. g.* THWAITES ex HOOK.). – Ceylon. – Mouth of the **Cor.**-tube 2.5–3.5 cm $\varnothing$, the indentations between the lobes projecting like pockets; the bud, seen from above, is 5-lobed and star-like.

**C. euryacme** SCHLTR. (§ VII/1a) (*C. decidua* BRUCE). – Kenya to Natal. – **St.** twining, slender, glabrous; **L.** usually fleshy, sometimes thin-membranous, usually ovate or oblong; **Cor.** glabrous outside, tube 7–19 mm long, hirsute in the middle part, lobes linear, widening to spatulate towards the tip, only the tips united, 4–6 mm long, with purple cilia inside.

**C. filiformis** (BURCH.) SCHLTR. (§ VIII/3) (*Systrepha f.* BURCH., *C. infundibuliformis* E. MEY.). – N.Cape. – **Roots** fleshy, tufted; **St.** twining, glabrous; **L.** narrow-linear to almost threadlike, **Infl.** on a short Ped., Cor. glabrous on both surfaces, tube 12–24 mm long, whitish or light green outside with purple bands and blotches, widening only slightly towards the mouth, lobes filiform, 12–20 mm long, dark purple, glabrous.

**C. fimbriata** E. MEY. (§ VIII/2) (*C. estelleana* R. A. DYER). – S.E.Cape. – **St.** very succulent, twining, glabrous; **L.** when present reduced to scales 1 cm long; **Fl.** usually solitary, Cor. glabrous outside, tube 3–5 cm long, inflated to spherical in the lowest 1/4, above that abruptly narrowed, widening again gradually towards the funnel-shaped mouth, glabrous inside, the Cor.-lobes widening to spatulate above, united to form an umbrella-shaped canopy (1–)2 cm across.

**C. floribunda** N. E. Br. (§ VIII/2) (*C. conrathii* Schltr.). – Botswana; Transvaal. – **St.** erect or twining, glabrous; **L.** fleshy, ovate, elliptical or lanceolate; **Infl.** often of many Fl., Cor. glabrous outside, tube 10–14 mm long, hirsute inside in the middle part, lobes linear, 6–8 mm long, purple, both surfaces glabrous.

**C. fusca** Bolle (§ III/2). – Canary Is.: Gran Canaria, Tenerife. – **St.** erect or trailing somewhat over bushes, to 2 m tall; **L.** very short-lived, linear, 2–4 cm long, 2–5 mm across; **Infl.** perennial, of many Fl., these being dark red-brown, tube 14–22 mm long, Cor.-lobes narrow-linear from a triangular base, frequently separating as the Fl. withers.

**C. galeata** H. Huber (§ VIII/2). – Kenya. – **St.** very succulent, twining, glabrous; **Cor.** glabrous outside, yellow-green or light ochre-yellow with brown or olive-green blotches, about 4 cm long, inflated to spherical in the lower half, above that abruptly constricted into the tube, and above that widening again to the funnel-shaped upper part, inside usually glabrous; Cor.-lobes with a triangular, laterally plicate base and above that broadly obovate, united roughly for the upper third to form a flattened canopy 2–2.5 cm across, the border with purple cilia.

**C. gemmifera** K. Schum. (§ V/1) (*C. nilotica* v. *simplex* H. Huber p. part.). – Ghana; Togo. – **St.** slightly succulent, slender, twining, with short, thick, readily detached branchlets of 1–4 Int. 1–3 cm long; **L.** ovate to ovate-lanceolate; **Infl.** of few Fl., Cor.-tube 2.5–3 cm long, the venter being without any constriction, the upper half glabrous inside, the lower part hirsute, the slender central tube-section very hairy in the lower half, the lobes triangular-ovate, extended into a short, broadly linear beak c. 1.5 cm long, with a large yellowish-white blotch at the base, a black transverse band in the middle, above that emerald-green, with purple hairs inside.

**C. hians** Svent. (Pl. 32/3) (§ III/2). – Canary Is.: La Palma. – **St.** branching freely from the ground, stiffly erect, 0.5–0.75 m tall; **L.** linear, 5–6 mm long, 3–4 mm across, with a petiole 3 mm long; **Infl.** perennial, of many Fl., tube yellowish-white, slightly bent, 2–2.5 cm long, Cor.-lobes yellow, separating, 1–1.5 cm long. – **C. hians** acc. H. Huber is hardly more than a ssp. of **C. dichotoma**.

**C.** × **hybrida** N. E. Br. (*C. meyeri-arthuri* Herter). – Garden hybrid: **C. elegans** × **C. sandersonii**. – **St.** robust; **L.** small, ovate, almost sessile; **Fl.**-tube 3–5 cm long, the venter glabrous inside its upper half, hirsute below, the middle and upper tube-sections hirsute towards the base and the mouth, the Cor.-lobes oblong, blunt, 2–2.5 cm long, hairy inside over the lower half. It seems rather dubious whether **C. elegans** is one parent; **C. nilotica** appears more probable.

**C. juncea** Roxb. (§ IV/1). – Peninsular India. – **St.** very succulent, twining; **L.** much reduced; **Cor.** 2–3.5 cm long, widened funnel-like at the mouth, lobes on a triangular base, then linear, with cilia towards the tip and hirsute on the inside, 1.5–2.5 cm long.

**C. krainzii** Svent. (§ III/2). – Canary Is.: Gomera. – **St.** stiffly erect, simple or little branched basally; **L.** very transient, linear, 1.5–3 cm long, 1.5 mm across, sessile; **Infl.** perennial, Fl. very numerous, tube yellowish-white, slightly curved, 2 cm long, Cor.-lobes yellow, separating slightly, 13 mm long.

**C. leroyi** Rauh et Marn.-Lap. (§ II/2a). – S.W.Madag. – Non-flowering **St.** Stapelia-like, up to 5 mm thick, prostrate, 5–10 cm long, finely tuberculate, flower-bearing St. much extended, twining and up to 0.5 m tall; **L.** linear-lanceolate, 2–2.5 cm long, the Cor.-tube much broadened in the lower 1/3, above that slender-terete, 2.5–3 cm long, lobes linear, 6–12 mm long.

**C. linearis** E. Mey. (§ VII/1c) (*C. caffrorum* Schltr.). – Natal; Moçambique. – **St.** slender, glabrous, twining; **L.** triangular-ovate or lanceolate or linear, fleshy; **Cor.** glabrous outside, the tube 12–15 mm long, light green outside with purple lengthwise stripes, the inside of the middle section hirsute, the lobes narrow-linear, slightly spatulately widening towards the tips, 6–8 mm long, dark purple-brown, the margins with purple ciliate hairs.

**C. linophyllum** H. Huber (§ VII/1b). – Ghana; S.Nigeria. – **St.** slender, twining, glabrous; **L.** rather fleshy, linear; **Cor.** glabrous outside, the tube 8–14 mm long, hirsute on the inside of the middle section, the lobes narrow-linear, 5–7 mm long, sparsely hairy inside.

**C. maiuscula** H. Huber (§ VII/1c). – Tanzania. – **St.** slender, twining, glabrous; **L.** linear, fleshy; **Fl.** tube 2–2.5 cm long, hairy on the inside of the middle section, Cor.-lobes linear, 12–15 mm long, very hirsute on the inside.

**C. multiflora** Bak. ssp. **multiflora** (§ VII/2). – N.Cape, Transvaal. – **St.** slender, twining, glabrous (f. **multiflora**), or with short, downy hairs (f. **pubescens** H. Huber); **L.** fleshy, rounded-ovate, elliptical, lanceolate or linear; **Infl.** of many Fl., Cor. greenish-white, glabrous outside, tube 12–26 mm long, hirsute on the inside of the middle section, lobes filiform from a triangular base, (5–)7(–15) mm long, hairy inside at the base, the lower half projecting obliquely outwards, the upper half directed almost horizontally inwards, the tips usually remaining united.

**C.** — ssp. **tentaculata** (N. E. Br.) H. Huber (*C. t.* N. E. Br.). – Angola; S.W.Afr.; Rhodesia. – Tube up to 26 mm long, lobes 7–15 mm long, the tips usually separating.

**C.** — ssp. — f. **puberula** H. Huber. – **St.** and **L.** with short, downy hairs.

**C.** — ssp. — f. **tentaculata**. – Glabrous form.

**C. nilotica** Kotschy (§ V/1) (*C. boussingaultifolia* Dtr., *C. constricta* N. E. Br., *C. mozambiquensis* Schltr., *C. plicata* E. A. Bruce). – Kenya; Sudan; trop. Afr. S. of the Equator. – **St.** somewhat succulent, twining, often 4-angled; **L.** usually ovate or rhombic, margins often finely dentate; **Infl.** cymose, Cor.-tube 2–4 cm long, greenish or yellowish-white with red-brown blotches, the venter transversely constricted, glabrous above, hirsute below, the lobes broad oblong-ovate or triangular, 6 to 12 mm long, with a large yellow or white blotch at the base, the middle usually having a purple-brown transverse band, above that green or red-brown, with fine and rather short purple hairs on the inside. – The following, of uncertain status, perhaps belong here: **C. decumbens** BALLY from Kenya, **St.** prostrate instead of twining, **L.** lanceolate, more fleshy; **C. grandis** E. A. BRUCE, Zululand, with a large **Fl.**, tube 4–6 cm long, lobes broad-ovate, broad-rounded at the tips, 16–18 mm long, green with a white transverse band at the base; in the N. of S.W. Afr. there is another related Ceropegia resembling **C. denticulata** v. **brownii,** having Cor.-lobes with clavate-thickened hairs 1 mm long.

**C. occulta** R. A. DYER (§ VII/2). – Cape: Worcester D. – **St.** slender, glabrous, twining; **L.** fleshy, ovate to lanceolate; **Infl.** of few Fl., Cor. glabrous outside, tube about 18 mm long, the middle section hirsute inside, the lobes obovate, widening towards the tip, 5–6 mm long, finely hairy on the inside, especially in the lower half, and on the margins.

**C. pachystelma** SCHLTR. ssp. **pachystelma** (§ VIII/1c) (*C. acacietorum* SCHLTR. ex DTR., *C. obscura* N. E. BR.) – Moçambique; Natal; Transvaal; S.W.Afr. – **St.** slender, twining, with fine, down-like hairs; **L.** fleshy, ovate, elliptical or oblong, both surfaces downy or becoming ± glabrous; **Cor.** with fine, short hairs outside, the tube occasionally becoming glabrous, 12–24 mm long, greenish outside, shortly hirsute inside in the middle section, the lobes narrow-linear, (5–)7(–12) mm long, light brown outside, red inside, both surfaces hairy or ± glabrous inside.

**C.** — ssp. **undulata** (N. E. BR.) H. HUBER (*C. u.* N. E. BR.). – Lobes 8–12 mm long, widening to slightly spatulate at the tip.

**C. patriciae** RAUH et BUCHLOH (§ XI). – S.Afr.; Transvaal. – **Caudex** spherical, 6 to 8 cm ⌀; **St.** scarcely 10 cm tall, not very succulent; **L.** ovate-lanceolate to lanceolate, glabrous on the upperside, densely short-hairy on the underside; **Infl.** of many Fl., Cor. glabrous outside, tube 5–6 mm long, long pitcher-shaped, the venter narrowing gradually to the mouth, grey-green outside with dark violet spots, glabrous inside, the lobes obovate, about 5 mm long, black-purple, glabrous, coarsely tuberculate on the inside, the tips initially connivent, later separating.

**C. racemosa** N. E. BR. ssp. **racemosa** (§ IV/2). – Trop. Afr. – With or without a small **caudex**; **roots** fleshy and thickened, tufted; **St.** slender, twining, not really succulent, shortly hairy; **L.** broad-ovate to linear; **Fl.** usually elongated, Cor. red-brown, glabrous outside, tube 1–2 cm long, evenly hairy on the inside, little widened at the mouth, lobes long-ovate, 5–12 mm long, sparsely white-haired inside.

**C.** — ssp. **glabra** H. HUBER. – Madag. – **St.** glabrous.

**C.** — ssp. **secamonoides** (S. MOORE) H. HUBER (*C. s.* S. MOORE, *C. cynanchoides* SCHLTR.). – Angola; Rhodesia; S.W.Afr. – **Cor.** initially with fine downy hair outside.

**C.** — ssp. **setifera** (SCHLTR.) H. HUBER (*C. s.* SCHLTR., *C. s.* v. *natalensis* N. E. BR.). – Transvaal; Natal. – Outer corona bristly.

**C. radicans** SCHLTR. v. **radicans** (§ V/1). – E.Cape: Komga D. – **St.** creeping, rooting from the nodes; **L.** ovate, elliptical or oblong; **Fl.** often solitary, tube 3.5–5 cm long, greenish-white with purple blotches, the venter not constricted, glabrous inside the upper half, below that hirsute, the slender middle tube-section hairy inside towards the base, the Cor.-lobes narrow-linear from a triangular base, 1.5–3 cm long, with a broad transverse band in the bottom third, above that a narrower dark purple-red band, the tip green, the upper part with long, dark purple, clavate cilia.

**C.** — v. **smithii** (HENDERS.) H. HUBER (*C. s.* HENDERS.). – Cape: East London D. – Lobes spatulate, obovate to almost cordate, c. 1.5 cm long.

**C. rendallii** N. E. BR. (§ VII/1a) (*C. galpinii* SCHLTR.). – Orange Free State, Transvaal. – **St.** slender, twining, or erect and remaining dwarf, glabrous or minutely downy; **L.** fleshy, ovate or linear; **Cor.** glabrous outside, the tube 12 to 20 mm long, hairy in the inside of the middle section, lobes narrow-linear in their lower half, widening above that to spatulate, to unite into an umbrella-shaped canopy 7–9 mm across, the margin being slightly 10-lobed and sparsely ciliate with purple hairs.

**C. rhynchantha** SCHLTR. (§ IV). – Upper Guinea. – **St.** slightly succulent, twining, glabrous; **L.** ovate or elliptical; **Cor.** glabrous outside, the tube 13–25 mm long, the venter glabrous inside, with little widening towards the mouth, the lobes abruptly widening from an ovate base, the inner surface white-hairy, especially towards the tips.

**C. robynsiana** WERDERM. (§ II/1). – Congolese Rep. – Closely related to **C. ballyana** and **C. succulenta;** Cor.-tube 2.5–3 cm long, lobes c. 2 cm long.

**C. × rothii** GÜRKE (§ V). – **C. radicans** SCHLTR. × **C. sandersonii** GÜRKE. – Garden hybrid. – Closely resembles **C. × hybrida** N. E. BR.; Cor.-lobes 12–15 mm long.

**C. rupicola** DEFL. (§ III/1). – S.Arabia. – **St.** erect or ascending, up to c. 1 m tall; **L.** broadly ovate; **Fl.** red-brown, tube 2.5–3.5 cm long, lobes 2 cm long.

**C. sandersonii** HOOK. f .(Pl. 33/3) (§ V/2) (*C. monteiroi* HOOK. f.). – Moçambique; S.Afr.; Transvaal. – **St.** robust, very succulent, twining; **L.** cordate; **Fl.** green, tube 3.5–5 cm long, hairy inside the venter, otherwise almost glabrous, the lobes united to form an umbrella-shaped canopy whose margin is shortly 10-lobed and covered with white motile hairs, the surface being covered with dark green blotches.

**C. sankuruensis** SCHLTR. (§ II/1). – Trop. Afr.: Sierra Leone to Tanzania. – **St.** slender, twining; **L.** slightly succulent, ovate to lanceolate, with a petiole over 1 cm in length; **Cor.** 2–2.5 cm long, the mouth funnel-shaped, 5–9 mm across, the lobes oblong-linear, 6–10 mm long.

**C. senegalensis** H. HUBER (§ VII/1c). – Senegal (near Dakar). – **St.** slender, twining, glabrous; **L.** usually linear, with minute cilia along the margins and on the underside along the M.rib; **Cor.** glabrous outside, the tube 12–24 mm long, hirsute inside the middle section, the lobes narrow-linear, almost filiform, 8–15 mm long, very hirsute on the inside.

**C. seticorona** E. A. BRUCE v. **seticorona** (§ IV/2) (*C. volubilis* v. *crassicaulis* H. HUBER). – Kenya; N.Tanzania. – **St.** very succulent, sparsely downy; **L.** ovate, cordate or lanceolate; **Infl.** of many Fl., Cor. shortly downy outside, the tube 2–2.5 cm long, glabrous inside the slightly swollen venter, extensively hairy in the central tube-section, the lobes spatulate from a broadly triangular base, narrowing somewhat in the middle, 8–10 mm long, margins glabrous. The word 'seticorona' refers to the five deeply 2-cleft sections of the outer subsidiary corona, every tooth of which bears a solitary long bristle-hair.

**C. —** v. **dilatiloba** BALLY. – Kenya; Uganda. – Tube much widened both at the base and at the mouth, lobes ovate, margins glabrous or sometimes with thin ciliate hairs.

**C. somalensis** CHIOV. f. **somalensis** (§ IV/2). – Somalia; Kenya. – **St.** slender, twining, slightly fleshy, glabrous; **L.** small, ovate, downy or becoming glabrous; **Cor.** with dense, short down outside, the tube 1–2 cm long, hairy inside the venter and the funnel-shaped mouth, glabrous in the middle section, the lobes narrowing from the ovate base to a threadlike stalk and widening to spatulate at the tip, in all almost as long as the tube, white-hairy inside.

**C. —** f. **erostrata** H. HUBER. – Lobes oblong-ovate, much shorter than the tube.

**C. stapeliiformis** HAW. v. **stapeliiformis** (Pl. 33/2, 4) (§ VI). – E.Cape. – **St.** stout, up to 1.5 m tall, very succulent, almost leafless, 1.75 cm thick below, twining, glabrous; **L.** reduced to small ovate scales; **Infl.** of several Fl., the Cor.-tube 2–4 cm long, whitish outside or greenish-white with black-purple blotches, the lobes linear from a triangular base, 2–5 cm long, free, spreading, white with the margins and tip greenish-yellow, black-purple or dark brown.

**C. —** v. **serpentina** (E. A. BRUCE) H. HUBER (*C. s.* E. A. BRUCE). – Lobes remaining united with one another towards the tip along part of their length, thus forming a tube which is closed above, but later tears open.

**C. stentiae** E. A. BRUCE (§ VII/1a). – Transvaal. – **St.** erect, c. 10 cm tall, the upper part with fine downy hairs; **L.** fleshy, narrow-linear, glabrous; **Infl.** with 1 or only very few Fl., Cor. glabrous outside, the tube 2–3 cm long, hairy inside the middle section, the lobes narrow-linear, as long as the tube or rather longer, hairy towards the base.

**C. subaphylla** K. SCHUM. (§ VII/1) (*C. botrys* K. SCHUM., *C. nuda* HUTCH. et E. A. BRUCE). – Somalia. – **Caudex** elongated, roots not thickened; **St.** twining, glabrous; **L.** when present lanceolate, up to 1 cm long; **Cor.** minutely downy outside, sometimes becoming glabrous, tube 15–35 mm long, whitish with red-brown blotches outside, the inside of the constricted section hairy above the venter, otherwise glabrous, widening slightly and gradually towards the mouth, the lobes narrow-linear from a triangular base, sometimes with some spatulate widening towards the tip, 1–1.5 cm long, hairy inside towards the base.

**C. succulenta** E. A. BRUCE (§ II/1). – Kenya; Tanzania; Uganda. – Closely resembles **C. ballyana**; the **Cor.**-tube 2–4 cm long, the lobes triangular, 1–2 cm long. **C. robynsiana** is nearly related to this species, whilst **C. evelynae** can hardly be differentiated from it. The classification and defining of this complex in relation to the closely related species **C. albisepta** and **C. viridis** is still very much in dispute.

**C. variegata** N. E. BR. (§ IX) (*C. tubulifera* DEFL.). – Arabia. – **St.** very succulent, twining or creeping, almost leafless; **Fl.** tinged with pale green or pink, spotted dark red-brown, Cor.-tube 3.5–4.5 cm long, with conical intermediate lobes at the mouth directed obliquely backwards, the lobes 1–1.5 cm long, longer than the intermediate lobes, with a triangular-ovate base above which the almost filiform sections are twisted helically together, to form a tip about 5 mm long.

**C. vignaldiana** A. RICH. (§ VII/1c) (*C. brosima* E. A. BRUCE et BALLY). – Ethiopia; Somalia; Kenya. – **St.** dwarf, erect or sometimes twining, slender; **L.** elliptical to linear and then often semi-terete; **Cor.** glabrous outside, the tube 12–18 mm long, hairy or glabrous inside, the lobes narrow-linear, with a slightly spatulate widening towards the tip, 5–8 mm long, with purple cilia along the margin and inside on the keel.

**C. viridis** CHOUX v. **viridis** (§ II/1) (*C. albisepta* v. *v.* (CHOUX) H. HUBER). – S.Madag. – Closely related to **C. albisepta;** more markedly succulent; **L.** 1–4 cm long, 5–9 mm across, usually lanceolate, with a rather short petiole, almost sessile; **Cor.** green, yellow or whitish, with or without purple-brown markings, the mouth of the tube widening to 1–1.5 cm, the lobes extended from a triangular base into a broad-linear blunt beak, sometimes as long as the tube but usually much shorter. The following are also referable here: **C. helicoides** CHOUX, **C. verrucosa** CHOUX and **C. decaryi** CHOUX, from Madagascar.

**C. —** v. **truncata** H. HUBER (*C. albisepta* JUM. et PERR. p. part.). – Lobes very much shortened, 3–5 mm long, without any beak.

**C. volubilis** N. E. BR. (§ IV/2). – Angola; Congolese Rep. – **St.** slightly succulent, twining, glabrous; **L.** ovate or elliptical, rounded at the base or with cordate indentations; **Infl.** of few Fl., Cor. purple or dark brown, glabrous outside, the tube 1.2–2 cm long, the only slightly swollen venter always glabrous inside, the middle tube-section either extensively hairy or with only a narrow ring of hairs at the base of the narrow section, the tube less often completely glabrous,

the mouth usually only slightly widened, the lobes long-ovate, 6–12 mm long, the margins with purple clavate ciliate hairs. (This species may be a hybrid of **C. distincta** or some related species.)
**C. woodii** SCHLTR. ssp. **woodii** (Pl. 33/3) (§ VII/1c) (*C. barbertonensis* N. E. BR., *C. hastata* N. E. BR., *C. linearis* ssp. *w.* (SCHLTR.) H. HUBER, *C. schoenlandii* N. E. BR., *C. tenuis* N. E. BR.). – From Rhodesia to E.Cape. – **St.** slender, glabrous, twining or creeping or pendulous, often forming spherical Tub. at the nodes; **L.** fleshy, rounded-reniform, cordate, triangular or lanceolate; **Cor.** glabrous outside, the tube 1–2 cm long, dull pink or light green outside, hirsute inside the middle section, the lobes narrow-linear, with a gradual spatulate widening towards the tip, 5–7 mm long, purple-brown, with purple ciliate hairs on the margin and along the keel inside. – The Tub. are useful as grafting stock for **Hoodia**, **Decabelone** and other difficult members of the Asclepiadaceae.
**C. —** ssp. **debilis** (N. E. BR.) H. HUBER (*C. d.* N. E. BR., *C. linearis* ssp. *d.* (N. E. BR.) H. HUBER). – Malawi; Rhodesia. – **L.** linear, semiterete; not forming small Tub. at the St.nodes.
**C. yorubana** SCHLTR. (§ IV/2). – Upper Guinea. – **St.** very twining, slightly succulent, glabrous; **L.** cordate-ovate; **Cor.** glabrous outside, the tube 17–26 mm long, all the lower third or half hairy inside, the mouth with a funnel-form opening, the lobes extended from a triangular base into a short, linear beak 1–1.5 cm long, with fine white hairs inside.
**C. zeyheri** SCHLTR. (§ VIII/1) (*C. patersoniae* N. E. BR.). – Cape: Oudtshoorn, Uitenhage D. – **Roots** fleshy, tufted; **St.** twining, glabrous, usually leafless; **L.** where present are small, ovate or lanceolate, scale-like, up to 1 cm long; **Cor.** with short downy hairs on the outside before the Fl. opens, less often glabrous, the tube 2–5 cm long, completely glabrous inside, the lobes linear, with a slightly spatulate widening towards the tip, 1.5–3 cm long, hirsute inside in the lower half.

**Chamaealoe** BGR. (*Bowiea* HAW. non HARV.). Liliaceae. – South Africa. – Greenhouse, warm. Propagation: seed, division.
**C. africana** (HAW.) BGR. (Pl. 34/1, 2) (*Bowiea a.* HAW., *Aloe bourea* SCHULT., *A. bowieana* S.D.). – S.Cape. – **Ros.** acauline, mat-forming; **L.** numerous, broadly-ovate at the base, narrow-linear to subulate, fleshy, grooved on the upperside, 9–11 cm long, 22 mm across at the base, pale green with distinct longitudinal lines, the underside with small whitish tuberculate blotches, the margins with small white T.; **Infl.** c. 20 cm tall, simple, Fl. 15–20, spreading on short pedicels, cylindrical, 1 cm long, greenish-white.

× **Chamaeleptaloe** ROWL. Liliaceae. – Intergeneric hybrid: **Chamaealoe** × **Leptaloe**. If REYNOLDS is followed, this would be classed as a × **Alchamaloe** since that author includes **Leptaloe** under **Aloe**. Acc. ROWLEY it would be more consistent if both genera were kept distinct, or merged under **Aloe,** in which case the hybrid would be listed under **Aloe.** – Cultivation as for the dwarf Aloes.
× **C. 1.** – **Chamaealoe africana** (HAW.) BGR. × **Leptaloe myriacantha** (HAW.) STAPF. – Intermediate between the parents; **Ros.** of slender, grass green, linear, finely serrulate **L.** to 25 cm long, concave and striped above, convex and spotted below. Hort. HORWOOD.

**Chamaegigas** DTR. Scrophulariaceae. – Southwest Africa, in dried-out water-holes, the dry season lasting from May to December or even longer. – Greenhouse, warm. Propagation: seed, division.
**C. intrepidus** DTR. (Pl. 34/3). – S.W.Afr. – **Caudex** fleshy, minute, scarcely the size of a pin-head; within a few minutes of first watering, subulate **L.** are produced 1 cm long; on the second day floating **L.** appear, these being 1 cm long, oval, in a Ros. of 4, from the centre of which the thin pedicel carries the pink-violet **Fl.** 8 mm ⌀.

**Chiastophyllum** (LEDEB.) STAPF (*Chiastophyllum* LEDEB. as Section of the Genus **Umbilicus** DC.). Crassulaceae. – Caucasus. – Culture as for **Umbilicus**, calcicole.
**C. oppositifolium** (LEDEB.) BGR. (Pl. 34/5) (*Umbilicus o.* LEDEB., *Cotyledon o.* LEDEB., *Sedum o.* (LEDEB.) HAMET). – W.Caucasus. – ♃ up to 30 cm tall; **St.** creeping, rooting, ending above in a small, rather lax, spicate and simple Rac., or the Infl. branches basally to become paniculate; **L.** 6–8, decussate, large, round-ovate, coarsely crenate-dentate, narrowing abruptly into the petiole, upper L. smaller; **Fl.** on short pedicels, older Fl. nodding, campanulate, 4 mm long, yellowish-white, bracts subulate.

**Chorisia** H. B. et K. Bombacaceae. – S.Am. – Greenhouse, warm. Propagation: seed.
**C. speciosa** A. SAINT HIL. (Pl. 34/7). – Brazil. – In age a tall ♄; **St.** clavate, branching, spiny; **L.** long-petiolate, pinnate, tapering-lanceolate, leaflets 5–7, margins entire or dentate; **Fl.** solitary from the L.axils, c. 10 cm ⌀, yellowish or reddish, Pet. hirsute on the underside.
**C. ventricosa** N.M. (Pl. 34/4). – ('Samuru tree'). – Argentine: dry forests. – Very large ♄; **St.** greatly swollen, branching, spiny; **Fl.** woolly outside, solitary or in clusters on old growth. Seldom cultivated.

**Chortolirion** BGR. Liliaceae. – S. and S.W. Afr. – Very dwarf bulbous plants with linear, dentate, slightly fleshy, almost grass-like, deciduous **L.** in a compact **Ros.**, dying back during the resting period to a ± underground bulb formed from the expanded amplexicaul L.bases; **Fl.** as for **Haworthia**. Acc. G. D. ROWLEY not separable from **Haworthia** Sect. V. **Fusiformes**. – Greenhouse, warm. Propagation: seed.
**C. angolense** (BAK.) BGR. (*Haworthia a.* BAK.). – Angola. – Bulb c. 35 mm long; **L.** c. 10, distinctly twin-jointed at the base, narrow-linear,

Plate 34. 1. **Chamaealoe** Infl.; 2. **C. africana** (HAW.) BGR.; 3. **Chamaegigas intrepidus** DTR. (Photo: H. STRAKA); 4. **Chorisia ventricosa** N. M. (Photo: W. RAUH); 5. **Chiastophyllum oppositifolium** (LEDEB.) BGR.; 6. **Coulterella capitata** VASEY et ROSE (Photo: R. MORAN); 7. **Chorisia speciosa** A. SAINT HIL.

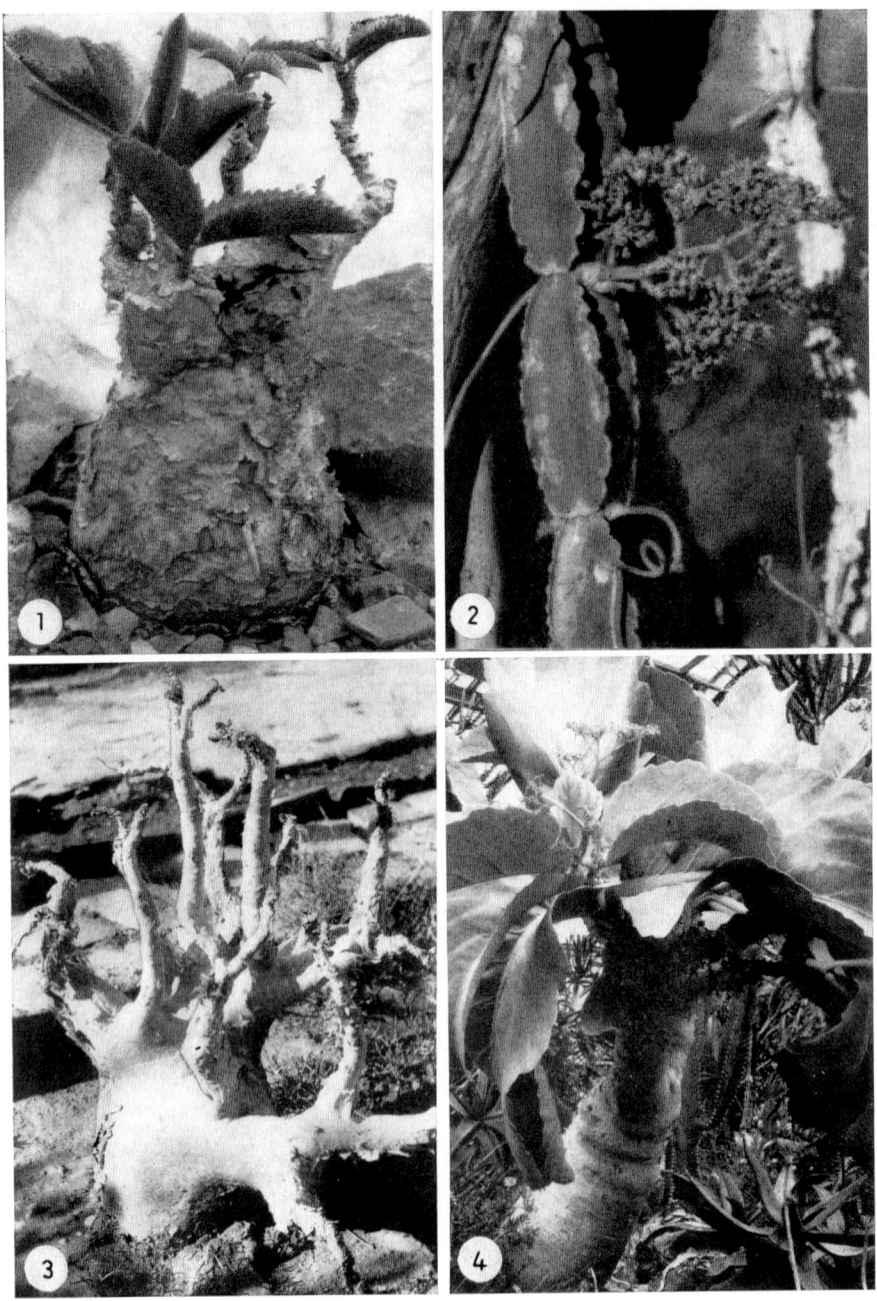

Plate 35. 1. **Cyphostemma bainesii** (Hook. f.) B. Desc.; 2. **Cissus quadrangularis** L. (Photo: P. R. O. Bally); 3. **Cyphostemma currori** (Hook. f.) B. Desc.; 4. **Cyphostemma juttae** (Dtr. et Gilg.) B. Desc. v. **juttae.**

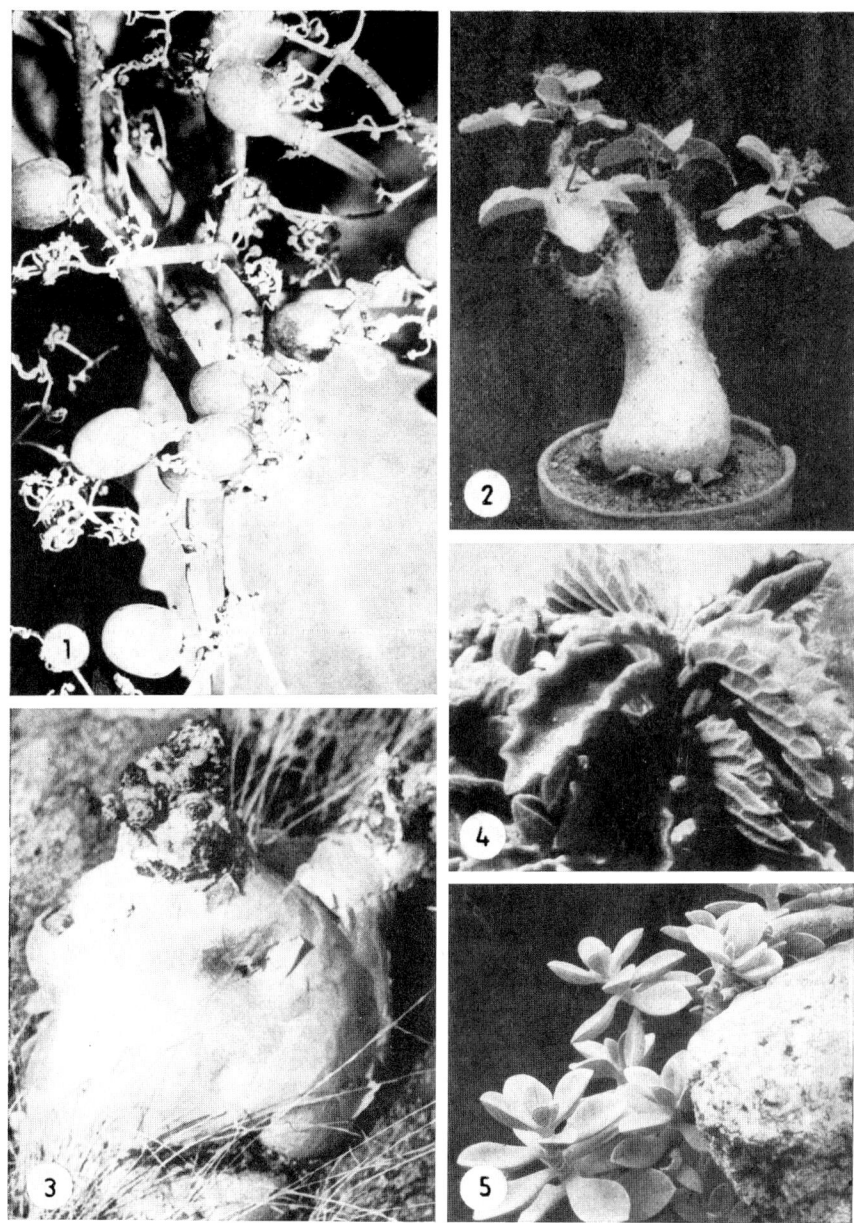

Plate 36. 1. **Cyphostemma** Fruits; 2. **C. betaeformis** CHIOV. (Photo: J. MARNIER-LAPOSTOLLE); 3. **C. seitziana** (GILG et BRANDT) B. DESC. (Photo: W. TRIEBNER); 4. **Coleus spicatus** BENTH.; **C. coerulescens** GÜRKE.

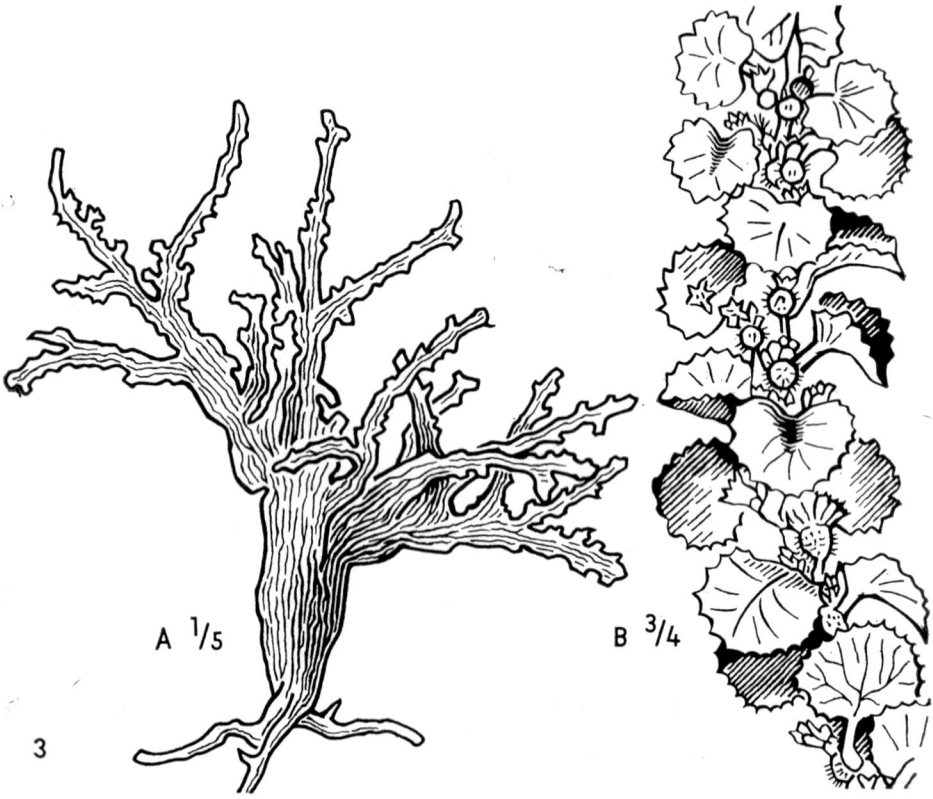

**Plate 37. 1. Commiphora spec.** (dulcis?) (Photo: K. H. HASENBALG); **2. Coreopsis gigantea** (KELLOG) HALL. (Photo: C. BACKEBERG); **3. Corallocarpus glomeruliflorus** SCHWEINF. (Repr.).

Plate 38. 1. **Cotyledon barbeyi** Schweinf. ex Penzig (Photo: P. R. O. Bally); 2. **C. buchholziana** Steph. et Schuldt (Photo: J. Marnier-Lapostolle); 3. **C. decussata** v. **hinrichseniana** Jacobs.; 4. **C. dinteri** Bak. f.; 5. **C. jacobseniana** v. Poelln.

Plate 39. 1. Cotyledon ladismithiensis v. POELLN.; 2. C. luteosquamata v. POELLN.; 3. C. orbiculata L. v. orbiculata; 4. C. orbiculata v. higginsiae JACOBS.; 5. C. orbiculata v. dinteri JACOBS.; 6. C. orbiculata v. oophylla DTR.

Plate 40. 1. Cotyledon paniculata L. f.; 2. C. reticulata Thunbg.; 3. C. schaeferana Dtr. (Photo: J. Marnier-Lapostolle); 4. C. racemosa E. Mey.; 5. C. pearsonii Schoenl.

Plate 41. 1. Cotyledon teretifolia THUNBG.; 2. C. undulata HAW. v. undulata; 3. C. tomentosa HAW. (Graf-Photo); 4. Crassula alstonii MARL. (Photo: E. HAHN).

Plate 42. 1. Crassula arborescens (MILL.) WILLD.; 2. C. sericea SCHOENL. (Photo: H. HERRE); 3. C. columnaris THUNBG.; 4. C. cotyledonis THUNBG.; 5. C. barbata THUNBG. (Photo: DE LAET); 6. C. columella MARL. et SCHOENL. (Photo: as 2).

Plate 43. 1. Crassula arta SCHOENL. (Photo: H. HERRE); 2. C. corallina THUNBG.; 3. C. cultrata L. v. cultrata; 4. C. scabra L. (Photo: E. HAHN); 5. C. cornuta SCHOENL. et BAK. f.; 6. C. hemisphaerica THUNBG.; 7. C. grisea SCHOENL.; 8. C. tomentosa THUNBG.

Plate 44. 1. Crassula ericoides HAW.; 2. C. cv. Jade Necklace (Repr.: Cact. Succ. J. Am. **XXXI**, 1959, 19, Photo: KIMNACH); 3. C. × **justi-corderoyi** JACOBS. et v. POELLN.; 4. C. **lactea** AIT.; 5. C. **lycopodioides** LAM. v. **lycopodioides**; 6. C. **lycopodioides** v. **purpusii** JACOBS. (Photo: J. MARNIER-LAPOSTOLLE).

Plate 45. 1. Crassula marnierana HUBER et JACOBS.; 2. C. orbicularis L.; 3. C. mesembryanthemopsis DTR.; 4. C. nodulosa SCHOENL. (Photo: E. HAHN); 5. C. argentea THUNBG.; 6. C. picturata BOOM.

Plate 46. 1. Crassula perfoliata L.; 2. C. portulacea LAM.; 3. C. perforata THUNBG.; 4. C. × pulverulenta BOOM.

Plate 47. 1. **Crassula quadrangula** (Eckl. et Zeyh.) Endl. ex Walp. (Repr.: Fl. Pl. Afr. **34**, 1961, Pl. 1334a, Drawing: C. Letty); 2. **C. quadrangula** (Repr.: Succulenta, **19**, 1937, 190. Photo: Bosch et Keuning, N. V. Baarn); 3. **C. pyramidalis** Thunbg.; 4. **C. reversisetosa** Bitter; 5. **C. sericea** Schoenl.; 6. **C. schmidtii** Rgl. v. schmidtii; 7. **C. × pulverulenta** Boom; 8. **C. rupestris** Thunbg.

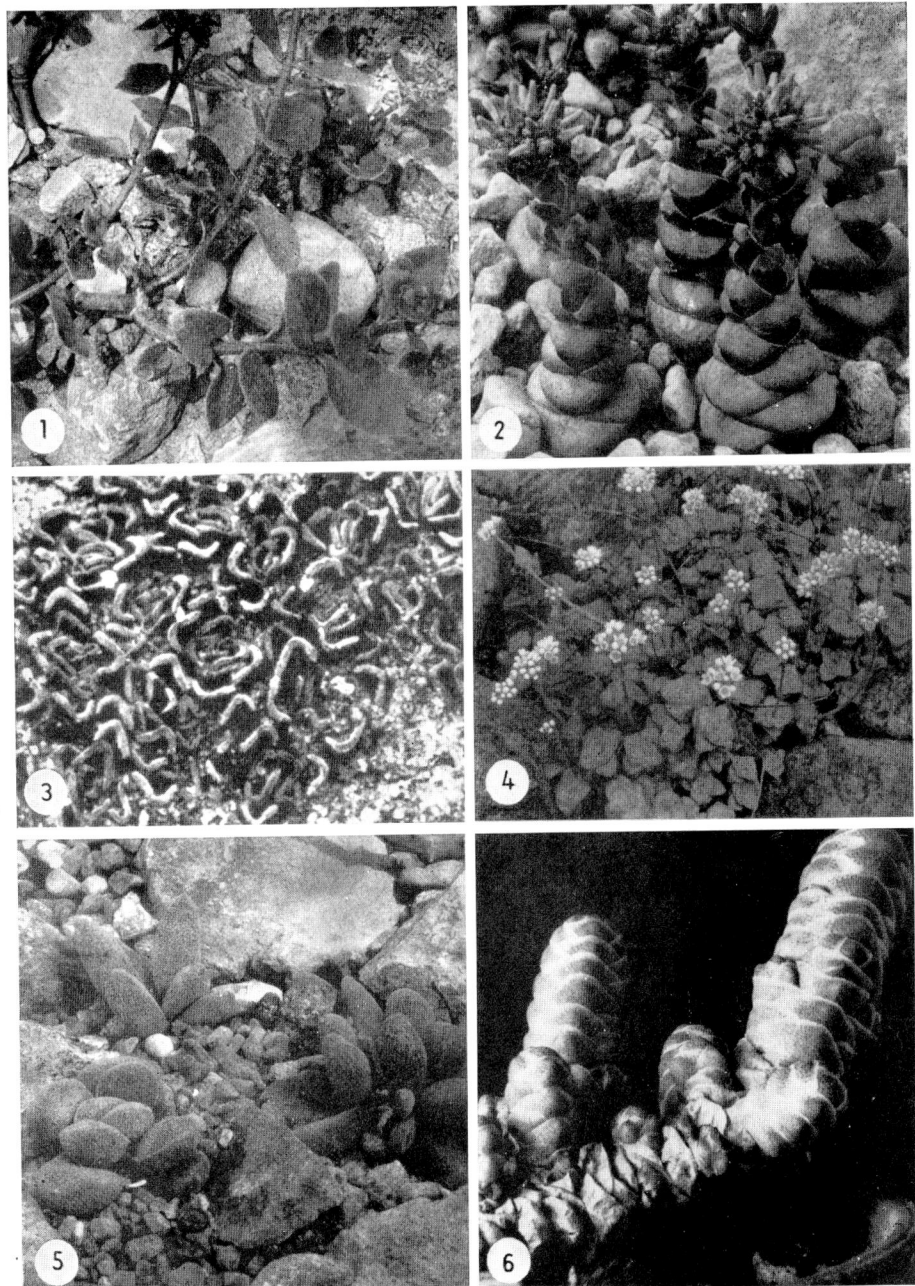

Plate 48. 1. C. sarmentosa Harv.; 2. C. semiorbicularis Eckl. et Zeyh.; 3. C. susannae Rauh et Friedr. (Photo: W. Rauh); 4. C. socialis Schoenl.; 5. C. tecta Thunbg.; 6. C. teres Marl.

Plate 49. 1. Crithmum maritimum L.; 2. Cussonia spicata Thunbg.; 3. Cynanchum perrieri Choux (Photo: J. Marnier-Lapostolle); 4. C. macrolobum Jum. et Perr. (Photo: W. Rauh); 5. C. messeri (Buch.) Jum. et Perr. (Photo: as 3).

5–7 cm long, grooved, margins dentate-hirsute; **Infl.** 20–25 cm long, Fl. 12 mm long.

**C. bergeranum** DTR. – S.W.Afr. – Bulb elongated, pink; **L.** 10–18, ± 20 cm long, linear, 5–6 mm across, blue-green, upperside slightly grooved, underside with oblong white blotches, margins with T. directed downwards; **Infl.** ± 50 cm tall, Fl. c. 20, 2 cm long, brownish-white.

**C. stenophyllum** (BAK.) BGR. (*Haworthia s.* BAK.). – Transvaal. – Bulb ovoid, c. 5 cm long; L. c. 9, jointed transversely above the ovate base, then narrow-linear, 10–12 cm long, hirsute at the base, veined, somewhat grooved, margins horny, dentate; **Infl.** 40–60 cm tall, Fl. 15 mm long, red.

**C. subspicatum** (BAK.) BGR. (*Haworthia s.* BAK.). – Transvaal. – Bulb spherical, 2.5 cm ⌀, pale; **L.** 6–10, narrow-linear, 3–5 cm long, 3 mm across, jointed at the base, margins with minute curved T.; **Infl.** 25–30 cm tall, Fl. c. 10, 12 to 13 mm long, white.

**C. tenuifolium** (ENGL.) BGR. (*Haworthia t.* ENGL.). – S.Botswana. – **L.** in a spiral Ros., linear, 10–20 cm long, 3–4 mm across, grooved, underside with small spots, margins horny, dentate; **Infl.** 20 cm tall, Fl. 15 mm long, white.

**Cissus** DC. Vitaceae (SG. Cissus, SG. Eucissus PLANCH.). – S.W.Afr.; Tanzania; Kenya. – G. containing many species, of which only succulent ones are given here. Climbing plants, with tendrils; sometimes caudiciform; **St.** 4-angled or rounded, fleshy; **L.** simple with entire margins, lobed or less often with 3–5 leaflets or digitate; **Infl.** corymbose, produced opposite a L., Cor. ovate or conical and without any constriction in the middle, stigma undivided; the **Fr.** is a berry, usually with a single seed. – Greenhouse, warm; in winter, keep dry during the resting period. Propagation: seed, cuttings.

**C. hypoleuca** HARV. (*C. quinata* DREGE). – Natal. – **St.** thick, soft, striate, arising from a tuberous caudex, minutely hairy, glandular or possibly glabrous; **L.** long-petiolate, 5-lobed, the leaflets stalked, broad-lanceolate, pointed or tapering, finely serrate, T. wide but shallow, sharp-pointed, L. underside densely felted; **Infl.** felted.

**C. nymphaeifolia** (WELW. ex BAK.) PLANCH. (*Vitis n.* WELW. ex BAK., *C. dinteri* SCHINZ). – S.W.Afr.: Gr. Namaqualand. – **Caudex** napiform, tuberous; **St.** several, fleshy, prostrate, 1 m long, Int. 2–3 cm long, grey; **L.** 10 cm long, almost stiffly erect, roundish-cordate, hirsute on the underside; **Fr.** greenish yellow.

**C. quadrangularis** L. (Pl. 35/2) (*Vitis q.* WALL, *C. cactiformis* GILG., *C. rotundifolia* VAHL, *C. succulenta* BURTT DAVY, *Vitis s.* GALPIN). – Somalia to Transvaal; Kenya; Tanzania; Arabia; peninsular India. – **St.** climbing by tendrils, very succulent, up to 5 cm ⌀, 4-angled, Br. thinner, much constricted at the nodes, Int. 8–10 cm long, angles wavily bent and projecting like wings, sharp, horny, dry; **L.** produced at the nodes, opposite a tendril, somewhat fleshy, cordate, 3-lobed, margins coarsely dentate, T. 2–4 cm long; **Fr.** black.

**C. quinquangularis** CHIOV. – Somalia. – Closely related to **C. quadrangularis** L. and probably not distinct; Int. 5-angled.

**C. rotundifolia** (FORSK.) VAHL (*Saelanthus r.* FORSK., *Vitis r.* DEFL., *V. crassifolia* BAK., *C. c.* (BAK.) PLANCH.). – S.Arabia to Moçambique and Transvaal. – Vigorous grower; **St.** green, 4-angled, angles sharp, corky, young St. and L. with brown spots; **L.** thick and fleshy, broadly ovate or round to reniform, margin almost entire or slightly wavy; **Fl.** green, small; **Fr.** red.

**Coleus** LOUR. Labiatae. – Trop. Afr.; Ethiopia; Somalia; Madag.; E.India; Malayan Archipelago. – Perennial ♄ or sub-♄ with variously-shaped **L.**; species from the E., e.g. **C. arabicus,** have smaller L. with fine felty hairs and a distinct tendency to succulence, species from Madag. are dwarf ♄ with rather small L., tending to fall in the dry season, and conspicuously succulent; **Fl.** in whorls, small, ± blue. Not all **Coleus** species are succulent. – Greenhouse, moderately warm. Propagation: seed, cuttings.

**C. arabicus** BENTH. – Arabia. – ♄, much branching, up to 1 m tall; **Br.** ascending, felted or with tufted hairs; **L.** ovate to ovate-oblong, with a blunt tip, margins crenate, thick, rough and felted or with tufted hairs, whitish to yellowish, 2.5–5 cm long, 2–2.5 cm across; **Fl.** light mauve. – Plant for the cold-house.

**C. coerulescens** GÜRKE (Pl. 36/5) (*C. coerulea* GÜRKE). – E.Ethiopia; Somalia. – Semi-succulent ♃; **St.** becoming prostrate, trailing over one another; **L.** ovate, blunt-tapering, 10–15 mm long, 8–10 mm across, silver-grey, slightly downy-hairy; **Infl.** of 1–5 Fl., Cor. blue with white.

**C. pachyphyllus** GÜRKE. – Ethiopia. – ♃; **St.** thick, fleshy, with minute downy hairs above; **L.** sessile, narrow-obovate to spatulate, margins coarsely dentate, blunt or possibly pointed, thick, fleshy, both surfaces very minutely downy-hairy, 5–10 cm long, 2–4 cm across; **Infl.** a dense spike with broadly ovate, pointed bracts 3–6 mm long, 2–3 mm across, margins with long, stiff ciliate hairs, Fl. violet.

**C. paniculatus** BENTH. ex WALL. – Peninsular India. – **St.** creeping, hispid; **L.** petiolate, broadly ovate, incised-dentate, thick, fleshy, roughly hairy; **Infl.** corymbose.

**C. pentheri** GÜRKE ex ZAHLBR. – Cape: Peddie D. – Semi-succulent plant; **St.** somewhat branching, 20 cm tall, hairy at the base, the tips hispid; **L.** shortly petiolate, cuneate-obovate, narrowing basally to the 3–8 mm long petiole, margins crenate, the tip either blunt or rounded, slightly thickened, hispid, 1–2 cm long, 5–15 mm across; **Infl.** of 4–6 Fl.

**C. spicatus** BENTH. (Pl. 36/4). – Peninsular India; ? Madag. – Mat-forming ♃; **St.** initially short; **L.** semi-succulent, with a petiole ± 1.5 cm long, obovate, pointed, margins notched, up to 6 cm long, 4.5 cm across, finely hirsute, especially on the upperside; **Infl.** c. 10 cm tall, **Cor.** c. 2 cm long, milky blue.

**Combesia** A. RICH. Crassulaceae. – Genus synonymous with **Crassula** L. **Sect.** I/5. This section disregarded here, since none of its species is succulent.

**Commiphora** JACQ. Burseraceae. – Tropical Africa: arid zones; Madagascar. – Trees or shrubs, some species dwarf; **St.** swollen or **Br.** thick, often thorny; **L.** sometimes simple, but more usually of 3 leaflets or imparipinnate, leaflets either sessile or petiolate, margins entire, crenate or serrate; **Fl.** small, in Pan. or clusters. – Greenhouse, warm. Propagation: seed.
**C. cervifolia** VAN DER WALT. – Cape: Lit. Namaqualand. – ♄ up to 2 m tall; **St.** fleshy, branching just above soil-level; **Br.** numerous, short, dense, glandular, bark grey-green to yellow-brown, with dark spots; **L.** with 3 leaflets, up to 1.5 cm long, leaflets small, cultrate, lobed; **Fl.** either in dichotomous cymes produced in the axils, or solitary, 2–3 mm long, yellow-green to brown.
**C. dulcis** ENGL. (Pl. 37/1). – ('Sugar-candy tree'). – S.W.Afr.: Namib. – Small, resiniferous, unpleasant-smelling ♄; **St.** caudiciform, usually tripartite, with Br. extending horizontally; **Br.** rapidly tapering to 5–8 mm thick at the tip, with L.scars; **L.** apparently short-lived; **Infl.** short, bracts lanceolate to cup-shaped, green, very finely hispid, 3-toothed.
**C. gracilifrondosa** DTR. ex VAN DER WALT. – S.W.Afr.; Cape: Lit. Namaqualand. – ♄ up to 3 m tall; **St.** repeatedly branching from just above soil-level, fleshy; **Br.** thin; **L.** with 3 leaflets, linear to cultrate, margins serrate-dentate, sparsely glandular; **Fl.** in axillary, dichotomously branching cymes, or solitary, yellow-green, up to 7 cm long.
**C. madagascariensis** JACQ. – Madag. – Branching ♃, about 1.2 m tall; **St.** round, about 2.5 cm thick; **L.** petiolate, lanceolate, glossy, margins serrate; **Fl.** white.
**C. saxicola** ENGL. – S.W.Afr.: Damaraland. – ♄; **Br.** short thick, initially red, later grey, with L. crowded at the tips; **L.** of 3 leaflets, 5–7 cm long, almost glabrous, leaflets almost orbicular, 10–11 mm ⌀, margins notched.
**C. virgata** ENGL. ('Soris-soris'.) – S.W.Afr. – ♄; **St.** caudiciform, thick, with a thin, papery bark, twigs thin, slender; **L.** with petiole 3–4 mm long, leaflets 5 mm long; **Fl.** sessile.

**Corallocarpus** WELW. Cucurbitaceae. – Afr.; Arabia. – **Caudex** tuberous; **St.** thin and climbing by tendrils, or St. and Br. succulent (**C. corallinus**); **L.** circular or lobed; ♂ **Fl.** in small **Infl.**, ♀ **Fl.** solitary or only few together; **Fr.** ovoid, usually rostrate. – Greenhouse, moderately warm. Propagation: seed.
**C. corallinus** (FENZL) COGN. (*Rhynchocarpa c.* NAUD., *C. fenzlii* HOOK. f.). – N. & S.Afr. – **Br.** stout, succulent, jointed at the nodes; **L.** with robust, fleshy petioles, lamina almost circular, digitate, 3–5-lobed, 8–12 cm long and across, somewhat rough; tendrils slender; **Fr.** 15 to 17 mm long, 8–9 mm thick.
**C. erostris** (SCHWEINF.) HOOK. f. (*Rhynchocarpa e.* SCHWEINF.). – N.Afr.: deserts. – **St.** robust, little branching, fleshy, furrowed-angular, slightly rough, jointed at the nodes; **L.** 2–3 cm long, with robust petioles, with dense tufted hairs, 2–3 cm long, lamina 5–8 cm long and across, rather thick, broadly ovate-cordate, round to 5-angled to slightly 5-lobed, margins slightly sinuate-dentate, hairy; tendrils robust; **Fl.** greenish-yellow; **Fr.** ovoid, yellow, 3.5 mm long, 3 mm thick.
**C. glomeruliflorus** SCHWEINF. (Pl. 37/3) (*Rhynchocarpa courbonii* DEFL., *Phialocarpus g.* DEFL.). – S.Arabia: Yemen. – Erect ♄, about 1 m tall; **St.** fleshy, short, the visible part being clavate, thick, jointed, branching from the apex; **Br.** stiff, furrowed, thickened basally, with dense wool in the L.axils; **L.** on petioles 1–2 cm long, almost circular, margins sinuate-dentate, somewhat wavy, lamina 1–2 cm long, 1.5–2 cm across, velvety-rough; **Fl.** unisexual; **Fr.** ovoid, furrowed, golden-yellow, 12–15 mm long, 8 to 10 mm thick, with 2–4 seeds.
**C. tenuissimus** BUSC. et MUSCHL. – Rhodesia. – Slender ♄, succulent, bluish, distinctly striate, jointed at the nodes; **L.** membranous, ovate to almost circular, at first with dense bristly hairs, digitate, 5-lobed; tendrils slender; **Fr.** with an almost filiform beak, 15–17 mm long, 8–9 mm thick.
**C. welwitschii** (NAUD.) HOOK. f. v. **welwitschii** (*Rhynchocarpa w.* NAUD.). – Zaire; Angola; S.W.Afr. – ♄ with fusiform or napiform caudex; **Br.** slender, fleshy, almost angular, clambering, up to 2 m tall, jointed at the nodes; **L.** 4–7 cm long and almost the same across, with petioles 2–4 cm long, lamina ovate, stiff, with short dense tufted hairs, palmate, 5-lobed; **Fl.** pale yellow; **Fr.** oblong, 17–20 mm long, 9–12 mm thick, scarlet, with 5–8 seeds.
**C. —** v. **subintegrifolius** COGN. – Cape. – **L.** triangular, almost sagittate, often slightly lobed.

**Coreopsis** L. Compositae. – USA: California. – Greenhouse, warm. Propagation: seeds, cuttings.
**C. gigantea** (KELLOG) HALL (Pl. 37/2) (*Leptosyne g.* KELLOG). – USA: Calif.: coast and Channel Is. – ♃; **St.** fleshy, robust, 60–200 cm tall, with clustered L. and corymbose Infl. at the tip; **L.** tripartite pinnate, each section filiform.

**Cotyledon** L. Crassulaceae. – S. and S.W.Afr.; Ethiopia; S.Arabia. – Succulent ♃, often compact and forming clumps or rarely perennials; **L.** decussate to alternate or spirally crowded, sessile or stalked, thick-fleshy; **Infl.** terminal, ± tall, branched, sometimes thorny and persisting; **Fl.** mostly pendent, rarely semi-erect, 5-partite, Cal. shorter than the tube of the Cor., tube ± wide-bellied to campanulate, cylindrical, faintly 5-angled, faintly contracted in the throat, lobes triangular to linear, reflexed or rolled backwards, turned spirally in the bud, yellowish to orange-coloured or red, often faint violet and finely puberulous. – Greenhouse, moderately warm. Propagation: seeds, cuttings.

## Division of the Genus Cotyledon into Sections according to Schoenland emend. v. Poellnitz et Jacobsen

Type species: C. orbiculata.

1. **Caulescent plants;** L. alternate or decussate.

    **Sect. I. Alternifoliae** JACOBS. (§ *Paniculatae* SCHOENL. p.part.). – **L.** alternate, crowded.

    **SSect. 1. Reticulateae** (SCHOENL.) JACOBS. (§ *Paniculatae* SCHOENL. Group *Reticulatae* SCHOENL). – **Infl.** persisting, thorny; **Fl.** nutant or pendent, ± bell-shaped. – Type species: C. reticulata. – Species: C. reticulata.

    **SSect. 2. Paniculateae** (SCHOENL.) JACOBS. (§ *Paniculatae* SCHOENL. Group *Paniculatae* SCHOENL.). – ♄ or small ♄ with thick, fleshy, rarely branched **St.**, bare in summer; **Fl.** erect or suberect. – Type species: C. paniculata. – Species: C. cacalioides, dinteri, eckloniana, fergusoniae, hirtifolia, paniculata, pearsonii, rubrovenosa, swartbergensis, wallichii.

    **SSect. 3. Ventricoseae** (SCHOENL.) JACOBS. (§ *Paniculatae* SCHOENL., Group *Ventricosae* SCHOENL.). – Small ♄, **St.** thick, fleshy; **L.** ± linear, mostly absent in the resting season; **Fl.** mostly suberect, tube ± conspicuously bellied. – Type species: C. ventricosa. – Species: C. chloroleuca, faucium, striata, ventricosa.

    **SSect. 4. Grandifloreae** (SCHOENL.) JACOBS. (§ *Paniculatae* SCHOENL. Group *Grandiflorae* SCHOENL.). – ♄ with thick and fleshy **St.**; **L.** absent in the flowering season; **Fl.** erect, not bellied. – Type species: C. grandiflora. – Species: C. grandiflora, racemosa.

    **Sect. II. Cotyledon** (*Decussatae* JACOBS.). – **L.** decussate.

    **SSect. 1. Velutineae** (SCHOENL.) JACOBS. (§ *Paniculatae* SCHOENL. Group *Velutinae* SCHOENL.). – ♄ 1–2 m high, much branched; **L.** large, flat; **Infl.** many branched, Fl. reddish, stamens much broadened at the base of the free part and without hairs. – Type species: C. velutina. – Species: C. mollis, velutina.

    **SSect. 2. Ramosissimeae** (SCHOENL.) JACOBS. (§ *Paniculatae* SCHOENL., Group *Ramosissimae* SCHOENL.). – Much branched ♄ about 1 m high, **St.** woody; **L.** rather thick, short, rather flat, obovate-cuneate, distant; **Fl.** solitary or few, reddish, stamens with hairs at the base of the free part. – Type species: C. salmiana (*C. ramosissima*). – Species: C. salmiana.

    **SSect. 3. Orbiculateae** (SCHOENL.) JACOBS. (§ *Paniculatae* SCHOENL. Group *Orbiculatae* SCHOENL.). – ± branched ♄, **St.** mostly procumbent, thick, branches ascending; **L.** large, bare; **Fl.** reddish or yellow, stamens not broadened at the base of the free part, here with hairs. – Type species: C. orbiculata. – Species: C. coruscans, decussata, galpinii, macrantha, ? obermeyerana, orbiculata, sturmiana, tricuspidata, undulata.

    **SSect. 4. Pillansieae** (SCHOENL.) JACOBS. (§ *Paniculatae* SCHOENL. Group *Pillansii* SCHOENL.). – Mostly stout ♄ as in Subsect. 3; tube of the **Fl.** usually shorter than the lobes and often other parts of the plants papillate, stamens also with hairs at the base of the free parts. – Type species: C. pillansii. – Species: C. barbeyi, deasii, pillansii, teretifolia, wickensii.

    **SSect. 5. Graciles** (SCHOENL.) JACOBS. (§ *Paniculatae* SCHOENL. Group *Graciles* SCHOENL.). – Small semi-♄, **St.** and Br. proportionally thin; **Infl.** few-flowered, mostly with a long, thin scape. – Type species: C. gracilis. – Species: C. glandulosa, gracilis, jacobseniana, muirii, papillaris.

    **SSect. 6. Tomentoseae** (SCHOENL.) JACOBS. (§ *Paniculatae* SCHOENL. Group *Tomentosae* SCHOENL.). – Branched semi-♄; **L.** hairy, thick, ovate-elongate; **Fl.** hairy outside. – Type species: C. tomentosa. – Species: C. ladismithensis, tomentosa.

    **SSect. 7. Zuluenses** (v. POELLN.) JACOBS. (§ *Paniculatae* SCHOENL. Group *Zuluensis* v. POELLN). – Appearance as Subsect. 3. **L.** finely felty-haired. – Type species: C. zuluensis. – Species: C. zuluensis.

**SSect. 8. Pygmaeae** JACOBS. − Dwarf, highly succulent **shrubs; Infl.** becoming thorny and persisting or Fl. stalks becoming thorny. − Type species: C. pygmaea. − Species: C. buchholziana, fragilis, luteosquamata, pygmaea, schaeferana, sinus-alexandrii. Acc. H.-CHR. FRIEDRICH the species of this SSect. belong to **SSect. I.: Reticulateae.** They all have L. easily deciduous and soon dropping, never opposite, partly persisting thorny Infl. and erect Fl.

## 2. Acauline plants.

**Sect. III. Rhizomatae** JACOBS. − **L.** developed from the rootstock. − Type species: C. singularis. − No further species.

Omitted from this classification are the following species: C. leucothrix, simulans, (both closely related to the G. **Adromischus**).

**C. barbeyi** SCHWEINF. ex PENZIG (Pl. 38/1) (§ II/4) (*Kalanchoe alternans* DEFL.). − Ethiopia; Arabia: Yemen; Somalia; Kenya. − Freely branching ♄; **L.** obcuneate or obovate, apiculate or acuminate, glabrous, fleshy, flat, deep grey-green, 6−14 cm long, 7.5 cm across; **Infl.** 60 cm tall, paniculate, Fl. numerous, nodding, with minute glandular hairs outside, 1−3 cm long, Cal. green, Cor. yellowish green, tinged with red or orange-coloured.

**C. buchholziana** STEPHAN et SCHULDT (Pl. 38/2) (§ II/8). − Cape: Lit. Namaqualand. − Low ♄, branching freely; **St.** terete, 6−12 mm ⌀, greygreen, at first with reddish scale-L., growingpoints spirally arranged round the Br., with conical depressions, lateral shoots dichotomous, often leafless; **L.** when present 1−4, linearlanceolate, 10−15 mm long, 3−5 mm across, upperside flat to grooved, subterete, fleshy; **Infl.** 10−12 cm tall, with a few almost thorny protuberances towards the base, Fl. solitary, 10 to 12 mm long, red-brown outside, lobes with involute margins, with several colourless papillae on the upperside. A cristate cultivar is grown also.

**C. cacalioides** L. f. (§ I/2). − ('Nenta'.) − Cape: Lit. Karroo, Gr. Karroo. − **St.** branching from the base, seldom exceeding 50 cm in height, slightly woody; **Br.** up to 20 cm long, sometimes longer, covered like the St. with spirally arranged, projecting L.bases; **L.** very narrow, terete, pointed, fleshy, crowded at the Br.tips, 5−10 cm long, grey-green; **Infl.** 30−60 cm tall, Cal. and Cor. with glandular hairs outside, Cor. 20−25 mm long, yellow, becoming rather darker. The plant is poisonous to goats.

**C. chloroleuca** DTR. ex FRIEDR. (§ I/3). − S.W.Afr.: Lüderitz South D. − Dwarf ♄, branching pseudo-dichotomously; **Br.** 2−6 cm long, 1−2 cm ⌀, initially rough, with tufts of glandular hair, bark pale yellowish; **L.** crowded, fleshy, broad-lanceolate, blunt to somewhat pointed, sessile, pale green, with glandular hairs, 2−5 cm long, 1−2.5 cm across, 2−3 mm thick; **Infl.** 5−10 cm tall, Fl. erect, with pedicels 1 to 1.2 cm long, roughly hairy, Cor. campanulate, pale green, hirsute outside.

**C. coruscans** HAW. (§ II/3) (*C. whitei* SCHOENL. et BAK. f., *Echeveria c.* v. ROED.). − Cape, Transvaal. − Erect ♄, less often prostrate, with a short St.; **L.** ± densely crowded, oblong, somewhat cuneate towards the tip, otherwise oblong-lanceolate, rather blunt to pointed, thick, very convex on the underside, concave on the upperside, the margins very thick and ± involute, initially very pruinose, 4−10 cm long, 8−20 mm across, 6−9 mm thick, glabrous; **Infl.** 20−45 cm tall, pruinose, reddish, Cor. slightly 5-angled, almost cylindrical, 2.5−3.5 cm long, orange-red or light yellow, green at the base.

**C. deasii** SCHOENL. (§ II/4). − Cape: Oudtshoorn D. − ♄; **St.** thick, prostrate, branching from the base; **L.** few, obovate-cuneate or oblanceolate-cuneate, pointed, ± amplexicaul, thick, almost flat or with the underside rounded, grey-green, 6−7 cm long, 1.5−2.5 cm across, glabrous; **Infl.** 30 cm tall, somewhat hirsute above, Fl. pendulous, Cor. with glandular hairs externally, yellowish red.

**C. decussata** SIMS v. decussata (§ II/3) (*C. flanaganii* et v. *karrooensis* SCHOENL. et BAK. f., *C. ungulata* LAM.). − Cape: Gr. Karroo to Lit. Namaqualand and S.W.Afr.: Gr. Namaqualand. − Erect or prostrate ♄ to 0.75 m tall; **L.** subterete or semi-terete, occasionally almost flat, linear to lanceolate to obovate-cuneate, the upperside sometimes concave, blunt to rather pointed, glabrous, pale to dark green, pruinose, ± red-tipped, 3.5−12 cm long, up to 1 cm across, 4−10 mm thick; **Infl.** 20−45 cm tall, Fl. numerous, ± nodding, Cor. 1−3 cm long, yellowish with 5 red longitudinal stripes.

**C. —** v. **dielsii** (SCHLTR. ms.) v. POELLN. (*C. dielsii* SCHLTR.). − **Fl.** smaller, 7 mm long, yellowish.

**C. —** v. **flavida** (FOURC.) v. POELLN. (*C. flavida* FOURC., *C. fourcadei* SCHOENL.). − **L.** linear, slightly concave, underside very rounded, glossy, dark green; **Fl.** about 3 cm long, light yellow.

**C. —** v. **hinrichseniana** JACOBS. (Pl. 38/3). − Lit. Namaqualand. − ♄ 20−30 cm tall; **L.** ovatecuneate, with a small red apiculus, sessile, up to 2 cm long, 12 mm across, narrowing teretely to the base, 5−6 mm thick, both sides convex, whitish-pruinose, margins red in the upper part.

**C. —** v. **rubra** v. POELLN. − Cape. − **L.** little flattened on the upperside, 5.5−7 cm long, 14 mm across, 11−12 mm thick; **Fl.** 2.5−3 cm long, red.

**C. dinteri** BAK. f. (Pl. 38/4) (descr. emend. v. POELLN.) (§ I/2). − S.W.Afr.: Gr. Namaqualand. − ♄ to 50 cm tall; **St.** short, brown; **Br.** 2.5−3 cm thick, the prominent and obliquely truncate L. bases (podaria) being arranged in a spiral; **L.** linear-lanceolate, rather pointed, terete, glabrous, 2.5−10 cm long, green or greygreen, falling in the flowering period; **Infl.** 30−40 cm tall, Fl. nodding, Cor. green or yellow-green, tube about 7 mm long, with small tuberculate papillae outside, lobes 4–5 mm long, hirsute on the upperside.

**C. eckloniana** Harv. (§ I/2) (*C. cacalioides* Eckl. et Zeyh., *C. maximiliana* R. Schltr.). – Cape: Lit. Namaqualand. – **St.** long, somewhat branching, with a light grey or brown cortex, the L.bases (podaria) arranged in a spiral, obliquely truncate and up to 15 mm long; **L.** falling in the flowering period, cylindrical, green, 5–6 cm long; **Infl.** 30–60 cm tall, Fl. numerous, nodding, glabrous, Cor. (?) reddish, tube 7 to 9 mm long. – Poisonous to cattle!

**C. faucium** v. Poelln. (§ I/3). – Cape: Sutherland. – Dwarf ♄; **St.** branching, up to 7.5 cm tall; **Br.** 1.5 cm thick, bearing L.scars; **L.** falling in the flowering period, obovate or obovate-lanceolate, the upperside more convex, covered with long white hairs; **Infl.** hirsute, Fl. erect, 14–16 mm long, glossy, green.

**C. fergusoniae** L. Bol. (§ I/2). – Cape: Lit. Karroo. – **Br.** and twigs with prominent L.scars, often 4–7 cm thick; **L.** narrowing towards the tip, rounded on the underside, 7–10 cm long, up to 1 cm thick; **Fl.** often pendulous, 3–3.5 cm long. (Acc. L. Bolus possibly a hybrid: **C. cacalioides × C. paniculata.**)

**C. fragilis** R. A. Dyer (§ II/8). – Cape: Strandfontein. – Small succulent plants 3–3.5 cm tall; **rootstock** tuberous, 1.5–6 cm long, 1–3 cm thick; **Br.** simple, fragile, 3–8 mm thick, glabrous, grey-green, the tip black pointed; **L.** some few at the ends of the Br., subcylindrical, 1.5–3 cm long, 3–6 mm ⌀, glabrous, dropping; **Sc.** of the Infl. 3–6 cm long, slender, subpersistent, 1–8-flowering, Fl.crowded, with stalks 3–8 mm long, Cal. c. 2.5 mm long, Cor. ± 1.5 cm long, outside glabrous, green-yellow, tube inside pilose, lobes c. 5 mm long, recurved.

**C. galpinii** Schoenl. et Bak. (§ II/3). – Cape: Queenstown. – Probably a natural hybrid: **C. decussata × C. orbiculata.** – ♄; **St.** prostrate; **L.** obovate-oblong or broad-oblanceolate, pointed or rather so, 5–7 cm long, 2–2.3 cm across, underside convex, pruinose, margins red above; **Fl.** 2 cm long, ± red and light yellow.

**C. glandulosa** N. E. Br. (§ II/5). – Zambia (?). – Glandular-hairy; **St.** 4–6 cm long, 2–3 mm thick, brownish, branching; **L.** 2–3 pairs at the Br.tips, 1–1.5 cm long, 5–7 mm across, 5–6 mm thick, pointed or blunt, grey-whitish, later olive-brown; **Infl.** 6 cm long, tube dirty green with red lines, 5.5 mm long, lobes 10 mm long, dirty green.

**C. gracilis** Haw. (§ II/5). – Cape: Lit. Namaqualand. – Small; **L.** linear, fleshy, 2–3 cm long, about 6 mm across; **Infl.** about 30 cm long, often prostrate, white-pruinose, with numerous bracts which are hirsute and leaf-like; Fl. red, 1.5 cm long. (Known only from the literature.)

**C. grandiflora** Burm. f. (§ I/4) (*C. tuberculosa* Lam., *C. curviflora* Sims, *C. purpurea* Haw. ? *C. interjecta* Haw.). – S.Cape: Cape, Swellendam D. – **St.** short, scarcely branching, nodose because of the persistent, light brown L.bases, 1–1.5 cm thick; **L.** linear to linear-lanceolate, sub-terete, pointed or rather blunt, (4–)7(–8) cm long, c. 12 mm across, shrivelling before the flowering period; **Infl.** 20–50 cm tall, Fl. erect, Ped., Cal. and Cor. hirsute outside, tube swollen, usually slightly bent, c. 2 cm long, Cor. orange-red, lobes 1.5–2 cm long.

**C. hirtifolira** Barker (§ I/2). – Cape: Lit. Namaqualand. – Up to 70 cm tall; **Br.** fleshy, 2.2 cm ⌀, glabrous, closely covered by hardened L.bases (podaria) which are prominent and 7 mm long; **L.** 5–8 cm long, 1.2–3 cm across, varying in appearance, young L. often semi-terete at the base, older L. oblanceolate or obovate, narrowing towards the base, the upperside grooved towards the tip, both sides hispid; **Infl.** terminal, with a few Br., glandular-hispid, up to 12 cm long, with several reduced L., Fl. pendulous, with pedicels 7 mm long, Sep. pointed, glandular-hispid, Cor. tube 14 mm long, 9 mm ⌀ above, yellowish-green outside, inside light green, lobes recurved.

**C. jacobseniana** v. Poelln. (Pl. 38/5) (§ II/5) (*C. gracilis* Bgr.). – Cape: Lit. Namaqualand. – Small, mat-forming semi-♄; **St.** prostrate, branching, up to 20 cm long, about 3 mm thick; **L.** crowded, 2–3 cm long, 5–7 mm thick, narrowing towards the tip and the base, terete, green, ± pruinose; **Infl.** erect, up to 13 cm tall, bracts glandular-hairy, Fl. nodding, tube bluntly 5-angled, 6–9 mm long, with glandular hairs outside, greenish, the angles red-brown, light green inside, lobes with reddish lines on both surfaces.

**C. ladismithensis** v. Poelln. (Pl. 39/1) (§ II/6) (*C. heterophylla* Schoenl.). – Cape: Ladismith D., Lit. Karroo. – Bushy, much branching, c. 30 cm tall; **Br.** woody below, leafless, fleshy above, softly hairy; **L.** in 4 pairs, oblong, narrowing teretely towards the base, thick, fleshy, softly hairy, the underside very convex, lower L. blunt or glandular-tipped, middle L. with 3 blunt, glandular T., upper L. with margins entire, about 5 cm long, 15–20 mm across, 10–12 mm thick; **Infl.** fleshy, hirsute, c. 9 cm long, Fl. c. 10, the Cor. being short-hairy on the outside.

**C. leucothryx** (C. A. Sm.) Fourc. (§ I/?) (*Adromischus l.* C. A. Sm.). – Cape: Ladismith D. – **Caudex** thickened; **St.** 4 cm tall, 3 cm ⌀, simple or branching a little from the base, glabrous; **L.** crowded at the St.tips, linear-oblong to oblanceolate-elliptical, narrowing towards both ends, up to 3.5 cm long, 5 mm across, semi-terete towards the base, deeply grooved on the upperside, very fleshy, deep green, all surfaces densely white-hirsute; **Infl.** with few Fl.

**C. luteosquamata** v. Poelln. (Pl. 39/2) (§ II/8). – Cape: Bushmanland. – Semi-♄ 8 cm tall, branching, bearing many desiccated Infl.; **L.** crowded at the St.tips, the upperside flat and furrowed, otherwise roundish, broad-oblong, 3–3.5 cm long, 3–4 mm across, green, with a grey coating and red-brown longitudinal stripes, somewhat tuberculate; **Infl.** 7 cm long, persistent, Fl. few, erect, 12–14 mm long, greenish.

**C. macrantha** L. v. **macrantha** (§ II/5). – Cape. – ♄ up to 1 m tall, robust, branching; **Br.** with light grey cortex and L.scars, 2–2.5 cm thick; **L.** thick, with margins entire, broadobovate, less frequently oblong, narrowed and shortly terete, rounded at the tip with an apiculus, fresh green, at first slightly pruinose, margins red above, up to 10 cm long, up to 8 cm across; **Infl.** up to 25 cm long, with many nodding Fl., Cor. vivid red, only the inside being yellow to yellowish-green.

**C.** — v. **virescens** (SCHOENL. et BAK. f.) v. POELLN. (*C. v.* SCHOENL. et BAK. f., ? *C. viridis* HAW.). – Cape: near Grahamstown. – **Br.** very fleshy, thick; **L.** often larger, green.

C. mollis DTR. is the soft-haired immature state of **C. paniculata** L.f.

**C. mollis** SCHOENL. (§ II/1). – Cape. – ♄ up to 1 m tall; young **Br.** hirsute; **L.** obovatecuneate, narrowing towards the base, almost flat, blunt or somewhat pointed at the tip, margins slightly wavy, c. 8 cm long, 2–3.5 cm across; **Infl.** hirsute, c. 35 cm long, Fl. pendulous, c. 2.5 cm long, Cor. with hairs outside.

**C. muirii** SCHOENL. (§ II/5). – Cape: Karroo. – **St.** thin, creeping, c. 4 mm thick, branching, the tip ascending; **twigs** c. 8 cm long, 5 mm thick, with fine glandular hairs above; **L.** linearoblong, narrowing teretely towards the base, usually somewhat oblique, grey-green, glabrous, margins red, with a pointed tip, 3–3.5 cm long, 7 mm across; **Infl.** 12–20 cm long, Fl. 3–5, nodding, Cor. pruinose, tube 10 mm long, 5-angled.

**C. obermeyerana** v. POELLN. (§ II/3?). – Cape: Queenstown Mts. – Glabrous ♃; **L.** obovate to oblong, c. 10 cm long, c. 5 cm across, pruinose; **Infl.** c. 30 cm tall, with c. 15 pendulous red Fl., tube 25 mm long, lobes 15 mm long.

**C. orbiculata** L. v. **orbiculata** (Pl. 39/3) (§ II/3) (*C. elata* HAW., *C. oblonga* HAW., *C. ovata* HAW., *C. ramosa* HAW.). – Cape: Natal; Transvaal; S.W.Afr.; ? Angola. – ♄ up to 1.5 m tall; **St.** usually thick, freely branching, prostrate or erect; **L.** conspicuously white or grey pruinosewaxy, margins usually smooth, less often wavy, upperside ± flat to convex, but sometimes concave, underside ± convex, broad-oblong or obovate-cuneate, only occasionally linear, blunt or almost pointed, 3.5–14 cm long, up to c. 6 cm across; **Infl.** up to 70 cm tall, Fl. few, pendulous, red or yellowish-red, Cor. up to 30 mm long.

**C.** — v. **ausana** (DTR.) JACOBS. (*C. ausana* DTR.). – S.W.Afr.: Gr. Namaqualand. – Up to 40 cm tall; **L.** with an attractive red margin; **Fl.** orange-red.

**C.** — v. **dinteri** JACOBS. (Pl. 39/5). – S.W.Afr.: Klinghardt area. – Differentiated from v. **oophylla** by the conspicuously larger L. which are c. 45 mm long, c. 20 mm across and almost as thick, grey-green, little pruinose, with crescent-shaped brown markings near the noticeably rounded tip.

**C.** — v. **engleri** (DTR. et BGR.) DTR. (*C. engleri* DTR. et BGR., *C. o.* v. *viridis* DTR.). – S.W.Afr.: Gr. Namaqualand. – **L.** sea-green and faintly pruinose.

**C.** — v. **higginsiae** JACOBS. (Pl. 39/4). – **L.** c. 35 mm long, 15 mm across in the upper part, 10 mm thick, apiculate, white-mealy pruinose, margins rounded and brown-red. Intermediate to v. **dinteri**.

**C.** — v. **oophylla** DTR. (Pl. 39/6). – S.W.Afr.: Gr. Namaqualand. – **St.** very short, up to 2 cm thick; **L.** 8–12, long-ovate, about 12–20 mm long, 12 mm across, 8–10 mm thick, ± curved, dark purple to grey-green, thickly bluish-white pruinose, with lunate markings at the rounded tip; **Fl.** shining orange-red.

**C. paniculata** L. f. (Pl. 40/1) (§ I/2) (*C. spuria* L., *C. fascicularis* AIT., *C. tardiflora* BONPL., *C. mollis* DTR.). – Cape: Karroo, Cape D.; S.W.Afr. – ('Botterboom'.) – Immature plants always softly hairy (*C. mollis*); **roots** slightly thickened; **St.** 30 cm to over 2 m in height, exceedingly soft, the thickness of a human torso, with a papery yellow-brown to yellowish bark; **Br.** few, short, thick, tuberculate; **L.** eventually falling, fleshy, numerous, initially hairy, obovate to lanceolate, blunt or pointed, margins entire, 5–11 cm long, 2.5–4 cm across, light grey-green, flat or with the upperside somewhat concave and the underside somewhat convex; **Infl.** 30–60 cm tall, Fl. nodding, Cor. hirsute outside, dark red to red-brown with yellow-green lines and stripes, lobes dark red.

**C. papillaris** L. f. v. **papillaris** (§ II/5) (? *C. cuneata* THUNBG., *C. meyeri* HARV., *C. angulata* v. *foliis-minoribus* E. MEY., *C. cuneata* E. MEY., *Adromischus c.* LEM.). – Cape: Karroo to Lit. Namaqualand. – **St.** prostrate, the tip ascending, bark brown or grey-brown, c. 15 cm long, 2–4 mm thick, branching; **Br.** often somewhat bent, initially papillose; **L.** often almost rosulate, dark green, little pruinose to powdery mealy-white, obovate-oblong or obovate-cuneate, pointed to bluntish, often almost rounded, often with a red apiculus, margins smooth, red above, the underside more convex than the upperside, 1.5–4.5 cm long, 6–14 mm across, 3–6 mm thick; **Infl.** up to 25 cm tall, Fl. pendulous, Cor. finely papillose outside, reddish to greenish yellow with 5 longitudinal stripes, lobes red to dark brown-red.

**C.** — v. **glutinosa** (SCHOENL.) v. POELLN. (*C. g.* SCHOENL., *C. p.* SCHOENL.). – **L.** densely hairy; **Infl.** Cal. and Cor. all with sticky hairs outside.

**C.** — v. **robusta** SCHOENL. et BAK. (*C. pseudogracilis* v. POELLN., *C. gracilis* HARV.). – Stouter, mat-forming; **L.** 2.5–6 cm long, 6 to 12 mm across, 5–9 mm thick.

**C.** — v. **subundulata** v. POELLN. – **L.** almost spatulate, 2–3 cm long, 7–10 mm across, 4 to 6 mm thick, margins in the upper third sharp and wavy, red-brown.

**C. pearsonii** SCHOENL. (Pl. 40/5) (§ I/2). – Cape: Lit. Namaqualand; S.W.Afr.: Gr. Namaqualand. – Freely branching ♄; **St.** almost woody, c. 2 cm thick; **Br.** c. 5 cm long, 1 cm thick, tuberculate towards the tip; **L.** very small, linear, deciduous; **Infl.** 7 cm long, becoming woody and thorny and persisting for several years, glandular-hairy, with 4–5 small and Sp.like bracts, Fl. few, pedicels 12–20 mm long, glandular-hairy, Cor. glandular-hairy outside, whitish with red stripes, tube funnel-shaped, 1 cm long, lobes 7 mm long.

**C. pillansii** SCHOENL. (§ II/4) (*C. cuneata* HARV.). – Cape: Lit. Namaqualand to the Karroo. – **St.** c. 35 cm tall, branching from the base, robust, 12–15 mm thick; **L.** obovate-oblong, cuneate towards the base, blunt or short-pointed, coarsely hairy to almost glabrous, green or grey-green, margins horny and red, often wavy, upperside ± concave, underside somewhat convex, 12 cm long, 10 cm across, the upper L. shorter and narrower; **Infl.** 30 to 60 cm tall, glandular hairy above, Fl. pendulous, Cor. glandular-hairy outside, yellowish-green.

**C. pygmaea** BARKER (§ II/8). – Cape: Lit. Namaqualand. – Extremely succulent, 4–5 cm tall; **St.** caudiciform with prostrate or erect Br., almost undivided, 4–5 mm ⌀, bark white; **L.** c. 2 pairs, sessile, ovate to round, narrowing below, upperside flat, underside convex, 8–10 mm long, 6–8 mm across, 3–4 mm thick, green, completely covered with whitish glossy papillae; **Fl.** small, inconspicuous, white.

**C. racemosa** E. MEY. descr. emend. N. E. BROWN (Pl. 40/4) (§ I/4). – Cape: Lit. Namaqualand. – **St.** 15–20 cm tall, c. 1.5 cm thick, erect, branching above, the greenish-brown bark tending to tear away; **Br.** 2–4 cm long, densely leafy, glabrous, green; **L.** linear-lanceolate or narrow-oblanceolate, pointed, narrowing towards the base, furrowed on the upperside, rather hirsute, 18–50 mm long, 5–10 mm across, 2–3 mm thick; **Infl.** 5–7 cm tall, Fl. 4–6, Cor. light green.

**C. reticulata** THUNBG. (Pl. 40/2) (§ I/1) (? *C. dichotoma* HAW., *C. juttae* DTR.). – Cape: Gr. Karroo, Lit. Namaqualand; S.W.Afr.: Gr. Namaqualand. – **St.** thick, short, dividing above into twiggy Br., very soft, spongy, the plant later becoming hemispherical and c. 30 cm tall; **L.** 4–8 on the cylindrical, tuberculate short spurs, subterete, rather grooved on the upperside, both sides narrowing or long-oval, glabrous, up to 16 mm long, with a brown tip; **Infl.** numerous, up to 8 cm long, becoming woody and persisting, Fl. erect, 8–10 mm long, greenyellow and glandular; the dried Cal. hangs loosely on the pedicel like a collar.

**C. rubrovenosa** DTR. (§ I/2). – S.W.Afr.: Gr. Namaqualand; Cape: Lit. Namaqualand. – Semi-♄ up to 25 cm tall; **St.** short, yellowbrown; **Br.** up to 1.5 cm long, thick, covered with ± projecting, thick, roundish L.bases; **L.** in clusters, subulate, ± terete, pointed, glabrous, bluish-white, soft, up to 5 cm long, deciduous; **Infl.** of about 20 nodding, glandular-hairy Fl., with glandular-hairy pedicels, Cal. and Cor., the last light brownish-white, lobes with fine reddish veins, glandular-ciliate.

**C. salmiana** v. POELLN. v. **salmiana** (§ II/2) (*C. ramosissima* S.D., ex HAW.). – S.Cape. – Freely branching, glabrous ♄ up to 1 m tall; **L.** obovate or obovate-cuneate, rounded above with reddish borders, often apiculate, both surfaces rather convex, green, somewhat pruinose, 2.5–5 cm long, 12–20 mm across, 4–5 mm thick, with tiny spots; **Infl.** up to 8 cm tall, Fl. (1–)3(–5), pendulous, Cor. greenish to greenish-red, 10–16 mm long, pruinose outside, lobes 12–18 mm long, dirty reddish.

**C. — v. woodii** (SCHOENL. et BAK.) v. POELLN. ↑(*C. woodii* SCHOENL. et BAK., *C. ramosissima* v. *w.* SCHOENL. et BAK.). – L. oblong or obovate-oblong, 15–20 mm long, 5–7 mm across, ± pruinose; **Fl.** solitary.

**C. schaeferana** DTR. (Pl. 40/3) (§ II/8) (*C. schaeferi* DTR., *C. hoerleiniana* v. *s.* DTR., *Adromischus s.* (DTR.) BGR., *C. hoerleiniana* DTR., *A. h.* (DTR.) v. POELLN., *A. keilhackii* WERDERM., *A. s.* v. *k.* (WERDERM.) v. POELLN.). – S.W.Afr.; Cape: Lit. Namaqualand. – **Root** ± tuberous, sometimes more than 1 cm thick; **St.** almost undeveloped or 3–4 cm long, 3–5 mm thick, branching, often caudiciform, producing aerial roots, bark light greenish-yellow; **L.** soon falling, solitary or several, irregularly spherical to obovate-cuneate, the upperside rather furrowed, at first with a few projecting glandular hairs, becoming glabrous, greenish-yellow, with numerous red lines, (7–)12(–20) mm long, and a half or two-thirds as wide; **Infl.** up to 4 cm tall, hirsute, Fl. 1–3, usually erect, Cor. 8 to 11 mm long, greenish, the angles having red lines, somewhat pruinose, lobes pink to grey-purple.

**C. simulans** SCHOENL. v. **simulans** (§ ?). – Cape: Transvaal: Pretoria. – **L.** oblong or linguiform, pointed, convex on both surfaces, glabrous, about 4.5 cm long, 12 mm across, 6 mm thick; **Infl.** c. 13 cm tall, Fl. 11 mm long, light red, powdery-white, lobes light red. (? **Adromischus**).

**C. — v. spathulata** SCHOENL. – **L.** spatulate, thick at the base, with sharp margins.

**C. singularis** R. A. DYER (§ III). – S.W.Afr.: Lüderitz D. – ♃; **caudex** tuberous; **L.** 1–2, produced from the caudex, short-petiolate, fleshy, ± circular and deeply cordate, 5–8 cm ⌀, light green and with prominent veins on the upperside, underside light chestnut-brown, both surfaces glandular-hairy, the L. dying off before the flowering period; **Infl.** solitary, the axis (in cultivation) being 15–35 cm tall, up to 5 mm thick, glandular-hispid, terminating in branching panicles, Fl. 1–3 together on pedicels 3 to 8 mm long, Cor. c. 2 cm long, greenish-yellow. Grow preferably in a moist, shady position.

**C. sinus-alexandrii** v. POELLN. (§ II/8). – Cape: Lit. Namaqualand. – Semi-♄, 2–3 cm tall, few Br. 2–3 mm thick with a light-coloured bark; **L.** crowded, round or almost oval, 7–10 mm long, 3–4 mm thick, bluish-green with minute red spots or lines, the underside purple, glabrous; **Fl.** 1–2, about 14 mm long, greenish.

**C. striata** HUTCHIS. (§ I/3). – Cape: Lit. Namaqualand. – **Caudex** short, gnarled, about 2 cm tall, the upper half projecting above the soil; **St.** not present or much shortened to 10 cm or little more in length, up to 1.5 cm ⌀, grey-white or yellow-brown, with short longitudinal striae, with a peeling bark; **L.** hirsute, viscous, linear, terete, 3–7 cm long, grooved on the upperside, later shrivelling and persisting; **Infl.** 20 cm or more tall, glandular-viscous, Fl. 10–15, erect, Cor. 22 mm long, green, tinged purple-brown, striate, the lobes with branching reddish lines.

**C. sturmiana** v. POELLN. (§ II/3). – Cape. – Branching ♄, at first glandular; **L.** oblong or linear-oblong, with a small red apiculus, rounded or with the upperside flat, somewhat glandular, 7–12 cm long, 6–9 mm across, 4 to 6 mm thick; **Infl.** c. 15 cm tall, red-brown, Fl. pendulous, c. 3 cm long, red.

**C. swartbergensis** v. POELLN. (§ I/2). – Cape: Ladismith D. – **St.** erect, fleshy, thick, 2 cm ⌀ at the base, branching above, Br. short, green, hirsute; **L.** narrow-oblong, bluntish to rather pointed, furrowed on the upperside, the underside convex to semi-terete, green, hairy, 3 to 5 cm long, 5–8 mm across, 3–4 mm thick; **Infl.** 5–7 cm tall, Fl. pendulous, c. 1 cm long, brownish-green.

**C. teretifolia** THUNBG. (Pl. 41/1) (§ II/4) (*C. t.* v. *subglaber* HARV., *C. campanulata* HARV.). – Cape: Karroo. – ♄; **St.** 20 cm long or more, with dense white hairs; **L.** linear, often widening towards the tip and narrowing towards the base, 3–12 cm long, 6–12 mm or more wide, c. 5 mm thick, rather flattened on the upperside, often furrowed, light to dark green, densely white-haired to almost glabrous; **Infl.** 30–45 cm tall, densely hairy, Cor. hirsute outside, green.

**C. tomentosa** HARV. (Pl. 41/3) (§ II/6). – Cape. – **St.** thin, laxly leafy, 10–15 cm long, 3–4 mm thick; **L.** ovate-oblong, narrowing to shortly terete, c. 2.5 cm long, 12 mm across, fleshy, thick; **Infl.** 10–20 cm tall, Fl. 4–6, Cor. nodding, red. The entire plant covered in dense felty hairs.

**C. tricuspidata** HARV. (§ II/3) (*C. papillaris* v. *t.* SALM ex DC.). – Cape: Queenstown D. – Probably a natural hybrid, related to **C. galpinii**; **L.** broad-grooved or oblong-cuneate, with 3 tips on the upper part, or subterete or semi-terete, or almost flat and obcuneate, margins smooth or wavy or irregularly dentate, these T. later falling.

**C. undulata** HAW. v. **undulata** (Pl. 41/2) (§ II/3). – Cape. – Erect ♄ 30–60 cm tall, immature growth white, pruinose; **L.** thick, fleshy, both surfaces convex, obovate-spatulate, usually truncate above, less often rounded, the margins wavy above midway, often becoming reddish, 8–12 cm long, up to 6 cm across, or often broader than long; **Infl.** up to 45 cm tall, Fl. on a ± long pedicel, nodding, tube up to 2.5 cm long, orange-coloured to red.

**C. —** v. **mucronata** (LAM.) v. POELLN. (*C. m.* v. POELLN., *Adromischus m.* LAM.). – **L.** 4 to 4.5 cm long, 2.5–3 cm across, with an apiculus 1–2 mm long.

**C. velutina** HOOK. f. v. **velutina** (§ II/1). – Cape. – ♄, 60–90 cm tall, L. and shoots covered with fine hairs; **L.** ± obovate, blunt, cordate at the base and amplexicaul, dark green, thick, margins entire, brown-red towards the tip, 7–12 cm long, 3–4.5 cm across, 3–4 mm thick; **Infl.** up to 20 cm tall, with many nodding Fl., Cor. 1.5–2 cm long, yellow above, light green below, the lobes having a red border.

**C. —** v. **beckeri** SCHOENL. (*C. b.* SCHOENL.). – **L.** 2–3 cm across, grey-green; **Fl.** dark red.

**C. ventricosa** BURM. f. v. **ventricosa** (§ I/3). – Cape: Karroo; S.W.Afr.: Gr. Namaqualand. – **St.** short, somewhat branching, covered with round projecting L.bases 3 mm long; **L.** crowded at the Br. tips, linear, terete, elongated, glabrous, 5–10 mm long, 2–4 mm thick; **Infl.** 25–30 cm tall, with many erect Fl., Cor. with fine red markings on both surfaces. Very poisonous.

**C. —** v. **alpina** HARV. – Cape: Uitenhage D. – **Infl.** 8–10 cm tall, Fl. fewer but larger.

**C. wallichii** HARV. (§ I/2). – S.Afr.: origin not known. – **St.** fleshy, branching, 30–35 cm tall, up to 3 cm thick; **L.** soon falling, subterete, upperside grooved, 5–10 cm long, grey-green, L.bases persisting and very prominent; **Infl.** 35–70 cm tall, glandular-hairy, with numerous nodding Fl., Cor. glandular-hairy outside, greenish-yellow. Very poisonous.

**C. wickensii** SCHOENL. v. **wickensii** (§ II/4). – N.Transvaal: Pietersburg D. – ♄ up to 2 m tall, branching basally; **L.** ovate-oblong to ovate-lanceolate, pointed or rounded, narrowing towards the base, 8–11 cm long, 2.5–6 cm across, convex on the underside, often carinate, fleshy, soft, green, pruinose; **Infl.** c. 30 cm tall, glandular-hairy, Fl. ± nodding, red.

**C. —** v. **glandulosa** v. POELLN. – **L.** more oblong, glandular, c. 16 cm long, c. 4 cm across; **Infl.** glandular, Fl. pendulous, glandular outside.

**C. —** v. **rhodesica** SCHOENL. – Rhodesia. – **L.** broader than in the type.

**C. zuluensis** SCHOENL. ex v. POELLNITZ (§ II/7). – Zululand. – Habit as for **C. orbiculata**; **L.** obovate to ovate-cuneate, blunt, 10 cm long, 7 cm across, finely felty; **Infl.** hirsute below, with many pendulous Fl., red or reddish.

**Coulterella** VASEY et ROSE. Compositae. – Mexico. – Medium-sized shrubs, of little horticultural importance; **L.** somewhat fleshy. Summer: in the open, Winter: unheated greenhouse. Propagation: seed.

**C. capitata** VASEY et ROSE (Pl. 34/6). – Mexico: S.Baja Calif., incl. the islands. – Glabrous ♄, up to 1 m tall, 2 m across, with an aromatic odour; **St.** up to 2 cm thick at the base; **L.** opposite to alternate, ovate to elliptical, pointed, usually dentate, bluish, 2–5 cm long, 1–4 cm across, 1.5–2 mm thick; **Fl.** solitary, Cor. yellow, lobes 3–4 mm long.

**Crassula** L. Crassulaceae. – S.Afr., trop. Afr.; Madag.; several annual and aquatic species distributed throughout the world. – Mostly succulent ♃ forming clumps sometimes with a tuberous rootstock, or semi-♄ with ± succulent Br. and twigs; **L.** opposite, often crowded into Ros., usually sessile, often united at the base, glabrous, hairy or scaly; **Fl.** fairly small, white, pink, more rarely yellow or greenish, mostly numerous, in terminal or apparently lateral cymes or thyrsoid **Infl.,** rarely solitary and from axils, Sep. and Pet. usually 5. – Summer: in the open, in winter: greenhouse, moderately warm. – Propagation: seeds, cuttings, leaf-cuttings. – Some species are scarcely succulent, these are omitted in the following.

## Division of the Genus Crassula into Sections acc. S. Schoenland and enumeration of the Species revised by H.-Ch. Friedrich

**Sect. I. Tillaeoideae** Schoenl. – Small ⊙ or ♃ **herbs,** or **semi-♄.**

**Ser. 1. Helophytum** (Eckl. et Zeyh.) Schoenl. – Disregarded Ser.

**Ser. 2. Vaillantii** Schoenl. – Disregarded Ser.

**Ser. 3. Aphylla** Schoenl. – Disregarded Ser.

**Ser. 4. Filicaulis** Schoenl. – **St.** often with elongated, more rarely short Int.; **Fl.** 5-partite, solitary or in cymes, on filiform pedicels, usually from the axils of the upper L., sometimes terminal. – S. and trop. E.Afr. – Partly included Ser. – Species: C. albicaulis, browniana, coleae, expansa, filicaulis, fragilis, humbertii, nakurensis, pyrifolia, tenuicaulis, thorncroftii, volkensii, woodii, zimmermannii.

**Ser. 5. Glomerata** Schoenl. – Disregarded Ser.

**Ser. 6. Muscosa** Schoenl. – Disregarded Ser.

**Ser. 7. Lycopodioides** Schoenl. – **Semi-♄,** densely leafy in 4 R.; **Br.** elongated, frequently with short shoots in the L.-axils; **Fl.** 5-partite, nearly sessile, very small, solitary or several together in the axils of the upper L. – S.W. to trop. Afr. – Partly included Ser. – Species: C. lycopodioides, phyturus, transvaalensis.

**Ser. 8. Umbellata** Schoenl. – Disregarded Ser.

**Ser. 9. Corallina** Schoenl. – Small ♃ **herbs; St.** single or with numerous short, erect or procumbent, densely leafy Br., rooting at the lower nodes; **Fl.** in terminal cymes, campanulate. – S.Afr. – Species: C. corallina, dasyphylla, deltoidea, vestita.

**Sect. II. Stellatae** Schoenl. – **St.** and Br. ± fleshy, distantly leafy; **L.** sessile or petiolate, ± flat, ovate or cordate-ovate or rarely sub-cylindrical, with entire or dentate margins, rarely serrate; **Infl.** usually terminal, paniculate or nearly umbellate, Fl. stellate.

**Ser. 1. Pellucida** Schoenl. – Spreading ♃ **herbs; St.** slender, Int. elongated; **L.** fleshy, flat, margins entire, glabrous or with soft hairs; **Fl.** on slender stalks, solitary or axillary or frequently several together in somewhat umbel-like, terminal Infl. – Trop. Afr.; S.Afr., E. to Transvaal; Madag. – Species: C. alsinoides, cordifolia, galunkensis, lineolata, marginalis, micans, nummulariifolia, pellucida, tysonii.

**Ser. 2. Spatulata** Schoenl. – Procumbent, inclined or erect ♃ **herbs** with long Int.; **L.** stalked, flat, fleshy; **Infl.** lax, paniculate, terminal. – S.E.Afr. to Natal. – Species: C. crenulata, cyclophylla, inandensis, latispatulata, multicava, sarmentosa, spatulata, streyi, wylei.

Ser. 3. *Lactea* Schoenl. is included in **Ser. 4.** now.

**Ser. 4. Arborescens** Schoenl. incl. *Lactea* Schoenl. – ♄ or ♄̄ up to 3–4 m high, freely branched, glabrous, **St.** and **Br.** very thick; **L.** flat, large; **Fl.** in terminal Pan., white or pink, not rarely 6-partite. – S.Afr.: Cape Prov. to Natal. – Species: C. arborescens, argentea, lactea, portulacea.

**Ser. 5. Cordata** Schoenl. – **Semi-♄** or ♄, mostly 30 cm high, **St.** and **Br.** slender, fleshy; **L.** stalked, flat, grey-green; **Fl.** in paniculate umbels. – S.E.Afr.: Uitenhage D. to Natal. – Species: C. cordata, glauca.

**Ser. 6. Peploides** Friedr. (*Galpinii* Schoenl.). – Succulent ♃ **herbs,** branched from the base; **L.** dentate at the apex; **Infl.** paniculate, stalked. – E.Cape: high mountains. – Species: C. peploides.

**Sect. III. Tuberosae** Schoenl. – Succulent ♃ **herbs,** with a tuberous, rarely elongated underground **caudex,** often also developing Tub. in the Stems; **L.** soft-fleshy, flat, margins entire or dentate, crenate or rarely deeply lobed, sessile or petiolate, the few pairs often densely crowded; **Fl.** 5–7-partite, stellate. – S.E. and S.W.Cape, partly to Natal. – Species not in cultivation omitted. – Species: C. bartlettii, capensis, flabellifolia, nemorosa, saxifraga, simulans, umbella.

**Sect. IV. Crassula** (*Campanutatae* Schoenl.). – Succulent ♃ **herbs** or **semi-♄**; **St.** fleshy or woody; **L.** ± fleshy; **Infl.** various, mostly cymose or paniculate, Fl. campanulate. – S. and trop. Afr.; Arabia.

**Ser. 1. Acutifolia** SCHOENL. – **Semi-♄,** glabrous, sparingly branched, St. and Br. leafy; **L.** dense, often imbricate, usually sub-cylindrical, always pointed; **Infl.** terminal, ± cymose, Fl. small. – S.Afr. – Species: C. acutifolia, connivens, planifolia, robusta, rudis, subsessilis, tetragona.

**Ser. 2. Perforata** SCHOENL. – **Semi-♄,** xerophytic, sparingly branched, **St.** woody; **L.** united at the base or perfoliate-amplexicaul, fleshy, ovate or oblong, rarely ovate to lanceolate, margins smooth or cartilaginous-ciliate or rarely papillose-ciliate; **Infl.** terminal, paniculate or thyrsoid, Fl. small, white, pink or yellow. – S.W.Afr.; S.Afr.: Lit.Namaqualand, Karroo. – Species: C. brevifolia, commutata, conjuncta, fusca, macowaniana, marnierana, montis-draconis, nealeana, perforata, rupestris, sladenii.

**Ser. 3. Biplanata** FRIEDR. (*Harveyi* SCHOENL. p.part.). – **Semi-♄,** loosely branched, xerophytic, **St.** woody or fleshy and with densely leafy Br.; **L.** sessile, united at the bases, glabrous or with a few hairs, ovate to lanceolate, ± flattened, subacute or acute, usually considerably less than 1.5 cm long, rarely longer; **Infl.** terminal, cymose, stalked or sessile. – S.Afr. – Species: C. biplanata, dependens, ericoides, griquaensis, kuhnii, parvisepala, punctulata, sarcocaulis.

*Ser. 4. Cymosa* SCHOENL. is included in **Ser. 5.** now.

**Ser. 5. Scabra** SCHOENL. incl. *Cymosa* SCHOENL. – **Semi-♄,** xerophytic, woody, rarely almost herbaceous; **St.** and **Br.** and L. ± papillose-bristly or strigose-hairy or glabrous; **L.** united or nearly so; **Fl.** in sessile or very short stalked, terminal, cymose Infl. on the Br. – S.W.Afr.; S.W.Cape. – Species: C. arenicola, burmanniana, dejecta, flava, pallens, petraea, pruinosa, pustulata, rubricaulis, rudolfii, scabra, undulata, whiteheadii.

**Ser. 6. Perfoliata** SCHOENL. – **Semi-♄,** stout; **St.** ± simple, erect, 50–90 cm high, leafy for the whole length; **L.** large and thick, united, grey green, triangular-lanceolate or falcate, with smooth margins; **Infl.** ± terminal, large, dense, cymose. – S.E.Cape to Natal. – Species: C. falcata, heterotricha, perfoliata.

**Ser. 7. Southii** SCHOENL. – **Semi-♄**; **St.** woody, fairly freely and dichotomously branched, Br. rather densely leafy; **L.** flat, sheathing at the base, ovate-lanceolate, with sharp tips, margins papillose-ciliate; **Fl.** white, in dense cymes at the ends of the Br. – S.E.Cape. – Species: C. southii.

**Ser. 8. Vaginata** SCHOENL. – ♃ **herbs,** root often fusiform, **shoots** simple, rarely more than 1 m high; **L.** sheathing at the base, lower L. more than 3 cm long, margins papilloseciliate; surfaces smooth or papillose. – S.E.Cape, through trop. Afr. to S.Arabia. – C. abyssinica, acinaciformis, alba, drakensbergensis, ellenbeckiana, goetzeana, natalensis, recurva, retrorsa, rubicunda, spectabilis, vaginata, wilmsii.

**Ser. 9. Ramuliflora** SCHOENL. – ♃ **herbs** with annual, simple or branched, slender stalks, rarely with decumbent **St.**; **L.** little united, flat and broad, margins papillose, surface often ± papillose-bristly; **Fl.** in elongated, spreading or capitate cymes. – S.W.Afr.; S.E.Cape to Natal, Transvaal and trop. Afr. – Species: C. bloubergensis, dregeana, globularioides, illichiana, lasiantha, liebuschiana, meyeri, nyikensis, peglerae, ramuliflora, reversisetosa, rubescens, tabularis.

**Ser. 10. Setulosa** SCHOENL. – ♃ **herbs,** rarely higher than 10 cm, mostly branched from the base; **St.** woody, leafy, with rough hairs or glabrous; lower **L.** sometimes crowded, upper more distant, free or a little united, fleshy, often flat, ciliate at the margins, surfaces glabrous or with soft or rough hairs, rarely covered with soft hairs; **Infl.** terminal, paniculate cymose, rarely the Fl. solitary or in pairs, white or red. – Mountains of S.W.Afr. to Transvaal, Natal and in S.E.Cape. – Species: C. barklyana, cooperi, curta, milfordae, picturata, schmidtii, sedifolia, setulosa.

**Ser. 11. Sediflora** SCHOENL. – **Semi-♄,** laxly branched; **St.** thin, Br. leafy, procumbent or ascending; **L.** flat, somewhat fleshy, margins rough or ciliate; **Infl.** terminal, lax-cymose, Fl. small, white. – S.E.Cape to Natal. – Species: C. amatolica, flanaganii, sediflora, tenuifolia.

**Ser. 12. Quadrangularis** SCHOENL. – Small ♃ **herbs,** freely branched from the base; **L.** usually densely crowded, sessile, united, ovate or obovate, often slightly folded, glabrous on both sides, margins papillose-ciliate; **Infl.** terminal, ± stalked, Fl. in capitate cymes or thyrsoid Infl., white. – S.E.Cape, Transvaal. – Species: C. compacta, montana, mossii, quadrangularis, socialis.

**Ser. 13. Rosularis** SCHOENL. – Succulent ♃ **herbs; L.** crowded, rosulate, glabrous, margins papillose-ciliate; **Infl.** stalked, paniculate or thyrsoid, more rarely capitate. – S.Afr., Swellendam to Natal. – Species: C. gillii, intermedia, orbicularis, rosularis.

**Ser. 14. Turrita** SCHOENL. – ☉ or ♃ succulent **herbs,** branched from the base; **L.** either basal and crowded in Ros. or along the St. and more distant, gradually diminishing in size towards the apex, glabrous or papillose, ciliate at the margins; **Infl.** thyrsoid, Fl. mostly white, rarely yellowish. – S.Afr., widely distributed. – Species: C. albanensis, barbata, brevistyla, broomii, capitella, corymbulosa, engleri, hemisphaerica, lettyae, luederitzii, maculata, nodulosa, nuda, pseudohemisphaerica, punctata, rhodogyna, sessilicymula, spicata, subbifaria, thyrsiflora, turrita.

**Ser. 15. Exilis** SCHOENL. – ⚁ **herbs** up to 6 cm high; **St.** fleshy, densely leafy, branched from the base; **L.** thick, fleshy, densely covered with small grey papillae; **Infl.** stalked, cymose or ± capitate, Fl. white. – S.Afr.: Lit.Namaqualand. – Species: C. ausensis, exilis, garibina, karasana, littlewoodii, mesembryanthemopsis.

**Ser. 16. Arta** SCHOENL. – Small, succulent ⚁ **herbs,** rarely more than 15 cm high, **St.** fleshy or more rarely woody, mostly branched from the base, densely leafy, with short Int.; **L.** thick, fleshy, glabrous, pulverulent, grey, white-hairy or grey-papillose; **Infl.** cymose or somewhat capitate on a slender stalk, Fl. small, whitish, yellow or pinkish. – S.Afr.: Lit.Namaqualand; S.W.Afr.: Gr. Namaqualand. – Species: C. alstonii, arta, bakeri, columella, cornuta, deceptor, densa, dinteri, elegans, globosa, grisea, humilis, namibensis, plegmatoides.

**Ser. 17. Argyrophylla** SCHOENL. – Sparingly branched, succulent ⚁ **herbs,** rarely **semi-**♄; **St.** fleshy or somewhat woody; **L.** united or nearly free, in Ros. or more distantly arranged, nearly always with soft hairs; **Fl.** in dense cymes, which are grouped into corymbose, paniculate or nearly spicate stalked or sessile Infl., Fl. whitish. – S.W.Afr.; E.Cape to Transvaal, Rhodesia, Moçambique and Malawi. – Species: C. argyrophylla, decidua, ernestii, hottentotta, klinghardtensis, lanuginosa, pachystemon, sericea, tecta, velutina, zombensis.

**Sect. V. Sphaeritis** (ECKL. et ZEYH.) SCHOENL. (*Sphaeritis* ECKL. et ZEYH. as genus). – Xerophytic **semi-**♄ with a woody St., or succulent ⚁ **herbs** of diverse habit; **Infl.** mostly stalked, corymbose-cymose or thyrsoid or capitate; **Fl.** white or slightly yellowish, Cor. urn-shaped. – S.W. & S.Afr.

**Ser. 1. Subulata** FRIEDR. (*Ramosa* SCHOENL.). – **Semi-**♄ up to 60 cm high; **Br.** leafy, the older ones somewhat woody and bare; **L.** numerous, usually densely and evenly spaced on the ultimate branchlets; **Infl.** terminal, capitate or cymose, sessile or short-stalked. – S.W.Cape up to Uitenhage in S.E. – Species: C. ciliata, fastigiata, leucantha, multiflora, muricata, rustii, subulata.

**Ser. 2. Cultriformis** FRIEDR. (*Clavifolia* SCHOENL. p. part.). – **Semi-**♄ more than 30 cm high, **St.** and older **Br.** somewhat woody and leafless; **L.** crowded at the base of the ultimate branchlets, fleshy, somewhat flat, always ± blunt, the margins glabrous or ciliate; **Infl.** paniculate or cymose, more rarely thyrsoid, with capitate cymes, more rarely a single cyme, mostly stalked, with distant, leaflike bracts. – S.W.Afr.; S.W. and W.Cape. – Species: C. anomala, cultriformis, watermeyeri.

**Ser. 3. Virgata** SCHOENL. – ♄, small; **L.**-pairs separated by distinct Int., thick, ± semi-cylindrical, glabrous or almost so; **cymes** somewhat dense, capitate, usually stalked, in paniculate or cymose Infl. or solitary. – S.W.Afr.: Gr. Namaqualand; central and W.Cape. – Species: C. clavifolia, incana, loganiana, puberula, purcellii, remota, rubella, serpentaria, smutsii, subaphylla, virgata.

**Ser. 4. Trachysantha** SCHOENL. – **Semi-**♄, sparingly branched, 20–30 cm high, with rather crowded L.-pairs; **L.** semi-cylindrical, acute or bluntish, gradually diminishing in size upwards, densely covered with acute, backwardly directed, bristly acuminate papillae; **Infl.** terminal, stalked, broad cymose. – S. and S.E.Cape. – Species: C. hispida, mesembryanthoides. –

**Ser. 5. Tomentosa** SCHOENL. – ⚁, ♄, often with woody bases; **L.** basal or sub-basal, ± flat, hairy, rarely only ciliate at the margins; **Infl.** moslty terminal stalked, usually with a few pairs of emty bracts, consisting of small cymes arranged in a thyrse. – S.W.Cape to S.W.Afr. – Species: C. glabrifolia, interrupta, scalaris, tomentosa.

**Ser. 6. Namaquensis** SCHOENL. – Succulent ⚁ **herbs,** branched from the base, never higher than 15 cm; **L.** thick, mostly with papillae or hairs; **Infl.** terminal, stalked, Fl. sessile or nearly so in capitate cymes, these solitary or several together at the tips of the Infl. – W.Cape. – Species: C. comptonii, hirtipes, hystrix, namaquensis, susannae.

**Sect. VI. Globulea** (HARV.) HARV. et SOND. (*Globulea* HARV. as genus). – Xerophytic, succulent ⚁ **herbs** or sub-♄, with leafy Br. or. **L.-Ros.; Fl.** in ± dense capitate cymes, mostly in paniculate, stalked Infl., Cor. white or yellowish, urn-shaped, Pet. with erect, globular appendages. – S.W.Afr.; Cape: Orange Free State; Lesotho; Natal; widely distributed. – Species: C. cephalophora, clavata, cotyledonis, cultrata, dewinteri, erosula, fergusoniae, herrei, hirta, inamoena, mollis, nudicaulis, obvallata, platyphylla, pubescens, radicans, rattrayi, rauhii, rogersii.

**Sect. VII. Pyramidalis** HARV. – Succulent ⚁ **herbs; L.**-pairs mostly close together, forming a Ros., column or pyramid; **Infl.** mostly capitate, Fl. white or yellowish. – S.W.Afr.; W. and Central Cape. – Species: C. alpestris, archeri, columnaris, congesta, cylindrica, laticephala, multiceps, pyramidalis, quadrangula, semiorbicularis, teres.

**C. abyssinica** A. RICH. (§ IV/8). – Ethiopia; trop. E.Afr. – ♃ with a fleshy, napiform taproot; **L.** at first congested in Ros., lanceolate, glabrous, with papillose-ciliate margins; flowering St. ± papillose-hairy with distant L. diminishing upwards; **Fl.** red in terminal flat-topped cymes.

**C. acinaciformis** SCHINZ (§ IV/8) (*C. aloides* N. E. BR.). – Transvaal; Rhodesia. – Monocarpic. ♃; **St.** 0.5–0.7 m, rarely up to 1 m high, thick below, glabrous; **L.** forming in 1 to several years a Ros. like an **Aloe**, acinaciform, margins with acute cilia; **Infl.** up to 40 cm ⌀, 1 m high.

**C. acutifolia** LAM. v. **acutifolia** (§ IV/1) (*C. bibracteata* HAW.). – Cape: coastal zones. – ♃ or semi-♄ forming mats 40 cm ⌀; **L.** subulate, 12 mm long; **Fl.** white, very small.

**C.** — v. **harveyi** SCHOENL. – **St.** more erect.

**C. alba** FORSK. (§ IV/8) (*C. abyssinica* A. RICH. p. part., *C. mannii* HOOK. f., *C. schweinfurthii* DE WILD.). – Trop. Afr. in mountains. – ♃, with a fleshy taproot; **L.** at first rosulate, fleshy, linear-subulate to lanceolate, margins very finely serrate; **Fl.** white.

**C. albanensis** SCHOENL. (§ IV/14). – S.E.Cape: near Grahamstown. – ♃, ± branching basally; **L.** subulate, rosulate, papillose; **Fl.** in a thyrse, white.

**C. albicaulis** HARV. (§ I/4). – S.W.Afr. – ♃; **St.** prostrate, ± angled, ± swollen at the nodes, white; **L.** ovate-oblong to linear-oblong; **Fl.** with ± elongated pedicels, white or with red tips.

**C. alpestris** THUNBG. (§ VII) (*C. massonii* BRITT. et BAK.?, *C. variabilis* N. E. BR.). – Cape: S.Lit. Namaqualand to Ceres Karroo. – **St.** ± branching at the base, with short, visible Int.; **L.** triangular at the base, tapering towards the tip, usually pointed; **Fl.** white in a single or rarely branched capitate cyme. (C. massonii seems to be a separate species with Fl.arranged in a thyrse.)

**C. alsinoides** (HOOK. f.) ENGL. (§ II/1) (*Tillaea a.* HOOK. f.). – Transvaal, Natal, trop. Afr.; Madag.? – **St.** prostrate, up to 20 cm long; **L.** fleshy, oblong, margins entire; **Fl.** from the upper leaf-axils, white.

**C. alstonii** MARL. (Pl. 41/4) (§ IV/16). – Cape: Lit. Namaqualand. – ♃ with a few Br. from the base; **St.** 8–10 cm tall; **L.** apparently in 2 rows, dense, virtually imbricate, round, somewhat involute, c. 2 cm across, the underside round, bluntly rounded at the tip, fleshy, grey-green; **Fl.** white.

**C. amatolica** SCHOENL. (§ IV/11). – S.E.Afr. – Related to **C. sediflora**; ♃; mat-forming; **St.** 40 cm tall; **L.** 5–10 mm long, lower ones ovate and pointed, the upper lanceolate.

**C. anomala** SCHOENL. et BAK. f. (§ V/2). – S.W.Cape. – Semi-♄ c. 30 cm tall; **St.** and **Br.** somewhat woody; **L.** ± crowded at the shoot tips, fleshy, somewhat flat, ± blunt, grey-green from ± densely set papillose hairs.

**C. arborescens** (MILL.) WILLD. (Pl. 42/1) (§ II/4) (*Cotyledon a.* MILL., *Crassula cotyledon* JACQ.). – Cape: Lit. Namaqualand to Natal. – ♄ 3–4 m tall or a small ♄; **St.** and **Br.** stout; **L.** rounded to obovate with a soft apiculus, 3.5–7 cm long, flat, glabrous, fleshy, grey-green, margins entire, often red, with red spots above; **Fl.** pink or white at first, becoming pink, up to 2 cm ⌀.

**C. archeri** COMPT. (§ VII). – Cape: Lit. Namaqualand. – Similar to **C. pyramidalis**, but larger, with short lateral Br.; **Fl.** pure white.

**C. arenicola** TOELKEN (§ IV/5) (*C. cymosa* sensu SCHOENL.). – Cape: Clanwilliam D., Cape Flats, Cape Peninsula to Hondeklip Bay. – Up to 35 cm tall, much branched at the base with the main **Br.** decumbent; **L.** elliptic-linear to strap-like, slightly tapering towards both ends, with a blunt apex, (1–)2(–3) cm long, 2–4 mm across, erect, flat, glabrous but with marginal cilia, green to yellowish-green, Sh. 1–2 mm long; **Infl.** terminal, erect, branched, 25 mm tall, with many Fl. c. 8 mm ⌀, light yellow or cream.

**C. argentea** THUNBG. (Pl. 54/5) (§ II/4) (*C. obliqua* AIT.). – S.Cape to Natal. – ♄ 1(–3) m tall; **St.** thick, freely branching; **L.** rather oblique-ovate, mostly short-pointed, 3–4 cm long. 2.5–3 cm across, fleshy, dark green with a silvery cuticle and with scattered darker spots; **Fl.** numerous, 1–1.5 cm ⌀, white or rarely pale pink.

**C. argyrophylla** DIELS (§ IV/17). – Transvaal; Swaziland; Rhodesia. – ♄ with few Br., up to 20 cm tall.

**C.** — v. **argyrophylla.** – **L.** ± flat, somewhat rosulate, cuneate-obovate, blunt, 15–36 mm long, 15–20 mm across, fleshy, softly hairy, dark green, often red, with ciliate margins; **Fl.** white.

**C.** — v. **ramosa** SCHOENL. – Freer-branching variety with smaller L. (Seems to be better considered as a separate species).

**C.** — v. **swaziensis** SCHOENL. (*C. s.* SCHOENL.). – **L.** glabrous apart from the margins.

**C. arta** SCHOENL. (Pl. 43/1) (§ IV/16) (*C. deltoidea* AUCT. non THUNBG., *C. cornuta* AUCT. non SCHOENL. et BAK. f.). – S.W.Afr.; Cape: Lit. Namaqualand. – ♃ up to 10 cm tall; **St.** simple or branching from the base, covered below with the remains of dry old L., 3–6 mm ⌀; **Br.** erect, covered with 8–15 pairs of L., the L.together forming a four-sided column 2.5 to 5 cm long, 1–2 cm thick; **L.** fleshy, united and perfoliate, triangular, the tip projecting, ± pointed, convex on the underside and almost carinate towards the tip, margins angular, bluish-green to grey, densely covered with minute spherical papillae, somewhat mealy-pruinose, with green spots on the outside, 8–15 mm long, 6–17 mm across, 4–8 mm thick; **Fl.** lemon to pale yellow or ± pink-tinged.

**C. ausensis** P. C. HUTCHIS. (§ IV/15) (*C. "ausiensis", C. hofmeyeriana* DTR. nom. nud.). – S.W.Afr.–Compact, branching ♃ up to 9 cm tall; **St.** robust, gnarled; **L.** densely rosulate or opposite and decussate, obovate to oblong, 9–17 mm long, 3–8 mm across, 3–6 mm thick, slightly hollow on the upperside, the underside convex, densely white-hairy; **Fl.** c. 12 mm ⌀, white.

**C. bakeri** SCHOENL. (§ IV/16). – Cape: Lit. Namaqualand. – Small, erect, fleshy, densely leafy; **L.** in 2–3 pairs, spherical-oblong, somewhat flattened on one side, 5–10 mm long, ash-grey, finely white-haired, persistent.

**C. barbata** THUNBG. (Pl. 42/5) (§ IV/14). – Cape: Calvinia D., Lit. Karroo. – ♃, branching basally; **L.** in a basal Ros., 3–4 cm ∅, convex and bent inwards, green, the margins with projecting hairs; **Fl.** small, white. Plants sometimes monocarpic.

**C. barklyana** SCHOENL. (§ IV/10). – S.E.Cape. – Succulent, dwarf ♃ resembling **C. cooperi**; **Fl.** solitary or in pairs.

**C. bartlettii** SCHOENL. (§ III). – Cape: S.Cape. – Underground **St.** up to 2.5 cm long, St.short; **L.** usually in 4 pairs, subpetiolate, spatulate, margins entire; **Fl.** white in a subumbellate Infl.

**C. biplanata** HAW. (§ IV/3). – Cape: S.-S.E.Cape. – Semi-♄, 8–12 cm high, ± branching; **Br.** brown, stiff; **L.** lanceolate-subulate, 1–1.5 cm long, grey-green, ± flattened; **Fl.** white, in a pedunculate cyme.

**C. bloubergensis** R. A. DYER (§ IV/9). – Transvaal: Pieterburg D. – ♃, branching basally, mat-forming; **L.** forming a basal Ros., pale olive-green with a brown tint, c. 1 cm long, 4 mm across, 2–2.5 mm thick; **Infl.** covered with minute papillae, Fl. pink or white.

**C. brevifolia** HARV. non SCHOENL. (§ IV/2) (*C. flavovirens* PILL., *C. pearsonii* SCHOENL.). – Cape: Lit. Namaqualand; S.W.Afr.: Lüderitz S. – Glabrous ♄, c. 20 cm tall; **St.** becoming woody, freely branching; **L.** oblong to obovate-oblong, blunt to pointed, 2–2.5 cm long, 7–8 mm ∅, fleshy, green, often turning yellowish-green; **Infl.** c. 2.5 cm tall, Fl. pale to lemon-yellow, rarely whitish.

**C. brevistyla** BAK. (§ IV/14). – Natal. – Closely related to typical **C. corymbulosa** and perhaps synonymous.

**C. broomii** SCHOENL. (§ IV/14). – Cape: Victoria West D. – ♃, branching basally, succulent; **L.** in a basal Ros., rather thick, spatulate or roundish, rather over 3 mm long; **Fl.** small, white.

**C. browniana** BURTT-DAVY (§ I/4) (*C. furcata* SCHOENL. nom. nud.). – Natal, Orange Free State, Transvaal; Rhodesia; ? trop. Afr. – Dwarf ♃ up to 5 cm tall; **St.** red, dividing dichotomously, downy, as are the L., Int. 6–15 mm long; **L.** fleshy, flat, obovate, blunt to spatulate, lower L. 8 mm long and 3–4 mm across; **Fl.** solitary. Acc. FRIEDRICH possibly identical with **C. zimmermannii** or **C. coleae**.

**C. burmanniana** D. DIETR. (§ IV/5). – S.Cape. – Related to **C. flava**; **Infl.** glabrous.

**C. capensis** (L.) BAILL. (§ III) (*Septas c.* L., *C. septas* THUNBG.). – S.W.Cape. – **Root** tuberous, ♃ with annual Br.; **L.** 1 cm or more in length, broader than long, more rarely round or ovate, margins sinuate, crenate or dentate, narrowing at the base towards the petiole.

**C. capitella** THUNBG. (§ IV/14). – Cape: S.-S.E.Cape. – ♃; **St.** short; **L.** in a basal Ros., subulate-linear to lanceolate, 4–7 cm long, c. 1 cm wide, margins papillose-ciliate; **Infl.** erect thyrsoid, Fl. white, in sessile clusters.

**C. cephalophora** THUNBG. (§ VI) (*C. canescens* HAW., *Globulea c.* HAW.). – Cape: Lit. Namaqualand; S. and S.W.Cape. – ♃, branching from the base; **St.** thick; **L.** oblong or lanceolate or almost linear, blunt to somewhat pointed, 2.5–7 cm long, the margins without cilia, the entire plant shortly and densely white-hairy; **Fl.** white.

**C. ciliata** L. (§ V/2). – S.W.Cape. – Small semi-♄; **St.** and Br.rather woody; **L.** oblong-ligulate, distinctly papillose-ciliate.

**C. clavata** N. E. BR. (§ VI). – Cape: Lit. Karroo. – Small ♃; **L.** clavate, 1 cm long, grey, spotted, glabrous.

**C. clavifolia** (E. MEY.) HARV. (§ V/3) (*Globulea c.* E. MEY.). – Cape: Lit. Karroo. – Erect semi-♄; **L.** oblong-spatulate, blunt, fleshy; **Fl.** small.

**C. coleae** BAK. (§ I/4). – Somalia. – Erect or prostrate ♃, branching basally; **St.** very brittle, with dense fine hairs; **L.** sessile, oblong, pointed, fleshy, with entire margins, 15–18 mm long, 4.5–6 mm across; **Fl.** white.

**C. columella** MARL. et SCHOENL. (Pl. 42/6) (§ IV/16). – Cape: Lit. Namaqualand. – Up to 15 cm tall; **St.** covered throughout by small compressed **L.** 15–25 mm long, 3–12 mm across, dark olive velvety green to brownish, minutely hirsute; **Fl.** greenish-white.

**C. columnaris** THUNBG. (Pl. 42/3) (§ VII). – Cape: Lit. Namaqualand, Gr. and Lit. Karroo; S.W.Afr. – **St.** up to 5 cm tall, monocarpic or proliferous; **L.** dense, arranged in 4 rows, fleshy, round to elliptical, the upperside mostly curved-concave, the underside rounded, 2–3 cm across, green to brownish, margins shortly ciliate; young plants mostly form a spherical Ros. which later develops into a ± 4-angled column; **Infl.** capitate, 3 cm ∅, Fl. white, scented.

**C. commutata** FRIEDR. (§ IV/2) (*C. brevifolia* AUCT. non HARV.). – S.W.Afr.: Lüderitz S.; Cape: Lit. Namaqualand. – ♄, 10–30 cm tall, branching; **Br.** fleshy, becoming woody, 2–5 mm thick; **L.** spreading horizontally, semi-terete, linear-lanceolate to oblong, pointed, the upperside flat, the underside convex and sometimes carinate, glabrous or with the margins minutely and sparsely papillose-ciliate, pale green, the upperside green spotted, 8–13 mm long, 3–6 mm across, 2–4 mm thick; **Fl.** white.

**C. compacta** SCHOENL. (§ IV/12) (*C. massonioides* DIELS). – Transvaal. – See supplement.

**C. comptonii** HUTCHINS. et PILL. (§ V/6). – Cape: Lit. Namaqualand. – ♃, acauline, mat forming; **L.** rosulate, fleshy, almost ovate, coarsely papillose-hirsute; **Fl.** pale to bright yellow.

**C. congesta** N. E. BR. (§ VII) (*C. pachyphylla* SCHOENL.). – Cape. – ♃; **St.** fleshy, with few Br.; **L.** ovate and blunt to lanceolate and rather pointed, underside very rounded, grey-green; **Fl.** in dense capitate Infl., white.

**C. conjuncta** N. E. BR. (§ IV/2). – E.Cape. – Related to **C. perforata**; **L.** perfoliate, amplexicaul, broadly ovate, tapering, bluish to blue, densely papillose-ciliate; **Fl.** white.

**C. connivens** SCHOENL. (§ IV/1). – Cape: Lit. Namaqualand, central Cape, Gr. Karroo. – Related to the more shrubby form of **C. acutifolia**.

**C. cooperi** RGL. f. **cooperi** (§ IV/10) (*C. bolusii* HOOK. f.?). – Transvaal; Lesotho. – Dwarf mat-forming ♃; **L.** crowded in a dense Ros., lanceolate-spatulate, the underside round-carinate, margins ciliate, 10–15 mm long, light green, the upperside pitted, spotted; **Fl.** c. 3 mm ⌀, white or pallid flesh-coloured.

**C.** — f. **macrophylla** JACOBS. – **L.** 15–20 mm long, up to 4 mm across.

**C. corallina** THUNBG. (Pl. 43/2) (§ I/9) (*C. 'Simiana'*). – S.W.Afr.: Gr. Namaqualand; Cape: Lit. Namaqualand, Karroo. – Resembles **C. dasyphylla**; **root** fusiform; **St.** always erect, c. 2–5 cm tall; **Fl.** with a pedicel rarely 5–6 mm long.

**C. cordata** THUNBG. (§ II/5) (*C. aitonii* BRITT. et BAK. f.). – S.E.Cape: Uitenhage D. to Natal. – ♃ up to 30 cm tall; **St.** erect, slender, freely branching; **L.** cordate or reniform with entire margins, glabrous, 18–20 mm long, densely white-pruinose, margins reddish; **Infl.** lax, often with adventitious buds, **Fl.** greenish-white.

**C. cordifolia** BAK. (§ II/1). – Madag.: Ankaratra Mts. – ♃; **St.** thin, angular, branching, hirsute at first, 7.5–15 cm tall; **L.** fleshy, green, glabrous, 6–8 mm long, amplexicaul; **Fl.** numerous. Acc. FRIEDRICH identical with **C. lineolata**.

**C. cornuta** SCHOENL. et BAK. f. (Pl. 43/5) (§ IV/16). – S.W.Afr.; Cape: Lit. Namaqualand. – Small, succulent ♃; **St.** fleshy, densely leafy; **L.** triangular-ovate, united, c. 10–20 mm long, 8–10 mm across, almost 10 mm thick, the upperside flat, the underside carinate, mealy-grey; **Fl.** small, white-yellowish.

**C. corymbulosa** LINK et OTTO v. **corymbulosa** (§ IV/14). – ♃; **St.** erect with few Br.; **L.** long-triangular, arranged in 4 slightly curved series, rather closely spaced, the lower L. up to 7 cm long, 2 cm across at the base, 2–3 mm thick, both sides convex, green, with dark green spots, the margins finely hairy, the St.later prolonged into an **Infl.** c. 30 cm tall; **Fl.** white.

**C.** — v. **cordata** SCHOENL. – **L.** short, ± cordate; **Infl.** short.

**C. cotyledonis** THUNBG. (Pl. 42/4) (§ VI) (*C. dubia* SCHOENL., *C. cephalophora* v. d. SCHOENL., *C. tayloriae* SCHOENL., *C. ceph.* v. *t.* (SCHOENL.) SCHOENL., *C. torquata* BAK., *C. rehmannii* BAK. f.). – Cape; S.W.Afr.: Lüderitz S. – ♃; **St.** simple, erect, almost 4-angled, leafless, felted, c. 30 cm tall; **L.** crowded on basal shoots, ovate, broadly oblanceolate-oblong, blunt, fleshy, the underside convex, margins entire, ciliate, somewhat white-felted; **Fl.** white, greenish-white or pale yellowish.

**C. crenulata** THUNBG. (§ II/2) (*C. telephioides* HAW.). – ♃; **St.** simple, densely leafy; **L.** elongate, variable in length and width, the upperside convex, the underside slightly carinate, margins dentate; **Fl.** white, in terminal cymes.

**C. cultrata** L. v. **cultrata** (Pl. 43/3) (§ VI). – Cape. – Small, succulent semi-♄; **St.** woody;

**L.** obliquely cultrate-obovate to spatulate, flat, rather stiff, 2–3 cm long, glabrous, sometimes ciliate; **Fl.** whitish.

**C.** — v. **ramosissima** SCHOENL. – Very freely branching; **L.** up to 2 cm long, very shortly and densely hirsute.

**C. cultriformis** FRIEDR. (§ V/2). – S.W.Afr.; Cape: Lit. Namaqualand. – Succulent semi-♄; **St.** woody and branching, ± mat-forming; **Br.** erect to prostrate, rooting from the nodes, fleshy at first, becoming woody, 5–10 cm long, 2–5 mm thick; **L.** ± crowded on the Br.tips, cultriform, spreading to incurved, margins almost angular, tinged red, otherwise green, glabrous, the upperside minutely spotted; **Fl.** white to pale yellow.

**C.** — ssp. **cultriformis**. – Up to 15 cm tall; **L.** oblong, ovate to broad-spatulate, 1–2.5 cm long, 10–15 mm across, 5–7 mm thick.

**C.** — ssp. **robusta** FRIEDR. – Up to 25 cm tall; **L.** oblong-linear to linear-lanceolate, 3.5–7 cm long, 8–15 mm across, 5–7 mm thick.

**C. curta** N. E. BR. (§ IV/10). – Natal; Lesotho: Drakensberge. – Mat-forming ♃, freely branched with many L.-Ros.; **L.** oblong-obovate, margins ciliate; **Fl.** white or red (v. **rubra** N. E. BR.) in terminal cymes. Closely related to **C. milfordae**.

**C. cyclophylla** SCHOENL. et BAK. f. (§ II/2). – S.E.Cape. – ♃; **St.** prostrate, with long Int.; **L.** petiolate, flat, fleshy, ovate, broad-ovate to circular, margins finely crenate-serrate; **Infl.** lax. Shade-loving plant.

**C. cylindrica** SCHOENL. (§ VII). – Cape. – Small succulent ♃; **St.** simple; **L.** broad-triangular, rather pointed, margins curved slightly inwards, the densely arranged L.forming a cylindrical column 8.5 cm tall; **Fl.** pale yellow.

**C. dasyphylla** HARV. (§ I/9) (*C. simiana* SCHOENL., *C. 'Corallina'*). – E.Cape; S.W.Afr.: Gr. Namaqualand. – Dwarf mat-forming ♃; **St.** with numerous Br.which root from the nodes; **L.** dense, compressed-spherical or elliptical, 5 mm long, 4 mm across, light green, with mealy-white powdering and green spots above; Fl. in terminal cymes, white.

**C. deceptor** SCHOENL. et BAK. f. (§ IV/16) (*C. deceptrix* SCHOENL.). – Cape: Lit. Namaqualand; S.W.Afr.: Lüderitz S. – Dwarf, succulent ♃, branching a little from the base; **St.** 4.5 cm and more in length; **L.** arranged in 4 dense rows, densely imbricate, thick, fleshy, c. 15 mm long and across, roundish-triangular, tips and the underside rounded, whitish-grey with raised tessellate-reticulate markings; **Fl.** small, cream to pale yellowish to pinkish.

**C. decidua** SCHOENL. (§ IV/17). – Cape: Karroo. – ♃; **St.** fleshy, becoming woody; **L.** obovate, blunt, tending to fall, very easely rooting, 1–2 cm long, 7–13 mm wide; **Fl.** white.

**C. dejecta** JACQ. (§ IV/5) (*Curtogyne d.* HAW., *Cr. undata* HAW., *C. obvallata* THUNBG., *Rochea albiflora* DC., *C. albiflora* SIMS). – S.Cape. – Semi-♄; **St.** and Br. ± woody,laxly leafy, prostrate and ascending; **L.** united, flat, oblong orbiculate, margins densely fringed with rounded papillae; **Fl.** numerous in terminal, freely branched cymes, white.

**C. deltoidea** THUNBG. (§ I/9) (*C. rhomboidea* N. E. BR.). – Cape: Lit. Namaqualand; S.W. Afr. – Up to c. 6 cm tall; **St.** and Br. fleshy; **L.** united, almost rhomboid, pointed, constricted towards the base, 16 mm long, up to 8 mm across, 4.5 mm thick, the underside blunt-carinate, the upperside broadly and deeply grooved, mealy-grey; **Fl.** small, dirty white.

**C. densa** N. E. BR. (§ IV/16). – S.W.Afr.: Lüderitz S.; Cape: Lit. Namaqualand. – Succulent ⚃, 2–5 cm tall; **L.** amplexicaul, the lower L. crowded, fleshy, obtuse, triangular to nearly globular, slightly tuberculate; **Fl.** capitate-crowded, the Cal. and pedicels beeing softly hairy.

**C. dependens** BOL. (§ IV/3) (*C. harveyi* BRITTEN et BAK. f. nom. prov., *C. h. v. d.* (BOL.) SCHOENL., *C. laxa* SCHOENL., *C. alpestris* AUCT. non THUNBG., *C. montis-moltkei* DTR.). – Cape: Orange Free State; Lesotho; E. Cape, Natal; S.W.Afr.: Windhoek D. – Semi-♄; **St.** dependent, ascendent or rarely erect, becoming ± woody, leafy; **L.** sessile, lanceolate, pointed, ± flattened, hairy, less than 15 mm long, green; **Fl.** white.

**C. dewinteri** FRIEDR. (§ VI) (*Sphaeritis biconvexa* ECKL. et ZEYH., *C. b.* (ECKL. et ZEYH.) HARV.). – S.W.Cape. – Resembles **C. rogersii** but more bristly haired and **L.** more clavate.

**C. dinteri** SCHOENL. (§ IV/16). – S.W.Afr.: Klinghardt Mts.; Cape: Lit. Namaqualand, Richtersveld. – **L.** seldom more than 1 cm long, scarcely half as wide, minutely densely tomentose, ashygrey-green.

**C. drakensbergensis** SCHOENL. (§ IV/8). – Cape Drakensberg Mts. – Resembles **C. vaginata**; Int. almost glabrous.

**C. dregeana** HARV. (§ IV/9). – E.Cape to S.Natal. – ⚃ with spreading Br., up to 52 cm tall; **Br.** and twigs ± herbaceous, slender, hairy; **L.** ovate or ovate-oblong, c. 12 mm long, c. 5 mm across and thick, green, bristly papillose; **Fl.** small, white.

**C. elegans** SCHOENL. et BAK. f. (§ IV/16). – Cape: Lit. Namaqualand. – Similar to **C. globosa**; **L.** green, glabrous; **Infl.** capitate, pedicels 2 cm long.

**C. ellenbeckiana** SCHOENL. (§ IV/8). – S.Ethiopia. – Similar to **C. alba**. The upper part of the St. with downwardsly pointing hairs.

**C. engleri** SCHOENL. (§ IV/14). – S.E.Cape. – ⚃, branching from the base, glabrous; **L.** crowded, rosulate, ovate, rather pointed, up to 3.5 cm long, spotted, glabrous; **Fl.** dirty yellowish, in a terminal thyrse.

**C. ericoides** HAW. (Pl. 44/1) (§ IV/3) (*C. jacobseniana* v. POELLN.). – S.Cape; Natal. – Dwarf semi-♄ up to 30 cm tall; **Br.** woody, with distant twigs; **L.** ± densely crowded, ovate, lanceolate, pointed, erect, margins slightly revolute, light green; **Infl.** sessile, Fl. white.

**C. ernestii** SCHOENL. et BAK. f. (§ IV/17). – Cape: Karroo. – Resembles **C. lanuginosa**; **L.** ovate-oblong, distant, c. 1 cm long.

**C. erosula** N. E. BR. (§ VI). – Cape: Lit. Namaqualand; S.W.Afr.: Lüderitz S. – ⚃; **Br.** prostrate; **L.** oblong-oblanceolate or linear-lanceolate, ± pointed; **Infl.** composed of densely capitate cymes,

**C. exilis** HARV. (§ IV/15). – Cape: Lit. Namaqualand. – ⚃ up to 6 cm tall; **St.** fleshy, branching basally; **L.** oblong, 12 mm long, thick, fleshy, densely grey-papillose; **Fl.** white.

**C. expansa** AIT. (§ I/4). – S.Afr. – ⊙ or ⚃; **St.** erect or ± prostrate, freely branched, thin; **L.** ± terete, linear to linear-oblanceolate, distant, green; **Fl.** small, greenish-white, with long thin pedicels.

**C. falcata** WENDL. (§ IV/6) (*C. perfoliata* sensu SCHOENL. p. part., *C. obliqua* ANDR., *Larochea f.* HAW., *Rochea f.* DC., *C. perfoliata* v. *miniata* TOELKEN, *C. falx* LDG. nom. inval., *C. johanniswinkleri* LDGR.). – E. Cape. – Up to 1 m tall, usually branching; **L.** lanceolate and caniliculate to falcate and flat, the lamina standing nearly vertical, (7–) 10 (-15) cm long, 3–4 cm across; **Fl.** scarlet-red to orange, also lighter.

**C. —** cv. Trifaria. – Cultivar from the garden of A. BAYNES, Bradford, England; **L.** arranged in 3 series.

**C. fastigiata** SCHOENL. (§ V/1). – S.W.Cape. – Semi-♄ 25–30 cm tall; **L.** numerous, evenly and densely arranged at the shoot tips, ovate-oblong, pointed, glabrous except for the margins with projecting cilia.

**C. fergusioniae** SCHOENL. (§ VI). – Cape. – Dwarf acaulescent succulent ⚃; **L.** obovate or oblong, c. 14 mm long, with soft whitish-grey hairs.

**C. filicaulis** HAW. (§ I/4) (*C. maritima* SCHOENL., *C. uniflora* SCHOENL.). – S.-S.E. Cape. – ⚃; **St.** thin, freely divaricately branched, ± prostrate, rooting; **L.** linear, subterete, subcanaliculate above, often flushed with red; **Fl.** solitary or few in the axils of the upper L., white, subcampanulate, with pedicels up to 5 mm long.

**C. flabellifolia** HAW. (§ III) (*C. weissii* N. E. BR.). – S.W. and central Cape. – ⚃, underground **St.** tuberous, annual shoots erect, 1.5 to 3 cm high; **L.** 2, fan-shaped, margins crenate; **Fl.** small, white, in ± branched panicles. The apical portion of the annual shoots forms a new tuber.

**C. flanaganii** SCHOENL. et BAK. f. (§ IV/11). – S.E.Afr. – Related to **C. sediflora**; **L.** up to 5 cm long, upper L. smaller, somewhat united, broad-lanceolate or oblanceolate.

**C. flava** L. (§ IV/5) (*Rochea f.* DC.). – S.Cape: Table Mt. – **St.** mostly simple; **L.** linear to linear-lanceolate, fringed, with papillose hairs; **Infl.** with white bristles on Ped. and bracts, Fl. creamy-white to pale yellow.

**C. fragilis** BAK. (§ I/4). – Madag. – Brittle ⊙ - ⊙; **St.** erect, dichotomously branching, hirsute, up to 10 cm tall, branching basally; **L.** fleshy, 3–4 mm long, hairy from midway down to the base; **Fl.** solitary, white to pale reddish.

**C. fusca** HERRE (§ IV/2). – Cape: Lit. Namaqualand; S.W.Afr.: Lüderitz S. – Glabrous ⚃; **St.** 20–30 cm tall, 5 mm thick, Int. 4–5 mm long, 1–2 mm long in the upper part; **L.** 2 to

**Crassula**

10 cm long, 9–15 mm across, amplexicaul, tapering, sometimes recurved at the tip, fleshy, slightly convex on the underside, glabrous but the margins covered with tiny white papillae, green or rust-brown; **Fl.** white to light pink, many in a terminal, pyramidal thyrse.

**C. galunkensis** ENGL. (§ II/1). – Kenya. – ♃; resembles **C. alsinoides**, but **Br.** ascending; **L.** more succulent; **Fl.** pink.

**C. garibina** MARL. et SCHOENL. (§ IV/15). – Cape: Lit. Namaqualand; S.W.Afr. – **Semi-♄**; **St.** thick, 10 cm high; **L.** oblong, 1.5 to 4 cm long, 10 mm across, grey-green, finely papillose-pubescent; **Fl.** in dense terminal cymes, white.

**C. gillii** SCHOENL. (§ IV/13). – S.E.Afr. – Related to **C. rosularis**; **Infl.** with 2–3 small terminal heads.

**C. glauca** SCHOENL. (§ II/5). – S.E.Cape to Natal. – ♃, up to 30 cm tall; **St.** and **Br.** slender; **L.** petiolate, flat, oblong-round or almost circular, ± cuneate at the base.

**C. glabrifolia** HARV. (§ V/5) (*C. eendornensis* DTR. nom. nud., *C. interrupta* v. g. (HARV.) SCHOENL.). – Cape: Lit. Namaqualand; S.W. Afr. – Resembling **C. scalaris**; **L.** glabrous except for the margins with long cilia.

**C. globosa** N. E. BR. (§ IV/16). – S.W.Cape. – Small ♃; **St.** fleshy with few **Br.**, densely leafy; **L.** spherical, somewhat flattened on one side, usually 3 pairs, tessellate-reticulate; **Fl.** white.

**C. globularioides** BRITTEN ex OLIVER (§ IV/9). – Mountains of trop. E Afr. – ♃, growing in tufts; **St.** short, ± shortly branched; **L.** ligulate, glabrous, margins fringed with cilia, ± densely set; **Infl.** terminal, Ped. 1–5 cm long, Fl. white.

**C. goetzeana** ENGL. (§ IV/8). – N.Malawi. – ♃ with a thickened root; **St.** simple, densely leafy at the base; **L.** sheathing, triangular-lanceolate, the margins finely papillose; **Infl.** much branched, rather dense, Fl. small, white. (Perhaps identical with **C. acinaciformis**.)

**C.** × **graeseri** BOOM (*C. hybrida* HORT.). – Hybrid: **C. rosularis** × **C. falcata**. – **L.** rosulate, decussate and opposite, spreading, 4–7 cm long, 0.5–3 cm across, oblong, pointed at the tip, grey-reddish, the underside convex, both surfaces with scattered pearl-like papillae; **Infl.** corymbose-paniculate, Fl. numerous, 6–8 mm ∅, pink to white.

**C. griquaensis** SCHOENL. (§ IV/3). – Cape: Griqualand. – Glabrous **semi-♄** with Br. and numerous short shoots in the L. axils; **L.** terete, pointed; **Fl.** white in shortly pedunculate terminal cymes.

**C. grisea** SCHOENL. (Pl. 43/7) (§ IV/16). – Cape: Lit. Namaqualand; S.W.Afr.: Lüderitz S. – Succulent ♃, scarcely 15 cm tall; **St.** fleshy, fragile, branching from the base or higher, not densely leafy; **L.** spreading almost horizontally, c. 3 cm long, lanceolate, 8–9 mm across, 5–6 mm thick, roundish-carinate on the underside, white-grey; **Fl.** small, white.

**C. hemisphaerica** THUNBG. (Pl. 43/6) (§ IV/14). – Cape: Calvinia D., Lit. Karroo. – Monocarpic or ♃; **St.** short, with 8–10 L. pairs forming a dense hemispherical Ros.; **L.** flat, rounded, finely white-ciliate; **Fl.** white in a terminal thyrse.

**C. herrei** FRIEDR. (§ VI). – Cape: Lit. Namaqualand. – **Semi-♄** 10–25 cm tall; **St.** woody, ± branching; **Br.** opposite, short, erect or prostrate, rooting at the nodes, 3–5 cm long, 3–5 mm thick, with a red-brown bark; **L.** oblong-ovate to almost clavate, the upperside convex to flat, the underside convex, blunt or minutely apiculate, up to 1 mm across and thick; **Fl.** numerous, the Cal. papillose-hirsute, the Cor. urn-shaped, white.

**C. heterotricha** SCHINZ (§ IV/6). – Resembles in habit **C. falcata**. ♄, somewhat stiff, much branched from the base; **L.** obliquely decussate, ± linear-falcate, 7–15 cm long, 1–1.5 cm broad, thick, falcately turned downwards, more green than grey, very finely and shortly hairy; **Infl.** up to 10 cm ∅, Fl. white.

**C. hirta** THUNBG. (§ VI). – Cape: Lit. Karroo. – Stemless ♃; **L.** in a basal Ros., linear-subulate, 2–9 cm long, ± densely white and rough haired; **Infl.** up to 30 cm tall, Fl. white.

**C. hirtipes** HARV. (§ V/6). – S.W.Cape. – Related to **C. hystrix**; dwarf ♃; **L.** globular-subclavate, glabrous or with few retrorse bristles.

**C. hispida** (HAW.) D. DIETR. (§ V/4) (*Globulea h.* HAW.). – S.E.Cape. – ♃; **St.** short, ± branched; **L.** subulate, in dense Ros., 3–7 cm long, covered with spreading coarse hairs, L. of the flowering St. distant, diminishing upwards; **Infl.** a dense, ± branched cyme, Fl. white to greenish-white.

**C. hottentotta** MARL. et SCHOENL. (§ IV/17) (*C. merxmuelleri* FRIEDR.). – S.W.Afr.: Lüderitz S.; Cape: Lit. Namaqualand. – Dwarf ♄ up to 10 cm tall; **St.** almost woody, covered with the remains of old L., 5–7 mm ∅, Br. short, Int. densely brown papillose-hairy; **L.** blue-green, hemispherical to almost spherical, round to almost rhomboid, margins somewhat angular, abruptly and teretely constricted at the base, densely covered with coarse roundish papillae, 1–2 cm long and across, c. 1.2 cm thick, the margins somewhat hairy at the base; **Fl.** white to yellowish.

**C. humbertii** B. DESC. (§ I/4). – S.Madag. – ♃, 3–5 cm tall, ± branching; **St.** terete, forking; **L.** united, fleshy, oblong-elliptical, pointed, with entire margins, 6.5–8.5 mm long, 2.5–4 mm across, pale green, sometimes reddish and with red spots on the upperside; **Fl.** solitary, white.

**C. humilis** N. E. BR. (§ IV/16) (*C. mesembrianthoides* SCHOENL. et BAK. f., *C. liquiritiodora* DTR. nom. nud., *C. schoenlandii* JACOBS., *C. corpusculariopsis* BOOM, *C. dinteri* AUCT.). – S.W.Afr.: Lüderitz Bay S.; Cape: Lit. Namaqualand. – Dwarf ♄; **St.** fleshy, branching basally, densely leafy; **L.** united, glabrous, green; **Fl.** whitish in lax or ± capitate cymes.

**C. hystrix** SCHOENL. (§ V/6). – Cape: Lit. Namaqualand. – ♃, 2–7 cm tall; **St.** becoming rather woody; **L.** ovate-elliptic to oblong, narrowing somewhat towards the base, 6–8 mm long, 4–6 mm across, 3–5 mm thick, very fleshy, grey-

Plate 50. 1. **Decabelone grandiflora** K. Schum. (Photo: H. Lang); 2. **Cyphostemma laza** B. Desc. v. laza (Photo: J. Bogner); 3. **Decaryia madagascariensis** Choux (Photo: as 2); 4. **Dendrosicyos socotrana** Balf. f. (acc. Schweinfurth in A. Engler, Pfl. welt Afr. I, 1910).

10/B Lexikon of Succulent Plants

Plate 51. 1. **Didierea madagascariensis** H. BAILL. (Photo: W. RAUH); 2. **D. trollii** CAPURON et RAUH (Photo: W. RAUH); 3. **Diopogon hirtus** ssp. **borealis** H. HUBER; 4. **Diopogon** Infl.

Plate 52. 1. Dorstenia crispa ENGL. v. crispa (Photo: P. R. O. BALLY); 2. D. bornimiana SCHWEINF. v. bornimiana (Photo: W. RAUH); 3. D. foetida (FORSK.) SCHWEINF. v. foetida (Photo: C. W. PITCHER); 4. D. hildebrandtii ENGL. (Photo: as 3).

Plate 53. 1. Drimia Infl.; 2. D. haworthioides BAK.; 3. × Dudleveria spiralis (DEL.) ROWL. (Photo: I. V. T. Wageningen, Netherlands); 4. Dolichos seineri HARMS (Photo: K. DINTER); 5. Diplocyatha ciliata (THUNBG.) N. E. BR.; 6. Dischidia pectinoides H. H. W. PEARSON (Photo: P. FISCHER).

Plate 54. 1. **Dudleya albiflora** ROSE (Photo: R. MORAN); 2. **D. attenuata** ssp. **orcuttii** (ROSE) MORAN; 3. **D. candida** BRITT. (Photo: as 1); 4. **D. farinosa** (LINDL.) BR. et R.

Plate 55. 1. Dudleya densiflora (ROSE) MORAN; 2. D. lanceolata (NUTT.) BR. et R. (Photo: J. MARNIER-LAPOSTOLLE); 3. D. formosa MORAN (Photo: R. MORAN); 4. D. hassei (ROSE) MORAN (Photo: as 3); 5. D. ingens ROSE; 6. D. rigida ROSE (Photo: as 3).

Plate 56. 1. **Huernia andreaeana** (RAUH) LEACH (Photo: D. ANDREAE); 2. **Duvalia parviflora** N. E. BR. (Photo: H. LANG); 3. **D. polita** v. **transvaalensis** (SCHLTR.) WHITE et SLOANE (Photo: as 2); 4. **D. procumbens** DYER (Repr.: Fl. Pl. Afr. XXXI ,1956, Pl. 1218, Drawing: C. LETTY); 5. **Dyckia sulfurea** C. KOCH.

Plate 57. 1. **Echeveria** Infl.: Cincinnus, raceme; 2. **E. agavoides** LEM. v. **agavoides**; 3. **E. elegans** ROSE; 4. **E. ciliata** MORAN (× 4) (Photo: R. MORAN); 5. **E. carnicolor** (BAK.) MORREN.

Plate 58. 1. Echeveria leucotricha J. A. Purp. (Photo: J. Marnier-Lapostolle); 2. E. derenbergii J. A. Purp.; 3. E. gibbiflora DC. cv. Carunculata; 4. E. globulosa Moran (Photo: R. Moran) 5. E. glauca Bak. v. glauca; 6. E. pulvinata Rose cv. Ruby.

Plate 59. 1. Echeveria nodulosa (BAK.) OTTO (Photo: R. MORAN); 2. E. setosa ROSE et PURP.; 3. E. peacockii (BAK.) MORREN; 4. E. cv. Hoveyii; 5. E. subrigida (ROBINS. et SEAT). ROSE.

Plate 60. 1. **Echidnopsis archeri** BALLY (Photo: J. BOGNER); 2. **E. cereiformis** HOOK. f. v. **cereiformis**; 3. **E. urceolata** BALLY (Photo: P. R. O. BALLY, enlarged); 4. **Edithcolea grandis** N. E. BR. v. **grandis** (Photo: as 3).

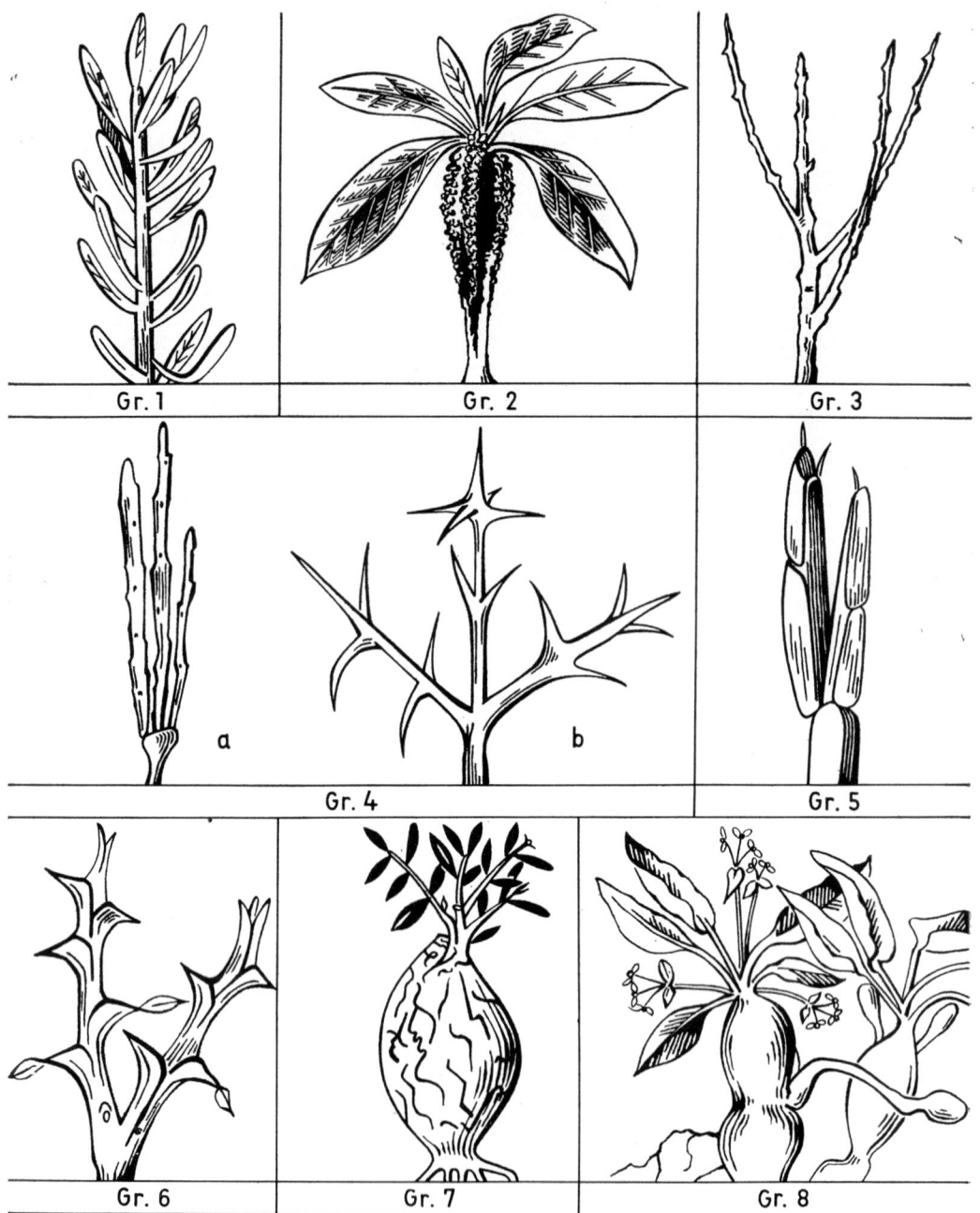

Plate 61. Typical species of the Groups 1-8 of the Genus Euphorbia (see pages 199, 200)
Gr. 1. Euphorbia atropurpurea Brouss; Gr. 2. E. leuconeura Boiss.; Gr. 3. E. frutescens N. E. Br.; Gr. 4a. E. dregeana E. Mey., b. E. brachiata E. Mey.; Gr. 5. E. sipolisii N. E. Br.; Gr. 6. E. hamata Sweet; Gr. 7. E. trichadenia Pax (acc. C. Letty); Gr. 8. E. tuberosa L. (acc. Burmann. Rar. Afr. Pl., Pl. 1778).

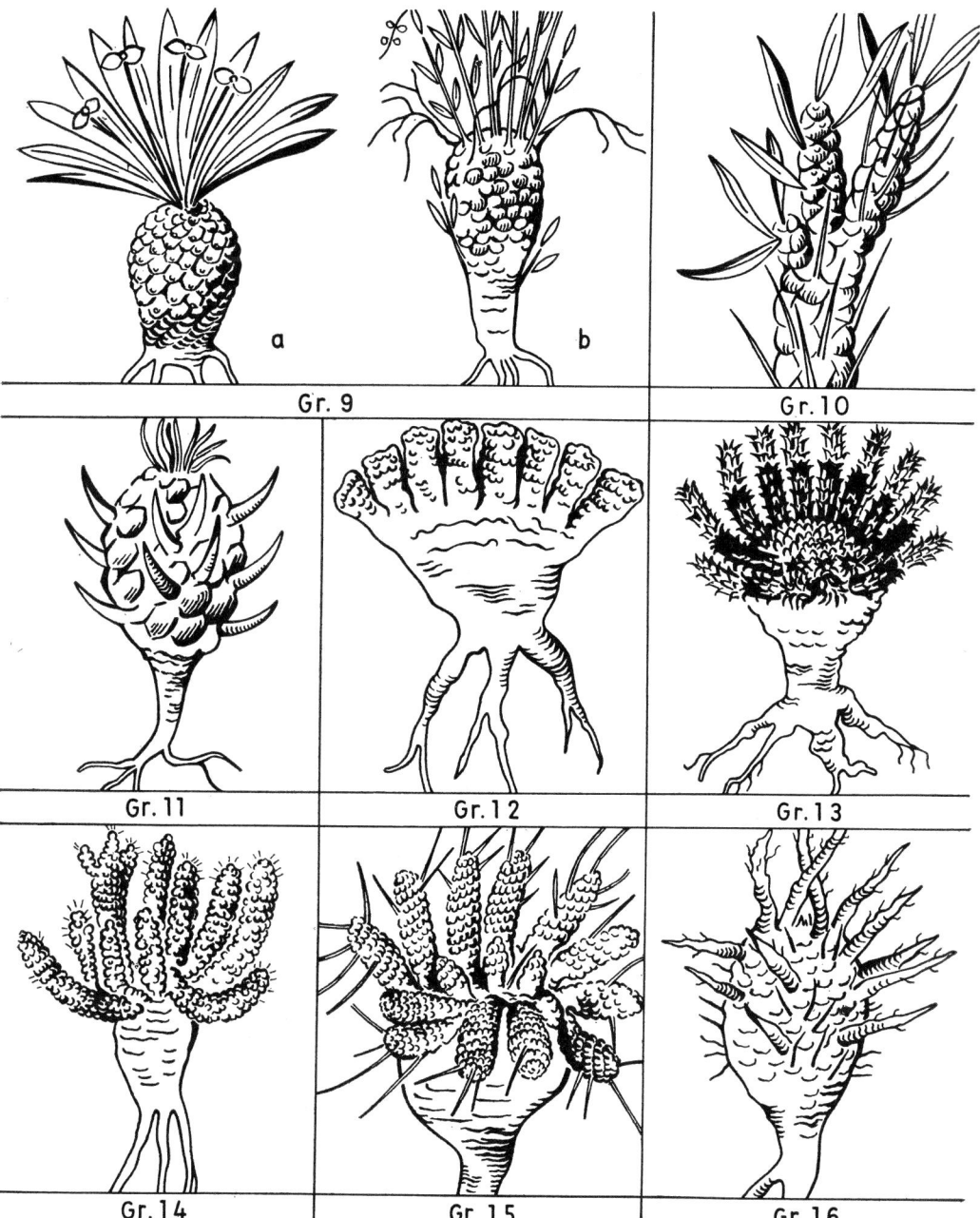

Plate 62. **Typical species of the Groups 9-16 of the Genus Euphorbia** (see page 200, 201)
Gr. 9 a. Euphorbia bupleurifolia JACQ., b. E. monteiri HOOK. f.; Gr. 10. E. loricata LAM.; Gr. 11. E. fasciculata THUNBG.; Gr. 12. E. clavarioides BOISS. v. clavarioides; Gr. 13. E. pugniformis BOISS.; Gr. 14. E. inermis MILL.; Gr. 15. E. pentops MARL.; Gr. 16. E. multiramosa NEL.

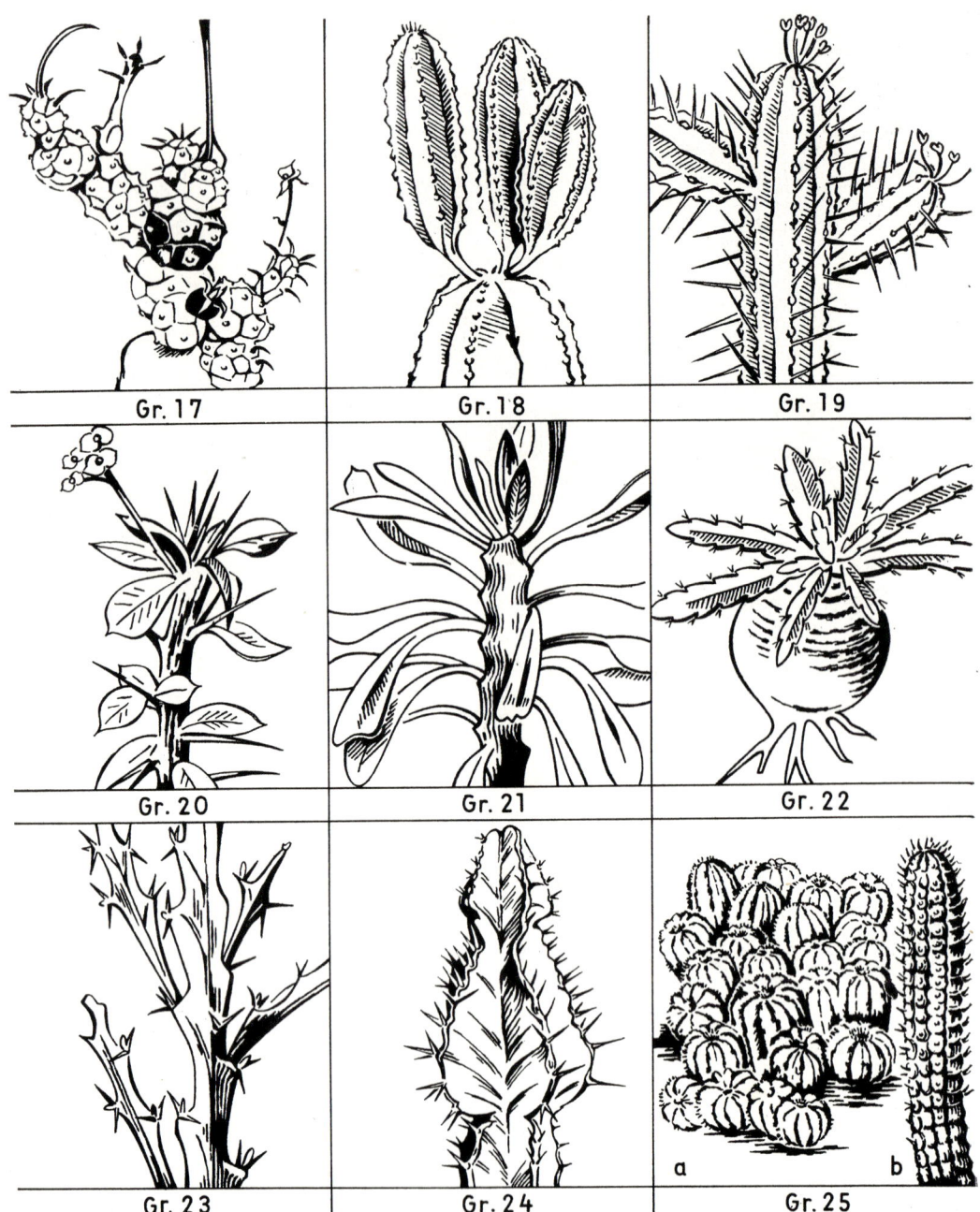

**Plate 63.** Typical species of the Groups 17–25 of the Genus Euphorbia; Groups 26–28 on **Plate 64** (see pages 201–203).
**Gr. 17.** Euphorbia globosa (Haw.) Sims; **Gr. 18.** E. tubiglans Marl.; **Gr. 19.** E. heptagona L.; **Gr. 20.** E. milii v. splendens (Boi. ex Hook.) Ursch et Leandri; **Gr. 21.** E. neriifolia L.; **Gr. 22.** E. stellata Wild.; **Gr. 23.** E. ramipressa Croiz.; **Gr. 24.** E. cooperi N. E. Br. v. cooperi; **Gr. 25 a.** E. multiclava Bally et Carter; **Gr. 25 b.** E. columnaris Bally.

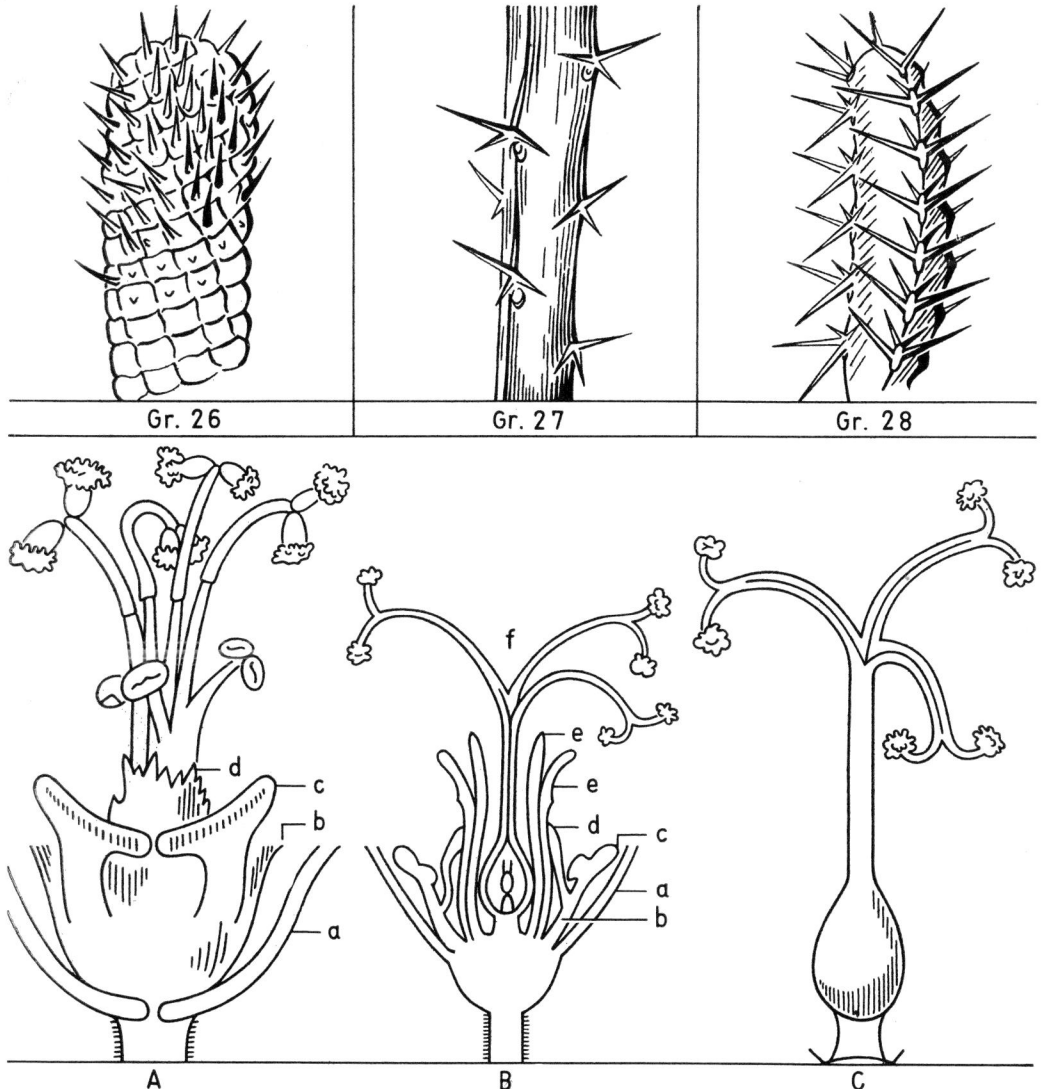

**Plate 64. Gr. 26. Euphorbia darbandensis** N. E. Br.; **Gr. 27. E. ballyana** Rauh; **Gr. 28. E. subsalsa** Hiern.

**Cyathia of Euphorbia.** After a drawing by J. G. Y. Meier from Succulenta, 1932.

**A. Male cyathium.**
a) bracts at the base of the involucre; b) the basal portion of the involucre; c) the involucral glands; d) the involucral lobes, with the male flowers above, two being in bud and four at maturity. Each consists of a pedicel below jointed to the filament above, which bears two anthers at the top. In most species there are many hair-like bracteoles amongst the male flowers.

**B. Bisexual cyathium.** (Vertical section.)
a) bracts under the involucre; b) base of the involucre; c) glands on the involucre; d) lobes of the involucre; e) male flowers surrounding one female flower which extends from the ovary below to the point of separation of the styles, the apex of the bifid tips forming the stigmatic surface or stigma.

**C. Female flower.**
No perianth; the single ovary is surmounted by a style and 3 bifid stigmas.

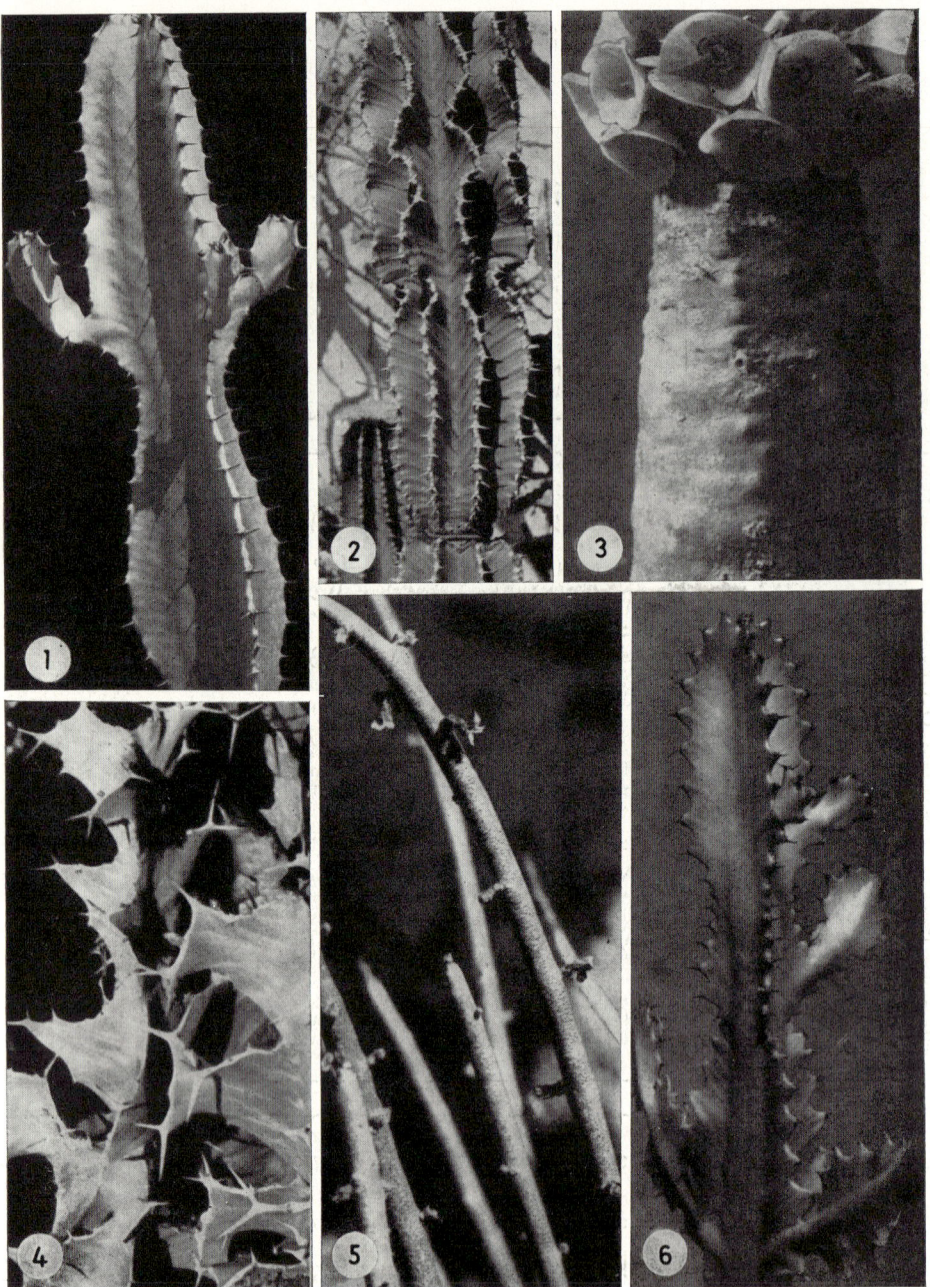

**Plate 65. 1. Euphorbia acrurensis** N. E. Br. (Photo: J. Marnier-Lapostolle); **2. E. abyssinica** Raeuschel (Photo: as 1); **3. E. ankarensis** P. Boit. (Photo: as 1); **4. E. angularis** Klotzsch (Photo: as 1); **5. E. antisyphilitica** Zucc. (Photo: as 1); **6. E. antiquorum** L. (Photo: as 1).

green, covered with white bristle-like hairs; **Fl.** white to pale yellowish, in a branched panicle.

**C. illichiana** ENGL. (§ IV/9). – Tanzania: W.Usambara. – ♃; **St.** short, branched from the base; **L.** oblong, hairy, blunt; **Fl.** red-violet or white.

**C. inamoena** N. E. BR. (§ VI). – S.W.Cape. – ♃, nearly stemless; **L.** in basal Ros., broad-linear or somewhat narrowed, margins without cilia; **Infl.** 10–20 cm high, Fl. white in 3–7 capitate cymes.

**C. inandensis** SCHOENL. et BAK. f. (§ II/2). – S.E.Cape to Natal. – Resembles **C. sarmentosa**; **L.** ovate or elliptical, narrowing towards the base, ± indistinctly dentate or the margins almost entire.

**C. incana** (ECKL. et ZEYH.) HARV. (§ V/3) (*Sphaeritis i.* ECKL. et ZEYH.). – Cape: Lit. Namaqualand, Lit. & Gr. Karroo; S.W.Afr. – Laxly branched **semi-**♄ with virgate laxly leaved **Br.**; **L.** oblong to ovoid-elliptic, grey-green, pubescent, easily dropping and rooting; **Fl.** white in terminal cymes.

**C. intermedia** SCHOENL. (§ IV/13). – S.E.Cape. – Related to **C. rosularis**; **L.** oblong-cuneate, ovate or obovate with a cuneate base; **Fl.** white with red anthers.

**C. interupta** E. MEY. (§ V/5). – S.W.Cape. – ♃, often becoming woody at the base and much shooting forming tufts; **L.** radical or almost so, ± flat, roundish, covered with whitish hairs, margins white-hairy; **Fl.** in small cymes, arranged in a thyrse 8–15 cm high.

**C. cv. Jade Necklace** (Pl. 44/2). – Hybrid: **C. falcata** × **C. marnierana** (HORT. Kimnach). – ♄; **St.** erect, later spreading, 1.5 cm thick, up to 20 cm long; **Br.** 3–5 mm thick, forking dichotomously several times; **L.** dense, perfoliate, the free margins being convolute, lamina broad-triangular, slightly navicular, 7 mm long, 15 mm across, c. 5 mm thick, with darker green spots, margins reddened, glabrous, but with a row of white papillae near the base; **Fl.** 8 mm ⌀, white tinged with delicate pink.

**C.** × **justi-corderoyi** JACOBS. et v. POELLN. ('Justus Corderoy') (Pl. 44/3) (*C. j.-c.* N. E. BR.). – Acc. N. E. BROWN a hybrid: **C. cooperi** × **C. falcata**. – A mat-forming ♃; **L.** numerous, lanceolate, 22 mm long, 8–10 mm across, thick-fleshy, the upperside flat, the underside roundish, dark green with red spots, with white hairs in distinct longitudinal rows; **Fl.** reddish.

**C. karasana** FRIEDR. (§ IV/15). – S.W.Afr. – ♃; **roots** somewhat fleshy; **St.** 5–10 cm tall, branching, Br. 2–3 cm long; **L.** 6–8, crowded, sessile, broadly and irregularly lanceolate to ovate-elliptical, short-pointed, with red angular margins, densely papillose-hirsute, 1–2 cm long, 5–12 mm across, 5–7 mm thick; **Fl.** white.

**C. cv. Klein Duimpje.** – Hybrid: **C. marnierana** × **C. rupestris.** – Compact **semi-**♄; **Fl.** numerous, white.

**C. klinghardtensis** SCHOENL. (§ IV/17). – S.W.Afr.: Klinghardt Mts. – Small succulent ♃; **St.** erect, up to 7 cm high; **L.** almost spherical or ovate, pointed, 1 cm long and across, 6 mm thick, ± densely set, ashy grey, densely papillose-hairy; **Fl.** white in dense capitate cymes.

**C. kuhnii** SCHOENL. (§ IV/3). – Cape. – Scarcely distinguishable from **C. punctulata**; Pet. 6 mm long. Both closely allied to **C. biplanata**

**C. lactea** AIT. (Pl. 44/4) (§ II/3) – E.Cape: Natal, Transvaal. – **Semi-**♄ 30–60 cm tall; **St.** and Br. thick, erect or curved, prostrate, with short Br.; **L.** ± rhombic-obovate, somewhat pointed or long-pointed, narrowing towards the base, fleshy, flat, glabrous, green, with white dots along the margins; **Fl.** white, scented, numerous.

**C. lanuginosa** HARV. (§V/17). – Cape: Karroo. – Mat-forming ♃ rarely 10 cm tall; **St.** and **Br.** thin, prostrate ascending rooting; **L.** semiterete, ovate, pointed or rather so, softly hairy; **Fl.** small, white, few in terminal cymes.

**C. lasiantha** E. MEY. (§ IV/9). – S.E.Afr. – ♃; **St.** prostrate; **L.** small, roundish or broad-ovate; **Fl.** small.

**C. laticephala** SCHOENL. (§ III). – Cape: Karroo. – Small, compressed ♃; **L.** fleshy, in 4 series, sharply reflexed, 3 cm long, 1.2 cm across the base, thick, subulate, acute, the upperside with a distinct keel, the underside slightly rounded, scurfy; **Fl.** in large heads, numerous, white.

**C. latispatulata** SCHOENL. et BAK. f. (§ II/2). – S.E.Cape to Natal. – Resembles a smaller form of **C. inandensis**; **L.** ovate or circular, narrowing towards the base, margins serrate-dentate.

**C. lettyae** PHILLIPS (§ IV/14). – S.Afr. – ♃; **L.** rosulate, 5–6 cm long, 2 cm across, oblong, margins ciliate. Resembles **C. barbata** with laxer Infl.

**C. leucantha** SCHOENL. et BAK. f. (§ V/1). – S.W.Cape. – **Semi-**♄ with leafy Br.; **L.** numerous, densely arranged on the non-flowering Br., oblong or oblong-ovate; **Fl.** cream, in ± flat-topped cymes.

**C. liebuschiana** ENGL. (§ IV/9). – W.Tanzania; Kenya. – ± mat-forming ♃; **St.** and Br. ascending, brown, laxly leafy beneath, densely leafy at the tip; **L.** 2–5 cm long, oblong, elliptical or ovate, glabrous, often flushed with red, margins ciliate; **Fl.** white in pedunculate cymes.

**C. lineolata** AIT. spec. aggreg. (§ II/1) (*C. involucrata* SCHOENL.). – Humid subtropical parts of S.Afr. – Very polymorphic ♃; **St.** spreading, slender, glabrous or hairy; **L.** sessile or narrowed below, glabrous or ± hairy, often with numerous lines (only visible in dried specimens); **Fl.** white.

**C. littlewoodii** FRIEDR. (§ IV/15). – S.W.Afr.: Keetmanshoop, Warmbad. – Dwarf ♄ 7 to 17 cm tall; **St.** branching; **Br.** erect, 5–10 cm tall, fleshy, 3–7 mm thick, hispid; **L.** fleshy, semi-terete, oblong-linear to oblong-lanceolate, almost pointed, perfoliate and united at the base, falcately involute, the upperside convex, the underside flat to slightly convex, margins angular towards the tip, the surface ± densely papillose-hairy, with tiny blisters, the tip often red, 1.5–3.7 cm long, 6–10 mm across, 3–6 mm thick; **Fl.** with campanulate Cor., white.

**C. loganiana** COMPT. (§ V/3). – Cape: Karroo, Whitehill. – Erect semi-♄ c. 30 cm tall, branching from the base; **Br.** thin, stiff, virgate, with dark brown rind; **L.** 2–7 mm long, 3–4 mm across, 3–4 mm thick, the upperside flat-convex, green, margins red at the tip, very easily dropping off and rooting; **Fl.** white. Closely allied to **C. subaphylla**.

**C. luederitzii** SCHOENL. (§ IV/14) (*C. triebneri* SCHOENL. ex JACOBS. nom. illeg.). – S.W.Afr. – ♃, 5–20 cm tall; **St.** undeveloped; **L.** rosulate, rhombic-oblanceolate, fleshy, the upperside almost flat, the underside rather convex, 1–2 cm long, 5–8 mm across, pale yellowis-hgreen, spotted with red, glabrous except for the papillose-ciliate margins; **Fl.** white, in thyrsoid Infl.

**C. lycopodioides** LAM. spec. aggreg. (§ I/7). – S.W.Afr.; S.Cape, S.E.Cape. – Semi-♄ up to 30 cm tall; **St.** slender, erect, branching irregularly, with the L.arranged ± densely in 4 series throughout; **L.** small, imbricate, triangular-ovate, short-pointed, green; **Fl.** minute, produced in the axils, greenish-white, yellow or the tips red-brown.

**C. — v. lycopodioides** (Pl. 44/5) (*C. muscosa* THUNBG., *C. anguinea* HARV., *C. imbricata* AIT., *Tetraphile litoralis* ECKL. et ZEYH., *T. l.* ECKL. et ZEYH., *T. propinqua* ECKL. et ZEYH., *T. polypodacea* ECKL. et ZEYH., *C. l.* v. *acuminata* JACOBS.). – **L.** densely arranged, 3–5 mm long, 2.4 mm across the base.

**C. —** cv. Monstrosa. – Monstrous cultivar. **St.** short, spreading, crest-like.

**C. — v. pseudolycopodioides** (DTR. et SCHINZ) WALTH. (*C. p.* DTR. et SCHINZ). – **St.** thicker and more robust; **L.** blunter, grey-green, with numerous short spurs from the axils.

**C. — v. — f. fragilis** HUBER. – Form with easily-detached short side-shoots.

**C. — v. — f. fulva** HUBER. – Brownish-yellow form.

**C. — v. purpusii** JACOBS. (Pl. 44/6) (*C. ericoides* HORT., *C. purpusii* HORT.). – Very large variety; **L.** curved, spreading, ovate-triangular, 5–6 mm long, 4 mm across, 2 mm thick, the underside very rounded.

**C. —** cv. Variegata (*C. l. v. v.* LAMB. et HALL). – Unusual cultivar having **L.** with a silvery sheen.

**C. — v. viridis** BGR. – Resembles the type; **L.** more pointed, light green.

**C. macowaniana** SCHOENL. (§ IV/2). – Cape: Lit. Namaqualand; S.W.Afr.: Lüderitz S. – Semi-♄; **L.** oblong-lanceolate to subulate, pointed, 4–15 cm long, glabrous; **Fl.** small, pink, many, in terminal panicles.

**C. — v. macowaniana** (*C. ramosa* HARV.). – **L.** subulate, up to 10 cm long, St. with few Br.

**C. — v. crassifolia** SCHOENL. – **L.** thicker, up to 15 cm long, St. more branched.

**C. maculata** SCHOENL. (§ IV/14). – S.E.Cape. – Succulent ♃, branching from the base; **L.** rosulate, oblong-cuneate, blunt or pointed, rather thick, light green with darker green spots; **Fl.** whitish.

**C. marginalis** SOL. (§ II/1) (*C. profusa* HOOK. f., *C. centauroides* HARV.). – Cape. – ♃; **St.** slender, erect; **L.** sessile, fleshy, flat, green, margins entire, with a row of darker spots.

**C. —** cv. Rubra. – **L.** purple-red.

**C. marnierana** HUBER et JACOBS. (Pl. 45/1) (§ IV/2). – Cape: Sutherland D. – ♃; **St.** ± erect, 4 cm and more in length, with forking Br.; **L.** broadly cordate and connate, forming a disc, glabrous, bluish, margins reddened, 3.5 mm long, 7 mm across, 2.5–3 mm thick; **Fl.** numerous, campanulate, white.

**C. mesembryanthemopsis** DTR. (Pl. 45/3) (§ IV/15) (*C. rapacea* SCHOENL. nom. nud.). – S.W.Afr., Gr. Namaqualand; Cape: Lit. Namaqualand. – ♃; **root** napiform; **St.** very short, with a Ros. of 4–5 L. pairs; **L.** cuneate, thick-fleshy, angular, with a truncate triangular-ovate blunt tip, whitish grey-green; **Fl.** white.

**C. mesembryanthoides** (HAW.) DIETR. (§ V/4) (*Globulea m.* HAW., *Sphaeritis trachysantha* ECKL. et ZEYH., *C. t.* (ECKL. et ZEYH.) HARV.). – S.E.Cape. – Semi-♄ up to 30 cm tall; **Br.** erect, slender but stiff, roughly hairy; **L.** terete, tapering, thickly covered with retrorse bristles; **Fl.** white-cream.

**C. meyeri** HARV. (§ IV/9). – S.E.Afr. – Resembles **C. ramuliflora**; **L.** oblong-lanceolate.

**C. micans** VAHL ex BAILL. (§ II/1). – Madag. – Glabrous ⊙ to ♄; **St.** slender, erect or prostrate, rooting from the base; **L.** obovate, tapering; **Fl.** 2 cm long on slender pedicels, pink.

**C. milfordae** BYLES (§ IV/10). – Lesotho. – Dwarf, only slightly succulent, mat-forming ♃; **St.** numerous, branching, thin, very slightly hairy, L.Ros. on the Br. ends; **L.** oblong to oblong-lanceolate, pointed, 8–10 mm long, 2.5–4 mm across, green or tinged with purple, margins ciliate; **Fl.** ivory-white. Almost hardy.

**C. mollis** THUNBG. (§ VI). – Cape: E.Lit. Karroo. – Semi-♄; **St.** woody, branching, brittle, slender, dainty; **L.** obliquely oblong-lanceolate, blunt or rather pointed, underside convex, c. 2 cm long, rather, thick, grey-green, pubescent; **Fl.** whitish.

**C. montana** THUNBG. (§ IV/12). – Cape: Mountains of the S.W. and bentral Cape. – Mat-forming ♃, branched from the base; **Br.** ending in a Ros. of dark green L.; **L.** obovate-cuneate, subacute, glabrous, margins ciliately fringed; **Infl.** pedunculate with 1 to several heads of white Fl.

**C. montis-draconis** DTR. (§ IV/2) (*C. fragillima* DTR.). – S.W.Afr.: Lüderitz S. – ♄ up to 40 cm tall, branched from the base; **L.** 2.5–4 cm long, 7–10 mm across, 8–10 mm thick, boat-shaped, margins distinct, underside semi-terete, blue-green pruinose, somewhat incurved or curved upwards; **Fl.** numerous, yellow.

**C.** cv. Morgan's Beauty. – Hybrid by Dr. MEREDITH MORGAN of Richmond, Calif.: **C. falcata** × **C. mesembryanthemopsis**. – Compact, mound-shaped plant, much branching, up to 20 cm ⌀ or more, 10 cm tall; **L.** 3.6 cm across, narrowing to the tip, in 4 slightly spiral rows, in almost opposite pairs, appressed, obliquely ovate-lanceolate, (1–)3(–4) cm long, (1–)2(–3) cm across, rarely somewhat pointed, upperside with a truncate surface near the tip, margins entire, green, completely covered with white papillae; **Fl.** deep crimson, scented.

**C. mossii** SCHOENL. (§ IV/12). – Transvaal. – **L.** broad-obovate, blunt or rather pointed.

**C. multicava** LEM. (§ II/2) (*C. quadrifida* BAK., *C. punctata* HORT.). – Natal, Transvaal. – Spreading ♃, freely branching, up to 30 cm tall; **L.** broadly oblong, attenuate at the base into a petiole, somewhat emarginate, almost as long as broad, grey-green or glossy green, minutely pitted and spotted; **Fl.** always 4-merous, pinkish-white, Infl. with adventitious buds in the axils.

**C.** — cv. Variegata. – **L.** white-mottled.

**C. multiceps** HARV. (§ VII). – S.W.Cape: Bokkeveld Karroo. – Dwarf ♃, 3–7 cm high, with short forking Br.; **L.** shortly triangular-subulate, densely set in whorls of 3; **Fl.** white in terminal, sessile cymes.

**C. multiflora** SCHOENL. et BAK. f. (§ V/1). – S.W.Cape. – Semi-♄ up to 70 cm high with yellow-brown Br.; **L.** oblong-lanceolate to linear, blunt or rather pointed, 3–5 cm long, 5 mm across, the upper L. smaller and narrower and more densely set; **Fl.** numerous, white.

**C. muricata** THUNBG. (§ V/1) (*C. divaricata* ECKL. et ZEYH.). – S.W. and S.E.Cape. – Semi-♄; **St.** and Br. woody, ± bristly papillose; **L.** oblong-ovate, 3–10 mm long, pointed, margins papillose; **Fl.** small, white.

**C. nakurensis** ENGL. (§ I/4). – Kenya. – ♃; **St.** prostrate, laxly branched; **L.** distant, obovate or subpetiolate, thinly pubescent; **Fl.** solitary, in the axils of the upper L., white with longer pedicels. (Resembles **C. coleae** or **C. browniana** and probably identical.)

**C. namaquensis** SCHOENL. et BAK. f. v. **namaquensis** (§ V/6). – Cape: Lit. Namaqualand. – ♃, branching from the base, up to 10 cm tall; **L.** thick, short, densely papillose; **Fl.** small, white, in capitate cymes.

**C.** — v. **lutea** SCHOENL. – W.Cape: Bokkeveld Karroo. – **Fl.** yellowish.

**C. namibensis** FRIEDR. (§ IV/16) (*C. mesembryanthemoides* DTR. et BGR., *C. alstonii* AUCT., *C. mesembrianthoides* AUCT., *C. schoenlandii* AUCT., *C. luederitzii* AUCT.). – S.W.Afr.: Lüderitz S. – Succulent ♃, 2.5–5 cm tall; **St.** almost woody at the base, short, compressed and mat-forming, (1–)1.5(–2) cm long, closely covered with (2–)4(–5) L. pairs; **L.** fleshy, hemispherical to triangular, boat-shaped, almost pointed to blunt, upperside flat, underside convex, margins angular, papillose-ciliate at the base, blue-grey or ± densely minutely papillose to rarely almost glabrous, 8–15 mm long, 5–12 mm across, 4 to 6 mm thick; **Fl.** Cor. campanulate, white.

**C. natalensis** SCHOENL. (§ IV/8). – Natal. – ♃; **St.** simple, up to 1 m tall; **L.** united into a sheath basally, ovate, blunt or rather pointed, cuneate below, margins papillose-ciliate; **Fl.** numerous in richly branched terminal, flat-topped cymes.

**C. nealeana** V. HIGGINS (§ IV/2) (*C. rupestris minor* HORT., *C. perfossa minor* HORT.). – Cape. – Semi-♄ branching from the base; **St.** erect, later creeping, woody at the base, with persistent dry L. pairs; **L.** usually touching one another, perfoliate-amplexicaul, broad-oval, pointed, fleshy, glabrous, ± flat, bluish-green or purple, bordered and spotted with red, the underside slightly carinate; **Fl.** pale yellow.

**C. nemorosa** ENDL. (§ III) (*C. nivalis* HARV.). – S.E.Cape. – Succulent ♃ with Tub. and annual Br.; **L.** cordate, petiolate, soft-fleshy, margins entire; **Fl.** terminal or axillary, white.

**C. nodulosa** SCHOENL. (Pl. 45/4) (§ IV/14) (*C. enantiophylla* BAK. f., *C. elata* N. E. BR., *C. pectinata* CONRATH, *C. mariae* R. HAMET?, *C. avasmontana* DTR., *C. guchabensis* MERXM.). – S.W.Afr.; Transvaal; Rhodesia; trop. E.Afr. – ♃, 8 cm tall; **St.** simple, stiff, hairy; **L.** rosulate, round-ovate, 2–7 cm long, up to 3 cm across, mostly acute, glabrous or ± papillose-hairy in the lower part except for the margins with translucent, pectinate bristle hairs, cauline L. diminishing upwards; **Fl.** numerous in sessile compact cymes.

**C. nuda** COMPT. (§ IV/14). – Cape: Karroo. – ♃ up to 20 cm tall, branching basally, glabrous; **St.** erect to prostrate; **L.** fleshy, green, ovate-triangular, somewhat pointed, the upper L. smaller, 12 mm long, 8–10 mm across, 4–5 mm thick, underside convex; **Cor.** campanulate, white.

**C. nudicaulis** L. (§ VI). – Central Cape and Cape Peninsula. – Succulent, a cauline ♃ with a thick taproot; **L.** semi-terete, subulate, ± deeply grooved on the upperside, glabrous or slightly pubescent; **Infl.** 15–30 cm high, Fl. white in capitate cymes.

**C. nummulariifolia** BAK. (§ II/1). – Central Madag. – Glabrous ☉ or short living ♃; **St.** very thin, creeping, rooting from the lower nodes; **L.** short-petiolate, fleshy, broad-ovate, almost round, blunt or somewhat pointed, 6–12 mm long and across; **Fl.** solitary.

**C. nyikensis** BAK. f. (§ IV/9) (*C. whyteana* SCHOENL.). – Trop. Afr. – ♃; **L.** imbricate, obovate-roundish, very blunt, finely papillose, margins ciliate; **Fl.** white in sessile terminal cymes.

**C. obvallata** L. (§ VI). – S.W.Cape. – Succulent ♃; **St.** thick, with a few Br. from the base; **L.** dense and almost rosulate, oblong-lanceolate or oblique-cultriform, blunt or rather pointed, green, both surfaces flat, glabrous but margins with cartilaginous ciliate hairs; **Fl.** small, white.

**C. orbicularis** L. (Pl. 45/2) (§ IV/13) (*C. sedioides* MILL.). – Cape: Swellendam D. to Natal. – Resembles **C. rosularis**; smaller, offsetting; **L.** spatulate to obovate; **Fl.** often reddish.

**C. pachystemon** SCHOENL. et BAK. f. (§ IV/17). – Cape: E.Karroo. – ± mat-forming ♃, at most 20 cm tall; **St.** slender, prostrate to ascending; **L.** ± flat, slightly convex on the underside, oblong, rather, blunt, 15–20 mm long, pubescent; **Fl.** white.

**C. pallens** SCHOENL. et BAK. f. (§ IV/5). – S.W.Cape. – Branching semi-♄; St. and Br. minutely strigose; L. ovate or lanceolate, pointed or rather blunt, very convex on the underside, 7–12 mm long, the upper L. 4 mm long, with a few minute papillae, margins smooth.

**C. parvisepala** SCHOENL. (§ IV/3) (*C. lignosa* BURTT-DAVY). – Cape; Transvaal. – Laxly branched semi-♄; St. and Br. fleshy; L. narrow-oblong, tapering to both ends, pointed, margins entire or finely serrate; Fl. white.

**C. peglerae** SCHOENL. (§ IV/9). – S.E.Cape to Natal, Transvaal. – Resembles **C. ramuliflora**; ♃ up to 15 cm tall; L. ovate or ovate-lanceolate, margins densely papillose.

**C. pellucida** L. (§ II/1) (*C. marginata* THUNBG.) – S.Cape and Cape Peninsula. – ☉ to ♃; St. and Br. slender, prostrate-ascending, glabrous, rooting; L. obovate, attenuated into a petiole, margins slightly crenate; Fl. white or pink in lax terminal cymes.

**C. peploides** HARV. (§ II/6) (*C. galpinii* SCHOENL.). – Mts. of E.Cape and Lesotho. – ♃, branching from the base; L. crowded, semi-terete, less often flat, margins dentate at the tip.

**C. perfoliata** L. (§ IV/6) (*Rochea p.* DC., *C. p.* v. *albiflora* HARV.). – S.E.Cape to Natal. – Semi-♄ up to 1 m tall, with few Br.; St. fleshy, erect; L. very thick and fleshy, decussate, united at the base, distinctly spreading, lanceolate acute, up to 10 cm long, rather dull grey-green; Fl. in a freely-branched cyme, white.

**C. perforata** THUNBG. (Pl. 46/3) (§ IV/2) (*C. perfossa* LAM., *C. perfilata* SCOP.). – Cape: Lit. Namaqualand, Karroo. – Semi-♄; St. becoming woody, with few Br.; L. united-amplexicaul, broadly ovate, short-tapering, 15–25 mm long, 9–13 mm across, light grey-green, with numerous minute red spots especially at the margins, with fine cartilaginous cilia; Fl. small, yellow. (*C. perfossa* has glabrous margins.)

**C. petraea** SCHOENL. (§ IV/5). – Cape: Lit. Namaqualand. – Almost glabrous semi-♄; L. oblong-lanceolate, somewhat rough at first, margins ciliate. (Probably identical with **C. whiteheadii**.)

**C. phyturus** MILDBR. (§ I/7) (*C. pentandra* v. *p.* (MILDBR.) HEDB., *C. parvifolia* E. A. BRUCE). – ♃ to ♄; St. and Br. thick, fleshy; L. in 4 straight vertical series, sessile, almost horizontal, semi-terete, linear-subulate, tapering or blunt, 6–7 mm long, 2 mm across, 1 mm thick, both surfaces convex, green, glossy.

**C. picturata** BOOM (Pl. 45/6) (§ IV/10) (*C. cooperi* v. *robusta* SCHOENL., *C. c.* v. *major* HORT.). – Cape: Steynburg D. – ♃; L. rosulate, 12–25 mm long, 8–10 mm across, the upper L. smaller, oval, pointed, the upperside with irregularly scattered red spots, glabrous but margins laxly white-ciliate; Fl. small, white.

**C. —** × **C. schmidtii** RGL. – Garden hybrid by A. GRÄSER, Nuremberg. – Intermediate between the parents.

**C. —** × **C. milfordae** BYLES. – Garden hybrid from the Plant Breeding Institute, Wageningen, Holland. – Fl. very numerous.

**C. planifolia** SCHOENL. (§ IV/1). – Cape: coasts. – Closely related to **C. tetragona**; L. almost flat, lanceolate, pointed, 22 mm long.

**C. platyphylla** HARV. (§ VI). – E.Cape to Orange Free State and Lesotho. – Related to **C. obvallata**; L. roundish and obovate, usually very blunt, without cilia.

**C. plegmatoides** FRIEDR. (§ IV/16) (*C. deltoidea* AUCT. non THUNBG. nec. HARV., *C. pseudocolumnaris* DTR. nom. nud., *C. arta* AUCT.). – S.W.Afr.; Cape: Lit. Namaqualand. – ♃, (5–)10(–15) cm tall; St. with few Br., covered at the base with dry old L., 4–6 mm thick; Br. erect, closely covered by the 6–8 pairs of imbricate L. which together form an almost 4-sided column (2.5–)5(–8) cm tall, 1–2 cm thick; L. very fleshy, united-perfoliate, hemispherical, almost triangular, involute at the tip, almost pointed, the underside very convex, densely covered with minute spherical papillae, blue-green, the upperside hollow toward the tip, margins angular, papillose-ciliate, 7–15 mm long and across, 4–8 mm thick; Fl. campanulate, white.

**C. portulacea** LAM. (Pl. 46/2) (§ II/4) (*C. lucens* GRAM., *Cotyledon ovata* MILL., *Cr. o.* (MILL.) DRUCE, *Cr. nitida* SCHOENL.). – Cape: S. Cape to Transvaal. – Closely resembles **C. argentea**; L. more rounded, glossy green with red borders.

**C. —** cv. Blue Bird (Blauer Vogel, Bluave Vogel, Oiseau Bleu). – L. rather oblong; Fl. pinkish. (Acc. B. K. BOOM possibly a hybrid: **C. portulacea** × **C. arborescens**).

**C. —** cv. Variegata. – L. white mottled.

**C. pruinosa** L. (§ IV/5) (*C. scabra* v. *minor* SCHOENL., *C. scabrella* HARV.). – S.W.Cape. – Semi-♄; St. and Br. woody, with a few, small papillose hairs; L. subulate, semi-terete, covered ± with rough papillose hairs, margins rough, 3–6 mm long.

**C. pseudohemisphaerica** FRIEDR. (§ IV/14). – S.W.Afr.: Lüderitz S.; Cape: Lit. Namaqualand. – ♃, ± mat-forming; L. densely rosulate, broad to round, apiculate, cuneate at the base, fleshy, almost flat, 1.5–2.5 cm long and across, green, margins with pectinate cilia, hairs white and directed downwards; Fl. small, white to yellowish.

**C. puberula** (ECKL. et ZEYH.) ENDL. ex WALP. (§ V/3) (*C. hispida* SCHOENL. et BAK. f.). – S.Cape. – Semi-♄ up to 60 cm tall; Br. virgate; L. numerous, ± densely arranged on the extreme tips of the Br., roughly hairy, otherwise like **C. subaphylla**.

**C. pubescens** THUNBG. (§ VI) (*C. fragilis* SCHOENL.). – S. and S.E.Cape: Lit. Namaqualand, Karroo. – Polymorphic semi-♄; St. branched from the base; Br. very fragile; L. oblanceolate-obovate, blunt, fleshy, ± pubescent to nearly glabrous; Fl. small, white, in terminal cymes.

**C.** × **pulverulenta** BOOM (Pl. 46/4; 47/7) (*C.* × *andegavensis* BOOM, *C. rubicunda* HORT.)

– Hybrid: **C. falcata** × **C. schmidtii.** – Erect ♃ up to 40 cm tall, densely hairy; **L.** lanceolate, pointed, margins densely white-ciliate, the upperside concave, the underside convex with numerous spherical papillae; **Fl.** numerous, 8 to 10 mm $\varnothing$, red.

**C. punctata** L. (§ IV/14). – W.Cape: Gr. Karroo. – ♃; **St.** short, with few Br. ; **L.** at first subrosulate, lanceolate, acute, thick, glabrous, with papillose-ciliate margins, spotted; flowering St. erect, with distant L.; **L.** diminishing upwards; **Fl.** white in shortly pedunculate cymes arranged on an elongated thyrse.

**C. punctulata** SCHOENL. et BAK. f. (§ IV/3). – Cape. – **Semi-**♄, rarely 10 cm tall, freely branching; **Br.** and twigs stiff, brown; **L.** rather crowded at the ends of the shoots, subulate, tapering, 5–10 mm long, c. 1 mm thick, glossy green, white-pruinose towards the tip; **Fl.** white.

**C. purcellii** SCHOENL. (§ V/3). – S.W.Cape. – **Semi-**♄ ; **Br.** virgate, 20–40 cm tall, laxly leafy; **L.** ± semi-terete, 1.5–4 cm long, 5–8 mm across and thick, grey-green; Br. and L. with a dense pubescence.

**C. pustulata** TOELKEN (§ IV/5) (*C. pruinosa* sensu HARV.). – Cape: Clanwilliam D. – Erect, often fastigiate, (10–)14(–20) cm tall, much branched semi-♄; lower **Br.** often rooting, Br. wiry, (1–)2(–3) mm thick at the base, hairy; **L.** linear-lanceolate, acute, 5–14 mm long, 1–2 mm across, almost terete, upper surface usually flat, erect or spreading, grey-green to brownish-green, Sh. 1–1.5 mm long; **Infl.** terminal, branching, flat-topped, up to 2 cm $\varnothing$, Fl. white to light yellow.

**C. pyramidalis** THUNBG. (Pl. 47/3) (§ VII). – Cape: E. Lit. Karroo. – ♃; **St.** 3–7 cm tall, ± dichotomously branching; **Br.** quadrangular; **L.** flat, crowded one above the other, broad-triangular, sides rounded, green, immature L. with pubescent margins, almost 1 cm across; Fl. in a capitate terminal cyme, white.

**C. pyrifolia** COMPT. (§ I/4). – Cape: Lit. Namaqualand. – **Semi-**♄; **Br.** terete, brittle, slightly swollen at the nodes; **L.** fleshy, bluntly obovoid to pyriform, narrowing towards the base, green, 8–12 mm long, 4–6 mm across and thick; **Fl.** reddish. (Closely allied to **C. albicaulis,** but Fl. solitary on slender pedicels.)

**C. quadrangula** (ECKL. et ZEYH.) ENDL. ex WALP. (Pl. 47/1; 2) (§ VII) (*Tetraphile q.* ECKL. et ZEYH.). – Cape: Calitzdorp. – ♃, monocarpic or (cultivated plants) with few detachable short offsets at the base; **St.** with densely set triangular L. forming an erect column up to 6 cm high; **Fl.** white, fragrant, in a dense terminal sessile cluster.

**C. quadrangularis** SCHOENL. (§ IV/12). – S. and Central Cape, mountains. – Mat-forming ♃, freely branched; **Br.** with subquadrangular L.Ros.; **L.** broadly ovate, ± pointed, slightly plicate, glabrous, with papillose-ciliate margins; **Fl.** white in pedunculate capitate cymes. (Probably identical with **C. montana.**) –

**C. radicans** D. DIETR. (§ VI) (*C. r.* HARV.). – Cape. – **Semi-**♄ up to 40 cm tall; **Br.** spreading or prostrate, rooting; **L.** 10–15 mm long, oblanceolate-spatulate, flat, usually glabrous; **Fl.** white.

**C. ramuliflora** LINK spec. aggreg. (§ IV/9). – S.E.Cape; S. Natal. – Extremely variable ♃; **St.** simple or branching, slender, erect, rarely prostrate; **L.** broad-ovate, ± pointed, or oblong or obovate, ± blunt, margins papillose; **Fl.** white, sometimes red. (Probably identical with **C. obovata** which would have priority.)

**C. rattrayi** SCHOENL. et BAK. f. (§ VI). – Cape: S.W. and Centr. Karroo. – Small ♃; **L.** spatulate or almost obovate, up to 3.5 cm long, pubescent. (Resembles small forms of **C. obvallata.**)

**C. rauhii** FRIEDR. (§ VI). – Cape: Clanwilliam D. – **Semi-**♄ up to 50 cm tall with a few **Br.** from the base, erect, 10–15 cm long, 5 to 9 mm thick, initially fleshy, red, laxly leafy; **L.** linear, strap-shaped, fleshy, blunt to almost pointed, the upperside flat to grooved, the underside convex, margins almost angular, 7–14 mm long, 7–12 mm across, 4–6 mm thick, glabrous, green, often tinged red; **Fl.** pale yellow to yellow-green, Cor. urn-shaped.

**C. recurva** N. E. BR. (§ IV/8) (*C. cernua* N. E. BR.). – Natal: Zululand. – ♃; **St.** densely leafy; **L.** grey-green, sheathing at the base, lanceolate, pointed, the lower L. over 2 cm long, the underside very convex; L., Ped. and pedicels white-hairy; **Fl.** deep pink.

**C. remota** SCHOENL. (§ V/3). – E.Gr. Karroo. – Semi-♄ with spreading Br.; **L.** ± oblong-ovate, 2 cm long, 12 mm across, 4 mm thick, rather obliquely pointed, grey-green, somewhat rough, puberulous, margins reddened above.

**C. retrorsa** HUTCHINS. (§ IV/8). – E.Transvaal. – ♃; **St.** erect, 15–20 cm tall, laxly leafy, white-haired; **L.** narrowly oblong-lanceolate, 2–3 cm long, c. 1 cm across, upperside glabrous, underside rather white-hairy, margins densely papillose-ciliate; **Fl.** numerous.

**C. reversisetosa** BITTER (Pl. 47/4) (§ IV/9) (*C. stachyera* ECKL. et ZEYH. v. *pulchella* HARV., *C. ramuliflora* v. *pulchella* (HARV.) SCHOENL.). – S.E.Cape; S.Natal. – ♃; **St.** red with white hairs; **L.** round to rhombic, 7–10 mm long and across, spotted, ciliate; **Fl.** small, white.

**C. rhodogyna** FRIEDR. (§ IV/14). – S. W. Afr.; Cape: Lit. Namaqualand. – Stemless ♃; **L.** in basal Ros., obovate, acute, glabrous, spotted, margins fringed with retrorse cilia, 1–1.5 cm long; flowering scape erect; **Infl.** thyrsoid, Fl. white.

**C. robusta** TOELKEN (§ IV/1) (*C. decussata* SALISB. nom. nud.). – E.Cape: near Grahamstown. – Resembles **C. tetragona**, but more robust; ♄, erect, up to 80 cm tall, glabrous, branching; **Br.** fleshy, up to 1.5 cm $\varnothing$, with flaking bark; **L.** lanceolate, subulate, (2–)3(–4) cm long, 5–7 mm across, usually terete, arched upwards, glabrous, green, sometimes yellowish-green, Sh. over 1 mm long; **Infl.** terminal, thyrsoid, Fl. numerous, Pet. up to 2 mm long, white or cream.

**C. rogersii** SCHOENL. (§ VI.) – Cape: E. Cape – Semi-♄ ; **St.** woody; **L.** obovate, blunt or rather pointed, both surfaces convex, distinctly pubescent.

**C. rosularis** HAW. (§ IV/13). – Cape: Swellendam to Natal. – Succulent ⚁; **St.** with basal Ros., offsetting laterally; **L.** densely crowded, linear-spatulate, rather tapering, 6–8 cm long, 1.5–2 cm across, slightly fleshy, glossy green, margins with minute cartilaginous cilia; **Fl.** white.

**C. rubella** COMPT. (§ V/3). – Cape: Karroo. – Small semi-♄; **St.** thin, red, stiff, hairy; **L.** 4–6 at the Br. tips, ovate-spatulate, underside asymmetrically carinate, minutely hairy, 1 cm long, 6 mm across, deciduous; **Fl.** white.

**C. rubescens** SCHOENL. et BAK. f. (§ IV/9). – S.E.Cape. – Certainly closely related to **C. ramuliflora.**

**C. rubicunda** E. MEY. ex HARV. (§ IV/8) (*C. milleriana* BURTT-DAVY). – S.E.Cape; Natal-Transvaal; Moçambique; Rhodesia. – ⚁ ,sometimes monocarpic; **St.** simple, up to 30 cm high, glabrous or hispidulous like the L., with a fleshy taproot; **L.** linear-lanceolate, subulate, ± canaliculate, green, often flushed or mottled with red, margins shortly papillose-ciliate, sheathing at the base, closely set beneath, up to 15 cm long, diminishing and more distant upwards; **Fl.** red. in a richly branched, flat-topped terminal cyme.

**C. rubricaulis** ECKL. et ZEYH. (§ IV/5). – S.E.Cape. – Semi-♄, 30 cm high; **St.** branched from the base; **Br.** forking, yellow-brown to ± reddish; **L.** obovate or ovate, sessile, papillose-ciliate; **Fl.** white in pedunculate terminal cymes.

**C. rudis** SCHOENL. et BAK. f. (§ IV/1). – Cape: Lit. Namaqualand. – Closely resembles **C. acutifolia** but less branched; **Br.** longer; **L.** grey-pruinose; **Pet.** 1.5 mm long.

**C. rudolfii** SCHOENL. et BAK. f. (§ IV/5). – S.W.Cape. – Erect **semi-♄**; **St.** covered with short, white hairs; **L.** lanceolate, glabrous but with ciliate margins, ± 1 cm long; **Fl.** white or pink.

**C. rupestris** THUNBG. (Pl. 47/8) (§ IV/2) (*C. punctata* AUCT. non LAM., *C. monticola* N. E. BR., *C. brevifolia* AUCT. non HARV. nec. V. HIGGINS). – Cape: Karroo area. – Glabrous, branching **semi-♄**; **St.** woody, Br. and twigs prostrate or erect; **L.** ± united at the base, ± ovate rather pointed, fairly thick, glabrous, upperside, flat to somewhat concave, underside convex-carinate, with margins entire and reddish, (5–)7(–10) mm long, (4–)6(–11) mm across, (2–)4(–5) mm thick, light green with a bluish tinge; **Fl.** small, white or pinkish. Variable species, with every intermediate form from those with thick, round L. (f. *typica* = *C. brevifolia* AUCT.), to flat and larger L. (*C. monticola*).

**C. rustii** SCHOENL. (§ V/1). – S. Cape. – Semi-♄, the **Br.** densely leafy; **L.** ovate-lanceolate, pointed, underside convex, with ciliate margins, 8 mm long, 2 mm across; **Fl.** white.

**C. sarcocaulis** ECKL. et ZEYH. (§ IV/3). – S.E.Cape (mountains) to Natal, Lesotho. – Semi-♄ ; **St.** fleshy, Br. forking; **L.** lanceolate-tapering; **Fl.** white or pink. Hardy.

**C. sarmentosa** HARV. (Pl. 48/1) (§ II/2). – S.E.Cape to Natal. – **St.** up to 1 m long, ascerding; **L.** ovate, pointed, coarsely serrate-dentate; **Fl.** stellate, white, 1 cm ⌀.

**C. saxifraga** HARV. (§ III). – S.W.Cape. to S.E.Cape. – ⚁ ; resembles in habit **C. capensis** but **Fl.** 5-partite, campanulate, Infl. mostly few-flowered.

**C. scabra** L. (Pl. 43/4) (§ IV/5) (*C. squamulosa* WILLD.). – S.W.Cape. – Semi-♄ ; **St.** and Br. woody, ± densely leafy, scaberulous from rough papillae; **L.** narrow-linear or linear-lanceolate, coarsely papillose-hairy; **Fl.** white.

**C. scalaris** SCHOENL. et BAK. f. (§ V/5). – S.W.Cape; S.W.Afr.: Lüderitz S. – ⚁ ; resembles **C. tomentosa,** but basal **L.** much broader, obovate to almost roundish, soft, densely pubescent, margins distinctly ciliate; **Fl.** white to pale yellowish in sessile cymes on an interrupted, spike-like thyrse.

**C. schmidtii** RGL. v. **schmidtii** (Pl. 47/6) (§ IV/10) (*C. gracilis* HORT., *C. hookeri* HORT., *C. impressa* N. E. BR.). – Centr. Cape to Transvaal, Natal. – Succulent, mat-forming ⚁; **St.** erect or ascending, 7–10 cm tall, green or ± red, with projecting hairs; **L.** crowded basally, united, linear-lanceolate, 3–4 cm long, 4–5 mm across, upperside flat, green, pitted and spotted, underside convex, reddish, margins white-ciliate; **Fl.** shining crimson or deep pink-red.

**C. —** v. **alba** BOOM. – **Fl.** white.

**C. sediflora** (ECKL. et ZEYH.) ENDL. (§ IV/11) (*Tetraphile s.* ECKL. et ZEYH.). – S.E.Cape. – Semi-♄ ; **Br.** thin, creeping, ascending, ± laxly leafy; **L.** flat, somewhat fleshy, 14–20 mm long, linear-lanceolate or oblong, pointed, margins rough; **Fl.** white.

**C. sedifolia** N. E. BR. (§ IV/10). – S.W.Afr.; Cape: Lit. Namaqualand. – Succulent ⚁, forming tufts, 2.5–5 cm tall, branching from the base; **L.** densely crowded, 1.5–7 mm long, almost as thick, sub-acute, apiculate, glabrous but the margins with some cilia and 3–5 red spots in a row. Related to **C. bolusii** but with a fleshy tap-root and more distinct Infl.

**C. semiorbicularis** ECKL. et ZEYH. (Pl. 48/2) (§ VII). – Cape: Lit. Namaqualand. – ⚁ ; monocarpic but often sprouting from the axils of the lower L.; **St.** erect, solitary, densely leafy; **L.** imbricate, perfoliate, semicircular, apiculate, margins horny and somewhat ciliate, upper L. of flowering plants more distant; **Fl.** yellowish. Unpleasantly scented.

**C. sericea** SCHOENL. (Pl. 42/2; Pl. 47/5) (§ IV/17) (*C. argyrophylla* v. *ramosa* AUCT., *C. dewinteri* FRIEDR. – S.W.Afr.; Cape: Lit. Namaqualand. – ⚁ up to 12 cm tall; **St.** ± branched from the base; **Br.** short, thick, covered with fine hairs which become brownish; **L.** ± densely set, obovate-clavate to sub-globose, velvety or silky-hairy; **Fl.** white in dense capitate pedunculate cymes.

**C. serpentaria** SCHOENL. (§ V/3). – W.Cape. – Unbranched, or with **Br.** over 40 cm long.

**C. sessilicymula** MOGG (§ IV/14). – Transvaal. – ⚁ ; **St.** 30–75 cm tall, 7.5 mm thick, branching basally, hairy above; **L.** sessile, across, perfoliate, lower L. 2.4 cm long, 1 cm almost 3 mm thick, middle L. 3 cm long, 7 mm across, 1 mm thick, upper L. more pointed, glabrous or ciliate; **Fl.** white.

**C. setulosa** HARV. spec. aggreg. (§ IV/10) (*C. deminuta* DIELS, *C.scheppigiana* DIELS, *C.schlechteri* SCHOENL.). – Transvaal; Natal. – Near **C. cooperi**; **L.** ovate-lanceolate, pointed, unspotted.

**C. simulans** SCHOENL. (§ III). – Centr. S.W.-Cape. – ♃; underground **St.** tuberous, annual St. 3.6 cm tall; **L.** in 2–3 pairs, obovate or oblong, crenate; **Fl.** white in a lax pedunculate cyme.

**C. sladenii** SCHOENL. (§ IV/2). – Cape: Lit. Namaqualand; S.W.Afr.: Lüderitz S. – Forming a semi-♄, 30 cm high; **L.** broadly ovate, blunt, somewhat grey; **Fl.** numerous.

**C. smutsii** SCHOENL. (§ V/3) (*C. margaritifera* (ECKL. et ZEYH.) HARV.). – S.W.Cape. – Related to **C. incana**; **L.** oblanceolate or linear, 27 mm long, 5–6 mm across.

**C. socialis** SCHOENL. (Pl. 48/4) (§ IV/12). – S.E.Cape. – Succulent ♃, branching basally and forming tufts or mats; **L.** dense, in small Ros., ovate-triangular, 6–7 mm long and wide, underside rounded, thick, smooth, light green, margins narrow, horny; **Fl.** white.

**C. southii** SCHOENL. (§ IV/7). – S.E.Cape. – Semi-♄; **St.** woody, fairly freely and dichotomously branching; **Br.** rather densely leafy; **L.** flat, sheathing at the base, ovate-lanceolate, apiculate, margins papillose-ciliate; **Fl.** white.

**C. spatulata** THUNBG. (§ II/2) (*C. cordata* LODD., *C. lucida* LAM.). – S.E.Cape to Natal. – Succulent ♃; **St.** slender, prostrate, slightly 4-angled; **L.** petiolate, broadly cordate, margins minutely crenate-dentate, short-tapering, glabrous, c. 1 cm long and wide, glossy green; **Fl.** numerous, flesh-coloured.

**C. spectabilis** SCHOENL. (§ IV/8). – Natal: Zululand. – Tall ♃, resembling **C. acinaciformis**; **L.** margins with roundish cilia.

**C. spicata** THUNBG. (§ IV/14). – S. and S.E. Cape. – Similar to **C. capitella** but the **L.** glabrous, margins not papillose-ciliate; **Infl.** an interrupted spike-like thyrse.

**C. streyi** TOELKEN (§ II/2). – See Supplement.

**C. subaphylla** (ECKL. et ZEYH.) HARV. (§ V/3) (*Sphaeritis s.* ECKL. et ZEYH.). – S.W. Cape. – Related to **C. incana**; **L.** ovate or oblong, semi-terete, blunt, glabrous or shortly pubescent

**C. subbifaria** SCHOENL. (§ IV/14). – S.E.Cape. – Succulent ♃, branching basally; **L.** densely rosulate, oblong-lanceolate, pointed, margins with retrorse papillose-ciliate hairs; **Fl.** white.

**C. subsessilis** W. F. BARKER (§ IV/1). – Cape: Laingsburg, Prince Albert D. – Semi-♄ up to 20 cm tall; **St.** freely branching, older Br. leafless, with rings of L. scars, flowering Br. all of the same height, with 10 L. pairs, the lower ones with short twigs sprouting from the axils; **L.** bluish-green, oblong-fusiform, very convex on the underside, the upperside ± flat, up to 1 cm long, 2 mm ∅; **Fl.** white.

**C. subulata** L. (§ V/1) (*C. sphaeritis* HARV., *C. cymosa* BERGIUS, *C. ramosa* THUNBG.). – S.W. and S.E.Cape. – Semi-♄, becoming woody; L. numerous, dense, narrow-lanceolate, pointed, glabrous apart from the margins; **Fl.** white.

**C. susannae** RAUH et FRIEDR. (Pl. 48/3) (§ V/6). – Cape: Lit. Namaqualand. – ♃; **root** napiform, fleshy, with a brown bark and L.Ros. above; **L.** 3–5, obliquely truncate, projecting, asymmetrical, hollow in the upper third, almost right-angled, 7 mm long, 5 mm across, 2 mm thick, with only the blunt edges covered with translucent papillae and projecting above soil-level; **Fl.** white.

**C. tabularis** DTR. (§ IV/9). – S.W.Afr. – ♃ with a napiform, fleshy **root**; **L.** in a cruciform Ros., imbricate, up to 3 cm long, ovate-lanceolate, pointed, 10–17 mm across, c. 2 mm thick, the upperside with closely set dark green spots, the spots on the underside scattered, margins with translucent cilia; **Fl.** small, white.

**C. tecta** THUNBG. (Pl. 48/5) (§ IV/17) (*C. decipiens* N. E. BR.). – W.Cape: Karroo. – ♃, branching basally and forming clumps; **L.** thick, semi-terete, green, densely covered with thick white papillae; **Fl.** in dense capitate clusters.

**C. tenuicaulis** SCHOENL. (§ I/4). – Cape. – ♃; **St.** spreading, slender; **L.** fleshy, flat, with entire margins, oblong or ovate-cuneate or narrowing, 4–6 mm long; **Fl.** solitary.

**C. tenuifolia** SCHOENL. (§ IV/11). – Natal. – Perhaps identical with **C. sediflora**; **L.** distinctly united, Sh. 2–3 mm long, linear-lanceolate, pointed.

**C. teres** MARL. (Pl. 48/6) (§ VII) (*C. barklyi* N. E. BR.). – Cape: Lit. Namaqualand. – Resembles **C. columnaris**; **St.** and L. together forming a short, slender column; **L.** with transparent margins.

**C. tetragona** L. (§ IV/1) (*C. fruticulosa* L., *C. caffra* L.). – E.Cape. – Erect semi-♄, up to 1 m tall, with few slender Br. up to finger-thick, roundish, glabrous; **L.** rather dense, 2.5–3 cm long, 7 mm across, 5 mm thick, subterete, tapering, green; **Fl.** white.

**C. thorncroftii** BURTT-DAVY (§ I/4). – Transvaal; Natal. – ♃; **St.** spreading, slender; **L.** flat, oblong-obovate or linear-oblanceolate, glabrous, 5–10 mm long; **Fl.** solitary on thin pedicels. Probably identical with **C. tenuicaulis**.

**C. thyrsiflora** THUNBG. (§ IV/14) (*C. corymbulosa* AUCT. non LINK. p. part.). – S.E.Cape. – ♃, branched from the base; **Br.** short; **L.** linear-lanceolate to broadly lanceolate, glabrous, ± spotted or flushed with red, margins fringed with papillose cilia, ± easily dropping off and rooting; flowering St. erect, with distant L. diminishing upwards; **Fl.** white, numerous in a terminal thyrse.

**C. tomentosa** THUNBG. (Pl. 43/8) (§ V/5). – S.W.Cape: Lit. Namaqualand; S.W.Afr.: Lüderitz S. – Succulent ♃; **L.** subrosulate, broadly obovate-roundish or broadly lanceolate, blunt, densely tomentose, margins ciliate, 2–7 cm long, 2–3 cm wide; flowering St. erect with distant L. diminishing upwards; **Fl.** white or yellowish in sessile dense cymes in the axils of the upper L.

**C. transvaalensis** O. KTZE. (§ I/7) (*Tillaea subulata* HOOK., *C. s.* (HOOK.) HARV., *C. selago* DTR.). – E. trop. Afr.; Angola; S.W. and S.Afr.;

Socotra. – ⚃ with a fusiform taproot; **St.** ± woody at the base, branching; **L.** subulate from an ovate base, lanceolate, tapering, small, imbricate, green; **Fl.** small.

**C. turrita** THUNBG. (§ IV/14). – Cape. – Succulent ♄; **L.** 4–6 cm long, 1.5–2 cm across, lanceolate to ovate, glabrous, in basal Ros.; flowering St. 60–70 cm tall, with L. diminishing upwards, with tiny clusters of minute white **Fl.** from the upper L. axils. (= **C. capitella**?).

**C. tysonii** SCHOENL. (§ II/1). – Cape. – Spreading ⚃; **Br.** slender, pubescent; **L.** fleshy, flat, with entire margins, sparsely hairy or almost glabrous.

**C. umbella** JACQ. (§ III) (*Septas u.* HAW.). – Cape: Lit. Namaqualand, Clanwilliam D. – ⚃; **root** tuberous, with annual Br.; **L.** united, forming a plate-shaped disc, margins crenate, 2–7 cm long; **Infl.** cymose.

**C. undulata** HAW. (§ IV/5). – S.W.Cape. – Semi-♄, branched from the base in a pseudo-dichotomous manner; **L.** ligulate or oblong, blunt, sheathing at the base, glabrous, margins ± undulate with blunt papillose cilia, 1–1.5 cm long; **Fl.** white in terminal cymes. Related to **C. dejecta** but smaller in all parts and much more branched.

**C. vaginata** ECKL. et ZEYH. (§ IV/8).– E. Cape. – Glabrous ⚃; **St.** simple, rarely up to 1 m tall; **L.** linear-lanceolate to subulate, sheathing at the base, flat, c. 5–10 cm long, margins entire and papillose-ciliate; **Fl.** yellowish.

**C. velutina** FRIEDR. (§ IV/17). – S.W.Afr.: Lüderitz S. – ⚃, 5–10 cm tall; **St.** almost woody at the base; **Br.** c. 3 cm long, 2–5 mm thick, with a brownish bark, densely hirsute; **L.** ovate, almost rhombic, broadly spatulate to almost circular, united at the base, underside convex, upperside flat to convex, blunt, green, minutely hairy, densely velvety-puberulous, (13–)30(to 35) mm long, (8–)25(–30) mm across, (2–) 6(–8) mm thick; **Fl.** pale yellow, Cor. campanulate.

**C. vestita** THUNBG. (§ I/9). – W.Cape. – Small ⚃; **St.** branching, weak, prostrate; **L.** dense, triangular or ovate, bluntish, underside very thickened, mealy-white, 4 mm long; **Fl.** in sessile terminal cymes.

**C. virgata** HARV. (§ V/3). – S.W.Cape. – Small semi-♄; **L.** thick, ± semi-terete, the entire plant glabrous, rarely delicately white-hairy.

**C. volkensii** ENGL. (§ I/4). – Kenya; Tanzania. – Glabrous ⚃; **St.** branching basally; **L.** lanceolate, rather blunt, 2.5–3 cm long; **Fl.** numerous, stellate, 6–10 mm ⌀, white, from the axils of the upper L.

**C. watermeyeri** COMPT. (§ V/2). – Cape: Nieuwoudtville. – Semi-♄; **St.** erect to prostrate, branching basally, minutely hairy; **L.** few, c. 4.2 cm long, c. 2 cm across, 4 mm thick, green, ovate-spatulate, blunt, underside convex, minutely hairy, margins red and short-ciliate; **Fl.** white.

**C. whiteheadii** HARV. (§ IV/5). – Cape: Lit. Namaqualand; S.W.Afr.: Lüderitz S. – Semi-♄; **St.** and Br. minutely bristly; **L.** sub-ovate to linear-lanceolate, carinate or triangular, margins with blunt cilia 3–5 mm long; **Fl.** white in lax terminal cymes.

**C. wilmsii** DIELS (§ IV/8). – Transvaal. – ⚃, monocarpic or with basal offsets. Resembles **C. rubicunda,** but **L.** more flat, linear-lanceolate; **Fl.** smaller, red.

**C. woodii** SCHOENL. (§ I/4). – Natal. – Spreading ⚃; **St.** slender; **L.** fleshy, flat, glabrous, margins entire; **Fl.** small, white, shortly pedicellate, in axillary clusters.

**C. wyliei** SCHOENL. (§ II/2). – S.E.Cape to Natal. – Resembles **C. inandensis**; **L.** ovate, cuneate at the base, pointed, margins sinuate or dentate, T. mealy-green.

**C. zimmermannii** ENGL. v. **zimmermannii** (§ I/4). – Tanzania; Usambara. – ⚃; **St.** 8 to 10 cm tall, branching ± symmetrically from the base; **L.** lanceolate, fleshy, lower L. up to 22 mm long, c. 5 mm across, slightly hairy; **Fl.** white.

**C.** — v. **uhligii** ENGL. – Kenya. – **L.** somewhat longer and narrower.

**C. zombensis** BAK. f. (§ IV/17). – Malawi; Moçambique. – Dwarf ⚃; **St.** and Br. short, somewhat woody at the base, bristly hairy; **L.** subulate, linear-oblong, 2–3 cm long, 4–7 mm wide, grey-green, densely hairy; **Fl.** white in terminal pedunculate cymes.

**Crithmum** L. Umbelliferae. – Temperate regions. – Summer: in the open; winter: unheated greenhouse. Propagation: seed, division.

**C. maritimum** L. (Pl. 49/1) (*Cachrys m.* SPR., *Crithmum canariense* CAVAN., *C. m.* v. *canariense* DC., *C. latifolium* BUCH.). – Mediterranean coasts; Canary Is.; Madeira; Portugal, northwards to Great Britain. – ⚃; **roots** hard, gnarled; **St.** 20–50 cm tall, terete, very slightly grooved, some Br. above; **L.** sea-green, glossy, almost tripinnatifid, leaflets 2.5–5 cm long, up to 6 mm across, slightly fleshy; **Fl.** in large umbels of 10–20 rays, small, yellowish or greenish-white. Hardy.

**Cucurbita** L. Cucurbitaceae. – North America. – Mostly annuals, some having a large caudex; **St.** clambering, roughly-hairy; **L.** longer than broad, in some species palmate; **Fl.** unisexual, Cor. yellow; **Fr.** spherical, hard-shelled, smooth. – Summer: in the open; winter: unheated greenhouse. Propagation: seed. The undermentioned species have a fleshy caudex.

**C. digitata** A. GRAY. – USA: S.W.New Mexico to S.E.Calif.; N.Mexico. – **L.** palmate with 5 lobes; **Fr.** pale yellow with longitudinal stripes.

**C. foetidissima** H. B. et K. – USA: Missouri and Nebraska to Texas, Arizona, S.Calif. to Mexico. – **St.** up to 6 m long, evil-smelling; **L.** triangular-ovate, longer than broad, lobes slightly angular, tapering, up to 30 cm long, upperside hirsute; **Fr.** about 10 cm ⌀.

**C. palmata** S. WATS. – USA: S.W.Arizona, S.Calif.; Mexico; Baja Calif. – **L.** up to 10 cm long, incised, lobes triangular, upperside light green, hairy.

**Cussonia** THUNBG. Araliaceae. – Tropical Africa; Madagascar; Comoro. – Genus of many species of evergreen ♄. Only those mentioned below have a thick caudex capable of storing water, and thick woody Br. and twigs; **L.** in clusters on the end of short shoots, bi- or tripartite; **Infl.** terminal. – Greenhouse, warm. Propagation: seed.

**C. myriacantha** BAK. – Central Madag. – Small ♃; **Br.** thick; **L.** with petiole 15 cm long, lobes 7.5–10 cm long, distinctly pointed, almost leathery; **Fl.** greenish.

**C. spicata** THUNBG. (Pl. 49/2). – Cape; Natal; Tanzania; Comoro. – **Caudex** large, swollen; **L.** digitate; **Fl.** densely crowded.

**C. thyrsiflora** THUNBG. – Cape: Table Mt., Uitenhage D. – **L.** simple, somewhat serrate or deeply pinatifid, often 3-lobed; **Fl.** dense, in a long Rac.

**Cyanotis** D. DON. Commelinaceae. – Trop. Afr. and Asia. – Dwarf ♃; **roots** fleshy; **shoots** erect or creeping; **L.** sessile, ± densely arranged, ovate to linear-lanceolate, ± hairy, ± fleshy; **Fl.** solitary or in single cincinni, pink to blue. Many species are inadequately known, so that only a few which qualify as succulents are mentioned here. – Greenhouse, warm and sunny. Propagation: seed, division.

**C. arachnoidea** C. B. CL. – Ghana; Liberia; Nigeria. – **St.** up to 45 cm long, freely branching, prostrate, forming loose mats on rocks, ± clothed with white hairs like silk or cottonwool; **L.** linear, up to 6 cm long, 3 mm or more across, succulent and covered with the same indumentum as the St.; **Fl.** pink to mauve, in single cincinni.

**C.** – v. **pilosa** BRENAN. – Nigeria. – Similar to the type, but **St.** and **L.** with dense, spreading hairs.

**C. lanata** BENTH. – Widespread in trop. Afr. – **St.** ○, up to 30 cm tall, often branching basally, usually ± woolly, sometimes subglabrous; **L.** linear, up to 6 cm long, succulent, hairy; **Fl.** in terminal and axillary clusters, Pet. purple, pink or white with a pink border, hairs on the filaments blue, pink or pinkish-purple.

**Cynanchum** L. Asclepiadaceae. – Madag.; trop. E.Afr. – Dwarf ♄, branching or climbing; **Br.** ± thick-fleshy, sometimes leafless; **L.** otherwise small, scale-like and soon falling, or larger and cordate; **Fl.** in small capitate Infl., small, the corona-lobes forming a pentagon, the angles arranged between the Sep. – Greenhouse, warm. Propagation: seed, cuttings.

**C. ampanihensis** JUM. et PERR. (*C. humbertii* CHOUX). – Madag. – **St.** creeping or somewhat climbing, 50–100 cm long, c. 2 mm ∅, subcylindrical, scarred; **Br.** numerous, fusiform, 3–8 cm long, 2–3 mm thick, reddish; **L.** scale-like, triangular, c. 1.5 mm long, 0.7 mm across; **Infl.** paired, each with 5 Fl. 1.7–3 mm ∅.

**C. aphyllum** (THUNBG.) SCHLTR. (*Asclepias a.* THUNBG., *Sarcocyphula gerrardii* HARV., *Sarcostemma tetrapterum* TURCZ., *C. sarcostemmatoides* K. SCHUM.). – S. and trop. Afr.; S.W.Madag. – **St.** scrambling, freely branching, rooting at the nodes; **Br.** 1.5–3 mm thick with nodes 15 cm apart; **L.** 3 mm long, 1 mm across; **Fl.** few in an umbel arising from a tuberculate projection, corona green with brown stripes.

**C. compactum** CHOUX v. **compactum**. – Madag. – Forming underground stolons; **St.** numerous, erect, 10–15 cm long, freely branching from the base, forming compact clumps, in cultivation often attenuated, cylindrical, 3 to 5 mm thick, often reddish-violet, slightly constricted at the nodes; **L.** scaly, 1–1.3 mm long; **Fl.** about 8 mm ∅, Cor. cup-shaped.

**C.** – v. **imerinense** DESC. – **St.** more slender; **Fl.** smaller.

**C. cucullatum** N. E. BR. – Madag. – Resembles **C. pycneuroides**, probably not in cultivation.

**C. decaisnianum** B. DESC. (*Decanema bojeriana* DCNE., *Cynanchum b.* (DCNE.) CHOUX, *Asclepias aphylla* BOJ. MS.). – W.Madag. – Erect branching ♄; **Br.** terete, leafless, jointed, joints 10 to 40 cm long; **Infl.** of few Fl.

**C. helicoideum** CHOUX. – Madag. – **Roots** tuberous or napiform; **St.** thin, dying off annually.

**C. implicatum** (JUM. et PERR.) JUM. et PERR. (*Sarcostemma i.* JUM. et PERR., *Vohemaria i.* (JUM. et PERR.) JUM. et PERR.). – Madag. – **St.** thin, minutely longitudinally striate.

**C. lineare** N. E. BR. – Madag. – **Roots** tuberous or napiform; **St.** thin, dying off annually.

**C. luteifluens** (JUM. et PERR.) DESC. (*Decanema l.* JUM. et PERR.). – W.Madag. – **St.** slender, striate; **Fl.** yellow, Cor. sinuate, lobes oblong.

**C. macrolobum** JUM. et PERR. (Pl. 49/4). – S.W.Madag. – Small ♄, branching irregularly from the base, up to 40 cm tall, becoming woody at the base, readily broken off at the nodes, up to 6 mm thick, the bark made very tuberculate by the peeling waxy coating, whitish-grey; **L.** scale-like, soon falling, L. cushions persisting; **Fl.** brownish, corona lobes dentate at the tip.

**C. madagascariense** K. SCHUM. – Madag.: Fort Dauphin. – **St.** slender, strongly twining, up to 1.5 mm ∅; **L.** almost rhombic to obovate, short-pointed, 1.5–2.5 cm long, 7–12 mm across, on a petiole 3–5 mm long; **Infl.** racemose, Fl. dense, 1–2 mm long including pedicel, Cor. 4 mm ∅, tube somewhat swollen, 2.5 mm long.

**C. madecassum** B. DESC. (*Voharanga madagascariensis* COST. et BOISS.). – Madag. – **St.** simple at first; **Br.** readily breaking off above, new growth ascending and elongated, c. 1 cm ∅; **L.** greatly reduced, 2–4 mm long, short-petiolate, appressed, soon falling, with 2 stipular glands; **Fl.** in umbel-like clusters at the constrictions, glabrous, 3 mm long, Cor. not very open, lobes triangular-lanceolate.

**C. mahafalense** JUM. et PERR. – S. and central Madag. – **St.** liana-like, twining, somewhat hirsute initially, becoming glabrous, with a grey waxy coating, 4–6 mm ∅, terete, constricted at the nodes, Int. 5–15 mm long, growth somewhat zig-zagging; **L.** scale-like, ± 2.5 mm long; **Fl.** terminal on short spurs, tube campanulate, cream-coloured, upperside very hirsute, with dark wine-red nerves.

**C. marnieranum** RAUH. – Central Madag. – Dwarf and much-branching ♄; **Br.** elongated, fleshy, usually creeping and rooting, sometimes with erect shoots, 5–7 mm ⌀, dark green, irregularly tuberculate and with appressed white hairs, Int. up to 5 cm long; **L.** deciduous, opposite and decussate, broadly elliptical, pointed, brownish-green, 1.5 mm long and across at the base; **Infl.** of 3–6 Fl. on pedicels 3–6 mm long, Sep. c. 1 mm long, Pet. narrow-lanceolate, 5–6 mm long, upperside light greenish-yellow, tips at first united to form a broadly curving canopy (as in **Ceropegia**), the Pet. later separating and recurved, rather widened at the base, with 2 membranous appendages, the underside of the Pet. brown with a yellowish-green border.

**C. messeri** (BUCH.) JUM. et PERR. (Pl. 49/5) (*Vohemaria m.* BUCH.). – Madag. – Clambering ♄; **Br.** in terminal whorls, woody, terete, wrinkled, reddish, waxy; **L.** small, scale-like; **Infl.** capitate, Fl. few, 5 mm ⌀, yellowish, Pet. with velvety hairs on the upperside.

**C. napiforme** CHOUX. – Madag. – **Roots** tuberous or napiform; **St.** thin, dying off annually.

**C. nodosum** (JUM. et PERR.) B. DESC. (*Mahafalia n.* JUM. et PERR.). – S.Madag. – **St.** clambering, cylindrical, 4–10 mm ⌀, very much thickened at the nodes, freely branching, glabrous, segments barely 1 cm long, with a thick white waxy coating; **L.** scale-like, c. 2 mm long; **Infl.** of few Fl., c. 2 mm ⌀.

**C. pachylobum** CHOUX. – Madag. – Clambering ♄; **St.** thick, at first pubescent; **L.** cordate, 5.5–7 cm long, 4–6 cm across, tapering to a point, with auricles 5–11 mm long, petioles 2–3 cm long, minutely pubescent; **Fl.** 8–13 in small umbels, about 5 mm ⌀.

**C. perrieri** CHOUX (Pl. 49/3). – Madag. – ♄ branching from the base, 80–160 cm tall; **St.** erect, round or slightly angular, 9–13 mm thick, with a few Br. above, smooth or slightly tuberculate, distinctly pinnate, slightly constricted at the nodes; **L.** scale-like; **Fl.** small, inconspicuous.

**C. pycneuroides** CHOUX. – Central Madag. – Creeping by means of underground stolons and re-rooting, with scale-like protective cataphylls; **St.** numerous, erect, with a green bark, rarely branching, 40–60 cm tall, 2.5 cm ⌀ at the base; **L.** numerous, long-acicular, arranged in 5–7 oblique rows, falling after 1–2 years leaving persistent scars, up to 6 cm long, up to 8 mm across, apiculate, M.nerve projecting on the underside and slightly carinate; **Fl.** small, greenish.

**C. rauhiana** DESC. – S.E.Madag. – Erect, scarcely branching ♄; **St.** 4-angled to almost cylindrical, 6–10 mm ⌀, glabrous, smooth; **Br.** with a waxy, flaking bark, constricted at the nodes, segments 4–5 cm long; **Fl.** 4 mm ⌀, lobes papillose inside.

**C. rossii** RAUH. – Central Madag. – Forming dense clumps c. 1 m ⌀; **St.** sharply 4-angled, dark green, covered with appressed white curly hairs, prostrate-creeping, rooting from the underside; **L.** deciduous, reduced to triangular scales; **Ped.** erect, ± round, Fl. solitary, rarely 2, pedicel scarcely developed and like the Sep. and the young Pet. woolly-hairy, Pet. recurved above midway, 4 mm long, 2.5 mm across at the base, upperside olive-green, underside yellow-green.

**C. rusillonii** HOCHR. – Madag. – Resembles **C. pycneuroides,** probably not in cultivation.

**Cyphostemma** (PLANCH.) ALSTON (*Cissus* D.C., SG *Cyphostemma* PLANCH. et AUCT. num.). – Vitaceae. – Afr.; Angola; Cameroon; Somalia; Madag. – G. with many species, but only succulent and some caudiciform species can be described here. – Some plants form trunks, these often being clavate, thick, tapering, not segmented (as in **Cissus**), very succulent, without tendrils when mature; or with or without a subterranean caudex; **St.** erect or prostrate, at first with rudimentary tendrils; **L.** clustered on the ends of the shoots, ± fleshy, glabrous or felted, deciduous, compound and pinnate or tripartite, less often with up to 5 leaflets or digitate; **Infl.** corymbose, with Fl. laxly arranged, Cor. almost cylindrical, constricted midway, stigma 2-lobed; **Fr.** (Pl. 36/1) a berry, usually with only one seed. – Greenhouse, warm. Winter: dry for resting period. Propagation: seed, cuttings.

**C. bainesii** (HOOK. f.) B. DESC. (Pl. 35/1) (*Vitis b.* HOOK. f., *Cissus b.* (HOOK. f.) GILG et BRANDT). – S.W.Afr. – **St.** spherical to bottle-shaped, 60 cm tall, up to 25 cm ⌀ at the base, often even more, frequently dividing above into 2 thick Br., bark light yellow to green, peeling away like paper; **L.** short-petiolate, usually tripartite, leaflets 12 cm or more long, 5 cm across, coarsely serrate, green, both surfaces covered in woolly hairs; **Fr.** coral-red.

**C. betiformis** CHIOV. (Pl. 36/2) (*Cissus b.* CHIOV.). – Somalia; Kenya. – **Caudex** tuberous or napiform, 5–10 cm long, 3–7 cm thick, passing over ± abruptly above into 1–3 thickened Br. 2.5–8.5 cm long, long-conical, 1–3 cm across at the base, blunt at the tips, bark brown; **L.** clustered at the Br.tips, minutely velvety, 2–3 mm thick, leaflets obovate, ± blunt-tipped, end-leaflet 5–8 cm long, 2–3.5 cm across, the lateral ones smaller, margins crenate-dentate; **Fr.** spherical-ovoid.

**C. cirrhosa** (THUNBG.) B. DESC. (*Vitis c.* THUNBG., *Cissus c.* (THUNBG.) WILLD.). – Cape: Sondags River. – **St.** succulent, compressed, grey, curved in different directions, with tufted hairs; **L.** petiolate, obovate, pointed, margins acutely serrate, bracts sessile or amplexicaul, tendrils simple; **Fl.** white; **Fr.** ovoid, green, hairy.

**C. cornigera** B. DESC. – Madag. – Climbing ♄; **St.** c. 2 m tall, 40 cm ⌀, fleshy, succulent, rich in sap, bark green; **Br.** glandular at first, becoming glabrous; **L.** with petiole 2–6 cm long, tripartite, leaflets almost round, with crenate margins, 4.5–7.5 cm long, 3.5 cm across, nerves puberulous on the underside; **Infl.** with scattered hairs on the Ped.; **Fr.** 1 cm long, 5 mm ⌀.

**C. coursii** B. Desc. – Madag. – **St.** creeping to scrambling and twining, 4-angled, bluish, thick and fleshy at the base; **L.** with petiole 2–4 cm long, almost sagittate, margins short-serrate, 5–9 cm long, 3.5–5 cm across, or up to 3 times this size, underside laxly hirsute; **Infl.** of many Fl.; **Fr.** 4-angled, 11 mm long, 6 mm ⌀, yellow.

**C. cramerana** (Schinz) B. Desc. (Pl. 1) (*Cissus c.* Schinz). – S.W.Afr. – ♄ up to 4 m tall; **St.** thick, straight, fleshy, bark light yellow, with short projecting Br.; **L.** tripartite, leaflets short-petiolate, ovate-oblong, irregularly and coarsely dentate, with a minutely felty coating, petiole 2 cm long; **Fl.** 5 mm ⌀; **Fr.** yellow-red.

**C. currori** (Hook. f.) B. Desc. (Pl. 35/3). – (Cissus c. Hook. f.) Cape: Elephant's Bay; Angola: Moçamedes. – **St.** thick conical, freely branching, in age reaching a height of 35–80 cm, tree-like; **Br.** thick, striate; **L.** 15 cm long, 10 cm across, pale green, tripartite, leaflets petiolate, ovate, covered with raised raphids, bracts 1.25 cm long, ovate; **Infl.** branching dichotomously, Fl. large; **Fr.** ovoid.

**C. echinocarpa** B. Desc. – S.E.Madag. – **St.** succulent, clambering; **L.** imparipinnate, 12 to 20 cm long, leaflets round to ovate, 3–7 cm long, 2–4.5 cm across, margins lobed, petiole 3–5 cm long; **Fl.** yellowish-green; **Fr.** elliptical, 8 mm long, 5 mm ⌀, prickly.

**C. elephantopus** B. Desc. – Madag. – **St.** 1 m tall, 20–40 cm ⌀ below, fleshy, with a parchment-like bark, producing a disc-shaped caudex 1–1.30 m ⌀ and 20–30 cm thick, with large-lobed margins; **Br.** prostrate to clambering, initially hairy; **L.** thick, fleshy, doubly pinnate; **Fl.** numerous; **Fr.** spherical, c. 2 cm long, black.

**C. fleckii** (Schinz) B. Desc. (*Cissus f.* Schinz, *C. marlothii* Dtr., *C. amboensis* Schinz). – S.W.Afr. – **Caudex** c. 20 cm ⌀, attenuated, gnarled, soft, 50–60 cm long; **St.** 4–5 cm thick, with a brown bark, rough; **Br.** thinner, with tendrils, often climbing, up to 4 m tall; **L.** palmate, 5-partite, leaflets narrow-lanceolate, 2 cm long, 4 cm across, thick, crenate, green.

**C. hereroensis** (Schinz) Desc. (*Cissus h.* Schinz). – S.W.Afr.: Hereroland. – ♃; **caudex** thick, tuberous, up to 70 cm long; **St.** 2–4, 30–50 cm long, fleshy, prostrate, usually ± zig-zagging, with very few Br., Int. 3–4 cm long, 3–7 mm thick, shortly grey-papillose; **L.** sessile, 5–7-partite, leaflets lanceolate to narrow-lanceolate, margins coarsely crenate-serrate, somewhat fleshy, 8–10 cm long, 1.5 cm across, underside grey-papillose; **Infl.** erect, branching and re-branching dichotomously, Fl. very small; **Fr.** 1.5 cm long, yellow-green.

**C. juttae** (Dtr. et Gilg) B. Desc. v. **juttae** (Pl. 35/4) (*Cissus j.* Dtr. et Gilg). – S.W.Afr.: Gr. Namaqualand. – **St.** 1–2 m tall, acute-conical, with several thick Br. above, bark yellowish-green, later peeling off like paper; **L.** sessile, oval-pointed, 10–15 cm long, 5–6 cm across, irregularly and coarsely serrate, glossy green, often reddish, waxy-pruinose, the ribs on the underside covered with translucent hairlets 2–3 mm long; **Fr.** dark red or light yellow.

**C.** – v. **ternatus** Jacobs. (Pl. 1) (*Cissus j.* v. t. Jacobs.). – S.W.Afr.: Otavi. – **L.** tripartite, leaflets oval-pointed, 30–35 cm long, 15 cm across, narrowing at the base to the 3 cm long petiole, margins pointed-serrate; **Fr.** scarlet.

**C. laza** B. Desc. v. **laza** (Pl. 50/2). – Madag. – **St.** 1–2 m tall, 0.7–1 m ⌀ at the base, fleshy, bark parchment-like, with 2–3 Br. thickened at the base, prostrate to clambering, 3–5 m long, initially white-felted; **L.** c. 16 cm long, 11 cm across, pinnatifid; **Fl.** green; **Fr.** ovoid-spherical, c. 1.5 cm long.

**C.** – v. **parvifolia** B. Desc. – Madag. – **L.** c. 7 cm long, c. 5 cm across.

**C. macropus** (Welw.) B. Desc. (*Cissus m.* Welw., *Vitis gastropus* Welw. ex Planch.). – Angola: Moçamedes. – **St.** thick, conical, in age tree-like and 35–80 cm tall; **L.** 3.5–5 cm long with petioles, simple, young L. hairy at first.

**C. migiurtinorum** (Chiov.) Chiov. (*Cissus m.* Chiov.). – Somalia. – **St.** 10 cm tall, 3 cm ⌀, smooth, with a brown bark; **Br.** 5–20 cm long, with a L.Ros. above; **L.** fleshy, simple, round or 3-partite, oval-cuneate, with very long hairs.

**C. montagnaci** B. Desc. – Madag. – **Caudex** large, napiform, cylindrical, conical at the tip, 50–70 cm long, 15–20 cm ⌀, bark black, leathery; **Br.** long, woody, initially fleshy and laxly leafy; **L.** thick, fleshy, 3-partite; **Fl.** green; **Fr.** 18–22 mm long, 14–26 mm ⌀, blackish.

**C. roseiglandulosa** B. Desc. – Madag. – **St.** 1–1.5 m tall, 35–50 cm ⌀ at the base; young **Br.** terete, grooved, densely puberulous, becoming glabrous; **L.** imparipinnate, 12–15 cm long; **Fl.** with red hairs; **Fr.** spherical-ovoid, c. 15 mm long, also with red hairs.

**C. rupicola** (Gilg et Brandt) B. Desc. (*Cissus r.* Gilg et Brandt). – Cameroon. – **St.** thick, fleshy; **L.** 5-partite with a short, thick petiole, leaflets fleshy, round or 3-partite, oval-cuneate, with very long hairs.

**C. seitziana** (Gilg et Brandt) B. Desc. (Pl. 36/3) (*Cissus s.* Gilg et Brandt). – S.W. Afr.: Granadilla Mts., Namib. – Resembles **C. bainesii**; **St.** almost spherical, unbranched; **L.** including the petiole 10–15 cm long, margins sharp-dentate, younger L. whitish-felted; **Infl.** on Ped. up to 45 cm long; **L.** 3-partite, rather fleshy, upperside glossy, dropping during the dry season.

**C. skalavensis** B. Desc. – Madag.: Ambongo D. – Sturdy ♄; **St.** 2–3 m tall; **Br.** terete, compressed, grooved, 5–7 mm ⌀, glabrous; **L.** imparipinnate, 20 cm long, glabrous, leaflets ± ovate, margins thickened, dentate, 6–8 cm long, 4–4.5 cm across; **Fr.** ovoid, 18–22 mm long, 12–14 mm thick, glabrous.

**C. uter** (Exell et Mendonca) B. Desc. (*Cissus u.* Exell et Mendonca). – Angola: Moçamedes. – **St.** almost spherical, fleshy, 7 to 30 cm ⌀, branching; **L.** leathery, with the petiole puberulous or glandular-hairy, 6 cm long, 5-partite, leaflets with petiolules 3 cm long, almost round, c. 8 cm ⌀, margins serrate, the underside with prominent ribs; **Fr.** elliptical, 11 mm long, 7 mm ⌀.

**Dasylirion** Zucc. Agavaceae. – USA: Arizona, Texas; Mexico. – ♃ with thick, erect, woody St.; **L.** clustered in Ros., long, stiff, the margins often thorny-dentate; **Infl.** lateral, tall, paniculate, Fl. campanulate, whitish. – Greenhouse, warm. Propagation: seed.

**D. acrotrichum** Zucc. (*D. a.* Bak., *D. gracile* Planch., *Bonapartea g.* Otto, *Roulinia g.* Brongn., *Yucca g.* Otto, *Y. a.* Schiede). – E. central Mexico. – **St.** short, thick; **L.** very numerous, up to 1 m long, c. 1 cm across, margins with closely-spaced minute T. and pale yellow, brown-tipped Sp. which are bent forwards, the tip having a tuft of fibres; **Infl.** 2–4 m tall.

**D. glaucophyllum** Hook. (*D. glaucum* Corr.) –. E. Mexico. – Forming a **St.**; **L.** bluish-green, more than 1 m long, c. 12 mm across, Sp. yellowish-white; **Fl.** greenish-yellow with red spots.

**D. graminifolium** Zucc. (*Yucca g.* Zucc.). – E. central Mexico. – Resembles **D. acrotrichum**; **L.** glossy green, 1 m long, 12 mm across, marginal Sp. very small, yellowish-white.

**D. longissimum** Lem. (*D. quadrangulatum* S. Wats., *D. juncifolium* hort.). – E. Mexico. – **St.** 1–2 m tall; **L.** very numerous, spreading gracefully, 1.2–1.8 m long, c. 6 mm across, gradually tapering, both surfaces almost angular-convex, in cross-section almost 4-angled; **Infl.** 1–2 m tall.

**D. serratifolium** Zucc. (*Yucca s.* Schultes, *Roulinia s.* Brongn., *D. laxiflorum* Bak.). – S.E. Mexico. – **L.** rough, 70–100 cm long, 2–3 cm across, Sp. fairly large, c. 2 cm apart, intervening margn minutely dentate.

**D.** × **vllarum** Winter. – Hybrid: **D. longissimum** × **D. graminifolium.** – Robust; **L.** 2 m long, green, curving over; **Infl.** 2–3 m tall.

**D. wheeleri** S.Wats. – USA: S.E. Arizona. – **St.** short; **L.** 1 m long, 25 mm across, fairly smooth, Sp. yellow, brown-tipped.

**Dasystemon** DC. Crassulaceae.
**D. calycinum** DC. is probably an abnormally developed **Crassula scabra** v. **minor** Schoenl.

**Decabelone** Decne. Asclepiadaceae. – Angola; S.W. Afr.; Botswana; Rhodesia; Transvaal; Orange Free State; Cape. – Succulent perennals; **St.** branching basally, cylindrical, grey-geen, with 6–12 ribs divided by sharp furrows, the angles dentate, the T. having 3 minute, bristle-like Sp.; **Fl.** several, at the base of young shoots or Br., short-pedicellate, large, conspicuous, Cal. 5-partite with pointed lobes, Cor. campanulate to funnel-shaped, somewhat obliquely widening towards the throat, 5-lobed; corona double. Outer corona united and annular at the base, with erect filiform extensions which are thickened at the tip. – Greenhouse, warm. Propagation: seed. Shoots can be grafted on **Stapelia** or the small Tub. of **Ceropegia woodii.**

**D. barklyi** T. Dyer (*Tavaresia b.* (T. Dyer) N. E. Br.). – Cape: Karroo, Lit. Namaqualand; Orange Free State; S.W. Afr. – Mat-forming; **St.** 7–12 cm tall, 2 cm or more ∅, with 10–12 acute dentate angles, the T. close together; **Fl.** 5–7 cm long, Cor. campanulate to funnel-shaped, somewhat curved, lobes triangular, abruptly tapering to a point, pale greenish outside, pale yellowish inside, with minute red spots, densely papillose, purple-red at the base.

**D. elegans** Decne (*Tavaresia angolensis* Welw., *Huernia tavaresii* Welw., *Stapelia digitaliflora* Pfersd., *D. sieberi* Pfersd.). – Angola. – Mat-forming; **St.** 10–15 cm tall, with 6–8 angles with projecting T.; **Cor.** campanulate to funnel-shaped, somewhat curved, up to 8 cm long, 3 cm ∅ at the throat, lobes triangular, narrowing abruptly to a short tip, dull yellowish outside, with minute brown-red spots and stripes, inside lighter, with circular or oblong red-brown markings and yellow ones in the centre, tube with numerous papillose hairs.

**D. grandiflora** K. Schum. (Pl. 50/1) (*Tavaresia g.* (K. Schum.) Bgr., *D. angolensis* N. E. Br. ex Jacobs., *Hoodia senilis* Jacobs.). – S.W. Afr.; Angola; Botswana; Transvaal, Cape. – **St.** very numerous, deep green, up to 20 cm tall, 1.5–2 cm ∅, with (10–)11(–14) densely tuberculate angles; **Fl.** several, Cor. campanulate to funnel-shaped, somewhat curved, 9–14 cm long, 4.5 cm ∅ at the throat, lobes broad-triangular, extending abruptly to a tip 7 mm long, glabrous or finely rough papillose outside, both surfaces light yellow with brown-red spots, with numerous papillae inside.

**D. meintjesii** (R. A. Dyer) Rowl. (*Tavaresia m.* R. A. Dyer). – Transvaal: Zoutpansberg. – **St.** 8–10 cm tall, 1.25–1.5 cm ∅, glabrous or very fine-hairy, branching, with 6–8 tuberculate angles, young tubercles with a projecting aciculate bristle; **Cor.** 7–7.5 cm long, tubular at the base, cream-coloured inside with blotches and chestnut-brown bands, margins brown, tube 3–3.5 cm long, 2.5 cm ∅ at the mouth, with long hairs, lobes triangular-lanceolate, c. 4 cm long, 1.5 cm across, the upperside minutely hirsute, margins with white to red ciliate hairs.

**Decaryia** Choux. Didiereaceae. – Madag. – Greenhouse, warm. Propagation: seed, cuttings.

**D. madagascariensis** Choux (Pl. 50/3). – S.W. Madag. – ♃, 6–8 m tall; **St.** straight; **Br.** spreading, twigs thorny, zigzagging, Th. 3 to 10 mm long, projecting horizontally; **L.** small, fleshy, obcordate, 5 mm long, 3 mm across, solitary below a pair of Th.; **Infl.** a single bipartite cyme, Fl. small.

**Dendrosenecio** Hauman. Compositae. – Trop. and E. Afr. – See **Senecio** L. (Gr. I). – Tree-like **Senecio** species; **St.** thick; **L.** in a large Ros.; **Infl.** large, terminal, capitula ± large. – The following species of **Senecio** fall within this category: **S. aberdaricus, adnivalis, battiscombei, brassica, brassiciformis, erici-rosenii, johnstonii, keniodendron.** For specialised literature see Olov Hedberg, Robert E. and Thore C. E. Fries.

**Dendrosicyos** Balf. f. Cucurbitaceae. – N. Afr.; Is. of Socotra. – Greenhouse, warm. Propagation: seed, cuttings.

**D. socotrana** BALF. f. (Pl. 50/4) (*D. jaubertiana* BAILL.) ('Cucumber Tree'). – Socotra; N.Afr.: deserts. – ⚶; **St.** thick, swollen, up to 6 m tall, often 1 m ⌀, bark white, crown sparsely branching; **Br.** somewhat pendulous, rather thin, rough, wrinkled, at first densely hispid; **L.** scattered, long-petiolate, round-cordate, ± 5-lobed, 7 cm ⌀ or more, rough with white bristles; **Fl.** grouped in the L. axils, unisexual, c. 2.5 cm ⌀; **Fr.** glandular-hairy.

**Dermatobotrys** H. BOLUS. Scrophulariaceae. – S.Afr. – Greenhouse, moderately warm. Propagation: seed, cuttings.
**D. saundersii** H. BOLUS. – Natal: Zululand. – ⚶; **caudex** c. 8 cm thick, with short, almost square Br. sprouting from the apex; **L.** clustered, stiff, fleshy, ovate or oblong, 5–15 cm long, 2.5–4 cm across, rather pointed, margins coarsely dentate; **Fl.** in whorls, Cor. tubular to trumpet-shaped, pink, lobes 5, ovate, yellow inside, tube hirsute inside; **Fr.** ovoid, green, 2.5 cm long, with many seeds.

**Diamorpha** NUTT. Crassulaceae. – N.Am. – Closely related to the G. **Sedum.** ☉ or ⊙ plants; **St.** slender, branching from the base; **L.** alternate, sub terete, oblong; **Fl.** small, few in a cyme. – Summer: in the open. Propagation: seed.
**D. cymosa** (NUTT.) BRITT. (*Tillaea c.* NUTT., *D. pusilla* NUTT.). – N.Am. – **St.** 8–10 cm tall; **L.** 2–5 mm long.

**Didierea** H. BAILL. Didiereaceae. – Madag. – Becoming tree-like, deciduous, thorny; **St.** simple or with Br. extending vertically or horizontally; **Br.** with tuberculate short shoots; **L.** rosulate, ± fleshy, linear to ovate, growing from the centre of a group of Th.; **Fl.** in many-flowered cymes, small, short-pedicellate, Cor. flat cup-shaped, Pet. 5 with attractively reticulate vein-markings, the Cal. later enclosing the Fr. – Greenhouse, warm. Propagation: seed, cuttings. W. RAUH of Heidelberg was successful in grafting on **Pereskiopsis.**
**D. madagascariensis** H. BAILL. (Pl. 51/1) (*D. mirabilis* H. BAILL.). – S.W.Madag. – ⚶, eventually 4–8 m tall; **St.** very woody, forming a cereoid column up to 40 cm thick, the tip sometimes curving over, densely covered with thorny short shoots, Th. very stout, arranged in false whorls of 4, 4–10 cm long; **L.** narrow-linear, 7–15 cm long, later falling; **Fl.** fairly large.
**D. trollii** CAPURON et RAUH (Pl. 51/2). – S.W.Madag. – Long shoots at first spreading horizontally, prostrate, forming bushes 50 cm tall and 2 m ⌀, with numerous, tangled twigs; from this immature form one or more erect long shoots emerge to form trunks with Br. projecting horizontally, the basal Br. eventually dying off; **St.** and Br. densely covered with short, convex tubercles, with short shoots whose tips always bear 5 thin Th. 2–4 cm long, arranged crosswise; **L.** usually 5, fleshy, ovate, elliptical or oblong, short-tipped, 1–2 cm long, 3 mm across, deciduous; **Fl.** greenish-yellow.

**Dinacria** HARV. Crassulaceae. – S.Afr. – Small, branching ☉; **L.** opposite, obovate-oblong, blunt, fleshy; **Fl.** pedicellate, whitish, in forking cymes. – Summer: in the open. Propagation: seed.
**D. filiformis** HARV. – Freely tripartite-branching, 5–8 cm tall; **Fl.** 3–4 mm long, in dense tufts.
**D. grammanthoides** SCHOENL. – Cape. – 3 to 4 cm tall; lower L.pair united, upper pair free; Fl. 3–4 mm long, 7–8 in a dichotomous cyme.
**D. sebacoides** SCHOENL. – Cape. – Resembles **D. grammanthoides; Fl.** larger.

**Diopogon** JORD. et FOURR. Crassulaceae. – Mountainous areas: E.Alps, Balkans, USSR: Caucasus. – *Jovibarba* OPIZ (1852) is the older name but since, according to H. A. FUCHS, it was not validly described, it is a nom. nud. Acc. G. D. ROWLEY *Jovibarba* is validly published and should be adopted in preference to **Diopogon,** as was done in Flora Europaea. See HOLUB & POUZAR in Folia Geob. Phyt. Praha 2: 408–409, 1967. The species previously belonged to the G. **Sempervivum** L. SG **JOVIBARBAE** DC., or Sect. **Jovibarbae** MERT. (The spelling *Jovisbarba* is incorrect.) – **Ros.**-forming ⚶, without offsets, propagated only by division, or Ros. with numerous adventitious Ros. on long, thin St. which are easily broken off; Ros.-L. oblong-spatulate, lanceolate to ovate or linear, ± hairy; **Infl.** simple, terminal with a flat or hemispherical cyme, Fl. 6-partite, campanulate to short-tubular, Pet. erect, fringed-laciniate, pale yellow or yellowish-white; carpels erect, not spreading at the tips. – Cultivation in the open. Propagation: seed, division, adventitious rosettes.
**D. heuffelii** (SCHOTT) H. HUBER (*Sempervivum h.* SCHOTT, *Jovibarba h.* (SCHOTT) A. et D. LÖVE, *S. hirtum* SIBTH. et SMITH, *S. brassenii* SCHUR.). – S.E.Eur.: Balkans. – **Ros.** without offsets, short-cauline, robust, with dichotomous branching to form clumps; Ros.L. oblong-spatulate; **Fl.** ventricose, Pet. obovate-lanceolate, tips with 3 T., underside carinate.
**D. —** v. **glabrum** (BECK. et SZYSZ) JACOBS. (*Sempervivum g.* BECK. et SZYSZ, *S. heuffelii* v. *g.* BECK. et SZYSZ, *Jovibarba h.* ssp. *g.* (BECK. et SZYSZ) HOLUB.). – Balkans. – Ros.L. glabrous on both surfaces, 2.5–3.5 cm long, 1.5–2 cm across; **Pet.** truncate, awned, yellow.
**D. —** v. **heuffelii.** – S.Eur. – **Ros.** open, 6–12 cm ⌀; Ros.L. awn-tipped, 5.5 cm long, 1.5 cm across, both surfaces with scattered glandular hairs, margins pectinate-ciliate, tips often reddened; **Fl.** yellowish-white to pale yellow.
**D. —** v. **kopaonikensis** (PANCIC) JACOBS. (*Sempervivum k.* PANCIC, *S. h.* v. *k.* (PANCIC) J. A. HUBER). – Bulgaria. – **Ros.** grey-green; Fl. somewhat more swollen.
**D. —** v. **patens** (GRISEB. et SCHENK) JACOBS. (*Sempervivum p.* GRISEB. et SCHENK, *S. h.* v. *p.* (GRISEB. et SCHENK) J. A. HUBER, *S. hirtum* ssp. *p.* STOJANOFF et STEFANOFF). – **Ros.L.** minutely hairy.

**D.** — cv. Glaucus (*Sempervivum h.* v. *p.* f. *glaucum* HORT.). – Ros.**L.** grey-green.

**D.** — v. **reginae-amaliae** (HELDR. et SART. ex BAK.) JACOBS. (*Sempervivum r.-a.* HELDR. et SART. ex BAK., *S. h.* v. *r.-a.* (HELDR. et SART. ex BAK.) BAK.). – Ros.**L.** purer green with reddened tips.

**D.** — v. **stramineus** (JORD. et FOURR.) JACOBS. (*D. s.* JORD. et FOURR., *Sempervivum s.* (JORD. et FOURR.) BAK., *S. h.* v. *s.* (JORD. et FOURR.) BAK.). – Bulgaria. – Ros.**L.** smaller, purer green and glossy (acc. H. CORREVON); **Fl.** deeper yellow.

**D. hirtus** (JUSLEN) H. P. FUCHS ex H. HUBER (*Sempervivum h.* JUSLEN, *Jovibarba h.* (JUSLEN) OPIZ, *S. adenophorum* BORB., *J. h.* v. *a.* (BORB.) LÖVE, *Sedum h.* LOUDON, *Semp. simonskaianum* DEGEN, *S. soboliferum* FLEISCH. et LINDEN, *S. tatrense* DOM., *J. h.* ssp. *t.* (DONN.) LÖVE). – S.W. and E.Alps; central and N.Eur. – **Ros.** medium-sized to small with numerous subsidiary Ros.; Ros.**L.** fleshy, spreading star-like or else curved inwards, lanceolate, ovate-lanceolate or obovate, pointed, 8–20 mm long, 2–12 mm across, margins glandular-ciliate, surfaces glabrous, glandular-hairy only in ssp. **allionii**; **Infl.** (Pl. 51/4) (8–)30(–40) cm tall, glandular-hairy above, Pet. with fibrous-ciliate margins, 12–17 mm long, pale yellow.

**D.** — ssp. **allionii** (JORD. et FOURR.) H. HUBER (*Diopogon a.* JORD. et FOURR., *Sempervivum a.* (JORD. et FOURR.) NYMAN, *Jovibarba a.* (JORD. et FOURR.) D. A. WEBB, *S. h.* ALL., *S. hirsutum* POLLINI, *D. arenarius* ssp. *a.* (JORD. et FOURR.) LEUTE). – S.W.Alps. – **Ros.** a tight sphere 2–4 cm ⌀; Ros.**L.** yellow-green, surfaces glandular-hairy; **Fl.** yellow.

**D.** — ssp. **arenarius** (KOCH) H. HUBER (*Sempervivum a.* KOCH, *Jovibarba a.* (KOCH) OPIZ, *S. h.* STERNBG., *S. h.* v. *pumilum* BERT., *S. h.* v. *glabriusculum* PARL., ? *S. soboliferum* HELDR. et SART., *S. hirtellum* SCHOTT, *S. kochii* FACCH., *D. arenarius* (KOCH) LEUTE). – E.Alps. – **Ros.** light-green, outer surfaces usually red-tinged, ± tightly spherical, 0.5–2 cm ⌀; Ros.**L.** lanceolate, 8–12 mm long, 3–5 mm across, surfaces glabrous, fresh green, with red-brown markings outside at the tip; **Pet.** 12–15 mm long.

**D.** — ssp. **borealis** H. HUBER (Pl. 51/3) (*Sempervivum soboliferum* SIMS, *Jovibarba s.* (SIMS) OPIZ, *Sedum s.* BREHM, *Semp. h.* WIMM. et GRAB., *S. globiferum* REICHENB., *D. g.* (L.) LEUTE). – N.Eur., E.Eur. to N.As. – **Ros.** tightly spherical, 1–3 cm ⌀; Ros.**L.** broadest in the upper third, the underside of the tip having a red mark; **Fl.** campanulate, yellow.

**D.** — ssp. — cv. Glaucus (*Sempervivum soboliferum* f. *glaucum* HORT.). – Ros.**L.** slightly grey-green.

**D.** — ssp. — f. **major** (MURRAY) JACOBS. (*Sempervivum soboliferum* f. *major* MURRAY). – **Ros.** larger than in ssp. **borealis**.

**D.** — ssp. **hirtus** (*Jovibarba hirta* ssp. *hirta*). – E.Alps to the Carpathians. – **Ros.** open-stellate, (3–)5(–7) cm ⌀; Ros.**L.** lanceolate, 15–20 mm long, (2–)5(–6) mm across, fresh green, glabrous, bracts on Ped. hirsute on the underside.

**D.** — ssp. — f. **austriacus** (JORD. et FOURR.) JACOBS. (*D. a.* JORD. et FOURR., *Sempervivum h.* v. *a.* JORD. et FOURR., *S. admontense* HORT. SÜNDERMANN nom. nud.?). – **Ros.** more open. (The following garden hybrids are included here: **S. hirtum** v. **austriacum** cv. Glaucum with a more grey-green Ros., and **S. h.** 'Major', with a larger Ros.)

**D.** — ssp. — f. **glabrescens** (SABR.) JACOBS. (*Sempervivum g.* SABR., *Jovibarba h.* ssp. *g.* (SABR.) SOO. et JAV.). – Glabrous form.

**D.** — ssp. — f. **hillebrandtii** (SCHOTT, NYM. et KOTSCHY) JACOBS. (*Sempervivum h.* SCHOTT, NYM. et KOTSCHY, *S. h.* v. *h.* (SCHOTT, NYM. et KOTSCHY) HAYEK). – Austria: Styria, Carinthia. – Ros.**L.** more grey-green, 8–12 mm across, bracts broader, becoming glabrous.

**D.** — ssp. — v. **hirtus** JUSLEN. – Ros.**L.** 5–6 mm across, bracts hairy on the underside.

**D.** — ssp. — v. **neilreichii** (SCHOTT, NYM. et KOTSCHY) HUBER (*Sempervivum n.* SCHOTT, NYM. et KOTSCHY, *S. h.* v. *n.* (SCHOTT, NYM. et KOTSCHY) HAYEK, *S. h.* ssp. *n.* (SCHOTT, NYM. et KOTSCHY) O. SCHWARTZ). – Lower Austria, Carinthia. – **Ros.** open: Ros.**L.** 2–3 mm across, deep green, bracts glabrous.

**D.** — ssp. — cv. Raripilus (*Sempervivum h.* v. *raripilum* HORT.). – Ped.-bracts less densely hirsute.

**D.** — ssp. **pseudohirtus** (LEUTE) JACOBS. (*D. arenarius* ssp. *p.* LEUTE, *Jovibarba a.* ssp. *p.* (LEUTE) HOLUB.). – Conspicuously rough-haired.

**Diplocyatha** N. E. BR. (*Diplocyathus* K. SCHUM.). Asclepiadaceae. – Related to the G. **Huernia**; easily recognizable by the campanulate tube from the base of which arises a second tube with a thickened border. – Greenhouse, warm. Propagation: seed, cuttings.

**D. ciliata** (THUNBG.) N. E. BR. (Pl. 53/5) (*Stapelia c.* THUNBG., *Podanthes c.* DON., *Tromotriche c.* SWEET). – Cape: Karroo. – Mat-forming; **St.** rooting from the base, ascending, 3–5 cm tall, having 4 angles with pointed T., glabrous, tinged with dirty red; **Fl.** with pedicels 15 mm long, 7–8 cm ⌀, tube short, campanulate to funnel-shaped, lobes spreading, ovate, pointed, almost white, roughly papillose, margins with long white capitate hairs, the tube-throat with a cup-shaped, roughly tuberculate annulus having a thickened border.

**Dischidia** R. BR. Asclepiadaceae. – Peninsular India to Taiwan, Philippines to New Guinea and Australia. – Epiphytic trailing plants; **L.** usually round, slightly fleshy, round-ovate or sometimes lanceolate, fleshy, in many species some transformed into large fleshy pitchers which collect rain-water and the moisture from transpiration, which is then taken up by the aerial roots growing in them; **Fl.** small, Cor. pitcher-shaped, lobes fleshy, corona with bifid lobes; **Fr.:** berries. – Greenhouse, warm and moist. Propagation: seed, cuttings. – About 46 species are known.

**D. merillii** BECC. – Resembles **D. pectinoides**: L. more vividly green, the saccate L. broadest midway, rather tapering above.

**D. pectinoides** H. H. W. PEARSON (Pl. 53/6). – Philippines. – **St.** slender, twining, rooting; **L.** oval-lanceolate, 18 mm long, pointed, about 9 mm across, rather thick, saccate L. mussel-shaped with a basal opening; **Fl.** carmine.

**D. rafflesiana** WALL. – Peninsular India to Austr. – **St.** thin, twining, rooting; **L.** short-petiolate, appressed, concealed under the L.roots, circular, saccate L. digitate, broadest in the middle, rounded at the tip, dark green; transitional L.forms are also present; **Fl.** small, white.

**Dolichos** L. Leguminosae. – S.W.Afr. – Greenhouse, warm. Propagation: seed.

**D. seineri** HARMS. (Pl. 53/4). – S.W.Afr.: N.Damaraland, Gr. Namaqualand. – **Caudex** very large, weighing up to 150 kg and storing much water; **St.** 3–5, robust, forming a bush 1 m tall; **L.** ternate, with silky glossy hairs; **Fl.** in erect Rac., blue; **Fr.** pod-like.

**Dorstenia** L. Moraceae. – Trop. to N.E. Afr.; Madag.; peninsular India; trop. Am. – Not all **Dorstenia** are succulent. – **Rhizome** in some species thick and tuberous; **St.** often fleshy or greatly thickened, with the L. on the upper nodes reduced to scales, but sometimes acauline with L. long-petiolate, arising from a thick caudex, less often ♄; **L.** petiolate, occasionally almost sessile, alternate, with margins entire or irregularly dentate, more rarely lobed, often ± tapering; Stip. narrow, usually falling, sometimes persistent; **Infl.** solitary, more rarely 2–3 on reduced St. from the upper L. axils, Ped. shorter than the L. or attenuated in the acauline species, the Fl. borne on a flat, saucer or even top-shaped receptacle (pseudanthium), the outline being very variable, almost cordate, stellate, oblong, rhombic or angular, with many or few bracts which vary in length from small-dentate to long-acicular. Both the ♂ and the ♀ Fl. are situated on the receptacle, all Fl. being without Pet.; the **Fr.** contains a single seed which, when ripe, is ejected from the receptacle. – Greenhouse, warm. Propagation: seed, cauline species also from cuttings.

**D. arabica** HEMSL. – S.E.Arabia. – **St.** fleshy, erect, simple, about 10 cm tall, 2.5 cm thick, tuberculate; **L.** somewhat fleshy, oblanceolate, with a petiole 4–6 cm long, tapering, sinuate-lobed, Stip. small, fleshy, pointed; receptacle circular, c. 1 cm ⌀, lobes c. 8, forming a star-shape, narrow, dentate.

**D. benguellensis** WELW. – Tanzania: Usambara. – **Caudex** disc-shaped or compressed-spherical, 2.5–4 cm ⌀; **St.** solitary, fleshy, unbranched or with a few short Br., 15–45 cm tall, puberulous at first; lower **L.** scale-like, those at the Br.tips oblong-lanceolate to elliptical, margins short-dentate, 2.5–6 cm long, 8–15 mm across, pubescent, margins short-dentate, Stip. c. 2.3 mm long, falling; **Infl.** solitary, Fl.head round, 5–7 mm ⌀, margin short-dentate, bracts 7–15, 15 mm long, thin, linear.

**D. bornimiana** SCHWEINF. v. **bornimiana** (Pl. 52/2). – Ethiopia; Congo: Kasindi; Arabia: Yemen. – **Caudex** compressed-spherical, 4 to 5 cm ⌀, 2.5 cm thick; **St.** with a few small Br.; **L.** long-petiolate, somewhat fleshy, short-cordate to cordate-ovate, blunt, margins wavy and blunt-dentate, 5–7.5 cm long, 5–6 cm across, the upper L. tripartite; receptacles standing vertically, oblong or narrow-oblong to linear, with a narrow border confluent with the bracts, the uppermost bract being longest, 2–3 cm long, c. 1 mm across.

**D. —** v. **angustior** ENGL. – Tanzania. – **L.** cordate, 2–2.5 cm long and across; **receptacles** narrow, c. 3 cm long, c. 3 mm across, the terminal bract 4–5 cm long.

**D. —** v. **ophioglossoides** (BUREAU) ENGL. (*D. o.* BUREAU). – Ethiopia. – **L.** cordate or scutate, c. 2.5 cm long and across; bracts narrow to linear, 6–8 mm long, pointed.

**D. —** v. **telekii** (SCHWEINF.) ENGL. (*D. t.* SCHWEINF.). – Kenya. – **L.** reniform to circular, 2 cm long and across; **receptacles** oblong, bracts linear, blunt, 4–6 mm long.

**D. —** v. **tropaeolifolia** (BUREAU) RENDLE (*D. t.* BUREAU, *Kosaria t.* (BUREAU) SCHWEINF., *D. peltata* ENGL.). – Ethiopia; Kenya. – **Caudex** 12.5–18 mm ⌀; **L.** solitary, scutate, 2.5 to 3 cm ⌀, with a petiole 12–18 mm long; bracts 5–10 mm long, pointed.

**D. braunii** ENGL. – Tanzania: Usambara. – Tuberous?; **St.** somewhat fleshy below, c. 10 mm ⌀, with few Br.; **L.** narrow-spatulate, some parts of the margin at intervals short-dentate, 4–6.5 cm long, 4–5 mm across, Stip. short-triangular; **Infl.** numerous, receptacles purple on the underside, elliptical, 9 mm across, bracts 7–8, c. 10 mm long.

**D. buchananii** ENGL. – Tanzania; Malawi. – **St.** succulent, c. 30 cm long, curving in different directions, white-hairy; **L.** elliptical, pointed, margins slightly sinuate-dentate, 5–7.5 cm long, 2–4.5 cm across, petiole up to 10 mm long, Stip. c. 3 mm long, falling; **receptacles** narrow-oblong, c. 2 cm long, 7 mm across, margins crenate, with a linear bract at each end, c. 5 cm long.

**D. caulescens** SCHWEINF. et ENGL. – N. equatorial Afr.: Sudan (Equatoria Prov.). – **Caudex** almost disc-shaped, c. 4 cm ⌀; **St.** 20–40 cm tall; **L.** with petiole c. 1 cm long, 8–10 cm long, 3–4 cm across; **receptacles** stellate, c. 1.5 cm ⌀, bracts 2–3.5 cm long, pointed.

**D. crispa** ENGL. v. **crispa** (Pl. 52/1). – Somalia; Kenya. – **St.** fleshy, (12–)30(–40) cm tall, erect, arising from a swollen base 4 cm thick, cylindrical, simple or with short Br. above, pale brown, tuberculate, bearing L. scars; **L.** narrow-oblong to linear-lanceolate, rather blunt to pointed, margins curly and dentate, 4–7 cm long, 1–2 cm across, initially with scattered hairs, Stip. small, bristly; **receptacles** round-stellate, 1.5–2 cm across, margin thickened, confluent with 6–10 subulate bracts variable in length between 12–25 mm, intermediate T. short.

**D. — v. lancifolia** RENDLE. – Kenya. – **St.** in old plants slender, more rarely slightly barrel-shaped, up to 25 cm long, up to 4 cm thick, unbranched, L.cushions arranged spirally; **L.** in an apical Ros., linear-lanceolate, pointed, 5 to 6 cm long, 7–10 mm across, with silver-grey spots on the upperside, margins curly, slightly sinuate-dentate.

**D. cuspidata** HOCHST. – Ethiopia. – **Rhizome** with thick, tuberous nodes; **St.** erect, 10–30 cm tall, usually solitary, thin; **L.** obovate and blunt or elliptical and pointed, often rhombic, margins slightly sinuate, shortly and broadly blunt-dentate, 5–6 cm long, 2–3 cm across; **receptacles** narrow-elliptical, 1 cm long, 2 mm across, with a bract at each end, upper receptacles stellate, 12 mm ∅, with 3–5 rays confluent with the 5–8 mm long bracts.

**D. ellenbeckiana** ENGL. – Ethiopia: Galla Highlands. – **Caudex** underground; **L.** somewhat fleshy, ovate or elliptical, margins wavy, minutely and bluntly dentate, c. 7.5 cm long, 3–4 cm across, hirsute; **Infl.** hairy, receptacles round, brown, c. 2.5 cm ∅, bracts numerous, about 10, usually c. 6 cm long, 1 mm across, but others linear-spatulate, 3–4 mm long, 1.5 mm across, 1–2 being less than 1 mm long.

**D. foetida** (FORSK.) SCHWEINF. v. **foetida** (Pl. 52/3) (*Kosaria f.* FORSK.). – S.Arabia. – **Caudex** flattened, up to 3 cm tall, 15 mm thick, Br. numerous, thick, up to 5 cm long; **L.** oblong-lanceolate; **receptacles** almost circular, bracts 8–10, up to 1 cm long, narrow.

**D. — v. obovata** ENGL. – Ethiopia; Arabia. – **Caudex** thick, fleshy; **St.** 1.5–2.5 cm long, 1 to 1.3 cm thick, with short Br.; **L.** deciduous, elliptical, ovate or obovate, 2–4 cm long, 2 to 3 cm across; **receptacles** roundish, c. 1 cm ∅, bracts 8–10, linear, uneven, 3–4 mm long.

**D. gigas** SCHWEINF. – Socotra; S.Arabia. – Tree-like; **St.** cylindrical, 1.2 m tall, up to 50 cm thick, tapering above and below, above with a crown of dichotomously dividing Br. and twigs; **L.** lanceolate, small, soon falling; **Infl.** inconspicuous, green.

**D. gypsophila** LAVR. – Somalia: Burao-Reg., Galkudal. – Extremely succulent plant to 120 cm tall, latex whitish having a smell of that of Tropaeolum; **St.** much thickened, columnar or globose, bark grey or whitish; **Br.** numerous, erect or ascending, bearing tessellate leaf scars; **L.** in Ros. on the apices of the St., up to 40 mm long and 35 mm broad, ovate, cordate or cuneate, minutely pubescent, margins undulate, crenate or dentate, petioles 5–30 mm long, Stip. minute; **receptacles** 4 mm ∅, produced singly in axils of the topmost L., Ped. 5 cm long, the margin denticulate to bearing 7–8 rays which are terete, unequal in length, up to 8 mm long, 0.5 mm ∅, ♂ and ♀ Fl. intermixed.

**D. hildebrandtii** ENGL. (Pl. 52/4). – Tanzania; Kenya. – **Caudex** small, partly above soil-level; **St.** several, with few Br. up to 2 cm thick at the base; **L.** oblong-oval, 4–6 cm long, margins often sinuate-crenate; **Infl.** from L.axils, receptacles roundish, dark purple, bracts 5–7, linear, purple.

**D. homblei** DE WILD. – Zaire. – **Caudex** 5–6.5 cm across, round or compressed the upperside depressed; **St.** solitary, cylindrical, up to 30 cm tall, with scale L. at the nodes; **L.** oblanceolate-elliptical, blunt, margins irregularly dentate, up to 6 cm long, c. 1.5 cm across, Stip. linear, up to 4 mm long; **receptacles** round, c. 1.5 cm ∅, bracts thin, linear.

**D. mirabilis** R. E. FRIES. – Rhodesia. – Caudex 7.5–12 cm ∅, 2–3 cm tall, compressed-spherical, depressed above, tuberculate; **St.** 1–3, fleshy, 15–20 cm tall, densely white-hairy, with pointed, fleshy, 4–6 mm long scales below; **L.** narrow-lanceolate, blunt, margins serrate, c. 3.5 cm long, 8–10 mm across, Stip. 3–5 mm long, persistent; **receptacles** round or irregularly round to 4-angled, 12–20 mm ∅, upperside depressed, dark purple margin narrow, with numerous dark purple bristles, bracts (4–)5(–6), the upper ones 12–15 mm long, the lower ones 8–10 mm.

**D. palmata** (SCHWEINF.) ENGL. (*Kosaria p.* SCHWEINF. MS.). – N. equatorial Afr.; Ghasal Spring reg. (??). – **Caudex** 3 cm ∅; **L.** stalk 10 cm long, blade cordate, 8–10 cm long, 7–8 cm across, margins entire or dentate, or the blade palmate with 3 oblong or 5–7 linear segments 10–12 mm across; **receptacles** 2–3 cm long, 7–10 mm across, with an apical bract 1.5 cm long and 2 mm across, and with rather shorter bracts at the base.

**D. philipsiae** HOOK. f. – Somalia. – **St.** rising from a 4 cm thick, conical caudex, 1.5 cm ∅ above, with short, thick, spreading Br. at the tip, with square L. scars; **L.** linear-oblong, pointed, 4–5 cm long, margins sharply dentate; **receptacles** broad-campanulate, bracts subulate, twisted or variously bent, 2.5 cm long.

**D. poggei** ENGL. – Zaire: on the Quango – Succulent ♄ up to 30 cm tall; **L.** fleshy, linear-lanceolate, 2–2.5 cm long, 5 mm across, margins short dentate; **receptacles** round, brown, with 10–12 linear bracts.

**D. quercifolia** R. E. FRIES. – Zaire. – **Rhizome** short, fleshy, 4–5 cm long, 7.5–15 mm thick; **St.** 1 or several, 8–15 cm tall, 5–6 mm thick, laxly hairy, with dark L.scars; **L.** rather fleshy, narrow-obovate, blunt, margins not dentate below, 4–6.5 cm long, 2.5 to 3 cm across, underside laxly hirsute, Stip. very small, falling; **receptacles** round, 5–8 mm ∅, margins with numerous T. 2–3 mm long and 8–10 narrow bracts up to 6 mm long.

**D. radiata** LAM. – Arabia. – **St.** c. 20 cm long and finger-thick, branching basally, tuberculate; **L.** scattered, with a petiole up to 5 cm long, lanceolate, lamina ± 5 cm long, margins wavy, almost without T., membranous, green; **receptacles** flat, round.

**D. rhodesiana** R. E. FRIES. – Rhodesia. – **Caudex** disc-shaped, c. 5 cm ∅, 9–12 mm tall; **St.** slightly fleshy, 20–25 cm tall, grey at the tip, covered with pointed, woolly scales 5 mm long; **L.** linear-lanceolate, almost pointed, margins thick-crenate, 5–7.5 cm long, 6–8 mm across, underside hairy, Stip. 2–3 mm long; **receptacles** round, 7–9 mm ∅, margins with a ring of T.

Plate 66. 1. **Euphorbia avasmontana** Dtr. v. **avasmontana**; 2. **E. atropurpurea** Brouss. (Photo: W. Rauh); 3. **E. ballyi** Carter (Repr.: Fl. Pl. Afr. 36, 1963, Pl. 1408, Drawing: M. Stones); 4. **E. balsamifera** Ait.

Plate 67. 1. **Euphorbia bravoana** SVENT. (Photo: W. RAUH); 2. **E. barnhartii** L. CROIZ. (Photo: J. MARNIER-LAPOSTOLLE); 3. **E. canariensis** L. v. **canariensis** (Photo: as 2); 4. **E. bupleurifolia** JACQ.

Plate 68. 1. Euphorbia caput-medusae L.; 2. E. cooperi N. E. Br. v. cooperi (Photo: J. Marnier-Lapostolle); 3. E. clavarioides v. truncata (N. E. Br.) White, Dyer et Sloane (Photo: G. C. Nel); 4. E. decaryi A. Guill. (Photo: as 2); 5. E. clandestina Jacq. (Photo: Bot. Gard. Königsberg).

11/B*

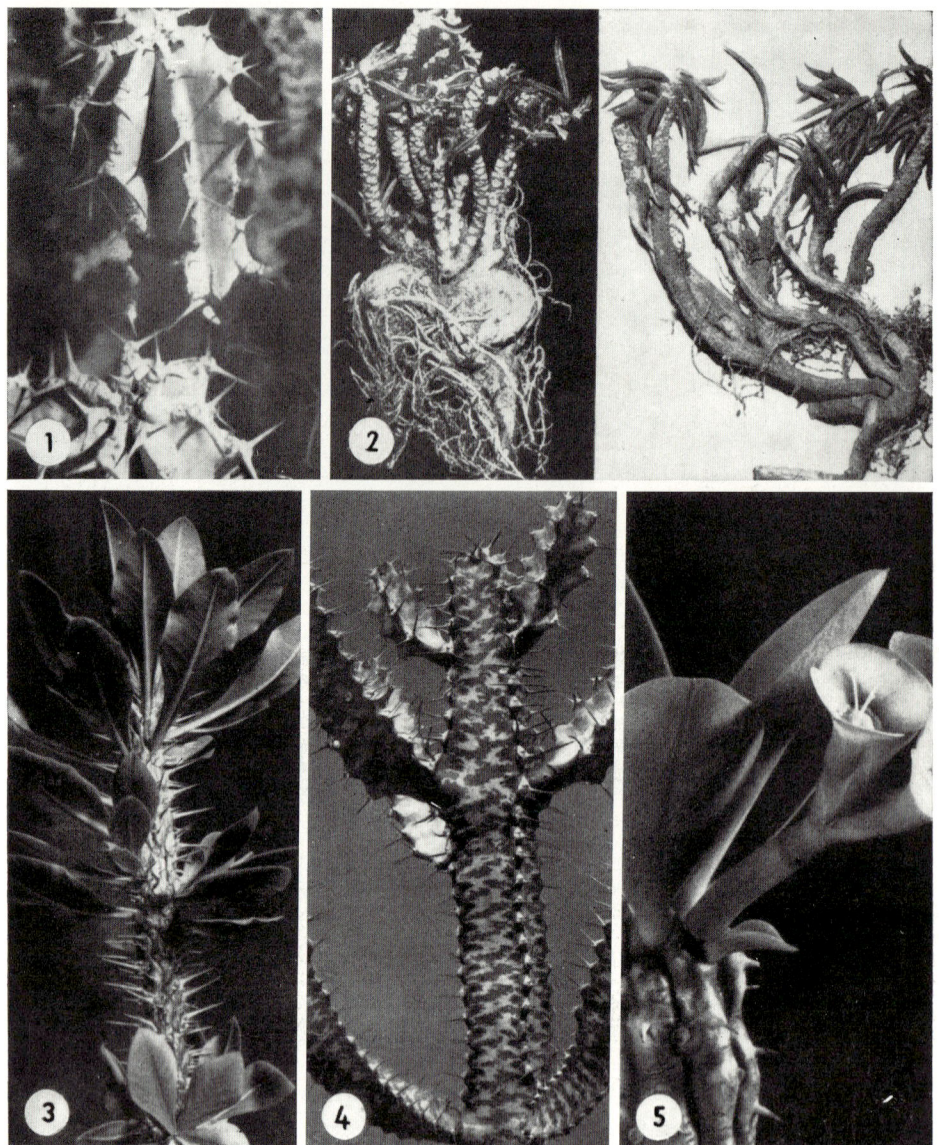

Plate 69. 1. **Euphorbia coerulescens** HAW. (Photo: J. MARNIER-LAPOSTOLLE); 2. **E. cylindrifolia** MARN.-LAP., left: ssp. **tuberifera** RAUH, right: ssp. **cylindrica** (Photo: W. RAUH); 3. **E. didieroides** M. DENIS ex LEANDRI (Photo: as 1); 4. **E. × doinetiana** A. GUILL. (Photo: J. DOINET); 5. **E. duranii** v. **ankaratrae** URSCH et LEANDRI (Photo: as 1).

Plate 70. 1. **Euphorbia echinus** Hook. f. et Coss. v. **echinus** (Photo: E. Hahn); **2. E. ferox** Marl. (Photo: J. Marnier-Lapostolle); **3. E. esculenta** Marl. (Photo: as 2); **4. E. fimbriata** Scop. (Photo: as 2).

Plate 71. 1. Euphorbia grandicornis GOEBEL ssp. grandicornis; 2. E. fruticosa FORSK. (Photo: J. MARNIER-LAPOSTOLLE); 3. E. filiflora MARL. (Photo: E. HAHN); 4. E. gariepina BOISS.; 5. E. globosa (HAW.) SIMS (Photo: as 2).

Plate 72. 1. **Euphorbia hamata** (Haw.) Sweet (Photo: J. Marnier-Lapostolle); 2. **E. horrida** Boiss v. **striata** White, Dyer et Sloane; 3. **E. heptagona** L. v. **heptagona** (Photo: H. Müller); 4. **E. guillauminiana** Boit. (Photo: as 1); 5. **E. handiensis** Burch. (Photo: O. Burchard).

Plate 73. 1. Euphorbia inconstantia R. A. Dyer (Photo: J. Marnier-Lapostolle); 2. E. horombensis Ursch et Leandri (Photo: as 1); 3. E. candelabrum Trem. v. candelabrum (Photo: as 1); 4. E. knuthii Pax ssp. johnsonii (N. E. Br.) Leach (Repr.: Cact. Succ. J. Gt. Brit. 25, 1963, 56, Photo: Nat. Herb. Pretoria); 5. E. intisy Drake v. intisy (Photo: as 1).

Plate 74. 1. Euphorbia juttae DTR. (Photo: K. DINTER); 2. E. leandriana P. BOIT. (Photo: W. KABEL); 3. E. lactea HAW. (Photo: J. MARNIER-LAPOSTOLLE); 4. E. lophogona LAM. (Photo: as 3); 5. E. mammillaris L. (Photo: as 3); 6. E. meloformis AIT.

Plate 75. 1. Euphorbia leucodendron Drake (Photo: J. Marnier-Lapostolle); 2. E. mauritanica L.; 3. E. milii v. bevilanensis (L. Croiz.) Ursch et Leandri (Photo: W. Rauh); 4. E. milii v. hislopii (N. E. Br.) Leandri; 5. E. milii v. splendens (Boj. ex Hook.) Ursch et Leandri.

Plate 76. 1. Euphorbia neriifolia L. (Photo: J. Marnier-Lapostolle); 2. E. neglecta N. E. Br. (Photo: as 1); 3. E. monteiri Hook. f.; 4. E. neohumbertii Boit. v. neohumbertii (Photo: as 1); 5. E. nivulia Haw. (Photo: as 1); 6. E. neutra Bgr. (Photo: as 1).

Plate 77. 1. **Euphorbia obovalifolia** A. Rich. (Photo: J. Marnier-Lapostolle) 2. **E. officinarum** L. v. **officinarum** (Photo: as 1); 3. **E. oncoclada** Drake; 4. **E. orthoclada** Bak. (Photo: as 1); 5. **E. obesa** Hook. f. (Photo: K. Schick).

Plate 78. 1. Euphorbia phillipsiae N. E. Br. (Photo: J. Marnier-Lapostolle); 2. E. phosphorea N. E. Br.; 3. E. pseudocactus Bgr.; 4. E. pachypodioides P. Boit. (Photo: W. Rauh); 5. E. lignosa Marl.

Plate 79. 1. Euphorbia pteroneura BGR. v. pteroneura; 2. E. pugniformis BOISS. (Photo: J. MARNIER-LAPOSTOLLE); 3. E. pulvinata MARL. (Photo: H. LANG); 4. E. ramipressa CROIZ.; 5. E. polyacantha BOISS. (Photo: as 2).

Plate 80. 1. **Euphorbia royleana** Boiss.; 2. **E. stellaespina** Haw. v. **stellaespina** (Photo: J. Marnier-Lapostolle); 3. **E. resinifera** Bgr. v. **resinifera**; 4. **E. rudis** N. E. Br. (Photo: as 2); 5. **E. subsalsa** Hiern v. **subsalsa** (Photo: as 2); 6. **E. sapinii** de Wild. (Photo: as 2).

Plate 81. 1. **Euphorbia squarrosa** Haw. (Photo: E. Hahn); 2. **E. tenuirama** Schweinf.; 3. **E. stenoclada** H. Baill.; 4. **E. valida** N. E. Br. (Repr.: White, Dyer et Sloane, Succ. Euph. fig. 619, Drawing C. Letty); 5. **E. tetragona** Haw. (Photo: J. Marnier-Lapostolle).

scarcely 1 mm long, bracts 8–10, linear, uneven, 10–12 mm long, 1–2 mm across.

**D. sessilis** R. E. FRIES. – Rhodesia. – **Caudex** compressed-spherical, c. 3 cm ⌀, fleshy; **St.** up to 30 cm long, 5 mm thick, very short-hairy; **L.** thin, oblong, broadly and shortly dentate, 10–12 cm long, 3–5 cm across, both surfaces sparsely hairy; **receptacles** flat, oblong, blunt, bract solitary, narrow-linear, blunt, 5 mm long, 1 mm across.

**D. vivipara** WELW. – Angola. – **Caudex** spherical, the size of a pea or walnut; **St.** soft, thin, eventually prostrate or ascending, rooting at the lower nodes, 15–20 cm tall, simple; **L.** long-petiolate, thin-membranous, blunt or rather pointed, margins wavy, dentate, 3.5–5 cm long, 2–2.5 cm across; **receptacles** top-shaped, 6–7 mm ⌀, upperside elliptical, bracts (5–)8(–9), 3–5 mm long.

**D. welleri** HEMSL. v. **welleri**. – Guinea; Malawi. – **Caudex** c. 4 cm ⌀; **St.** simple, fleshy, short-hairy, 20–40 cm tall, with minute reduced L. at the lower nodes; **L.** crowded on the upper part of the St., elliptical, blunt, margins slightly wavy, 8–14 cm long, 2.5–5 cm across, Stip. linear-linguiform, 2–5 mm long; **receptacles** stellate, with 4–7 rays, up to 2.5 cm ⌀, margin crenate, each ray terminating in a narrowly triangular bract 2–7.5 cm long, with a filiform tip.

**D. —** v. **minor** RENDLE (*Dorstenia unyika* ENGL. et WARB.). – Congolese Rep. – **St.** 10 to 15 cm tall; **L.** elliptical to narrow-ovate, c. 5 cm long; **receptacles** 3–7 mm ⌀.

**D. zanzibarica** OLIV. – Tanzania: Usambara, Zanzibar. – **Caudex** fleshy; **St.** succulent, branching, up to 90 cm tall, densely leafy above; **L.** elliptical, margins dentate above the cuneate base, 5–10 cm long, 2.5–3 cm across, Stip. 1 to 2 mm long, falling; **receptacles** triangular, less often 4 or 5-sided, c. 1 cm ⌀, margin narrow, at each angle with a linear blunt bract 6–9 mm long.

**Drimia** JACQ. Liliaceae. – Greenhouse, warm. Propagation: seed.

**D. haworthioides** BAK. (Pl. 53/2). – Cape. – Succulent ⚄; **root** fleshy, with a few offsets from ground-level; **L.** densely rosulate, the bases thick and fleshy, so close together as to form a bulb 5–6 cm ⌀, almost entirely below soil-level; **L.** oblong-spatulate, 16–20 mm long, 6–7 mm across, 2 mm thick, pointed above, underside rounded, purple-grey, virtually without chlorophyll, the tips usually drying up; in summer after the rest-period 3–4 soft thin L. are produced from the centre, these being oblong lanceolate, fresh green, c. 10 cm long, 1 cm across and deciduous; **Fl.** 10–20 in a long Rac. (Pl. 53/1), tube 2–3 mm long, Pet. completely recurved, greenish.

× **Dudleveria** ROWL. Crassulaceae. – Hybrid: **Dudleya** × **Echeveria**. – Greenhouse, warm. – Propagation: division. Apart from the following there are probably further hybrids.

× **D. spiralis** (DEL.) ROWL. (Pl. 53/3) (*Echeveria* × *spiralis* DEL. ex MORREN). – Hybrid: **Echeveria decipiens** × **Dudleya caespitosa**. – Branching basally; **St.** forming an extended Ros. 20–25 cm tall, densely spirally leafy; **L.** ± triangular, tapering, 3 cm long, underside rounded, with 2 minute spurs at the base, light green, often pink-tinged; **Infl.** lateral, Ped. with appressed L., Fl. 2–5, Pet. ± 8 mm long, 2.5 mm across, white, underside with a red keel. – Parentage doubtful; perhaps **Villadia** and not **Dudleya** was one parent.

**Dudleya** BRITT. et ROSE. (*Stylophyllum* BR. et R., *Hasseanthus* ROSE). Crassulaceae. – USA; Mexico. – ⚄, glabrous, succulent; **St.** simple or mostly forked; **L.** in Ros., semi-amplexicaul, veins several, parallel; **Infl.** from the L.-axils, with sessile bracts, cymose, the final Br. Cinc.; **Fl.** 5-partite, Sep. erect, Pet. erect or spreading, yellowish, white or red. – Type species: D. lanceolata. –Greenhouse, warm. – Text abridged from a contribution by R. MORAN.

## Division of the Genus Dudleya into Subgenera by R. Moran

A. Main **St.** overground, often elongate, often branched; **Ros.-L.** mostly evergreen, mostly 1–5 mm thick at the very base.

B. Pet. variously spreading from near the middle; receptacle conical, follicles patent to wide-spreading before dehiscence.

    SG. I. **Stylophyllum** (BR. et R.) MORAN (*Stylophyllum* BR. et R. as a G.; *Echeveria* DC. 5. Sect. *Stylophyllum* (BR. et R.) BGR.). – Type-species: D. edulis. – Species: D. anomala, attenuata, densiflora, edulis, formosa, hassei, traskiae, virens, viscida.

B. Pet. erect, pressed together in the form of a tube, only the tips sometimes spreading; receptacle flat, the follicles mostly erect and appressed before dehiscence.

    SG. II. **Dudleya** (*Echeveria* DC. 4. Sect. *Dudleya* (BR. et R.) BGR.). – Type-species: D. lanceolata. – Species: D. abramsii, acuminata, albiflora, anthonyi, bettinae, brevipes, brittonii, caespitosa, candelabrum, candida, cultrata, cymosa, elongata, farinosa, gatesii, greenei, guadalupensis, ingens, lanceolata, linearis, nubigena, palmeri, parva, pulverulenta, rigida, rigidiflora, rubens, saxosa, stolonifera.

A. Main **St.** underground, caudex-like, not branched; **Ros.-L.** dying off before the flowering time, less than 0.5 mm thick at the very base; Pet. and follicles mostly widespreading.
   **SG. III. Hasseanthus** (Rose) Moran (*Hasseanthus* Rose as a G.). – Type-species: D. variegata. – Species: D. blochmaniae, multicaulis, nesiotica, variegata.

**D. abramsii** Rose (§ II). – **St.** short, simple or branching; **Ros.** usually 2–6 cm ⌀; **L.** 12–20, pruinose, oblong-lanceolate, pointed, 4–60 mm long, 1.5–12 mm across, 2–4 mm thick; **Infl.** 2–3, simple or bipartite, Pet. pale yellow with red lines.

**D.** — ssp. **abramsii** (*D. tenuis* Rose, *Cotyledon t.* (Rose) Fedde, *C. a.* (Rose) Fedde, *Echeveria a.* (Rose) Bgr., *E. t.* (Rose) Bgr. non Rose). – USA: Calif.; Mexico: Baja Calif. – **St.** 1–1.5 cm ⌀; lower cauline **L.** usually 4 to 15 mm long, 1.5–4 mm across; **Pet.** 2–2.5 mm across, united for 2–4.5 mm.

**D.** — ssp. **murina** (Eastw.) Moran (*D. m.* Eastw.). – USA: Calif. – **St.** 1–3 cm ⌀; lower cauline **L.** 1–3 cm long, 5–11 mm across; **Pet.** 3–3.5 mm across, united for 1.5–3 mm.

**D. acuminata** Rose (§ II) (? *D. brandegei* Rose, *Cotyledon a.* (Rose) Fedde, ? *C. b.* (Rose) Fedde, *Echeveria a.* (Rose) Bgr., ? *E. b.* (Rose) Bgr.). – Mexico: central Baja Calif. – **St.** short, with few or no Br., 1–2.5 cm ⌀; **Ros.** 3–10 cm ⌀; **L.** 10–20, pale green, ovate to triangular-lanceolate, pointed or tapering, 2 to 10 cm long, 1.5–3 cm across, 2–5 mm thick; **Infl.** 5–40 cm tall, with 2–3 Br., Pet. yellow, 10–14 mm long, acuminate.

**D. albiflora** Rose (Pl. 54/1) (§ II) (*Cotyledon a.* (Rose) Fedde non Hemsl., *Echeveria a.* (Rose) Bgr., ? *D. moranii* D. A. Johansen). – Mexico: central and S.Baja Calif. – **St.** 1–2 cm ⌀, up to 30 cm long, usually branching to form clumps up to 70 cm ⌀; **Ros.** 2–10 cm ⌀; **L.** 10–25, green to pruinose, slightly broader midway, tapering, 2–6 cm long, 5–15 mm across, 2–6 mm thick, terete above; **Infl.** up to 45 cm tall, with 2–3 Br., Pet. white, 10–15 mm long.

**D. anomala** (Davids.) Moran (§ I) (*Stylophyllum a.* Davids., *S. insulare* Froed. non Rose, *S. coronatum* Froed.). – Mexico: Baja Calif. – **St.** up to 50 cm long, 5–15 mm thick, branching to form dense cushions of 60 cm ⌀; **Ros.L.** 20–30, strap-shaped to lanceolate, pointed, 1.5–8 cm long, 5–10 mm across, 2.5 to 7 mm thick, rather viscous; **Infl.** up to 25 cm tall, with 2–3 Br., Pet. white, elliptical, pointed, 7–10 mm long.

**D. anthonyi** Rose (§ II) (*Cotyledon a.* (Rose) Fedde, *Echeveria a.* (Rose) Bgr.). – Mexico: Baja Calif. – **St.** 4–9 cm ⌀, up to 80 cm long, unbranched; **Ros.** 15–50 cm ⌀; **L.** 35–90, chalky-pulverulent, oblong, acute to gradually tapering, 8–25 cm long, 3–7 cm across, 5–10 mm thick; **Infl.** up to 1 m tall, Cinc. 3–10, Cor. red, 12–17 mm long, lobes 5–10 mm long. – Hybridises with **D. cultrata**.

**D.** — × **D. cultrata** Rose (nat. hybr. Moran). – Mexico: Baja Calif., Is. of San Martín. – **St.** 1.5–4 cm thick, up to 7 cm long, with few Br.; **Ros.** about 10 cm ⌀; **L.** 30–40, somewhat pruinose, oblong, tapering to a point, 10–20 cm long, 2–4 cm across, 3–5 mm thick; **Infl.** up to 55 cm tall, with 3 simple or forking Br., Pet. yellow with red stripes, 12–14 mm long.

**D. attenuata** (S. Wats.) Moran (§ I). – USA: Calif.; Mexico. – **St.** 3–15 mm ⌀, up to 30 cm tall, branching to form clumps up to 40 cm ⌀; **Ros.** 2–5 cm ⌀; **L.** 5–20, pruinose, linear to linear-oblanceolate, pointed, terete, 2–10 cm long, 2–5 mm across; **Infl.** up to 25 cm tall, with 1–3 simple Cinc., Pet. elliptical, pointed.

**D.** — ssp. **attenuata** (*Cotyledon a.* S. Wats. non Lindberg f., *Stylophyllum a.* (S. Wats.) Br. et R., *Cotyledon edulis* v. *a.* (S. Wats.) Jeps., *Echeveria a.* (S. Wats.) Bgr., *E. e. v. a.* Jeps., *D. a.* ssp. *typica* Moran). – Mexico: Baja Calif. – **Pet.** with yellowish-red lines, 5–8 mm long.

**D.** — ssp. **orcuttii** (Rose) Moran (Pl. 54/2) (*Stylophyllum o.* Rose, *S. parishii* Britt., *Cotyledon o.* (Rose) Fedde, *C. p.* (Britt.) Fedde, *Echeveria o.* (Rose) Bgr., *E. palensis* Bgr.). – Mexico: Baja Calif. and islands. – **Pet.** white, often tinged with pink, with red lines, 6–10 mm long. Hybridizes with **D. brittonii, candida, formosa** and **variegata**.

**D.** — ssp. — × **D. formosa** Moran (nat. × Moran). – Mexico: Baja Calif. – Intermediate hybrid. – **St.** 7–20 mm ⌀, up to 15 cm long, branching, forming cushions of 10–12 Ros.; **Ros.** 4–5 cm ⌀; **L.** 12–20, pruinose, linear to linear-triangular, acuminate, 6–9 mm long, 6 to 8 mm across below, 3–4 mm thick; **Infl.** up to 15 cm tall, with 3–4 Br., Pet. pink, the keel red-lined, 8–10 mm long.

**D.** — ssp. — × **D. variegata** (Rose) Moran (nat. × Moran). – Mexico: Baja Calif. – Intermediate hybrid. – Principal **St.** with few Br., 1–4 cm long, 3–8 mm thick; **L.** rosulate, often withering before flowering, pruinose, linear-oblanceolate, pointed, 4–7 cm long, 1.5–2.5 cm across, terete above, flat below; **Infl.** up to 18 cm tall, with 2 Br., Pet. yellowish-white.

**D. bettinae** R. Hoover (§ II). – USA: Calif. – **St.** 10–18 mm ⌀, branching very freely to form mounds; **Ros.** 3–5 cm ⌀; **L.** c. 15, pruinose, terete or semi-terete, 2–7 cm long, 3–7 mm across; **Infl.** 15–25 cm tall, Cinc. few, Pet. straw-coloured, 8–12 mm long.

**D. blochmaniae** (Eastw.) Moran (§ III). – USA: Calif. – **L.** linear-oblanceolate to spatulate with a narrow petiole, pointed to rounded, 7 mm to 6 cm long, 1–4 mm across; **Infl.** up to 22 cm tall, with 2–3 Br., Pet. white, elliptical, pointed, 5–10 mm long.

**D.** — ssp. **blochmaniae** (*Sedum b.* Eastw., *Hasseanthus b.* (Eastw.) Rose, *H. kessleri* Davids., *H. variegatus* v. *b.* (Eastw.) Jeps., *S. gertrudianum* Eastw.). – USA: Calif.: coastal reg.; Mexico: Baja Calif. – **Caudex** hemispherical to fusiform, 7–25 mm long, 4–15 mm thick; **L.** 1–6 cm long, 3–8 mm across, 2–4 mm thick. Hybridizes with **D. edulis**.

**D.** — ssp. **brevifolia** MORAN (*Hasseanthus b.* v. *b.* MORAN). – USA: Calif. – **Caudex** 1.3–3.5 cm long, 1–6 mm ⌀; **L.** 7–15 mm long, 2.7 mm across, 2–4 mm thick. Hybridizes with **D. edulis.**

**D.** — ssp. **insularis** MORAN (*Hasseanthus b.* v. *i.* MORAN). – USA: Santa Rosa Is. (Calif.). – **Caudex** almost spherical, 1–2 cm long, 5–20 mm ⌀; **L.** 15–30, 1–3.5 cm long, 2–7 mm across, 1–3 mm thick.

**D. brevipes** ROSE (§ II) (*Cotyledon b.* FEDDE sphalm., *Echeveria b.* (ROSE) BGR.). – Mexico: Baja Calif. – Probably to be classified as **D. acuminata.**

**D. brittonii** D. A. JOHANSEN (§ II). – Mexico: Baja Calif. – **St.** 2–10 cm ⌀, usually unbranched; **Ros.** 10–50cm ⌀; **L.** 40–100, oblonglanceolate, widening and thickening towards the base, 7–25 cm long, 3.5–8.5 cm across, 4 to 11 mm thick, pointed to tapering, green or very white-pruinose; **Infl.** up to 1 m tall, with 3–6 Br., Pet. pale yellow. Hybridizes with **D. attenuata** ssp. **orcuttii** and **D. formosa.**

**D.** — × **D. formosa** MORAN (nat. × MORAN). – Mexico: N.Baja Calif. – Intermediate hybrid. – **St.** 2–4.5 cm thick, forming clumps of up to 20 **Ros.** which are 10–25 cm ⌀; **L.** 25–40, green, not pruinose, oblong, short-tapering, 5–12 cm long, 2.5–5 cm across, 3–4 mm thick; **Infl.** up to 40 cm tall, 5-branched, Pet. white or pink.

**D. caespitosa** (HAW.) BR. et R. (§ II) (*Cotyledon c.* HAW., *Sedum cotyledon* JACQ., *C. linguaeformis* R. BR., *C. reflexa* WILLD., *Echeveria c.* (HAW.) DC., *E. laxa* LINDL., *C. californica* BAK., *E. c.* BAK., *C. l.* (LINDL.) BREW. et WATS., *D.l.* (LINDL.) BR. et R., *D. helleri* ROSE, *D. cotyledon* (JACQ.) BR. et R., *C. h.* (ROSE) FEDDE, *E. cotyledon* (JACQ.) NELS. et MACBR., *E. h.* (ROSE) BGR.). – USA: central Calif. – **St.** 1.5–4 cm ⌀, up to 20 cm long, branching; **Ros.** 5–15 cm ⌀; **L.** 15–30, green or pruinose, oblong-oblanceolate, pointed, 5–15 cm long, 1–2 cm across, 3–8 mm thick; **Infl.** up to 60 cm tall, Pet. yellow, pointed.

**D. candelabrum** ROSE (§ II) (*Cotyledon c.* (ROSE) FEDDE, *Echeveria c.* (ROSE) BGR.). – USA: Calif.: Is. of Santa Cruz and Santa Rosa. – **St.** 2–8 cm ⌀, up to 20 cm long; **Ros.** 10 to 30 cm ⌀; **Ros.** 7–21 cm ⌀; **L.** 20–45, green, obovate to oblong-oblanceolate, tapering, 6–17 cm long, 3–7 cm across; **Infl.** up to 35 cm tall, 3-branched, Pet. pale yellow.

**D. candida** BRITT. (Pl. 54/3) (§ II) (*D. bryceae* BRITT., *Cotyledon c.* (BRITT.) FEDDE, *C. b.* (BRITT.) FEDDE, *Echeveria c.* (BRITT.) BGR., *E. b.* (BRITT.) BGR.). – Mexico: Baja Calif. – **St.** 2–6 cm ⌀, branching to form clumps up to 80 cm ⌀; **Ros.** 7–21 cm ⌀; **L.** 30–70, pruinose or green, tri-angular-ovate to oblong-oblanceolate, 5–11 cm long, 1–2 cm across; **Infl.** up to 50 cm tall, with 3–5 Br., Pet. pale yellow.

**D. cultrata** ROSE (§ II) (*Cotyledon c.* (ROSE) FEDDE, *Echeveria c.* (ROSE) BGR.). – Mexico: Baja Calif., islands. – **St.** 2–4 cm ⌀, up to 20 cm long, branching; **Ros.** 3–8 cm ⌀; **L.** 20–30, green, oblong, pointed, 5–13 cm long, 1–1.5 cm across, 3–5 mm thick; **Infl.** up to 40 cm tall, with c. 3 Br., Pet. pale yellow.

**D. cymosa** (LEM.) BR. et R. (§ II) (*Echeveria laxa* AUCT. non LINDL.). – USA: Calif. – **St.** short, simple or with few Br., 2–30 mm ⌀; **Ros.** usually 6–15 cm ⌀; **L.** 10–25, mostly evergreen, green to pruinose, usually oblanceolate to spatulate, pointed; **Infl.** with 2–4 Br., Pet. yellow to red, 7–14 mm long.

**D.** — ssp. **cymosa** (*Echeveria c.* LEM. *Cotyledon c.* (LEM.) BAK., *C. nevadensis* S. WATS., *C. purpusii* K. SCHUM., *Echeveria p.* (K. SCHUM.) WITTM., *C. plattiana* JEPS., *Sedum p.* (K. SCHUM.) KUNTZE non ROSE, *D. angustiflora* ROSE, *D. n.* (S. WATS.) BR. et R., *D. sheldonii* ROSE, *D. p.* (JEPS.) BR. et R., *D. purpusii* (K. SCHUM.) BR. et R., *C. a.* (ROSE) FEDDE, *C. s.* (ROSE) FEDDE, *E. n.* (S. WATS.) NELS. et MACBR., *E. p.* (JEPS.) NELS. et MACBR., *C. laxa* v. *cymosa* (LEM.) JEPS., *C. l.* v. *n.* (S. WATS.) JEPS., *E. a.* (ROSE) BGR., *E. s.* (ROSE) BGR., *E. l.* v. *c.* (LEM.) JEPS., *E. l.* v. *n.* (S. WATS.) JEPS.). – USA: Calif. – **L.** 3–12 cm long, 1–5.5 cm across; **Infl.** up to 25 cm tall, Pet. light yellow to red.

**D.** — ssp. **gigantea** (ROSE) MORAN (*D. g.* ROSE, *Cotyledon g.* (ROSE) FEDDE, *Echeveria amadorana* BGR., *E. lanceolata* v. *incerta* JEPS.). – USA: Calif. – **L.** pruinose, lanceolate, pointed to tapering, 4–17 cm long, 1.5–6 cm across; **Infl.** up to 45 cm tall, Pet. red.

**D.** — ssp. **marcescens** MORAN. – USA: Calif. – **St.** 2–7 mm ⌀; **L.** green, oblanceolate, 1.5–4 cm long, 5–12 mm across; **Infl.** up to 10 cm tall, Pet. pale yellow.

**D.** — ssp. **minor** (ROSE) MORAN (*D. m.* ROSE, *D. pumila* ROSE, *D. bernardina* BRITT., *D. goldmanii* ROSE, *Cotyledon p.* (ROSE) FEDDE, *C. b.* (BRITT.) FEDDE, *C. m.* (ROSE) FEDDE, *C. roseana* FEDDE, *Echeveria parva* BGR., *E. b.* (BRITT.) BGR., *E. g.* (ROSE) BGR. non ROSE, *E. m.* (ROSE) BGR., *E. laxa* v. *m.* (ROSE) JEPS., *D. nevadensis* ssp. *m.* (ROSE) ABRAMS). – USA: Calif. – **L.** green or pruinose, rhombic-oblanceolate to spatulate, short-pointed, 1.5–5 cm long, 1–3 cm across; **Infl.** up to 25 cm tall, Pet. light yellow to red.

**D.** — ssp. **ovatifolia** (BRITT.) MORAN (*D. o.* BRITT., *Cotyledon o.* (BRITT.) FEDDE, *Echeveria o.* (BRITT.) BGR.). – USA: Calif. – **St.** 1–1.5 cm ⌀; **L.** green, ovate to elliptical, 2–5 cm long, 1.5 to 2.5 cm across; **Infl.** up to 15 cm tall, Pet. light yellow.

**D.** — ssp. **setchellii** (JEPS.) MORAN (*Cotyledon laxa* v. *s.* JEPS., *C. caespitosa* v. *paniculata* (JEPS.) BR. et R., *D. p.* (JEPS.) BR. et R., *D. humilis* ROSE, *C. s.* (JEPS.) FEDDE, *C. p.* (JEPS.) FEDDE non THUNBG., *C. h.* (ROSE) FEDDE non MARL., *E. s.* (JEPS.) NELS. et MACBR., *E. jepsonii* NELS. et MACBR., *C. l.* v. *p.* JEPS., *E. diaboli* BGR., *E. l.* v. *p.* JEPS., *E. l.* v. *s.* JEPS.). – USA: Calif. – **St.** 1–2 cm ⌀; **L.** pruinose, oblong-oblanceolate to triangular, 3 to 10 cm long, 0.5–2 cm across; **Infl.** up to 25 cm tall, Pet. pale yellow.

**D. densiflora** (ROSE) MORAN (Pl. 55/1) (§ I) (*Cotyledon nudicaule* ABRAMS non LAM., *Stylophyllum d.* ROSE, *C. d.* (ROSE) FEDDE, *Echeveria d.* (ROSE) BGR., *E. n.* (ABRAMS) MUNZ, *D. n.* (ABRAMS) MORAN, *S. n.* (ABRAMS) ABRAMS). – USA: Calif. – **St.** 1–2.5 cm ⌀, up to

10 cm long, branching; **Ros.** 7–25 cm ⌀; **L.** 20–40, pruinose, linear, apiculate, terete above, 6–15 cm long, 6–12 mm across, 5–8 mm thick; **Infl.** up to 30 cm tall, branching, Pet. whitish to pink, 5–10 mm long.

**D. edulis** (NUTT.) MORAN (§ I) (*Sedum e.* NUTT., *Cotyledon e.* (NUTT.) BREW. et WATS., *Stylophyllum e.* (NUTT.) BR. et R., *Echeveria e.* (NUTT.) BGR.). – USA: S.W.Calif.; Mexico: N.W.Baja Calif. – **St.** 1.5–4.5 cm ⌀, up to 15 cm long, branching to form clumps of 40 cm ⌀; **Ros.** 5–10 cm ⌀; **L.** 15–25, pale green, linear, pointed, apiculate, terete above, 8–20 cm long, 4–10 mm ⌀; **Infl.** up to 50 cm tall, branching, Pet. white. Hybridizes with **D. blochmaniae** and **D. stolonifera**.

**D.** — × **D. blochmaniae** ssp. **blochmaniae** (nat. × MORAN). – USA: Calif. – Intermediate hybrid. – **St.** spherical to oblong, 6–15 mm long, 6–13 mm thick; **L.** linear-lanceolate or almost oblanceolate, pointed, then 3–7 cm long, 5–10 mm across at the base, then 2–6 mm across and 1.5–2 mm thick, 5–8 mm across above and 2–4 mm thick; **Infl.** 11 cm tall, with 1–3 Br., Pet. white with purple-spotted M.rib.

**D.** — × **D. blochmaniae** ssp. **brevifolia** MORAN (nat. × MORAN). – USA: Calif. – Intermediate hybrid. – **St.** oblong, 1–2.5 cm long, 5–12 mm thick; **L.** 5–15, usually linear-oblong, pointed, 1–6 cm long, 5–10 mm across at the base, then 1.5–6 mm across, broader above and 1–2 mm thick, subterete towards the tip, rarely a few L. are spatulate; **Infl.** up to 25 cm tall, with several Br., Pet. white with red-striped keel.

**D.** — × **D. stolonifera** MORAN (nat. × MORAN). – USA: Calif. – **St.** 2–2.5 cm ⌀, branching dichotomously, forming small clumps; **Ros.** 6–12 cm ⌀; **L.** 25–30, broad-linear, somewhat tapering, 5–11 cm long, 9–15 mm across, 3–4 mm thick, not pruinose, margins rounded; **Infl.** up to 30 cm tall, Pet. pale yellow.

**D. elongata** ROSE (§ II) (*Cotyledon e.* (ROSE) FEDDE, *Echeveria e.* (ROSE) BGR.). – USA: Calif. – Resembles **D. lanceolata**; **St.** up to 20–40 cm long, attenuated.

**D. farinosa** (LINDL.) BR. et R. (Pl. 54/4) (§ II) (*Echeveria f.* LINDL., *Cotyledon f.* (LINDL.) BAK., ? *C. lingula* S. WATS., *D. compacta* ROSE, *D. eastwoodiae* ROSE, *D. septentrionale* ROSE, ? *D. l.* (S. WATS.) BR. et R., *C. c.* (ROSE) FEDDE, *C. e.* (ROSE) FEDDE, *S. s.* (ROSE) FEDDE, ? *E. l.* (S. WATS.) NELS. et MACBR., *E. c.* (ROSE) BGR., *E. e.* (ROSE) BGR., *E. s.* (ROSE) BGR., *E. farinulenta* LEM.). – USA: coasts of N. and central Calif. – **St.** 1–3 cm ⌀, often attenuated, freely branching; **Ros.** 4–10 cm ⌀; **L.** 15–30, green to pruinose, ovate-oblong, pointed, 2.5–6 cm long, 1–2.5 cm across, 3–6 mm thick; **Infl.** up to 35 cm tall, with 3–5 Br., Pet. pale yellow, 10–14 mm long.

**D. formosa** MORAN (Pl. 55/3) (§ I). – Mexico: Baja Calif. – **St.** 0.5–2.5 cm ⌀, up to 50 cm long, branching, forming large clumps; **Ros.** 4 to 13 cm ⌀; **L.** 10–20, green with red tips, oblong to oblong-ovate, pointed to blunt, 2–8 cm long, 1–3 cm across, 3–6 mm thick; **Infl.** up to 15 cm tall, branching, Pet. white with red. Hybridizes with **D. attenuata** ssp. **orcuttii** and **D. brittonii**.

**D. gatesii** D. A. JOHANSEN (§ II). – Mexico: Baja Calif., Cedros Is. – **St.** 2–5 cm ⌀, up to 30 cm long, simple or branching; **Ros.** 10 to 25 cm ⌀; **L.** 20–50, pale green or slightly pruinose, oblong, pointed, 5–15 cm long, 1.5 to 6 cm across, 5–10 mm thick; **Infl.** up to 70 cm tall, with 2–4 Br., Pet. white or pink.

**D. greenei** ROSE (§ II) (*Cotyledon g.* (ROSE) FEDDE, *Echeveria g.* (ROSE) BGR., *D. hoffmannii* D. A. JOHANSEN, *D. regalis* D. A. JOHANSEN, *D. echeverioides* D. A. JOHANSEN). – USA: Calif., islands. – **St.** 2–5 cm ⌀, often attenuated, branching to form clumps of 1.5 m ⌀; **Ros.** 5–15 cm ⌀; **L.** 15–50, green to pruinose, oblong-oblanceolate to ovate, pointed, 3–11 cm long, 1–3.5 cm across, 4–8 mm thick; **Infl.** up to 40 cm tall, with 3–5 Br., Pet. pale yellow or whitish, 8–12 mm long.

**D. guadalupensis** MORAN (§ II). – Mexico: Guadalupe Is. – **St.** 1.5–3.5 cm ⌀, up to 15 cm long, freely branching to form clumps up to 50 cm ⌀; **Ros.** 3–10 cm ⌀; **L.** 35–70, green to pruinose, oblanceolate, apiculate, pointed, 2.5 to 7 cm long, 8–17 mm across, 2–4 mm thick; **Infl.** up to 30 cm tall, with 2–3 Br., Pet. whitish.

**D. hassei** (ROSE) (Pl. 55/4) (§ I) (*Stylophyllum h.* ROSE, *Cotyledon h.* (ROSE) FEDDE, *Echeveria h.* (ROSE) BGR.). – USA: Calif.: Is. Santa Catalina. – **St.** 1–3 cm ⌀, up to 30 cm long, freely branching; **Ros.** 6–10 cm ⌀; **L.** 15–30, pruinose, linear-lanceolate, blunt, 3–10 cm long, 5–15 mm across, 2–4 mm thick; **Infl.** up to 30 cm tall, with 2–4 Br., Pet. white, 8–10 mm long.

**D. ingens** ROSE (Pl. 55/5) (§ II) (*Cotyledon rugens* FEDDE sphalm., *Echeveria i.* (ROSE) BGR., ? *D. viridicata* D. A. JOHANSEN, *D. eximia* D. A. JOHANSEN). – Mexico: Baja Calif. – **St.** 1.5–6 cm ⌀, up to 40 cm long, simple or somewhat branching; **Ros.** 5–40 cm ⌀; **L.** 20–70, green or slightly pruinose, oblong, tapering, 7–25 cm long, 1.5–5 cm across, 3–11 mm thick; **Infl.** up to 90 cm tall, with 3–4 Br., Pet. pale yellow to white, 10–15 mm long, pointed.

**D. lanceolata** (NUTT.) BR. et R. (Pl. 55/2) (§ II) (*Echeveria l.* NUTT., *Cotyledon l.* (NUTT.) BREW. et WATS. non FORSK. nec BLANCO, *D. hallii* ROSE, *D. robusta* BRITT., *D. parishii* ROSE, *D. lurida* ROSE, *D. brauntonii* ROSE, ? *D. congesta* BRITT., *C. h.* (ROSE) FEDDE, *C. r.* (BRITT.) FEDDE, *C. p.* (ROSE) FEDDE non (BRITT.) FEDDE, *C. l.* (ROSE) FEDDE, *C. b.* (ROSE) FEDDE, ? *C. c.* (BRITT.) FEDDE, *D. reflexa* BRITT., *E. h.* (ROSE) NELS. et MACBR., *E. p.* (ROSE) BGR., *E. refl.* (BRITT.) BGR., *E. rob.* (BRITT.) BGR., *E. monicae* BGR., ? *E. c.* (BRITT.) BGR., *E. b.* (ROSE) BGR., *E. l.* v. *lurida* (ROSE) MUNZ). – USA: Calif.; Mexico: Baja Calif. – **St.** 1–3 cm ⌀, short, simple or with few Br.; **Ros.** 3–25 cm ⌀; **L.** 10–25, green to pruinose, oblong-lanceolate, pointed, 5–20 cm long, 1–3 cm across, 1.5–6 mm thick; **Infl.** up to 75 cm tall, with 2–3 Br., Pet. yellow to (more usually) red, 10–16 mm long, pointed.

**D. linearis** (GREENE) BR. et R. (§ II) (*Cotyledon l.* GREENE, *Echeveria l.* (GREENE) BGR.). – Mexico: Baja Calif., Is. San Benito. – **St.** 1–2 cm ⌀, freely branching, mat-forming; **Ros.** 2–5 cm ⌀; **L.** 20–40, green, oblong-oblanceolate, almost pointed to tapering, 2.5–6 cm long, 5–10 mm across, 2–3 mm thick; **Infl.** up to 17 cm tall, with 2–3 Cinc., Pet. yellow, 7–12 mm long.

**D. multicaulis** (ROSE) MORAN (§ III) (*Hasseanthus m.* ROSE, *H. elongatus* ROSE, *Sedum e.* (ROSE) FEDDE non LEDEB., *S. m.* (ROSE) FEDDE non WALL., *H. variegatus* v. e. (ROSE) JOHNST., *S. oblongirhizum* BGR., *S. sanctae-monicae* BGR.). – USA: Calif.: coasts. – **Caudex** 1.5–5 cm long, 3–18 mm ⌀; **Ros.** 3–10 cm ⌀; **L.** 5–15, linear-tapering, terete above, 4–15 cm long, 2–6 mm across; **Infl.** up to 35 cm tall, branching twice or more, Pet. yellow, 5–9 mm long, pointed.

**D. nesiotica** MORAN (§ III) (*Hasseanthus n.* MORAN). – USA: Calif.: Is. Santa Cruz. – **Caudex** almost spherical, 1–3 cm long, 7–20 mm ⌀; **Ros.** 5–10 cm ⌀; **L.** oblanceolate, sometimes spatulate, pointed or blunt, 2.5–5 cm long, 0.5 to 2.5 cm across, 2–5 mm thick; **Infl.** up to 10 cm tall, with 2 Cinc., Pet. white.

**D. nubigena** (BRANDEG.) BR. et R. ssp. **nubigena** (§ II) (*Cotyledon n.* BRANDEG., *D. xantii* ROSE, *C. x.* (ROSE) FEDDE, *Echeveria n.* (BRANDEG.) BGR., *E. x.* (ROSE) BGR.). – Mexico: Baja Calif. – **St.** 1–2 cm ⌀, up to 6 cm long, simple; **Ros.** 4–18 cm ⌀; **L.** 10–20, usually pruinose, oblong-lanceolate, pointed, 4–11 cm long, 1–3 cm across, 2–3 mm thick; **Infl.** up to 30 cm tall, with 2–3 Cinc., Pet. yellow, usually with red.

**D.** — ssp. **cerralvensis** MORAN. – S.W.side, Is. Cerralvo. – **L.** green to slightly bluish, not pruinose; Cor. pure yellow.

**D. palmeri** (S. WATS.) BR. et R. (§ II) (*Cotyledon p.* S. WATS., *Echeveria p.* (S. WATS.) NELS. et MACBR. non ROSE). – USA: Calif. – **St.** 2–4 cm ⌀, up to 20 cm long, laxly branching; **Ros.** 5–20 cm ⌀; **L.** 15–25, green or reddish, oblong-lanceolate, pointed, 5–20 cm long, 1.5 to 5 cm across, 3–8 mm thick; **Infl.** up to 65 cm tall, with ± 3 Br., Pet. yellow with red, pointed.

**D. parva** ROSE et DAVIDSON (§ II). – USA: Calif. – **St.** 1–7 mm ⌀, up to 5 cm long, often branching; **Ros.** 1–3 cm ⌀; **L.** 5–10, slightly pruinose, linear to oblanceolate, pointed, 1.5 to 4 cm long, 3–6 mm across, 1.5–2 mm thick; **Infl.** up to 18 cm tall, with 1–2 Cinc., Pet. pale yellow, pointed.

**D. pauciflora** ROSE (§ II) (*Cotyledon p.* (ROSE) FEDDE, *Echeveria p.* (ROSE) BGR.). – Mexico: Baja Calif. – **St.** short, 1–2.5 cm ⌀, mat-forming; **Ros.** 3–7 cm ⌀; **L.** 15–30, green or pruinose, triangular to oblong, pointed, 2 to 7.5 cm long, 7–15 mm across, 2–5 mm thick; **Infl.** up to 30 cm tall, with 2–3 Br., Pet. yellow with red, pointed.

**D. pulverulenta** (NUTT.) BR. et R. (§ II). – USA: Calif.; Mexico. – **St.** unbranched, the entire plant mealy-pruinose; **Infl.** with 3–30 Br., Cinc. twisted at the base so that the Fl. are ± on the underside.

**D.** — ssp. **arizonica** (ROSE) MORAN (*D. a.* ROSE, *Echeveria lagunensis* MUNZ, *D. l.* (MUNZ) E. WALTH., *E. a.* (ROSE) KEARN et PEEB. non HORT. ex BGR., *E. p.* ssp. *a.* (ROSE) CLOKEY). – USA: S.Nevada, Calif., W.Arizona; Mexico: Baja Calif., N.W.Sonora. – **St.** 1–4 cm ⌀, up to 15 cm long; **Ros.** 3–30 cm ⌀; **L.** 10–35, oblong to oblong-obovate, gradually tapering, 3 to 17 cm long, 1–5 cm across, 2–4 mm thick; **Infl.** up to 60 cm tall, Cinc. 3–6, Cor. red to yellow.

**D.** — ssp. **pulverulenta** (*Echeveria p.* NUTT., ? *E. argentea* LEM., *Cotyledon p.* (NUTT.) BAK., *D. p.* ssp. *typica* MORAN). – USA: Calif. to Mexico: Baja Calif. – **St.** 4–10 cm ⌀, up to 60 cm long; **Ros.** 25–55 cm ⌀; **L.** 30–80, oblong, tapering to pointed, 8–27 cm long, 3–10 cm across, 3–10 mm thick; **Infl.** up to 1.5 m tall, Cinc. 3 or more, Cor. red.

**D. rigida** ROSE (Pl. 55/6) (§ II) (*Cotyledon r.* (ROSE) FEDDE, *Echeveria r.* (ROSE) BGR.). – Mexico: Baja Calif. – **St.** 1–3.5 cm ⌀, short, branching to form clumps of 25 cm ⌀; **Ros.** 6–15 cm ⌀; **L.** 10–25, usually slightly pruinose, triangular-ovate to triangular-lanceolate, tapering, 5–8 cm long, 2.5–4 cm across, 6–10 mm thick; **Infl.** up to 50 cm tall, with 2–3 Cinc., Pet. yellow with red markings.

**D. rigidiflora** ROSE (§ II) (*Cotyledon r.* (ROSE) FEDDE, *Echeveria r.* (ROSE) BGR.). – Mexico: Baja Calif. – **St.** 1.5–4 cm ⌀, up to 20 cm long, simple or branching to form cushions up to 1 m ⌀; **Ros.** 5–15 cm ⌀; **L.** 20–60, pale green, slightly pruinose, linear, narrowing from the base, pointed, 8–15 cm long, 1–2 cm across; **Infl.** up to 40 cm tall, with several Br., Pet. white, pointed.

**D. rubens** (BRANDEG.) BR. et R. (§ II) (*Cotyledon r.* BRANDEG., *Echeveria r.* (BRANDEG.) BGR.). – Mexico: Baja Calif. – **St.** 1–2.5 cm ⌀, up to 12 cm long, usually with a few Br.; **Ros.** 5–15 cm ⌀; **L.** 10–25, pruinose, oblong-lanceolate to oblanceolate, pointed, 4–10 cm long, 1–3.5 cm across, 2–4 mm thick; **Infl.** up to 30 cm tall, with 2–4 Br., Cor. dull red.

**D. saxosa** (M. E. JONES) BR. et R. (§ II). – USA: Calif., Arizona. – **St.** short, simple or branching, not usually forming clumps; **Ros.** 3–12 cm ⌀; **L.** 10–25, initially pruinose, oblong-lanceolate, pointed; **Infl.** with 2–3 Br., Pet. yellow, pointed.

**D.** — ssp. **aloides** (ROSE) MORAN (*D. a.* ROSE, *D. grandiflora* ROSE, ? *D. delicata* ROSE, *Cotyledon a.* (ROSE) FEDDE, *C. g.* (ROSE) FEDDE, ? *C. d.* (ROSE) FEDDE, *E. a.* (ROSE) BGR., *E. g.* (ROSE) BGR., ? *E. d.* (ROSE) BGR., *E. lanceolata* v. *a.* (ROSE) MUNZ, *E. l.* v. *compacta* JEPS.). – USA: S.Calif. – **St.** 1–3 cm ⌀; **L.** 4–15 cm long, 6–20 mm across, 2–5 mm thick; **Infl.** up to 35 cm tall.

**D.** — ssp. **collomiae** (ROSE) MORAN (*D. c.* ROSE, *Echeveria c.* (ROSE) KEARN. et PEEB.). – USA: central Arizona. – **St.** 1.5–3 cm ⌀; **L.** 6–15 cm long, 1–2 cm across, 2–6 mm thick; **Infl.** up to 40 cm tall.

**D.** — ssp. **saxosa** (*Cotyledon s.* M. E. JONES, *E. s.* (M. E. JONES) NELS. et MACBR., *C. lanceolata* v. *s.* (M. E. JONES) JEPS., *E. l.* v. *s.*

(M. E. Jones) Jeps.). – USA: Calif. – **St.** 1–1.5 cm ⌀; **L.** 3–9 cm long, 0.5–1.5 cm across, 1.5–3 mm thick; **Infl.** up to 20 cm tall.

**D. × semiteres** (Rose) Moran (§ I, § II) (*Stylophyllum s.* Rose, *Cotyledon s.* (Rose) Fedde, *Echeveria s.* (Rose) Bgr.). – Mexico: N.Baja Calif. and islands. – Nat. hybrid: **D. attenuata** ssp. **orcuttii** × **D. candida**. – **St.** 1–3 cm ⌀, up to 15 cm tall, with few Br.; **Ros.** 3–10 cm ⌀; **L.** 15–40, green or pruinose, narrowing, pointed, 2–18 cm long, 6–15 mm across, 3–6 mm thick; **Infl.** up to 50 cm tall, with 2–4 Br., Pet. white or yellowish, pointed.

**D. stolonifera** Moran (§ II). – USA: Calif. – **St.** 1.5–3 cm ⌀, up to 10 cm long, branching by means of offsets; **Ros.** 5–12 cm ⌀; **L.** 15–25, green becoming marked with red, oblong-obovate, short-tapering, 3–7 cm long, 1.5 to 3 cm across, 3–4 mm thick; **Infl.** up to 25 cm tall, with 2 or more Cinc., Pet. yellow, pointed. Hybridizes with **D. edulis.**

**D. traskiae** (Rose) Moran (§ I) (*Stylophyllum t.* Rose, *Cotyledon t.* (Rose) Fedde, *Echeveria t.* (Rose) Bgr.). – USA: Calif.: Is. Santa Barbara. – **St.** 1–3 cm ⌀, branching; **Ros.** 10–20 cm ⌀; **L.** 25–35, pruinose, oblong-oblanceolate, pointed, 4–15 cm long, 1–4 cm across, 4–6 mm thick; **Infl.** up to 30 cm tall with c. 3 Br., Pet. yellow.

**D. variegata** (S. Wats.) Moran (§ III) (*Sedum v.* S. Wats., *Hasseanthus v.* (S. Wats.) Rose). – USA: San Diego, Calif.; Mexico: Baja Calif. – **Caudex** almost spherical to oblong, 1–3 cm long, 3–15 mm ⌀; **Ros.** 2–5 cm ⌀; **L.** 4–12, oblanceolate to spatulate, pointed to blunt, 1–7 cm long, blade 3–11 mm across, 1.5–4 mm thick, with a petiole 0.5–3 mm across; **Infl.** up to 20 cm tall, with 2–3 Cinc., Pet. yellow. Hybridizes with **D. attenuata** ssp. **orcuttii**.

**D. virens** (Rose) Moran (§ I) (*Stylophyllum v.* Rose, *S. albidum* Rose, *S. insulare* Rose, *Cotyledon v.* (Rose) Fedde, *C. a.* (Rose) Fedde, *C. i.* (Rose) Fedde, *C. viscida v. i.* (Rose) Jeps., *Echeveria v.* (Rose) Bgr., *E. a.* (Rose) Bgr., *E. i.* (Rose) Bgr., *E. visc. v. i.* (Rose) Jeps.). – USA: Calif. and islands; Mexico: Baja Calif., Is. Guadalupe. – **St.** 1–4 cm ⌀, branching to form clumps up to 1 m ⌀; **Ros.** 10–25 cm ⌀; **L.** 20–45, green or pruinose, oblong, pointed, 8–20 cm long, 1.5–3 cm across; **Infl.** up to 60 cm tall, with several forking Br., Pet. white, pointed.

**D. viscida** (S. Wats.) Moran (§ I) (*Cotyledon v.* S. Wats., *Stylophyllum v.* (S. Wats.) Br. et R., *Echeveria v.* (S. Wats.) Bgr.). – USA: S.Calif. – **St.** 1–4 cm ⌀, short-branching and mat-forming; **Ros.** 10–30 cm ⌀; **L.** 15–40, dark green, viscous, resin-scented, linear-triangular, 6–15 cm long, 0.5–1.5 cm across, 3–5 mm thick; **Infl.** up to 40 cm tall, with 3 or more Br., Pet. white with red lines.

**Duvalia** Haw. Asclepiadaceae. – Cape; S.W. Afr.; E.Afr.; Arabia. – Succulent ♃; **St.** prostrate, scarely erect, mat-forming, short and thick, with 4–6 angles, these being obtuse, short-dentate, with protuberances often divided into tubercles by transverse furrows; **L.** insignificant; **Fl.** solitary or several on young shoots, with pedicels 1.5–3 cm long, small or medium-sized, the tube annular, fleshy, the margin deeply 5-cleft, lobes 5, the borders and the indentations usually very recurved, often hirsute inside, variously coloured, the corona united to form a somewhat 5-angled fleshy Bo. – Greenhouse, warm. Propagation: seed, cuttings.

**D. angustiloba** N. E. Br. – Cape: Karroo. – Mat-forming; **St.** almost spherical, 1.5–2 cm tall and thick, angles 4–5 with tuberculate pointed T.; **Fl.** 5–20, Cor. 2 cm ⌀, lobes 8–10 mm long, very narrow-lanceolate, dark brown, very inconspicuously hairy at the base, annulus very slightly prominent.

**D. caespitosa** (Mass.) Haw. (*Stapelia c.* Mass.). – Cape: Karroo. – **St.** 2–4 cm long, 8–11 mm ⌀, ovoid to oblong, roundish 4-angled, tuberculate-dentate; **Fl.** 1–2, Cor. 1.5–2 cm ⌀, greenish outside, brown inside, glossy, the fleshy annulus very prominent, somewhat 5-angled, 3–5 mm ⌀, pubescent, lobes 5–6 mm long, margins ciliate.

**D. compacta** Haw. (*Stapelia mastodes* Jacq., *S. c.* Schultes, *D. m.* Sweet). – Cape: Lit. Namaqualand. – **St.** ± ovate-spherical, (1.5–) 2.5–4(–5) cm long, 2 mm ⌀ or more, with 4–6 angles, often brownish, with roundish tubercles and small T.; **Fl.** 1–5, Cor. c. 2 cm ⌀, dark brown, lobes very spreading, 6–7 mm long, lanceolate-pointed, the annulus glabrous or pubescent, with ciliate hairs at the lobe-bases.

**D. corderoyi** (Hook. f.) N. E. Br. (*Stapelia c.* Hook. f.). – Cape: Karroo. – **St.** roundish, 1.5–3 cm long, 2 cm thick, with 6 indistinctly tuberculate ribs, often purple; **Fl.** 2–4, Cor. 3–5 cm ⌀, lobes 2 cm long, dull olive-green, annulus 10–12 mm ⌀, densely covered with long purple hairs.

**D. eilensis** Lavr. – Somalia: Bosaso Reg. – **St.** 4-angled, 20–30 cm long, up to 10 mm thick, greenish or brownish with a few dark spots, erect; **Fl.** solitary from the base of the St., pedicels 4–4.5 cm long, Sep. deltoid, acute, 3 mm long, Cor. rotate, 22–24 mm ⌀, annulus thick, fleshy, 10–13 mm ⌀, lobes replicate, very slightly ascending, 7 mm long, 3.5 mm broad at base, outside green, glabrous, inside yellowish, densely covered with brown-purple spots, densely tuberculate-papillose, papillae on the lobes bearing stiff hairs throughout, corona 6 mm ⌀, stipitate.

**D. elegans** (Mass.) Haw. v. **elegans** (*Stapelia e.* Mass., *S. radiata* Jacq., *S. jacquiniana* Schult., *D. j.* Sweet). – Cape: Karroo. – **St.** prostrate, mat-forming, 2–4 cm long, 1–1.5 cm thick, angles blunt-rounded, dentate, reddish; **Fl.** 1–3, Cor. 2 cm ⌀, rather flat, lobes triangular-ovate, pointed, 5–6 mm long, 4.5 mm across, black to violet inside, with long purple hairs, smooth and glabrous outside.

**D. — v. namaquana** N. E. Br. – Cape: Lit. Namaqualand. – Lobes 7–10 mm long.

**D. — v. seminuda** N. E. Br. – **St.** 4–10 cm long; lobes glabrous in the upper half.

**D. emiliana** WHITE. – Origin unknown. – **St.** 2–3 cm long, grey-green, with 4–6 dentate angles; **Fl.** 1–2, Cor. 2.5–3 cm $\varnothing$, yellow-green, lobes spreading, 1 cm long, with a few red hairs at the base, purplish green at the tip, annulus roundish to 5-angled, minutely papillose, with minute brown spots, with lighter spots inside.

**D. maculata** N. E. BR. v. **maculata.** – Cape: Karroo, Lit. Namaqualand; S.W.Afr.: Gr. Namaqualand. – Mat-forming; **St.** prostrate, 1.5–3 cm long, 8–12 mm $\varnothing$, oblong, dark green, with 4–5 blunt angles and very pointed T. 4–5 mm long; **Fl.** 4–8, Cor. 15–20 mm $\varnothing$, lobes 5–7 mm long, the sides vertically folded back, very pointed, glabrous, olive or red-brown, finely ciliate at the base, annulus 5–7 mm $\varnothing$, slightly 5-angled, pubescent, with red-brown spots.

**D. –** v. **immaculata** LUCKH. – **Cor.** deep purple, without spots.

**D. minuta** NEL. – S.W.Afr.: Gr. Namaqualand. – Clump-forming; **St.** almost spherical or cylindrical, 10–18 mm long, 10–12 mm $\varnothing$, green, with 5 distinctly tuberculate angles and pointed triangular T.; **Fl.** 2–4, Cor. 15 mm $\varnothing$, lobes 5 mm long, linear, apiculate, both surfaces glabrous, yellowish-green with light purple blotches, margins with short white cilia, annulus 5 mm $\varnothing$, roundish to 5-angled, border thickened. white with purple blotches, with white hairs.

**D. modesta** N. E. BR. – Cape: Karroo. – **St.** 2–2.5 cm long, 8–12 mm $\varnothing$, ovate or oblong, glabrous, with 4–5 obtuse dentate angles; **Fl.** 2–3, Cor. 12–18 mm $\varnothing$, dark brown, annulus c. 5 mm $\varnothing$, bluntly 5-angled, glabrous, lobes ovate-pointed, 5–6 mm long, margins sharply recurved, hairy up to midway.

**D. parviflora** N. E. BR. (Pl. 56/2). – Cape: Karroo. – **St.** c. 25 mm long, 12 mm $\varnothing$, oblong, with 5–6 very obtuse, dentate angles, light green spotted with red or slightly grey-green; **Fl.** 4–5, Cor. 10–14 mm $\varnothing$, cream-coloured, lobes 4 to 5 mm long, vertically folded back at the sides, pale red above, with a few light cilia below.

**D. pillansii** N. E. BR. v. **pillansii.** – Cape: Karroo. – **St.** numerous, 2–2.5 cm long, c. 1 cm thick, almost cylindrical, green, somewhat reddish, the 4 angles having a few stout T.; **Fl.** numerous, Cor. c. 2 cm $\varnothing$, lobes triangular, deep-furrowed on the underside, greenish outside, velvety purple-brown inside, the annulus yellowish, the margin with reddish tubercles up to midway.

**D. –** v. **albanica** N. E. BR. – Lobes spreading to erect.

**D. polita** N. E. BR. v. **polita** (*D. dentata* N. E. BR., *Stapelia p.* HORT. angl. ex N. E. BR., *S. echinata* HORT. angl. ex N. E. BR.). – Botswana; S.W.Afr.: Damaraland; Angola; Transvaal; Moçambique. – **St.** prostrate, rooting, 6–8 cm long, up to 2 cm thick, 6-angled, furrowed, the angles obtuse, rounded, dentate, with a transverse furrow above the T., dark green to brownish; **Fl.** Cor. 2.5–3 cm $\varnothing$, lobes broad-triangular, smooth, dark brown, greenish outside, ciliate in the indentations, annulus paler, hirsute.

**D. –** v. **transvaalensis** (SCHLTR.) WHITE et SLOANE (Pl. 56/3) (*D. t.* SCHLTR., *D. t.* v. *parviflora* L. BOL.). – Transvaal; Botswana: Kalahari. – **St.** 7 cm tall, 1 cm $\varnothing$; **Fl.** several, Cor. yellowish-green, upperside with bluish markings, underside pale green, 3 cm $\varnothing$, lobes 11 mm long, 8 mm across, annulus 12 mm $\varnothing$, purple-brown.

**D. pubescens** N. E. BR. v. **pubescens.** – Cape: Lit. Namaqualand; S.W.Afr.: Gr. Namaqualand. – **St.** prostrate, 2–5 cm long, 8–16 mm $\varnothing$, the 4–5 obtuse angles having T. 2–4 mm long; **Fl.** 2–4, Cor. c. 2.5 cm $\varnothing$, all inner surfaces pubescent, especially the annulus, dark chocolate-brown, lobes sharply projecting, margins vertically folded back, annulus 8 mm across, 2–2.5 mm tall, slightly 5-angled.

**D. –** v. **major** N. E. BR. – Cape: Lit. Namaqualand. – **Cor.** c. 3 cm $\varnothing$, lobes sparsely hirsute at the base, glabrous above.

**D. radiata** (SIMS) HAW. v. **radiata** (*Stapelia r.* SIMS, *S. replicata* JACQ., *D. repl.* SWEET). – Orange Free State; Cape: Karroo. – **St.** rather prostrate, c. 4–5 cm long, 2–2.5 cm $\varnothing$, the 4–5 angles having stout T.; **Fl.** 1–2, Cor. 2.5 to 3 cm $\varnothing$, lobes curved and projecting-erect, the margins sharply folded back, especially those in the indentations, glossy brown inside, the annulus large, fleshy and glabrous.

**D. –** v. **hirtella** (JACQ.) WHITE et SLOANE (*Stapelia h.* JACQ., *S. caespitosa* DC., *S. reclinata* SIMS, *D. h.* SWEET). – Orange Free State; Cape: Karroo. – **Cor.** green outside, black-purple inside, the annulus large, fleshy, all surfaces pubescent, lobes ciliate to midway along the margins.

**D. –** v. **minor** (N. E. BR.) WHITE et SLOANE (*D. hirtella* v. *m.* N. E. BR.). – Lobes 7 mm long, ciliate.

**D. –** v. **obscura** (N. E. BR.) WHITE et SLOANE (*Duvalia hirtella* v. *o.* N. E. BR.). – Cape: Karroo. – Lobes ciliate, annulus glabrous to softly hairy.

**D. reclinata** (MASS.) HAW. v. **reclinata** (*Stapelia r.* MASS., *S. radiata* LINK, *D. propinqua* BGR.). – Cape: Karroo. – **St.** ± erect, ascending or recurved, (2.5–)4(–10) cm long, green, glabrous, with 4–6 sharply dentate angles; **Fl.** 3–4, Cor. 2–2.5 cm $\varnothing$, dirty green outside, smooth inside, glossy, dark brown, the annulus large, fleshy, pubescent, its margins strongly folded back between the indentations, lobes curving and projecting-erect, pointed, borders completely folded back and ciliate.

**D. –** v. **angulata** N. E. BR. – Annulus bluntly 5-angled, pubescent, lighter brown than the lobes, having 5 straw-coloured blotches in the middle.

**D. –** v. **bifida** N. E. BR. – Annulus pubescent, with purple hairs on the rim.

**D. somalensis** LAVR. – Somalia: N.reg. – Forming dense groups; **St.** clavate, ascending or erect, bluish-green with purple blotches, the 4 angles with projecting T., leafless, 3–5 cm long, c. 15 mm $\varnothing$; **Fl.** 3–4 at the base of the shoots, pedicels 3 cm long and spreading horizontally, Sep. 3 mm long, pointed; Cor. glabrous, round, campanulate, 3–3.6 cm $\varnothing$, annulus raised, red-brown, minutely tuberculate, tube 2 mm long, lobes ovate-triangular, pointed, c. 15 mm long, with longitudinal furrows, yellowish with large, convex, red-brown blotches and lines.

**D. sulcata** N. E. BR. v. **sulcata**. – Arabia. – St. prostrate, 2.5–6.5 cm long, with 4 angles, 1.2–1.5 cm thick, glabrous, whitish green, purple-spotted, T. 6–10 mm long; Fl. 1–3, Cor. c. 4.5 cm $\varnothing$, brownish red, the annulus densely covered with long, pale red hairs, lobes 1.6 to 1.8 cm long, 1–2 cm across, ovate, pointed, with 5 furrows, tips rough, with motile clavate hairs at the base, annulus 1.2–1.4 cm $\varnothing$.

**D. —** v. **seminuda** LAVR. – S.W.Arabia. – Annulus completely glabrous, but the lobe margins have purple clavate hairs.

**Dyckia** SCHULT. Bromeliaceae. – S.Am. – Plants acauline, forming clumps; L. in a dense Ros., rather thick, the margins thorny-serrate, rather hard; Infl. always lateral, spicate, Fl. yellow or orange-coloured. – Greenhouse, moderately warm. Propagation: seed, division.

**D. altissima** LINDL. (*D. princeps* LEM., *D. gigantea* C. KOCH, *D. ramosa* HORT., *D. laxiflora* MART.). – Brazil. – L. 25 cm long or more, margins with Sp. 5 mm long; Infl. branching, hoary, of many brilliant yellow Fl.

**D. brevifolia** BAK. (*D. princeps* HORT.). – Brazil. – L. up to 20 cm or more long, underside with light lines, with whitish scales, marginal Sp. 2 mm long; Infl. with fairly numerous yellow Fl.

**D. cinerea** MEZ. – Brazil. – With a few offsets; L. oblong-triangular, tapering, 6–7 cm long, c. 4 cm across at the base, 5–6 mm thick, both surfaces dark green with dense whitish-silvery scales, margins with horny soft T. 2 mm long and directed backwards.

**D. rariflora** SCHULT. – Brazil. – Resembles **D. altissima**; L. smaller, marginal Sp. shorter; Infl. ± mealy-pruinose, Fl. few, orange-yellow.

**D. remotiflora** OTTO et DIELS. – Uruguay. – L. up to 25 cm long, margins with Sp. 1–3 mm long; Infl. hirsute towards the tip, Fl. deep orange-yellow.

**D. rubra** WITTM. – S.Brazil. – L. c. 1 m long, margins very spiny, smooth above, the underside having light-coloured scales; Infl. branching, conspicuous, Sep. with rust-coloured hairs, Pet. 12 mm long, red.

**D. sulfurea** C. KOCH (Pl. 56/5). – Brazil. – Closely resembles **D. brevifolia**; Fl. more widely spreading, sulphur-yellow.

**Echeveria** DC. (*Courantia* LEM., *Cotyledon* L. SG *Echeveria* (DC.) BENTH. et HOOK.; *Cotyledon* L. Sect. *Echeveria* (DC.) SCHOENL.; *Oliverella* ROSE; *Oliveranthus* ROSE; *Echeveria* DC. Sect. *Euecheveria* BGR.). – Crassulaceae. – Mostly Mexico, extending northwards to USA: Texas and southwards to Centr. Am. and through the Andes. – Succulent glabrous or hairy ♃ or semi-♄; L. alternate, sparse or mostly in Ros., flat and attenuated to the base; Infl. axillary, always annual, with sessile, often spurred bracts, a spike or Rac. (Pl. 57/1), or consisting of one or several Cinc., more rarely reduced to one Fl.; Fl. 5-partite, Sep. erect to reflexed, equal to strongly unequal, Pet. erect, united at the base, mostly imbricate, thin or usually very thick. – Type species: E. coccinea. For hybrids with **Dudleya** s. × **Dudleveria**; with **Graptopetalum** s. × **Graptoveria**; with **Pachyphytum** s. × **Pachyveria**; with **Sedum** s. × **Sedeveria**. Also crosses with **Thompsonella**. – This treatment of **Echeveria** is abridged from that drawn up by R. MORAN. – Summer: most species in the open, winter: cold-house. Propagation: seed, cuttings, also leaf cuttings. (Lit.: WALTHER, E.: Echeveria. San Francisco 1972).

## Division of the Genus Echeveria into Series by R. Moran

Some of the series are ill-defined, and various species are still too little known for proper placement.

**A.** Fl. in a spike or Rac. (attached singly on all sides of the St.; (see Pl. 57/1) or with a few lower branches with more than one Fl.; each pedicel with two bracteoles.

  **B.** Plants hairy.

    **Ser. 1. Echeveria** (*Oliverella* ROSE = *Oliveranthus* ROSE as genus; *Echeveria* DC. Sect. *Oliveranthus* (ROSE) BGR.; *Echeveria* DC. Ser. *Vestitae* E. WALTH.). – St. short or commonly elongate; Ros. small to medium-sized, of 10–25 L., often very lax; Infl. a spike or Rac., rarely reduced to one Fl., Sep. subequal, erect to spreading; Cor. 1–3 cm long, the sides channelled. – Mexico: Jalisco and Hidalgo to Guatemala, c. 800–2500 m. – Type-species: E. coccinea. — Species: E. amphoralis, carminea, coccinea, harmsii, leucotricha, macrantha, pringlei, pulvinata.

  **B.** Plants not hairy, at most papillose.

    **C.** Ros. long-stemmed, often lax, or the L. scattered; the plants subshrubby.

      **Ser. 2. Nudae** E. WALTH. (Ser. *Australes* E. WALTH., Ser. *Elatae* E. WALTH., Ser. *Bracteolatae* E. WALTH.). – Ros. small or medium-sized, of 10–25 L.; foliage smooth or papillose; **Infl.** mostly a Rac., rarely spicate; **Sep.** subequal, mostly ascending to reflexed; Cor. 2–2.5 cm long, the sides mostly channelled. – Mexico: Puebla to; Peru, c. 900–3500 m. Perhaps an artificial group, but several of the species are still too little known. – Type-species: E. nuda. – Species: E. aequatorialis, alata, australis, bicolor, bracteolata, buchtienii, colombiana, globuliflora (?), goldmanii, gracilis, guatemalensis, johnsonii, macdougallii, maxonii, montana, multicaulis, nodulosa, nuda, pachanoi, procera, quitensis, sedoides, spectabilis, sprucei, viridissima, waltheri.

**Ser. 3. Spicatae** (BAK.) BGR. (*Courantia* LEM. as genus; *Cotyledon* L. Sect. *Spicatae* BAK., *Echeveria* DC. Sect. *Courantia* (LEM.) BGR.). – The 20–50 **L.** often scattered and scarcely rosulate; **Infl.** a dense spike or Rac.; **bracts** and **Sep.** often coloured, sometimes exceeding the corolla; **Cor.** 10–13 mm long, pentagonal with flat sides; pistils long-styled; seeds linear, often more than 1 mm long. – Mexico: Tamaulipas and Jalisco to Costa Rica, c. 1100–2500 m. – Type-species: E. rosea. – Species: E. chiapensis, omiltemiana, pittieri, rosea.

C. Ros. mostly subsessile (or if caulescent, the pedicels 1 mm thick or less), the L. mostly crowded.

**Ser. 4. Racemosae** (BAK.) BGR. (*Cotyledon* L. Sect. *Racemosae* BAK.). – **Ros.** small to large, often of 15–50 L., the foliage smooth or papillose; **Infl.** racemose, the pedicels commonly slender; **Sep.** unequal or subequal, erect to reflexed; **Cor.** 7–20 mm long, the sides channelled to slightly convex. – Mexico: Hidalgo; to the Andes, c. 1000 to 4300 m. – Probably not a natural group, but some of the species still little known. – Type-species: E. racemosa. – Species: E. atropurpurea, ballsii, bella, canaliculata, carnicolor, chiclensis, chilonensis, cuencaensis (?), eurychlamys, excelsa, megacalyx, penduliflora, peruviana, racemosa, rauschii, sessiliflora, steyermarkiana, vanvlietii.

**Ser. 5. Mucronatae** E. WALTH. – **Roots** thickened; **Ros.** medium-sized, of 15–25 L.; **L.** often withering after anthesis, the margins acute, often hyaline or papillose; **Infl.** spicate or subracemose, the pedicels to 2 mm long; **Sep.** unequal, erect to spreading; **Cor.** 6–18 mm long, the sides convex to channelled. – Mexico: Durango to Guatemala, c. 1400–3300 m. – Type-species: E. mucronata. – Species: E. corallina, crassicaulis, huehueteca, mucronata, pinetorum.

A. Fl. in one or more Cinc. (i.e. in two rows on one side of each Br., the Br. ± zig-zag and curled before flowering; see Pl. 57/1); each pedicel with one bracteole or mostly none.

D. Fl. and Infl. pubescent; Ros.-L. pubescent or at least ciliate.

**Ser. 6. Ciliatae** MORAN. – **Ros.** subsessile, medium-sized, of 25–170 L.; **Cinc.** 1 to c. 10; **Sep.** subequal, ascending; **Cor.** 10–16 mm long, the sides channelled. – Mexico: near the Puebla-Oaxaca border, c. 2000–2200 m. – Type-species: E. ciliata. – Species: E. ciliata, pilosa, setosa.

D. Fl. and Infl. glabrous; Ros.-L. mostly glabrous, rarely puberulent or ciliate.

E. Cor. terete or subpentagonal with ± rounded sides.

**Ser. 7. Paniculatae** BGR. (Ser. *Amoenae* E. WALTH.). – **Ros.** subsessile, small, of 25–40 L.; **L.** clavate with rounded margins, both Ros.- and St.-L. easily detached and rooting; **Cinc.** 2 to several, the pedicels slender; **Sep.** subequal, short; **Cor.** 50–10 mm long, tubular, the **Pet.** thin. – Mexico: near the Puebla-Veracruz border, c. 2400 m. – Type-species: E. amoena. – Species: E. amoena, expatriata, microcalyx, pulchella.

**Ser. 8. Urceolatae** E. WALTH. (*Urbinia* BR. et ROSE as genus; *Echeveria* DC. Sect. *Urbinia* (BR. et ROSE) BGR., Ser. *Urbiniae* E. WALTH.). – **Ros.** subsessile, small to medium-sized, of 20–110 crowded L.; **Cinc.** mostly 1–2, the pedicels slender; **Sep.** unequal, ascending to reflexed; **Cor.** 7–14 mm long, urceolate, the sides flattened to rounded. – Mexico: Chihuahua to Oaxaca, c. 1200–2800 m. – Type-species: E. agavoides. – Species: E. agavoides, albicans, chihuahuensis, elegans, × gilva, goldiana, halbingeri, humilis, hyalina, parrasensis, potosina, purpusorum, simulans, tobarensis, tolimanensis.

**Ser. 9. Longistylae** E. WALTH. – **Ros.** short-stemmed, medium-size, of c. 30 L.; **Cinc.** 1–2, the pedicels rather long; **Sep.** subequal, spreading; **Cor.** 3 cm long, bicoloured, rounded-pentagonal; styles 2–3 times longer than the ovaries. – Mexico: Southern Puebla, c. 2100 m. – Type-species: E. longissima. – Only species.

E. Cor. pentagonal, with flat or channelled sides.

F. Cinc. (one-sided floral Br.) mostly 1–2.

**Ser. 10. Valvatae** MORAN. – **Ros.** subsessile, medium-sized, of 15–45 L.; **L.** rather thin, with acute margins; **Cinc.** mostly solitary, with imbricate bracts and short pedicels; **Sep.** erect or ascending, about equalling the corolla; **Cor.** 7–11 mm long, pentagonal with flat sides, the Pet. valvate. – Mexico: Michoacan and the State of Mexico, c. 1200–2000 m. – Type-species: E. valvata. – Species: E. calycosa, valvata.

**Ser. 11. Secunda** (BAK.) BGR. (*Cotyledon* L. Sect. *Secunda* BAK.). – **Ros.** subsessile, often clustered, small to medium-sized, of 25–60 crowded L.; **Cinc.** 1–2, the pedicels rather long; **Sep.** unequal, ascending to spreading; **Cor.** 8–15 mm long, ovoid to pyramidal, the sides channelled, the Pet. more than twice as wide as thick, obtusely keeled. – Mexico: Coahuila to Oaxaca, c. 1900–4300 m. – Type-species: E. secunda. – Species: E. cuspidata, derenbergii, globulosa, meyraniana, minima, secunda, subalpina, tolucensis (?), turgida. – E. alpina, byrnesii, elatior, glauca & tolucensis are perhaps all vars. or synonyms of E. secunda.

**Ser. 12. Chloracanthae** MORAN (*Pachyphytum* LK., KLOTZSCH et OTTO Sect. *Echeveriopsis* E. WALTH.). – **Ros.** subsessile, medium-sized, of 40–70 L.; **Cinc.** mostly solitary, the pedicels short and stout; **Sep.** markedly unequal, ascending; **Cor.** 8–11 mm long, pentagonal with flat sides, opening slowly over several days, the Pet. more than twice as wide as thick, obtusely keeled, green. – Mexico: Near the Puebla-Veracruz boundary, c. 2100 m. – Type-species: E. heterosepala. – Only species.

**Ser. 13. Pruinosae** E. WALTH. – **Ros.** subsessile, medium-sized, of 45–55 L.; **L.** strongly pruinose, flat or with crenulate margins, broadly acute; **Cinc.** mostly solitary, the pedicels short; **Sep.** unequal, ascending; **Cor.** 8–14 mm long, the sides channelled, the Pet. triquetrous, acutely keeled. – Mexico: Tamaulipas to Puebla, c. 2100 m.? – Type-species: E. peacockii. – Species: E. peacockii, runyonii (?), shaviana.

**Ser. 14. Angulatae** E. WALTH. – **Roots** often thickened; **Ros.** subsessile, medium-sized, of 10–50 L., the **L.** often channelled, often narrowly acute; **Cinc.** 1–3, the pedicels very short to long; **Sep.** unequal, appressed to slightly reflexed; **Cor.** 10–17 mm long, the sides channelled, the Pet. triquetrous, acutely keeled. – U.S.A.: Texas to Mexico: Hidalgo, c. 1000–2400 m. – Type-species: E. teretifolia. – Species: E. bifida, lutea, strictiflora, tenuis, teretifolia, walpoleana.

F. Cincinni mostly 3 to several (sometimes 2 in Ser. 18).

**Ser. 15. Induplicatae** MORAN. – **St.** elongate; **Ros.** medium-sized, of 25–35 turgid L.; **Infl.** a pendent thyrse or apically racemose, the axis zigzag with a protrusion at each node, the Br. reflexed, each a Cinc.; **Sep.** erect, about equalling the Cor.; **Cor.** 7–10 mm long, the Pet. thin, induplicate-valvate. – Range unknown. – Type-species: E. linguifolia. – Only species.

**Ser. 16. Occidentales** MORAN. – **Ros.** subsessile, medium-sized, of 20–50 L.; **Infl.** of several mostly bifurcate Br., the ultimate Br. Cinc.; pedicels rather long; **Sep.** subequal or ascending; **Cor.** 9–13 mm long, pentagonal with flat sides, the Pet. rather thin. – Mexico: Chihuahua to Durango and Sinaloa, c. 2100–2300 m. – Type-species: E. affinis. – Species: E. affinis, craigiana.

**Ser. 17. Thyrsiflorae** MORAN. – **Roots** thickened; **Ros.** sessile, medium-sized to large, of 12–25 L.; **L.** often withering after anthesis, the margins acute, sometimes hyaline or papillose; **Infl.** thyrsoid or subracemose but determinate, the Cinc. few-flowered, with the terminal Fl. often long-pedicellate; **Sep.** unequal, ascending; **Cor.** 10–16 mm long, the sides channelled. – Mexico: Chihuahua to Hidalgo and Guerrero, c. 2100 to 2600 m. – Type-species: E. maculata. – Species: E. longipes, maculata, paniculata, platyphylla.

**Ser. 18. Gibbiflorae** (BAK.) BGR. (*Cotyledon* L. Sect. *Gibbiflorae* BAK., Ser. *Grandes* E. WALTH.). – **Ros.** sessile to tall-stemmed, medium-sized to large, of 12–60 L.; **Cinc.** 2 to several; pedicels short to long; **Sep.** unequal, ascending to spreading; **Cor.** 1–3 cm long, pentagonal with flat to channelled sides. – Mexico: Chihuahua?, Durango to Chiapas, c. 100–2900 m. – Type-species: E. gibbiflora. – Species: E. acutifolia, crenulata, fimbriata, fulgens, gibbiflora, gigantea, grandifolia, grisea, juarezensis, lozanii (?), longiflora, obtusifolia, pallida, rubromarginata, scheeri, semivestita, subrigida, violascens.

**E. acutifolia** LINDL. (§ 18) (*Cotyledon a.* BAK., *C. devensis* N. E. BR., *E. d.* (N. E. BR.) M. G. GREENE, *E. holwayi* ROSE, ? *E. calophana* HORT. angl. ex LEM.). – Mexico: Oaxaca. – Caulescent; **Ros.** densely leafy; **L.** ± rhombic, usually pointed, green, ± pruinose, margins often wavy, often tinged with red, 10–19 cm long, 5–9 cm across; **Infl.** sometimes over 1 m long, Fl. 9 to 15 mm long, red, yellow above.

**E. aequatorialis** ROSE ex v. POELLN. (§ 2). – Ecuador. – Acauline; **Ros.** laxly leafy; **L.** obovate-cuneate, rounded or blunt above, apiculate, 3–4 cm long, 1.5–2 cm across, margins reddish; **Fl.** (?) red.

**E. affinis** E. WALTH. (§ 18). – Mexico. – **St.** very short, usually simple; **Ros.** densely leafy; **L.** lanceolate, short-pointed, up to 5 cm long, 2 cm across, very concave below, flat above, brownish olive-green; **Fl.** 10 mm long, 8 mm ⌀, scarlet.

**E. agavoides** LEM. v. **agavoides** (Pl. 57/2) (§ 8) (*Cotyledon a.* BAK., *E. yuccoides* MORR., *Urbinia a.* ROSE, *E. obscura* (ROSE) BGR.). – Mexico: San Luis Potosí. – Acauline, often offsetting from the base to form large clumps; **Ros.** firm and dense; **L.** 15–25, ovate, pointed, brown-tipped, stiff, thick, light apple-green, 3–9 cm long, 2.5–5 cm across, often becoming reddish; **Infl.** up to 50 cm tall, Fl. reddish, dark yellow above.

**E. — v. corderoyi** (MORR.) v. POELLN. (*E. c.* MORR., *Urbinia c.* ROSE). – Mexico: San Luis Potosí. – **L.** 60–70; **Fl.** up to 18 mm long, red, yellow above.

**E. — cv. Cristata.** – Monstrous garden form.

**E. alata** ALEXANDER (§ 2). – Mexico: Oaxaca. – ♄, branching from the base; **L.** not all rosulate, oblanceolate, short-pointed, 5–6 cm long, 2 cm across, thick, fleshy, with a red border; **Infl.** up to 20 cm tall, Fl. 20–22 mm long, light scarlet outside, creamy-yellow inside.

**E. albicans** E. WALTH. (§ 8) (*E. elegans* v. *kesselringiana* v. POELLN., *E. alba* HORT. Calif.). – Mexico. – **Ros.** acauline, later offsetting; **L.** densely crowded, obovate-oblong, 3–5 cm long, 15–25 mm across, thick and swollen, thickest below the tip, curved upwards, blunt to truncate, with a small whitish apiculus, pale olive-green, margins a little translucent; **Infl.** up to 25 cm tall, Fl. 14–18 mm long, light pink.

**E. alpina** E. WALTH. (§ 11). – Mexico: Puebla. – Acauline, mat-forming; **L.** numerous, long-spatulate, 7 cm long, 3 cm across, truncate at the tip, pointed, an attractive blue colour; **Infl.** 12 cm tall, Fl. 15 mm long, red to yellow. (Acc. R. MORAN possibly a variety of, or synonymous with, **E. secunda.**)

**E. amoena** L. DE SMET (§ 7) (*E. pusilla* BGR.). – Mexico: Vera Cruz. – Acauline or short-caulescent, pruinose, with many offsets; **L.** densely rosulate, oblong-spatulate, pointed or rather blunt, sometimes tapering, thick, 2 to 2.5 cm long, 6–8 cm across; **Infl.** up to 20 cm tall, Fl. 6–10 mm long, yellow-red.

**E. amphoralis** E. WALTH. (§ 1). – Mexico: Oaxaca. – Mealy-pruinose semi-♄ up to 20 cm or more in height, with numerous Br.; **L.** numerous, almost rosulate or lax, ovate-cuneate, apiculate, up to 3.5 cm long, 2 cm across, rather thick, underside indistinctly carinate; **Infl.** several, up to 20 cm tall, Fl. c. 24 mm long, pitcher-shaped, red, lemon-yellow along the outer angles and inside.

**E. × atropulla** ROLLISON. – Hybrid: **E. rosea × E. glauca.** – **Ros.** compact; **L.** large, spatulate, tapering towards the tip, wavy, deep green with a bluish sheen, margins red; **Fl.** red and yellow.

**E. atropurpurea** BAK. (§ 4) (*Cotyledon a.* BAK., *E. sanguinea* MORR.). – Mexico. – **St.** 10–20 cm tall; **L.** densely rosulate, obovate to spatulate, pointed or tapering, dark red, grey-pruinose, 10–14 cm long, 3–5 cm across; **Infl.** up to 60 cm tall, Fl. 10–12 mm long, red.

**E. australis** ROSE (§ 2). – Costa Rica to Panama. – **St.** branching, 20–30 cm tall; **L.** densely rosulate, broadly spatulate to obovate, ± rounded above, ± pointed, grey-pruinose, often reddish, cochleariform, 2.5–7 cm long, up to 3 cm across; **Infl.** 30–50 cm tall, Fl. 11 to 14 mm long, brilliant red.

**E. ballsii** E. WALTH. (§ 4). – Colombia. – **St.** short, several; **L.** thick, swollen above, oblong-ovate, rather pointed, up to 3.5 cm long, 1 cm across, green; **Infl.** 25–30 cm tall, Fl. 12 mm long, scarlet, yellow inside.

**E. bella** ALEXANDER (§ 4). – Mexico: Chiapas. – Mat-forming; **Ros.** 2–4 cm ⌀, densely leafy; **L.** 12–18 mm long, 2–4 mm across, oblanceolate, pointed, light yellowish green; **Infl.** 12–20 cm tall, Fl. 8–10 mm long, orange-yellow.

**E. bicolor** (H. B. et K.) E. WALTH. v. **bicolor** (§ 2) (*Sedum b.* H. B. et K., *Cotyledon subspicata* BAK., *E. s.* (BAK.) BGR., *E. b. v. s.* (BAK.) E. WALTH., *E. venezuelensis* ROSE). – Venezuela; Colombia. – **St.** up to 30 cm tall, up to 2 cm ⌀, branching; **L.** 20–30 in an extended Ros., lanceolate-spatulate or broad-spatulate, rounded above and scarcely tapering, light green, glossy, 6.5–8 cm long, up to 18 mm across, underside convex; **Infl.** up to 60 cm tall, Fl. 10–12 mm long, red.

**E. — v. turumiquirensis** STEYERM. – Mexico. – **L.** leathery-fleshy, pale green; **Fl.** deep red.

**E. bifida** SCHLTD. (§ 14) (*Cotyledon b.* HEMSL., *E. trianthina* ROSE). – Mexico: Hidalgo. – **Ros.** caulescent, offsetting freely; **L.** oblanceolate, upperside concave, young L. bright red and pointed, becoming greenish, 6–12 cm long, 10 to 16 mm across; **Infl.** 30–40 cm tall, Fl. numerous, 10–12 mm long, reddish.

**E. cv. Blue Spur.** – A seedling of cv. Edna Spencer. – **Ros.** with many offsets; **L.** with fewer tubercles, some of these spur-shaped.

**E. bracteolata** LK., KLOTZSCH et OTTO (§ 2) (*Cotyledon b.* (LK., KLOTZSCH et OTTO) BAK.). – Venezuela. – Semi-♄ with few erect Br. as thick as a finger; **L.** broad-spatulate, short-tipped, fleshy, 5–8 cm long, 2–3 cm across, upperside light green, minutely pruinose, underside roundish-carinate, whitish sea-green; **Infl.** 25–30 cm tall, Fl. oblong to urn-shaped, yellow to brick-red.

**E. buchtienii** v. POELLN. (§ 2). – Bolivia. – **St.** branching basally, prostrate, rooting, up to 20 cm long, 1–1.5 cm ⌀; **L.** in a crowded Ros., obovate-spatulate, somewhat pointed, red-green, underside convex, up to 5 cm long, up to 18 mm across; **Infl.** up to 35 cm tall, Fl. 16–20 mm long, red.

**E. byrnesii** ROSE (§ 11) (*E. secunda* v. *b*. (ROSE) v. POELLN.). – Mexico: Mexico D. – Resembles **E. secunda; L.** obovate to oblanceolate, pointed, 4–5 cm long, c. 2 cm across, light green, often reddish, not grey-pruinose. (Acc. R. MORAN possibly a variety of, or synonymous with, **E. secunda.**)

**E. cv. Callosa.** – Hybrid: **E. 'Van-celstii'** × **E. atropurpurea.**

**E. calycosa** MORAN (§ 10). – Mexico: Michoacán. – **St.** rarely 4 cm long, up to 1.5 cm thick; **Ros.** 5–10 cm ⌀; **L.** 15–25, spatulate, rounded at the tip, apiculate, light green, (2.5–)5(–9) cm long, (1.5–)2.5(–3.5) cm across, with a pronounced midrib and 2 side veins, margins narrow, membranous; **Infl.** up to 20 cm tall, Fl. 7.5 mm long, yellow above, tinged pink.

**E. cv. Cameo.** – Hybrid: **E. gibbiflora** v. **carunculata** × **E.** 'Edna Spencer'. – Resembles **E. gibbiflora** v. **carunculata;** L.-protuberances more prominent.

**E. canaliculata** HOOK. (§ 4) (*Cotyledon c.* BAK.). – Mexico: Hidalgo. – **St.** 10–20 cm tall, 2.5–3 cm thick; **L.** densely rosulate, 10–20 cm long, 2.5–3 cm across, narrow-oblong or lorate to oblong, tapering, fleshy, upperside very concave, underside rounded, grey-green, tinged purple; **Infl.** up to 50 cm tall, Fl. 22–25 mm long, red outside, yellow to orange inside.

**E. cv. Carinata** (*E.* × *undulata* HORT.). – Hybrid: **E. gibbiflora** v. **metallica** × **E. atropurpurea.**

**E. carminea** ALEXANDER (§ 1). – Mexico: Oaxaca. – **St.** up to 70 cm tall, densely papillose-pulverulent; **L.** almost rosulate, deeply grooved, 10–12 cm long, 2.5–3 cm across, deep green, margins blackish-purple, underside bluish, dusted with silver, short-pointed, M.rib distinctly raised only at the cuneate base; **Infl.** 40–50 cm tall, Fl. 22–24 mm long, very acutely angled, crimson with yellow tips.

**E. carnicolor** (BAK.) MORR. (Pl. 57/5) (§ 4) (*Cotyledon c.* BAK.). – Mexico: Vera Cruz. – **Ros.** 6–8 cm ⌀, producing many offsets; **L.** oblanceolate-spatulate, underside convex, very fleshy, flesh-coloured, slightly pruinose, rough, with a metallic sheen; **Infl.** 15–20 cm tall, Fl. orange-red.

**E.** × **carnitricha** WERD. – Hybrid: **E. carnicolor** × **E. leucotricha.**

**E. chiapensis** ROSE ex v. POELLN. (§ 3). – Mexico: Chiapas. – Caulescent, branching; **L.** not rosulate, obcuneate, narrowing teretely to the base, pointed or rather so, grey-pruinose, up to 4.5 cm long, 1.5 cm across; **Infl.** 20–25 cm tall, Fl. reddish. (May be identical with **E. goldmanii.**)

**E. chiclensis** (BALL.) BGR. (§ 4) (*Cotyledon c.* BALL., *E. backebergii* v. POELLN., *E. neglecta* v. POELLN.). – Peru: Andes. – Acauline, offsetting from a napiform **root**; **L.** densely rosulate, fairly numerous, lanceolate or oblong-lanceolate, blunt or rather pointed, often with a minute white apiculus, up to 6 cm long, 6–18 mm across, dirty green, somewhat rough, covered with minute whitish glossy papillae, underside convex and little carinate, margins papillose; **Infl.** up to 30 cm tall, Fl. 15 mm long, intense red to yellowish.

**E. chihuahuaensis** v. POELLN. (§ 8). – Mexico: Chihuahua. – Acauline; **L.** densely rosulate, obovate-spatulate, tapering, pointed, with red margins, grey-pruinose, up to 4 cm long, 2 cm across; **Infl.** c. 20 cm long, Fl. c. 11 mm long, red (?).

**E. chilonensis** (O. KTZE.) E. WALTH. (§ 4) (*Sedum c.* O. KTZE., *E. whitei* ROSE). – Bolivia. – Caulescent, branching, dwarf, Br. c. 5 cm long; **L.** in a lax Ros., oblanceolate, tapering, 3–4 cm long, 1–1.5 cm across; **Infl.** 10–12 cm tall, Fl. 10–12 mm long, yellow.

**E. ciliata** MORAN (Pl. 57/4) (§ 6). – Mexico: Oaxaca. – **Ros.** solitary, almost sessile; **St.** occasionally up to 7 cm long, 1–2 cm thick, hirsute, usually unbranching, sometimes with basal Br. below the dense, hemispherical or cup-shaped Ros. 5–12 cm ⌀; **L.** 30–80, deep green, often with a narrow red border, ovate-cuneate, initially tapering, becoming blunt, apiculate, 2.5–5 cm long, 1.5–3.5 cm across, 5–9 mm thick, underside bluntly carinate, upperside broadly grooved, the keel and the margins with hairs 1–3 mm long almost to the base, upperside glabrous with a few scattered hairs; **Infl.** 6 to 13 cm tall, all surfaces hirsute, Fl. 10–16 mm long, green at first, becoming scarlet.

**E. coccinea** (CAV.) DC. (§ 1) (*Cotyledon c.* CAV., *E. longifolia* HORT. Kew.). – Mexico: Hidalgo. – Hirsute ♄ 30–70 cm tall; **L.** 8–12 towards the Br.tips, lanceolate-spatulate or obovate-cuneate, pointed, terete at the base, upperside usually concave, underside convex, up to 10 cm long, up to 2.5 cm across, green, pubescent, older L. reddish; **Infl.** up to 35 cm tall, Fl. 10 mm long, reddish-yellow.

**E. cv. Cochlearis.** – Hybrid: **E. linguifolia** × **E. atropurpurea.**

**E. columbiana** v. POELLN. (§ 2). – Colombia: E.cordillera. – **St.** up to 15 cm tall, very fleshy; **L.** in a dense or lax Ros., obovate-oblong or obovate-cuneate, mostly rounded above or blunt, pointed, 2.5–3.5 cm long, 1.5–2 cm across; **Infl.** c. 35 cm tall, Fl. 7 mm long, yellow.

**E. corallina** ALEXANDER (§ 5). – Mexico: Chiapas. – **St.** 2–3 cm tall, sometimes branching basally; **L.** in a lax Ros., oblanceolate, short-pointed, 6–8 cm long, 15–16 mm across, pale green, margins brownish-purple, with a red apiculus, deep blue tinged brownish-red; **Infl.** of 20–25 Fl. 14 mm long, coral-red.

**E. cv. Corymbosa** (*E.* × *corymbosa* GOSSOT). – Hybrid: Acc. VAN KEPPEL probably **E. agavoides** × **E. derenbergii.** – **Ros.** dense, acauline, c. 12 cm ⌀, sometimes producing offsets from undeveloped Infl.; **L.** glabrous, thick, fleshy, obovate-oblong, apiculate, both surfaces green, the underside often reddened in winter, epidermis crystalline; **Infl.** 12–15 cm tall, Fl. 12 to 15 mm long.

**E. craigiana** E. WALTH. (§ 16). – Mexico: Chihuahua. – Acauline or short-stemmed, mature plants branching; **Ros.** dense; **L.** 30–40, semi-terete, linear-oblong, 8–11 cm long, up to 2 cm across, underside rounded and carinate towards the tip, which is tapering and shortly subulate-bristly, bottle green at the base otherwise light brown and somewhat bluish; **Infl.** erect, Fl. 11 mm long, pink outside, red inside.

**E. crassicaulis** E. WALTH. (§ 5). – Mexico: Fed. D. – **St.** subterranean, short; **L.** rosulate, dark green, oblong to oblanceolate, rhombic or circular, rounded above, tapering, 5–8 cm long, 2–2.5 cm across; **Infl.** up to 60 cm tall, Fl. 15 mm long, yellow.

**E. crenulata** ROSE (§ 18). – Mexico: Morelos. – **St.** ± short, thick; **L.** broad-obovate, rounded above, distinctly narrowed and terete at the base, pale green, somewhat pruinose, margins wavy, red, up to 30 cm long, 15 cm across; **Infl.** up to 1 m tall, Fl. 15 mm long, yellowish-red.

**E.** × **crinita** GOSSOT. – Hybrid: **E. pilosa** × **E. setosa.**

**E. cuencaensis** V. POELLN. (§ 4?) (*E. ingens* ROSE). – Ecuador. – Acauline (?); **L.** oblong or obovate-oblong, pointed, grey-pruinose, up to 7 cm long, 2.5 cm across, margins red; **Infl.** extended, Fl. 15 mm long, red.

**E. cuspidata** ROSE (§ 11). – Mexico: Coahuila. – Acauline; **Ros.** dense; **L.** often more than 100, triangular-tapering, very pointed, very pruinose, somewhat reddish-bronze, 6–8 cm long, 3–5 cm across; **Infl.** 24–40 cm tall, Fl. 10–12 mm long, purple-red.

**E.** × **dasyphylla** DEL. – Hybrid: **E. amoena** × **E. agavoides.** – **St.** short; **L.** dense, cuneate, imbricate, blunt-tipped, brilliant green with a red tip.

**E.** cv. Decora. – Somatic mutation of **E.** cv. Metallica. – **St.** 20–60 cm tall, 2–3 cm thick; **Ros.** dense, 10–20 cm ⌀; **L.** 10–15, fleshy, 4–10 cm long, 2–5 cm wide, apiculate, upperside often deeply chanelled, colour marbled green, the irregular margins whitish and pinkish, sometimes striped; **Cor.** pinkish outside, yellowish inside.

**E. derenbergii** J. A. PURP. (Pl. 58/2) (§ 11). – Mexico: Oaxaca. – Mat-forming, acauline or caulescent; **Ros.** spherical to somewhat cylindrical, 3–6 cm ⌀; **L.** numerous, broad-spatulate, rounded above to roundish-truncate, red-tipped, light green, very whitish-grey pruinose, 2.5 to 4 cm long, 2–2.5 cm across, margins ± reddish, underside convex; **Infl.** c. 8 cm tall, Fl. 11 to 15 mm long, reddish-yellow or almost yellow.

**E.** × **derosa** V. ROED. (*E. derenbergii major* HORT., *E.* × *dasyphylla* HORT.). – Hybrid: **E. derenbergii** × **E. setosa.** – **Ros.** of many L. curving up to form a cup-shape up to 16 cm ⌀, with lateral Ros. between the L.; **L.** lanceolate-spatulate, bluntly rounded or somewhat pointed at the tip, narrowing gradually towards the base, up to 8 cm long, 2.3 cm across near the tip, both surfaces carinate, bluish, the margins with minute irregular papillae, soft green, the tip scarlet; **Infl.** of few Fl., yellow inside, orange-coloured outside.

**E.** × — cv. Worfield Wonder. – Very compact, 10 cm tall.

**E.** cv. Edna Spencer. – Cultivar of uncertain origin, certainly related to **E. gibbiflora** v. **metallica.** – **Ros.** large, offsetting; **L.** slightly rolled and twisted, upperside ± tuberculate.

**E.** cv. Edna's Giant. – A seedling of the preceding. – **Ros.** c. 40 cm ⌀ with numerous offsets; **L.** exceedingly tuberculate.

**E. elatior** E. WALTH. (§ 11). – Mexico: Hidalgo. – Acc. E. WALTHER probably only a variety of **E. secunda**; **Ros.** numerous, dense; **L.** crowded, swollen, ovate-cuneate, 4–5 cm long, 2.5 cm across, upperside concave, with an acutely pointed tip, dark blue-green; **Infl.** 30 cm tall, Fl. 12 mm long, red.

**E. elegans** ROSE (Pl. 57/3) (§ 8) (*E. perelegans* BGR.). – Mexico: Hidalgo. – Acauline, mat-forming, with daughter-Ros. on long stolons; **L.** numerous, densely rosulate, obovate, rounded above, young L. pointed, fairly thick, 3–6 cm long, very pruinose, alabaster-white, margins often reddish, often also ± translucent; **Infl.** 10–25 cm tall, Fl. 10–12 mm long, yellow above.

**E.** cv. Erecta. – Hybrid: **E. coccinea** × **E. atropurpurea.**

**E. eurychlamys** (DIELS) BGR. (§ 4) (*Cotyledon e.* DIELS). – Peru: Cajamarca. – Acauline; **L.** rosulate, elliptical-ovate, pointed, 3–3.5 cm long, up to 2 cm across; **Infl.** 25–30 cm tall, Fl. 13–17 mm long, flesh-coloured.

**E. excelsa** (DIELS) BGR. (§ 4) (*Cotyledon e.* DIELS). – Peru. – Acauline; **L.** densely rosulate, obovate-cuneate, pointed, 12–15 cm long, up to 4 cm across; **Infl.** up to 1 m tall, Fl. 15–20 mm long, scarlet.

**E. expatriata** ROSE (§ 7). – Mexico (?). – **St.** up to 10 cm tall, branching basally, weak; **L.** rosulate, oblanceolate, thick, narrow, both surfaces convex, pointed, grey-pruinose, 2–2.5 cm long; **Infl.** slender, Fl. 5–6 mm long, reddish.

**E.** × **fallax** GOSSOT. – Hybrid: **E. derenbergii** × **E. elegans.**

**E.** cv. Ferrea. – Hybrid: **E. scheeri** × **E. acutifolia.**

**E. fimbriata** C. H. THOMPS. (§ 18). – Mexico: Morelos. – Resembles **E. scheeri**; **L.** margins fringed.

**E.** × **fulgens** HAAGE jr. Acc. VAN KEPPEL may possibly be a hybrid of **E. elegans**, also called **E.** × **splendens** HAAGE jr.

**E. fulgens** LEM. (§ 18) (*E. retusa* LINDL., *Cotyledon r.* BAK., *C. f.* BAK.). – Mexico: Michoacan. – **St.** short or 10–15 cm tall, often branching; **L.** 8–20 in a dense Ros., spatulate to obovate-spatulate, rounded above, less often emarginate, indistinctly pointed, broad at the base, light grey-green, a little pruinose, margins often reddish, up to 10 cm long, 4–5 cm across; **Infl.** 30–50 cm tall, Fl. 12–15 mm long, yellow.

**E. gibbiflora** DC. v. **gibbiflora** (§ 18). – Mexico: S. of Mexico City. – **St.** simple or little branching, mature plants forming new Ros. at the base; **L.** 12–20, rosulate, obovate-spatulate to oblong-spatulate, somewhat triangular or cordate above, blunt-tapering, carinate on the underside at the base, upperside convex, light grey-green, often tinged reddish-brown, 13 to 30 cm long, 7–14 cm across; **Infl.** more than 60 cm tall, Fl. up to 25 mm long, light red, yellow inside, pruinose outside; often with adventitious buds in the Infl. and the axils of the St.L.

E. — cv. Carunculata (Pl. 58/3). – **L.** upperside with tuberculate or vesicular protuberances.

E. — v. **crispata** BAK. - Margins wavy.

E. — v. **gibbiflora** (*Cotyledon g.* Moc. et SESSE ex DC., *E. grandis* MORR., *E. bernhardiana* FOERST. ex LEM., *E. g.* v. *typica* BGR.). – **L.** narrow, shortly and bluntly triangular-tipped.

E. — v. **metallica** (LEM.) BAK. (*E. m.* LEM.). – **L.** much rounded towards the tip, bronze-coloured, more pruinose, margins white or reddish.

E. — cv. Monstruosa. - **St.** fasciated; **L.** smaller.

E. **gigantea** ROSE et PURP. (§ 18). – Mexico: Puebla. - Clump-forming; **St.** 20–30 cm tall; **L.** rosulate, oblanceolate, narrowing at the base into a thick, fleshy petiole, light green, somewhat grey-pruinose, margins deep red, smooth, up to 25 cm long, up to 15 cm across, the tip rounded and somewhat emarginate, always apiculate; **Infl.** up to 2 m tall, Fl. 12–14 mm long, flesh-coloured.

E. — cv. **Crispata**. – **L.** margins wavy.

E. × **gilva** E. WALTH. (§ 8) cv. Gilva. – Hybrid: E. agavoides × E. elegans. – Acauline; **Ros.** densely leafy, 15–20 cm ⌀, later with many side-shoots; **L.** numerous, thick fleshy, obovate-oblong, blunt, reddish, apiculate, 5 to 8 cm long, 2–3 cm across, underside slightly carinate, surface crystalline; **Infl.** 30 cm tall, Fl. 1 cm long, pinkish red, yellow inside.

E. × — cv. Blue Surprise. – Somatic mutation of cv. Gilva; **L.** blue, tinged reddish.

E. **glauca** BAK. v. **glauca** (§ 11) (Pl. 58/5) (*Cotyledon g.* BAK., *E. secunda* BOOTH. v. *glauca* OTTO, ? *E. globosa* HORT. ex MORR.). – Mexico: Mexico D. - Acauline, rarely caulescent, with numerous offsets; **Ros.** 10 cm ⌀; **L.** 40–50, broadly obovate-spatulate, rounded above or often truncate, ± purple, tapering, both surfaces pruinose, margins becoming reddish, 5 cm long, 2.5 cm across; **Infl.** 20–30 cm long, Fl. 8–10 mm long, intense red. Acc. R. MORAN possibly only a variety or synonym of **E. secunda.**

E. — v. **pumila** (SCHLTD.) v. POELLN. (*E. p.* SCHLTD., *Cotyledon p.* BAK., *E. s.* v. *p.* MORR., *E. s.* v. *p.* OTTO). – **L.** rather narrower, often longer; **Fl.** often more yellow, red below.

E. **globuliflora** E. WALTH. (§ 7). – Mexico: Oaxaca. - **St.** simple, 10 cm tall; **L.** numerous, almost rosulate, oblong-lanceolate to obovate-cuneate, apiculate, c. 5 cm long, 15 mm across, thin, curving upwards, somewhat oblique, underside carinate, lettuce-green, margins and keel reddened; **Infl.** up to 25 cm tall, Fl. up to 1 cm long, 8 mm ⌀, peach-coloured, margins orange-yellow.

E. **globulosa** MORAN (Pl. 58/4) (§ 11). – Mexico: Oaxaca. - **St.** short, up to 9 mm thick, offsetting from the base; **Ros.** compressed-spherical, densely leafy, 4–4.5 cm ⌀; **L.** 50–60, cuneate-spatulate, blunt to somewhat truncate, apiculate, 2–2.5 cm long, 9–12 mm across above, 4–4.5 mm thick, bluish, often with a red margin, underside often red-spotted, carinate below, margins sharp; **Infl.** 6–8 cm tall, Fl. up to 12 mm long, yellow.

E. **goldiana** E. WALTH. (§ 8). – Mexico: Mexico D. - Acauline, simple, or eventually with lateral offsets; **L.** densely rosulate, 40 or more, broadly obovate-cuneate, much swollen, underside rounded, upperside slightly convex, apiculate, up to 4 cm long, c. 15 mm across, green, glossy; **Infl.** up to 40 cm tall, Fl. pitcher-shaped, 13 mm long, begonia-pink, greenish-yellow towards the tip.

E. **goldmanii** ROSE (§ 2). – Mexico: Chiapas. - **St.** up to 20 cm tall, becoming prostrate and rooting; **L.** laxly arranged, narrow-oblong, pointed or rather blunt, narrowing teretely towards the base, light green, margins red, 2–3 cm long; **Infl.** up to 20 cm tall, Fl. up to 11 mm long, reddish.

E. **gracilis** ROSE (§ 2). – Mexico: Puebla. – **St.** very short; **L.** scattered or in a lax Ros., thick, obovate-oblong, pointed, underside rounded, up to 3 cm long, 12 mm across, 5 mm thick, green, apiculus red; **Infl.** up to 20 cm tall, Fl. 10 mm long, light scarlet.

E. **gracillima** MÜHLENPF. – Probably a small-flowered form of **E. secunda,** or possibly a hybrid.

E. × **graessneri** VAN KEPPEL (*E.* × *haageana* HORT. ex WALTH. nom. illeg.). – Hybrid: E. derenbergii × E. pulvinata. – **St.** 10–15 cm tall, with some Br. from the base; **L.** crowded into a dense Ros., overlapping unevenly, underside rounded and indistinctly carinate, apiculate, ovate-spatulate, up to 6 cm long, 3 cm across, greenish-blue, margins sometimes reddish, surface rough granular; **Infl.** up to 12 cm tall, Fl. 2 cm long, salmon to orange-coloured, scarlet above, becoming yellow.

E. **grandifolia** HAW. (§ 18) (*E. campanulata* KUNZE). – Mexico: Guerrero. - Caulescent; **L.** rosulate, spatulate, thick at the base, narrowing teretely, blunt, tapering, upperside convex or concave, very grey-pruinose, margins often light pink, up to 8 cm long; **Infl.** up to 70 cm tall, Fl. 15 mm long, reddish, yellow inside.

E. cv. Grandisepala. – Hybrid: E. gibbiflora v. metallica × E. rosea.

E. **grisea** E. WALTH. (§ 18). – Mexico: Guerrero. - **St.** 2–3 cm thick, short; **L.** few, laxly rosulate, 10–15 cm long, 5–8 cm across, broadly obovate-spatulate, rounded at the tip, apiculate, narrowing gradually towards the base, margins often distinctly wavy, larkspur-green, sometimes with purple to wine-coloured spots; **Infl.** up to 50 cm tall, Fl. 13 mm long, pink, yellow at the tip, pale pink inside.

E. **guatemalensis** ROSE (§ 2). – Guatemala. - Mat-forming; **St.** 15 cm tall, branching; **L.** not rosulate, spatulate, rounded above, pointed, broadly petiolate, fleshy, light greenish, somewhat grey-pruinose, margins reddish; **Infl.** 20 to 30 cm tall, Fl. 1 cm long, reddish-yellowish.

**E. × haageana** F. A. HAAGE jr. emend. VAN KEPPEL. – Hybrid: **E. agavoides × E. derenbergii.** – **Ros.** acauline, regular, dense, 10 to 15 cm ⌀, freely branching from the base; **L.** numerous, oblong, pointed, apiculate, 5 cm long, 2 cm across, bluish-green, tinged reddish, underside slightly carinate; **Infl.** 10–15 cm tall, Fl. 12 mm long, pink-red, yellow inside.

**E. halbingeri** E. WALTH. (§ 8). – Mexico. – Acauline, mat-forming; **L.** c. 50, densely rosulate, thick, swollen, up to 2.5 cm long, 13 mm across, almost triangular in the upper half, blunt, apiculate, light green, tinged bluish; **Infl.** up to 12 cm tall, Fl. c. 12 mm long, orange to deep yellow, pale orange inside.

**E. harmsii** F. MCBRIDE (§ 1) (*Oliverella elegans* ROSE, *Oliveranthus e.* ROSE, *Cotyledon e.* N. E. BR., *E. e.* BGR.). – Mexico. – Small and softhaired semi-♄; **L.** crowded on the Br.tips, lanceolate-spatulate, pointed, green, 2–3 cm long, c. 1 cm across, in a lax Ros.; **Infl.** 10 to 20 cm tall, Fl. 2–3 cm long, red with light tips.

**E. heterosepala** ROSE (§ 12) (*Pachyphytum h.* ROSE) E. WALTH., *E. viridiflora* ROSE nom. nud., *P. chloranthum* E. WALTH.). – Mexico: Puebla. – **St.** up to 5 cm long; **L.** c. 25, rosulate, rhombic-lanceolate, 4–7 cm long, 1.5–2.5 cm across, c. 3 mm thick, upperside flat, yellowishgreen; **Infl.** up to 45 cm tall, Fl. light green, pink to red.

**E. cv. Hoveyi** (Pl. 59/4) (*E. h.* ROSE, *E. zahnii* cv. Hoveyi). – Acc. H. HALL a mutation (probably chimerical) of **E.** cv. Zahnii; **Ros.** lax; **L.** long-spatulate, mucronate, c. 5 cm long, c. 8 mm across at the base and 22 mm across above, grey-green with irregular pink stripes, finely waxy-pruinose.

**E. huehueteca** STANDL. et STEYERM. (§ 5). – Guatemala. – Acauline; **Ros.** dense; **L.** oblong-lanceolate to almost elliptical, broadly cuneate or obovate-oblong, 2.5 cm long, 1–2.5 cm across, rather tapering to blunt at the tip, apiculate, both surfaces grass-green, sometimes tinged with purple; **Infl.** 20–30 cm tall, Fl. 8–10 mm long, scarlet.

**E. humilis** ROSE (§ 8). – Mexico: San Luis Potosí. – Acauline or short-stemmed; **L.** densely rosulate, lanceolate, rather thick, underside convex, short-tapering, 3–6 cm long; **Infl.** about 10 cm tall, Fl. 8–9 mm long, reddish.

**E. hyalina** E. WALTH. (§ 8). – ? Mexico. – **Ros.** acauline, eventually mat-forming; **L.** numerous, ovate-cuneate, up to 6 cm long, 3.5 cm across, whitish-crystalline, thin, margins thin and translucent, pale green, often somewhat bluish; **Infl.** up to 30 cm tall, Fl. c. 11 mm long, pink to flesh-coloured, green inside.

**E. cv. Imbricata** (*E. × glauco-metallica* HORT.). – Hybrid: **E. glauca × E. gibbiflora v. metallica.** – **Ros.** saucer-shaped; **L.** round-obovate, round at the tip, apiculate, underside carinate, grey-green or almost white, often reddened.

**E. × indecorata** GOSSOT. – Hybrid: Parents unknown.

**E. johnsonii** E. WALTH. (§ 2). – Ecuador. – **St.** up to 10 cm tall, branching, erect or prostrate; **L.** not rosulate, clavate to linear-oblong, semi-terete, blunt, apiculate, c. 3.5 cm long, 9 mm thick, light green, margins with red lines towards the tip; **Infl.** c. 10 cm tall, Fl. c. 18 mm long, yellow with red tips.

**E. juarezensis** E. WALTH. (§ 18). – Mexico: Oaxaca. – **St.** up to 8 cm tall; **Ros.** of c. 30 L.; **L.** obovate-cuneate, pointed, apiculate, thick, stiff, upperside very concave, underside rounded to carinate, 5 cm long, 3 cm across, green; **Infl.** up to 20 cm tall, Fl. c. 12 mm long, pitcher-shaped, scarlet, light orange coloured inside.

**E. × lancifolia** DEL. – Hybrid: **E. rosea × E. glauca.** – **Ros.** compact; **L.** numerous, lanceolate, lush blue-green, tinged pink-violet.

**E. leucotricha** J. A. PURP. (Pl. 58/1) (§ 1). – Mexico: Puebla. – Semi-♄ covered with dense white felty hairs; **Ros.** lax, 10–15 cm ⌀; **L.** lanceolate, rather blunt, pointed, up to 10 cm long, 2–2.5 cm across, thick, densely white-hairy but the tips brown-haired; **Infl.** 30–40 cm tall, Fl. 2 cm long, vermilion.

**E. linguifolia** LEM. (§ 15) (*Talinum l.* LEM., *Anacampseros l.* LEM., *Cotyledon l.* LEM., *Pachyphytum linguae* HORT. ex BAK.). – Mexico?. – **St.** branching, 20–30 cm tall; **L.** densely rosulate, broad-spatulate to obovate, ± rounded above, indistinctly pointed, grey-pruinose, often somewhat reddish, spoon-shaped, 2.5–7 cm long, up to 3 cm across; **Infl.** 30–50 cm tall, Fl. up to 14 mm long, hirsute.

**E. × longicaulis** GOSSOT. – Hybrid: Parents unknown; probably a × **Graptoveria.**

**E. longiflora** E. WALTH. (§ 18). – Mexico: Guerrero. – **St.** simple, short; **L.** few, over 15 cm long, over 8 cm across, obovate-round, obliquely-truncate at the tip, narrowing at the base into the broad grooved alate petiole, green, ± pruinose, tinged light wine-red; **Infl.** up to 75 cm tall, Fl. up to 22 mm long, pale wine-red to mauve at the base, pink above.

**E. longipes** E. WALTH. (§ 17). – Mexico: Hidalgo. – **St.** subterranean, short; **L.** rosulate, linear-oblong, very pointed, 8–10 cm long, 15 mm across, light elm-green, somewhat grey-pruinose; **Infl.** usually of 1 Fl. 14 mm long, flesh to ochre-coloured, yellow inside.

**E. longissima** E. WALTH. (§ 9) (*E. harmsii* v. *multiflora* E. WALTH.). – Mexico: between Puebla and Oaxaca. – Usually acauline; **L.** in a closed Ros., broad-ovate, pointed, thickish, underside rounded, 2.5–4.5 cm long, 1.5–2 cm across, underside ± papillose and reddish, upperside apple-green; **Fl.** 3 cm long, reddish-green.

**E. lozanii** ROSE (§ 18). – Mexico: Jalisco. – Acauline; **L.** densely rosulate, lanceolate or often oblanceolate-oblong, thick, flat, pointed, 12 to 20 cm long, 2–5 cm across, coppery; **Infl.** 20–50 cm tall, Fl. 10–15 mm long, coppery.

**E. lutea** ROSE (§ 14). – Mexico: San Luis Potosí, Hidalgo. – Acauline; **L.** rosulate, ± oblong-lanceolate, tapering, tip curved upwards like a horn, stiff, margins curved upwards,

7–10 cm long, green to yellowish-green; **Infl.** 20–30 cm tall, Fl. 12–17 mm long, lemon-yellow.

**E. macdougallii** E. WALTH. (§ 2). – Mexico: Oaxaca. – Semi-♄ with numerous Br. up to 12 cm long; **L.** numerous, almost rosulate, oblong-ovate to ovate, cuneate below, thick clavate, ± semi-terete or angular, tapering, indistinctly papillose, 3 cm or more long, 1 cm thick, green, often reddish along margins and at the tip; **Infl.** up to 10 cm long, Fl. up to 18 mm long, red, yellow inside.

**E. macrantha** STANDL. et STEYERM. (§ 1). – Guatemala: Jalapa. – Semi-♄ with few Br.; **L.** densely rosulate, sessile, round-cuneate, c. 3 cm long, 2–2.8 cm across, broadly rounded to almost truncate at the tip, short-apiculate, thick, fleshy, sparsely hirsute, pale yellowish-green with pink margins; **Infl.** 4–5 cm long, of c. 3 Fl., Cor. densely puberulous outside, segments c. 2 cm long, free.

**E. maculata** ROSE (§ 17). – Mexico: Hidalgo. – **Ros.** dense, acauline; **L.** oblong-lanceolate, thickish, c. 10 cm long, 1.5–2 cm across, pointed, dark green, somewhat mottled; **Infl.** 60–80 cm tall, Fl. 1 cm long, lemon-yellow.

**E. maxonii** ROSE (§ 2). – Guatemala. – Semi-♄; **Br.** erect, eventually prostrate, 60–80 cm long; **L.** rather lax, spatulate, blunt or rather pointed, narrowing distinctly and teretely below, 3 to 10 cm long, upperside very concave, underside rounded, margins reddish; **Infl.** 20–60 cm tall, Fl. 1 cm long, salmon-pink.

**E. megacalyx** E. WALTH. (§ 4). – Mexico: Oaxaca. – **St.** short, eventually offsetting laterally; **L.** numerous, densely rosulate, thin, almost flat, oblong-spatulate, up to 10 cm long, 2.5 cm across, mucronate, grass-green to dark bluish, minutely spotted, margins at first often incised; **Infl.** up to 45 cm tall, Cor. pitcher-shaped, up to 8 mm long, dull greenish-yellow.

**E. cv. Metallica** (*E.* × *metallica* HORT.). – Hybrid: Parentage unknown. – Resembles **E. gibbiflora** v. **metallica**; **L.** narrower; **Fl.** lighter coloured, anthers sterile.

**E. meyraniana** E. WALTH. (§ 11). – Mexico: Puebla. – Acauline, mat-forming; **L.** numerous, densely rosulate, oblong-lanceolate, up to 5.6 cm long, 2 cm across, tapering, underside rounded, somewhat carinate, green, ± pruinose; **Infl.** up to 15 cm tall, Fl. c. 1 cm long, begonia-pink, chrome-yellow inside.

**E. microcalyx** BR. et R. (§ 7) (*E. purpusii* BRITT.). – Mexico: Vera Cruz, Pueblo. – **St.** about 10 cm tall, often branching and rooting; **L.** laxly rosulate, blunt, tapering, thick, 2–3 cm long, 1–1.5 cm across; **Infl.** of c. 5 Fl. 6–8 mm long, yellowish-red.

**E. minima** MEYRAN (§ 11). – Mexico: Hidalgo D. – Glabrous, simple; **Ros.** sessile, densely leafy, c. 3.5 cm ⌀; **L.** cuneate, apiculate, bluish, margins and apiculus red, c. 15 mm long, 7–9 mm across; **Infl.** 1–4, up to 3.5 cm tall, with 4–6 L., Sep. 4–5 mm and 6–9 mm long, Cor. 8.5–11 mm long, 5–7 mm ⌀, yellow, pink at the base.

**E. montana** ROSE (§ 2) (*E. nuda* v. *m.* (ROSE) v. POELLN.). – Mexico: Oaxaca. – **St.** erect, ± branching, 20–50 cm tall, later with L.scars; **Ros.** flat, 10–25 cm ⌀; **L.** spatulate to obovate-spatulate, blunt to rounded, apiculate, 4 to 12 cm long, 2.5–4.5 cm across, 5–8 mm thick, underside rounded and slightly carinate, upperside usually grooved, green to reddish, margins sharp, narrow, membranous, often red, papillose, sometimes irregularly crenate, minutely wavy or almost fringed; **Infl.** up to 55 cm tall, Fl. up to 15 mm long, yellow with red.

**E. mucronata** (BAK.) SCHLTD. (§ 5) (*Cotyledon m.* BAK.). – Central Mexico. – Acauline; **L.** rosulate, narrowly obovate-spatulate, tapering, pointed, green, 7–10 cm long, 2.5 cm across; **Infl.** c. 80 cm tall, Fl. up to 18 mm long, reddish-yellowish.

**E. multicaulis** ROSE (§ 2). – Mexico: Guerrero. – **St.** 10–20 cm tall, branching, rough below; **L.** in a dense Ros., broadly obovate-spatulate, somewhat rounded to almost truncate above, fairly flat, dark green, margins reddened, 2 to 3 cm long, 1.5–2 cm across; **Infl.** 40 cm tall, Fl. 12–13 mm long, reddish, yellowish inside.

**E. cv. Mutabilis.** – Hybrid: **E. carnicolor** × **E. linguifolia.**

**E. nodulosa** (BAK.) OTTO (Pl. 59/1) (§ 2) (*Cotyledon n.* BAK., *E. misteca* L., *C. bicolor* HORT. Herb.Kew., *E. discolor* L. DE SMET, *E. sturmiana* v. POELLN.). – Mexico: Oaxaca, Puebla. – Overall minutely prickly-papillose; **St.** simple or branching; slow-growing, eventually up to 50 cm long, becoming prostrate, often with aerial roots, bearing scars of old L. bases; **Ros.** 5–13 cm ⌀, dense or more usually lax; **L.** 10–25, oblanceolate-spatulate, broadly pointed or rounded, apiculate, 3–8 cm long, 1–3.5 cm across above, 5–10 mm across below, 3–8 mm thick, appearing whitish because of the papillae, upperside somewhat asymmetrically grooved and with 3 asymmetric ridges, margins and keel red, both surfaces often asymmetrically marked; **Infl.** 15–60 cm tall, Fl. 12–17 mm long, angular, red below, yellow above.

**E. nuda** LINDL. (§ 2) (*Cotyledon n.* BAK., *E. navicularis* L.). – Mexico: Vera Cruz (Orizaba). – **St.** 10–20 cm tall; **L.** arranged laxly, obovate-spatulate, rounded or tapering above, with red margins, 4–7 cm long, 2.5–3 cm across; **Infl.** 15–30 cm tall, Fl. 10–12 mm long, red below, yellow above.

**E. obtusifolia** ROSE (§ 18) (*E. scopulorum* ROSE, *E. o. v. s.* (ROSE) v. POELLN.). – Mexico: Morelos. – **Ros.** acauline or short-stemmed; **L.** obovate-lanceolate, rounded above, not tapering or scarcely so, much narrowed towards the base, light green, becoming reddened, rather thin, 7.5–10 cm long, 3.5 cm across; **Infl.** 20–35 cm tall, Fl. 8–12 mm long, red. Possibly a form of **E. fulgens.**

**E. omiltemiana** MATUDA (§ 3). – Mexico: Guerrero. – Possibly identical with **E. rosea**; **St.** 20–30 cm long; **L.** 14 or more in a lax Ros., lanceolate-ovate, cuneate, tapering, green, at first tinged pink, 5.5–6.5 cm long, 1.5–2.5 cm across; **Infl.** 15–25 cm tall, Fl. 12 mm long, pitcher-shaped, yellowish-white.

**E.** × **opalina** ROLLIS. – Hybrid: **E. secunda** × **E. scheeri**. – **L.** broad, flat, ovate, tapering, blue-green, tinged metallic pink; **Fl.** red and yellow.

**E.** cv. **Ovata**. – Hybrid: **E. scheeri** × **E. gibbiflora**. – Intermediate between the parents.

**E. pachanoi** ROSE ex v. POELLN. (§ 2). – Ecuador. – Caulescent, branching; **L.** densely rosulate, obovate, cuneate towards the base, rounded above, pointed, grey-pruinose, 1 to 1.5 cm long, 7–10 mm across; **Infl.** 10 cm tall, Fl. 1 cm long.

**E. pallida** E. WALTH. (§ 18). – Origin unknown. – **St.** 3 cm thick, simple; **L.** in a lax Ros., ovate-spatulate, c. 15 cm long, 9 cm across, rounded and tapering above, narrowed teretely below, the upperside deeply grooved below, the underside sharply carinate, pale green, margins reddish; **Infl.** 50 cm tall, Fl. 16 mm long, pink.

**E. paniculata** A. GRAY (§ 17) (*Cotyledon grayii* BAK., *E. g.* BAK. also MORR., *E. schaffneri* ROSE). – Mexico: Chihuahua, Durango, Hidalgo. – **Ros.** acauline, dense; **L.** 15, obovate-linguiform or narrow-lanceolate or somewhat oblong, pointed, tapering, 6–10 cm long, 1.5–2 cm across, green, often somewhat blotched; **Infl.** 40–80 cm tall, Fl. reddish or pale yellow.

**E. parrasensis** E. WALTH. (§ 8) (*E. cuspidata* J. A. PURP.). – Mexico: Coahuila. – **St.** up to 8 cm tall, 2 cm thick, with few Br.; **L.** 40–100 or more, obovate-cuneate to oblong-lanceolate, 5–6 cm long, 2–3 cm across, 12 mm across at the base, underside rounded, light green, powdery; **Infl.** up to 25 cm tall, Fl. c. 11 mm long, pitcher-shaped, scarlet to pink.

**E. peacockii** (BAK.) MORREN (Pl. 59/3) (§ 13) (*Cotyledon p.* BAK., *E. desmetiana* L. DE SMET, *C. d.* HEMSL., *E. subsessilis* ROSE). – Mexico. – **Ros.** acauline or short-stemmed, densely leafy; **L.** almost ovate to ± oblong or almost obovate, blunt, somewhat tapering, very conspicuously blue-white pruinose, margins and tip often reddish, 3–7 cm long, 2–4 cm across; **Infl.** 15–35 cm long, Fl. 9–12 mm long, intense red.

**E.** — cv. Cristata. – Monstrous cultivar.

**E. penduliflora** E. WALTH. (§ 4). – Mexico: Oaxaca. – **St.** usually simple, up to 30 cm tall; **L.** almost rosulate, thin, oblong-lanceolate, up to 14 cm long, 4 cm across, narrowing below to the carinate petiole 2 cm across, apiculate, green, underside slightly carinate; **Infl.** up to 30 cm tall, Fl. pendulous, 13 mm long, geranium-pink.

**E.** cv. Perle von Nürnberg (HORT. GRAESER). – Hybrid: **E. gibbiflora** × **E. potosina**. – **L.** coloured conspicuously purple.

**E. peruviana** MEYEN (§ 4) (*Cotyledon v.* BAK.). – Peru to N.Chile; N.Argentine. – **Ros.** acauline; **L.** dense, obovate-spatulate, tapering, pointed, green, very grey-pruinose, 5–7.5 cm long, 2 to 2.5 cm across; **Infl.** up to 35 cm tall, Fl. 12 to 15 mm long, red.

**E. pilosa** J. A. PURP. (§ 6). – Mexico: Puebla. – Plant densely covered with projecting hairs; **St.** short, later branching, red-brown felted; **L.** in a lax Ros., ± long spatulate-cuneate, somewhat pointed, thick, 7–8 cm long, 3 cm across, densely white-hairy, upperside somewhat concave, with brown hairs at the tip; **Infl.** c. 30 cm tall, Fl. 9–13 mm long, dull orange-red.

**E. pinetorum** ROSE (§ 5) (*E. sessiliflora* v. p. (ROSE) v. POELLN.). – Mexico: Chiapas. – **L.** narrowly oblanceolate, 1–1.5 cm across, light green with red margins.

**E. pittieri** ROSE (§ 3). – Guatemala. – Semi-♄ 10 cm or more tall; **L.** in a lax Ros., oblanceolate, somewhat tapering, narrowing teretely towards the base, green, turning pink, little pruinose, 5–8 cm long, 2 cm across; **Infl.** 10–20 cm tall, Fl. numerous, 10–13 mm long, reddish-yellow, grey-pruinose.

**E. platyphylla** ROSE (§ 17). – Mexico: Hidalgo, Mexico D. – **Ros.** acauline; **L.** dense, oblong-rhombic, narrowing towards the base, tapering, rather thin, mucronate, pale green, 4–8 cm long, 2–3 cm across; **Infl.** 20–40 cm tall, Fl. 10–12 mm long, reddish-yellow.

**E. potosina** E. WALTH. (§ 8). – Mexico: San Luis Potosí. – **Ros.** acauline, densely leafy, with few offsets; **L.** obovate-cuneate, truncate, pointed, very thick, swollen, light grey-green, marbled, purple towards the tip, 4–6 cm long, 2–3 cm across; **Infl.** up to 30 cm tall, Fl. 13 mm long, pitcher-shaped, pink.

**E. pringlei** (S. WATS.) ROSE (§ 1) (*Cotyledon p.* S. WATS.). – Mexico: Jalisco. – Plant whitish-hairy; **St.** c. 30 cm long, prostrate; **L.** usually in a lax Ros., oblanceolate, pointed, 5–10 cm long; **Fl.** c. 16 mm long, intense red.

**E. procera** MORAN (§ 2). – Mexico: Oaxaca. – **St.** erect to prostrate, up to 2 m long, with few Br., covered with numerous rhombic L.-bases; **Ros.** 15–25 cm ⌀; **L.** 20–30, oblong-obovate, rounded to slightly truncate at the tip, apiculate, 7–10 cm long, 13 mm across at the base and 7–10 mm thick, upperside deeply grooved, underside roundish-carinate, green, with narrow red borders; **Infl.** 30–110 cm tall, Fl. 11–12 mm long, almost red outside, light yellow inside.

**E.** cv. Pruinosa. – Hybrid: **E. linguifolia** × **E. coccinea**. – Intermediate between the parents.

**E.** × **pseudoagavoides** GOSSOT. – Hybrid: Parents not known.

**E.** cv. Pseudolancifolia (*E. p.* GOSSOT, *E. lanceolata* GOSSOT non NUTT., *E. gossotii* ROWL.). – Hybrid by P. GOSSOT of **E. agavoides** × **E. lancifolia**. Intermediate between the parents.

**E. pubescens** SCHLTD. (§ 1) (*Cotyledon p.* BAK.). – E.Mexico. – Hairy ♄ 50–70 cm tall; **L.** 8–15 in a lax Ros., obovate-spatulate, up to 9 cm long, up to 2.5 cm across, pointed or rather so, tapering, almost terete at the base, pale green, becoming reddish; **Infl.** 20–50 cm tall, Fl. c. 15 mm long, intense red, yellow-tipped. Acc. R. MORAN possibly identical with **E. coccinea**.

**E. pulchella** BGR. (§ 7). – Mexico. – Acauline; **L.** densely rosulate, obovate-spatulate, short-pointed, underside somewhat carinate or very convex, lush green, scarcely pruinose, 4–5.5 cm long, 1.5 cm across, 4–6 mm thick; **Infl.** 30 to 40 cm tall, Fl. 9 mm long, light red.

**E. × pulvi-carn** E. WALTH. – Hybrid: **E. carnicolor × E. pulvinata.** – St. up to 4 cm tall; L. numerous, oblong-lanceolate, up to 7 cm long, 2.5 cm across, 8 mm thick, underside rounded, pointed, covered like the rest of the plant with dense minute papillose hairs; **Infl.** up to 60 cm tall, Fl. up to 17 mm long, peach-coloured.

**E. pulvinata** ROSE (§ 1) (*Cotyledon p.* HOOK. f.). – Mexico: Oaxaca. – Small silvery semi-♄, becoming brownish-felted; **Ros.** lax; L. obovate, blunt above, mucronate, 4–5 cm long, 2–2.5 cm across, 1 cm thick, soft and white-hairy; **Infl.** branching horizontally, Fl. up to 2 cm long, red or yellow-red.

**E. —** cv. Ruby (Pl. 58/6). – L. narrower; L. and Sep. red-haired along the margins and at the tips; **Pet.** golden-yellow.

**E. purpusorum** BGR. (§ 8) (*Urbinia p.* ROSE). – S.Mexico. – Acauline; L. 8–12, densely rosulate, broadly ovate or triangular-ovate, underside carinate, grey-green, spotted and tinged with brown, 3–4 cm long, 3 cm across at the base, pointed; **Infl.** c. 30 cm tall, Fl. 10–12 mm long, red, yellowish above.

**E. quitensis** (H. B. et K.) LINDL. (§ 2) (*Sedum q.* H. B. et K., *Cotyledon q.* BAK., *E. q. v. gracilior* SOD.). – Ecuador, Colombia, Bolivia: mountains. – St. often up to 1 m tall, branching, bushy; L. 20–30 in a lax Ros., oblanceolate to oblong, rounded or rather pointed above, mucronate, light green, somewhat grey-pruinose, becoming reddish, 2–6 cm long, 8–12 mm across; **Infl.** 10–24 cm long, Fl. 11–15 mm long, reddish-yellow.

**E. racemosa** SCHLTD. et CHAM. v. **racemosa** (§ 4) (*E. lurida* HAW., *Cotyledon l.* BAK.). – Mexico: Vera Cruz. – Ros. acauline, dense; L. lanceolate to narrow-oblong, pointed, with narrow cartilaginous margins, upperside somewhat concave, grey-pruinose, conspicuously tinged red or deep red, 5–10 cm long; **Infl.** 20–50 cm tall, Fl. 10–15 mm long, red or orange-red.

**E. —** v. **lucida** STEUD. (*E. l.* STEUD.). – L. with a conspicuous sheen.

**E. rauschii** VAN KEPPEL (§ 4). – Bolivia: N.W. of Sucre. – St. c. 5 cm tall, 1–2 cm ⌀, erect or very thin, longer and prostrate, branching basally; **Ros.** dense, 5–12 cm ⌀; L. 10–15, fleshy, oblong-lanceolate to ovate-triangular, pointed, with a red apiculus, upperside concave and flat towards the tip, underside convex, slightly carinate, fresh green, with red margins, 4–7 cm long, 8–15 mm across; **Infl.** with axis 10–25 cm tall, Rac. 10 cm long, Fl. 7–20, with pedicels up to 2 cm long, Cor. orange to orange-red, borders yellow, lobes 10 mm long, 6 mm across at the base.

**E. cv. Retusa Autumnalis.** – Intermediate hybrid: **E. glauca × E. fulgens** (syn. *E. retusa*).

**E. cv. Retusa Hybrida.** – Hybrid: **E. fulgens** (syn. *E. retusa*) **× E. gibbiflora v. metallica.** – L. blue-green; **Fl.** brilliant orange-red.

**E. × rosacea** LINDL. et ANDR. – Intermediate hybrid: **E. glauca × E. secunda.**

**E. rosea** LINDL. (§ 3) (*Courantia echeverioides* LEM., *C. rosea* ROSE, *C. roseata* BAK., *Cotyledon roseata* BAK.). – Mexico: San Luis Potosí to Oaxaca and Vera Cruz. – ♄ up to 30 cm tall; L. oblanceolate-spatulate, almost terete below, 6–9 cm long, c. 2 cm across, upper L. in a lax Ros., ± grey-green, becoming reddish; **Infl.** 10 cm and more tall, Fl. 10–12 mm long, light yellow.

**E. × rubescens** DEL. – Hybrid: **E. × imbricata × E. atropurpurea.** – L. oblong, large and broad, margins wavy-serrate, green, intensely tinged with purple-red; **Fl.** deep pink.

**E. rubromarginata** ROSE (§ 18) (*E. gloriosa* ROSE). – Mexico: Vera Cruz, Puebla. – **Ros.** acauline or short-stemmed; L. 12–15, oblanceolate to ± obovate-round, blunt, pointed, narrowing teretely towards the base, grey-pruinose, with red margins, 6–12 cm long, up to 5 cm across; **Infl.** 50–120 cm tall, Fl. 10–12 mm long, light pink.

**E. runyonii** ROSE ex WALTH. (§ 13?). – Origin unknown. – **Ros.** acauline or short-stemmed; L. spatulate-cuneate, truncate or blunt above, deep bluish, 6–8 cm long, 3–4 cm across, almost flat; **Infl.** 15–20 cm tall, Fl. 2 cm long, pink.

**E. —** v. **macabeana** E. WALTH. – L. pointed, deep green to bluish.

**E. × scaphylla** DEL. (*E. × scaphophylla* BGR.). – Hybrid: **E. agavoides × E. linguifolia.** – St. short, thick; Ros. dense, up to 25 cm ⌀; L. 10–40, obovate to lanceolate, tapering, upperside very concave, underside convex and carinate, 8–15 cm long, 3 cm across, 5 mm thick, fresh grass-green; **Infl.** curving downwards, Fl. 1 cm long, light yellow.

**E. scheeri** LINDL. (§ 18) (*Cotyledon s.* BAK.). – Mexico: Chihuahua. – Related to **E. gibbiflora**; St. up to more than 20 cm tall, freely branching; L. rosulate, oblong-spatulate, tapering, narrowing towards the base, with a long petiole, upperside concave, c. 20 cm long, 5–7 cm across; **Infl.** of many Fl. 12–18 mm long, orange-red or red, yellow above.

**E. secunda** BOOTH ex LINDL. (§ 11) (*Cotyledon s.* BAK., *E. spilota* KUNZE). – Mexico: Hidalgo, Puebla. – **Ros.** offsetting freely, acauline or short-stemmed; L. numerous, broadly obovate-cuneate, rounded above, distinctly mucronate, light green, initially somewhat pruinose, reddish towards the tip; **Infl.** c. 30 cm tall, Fl. 9–12 mm long, red, yellow inside.

**E. —** cv. Callosa. – L. with prominent, warty callosities.

**E. —** cv. Major. – Larger cultivar.

**E. sedoides** E. WALTH. (§ 2). – Mexico: Oaxaca. – ♄ up to 25 cm tall; L. lax, clavate, oblong-ovate, almost triangular to subterete, c. 2.5 cm long, 8–10 mm across, 6–8 mm thick, green; **Infl.** branching, Fl. 16 mm long, scarlet, yellow inside. Acc. R. MORAN possibly identical with **E. macdougallii.**

**E. semivestita** MORAN (§ 18). – Mexico: Hidalgo. – St. c. 20 cm tall, green, puberulous, branching basally; lower L. more scattered, upper ones more rosulate, lanceolate, pointed, underside slightly carinate, upperside grooved, 11–14 cm long, 1.5–3 cm across, 3–4 mm thick, dark green, often with purple margins, terete below with a spur 1–2 mm long; **Infl.** 10 cm tall, Fl. 12–13 mm long, bluish, yellowish inside.

**E. — v. floresiana** E. WALTH. – All parts completely glabrous.

**E. — v. semivestita.** – Plant hirsute all over.

**E. sessiliflora** ROSE (§ 4). – Mexico: Chiapas. – **Ros.** acauline, dense; **L.** lanceolate, pointed, light bluish-green, 2–5 cm long, 1 cm across; **Infl.** 10–30 cm tall, Fl. sessile, 1 cm long, reddish.

**E. × setorum** VAN KEPPEL cv. Victor (× *Urbino-Echeveria angustata* VAN LAREN). – Hybrid: **E. setosa** × **E. purpusorum** (*Urbinia purpusii*). – **Ros.** acauline, solitary, 10 cm ⌀; **L.** dense, erect to spreading, oblong, narrowing towards the tip, terminating in a thin apiculus, 4–8 cm long, up to 1.5 cm across, grey-green, keel and underside with red spots, margins and keel ciliate; **Infl.** 2, up to 20 cm tall, Rac. simple or bipartite, Fl. 7 to 14, 1.5 cm long, with pedicels 1–3 cm long, Sep. green, thick, Pet. red with yellow tips, yellow inside.

**E. setosa** ROSE et PURP. (Pl. 59/2) (§ 6). – Mexico: Puebla. – **Ros.** acauline, densely leafy, spherical or rather flatter; **L.** clavate-spatulate, ovate above, both surfaces convex, obtuse at the tip, 7–8 cm long, both surfaces covered with dense white bristles; **Infl.** 20–30 cm tall, Fl. numerous, red-yellow.

**E. — cv. Cristata.** – Monstrous cultivar.

**E. × set-oliver** E. WALTH. – Hybrid: **E. harmsii × E. setosa.** – **Ros.** numerous; **L.** hirsute; **Infl.** 40 cm tall, Fl. 21 mm long, red-yellow.

**E. shaviana** WALTHER (§ 13) (*E. 'Shazan', E. 'Pink Shaviana'* HORT. angl.). – Mexico. – Acauline or short-stemmed, with some offsets from the base; **Ros.** 7–10 cm ⌀; **L.** 20–40, obovate-spatulate, 4.5 cm long, 8 mm across at the base and 2.5 cm at the tip, upperside concave, underside convex, glabrous, margins pink to white, wavy-crenate above, grey-green, very blue-pruinose, slightly pink-tinged, the tip sharp and curved upwards; **Infl.** erect, 30 cm tall, pruinose, a Cinc. of 20–30 Fl. with pedicels 2–3 mm long, Pet. sharply carinate on the underside, 12 mm long, pink outside, orange inside.

**E. simulans** ROSE (§ 8) (*E. elegans* v. *simulans* (ROSE) v. POELLN.). – Mexico: Nuevo Leon. – Resembles **E. elegans**; **Ros.** somewhat flat; **L.** somewhat longer, rather pointed; **Infl.** up to 40 cm tall.

**E. spectabilis** ALEXANDER (§ 2). – Mexico: Oaxaca. – Branching ♄ up to 60 cm tall, finely papillose; **L.** distinctly petiolate, almost rosulate but also laxly arranged in the lower part, 4–7 cm long, 2–3 cm across, deep yellow-green with red margins, short-pointed; **Infl.** 25–70 cm tall, Fl. 24 mm long, lemon-yellow.

**E. sprucei** BAK. (§ 2) (*Cotyledon s.* BAK., *E. quitensis* v. *s.* (BAK.) v. POELLN., *E. s.* (BAK.) BGR.). – Ecuador: Andes. – Related to **E. quitensis**; **L.** only 1.5 cm long, 6 mm across, pointed; **Fl.** red.

**E. steyermarkiana** STANDL. (§ 4). – Guatemala. – **Ros.** acauline, simple or mat-forming; **L.** numerous, narrowly to broadly lanceolate-spatulate, rounded to rather blunt at the tip, with a short blunt apiculus, fleshy, 6.5 cm long, 1–2 cm across, green, often reddish; **Infl.** 5–20 cm tall, Fl. c. 2 cm long, red to pink or reddish-yellow, Pet. with scarlet margins.

**E. × stolonifera** (BAK.) OTTO (*Cotyledon s.* BAK., *E. pfersdorfii* HORT. ex MORR.). – Hybrid: **E. glauca × E. grandifolia.** – **Ros.** acauline or very short-stemmed, offsetting freely; **L.** 30–45 in a dense Ros., broadly obovate to spatulate, somewhat triangular at the tip, distinctly pointed, somewhat pruinose, 5–7 cm long, 2.5 to 3.5 cm across; **Infl.** 15–20 cm tall, Fl. 12 mm long, yellowish.

**E. strictiflora** A. GRAY (§ 14) (*Cotyledon s.* BAK.). – Mexico; USA: W.Texas. – **Ros.** acauline, mat-forming; **L.** c. 20, rhombic or oblanceolate, tapering, 7–9 cm long, 1.5–2 cm across, c. 6 mm thick below, upperside somewhat hollow, underside indistinctly carinate, dark olive-green to brown; **Infl.** 15–20 cm tall, Fl. 15 mm long, pink, very short-pedicellate.

**E. subalpina** ROSE et PURP. (§ 11) (*E. akontiophylla* WERD.). – Mexico. – **Ros.** acauline or almost so; **L.** densely rosulate, ± linear-lanceolate, narrowing towards the base, upperside ± flat, underside convex, with a dark brown apiculus, hard, very grey-pruinose, often reddish, 4–10 cm long, up to 2 cm across; **Infl.** up to 30 cm tall, Fl. 12–15 mm long, reddish or crimson.

**E. subrigida** (ROBINS. et SEAT.) ROSE (Pl. 59/5) (§ 18) (*Cotyledon s.* ROBINS. et SEAT., *E. palmeri* ROSE, *E. rosei* NELS. et MACBR., *E. angusta* v. POELLN.). – Mexico: San Luis Potosí, Mexico D. – **Ros.** acauline, dense; **L.** ovate to rhombic or oblanceolate, pointed, flat, bluish green, margins often intensely red, ± strongly grey-pruinose, 7.5–20 cm long, up to 12 cm across; **Infl.** over 60 cm tall, Fl. 15–23 mm long, reddish or pink.

**E. tenuis** ROSE (§ 14). – Mexico: Zacatecas. – Acauline; **L.** in a flat Ros., numerous, fleshy, oblong, 4–5 cm long, much narrowed towards the base, pointed; **Fl.** 9 mm long.

**E. teretifolia** DC. v. **teretifolia** (§ 14) (*Sedum t.* MOC. et SESSE, *Cotyledon subulifolium* BAK., *E. s.* MORR.). – Mexico: Hidalgo. – **Ros.** caulescent; **L.** oblong-cuneate to lanceolate-tapering, flat to concave, ± light green, often reddish, 5–7 cm long, 1.5–2 cm across; **Infl.** 30–35 cm tall, Fl. 10–12 mm long, red.

**E. — v. bifurcata** (ROSE) E. WALTH. (*E. b.* ROSE). – Infl. consisting of 2 Cinc.

**E. — v. schaffneri** (S. WATS.) E. WALTH. (*Cotyledon s.* S. WATS., *E. s.* (S. WATS.) E. WALTH.). – **L.** narrow-lanceolate, tapering, 7.5–10 cm long, 12 mm across; **Fl.** 12–16 mm long, reddish and yellow.

**E. tobarensis** BGR. (§ 8) (*Urbinia lurida* ROSE). – Mexico: Durango. – **Ros.** acauline, dense; **L.** ovate, tapering, purple or dull red,

3–4 cm long, 1.5–2.5 cm across; **Infl.** 25 cm tall, Fl. 6–7 mm long.

**E. tolimanensis** E. MATUDA (§ 8). – Mexico: Hidalgo. – **Ros.** acauline, dense; **L.** sessile, fleshy, oblong, tapering, ± apiculate, upperside somewhat concave, pruinose, 4–8.5 cm long, 2–2.5 cm across, 1–1.5 cm thick; **Infl.** 20–26 cm tall, Fl. yellow or golden-yellow.

**E. tolucensis** ROSE (§ 11) (*E. glauca* v. *r.* (ROSE) v. POELLN.). – Mexico: Mexico D. – Related to **E. glauca**; **Ros.** laxer; **L.** more lanceolate.

**E.** × **translucida** ROLLIS. – Hybrid: **E. gibbiflora** v. **metallica** × **E. peacockii.** – **Ros.** large; **L.** large, with serrate margins, metallic pink, slightly powdery; **Fl.** deep red.

**E. turgida** ROSE (§ 11). – Mexico: Coahuila. – **Ros.** acauline, dense; **L.** oblong-spatulate, blunt, acuminate, thickened, grey-green, little pruinose, margins and tip reddish, 3–8 cm long; **Infl.** 10 cm or more tall, Fl. 10–14 mm long, crimson.

**E. valvata** MORAN (§ 10). – Mexico: Mexico D. – **Root** napiform; **Ros.** simple, sessile, (8–)15(–20) cm ⌀; **L.** 20–45, cuneate-spatulate to narrowly obovate or somewhat rhombic, broad-pointed or rounded at the tip, often apiculate, light green or purple or reddish, 4–10 cm long, 2–3.5 cm across, 5–12 mm across at the base and 2–5 mm thick, flat to rounded-grooved, underside slightly blunt-carinate, margins narrow, membranous, often irregularly papillose-dentate, often wavy; **Infl.** 10–35 cm tall, Fl. 7 to 11 mm long, dark red.

**E.** cv. Vanbreen. – Acc. VAN KEPPEL a hybrid: **E. derenbergii** × **E. carnicolor.** – **Ros.** acauline, 9–12 cm ⌀, offsetting basally; **L.** numerous, crowded, swollen-spatulate to obovate-oblong, 4–6 cm long, 2–3 cm across, long-apiculate, bluish-green, both surfaces flat to convex, scarcely carinate, margins and apiculus often reddened; **Infl.** of 1–3 Rac., Fl. 1–1.5 cm long, reddish outside, becoming salmon to orange-coloured, orange-yellow inside, Pet. bordered with red above.

**E. vanvlietii** VAN KEPPEL (§ 4). – Bolivia: Chica-Chica Mts. – **St.** up to 5 cm or more tall, 1–2 cm thick, erect, sometimes prostrate; **Ros.** dense, 8–12 cm ⌀; **L.** fleshy, 20–25, spreading, oblong-lanceolate, 4–8 cm long, 1–2 cm across, deeply grooved above, apiculate, underside convex, somewhat carinate, pale grey-green, tinged purple-bronze, margins paler, less often reddish; **Infl.** with axis 20–40 cm tall, Rac. 10–20 cm long, Fl. 15–40, on pedicels 5–10 mm long, Cor. creamy-white.

**E. violascens** E. WALTH. (§ 18) (*E. gibbiflora* v. *metallica* HORT.). – Mexico. – **St.** up to 60 cm tall, later branching; **L.** 10–15, laxly rosulate, ovate-spatulate, rounded at the tip, apiculate, narrowing to 2 cm across at the petiole, the blade folded upwards, margins wavy, green with a mauve tint; **Fl.** 12–14 mm long, geranium pink to dark pink.

**E. viridissima** E. WALTH. (§ 2). – Mexico: Oaxaca. – Semi-♄ with numerous spreading Br. up to 20 cm and more in length; **L.** almost rosulate, obovate to cuneate, mucronate, underside somewhat carinate, up to 10 cm long, 6 cm across, green, margins and tip red on the underside; **Infl.** 35 cm tall, Fl. up to 16 mm long, red outside, orange-yellow inside.

**E. walpoleana** ROSE (§ 14). – Mexico: San Luis Potosí. – **Ros.** acauline or later short-stemmed; **L.** dense, almost oblong or oblanceolate and tapering, pointed, light-green, margins and blotches later reddish, upperside very concave, 6–8 cm long, 2–2.5 cm across; **Infl.** 30 to 40 cm tall, Fl. 12–15 mm long, dark orange-coloured.

**E. waltheri** MORAN et MEYRAN (§ 2). – Mexico: Mexico D. – **St.** prolonged to prostrate, with 1–2 Br., c. 9 cm long; **L.** 12–25 in a lax Ros., spatulate to rhombic, apiculate, 2.5–8 cm long, 1–3 cm across, almost cylindrical at the base, margins obtuse, the surface minutely papillose; **Infl.** 20–25 cm tall, Fl. 11–16 mm long, white and red.

**E.** cv. Zahnii. – **Ros.** regular; **L.** grey-green, upperside with scattered pink blotches. Hort. Cambridge.

**Echidnopsis** HOOK. f. Asclepiadaceae. – S. Arabia; Socotra; trop. and S.Africa. – Succulent ♃; **St.** spreading and branching, ascending, continued growth from the tips, with 6–20 angles, the ribs being divided into long, hexagonal tubercles, with small deciduous **L.**, grey-green; **Fl.** small, 2–4 together produced from the furrows at the St.-tips, Cor. rotate or very short-tubular, sometimes pitcher-shaped, having 5 lobes which are triangular-ovate or linear, campanulate-erect or rotate and spreading, corona simple or double, variously shaped, yellow, brown or red in colour. – Greenhouse, warm. Propagation: seed, cuttings. (Lit.: BALLY, P.R.O.: The Genus Echidnopsis in Tropical East Africa. In Cact. Succ. Journ. Gt. Brit. **18**: 107–109, 1956.)

**E. angustiloba** BRUCE et BALLY. – Kenya. – **St.** freely branching, with 11–13 angles, grey to olive-green; **Cor.**lobes thin, pale lemon-yellow, the Fl. centre forming a tiny brown star.

**E. atlantica** DTR. – S.W.Afr. – **St.** c. 40 cm long, 34 mm thick, ± prostrate, with 10 angles, the ribs divided up into tubercles 7–8 mm high, with a hard Sp. 1.5 mm long, indistinctly marbled.

**E. archeri** BALLY (Pl. 60/1). – Kenya. – **St.** cylindrical, reticulate, 8-angled, up to 35 cm long, 15 mm thick, creeping, somewhat branching, often forming thickened nodes; **L.** scale-like, persistent; **Fl.** paired, Cor. cup-shaped, lobes triangular, pointed, crimson and minutely papillose inside, pale pink and glabrous outside.

**E. ballyi** (J. MARN.-LAP.) BALLY (*Stapeliopsis ballyi* J. MARN.-LAP.). – N.Somalia. – **St.** creeping, clump-forming, 5–7 cm long, 1.5–2 cm thick, with 6–7 tuberculate angles, the tubercles with a somewhat recurved tip, bluish-green; **Fl.** solitary, Cor. urn-shaped, almost pyriform, up to 15 mm long, 8–10 mm ⌀, both surfaces dark purple, wrinkled inside, lobes triangular, pointed, 3 mm long, 2 mm across, dark purple outside, paler inside.

**E. bentii** N. E. Br. – S.Arabia. – **St.** up to 15 cm tall, branching, with 7–8 obtuse angles, divided into long hexagonal tubercles; **Fl.** paired, c. 12 mm $\varnothing$, Cor. broad-campanulate, almost rotate, lobes ovate, pointed, greenish outside with brown spots, wine-red inside and minutely papillose-tuberculate.

**E. bihendulensis** Bally. – Somalia. – **St.** as for **E. planifolia**, but the scale-L. and Sep. rough; Cor. cup-shaped, lobes erect, margins sharply recurved, both surfaces minutely papillose, velvety, sulphur-yellow to chestnut-brown.

**E. cereiformis** Hook. f. v. *cereiformis* (Pl. 60/2) (*Stapelia cylindrica* Hook. f., *Apteranthes tessellata* Dec., *Piaranthus fascicularis* hort., *E. t.* K. Schum.). – Ethiopia; Somalia; S.Arabia. – **St.** erect or bent and prostrate, often re-rooting, 15–30 cm long, 15–25 mm thick, simple or irregularly branching, ribs obtuse and divided by transverse furrows into 4–6-sided tubercles, dull dark green or brownish, finely papillose-tuberculate, with white spots; **Fl.** 2–4 together, 1 cm $\varnothing$, lobes campanulate to curved inwards, ovate-lanceolate to ovate, pointed, brownish-yellow outside, rather rough, light yellow inside.

**E. — v. brunnea** Bgr. (*E. dammaniana* v. *b.* Damm.). – **St.** thinner, with acuter angles; **Fl.** yellowish-brown to purple-brown.

**E. — v. obscura** Bgr. (? *Apteranthes cylindrica* Dec., ? *E. c.* K. Schum.). – **Fl.** intense yellow-brown.

**E. chrysantha** Lavr. – Somalia: N.reg. – Offsetting to form small groups; **St.** erect or ascending, partly subterranean, dark green, with 8–9 angles, 3–10 cm long, 7–14 mm $\varnothing$, deeply tuberculate; **L.** linear, pointed, 1.5 to 2.5 mm long, margins with green T. drying and persisting as white bristles; **Fl.** paired, borne on the upper parts of the St., with pedicels 2 cm long, Sep. 1.8 mm long, narrowly triangular and, like the pedicels, minutely tuberculate, Cor. round-campanulate, 9 mm $\varnothing$, golden-yellow, both surfaces sparsely tuberculate, tube 2 mm long, broadly campanulate, lobes ascending-spreading, broadly triangular, c. 3 mm long.

**E. ciliata** Bally. – Somalia. – **St.** green, cylindrical, creeping, with 8 tuberculate angles, the few Br. easily broken off and rooting; **Fl.** paired, c. 11–12 mm $\varnothing$, glabrous outside, dark purple and minutely papillose inside, tube 3.2 mm long, lobes broad-triangular, pointed, margins revolute and hirsute.

**E. columnaris** (Nel) Dyer et Hardy (*Trichocaulon c.* Nel). – Cape: Lit. Namaqualand. – **St.** erect, greenish-grey, simple or with 2–3 shoots from the base, columnar, cylindrical-clavate, 15–18 cm tall, 2–2.5 cm $\varnothing$, with 8 angles, the furrows c. 1 cm across, ribs divided transversely, with small T. curved downwards; **Fl.** 10–15, 4–8 mm $\varnothing$, lobes 4 mm across, 2 mm long, ovate, pointed, minutely spotted and with red blotches outside, white-hairy inside, yellowish-green with reddish spots.

**E. dammaniana** Spreng. (? *Boucerosia cylindrica* Brong., *E. somalensis* N. E. Br.). – Ethiopia. – **St.** up to 20 cm tall, 1–2 cm $\varnothing$, angles divided sharply into small, irregular tubercles; **Fl.** 2–5 together, c. 9 mm $\varnothing$, Cor. rotate, lobes projecting, ovate-pointed, rough tuberculate, Cor. yellow with purple-brown blotches to concolorous purple-brown.

**E. ericiflora** Lavr. – Kenya: Coast Prov. – **Roots** fibrous; **St.** congested, forming dense mats, rooting, 6-angled, green, up to 20 cm long, 4–6 mm thick, the tessellae oblong-hexagonal, 2–3 mm long, c. 1.8 mm broad; **L.** deltoid, soon drying; **Fl.** in pairs, stalked 1.5 mm long, Cor. wine-red, shortly urceolate, c. 5 mm long, tube much inflated to 4 mm $\varnothing$, narrowing to 2 mm below the mouth, lobes broadly deltoid, c. 1 mm long and as broad; corona biseriate, 2–2.5 mm $\varnothing$.

**E. framesii** White et Sloane (*Caralluma tessellata* Pill.). – Cape: Van Rhynsdorp D. – **St.** prostrate, later erect, up to 10 cm tall, 10–13 mm $\varnothing$, somewhat curved, with a few Br. from the base, rounded to 6-angled, the angles divided into 5–6-angled tubercles; **Fl.** 2–5, Cor. 8–9 mm $\varnothing$, lobes 3–4 mm long, ovate, pointed, slightly recurved, purple, papillose apart from margins and tip, underside glabrous.

**E. insularis** Lavr. – Socotra. – **St.** ascending or prostrate, 2–6 cm long, c. 5 mm $\varnothing$, reticulate, brown or brownish-green; **L.** minute, soon falling; **Fl.** solitary or 2, on upper parts of the St., on pedicels 2.5–3 mm long, Cor. very fleshy, 10–11 mm long, 4 mm $\varnothing$ at the base, 3 mm $\varnothing$ at the mouth, cup-shaped, greenish-yellow outside, glabrous, tube greenish-yellow inside with thin, purple, longitudinal lines 7–8 mm long, lobes erect, straight, pointed, triangular, 3 mm long; outer Cor. 3 mm $\varnothing$, cup-shaped, yellow.

**E. leachii** Lavr. – Tanzania: Ruaha George. – **St.** congested, rooting, ascending, tessellate, markedly rugose, 6-seriate, green or brownish, up to 15 cm long, 8–10 mm thick, the tessellae oblong, 6-angled; **Fl.** in pairs from the apical part of the St., short stalked, Sep. deltoid, acute, Cor. campanulate, 4 mm long, 4.5 mm $\varnothing$, outside dark purple-pink, glabrous, tube 2 mm long, glabrous, yellowish inside, lobes rather steeply ascending, ovate deltoid, acute, their apices slightly reflexed, bright purple-pink, 2.2 mm long, 1.8 mm broad at base, glabrous; corona bright yellow.

**E. mijerteina** Lavr. – Somalia: Mijertein Prov. – Forming dense clumps; **St.** with 8 or 9 angles, dull green, up to 15 cm long, 8 to 12 mm $\varnothing$, prostrate, conspicuously tuberculate; **L.** narrow, usually linear, 1–2 mm long, persisting for a short time as a dry bristle; **Fl.** solitary or in pairs from the sides of young shoots, pedicel 6 mm long, Sep. linear, pointed, 3 mm long; St., L., pedicels and Sep. all densely and minutely papillose; Cor. narrow-tubular, curved, 25 mm long, 4–5 mm $\varnothing$, lobes very shortly triangular, 1–2 mm long, erect, white outside, densely short-papillose, wine-red to purple inside, with 5 raised nerves, transversely ribbed, glabrous.

**E. montana** (R. A. Dyer et E. A. Bruce) Bally (*Caralluma m.* R. A. Dyer et E. A. Bruce). – S.Ethiopia. – Branching basally to form clumps; **St.** erect to creeping, rooting, up to 15 cm long, 7–12 mm $\varnothing$, with 6 tuberculate angles, the tubercles 1–2 mm high; **Fl.** few, Cor. with linear, spreading, revolute lobes, yellow-green at the base, pale purple above, pointed, 9–11 mm long, 11 mm across at the base.

**E. nubica** N. E. Br. (*E. dammaniana* Schweinf.). – Sudan; Ethiopia; S.Arabia. – **St.** usually with 8 angles, 20 cm and more long, 15–25 mm ⌀, branching, erect or variously curving, tubercles with 4–6 sides; **Fl.** 1–3 together, Cor. rotate, dull purple outside, green inside, densely papillose, with brown-purple blotches, the tip brown-purple.

**E. planiflora** Bally. – Ethiopia. – **St.** with many angles, tuberculate, thick, erect; **L.** persisting, hairlike, on the tubercles; **Fl.** almost sessile, Cor. 7.5 mm ⌀, glabrous, cup-shaped, dull pale yellowish-green, the edges reddish-brown, the lobe-tips revolute, darker.

**E. repens** R. A. Dyer et J. C. Verd. – Tanzania. – **St.** with few Br., creeping and rooting, almost cylindrical, 6–9 mm thick, with 8 to 10 angles divided into tessellate tubercles; **Cor.** 9 mm ⌀, cup-shaped below, glabrous on the underside, wine-red inside, with a few hairs in the middle, lobes ovate, 3–3.5 mm long, margins recurved and hirsute, the indentations particularly hairy.

**E. scutellata** (Defl.) Bgr. (*Caralluma s.* Defl.). – S.Arabia. – **St.** 10–30 cm long, 1 cm ⌀, ascending and prostrate, green to slightly grey-green, freely branching, 8-angled, the ribs divided into hexagonal prominent tubercles; **Fl.** 1–2 together, Cor. with a broad, campanulate tube 3.5 mm long, dirty green outside, yellowish and minutely papillose inside, the throat red-spotted, lobes triangular, 3 mm long, projecting, the tips reddened.

**E. seibanica** Lavr. – S.Arabia. – **St.** almost subterranean, or branching above soil-level, prostrate, ascending, also pendulous, 5–15 cm long, 6–9 mm ⌀, 6-angled, green or bluish-green, almost cylindrical, tubercles hexagonal; **Fl.** in pairs, Cor. campanulate, fleshy, 2.5 mm long, cream-coloured outside, minutely tuberculate, the tube with a few reddish blotches, sulphur-yellow with minute red spots inside, minutely tuberculate, lobes triangular, later spreading, 2.5 mm long.

**E. serpentina** (Nel) White et Sloane (*Caralluma s.* Nel). – Cape: Van Rhynsdorp D. – **St.** almost cylindrical, creeping, branching, 3–10 cm long, 10–12 mm ⌀, with 8 angles divided into hexagonal tubercles, greenish-purple; **Fl.** 8–9 together at the St.-tips, on pedicels 3–4 mm long, Cor. 8 mm ⌀, lobes ovate, terminating in an acute white tip, white outside, granular, pink-purple inside, white-bristled, with a narrow yellow ring in the middle.

**E. sharpei** White et Sloane. – Kenya. – **St.** branching, c. 14 cm long, 1–1.5 cm ⌀, erect or creeping, cylindrical, tessellate, 8-angled; **Fl.** 1–2 together, Cor. 1 cm ⌀, glabrous outside, pale red below, grey-green with reddish lines above, velvety-red inside, glabrous, lobes 3 mm long, ovate, with a few reddish ciliate hairs.

**E. squamulata** (Decne.) Bally (*Ceropegia s.* Decne.). – S.Arabia: Yemen. – **St.** 7–8-angled, tuberculate-reticulate, creeping, forming dense cushions, up to 15 cm long, 8 mm ⌀, grey-green, somewhat brownish, densely and minutely hairy; **Fl.** in pairs, Cor. pitcher-shaped, distinctly 5-angled, 14 mm long, 5 mm ⌀, 2 mm ⌀ at the mouth, tube reddish brown outside, glabrous, reddish brown inside, lobes broadly triangular, pointed, 2.5 mm long, reddish brown inside, greenish-yellow outside with reddish brown blotches.

**E. urceolata** Bally (Pl. 60/3). – Kenya: N.Frontier reg. – **St.** erect, solitary or up to 6, up to 8 cm tall, 2.5 cm ⌀, cylindrical, with 18–20 angles, densely tuberculate, tubercles 1 mm across and high, with persistent, filiform L. 3–3.5 mm long; **Fl.** solitary, Cor. up to 10.5 mm long, 6–7 mm ⌀, tube urn-shaped, 9 mm long, 6–7 mm ⌀ in the middle, up to 4 mm ⌀ at the mouth, glabrous outside, pale green to purple, transversely rough inside, purple-red, lobes broadly triangular, 3 mm long and across, both surfaces pale green.

**E. virchowii** K. Schum. (*Virchowia africana* Vatke). – Tanzania. – **St.** erect, branching, 10–12 mm ⌀, with 6 straight angles, green or brown, minutely tuberculate; **Fl.** 2–4, 6 to 7 mm ⌀, Cor. circular, minutely tuberculate outside, dull green to dull purple, minutely papillose inside, yellow-green with purple blotches, lobes purple-brown.

**E. watsonii** Bally. – N.Somalia: Borama D. – **St.** creeping, with few Br., up to 20 cm long; **Br.** shorter, thinner, indistinctly 8–12-angled, tuberculate, tubercles reticular, 4–5-angled, 3–4 mm long and across, 1–2 mm high; **L.** narrow-lanceolate, becoming bristly when dry; **Fl.** 1 or few together, Cor. pyriform, c. 7 mm long, 4–5 mm ⌀ at the base, 3 mm ⌀ at the mouth, both surfaces glabrous and dark purple-red, lobes spreading, broad-linear, tapering, 5 to 7.5 mm long, 1.5 mm across, dark purple-red outside, yellow inside, rough.

**Echinocystis** Torr. et Gray (*Marah* Kellog). Cucurbitaceae. – N.Am. – Mostly ☉ plants, not succulent. There is only one species with a tuberous caudex. – Greenhouse, warm. Propagation: seed.

**E. gilensis** Greene (*Marah g.* Greene). – USA: S.W.New Mexico, Arizona. – ♃; **caudex** very large and tuberous; **St.** succulent, climbing; **L.** deeply cleft, lobes triangular or oblong-lanceolate; **Fl.** up to 10 cm ⌀, whitish; **Fr.** spherical, 2–3 cm ⌀, prickly.

**Edithcolea** N. E. Br. Asclepiadaceae. – Socotra; Somalia; Tanzania. – Related to the G. **Caralluma**; Fl. conspicuously large. – Greenhouse, warm. Propagation: seed.

**E. grandis** N. E. Br. v. **grandis** (Pl. 60/4) (*E. sordida* N. E. Br.). – Somalia; Kenya; Socotra. – **St.** ascending, up to 30 cm tall, 2.5 cm thick, glabrous, angles armed with very sharp thorny T.; **Fl.** solitary, almost at the St.tips, on pedicels 14–18 mm long, ± 10 cm ⌀, the centre plate-shaped, 5-cleft up to the middle, lobes triangular-ovate, 5 cm long, 2.5–3 cm across, pale yellow, with red-brown blotches, the throat round the 4 mm deep tube tuberculate-callused, with clavate hairs in 5 lines from the base to the indentations, lobe-margins hairy, revolute above, brown.

E. — v. **baylissiana** LAVR. et HARDY. – Tanzania. – **St.** 20–75 cm long, creeping and rooting, 4-angled; lobes of the **Cor.** shorter and broader.

**Elaeophorbia** O. STAPF. Euphorbiaceae. – W. trop. Afr.; Guinea; Angola; Ghana; Togo; Dahomey. – ♄ or ± tall ♄, succulent, leafy, thorny, with milky sap. In Journ. Arn. Arb. **48**, 357, 1967, WEBSTER transfers this G. as Sect. Elaeophorbia (O. STAPF), WEBSTER to the G. **Euphorbia** L. – Greenhouse, warm. Propagation: seed, cuttings.

E. **beillei** (CHEV.) JACOBS. (*Euphorbia b.* CHEV.). – W.Afr. – Branching ♄; **Br.** with 4 almost alate angles, Th. short; **L.** oblong-spatulate, narrowing below to the petiole. Acc. P. CHEVALIER possibly identical with E. **leonensis**.

E. **drupifera** (THONN. ex SCHUM.) STAPF v. **drupifera** (*Euphorbia d.* THONN. ex SCHUM., *E. grandifolia* HAW.). – Guinea; Sierra Leone. – ♄; **St.** cylindrical, thick at the base; **Br.** with prominent L.cushions (podaria) in 5 spiral R., thus slightly 5-angled, Th. straight, scarcely 4 mm long; **L.** obovate-oblong, 15–23 cm long, narrowing below to the petiole, blunt, the underside with a carinate M.nerve.

E. — v. **elastica** (POISS.) JACOBS. (*Euphorbia e.* POISS., *E. d. v. e.* (POISS.) CHEV.). – **L.** ovate-oblong, with margins entire and wavy, shortly petiolate.

E. **leonensis** (N. E. BR.) JACOBS. (*Euphorbia l.* N. E. BR.). – Sierra Leone. – ♄; **St.** and **Br.** 4-angled, 12 mm ⌀, Th.pairs 2–3 mm long, along the angles; **L.** 6–10 cm long, 2–3 cm across, oblong-oval, blunt, narrowing below to the short petiole, margins entire, leathery or somewhat fleshy; **Infl.** sessile. Acc. P. CHEVALIER possibly a variety of E. **drupifera**.

E. **hiernii** L. CROIZ. – Angola. – ♄; **St.** 5–6 m tall, branching; **Br.** 5-angled, Th. in pairs, projecting, 3–4 mm long, blackish-brown; **L.** somewhat spatulate, c. 19 cm long, 5 cm across, petiole 1 cm long, margins curved somewhat upwards, dark green, M.nerve thin-triangular.

**Espeletia** MUTIS ex HUMB. et BONPL. Compositae. – S.A.: from Venezuela to Colombia. – Rather tree-like ♄, not very tall, some being succulent in character; unbranched or with only few **Br.**; **L.** in clusters at the Br.tips, usually lanceolate, long; **Infl.** spicate or racemose. – Greenhouse, warm. Propagation: seed. Seldom in cultivation.

E. **grandiflora** HUMB. et BONPL. – Venezuela to Colombia. – Tree-like; **St.** 1.2–6 m tall, thick, densely covered with dry L. remains; **L.** ensiform, densely hairy, grey-white.

E. **insignis** CUATR. – Venezuela. – ♄ 1.5 to 2 m tall.

E. **killipii** CUATR. – Venezuela. – **St.** thick, short, unbranching; **L.** in an erect Ros.

E. **lopezii** CUATR. – Venezuela. – **St.** almost 2 m tall.

E. **rositae** CUATR. – Venezuela. – **St.** very short, thick.

**Euphorbia** L. Euphorbiaceae. – S. and E.Afr.; Ethiopia; Somalia; Morocco; Madag.; E.India; Ceylon; Canary Is.; America. – Succulents of very variable habit, some similar to cacti, ♄-like or bushy; milky sap ± poisonous. **St.** partly columnar, with many or few ribs, some ± strongly branched, some with terete St., some much reduced and ± spherical, with a spherical-clavate caudex and spreading **Br.**, there being many intermediates between these extremes; only few species are conspicuously leafy, mostly the **L.** are deciduous or small, soon dropping; many species with strong Th. at the L. bases, many with Infl. which develop into Th.; **Infl.** (cyathium) often with conspicuous or coloured bracts (cyathophylls), Fl. all unisexual, plants either monoecious or dioecious. – Greenhouse, warm. Some species during summer in the open. Propagation: seed, cuttings.

# Division of the Genus Euphorbia into Sections and Groups according to growth-forms

**Sect. I. Pedunculacanthae** JACOBS. – Succulent plants, thornless or with Th. formed from the persistent, hardened remains of Ped.

**A.** At first succulent, later woody, leafy bushes and succulent ♄, in one case a ♄ (Gr. 4), Br. slender, virgate, jointed, round, flat or angled, mostly without tubercles, a few species with distinct, separate tubercles.

**Group 1.** (Sect. *Tithymalus* BOISS.). – **Bushes**, when young with succulent, cylindrical Br., densely spirally leafy; **L.** sometimes deciduous; **Infl.** umbellate, surrounded by whorled or opposite, red, violet or yellowish bracts. (Type s. Pl. 61). – Species: E. anachoreta, atropurpurea, balsamifera, barbicollis, berthelotii, bravoana, dendroides, lambii, mellifera, noxia, orthoclada, piscatoria, pseudograntii, punicea, quadrata, tuckeyana.

**Group 2.** (Sect. *Goniostema* BAILL.). – ♃, ± succulent, branched ± in whorls, or small ♄; **Br.** thickened towards the apex, slightly angled, covered with comb-like stipules on the edges; **L.** large, stalked, pinnately veined (Type s. Pl. 61). – Species: E. boissieri, commersonii, epiphylloides, ? hedyotoides, leuconeura, lophogona, neohumbertii, viguieri.

**Group 3.** (Key 1 according to WHITE, DYER and SLOANE). – Woody ♃; **Br.** glabrous or rough from the ± projecting cushion-like remains of small, reduced Infl., thornless or the tips of the Br. becoming thorny. (Type s. Pl. 61). – Species: E. espinosa, frutescens, giumboensis, grosseri, guerichiana, sacchii, scheffleri, somalensis.

**Group 4.** (Sect. *Arthrothamnus* BOISS.; Sect. *Lycopsis* BOISS.; Sect. *Tirucallii* BOISS.; Key 2 according to WHITE, DYER and SLOANE). – ♃, mostly dwarf or taller and more slender, in one case a ♄; **Br.** numerous, succulent, rarely becoming woody, usually cylindrical, virgate or jointed or more rarely laterally compressed, dichotomous, two or three times branched, thornless or more rarely the end becoming thorny, smooth or rough, without tubercles, more rarely with evidence of tubercles; **L.** minute, opposite, at the Br.-joints, or somewhat larger, spirally arranged, soon falling (Type s. Pl. 61). – Species: E. aequoris, alata, amarifontana, angrae, antisyphilitica, aphylla, arbuscula, arceuthobioides, aspericaulis, brachiata, burmannii, cameronii, cassythoides, carunculifera, caterviflora, cerifera, chersina, cibdela, congestiflora, corymbosa, decussata, dregeana, ephedroides, fiherensis, fragiliramosa, gentilis, gossypina, gregaria, gummifera, herrei, intisy, juttae, karroensis, lactiflua, laro, lateriflora, leucodendron, lignosa, macella, mauritanica, muricata, mundii, nubica, obtusifolia, oncoclada, paxiana, perpera, platyclada, pseudobrachiata, plagiantha, rectirama, rhombifolia, rudolfii, sarcostemmatoides, schimperi, silicicola, spartaria, spicata, spinea, stapelioides, stenoclada, stolonifera, tenax, tirucallii, transvaalensis, verruculosa, xylophylloides.

**Group 5.** (Sect. *Pteroneurae* BGR.). – ♃; with succulent, angled **Br.**, angles formed by the decurrent L. bases; **L.** distant, of medium size or smaller, soon falling. (Type s. Pl. 61). Only from America. – Species: E. phosphorea, pteroneura, sipolisii, weberbaueri.

**Group 6.** (Sect. *Treisia* HAW. p.part.; Key 3 according to WHITE, DYER and SLOANE). – ♃ bushy, many-branched from the base; **Br.** succulent, with mostly projecting, solitary, alternate tubercles (raised L.-bases or podaria); **L.** small, soon falling; peduncle deciduous. (Type s. Pl. 61). – Species: E. gariepina, halleri, hamata, peltigera, schaeferi.

**B.** Low plants, caudex tuberous or thickened and transitional to the main shoot, with serpentine subterranean Br., with herbaceous, leafy Br. or with a short leafy St.

**Group 7.** (Key 4 according to WHITE, DYER and SLOANE). – Plants consisting of a tuberous or thickened **caudex** which is completely hidden in the ground, simple or with two or several serpentine subterranean Br. which develop a number of herbaceous leafy **Br.** above ground. (Type s. Pl. 61). – Species: E. gueinzii, ledermanniana, multifida, pseudotuberosa, trichadenia.

**Group 8.** (Key 5 according to WHITE, DYER and SLOANE). – Plants consisting of a tuberous or thickened **caudex** which is completely hidden in the ground, forming a short St. above, simple or branched, with a number of deciduous **L.** above the ground. (Type s. Pl. 61). – Species: E. crispa, ecklonii, pseudohypogaea, rubella, silenifolia, tuberosa.

**C.** Dwarf or sometimes slender ♃; **St.** and **Br.** above the ground mostly intensely succulent, with ± distinct, flat or projecting crowded tubercles, often angled, with or without persistent Ped.

**Group 9.** (Sect. *Treisia* HAW. p.part.; Sect. *Pseudoeuphorbium* PAX; Key 6 according to WHITE, DYER et SLOANE). – Dwarf or shrubby succulents, one species sometimes arborescent; **caudex** succulent, densely covered with spirally arranged L.-bases (raised L.-cushions); **L.** crowded in tufts above or on long herbaceous shoots from the axils of the L.-bases, these often umbellately branched above; umbel-Br. often forked, dying off, alternately leafy; **Ped.** from the axils of the L.-bases, drying and persisting. (Type s. Pl. 62). – Species: E. bubalina, bupleurifolia, clandestina, clava, cylindrica, grantii, hallii, longetuberculosa, montieri, oxystegia, platycephala, pubiglans, wildii.

**Group 10.** (Key 7 according to WHITE, DYER and SLOANE). – Dwarf or somewhat shrubby succulents; **St.** succulent, densely covered with spirally arranged L.-bases, ± branched; **L.** mostly linear, plicate, several or all Ped. persisting and becoming woody to slender or curved thorns. (Type s. Pl. 62). – Species: E. eustacei, loricata, multifolia.

**Group 11.** (Key 8 according to WHITE, DYER and SLOANE). – Erect succulent plants, dwarf or up to 1.20 m high; the main **shoot** not or little branched, densely covered with rather raised tubercles; **tubercles** with a hollow or pit on their upper slanting face from which the Fl.-stalks arise, or the tubercles without such a hollow or pit, then the Fl.-stalks arising from the axils of the tubercles; **Ped.** persisting and becoming thorny. (Type s. Pl. 62). – Species: E. fasciculata, restituta, schoenlandii.

**Group 12.** (Sect. *Treisia* HAW. p.part.; Key 9 according to WHITE, DYER and SLOANE). – Dwarf succulents, the **caudex** usually subterranean, with densely arranged short Br. near the apex raised above the ground, the truncate tips of which form together a dense cushion; **Br.** covered with distinct tubercles; **Ped.** falling off. (Type s. Pl. 62). – Species: E. clavarioides.

**Group 13.** (Sect. *Pseudomedusae* BGR.; Key 10 according to WHITE, DYER and SLOANE). – Dwarf succulents; the **caudex** clavately thickened above, sometimes completely hidden in the ground, with several radiating rows of Br. which crown the tip of the shoot; the **Br.** covered with distinct tubercles, dying off when old; **Infl.** sessile or short-stalked, Ped. along the St. or along the Br., falling. (Type s. Pl. 62). – Species: E. ernestii, flanaganii, franksiae, gatbergensis, gorgonis, pugniformis, woodii.

**Group 14.** (Sect. *Medusae* HAW. p.part.; Key 11 according to WHITE, DYER and SLOANE). – Dwarf succulents, the **caudex** nearly completely subterranean, mostly with various spirally and densely arranged rows of serpentine elongated Br. which crown the apex, or the caudex completely subterranean with few to several Br. above the ground; **Br.** densely covered with spirally arranged tubercles; **L.** small, deciduous; **Ped.** few at the tips of the Br., hardly more than 10 mm long, deciduous or persisting. (Type s. Pl. 62). – Species: E. bergeri, bolusii, caput-medusae, colliculina, confluens, davyi, duseimata, esculenta, fortuita, hypogaea, inermis, maleolens, marlothiana, muirii, pseudoduseimata, ramiglans, superans, tuberculata, tuberculatoides.

**Group 15.** (Sect. *Medusae* HAW.; Key 12 according to WHITE, DYER and SLOANE). – Dwarf succulents, the **caudex** globose or subcylindrical, above the ground or usually projecting from the ground, with stiff Br., or the caudex dividing below into two or several St.-like Br. and again shortly rebranched, often the Br. so numerous that they form a dense cushion together; **Br.** densely covered with spirally arranged tubercles; **Ped.** solitary or several together, up to 7.5 cm long, mostly near the apices of the Br., persisting. Some species, e.g. E. arida, are transitional to the next group. (Type s. Pl. 62). – Species: E. albertensis, argillicola, arida, baliola, bergii, brakdamensis, braunsii, brevirama, crassipes, decepta, filiflora, friedrichiae, hopetownensis, inornata, marientalensis, melanohydrata, multiceps, namibensis, nelii, orabensis, pentops, rangeana, rudis.

**Group 16.** (Key 13 according to WHITE, DYER and SLOANE). – Dwarf succulents; the **caudex** thick, growing above the ground, usually conical or cylindrical, covered with conspicuous tubercles; **Br.** short, nearly conical when young, the upper part attenuated when older; **Ped.** from the caudex or from the Br., very short or up to 12 mm long, often similar to slender Br., persisting. (Type s. Pl. 62). – Species: E. namaquensis, multiramosa.

**Group 17.** (Sect. *Dactylanthes* HAW.; Key 14 according to WHITE, DYER and SLOANE). – Dwarf succulents often forming clumps; **main shoot** subterranean, with globose or subglobose Br. and shoots and with globose, clavate or cylindrical, jointed Br. above the ground or with a tuberous root with elongated rhizomes and numerous crowded, clavate, cylindrical or also ± conical Br.; **Br.** covered with spirally arranged tubercles; **Infl.** sessile or stalked; Ped. often persisting. (Type s. Pl. 63). – Species: E. globosa, ornithopus, planiceps, polycephala, tridentata, wilmanae.

**Group 18.** (Sect. *Meleuphorbia* BGR.; Key 15 according to WHITE, DYER and SLOANE). – Dwarf succulents; **main-St.** half below or above ground, branched; **Br.** arising at or above the base, like the main St. globular or more columnar; **tubercles** confluent to form continuous ribs, the plant body hence angled; **angles** 5–16; **tubercles** often protracted and bent downwards or scarcely discernible; **Infl.** solitary, almost sessile or branching; **Ped.** often woody and persistent. (Type s. Pl. 63). – Species: E. jansenvillensis, juglans, meloformis, obesa, pseudoglobosa, susannae, symmetrica, tubiglans, turbiniformis, valida.

**Group 19.** (Sect. *Anthacantha* LEM.; Sect. *Florispinae* HAW.; Key 16 according to WHITE, DYER and SLOANE). – Dwarf or shrubby succulents, partly with tuberous roots; the **main shoot** as well as the Br. cylindrical with tubercles (L.-bases) arranged in longitudinal rows often forming areolate angles; **angles** 6–18, ± prominent; sterile **Ped.** persistent as Th., rarely falling. (Type s. Pl. 63). – Species: E. aggregata, anoplia, cereiformis, cucumerina, cumulata, enopla, ferox, fimbriata, heptagona, horrida, inconstantia, leviana, mammillaris, nesemannii, pentagona, pillansii, poissonii, polygona, pulvinata, stellaespina, submammillaris.

**Sect. II. Euphorbia** (*Stipulacanthae* JACOBS.). – Plants with **Th.** mostly in pairs, but also solitary or three together, rarely four of which 3 are from Stip., or developed from the petiole or of unknown origin; occasionally also undeveloped or indistinct; the **bases** of the L. isolated or confluent along the St. and Br. to form fleshy continuous angles.

**SSect. A. Euphorbia** (Sect. *Diacanthium* BOISS.). – **Th.** in pairs.

a) **Teretes.** – **Br.** round or only imperfectly ribbed, L.-bases not completely confluent; **L.** well developed.

**Group 20.** (Sect. Euphorbia. – 1. *Splendentes* BGR., 1. *Milii* JACOBS.). – ♄ with thin, slender, not very succulent Br., curved this way and that, often with thicker, fleshy Br.; **Br.** round or only imperfectly ribbed, L.-bases not completely confluent; **L.** on the short shoots or crowded in tufts at the tips of the Br., 1–18 cm long, ± falling in the resting period; **Infl.** mostly at the ends of the shoots, Ped. long, later deciduous; **bracts** red or yellow, white or pink. (Type s. Pl. 63). – Species: E. beharensis, boiteaui, brachyphylla, capuronii, delphinensis, denisiana, duranii, fianarantsoae, genoudiana, guillemetii, horombensis, isaloensis, leandriana, mahafalensis, mangokyensis, milii, quartzicola, razafinjohanii, rossii, tardieuana, tzimbazazae, zakamenae.

**Group 21.** (Sect. *Euphorbia*. – 2. *Grandifoliae* BGR. p.part.). – Succulent ♄ or ♄ with fleshy St. and Br.; **L.-cushions** not completely confluent to form ribs, therefore the shoots only ± distinctly angled; **L.** large and stout, with pinnate transverse veins and a thick median vein; **Infl.** short-stalked; **bracts** not brilliantly coloured. (Type s. Pl. 63). – Species: E. desmondii, neriifolia, nivulia, royleana, sudanica, teke, trapiifolia, undulatifolia.

b) **Angulatae.** – **Br.** clearly ribbed; **ribs** 2–13; **L.** often very much reduced.

**Group 22.** (Sect. *Euphorbia*. – 3. *Scolopendriae* BGR.; Key 17 according to WHITE, DYER and SLOANE). – Dwarf succulents not or rarely more than 30 cm high; **caudex** short and thick, above the ground or semi-subterranean, often in connection with a thickened **root**; **St.** from the apex, radiate or ascending, with or without lateral Br. which are 2–4-angled, often spirally twisted, the leaf bases with thorn pairs upon small Th.-shields which are solitary or confluent into horny bands; **Infl.** from the ends of the St. (Type s. Pl. 63). – Species: E. aeruginosa, clavigera, clivicola, decidua, dekindtii, enormis, fanshawei, groenewaldii, imitata, knuthii, micracantha, persistens, restricta, schinzii, squarrosa, stellata, tortirama, tortistyla, vandermerwei.

**Group 23.** (Sect. *Euphorbia*. – 4. *Compressa* BGR.). – Caulescent, arborescent succulents; **St.** from base with elongated, angled **Br.** which are numerous, often arranged in whorls, compressed to 2-angled and flat, often pinnately branched. (Type s. Pl. 63). – Species: E. alcicornis, dawei, ramipressa.

**Group 24.** (Sect. *Euphorbia*. – 5. *Trigonae* BGR. and 6. *Polygonae* BGR.; Keys 18 and 19 according to WHITE, DYER and SLOANE). – Shrubby and arborescent succulents 30 cm up to more than 3 m high; Br. developed at the base of a ± shortened **main St.**, with or without side-shoots, or with an erect trunk, often divided into two or more main St.; **Br.** with or without flowering Br. and lateral shoots which are 3–13-angled, trunk always many-angled, ± roundish when old; **Th.** short or up to 5 cm long; **Th.-shields** with pairs of Th. solitary or confluent; **L.** large, later falling. (Type s. Pl. 63). –Species: E. abyssinica, acrurensis, ambroseae, angularis, antiquorum, atrocarmesina, avasmontana, ballyi, barnardii, barghartii, baylissii, × bothae, bougheyi, breviarticulata, buruana, cactus, canariensis, candelabrum, carterana, coerulescens, complexa, confertiflora, confinalis, conspicua, contorta, cooperi, curvirama, cussonioides, dawei, decliviticola, deightonii, disclusa, × doinetiana, echinus, eduardoi, erlangeri, evansii, excelsa, fortissima, franckiana, golisana, gracilicaulis, grandialata, grandicornis, grandidens, graniticola, griseola, halipedicola, handiensis, heterochroma, hottentota, hubertii, inarticulata, inculta, ingenticapsa, intercedens, jubata, kamerunica, keithii, kibwezensis, knobelii, lactea, ledienii, lemaireana, letestui, lividiflora, longispina, lydenbergensis, macroglypha, malveola, mbaluensis, memoralis, migiurtinorum, mlanjeana, neglecta, neutra, nigrispina, obovalifolia, officinarum, opuntioides, paganorum, parciramulosa, perangusta, persistentifolia, phillipsiae, polyacantha, proballyana, qarad, quinquecostata, ramulosa, reinhardtii, resinifera, robecchii, rowlandii, semperflorens, sekukuniensis, seretii, spiralis, strangulata, tanaensis, tenuirama, tetragona, thi, tortilis, triangularis, trigona, venenata, virosa, volkmanae, wakefieldii, waterbergensis, williamsonii, winkleri, zoutpansbergensis.

c) **Costatae.** **Br.** with up to 16 ribs; **ribs** warty or dentate; **L.** much reduced.

**Group 25 a.** – **St.** with short or longer branches; **ribs** up to 13, warty; **Th.-shields** solitary, thorns in pairs, short; **L.** scale-like. (Type s. Pl. 63). – Species: E. cryptospinosa, fruticosa, multiclava.

**Group 25 b.** – **St.** always unbranched, with 13–16 dentate **ribs**; **Th.-shields** confluent, Th. bifurcate, strong; **L.** scale-like; **Infl.** up to 30 above a Th.-pair. (Type s. Pl. 63). – Species: E. columnaris.

**SSect. B. Monacanthium** CHEV. – **Th.** solitary.

**Group 26.** – **Shrubby** succulents; **Br.** round, with tubercles arranged in spiral rows, the **tubercles** with one **Th.** (Type s. Pl. 64). – Species: E. darbandensis, immersa, poissonii, sapinii, unicornis, unispina, venenifica. (**E. monacantha** belongs morphologically to SSect. Triacanthium. The two lateral Stip. are reduced and easely overlooked).

**SSect. C. Triacanthium** JACOBS. – **Th.** three together.

**Group 27.** – **Shrubby** succulents; **Br.** 5–7-angled; **Th.-shields** with one central and two lateral smaller Th. (Type s. Pl. 64). – Species: E. ballyana, glochidiata, graciliramea, monacantha, schizacantha, triaculeata.

**SSect. D. Tetracanthium** JACOBS. – **Th.** four together.

**Group 28.** – Dwarf or taller ♄; **Br.** 4-angled to irregularly angled or cylindrical; **Th.-shields** with two Th.-pairs. (Type s. Pl. 64). – Species: E. angustiflora, coerulans, ellenbeckii, inaequispina, isacantha, ndurumensis, nyassae, quadrangularis, subsalsa, tetracantha, tetracanthoides, uhligiana, whellanii.

## Division of the Madagascan Euphorbias into Groups according to J. LEANDRI

**Complex M.** – Madagascan species which can as yet only be partly inserted into the previous Groups.

**M. I.**   **E. stenoclada Group.** – (These belong to Group 4 of the preceding Sect. I.).

**M. II.**  **E. lophogona Group.** – Species: E. boissieri, ? hedyotoides, leuconeura, lophogona, neohumbertii, pyrifolia, viguieri (these belong to Group 2 of the preceding Sect. I.). –

**M. III.** **E. perrieri Group.** – **Br.** thick, cylindrical; **Th.** many, often decurrent, consisting of transformed stipules sitting at the bases of old L., often with secondary small Th., but the L.-bases forming small, connected, winged angles; **L.** elongate-lanceolate or ovate-spatulate, large or rather small; **Infl.** at the end of the shoots. – Species: E. caput-aureum, croizatii, didieroides, guillauminiana, pauliana, perrieri.

**M. IV.**  **E. pedilanthoides Group.** – **Th.** developed from stipules, near the L. or along the old L.-scars, often decurrent or accompanied by more slender Th.; **Br.** thick, cylindrical, the Th.-bases not forming confluent, decurrent angles; **L.** linear or narrow; **Infl.** at the ends of the Br. – Species: E. biaculeata, pedilanthoides.

**M. V.**   **E. pachypodioides Group.** – Plant small, thorny, very fleshy; **L.** crowded at the end of the St. – Species: E. pachypodioides.

**M. VI.**  **E. milii Group.** – **St.** and **Br.** cylindrical; **Th.** decurrent or with two lateral Th., but these never united to form an angle; **cyathophylls** obovate and mostly broadened at the end or with a point. – Species: E. beharensis, brachyphylla, capuronii, delphinensis, duranii, fianarantsoae, genoudiana, guillemetii, horombensis, leandriana, mangokyensis, milii, quartzicola, razafinjohanii, tardieuana, tsimabazazae, zakamenae. – Possible hybrids, cult. at the Botan. Garden at Tsimbazaza, Madag. are: E. × ambohipotsiensis, × andrefandrovana, × ingezalahiana, × mitsimbinensis, × soanieranensis, × zanaharensis. (These belong to Group 20 of the preceding Sect. II.).

**M. VII.** **E. ankarensis Group.** – Stip.-Th. inserted on the top of the L.-attachment, sometimes ± undeveloped, but not united to form a wing-like border; **St.** mostly thick-cylindrical; **Cy.** pendent or horizontal, the **cyathophylls** ending in a sharp point. – Species: E. ankarensis, boiteaui, cap-saintemariensis, cylindrifolia, decaryi, francoisii, millotii, moratii, primulifolia.

**M. VIII.** **E. bosseri Group.** – (Temporarily inserted here, but not by LEANDRI). – Without Th.; **St.** terete or partly flattened; **L.** very small and soon dropping. – Species: E. bosseri, platyclada.

**E. abdelkuri** BALF. f. (Gr. ?). – Socotra Archipelago: Abd-al-Kuri. – **Br.** numerous, candelabrum-like from a common axis, cylindric, slightly constricted at intervals, up to 2 m high, grey-green, leafless, thornless, only young St. with tiny rudimentary L.

**E. abyssinica** RAEUSCHEL (Pl. 65/2) (Gr. 24) (*E. grandis* LEM.). – Ethiopia. – ♄; **Br.** with 8 almost alate angles, c. 5 cm high, with a grey horny margin, Th. in pairs, c. 1 cm long; **L.** numerous, linear-lanceolate, 4–5 cm long, 1 cm across, deciduous.

**E. acrurensis** N. E. BR. (Pl. 65/1) (Gr. 24) (*E. abyssinica* v. *tetragona* SCHWEINF.). – Ethiopia. – ♄ up to 10 m tall, with Br. ± in whorls; **Br.** with 4 angles and joints up to 30 cm long, the angles alate, 5–6 cm across, dark green with distinct callused veins, often wavy, curving and sinuate, Th. c. 3 cm apart, c. 3 mm long, in pairs, black.

**E. aequoris** N. E. Br. (Gr. 4). – Cape: S.W. Afr. – **Caudex** round; **St.** simple, seldom branching, up to 20 cm tall, Th. developing eventually, L.cushions long and rather prominent; **L.** linear-lanceolate, small.

**E. aeruginosa** Schweick. (Gr. 22). – Transvaal. – **Caudex** arising from the main root to form a rather thickened Bo., usually subterranean, freely branching from the base; **Br.** c. 15 cm tall, 5–7.5 mm thick, with 4–5 indistinct angles, often contorted, coppery-green, Th.shi. prominent, 5–7 mm long, Th. in pairs, 2 cm long, usually with smaller Th. above them.

**E. aggregata** Bgr. (Gr. 19) (*E. enneagona* Bgr.). – Cape: Karroo, Orange Free State. – Thorny ♄ 5–75 cm tall, freely branching from the base; **St.** short, Br. 3–3.5 cm thick with 8–9 angles, divided by sharp furrows, scarcely dentate; thorny **Infl.** numerous.

**E. —** cv. Cristata. – **St.** very much fasciated.

**E. alata** Hook. (Gr. 4). – Jamaica. – **Br.** virgate, alate. Phosphorescent.

**E. albertensis** N. E. Br. (Gr. 15) (*E. crassipes* Marl. p. part.). – Cape: Prince Albert D. – **Caudex** 10 cm long, 4 cm thick, tuberous, cylindrical, with numerous **Br.** above, some erect, others ± spreading, c. 2 cm long, 6–8 mm thick, with small, rhombic, slightly raised tubercles, thornless, but with the persistent remains of the 15 mm long Ped.

**E. alcicornis** Bak. (Gr. 23). – Madag. – Thornless leafless ♄, branching and re-branching dichotomously; main **St.** 1–6 cm ∅; **Br.** 2–3 cm long. Insufficiently known species. (*E. a.* hort. = **E. ramipressa** L.)

**E. amarifontana** N. E. Br. (Gr. 4). – Cape: Van Rhynsdorp D. – Succulent thornless ♄, 30 cm or more in height, related to **E. chersina**.

**E. × ambohipotsiensis** Ursch et Leandri. – Hybrid: **E. milii** v. **milii** × **E.** ? **viguieri**. – Cult. Bot. Gard. Tsimbazaza (Madag.).

**E. ambroseae** Leach (Gr. 24). – Moçambique. – Thornless ♄, 1.8–2.5 m tall, but also often dwarf; **caudex** up to 4 cm ∅, branching basally and re-branching; **Br.** virgate with 4 (or rarely 5) angles, up to 2 cm ∅, sides usually flat, angles sinuate-crenate, Th.shi. small, often merging into a narrow border, main Th. not present, subsidiary Th. 1–2 mm long or missing; **L.** thick, fleshy, round, soon falling.

**E. anachoreta** Svent. (Gr. 1). – Canary Is.: Salvajita Is. – Compact branching ♄, 30–40 cm tall; **St.** robust, thickened, bark white, young shoots green, L. in a dense Ros. at the tip; **L.** linear-elliptical, almost truncate-blunt at the tip, sessile, densely white-farinose, Ros. flat; **Infl.** simple, sessile, corymbose, floral bracts 5, elliptical, yellowish, Cy. spherical, with Ped. 10 mm long.

**E. angrae** N. E. Br. (Gr. 4). – S.W.Afr.: Gr. Namaqualand. – Compact, succulent, thornless ♄ branching densely from the base and above; **Br.** usually forking, joints 3–25 mm long, 6 mm thick, terminal joints 2 mm thick, almost cylindrical; **L.** 13–15 mm long, 2–2.5 mm across, tapering, thickish, upperside minutely velvety.

**E. angularis** Klotzsch (Pl. 65/4) (Gr. 24) (*E. abyssinica* v. *mozambicensis* Boiss., *E. cactus* sensu Schweinf. p. part.). – Moçambique, Goa Is. – Small ♄ or large ♄ 3–5 m tall; **St.** short, Br. numerous, dividing, joints 7.5–20 cm long, 5–11 cm thick, with 3–4 broad alate angles 2.5–6 cm tall, irregularly sinuate-dentate, with a grey horny margin, Th.pairs 4–10 mm long, projecting, grey.

**E. angustiflora** Pax (Gr. 28). – Malawi; Tanzania. – Cushion-forming ♄ up to 30 cm tall; **Br.** short, shoots 1 cm ∅, with 4 obtuse sinuate-dentate angles, Th.cushions in vertical groups of 2 pairs, the lower Th. 4–7 mm long, the upper ones 1–2 mm long.

**E. ankarensis** P. Boit. (Pl. 65/3) (Gr. 20; M. VII). – N.W.Madag. – **St.** up to 20 cm tall, unbranched, broader at the tip, 4 cm ∅, fleshy, the corky covering tearing and separating; **L.** produced at the end of the flowering period in a cluster of 5–9, oblong-oval, tapering, 5–7 cm long, 2–3 cm across, both surfaces softly hairy, narrowing below towards the 5–10 mm long hirsute petiole, bracts transformed into short stout T. on oblong tubercles; **Infl.** almost terminal, Cy. pendulous, cyathophylls pale flesh-coloured, margins red, long-pointed.

**E. anoplia** Stapf (Gr. 19). – ? Cape: Uniondale D. – ♄ c. 18 cm tall with a few Br. from the base; **St.** 5 cm thick, with 7–9 angles separated by deep furrows, the angles with white pointed tubercles 4–6 mm apart; **Br.** 5-angled, shorter, the tubercles less prominent.

**E. antiquorum** L. (Pl. 65/6) (Gr. 24). – Peninsular India. – Thorny succulent ♄ 3–4 m tall; **St.** 4–5-angled; **Br.** with 3 sinuate angles, constricted to form segments, 3–5 cm across, sides flat or slightly depressed, Th.pairs 2–3 cm apart, 4–6 mm long, Th.shi. roundish, small.

**E. antisyphilitica** Zucc. (Pl. 65/5) (Gr. 4). – Mexico. – Resembles **E. cerifera**; **St.** somewhat thicker, with short shoots with red Infl.

**E. aphylla** Brouss. (Gr. 4). – Canary Is. – Freely branching, succulent ♄; **Br.** segmented, cylindrical, dividing dichotomously or in whorls, curved upwards, 5–8 cm long, 5–6 mm thick, terete, with L.scars.

**E. —** × **E. atropurpurea** Brouss. – Intermediate hybrid.

**E. arbuscula** Balf. f. (Gr. 4). – Socotra. – ♄ or ♄ 6–7 m tall; **Br.** terete, dividing alternately or in whorls, shoots terete or rather flat, segmented; **L.** scale-like, small.

**E. arceuthobioides** Boiss. (Gr. 4) (*E. tirucallii* Thunbg. p. part.). – Cape. – Thornless ♄, 22.5–30 cm tall from a thickened **caudex**, branching freely from the base and re-branching, 2–4 mm thick, shoots 2 mm thick, often variously curved, minutely rough.

**E. argillicola** Dtr. (Gr. 15). – S.W.Afr.: Namaqualand. – **Taproot** up to 30 cm long, 12–15 cm thick; **Br.** segmented, 5–9 cm long, 1.5–2 cm thick; **L.** in dense clusters, blue-green, 2–6 cm long, 1.5–2 cm across, longitudinally folded, margins with well-spaced bristly T.; **Infl.** 3–4-partite, sometimes developing Th. 1–3 cm long.

**E. arida** N. E. BR. (Gr 15). – Cape, Orange Free State. – **Caudex** cylindrical, projecting only 5 cm above the soil, 4.5–5 cm thick; **Br.** numerous, erect or spreading, thornless but with the persistent remains of the 6–8 mm long Ped., and with 5–6-angled, slightly convex zones 8 mm ⌀, shoots 2.5–3.7 cm long, 8–10 mm thick, cylindrical, tuberculate, later falling; **L.** linear-lanceolate, green, deciduous.

**E. aspericaulis** PAX (Gr. 4). – Cape: Calvinia D. – Thornless ♄, 30–45 cm tall with a **caudex**, branching from the base or higher; **Br.** ± alternate, 2–3 mm thick, 6-angled, furrowed, minutely rough.

**E. atrocarmesina** LEACH (Gr. 24). – Angola: Cuanza Sul, Huambo. – Succulent, thorny ♄ branching basally; **Br.** spreading, ascending-erect, jointed, with 4–6 broad, alate angles, deeply constricted into almost round, elliptical, ovate or occasionally oblong segments, (5–)8(–15) cm long, 2–6 cm across, the angles slightly or very crenate, continuous, horny, brown, Th. 2, spreading, straight or somewhat curved; **L.** ovate-triangular, c. 2 mm long, 2.5 mm across, deciduous; stipular Th. minute; **Infl.** 1–3, short-pedunculate, Cy. deepest pink with yellow pollen.

**E. atropurpurea** BROUSS. (Pl. 61/1; 66/2) (Gr. 1). – Canary Is.: Tenerife. – ♄ branching dichotomously or tripartitely; **St.** thickened; **Br.** c. 2 cm thick, fleshy, becoming woody, with L. scars, young Br. densely leafy; **L.** obovate-lanceolate, bluntish, 5–9 cm long; **Cy.** in 5–10-rayed umbels, bracts brown to violet.

**E. —** v. **modesta** SVENT. – Canary Is.: Tenerife. – **L.** striped green to purple; **Infl.** with ovate-spatulate bracts.

**E. avasmontana** DTR. v. **avasmontana** (Pl. 66/1) (Gr. 24) (*E. kalaharica* MARL., *E. karasmontana* DTR. nom. nud.). – S.W.Afr.: Gr. Namaqualand; Cape. – Thorny succulent ♄, c. 2 m tall; **caudex** much reduced, branching freely from the base; **Br.** occasionally dividing, 5–7 cm thick, 5–7-angled, with joints 5–13 cm long, greenish-yellow to bluish, angles not very prominent, Th.shi. merging into the broad horny border, Th. in pairs, 1 cm apart, sharp, rigid, 1–2 cm long.

**E. —** v. **sagittaria** (MARL.) WHITE, DYER et SLOANE (*E. s.* MARL.). – Cape. – Up to 1.7 m tall, with 20–30 4–5-angled **Br.** clearly divided into short symmetrical segments.

**E. baga** A. CHEV. (Gr. 8). – Sudan; Ghana; Upper Volta; Nigeria. – **Caudex** ovoid, 10 to 15 cm tall, 4 cm ⌀; **Br.** very short; **L.** 1–2, linear-lanceolate, 10–20 cm long, 2–4 cm across, tapering, narrowing to the short petiole, somewhat fleshy, margins minutely wavy, pink.

**E. baliola** N. E. BR. (Gr. 15). – S.W.Afr.: Gr. Namaqualand. – Related to **E. pentops**; only one plant hitherto discovered.

**E. ballyana** RAUH (Pl. 64/27) (Gr. 27). – Kenya: Nairobi highlands. – Dwarf ♄; **root** napiform, up to 10 cm long, 4 cm thick; **caudex** erect, with a few Br. above, up to 30 cm tall; **Br.** 5–10 mm thick, obtusely 4-angled, grey-green to yellowish-green, with confluent darker podaria, angles almost straight; **L.** scale-like, deciduous, Th. 3, very thin, sub-aciculate, main Th. 10–15 mm long, lateral Th. 2–3 mm long; **Cy.** 3.

**E. ballyi** CARTER (Pl. 66/3) (Gr. 24). – Somalia. – Closely resembles **E. grandicornis**; thorny ♄ up to 1.2–1.5 m tall, 1.8 m ⌀, branching from the base, with some lateral shoots, Br. and shoots erect, with 4–6 broadly alate angles, the wings constricted at intervals to form segments which are 3-angled, 1.5–3.5 cm long, up to 2.5 cm across, the upper margin dentate, the Th.shi. forming a continuous border, Th. 2, up to 3.5 cm long; **L.** minute, scale-like and deciduous; **Infl.** solitary, forking.

**E. balsamifera** AIT. (Pl. 66/4) (Gr. 1). – Canary Is.; W.Afr.; S.Arabia; Somalia. – Freely branching ♄, the St. often thickened, Br. grey; **L.** crowded at the Br.tips, rather thick, linear-lanceolate or ovate-oblong, green or bluish, 18–24 mm long, 4–5 mm across; **Cy.** solitary between the L.Ros.

**E. —** ssp. **adenensis** (DEFL.) BALLY (*E. a.* DEFL.). – S.Arabia: Aden; Somalia. – More compact in growth than ssp. **balsamifera**; **L.** shorter, obovate-oblong, bluer.

**E. —** ssp. **balsamifera.** – Canary Is.; W.Afr. – **L.** linear-lanceolate.

**E. barbicollis** BALLY (Gr. 1). – Somalia. – **Root** napiform, 10–12 cm long, 3–4 cm thick; **St.** few, clambering, little branching, up to 1 m long, c. 5 mm thick, short shoots 2 mm thick, sparsely hairy; **L.** rosulate on Br.tips, linear, up to 3.7 cm long, 3 mm across, sessile, fleshy, upperside deeply grooved, both surfaces with scattered hairs; **Infl.** terminal.

**E. barnardii** WHITE, DYER et SLOANE (Gr. 21). – Transvaal. – Thorny succulent ♄ c. 60 cm tall, with a somewhat tuberous main root, branching freely below; **caudex** much reduced; **Br.** distinctly jointed, segments c. 10 cm long, 4–7 cm thick, thickest at the base, with 6 much compressed angles 1.5–3 cm high, Th.shi. decurrent into a horny grey border, Th. in pairs, 1 cm long, on prominent tubercles.

**E. barnhartii** L. CROIZ. (Pl. 67/2) (Gr. 24) (*E. trigona* ROXB.). – Peninsular India. – ♄; **Br.** triangular, the angles compressed, deeply arcuate-sinuate, light green, L.cushions dentate, prominent, Th. 2–4, c. 10 mm long; **L.** 6 cm long, 2.5–3 cm across, ovate-spatulate, soon falling.

**E. baylissii** LEACH (Gr. 24). – Moçambique. – Thorny succulent ♄, 30–180 cm tall, usually with a single St., few Br. or twigs; **St.** and **Br.** constricted to form segments, angles very sinuate, dentate, alate, with dark green and usually white longitudinal stripes, wings 1.2 cm across, c. 1 cm thick, T. triangular, up to 4.5 mm high, 8–17 mm apart, Th.shi. on the tips of the T., Th. 2, aciculate, (1.5–)4(–6) mm long, with minute stipular Th.; **L.** minute, soon falling.

**E. beharensis** J. LEANDRI (Gr. 20; M. VII). – S.W.Madag. – Thorny compact ♄; **Br.** 3–5 mm thick, reddish-grey, Th. 10–15 mm long, 2 to 10 mm apart, stipular Th. present; **L.** small, soon falling.

**E. bergeri** N. E. Br. (Gr. 14) (*E. caput-medusae* Lam., *E. fructuspina* Sweet p. part., *E. parvimamma* Bgr.). – ? Cape. – Resembles **E. caput-medusae** L.; **St.** short, almost spherical; **Br.** numerous, 7.5–22.5 cm long, 8–17 mm thick, usually simple, cylindrical, tuberculate, curved, smooth, green, L.cushions rhombic or long-hexagonal, 5 mm across.

**E. bergii** White, Dyer et Sloane (Gr. 15). – Orange Free State, Cape. – **Caudex** arising from the somewhat thickened main root, partly below soil-level, with Br. from the base dividing 2–3 times and forming spherical clumps of 20–25 cm ⌀; **Br.** up to 15 cm long, up to 12 mm thick, cylindrical, tuberculate, the tubercles being rhombic, 4–5 mm long and across and 1–2 mm high, with L.scars; **L.** in clusters at the Br.tips, deciduous, 4–6 mm long, linear-lanceolate, plicate.

**E. berthelotii** C. Bolle (Gr. 1). – Canary Is.: Gomera. – Small ♄ 2–2.5 m tall, crown 2 m across, Br. grey; **L.** sessile, broadly linear, c. 5 cm long, 10–12 mm across, grey-green.

**E. biaculeata** M. Denis (Gr. M. IV). – S.W. Madag. – ♄ with few Br., 1–1.5 m tall, thorny, not very fleshy, Th. in pairs, 1 cm long; **L.** sessile, oblong-linear, 6 cm long, 3 mm across.

**E. boissieri** H. Baill. (Gr. 2; M. II). – E. Madag. – Resembles **E. lophogona**; **L.** long-linear, 30 cm long, 5 cm across. – Insufficiently known species.

**E. boiteaui** Leandri (Gr. 20; M. VII). – S.E. Madag. – Acc. Ursch and Leandri possibly only a variety of **E. ankarensis**; **St.** up to 15 cm tall, up to 12 mm ⌀, stipular Th. arranged in 5 R., 2–3 mm long; **L.** few, rosulate, elliptical-tapering, sessile, c. 2 cm long, 8 mm across; **Cy.** smaller.

**E. bolusii** N. E. Br. (Gr. 14). – ? Transvaal. – Related to **E. maleolens**; a still unclarified species.

**E. bosseri** Leandri (Gr. M. VIII). – Madag. – Fleshy, ± creeping semi-♄ c. 30 cm tall, with few Br.; **Br.** terete to ± angular, 3–4 mm thick, constricted at the base, grey, pink-tipped with green lines and blotches; **L.** deciduous, membranous, 1 mm long; **Cy.** terminal or axillary.

**E. × bothae** Lotsy et Godd. (Gr. 24). – Cape: Albany, Uitenhage, Alexander D. – Nat. hybrid: **E. coerulescens** × **E. tetragona**. – Thorny succulent ♄ with a rudimentary main St., **Br.** 3–7-angled, joints variable in length, yellowish-green, blue-green or grey, angles ± compressed, ± sinuate-dentate, Th.shi. confluent into a broad horny border, Th. in pairs.

**E. ×** — nm. **anticaffra** (Lotsy et Godd.) Rowl. (*E.* × *a.* Lotsy et Godd.). – Always tree-like, with a trunk up to 1 m tall, c. 14 cm thick.

**E. bougheyi** Leach (Gr. 24). – Moçambique. – ♄ up to 7 m tall, thorny; **St.** cylindrical, rather thin, thorny above, with up to 9 angles; **Br.** curving-ascending to form an umbrella-like crown, with joints of varying lengths, and 2–5 rather thin wings 2.5–5 cm across, curled and undulating, youngest twigs often with only 2 wings, Th.shi. normally solitary, sometimes confluent with a very narrow horny border; **L.** ovate-triangular, pointed, later falling, leaving minute stipular Th.; **Infl.** short-pedunculate, Cy. 3 together.

**E. brachiata** E. Mey. (Pl. 61/4) (Gr. 24) (*E. muricata* Thunbg.). – Cape: Van Rhynsdorp D.; S.W.Afr. – Thornless succulent ♄, with many Br. from the base; **Br.** usually projecting horizontally. Resembles **E. lignosa**, but daintier.

**E. brachyphylla** M. Denis (Gr. 20; M. VI). – Madag. – ♄ with few Br. from the base, 1 to 1.2 m tall; **Br.** c. 1 cm thick, densely covered with 7 mm long main Th., initially densely white-hairy and flattened almost ribbon-like, beside them shorter subsidiary Th.; **L.** on small tuberculate short shoots between the Th., linear to narrow-linear, 5–8 cm long, 3–5 mm across, short-pointed; **Infl.** with a long Ped., the axis with dense bristly hairs, cyathophylls broad, slightly hairy on the underside.

**E. brakdamensis** N. E. Br. (Gr. 15). – Cape: Lit. Namaqualand. – **Caudex** short, c. 20 cm tall; **Br.** numerous, forming smaller or larger clumps, thornless but covered with the remains of the 2.5–5 cm long Ped., simple or dividing, tuberculate, 5.5–12 cm long, 8–10 mm thick, tubercles 5–8 mm long, 2 mm across, 2–3 mm high, oblong-rhombic; **L.** 12–15 mm long, 7 mm across, linear, pointed, plicate, bluish.

**E. braunsii** N. E. Br. (Gr. 15). – Cape: Karroo. – **Caudex** arising from the thickened main root, partly below soil-level; freely branching and re-branching, forming clumps 5–15 cm tall and up to 25 cm ⌀; **Br.** 5–10 cm long, up to 2 cm thick, cylindrical, tuberculate, grey-green, tubercles 4–6-sided, 5–6 mm ⌀, flat with projecting white podaria, thornless but with thorny Ped.

**E. bravoana** Svent. (Pl. 67/1) (Gr. 1). – Canary Is.: Gomera. – ♄; **St.** fleshy, 1–2 m tall, 2–4 cm and more ⌀, laxly branched, candelabra-like, bark green-yellow, becoming black, with L.scars; **L.** 30–40, rosulate, sessile, linear-lanceolate, blunt, stiff, fleshy, bluish-violet, margins whitish, upper L. deep violet-purple during flowering; **Infl.** a terminal umbel.

**E. breviarticulata** Pax (Gr. 24) (*E. grandicornis* sensu N. E. Br. p. part.). – Tanzania; Somali Repub.; Kenya. – Succulent ♄ up to 1.20 m tall, branching freely from the base; **Br.** erect to projecting, much constricted to form almost sagittate, ovate or reniform segments 5–8 cm long, 8–9 cm ⌀, triangular, central part 2.5 cm thick, glabrous, angles much compressed, 2.5–3 cm across, 4–6 mm thick, with continuous, horny, greyish white, wavy borders; **L.** scale-like, small, deciduous, Th. stout, 1.5–6 cm long, in pairs; **Infl.** of several umbels, short-pedunculate, often with only 1 ♂ calyculus.

**E. brevirama** N. E. Br. (Gr. 15). – Cape: Jansenville D. – **Caudex** conical, 5–6 cm ⌀, flat above, the middle somewhat depressed, divided into octagonal tubercles, the crown-margin bearing 3 R. of short, fleshy green Br., the outer Br. c. 8 cm long, the inner ones shorter, 5–6 mm thick, tuberculate; **L.** very short, deciduous; **Infl.** at the Br.tips, 4–6 mm long, becoming thorny.

**E. brevis** N. E. Br. (Gr. 22). – Angola. – **Caudex** 5–8 cm ⌀; **Br.** numerous, erect, 5 to 7 cm tall, 4–5 mm ⌀, triangular, glabrous, yellowish-green, the angles pointed, ± sinuate-dentate; **L.** rudimentary, scale-like, scarcely 1 mm long, Th. 2–3 mm long, in pairs 8–10 mm apart, Th.shi. with hemispherical L.scars; **Infl.** with Ped. 6–9 mm long.

**E. brevitorta** Bally (Gr. 22). – Kenya: S.Prov. – **Caudex** clavate, simple, semi-subterranean; **Br.** with 2, more rarely 3 angles, contorted, tuberculate, (5–)8(–15) cm long, constricted into segments 1–2 cm long, podaria tipped with a T., Th. 2, c. 7 mm long, stipular Th. small, deciduous; **L.** triangular, small, deciduous; **Infl.** solitary, with 3 Cy.

**E. bubalina** Boiss. (Gr. 9) (*E. clava* E. Mey., *E. oxystegia* Bak., *E. laxiflora* O. Ktze.). – Cape. – **St.** erect, with few Br., up to 1.3 m tall, cylindrical, 2 cm thick at the base, thickened above, green, becoming grey, L.cushions long, flat, rhombic; **L.** at the Br.tips, 7–10 cm long, lanceolate, blunt, mucronate, broadly petiolate, soft, thin, light green, deciduous.

**E. bupleurifolia** Jacq. (Pl. 62/9; 67/4) (Gr. 9) (*Tithymalus b.* Haw., *E. proteifolia* Boiss., *E. squamosa* Mass. ex Bgr.). – Cape: Komga D., Natal. – **Caudex** simple, rarely with a few Br., very thick, somewhat ovoid, 10–12 cm tall, 7–8 cm thick, L.cushions 4-sided, brown, scale-like, in a double spiral; **L.** from the St.tip, 10–15 cm long, lanceolate, tapering, narrowing below to the long petiole, light green; **Infl.** long-pedunculate.

**E. burmannii** E. Mey. (Gr. 4) (*E. viminalis* Burm., *E. tirucallii* Thunbg. p. part., *E. biglandulosa* Willd.). – Cape: widely distributed. – Thornless ♄ branching basally, with opposite Br. above, succulent at first, becoming woody, forming clumps of 30–70 cm ⌀, Br. c. 5 cm thick, shoots thinner, with large, persistent, dark brown L.bases at the nodes.

**E. buruana** Pax (Gr. 24). – Uganda; Tanzania. – Leafless ♄ 45–60 cm tall; **Br.** 10–13, 25–35 mm ⌀, thickened below, the upper part with 2–3 narrow neck-like constrictions between 2–3 widened roundish segments, with 5 alate angles which are sometimes wavy and sinuate-dentate, Th. dissimilar, 3–12 mm long, in pairs 8–10 mm apart, Th.shi. horny, solitary; **L.** rudimentary, minute.

**E. cactus** Ehrenb. ex Boiss. v. **cactus** (Gr. 24). – S.Arabia. – Thorny ♄ 1.5–3 m tall; **St.** 3–4-angled with candelabra-like Br., St. and Br. long-segmented, segments 10–30 cm long, 7 to 10 cm ⌀, sides initially grooved, later flat, angles compressed, somewhat wavy-curved, with a broad, light green, horny margin, Th.pairs 15 mm apart, 1–4 mm long.

**E. —** v. **aureo-variegata** Schweinf. – **Br.** mottled pale yellow, with a long, pale yellow M.band along the furrows, and double, curved lines towards the angles.

**E. —** v. **tortirama** Rauh et Lavr. – **Br.** contorted.

**E. cameronii** N. E. Br. emend. Bally (Gr. 4). – N.Somalia. – Freely branching ♄ up to 3 m tall, 3.5 m ⌀, with Br. in a dense conical crown; **St.** cylindrical, 1.5–3 cm thick, Br. dense, 3.5–20 cm long, 7–9 mm thick, lateral shoots 3–6 cm long, 6–7 mm thick, L.scars arranged spirally, slightly raised; **L.** on the St.tips, fleshy, obovate, deciduous; Cy. produced from a whorl of 3–5 white bracts.

**E. canariensis** L. v. **canariensis** (Pl. 67/3) (Gr. 24). – Canary Is. – ♄ branching from the base, up to 12 m tall; **Br.** numerous, ascending, (4–)5(–6)-angled, fresh green, sides flat, angles acute, sinuate-tuberculate, Th.pairs c. 14 cm apart, 4–5 mm long, thin.

**E. —** v. **spiralis** Bolle. – **Br.** reddish and with angles turned in a spiral line.

**E. —** v. **viridis** Kunkel. – Is. Gomera. – Differs in having green **Fr.**

**E. candelabrum** Trem. v. **candelabrum** (Pl. 73/3) (Gr. 24) (*E. ammak* Schweinf., *E. officinarum* v. *arboreum* Forsk., *E. ingens* E. Mey., *E. grandidens* Adlam, *E. cooperi* Bgr., *E. similis* Bgr., *E. natalensis* Hort. ex Bgr.). – S.Afr. to Somalia. – Thorny succulent ♄ up to 10 m tall; **St.** branching and re-branching to form a broad round crown, triangular, becoming 4–5-angled; **Br.** erect or ascending, candelabra-like, 4–5-angled, deeply alate with prominent wings, divided into segments 2.5–15 cm long, angles 2–3 cm high, 5–7.5 mm thick, undulating or sinuate, deep green, Th.Shi. (2–)3(–10) mm long, obovate, horny, projecting, Th. in pairs, very variable, 2–18 mm long, all are aciculate; **L.** very rudimentary, minute.

**E. —** v. **erythraea** Bgr. – Ethiopia. – **St.** 3-angled; **Br.** with 4 rather fleshy angles, sides darker green, Th.shi. rounder and more closely spaced.

**E. cap-saintemariensis** Rauh (Gr. M. VII). – Madag.: Cape Ste. Marie. – **Caudex** napiform, branched, up to 30 cm long, 10 cm ⌀, bark thick, silver-grey, with a dense crown of Br.; **Br.** decumbent and resting on the soil, or ascending, 5–10 cm long, 5–10 mm across, terete but slightly angled near the apex, with the scars of the deciduous L.; **L.** in a spreading terminal Ros., succulent, reddish-green, 2–2.5 cm long, 5–8 mm wide, margins curved upwards and undulate, the base of the short succulent petiole surrounded by many short, dissected, silver-grey, stiff bristles inserted on a succulent, acutely-angled podarium; **Infl.** few, subterminal, with 2–4 erect Cy., cyathophylls broad-ovate, olive-green, 2–5 mm across.

**E. capuronii** Ursch et Leandri (Gr. 20; M. VI). – S.W.Madag. – ♄ up to 1 m tall, freely branching from the base; young **Br.** ± 1 cm thick, becoming corky; **L.** in terminal Ros. on numerous short shoots, oblong-lanceolate, tapering, narrowing at the base to the short petiole, (3–)5(–7) cm long; Th. 1.5–2 cm long, from a broad base from which 1 or several other Th. arise; **Infl.** several, subterminal.

**E. caput-aureum** M. Denis (Gr. M. III). – W.Madag. – Small ♄; **St.** with 5 spirally contorted angles, simple or ± branching, Th. solitary, 1 cm long, with much smaller Th. at the base; **L.** obovate-spatulate, gradually tapering towards the base, mucronate, 3–12 cm long, 2–3.5 cm across; **Infl.** with 10–20 yellow Cy.

**E. caput-medusae** L. (Pl. 68/1) (Gr. 14) (*E. fructuspini* MILL., *E. medusae* THUNBG., *E. commelinii* DC., *E. tessellata* SWEET). – Cape: Cape D. – **Caudex** short, thickened above, up to 20 cm ⌀; **Br.** numerous, extending ray-like, serpentine, 3–5 cm thick, up to 75 cm long, greygreen, L.cushions 4–5-angled, separated by sharp furrows; **L.** 15–25 mm long, linearlanceolate, later falling; **Infl.** numerous on the tips of young Br., pedunculate, Ped. persisting.

**E. carterana** BALLY (Gr. 24). – Somalia. – ♄ up to 2 m tall, 2–3 m ⌀; **Br.** spreading, with 3–4 irregularly alate and dentate angles, wings 5–34 mm across, T. 0.5–4 mm apart, Th.shi. virtually merging, Th. in pairs, up to 1.5 cm long, subsidiary Th., 2, small; **L.** unknown; **Infl.** terminal.

**E. carunculifera** LEACH (Gr. 4). – Angola: Moçamedes D. – Thornless ♄ up to 2.5 m tall, monoecious (?), branching from the base; **Br.** virgate, bluish, glabrous or often minutely tawny or orange felted towards the tip, or with minute tuberculate white spots, fleshy at first, becoming woody, irregularly longitudinally wrinkled, the lower Br. dividing, ascending, later branching dichotomously and almost erect and ± straight; **L.** unknown; **Infl.** terminal, ♂ Cy. in dense sessile clusters, ♀ Cy. terminal, in a few short-pedunculate clusters (?).

**E. cassythoides** BOISS. (Gr. 4). – Cuba. – Resembles **E. pendula**; **Br.** and twigs appreciably thinner and daintier. Phosphorescent.

**E. caterviflora** N. E. BR. (Gr. 4) (*E. hastisquamata* N. E. BR., *E. mundtii* R. A. DYER). – Cape: Karroo; Orange Free State. – ♄ 15 to 30 cm tall, thornless, the main root tuberously thickened, with numerous bent, irregularly shaped **Br.**; terminal shoots very short, ending umbellately, slightly 6-angled, somewhat furrowed, smooth or minutely rough, green with reddish blotches.

**E. cereiformis** L. (Gr. 19) (*E. erosa* WILLD., *E. odontophylla* WILLD., *E. echinata* S. D.). – Cape. – Thorny succulent ♄; **St.** up to 90 cm tall, branching from the base, re-branching above, shoots erect, 2.5–5 cm thick, dark green, ribs (9–)11(–15), usually vertical, sometimes somewhat spirally contorted, the ribs separated by sharp furrows, dentate, the T. directed slightly downwards; **Infl.** with solitary, 5 to 10 mm long, spreading, thorny Ped.

**E. cerifera** ALC. (Gr. 4). – Mexico. – ('Candelilla', the Candle-Euphorbia). – ♄ up to 1.5 m tall, roots woody; **Br.** erect, 5 mm thick, cylindrical, almost glabrous, ash-grey, waxypruinose; **L.** lanceolate, 3–5 mm long, 1 mm across, dark red, with a few tiny hairs, soon falling; **Cy.** in groups, yellowish-white, almost 1 cm ⌀.

**E. chersina** N. E. BR. (Gr. 4). – S.W.Afr.: Gr. Namaqualand; Cape: Lit. Namaqualand. – Thornless ♄ up to 60 cm tall, repeatedly branching from the base, each **Br.** bearing several umbels of shoots 12–15 cm ⌀, Br. c. 15 cm long, 5.5–7 mm ⌀.

**E. cibdela** N. E. BR. (Gr. 4). – S.W.Afr.: Gr. Namaqualand. – Thornless ♄, 0.7–1 m tall, branching from the base or higher, becoming woody, green, secondary shoots opposite, 2 to 4 mm thick, 18–30 mm long.

**E. clandestina** JACQ. (Pl. 68/5) (Gr. 9). – Cape: Karroo. – Thornless erect ♄ up to 60 cm tall, with few Br.; **Br.** 25–37 mm ⌀, cylindricalclavate, with numerous spirally arranged short-conical tubercles, somewhat curved at the tip, 4–8 mm long; **L.** at the Br.tips, linearlanceolate, 2–4 cm long, 2–4 mm across, numerous, ± plicate, minutely hirsute.

**E. clava** JACQ. (Gr. 9) (*E. caput-medusae* L. p. part., *E. canaliculata* LAM., *E. clavata* SALISB., *E. coronata* THUNBG., *E. radiata* BOISS., *Tithymalus aizoides* COMM.). – Cape: Karroo. – St. thornless, 30–100 cm tall, cylindrical or clavate, often thickened above, usually without Br., L.cushions little prominent, topped with a few **L.** 10–12 cm long, 5–10 mm across, linear, tapering, grooved, light green, later falling; **Infl.** with Ped. shorter than the L., becoming woody, persisting.

**E. clavarioides** BOISS. v. **clavarioides** (Pl. 62/12) (Gr. 12) (*E. basutica* MARL.). – Cape: Karroo, Orange Free State, Transvaal, Natal, Lesotho. – **Caudex** dwarf, with very numerous 2–7 cm long Br. above, forming a very dense cushion 10 to 30 cm ⌀; young **Br.** 8–17 mm thick, spherical at first, becoming cylindrical and clavate, usually thicker at the tip, rounded above, tuberculate, green to reddish, tubercles 3–4 mm ⌀, 1 mm tall, rhombic or hexagonal, broad-conical or pointed; **L.** 1–2 mm long, lanceolate, plicate.

**E.** – v. **truncata** (N. E. BR.) WHITE, DYER et SLOANE (Pl. 68/3) (*E. t.* N. E. BR.). – Transvaal; Botswana. – **Br.** short, truncate, the tops forming an almost flat surface up to 30 cm ⌀.

**E. clavigera** N. E. BR. (Gr. 22). – Swaziland. – **Caudex** produced from the thickened main root, forming with the St. a large tuberous Bo. completely concealed below the soil, crowned with a number of short thornless **Br.** which are also hidden in the ground, with a few erect shoots above, these 7.5–15 cm long, often somewhat clavate, 2–2.5 cm thick, triangular, glabrous, green, the angles very compressed, alate, deeply tuberculate, tubercles 1–1.8 cm apart, 4–5 mm high, triangular, Th.shi. solitary, small, Th. in pairs, 6–10 mm long, pale brown; **L.** deciduous, 5–10 mm long.

**E. clivicola** R. A. DYER (Gr. 22). – Transvaal. – **Root** tuberous, producing the thickened main St., to form a large tuberous caudex 15 cm long, 2–3 cm thick, branching below soillevel, young Br. above-ground; **Br.** numerous, dense, yellowish-green, 2–3 cm long, 1.5 cm thick, thinner above, tuberculate, indistinctly 4-angled because of the opposite and decussate tubercles and their pairs of Th., Th. up to 5 mm long; **L.** rudimentary, soon falling.

**E. coerulans** PAX (Gr. 28). – Angola: Benguela, Moçamedes. – ♄; **Br.** c. 8 mm ⌀, irregularly angular; Th.cushions solitary or in

Plate 82. 1. **Euphorbia trigona** Haw.; 2. **E. tirucallii** L. v. **tirucallii** (Photo: E. Hahn); 3. **E. tubiglans** Marl. (Repr.: White, Dyer et Sloane, Succ. Euph. Pl. XIII, Drawing: C. Letty); 4. **E. viguieri** M. Denis v. **viguieri** (Photo: J. Marnier-Lapostolle); 5. **E. virosa** Willd. ssp. **virosa**.

Plate 83. 1. **Euphorbia venenata** Marl. (Photo: J. Marnier-Lapostolle); 2. **E. xylophylloides** Ad. Brogn. ex Lem. (Photo: as 1); 3. **Fockea crispa** (Jacq.) K. Schum.; 4. **Ficus palmeri** S. Wats. (Photo: C. Backeberg).

Plate 84. 1. **Fockea multiflora** K. SCHUM. (Photo: WALTER); 2. **F. crispa** (JACQ.) K. SCHUM. Br. with Fl.; 3. **Fouquieria fasciculata** (HUMB. ex ROEM. et SCHULT.) NASH (Repr.: Cact. y Succ. Mex. **VIII,** 1963, 65, Photo: S. MATUDA); 4. **F. diguetii** (VAN TIEGHEM) J. M. JOHNSTON (Photo: J. MARNIER-LAPOSTOLLE).

Plate 85. 1. **Furcraea selloa** C. Koch v. **selloa;** 2. The same, Br. of Infl. with Fl.; 3. The same, plantlets in Infl.; 4. × **Gasterhaworthia bayfieldii** (S. D.) Rowl. (Photo: G. D. Rowley); 5. **Gasteria armstrongii** v. Poelln. Juvenile form.

Plate 86. 1. Gasteria × cheilophylla BAK.; 2. G. laetipuncta HAW.; 3. Gasteria Infl.; 4. G. batesiana ROWL. (Photo: A. J. A. UITEWAAL); 5. G. ernesti-ruschii DTR.; 6. G. disticha (L.) HAW. v. disticha.

Plate 87. 1. Gasteria poellnitziana Jacobs.; 2. G. marmorata Bak. (Photo: J. Marnier-Lapostolle); 3. G. liliputana v. Poelln.; 4. G. verrucosa (Mill.) Duv. v. verrucosa; 5. G. trigona Haw. v. trigona.

Plate 88. 1. × **Gastrolea lapaixii** (Radl) Jacobs.; 2. × **G. beguinii** (Radl) E. Walth. (Photo: J. Marnier-Lapostolle); 3. **Graptopetalum paraguayense** (N. E. Br.) E. Walth.; 4. **G. filiferum** (S. Wats.) Whitehead (Photo: J. Bogner); 5. **Gerrardanthus lobatus** (Cogn.) C. Jeffrey (Photo: P. R. O. Bally).

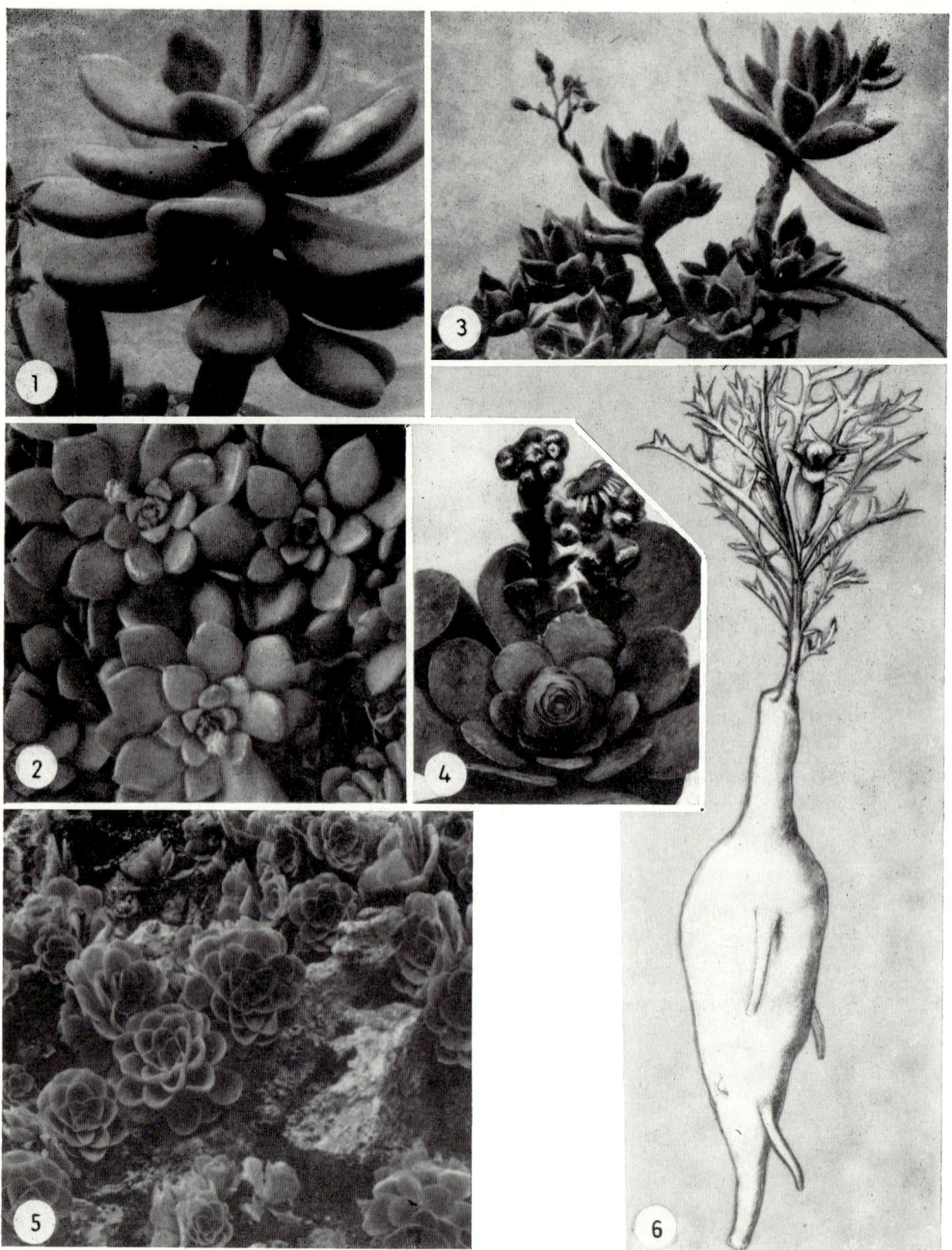

Plate 89. 1. **Graptopetalum amethystinum** (Rose) E. Walth.; **2. G. rusbyi** (Greene) Rose (Photo: R. Moran); **3. × Graptoveria** cv. Calva (Repr.: Succulenta, 1964, 166, Photo: van Arkel); **4. Greenovia dodrantalis** (Willd.) Webb et Berth. (Photo: E. Hahn); **5. G. aurea** (Sm.) Webb et Berth. (Photo: J. Vatrican); **6. Pterodiscus speciosus** Hook. (Repr.: Bot. Jahrb. **X,** 1889, Pl. VII (as *Harpagophytum pinnatifidum* Engl.).

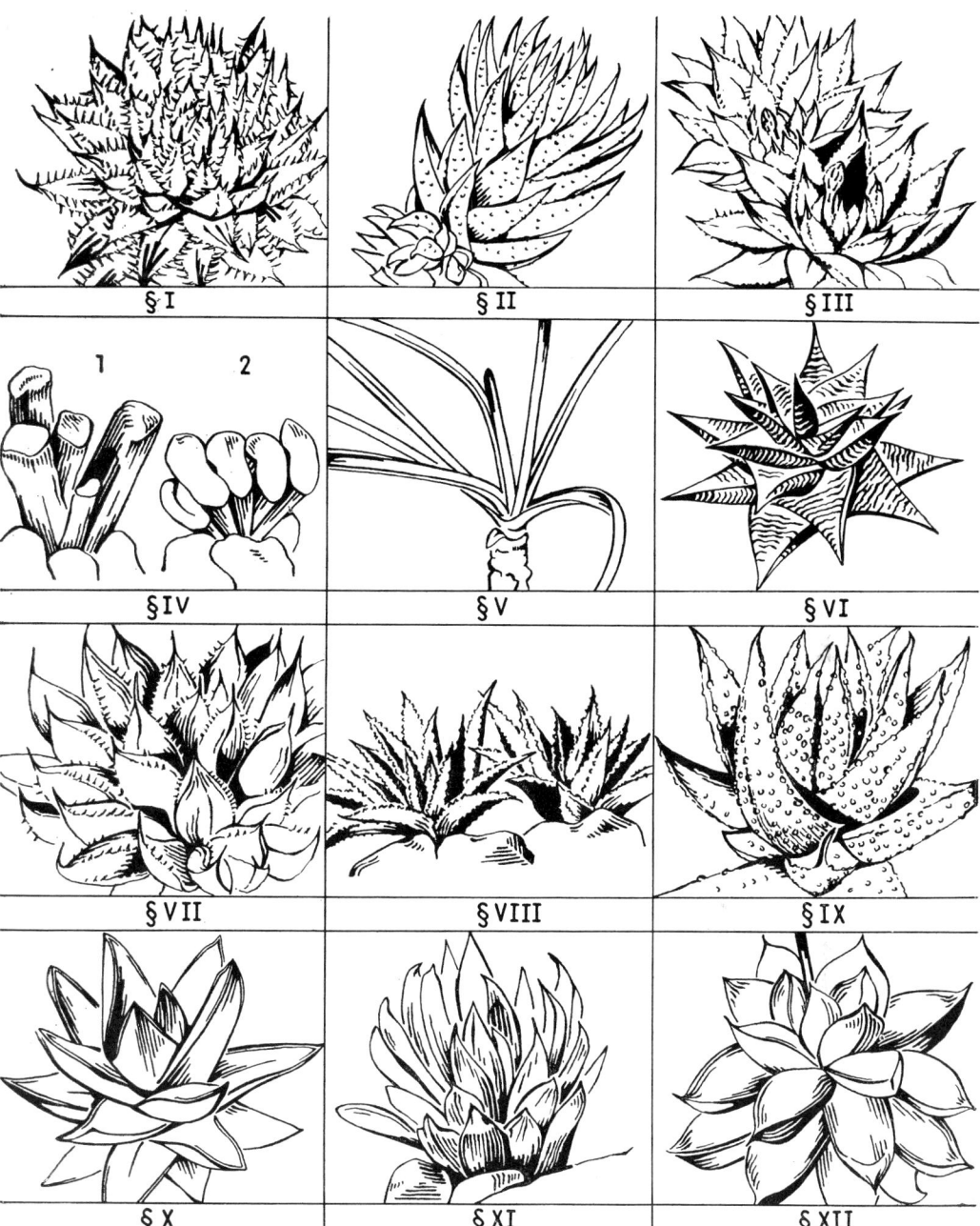

Plate 90. Typical species of the Sections I–XII of the Genus Haworthia (see pages 249, 250).
§ I. Haworthia setata v. major Haw.; § II. H. coarctata (Salm) Haw. v. coarctata; § III. H. altilinea v. denticulata (Haw.) v. Poelln.; § IV.1. H. maughanii v. Poelln. 2. H. truncata Schoenl.; § V. H. blackburniae Barker; § VI. H. limifolia Marl. v. limifolia; § VII. H. cooperi Bak.; § VIII. H. angustifolia Haw. v. angustifolia; § IX. H. margaritifera (L.) Haw. v. margaritifera; § X. H. marginata v. virescens (Haw.) Uitew.; § XI. H. incurvula v. Poelln.; § XII. H. cymbiformis v. obesa v. Poelln.

Plate 91. Typical species of the Sections XIII-XX of the Genus Haworthia (see page 250).
§ XIII. Haworthia planifolia Haw. v. planifolia; § XIV. H. retusa (L.) Haw.; § XV. H. rigida (Lam.) Haw.; § XVI. H. sordida Haw.; § XVII. H. subregularis Bak.; § XVIII. H. tessellata Haw. v. tessellata; § XIX. H. tortuosa Haw.; § XX. H. viscosa v. concinna (Haw.) Bak.

Plate 92. 1. Haworthia Infl.; 2. H. attenuata v. clariperla (Haw.) Bak. (Photo: J. Brown); 3. H. asperiuscula v. Poelln. v. asperiuscula; 4. H. blackbeardiana v. Poelln. v. blackbeardiana; 5. H. bilineata v. affinis (Bak.) v. Poelln. (Photo: as 2); 6. H. cassytha Bak. (Photo: as 2); 7. H. bolusii v. semiviva v. Poelln. (Photo: as 2); 8. H. aegrota v. Poelln. (Photo: as 2); 9. H. caespitosa f. caespitosa (Photo: J. Marnier-Lapostolle).

Plate 93. 1. Haworthia cuspidata HAW. (Photo: E. HAHN); 2. H. fasciata (WILLD.) HAW. v. f. (Photo: as 1); 3. H. cymbiformis v. translucens TRIEBN. et v. POELLN.); 4. H. glabrata v. concolor SALM (Photo: J. BROWN); 5. H. cordifolia HAW.; 6. H. ferox v. POELLN. v. ferox (Photo: as 4); 7. H. gracilis v. POELLN. (Photo: as 4); 8. H. glauca BAK.

Plate 94. 1. Haworthia greenii v. minor Res. (Photo: J. Brown); 2. H. haageana v. Poelln. v. haageana; 3. H. helmae v. Poelln. (Photo: as 1); 4. H. hybrida (Salm) Haw. (Photo: as 1); 5. H. herbacea (Mill.) Stearn (Photo: J. A. Huber); 6. H. isabellae v. Poelln. (Photo: as 1); 7. H. heidelbergensis G. G. Smith (Photo: as 1); 8. H. herrei v. Poelln. v. herrei.

Plate 95. 1. Haworthia lockwoodii ARCH. (Photo: J. BROWN); 2. H. margaritifera v. maxima (HAW.) UITEW. sv. maxima (Photo: J. MARNIER-LAPOSTOLLE); 3. H. marginata (LAM.) STEARN v. marginata (Photo: K. POLLARD); 4. H. mucronata HAW. v. mucronata; 5. H. nigra (HAW.) BAK. v. nigra (Photo: as 1); 6. H. nitidula v. POELLN. v. nitidula.

Plate 96. **1.** Haworthia papillosa (Salm) Haw. v. papillosa (Photo: J. Brown); **2.** H. otzenii G. G. Smith (Photo: as 1); **3. H. obtusa** v. pilifera f. truncata Jacobs.; **4. H. planifolia** v. setulifera v. Poelln.; **5.** H. pallida Haw. v. pallida (Photo: as 1); **6.** H. pearsonii C. H. Wright (Photo: as 1).

Plate 97. 1. Haworthia reinwardtii (S. D.) Haw. v. reinwardtii (Photo: E. Hahn); 2. H. setata v. bijliana v. Poelln. (Photo: J. Brown); 3. H. ryderana v. Poelln. (Photo: as 2); 4. H. radula (Jacq.) Haw.; 5. H. reticulata Haw. v. reticulata; 6. H. subfasciata (Salm) Bak. v. subfasciata (Photo: as 2).

distinct longitudinal R., Th. in 2 pairs, the upper Th. smaller than the lower ones; **Cy.** distinctly pedunculate, golden yellow.

**E. coerulescens** HAW. (Pl. 69/1) (Gr. 24) (*E. virosa* BOISS. p. part., *E. v.* v. *c.* BGR.). – Cape: Jansenville, Steytlerville, Uitenhage D. – Thorny, succulent ♄, spreading below ground and with numerous erect St. up to 1.5 m tall, forming broad bushes; **St.** 3–4 cm thick, often branching above, segments round, 4–6-angled with very concave sides, grey-blue, angles sinuate-tuberculate with a brownish horny band. Th. in pairs, 6–12 mm long, stiff, dark brown.

**E. colliculina** WHITE, DYER et SLOANE (Gr. 14). – Cape: Oudtshoorn D. – The conical or subcylindrical main root and the main St. forming a **caudex** almost completely buried in the soil; **Br.** rosulate around a depressed, tuberculate, branchless, central surface, Br. 3–12 cm long, 4–5 mm thick, cylindrical, tuberculate, glabrous, tubercles c. 5 mm long and across, oblong or hexagonal; **L.** deciduous, c. 1.5 mm long, grooved; **Ped.** 1–2 mm long, persisting.

**E. columnaris** BALLY (Pl. 63/25b) (Gr. 25b). – N.Somalia. – **St.** solitary, erect, columnar, unbranching, with 13–16 ribs, up to 1.3 m tall, up to 8 cm ∅, deeply pitted between the ribs which are often dentate, Th.shi. often forming a continuous border, Th. in pairs, 11–18 mm long; **L.** scale-like, 1.5 mm long, deciduous; **Cy.** very numerous.

**E. commersonii** (H. BN.) M. DENIS (Gr. 2) (*E. pyrifolia* LAM. ex H. BN. p. part., *E. spathulata* COMMERSON nom. nud.). – Madag.: coastal zone. – Small ♄ related to **E. leuconeura**, with succulent Br., the shoots thickened at the tip.

**E. complexa** R. A. DYER (Gr. 24). – Transvaal. – Succulent thorny ♄ c. 40 cm tall; **caudex** much shortened, branching from the base; **Br.** simple or branching above, usually 1 cm thick, usually with 4 angles and depressed sides, angles sinuate-tuberculate, Th.shi. solitary, Th. in pairs, 4–5 mm long, with 2 lateral Th. 1.5–2 mm long.

**E. confertiflora** VOLK. (Gr. 24). – Tanzania. – Thorny succulent ♄ 1 m tall; **Br.** erect with 4 alate angles 2 cm high, Th.shi. 1 cm apart, Th. spreading, 5 mm long. Insufficiently known species.

**E. confinalis** R. A. DYER (Gr. 24). – Transvaal; Moçambique. – ♄ up to 8 m tall, erect, unbranched or with 1–2 Br., with a crown of curved, 3–6-angled St. constricted to form segments 5–20 cm long, sides ± parallel, angles c. 5 mm thick, 3–3.5 cm tall, Th.shi. 1–2 cm apart, Th. in pairs, 0.5–8 mm long, on a continuous horny base; **Infl.** each of 3 Cy.

**E. –** ssp. **confinalis.** – St. at first with (3–)4(–5) angles, simple or with 1–2 Br.

**E. –** ssp. **rhodesica** LEACH. – Rhodesia. – **St.** at first with 5–6 angles, with 1 or more Br., each with a crown of 4–5-angled twigs with stout Th.

**E. confluens** NEL (Gr. 14). – Cape: Lit. Namaqualand. – **Caudex** obconical, partly hidden in the soil, the upper half completely covered with spreading shoots with lateral Br., 17 to 18 cm long, 2–2.5 cm thick, cylindrical, with very prominent hexagonal podaria 8–10 mm across, with white L.scars; **L.** on Br. ends, 15–20 mm long, 2 mm across, eventually falling; **Ped.** up to 35 mm long, persisting for a long time.

**E. congestiflora** LEACH (Gr. 4). – Angola: Moçamedes D. – Semi-succulent, virgate ♄ 1.2–2 m tall, sparsely branching; **Br.** somewhat whip-like, rather long, ± straight, 3–3.5 mm thick, light to dark brown, with numerous, minute, white blotches, minutely striate and waxy, St. and short shoots alternate or sometimes from the L.axils, 1.5–2 mm thick; L.scars transverse, somewhat lunate, often with 2 minute lateral glands; **Infl.** terminal, a sessile or sub-sessile corymb or cluster, Cy. fairly crowded, Ped. with 2 hemispherical bracts, often bifid.

**E. conspicua** N. E. BR. (Gr. 24) (*E. candelabrum* WELW.). – Port. Guinea. – Succulent, leafless, thorny ♄ c. 4.5 m tall; **St.** 30–70 cm tall; **Br.** and shoots in whorls, spreading radially, ascending, curving, with 3–8 angles, tapering towards the constrictions, shoots mostly 3-angled, the sides 2.5–3.75 cm across, the angles alate, sinuate-dentate, T. 1–3 cm apart; **Cy.** 1–3, calyculus 5–6 mm ∅, red.

**E. contorta** LEACH (Gr. 24). – Moçambique. – Thorny succulent ♄ c. 1 m tall; **Br.** spreading, sometimes prostrate and climbing, spirally contorted and curving, (4–)7(–9)-angled, sometimes constricted to form segments up to 30 cm long, 2.5–3 cm ∅, the sides between the angles deeply pitted, angles sinuate-tuberculate, main Th. in pairs on the tips of the tubercles, stout, 3 to 10 mm long, red, becoming grey, 8–20 mm apart, subsidiary Th. 3 mm long, Th.shi. not confluent; **L.** sessile, round, pointed, 8–22 mm long, 7–12 mm across; **Cy.** 3.

**E. cooperi** N. E. BR. (Gr. 24) (*E. angularis* sensu ENGLER, EYLES, WHITE, DYER et SLOANE, *E. grandidens* sensu BURTT DAVY). – E.Afr. – ♄ 3–5 m or more in height; **St.** cylindrical, 15–20 cm thick, much branching above; **Br.** curving-ascending, 4–6-angled, divided by conspicuous constrictions into ovate-cuneate or cordate segments 5–15 cm long, 3.7–7.5 cm ∅, smooth, angles alate, separated by a groove 2–3.7 cm deep, sides green with numerous dark transverse lines, margins grey and horny, Th. in pairs, 6–18 mm apart, 3–10 mm long, projecting, grey, black-tipped.

**E. –** v. **calidicola** LEACH (*E. spec.* 1 WHITE, *E. angularis* sensu BOUGHEY). – Rhodesia; Zambia; Moçambique. – Segments very variable in shape, wings narrower, margins 3 mm thick, ± sinuate-dentate or crenate, often variously curved and wavy, Th. very variable, usually weak and often missing, less often stout as in **E. grandicornis.**

**E. –** v. **cooperi** (Pl. 63/24; 68/2). – Natal; Swaziland; Transvaal; Rhodesia; S.Moçam-

14 Lexicon of Succulent Plants

bique. – **Br.** 4–6-angled, segments $\pm$ conical-ovate, wings fairly stout, angles $\pm$ straight or slightly sinuate-dentate, with a continuous horny band c. 5–6 mm across, Th. rather stout, usually less than 10 mm long.

**E. —** v. **ussanguensis** (N. E. BR.) LEACH (*E. u.* N. E. BR., *E. nyikae* PAX, *E. angularis* sensu PAX p. part., *E. strangulata* sensu HUTCHINSON, *E. bussei* PAX). – Tanzania; Zambia; Malawi. – **Br.** segments normally more regular in shape, usually broad or almost circular in outline, angles alate and stout, the horny band broader (up to 10 mm), wings often more numerous.

**E. corymbosa** N. E. BR. (Gr. 4). – Cape: Riversdale D. – Thornless $\hbar$ 30 cm or more in height, branching basally or higher up; **Br.** 4–5 mm thick, twigs opposite or alternate, terete, glabrous.

**E. crassipes** MARL. (Gr. 15). – Cape: Karroo. – Main St. and thickened main root forming a spherical to cylindrical **caudex** 10–15 cm long, almost as thick, usually completely hidden in the soil, flat above, bearing a Ros. of erect Br., 15–30 cm $\varnothing$; **Br.** 4–6 cm long, 1–1.5 cm thick, cylindrical, densely covered with small tubercles **Ped.** becoming thorny.

**E. crispa** (HAW.) SWEET (Gr. 8) (*Tithymalus c.* HAW., *E. elliptica* v. *undulata* BOISS.). – Cape: Karroo. – Conical taproot and St. forming an elliptical or spherical **caudex,** with 2 or more cylindrical St.-like Br. above; **L.** with a petiole 12–15 mm long, blade 12–50 mm long, 5–10 mm across, lanceolate-elliptical, tapering, plicate, smooth or minutely velvety-hairy, margins curly-wavy.

**E. croizatii** J. LEANDRI (Gr. M. III). – S.W. Madag. – Thorny $\hbar$ 50–75 cm tall; **Br.** projecting, somewhat fleshy, Th. in 5 longitudinal rows, pointed, arranged along the angles, 1 cm long, normally in 3's, the upper Th. larger, bark brown-red to brown-grey; **L.** obovate to almost circular, 8–10 mm long, 5–6 mm across, grey-hairy.

**E. cryptospinosa** BALLY (Gr. 25a). – Kenya; Somalia. – Climbing $\hbar$ up to 2 m tall; **St.** cylindrical, somewhat branching, terete, with 5–7 longitudinal furrows; **Br.** alternate, opposite or clustered, 15–25 cm long, c. 3 mm thick, furrowed, green, Th. in pairs, up to 1.5 mm long, often missing; **L.** triangular, up to 1.5 mm long, deciduous; **Cy.** in dichotomously-branched Infl. in concentric rings on young shoots, light red or sometimes cream-coloured.

**E. cucumerina** WILLD. (Gr. 19). – Cape: Lit. Namaqualand. – Known only from the Lit., classification in Gr. 19 is uncertain.

**E. cumulata** R. A. DYER (Gr. 19). – Cape: Albany D. – **St.** numerous from underground stolons, erect, unbranched, 2–3 cm thick, up to 30 cm tall, 7–9-angled, smooth, grey, angles 2–4 cm high, separated by sharp furrows; thorny **Ped.** solitary, spreading, 2–4 cm long, grey, persisting.

**E. curvirama** R. A. DYER (Gr. 24). – Cape: Karroo. – Thorny succulent $\hbar$ 5–6.5 m tall; **St.** cylindrical, 20–30 cm thick, simple or with 1–3 Br. above, each producing a round crown of mostly curving and then projecting Br.; **Br.** 1.3–2 m long, 5–7.5 cm thick, with 3–5 angles, with deep constrictions forming segments 5 to 15 cm long with curving outlines, angles alate, 3–4 cm tall, margins straight or sinuate-tuberculate, Th.shi. confluent into a broad horny border, Th. in pairs, 9–18 mm apart, 5–15 mm long.

**E. cussonioides** BALLY (Gr. 24). – Kenya. – Fleshy $\hbar$ 20–25 m tall; **trunk** erect, 12 m tall; main **Br.** few, spiralled, round, sparsely branched with numerous 3–4-angled, fleshy short shoots, with segments up to 20 cm long, rebranching from the constrictions, 3–4-angled, the angles much compressed, 2–3.5 cm across, margins wavy or blunt-dentate, Th.shi. 5 mm long, Th. in pairs, thin, up to 1.5 mm long, often missing; **L.** 9.5 mm long, 10 mm across, fleshy, soon falling.

**E. cylindrica** WHITE, DYER et SLOANE (Gr. 9). – Cape: Calvinia D. – **Caudex** simple or with 1–3 small Br., 4–5 cm $\varnothing$, cylindrical or usually clavate-cylindrical, with tubercles in 9–10 spiral R., conical, 5 mm long, tips projecting horizontally; **L.** on the Br.ends, elliptical-oblong, 3–6 cm long, 5–9 mm across, margins wavy, narrowing to the 1 cm long petiole.

**E. cylindrifolia** J. MARN.-LAP. et RAUH (Gr. M. VII). – S.E.Madag. – Forming loose mounds; underground stolons c. 5 mm thick, freely branching, covered with whitish-brown scale-L., or with an underground Tub.; **L.** at first in a Ros. lying on the soil, later 10–25, crowded at the tips of shoots which become 15 cm long and 5 mm thick, the lower St. then leafless, with L.scars surrounded by the raised cushions of the deciduous Stip., L. sessile or short-petiolate, succulent, cylindrical, (2–)2.5(–3) cm long, 3–5 mm thick, apiculate, grey-green to reddish-violet, rough papillose, the upperside narrowly to broadly grooved; the Stip. reduced to numerous corky, whitish cilia with entire or dentate margins, 1–2 mm long, later falling, their succulent bases forming an annular swelling; **Infl.** solitary or several, with Ped. 1–3 cm long.

**E. —** ssp. **cylindrifolia** (Pl. 69/2). – Rooting from long, freely branching, yellowish-brown stolons to form large mats.

**E. —** ssp. **tuberifera** RAUH (Pl. 69/2). – Without stolons, with an underground caudex 5–10 cm long, flattened above and passing over below into the freely branching main root; **shoots** numerous, 5–15 cm long, not rooting. Plant always solitary.

**E. darbandensis** N. E. BR. (Pl. 64/26) (Gr. 26). – W.Afr. – **St.** cylindrical, 4–5 cm $\varnothing$, Br. where present produced basally, 15–60 cm tall, smooth below, with transverse furrows, with rhombic tubercles above in spiral R., with a Th.shi. and 1 flattened Th. 12–15 mm long; **L.** few on the Br.tips, ovate, fleshy, c. 4 mm long, 1.5 mm across, deciduous; **Infl.** solitary from the axils of the L.scars, 25 mm long, with 3 Cy.

**E. davyi** N. E. BR. (Gr. 14). – Transvaal; Botswana. – **Caudex** hemispherical, long-obconical or obovate, arising from the thick main root, mostly below soil-level, 5.5 cm or more in ⌀, truncate above, covered with large tubercles; the central zone bearing a crown of Br. in 2–3 circular R., tubercles rhombic, 10–18 mm long, 6–12 mm across, 6–9 mm thick, Br. 5–15 cm long, 12–20 mm thick, cylindrical, tuberculate.

**E. dawei** N. E. BR. (Gr. 23). – Uganda. – Small ♄ 6–9 m tall, almost leafless, thorny, the loss of the lower Br. giving the plant a palm-like appearance; **Br.** succulent, flat and thin, also 3-winged, constricted to form oblanceolate-oblong segments 10–20 cm long, 4–6 cm across, with side-Br. narrowing teretely below, the angles sinuate-dentate, bluish; **L.** rudimentary, 3–4 mm long, soon falling, Th. in pairs, 3–4 mm long, 8–12 mm apart, Th.shi. usually solitary, but often confluent with the horny border.

**E. decaryi** A. GUILL. (Pl. 68/4) (Gr. M. VII). – S.E.Madag. – Subterranean stolons long, 5 mm thick, whitish to whitish-brown, covered with scales; **St.** curving or partly prostrate, up to 12 cm long, 8–12 mm thick, slightly angular; **L.** in Ros. at the St.tips, arranged in 5 straight or slightly spiral R., petiole 2–5 mm long, semiterete, upperside grooved, reddish, lamina succulent, lanceolate to oval, (3–)5(–7) cm long, c. 1 cm across, margins very wavy, upperside silver-grey to green or reddish, underside silver-grey, both surfaces rough papillose, Stip. reduced to several T. with a succulent tubercular base sometimes terminating in a fringed tip, becoming corky, light grey, later falling, eventually uniting with the alate angles of the St.axis; **Infl.** with Ped. 1.5 cm long, with 2 Cy., cyathophylls broadly triangular, greenish-yellow with a red border.

**E. decepta** N. E. BR. (Gr. 15) (*E. caput-medusae* E. MEY.). – Cape: Karroo. – **Caudex** spherical, partly below soil-level, with numerous projecting and spreading Br. produced from the sides and above, the central part of the tip being without Br., 6–10 cm thick, covered with impressed lines forming flat zones 1.2 cm ⌀, and in the centre of the zones are small, compressed-conical olive-green, often brownish tubercles; **Br.** 12–37 mm long, 7–8 mm thick, cylindrical, with rhombic tubercles, 2–5 mm long and across and slightly raised; **L.** deciduous, small; **Ped.** numerous, persisting, 4–15 mm long.

**E. decidua** BALLY et LEACH (Gr. 22). – Angola; Rhodsia; Malawi; Zambia. – **Caudex** tuberous; **St.** producing from its upper surface a cluster of **Br.** which are simple or rarely dichotomously dividing, erect, up to 12 cm tall, up to 6 mm thick, triangular, deciduous, the T. along the angles up to 2.5 mm high, 2–10 mm apart, with a round Th.shi. above, Th. in pairs, 1.5–4.5 mm long, stipular Th. 2, recurved; **L.** up to 3 mm long, pointed; **Infl.** borne along the angles after the fall of the Br. from the subterranean stolons, Infl. up to 3 cm tall.

**E. decliviticola** LEACH (Gr. 24). – Moçambique. – Small ♄ or tree-like ♄, up to ± 3 m tall, trunk stout, eventually multi-angled, simple, clearly marked with rings of scars of the fallen Br., crowned with a head of crowded, verticillately arranged, mostly simple, winged, spiny Br.; **Br.** spreading, arcuate ascending, ± 60 cm long, constricted into ± 2.5–15 cm long segments, generally ± 6 cm thick with 4 to 6 wing-like angles, wings 2–2.5 cm broad with a continuous, whitish, horny margin 1.5–3 mm wide, Th. in divergent pairs, ± horizontally spreading, ± 6 mm long; **L.** fleshy, caducous, up to 7 mm long, with inconspicuous flanking prickles; **Infl.** cymose, with 1–3 shortly pedunculate cymes, each bearing 3 Cy.

**E. decussata** E. MEY. (Gr. 4) (*E. tirucalli* THUNBG. p. part., *E. indecora* N. E. BR.). – Cape; S.W.Afr. – ♄ 70–100 cm tall, branching basally, re-branching freely above, old succulent St. with a thorny tip; **Br.** 6–8 mm thick, twigs 2–5 mm thick, stiff.

**E. deightonii** CROIZ. (Gr. 24). – Sierra Leone: Mario D.; S.Ghana; Nigeria. – Freely branching ♄ or ♄ up to 6 m tall, c. 12 m ⌀; **Br.** triangular, constricted-articulate, 4.5 cm ⌀, angles alate, sinuate-dentate, T. 2.5 cm high, indentations 3–5 cm long, angles 2–3 mm thick, Th. in pairs, projecting horizontally; **L.** obovate, fleshy, 3 mm long, 2.5 mm across; **Infl.** with campanulate Cy.

**E. dekindtii** PAX emend. LEACH et CANNELL (Gr. 22) (*E. polyacantha* sensu HIERN p. part., *E. fraterna* N. E. BR. p. part.). – Angola. – Succulent dwarf ♄; **caudex** semi-subterranean, often forming a very short, gnarled Bo. above soil-level; **Br.** simple, segmented, with (3–)6(–7) wings, up to 20 cm long, segments ± circular or elliptical, 2–4 cm long, up to 4.5 cm ⌀, border continuous, crenate-dentate, horny, grey, Th. in pairs at the tips of the projecting wing-sections, 1.5–10.5 mm long; **L.** soon falling, leaving a distinct scar, often with 2 fleshy lateral stipular Th.; **Infl.** with short Ped., normally with 1 bisexual Cy.

**E. delphinensis** URSCH et LEANDRI (Gr. 20; M. VI). – Madag. – Thorny, freely branching ♄; **Br.** 8 mm thick, Th. 10–18 mm long, 2–10 mm apart, short shoots with 1–3 almost leathery L. 2 cm long, 12 mm across, ovate, with slightly wavy margins; **Infl.** 10–30 mm long, Cy. 5–8.

**E. dendroides** L. (Gr. 1) (*E. divaricata* JACQ.). – Medit. area. – ♄ or small ♄ 2–3 m tall, forked or tripartitely branched; **Br.** reddish-brown, pencil-thick; **L.** sessile, linear-lanceolate, up to 5 cm long, 7 mm across; **Infl.** a 3–10-partite umbel with yellowish bracts.

**E. denisiana** GUILL. (Gr. 21; M. VI). – Madag. – **St.** erect, 10–15 cm tall, c. 2 cm ⌀, tapering above and below, somewhat 8-angled, the angles almost imperceptible, somewhat tuberculate, Th. probably rudimentary.

**E. desmondii** KEAY et MILNE-REDHEAD (Gr. 21). – Nigeria; Cameroon. – Tree-like, erectly branching ♄ 4–6 m tall; **St.** round, c. 10 cm ⌀; **Br.** 3–4-angled, 1.5–3.5 cm ⌀, angles parallel and somewhat wavy, segmented, the Th. on the angles robust, red-brown, 3–7 mm long, with Shi. 18 mm long; **L.** deciduous, arranged on the Br.tips, leathery-fleshy, spatulate-obcordate, c. 12 cm long, 6 cm across, pale green.

**E. didieroides** M. DENIS et LEANDRI (Pl. 69/3) (Gr. M. III). – Central Madag. – ♄ up to 2.5 m tall, less often a single-stemmed, small ♄, freely branching from the base; **Br.** becoming very woody, up to 15 cm ⌀ below, 2–3 cm ⌀ above, with or without short shoots all along; **L.** periodically deciduous, few, rosulate at the Br.tips, short-petiolate, oblong-oval, 2.5–4 cm long, 1–1.5 cm across, folded inwards above, grey-green with a red border, the underside hairy, the short shoots with 3–5 smaller L., stipular Th. numerous, Th. up to 5 on prominent L.bases, stout, hard, sharp, up to 1 cm long; **Infl.** with repeated dichotomous branching, Ped. 5–15 cm long, wine-red, with (15–)32(–100) Cy., cyathophylls yellowish-green to orange-coloured.

**E. disclusa** N. E. BR. (Gr. 24). – Ethiopia. – ♄; **St.** 5-angled; **Br.** 4-angled, not very conspicuously segmented, 10 cm ⌀, angles flat-alate, dentate-lobed at 2 cm intervals, Th. on cordate Shi. 4–8 mm long, black-brown.

**E. × doinetiana** A. GUILL. (Pl. 69/4) (Gr. 24). – Hybrid: **E. pseudocactus × E. franckiana.** Hort. M. J. DOINET, Glain-lez-Liège, Belgium. – **Br.** spreading-erect, somewhat contorted, dark green, with silver-white angular mottling in 2 R. on the surfaces, Th. 5 mm long, light reddish-brown.

**E. × — nm. lyttoniana** FRICK. – Mottling not angular, but circular around the Th.shi., Th. undeveloped.

**E. dregeana** E. MEY. (Pl. 61/4) (Gr. 4) (*E. elastica* MARL.). – Cape: Lit. Namaqualand; S.W.Afr.: Gr. Namaqualand. – Thornless succulent ♄ 1–2 m tall, several m across; **trunk** 2.5–5 cm thick at the base, erect, other Br. 8–12 mm thick, cylindrical, with convex L.scars, younger Br. minutely velvety, becoming smooth, whitish-green.

**E. duranii** URSCH et LEANDRI v. **duranii** (Gr. 20; M. VI). – Madag. – Thorny ♄ 20 to 40 cm tall; **St.** 8-angled, branching; **Br.** 1.5 to 2 cm ⌀, 8–10 mm ⌀ below, Th. simple, 10 to 16 mm long; **L.** arranged on the Br.tips, glabrous, 3–5 cm long, 14–35 mm across, ovate-lanceolate, apiculate, with petioles 2–3 mm long; **Infl.** with 2–4 Cy. Acc. W. RAUH very closely related to **E. fianarantsoae** and possibly identical.

**E. — v. ankaratrae** URSCH et LEANDRI (Pl. 69/5). – Smaller variety; **St.** 12 mm ⌀, Th. 5–8 mm long; **L.** 3.5 cm long, 13–15 mm across.

**E. duseimata** R. A. DYER (Gr. 14) (? *E. fusca* PHILL.). – Botswana; Cape: Vryburg D. – **Caudex** with 1–2 underground St., each with 6–9 Br. from the St.sections above-ground, ± cylindrical, 1.2–2.5 cm thick; **St.** up to 7 cm long, 5–20 mm thick; **Br.** cylindrical, tuberculate, 4–6 cm long, 2.5–4 mm thick, periodically dying off, tubercles projecting c. 1 mm; **L.** deciduous.

**E. echinus** HOOK. f. et COSS. v. **echinus** (Pl. 70/1) (Gr. 24). – S.Morocco. – Variously shaped ♄ 10–100 cm tall, forming dense, hemispherical clumps; **St.** numerous, crowded, 4 to 5 cm thick; **Br.** ascending, with (5–)8–11(–13) angles, the sides not much furrowed, dull light green, the angles scarcely dentate, Th. in pairs, 5–7 mm apart, 5–20 mm long, reddish, becoming light grey, on confluent Th.shi.; **Cy.** brown-red.

**E. — v. brevispina** HOOK. f. et COSS. – Th. shorter.

**E. — v. chlorantha** MAIRE. – **Cy.** green.

**E. — v. cristata** J. GATTEFOSSE. – **St.** with terminal crests.

**E. — v. hernandez-pachecoi** (CABALLERO) MAIRE (E. h.-p. CABALLERO). – No diagnosis available.

**E. ecklonii** (KLOTZSCH et GARCKE) HÄSSL. (Gr. 8) (*Tithymalus e.* KLOTZSCH et GARCKE, *E. pistiaefolia* BOISS.). – Cape. – Resembles **E. tuberosa; L.** lying on the soil-surface, forming a Ros. on the shortened main St.

**E. eduardoi** LEACH (Gr. 24). – Angola: Moçamedes D.; S.W.Afr.: Kaokoveld. –Thorny, fleshy, candelabra-shaped ♄ up to 10 m tall; **St.** cylindrical or 5–6-angled, unbranched, with a fairly small crown of ± curving-ascending or ascending-erect Br. arranged ± in a whorl; **Br.** usually simple, slightly constricted into ± elliptical segments of 7.5 cm ⌀ and variable length, blunt-tipped and with 4–5 (usually 5) angles, slightly wavy and alate only on younger growth, the wings then 2–2.5 cm across, 8 mm thick, older growth with ± flat to slightly concave sides, Th.shi. reddish-brown, becoming corky and grey, c. 12 mm long, 8 mm across, at first solitary, later usually confluent, Th. in pairs, 10–15 mm long, straight or slightly curved; **Infl.** of 3 Cy.

**E. ellenbeckii** PAX (Gr. 28). – Ethiopia. – Thorny succulent plant, with **Br.** and shoots 5–8 cm long, 1 cm thick, probably cylindrical, with Th.shi. not confluent in 6 spiral R., on very prominent tubercles, Th. 4, almost forming a transverse line, the 2 outer ones 2–3 mm long, the inner ones 6–18 mm long.

**E. enopla** BOISS. v. **enopla** (Gr. 19). – Cape: Karroo. – Thorny succulent ♄ 30–100 cm tall, with many shoots from the base, branching above; **Br.** 12–22 cm long, 22–30 mm thick, with 6–7 angles broader than high, furrows acute, straight or slightly wavy, margins straight or slightly crenate, with 18–20 mm long, orange-red, thorny Ped. 5–6 mm apart.

**E. — v. viridis** WHITE, DYER et SLOANE. – Th. yellowish-green.

**E. enormis** N. E. BR. (Gr. 22). – Transvaal. – Main St. arising from the thickened root to form a conical or napiform **caudex** 7.5–10 cm ⌀, entirely below soil-level, dividing to form several underground Br., each with 10–15 branchlets above; **branchlets** almost erect, simple or re-branching several times, 15–20 cm long, up to 2.5 cm thick, the shorter ones often clavate, the others segmented, having 4 (or 3) compressed and irregularly dentate angles, green with whitish markings, tubercles 4–10 mm apart, 2–6 mm high, Th.shi. solitary, Th. in pairs on the tips of the tubercles, 4–8 mm long, with a pair of small Th. above them; **L.** small, deciduous.

**E. ephedroides** E. MEY. (Gr. 4). – Cape: Lit. Namaqualand. – Thornless ♄ 40 cm tall, branching from the base or higher; **Br.** 4 mm or more thick, twigs opposite, often tangled, 2–7 cm long, c. 2 mm thick, the tips of Br. and twigs usually drying up.

**E. epiphylloides** KURZ (Gr. 2). – India: Andaman Is. – ♄ with few side-shoots; **St.** flattened; **L.** ovate, 16 mm long, 10–12 mm across, short-petiolate, stout, with 2 short fleshy Stip. laterally on the petiole.

**E. erlangeri** PAX (Gr. 24). – Ethiopia. – Leafless thorny ♄ 2–3 m tall; **Br.** opposite or alternate, not more than 3 mm thick, cylindrical or slightly angular (?), with 4–5 flat furrows, bark dividing into 4–5 rib-like stripes with scattered Th.pairs without Shi., Th. 2–3 mm long; **L.** rudimentary, minute.

**E. ernestii** N. E. BR. (Gr. 13). – Cape: Queenstown D. – The thick main St. and the cylindrical main root forming a **caudex** up to 6 cm thick; **St.** with numerous, short, usually radially-spreading Br., the inner ones 3–6 mm long and almost spherical, the outer ones up to 6 cm long, 6–8 mm thick, cylindrical, tuberculate, smooth, green, tubercles 3–4 mm ⌀, rhombic or hexagonal, with whitish L.bases.

**E. esculenta** MARL. (Pl. 70/3) (Gr. 14) (*E. inermis* v. *laniglans* N. E. BR.). – Cape: Karroo. – Root and main St. forming an obconical **caudex** 10–20 cm ⌀, of which only the tip projects above-ground; **Br.** thick, numerous, rosulate, tuberculate, 5–20 cm long, 1.5–2 cm thick, the inner Br. 4 cm thick, cylindrical, tubercles 3–12 mm long, 2–5 mm across, up to 4 mm high, oblong to hexagonal, with a small L.scar. Used as cattle-fodder.

**E. espinosa** PAX (Gr. 3) (*E. gynophora* PAX). – Tanzania; Rhodesia. – ♄ up to 3 m or more in height, branching from the base and higher up, bark brown; **Br.** alternate, ± verticillate, 2–4 mm thick, with small cushion-like Ped. at the nodes, but not thorny.

**E. eustacei** N. E. BR. (Gr. 10) (*E. hystrix* MARL.). – Cape: Laingsburg D. – Dwarf thorny succulent ♄; main **St.** with numerous crowded Br., forming dense hemispherical clumps 11 to 15 cm tall and up to 30 cm ⌀; **Br.** 6–11 cm long, 17–19 mm thick, cylindrical, tapering, divided by reticulate lines into hexagonal or rhombic zones 4 mm long and 4–6 mm across, glabrous, light green, the thorny Ped. solitary, 2–5 cm long, stiff, white; **L.** deciduous, 17–37 mm long, 4–9 mm across, plicate, very minutely papillose.

**E. evansii** PAX (Gr. 24). – Transvaal. – Thorny succulent ♄ up to 10 m tall; **St.** cylindrical or somewhat angular, with 1 or several erect thick Br. with spreading or projecting twigs forming a short crown; branchlets verticillate, c. 1.5 cm thick, pointed, sometimes segmented, 3–4-angled, the sides ± deeply grooved, the angles sinuate-dentate, tubercles 1.5–2 cm apart, with Th.shi., Th. in pairs, 6 cm long, projecting.

**E. excelsa** WHITE, DYER et SLOANE (Gr. 21). – Transvaal. – Thorny ♄ (7–)10(–15) m tall; trunk erect, with a crown of projecting-spreading St., up to 25 cm thick, dividing above into round **Br.** usually in whorls, often dividing, 1 m or more long, usually 4-angled, 2.5–3 cm thick, often with segments 8–15 cm long, shoots up to 2 cm thick, segments 5–8 cm long, sides mostly flat, bluish, angles with a long brown margin, Th. in pairs, 5–10 mm apart, c. 8 mm long. – Host-plant to the parasitic mistletoe **Viscum crassulae** ECKL. et ZEYH.

**E. fanshawei** LEACH (Gr. 22). – Zambia: Kawambwa D. – Perennial ♃ with turnip-shaped or ± oblate tuberous **root**, with succulent spiny Br. radiating from the apex of the much reduced subterranean St. which merges imperceptibly with the root; **Br.** usually numerous, simple, somewhat clavate, initially erectly spreading, up to 12.5 cm long, (4–)5(–6)-angled, with prominent fleshy tubercle T. up to 7 mm high along the angles, Th. in divergent pairs, up to 4 mm long, Th. shields subquadrate at the apex of the T., erect, up to 2.5 mm long; **Infl.** axillary, sometimes cymose or often comprising a shortly pedunculate, solitary, bisexual Cy.

**E. fasciculata** THUNBG. (Pl. 62/11) (Gr. 11). – Cape: Lit. Namaqualand. – **Caudex** thick, clavate, the part above-ground being ± fist-sized, plants 100 years old being up to 70 cm tall, the surface divided into hexagonal, projecting, spiralled zones with projecting L.cushions resembling stout Sp.; **L.** 4 cm long at the start of the growing season; Ped. transformed into very stout Th., 3 cm long, ± wavy and curving.

**E. ferox** MARL. (Pl. 70/2) (Gr. 19) (*E. captiosa* N. E. BR.). – Cape: Karroo. – Thorny succulent c. 15 cm tall, branching below soil-level, forming clumps of 20–60 cm ⌀; **Br.** 7.5–25 cm long, 2/3 hidden in the soil, 3–4.5 cm thick, 9–12-angled, light green, angles 3–5 mm high and rounded; Ped. along the angles, 12–30 mm long, 3–6 mm apart, stiff, becoming thorny, brown.

**E. fianarantsoae** URSCH et LEANDRI (Gr. 20, M. VI). – Madag. – Very free-branching, forming a compact spherical ♄; **St.** dense, branching or not, Br.tips all reaching the same height, bark corky, 1.5–2 cm thick, with 8 R. of hard thin Th. 5–15 mm long and projecting horizontally; **L.** in a terminal Ros., very short-petiolate, 2–3 cm long, 5–10 mm across, pointed, borders red; **Infl.** almost sessile, Cy. 2–4, cyathophylls pale yellow, becoming red.

**E. fiherensis** H. POISS. (Gr. 4). – ? Madag. – ♄ 6–10 m tall; **Br.** virgate, leafless.

**E. filiflora** MARL. (Pl. 71/3) (Gr. 15). – Cape: Lit. Namaqualand; S.W.Afr. – **Caudex** clavate, branching above, 7.5–10 cm thick, with tubercles in spiral R., St. tubercles conical, those on the Br. rhombic, projecting or slightly curved; Ped. numerous, ± curving, becoming ± thorny; **L.** on the Br.tips, linear, 2–3 cm long.

**E. fimbriata** SCOP. (Pl. 70/4) (Gr. 19) (*E. enneagona* HAW., *E. polygona* MARS. ex SCOP., *E. mammillaris* HAW. p. part., *E. scopoliana* STEUD., *E. cereiformis* K. SCHUM., *E. m.* v. *spinosior* BGR., *E. erosa* BGR., *E. latimammillaris* CROIZ. p.part., *E. platymammillaris* CROIZ.). –

Cape: Karroo. – Main **St.** erect and branching, 30 cm tall, often taller, or reduced with many shoots from near ground-level, St. and Br. 2–4 cm thick, cylindrical, often thickened, branching laterally and irregularly, ribs 7–12, flat, divided into hexagonal zones, each with a white L.scar, Ped. thorny arising from the transverse furrows in vertical R.

**E. flanaganii** N. E. BR. (Gr. 13). – Cape. – **Caudex** almost cylindrical or cylindrical-conical, rising 2.5 5 cm above the ground, 3.7–5 cm thick, with Br. in 2–3 R. in a crown around the tuberculate head; **Br.** almost erect, 1.2–3.7 cm long, sometimes longer, 5–6 mm thick, tuberculate, tubercles often like tiny T.; **L.** deciduous, 6–10 mm long, up to 1 mm across, linear-tapering.

**E. fortissima** LEACH (Gr. 24). – Rhodesia; Zambia. – Related to **E. cooperi** and can be considered as intermediate to **E. candelabrum**; Candelabra-like ♄ 5–7 m tall; **St.** 22–30 cm thick; **Br.** numerous, projecting-ascending, alate, simple or sometimes branching above midway, extending far above the St., with 3–4 wings, up to 5 m long, constricted to form broadly ovate or elliptical joints 4.5–9 cm across, 3–10 cm long, the angles with a horny margin 3–8 mm across, Th. in very divaricate pairs, (3–)7(–10) mm long, apical shoots 3–5-angled, usually very short; **L.** minute, ovate-pointed, soon falling; **Infl.** cymose.

**E. fortuita** WHITE, DYER et SLOANE (Gr. 14). – Cape: Ladismith D. – Main root tuberous, passing over into the main St. to form an almost cylindrical **caudex**, with the scars of numerous fallen Br.; main **St.** 5–16 cm thick; **Br.** rosulate, 3–13 cm long, up to 13 mm thick, cylindrical, tuberculate, glabrous, tubercles up to 7 mm long and more in width, fleshy, slightly raised, almost hexagonal, with a small L.scar; **L.** deciduous, 1 mm long, fleshy.

**E. fragiliramosa** LEACH (Gr. 4). – Angola: Moçamedes D. – Dwarf, thornless, semi-succulent ♄, forming cushions ± 40 cm ⌀; **Br.** and St. opposite, wide-spreading, bluish, terete, easily broken off at the constrictions and falling, especially the Infl.; **Infl.** terminal, repeatedly forking, the branchlets diminishing in size upwards.

**E. franckiana** BGR. (Gr. 21). – Origin unknown, probably Cape: Karroo. – Thorny ♄ 60–100 cm tall; **St.** erect, branched, 3–4-angled, segments 2.5–7.5 cm long, 2.5–3 cm thick, light green, becoming grey-green, angles acute, the sides flat in triangular shoots but deeply grooved in young 4-angled ones, angles ± sinuate-dentate, tubercles 6–12 mm apart, Th.shi. forming a ± continuous horny grey border, Th. in pairs, 4–8 mm long.

**E. francoisii** J. LEANDRI (Gr. M. VII). – S.E. Madag. – Related to **E. decaryi**; ? with underground stolons; L.Ros. at soil-level, L. on a petiole 5–10 mm long, blade succulent, oblong-linear, oval to rhombic, 2–6 cm long, 5–8 mm across, margins somewhat wavy, distinctly veined, Stip. reduced to numerous bristly, silver-grey T. around the L.base; **Cy.** (1–)2(–4), cyathophylls ± reniform, greenish-yellow.

**E. franksiae** N. E. BR. v. **franksiae** (Gr. 13). – Natal. – Thornless, succulent, dwarf ♄; **caudex** usually subterranean, forming a continuation of the thick main root which is mostly cylindrical or slightly conical, 2.5–3.5 cm thick, covered with closely-spaced, small tubercles 1.5 mm high; **Br.** in 2–3 R. at the tip of the St., 1.5 to 2.5 cm long, 5 mm thick, cylindrical, tuberculate, green, tubercles rhombic-oblong, c. 4 mm long, 2.5–3 mm across; **L.** deciduous, 2–6 mm long, oblong, grooved, glabrous.

**E. —** v. **zuluensis** WHITE, DYER et SLOANE. – **Caudex** contracted towards the apex, unbranched or with a few stout short Br.; involucral glands greenish.

**E. friedrichiae** DTR. (Gr. 15). – Southern S.W.Afr. – **Caudex** conical, 15–20 cm tall above soil-level, with dense, repeatedly dividing Br. above; **L.** up to 4 cm long, 2 cm across, strap-like, margins red, soft-dentate, those on the youngest shoots smaller; Ped. becoming thorny.

**E. frutescens** N. E. BR. (Pl. 61/3) (Gr. 5). – S.W.Afr.: Gr. Namaqualand. – Thornless ♄ up to 1.5 m tall, freely branching; **Br.** with projecting shoots 7.5–25 cm long, 2–2.7 mm thick, stiff, with numerous, opposite remains of old Ped.

**E. fruticosa** FORSK. (Pl. 71/2) (Gr. 25a). – S.Arabia: Yemen. – Thorny ♄ with a few Br. from the base, forming clumps 50 cm tall; **Br.** erect, scarcely segmented, cylindrical or somewhat clavate, grey-green, 5–7 mm ⌀, 10–13-angled, ribs narrowly triangular in cross-section, furrows acute, angles with closely spaced T., somewhat tuberculate, Th.shi. narrow-confluent, Th.pairs of unequal length, 15–20 mm long, projecting at right-angles.

**E. fusca** MARL. (Gr. 15) (*E. eendornensis* DTR.). – Cape: Karroo; S.W.Afr.: Gr. Namaqualand. – **Caudex** clavate, up to 30 cm tall, 20 cm ⌀, a continuation of the thick root, the tuberculate head producing a circle of dense, spreading Br.; **Br.** 2–15 cm long, 9–10 mm thick, cylindrical, smooth, tuberculate, tubercles rhombic-hexagonal with a whitish L.scar, little prominent, thornless or Ped. becoming thorny; **L.** rudimentary, short.

**E. gariepina** BOISS. (Pl. 71/4) (Gr. 6) (*E. bergeriana* DTR., ? *E. schaeferi* DTR., ? *E. halleri* DTR.). – Cape: Lit. Namaqualand to Karroo; S.W.Afr.: Gr. Namaqualand, Damaraland. – Thornless ♄ 15–65 cm tall, branching dichotomously from the base, with few Br. above, forming compact round clumps; **Br.** 1 cm ⌀, tapering above, with thornlike, projecting, ovate L.cushions 6–8 mm long, 4–5 mm across, tipped with a small Th. ♀ plants branch more than the ♂.

**E. gatbergensis** N. E. BR. (Gr. 13). – Cape: Transkei, W.Griqualand. – **Caudex** very short, conical, up to 10 cm thick, a continuation of the napiform main root, with a circle of Br. above; **Br.** 5 cm long, 1 cm thick, almost cylindrical, densely covered with raised tubercles, 3 mm ⌀, grey.

**E. genoudiana** URSCH et LEANDRI (Gr. 20; M. VI). – S.W.Madag. – Closely related to **E.**

**capuronii**, all parts being much less robust; **St.** simple (acc. W. RAUH branching in cultivation), up to 25 cm tall; **L.** in clusters of 4, linear, 4 cm long, 5–6 mm across; **Th.** 5–15 mm long, solitary, projecting; **Infl.** of 4–8 Cy., cyathophylls vivid green.

**E. gentilis** N. E. BR. (Gr. 4). – Cape: Van Rhynsdorp, Calvinia D. – Thornless ♄ 7.5 to 15 cm tall, branching and re-branching from the base and sometimes higher up; **twigs** opposite and alternate, densely crowded, 5–6 mm thick; **L.** very small.

**E. giumboensis** HAESSL. (Gr. 3). – Somalia. – Freely branching ♄ or ⧠; **Br.** 1 cm thick, terete, short shoots numerous, c. 2 mm thick, round to ± angular; **L.** clustered at the tips of the short shoots, on a petiole 1–5 mm long, lamina 2–2.5 cm long, 1.5–2.5 cm across, truncate-rounded at the tip, apiculate, falling at the flowering period; **Infl.** of 3 Cy.

**E. globosa** (HAW.) SIMS (Pl. 63/17; 71/5) (Gr. 17) (*Dactylanthes g.* HAW., *E. glomerata* HORT. ex BGR.). – Cape: Karroo. – Dwarf semi-♄ with a tuberous main root; **Br.** with spherical segments, lower ones 1.5–2.5 cm long, younger ones more ovate, up to 4 cm long, furrowed, L.cushions rather prominent and in spirals; Ped. forking, up to 10 cm long, woody, persisting for a long time.

**E. glochidiata** PAX (Gr. 27). – Kenya; Somalia; Ethiopia. – ♄ with spreading Br., up to 30 cm tall from a fleshy napiform caudex; **St.** 10–15 mm thick, light-grey, the broad, confluent L.bases forming 4 narrow, alate angles; **L.** reduced to deciduous scales, Th. 3, very stout, up to 15–20 mm long, forking at the tip, Th.shi. uniting into a horny border, stipular Th. up to 5 mm long; **Cy.** in shortly-pedunculate Infl.

**E. golisana** N. E. BR. (Gr. 24). – Somalia. – **St.** erect, branching, crowded, c. 8 cm tall, 12–18 mm thick, variously angled, angles with a continuous, horny border, with closely-spaced small tubercles, each with a pair of aciculate, dark brown Th. 8–12 mm long; **L.** either absent or very minute.

**E. gorgonis** BGR. (Gr. 13). – Cape: Karroo. – **Caudex** spherical-obconical, 5–10 cm ∅, with only the tip above soil-level, with numerous 4–6-angled prominent L.cushions covered with white L.scars; **Br.** in 3–5 circles around the depressed crown, 8–25 mm long, 6–10 mm thick, with L.cushions in spirals, green, often reddish, tubercles 2.5–4 mm ∅, 5–6-angled, pointed, conical; sessile **Infl.** only on the caudex.

**E. gossypina** PAX v. gossypina (Gr. 4). – Tanzania. – ♄ (?); youngest **St.** 8–10 cm long, 3 mm thick, blue-green, covered with the scars of fallen L.; **L.** 12 mm long, 10 mm across, soon falling; **Cy.** 1 cm ∅, with dense, white, woolly hairs.

**E. — v. coccinea** PAX. – **St.** 3–5 cm long; involucral bracts red.

**E. gracilicaulis** LEACH (Gr. 24). – Angola: Benguela, Huila D. – Thorny succulent ♄ or small ♄ 3.5–5 m tall; **St.** thin, at first 5–6-angled, becoming ± cylindrical; **Br.** in a crown above, 5–6-angled, c. 3 cm ∅, margins ± parallel, up to 3 cm long, spreading, curving-ascending, becoming erect, somewhat constricted into segments of varying length, the lower deciduous, rarely with apical shoots, Th.shi. confluent forming a brown border which soon becomes grey, widening at the Infl., Th. horizontally divaricate, c. 6 mm long, 5–12 mm apart; **L.** rudimentary, minute, soon dropping, with recurved stipular Th. at the side of the L.scar; **Infl.** corymbose with a short Ped.

**E. graciliramea** PAX (Gr. 27). – Kenya. – Dwarf ♄ with a long thick napiform **caudex**; **Br.** erect, prostrate or curved and bent, 10 to 15 cm long, grey-green or yellowish-striate, 5 to 10 mm thick, made sinuate by the tuberculate projections of the almost continuous L.bases, narrowly 4-angled and alate, occasionally also roundish, main Th. 10–15 mm long, stipular Th. 1–2 mm long; **L.** scarcely 2 mm long, cylindrical-tapering, soon falling; **Cy.** numerous.

**E. grandialata** R. A. DYER (Gr. 24). – Transvaal: Lydenburg D. – Thorny succulent ♄; main **St.** much reduced, branching freely from the base; **Br.** curving-projecting, seldom re-branching, forming large clumps 2 m tall and 2–2.5 m across, Br. with 4 (sometimes 3) angles, constricted to form pyramidal-cordate segments 7–15 cm long, 7–13 cm thick, the angles much compressed, alate, 5–8 cm high, scarcely 7 mm thick, somewhat sinuate-tuberculate, the sides green with yellow-green curving bands, angles with a grey horny border, Th. usually in 2 pairs, 15–25 mm or 3–5 mm long.

**E. grandicornis** GOEBEL ex N.E.BR. (Gr. 24) (*E. grandidens* sensu GOEBEL). – E.Afr. – Thorny succulent ♄; main **St.** much reduced, with numerous Br., triangular; **Br.** dividing in tiers, erect-projecting, broadly alate, constricted to form ± long segments, angles wavy-curving, with a horny border, green, Th. in pairs, very robust, 2–5 cm long, light brown, with or without a pair of minute stipular Th. at the base; **Fr.** coral-red.

**E. — ssp. grandicornis** (Pl. 71/1). – Moçambique; Swaziland; Natal. – Usually forming dense colonies; **Br.** mostly triangular, in some parts of Maputoland also often 4-angled.

**E. — ssp. sejuncta** LEACH. – Moçambique. – Smaller than the type, sometimes creeping, up to 1 m tall; **Br.** with 2–3 wings, Th. variable in length.

**E. grandidens** HAW. (Gr. 24) (*E. magnidens* HAW. ex S. D., *E. arborescens* S.D.). – Cape: coastal zones. – Thorny succulent ♄ 10–16 m tall; **St.** roundish to 6-angled; **Br.** in whorls, arranged in tiers, 3–4-angled, often spirally contorted, erect to projecting, later rather pendulous, fairly long, sinuate-dentate, Th. in pairs on small Shi., 4–7 mm long, brownish.

**E. grandilobata** CHIOV. is the mottled form of **E. grandicornis**, the sides of the shoots having U-shaped, yellowish-grey markings.

**E. graniticola** LEACH (Gr. 24). – Moçambique. – Related to **E. memoralis**; thorny succulent ♄ or small ♄ up to 2 m tall, acauline or with a truncate St. 12 cm ∅, very rarely with 1 or more trunks; **Br.** numerous, 4–6-angled, ± 1 m

**Euphorbia**

long, with segments 25 cm long and tapering above, 6 cm ⌀ below and 4 cm ⌀ above, angles with wings 3 cm across, margins horny, white to whitish grey, sinuate-dentate, 3 mm across, Th. in pairs, 5–15 mm apart, spreading horizontally, with minute stipular Th.; **L.** sessile, fleshy, linear, pointed, up to 34 mm long, 4 mm across, soon falling; **Infl.** with 3 Cy.

**E. grantii** OLIV. (Gr. 9). – Congo to Tanzania, Uganda. – ♄ up to 2 m tall; **St.** erect, terete, Br. bluish at their tips; **L.** oblong-linear, tapering, lower L. broadly ovate-lanceolate, 10–15 cm long, 1–2 cm across; **Infl.** cymose.

**E. gregaria** MARL. (Gr. 4). – Cape: Lit. Namaqualand; S.W.Afr.; Gr. Namaqualand. – Thornless ♄ branching basally and above to form large clumps 1–6 m ⌀ and 1–2 m high; **Br.** 3–5 cm thick, shoots of pencil-thickness, roundish, grey, waxy.

**E. griseola** PAX (Gr. 24) (*E. g.* v. *robusta* PAX ex ENGLER). – Botswana; Malawi; Rhodesia; Moçambique; Transvaal. – ♄ with a shortened **caudex**, branching freely from the base and above, forming ± dense clumps 50–75 cm high, otherwise ♄-like with a thin main St. up to 3.5 m tall; **Br.** and shoots 1–1.5 cm thick, (4–)5–9(–12)-angled, green with lighter markings in the furrows, Br. ± constricted into segments, angles somewhat compressed, sinuate-tuberculate, tubercles 8–15 mm apart, Th.shi. ± confluent, forming a grey horny border, or solitary, Th.pairs 5–7.5 mm long, thin, often with a lateral pair of small Th.

**E. —** ssp. **griseola**. – Botswana; Transvaal; Rhodesia. – Up to 1 m tall; **Br.** and twigs 4–6-angled, constrictions missing or widely spaced, with a continuous horny margin.

**E. —** ssp. **mashonica** LEACH. – Rhodesia; Malawi; Moçambique. – ♄-like, **St.** 3.5 m tall, 9–12-angled.

**E. —** ssp. **zambiensis** LEACH. – Zambia. – ♄ with the main **St.** greatly reduced or absent; **Br.** distinctly segmented, Th.shi. distinctly separated.

**E. groenewaldii** R. A. DYER (Gr. 22). – Transvaal: Pieterburg D. – Main St. arising from the thick main root, forming a **caudex** 18 cm long and 7 cm thick, mostly subterranean, with 3–7 Br. above; **Br.** 2.5–7 cm long, 12–30 mm thick, often with 1 or 2 sideshoots, 3-angled, spirally contorted, angles broken up into very prominent tubercles 5–10 mm long, pointed, blue-green with lighter markings, Th.shi. with a pair of Th. 3–10 mm long.

**E. grosseri** PAX (Gr. 3). – Ethiopia; Somalia. – ♄ 4–5 m tall, branching from the base; **L.** 6–8 cm long, 3–4 cm across, rather ovate-spatulate towards the base, blunt-tapering, scarcely petiolate.

**E. gueinzii** BOISS. (Gr. 7). – Natal, Orange Free State, Transvaal; Botswana; Swaziland. – **Caudex** producing 1 or several subterranean St. above, from which herbaceous Br. arise; branchlets 5–15 cm tall, simple or re-branching, thin or thick, hairy or glabrous; **L.** almost sessile or short-petiolate, ± numerous, 6–30 mm long, 2–11 mm across, linear, lanceolate or elliptical, tapering, rarely with a few long hairs.

**E. guerichiana** PAX (Gr. 3) (*E. commiphora* DTR.). – S.W.Afr.: Gr. Namaqualand; Cape: Lit. Namaqualand, Transvaal. – Thornless woody ♄ over 2 m tall, often ♄-like with the trunk then 30 cm thick, with a papery bark, laxly branched; **Br.** ± erect; **L.** lanceolate, tapering, very small, soon falling.

**E. guillauminiana** BOIT. (Pl. 72/4) (Gr. M. III). – W.Madag. – ♄ up to 1 m tall, St. 1 cm thick; **Br.** projecting ± horizontally, Th. arranged in 8 R., 2–3 cm long; **L.** sessile, crowded at the Br.tips, with a pink border; **Cy.** yellow or red.

**E. guillemetii** URSCH et LEANDRI (Gr. 20; M. VI). – Madag. – Small thorny ♄, Th. 3 to 8 mm long, thin, arranged irregularly; short shoots with 4–5 **L.** which are thick, 7–12 mm long, c. 4 mm across, pointed, margins undulating; **Infl.** with 2 (?) Cy., cyathophylls greenish-yellow.

**E. gummifera** BOISS. (Gr. 4). – Cape: Lit. Namaqualand; S.W.Afr.: Gr. Namaqualand. – Thornless ♄, branching basally, forming clumps up to 1.3 m tall; **Br.** fleshy at first, becoming woody, 5–10 mm thick, often with angles formed by the line of decurrent L.cushions, smooth, ± covered with dry, rubbery exudations.

**E. hadramautica** BAK. (Gr. 9) (*E. oblongicaulis* BAK., *E. napoides* PAX). – Arabia: Hadramaut; Ethiopia; N. and S.Somalia; Socotra. – **St.** erect or prostrate, fleshy, napiform to clavate, 3–12 cm long, 1–3 cm thick, glabrous, covered with spiralled L.scars; **L.** crowded at St.tips, petiolate, linear to broadly ovate, 3–9 cm long, 0.5–1.5 cm across, glabrous or densely or sparsely bristly, blunt at the tip; **Infl.** ± bristly, Cy. with Ped. 1–2 mm long, campanulate, densely bristly.

**E. halipedicola** LEACH (Gr. 24) (*E. nyikae* sensu REYNOLDS). – Moçambique. – Thorny succulent ♄ or small ♄, (4–)5(–10) m tall; **St.** stout, somewhat constricted and 6-angled to cylindrical; **Br.** spreading, often at first curving and ascending, becoming erect, lower Br. often falling, giving the plant the appearance of a ♄ with a circular crown of Br. arranged ± in whorls, with a few **shoots** with 3–4 (usually 4) wings, sharply constricted, the lower segments terete, secondary branchlets usually 3-winged and arising from the base of the segments which are usually slightly tapering, up to 32–40 cm long, 20 cm across, sometimes broad-triangular or ± ovoid, wings broad, margins horny, crenate-dentate, very undulating, up to 2 mm thick; **L.** ± circular or ovate, 3 mm long, dropping, Th. spreading, up to 15 mm long, with 2 small stipular Th. at the side of the L. scar; **Infl.** with 3 Cy.

**E. halleri** DTR. (Gr. 5). – S.W.Afr. – Possibly identical with **E. gariepina**.

**E. hallii** R. A. DYER (Gr. 4). – Cape: Calvinia D. – **Roots** thickened, spreading; main **St.** 2–3 cm ⌀, becoming straight and thinner above; **caudex** simple or with a few Br., thinner above, cylindrical, tuberculate; **Br.** 1–2 cm thick, with hexagonal tubercles, young tubercles bearing at their tips linear-lanceolate, grooved L. up to 1 cm long and soon falling; **Infl.** with 1 or several Ped., umbels with 1–3 Cy.

**E. hamata** (Haw.) Sweet (Pl. 61/6; 72/1) (Gr. 6) (*Dactylanthes h.* Haw., *E. antiquorum* E. Mey., *E. cervicornis* Boiss.). – Cape: Karroo, Lit. Namaqualand; S.W.Afr.: Gr. Namaqualand. – Thornless succulent ♄ up to 45 cm tall; **caudex** short, 2.5–5 cm thick, with tuberous roots; **St.** numerous, ± branching above, forming dense clumps 45 cm ⌀, shoots 6–13 mm thick, distinctly triangular, smooth, green, becoming grey, L.cushions prominent, 6–16 mm long, pointed, curved; **L.** oval-pointed, 18 mm long, 14 mm across. Used as cattle-fodder.

**E. handiensis** Burch. (Pl. 72/5) (Gr. 24). – Canary Is.: Fuerteventura. – Cactiform; branching freely from the base; **St.** and Br. erect, 80–100 cm and more tall, 6–8 cm thick, with 8–12 angles separated by deep furrows, vivid green, with prominent, white, cordate Th.shi. closely-spaced along the ribs, Th. in pairs, projecting, pointed, 2–3 cm long.

**E. hedyotoides** N. E. Br. (? Gr. II; M. II) (*E. decaryiana* Croiz.). – S.Madag. – Woody ♄ 50 to 100 cm tall; **root** napiform, 20 cm long; **St.** few, up to 2 cm thick, freely branching above; **Br.** with short spurs above, 0.5–1.5 cm long; **L.** 3–5 on the short spurs, spreading radially, deciduous, sessile, narrow-linear, (1.3–)3(–5) cm long, 2–3 mm across, with 2 ephemeral Stip. at the base.

**E. heptagona** L. v. **heptagona** (Pl. 63/19; 72/3) (Gr. 19) (*E. enoplea* Bgr., ? *E. morinii* Bgr.). – Cape: Karroo. – Thorny succulent ♄ (25–)30(–100) cm tall, branching basally; **Br.** 7–10-angled, twigs 3–5 cm long, ± clavate, with 7 (more rarely 5–6) smooth or somewhat undulating angles, divided into tubercles; thorny Ped. solitary, up to 3 cm long. (Acc. J. A. Janse this species is not identical with E. morinii because of different floral characteristics.)

**E. —** v. **dentata** (Bgr.) N. E. Br. (*E. enoplea* v. *d.* Bgr.). – **Shoots** with the angles tuberculate-dentate between the Th.

**E. herrei** White, Dyer et Sloane (Gr. 4) (*E. stapelioides* Boiss.). – Cape: Lit. Namaqualand. – Thornless ♄ scarcely 5–8 cm tall; **caudex** hemispherical, branching; **Br.** ± conspicuously constricted, Int. 1–2 cm long, 1 cm thick below, 5 mm thick above, roundish to somewhat compressed, often with 2 ribs, or angular, bluish.

**E. heterochroma** Pax (Gr. 24) (*E. stuhlmanii* Schweinf. ex Volkens, *E. stapfii* Bgr., *E. impervia* Bgr.). – Tanzania; Kenya; Uganda. – Thorny succulent ♄ 0.5–2 m tall, branching from the base, often re-branching; **Br.** 12–25 mm thick, 4–5-angled, sides at first rounded, distinctly segmented, smooth, green, angles ± distinctly sinuate-tuberculate, Th.shi. brown, ± merging into a horny border, Th. [in pairs, 5–6 mm long, 6–20 mm apart, grey.

**E. hopetownensis** Nel (Gr. 15). – Cape: Gr. Karroo. – **Caudex** c. 5 cm thick; **Br.** in 3 R., ± erect, around the somewhat depressed centre, 2.5 cm long, 1 cm thick, inner Br.shorter, densely covered with small podaria in distinct R.; **L.** small; **Infl.** Ped. 1 cm long at the Br.tips, persistent.

**E. horombensis** Ursch et Leandri (Pl. 73/2) (Gr. 20; M. VI). – Madag. – Thorny ♄ 0.2–1 m tall, branching; **St.** 2–3 cm ⌀, 3-angled, Th. solitary in vertical R., 3–5 mm apart, 8–15 mm long, stiff, on a widened base; **L.** crowded at the Br.tips, 4–8 cm long, 18–30 mm across, with a red border; **Infl.** with 20–40 Cy.

**E. horrida** Boiss. v. **horrida** (Gr. 19). – Cape: Karroo. – **St.** usually dwarf, but sometimes up to 1 m tall, with some Br. from the base, forming clumps, 10–15 cm thick, with 12–14 or more narrow ribs divided by furrows, angles dentate, bearing thorny Ped., usually 3 together, 1–2 cm long. Host-plant for the mistletoe **Viscum minimum.**

**E. —** v. **striata** White, Dyer et Sloane (Pl. 72/2). – Angles very undulating, with white stripes.

**E. hottentota** Marl. (Gr. 24). – Cape: Lit. Namaqualand; S.W.Afr.: Gr. Namaqualand. – Thorny succulent ♄ up to 2 m tall; **caudex** much reduced, with numerous Br. from the base, curved outwards or erect, rarely re-branching; **Br.** 4 cm thick, 5–6-angled, indistinctly segmented, sides somewhat depressed, bluish-green, shoots short, 4-angled, Th.shi. confluent, forming a horny border, Th. in pairs, 3–5 mm long.

**E. hubertii** Pax (Gr. 24). – Tanzania. – Related to E. **nyikae,** but Br. less deeply constricted, wings narrower, shoots with 3 to 4 wings. – ♄ 3–4 m tall, round below; lower **Br.** projecting horizontally, grey-bluish, angles undulating and slightly lobed, Th. 6 mm long, Th.shi. confluent.

**E. hypogaea** Marl. (Gr. 14). – Cape: Beaufort West D. – **Caudex** long and tuberous, almost entirely subterranean; primary **Br.** also below soil-level, secondary Br. 25 mm long, 12 mm thick, clavate or more cylindrical, glabrous, tuberculate, tubercles cylindrical, conical, 3 to 5 mm high; **L.** deciduous, 3–4 mm long, 0.7 mm broad, linear, pointed, plicate; Ped. 8–15 mm long, persistent.

**E. imitata** N. E. Br. (Gr. 22). – Angola. – **Caudex** tuberous, spherical; **St.** numerous, erect, 5–8 cm tall, 3–4 mm thick, triangular above, glabrous, yellowish-green, angles acute, sinuate-dentate, Th. 3–5 mm long, in pairs, swollen below; **L.** rudimentary scale-like.

**E. immersa** Bally et Carter (Gr. 26). – Somalia. – **Root** napiform, passing over into the main St. to form a short, thick, semi-subterranean caudex; **Br.** 5–15 from the flattened head, terete, 0.4–2.4 cm long, 6–8 mm thick, bluntly dentate or tuberculate, tubercles tipped with a triangular Th.shi., Th. solitary, 1.5 to 2.5 mm long; **L.** minute, scale-like, deciduous; **Infl.** forking once. The milky sap is used by the natives as chewing gum.

**E. inaequispina** N. E. Br. (Gr. 28). – Somalia. – Resembles **E. ellenbeckii**, but this species is incompletely documented; probably a dwarf branching ♄ with many Th.; **Br.** cylindrical, about 1 cm thick, Th.shi. 2–3 mm long, obovate, horny, Th. in pairs, very variable, some not more than 3 mm while others are up to 18 mm long, all are acicular; **L.** very rudimentary, minute.

**E. inarticulata** Schweinf. (Gr. 24). – S.Arabia: Yemen. – Erect ♄ 1–1.5 m tall; **Br.** slender, 10–13 mm across, triangular, green, not segmented, sides flat, angles somewhat sinuate at first, with narrow, white, confluent Th.shi., Th. in pairs, 5–10 mm long, 5–10 mm apart, thin.

**E. inconstantia** R. A. Dyer (Pl. 73/1) (Gr. 19). – Cape: Albany D. – Trunk simple, often with many **St.**, little branched above, 30–170 cm tall; main St. and Br. 3.7–7.5 cm ∅, Br. often curving down and after that erect, twigs 12 to 50 mm thick, with 7–10 acute angles, 6–10 mm high, slightly crenate-tuberculate; thorny Ped. 2–3 together along the angles (in ♀ plants usually solitary), 6–20 mm long, reddish, becoming grey.

**E. inculta** Bally (Gr. 24). – N.Somalia. – Stout thorny succulent ♄ 2–3 m tall; **St.** 2–3 cm thick; **Br.** irregular, erect, spreading, short, re-branching little or not at all, green to bluish, Th. in pairs, up to 15 mm long, 6–10 mm apart, Th.shi. rarely confluent; **Infl.** forking once or twice.

**E. inermis** Mill. v. **inermis** (Pl. 62/14) (Gr. 14) (*E. viperina* Bgr., *E. serpentina* hort.). – Cape: Karroo. – Tuberous **root** merging into the short main St. to form a **caudex**; **Br.** very numerous, in 3 or more circular R. around the often depressed crown, 13–15 mm thick, up to 30 cm or more long, often serpentine, L.cushions longhexagonal, in 6–7 spiral R.; **L.** small, soon falling; Ped. often becoming thorny.

**E. —** v. **huttonae** (N. E. Br.) White, Dyer et Sloane (*E. h.* N. E. Br.). – Only floral characteristics differentiate it from the v. inermis.

**E. ingenticapsa** Leach (Gr. 24). – Angola. – ♄, robust, succulent, strongly armed, erect, c. 1.8 up to ± 2.5 m high, branched from the base; **trunk** much reduced, stout, 3–5-angled; **Br.** spreading-ascending, mostly simple, very rigid, deeply constricted into segments, (4–)5(–7)-winged, wings broad, much compressed, ± 5 mm thick, segments ± trullate or elliptic or sometimes subcircular, ± 6–12 cm long, 9–12 cm broad, margins continuous, hard, horny, sinuate-dentate, often wavy, pale brown, strongly armed, **Th.** stout, up to 18 mm long, in spreading divergent pairs; **L.** thick, fleshy, up to 6 mm long, caducous, flanked by a pair of short stout prickles.

**E. × ingezalahiana** Ursch et Leandri. – Hybrid: parents unknown, cult. Bot. Gard. Tsimbazaza, Madag.

**E. inornata** N. E. Br. (Gr. 15). – Cape: W. Griqualand, Kimberley, Hopetown D. – **Caudex** almost spherical, c. 10 cm ∅, smooth, olive-green to brown; **Br.** dense, in 5 R. around the 3–4 cm ∅ central surface at the tip of the caudex, spreading, c. 7.5 cm long, 1 cm thick, cylindrical, tuberculate, green, often reddish, tubercles dense on the caudex, pointed, conical, those on the Br. rhombic, 3–8 mm long, 2–4 mm across, 1.3 mm high, with a white L.scar; **Infl.** Ped. becoming thorny.

**E. intercedens** Pax (Gr. 24). – Moçambique. – Succulent, spiny, leafless ♄ 4–8 m tall with a trunk 1.5–3 m high; **Br.** 3–5 cm ∅, 4-angled, the angles compressed, winglike and shallowly sinuate-dentate; Sp. in pairs, 2–8 mm long, grey. Incompletely described.

**E. intisy** Drake v. **intisy** (Pl. 73/5) (Gr. 4). – Madag. – ♄ or ♅ 6–7 m tall; **Br.** and shoots terete, very thin, usually forking, grey-green, covered with white spots in irregular transverse lines; **L.** on gibbose projections, giving the Br. a somewhat nodular appearance. Provides rubber. (Acc. Drake long "water-roots" are present.)

**E. —** v. **maintyi** (Decorse) E. H. Poiss. (*E. m.* Decorse). – Robuster variety. (Acc. Swingle a form of **E. laro**.

**E. isacantha** Pax (Gr. 28). – Tanzania. – Cushion-forming ♄; **Br.** with short shoots c. 1 cm ∅, Th.cushions closely-spaced, with 2 Th.pairs, upper Th. 6–7 mm long, the lower ones curved downwards and 6–7 mm long.

**E. isaloensis** Drake (Gr. 20; M. VI) (*E. mainiana* H. Poiss., *E. splendens* v. *m.* (H. Poiss.) Leandri, *E. isalensis* Leandri). – Madag. – ♄; **Br.** 5–7 mm ∅, reddish to grey, Th. close together, 15 mm long, projecting or directed downwards; **L.** at the twig-tips, obovate oblong, tapering, apiculate, 2–2.5 cm long, 6 to 7 mm across; **Infl.** c. 3 cm long, cyathophylls yellow.

**E. jansenvillensis** Nel (Gr. 18). – Cape: Jansenville, Uitenhage D. – **Caudex** erect with many shoots from the base, 8–16 cm tall, 1.5 cm thick, with 5 ribs, angles ± projecting above, with short acute L.cushions, glabrous, greengrey.

**E. johnsonii** N. E. Br. (Pl. 73/4) (Gr. 22). – Moçambique. – **Caudex** with 1 or several necklike shoots above, each with a number of thorny Br. produced at or just below soil-level; **Br.** variously curving, spreading-ascending, usually triangular, with 2 angles below, less often with 2 angles throughout the entire length, pale green to dark red-brown, with lighter or whitish longitudinal stripes, 8–24 cm long, 5–8 mm ∅, with tubercles 1–3 cm apart along the angles, the tubercles triangular or ± truncate, 3–5 mm high, more step-like below, Th.shi. horny, decurrent but not forming a horny border, Th. in pairs, aciculate, 2–9 mm long, often with 2 minute Th. at the base of the L.scars; **L.** lanceolate, pointed, 8–15 mm long, 3–5 mm across, fleshy, soon falling; **Infl.** with 3 Cy. (Acc. L. C. Leach may possibly be identical with **E. knuthii**.)

**E. × jubaephylla** Svent. – Canary Is.: Tenerife. – Nat. hybrid: **E. regis-jubae** × **E. aphylla**. – Hemispherical compact ♄ 1–1.7 m tall; **Br.** terete, ± whorled, at first somewhat clavate, up to 5 mm ∅, bark becoming corky, brown; **L.** linear-spatulate to linear, 2–3 cm long, 3–4 mm across, soon falling; **Infl.** at the Br.tips, in corymbs of 3–5 Cy.

**E. jubata** LEACH (Gr. 24). – Zambia. – Thorny succulent ♄ c. 15 cm tall, densely branching and re-branching, somewhat pulverulent; **Br.** and shoots with 4(–5) angles, c. 1.24 cm thick, with acute longitudinal furrows, usually somewhat constricted into segments, Th.shi. on the tubercle T. sometimes confluent, in part forming a horny border, pale brown, Th.pairs 4–7 mm long, often with minute hooked stipular Th.; **L.** sessile, fleshy, semi-terete to oval, pointed, c. 3.8 mm long, soon falling; **Infl.** of 3 Cy.

**E. juglans** COMPT. (Gr. 18). – Cape: Ladismith D. – **Caudex** 2–3 cm thick, sub-cylindrical, the major part underground; **Br.** 1–5, 2–5 cm long, 1.5–2.5 cm thick, erect, with 6–9 angles, the sections above ground hemispherical, thinner below, angles obtuse, with the small scars of old Ped. at intervals of 2–3 mm; **L.** rudimentary.

**E. juttae** DTR. (Pl. 74/1) (Gr. 4). – S.W.Afr.: Gr. Namaqualand. – Thornless ♄ 10–15 cm tall, branching basally, with numerous shoots above, Int. angular, 6–15 mm long, often thickened above, blue-green.

**E. kamerunica** PAX (Gr. 24) (*E. barteri* N. E. BR.). – Upper Guinea. – ♄ up to 10 m tall, up to 50 cm ⌀ at the base, branching candelabra-like; **Br.** 5-angled, shoots 4-angled, 6 cm ⌀, constricted into segments, angles crenate, T. with Th.pairs, 5–6 mm long.

*E. karasmontana* DTR. is probably an ecotype of *E. avasmontana* DTR.

**E. karroensis** (BOISS.) N. E. BR. (Gr. 4) (*E. burmannii* v. *k.* BOISS.). – Cape: Karroo. – Resembles **E. burmannii**; **St.** more erect.

**E. keithii** R. A. DYER (Gr. 24). – Swaziland. – ♄ or ♄ 2–6 m tall; when old **Br.** wither and fall, dark green, projecting to spreading, 1–2 m long, 3–6-angled, constricted into segments up to 25 cm long, 3–4 cm across, sides ± parallel, angles alate, 0.7–1.5 cm high, slightly tuberculate, Th.shi. confluent, Th.pairs 5–8 mm long; **L.** ovate-cordate, c. 5 mm long, soon falling; **Infl.** with 3 Cy., almost sessile.

**E. kibwezensis** N. E. BR. (Gr. 24). – Uganda. – ♄ up to 10 m tall; **Br.** 5–10 cm ⌀, with 3 to 4 alate angles, wings 2–5 cm across, 2–3 mm thick, ± sinuate and somewhat undulating, borders continuous, horny, light brown; Th.-pairs 4–6 mm long, with or without a pair of minute stipular Th. at the base; **L.** rudimentary, scale-like, semi-circular; **Infl.** almost sessile, forming clusters along the angles of the terminal Br.

**E. knobelii** LETTY (Gr. 24). – Transvaal. – Thorny succulent ♄ c. 1 m tall; trunk reduced, mainly hidden underground; **St.** numerous, projecting, rarely branching, 5-angled, with distinct segments, 2–2.5 cm thick at their broadest, yellowish-green with dark green markings, angles compressed, sinuate-tuberculate, sides concave, Th.shi. solitary, rarely confluent to form a border 3 mm across, Th. in pairs, 1 cm long.

**E. knuthii** PAX ssp. **knuthii** (Gr. 22). – Moçambique; S.Afr.: Transvaal, Natal; Swaziland. – **Caudex** serpentine, branching from the base, often with numerous subterranean offsets; **St.** 5–15 mm thick, often branching above, 3–4-angled, light green with light grey-green stripes, angles sinuate-tuberculate, tubercles 6–12 mm apart, 2–4 mm high, triangular, each with 2 light brown Th. 4–5 mm long.

**E. —** ssp. **johnsonii** (N. E. BR.) LEACH (Pl. 73/4) (*E. j.* N. E. BR.). – Roots tuberous, simple.

**E. lactea** HAW. (Pl. 74/3) (Gr. 24). – ? Peninsular India; Molucca Is.; Ceylon; Caribbean Is. and the USA: naturalised in Florida and Cuba (*E. habanensis* HORT.). – ♄ or ♄; **St.** and Br. 3–4-angled, 3–5 cm ⌀, sides flat, dark green, with a whitish band down the middle, angles sinuate with whitish curved lines out to the indentations, Th. thick, c. 5 mm long, brown.

**E. —** cv. Cristata. – Monstrous cultivar.

**E. lactiflua** PHIL. (Gr. 4). – Chile: Atacama desert. – Related to **E. tirucallii**.

**E. lagunillarum** CROIZ. – S.W.Venezuela. – Semi-succulent ♄, of little interest in cultivation (see Cact. Succ. Journ. Am. **39**, 144, 1967).

**E. lambii** SVENT. (Gr. 1). – Canary Is.: Gomera. – Erect ♄ 1–2.5 m tall; **St.** at first simple, branching dichotomously above, then tripartitely; **Br.** erect to projecting, c. 1 cm ⌀, fleshy, stiff, glabrous, leafy at the tips, bark corky and splitting, brown, with L.scars; **L.** in a lax Ros., sessile, thickish, linear to elliptical, bluntish, bluish to yellowish, 4–7 cm long, 8–10 mm across; **Infl.** with many Cy., terminal, 10–12 cm ⌀, with yellow bracts.

**E. laro** DRAKE (Gr. 4). – Madag. – ♄; **Br.** with shoots in whorls, 5–6 mm thick, round, shoots curving to projecting, dark green, furrowed longitudinally, with light spots and stripes in the furrows.

**E. lateriflora** SCHUM. et THONN. (Gr. 4). – Trop. W.Afr. – ♄; **St.** c. 1 m tall, climbing, fleshy, terete; **Br.** laxly pendulous, with L.scars, glabrous, bluish, alternate or sometimes clustered, 2–7 mm thick; **L.** alternate, 2–7 mm apart, with a whorl of L. at the base of the Infl., sessile, 2.5–3.5 cm long, 2–6 mm across, pointed; **Infl.** a terminal umbel.

**E. leandriana** P. BOIT. (Pl. 74/2) (Gr. 20; M. VI). – Madag. – Thorny ♄; **St.** robust, branching, Th. compressed, curving from the base; **L.** oblong-spatulate, sessile, glabrous; **Infl.** subterminal, cyathophylls cream or orange-coloured.

**E. ledermanniana** PAX et K. HOFFM. (Gr. 7). – Nigeria; Ghana. – **St.** fleshy, up to 30 cm long, arising annually from the perennial woody caudex; **L.** broadly linear to elongate-elliptical, narrowing to an obtuse, subsessile base, 5 to 14 cm long, 1–2 cm across; involucres solitary, 6–9 mm ⌀, with 5 broad flap-like glands, transversely elliptical with crenate margins; Ped. 2–6 mm long.

**E. ledienii** BGR. (Gr. 24). – Cape: Uitenhage, Humansdorp, Albany D. – Thorny succulent ♄ up to 2 m tall; **Br.** ascending, 5-angled, grey-green, with segments 7–20 cm long, 4–6 cm ⌀, tapering above, angles somewhat compressed, somewhat sinuate-dentate, sides rather grooved, later flatter, Th.pairs up to 18 mm apart, 8 mm long, brown, Th.shi. ± confluent to form a horny border; cyathophylls yellow.

**E.** — v. **dregei** (N. E. BR.) N. E. BR. (*E. d.* N. E. BR., ? *E. canariensis* THUNBG.). – Differentiated from the type only by the shape of the involucre.

**E. lemaireana** BOISS. (Gr. 24) (*E. fimbriata* HORT. ex LEM., *E. crispata* LEM., *E. grandicornis* sensu BGR. p.part., *E. angularis* sensu N. E. BR. p.part., *E. a.* sensu WHITE, DYER et SLOANE p.part., *E. a.* sensu JACOBS. p.part., *E. nyikae* WERTH.). – Zanzibar. – Small, very succulent ♄ or large ♄ 3–5 m tall; **St.** thick, short; **Br.** numerous, surrounding the St., projecting, branching, shoots deeply divided into segments 7.5–20 cm long and 5–11 cm thick, broadly 3–4-angled, upper ones always triangular, middle section thin, angles alate, 2.5–6 cm across, irregularly lobed or sinuate, tubercles 1.2–2.5 cm apart, border continuous, grey, horny, Th. in pairs, 4–10 mm long, grey; **L.** scale-like, stiff, 2–3 mm long, deciduous, leaving 2 small, stiff, sometimes recurved tubercles between the L.bases; **Infl.** 1–2, with 3 small involucres.

**E. letestui** J. et A. RAYNAL (Gr. 24). – Cameroon. – Related to **E. trigona** and **E. kamerunica**; succulent, hemispherical, erect ♄ 2–4 m tall; **St.** cylindrical, c. 20 cm thick; **Br.** dense, erect to projecting or recurved, the tips bent back, with 3 wings, divided into oblong-elliptical to oblanceolate segments, 10–30 cm long, 5–10 cm across, side-Br. cylindrical, ± triangular, 4 to 10 mm thick, leafy when young, L.cushions evenly spaced, 10–24 mm apart, with two 2–3 mm long and divaricate Th.; **L.** obovate to oblanceolate, 3.5–8 cm long, 1.2–3 cm across, blunt or rounded at the tip, fleshy, with a petiole 5–15 mm long; **Infl.** with few Cy.

**E. leucochlamys** CHIOV. (Gr. ?). – S.Somalia. – ♄ up to 2 m tall; **Br.** fleshy, cylindrical, with L.scars, bark reddish-grey to coppery red; **L.** crowded at the Br.tips, almost sessile, obovate, 1–2 cm long, 7–15 mm across, slightly fleshy, light sap-green; **Infl.** with 1–3 Cy.

**E. leucodendron** DRAKE (Pl. 75/1) (Gr. 4) (*E. alluaudii* DRAKE). – Madag. – ♄ with cylindrical virgate **Br.,** procumbent from the base.

**E. leuconeura** BOISS. (Pl. 61/2) (Gr. 2; M. II) (*E. fournieri* HORT., *E. madagascariensis* COMM. MS.). – Related to **E. lophogona**; ♄; **St.** with 4 laciniate-fringed angles; **L.** crowded on the St.tips, obovate-oblong, tapering, narrowing to the petiole, 10–12 cm long, 3.5–4 cm across; **Infl.** of 3 sessile Cy.

**E. leviana** L. CROIZ. (Gr. 19). – L. CROIZAT puts forward this name for the plants of **E. cereiformis** L. found in cultivation, since they do not conform to the species description of LINNAEUS. (L. CROIZAT, De Euph. Ant. atque Off.)

**E. lignosa** MARL. (Pl. 78/5) (Gr. 4). – S.W. Afr.: Gr. Namaqualand. – Spherical semi-♄, little more than 50 cm tall; **trunk** very short and compressed; **St.** numerous, finger-thick, rebranching in 3's or 4's, whitish-grey to green, shoot tips becoming thorny.

**E. lividiflora** LEACH (Gr. 24). – Moçambique. – Almost thornless ♄ c. 4–10 m tall and freely branching; **St.** woody, cylindrical, 12.5–25 cm thick; main **Br.** in whorls, likewise the shoots, Br. 5-angled, with segments 7.5–18 cm long, 4.5 cm ⌀ below, side-Br. and shoots 3–4-angled, segmented, angles prominently sinuate-tuberculate, Th. in pairs at the tips of the tubercles, up to 5 mm long, often missing, Th.shi. solitary, often confluent, with 2 minute stipular Th. at the side of the L.scars; **L.** round, up to 13 mm long, sessile, fleshy, with margins curved upwards, soon falling; **Infl.** a corymb with 3 Cy.

**E. longetuberculosa** HOCHST. ex BOISS. (Gr. 9). – Arabia; Somalia; Kenya; Ethiopia. – **St.** short, fleshy, ovate, tapering below; podaria conical to rather long; **Br.** numerous, thin, dividing dichotomously, covered with short tubercles; **L.** at the tips of younger Br., linear-oblong, short-petiolate; **Infl.** with Ped. sometimes branching dichotomously.

**E. longispina** CHIOV. (Gr. 24). – Somalia. – Much-branched ♄ c. 1.2 m tall, 1.5 m ⌀; **Br.** 5–6-angled, Th. on extended, horny, partly confluent grey Shi.; Th.pairs alternately 4–5 cm or 6–20 mm long, acicular; **Infl.** with 2 Cy.

**E. lophogona** LAM. (Pl. 74/4) (Gr. 2; M. II). – Madag. – ♄ up to 50 cm tall, with Br. ± in whorls; **St.** and Br. thickened above, the reddish angles set with pectinate Stip.; **L.** at the Br.tips, narrowing to the long reddish petiole, 12 cm long, 5 cm across, mucronate, emerald green with white pinnate veins; **Cy.** on Ped. 4–5 cm long, cyathophylls open, white or pink. For the heated greenhouse.

**E. loricata** LAM. (Pl. 62/10) (Gr. 10) (*E. hystrix* JACQ., *E. armata* THUNBG.). – Cape: Karroo. – Thorny ♄ 30–100 cm tall, branching from the base or higher; **Br.** 8–12 mm thick, cylindrical, with indistinct tubercles in spiral R., glabrous, thorny Ped. solitary, numerous, 12–50 mm long; **L.** at the Br.tips. 2.5–7.5 cm long, 3–6 mm across, tapering.

**E. lutzenbergerana** CROIZ. – Colombia; Venezuela. – Semi-succulent ♄, of little importance in cultivation (see Cact. Succ. Journ. Am. 39, 142, 1967).

**E. lydenburgensis** SCHWEICK. et LETTY (Gr. 24). – Transvaal. – Thorny succulent ♄ up to 1.5 m tall; **St.** shortened, branching from the base; **Br.** dividing near the tip, up to 1.5 cm thick, with 4–5 angles, sides slightly depressed, pale greenish yellow, angles rather sinuate-tuberculate, Th.shi. c. 1 cm long, usually forming a continuous, brown or grey, horny border, Th. in pairs, 5–7 mm long, with 2 lateral Th. 1 mm long.

**E. macella** N. E. Br. (Gr. 4). – Cape: Mossel Bay D. – Related to **E. karroensis.**

**E. macroglypha** Lem. (Gr. 24). – **St.** several m tall, 4–5-angled, with stout Th. in pairs, 3–4 cm apart, angles sinuate; **Br.** in whorls, triangular, deeply furrowed, Th. as on St.

**E. mahafalensis** M. Denis v. **mahafalensis** (Gr. 20; M. VII). – S.W.Madag. – Small ♄ with many Br.; outer **Br.** thorny; **L.** at the Br.tips, oblanceolate, tapering, 1–2 cm long, 6 mm across, softly hairy; **Infl.** sessile, cymose.

**E.** — v. **xanthadenia** (Denis) Leandri (*E. x.* Denis). – Thorny ♄ up to 1 m tall; **Br.** with few twigs, 1 cm ⌀, yellowish, Th. 1 cm long with smaller subsidiary Th.; **L.** up to 25 mm long, 8–14 mm across.

**E. maleolens** Phill. (Gr. 14). – Transvaal. – **Caudex** sub-ovate or sub-spherical, mostly underground, dividing above into 2 St., 3–9 cm thick, with numerous Br. up to midway and large rhombic tubercles 1–1.5 cm long, 5–12 mm across, compressed above; **Br.** 10–20 cm long, c. 1 cm thick, projecting, later curved upwards, tuberculate, glabrous, tubercles up to 1 cm long, 5 mm across, with a white L.scar at the tip; **L.** at the Br.tips, 12 mm long, linear-lanceolate, plicate, later falling.

**E. malevola** Leach (Gr. 24). – E.Afr.: Zambesi area. – Related to **E. complexa**; thorny succulent ♄, branching basally, up to 1.5 m tall; **Br.** spreading to ascending, 4–5-angled, 1–2 cm thick, sides concave or flat, angles sinuate-tuberculate, Th.shi. confluent but not forming a continuous border, dark purple or purple-brown, Th. in pairs, aciculate, 5–7 mm long, stipular Th. c. 2 mm long; **L.** broadly oval, pointed, c. 1.5 mm long, soon dropping; **Infl.** with 3 Cy. in an umbel.

**E.** — ssp. **bechuanica** Leach. – Botswana. – **Infl.** sessile.

**E.** — ssp. **malevola.** – Rhodesia; Zambia; Malawi. – **Infl.** very shortly pedunculate.

**E. mammillaris** L. (Pl. 74/5) (Gr. 19). – Cape: Riversdale, Oudtshoorn D. – **Trunk** rudimentary, branching basally; **St.** erect, with few Br., cylindrical, 4–6 cm thick, barely 20 cm tall, 7–17-angled, angles tessellate-tuberculate, tubercles cubical, separated along the angles by horizontal furrows, broad, hexagonal, flat to conical, each with a white L.scar; thorny Ped. solitary, 6–10 mm long, grey.

**E. mangokyensis** M. Denis (Gr. 20; M. VI). – Madag. – Thorny ♄; **Br.** numerous, warty, the lower segments angular, main Th. 1 cm long, hooked and recurved at the tip, with 2 stipular Th. which soon fall; **L.** bristly-hairy on both surfaces, ovate, 1.5–2 cm long, c. 1 cm across, margins very undulating; **Infl.** an umbel.

**E. marientalensis** Dtr. (Gr. 15) (*E. marientalii* Dtr. ex Range, Fl. S.W.Afr.). – S.W.Afr.: Gr. Namaqualand. – Related to **E. rangeana**; **St.** 3–5, clavate, 2–3 cm thick, rarely branching, greenish brown.

**E. marlothiana** N. E. Br. (Gr. 14) (*E. marlothii* N. E. Br.). – Cape: Cape, Wynberg D. – **Caudex** oblong-ovoid with a spherical thickening above; **Br.** from the crown, erect or spreading, 7.5–40 cm long, up to 10 mm thick, tuberculate, often eventually re-rooting and then clavately thickened, tubercles rhomboid or oblong, not very prominent; Ped. becoming ± thorny.

**E. mauritanica** L. (Pl. 75/2) (Gr. 4) (*E. tirucallii* Thunbg. p.part.). – Cape; SW.Afr. to Natal. – ('Milk tree' of the Boers.) – Thornless ♄ branching freely from the base, 1–1.5 m tall; **Br.** erect, elongated, pencil-thick, L.cushions rather prominent; **L.** linear-lanceolate, 12 to 15 mm long, soon dropping. – There are several ± distinct ecotypes.

**E. mbaluensis** Pax (Gr. 24). – Tanzania: W.Usambara. – ♄ or candelabra-like ♄; **Br.** triangular, constricted, segmented, angles alate, fleshy, undulating and sinuate, 2.3–3.5 cm across, Th.shi. confluent, forming a grey border, Th. in pairs, 5–6 mm long, with minute stipular Th. at the base; **Cy.** in crowded clusters at the Br.tips.

**E. melanohydrata** Nel (Gr. 15). – Cape: Lit. Namaqualand. – **Caudex** sub-cylindrical, 8 cm long and thick, mostly underground, often forking into 2 thick trunks above, with cylindrical, 1 cm long, tuberculate, bluish-green **St.**, tubercles hemispherical or hexagonal, tipped with a white L.scar; Ped. becoming thorny, numerous, stout.

**E. mellifera** Ait. v. **mellifera** (Gr. 1). – Is. Madeira. – Arborescent ♄ with few Br.; **St.** erect, straight; **Br.** with L.scars; **L.** long-lanceolate, tapering towards both ends, almost sessile, 12–20 mm long, 2–2.5 mm across; **Cy.** in alternate paniculate Br., with a strong scent of honey.

**E.** — v. **canariensis** Boiss. – Canary Is.: Tenerife. – ♄ up to 10 m tall; **Br.** rather shorter.

**E. meloformis** Ait. (Pl. 74/6) (Gr. 18) (*E. pomiformis* Thunbg., *E. meloniformis* Link., *E. infausta* N.E. Br., *E. pyriformis* N. E. Br., *E. falsa* N. E. Br.). – Cape: Karroo. – **Bo.** spherical, usually solitary but sometimes branching from the base, often 8–10 cm ⌀, usually broader than tall, with a thick taproot, ribs (8–)10(–12), vertical or slightly spiral, ± deeply furrowed, crown depressed, green or grey-green, the sides of the angles with lighter or reddish transverse bands, the angles with roundish L.scars; Ped. becoming woody, forking several times, eventually falling or persistent (*E. falsa*). – Hybridises with **E. obesa.**

**E. memoralis** R. A. Dyer (Gr. 24). – Rhodesia. – Thorny succulent ♄ up to 3 m tall; **St.** cylindrical, 5–7-angled, c. 10 cm ⌀, branching and re-branching; **Br.** 4–6-angled, ± 1 m long, 1.5–3 cm ⌀, with segments 5–15 cm long, angles with twin-thorned horny Shi., Th. 6 mm long; **L.** up to 2 cm long, lanceolate and 6–7 mm across, or triangular, up to 2 cm long; **Infl.** an umbel with 5–7 Cy.

**E. micracantha** Boiss. (Gr. 22) (*E. tetragona* Bak., *E. gilbertii* Bgr., *E. lombardensis* Nel). – Cape: Karroo. – **Caudex** short, conical, forming a continuation of the thick taproot, projecting only slightly above the soil, 12.5–15 cm long, 3–7 cm thick, with ± radially projecting Br. above, ± prostrate, contorted or spiral,

4–12 cm long, 6–15 mm thick, with 3–4 sinuate angles, with tubercles 4–8 mm apart, green, sides slightly grooved, Th. in pairs on the tubercle-tips, 3–6 mm long, thin, pointed, grey-brown.

**E. migiurtinorum** Chiov. (Gr. 24). – Somalia: coasts. – Resembles E. erlangeri; ♄; **Br.** and shoots alternate to opposite, extended, obtusely 4-angled, up to 6 mm thick, angles indistinct, Th.shi. confluent, Th. in 2 pairs, the lower Th. 4–6 mm long, projecting horizontally, the upper ones scarcely 1 mm long; **Cy.** solitary.

**E. milii** des Moulin (Gr. 20; M.VI). – Madag. – Thorny xerophytic ♄, very variable in shape; **L.** variable in shape and size, oblong to circular; **Cy.** ± large with coloured cyathophylls.

**E. —** v. **betsileana** Leandri (*E. splendens* v. *b.* Leandri). – Related to v. **milii**; **L.** ovate-rhombic.

**E. —** v. **bevilaniensis** (L. Croiz.) Ursch et Leandri (Pl. 75/3) (*E. b.* L. Croiz., *E. splendens* v. *b.* (L. Croiz.) Leandri). – S. Madag. – Up to 1.5 m tall; **Br.** round, smooth, brown, 1.5 cm ⌀, Th. well spaced, not swollen at the base, c. 1 cm long; **L.** opposite or in whorls, apiculate, 25 mm long, 20 mm across; cyathophylls red.

**E. —** v. — f. **rubro-striata** Drake et Castillo (*E. splendens* v. *b.* f. *r.-s.* Drake et Castillo). – Cyathophylls yellowish-red striate.

**E. —** v. **bosseri** Rauh. – E. central Madag. – ♄ c. 30 cm tall, branching freely from the base, 30–40 cm ⌀; **caudex** thickened, napiform, with a grey-brown bark, older plants with a distinct trunk; **Br.** spreading horizontally, 3–10 mm thick, green, becoming silvery-green; **L.** deciduous, lanceolate to linguiform, mucronate, 3–5 cm long, 3–5 mm across, narrowing below to the short petiole, with a narrow, membranous, shortly dentate border; **Th.** in pairs, thin, sharp; many short Br. 1–2 cm long arise from the upper L.axils on older shoots; **Infl.** subterminal, with 2–4 Cy., cyathophylls triangular, 3 × 3 mm, light green.

**E. —** v. **breonii** (L. Nois.) Ursch et Leandri (*E. b.* L. Nois., *E. neumannii* Hort., *E. splendens* v. *b.* (L. Nois.) Leandri). – **L.** in a terminal Ros., 10–15 cm long, 2–3 cm across, the tip rounded and mucronate, margins bicoloured, minutely wavy; cyathophylls brilliant red.

**E. —** v. **hislopii** (N. E. Br.) Ursch et Leandri (Pl. 75/4) (*E. h.* N. E. Br., *E. splendens* v. *h.* (N. E. Br.) Leandri). – ♄ up to 2 m tall, ± branching; main **St.** 3–6 cm thick, 8–10-angled, olive-green becoming grey, angles very thorny, Th. 12–24 mm long, with a broadly conical base; **L.** at the Br.tips, ovate-lanceolate, 12–18 cm long, 4–5 cm across, becoming brilliant red before falling; cyathophylls pink or red, 1 cm across.

**E. —** v. **imperatae** (Leandri) Ursch et Leandri (*E. splendens* v. *i.* Leandri). – Rather small and thinly-branched representative of the **milii**-complex. ♄ 30–50 cm tall, freely branching; **Br.** c. 5 mm thick, curving and pendulous, rather densely thorny, Th. 1–1.5 cm long, slender, with a broadened and flattened base, tips dark brown; **L.** leathery, variable in shape, oblong-oval and 1 cm long, apiculate, L. on short shoots broadly oval; **Infl.** with 2–4 Cy., cyathophylls crimson, broadly oval, 1 cm across.

**E. —** v. — f. **lutea** Rauh. – **Br.** still thinner, often bent and zigzagging, densely thorny; cyathophylls 5 mm long, 7 mm across, apiculate.

**E. —** v. **longifolia** Rauh. – Madag. – Branching basally; **St.** 80–100 cm tall, almost uniform in thickness throughout, 1.5–2 cm ⌀, round, with numerous short shoots up to 3 cm long; **L.** on longer St. very short-petiolate, narrow-lanceolate, 1–2 cm long, 8–10 mm across, mucronate, ± folded inwards, L. on short shoots rosulate, 3–4 cm long, c. 5 mm across, Th. simple; **Infl.** 1 or several, with viscous Ped., Cy. 4–30, cyathophylls pale sulphur-yellow, broadly oval, 5 mm long.

**E. —** v. **milii** (*E. bojeri* Hook., *E. splendens* v. *b.* Cost et Gallaud, *E. s.* ssp. *b.* M. Denis, *Sterigmanthe b.* Klotzsch et Garcke). – W.Madag. – **St.** and Br. slender, 8–10 mm ⌀, Th. well spaced, thin, scarcely wider at the base, c. 1 cm long, 2 mm thick at the base; **L.** somewhat leathery, obovate, apiculate; cyathophylls 7–8 mm across, intense red.

**E. —** v. **roseana** J. Marn.-Lap. – ♄ 80–100 cm tall; **Br.** erect, later more projecting, cylindrical, thorny at first, becoming smooth, grey-yellowish, Th. simple, projecting horizontally, 12–13 mm long, thin, soft, grey, later falling; **L.** oblong-lanceolate, 3–9 cm long, 1.5–2.5 cm across, apiculate, margins red; **Infl.** with Ped. viscous, with 4 Cy., cyathophylls rounded, 5.5 cm long, apiculate, whitish-yellow.

**E. —** v. **splendens** (Boj. ex Hook.) Ursch et Leandri (Pl. 63/20; 75/5) (*E. s.* Boj. ex Hook., *E. s.* v. *typica* Leandri, *Sterigmanthe s.* Klotzsch et Garcke). – ♄ up to 2 m tall; **Br.** spreading, ± angular, slightly furrowed, 1 cm ⌀ ,Th. numerous, 2 cm long, 5 mm across at the base, ± compressed, black; **L.** usually at the Br.tips, obovate-oblong, tapering, 4–5 cm long, 2 cm across, leathery, light green; **Cy.** in repeatedly branching Infl., cyathophylls c. 1 cm long, brilliant intense red.

**E. —** v. — f. **platyacantha** Leandri (*E. splendens* v. *typica* f. *p.* Leandri). – Th. conspicuously broad.

**E. —** v. **tananarivae** Leandri (*E. splendens* v. *t.* Leandri). – Resembles v. **hislopii**; **Br.** few, 2–3 cm thick; **L.** 7–10 cm long, 3–4 cm across, rounded at the tip, petiole 2–3 mm long, underside puberulous; **Infl.** with 4–8 Cy., cyathophylls yellow with a red border.

**E. —** v. **tulearensis** Ursch et Leandri. – ♄ 60–70 cm tall; **St.** 2–3 cm ⌀, subcylindrical, irregularly angular, Th. not numerous, thin, soft, 5–10 mm long, curved downwards; **L.** persisting at the Br.tips, 1–5 mm long, petiolate, lanceolate, apiculate, 3–4 cm long, 10–14 mm across; **Infl.** with 2–16 Cy., cyathophylls round, pale red.

**E.** — v. **vulcanii** LEANDRI (*E. splendens* v. v. LEANDRI). – Central Madag. – **Br.** 1 cm and more thick, Th. grouped longitudinally, simple, fleshy at the base, 1 cm long, narrowing abruptly to the sharp 5–6 mm long tip; **L.** in a terminal Ros., 15–20 cm long, 4–5 cm across, rounded at the tip, mucronate; cyathophylls red.

**E. millotii** URSCH et LEANDRI (Gr. 20; M.VII). – S.E.Madag. – **St.** fleshy, almost thornless, with L.scars, grey, 10–15 mm $\varnothing$, red at first, branching from the base, becoming densely bushy; **L.** deciduous, few in a terminal Ros., glossy, both surfaces reddish-green, nerves light red, c. 3.5 cm long, 1.2 cm across, petiole 8–12 mm long, flattened, Stip. transformed into minute short T. on projecting tubercles; **Infl.** with 2–3 nodding Cy.

**E.** × **mitsimbinensis** URSCH et LEANDRI. – Hybrid: **E. brachyphylla** × **E. spec.** – Hort. Tzimbazaza.

**E. mlanjeana** LEACH (Gr. 24). – Malawi: Mlanje Mt. – Often tree-like ♄ up to ± 1 m high, acauline or with a cylindric nude short **trunk** ± 10 cm $\varnothing$, with a crown of crowded, almost straight or slightly curved, spreading, ascending Br.; **Br.** up to 60 cm long, ± 3 cm $\varnothing$, sometimes slightly constricted into 2–3 relatively long segments, (3–)5(–6)-winged, wings thin, up to 1.2 cm wide, 1.5–2 mm thick, margins weakly armed, sinuate toothed, Th. in pairs, widely divergent, horizontally spreading, 4–4.5 mm long, ± 10 mm apart along the angles, Th.shields usually becoming confluent into a continuous or sub-continuous horny grey margin with age; **L.** sessile, 3–4 mm long, ± 1 mm wide, caducous; **Infl.** glabrous, yellow, comprising a single, very short pedunculate cyme of 3 horizontally arranged Cy.

**E. monacantha** PAX (Gr. 27; morphologically should not be incl. in this Gr., since the 2 lateral stipular Th. are much reduced.). – Ethiopia; N.Somalia. – Leafless semi-♄ 10–20 cm tall; **Br.** tuberculate, ± 1 cm thick, Th.cushions grey, not confluent, with a stout, conspicuously solitary Th. 1 cm long or more and with 1 much reduced stipular Th. on each side; **Cy.** 2, yellow.

**E. montieri** HOOK. f. (Pl. 62/9; 76/3) (Gr. 9) (*E. baumii* PAX, *E. marlothii* PAX, *E. longibracteata* PAX). – Angola; S.W.Afr.: Gr. Namaqualand; Botswana. – **Caudex** clavate, sometimes ± cylindrical, stout, tuberculate, with dried persistent erect and incurved Ped. at the tip, up to 10 cm $\varnothing$, usually 30 cm (rarely up to 1 m) tall, usually simple, occasionally with a few trunks from the base or higher, with tubercles in dense spirals, disappearing in age; **L.** solitary from the tubercle-tips, linear, spatulate-oblong or narrow-elliptical, pointed or blunt, usually apiculate, narrowing cuneately to the 1 cm long petiole, up to 21 cm long, 6–30 mm across; **Infl.** an umbel with Ped. 20 cm long, bracts in whorls, up to 15 cm long and gradually tapering.

**E.** — ssp. **ramosa** LEACH. – Botswana; Transvaal. – More freely branching; **Br.** rather thin, tubercles in laxer spirals; lower part of the **Infl.** often persisting and developing further Infl.

**E. moratii** RAUH (Gr. M.VII). – Madag.: Antsingy near Bekopaka. – **Caudex** napiform, thick, fleshy, ± 10 cm long, 2–4 cm thick, neck up to 2 cm long, 1 cm thick, St.axis covered with L.scars; **L.** 10–12 in a dense basal Ros., with a 1 cm long, recurved petiole, wine-red, upperside grooved, with 2 parchment-like laciniate Stip. at the base, lamina lanceolate, up to 9 cm long, 2 cm across, rounded below, pointed above, with a narrow, red, undulating border, upperside dark green with white blotches, underside reddish-green, nerves prominent; **Infl.** several, forking, Cy. usually 2, ⚥, cyathophylls leather-brown.

**E. muirii** N. E. BR. (Gr. 14). – Cape: Karroo. – Thickened main root and main St. forming a **caudex**; **St.** few from the base, up to 18 cm long, 6–15 mm thick, thicker above, often branching above, tuberculate, smooth, spreading and erect again at the tip, shoots often contorted, up to 3.5 cm long, tubercles rhombic, 5–13 mm long, slightly raised, with a L.scar at the tip; **Ped.** becoming thorny.

**E. multiceps** BGR. (Gr. 15). – Cape: Karroo. – **Caudex** conical with numerous dense, horizontal Br. below, these being 7.5 cm long, 2–3 cm thick, clavate-cylindrical, covered with rhombic or hexagonal tubercles 2–4 mm tall in spirals, smooth, green; Ped. becoming thorny, up to 7 cm long, ± curved, angular.

**E. multiclava** BALLY et CARTER (Pl. 63/25a) (Gr. 25a). – N.Somalia. – **St.** simple, 6–12 cm long, 2–3 cm thick, with 10–16 ribs; **Br.** forming clumps, 2–7 cm long, 1–2.5 cm thick, shortly bifurcate, with up to 16 ribs, tubercles 1–2 mm high, confluent, with Th.shi. above, with 1 pair of Th. 0.5–5 mm long; **L.** scale-like, very small, soon dropping.

**E. multifida** N. E. BR. (Gr. 7). – ? Natal. – Insufficiently known species, related to **E. gueinzii**.

**E. multifolia** WHITE, DYER et SLOANE (Gr. 10). – Cape: Laingsburg D. – Main **St.** with numerous curved Br. forming dense cushions 15 cm tall; **Br.** 13 cm or more long, 2.5–3 cm thick, ± cylindrical, with small, densely tuberculate shoots; **L.** numerous at the Br.tips, linear, plicate, 4 cm long, 2–3.5 cm across; **Ped.** at the Br. tips, becoming thorny.

**E. multiramosa** NEL (Pl. 62/16) (Gr. 16). – Cape: Lit.Namaqualand. – **Caudex** sub-cylindrical, 20 cm tall, c. 12 cm $\varnothing$, with numerous Br. densely arranged around a central zone of 3 cm $\varnothing$; **Br.** 4–11 cm long, 5–15 mm thick, simple or branching, tuberculate, prolonged into a long, Th.-like tip which tends to wither, tubercles triangular, with white T. at their tips; **Ped.** becoming thorny.

**E. mundtii** N. E. BR. (Gr. 4) (*E. decussata* BOISS. p.part.). – Cape: Karroo. – Succulent ♄ 45 cm and more tall, branching from the base; **Br.** with shoots 3–6 mm thick, sideshoots spreading horizontally, segmented, without Th.; **L.** 1–2 mm long and across, brown and reddish.

**E. muricata** THUNBG. (Gr. 4) (*E. brachiata* BOISS. p. part.). – Cape: Clanwilliam D. – Thornless ♄ 45–60 cm tall, with a tuberous root, branching freely from the base and higher, sideshoots often opposite, glabrous, smooth.

**E. namaquensis** N. E. Br. (Gr. 16). – Cape: Lit. Namaqualand.; S.W.Afr.: Gr. Namaqualand. – **Caudex** obconical or cylindrical, main **St.** 4–6 cm ⌀, smooth, green; **Br.** numerous, erect, 2.5–12.5 cm long, 4–7 mm thick, cylindrical and rounded above, often prolonged into a thorny tip, tuberculate, smooth, tubercles in spirals, 1–4 mm tall, cylindrical-conical, often with a whitish hard tip, often curved; **Ped.** almost 5 cm long, becoming thorny.

**E. namibensis** Marl. (Gr. 15). – S.W.Afr.: Gr. Namaqualand. – **Caudex** thick-clavate, arising from the thick napiform taproot, 30–35 cm tall, weighing up to 1 kg, crown up to 15 cm ⌀, densely covered with erect or spreading, 3 cm long **Br.** of pencil-thickness, whitish-grey and smooth; **L.** in a crowded cluster, 3–4 cm long, c. 2 mm across, margins with widely spaced bristly **T.**; **Ped.** becoming thorny, 12 mm long, unbranched.

**E. ndurumensis** Bally (Gr. 28) (*E. taitensis* Pax non Boiss.). – Kenya; Tanzania. – Leafless thorny ♄ up to 2 m tall; **St.** and **Br.** thin, 4–8 mm ⌀, shoots opposite or alternate, 4-angled, becoming terete, the angles made prominent by the short triangular **T.**; **L.** much reduced, minute, scale-like; **Th.** thin, blackish, in pairs, 2–5 mm long, with or without a pair of minute stipular Th., Th.shi. not confluent.

**E. neglecta** N. E. Br. (Pl. 76/2) (Gr. 24). – Ethiopia. – ♄ up to 10 m tall; **St.** with **Br.** in whorls; **Br.** divided into segments 10–30 cm long, light or dark green, with 5–8 broadly alate angles, compressed, little sinuate, somewhat undulating and curving, sides much depressed, with double horny nerves, Th.pairs c. 3 cm apart, 3 mm long, almost black, stout, situated deep in the indentations; **L.** small, soon dropping.

**E. nelii** White, Dyer et Sloane (Gr. 15) (*E. meyeri* Nel). – Cape: Lit. Namaqualand. – **Caudex** erect, tuberculate, 15–18 cm tall, 1.5 to 3 cm thick, with numerous, erect, 15–30 mm long, clavate Br. above, tubercles small, curved; **Ped.** usually 4, up to 3.5 cm long, soon dropping.

**E. neohumbertii** P. Boit. v. **neohumbertii** (Pl. 76/4) (Gr. 2; M.II). – N.Madag. – **St.** simple or with few Br., thickened above, 5-angled, vivid dark green, often with lighter horizontal bands, grey-corky below, 20–25 cm tall, c. 5 cm ⌀ in the upper thickened part; **L.** developing at the tip after the flowering period, deciduous, rosulate, oval, pointed, c. 10 cm long, 6.5 cm across, narrowing towards the short broad petiole, upperside bluish-green, lighter on the underside, stipular Th. stout, leather-brown, up to 1.5 cm long, accompanied by small, bristle-like, generally alate Th. 2–5 mm long uniting to form a border on the older St.-sections; **Infl.** with 4–8 Cy., cyathophylls green with vivid crimson tips.

**E. —** v. **aureo-viridiflora** Rauh. – Cyathophylls yellow-green.

**E. neriifolia** L. (Pl. 63/21; 76/1) (Gr. 21). – Peninsular India. – Becoming a ♄; **St.** slightly 5-angled, becoming round; **Br.** ± in whorls, with 5 slightly spiralled angles, thick, light green, becoming grey, Th. short and black; **L.** 7–12 cm long, obovate, tapering, fleshy to leathery, light green, deciduous.

**E. —** cv. Cristata. – Monstrous cultivar.

**E. nesemannii** R. A. Dyer (Gr. 19) (*E. fimbriata* N. E. Br., *E. latimammillaris* Croiz. p.part.). – Cape: Robertson, Worcester D. – Short main St. and tuberous root forming a ± cylindrical **caudex**, branching basally, often with 5 or more short St. above, 8–40 cm long, up to 3 cm thick, usually thicker above, with 6–14 contorted angles 2–3 cm high, divided into tubercles by horizontally impressed lines; Ped. along the angles, thorny, pointed.

**E. neutra** Bgr. (Pl. 76/6) (Gr. 24). – Origin unknown. – Resembles **E. abyssinica**; St. and Br. with 5–6 angles, with distinct segments narrowing above, the angles alate, rather robust, little sinuate, Th. pairs 1–1.5 cm apart, Th. 4–5 mm long.

**E. nigrispina** N. E. Br. (Gr. 24). – Somalia. – Thorny leafless ♄; **Br.** 8–10 mm ⌀, the 4 angles with a continuous, horny, brown border, Th. 8–16 mm long, aciculate, in pairs, 8–16 mm apart, with 2 minute stipular Th.

**E. nivulia** Ham. (Pl. 76/5) (Gr. 21) (*E. varians* Haw., *E. helicothele* Lem.). – Peninsular India. – Branching ♄; **St.** and **Br.** terete; **Br.** in whorls of 4, c. 2.5 cm ⌀, L.cushions rather conical, compressed laterally, Th. directed downwards, short, blackish; **L.** obovate, tapering, narrowing below towards the short petiole, otherwise resembling **E. neriifolia**.

**E. noxia** Pax (Gr. 1). – Somalia. – ♄ up to 3 m tall; **Br.** rather fleshy; **L.** crowded at the Br.tips, up to 15 cm long, 4 cm across, sessile, narrowly ovate to oblong, apiculate; **Infl.** an umbel. Sap used to poison arrow-heads.

**E. nubica** N. E. Br. (Gr. 4). – Sudan: Nubia; Ethiopia. – Succulent, leafless, thornless ♄ up to 2 m tall, short shoots alternate, ± clustering, very little divergent, 6–30 cm long, 2–3 mm thick, terete, with L.scars; **L.** unknown, early deciduous; **Infl.** an umbel.

**E. nyassae** Pax (Gr. 28) (*E. tetracantha* Pax). – Malawi; Tanzania. – Thorny, leafless ♄ up to 30 cm tall; **Br.** and shoots 8–13 mm ⌀, sub-cylindrical, slightly angular, tubercles arranged in spirals, not very prominent, each with a horny grey Th.shi. with 2 Th.pairs, the lower Th. 3–4 mm long, the upper ones 1–2 mm long; **L.** very rudimentary, scale-like; **Infl.** solitary, with 2 yellowish-red cyathophylls.

**E. obesa** Hook. f. (Pl. 77/5) (Gr. 18). – Cape: Graaff Reinet D. – **St.** simple, unbranched, spherical or rather taller than broad, 8–12 cm ⌀, with 8 broad vertical ribs, with very small, brownish, blunt T., furrows distinct, these and the crown scarcely depressed, light grey-green, with reddish-brown longitudinal and transverse stripes. Dioecious.

**E. —** × **E. meloformis** Ait. – Garden hybrid.

**E. —** × **E. ferox** Marl. – Natural hybrid. – Cape: Graaff Reinet D. – Almost intermediate; **Bo.** somewhat more slender than in **E. obesa**,

Plate 98. 1. Haworthia tuberculata v. acuminata v. POELLN.; 2. H. translucens HAW. (Photo: J. BROWN); 3. H. turgida v. suberecta v. POELLN. (Photo: as 2); 4. H. variegata L. BOL.; 5. H. viscosa v. torquata (HAW.) BAK. (Photo: J. MARNIER-LAPOSTOLLE); 6. H. vittata BAK.; 7. H. zantneriana v. POELLN.

Plate 99. 1. **Hechtia argentea** Bak.; 2. **Hesperaloe parviflora** Coult. (Photo: J. Marnier-Lapostolle); 3. **Hoya carnosa** (L.) R. Br.; 4. **Hoodia macrantha** Dtr. (Photo: W. Triebner); 5. **Hoodiopsis triebneri** Luckh. v. **triebneri** (Photo: J. Luckhoff).

Plate 100. 1. Huernia brevirostris N. E. Br. v. brevirostris (Photo: H. Lang); 2. H. nouhuysii Verd. (Photo: as 1); 3. H. insigniflora C. A. Maass (Photo: as 1); 4. H. oculata Hook. f. (Photo: as 1); 5. H. primulina N. E. Br. v. primulina (Photo: as 1); 6. H. zebrina N. E. Br. v. zebrina (Photo: as 1).

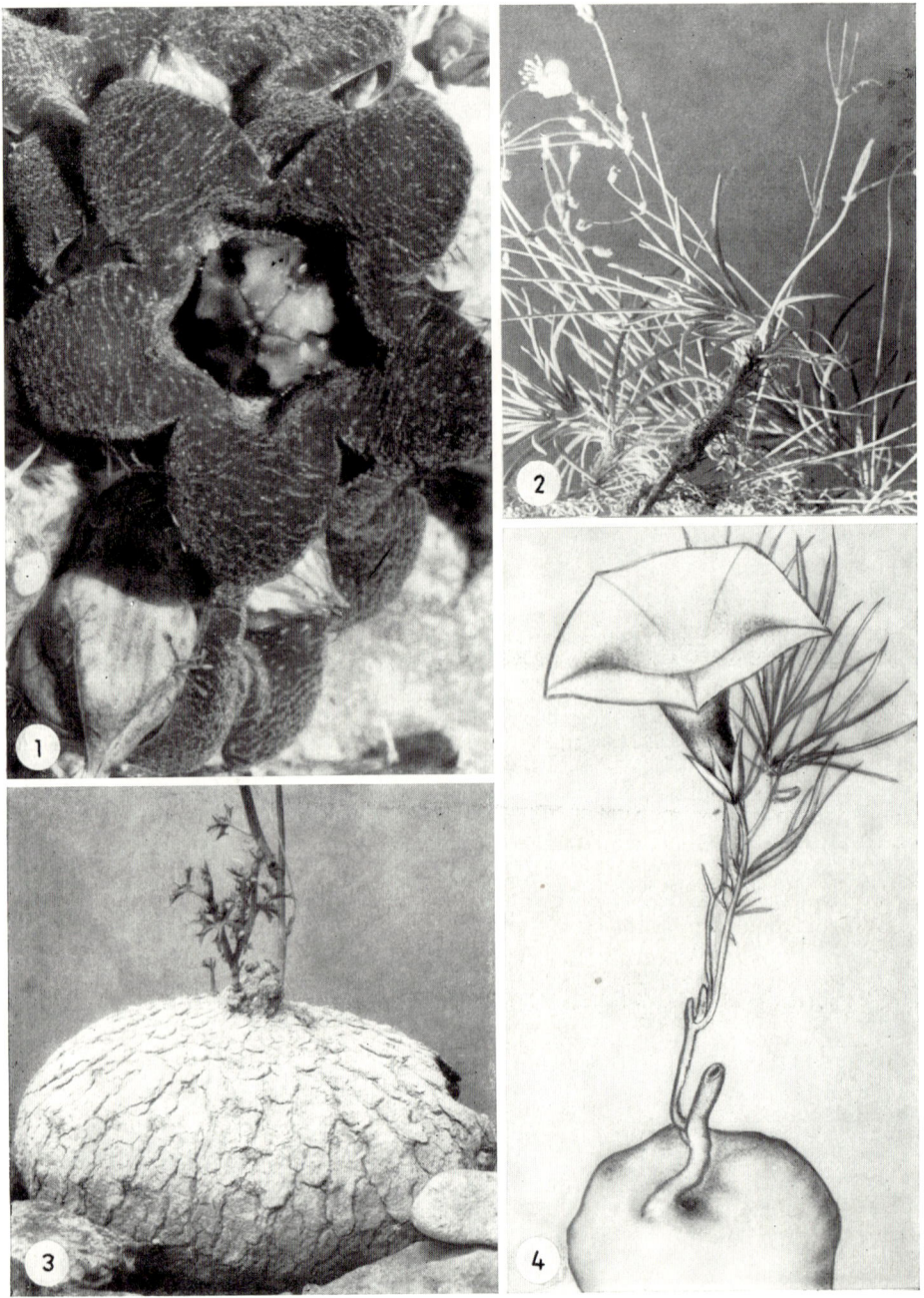

Plate 101. 1. **Huerniopsis decipiens** N. E. Br. (Photo: H. Lang); 2. **Hypertelis verrucosa** (Eckl. et Zeyh.) Fenzl v. **verrucosa** (Photo: J. Marnier-Lapostolle); 3. **Ibervillea sonorae** (S. Wats.) Greene; 4. **Ipomoea angustisecta** Engl. (Repr.: Bot. Jahrb. 1889, Pl. VII).

Plate 102. 1. **Idria columnaris** KELLOG (Photo: C. BACKEBERG); 2. **Impatiens tuberosus** PERR.; 3. **Jatropha berlandieri** TORR.; 4. **J. podagrica** HOOK.

Plate 103. 1. Kalanchoe beauverdii HAMET v. beauverdii; 2. K. beharensis DRAKE DEL CASTILLO v. beharensis; 3. Left: K. daigremontiana HAMET et PERR., right: K. tubiflora (HARVEY) HAMET; 4. K. fedtschenkoi HAMET et PERR.

Plate 104. 1. **Kalanchoe eriophylla** HILSENB. et BOJ. (Photo: J. MARNIER-LAPOSTOLLE); 2. **K. farinosa** BALF. f. (Photo: J. BOGNER); 3. **K. grandiflora** WIGHT et ARN.; 4. **K. jongmansii** HAMET et PERR. v. **jongmansii** (Photo: as 1).

Plate 105. 1. Kalanchoe longiflora v. coccinea Marn.-Lap.; 2. K. millotii Hamet et Perr. (Photo: J. Marnier-Lapostolle); 3. K. rhombopilosa Mann. et Boit. (Photo: as 2); 4. K. prolifera (Bowie) Hamet; 5. K. marmorata Bak.

simple or branching a little from the base, with 6–9 rather more prominent angles, with small thorny Ped. along the angles, with horizontal bands.

**E. — × E. submammillaris** L. – Garden hybrid. – **Bo.** as **E. obesa,** more deeply furrowed, with several side-shoots.

**E. obovalifolia** A. RICH. descr. BALLY (Pl. 77/1) (Gr. 24) (*E. winkleri* PAX). – Ethiopia to Tanzania. – Thorny ♄; **Br.** fleshy, with segments 5–11 cm long, 2.5–6 cm across, triangular, angles alate and sinuate-dentate, T. 6–12 mm apart, 1.4–5 mm long; **L.** petiolate, 2–12.5 cm long, 1.2–4.5 cm across, oblong to elliptical or long-ovate, rounded above, pointed, deciduous, Th. in pairs, 1.5–4 mm long, brown, often missing, Th.shi. little or shortly decurrent; **Infl.** sessile, with 3 Cy.

**E. obtusifolia** POIR. v. **obtusifolia** (Gr. 4). – Canary Is.: Tenerife. – ♄ 1–2 m tall, branching in whorls of 3 or more; **Br.** virgate, fairly densely leafy; **L.** 5–7 cm long, 4–6 mm across, sessile, linear, blunt or short-pointed; **Cy.** 4–6 in a terminal umbel.

**E. — v. regis-jubae** (WEBB et BERTH.) MAIRE (*E. r.-j.* WEBB et BERTH.). – Canary Is.; Morocco. – **Br.** 3–5 in a whorl; **L.** narrow-linear; umbel with 6–10 rays. A moisture-loving form from HUBERT MÜLLER, Tunis, is much stouter, with the **L.** appreciably broader; **Fl.** and **Fr.** larger. Habitat: Cape Verde Is.

**E. — v. — sv. pseudodendroides** RIKLI. – S.W.Morocco. – Arborescent sub-variety.

**E. officinarum** L. v. **officinarum** (Pl. 77/2) (Gr. 24). – N.Afr. – Thorny succulent ♄; **St.** erect, up to 1 m tall, with few Br., 6–8 cm ⌀, angles (9–)11(–13), acute, almost straight, little sinuate, not deeply furrowed, with a horny border, Th. in pairs, very uneven, 5–15 mm long, whitish-grey, often directed downwards. – This species, reputedly discovered by King JUBA, is said to have been named after his doctor EUPHORBIUS.

**E. — v. beaumerana** (HOOK. f. et COSS.) MAIRE. (*E.b.* HOOK. f. et COSS.). – Morocco. – 25 cm–2 m tall; **St.** clavately thickened above; **Br.** numerous, usually higher than the St., with 9–10 straight angles, Th. on confluent Shi. Some populations have Th. 2 cm long, or red in colour, in others the Th. are quite short or completely absent.

**E. oncoclada** DRAKE (Pl. 77/3) (Gr. 4). – Madag. – Thornless succulent ♄ with a few erect Br.; **St.** almost 2 cm thick, with segments 2–4 cm long, tapering towards both ends, covered with. L.scars in spirals, grey-green; **L.** 2–3 mm long, 1–2 mm across, linear, pointed, plicate, soon dropping.

**E. opuntioides** N. E. BR. (Gr. 24). – Angola: Malanje D. – ♄; **St.** low, branching basally; **Br.** ascending or prostrate, fleshy to woody, flattened-compressed, c. 15 cm long, leafless, remarkably like an **Opuntia,** segments (2.5–)3(–8)cm long and across, margins undulating, coarsely crenate or sinuate, with a few thin Th. 3–4 mm long; **Cy.** 3 in an Infl.

**E. orabensis** DTR. (Gr. 15). – S.W.Afr.: Gr. Namaqualand. – **Caudex** slender-clavate with a long taproot, usually c. 15 cm tall with the crown 5 cm ⌀, with St. 3 cm long, 5–6 mm thick, branching; **L.** green, 3–4 cm long, 2 mm across, curved inwards to form a groove; without thorny Ped.

**E. ornithopus** JACQ. (Gr. 17) (*Dactylanthes patula* HAW., *E. p.* SWEET). – Cape: Karroo. – **Caudex** forming a continuation of the tuberous root, rhizomes cylindrical, running horizontally, producing small Br. 1–3 cm long, 6–10 mm thick, thickened above, with 3–5 prominent conical tubercles; **Cy.** on extended, dichotomously-branching Ped.

**E. orthoclada** BAK. (Pl. 77/4) (Gr. 1). – Central and S. Madag. – Scrambling or prostrate ♄, branching freely; **Br.** terete, fleshy, becoming woody; **L.** oblong-lanceolate, falling; **Cy.** at the Br.tips.

**E. — ssp. orthoclada** (*E. lohaensis* H. BN., *E. cynanchoides* DRAKE). – Erect to prostrate; **Cy.** up to 3 in an umbel.

**E. — ssp. vepretorum** (DRAKE) J. LEANDRI (*E. v.* DRAKE). – Climbing; **L.** up to 2 cm long, 2–4 mm across; **Cy.** 2–3.

**E. oxystegia** BOISS. (Gr. 9) (*E. bupleurifolia* E. MEY., *E. stegmatica* NEL). – Cape: Lit. Namaqualand. – **Caudex** succulent, 15–20 cm tall, irregularly branching, Br.thornless, up to 16 mm thick, cylindrical, grey-green, with several tubercles in spirals; **L.** in crowded Ros. at the Br.tips, lanceolate, 3–10 cm long, 6–14 mm across, short-petiolate, often plicate; **Ped.** becoming thorny.

**E. pachypodioides** P. BOIT. (Pl. 78/4) (Gr. M.V.) (*E. antankara* J. LEANDRI). – N.Madag. – **Caudex** columnar, unbranched, up to 50 cm tall, up to 5 cm ⌀, with a grey bark, densely covered with prominent, tuberculate L.scars, arranged in 8–12 spiral R., flanked by stout stipular Th. up to 0.5 mm long; **L.** in a terminal Ros., 10–12 cm long, 2–5 cm across, apiculate, narrowing towards the short, thick petiole, vivid green, deciduous; **Infl.** a cyme with 20–40 Cy. with long Ped., cyathophylls dark purple-red.

**E. paganorum** A. CHEV. (Gr. 24). – Sudan; Port. Guinea; Nigeria. – Candelabra-like ♄ 1–1.5 m tall; **Br.** verticillate, cylindrical, 3 to 4 cm ⌀, with well-spaced tubercles arranged in spirals, 12–15 mm long, 8–10 mm across, with a small Th.shi., Th. in pairs, 8–12 mm long; **L.** clustered at the Br.tips, 6 cm long, 3–5 cm across, apiculate, narrowing towards the very long petiole; **Infl.** small, with 2 Cy.

**E. parciramulosa** SCHWEINF. (Gr. 24) (? *E. canariensis* FORSK.). – S.Arabia: Yemen. – Short-stemmed ♄ over 3 m tall; **Br.** dense, almost unbranched, straight and ascending, not segmented, dark green, with 3–4 slightly sinuate angles, sides flat, 4–5.5 cm across, Th.shi. confluent to form a horny border, Th. in pairs, thick, 4–5 mm long; **L.** oblong-ovate, up to 8 mm long.

**E. pauliana** URSCH et LEANDRI (Gr. M.III). – Madag. – St. unbranched, up to 40 cm tall, up to 4 cm thick, with a grey-green bark, covered with the broadly oval scars of fallen L., with 0.5 cm long and broadly-based stipular Th. arranged in spiral R.; **L.** rosulate at the Br.tips, oblong-oval, 18–25 cm long, up to 4 cm across; **Infl.** repeatedly forking, with up to 300 Cy.

**E. paxiana** Dtr. (Gr. 4). – S.W.Afr. – ♄, little branched with few St., up to 70 cm tall; **Br.** 5 mm thick, slender, erect, soft, blue-green; **L.** at the Br.tips, c. 4 cm long, 8 mm across, soft.

**E. pedilanthoides** M. Denis (Gr. M.IV). – W.Madag. – Thorny ♄ 50–100 cm tall, branching basally, laxly leafy, Th. in pairs, 1 cm long, often with 2 other smaller Th.; **L.** small.

**E. peltigera** E. Mey. (Gr. 5). – Cape: Lit. Namaqualand. – Thornless succulent ♄ forming large clumps, c. 30–40 cm tall, with a tuberous root and short **St.**, branching basally, re-branching above; **Br.** 6–12 mm thick, minutely velvety at first, with opposite, conical, spreading, prominent and confluent L.cushions.

**E. pendula** Boiss. (Gr. 4) (*Tithymalus pendens* Haw.). – Cape: Cape D. – Freely branching ♄; **Br.** long, pendulous, 3–5 mm thick, dividing dichotomously, terete, dull green, with minute white spots, with segments 5–8 cm long; **L.** small, scale-like, soon dropping.

**E. pentagona** Haw. (Gr. 19) (*E. heptagona* Bgr., *E. tetragona* Sim. p.part.). – Cape: Karroo. – ♄ up to 3 m tall; **St.** erect, 3–4 cm thick, with Br. in irregular whorls; **Br.** erect, with 5–8 angles, 5–6 mm high, separated by sharp furrows, margins with downward-pointing T. or slightly crenate, with small transverse furrows in the indentations, from which the stout, 15–19 mm long, yellowish, thorny Ped. are produced at intervals.

**E. pentops** Marl. descr. White, Dyer et Sloane (Pl. 62/15) (Gr. 15). – Cape: Lit. Namaqualand. – **Caudex** compact, partly underground, with a number of **Br.** above, covered with conical tubercles in spirals; **L.** at the Br.tips, 15–20 mm long, linear, deciduous; **Ped.** becoming thorny.

**E. perangusta** R. A. Dyer (Gr. 24). – Transvaal. – Succulent thorny ♄ up to 1 m tall; main **St.** very short, with several St.-like Br. bearing numerous flowering shoots, constricted to form segments 1.5–9 cm long and thickest below, 5 cm ⌀ above, with alate angles, Th.shi. confluent to form a horny border, Th. in pairs, 13 mm long.

**E. perpera** N. E. Br. (Gr. 4) (*E. brachiata* E. Mey. p.part.). – Cape: Lit. Namaqualand. – Compact succulent thornless ♄, branching from the base or higher; **Br.** and shoots opposite or forked, projecting, fleshy; L. 1 mm long, broadly triangular.

**E. perrieri** Drake v. **perrieri** (Gr. M.III). – S.W.Madag. – Branching ♄ up to 2 m tall; **Br.** deciduous, Th. large, very closely spaced, arranged in spirals and thus forming angles; L. in crowded clusters at the Br.tips, oblong-lanceolate, 25 cm long, 4 cm across; **Infl.** a cyme, with 15–30 Cy.

**E. —** v. **elongata** M. Denis. – Robuster variety; **Br.** 5 cm ⌀.

**E. persistens** R. A. Dyer (Gr. 22). – Moçambique. – **Caudex** napiform, 30 cm long, 15 cm thick at the crown, divided into thick trunks above, completely buried in the ground, with many erect Br. above ground; **Br.** 10–20 cm long, at first 13–20 mm thick, later constricted into uneven segments and 2–3 cm thick, with (3–)4(–5) irregularly tuberculate angles, with furrowed sides, Th.shi. solitary, Th. in pairs, up to 15 mm long, brown, with 2 small lateral Th.

**E. persistentifolia** Leach (Gr. 24). – Rhodesia; Zambia. – Stout thorny succulent ♄ up to 2 m tall; main **St.** stout, angular; **Br.** verticillate, erect; less often ♄ up to 3.5 m tall, with the spreading Br. in a crown; **Br.** stout, 2–2.5 cm ⌀, with 4–5 sinuate-tuberculate and more rarely alate angles, usually with a whitish horny border, sides flat or concave, divided into 40 cm long segments, Th. in pairs, (3–)7(–15) mm long, 1–2 cm apart, with a smaller pair of Th. at the base; **L.** up to 10 cm long, 2.5 cm across, somewhat fleshy, pointed; **Infl.** cymose.

**E. phillipsiae** N. E. Br. (Pl. 78/1) (Gr. 24). – Somalia; S.Arabia. – ♄; **St.** and Br. 2–3 cm thick, with 9 angles separated by sharp furrows, 4 mm high, somewhat tuberculate, Th.shi. almost confluent, 2–5 mm long, brown, Th. in pairs, 8–16 mm long, slender, brown; **L.** minute, triangular, deciduous.

**E. phosphorea** Mart. (Pl. 78/2) (Gr. 5). – Brazil: Bahia. – Succulent ♄; **Br.** erect, with at least 6 longitudinal angles, segmented; **L.** minute; appearance otherwise as **E. pteroneura.** Phosphorescent.

**E. pillansii** N. E. Br. (Gr. 19) (*E. stellaespina* Phill.). – Cape: Ceres Karroo. – **St.** up to 30 cm tall, branching basally or higher; **Br.** 3–5 cm thick, roundish, with 7–9 angles, often with only 5 in young Br., with tubercles 1–1.3 mm high, sides flat, with light and dark green transverse bands; **Ped.** 8–14 mm long, becoming thorny.

**E. piscatoria** Ait. (Gr. 1). – Madeira. – ♄ or ♄ 2–3 m tall and across; **St.** over 20 cm ⌀, with grey-brown, granular bark; **Br.** erect to projecting, light grey, with brown L.scars running transversely, shoots in whorls of 6–12, curving upwards.

**E. plagiantha** Drake (Gr. 4). – Madag. – ♄ 5 m tall, with succulent Br.

**E. planiceps** White, Dyer et Sloane (Gr. 17). – Cape: W.Griqualand, Hay D. – **Root** tuberous and often dividing into carrot-like or longer rootlets passing over into the main St., with numerous dense St. from the crown and often rebranching several times, generally forming a flat or plate-shaped cushion 30 cm ⌀; **Br.** 2–2.5 cm long and 1 cm ⌀, with almost cylindrical tubercles 3 mm high; **L.** small, dropping; **Ped.** becoming thorny, 1.5–2.5 cm long.

**E. platycephala** Pax (Gr. 26). – Tanzania; W.Afr. – ♃; **St.** rather fleshy; **L.** sessile, linear-oblong, pointed, 5–6 cm long, up to 1 cm across.

**E. platyclada** Rauh v. **platyclada** (Gr. 4; M.VIII). – Madag. – Small ♄ with spreading

Br., up to 50 cm tall, shoots prostrate, creeping to ascending or erect, succulent, with thin, freely branching, adventitious roots, old St. roundish to slightly flattened, c. 5 mm ⌀, with a minutely scaly, peeling, waxy layer, with thickened L.bases and with horizontal to curving, rooting shoots from the base of the axils of erect shoots, these shoots up to 1 cm thick below, thereafter forming flat, grey to violet shoots, up to 1 cm across, 2 mm thick and (5–)10(–15) cm long, with dark or very dark red lines, at first with solitary, bent bristles and tubercles; **L.** distichous on long shoots, scarcely 1 mm long, with bristles, ephemeral, L.bases swollen and continuing as a raised line to the next node; **Cy.** 1–5, terminal, on the flat shoots.

**E. — v. hardyi** RAUH. – E.Tuléar. – Slighter and less freely branching; **Br.** up to 40 cm long, ascending shoots ± round or made slightly angular by the decurrent L.bases, 2–3mm⌀, leaflike segments rhomboid, 2–4 cm long, 1.5 cm across, 1.5–2 mm thick, grey-green, the raised L.bases continuing as a swollen band, with an irregular network of dark green tubercles.

**E. poissonii** PAX (Gr. 26). – Ghana; Togo; N.Nigeria; Dahomey; Ivory Coast. – ♄ 1.2 to 1.6 m tall; **Br.** cylindrical, 3–3.5 cm ⌀, often indistinctly tessellate, thornless; **L.** 5–6 at at the Br.tips, obovate to oblong, truncate above with 2 distinc tlobes 2–14 cm long, 4–6 cm across, fleshy; **Infl.** a cyme.

**E. polyacantha** BOISS. (Pl. 79/5) (Gr. 24). – Ethiopia. – Dwarf, branching ♄; **Br.** ascending, grey-green, somewhat segmented, with 4–5 angles having a continuous, brown, horny border, sides flat, with indistinct curved nerves to the Th.pairs, Th. 5 mm apart, divaricate, almost pectinate, projecting, grey with black tips.

**E. polycephala** MARL. (Gr. 17). – Cape: Cradock D. – Main St. and tuberous root forming a compact **caudex,** dividing below into several carrot-like roots, with numerous, short, dense Br. above, forming a clump 30–70 cm ⌀; **Br.** ± conical, tuberculate.

**E. polygona** HAW. (Gr. 19). – Cape: Karroo. – Succulent ± thorny ♄; **St.** 50–70 cm tall, 7–10 cm thick, branching basally, with 12 to 30 ribs, the angles straight or spirally contorted, c. 15 mm high, slightly dentate, the furrows sharp; **Ped.** few, 1 cm long, becoming thorny. Host plant of the mistletoe **Viscum minimum.**

**E. primulifolia** BAK. (Gr. M.VII) (*E. subapoda* H. BAILL.). – Madag. – Developing a **caudex** 10–15 cm long, 5–7 cm thick, with a thick, brown, fissile, corky covering, tapering above to the shortened main **St.**; **L.** 6–12 in a radical Ros., oblong-oval, margins often rather undulating or crenate, rather thick, dull green, narrowing towards the flattened petiole; **Infl.** from the axils of bracts, with Ped. 4 cm long, with 4 Cy., cyathophylls white, pink to violet.

**E. proballyana** LEACH (Gr. 24). – Tanzania. – Densely branching, rather flat-topped ♄ up to 1.3 m tall and 1.5 m ⌀; **Br.** in whorls, with 4–5 sinuate-dentate angles, T. up to 3 mm high, sides ± parallel, not very constricted, segments 6–20 cm long, 1–3 cm thick, Th.shi. broadly decurrent, Th. in pairs, 8–10 mm long, grey with black tips.

**E. pseudobrachiata** DTR. (Gr. 4) (*E. brachiata* E. MEY.). – S.W.Afr. – Semi-♄ with a tuberous root; **Br.** forking or tripartitely branched, thick as a quill, grey-green, side shoots becoming ± thorny; **L.** very small, triangular.

**E. pseudocactus** BGR. (Pl. 78/3) (Gr. 24). – Natal. – Thorny succulent ♄ with the reduced main St. usually hidden in the soil; **Br.** numerous, suberect, less often branching, usually with 3 angles below, with 4–5 coarsely dentate angles above, with a distinct horny border, sides with broad yellowish-green U-shaped curving lines, segments 10–15 cm long, up to 5 cm thick, Th.pairs up to 12 mm long, brown.

**E. pseudoduseimata** WHITE, DYER et SLOANE (Gr. 14). – S.W.Afr. – Related to **E. duseimata** and doubtfully distinct. Ped. persistent.

**E. pseudoglobosa** MARL. (Gr. 18) (*E. frickiana* N. E. BR.). – Cape: Lit. Karroo. – Taproot thick; **St.** numerous, oblong to round, c. 22 cm long, 15 mm thick, almost smooth, with 5–6 furrows towards the tip, the angles indicated by a few L.scars, brownish green.

**E. pseudograntii** PAX (Gr. 1). – Tanzania. – Semi-♄ 1–2 m tall; **L.** with a petiole 2 cm long, passing over into the blade, 25 cm long, 6 cm across, L. on the Ped. round-cordate or shortly caudate-tapering; **Infl.** an umbel, Cy. broadly cup-shaped.

**E. pseudohypogaea** DTR. (Gr. 8). – S.W.Afr. – Root light-coloured, soft, looking like a horse-radish, producing from the thickened neck 2–5 clavate Br. 10 cm long, up to 3 cm ⌀, dark green; **L.** clustered at the Br.tips, c. 5 cm long, narrow-lanceolate.

**E. pseudotuberosa** PAX (Gr. 7). – Transvaal, Cape; Botswana. – ♃ with 1 or more extended, fleshy, underground shoots from the tuberous **caudex**; **Br.** several, herbaceous, leafy, 1–5 cm tall; **L.** 12–37 mm long, 1.5–4 mm across, linear-lanceolate, pointed, grooved, glabrous, bluish-green.

**E. pteroneura** BGR. v. **pteroneura** (Pl. 79/1) (Gr. 5). – ? Mexico. – Succulent ♄ seldom over 50 cm tall; **Br.** ± pencil-thick, with 5–6 angles formed by the 3 decurrent angles from each L.base; **L.** alternate, ovate-lanceolate, short-petiolate, 2–4 cm long, 1–2 cm across, deciduous.

**E. — v. macdougallii** CROIZ. – No diagnosis available.

**E. pubiglans** N. E. BR. (Gr. 9). – Cape: Karroo. – Related to **E. bupleurifolia**; **caudex** up to 30 cm tall, rarely branching, c. 3.5 cm ⌀ above, with closely spaced, grey, rhomboid or hexagonal tubercles in 13 spiral R.; **L.** 20–35 mm long, linear, almost completely folded together; cyathophylls up to 25 mm ⌀.

**E. pugniformis** BOISS. (Pl. 62/13; 79/2) (Gr. 13) (*E. caput-medusae* v. δ L., *E. procumbens* SWEET, *E. p.* MILL. ?).–Cape: Karroo. – Main St. and thick taproot together forming a hemispherical **caudex** 5–8 cm ⌀, partly buried, truncate above, with a crown of Br. around the slightly depressed and densely tuberculate

**Euphorbia**

centre; **Br.** in 2–3 R., spreading or curved slightly upwards, 12–20 cm long, 6–8 mm thick, with rhombic or slightly hexagonal tubercles, 2–3 mm ⌀.

**E. pulvinata** MARL. (Pl. 79/3) (Gr. 19). – Cape, Natal, Orange Free State, Transvaal; Lesotho. – Branching basally; **Br.** numerous, of equal length, forming a low and slightly convex cushion up to 1.5 m ⌀, 3–6 cm tall, 3–4 cm ⌀, with (6–)7(–8) slightly crenate angles 7–9 mm high; **Ped.** numerous along the angles, wine-red or purple-brown, becoming thorny, 10–15 mm long.

**E. punicea** SWARTZ (Gr. 1). – Jamaica; Cuba; Bahamas. – ♄ up to 10 m tall; **St.** tripartitely branching; **Br.** with L.scars; **L.** almost rosulate towards the Br.tips, oblanceolate, blunt, mucronate, narrowing below towards the short petiole; **Infl.** an umbel with 2 bracts together, red.

**E. pyrifolia** LAM. (Gr. M. II). – Madag.; Reunion; Mauritius; Seychelles. – ♄; **Br.** thickened at the tips. Related to **E. lophogona**.

**E. qarad** DEFL. (Gr. 24). – Arabia. – Thick fleshy ♄ 2–2.5 m tall, 12–20 cm ⌀ at the base; **St.** simple below, branching above; **Br.** 1.5 to 2 cm ⌀, greenish to bluish, curved inwards or ascending, with 6 alate angles, stipular Th. short, divaricate, 3–4 mm long, Th.cushions conical, tuberculate, hexagonal; **L.** few along the angles, dropping.

**E. quadrangularis** PAX (Gr. 28). – Tanzania. – Related to **E. heterochroma**; ♄ with succulent, leafless, thorny St. and Br.; **Br.** 8–10 mm thick, white to bluish-pruinose, with usually 4 much compressed and dentate angles, T. c. 2 mm long, broadly triangular, Th. 2–4 mm long, 6–8 mm apart, with 2 minute stipular Th., Th.shi. confluent to form an almost 1 mm wide continuous margin; **L.** rudimentary, 1.5 mm long, soon dropping; **Infl.** sessile, with 3 Cy.

**E. quadrata** NEL (Gr. 1). – Cape: Lit. Namaqualand. – Succulent to woody ♄ with a tuberous root; **Br.** few, erect, up to 60 cm tall, 3–6 mm thick, tapering above, almost reddish-purple, white-tomentose, striate below; **L.** 3–6 cm long, 4–12 mm across, lower L. spatulate, upper ones linear-spatulate, somewhat grooved.

**E. quartziticola** LEANDRI (Gr. 20; M.VI). – Madag. – Related to **E. primulifolia**, possibly only an ecotype of this; eventually developing a napiform root dividing into several thick lateral roots, with several heads above; **L.** ± stout and leathery, glossy green, petiolate, lamina ovate to oblong-oval, 3–5 cm long, 1 to 3.5 cm across, short-pointed, narrowing below towards the broad petiole 0.5–4 cm long, margins red, Stip. 3.5 mm long, thin, bristly, dropping; **Infl.** short, with 2–4 Cy.

**E. quinquecostata** VOLKENS (Gr. 24). – Tanzania. – ♄ 3–5 m tall; **St.** thick as a man's thigh, Br. forming a spherical crown; **Br.** bent inwards, ± 2 cm ⌀, with 5 thickened angles with a continuous, horny border, Th.pairs c. 1 cm apart, 1–2 mm long.

**E. ramiglans** N. E. BR. (Gr. 14). – Cape: Lit. Namaqualand. – Related to **E. bergeri**; insufficiently known species.

**E. ramipressa** L. CROIZ. (Pl. 63/23; 79/4) (Gr. 23) (*E. alcicornis* HORT. non BAK.). – Madag. – Thorny succulent ♄, becoming ± ♄-like; **St.** up to 7 cm thick, bark becoming grey, often branching from the base, with 5 obtuse angles, with 3–4 mm long, dark Th.; **Br.** forming 2–3 R. or pinnately arranged, dark green, often slightly striate, terete below, then triangular, flat above and 2-angled, angles sinuate-dentate, Th. 2–3 mm long, brown.

**E. ramulosa** LEACH (Gr. 24). – Moçambique. – Thorny, succulent, dwarf ♄, usually less than 15 cm tall, cushion-forming; **Br.** and shoots erect, stiff, crowded, with 4 angles, somewhat compressed above, tuberculate-dentate, c. 1 cm thick towards the truncate tip, not constricted to form segments, with tuberculate triangular T. 2.5 mm high, 8–12 mm apart, Th.shi. forming a continuous horny border 1.5 mm across, Th. in pairs, stout, up to 5–8 mm long; **L.** minute, soon dropping, with or without 2 stipular Th.; **Infl.** pedunculate, with 3 Cy.

**E. rangeana** DTR. (Gr. 15). – S.W.Afr.; Gr. Namaqualand. – **Caudex** thick, napiform; **Br.** 5 cm long, green, brown or olive-coloured, 15 mm thick, L.cushions 5–6-angled, very prominent, with round L.scars; **L.** 15 mm long, 3 mm across, short-pointed; **Ped.** solitary, becoming thorny, 1 cm long.

**E. razafinjohanii** URSCH et LEANDRI (Gr. 20; M.VI). – Madag. – Thorny ♄; **St.** branching, 3 cm ⌀; **Br.** narrower at the base, with Th. in irregular R., 5–20 mm long, lateral Th. 1–3; **L.** clustered at the Br.tips, oblong-lanceolate, pale green, with a red border, 6–11 mm long, 2–3 mm across, with a petiole 3 mm long; **Infl.** with 4–8 Cy.

**E. rectirama** N. E. BR. (Gr. 4) (*E. ephedroides* BOISS. p.part.). – Cape, Orange Free State. – Thornless ♄ up to 1 m tall, branching basally or higher up; **Br.** 4–6 mm thick below, shoots opposite, terete, glabrous, Int. 2–7.5 cm long; **L.** 1.5–2 mm long, spatulate.

**E. reinhardtii** VOLKENS (Gr. 24). – Tanzania. – ♄ with a trunk as thick as a man, c. 3 m tall; **St.** candelabra-like, in a flat crown above, 12–15 m tall; **Br.** ascending, 20–30 cm long, slightly segmented, terminal segments shorter and somewhat tapering towards the tip, with 4–5 alate, compressed, straight angles, thick, c. 8 cm across, with 2 opposite angles differing in size from the others, sides deeply furrowed, Th. pairs 2–3.5 cm apart, Th. 1 cm long, divaricate, 3 mm thick at the base, Th.shi. triangular to round.

**E. resinifera** BGR. v. **resinifera** (Pl. 80/3) (Gr. 24) (*E. sansalvador* HORT., *E. mogadorensis* HORT.). – Morocco. – ♄ branching irregularly and freely from the base; **Br.** ascending, c. 50 cm tall, light grey-green, the 4 angles not dentate, the sides somewhat grooved at first, becoming flat, Th. in pairs, 5–10 mm apart 5–6 mm long, brownish. Source of the drug Euphorbium, in use since ancient times.

**E. —** v. **chlorosum** CROIZ. – **Br.** and shoots slighter, green, Th. longer.

**E. restituta** N. E. Br. (Gr. 11) (*E. radiata* E. Mey., *E. graveolens* N. E. Br.). – Cape: Lit. Namaqualand. – Succulent ♄ 24 cm or more tall; main St. with a few **Br.** from the base, re-branching above, cylindrical, c. 6 cm thick, with tubercles in 20 ± spiral R. ± conical, c. 1.5 mm high, decurved at the tip; **L.** 2.5 cm long, lanceolate-spatulate, deciduous; **Ped.** 7–12 cm long, persisting.

**E. restricta** R. A. Dyer (Gr. 22). – Transvaal. – Main root and reduced St. combined in a caudex, dividing above into 2 or several Br., each with a tuft of aerial flowering Br.; **caudex** 4–8 cm ⌀, irregularly tuberculate, tubercles 0.5 mm high; **Br.** up to 16 cm long, usually simple, deciduous, c. 3 cm ⌀, constricted to form segments 1–2 cm long, with 4–5 acute, alate, horny angles, Th.pairs up to 1 cm long, stipular Th. minute; **Infl.** solitary, with 3 Cy.

**E. rhombifolia** Boiss. (Gr. 4) (*E. racemosa* E. Mey. and v. *triceps*, v. *cymosa* and v. *laxa* N. E. Br.). – Cape, Natal, ? Transvaal. – Thornless succulent ♄ 30–70 cm tall, branching from the base or higher up; **Br.** 3–6 mm thick, simple or with 1–5 pairs of thin, opposite, 12.5 to 45 cm long secondary Br., round or angular, smooth or rough, any roughness usually along the angles, which are often rib-like and compressed; **L.** 1–2 mm long, spatulate.

**E. robecchii** Pax (Gr. 24) (*E. pimeleodendron* Pax, *E. ruspolii* Chiov.). – Trop. E.Afr. – In the seedling stage resembling **E. candelabra**; erect, unbranched, sharply 4-angled, dark green with pale green blotches, Th.shi. confluent along the angles, Th. hamate and curved downwards; later with thin lateral **Br.**, with 3 rounded angles, Th.shi. not confluent; eventually forming a ♄ up to 10–15 m tall; **trunk** erect, 2–3 m tall, up to 65 cm ⌀; **Br.** in a ± dense, tangled crown, spreading to ascending, 5-angled, short shoots completely smooth, thornless. Wood used for crates.

**E. rossii** Rauh et Buchloh f. **rossii** (Gr. 20; M.VI). – Madag. – ♄ branching freely from the base, up to 1 m tall; **root** napiform, 30 cm long, 5 cm thick; **Br.** numerous, 3 cm thick below, 1–2 cm thick above, glossy grey to olive-green, with numerous Th. on the thickened podaria, main Th. 1.5–2 cm long, hairy at the tip, with smaller lateral stipular Th., short shoots several mm long, developing from axillary buds; **L.** narrow-linear, c. 4 cm long, 2–3 mm across; **Infl.** short-pedunculate, Cy. almost capitate, cyathophylls olive-green to dirty wine-coloured, pruinose, underside hirsute.

**E. —** f. **glabra** Rauh et Buchloh. – Br. of the Infl., bracts and cyathophylls completely glabrous.

**E. rowlandii** R. A. Dyer (Gr. 24). – Transvaal. – Succulent ♄ 1–3 m tall; main **St.** rather short, branching just above the base; **Br.** numerous, spreading to erect, very rarely re-branching, constricted to form segments 7–15 cm long, 3–5 cm across below, narrowing above to 2 cm across, with 5–7 acute, alate angles with a narrow, horny margin, Th.pairs 1 cm apart, 5–10 mm long, with or without rudimentary stipular Th.; **Infl.** solitary, with 3 Cy.

**E. royleana** Boiss. (Pl. 80/1) (Gr. 21) (*E. pentagona* Royle). – Peninsular India. – ♄ 6–8 m tall; **St.** 40–50 cm ⌀; **Br.** ascending, 5–7 cm ⌀, with 5 slightly sinuate, ± straight angles, Th.shi. 12–15 mm apart, Th.pairs 4–5 mm long, directed downwards.

**E. rubella** Pax (Gr. 8). – N.E.Afr. –Dwarf, glabrous; **caudex** subterranean, cylindrical, tuberculate, rough, up to 5 cm tall, 0.5–3 cm thick, only the tip projecting above ground; **L.** from the St.tip, rosulate, resting on the ground, narrow-elliptical, narrowing towards the short petiole, 2 cm long, 1 cm across. vivid green, underside very dark crimson; **Infl.** pedunculate, forking once or twice, cyathophylls white, tinged pink, underside pink, glands brownish-yellow.

**E. —** v. **brunellii** (Chiov.) Bally (*E. b.* Chiov.). – Ethiopia; Kenya; Uganda. – **Caudex** ± conical, 2–5 cm tall, 1.5–3 cm ⌀ at the base, 0.5–1 cm thick above, tip roundish, bark ± rough, dark; **L.** inconspicuous, often missing; **Infl.** 1–2.

**E. —** v. **rubella**. – Ethiopia; Sudan. – **Caudex** 3 cm long, ± 2 cm thick, tapering below, tuberculate.

**E. rudis** N. E. Br. (Pl. 80/4) (Gr. 15) (? *E. rangeana* Dtr.). – S.W.Afr.: Gr. Namaqualand. – Main root passing over into the main St. to form a fleshy **caudex** ± hidden in the soil; **St.** often shortly branched above, usually all of equal length and together forming a convex clump of 20 cm ⌀, Br. 2.5–15 cm long, 8–17 mm thick, cylindrical, with rhombic, conical tubercles c. 4 cm long, 2–3 mm ⌀, 2 mm high; **L.** narrow-linear, c. 10 cm long, rather pointed, longitudinally plicate; **Ped.** becoming thorny.

**E. rudolfii** N. E. Br. (Gr. 4). – Cape: Van Rhynsdorp D. – Thornless ♄ 30 cm or more tall; **Br.** and shoots terete, 3–5 mm thick, Int. 2.5–6 cm long; **L.** reduced; **Infl.** a cyme, 10 to 20 cm ⌀.

**E. sacchii** Chiov. (Gr. 4). – Somalia. – Small ♄; **Br.** thick, flowering Br. plicate-compressed, 5–6 mm thick, brown, glabrous; **L.** leathery, 5–6 in a Ros. on the flowering Br., elliptical-obovate, narrowing below towards the 3–5 mm wide petiole, 7–10 cm long, 3–4.5 cm across; **Fl.** at the Br.tips.

**E. sapinii** de Wild (Pl. 80/6) (Gr. 26). – N. equatorial Afr. – Thorny succulent ♄: **St.** branching from the base or above, St. and Br. round, densely covered with numerous spiralled, 4-angled, flat tubercles, with small blackish-grey Th., older Th. appressed; **L.** in crowded clusters at the Br.tips, linear-oblong, almost 10 cm long, c. 1 cm across, mucronate, with a reddish margin, deciduous; **Infl.** with a long Ped.

**E. sarcostemmatoides** Dtr. (Gr. 4). – S.W.Afr.: S.E. reg. – ♄ branching basally; **Br.** 5–6 mm thick, light green, Int. c. 15 cm long; **L.** opposite, scale-like, soon shrivelling.

**E. schaeferi** DTR. (Gr. 6). – S.W.Afr.: Lit. Karas Mts. – Probably identical with **E.gariepina**.

**E. scheffleri** PAX (Gr. 3). – Tanzania. – Small ♄, soft-woody, with a grey bark; **L.** with a petiole 1–1.5 cm long, 10–11 cm long, 6 cm across; **Cy.** light green, on thick Ped. almost 2 cm long.

**E. schimperi** PRESL. (Gr. 4). – S.Arabia: near Aden. – Laxly branching succulent ♄ up to 2 m tall; **Br.** extended, terete, pencil-thick, erect at first, later ± prostrate, with L.scars; **L.** 3–15 mm long, triangular-oblong, tapering, deciduous.

**E. schinzii** PAX (Gr. 22). – Transvaal; Rhodesia. – Slightly thickened main root passing over into the main St. to form an underground **caudex** with many subterranean offsets and Br., forming a compact clump; **St.** often branching above, 10–15 cm long, 8–10 mm thick, usually with 4 angles with roundish tubercles, sides flat or slightly grooved, green, Th. in pairs, 10 to 12 mm long.

**E. schizacantha** PAX (Gr. 27). – Somalia: Ethiopia. – **St.** clavate, 1.2 m tall, 5 cm ⌀, fleshy; **Br.** numerous, dense, 5–8 cm long, 0.5 cm thick; Th.cushions ± decurrent with 3 grey Th., the central one 2–3 cm long with 2 lateral points, the 2 lateral stipular Th. 5 mm long.

**E. schoenlandii** PAX (Gr. 11) (*E. fasciculata* N. E. BR. p.part.). – Cape: Clanwilliam, Van Rhynsdorp D. – Main **St.** thick, erect, seldom branching, up to 1.3 m tall, 20 cm thick, covered with hexagonal spiralled tubercles tipped with a small Th.; **Ped.** up to 5 cm long, becoming thorny, curved inwards.

**E. sekukuniensis** R. A. DYER (Gr. 24). – Transvaal. – Thorny succulent ♄ 2–7 m tall; **St.** roundish, 8–10-angled, simple or with a trunk, with a crown of Br. above, 2 m ⌀; **Br.** spreading, 1 m long, 15–20 mm thick, shoots 1–1.5 cm thick, usually triangular, the angles with a horny border, Th. in pairs, 1–1.5 cm apart, 5–10 mm long, stiff.

**E. semperflorens** LEACH (Gr. 24). – Angola: Benguela, Moçamedes D. – Thorny erect succulent ♄ branching from the base, up to ± 1 m tall; **Br.** stiff, bluish, the segments with 3–4 alate angles, the upper shoots with 3 wings, segments variously shaped, usually ± trullate, 4–10 cm long, 4–7.5 cm across, the angles hard, with a continuous horny border, usually variously curving, undulating to dentate, brown, becoming grey; **Th.** in pairs on the marginal T., very divaricate to strongly curved, 5–15 mm long; **L.** scale-like, slightly plicate, stiff, broadly oval, 5 × 5 mm, soon dropping, with minute stipular Th. above the L.scars; **Infl.** 1–4, subsessile, Cy. unisexual, bracts fleshy.

**E. seretii** DE WILD. v. seretii (Gr. 24). – Congo: N.E.Provinces, Ngoa. – Candelabrashaped ♄ ± 1.5 m tall; **Br.** thorny, with 4–5 wings up to 2 cm across, the angles slightly undulating, with a horny border, segments almost round to ovoid, 5–7 cm long, Th. spreading horizontally, in pairs, 2–11 mm long; **L.** small, deciduous, with minute stipule Th.

**E. —** ssp. **variantissima** LEACH. – Up to 3 m tall, branching from the base, almost tree-like; **Br.** erect, c. 1.6 m long, with 3–6 wings, segments variable, c. 6 or 10 cm long, shoots with 3 wings.

**E. silenifolia** (HAW.) SWEET (Gr. 8) (*Tithymalus* s. HAW., *E. elliptica* THUNBG.). – Cape. – ♃ with a thick, fusiform main **caudex** 2–5 cm thick; **St.** 1 or several, very short, simple or branching, with L. and Fl. borne at the tips; **L.** with a petiole 1.2–10 cm long, lamina 2.5 to 10 cm long, 2–12 mm across, linear to elliptical-lanceolate, usually tapering, somewhat plicate.

**E. silicola** DTR. (Gr. 4). – S.W.Afr. – Closely resembles **E. juttae** and may be identical.

**E. sipolisii** N. E. BR. (Pl. 61/5) (Gr. 5). – Brazil: Minas Gerais. – Succulent ♄; **Br.** erect, leafless, with segments c. 11 cm long, with 4 acute, rather convex angles, sides 1 cm across, somewhat concave, grey-green; **L.** minute, triangular, pointed, soon dropping.

**E. somalensis** PAX (Gr. 3). – Somalia. – Thornless, glabrous, leafless ♄; **Br.** alternate, with a brown bark; **Cy.** 6–8 mm ⌀, few, at the Br.tips.

**E.** × **scanieranensis** URSCH et LEANDRI. – Hybrid: parents not named, Bot.Gard. Tsimbazaza, Madag.

**E. spartaria** N. E. BR. (Gr. 4). – S.W.Afr.; Cape: Lit. Namaqualand. – Semi-♄ 20–30 cm tall, branching basally, little rebranched; shoots in 2–3 pairs, 2–3 mm thick, bluish-green, Int. 3–4 cm long; **L.** opposite, minute.

**E. spicata** E. MEY. (Gr. 4). – Cape: Lit. Namaqualand. – Thornless succulent ♄ 5 to 15 cm tall, branching freely from the base or above, rebranching; **Br.** opposite or alternate, 2.5–10 cm long, c. 5 mm thick, terete, smooth or rough.

**E. spinea** N. E. BR. (Gr. 4). – S.W.Afr. – Rigid ♄ 15–25 cm tall, with numerous sharp-tipped **Br.** forming ± hemispherical clumps of 1 m ⌀; shoots alternate, all terminating in a thorny tip, smooth, bluish.

**E. spiralis** BALF. f. (Gr. 24). – Socotra. – Candelabra-like small ♄, branching basally, 40–80 cm tall; **Br.** with 5–7 compressed, spirally contorted, sinuate angles, Th. small, Shi. solitary.

**E. squarrosa** HAW. (Pl. 81/1) (Gr. 22) (*E. mammillosa* LEM.). – Cape: Albany, Bathurst, Fort Beaufort, King Williamstown D. – Thick main root and main St. forming together an ovoid-oblong **caudex** 10 cm thick, usually entirely underground, with numerous prostrate or ± erect, radiating **Br.** above, 4–15 cm or more long, 1–2.5 cm thick, ± re-branching, dark green, with (2–)3(–5) angles divided into cylindrical-conical tubercles up to 10 mm long, tipped with a pair of Th. 1–6 mm long.

**E. stapelioides** BOISS. (Gr. 4). – Cape: Lit. Namaqualand; S.W.Afr. – **St.** thornless, c. 8 cm tall, rising from a tuberous root 10 cm long, branching somewhat from the base; **Br.** constricted into solitary, somewhat flattened segments scarcely 10 cm long, 5-10 mm thick, green, somewhat reddish.

**E. stellispina** HAW. v. **stellispina** (Pl. 80/2) (Gr. 19). – Cape: Lit. Namaqualand. – **St.** cylindrical, usually forming a truncated cone, branching from the base, forming dense clumps, 5-7 cm thick, with (10-)14(-16) not very prominent and transversely notched ribs, L.bases hamate, curved downwards; **L.** up to 1 cm long, narrow, deciduous; thorny Ped. usually branching 4-fold, arranged in zones, 4-10 mm long, grey.

**E.** — v. **atrispina** (N. E. BR.) WHITE, DYER et SLOANE (*E. a.* N. E. BR.). – Very densely armed with subsessile or sessile, stout, stellate, waxy-blue or even black Th.

**E.** — v. — f. **viridis** (WHITE, DYER et SLOANE) JACOBS. (*E. a.* v. *v.* WHITE, DYER et SLOANE). – **Br.** 6-8-angled, Th. yellowish-green.

**E. stellata** WILLD. (Pl. 63/22) (Gr. 22) (*E. procumbens* MEERBG., *E. uncinata* DC., *E. radiata* THUNBG., *E. scolopendria* DONN., *E. scol.* HAW.) – Cape: Uitenhage, Port Elizabeth, Peddie, Albany D. – **Caudex** small, conical, truncate, with a thick napiform root; **Br.** projecting radially from the crown, prostrate, curving, flat, with 2 sinuate, tuberculate and dentate angles, with segments up to 30 cm long, 2 cm across, upperside grooved, sides green with lighter striae, Th.pairs grey-brown, 2-4 mm long.

**E. stenoclada** H. BAILL. (Pl. 81/3) (Gr. 4; M.I) (*E. cirsioides* COST. et GALL., *E. insulae-europae* PAX). – Madag. – ♄ 50-100 cm tall; **St.** round; **Br.** flat, with or without Th., leafless.

**E. stolonifera** MARL. (Gr. 4). – Cape: Laingsburg D. – Acauline succulent ♄ 20-60 cm tall, usually wider than tall, arising from a thickened root; **Br.** subcylindrical, 5-9 mm thick, often thickest midway, simple or with a few branchlets, dark green to bluish, often with a waxy coating.

**E. strangulata** N. E. BR. (Gr. 24) (*E. polyacantha* sensu HIERN, *E. fraterna* N. E. BR. p.part.). – Angola. – Thorny succulent ♄ branching from the base, 15-100 cm tall; **Br.** spreading to ascending, deeply constricted, segments round-ovoid or shortly conical-ovoid, (1.5-)4(-7.5) cm long, 2.5 cm ⌀, light green, with alate, slightly sinuate-dentate angles 1.5-2 cm high, 1.25 cm thick, with a continuous horny border, Th. in pairs, 3-7 mm long, 6-12 mm apart, stipular Th. rudimentary; **L.** scale-like, 1.5 mm long and across, reddish-brown, soon dropping.

**E.** — ssp. **deminuens** LEACH – Malanje D. – (15-)20(-30) cm tall, forming clumps; **St.** erect, spreading or sometimes prostrate with ascending tips, usually 10-15 cm long, at first with 2-3 angles, deeply constricted into almost spherical, elliptical or ovoid segments 2-5 cm long, (2.5-)3.5(-5) cm across.

**E.** — ssp. **strangulata**. – Malanje D. – Up to 60 cm tall, main **St.** reduced; **Br.** spreading, projecting, simple, with (4-)5(-6) wings initially 3-4-angled, sometimes constricted, with ± oblong segments of very variable size and shape, mostly ± spherical or ovoid, up to 7.5 cm long, 5 cm across.

**E. submammillaris** (BGR.) BGR. (Gr. 19) (*E. cereiformis* BGR. and v. *s.* BGR., *E. pfersdorfii* HORT.). – Cape. – **St.** irregularly branching from the base, c. 3 cm thick, rather grey, roundish, thickened and 9-10-angled above; **Br.** rather thinner, with 5-8 angles 4-5 mm high, separated by sharp furrows, divided horizontally into zones c. 7 mm across, prominent and dentate, with a small L.scar; **Ped.** thorny, up to 2 cm long, slender, red, becoming brown.

**E. subsalsa** HIERN v. **subsalsa** (Pl. 64/28; 80/5) (Gr. 28) (*E. heteracantha* PAX). – Angola; S.W.Afr. – Thorny succulent ♄ up to 1.2 m tall; **Br.** numerous from the base, 8-12 mm thick, glabrous, light green, with 4 sinuate-tuberculate angles 8-22 mm apart, Th.shi. closely spaced, long-decurrent, Th. in pairs, 4-12 mm long, with small lateral Th.

**E.** — v. **kaokoensis** WHITE, DYER et SLOANE. – **Br.** with 5-6 angles.

**E. sudanica** A. CHEV. (Gr. 21). – Sudan; Niger valley; Senegal; Mali. – Thorny succulent ♄ 1-2 m tall; **Br.** spreading, eventually whitish, at first cylindrical, not angular, c. 2 cm ⌀, L.cushions in 5-8 spiral R., Th.shi. grey, Th. in pairs on the tubercles, 5-10 mm long, divaricate, often with a pair of very thin stipular Th. above; **L.** clustered at the Br.tips, oblong, triangular, finely dentate, 5-8 cm long, 1.5-2 cm across, dropping.

**E. superans** NEL (Gr. 14). – E.Cape. – Resembles **E. inermis**; **Br.** acute at the apex; **Infl.** white (not yellow).

**E. suzannae** MARL. (Gr. 18). – Cape: Ladismith D. – **St.** short, flat-spherical, up to 7 cm thick, or obovoid, with a napiform root, bluish-green, with 16 vertical, somewhat prominent ribs separated by distinct furrows, L.cushions prominent, up to 1 cm long, thick-hamate and curving downwards.

**E. symmetrica** WHITE, DYER et SLOANE (Gr. 18). – Cape: Willowmore, Beaufort West D. – Resembles **E. obesa**; **St.** usually 6 cm tall, 7 cm thick, usually 8-angled, grey-green, with deep purple, horizontal bands, the upper surface with pitted lines crossing the bands, the 8 ribs somewhat raised midway, with small tubercles 15-20 mm apart towards the tip.

**E. tanaensis** BALLY (Gr. 24). – Kenya. – Succulent ♄ up to 18 m tall; trunk cylindrical, up to 40 cm ⌀, 16 m tall, **St.** in a crown 1.8 m tall, sometimes with further superimposed crowns of Br.; **Br.** up to 2.7 m long, spreading-ascending, with 5-6 alate and slightly undulating angles, the wings up to 3 cm across, constricted into segments 35 cm long, Th.shi. 2-3 cm apart, with 2 weak divaricate Th. 3-5 mm long; **L.** on the upper St.segments, 8-12 mm long, 4-6 mm across, elliptical, pointed, fleshy; **Infl.** dichotomously branched.

**E. tardieuana** J. LEANDRI (Gr. 20; M.VI). – E.Madag. – Succulent branching ♄ c. 1 m tall; St. and **Br.** terete, bark brownish red, stipular Th. small, soon dropping; **L.** at the Br.tips, obovate-oblong, somewhat apiculate, 3-4 cm long, 10-14 mm across.

**E. teke** Schweinf. (Gr. 21). – N.Congo (Kinshasa). – Resembles **E. neriifolia**; succulent ♄ with a round St., each prominent Th.shi. with 2 small, 2–3 mm long, blackish-brown Th.; **L.** ± fleshy, ± ovate, tapering, 3–4 cm long, 2 cm across, sometimes up to 10 cm long, 5 cm across.

**E. tenax** Burch. (Gr. 4). – Cape. – Related to **E. caterviflora**. Known only from herbarium material.

**E. tenuirama** Schweinf. (Pl. 81/2) (Gr. 24). – Probably Arabia. – Erect ♄; **St.** 5-angled; **Br.** with 3 angles separated by slight indentations 10 mm across, often with 2–3 Br. immediately above one another on an angle, Th. in pairs, 5–6 mm long, slender, light grey, widely divaricate.

**E. tetracantha** Rendle (Gr. 28). – Somalia. – ♄ up to 15 cm tall; **Br.** from the shortened trunk, the 4 angles with hard, decurrent Th.-cushions, each with 4 stiff, prominent, sharp-tipped Th., the 2 lower ones up to 15 mm long, the upper ones thinner and c. 8 mm long; **Infl.** with 3 Cy.

**E. tetracanthoides** Pax (Gr. 28). – Tanzania. – ♄ 20–30 cm tall, up to 60 cm ⌀; **Br.** blue-green, shoots short, c. 1 cm ⌀, with 4 distinct, slightly sinuately lobed angles, Th.cushions 6–8 mm apart, solitary, with 2 Th.pairs, the lower Th. 5–10 mm long, the upper ones shorter.

**E. tetragona** Haw. (Pl. 81/5) (Gr. 24). – Cape. – Thorny succulent ♄ up to 15 m tall; **St.** up to 15 cm ⌀, 6–7-angled, constricted to form short segments; **Br.** in whorls, in layers, curving upwards, 2–3 cm ⌀, with 4–6 almost straight and flat-sided angles, Th. in pairs, c. 15 mm apart, 1 cm long, thin, brown.

**E. thi** Schweinf. v. **thi** (Gr. 24). – Southern Egypt, northern Sudan. – ♄ 80–150 cm tall, with candelabra-like **Br.**, 3 together, with 5–6 crenate angles, terminal Br. with 4 angles, 2.5–3 cm across, with flat sides, Th.shi. confluent, Th. up to 2 cm long.

**E. —** v. **subinarticulata** N. E. Br. – Eritrea; Ethiopia. – **Br.** 4–5-angled, more slender; Th.-smaller.

**E. tirucallii** L. v. **tirucallii** (Pl. 82/2) (Gr. 4) (*E. viminalis* Mill.). – Trop. E. and S.Afr.; growing wild in peninsular India. – ♄ with **Br.** forking or ± verticillate, light green with lighter longitudinal lines, segments 7–10 cm long, round, c. 6 mm thick; **L.** linear-lanceolate, 4–15 mm long, 1–3 mm across, soon dropping.

**E. —** v. **rhipsaloides** (Welw.) A. Chev. (*E. r.* Welw., *E. r.* Glaz.). – Angola; Congo: St. Thomas Is. – **Br.** cylindrical, eventually dropping, young Br. erect, clustered, simple or forked; **L.** 15–18 mm long, 2–3 mm across.

**E. tortilis** Rottler (Gr. 24). – India: near Madras. – **St.** erect, branching; **Br.** with segments 20–25 cm long, 4–5 cm across, with 3 spirally contorted, sinuate angles, with somewhat depressed sides, Th.shi. 3–4 cm apart, roundish, with 2 erect to divaricate Th. c. 1 cm long.

**E. tortirama** R. A. Dyer (Gr. 22). – Transvaal. – Thickened main root and main St. combining to form a napiform **caudex** 30 cm long, 15 cm thick, usually below ground, usually simple; **Br.** 20–50 or more from the crown, 6–30 cm long, 2–4.5 cm thick, becoming constricted to form segments, with 3 irregularly tuberculate, spirally contorted, compressed angles (at first rarely only 2 angles), tubercles 9 mm apart, Th.shi. confluent to form a horny border, Th. in pairs, c. 2 cm long.

**E. tortistyla** N. E. Br. (Gr. 22). – Rhodesia. – Related to **E. aeruginosa**; insufficiently known species.

**E. transvaalensis** Schltr. (Gr. 4) (*E. galpinii* Pax, *E. ciliolata* Pax, *E. goetzei* Pax). – Transvaal; Moçambique; S.W.Afr. – Thornless ♄ 0.6–1.5 m tall, branching basally; **St.** forking, branching above, terminal Br. 6–11 cm long, 4–5 mm thick; **L.** usually along the upper Br., several in a whorl, on a petiole 6–30 mm long, lamina 10.5 cm long, 15–50 mm across, oblong-lanceolate or elliptical-ovate, tapering, usually glabrous; **Infl.** at the tips of the upper Br.

**E. trapifolia** A. Chev. (Gr. 21) (incorrectly *E. trapaefolia* and *trapaeifolia*). – W.Sudan: Gourma. – Woody ♄, fleshy at first, 1–2 m tall; **St.** freely branching above; **Br.** 2 cm ⌀, with prominent tubercles in 5–8 spiral R., Th.shi. grey, with 1–2 pairs of variously large Th., main Th. 4–8 mm long, sharp-pointed, projecting horizontally; **L.** in crowded clusters at the Br.tips, fleshy, obovate-triangular or fan-shaped, 2–3 cm long, 1.5–2 cm across above, narrowing below towards the petiole, tip bifid, fan-like.

**E. triaculeata** Forsk. v. **triaculeata** (Gr. 27) (*E. infesta* Pax). – Arabia; Ethiopia; Somalia; Kenya. – ♄; **Br.** with 5–7 angles, Th. cushions with 1 thin central Th. 2–4 cm long and 2 lateral stipular Th. 3–5 mm long.

**E. —** v. **triacantha** (Ehrenb.) N. E. Br. (*E. triacantha* Ehrenb.). – Ethiopia: Eritrea. – Acc. P. R. O. Bally differs in appearance from the type; **Br.** with 3–5 angles.

**E. triangularis** Desf. (Gr. 24) (*E. grandidens* Sim., *E. evansii* N. E. Br.). – Cape, Natal. – Thorny succulent ♄ 9–18 m tall; **St.** 6-angled at first, becoming round; **Br.** in well-spaced tiers and whorls, projecting-ascending, green, segmented, re-branching, with 2–5 alate, compressed, little sinuate angles with a continuous grey or brown horny border, or the Th.shi. solitary, Th. in pairs, 8–20 mm apart, 1 cm long.

**E. tribuloides** Lam. – Acc. L. Croizat this is the valid name for **E. handiensis** Burch. (L. Croizat: De Euph. Ant. atque Offic.).

**E. trichadenia** Pax (Pl. 61/7) (Gr. 7) (*E. benguelensis* Pax, *E. subfalcata* Hiern, *E. gossweileri* Pax). – Angola; Transvaal; Natal; Rhodesia. – ♃ with a **caudex** 10 cm long, 6 cm thick, with a corky covering, with several underground St. above; **Br.** 3–10 cm tall; **L.** linear-lanceolate, up to 6 cm long, up to 5 mm across, blue-green, margins horny, often dentate.

**E. tridentata** Lam. (Gr. 17) (*E. anacantha* Ait., *Dactylanthes a.* Haw.). – Cape. – Distinguished from **E. ornithopus** by the elongated **Br.** and **Infl.** short-pedunculate or sessile, at the Br.tips.

**E. trigona** Haw. (Pl. 82/1) (Gr. 24) (*E. hermentiana* Lem.). – Trop. S.W.Afr. – ♄ or ♄; **Br.** erect, 4–6 cm ⌀, with segments 15–25 cm long, dark green with irregular whitish mottling, with 3–4 almost alate and compressed, sinuate, shortly dentate angles, Th. 4–5 mm long, reddish-brown; **L.** 3–5 cm long, spatulate, short-pointed, persisting for a long time.

**E. tsimbazazae** J. Leandri (Gr. 20; M.VI). – Possibly a hybrid: **E. milii** v. **splendens** × **E. spec.**, Bot. Gard. Tsimbazaza, Madag. – **Br.** dark green, c. 1 cm thick, with 2 stipular Th., usually double, variable in size, larger ones normally 1 cm long, robust; **L.** sessile, narrow-oblong, 6–7 cm long, 7–8 mm across.

**E. tuberculata** Jacq. (Gr. 14) (*Dactylanthes t.* Haw.). – Cape: Lit. Namaqualand. – **St.** combined with the main root to form a conical, stout **caudex**, usually underground; **Br.** numerous, erect or prostrate, forming clumps up to 75 cm tall, often 4 cm ⌀, thicker above, green, often whitish, with rhombic, very prominent tubercles in spiral R.; **L.** up to 4 cm long, 15 mm across, linear, with reddish margins; **Ped.** becoming thorny.

**E. —** 'Cristata'. – St.fasciation found by Christian Löri in a large old plant near Eendekuil, Malmesbury D., Cape.

**E. tuberculatoides** N. E. Br. (Gr. 14). – Cape: Malmesbury D. – Thickened main root combined with the shortened main St. to produce a **caudex**, usually completely hidden in the soil; **Br.** numerous from the crown, together forming a clump 45 cm tall, 1.2–1.7 cm ⌀, cylindrical, glabrous, with rhombic, short-conical tubercles in several spiral R., 6–8 mm long, 4–6 mm across, 2–3 mm high; **L.** 8–10 mm long, up to 1 mm across, deciduous; **Ped.** becoming thorny.

**E. tuberosa** L. (Pl. 61/8) (Gr. 8). – Cape: Stellenbosch, Tulbagh, Van Rhynsdorp D. – ♃, the tuberous root forming an extended and often constricted **caudex** of 8–24 mm ⌀; **Br.** short; **L.** few, with petiole almost 4 cm long, lamina 2–5 cm long, 1–2.5 cm across, pointed above, margins undulating.

**E. tubiglans** Marl. (Pl. 63/18; 82/3) (Gr. 18). – Cape: Karroo. – **Caudex** short, thick, passing over below into the napiform root; **Br.** erect, up to 2 cm ⌀, bluish-waxy, narrowing teretely below, this section 4–15 mm long, with 5 slightly dentate angles separated by deep furrows.

**E. tuckeyana** Steudel (Gr. 1). – Cape Verde Is. – Leafy ♄, probably not found in cultivation.

**E. turbiniformis** Chiov. (Gr. 9?). – S.Ethiopia; Somalia. – **Caudex** sub-spherical or turbiniform, buried in the soil, not exceeding 4 cm ⌀, surface tessellate-tuberculate, with only 1 growing point (rarely 2 or 3); **L.** not developed; **Cy.** 4–7 in umbellate cymes, Ped. stout, 5 mm long and branched, the apical Cy. ♂, 1.5 mm ⌀, cyathophylls 2, opposite, minute, involucre cup-shaped, yellow.

**E. uhligiana** Pax (Gr. 22). – Tanzania. – Insufficiently known species; **Br.** grey-green, extended, less than 1 cm ⌀, angles grey, ± alate, Th.shi. confluent, lower Th. 8 mm long, upper ones 2–3 mm long.

**E. undulatifolia** Janse (Gr. 21) (*E. quercifolia* hort., *E. helicothele* hort., *E. gardenifolia* hort.). – Resembles **E. nivulia**; **St.** ± 5-angled; **L.** large, on the tips of young Br., margins undulating.

**E. unicornis** R. A. Dyer (Gr. 26). – Moçambique. – Succulent branching ♄ c. 30 cm tall; **Br.** cylindrical, c. 1 cm thick, very inconspicuously constricted into segments, usually with 2 shoots arising from a common base, ± cylindrical or with 6–7 tuberculate angles, with 6–7 grooves 1 mm across, tubercles little raised, cushion-like, with a solitary 4–6 mm long Th., and above this 2 stipular Th. c. 1.5 mm long; **Infl.** solitary, with 3 Cy.

**E. unispina** N. E. Br. (Gr. 26). – Sudan; Congolese Republic; Togo; Nigeria. – Succulent, compact, thorny ♄ up to 4 m tall; **Br.** 16 to 22 mm or more thick, cylindrical, with several spiral R. of Th., these merging abruptly into convex, horny Shi. 4–5 mm long; **L.** few, clustered at the Br.tips, 4–5 cm long, 7–20 mm across at the tip, cuneate to linear-cuneate, the tip often roundish or 2-lobed, margins somewhat curly, fleshy; **Infl.** small.

**E. valida** N. E. Br. (Pl. 81/4) (Gr. 18). – Cape: Jansenville, Somerset East, Steytlerville D. – Resembles **E. meloformis**, but without a napiform root; **Bo.** spherical at first, becoming cylindrical, up to 30 cm tall, up to 12 cm thick, simple or branching basally, deep green, with ± distinct blue-green bands, angles often spirally contorted; **Infl.** very woody, very persistent.

**E. —** 'Viridis'. – **Bo.** grass-green, with no markings.

**E. vandermerwei** R. A. Dyer (Gr. 22). – Transvaal. – **Caudex** with thickened main root usually hidden in the soil, dividing above, usually into 2 thick St., with numerous Br. above, thus forming clumps 20 cm ⌀; **Br.** up to 30 cm long, 15–20 mm thick, with 4–5 ± curving, contorted, somewhat spreading, pointed-tuberculate angles, tubercles 5–15 mm apart, 2.5–5 mm high, Th.shi. on the tubercle-tips, Th. in pairs, 1 cm long.

**E. venenata** Marl. (Pl. 83/1) (Gr. 24). – S.W. Afr. – Thorny succulent ♄; main **St.** much reduced, with simple Br. from the base, these rarely re-branching above, forming large bushes 2 m tall and 1.3–2 m across; **Br.** distinctly segmented, segments broader below than above, with 6 very compressed angles, separated by deep furrows, Th.shi. merging into a broad, horny border, Th. in pairs, 5 mm long.

**E. venenifica** Trem. et Kotschy (Gr. 26). – W.Afr.: Ivory Coast. – Succulent ♄ 1–2 m tall, branching basally; **Br.** cylindrical, 2–3 cm ⌀, with tubercles in many spiral R., Th. solitary, c. 1 cm long, with no Shi.; **L.** clustered at the Br.tips, oblanceolate, triangular, or notched at the tip, often with an apiculus in the notch, 4.5–7 cm long, 1.8–3 cm across, with an alate petiole.

**E. verruculosa** N. E. Br. (Gr. 4). – S.W.Afr.: Gr. Namaqualand. – Small ♄ with a woody root c. 12 mm thick; **St.** short; **Br.** numerous, re-branching many times, 2.5–7 cm long, 6 to

7 mm thick, densely covered with ± spiralled tubercles.

**E. viguieri** M. DENIS v. **viguieri** (Pl. 82/4) Gr. 20; M.VI). – W.Madag. – ♄ 20–150 cm tall; **St.** clavate, thick, 20–30 mm ∅ above, with 6 angles consisting of broadly triangular L.-cushions with a light brown stipular Th. with several tips, 5 mm long; **L.** in crowded clusters at the Br.tips, long-oval, 9–50 cm long, 3–11 cm across, narrowing towards the short petiole, emerald green, M.rib and lateral veins distinctly whitish, the L.base brilliant red; **Cy.** on Ped. 4 cm long, cyathophylls red or yellowish green.

**E. — v. ankarafantsiensis** URSCH et LEANDRI. – **St.** simple or with few Br., Th. laciniate on a thick, tuberculate base, 2 cm long, secondary Th. longer or shorter; **Cy.** 8–24, bracts yellowish-green with a red border.

**E. — v. capuroniana** URSCH et LEANDRI. – 30–40 cm tall; angles 2–2.5 cm deep, 2–8 mm across, lateral Th. short, projecting horizontally, symmetrical; **L.** 10 cm long, 5 cm across, apiculate; **Cy.** 4–8, bracts yellowish green, borders reddish-orange.

**E. — v. tsimbazazae** URSCH et LEANDRI. – 40–150 cm tall; **St.** with several angles, Th. laciniate; **L.** on a petiole 25–30 cm long, 25 cm long, 8 cm across; **Cy.** 12–32, bracts red.

**E. — v. vilanandrensis** URSCH et LEANDRI. – **St.** simple, 20–60 cm tall, 2–3 cm ∅, Th. 5 to 10 mm apart; **L.** with a short, thick petiole, 18 cm long, lamina up to 11 cm across, with 35–40 oblique secondary veins; **Cy.** 6–16, bracts small, green, bordered with pale red.

**E. virosa** WILLD. (Gr. 24). – Cape; S.W.Afr.; Angola. – Thorny, succulent, ± tree-like or low ♄; **St.** 4–5-angled; **Br.** numerous in whorls, projecting and ascending, 2.5 cm ∅, grey green, with segments 4 cm long, 3-angled below, angles often spirally contorted, with 5–8 angles above, irregularly and almost rectangularly sinuate, curving, with a horny border, sides ± flat, Th. in pairs, stout, 11–13 mm long.

**E. — ssp. arenicola** LEACH. – Angola: Moçamedes D. – Low and spreading ♄ with a ∅ much exceeding its height; **Br.** horizontally spreading and becoming erect only towards the apex, more freely rebranched than those of ssp. **virosa**.

**E. — ssp. virosa** (Pl. 82/5) (*E. dinteri* BGR., *E. bellica* HIERN). – Cape: Lit. Namaqualand. – ♄ nearly tree-like, forming big clumps of 1.30 to 2 m ∅, more rarely up to 2.7 m high, 1.5 to 3 m ∅.

**E. — ssp. — f. caespitosa** JACOBS. – S.W.Afr.: Gr. Namaqualand. – With many crowded **St.**, very compact; **Br.** scarcely 50 cm tall.

**E. — ssp. — f. striata** JACOBS. – S.W.Afr.: Gr. Namaqualand. – Sides of the Br.angles with curving yellowish lines.

**E. volkmanniae** DTR. (Gr. 24). – S.W.Afr. – Insufficiently known species, probably identical with **E. avasmontana**.

**E. wakefieldii** N. E. BR. (Gr. 24). – Kenya; Tanzania. – Palm-like ♄, 7–15 m tall; **St.** fibrous, cylindrical, with Br.scars, with a grey bark and Br. in a circular crown; **Br.** few, short, thin, with alate angles, shoots few, clustered, 5 to 25 cm long, c. 2.5 cm ∅, with 3–4 distinctly dentate angles, T. ± well spaced and up to 5 mm high, with a 4–10 mm long Th.shi. above, usually with 2 minute stipular Th., or these missing; **L.** at the Br.tips, minute, scale-like, dropping; **Infl.** subsessile, forked.

**E. waterbergensis** R. A. DYER (Gr. 24). – Thorny succulent ♄ 1.5 m tall; main **St.** somewhat compressed, branching basally; **St.** deep green, little branching, ± erect, pendulous in age, 2–2.2 cm ∅, constricted into segments 2–20 cm long, sides parallel, with 6–8 horny-bordered angles, Th.-pairs 7–15 mm apart, c. 5 mm long, spreading horizontally, with rudimentary and deciduous lateral Th.

**E. weberbaueri** MANSF. (Gr. 5). – Peru. – Closely resembles **E. phosphorea**; laxly branched ♄; **Br.** and shoots erect, 5–8 mm ∅, becoming woody, with blunt longitudinal ribs or keels, with oblong-oval segments; **L.** soon dropping, scale-like, 1.5 mm long, 1 mm across; **Cy.** subsessile.

**E. whellanii** LEACH (Gr. 28). – Zambia. – Thorny, succulent, very dwarf ♄, branching densely from the base; **Br.** usually simple, sub-cylindrical, up to 17.5 cm long, up to 8 mm thick, with tubercles up to 1 mm high, usually in 6 vertical R., but sometimes in 5 spirally contorted R., Th.shi. 1–2 mm long, with 4 Th. 2.5–3.5 mm long, thin, whitish, the pairs of variable length; **Infl.** with 3 Cy.

**E. wildii** LEACH (Gr. 9) (*E. montieri* sensu WHITE, DYER et SLOANE, Succ. Euph. **1**, 266, Pl. 252). – Rhodesia. – Succulent ♄, sometimes ♄-like and up to 3 m tall; **root** fleshy, thickened, almost tuberous; **St.** cylindrical, up to 10 cm ∅, with few Br., tuberculate; **Br.** ± ascending-spreading, c. 5 cm ∅, with tubercles in 5 spiral R.; **L.** at the Br.tips, narrow-ovate, pointed and apiculate, c. 12 cm long, 4 cm across, ± plicate; **Ped.** 3–10 cm long, becoming thorny.

**E. cv. William Denton** (*E.* × *dentonii* HORT.). – Hybrid: **E. obesa** × ? **E. mammillaris**. – Hort. WILLIAM DENTON, London.

**E. williamsonii** LEACH (Gr. 22). – Zambia: N.Prov. – Thorny ♄ up to 1 m tall; **Br.** numerous from a ± caudiciform St., erect, simple, triangular, constricted into variously shaped segments, ± oblong, ± roundish, 2.5–5 cm long, 2.5–5.25 cm across, sides flat or slightly concave, shoots when present very short, Th.shi. ± obovate, decurrent and solitary, or confluent to form an undulating, horny border, Th. in pairs, up to 18 mm apart, up to 1 cm long; **L.** rudimentary, soon dropping, with minute stipular Th.

**E. wilmaniae** MARL. (Gr. 17). – Cape: Lit. Namaqualand. – Main St. and tuberous main root forming an irregular **caudex**, almost completely subterranean, roots napiform, with numerous Br. above; **Br.** extended-cylindrical, those above soil-level being short, blunt, clavate, with pointed tubercles.

**E. woodii** N. E. BR. (Gr. 13) (*E. pugniformis* BAK., *E. procumbens* N. E. BR., *E. passa* N. E. BR., *E. discreta* N. E. BR.). – Natal. – **Caudex** obconical, usually underground, 12 to

15 cm thick, with long roots below, with 20 to 40 Br. above, arranged in 2 or more R. around a tuberculate zone; **Br.** c. 20 cm long, 10 to 15 mm thick, light green, with rhombic tubercles, 1 cm long, rather prominent; **L.** small, deciduous.

**E. xylophylloides** AD. BROGN. ex LEM. (Pl. 83/2) (Gr. 4) (*E. enterophora* DRAKE). – Madag. – ♄ or ♄; **St.** 1–2 m tall, branching from the base or higher up, with spreading Br. arranged in whorls; **St.** and erect Br. light green, very much flattened, with 2 angles, slightly crenate at intervals, with segments c. 15 cm long, 10 to 12 cm across, 3 mm thick midway.

**E. zakamenae** J. LEANDRI (Gr. 20?; M.VI). – Central Madag. – Dichotomously branching ♄; **Br.** erect, green at first, becoming brown and rather fleshy, stipular Th. scarcely visible; **L.** crowded at the Br.tips, oblong-obovate, 6 to 10 cm long, 3–4 cm across, leathery, margins involute, petiole 1 cm long.

**E.** × **zanaharensis** URSCH et LEANDRI. – Hybrid: parents not named. Bot. Gard. Tsimbazaza, Madag.

**E. zoutpansbergensis** R. A. DYER (Gr. 24). – Transvaal. – Small, thorny, succulent ♄ up to 5 m tall; **St.** with many spreading, projecting Br.; **Br.** little re-branched, with segments 5–10 cm long, 2–3.5 cm thick, with (5–)6(–8) alate angles, mostly with a continuous grey horny border, Th. in pairs, 10–17 mm apart, 1 cm long, very sharp.

**Ficus** L. Moraceae. – N.Am. – Only one species need be described here. – Greenhouse, warm. Propagation: seed, cuttings.

**F. palmeri** S. WATS. (Pl. 83/4). – Mexico: Baja Calif., San Pedro Martin Is. – ♄, 3–4 m tall; **St.** much swollen at the base, branching basally; young **Br.** with white, velvety hairs; **L.** at first white-felted on the underside, eventually both sides green and upperside minutely hairy, rather thick, ovate, pointed, cordate at the base, on a petiole 2 cm long, lamina 6 cm long, 4–5 cm across; **Fl.** unisexual, small; compound **Fr.** (= figs) 2 together in the L.axils, thick, fleshy, 1 cm ∅.

**Fockea** ENDL. Asclepiadaceae. – Succulent velds of Afr., from Angola to the dry Karroo velds. – Caudiciform succulents; **St.** tuberous or napiform; **Br.** often thin, twining or erect; **L.** oblong, flat or with undulating margins; **Fl.** moderately large, axillary, solitary or several together in dense clusters. Dioecious. – Greenhouse, warm. Propagation: seed.

**F. angustifolia** K. SCHUM. – Cape: Lit. Namaqualand. – **Caudex** large; **St.** several, erect, twining or climbing, 50–70 cm long, minutely hairy; **L.** 1.5–10 cm long, 2–6 mm across, linear; **Fl.** in clusters of 2–6, green.

**F. crispa** (JACQ.) K. SCHUM. (Pl. 83/3; 84/2) (*Cynanchum c.* JACQ., *F. capensis* ENDL.). – Cape: Karroo. – **Caudex** napiform, up to 3 m ○, almost completely underground; **St.** thin, ± twining or prostrate; **L.** oval-pointed, 2–3 cm long, 1–2 cm across, margins undulating; **Fl.** 2–3 together, 3–4 cm ∅, greenish grey, with small brown blotches; the entire plant puberulous. (A specimen of this plant has been in cultivation at Schönbrunn near Vienna since 1799!)

**F. dammarana** SCHLTR. – S.W.Afr.: Hereroland. – ♄; **caudex** ± thickened; **Br.** laxly leafy, at first rather felty-haired; **L.** linear-lanceolate, pointed, margins slightly undulating, 12–23 mm long, 3–6 mm across; **Fl.** few.

**F. edulis** (THUNBG.) K. SCHUM. (*Chymocormus e.* (THUNBG.) HARV., *F. glabra* DCNE.). – **Caudex** large, edible; **L.** oblong to elliptical; **Fl.** solitary from the L.axils.

**F. multiflora** K. SCHUM. (Pl. 84/1). – S.Angola; Tanzania. – **Caudex** large; **Br.** stout, ± gnarled; **L.** small, flat, white-felted on the underside; **Fl.** numerous, in clusters. The large quantity of milky sap contains 12% rubber.

**F. schinzii** N. E. BR. – S.W.Afr.; Angola. – Insufficiently described species.

**Folotsia** COST. et BOIS. Asclepiadaceae. – Madag. – Plant smooth, climbing or shrubby; **St.** segmented, round; **Infl.** terminal, with numerous Fl.; corona-lobes large, united at the base, forming a pentagon. – Greenhouse, warm. Propagation: seed, cuttings.

**F. aculeatum** (B. DESC.) B. DESC. (*Prosopostelma a.* B. DESC.). – S.Madag. – **St.** trailing, long, thin, 1.5–2 mm ∅, branching basally, 30–40 cm long, white-powdery; **Infl.** of few Fl., Cor. large, papillose inside, white, tube c. 3.5 mm long, lobes triangular-oblong, bifid.

**F. floribundum** B. DESC. (*Prosopostelma grandiflorum* CHOUX). – N.Madag. – Freely-branching ♄ up to 1.5 m tall; **Br.** erect, prostrate or curving and pendulous, intricately tangled, or climbing, up to 3 m tall, distinctly segmented; **Infl.** of many, fairly large, delicately scented Fl.

**F. grandiflorum** (JUM. et PERR.) JUM. et PERR. (*Decanema g.* JUM. et PERR.). – Madag. – Trailing plant; **Br.** numerous, green, becoming whitish; **Fl.** c. 20, in sessile umbels at the nodes, Cor. round, large, lobes 7 mm long, triangular, 2 mm across, pointed, with tiny pockets in the indentations.

**F. madagascariensis** (JUM. et PERR.) B. DESC. (*Prosopostelma m.* JUM. et PERR.). – Madag. – Leafless liana; **St.** 1 cm ∅; **Fl.** 3.5–4 mm ∅, white.

**F. sarcostemmatoides** COST. et BOIS. – Madag. – **St.** fleshy, constricted into segments, c. 15 mm ∅; **Infl.** an umbel with numerous Fl. in stories, one above the other, from the nodes, Cor. white, lobes triangular, somewhat fleshy.

**Fouquieria** H. B. et K. Fouquieriaceae. – Mexico: Baja California. – ± succulent ♄ or ♄; **St.** swollen, branching, with numerous thorny shoots; **L.** rather small, oval, later dropping; **Fl.** conspicuously large, usually red. – Greenhouse, warm. Propagation: seed, cuttings. Some species are decorative shrubs. (Lit.: HENRICKSON, J.: A taxonomic revision of the Fouquieriaceae. In Aliso **7**: 439–537, 1972).

**F. campanulata** NASH. – Mexico. – ♄; main St. short; **Fl.** campanulate.

**F. diguetii** (VAN TIEGHEM) J. M. JOHNSTON (Pl. 84/4) (*Bronnia d.* VAN TIEGHEM, *B. thiebautii* VAN TIEGHEM, *F. peninsularis* NASH). – Mexico: Baja Calif. – **Caudex** well developed; main **St.** spreading, stout, freely branching, bent, with a brown bark; **Fl.** in corymbs, scarlet.

**F. fasciculata** (HUMB. ex ROEM. et SCHULT.) NASH (Pl. 84/3) (*Cantua f.* HUMB. ex ROEM. et SCHULT., *F. spinosa* H. B. et K., *Brownia s.* H. B. et K.). – Mexico: Hidalgo, Durango. – ♄; **St.**-bases thickened; **Infl.** clustered, Cor. red.

**F. formosa** H. B. et K. – Mexico: Puebla. – Thorny freely branching ♄ 1–2 m tall; **Fl.** in terminal spikes, scarlet.

**F. macdougallii** NASH. – Mexico: both sides of the Gulf of Calif. – ♄ or small ♄ with several main repeatedly forking **Br.**, bark green; **Fl.** in corymbs, red.

**F. purpusii** BRANDEG. – Mexico: Oaxaca. – ♄ with 1 or more side-St. from the base, with a conical swelling below, tapering above, bark thick and tearing transversely; **St.** numerous, re-branching, thorny; **L.** small, oval, soon dropping; **Fl.** very numerous, white.

**F. splendens** ENGELM. (*F. spinosa* TORREY). – ('Ocotilla'.) – N.Mexico to USA: S.Calif. – Thorny succulent ♄ c. 2 m tall, freely branching from the base; **St.** erect, spreading, little branching, round, with L.scars, thorny, dark green with whitish stripes, Th. 1 cm long; **L.** small, oval, soon dropping; **Fl.** numerous, brilliant red. Source of ocotilla wax, used in medicine.

**Furcraea** VENT. (*Fourcroya* SCHULT.). Agavaceae. – Mexico; Caribbean Is. – Acauline or caulescent plants; **L.** clustered or in Ros. at the St.tips, lanceolate, margins minutely dentate or spiny, with a carinate, almost fleshy midrib; **Infl.** terminal, in a large pyramidal panicle, Fl. 1–3 in the bract-axils, slender-pedicellate, pendulous, Pet. 6, elongated-round, broadcampanulate and spreading, white inside, greenish outside, with small adventitious bulbils or plantlets between the Fl. – Greenhouse, light. Propagation: seed, offsets.

**F. bedinghausii** C. KOCH (*Beschorneria multiflora* HORT., *Yucca pringlei* GREENM., *Roezlia bulbifera* HORT., *Y. b.* HORT.). – Central Mexico. – Similar to **F. roezlii** but smaller in all its parts; **St.** c. 1 m tall; **L.** 60 cm long, 5–7 cm across; **Infl.** 3–4 m tall.

**F. cubensis** VENT. (*Agave c.* JACQ., *F. hexapetala* URBAN). – Cuba; Haiti. – **St.** short or undeveloped; **L.** lanceolate, ensiform, the underside rounded, distinctly carinate, with several distinct longitudinal folds and ribs above midway, rough, marginal T. 2–3 mm long, 2–2.5 cm apart; **Infl.** up to 5 m tall, Fl. 5 cm long, milky-white inside.

**F. gigantea** VENT. v. **gigantea** (*Agave foetida* L., *F. f.* HAW., *F. madagascariensis* HAW.). – S.Brazil. – Acauline or caulescent; **L.** broad, oblanceolate, ± flat, margins somewhat wavy, underside rather rough, up to 2.5 m long, up to 20 cm across, margins entire, with a few triangular, hamate T. at the base.

**F. —** v. **medio-picta** TREL. (*F. variegata* HORT., *F. watsoniana* HORT.). – **L.** with a creamy-white longitudinal band.

**F. —** v. **willemetiana** TREL. (*F. commelinii* SALM). – **Infl.** 8–10 m tall, freely branching, Fl. 5 cm long, greenish-white.

**F. macdougallii** MATUDA. – Mexico: Oaxaca. – **St.** simple, 6–7 m tall, c. 20 cm ∅, densely covered with withered L.; **L.** becoming pendulous, fleshy, linear-oblong, narrowing-subulate, Sp.-tipped, margins spiny-dentate, 2–2.5 m long, 7 cm across; **Infl.** 8 m tall, branching, Fl. pendulous, whitish, dirty green inside.

**F. pubescens** TOD. – ? Mexico. – Acauline; **L.** numerous, lanceolate, long-tapering, thickened at the base, up to 1.3 m long, 7 cm across, margins with triangular, forwardly directed T.; **Infl.** up to 7 m tall, panicles freely branching, up to 5.5 m long, Fl. greenish-yellow, panicle-Br. and Fl. pubescent.

**F. roezlii** ANDRE (*Agave argyrophylla* HORT., *Yucca a.* HORT., *Y. parmentieri* HORT., *Y. toneliana* HORT., *A. t.* HORT., *Beschorneria floribunda* HORT., *Lilia regia* HORT., *Roezlia bulbifera* HORT., *R. r.* HORT.). – S.Mexico. – **St.** 3–4 m tall, 30 to 35 cm thick, covered with remnants of old L.; L.Ros. 2–3 m across, L. 1–1.2 m long, up to 8 cm across, ensiform, long-tapering, leathery-fleshy, margins rather thin and curved upwards, minutely dentate; **Infl.** 4–5 m tall, panicle freely branching, pyramidal, Fl. pubescent, white inside, 5–6 cm long.

**F. selloa** C. KOCH v. **selloa** (Pl. 85/1, 2, 3) (*F. flavoviridis* HOOK.). – Colombia. – **St.** up to 1 m tall; **L.** numerous, narrow-lanceolate, ensiform, much narrowed towards the base, glossy dark green, rough, over 1 m long, 7–10 cm across, marginal T. large, widely-spaced, hamate, brown; **Infl.** up to 6 m tall, laxly branching, Fl. 6.5 cm long, whitish.

**F. —** v. **marginata** TREL. (*F. lindenii* JACOBI, *F. cubensis* v. *l.* HORT., *Agave c.* v. *striata* HORT.). – **L.** border white at first, becoming yellow.

**F. —** v. **—** f. **edentata** (TREL.) JACOBS. (*F. s.* v. *e.* TREL.). – **L.** pink, margins without T.

**F. tuberosa** AIT. (*Agave t.* MILL., *F. spinosa* TARZ.-TOZZ., *F. interrupta* HORT.). – Cuba; Haiti. – Acauline or almost so; **L.** broad-lanceolate, 20–25 cm across midway, margins curved upwards towards the tip, T. 5–9 mm long, usually 2–3 mm apart, often missing at the base and towards the tip, reddish-brown, hamate; **Infl.** 5–8 m tall, panicle elongated.

**F. undulata** JACOBI (*F. pubescens* BAK.). – ? Mexico. – Acauline. – **L.** narrow-lanceolate, gradually tapering, with a blunt, brown terminal Sp., 45 cm long, 5 cm across midway, olive-green, with a carinate midrib on the underside, margins undulating, with triangular, brown, bent T.; **Infl.** 3–4 m tall, panicle-Br. and Fl. pubescent.

× **Gasterhaworthia** GUILL. (× *Haworthiogasteria* KONDO et MEGATA). Liliaceae. – Intergeneric hybrids: **Gasteria** × **Haworthia**. – Greenhouse, warm. Propagation: division.

× **G. bayfieldii** (S. D.) ROWL. (Pl. 85/4) (*Aloe b.* S. D., *Gasteria b.* (S. D.) BAK.). – Parents unknown. – **St.** simple, foliate, branching basally, 10–30 cm tall; **L.** numerous, densely spiralled, forming Ros. with many R., 10–14 cm long, 2.5–3 cm across at the base, the tip acute and mucronate, concave-grooved on the upperside, convex and distinctly carinate at the tip on the underside, with numerous raised spots, often in distinct transverse lines, margins hornytuberculate; **Infl.** 60–70 cm tall, Fl. 13 mm long, spherical at the base, pale red.

× **G. holtzei** (RADL) GUILL. (*Aloe h.* RADL., *Gasteria × h.* (RADL) BGR.). – Hybrid: **Gasteria verrucosa** v. **intermedia** × **Haworthia radula**. – **L.** distichous, later spiralled, lanceolate-triangular, shortly tapering, 9 cm long, 3 cm across, upperside grooved, underside simply or doubly carinate, green with white tubercles, ± merging on the underside into transverse R.

× **G.** cv. Royal Highness. – Parents unknown. – Appearance as **Haworthia margaritifera** or **H. papillosa,** robuster; **Ros.** compact, spiralled, offsetting from the base; **L.** stout, brittle, triangular, up to 10 cm long, 2.5–3.5 cm across at the base, with a soft pointed tip, overall with evenly distributed, large, hemispherical, solitary, white tubercles of 1 mm ⌀; **Infl.** over 1 m tall, Fl. 16–20 mm long, whitish-green.

**Gasteria** DUVAL. Liliaceae. – Cape to S.W. Afr. – ♃, usually acauline, offsetting, matforming; **L.** usually distichous, less often rosulate, linguiform or elongate, with entire margins, firm in texture, fleshy, dark green, often reddish, with white spots or tubercles; **Fl.** in lax Rac. or Pan. (Pl. 86/3), tubular, swollen at the base, usually pendulous, red, often with green tips. – Greenhouse, moderately warm. Propagation: seed, cuttings, leaf-cuttings. (For a critical appraisal of various Gasteria complexes, see E.A.C.L.E. SCHELPE, Gasteria, A Problem Genus of S.Afr. Succ. Plants, in J. Bot. Soc. S.Afr. XLIV, 1958.)

**G. acinacifolia** (JACQ.) HAW. v. **acinacifolia** (*Aloe a.* JACQ., *G. pluripunctata* HAW., *A. p.* ROEM. et SCHULT., *G. a.* v. *p.* BAK.). – Cape. – **L.** 15 in a large Ros., ensiform, 35 cm long, 5 cm across, trigonous-rounded at the tip, mucronate, one side flattened, the keel marginate, upperside concave, with irregularly arranged 2–4 mm long spots, margins small-tuberculate below, hornyserrate above.

**G.** — v. **ensifolia** (HAW.) BAK. (*G. e.* HAW., *Aloe e.* ROEM. et SCHULT.). – **L.** more numerous, trigonous-tapering at the tip, blotches scattered and more numerous.

**G.** — v. **nitens** (HAW.) HAW. (*G. n.* HAW., *Aloe n.* ROEM. et SCHULT., *G. a.* v. *patula* HAW., *A. candicans* v. *n.* SALM). – **L.** longer, less tapering, spots numerous, pale, mottled.

**G.** — v. **venusta** (HAW.) BAK. (*G. v.* HAW., *Aloe v.* ROEM. et SCHULT., *A. a.* v. *v.* SALM). – Markings in decorative transverse rows.

**G.** cv. Amoena. – Cultivar from DE LAET, similar to **G. verrucosa**.

**G. angulata** (WILLD.) HAW. v. **angulata** (*Aloe a.* WILLD., *G. disticha* v. *a.* BAK., *A. lingua* v. *longifolia* HAW., *A. l.* v. *latifolia* WILLD., *G. longifolia* HAW.). – S.Cape. – Acauline, offsetting; **L.** distichous, 20–25 cm long, 5 cm across, margins angular, blunt-tipped, green, with depressed spots in transverse R.

**G.** — v. **truncata** (WILLD.) BGR. (*A. a.* v. *t.* WILLD.). – **L.** rather shorter, oblique-angular, blotches numerous, well spaced on the upperside, in transverse bands on the underside.

**G. angustiarum** v. POELLN. – Cape: between Ladismith and Calitzdorp. – Acauline, offsetting; **L.** 5–8, distichous, 8–9 cm long, 2 to 2.5 cm across, 7–9 mm thick midway, dark green, often reddish, with slight longitudinal furrows midway on the upperside, underside convex, with a brownish tip, margins roundish, tuberculate towards the tip, often with a white horny border, blotches often ± merging, often in transverse bands on the underside.

**G. angustifolia** (AIT.) HAW. (*Aloe lingua* v. *a.* AIT., *A. a.* SALM, *G. disticha* v. *a.* BAK.). – S.Cape. – Mat-forming; **L.** 8–10, distichous, linguiform, 20–22 cm long, c. 3.5 cm across, fleshy, firm, upperside flat, plicate above midway, margins tapering at the tip, dark green, both surfaces spotted, margins rough tuberculate, exeedingly sharp, dentate towards the tip.

**G.** — v. **laevis** (SALM) HAW. (*Aloe l.* SALM, *A. lingua* LINK, *A. a.* v. *l.* SALM, *G. l.* HAW.). – **L.** smoother, spots more confluent.

**G.** × **apicroides** BAK. – Cape. – Probably a natural hybrid between **Gasteria** × **Astroloba**. – **St.** laxly leafy, c. 20 cm tall; **L.** 20–30, in many R., 10–15 cm long, 3 cm across, triangular-lanceolate, tapering, underside sharply carinate, upperside grooved, green to reddish, with numerous small, whitish spots in distinct transverse R., margins with cartilaginous tubercles, whitish.

**G. armstrongii** SCHOENL. (Pl. 85/5). – Cape: Humansdorp D. – Acauline or caulescent; **L.** at first distichous, eventually forming a spiralled Ros., older L. resting on the soil; juvenile L. 3–5 cm long, 3 cm across, linguiform, the tip rounded and apiculate, upperside very concave, tuberculate, margins slightly tuberculate towards the tip, ovate-triangular, tapering, 5 to 6 cm long, 3–4 cm across at the base, 1 cm thick, the underside sharply carinate towards the tip, tuberculate-rough to smooth.

**G. batesiana** ROWL. (Pl. 86/4) (*G. carinata* ROWL., *G. subverrucosa* v. *marginata* HORT.). – Zululand. – Ros. acauline, with few offsets; **L.** in spirals, triangular to linguiform-lanceolate, stiff, somewhat grooved towards the tip, oblique-trigonous, underside sharply and obliquely carinate, deep olive-green, rough with numerous larger green and smaller white spots in irregular transverse bands in the lower half.

**G. beckeri** Schoenl. – Cape: Albany D. – Acauline; L. 6–8 in a spiralled Ros., 12 cm long, 5 cm across at the base, with slightly raised, elongated tubercles.

**G. bicolor** Haw. (*Aloe b.* Roem. et Schult.). – S.Cape. – **St.** leafy, 10–15 cm tall; L. 12–16, distichous, 15–23 cm long, 2–3 cm across, 10 to 12 mm thick below, tapering above, dull light green, with small white spots, especially at the base, margins cartilaginous.

**G. biformis** v. Poelln. – Cape: Cradock D. – **St.** very extended, up to 14 cm long; L. distichous, 20–25 cm long, 4.5–5 cm across, grooved, gradually tapering, trigonous and mucronate above, smooth, green, spotted, spots roundish, usually in indistinct transverse bands.

**G. brevifolia** Haw. (*A. b.* Roem. et Schult., *A. brachyphylla* Salm). – S.Cape. – Acauline; L. 12–14, ± distichous, 8–15 cm long, 3.5 to 5 cm across, thick, linguiform, rounded and mucronate above, dark green, with ± conspicuous white blotches in distinct transverse bands, rather rough tuberculate, margins with cartilaginous T.

**G. caespitosa** v. Poelln. – Cape: Somerset East D. – Freely offsetting, soon forming clumps; **St.** 1–1.5 cm long; L. distichous, 10 to 14 cm long, 2 cm across, gradually tapering above, the tips trigonous, pointed, mucronate, concave below on the upperside, margins blunt, with a tuberculate horny border, the upperside with ± numerous, light green spots, usually in transverse R.

**G. candicans** Haw. (*Aloe c.* Roem. et Schult., *G. linita* Haw., *A. l.* Roem. et Schult., *A. c.* v. *l.* Salm). – S.W.Cape. – **Ros.** large, with offsets; L. 12–20, in many series, broadly ensiform, 25–30 cm long, 7 cm across, concave on the upperside, obliquely carinate on the underside, glossy green, with minute spots in indistinct transverse R.

**G. carinata** (Mill.) Haw. v. **carinata** (*Aloe c.* Mill. and v. *acinaciformis* Salm). – Cape. – Freely offsetting; L. in a spiralled Ros., lanceolate, trigonous, blunt or pointed, 12–15 cm long, upperside furrowed, underside obliquely carinate, dark green with raised white spots in transverse R., margins rough.

**G.** – v. **falcata** Bgr. – L. longer, narrower, spots less numerous.

**G.** – v. **latifolia** Bgr. – L. 11 cm long, 5 cm across, underside broadly keeled, this third face 2 cm across.

**G.** – v. **strigata** (Haw.) Bgr. (*G. s.* Haw., *Aloe s.* Roem. et Schult., *A. c.* v. *laevior* Salm). – Robuster; L. 20–25 cm long, smoother, spots less distinctly in R.

**G. chamaegigas** v. Poelln. – Cape: Port Elizabeth D. – Related to **G. maculata**; **St.** 8 cm long, leafy; L. crowded, in distichous spirals, 6–7 cm long, 2.5 cm across, linguiform, rounded and mucronate above, dark green, with light greenish spots in transverse bands, the underside with a distinct, angular keel.

**G.** × **cheilophylla** Bak. (Pl. 86/1). – Hybrid: **G. verrucosa** × **G. pulchra**; L. in a spirally contorted Ros., ensiform, 20–25 cm long, 2 to 2.5 cm across at the base, glossy dark green, with numerous white, ± indistinct dots, especially on the underside.

**G. colubrina** N. E. Br. – Cape: Uitenhage D. – **St.** leafy, 7–15 cm tall; L. 8–10, distichous, later more spiralled, ligulate, 23–35 cm long, 3 cm across, 8–12 mm thick, trigonous-tapering above, convex on the upperside, one face cuneate, dark green to purple, spotted.

**G. conspurcata** (Salm) Haw. (*Aloe c.* Salm and v. *truncata* Salm, *G. disticha* v. *c.* Bak.). – S.Cape. – **L.** distichous at first, later more spiralled, broad-linguiform, sharply trigonous, 30 cm long, 4 cm across, blunt and mucronate above, with rough tubercles in distinct transverse R.

**G. croucheri** (Hook. f.) Bak. (*Aloe c.* Hook. f., *G. natalensis* Bak.). – Cape, Natal. – **Ros.** 60 cm $\varnothing$, 30 cm tall; L. c. 17, crowded, in many series, long-tapering, blunt above, 30–40 cm long, 7–9 cm across at the base, broadly concave on the upperside, doubly carinate on the underside, the 3 surfaces 2–2.5 cm across, green, smooth, with white spots, margins and keel rough tubercled, dentate at the tip.

**G.** — cv. Spathulata. – L. 50 cm long, 6–7 cm across at the base, more grooved on the upperside, spots less numerous, greenish. Hort. Kew.

**G. decipiens** Haw. (*Aloe d.* Roem. et Schult., *Haworthia nigricans* Haw.). – Acauline; L. 10–15, in a spiralled Ros., triangular-tapering, c. 10 cm long, 4 cm across at the base, thick, upperside very concave, underside carinate, the tip sharp, dark green, with dirty white spots, margins blunt.

**G. dicta** N. E. Br. (*G. subnigricans* v. *torta* Bak.). – Hort. Kew. – **St.** leafy, 5–7-angled, 5–7 cm tall; L. 12–14, spiralled-distichous, linguiform, 10–13 cm long, 3 cm across, 10 to 12 mm thick, underside with 2 keels, with a trigonous tip, green with tuberculate spots, margins tuberculate and reddish.

**G. disticha** (L.) Haw. v. **disticha** (Pl. 86/6) (*Aloe d.* L., *A. lingua* Salm and v. *angustifolia* Salm, *G. denticulata* Haw., *A. d.* Roem. et Schult., *G. lingua* Bgr.). – Cape: Clanwilliam D. – Acauline, offsetting basally; L. 10–12, distichous, linguiform, 20–25 cm long, c. 5 cm across, margins sharp and curved, rounded and pointed above, dark green, with roundish white spots in transverse bands.

**G.** — v. **latifolia** (Salm) v. Poelln. (*Aloe lingua* v. *l.* Salm, *G. d.* v. *major* Haw., *A. d.* v. *l.* Kunth). – L. larger than in the type.

**G.** — v. **minor** Bak. – Smaller variety.

**G. elongata** Bak. – Cape. – L. 6–8, distichous, 15–25 cm long, 2.5–3 cm across, upperside flat, underside rounded, margins often double, cartilaginous-dentate below, trigonous-tapering above, with numerous small white spots.

**G. ernestii-ruschii** Dtr. (Pl. 86/5). – S.W.Afr.: Orange R. – Offsetting from the base; **St.** 2.5 cm long, leafy; L. c. 12, very densely distichous, rough with minute papillae, very thick, linguiform, 5–7 cm long, c. 3 cm across, dark green, with a slight longitudinal furrow midway, margins with a light-coloured horny border or small tubercles, both surfaces with greenish spots, tending to merge.

**G. excavata** (WILLD.) HAW. (*A. e.* WILLD., *A. obscura* WILLD., *G. latifolia* HAW.). – S.Cape. – Acauline; **L.** 12–16, spiralled, lanceolate, bluntly rounded and mucronate above, upperside boat-shaped and concave, 10–13 cm long, 2.5–3.5 cm across, with well-spaced whitish round spots.

**G. fasciata** (SALM) HAW. v. **fasciata** (*Aloe nigricans* v. *f.* S.D., *A. vittata* ROEM. et SCHULT. and v. *latifolia* SALM). – S.Cape. – Acauline; **L.** 10–20, initially distichous, later in spirals, narrow, rounded above, acuminate, 15–17 cm long, 2.5–3 cm across, underside rounded, margins tuberculate above, green, glossy, underside mottled, with blotches in transverse bands.

**G. —** v. **laxa** HAW. (*Aloe vittata* v. *l.* ROEM. et SCHULT.). – **L.** 4 cm across, ± imbricate.

**G. —** v. **polyspila** (BAK.) BGR. (*G. nigricans* v. *p.* BAK.). – **L.** margins smooth, blotches small, greenish.

**G. fuscopunctata** BAK. (*G. excelsa* BAK.). – Cape: King Williamstown D. – **Ros.** 35 to 40 cm ∅; **L.** 25–27, tapering from the base, triangular and mucronate above, c. 30 cm long, 8–9 cm across, 2–2.5 cm thick, upperside grooved-concave, underside with 2 keels, green, often reddish, with isolated brown, roundish spots, margins and keels cartilaginous-dentate.

**G. glabra** HAW. (*Aloe g.* SALM and v. *minor* SALM and v. *major* KUNTH, *A. carinata* KER.). – S.Cape. – Acauline, offsetting; **L.** 10–18 in a spiralled Ros., triangular-lanceolate, roundish-triangular and mucronate above, 17–20 cm long, 5–6 cm across, upperside concave, green, white-spotted, margins horny.

**G. gracilis** BAK. – Natal. – Acauline; **L.** in a dense Ros., lanceolate, 7–10 cm long, 2 to 2.5 cm across, with a roundish tip, margins horny, underside carinate, green, with numerous white spots, keel and margins rough tuberculate.

**G. herreana** v. POELLN. – Cape. – **St.** 2.5 to 3 cm tall, leafy; **L.** 9–10, distichous, 12–14 cm long, 2–2.5 cm across, 5–8 mm thick at the base, trigonous and mucronate at the tip, brownish above, both surfaces with numerous merging spots, margins serrate above.

**G. huttoniae** N. E. BR. – Cape: Stutterheim D. – **L.** rosulate, c. 20 cm long, ± ovate-triangular, c. 8 cm across at the base, narrowing towards the tip, blunt, dark green, both surfaces with large, light green blotches mostly in distinct transverse R., one face carinate, the third face 2.5 cm across, the keel with a serrate horny border, margins with several small lighter-coloured tubercles.

**G. humilis** v. POELLN. – Cape: Brak R. – Acauline, c. 14 cm ∅; **L.** 8–12, ± triangular, usually rather blunt above, acuminate, upperside very concave, underside convex and very obliquely carinate, 8–10 cm long, 2–4 cm across below, glossy dark green, with whitish spots, solitary or merging to form transverse bands, margins and keel tuberculate.

**G. inexpectata** v. POELLN. – Cape. – **L.** lanceolate, pointed, mucronate, 14–15 cm long, 3–4 cm across at the base, 8 mm thick, very deep green, rounded, underside very rounded and carinate, keel angular, keel and margins rough tuberculate, horny towards the tip, both surfaces blotched, markings on the upperside greenish-white, roundish, tending to merge, those on the underside tending to form transverse bands.

**G. joubertii** v. POELLN. – Cape: Ladismith D. – ± acauline; **L.** distichous, linguiform, very blunt above, trigonous or trigonous-rounded, acuminate, 10 cm long, 3.5–4 cm across at the base, dull, smooth, green, with greenish-white spots, ± merging into transverse bands, the upperside with a M.groove, underside convex, also with a M.groove, the margins with lighter-coloured small tubercles towards the tip and a horny border.

**G. × kewensis** BGR. – Hybrid: **G. verrucosa** v. **latifolia** × **G. brevifolia**. – Hort. Kew. – **L.** distichous, c. 10, c. 10 cm long, 4 cm across, thick, stiff, the margins especially thick, apiculate, with white tubercles 1 mm ∅ in irregular, ± conspicuous, transverse bands.

**G. kirsteana** v. POELLN. – Cape: N. of Adelaide. – **St.** c. 5 cm tall; **L.** distichous, in spirals, linguiform-ligulate, roundish-trigonous above with a large point, underside rounded, truncate on one side, margins with light green tubercles, not horny, blackish green with whitish-green spots 4–8 mm long, solitary or tending to merge, 20–22 cm long, 6–7 cm across below, 1–1.5 cm thick midway.

**G. laetipunctata** HAW. (Pl. 86/2) (*Aloe laetepunctata* ROEM. et SCHULT., *G. l.* BAK.). – S.Cape. – Resembles **G. carinata** but smaller; **L.** narrower, tapering, tubercles scattered, white, in distinct bands.

**G. × lauchei** (RADL) BGR. (*Aloe × l.* RADL). – Hybrid: **G. pulchra** × **G. scaberrima**. – **L.** 25 cm long, 2 cm across, dark green, with conspicuous, white, pear-like tubercles.

**G. liliputana** v. POELLN. (Pl. 87/3) (*G. pulchra* PHILLIPS, *G. minima* HORT.). – Cape: Peddie D. – **St.** very short, mat-forming; **L.** spirally rosulate, 3.5–6 cm long, 1–1.5 cm across, dark green, glossy, lanceolate, pointed, mucronate, underside roundish and distinctly carinate, keel tuberculate towards the base, margins serrate, white or greenish, with spots 1–2 mm ∅, ± in transverse R.

**G. loeriensis** v. POELLN. – Cape: near Loeriesfontein. – **St.** up to 3 cm tall; **L.** in a spiralled Ros., lanceolate, gradually tapering, mucronate, with a keel forming the border, dark green, both surfaces with ± whitish spots, those on the underside rarely in transverse R. but often merging, both surfaces with ± long, broad longitudinal bands, 16–20 cm long, 2.5 cm across below.

**G. longiana** v. POELLN. – Cape: Grahamstown D. – Caulescent, offsetting; **St.** 12 cm long; **L.** numerous, spiralled, ensiform-lanceolate, very tapered, with a long brown tip, 15–25 cm long, 2–4 cm across below, 5–8 mm thick, upperside rather concave, underside carinate, the keel angular towards the tip, the side of the angle 1–1.5 cm across at the base, margins slightly serrate-bristly, spots numerous, greenish-white, 4–10 mm long, usually merging to form broad transverse bands.

**G. longibracteata** v. POELLN. – Cape: Uitenhage D. – **L.** ligulate, roundish or roundish-trigonous above, mucronate, 22–30 cm long, 4–5 cm across below, 5–6 mm thick, smooth, dark green, both surfaces with whitish spots 3–5 mm long, merging into indistinct transverse bands, both surfaces roundish, margins sharply angular towards the tip with white tubercles continuing into the horny border.

**G. lutzii** v. POELLN. (*G. vroomii* HORT.). – Cape: Grahamstown D. – **L.** in a dense, spiral Ros. 45–50 cm ⌀, c. 24 cm long, c. 7 cm across below, often ± falcate, narrowing towards the tip, ± acuminate, always truncate at one side, greenish-reddish, very slightly glossy, blotches large, few, margins horny with small confluent tubercles.

**G. maculata** (THUNBG.) HAW. v. **maculata** (*Aloe m.* THUNBG., *A. obliqua* HAW., *G. o.* HAW., *A. maculata* v. *o.* AIT., *A. lingua* KER., *G. latifolia* BAK., *G. nigricans* v. *platyphylla* BAK.). – Cape: George, Uitenhage D. – **St.** leafy, offsetting from the base; **L.** c. 34, in spirals or distichous, firm, thick, linguiform, 16–20 cm long, 4.5–5 cm across, underside convex and with 2 angles, the third face 7 mm across, margins and keels horny-rough, tip blunt-rounded to tri-angular-tapering, dark green, glossy, with confluent white blotches 4–5 mm long, forming bands towards the tip.

**G. — v. dregeana** BGR. – **L.** 25 cm or more long, keel and margins with tubercles or white, horny and serrate towards the tip, the spots ± white on both surfaces.

**G. — v. fallax** HAW. (*Aloe obliqua* v. *f.* ROEM. et SCHULT., *A. m.* v. *angustior* S.D.). – Smaller than the type in all parts; **L.** narrower and more whitish.

**G. × margaritifera** BGR. – Hybrid: **G. verrucosa** × **G. spec.** ?. – Offsetting, mat-forming; **L.** 7–10, in spirals, narrow-triangular, underside with 2 angular keels, dirty green with scattered white spots, these more numerous on the underside, often in transverse bands.

**G. marmorata** BAK. (Pl. 87/2). – S.Cape. – **St.** leafy, 15–25 cm tall; **L.** 20–30 in many series, lanceolate, rounded-tapering above, 13–15 cm long, 3–4 cm across at the base, dark green with darker mottling.

**G. × metallica** BGR. – Hybrid: **G. acinacifolia** v. **nitens** × **G. subcarinata** ?. – **L.** in spirals, 16 cm long, 2.5 cm across below, metallic brownish-green, with indistinct whitish tubercles, margins with small whitish T.

**G. mollis** HAW. (*Aloe m.* ROEM. et SCHULT.). – S.Cape. – Acauline; **L.** 6–8, distichous, linguiform or triangular-tapering, mucronate, 7 to 10 cm long, 3–3.5 cm across, green with some white spots, fleshy, soft, upperside slightly velvety, margins rounded, horny.

**G. multiplex** v. POELLN. – Cape: Bruinjes Hoogte. – **St.** up to 28 cm long; **L.** distichous, gradually narrowing towards the tip, upperside rather concave to flat, underside somewhat rounded, not carinate, 14–22 cm long, 3–4.5 cm across, pointed to bluntish, apiculate, ± glossy, green to dark green, ± light-spotted, indistinctly striate.

**G. neliana** v. POELLN. – Cape: Lit. Namaqualand. – **L.** 4–6, distichous, linguiform, 18–20 cm long, 4–4.5 cm across at the base, 6–8 mm thick, tip blunt-triangular and apiculate, both surfaces slightly convex, rough, emerald-green, with ± confluent whitish-green blotches in ± transverse R., margins 1–2 mm across, minutely dentate, pale, cartilaginous.

**G. nigricans** HAW. v. **nigricans** (*Aloe n.* HAW., *A. obliqua* JACQ.). – Cape: Port Elizabeth D. – **St.** very short, with few offsets; **L.** up to 20, distichous or somewhat spiralled, 6–20 cm long, 4–5 cm across, linguiform, the tip round-trigonous and apiculate, dark green, glossy, both surfaces convex and often with a M.groove, margins sheath-like, spots white or greenish-white, often indistinct, often in irregular transverse bands.

**G. — v. crassifolia** (AIT.) HAW. (*Aloe lingua* v. *c.* AIT., *A. n.* v. *c.* SALM, *G. c.* HAW.). – **L.** narrower and thicker.

**G. — v. marmorata** HAW. (*Aloe n.* v. *m.* SALM). – **L.** 6 cm long, 4 cm across, flat, both surfaces mottled.

**G. nitida** (SALM) HAW. v. **nitida** (*Aloe n.* SALM). – Cape: Alicedale. – **St.** short; **L.** 12–15, in several series in a spiralled Ros., broadly triangular-lanceolate, upperside grooved, underside carinate, with a cartilaginous border, smooth, glossy green, indistinctly white-spotted, the spots on the underside in indistinct transverse bands.

**G. — v. grandipunctata** (SALM) HAW. (*Aloe n.* v. *g.* SALM). – Spots considerably larger, in transverse bands.

**G. — v. parvipunctata** (SALM) HAW. (*Aloe n.* v. *p.* SALM, *A. n.* KER.). – Spots less numerous, smaller.

**G. obtusa** (SALM) HAW. (*Aloe nitida* v. *o.* SALM and v. *brevifolia* SALM, *A. o.* ROEM. et SCHULT.). – Cape: Douglas. – **St.** leafy; **L.** 11–13 in a spiralled Ros., 10–15 cm long, 3.6 cm across below, gradually tapering, roundish-trigonous at the tip, mucronate, underside carinate-triangular, keel angular, green, glossy, with numerous indistinct blotches in transverse bands, margins horny-dentate.

**G. obtusifolia** (SALM) HAW. (*Aloe o.* SALM, *A. lingua* v. *brevifolia* SALM). – Cape: Bonnievale. – Acauline; **L.** 12–14, crowded-distichous, 15–18 cm long, 5–6 cm across, linguiform, rounded-truncate at the tip, underside convex, with greenish-white spots in transverse bands, margins horny, serrate towards the tip.

**G. pallescens** BAK. – Cape: around Algoa Bay. – **St.** leafy, 5 cm tall; **L.** 8–10, crowded, spiralled-distichous, ensiform, 15 cm long, 2.5 cm across below, 10–12 mm thick, upperside grooved, underside roundish, triangular-pointed at the tip, green, with indistinct whitish-green spots.

**G. parvifolia** BAK. (*G. bijliae* v. POELLN., ? *G. parva* HAW.). – Cape: Oudtshoorn D. – Acauline, offsetting from the base; **Ros.** 12–14 cm ⌀; **L.** initially distichous, later rosulate, 7–8 cm long, 3–4 cm across below, tapering, blunt triangular or oblique-triangular at the tip, with a whitish mucro, the left margin truncate, the resulting face 10–13 mm across, underside somewhat convex, glossy green, both surfaces with tuberculate, whitish, roundish spots 1–1.5 mm ⌀, those on the underside merging into 7–8 distinct transverse bands, those on the upperside less so, margins and keel rough and horny towards the tip.

**G. patentissima** v. POELLN. – Cape: Brak R. – **St.** c. 2 cm tall; **L.** distichous, ligulate, trigonous and mucronate at the tip, 18 cm long, 3–3.5 cm across, 5 mm thick, glossy, smooth, dark green, underside somewhat convex, margins serrate-horny above, both surfaces with numerous, ± confluent spots 3 mm ⌀, in distinct transverse bands.

**G. picta** HAW. v. **picta** (*Aloe bowieana* ROEM. et SCHULT.). – S.Cape. – Eventually caulescent; **L.** 12–20 in a distichous, spirally contorted Ros., ligulate-lanceolate, 25–35 cm long, 4–5 cm across, slightly rounded on the upperside, obliquely convex on the underside, the triangular tip with dentate margins, blackish-green, glossy, with white confluent spots in transverse bands.

**G.** – v. **formosa** (HAW.) BAK. (*G. f.* HAW., *Aloe f.* ROEM. et SCHULT., *A. marmorata* HAW., *A. bowieana* v. *f.* SALM). – **L.** c. 2 cm across, with thinner margins.

**G. pillansii** KENSIT. – Cape: Clanwilliam D. – Acauline; **L.** distichous, linguiform, exceedingly obtuse, apiculate, margins tuberculate, both surfaces dirty green with numerous whitish spots, 10–17 cm long, 4–4.5 cm across.

**G. planifolia** BAK. (*Aloe p.* BAK.). – Cape: vicinity of Algoa Bay. – **St.** leafy, 15–25 cm tall; **L.** 12–20, distichous, long-ensiform, 15–25 cm long, 2 cm across, flat-grooved to concave, smooth, glossy, green, with large white spots, more confluent on the underside, rarely in transverse bands.

**G. poellnitziana** JACOBS. (Pl. 87/1). – Cape. – With few offsets or none; **L.** 8–10 in a spiralled Ros., narrow-linear, long-tapering, c. 20 cm long, 15–18 mm across at the base, 5–6 mm thick, emerald-green, upperside concave up to midway, underside obliquely and obtusely carinate, with more numerous, ± prominent, white tubercles c. 1 mm long, ± merging into groups, keel and angles white-tuberculate.

**G. porphyrophylla** BAK. – S.Cape. – **St.** leafy, 4–5 cm tall; **L.** 8–10, distichous-spiralled, 15 to 20 cm long, 2.5 cm across below, ligulate, margins often double, triangular-tapering at the tip, dirty purple with white, confluent blotches.

**G. prolifera** LEM. (*G. carinata* HORT. ex LEM.). – S.Cape. – Offsetting freely; **Ros.** large; **L.** thick, triangular, 20–30 cm long, 13–15 mm across below, upperside concave, dirty green with small white spots.

**G. pseudonigricans** (SALM) HAW. v. **pseudonigricans** (*Aloe p.* SALM, *G. subnigricans* HAW., *A. s.* HAW., *A. s.* KUNTH, *G. nigricans* v. *subnigricans* BAK.). – Cape. – **St.** leafy, offsetting; **L.** 14–16, distichous, 15–20 cm long, 3.5–4 cm across below, narrow-ligulate, compressed at the base, with a triangular-acuminate tip, green, glossy, with indistinct white spots, angles tuberculate, sharp and horny at the tip.

**G.** – v. **canaliculata** (SALM) JACOBS. (*Aloe subnigricans* v. *c.* SALM, *G. s.* v. *c.* SALM). – **L.** 20–25 cm long, upperside grooved-concave.

**G.** – v. **glabrior** (HAW.) JACOBS. (*G. subnigricans* v. *g.* HAW., *Aloe guttata* SALM, *G. nigricans* v. *guttata* BAK.). – **L.** grooved below, dark green, with larger spots in lateral R., glossy, smooth, margins horny-dentate.

**G. pulchra** (AIT.) HAW. (*Aloe maculata* v. *p.* AIT., *A. m.* KER., *A. p.* JACQ., *A. obliqua* DC.). – Cape: Ugie. – **St.** leafy, 15–30 cm tall, offsetting basally; **L.** 15–20, ± distichous, linear, ensiform, 3-angled with unequal sides, long-tapering, often doubly carinate, upperside concave, 20–30 cm long, 2.5 cm across below, dirty green, glossy, with numerous white, confluent blotches in irregular transverse bands, margins white, horny.

**G. radulosa** BAK. – Cape. – **St.** leafy, 4 cm tall; **L.** c. 6, ligulate-ensiform, 15–20 cm long, 4 cm across, 4–5 mm thick, curved, almost flat, the tip rounded and apiculate, dark green, the underside and margins dentate, rough, with numerous, whitish, raised tubercles.

**G.** × **repens** HAW. (*G. intermedia* v. *r.* SALM, *Aloe r.* ROEM. et SCHULT.). – Hybrid: **G. verrucosa** v. **intermedia** × **G. carinata**. – Acauline; **L.** 8–10, distichous, linguiform, 6–8 cm long, 16–18 mm across, 3 mm thick, rounded to triangular-tapering at the tip, green, with small tubercles in transverse bands, margins horny, white.

**G. retata** HAW. (*G. dyctioides* ROEM. et SCHULT.). – Cape. – **St.** leafy, 5–8 cm tall; **L.** 10–12, spiralled-distichous, ± ensiform, 25 to 30 cm long, 3–3.5 cm across below, 10–12 mm thick, often forming a right-angled triangle in cross-section, upperside flat, underside rounded, margins horny-tuberculate, dentate towards the tip, green, with roundish, white, confluent spots forming indistinct transverse bands towards the tip.

**G.** × **rufescens** BGR. – Cape. – A hybrid, probably **G. nigricans** × **G. spec.** ?. – Acauline; **L.** c. 12, distichous to spiralled, 8–10 cm long, linguiform, tapering from the base, 5–6 cm across, rounded above, margins rounded and often double, tuberculate-horny, green to reddish with a metallic sheen, spots 2 mm long in irregular transverse bands, more numerous on the underside.

**G. salmdyckiana** v. POELLN. – Cape: Alexandria D. – **L.** laxly distichous-spiralled, long-ensiform or triangular, mucronate, smooth, glossy, dark green, both surfaces with light green spots, in ± transverse bands, 20–30 cm long, 4–5 cm across, 5–7 mm thick, bluntly carinate at the margins, keel angular and 1.5 cm across, margins and keel horny, brown.

**G. schweickerdtiana** v. POELLN. – Cape: Brak R. – Forming low clumps; L. 8–12, initially distichous, later spiralled, rigid, 14–18 cm long, 3.5–4 cm across below, narrowing towards the tip, rounded to triangular-tapering above, with a brownish apiculus, upperside concave, underside convex, light green, glossy, with white spots, usually solitary on the upperside, merging on the underside into transverse bands.

**G. spiralis** BAK. v. **spiralis**. – S.Cape. – St. leafy, 10–15 cm tall; L. 16–18, spiralled-distichous, rigid, linguiform, 10–15 cm long, 2.5–3 cm across, 6 mm thick, upperside flat, underside rounded, margins often double, glossy, green to reddish, tuberculate with ± indistinct, white spots, tip triangular-tapering.

**G. —** v. **tortulata** BAK. – L. narrower towards the tip, conspicuously twisted from the base.

**G. × squarrosa** BAK. – Hybrid: **Gasteria × Haworthia**, or **Gasteria × Astroloba**. – St. leafy, c. 15 cm tall; L. 20–30, in many R., widely spreading, 10–13 cm long, 2.5 cm across below, upperside concave, underside rounded, green with solitary, indistinct, whitish tubercles.

**G. stayneri** v. POELLN. – Cape: Port Elizabeth D. – Acauline, freely offsetting, forming clumps; L. ovate-oblong to linguiform, the tip broadly rounded to triangular-rounded and mucronate, dark green, underside convex, with numerous dark green tubercles, margins tuberculate.

**G. subcarinata** (SALM) HAW. v. **subcarinata** (*Aloe s.* SALM, *A. pseudoangulata* SALM). – Cape. – Acc. BAKER possibly a hybrid: **G. carinata × G. disticha**. – Short-stemmed, offsetting; L. 10–15 in several series, almost rosulate, lanceolate, forming a triangle with unequal sides, upperside concave, underside convex, the tip obtuse, somewhat oblique, 10–15 cm long, 2.5 to 3 cm across, green, with flat, white tubercles, margins cartilaginous-serrate.

**G. —** v. **striata** (WILLD.) HAW. (*A. angulata* v. *striata* WILLD., *Aloe pseudoangulata* v. *s.* SALM, *G. s.* HAW.). – L.margins attractively striped.

**G. subverrucosa** (SALM) HAW. v. **subverrucosa** (*Aloe s.* SALM and v. *grandipunctata* SALM, *G. s.* v. *g.* HAW.). – Cape: vicinity of Algoa Bay. – L. 8–10, distichous, ligulate, 20–25 cm long, c. 3 cm across, both surfaces convex, with irregular white spots in distinct transverse bands, margins thickened, tuberculate, the tip rounded, acuminate, horny.

**G. —** v. **marginata** BAK. – L. shorter, margins horny.

**G. —** v. **parvipunctata** (SALM) HAW. (*Aloe s.* v. *p.* SALM). – L. longer, ensiform, spots smaller and more numerous.

**G. sulcata** (SALM) HAW. (*Aloe s.* SALM, *A. linguiformis* DC., *A. lingua* v. *angulata* HAW., *G. a.* HAW.). – S.Cape. – L. linguiform, distichous, 10–11 cm long, 3 cm across, 6–8 mm thick, dark green, deeply furrowed, somewhat spotted, margins narrow, white.

**G. thunbergii** N. E. BR. – Cape. – L. precisely distichous, 7.5–17.5 cm long, 1.5–2 cm across, apiculate, both surfaces with numerous, small, white, prominent tubercles along the margins.

**G. transvaalensis** BAK. – Transvaal. – Related to **G. nigricans**; Ros. distichous or slightly oblique; L. 8, ligulate, dark green, glossy, 10–12 cm long, 2.5 cm across, 6–7 mm thick, with triangular T. along the white, horny margin, spots greenish-white, merging into transverse bands.

**G. triebneriana** v. POELLN. – Cape: Karroo. – St. short; L. few, 15–20 cm long, 4–5 cm across, distichous, linguiform, mucronate, smooth, dark green, older L. much curved inwards, upperside with a distinct M.rib, underside convex, margins and keel smooth towards the tip or with very small tubercles towards the base, margins serrate and horny above, spots whitish, solitary or merging into indistinct transverse bands.

**G. trigona** HAW. v. **trigona** (Pl. 87/5) (*Aloe acinacifolia* v. *angustifolia* and v. *laetevirens* SALM, *A. elongata* SALM, *A. t.* v. *e.* SALM and v. *minor* ROEM. et SCHULT., *G. t.* v. *e.* and v. *m.* HAW.). – Cape. – Acauline, offsetting; L. 9–11, in many series, ensiform to lanceolate, acute at the tip, 15–20 cm long, 3–3.5 cm across, upperside concave, underside carinate, fresh green, with white spots in bands, margins and keel horny, tuberculate-dentate.

**G. —** v. **kewensis** BGR. – L. swollen on the upperside, furrowed at the base and midway, blue-green, spots larger, translucent.

**G. variolosa** BAK. – Cape: vicinity of Algoa Bay. St. leafy, 4–5 cm tall; L. 15–18, spiralled-distichous, 20–27 cm long, 4 cm across, 6–8 mm thick, firm, ligulate, tapering above, with a triangular, tapering tip, margins tuberculate, upperside concave, underside convex, dirty green, with whitish confluent spots.

**G. verrucosa** (MILL.) DUV. v. **verrucosa** (Pl. 87/4) (*Aloe v.* MILL., *A. disticha* THUNBG., *A. v.* MED., *A. acuminata* LAM., *A. racemosa* LAM.). – Cape: near King Williams Town. – Offsetting from the base to form a mat; L. 6–10, distichous, 10–15 cm long, 2 cm across below and 1.5 cm across midway, long-tapering, acuminate, upperside grooved, underside convex, dirty green, with numerous, irregular, ± confluent, white tubercles, margins truncate, thickened, 4–5 mm thick, tuberculate, horny and smooth towards the tip.

**G. —** v. **asperrima** (SALM) v. POELLN. (*Aloe intermedia* v. *a.* SALM, *G. i.* v. *a.* HAW., *A. scaberrima* SALM ex ROEM. et SCHULT., *G. v.* v. *s.* (S. D.) BAK.). – L. ligulate, thick, curved-ensiform at the tip, tubercles larger.

**G. —** v. **intermedia** HAW. (*A. i.* HAW., *G. i.* HAW.). – L. considerably thicker, longer and broader, margins blunter, tubercles more numerous, with a white sheen, wider-spaced.

**G. —** v. **latifolia** HAW. – Considerably more robust; L. up to 30 cm long.

**G. —** v. **striata** (SALM) v. POELLN. (*A. v.* v. *s.* SALM). – Tubercles in ± distinct longitudinal lines.

**G. vlaaktensis** v. POELLN. – Cape: Ghewanie Vlaakte. – **St.** very short; **L.** distichous, linguiform, bluntly rounded and acuminate above, glossy, smooth, dark green, 11–12 cm long, 3–3.5 cm across, 4–5 mm thick, upperside with a flat M.groove, underside somewhat convex, margins with small tubercles, lighter midway, serrate towards the tip, both surfaces with numerous, whitish spots 2–3 mm long, those on the upperside merging into indistinct transverse bands.

**G. zeyheri** (SALM) BAK. (*A. z.* SALM). – Cape. – **St.** leafy, c. 15 cm tall; **L.** 10–12, spiralled-distichous, almost ensiform, 25–30 cm long, 3–3.5 cm across below, often forming a right-angled triangle in cross section, thickened, underside convex, the tip tapering, margins horny tuberculate, serrate towards the tip, with roundish spots in irregular transverse bands.

× **Gastrolea** E. WALTH. Liliaceae. – Intergeneric hybrids: **Gasteria** × **Aloe**. – Usually mat-forming, acauline **Ros.**; **L.** arranged spirally, very variable in shape, spotted or tuberculate, margins mostly horny-dentate, underside mostly triangular-carinate; **Infl.** tall, simple or branching, laxly racemose, Fl. pedic. llate, Per. sometimes slightly swollen at the base, often ± curved, reddish at the base, greenish at the tips. – Greenhouse, warm. Propagation: division, or adventitious plantlets from the Infl.

× **G. bedinghausii** (RADL) E. WALTH. (*Aloe* × *bedinghausii* RADL). – Hybrid: **Aloe aristata** × **G. nigricans**. – Like Aloe aristata in appearance; **Ros.** up to 25 cm ⌀; **L.** robust, with thick, pearl-like tubercles, margins white-dentate with only a rudimentary awn at the tip.

× **G. beguinii** (RADL) E. WALTH. (Pl. 88/2) (*Aloe* × *b.* RADL). – Hybrid: **Aloe aristata** × **G. verrucosa**. – Ros. 20 cm ⌀, densely leafy; **L.** leathery, dark green, with a reddish tip, triangular-tapering, 7–8 cm long, less than 3 cm across below, triangular above midway, apiculate, the upperside with several horny transverse bands, the underside with numerous spots in transverse bands, margins and keel with tuberculate T.

× **G. — nm. chludowii** (RADL) ROWL. (*Aloe* × *c.* RADL, × *G. c.* (RADL) E. WALTH.). – **L.** 12–16 cm long, ligulate, carinate, with numerous gibbose tubercles, margins slightly dentate.

× **G. — nm. perfectior** (RADL) ROWL. (*Aloe* × *beguinii* v. *p.* RADL, *A.* × *p.* BGR., × *G. p.* (BGR.) E. WALTH.). – Resembles × **G. beguinii**, up to 20 cm tall; **L.** considerably longer, light green.

× **G. derbetzii** (HORT. BGR.) E. WALTH. (*Aloe* × *d.* HORT. BGR.). – Hybrid: **Aloe striata** × **G. acinacifolia**. – Intermediate between the parents.

× **G. imbricata** (BGR.) E. WALTH. (*Aloe* × *i.* BGR.). – Hybrid: **Aloe variegata** or **A. serrulata** × **G. spec. ?** – Ros. c. 13 cm ⌀; **L.** in 5 spirals, lanceolate-triangular, 6 cm long, 2 cm across, very thick, green, margins horny-tuberculate, apiculate.

× **G. lapaixii** (RADL) JACOBS. (Pl. 88/1) (*Aloe* × *l.* RADL). – Hybrid: **A. aristata** × **G. maculata**. – Mat-forming; **L.** in a Ros. in many series, triangular-lanceolate, slightly tapering, 10 cm long, 2 cm across, 6–8 mm thick, underside somewhat obliquely carinate, with a cartilaginous tip, dull green, upperside with spots in indistinct transverse bands, underside with more numerous spots, margins dentate, serrate, T. triangular, cartilaginous.

× **G. lynchii** (BAK.) E. WALTH. (*Aloe* × *l.* BAK.). – Hybrid: **A. striata** × **G. verrucosa**. – Almost acauline or very short-stemmed; **L.** 20 to 24 cm long, 5 cm across below, long-tapering, underside convex, with 2 keels, pale green, with numerous, white, somewhat tuberculate spots, margins with minute, cartilaginous T.

× **G. mortolensis** (BGR.) E. WALTH. (*Aloe* × *m.* BGR.). – Hybrid: **A. variegata** × **G. acinaciformis**. – **St.** 40–60 cm long, leafy, creeping, offsetting basally, mat-forming; **L.** in a spiralled Ros., triangular-lanceolate, 25 cm long, 9 cm across below, upperside broadly concave, underside convex and somewhat obliquely carinate, dull green, the upperside with a few roundish spots, these more numerous and in transverse bands on the underside, margins horny, with T. 2 mm long.

× **G. nowotnyi** (RADL) E. WALTH. (*Aloe* × *n.* RADL). – Hybrid: ? **Aloe aristata** × **G. nigricans**. – Ros. c. 15 cm ⌀; **L.** with white tubercles and small, whitish T., light green, somewhat longer than in × **G. bedinghausii** and not as broad.

× **G. peacockii** (BAK.) E. WALTH. (*Aloe* × *p.* BAK.). – Hybrid: **A. heteracantha** × **G. acinacifolia** v. **ensifolia**. – Intermediate between the parents.

× **G. pethamensis** (BAK.) E. WALTH. (*Gasteria* × *p.* BAK.). – Hybrid: **Gasteria verrucosa** × **A. variegata**. – Closely resembles **G. verrucosa**; **L.** shorter, broader and thicker, tubercles less numerous and thicker.

× **G. pfrimmeri** (GUILL.) E. WALTH. ( × *Gasteraloe p.* GUILL.). – Hybrid: **Aloe variegata** × **Gasteria spec. ?** – **L.** in a spiralled Ros., hard, c. 10 cm long, 2.5–3 cm across below, long triangular-tapering, upperside grooved-concave, underside sharply carinate, both surfaces with well-spaced tuberculate spots, margins tuberculate-horny.

× **G. prorumpens** (BGR.) E. WALTH. (*Aloe* × *p.* BGR.). – Hybrid: **Aloe aristata** × **G. spec. ?** – Acauline, simple; **L.** c. 40 in a dense Ros., triangular, gradually tapering, 12 cm long, 4.5 cm across, dark green, upperside flat with a few spots, underside convex, with more numerous spots, carinate at the tip, often doubly carinate with white gibbose tubercles, margins with a few horny T.

× **G. quehli** (RADL) E. WALTH. (*Aloe* × *q.* RADL). – Hybrid: **Aloe aristata** × **G. spec**. – Ros. compact; **L.** dark green with white spots.

× **G. rebutii** (BGR.) E. WALTH. (*Aloe* × *r.* BGR.). – Hybrid: **Aloe variegata** × **G. spec. ?** – **L.** in 5 spiral series, fleshy, dull green, 13–14 cm long, 2.5–3.5 cm across below, long-tapering, upperside very concave, underside obliquely carinate-triangular, with numerous spots in ± distinct transverse bands, margins and keel cartilaginous-tuberculate.

× **G. sculptilis** POIND. – Hybrid: **Aloe variegata** × **G.** × **cheilophylla.** – Acauline or short-stemmed; L. c. 18 cm long, 8.5 cm across, thick, fleshy, upperside broadly grooved, underside keeled, margins and keel bordered white and irregularly rough-dentate, olive-green, with whitish spots in ± distinct transverse bands.

× **G. simoniana** (DEL.) GUILL. (*Aloe* × *s.* DEL., × *G. s.* (DEL.) JACOBS.). – Hybrid: **Aloe aristata** × **G. disticha.** – Ros. 30–40 cm ⌀; L. 15–20 cm long, 5–6 cm across below, tapering, with a triangular tip, dull green, spotted, margins cartilaginous.

× **G. smaragdina** (BGR.) E. WALTH. (*Aloe* × *s.* BGR.). – Hybrid: **Aloe variegata** × **G. candicans** ?. – Offsetting basally, mat-forming; L. c. 15, distichous or rosulate, fleshy, lanceolate-triangular, 20 cm or more long, 7 cm across below, underside carinate towards the tip, pale green, glossy, with numerous whitish-green spots in transverse bands, margins cartilaginous, minutely serrate-dentate.

× **Gastrolirion** E. WALTH. Liliaceae. – Intergeneric hybrids: **Gasteria** × **Chortolirion.** – Intermediate. – Greenhouse, warm. Propagation: division.

× **G. orpetii** E. WALTH. – Acauline; L. in a spiralled Ros., the broad bases overlapping like a bulb, rather thin and limp, 30–40 cm long, 3 cm across below, 1 cm across at midway, upperside deeply grooved, margins acute, horny with numerous T., dark green, with indistinct white spots; Fl. in a lax Rac., resembling **Gasteria,** somewhat paler.

**Gerrardanthus** HARV. Cucurbitaceae. – Afr. – ♃ climbing plants; **caudex** disc-like or spherical, partly underground; St. up to 6 m long, with tendrils; L. thin, glabrous, petiolate, ovate-reniform or narrow-cordate, ± conspicuously lobed, lobes tapering; Fl. browny-orange to browny-green, unisexual, ♂ Fl. in pedunculate spikes, ♀ Fl. solitary; Fr. somewhat angular, veined, c. 6 cm long, 2 cm ⌀, seeds fusiform and alate. Greenhouse, warm. Propagation: seed.

**G. lobatus** (COGN.) C. JEFFREY (Pl. 88/5) (*G. grandiflorus* v. *l.* COGN., *G. macrorhizus* HARV. ex JACOBS.). – Uganda; Kenya; Tanzania; Malawi; Nigeria. – **Caudex** dome-shaped, 50 cm or more ⌀; St. ± succulent; L. ovate to reniform-cordate, deeply 5-lobed; Fr. 4.5–6 cm long, up to 2 cm ⌀.

**G. macrorhizus** HARV. (*G. portentosus* NAUD. et DURIEU). – E. and S.Afr.: deserts. – **Caudex** disc-shaped, 30–60 cm tall; St. woody; L. broadly ovate to narrowly triangular, with 5–7 angular lobes; Fr. straw-coloured.

**Graptopetalum** ROSE (*Byrnesia* ROSE; *Sedum* L. Sect. *Graptopetalum* (ROSE) BGR.; *Echeveria* DC. Sect. *Graptopetalum* (ROSE) KEARN. et PEEBLES). – Crassulaceae. – USA: Arizona to Mexico: Oaxaca, up to 2 300 m above sea-level. – Glabrous, succulent, ♃ plants or semi-♄, resembling **Echeveria**, but the Pet. thin, ascending or spreading above midway and irregularly spotted or transversely banded with dark red; Infl. a lax cyme, normally consisting of 1–10 lax Cinc., Fl. pedicellate, 5–7-partite, Sep. erect, usually almost equal, Pet. united below, imbricate in the bud-stage, pointed, anthers eventually recurved. – Greenhouse, cool. Propagation: seed, cuttings. – Type species: **G. pusillum.** Text abbreviated from a contribution by **R. Moran.**

**G. amethystinum** (ROSE) E. WALTH. (Pl. 89/1) (*Pachyphytum a.* ROSE, *Echeveria a.* HORT. ex JACOBS.). – Mexico: Durango, Jalisco. – St. prostrate, up to 40 cm long, 8–13 mm thick; Ros. 10–15 cm ⌀; L. 12–15, obovate, blunt to rounded, 3–7 cm long, 2.5–4 cm across, 10 to 18 mm thick, margins rounded, bluish, reddish; Infl. 5–15 cm tall, with 3–10 Cinc., pedicels 5–8 mm long, Cor. 15–18 mm long, tube 4 mm long, lobes with dense, dark red, transverse bands.

**G. bartramii** ROSE (*Echeveria b.* (ROSE) KEARN. et PEEBLES). – USA: Arizona; Mexico; Chihuahua. – Ros. almost sessile, solitary or several (?), 8–15 cm ⌀; L. 20–30, ovate to elliptical, tapering, 4–7 cm long, c. 2 cm across, 3 mm thick, light green and bluish; Infl. 15 to 30 cm long, Cinc. 8–15, of 1–3 Fl., Cor. 19 to 28 mm long, tube 3 mm long, lobes with dark red transverse bands.

**G. filiferum** (S. WATS.) WHITEHEAD (Pl. 88/4) (*Sedum f.* S. WATS.). – Mexico: Chihuahua. – Ros. almost sessile, clump-forming, 2–6 cm ⌀; L. 75–200, cuneate-spatulate, blunt, 1–3 cm long, 3–12 mm across, 1–4 mm thick, upperside papillose towards the tip, terminating in a thin brown bristle up to 15 mm long; Infl. 4–8 cm tall, each Br. with few Fl., pedicels 0.5–1.5 cm long, Cor. 15–17 mm long, tube 3 mm long, lobes with dense, dark red, transverse bands.

**G. fruticosum** MORAN. – Mexico: Jalisco. – ♄ c. 40 cm tall and ⌀, branching from the base; L. subrosulate, cuneate-spatulate, blunt, almost apiculate, 2–4 cm long, 8–20 mm across, 2.5 to 4 mm thick; Infl. a lax cyme, branching laterally, pedicels 5–18 mm long, Cor. 16–21 mm across, yellow, lobes 7–9 mm long, with intermittent brownish-red transverse bands.

**G. grande** ALEXANDER. – Mexico: Oaxaca. – Semi-♄ 80 cm or more tall; L. 15–25, rosulate to rather scattered, cuneate-spatulate, rounded to ± truncate, 5–8.5 cm long, 2–3.5 cm across, c. 4 mm thick, slightly bluish; Infl. 30–40 cm tall, with 8–10 simple or forked Br., pedicels 5–8 mm long, Cor. 15–23 mm ⌀, tube 1.5 to 2 mm long, lobes with c. 10 red transverse bands.

**G. macdougallii** ALEXANDER. – Mexico: Oaxaca. – St. short, erect, 8–17 mm ⌀, with long offsets; Ros. dense, 2–7 cm ⌀; L. 25–30, oblong-cuneate, short-tapering, 2–4.5 cm long, 1–1.5 cm across, 3–4 mm thick, deep bluish; Infl. slight, 5–15 cm long, with 1–3 Cinc. of few Fl., pedicels 0.5–2 cm long, Cor. 2–3 cm ⌀, tube 3–5 mm long, lobes with dense red lines.

**G. occidentale** ROSE. – Mexico. – **Ros.** sessile, dense, c. 3 cm ⌀; **L.** oblanceolate to spatulate, tapering, 1–1.5 cm long; **Infl.** c. 7 cm tall, Cinc. with c. 5–6 Fl., pedicels 5–8 mm long, Cor. 7–8 mm long.

**G. pachyphyllum** ROSE (*Sedum atypicum* BGR., *Echeveria minutifoliolata* v. POELLN.). – Mexico: San Luis Potosí, Querétaro, Hidalgo. – Mat-forming; **Ros.** dense, or St. up to (?) 10 cm tall, with ± scattered L., offsetting; **L.** clavate-oblanceolate to clavate-obovate, broad-pointed to blunt, apiculate, 1–1.5 cm long, 4–7 mm across, 2–6 mm thick, blue to bluish, often red-tipped; **Infl.** 2–4 cm tall, usually with 1–2 Cinc. of 2–5 Fl., pedicels 0.5–2 cm long, Cor. 15 to 20 mm long, tube 4 mm long, lobes with a few red spots or somewhat transverse-banded.

**G. paraguayense** (N. E. BR.) E. WALTH. (Pl. 88/3) (*Cotyledon p.* N. E. BR., *Byrnesia weinbergii* ROSE, *E. w.* HORT. ex ROSE, *E. arizonica* HORT. ex ROSE, *Sedum w.* (ROSE) BGR., *G. w.* (ROSE) BGR., *G. w.* (ROSE) E. WALTH., *G. byrnesia* E. WALTH. sphalm., *E. p.* v. POELLN., *S. p.* (N. E. BR.) BULLOCK). – Probably Mexico. – **St.** prostrate; **Ros.** 10–15 cm ⌀; **L.** obovate-spatulate, pointed or tapering, 5–8 cm long, 1.5–4 cm across, up to 1 cm thick, flattened, with margins and keel blunt, grey-green, tinged with purple; **Infl.** 5–10 cm tall, Cinc. 2–6, with 3–10 Fl., pedicels 0.5–1.5 cm long, Cor. 1.5 to 2 cm ⌀, white, tube 4 mm long, lobes with a few red spots.

**G. pusillum** ROSE (*Sedum graptopetalum* BGR.). – Mexico: Durango. – **Ros.** subsessile, 3–5 cm ⌀; **L.** 25–35, elliptical, tapering, apiculate, 2–3 cm long, c. 8 mm across, 3 mm thick, slightly bluish; **Infl.** 10 cm tall, with 1–2 Cinc. of few Fl., pedicels 4–8 mm long, Cor. c. 17 mm ⌀, tube 2–3 mm long, lobes with red transverse bands.

**G. rusbyi** (GREENE) ROSE (Pl. 89/2) (*Cotyledon r.* GREENE, *Dudleya r.* (GREENE) BR. et R., *Echeveria r.* (GREENE) NELS. et MACBR., *G. orpetii* E. WALTH.). – USA: Arizona; Mexico: Chihuahua. – **Ros.** subsessile, 2–10 cm ⌀; **L.** 15–35, oblanceolate to spatulate, blunt, apiculate, 1–5 cm long, 3–17 mm across, 2–4 mm thick, slightly bluish; **Infl.** 5–15 cm tall, with 2–3 Cinc. of 2–5 Fl., pedicels 1–6 mm long, Cor. 13–16 mm ⌀, tube 3 mm long, lobes with red transverse bands.

× **Graptoveria** GOSSOT (× *Graptoveria* ROWL., × *Echepetalum* GOSSOT). – Crassulaceae. – Intergeneric hybrids: **Graptopetalum × Echeveria**. – Caulescent, branching plants; **Ros.** short or extended, at the Br.-tips, usually offsetting freely from the base or higher; **L.** obovate-round to lanceolate, tapering, greenish, with irregular reddish blotches or blue-green pruinose, margins reddish; **Infl.** usually numerous, racemose, with bracts, Fl. 10 or more, with pedicels 1–3 cm long, usually reddish-yellow to yellow, often with irregular red markings, corona lobes almost erect, not projecting radially, Sep. appressed, short. – Greenhouse, moderately warm. Propagation: cuttings, also leaf-cuttings.

× **G.** cv. Acaulis (*Echeveria a.* GOSSOT). – Hybrid: **Graptopetalum paraguayense × E. amoena**. – Short-stemmed; **Ros.** dense, compact; **L.** obovate-round to oblong, 1–3 cm long, up to 1.5 cm across, rather thick, bluish-green, waxy-pruinose, with reddish spots; **Fl.** light orange, spotted or not, Cor. 1 cm long.

× **G.** cv. Caerulescens (*Echeveria × c.* GOSSOT). – Hybrid: **Graptopetalum paraguayense × E. elegans**. – Caulescent; **Ros.** lax, 10 to 15 cm ⌀, later with offsets from the L.axils; **L.** obovate-lanceolate, pointed, ± 7 cm long, upperside flat-concave, underside convex, carinate, bluish, white-pruinose, margins translucent, white, becoming pink; **Infl.** forked, Fl. ± 13 mm long, yellow outside, otherwise pink, unspotted.

× **G.** cv. Calva (Pl. 89/3) (*Echeveria × c.* GOSSOT). – Hybrid: **Graptopetalum paraguayense × E. pulvinata**. – Caulescent; **Ros.** lax, 15 cm ⌀, with offsets between the L.; **L.** obovate-lanceolate, pointed, ± 7 cm long, upperside concave to flat, underside very convex, carinate, thick, sea-green, margins lighter, glabrous and glossy, pruinose; **Infl.** with large bracts, Fl. yellow with red spots.

× **G.** cv. Haworthioides (*Echeveria × h.* GOSSOT). – Hybrid: **Graptopetalum paraguayense × E. agavoides**. – Ros. as for E. agavoides, flatter, could be mistaken for a **Haworthia**; L. as for **G. p.** but remarkable for the pale yellow colour, unlike that of either parent.

× **G.** cv. Titubans (*Echeveria × t.* GOSSOT). – Hybrid: **Grapoptetalum paraguayense × E. derenbergii**. – **Ros.** densely leafy, 5–10 cm ⌀, at the tip of the St. which has few offsets, and later becomes thickened; **L.** obovate-round, acutely pointed, upperside flat, underside convex, slightly carinate, up to 6 cm long, 3 cm across, 0.5–1 cm thick, light green, grey-blue pruinose; **Infl.** of 9–22 Fl., 1 cm long, yellow inside with some spots, orange-yellow outside.

× **Greenonium** ROWL. Crassulaceae. – Intergeneric hybrids: **Greenovia × Aeonium**. – Greenhouse, cool. Propagation: division.

× **G. bramwellii** ROWL. – Canary Is.: Acc. PRAEGER a spontaneous hybrid: **Greenovia dodrantalis × Aeonium ? spathulatum**. – Bright green tufted plant the size of **G. dodrantalis** with many flattish open Ros. 3–4 cm ⌀; L. obovate-spatulate, 20–25 mm long, 10–15 mm broad, apiculate, glabrous; margins green, slightly cartilaginous, extremely erose, with irregularly disposed pellucid beads; Fl. yellow.

× **G. rowleyi** BRAMW. – Canary Is.: Tenerife. – Acc. PRAEGER a spontaneous hybrid: **Greenovia dodrantalis × Aeonium haworthii**. – Habit of **G. dodrantalis** but larger; St. short, rather slender, with short patent Br.; Ros. flattish globular, rather dense, 6–8 cm ⌀; L. ovate-spatulate, 4–5 cm long, 15–25 mm wide, fresh green, glabrous, cartilaginous, margins whitish, strongly erose, ± ciliate and glandular; **Infl.**

30–40 cm tall, 10–12 cm broad, intermediate between those of the parents; Fl. 2 cm ⌀, pale yellow.

× **G. 1.** – Dubious hybrid, not re-collected: **Greenovia aurea** × **A. glutinosum.** Acc. Haworth cultivated in the Physic Garden, Chelsea, London.

**Greenovia** Webb et Berth. Crassulaceae. – Canary Is.: mountainous areas. – ⚳ dwarf ♄, acauline or caulescent, offsetting; **L.** forming spherical or cup-shaped Ros. which die off after the flowering period, broad, spatulate, thin, usually blue-green, eventually becoming an attractive reddish colour; **Infl.** densely leafy, branching dichotomously, Fl. glossy yellow. – Summer: outdoors; winter: cold greenhouse. Propagation: seed, division.

**G. aizoon** Bolle (*G. quadrantalis* Webb, *Sempervivum a.* Christ). – Canary Is.: Tenerife. – Very freely branching; **Ros.** 5–6 cm ⌀; **L.** broadly roundish-spatulate or ± square, light green, with dense white hairs.

**G. aurea** (C. Sm.) Webb et Berth. (Pl. 89/5) (*Sempervivum a.* C. Sm., *S. calyciforme* Haw.). – Canary Is.: Hierro, Gomera, Tenerife, Gran Canaria. – Dwarf prostrate ♄; **Ros.** up to 40 cm ⌀, cup-shaped; **L.** obovate-round, spatulate, bluntly-rounded, blue-green pruinose; **Fl.** dark yellow.

**G. × aureozoon** Bramw. et Rowl. – Canary Is.: Tenerife; hybrid swarms abundant where the parents grow together. Acc. Praeger a spontaneous hybrid: **G. aizoon** × **G. aurea.** – Variable; all ± intermediate between the parents.

**G. diplocycla** Webb (*Sempervivum d.* Burch.). – Canary Is.: Hierro, La Palma, Gomera. – **Ros.** seldom offsetting, cup-shaped; **L.** with ciliate, cartilaginous margins, without T.

**G. dodrantalis** (Willd.) Webb et Berth. (Pl. 89/4) (*Sempervivum d.* Willd., *G. gracilis* Bolle, *S. g.* Christ). – Canary Is.: Tenerife. – Small, slender, prostrate ♄ forming cushions; **Ros.** small, numerous; **L.** broad, saucerlike, roundish-spatulate, vivid blue-green with a waxy surface; **Fl.** yellow.

**Gynura** Cass. Compositae. – E.Afr. – ♄ with a swollen **St.,** sometimes clambering; **L.** pinnately compound; **Fl.** in corymbose heads. – Greenhouse, warm. Propagation: seed.

**G. scandens** O. Hoffm. – Tanzania: Kenya. – **St.** swollen, fleshy; **Br.** climbing; **L.** ovate-pointed, somewhat fleshy, margins ± incised, upperside densely hairy; **Fl.**heads flat, **Fl.** orange-yellow.

**G. valeriana** Oliv. – Tanzania; Kenya. – ♄ 60–80 cm tall; **St.** very fleshy; **L.** fleshy, pinnately compound, pubescent; **Fl.** orange-red.

**Haemanthus** L. Amaryllidaceae. – Mostly attractive flowering herbs, only a few species can be considered as succulent. – Greenhouse, warm. Propagation: seed, division.

**H. albiflos** Jacq. v. **albiflos.** – S.Afr. – Compressed bulb consisting of fleshy, distichous scales, eventually producing 2–4 L. at the same time as the Infl.; **L.** thick-fleshy, 15–20 cm long, 6–12 cm across, deep green, margins ciliate; **Fl.**scape with a round umbel of many Fl., with yellow anthers projecting on white filaments.

**H. —** v. **pubescens** Bak. – **L.** hairy on the upperside.

**H. canaliculatus** M. R. Levyns. – Cape. – Bulb c. 9 cm long; **L.** distichous, 2(–3), narrow-linear, fleshy, upperside very grooved, up to 36 cm long; **Fl.**scape thick, pink, bracts pointed, red, Fl. numerous, crowded, pink.

**H. cv. Clarkei.** – Hybrid: **H. albiflos** × **H. coccineus.** – Resembles **H. coccineus**; **Fl.** dull green.

**H. coccineus** L. – S.Afr. – Bulb compressed, up to 10 cm ⌀; **L.** 2, linguiform, 45–60 cm long, 15–20 cm across, green, smooth; **Fl.**scape 15 to 25 cm tall, umbel dense, 6–8 cm ⌀, with 6 to 8 imbricate red bracts below the Fl.head, Fl. brilliant red, 3 cm long.

**Harpagophytum** Cd. ex Meissn. (*Uncaria* Burch.). Pedaliaceae. – S.Afr. – Prostrate ⚳; roots fleshy, lateral roots with tuberous swellings; **L.** opposite, ± sinuate or lobed, somewhat fleshy; **Fl.** solitary, from the L.axils, tube widening above, Cor.-lobes 5; **Fr.** hamate.

The species **H. procumbens** (Burch.) DC. ex Meissn. and **H. zeyheri** Decne., both from the Transvaal, are only to some extent succulent. Of little importance in cultivation. – *H. pinnatifidum* Engl. is **Pterodiscus speciosus** Hook. (Cf. Ihlenfeldt in Mitt. Bot. Staats. München, VI, 605, 1967.)

**Haworthia** Duval (*Apicra* Willd.). Liliaceae. – S. and S.W.Afr. – Low ⚳ with short or ± elongated **St.** often with shoots from the base forming clumps; **L.** in Ros. or the shoot elongated and the L. densely imbricate in several series, rarely distichous, short or oblong-lanceolate, blunt, acute or truncate, fleshy, often covered with pearly tubercles or ± transparent, margins often dentate or ciliate and the tip with a bristle point; **Infl.** simple or up to 5-branched; **Fl.** in loose Rac., faintly zygomorphic (Pl. 92/1), the 6 tepals united at the base to form a tube, the free tips above curved into 2 lips, whitish-green, rarely pink. – Greenhouse. Propagation: seeds, division of cushions, also leaf-cuttings. Type: **H. arachnoidea** (L.) Duval.

The division into Sections is based on position and form of L. (acc. A. Berger emend. K. v. Poellnitz.

In Cact. a. Succ. J. of Gr.Brit. **34**, No. 2, p. 35, 1972 M. B. Bayer from the Karroo Garden, Worcester, S.Afr. undertakes an exhaustive examination of many species of **Haworthia** and proposes a revision of the nomenclature. It has not been possible to incorporate this work in this edition of the Lexicon.

## Key to the Sections of the Genus Haworthia according to K. v. Poellnitz emend. G. D. Rowley

A. Plants with elongated shoots.
- **1.** L. in 3 straight or twisted vertical rows .......................................... **2.**
- **1a.** L. in 5 or more spiral series, juvenile L. flattened or convex on the upper surface ... **3.**
- **2.** L. in 3 straight or slightly twisted vertical rows .................... **§ XX. Trifariae**
- **2a.** L. in 3 strongly twisted vertical rows .......................... **§ XIX. Tortuosae**
- **3.** L. spreading, tuberculate below or on both sides ..................... **§ XV. Rigidae**
- **3a.** L. ± curved inwards at the tip, with ± raised lines and often tuberculate below ....... ........ **§ II. Coarctatae**

A. Plants with short shoots.
- **1.** Roots a cluster of fusiform Tub.; L. long and slender from broad, swollen bases ....... ........ **§ V. Fusiformes**
- **1a.** Roots not fusiform; L. not as above ................................................. **2.**
- **2.** L. rigid, stiff, stout, unicolorous, with or without marginal T. ...................... **3.**
- **2a.** L. especially towards the tip usually paler or striped, usually with T. at the margins ... ........ **8.**
- **3.** L. with T. at the margins ......................................... **§ VIII. Loratae**
- **3a.** L. without T. at the margins ..................................................... **4.**
- **4.** L. smooth on both sides .......................................................... **5.**
- **4a.** L. with very small to very large tubercles or warts ............................ **6.**
- **5.** L. whitish, ovate-triangular, 7 cm long, 4–4.5 cm wide ............... **§ X. Marginatae**
- **5a.** L. green, ovate or lanceolate-triangular, gradually tapering to the tip, c. 8 cm long, 1.5 cm wide ........................................................... **§ XV. Scabrae**
- **6.** Tubercles small ..................................................... **§ XV. Scabrae**
- **6a.** Tubercles large to very large, ± confluent into transverse rows ..................... **7.**
- **7.** Plants stoloniferous; L. file-like, with horizontal ridges which are the same colour as the L. or lighter ........................................................ **§ VI. Limifoliae**
- **7a.** Plants not stoloniferous; L. not so ridged .......................... **§ IX. Margaritiferae**
- **8.** L. rigid, stiff, stout ........................................................... **9.**
- **8a.** L. less rigid, rather soft ..................................................... **12.**
- **9.** L. truncate at right angles at the tip .......................................... **10.**
- **9a.** L. oblique or recurved at the tip ............................................. **11.**
- **10.** Truncated upper surface round or ovate, without green longitudinal stripes; L. distichous or spiral ......................................................... **§ IV. Fenestratae**
- **10a.** Truncated upper surface ± triangular with green lines; L. spiral ..... **§ XVI. Retusae**
- **11.** L. broad- or lanceolate-triangular, upper surface mostly flat, recurved near the tip, colourless, with green lines which are mostly united into a network, dorsally strongly convex, rough or tuberculate, ± toothed at the margins ......... **§ XVIII. Tessellatae**
- **11a.** L. ± triangular to truncate and recurved from the middle up or at the tip, so that often a pale end surface is present with ± dark lines ..................... **§ XIV. Retusae**
- **12.** L. not truncated, mostly somewhat recurved or swollen .......................... **15.**
- **12a.** L. horizontally truncate, the upper end not triangular and without green lines .......... ........ **§ IV. Fenestratae**
- **12b.** L. ± triangular to truncate and recurved at the top or from the middle up; end surface mostly paler and with ± long green lines ..................................... **13.**
- **13.** L. truncated at right angles or obliquely, upper end surface distinct ... **§ XIV. Retusae**
- **13a.** L. not obviously truncate, upper end surface less distinct ........................ **14.**
- **14.** Margins of L. usually entire, smooth; L. obovate, 2 cm wide ........ **§ XII. Obtusatae**
- **14a.** Margins of L. mostly dentate, L. narrower ............................ **§ XI. Muticae**
- **15.** L. oblanceolate, obovate-lanceolate or triangular-subulate ....................... **18.**
- **15a.** L. broader ..................................................................... **16.**
- **16.** L. with small tubercles and minute T. arranged in longitudinal series towards the tip ... ........ **§ XVII. Subregulares**
- **16a.** L. without tubercles on either side ........................................... **17.**
- **17.** L. indistinctly truncate above or recurved (See further 14, 14a) ....... **§ XII. Obtusatae**
- **17a.** L. not recurved, broad ovate, obtuse, mostly flat, rather lighter towards the tip, with reticulate lines .................................................... **§ XIII. Planifoliae**
- **18.** Lighter part of the L. sharply defined from the green part, with ± numerous dark lines, L. rather thick ..................................................... **§ VII. Limpidae**
- **18a.** Lighter parts of the L. not so sharply defined from the green part which is thinner here with paler spots or stripes ...................................................... **19.**

19. L. proportionately wide, lighter above, with a short tip . . . . . . . . . . . § III. **Denticulatae**
19a. L. narrower, or if broad, not lighter above . . . . . . . . . . . . . . . . . . . . . . . . . . . . . . . 20.
20. L. triangular-subulate, with or without a very small tip, with minute T. on the margins . . .
. . . . . . . . § VIII. **Loratae**
20a. L. ± narrow-lanceolate, often with a long terminal bristle . . . . . . . . § I. **Arachnoideae**[1])

C. A. E. PARR founded the Sect. Quinquefariae in which he included the species of the genera **Astroloba** UITEW. and **Poellnitzia** UITEW. transferred to **Haworthia** DUV. (see A.S.P.S. Bull VI: 145–150, 195–197, 1971). This revision is not accepted here. (See also: M. B. BAYER in Nat. Cact. & Succ. J. **27**, 1972: 77–79).

## The Sections of the Genus Haworthia and their Species

**Sect. I. Arachnoideae** HAW.[1]) (Pl. 90). – Ros. stemless; L. uniformly coloured, gradually becoming nearly transparent towards the apex, narrow, often with a short terminal bristle, margins and keel with small bristles or T., not truncate-recurved towards the tip. – Species: H. aegrota, arachnoidea, bolusii, decipiens, ferox, gracilis, guttata, helmae, herbacea, isabellae, luteorosea, marumiana, pallida, pearsonii, setata, stiemei, submaculata, tenera, translucens.

**Sect. II. Coarctatae** BGR. (Pl. 90). – St. elongated, densely spirally leafy; L. ± erect, mostly somewhat curved inwards, smooth or tuberculate on both sides or only on the upper surface. – Species: H. armstrongii, baccata, carrissoi, cassytha, coarctata, coarctatoides, eilyae, fulva, greenii, henriquesii, herrei, jacobseniana, jonesii, kewensis, lisbonensis, musculina, peacockii, reinwardtii, resendeana, revendettii, rubrobrunnea, sampaiana.

**Sect. III. Denticulatae** BAK. (Pl. 90). – Ros. stemless, spirally leafy; L. oblanceolate or oblanceolately tapering, rather soft, margins glabrous or dentate, gradually colourless-transparent towards the tip. – Species: H. altilinea, globosiflora, janseana, laetevirens, lockwoodii, mucronata, nortieri.

**Sect. IV. Fenestratae** V. POELLN. (Pl. 90). – L. distichous or spiral, ± erect, ovate-triangular or ovate-elongate, horizontally truncate above, the truncate portion transparent, covered with very numerous, tiny, transparent tubercles. – Species: H. maughanii, truncata.

**Sect. V. Fusiformes** BARKER (Pl. 90). – Roots thick, fusiform; St. short; L. linear, acute, erect or spreading from a broad base, firm, glabrous, green, margins with minute, horny T. – Species: H. blackburniae, graminifolia (c.f. **Chortolirion**).

**Sect. VI. Limifoliae** G. G. SMITH (Pl. 90). – Plant stoloniferous, forming offsets; L. in a stemless Ros., spirally arranged, ovate-lanceolately tapering, uniformly coloured, set with transverse or longitudinal, confluent or solitary, similarly coloured or lighter tubercles. – Species: H. koelmaniorum, limifolia.

**Sect. VII. Limpidae** BGR. (Pl. 90). – L. in a stemless Ros., spirally arranged, lower portion green, upper part sharply defined, wholly transparent, with a few green, longitudinal stripes in the lighter part. – Species: H. bilineata, blackbeardiana, cooperi, habdomadis, leightoniae, obtusa, sessiliflora, setata.

**Sect. VIII. Loratae** (SALM) BGR. (Pl. 90). – L. spirally arranged in a stemless Ros., narrow, triangular-subulate or ovate-lanceolate, not truncate above, mucronate, rather firm, ± erect, margins glabrous, often armed with minute T. – Species: H. angustifolia, chloracantha, floribunda, mclarenii, monticola, variegata, venteri, wittenbergensis, zantneriana.

**Sect. IX. Margaritiferae** HAW. (Pl. 90). – Stemless or very short-stemmed; L. spirally arranged, firm, nearly lanceolate or ovate triangular, tuberculate on both sides or only the lower surface. – Species: H. attenuata, browniana, fasciata, glabrata, icosiphylla, longiana, margaritifera, mutabilis, papillosa, poellnitziana, radula, rugosa, semiglabrata, smithii, subattenuata, subfasciata, subulata, tisleyi, tuberculata.

**Sect. X. Marginatae** UITEW. (Pl. 90). – Ros. stemless, little suckering; L. not numerous, spirally arranged, ± inclined, ovate-lanceolate, gradually tapering, stiff, rigid, mucronate, back surface sharply keeled, glabrous, dark green, coated with a whitish skin, margins and keel with a ± distinct white edge. – Species: H. marginata, uitewaaliana.

**Sect. XI. Muticae** BGR. (Pl. 90). – L. spirally arranged in a stemless Ros., upper surface flat towards the base, somewhat inflated towards the apex and thus appearing recurved, equally coloured or lighter towards the apex, or half-transparent, margins glabrous or with minute T., often with a very short terminal bristle, more rarely without that. – Species: H. batesiana, baylissii, caespitosa, haageana, hurlingii, incurvula, integra, intermedia, reticulata, umbraticola.

---

[1]) = Sect. Haworthia

**Sect. XII. Obtusatae** BGR. (Pl. 90). – Stemless, freely suckering; **L.** numerous, broadly obovate, shortly tapering, thickened above and somewhat truncate, upper surface very much convex, keeled towards the apex, often with somewhat connected longitudinal lines, margins and keel glabrous or with tiny T. – Species: H. cymbiformis, lepida, ramosa.

**Sect. XIII. Planifoliae** HAW, BGR. (Pl. 91). – **L.** spirally arranged in a stemless Ros., fairly soft, not recurved at the apex, upper surface ± flat, broadly ovate or more narrow, margins and keel glabrous or with minute T., unicoloured or lighter towards the apex and with darker, connected longitudinal lines, mucronate or with a short or longer terminal bristle. – Species: H. aristata, perplexa, planifolia.

**Sect. XIV. Retusae** HAW. (Pl. 91). – **L.** spirally arranged in a stemless Ros., rather firm, ± erect, mostly with minute T. on the edge, more rarely smooth, truncate-recurved above, the terminal area thus produced being somewhat transparent, with few or several green stripes, glabrous, tuberculate or rarely with minute T. – Species: H. asperula, atrofusca, badia, comptoniana, correcta, cuspidata, dekenahii, emelyae, fouchei, geraldii, heidelbergensis, longibracteata, magnifica, maraisii, mirabilis, mundula, mutabilis, nitidula, otzenii, paradoxa, parksiana, picta, pubescens, pygmaea, retusa, rossouwii, ryderana, schuldtiana, springbokvlaktensis, sublimpida, triebnerana, turgida, willowmorensis.

**Sect. XV. Rigidae** HAW. emend. BGR. (Pl. 91). – **St.** elongated; **L.** in 5 or more spiral series, spreading above, with small tubercles on both sides or only on the lower surface. – Species: H. hybrida, rigida.

**Sect. XVI. Scabrae** BGR. (Pl. 91). – **L.** spirally arranged, ovate-lanceolate or nearly triangular, long or short-tapering, firm, unicoloured, somewhat rough or minutely tuberculate, margins and keel without T. or bristles. – Species: H. granulata, lateganae, morrisiae, pseudogranulata, scabra, sordida, starkiana.

**Sect. XVII. Subregulares** BGR. (Pl. 91). – **L.** spirally arranged in a stemless Ros., broad, thickened in the middle, with darker stripes and regularly arranged tubercles on both sides, minutely dentate and minutely mucronate. – Single species: H. subregularis.

**Sect. XVIII. Tessellatae** (SALM) BAK. (Pl. 91). – Plants forming clumps, ± buried in the ground except the upper surface of the L.; **L.** spirally arranged in a stemless Ros., fleshy, firm, triangular or lanceolate-triangular, spreading, recurved, often somewhat erect during the resting period, upper surface somewhat transparent and with longitudinal lines, mostly less perceptible in the native country than in cultivated plants, with the exception of **H. recurva**, ± tessellately connected, L.-margin dentate, lower surface ± tuberculate. – Species: H. recurva, tessellata, venosa, wooleyi.

**Sect. XIX. Tortuosae** HAW. emend. BAK. (Pl. 91). – **St.** elongated, leafy; **L.** firm, thick, in three fairly strongly spirally twisted longitudinal series, tuberculate on both sides or only on the lower surface. – Single species: H. tortuosa.

**Sect. XX. Trifariae** HAW. (Pl. 91). – Shoots elongated, forming **St.**; **L.** in 3 somewhat twisted longitudinal series, thick, firm, fleshy, dark green, rough, covered with tubercles. – Species: H. asperiuscula, cordifolia, nigra, viscosa.

**H. aegrota** v. POELLN. (Pl. 92/8). (§ I). – Cape: Worcester, Swellendam, Caledon, Bredasdorp D. – **L.** 3–4 cm long, 7–10 mm across, ± erect, oval-lanceolate or lanceolate-tapering, the underside with 2 keels, not glossy, whitish-green, often reddish-green, both surfaces with reddish longitudinal lines, margins and keels dentate, terminal bristle 3–4 mm long.

**H. altilinea** HAW. (§ III) – Cape.

**H.** — v. **altilinea** (*Aloe a.* ROEM. et SCHULT., *H. a.* v. *typica* v. POELLN.). – Stockenstroom, Uitenhage, Prince Albert D. – **L.** 3–6 cm long, 1–2 cm across, terminal bristle simple, 5–10 mm long, M.rib prominent, with 4–8 ± connected, green, longitudinal lines, T. light-coloured, very small.

**H.** — v. **denticulata** (HAW.) v. POELLN. (Pl. 90/III) (§ III) (*H. d.* HAW., *Aloe a.* v. *d.* SALM). – Lit. Karroo and southern D. – **L.** 9–18 mm across, T. well spaced on margins and keel, c. 1 mm long.

**H. angustifolia** HAW. (§ VIII). – Cape. – Ros. 2–6 cm ⌀; **L.** lanceolate, tapering, 2 to 10 cm long, 10–14 mm across at the base, older L. recurved, ± rigid, terminating in a white bristle, upperside with a broad M.line and 5–7 indistinct longitudinal lines, underside triangular towards the tip, with 5 indistinct longitudinal lines, margins and keel with transparent T.

**H.** — v. **albanensis** (SCHOENL.) v. POELLN. (*H. a.* SCHOENL.). – Grahamstown, Calitzdorp, Riversdale D. – **Ros.** 2–3 cm ⌀; **L.** 3.5–4 cm long, 1 cm across at the base, with dark lines, slighthly rough towards the tip.

**H.** — v. **angustifolia** (Pl. 90/VIII) (*Aloe stenophylla* ROEM. et SCHULT.). – Steytlerville, Bredasdorp, Calitzdorp D. – **Ros.** 5–6 cm ⌀; **L.** 4–5 cm long, 10 mm across.

**H.** — v. **denticulifera** v. POELLN. – Calitzdorp, Riversdale, Montagu D. – **L.** with marginal T. c. 1 mm long, upperside with M.line often tuberculate or dentate, tip 1–4 mm long.

**H. — v. grandis** G. G. SMITH. - Albany D. - L. 6–10 cm long, 14 mm across below, 3.5 mm thick.

**H. — v. liliputana** UITEW. - Mossel Bay. - L. 2 cm long.

**H.—v. paucifolia** G. G. SMITH.-Grahamstown D. - **L.** c. 12, c. 5 cm long, 4.5 mm thick below, terminal bristle 2 mm long.

**H. — v. subfalcata** v. POELLN. - Robertson D. - **L.** ± laterally curved (like a sickle).

**H. arachnoidea** (L.) DUV. (§ I) (*Aloe pumila* v. *a.* L., *A. a.* LAM. and v. *pellucens* SALM, and v. *klugii* SALM, *Catevala a.* MEDIC., *Apicra arachnoides* WILLD., *H. a.* (AIT.) HAW.) – Cape: Montagu D. - **Ros.** 3 cm ∅; **L.** c. 15 mm long, oblong, with a terminal bristle 6–7 mm long, both surfaces transparent with continuous, darker lines, keel and margins with bristles 1.5 mm apart, 2 mm long.

**H. aristata** HAW. (§ XIII) (*H. altilinea* AUCT. p.min.part., *Aloe aristata* ROEM. et SCHULT., *H. setata* v. *subinermis* v. POELLN., ? *H. unicolor* v. POELLN.). - Cape: Barrydale, Montagu, Ladismith D. - Mat-forming; **Ros.** leafy, c. 4 cm ∅; **L.** ovate-lanceolate, abruptly or gradually tapering, 4–6 cm long, c. 10–15 mm across, dark to light green with dark longitudinal lines, 4–6 cm long, 10–15 mm across, upperside swollen towards the tip, underside very convex, keel small-tuberculate, terminal bristle 6–15 mm long.

**H. armstrongii** v. POELLN. (§ II). - Cape: Adelaide. - **St.** branching basally, over 10 cm long; **L.** densely spiralled, 3–4 cm long, c. 9–12 mm across below, lanceolate, margins rather thickened, upperside slightly concave, underside obliquely carinate towards the tip, dark green, ± pruinose, apiculate, projecting keel and margins tuberculate.

**H. asperula** HAW. (§ XIV). – Cape: Barrydale, Oudtshoorn, Union D. - **Ros.** 7–8 cm ∅; **L.** c. 10, ovate-triangular, 3–3.5 cm long, the tip on the upperside flat to rectangularly recurved to form a triangular-roundish end-surface, transparent, rough-papillose, the underside very convex and carinate, green.

**H. asperiuscula** HAW. v. **asperiuscula** (Pl. 92/3) (§ XX) (*Aloe a.* ROEM. et SCHULT.). – Cape: Laingsburg, Robertson, Prince Albert D. - **Ros.** 4–5 cm tall, 2 cm ∅; **L.** in 3 slightly contorted R., triangular-pointed, 14 mm long, 12 mm across below, upperside concave, underside sharply carinate, dark green, ± viscous, minutely rough.

**H. — v. patagiata** G. G. SMITH. – Cape: Willowmore D. - **L.** more pointed, margins horny, glossy.

**H. — v. subintegra** G. G. SMITH. – Cape: Ladismith D. - **L.** margins and keel with horny concolorous T.

**H. atrofusca** G. G. SMITH (§ XIV). – Cape: Riversdale D. - **Ros.** acauline, c. 6.5 cm ∅, with few offsets; **L.** rectangularly recurved above, firm, c. 4 cm long, 17 mm across and 10 mm thick below, long-tapering to triangular, green to reddish-brown, end-surface rather swollen, tuberculate, 17 mm long and across, with 3–5 reddish-brown lines and several small tubercles, margins with a few T.

**H. attenuata** HAW. (§ IX). – Cape. – Polymorphic. – **Ros.** acauline or short-stemmed; **L.** numerous, oblong-triangular, dark green, very tuberculate, tubercles sometimes merging into transverse bands. Many "varieties" acc. G. D. ROWLEY are only cultivars unworthy of naming.

**H. — v. argyrostigma** (BAK.) BGR. f. **argyrostigma** (*H. subfasciata* f. *a.* BAK., *H. att.* v. *arg.* f. *typica* (BAK.) BGR.). – Caulescent; **L.** 3 cm long, 15 mm across, upperside with small colourless tubercles, underside with tubercles white and in transverse R.

**H. — v. attenuata** (*Aloe a.* HAW., *A. radula* KER., *H. a.* v. *typica* HAW.). – Cape: Graaff Reinet, Matjesfontein, Laingsburg D. - **L.** 30–40, oblong-triangular, tapering, 6–7.5 cm long, 15 mm across, glossy dark green, upperside with a M.line of white tubercles, those on the underside in transverse bands.

**H. — v. britteniana** v. POELLN. (*H. b.* v. POELLN., *H. a.* v. *b.* f. *typica* v. POELLN.). – Cape: Karroo. – **L.** 6–7 cm long, 15 mm across, upperside with tubercles not clearly in a M.row, those on the underside minute, with larger ones forming transverse bands. – FARDEN names 5 other forms in which the L. have ± larger or smaller, colourless or white tubercles.

**H. — v. caespitosa** (BGR.) FARDEN (*H. fasciata* v. *c.* BGR.). – **L.** 6–7 cm long, 15 mm across below, upperside with small tubercles, those below in distant transverse R.

**H. — v. clariperla** (HAW.) BAK. (Pl. 92/2) (*H. c.* HAW., *Aloe c.* ROEM. et SCHULT. *H. a.* v. *minor* SALM, *Aloe a.* v. *c.* SALM ex ROEM. et SCHULT.). – Cape: Port Elizabeth D. - **L.** 6 cm long, 15 mm across, upperside wholly covered with small white tubercles, underside with larger tubercles ± merging into R.

**H. — v. deltoidea** FARDEN (*H. att.* f. *typica* FARDEN). – **L.** 3–3.5 cm long, 12–13 mm across, underside with tubercles irregular, not in distinct R., tubercles on the upperside white or colourless, with a few white ones along the M.line.

**H. — v. inusitata** FARDEN. – **L.** 3–5 cm long, 15 mm across, tubercles on the underside very irregular, sometimes in transverse R., solitary and minute ones in between, 2 mm long tubercles along the keel, with tubercles along the M. line on the upperside.

**H. — v. linearis** FARDEN. – **L.** 6 cm long, 8 mm across, long-tapering, tubercles on the underside white, small, not in distinct transverse R., those on the upperside in longitudinal R.

**H. — v. minissima** FARDEN. – **L.** 4–5 cm long, 10–12 mm across, very pointed, upperside concave, very deep green, tubercles small on the underside, still tinier on the upperside.

**H. — v. odonoghueana** FARDEN. – **L.** 3 to 4 cm long, 14 mm across, flat, light green, tubercles on the underside forming 2 mm wide bands, upperside with a M.row of small whitish tubercles.

**H. — v. uitewaaliana** FARDEN. – **L.** 3–4 cm long, 2 cm across, light green, tubercles in R.

2 mm apart on the underside, keel-line conspicuously tuberculate, upperside with an indistinct M.line but otherwise minutely tuberculate.

**H. baccata** G. G. SMITH (§ II). – Cape: Stutterheim D. – **St.** erect, simple, c. 10 cm tall, leafy, offsetting from the base to form clumps; **L.** in many spiralled R., green, c. 3 cm long, 16 mm across at the base, 5 mm thick midway, ovate, tapering, upperside flat to concave, with transverse-oblong white tubercles in the upper third, underside rounded, with large tubercles in transverse R.

**H. badia** v. POELLN. (§ XIV). – Cape: Bredasdorp D. – **Ros.** acauline, 7–10 cm $\varnothing$; **L.** 4–5 cm long, 2–2.5 cm across, stiff, erect, ovate-oblong, green to brownish, upperside $\pm$ flat, obliquely truncate above, underside very convex or with a second roundish keel, tuberculate towards the tip, with light-coloured small T. below, tip-area truncate at 45°, semi-translucent, often tuberculate, abruptly tapering, with several longitudinal lines, terminal bristle light-coloured.

**H. batesiana** UITEW. (§ XI). – Cape: Graaff Reinet D. – **Ros.** 4–5 cm $\varnothing$, offsetting basally to form a cushion; **L.** ovate-lanceolate, tapering, with a 2 mm long bristle-tip, 2–2.5 cm long, 8–10 mm across, green, with 4–5 indistinct, darker, longitudinal lines becoming reticulate at the tip, underside short-carinate, keel and margins with irregularly spaced, minute T. arranged on a light translucent line.

**H. baylissii** C. L. SCOTT (§ XI). – Cape: Somerset East D. – **Ros.** acauline, up to 7 cm $\varnothing$, offsetting basally, forming a cushion; **L.** c. 25, in 8 spiralled R., older L. much recurved, $\pm$ stout, lanceolate, c. 4 cm long, up to 13 mm across, 4 mm thick, tapering, dull, dark green, upperside slightly rounded, with a prominent M.line, green, with 10 indistinct, dirty green lines, with some white spots, underside rounded, with 8 indistinct lines, carinate, keel white translucent T., terminal bristle 5 mm long, white, dropping.

**H. bilineata** BAK. v. **bilineata** (§ VII). – Cape: Origin unknown. – **Ros.** of c. 15 oblong-lanceolate, tapering **L.** 3–5 cm long, 12 mm across, 8 mm thick, upperside with 1–2 longitudinal lines, flat, underside carinate above, with a bristly tip, margins and keel with ciliate T.

**H. —** v. **affinis** (BAK.) v. POELLN. (Pl. 92/5) (*H. affinis* BAK.). – **L.** shorter, T. on margins and keel shorter.

**H. —** v. **gracilidelineata** v. POELLN. (*H. g.* v. POELLN.). – Lines much more distinct and more numerous.

**H. blackbeardiana** v. POELLN. v. **blackbeardiana** (Pl. 92/4) (§ VII). – Cape: Grahamstown, Dornburg, Cradock D. – **Ros.** acauline, with many L., c. 5 cm $\varnothing$; **L.** with the thickened tip curving inwards, c. 3 cm long, 10–12 mm across, pale green, translucent in the upper third, with 4–6 confluent longitudinal lines, underside indistinctly keeled, margins serrate towards the base, margins and keel with 2 mm long bristles towards the tip, terminal bristle 1 cm long, $\pm$ curved, white.

**H. —** v. **major** v. POELLN. – **L.** 5–7 cm long, 12–15 mm across, terminal bristle up to 12 mm long, margins reddish, with white, 3–5 mm long bristles.

**H. blackburniae** BARKER (Pl. 90/V) (§ V). – Cape: Oudtshoorn D. – **Roots** fusiform, up to 12 mm thick; **St.** short, simple, up to 1 cm thick; **L.** at first 4–6, $\pm$ distichous, later 10, multifariously arranged, linear, pointed, almost erect, firm, smooth, up to 15 cm long, 5 mm across, 1 mm thick, upperside grooved, underside convex to very bluntly carinate, margins with horny T. curving inwards.

**H. bolusii** BAK. v. **bolusii** (§ I). – Cape: Beaufort West, Laingsburg, Graaff Reinet D. – **Ros.** 5–7 cm $\varnothing$; **L.** numerous, curved inwards, oblong-lanceolate, long-tapering, 2–4 cm long, 22–25 mm across, pale green, tipped with a 12 mm long bristle, upperside slightly concave, translucent, with several green longitudinal lines, underside carinate above, margins and keel with variously curved, fibrous T. c. 5 to 6 mm long, so that the plant appears covered with a web of white bristles. – The L. close up tightly in the resting period.

**H. —** v. **aranea** BGR. – **L.** margins with snow-white, minute, very fine bristle-hairs up to 5 mm long, terminal bristle 1 cm long.

**H. —** v. **semiviva** v. POELLN. (Pl. 92/7). – **L.** tip distinctly translucent, drying up and becoming paper-thin in the resting period.

**H. browniana** v. POELLN. (§ IX). – Cape: Uitenhage D. – **St.** up to 2 cm long, offsetting basally; **L.** numerous, spiralled, ovate-lanceolate, very tapered, mucronate, 5–7.5 cm long, 2–2.5 cm across, light grey-green, upperside convex, smooth, with 1–4 indistinct longitudinal lines, underside blunt-carinate above, with large tubercles, solitary or in transverse bands, interspersed with small, greenish-white tubercles.

**H. caespitosa** v. POELLN. (§ XI). – Cape: Uitenhage D. – **Ros.** acauline, 3–4 cm $\varnothing$, $\pm$ freely offsetting, mat-forming; **L.** elongate-tapering, 1–4 cm long, 6–8 mm across, translucent towards the tip, pallid green, with 3–4 rather dark longitudinal lines, often minutely tuberculate, underside carinate above, often with light spots and longitudinal lines, margins with 0.3 mm long T., the very extended L.tip with T. 3–4 mm long.

**H. —** f. **caespitosa** (Pl. 92/9) (*H. c.* f. *typica* v. POELLN.). – **L.** 1–2 cm long.

**H. —** f. **subproliferans** v. POELLN. – Usually with only 2 lateral shoots; **L.** up to 4 cm long, markings indistinct, often without lines.

**H. —** f. **subplana** v. POELLN. – **L.** 2–3 cm long, markings more pronounced, T. longer.

**H. carrissoi** RES. (§ II). – Cape: Suurberge. – **St.** extended, erect, c. 15 cm tall, branching basally; **L.** spiralled, almost appressed, ovate-lanceolate, tapering, apiculate, c. 3 cm long, 16 mm across below, dark green, rather glossy, upperside somewhat concave, underside carinate above, with rounded tubercles in longitudinal R.

**H. cassytha** BAK. (Pl. 92/6) (§ II). – Acc. RESENDE possibly a hybrid: **H. lisbonensis × H. tortuosa.**

**H. chloracantha** HAW. v. **chloracantha** (§ VIII) (*Aloe c.* (HAW.) ROEM. et SCHULT.). – Cape: Riversdale D. – **Ros.** 6–7 cm ∅, offsetting basally; **L.** 20–25, ovate-lanceolate, tapering, 3.5–4 cm long, 12–18 mm across below, rather stiff, dark green to purple, underside convex-triangular to sharply keeled, apiculate, margins and keel fringed-serrate, minutely dentate.

**H. — v. subglauca** v. POELLN. – **L.** blue-green, underside with solitary T., margins bristly below, terminal bristle 1 mm long.

**H. coarctata** (SALM) HAW. (§ II) (*Aloe c.* SALM). – Cape: Graaff Reinet, Uitenhage D.

**H. — v. coarctata** (Pl. 90/II) (*H. c. v. haworthii* RES. and f. *major* RES.). – **Ros.** extended St.-like up to 20 cm long, offsetting basally; **L.** in a dense spiral, curved inwards, lanceolate-triangular, long-tapering, 4–6 cm long, 15 mm across, upperside smooth or with a few greenish-white tubercles in longitudinal or sometimes transverse R.

**H. — v. krausii** RES. – **L.** rather shorter and broader, broad-triangular, tapering.

**H. coarctatoides** RES. et VIVEIROS (§ II). – After research into its chromosomes, the authors regard this as a hybrid: **H. coarctata v. coarctata × H. reinwardtii.**

**H. comptoniana** G. G. SMITH (§ XIV). – Cape: Willowmore D. – **Ros.** acauline, up to 8.5 cm ∅, with or without basal offsets; **L.** spreading, curved inwards at the tip, almost 4.5 cm long, the tip-area recurved at 90°, 2 cm across, 16 mm thick, ovate-triangular, terminal bristle white, 1.5 mm long, upperside concave, flat to swollen above, 24 mm long, smooth or with concolorous tubercles and white spots, translucent green, with 5–7 reticulate lines, underside carinate above, with longitudinal lines, keel with small light green T. above, the margins similarly.

**H. cooperi** BAK. (Pl. 90/VII) (§ VII). – Cape: Gr. Karroo. – Resembles **H. obtusa v. pilifera; L.** somewhat longer.

**H. cordifolia** HAW. (Pl. 93/5) (§ XX) (*Aloe c.* ROEM et SCHULT.). – Origin unknown. – Possibly only a variety of **H. viscosa,** which is very variable.

**H. correcta** v. POELLN. (§ XIV) (*H. blackburniae* v. POELLN.). – Cape: Calitzdorp D. – **Ros.** acauline, with few L., c. 4 cm ∅; **L.** c. 3 cm long, 12–15 mm across, upperside flat or ± concave, often with a rather prominent M.line, the tip truncate almost at right-angles or somewhat obliquely, underside convex, almost carinate midway, with flesh-coloured, ±translucent spots, terminal bristle whitish, 1 mm long, dropping, margins and keel rarely indistinctly dentate, tip-area triangular, 12–13 mm long, 12–15 mm across, pointed, margins or tip swollen, with 3–4 longitudinal lines and roundish tubercles.

**H. cuspidata** HAW. (Pl. 93/1) (§ XIV) (*Aloe c.* ROEM. et SCHULT., *H. cymbiformis* BGR.). – Cape: George D. – **Ros.** acauline, mat-forming, c. 5–6 cm ∅; **L.** 30–40, ovate-lanceolate, recurved at the tip, mucronate, 2.5 cm long, 13 mm across, pale green, translucent and with dendritic lines towards the tip, underside rounded and carinate above, margins and keel minutely serrate-dentate.

**H. cymbiformis** (HAW.) DUV. (§ XII) (*Aloe c.* HAW., *A. cymbaeformis* SCHRAD., *Apicra c.* WILLD., *H. concava* HAW.). – Cape. – **Ros.** freely offsetting, mat-forming; **L.** usually numerous, obovate, with translucent dark lines.

**H. — v. angustum** v. POELLN. – Cape: Port Elizabeth D. – **L.** not very numerous, 2–3 cm long, c. 1 cm across, tip thickened, lined and spotted.

**H. — v. — f. subarmata** v. POELLN. – **L.** with a minute terminal bristle, margins and keel minutely dentate.

**H. — v. brevifolia** TRIEBN. et v. POELLN. – Cape: Uitenhage D. – **L.** 1–2 cm long and across, dull light green, tip semi-translucent, with a dark line.

**H. — v. compacta** TRIEBN. – **L.** 10–15, very obtuse, 2–3 cm long, 1.5–2 cm across, grey-green, margins rounded and thickened, translucent, with dark green spots and blotches, margins and keel smooth.

**H. — v. cymbiformis** (*H. c. v. typica* TRIEBN. et v. POELLN.). – **Ros.** acauline, 7–10 cm ∅; **L.** numerous, 3–4 cm long, 2–2.5 cm across, obovate, short-tapering, thickened and somewhat truncate above, upperside concave, underside very convex, carinate above, grey-green, smooth, translucent at the tip, with somewhat interconnected longitudinal stripes.

**H. — v. multifolia** TRIEBN. – Cape: Uitenhage D. – **L.** 30–40, 2–2.5 cm long, 10–15 mm across, terminal bristle 1–1.5 mm long, usually dropping.

**H. — v. obesa** v. POELLN. (Pl. 90/XII). – **Ros.** 3–4 cm ∅ with 20–40 L.; **L.** c. 2 cm long, c. 1.5 cm across, keel with 0.5 mm long T., terminal bristle dentate, 4–5 mm long.

**H. — v. translucens** TRIEBN. et v. POELLN. (Pl. 93/3). – Cape: Uniondale D. – **L.** 1.5–2.5 cm long, 8–15 mm across, translucent from the base, with 8–10 stripes, usually brownish-red, margins minutely dentate.

**H. decipiens** v. POELLN. (§ I). – Cape: Prince Albert D. – **Ros.** c. 7 cm ∅; **L.** numerous, oblong-lanceolate, tapering, 2–4 cm long, 10 to 12 mm across, terminal bristle 1 cm long, underside convex and carinate, dentate, with reticulate lines, margins with curved T. 2 to 2.5 mm long.

**H. dekenahii** G. G. SMITH (§ XIV). – Cape: Riversdale D. – **Ros.** acauline, c. 8 cm ∅; **L.** c. 15, firm, 4.8 cm long, 20–25 mm across, 14 mm thick, pointed, apiculate, the recurved tip-area triangular-pointed, 28 mm long, 23 mm across, light green, translucent, tuberculate, with up to 10 greenish-white lines, underside ± oblique-carinate, keel dentate, margins acute, dentate.

**H. — v. argenteo-maculosa** G. G. SMITH. – Cape: Mossel Bay D. – Tip-area 15 mm long and across, indistinctly tuberculate, with a whitish spot and 5 lines, underside light green, with c. 10 lines, margins horny, translucent, white and papery above.

**H. eilyae** v. POELLN. (§ II). – Cape: Steytlerville D.

**H. — v. eilyae** (*H. e.* v. *poellnitziana* RES.). – St. 20 cm long, offsetting basally; **L.** densely spiralled, lanceolate, 4–5 cm long, 9–13 mm across below, upperside concave, the tip flat, with raised longitudinal lines, green, little pruinose, tuberculate, tubercles in a longitudinal line along the margins, underside obliquely keeled, with 8–12 gibbose longitudinal lines, margins and keel tuberculate.

**H. — v. zantneriana** v. POELLN. – **L.** smaller, broader and thicker at the base.

**H. emelyae** v. POELLN. v. **emelyae** (§ XIV). – Origin unknown. – **Ros.** acauline, c. 4 cm $\varnothing$; **L.** oblong, c. 3–4 cm long, c. 1.5 cm across, upperside rather concave, tip obliquely truncate, underside very convex, obliquely carinate, dark green, with a few small tubercles above, margins and keel with small confluent T., tip-area triangular, 1.5 cm long and across, with green, glistening tubercles and lines.

**H. — v. beukmannii** v. POELLN. – Caledon D. – **L.** 4–5 cm long, 1.5–2 cm across, both surfaces brownish and tuberculate above.

**H. fasciata** (WILLD.) HAW. (§ IX) (*Apicra f.* WILLD., *Aloe f.* SALM). – Cape: Uitenhage, Albany D. – **Ros.** acauline, 5–7 cm $\varnothing$; **L.** numerous, 3–8 cm long, 10–20 mm across, triangular-lanceolate, tapering, underside very convex, carinate towards the tip, smooth, firm, green, with large white tubercles merging into transverse bands.

**H. — v. concolor** SALM. – Completely concolorous.

**H. — v. fasciata** (Pl. 93/2) (*Apicra f.* WILLD., *Aloe f.* SALM.) – Uitenhage, Albany D. – **L.** 3–4 cm long, up to 13 mm across.

**H. — v. — f. major** (SALM) v. POELLN. (*H. m.* SALM, *H. f.* v. *m.* (SALM) BGR., *Aloe f.* v. *m.* SALM ex ROEM. et SCHULT.). – Humansdorp D. – **L.** 5–8 cm long, 2 cm across, underside as for f. **subconfluens**.

**H. — v. — f. ovato-lanceolata** v. POELLN. – **L.** ovate-oblong, 4–5 cm long, c. 2 cm across.

**H. — v. — f. sparsa** v. POELLN. – **L.** 3–5 cm long, 10–15 cm across, tubercles on the underside more scattered.

**H. — v. — f. subconfluens** v. POELLN. (*H. fasciata* v. *s.* v. POELLN.). – **L.** 3–3.5 cm long, c. 15 mm across below, tubercles $\pm$ translucent.

**H. — v. — f. vanstaadensis** v. POELLN. – **L.** 4–5 cm long, 10–15 mm across, tubercles on the underside mostly solitary, not numerous, usually in longitudinal lines.

**H. — v. — f. variabilis** v. POELLN. – **L.** 5 cm long, 10–15 mm across below, dark green, upperside smooth, underside with a few scattered tubercles midway only on some L., mostly in longitudinal lines.

**H. — v. perviridis** SALM. – **L.** deep green.

**H. ferox** v. POELLN. v. **ferox** (Pl. 93/6) (§ I). – Cape: Graaff Reinet D. – **Ros.** c. 7 cm $\varnothing$; **L.** many, c. 5 cm long, scarcely 1 cm across, lanceolate, rather stiff, tapering, with a terminal bristle up to 13 mm long, whitish, curved, simple or dentate, underside convex to carinate, dark blackish-green, both surfaces with dentate tubercles, margins and keel white-dentate.

**H. — v. armata** v. POELLN. – Cape: Oudtshoorn D. – **L.** 8 cm long, 15 mm across, upperside with 3–4 mm long dentate tubercles, underside with blunt tubercles in 5 longitudinal lines, margins with 4 mm long bent T., terminal bristle 4 mm long.

**H. floribunda** v. POELLN. (§ VIII). – Cape: Swellendam D. – **Ros.** acauline, 6–7 cm $\varnothing$; **L.** few, rosulate or often distichous, 3.5 cm long, ovate-lanceolate or tapering from base to tip and wider midway, the tip rounded or blunt, upperside often somewhat grooved, underside carinate, dull, green, indistinctly striate, margins and keel fine-dentate towards the tip; **Fl.** numerous.

**H. fouchei** v. POELLN. (§ XIV). – Cape: Riversdale D. – **Ros.** acauline, with few basal offsets; **L.** oblong-lanceolate, $\pm$ tapering, 7–8 cm long, 15–20 mm across, flattened below the tip at 30–45°, underside smooth, green, somewhat spotted below the tip, tip-area 2.5–3 cm long, 15–20 mm across, ovate-triangular, somewhat wrinkled, semi-translucent, with a few lines, terminal bristle 4 mm long.

**H. fulva** G. G. SMITH (§ II). – Cape: Bathurst D. – **St.** simple, c. 10 cm tall, offsetting basally to form clumps, densely leafy; **L.** spreading, multifarious, c. 3 cm long, c. 10 mm across below, c. 5 mm thick midway, sublanceolate tapering, green to brownish-green, some L. with whitish tubercles along a M.line, underside indistinctly carinate, with raised tubercles forming c. 12 transverse rows.

**H. geraldii** C. L. SCOTT (§ XIV). – Cape: Riversdale D. – **Ros.** acauline, up to 10 cm $\varnothing$, offsetting from the base to form a cushion; **L.** c. 18, up to 6 cm long, pale green, with 8–12 lines and white spots, the tip truncate at 80°, the tip-area 25 mm thick at the base, with small transparent tubercles, the underside with a dentate keel above, terminal bristle 3–6 mm long, red-brown to white.

**H. glabrata** (SALM) BAK. v. **glabrata** (§ IX) (*Aloe g.* SALM). – Cape: origin unknown. – **Ros.** 18 cm $\varnothing$, often dichotomously branched; **L.** 35–45, erect-projecting, 10–15 cm long, 2.5 cm across below, triangular-tapering, bluish-green, the underside carinate towards the tip, with well-spaced tubercles in distant R., margins and keel with translucent tubercles.

**H. — v. concolor** SALM (Pl. 93/4). – Garden form (Kew Gardens). – Tubercles concolorous-green, whitish on the margins and keel.

**H. — v. perviridis** SALM. – Garden form. – **L.** more numerous, green to bluish, tubercles larger.

**H. glauca** BAK. (Pl. 93/8) (§ II). – Orange Free State. – **St.** 5–8 cm long, densely leafy;

**L.** multifarious, ascending, long-lanceolate, 2–2.5 cm long, 8 mm across, 4 mm thick, pale green to bluish, often reddish, underside rounded to carinate, with 5–7 raised, darker green stripes.

**H. globosiflora** G. G. SMITH (§ III). – Cape: Calvinia D. – **Ros.** acauline, c. 8 cm ∅, with very few offsets basally; **L.** c. 40, ovate-tapering, c. 4.5 cm long, 1 cm across below, 6 mm thick towards the tip, soft, smooth, with 1 mm long, translucent, raised spots, some spots with T., with 5–6 translucent reticulate lines in the upper part, underside triangular-convex, spotted similarly, margins acute, dentate, keel often double, dentate, terminal bristle 4 mm long, dentate; **Fl.** spherical.

**H. gracilis** v. POELLN. (Pl. 93/7) (§ I). – Cape: Albany, Graaff Reinet, Willowmore D. – **Ros.** acauline, 6–7 cm ∅; **L.** many, oblong to ± ovate-oblong, very tapered, c. 3.5–4 cm long, 8–9 mm across, bluish, rather glossy, upperside with 3–5 lines, often connected, underside with 5–7 longitudinal lines, carinate towards the tip, margins and to a lesser extent the keel with 2 mm long, curved, widely spaced T., terminal bristle 6–8 mm long, simple or dentate.

**H. graminifolia** G. G. SMITH (§ V). – Cape: Oudtshoorn D. – **Roots** fusiform, up to 12 mm ∅, c. 8 cm long; **St.** simple, up to 4 cm long, c. 1 cm thick, covered with dry old L.bases; **L.** up to 14, ± linear, grass-like, subulate, firm, c. 45 cm long, the final 10 cm often dried, 2 cm across at the base, V-shaped in cross-section, smooth, dark green, underside with 4–5 longitudinal lines with white tubercles along them, some tubercles dentate, margins rather acute, fine-dentate.

**H. granulata** MARL. (§ XVI). – Cape: Roggeveld Mts. – **Ros.** 3–4 cm ∅, offsetting basally; **L.** not numerous, ovate-lanceolate, short-tapering, 2–3 cm long, 10–15 mm across, upperside concave, underside very convex, grey-green, with numerous grey-green tubercles arranged in transverse R.

**H. greenii** BAK. (§ II). – Cape: Grahamstown D.

**H. — v. greenii** (*H. g. f. bakeri* RES.). – **St.** simple, 15–20 cm tall, prostrate in age, offsetting basally to form clumps; **L.** crowded, multifarious, stout, 3–4 cm long, 15–20 mm across below, upperside concave below, slightly convex towards the tip, underside with 6–8 raised longitudinal lines, both surfaces with solitary, whitish, oblong, transversely situated tubercles.

**H. — v. minor** RES. (Pl. 94/1). – **L.** 2.5 cm long, 1.5 cm across, tubercles more conspicuous.

**H. — v. pseudocoarctata** v. POELLN. (*H. reinwardtii* v. *p.* v. POELLN., *H. coarctata* v. *p.* (v. POELLN.) RES. – **L.** 2.5 cm long, 10–15 mm across, with numerous white tubercles, merging into longitudinal lines and indistinct transverse R.

**H. — v. silvicola** G. G. SMITH. – **St.** up to 15 cm long; **L.** up to 3 cm long, 1 cm across below, upperside with raised longitudinal lines, underside with slightly raised, white or concolorous tubercles, with a ± distinct line along the rather oblique keel, margins tuberculate.

**H. guttata** UITEW. (§ I). – Cape: Robertson D. – **Ros.** acauline, c. 3 cm ∅; **L.** not numerous, spiralled, 2.5–3.5 cm long, 7–8 mm across, oblong-lanceolate, dark grey-green to reddish, translucent towards the tip, with 3–4 reticulate longitudinal lines and spots, underside often doubly carinate, often with tuberculate spots, margins and keels with 0.5 mm long T., terminal bristle 1 mm long.

**H. haageana** v. POELLN. v. **haageana** (Pl. 94/2) (§ XI). – Cape: Oudtshoorn, Robertson, Steytlerville D. – **Ros.** c. 7 cm ∅, not offsetting; **L.** numerous, rather firm, short-lanceolate to roundish-triangular, 3 cm long, 1 cm across, 5 mm thick, light green to reddish, often mottled, underside rounded, acutely carinate, keel ± fine-tuberculate, surface with fine, reticulate, longitudinal lines; **Fl.** greenish-white.

**H. — v. subreticulata** v. POELLN. – Cape: near Grahamstown (?). – The darker lines at the tip rather more distinctly prominent, margins and keel less conspicuously dentate, terminal bristle 2 mm long.

**H. habdomadis** v. POELLN. (§ VII). – Cape: Sevenweekspoort. – **L.** oblong or obovate, rather tapering, tip somewhat incurved, 2.5 to 3.5 cm long, 9–12 mm across, light grey-green, both surfaces semi-translucent, with several dark reticulate lines, terminal bristle 3–4 mm long.

**H. heidelbergensis** G. G. SMITH (Pl. 94/7) (§ IV). – Cape: Swellendam, Riversdale D. – **Ros.** acauline, c. 7 cm ∅, offsetting basally; **L.** c. 35, up to 3.5 cm long, 8 mm across, 6 mm thick, the tip-area ovate-oblong, up to 16 mm long, 8 mm across, bluish-translucent, with 3(–4) light green lines, underside convex, with mostly dentate tubercles in several R. midway and with several spots in the upper half, dentate-keeled in the upper third, margins acute, dentate, terminal bristle 3–5 mm long, whitish-translucent.

**H. helmae** v. POELLN. (Pl. 94/3) (§ I). – Cape: Worcester, George D. – **Ros.** mat-forming; **L.** oblong or lanceolate, tapering, 3–4 cm long, 6–9 mm across, upperside somewhat convex, little swollen, smooth or rough, with a longitudinal tuberculate line, underside singly or doubly carinate, both surfaces green, translucent, margins and keels with whitish T. 0.75 mm long, terminal bristle 1–4 mm long.

**H. henriquesii** RES. (§ II). – Origin unknown. – Acauline; **L.** densely crowded, multifarious, triangular-ovate, tapering, the tip incurved, 3–3.5 cm long, 10–15 mm across, 4–5 mm thick, underside convex, with 10–15 keel-like longitudinal lines, tubercles in conspicuous transverse bands.

**H. herbacea** (MILL.) STEARN (Pl. 94/5) (§ I) (*Aloe h.* MILL., *A. atrivirens* DC., *H. a.* (DC.) HAW., *H. pumila* HAW.). – Cape: Origin unknown. – **Ros.** acauline, mat-forming, 2–5 cm ∅; **L.** numerous, 15-20 mm long, 6–8 mm across, lanceolate, long-tapering, green, dark green towards the tip, with translucent spots, with rather darker longitudinal and transverse lines, upperside convex, underside carinate, whitish-tuberculate, margins and keel cartilaginous-dentate.

**H. herrei** v. POELLN. (§ II). – Cape: Steytlerville, Jansenville, Somerset East D.

**H. — v. depauperata** v. POELLN. – **L.** with 1–5 ± tuberculate longitudinal lines on the underside.

**H. — v. herrei** (Pl. 94/8) (*H. h.* v. *poellnitzii* RES.). – **St.** spirally leafy, offsetting from the base, forming small clumps, up to 10 cm long, ascending-erect; **L.** lanceolate-triangular, long-tapering, shortly acuminate, grey-green, somewhat pruinose, 4–5 cm long, 10–15 mm across below, upperside almost flat, with up to 2 slightly raised, tuberculate, longitudinal lines, underside convex, with 5–10 gibbose longitudinal lines.

**H. hurlingii** v. POELLN. v. **hurlingii**(§ XI). – Cape: Worcester, Robertson D. – **Ros.** 3 cm ⌀, mat-forming; **L.** nearly ovate or ovate-oblong, ± mucronate, minutely apiculate, 2 cm long, 8 mm across, upperside slightly swollen at the base, pallid green, with 5–7 darker longitudinal lines, underside rounded, carinate above, also with longitudinal lines, margins and keel soft or with small T. at the base.

**H. — v. ambigua** TRIEBN. et v. POELLN. – Cape: Montagu D. – **Ros.** 2–3; **L.** 12 mm across, more distinctly carinate, longitudinal lines clearer.

**H. hybrida** (SALM) HAW. (Pl. 94/4) (§ XV) (*Aloe h.* SALM, *A. h.* v. *asperior* SALM). – Possibly a hybrid: **H. margaritifera** × **H. pseudorigida** (acc. SALM-DYCK), or **H. rigida** × **H. radula** (acc. BAKER), or **H. rigida** × **H. tortuosa** (acc. BERGER).

**H. icosiphylla** BAK. (§ IX). – Origin unknown. – **Ros.** 7–8 cm ⌀; **L.** c. 20, spiralled, lanceolate-triangular, 3–4 cm long, 2 cm across in the middle, 3 mm thick, upperside concave, underside convex and carinate, green to reddish, tuberculate-rough.

**H. incurvula** v. POELLN. (Pl. 90/XI) (§ XI; acc. G. G. SMITH § XII). – Cape: Albany D. – **Ros.** up to 6 cm ⌀, offsetting basally; **L.** 7–40, lax, soft, broadly ovate-oblong, blunt-tipped, 2 cm long, 12 mm across, underside convex, ± carinate, upperside glassy-transparent, both surfaces with dark green longitudinal lines.

**H. integra** v. POELLN. (§ XI). – Cape: Lit. Karroo. – **Ros.** offsetting basally, mat-forming; **L.** numerous, ± oblong or ovate-oblong, c. 4 cm long, 13 mm across midway, blue-green, both surfaces with reticulate longitudinal lines, upperside swollen in the middle, underside with 1–2 distinct keels and swollen at the tip, terminal bristle 8 mm long.

**H. intermedia** v. POELLN. (§ XI). – Cape: near Port Elizabeth. – **Ros.** acauline, offsetting basally, 5–6 cm ⌀; **L.** numerous, oblong-lanceolate, 3.5–4 cm long, c. 1 cm across, light green or light grey-green, both surfaces ± translucent towards the tip, upperside rather swollen towards the tip, often with a prominent longitudinal line and 7–10 ± connected, very irregular longitudinal lines, underside rather convex and bluntly keeled, also with indistinct, connected, longitudinal lines, terminal bristle 1 mm long, margins somewhat dentate towards the base.

**H. isabellae** v. POELLN. (Pl. 94/6) (§ I). – Cape: near Port Elizabeth. – **Ros.** acauline, c. 7 cm ⌀; **L.** numerous, narrow-lanceolate, tapering, 3.5–4.5 cm long, 8–10 mm across, upperside flat to grooved-concave, underside convex-carinate, grey-green, dull, both surfaces with several darker lines, margins and keel bristly, terminal bristle 3 mm long.

**H. jacobseniana** v. POELLN. (§ II). – Cape: Jansenville D. – **St.** c. 10 cm long, offsetting from the base; **L.** 15–20 mm long, 7–8 mm across below, upperside slightly convex at the base, underside convex, carinate, with 6–7 ± conspicuous longitudinal lines, margins and keel tuberculate, serrate towards the tip, the L. otherwise green, very pruinose, tubercles concolorous, ± merging, rarely in transverse R.

**H. janseana** UITEW. (§ III). – Origin unknown. – **Ros.** acauline, 4.5–5 cm ⌀, offsetting basally, mat-forming; **L.** numerous, ovate-lanceolate, tapering, 2.5 cm long, 8–10 mm across, pale grey-green, with many longitudinal lines, upperside grooved, underside convex-carinate, margins and keel irregularly and minutely dentate, T. 0.6 mm long, terminal bristle 2 mm long.

**H. jonesiae** v. POELLN. (§ II). – Cape: Steytlerville D. – **St.** spirally leafy, over 20 cm long, offsetting basally, less often forking dichotomously higher up; **L.** lanceolate, 1.5–2.5 cm long, 1.5–2.5 cm across below, dark green, very pruinose, with a brownish apiculus, upperside with thickened margins and slightly concave midway, underside convex, with inconspicuous longitudinal lines and a distinct keel with dark tubercles, margins rough.

**H. kewensis** v. POELLN. (§ II) (*H. peacockii* Hort. Kew non BAK.). – Origin unknown. – **St.** extended, offsetting basally; **L.** spiralled, stiff, straight, tapering, c. 3 cm long, 1.5–2 cm across below, up to 10 mm thick, dark green, dull, upperside with solitary tubercles in indistinct longitudinal and transverse R., underside sharply keeled, often with 2 keels, tubercles in longitudinal R., margins and keels tuberculate-dentate, terminal bristle small, soon dropping.

**H. koelmaniorum** OBERM. et HARDY (§ VI) (*H. koelmaniora* OBERM. et HARDY err. est). – Transvaal: Groblersdal D. – **Root** 5–9 mm thick, seldom branching; **L.** 18–30 in a dense, stiff, erect Ros. in a treble spiral, fleshy, linguiform, up to 7 cm long, 2 cm across below, tapering, the margins above folded inwards, smooth at the base, pale, otherwise dark purple-brown, upperside with small, glossy, translucent tubercles in irregular wavy transverse bands, margins indistinct.

**H. cv. Krausiana** – Garden plant, probably a hybrid: **H. glabrata** v. **perviridis** × **H. tortuosa**. Hort. HAAGE and SCHMIDT.

**H. laetevirens** Haw. (§ III) (*Aloe l.* Salm.). – Cape: Sevenweekspoort. – Ros. 5–7 cm ∅; L. 20–30, oblong-lanceolate, tapering, 2.5–3 cm long, 12–16 mm across, underside rounded, often thicker at the tip, pale green, both surfaces translucent towards the tip, with 3–7 pale reticulate lines, with spots with light T., margins and keel dentate, terminal bristle small.

**H. lateganae** v. Poelln. (§ XVI). – Cape: Oudtshoorn D. – St. 5 cm long; L. numerous, in spirals, ovate-lanceolate, very tapered, apiculate, 5.5–6 cm long, 16–20 mm across below, grey-green, slightly pruinose, slightly rough, upperside very concave, underside very rounded and obliquely carinate, margins smooth, darker green, curved horizontally inwards above the middle.

**H. leightoniae** G. G. Smith (§ VII). – Cape: East London D. – Ros. acauline, 7 cm ∅, offsetting basally, forming large clumps; L. c. 45, c. 43 mm long, c. 16 mm across below, 5 mm thick towards the tip, oblong, tapering, acuminate, upperside flat to rather concave, slightly swollen and translucent towards the tip, which has 5 red reticulate lines, otherwise green, dark red in the upper part, underside convex, also with lines at the tip, obliquely carinate above, dentate, terminal bristle 5–6 mm long.

**H. lepida** G. G. Smith (§ XII). – Cape: Albany D. – Ros. acauline, 6 cm ∅, offsetting from the base, forming clumps; L. c. 40, ovate-oblong, broad-tapering, soft, 2–3 cm long, 12–13 mm across below, 6–8 mm thick midway, upperside rounded, swollen towards the tip, with several translucent spots along the margins and 7 indistinct reticulate lines, light green, underside rounded, swollen towards the tip, spotted, keel obtuse, oblique, margins translucent above, terminal bristle 0.75 mm long.

**H. limifolia** Marl. (§ VI). – Cape: Barberton D., Transvaal. – Ros. with or without offsets from the base, 6–12 cm ∅; L. almost horizontal to inclined, broadly ovate-triangular at the base, tapering, upperside ± concave, underside very rounded, keel roundish, dark brownish green, both surfaces with 15–30 transverse callosities.

H. — v. **gigantea** M. B. Bayer. – Natal: Nongoma D. – St. 4 cm thick, offsetting; L. c. 70 in a compact Ros., up to 10 cm long, 4 cm across midway, ovate-lanceolate or tapering, purplish-pink below, dark green towards the tip, the upper ³⁄₄ with light green tubercles merging into irregular transverse R. 4 mm long, underside convex, tubercles as for upperside, keel oblique in the upper third; **Infl.** up to 75 cm tall with c. 50 Fl.

H. — v. **keithii** G. G. Smith. – Swaziland. – Ros. 11 cm ∅, offsetting; L. up to 7 cm long, upperside with 8 dark red, translucent, longitudinal lines, with solitary tubercles in wavy transverse R. and a few wrinkles above, pinkish-brown below, olive-green above, underside carinate, also tuberculate-wrinkled, margins ± acute, concolorous.

H. — v. **limifolia** (Pl. 90/VI) (f. *tetraploidea*) (*H. l.* v. *marlothiana* Res.). – **St.** not offsetting; Ros. 6–11 cm ∅; L. many, triangular, sharply tapering, margins curved upwards, 3–5 cm long, 2–4 cm across below, 2–6 mm thick, concave on the upperside, both surfaces with 15–25 transverse callosities.

H. — v. **schuldtiana** Res. (f. *diploidea*). – St. not offsetting; Ros. 9–12 cm ∅; L. lanceolate-tapering, 4–6 cm long, 15–27 mm across below, margins ± involute, with 20–30 transverse callosities, these often consisting of tubercles.

H. — v. **stolonifera** Res. – With long offsets.

H. — v. — f. **major** Res. – L. 5–8 cm long, triangular-lanceolate.

H. — v. — f. **pimentellii** Res. – L. c. 5 cm long, 4 cm across, ovate-lanceolate towards the tip, tapering.

H. — v. **ubomboensis** (Verd.) G. G. Smith (*H. ubomboensis* Verd.). – Swaziland. – Almost acauline, offsetting; Ros. with c. 20 L., ovate-lanceolate, upperside with dark longitudinal lines, not tuberculate or only slightly so, tubercles in longitudinal lines, margins horny, tip pointed.

**H. lisbonensis** Res. (§ II; acc. G. G. Smith § XV). – Origin unknown. – St. c. 12 cm long, branching from the base; L. in spiral R., ovate-lanceolate, tapering, 3–4.5 cm long, 1–2 cm across, 2–6 mm thick, both surfaces with 2 distinct keels and minutely rough tuberculate, dark green, tip lighter.

**H. lockwoodii** Arch. (Pl. 95/1) (§ III). – Cape: Laingsburg D. – Resembles **H. altilinea**; L. broader and thicker, without marginal T., tips of old L. white, papery.

**H. longibracteata** G. G. Smith. (§ XIV). – Cape: Riversdale D. – Ros. acauline, 10 cm ∅, offsetting basally, forming large clumps; L. c. 34, oblong-ovate, pointed-tapering, firm, 5.5 cm long, 17 mm across, 1 cm thick, upperside convex, tip area recurved at 40°, 17 mm long, 16 mm across, smooth, spotted, light green, translucent, with 4–7 lines, underside oblique-carinate, angular, dentate, margins roundish, dentate, terminal bristle 2–3 mm long.

**H. longiana** v. Poelln. v. **longiana** (§ IX). – Cape: Humansdorp D. – Ros. acauline; L. few, ovate-triangular at the base, straight towards the tip, 2–2.5 cm long, 15 mm across, upperside flat with an inconspicuous M.line, underside rounded, both surfaces with tubercles often in indistinct longitudinal lines, terminal bristle brownish.

H. — v. **albinota** G. G. Smith. – Cape: Humansdorp D. – L. lanceolate, tapering-subulate, up to 3 cm long, 28 mm across below, upperside indistinctly tuberculate, underside with tubercles 2 mm ∅, bluish-white, usually in irregular transverse bands.

**H. luteorosea** Uitew. (§ I). – Cape: probably near Grahamstown. – Ros. almost acauline, offsetting basally, 4–5 cm ∅; L. numerous, obovate to oblong-lanceolate, tapering, 2–3 cm long, 7–9 mm across, pale blue-green, upperside with 3–4 dark longitudinal lines, with translucent spots towards the tip, underside convex and

Plate 106. 1. **Kalanchoe tomentosa** BAK.; 2. **K. thyrsiflora** HARV.; 3. **Kinepetalum schultzei** SCHLTR. (Photo: K. DINTER); 4. × **Rocheassula langleyensis** (VEITCH) ROWL. (Photo: E. HAHN); 5. **Kedrostis nana** (LAM.) COGN. v. nana; 6. **Karimbolea verrucosa** DESC. (Repr.: Cactus (France) No. 68/69, 1960, 79).

Plate 107. 1. **Lenophyllum guttatum** Rose (Photo: A. J. A. Uitewaal); 2. **Lewisia rediviva** Pursh v. rediviva (Repr.: Parey's Blumengärtnerei); 3. **Litanthus pusillus** Harvey; 4. **Lobelia keniensis** R. E. et Th. Fries (Repr.: Afroalp. Vasc. Pl., 1957, Pl. 6); 5. **Lomatophyllum citreum** Guill. (Photo: J. Marnier-Lapostolle).

Plate 108. 1. Meterostachys sikokianus (MAKINO) NAKAI (Photo: R. MORAN); 2. Momordica rostrata A. ZINN.; 3. Monadenium ellenbeckii N. E. BR. (Photo: P. R. O. BALLY); 4. M. lugardae N. E. BR. (Photo: H. LANG).

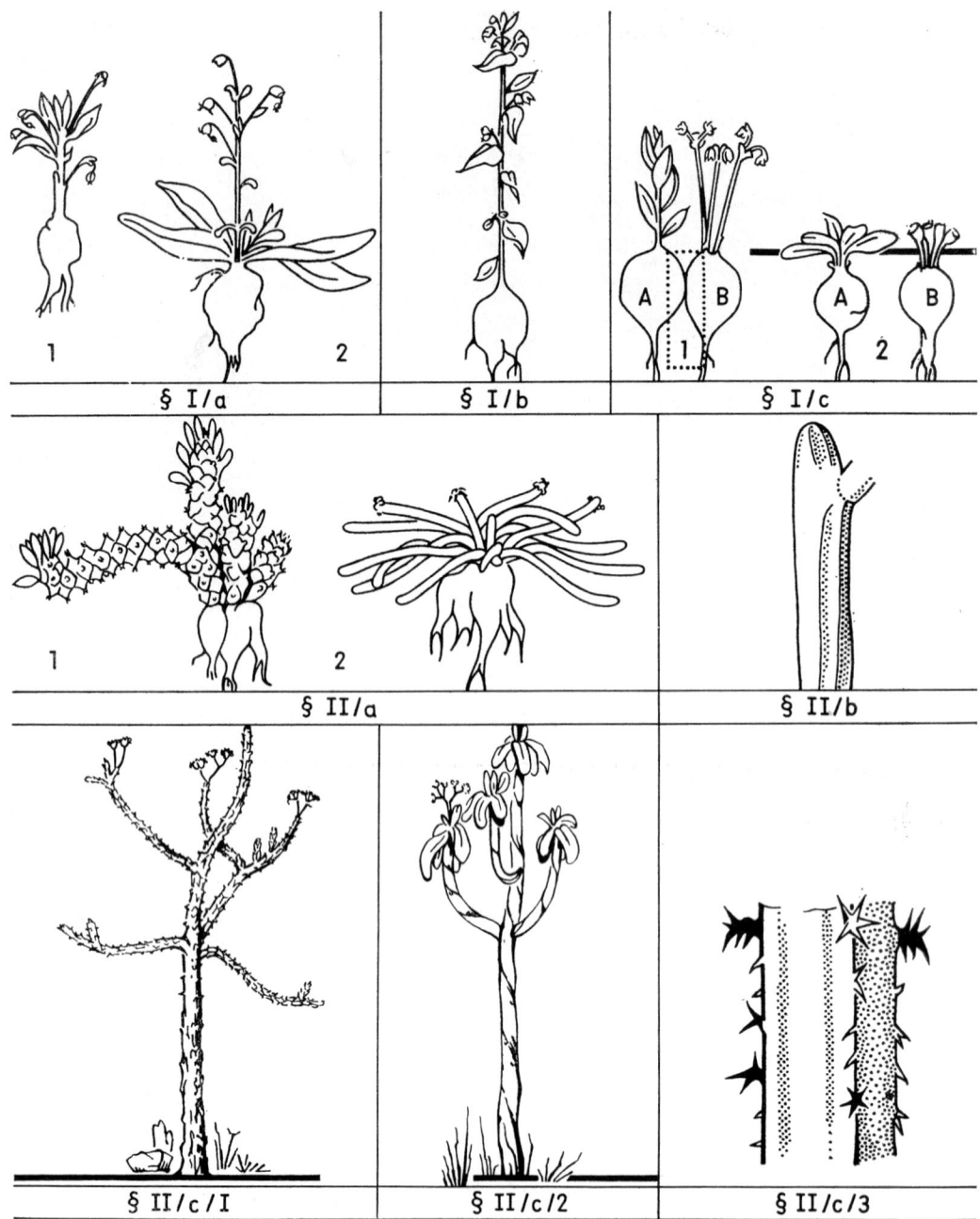

Plate 109. Division of the Genus Monadenium according to growth form.
§ I. Perennials. a–c. Dwarf, semisucculent geophytes.
§ II. Succulents.
a) Dwarf succulents. b) Shrubby plants. c) Trees.
Details see page 296.
Drawings acc. P. R. O. Bally.

Plate 110. 1. **Monadenium erubescens** N. E. Br. (Drawing P. R. O. Bally); 2. **M. guentheri** Pax v. **guentheri** (Drawing as 1); 3. **M. stapelioides** Pax v. **stapelioides** (Photo: P. R. O. Bally); 4. **Monolena primuliflora** Hook. f.; 5. **Monanthes polyphylla** (Webb) Haw. (Photo: J. A. Huber); 6. **M. anagensis** Praeg.; 7. **M. subcrassicaulis** (O. Ktze.) Praeg.

Plate 111. 1. **Moringa hildebrandtii** Engl. (Photo: W. Rauh); 2. **Mucizonia hispida** (Lamk.) Bgr.; 3. **Myrmecodia platyrea** Becc. (Photo: P. Fischer); 4. **Nolina recurvata** Lem. Young plant.

Plate 112. **1. Neoalsomitra podagrica** VAN STEENIS (Repr.: Nat. Cact. Succ. J. **22,** 1957, 46, Photo: F. K. HORWOOD); **2. Orostachys iwarenge** (MAKINO) HARA (Repr.: Succulenta, 1962, 103, Photo: I. V. T.); **3. O. spinosus** (L.) BGR.; **4. Othonna capensis** L. H. BAILEY; **5. O. clavifolia** MARL.

Plate 113. 1. Oxalis succulenta Barn.; 2. Othonna herrei Pill. (Photo: E. Hahn); 3. Oxalis carnosa Mol.; 4. Othonna dentata L.

similarly lined, lines ± reticulate, margins and keel on the underside bristly-dentate, terminal bristle dentate; Fl. yellowish pink.

**H. magnifica** v. POELLN. (§ XIV). – Cape: Riversdale, Robertson D. – Ros. with few L. c. 3.5 cm long, stiff, tip perpendicularly recurved, tip-area triangular, 18 mm long, 9 mm across at the base, somewhat swollen, furrowed in the middle, translucent, acute-dentate, with 4–5 lines, underside roundish-carinate, margins and keel with whitish T., with a white apiculus.
⚹ **H.** × **mantelii** UITEW. – Intermediate hybrid: **H. truncata** × **H. cuspidata.**

**H. maraisii** v. POELLN. (§ XIV). – Cape: Swellendam, Robertson D. – Ros. up to 8 cm ⌀, offsetting basally; L. ovate-lanceolate, 3.5 cm long, 2 cm across, light green, underside very rounded-carinate, upperside convex, with the upper half almost horizontally truncate, tip-area roundish-triangular, with 3 usually short longitudinal lines, irregularly tuberculate, L.-margins ± distinctly dentate.

**H. margaritifera** (L.) HAW. (§ IX). – Cape: W. reg., widespread. – Type s. v. **margaritifera.**

**H.** — v. **corallina** BAK. – Cape: Ceres D. – L. 8–10 cm long, ovate-oblong, long-tapering, 2–2.5 cm across, dark green, upperside with green tubercles ± in longitudinal lines, tubercles on the underside white, numerous, in indistinct longitudinal R., c. 1 mm thick, at the tip in wavy transverse bands, margins horny, whitish, keel with a white, horny border towards the tip.

**H.** — v. **margaritifera** (Pl. 90/IX) (*H. m.* v. *typica* BGR., *Aloe pumila* v. *m.* α L., *A. p.* v. *m.* THUNBG. p.part., *A. m.* MILL. v. *major* DC., *H. major* DUV., *Apicra m.* WILLD. and v. *major* WILLD., *A. m.* v. *major* AIT., *A. m.* v. *m.* WILLD.). – Ros. acauline, offsetting freely, up to 15 cm ⌀; L. triangular-ovate, thick and firm-fleshed, 7–8 cm long, up to 3 cm across at the base, upperside flat to somewhat convex, underside carinate towards the tip, dark green, both surfaces with large, roundish pearl-like tubercles; Infl. 5-branched, Cor. segments stellate-spreading.

**H.** — v. **maxima** (HAW.) UITEW. sv. **maxima** (Pl. 95/2) (*Aloe margaritifera* v. *m.* HAW., *Aloe semimargaritifera* SALM, *H. maxima* DUV., *Aloe s.* v. *maxima* SALM, *Haw. semimargaritifera* HAW., *H. s.* v. *maxima* HAW., *H. marg.* v. *s.* SALM, *H. marg.* v. *maxima* sv. *maxima* HAW., *H. marg.* v. *semimargaritifera* (SALM) BAK.). – Cape: Worcester D. – L. long-ovate, 7–10 cm long, upperside with few tubercles or none, tubercles on the underside large and very prominent.

**H.** — v. — sv. **major** (SALM) UITEW. (*Aloe semimargaritifera* v. *major* SALM, *H. marg.* v. *maxima* sv. *major* HAW., *H. semimarg.* v. *major* HAW., *H. marg.* sv. *major* (SALM) BGR.). – Larger than the preceding sv.

**H.** — v. — sv. **multiperlata** (HAW.) UITEW. (*H. semimargaritifera* v. *m.* HAW., *Aloe sem.* v. *m.* SALM, *A. sem.* v. *m.* ROEM. et SCHULT., *H. marg.* v. *max.* sv. *multipapillosa* SALM,

*H. marg.* v. *s.* sv. *m.* (HAW.) v. POELLN.). – Pearly tubercles more numerous.

**H.** — v. **minima** (AIT.) sv. **minima** (*Aloe marg.* v. *m.* AIT., *H. marg.* v. *granata* (WILLD.) BAK., *H. minima* DUV., *Apicra granata* WILLD., *Aloe pumila* v. *marg.* γ. L., *A. pumila* v. *marg.* THUNBG. p.part., *A. marg.* v. β. LAM. p.min.part., *H. minima* HAW., *H. brevis* HAW., *A. marg.* SPRENG., *H. granata* HAW., *A. g.* ROEM. et SCHULT., *A. brevis* ROEM. et SCHULT., *A. granata* v. *major* SALM). – L. ovate-triangular, 5–6 cm long, c. 2 cm across, tubercles on both surfaces in irregular transverse R., ± confluent.

**H.** — v. — sv. **laetevirens** (SALM) UITEW. (*Aloe erecta* v. *l.* SALM., *A. granata* v. *l.* SALM, *H. marg.* v. *granata* sv. *l.* (SALM) BGR.). – L. fresh green.

**H.** — v. — sv. **polyphylla** (HAW.) UITEW. (*H. granata* v. *p.* HAW., *Aloe g.* v. *minor* SALM, *H. marg.* v. *m.* SALM, *A. marg.* DC., *A. granata* v. *p.* HAW., ? *A. g.* v. *minima* SALM, *H. marg.* v. *g.* sv. *minor* BGR., *H. marg.* v. *g.* sv. *p.* (HAW.) v. POELLN.). – Cape: Albany D. – Smaller than sv. **minima**; L. more numerous, tubercles smaller, closer together.

**H.** — v. **minor** (AIT.) UITEW. (*Aloe marg.* v. *m.* AIT., *A. marg.* v. *media* AIT., *H. marg.* v. *erecta* (HAW.) BAK., *H. m.* DUV., *H. erecta* HAW., *A. pumila* v. *marg.* β. L., *A. p.* v. *marg.* THUNBG. p.part., *A. marg.* v. β. LAM. p.max.part., *A. marg.* v. *m.* WILLD., *Apicra marg.* v. *m.* WILLD., *A. m.* ROEM. et SCHULT., *A. erecta* SALM, ? *A. mamillaris* v. *major* SALM, *A. erecta* v. *minor* SALM). – L. long-triangular, 7–8 cm long, c. 2 cm across, tubercles on both sides scattered, not confluent.

**H.** — v. **subalbicans** (SALM) BGR. (*Aloe s.* SALM, ? *A. semi-margaritifera* v. *glabrata* SALM). – Variety with a somewhat whitish sheen.

**H.** — v. — sv. **acuminata** (SALM) BGR. (*Aloe s.* v. *a.* SALM). – L. long-tapering.

**H.** — v. — sv. **laevior** (SALM) BGR. (*Aloe s.* v. *l.* SALM). – L. rather smoother.

**H. marginata** (LAM.) STEARN v. **marginata** (Pl. 95/3) (§ X) (*Aloe m.* LAM., *H. albicans* HAW., *Aloe a.* HAW.). – Cape: Riversdale D. – Ros. acauline, c. 12 cm ⌀; L. not numerous, ovate-lanceolate, long-tapering, apiculate, rigid, underside rounded and sharply carinate, both surfaces dark green, covered with a fine, papery, white membrane, margins and keel with a white margin made up of ± confluent white tubercles, underside with solitary tubercles.

**H.** — v. **laevis** (HAW.) JACOBS. (*H. a.* v. *l.* HAW.). – L. ovate, long-tapering, with a thin white border.

**H.** — v. **ramifera** (HAW.) JACOBS. (*H. a.* v. *r.* HAW.). – Ros. offsetting basally, and overlapping; L. ovate-tapering, with a few scattered tubercles.

**H.** — v. **virescens** (HAW.) UITEW. (Pl. 90/X) (*H. v.* HAW., *H. a.* v. *v.* (HAW.) BAK.). – L. more ovate-tapering, less conspicuously emarginate, with few scattered, white, shining tubercles.

**H. marumiana** UITEW. (§ I). – Cape: Ladismith D. – Ros. acauline, 1.5–2 cm ⌀, offsetting basally; **L.** numerous, ovate to elongate-lanceolate, tapering, 12–20 mm long, 5–8 mm across, with a 2 mm long bristly tip, upperside with 2–4 reticulate, translucent lines and tuberculate spots, underside rounded, lined and spotted like the upperside, margins and keel with T. forming a light border.

**H. maughanii** v. POELLN. (Pl. 90/IV/1) (§ IV) (*H. truncata* v. *m.* (v. POELLN.) FEARN). – Cape: Calitzdorp D. – **L.** rosulate, c. 25 mm long, ± semi-terete, 15 mm across below, grey-green to reddish-brown, rough, tip truncate as if cut off cleanly, 10–12 mm across and 8 to 10 mm thick, with a ± distinct Wi., in habitat only the tip-area protruding above the soil. – See note under **H. truncata**.

**H. mclarenii** v. POELLN. (§ VIII). – Cape: Tulbagh D. – **Ros.** acauline, 7–8 cm ⌀; **L.** many, spiralled, oblong-lanceolate or ± ovate-lanceolate, tapering, 3.5–4 cm long, 8–10 mm across, upperside flat with longitudinal lines, underside rounded-carinate, surface dark green to red-brown, somewhat pruinose, margins and keel irregularly dentate, T. 1 mm long, terminal bristle 10–15 mm long, with small T. below.

**H. mirabilis** HAW. (§ XIV) (*Aloe m.* SALM, *Apicra m.* WILLD., *H. multifaria* HAW., *A. m.* ROEM. et SCHULT.). – Origin unknown. – **Ros.** c. 5 cm ⌀, offsetting basally; **L.** 18–20, spiralled, 3–3.5 cm long, the upper half recurved, almost horizontal, this tip-area long-triangular, convex, translucent, with 3–5 pallid green lines, underside triangular-rounded, sharply keeled, tapering above, mucronate, margins and keel dentate.

**H. monticola** FOURC. (§ VIII) (*H. angustifolia* BAK. non HAW., *Aloe stenophylla* SCHULTES). – Cape: George, Uniondale D. – **St.** short; **Ros.** c. 7.5 cm ⌀; **L.** c. 20, erect, straight or incurved, lanceolate, acuminate, 4–5 cm long, flat above, convex below with 1–3 faint keels above, dark green above, translucent in the lower half, minutely ciliate along the margins.

**H. morrisiae** v. POELLN. (§ XVI). – Cape: Oudtshoorn D. – **Ros.** acauline, 3–4 cm ⌀; **L.** few, ovate-lanceolate, 3–3.5 cm long, 12 to 15 mm across, mucronate, upperside flat, underside keeled, both surfaces with small, often confluent, concolorous tubercles in indistinct transverse R., margins and keel rough-tuberculate, surface dark green.

**H. mucronata** HAW. v. **mucronata** (Pl. 95/4) (§ III) (*A. altilinea* v. *m.* (HAW.) v. POELLN., *Aloe m.* SCHULT, *H. altilinea* BGR. p.min.part., *H. a.* v. *brevisetata* v. POELLN.). – Cape: Lit. Karroo. – **L.** 1.5–6 cm long, 4–16 mm across, ± long-tapering, upperside flat or rather thickened towards the tip, underside slightly carinate, margins small-dentate, the tip usually translucent.

**H. — v. bicarinata** (TRIEBN.) v. POELLN. (*H. altilinea* v. *b.* TRIEBN.). – **L.** 3–5 cm long, 6–8 mm across, with 4–6 darker lines in the light part, underside with 2 keels, with 2 mm long T.

**H. — v. limpida** (HAW.) v. POELLN. (*H. l.* HAW., *Aloe l.* SCHULT., *H. altilinea v. l.* (HAW.) v. POELLN.). – **L.** 3–4 cm long, 10–16 mm across, upperside translucent, with 7–16 longitudinal lines, underside with 2 keels and 11–13 longitudinal lines, terminal bristle 3–8 mm long.

**H. — v. — f. acuminata** v. POELLN. (*H. altilinea v. l. f. a.* v. POELLN.). – **L.** conspicuously long and gradually tapering.

**H. — v. — f. inconfluens** v. POELLN. (*H. altilinea v. l. f. c.* v. POELLN.). – **L.** with 8–16 longitudinal lines on the upperside in the light part, margins often slightly dentate.

**H. — v. — f. inermis** v. POELLN. (*H. altilinea v. l. f. i.* v. POELLN., *H. a. v. i.* v. POELLN., *H. i.* v. POELLN.). – **L.** not conspicuously long-tapering, margins not dentate.

**H. — v. — f. limpida** (*H. m. v. l. f. typica* v. POELLN.). – **L.** long ovate-lanceolate, tapering, lines irregularly interconnected in the light part, terminal bristle c. 6 mm long.

**H. — v. morrisiae** v. POELLN. (*H. altilinea v. m.* v. POELLN.). – **L.** yellowish-green, margins and keel smooth, upperside with 8–14 irregularly interconnected longitudinal lines, 2–5 cm long, pointed tip scarcely 1 mm long.

**H. — v. — f. subglauca** v. POELLN. (*H. altilinea v. m. f. s.* v. POELLN.). – **L.** almost grey-green, usually smooth.

**H. — v. polyphylla** (BAK.) v. POELLN. (*H. p.* BAK., *H. altilinea v. p.* (BAK.) v. POELLN.). – **L.** numerous, 4–5 cm long, underside with 2–3 keels, keels and margins with T. 2–4 mm long, terminal bristle 5–8 mm long.

**H. — v. — f. minor** (TRIEBN.). v. POELLN. (*H. altilinea v. p. f. m.* TRIEBN.). – **L.** 1.5 to 2.5 cm long, 4–7 mm across, T. numerous, closer together, up to 3 mm long.

**H. — v. setulifera** (TRIEBN. et v. POELLN.) v. POELLN. (*H. altilinea v. s.* TRIEBN. et v. POELLN.). – **L.** yellow-green, margins distinctly banded, margins and keel with 2 mm long bristles, directed backwards.

**H. mundula** G. G. SMITH (§ XIV). – Cape: Bredasdorp D. – **Ros.** acauline, 6 cm ⌀, branching basally; **L.** oblong, tapering, firm, almost 4 cm long, 14 mm across, the tip-area recurved, 10 mm thick, 13 mm across, 17 mm long, ± convex, smooth or with a few tubercles in 1–2 longitudinal lines, greenish-translucent, underside convex to keeled, with lines and whitish spots, keel and margins angular and dentate, terminal bristle 3 mm long, soon dropping.

**H. musculina** G. G. SMITH (§ II). – Cape: Bathurst D. – **St.** leafy, up to 20 cm tall, erect, becoming prostrate, offsetting basally, forming large clumps; **L.** in many R., ovate-lanceolate, pointed, 4 cm long, 2 cm across below, almost 1 cm thick, firm, leathery, light green to bronze-coloured, smooth, with a concolorous M.line and 2 lateral lines, underside rounded, with c. 15 rows of whitish, hard, almost 1 mm long tubercles, carinate above.

**H. mutabilis** v. POELLN. (§ IX). – Cape: Bredasdorp D. – **Ros.** acauline; **L.** spirally arranged, ovate at the base, then straight to lanceolate, stiff but supple, leathery, with a long triangular tip, green to brownish-green, upperside almost flat, with a conspicuous M.line, rarely with 1–3 tubercles, underside convex-carinate, tuberculate, margins and keel with T.-like tubercles.

**H. nigra** (HAW.) BAK. v. **nigra** (Pl. 95/5) (§ XX) (*Apicra n.* HAW., *Aloe n.* ROEM. et SCHULT., *H. ryneveldiae* v. POELLN.). – Cape: Bedford, Somerset East D. – **St.** 6–12 cm tall; **L.** in 3 indistinct R., ovate, pointed, spreading horizontally, 1–6 cm long, 10–18 mm across, 4–6 mm thick, upperside concave below, flat-convex above, underside rounded and obliquely keeled, tip somewhat pungent, blackish-green, both surfaces with fairly large, closely spaced tubercles, margins tuberculate.

**H. —** v. **angustata** (v. POELLN.) UITEW. (*H. schmidtiana* v. *a.* v. POELLN.). – Cape: Ladismith D. – **L.** laxer, in general less robust than v. **schmidtiana**.

**H. —** v. **diversifolia** (v. POELLN.) UITEW. (*H. d.* v. POELLN., *H. schmidtiana* v. *d.* v. POELLN.). – Cape: Prieska D. – **St.** 10 cm long; **L.** erect-appressed, upper L. horizontally spreading, ovate-oblong, c. 2 cm long, 12 mm across, both surfaces tuberculate, underside with tubercles in distinct R. or irregular Gr.

**H. —** v. — f. **nana** (v. POELLN.) UITEW. (*H. schmidtiana* v. *d.* f. *n.* v. POELLN.). – Cape: Beaufort West D. – Smaller than f. **diversifolia**; **L.** 13–15 mm long, 12–13 mm across, tip recurved.

**H. —** v. **elongata** (v. POELLN.) UITEW. (*H. schmidtiana* v. *e.* v. POELLN.). – Cape: Somerset East D. – **St.** 6–8 cm long; **L.** 3.5–6 cm long, 10–18 mm across, almost lanceolate, tubercles oblong, often in longitudinal R.

**H. —** v. **pusilla** (v. POELLN.) UITEW. (*H. schmidtiana* v. *p.* v. POELLN.). – Aberdeen D. – **L.** 10–15 mm long, 8–11 mm across.

**H. —** v. **schmidtiana** (v. POELLN.) UITEW. (*H. s.* v. POELLN., *H. marginata* HORT.). – Cape: Somerset East, Ladismith, Cradock, Willowmore D. – **St.** rarely offsetting, 7–8 cm tall; **L.** ovate-lanceolate, 14–30 mm long, 15 mm across below, spreading horizontally, surface dark green, somewhat viscous, with thick tubercles, those on the underside in dense transverse R., margins and often the keel with 1–2 R. of ± confluent tubercles.

**H. —** v. **suberecta** (v. POELLN.) UITEW. (*H. schmidtiana* v. *s.* v. POELLN.). – **L.** almost erect, lanceolate, less ovate at the base, 3–4 cm long, 10–15 mm across, the upperside with tubercles merging in fairly long, longitudinal rows, underside as v. **schmidtiana**.

**H. nitidula** v. POELLN. v. **nitidula** (Pl. 95/6) (§ XIV). – Cape: Worcester, Swellendam, Caledon, Bredasdorp D. – **Ros.** acauline, 5 cm ∅, with few offsets from the base; **L.** many, long-lanceolate, tapering, 5–6 cm long, 8 to 10 mm across, surface light green, tip-area recurved at 30–45°, 25 mm long, smooth, with 6–8 green longitudinal lines, underside rounded, indistinctly carinate, with a few light spots or lines towards the tip, margins and keel with light T. towards the tip, terminal bristle 5–6 mm long.

**H. —** v. **opaca** v. POELLN. – Darker variety.

**H. nortieri** G. G. SMITH v. **nortieri** (§ III). – Cape: Van Rhynsdorp D. – **Ros.** acauline, 6.5 cm ∅, with few or no offsets from the base; **L.** c. 40, obovate, lanceolate-tapering, ± firm, 37 mm long, 12 mm across, 6 mm thick above midway, upperside convex, swollen, with translucent spots and greenish, reticulate lines, reddish-green to brownish, underside convex, spotted and with reticulate lines, margins angular, with 1.25 mm long, whitish, translucent T., keel similarly dentate.

**H. —** v. **giftbergensis** G. G. SMITH. – Smaller variety, terminal bristle shorter, both surfaces with fewer lines.

**H. —** v. **montana** G. G. SMITH. – **L.** shorter, similarly the T. on keel and margins, lines fewer on both surfaces.

**H. notabilis** v. POELLN. (§ XIV). – Cape: Robertson D. – **Ros.** acauline; **L.** numerous, oblong-lanceolate, somewhat broader at the base, 4–5 cm long, 7–10 mm across, the tip truncate-recurved, tip-area triangular, somewhat convex, 13–18 mm long, 7–9 mm across, translucent, smooth or less often with a few or many T. and with 3 ± confluent, green, longitudinal lines, underside obliquely carinate, often double-keeled, spotted or with small tubercles in longitudinal R.

**H. obtusa** HAW. emend. UITEW. v. **obtusa** (§ VII) (*H. cymbiformis* v. *o.* (HAW.) BAK.). – Cape: King Williams Town, Somerset East, Beaufort West D. – **Ros.** acauline; **L.** dirty green, oblong to ovate-oblong, somewhat thickened at the tip, abruptly truncate and long tapering, 2–3.5 cm long, upperside somewhat convex at the tip, underside carinate towards the tip, the upper tip-area translucent, with darker longitudinal lines, terminal bristle 2–6 mm long or absent.

**H. —** v. **columnaris** (BAK.) UITEW. (*H. c.* BAK., *H. pilifera* v. *c.* (HAW.) v. POELLN.). – Cape: Oudtshoorn D. – **L.** darker, erect, up to 3.5 cm long, upperside with green lines in the lighter area, terminal bristle long.

**H. —** v. **dielsiana** (v. POELLN.) UITEW. (*H. d.* v. POELLN., *H. pilifera* v. *d.* v. POELLN., *H. reticulata* TROLL). – **L.** numerous, 2–3 cm long, tip 5 mm thick, green to reddish, upperside with c. 12 translucent lines, margins and keel with small white T., terminal bristle 3 mm long.

**H. —** v. — f. **acuminata** (v. POELLN.) UITEW. (*H. pilifera* v. *d.* f. *a.* v. POELLN.). – **L.** very tapering.

**H. —** v. **gordoniana** (v. POELLN.) UITEW. (*H. g.* v. POELLN., *H. pilifera* v. *g.* v. POELLN.). – With 4–6 lines in the light area, the tip-section 4–5 mm thick, terminal bristle 2 mm long, with small lateral bristles, margins and keel with 1 mm long, spreading bristles.

**H. — v. pilifera** (BAK.) UITEW. (*H. p.* BAK.). – **L.** 2–2.5 cm long, both surfaces convex, short-tapering, both surfaces nearly pellucid in the upper third, with darker simple lines, margins and keel finely dentate, terminal bristle 5–6 mm long.

**H. — v. — f. truncata** JACOBS. (Pl. 96/3). – **L.** tips very truncate, translucent, margins with very reduced T. or none, no terminal bristle.

**H. — v. salina** (v. POELLN.) UITEW. (*H. stayneri* v. *s.* v. POELLN., *H. pilifera* v. *s.* v. POELLN.). – **L.** rather blunt, upperside with 4–5 freely tessellate lines, terminal bristle simple, 2 mm long, margins and keel with large T.

**H. — v. stayneri** (v. POELLN.) UITEW. (*H. s.* v. POELLN., *H. pilifera* v. *s.* v. POELLN.). – **L.** oblong to long-ovate, 2–3 cm long, 10–13 mm across, gradually tapering, blunt, upperside with 3–4 longer and 4–5 shorter lines, underside very convex, blunt-carinate and thickened to 5–6 mm towards the tip, with 5–7 longer lines with shorter ones in between, keel smooth, margins very shortly dentate.

**H.** cv. **Ollasonii**. – Hybrid raised by L. O. OLLASON, Sydney: **H. retusa** × **H. obtusa** v. **dielsiana**. – **Ros.** scarcely offsetting; **L.** convex on both surfaces, c. 4 cm long, c. 2.5 cm across, c. 13 mm thick, margins and keel with minute, translucent T., tip with very translucent Wi. having 6–8 dark green lines on the upperside, and 14 lines on the underside, bristle-tip 1 mm long.

**H. otzenii** G. G. SMITH. (Pl. 96/2) (§ XIV). – Cape: Bredasdorp D. – **Ros.** acauline, 5 cm ⌀, with few or no basal offsets; **L.** c. 15, firm, c. 4 cm long, the tip-area recurved at 60°, 14 mm long, 11 mm across, ± convex, smooth, translucent-green, with 5–7 parallel, pale green lines, underside convex, spotted, obliquely carinate at the tip, margins and keel smooth, terminal bristle dropping.

**H. pallida** HAW. v. **pallida** (Pl. 96/5) (§ I) (*A. p.* ROEM. et SCHULT.). – Cape: Ceres, Riversdale, Robertson D. – **Ros.** acauline, 3–7 cm ⌀, mat-forming; **L.** many, lanceolate, 2–4 cm long, 6–10 mm across, grey-green, both surfaces with dark longitudinal lines, with light spots towards the tip, often with several tubercles, underside doubly carinate above, margins and keels dentate, terminal bristle 1–2 mm long.

**H. — v. paynii** v. POELLN. (*H. p.* v. POELLN.). – Cape: Robertson D. – **L.** with ± distinct spots and a few small tubercles on the upperside, underside often with 3 keels, often with several T.

**H. papillosa** (SALM) HAW. v. **papillosa** (Pl. 96/1) (§ IX) (*Aloe p.* SALM., *Apicra margaritifera* v. *maxima* WILLD., *Aloe p.* v. *m.* SALM). – Cape: Swellendam, Riversdale D. – **Ros.** acauline, with few offsets, 8–10 cm ⌀; **L.** not numerous, ovate-lanceolate, long-tapering, with a slender tip 2–3 mm long and 7–8 mm across, upperside flat or slightly concave, underside very rounded, both surfaces with round, white tubercles in ± distinct transverse R., margins and keel tuberculate.

**H. — v. semipapillosa** HAW. (*Aloe p.* v. *s.* ROEM. et SCHULT., *A. papillosa* v. *minor* SALM). – Tubercles less numerous.

**H. paradoxa** v. POELLN. (§ XIV). – Cape: Riversdale D. – **Ros.** acauline, offsetting basally, 3.5–4 cm ⌀; **L.** ± ovate or ovate-oblong, c. 3 cm long, c. 1 cm across, underside very convex, distinctly keeled, dark green, with indistinct longitudinal lines, spotted, upperside with a triangular tip-area 10–14 mm long, 9–12 mm across at the base, translucent, glossy, tuberculate, with 5–9 often broken, longitudinal lines, and T. in 3 divergent lines, terminal bristle 1 mm long.

**H. parksiana** v. POELLN. (§ XIV). – Cape: Brak R. – **Ros.** acauline, offsetting from the base, 2–2.5 cm ⌀; **L.** many, ovate-oblong, pointed or rather so, tip recurved, tip-area triangular, falcate, 6–8 mm long, 4–5 mm across, glossy dark green, tuberculate, underside rounded-carinate.

**H. peacockii** BAK. (§ II). – Origin unknown. – **St.** 15–23 cm long; **Ros.** 5–7 cm ⌀; **L.** multifarious, deltoid, ascending, c. 2.5 cm long and broad, 4–6 mm thick, concave above, rounded below with a slightly eccentric keel in the upper half, covered with small, round, immersed, greenish-white spots; margins slightly scabrous.

**H. pearsonii** C. H. WRIGHT (Pl. 96/6) (§ I). – Cape: origin unknown. – **L.** c. 80, obovate to lanceolate-oblong, 2.5 cm long, 8–10 mm across, light green, both surfaces translucent in the upper half, upperside with 5 ± translucent, longitudinal lines, underside with 9–11 lines, margins and keel with tiny T. 2–3 mm long.

**H. perplexa** v. POELLN. (§ XIII). – Cape: Grahamstown D. – **Ros.** acauline, offsetting freely from the base, 9 cm ⌀; **L.** long-ovate or lanceolate, tapering, soft, 5–6 cm long, 1.5 to 2 cm across, upperside flat, with 2 confluent, longitudinal lines, underside convex, with 2–3 keels, with small dentate tubercles, grey-green, the tip lighter and ± translucent, with or without a terminal bristle.

**H. picta** v. POELLN. (§ XIV). – Cape: Lit. Brak R. – **Ros.** acauline, c. 6 cm ⌀; **L.** few, ± ovate-lanceolate, tapering, 3.5–4.5 cm long, 1.5–2 cm across, horizontally truncate at the tip, the tip-area 2.5 cm long, 1.5–2 cm across, triangular, with 5–8 transparent, longitudinal lines, underside very convex-keeled, with spots which are rarely tuberculate, margins and keel rough-dentate, terminal bristle 3 mm long.

**H. planifolia** HAW. (§ XIII) (*Aloe p.* ROEM. et SCHULT., *H. cymbiformis* v. *p.* BAK.). – Cape. – Type s. v. **planifolia** f. **planifolia**.

**H. — v. exulata** v. POELLN. – Origin unknown. – **L.** flatter and broader than those of v. **planifolia** f. **p.**, ± prostrate, 4–5 cm long, 1.5–2 cm across, the keel minutely dentate above, terminal bristle 3–5 mm long, dentate, light, eventually dropping.

**H. — v. incrassata** v. POELLN. – Cape: Graaff Reinet D. – **L.** 2.5–3 cm long, 10–15 mm across, distinctly thickened below the tip to 5–6 mm, ovate-oblong, tapering, margins and keel smooth, mucronate.

**H. – v. longifolia** TRIEBN. et v. POELLN. – L. 5–9 cm long, 1.5–2 cm across, grey-green, the point 1 mm long.

**H. — v. — f. calochlora** TRIEBN. et v. POELLN. – Cape: Port Elizabeth. – L. light green, ovate-lanceolate to oblong-lanceolate, underside acutely carinate, keel and margins smooth, only the lower part of the keel minutely dentate, terminal bristle 3 mm long.

**H. — v. planifolia f. agavoides** TRIEBN. et v. POELLN. – Cape: Fort Beaufort D. – More robust than f. **planifolia; L.** 4–5 cm long, 2–3 cm across, ovate-oblong, fairly gradually tapering, terminal bristle 2 mm long.

**H. — v. — f. alta** TRIEBN. et v. POELLN. – Cape: Grahamstown D. – **St.** up to 2 cm long, with up to 20 lateral Ros.; **L.** with a broad, dark border, otherwise as f. **planifolia**.

**H. — v. — f. olivacea** TRIEBN. et v. POELLN. – Cape: Quagga W. – **Ros.** 10 cm ⌀, compressed; **L.** olive-green, underside acute-carinate, otherwise as f. **planifolia**.

**H. — v. — f. planifolia** (Pl. 91/XIII) (*H. p.* v. *typica* TRIEBN. et v. POELLN.). – Cape: Albany D. – **Ros.** acauline, freely offsetting below, 5–10 cm ⌀; **L.** ovate or ± ovate-oblong, short-tapering, upperside usually flat with 1–2 ± distinct longitudinal lines, underside convex, ± distinctly and obliquely carinate towards the tip, dull, light green, tip translucent, particularly towards the margins, 2.5–3.5 cm across, with a small point usually 3 mm long, soon dropping.

**H. — v. — f. robusta** TRIEBN. et v. POELLN. – Cape: near Port Elizabeth. – Very vigorous in growth; **L.** 3.5–6 cm long, 2–3 cm across, rather long-tapering, margins and keel smooth, light green, point 3 mm long.

**H. — v. poellnitziana** RES. – Origin unknown. – **Ros.** 9–14 cm ⌀; **L.** very broad, short-tapering or almost rounded, 6–6.5 cm long, 2–2.5 cm across, somewhat convex on the upperside towards the tip, with 2 shallow longitudinal grooves, underside often with 2 indistinct, oblique keels, these smooth or with small, green, widely-spaced T., tip-area with translucent spots and a minute point.

**H. — v. setulifera** v. POELLN. (Pl. 96/4). – Cape: East London. – **L.** 5–5.5 cm long, 2–2.5 cm across, margins and keel smooth, terminal bristle 5–7 mm long.

**H. — v. sublaevis** v. POELLN. – Cape: Albany D. – **L.** long-tapering, 4–4.5 cm long, 1–1.5 cm across, margins smooth or with dentate protuberances, terminal bristle simple.

**H. — v. transiens** v. POELLN. – Cape: Prince Albert D. – Possibly a hybrid: **H. planifolia × H. cymbiformis; L.** narrow-oblong, rather tapered, upperside concave, 2.5 cm long, 1–2 cm across, margins and keel smooth, tip minute.

**H. — × H. altilinea** HAW. – Nat. hybrid. – Cape: Grahamstown D. – **Ros.** acauline, c. 9 cm ⌀, offsetting freely from the base; **L.** numerous, ± ovate-oblong, tapering, 5 to 6 cm long, 16–20 mm across, upperside flat, with raised longitudinal lines above, underside with (1–)2(–3) keels, longitudinally striped, keels and margins ± smooth, or slightly rough with minute T.

**H. poellnitziana** UITEW. (§ IX). – Cape: Swellendam D. – **Ros.** acauline; **L.** somewhat curved, 7–10 cm long, 1.5–2 cm across, c. 7 mm thick, usually ovate at the base, narrowing towards the tip which is long and subulate, underside convex-carinate, both surfaces with numerous thick, parchment-like, white, glossy tubercles 1–1.5 mm long, in ± distinct R., keel and margins similarly tuberculate, terminal bristle 5 mm long.

**H. pseudogranulata** v. POELLN. (§ XVI). – Cape: Oudtshoorn D. – **Ros.** acauline; **L.** few, ovate or ovate-oblong, tapering, round to truncate, rather dark green, c. 2 cm long, 1.5 cm across, upperside concave, the tip swollen, underside rounded, obliquely keeled, both surfaces with concolorous tubercles, terminal bristle small.

**H. pubescens** M. B. BAYER (§ XIV). – Cape: Worcester D. – **Ros.** acauline, 2.5–4 cm ⌀; **roots** thick, firm, white-fleshed; **St.** thick, rarely elongated and offsetting; **L.** up to 50, erect, spreading, incurved at the tips, firm, up to 30 mm long, 8 mm across, 4 mm thick, ovate-lanceolate, acuminate, with a bristle up to 2 mm long, very dark green, hoary-looking, with ± distinct longitudinal lines, upperside flat to concave at the base, convex-turgid towards the middle and tip, slightly to moderately muricate, underside convex and spiny, often with a second keel, margins subacute, spines white, up to 0.5 mm long; **Infl.** simple, with 8–10 Fl.

**H. pygmaea** v. POELLN. (§ XIV). – Cape: Oudtshoorn, Beaufort West D. – **Ros.** usually simple, c. 5 cm ⌀; **L.** 9–10, long-oval, 2.5 to 3 cm long, upperside flat to concave, the upper third abruptly truncate at 30–40°, the tip-area triangular-roundish, 12 mm across and long, ± translucent, with 4–5 light, longitudinal stripes, underside very rounded and carinate, glassy dark green, slightly granular.

**H. radula** (JACQ.) HAW. (Pl. 97/4) (§ IX) (*Aloe r.* JACQ., *Apicra r.* WILLD., *Aloe r.* v. *media* SALM, *H. r.* v. *pluriperlata* HAW.). – Cape: Humansdorp D. – **Ros.** acauline, mat-forming; **L.** 6–8 cm long, 2 cm across below, abruptly narrowing and long-tapering above, underside often carinate, both surfaces green, rough with minute white pearly tubercles.

**H. ramosa** G. G. SMITH (§ XII). – Cape: Peddie D. – Short-stemmed; **St.** 5 cm tall, 7 mm thick, with 4 cm long, rooting, prostrate Br., forming dense clumps; **Ros.** 5 cm ⌀; **L.** ovate to round, short-tapering, 3.5 cm long, 22 mm across, 6 mm thick near the tip, upperside flat to concave, with a transparent M.line and other reticulate lines and spots near the tip, underside rounded, with c. 20 reticulate lines and spots, light green, often with 1–2 keels, margins greenish-white, terminal bristle 1 mm long, reddish.

H. **recurva** HAW. (§ XVIII) (*Aloe r.* HAW., *A. anomala* HAW., *Apicra r.* WILLD., *A. anomala* WILLD.). – Cape: Lit. Karroo. – **Ros.** acauline, 5–7 cm ⌀, offsetting basally, matforming; **L.** 12–15, in many spirals, triangular-ovate, tapering, 3–4 cm long, 2–2.5 cm across below, rigid, fleshy, olive-green to dirty yellowish, upperside flat, with 3–5 pale longitudinal lines, underside convex to obliquely keeled, margins and keel horny-tuberculate.

H. **reinwardtii** (S.D.) HAW. (§ II) (*Aloe r.* S.D.). – Cape. – Type s. v. **reinwardtii**. **Ros.** stemlike, elongated, up to 20 cm or more tall, with **L.** in a dense spiral; **L.** erect to recurved, ovate-lanceolate, pointed, usually dark green, with ± numerous tubercles, arranged ± in R.

H. — v. **adelaidensis** v. POELLN. – Cape: Adelaide. – **L.** up to 3 cm long, 12 mm across, tubercles 0.5 mm ⌀, often merging on the underside into short longitudinal R.

H. — v. **archibaldiae** v. POELLN. – Cape: Peddie D. – **L.** 4 cm long, 12 mm across, upperside with white, large tubercles, underside with white tubercles 1 mm thick, mostly solitary, in 10–12 longitudinal R. and ± in transverse R.

H. — v. **bellula** G. G. SMITH. – Cape: Albany D. – **St.** 5 cm tall; **L.** 15 mm long, 7 mm across, 3.5 mm thick, ovate, pointed, upperside with 3 indistinct lines, underside convex, with whitish tubercles 0.75 mm long in 8–10 transverse R. and in c. 10 indistinct longitudinal R., keel ± acute towards the tip.

H. — v. **brevicula** G. G. SMITH. – Cape: Albany D. – **St.** 8 cm tall; **L.** 26 mm long, 8 mm across, 4.5 mm thick, lanceolate, tapering, upperside with small, raised, white tubercles in 3–4 longitudinal R., the tubercles greenish below, underside with roundish tubercles 1 mm thick in 12–13 transverse R., and in 9 indistinct longitudinal R.

H. — v. **chalumnensis** G. G. SMITH. – Cape: East London D. – **St.** simple, c. 13 cm tall; **L.** 4–5 cm long, 13–16 mm across, 4.5–6.5 mm thick, lanceolate, tapering, upperside with indistinct, raised, longitudinal lines, in the upper half with roundish, whitish tubercles 0.75 mm long, underside rounded to rather blunt-keeled, with 12–14 ± wavy transverse R. of pure white, raised, truncate tubercles, mostly merging into bands.

H. — v. **chalwinii** (MARL. et BGR.) RES. (*H. chalwinii* MARL. et BGR.). – Cape: Graaff Reinet D. – **St.** rather long, erect, later creeping; **L.** ovate-triangular, tapering, 2.5 cm long, 2 cm across, upperside smooth, underside very convex-carinate, with 12–14 regular R. of white, ± confluent tubercles.

H. — v. **committeesensis** G. G. SMITH. – Cape: Albany D. – **St.** 17 cm long; **L.** 44 mm long, 14 mm across, 6 mm thick, lanceolate, tapering, upperside smooth or with a few whitish tubercles along a prominent M.line, light green, underside rounded, with 12 R. of raised, white tubercles in transverse bands, and the same number of indistinct longitudinal R., keel oblique.

H. — v. **conspicua** v. POELLN. – Cape: Port Elizabeth D. – **St.** over 20 cm long; **L.** 5–7 cm long, 16–18 mm across, upperside smooth at the base, otherwise with whitish tubercles in 6–8 longitudinal R., white tubercles on the underside in 10–12 longitudinal R., often also in confluent transverse R.

H. — v. **diminuta** G. G. SMITH. – Cape: Albany D. – **St.** erect, c. 5.5 cm tall; **L.** 2 cm long, 6 mm across, 3 mm thick, lanceolate, tapering, M.line on the upperside with small, white tubercles, often with 2 lateral R., underside rounded, bluntly carinate, with prominent white tubercles in 12 transverse R.

H. — v. **fallax** v. POELLN. (*H. f.* v. POELLN.). – Cape: Grahamstown, East London D. – **L.** up to 5 cm long, 12 mm across, upperside smooth, underside with larger and smaller tubercles in 5–12 longitudinal R. and often in confluent transverse R. Resembles v. **reinwardtii** and is transitional to this variety.

H. — v. **grandicula** G. G. SMITH. – Cape: Peddie D. – **St.** 13 cm long; **L.** lanceolate, tapering, 38 mm long, 11 mm across, 4.5 mm thick, upperside smooth, light green, underside rounded-carinate, with raised, white tubercles c. 1 mm thick, in ± wavy transverse R. and indistinct longitudinal R., tubercles oblong along margins and keel.

H. — v. **huntsdriftensis** G. G. SMITH. – Cape: Albany D. – **St.** 17 cm long; **L.** dark green, c. 38 mm long, 14 mm across, upperside flat to rounded, with prominent longitudinal lines and 2 lateral lines, light to dark green, underside rounded, obliquely carinate above, with raised, white, mostly small tubercles in 11 transverse R., and sometimes transversely confluent.

H. — v. **kaffirdriftensis** G. G. SMITH. – Cape: Peddie D. – **St.** 12 cm long; **L.** lanceolate, pointed and tapering, upperside flat-convex, with 1–2 prominent lines, smooth or with some white tubercles, green, underside rounded-carinate, with pure white, round tubercles in 5–8 indistinct longitudinal R., tubercles smaller towards the base.

H. — v. **major** BAK. (*H. r.* v. *pulchra* v. POELLN.). – Cape: Bathurst D. – Like v. **reinwardtii** but generally larger; **L.** with tubercles on the underside c. 1 mm long in a few transverse R.

H. — v. **minor** BAK. – **L.** 2–2.5 cm long.

H. — v. **olivacea** G. G. SMITH. – **L.** olive-green.

H. — v. **peddiensis** G. G. SMITH. – Cape: Peddie D. – **St.** 13 cm long, erect; **L.** lanceolate, tapering, 28 mm long, 12.5 mm across, 4.5 mm thick, green, becoming brownish-green, upperside flat to convex, with very small, light tubercles, mostly in 3 longitudinal R. in the upper half, underside rounded, obliquely carinate, with larger tubercles in 10–12 ± wavy transverse R. and 10–12 indistinct longitudinal R.

H. — v. **reinwardtii** (Pl. 97/1) (*H. r.* v. *typica* v. POELLN.). – Cape: Grahamstown, East London, Cradock, Port Alfred, Port Elizabeth, Peddie D. – **L.** 2–5 cm long, 12–15 mm across below, upperside midway with greenish or whitish small tubercles in 1–3 indistinct R., underside with numerous tubercles, solitary or in 8–15 longitudinal R. or in regular transverse bands.

**H. — v. riebeekensis** G. G. SMITH. – Cape: Albany D. – **St.** 8 cm long; **L.** 2 cm long, up to 10 mm across, 6 mm thick, ovate-lanceolate, tapering, upperside light green, smooth, underside with small, whitish tubercles in 10–11 longitudinal R. and 9–12 transverse R., keel ± acute.

**H. — v. tenuis** G. G. SMITH. – Cape: Alexander D. – **St.** simple, prostrate, 20 cm long; **L.** 27 mm long, 1–2 cm across, lanceolate, pointed, tapering, upperside concave, reddish-green, with a few, solitary, concolorous tubercles in 5–7 longitudinal R., underside rounded-carinate, with small white tubercles in c. 12 transverse R. and c. 14 indistinct longitudinal R.

**H. — v. triebneri** RES. – Origin unknown. – **L.** 3–4 cm long, 8–10 mm across, 2–3 mm thick, underside with numerous confluent tubercles in transverse bands, upperside smooth.

**H. — v. valida** G. G. SMITH. – Cape: Peddie D. – **L.** lanceolate, tapering, with a tiny point, upperside smooth or with several small, concolorous tubercles along the M.line, light green, underside rounded-carinate, with large white tubercles in 9–10 transverse bands and 10–11 indistinct longitudinal lines.

**H. — v. zebrina** G. G. SMITH. – Cape: Peddie D. – **St.** up to 14 cm tall; **L.** 5 cm long, 12 mm across, 6 mm thick, lanceolate, tapering, upperside light to dark green, rounded, with a raised M.line and 2 lateral lines, with a few whitish to concolorous tubercles above, underside rounded, with silvery-white tubercles merging into 12–15 transverse R., tubercles towards the tip becoming smaller and solitary.

**H. resendeana** v. POELLN. (§ II). – Cape: Origin unknown. – **St.** extended, erect to prostrate; **L.** arranged spirally, 1.5–2 cm long, 10–13 mm across below, ovate or ovate-triangular, tapering, dark green, upperside smooth, underside convex, often with 2 tuberculate keels, tubercles solitary, mostly lying transversely, L.tip with a horny border.

**H. reticulata** HAW. v. **reticulata** (Pl. 97/5) (§ XI) (*Aloe pumilio* JACQ., *A. herbacea* DC., *A. arachnoidea* v. *r.* KER., *Apicra p.*, *A. r.* WILLD.). – E.Cape. – Resembles **H. altilinea**; **L.** very thick, especially above, underside carinate, with a terminal bristle c. 1 cm long, margins cartilaginous-dentate, fresh green, the upper third ± pellucid, with a few thin, green longitudinal lines.

**H. — v. acuminata** v. POELLN. – Cape: Robertson D. – **L.** very long-tapering.

**H. retusa** (L.) HAW. v. **retusa** (Pl. 91/XIV) (§ XIV) (*Aloe r.* L., *Catevala r.* MED., *Apicra r.* WILLD.). – Cape: Uniondale, Willowmore, Riversdale D. – **Ros.** acauline, 7–9 cm ⌀, offsetting basally, mat-forming; **L.** 15–20, 3–5 cm long, ovate-triangular, recurved horizontally in the upper half, tip-area ± triangular-tapering, rounded, ± translucent, with 5–8 pale lines, underside very rounded, margins with isolated T., surface pallid green.

**H. — v. densiflora** G. G. SMITH. – Cape: Riversdale D. – Related to v. **solitaria**; **L.** 4.5 cm long, 23 mm across, 14 mm thick; **Infl.** dense.

**H. — v. multilineata** G. G. SMITH. – Cape: Riversdale D. – **L.** larger than in the type; the tip-area with 15–22 lines, terminal bristle 5 mm long.

**H. — v. mutica** (HAW.) BAK. (*H. m.* HAW., *Aloe m.* ROEM. et SCHULT., *A. retusa* v. *m.* SALM). – Cape: Worcester, Willowmore D. – **L.** margins without any T.

**H. — v. solitaria** G. G. SMITH. – Cape: Riversdale D. – Very flat-growing; **Ros.** 8 cm ⌀, with few basal offsets; **L.** as in the type, tip-area rough with minute tubercles, lines here more numerous, margins and keel with numerous small T.

**H. revendettii** UITEW. (§ II) (*H. r.* HORT.). – S.Afr.: origin unknown. – **St.** 10–18 cm long, offsetting basally; **L.** in 5 R., ovate-triangular, tapering, 4.5–5.5 cm long, ± 2 cm across, thick, rigid, rather glossy, tip sharp and ± recurved, upperside slightly convex, grooved below, ± pale green, with indistinct longitudinal lines, underside rounded-carinate, ± green, with ± colourless tubercles ± merging into horizontal lines, margins with small tuberculate T.

**H. rigida** (LAM.) HAW. v. **rigida** (Pl. 91/XV) (§ XV) (*Aloe cylindrica* v. *r.* LAM., *A. r.* DC: *A. expansa* v. *paulo major* HAW., *Apicra expansa* WILLD., *H. e.* v. *major* HAW.). – S.Afr.: origin unknown. – **St.** 7–12 cm long, erect, leafy, offsetting basally, mat-forming; **L.** ± in 5 R., ovate-lanceolate, tapering, thick, rigid, vivid green, often brown, upperside flat, smooth, underside convex with a triangular keel or with several keels, margins and keels horny-crenate.

**H. — v. expansa** (HAW.) BAK. (*Aloe e.* HAW., *Apicra patula* WILLD., *H. e.* HAW., *Aloe r.* SALM (?), *A. r.* v. *e.* SALM, *A. r.* v. *minor* SALM). – Rather smaller than the type; **L.** shorter, c. 5 cm long, 15 mm across, rough and glossy.

**H. rossouwii** v. POELLN. (§ XIV). – Cape: Bredasdorp D. – **Ros.** 5–6 cm ⌀, acauline; **L.** many, oblong-lanceolate, tapering, 5–6 cm long, 8–10 mm across, light green, dull, tip-area recurved and truncate above midway at an angle of 20°, triangular to long-tapering, 2 cm long, 7–9 mm across, semi-translucent, smooth, rarely tuberculate, with 3 longitudinal lines midway, underside carinate, margins and keel dentate, terminal bristle 5–7 mm long, light.

**H. rubrobrunnea** v. POELLN. (§ II). – S.Afr.: origin unknown. – **St.** extended, freely offsetting from the base; **L.** arranged spirally, ovate-lanceolate, mucronate, c. 3 cm long, 8–10 mm across below, rather dull, reddish-brown, underside convex, with 1–2 blunt keels, both surfaces with little prominent, concolorous tubercles, often merging into ± irregular transverse R., margins and keel with lighter tubercles like small T.

**H. rugosa** (SALM) BAK. v. **rugosa** (§ IX) (*Aloe r.* SALM ex ROEM. et SCHULT., *A. radula* v. *minor* SALM, *H. radula* v. *asperior* HAW.). – Cape: Riversdale D. – Ros. acauline, with few offsets, often branching dichotomously; L. numerous, triangular-lanceolate, long-tapering, with a triangular tip, 7–10 cm long, 20–25 mm across, upperside flat-convex, dark green, with few tubercles interspersed with larger, ± translucent ones.

**H. — v. perviridis** (SALM) BGR. (*Aloe r.* v. *p.* SALM). – L. lighter green.

**H. ryderana** v. POELLN. (Pl. 97/3) (§ XIV). – S.Afr.: origin unknown. – Ros. acauline, 4–6 cm ⌀; L. many, ovate-triangular, tapering, 3–4 cm long, c. 1.5 cm across, dark brownish- or bluish-green, upperside often with a distinct M.nerve, the tip recurved and truncate at 40°, tip-area triangular or triangular-tapering, 10–13 mm long, 15 mm across, flat or slightly swollen, ± translucent, almost smooth, with a few indistinct tubercles in longitudinal lines and 8–10 green, ± confluent, reticulate, longitudinal lines, underside convex-carinate, spotted, margins and keel with small, light T., terminal bristle curved, translucent, 5–6 mm long.

**H. sampaiana** RES. v. **sampaiana** (§ II) (*H. coarctata* v. *s.* RES.). – S.Afr.: origin unknown. – Resembles **H. coarctata**; L. 4–8 cm long, 2–2.5 cm across, upperside concave, tubercles on the underside usually in a zigzag, often in indistinct transverse R.

**H. — v. broterana** (RES.) RES. et PINTO-LOPES (*H. b.* RES.). – L. more numerous, more pointed, tubercles on the underside closer, larger, almost white.

**H. scabra** (ROEM. et SCHULT.) HAW. (§ XVI) (*Aloe s.* ROEM. et SCHULT.). – S.Afr.: origin unknown. – Ros. acauline; L. few, triangular-ovate, 5 cm long, 2–2.5 cm across below, dark green, with minute transverse wrinkles and tubercles.

**H. schuldtiana** v. POELLN. v. **schuldtiana** (§ XIV). – Cape: Ladismith D. – Ros. acauline, 3.5–4 cm ⌀; L. oblong-lanceolate, 2.5–6 cm long, 6–11 mm across, stiff, brown- to reddish-green, lower part of the upperside not translucent, smooth or with a few, small, blunt, light tubercles, the upper part bluntly recurved, tip-area 10–14 mm long, tuberculate and small-dentate, with 3–5 green longitudinal lines, underside rounded-carinate, with a number of darker lines and small blunt tubercles, margins and keel whitish-dentate.

**H. — v. erecta** TRIEBN. et v. POELLN. – Cape: Robertson D. – L. erect, up to 3 cm long, 7–9 cm across, dark grey-green, tip-area with only a few small tubercles.

**H. — v. maculata** v. POELLN. – Cape: Worcester, Swellendam, Bredasdorp, Caledon D. – L. light green, with translucent spots.

**H. — v. major** G. G. SMITH. – Cape: Riversdale D. – Larger than the type; T. and tubercles more numerous and larger, terminal bristle long.

**H. — v. minor** TRIEBN. et v. POELLN. – L. smaller.

**H. — v. robertsonensis** v. POELLN. – Cape: Robertson D. – L. more numerous, grey-green, up to 6 cm long, the tip little translucent, tip-area abruptly tapering, with 3 lines and blunt, dentate tubercles, T. bent downwards on margins and keel.

**H. — v. simplicior** v. POELLN. – Cape: Swellendam D. – L. as the type, tip-area with one or often 2 very short, almost imperceptible lines.

**H. — v. sublaevis** v. POELLN. – L. little truncate, up to 3 cm long, 6–8 mm across, tip-area smooth or with a few small tubercles in longitudinal lines.

**H. — v. subtuberculata** v. POELLN. – L. 2.5 cm long, tip-area 12–14 mm long, 6–8 mm across, with numerous small tubercles and 3 lines.

**H. — v. whitesloaneana** v. POELLN. (*H. w.* v. POELLN.). – Cape: Robertson D. – L. ovate-lanceolate, 3.5–4 cm long, 9–11 mm across, tip-area truncate at 45°, the numerous tubercles with small bristles, with green stripes c. 3 mm long, margins and keel dentate.

**H. semiglabrata** HAW. (§ IX) (*Aloe s.* ROEM. et SCHULT.). – Cape: Riversdale D. – Ros. acauline, c. 13 cm ⌀, often branching dichotomously; L. 30–40, lanceolate-triangular, tapering, 7–8 cm long, 2 cm across, upperside flat, ± tuberculate, green, underside ± convex, triangular-carinate, with a horny tip, tubercles pearl-like in indistinct transverse lines; Infl. 5-branched, Per. segments spreading stellately.

**H. sessiliflora** BAK. (§ XII) – Origin unknown. – Ros. acauline, 6–8 cm ⌀; L. c. 20, ovate, 2.5–3 cm long, 15–18 mm across, 8 mm thick, much recurved, fresh green, colourless in the upper third, with 7 green longitudinal lines, underside shortly and acutely carinate, with light spots, margins and keel dentate, terminal bristle 4 mm long; Fl. sessile.

**H. setata** HAW. (§ I) (*Aloe setosa* ROEM. et SCHULT.). – Cape: S.W. reg. – Ros. acauline, 5–12 cm ⌀; L. 30–40, oblong-lanceolate, 2 to 6 cm long, 3–12 mm across, terminating in a long, translucent bristle, underside with 1–2 keels, deep green, margins and keels with 2–5 mm long, snow-white bristles.

**H. — v. bijliana** v. POELLN. (Pl. 97/2) (*H. b.* v. POELLN., *H. fergusoniae* v. POELLN.). – Cape: Lit. Namaqualand. – L. lanceolate, c. 35 mm long, 3–10 mm across, grey-green to dark green, with ± translucent spots, tip brownish, with a 7–8 mm long, white, ramose, terminal bristle, margins and keels with 2 mm long bristles.

**H. — v. gigas** v. POELLN. (*H. g.* v. POELLN.). – Cape: Lit. Karroo. – Ros. 10–12 cm ⌀; L. many, soft, fleshy, 6 cm long, 10 mm across, long-tapering, terminal bristle 1 cm long, thin, margins and keels with 5 mm long, opalescent, crescent-shaped bristles.

**H. — v. joubertii** v. POELLN. (*H. bijliana* v. *j.* v. POELLN.). – Cape: Ladismith D. – L. upperside with some bristles towards the tip, bristles on the underside fairly long, those on keel and margins 4–5 mm long, terminal bristle 10–15 mm long.

**H. — v. major** Haw. (Pl. 90/I) (*Aloe setosa* v. *major* Roem. et Schult., *H. arachnoides* v. *minor* Haw.). – Larger variety.

**H. — v. media** Haw. (*Aloe setosa* v. *m.* Roem. et Schult.). – Intermediate in size between the type and v. major.

**H. — v. nigricans** Haw. (*Aloe setosa* v. *n.* Roem. et Schult.). – Darker than the type.

**H. — v. setata** (*Aloe setosa* Roem. et Schult.). – **Ros.** c. 5 cm $\varnothing$; **L.** 2–2.5 cm long, 10–12 mm across, bristles 2–3 mm long.

**H. — v. xiphiophylla** (Bak.) v. Poelln. (*H. x.* Bak., *H. longiaristata* v. Poelln.). – Cape: Uitenhage D. – **Ros.** 5 cm $\varnothing$; **L.** narrowly oblong-lanceolate, very tapered, 4–5 cm long, 6 mm across, concolorous, rarely with a few, inconspicuous, light spots, both surfaces with a few longitudinal lines, margins and keel with $\pm$ distant T., terminal bristle 7–10 mm long, dentate.

**H. skinneri** (Bgr.) Res. (Mem. Brot. 1953, Pl. p. 76) is certainly not *Apicra skinneri* Bgr. = × **Astroworthia bicarinata** (Haw.) Rowl. nm. **skinneri** (Bgr.) Rowl.

**H. smithii** v. Poelln. (§ IX). – Cape: Oudtshoorn D. – **Ros.** acauline or short-stemmed, c. 10 cm $\varnothing$; **L.** very numerous, 6–8 cm long, 3 to 4 cm across, ovate-triangular, tapering, mucronate, upperside with angles curving inwards towards the tip, $\pm$ concave, underside convex, carinate, with $\pm$ confluent, dark green tubercles 1 mm long.

**H. sordida** Haw. (Pl. 91/XVI) (§ XVI). – Cape: Steytlerville D. – **Ros.** acauline; **L.** few, c. 10 cm long, 3 cm across, triangular-tapering, fleshy, rigid, dirty green, minutely rough tuberculate, upperside flat or $\pm$ concave, underside somewhat convex-carinate.

**H. — v. agavoides** (Zant. et v. Poelln.). G. G. Smith (*H. a.* Zant. et v. Poelln.). – Origin unknown. – **Ros.** with very few basal offsets; **L.** stiff, 9–10 cm long, 10–15 mm across, narrowly triangular, tapering towards the tip, acuminate, dull, dirty grey-green, upperside very concave, $\pm$ grooved, rough, underside very convex, acute-carinate, with somewhat darker, solitary tubercles, those towards the tip larger, often merging into wavy, transverse bands, margins tuberculate.

**H. springbokvlaktensis** C. L. Scott (§ XIV). – Cape: Uitenhage D. – **Ros.** acauline, c. 7 cm $\varnothing$; **L.** c. 37 mm long, up to 18 mm across, tip-area c. 17 mm long, 18 mm across, convex, with minute, concolorous tubercles, with 4 long and 5 short, branching lines, underside minutely tuberculate, margins rounded, with 0.5 mm long, translucent T., keel rounded.

**H. starkiana** v. Poelln. (§ XVI) (*H. taylori* Barker MS.) – Cape: Oudtshoorn D. – **Ros.** acauline, offsetting basally, mat-forming, 12 to 15 cm $\varnothing$; **L.** numerous, ovate-lanceolate or triangular-lanceolate, tapering, often flat and laterally curved, c. 7 cm long, 2 cm across, upperside slightly convex, underside very convex-carinate, green to brownish-green, $\pm$ glossy, brown-tipped.

**H. stiemei** v. Poelln. (§ I). – Cape: Uitenhage D. – **Ros.** acauline; **L.** many, 3.5–4 cm long, 7–9 mm across, narrow-oblong, tapering, dull, dark grey-green, slightly translucent towards the tip which has 3 longitudinal lines on the upperside, underside with numerous longitudinal lines, carinate above midway, margins and keel with 1 mm long T., tip smooth.

**H. subattenuata** (Salm) Bak. (§ IX) (*Aloe s.* Salm, *A. semimargaritifera* v. *minor* Salm, *H. s.* v. *m.* Haw., *H. radula* v. *magniperlata* Haw., *Aloe r.* v. *margaritacea* Salm ex Haw.). – Origin unknown. – **Ros.** acauline, offsetting basally, c. 15 cm $\varnothing$; **L.** lanceolate-triangular, linear-tapering, 5–7 cm long, 15–20 mm across, deep green, often reddish, upperside slightly convex, smooth, with small green tubercles along the M.line, underside very rounded to acute-carinate, tubercles numerous, white, solitary, in distinct R., L. often also with white stripes and border.

**H. subfasciata** (Salm) Bak. v. **subfasciata** (Pl. 97/6) (§ IX) (*Aloe s.* Salm ex Roem. et Schult., *A. fasciata* v. *major* Salm, *H. f.* v. *m.* Haw.). – Cape: Uitenhage D. – **Ros.** offsetting little from the base, often branching dichotomously; **L.** numerous, lanceolate-triangular, tapering, dirty green, 12–13 cm long, 25–30 mm across, upperside convex, smooth, glossy, underside convex-carinate, with a few horny, translucent tubercles in transverse bands, margins tuberculate.

**H. — v. kingiana** v. Poelln. (*H. k.* v. Poelln.). – **L.** ovate-tapering, 4.5–5.5 cm long, 3 cm across, both surfaces tuberculate, tubercles on the upperside numerous, 1 mm long, in curves or transverse lines.

**H. sublimpida** v. Poelln. (§ XIV). – Cape: Swellendam D. – **Ros.** acauline, 3.5–5 cm $\varnothing$; **L.** few, $\pm$ ovate-lanceolate, 2.5–5 cm long, 8 mm across, intense green, glossy, upperside $\pm$ swollen below, often with a conspicuous M.line, the upper part recurved, the tip-area triangular-ovate, 11–12 mm long, 5–7 mm across, slightly grooved, with 2–3 short, green lines and small light tubercles, underside very rounded, distinctly 2–3-keeled, tuberculate.

**H. submaculata** v. Poelln. (§ I). – Cape: Worcester, Swellendam, Caledon, Bredasdorp D. – **Ros.** acauline. c. 6 cm $\varnothing$; **L.** numerous, ovate-lanceolate, tapering, dark grey-green to reddish, c. 4 cm long, 10–13 mm across, upperside with an indistinct M.line or solitary, indistinct tubercles, underside very rounded, often with 2 keels, with numerous longitudinal lines and light spots, margins and keels dentate, terminal bristle 1–2 mm long.

**H. subregularis** Bak. (Pl. 91/XVII) (§ XVII). – S.E.Cape. – **Ros.** 6–7 cm $\varnothing$; **L.** 20–30, densely crowded, fleshy, broad-lanceolate to $\pm$ rhombic, acuminate, 3–3.5 cm long, 1.5 cm across midway, underside carinate below the tip, pale blue-green, upperside with 5–6 R. of whitish, tuberculate spots, margins minutely dentate.

**H. subulata** (Salm) Bak. (§ IX) (*Aloe s.* Salm, *A. radula* v. *major* Salm, *H. r.* v. *laevior*

SALM ex HAW.). – Cape: Peddie D. – **Ros.** c. 15 cm $\varnothing$, offsetting basally, often dichotomously divided; **L.** triangular-lanceolate, 10 to 12 cm long, 25 mm across, tapering, with a long, triangular tip, green, upperside flat, with a prominent, carinate M.line, rough with small, concolorous tubercles, underside convex, with small, pearly tubercles $\pm$ in transverse R., margins tuberculate-rough.

**H. tenera** v. POELLN. (§ I). – Cape. – **Ros.** acauline, closing up in the resting period, 1–3 cm $\varnothing$; **L.** many, narrow-lanceolate or narrowly ovate-lanceolate, tapering, 1.5–3.5 cm long, 4–10 mm across, dull, grey-green, both surfaces light or light-spotted towards the tip, underside with 1–2 keels towards the tip, margins and keels with $\pm$ confluent, longitudinal lines, the underside similar, terminal bristle 2–10 mm long, dentate.

**H. — v. confusa** (v. POELLN.) UITEW. (*H. c.* v. POELLN., *H. minima* v. *c.* v. POELLN.). – ? Willowmore D. – **L.** very tapered, reddish-green, up to 3.5 cm long, 6–10 mm across, both surfaces with a light spot and several $\pm$ connected longitudinal lines towards the tip, margins and keel with numerous T., terminal bristle 5–10 mm long.

**H. — v. major** (v. POELLN.) UITEW. (*H. minima* v. *major* v. POELLN.). – Oudtshoorn D. – **L.** long-tapering, 3–3.5 cm long, 6–8 mm across, margins and keels with 1 mm long, triangular T., terminal bristle dentate, 4 to 6 mm long.

**H. — v. tenera** (*H. minima* BAK.). – Albany D. – **L.** 1.5–2 cm long, 4–7 mm across, terminal bristle 2–3 mm long.

**H. tessellata** (SALM) HAW. (§ XVIII) (*Aloe t.* SALM). – Cape; S.W.Afr. – The many varieties $\pm$ transitional to the type, v. **tessellata**. – **Ros.** acauline, offsetting basally, mat-forming, less often solitary; **L.** up to 15, spiralled, usually broadly ovate-triangular, mucronate, recurved-spreading, fleshy, rigid, dark green to reddish-brownish, smooth, flat, usually translucent, often with reticulate, branching, longitudinal lines, underside very rounded, usually rough with tiny tubercles, margins usually with recurved white T.

**H. — v. coriacea** RES. et v. POELLN. – Origin unknown. – **L.** leathery, 3–7 cm long, 2.5–3 cm across, with 2–5 branching, longitudinal lines, underside with 2 keels, tuberculate, marginal T. 1 mm long.

**H. — v. — f. brevior** RES. et v. POELLN. – **L.** ovate-triangular, 3–3.5 cm long, 2–2.5 cm across.

**H. — v. — f. longior** RES. et v. POELLN. – **L.** 5 cm long.

**H. — v. elongata** v. WOERDEN. – S.W.Afr. – Offsetting, St. usually 2–3, 10–12 cm long; **L.** in 5 spiral R., lanceolate-triangular, upperside boat-shaped, concave, semi-translucent, with 2 interconnected, longitudinal lines, underside grey-green to reddish-brown, rough-tuberculate, margins and keel with small, white tubercles.

**H. — v. engleri** (DTR.) v. POELLN. (*H. e.* DTR.). – S.W.Afr.: Gr. Namaquland. – **Ros.** 3–4.5 cm $\varnothing$; **L.** up to 9, 2–3 cm long, 2 cm across, 8–9 mm thick, underside with small tubercles in continuous transverse R., upperside smooth, dark green, with 2–3 greenish-white, longitudinal lines and transverse, greenish-white veins, margins with white, horny, blunt T.

**H. — v. inflexa** BAK. (*H. pseudotessellata* v. POELLN.). – Cape: Prieska, Beaufort West D. – **L.** upperside concave, with 5–7 longitudinal lines.

**H. — v. luisieri** RES. et v. POELLN. – **L.** ovate-lanceolate, 3 cm long, 1.5 cm across, upperside grooved, with 3–5 interconnected, longitudinal lines, underside with irregularly arranged tubercles, margins minutely dentate.

**H. — v. minutissima** (v. POELLN.) VIVEIROS (*H. m.* v. POELLN.). – Cape: Cradock D. – **Ros.** c. 1.5 cm $\varnothing$, mat-forming; **L.** few, ovate-triangular, usually tapering, dark grey-green or $\pm$ blackish-green, often reddish-green, 10 to 13 mm long, underside very rounded, with $\pm$ numerous tubercles towards the tip, margins and keel with small colourless T., the tip-area semi-translucent, rather rough.

**H. — v. obesa** RES. et v. POELLN. – Origin unknown. – **Ros.** 10 cm $\varnothing$; **L.** up to 5 cm long, 2–2.5 cm across, ovate, long-tapering, upperside with 4–8 reticulate, interconnected longitudinal lines, underside with a few small tubercles, margins with translucent small T.

**H. — v. palhinhae** RES. et v. POELLN. – In Bot. Gard., Lisbon. – **L.** 5 cm long, 2 cm across, upperside with 3 reticulate, interconnected, longitudinal lines, underside with glossy, irregularly arranged tubercles.

**H. — v. parva** (HAW.) BAK. (*H. p.* HAW., *Aloe p.* ROEM. et SCHULT.). – Cape: Aliwal North D. – In general smaller than v. **tessellata**.

**H. — v. simplex** RES. et v. POELLN. – In Bot. Gard., Lisbon. – **Ros.** without basal offsets; **L.** 3–4 cm long, 5–10 mm across, somewhat swollen below, with 7–9 translucent, longitudinal lines, margins and tip somewhat tuberculate on the underside.

**H. — v. stephaneana** RES. et v. POELLN. – Origin unknown. – **Ros.** c. 6 cm $\varnothing$, offsetting basally, mat-forming; **L.** lanceolate, c. 4 cm long, 1.5 cm across, upperside flat, $\pm$ glossy, dark green, semi-translucent, with 3 reticulately branching longitudinal lines, underside with several tubercles, margins irregularly dentate.

**H. — v. tessellata** (Pl. 91/XVIII) (*H. t.* v. *typica* HAW., *H. t.* v. *haworthii* RES.). – Cape: Victoria West, Carnarvon to Graaff Reinet D.; S.W.Afr.: Gr. Namaquland. – **Ros.** 6–10 cm $\varnothing$; **L.** 10–15, 3–5 cm long, 2–2.5 cm across, broadly ovate-triangular, short-tapering, mucronate, dark green, underside often reddish, rough with small white tubercles, upperside with 5–7 longitudinal lines, usually reticulate-branching, margins with small, white, recurved T., smooth towards the tip.

**H. — v. tuberculata** v. POELLN. – Cape: ? Oudtshoorn D. – **L.** upperside with 4–7 longitudinal lines, tessellate-reticulate, underside with large tubercles in transverse bands, margins with large and small concolorous tubercles.

**H. — v. velutina** RES. et v. POELLN. – Origin unknown. – **L.** 3.5 cm long, 1.5 cm across, dark green, velvety, with 3–8 longitudinal lines and numerous tubercles in transverse R., margins and keel translucent-dentate.

**H. tisleyi** BAK. (§ IX). – Cape: ? Oudtshoorn D. – **Ros.** 7–8 cm $\varnothing$; **L.** 30–40, lanceolate-triangular, 2.5–3 cm long, 15 mm across, 6 mm thick, upperside flat, underside convex and obliquely carinate, dark green, often reddish, ± rough, with tiny concolorous tubercles.

**H. tortuosa** HAW. v. **tortuosa** (Pl. 91/XIX) (§ XIX) (*Aloe t.* HAW., *A. rigida* KER., *Apicra t.* WILLD. p.part., *Aloe subtortuosa* SPRENG.). – Origin unknown. – **St.** up to 13 cm long, offsetting basally; **L.** in 3 spiral R., densely imbricate, erect to projecting, ovate-lanceolate, pointed, 2.5–4 cm long, upperside concave, underside convex, carinate towards the tip, green, rough, margins and keel tuberculate, rough.

**H. — v. curta** HAW. (*H. c.* HAW., *Aloe c.* HAW.). – Probably a cultivar. – **St.** scarcely 5 cm tall; **L.** blackish-green, 2.5 cm long, upperside minutely tuberculate.

**H. — v. major** (SALM) BGR. (*Aloe t.* v. *m.* SALM). – Larger, robuster variety.

**H. — v. pseudorigida** (SALM) BGR. (*Apicra p.* SALM, *Aloe rigida* JACQ., *Apicra r.* WILLD., *Aloe subrigida* ROEM. et SCHULT., *H. s.* BAK.). – **L.** thicker, less grooved.

**H. — v. tortella** (HAW.) BAK. (*H. t.* HAW.). – **St.** scarcely 7 cm tall.

**H. translucens** HAW. (Pl. 98/2) (§ I) (*Aloe t.* HAW., *Apicra t.* WILLD., *Aloe arachnoides* v. *t.* KER., *H. pellucens* HAW.). – Cape: Bredasdorp, Riversdale D. – **Ros.** acauline, offsetting freely from the base, 4–6 cm $\varnothing$; **L.** many, ± oblong-lanceolate, long-tapering, c. 3 cm long, 6 mm across, fresh light green, rather soft, with reticulate, connected, dark green lines, both surfaces towards the tip with a transparent, bristle-tipped spot, underside often with 2 keels, margins and keels with ± translucent T., terminal bristle slender. In the resting period the Ros. closes up bulb-like.

**H. — v. delicatula** (BGR.) v. POELLN. (*H. pellucens* v. *d.* BGR.). – T. on L.margins less numerous and rather longer than in the type.

**H. triebnerana** v. POELLN. (§ XIV). – Cape. – **Ros.** acauline, c. 6 cm $\varnothing$; **L.** thick, ± lanceolate, up to 8 cm long, 4–25 mm across, upperside ± flat, the tip recurved at c. 30°, tip-area somewhat translucent, triangular, with a green M.line and 2 shorter lines, underside obliquely carinate, with small light tubercles, often with white T., margins with white T., terminal bristle up to 10 mm long.

**H. — v. depauperata** v. POELLN. – Robertson D. – **L.** grey or dark green, margins with minute, light, bristly T., underside smooth or with a few small tubercles with minute, light bristles.

**H. — v. diversicolor** TRIEBN. et v. POELLN. – **L.** 4–6 cm long, 4–6 mm across, long-tapering, grey-green to yellow, pink or red, tip-area with dark longitudinal lines or with small tubercles in lines, each with a 1 mm long, light bristle.

**H. — v. lanceolata** TRIEBN. et v. POELLN. – Robertson D. – **L.** 3–5 cm long, 8–10 mm across, long-tapering, upperside dark grey to dark green, dull, underside with many concolorous tubercles, tip-area only little transparent, with a dark M.line, terminal bristle up to 6 mm long, soon dropping.

**H. — v. multituberculata** v. POELLN. – **L.** 6–7 cm long, 1.5–2.5 cm across, light green, tip-area glossy, translucent, with glossy, variegated tubercles, margins and keel with T. up to 2 mm long, terminal bristle 4–10 mm long, dentate, underside with numerous, minute tubercles towards the tip.

**H. — v. napierensis** TRIEBN. et v. POELLN. – **L.** 2.5–3 cm long, 18–20 mm across, grey to dark green, tip-area translucent, with 1–2 long and 2–6 shorter longitudinal lines, dull, underside with numerous distinct tubercles, margins and keel with 2 mm long T., terminal bristle up to 3 mm long, often missing.

**H. — v. nitida** v. POELLN. – **L.** dark green, glossy, tip-area with very few tubercles or smooth, with 1–2 shorter and 2 medium-long lines, underside smooth or with few tubercles.

**H. — v. pulchra** v. POELLN. – Robertson D. – **L.** dark grey-green, dull, 2.5–3 cm long, 6–8 mm across, tip-area semi-translucent, with very numerous tubercles and 3 green longitudinal lines, margins and keel with light T.

**H. — v. rubrodentata** TRIEBN. et v. POELLN. – **L.** 6–8 cm long, 12–15 mm across, green, upperside pale green and semi-translucent towards the tip, with 1–2 light, ± brown lines and with up to 8 light lines, underside reddish, with numerous light tubercles, margins and keel with reddish-brown T. 1–2 mm long, each with a white bristle 1 mm long.

**H. — v. sublineata** v. POELLN. – Bredasdorp D. – **L.** with a varnished appearance, dark green, tip-area glossy, translucent, with large tubercles and mostly short longitudinal lines, underside with numerous tubercles.

**H. — v. subtuberculata** TRIEBN. et v. POELLN. – Caledon D. – **L.** 18–25 mm long, 10–15 mm across, dark green, tip-area dull, semi-translucent, with concolorous, somewhat glossy tubercles, underside with numerous tubercles, T. green with a white tip.

**H. — v. triebnerana**. – Karroo. – **L.** 3–3.5 cm long, 8–12 mm across, terminal bristle 3–4 mm long.

**H. — v. turgida** TRIEBN. – **L.** 3–5 cm long, almost 2 cm across, upperside light green, very convex, tip-area convex, with 5 lines, underside acute-carinate, T. the same colour as the L.

**H. truncata** SCHOENL. (Pl. 90/IV/2) (§ IV). – Cape: Oudtshoorn, Calitzdorp D. – **L.** 7–13, distichous, ± erect, curved inwards, linear, c. 2 cm long, 15–25 mm across, 3–12 mm thick, sharply truncate above as if sliced away, dark green to brownish, both surfaces rough-tuberculate, the truncate tip ± translucent at first. Plant in habitat buried up to the translucent L.tips. – Acc. B. FEARN (Nat. Cact. Succ. Journ. **21,** 1966, p. 29), this species is an arrested juvenile form of **H. maughanii,** but this is only a supposition. – **H. truncata** has been crossed with **H. setata** and **H. retusa.**

**H.** — f. **crassa** v. POELLN. – Cape: Calitzdorp D. – **L.** at least 7, 15–20 mm across, 9–12 mm thick, often with short, white, radiating lines on the truncate surface.

**H.** — f. **tenuis** v. POELLN. – Cape: Oudtshoorn D. – **L.** up to 13, 20 mm across, 3–5 mm thick.

**H.** — f. **truncata** (*H. t.* f. *normalis* v. POELLN.). – Cape: Oudtshoorn, Ladismith D. – **L.** up to 13, 25 mm across, 3–6 mm thick.

**H. tuberculata** v. POELLN. (§ XIV). – Cape. – Ros. acauline or caulescent; **L.** appearing as if verticillate, up to 10, ovate-lanceolate, blunt above, tip falcate, 2–8 cm long, 8–25 mm across, hard, fleshy, upperside flat or slightly concave, underside round, carinate, both surfaces dark brownish-green, with closely and irregularly arranged, large, concolorous tubercles; **Fl.** glossy whitish-yellow.

**H.** — v. **acuminata** v. POELLN. (Pl. 98/1). – Uniondale, Oudtshoorn D. – **Ros.** always acauline; L. 2–4 cm long, c. 1 cm across, very tapered, margins tuberculate up to the tip.

**H.** — v. **angustata** v. POELLN. – Oudtshoorn D. – **St.** short; **L.** 7–8 cm long, 8–12 mm across, ± falcate, green, underside with small tubercles, margins and keel tuberculate.

**H.** — v. **subexpansa** v. POELLN. – Ladismith D. – **L.** c. 7 cm long, straight, elongateacute, later somewhat curved upwards, upperside very concave, margins and keel tuberculate.

**H.** — v. **sublaevis** v. POELLN. – Oudtshoorn D. – **L.** numerous, curved inwards, usually very acute, underside sharply carinate, margins and keel horny and rough above midway, without tubercles, the tip with a horny border, both surfaces conspicuously tuberculate.

**H.** — v. **tuberculata.** – Oudtshoorn, Ladismith, Calitzdorp D., Lit. Karroo. – **L.** 6–7 cm long, c. 2.5 cm across.

**H. turgida** HAW. (§ XIV). – Cape. – **Ros.** almost acauline, offsetting basally, 5–6.5 cm ⌀; **L.** rather stiff, 1.5–4 cm long, c. 8–10 mm across and thick, upperside swollen above, terminating in a stiff point, the convex parts translucent, with 3–7 longitudinal lines, underside rounded and carinate, with reticulate, connected, longitudinal lines above, pellucid below the tip, keel and margins smooth, more rarely rough, the rest of the L. pale green.

**H.** — v. **pallidifolia** G. G. SMITH. – Riversdale D. – **L.** longer and broader, thicker and more tapering than in v. **turgida,** the tip-area with numerous spots.

**H.** — v. **suberecta** v. POELLN. (Pl. 98/3) (*H. cuspidata* v. POELLN.). – **L.** ± dull, deep greygreen, more erect, 3–4 cm long, tip-area less swollen, underside without light spots towards the tip, pointed tip almost missing, margins and keel almost smooth.

**H.** — v. **subtuberculata** v. POELLN. – Prince Albert D. – **L.** (2–)2.5(–3) cm long, fresh or dull green, tip-area with minute, semi-translucent tubercles, keel and margins minutely dentate.

**H.** — v. **turgida** (*Aloe t.* ROEM. et SCHULT.). – Swellendam, Oudtshoorn D., Lit. Karroo. – **L.** 15–20 mm long.

**H. uitewaaliana** v. POELLN. (§ X) (*H. albicans* var.? ZANTNER). – Cape: Lit. Namaqualand. – **Ros.** acauline, ± 15 cm ⌀, offsetting basally; **L.** broadly ovate-lanceolate, long-tapering, somewhat incurving above, c. 5 cm long, 1.5–2 cm across, with a mucro 2 mm long, upperside concave, underside rounded and obliquely carinate, bluish-green, dull, margins above midway and keel with thick, white, irregular, ± confluent tubercles, those towards the tip uniting to form a horny border.

**H. umbraticola** v. POELLN. v. **umbraticola** (§ XI). – Cape: Alicedale D. – **Ros.** acauline, offsetting basally, 5–6 cm ⌀; **L.** many, obovateoblong, blunt or rather so, c. 3 cm long, 11 to 13 mm across above midway, light green, ± glossy, both surfaces translucent at the tip, with darker, irregularly connected, longitudinal lines, upperside rather convex, somewhat thickened at the tip, underside blunt-carinate, thickened below the tip and 6–8 mm thick, margins and keel with minute light T., deciduous terminal bristle c. 1 mm long.

**H.** — v. **hilliana** v. POELLN. (*H. v.* v. POELLN.). – Ros. c. 2.5 cm ⌀; **L.** not numerous, up to ± 2 cm long, 7–9 mm across towards the tip, rather blunt, short-tapering, pale green, margins and keel smooth or minutely dentate.

**H. variegata** L. BOL. (Pl. 98/4) (§ VIII). – Cape: Riversdale D. – **Ros.** acauline, 3.5 cm ⌀; L. 20–30, linear-lanceolate, long-tapering, 5.5 cm long, 1 cm across, deep brown with pale spots, upperside with a distinct M.line, underside carinate above, margins and keel with small sharp T.

**H. venosa** (LAM.) HAW. (§ XVIII) (*Aloe v.* LAM., *A. tricolor* HAW., *Apicra r.* WILLD., *H. distincta* N. E. BR.). – Cape: Graaff Reinet, Swellendam D. – **Ros.** 8–10 cm ⌀; **L.** 12–15, in 5 R., lanceolate-triangular, tapering, apiculate, 5–7 cm long, c. 22 mm across, fleshy, stiff, upperside slightly convex towards the tip, green, somewhat purple, the tip with 5–6, often reticulately connected, longitudinal lines, underside very rounded-carinate, very rough, margins with irregular small white T.

**H. venteri** v. POELLN. (§ VIII). – Cape: Worcester, Swellendam, Caledon, Bredasdorp D. – **Ros.** acauline, with few offsets basally; **L.** numerous, stiff, 5–6 cm long, straight and longtapering, with a small, curved, rather lighter tip 10 mm long, grey-green to brownish-green, with a few inconspicuous longitudinal lines, underside very rounded and blunt-carinate, often with slight keels near the margins, margins and keels with minute tubercles like T.

**H. viscosa** (L.) HAW. (§ XX) (*Aloe v.* L., *A. triangularis* MEDIC., *Apicra v.* WILLD., *Aloe concinna* SPRENG.). – Cape: S.W.reg., widespread. The many varieties are transitional to the type, v. **viscosa,** and to one another. – **St.** extended, offsetting basally; **L.** in 3 R., imbricate, erect to projecting, lanceolate, triangular, tapering, sharp-tipped, firm, dark green, both surfaces often tuberculate, upperside concave, underside carinate.

**H. —** v. **caespitosa** v. POELLN. – Ladismith D. – **St.** c. 6 cm tall, freely offsetting from the base to form clumps; **L.** 10–12 mm across, the sharp tip slightly recurved.

**H. —** v. **coegaensis** G. G. SMITH. – Willowmore D. – **St.** 9 cm tall; **L.** densely imbricate, 21 mm long, 14 mm across, 4 mm thick, ovate-pointed, older L. recurved at 90°, upperside triangular-concave, with small lighter papillae in the upper third, pink below, light green above, underside convex, minutely rough in the upper third, margins acute.

**H. —** v. **concinna** (HAW.) BAK. (Pl. 91/XX) (*H. c.* HAW., *Aloe c.* v. *major* SALM, *Aloe c.* ROEM. et SCHULT.). – **L.** in 3 straight longitudinal R., lanceolate-triangular.

**H. —** v. **indurata** (HAW.) BAK. (*H. i.* HAW., *Aloe i.* ROEM. et SCHULT., *A. v.* v. *i.* SALM). – Ladismith D. – **L.** in 3 straight longitudinal R., ovate-triangular.

**H. —** v. **pseudotortuosa** (SALM) BAK. (*H. p.* SALM, *Apicra tortuosa* WILLD. p.part., *A. p.* SALM, *A. subtortuosa* ROEM. et SCHULT.). – Prince Albert D. – **L.** in spirally contorted longitudinal R., triangular.

**H. —** v. **quaggaensis** G. G. SMITH. – Humansdorp D. – Compact; **L.** 27 mm long, 12 mm across, 6.5 mm thick, ovate-lanceolate, pointed, upperside triangular-concave, minutely rough in the upper half, underside rounded, triangular towards the tip, minutely rough, with 11 reticulate lines, margins horny, light green.

**H. —** v. **subobtusa** v. POELLN. – Ladismith D. – **L.** in 3 straight or somewhat contorted R., blunt above with a minute apiculus, margins reddish-brown.

**H. —** v. **torquata** (HAW.) BAK. (Pl. 98/5) (*H. t.* HAW., *Aloe pseudotortuosa* v. *elongata* SALM, *A. t.* ROEM. et SCHULT., *A. t.* v. *laevior* SALM). – Prince Albert D., Oudtshoorn, Steytlerville D. – **St.** 20 cm tall, more rarely dichotomously branched; **L.** very concave, crowded-imbricate, lanceolate-triangular, long-tapering, 5 cm long, up to 15 mm across, very thick, upperside very grooved, underside very carinate, keel and margins yellowish-green, glabrous.

**H. —** v. **viridissima** G. G. SMITH. – Steytlerville D. – **St.** 10 cm tall; **L.** imbricate, 22 mm long, 13 mm across, 3.5 mm thick, ovate-tapering, upperside concave, triangular in the upper quarter, rough, with 2 darker longitudinal lines, glossy light green above, underside rounded, pointed-triangular towards the tip, minutely rough, keel horny, margins sharp, horny.

**H. —** v. **viscosa** (*H. v.* v. *typica* BGR.). – Cape: S.W.reg., widespread. – **St.** 20 cm tall; **L.** upperside concave, pointed to aciculate, dirty green, minutely rough.

**H. vittata** BAK. (Pl. 98/6) (§ VII). – Cape: Port Elizabeth D. – **Ros.** 5–7 cm ⌀; **L.** 20–30, oblong-lanceolate, tapering, 3.5–4 cm long, 12 to 15 mm across, 6 mm thick, upperside flat, with 3–5 short, pellucid stripes on a brown ground, underside convex and swollen, carinate, green, margins with small ciliate T., awn-tip 3.5–4 cm long.

**H. willowmorensis** v. POELLN. (§ XIV). – Cape: Willowmore D. – **Ros.** acauline, 3.5 to 4 cm ⌀; **L.** few, stiff, c. 3 cm long, 10–12 mm across, upperside green, smooth, almost translucent above and obtusely recurved, underside greenish-red, rounded, obliquely carinate towards the tip, with pallid tubercles in longitudinal R., tip-area ovate-triangular, tapering, c. 16 mm long, c. 10 mm across, ± rounded, with minute, concolorous tubercles and several translucent longitudinal lines, margins with small T., terminal bristle 1–2 mm long.

**H. wittebergensis** BARKER (§ VIII). – Cape: Laingsburg D. – **St.** simple, 1–2 cm long, 1 cm ⌀, covered with old L.bases; **L.** in 9 or more R., ± lanceolate, tapering, leathery, deep purplish-green, up to 10 cm long, 14 mm across, upperside flat to slightly grooved, underside indistinctly keeled, with several tuberculate, longitudinal ribs near the base, margins and keel with numerous white T. in the upper part, L.tips soon drying and dropping.

**H. wooleyi** v. POELLN. (§ XVIII). – Cape: Steytlerville D. – **Ros.** acauline, 7–10 cm ⌀, freely offsetting from the base, mat-forming; **L.** many, ovate-lanceolate, very tapered, curved back ± horizontally towards the tip, 6–8 cm long, 12–20 mm across, upperside somewhat concave, light green, with several dark green, broken lines, underside dark green, very rounded and carinate, with tubercles in wavy, intersecting lines, margins with 1 mm long T., terminal bristle small.

**H. zantneriana** v. POELLN. (Pl. 98/7) (§ VIII). – E.Cape. – **Ros.** acauline, offsetting basally, mat-forming; **L.** not numerous, lanceolate-subulate, 5–7 cm long, 6–8 mm across, rather pointed, not dentate, upperside ± flat, underside roundish with a ± lateral, blunt keel, dull, rather greyish-green, with large irregular, translucent blotches and longitudinal stripes, margins often with pallid stripes.

**Hechtia** KLOTZSCH. Bromeliaceae. – Mexico; USA: Texas. – Dioecious plants; **L.** usually in a dense, acauline Ros., long, very prickly, dentate; **Fl.** Sc. terminal, very long, simple or little branching, Fl. rather small, in clusters along the axis, sometimes extending to short spikes. – Greenhouse, warm. Propagation: seed, division.

**H. argentea** BAK. (Pl. 99/1). – Mexico. – **L.** 30 cm long, silvery on both surfaces, marginal Sp. 7 mm long, sharp; **Fl.** Sc. rather scaly, 20 cm long, with many inconspicuous, greenish-white Fl.

**H. desmetiana** (BAK.) MEZ (*Dyckia d.* BAK.). – Mexico. – **L.** 40 cm long, rather thick, fleshy.

**H. ghiesbreghtii** LEM. – Mexico: San Luis Potosi. – **L.** 15 cm long, **Sp.** with dense silvery scales.

**H. glomerata** ZUCC. – Mexico: valley of Mexico City. – **L.** 30 cm long, contorted, underside with silvery scales, upperside brownish-red in full sun, **Sp.** 6 mm long, stout; **Infl.** 40 cm tall.

**Hertia** LESS. Compositae. – N.Afr., S.W.Afr. – Small sub-♄; **Br.** woody; **L.** ± fleshy, ± dentate; **Fl.** terminal, yellow. – Of little interest in cultivation. Summer: in the open, winter: coldhouse. Propagation: seed, cuttings.

**H. cheirifolia** (L.) O. KTZE. (*Othonna ch.* L., *O. crassifolia* L., *O. calthoides* MILL., *H. crassifolia* LESS., *Othonnopsis ch.* (L.) BENTH. et HOOK.). – Algeria, Tunisia. – Prostrate; **Br.** creeping, ascending at the tip, glabrous, greygreen; **L.** alternate but distichous in young plants, sessile, ± fleshy, lanceolate, 3-veined, margins entire, scarcely dentate, green to purple, up to 5 cm long; **Fl.** heads solitary or in a panicle, each head with c. 12 yellow ray-florets.

**H. ciliata** (HARV.) O. KTZE. (*Othonna dinteri* MUSCHL. ex DTR., *O. macrocephala* MUSCHL.). – S.W.Afr. – **Roots** woody; **St.** 5–10, up to 40 cm long, little branching; **L.** ± dense, 2–3.5 cm long, up to 1.5 cm across, with a few coarse **T.**, blue-green, somewhat fleshy; **Fl.**heads 2 cm ⌀, terminal.

**Hesperaloe** ENGELM. Agavaceae. – Mexico; USA: Texas. – Acauline plants; **L.** long, linear, firm, leathery, grooved, with fibrous margins; **Infl.** 1–2.5 m tall, simple or ± branched, Fl. long-campanulate, c. 15 mm ⌀, reminiscent of **Aloe**. – Greenhouse. Propagation: seed, division.

**H. funifera** TREL. (*Yucca f.* KOCH, *Agave f.* LEM., *Hesperoyucca engelmannii* BAILL., *Hesperoyucca davyi* BAK.). – N.Mexico. – Fl.Sc. 2.5 m tall, slender, branching, Fl. greenish.

**H. parviflora** COULT. (Pl. 99/2) (*Yucca p.* TORR., *H. yuccaefolia* ENGELM., *Aloe y.* GRAY). – Fl. 23–35 mm long, pinkish-red.

**Heurnia** R. BROWN corr. SPRENG. See **Huernia** R. BROWN.

**Heurniopsis** N. E. BR. corr. K. SCHUMANN. See **Huerniopsis** N. E. BR.

**Holubia** OLIV. Pedaliaceae. – N.Transvaal. – Only one species known, not found in cultivation.

**H. saccata** OLIV. – Erect ☉ plants; **L.** opposite, large, ± fleshy; **Fl.** solitary, from the upper L.axils, tube narrow, long, with a large saccate spur at the base, the 5 Cor.lobes ± revolute; **Fr.** a disc-shaped schizocarp.

**Hoodia** SWEET. Asclepiadaceae. – Angola; S.W.Afr.; Cape. – Succulent ♃ plants, resembling **Stapelia**; **St.** up to 80 cm tall, robust, grey-green, the numerous angles with many thornlike hard T.; **Fl.** 1–5, with pedicels 2 cm long, from the furrows higher up the St., Cor. ± rotate and actinomorphic, ± circular but with 5 points, campanulate or flat, yellowish or brownish, glabrous or hairy, corona double, the outer corona cup-shaped. – Greenhouse, warm. – Propagation: seed, grafting sections on to **Stapelia** or Tub. of **Ceropegia**.

**H. albispina** N. E. BR. – Cape: Carnarvon D. – **St.** 15-angled, each tubercle tipped with a white Sp.; **Cor.** 7.5–9 cm ⌀, c. 2.5 cm deep, lobes very broad, with a tip 3–4 mm long, rough-papillose in the middle.

**H. bainii** DYER v. **bainii**. – Cape: Prince Albert, Carnarvon, Clanwilliam, Fraserburg, Kenhardt, Hay D., Bushmanland; S.W.Afr. – **St.** cylindrical, tubercles compressed, spiralled, merging to form longitudinal ribs, terminating in a Sp. 1 cm long; **Fl.** 1–3, **Cor.** 6–7 cm ⌀, deep-campanulate, slightly 5-lobed, dull yellow.

**H. —** v. **juttae** (DTR.) H. HUBER (*H. j.* DTR.). – S.W.Afr.: Lit. Karas Mts. – **Cor.** quite flat, 2.2–5.5 cm ⌀, almost circular, indentations very shallow, light yellow to brownish, greenish to yellowish-brown to dark reddish-brown inside, with dark veins.

**H. barklyi** DYER. – Cape: Karroo. – **St.** manyangled, over 20 cm tall, 4 cm thick, branching, with stout Sp.; **Cor.** 5–6 cm ⌀, sinuate, lobes obtuse, with a short point, yellowish, with red spots at the base.

**H. burkei** N. E. BR. (*Stapelia gordonii* HOOK., *Scytanthus g.* HOOK.). – Cape: Beaufort West, Prince Albert D. – **St.** 30 cm tall, with 12 to 14 longitudinal ribs, tubercles with a 6–8 mm long Sp.; **Cor.** 9–10 cm ⌀, flat, almost circular, slightly crenate, lobes with a point 6 mm long, dark brown.

**H. currori** (HOOK.) DECNE (*Scytanthus c.* HOOK, *S. burkei* HOOK.). – Angola; S.W.Afr.: Gr. Namaqualand, Damaraland. – **St.** branching basally, 30–60 cm tall, 5–6 cm ⌀, many-angled, Sp. sharp, pointed; **Cor.** 9–12 cm ⌀, campanulate-rotate, 5-lobed, green to yellow or pink, becoming yellowish-pink, with dense violet hairs, the tube orange to yellowish-red, densely hairy.

**H. dregei** N. E. BR. – Cape: Prince Albert, Beaufort West D. – **St.** 4 cm thick, 20–24-angled, tubercles conical, pointed, with a bristle 5 to 6 mm long; **Cor.** 3–3.5 cm ⌀, ± cup-shaped, with 5 indistinct, broad lobes, light brown inside, with dense white hair, glabrous outside, lobes with a subulate tip up to 3 mm long.

**H. gibbosa** NEL. – S.W.Afr.: E. of Swakopmund. – Caespitose; **St.** 30 cm tall, up to 5 cm ⌀, cylindrical, with 14 very prominent angles, tubercles with a stiff Sp. 7 mm long; **Cor.** 8 cm ⌀, slightly concave, ± round, margins slightly triangular, lobes with a 5 mm long point, glossy grey-green outside, purple to wine-coloured inside, covered with purple hairs, tube-mouth orange-coloured.

**H. gordonii** (MASS.) SWEET (*Stapelia g.* MASS., *Gonostemon g.* SWEET, *Monothylaceum g.* G. DON.). – Cape: Lit. Namaqualand, Prieska, Herbert D.; S.W.Afr. – Plant with a taproot; **St.** numerous, cylindrical, with 12–14 longitudinal ribs, tubercles with a woody Sp. 1 cm long; **Cor.** saucer-shaped, ± circular, lobes little

**H. husabensis** NEL. – S.W.Afr. – **St.** 60–70 cm tall, c. 4 cm ⌀, 16–20-angled, tubercles oblong, with a bristle 7 mm long; **Cor.** 7 cm ⌀, ± circular, lobes very rounded, with a point 1 mm long, pinkish-violet to greyish-violet, with many light longitudinal stripes and a zone of blackish-purple hairs 1 cm wide in the middle.

**H. longii** OBERM. et LETTY. – Cape: W.reg.; Botswana. – **St.** c. 40 cm tall, with many subsidiary shoots from the base, c. 3 cm ⌀, 14-angled, tubercles 4–8 mm long; **Cor.** 5 cm ⌀, cup-shaped, pinkish-brownish, lobes very rounded, with a point 3 mm long, densely covered inside with hairs 3 mm long.

**H. lugardii** N. E. BR. – Cape; Botswana; S.W.Afr.: Gr. Namaqualand. – **St.** up to 45 cm tall, with c. 16 angles, tubercles small, spiny; **Cor.** campanulate-rotate, slightly 5-lobed, lobes abruptly narrowing into a subulate point 7 to 10 mm long, brick-red, sparsely hairy.

**H. macrantha** DTR. (Pl. 99/4). – S.W.Afr. – **St.** c. 20, up to 80 cm tall, up to 8 cm ⌀, forming clumps; **Cor.** rotate, flat, up to 20 cm ⌀ including the lobe-tips, 10–12 cm ⌀ across the indentations, lobes with a tip 6–8 mm long, light purple, yellowish along the veins, covered with purple hairs 2–3 mm long, tube crater-like, orange-yellow, annulus 5-partite, yellow.

**H. montana** NEL. – S.W.Afr. – **St.** 20 cm tall, 2–4.5 cm ⌀, 20–24-angled, tubercles ovate to elliptical with a stiff bristle 5–10 mm long; **Cor.** 9 cm ⌀, ± circular, somewhat concave, sinuate between the lobes, lobes with a point 5–7 mm long, light yellow with purple hairs, tube 1 cm ⌀, deep purplish-red, with purple hairs in the upper part, around the mouth an orange-yellow zone 2–3 mm across, with long purple hairs.

**H. parviflora** N. E. BR. – Angola. – **St.** c. 40 cm tall, many-angled, c. 2 cm ⌀, Sp. stout, 5–8 mm long, horny; **Cor.** 3 cm ⌀, slightly 5-lobed, lobes with a subulate tip 2–3 mm long, glabrous outside, darker and hairy inside.

**H. pillansii** N. E. BR. – Cape: Prince Albert D. – **St.** 12–15 cm tall, branching bush-like; **Br.** up to 3.5 cm long, with 15–18 tuberculate angles, tubercles with a spine-like tip 6–9 mm long; **Cor.** 6–6.5 cm ⌀, margin slightly 5-angled, lobes with a point 3–4 mm long, base and middle minutely rough-papillose, salmon-coloured, peach-coloured in the middle.

**H. rosea** OBERM. et LETTY. – Cape: Botswana, W.reg. – **St.** c. 30 cm tall, with c. 14 angles, tubercles with Sp. 4–8 mm long; **Cor.** 7 cm ⌀, cup-shaped, indistinctly 5-angled, lobes with a slightly recurved point, also with a minute point between the lobes, light pink, sparsely hirsute inside.

**H. ruschii** DTR. – S.W.Afr.: Gr. Namaqualand. – With many St. when old; **St.** up to 50 cm tall, c. 4 cm ⌀, with 22–24 ribs, Sp. 6 mm long; **Cor.** broad-campanulate, c. 4 cm ⌀, lobes incised to c. 6 mm, reddish-brown inside, yellow in the middle, with minute papillae having tiny hairs c. 0.5 mm long.

**H. triebneri** SCHULDT. – S.W.Afr. – **St.** compressed, much branched from the base, 10 to 20 cm tall, with c. 12 angles, tubercles spine-tipped; **Cor.** c. 4 cm ⌀, flat saucer-shaped, lobes broad-tapering, greenish-pink, glabrous, smooth.

**Hoodiopsis** LUCKH. Asclepiadaceae. – Possibly intergeneric hybrid: **Caralluma** × **Hoodia**. – Greenhouse, warm. Propagation: seed, grafting as for **Hoodia**.

**H. triebneri** LUCKH. v. **triebneri** (Pl. 99/5). – S.W.Afr.: Gr. Namaqualand. – Forming clumps up to 30 cm ⌀; **St.** erect, branching basally, 12–18 cm tall, with 7–9 prominent angles having raised, compressed tubercles with minute T., light green, purple-striate; **Fl.** solitary, from midway along the St., pedicel 2.5 cm long, Cor. 10.5 cm ⌀, tube 5 cm ⌀, lobes spreading, 2.5–3 cm long, somewhat laterally curved, margins recurved, green or light, pale pink outside, rough papillose inside, lobes with 5 longitudinal furrows and ridges, deep wine-coloured, inner corona-lobes with 2 horns.

**H. —** v. **ciliata** LUCKH. – Robuster; **Cor.** more rough-papillose, lobes more tapering, ciliate.

**Hoya** R. BR. Asclepiadaceae. – China, peninsular India, Malayan archipelago, Austr. – Climbing or lax ♄, usually rooting from the Br.; **L.** opposite, fleshy or leathery, evergreen, glossy; **Fl.** in axillary cymes, 5-angled, white, yellowish, pink or red. – Greenhouse, warm. Propagation: cuttings, sometimes from seed. (Not all species of **Hoya** are succulent.)

**H. australis** R. BR. (*H. bicarinata* A. GRAY, *H. dalrympleana* F. v. MUELL.). – Austr. – **L.** short-petiolate, ovate, obovate or ± circular, thick, fleshy, 5–8 cm long; **Fl.** in simple umbels, c. 1 cm ⌀, white with a red centre.

**H. bella** HOOK. (*H. paxtonii* HORT.). – Java. – Bush, not climbing; **St.** branching, very leafy; **L.** small, ovate-lanceolate, tapering, fleshy, ± convex; **Infl.** of shortly pedunculate umbels, Fl. 15 mm ⌀, white, with pointed, 5-angled lobes. Best grafted on **H. carnosa**.

**H. carnosa** (L.) R. BR. (Pl. 99/3) (*Asclepias c.* L., *Schottia crassifolia* JACQ.). – China, Austr. – Twining ♄; **L.** ovate-cordate, blunt to short-tapering, thick and fleshy, 5–8 cm long, 4–5 cm across, short-petiolate; **Fl.** in fairly large umbels, fleshy, 15 mm ⌀, white or pale flesh-coloured, with a red centre.

**H. —** cv. Compacta. – **L.** shorter, plicate, ± stunted; **Fl.** smaller.

**H. —** cv. Marmorata. – **L.** yellow-mottled.

**H. —** cv. Variegata. – **L.** yellowish-white, often with a red border.

**H. englerana** HOSS. – Siam. – Epiphytic; **L.** 1.5 cm long, 4 mm across, upperside convex; **Fl.** 4 together, terminal, 15 mm ⌀, white and violet.

**H. fraterna** BL. – Java. – **L.** very large and thick, broad-elliptical; **Fl.** pinkish-red with a yellow centre.

**H. imperialis** Lindl. – Peninsular India. – Vigorous, climbing plant, almost entirely covered in felt; **L.** leathery, oblong, slender-tipped, slightly cordate at the base, smooth; **Fl.** in pendent umbels, 5–7 cm ∅, dark purple, greenish outside.

**H. —** cv. Rauschii. – **Fl.** lacquer-red.

**H. linearis** Wall. – Subtrop. Himalayas. – Resembles **H. bella**; **L.** cylindrical, narrow, hairy; **Fl.** white.

**H. longifolia** Wall. – Trop. Himalayas. – Resembles **H. bella**; **L.** larger, much elongated, linear-lanceolate.

**H. macrophylla** Bl. – Java. – Twining ♄; **Br.** subcylindrical; **L.** elliptical-lanceolate, tapering, light green, with 3 longitudinal veins, fleshy; **Fl.** in ± spherical umbels, fleshy, white, papillose-hirsute inside.

**H. pallida** Lindl. – China. – **St.** twining; **L.** pale green, ovate-lanceolate, tapering, fleshy, thick; **Fl.** c. 1 cm ∅, pale yellow, scented.

**Huernia** R. Brown (*Heurnia* R. Brown corr. Spreng., *Decodontia* Haw.). – Asclepiadaceae. – S. and E.Afr.; Ethiopia; Arabia. – ♃; **St.** low, hardly 10 cm high, with (4–)7 (rarely more) angles, fleshy, edges prominent, armed with large T., grey-green, often reddish, glabrous, branched from the base, forming clumps; **Fl.** in short-stalked umbels at the base of the young St., 2–3 cm ∅, fleshy, ± tuberculate, variously coloured and spotted, smelling slightly of carrion, in summer and autumn, Cor. campanulate at the base, margin 5-angled or broadly 5-cleft, with another small lobe in the sinuses of the lobes, and a ring-shaped corona at the centre. – Greenhouse, warm. Propagation: seeds, cuttings.

## Division of the Genus Huernia into Groups acc. White and Sloane

I. St. not more than 7-angled; Fl. with both an outer and inner corona.
  A. Cor. tube not deeper than 1 cm.

  **Gr. I. Macrocarpa-Brevirostris Group.** – The annulus absent or indistinct. – Species: H. albomaculata, aspera, brevirostris, concinna, erinacea, hadhramautica, herrei, hystrix, inornata, keniensis, kennedyana, lodarensis, loesenerana, macrocarpa, marnierana, montana, namaquensis, nouhuysii, oculata, piersii, quinta, schneiderana, similis, stapelioides, stricta, thuretii, transmutata, verekeri, vogtsii, volkartii, whitesloaneana, witzenbergensis.

  **Gr. II. Somalica-Guttata Group.** The annulus distinctly perceptible or raised. – Species: H. andreaeana, confusa, guttata, humilis, insigniflora, ocellata, praestans, procumbens, reticulata, somalica, tanganyikensis, transvaalensis, venusta, zebrina.

  A. Cor. tube more than 1 cm deep.

  **Gr. III. Longituba Group.** – Tube not contracted from the outside, so that the Fl. appears subcylindrical. – Species: H. decemdentata, leachii, levyi, longituba, pendula.

  **Gr. IV. Campanulata Group.** – Tube contracted from the outside, thus the Fl. forming a wide-mouthed bell. – Species: H. barbata, campanulata, clavigera, erectiloba, hallii, hislopii, kirkii, occulta.

II. St. or Fl. not as above.

  **Gr. V. Miscellaneous Group.** – Species: G. distincta, pillansii, simplex.

**H. albomaculata** White et Sloane (§ I). – Origin unknown. – **St.** acutely dentate; **Cor.** 18 mm ∅, tube campanulate, 8 mm ∅, 6 mm deep, red, white-spotted inside, white-papillose around the mouth, lobes 6 mm across and long, purple with white spots, white-papillose, tips purple-red.

**H. andreaeana** (Rauh) Leach (Pl. 56/1) (§ II) (*Duvalia a.* Rauh). – Kenya. – **St.** prostrate, creeping, with a few Br. up to 1 m long, rooting from the underside, 7–10 mm thick, grey-green, often with wine-coloured spots, with 4–5 ribs; **L.** 2–3 mm long, persisting as scales for a long time; **Fl.** solitary, Cor. 2.5–2.7 cm ∅, tube 1–1.2 cm ∅, annulus 7 mm ∅, pale wine-coloured, lobes triangular, with an acute tip, c. 1 cm long, underside pale ochre-coloured, upperside with scattered blackish-violet, clavate hairs.

**H. aspera** N. E. Br. (§ I). – Probably Tanzania or Zanzibar. – **St.** prostrate, 10–15 cm long, almost cylindrical, T. short, brown; **Cor.** broad-campanulate, 2–2.5 cm ∅, reddish-brown outside, lobes triangular, deep purple-brown inside, both surfaces with whitish spots.

**H. barbata** (Mass.) Haw. (§ IV) (*Stapelia b.* Mass., *H. crispa* Haw., *H. b. v. c.* Loud.). – Orange Free State: Fauresmith; Cape: Prieska to Albany D. – **St.** numerous, erect, 2–6 cm long, 1.5–2 cm thick, with 4–5 acute, sinuate-dentate angles; **Cor.** campanulate, c. 5 cm ∅, tube 15 to 20 mm long, lobes triangular, long-tapering, smooth outside, pale sulphur-yellow or brownish with blood-red spots inside, with long blood-red hairs.

**H. —** v. **griquensis** N. E. Br. – Cape: W. Griqualand, Hay D. – Lobes 10 mm long, narrow-triangular, very pointed, hairs more scattered.

Plate 114. 1. **Pachycormus discolor** (BENTH.) COVILLE v. **discolor** (Repr.: Des. Pl. Life, **11**, 1939, 10, Photo: G. LINDSAY); 2. **Pachyphytum glutinicaule** MORAN (Photo: R. MORAN); **3. P. viride** E. WALTH.; **4. P. oviferum** J.A. PURP. (Photo: W. BORWIG); **5. P. compactum** ROSE.

17/B  Lexicon of Succulent Plants

Plate 115. 1. **Pachypodium geayi** Cost. et Bois (Photo: W. Kabel); **2. P. rosulatum** v. **horombense** (H. Poiss.) Rowl.; **3. P. brevicaule** Bak. (Photo: J. Marnier-Lapostolle); **4. P. lealii** Welw. ssp. **lealii** (Photo: Walter); **5. P. lealii** Welw. ssp. **lealii**; **6. P. namaquanum** Welw. (Photo: W. Triebner).

Plate 116. 1. × **Pachyveria** cv. Clavata; 2. × **P.** cv. Scheideckeri; 3. × **P.** cv. Glauca; 4. **Pedilanthus tithymaloides** ssp. **smallii** (MILLSP.) DRESSLER; 5. **P. tithymaloides** POIT. ssp. **tithymaloides** cv. Nanus; 6. **Pectinaria asperiflora** N. E. BR. (Photo: E. HAHN).

Plate 117. 1. **Pelargonium carnosum** (L.) AIT.; 2. **P. ceratophyllum** L'HER.; 3. **P. crithmifolium** J. E. SMITH; 4. **P. tetragonum** L'HER.

Plate 118. 1. Peperomia dolabriformis v. tenuis JACOBS.; 2. P. galioides H. B. et K.; 3. P. dolabriformis KUNTH. v. dolabriformis; 4. P. nivalis MIQ. v. nivalis (Photo: W. RAUH); 5. Phytolacca dioica L.; 5 year old seedling.

Plate 119. 1. **Piaranthus comptus** N. E. Br. v. **comptus** (Photo: H. Lang); 2. **P. decorus** (Mass.) N. E. Br. (Photo: as 1); 3. **Pilea serpyllacea** (H. B. et K.) Wedd.; 4. **Pistorinia breviflora** ssp. **salzmannii** (Boiss.) Jacobs.

Plate 120. 1. **Plectranthus prostratus** Gürke; 2. **Plumeria rubra** L.; 3. **Poellnitzia rubriflora** (L. Bol.) Uitew. v. **rubriflora**; 4. **Portulaca jacobseniana** Rowl.; 5. **P.** cv. Grandiflora Flore Pleno (Photo: E. Hahn); 6. **Pseudopectinaria malum** Lavr. × 1.6 (Photo: Bot. Res. Inst. Pretoria).

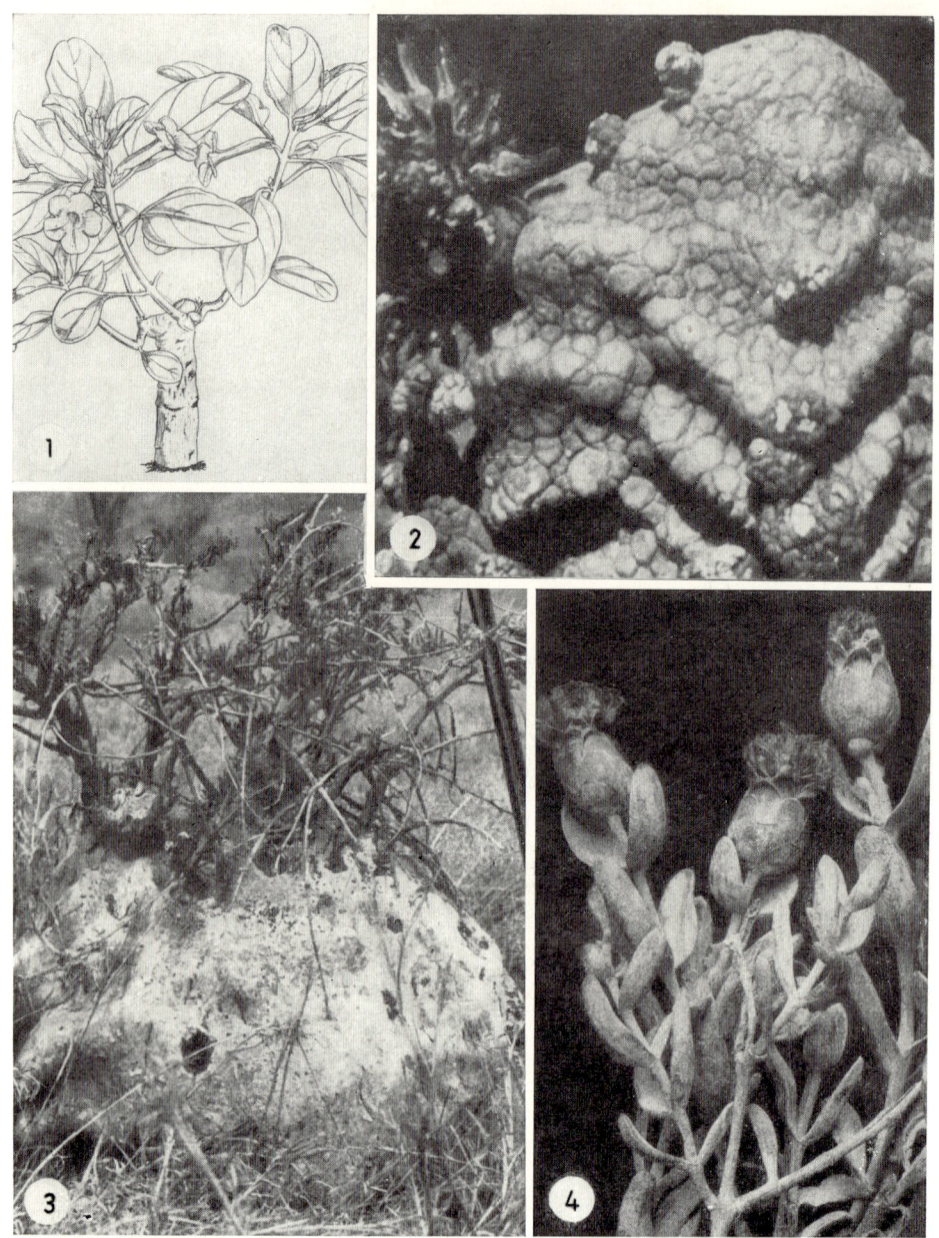

Plate 121 1. **Pterodiscus ruspolii** ENGL. (Repr.: Fl. Trop. E. Afr. Pedal. Fig. 3, 1953); 2. **Pseudolithos cubiforme** (BALLY) BALLY (Photo: J. BOGNER); 3. **Pyrenacantha malvifolia** ENGL. (Photo: P. R. O. BALLY); 4. **Pteronia succulenta** THUNBG. (Photo: Heimatkundl. Ver. Lüderitzbucht).

**H. brevirostris** N. E. Br. v. **brevirostris** (Pl. 100/1) (§ I). – Cape: Graaff Reinet, Swellendam, Cradock D. – **St.** numerous, 4 cm long, 2–2.5 cm ⌀, with 4–5 thick, sinuate-dentate angles, with brownish spots, T. pointed; **Cor.** waxy, fleshy, 3.5 cm ⌀, tube campanulate, lobes triangular, tapering, pallid green outside, with some red spots, whitish inside, minutely papillose, red-spotted, throat blood-red, not evil-smelling.

**H.** — v. **ecornuta** (N. E. Br.) White et Sloane (*H. scabra* v. *e*. N. E. Br.). – Cape: Victoria West, Willowmore D. – **Cor.** 33 to 37 mm ⌀, pale canary-yellow with light red spots, rough papillose.

**H.** — v. **histronica** White et Sloane. – Cape: Graaff Reinet D. – **Cor.** sulphur-yellow inside, with large purple blotches.

**H.** — v. **immaculata** (N. E. Br.) White et Sloane (*H. scabra* v. *i*. N. E. Br.). – Cape: Victoria West, Cradock D. – **Cor.** 37–43 mm ⌀, primrose-yellow, unspotted, papillose.

**H.** — v. **intermedia** N. E. Br. – Cape: Graaff Reinet D. – **Cor.** wrinkled, primrose-yellow inside, with purple-tipped papillae, tube paler, purple at the base.

**H.** — v. **longituba** (N. E. Br.) White et Sloane (*H. scabra* v. *l*. N. E. Br.). – Cape: Beaufort West, Willowmore D. – **Cor.** 25 to 31 mm ⌀, lobes 9–10 mm long, 10–11 mm across, yellow with small furrows, rough.

**H.** — v. **pallida** (N. E. Br.) White et Sloane (*H. scabra* v. *p*. N. E. Br.). – Cape: Victoria West D. – **Cor.** 32–40 mm ⌀, border with an annular thickening, lobes very pointed, ± canary-yellow inside, spotted light red, tube-base purplish-red.

**H.** — v. **parvipunctata** White et Sloane. – Cape: Graaff Reinet D. – **Cor.** with smaller spots inside than in the type.

**H.** — v. **scabra** (N. E. Br.) White et Sloane (*H. s.* N. E. Br.). – Cape: Victoria West D. – **Cor.** 32–40 mm ⌀, tube 10 mm ⌀, border with an annular thickening, lobes very pointed, pale flesh-coloured to canary-yellow inside, minutely papillose, with light red spots, annulus with large papillae.

**H. campanulata** (Mass.) R. Br. (§ IV) (*Stapelia c.* Mass.). – Cape: Prince Albert, Ladismith, Oudtshoorn D. – **St.** numerous, erect, 5–10 cm tall, almost 2 cm ⌀, red-spotted, with 4–5 acute, sinuate-dentate angles, T. pointed; **Cor.** campanulate, constricted below midway, lobes broad-triangular, pointed, whitish outside, with reddish spots, ± sulphur-yellow inside, with larger blackish-purple blotches, tube with long purple hairs, with blackish-red transverse lines.

**H. clavigera** (Jacq.) Haw. (§ IV) (*Stapelia c.* Jacq., *St. campanulata* Sims., *St. clavata* Decne). – Cape: Swellendam, Laingsburg, Prieska D., Lit. Namaqualand. – Mat-forming; **St.** 5–8 cm tall, c. 2 cm ⌀, red-spotted, with 4–5 deep, acute, dentate angles, T. 10–13 mm apart; **Cor.** campanulate, with a border 5 mm wide, lobes broad-triangular, tapering, glabrous outside, pale dirty green, inside dirty yellow to dark yellow, everywhere with red spots and papillae and blood-red hairs, the tube with minute striae, the base blood-red.

**H. concinna** N. E. Br. (§ I) (*H. macrocarpa* Taubert). – Ethiopia. – Mat-forming; **St.** 2.5 to 5 cm long, up to 1 cm ⌀, with 5 sinuate-dentate angles, T. pointed; **Cor.** campanulate, over 2 cm ⌀, pale and smooth outside, inside yellowish, with minute red spots, with yellow papillae, lobes triangular, tapering.

**H. confusa** Phillips (§ II) (*H. humilis* Schltr.). – Transvaal: Pietersburg D. – **St.** c. 6 cm tall, the 4–5 angles with triangular-pointed T.; **Cor.** 3 cm ⌀, tube cup-shaped, smooth, yellowish-pink, the mouth with a broad, yellowish-pink, white-spotted annulus, lobes pale greenish-white, with red markings, minutely papillose.

**H. decemdentata** N. E. Br. (§ III). – S.W. Afr. – **St.** probably short; **Cor.** campanulate, c. 18 mm ⌀, minutely rough-papillose outside at the tip, smooth and glabrous inside at the base, above that covered with clavate hairs, lobes 10.

**H. distincta** N. E. Br. (§ V). – Cape: Laingsburg D. – Forming dense cushions; **St.** 4–6 cm tall, 10–15 mm ⌀, with 8–9 obtuse angles, indistinct at first, T. 3 mm long; **Cor.** c. 2.5 cm ⌀, tube campanulate, 1 cm ⌀ and deep, lobes spreading, 8–10 mm long and across, very pointed, papillose inside, pale yellow with intricate pink markings.

**H. erectiloba** Leach et Lavr. (§ IV). – Moçambique. – **St.** 4-angled, (4–)8(–20) cm tall, 6–11 mm thick, T. up to 5 mm long; **Cor.** bicampanulate, 22–25 mm long, almost 30 mm ⌀, lobes erect, ± rough outside, cream-coloured, with red spots and prominent veins, tube cylindrical, slightly constricted at the mouth, with an annular crest, deep blood-red inside below, above that with vesicular papillae, with deep blood-red convergent lines, the tube hirsute in the middle, with its margin spreading horizontally, cream-coloured with blood-red spots, lobes deep blood-red, minutely callused with tubercles.

**H. erinacea** Bally (§ I). – Kenya. – **St.** prostrate, 20–60 cm long, 10–14 mm thick, with 5 obtuse angles, with dentate tubercles 2–3 mm long, with scale-like, deciduous **L.** 1 mm long; **Cor.** fleshy, ± constricted at the base of the lobes, 16–17 mm ⌀, lobes pointed-triangular, 2 cm long, upperside with papillae up to ¾ mm long, yellow, with blackish-purple blotches.

**H. guttata** (Mass.) R. Br. (§ II) (*Stapelia g.* Mass., *St. lentiginosa* Curtis, *St. venusta* Jacq., *H. lentiginosa* Haw.). – Cape: Bushmanland, Clanwilliam, Van Rhynsdorp D. – **St.** erect, 6–8 cm tall, somewhat 4–5-angled, the angles sinuate with projecting T.; **Cor.** campanulate-hemispherical, constricted at the throat, with a prominent annulus, lobes broad-triangular, tapering, papillose, pallid outside, sulphur-yellow inside, with minute, blood-red blotches and spots, tube with minute, pointed papillae at the throat.

**H. hadhramautica** Lavr. (§ I). – S.Arabia. – **St.** robust, 5-angled, 4–6 cm long, 22 mm thick, T. very truncate, 12 mm long, very pointed; **Cor.** broad-campanulate, 10 mm long, 25 mm ⌀, cream-coloured and densely papillose outside, inside dark purplish-red, papillose, the tube pink below, lobes broad-triangular.

**H. hallii** E. et B. M. Lamb (§ IV). – S.W.Afr.: Grünau. – **St.** erect, 3–4 cm tall, 10–15 mm ⌀, with (4–)5(–6) obtuse angles, T. 5 mm long; **Cor.** 2–2.5 cm ⌀, milk-coloured outside, white inside, yellow towards the lobes, spotted towards the tube, tube 7.5 cm tall and wide, ± campanulate, transversely striate, minutely papillose, lobes triangular, pink.

**H. herrei** White et Sloane v. **herrei** (§ I). – Cape: Lit. Namaqualand. – **St.** 3–5 cm tall, 1–1.5 cm ⌀, with 5 roundish angles, T. curved downwards, compressed; **Cor.** 22–25 mm ⌀, tube 8 mm deep, 10 mm across, lobes triangular-pointed, whitish-yellow inside, with purple spots, the tube-mouth with large papillae.

**H. —** v. **immaculata** White et Sloane. – **Cor.** overall whitish-yellow, unspotted.

**H. hislopii** Turrill (§ IV). – Rhodesia; Moçambique. – **St.** branching from the base, 5 cm tall, 1 cm thick, 5–7-angled, T. 3 mm long; **Cor.** c. 25 mm ⌀, tube 16 mm long, swollen below, 11 mm ⌀ at the mouth, hirsute above, lobes triangular-tapering, cream-coloured, with blood-red spots and dark blood-red lines inside.

**H. —** ssp. **hislopii**. – Rhodesia: Bikita D.; Moçambique: Revué D. – **St.** always 5-angled; the spherically widened **tube-**base with distinct concentric lines inside, lobes longer than wide.

**H. —** ssp. **robusta** Leach et Plowes. – Rhodesia: Gokwe D. – Very robust; **St.** 5- to 7-angled; **Cor.**tube little widened, very indistinctly lined inside, lobes as long as they are broad, or little longer.

**H. humilis** (Mass.) Haw. (§ II) (*Stapelia h.* Mass.). – Cape: Beaufort West, De Aar D. – Caespitose; **St.** 10 cm tall, 1.5–2.5 cm ⌀, tapering above, with 4–5 sinuate angles, T. 2–3 mm long; **Cor.** with a short, broadly campanulate tube, limb spreading c. 3 cm ⌀, the annular thickening around the throat blackish-purple with wavy white blotches, the throat purple at the base, the lobes triangular-pointed, pale outside, sulphur-yellow inside, with purple spots.

**H. hystrix** (Hook. f.) N. E. Br. v. **hystrix** (§ I) (*Stapelia h.* Hook. f.). – Natal: Zululand, Orange Free State, Transvaal; Moçambique; S.Rhodesia. – **St.** 5–12 cm tall, c. 9 mm thick, with 5 obtuse, dentate angles, T. stout; **Cor.** 3 to 4 cm ⌀, campanulate, pale green outside, lobes triangular-tapering, with ± revolute margins, yellow inside, with red transverse lines, with dense, red, pointed papillae 3–5 mm long.

**H. —** v. **appendiculata** (Bgr.) White et Sloane (*H. a.* Bgr.). – **St.** more creeping, T. sharper; **Cor.** lobes broader than long, sulphur-yellow below, the markings and the papillae more conspicuous.

**H. inornata** Oberm. (§ I). – Cape: Albany D. – **St.** c. 8 cm tall, 12 mm ⌀, roundish to 4-angled, T. recurved, small; **Cor.** 8 mm ⌀, cylindrical, 12 mm long, glabrous, red outside, with dark red spots, dark brown inside with wine-coloured blotches, tube reddish inside, lobes 5 mm long and across, pointed, spotted.

**H. insigniflora** C. A. Maass (Pl. 100/3) (§ II). – Cape: ? Graaff Reinet D. or Transvaal. – **St.** 5–10 cm tall, 15–20 mm ⌀, with 4 slightly compressed angles, T. pointed, 5–6 mm long; **Cor.** 4–4.5 cm ⌀, rotate, tube 4 mm ⌀, limb spreading, margin dark purple-brown, lobes 10 to 12 mm long, 12 mm across, with a raised rib in the middle, minutely papillose outside, glabrous inside, with pale pink markings, intermediate lobes much reduced.

**H. keniensis** R. E. Fries v. **keniensis** (§ I). – Kenya: Mt. Kenya. – **St.** prostrate, irregularly branching, 5–12 cm long, 10–17 mm thick, angles roundish, sides slightly furrowed; **Cor.** campanulate, 25–28 mm long, 25–30 mm ⌀ above, rough-tuberculate and reddish-purple outside, paler and tuberculate below, deep purple and minutely papillose inside, lobes 4–5 mm long, 12–14 mm across, intermediate lobes tooth-like.

**H. —** v. **grandiflora** Bally. – Samburu D. – **Cor.** 37 mm long, 5 cm ⌀, whitish outside, ± glabrous, with few veins, very dark blackish-purple inside, densely covered with minute flat papillae.

**H. —** v. **molonyae** White et Sloane. – Kenya: near Nairobi. – **Cor.** 37 mm long, 35 mm ⌀, less tuberculate outside, inside deep dark purple or almost dark blue, little papillose, lobes 4.5 mm long, 15 mm across.

**H. —** v. **nairobiensis** White et Sloane. – Kenya: near Nairobi. – **Cor.** more cup-shaped, 3.5 cm ⌀, reddish-purple outside, with prominent yellowish veins, rough, inside deep reddish-purple, minutely papillose, lobes 5 mm long, 16 mm across, intermediate lobes very small.

**H. kennedyana** Lavr. (§ I). – Cape: Cradock D. – **St.** spherical, with 6–8 angles, c. 2 cm long and thick, reticulate-tuberculate; **Cor.** 22 mm ⌀, 10 mm long, tube campanulate, 6 mm long, lobes 7 mm across, 6 mm long, pointed, intermediate lobes 1 mm long, pink to cream-coloured outside, with scattered purplish-brown blotches, with 5 prominent veins, cream-coloured inside, with reddish-brown blotches and short transverse bands, glabrous below, otherwise covered with 2.5 mm long, fleshy, whitish papillae.

**H. kirkii** N. E. Br. (§ IV) (*H. bicampanulata* Verdoorn). – Transvaal; Mozambique. – **St.** 2.5–4 cm tall, 12–18 mm ⌀, with 4 acute angles, T. 2–2.5 mm long, pointed; **Cor.** 4–5 cm ⌀, tube 12–14 mm long, 25 mm ⌀, limb spreading and lobed to midway, lobes broader than long, intermediate lobes scarcely 1.5 mm long, corona glabrous outside, the tube with indistinct concentric lines inside, dark purple, papillose-bristly, lobes pale yellow, with irregular red spots.

**H. leachii** LAVR. (§ III). – Moçambique. – **St.** prostrate, up to 1.50 m long, 5–8 mm $\varnothing$, indistinctly 4-angled; **L.** 2–2.5 mm long, tooth-like; **Cor.** campanulate, 15 mm long, c. 20 mm $\varnothing$, lobes triangular-tapering, 8–9 mm long, 10 mm across, with 3 veins, intermediate lobes scarcely present, corona cream-coloured outside, $\pm$ rough, with small purplish-brown spots and densely white papillose, inside with dark purplish-brown transverse bands, with bristly, dark brown papillae.

**H. levyi** OBERM. (§ III). – Rhodesia: Wankie; Zambia: near Broken Hill. – **St.** 7 cm tall, 3–4 cm $\varnothing$, 4–5-angled, sides deeply furrowed, T. 1 cm long; **Cor.** flat, 3.5 cm $\varnothing$, tube cylindrical, 3 cm long, 12 mm $\varnothing$ at the base and 17 mm $\varnothing$ above, lobes triangular-tapering, rough outside with reddish spots, inside velvety below, dark chestnut-brown, with a small annulus, densely papillose, bristly-papillose above the ring, creamy-yellow above, with red spots and papillose.

**H. lodarensis** LAVR. (§ I). – Arabia: S.Yemen: Audhali. – **St.** very compressed, 5–7 cm tall, 2.5 cm thick, with purple blotches, 5-angled, T. pointed and spreading horizontally; **Cor.** 15 mm long, 32 mm $\varnothing$, cream-coloured outside, minutely tuberculate, inside cream-coloured with a narrow purplish-brown border, with purplish-brown spots, papillos, tube campanulate, 11 mm long and across, lobes spreading horizontally, very pointed, 9 mm across and long, with 5 prominent veins outside.

**H. loesenerana** SCHLTR. (§ I). – Transvaal: Pretoria, Belfort, Middelburg D. – **St.** 3–4 cm tall, 10–12 mm $\varnothing$, with 4 acute angles, T. 2 to 3 mm long; **Cor.** campanulate, 25 mm $\varnothing$, lobes triangular, very pointed, 5 mm long, minutely rough outside, reddish-brown, inside bluntly papillose from midway to the tip, dull yellow with reddish-brown transverse stripes and spots, tube 8 mm long, dark reddish-brown at the base.

**H. longituba** N. E. BR. v. **longituba** (§ III) (*H. bicampanulata* sensu BREMEK. et OBERM.). – Cape: Botswana. – Forming dense clumps; **St.** 2–5 cm tall, 1–2 cm $\varnothing$, acutely 4–6-angled; **Cor.** campanulate, glabrous outside with 20 prominent veins, tuberculate inside, creamy-yellow with red spots, glabrous at the base, smooth, whitish with dark red transverse lines, tube 18–21 mm long, lobes triangular, very pointed, 10 mm long.

**H. —** ssp. **cashelensis** LEACH et PLOWES. – Rhodesia: Melsetter D. – **St.** 5–6-angled, smaller than in ssp. *longituba*; **Fl.** smaller, tube slightly widened below, papillae pointed, stiff-bristly, with orange-red blotches.

**H. —** ssp. **longituba.** – Cape: Barkly West, Herbert, Kuruman, Vryburg, Fauresmith D.; Transvaal: Lichtenberg D.; Botswana. – **St.** 4–5-angled; **Fl.**tube not always widened below, papillae pointed, short-apiculate, without a bristle.

**H. macrocarpa** (A. RICH.) SPRENG. v. **macrocarpa** (§ I) (*Stapelia m.* A. RICH., *H. penzigii* v. *schimperi* BGR.). – Ethiopia. – **St.** 4–5-angled, c. 9 cm long, $\pm$ thick, pointed above, angles sinuate-dentate, T. 7–8 mm long, pointed; **Cor.** campanulate, 1 cm long, 11–20 mm $\varnothing$, lobes broadly triangular, tapering, greenish-yellow outside, lighter yellow inside, with concentric, minute, brownish transverse stripes.

**H. —** v. **arabica** (N. E. BR.) WHITE et SLOANE (*H. a.* N. E. BR., *H. penzigii* v. *a.* BGR.). – Arabia: S.Yemen. – **St.** slender, 4-angled; **Cor.** 11 mm $\varnothing$, campanulate, rough-papillose inside, lobes 5 mm long.

**H. —** v. **cerasina** WHITE et SLOANE. – **Cor.** chestnut-brown and light yellow inside.

**H. —** v. **penzigii** (N. E. BR.) WHITE et SLOANE (*Stapelia p.* N. E. BR., *H. p.* N. E. BR., *St. macrocarpa* BGR., *H. m.* BGR.). – Ethiopia. – **Cor.** broad-campanulate, c. 2 cm $\varnothing$, paler outside or dirty white, inside deep blackish-red, papillose-tuberculate, lobes acutely pointed, intermediate lobes small, $\pm$ plicate.

**H. —** v. **schweinfurthii** (BGR.) WHITE et SLOANE (*H. penzigii* v. *s.* BGR.). – Ethiopia: Eritrea. – Inner corona lobes dark reddish-brown.

**H. marnierana** LAVR. (§ I). – S.W.Arabia. – **St.** erect, 4–6 cm long, 14–17 mm $\varnothing$, 5-angled, brown-spotted, T. compressed, very pointed, up to 8 mm long; **Cor.** rather fleshy, flat, broadly cup-shaped, 33 mm $\varnothing$, 2–5 mm high, cream-coloured outside, lobe-tips brownish, with 5 prominent veins, whitish inside, tube pink below, with a narrow red border, red-tuberculate overall, very flat, c. 4 mm high, lobes 12 mm across, 7 mm long, pointed, intermediate lobes 1 mm long or missing.

**H. montana** KERS. (§ I). – Angola: Huila D. – **St.** erect, branching freely from the base; **Br.** 2–5 cm tall, up to 1.5 cm across, with 5 acute angles, $\pm$ dull green and dull violet, angles passing over into whitish, sharp T.; **Fl.** (1–)2(–3) in the lower part of the shoot, pedicels 1 cm long, Sep. 5 mm long, subulate, corona broad-campanulate, tube 7 mm long, widening above to the spreading 7 mm long lobes, tube dark brown, minutely rough above, with 2–3 mm long papillae, white below, tipped and banded dark brown.

**H. namaquensis** PILLANS (§ I). – Cape: Lit. Namaqualand. – **St.** 3.5–6 cm tall, 10–15 mm $\varnothing$, 4–5-angled, T. 3–4 mm long, short-pointed; **Cor.** 24–26 mm $\varnothing$, tube 6 mm deep, 8 mm $\varnothing$ and papillose at the mouth, pale yellow outside, whitish-yellow inside, somewhat purple-spotted, limb spreading, whitish-yellow, purple-spotted, papillose, lobes triangular, tapering, 5 mm long, plicate towards the tip, intermediate lobes 1 mm long, broad-triangular.

**H. nouhuysii** VERDOORN (Pl. 100/2) (§ I). – Transvaal: Soutpansberg D. – **St.** 8–20 cm tall, branching, pointed, with 4–6 spirally contorted, dentate angles, T. 5 mm long, deltoid-tapering; **Cor.** campanulate, lobes triangular, tapering, 5 mm long, greenish inside, with red spots and lines, with small, red-tipped, conical protuberances up to the lobe-tips, intermediate lobes small, pointed.

**H. occulta** LEACH et PLOWES (§ IV). – Rhodesia: Victoria D. – **St.** thin, creeping, rooting, 5-angled, deeply furrowed, up to 32 cm long, 3–9 mm ⌀, T. 1–3 mm long, pointed; **Cor.** bicampanulate, outside slightly rough below, the tube glossy, blackish-purple inside below, otherwise cream-coloured with flame-coloured spots, papillose above, lobe-margins minutely tuberculate, tube cylindrical, slightly constricted at the mouth, c. 7.5 mm long, 10 mm ⌀, the limb cup-shaped and spreading, the lobes triangular-tapering, ± erect, c. 13 mm long, 12 mm across, with 5 veins inside.

**H. ocellata** (JACQ.) SCHULT. (§ II) (*Stapelia o.* JACQ.). – Cape: Somerset East D. – **St.** 5–10 cm tall, rather thick, with 4–5 sinuate-dentate angles, T. small, projecting; **Cor.** with a campanulate tube and a very spreading limb 5 mm across, and a large, broad, smooth annulus, lobes broad-triangular, pointed, intermediate lobes small, corona glabrous outside, pale, lobes and limb densely and minutely papillose, yellow, with small blood-red spots, often tuberculate, the tube often hirsute in the upper part.

**H. oculata** HOOK. f. (Pl. 100/4) (§ I) (*H. rogersii* R. A. DYER). – S.W.Afr.: Damaraland. – Caespitose; **St.** numerous, 8–12 cm tall, with 5 sinuate-dentate, laterally compressed angles, T. large, thorn-like, soft; **Cor.** 2.5 cm ⌀, 10-lobed, the larger lobes short-deltoid, tapering, intermediate lobes very small, limb little spreading, dark reddish-brown, sharply contrasting with the white throat, minutely papillose inside, greenish outside, with prominent veins.

**H. pendula** E. A. BRUCE (§ III). – Cape: Bolo Native Reserve. – **St.** pendulous, 45–150 cm long, branching or simple, ± cylindrical or slightly 4-angled, the angles blunt or not very prominent, with tubercles in opposite or alternate pairs; **Br.** spreading ± at right angles to the stem, (3–)4(–9) cm long, slightly tapering towards the tip; **Cor.** campanulate, 15 mm ⌀, lobes 4 mm long, tuberculate inside, with longitudinal ribs, intermediate lobes T.-like, small, tube 6–8 mm long, 10 mm ⌀, biscuit-coloured and slightly purple outside, longitudinally ribbed, deep chestnut-brown inside and densely and minutely tuberculate.

**H. piersii** N. E. BR. (§ I). – Cape: Wodehouse D. – Caespitose; **St.** 2–3 cm tall, 12 mm ⌀, with 4 acute angles, T. projecting, pointed, 2 mm long; **Cor.** 2.5–3 cm ⌀, lobes abruptly and widely spreading, very sharply tapering, 8 to 9 mm across and long, minutely papillose inside, otherwise smooth, deep ochre-yellow with red spots merging into transverse bands, tube campanulate, 6 mm long, 2 cm ⌀, cream-coloured inside, with purple transverse lines, throat red, with sparse, reddish-brown hairs.

**H. pillansii** N. E. BR. ssp. **pillansii** (§ V). – Cape: Laingsburg, Prince Albert, Oudtshoorn, Calitzdorp D. – **St.** 2–4 cm tall, 1.5–2 cm ⌀, almost spherical at first, with tubercles in dense, spiral, longitudinal R., with hair-like tips; **Cor.** campanulate, tube 8 mm long, 6 mm ⌀, glabrous inside below, reddish to cream-coloured, with red spots, lobes 10–12 mm long, 6–7 mm across, recurved, long-tapering, glabrous outside, inside reddish papillose and pale yellow, with small red spots, intermediate lobes short, subulate.

**H. —** ssp. **echidnopsioides** LEACH. – Cape: Steytlerville D. – Caespitose, with numerous underground rhizomes, **St.** above ground ± cylindrical, up to 16 cm long, 1 cm ⌀, tubercles reticulate, in 8–10 spiral R., 1–2 mm high, with a button-like protuberance at the tip, with deep furrows between the R.

**H. praestans** N. E. BR. (§ II). – Cape: Laingsburg, Riversdale, Ladismith D. – **St.** 3–5 cm tall, c. 12 mm ⌀, red-spotted, the 4 acute angles with projecting T. 3–4 mm long; **Cor.** 4 cm ⌀, tube campanulate, 6 mm long, 12 mm ⌀, throat projecting horizontally, much thickened and somewhat 5-angled, lobes triangular-pointed, c. 12 mm long, smooth outside, glabrous, cream-coloured to reddish, with red veins and ribs, densely and minutely papillose inside, with dark red, short hairs and cream-coloured papillae on the annulus and in 5 triangles, annulus and tube cream-coloured with red spots, the base glossy, deep red.

**H. primulina** N. E. BR. v. **primulina** (Pl. 100/5) (§ I). – Cape: Albany, Uitenhage, Somerset East, Cradock, Tarka D. – **St.** crowded, 3–8 cm tall, 10–12 mm thick, with 4–5 sinuate-dentate angles; **Cor.** with a campanulate, pale reddish, waxy, fleshy tube 6 mm long, 8 mm broad, lobes triangular-tapering, intermediate lobes plicate, yellowish-white inside, with some reddish spots on the blackish-brown subsidiary corona, base-colour often golden-yellow, unspotted.

**H. —** v. **rugosa** N. E. BR. (*H. flava* N. E. BR.). – **Cor.** limb tuberculate.

**H. procumbens** (R. A. DYER) LEACH (§ II) (*Duvalia p.* R. A. DYER). – Transvaal. – **St.** prostrate or often pendulous, 10–15 cm long, 7–12 mm ⌀, dull green with a purple sheen, with 5 blunt, tuberculate angles; **Fl.** 1–2 together, Cor. 2.5–3 cm ⌀, annulus 8–9 mm ⌀, chestnut-brown, 3 mm high, lobes narrowing, lanceolate, 1.5 cm long, 4–5 mm across below, parchment-coloured, upperside with short brownish hairs, the terracotta-coloured margins folded upwards towards the tip.

**H. quinta** (PHILLIPS) WHITE et SLOANE (§ I) (*H. scabra* v. *q.* PHILLIPS). – Transvaal: Potgietersrust D. – **St.** 7 cm tall, with 4 acute angles, T. horn-tipped; **Cor.** 27 mm ⌀, tube campanulate, 4 mm long, 9 mm across, bristly-papillose and with dark red bands at the mouth, the limb spreading horizontally, lobes ovate, short-tapering, 8 mm long, 9 mm across, papillose.

**H. reticulata** (MASS.) HAW. (§ II) (*Stapelia r.* MASS., *St. crassa* J. DONN., *H. barbata* SCHLTR.). – Cape: Clanwilliam, Van Rhynsdorp D. – Caespitose; **St.** 5–10 cm tall, 15–25 mm ⌀, red-spotted, with 5 acute, sinuate-dentate angles, T. pointed; **Cor.** broad-campanulate, the limb bent upwards or plate-like, c. 5 cm ⌀, with an annular thickening which is blackish-red and glossy, tube campanulate, fleshy, c. 8 mm long, 10 mm ⌀, glossy blood-red inside, hairy, lobes broad-triangular, short-pointed, yellowish, irregularly spotted so that the base-colour appears reticulate, intermediate lobes small, pointed.

**H. schneiderana** BGR. (§ I). – Malawi: near Kimbila; Moçambique: Mangulane. – **St.** numerous, slender, 20 cm tall, with 5–7 slightly dentate angles; **Cor.** campanulate, c. 3 cm $\varnothing$, brownish outside, velvety-black inside, margins reddish to flesh-coloured, with a brownish-red, papillose border, lobes recurved.

**H. similis** N. E. BR. (§ I). – Angola: Pungo Ndongo. – **St.** extended, over 20 cm long, over 1 cm $\varnothing$, cylindrical, with 5 obtuse angles with small T.; **Cor.** c. 1.5 cm $\varnothing$, 7 mm long, campanulate, minutely papillose inside, glabrous outside, lobes projecting, spreading, triangular, tapering.

**H. simplex** N. E. BR. (§ V). – Cape: Victoria West D. – **St.** 3–4 cm tall, with 4–5 acute angles, T. spreading, 1.5–2 mm long; **Cor.** c. 2 cm $\varnothing$, smooth outside, lobes 7 mm long and across, pointed, spreading, minutely papillose inside, yellow, with pink spots in the middle, tube c. 7 mm long, 11 mm $\varnothing$, campanulate.

**H. somalica** N. E. BR. (§ II). – Somalia. – Caespitose; **St.** 4–5 cm long, 15–20 mm $\varnothing$, brownish-spotted, rounded, the 5 blunt angles with large, pointed, deltoid T.; **Cor.** deep campanulate, with an annular, thickened rim 3.5 to 4.5 mm broad around the throat, pale outside, smooth and glossy reddish-brown inside, tube c. 1 cm broad and deep, lobes broadly triangular, paler, almost flesh-coloured, with minute, spiny tubercles, intermediate lobes small.

**H. stapelioides** SCHLTR. (§ I). – Transvaal: Pietersburg D. – **St.** c. 4 cm tall, 12–13 mm $\varnothing$, 4-angled, T. spreading; **Cor.** broad-campanulate, almost rotate, c. 3 cm $\varnothing$, glabrous outside, inside densely tuberculate, lobes triangular, very tapered, intermediate lobes T.-like.

**H. striata** OBERM. (§ I). – S.W.Afr.: Gr. Namaqualand. – **St.** 5–8 cm tall, 10–15 mm $\varnothing$, roundish, with 4–6 angles, T. 2–4 mm long; **Cor.** c. 27 mm $\varnothing$, tube 6 mm long, 4–6 mm $\varnothing$, pale green or $\pm$ maroon outside, pale yellow inside, with thin, dark, transverse stripes, the tube with a flat border 14–17 mm $\varnothing$, this and the lobes light yellow with darker, transverse stripes, minutely papillose towards the margins, lobes 5 mm long, triangular, pointed.

**H. tanganyikensis** (BRUCE et BALLY) LEACH (§ II) (*Duvalia t.* BRUCE et BALLY). – Tanzania. – **St.** creeping, little branching, rooting from the prostrate sections, T. small, solitary, sharp; **Fl.** 2 together at the St.tips, Cor. with a distinct annulus, lobes tapering, salmon-pink.

**H. thuretii** CELS. (§ I) (*Stapelia t.* CROUCH.). – Cape: Transkei, Albany D. – **St.** 3–5 cm tall, the 5 acute angles with sharp T. 2–3 mm long; **Cor.** $\pm$ hemispherical-campanulate, tube c. 8 mm long and across, blood-red at the base, somewhat constricted at the throat, limb spreading, lobes broad-triangular, very pointed, 8 mm long, 9 mm across, with blood-red spots, both surfaces smooth, dull ochre-yellow inside, the annulus with blood-red blotches.

**H. transmutata** WHITE et SLOANE (§ I). – Origin unknown. – **St.** 5 cm tall, 4–5-angled, T. short; **Cor.** 3 cm $\varnothing$, tube campanulate, 12 mm $\varnothing$, 5 mm deep, maroon, with scattered, cream-coloured spots, lobes projecting-spreading, 8 mm long, 8–10 mm across, tapering, dark maroon, with yellow transverse lines near the tip, papillose and with light yellow blotches near the base.

**H. transvaalensis** STENT. (§ II). – Transvaal: Pretoria, Rustenberg D. – **St.** 4–6 cm tall, 15 to 18 mm $\varnothing$, 4–5-angled, T. 4–7 mm long; **Cor.** 4–5 cm $\varnothing$, smooth outside, suffused with purple, tube campanulate, 7–8 mm long, 6–7 mm $\varnothing$, reddish-papillose at the mouth, limb spreading, annulus convex, dark purple, very glossy, lobes 15 mm long and across, tapering, reddish-purple, with pale yellow spots, intermediate lobes very small.

**H. venusta** (MASS.) R. BR. (§ II) (*Stapelia v.* MASS.). – Cape: Karroo. – **St.** 12 cm or more tall, 20–25 mm $\varnothing$, with 4–5 sinuate-dentate angles; **Cor.** c. 5.5 cm $\varnothing$, campanulate, limb spreading, 5-partite, sulphur-yellow, with red spots, glabrous, lobes triangular, rather narrow-tapering, tube glabrous inside, with a prominent, lighter, conspicuously blotched annulus at the throat, intermediate lobes short.

**H. verekeri** STENT. (§ I). – S.W. and C.Afr. – Type: v. **verekeri**.

**H. — v. pauciflora** LEACH. – Moçambique. – **St.** longer then in v. verekeri, creeping, less branched, angles rounded, T. smaller, less distant; **Fl.** less numerous.

**H. — v. verekeri** (*H. v.* v. *stevensoniae* WHITE et SLOANE). – Rhodesia: Nyamandhlovu D.; Malawi; Zambia; Moçambique; S.W.Afr. – Dwarf, clump-forming, glabrous, usually with many Fl.; **St.** up to 10 cm long, 1.25 cm $\varnothing$, 5–7-angled, sides sharply furrowed, T. spreading, deltoid, tapering; **Fl.** with pedicels 10 mm long, Cor. c. 3.5 cm $\varnothing$, glabrous outside, tube $\pm$ hemispherical, c. 3 mm deep, white tinged with red, annulus 5-angled, 6.5–8 mm $\varnothing$, lobes greenish-yellow with red hairs, 15 mm long, tapering, intermediate lobes small.

**H. vogtsii** PHILLIPS (§ I). – Transvaal: Pretoria, Lydenburg, Waterberg, Pietersburg D.; Swaziland. – **St.** branching; **Br.** 4–9 cm long, 12–20 mm $\varnothing$, with 4 acute angles, T. $\pm$ sharp; **Cor.** 37 mm $\varnothing$, tube 5 mm deep, 9 mm $\varnothing$, minutely spotted outside, glabrous inside below, densely papillose above, lobes ovate, tapering, with papillae which are pale yellow in the lower part, chestnut-brown above, base-colour of the Cor. pale yellow, with chestnut-brown transverse bands, intermediate lobes small.

**H. volkartii** WERDERM. et PEITSCH (§ 1) (*H. v.* GOSSW.). – Angola; Nigeria; Moçambique. – Type: v. **volkartii**.

**H. — v. nigeriana** (LAVR.) LAVR. (*H. nigeriana* LAVR.). – Nigeria. – **St.** c. 7 cm tall, 5-angled; **L.** 2 mm long, tapering, projecting, persisting for 2 years; **Cor.** campanulate, tube densely tuberculate-spotted outside, cream-coloured with pink spots inside, the tube-mouth with papillae 1 mm long.

**H. — v. repens** (LAVR.) LAVR. (*H. r.* LAVR.). – Port. Moçambique; Rhodesia. – **St.** usually prostrate, creeping, 30–50 cm long, 7–9 mm thick, 5-angled; **Fl.** 22 mm ⌀, otherwise like the type.

**H. — v. volkartii.** – Angola. – **St.** creeping to ascending, 4–5-angled; **Cor.** 22 mm ⌀, tube cup-shaped, rough tuberculate outside, smooth inside, creamy-white with pink transverse lines, lobes spreading, densely covered with creamy-white papillae.

**H. whitesloaneana** NEL. (§ I). – Transvaal: Soutpansberg D. – Caespitose; **St.** 5 cm tall, greenish, purple above, with 4–5 prominent, ± spirally contorted angles, each tubercle with a deltoid T.; **Cor.** campanulate, 12 mm ⌀, 14 mm long, with purple blotches outside, distinctly ribbed, tube 1 cm long, light blood-red inside, with 5 concentric, reddish-purple lines in the middle, the upper part papillose, lobes 4 mm long, pointed, with purple spots and large papillae.

**H. witzenbergensis** LUCKH. (§ I). – Cape: Tulbagh D. – **St.** 5–7 cm tall, 1–2 cm ⌀, with 4–5 acute angles, purple-spotted, T. 5–6 mm long, tapering; **Cor.** c. 4 cm ⌀, glabrous outside with distinct veins, tube 6 mm deep, 12 mm ⌀, campanulate, slightly constricted at the mouth, limb spreading, lobes 13 mm long, 15 mm across, tapering, the tube glabrous inside, all other surfaces covered with watery papillae, those at the mouth tipped with a short hair, tube and annulus light sulphur-yellow, lobes black.

**H. zebrina** N. E. BR. v. **zebrina** (Pl. 100/6) (§ II). – Natal: Zululand, Transvaal; Botswana; S.W.Afr.: Gr. Namaqualand. – **St.** 6–8 cm tall, up to 2.5 cm thick, tapering above, T. projecting, 4–5 mm long; **Cor.** 3.5–4 cm ⌀, tube short-campanulate, narrowed at the throat, 6 mm ⌀, limb spreading horizontally, with a much-thickened annulus, lobes triangular, very pointed, 10 mm long, 12 mm across, glabrous outside, softly hairy inside, pale greenish-yellow or sulphur-yellow, with broken, reddish-brown, transverse bands, passing over to blotches towards the annulus, intermediate lobes small.

**H. — v. magniflora** PHILLIPS (*H. blackbeardae* R. A. DYER). – Transvaal: Potgietersrust D. – **Cor.** 6.5–7 cm ⌀, yellowish-white, transverse bands light purple or crimson, ± minutely hairy, lobes 22–24 mm long.

**Huerniopsis** N. E. BR. (*Heurniopsis* N. E. BR. corr. K. SCHUM.). – Asclepiadaceae. – S.W.Afr.; Botswana; Cape. – Succulent ♃; **St.** with 4 to 5 dentate angles; **Fl.** from the base, the middle or the tip of the St., tube campanulate, lobes spreading or recurved, with or without a small T. between the lobes, outer corona missing, corona-lobes thick, erect, extending beyond the anthers. – Greenhouse, warm. Propagation: seed, cuttings, or grafting on **Stapelia** or Tub. of **Ceropegia**.

**H. atrosanguinea** (N. E. BR.) WHITE et SLOANE (*Stapelia a.* N. E. BR., *Caralluma a.* N. E. BR.). – Botswana: N. Kalahari; Transvaal: Pretoria, Soutpansberg D. – **St.** as **H. decipiens**; **Cor.** 4 cm ⌀, glabrous, suffused dull red outside, inside an even, deep, blackish-red, unspotted, tube 7 mm deep, 15 mm ⌀, campanulate, lobes almost 2 cm long, 1 cm across, projecting, ovate-tapering, with a small pointed tip, margins slightly revolute, rough with minute callosities.

**H. decipiens** N. E. BR. (Pl. 101/1). – S.W.Afr.: Gr. Namaqualand; Botswana; Cape: widely distributed. – **St.** prostrate, 3–7 cm long, 1 to 1.5 cm thick, offsetting to form clumps, with 4–5 rounded angles, T. sharp, projecting, with a thorny tip, with 2 subsidiary T.; **Cor.** 2.5 cm ⌀, tube campanulate, lobes triangular, recurved, glabrous, greenish outside, slightly spotted, dull purple inside, with some yellowish blotches, the indentations somewhat ciliate, with a strong, unpleasant smell in the evening.

**H. gibbosa** NEL. – Botswana: Lobatsi. – Caespitose; **St.** prostrate to ± erect, 5 cm long, with 4 prominent angles, T. 3 mm long, with 2 subsidiary T. at the base; **Cor.** c. 4 cm ⌀, tube very short, very minutely rough, lobes 18 to 25 mm long, 9 mm across, ovate-pointed, greenish-yellow inside above, dark purple below, rough-tuberculate, with fine papillae around the mouth of the tube.

**H. papillata** NEL. – Botswana: Debeeti. – **St.** 3–5 cm long, 5–15 mm ⌀, prostrate, the tip ± ascending, 4-angled, with conspicuous tubercles terminating in a deltoid T. with 2 small denticles at the tip; **Cor.** 32 mm ⌀, tube campanulate, purple-spotted above and with numerous white hairs, lobes 15 mm long, 9 mm across, tapering, somewhat tuberculate, white with reddish-purple spots, densely papillose, white-hairy, lobes ciliate at the base.

**Hydnophytum** JACK. Rubiaceae. – For culture, see **Myrmecodia**.

**H. formicarum** JACK. – New Guinea. – Resembles **Myrmecodia**; **caudex** spineless, with several shoots; **L.** smaller. Rarely seen in cultivation.

**Hymenanthera** R. BR. Violaceae. – New Zealand, Austr. – Only one of the c. 6 species is succulent. – Coldhouse; summer: in the open. Propagation: cuttings.

**H. crassifolia** HOOK. f. – New Zealand. – Stiff ♄ up to 1 m tall; **Br.** prostrate, light green, soft, hairy; **L.** alternate, small, fleshy; **Fl.** inconspicuous, small, yellowish-white.

**Hypagophytum** BGR. Crassulaceae. – Ethiopia. – Related to the genera **Sempervivum** and **Sedum**. – **Roots** tuberous; **L.** opposite. – Summer: in the open, winter: cold-house. Propagation: easy, by division.

**H. abyssinicum** (HOCHST.) BGR. (*Sempervivum a.* HOCHST. ex A. RICH., *Sedum a.* (HOCHST.) HAMET, *Sed. malladrae* CHIOV., *Crassula m.* CHIOV.). – Ethiopia. – Dainty ♃; **roots** tuberous, swollen; **St.** slender, terete, simple, c. 15 cm long, glabrous, reddish, thickened at the L.nodes, Int. 7–35 mm long; **L.** opposite, fleshy, ovate to ovate-spatulate, blunt, margins entire, striate, red-spotted; **Infl.** a flat cyme, Fl. many, pedicellate, white with red spots.

**Hypertelis** E. MEY. ex FENZL. (*Hyperstelis* PAX). Molluginaceae. – S.Afr. – ⊙, ⊙ or small sub-♄; **Br.** usually prostrate; **L.** often in clusters, filiform, fleshy; **Fl.** in umbels, Sep. large, Pet. not present. Cold-house. Propagation: seed, cuttings.

**H. angrae-pequenae** FRIEDR. – S.W.Afr.: Gr. Namaqualand. – Spreading ♄ c. 20 cm tall; **L.** ± opposite, oblong-spatulate, blunt, semifleshy, 10–15 mm long, 2–3 mm across, stipules 3–4 mm long; **Infl.** of 1–2 Fl., Sep. blue, with a white border.

**H. arenicola** SOND. – Cape: Cape D. – ⊙–⊙; **St.** creeping, branching; **L.** opposite or in clusters, fleshy, linear-cylindrical; **Infl.** of 2–3 Fl., Sep. very soft.

**H. bowkerana** SOND. – Cape: Albany D. – ⊙, bluish; **St.** creeping, branching, Br. whitish; **L.** alternate or in clusters, linear-cylindrical, 22 mm long; **Infl.** of 3–4 Fl., Cal. rather tuberculate or smooth.

**H. caespitosa** FRIEDR. – S.W.Afr.: Damaraland. – ⊙–⊙, caespitose; **St.** creeping, branching; **L.** linear, narrow, 1–2 cm long, 1–2.5 mm across. Scarcely succulent.

**H. spergulacea** E. MEY. – Cape: Cape D. – Sub-♄, c. 40 cm tall; **roots** white; **St.** simple or forked; **Br.** stiff, filiform; **L.** at the St.-base fleshy, 8–10 mm thick, L. on the Br. in clusters of 5–10 from the swollen nodes, 6–11 mm long; **Infl.** ± an umbel.

**H. verrucosa** (ECKL. et ZEYH.) FENZL v. **verrucosa** (Pl. 101/2) (*Pharnaceum v.* ECKL. et ZEYH., *P. salsaloides* BURCH.). – Cape: Swellendam, Graaff Reinet D. – Sub-♄; **roots** woody; **Br.** and twigs whitish, Br. short, shoots 20 to 60 cm long; **L.** alternate or in clusters, curving, filiform, cylindrical, fleshy, up to 10 cm long, 1–2 mm thick, bluish-white; **Infl.** with pedicels and Cal. tuberculate, Fl. 5 mm ⌀.

**H. — v. laevigata.** AUCT.? – **St.** extended, branching; **Infl.** and pedicels smooth.

**Ibervillea** GREENE (*Maximowiczia* COGN., 1881, non M. RUPR. 1856). Cucurbitaceae. – USA: W.Texas to S.E.Arizona; N.Mexico. – Dioecious climbing plants; **caudex** thickened, partly above ground; **St.** thin, glabrous, Int. short, tendrils simple; **L.** with 3–5 deep incisions, the lobes variously further incised, linear to ovate, glabrous to roughly puberulous; ♂ **Fl.** in Rac., usually 5-merous, Cal. tubular to narrow-campanulate, lobes pointed, Cor. narrow-campanulate, lobes with margins entire to bifid, undulating, hirsute, ♀ Fl. solitary, with a round Cor., tube and lobes narrower than in the ♂ Fl.; **Fr.** a spherical or ovoid berry, red to yellowish, seeds numerous, flat. – Greenhouse, warm. Propagation: seed, ripened woody cuttings.

**I. insularis** (BRANDEG.) JACOBS. (*Max. i.* BRANDEG., *M. sonorae* v. *brevicaulis* I. M. JOHNSTON). – Mexico: Lower Sonora reg., Baja Calif. and offshore Is. – **Caudex** swollen; **St.** woody, 5–15 cm long, curved in various directions, whitish to bluish, ± puberulous, tendrils slender; **L.** deeply incised, primary lobes 4–7, broad-spatulate, crenate, dentate or more often again incised, roughly hairy; **Fl.** light green, hirsute inside; **Fr.** spherical, 2.5–3 cm long, orange-red.

**I. sonorae** (S. WATS.) GREENE (Pl. 101/3) (*Max. s.* S. WATS.). – Mexico. – **Caudex** irregularly spherical or bottle-shaped, often weighing many kg., bark fissured, corky, greyish white, roots fibrous; **St.** several, 2–3 m long, with glabrous tendrils; **L.** round to broad-ovate, deeply incised to form 3 primary lobes, coarsely and deeply or flat-dentate, glabrous or the underside sometimes with rough white hairs, 4–10 cm long; **Fl.** hirsute on both surfaces; **Fr.** ovoid, 3–5 cm long, red when ripe.

**I. — v. sonorae** (*Max. s.* v. *s.*). – Mexico: Lower Sonora desert. – **L.** bluish, the underside sometimes with rough white hairs.

**I. — v. peninsularis** (I. M. JOHNSTON) JACOBS. (*Max. s.* v. *p.* I. M. JOHNSTON). – Mexico: lower Sonora reg., central Baja Calif. – **St.** longer; **L.** not bluish or rough-haired.

**I. tenuisecta** SMALL (*Sicydium lindheimeri* v. *t.* A. GRAY., *Max. l.* v. *t.* (A. GRAY) COGN.). – USA: S.Arizona to E.Texas; Mexico. – **St.** thin, glabrous, bluish, minutely striate, tendrils very small, divided; **L.** palmate, 3–5-partite, again divided, the secondary lobes linear, coarsely dentate, upperside glabrous, underside pockmarked, hirsute; **Fl.** yellow; **Fr.** spherical, 1.5 to 3 cm long, reddish.

**Idria** KELLOG. Fouquieriaceae. – Greenhouse, warm. Propagation: seed. Acc. G. D. ROWLEY not generically distinct from **Fouquieria**.

**I. columnaris** KELLOG (Pl. 102/1) (*Fouquieria c.* KELLOG, *F. gigantea* ORCUTT). – Mexico: Baja Calif. to Sonora. – ♄-like, succulent; **caudex** soft, spongy, up to 25 m tall, much thickened at the base, up to 90 cm ⌀, usually simple, sometimes with 2–3 main St., bark whitish, with numerous small, simple Br. in spiral R., 30–60 cm long, 5–7 mm thick; **L.** dimorphous, those on the lateral Br. alternate, narrow-ovate, 15–20 mm long, 5–6 mm across, with a petiole 1 cm long, the petiole and part of the M.rib of the dead L. persisting and becoming thorny, other L. smaller, 2–3 together in the axils of the large L., 5 mm long, 1 mm across; **Fl.** in corymbs at the tips of the upper Br., Sep. 5, Cor. 6–8 mm long, narrow-campanulate, lobes round, yellow.

**Impatiens** L. Balsaminaceae. – Summer-flowering, with ± fleshy **St.**, in general not classed as succulent plants except for the following species. Greenhouse, warm. Propagation: seed.

**I. mirabilis** HOOK. f. – Is. of Langkawi, E. coast of Sumatra. – **St.** simple, much swollen and caudiciform below, columnar, cylindrical above, up to 2 m tall, 20–40 cm thick, very fleshy, light green, becoming greyish-brown; **L.** in crowded clusters at the St.tips, ovate to oblong, 20 to 25 cm long, 6–7 cm across; **Fl.** numerous, large, yellow.

**I. tuberosus** PERR. (Pl. 102/2). – W.Madag. – **Caudex** 30–40 cm ⌀; **St.** erect, 40–60 cm tall, simple, fleshy; **L.** alternate, 3–10 cm long, petiolate, ovate-oblong to elliptical, pointed, 10 to 15 cm long, 4–6 cm across, finely dentate; **Fl.** red.

**Inula** L. Compositae. – Summer: in the open, winter: in S. latitudes in the open, otherwise in the cold-house.
**I. crithmoides** L. (*I. crithmifolia* WILD., *Senecio succulentus* FORSK.). – France: S. coast. – Erect, branching ♄ 90–120 cm tall; **L.** linear, fleshy, very numerous; **Fl.** solitary on the branchlets, terminal, yellow.

**Ipomoea** L. Convolvulaceae. – S.W·Afr.; Cape. – A few species are caudiciform plants with leafy Br. and large Fl. – Greenhouse, warm. Winter: dry for the resting period. Propagation: seed.
**I. angustisecta** ENGL. (Pl. 101/4). – Cape: W. Griqualand. – **Caudex** c. 5 cm ⌀; **St.** branching from the base, 10–15 cm long, Int. c. 1.5 cm long; **L.** thickish, with a petiole c. 1 cm long, palmately incised, middle segments 5–7 cm long, lateral ones 1.5 cm long, 1.5 mm across; **Cor.** 3–3.5 cm long, 1 cm across at the base, 6 cm ⌀ at the mouth, pinkish-purple, with purplish-red veins at the mouth.
**I. inamoena** PILG. v. **inamoena**. – S.W.Afr.: Damaraland. – **Caudex** head-sized, with blackish-brown bark; **St.** 2–3, up to 50 cm long, prostrate; **Br.** curved upwards; **L.** directed upwards, on a petiole 1 cm long, broad-lanceolate, with margins and veins ciliate on both surfaces, 7–8 cm long, up to 3.5 cm across; **Fl.** solitary, pedicellate, 7.5 cm long, campanulate, 7 cm ⌀ above, whitish-pink, violet to pink inside.
**I. —** v. **trisecta** DTR. – **L.** deeply 3-cleft.

**Jatropha** L. Euphorbiaceae. – Trop. N., central and S.Am.; trop. Afr.; Madag. – ♄ or more rarely ♄ or herbaceous with thick, perennial and often tuberous rhizomes, variable in appearance, some with a thickened, fleshy caudex with milky sap, the St. often tipped during the growing period with a cluster of petiolate L., with few cymose Infl.; **L.** usually petiolate, margins entire or lobed, Stip. laciniate, often hardened, sometimes glandular or very reduced or missing; **Fl.** usually in dichotomous Infl., unisexual or bisexual, ♂ Fl. lateral, 5-partite with conspicuous Pet., ♀ Fl. solitary at the Ped. tips .(See J. PAX, **Jatropha**, in A. ENGLER, Pflanzenreich, Vol. 42, **IV**, pp. 21–113 for further species). – Heated greenhouse, dry in winter during the resting period. Propagation: seed.
**J. baumii** PAX. – S.W.Afr.: Kunene reg. – **Caudex** tuberous, spherical, 5–6 cm ⌀; **St.** 5–6 cm tall; **L.** lanceolate, margins undulating, curly, rather cartilaginous, 1–2 cm long, 3–4 mm across, blue, Stip. minute, bristly-filiform; **Infl.** of many Fl.
**J. berlandieri** TORR. (Pl. 102/3) (*J. cathartica* TERAN et BERLAND). – USA: Texas, valley of the Rio Grande; Mexico: Nuevo Leon, San Juan. – **Caudex** spherical, 10–20 cm ⌀ or more; **St.** short, several; **L.** with petiole ± 10 cm long, blade 6 cm across, 7-partite, lobes tapering, pinnate-tipped, underside bluish, Stip. subulate and 3-cleft, 3–4 mm long, pale; **Infl.** long-pedicellate, Fl. numerous, red.
**J. brachydenia** PAX et HOFFM. – Moçambique: Chiloane. – ♃; **St.** simple, 10–50 cm tall, rhizome thick, with tufted hairs; **L.** sessile, cuneate at the base, margin entire to slightly 3-lobed, 5–6 cm long, 3–3.5 cm across, glandular-dentate, with tufted hairs, Stip. laciniate, 3–4 mm long; **Infl.** tufted, Fl. pale yellow.
**J. ciliata** MUELL. – Argentine; Peru; Huancabamba. – ♄; **Br.** stout, fleshy, partly glabrous; **L.** with petiole 6–10 cm long, lamina 10–12 cm across, cordate below, shortly 5-lobed, lobes tapering, ciliate-dentate, Stip. glandular, sessile; **Infl.** short.
**J. dissecta** (CHODAT. et HASSLER) PAX (*J. gossypifolia* ssp. *heterophylla* v. *d.* CHODAT et HASSLER). – Paraguay. – ♃; rhizome 4–5 cm ⌀, extended; **St.** up to 1 m tall, simple or somewhat branched, rough-hairy; **L.** having petioles up to 2 cm long, roughly hairy, with distichous laciniate glands above, lamina 10–13 cm long and across, 3–5-lobed from the base, outer lobes linear, pointed, margins glandular, densely ciliate, veins prominent; **Infl.** short-pedunculate, Fl. numerous, dark purplish-red to reddish-brown or yellowish, Pet. often with a pale yellow margin, underside green.
**J. elliptica** (POHL) MUELL. ARG. (*Adenoropium e.* POHL, *J. officinalis* MART., *J. opifera* MART., *J. lacerti* SILVA MANSO). – S.Brazil: Minas Gerais, São Paulo. – ♃; rhizome thick, fleshy, becoming woody; **St.** 30–80 cm tall, herbaceous; **L.** short-petiolate, 10–15 cm long and across, oblong-elliptical, pointed, stiff, membranous, glabrous, margins densely glandular-ciliate, Stip. persistent, 5–6 mm long, laciniate; **Infl.** hairy, Fl. pale.
**J. erythropoda** PAX et K. HOFFM. – S.W.Afr.: Neitsas. – ♃; **caudex** thick, 10 cm long, 3 to 3.5 cm ⌀, with many heads, red; **St.** 7–12 cm tall, branching, bluish, puberulous; **L.** sub-sessile, glabrous, bluish, stiff, 5–6 cm long, irregularly laciniate, lobes linear, serrate-incised, Stip. Th. bristly, twin-tipped; **Infl.** glabrous.
**J. macrantha** MUELL. ARG. – S.Peru: Huánuco. – Spreading ♄ c. 1 m tall; **Br.** thick, ± fleshy, with L.scars; **L.** on petioles 5–8 cm long, blade glabrous, 10–12 cm across, 9–10 cm long, 3-lobed above midway, lobes broad-ovate, pointed, margins entire, Stip. reduced to sessile glands; **Infl.** capitate, with few large, scarlet Fl., Pet. 2 cm long.
**J. macrocarpa** GRISEB. – Argentine: Catamarca. – ♄ 2 m tall; **Br.** thick, fleshy, with L.scars; **L.** on a petiole 9 cm long, blade 9 cm long, 10–12 cm across, 5-lobed, lobes broadly oblong-ovate to rhombic, pointed, margins entire, ± cartilaginous, Stip. widened, deciduous; **Infl.** of few large Fl.
**J. macrorrhiza** BENTH. – Mexico; USA: S.W. Texas. – Very poisonous ♃; rhizome fleshy, 6–7 cm ⌀; **St.** c. 13 cm tall, simple, old petioles and L.veins puberulous; **L.** on a petiole 1–2 cm long, blade 12–15 cm long and across, with 3 or

5 ovate to lanceolate, pointed, dentate lobes, Stip. 3–5 mm long, bristly, laciniate; **Infl.** with few Fl.

**J. multiflora** PAX et K. HOFFM. – Argentine: Jujuy-Santa Clara. – ♄ up to 3 m tall; **Br.** thick, fleshy, with conspicuous L.scars; **L.** petiolate, firm, with 5 lanceolate, pointed lobes, margins entire; **Infl.** laxly branched, Fl. crowded.

**J. nudicaulis** BENTH. – Colombia; Ecuador. – Treelike, up to 2 m tall; **St.** thick, rough, with an apical crown of L. and Fl.; **L.** with a long petiole, pedate, with 5 ovate lobes, margins entire, Stip. short, stiff, bifurcated; **Infl.** with many Fl.

**J. pachypoda** PAX. – S.Bolivia. – ♄ up to 1.5 m tall; **St.** thick, fusiform, fleshy, with scarlet sap, bark brown, papery; **Br.** with L.scars; **L.** on a petiole 1 cm long, blade bluish, glabrous, 10 cm long, 15 cm across, with 5 ovate, pointed, glandular-serrate lobes above midway, Stip. reduced to sessile glands; **Infl.** dense, with many Fl.

**J. peltata** H. B. et K. (*Adenoropium p.* POHL). – Peru; Brazil. – ♄ up to ± 2 m tall; **Br.** thick; **L.** petiolate, blade 3–5 cm long and across, pedate, glabrous, bluish, with 3–7 short-ovate, pointed, glandular-ciliate lobes; **Infl.** corymbose, Fl. scarlet.

**J. podagrica** HOOK. (Pl. 102/4). – Guatemala; Nicaragua; Costa Rica; Panama. – **St.** 40 to 150 cm tall, thickened at the base to an ovate-clavate shape, dichotomously branched, bark grey, later ± peeling; **L.** in clusters of 6–8 at the Br.tips, long-petiolate, 3-lobed, 15–18 cm long and across, thin, green, glabrous, slightly waxy-pruinose, Stip. branched, thornlike; **Infl.** repeatedly forking, Fl. brilliant red.

**J. seineri** PAX. – S.W.Afr.: Caprivi Strip. – Dwarf ♃; rhizome thickened, tuberous, softly hairy; **St.** prostrate, 12–15 cm long; **L.** on a petiole 5 cm long, narrow-oblong, tapering, margins undulating, 4–5 cm long, 1.5–2 cm across, puberulous, Stip. minute to missing; **Infl.** lax, puberulous.

**J. texana** MUELL. ARG. (*Cnidoseolus stimulosus* ENGELM. et GRAY). – USA: Texas. – **Caudex** 10 cm or more in ⌀, with a short rhizome-like St.segment above, producing 10–20 shoots, glabrous or with a few stinging hairs; L.having petioles 3–5 cm long, blade 8–15 cm across, with 3–5 angular to shallow, sinuate-dentate, tapering lobes, glabrous or with stinging hairs, Stip. small, dentate; **Infl.** of few red Fl., with stinging hairs.

**J. tuberosa** PAX. – Sudan: S. Prov. – ♃; rhizome thick, napiform, 3–4 cm ⌀; **St.** simple, 40–75 cm tall, glabrous, leafy at the tip; L. 8–15 cm long, 2–2.5 cm across, subsessile, lanceolate, bristly-serrate, pointed, stipules 3 mm long, divided; **Infl.** crowded.

**J. woodii** O. KTZE. v. **woodii.** – Natal. – ♃ with a thick rhizome; **St.** simple, 8–50 cm tall, hairy; **L.** hairy, sessile, laciniate, with 5–7 lanceolate lobes, ± 5 cm long, 1.5 cm across, with glandular-bristly margins, Stip. very much divided, glandular, 5–6 mm long; **Infl.** short-pedunculate.

**J. —** v. **vestita** PAX. – Natal: Ladysmith. – **St.** 8–15 cm tall, densely puberulous; **L.** c. 4 cm long, margins ± entire or slightly lobed.

**J. —** v. **woodii** (*J. woodii* v. *kuntzei* PAX). – Ladysmith. – Robust, up to 50 cm tall; **L.** 10–12 cm long.

**Kalanchoe** ADANS. – Crassulaceae. – S. and trop. Afr.; Madag. and neighbouring islands; Socotra; Cyprus; E.India; Indochina; Malayan Peninsula; Java; Taiwan; trop. Am. – Glabrous or hairy ± succulent sub-♄ or ♄, ♃, ⊙ or ⊙ herbs, some like lianas; **L.** in some species with adventitious buds or decussate or 3 together in whorls, often with an amplexicaul petiole or base, ± fleshy, simple, entire or dentate, lobes pinnatisect or pinnate; the St. passing over into the Infl. above or the Infl. axillary, mostly 3-partite, paniculate-cymose, sometimes with reduced bracts, many-flowered, often with adventitious buds in the axils of the Fl.; **Fl.** erect or pendent, ± large, white, pink, violet, greenish, yellow or red; Sep. nearly free at the base or ± connate, often ± producing a tube, usually shorter than the Cor., fleshy, Cor. connate into a 4-angled tube, this cylindrical or expanded and urn-shaped at the centre or more frequently at the base, often narrowed below the lobes which are usually shorter than the tube, triangular or rounded or ovate, spreading or recurved, stamens up to 8 in 2 whorls, Carp. erect, separate or connivent; **follicles** with many small seeds. – Summer: out of doors, winter: cold-house. Propagation: seeds, cuttings, adventitious buds. (Lit.: RAYMOND HAMET and MARNIER-LAPOSTOLLE, J.: Le Genre Kalanchoe au Jardin Botanique 'Les Cedres'. in Arch. Nat. Mus. Hist., ser. 7, 8: 1–110, Paris 1964).

## Division of the Genus Kalanchoe into Sections according to P. Boiteau and O. Mannoni

**Sect. I. Kitchingia** (BAK.) BOIT. et MANN. (*Kitchingia* BAK. as a G.). – ♃ plants or epiphytes with opposite, dentate **L.**; **Infl.** terminal; **Fl.** usually large, pendent, coloured, in panicles of 3–4-partite cymes, Sep. small, triangular or roundish, ± long connate at the base, much smaller than the Cor. which is elongated, tubular-campanulate, with short, ovate or semicircular lobes. – Species: K. ambolensis, campanulata, gracilipes, peltata.

**Sect. II. Bryophyllum** (SALISB.) BOIT. et MANN. (*Bryophyllum* SALISB. as a G.). – ♃ or ⊙, terrestrial or epiphytic plants, some rambling and twining with St. thin and often several metres long; **L.** opposite or mostly in whorls of 3, simple, pinnatisect or pinnate; Infl. terminal, occasionally also axillary, 3-partitely branched, often with adventitious buds in the crenatures of the leaves;

Fl. large, mostly pendent, conspicuously coloured, as are often parts of the Infl. and the Cal., Sep. often free, usually however ± connate at the base into a tube, campanulate or vesicularly inflated, Cor. of various shapes, mostly with a 4–8-angled, constricted, straight or somewhat expanded tube, the 4 lobes shorter, rarely longer than the tube, roundish or ovate or triangular, spreading or recurved. – Species: K. adelae, beauverdii, bergeri, bouvetii, daigremontiana fedtschenkoi, × hybrida, jongmansii, laxiflora, macrochlamys, manginii, marnierana, miniata, pinnata, poincarei, porphyrophylla, prolifera, pseudocampanulata, rolandibonapartei, rosei, rubella, schizophylla, serrata, streptantha, suarezensis, tubiflora, uniflora, waldheimii.

**Sect. III. Kalanchoe** (*Eukalanchoe* BOIT. et MANN.). – Succulent ⁀ herbs, sub-♄ or ♄; L. opposite, hairy or glabrous, often with an amplexicaul petiole or base, fleshy, entire or dentate or pinnatisect; the St. passing above into the **Infl.** which is mostly 3-partite, paniculate, cymose, with reduced bracts, mostly many-flowered, the Fl. mostly erect, white, yellow, reddish or red, Sep. free or ± connate at the base, of various lengths, ± triangular or lanceolate, mostly shorter than the tube of the Cor., Pet. united into a ± 4-angled tube which is regular or urn-shaped, expanded at the centre or oftener at the base, often narrowed below the lobes, lobes mostly shorter than the tube, spreading or recurved, usually narrowed at the base. – Species: K. abrupta, adelae, adolphi-engleri, arborescens, aromatica, aubrevillei, ballyi, baumii, beharensis, bentii, bipartita, bitteri, blossfeldiana, boisii, brachygloba, bracteata, brasiliensis, brittenii, 'Cantabrigiensis', citrina, craibii, dangeardii, decumbens, densiflora, dicksoniana, ebracteata, elizae, elliptica, eriophylla, exellii, farinosa, faustii, fernandesii, figueiredoi, flammea, floribunda, gastonis-bonnieri, gentyi, globulifera, gloveri, grandidieri, grandiflora, hametorum, heckelii, hemsleyana, hildebrandtii, hirta, humbertii, humilis, integra, integrifolia, laciniata, lanceolata, linearifolia, longiflora, lugardii, luciae, migiurtinorum, marmorata, millotii, mitejea, mocambicana, montana, mortagei, nadyae, oblongifolia, obtusa, orgyalis, peteri, petitiana, platysepala, pubescens, pumila, quartiana, rhombopilosa, robusta, rosulata, rotundifolia, salazarii, scapigera, schimperana, schumacheri, scilleana, somaliensis, stearnii, synsepala, teretifolia, tetraphylla, thyrsiflora, tomentosa, trichantha, tuberosa, usambariensis, vadense, varians, vatrinii, velutina, viguieri, zimbabwensis.

**K. abrupta** BALF. (§ III). – Socotra. – Glabrous ♄; L. with a petiole 5–10 cm long, obovatespatulate, shortly acuminate, 5–9 cm long, 2.5–4 cm across, margins entire; **Cor.** 12–18 mm long, lobes 7–10 mm long.

**K. adelae** HAMET (§ III) (*Bryophyllum a.* (HAMET) BGR.). – Comoro. – Resembles **K. floribunda**; Sep. much longer.

**K. adolphi-engleri** HAMET (§ III). – S.E. Madag. – St. robust, glabrous; L. sessile, oblong, acute, margins crenate, glabrous; **Infl.** lax, Fl. large, Cor. long urn-shaped.

**K. ambolensis** HUMBERT (§ I). – E. and central Madag. – ♄ 50–80 cm tall, **St.** erect, bluish; L. fleshy, glabrous, with a petiole 5–7 cm long, oblong, blunt, 10–15 cm long, 3–6 cm across, dentate, crenate, simple, basal L. tripartite; Infl. with small adventitious buds, Fl. 9 mm long, red.

**K. arborescens** H. HUMB. (§ III). – Madag. – ♄-like, up to 4 m tall; **St.** simple, 5–8 cm ∅; Br. forming a conspicuous, capitate crown, 5–10 cm long, thick; L. 3 together in whorls or alternate, narrowing teretely, blade ± round, 1.5–2.5 cm long, apiculate; **Infl.** corymbose to pyramidal, Fl. pendulous, Cal.-tube cylindricalcampanulate, Cor. thick, fleshy, pale green, Cor.-tube slightly urn-shaped, lobes pointed.

**K. aromatica** PERR. (§ III). – Madag. – ⁀ with many St., viscous, aromatic, 30–60 cm tall; L. on a petiole 15–20 mm long, oblong-lanceolate, tapering, up to 13 cm long, 1.5–5 cm across, with large T.; **Infl.** corymbose, Cor. glandular, cylindrical, up to 8 mm long, lobes pointed.

**K. aubrevillei** HAMET ex CUF. (§ III). – N.Tanzania. – Glabrous plant, ascending obliquely up to 1.5 m tall; **St.** simple, c. 2 cm thick below; L. opposite, pruinose, broad-ovate to round, 18 cm long, 15 cm across, margins irregularly lobed or coarsely crenate, petiole 10 cm long, grooved, semi-amplexicaul; **Infl.** corymbose, c. 20 cm tall and across, with c. 40 white Fl., bracts subulate, pedical c. 1 cm long, tube c. 6 cm long, lobes ovatelanceolate.

**K. ballyi** HAMET ex CUF. (§ III). – Kenya; Tanzania. – Glabrous ⁀ (?), ± creeping, up to 1.20 m tall; **St.** simple, erect, terete, with black lenticels, 15 mm thick below, Int. 3 cm or more long; L. opposite, green, fleshy, sessile, oblong-spatulate, c. 20 cm long, 6 cm across, upper L. more pointed and smaller, margins entire to irregularly undulating; **Infl.** with branchlets c. 12 cm long, bracts filiform, c, 3 mm long, Sep. thick, tube wider at the base. up to 12 mm long, lobes 5–6 mm long, 3.5 mm across, Fl. brick-red or coral-red.

**K. baumii** ENGL. et GILG (§ III) (*K. prasina* N. E. BR.). – Zambesi reg. – Related to **K. floribunda**; L. oblong-lanceolate, blunt, margins entire; Fl. tube 6–11 mm long, lobes 2–3 mm long.

**K. beauverdii** HAMET (§ II) (*Bryophyllum b.* HAMET, *B. costantinii* HAMET, *K. jueli* HAMET et PERR., *Br. j.* (HAMET et PERR.) BGR., *K. scandens* PERR., *Br. s.* (PERR.) BGR.). – Madag. – Climbing plant; **St.** very thin, 2–3 m long, 2–3 mm thick; L. linear-lanceolate, 2 to 8 cm long, 7–15 mm across, thick, curved, reflexed, T. 5–6, with adventitious buds; **Infl.** lax, with many spotted, blackish-violet Fl. 11–16 mm long.

**K. — v. beauverdii** (Pl. 103/1) (*K. b.* v. *typica* BOIT. et MANN.). – **L.** 20–25 cm long; **Fl.** broad-campanulate, tube 14–15 mm long.

**K. — v. guignardii** HAMET. (*K. g.* HAMET et PERR.). – **Fr.** different from the type.

**K. — v. parviflora** MANN. et BOIT. – **Fl.** 11–12 mm long.

**K. beharensis** DRAKE et CASTILLO v. **beharensis** (Pl. 103/2) (§ III) (*K. vantieghemi* HAMET). – S.Madag. – **St.** slender, nodal, 2–3 m tall, glabrous; **L.** on a petiole 3–4 cm long, blade broadly ovate-lanceolate, fleshy, 10 to 20 cm long, 5–10 cm across, glabrous, older L. concave on the upperside, underside very convex, with doubly crenate margins, olive-green, grey-waxy; **Infl.** 50–60 cm tall, a large corymb, Cor. urn-shaped, 7 mm long.

**K. — v. aureo-aeneus** JACOBS. – **St.** woolly-hairy; **L.** with dense, woolly, golden-bronze to brownish hairs on both surfaces, the tips of the marginal T. with brown hairs, St. and young L. with white hairs.

**K. — v. subnuda** JACOBS. – **L.** underside sparsely white-haired, upperside ± glabrous.

**K. bentii** C. H. WRIGHT (§ III) (*K. teretifolia* DEFL., *K. deflersii* GAGNEPAIN). – S.Arabia. – Sub-♄ 1–1.2 m tall; **St.** 4–4.5 cm ⌀ below, glabrous, green, becoming corky, erect, leafy; **L.** cylindrical to ± triangular, thick, upperside ± grooved, curved, tapering above midway, 30–40 cm long, blunt-truncate at the tip; **Infl.** thyrsoid, Cor. 2–3 cm long, whitish, tube swollen below, lobes glabrous or papillose inside.

**K. — v. bentii.** – S.Yemen. – **Cor.** glabrous inside.

**K. — v. somalica** CUF. – N.Somalia. – **Cor.** glandular-papillose inside.

**K. bergeri** HAMET (§ II). – Central Madag. – ♃ 10–20 cm tall; **St.** ascending at the tip, little branched; **L.** on a petiole 5–15 mm long, blade ovate-oblong, rounded above, margins crenate, 1–4 cm long, 7–24 mm across; **Infl.** of few, pale yellow Fl., Cor. tube 10–12 mm long and across.

**K. — v. bergeri** (*K. b.* v. *typica* BOIT. et MANN.). – Plant hairy.

**K. — v. glabra** MANN. et BOIT. – Plant glabrous.

**K. bipartita** CHIOV. (§ III). – S.Somalia. – **Infl.** consisting of 2–3 thin monochasia with red Fl.

**K. bitteri** HAMET et PERR. (§ III). – Central Madag. – Glabrous plant; **L.** cylindrical, upperside slightly furrowed, up to 11 cm long, 10 to 13 mm across; **Infl.** with Ped., Cal. and Cor.-tube glandular-hairy, tube urn-shaped, 5 mm long, lobes 6 mm long, Fl. white.

**K. blossfeldiana** v. POELLN. (§ III) (*K. globulifera* v. *coccinea* PERR.). – Madag. – Compact ♃; **St.** erect, little branching, c. 30 cm tall, glabrous, smooth; **L.** on a petiole 2.5 cm long, blade oblong or ovate-oblong, c. 7 cm long, 4 cm across, dark green, glossy, margins red, ± sinuate-crenate above midway; **Infl.** often capitate, with many scarlet Fl. 9–10 mm ⌀, from Jan. to April. – Well-known as a house-plant; by applying a short-day regime during several weeks in summer, the flowering period can be advanced to October. Many cvs. have been obtained by selective breeding and crossing.

**K. boisii** HAMET et PERR. (§ III). – Madag. – Plant with long glandular hairs, up to 27 cm tall, possibly ⊙; **L.** ovate, c. 2 cm long, coarsely dentate; **Fl.** 12–13 mm long, golden-yellow.

**K. bouvetii** HAMET et PERR. (§ II) (*Bryophyllum b.* HAMET et PERR.) BGR. – Madag. – Softly hairy ♄ up to 40 cm tall; **L.** petiolate, 6–7 cm long, 6–22 mm across, oblong, bluntish, coarsely dentate; **Cal.** campanulate, hirsute, Cor. 12–22 mm long, delicate pink, glandular-hairy.

**K. brachyloba** WELW. ex J. BRITTEN (§ III) (*K. pruinosa* DTR., *K. multiflora* SCHINZ). – S.W.Afr. – ⊙; **St.** 2–5 cm long, 2–2.5 mm thick; **L.** in Ros. of 2–3 pairs, 12–15 cm long, 6–7 mm across, long-elliptical, coarsely crenate, whitish blue-green, curved perpendicularly upwards midway, 2 mm thick; **Infl.** c. 70 cm tall, Fl. 15 mm long, yellowish-green, lobes orange-yellow.

**K. bracteata** SCOTT-ELLIOT (§ III). – S.Madag. – Plant glabrous; **L.** sessile, ovate, margins entire, 2.5–3 cm long, 2 cm across, rather pointed; **Cor.** 7–8 mm long.

**K. brasiliensis** ST. HIL. (§ III). – Brazil. – ♄; **Br.** cylindrical, short-haired; **L.** short-petiolate, ovate-lanceolate, tapering, crenate-serrate, upper L. linear, smaller, margins entire; **Infl.** a dense corymb, Fl. pink, tube widened at the base, lobes tapering.

**K. brittenii** HAMET (§ III). – E.Afr. – Related to **K. usambariensis**; **L.** dentate above; **Infl.** glandular.

**K. campanulata** (BAK.) BAILL. v. **campanulata** (§ I) (*Kitchingia c.* BAK., *Kit. parviflora* BAK., *Kal. p.* (BAK.) BAILL., *Kit. panduriformis* BAK., *Kal. p.* (BAK.) BAILL., *Kit. amplexicaulis* BAK., *Kal. c.* (BAK.) BAILL.). – Central and S.E. Madag. – ♃; **St.** up to 1 m tall; **L.** ± fleshy, usually sessile, linear-oblong, blunt-dentate, up to 12 cm long, light green, white-pruinose, somewhat spotted; **Fl.** campanulate, 15 mm long, 1 cm ⌀, brilliant red.

**K. — v. orthostyla** MANN. et BOIT. – **L.** smaller, narrower; **Fl.** narrower.

**K. cv. Cantabrigiensis** (§ III). – Hybrid: Bot.Gard. Cambridge, parents unknown. – Glabrous ♄; **L.** on a petiole 2.5 cm long, oval, 10 cm long, 8 cm across, margins crenate; **Infl.** corymbose, Cor. 13 mm long, yellow, lobes broad-oval.

**K. cassiopeja** DAMMAN is probably a form or hybrid of **K. blossfeldiana**; **L.** blue-green; **Fl.** orange.

**K. citrina** SCHWEINF. (§ III) (*K. c.* v. *erythraea* SCHWEINF.). – Arabia; Yemen; N.Somalia; Ethiopia; Kenya. – Plant hirsute; **L.** sessile, ovate-lanceolate, up to 6 cm long, 12 mm across, margins with 5–7 large T.; **Fl.** yellow.

**K. craibii** HAMET (§ III). – Thailand. – **St.** erect, robust, simple, glabrous, hirsute above; **L.** opposite and decussate, trebly incised and widened at the base, petiole short, slender; **Infl.** with no distinct Ped., corymbose, umbels almost simple, Fl. short-pedicellate, Cal. campanulate, hairy, Cor. ± tubular.

**K.** cv. Crenodaigremontiana (*Bryophyllum* × *c.* RES. et VIANA). – Artifical hybrid from the Bot. Garden in Lisbon. – Parents: **K. laxiflora** (*Bryophyllum crenatum*) × **K. daigremontiana** (*B. d.*). (Lit. s. RESENDE et VIANA).

**K. daigremontiana** HAMET et PERR. (Pl. 103/3) (§ II) (*Bryophyllum d.* (HAMET et PERR.) BGR.). – S.W.Madag. – Robust, 50–100 cm tall; **St.** simple, erect; **L.** on petiole up to 5 cm long, blade long-triangular, long-tapering, 15 to 20 cm long, 2–3 cm across, often with cup-shaped protuberances at the base, rather fleshy, with brownish-red spots, margins crenate, with adventitious buds; **Infl.** branching, Cor. 16–19 mm long, grey-violet.

**K. dangeardii** HAMET (§ III). – Angola. – Related to **K. teretifolia**; **L.** up to 40 cm long, cylindrical; **Cor.** lobes long-tapering.

**K. decumbens** COMPTON (§ III). – Swaziland. – **Taproot** perennial; **Br.** spreading horizontally, finally erect; **L.** scattered, rarely in pairs, narrow lanceolate-clavate, blunt, up to 3.5 cm long, 5 mm across; **Infl.** up to 30 cm tall, with a ± capitate corymb, Cor. c. 1 cm long, crimson.

**K. densiflora** ROLFE (§ III). – Ethiopia; Sudan; Uganda; Tanzania. – Plant up to 40 cm tall; **St.** terete below, 4 mm ⌀, ribbed to 4-angled and thinner above; **L.** lanceolate and narrowing cuneately into the 3 cm long petiole, blade 6 cm long, 3 cm across, upper L. linear-spatulate and ± sessile, often dirty red, margins crenate-serrate below; **Infl.** compact, paniculate-cymose, Cor. 11 mm long, lobes broad-ovate, dirty red.

**K.** — v. **densiflora** (*K. glaberrima* VOLKENS, *K. bequaertii* DE WILD, *K. petitiana* ANDREWS). – **Infl.** and Cal. glabrous.

**K.** — v. **subpilosa** CUF. (*K. brachycalyx* ENGL., *K. lentiginosa* CUF.). – Ethiopia. – **Infl.** and Cal. sparsely hirsute.

**K. dixoniana** HAMET (§ III). – Thailand. – **St.** erect, robust, glabrous; **L.** opposite and decussate, with a petiole 2–3 cm long, obovate, blunt, 11 cm long, 3 cm across, margins slightly crenate; **Infl.** ± sessile, corymbose, umbels little branching, Fl. short-pedicellate, Cal. hairy, tube 12 mm long.

**K. ebracteata** SCOTT-ELLIOT (§ III). – Madag. – ♄ 40–60 cm tall; **Br.** densely velvet-felted; **L.** with a petiole 5–7 mm long, 2–3 cm long, 1–1.5 cm across, ovate, tapering, fleshy, white-felted; **Infl.** of many Fl., Cor. c. 1 cm long.

**K. elizae** BGR. (§ III) (*Cotyledon insignis* N. E. BR.). – Malawi. – Plant glabrous, red-tinged; **St.** simple, c. 20 cm tall; **L.** obovate-oblong, blunt-tapering, fleshy, with entire margins, c. 9 cm long, 4 cm across; Fl. in clusters in the axils of the leaf-like bracts, finally in a thyrsoid panicle, Cor. c. 2 cm long, bent, 4-angled, reddish-brown, lobes linear, 12 mm long.

**K. elliptica** E. RAADTS (§ III). – Somalia: Afra and Issa; between Erivago and Mait. – ♃ glabrous herb 35–90 cm tall; **St.** short, 2-costate, 1.2 cm thick at the base; **L.** 6–8 cm long, 3.5–6 cm wide, opposite, stalked, fleshy, green, blade elliptical to broad-elliptical, obtuse, attenuated into the petiole, margins entire to minutely crenate, M.-rib robust, petiole 2–3 cm long and widened at the base, upper L. smaller, narrowly elliptical; **Infl.** paniculate, bracts minute, Fl. erect, pedicels 3–5 mm long, Cor. c. 12 mm long, greenish-yellow.

**K. eriophylla** HILSENB. et BOJER (Pl. 104/1) (§ III) (*Cotyledon pannosa* BAK.). – Madag. – **St.** slender, woolly towards the tip; **L.** ovate-oblong, sessile, blunt, woolly, 15–20 mm long, margins entire; **Cor.** up to 6.5 mm long.

**K. exellii** HAMET (§ III). – Angola. – **St.** rhizome-like, c. 3.5 mm thick, simple to tripartite, somewhat hairy; **L.** several, at the St.tip, 3 mm long, 0.8 mm across; **Infl.** slender, leafy, these L. narrow-oblong, ± pointed, margins sinuate, 18 mm long, 5 mm across; **Fl.** short-pedicellate, Cal. hirsute, Cor. ± urn-shaped, c. 13 mm long, lobes 5 mm long.

**K. farinacea** BALF. f. (Pl. 104/2) (§ III). – Socotra. – Compact ♄, **Br.** few, white-farinose; **L.** crowded at the Br. tips, sessile, obovate, 2–3 cm long, rather thick, pale green, farinose-pruinose, margins entire, pink; **Infl.** almost corymbose, Cor. campanulate, yellow, lobes light scarlet.

**K. faustii** FONT Y QUER (§ III). – S.Morocco. – **St.** 2–2.5 cm thick, 30–70 cm tall, branching basally; **L.** sessile, 6–8 cm long, 3–4 cm across, ovate, pointed, margins entire, irregularly curly, upper L. linear-lanceolate, ± reddish; Fl. in cymes, 12–15 mm long, golden-yellow.

**K. fedtschenkoi** HAMET et PERR. (Pl. 103/4) (§ II). – Madag. – Dense ♄; **Br.** initially prostrate; finally ascending, forming stiff proproots up to 30 cm long; Int. 1–2 cm long; **L.** dense, oval to oblong, flat, fleshy, rounded above, 1.5–5 cm long, 1–2.5 cm across, with petiole 1–6 mm long, bluish, margins crenate below and brownish at the notches, with adventitious buds; **Infl.** 15–20 cm tall, Fl. numerous, 17–20 mm long, brownish-pink.

**K.** — v. **fedtschenkoi** (*K. f.* v. *typica* BOIT. et MANN.). – **L.** fairly large, with 5–8 irregular notches, pointed.

**K.** — v. **isalensis** MANN. et BOIT. – **Br.** violet; **L.** very crowded; Fl. smaller, more tubular, deeper red.

**K.** — cv. Variegata. – **L.** yellow-mottled.

**K.** × **felthamensis** HAMET. – Hybrid: **K. flammea** × **K. velutina**. Hort. VEITCH.

**K. fernandesii** HAMET (§ III). – Moçambique. – **St.** erect, robust, glandular-hairy; **L.** ± sessile, ovate, blunt, upperside hairy, margins crenate above; **Infl.** paniculate to thyrsoid, with many Fl., Cal. and Cor. hairy.

**K. figueiredoi** CROIZ. (§ III). – Moçambique. – **Taproot** fleshy, creeping, 1 cm thick; **St.** erect, 15–20 cm long, simple or branching; **L.** in whorls of 6–8, obovate to spatulate, 3–6 cm long, 1.5–3 cm across, minutely tuberculate, with purple spots, margins irregularly crenate towards the tip; **Infl.** freely dichotomously branched, 20–30 cm tall, Fl. pale green, Cor.lobes whitish to pink or scarlet.

**K. flammea** STAPF (§ III). – Somalia. – ♃; **St.** with few Br., 30–40 cm tall; **L.** obovate, blunt-rounded above, 6–7 cm long, 2.5–3 cm across, margins entire or sinuate-dentate; **Fl.** many in a cyme, c. 2 cm ⌀, brilliant orange-coloured.

**K. floribunda** TUL. (§ III). – Comoro. – Dwarf ♃; **L.** oblong, 7–8 cm long, 2.5–3 cm across, margins dentate; **Cor.** tube 5–6.5 mm long, lobes 3.5 mm long.

**K. gastonis-bonnieri** HAMET et PERR. (§ III). – ? Madag. – **St.** robust, 12 mm ⌀ below, up to 50–60 cm tall; **L.** broad-amplexicaul, ovate-lanceolate, 13.5–16.5 cm long, 4.5–5.5 cm across, rounded above, surface whitish to farinose-pruinose, margins coarsely roundish-crenate; **Infl.** corymbose, up to 30 cm tall, Cal. campanulate, Cor. pale pink.

**K. gentyi** HAMET et PERR. (§ III). – Madag. – Plant glandular-hairy; **L.** oblong-lanceolate, irregularly and distantly acute-dentate above midway, 6–11.5 cm long, 3–4 cm across; **Fl.** with tube 3.5–4 mm long.

**K. glaucescens** BRITT. Acc. A. BERGER = **K. laciniata** (L.) DC.; acc. G. CUFODONTIS to be maintained as a species, with the following synonyms: *K. holstii* ENGL., *K. magnidens* N. E. BR., *K. marinellii* PAMPANINI, *K. beniensis* DE WILD, *K. crenata* BROUN. et MASSEY, and ssp. *glaucescens* and ssp. *arabica* CUF.

**K. globulifera** PERR. (§ III). – Madag. – Dwarf ♃ branching basally; **St.** glandular-hairy, Br. projecting; **L.** in a Ros. of 3–4 pairs at the Br.tip, ± sessile, glabrous, ovate-spatulate, rounded, 14–18 mm long, 13–18 mm across, margins ± dentate above; **Fl.** 7–10, yellow, in ± capitate clusters forming a cyme 10–15 cm tall.

**K. gloveri** CUF. (§ III). – N.Somalia. – **St.** up to 1 m tall, younger sections soft-woolly; **L.** known only from fragments, narrowing towards the petiole, lobed, glandular-puberulous; **Infl.** densely corymbose, Cor. 15–17 mm long, yellow, lobes lanceolate, pointed.

**K. gracilipes** (BAK.) BAILL. (§ I) (*Kitchingia g.* BAK.). – Central Madag. – Dwarf epiphytic ♄; **St.** slender, rooting, up to 60 cm long; **L.** petiolate, obovate-oblong, blunt, crenate-serrate up to 3 cm long, 2 cm across; **Fl.** solitary, urn-shaped to campanulate. For the heated greenhouse.

**K. grandidieri** BAILL. (§ III). – S.Madag. – ♄ 1–2 m tall; **L.** obovate, mucronate, 9–15 cm long; **Fl.** nodding, Cor. 12–13 mm long, 4-angled, violet, lobes up to 7.5 mm long.

**K. grandiflora** WIGHT et ARN. (Pl. 104/3) (§ III) (*K. nyikae* ENGL.). – E.Afr.; peninsular India. – **St.** erect; **L.** sessile, obovate, 4–6 cm long, 3–3.5 cm across, crenate-dentate, glabrous, bluish-violet, waxy-pruinose; **Fl.** in cymes, tube 12–14 mm long, yellow.

**K. hametorum** HAMET (§ II). – Moçambique. – **St.** erect, slender, usually simple, hirsute; **L.** hirsute, the blade narrowing towards the petiole, narrowly linear to ovate-oblong, ± pointed, 6.5–8.5 mm long, 4–5 mm across, margins entire to crenate; **Infl.** corymbose, Fl. few, Cor. tubular to urn-shaped, salmon-pink.

**K. heckelii** HAMET et PERR. (§ III). – Central Madag. – Plant glabrous; **L.** obovate; **Fl.** with sparsely hairy Cal. lobes.

**K. hemsleyana** CUF. (§ III) (*K. nyikae* AUCT. non ENGL.). – Kenya: Kilifi D. – Plant glabrous, up to 1 m tall; **St.** 3 mm thick below; **L.** opposite, with margins entire, bluish, pruinose, ovate to ± round, up to 18 cm long, 16 cm across, distinctly peltate, petiole 3–4 cm long; **Infl.** lax, c. 20 cm tall and across, corymbose, irregular, branchlets opposite or solitary, bracts leaflike, Fl. with pedicels 1 cm long, Cor. pale pink, tube widened basally, cylindrical, c. 16 mm long, lobes ovate-lanceolate.

**K. hildebrandtii** BAILL. (§ III). – Central Madag. – Rather treelike ♄ 1.75–5 m tall; **L.** roundish to obovate-round, abruptly petiolate, 16–40 mm long, 13–35 mm across, hairy like the rest of the plant; **Cor.** whitish, sparsely hirsute.

**K. hirta** HARV. (§ III). – Cape: Natal. – Plant hairy; **L.** oblong, blunt-dentate, 7–8 cm long, 4.5 cm across, rough-haired.

**K. humbertii** GUILL. (§ III). – Angola. – ♄ with ovoid root-Tub.; **St.** erect, with L.scars; **L.** sessile, spatulate, 6–7 cm long, 2.5–3 cm across, tip rounded, upper L. linear; **Infl.** corymbose, Fl. yellowish, Cor. 2 cm long, lobes tapering, 7.5 mm long.

**K. humilis** BRITTEN (§ III). – Moçambique. – Plant glabrous; **root** creeping; **St.** thin, erect, 8–10 cm tall; **L.** spatulate, ± sessile, 2 cm long, 8 mm across, dentate above; **Infl.** of few Fl., Cor. 4.5 mm long, lobes 2.5 mm long.

**K. cv. Hybrida** (§ II) (*Bryophyllum h.* HORT.). Intermediate hybrid: **K. daigremontiana** × **K. tubiflora**. Usually sterile. Acc. F. RESENDE K. poincarei HAMET seems to be identical.

**K. integra** (MED.) O. KTZE. (§ III) (*Cotyledon i.* MED., *C. deficiens* FORSK., *K. d.* (FORSK.) ASCHERS. et SCHWEINF., *K. glaucescens* v. *d.* (FORSK.) SENN., *C. nudicaulis* MURR., *Verea n.* (MURR.) SPRENG., *C. aegyptiaca* LAM., *K. ae.* (LAM.) DC.). – Yemen; Ethiopia (cult.); W.Kenya; Tanzania; Congo; Malawi (?); Moçambique; Rhodesia; Angola; S.E.Cape; Is. St. Tomé and Annobon; Dahomey; Ivory Coast; Sierra Leone; Guinea; Mali; introduced to E. Brazil, now growing wild, similarly to the W.Indies: Cuba, Andros and the Virgin Is. – Plant completely glabrous or not so in all parts; **St.** up to 90 cm tall; **L.** long-petiolate, oblong-lanceolate, doubly dentate and spotted; **Infl.** partly glabrous, hairy or glandular, ultimate branchlets scorpioid, pedicels short or undeveloped, Fl. yellow, reddish or scarlet to golden-yellow or reddish-brown.

Intermediates have been observed between the following varieties: between v. **integra** and v. **verea** in Tanzania, Ghana, Sierra Leone and Mali; between v. **integra** and v. **crenato-rubra** in E. Tanzania; and between v. **verea** and v. **crenata** in Tanzania.

**K.** — v. **crenata** (ANDR.) CUF. (*Verea c.* ANDR., *K. c.* (ANDR.) HAW., *K. afzeliana* BRITT., *K. c. v. c., K. laciniata* AUCT. non (L.) DC., *K. i.* AUCT. non O. KTZE.). – Tanzania; Rhodesia; Cape. – Plant completely glabrous in all parts; **Fl.** lighter or darker yellow.

**K.** — v. **crenato-rubra** CUF. – Tanzania. – Plant completely glabrous in all parts; **Fl.** lighter or darker red.

**K.** — v. **integra** (*K. coccinea* WELW. ex BRITT., *K. crenata* (ANDR.) HAW. v. *cocc.* CUF.). – Ethiopia; Kenya; Tanzania; Angola; Sierra Leone. – Plant not completely glabrous in all parts; **Fl.** distinctly pedicellate, lighter or darker red.

**K.** — v. **subsessilis** (BRITT.) CUF. (*K. coccinea* v. *s.* BRITT., *K. kirkii* N. E. BR.). – Moçambique. – Plant not completely glabrous in all parts; **Fl.** sessile or almost so, lighter or darker red.

**K.** — v. **verea** (JACQ.) CUF. (*Cotyledon v.* JACQ., *K. crenata* v. *v.* (JACQ.) CUF., *C. c.* (ANDR.) VENT., *K. c.* (ANDR.) TRATT., *C. brasilica* VELL., *K. b.* (VELL.) STELLFELD, *K. brasiliensis* CAMB., *K. coccinea* AUCT. non WELW. ex BRITT.). – Uganda; Tanzania; Congo; Moçambique; Is. Annobon; S.Nigeria; Ivory Coast; Brazil. – Plant not glabrous in all parts; **Fl.** axis, upper L. and Fl. densely glandular-hairy, Infl. branchlets extended, Fl. lax, lighter or darker yellow.

**K. integrifolia** BAK. (§ III). – Central Madag. – $2\!\!\!|$; **St.** up to 30 cm tall, erect; **L.** sessile, glabrous, very thick, oblong-cuneate, margins entire, c. 3 cm long, 8–12 mm across; **Infl.** dense, branchlets hairy, Cal. soft-haired, Cor. red, up to 4 mm long, lobes 5 mm long.

**K. jongmansii** HAMET et PERR. (§ II) (*K. j.* HUMBERT). – Madag. – Small $2\!\!\!|$; **Br.** $\pm$ woody, 9–20 cm long; **L.** sessile, rather fleshy, oblong, rounded above, 16–40 mm long, up to 10 mm across, margins entire or crenate-serrate above; **Infl.** glandular-hairy, Fl. few, up to 3 cm long, yellow.

**K.** — v. **ivohibensis** HUMBERT. – **Fl.** 3 cm long.

**K.** — v. **jongmansii** (Pl. 104/4). – **Fl.** 20 to 25 mm long.

**K.** × **kewensis** THISELTON-DYER. – Hybrid: **K. teretifolia** DEFL. × **K. flammea** STAPF. – Intermediate in habit but **L.** pinnatisect and **Fl.** pink.

**K. laciniata** (L.) DC. (§ III) (*Cotyledon l.* L., *K. acutifolia* HAW., *K. aegyptiaca* DC., *K. alternans* PERSOON, *K. angustifolia* A. RICH., *K. carnea* N. E. BR., *K. ceratophylla* HAW., *K. coccinea* WELW., *K. diversa* N. E. BR., *K. floribunda* WIGHT, *K. glaucescens* BRITT., *K. integra* KUNTZE, *K. magnidens* N. E. BR., *K. ndorensis* SCHWEINF., *K. rohlfsii* ENGL., *K. rosea* CLARKE, *K. schweinfurthii* PENZIG, *K. spathulata* DC., *K. stenosiphon* BRITT.,

*K. welwitschii* BRITT.). – W., E. and S.Afr.; E.As.: Sunda Is.; peninsular India; Yemen; Brazil. – Succulent $2\!\!\!|$; **St.** erect; **L.** petiolate, glabrous or hairy, simple or pinnatifid with linear segments; **Fl.** yellowish.

**K. lanceolata** (FORSK.) PERSOON (§ II). – E.Afr.; Congo Rep.; Angola; peninsular India. – Plant minutely hirsute, 25–35 cm tall; **L.** ovate to lanceolate, margins serrate above, 3–7.5 cm long, 10–26 mm across, $\pm$ appressed to the St.; **Fl.** in double cymes from the L.axils, Cal. spherical, tube 1 cm long, orange, often tinged red.

**K.** — v. **lanceolata** (Synonyms acc. G. CUFO-DONTIS: *Cotyledon l.* FORSK., *K. pubescens* R. BR., *K. floribunda* WIGHT et ARN., *Verea f.* (WIGHT et ARN.) DIETR., *K. brachycalyx* RICH., *K. modesta* KOTSCHY et PEYR., *K. glandulosa* HOCHST. ex RICH. v. *bengulensis* ENGL., *K. crenata* v. *collina* ENGL. p.max.part., *K. goetzei* ENGL., *K. diversa* N. E. BR., *K. brachycalyx* v. *erlangeri* ENGL., *K. ellacombei* N. E. BR., *K. homblei* DE WILD, *K. h.* v. *reducta* DE WILD, *K. laciniata* v. *brachycalyx* (RICH.) CHIOV., *K. gregaria* DTR., *K. glandulosa* HOCHST. ex RICH. v. *rhodesiaca* BAK. f. ex RENDLE; A. BERGER adds: *K. heterophylla* WIGHT, *K. pilosa* BAK.). – Peninsular India; Yemen; Somalia; Ethiopia; Sudan; Kenya; Uganda; Tanzania; Moçambique, Malawi; Congo Rep.; Zambia; Rhodesia; ? Transvaal; Angola; S.W.Afr.; Nigeria; Ghana; Mali. – The united part of the Cal. much shorter than the lobes.

**K.** — v. **glandulosa** (HOCHST. ex RICH.) CUF. (*K. glandulosa* HOCHST. ex RICH., *K. ritchieana* DALZELL). – Peninsular India; Ethiopia; Kenya; Uganda; Tanzania. – The united part of the Cal. as long as the lobes, or longer.

**K. laxiflora** BAK. (§ II) (*Kitchingia l.* BAK., *Bryophyllum l.* BAK., *K. crenata* HAMET, *K. tieghemii* HAMET). – Madag. – Glabrous $2\!\!\!|$; **Br.** 50 cm tall, resting on the ground and rooting; **L.** 2–6 cm long, petiolate, blade variable, margins crenate, $\pm$ distinctly auriculate at the base, with adventitious buds in the notches; **Infl.** branching, Fl. pink, orange or red.

**K.** — v. **stipitata** MANN. et BOIT. – **L.** brown-spotted, auricles much reduced; **Cor.** distinctly pedicellate, red or yellowish-orange.

**K.** — v. **subpeltata** MANN. et BOIT. – **L.** margins with conspicuous red blotches in the notches, auricles large; **Cor.** orange-red.

**K.** — v. **violacea** MANN. et BOIT. – **L.** margins red, auricles little developed; **Cal.** violet, Cor. red or pink.

**K. linearifolia** DRAKE DEL CASTILLO (§ III). – S.Madag. – Plant glabrous; **L.** sessile, cylindrical, acute, 3–13 cm long, 4–10 mm across, upperside furrowed; **Cor.** 7–9 mm long, urn-shaped, red.

**K. longiflora** SCHLTR. v. **longiflora** (§ III) (*K. petitiana* v. *salmonea* HORT.). – Robust, very leafy; **St.** simple, distinctly 4-angled; **L.** sessile, ovate-oblong, blunt, margins serrate above, 5–6 cm long and across, light grey-green to yellowish to slightly orange-coloured; **Fl.** tube 14–16 mm long, lobes 4 mm long.

**K.** — v. **coccinea** MARN.-LAP. (Pl. 105/1) (*K. petitiana* HORT.). – **L.** plicate, margins coarsely dentate, salmon-coloured to scarlet in full sun.

**K. luciae** HAMET (§ III). – Cape, Transvaal. – Plant glabrous; **L.** sessile, obovate, blunt-rounded, margins entire; **Infl.** paniculate, Cor.-tube urn-shaped.

**K. lugardii** BULLOCK (§ III) (*K. robynsiana* HAMET). – Ethiopia; Kenya; Uganda; Tanzania. – Robust ♄; **St.** erect, up to 2.25 m tall, glabrous; **L.** little petiolate, broad-spatulate and sessile, coarsely dentate to pedate and thin-petiolate with intermediate forms with auriculate or cordate blades, often with darker stripes and blotches in the indentations of the margins; **Infl.** corymbose, with many Fl. up to 2.5 cm ∅, white and yellow.

**K. macrochlamys** PERR. (§ II) (*Bryophyllum m.* (PERR.) BGR.). – Madag. – Glabrous ♄ 80 to 120 cm tall, root tuberous; **L.** sessile, simple to pinnatifid, auriculate at the base, 30 to 40 cm long, 10 cm across, segments oblong, up to 20 cm long, crenate-dentate; **Infl.** cymose, Cal. much inflated, yellowish, tube spherical, hirsute, lobes pointed.

**K. manginii** HAMET et PERR. (§ II) (*Bryophyllum m.* (HAMET et PERR.) NOTHDURFT). – S.Madag. – **Br.** numerous, woody, flowering Br. 10–30 cm long, simple; **L.** ovate-spatulate, very fleshy, up to 8 mm thick, 17–30 mm long, 9 to 14 mm across, at first minutely hairy; **Infl.** with adventitious buds and few Fl., Cor.tube up to 24 mm long, red.

**K.** — v. **triploidea** BOIT. et MANN. (*Br. m.* v. *t.* (BOIT. et MANN.) NOTHDURFT). – Natural triploid variety, **Fl.** larger.

**K. marmorata** BAK. (Pl. 105/5) (§ III) (*K. macrantha* BAK., *K. grandiflora* A. RICH., *K. somaliensis* BAK., *K. kelleriana* SCHINZ, *K. grandiflora* v. *angustipetala* ENGL., *K. m.* v. *maculata* TERR., *K. m.* f. *somaliensis* (BAK.) PAMPANINI, *K. rutshurensis* LEBRUN et TOUSSAINT). – Somalia; Ethiopia; Sudan; Kenya; Congo Rep. – ♃ branching basally; **St.** erect to prostrate; **L.** sessile, obovate, c. 10 cm long, 6–8 cm across, green, grey-pruinose, both surfaces with large brown blotches, margins sinuate-dentate; Fl. 6–8 cm long, white.

**K. marnierana** JACOBS. (§ II) (*K. humbertii* MANN. et BOIT. non GUILL.). – S.W.Madag. – Woody glabrous sub-♄; **Br.** creeping at the base, up to 30 cm long, with stiff roots; **L.** ± oval, rounded above, 3 cm long, 2.3 cm across, not sinuate or crenate, in the natural short-days (winter) with numerous adventitious buds along the margins; **Fl.** numerous, large, pink.

**K. migiurtinorum** CHIOV. (§ III). – N.E.Somalia. – Plant glabrous, 60 cm tall; **St.** terete; **L.** sessile, oblong-lanceolate to ± broad-spatulate, thick, leathery, with entire margins, up to 9 cm long, up to 12 mm across; **Infl.** a dense cyme, Cor. white to yellowish, lobes 6–7 mm long.

**K. millotii** HAMET et PERR. (Pl. 105/2) (§ III). – Madag. – Much-branched, minutely soft-haired ♄; **L.** with a grooved petiole, ovate-pointed, margins sinuate-crenate, 3–4 cm long and across.

**K. miniata** HILSENB. et BOJER (§ II). – Madag. – ♃; **Br.** rooting from the base, 50 to 80 cm tall, glabrous; **L.** ± petiolate, long-triangular to tripartite or roundish, 2.5–8 cm long; crenate to dentate, auriculate, amplexicaul, upperside with red blotches; **Infl.** of campanulate Fl. up to 2.5 cm long.

**K.** — v. **andringitrensis** PERR. – **Infl.** large, Fl. numerous.

**K.** — v. **anjirensis** PERR. – Plant 50–60 cm tall; **L.** crenate, with 2 unequal wings at the petiole.

**K.** — v. **confertifolia** PERR. – **L.** ovate-roundish, margins undulating, with few red blotches; Fl. red.

**K.** — v. **miniata** (*Kitchingia m.* BAK., *Bryophyllum m.* (HILSENB. et BOJ.) BGR., *K. subpeltata* BAK., *K. m.* v. *typica* PERR.). – **L.** with red blotches on the upperside; **Fl.** glossy red.

**K.** — v. **peltata** BAK. – **L.** small, ± scutate; Fl. red.

**K.** — v. **sicaformis** MANN. et BOIT. – **L.** tripartite, stiletto-shaped, with minute red spots.

**K.** — v. **subsessilis** PERR. – **L.** ± sessile.

**K. mitejea** LEBL. et HAMET (§ III). – E.Afr. – Related to **K. rotundifolia**; **L.** usually auriculate at the base, 16 mm long, 5.5 cm across, the lower half slightly dentate; **Fl.** sulphur-yellow.

**K. mocambicana** RES. et SOBRINO (§ III). – Moçambique. – Glabrous ♃; **St.** up to 4, dying after flowering, erect, up to several m tall; **L.** broad-elliptical, 9–15 cm long, 5–8 cm across, irregularly dentate-crenate, petiole up to 3 cm long, grooved on the upperside; **Infl.** corymbose, forking, with many Fl., tube cup-shaped, greenish-yellow, lobes up to 3 mm long.

**K. montana** COMPTON (§ III). – Swaziland. – Robust plant, with a Ros. of L. at the base, producing a prostrate St. with a terminal Infl.; **St.** fleshy, minutely hairy; **L.** simple, obovate to lanceolate, rounded above, softly fleshy, margins often slightly undulating, surface minutely puberulous, up to 18 cm long 7 cm across; **Infl.** very large, Fl. erect to projecting, Cor. greenish-yellow, slightly asymmetrical, lobes pointed.

**K. mortagei** HAMET et PERR. (§ III). – N.Madag. – **St.** erect, simple, glabrous, 30 to 60 cm tall; **L.** petiolate, glabrous, narrowly ovate-reniform, blunt, 7–10 cm long, up to 4 cm across, margins crenate, red; **Infl.** ± racemose, Fl. 3 cm long, red.

**K. nadyae** HAMET (§ III). – Central Madag. – **St.** erect, robust, branching, glabrous; **L.** circular to ovate, with entire margins, 20 to 27 mm long and across, petiole 7–9 mm long; **Infl.** corymbose, Cor.tube urn-shaped, lobes ovate.

**K. oblongifolia** HARV. (§ III). – Cape: Hopetown D. – **St.** robust, glabrous; **L.** obovate-oblong; **Fl.** yellow. Insufficiently known.

**K. obtusa** ENGL. (§ III). – Tanzania. – Up to 10 cm tall; **L.** ± circular, c. 22 mm ⌀, fleshy; **Infl.** of many Fl.

**K. orgyalis** BAK. (§ III) (*K. antonasyana* DRAKE DEL CASTILLO). – Madag. – Erect, branching ♄ up to 1 m tall, with woolly hairs; **L.** ovate-spatulate, tapering towards the petiole, tapering above, margins curved upwards, 5 to 7 cm long, 3–4 cm across, minutely hairy, bronze-coloured, underside sometimes with adventitious buds.

**K. peltata** (BAK.) BAILL. v. **peltata** (§ I) (*Kitchingia p.* BAK.). – Central Madag. – Sub-♄ 1–2 m tall; **L.** scutate, petiole 2 to 10 cm long, blade ovate, blunt, coarsely sinuate-dentate, 3–7 cm long, 2–6 cm across; **Fl.** pendulous, campanulate-tubular, pink.

**K. — v. mandrakensis** PERR. (*K. mandrakensis* PERR., *Kitchingia m.* (PERR.) BGR.). – **L.** broadly ovate-triangular, dentate, 9 cm long, 7 cm across; **Fl.** red.

**K. — v. stapfii** (HAMET et PERR.) PERR. (*K. s.* HAMET et PERR., *Kitchingia p. v. s.* (HAMET et PERR.) BGR.). – More robust than the type; **L.** more dentate, less scutate; **Fl.** longer.

**K. peteri** WERD. (§ III). – Tanzania. – Plant 40–60 cm tall; **L.** petiolate, oblong to ovate, 19 cm long, 10 cm across, rounded above, auriculate below, reddish-blue to flesh-coloured, margins ± crenate; **Infl.** terminal, Cor. 33 mm long, lobes 16–17 mm long, pale yellow.

**K. petitiana** A. RICH. (§ III). – Ethiopia. – Glabrous or glandular ♄; **St.** branching from the tip; **L.** ovate-oblong, narrowing towards the base, bluntish above, margins serrate above; **Infl.** tripartite, Fl. yellow, Cor. widened below, lobes bluntish.

**K. — v. neumannii** (ENGL.) CUF. (*K. n.* ENGL.). – Plant ± short-glandular.

**K. — v. petitiana** (*K. petitiaesii* A. RICH., *K. quartiana* v. *micrantha* PAMPANINI). – **Fl.** with Cor. 3 cm long.

**K. pinnata** (LAM.) PERSOON v. **pinnata** (§ II) (*Cotyledon p.* LAM., *Bryophyllum calycinum* SALISB., *B. p.* (LAM.) KURZ, *Sedum madagascaricum* CLUS.). – Trop. Afr.; Cape Verde Is.; Madeira; Reunion, Mauritius, Rodriguez, Comoro; trop. As.; Austr.; trop. N. and S.Am. – **St.** stout, erect, fleshy, up to 1 m tall; **L.** at first simple, ovate, crenate-serrate, younger L. pinnate with oblong-round segments, crenate-dentate, up to 13 cm long, leathery-fleshy, with young plants forming in the indentations of the T.; **Fl.** up to 3.5 cm long, greenish-whitish, reddish, Cal.tube campanulate.

**K. — v. calcicola** PERR. (*Bryophyllum p. v. c.* PERR.). – **L.** ternate.

**K. platysepala** WELW. (§ III). – Angola; Zambesi reg. – **St.** hirsute; **L.** linear-lanceolate, over 3 mm across, margins entire or sinuate.

**K. poincarei** HAMET (§ II). – S.E.Madag. – **St.** long, thin, whitish, with minute red spots, 3–3.5 m long, 4–5 mm ⌀, rooting from the lower part; **L.** ovate-lanceolate, the upper L. petiolate, margins acute-dentate, 3.5–4 cm long, 14–19 mm across; **Fl.** pink-purple.

**K. porphyrocalyx** (BAK.) BAILL. (§ II). – Central Madag. – Epiphytic, rarely terrestrial and growing in moss; **St.** erect, stiff, fleshy, up to 35 cm long; **L.** variable in shape, sessile, 2.5–5.5 cm long, 5–40 mm across, margins irregularly crenate; **Infl.** with up to 9 red to yellow Fl., Cal. violet.

**K. — v. porphyrocalyx** (*K. p. v. typica* BOIT. et MAN., *Kitchingia p.* BAK., *Bryophyllum p.* (BAK.) BGR., *Kit. sulphurea* BAK., *B. s.* (BAK.) BGR., *K. s.* (BAK.) BAK.). – **L.** obovate; **Fl.** with Cor. 25 mm long, 12 mm ⌀.

**K. — v. sambiranensis** HUMBERT. – **St.** 4–5 cm long; **L.** linear, 25–32 mm long, 5 to 6 mm across; **Cor.** 20–22 mm long, red, lobes orange.

**K. — v. sulphurea** BAK. – **St.** up to 20 cm long; **L.** more oblong, irregularly crenate, up to 45 mm long, up to 14 mm across; **Cor.** 28 mm long, 10 mm ⌀.

**K. prolifera** (BOWIE) HAMET (Pl. 105/4) (§ II) (*Bryophyllum p.* BOWIE). – Central Madag. – **St.** 1–3 m tall, 4-angled; **L.** petiolate, entire or pinnate, segments oblong-lanceolate or ovate-oblong, dentate or pinnatifid, 3–5 cm long, sometimes with adventitious buds in the notches of entire L. and along the M.ribs; **Infl.** also with adventitious buds, Cal.tube 14 mm long, Cor. 16 mm long, lobes ovate, 3–4 mm long, Fl. yellow.

**K. pseudocampanulata** MANN. et BOIT. (§ II) (*K. miniata* v. *decaryana* PERR.). – E.Madag. – ♃; **St.** up to 1 m tall; **L.** rather fleshy, sessile, violin-shaped at the base, amplexicaul, margins fine-serrate; **Infl.** with numerous adventitious buds, Fl. many, red, Cor. campanulate.

**K. pubescens** BAK. (§ II). – Central, N. and E.Madag. – **St.** up to 1 m tall, ± glabrous to long glandular-hairy; lower **L.** ± sessile, upper ones petiolate, 16–43 mm long, 6–30 mm across, bluntly roundish-dentate, often auriculate at the base, minutely hairy; **Infl.** of many campanulate, yellow to red Fl., with numerous adventitious buds.

**K. — v. alexiana** MANN. et BOIT. – Madag. – Fl. yellowish-red, lobes with 5 red lines.

**K. — v. brevicalyx** MANN. et BOIT. – Hairs red; Cal. small.

**K. — v. decolorata** MANN. et BOIT. – **Fl.** yellow or pink, with light stripes.

**K. — v. grandiflora** MANN. et BOIT. – **Fl.** 3 cm long, lobes 13 mm long.

**K. — v. pubescens** (*K. aliciae* HAMET, *Bryophyllum a.* (HAMET) BGR., *K. miniata* v. *p.* PERR., *K. m.* v. *glandulosa* PERR., *K. m.* v. *tsinjoarivensis* PERR., *K. p. v. typica* MANN.). – Fl. 2 cm long, lobes 5 mm long, yellowish-orange, with reddish lines.

**K. — v. — f. reducta** HUMBERT (*K. p. v. typica* f. *r.* HUMBERT). – Plant 20 cm tall; **L.** 16 mm long, 6 mm across; **Fl.** as f. **pubescens**.

**K. — v. subglabrata** MANN. et BOIT. – ± glabrous variety.

**K. — v. subsessilis** MANN. et BOIT. – Plant red-hairy; **L.** sessile or nearly so; **Fl.** as v. **brevicalyx.**

**K. pumila** BAK. (§ III) (*K. brevicaulis* BAK.). – Central Madag. – Plant 10–20 cm tall; **L.** sessile, ovate, narrowing cuneately towards the petiole, 2–2.5 cm long, 15 mm across, margins crenate above, minutely farinose-pruinose; **Fl.** numerous, in a botryoidal panicle, reddish-violet, tube urn-shaped.

**K. quartiana** A. RICH. (§ III) (*K. dyeri* N. E. BR.). – Malawi; Ethiopia. – ♃; **L.** petiolate, ovate-oblong, blunt, dentate, 13–18 cm long; **Fl.** up to 4.5 cm long.

**K. rhombopilosa** MANN. et BOIT. (Pl. 105/3) (§ III). – Madag. – Small glabrous ♄ with few Br.; **L.** very short-petiolate, narrowing cuneately towards the petiole, broad-triangular, rounded above, irregularly crenate-sinuate, green-grey with silvery scales and red spots, 2 cm long and across, 2 mm thick; **Infl.** very tall, branching, **Fl.** small. Fallen L. root easily.

**K. robusta** J. B. BALFOUR (§ III). – Socotra. – Robust; **L.** scarcely petiolate, elliptical, blunt, 4.5–6.5 cm long, 3–3.5 cm across, margins entire; **Fl.** up to 3.5 cm long, lobes 8–10 mm long, 3.5 mm across.

**K. rolandi-bonapartei** HAMET et PERR. (§ II) (*K. tsaratananensis* PERR., *Bryophyllum t.* (PERR.) BGR., *K. t.* FRANCOIS). )– Central Madag. – Robust ♃; **St.** up to 2 m tall, much branching, minutely hairy; **L.** petiolate, ovate-lanceolate, 15–25 cm long, 5–7 cm across, doubly crenate-dentate; **Infl.** paniculate, Fl. orange-coloured.

**K. rosei** HAMET et PERR. (§ II). – Madag. – ♃, ☉ or ☉; **St.** erect, up to 1.80 m tall, unbranched; **L.** usually lanceolate, tripartite, 4–17 cm long, 5–20 mm across, irregularly sinuate-dentate, slightly spotted, terminal lobe triangular, again tripartitely divided, with adventitious buds in the notches; **Infl.** branching, Fl. deep pink, Cor. c. 3 cm long, lobes 6 mm long.

**K. — v. rosei** (*Bryophyllum r.* (HAMET et PERR.) BGR., *K. r. v. typica* HUMBERT). – L. tripartite-hastate or slightly pinnatifid.

**K. — v. seyrigii** MANN. et BOIT. – L. linear-lanceolate, 4–5 cm long, 5–6 mm across, margins serrate-dentate; Fl. large.

**K. — v. variifolia** GUILL. et HUMBERT. – L. large, linear or linear-lanceolate, 6–12 cm long, 1–2 cm across, more dentate than v. rosei; **Fl.** ± deeply incised, petiole 2 cm long.

**K. rosulata** E. RAADTS (§ III) (*K. heimii* nom. nud.). – Arabia: S.Yemen, Audhali Plateau. – ♃ glabrous herbs (20–)50(–75) cm tall, roots thick; **St.** terete, 1–1.2 cm thick at the base, 3–4 mm thick above; **L.** apparently rosulate but opposite, sessile, lower L. approximate, Int. 2–3 mm long, blade fleshy-leathery, entire to nearly entire, ovate, obtuse to acute, attenuated at the base, to 8 cm long, 5 cm wide, dark green to red, upper L. smaller; **Infl.** laxly paniculate, bracts minute, Fl. c. 15 mm long, pale red to yellowish-green.

**K. rotundifolia** HAW. (§ III). – Cape; Socotra. – ♃; **St.** slender, up to 1.5 m tall, densely leafy; L. obovate, rarely 3-lobed, blunt-tipped, narrowing towards the short petiole, 2.5–3 cm long, 1.5–2.5 cm across, margins entire; **Fl.** c. 13 mm long, crimson.

**K. rubella** (BAK.) HAMET (§ II) (*Bryophyllum r.* BAK.). – Madag. – Related to **K. prolifera**; **St.** robust, glabrous, mottled; **L.** opposite and decussate, lower L. simple, petiolate, ovate, bluntish, margins crenate, middle and upper L. with 3–9 pinnae, 6–12 cm long, 2.5–5 cm across, leaflets crenate; **Infl.** corymbose.

**K. salazari** HAMET (§ III). – Angola. – **St.** erect, slender, simple, glabrous, fairly tall; **L.** sessile, lanceolate, pointed, 22–33 mm long, 8–10 mm across, margins entire; **Infl.** corymbose, with many Fl., Cor. ± urn-shaped.

**K. scapigera** WELW. (§ III). – Angola; Socotra. – Plant stout, glabrous, up to 40 cm tall; **L.** crowded at the St.tips, sessile, obovate, with entire margins, 20–27 mm long, 1–2 cm across, very waxy-pruinose; **Fl.** brilliant red.

**K. schimperana** A. RICH. (§ III) (*K. neumannii* ENGL., *Cotyledon deficiens* HOCHST. et STEUDEL). – Ethiopia. – Upright plant; **L.** petiolate, hairy, round-ovate, blunt, 5–6 cm long and across; **Cor.** 4–8 cm long.

**K. schizophylla** (BAK.) BAILL. (§ II) (*Kitchingia s.* BAK., *Bryophyllum s.* (BAK.) BGR.). – Central Madag. – ♃; **St.** woody, climbing, 6–8 m long; **L.** petiolate, pinnatifid, 8–10 cm long, with 6–8 pairs of recurved, subentire leaflets; **Infl.** with 4 Br., Cor. 15 mm long.

**K. schumacheri** KOORD. (§ III). – Origin unknown. – Erect plant c. 50 cm tall, glabrous, blue; **L.** on a petiole up to 4 cm long, cuneate at the base, 10 cm long, 5 cm across, margins doubly serrate, more rarely sinuate and lobed, upper L. more linear; **Infl.** terminal or axillary, Fl. yellow, Cor. up to 2 cm long.

*K. schweinfurthii* PENZIG = **K. laciniata** (L.) DC. (G. CUFODONTIS separates this species from K. laciniata, but adds the following synonym: *K. laciniata* WIGHT.) – E.Indies; Ethiopia; Angola.

**K. scilleana** HAMET (§ III). – Cape: Prieska D. – Related to **K. laciniata**; **L.** oblong, margins entire or slightly sinuate.

**K. serrata** MANN. et BOIT. (§ II). – Central Madag. – Glabrous sub-♄ forming dense bushes; **Br.** creeping at the base, then erect, 30–60 cm tall; **L.** elliptical to oval, 4–6 cm long, 2.5–4.5 cm across, rounded above, with small wings below, with serrate-dentate margins, 3 mm thick, ± sessile, bluish, minutely whitish-pruinose, with a few reddish blotches; **Infl.** up to 30 cm tall, Fl. tubular, 3 cm long, reddish-orange.

**K. somaliensis** HOOK. f. (§ III). – E.Afr. – ♃ branching from the base; **St.** densely leafy; **L.** obovate, narrowing to the base which is auriculate, 16 cm long, up to 10 cm across, with serrate-dentate margins, grey-white pruinose, with a few brown spots; **Fl.** c. 8 cm long, white.

**K. stearnii** HAMET (§ III). – Origin unknown. – **St.** erect, ± creeping in the lower part, slender, glabrous; **L.** opposite and decussate, thick-petiolate, glabrous, broad-obovate, bluntish, margins crenate above; **Infl.** shortly pedunculate, Fl. few, in corymbs, short-pedicellate, Cor. ± urn-shaped.

**K. streptantha** BAK. (§ II) (*Kitchingia s.* BAK., *Bryophyllum s.* (BAK.) BGR.). − Central and W.Madag. − ♃; **St.** robust, up to 1.20 m tall, branching basally, with L. in a crowded Ros. above; **L.** ovate-lanceolate, 4−13 cm long, 1−7.5 cm across, tapering, with margins entire or slightly sinuate, petiolate, bluish; **Infl.** corymbose, Fl. up to 3.5 cm long, yellow.

**K. suarezensis** PERR. (§ II) (*Bryophyllum s.* (PERR.) BGR.). − N.Madag. − ♄; **St.** 40−60 cm tall; **L.** dense, 12−15 cm long, with large and irregularly spaced T., grey-green, with adventitious buds at the tip; Cal. reddish-violet, tube 15 mm long, lobes 1 cm long.

**K. synsepala** BAK. (§ III). − Central Madag. − **St.** short, erect, with adventitious buds from the L.axils; **L.** few, ovate-spatulate, narrowing to the petiole which is short or up to 5 cm long, rounded above, 6−14 cm long, 4−5 cm across, margins undulating-dentate; **Infl.** long, slight, capitate, Fl. pendulous, Cor.tube ± hirsute, 8−9.5 mm long, lobes up to 5 mm long.

**K. teretifolia** HAW. (§ III). − Peninsular India. − Very stiff, erect ♄ up to 2 m tall; **St.** fleshy, simple, leafy, purple-red at the base, with L.scars; **L.** 3-pinnatifid, segments again pinnatifid and linear to semi-terete, pointed, terminal segment extended and caudate, petiole thick, fleshy; **Infl.** corymbose, Cor. plate-shaped, jasmine-like, light yellow.

**K. tetraphylla** PERR. (§ III). − Central Madag. − **St.** up to 20 cm tall; **L.** 4 or more, 13−14 cm long, softly hairy, indistinctly dentate, becoming glabrous; **Infl.** corymbose, axillary, Fl. pendulous.

**K. thyrsiflora** HARV. (Pl. 106/2) (§ III). − Cape: Transvaal. − **St.** up to 60 cm tall, densely leafy, dying after flowering; **L.** initially rosulate, 10−15 cm long, 5−7 cm across, blunt-rounded above, like the St. densely white-pruinose, margins often reddish; **Infl.** paniculate, Fl. 1.5 cm long, yellow, Cor. tubular to urn-shaped.

**K. tomentosa** BAK. (Pl. 106/1) (§ III) (*K. pilosa* HORT.). − Central Madag. − ♄ up to 50 cm tall, the entire plant densely white-felted; **St.** branching from the base, densely leafy; **L.** sessile, long-oval, blunt, 7 cm long, c. 2 cm across, grooved-concave above, underside carinate, margins rounded, coarsely serrate above, L.tips with brown spots at the T.; **Infl.** very tall.

**K. trichantha** BAK. (§ III) (*K. brachycalyx* BAK.). − Madag. − **St.** very thin, erect; **L.** with a petiole up to 5.5 cm long, fleshy, long-spatulate, 12−15 cm long, 4−7 cm across, green, glabrous, margins sharply serrate in the bottom third; **Infl.** paniculate-botryoidal, Fl. 3−5 cm ⌀, with a long tube.

**K. tuberosa** PERR. (§ III). − Madag. − **Root** woody-tuberous, the entire plant farinose-pruinose; **L.** 2.5−3 cm long, sinuate-dentate; **Infl.** cymose, Fl. 4 cm long, pink.

**K. tubiflora** (HARVEY) HAMET (Pl. 103/3) (§ II) (*Bryophyllum t.* HARVEY, *K. dolagoensis* ECKL. et ZEYH., *B. d.* (ECKL. et ZEYH.) H. SCHINZ, *K. verticillata* SCOTT-ELLIOT, *B. v.* (SCOTT-ELLIOT) BGR., *Geaya purpurea* CONST. et POISS.). − S.Madag. − Robust, glabrous plant up to 1 m tall; **St.** erect, little branching; **L.** ± cylindrical, reddish-brown spotted, upper-side somewhat grooved, up to 10 cm long, 4 to 6 mm across and thick, the tip with 2−8 notches with adventitious buds; **Infl.** compact, with many red to purple Fl., Cal. campanulate, Cor. 20−24 mm long, lobes 7−10 mm long.

**K. uniflora** (STAPF) HAMET (§ II). − Madag. − Epiphyte; **Br.** creeping and rooting, 2 mm thick, sometimes extended; **L.** ± sessile, orbicular to obovate-oblong, glabrous, 4−35 mm long, 4 to 15 mm across, very thick; **Infl.** hairy, with few red, campanulate to urn-shaped Fl.

**K. —** v. **brachycalyx** MANN. et BOIT. − **L.** oblong, 35 mm long, 14 mm across; Cal. short, campanulate, Cor. 10−13 mm long, 6 mm ⌀.

**K. —** v. **uniflora** (*Kitchingia u.* STAPF, *Bryophyllum u.* (STAPF) BGR., *K. ambrensis* PERR., *Br. a.* (PERR.) BGR., *K. u.* v. *typica* MANN. et BOIT.). − **L.** ± circular or oval, 4−18 mm long, 4−15 mm across; Cor. 11−19 mm long.

**K. usambariensis** ENGL. et HAMET (§ III). − Tanzania: Usambara. − Low-growing; **L.** 4 to 10 cm long, up to 4.5 cm across, margins entire or slightly sinuate; **Infl.** glandular, Cor. 1 cm long, red.

**K. × vadensis** BOOM et ZEILINGA (§ III). − Amphidiploid from colchicine-treatment of the hybrid **K. blossfeldiana × K. grandiflora** (Inst. of Plant Breeding, Wageningen). − Resembles **K. grandiflora**, up to 1 m tall; **L.** circular or broad-oval, 8−12 cm long and across, ± 3 mm thick, tip rounded, narrowing below towards the 3−4 cm long petiole; **Infl.** forking, Fl. with pedicel 1 cm long, Cor. 2.5−3 cm long, dark crimson.

**K. varians** HAW. (§ III) (*K. subamplectens* nom. nud.). − Nepal. − Robust sub-♄ 1−1.20 m tall; **St.** cylindrical, finger-thick below, purple, branching above the base; **L.** thick, fleshy, bluish, blunt-serrate, lower L. ovate, 10−12 cm long, others tripartite, lateral segments lanceolate, tapering, disparate, upper L. narrow-lanceolate to linear, grooved, crenate, petiole thick, amplexicaul, 2.5−5 cm long; **Infl.** ± cymose, bracts tripartite, upper ones linear, top ones scale-like, Cor. 4-partite, yellow.

**K. vatrinii** HAMET (§ III). − E.Afr.: Zambesi reg. − Related to **K. crenata**; **L.** oblong, margins dentate or incised above; **Infl.** glabrous or hairy.

**K. velutina** WELW. (§ III) (*K. lateritia* ENGL., *K. cuisinii* DTR., *K. kirkii* N. E. BR., *K. angolensis* N. E. BR., *K. coccinea* v. *subsessilis* BRITT.). − Angola; Congo Rep.; Malawi; Tanzania; Zanzibar. − Rhizome thick, fleshy, creeping, sprouting, the entire plant shortly hairy, including the Fl.; lower **L.** oblong to ovate-lanceolate, sinuate, blunt to rather pointed, crenate above, rounded, 11 cm long, narrowing towards the short, amplexicaul petiole, up to 8 cm across, upper L. linear-lanceolate, rounded; **Fl.** 12 mm long, yellow.

**K. viguieri** HAMET et PERR. (§ III). – Madag. – White-woolly ♄ 1–1.20 m tall; **L.** petiolate, ovate, with entire margins, 10–16 mm long; **Cor.** 15–20 mm long.

**K. waldheimii** HAMET et PERR. (§ II). – Central Madag. – Glabrous sub-♄; **Br.** numerous, somewhat creeping at the base, 20–40 cm long; **L.** flat, with small auricles at the base, oval, rounded above, 10 cm long, 2–4.5 cm across, bluish, with a few red spots, notches large, roundish, with few adventitious buds; **Infl.** divided, Fl. 3 cm long, pink.

**K. zimbabwensis** RENDLE (§ III). – Rhodesia. – Erect, papillose-hairy ♃ or ⊙; **St.** fairly stout, simple; **L.** ovate-roundish, ± sessile, c. 5 cm long, 4 cm across, margins indistinctly sinuate, tip rounded, upper L. ± lanceolate, 4 cm long, 1–3 cm across, with a petiole 7 mm long; **Infl.** of few golden-yellow Fl., Cal. densely papillose-hairy, Cor.tube hairy, 8.5 cm long, lobes 4 to 5 mm long.

**Karimbolea** DESC. Asclepiadaceae. – Greenhouse, warm. Propagation: seed.

**K. verrucosa** DESC. (Pl. 106/6). – Madag.: Prov. Tuléar. – Low, leafless sub-♄, c. 15 cm tall; **Br.** flexible, initially prostrate, becoming stiff, erect, 5–10 cm long, glabrous, brown, becoming grey, tuberculate; **L.** oblong-ovate, 1–1.5 mm long, soon dropping; **Infl.** lateral, of 3–6 Fl., Pet. lanceolate, thick, c. 9 mm long, pale pink to violet outside.

**Kedrostis** MEDIC. Cucurbitaceae. – Trop. Afr.; peninsular India. – Monoecious, more rarely dioecious, climbing or creeping plants, glabrous or hairy or rough, sometimes ± fleshy; **caudex** perennial, often tuberous and fleshy; **L.** with margins entire, dentate or compound; tendrils simple, or less often bipartite; **Fl.** small to minute, yellow to green, ♂ Fl. in a Rac. or corymb, ♀ Fl. solitary or few together; **Fr.** small, ovoid, berry-like. – Greenhouse, warm. Propagation: seed. – 6 new species from Madag. described by M. KEDRAUDEN have no characteristics of succulence.

**K. africana** (L.) COGN. (*Bryonia a.* L., *B. dissecta* THUNBG., *B. multifida* E. MEYER, *B. pinnatifida* BURCH., *Coniandra a.* SOND., *C. dissecta* SCHRAD., *C. pinnatifida* SCHRAD., *Rhynocarpa a.* ASHERS., *R. dissecta* NAUD.). – S. and S.W.Afr. – Monoecious; **caudex** white, fleshy, succulent, brittle, up to 0.5 m long, simple or branched; **St.** 4–6 m long; **L.** with a sparsely hairy petiole 4–12 mm long, blade triangular, slightly rough, 6–10 cm long, 3–5-partite, segments laciniate, dentate; tendrils filiform; **Fl.** dirty green, Pet. striate; **Fr.** orange-red, 12–15 mm long.

**K. crassirostrata** BREMEK. – Transvaal. – Monoecious; **St.** slender, furrowed; **L.** ± sessile, blade ± round, upperside white-spotted, 2.5 cm long and across, digitate or 3-partite, the segments pinnatifid; tendrils simple; **Fl.** 3 mm ⌀; **Fr.** red, rostrate, 14–17 mm long.

**K. glauca** (SCHRAD.) COGN. v. **glauca** (*Coniandra g.* SCHRAD., *Bryonia grossulariaefolia* E. MEYER, *C. g.* ARN., *K. g.* PRESL.). – S.Afr. – Monoecious; **caudex** partly above ground, ovoid to ± spherical, tuberculate below, green; **St.** ± filiform, curved in various directions, furrowed-angular, subglabrous; **L.** with a petiole 2–7 mm long, blade ± circular, usually glabrous, 3–4 cm long and across, 3–5-partite, segments ovate-cuneate, dentate to lobed, apiculate; **Fl.** with thin pedicels 1–3 cm long, Pet. densely papillose; **Fr.** ovoid, 10–12 mm long, 6–8 mm thick.

**K.** — v. **dissecta** (SOND.) COGN. (*Coniandra g.* v. *d.* SOND.). – **L.** segments distinctly dentate to incised-serrate.

**K. gracilis** R. FERNANDES. – Angola: Moçamedes D. – Dioecious; **caudex** thick, cortex splitting, scaly, corky; **St.** thickened and woody at the base; **Br.** numerous, slender, c. 25 cm long, angular-furrowed, hairy at first; **L.** 1 cm long and across, ovate, ± 3-lobed, blunt, margins crenate-curly, with dense, tufted hairs; tendrils simple, filiform; **Fl.** small; **Fr.** unknown.

**K. hirtella** (HOCHST.) COGN. – Ethiopia. – **Caudex** fleshy; **St.** climbing or trailing, to 2 m or more, with spreading hairs; **L.** broad ovate to reniform or cordate, scabrid, 3–10.5 cm long, 3.5–13 cm broad, usually ± 3–5-lobed, petiole 1.5–6.5 cm long; **Fl.** greenish-white to greenish-yellow.

**K. mollis** (KUNTZE) COGN. (*Cyrtonema m.* KUNTZE, *Coniandra m.* SOND.). – Cape: Uitenhage D. – Monoecious; **caudex** ± cuneate-ovoid, only the tip aboveground, bipartite at the base, with scattered tubercles; **St.** slender, angular-furrowed, ± simple, with scattered hairs; **L.** with stout, densely woolly petioles 2–3.5 cm long, blade fleshy, 3–3.5 cm long, 4–7 cm across, reniform to ± circular, 5-angled, margins crenate-dentate, upperside greenish-grey, with long, soft felt, underside whitish-grey, felted; tendrils slender; **Fl.** small; **Fr.** yellowish to reddish, ovoid, rostrate, up to 2 cm long.

**K. nana** (LAM.) COGN. (*Bryonia n.* LAM., *Sicyos angulata* BERG., *B. triloba* THUNBG., *B. trilobata* DER., *Cyrtonema trilobata* SCHRAD., *C. thunbergii* SOND.). – Cape. – Monoecious; **St.** slender, branching, furrowed, sparsely hairy, becoming glabrous; **L.** with petioles 1–4 cm long, slender, short-haired, blade leathery, broadly reniform to cordate, simple or with 3–5 small lobes, margins entire or slightly undulating, pale green, glabrous to sparsely short-haired, 2–6 cm long and across; tendrils filiform, short; **Fl.** greenish-yellow; **Fr.** ovoid, 15–17 mm long, 7–8 mm thick, red. – The plants in cultivation have a thick **caudex,** but this is not mentioned in the original description.

**K.** — v. **latiloba** (SCHRAD.) COGN. – **L.** 3–5-lobed from midway or higher up.

**K.** — v. **nana** (Pl. 106/5). – **L.** reniform to cordate, angular to scarcely lobed.

**Kinepetalum** SCHLTR. Asclepiadaceae. – S.Afr. – Related to the G. **Ceropegia.** Greenhouse, warm. Propagation: seed, cuttings.

**K. schultzei** SCHLTR. (Pl. 106/3) (*Ceropegia phalangium* DTR.). – S.W.Afr.: Gr. Namaqualand. – **Caudex** smooth, 8 cm ⌀, 4–5 cm tall; **St.** numerous, erect, up to 1 m tall, forked, terete, minutely hairy, light grey to violet, Int. 2–5 mm long; **L.** narrow-linear, curved lengthwise to form a groove, 4–6 cm long, 1–2 mm across, bluntly acuminate, minutely hairy, green; **Fl.** 1–2 together, pedicellate, tube campanulate, 3 mm long and ⌀, long-hairy inside, whitish with green spots, terminating in filiform, grey-green, projecting lobes 3 cm long.

**Lenophyllum** ROSE. Crassulaceae. – Mexico; USA: Texas, Calif. – Glabrous, caespitose ♃ herbs; **L.** opposite, rosulate, fleshy; flowering St. terminal, ± densely leafy, **Fl.** solitary or a few together, yellow, in one-sided Rac. or spikes. – Summer: in the open, winter: cold-house. Propagation: seed, L.-cuttings.

**L. acutifolium** ROSE. – USA: Calif. – Up to 10 cm tall; **L.** lanceolate, pointed, upperside grooved; **Fl.** greenish-yellow, numerous.

**L. guttatum** ROSE (Pl. 107/1) (*Sedum g.* ROSE). – Mexico. – **L.** broadly ovate-lanceolate, acute, 2–3 cm long, c. 1 cm across, with a blunt tip, upperside broadly grooved, grey-green with brown spots; **Infl.** 3–4-branched, Sep. very thick, Pet. light yellow.

**L. pusillum** ROSE. – Mexico. – **L.** 8–16 mm long, narrow, pointed, upperside furrowed, underside carinate, fleshy, dirty reddish green, readily dropping and rooting; **Fl.** solitary, yellow, pedicels 4–5 cm long.

**L. reflexum** WHITE. – Mexico. – **L.** broad-oval to elliptical, up to 4.5 cm long, 3 cm across, pointed, upperside flat or convex, mature L. tinged purple, otherwise light green; **Infl.** corymbose, with many yellow Fl. c. 12 mm ⌀.

**L. texanum** (J. G. SMITH) ROSE (*Sedum t.* J. G. SMITH, *Villadia t.* ROSE). – USA: Texas. – Up to 10 cm tall; **L.** obovate to ovate-lanceolate, thick; **Fl.** numerous, reddish.

**L. weinbergii** BRITT. – Mexico. – **L.** very blunt and wide, constricted and terete below, broadly grooved, 15 mm long and across.

× **Leptaloinella** ROWL. Liliaceae. – Intergeneric hybrid: **Leptaloe** × **Aloinella.** REYNOLDS includes both these genera in **Aloe,** and is followed here, so that the following would be listed under **Aloe.** – Cultivation as for the dwarf Aloes.

× **L. 1.** – **Leptaloe albida** STAPF × **Aloinella haworthioides** (BAK.) LEMÉE. Intermediate between the parents. HORT. Rowley.

× **Leptauminia** ROWL. Liliaceae. – Intergeneric hybrid: **Leptaloe** × **Guillauminia.** REYNOLDS includes both these genera in **Aloe.** – Cultivation as for the dwarf Aloes.

× **L. 1.** – **Leptaloe albida** × **Guillauminia albiflora** (GUILL.) A. BERTR. Intermediate between the parents. Hort. Makin.

**Lewisia** PURSH. Portulacaceae. – Western N.Am.: mts.; Central Am.; Bolivia. – ♃ of a xerophytic character; **caudex** fleshy, thick or tuberous, with ± dense Ros. of ± fleshy or leathery L. above; **L.** persistent in some species, otherwise dying back annually; **Infl.** with several ± large, white to pink Fl., Pet. 5–10. – Cultivated in the open, not completely frost-resistant. – Propagation: seed, division. – See R. C. ELLIOT: The Genus Lewisia. Alp. Gard. Soc. 1966.

## Key to the commonly cultivated species of Lewisia by G. D. Rowley

1. Plants evergreen (exc. some **cotyledon** cvs.); Ped. elongated .............................. **2.**
1a. Plants deciduous; Ped. (exc. **oppositifolium**) short, with Fl. ± in amongst the L. ........ **8.**

2. Fl. 1–4 together on a Ped., c. 5 cm across; plant robust ...................... **tweedyi**
2a. Fl. more together, in panicles or cymes ............................................ **3.**

3. L. narrow, grass-like, terete or nearly so; Fl. very small, c. 6 mm ⌀ ............. **leeana**
3a. L. broader, not grass-like, spatulate or obovate, flattened; Fl. larger ................ **4.**

4. Small plants with offsets; Ros. 5–10 cm ⌀; Fl. to 18 mm ⌀ ....................... **5.**
4a. Larger plants rarely offsetting; Ros. 10–15 cm ⌀; Fl. 18–38 mm ⌀ ................... **6.**

5. Fl. white or flesh-coloured ....................................... **columbiana** v. **columbiana**
5a. Fl. bright pink ............................................... **columbiana** v. **rosea**

6. L. flat, entire ................................................. **cotyledon** v. **cotyledon**.
   ('Finchae', 'Purdyi' and many other cvs. belong here)
6a. L. with crisped or toothed margins .............................................. **7.**

7. L. crisped, rounded at the apex; Pet. oblanceolate, pink with a carmine centre streak .....
   ........ **howellii**
7a. L. serrate, acutish or ± obtuse; Pet. linear-oblong, rose, not streaked ..................
   ........ **cotyledon** v. **heckneri**

8. Ped. elongated, well clear of the L. ............................................ **oppositifolia**
8a. Ped. short ...................................................................... **9.**

9. Stamens few, 5–12; stigmas 3–6 ............................................... **10.**
9a. Stamens many, 10–50; stigmas 5–8 ............................................ **11.**

10.  L. linear, strap-like; bracts short, below 1 cm; Sep. short, 5 mm or less ........ **pygmaea**
10a. L. subulate; bracts and Sep. longer ............................... **nevadensis**
11.  L. lanceolate, flat, persisting during flowering; Sep. 2, short, 6 mm long or less ..........
     ........ **brachycalyx**
11a. L. subulate, terete, dying before Fl. open; Sep. 6–9, 15–25 mm long ................ **12.**
12.  L. linear-oblong, subterete; Pet. 17–35 mm long.................... **rediviva v. rediviva**
12a. L. clavate to narrow oblanceolate; Pet. shorter .................... **rediviva v. minor**

**L. brachycalyx** ENGELM. ex GRAY. – Western N.Am. – **Caudex** firm, fleshy; **L.** oblong to spatulate, 4–6 cm long, blunt, often bluish-green, somewhat fleshy; **Fl.** solitary, c. 4 cm ⌀, white.

**L. columbiana** ROBINS. f. **columbiana.** – USA: Columbia R.; Oregon. – **L.** numerous, linear-spatulate, 2.5–5 cm long, flat, green; **Fl.** in panicles, Pet. 4–7, often dentate, white or pink with red veins.

**L. — f. rosea** JACOBS. – **Fl.** dark pink.

**L. cotyledon** ROBINS. v. **cotyledon** – USA: N.Calif. – **Ros.** dense, 10–12 cm ⌀; **L.** numerous, ovate-spatulate; **Fl.** in panicles 10 cm tall, Pet. 7–10, white with red veins.

**L. — v. heckneri** (MORTON) MUNZ (*L. heckneri* MORTON). – **L.** fleshy, very dentate, 6.5–13 cm long; **Fl.** pale to deep pink, sometimes with pale pink stripes.

**L. cv. Finchae.** – Cultivar of **L. cotyledon**; **L.** persisting; **Fl.** with delicate pink, conspicuously veined Pet.

**L. howellii** ROBINS. – USA: Oregon. – **Ros.** c. 15 cm ⌀; **L.** oblong to obovate, margins curly; **Fl.** deep pink.

**L. leeana** ROBINS. – USA: N.Calif. – **L.** linear, usually rounded, often widening towards the tip, 4–6 cm long, bluish; **Fl.** rather small, red to white, with darker veins.

**L. nevadensis** ROBINS. – USA: Nevada. – **Root** napiform; **L.** thin, filiform, soon dropping; **Fl.** white.

**L. oppositifolia** ROBINS. – USA: Calif., Oregon. – **L.** oblanceolate, 5–10 cm long, blunt, green, glossy; **Fl.** 5 cm ⌀, Pet. pearly white to pink.

**L. cv. Purdyi.** – Cultivar of **L. cotyledon**; **L.** persistent; **Fl.** large, pink to apricot-coloured.

**L. pygmaea** ROBINS. – N.Am.: Rocky Mts. – **L.** rather short, fleshy, broad-linear to oblong-lanceolate; **Fl.** numerous, small, Pet. 6–8, pinkish-red.

**L. rediviva** PURSH. v. **rediviva** (Pl. 107/2) (*L. alba* KELLOG). – Western N. Am.: Rocky Mts. – **Roots** tapering, fleshy, edible; **L.** linear, blunt, fleshy, smooth; **Fl.** solitary, large, white.

**L. - v. minor** (RYDBERG) HOLMGREEN (*L. minor* RYDBERG). – Smaller variety.

**L. tweedyi** ROBINS (*L. tweedei* A. GRAY).–USA: State of Washington- – **L.** numerous, broad-oval or obovate, blunt, 5–7.5 cm long; **Infl.** of 1–2 flesh-pink Fl. 5–7.5 cm ⌀.

**Litanthus** HARVEY. Liliaceae. – Greenhouse, warm. Resting period: summer. Propagation: seed, subsidiary bulbs.

**L. pusillus** HARVEY (Pl. 107/3). – S.Afr. – ♃; bulbs spherical, c. 1 cm ⌀; the thin, simple **Fl.** scape 2–3 cm tall develops first, and bears 1 small, nodding, white Fl.; **L.** appear later, linear to filiform, 3–4 cm long. (Not a true succulent plant).

**Lobelia** L. Campanulaceae. – G. with numerous species, of which only 2 from E.Afr. can be described here; even these are scarcely found in cultivation and not true succulent plants.

**L. keniensis** R. E. FRIES et TH. FRIES f. (Pl. 107/4) (*L. gregoriana* BAK. f. nom. conf.). – Kenya: Mt. Kenya. – Up to 1.80 m tall, acauline or almost so; **L.** in a dense Ros., 18.5–21.5 cm long, 4.8–5.2 cm across, sessile, papery, narrow-lanceolate, broadest midway, upperside glabrous, underside with tufted hairs along the thick M.nerve, margins entire, densely ciliate; **Infl.** c. 1 m tall, terminal, very dense, bracts 6.5–8 cm long, 4–5 cm across, ovate, tapering, Fl. on pedicels 1 cm long, Cor. c. 2.7 cm long, bluish-violet.

**L. rhynchopetala** (HOCHST.) HEMSL. (*Tapa r.* HOCHST., *Rhynchopetalum montanum* FRES.). – Tanzania: on Mt. Klimanjaro; ? Ethiopia. – **St.** 3–4 m tall, with the persistent L. bases in 5–8 years attaining the thickness of a man's thigh, glabrous, with a large crown of L. above; **L.** long, broad-linear; **Infl.** 4–5 m tall, bracts narrow-lanceolate, Pet. rostrate; **Fr.** 2.5 cm long.

× **Lomataloe** GUILL. Liliaceae. – Intergeneric hybrid: **Lomatophyllum** WILLD. × **Aloe** L. – Culture: as for **Aloe.** Propagation: cuttings.

× **L. hoyeri** (RADL) GUILL. (*Aloe* × *hoyeri* RADL). – Hybrid: **Lomatophyllum purpureum** × **Aloe serrulata.** – **Ros.** elongated, 10 cm tall, 15 cm ⌀; **L.** in 5 spiral R., 10–16 cm long, 3–4 cm across, fleshy below, narrowing abruptly and long-tapering, dull green, indistinctly striate, margins cartilaginous, minutely dentate.

× **Lomateria** GUILL. Liliaceae. – Intergeneric hybrid: **Lomatophyllum** WILLD. × **Gasteria** DUVAL. – Culture as for **Gasteria.** – Propagation: cuttings.

× **L. gloriosa** (RADL) GUILL. (*Aloe* × *gloriosa* RADL). – Hybrid: **Lomatophyllum purpureum** × **Gasteria maculata.** – **L.** c. 40 cm long, c. 5 cm across, dark green, with large white spots, margins ± imperceptibly armed.

**Lomatophyllum** WILLD. Liliaceae. – Madag.; Mauritius. – Related to the G. **Aloe;** ♄ or ♄; **L.** in dense Ros. at the Br.tips, ± fleshy, aloe-like; **Infl.** simple or branching, racemose, with many Fl.; **Fr.** berry-like. – For the warm-house. Propagation: seed, cuttings.

**L. antsingyense** LEANDRI. – W.Madag. – **St.** 40–50 cm tall, c. 1 cm thick; **L.** not numerous, 2–3 cm apart, 30–50 cm long, 2 cm across, Sh. with c. 20 large, very prominent ribs, white-spotted, marginal T. up to 1 mm long, 10 mm apart; **Fl.** red.

**L. citreum** GUILL. (Pl. 107/5). – Madag. – Acauline; **L.** c. 16, 28 cm long, thin, dark green, marginal T. triangular, c. 3 mm long, 1–2 mm apart, sharp; **Fl.** yellowish-green.

**L. macrum** (HAW.) SALM ex ROEM. et SCHULT. (*Aloe m.* HAW.). – Madag.; Mauritius. – ♄; **St.** 30 cm tall; **L.** numerous, narrow-ensiform, long-tapering, 30–35 cm long, 3 cm across, upperside concave, green, T. small, horny, red.

**L. occidentale** PERR. – W.Madag. – Acauline or short-stemmed, up to 1 m tall, 6–10 cm ⌀; **L.** 15–20, stiff, recurved, 80–100 cm long, 10–12 cm across, upperside concave, margins with 4 mm long, triangular T. 6–25 mm apart; **Infl.** 2–5 dense Rac. of 50–80 Fl.

**L. oligophyllum** (BAK.) PERR. (*Aloe o.* BAK.). – Central Madag. – **St.** thin, extended; **L.** 2–4 in a Ros., elongated, 40–50 cm long, 12–15 mm across, tapering, margins with 1–3 mm long green T. 12–15 mm apart; **Infl.** 15–20 cm tall, Fl. not known.

**L. orientale** PERR. – E.Madag. – Resembles **L. occidentale**; **L.** denser, more tapering, tip stouter, more rounded.

**L. prostratum** PERR. – W.Madag. – Clump-forming; **Ros.** dense, lying close to the ground; **L.** 12–20, 15–20 cm long, 15–20 mm across, thin, tapering, blackish with white spots, margins dentate, T. white, 5–15 mm apart; **Infl.** of few Fl.

**L. purpureum** (LAM.) TH. DUR. (*Aloe p.* LAM., *Dracaena marginata* AIT., *Aloe m.* WILLD., *L. m.* HOFFMGG., *A. marginalis* DC., *L. borbonicum* WILLD., *L. aloiflorum* NICH., *Phylloma a.* KER., *P. borbonicum* HAW., *Aloe dentata* PERS.). – Madag.; Mauritius; Aldabra Is. – Almost tree-like; **St.** erect, becoming woody, up to 2 m tall; **L.** numerous, rosulate, up to 80 cm long, 8–10 cm across, linear-lanceolate, upperside grooved, fleshy-leathery, deep green, with red, horny, dentate margins; **Infl.** 50 to 60 cm tall, Fl. yellowish-red.

**L. roseum** PERR. – Central and E.Madag. – Acauline; **Ros.** of 12–15 L. 30–45 cm long, 2.5–4 cm across, thin, gibbose, margins sinuate-dentate, T. 5 mm long, 6–18 mm apart; **Infl.** 25–30 cm tall, with 25–30 pink Fl.

**L. rufocinctum** (HAW.) SALM ex ROEM. et SCHULT. (*Aloe r.* HAW.). – Mauritius. – Resembles **L. purpureum**; **L.** erect, decurved, ligulate-lanceolate, tapering, green, upperside concave, T. horny, pink.

**L. sociale** PERR. – W.Madag. – **St.** numerous, prostrate, 30 cm long; **L.** 14–16, partly rosulate, stiff, thick, flat, 30–40 cm long, 10–15 mm across, roundish-sinuate, dentate; **Infl.** 15–25 cm tall, Fl. crimson.

**L. viviparum** PERR. – Madag. – **St.** short, prostrate; **Ros.** of 12–15 L. 45–55 cm long, 2.5–3.5 cm across, margins sinuate, with 3 Sp. 1 mm long in the indentations, the central Sp. longest; **Infl.** 50–80 cm tall, with adventitious buds.

**Luckhoffia** WHITE et SLOANE. Asclepiadaceae. – Related to **Hoodiopsis**; inner corona lobes with a single horn. – Greenhouse, warm. Propagation: seed.

**L. beukmanii** (LUCKH.) WHITE et SLOANE (*Stapelia b.* LUCKH.). – Cape: Clanwilliam D. – **St.** erect, sprouting from the base, up to 75 cm tall, 3 cm ⌀, with 9 angles resolved into compressed tubercles, these at first with a short Sp., grey-green, minutely hairy; **Fl.** 2–3 together from the St.tip, pedicels 3 cm long, Cor. 6–6.5 cm ⌀, 5-cleft to midway, flat, glabrous outside, green to light pink, papillose inside, with horizontal, black, minute hairs, lobes with 3 longitudinal ribs, the united part brown, with yellow blotches, the upper half concolorous brown, lobes 18 mm long, tapering, margins recurved, with some ciliate hairs.

**Mesembryanthemaceae** with their genera, species etc., see **Part II.** and **Invalid Designations**.

**Meterostachys** NAKAI. Crassulaceae. – E.As. – ☉, ☉ or ♃, hardy. Propagation: seed, winter buds.

**M. sikokianus** (MAKINO) NAKAI (Pl. 108/1) (*Cotyledon s.* MAKINO, *Sedum s.* (MAKINO) HAMET, *Orostachys s.* (MAKINO) OHWI, *S. leveilleanum* HAMET, *S. orientoasiaticum* MAKINO). – Central and S.Korea; Japan: Shikoku, W.Honshu. – **L.** rosulate, fleshy, with a cartilaginous apiculus, L. in the sterile Ros. linear-lanceolate, 0.5–3 cm long, 10–13 mm across, drying in autumn, leave a compact winter bud of small horny L., this bud developing from the base in spring to form the outer L. of the Ros.; **Infl.** corymbose, terminal and lateral, erect, 2–8 cm tall, leafy, dichotomous, Rac. one-sided, Fl. 2–8, pure white, pedicels c. 5 mm long.

**Microcnemum** UNGERN-STERNBERG. Chenopodiaceae. – Halophyte, probably not in cultivation.

**M. coralloides** (LOSCOS et PARDO) FONT Y QUER (*Arthocnemum c.* LOSCOS et PARDO). – Spain: salt-lake of Caspe, Iberian savannah grasslands in Aragon. – ☉ with segmented branchlets, the purple colouring of the fertile spikes giving the plant a resemblance to coral.

**Momordica** L. Cucurbitaceae. – Trop. climbing plants, usually ☉, with one ♃ species known to be succulent. – Greenhouse, warm. Propagation: seed.

**M. rostrata** A. ZIMM. (Pl. 108/2). – Kenya. – Dioecious plant; **caudex** partly hidden underground, fleshy, up to 25 cm tall, 25 cm ⌀ below, tapering above, splitting irregularly, longitudinally furrowed, with several thin, liana-like, very leafy St. above; **L.** petiolate, 5-lobed, dropping in the resting period; **Fl.** vivid yellow with a dark brown throat.

**Monadenium** Pax. Euphorbiaceae. – E. and S. trop. Afr.; Ethiopia; Somalia; Congo Rep.; Uganda; Kenya; Tanzania; Rhodesia; Malawi; Angola; S.W.Afr.; Botswana; Transvaal, Natal. – ♃ with annual, usually unbranched St. from an underground **caudex,** or plants with ♃, fleshy, cylindrical or angular, often tuberculate St., with or without Th.; **St.** creeping or erect, in some species tree-like, several m tall and sparsely branched; **L.** alternate, ± fleshy, often soon dropping, simple, with margins entire or serrate or undulating; **Infl.** from the L. axils, or subterminal or terminal, in some species repeatedly forking and producing cymes of many Fl. heads, often vividly coloured; primary Cy. usually solitary, developing a pair of lateral Cy.; bracts irregular, cup-shaped, blunt, open completely or half-way on one side, with a marginal gland. (For further details see P. R. O. Bally, **The Genus Monadenium,** published by Benteli, Bern, 1961.) – All species of **Monadenium** have a milky sap. – Greenhouse, warm. Propagation: seed, cuttings.

## Division of the Genus Monadenium according to the Form of Growth by P. R. O. Bally

§ I. ♃ **herbs.** – Roots tuberous, above-ground St. withering away annually.

 a). Dwarf subfleshy.

  1. L. terminally crowded. – Species: M. chevalieri, erubescens, fwambense, gracile, letestuanum, majus, nervosum, rhizophorum, trinerve.

  2. L. rosulate. – Single species: M. pseudoracemosum.

 b). Herbs to 60 cm high, L. equidistant. – Species: M. capitatum, crenatum, descampsii, discoideum, echinulatum, fanshawei, goetzei, herbaceum, hirsutum, intermedium, invenustum, kaessneri, laeve, montanum, parviflorum.

 c). Geophytic, hysteranthous plants; caudex tuberous, St. growing subterraneously, sterile St. short.

  1. St. developing immediately from the caudex, sterile, erect, withering before the Infl. develops. – Species: M. friesii, nudicaule.

  2. St. reduced to a short, subterranean axis; L. withering before the Infl. develops. – Species: M. angolense, orobanchoides, simplex.

§ II. **Succulents.**

 a). Dwarf succulents.

  1. St. tuberculate, warts wholly flat (M. lugardae) to very thin and stalk-like (M. reflexum, stellatum). – Species: M. guentheri, heteropodum, lugardae, reflexum, ritchiei, schubei, stapelioides, stellatum, yattanum.

  2. St. cylindrical, tessellate. – Single species: M. ellenbeckii.

 b). Shrubs: St. and Br. weakly angled, smooth. –

  1. Infl. singly branched. – Single species: M. virgatum.

  2. Infl. repeatedly branched. – Single species: M. coccineum.

 c). Trees.

  1. St. and Br. cylindric, warty, strongly thorny. – Single species: M. spinescens.

  2. St. and Br. 5-angled, smooth. – Single species: M. arborescens.

  3. Angles set with branched Th. – Single species: M. magnificum.

**M. angolense** Bally (§ Ia/2). – Angola. – ♃, mostly subterranean, with an underground **caudex; St.** much reduced, 2.5–4 cm tall, erect, fleshy, up to 1 cm thick; **L.** scale-like on flowering St., 8–11 mm long, 4–6.5 mm across; **Infl.** cymose with terminal clusters, bract-cup light red, oblique-elliptical, Cy. barrel-shaped.

**M. arborescens** Bally (§ IIc/2). – Tanzania: Kilosa D. – Small succulent ♄ up to 3.6 m tall; **St.** simple, up to 10 cm thick, with 5 somewhat spiralled angles; **Br.** few, up to 1 m long, 3.5–4 cm ⌀, 5-angled; **L.** terminal, sessile, fleshy, obovate, pointed, cordate, 7–19 cm long, 5.6–11 cm across, L.bases with a solitary, 1.5–5 mm long stipular Th. to the right; **Infl.** axillary, bract-cup glabrous, oblong-ovate, bluntly bifid, dull red, Cy. greenish-yellow.

**M. capitatum** Bally (§ Ib). – Tanzania: S.Highlands Prov. – Erect, fleshy ♃ up to 60 cm tall, with a **caudex; St.** solitary or few, cylindrical, densely papillose; **L.** fleshy, oblanceolate, up to 6.6 cm long, 17 mm across, glabrous, short-petiolate, apiculate, margins a little dentate at the base, underside with a prominent M.rib; **Infl.** almost sessile.

**M. chevalieri** N. E. Br. (§ Ia/1).

**M.** — **v. chevalieri.** – W. equatorial Afr. – Semi-succulent ♃; caudiciform; **St.** solitary or several, 5–24 cm tall, herbaceous; **L.** linear to narrow-elliptical, 18–25 mm long, 3–7 mm across, longitudinally plicate; **Infl.** rarely branching, bract-cups elliptical, pointed.

**M. —** v. **filiforme** BALLY. – Zambia. – **Caudex** conical, 2 cm tall and ∅; **St.** solitary, fleshy, 3–3.5 mm thick; **L.** scale-like to narrow-lanceolate, upper L. crowded, filiform, pointed, margins curved over; Cy. white to pink.

**M. —** v. **spathulatum** BALLY. – Malawi: Rumpi D. – Caudiciform; **St.** reduced to an underground axis 1–2 cm long, up to 4 mm thick; **L.** scale-like, broad-lanceolate, blunt, 4–6 mm long; bract-cups spatulate, 3.5–5 mm long, 1.5 mm across.

**M. coccineum** PAX (§ IIb/2). – Tanzania: Masai D. – Erect, semi-climbing plant; **St.** 1–2, simple or sometimes branching, 30–130 cm long, 12–14 mm thick, glabrous, 5-angled; **L.** glabrous, fleshy, ± sessile, oblanceolate to obovate, pointed, M.rib carinate, spiny, margins serrate; **Infl.** several, terminal, 5–6-forked, brilliant red, forming heads up to 4 cm ∅, bract-cups glossy scarlet.

**M. crenatum** N. E. BR. (§ Ib). – Rhodesia; Moçambique. – ♃; **St.** simple, glabrous, 7 to 12 cm tall, 6–7 mm thick; **L.** linear-lanceolate, pointed, 4–6 cm long, 7–12 mm across, margins curly and undulating, underside with the M.rib prominent; **Infl.** terminal, pedunculate, 1–2-forked, bract-cups oblique, ovate-lanceolate, tapering, c. 5 mm long.

**M. crispum** N. E. BR. (§ Ib). – Tanzania: Tanga, Mpwapwa, Shinyanga D. – Caudiciform ♃; **St.** erect, up to 32 cm tall, with L.scars; **L.** ovate-lanceolate, pointed, narrowing cuneately towards the base, 4–5.5 cm long, 1–1.5 cm across, little petiolate, slightly fleshy, margins serrate, undulating or curly, underside with a prominent M.rib; **Infl.** axillary, 1–2-forked, bract-cups oblique-ovate, pointed.

**M. descampsii** PAX (§ Ib). – Congo: Katanga. – Fleshy ♃, probably caudiform; **St.** simple, c. 25 cm tall; **L.** sessile, linear-lanceolate, very fleshy, c. 3 cm long, 5 mm across, pale green with margins entire, slightly undulating towards the tip; **Infl.** solitary from the L.axils, bract-cups ± round, pointed, 18 mm long and across, fleshy.

**M. discoideum** BALLY (§ Ib). – Zambia. – Slightly fleshy ♃; **St.** simple, solitary, up to 15 cm tall, bristly; **L.** sessile or short-petiolate, elliptical, cuneate, pointed, up to 4.2 cm long, 1.4 cm across, margins entire, slightly undulating, both surfaces bristly; **Infl.** solitary from the L.axils, forked, bristly, bract-cups transversely ovate, usually round, 9–11 mm long, 11–18 mm across.

**M. echinulatum** STAPF v. **echinulatum** (§ Ib) (*M. aculeolatum* PAX, *M. asperrima* PAX). – Tanzania: N. and Central Prov. – Fleshy ♃; **St.** simple, 30–70 cm tall, glabrous, spiny; **L.** almost sessile or short-petiolate, 2.5–12 cm long, 1–6.5 cm across, elliptical to ± ovate to lanceolate, rather pointed, both surfaces glabrous or the underside with small soft Sp., M.rib sharply carinate, margins minutely serrate; **Infl.** axillary, 1–5-forked, bract-cups oblique,

sinuate, 6–14 mm long, 10–21 mm across, slightly rough or prickly, pale green, often slightly pink, with dark green veins, margins white.

**M. —** v. **glabrescens** BALLY (*M. pedunculatum* MANSF. Ms.). – Tanzania: Kilwa D. – Caudiciform; **St.** 20 cm or more tall, L.scars reniform with 2 lateral stipular Th.; **L.** 5–9 cm long, up to 5.5 cm across, both surfaces glabrous.

**M. ellenbeckii** N. E. BR. f. **ellenbeckii** (Pl. 108/3) (§ IIa). – Ethiopia: Sidamo; Kenya: N.Frontier Prov. – Sparsely-branching ♄ up to 1 m tall; **St.** one to several, erect, cylindrical, 12–25 mm thick, indistinctly tessellate-reticulate, longitudinally pitted, glabrous, green; **Br.** few, short; **L.** few, terminal, soon dropping, 6–10 mm long, 7–8 mm across, fleshy, stiff, minutely hairy; **Infl.** axillary, 1–2-forked, bract-cups oblique, 6 mm long, 10 mm ∅, yellowish-green.

**M. —** f. **caulopodium** BALLY (*M. zavatteri* CHIOV.). – **St.** weak, creeping or prostrate, numerous, very rarely branching, scarcely over 10 cm tall.

**M. erubescens** (RENDLE) N. E. BR. (Pl. 110/1) (§ Ia/1) (*Lortia e.* RENDLE). – Somalia. – Caudiciform; **St.** ☉, usually 2 together, 3 to 14 cm tall, cylindrical, glabrous; **L.** stiff, flat, fleshy, ± sessile, round-rhombic, pointed, up to 5 cm long, 2.5 cm across, margins curly, both surfaces puberulous, upperside dark green to purple, with distinctly paler M.rib and veins; **Infl.** solitary from the upper L.axils, bract-cups capitate, obliquely urn-shaped, sinuate, white to somewhat pink.

**M. fanshawei** BALLY (§ Ib). – Zambia: W. and N.Prov. – Caudiciform ♃; **St.** solitary, up to 30 cm tall, simple, 2–4 mm thick; **L.** narrow-elliptical to lanceolate, pointed, narrowing cuneately to the ± petiolate base, 4–8 cm long, 1–2.4 cm across, margins minutely dentate, M.rib prominent, sometimes carinate, upperside dark green, underside bluish; **Infl.** solitary from the L.axils, 1–2-forked, fleshy, bract-cups 2.5–3 mm across, green, rather fleshy.

**M. friesii** N. E. BR. (§ Ic/1). – Zambia; Tanzania. – Caudiciform; **St.** solitary, ☉, cylindrical, 4–4.5 cm tall, 4.5 mm thick, densely spiny, sterile; **L.** 5, crowded at the tip of the St., elliptical, pointed, scarcely petiolate, both surfaces and the margins spiny; **Infl.** 1–3 from the caudex, forked 3 times or more, erect, 5–8.5 cm tall, spiny, bract-cups 9 mm long, 16 mm across, thin-fleshy, whitish-green, with dark green veins, with scattered prickly hairs.

**M. fwambense** N. E. BR. (§ Ia/1). – Zambia. – Slightly succulent ♃; **St.** 1–3, up to 6 cm tall, erect, cylindrical, glabrous; **L.** lanceolate or linear-lanceolate, pointed, ± sessile, M.rib acute-carinate, margins entire; **Infl.** borne on the upper St. portions, bract-cups pointed, bifid, 6.5 mm long, 4.5 mm across.

**M. goetzei** PAX (§ Ib). – Tanzania: S.Highland Prov. – Semi-succulent ♃; caudiciform; **St.** up to 50 cm tall; **L.** up to 14 cm long, 1.5 cm across; bract-cups dark violet to green, reticulate-veined, 7–8 mm long, 10–18 mm across, margin broadly rounded, crenate or curly.

**M. gossweileri** N. E. Br. – Angola. – Insufficiently known species; acc. P. R. O. Bally more closely related to the G. **Synadenium.** – See Supplement: **Endadenium** Lerch.

**M. gracile** Bally (§ Ia/1). – Tanzania: Central Prov. – Dwarf ⚁; **St.** erect, up to 5 cm tall, solitary or few, little branching, glabrous; **L.** linear, falcate, longitudinally plicate down the middle, 3–14 mm long, crowded, terminal; **Infl.** axillary, capitate, bracts divided, lanceolate, slightly fleshy, with undulating margins, membranous, white, 6.5 cm long, 3 mm across, as long as the involucre.

**M. guentheri** Pax (§ IIa/1). – E. Afr.
**M. — v. guentheri** (Pl. 110/2). – Kenya: Taita D. – **St.** 15–20 cm tall, longer Br. up to 90 cm long, rampant, simple, fleshy, cylindrical, with tubercles hexagonal at the base, up to 6 mm high, slightly recurved, with 1–3 sharp Sp. around the L.base; **L.** sessile, 1–4 cm long, 3–20 mm across, linear-lanceolate to obovate, very fleshy, margins entire, slightly curly at the tip; **Infl.** terminal, short-pedunculate, bract-cups 6–8 mm long, with 2 short, pointed lobes, glabrous, greenish-white with purple mottling.

**M. — v. mammillare** Bally. – Tanzania: Tanga. – Tubercles cup-shaped, sap-green, Sp. less well developed.

**M. herbaceum** Pax (§ Ib). – Congo (Kinshasa): Katanga D. – Caudiciform ⚁; **St.** erect, 25–40 cm tall, thin, simple, glabrous; **L.** oblong to lanceolate, smaller and linear-lanceolate towards the tip, pointed, apiculate, leathery, both surfaces glabrous, underside with the M.rib acute-carinate; **Infl.** from the upper L.axils, bract-cups up to 8 mm long, 6.5 mm across, elliptical to ± round, underside with a pectinate, undulating keel, pale green to somewhat pink.

**M. heteropodum** (Pax) N. E. Br. (§ IIa/1) (*Euphorbia h.* Pax). – Tanzania: Tanga. – Caudiciform; **St.** erect, up to 35 cm tall, creeping if longer, cylindrical, up to 3 cm ∅, with tubercles which are rhomboid to square below or, during rapid growth, narrowly hexagonal, Sp. very small, soft, 2–3; **L.** very fleshy, soon dropping, rhombic-spatulate, ± petiolate, up to 3.5 cm long, 1.25 cm across; **Infl.** consisting of one central and 2 lateral involucres, bract-cups 13 mm long, 10 mm across, greenish-white.

**M. hirsutum** Bally (§ Ib). – Zambia. – Caudiciform ⚁; **St.** solitary, erect, up to 25 cm tall, cylindrical, with some hairs; **L.** lanceolate or narrow-elliptical, up to 6 cm long, 1.5 cm across, narrowing towards the ± petiolate base, both surfaces hairy, margins entire, densely hirsute; **Infl.** axillary, densely hairy, bract-cups 8.5 mm long, 10 mm across, membranous, glabrous.

**M. intermedium** Bally (§ Ib). – Tanzania: Tanga. – Fleshy ⚁; **St.** cylindrical, erect, up to 35 cm tall, glabrous; **L.** short-petiolate, broad-elliptical, pointed, margins somewhat curly, undulating; **Infl.** solitary, from the upper L.axils, forking once, bracts light green with dark green veins, 5.5 mm long, 4.5 mm across.

**M. invenustum** N. E. Br (§ Ib). – Kenya.
**M. — v. angustum** Bally. – Kenya: central reg. – Differentiated from v. **invenustum** as follows; **L.** narrow-lanceolate, sessile, margins not curly; bract-cups narrower.

**M. — v. invenustum.** – Kenya: central reg. – Succulent, caudiciform ⚁; **St.** erect, 25–80 cm tall, cylindrical, glabrous, bluish-green; **L.** fleshy, round-ovate to almost rhombic-ovate, 1.7–4 cm long, 1.3–3.5 cm across, margins sometimes crenate-curly, upperside dark bluish-green with pale green veins, underside concolorous pale green; **Infl.** axillary, on the upper St., consisting of 3 bracteate heads.

**M. kaessneri** N. E. Br. (§ Ib). – Congo: Kundelungu Mts. – Caudiciform ⚁; **St.** erect, herbaceous, glabrous, 54 cm tall, simple; **L.** somewhat fleshy, elliptical, lanceolate at first, blunt or rather pointed, apiculate, both surfaces glabrous, 7.5–11.5 cm long, 2.5–5 cm across, the M.rib prominent and carinate on the underside, veins indistinct; **Infl.** axillary, nodding, involucral bracts oblique-elliptical, 13 mm long, 18 mm across, green.

**M. laeve** Stapf f. **laeve** (§) Ib. – Malawi: Tanzania. – Semi-succulent, caudiciform ⚁; **St.** 45 cm or more tall; **L.** oblanceolate to elliptical-obovate, 6–17 cm long, 2 cm across, narrowing below midway into a petiole 7–18 mm long, shortly hairy above, the M.rib prominent on the underside and winged; **Infl.** axillary, forking up to 3 times, bract-cups 6–14 mm long, 6–16 mm across, membranous, reticulate-veined, margins undulating, crenate.

**M. — f. depauperata** Bally. – Tanzania: Highlands Prov. – **St.** thinner, shorter, bristly or spiny; **L.** smaller, ± densely bristly; bract-cups smaller, narrower.

**M. letestuanum** Denis (§ Ia/1). – W. equatorial Afr.

**M. — v. letestuanum.** – Central Afr. Rep.: above Ubangi. – With a ⚁ caudex; **St.** ⊙ or rarely seasonal, erect, cylindrical, 2–3 cm long, glabrous; **L.** 4–8, crowded at the St.tips, leathery, lanceolate or oblanceolate, apiculate, margins entire, petiole 3–5 mm long, blade 2–4 cm long, 1–1.5 cm across, underside often violet; **Infl.** solitary from the L.axils, glabrous, up to 3 cm long, bract-cups triangular, pointed 4 mm long.

**M. — v. rotundifolium** Bally. – Central Afr. Rep. – **L.** broad-spatulate, blunt; **Infl.** puberulous.

**M. lugardae** N. E. Br. (Pl. 108/4) (§ IIa/1). – S.W.Afr.; Transvaal, Natal; Rhodesia. – Caudiciform; **St.** thick, cylindrical, without Sp., glabrous, sometimes puberulous, 12–60 cm long, 1.5–3 cm thick, with pentagonal or hexagonal tessellation; **L.** crowded at the Br.tips, fleshy, 1.5–9 cm long, 0.5–4 cm across, obovate or spatulate, blunt or ± pointed, scarcely petiolate, margins curly-serrate towards the tip; **Infl.** from the upper L.axils, simply forked, bract-cups oblique, 6–7 mm long, bicarinate, crenate on the underside, pale green.

**M. magnificum** E. A. Bruce (§ IIc/3). – Tanzania: central prov. – Succulent ♄ up to 1.5 m tall, or ♄ up to 3.5 m tall; **St.** solitary or several, up to 4 cm thick at the base, with few Br., St. and Br. with 4–5 acute, continuous angles set with reddish-brown spiny clusters of c. 3 mm $\varnothing$; **L.** fleshy, glabrous, broad-elliptical to obovate to $\pm$ cordate, blunt, apiculate, the M.rib acute-carinate on the underside, spiny, margins serrate, undulating in the upper half; **Infl.** 5–7, axillary along the upper parts of the angles, forking up to 8 times, Ped. acute-angled, prickly, vivid scarlet, bract-cups light scarlet, 8–14 mm long, with dentate keels.

**M. majus** (Pax) N. E. Br. f. **majus** (§ Ia/1) (*Lortia m.* Pax). – Ethiopia: Harar. – Succulent, caudiciform ♃; **St.** ☉, solitary, erect, glabrous, green, c. 10 cm tall, usually climbing and up to 40 cm tall, 8–10 mm thick; **L.** fleshy, subpetiolate, cuneate at the base, apiculate, folded along the M.rib, sharply carinate, up to 9.5 cm long, 4.5 cm across, glabrous or puberulous, margins serrate, often curly towards the tip; **Infl.** from the upper L.axils, 1–2-forked, corymbose, with 2–14 cymes forming a head 10 cm tall and 8 cm $\varnothing$, bract-cups 3.3 mm $\varnothing$, cherry-red outside, inside white, with green veins.

**M.** — f. **floribundum** Bally. – Ethiopia: Sidamo. – **St.** numerous, forming large, dense bushes; **L.** smaller, M.rib rounded; bract-cups white to pinkish-red.

**M. montanum** Bally (§ Ib). – Tanzania; Kenya.

**M.** — v. **montanum.** – Tanzania: N.prov. – Small, succulent, caudiciform ♃; **St.** solitary, erect or prostrate, cylindrical; **L.** sessile, narrow to broad-elliptical, very fleshy, 24–44 mm long, 6–25 mm across, green, often spotted with purple, rough; **Infl.** solitary from the L.axils, bract-cups grey-green, with 2 keels, 5 mm long, sinuate.

**M.** — v. **rubellum** Bally. – Kenya: Central prov. – **L.** completely glabrous, narrower, usually linear, usually red-spotted; **Infl.** pink, bract-cups longer and narrower.

**M. nervosum** Bally (§ Ia/1). – Tanzania: S.prov.; Zambia. – Dwarf, caudiciform ♃; **St.** solitary, erect, simple, cylindrical, up to 2.5 cm tall; **L.** 4–5, forming a dense Ros. at the St.tip, oblanceolate to obovate, cuneate at the base, blunt, up to 9 cm long, 3 cm across, green, sometimes purple-spotted on the upperside, underside pale green, M.rib and veins prominent; **Infl.** solitary from the L.axils, puberulous, bract-cups 2.5 mm long, 2–2.5 mm across, slightly dentate.

**M. nudicaule** Bally (§ Ia/1). – Tanzania: S. Highland prov. – Succulent, caudiciform ♃; **St.** ☉, erect, glabrous, 1.5–5 cm tall, fleshy, cylindrical, with few Br.; **L.** scale-like, 4 mm long, 3 mm across, soon dropping; **Infl.** from the upper L.axils, 1–3-forked, bract-cups glabrous, 6 mm long, 7 mm across, bicarinate, pointed, bifid, crenate, whitish-green, with purple blotches.

**M. orobanchoides** Bally (§ Ic/2). – Angola; Tanzania.

**M.** — v. **calycinum** Bally. – Angola: Benguela D. – **Caudex** 2–2.5 cm $\varnothing$; flowering **St.** 1–6, erect, 4–5 cm long, 2–4 mm $\varnothing$, thinly puberulous; **L.** scale-like, 6–8 mm long, 4–4.5 mm $\varnothing$, carinate, fleshy, margins serrate in the upper third; **Infl.** reduced to pedunculate heads, solitary from the L.axils, bract-cups broad-oval, pointed, 7 mm long, 5.5 mm across, carinate, thinly puberulous outside.

**M.** — v. **orobanchoides.** – Tanzania. – Dwarf ♃ with a persistent, napiform **caudex**; **L.** 4–5 in a Ros., developing in the growing period, ovate-lanceolate, up to 6 cm long, 2.5 cm across, blunt-tipped, margins entire, upperside dark green, underside pale green, M.rib prominent; flowering shoots from the caudex, with scale-like L. 4–4.5 mm long, 3 to 5 mm across; **Infl.** axillary, crowded at the St.tips, bract-cups in pairs, 3.5–4 mm long, 3 mm across.

**M. parviflorum** N. E. Br. (§ Ib). – Malawi. – ♃, probably with a single St. from a caudex; **St.** 23 cm or more long, cylindrical, glabrous, with L.scars; **L.** not known; **Infl.** axillary, crowded at the St.tip, 3–5-forked, forming compact heads, glabrous, bract-cups bifid to midway, 3 mm long, 3.5–4 mm $\varnothing$, green or reddish, with darker veins.

**M. petiolatum** Bally (§ Ib). – Tanzania: Central prov. – Glabrous, caudiciform ♃; **St.** erect, simple, solitary, up to 17 cm tall; **L.** obovate, narrowing towards the base, distinctly petiolate, blade 6–7.5 cm long, 2–2.5 cm across, the M.rib prominent on the underside; **Infl.** solitary from the L.axils, short-pedunculate, bract-cups divided, oblique-ovate, apiculate, 3.5 mm long, 3 mm across.

**M. pseudoracemosum** Bally (§ Ia/2). – Zambia; Tanzania.

**M.** — v. **lorifolium** Bally. – Zambia: Abercorn D. – **St.** up to 9 cm tall; **L.** up to 9 cm long, 3 cm across, oblanceolate, cuneate below, pointed above; **Infl.** sometimes forked.

**M.** — v. **pseudoracemosum.** – Tanzania: S. Highland prov. – Small, caudiciform ♃; **St.** solitary cylindrical, 2–4 cm tall, 2–3 mm thick; **L.** 3–4, rosulate, spreading on the ground, slightly fleshy, glabrous, subpetiolate, obovate, 4–6 cm long, 2–2.8 cm across, cuneately narrowed below, pale green, M.rib prominent, L. in the extended upper St.-section 4–12 mm apart, narrow-lanceolate, pointed, sessile, 6–20 mm long, 2–6 mm across; **Infl.** only from the upper L.axils, capitate, bract-cups crenate, glabrous, 5 mm long, 9 mm across, bicarinate.

**M. reflexum** Chiov. (§ IIa/1). – Ethiopia: Sidamo; Kenya; N. Frontier reg. – **Roots** fibrous; **St.** up to 35 cm tall, up to 6 cm $\varnothing$, Br. not present or few, cylindrical, ascending, densely covered with recurved, narrowed tubercles 6–21 mm long, with a shallow longitudinal furrow above, basal tubercles reduced to $\pm$ rhombic areas, tipped with a L.scar, with 2 lateral, minute stipular Th.Shi.; **L.** terminal at the tubercle-tips, very soon dropping, fleshy, up to 24 mm long, 8 mm across, spatulate, very minutely puberulous, margins entire, curly at the tip; **Infl.** crowded at the St.tips, 1–2-forked, bract-cups obliquely cup-shaped, underside crenate, bicarinate, yellowish-green, often with a pink tinge.

**M. rhizophorum** BALLY (§ Ia/1). – Kenya.

**M. — v. rhizophorum.** – Kenya: S. reg. – Small glabrous ♃; **roots** thin, tuberous, forming a long rhizome; **St.** erect, cylindrical, simple, 4–10 cm long, 5–7 mm thick, solitary and 6–10 cm apart along the rhizome, L.scars rather prominent; **L.** crowded at the St.tips, 10 to 36 mm long, 6–22 mm across, very fleshy, obovate to spatulate, sessile, green, often purple between the veins, margins curly; **Infl.** axillary, involucres usually solitary, bract-cups oblique, up to 7 mm long, 5 mm ⌀, rather fleshy, whitish-green with dark green lines and spots, often with red blotches.

**M. — v. stoloniferum** BALLY. – Kenya: S. reg. – Caudiciform; **St.** 1 or several, up to 5 mm ⌀ but usually thinner, 6 cm or more long, erect at first, eventually prostrate and rooting from the nodes and forming secondary caudices.

**M. ritchei** BALLY (§ IIa/1). – Kenya: N. Frontier reg. – Caudiciform; **St.** solitary or several, little branching, creeping, up to 40 cm long, cylindrical, 1.5–2.5 cm thick, with tubercles up to 7 mm high, with a rhombic base and 3 to 5 Sp.-like Th. above; **L.** fleshy, stiff, ovate, up to 27 mm long, 20 mm across, puberulous, upperside dark green, underside pale green, margins serrate; **Infl.** solitary from the upper L.axils, bract-cups oblique, 6–8 mm long, bicarinate, white, with pale pink veins, gland light red.

**M. schubei** (PAX) N. E. BR. (§ IIa/1). – Tanzania; Rhodesia.

**M. — v. formosum** BALLY. – Tanzania: Tanga. – **Shoots** rather thinner than in the type, less erect, always unbranched; **Infl.** densely crowded, pure white.

**M. — v. schubei** (*Euphorbia s.* PAX). – Tanzania: central prov., Tanga; Rhodesia. – Plant stout, succulent; **shoots** solitary or several, erect, up to 45 cm tall, creeping if longer, little branched, up to 4 cm ⌀, tubercles rectangular or hexagonal below, 8–12 mm ⌀, 8–10 mm high, tapering to a recurved tip with a L.scar, Th.shi. below this, with 3–5 stout Th.; **L.** early deciduous, crowded at the shoot-tips, thick-fleshy, oblanceolate, 38–63 mm long, 18–20 mm across, often subulate at first, sessile, with ± curly margins; **Infl.** axillary, 1–2-forked, bract-cups oblique, bluntly bicarinate, pale green, with a white border and dark green veins, often suffused pink.

**M. simplex** PAX (§ Ic/2). – Zambia; Angola.

**M. — v. pubibundum** (BALLY) BALLY (*M. p.* BALLY). – Zambia: Mwinilunga D. – **L.** rosulate, spreading horizontally, spatulate, apiculate, 3.2 cm long, 1.5 cm across, margins entire, curly, fringed; **Infl.** produced after the L. have dropped, from the axils of scale-like bracts along the short, underground shoot, bract-cups oblique, funnel-shaped, yellowish-green to white, 8 mm long, 4 mm ⌀.

**M. — v. simplex.** – Angola: Huila. – Dwarf ♃ with a tuberous, napiform **caudex** c. 3 cm long, 2.5 cm ⌀; flowering **shoots** (only these being known) reduced to a semi-underground axis a few mm long, with a few scale-like, lanceolate, acute, membranous L. 3.5 mm long and 1.5 mm across; **Infl.** axillary, bract-cups 4–5 mm long, 10 mm across, thin-fleshy, bicarinate, cordate, crenate.

**M. spinescens** (PAX) BALLY (§ IIc/1) (*Stenadenium s.* PAX). – Tanzania: S. Highlands and central prov. – Somewhat fleshy ♄ up to 6 m tall; **trunk** firm, with few Br., up to 17 cm ⌀, bark yellowish-brown, peeling away in papery layers; **Br.** with tubercles with 3 stout, spreading Th., central one 6–14 mm long, slightly recurved, the lateral Th. 2–7 mm long, straight; **L.** on the upper part of the Br., fleshy, glabrous, oblanceolate to obovate, acute, up to 9 cm long, 4 cm across, dark green, the upperside with pale green veins, the underside pale green, with a carinate M.rib, prickly, margins serrate, curly; **Infl.** axillary, forked 7 times or more, forming dense heads 7 cm ⌀, branchlets angular, red, bract-cups pale green, somewhat suffused pink, cleft, up to 1 cm long, 7 mm across.

**M. stapelioides** PAX f. **stapelioides** (Pl. 110/3) (§ IIa) (*M. succulentum* SCHWEICK.). – Tanzania; Kenya. – **Caudex** tuberous; **shoots** solitary to numerous, erect, cylindrical, 3 to 15 cm tall, often longer and prostrate, simple, 1–1.75 cm thick, with rhombic or hexagonal tubercles in spirals, with a thornless L.scar above; **L.** always at the shoot-tips, sessile, obovate or rhombic-spatulate, acute, narrowing cuneately towards the base, 2–3 cm long, 1 to 1.75 cm across, thick-fleshy, glabrous, M.rib very prominent and round, margins serrate; **Infl.** crowded at the shoot-tips, bract-cups oblique, bifid, bicarinate, crenate, white to greenish-white.

**M. — f. congestum** BALLY. – Tanzania; Kenya; Uganda. – **St.** thicker, tubercles more numerous, smaller; **L.** narrower, very densely crowded, ± sessile.

**M. stellatum** BALLY (§ IIa/1) (*Euphorbia ellenbeckii* CHIOV., *M. guentheri* CHIOV.). – Somalia. – Succulent, semi-woody plant, up to 2.1 m tall; **St.** simple, climbing, cylindrical, with few Br., up to 5 cm thick below, with longitudinal furrows and tubercles bluntly-rhombic at the base or long-hexagonal, thin, stalk-like, spreading horizontally, 4–8 mm long, with a L.scar above, with 2 lateral Th.shi., each with a group of 4–5 Th.; **L.** fleshy, borne at the St.tips, soon dropping, spatulate, tapering, up to 15 mm long, up to 10 mm across, margins entire, in the upper part undulating or curly; **Infl.** crowded at the St.tips, 1–2-forked, bract-cups 2.3 mm long, 2 mm across, crenate, bicarinate, whitish-green.

**M. subuliferum** CHIOV. – Somalia. – Acc. P. R. O. BALLY, this appears from the herbarium material to be a shoot of **Senecio**, resembling **S. pendulus.**

**M. trinerve** BALLY (§ Ia/1). – Kenya: central reg. – Caudiciform; **St.** solitary, erect, ○, up to

15 cm tall, 7 mm thick, green, smooth, glabrous; **L.** few, very fleshy, broadly elliptical to rhombic, ± sessile, pointed above, up to 3.5 cm long, 2.5 cm across, M.rib on the upperside paler green, the underside often purple-spotted, margins entire, often curly; **Infl.** from the L.axils, bract-cups rather fleshy, 7 mm long, 6 mm ⌀, bicarinate, notched, greenish-white, with darker green or purple blotches.

**M. virgatum** BALLY (§ IIb/1). – Kenya: coastal reg. – Glabrous, laxly branching ♄; **Br.** virgate, numerous, up to 2 m tall, fleshy, cylindrical, with a few shoots 4–7 cm long, bluish, 7 mm ⌀, with smooth, very slightly raised, longitudinal lines or veins c. 2 mm apart, L.scars prominent, 2 mm across; **L.** fleshy, deciduous, borne at the tips of Br. and shoots, narrow-elliptical, pointed, apiculate, 14–18 mm long, 7–8.5 mm across, the M.rib carinate on the underside, margins entire; stipular Th. paired, minute; **Infl.** solitary, often doubly forked, bract-cups triangular, pointed, fleshy, pale yellow or reddish-brown, 7.5 mm long, 5–6 mm across, acute-carinate, apiculate.

**M. yattanum** BALLY (§ IIa/1). – Tanzania; Kenya.

**M. — v. gladiatum** BALLY. – Tanzania: N. prov. – **St.** ± branching, somewhat climbing, tubercles slightly recurved and longer than in the type; **L.** narrow-oblanceolate.

**M. — v. yattanum.** – Kenya: central reg. – Caudiciform; **St.** numerous, little branching, c. 20 cm long, prostrate if over 15 cm long, 10 to 14 mm thick, cylindrical, fleshy, reticulate or slightly tuberculate, tubercles at the St.base elongated, up to 27 mm long, 8 mm across, those above being 4.5 mm long, 6 mm across, L.scars 1.5 mm high, 4 mm across, with 2 rudimentary, deciduous, linear Stip.; **L.** crowded higher up, very fleshy, oblanceolate, pointed, underside light green, margins serrate, curly towards the tip; **Infl.** axillary, solitary, forking, capitate, bract-cups oblique, 6.5 mm long, 8 mm across, tapering, bicarinate, with purple longitudinal stripes.

**Monanthella** BGR. Crassulaceae. – Acc. A. BERGER **Sedum jaccardianum** MAIRE et WILCZEK, with its 7–10-partite Fl., should be included in this G.; acc. FRÖDERSTRÖM **Sedum atlanticum** MAIRE should also be included; E. R. SVENTENIUS considers both these species to be intermediate between **Sedum** L. and **Monanthes** HAW. and would be prepared to admit this G., but living material is not available for study. In the meantime, both the above-named species are included in Sect. 8. **Monanthoidea** BATT. et TRABUT of the G. **Sedum** L.

**Monanthes** HAW. Crassulaceae. – Canary Is., Salvage Is.; N.Afr. – ☉ plants, ♃ herbs or low shrubs of intricate growth, caespitose; **L.** ± rosulate, crowded at the end of the Br., fleshy, thick, ovate to subcylindrical, tuberculate; **Fl.** small and inconspicuous upon thin pedicels, often in small Rac., pink, yellow or whitish-green, in summer. – Summer: in the open, winter: cold-house. Propagation: seeds, cuttings.

## Division of the Genus Monanthes into Sections by E. R. Sventenius

I. ☉ plants; L. alternate, nearly sessile; Infl. cymose, terminal; Fl. with 6 Pet.

    **Sect. I. Annua** SVENT. – Species: M. icterica.

II. ♃ plants.

    A. Plants semishrubby, intricately branched; L. subopposite to alternate, sessile, globose to spindle-shaped; Fl. with 5–8 Pet.

    **Sect. II. Sedoidea** SVENT. – Species: M. anagensis, laxiflora.

    A. Plants ± caespitose, branched or simple; L. arranged in Ros.

        a) Plants caespitose, many-headed: Ros. globose to elongated; L. spatulate, cuneate, nearly sessile; Infl. from the middle, with or without bracts, Fl. with 6–9 Pet.

    **Sect. III. Monanthes** (Polyclada SVENT.). – Species: M. amydros, atlantica, muralis, polyphylla, subcrassicaulis.

        a) Plants small, simple or branched from the base, stemless or nearly so, with ± flat Ros., compressed; L. narrowed into the petiole; Infl. lateral from the base, with bracts; Fl. with 6–8 Pet.

    **Sect. IV. Petrophylla** SVENT. – Species: M. adenoscepes, brachycaulon, dasyphylla, minima, niphophila, pallens, praegeri, purpurascens, silensis.

**M. adenoscepes** SVENT. (§ IV). – Canary Is.: Tenerife. – Acauline, ♃, with fusiform **roots**; **Ros.** 3–4 cm ⌀, compressed, densely leafy; **L.** thick, rhombic to lanceolate-spatulate, bluntish, narrowing towards the filiform petiole, greenish-brown, with glandular hairs; adventitious buds spherical; **Fl.** 6 mm ⌀.

**M. amydros** SVENT. (§ III). – Canary Is.: Gomera. – Mat-forming; **Ros.** numerous, 10 to 15 mm ⌀, hemispherical to conical; **L.** densely crowded, fleshy, 6–9 mm long, 3–3.5 mm across, compressed cylindrical to spatulate, upperside ± triangular, tuberculate; **Fl.** with stiff, hirsute pedicels, 8–10 mm ⌀, yellow.

**M. anagensis** PRAEG. (Pl. 110/6) (§ II). – Canary Is.: Tenerife. – ♃ herbs or sub-♄, spreading, branching dichotomously; **L.** almost cylindrical, pointed, upperside furrowed, glabrous, green, often reddish; **Fl.** yellowish-green.

**M.** × **anagiflora** BRAMW. et ROWL. – Canary Is.: Tenerife. – Acc. PRAEGER a spontaneous hybrid: **M. anagensis** × **M. laxiflora**. – Subshrubby, mostly smaller than **M. anagensis**; **L.** alternate, rarely opposite, narrowly elliptic or linear-elliptic, green or purplish, subterete, flattish above; otherwise intermediate between the parents.

**M. atlantica** BALL. (§ III) (*M. muralis* HOOK. f., *Sedum surculosum* COSSON, *S. a.* (BALL.) MAIRE). – N.Afr.: High Atlas. – Mat-forming; **Ros.** with 20–30 spatulate, smooth, dark green **L.**; **Fl.** yellow, tinged reddish-brown.

**M. brachycaulon** (WEBB et BERTH.) LOWE v. **brachycaulon** (*Petrophyes b.* WEBB et BERTH.). – Canary Is.: Tenerife, Gran Canaria; Salvage Is. – Resembles **M. purpurascens**; **Ros.** laxer, smaller, without offsets; **L.** clavate-spatulate; **Infl.** minutely hairy, with 5–7 Fl.

**M.** — v. **adenopetala** SVENT. – Canary Is.: Tenerife. – **Ros.** dense; **L.** shortly oblong-spatulate at the tip, narrowing below towards the long petiole; flowering Br. horizontal, bent, c. 5 cm long, with densely rosulate, lanceolate **L.**; **Fl.** with narrow-linear, glandular-hairy Pet.

**M.** — v. **nivata** SVENT. – Canary Is.: Tenerife. – **Ros.** dense; **L.** cylindrical-spatulate, stiff, angular at the tip; Ros. of offsets compressed-spherical.

**M.** — v. **ramosa** PRAEG. – Grand Canary, Salvage Is. – **St.** with radiating, horizontal Br. 2 cm long, with Ros. bearing Infl. and further similar Br.

**M.** × **burchardii** BRAMW. et ROWL. – Canary Is.: Tenerife, Gomera. – Acc. PRAEGER a spontaneous hybrid: **M. laxiflora** × **M. silensis** (*M. l.* × *M. pallens* v. *silensis*). – Habit of **M. laxiflora** but smaller and greyer, 8 cm tall, much branched; **L.** laxly rosulate; **Fl.** few.

**M. dasyphylla** SVENT. (§ IV). – Canary Is.: Tenerife. – **Roots** short-cuneate; **Ros.** 2 to 5 cm ⌀; **L.** dense, compressed, somewhat depressed midway, clavate, thick, ovate-cuneate at the tip, blunt, narrowing towards the filiform petiole, 15–35 mm long, 2–4 mm across, inner L. rather shorter, incurved, reddish-green to dark green, grey-hairy, with purple spots and stripes; **Infl.** papillose, red, with 2–5 Fl. 6–8 mm ⌀, Pet. yellowish with red, longitudinal stripes.

**M.** × **hybrida** BRAMW. et ROWL. – Canary Is.: Tenerife. – Acc. PRAEGER a spontaneous hybrid: **M. brachycaulon** × **M. pallens**. – Intermediate between the parents; **St.** short, unbranched, with rather large, solitary **Ros.**; **L.** 15 mm long, purplish; **Infl.** axillary.

**M. icterica** (WEBB) PRAEG. (§ I) (*Petrophyes i.* WEBB, *Aichryson mollii* PIT.). – Canary Is.: Tenerife, Gomera. – ⊙, pallid green; **Br.** spreading, 25 mm tall; **L.** ± rosulate, ovate, glabrous; **Fl.** numerous, greenish-yellow or brownish; with glandular, softly hairy pedicels.

**M.** × **intermedia** BRAMW. et ROWL. – Canary Is.: Tenerife. – Acc. PRAEGER a spontaneous hybrid: **M. brachycaulon** × **M. polyphylla**. – Like a green, lax, broad-leaved **M. polyphylla**; **Infl.** terminal, short, irregularly branched, leafless, with 5–10 **Fl.** on long Ped.

**M. laxiflora** (DC.) BOLLE (§ II) f. **laxiflora** (*Sedum l.* DC., *Sempervivum agriostaphys* O. KTZE., *Petrophyes a.* WEBB et BERTH., *M. a.* CHRIST). – Canary Is.: Tenerife, Gomera, La Palma, Gran Canaria, Fuerteventura, Lanzarote. – ♃ herb or dwarf ♄; **L.** rather more crowded at the Br.tips, thick, oblong to round, reticulate-rough and wrinkled, green or brownish; **Infl.** of 3–5 minutely red-spotted Fl.

**M.** — f. **foliis-aureis** PRAEG. – Form with yellowish **L.**

**M.** — v. **minor** PRAEG. – Dwarf variety; **Br.** erect; **L.** rather smaller, ovate; **Infl.** and Cal. with longer hairs.

**M.** — v. **chlorotica** (BORNM.) PRAEG. (*M. c.* BORNM.). – **L.** pale green, glabrous.

**M.** — v. **eglandulosa** BORNM. – **Infl.** glabrous.

**M. microbotrys** (BOLLE et WEBB) JACOBS. (*Petrophyes m.* BOLLE et WEBB). – Fuerteventura. – Probably a natural hybrid.

**M. minima** (BOLLE) PRAEG. (§ IV). – Canary Is.: Tenerife. – **Root** tuberous, napiform; **Ros.** dense, offsetting; **L.** spatulate, blunt to rounded, margins short-ciliate; **Infl.** with minute reddish hairs, Fl. few, small.

**M. muralis** (WEBB) CHRIST (§ III) (*Petrophyes m.* WEBB). – Canary Is.: Hierro, La Palma, Gomera. – Small sub-♄, tree-like in appearance, brownish-red; **Br.** ± erect, greyish-red; **L.** obconical, papillose, tuberculate, with grey and red spots, upperside slightly furrowed, 5 to 10 mm long; **Fl.** whitish-pink.

**M. niphophila** SVENT. (§ IV). – Canary Is.: Tenerife. – **Roots** fusiform, simple, thickened and then constricted intermittently, 4–5 cm long; **Ros.** 3–5 cm ⌀; **L.** leathery-fleshy, ± spreading, blistery-papillose at the tip, ± rhombic, blunt, angular-tuberculate above, dirty greenish-blue, with minute purple spots and stripes; **Infl.** with 3–7 yellowish-green Fl. 1 cm ⌀.

**M. pallens** (WEBB) CHRIST v. **pallens** §(IV) (*Petrophyes p.* WEBB). – Canary Is.: Gomera, Hierro; Tenerife. – **Roots** fibrous; **Ros.** 3–5 cm ⌀; **L.** very soft, short-petiolate, roundish or ovate, somewhat narrowed below.

**M. × polycaulis** BRAMW. et ROWL. – Canary Is.: Gomera. – Acc. PRAEGER a spontaneous hybrid: **M. polyphylla** × **M. subcrassicaulis.** – Intermediate between the parents; **St.** creeping as in **M. polyphylla** but less slender and less branched, ascending; **L.** dark purplish-green, obovate, subsessile, 10–15 mm long, 5–6 mm wide; **Infl.** short, central.

**M. polyphylla** (WEBB) HAW. (Pl. 110/5) (§ III) (*Petrophyes p.* WEBB, *Sempervivum monanthes* AIT.). – Canary Is.: Tenerife, Gran Canaria, Gomera, La Palma. – Forming dense and dainty cushions; **St.** filiform, prostrate; **Ros.** dense, 1–2 cm ⌀; **L.** many, cylindrical, bluntly truncate, light green, glabrous, papillose at the tip; **Fl.** 1–4, reddish.

**M. praegeri** BRAMWELL (§ IV). – Canary Is.: coast of N.E.Tenerife. – **Roots** filiform or fibrous; **St.** acauline or caulescent, unbranched, (1–)2(–3) cm ⌀; **L.** dense, clavate, thick, blunt, narrowing teretely below, glabrous, green, ± tinged reddish-purple, papillose-glandular towards the tip, 6–12 mm long, 1.5–2.5 mm across; **Infl.** 3–4 cm tall, from the L.axils, with basal Ros. of leaf-like bracts 4–7 mm long, 2–5 mm across, Fl. 3–6, 5.5–6.5 mm ⌀, Pet. yellowish on the upperside, with red stripes; Ped., pedicels, Cal. and Pet. all glandular-hairy on the underside.

**M. × pumila** BRAMW. et ROWL. – Canary Is.: Tenerife. – Acc. PRAEGER a spontaneous hybrid: **M. brachycaulis** × **M. silensis (M. b.** × *M. pallens* v. *silensis*). – Intermediate between the parents; plants small, compact, grey; **St.** short, unbranched; **L.** 1 cm long; **Infl.** axillary.

**M. purpurascens** (BOLLE et WEBB) CHRIST (§ IV) (*Petrophyes p.* BOLLE et WEBB). – Canary Is.: Gran Canaria. – **Ros.** dense, small, with numerous offsets; **L.** long-spatulate, long-petiolate, smooth, papillose; **Infl.** with dense bracts, Fl. 8, densely cobweb-hairy.

**M. silensis** (PRAEG.) SVENT. (*M. pallens* v. *s.* PRAEG.). – Canary Is.: W.Tenerife. – Similar to **M. polyphylla,** but with a distinct taproot; **Ros.** small, glaucous, convex; **L.** densely imbricate, up to 1 cm long, clavate with a convex, truncate apex, papillose to verrucose; Infl. 3–5-flowered; pedicels ± shortly pilose-glandular; Pet. 2.5 mm long.

**M. × silophylla** BRAMW. et ROWL. – Canary Is.: Tenerife. – Acc. PRAEGER a spontaneous hybrid: **M. silensis** × **M. polyphylla** (*M. pallens* v. *silensis* × **M. polyphylla**). – Intermediate between the parents; **St.** short, less creeping and less branched than that of **M. polyphylla**; **Ros.** like those of **M. silensis** but smaller, flat, crowded.

**M. subcrassicaulis** (O. KTZE.) PRAEG. (Pl. 110/7) (§ III) (*Sempervivum monanthes* v. *s.* O. KTZE.). – Canary Is.: Gomera, La Palma, Tenerife. – Mat-forming; **Br.** prostrate, 1–2 cm long, green; **L.** dense, 10–12 mm long and across, clavate, truncate, semi-terete, upperside flat, red at the base; **Infl.** of 1–5 Fl., roughly hairy, including the Pet.

**M. × sventenii** BRAMW. et ROWL. – Canary Is.: Tenerife, Gomera. – Acc. PRAEGER a spontaneous hybrid: **M. laxiflora** × **M. pallens.** – Habit of **M. laxiflora,** but **L.** alternate, crowded, flattish, broadest near the apex; **Infl.** branched; **Fl.** larger, Pet. narrower.

**M. × tilophila** (BOLLE) CHRIST (*Petrophyes* × *t.* BOLLE). – Canary Is.: Gran Canaria, Tenerife. – Acc. PRAEGER a spontaneous hybrid: **M. brachycaulon** × **M. laxiflora.** – Canary Is.: Gran Canaria, Tenerife. – **Br.** erect or ascending from a creeping St.; **L.** numerous, scattered or subrosulate, alternate, oblanceolate to obovate with a cuneate base or rhombic-elliptic, blunt, purplish, 9–12 mm long, 4–6 mm broad, 2–4 mm thick; **Fl.** intermediate between those of the parents, in racemes of 4–6.

**Monolena** TRIANA. Melastomataceae. – S.Am. – Glabrous, fleshy ♃ with a caudiciform **taproot**; **L.** conspicuous; **Fl.** large, pink. – Culture: Greenhouse, warm. Propagation: seed.

**M. primuliflora** HOOK f. (Pl. 110/4). – Colombia. – **Caudex** thick, fleshy, smooth, tuberous, knotty from the scars of old L., Stip. persisting as soft Th., up to 15 cm ⌀, brownish-green; **L.** smooth, leathery, broad-elliptical, margins dentate and ciliate, with 3–5 veins, up to 15 cm long, 6 cm across, upperside metallic, glossy green, veins red, underside red, petiole 2.5–3 cm long, red; **Infl.** of 3 Fl. 2.5 cm ⌀, pink with a white eye.

**Moringa** BURM. Moringaceae. – Afr.; Madag. – ♄, sometimes with a thick trunk and smooth bark; **L.** large, readily dropping, alternate, simply imparipinnate, Stip. missing or small and developing as glands at the bases of petioles and leaflets; **Fl.** medium-large, white or red, in axillary panicles; **Fr.** long, narrow, pod-like, with numerous conical or ovoid, winged or unwinged seeds. – Greenhouse, warm. Propagation: seed.

**M. drouhardii** JUM. – S.Madag. – **St.** c. 10 m tall, very thick, reminiscent of **Adansonia**; **L.** 20–30 cm long, doubly or trebly pinnate, segments very thin; **Fr.** 30–50 cm long.

**M. hildebrandtii** ENGL. (Pl. 111/1). – Madag.; trop. Afr. – **Trunk** 20–25 m tall, much thickened, especially at the base; **L.** up to 60 cm long, doubly pinnate.

**M. longituba** ENGL. (*Hyperantha longituba* (ENGL.) CHIOV.). – W.Somalia. – **Caudex** large, tuberous; **St.** few; Cal.tube 2–3 cm long.

**M. ovalifolia** DTR. et BGR. (*M. ovalifoliolata* DTR. et BGR.). – S.W.Afr.: Damaraland, Gr. Namaqualand. – **Trunk** with few Br., 2–6 m tall, up to 1 m thick, wood spongy, bark smooth, light grey; **L.** 50–80 cm long, 40–60 cm across, doubly pinnate; **Fl.** small, white, in much-branched panicles 40–50 cm long.

**Mucizonia** (DC.) BGR. (Umbilicus DC. Sect. *Mucizonia* DC.). Crassulaceae. – Iberian peninsula; N.Afr.; Canary Is. – ⊙, reminiscent of **Pistorinia.** – Summer: in the open. – Propagation: seed.

**M. hispida** (LAMK.) BGR. (Pl. 111/2) (*Cotyledon h.* LAMK., *Umbilicus h.* (LAMK.) DC., *Cot. mucizonia* ORTEGA, *Sedum m.* (ORTEGA) HAMET, *S. m.* (ORTEGA) ROTHM.). – Spain; Algeria; Morocco; Canary Is. – 5–10 cm tall, branching basally; **L.** alternate, ± cylindrical, blunt, glabrous, 13 mm long, red-spotted, soon dropping; **Fl.** numerous, pedicellate, reddish, in Cinc. in a dichotomous cyme, Ped. with long, projecting, bristly hairs.

**M.** — v. **glabra** BRAUN-BLANQUET et MAIRE. – Morocco. – Glabrous in all parts.

**M. lagascae** (PAU) LAINZ (*Sedum l.* PAU, *S. villosum* v. ? *campanulatum* WK., *S. hispanicum* LAG.). – Spain. – Similar to **Sed. villosum** L., but only glandulose-pubescent; **Infl.** corymbose-ramose.

**Myrmecodia** JACK. Rubiaceae. – Malayan Archipelago; Austr. – Epiphytes, for the heated greenhouse. Propagation: seed.

**M. echinata** MIQ. (*M. tuberosa* JACK., *M. inermis* DC., ? *M. antonii* BECC.). – Malayan Archipelago; Austr. – Small sub-♄ with a spiny tuberculate **caudex** up to 10 cm thick, with peculiar internal cavities which are inhabited by ants; **St.** several, fleshy, thick, 5–10 cm or more tall, produced from the tip of the caudex; **L.** several, leathery, ± fleshy, arising at the St.tips from a spiny cushion; **Fl.** small, tubular, white; **Fr.** berry-like, oblong, red.

**M. platyrea** BECC. (Pl. 111/3). – Resembles **M. echinata**; caudex flatter, **L.** on a very short **St.** c. 5 mm long.

**Neoalsomitra** HUTCH. Cucurbitaceae. – Peninsular India to Polynesia; Austr. – Some 23 species of this G. are known, mostly non-succulent climbing plants; only the following have some characteristics of succulence. – Greenhouse. Propagation: seed, cuttings.

**N. podagrica** VAN STEENIS (Pl. 112/1). – ? Celebes, Lit. Sunda Is. – **St.** with fusiform thickening at the base, often with side shoots also thickened basally, these swellings green, fleshy, often covered with obliquely projecting, stout, green, hard Th. up to 1.5–5 cm long, St.-climbing, up to 30 m long, the bottom 3–5 m also thorny, tendrils only on the youngest shoots, 5–15 cm long, shortly bipartite; **L.** petiolate, 5-partite, upper L. tripartite, 6–11 cm long, 4–7 cm across; ♂ **Fl.** 2.5–3.5 mm ⌀, greenish-yellow, many in lateral, branching, axillary, pyramidal Infl.; ♀ **Fl.** 4–5 mm ⌀, in lateral Rac.; **Fr.** tubular to cup-shaped, 1.5 to 2 cm long, c. 8 mm ⌀.

**N. sarcophylla** (ROEM.) HUTCH. (*Alsomitra s.* ROEM., *Zazonia s.* WALL.). – Burma; Thailand; Philippines. – **St.** thin, much-branching, climbing; **Br.** pendulous; **L.** 5–7.5 cm long, elliptical-ovate or oblong or ovate-lanceolate, blunt, apiculate, with entire margins, very succulent, c. 2.5 mm thick, upperside light green, with 3 veins, underside paler and reticulate, tendrils unbranched; **Fl.** many, greenish-yellow; plant dioecious.

**Neorautenenia** SCHINZ. Leguminosae. – S. trop. Afr. – Dwarf ♃; **caudex** thick, fleshy; **L.** somewhat fleshy (?), pubescent; **Fl.** in axillary Rac., yellowish. – Greenhouse, warm, moist.

**N. ficifolia** (BENTH.) C. A. SMITH (*Rhynchosia f.* BENTH.). – Transvaal. – **Caudex** compressed, angular, up to 70 cm long, 15 cm across, 6 cm thick, attenuated towards both ends, upper surface rough, brown; **L.** short-petiolate, tripartite, leaflets cuneate-obovate (resembling a fig), mostly 3–5-lobed, the lateral lobes oblique, 6–10 cm long, 5–6 cm across, rather rigid, softly pubescent and very pale green, reticulate above, with prominent ribs and veins beneath; Stip. lanceolate; **Infl.** racemose, Ped. elongate, Fl. yellow.

**Nolina** MCHX. Agavaceae. – Mexico. – Small ♄; **trunk** thickened, succulent, slender above; **L.** clustered at the top of the trunk, long, pendulous, narrow; **Fl.** small, in large, terminal panicles. – Greenhouse. Propagation: seed.

**N. bigelovii** S. WATS. (*Dasylirion b.* TORR., *Beaucarnea b.* BAK.). – USA: W.Arizona; Mexico: Sonora. – **Trunk** up to 3 m tall, clavately thickened and 70 cm ⌀ at the base, much tapered above, sparsely branching, with a rough bark; **Br.** very short; **L.** with a broad-triangular base, 2 cm across above the base, linear-tapering, up to 1 m long, margins rough.

**N. longifolia** HEMSL. (*Yucca l.* SCHULT., *Dasylirion l.* ZUCC., *Beaucarnea l.* BAK.). – S.Mexico. – **Trunk** over 2 m tall, base very rounded and widened, tapering above, bark thick and corky; **L.** in a dense crown, hanging down for over 2 m, over 2.5 cm across, very gradually tapering, thin, firm, green, margins minutely rough; **Infl.** 2 m tall, Fl. white.

**N. recurvata** LEM. (Pl. 111/4) (*Beaucarnea r.* LEM., *B. tuberculata* ROEZL., *N. t.* HORT., *Pincenectitia t.* LEM.). – S.E.Mexico. – **Trunk** spherically swollen at the base, often 50 cm or more ⌀, above that slender and 4–6 m tall, with a few Br. above; **L.** thin, linear, long-tapering, with entire margins, green, elegantly recurved, 1 m long, c. 2 cm across; **Fl.** inconspicuous.

**Orostachys** FISCH. Crassulaceae. – USSR to China and Korea; Japan. – Glabrous ♃, appearance variable, 5–30 cm tall, perhaps more; **L.** crowded in an acauline Ros., often with offsets, usually ± linear, terminating in a cartilaginous apiculus, or broader and not apiculate; **Infl.** terminal, the Ros. dying after fruiting, sometimes with smaller lateral Infl., ±bracteate, often branched, Fl. ± short-pedicellate, very numerous and dense in a long, cylindrical, spicate Rac. or Pan., or from the axils of the cauline L. in small Rac., white, whitish-yellow or reddish, Pet. often red-spotted. Flowers in summer, in the second or third year. In the open. Propagation: seed, offsets.

**O. boehmeri** (MAKINO) HARA (*Cotyledon malacophylla* v. *b.* MAKINO, *Cot. b.* (MAKINO) MAKINO, *Sedum b.* (MAKINO) MAKINO, *O.*

*aggregatus* v. *b.* (MAKINO) OHWI). – Japan: Honshu, Hokkaido. – Small counterpart of **O. malacophyllus,** with elongated offsets.

**O. chanetii** (LEV.) BGR. (*Sedum c.* LEV., *S. pyramidalis* PRAEG.). – China: Kansu. – Stolon-Ros. with L. of 2 different lengths; **L.** linear, underside convex, grey-green; **Fl.** in a pyramidal Pan., very numerous, white, reddish outside.

**O. erubescens** (MAX.) OHWI v. **erubescens** (*Umbilicus e.* MAX., *Cotyledon e.* (MAX.) FRANCH. et SAVAT, *O. spinosus* v. *e.* (MAX.) BGR., *Sedum e.* (MAX.) OHWI). – W.Japan; E.As. – Winter-L. narrow-spatulate, fleshy, 1.5–3 cm long, 4 to 6 mm across, somewhat flat, the hard tip with several T. and a short apiculus; **St.-L.** (summer-L.) soft, with a short apiculus; **Ped.** 6–15 cm tall, with many aciculate-lanceolate L., **Fl.** in a spike 4–10 cm long, 1.5–2 cm ∅, pedicellate, whitish-pink, anthers red at first, becoming darker.

**O.** — v. **japonicus** (MAX.) OHWI (*Cotyledon j.* MAX., *O. j.* (MAX.) BGR., *Sedum e.* v. *j.* (MAX.) OHWI, *S. japonicola* MAKINO, *S. spinosum* THUNBG.). – Ros.-L. ovate, oblong or narrow-oblong, sometimes 10 mm across; St.-L. oblong-linear.

**O.** — v. **polycephalus** (MAKINO) OHWI (*Cotyledon p.* MAKINO, *Sedum p.* (MAKINO) MAKINO, *O. p.* (MAKINO) HARA, *S. e.* v. *p.* (MAKINO) OHWI. – Caespitose through offsets from the St.-base.

**O. fimbriatus** (TURCZ.) BGR. v. **fimbriatus** (*Cotyledon f.* TURCZ., *U. f.* (TURCZ.) TURCZ., *Sedum f.* (TURCZ.) FRANCH.). – Mongolia; China. – Ros. offsetting; L. with a white apiculus, broader at the base, passing over into a white, crescent-shaped appendage clasping the L.tip, the bracts and Sep. apiculate; **Fl.** white.

**O.** — v. **ramosissimus** (MAXIM.) BGR. (*Umbilicus r.* MAXIM.). – Offsetting very freely from the base; **Infl.** broad-conical, Fl. reddish.

**O.** — v. — sv. **limunoides** (PRAEG.) JACOBS. (*Sedum l.* PRAEG.). – **Infl.** laxer, sometimes axillary.

**O. iwarenge** (MAKINO) HARA (Pl. 112/2) (*Cotyledon i.* MAKINO, *Sedum i.* (MAKINO) MAKINO, *S. malacophyllum* v. *i.* (MAKINO) FROED.). – Japan. – R. MORAN considers this species to be a large, farinose form of **O. malacophyllus,** the Ros. persisting in winter, with lax winter-buds. Acc. FROEDERSTROEM it is differentiated from **O. malacophyllus** by the Sep. being obtuse, those in **O. m.** being acute.

**O.** — cv. Fuji. – Cultivar from Japan; **L.** with attractive mottling.

**O. malacophyllus** (PALLAS) FISCH. (*Cotyledon m.* PALLAS, *Sedum m.* (PALLAS) STEUD., *Umbilicus m.* (PALLAS) DC., *U. stamineus* LEDEB., *U. inermis* MIQ., *Cot. m.* v. *japonicus* FRANCH. et SAVAT, *Cot. aggregata* MAKINO, *Cot. filifera* (MAKINO) MAKINO, *O. m.* v. *japonicus* (FRANCH. et SAVAT) BGR., *O. aggregatus* (MAKINO) HARA, *O. filifera* (NAKAI) NAKAI, *O. furusei* OHWI, *O. aggregatus* v. *roseus* OHWI). – E.Siberia to S.Korea; Japan. – **Ros.** solitary or grouping, 1–13 cm ∅; **L.** 10–65, oblong-ovate to oblanceolate, rounded or blunt, inner L. pointed, 1–9 cm long, 0.5–4 cm across, 1.5 to 2.5 mm thick, green or bluish, margins entire to papillose, ± translucent; winter-buds ± spherical, 2–15 mm ∅, usually compact, with 8–20 erect L., often covered with 1–6 imbricate, longer L.; floral St. up to 35 cm tall, **Infl.** cylindrical, up to 19 cm long, Fl. 5–12 mm ∅, white.

**O. minutus** (KOMAROV) BGR. (*Cotyledon minuta* KOMAROV, *O. kanboensis* OHWI). – Southern N.E.China to S.Korea. – Usually caespitose, Ros. (0.5–)2.5(–8) cm ∅; **L.** 14–20, outer ones elliptical-oblong, (0.5–)2.5(–3.5) cm long, (2–)6(–10) mm across, (1–)2(–3) mm thick, with a rhombic swelling 2–6 mm across, green, later with red markings, margins entire, terminal Sp. 0.5–1 mm long, rigid, inner L. triangular-lanceolate, tapering, Sp. soft; winter-buds 2–9 mm ∅, L. 8–40, slightly incurving, ovate, wrinkled; floral St. (2.5–)8(–12) cm tall, **Infl.** cylindrical, up to 8 cm long, Fl. 5–8 mm ∅, white, Pet. with glandular blotches above which eventually turn pink to red.

**O. roseus** (LESS.) BGR. (*Cotyledon r.* LESS.). – S.Ural Mts. – **L.** triangular, spiny tip long, whitish; **Fl.** pale pink.

**O. spinosus** (L.) BGR. (Pl. 112/3) (*Cotyledon s.* L., *Umbilicus s.* (L.) DC.). – E.As.: Siberia, Mongolia, Altai, Tienshan, W.Tibet. – Caespitose; **L.** rosulate, oblong, ± cuneate, usually of 2 different lengths, spiny tip white, soft; **Fl.** in a spicate panicle, yellow.

**O. thyrsiflorus** FISCH. (*Cotyledon t.* (FISCH.) MAXIM., *Umbilicus t.* (FISCH.) DC., *U. leucanthum* LEDEB.). – Urals, Mongolia. – Related to **O. fimbriatus,** appendage not dentate; Fl. white.

**Othonna** L. Compositae. – S. and S. W. Afr. – Small succulent ♃ or sub-♄, sometimes caudiciform; **St.** glabrons or somewhat hairy, or with woolly hair from the L. axils; **L.** alternate, opposite or basal, cylindrical or flat, ± fleshy, often thick; **Fl..** heads long-pedicellate, solitary or in a paniculate cyme, yellow, rarely white or purple. – Not all species of **Othonna** are succulent. – Greenhouse, winter in the cold-house. – Propagation: seed, cuttings.

**Plate 122. 1. Rhodiola crassipes** (WALL.) JACOBS. (Photo: J. A. HUBER); **2. R. rosea** L. ssp. **rosea** (Photo: as 1); **3. Rochea coccinea** (L.) DC.; **4. Rhytidocaulon subscandens** BALLY (Drawing: P. R. O. BALLY).

Plate 123. 1. Salicornia fruticosa L.; 2. Left: Sansevieria stuckyi GODEFR.-LEB., right: S. trifasciata PRAIN; 3. Salicornia europaea f. stricta (DU MORTIER) JACOBS.; 4. 1. Sansevieria trifasciata cv. Hahn'i, 2. cv. Golden Hahnii, 3. cv. Silver Hahnii (Photo: H. PFENNIG).

Plate 124. 1. **Sarcocaulon multifidum** R. KNUTH (Photo: C. BACKEBERG); 2. **S. burmannii** SWEET (Photo: K. DINTER); 3. **Sarcostemma viminale** R. BR. (Photo: J. MARNIER-LAPOSTOLLE); 4. × **Sedeveria hummelii** E. WALTH.

Plate 125. 1. Sedum album L. ssp. teretifolium (LAMK.) SYME v. teretifolium; 2. S. bellum ROSE; 3. S. coeruleum VAHL; 4. S. brevifolium DC. v. brevifolium; 5. S. hintonii R. T. CLAUSEN (Repr.: Cact. Succ. J. Am. XXV, 1953, 97. Photo: M.KIMNACH); 6. S. lineare v. lineare f. variegatum PRAEG.

Plate 126. 1. a) Sedum pachyphyllum Rose, b) S. nussbaumeranum Bitter, c. S. allantoides Rose; 2. S. morganianum E. Walth.; 3. S. oaxacanum Rose; 4. S. oreganum Nutt.; 5. S. pilosum Marsch. Bieb.; 6. S. potosinum Rose.

Plate 127. 1. Sedum sempervivoides FISCH.; 2. S. spectabile BOREAU; 3. S. sediforme (JACQ.) C. PAU v. sediforme; 4. S. treleasei ROSE.

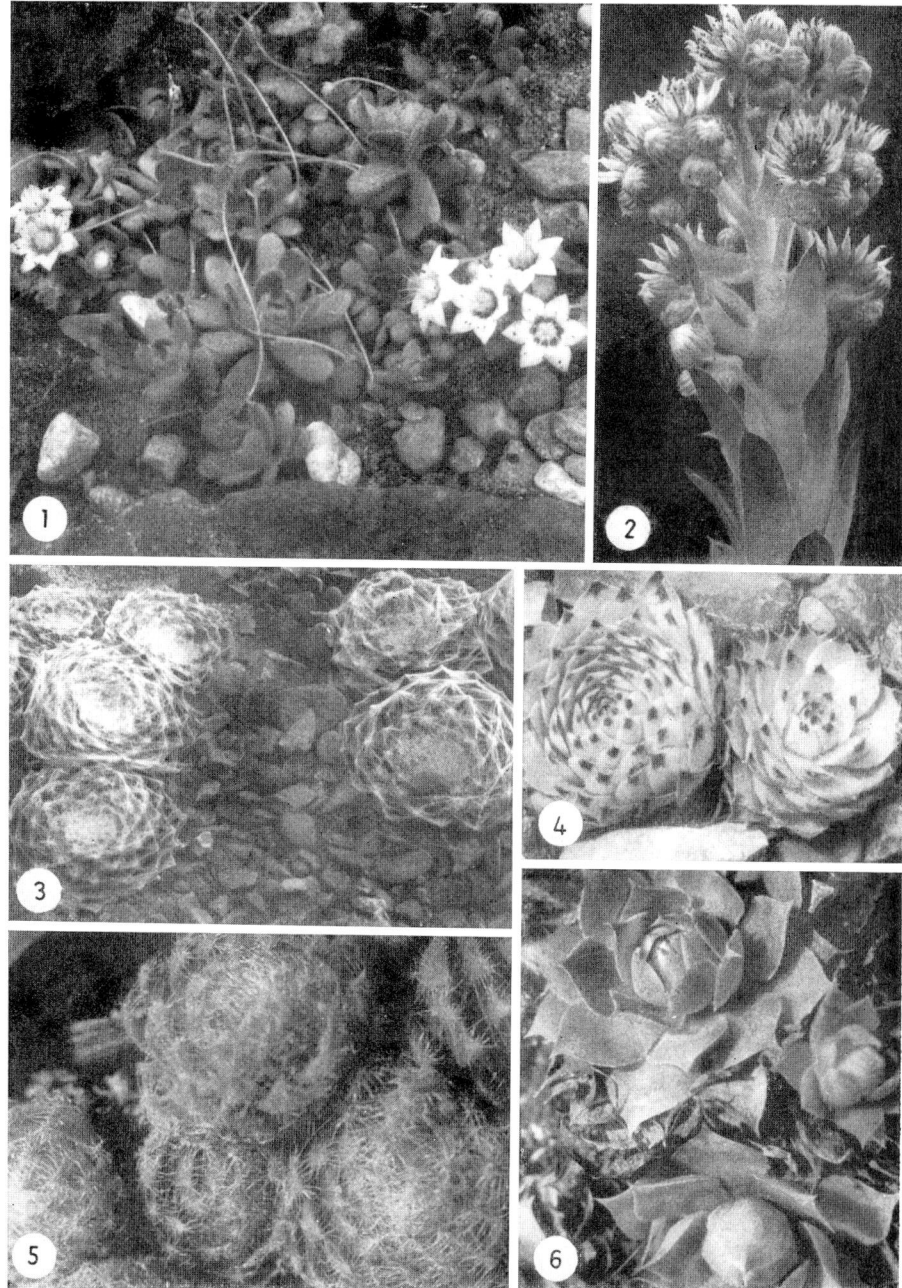

Plate 128. 1. **Sempervivella alba** (EDGEW.) STAPF; 2. **Sempervivum tectorum** ssp. **alpinum** (GRISEB. et SCHENK) WETTST. Infl.; 3. **S. arachnoideum** L. v. **arachnoideum**; 4. **S. tectorum** ssp. **calcareum** (JORD.) CARIOT et ST. LAGER; 5. **S. ciliosum** CRAIB f. **ciliosum** (Photo: J. A. HUBER); 6. **S. wulfenii** HOPPE (Photo: as 5).

Plate 129. 1. Senecio antandroi Scott Elliot; 2. S. amaniensis (Engl.) Jacq. (Photo: P. R. O. Bally; 3. S. aizoides (DC.) Sch. Bip.; 4. S. citriformis Rowl.

## Synopsis of the Genus Othonna by G. D. Rowley

Only species mentioned in the Lexicon are included here. The key should be regarded as provisional only, until the plants have been more fully studied in the field.

**Gr. I.** Soft, prostrate to decumbent **herbs** with ± terete evergreen L.

| | | |
|---|---|---|
| 1. | Plant stoloniferous, forming Tub. at the St. tips | rechingeri |
| 1a. | Plant not forming Tub. at the St. tips | 2. |
| 2. | St. shrubby and fleshy; L. spiralled | 3. |
| 2a. | St. slender and limp; L. sometimes whorled, up to 3 cm long | capensis |
| 3. | L. crowded at the Br. tips, 2.5–5 cm long | carnosa |
| 3a. | L. ± throughout the Br., 6–10 cm long | opima |

**Gr. II. Shrubby** or **long-caudiciform** with slender non-succulent Br.

| | | |
|---|---|---|
| 1. | Long-caudiciform with a dark peeling bark | 2. |
| 1a. | Shrubby, not as above | 5. |
| 2. | L. linear | 3. |
| 2a. | L. broader, ± lanceolate, oblanceolate or spathulate | 4. |
| 3. | L. bearded at the base | obtusiloba |
| 3a. | L. bare at the base | protecta |
| 4. | L. minute, to 6.5 mm long | pachypoda |
| 4a. | L. larger, c. 5 cm long | retrofracta |
| 5. | L. terete, subterete or semiterete | 6. |
| 5a. | L. flat | 8. |
| 6. | L. minute, terete, 6–14 mm long | sedifolia |
| 6a. | L. larger, 1–8 cm long | 7. |
| 7. | Ped. 1–4; ray florets present | cylindrica |
| 7a. | Ped. solitary; no ray florets | sparsiflora |
| 8. | L. tapered below, sessile or short petiolate | 9. |
| 8a. | L. abruptly narrowed into a slender petiole 7–8 mm long | cyclophylla |
| 9. | L. pinnatifid or laciniate | graveolens quercifolia |
| 9a. | L. entire or with up to 3 lobes | 10. |
| 10. | Ray florets absent | furcata |
| 10a. | Ray florets present | 11. |
| 11. | L. bearded at the base | arborescens macrosperma quinqueradiata |
| 11a. | L. bare at the base | 12. |
| 12. | L. quite entire | triplinervia |
| 12a. | L. dentate | dentata |
| 12b. | L. with up to 3 teeth | osteospermoides othonnites |

**Gr. III. Geophytes** with the caudex more than half below ground

| | | |
|---|---|---|
| 1. | Plants with deciduous Br. and Infl. | 2. |
| 1a. | Plants with deciduous Infl. only and radical L. | 6. |
| 2. | Plant with underground stolons rooting at the nodes | filicaulis |
| 2a. | Plant without underground stolons | 3. |
| 3. | Radical L. linear, 15–25 cm long | stenophylla |
| 3a. | Radical L., when present, shorter, petiolate with a flat blade | 4. |
| 4. | Cauline L. ± amplexicaul | 5. |
| 4a. | Cauline L. not amplexicaul | gracilipes |
| 5. | St. elongated, laxly leafy | amplexifolia |
| 5a. | St. short, flexuous, closely leafy | tuberosa |
| 6. | L. bearded at the base | retrorsa |
| 6a. | L. bare at the base | 7. |
| 7. | L. petiolate | hederifolia |
| 7a. | L. sessile | primulina |

**Gr. IV. Short-caudiciform,** more than half the caudex above ground, without slender deciduous leafy shoots

| | | |
|---|---|---|
| 1. | Whole plant body covered in imbricate scales from the persistent L. bases | lepidocaulis |
| 1a. | Plant body not covered in imbricate scales | 2. |
| 2. | L. bearded at the base | 3. |
| 2a. | L. bare at the base | 4. |

3. L. thick and fleshy .................................................... **hallii**
3a. L. thin ............................................................. **cacalioides**
　　　　　　　　　　　　　　　　　　　　　　　　　　　　　　　　　　**intermedia**
4. Plant spiny ....................................................... **euphorbioides**
4a. Plant unarmed ............................................................. 5.
5. Caudex strongly tuberculate ............................................ **herrei**
5a. Caudex not tuberculate ......................................................... 6.
6. L. clavate ......................................................... **clavifolia**
6a. L. thin, flat ..................................................... **spektakelensis**

**O. amplexifolia** DC. (Gr. III). – Cape: Lit. Namaqualand; S.W.Afr.: Klinghardt Mts. – Herbaceous; caudiciform; **St.** weak; **L.** broadly cordate-amplexicaul; **Fl.**heads solitary, long-pedicellate, ray-florets linguiform.

**O. arborescens** L. (Gr. II) (*O. coronopifolia* THUNBG.). – Cape: Cape Flats. – Erect, fleshy, glabrous ♄ with very tangled Br.; **L.** ovate-oblong, margins entire or blunt-dentate.

**O. cacalioides** L.f. (Gr. IV) (*O. minima* DC., *O. pillansii* HUTCHINS., *O. pygmaea* COMPT.). – Cape: Giftberg, Van Rhynsdorp D.; S.W.Afr.: Lüderitz Bay. – **Caudex** compressed-spherical, 2–3 cm ∅, woolly at the tip, roots fleshy; **L.** rosulate, lanceolate-spatulate, margins entire or with 1–2 T., 10–15 mm long, 4–6 mm across, leathery, glabrous; **Infl.** of one yellow Fl. head.

**O. capensis** L. H. BAILEY (Pl. 112/4) (Gr. I) (*O. crassifolia* HARV., *O. filicaulis* ECKL.). – Cape: Uitenhage D. – ♃ herb or sub-♄; **St.** short, freely and irregularly branching; **Br.** often over 1 m long, thin, prostrate or pendulous; **L.** crowded at the forks, ± cylindrical, 2–3 cm long, 5–7 mm thick, blunt or with a soft mucro, upperside slightly furrowed, soft-fleshy, light green, slightly waxy-pruinose; **Fl.** in terminal corymbs of few yellow Fl., summer.

**O. carnosa** LESS. (Gr. I) (*Kleinia crassulaefolia* BAK.). – Cape: Uitenhage, Albany D. – Small sub-♄; **Br.** cylindrical, 10–12 cm long, fairly densely leafy, often dying off in winter; **L.** 5–7 cm long, 5–6 mm thick, sessile, short-tapering, green, reddish at the tip; **Infl.** terminal, 15–20 cm long, forking, with 3–5 yellow to creamy-white Fl. heads.

**O. clavifolia** MARL. (Pl. 112/5) (Gr. IV). – S.W.Afr.: Lüderitz D. – Very dwarf ♄; **caudex** short, with a napiform thickening below; **Br.** short, thick; **L.** at the Br.tips, up to 24 mm long, cylindrical, thickening above to 9 mm, thick-clavate or almost berry-shaped, mucronate, light green, becoming greyish-purple in the resting period; **Infl.** with long Ped. bearing 1–2 Fl.heads with yellow ray-florets.

**O. cyclophylla** MERXM. (Gr. II). – Cape: Lit. Namaqualand; S.W.Afr.: Warmbad D. – Small, fairly succulent ♄ up to 60 cm tall, 3–4 cm thick at the base, branching above; **Br.** 5 mm thick, flexible, eventually rigid, bark whitish-grey, granular, Br.tips and short shoots floccose-felted, the plant otherwise glabrous; **L.** crowded at the Br.tips and on the lateral short shoots, rather fleshy, flat, circular, 1–2 cm ∅, petiole 7–8 mm long, margins slightly crenate-dentate, blue-green; **Fl.**heads solitary, Ped. c. 2 cm long, involucral bracts 11–13 mm long.

**O. cylindrica** (LAM.) DC. (Gr. II) (*Cacalia c.* LAM., *O. tenuissima* HAW.). – Cape: Clanwilliam D.; S.W.Afr.: coast. – ♄; **Br.** lax, verticillate, cylindrical, fleshy; **L.** semi-cylindrical, linear, 3–8 cm long, 2–3 mm thick, fleshy, grey-green, tapering; Ped. 10–15 cm long, terminal, Fl.heads (1–)2(–4) in paniculate false umbels, ray-florets yellow.

**O. dentata** L. (Pl. 113/4) (Gr. II) (*O. bulbosa* β L.). – Cape: Caledon D. – Glabrous, fleshy ♄, **L.** rosulate, crowded at the Br.tips, ± sessile, obovate, acute-dentate.

**O. euphorbioides** HUTCHINS. (Gr. IV). – Cape: Lit. Namaqualand. – Dwarf, succulent ♄ scarcely 10 cm tall, with few Br. from the base; **Br.** terete, 6–8 cm long, 15–20 mm thick, bark yellowish-grey; **L.** in dense clusters at the St.-tips, oblong-spatulate, narrowing below towards the petiole, c. 22 mm long, 4 mm across above, with a rounded tip, light green, waxy-pruinose, with numerous, much-forked, thin Th. almost as long as and between the L.; **Fl.**heads solitary, yellowish, with Ped. 2.5 cm long.

**O. filicaulis** JACQ. (Gr. III). – Cape: Cape peninsula. – Very variable, small with few Br., or more luxuriant, with fleshy underground stolons, rooting from the nodes, with small caudices 1–2 cm ∅; aerial **shoots** thin, filiform, ± erect, 5–10 cm tall; **L.** very variable, ± large, sessile, ovate-cordate or lanceolate or long-petiolate and subcircular, mucronate, c. 2 cm long, 1 cm across; **Fl.**heads with Ped. 2 cm long.

**O. furcata** (LINDL.) DRUCE (Gr. II) (*Ceradia f.* LINDL., *O. aeonioides* DTR., *Doria ceradia* (LINDL.) HARV.). – S.W.Afr.: Gr. Namaqualand, Ichabo Is., Klinghardt Mts. – Hemispherical ♄ up to 60 cm tall, with a white bark; **St.** up to 6 cm thick, dividing c. 20 cm above the base into 2–4 ascending and repeatedly forking Br., terminal shoots with spiralled L. c. 3–4 cm long, 1.5 cm thick, broad-spatulate, fleshy, sap-green, margins entire, petiole short and thick; **Fl.**heads terminal, 7–8, ray-florets missing.

**O. gracilipes** HUTCHINS. (Gr. III). – Cape: Prince Albert D. – **Caudex** short, c. 3 cm long, constricted at intervals, the bark longitudinally plicate, the tip with densely woolly, light brown, floccose hairs; **shoots** solitary, very thin, probably creeping, Int. 7–9 cm long, thinnest midway; **L.** broad-ovate or flat-circular, abruptly cuneate at the base, petiolate, rounded above, apiculate, fleshy, large L. digitate, glabrous, upperside green, underside purple; **Infl.** up to 10 cm tall, with violin-shaped bracts; Fl.heads disc-shaped, lemon-yellow.

**O. graveolens** O. HOFFM. (Gr. II) (*Senecio cactaeformis* KLATT). – S.W.Afr.: Gr. Namaqualand. – ♄ c. 60 cm tall, unbranched or sparsely branched above; **Br.** c. 1 cm ⌀, fleshy, with a thick coating of light brown resin; **L.** in crowded clusters at the Br.tips, broad-obovate to spatulate, narrowing below towards the petiole, ± laciniate, 3.5 cm long, 1 cm across above, light green; **Fl.**heads small, yellow, in dense clusters.

**O. hallii** B. NORD. (Gr. IV). – Cape: Lit. Namaqualand. – **Caudex** short, obconical or napiform, hard and woody, up to 4 cm ⌀, narrowing abruptly below towards the vertical main root; **L.** rosulate at the tip of the caudex, with wool between the L.bases, numerous, glabrous, oblanceolate or spatulate, 2.8 cm long, 0.3–2 cm across, thickish and rather fleshy, dull green or slightly bluish, margins smooth or minutely and distantly dentate, the tip pointed or rounded; **Infl.** solitary or several from the tip of the caudex, paniculate, 12–50 cm tall, Fl.heads 5–7, yellowish.

**O. hederifolia** B. NORD. (Gr. III). – Cape Prov., Calvinia D. – Dwarf, caudiciform, with a fusiform root below 2–10 cm long and 1–2.5 cm thick; **L.** all radical, with a flat, broadly cuneate blade with c. 5 sinuate lobes at the tip, on a slender petiole 2–7 cm long; Ped. simple, 3–17 cm long; **Fl.** yellow, rayed.

**O. herrei** PILL. (Pl. 113/2) (Gr. IV). – Cape: Lit. Namaqualand. – **St.** short, 7–20 cm tall, 1.5–3 cm thick, little branching, the persistent L.bases giving it a nodal appearance; **L.** at the Br.tips, 5–6 cm long, c. 3 cm across, irregularly ovate, narrowing towards the petiole, margins ± undulating and curved, ± smooth or irregularly thorny-dentate, green, bluish-pruinose; **Fl.**heads usually several on a Ped. 5–7 cm long, 12 mm ⌀, yellow.

**O. —** cv. Cristata. – Acc. H. HERRE a St.-fasciation found by R. RAWÉ on a plant at Stinkfonteinberg, Lit. Namaqualand, Cape.

**O. intermedia** COMPT. (Gr. IV). – Cape: Van Rhynsdorp D. – **Caudex** small, dark, tuberous, resiniferous, radish-like; **L.** from the upper surface, with white woolly hairs at the base.

**O. lepidocaulis** SCHLTR. (Gr. IV). – Cape: Lit. Namaqualand. – Dwarf, glabrous ♄ 15–25 cm tall; **Br.** short, thick, densely covered with smooth, glossy, horny L.bases, giving the appearance of a pine-cone; **L.** in a small Ros. from the woolly Br.tips, linear-obtuse to oblanceolate, thickish, 3.5–6 cm long, 5 mm across, the petiole widening abruptly at the base into a thick scale; **Infl.** subumbellate, of up to 8 Fl.heads, ray-florets yellow, rarely absent.

**O. macrosperma** DC. (Gr. II). – Cape: Olifants R. – Fleshy, glabrous ♄; **Br.** ascending, hirsute; **L.** oblong, flat, blunt, narrowing towards the base, fleshy, c. 22 mm long, 3–5 mm across, margins entire or with a small lobe midway; flowering Br. striate, with few Ped.

**O. obtusiloba** HARV. (Gr. II). – Cape: Knakerberge, Komkans, Bitterfontein. – Fleshy sub-♄; **Br.** glabrous below, L.scars woolly, shoots leafy; **L.** linear, fleshy, blunt, 4–5 cm long, 2.5–5 mm across, usually with 1–2 opposite, linear, very blunt lobes 1.5–2 cm long midway, these lobes sometimes having a further pair of lobes or T., margins revolute; **Infl.** terminal, forked, with 3–5 heads.

**O. opima** MERXM. (Gr. I). – S.W.Afr.: Lüderitz D.; Cape: Lit. Namaqualand. – Succulent ♄ 30–60 cm tall, with few Br., partly glabrous and ± pruinose; **St.** and Br. cylindrical, c. 8–10 mm ⌀; **L.** fleshy, cylindrical, c. 6–10 cm long, (8–)10(–15) mm thick, slightly curved, blunt to ± pointed above; **Infl.** several-headed, ray-florets yellow.

**O. osteospermoides** DC. (Gr. II) (*O. barkerae* COMPT.). – Cape: Ladismith D. – Erect, softly woody ♄ up to 1 m tall; **L.** lanceolate, rather fleshy, 8–10 cm long, 1.5–2.5 cm across, dentate; **Infl.** terminal or almost so, **Fl.** heads large.

**O. othonnites** (L.) DRUCE (Gr. II) (*O. frutescens* L.). – Cape: Stellenbosch, Piquetberg D. – Erect, glabrous ♄; **L.** subpetiolate, obovate, with 2 large T. at the tip, L. on the floral St. longer.

**O. pachypoda** HUTCHINS (Gr. II). – Cape: Van Rhynsdorp D. – **Caudex** dwarf, robust, c. 15 cm tall, 3 cm ⌀; **Br.** numerous, slender, pendulous, up to 60 cm long, terete, with a pale brown bark, glabrous; **L.** scattered, linear-spatulate to linear-lanceolate, blunt to rounded at the tip, 3.5–6.5 mm long, 4.8 mm across, somewhat fleshy, glabrous, margins entire; **Fl.**heads yellow, 3–4 in a corymbose Infl.

**O. primulina** DC. (Gr. III). – Cape. – Dwarf, glabrous ♄ resembling **Primula auricula**; **L.** crowded at the St.tips, obovate, obtuse, sessile, narrowing towards the base, denticulate, 13 to 18 mm long, 6–8 mm wide; Ped. one-headed.

**O. protecta** DTR. (Gr. II) (*O. crassicaulis* COMPT.). – S.W.Afr.: Gr. Namaqualand. – Small ♄; **caudex** slender bottle-shaped, up to 30 cm tall; **Br.** few, erect, often liana-like and climbing; **L.** crowded at the Br.tips, sessile, widened at the base, curving-ascending, 4 to 10 cm long, ± cylindrical and narrow-linear, 2.5–3 mm thick, upperside flat to furrowed, glabrous, mucronate, dark blue-green to greenish-purple; flowering Br. purple, blue-pruinose, with 2–3 large Fl.heads, ray-florets pale.

**O. quercifolia** DC. (Gr. II). – Cape: Clanwilliam D. – Erect, glabrous ♄; **St.** fleshy; **L.** 6–7 cm long, c. 2.5 cm across, terete below, cuneate-tapering, blunt, pinnatifid, segments ovate-lanceolate, slightly apiculate, blades 3-veined.

**O. quinqueradiata** DC. (Gr. II). – Cape. – Erect, glabrous ♄; **St.** subcylindrical, fleshy, with few Br.; **L.** crowded at the Br.tips, with wool in the L.axils, oblong, narrowing towards the base, blunt, submucronate, flat, margins entire. Doubtfully distinct from **O. arborescens**.

**O. rechingeri** B. NORD. (Gr. I). – Cape Prov., Calvinia D. – Glabrous ♃ herb with erect flowering St. from a basal Tub.; **Tub.** 1–2 cm ⌀, emitting lateral stolons which form further Tub. at the tip; **L.** alternate, linear, terete, soft, 1–10 cm long, 1.5–4 mm thick, rounded or mucronulate at the tip; Ped. 10–30 cm long, with 1–3 Br.; **Fl.** yellow, rayless.

**O. retrofracta** Jacq. (Gr. II) (*O. lamulosa* Schinz, *Doria lasiocarpa* DC., *O. l.* (DC.) Sch. Bip., *O. litoralis* Dtr., *O. schaeferi* Muschler ex Dtr. nom. nud., *O. surculosa* Muschler ex Dtr. nom. nud.). – S.W.Afr.; Cape: Lit. Namaqualand. – Small ♄; **St.** up to 40 cm tall, usually thickened below and branching from there; **Br.** spreading, tangled; along the coast rarely with the St. caudiciform, bottle-shaped and only 10 cm tall, with only a few short, spreading Br.; bark grey to reddish to glossy reddish-brown at first, eventually dark grey, peeling and papery, often with a resinous coating; **L.** usually oblanceolate to narrow-lanceolate with a long cuneate base, more rarely ovate-spatulate, margins entire, with a few short or longer, pointed T., or ± pinnately lobed, the lobes pointed; **Fl.** heads rarely solitary, usually several, the Ped. usually very reduced.

**O. retrorsa** DC. (Gr. III) v. **retrorsa**. – Cape: Lit. Namaqualand. – **Caudex** erect, 5–6 cm tall, 2–3 cm thick, woolly, densely covered with the remains of dried L.; **L.** in crowded rosulate clusters at the tips of the shoots, oblong-spatulate or linear-spatulate, narrowing towards the base, rather pointed, thin, the nerves distinct on both surfaces, margins with spreading hairs; Ped. erect, **Fl.** yellow.

**O.** — v. **linearifolia** Harv. et Sond. – **L.** linear.

**O. sedifolia** DC. (Gr. II) (*Euryops schenkii* O. Hoffm., *O. papillosa* Dtr., *O. papulosa* Dtr. nom. nud.). – Cape: Lit. Namaqualand; S.W. Afr.: Gr. Namaqualand. – ♄ up to 60 cm tall, 50–60 cm across; **St.** up to 4 cm thick, laxly branched, with a light bark; **L.** laxly spiralled on the terminal shoots, constricted abruptly below into the 1–2 mm long, broad-based petiole, ± spherical to short-cylindrical, 5 to 10 mm long, 4 mm thick, upperside rather flat, blue-green, with rather prominent papillae or tubercles; **Infl.** almost always single-headed, ray-florets yellow.

**O. sparsiflora** (Sp. Moore) B. Nord. (Gr. II) (*Euryops s.* Sp. Moore, *O. floribunda* Schltr. ex Range). – S.W.Afr.: Gr. Namaqualand; Cape: Karree Mts. – Branching, erect ♄ up to 3 m tall, young shoots thick, leafy; **L.** sessile, narrow-linear, (10–)13(–30) mm long, 1–5 mm across, blunt, thick, fleshy; **Fl.** heads solitary, disc-shaped, c. 1 cm ⌀, on Ped. up to 7 cm long.

**O. spektakelensis** Compt. (Gr. IV) – Cape: Lit. Namaqualand. – **Caudex** obconical, woody, with a long taproot, with densely leafy, extended shoots above; **L.** very variable, up to 6 cm long, 1.5 cm across, lanceolate, narrowing towards the base, thin, glabrous, margins dentate, T. variable, horny, c. 8 to the cm, veins prominent, L. bases woolly; **Infl.** axillary.

**O. stenophylla** Levyns. (Gr. III) (*O. linifolia* L. f. non L.). – Cape. – **St.** erect, dichotomous; **Br.** ± glabrous; **L.** linear, narrowing towards the base, margins entire, wrinkled.

**O. triplinervia** DC. (Gr. II). – Cape. – ♄ with a few Br. from the base; **L.** rosulate on young shoots, obovate-lanceolate, 5–11 cm long, 2.3 to 5 cm across, blunt, with 3 distinct longitudinal veins, margins usually sinuate; **Fl.** yellow, usually 8 in a tall corymb.

**O. tuberosa** Thunbg. (Gr. III) (*O. bulbosa* x L., *O. b.* Willd.). – Cape. – Caudiciform; **shoots** herbaceous, rather hairy below, ascending, simple, more rarely bipartite, leafy below; **L.** petiolate, broadly ovate, blunt, cordate below, irregularly crenate.

**Oxalis** L. Oxalidaceae. – Mostly ☉ or ♃, but including several succulent ♄ from S.Am. – **L.** tripartite, petiole often much thickened. – Greenhouse. Propagation: seed, cuttings.

**O. carnosa** Mol. (Pl. 113/3). – Chile, Bolivia: coasts. – Small ♄ with a tuberous root; **St.** thick, fleshy, becoming woody, with few Br., c. 10 cm tall, somewhat gnarled, 18–20 mm thick; **L.** in crowded terminal clusters, petiole 3–4 cm long, leaflets 3, green, rather fleshy; **Infl.** borne between the L., Fl. 3–4, yellow.

**O. paposana** Phil. – Chile; Peru; Bolivia. – **St.** thick, fleshy, c. 10 cm tall, forked above, with small short shoots; **L.** in crowded clusters on the short shoots, petiole 3–4 cm long, leaflets 3, green; **Infl.** of 4–5 yellow Fl.

**O. paucartambensis** R. Knuth. – S. and W.Peru: Andes. – **St.** fleshy, covered with the remains of old, persistent petioles; **L.** numerous at the shoot-tips, with 3 obcordate leaflets 13 mm long, 14 mm across, green, the petiole 4–6 times longer, glabrous, rather thick; **Infl.** long-pedunculate, Fl. yellow.

**O. sepalosa** Diels. – Peru; Bolivia. – Succulent ♄ up to 40 cm tall; **St.** c. 2 cm thick, swollen at the base, with short Br. above; **L.** numerous, crowded at the Br. tips, long-petiolate, tripartite, green; **Infl.** of several yellow Fl.

**O. succulenta** Barn. (Pl. 113/1). – Chile; Peru. – **St.** short, somewhat branched, thick, scaly; **L.** with a fleshy petiole c. 25 mm long, leaflets broad-obcordate, upperside glabrous, underside hairy or glabrous; **Infl.** forking above, with many yellow, long-pedicellate Fl.

**Pachycormus** Coville. – Anacardiaceae. – Mexico. – Succulent ♄ with spongy wood and white, milky sap; **L.** pinnatisect; **Fl.** in corymbs, pink or light red.

**P. discolor** (Benth.) Coville (*Schinus d.* Benth.). – ('Elephant tree'.) – Mexico: Baja Calif. – **St.** several, compact, up to 4 m tall; **L.** pinnatisect, deciduous; **Fl.** numerous, white to red, often yellow.

**P.** — v. **discolor** (Pl. 114/1). – Mexico: Magdalena Bay, Is. of Magdalena and Santa Margarita. – **L.** up to 8 cm long, apical leaflet obovate to cuneate, dentate or lobed above midway, ± hairy; **Fl.** pink or red, in a lax or dense Rac.

**P.** — v. **pubescens** (S. WATS.) GENTRY (*Bursera p.* S. WATS., *Veatchia d.* v. *p.* I. M. JOHNSTON). – Mexico: from El Marmol and San Augustin southwards. – **L.** smaller, apical leaflet less dentate; Rac. pyramidal, lax, Fl. pale pink or white.

**P.** — v. **veatchiana** (KELL.) GENTRY (*Rhus v.* KELL., *Veatchia cedrosensis* A. GRAY, *B. d.* v. *v.* I. M. JOHNSTON). – Mexico: Cedros Is. – **L.** as v. **pubescens**; **Fl.** in a dense, cylindrical corymb, Sep. deep pink, Pet. light pink or red.

**Pachycornia** HOOK. f. Chenopodiaceae. – Austr. – Halophytes, resembling **Salicornia**, of little interest in cultivation. – Leafless, segmented, shrubby plants. Known species include: **P. triandra** (F. V. MUELL) BLACK, **P. tenuis** (BENTH.) BLACK.

**Pachyphytum** LINK, KLOTZSCH et OTTO. Crassulaceae. – Mexico. – Glabrous succulent ♃ plants similar to **Echeveria**; **St.** mostly decumbent with age, somewhat branching; **L.** crowded to scattered, turgid, with margins broadly rounded; **Infl.** a single Cinc., Sep. erect, equal or unequal, Pet. erect throughout or outcurved to spreading from the middle, free nearly to the base and separated below, rather thin, each with 2 scales reaching about to the middle of the inner surface. (Abridged from a text contributed by **R. Moran**.) – Cold-house. Propagation: seed, cuttings, leaf-cuttings. – For hybrids with **Echeveria** see × **Pachyveria** HAAGE et SCHMIDT.
Type species: P. bracteosum.

## Division of the Genus Pachyphytum into Sections by R. Moran

**A.** Pet. shorter than the Cal., each with a full-width ± round dark red spot ventrally just below the apex; Sep. markedly unequal, the larger imbricate at least nearly to and after anthesis; bracts of the young Cinc. imbricate.

**Sect. 1. Pachyphytum** (*Eupachyphytum* BGR.). – Type species: P. bracteosum. – Species: P. bracteosum, kimnachii, longifolium, oviferum, viride, werdermannii.

**A.** Pet. equalling or exceeding the Cal., at least in later Fl., variously coloured but without a dark red spot; Sep. equal or unequal, mostly open well before anthesis.

**B.** L. elliptic to obovate, mostly two or more times wider than thick; bracts of the young Cinc. imbricate; Pet. outcurved in late anthesis; Sep. markedly unequal, the 3 largest imbricate in early bud or nearly until anthesis; St. glutinous or not.

**Sect. 2. Ixiocaulon** MORAN. – Type species: P. glutinicaule. – Species: P. fittkaui, glutinicaule.

**B.** L. fusiform to subcylindric, scarcely wider than thick; bracts not imbricate; Pet. remaining erect or nearly so; Sep. equal in length but not in breadth, open; St. not glutinous.

**Sect. 3. Diotostemon** (S. D.) E. WALTH. (*Diotostemon* S. D. as a G.). – Type species: P. hookeri. – Species: P. coeruleum, compactum, hookeri. Doubtfully placed here: P. brevifolium.

(The Sect. *Echeveriopsis* E. WALTH. is returned to Echeveria D. C. as Ser. **Chloranthae** MORAN.)

**P. bracteosum** LINK, KLOTZSCH et OTTO (§ 1) (*Echeveria b.* (LINK, KLOTZSCH et OTTO) LINDL. et PAXT., *Cotyledon pachyphytum* BAK., *Ech. p.* (BAK.) MORREN). – Mexico: Hidalgo. – **L.** 15–30, obovate to spatulate, 4–11 cm long, 2.5–5 cm across, 3–10 mm thick, pruinose; **Fl.** 10–28, red.

**P. brevifolium** ROSE (§ 3 ?). – Mexico: Guanajuata. – **L.** c. 30, elliptical to narrow-obovate, round or broad and blunt, 1–2 cm long, 6–9 mm across; Fl. 5. – Insufficiently known.

**P. coeruleum** MEYRAN (§ 3). – Mexico: Querétaro or Hidalgo. – **L.** c. 30, crowded, fusiform, ± pointed, apiculate, 2–3 cm long, 5 to 12 mm across, 4–10 mm thick, pruinose; Cinc. of 6–10 Fl., Pet. light yellow with green spots.

**P. compactum** ROSE (Pl. 114/5) (§ 3). - Mexico: Hidalgo. – **L.** 30–60, usually crowded, oblong-lanceolate, narrowing above, with a broad apiculus, pruinose, 2–4 cm long, 12 to 16 mm across, 9–12 mm thick, surface rather angular, keel asymmetrical, with lines along the margins; **Infl.** of Cinc. with 3–10 Fl., Cal. deep pink, Pet. orange, tips bluish.

**P.** — cv. **Weinbergii**. – Cultivar of more robust growth, Cal. and Fl. deeper red.

**P.** — cv. **Weinbergii Cristata**. – **St.** fasciated.

**P.** — × **P. viride** E. WALTH. – Natural hybrid. – Mexico: Querétaro, Cerro Mexicano. – **St.** erect, up to 8 cm tall, 2 cm ⌀; **Ros.** up to 13 cm ⌀; **L.** c. 45, rather crowded, oblong-lanceolate, the tip bluntly rounded and ± apiculate, semi-cylindrical, 4.5–6 cm long, 12 to 17 mm across, up to 10 mm thick, upperside flat, underside rounded and angular, with markings as **P. compactum**; Fl. shoots 30–40 cm tall, Cinc. c. 11 cm long, of 12–17 Fl., bracts lanceolate, c. 12 mm long, Cal. 11–13 mm long, pink, Pet. apiculate, with a deep pink spot inside, tips yellow.

**P. fittkaui** MORAN (§ 2). – Mexico. – **St.** erect, up to 50 cm tall, or prostrate to pendulous, up to 1 m long, up to 3.5 cm thick, branching from the base, L.scars round to elliptical; **Ros.** lax, 10–20 cm ⌀; **L.** 10–25, deep green, tinged ± purplish-red, elliptical to narrow-obovate, rounded above, apiculate, 3–9.5 cm long, 2–4 cm across, up to 15 mm thick, upperside slightly concave; **Infl.** up to 35 cm tall, bracts deciduous, leaflike, Cinc. 10–20 cm long, with 12–25 Fl. on pedicels 2–6 mm long, Sep. unequal, Cor. 12 to 17 mm long, 12–15 mm ∅, deep pink inside, underside paler.

**P. glutinicaule** MORAN (Pl. 114/2) (§ 2). – Mexico: Querétaro, Hidalgo. – **St.** viscous at first; **L.** 20–35, obovate, blunt to rounded, apiculate, 2.5–6.5 cm long, 1.5–3.5 cm across, 3–15 mm thick, pruinose; Cinc. with 6–20 **Fl.**, Pet. light red.

**P. hookeri** (S.D.) BGR. (*Diotostemon h.* S.D., *Echeveria h.* (S.D.) LAM., *Cotyledon adunca* BAK., *P. roseum* HORT. ex BAK., *E. adunca* (BAK.) OTTO, *P. uniflorum* ROSE, *P. aduncum* (BAK.) ROSE). – Mexico: San Luis Potosí. – **L.** 25–40, scattered, clavate-fusiform to cylindrical, blunt to rounded, apiculate, 2.5–5 cm long, 6–16 mm across, 5–11 mm thick, green to pruinose; Cinc. with 5–15 Fl., Sep. pink, green-tipped, Pet. light red.

**P. kimnachii** MORAN (§ 1). – Mexico: San Luis Potosí. – **Ros.** lax, 15–20 cm ∅; **L.** 12–25, pruinose, often purple at first, elliptical-oblong, blunt, apiculate, 5–10 cm long, 17–25 mm across, 7–13 mm thick, margins broad; **Cinc.** with 12–18 Fl., Cor. 5-angled, Pet. pink with a crimson blotch.

**P. longifolium** ROSE (§ 1). – Mexico: Hidalgo. – **L.** 20–60, oblanceolate, pointed to blunt, apiculate, 6–11 cm long, 1.5–2.5 cm across, 4 to 10 mm thick, pruinose, underside usually grooved; Cinc. with 10–45 **Fl.**

**P. oviferum** J. A. PURP. (Pl. 114/4) (§ 1). – Mexico: San Luis Potosí. – **L.** 12–20, obovate, rounded, 3–5 cm long, 18–30 mm across, 10 to 16 mm thick, pruinose; Cinc. of 10–15 greenish-white **Fl.**

**P. viride** E. WALTH. (Pl. 114/3) (§ 1). – Mexico: Querétaro. – **L.** 12–40, scattered, elliptical-oblong, rather blunt to short-rounded, 6–14 cm long, 15–28 mm across, 12–16 mm thick, green to purplish-red; Cinc. of 10–22 **Fl.**

**P. werdermannii** V. POELLN. (§ 1). – Mexico: Tamaulipas. – **St.** often creeping, up to 1 m long; **L.** 10–35, rather scattered, elliptical to oblong, blunt to narrow-rounded, 4–10 cm long, 1.5–3.5 cm across, 5–12 mm thick, light green, pruinose, margins rounded; Cinc. of 10–22 **Fl.**, Pet. light pink with a deep crimson blotch.

G. D. ROWÄEY **P. succulentum** by root cuttings.

**Pachypodium** LDL. Apocynaceae. – S.Afr., S.W.Afr., Angola, Madag. – St.-succulents, slightly woody, soft-fleshy; **trunk** barrel- or bottle-shaped, up to 10 m high, or shrubby or caudiciform, thorny; **L.** often few in a Ros., short-lived; **Fl.** short-stalked in a few-flowered Infl., Cor. turbinate with 5 free, mostly spreading lobes, white or striped with red, pink or yellow. – Greenhouse, warm. Propagation: seed. Acc. to W. RAUH **P. bispinosum** also propagates by cuttings, and acc. G. D. ROWLEY **P. succulentum** by root-currings.

## Division of the Genus Pachypodium into Subgenera and Sections by M. Pichon

**SG. I. Pachypodium** (SG. *Chionopodium* M. PICHON). – Fl. white, striped whitish-red or dirty brown-red.

a) **Sect. Adeniopsis** BENTH. et HOOK. – S.Afr., S.W.Afr., Angola. – Fl. at the ends of leafy and thorny Br.; Cor.-lobes broad, strikingly asymmetrical, white or striped whitish-red. – Species: P. lealii.

b) **Sect. Pachypodium** (Sect. *Eupachypodium* BENTH. et HOOK.). – S.Afr. – Fl. at the ends of leafy and thorny Br.; Cor.-lobes narrow or broad, less asymmetrical; caudex clavate, subterranean. – Species: P. bispinosum, succulentum.

c) **Sect. Erianthum** M. PICHON. – S.Afr., S.W.Afr. – Fl. numerous from the L.-axils, velvety, reddish-brown; trunk ± cylindrical, thorny in the upper 2/3. – Species: P. namaquanum.

d) **Sect. Leucopodium** M. PICHON. – Madag. – Mostly tree-like, only P. ambongense and P. decaryi with a short trunk; Infl. dense, Fl. large, white or partly pink. – Species: P. ambongense, decaryi, lamerei, rutenbergianum, sofiense.

e) **Sect. Cladanthum** M. PICHON. – Madag. – Infl. more lax, Fl. small, white, Cor.-tube very short, from which the filaments freely extend. – Species: P. geayi.

**SG. II. Chrysopodium** M. PICHON (Sect. *Velutinii* COST. et BOIS; Sect. *Gymnopus* K. SCHUM.). – Madag. – Fl. bright yellow; dwarf caudiciform; Infl. and Fr. ± woolly-hairy. – Species: P. brevicaule, densiflorum, rosulatum.

**SG. III. Porphyropodium** M. PICHON. – Madag. – Fl. red; caudiciform. – Species: P. baronii.

## Key to the Species of the Genus Pachypodium by G. D. Rowley

1. Caudex wholly or almost wholly above ground ............................................. **2.**
1a. Caudex mainly below ground; aerial Br. relatively slender ............................. **18.**
2. Th. acicular, in threes, the 2 laterals usually longer and spreading, the central sometimes undeveloped ............................................................................... **3.**
2a. Th. ± conical, solitary or in pairs .......................................................... **8.**
3. L. less than 10 times as long as wide ...................................................... **4.**
3a. L. linear, 10 or more times as long as wide ............................................. **6.**
4. Plant with a single erect main St. and 0–few thick erect side branches .... **namaquanum**
4a. Caudex variously shaped but with many relatively thin twiggy Fl. shoots ............ **5.**
5. L. tomentulose on both surfaces ................................................... **lealii ssp. lealii**
5a. L. glabrous exc. for pubescent midrib and fimbriate margin ........ **lealii ssp. saundersii**
6. Young St. and L. lanate ................................................................... **geayi**
6a. Young St. and L. not lanate ................................................................ **7.**
7. L. not mucronate, rarely glabrous on both surfaces ..................... **lameri v. lameri**
7a. L. mucronate; whole plant glabrous .......................................... **lamerei v. ramosum**
8. Tree-like to 8 m tall, with a gradually tapering trunk .................................. **9.**
8a. Shrub with a swollen caudex abruptly narrowed into Br. ........................... **11.**
8b. Dwarf geophyte with a flattened caudex many times wider than tall .......... **brevicaule**
9. L. oblong to linear, up to 16 cm long and 4–4.5 cm wide; Cor. tube less than 4 cm long . **10.**
9a. L. obovate, shortly petiolate, up to 12 cm long and 5–6 cm wide; Cor. tube 4–6 cm long .... **sofiense**
10. Lateral veins 25–60 pairs; Cor. 6–7 cm ⌀; tube 25–35 mm long .......................
     **rutenbergianum v. rutenbergianum**
10a. Lateral veins 60–100 pairs; Cor. 3.5–4.5 cm ⌀; tube 17–22 mm long .....................
     **rutenbergianum v. meridionale**
11. Th. minute, black, deciduous, in pairs at the L.bases; Fl. whitish .............. **decaryi**
11a. Th. conspicuous, persistent; Fl. coloured ............................................... **12.**
12. Stamens exserted; Fl. orange ............................................................. **13.**
12a. Stamens included within Cor. tube; Fl. yellow or red .............................. **14.**
13. L. not slender petiolate; Fl. 1.5–3 cm ⌀, glabrous within; Sep. up to 8 mm long .........
     **densiflorum v. densiflorum**
13a. L. slender petiolate; Fl. smaller, pubescent within; Sep. up to 4 mm long ...........
     **densiflorum v. brevivalyx**
14. Fl. pale yellow, with narrow Sep. 6–10 mm long ..................................... **15.**
14a. Fl. ± red, with triangular Sep. 3 mm long ............................................ **17.**
15. Cor. tube cylindric, c. 3 times as long as wide ........................................ **16.**
15a. Cor. tube campanulate, almost as wide as long ................. **rosulatum v. horombense**
16. Br. thick, with stout brown Th. .................................... **rosulatum v. rosulatum**
16a. Br. thin, with almost needlelike red-brown Th. ................ **rosulatum v. gracilius**
17. L. up to 15 cm long; Ped. 4–25 cm long; Fl. red ..................... **baronii v. baronii**
17a. L. 5–6 cm long; Ped. almost undeveloped; Fl. yellowish with red margins ...............
     **baronii v. windsorii**
18. Cor. tube narrow cylindric, with narrow obovate to oblanceolate lobes ....... **succulentum**
18a. Cor. tube funnelform-campanulate, with broadly ovate lobes................. **bispinosum**

**P. baronii** Cost. et Bois (§ III). – N.Madag. – Caudiciform.
   **P. —** v. **baronii** (*P. b.* v. *erythreum* H. Poiss., *P. b.* v. *typicum* M. Pichon). – Madag.: Majunga. – **Caudex** large, stout; **St.** up to 3 m tall; **L.** oblong to oval, tapering, up to 15 cm long; **Fl.** large, red.
   **P. —** v. **windsori** (H. Poiss.) M. Pichon (*P. w.* H. Poiss.). – Madag.: near Diégo-Suarez. – **Caudex** spherical, up to 20 cm ⌀, tapering above into the St. which is c. 80 cm long, 2–3 cm thick, L.scars present, Th. 5 mm long; **L.** oval, 5–6 cm long, 3–5 cm across, underside felted; **Fl.** with Cor. up to 5.5 cm ⌀, yellowish, hairy inside, lobes brilliant red.
   **P. bispinosum** (L. f.) A. DC. (§ I/b) (*Echites b.* L. f., *Belonites b.* (L. f.) E. Mey., *P. glabrum* G. Don. nom. ill., *P. tuberosum* sensu Lodd.,

***P. t.*** LINDL. v. *loddigesii* A. DC.). – Cape: widely distributed. – Succulent ♄; **caudex** large, up to 18 cm ∅, usually more than half-buried in the soil; **St.** ± numerous, thin, erect, scarcely branching, up to 45 cm tall, 5–10 mm ∅; Th. in pairs, 1–2 cm long, often with short, intermediate stipular Th.; **L.** lanceolate to linear-lanceolate, 2–4 cm long, 2–7 mm across, upperside rough, underside hairy; **Fl.** few in an Infl., purple to pink, tube hirsute inside, purple, lobes usually paler, rarely white. Can also be propagated by cuttings.

**P. brevicaule** BAK. (Pl. 115/3) (§ II). – Madag.: Itremo reg. – **Caudex** flattened, shapeless, up to 60 cm ∅, with a silvery-grey bark; **L.** few, ovate, 1–3 cm long, up to 1 cm across, both surfaces hairy; Th. white-felted; **Infl.** usually of paired Fl., Cor.tube c. 1.5 cm long, lobes flat, slightly asymmetrical, lemon-yellow.

**P. decaryi** H. POISS. (§ I/d). – N.Madag. – **Caudex** short tuberous, ± round to pyriform, up to 40 cm tall and ∅, tapering abruptly above, with several thin St., thornless; **L.** 6–24, oblong-oval, 5–6 cm long, 4–5 cm across, margins softly hairy; **Fl.** with a fleshy Cor.tube 5–8 cm long, greenish-white, lobes pure white with a yellow blotch at the base.

**P. densiflorum** BAK. (§ II). – Central, S.W. and S.E.Madag. – **Fl.**-size variable, very small or up to 3 cm ∅.

**P. — v. brevicalyx** PERR. (*P. brevicalyx* (PERR.) M. PICHON). – S.E.Madag. – **L.** very slender-petiolate; **Fl.** very small, Cor.tube slightly hairy inside, otherwise as the type.

**P. — v. densiflorum.** – Central, S.W.Madag. – **Caudex** small at first, eventually much thickened, thorny; **L.** 3–5, oblong-oval, up to 10 cm long, 5 cm across, underside very felted; **Infl.** up to 40 cm tall, Fl. numerous, up to 3 cm ∅, lobes flat, slightly asymmetrical, brilliant yellow.

**P. geayi** COST. et BOIS (Pl. 115/1) (§ I/e). – S.W.Madag. – Tree-like, up to 8 m tall; **caudex** swollen; **Br.** forming a ± spherical, lax crown, thorny; **L.** long-linear, underside with silvery-grey hairs, up to 40 cm long, 16 mm across; **Infl.** lax, of many, very small Fl., lobes up to 19 mm long, white.

**P. lamerei** DRAKE (§ I/d). – S. and S.W. Madag. – **Caudex** barrel-shaped, up to 60 cm ∅, 1–6 m tall, with a few Br. from the crown, Th. 3 together; **L.** variable, up to 40 cm long, 2–11 cm across, underside hairy or glabrous; **Fl.**-tube 3–5 cm long, widened above, glabrous, white, lobes very asymmetrical, not contorted.

**P. — v. lamerei** (*P. l.* v. *typicum* DRAKE). – **Caudex** up to 5 m tall, Th. stout, up to 2 cm long; **L.** 12–40 cm long, 2–4 cm across; **Fl.** white.

**P. — v. ramosum** (COST. et BOIS) M. PICHON (*P. r.* COST. et BOIS, *P. menabeum* LEANDRI). – **L.** not hirsute on the underside, broad-oval, terminating abruptly in a sharp tip. – **P. champenoisianum** with yellowish Fl. is regarded by H. PICHON as a form of this species.

**P. lealii** WELW. (§ I/a). – W.Afr. and Zululand. – ♄ or small ♄ up to 6 m tall; **caudex** irregular in shape, sometimes with erect Br., 40–100 cm ∅ at the base, tapering evenly above or ± bottle-shaped, fleshy, bark thin, papery, with transverse L.scars; **L.** obovate-oblong to oblong-lanceolate, tip blunt, apiculate, cuneate below, 2.5–8 cm long, 1.6–4 cm across, Stip. thorny, on small cushions, Th. in divaricate pairs on the long shoots, 1.5–3 cm long, often with 3 stipular Th.; **Infl.** terminal, corymbose, Fl. few, with pedicels 2–4 mm long, Cor. large, scented.

**P. —** ssp. **lealii** (Pl. 115/4, 5) (*P. giganteum* ENGL.). – Angola; Lit. Namaqualand. – **Caudex** up to 5 m tall, bottle-shaped, up to 40 cm ∅ at the base, branching 2–3-partitely above; **Fl.** conspicuous, Cor. 7–8 cm long, white.

**P. —** ssp. **saundersii** (N. E. BR.) ROWL. (*P. s.* N. E. BR.). – Zululand. – **Caudex** up to 6 m tall, tuberous below, up to 1 m ∅, with 1 or more erect Br. 5–10 cm thick; **Fl.** white with reddish stripes, lobes tinged with purple at the base and near the tube.

**P. namaquanum** (WYLEY ex HARV.) WELW. (Pl. 115/6) (§ I/c) (*Adenium n.* WYLEY ex HARV.). – ('Club-foot', 'Elephant's trunk'.) – Cape: Lit. Namaqualand; S.W.Afr.: Gr. Namaqualand. – **Caudex** 1.5–2.5 m tall, usually unbranched, the upper two-thirds with dense brown Th. up to 5 cm long in spiralled R. on the tuberculate L.scars; **L.** clustered at the tips of the shoots, 8–12 cm long, 2–6 cm across, both surfaces hairy, margins undulating-curly; **Fl.** numerous, velvety reddish-brown, with yellow stripes inside. Tip of caudex always inclined to the N.

**P. rosulatum** BAK. (§ II). – N. to S.Madag. – Variable species because of wide distribution in differing ecological conditions.

**P. — v. gracilius** PERR. – Madag.: Isalo Mts. – More slender than v. **rosulatum**; **caudex** usually round, up to 40 cm ∅, lateral St. little branched, Th. very thin, up to 10 mm long, reddish-brown; **L.** oblong-linear, up to 8 cm long, 2 cm across, underside very felted, upperside sparsely hairy; **Fl.** numerous, Cor.lobes ± roundish, 12–15 mm long, flat.

**P. — v. horombense** (H. POISS.) ROWL. (Pl. 115/2) (*P. h.* H. POISS.). – Central and S.Madag. – **Caudex** irregular; Br. forming a bushy crown up to 1.5 m tall, much thickened, Th. robust, closely spaced; **L.** oval, 3–5 cm long, up to 2 cm across, underside densely felty, upperside slightly hairy; **Infl.** an umbel up to 40 cm tall, Cor.tube broad-campanulate, lobes asymmetrical, yellow.

**P. — v. rosulatum** (*P. cactipes* K. SCHUM., *P. r.* v. *typicum* COST. et BOIS, *P. drakei* COST. et BOIS, *P. r.* v. *delphinense* PERR.). – 50 to 150 cm tall; **caudex** irregular, up to 1 m ∅; **St.** in a much branched crown, Th. very firm, 5–10 mm long; **L.** variable, margins rolled slightly under, underside felty-haired, upperside green, sparsely hairy; **Infl.** up to 30 cm tall, hirsute, Fl. numerous, Cor. up to 7 cm long, hairy outside, lobes ± round, ± asymmetrical, flat, yellow.

**P. rutenbergianum** VATKE (§ I/d). – N., central and W.Madag. – ♄ with trunk up to 60 cm ⌀, with sparse Br. from the crown and dense, brown Th. up to 1.5 cm long or absent; **L.** with a thick petiole 3 cm long, blade lanceolate, pointed, 16 cm long, 4.5 cm across, glabrous; **Infl.** cymose, Cor. tube angularly furrowed, up to 3.5 cm long, whitish-green, lobes very asymmetrical, white, 2–4 cm long, tapering, margins undulating, often spirally contorted.

**P.** — v. **meridionale** (M. PICHON) PERR. (*P. meridionale* M. PICHON). – Central to S.W. Madag. – ♄ 6–8 m tall; **trunk** up to 60 cm ⌀, smooth, thornless, branching freely at a height of 2–3 m; **L.** with 60–100 pairs of side-veins; **Fl.** with Cor. 3.5–4.5 cm ⌀, tube 17–22 mm long, Pet. reddish on the underside.

**P.** — v. **rutenbergianum** (*P. r.* v. *typicum* PERR.). – N.Madag. – **L.** with 25–60 pairs of side-veins; Cor. 6–7 cm ⌀, tube 2.5–3.5 cm long.

**P. sofiense** (H. POISS.) PERR. (§ I/d) (*P. rutenbergianum* v. *s.* H. POISS., *P. r.* v. *perrieri* H. POISS.). – N. and N.W.Madag. – Appearance and Th. as **P. rutenbergianum**; tree-like, Th. stout; **L.** short-petiolate, blade obovate, up to 12 cm long, up to 5 cm across, both surfaces glabrous, with 10–15 pairs of side-veins; **Infl.** short-pendunculate, Cor. tube 4–6 mm long, lobes blunt, broad, flat, white; **Fr.** conspicuously thick, up to 16 cm long, 6.5 cm ⌀.

**P. succulentum** (L. f.) A. DC. (§ I/b) (*Echites s.* L. f., *P. tuberosum* LINDL., *P. tomentosum* G. DON nom. ill., *P. griquense* L. BOL., *Belonites s.* (L. f.) E. MEY., *Barleria rigida* SPRENG. ex SCHLTR. nom. nud., *P. jasminiflorum* L. BOL.). – Cape: widely distributed. – Succulent ♄; **caudex** half-hidden in the soil, passing over into the napiform root, up to 15 cm ⌀; **St.** several to numerous, simple or sparsely branching, 15–60 cm tall, Th. dense, paired, 2–2.5 cm long, often with a third stipular Th.; **L.** sessile, linear to linear-lanceolate, 1.7–4.5 cm long, 2–8 mm across, both surfaces hirsute; **Fl.** few in a corymb, small, crimson or delicate pink, rarely white (*P. jasminiflorum*, *P. tuberosum*), tube hairy, lobes bordered with white, with a red M.stripe.

× **Pachysedum** ROWL. Crassulaceae. – Intergeneric hybrids: **Pachyphytum** × **Sedum**.

× **P.** cv. Rose Tips. – Hort. Johnson, California, 1954. – **L.** plump, powdered white over dove grey, suffused with red; **Fl.** bright lemon yellow. – Parentage not recorded.

× **Pachyveria** HAAGE et SCHMIDT (× *Echephytum* GOSSOT, × *Pachyrantia* E. WALTH., × *Urbiphytum* GOSSOT). Crassulaceae. – Intergeneric hybrids: **Echeveria** × **Pachyphytum**. – Usually caulescent; L.Ros. often extended; **L.** oblong-spatulate, sometimes ± cylindrical, underside convex, upperside usually concave, green, somewhat bluish-pruinose, often with a reddish tinge; **Infl.** from the L.axils, not strongly developed, bracts sparse, deciduous, or Infl. 2–3-partite with persistent bracts, Sep. appressed, Cor. campanulate, light red, reddish-orange or red, often spotted or blotched, Pet. with or without a scale-like appendage inside below. – Greenhouse. Propagation: cuttings.

× **P. albomucronata** (GOSSOT) ROWLEY cv. Albo-mucronata (× *U. a.-m.* (GOSSOT) JACOBS., × *E. a.-m.* (GOSSOT) JACOBS.). – Hybrid: **E. purpusorum** × **P. uniflorum.** – Not offsetting; **L.** small, thick, fusiform, mucronate, bluish, pruinose; **Fl.** reddish-orange.

× **P. clavata** E. WALTH. cv. Clavata (Pl. 116/1) (*E. clavifolia* HORT. ex E. WALTH.). – Hybrid: **P. bracteosum** × **E. spec. ?**. – Caulescent; **L.** c. 10 cm long, 3 cm across above, spatulate, grey-green; **Fl.** reddish.

× **P.** — cv. Cristata (*E. clavifolia* v. *cristata* HORT.). – Monstrous cultivar.

× **P. clavifolia** (BGR.) E. WALTH. cv. Clavifolia (*E. clavifolia* DEL. ex MORREN, 1874, *Sedum c.* BGR., *Cotyledon c.* BGR., *E. c.* BGR., *Diostemon c.* BGR., *E. c.* HORT. ex BGR., × *Pachyrantia c.* (BGR.) E. WALTH., × *Pachyrantia echeverioides* (BGR.) E. WALTH., × *P. c.* (BGR.) JACOBS., *E. brachyantha* PRAEG.). – Hybrid: **P. bracteosum** × **E. rosea.** – Caulescent; **L.** oblong-clavate, c. 5 cm long, 1.5 cm across, bluish-green; **Fl.** with crimson Pet., silvery outside.

× **P.** cv. Clevelandii (*E. magnifica* HORT., *E. nobilis* HORT.). – Hybrid: **P. bracteosum** × **E. secunda.**

× **P.** cv. E. O. Orpet. – Hybrid: **P. bracteosum** × **E. spec.?** –

× **P. fucifera** (ROLL.) ROWL. cv. Fucifera (*E. fucifera* ROLL.). – Hybrid: **E. hybrida** (**E. gibbiflora** v. **metallica** × **E. glauca**) × **P. hookeri.** – **L.** large, fuciform, upperside grooved, bluish pea-green.

× **P. glauca** HAAGE et SCHMIDT cv. Glauca (Pl. 116/3) (*Diotostemon g.* HORT., × *P. haagei* HORT., *E.* × *fruticosa* ROLL.). – Hybrid: **P. compactum** × **E. spec.?** (acc. KEPPEL). – Acauline; **Ros.** dense; **L.** oblong, semi-cylindrical, bluish-pruinose, with reticulate markings, ± 6 cm long, 1.5 cm across, 1 cm thick; **Pet.** yellow with red tips.

× **P. glossoides** GOSSOT cv. Glossoides. – Parents unknown. – **Ros.** 30 cm long; **L.** 4 to 6 cm long, tapering, light at the tip, blue-green, pruinose; **Fl.** yellow to reddish.

× **P.** cv. Guichnettii. – A cultivar of unrecorded parentage.

× **P.** cv. La Rochette. – A cultivar recorded by ERIC WALTHER as **P. bracteosum** × **E. flammea?**

× **P.** cv. Lesliei. A cultivar recorded in Johnson's 1954 catalogue without parentage. – **L.** thick, pale, silvery green tipped with red.

× **P. mirabilis** (DEL.) E. WALTH. cv. Mirabilis (*E. mirabilis* DEL.). – Hybrid: **P. bracteosum** × **E. scheeri.**

× **P. morreniana** (DEL.) E. WALTH. cv. Morreniana (*E. m.* DEL. 1874). – Hybrid: **P. bracteosum** × **E. peacockii.** – Possibly identical with × **P. mirabilis** cv. Mirabilis.

× **P.** cv. Mrs. Scannavino. – Hybrid: **P. bracteosum** × **E. spec.?**

× **P. muelleri** JACOBS. cv. Muelleri. – Hybrid: **P. oviferum** × **E. derenbergii**. – **Ros.** 10 cm ∅; **L.** 4.5 cm long, 15–17 mm across, 1 cm thick, blue-green, slightly waxy-pruinose.

× **P. pachyphytoides** (L. DE SMET) E. WALTH. cv. Pachyphytoides (*E. p.* L. DE SMET ex MORREN, *P. p.* (L. DE SMET) BGR.). – Hybrid: **P. bracteosum** × **E. gibbiflora** v. **metallica**. – Caulescent; **L.** oblong-ovate to spatulate, 12 cm long, 5 cm across above, blue-green, somewhat reddish; **Fl.** light red.

× **P. paradoxa** (GOSSOT) ROWL. cv. Paradoxa (*E.* × *p.* GOSSOT, 1936, × *Echephytum p.* (GOSSOT) JACOBS.). – Hybrid: **E. setosa** × **P. oviferum**. – **Ros.** usually solitary, dense, regular; **L.** oblong-lanceolate, thick, fleshy, glabrous, green, not pruinose; **Fl.** reddish-green.

× **P. pfitzeri** W. MUELLER cv. Pfitzeri. – Hybrid: **P. oviferum** × **E. elegans**. – **Ros.** dense, with numerous daughter-Ros.; **L.** 6 cm long, alabaster-white; **Pet.** without a blotch at the base.

× **P. scheideckeri** (DE SMET) E. WALTH. cv. Scheideckeri (Pl. 116/2) (*E.* × *scheideckeri* DE SMET, 1877). – Hybrid: **P. bracteosum** × **E. secunda**. – Short stemmed; Ros.L. numerous, oblong-obovate, 6–7 cm long, 1–2 cm across, 5–7 mm thick, blue-green, whitish-grey pruinose, both surfaces white-striate; **Fl.** orange-red.

× **P.** — cv. Albocarinata (*E. albocarinata* HORT., × *P. s.* v. *a.* E. WALTH., 1934, *E. s.* v. *striata* HAAGE jr. Cat. 1935 nom.). – Mutation of 'Scheideckeri'. – Differences as follows; **L.** irregular, ± cylindrical, somewhat deformed, white-pruinose, often pink; **Fl.** lighter.

× **P. sempervivoides** (GOSSOT) ROWL. cv. Sempervivoides (× *E. s.* GOSSOT, 1938). – Hybrid: **P.** 'linguifolium' × **E. elegans**. – **Ros.** solitary; **L.** not numerous, small, glabrous, thick, fleshy, ovate-oblong, apiculate, green; **Fl.** reddish-yellow. – Probably an **Echeveria** hybrid: **E. linguifolia** × **E. elegans**.

× **P. sobrina** (BGR.) E. WALTH. cv. Sobrina (*E.* × *sobrina* BGR., *E. fusifera* HORT.). – Hybrid: **P. bracteosum** × **E. spec.?**

× **P. sodalis** (BGR.) E. WALTH. cv. Sodalis (*E.* × *s.* BGR., *P. s.* (BGR.) ROSE, *E. bergeriana* HORT.). – Hybrid: **P. bracteosum** × **E. spec.?**

× **P.** cv. Spathulata (*E.* × *s.* DEL. ex MORREN). – Hybrid: **P. bracteosum** × **E. grandifolia**. – **Ros.** as large as **E. grandifolia**; **L.** rather narrower, very fleshy, grey-green, densely white-pruinose; **Fl.** red.

× **P.** cv. White Nun. – Parentage not recorded; Hort. Johnson 1954. – Densely leafy; **L.** glaucous, whitish.

**Pagella** SCHOENL. Crassulaceae. – S.Afr. – ☉ succulent plant. Propagation: seed.

**P. archeri** SCHOENL. – Cape: Montagu D. – **St.** at first obconical, eventually flat, to 20 mm ∅; **L.** alternate and decussate, eventually rosulate, crowded, broad below, narrowed teretely above, deep green, 3 mm long; **Fl.** small, red to white. – Of botanical import only.

**Pectinaria** HAW. Asclepiadaceae. – Cape. – St. 4–6-angled or tessellate-tuberculate, ± prostrate, the tips penetrating into the soil; **Fl.** small, solitary or several together from the furrows on young shoots, Cal. 5-partite, lobes pointed, Cor. with a short and broadly campanulate tube and 5 triangular, pointed lobes united at their tips, leaving only small slits open in between, corona double. – Greenhouse, warm. Propagation: seed, cuttings.

**P. arcuata** N. E. BR. – Cape: Bedford, Cradock D. – **St.** obtusely 4-angled, 5–10 cm or more long, 6–10 mm thick, curving, prostrate, glabrous, green; **Fl.** 1–3 together, tapering ovoid, tube 3 mm long, hemispherical, dark red, lobes 7 mm long, yellowish outside, smooth, creamy-white inside, with red spots.

**P. articulata** (AIT.) HAW. v. **articulata** (*Stapelia a.* AIT.). – Cape: Calvinia D. – **St.** with spreading and projecting Br., segments 3–5 cm long, finger-thick. 5–6-angled, grey-green, often reddish; **Fl.** solitary, Cor. 6–8 mm across, green outside, papillose, lobes triangular, blackish inside, minutely papillose.

**P.** — v. **namaquensis** N. E. BR. – Cape: Lit. Namaqualand. – **St.** 6–12 mm thick, with tubercles 1–2 mm long; **Fl.** with Cor. c. 4 mm long, 5–6 mm across, greenish outside, inside papillose-tuberculate.

**P. asperiflora** N. E. BR. (Pl. 116/6). – Cape: Laingsburg D. – **St.** numerous, curved, 2–8 cm long, 13–20 mm thick, 6–8-ribbed, dark purple, glabrous; **Fl.** pendulous, 5-angled, campanulate, the lobes purplish browny-red outside, with few papillae, inside white with purple spots, densely papillose.

**P. breviloba** R. A. DYER. – Cape: Worcester D. – **St.** produced from thin rhizomes, erect, 5 cm tall, 7–9 mm thick, bluntly 4-angled, with pointed tubercles 1.5–2 mm high; **Cor.** chestnut-brown, tubular, ± ellipsoid, 1–1.3 cm long, 5–6 mm ∅, smooth outside, glabrous, inside with several long hairs in the upper half, tuberculate or papillose towards the base, tube almost barrel-shaped, 8–10 mm long, lobes 2.5–3 mm long.

**P. pillansii** N. E. BR. – Cape: Somerset East D. – **St.** acutely 4-angled, 15 cm or more long, prostrate, glabrous, dark green, with T. 3 to 4 mm long, brownish; **Fl.** depressed-pyriform, 7 mm ∅, glabrous and smooth outside, inside light red with watery papillae. Fl. said to develop underground.

**P. saxatilis** N. E. BR. – Cape: Laingsburg, Oudtshoorn, Willowmore D. – **St.** prostrate, 3–5 cm long, 12–20 mm thick, angles 4, acute, compressed, T. 2–3 mm long, projecting horizontally; **Fl.** 8–10 mm ∅, broad-ovoid, 9 to 11 mm long, pointed, blackish-purple, minutely hairy.

**P. tulipiflora** LUCKH. – Cape: Van Rhynsdorp D. – **St.** creeping, 2–6 cm long, 1 cm ∅, with 4 acute angles and flat sides, with reddish spots, T. pointed, 1.5 mm long; **Fl.** ± ovoid, tube cup-shaped, 6.5 mm long and across, deep purplish-red, glabrous outside, inside minutely papillose, lobes 5.5 mm long, the tips often free.

**Pedaliodiscus** IHLENF. Pedaliaceae. – Tanzania, Kenya. – Only one species known, of little importance for cultivation.

**P. macrocarpus** IHLENF. – Tanzania, Kenya. – Prostrate or ascending ⊙; **L.** rather fleshy, opposite, blade oblong-ovate, sinuate, 6–7 cm long, 1.5–2.5 cm across, petiole 2.5–2.5 cm long; **Fl.** solitary in the upper L.axils, 10 to 15 mm long, yellowish-white; **Fr.** a pendulous, thorny achene.

**Pedalium** ROYEN. Pedaliaceae. – Peninsular India; Socotra; E.Afr.; Madag. – ⊙ succulent herbs; **L.** coarsely dentate or laciniate; **Fl.** axillary, tube-shaped, with 5 rounded lobes; **Fr.** angular, prickly. Of little importance in cultivation. Propagation: seed.

**P. murex** L. – Kenya: Mombasa D.; Tanzania: Rufiji D.; Zanzibar; Ghana; Somalia; Ethiopia; Socotra; Madag.; peninsular India; Ceylon. – Plants of sandy places, near coasts. – Simple or branching; **St.** succulent, erect, spreading or ± creeping, little glandular, 12–75 cm tall; **L.** with a petiole up to 3.5 cm long, elliptical, obovate or oblong, 1.5–5 cm long, 0.8–3.5 cm across, margins entire or coarsely dentate; **Fl.** solitary from the L.axils, pale pink to yellow, tube 2–2.5 cm long, glandular, lobes ± round.

**Pedilanthus** NECKER. Euphorbiaceae. – Mexico; USA: Florida; Central Am.; Caribbean Is.; S.Am. – Small or larger ♄, often branching basally to form clumps, or ♄ up to 3 m tall; **roots** usually woody and freely branched; **St.** sometimes woody or ± succulent, sometimes underground, with a poisonous, milky sap; **L.** opposite, the uppermost L. sometimes alternate, palmate to lanceolate or ovate, short-petiolate, variable in size, ± succulent, sometimes deciduous, the M.rib on the underside usually prominent and carinate, often rather undulating, the blade green, sometimes ± puberulous, Stip. rudimentary or missing, often barely 1 mm long, rounded or hemispherical, usually conical, spur-like; **Infl.** cymose with a forked main axis, with flower-like bract-cups, terminal or axillary or both, differentiated from the related G. **Euphorbia** by the oblique, irregular, flower-like Cy. with a unilateral, spur-like appendage. – Greenhouse, warm. Propagation: seed, cuttings. (See ROBERT L. DRESSLER, The Genus **Pedilanthus** in Contr. GRAY Herb. HARVARD Univ. CLXXXII, Cambridge, Mass., USA 1957). The numerous synonyms of the invalid G. *Tithymalus* BOISS. and of the G. *Cropidaria*, *Cubanthus* and *Diadenaria* have not been included in the following. – Only the more succulent species are described here.

**P. bracteatus** (JACQ.) BOISS. (*Euphorbia b.* JACQ., *P. articulatus* (KL. et GK.) BOISS., *P. involucratus* (KL. et GK.) BOISS., *P. pavonis* (KL. et GK.) BOISS., *P. rubescens* BRANDEG., *P. spectabilis* ROBINS., *P. greggii* MILLSP.,

*P. olsson-sefferi* MILLSP.). – Mexico. – ♄ 1 to 3 m tall, branching basally; **St.** semi-succulent or very succulent, puberulous at first; **L.** densely puberulous, ovate to oblanceolate, (2.5–)10(–17) cm long, (2.5–)6(–7.5) cm across, tip blunt, apiculate, the M.rib carinate on the underside, Stip. up to 1 mm long and across, brown; **Infl.** terminal, Cy. fleshy, green, bract-cup 10–16 mm long.

**P. macrocarpus** BENTH. – Mexico: Sonora Desert, Baja Calif. – ♄ 0.5–1.50 m tall, branching basally; **St.** 5–10 mm thick, fleshy, hispid-hairy at first, axillary buds very woolly-hairy; **L.** deciduous, spatulate to ovate, the blade often involute and fleshy, 3–10 mm long, 1.5 to 4 mm across, pointed or blunt, Stip. missing; **Infl.** of few densely hairy Cy., bracts red, 5 to 11 mm long, pointed.

**P. millspaughii** PAX et K. HOFFM. – Costa Rica. – **St.** thick and fleshy; **L.** glabrous, ovate, pointed, 9–11 mm long, 4–7 mm across; **Infl.** short, bracts small, Cy. 10–12 mm long. Insufficiently known.

**P. tehuacanus** BRANDEG. – Mexico: Tehuacàn. – ♄; **St.** thick, succulent, sparsely puberulous, axillary buds densely woolly; **L.** with both surfaces puberulous or densely so, fleshy, probably readily deciduous, narrow-obcuneate, the M.rib thick but not carinate on the underside, petiole 1–3 mm long and across, Stip. spur-like, c. 0.5 mm long; **Infl.** often branched, bracts ovate to round-ovate, pointed, conspicuously hood-shaped, puberulous, 12–20 mm long, 11–15 mm across, Cy. green, puberulous.

**P. tithymaloides** POIT. – Type: ssp. **tithymaloides.**

**P. — ssp. angustifolius** (POIT.) DRESSLER (*P. a.* POIT.). – From E. Cuba through the Greater Antilles to St. John and St. Croix .– **St.** 0.8 to 3 m tall, thin to rather thick and fleshy, sparsely puberulous at first; **L.** linear to ribbon-like or narrow-lanceolate, 2–9 cm long, 2.5 to 16 mm across, blunt to broadly tapering, the M.rib carinate on the underside, petiole 1 to 5 mm long, curly-puberulous, Stip. hemispherical, brown, 0.5 mm across; **Infl.** puberulous, involucral tube green below, 6–9 mm long, hairy.

**P. — ssp. bahamensis** (MILLSP.) DRESSLER (*P. b.* MILLSP.). – Bahamas. – Erect ♄ 0.7–1.5 m tall, branching basally, Br. ± curving; **L.** lanceolate, 1.5–3 cm long, 3.5–6.5 mm across, the M.rib very alate-carinate on the underside, stipules hemispherical; involucres red, tube 7.5–12 mm long, sparsely hirsute outside.

**P. — ssp. — cv. Variegata. – St.** green with white stripes; **L.** white-mottled.

**P. — ssp. jamaicensis** (MILLSP. et BRITT.) DRESSLER (*P. j.* MILLSP. et BRITT., *P. grisebachii* MILLSP. et BRITT.). – Jamaica: coast. – ♄ (0.4–)2(–4) m tall; **St.** thin, at first puberulous; **L.** narrow-lanceolate to elliptical or oblanceolate, 1 – c. 10 cm long, 2–40 mm across, pointed or blunt, the M.rib somewhat or very carinate on the underside, Stip. 0.3–1 mm ∅; involucres red, glabrous to hairy-puberulous outside.

**P. — ssp. padifolius** (L.) DRESSLER (*Euphorbia t. β padifolius* L., *E. anacampseroides* LAM., *P. a.* (LAM.) KL. et GK.). – Lesser Antilles. – ♄ 1–3 m tall; **St.** thin, moderately succulent; **L.** ovate to elliptical, 6.5–12.5 cm long, 2.5 to 2.8 cm across, underside with the M.rib thickened and often carinate, Stip. spur-like; **Infl.** fairly large, involucres red, involucral tube yellowish-green or red.

**P. — ssp. parasiticus** (KL. et GK.) DRESSLER (*P. p.* KL. et GK., *P. itzaeus* MILLSP., *P. latifolius* MILLSP. et BRITT., *P. ramosissimus* BOISS.). – Greater Antilles. – ♄ 0.8–2 m tall; **St.** thick, fleshy, rather curly-puberulous, eventually glabrous, straight or ± zigzagging; **L.** triangular-ovate or broad-ovate, pointed above, 3 to 9 cm long, 2–6.5 cm across, underside with the M.rib slightly carinate, Stip. blunt or spur-like; **Infl.** with red bracts, involucre pink.

**P. — ssp. retusus** (BENTH.) DRESSLER (*P. r.* BENTH.). – S.Am.: Amazon reg. – ♄ 0.7 to 1.7 m tall; **St.** initially sparsely puberulous; **L.** ovate to broad-elliptical, 3.2–7.5 cm long, 1.5–4.8 cm across, blunt, underside with the M.rib carinate, Stip. hemispherical.

**P. — ssp. smallii** (MILLSP.) DRESSLER (Pl. 116/4) (*P. s.* MILLSP.). – USA: S.Florida; Cuba. – ♄ 0.7–2 m tall; **St.** at first puberulous, distinctly zigzagged; **L.** 2.5–7 cm long, 1.3–3.2 cm across, lanceolate-ovate, acute, underside sparsely hairy to puberulous, with M.rib carinate or alate, Stip. spur-like, brown.

**P. — ssp.** — cv. Silver Slipper. – **St.** green, white-striate; **L.** white-mottled.

**P. — ssp. tithymaloides** (*Euphorbia t.* L. and v. *myrtifolia* L., *E. canaliculata* LODD., *P. c.* (LODD.) SWEET, *P. houlletii* BAILL., *P. fendleri* BOISS,. *P. pringlei* ROBINS., *P. gritensis* ZAHLBR., *P. deamii* MILLSP., *P campester* BRANDEG., *P. petraeus* BRANDEG., *P. irensis* BRITT., *P. camporum* STANDL. et STEYERM., ? *E. carinata* LODD., ? *P. c.* SPRENG.). – S.Mexico to Central Am. to S.Columbia. – ♄ 0.4–3 m tall; **St.** thin and woody or thick and conspicuously succulent, puberulous at first; **L.** deciduous or evergreen, ovate, elliptical or ovate-lanceolate, glabrous to puberulous, 1–16 cm long, 0.8 to 10 cm across, the tip pointed, bluntly tapering or truncate, usually ± tapering, cuneate below, the M.rib carinate or very undulating on the underside, petiole 2–12 mm long, Stip. dark brown, spur-like, 0.3–1.2 mm ∅; **Infl.** terminal or axillary, Int. glabrous or puberulous, 1–3 mm long, bracts red, glabrous to curly-puberulous (4–)12(–14) mm long, 2 5 mm across, involucral tube red on the upperside, green or greenish-yellow at the base, glabrous or hairy, 7–14.5 mm long.

**P. — ssp. — cv. Nanus** (Pl. 116/5). – Dwarf form; **St.** with segments c. 10 mm long, ± spherical, 8–10 mm thick; **L.** 2–2.5 cm long, 6–8 mm across.

**P. — ssp. — cv. Variegatus.** – **L.** with white or pink borders.

**P. tomentellus** ROBINS. et GREENM. – Mexico: Oaxaca valley. – ♄ 2–4 m tall; **St.** rather succulent, at first densely red-hairy, later sparsely puberulous; **L.** rather fleshy, underside more puberulous, ovate to ovate-elliptical, 3.8–7.5 cm long, 1.5–4.5 cm across, pointed or blunt above, the M.rib slightly carinate on the underside, petiole with red hairs, 2–6 mm long, Stip. dark brown, spur-like; **Infl.** terminal, forking, Int. with red hairs, bracts red, involucres puberulous all over, green.

**Pelargonium** L'HER. Geraniaceae. – S.Afr., S.W.Afr., Madag., As. Minor. – Some 240 shrubby species are known; mention can be made here only of the ± distinctly succulent species. – Usually low ♄ branching ± dichotomously, or caudiciform, variable in height; **L.** petiolate, with the blade cordate or truncate below, not or only slightly decurrent, blade ± lobed (Sect. **Cortusine** acc. H. MERXMÜLLER), or never cordate at the base but pinnatisect (Sect. **Otidia** acc. MERXMÜLLER), dropping in the resting period; **Fl.** usually in irregularly-branched umbels, the uppermost Sep. spur-like, Pet. usually unequal in size, the upper two usually very large, white or pink, often with purple blotches and stripes, or else scarlet or brownish-yellow; **Fr.** rostrate, 5-partite, the part-Fr. with a flexible awn which rolls back when the Fr. is ripe. – Greenhouse, warm. Propagation: seed, cuttings.

**P. alternans** WENDL. – Cape: Kraus Valley. – **Caudex** branching; **Br.** short, gnarled; **L.** laciniate, white-haired, lobes tridentate; **Infl.** an umbel of 3–4 Fl.

**P. caylae** HUMB. – Madag. – Laxly-branched ♄; **Br.** fleshy to woody, erect, firm, cylindrical, 5–7 mm thick, initially minutely hairy, becoming bluish, glabrous; **L.** ± fleshy, pale blue, cordate below, 5-angled, c. 3 cm long, 5–7-lobed, lobes crenate-serrate, semi-ovate, minutely whitish-hairy; **Infl.** of 4–6 red Fl. c. 3 cm ∅.

**P. carnosum** (L.) AIT. (Pl. 117/1) (*Octidia c.* L., *P. sisonifolium* BAK., *P. ferulaceum* v. *polycephalum* E. MEY.). – S.W.Afr.: Kahanstal. – **Caudex** up to 30 cm tall, up to 5 cm thick, green, smooth; **Br.** erect, thick, L.bases persisting as short Th.; **L.** with a petiole 3 to 6 cm long, blade 7–14 cm long, 3–6 cm across, with broad-ovate, lobed and crenate-dentate segments, green, hairy, especially along the margins; **Ped.** and pedicels hairy, capitate, with 5–12 whitish-pink Fl.

**P. ceratophyllum** L'HER. (Pl. 117/2) (*P. ferulaceum* AUCT. non (BURM. f.) WILLD., *P. crithmifolium* AUCT. non J. E. SMITH). – S.W.Afr.: S.Namib, Lüderitz D. – **Caudex** up to 20 cm tall, 2–4 cm thick, usually branching only from the base; **Br.** thinner, bark wrinkled, rough, with persistent, white L.-bases; **L.** few, with a petiole 5 cm long, blade 5 cm long, 1.5 to 3 cm across, simply pinnatifid, segments broad-linear, undivided or short-dentate; **Infl.** little branched, Ped. often becoming woody and persisting, Fl. numerous, whitish. Like **Sarcocaulon rigidum**, the St. and Br. are thickly resin-coated.

**P. cortusifolium** L'HER. (*P. monsoniaefolium* DTR.). – S.W.Afr.: S. Namib, Lüderitz D. – **Caudex** thick, sparsely branching, up to 25 cm tall, sand-coloured, pointed-tuberculate and wrinkled because of the woody bases of petioles and stipules; **L.** with a petiole c. 4 cm long, blade 3 cm long, 2.5 cm across, roundish-ovate, cordate at the base, plicate along the veins, lobed, irregularly dentate, both surfaces with glossy, long, silken hairs or felt; **Infl.** with long tufts of hairs, Fl. many, white to pinkish-red, with purple markings.

**P. cotyledonis** (L.) L'HER. – St. Helena. – Dwarf ♄ with short, thick, sparingly branched succulent **St.**; **L.** very broadly ovate, cordate below, slightly lobed and denticulate, petiolate, tomentose underneath at first, deciduous; Ped. few-flowered, **Fl.** small, white.

**P. crassicaule** L'HER. (*P. mirabile* DTR.). – S.W.Afr.: Gr. Namaqualand. – **Caudex** smooth, glossy, dark brown, unbranched or with only few Br., 10–15 cm tall, up to 3 cm thick and mores or (in mountainous areas) St. thinner, freely branching and forming spherical cushions, young St. segments covered with stipular Th.; **L.** with a petiole 3–4 cm long, blade 4–6 cm long, 5–6 cm across, roundish-cordate, lobed, margins bluntly dentate, grey-green, both surfaces short-felty; **Infl.** with long tufted hairs, Fl. in umbels of 4–5, c. 12 mm ⌀, white, purple-spotted in the centre.

**P. crassipes** HARV. – Cape. – **Caudex** short, fleshy, densely covered with dry, old L.-bases; **L.** with thick petioles; **Infl.** umbels of many small, dark purple Fl.

**P. crithmifolium** J. E. SMITH (Pl. 117/3) (*P. alternans* AUCT. non WENDL.). – Cape: Lit. Namaqualand; S.W.Afr.: Gr. Namaqualand. – **Caudex** 30–40 cm tall, up to 5 cm thick, usually much branched, smooth, greenish; **L.** with a petiole 10 cm long, blade 16 cm long, 9–10 cm across, sometimes doubly pinnatifid, rachis narrow, L.segments cuneate, dentate only at the ends, very shortly hairy; **Infl.** with many, closely-spaced branchlets, forking pseudo-dichotomously, each section of the Infl. with 2–5 white Fl., often with red markings.

**P. echinatum** CURT. v. **echinatum** (*Geraniospermum e.* O. KTZE., *Geranium aculeatum* PAT., *G. e.* THUNBG., *P. hamatum* JACQ.). – S.Afr. – **Caudex** erect, with few Br., dwarf, fleshy, c. 6 mm thick, covered with fleshy, spiny Stip. 4–5 mm long; **L.** eventually dropping, long-petiolate, blade cordate-oval, 3–5-lobed, margins crenate, white-hairy; **Fl.** purplish-lilac.

**P. —** v. **albiflorum** SCHLTR. – Fl. white.

**P. endlicheranum** FENZL. – S.As. Minor. – Caudex short, thick, fleshy, somewhat branching; basal **L.** cordate-auriculate, broadly 5-lobed, crenate-dentate, blue-green, with appressed hairs, long-petiolate; **Infl.** of many pink Fl., the 2 upper Pet. large, with a red border.

**P. ferulaceum** (BURM. f.) WILLD. (*Otidia f.* BURM. f.). – Cape: Lit. Namaqualand; S.W.Afr. ? – **Caudex** thick, semi-woody, 6–16 cm tall, 5–8 mm thick, almost unbranched; **L.** long-petiolose, usually deeply pinnatifid, rather thick, both surfaces with bristly of soft hairs; **Fl.** 2–5 in an umbel, Sep. white-bordered, Pet. white with pink blotches.

**P. fulgidum** (L.) AIT. (*Otidia f.* L., *P. chelidonifolium* SALISB.). – S.W.Cape. – **Caudex** semi-woody, fleshy, 1 cm thick, hairy, up to 75 cm tall; **L.** tripartite-pinnatifid, both surfaces with silky hairs, the lateral segments 3-cleft, deeply dentate, terminal lobes oblong, deeply incised; **Infl.** of many small, brilliant scarlet Fl.

**P. gibbosum** L'HER. – S.W.Cape, near coasts. – ♄ with spreading brittle Br. 2–3 mm thick, much thickened and often 2 cm ⌀ at the nodes; **L.** long-petiolate, pinnatifid, lobed and re-lobed, green, minutely hairy.

**P. juttae** DTR. – S.W.Afr.: Lüderitz D. – Sub-♄; **caudex** short, tuberous, up to 10 cm tall, 15–20 mm thick, with a light brown bark, with a few short Br.; **L.** crowded at the Br.tips, petiole 8–10 cm long, blade doubly pinnatisect, light green, very minutely and densely hairy.

**P. klinghardtense** R. KNUTH (*P. paradoxum* DTR., *P. jacobi* R. A. DYER). – S.W.Afr.: Gr. Namaqualand, Klinghardt Mts.; Cape: Lit. Namaqualand. – **Caudex** up to 70 cm tall, up to 6 cm thick, sparsely branched, yellowish-green; **L.** fleshy, thickish, spatulate-obovate, distinctly and gradually tapering towards the petiole, margin dentate to lobed, grey-green, minutely hairy, blade (3–)4(–6) cm long, c. 2–4 cm across; **Infl.** stout, minutely hirsute, the thick branchlets becoming woody and persisting, with many inconspicuous Fl., Sep. light green, Pet. concolorous-white.

**P. longifolium** JACQ. – W.Cape. – **Caudex** c. 3 cm tall, 2 cm thick, with a thick taproot; **L.** oblong-lanceolate, narrowing towards the 1 cm long petiole, 2 cm long, 6–8 mm across; Fl. numerous, whitish-pink.

**P. ovato-stipulatum** KNUTH. – Cape: Lit. Namaqualand. – **Caudex** 4–5 cm long, little branching, gnarled, 5–6 mm thick, covered with dry, ovate stipules, bark flaking; during the growing period clusters of small, obovate Stip. c. 6 mm long, 3–4 mm across, with sinuate margins are produced at the St.tips; **L.** with a petiole 1.5 cm long, 3-lobed, the central lobe 1.5–2 cm long, the two side-lobes 5–6 mm long, 2–3 mm across, green, deciduous; **Infl.** bipartite, Pet. white, with pink veins.

**P. paniculatum** JACQ. – S.W.Afr.: Gr. Namaqualand. – **St.** 50–120 cm tall, almost unbranched, up to 6 cm thick, with conspicuous, white, persistent L.-bases; **L.** doubly or trebly pinnatifid, petiole 5–8 cm long, blade 15–16 cm long, 10–18 cm across, segments broad-linear, dentate only in front, with thin, short hairs; **Infl.** pyramidal, with very numerous, white Fl.

**P. roessingense** DTR. (*P. damarense* R. KNUTH). – S.W.Afr.: Gr. Namaqualand. – Hemispherical, brittle sub-♄ c. 25 cm tall; **L.** with a petiole 2.5 cm long, rather fleshy, green, minutely hairy, cordate, 2–2.5 cm long, 6 cm across, slightly 5-lobed, lobes with 2–3 blunt T.; **Fl.** on pedicels 6–9 mm long.

**P. saffrum** ECKL. et ZEYH. – Cape: Winterberg. – **Caudex** short, fleshy; **L.** long-petiolate, digitate, segments linear, laciniate, slightly hairy; **Infl.** a many-flowered umbel.

**P. sericeum** E. MEY. – Cape: Lit. Namaqualand. – Low ♄; **caudex** fleshy, with short, thick Br. from the base; **L.** short-petiolate, laciniate, 3-lobed, c. 2 cm long, 1 cm across; **Fl.** an attractive red, with pedicels c. 6 cm long.

**P. sibthorpiifolium** HARV. (*P. graniticum* R. KNUTH, *P. eberlanzii* DTR., *P. amabile* DTR.).. – S.W.Afr.: Gr. Namaqualand. – **Caudex** short, underground, 6–7 cm long, 1 cm thick, with roots up to ½ m long, with 3–4 small Tub.; **L.** 3–7, rosulate, with petioles 3–4 cm long, 1.5 mm thick, sparsely hairy, resting on the ground, blade reniform to cordate, 12–30 mm long, 20–40 mm across, crenate, dark green, hirsute; **Infl.** an umbel 2–4 cm tall, of 4–7 pink to deep pink Fl. 2 cm ∅, with red blotches and lines, Cal. with grey hairs.

**P. tetragonum** (L. f.) L'HER. (Pl. 117/4) (*Geranium t.* L. f.). – S.Afr. – Erect ♄ up to 70 cm tall, dichotomously branched; **Br.** 3–4-angled, c. 7 mm thick, fleshy, pale green, smooth; **L.** broad-cordate, 5-lobed, crenate-dentate, dropping in autumn; **Fl.** usually 3 together, long-pedicellate, Pet. pink, with purple veins.

**P. triste** (L.) AIT. v. **triste** (*Otidia t.* L.). – S.W.Cape. – **Caudex** very short, fleshy; **L.** large, 2–3, ± pinnatisect, hirsute, outer lobes glandular; **Infl.** hairy, of 7–14 Fl., Pet. ligulate, blackish-purple, with a yellow or brownish-yellow border and brown blotches.

**P. —** v. **daucifolium** HARV. (*P. millefoliatum* SWEET). – **L.** minutely pinnatisect (like a carrot).

**P. —** v. **filipendulifolium** SIMS. – **L.** coarsely pinnate.

**P. —** v. **laxum** HARV. – **L.** very laxly pinnatisect.

**P. xerophytum** SCHLTR. – SW.Afr.: Gr. Namaqualand. – Sub-♄ forming large, flat cushions; **Br.** woody-fleshy, prostrate, 1 cm thick, bark grey, shoots short, with short Sp.; **L.** at the shoot tips, short-petiolate, broad-spatulate, c. 1 cm long and across, later dropping, margins crenate-serrate; **Fl.** usually solitary, pedicel 7 cm long, Pet. white with red markings.

---

**Peperomia** RUIZ. et PAV. Piperaceae. – Trop. S.Am.; some species in Afr. – More than 500 species are known. Mostly ♃, some species with a **caudex**, bearing mostly peltate L., others with compact or elongated, ± fleshy leafy stems, some also low and semi-epiphytic in growth; L. ± fleshy, alternate or verticillate, of various sizes, often beautifully marked; **Fl.** in terminal or axillary Rac., inconspicuous. – Not all **Peperomia** species are succulent plants. (See W. RAUH, Kakt. u. a. Sukk. **11**, 116, 1960.) – Greenhouse, warm, moist. Propagation: Seed, some by cuttings.

## Key to the species of Peperomia described below by G. D. Rowley

| | | |
|---|---|---|
| 1. | L. developing from a basal Tub. | **2.** |
| 1a. | L. on ± branched, erect or creeping St. | **5.** |
| 2. | L. usually less than 3 cm long, obtuse; Rac. mostly felty | **3.** |
| 2a. | L. 3–10 cm long or longer, acute; Rac. fleshy | **hadrostachys** |
| 3. | L. seldom as much as 1 cm long; Rac. 1–2 cm long | **parvifolia** |
| 3a. | L. mostly larger; Rac. 2 cm long or longer | **4.** |
| 4. | L. shield-shaped (peltate) from 1/3 above the base | **cyclaminoides** |
| 4a. | L. shield-shaped from the middle | **campylotropa peruviana** |
| 5. | L. alternate | **6.** |
| 5a. | L. in whorls | **9.** |
| 6. | L. windowed on the upper surface | **7.** |
| 6a. | L. not windowed | **balansana** |
| 7. | St. erect; L. hatchet-shaped (dolabriform) | **dolabriformis** |
| 7a. | St. creeping | **8.** |
| 8. | L. boat-shaped | **nivalis** |
| 8a. | L. round or elliptical | **rotundifolia** |
| 9. | St. erect | **10.** |
| 9a. | St. creeping; plants epiphytic | **11.** |
| 10. | L. small, transparent-dotted | **galioides** |
| 10a. | L. large, velvety | **doellii** |
| 11. | L. villose, thick and fleshy | **carnifolia** |
| 11a. | L. glabrous, impressed-dotted | **reflexa** |

**P. balansana** C. DC. – Paraguay; Argentine; Bolivia. – Glabrous, ± erect, succulent, moderately large; **St.** at least 3–5 mm thick, Int. up to 6 cm long; **L.** round-ovate, tip pointed to rounded, 2.5–5.5 cm long, 2.5–5 cm across, with 5–7 veins, petiole 1.5–3 cm long; **Fl.** spike c. 6 cm long, lax, bracts round-scutate.

**P. campylotropa** A. W. HILL. (*P. umbiciliata* KUNTH., *P. u.* R. et P., *P. u.* v. *macrophylla* C. DC.). – Mexico: cooler reg. – **Caudex** subterranean, disc-shaped to spherical, c. 1–2 cm ∅, 8–12 mm tall; L.rosulate, petiole (3–)8–15(–20) cm long, blade ± circular, 1.4–3.5 cm ∅, membranous to leathery-membranous; **Fl.** spike ± dense, Ped. (8–)20(–30) cm long, bracts ovate, tapering, 15–17 mm long, 8–9 mm across.

**P. carnifolia** YUNCKER. – Bolivia: Cochabamba. – Epiphyte; **St.** at least 5 mm thick, 1 m or more long, laxly branching, tufted, Int. up to 12 cm long; **L.** 3–4 at each node, elliptical to elliptical-ovate, blunt, 1–2 cm long, 8–14 mm across, 3–5 mm thick, 3-veined, both surfaces rather villose, petiole 1–2 mm long, densely puberulous; **Fl.** spike 1.5 cm long.

**P. cyclaminoides** HILL. – Bolivia: near Tarija. – **Caudex** 1–2 cm long; **L.** round-ovate, blunt, upper 2/3 scutate, 1.2 cm long, 1–1.5 cm across, with a petiole 3–7 cm long; **Fl.** spikes several, 2–4 cm long, bracts round-ovate, pointed.

**P. dolabriformis** KUNTH. v. **dolabriformis** (Pl. 118/3). – Peru: along the R. Huancabamba, near San Felipe. – Sparsely-branched ♄ 10 cm or more tall; **L.** alternate, sometimes in whorls, crowded, on finger-thick St., hatchet-shaped, c. 6 cm long, at the broadest point c. 17 mm across, c. 6 mm thick, with a tiny subulate tip, narrowing below towards the clavate petiole, the narrow, 6 mm thick angle being the transparent upperside, the broad sides are the underside, grey-blue or ± powdery grey-blue; **Fl.** spikes very long, branching.

**P.** — v. **tenuis** JACOBS. (Pl. 118/1). – **L.** thinner, with the transparent strip only 1 mm across.

**P. doellii** PHILIPP. – Chile. – **St.** erect, up to 20 cm tall, simple or with a few Br. above, usually with several Br. forming a flat cushion, very velvety-puberulous, Int. 1–2 cm long; **L.** (3–)4(–5) in whorls, ovate, elliptical-obovate, ± spatulate or even circular, tip rounded, blunt or slightly emarginate, cuneate at the base, very thick and fleshy, upperside sparsely puberulous, underside reddish and glabrous, 7–10 mm long, 3–8 mm across, 3-veined, petiole 1 mm long; **Fl.** spikes 6–14 cm long, lax, bracts scutate.

**P. galioides** H. B. et K. (Pl. 118/2). – Colombia; Bolivia. – ± erect, branching, succulent plants, either epiphytic or terrestrial; **St.** 2.5–8 mm thick, up to 1 m long, Br. bi-partite, tri-partite or in whorls, often thin and often whip-like, Int. (1–)5(–10) cm long, with fairly dense, tufted hairs; **L.** 3–8 at the nodes, elliptical, elliptical-oblong, ± spatulate or oblanceolate, blunt-tipped, 3–30 mm long, 2 to 5 mm across, indistinctly 3-veined, ciliate at the tip, with transparent spots, underside with yellow glands, the petiole short-bristly; **Fl.** spikes 4–7 cm long, lax, bracts round-scutate to long-ovate, thin.

**P. hadrostachys** YUNCKER. – Argentine: Salta. – Acauline, tuberous plant; **L.** round-ovate, scutate for 1–3 cm above the rounded, blunt to slightly cordate base, pointed, 5 to 12 cm long, 4–9 cm across or possibly more, with 9–11 veins, and a petiole 10–15 cm long; **Fl.** spike thick, fleshy, 5–12 cm long, bracts round-scutate.

**P. nivalis** MIQ. v. **nivalis** (Pl. 118/4). – Peru. – Low-growing; **St.** creeping and forming a mat, or up to 10 cm tall and ± erect; **L.** alternate, densely crowded at the St.tips, sessile, pendulous, navicular, 12 mm long, 4 mm across, 3–4 mm thick across the upperside of the L. which is grooved, the 4 mm width representing the carinate underside; sap with a strong scent of aniseed; **Fl.** spikes very compressed, thick, bracts linear, thick.

**P.** — v. **crassa** JACOBS. – **L.** 6–7 mm thick.

**P.** — v. **minor** JACOBS. – **L.** 5–7 mm long, 2–3 mm across, 1–2 mm thick.

**P. parvifolia** C. DC. – Peru; Bolivia; N.Argentine. – **Caudex** tuberous, 1–2 cm ⌀; **L.** round or ± triangular, scutate above midway, 5 to 10 mm long, 7-veined, petiole 1.5–2.5 cm long; **Fl.** spike c. 2 cm long, bracts ± elliptical-scutate.

**P. peruviana** (MIQ.) DAHLST. v. **peruviana** (*Tildenia p.* MIQ.). – Peru; Argentine; Bolivia. – Caudiciform; **L.** round, usually 1–2 cm ⌀, scutate above midway, petiole 2.5 cm long; **Fl.** spike 2–6 cm long, bracts round-ovate, rather pointed.

**P.** — v. **major** HILL. – Peru; Argentine; Bolivia. – **L.** 2.5–3.5 cm ⌀, petiole 6–14 cm long; **Fl.** spike 5–8 cm long.

**P. reflexa** (L. f.) A. DIETR. (*Piper r.* L. f.). – Cape: Cape D.; Argentine; Bolivia. – Tending to form a mat, or creeping, branching, epiphytic; **St.** at least 1–2 mm thick, fertile St. ascending, up to 5 cm long, Int. 1–4 cm long, with minute, short bristles; **L.** usually in whorls of 4, rhombic-ovate to ± round, (8–)10(–15) mm long, 4–12 mm across, with a rounded tip, upperside glabrous with impressed spots, underside puberulous to glabrous, indistinctly 3-veined, petiole 2–3 mm long; **Fl.** spike 1–3 cm long, dense.

**P. rotundifolia** (L.) H. B. et K. v. **rotundifolia** (*Piper r.* L., *Acrocarpidium nummularifolium* MIQ., *Piper n.* Sw., *Pep. n.* H. B. et K.). – Probably Is. of Martinique; Argentine; Bolivia. – Creeping, often epiphytic; **St.** filiform, at least 1 mm thick; **Br.** erect, short, fertile, rather curly-puberulous or ± bristly; **L.** alternate, round, fleshy, c. 5 mm ⌀, or round-elliptical and up to 12 mm long, 10 mm across, rounded at the tip, the base round to almost scutate, the lower margin of the blade projecting slightly over the petiole, or the L. ± pointed, curly-puberulous, ± ciliate, 3-veined, petiole (1–)5(–10) mm long; **Fl.** spike solitary, bracts round-scutate.

**P.** — v. **pilosior** (MIQ.) C. DC. (*Acrocarpidium nummularifolium* f. *pilosior* MIQ., *P. n.* v. *pubescens* C. DC.). – Bolivia. – Young St. and L. densely curly-puberulous or with bristly hairs; **L.** lentiform; **Fl.** spikes short.

**Phytolacca** L. Phytolaccaceae. – Tropics. – ♄, ♄ or herbaceous plants; c. 10 species are known, but only the under-mentioned species is succulent. – Summer: in the open, winter: cold-house.

**P. dioica** L. (Pl. 118/5). – S.Am. – ♄ 8–10 m tall; **trunk** thick, fleshy; **Br.** thick; in S.Eur., where it is often planted by the wayside, its thick, fleshy roots tend to break up the road-surface; bark rough, splitting, grey, young shoots olive-green, with white lenticels 10 mm long; **L.** spiralled, crowded at the tips of the shoots, with a petiole 7–10 cm long, ovate, with an elongated tip 1–2 cm long, with entire margins, 18–25 cm or more long, 10–15 cm across, green, petiole and veins red; Fl. inconspicuous, in a long-pedunculate, pendulous Rac.; Fr. reddish-black berries.

**Piaranthus** R. Br. Asclepiadaceae. – S. Afr., S.W.Afr. – Small, succulent ♃; shoots short, oblong, usually 4-angled, angles blunt, with pointed T. flanked below by 2 subsidiary T.; Fl. small to medium-sized, several together towards the St.tip, erect, pedicellate, corona rotate, tube campanulate or flat, lobes triangular-pointed, velvety-hairy inside, variously coloured and spotted, corona-lobes with pectinate protuberances on the underside. – Greenhouse, warm. Propagation: seed, cuttings.

**P. comptus** N. E. Br. v. comptus (Pl. 119/1) (*Caralluma c.* Schltr.). – Cape: Beaufort West, Prince Albert, Hay D., W.Griqualand, Lit. Namaqualand. – St. crowded, 2–3 cm long, 8–14 mm thick, dull grey-green, bluntly 4-angled, T. short, pointed; Fl. 1–4 together, Cor. ± rotate, 16–18 mm ⌀, lobes 5–9 mm long, margins slightly recurved, glabrous outside, brownish-green, inside with dense, short, soft hairs, with dark, reddish-brown blotches.

P. — v. **ciliatus** N. E. Br. – Cape: Prince Albert, Laingsburg D. – Lobes ciliate.

**P. cornutus** N. E. Br. v. cornutus – Cape: Lit. Namaqualand. – St. prostrate to ascending, 1.5–3.5 cm long, 12–16 mm thick, angles very obtuse, with 3–5 tuberculate T.; Fl. 2 together, Cor. very deeply 5-cleft, with no tube, lobes 10 mm long, tapering, glabrous outside, velvety inside, softly hirsute, very pale yellow or whitish, with red blotches.

P. — v. **grandis** N. E. Br. – Cape: Victoria West D. – Lobes 12–14 mm long.

**P. decorus** (Mass.) N. E. Br. (Pl. 119/2) (*Stapelia d.* Mass., *S. serrulata* Jacq., *Obesia d.* Haw., *O. s.* Sweet, *P. s.* N. E. Br., *Caralluma d.* Schltr., *C. s.* Schltr.). – Cape: Swellendam, Oudtshoorn, Ladismith D. – St. crowded, prostrate, ovoid-oblong, 2.5–3.5 cm long, 15 mm thick, grey-green or light green, angles round, with small pointed T.; Fl. 2 together, Cor. 30–32 mm ⌀, deeply 5-cleft, rotate, glabrous outside, pale, lobes 10–12 mm long, pointed, minutely velvety-hairy inside, yellow, with dense, blackish-red blotches, borders slightly recurved.

**P. disparilis** N. E. Br. v. disparilis – Cape: Laingsburg, Oudtshoorn D. – St. ± cylindrical; Fl. on pedicels 6–8 mm long, Cor. rotate, tube indistinct, lobes 8 mm long, tapering, margins slightly recurved, glabrous outside, inside with minute, soft, velvety hairs, light red, with fine, pale yellow, transverse lines.

P. — v. **immaculatus** Luckh. – Fl. concolorous.

**P. foetidus** N. E. Br. v. foetidus – Cape: Bedford, Graaff Reinet, Van Rhynsdorp, Cradock D., W. Griqualand; reputedly also in S.W.Afr. – St. oblong-ovoid, almost spherical, up to 4 cm long, with 4–5 tuberculate-dentate angles, green to grey-green; Fl. usually 2 together, Cor. rotate, 14–22 mm ⌀, lobes ovate-lanceolate, pointed, margins rather recurved, greenish-reddish outside, glabrous, inside softly hairy, yellow, with red transverse lines and blotches, evil-smelling.

P. — v. **diversus** N. E. Br. – Lobes dark purplish-red with creamy-yellow, rough lines and blotches, with short, grey hairs below, above with longer, dark purple hairs.

P. — v. **multipunctatus** N. E. Br. – Lobes all over with small, rough, round, purplish-red blotches.

P. — v. **pallidus** N. E. Br. – Lobes pale purple with creamy-white, rough lines and blotches below.

P. — v. **purpureus** N. E. Br. – Lobes lighter purple, with yellow, rough markings in the lower part.

**P. framesii** Pill. – Cape: Van Rhynsdorp D. – St. ascending, 3–4 cm long, 12–14 mm thick, with 4–5 tuberculate-dentate angles, bluish-green; Fl. 1–2 together, Cor. 17–20 mm ⌀, tube campanulate, glabrous outside, whitish inside with reddish blotches, with horizontal lines, lobes 7–8 mm long, tapering, wrinkled, purple, with transverse lines, minutely papillose above, margins with purple ciliate hairs.

**P. geminatus** (Mass.) N. E. Br. (*Stapelia g.* Mass., *Podanthes g.* Nich., *Caralluma g.* Schltr.). – Cape: Jansenville D. – St. prostrate, almost ovoid, 2.5–4.5 cm long, 12–20 mm thick, very obtusely 4-angled, T. few, projecting; Fl. usually 2 together, Cor. rotate, 2.5–3 cm ⌀, deeply 5-cleft, lobes lanceolate, 10–14 mm long, margins somewhat revolute, greenish outside, suffused with red, dull yellow to orange-yellow inside, with minute red spots, densely and minutely hairy.

**P. globosus** White et Sloane. – Origin unknown. – St. creeping or ascending, c. 2 cm long, 12 mm thick, ovoid to spherical, very indistinctly 4-angled, angles with 2–4 small T., glabrous, light green; Fl. 1–2 together, Cor. rotate, 13 mm ⌀, lobes very spreading, 7 mm long, margins somewhat recurved, underside glabrous, velvety-hairy inside, light greenish-yellow, with reddish or mauve blotches.

**P. grivanus** N. E. Br. (*Caralluma g.* Schltr.). – Cape: W.Griqualand. – St. 2–5 cm tall, with 4 blunt angles and pointed T. 4–6 mm long, those above being dry; Fl. with Cor. 2.5 cm ⌀, tube short and broadly campanulate, lobes triangular-ovate, green outside, with darker veins, wrinkled and blackish-brown inside, both surfaces glabrous.

**P. mennellii** Luckh. – Cape: Kenhardt D. – St. hemispherical or oblong, roundish to 4-angled, with small T., glabrous, green; Fl. 2–6 together, Cor. without any visible tube, underside glabrous, inside velvety with white or purple hairs, pale yellow with scattered purple blotches, lobes 6–7 mm long, margins recurved.

**P. pallidus** Luckh. – Cape: Kenhardt D. – St. ± spherical or oblong, with 4 very rounded angles, glabrous, green; Fl. 2–4 together, Cor. rotate, tube indistinct, lobes c. 13 mm long, tapering, margins recurved, underside glabrous, velvety-hairy inside, pale yellow.

**Plate 130. 1. Senecio ficoides** (L.) Sch. Bip.; **2. S. haworthii** (Haw.) Sch. Bip. (Photo: G. D. Rowley); **3. S. herreanus** Dtr.; **4. S. hallianus** Dtr.; **5. S. fulgens** Nich.

20/B  Lexicon of Succulent Plants

Plate 131. 1. Senecio johnstonii Oliv. (Photo: G. Klatt); 2. S. kleiniiformis Suesse NG.; 3. S. jacobsenii Rowl.; 4. S. longiflorus (DC.) Sch. Bip. v. longiflorus; 6. S. klinghardtianus Dtr.

Plate 132. 1. Senecio medley-woodii Hutchis. (Photo: F. Riviere de Caralt); 2. S. rowleyanus Jacobs.; 3. S. spiculosus (Sheph.) Rowl.; 4. S. radicans (L. f.) Sch. Bip.

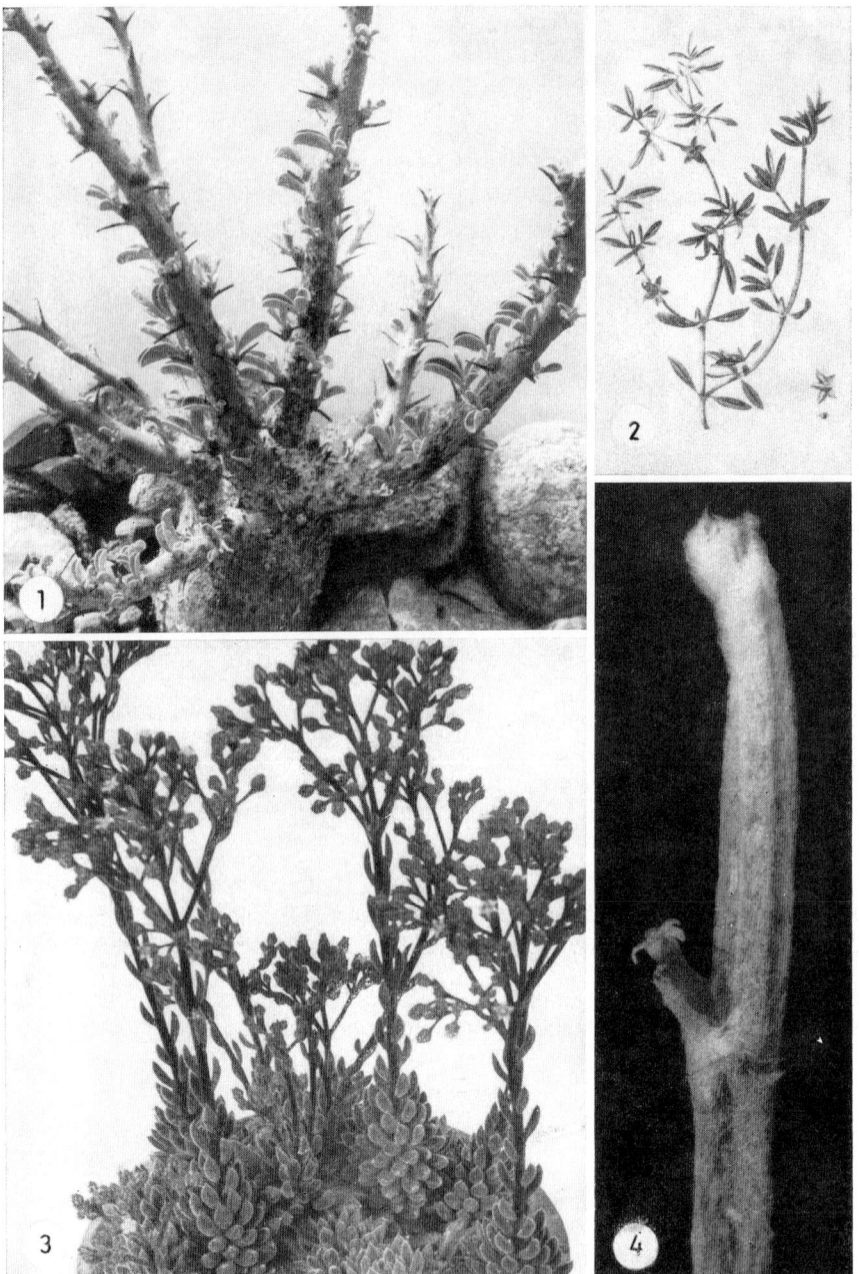

Plate 133. 1. Sesamothamnus lugardii N. E. BR.; 2. Sesuvium portulacastrum L. (From an unpublished plate by REDOUTÉ: (Photo: G. D. ROWLEY); 3. Sinocrassula yunnanensis (FRANCH.) BGR. (Photo: E. HAHN); 4. Seyrigia humbertii KERAUDREN, enlarged (Photo: J. MARNIER-LAPOSTOLLE).

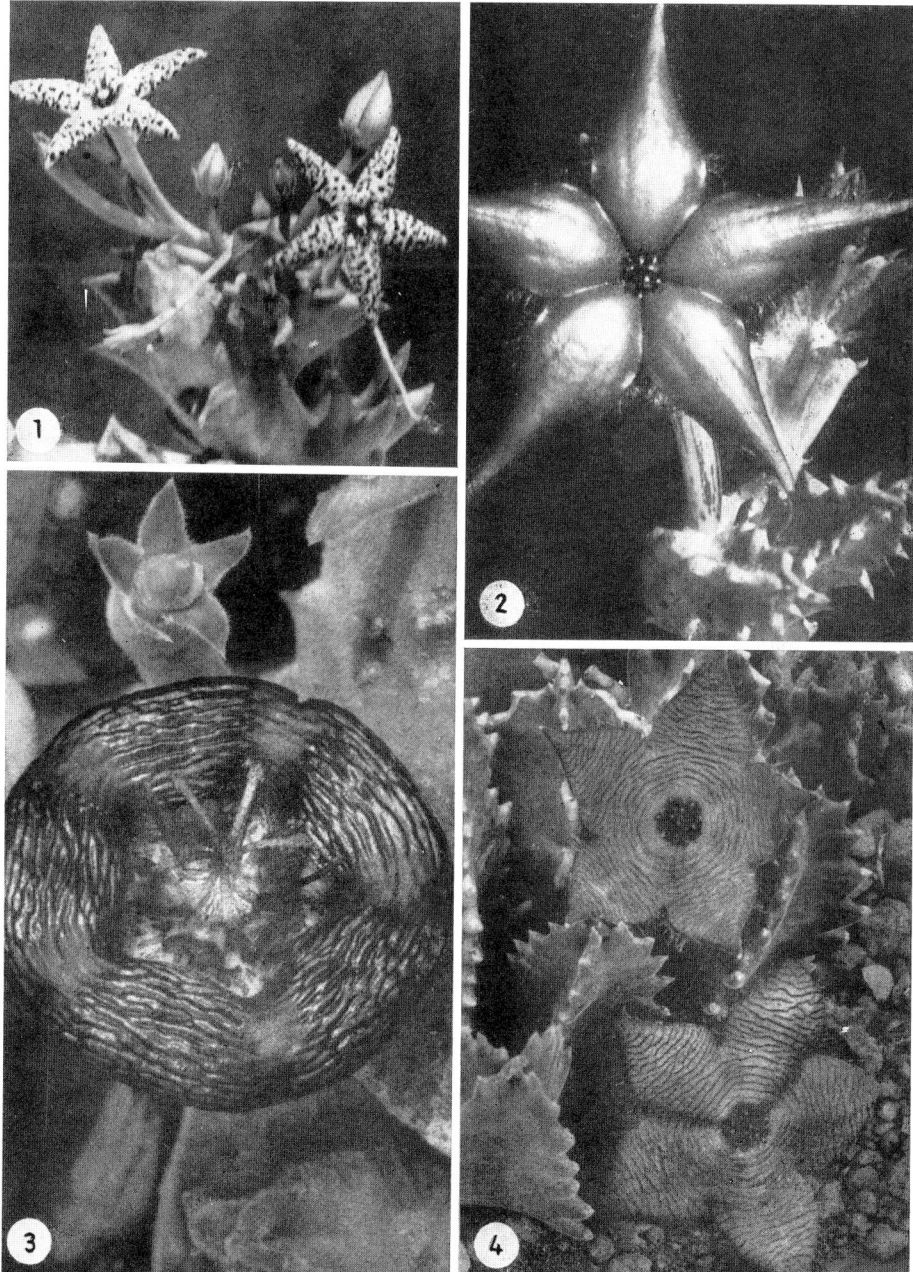

Plate 134. 1. **Caralluma albo-castanea** (MARL.) LEACH (Photo: K. DINTER); 2. **Stapelia bergerana** DTR. (Photo: as 1); 3. **S. englerana** SCHLTR. (Photo: H. LANG); 4. **S. clavicorona** VERD. (Photo: as 3).

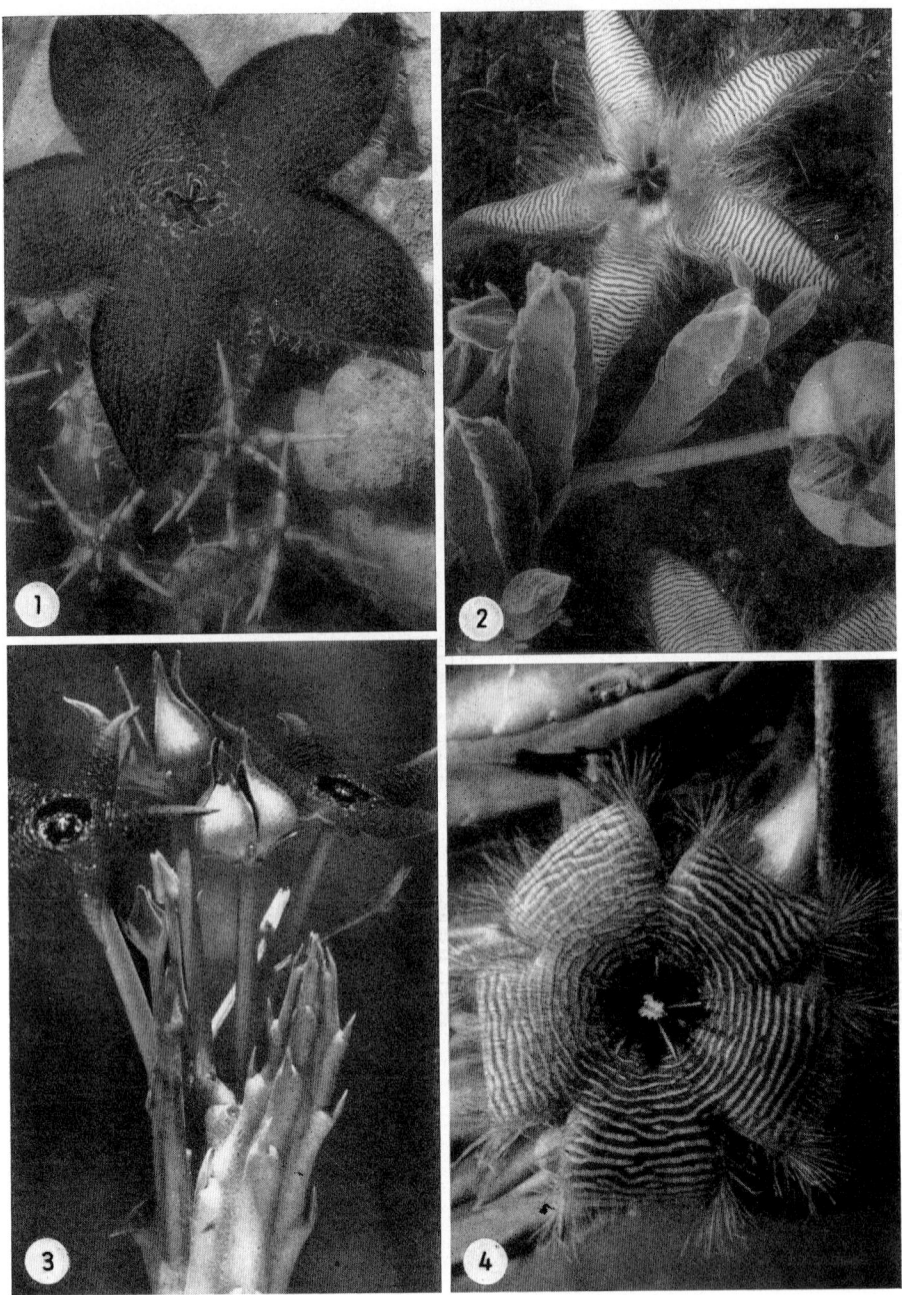

Plate 135. 1. Stapelia gemmiflora MASS. v. gemmiflora (Photo: H. LANG); 2. S. gettleffii POTT. (Photo: as 1); 3. S. kwebensis v. longipedicellata BGR. (Photo: K. DINTER); 4. S. grandiflora MASS.

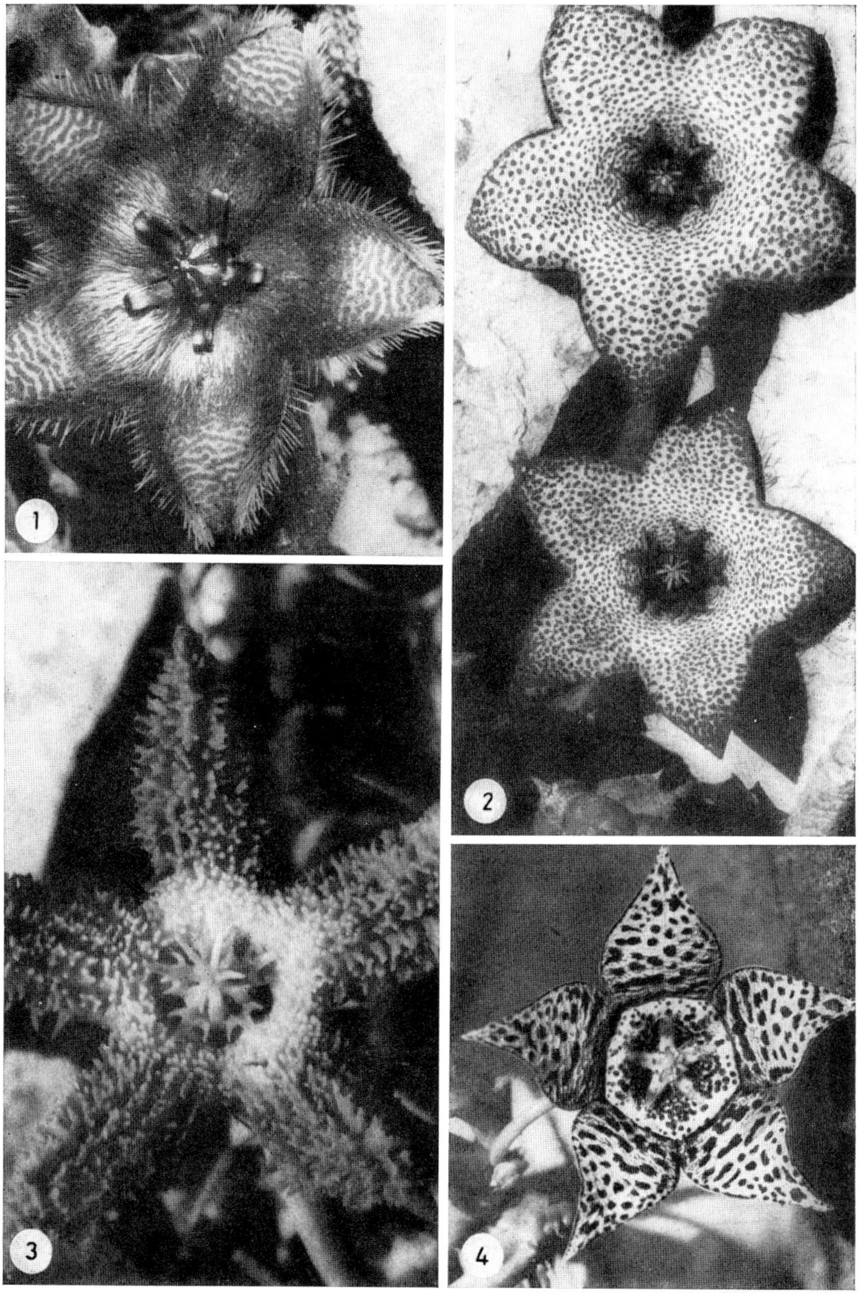

**Plate 136.** 1. **Stapelia nouhuysii** Phill. (Photo: H. Lang); 2. **S. parvipuncta** N. E. Br. (Photo: as 1); 3. **S. virescens** N. E. Br. (Photo: as 1); 4. **S. variegata** v. **curtisii** (Haw.) N. E. Br. (Photo: G. Kaiser).

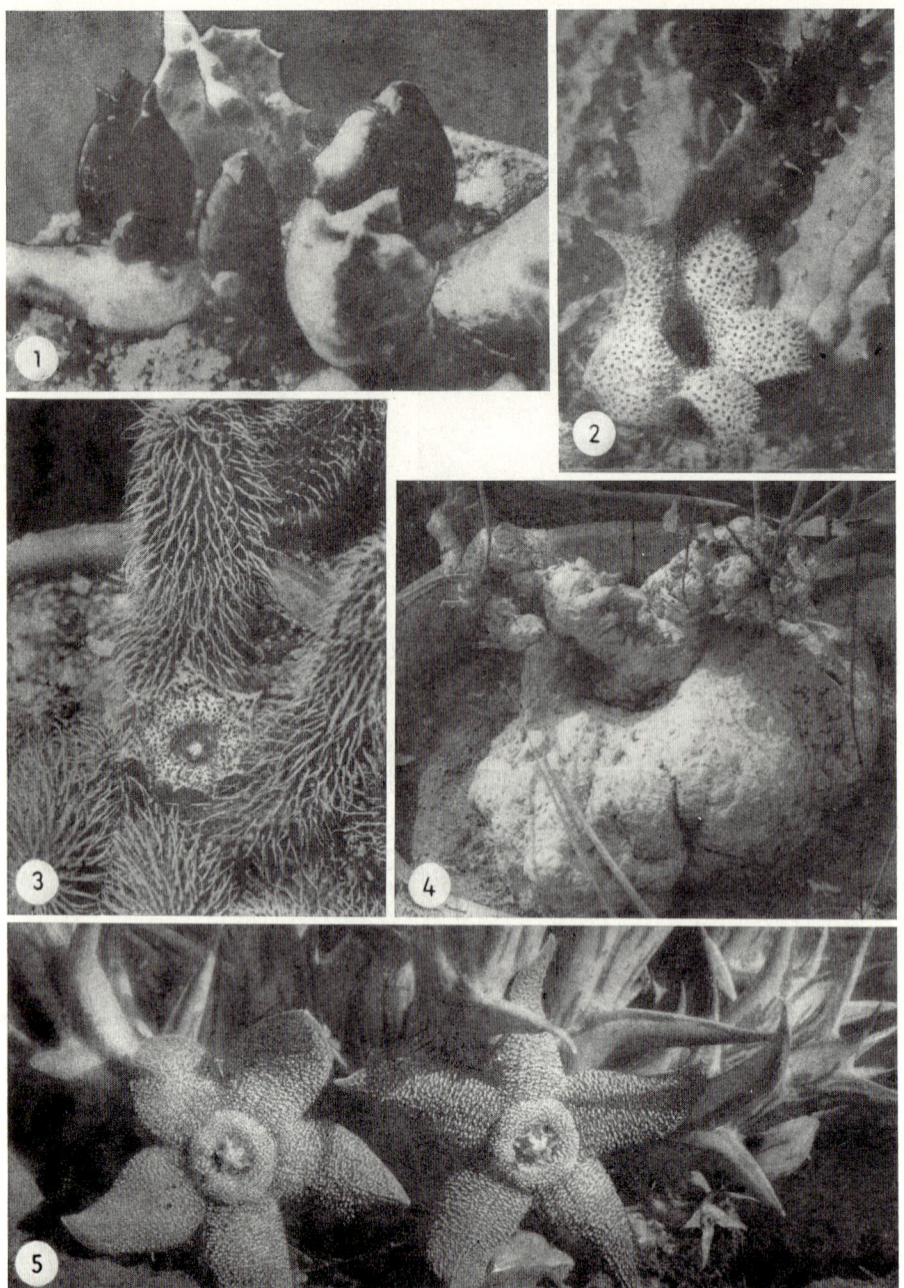

Plate 137. 1. **Stapeliopsis urniflora** Lavr. (Repr.: J. S. Afr. Bot. **XXXII,** 1966, Photo: J. J. Lavranos); 2. **Stapelianthus decaryi** Choux (Photo: W. Rauh); 3. **Stapelianthus pilosus** (Choux) Lavr. et Hardy (Photo: as 2); 4. **Stephania rotunda** Lour. (Photo: G. D. Rowley); 5. **Stultitia tapscottii** (Verd.) Phill. (Photo: H. Lang).

**P. parvulus** N. E. BR. – Cape: Laingsburg, Prince Albert D. – **St.** prostrate or ascending, 2–4.5 cm tall, 8–14 mm thick, oblong or ovoid-oblong, with 4 obtuse angles and tuberculate T. 3 mm long, grey-green, with pale red blotches; **Fl.** up to 12, Cor. rotate, 10–12 mm ⌀, tube indistinct, lobes triangular-lanceolate, pointed, glabrous outside, velvety-hairy inside, straw-coloured.

**P. pillansii** N. E. BR. v. **pillansii.** – Cape: Oudtshoorn, Willowmore, Riversdale, Ladismith D. – **St.** prostrate, 3–4 cm long, 10–15 mm thick, ± clavate, very obtusely angled, light green, suffused reddish; **Fl.** 2–6, Cor. 3–3.5 cm ⌀, rotate, 5-cleft almost to the base, lobes narrow-lanceolate, pointed, margins rather recurved, glabrous outside, shortly soft-hairy inside, yellowish or greenish.

**P. — v. fuscatus** N. E. BR. – **Cor.** dark red inside, with minute, intricate, greenish-yellow, transverse lines.

**P. — v. inconstans** N. E. BR. – **Cor.** ochre-yellow inside, with dense, light red, transverse blotches or spots, often so densely spotted as to appear ± concolorous reddish-brown.

**P. pulcher** N. E. BR. v. **pulcher.** – Cape: ? Colesberg D. – **St.** 15–25 mm long, 6–12 mm thick, oblong to spherical, the 4 obtuse angles with tuberculate T., grey-green; **Fl.** 1–4 together, Cor. ± rotate, tube distinct, 2 cm ⌀, underside glabrous, dull greenish to brown, minutely velvety inside with white or red hairs, light greenish-yellow, with dark reddish-brown blotches all over, lobes erect to spreading, 9 to 11 mm long, narrow.

**P. — v. nebrownii** (DTR.) WHITE et SLOANE (*P. n.* DTR.). – S.W.Afr.: Lit. Karas Mts.; Cape: Lit. Namaqualand. – Lobes ascending, with recurved tips, with purple borders.

**P. punctatus** (MASS.) R. BR. (*Stapelia p.* MASS., *Obesia p.* HAW., *Caralluma p.* SCHLTR.). – Cape: Lit. Namaqualand. – **St.** prostrate, 3.5 to 5 cm long, 15–20 mm thick, bluntly 4-angled, T. short, thick; **Fl.** 2–4 together, Cor. campanulate below, lobes tapering, spreading, glabrous outside, pale, minutely tuberculate inside, whitish, with blood-red spots.

**P. ruschii** NEL. – S.W.Afr.: Pockenbank. – **St.** creeping or erect, 10–18 mm long, 8–12 mm thick, ± spherical or oblong, roundly 4-angled, tuberculate-dentate, grey-green, **Fl.** 2–3 together, Cor. rotate, lobes tapering, 11 mm long, greenish-yellow inside, with numerous blackish-brown blotches, white-hairy.

**Pilea** LINDL. Urticaceae. – S.Am.; Mexico; USA: Florida. – ⊙, ± herbaceous or shrubby ♃; **St.** ± fleshy; **L.** sometimes thickish; **Fl.** in small axillary cymes, minute, unisexual or bisexual. – In general, plants of trop. or subtrop. shade-forests, some suitable for cultivation. – Greenhouse, warm. Propagation: seed, cuttings.

**P. alsinifolia** WEDD. – Colombia. – ♄; **Br.** thick, fleshy, glabrous, bearing numerous cystoliths; **L.** round-ovate, 1–2.5 cm long, 1–1.5 cm across, tip blunt or ± pointed, thick. Dioecious.

**P. buchtienii** KILLIP. – Bolivia. – **St.** creeping at first, later erect, simple, c. 20 cm tall, fleshy; **L.** broad-ovate, 8–12 cm long, 3–6 cm across, the tip pointed or caudate-tapering, doubly crenate, fleshy. Monoecious.

**P. crugeriana** WEDD. – Venezuela. – Succulent sub-♄; **St.** glabrous, densely covered with dark, linear cystoliths; **L.** with a caudate-tapering tip, crenate-serrate, 3-veined, up to 12 cm long, 3 cm across. Monoecious.

**P. cushiensis** KILLIP. – Central Peru. – Succulent ♃ 1–1.5 m tall; **L.** elliptical-ovate, 7–15 cm long, 3.5–6 cm across, tapering above, crenate-serrate. Dioecious.

**P. dombeyana** WEDD. – Central Peru. – Dwarf, glabrous, succulent; **L.** at the St.tips circular, up to 2.5 cm long and across, often with thickened margins. Monoecious or dioecious.

**P. goudotiana** WEDD. – Colombia. – Glabrous, succulent; **St.** creeping, Br. erect, up to 1.5 m tall; **L.** oblong or ovate-oblong, 5–9 cm long, 2–4.5 cm across, abruptly tapering at the tip, thick, upperside papillose. Dioecious.

**P. microphylla** (L.) LIEBM. (*Parietaria m.* L.). – Mexico; USA: S.Florida, to Peru and N.Brazil. – Glabrous, succulent, 4–30 cm tall; **L.** with the tip blunt or rather pointed and margins entire, in disparate pairs, obovate and up to 10 cm long, or circular and up to 3 cm long, the upperside with linear cystoliths in transverse stripes. Monoecious.

**P. serpyllacea** (H. B. et K.) WEDD. (Pl. 119/3) (*Urtica s.* H. B. et K., *U. thymifolia* H. B. et K., *U. globosa* PAVON MS., *P. g.* WEDD., *P. thymifolia* BLUME, *P. subcrenata* WEDD.). – Peru: W. of Cuzco. – Low ♄ 5–10 cm tall, branching basally; **Br.** fleshy, becoming woody, 2–2.5 mm thick, brownish-red, laxly leafy; **L.** rather crowded at the Br.tips, hemispherical to ± spherical, 3–4 mm long and almost as broad, 2–3 mm thick, upperside hemispherical, purplish-red, underside less rounded, pellucid; ♀ **Fl.** very small, purplish-red, ♂ Fl. c. 3.5 mm ⌀, perianth 4-partite, segments much widened at the tip, white.

**Pisosperma** SOND. Cucurbitaceae. – E. and S.Afr. – Climbing, caudiciform plants. – Greenhouse, warm. Propagation: seed, cuttings.

**P. capense** SOND. – Deserts of E. and S.Afr.: Orange Free State. – **Caudex** tuberous; **St.** erect, 2–4 cm long; **Br.** slender, 20–30 cm long, re-branching, angular, striate, roughly hairy; **L.** with short, tufted hairs, with petioles 4 to 15 mm long, blade rigid, green-grey, oblong to ovate-oblong, with linear, pointed lobes and scattered T. or further small lobes; tendrils short or missing; ♀ Fl. few or many, ♂ Fl. in Rac., Cor. 5-partite, hairy, pale yellow; **Fr.** ± spherical, with 6–12 seeds, 1.5–2 cm thick, seeds pea-shaped.

**Pistorinia** DC. Crassulaceae. – Iberian peninsula; N.Afr. – Glandular-hairy, erect ⊙ herbs; **L.** alternate, linear, ± terete, thickish, obtuse, sessile, green, soon dropping; **Fl.** numerous, corymbose, erect, short-pedicellate, ± red and yellow, Cor. tube glandular-hairy outside. – In the open. Propagation: seed.

**P. breviflora** BOISS. ssp. **breviflora** (*Cotyledon b.* (BOISS.) MAIRE, *C. cossoniana* BAILL., *P. b.* Coss. et BAL.). – S.Spain; Morocco; Algeria. – Cor. tube 12 mm long, c. 6 times longer than the Sep., red-striate, Pet. ovate, pointed, golden-yellow with reddish-brown tips and minute spots.

**P. —** ssp. **intermedia** (BOISS. et REUT.) JACOBS. (*P. i.* BOISS. et REUT., *Cotyledon b.* ssp. *i.* (BOISS. et REUT.) MAIRE). – Fl. smaller than in ssp. **salzmannii**.

**P. —** ssp. **—** v. **rubella** (BATT.) JACOBS. (*Cotyledon b.* ssp. *r.* BATT.). – Fl. red.

**P. —** ssp. **salzmannii** (BOISS.) JACOBS. (Pl. 119/4) (*P. s.* BOISS., *Cotyledon s.* SCHOENL., *C. b.* ssp. *s.* (BOISS.) MAIRE). – Cor. 12 mm long, red-striate.

**P. —** ssp. **—** v. **flaviflora** (BATT.) JACOBS. (*Cotyledon b.* ssp. *s.* v. *f.* BATT.). – Fl. yellow.

**P. —** ssp. **—** v. **rhodantha** (MAIRE) JACOBS. (*Cotyledon b.* ssp. *s.* v. *r.* MAIRE). – Fl. pink.

**P. hispanica** (L.) DC. v. **hispanica** (*Cotyledon h.* L., *Cot. pistorinia* ORT., *Sedum h.* HAMET). – Central and S.Spain; Portugal; N.Afr.: Atlas. 10–15 cm tall; Fl. 15–20 mm long, constricted below the throat, reddish or brownish, segments ovate-lanceolate, reddish-yellow, yellow in the throat.

**P. —** v. **flaviflora** (MAIRE) JACOBS. (*Cotyledon h.* v. *f.* MAIRE). – Morocco. – Fl. yellow.

**P. —** v. **maculata** (MAIRE) JACOBS. (*Cotyledon h.* v. *m.* MAIRE). – Morocco. – Fl. spotted.

**P. —** v. **purpurea** (MAIRE) JACOBS. (*Cotyledon h.* v. *p.* MAIRE). – Morocco. – Fl. purplish-red.

**Plectranthus** L'HER. Labiatae. – Trop. and sub-trop. reg. of As., Afr., Austr. – Some 120 species are known, of which a few have succulent L. or Ped. Greenhouse, warm. Propagation: seed, cuttings.

**P. fischeri** GÜRKE. – Tanzania: Kilimanjaro. – Matforming; St. erect, puberulous above; L. obovate to ± circular, passing cuneately over into the short petiole, thick-fleshy, puberulous, 15–35 mm long, 10–15 mm across, tip obtuse to apiculate, margins entire or crenate; Infl. densely spicate, terminal or axillary, Fl. pedicellate, Cal. pubescent, Pet. light blue.

**P. prostratus** GÜRKE (Pl. 120/1). – Tanzania: Kilimanjaro reg. – Mat-forming; St. somewhat pubescent, with side-shoots 3–5 cm long, rooting from the nodes; L. thick-fleshy, the blade passing over into the petiole, c. 15 mm long, 10 mm across, obtuse, margins crenate or undulating, L. on the side-shoots smaller, passing over into the bracts which are scarcely 1 mm long; Infl. a lax, hirsute Rac., Fl. violet.

**Plumeria** L. Apocynaceae. – Trop. Am. – ♄ or ♃; trunk fleshy; Br. thick, fleshy; L. usually long-petiolate, spiralled, pinnately veined; Fl. fairly large, perfumed, in terminal panicles. – Planted for decoration in the tropics. – Propagation: seed, cuttings.

**P. acuminata** AIT. (*P. acutifolia* POIR.). – Mexico; trop. As. (cultivated). – L. large, oblong; Fl. buds white outside, Fl. white, red inside.

**P. alba** L. – W.Indies. – L. small, lanceolate; Fl. white.

**P. lutea** RUIZ. et PAV. – L. large, oblong; Fl. buds red and yellow striped, Fl. whitish-yellow inside.

**P. purpurea** RUIZ. et PAV. – Peru. – Fl. red, yellow and hairy inside.

**P. rubra** L. (Pl. 120/2). – Mexico; Venezuela. – Fl. red. with a yellow throat.

× **Poellneria** ROWL. Liliaceae. – Intergeneric hybrid: **Poellnitzia** × **Gasteria**. – Cultivation as for **Gasteria**.

× **P. 1.** – **Poellnitzia rubriflora** (L. BOL.) UITEW. × **Gasteria** sp. – Habit dwarf, similar to that of **Aloe variegata**; L. acute, dark green with copious paler spots all over, 10 cm long, concave above, convex below, with thick broad bases. Hort. Kimnach.

**Poellnitzia** UITEW. Liliaceae. – S.Cape. – Resembles **Astroloba**; Fl. like **Aloe** or **Gasteria**. (Classified by C. A. PARR as **Haworthia** § *Quinquefariae*). – Greenhouse, warm. Propagation: seed, division.

**P. rubriflora** (L. BOL.) UITEW. v. **rubriflora** (Pl. 120/3) (*Apicra r.* L. BOL., *Haworthia r.* (L. BOL.) PARR). – S.Cape. – Dwarf ♃; St. elongated, 15–20 cm tall; L. in 5 dense series, ovate-triangular, tapering, terminating in a sharp point, 4 cm long, 2.5 cm across, rigid, blue-green, minutely rough, underside very convex and obliquely keeled, margins and keel minutely rough; Infl. c. 55 cm long, Fl. on pedicels 5 mm long, 2.5 cm long, dark orange-red, lobes recurved, 3–4 mm long.

**P. —** v. **jacobseniana** (V. POELLN.) UITEW. (*Apicra j.* v. POELLN., *Haworthia r.* v. *j.* (v. POELLN.) PARR. – Cape: Worcester D. – St. up to 6.5 cm long; L. projecting horizontally, up to 3.5 cm long, 1.5 cm across, upperside deeply grooved, dirty bluish-green, little tuberculate, margins and keel rough-tubercluate.

**Portulaca** L. Portulacaceae. – S.Am.; S. and E.Afr.; Austr.; Pacific Is. – ☉ or ♃ succulent plants, roots fibrous or thickened; St. erect, often woody at the base, ± branching and re-branching; L. alternate or ± alternate, the upper ones often embracing the Fl., Stip. dry, membranous or a cluster of bristles; Fl. usually terminal, solitary or in Cinc., small, not opening, or large and conspicuous, red, yellow or white, Sep. 2, variable in length, Pet. 4–6; Fr. a membranous capsule, opening with a lid, seeds many, often gibbose or tuberculate. Many of the numerous species are insignificant weeds. – In the open; ♃ species over-wintered in the cold-house. Propagation: seed, sometimes from cuttings.

**P. boniensis** TUYAMA. – Japan. – ♃; roots thick, fleshy, brown; St. thick and fleshy below, rough, branching basally, 10–15 mm thick;

**Br.** thick, ascending or creeping; **L.** crowded at the **St.** tips, 3–5 mm long, 1.5–2.5 mm thick, elliptical, upperside furrowed, acute to obtuse, with white hairs 2–3 mm long below; **Fl.** yellow.

**P. cinerea** v. POELLN. – S.Am. – ☉; **roots** slender; **St.** erect, branching, 12 cm long; **L.** linear, cylindrical, obtuse, c. 18 mm long, c. 2.2 mm across, with few, white, axillary hairs c. 3 mm long; **Fl.** terminal.

**P. collina** DTR. – S.W.Afr. – ♃ with a napiform taproot; **St.** ± segmented, 5–7 cm tall; **L.** lanceolate, acute, margins curved downwards, 8 mm long, with fishbone-like markings, the base encircled by hairs; **Fl.** 8 mm ⌀, golden-yellow.

**P. conico-operculata** v. POELLN. – Brazil. – ☉, **roots** numerous, thickened; **St.** ± creeping, c. 12 cm long, numerous, branching; **L.** numerous, linear, ± obtuse, 3–3.5 mm long, 1 mm across, the margins membranous below, with numerous axillary hairs; **Fl.** usually not opening; capsule with a conical lid.

**P. conzattii** E. J. ALEXANDER. – S.Mexico. – **St.** branching, with shaggy wool; **L.** ligulate, acute; **Fl.** 2 cm ⌀, pale yellow.

**P. cyanosperma** EGLER (*P. villosa* SKOTTSB.). – Hawaiian Is.: Kauai. – ♃ with a thickened taproot; **St.** creeping, fleshy, mat-forming, cylindrical, up to 3 mm thick, becoming red; **L.** with axillary hairs up to 7 mm long, cylindrical to subulate, c. 10 mm long, 2 mm across, 0.5 mm thick, greenish to red; **Fl.** dark pink; **seeds** bluish.

**P. decipiens** v. POELLN. – Austr.: Fanny Bay. – **St.** 9 cm tall, branching; **L.** linear, subcylindrical, c. 13 mm long, c. 0.5 mm across, axillary hairs brownish, 4 mm long; **Fl.** not known.

**P. denudata** v. POELLN. – Venezuela. – ☉, c. 13 cm tall, little branched; **L.** oblong-ovate, subpetiolate, c. 22 mm long, c. 7 mm across, obtuse, with few, grey axillary hairs; **Fl.** yellow.

**P. depauperata** v. POELLN. – Uruguay. – ☉; **St.** branching, c. 20 cm long, spreading-ascending, **root** tuberous; **L.** narrow-oblong, obtuse, margins membranous below, subpetiolate, c. 5 mm long, c. 1.5 mm across, axillary hairs few, short, white; **Fl.** solitary.

**P. diversifolia** v. POELLN. – Brazil. – ♃, c. 13 cm tall, **roots** fusiform, 8 cm long; **St.** erect to spreading, branching; **L.** flat, ovate-oblong, tip rounded, mucronate, c. 17 mm long, c. 5 mm across, axillary hairs numerous, curly, white; **Fl.** unknown.

**P. eitensis** LEGR. – Brazil. – ♃, **rootstock** thin-branched c. 5 cm long, shoots divided, 6–9 cm long; **L.** alternate, remote, linear-elongate, subpetiolate, the tip obtuse, rounded, 3–7 mm long, 1–1.5 mm wide, axillary hairs few, up to 1 mm long; **Fl.** terminal, cleistogamous.

**P. foliosa** KER. – Trop. W. and E.Afr. – Robust ♃; **St.** creeping or erect; **L.** sausage-shaped, 2 cm long, 5–6 mm across; **Fl.** 12 mm ⌀, light yellow.

**P.** cv. Grandiflora (non **P. grandiflora** HOOK.). – Long known in cultivation. – ☉; **St.** prostrate, spreading, round; **L.** cylindrical, fleshy, glabrous or hairy in the L.axils, 2 cm long; **Fl.** terminal, 2–4 cm ⌀, crimson to purplish-red with a white centre; there are also single to double cultivars with Fl. white, pink, red, yellow, orange etc.

**P.** cv. Grandiflora Flore Pleno (Pl. 120/5). – Petals numerous.

**P. hainanensis** CHUM. et How. – China: Yai Hsien (Hainan). – ♃ branching basally, spreading; **St.** 1–1.5 mm ⌀, **roots** stout and fleshy, 4–8 mm thick; **L.** thickish, obovate to linear-spatulate, 5–9 mm long, 2–4 mm across, obtuse, with long, yellow hairs in the L.axils; **Fl.** solitary, c. 5 mm ⌀, Cal. hairy.

**P. hatschbachii** LEGR. – Brazil. – **Taproot** 3–6 mm thick; **St.** numerous, simple to branched, thick, 3–4 cm long; **L.** crowded, oblong-linear, tapering, apiculate, 4–6 mm long, 1–1.5 mm across, with dense, woolly axillary hairs; **Fl.** solitary or few together in clusters, 6 mm ⌀, purple.

**P. hawaiiensis** DEGENER. – Hawaii. – Spreading ♃ with fleshy **roots**; **St.** 10–20 cm long, fleshy; **L.** ovate-oblong, 5–15 mm long, 3–4 mm across, 2 mm thick, obtuse, L.axils with flame-coloured hairs 1–3 mm long; **Fl.** 2 cm ⌀, white and red.

**P. immerso-stellulata** v. POELLN. – Argentine. – ♃ with long **roots** c. 1 cm thick; **St.** solitary to numerous, branching basally, c. 15 cm long, erect to spreading; **L.** cylindrical, linear, acute, margins membranous below, axillary hairs numerous, curly, stiff, brownish; **Fl.** terminal; **seeds** black, with a starlike pattern.

**P. insularis** HOSOKAWA. – Japan: Riu-Kiu Is. – Spreading ♃; **St.** branching, fleshy, glabrous; **L.** subsessile, obovate to circular to ovate, often thickened, 3–7 mm long, 1–3 mm across, with axillary hairs; **Fl.** solitary, red.

**P. jacobseniana** ROWL. (Pl. 120/4) (*P. poellnitziana* WERDERM. et JACOBS. non LEGRAND). – Brazil. – ☉ up to 20 cm tall, **roots** fibrous; **St.** erect, simple or branching basally, covered with white hairs; **L.** sessile, up to 18 mm long, 2 mm across, fleshy, cylindrical, with numerous white axillary hairs c. 10 mm long; **Fl.** 10 to 12 mm ⌀, lemon-yellow.

**P. longiusculo-tuberculata** v. POELLN. – Bolivia. – **Roots** tuberous; **St.** branching from the base, c. 25 cm long, suberect, c. 5 mm thick; **L.** obovate, rather obtuse or acute, c. 2 cm long, 8 mm across, axillary hairs numerous, curly, white, 8–10 mm long; **Fl.** terminal; **seeds** black, tuberculate.

**P. lutea** SOLAND. – Hawaii. – Thick, fleshy ♃ with thick **roots** up to 30 cm long; **St.** numerous, creeping, Br. becoming ± woody, very brittle; **L.** c. 1 cm long, 1.5 cm across, glabrous; **Fl.** 2 cm ⌀, light yellow.

**P. neumannii** ENGL. ex v. POELLN. – Ethiopia. – **St.** numerous, branching, spreading and creeping like irregular rays, c. 20 cm long; **L.** linear, c. 2 cm long, c. 2 mm thick, rather obtuse, with short axillary hairs; **Fl.** yellow.

**P. obtusa** v. POELLN. – Argentine. – **St.** erect, branching, c. 20 cm tall; **L.** oblong to ovate-cuneate to ovate-spatulate, short-petiolate, obtuse, c. 1 cm long, 2–2.5 mm across, flat, axillary hairs c. 6 mm long, curly, brownish; **Fl.** white.

**P. okinawensis** WALKER et TAWADA. – Japan: Okinawa Is. – Mat-forming ♃, 5 to 10 cm tall; **St.** numerous, herbaceous, green, becoming grey; **L.** at the Br.tips, subsessile, thick-fleshy, elliptical-ovate to oblong, 2 to 4 mm long, with a rounded tip; **Fl.** c. 16 mm ⌀, yellowish to reddish.

**P. oleracea** L. – Cosmopolitan. – ☉; **St.** glabrous, creeping, fleshy, with spreading Br. 10–20 cm long; **L.** 1–2 cm long, spiralled or ± opposite, lower L. ovate-oblong, fleshy, glossy, green; **Fl.** 8–12 mm ⌀, yellow.

**P.** — cv. Sativa. – Luxuriant, erect cultivar used as a pot-herb. 'Purslane'.

**P. pilosa** L. – Trop. Am.; trop. Afr. – Creeping, densely hairy; **L.** very weak, fleshy; **Fl.** small, light red to yellowish.

**P. quadrifida** L. (*P. meridiana* L. f., *P. microphylla* A. RICH.). – Trop. As.; W.Austr.; Antilles. – Spreading, prostrate; **L.** flat, elliptical-oblong, fleshy, with woolly axillary hairs; **Fl.** small, yellow.

**P. samoensis** v. POELLN. – Samoa. – ☉; **St.** branching, spreading to erect, c. 14 cm long; **L.** fleshy, oblong to ovate-oblong, rather obtuse, c. 9 mm long, 2 mm across, axillary hairs 4 mm long, brownish; **Fl.** solitary, yellow.

**P. saxifragoides** WELW. ex OLIVER. – Guinea; Angola. – ☉, sometimes ♃ from a persistent rhizome; **St.** spreading or projecting, 2–6 cm tall, densely woolly at the nodes; **L.** fleshy, ovate, flat to convex, obtuse, 2–3 mm long; **Fl.** whitish.

**P. sedoides** WELW. ex OLIVER. – Guinea; Angola. – Fleshy, erect or tangled, di- or trichotomous, 2–4 cm tall, pinkish-purple or green; **L.** fleshy, lanceolate or oval, 4–8 mm long; **Fl.** pink, in dichotomous Infl.

**P. striata** v. POELLN. – Brazil. – ♃; **St.** c. 12 cm long, with spreading Br.; **L.** linear to linear-oblong, acute or rather obtuse, c. 9 mm long, 1.5 mm across, axillary hairs numerous, curly, brownish, 9 mm long; **Fl.** pink.

**P. stuhlmannii** v. POELLN. – E.Afr. – ♃ with fleshy **roots**; **St.** branching, 12 cm long, brown, furrowed; **L.** linear, obtuse, subcylindrical, c. 15 mm long, 2 mm across; **Fl.** solitary.

**P. tuberculata** LEON. – Cuba. – Low-growing, insignificant; **St.** creeping, 4–7 cm long; **L.** oblong-ovate to ± round, 1–3.5 mm long, 0.8–1.3 mm across; **Fl.** from the L.axils, red.

**P. werdermannii** v. POELLN. – Brazil. – ♃ with a woody **taproot**; **St.** creeping or ±erect, 13 cm long, somewhat branched; **L.** linear-oblong, cylindrical, obtuse, 10 mm long, 2 mm across, with numerous, white axillary hairs; **Fl.** purple-violet.

**Portulacaria** JACQ. Portulacaceae. – S.Afr. – Succulent ♄. – Summer: in the open, winter: cold-house. Propagation: seed, cuttings.

**P. afra** JACQ. v. **afra** (*Crassula portulacaria* L., *Portulaca fruticosa* THUNBG.). – S.Afr. – Fleshy ♄; **St.** with a grey bark, Br. projecting ± horizontally, nodally segmented, cylindrical; **L.** opposite, sessile, obovate, with or without a small pointed tip, upperside flat, underside slightly convex, 12 mm long, 10–21 mm across, 2 mm thick, smooth, glossy green; **Fl.** c. 1 mm ⌀, delicate pink, inconspicuous.

**P.** — v. **foliis-variegatis** JACOBS. – L. yellow-mottled.

**P.** — v. **macrophylla** JACOBS. – L. 25 mm long, 17 mm across.

**P.** — v. **microphylla** JACOBS. – L. circular, 6 mm long and across.

**Pseudolithos** BALLY (*Lithocaulon* BALLY). Asclepiadaceae. – Somalia. – Highly succulent, unbranched, leafless plants; **St.** simple, erect, hemispherical or distinctly 4-angled, surface tessellate-tuberculate, green to grey-green; **Fl.** numerous, in umbels from hemispherical Ped., lobes of the Cor. tube ± hairy, with motile, clavate hairs at the tips. – Greenhouse, warm. Propagation: seed, or grafting on Tub. of **Ceropegia woodii**.

**P. cubiforme** (BALLY) BALLY (Pl. 121/2) (*Lithocaulon c.* BALLY). – Somalia. – **St.** 3 to 5 cm tall, 4–6 cm ⌀, with 4 blunt angles, pale green, tubercles flat, blunt, 3–6-angled, 2–3 mm high, those along the angles cordate, 4–6 mm ⌀, those at the tip crest-like; **Infl.** up to 4.3 cm ⌀, Cor. tube spherically widened, 3 mm tall, 5 mm ⌀, both surfaces glabrous, pale green outside, brownish inside, lobes 9 mm long, pointed, margins revolute, grey-green, with stiff, pink hairs inside, and 2–3 motile, clavate hairs at the tips.

**P. migiurtinorum** (CHIOV.) BALLY (*Whitesloanea m.* CHIOV., *Lithocaulon sphaericum* BALLY, *P. s.* (BALLY) BALLY). – Somalia. – **St.** simple, hemispherical, up to 65 mm ⌀, somewhat compressed above, tubercles blunt, many-angled, 2–8 mm ⌀, the pattern broken up by 4 vertical R. of large tubercles, which are often transverse-elliptical; **Ped.** hemispherical, up to 10 mm long, 7.5 mm ⌀, Cor. tube 4 mm long, 5 mm ⌀, cup-shaped, glabrous, both surfaces pale green, lobes spreading, 3 mm long, triangular, glabrous outside, hairy below, purplish-brown, margins recurved, bristly-haired, tips with a cluster of motile hairs 2.8 mm long.

**Pseudopectinaria** LAVR. Asclepiadaceae. – Somalia. – Succulent plants; **St.** creeping, with laterally compressed angles; **L.** rudimentary, soon dropping; **Fl.** fleshy, Cor. saccate, the tube constricted at the opening, lobes curved inwards, touching one another laterally, with the tips united; outer corona cup-shaped. – Culture as for **Stapelia**.

**P. malum** LAVR. (Pl. 120/6). – Somalia: N. reg. – **St.** 3–15 cm long, 8–10 mm ⌀, creeping, rooting throughout its length, indistinctly long-tessellate; **L.** tapering, 1 mm long, soon dropping; **Fl.** solitary from the upperside of the St., pedical 15 mm long, Sep. linear, short, Cor. broad ellipsoidal, 14–16 mm long, 13–15 mm ⌀, minutely and densely papillose outside, deep purplish-red inside, the spherical section with white hairs 1.5 mm long at the base and at the tip, the tube-mouth star-shaped, constricted to c. 5 mm, incurved, lobes triangular, 3–4 mm long, touching above, tapered inwards, glabrous inside; outer corona cup-shaped, 4–5 mm long, c. 4 mm ⌀, golden-yellow.

**Pseudosedum** (BOISS.) BGR. (*Pseudosedum* BOISS. as Sect. of **Umbilicus** DC.). Crassulaceae. – Central As. – Sedum-like ♃; **rootstock** stout, buried obliquely in the ground or ± creeping; **St.** dying back annually, 10–30 cm tall, rigidly erect; **L.** densely alternate, ± linear-cylindrical, broader at the base, with an obtuse or lobed spur, soon dropping; **Infl.** a crowded, much-forked, flat cyme, Fl. small. – In the open, frost-resistant. Propagation: seed, division.

**P. affine** (SCHRANK.) BGR. (*Umbilicus a.* SCHRANK., *Cotyledon a.* (SCHRANK.) MAX., *Sedum a.* (SCHRANK.) HAMET, *S. alberti* REGL.). – USSR: Central As.; China: Sinkiang. – **Fl.** 6 mm long, white.

**P. kuramense** BORISS. – Central As. – **Rootstock** 1–3 or many, clustered, c. 10 cm long, thickened, not tuberous; flowering St. 4–12, erect, densely leafy, 10–20 cm tall; **L.** numerous, alternate, linear, 5–10 mm long, c. 1 mm across, acute; **Infl.** corymbose-capitate, 2.5–4.5 cm tall, Fl. numerous, dense, short-pedicellate, Cor. 8 to 10 mm long, pink to violet (?).

**P. lievenii** (LEDEB.) BGR. (*Umbilicus l.* LEDEB., *Cotyledon l.* (LEDEB.) MAX., *Sedum l.* (LEDEB.) HAMET). – Central As. – **St.** 30 cm tall; **Infl.** a cyme 3–6 cm ⌀, Fl. over 12 mm long, ± pink.

**P. longidentatum** BORISS. – C.As. – Resembling **P. kuramense**; ♃; **roots** tuberous; St. 1–3, 25–40 cm high; **L.** with long T.

**Pterodiscus** HOOK. Pedaliaceae. – Trop. E.Afr.; Cape; S.W.Afr.; Angola. – Small ♃ herb or more usually small ♄, seldom over 30 cm tall, often semi-succulent, with a swollen **caudex** and tuberous roots; **St.** solitary or several, simple or branched; **L.** variable in shape, margins entire, undulating, dentate or laciniate; **Fl.** solitary from the L.axils, variously coloured, Cal. small, Cor. tube funnel-shaped, often slightly tuberculate below, the margin spreading, somewhat 2-lipped, lobes unequal, ovate, circular or transverse-elliptical. – Greenhouse, warm. Summer: dry during resting period. Propagation: seed.

**P. angustifolius** ENGL. (*Pedaliophytum busseanus* ENGL., *Pedalium b.* (ENGL.) STAPF). – Tanzania: Mwanza, Shinyanga D. – Branching basally; **Br.** spreading, ± fleshy, 9–20 cm long, purple, glabrous; **L.** dense, ± fleshy, with a petiole up to 2.5 cm long, oblong-lanceolate, 2.5–13 cm long, 6–12 cm across, slightly glandular at first, tip obtuse or somewhat pointed, dark green, margins entire or undulating, rarely dentate towards the tip; **Fl.** yellow or orange, often with purple blotches in the tube, lobes ciliate, hairy inside.

**P. aurantiacus** WELW. – Angola; S.W.Afr.: Gr. Namaqualand. – **Caudex** bottle-shaped, up to a foot high, with several thick Br. from the apex; **L.** oblong-lanceolate or ovate-spatulate, margins emarginate-sinuate, smooth, bluish; **Fl.** brilliant red.

**P. coeruleus** CHIOV. – Somalia; Kenya. – St. simple or branched, 5–20 cm tall; **L.** with a petiole 4–25 mm long, cuneate at the base, 13–40 mm long, 4–15 mm across, margins undulating, underside sparsely glandular; **Fl.** white, or white suffused mauve in the tube-throat, lobes with red veins.

**P. kelleranus** SCHINZ (*P. heterophyllus* STAPF). – Somalia. – **Caudex** fleshy, water-storing, edible; basal **L.** ± elliptical, with undulating margins, upper ones very narrowly lanceolate, usually with margins entire or somewhat incised, sometimes distinctly pinnatisect.

**P. luridus** HOOK. – Cape: Kalahari, W. Griqualand; S.W.Afr.: Gr. Namaqualand. – **Caudex** conical below, fleshy, c. 50 cm tall, 7–8 cm thick below, bark smooth, grey; **Br.** spreading, 15–20 cm long, pruinose; **L.** ± numerous, oblong, spatulate below, ± laciniate above, 7–8 cm long, 2.5 cm across, upperside dark green, underside whitish or bluish, lobes linear, pruinose all over; **Fl.** yellow, dotted with red outside.

**P. ruspolii** ENGL. (Pl. 121/1) (*P. somaliensis* (BAK.) STAPF, *P. welbyi* STAPF). – Kenya: Is. in L. Rudolf, Meru and Turkana D.; Ethiopia; Sudan; Somalia. – **Caudex** thick below, fleshy, 4–8 cm long, 0.5–2 cm thick; **St.** c. 20, erect, 4–20 cm long; **L.** usually with a petiole 0.5 to 3.5 cm long, obovate to elliptical, 1.5–6.5 cm long, 8–35 mm across, rounded at the tip, underside glandular, margins entire or undulating; **Fl.** light yellow to orange, often with red or purple blotches in the centre, lobes often ciliate.

**P. speciosus** HOOK. (Pl. 89/6) (*Harpagophytum pinnatifidum* ENGL.). – Cape: W.Griqualand, Transvaal. – **Caudex** spherical below, c. 15 cm tall, 6 cm ⌀; **L.** ± numerous, linear to linear-oblong, irregularly dentate or shortly incised, 3–6 cm long, 5–10 mm across; **Fl.** light purplish-red.

**Pteronia** L. Compositae. – S.Afr. – Slightly succulent, very large ♄ resembling **Senecio**, not found in cultivation. – Propagation: seed, cuttings.

**P. scariosa** L. – Cape; S.W.Afr.: Gr. Namaqualand. – Stout, branching ♄; **L.** ovate-oblong, pointed, fleshy, 3–5 mm long, 2–3 mm across, narrowing towards the base; **Infl.** terminal, with many yellow Fl.

**P. succulenta** THUNBG. (Pl. 121/4) (*P. carnosa* MUSCHLER). – Cape: Karroo; S.W.Afr. – Smooth, branching ♄; **L.** opposite, linear-triangular, tapering, fleshy, c. 2 cm long, spreading or recurved, margins somewhat thickened; **Infl.** of many yellow Fl. heads, with a strong scent of aniseed.

**Pyrenacantha** WIGHT. Icacinaceae. – Trop. Afr. – Caudiciform plants with a thickened St. and climbing shoots.

**P. malvifolia** ENGL. (Pl. 121/3). – Tanzania: Kilimanjaro. – **Caudex** irregular, swollen, 1 m tall and $\varnothing$, looking like a smooth rock, bark light leather-coloured; during the growing season producing twining shoots up to 2 cm thick; **L.** mallow-like.

**P. ruspolii** ENGL. – Somalia. – Resembles **P. malvifolia**, possibly identical.

**P. vitifolia** ENGL. – Kenya. – **Caudex** large, semi-underground; **Br.** not very fleshy, semi-climbing, bark leathery, green; **L.** rounded, deeply lobed, deciduous; **Fl.** small, in a terminal spike; **Fr.** rhomboid, orange-coloured, edible. Dioecious.

**Raphionacme** HARV. Asclepiadaceae. – Eastern S.Afr.; trop. W.Afr. – **Roots** clustered, fleshy, fusiform or plant tuberous, developing short, branched or simple shoots, caudices variable in shape and size, weighing up to 2.8 kg (**R. hirsuta**), usually flattened above or $\pm$ bottle-shaped, with a milky sap; **Br.** thin, simple, more rarely twining; **L.** narrow, lanceolate or ovate, grey-green, often purple; **Fl.** small, green, violet or purple, in cymes. – Greenhouse, warm. Propagation: seed.

**R. brownii** SC. ELLIOT. – Ghana; Nigeria; Guinea; Sierra Leone. – **Caudex** fleshy, tuberous; **shoots** up to 30 cm tall; **L.** linear, up to 20 times as long as broad; **Fl.** in lax terminal or lateral cymes, with pedicels 8–14 mm long, Cor. tube 3 mm long, lobes 5–7 mm long, oblong, spreading or reflexed, pink; corona lobes erect, filiform, 5–6 mm long.

**R. daronii** BERHAUT. – Senegal; Ghana. – **Caudex** tuberous, 5–10 cm long, 3–7 cm thick; **shoots** herbaceous, glabrous; **L.** alternate, elliptic to lanceolate, 8–12 cm long, 1–3 cm across, petiole 2–6 mm long; **Fl.** 2–3 together in the axils, pedicellate, Cor. 15–16 mm $\varnothing$, violet.

**R. galpinii** SCHLTR. – Transvaal. – **Caudex** 8 cm $\varnothing$, 14 cm long, with a neck 3 cm long, 1.5 cm $\varnothing$; **shoots** 7 cm long; **L.** lanceolate, 2–4 cm long, hairy; **Infl.** a cyme of many greenish Fl. 1 cm $\varnothing$.

**R. hirsuta** (E. MEY.) DYER (*Brachystelma h.* E. MEY., *R. divaricata* HARV.). – Transvaal, Orange Free State, Natal, E. Cape. – Resembles **R. galpinii**, more branching; **Fl.** purple.

**R. keayi** BULLOCK. – Nigeria. – **Shoots** simple, numerous, erect, up to 1 m tall, from a woody caudex; **L.** short-petiolate, oblong-oblanceolate, up to 12 cm long and 2 cm across, apex and base acute, glaucous below and crispate-pubescent on both surfaces; **Fl.** numerous, in lateral cymes, pedicels 4–6 mm long; Cor. tube 2 mm long, lobes triangular, 5 to 6 mm long, spreading, green with a deep red zone at the base, minutely papillate inside; corona lobes rectangular at the base, with a flexuose, whip-like, apical appendage, intertwined and connivent over the style and stamens.

**P. vignei** BRUCE. – Ghana. – **Shoots** simple, erect, 6–16 cm tall, puberulous, from a fleshy, disciform, edible **caudex**; **L.** obovate-oblanceolate, the uppermost largest, 4–10 cm long, 1 to 3 cm across, glabrescent above, scabrid to puberulous beneath, apex rounded, cuneate at the base and shortly petiolate; **Fl.** in branched terminal cymes, on pedicels 1–2 mm long, Cor. lobes ovate-lanceolate, c. 3.5 mm long, the tube cylindric to campanulate, 2 mm long, pale green; corona lobes filiform, 5 mm long.

**Rhodiola** L. (*Sedum* L. Sect. 1, *Rhodiola* (L.) SCOP.; *Sedum* L. Sect. 2, *Pseudorhodiola* DIELS; *Sedum* Sect. 17. *Seda Genuina* KOCH, Series 4, *Stapfiana* BGR. p.part.; *Chamaerhodiola* (FISCH. et MEY.) NAKAI as a genus). Crassulaceae. – N. temp. reg. to Arctic. – Dwarf $\hbar$ with branched St., with crowded L. in Ros., or herbs with a perennial rootstock and a tuft of green L., or set with dry scale L.; Fl.-Br. from the axils of the Ros.-L. or scales, unbranched, dying off yearly; **L.** somewhat succulent, alternate, flat, dentate, crenate or entire; **Infl.** densely crowded, flat or semiglobose cymose, Fl. in some species ⚥, but mostly unisexual, 4(–5)-partite, Sep. free or shortly united, Pet. narrow and rotate or $\pm$ erect, campanulate. Hardy. Propagation: seeds, division. Some of the following species are only subspecies or geographical races.

## Division of the Genus Rhodiola into Sections acc. Fu Shu-Hsia

**Sect. 1. Primuloides** (PRAEG.) FU (*Sedum* L. Sect. 1. *Rhodiola* (L.) SCOP. Ser. II. *Primuloides* PRAEG.; *Sedum* L. Sect. 1. *Rhodiola* (L.) SCOP. SSect. II. *Primuloides* (PRAEG.) BGR.; *Sedum* L. Sect. 1. *Rhodiola* (L.) SCOP. Gr. 1. *Primuloides* (PRAEG.) FROED.). – Rootstock with large Ros. of true **L.**, these with thick bases narrowed above into petioles; **flowering Br.** from the axils of the often already fallen L. – Type-species: R. primuloides. – China, Afghanistan. – Species: R. balfourii, durisii, hobsonii, humilis, karpelesae, modesta, nuristanica, petiolata, pleurogynantha, primuloides, sangpo-tibetiana, tuberosa.

**Sect. 2. Chamaerhodiola** FISCH. et MEY. (*Sedum* L. Sect. 1. *Rhodiola* (L.) SCOP. Ser. 1. *Rhodiolae* PRAEG., Gr. 2. *Himalense* PRAEG.; *Sedum* L. Sect. 1. *Rhodiola* (L.) SCOP. Ser. 1. *Rhodiolae* PRAEG. Gr. 1. *Rosae* PRAEG.; *Sedum* L. Sect. 1. *Rhodiola* (L.) SCOP. Ser. 2. *Crassipedes* PRAEG. p.min.part.; *Sedum* L. Sect. 1. *Rhodiola* (L.) SCOP. Gr. 2. *Chamaerhodiola* (FISCH. et MEY.) FROED.; *Chamaerhodiola* (FISCH. et MEY.) NAKAI as a genus). – **Rootstock** thick, set with the remains of old scale-**L.**, these at least initially green and fleshy, occasionally with a narrow, pointed lamina; **flowering Br.** slender, ultimately withering and remaining. – Type-species: R. quadrifida. –

**Ser. 1. Dumulosae** Fu (**Sedum** L. Sect. 1. *Rhodiola* (L.) Scop. Gr. 2. *Chamaerhodiola* Fisch. et Mey. Sub-Gr. 1. *Dumulosum* Froed. nom. nud.). – **Rootstock** short, branched; **St.** many, current season's St. 15 cm long; **L.** linear to elongate linear; **Infl.** many-flowered, compact; **Fl.** large. – Type-species: R. dumulosa. – China. – No further species.

**Ser. 2. Quadrifidae** (Froed.) Fu (**Sedum** L. Sect. 1. *Rhodiola* (L.) Scop. Gr. 2. *Chamaerhodiola* (Fisch. et Mey.) Froed. Sub-Gr. 2. *Quadrifidum* Froed.). – **Rootstock** of older plants short, thick, branched; older **St.** remaining, many (often more than 100), graceful, surrounded by flowering St.; annual **flowering St.** short, to 10 cm long; **L.** small, elongate to elongate linear; **Infl.** few-flowered; **Fl.** small. – Type-species: R. quadrifida. – China. – Species: R. brevipetiolata, juparensis, likiangensis, quadrifida.

**Ser. 3. Fastigiatae** (Froed.) Fu (**Sedum** L. Sect. 1. *Rhodiola* (L.) Scop. Gr. 2. *Chamaerhodiola* (Fisch. et Mey.) Sub-Gr. 3. *Fastigiata* Froed.). – The overground **rootstock** elongated up to 20 cm; **St.** many to few; **flowering St.** fleshy, erect; **L.** lanceolate; **Infl.** compact, many-flowered, cymose; **Fl.** large. – Type-species: R. fastigiata. – Nepal, USSR, China, Korea. – Species: R. algida, alsia, concinna, fastigiata, himalensis, recticaulis, tachoensis, tibetica, venusta.

**Ser. 4. Megalantha** (Froed.) Fu (**Sedum** L. Sect. 1. *Rhodiola* (L.) Scop. Gr. 2. *Chamaerhodiola* Fisch. et Mey. Sub-Gr. 4. *Megalacanthum* Froed.; **Sedum** L. Sect. 1. *Rhodiola* (L.) Scop. Gr. 1. *Primuloides* (Praeg.) Froed. p.min.part.; **Sedum** L. Sect. 17, *Seda Genuina* Koch. Row 4. *Stapfiana* Bgr. p.part.). – The overground **rootstock** short; **St.** few, graceful; **flowering St.** fleshy, erect; **L.** large, elongate to round; **Infl.** cymose, compact; **Fl.** large or small. – Type-species: R. rotundata. – E.China. – Species: R. euryphylla, megalophylla, rotundata, smithii, stapfii.

**Sect. 3. Rhodiola** (**Sedum** L. Sect. 1. *Rhodiola* (L.) Scop. Gr. 3. *Eurhodiola* Fisch. et Mey.; **Sedum** L. Sect. 1. *Rhodiola* (L.) Scop. Ser. 1. *Rhodiolae* Praeg. Gr. 1. *Roseae* Praeg.; **Sedum** L. Sect. 1. *Rhodiola* (L). Scop. SSect. 1. *Eurhodiola* (Praeg.) Bgr. Ser. *Rosea* (Praeg.) Bgr.). – **Rootstock** set with brown or greenish **scale-L.** – Type-species: R. rosea.

**Ser. 1. Roseae** (Praeg.) Fu (**Sedum** L. Sect. 1. *Rhodiola* (L.) Scop. Gr. 3. *Eurhodiola* Fisch. et Mey.; **Sedum** L. Sect. 1. *Rhodiola* (L.) Scop. Ser. 1. *Rhodiolae* Praeg. Gr. 1. *Roseae* Praeg.). – **Carpels** lanceolate, thickened at the base, 3 times longer than wide. – Type-species: R. rosea. – N. temp. and arct. reg. – Species: R. heterodonta, irmelica, kirilowii, komarovii, longicaulis, robusta, rosea, sexifolia, subopposita.

**Ser. 2. Bupleuroides** Fu (**Sedum** L. Sect. 1. *Rhodiola* (L.) Scop. Gr. 3. *Eurhodiola* Fisch. et Mey. Sub-Gr. 2. *Bupleuroides* Froed. nom. nud.). – **Carpels** short to elongate, thick at base, twice as long as wide. – Type-species: R. bupleuroides. – Nepal, Sikkim, China. – Species: R. bhutanensis, bupleuroides, discolor, hookeri, purpureo-viridis.

**Ser. 3. Crassipes** Fu (**Sedum** L. Sect. 1. *Rhodiola* (L.) Scop. Gr. 1. *Crassipes* Froed. nom. nud.). – **Carpels** narrow-ovate to lanceolate, substipulate at the base. – Type-species: R. crassipes. – Pakistan, Nepal, Sikkim, China, USSR: Siberia, Japan. – Species: R. crassipes, eurycarpa, kansuensis, macrocarpa, macrolepis, semenovii.

**Ser. 4. Yunnanenses** Fu (**Sedum** L. Sect. 1. *Rhodiola* (L.) Scop. Gr. 4. *Yunnanensis* Froed. nom. nud.; **Sedum** L. Sect. 2. *Pseudorhodiola* Diels). – **Carpels** erect at maturity, the tip suddenly recurved, the style then spreading. – Type-species: R. yunnanensis. – S.W. and Central China. – Species: R. papillocarpa, rotundifolia, sinica, yunnanensis.

**Sect. 4. Trifida** (Froed.) Fu (**Sedum** L. Sect. 1. *Rhodiola* (L.) Scop. Gr. 4. *Trifida* Froed.). – **Rootstock** as in Sect. *Rhodiola*, rather short and thick; annual **St.** several, erect or spreading, thin, glabrous; **L.** alternate, usually crowded, mostly verticillate, upper ones petiolate, lamina dentate or deeply lobed; **Infl.** a lax cyme; **Fl.** large, usually 5-partite, Pet. purple-red or white, 6–12 mm long. – Type-species: R. linearifolia. – China, N.Am. – Species: R. chrysanthemifolia, dielsiana, liciae, linearifolia, ovatisepala, rhodantha, sacra, sinuata, tieghemii, trifida, tsuiana.

**R. algida** (Ledeb.) Fisch. et Mey. (§ 2/3) (*Sedum a.* Ledeb.). – USSR: E. Pamir, Altai; N.China; Mongolia. – **Taproot** fleshy; **St.** over 15 cm tall; **L.** flat, ± linear, 8–12 mm long, 2 mm across, margins entire.

**R.** — v. **algida**. – Plants dioecious.

**R.** — v. **altaica** (Max.) Jacobs. (*Sedum a.* v. *a.* Max.). – USSR: Altai. – **Fl.** ⚥.

**R.** — v. **jeniseensis** (Max.) Jacobs. (*Sedum a.* v. *j.* Max.). – **L.** broader than in v. **algida**; dioecious.

**R.** — v. **tangutica** (Max.) Jacobs. (*Sedum a.* v. *t.* Max.). – E. Mongolia; N.China. – **L.** narrow-linear; dioecious.

**R. alsia** (Froed.) Fu (§ 2/2) (*Sedum a.* Froed.). – China. – Dioecious; **caudex** thick, underground, with a short section above ground, simple to slightly branching; sterile **shoots** few, 10–17 cm long, flowering shoots 16–19 cm long; **L.** long-lanceolate, with a few T. at the tip, 8–15 mm long; **Infl.** of many yellow and red Fl. c. 1 cm long.

**R. balfourii** (HAMET) FU (§ 1) (*Sedum b.* HAMET, *S. mossii* HAMET, *S. orichalcum* W. W. SM., *S. balanense* LIMPR.). – N.China. – **Caudex** short, basal **L.** oblong-lanceolate, rather pointed; cauline L. oblong, acute; **Infl.** forking, Fl. yellow.

**R. bhutanensis** (PRAEG.) FU (§ 3/2) (*Sedum b.* PRAEG., *S. cooperi* PRAEG.). – Himalayas; China: Yunnan. – **Caudex** large; **shoots** erect, 30–60 cm tall, up to 3 mm thick; **L.** sessile or short-petiolate, ovate to elliptical, slightly dentate, 35 mm long, 18 mm across, dark green, underside whitish; **Infl.** lax, Fl. purple.

**R. brevipetiolata** (FROED.) FU (§ 2/2) (*Sedum b.* FROED.). – China. – Dioecious; **caudex** thickened, freely branching from the aerial portion; sterile **shoots** numerous, c. 3 cm long, flowering shoots 3–5 cm long; **L.** ovate, tapering, 6–7 mm long; **Infl.** of few Fl., ♂ Fl. yellow, ♀ ones pale.

**R. bupleuroides** (WALL. ex HOOK. f. et THOMS.) FU (§ 3/2) (*Sedum b.* WALL. ex HOOK. f. et THOMS.). – Himalayas. – **Caudex** large; **shoots** erect, slender, 20–30 cm tall; **L.** cordate to ovate-lanceolate, acute, 12–25 mm long, 6 to 12 mm across, green, often red-spotted; **Infl.** flat, lax, Fl. dark purple.

**R. chrysanthemifolia** (LÉVL.) FU (§ 4) (*Sedum c.* LÉVL., *S. trifidum* v. *balfourii* HAMET, *S. linearifolium* v. *forrestii* HAMET, *S. l.* v. *balfourii* HAMET). – W.China. – Possibly a variety of **R. linearifolia; L.** crowded towards the tips of the shoots, ovate-oblong to circular, ± deeply lobed.

**R. concinna** (PRAEG.) FU (§ 2/3) (*Sedum c.* PRAEG.). – China: Yunnan. – **Caudex** fleshy, scaly; **L.** ± lanceolate; **Infl.** 2–3 cm long.

**R. crassipes** (HOOK. f. et THOMS.) BORISS. (Pl. 122/1) (§ 3)3 (*Sedum c.* HOOK. f. et THOMS., *S. c.* WALL. nom. nud., *S. asiaticum* WALL. p. part. et C. B. CLARKE, *S. wallichianum* HOOK., *R. w.* (HOOK.) FU). – W.Himalayas to W. China: Tibet. – **Caudex** often creeping, underground, with short offsets aboveground bearing sterile shoots; ⊙ shoots usually numerous, clustered, up to 30 cm long, papillose above; **L.** dentate, oblanceolate, ± obtuse; **Infl.** with bracts, Fl. usually bisexual, Pet. pale yellow.

**R. — v. cholaensis** (PRAEG.) JACOBS. (*Sedum c.* v. *c.* PRAEG., *S. crassipes* AUCT., *R. wallichiana* v. *cholaensis* PRAEG., *S. crassipes* AUCT., *R. wallichiana* v. *cholaensis* (PRAEG.) FU). – E. As.: E. Sikkim. – Larger in all parts than v. **crassipes,** greyer; L. 25–35 mm long, 5–6 mm across.

**R. — v. crassipes.** – Fl. ⚥.

**R. — v. cretinii** (HAMET) JACOBS. (*Sedum c.* HAMET, *S. c.* v. *c.* (HAMET) FROED.). – Sikkim. – Underground **caudex** very thin, creeping; ⊙ **shoots** few, short (2–10 cm), erect; **L.** oblong, small, margins ± entire; **Fl.** unisexual.

**R. — v. stephanii** (CHAM.) JACOBS. (*Sedum s.* CHAM., *S. c.* v. *s.* (CHAM.) FROED.). – Central As. – **Caudex** fleshy, without offsets; ⊙ **shoots** glabrous; **L.** broadly oblanceolate, deeply dentate; **Fl.** usually unisexual.

**R. dielsiana** (LIMPR. f.) FU (§ 4) (*Sedum d.* LIMPR. f., *S. linearifolium* v. *d.* (LIMPR. f.) HAMET). – China: Szechuan. – Possibly a variety of **R. linearifolia; L.** in crowded whorls at the tips of the shoots, deeply divided, segments narrow-linear, margins entire or incised.

**R. discolor** (FRANCH.) FU (§ 3/2) (*Sedum d.* FRANCH., *S. bupleuroides* v. *d.* (FRANCH.) FROED.). – Sikkim; China: Yunnan. – Perhaps a variety of **R. bupleuroides; caudex** fleshy, slender, creeping, underground; **L.** obovate; **Infl.** of few Fl.

**R. dumulosa** (FRANCH.) FU f. **dumulosa** (§ 2/1) (*Sedum d.* FRANCH., *S. rariflorum* N. E. BR.). – N.China. – **Caudex** thick, without underground offsets; **shoots** 10–15 cm tall; **L.** alternate, linear, acute, 12–25 mm long, 2–3 mm across, fleshy, with entire margins; **Infl.** very dense, of 6–12 campanulate, white Fl.

**R. — f. farreri** (W. W. SMITH) JACOBS. (*Sedum f.* W. W. SMITH). – E.As. – More robust.

**R. durisii** (HAMET) FU (§ 1) (*Sedum d.* HAMET). – Central As. – **Caudex** short, thick; basal **L.** hairy, broad; **Infl.** of several Fl. – Close to **R. barnesiana,** possibly only a variety.

**R. eurycarpa** (FROED.) FU (§ 3/3) (*Sedum e.* FROED., *S. progressum* DIELS). – China: N. Szechuan, Sung-p'an. – Dioecious; **caudex** robust to slender, erect to ± creeping, offsets 3.5 to 13 cm long, densely leafy; flowering shoots 2–7, slender, rough above, c. 24 cm long; cauline **L.** alternate, linear-lanceolate to narrow-spatulate, obtuse, 5–20 mm long, 1–3 mm across, margins dentate midway, greenish-yellow; **Infl.** dense, of 5–20 greenish-yellow Fl.

**R. euryphylla** (FROED.) FU (§ 2/4) (*Sedum e.* FROED.). – China: N.W.Yunnan. – **Caudex** short, above ground, robust, covered with scale-L.; **shoots** erect, 9–15 cm long, flowering shoots rather stout, 14–17 cm tall; cauline **L.** imbricate, obovate to round, margins entire or minutely crenate, tip obtuse to apiculate, 17–26 mm long, 12–18 mm across; **Infl.** a dense, broad corymb, Fl. pink.

**R. fastigiata** (HOOK. f. et THOMS.) FU (§ 2/3) (*Sedum f.* HOOK. f. et THOMS., *S. taliksiense* FROED., *S. doratocarpum* FROED.). – Himalayas; China: Tibet, W.Yunnan. – **Caudex** fleshy, thick, elongated; **shoots** erect, 7–15 cm tall; **L.** narrowly linear-lanceolate, fleshy, dark green or grey; **Infl.** of few, white to purple or pale yellow, unisexual or ⚥ Fl.

**R. — v. fastigiata.** – **L.** narrowly linear-lanceolate, glossy dark green.

**R. — v. gelida** (LEDEB.) JACOBS. (*Sedum g.* LEDEB.). – **L.** often dentate; **Fl.** bisexual.

**R. heterodonta** (HOOK. f. et THOMS.) BORISS. (§ 3/1) (*Sedum h.* HOOK. f. et THOMS.). – W.Himalayas; China: Tibet; Afghanistan. – **Caudex** elongated; **shoots** erect, 30–45 cm tall; **L.** triangular to ovate, 12–18 mm long, 12 mm across, acutely dentate, green or grey-green; **Infl.** dense, Fl. yellowish or reddish.

**R. himalensis** (PRAEG.) FU (§ 2/3) (*Sedum h.* PRAEG.). – Himalayas to China. – **Caudex** fleshy, thickened and elongated; **shoots** rough, 15 to

30 cm tall, reddish; **L.** ovate-oblong to ovate-lanceolate, 15–25 mm long, 5–8 mm across, ± acute, margins entire, ± papillose, dark green; **Infl.** of few, dark red or yellow Fl.

**R. — v. bouvieri** (HAMET) JACOBS. (*Sedum b.* HAMET). – Nepal. – **Shoots** and **L.** covered with long papillae.

**R. — v. himalensis.** – **L.** minutely papillose.

**R. — v. ishidae** (MAKINO) JACOBS. (*Sedum i.* MAKINO). – **L.** light green; **Fl.** yellow.

**R. hobsonii** (HAMET) FU (§ 1) (*Sedum h.* HAMET, *S. praegerianum* W. W. SMITH). – China: Tibet. – **Caudex** very short, thick; basal **L.** linear-oblong to rhombic-oblong; **Infl.** forked, 8–15 cm long, Fl. pink.

**R. hookeri** FU (§ 3/2) (*Sedum elongatum* WALL. nom. nud.). – Himalayas. – **Caudex** fleshy, thick, branching and projecting above the soil; **L.** short-petiolate, obovate, obtuse, 5 cm long, 2–2.5 cm across, green, underside pale; **Infl.** large, slightly hairy, Fl. almost 1 cm ⌀, purplish-red.

**R. humilis** (HOOK. f. et THOMS.) FU (§ 1) (*Sedum h.* HOOK. f. et THOMS., *S. levii* HAMET, *S. barnesianum* PRAEG., ? *S. prainii* PRAEG.). – Sikkim; China: Tibet. – **Caudex** short, thick; basal **L.** long-petiolate, oblanceolate, with entire margins, ± acute, ☉ **shoots** 2–3 cm long, with oblong, ± acute L.; **Fl.** solitary.

**R. irmelica** BORISS. (§ 3/1). – USSR: Urals. – Probably only an ecotype of **R. rosea.**

**R. juparensis** (FROED.) FU (§ 2/2) (*Sedum j.* FROED.). – China: E.Tibet. – **Caudex** very fleshy, up to 25 cm long, the aerial part divided, c. 1 cm thick, with triangular, obtuse scale-L. and the dry remains of numerous shoots; sterile **shoots** 1.5–3 cm long, flowering **shoots** numerous, spreading like a fan, 2–4 cm long; **L.** linear-lanceolate, with entire margins, bristle-tipped, 3–4 mm long; **Infl.** compact, of few Fl. with yellow Pet.

**R. kansuensis** (FROED.) FU (§ 3/3) (*Sedum semenovii* v. *k.* FROED.). – China. – Close to **R. semenovii**; sterile **shoots** 2–3 cm long, densely leafy, flowering shoots 7–8 cm long; **L.** linear, up to 10 mm long; **Infl.** a dense corymb of few Fl.

**R. karpelesae** (HAMET) FU (§ 1) (*Sedum k.* HAMET). – China: Tibet. – **Caudex** short, thick; cauline **L.** ± verticillate; **Fl.** solitary.

**R. kirilowii** (RGL.) FU (§ 3/1) (*Sedum k.* RGL.). – USSR: Central As.; Himalayas; N.W. China; Mongolia. – **Caudex** fleshy, elongated; **shoots** erect, unbranched, 30 cm or more tall, glabrous; **L.** ± lanceolate to linear, acute, 25–40 mm long, 6–15 mm across, margins rarely entire; **Infl.** dense, Fl. greenish-yellow to orange or red.

**R. — v. aurantiacum** (RGL.) JACOBS. (*Sedum k.* v. *a.* RGL.). – Fl. orange.

**R. — v. kirilowii.** – Fl. greenish-yellow.

**R. — v. latifolia** FU. – **L.** 12–15 mm across.

**R. — v. rubra** (PRAEG.) JACOBS. (*Sedum k.* v. *r.* PRAEG.). – Robust variety; Fl. brownish-red.

**R. komarovii** BORISS. (§ 3/1) (*Sedum k.* (BORISS.) CHU, *S. polytrichoides* KOM.). – E.USSR: Ussuri reg. – **Roots** thin; **caudex** long, thin, 5–7 mm thick, scale-L. imbricate, 13 mm long, 2–3 mm across; **shoots** 3–4, glabrous, 2–10 cm long; **L.** dense, linear, ± cylindrical, fleshy, margins usually entire, occasionally with 1–2 T., rather blunt above, 1–2 cm long, 1–2 mm across, green; **Infl.** with pale yellow Fl., solitary, or up to 5 in a corymb.

**R. liciae** (HAMET) FU (§ 4) (*Sedum l.* HAMET). – China: Yunnan. – Close to **R. linearifolia**; **L.** alternate, long-petiolate, circular, slightly dentate, very obtuse; **Infl.** fairly dense, Pet. very long-apiculate, white, 8–11 mm long.

**R. likiangensis** (FROED.) FU (§ 2/2) (*Sedum l.* FROED.). – China: Yunnan (Likiang). – **Roots** thin, c. 10 cm long; aerial **caudex** divided, c. 1 cm thick, covered with the remains of dry shoots; sterile **shoots** spreading, fan-like, 2 to 3 cm long, flowering shoots 2–3.5 cm long; **L.** linear-lanceolate, with entire margins, tapering, 4–7 mm long; **Infl.** with reddish Fl., solitary or few.

**R. linearifolia** (ROYLE) FU (§ 4) (*Sedum l.* ROYLE, *S. mucronatum* EDGEW., *S. paucifolium* EDGEW.). – ♃; **L.** usually only crowded towards the top of the shoots, often ± verticillate, 4–9 cm long, with entire margins, very variable in shape; **Fl.** large, red.

**R. longicaulis** (PRAEG.) FU (§ 3/1) (*Sedum l.* PRAEG., *S. kirilowii* v. *altum* FROED.). – E. and Central As. – **Caudex** elongated, fleshy; **shoots** 60–90 cm tall, erect; **L.** narrow-linear, 5–9 cm long, 6 mm across, dark green; **Infl.** dense, Fl. greenish.

**R. macrocarpa** (PRAEG.) FU (§ 3/3) (*Sedum m.* PRAEG.). – China. – **Caudex** fleshy; **L.** ± verticillate, ± lanceolate to linear, dark green, like the entire plant papillose; **Fl.** yellow or reddish. – Possibly only a variety of **R. kirilowii.**

**R. macrolepis** (FRANCH.) FU (§ 3/3) (*Sedum m.* FRANCH.). – China: E.Tibet. – **Caudex** thick, short; **shoots** few, erect, thin, papillose above, 10–15 cm long; **L.** linear-lanceolate, with ± entire margins, glabrous, very acute, 15–20 mm long; **Infl.** fairly dense, of few, yellow Fl.

**R. megalophylla** (FROED.) FU (§ 2/4) (*Sedum m.* FROED.). – China. – Dioecious; **caudex** thick, with a short, divided aerial section, Ros.L. oblanceolate to broad, tapering, 10–20 cm long; **shoots** few, slender, 3–5 cm long, sterile shoots robust, 4–8 cm long; **L.** ± imbricate, acute, 17–25 mm long; **Infl.** few flowered.

**R. modesta** (BORNM.) PARSA (§ 1) (*Cotyledon m.* BORNM.). – S.E.Iran. – **Caudex** thick, ± tuberous; **Ros.** dense, 2–3 cm across, sparsely rough-hairy; **L.** linguiform-spatulate, short-tapering, 6–15 mm long, 2–4 mm across, with entire margins, L. on the flowering shoots 4–6 mm long, up to 5 mm across; **Infl.** of few, pink Fl.

**R. nuristanica** (KITAMARU) JACOBS. (§ 1) (*Sedum n.* KITAMARU). – Afghanistan. – **Caudex** thickened, c. 5 mm ⌀, roots branching obliquely; basal **L.** rosulate, oblong, 9–10 mm long, 5 mm across, acute, sessile, with entire margins, sparsely papillose to glabrous, thickish; flowering shoots 4, 3.5–4 cm tall; **L.** narrow-oblong, obtuse or apiculate; **Infl.** a cyme of 5–8 Fl., Pet. and anthers purplish-pink.

**R. ovatisepala** (HAMET) FU v. **ovatisepala** (§ 4) (*Sedum linearifolium* v. *o.* HAMET, *S. trifidum* HOOK. f., CLARKE et GRAY p.part.). – E.Himalayas. – Probably only a variety of **R. linearifolia**; **L.** crowded towards the tips of the shoots, oblong or ovate-oblong, ± deeply lobed.

**R.** — v. **chingii** FU. – China: Yunnan. – Sep. 7–8 mm long, 1 mm across.

**R. papillocarpa** (FROED.) FU (§ 3/4) (*Sedum yunnanense* v. *p.* FROED.). – China: Yunnan. – Possibly only a variety of **R. yunnanensis**; flowering **shoots** several, erect, 20–35 cm long; **L.** oblong, dentate or almost lobed, acute, 10 to 30 mm long; ♀ **Fl.** 1 mm long; **Fr.** rough-tuberculate.

**R. petiolata** (FROED.) FU (§ 1) (*Sedum p.* FROED.). – China: Tibet. – **Caudex** short, thick, broad, covered with scale-L. which are triangular-linear, tapering, c. 17 mm long, or broad-linear, obtuse, 6–7 mm long; sterile **shoots** not observed; flowering shoots 3–4 cm long, simple, slender; **L.** narrowed teretely below for 15 mm, the blade reniform to circular, obtuse, c. 15 mm long; **Infl.** a corymb of white Fl. 12–15 mm ⌀.

**R. pleurogynantha** (HAND.-MAZZ.) FU (§ 1) (*Sedum p.* HAND.-MAZZ., *S. primuloides* v. *p.* (HAND.-MAZZ.) FROED.). – China: S.W.Szechuan. – **Roots** moniliform, thickened; **caudex** short, erect, thick, simple to many-headed; **shoots** persisting, thin, stiff, with an apical Ros. of many L.; **L.** rhombic to ± linear-lanceolate, 6 mm long and 2.5 mm across, or 11 mm long and 1 mm across, obtuse, margins ± erose; pedicels spreading radially from the Ros., 7–20 mm long, densely leafy, each with a solitary white Fl. c. 1 cm ⌀.

**R. primuloides** (FRANCH.) FU (§ 1) (*Sedum p.* FRANCH.). – China: Yunnan. – **Caudex** fleshy, elongated; **shoots** up to 10 cm tall; **L.** rosulate at the tips of the shoots, ovate to obovate, 9 mm long, 4–6 mm across, fleshy, flat, with entire margins, green; flowering shoots slender; **Infl.** of 1–3 white Fl.

**R.** — v. **pachyclada** (AITCH. et HEMSL.) JACOBS. (*Sedum p.* AITCH. et HEMSL.). – Afghanistan. – **L.** obovate.

**R.** — v. **primuloides.** – **L.** ovate.

**R. purpureoviridis** (PRAEG.) FU (§ 3/2) (*Sedum p.* PRAEG.). – W.China: Yunnan. – **Caudex** thick, branched; **shoots** densely hairy, 20–30 cm tall, erect, whitish, simple; **L.** 10 mm long, 6 mm across, narrow-oblong to lanceolate, slightly dentate, glandular-hairy on the underside; **Infl.** densely leafy, Fl. greenish.

**R. quadrifida** (PALL.) FISCH. et MEY. (§ 2/2) (*Sedum q.* PALL., *S. coccineum* ROYLE, *S. asiaticum* WALL. nom. nud., *R. a.* D. DON., *S. a.* DC., *S. humile* HOOK. f. et THOMS., *S. nobile* FRANCH., *S. scabridum* FRANCH., *R. s.* (FRANCH.) FU, *S. brachystelma* FROED., *S. atunsuense* PRAEG., *R. a.* (PRAEG.) FU, *S. horridum* PRAEG.). – Himalayas; China: Tibet; USSR: Siberia. – Aerial **caudex** elongated or branched and clustered, densely covered with up to 100 dead shoots, ⊙ shoots short, spreading, often rough and papillose; **L.** with entire margins, oblong to lanceolate, often rough, ± acute; **Infl.** of 1 or few, yellow or purple Fl.

**R. reticaulis** BORISS. (§ 2/3) (*Sedum r.* (BORISS.) WENDELBO). – USSR: Pamir; Afghanistan. – **Roots** thick, woody; **caudex** (3–)4(–6) cm thick, with Br. 1.5 cm thick, covered with scale-like, imbricate, membranous, brown, triangular, obtuse L. over 1 cm long and across; **shoots** numerous, covered with dead L., (8–)12(–15) cm tall, 1.5–2 mm thick, slightly furrowed; **L.** elliptical to elliptical-oblong, coarsely dentate, tapering, 8–10 mm long, 2–3 mm across, dark green; **Infl.** dense, capitate-corymbose, Fl. small, yellow, bisexual.

**R. rhodantha** (A. GRAY) JACOBS. (§ 4) (*Sedum r.* A. GRAY, *Clementsia r.* ROSE). – USA: Montana to Arizona, Rocky Mts. – **Caudex** thick, fleshy, elongated; **shoots** up to 30 cm tall; **L.** sessile, linear-oblong, margins entire or indistinctly dentate, 2.5 cm long, 6 mm across; **Infl.** racemose, Fl. pinkish-red.

**R. robusta** (PRAEG.) FU (§ 3/1) (*Sedum r.* PRAEG.). – China: Tibet, Yunnan. – **Caudex** fleshy; **L.** linear. Possibly an ecotype of **R. longicaulis**.

**R. rosea** L. (§ 3/1) (*Sedum r.* (L.) SCOP., *S. rhodiola* DC.). – Central Eur. mountains, incl. Pyrenees, N.Eur.; As.; N.Am. – **Caudex** fleshy, rather scaly; **shoots** 15–30 cm tall; **L.** sessile, broad-linear to ovate, ± acute, 5–30 mm long, flat, fleshy, green to grey-green, ± dentate above, glabrous; **Infl.** dense, Fl. yellow, green or red.

**R.** — ssp. **alaskana** (ROSE) JACOBS. (*Sedum a.* ROSE, *S. r. v. a.* (ROSE) MAX.). – USA: Alaska. – **L.** acute, very dentate above; **Fl.** reddish.

**R.** — ssp. **arctica** (BORISS.) LÖVE (*R. a.* BORISS., *Sedum a.* (BORISS.) RONNING). – USSR: Kola peninsula. – Dioecious. **Caudex** thick, moniliform, with "beads" 4–5 cm long and thick; **L.** imbricate, dense, round-ovate, margins entire above or with a few T.; **Fl.** yellow.

**R.** — ssp. **atropurpurea** (TURCZ.) JACOBS. (*Sedum a.* TURCZ., *S. r. v. a.* (TURCZ.) PRAEG.). – N.E.As.; N.W.Am. – **L.** elliptical-spatulate or oblong-lanceolate, dentate above; **Fl.** deep red.

**R.** — ssp. **elongata** (LED.) JACOBS. (*Sedum e.* LED., *S. r. v. e.* (LED.) MAX.). – USSR: Siberia. **L.** ± elliptical, spatulate or oblong-lanceolate, acute, dentate towards the tip; **Fl.** pink.

**R.** — ssp. **integrifolia** (RAF.) JACOBS. (*R. i.* RAF., *Sedum r. v. i.* (RAF.) MAX.). – USA: Rocky Mts. – **L.** obovate, acute; **Fl.** dark red.

**R.** — ssp. **leedyi** ROSENDALE et MOORE. – N.Am.: S.E.Minnesota. – **L.** many times longer than broad.

**R.** — ssp. **microphylla** (FROED.) FU (*Sedum r. v. m.* FROED.). – China. – **L.** oblanceolate to oblong, dentate, 8–10 mm long; **Fl.** small, yellow.

R. — ssp. **neo-mexicana** (BR.) JACOBS. (*R. n.-m.* BR., *Sedum r. v. n.-m.* (BR.) MAX.). — USA: New Mexico. — L. linear-oblong, acute at both ends, with entire margins; Fl. yellow or yellowish-green.

R. — ssp. **polygama** (RYDB.) JACOBS. (*Sedum p.* RYDB., *R. p.* BR. et R., *S. r. v. p.* (RYDB.) MAX.). — USA: N.Carolina. — L. oblanceolate, minutely dentate or with entire margins; Fl. dark red.

R. — ssp. **roanensis** (BR.) JACOBS. (*Sedum r.* BR., *S. r. v. r.* (BR.) MAX., *R. roanensis* BR. et R.). — USA: N.Carolina. — L. oblanceolate, margins entire or with 1–2 T.; Fl. red or reddish.

R. — ssp. **rosea** v. **rosea** (Pl. 122/2) (*Sedum r. v. r.*, *S. r. v. vulgare* MAX.). — Eur. mountains. — L. ± elliptical, acute, serrate above; Fl. yellow.

R. — ssp. — v. **continentalis** (MAX.) JACOBS. (*Sedum r. v. c.* MAX.). — Eur. mountains. — L. rather more elongated than in v. **rosea**.

R. — ssp. — v. **maritima** (MAX.) JACOBS. (*Sedum r. v. r. sv. maritima* MAX.). — Arctic Eur. — Plant conspicuously grey; L. broader than in v. **rosea**.

R. — ssp. **sino-alpina** (FROED.) JACOBS. (*Sedum r. v. s.-a.* FROED.). — China: N.W. Yunnan. — L. oblong-spatulate, noticeably petiolate, 5–8 mm long, obtuse, with entire margins, tuberculate; Infl. of few bisexual Fl.

R. — ssp. **tachiroi** (FRANCH. et SAV.) JACOBS. (*R. t.* FRANCH. et SAV., *Sedum r. v. t.* (FRANCH. et SAV.) MAX.). — Japan. — L. elliptical, with entire margins, upper L. linear-spatulate, dentate towards the tip; Fl. yellow.

R. **rotunda** (HEMSL.) FU (§ 2/4) (*Sedum r.* HEMSL., *S. r. v. oblongum* MARQ. et SHAW, *S. megalanthum* FROED.). — Himalayas; China: Yunnan. — Caudex very thick, blackish; shoots robust, 15 cm tall; L. broad-ovate to circular, very short-petiolate, 30 mm long, 25 mm across, dark green to silvery; Infl. lax, of few dark red Fl.

R. **rotundifolia** (FROED.) FU (§ 3/4) (*Sedum yunnanense v. r.* FROED.). — China. — Close to R. **yunnanensis**, possibly only a variety of this; flowering shoots simple, erect, 46 cm tall; L. round, very obtuse, 2–4 mm long and across, dentate; Infl. paniculate, with few, yellowish Fl.

R. **sacra** (PRAIN ex HAMET) FU (§ 4) (*Sedum s.* PRAIN ex HAMET, *S. linearifolium v. s.* (PRAIN ex HAMET) HAMET). — China: Tibet. — Possibly only a variety of R. **linearifolia**; L. obovate or obovate-lanceolate or oblong, little lobed.

R. **sangpo-tibetana** (FROED.) FU (§ 1) (*Sedum s.-t.* FROED.). — China: Tibet. — Roots fibrous; caudex short and broad, with triangular scale-L. at the apex; some L. scale-like, others linear-lanceolate, ± acute, shortly spurred below, 10–15 mm long; Infl. a corymb of yellow Fl.

R. **semenowii** (RGL. et HERD.) BORISS. (§ 3/3) (*Umbilicus s.* RGL. et HERD., *Sedum s.* (RGL. et HERD.) MAST., *U. linifolius* OST. SACK., *U. linearifolius* FRANCH., *Cotyledon s.* O. et B. FEDTSCH.). — USSR: Central As. — Caudex thick; shoots erect, 30–60 cm tall; L. linear,

with entire margins, 25–30 mm long, 1–2 mm across, green; Infl. dense, with ⚥, greenish-white Fl.

R. **sexifolia** FU (§ 3/1). — China: Tibet. — Caudex not known; shoots ascending, 17.5 cm tall; L. in whorls up to 6, elliptical-oblong 4 cm long, 3.5 cm across, tuberculate, laciniate to lobed, lobes with margins entire or distantly serrate, obtuse, sessile; Infl. dense, bracts tuberculate, Fl. c. 15 mm ∅.

R. **sinica** (DIELS) JACOBS. (§ 3/4) (*Sedum s.* DIELS). — China. — Caudex thick, tuberous; shoots erect, 30–40 cm tall; L. verticillate, rhombic or ± square, up to 3.5 cm long, up to 3 cm across; Infl. compressed.

R. **sinuata** (ROYLE ex EDGEW.) FU (§ 4) (*Sedum s.* ROYLE et EDGEW., *S. linearifolium v. s.* (ROYLE et EDGEW.) HAMET, *S. trifidum* HOOK. f., CLARKE et PRAEG. p.part.). — W. Himalayas. — Possibly only a variety of R. **linearifolia**; L. deeply pinnatisect, segments linear.

R. **smithii** (HAMET) FU (§ 2/4) (*Sedum s.* HAMET). — Possibly only an ecotype of R. **fastigiata**.

R. **stapfii** (HAMET) FU (§ 2/4) (*Sedum s.* HAMET). — Caudex creeping; L. pointed; pedicels slender. — Very close to **Sedum filipes**.

R. **subopposita** (MAX.) JACOBS. (§ 3/1) (*Sedum s.* MAX.). — W.China. — Caudex fleshy; shoots numerous, weak; L. scattered, 2–3 cm long, 10–15 mm across, short-petiolate or sessile, broadly elliptical to ovate, margins almost entire or irregularly divided; Infl. densely leafy, Fl. bisexual, yellow.

R. — v. **subopposita**. — L. broadly elliptical to ovate, irregularly crenate, 2 cm long, 1 cm across.

R. — v. **telephioides** (MAX.) JACOBS. (*Sedum s. v. t.* MAX.). — Grey-green variety; L. ± round, 25–30 mm long, 15 mm across, tapering, margins ± entire.

R. **tachoensis** FU (§ 2/3). — China. — Caudex simple to slightly branched; shoots few, erect; ♂ plants with flowering shoots erect, puberulous, straw-coloured, 10–13 cm long; L. laxly imbricate, long-lanceolate, (7–)8(–10) mm long, 1.6 to 1.8 mm across, subacute, with entire margins, puberulous; pedicels puberulous, short, Fl. c. 3 mm ∅, dirty green. ♀ plant not known.

R. **tibetica** (HOOK. f. et THOMS.) FU (§ 2/3) (*Sedum t.* HOOK. f. et THOMS.). — Himalayas. — Possibly only a variety of R. **fastigiata**; L. dentate, grey-green, 18 mm long; Fl. dark red.

R. **tieghemii** (HAMET) FU (§ 4) (*Sedum t.* HAMET, *S. linearifolium v. t.* HAMET). — China: Szechuan. — Possibly only a variety of R. **linearifolium**; L. ovate or ovate-oblong, little lobed; Pet. fringed-ciliate.

R. **trifida** (WALL.) JACOBS. (§ 4) (*Sedum t.* WALL.). — Himalayas. — Caudex thick, often elongated; shoots erect, 15–20 cm tall, slender; L. linear below, ovate above, deeply dentate, 3.5–7.5 cm long; Infl. very leafy, papillose, Fl. 15 mm ∅, purplish-red.

**R. tsuiana** Fu (§ 4). – China: E.Tibet. – **Caudex** thick; plant hairy; flowering **shoots** erect, simple, 9–19 cm long; **L.** sessile, rhombic to ovate, 15–17 mm long, 9–13 mm across, acute, margins coarsely dentate; **Infl.** dense, bracts with entire margins, Fl. 15–18 mm ∅.

**R. tuberosa** (Coss. et Letourn.) Jacobs. (§ 1) (*Sedum t.* Coss. et Letourn.). – Algeria; Tunisia. – **Caudex** large, ovoid to spherical, without offsets; **Ros.** 1–3; **L.** fleshy, obovate-oblong to linear, teretely narrowed, obtuse, margins entire, with translucent papillae; **Infl.** 8–15 cm tall, cauline L. linear-oblong, with basal spurs, cyme branchlets each with 2 to 3 yellow Fl.

**R. venusta** (Praeg.) Fu (§ 2/3) (*Sedum v.* Praeg.). – China: Tibet, Yunnan; Burma. – Possibly only a variety of **R. fastigiata**; **L.** oblong-ovate; Fl. unisexual.

**R. yunnanensis** (Franch.) Fu v. **yunnanensis** (§ 3/4) (*Sedum y.* Franch., *S. mengtzeanum* Ulbr., *S. y.* v. *oxyphyllum* Froed.). – W.China: Yunnan, Szechuan, Shansi, W.Hupeh. – **Caudex** thick, tuberous; **shoots** solitary or few, erect, up to 1 m tall, glabrous; **L.** 2–10 cm long, margins ± entire to slightly lobed, glabrous, rather obtuse; **Infl.** paniculate, Fl. usually unisexual, ♂ Fl. yellowish-green to reddish-purple.

**R. —** v. **forrestii** (Hamet) Jacobs. (*Sedum y.* v. *f.* Hamet, *R. f.* (Hamet) Fu, *S. y.* v. *muliense* Froed., *S. y.* v. *strictum* Froed.). – China: Likiang. – **L.** in whorls of 4, broad-lanceolate, ± lobed; **Infl.** of few Fl.

**R. —** v. **henryi** (Diels) Jacobs. (*S. h.* Diels, *S. y.* v. *h.* (Diels) Hamet, *R. h.* (Diels) Fu). – **L.** alternate, or up to 3 in whorls, ovate to rhombic, dentate; **Infl.** short-cylindrical.

**R. —** v. **valerianoides** (Diels) Jacobs. (*Sedum v.* Diels). – **L.** usually 3 in a whorl, oblong to ovate, margins ± entire; **Infl.** tall, slender.

**Rhytidocaulon** Bally. Asclepiadaceae. – Ethiopia; Somalia; S.Arabia. – Close to the G. **Echidnopsis**. – **St.** and **Br.** fleshy, cylindrical, up to 110 cm tall, up to 15 mm ∅, 4–6-angled or furrowed, indistinctly reticulate, with oblong or rectangular zones 7–12 mm long and 3–6 mm across, all parts rough, papillose, grey or dull green; **L.** crowded at the Br.tips, small, fleshy, with a prominent M.rib, rough, papillose, 2.5 to 3 mm long, acute, soon dropping; **Fl.** always solitary, scattered along the St. and Br., often very numerous, solitary from the L.axils, ± sessile or short-pedicellate, Cor. c. 9 mm ∅, 5-lobed, lobes spreading, triangular, corona variable. – Greenhouse, warm. Propagation: seed, cuttings.

**R. fulleri** Lavr. et Mort. – S.Arabia: Muscat and Oman, coast of Dhofer. – Main **St.** up to 15 cm tall, sparsely branched, 2 cm ∅, lateral Br. ascending, c. 15 mm ∅, all shoots cylindrical, with indistinct longitudinal reticulate markings, indistinctly 4- or 5-angled, rough, papillose, with a thick waxy coating, green at first, eventually grey; **L.** fleshy, rough, papillose, ovate-triangular, up to 3 mm long, soon dropping; **Fl.** solitary or paired, scattered on the younger shoot sections, Cor. c. 1 cm long, 4.5 mm ∅ at the base, lobes subulate, free below, united at the tips, margins hairy and recurved, minutely hairy outside, glabrous inside, both surfaces yellow, with brownish-red blotches and transverse lines.

**R. macrolobum** Lavr. – S.W.Arabia. – **St.** erect or ± climbing, at least 25 cm long, 20 mm ∅, round, indistinctly 4-angled, longitudinally tessellate, sparsely branching; **Br.** rather short, not jointed, St. and Br. grey with a waxy coating, young shoots grey-green, rough, papillose; **L.** fleshy, rough, papillose, ovate, 3 mm long, soon dropping; **Fl.** solitary, axillary, scattered, pedicels 3 mm long, Sep. triangular, acute, carinate, 1.2 mm long; Cor. broadly campanulate, very deeply lobed, 15 mm ∅, whitish outside below, lobes pale yellow, with dark purple blotches and borders, yellowish-white inside below, the lower half of the lobes white, with dark purple transverse bands and borders, the upper part concolorous dark purple, narrowly ovate to triangular, 6.5–8 mm long, 2 mm across below, margins minutely ciliate, with a cushion-like thickening inside in the upper half with a cluster of 3–5 motile, simple hairs 3 mm long, tube not present.

**R. paradoxum** Bally. – Ethiopia. – **St.** and Br. dark green; **Fl.** pinkish-brown, Cal. fleshy, with scattered papillae outside, Cor. 9 mm ∅, ± cup-shaped below, lobes patent, acute, 2.9 mm long, 3.2 mm across, corona 4 mm ∅, round, outer corona-lobes crowded together.

**R. piliferum** Lavr. – Somalia: N. reg. – **St.** and Br. erect or ascending, rigid, cylindrical, up to 40 cm long, 5–12 mm ∅, 4–6-angled or furrowed, the angles indistinctly tessellate, rough all over, waxy; **L.** ovate-triangular, 1 to 3 mm long, deciduous; **Fl.** axillary, 1–4, with pedicels 2 mm long, Sep. fleshy, triangular, densely papillose, Cor. round-campanulate, c. 10 mm ∅, glabrous, lobes usually subulate, 4 mm long, whitish below, with dark purple transverse bands, dark purple at the tips, with a cluster of dark, motile hairs 1 mm long, tube 1 mm long, 2 mm ∅.

**R. subscandens** Bally (Pl. 122/4). – Somalia; Ethiopia. – **Fl.** on pedicels 0.5 mm long, Cal. lobes lanceolate, 2.5 mm long, fleshy, minutely papillose outside, Cor. shortly cup-shaped below, lobes patent, triangular, acute, 3 mm long and across, papillose outside, green, glabrous inside, whitish, with red blotches, corona-lobes shortly bifid. – **St.** as in G.-description.

**Rochea** DC. Crassulaceae. – S.Afr. – Succulent ♃ herbs or sub-♄; **L.** opposite and decussate, united in pairs at the base, margins hairy; **Fl.** in capitate cymes, white, yellow or red. – Greenhouse, moderately warm. – Propagation: seed, cuttings.

**R. coccinea** (L.) DC. (Pl. 122/3) (*Crassula c.* L., *Kalosanthes c.* Haw.). – S.Afr. – Sub-♄ 30–60 cm tall; **St.** densely leafy; **L.** imbricate, obovate, acute, green, up to 25 mm long, 10 mm across; **Fl.** numerous, scarlet, crimson or white in summer.

**R.** — cv. Bicolor. – **Fl.** red and white.
**R.** — cv. Flore Albo. – **Fl.** white.
**R. jasminea** (SIMS) DC. (*Crassula j.* SIMS, *Kalosanthes j.* HAW., *R. microphylla* E. MEY.). – S.Afr. – Sub-♄; **St.** erect or prostrate, densely leafy; **L.** 2 cm long, spatulate, obtuse, with ciliate margins, upperside green, underside red; **Fl.** white, becoming red.
**R. odoratissima** (ANDR.) DC. (*Crassula o.* ANDR., *C. capitata* HORT., *Kalosanthes o.* HAW.). – S.Afr. – **St.** rough, Br. erect, laxly leafy; **L.** united and sheathing at the base, erect to projecting, linear, acute, grooved, 2–4 cm long, 2–4 mm across, green; **Fl.** numerous, pale yellow or pink.
**R. versicolor** (BURCH.) DC. (*Crassula v.* BURCH., *Kalosanthes v.* HAW.). – S.Afr. – Sub-♄ 30–60 cm tall; **St.** very fleshy, densely leafy; **L.** united and sheathing at the base, oblong-lanceolate, acute, margins rather cartilaginous, dark green; **Fl.** numerous, white with the outside often glossy red, pink, yellowish, or all over red.

× **Rocheassula** ROWL. ( × *Kalorochea* VEITCH). Crassulaceae. – Intergeneric hybrid: **Rochea** DC. (*Kalosanthes* HAW.) × **Crassula** L. – Greenhouse, warm. Propagation: cuttings.

× **R. langleyensis** (VEITCH) ROWL. (Pl. 106/4) (× *Kalorochea l.* VEITCH). – Hybrid: **Rochea coccinea** ♀ × **Crassula falcata** ♂. – Intermediate hybrid.; Fl. small, colour as **C. falcata**. – This hybrid raised in c. 1898 by SEDEN; the same cross but reversed was repeatd by R. GRÄSER of Nuremberg and by FRANK REINELT of Capitola, Calif. as cv. Capitola.

**Rogeria** J. GAY ex DELILE. Pedaliaceae. – Trop. and S.W.Afr. – Erect, ⊙, slightly succulent herbs; **L.** opposite, largely sinuate to lobed, ± fleshy; **Infl.** axillary, Fl. solitary or few together, short-pedicellate, violet, red or white; **Fr.** curved, prickly.
The following species are ± succulent: – **R. adenophylla** J. GAY ex DEL., Senegal, Sudan, S.Angola, northern S.W.Afr.; **R. bigibbosa** ENGL., S.Angola, central S.W.Afr.; **R. longiflora** (ROYEN in L.) DC., S.W.Afr. and **R. petrophila** DE WINTER, Cape; S.W.Afr. – Of little interest for cultivation.

**Rosularia** (DC.) STAPF (*Rosularia* DC. as Sect. of **Umbilicus** DC.). Crassulaceae. – As. Minor to the Himalayas. – Hardy ♃, offsetting, with crowded L.-**Ros.**; **L.** flat, ± spatulate, glabrous or hairy; **Fl.** in terminal or axillary Pan., reddish or yellowish-white, summer. – In the open. – Propagation: seed, division. (Acc. SAMUELSSON, Arkiv. Bot. Stockholm, **5**, 1960, p. 186, the nomenclature is exceedingly confused.)

**R. aizoon** (FENZL) BGR. (*Umbilicus a.* FENZL, *Cotyledon a.* (FENZL) SCHOENL., *Sedum chrysanthum* v. *a.* HAMET). – Turkey. – **Ros.-L.** dense, linguiform-oblong, ciliate, pubescent, 8–10 mm long, 4–10 mm across; **Infl.** of few yellow Fl.

**R. elymaitica** (BOISS. et HAUSSKN.) BGR. (*Umbilicus e.* BOISS. et HAUSSKN., *Sedum e.* (BOISS. et HAUSSKN.) HAMET). – S.W.Iran. – **Ros.-L.** long-spatulate, glabrous, 25 mm long; **Infl.** broadly pyramidal, ± glandular-hairy, 15–20 cm tall, Fl. numerous, flesh-coloured.

**R. glabra** (RGL. et WINKLER) BGR. (*Umbilicus g.* RGL. et WINKLER). – N.W.China. – Glabrous; **Ros.-L.** elliptical-spatulate, obtuse, mucronate, vivid green, cauline L. oblong; **Fl.** pale yellowish or greenish.

**R. globulariifolia** (FENZL) BGR. (*Umbilicus g.* FENZL, *Sedum g.* (FENZL) HAMET). – As. Minor. – **L.** spatulate, 3–4 cm long; **Infl.** glandular, 25 to 30 cm tall, Fl. urn-shaped, reddish.

**R. haussknechtii** (BOISS. et REUT.) BGR. (*Umbilicus h.* BOISS. et REUTT.). – Iraq. – **L.** linear-spatulate, 3–4 mm across, cauline L. linear, very long, acute, with a spur below; **Infl.** up to 12 cm tall, Fl. small.

**R. libanotica** (L.) SAM. v. **libanotica** (*Sedum l.* L., *Cotyledon l.* (L.) LABILL, *Umbilicus l.* (L.) DC.). – USSR: Caucasus; N.W.Iran; As. Minor to Israel and Jordan. – **Ros.-L.** spatulate, with minute glandular hairs (acc. A. BERGER), or glabrous (acc. SAMUELSSON ex LINNAEUS); **Infl.** of many purplish-pink Fl.

**R.** — v. **pubescens** FROED. (*R. parvifolia* FROED. et SAM.) RECH. f. – Syria. – Plant hairy all over; **L.** linear below ± spatulate at the tip, apiculate, 2–4 cm long; **Infl.** 9–20 cm tall, cauline L. oblong-spatulate, **Fl.** red.

**R. lineata** (BOISS.) BGR. (*Umbilicus l.* BOISS.). – Jordan, Israel. – **L.** linguiform-spatulate, obtuse, long-ciliate, dentate; **Infl.** 10–15 cm tall, forking, very coarsely hairy, Fl. pink.

**R. modesta** (BORNM.) PARSA (*Cotyledon m.* BORNM.). – S.E.Iran: Prov. Keman. – **Caudex** thick, ± tuberous; **Ros.** 2–3 cm across, dense; **L.** linguiform-spatulate, short-tapering, with sparse, long hairs, 6–15 mm long, 3–4 mm across, margins entire and glabrous; **Fl.**scape 2–3, with long-linear L., **Infl.** velvety-papillose, of few pink Fl.

**R. pallida** (SCHOTT et KOTSCHY) STAPF (*Umbilicus p.* SCHOTT et KOTSCHY, *U. chrysanthus* BOISS., *Sedum c.* HAMET, *R. c.* (BOISS.) HAMET). – Turkey. – Resembles **Sempervivum**; **Ros.-L.** oblong-spatulate, obtuse, 12–18 mm long, 6 mm across, with dense, grey hairs and cilia; **Infl.** glandular, cauline L. elliptical, acute, Fl. large, whitish-yellow.

**R. paniculata** (RGL. et SCHMALH.) BGR. (*Umbilicus p.* RGL. et SCHMALH., *U. platyphyllus* BOISS., *Sedum radicosum* BOISS.). – Glabrous; **L.** broad-spatulate, 4 cm long; **Infl.** up to 50 cm tall, terminal, Fl. whitish to pale red.

**R. persica** (BOISS.) BGR. (*Umbilicus p.* BOISS., *U. libanoticus glaber* BOISS., *Sedum sempervivum* v. *glabrum* HAMET). – Iran, Lebanon. – **Ros.-L.** linguiform-spatulate, with cartilaginous marginal T.; **Infl.** simple or bipartite, cauline L. linear, glabrous, Cor. with pink lines.

**R. pestalozzae** (BOISS.) SAM. (*Umbilicus p.* BOISS., *Cotyledon sempervivum* MARSCH. BIEB., *Sedum s.* (MARSCH. BIEB.) HAMET, *R. s.*

(MARSCH. BIEB.) BGR.). – As. Minor. – Densely hairy; **Ros.-L.** slightly spatulate below, obtuse and broadly apiculate above, up to 4 cm long, offsetting from the base of the St. where the L. are crowded, oblong, obtuse, 12–18 mm long; **Fl.** with Cor. 7 mm long.

**R. platyphylla** (SCHRENK.) BGR. (*Umbilicus p.* SCHRENK.). – USSR: Altai. – **L.** obovate-spatulate, obtuse or subacute, glandular-hairy, margins with longer hairs; **Infl.** hairy, **Fl.** whitish-yellow.

**R. serrata** (L.) BGR. (*Cotyledon s.*, *Umbilicus s.* (L.) DC., *Cotyledon saminum* URV., *U. saminus* (URV.) DC.). – Greek Is.; As. Minor. – Glabrous; **Ros.-L.** oblong-spatulate, 25–35 mm long, 6 to 8 mm across, short-tapering, margins with minute, blunt, cartilaginous T.; **Fl.** red.

**R. turkestanica** (RGL. et WINKLER) BGR. (*Umbilicus t.* RGL. et WINKLER). – USSR: Central As., S.Kazakhstan (Ala-tau). – **Ros.-L.** lanceolate or oblong-lanceolate, short-tapering, with minute rough hairs; **Infl.** glabrous, Pet. white, ± with red lines outside.

**Salicornia** L. Chenopodiaceae. – Distribution ± worldwide, not found in Austr., trop. Afr., S.Am. – Halophytes. Succulent, apparently leafless ☉ or ♃ herbs or ♄, with jointed Br. and shoots, fertile shoots with honeycomb-like cavities after the Fr. have dropped; **Fl.** minute. – Unimportant in cultivation. In the open; ♄ in winter: cold-house. Propagation: seed, in some instances cuttings.

**S. europaea** L. f. **europaea** (*S. herbacea* L.). – Eur., Afr., coasts. – ☉; **Br.** erect or prostrate, jointed, fleshy, 20–30 cm tall, up to 5 mm thick, leafless; **Fl.** in fleshy spikes.

**S.** — f. **brachystachya** (MEYER) JACOBS. et ROWL. (*S. b.* MEYER, *S. herbacea* f. *b.* (MEYER) JACOBS., *S. patula* DUVAL-JOUWE). – Prostrate form; **Fl.** spikes short.

**S.** — f. **stricta** (DUMORTIER) MEYER (*S. s.* DUMORTIER, *S. herbacca* f. *s.* (DUMORTIER) JACOBS.). – Erect form; **Fl.** spikes long. (Planted along the Friesian coast to bind flood-prone land.)

**S. fruticosa** L. (Pl. 123/1). – S.Eur.; S.Afr.; S.W.Afr. – Erect ♄; **Br.** 3 mm thick, jointed, grey-green, fleshy, Int. 8 mm long, leafless; **Fl.** solitary, small.

**Salsola** L. Chenopodiaceae. – Coasts and other saline soils: Eur., As., S.Afr., Am., Austr. – ☉, ☉ or ♃ halophytes. – **L.** alternate, rarely opposite, sessile, subulate, narrow, sometimes scale-like, sometimes fleshy, usually hairy, thorn-tipped; **Fl.** solitary or several together, axillary, Cal. usually 5-partite, membranous, small. – Not important for cultivation.

The following species are widespread: **S. kali** L., ± worldwide on saline soils; **S. soda** L., Medit. basin, Central As.

**Samuela** TREL. Agavaceae. – Mexico; USA: Texas. – G. closely related to **Yucca**; **Pet.** united basally to form a tube. – Cold-house. Propagation: seed.

**S. carnerosana** TREL. – USA: S.W.Texas; Mexico. – **St.** up to 6 m tall, up to 70 cm thick, simple or branching; **L.** as for the undermentioned species; **Infl.** freely branched.

**S. faxoniana** TREL. – USA: S.W.Texas; N.Mexico. – **St.** 1.5–5 m tall, 30–60 cm thick, very little branched; **L.** in crowded apical clusters, 1–1.25 m long, 5–7.5 cm across, grooved, firm, short-apiculate, margins with stout white threads; **Infl.** erect, broadly pyramidal, bracts large, **Fl.** large, white.

**Sansevieria** THUNBG. Agavaceae. – Trop. and S.Afr.; Madag.; Arabia; India. – (Abbreviated from a contribution by **Horst Pfennig**.) – Acauline ♃ plants with a creeping rhizome, or caulescent and branching near the base; **L.** ± leathery or fleshy, ± lanceolate and flat or semicylindrical or cylindrical, flexible or stiff, often with lighter blotches or transverse bands, some cultivars with a yellow border; **Infl.** extended or capitate-racemose, or paniculate, Cor. developing a tube, **Fl.** mostly whitish, inconspicuous, perfumed. – Greenhouse, warm. Propagation: division, leaf-cuttings, seed.

**S. aethiopica** THUNBG. (*S. zeylanica* REDOUTÉ, *S. thunbergii* MATTEI). – Trop. and S.Afr. – Acauline; **L.** up to 30 on a shoot, ± linear, 15 to 40 cm long, 1–1.5 cm across, smooth, with a dry tip up to 3 cm long, blue-green, at first with lighter transverse bands, margins reddish.

**S. arborescens** CORNU. – Trop. E.Afr. – **St.** erect, up to 1.50 m tall, 2.5 cm thick, densely leafy; **L.** lanceolate, flat, 20–45 cm long, 2 to 4.5 cm across, 5 mm thick, smooth, grass-green, tip brownish, hard, margins whitish or reddish.

**S. burmanica** N. E. BR. – Burma. – Acauline, with a creeping rhizome; **L.** 8–13, erect, dense, linear or narrowly lanceolate, 45–75 cm long, 1.3–3.2 cm across, 3–4 mm thick, ± flat, the subulate tip 3–10 cm long, with light green transverse markings, underside with 6–9 darker green longitudinal lines.

**S. caespitosa** DTR. – S.W.Afr. – Rhizome creeping, c. 6 mm thick, scaly; **L.** up to 20, densely rosulate, linear-lanceolate, flat, erect to recurved, 10–20 cm long, 1–2 cm across, smooth, with a subulate tip up to 5 cm long, dark green, margins green.

**S. canaliculata** CARR. – Trop. Afr.; Madag. – Rhizome creeping; **L.** 1–2, erect, cylindrical, with 5–6 longitudinal furrows, 15–70 cm long, 1–2 cm thick, dark green; **Fl.** scape 9–16 cm tall.

**S. caulescens** N. E. BR. – Trop. E.Afr. – **St.** erect, up to c. 60 cm tall, 2.5–3.8 cm thick, densely leafy; **L.** spreading, grooved, concave above, 45–80 cm long, 1.9–3.4 cm across below, 1.3–1.8 cm thick, roundish above, with a light brown tip, deep green, at first with indistinct transverse bands, underside with 9–12 darker green longitudinal lines, eventually with a whitish margin.

**S. cylindrica** BOJ. v. **cylindrica**. – Trop. W.Afr. – Acauline; **L.** 2–4 on a shoot together with several basal scales, distichous, erect, cylindrical, 0.6–1.5 m long, 2–3 cm thick below, tapering above into a short, hard tip, green, with dark green transverse bands; Fl. Sc. 60 to 90 cm tall, Fl. c. 3.5 cm long.

**S. — v. patula** N. E. Br. – L. 3–6 on a shoot, sharply recurved from the base.

**S. dooneri** N. E. Br. – Trop. E.Afr. – Rhizome creeping, 6–8 mm thick, scaly; L. up to 20 in a dense Ros., linear-lanceolate, flat or slightly concave, sharply recurved, smooth, with a subulate tip up to 5 cm long, dark green, the underside lighter, both surfaces with lighter transverse bands, margins green, 10–40 cm long, 1.5–3 cm across; Fl. Sc. c. 35 cm tall.

**S. ehrenbergii** SCHWEINF. – Trop. E.Afr. – St. up to 25 cm tall; L. 5–9 on one shoot, distichous, arranged like a fan, up to 1.5 m long, the upperside with a triangular groove, dark green, the tip rigid and spiny, margins reddish-brown and white, underside with 5–12 darker green, longitudinal lines, in seedling plants the L. are short and not at first distichous.

**S. gracilis** N. E. Br. – Trop. E.Afr. – St. 2–8 cm tall, surrounded by L. sheaths, offsets 15–90 cm long and c. 8 mm thick, produced above soil-level, covered with scales 12–25 mm long, passing over into the L.; L. 8–12 on a shoot, dense, 25–80 cm long, with 5–12.5 cm long grooves below, cylindrical above, 6–9 mm thick, with a spiny tip, smooth, but in older plants becoming grooved.

**S. grandicuspis** HAW. – Origin unknown. – Acauline, with a creeping rhizome; L. 5–15 on one shoot, linear-lanceolate, ± grooved, 18 to 50 cm long, 13–38 mm across, erect, smooth, with a subulate, rigid tip 1.7–5 cm long, both surfaces with dark and light green transverse bands, underside with 5–7 darker green, impressed lines or shallow furrows, margins green.

**S. grandis** HOOK. f. v. **grandis.** – ? S.Afr. – Acauline, rhizome creeping, scaly, up to 60 cm long, 3.5 cm thick, terminating in 2–5 elliptical or broadly lanceolate L. 20–60 cm long, 8 to 15 cm across, ± smooth, dark or bluish-green, with light transverse bands at first, margins becoming hard and reddish-brown.

**S. — v. zuluensis** N. E. Br. – S.Afr. – L. more oblong, 5–12 cm across, light green initially, with slightly reddish, horizontal bands below, rather rough.

**S. humbertiana** GUILL. – Trop. E.Afr. – Acauline; L. up to 10 on one shoot, up to 33 cm long, 1.3 cm thick, cylindrical, long-tapering, dark green, with indistinct transverse bands, upperside grooved, with a subulate white tip, margins acute below midway, whitish to reddish, surface rough, little furrowed.

**S. intermedia** N. E. Br. – Trop. E.Afr. – Acauline; L. up to 7 on one shoot, ± erect, the outer L. semicylindrical, the inner ones cylindrical, upperside grooved, 45–120 cm long, 1.3 to 1.9 cm thick, long-tapering, dark green, with indistinct transverse bands, the tip ± spiny, whitish, margins acute below midway, whitish, surface rough, with numerous impressed or slightly furrowed longitudinal lines.

**S. kirkii** BAK. v. **kirkii.** – Trop. E.Afr., Zanzibar. – Acauline, rhizome creeping; L. 1–3 on one shoot, the upper part sometimes pendulous, linear-lanceolate or broadly ligulate, up to 1.80 m long, up to 9 cm across, smooth, green, with whitish or whitish-green blotches, margins undulating, reddish-brown; Fl. Sc. 35–60 cm tall, Fl. capitately arranged.

**S. — v. pulchra** N. E. Br. – Zanzibar. – L. markings dull brown or reddish.

**S. liberica** GÉR. et LABR. (*S. chinensis* GENTIL, *S. gentili* MATTEI). – Trop. W.Afr. – Acauline, rhizome creeping; L. 1–6 on a shoot, erect at first, becoming more spreading, stiff and leathery, ligulate to lanceolate, flat, 45–100 cm long, 5–11.5 cm across, usually passing over below midway into a concave, grooved stalk, both surfaces with alternating light and dark green transverse bands, the subulate tip green at first, turning whitish-brown, 2–12 mm long, margins slightly undulating, whitish to brownish; Fl. Sc. 60–80 cm tall, Fl. c. 5 cm long.

**S. longiflora** SIMS. – Trop. Afr. – Acauline, rhizome creeping; L. 4–6 on one shoot, lanceolate, up to 1.5 m long, 9 cm across, grooved below, flat above, dark green, with lighter blotches or irregular bands, margins reddish-brown; Fl. Sc. c. 25 cm long, Infl. capitate or densely racemose, 8–35 cm long, Fl. c. 12 cm long.

**S. metallica** GÉR. et LABR. v. **metallica** (*S. guineensis* BAK.). – Trop. Afr. – Acauline, rhizome 2.5–4 cm thick, creeping; L. (1–)3(–4) on a shoot, long-lanceolate or broad-ligulate, 45 to 150 cm long, the bottom 10–60 cm forming a grooved petiole, 5–12.5 cm across, very flexible, with a soft, subulate tip, dull dark green, the upperside with pale green transverse markings at first, margins green, becoming whitish or slightly reddish-brown; Fl. Sc. 45–120 cm tall, the upper half bearing Fl. 3.3 cm long.

**S. — v. longituba** N. E. Br. – Fl. c. 5.6 cm long.

**S. parva** N. E. Br. – Trop. E.Afr. – Rhizome creeping, 8 mm thick, scaly; L. 6–14 on a shoot, linear or linear-lanceolate, concave or deeply grooved, 20–45 cm long, 8–14 mm across, smooth, the subulate tip up to 7.5 cm long, both surfaces at first with distinct light and dark green transverse bands, margins green; Fl.Sc. c. 30 cm tall.

**S. pinguicula** BALLY. – Kenya. – L. 5–7, rosulate, 12–30 cm long, thick-fleshy, 2.8 to 3.5 cm thick below midway, upperside very concave, tapering gradually towards the spiny tip; Infl. paniculate.

**S. raffillii** N. E. Br. v. **raffillii.** – Trop. E.Afr. – Acauline, with a rhizome 2–5 cm thick; L. 1–2, surrounded by several scales below, long-lanceolate or broad-ligulate, erect, stiff, 60 to 150 cm long, 5.5–12.5 cm across, passing over below midway into a concave petiole, the conspicuous marking consisting of large, yellowish-green, oblong or oval blotches or irregular transverse bands on a darker green ground, slightly pruinose, margins reddish-brown; Fl.Sc. 90–115 cm tall, Fl. c. 3.5 cm long, pedicels 4–6 mm long.

**S. — v. glauca** N. E. Br. – **L.**-markings green on a bluish-green ground, bluish-pruinose.

**S. roxburghiana** Schult. (*S. zeylanica* Roxb.). – Peninsular India. – Acauline, rhizome creeping; **L.** 6–24, 20–60 cm long, 1.3–2.5 cm across, 3–4 mm thick, the upperside deeply grooved, the tip subulate, 0.8–5 cm long, the transverse markings dark green, the underside with 6–11 dark green, longitudinal lines, the upperside with 1–3 lines.

**S. scabrifolia** Dtr. – S.W.Afr. – Acauline, rhizome creeping, 2 cm thick; **L.** (6–)7(–20) on a shoot, linear-lanceolate to linear, grooved, concave, 10–30 cm long, 1.5–2.5 cm across, rough, the 2.5 cm long tip soon becoming dry, dark green, at first with lighter transverse bands, margins broad, reddish-brown.

**S. senegambica** Bak. (*S. cornui* Gér. et Labr.). – Trop. W.Afr. – Acauline, rhizome creeping, 12–18 mm thick; **L.** 2–4 on a shoot, linear-lanceolate to lanceolate, 30–60 cm long, 3–6 cm across, tapering gradually above midway into a subulate tip up to 13 mm long, narrowing below to 1 cm across and grooved, concave, otherwise ± flat, upperside dark green, with indistinct transverse bands, the underside lighter, with more distinct transverse bands, margins green.

**S. singularis** N. E. Br. – Trop. E.Afr. – Rhizome creeping, up to 4.5 cm thick; **L.** solitary, cylindrical, erect, stiff, slightly rough, 0.5–2.4 m long, 2–4.5 cm thick below, with a hard, whitish tip, with grooves 3–6 mm across and with 4–6 impressed, longitudinal lines which later become furrowed, dark greyish or bluish-green, often suffused brownish, at first with light green markings; **Infl.** capitate, scarcely raised above the soil.

**S. stuckyi** Godefr. (Pl. 123/2). – Trop. E.Afr. – Rhizome creeping, up to 5 cm thick; **L.** 1–2 on a shoot, erect, stiff, ± smooth, 80–270 cm long, 3–8 cm thick, with a groove 8–30 mm across below, narrowing above, with up to 20 dark green longitudinal lines or shallow furrows, dark green, at first with lighter transverse markings.

**S. suffruticosa** N. E. Br. – Trop. E.Afr. – **St.** up to 30 cm tall, surrounded by L.sheaths, with stolons 10–15 cm long and 12–18 mm thick, produced above soil-level, covered with pointed scales 18–37 mm long, passing over into the L. above; **L.** 7–18 on one shoot, variously directed or distichous, 15–60 cm long, shortly concave and grooved below, then cylindrical and 12 to 19 mm thick, gradually tapering above into the spiny tip, rough, dark green, at first with light bands, with dark longitudinal lines which later become furrowed.

**S. thyrsiflora** Thunbg. (*S. guineensis* Willd.). – S.Afr. – Acauline, rhizome creeping; **L.** 2–5 on one shoot, lanceolate, flat, 15–45 cm long, 2.5–8 cm across, passing over below midway into a concave, grooved petiole, smooth, with a whitish tip 1–16 mm long, dark green, the lighter transverse bands initially rather indistinct, the whitish or reddish-brown margins becoming hard; Sc. 45–75 cm tall, **Infl.** a thyrse, Fl. 18 mm long.

**S. trifasciata** Prain v. **trifasciata** (Pl. 123/2) (*S. guineensis* Gér. et Labr.). – Trop. W.Afr. – Acauline, rhizome creeping; **L.** 2–6 on one shoot, stiff, linear-lanceolate, flat, 30–160 cm long, 2.5–8 cm across, with a green, subulate tip 0.4–3.5 cm long, passing over below midway into a concave, grooved petiole, both surfaces with whitish to light green transverse bands alternating with deep grass-green to blackish-green ones, margins green; Fl.Sc. 30–75 cm tall, Fl. c. 2.5 cm long.

**S. — cv. Hahnii** (Pl. 123/4/1) (*S. hahnii* Hort.). – **L.** c. 15, rosulate, or up to 60 when the plant is caulescent, elliptical, up to 15 cm long, 8 cm across, markings as for the type.

**S. — cv. Golden Hahnii** (Pl. 123/4/2). – Like cv. Hahnii but with yellowish markings.

**S. — cv. Silver Hahnii** (Pl. 123/4/3). – Like cv. Hahnii but with markings like cv. Silver Cloud.

**S. — v. laurentii** (de Wild.) N. E. Br. (*S. l.* de Wild.). – Congo Rep. – **L.** margins with ± broad, whitish or yellowish longitudinal stripes.

**S. — v. —** (?) **cv. Craigii.** – Longitudinal markings much broader than in v. **laurentii**, extending over almost the entire **L.**, which is shorter and narrower; slow-growing.

**S. — cv. Plumbea.** – Similar to cv. Silver Cloud but L.more greyish.

**S. — cv. Silver Cloud.** – **L.** whitish-green at first, becoming dark green, transverse markings almost completely absent.

**S. zeylanica** Willd. – Ceylon. – Acauline, rhizome creeping; **L.** 5–11 on one shoot, linear or narrow-lanceolate, 45–75 cm long, 8–20 mm across, 5–8 mm thick, folded longitudinally to form a groove, with a subulate tip 13–38 mm long, dark green, transverse markings light green, underside with 4–7 darker green, longitudinal lines.

**Sarcocaulon** (DC.) Sweet (*Monsonia* DC.). Geraniaceae. – Angola; S.W.Afr. to Cape: Lit. Namaqualand. – Mostly low-growing or small ♄ with spreading Br., sometimes thorny, St. and Br. with a resin coating; **L.** opposite and decussate, variable in size, early deciduous; **Fl.** solitary, white, yellowish, red or pink. – Greenhouse, warm. Growing period short, in winter. – Propagation: seed.

**S. burmannii** (DC.) Sweet emend. Rehm. (Pl. 124/2) (*Monsonia b.* DC.). – Cape: Bushmanland. – ♄ 10–30 cm tall, with a grey bark; **Br.** 10–15 cm long, 5–10 mm thick, glabrous, Th. 2–3 cm long; **L.** with entire or irregularly dentate margins, obcordate, 10–15 mm long, 10–13 mm across, leathery, petiole 2 mm long; Fl. white to pale pink, Pet. 2 cm long, 15 to 18 mm across.

**S. crassicaule** Rehm. – Cape: Lit. Namaqualand. – ♄ 20–30 cm tall; **Br.** up to 15 cm long, 15–20 mm thick, tuberculate, brown, bark ± splitting, Th. stout, recurved, whitish-grey; **L.** ± leathery, bluish, sinuate, 15–19 mm long, 14–19 mm across, petiole 1 mm long; **Fl.** on shortly hairy pedicels c. 4 cm long, Pet. 25 mm long, 22 mm across, yellowish to white.

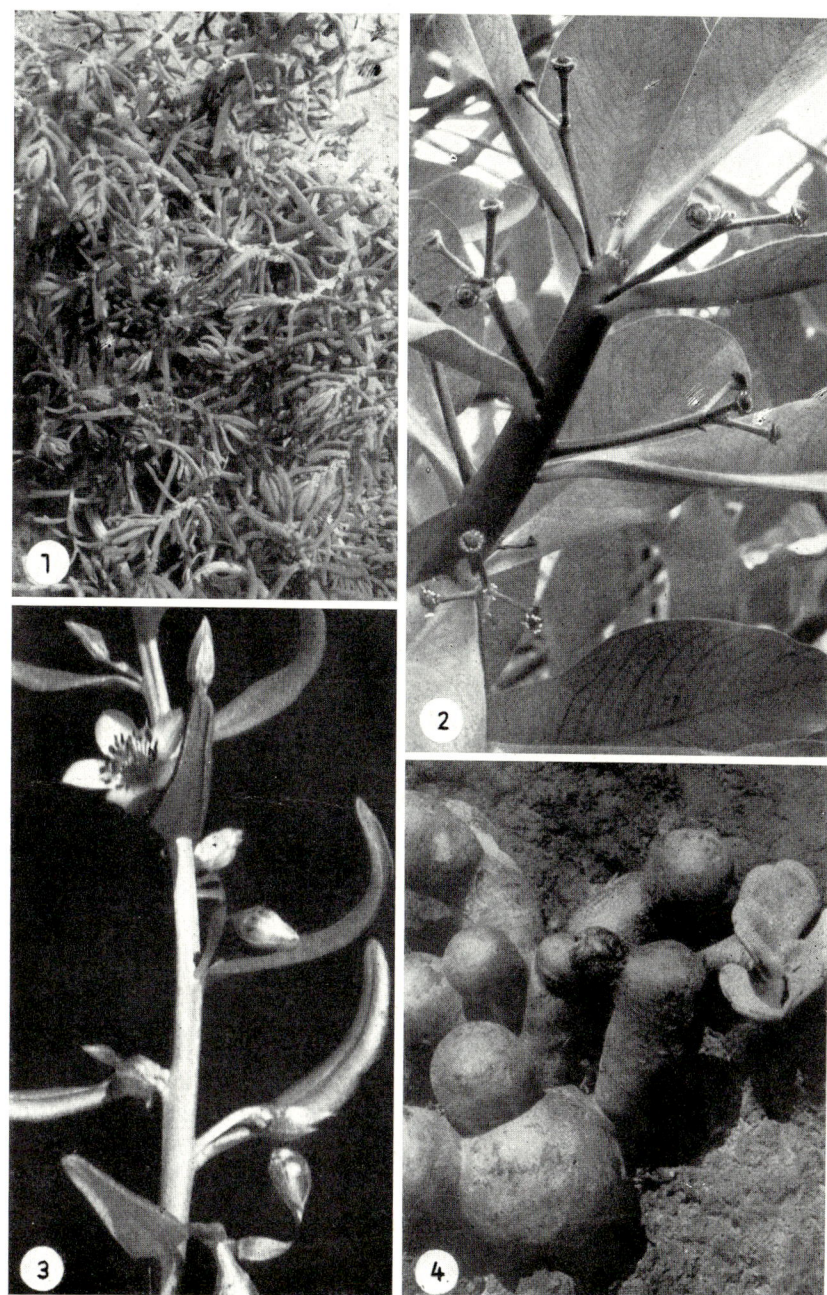

Plate 138. 1. Suaeda maritima (L.) Dum.; 2. Synadenium grantii Hook. f.; 3. Talinum esculentum Dtr. et Schellenbg. (Photo: K. Dinter); 4. Talinum guadalupense Dudl. (Repr.: Cact. Succ. J. Am. 1951/2, Photo: G. Lindsay).

Plate 139. 1. Testudinaria elephantipes (L'Her.) Lindl. f. elephantipes; 2. Thompsonella platyphylla Rose (Photo: R. Moran); 3. Trematosperma cordatum Urb. (Photo: P. R. O. Bally); 4. Tradescantia navicularis Ortg.

Plate 140. 1. **Trichocaulon clavatum** (WILLD.) H. HUBER; 2. **T. halenbergense** DTR. (Photo: E. RUSCH.); 3. **Umbilicus rupestris** (SALISB.) DANDY v. rupestris; 4. **Vauanthes dichotoma** (L.) KTZE. (Photo: H. HERRE).

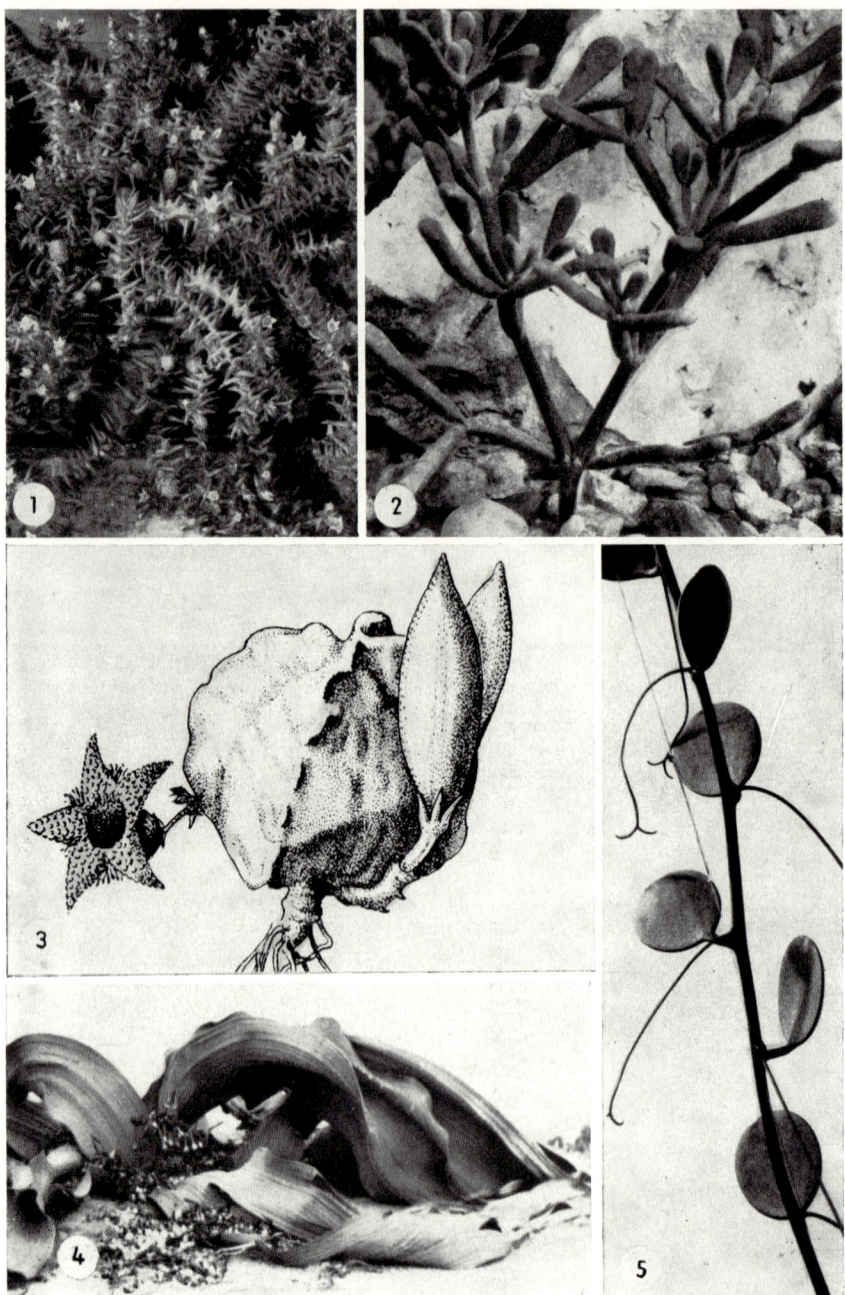

Plate 141. 1. **Villadia guatemalensis** Rose; 2. **Zygophyllum fontanesii** Webb et Berth.; 3. **White-sloanea crassa** (N. E. Br.) Chiov. (Drawing: P. R. O. Bally acc. Fl. Pl. Afr. 34, Pl. 1338, 1961); 4. **Welwitschia mirabilis** Hook. f., female plant (Photo: W. Triebner); 5. **Xerosicyos danguyi** Humbert.

CR = Chamber-roofs
VW = Valve-wings
C = Calyx
F = Funicle
P = Placenta
T = Placental-tubercle
R = Placental-rampart(-wall)
PW = Partition wall
× = acc. S. Dupont
+ = acc. J. A. Huber
* = acc. G. Schwantes

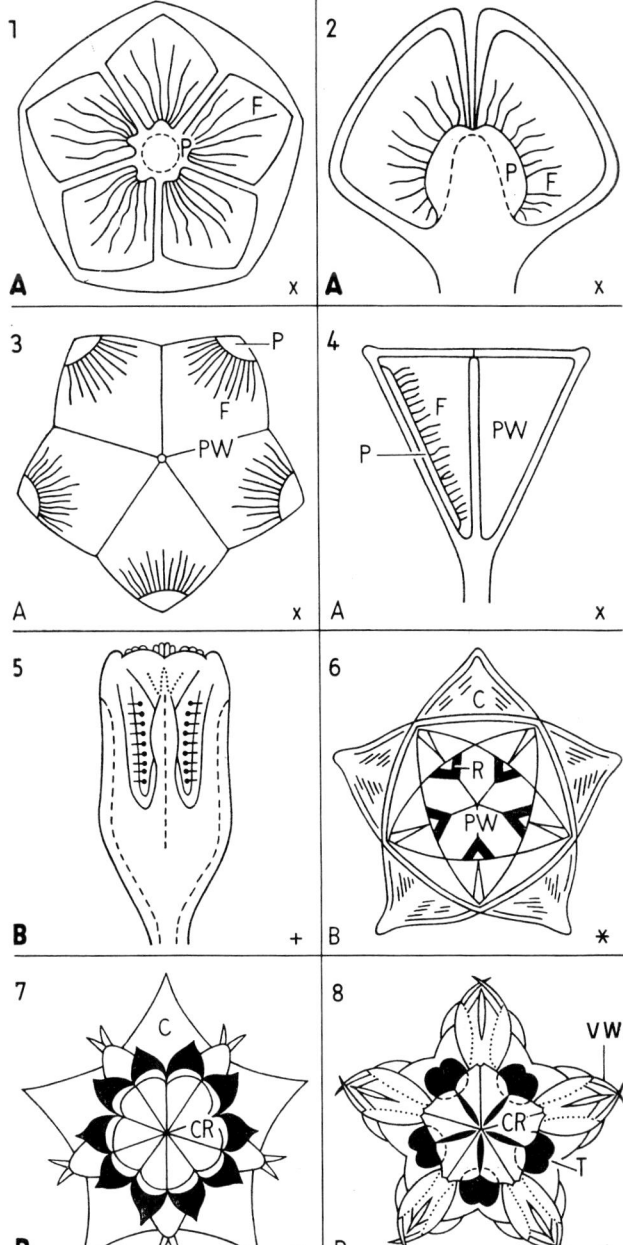

**Plate 142.** For Part II: Key to the genera of the Family Mesembryanthemaceae, (p 398)

1. **Mesembryanthemum crystallinum**
   Cross-section of capsule
2. **Mesembryanthemum crystallinum**
   Longitudinal-section of capsule
3. **Aethephyllum pinnatifidum**
   Cross-section of capsule
4. **Aethephyllum pinnatifidum**
   Longitudinal-section of capsule
5. **Carpobrotus aequilaterus**
   Longtudinal-section of fruit
6. **Skiatophytum tripolium**
   Opened capsule
7. **Dorotheanthus apetalus**
   Opened capsule
8. **Rhombophyllum nelii**
   Opened capsule

Plate 143. 1. **Acrodon bellidiflorus** (L.) N. E. Br. v. **bellidiflorus**; 2. **Aethephyllum pinnatifidum** N. E. Br. (Photo: G. D. Rowley); 3. **Aloinopsis rosulata** (Kensit) Schwant. (Photo: G. Schwantes).

Plate 144. 1. Aloinopsis lodewykii L. Bol. (Photo: G. Kaiser); 2. A. rubrolineata (N. E. Br.) Schwant. (Photo: Kakteen-Haage, Erfurt); 3. Apatesia helianthoides (Ait.) N. E. Br.; 4. Aloinopsis villetii (L. Bol.) L. Bol. (Photo: G. C. Nel); 5. Antegibbaeum fissoides (Haw.) Schwant. ex C. Weber; 6. Aloinopsis schoonesii L. Bol. v. schooneesii (Photo: G. Schwantes).

Plate 145. 1. Arenifera pillansii (L. Bol.) Herre (Photo: H. Herre); 2. Argyroderma delaetii Maass (Photo: J. Derenberg; 3. A. fissum (Haw.) L. Bol.; 4. Aptenia cordifolia (L. f.) Schwant. (Photo: G. Schwantes).

**S. flavescens** REHM. – S.W.Afr.: Gr. Namaqualand. – ♄ 15–20 cm tall; **Br.** short, c. 1 cm thick, Th. 1–2 cm long, thin; **L.** round, bluntly dentate, shortly hairy, 10–14 mm long, 8–16 mm across; **Fl.** with a pubescent pedicel 11–14 mm long, Pet. 10–20 mm long, 15–17 mm across, yellowish.

**S. herrei** L. BOL. – Cape: Richtersveld. – Small, branching ♄; **Br.** thorny, up to 25 cm long, Th. 25 mm long, L.scars 8 mm ⌀, herbaceous parts silky; **L.** ± round to triangular, doubly or trebly incised, densely woolly at the tips of the segments, 18 mm long, 12 mm across; Fl. 22 mm ⌀, Pet. white, lemon-yellow below.

**S. inerme** REHM. – S.W.Afr.: Gr. Namaqualand. – ♄ 20–30 cm tall; **Br.** up to 20 cm long, 7–14 mm thick, rough, brown, thornless; **L.** thin, ovate, margins crenate-dentate, rather hairy or felted, 15–29 mm long, 11–17 mm across, petiole 2–5 mm long; **Fl.** with a hairy pedicel 3.5 cm long, Pet. up to 25 mm long, 17 mm across, pale violet.

**S. l'heritieri** (DC.) SWEET v. **l'heritieri** (*Monsonia l'h.* DC.). – S.W.Afr.: Namib. – Erect ♄; **Br.** slender, with some short shoots, up to 30 cm long, all green parts waxy-pruinose, glabrous, with sharp Th. 7 mm long in 4 series; **L.** long-cordate, 2–3 cm long; **Fl.** 22 mm ⌀, yellow.

**S. —** v. **brevimucronatum** SCHINZ. – **L.** apiculate.

**S.** **mossamedense** (WELW.) HIERN (*Monsonia m.* WELW., *S. marlothii* ENGL.). – S.W. Afr. – ♄ (30–)50(–150) cm tall; **Br.** 1 cm thick below, dividing above, covered with sharp, thorny pedicel-bases; **L.** cordate, entire, c. 1 cm long and across, margins minutely dentate, petiole 3–4 mm long; **Fl.** on a pedicel 3 cm long, 2 cm ⌀, red.

**S. multifidum** R. KNUTH (Pl. 124/1) (*Monsonia m.* E. MEY.). – S.W.Afr.: Orange R. – **Br.** horizontal, finger-thick, thornless, 8–10 cm long; **L.** in clusters on the tubercles which are arranged in 2 series, 7–10 mm long, fine-laciniate, woolly, on a petiole 8–10 mm long; **Fl.** with pink Pet. with a red blotch below.

**S. patersonii** (DC.) ECKL. et ZEYH. – Type: ssp. **patersonii**.

**S. —** ssp. **badium** REHM. – Cape: Lit. Namaqualand. – **Br.** cylindrical, glabrous, brown, 15 cm long, 5–8 mm thick, Th. 12–28 mm long, brown; **L.** 7 mm long, margins glabrous or slightly hairy below; **Fl.** pink.

**S. —** ssp. **curvatum** REHM. – S.W.Afr.: Namib. – **Br.** cylindrical, glabrous, gnarled, brown, 2.5–6 mm thick, 7–10 cm long, curved downwards, Th. 5–13 mm long, recurved, thin, brittle; **L.** 3–6 mm long, margins hairy below, narrow-cuneate to cordate; **Fl.** pink.

**S. —** ssp. **patersonii** (*Monsonia p.* DC., *S. p.* ssp. *typicum* REHM). – S.W.Afr.: Gr. Namaqualand. – Dwarf ♄; **Br.** wrinkled, light grey, up to 20 cm long, 5–8 mm thick, Th. light, straight, 12–30 mm long, decurrent for 6–11 mm along the Br.; **L.** (8–)10(–11) mm long, margins shortly hairy; **Fl.** on a pedicel 7–9 mm long, Pet. 12 to 16 mm long, red.

**S. rigidum** SCHINZ. – Type: ssp. **rigidum**.

**S. —** ssp. **glabrum** REHM f. **glabrum**. – S.W. Afr.: Gr. Namaqualand. – **Br.** ascending; **L.** with a distinct waxy coating, bluish-green; **Pet.** cuneate, rounded at the tip, 20–22 mm long, 18 mm across.

**S. —** ssp. **—** f. **parviflorum** REHM. – Pet. 13–18 mm long, 11–15 mm across.

**S. —** ssp. **rigidum** (*S. r.* v. *typicum* REHM). – S.W.Afr.: Lüderitz Bay, coasts. – Dwarf ♄ with a water-storing St. 3–4 cm thick; **Br.** horizontal, shoots of finger-length and thickness, terete, ± thorny, Br. and shoots with a thick, inflammable, resinous coating ('Bushman's candle'); **L.** few, c. 8 mm long, bilobed, vivid green, without any waxy coating; **Fl.** brilliant red, Pet. 18–20 mm long, 10 mm across, narrow-cuneate.

**S. spinosum** (BURM.) O. KTZE. emend. REHM. v. **spinosum** (*Monsonia s.* BURM.). – S.W.Afr.: Gr. Namaqualand; Cape: Lit. Namaqualand. – **Br.** up to 15 cm long, 9–15 mm thick, with Th. 2.5–7 cm long, 2–3 mm thick below; **L.** 1 to 2.5 cm long, 14 mm across, dentate, densely hairy, variable in shape; **Fl.** on pedicels 2–7 cm long, Pet. 19–24 mm long, 14–17 mm across, yellowish.

**S. —** v. **hirsutum** REHM. – S.W.Afr.: Gr. Namaqualand. – **Th.** 3–5 cm long, 3 mm thick below; Sep. with silky, projecting hairs, Pet. 23–32 mm long, 16–18 mm across.

**S. vanderietiae** L. BOL. – Cape: Albany D. – Compact ♄ 12 cm tall, 25 cm ⌀; **St.** 1 cm thick, Br. short, 5–8 mm thick, Th. thin, straight, up to 3.5 cm long, on a round, scutate base; **L.** thick, glabrous, glossy, emarginate, 6–8 mm long and across; **Fl.** 4 cm ⌀, pale pink.

**Sarcostemma** R. BR. Asclepiadaceae. – Trop. and subtrop. As.; Afr.; Austr.; Madag. – Succulent sub-♄; **Br.** usually only pencil-thick, ± jointed, green, leafless; **Fl.** in shortly pedunculate umbels, small, whitish, Cor. rotate, with projecting or recurved lobes, corona double. – Greenhouse, warm. Propagation: seed, cuttings.

**S. andongense** HIERN. – S.W.Afr.: Klinghardt Mts. – **Br.** greenish-white, pencil-thick, often rooting.

**S. aphyllum** (THUNBG.) R. BR. (*Asclepias a.* THUNBG.). – Cape. – Erect ♄ with climbing, leafless **St.**, alternately branched, glabrous and faintly 4-angled; Ped. 1–4 in lateral umbels; involucre as long as Ped.

**S. australe** R. BR. – Austr. – Erect ♄ 1–2 m tall; **St.** cylindrical, freely and densely branched; **Br.** and shoots cylindrical, 5–10 mm thick, fleshy, succulent, green to grey-green, often whitish and pruinose, erect or climbing up other bushes; **L.** scale-like, early deciduous; **Fl.** (6–) 8(–10) in nodal clusters.

**S. brunonianum** WIGHT et ARN. – Peninsular India. – Twining or prostrate ♄, distantly dichotomous, with joints 5–6 cm long, 5–6 mm thick, cylindrical, green; **Fl.** 8–12 in umbels, whitish.

**S. decorsei** Cost. et Gall. (*Drepanostemma luteum* Jum. et Perr.). – Madag. – **Shoots** 50 to 100 cm long, branched, cylindrical, c. 4 mm thick, distinctly nodal, bluish; **L.** reduced ± to scales; **Infl.** of 4–10 Fl., Cor. 5 mm ⌀, lobes 2–2.5 mm long.

**S. insigne** (N. E. Br.) B. Desc. (*Platykeleba i.* N. E. Br.). – Madag. – Erect ♄; **shoots** numerous, branched, 30–50 cm tall, cylindrical, 2.5 to 3.5 mm ⌀, minutely longitudinally striate, glabrous, pale red to green; **L.** scale-like, triangular, ± 1 mm long; **Infl.** of 2–4 Fl. c. 15 mm ⌀, Cor. campanulate, lobes broadly triangular.

**S. madagascariense** B. Desc. (*Decanemopsis aphylla* Cost. et Gall.). – Madag.: Cape Sainte Marie. – Liana-like, with milky sap; leafless, or **L.** much reduced; **Fl.** in groups of 2–3, terminal or along the shoots, Cor. fleshy.

**S. socotranum** Lavr. – Socotra. – Small ♄; **St.** and **Br.** up to 1.5 m long, 2–5 mm ⌀, trailing or hanging, terete, the nodes much constricted, Int. 2–5 cm long; **L.** deltoid, minute, soon drying but persistent; **Infl.** umbellate with 2–8 Fl., pedicels c. 3 mm long, **Cor.** glabrous, campanulate, greenish, lobed ± to the base, soft, 3.5 mm long, 1.2 mm across; corona double, c. 2 mm ⌀, white.

**S. viminale** R. Br. (Pl. 124/3) (*Euphorbia v.* L., *Cynanchum aphyllum* L.). – Trop. and subtrop. Afr.; S.W.Afr.; Cape. – **Br.** erect or pendulous, cylindrical, 4–5 mm thick, jointed, dichotomous, with triangular scale-L. on the joints, light grey-green, often brownish, with a milky sap; **Fl.** several together, in shortly pedunculate umbels, c. 1 cm ⌀, white.

× **Sedeveria** E. Walth. Crassulaceae. – Hybrids intermediate between the parents: Sedum × Echeveria. – Summer: in the open, winter: cold-house. Propagation: cuttings.

× **S. hummelii** E. Walth. (Pl. 124/4). – Hybrid: **Sedum pachyphyllum** × **Echeveria derenbergii.** – Appearance as **S. pachyphyllum,** more luxuriant; sub-♄ up to 10 cm tall, branching from the base; **L.** numerous, ± in a spiralled Ros., obovate-oblong to cuneate, rounded or tapering above, thick, up to 3.5 cm long, sea-green, reddened at the tip; **Infl.** several, axillary, Ped. thin, orange-brown, branchlets with 5 or more Fl., Pet. spreading, acute, thick, carinate, 10 mm long, brilliant yellow.

**Sedum** L. Crassulaceae. – N. temp. and cool reg.; several species in Central Afr., Madag.; S.Am. – ☉, ☉, ♃, sub-♄ and ♄ of various shapes; **rootstock** sometimes woody or caudiciform, fleshy; roots often fleshy and turnip-shaped; **St.** ± fleshy, in several species becoming woody and perennial or in part dying off annually, mostly dichotomously branched, erect or procumbent or creeping and rooting; **L.** mostly alternate, more rarely opposite or verticillate, flat, entire or dentate or deeply incised or ± cylindrical, often elongated spur-like over the base, mostly more crowded on the sterile shoots than on the fertile where they diminish towards the apex and pass into bracts; **Infl.** mostly cymose, the Br. with scorpioid cymes and bracts; more rarely the Br. short and the Fl. more crowded; **Fl.** ± stalked or sessile, (5–)6–7(–9)-partite, mostly conspicuously coloured, white or yellow, more rarely reddish, violet or red, still more rarely blue. – Partly hardy. Species from the S. hemisphere also some N.American species in summer outdoors, winter: cold-house. – Propagation: seeds, cuttings, division. (Lit.: Froederstroem, H.: The Genus Sedum L. In Acta Hort. Goth. **5:** 3–75; **6:** 3–11; **7:** 3–126; **10:** 2–262, Goteborg 1930–1935. Praeger, R. L.: An Account of the Genus Sedum as found in Cultivation. In Jour. Roy. Hort. Soc. **46:** 1–314, London 1921. Reprinted London 1967).

### Division of the Genus Sedum into Sections after A. Berger[1])

Sect. 1. *Rhodiola* (L.) Scop. = **Rhodiola** L.

Sect. 2. *Pseudorhodiola* Diels = **Rhodiola** L.

Sect. 3. **Telephiastrum** S. F. Gray. – ♃; **rootstock** short, thick or with tuberous roots; **St.** ± erect, withering in autumn, arising from lateral buds of the St. of the last year; **L.** flat, broad, alternate, opposite or verticillate; **Fl.** in flat or spherical corymbs, 5-partite, ⚥, white, reddish or rose-red or greenish. Hardy species. – N. temp. reg. – Species: S. alboroseum, anacampseros, cauticolum, cyaneum, ewersii, hidakanum, maximum, ningjinianum, pallescens, pluricaule, rosthornianum, shimizuanum, sieboldii, sordidum, spectabile, taquetii, tatarinowii, telephioides, telephium, tsugarense, verticillatum.

Sect. 3a. **Sinosedum** Fu. – **Roots** fibrous; **L.** not spurred, sessile, alternate, thick-fleshy; **Infl.** compact, cymose, Fl. large, Sep. not spurred, Pet. free at the base, white, carpel not constricted at the base. – China. – Species: S. paochingense.

---

[1]) in A. Engler, Natürl. Pflanzenf. Ed. II, 18a: 436–462, 1930. – Sect. 3a and Sect. 11a are included later.

**Sect. 4. Sedastrum** (ROSE) BGR. (*Sedastrum* ROSE as a genus). – Evergreen plants; **rootstock** ± fleshy, thickened, roots fibrous; basal **L.** rosulate, thick; **St.** usually flowering in the second year, ± hairy, erect, 10–30 cm tall or more, then dying off and renewed by new lateral shoots; **Fl.** 5-partite, white. – Mexico. – Species: S. ebracteatum, glabrum, hemsleyanum, hintonii, rubricaule.

Sect. 5. *Hasseanthus* (ROSE) BGR. (*Hasseanthus* as a genus) = **Hasseanthus** (ROSE) MORAN Sub-Genus 3 of the genus **Dudleya** BR. et R.

**Sect. 6. Lenophyllopsis** BGR. – Low ♃ with small Tub. or tuberous **rootstock; L.** partly rosulate, alternate and elongate; **cyme** few-flowered. – Mexico; USA: California. – Species: S. cockerellii, flaccidum, lenophylloides, minimum, napiferum, niveum, pinetorum, pringlei.

Sect. 7. *Graptopetalum* (ROSE) BGR. (*Graptopetalum* ROSE as a genus) = **Graptopetalum** ROSE.

**Sect. 8. Monanthoidea** BATT. et TRABUT. (*Monanthella* BGR. perhaps as a genus)[1]. – ♃ with a **taproot** and dense L.-Ros., from the L. axils from which offsets arise; **L.** fleshy, obovate-spatulate, subobtuse, suddenly narrowed into a long petiole below, glandularly soft-hairy all over, especially at the margins, offsets elongated, ending in a Ros., blooming in the first or second year, L. of the flowering St. hardly petiolate; **Fl.** in a bracted scorpioid cyme, 7–10-partite. – N.Afr. – Species: S. atlanticum, jaccardianum.

**Sect. 9. Populisedum** BGR. – Sub-♄ with erect branches 20–40 cm high; **L.** alternate, long-stalked, ovate or ovate-cordate, irregularly large and obtusely dentate; cyme flat and rounded; **Fl.** white to pale pink, scented. Hardy species. – USSR: Siberia. – Species: S. populifolium.

**Sect. 10. Perrierosedum** BGR. – ♄, freely branched, Br. 4-angled; **L.** opposite, oblong-spatulate, sessile, rounded, very minutely toothed, narrowed towards the base; **Fl.** 5–10 in a short-stalked cyme, on pedicels, 6-partite. – Madag. – Species: S. madagascariense.

**Sect. 11. Pachysedum** BGR. – Sub-♄ or ♄ with a fleshy St.; **L.** thick, subcylindrical or cylindrical or much thickened on the back surface; **Infl.** lax; **Fl.** white or yellow. Tender species. – Mexico. – Species: S. adolphii, cremnophila, craigii, eichlamii, guatemalense, lucidum, luteoviride, morganianum, nussbaumeranum, pachyphyllum, rubrotinctum, treleasei, viride.

**Sect. 11a. Centripetala** ALEXANDER. – Succulent ♄, glabrous; **L.** alternate, terete or flattened, round ovate; **Infl.** paniculate; **Fl.** centripetal. – Mexico. – Species: S. allantoides, platyphyllum.

**Sect. 12. Dendrosedum** BGR. – Sub-♄ or ♄ with a fleshy St.; **L.** flat, thickish; **Infl.** terminal or lateral; **Fl.** deep yellow. – Mexico. – Species: S. aoikon, botteri, compressum, confusum, dendroideum, palmeri, purpusii, torulosum.

**Sect. 13. Frutisedum** BGR. – Sub-♄ or ♄ with a fleshy St.; **L.** flat, rather narrow and thin; **Fl.** white, reddish or yellowish. Tender or almost hardy. – Mexico. – S. amecamecanum, chloropetalum, conzattii, cuspidatum, frutescens, hultenii, oxypetalum, quevae, pulvinatum, retusum, tortuosum, tuberculatum.

**Sect. 14. Leptosedum** BGR. – Small ♄ or sub-♄ with a slender but fleshy St.; **L.** subulate or narrow-linear; **Fl.** white or greenish-white. – Mexico. – Species: S. bourgaei, brandtianum, diffusum, griseum, guadalajarum.

**Sect. 15. Afrosedum** BGR. – Erect or spreading small ♄ or **sub-♄**, mostly freely branched; **L.** alternate, short, ± cylindrical or flattened on the upper surface, obtuse or subacute; **cyme** short-stalked, with 1–3 simple Br.; **Fl.** yellow. – High mountains of Afr. up to the Canary Is. and Madeira. – Species: S. brissemoretii, churchillianum, epidendrum, fusiforme, lancerottense, multiceps, nudum, ruwenzoriense.

**Sect. 16. Aizoon** KOCH. – ♃; **rootstock** woody, often thickened, with fibrous roots; **St.** dying each year, the young shoots often appearing already in autumn (those of **S. hybridum** evergreen); **L.** alternate, flat, mostly dentate towards the apex; **Fl.** yellow. Hardy species. – N.Asia. – Species: S. aizoon, floriferum, hybridum, kamtschaticum, kurilense, maximowiczii, selskianum, sichotense, yabeanum.

**Sect. 17. Sedum** (Seda Genuina KOCH). – ♃ without a real rootstock (also ⊙), mostly evergreen and usually with caespitose, creeping and ascending sterile St.

    **Group 1. Involucrata** MAX. – ♃ with creeping sterile St. which produce ± dense clumps, often with rosulate L.; upper L. often alternate, with a cuneate base, margins papillately ciliate, the papillae hair-like towards the petiole; **cyme** dichotomous, flat; **Fl.** short-stalked. Hardy species. – USSR: Caucasus; China. – Species: S. baileyi, crenatum, involucratum, spurium, stevenianum, stoloniferum.

---

[1] See Berger's observations under **Monanthella** BGR., l.c. 446.

**Group 2. Propontica** BGR. – ♃; **rootstock** creeping, with erect or ascending St., or mostly subterranean, very short, small, bearing tuberous buds the size of a pea or hazelnut, which consist of shortened and thickened L.; these buds produce an overground **L.-Ros.** in autumn and later the flowering St.; **L.** opposite, the upper alternate, ± entire, margin minutely papillose; **Infl.** a 3- to many-branched cyme with bracts. – As. Minor. – Species: S. bornmuelleri, listoniae, obtusifolium.

**Group 3. Ternata** BGR. – ♃; **L.** flat, in whorls of 3–4; **Fl.** whitish or red. Hardy species. – N.Am. – Species: S. nevii, rhodocarpum, ternatum.

**Group 4. Filipes** (FROED.) FU (Ser. *Stapfiana* BGR. p.part.). – ♃; **rootstock** somwhat thickened; **St.** very slender; lower **L.** opposite, upper L. larger, in whorls of 2–4, crowded; **Infl.** shortly forked, few-flowered; Fl. very small. – E.As. – Species: S. filipes.

**Group 5. Americana** BGR. – ♃; **L.** flat, rosulately crowded on the sterile shoots; **Fl.** white. – N.Am.; Peru. – Species: S. backebergii, bellum, burhamii, caducum, californicum, chontalense, eastwoodiae, griffithsii, laxum, longipes, lumholtzii, madrense, moranii[1]), puberulum, purdyi[1]), versadense, wootonii, wrightii.

**Group 6. Alsinefolia** BGR. – Dainty shade- or mountain-plants; **L.** ± rosulately crowded on the sterile shoots, ± spatulate or elongate; **Fl. Sc.** and the lax Infl. slender, glandularly hairy; **Fl.** stalked. – Eur.; As. – Species: S. adenotrichum, alsinefolium, rosulatum.

**Group 7. Compacta** BGR. – ♃, small, tufted; **L.** roundish, 3–5 mm long, in short or hardly elongated Ros.; **Fl. Sc.** short, few-flowered. – N.Am. – Species: S. compactum.

**Group 8. Alamosana** PRAEG. – Low ♃; sterile **shoots** catkinlike, densely leafy; **L.** 6 mm long, papillose; **Fl.** only few on laxly leafy, long stalks. – N.Am. – Species: S. alamosanum.

**Group 9. Moranense** BGR. – Small ♃ branched from the base, habit like **S. acre,** but with few, white **Fl.** Hardy species. – N.Am. – Species: S. furfuraceum, liebmannianum, moranense, pentastamineum.

**Group 10. Pulchella** BGR. – Low ♃; **L.** narrow-linear, densely arranged on the shoot; **Infl.** dense. – N.Am. – Species: S. calcaratum, mellitulum, oxycoccoides, potosinum, pulchellum, semiteres, stelliferum.

**Group 11. Alba** BGR. – Low, caespitose ♃; **L.** in whorls of up to 4, alternate or opposite on the leafy St., thick, cuneate-ovate or globose, clavate to cylindrical or somewhat flattened or channelled. Hardy species. – Eur.; As. – Species: S. acutifolium, album, anglicum, brevifolium, dasyphyllum, erythraeum, farinosum, gracile, hirsutum, kostovii, lydium, magellense, monregalense, polystriatum, serpentinii, tenellum, tristriatum.

**Group 12. Stahliana** BGR. – **St.** thread-like, minutely hairy; **L.** opposite, oblong-spherical or ovate; **cyme** 2–3-branched; **Fl.** yellow. Tender species. – Mexico. – Species: S. obcordatum, stahlii.

**Group 13. Oaxacana** BGR. – **St.** freely branched; **Br.** spreading, rooting, minutely rough tuberculate; **L.** scattered along the St., sessile with a broad base; **Fl.** yellow. Tender species. – Mexico. – Species: S. oaxacanum.

**Group 14.** Diversifolia BGR. – Not maintained as a Series.

**Group 15. Humifusa** BGR. – Low plant; **L.** small or minute, ± flat or cylindrical, ciliate at the margin, densely imbricate on the procumbent, catkin-like St.; **Fl.** yellow. Tender species. – Mexico. – Species: S. humifusum.

**Group 16. Alpestria** BGR. – Tiny, low ♃; **shoots** not catkin-like; **L.** broader towards the apex, glabrous; **Fl.** yellow. Hardy species. – Eur. – Species: S. alpestre, horakii.

**Group 17. Acria** BGR. – Low, caespitose ♃, dichotomously branched at the base; **L.** ± imbricated, more lax on the leafy St., triangular-ovate, broadest at the base, appressed, bluntish; **cyme** forked; **Fl.** yellow. Some hardy, some not so. – N. temp. reg. – Species: S. acre, cupressoides, grandipetalum, greggii, krajonae, muscoideum, novakii, rohlenae, sartoranum, zlatiborense.

**Group 18. Mitis** BGR. – Low, caespitose ♃; **L.** ± cylindrical. Hardy species. – Eur. – Species: S. borissovae, laconicum, mite.

**Group 19. Hametiana** BGR. – Small ♃; **L.** alternate, small, ± linear or subulate; **Fl.** yellow. – E.As. – Species: S. barbeyi, beauverdii, celiae, daigremontianum, dugveyi, feddei, gagei, heckelii, multicaule, pampaninii, platysepalum, rosei, susannae, trullipetalum.

---

[1]) R. T. CLAUSEN arranged these into Subgenus *Gormania* (BRITT.) CLAUSEN of the Genus **Sedum** L.

**Group 20. Rupestria** BGR. – ♃ with numerous sterile shoots; **L.** alternate, ± cylindrical, acute, softly mucronate, spurred at the base, diminishing towards the apex of the flowering St.; **Fl.** yellow. Mostly hardy species. – Eur.; N.Am. – Species: S. forsteranum, lanceolatum, montanum, ochroleucum, pruinatum, reflexum, sediforme, stenopetalum, tenuifolium, victorianum.

**Group 21. Occidentalia** BGR. – ♃ with the habit of Gr. 20. **Rupestria; L.** flat on the upper surface and more lanceolate, often spurred and with small shoots in the axils; **Fl.** yellow. – N.Am. – Species: S. shastense.

**Group 22. Orientalia** BGR. – ♃; **L.** mostly alternate, sometimes also opposite, from linear-elongate to spatulate, obtuse, often spurred at the base; **Fl.** short-stalked, yellow. – China, Taiwan; Japan. – Species: S. alfredii, engleri, giajai, japonicum, makinoi, margaritae, morotii, pseudosubtile, shaeri, subtile, variicolor.

**Group 23. Rosulata** BGR. – ♃; **L.** flat, ± spatulate, alternate, rosulately crowded along the sterile, often offset-like shoots; **Infl.** laxly leafy, cyme usually flat. Hardy or nearly so. – Western N.Am. – Species: S. anomalum, debile, hallii, leibergii, obtusatum, oreganum, pruinosum, spathulifolium, watsonii, woodii, yosemitense.

**Group 24. Divergentia** BGR. – ♃, habit similar to that of the Gr. 23. **Rosulata,** but the **L.** much thicker and mostly opposite. Fairly hardy species. – Western N.Am. – Species: S. divergens, rubroglaucum.

**Group 25. Chinensia** BGR. – ♃; **St.** erect or procumbent; **L.** in whorls, linear to lanceolate, acute or obtuse, mostly spurred at the base, often alternate on the flowering shoots; **Fl.** in dichotomous cymes. Tender to almost hardy. – China; Japan. – Species: S. bergeri, bracteatum, chauveaudii, chrysastrum, concarpum, lineare, lungtsuanense, quaternum, sarmentosum, shigatsense, wenchuanense, yvesii, zentaro-tashiroi.

**Group 26. Galioidea** BGR. – ♃; **St.** ascending and erect, slender and delicate, glabrous, fresh green, 10 to 20 cm tall; **L.** mostly in whorls of 3–5, upper L. alternate, narrow-linear, subcylindrical or somewhat compressed above, 6–15 mm long, not spurred at the base; **cyme** flat, forked; **Fl.** yellow. Tender species. – Mexico. – Species: S. mexicanum.

**Group 27. Clavifolia** BGR. – Densely caespitose ♃; **St.** 1 to 3 cm high; **L.** 1–3 cm long, thick, upper surface flat, obtuse, contracted towards the base into a long, cylindrical stalk; **cymes** few-flowered, pedicels 5–8 mm long, **Fl.** pale greenish-yellow. – Mexico. – Species: S. clavifolium, gormiferum.

**Sect. 18. Prometheum** BGR. – Biennial to triennial plants, with many-leaved **Ros.** similar to **Sempervivum; Infl.** terminal, densely covered with leaf-like bracts up to the broad, somewhat capitate Infl.; **Fl.** 5-partite. Hardy species. – As. Minor to USSR: Caucasus. – Species: S. pilosum, sempervivoides.

**Sect. 19. Cyprosedum** BGR. – Biennial to triennial plants with lax **L.-Ros.; Infl.** terminal, a long, often narrow, many-flowered panicle upon a short, leafless scape, ± glandularly hairy; **Fl.** 5-partite, yellowish or reddish. – Cyprus; Greece: Crete. – Species: S. creticum, lampusae, microstachyum, tympheum.

**Sect. 20. Epeteium** BOISS. – ☉ or ☉☉ herbs, occasionally longer lived; **L.** usually alternate, mostly cylindrical or flat; **Infl.** ± branched, cymose, often corymbose; **Fl.** (4–)5(–9)-partite, white, yellow, red or blue. Hardy species. – N. and S. temp. reg.

A. **L.** flat and broad.
  a) **Fl.** white.
     Gr. 1. – Species: S. cepaea, jaliscanum.
  b) **Fl.** yellow.
     Gr. 2. – Species: S. drymarioides, esquirolii, formosum, meyeri-johannis, roborowskii, somenii, stellariifolium, triactinum, viscosum.
  c) **Fl.** red. Hardy.
     Gr. 3. – Species: S. stellatum.

A. **L.** ± cylindrical or thick and somewhat flattened.
  d) **Fl.** blue. Hardy ☉ species.
     Gr. 4. – Species: S. coeruleum.
  e) **Fl.** white or red. Hardy.

**Gr. 5.** – Species: S. aetnense, andegavense, atratum, confertiflorum, callichroum, candollei, constantinii, crassularia, forreri, hispanicum, jahandiezii, kotschyanum, longibracteatum, nevadense, pallidum, pedicellatum, porphyreum, pusillum, rubens, rubrum, sanguineum, steudelii, villosum.

f) **Fl.** yellow. Hardy species.

**Gr. 6.** – Species: S. annuum, assyricum, flexuosum, fui, grisebachii, hispidum, leblancae, littoreum, nanum, nuttallianum, oreades, palaestinum, przewalskii, versicolor.

**Sect. 21. Sedella** (BR. et R.) BGR. (*Sedella* BR. et R. as a genus). – Minute ☉ herbs; **L.** ovate to ovate-oblong; **Fl.** ± sessile, small, yellow, 5-partite. Tender species. – USA: California. – Species: S. congdonii, pumilum.

**Sect. 22. Telmissa** (FENZL) SCHOENL. (*Telmissa* FENZL as a genus). – Minute ☉ herbs; **L.** cylindrical, blunt; **Fl.** in a one-sided spike, sessile, 3–5-partite, whitish. – As. Minor. – Species: S. microcarpum.

Note: In Acta Hort. Gothob. **6,** App. 77 (1931) FROEDERSTROEM founded the **Section Asiatica Genuina Kyphocarpia** which was accepted by FU SHU-HSIA in Acta Phytotaxonomica Sinica Add. I. Dec. 1965, p. 115. FU added the following **Series : 1. Aizoonta** (MAX.) FU; **2. Japonicae** (MAX.) FU; **3. Bracteatae** (FROED.) FU. The new species published therein are arranged here according to A. BERGER's concept of the genus.

**S. acre** L. (§ 17/17) v. **acre** ('Biting Stonecrop', 'Wall-Pepper'). – Eur.; As. Minor; N.As. – Dwarf, cushion-forming ♃; **L.** ± imbricate, densely appressed, ovate-rhombic, obtuse, 1.5 mm long, glabrous, green; **Fl.** solitary or 2–3 together, light yellow. Hardy.

**S.** — v. **acutifolium** STEFANOFF. – **L.** narrow-elliptical.

**S.** — cv. Aureum ('*Variegatum*'). – **Shoot** tips pale yellow in early summer.

**S.** — v. **confertum** DOM. – **L.** crowded, even on flowering St., but not imbricate.

**S.** — v. **degenianum** DOM. – Hungary. – Forming lax cushions; **L.** broad-ovate; **Fl.** 3–6 together, fairly large.

**S.** — v. **fastigiatum** BECK. – Lower Austria. – **L.** distant, small, spherical; **Fl.** shoots straight, erect.

**S.** — v. **glaciale** (CLARION) DUBY (*S. g.* CLARION). – French Alps; Pyrenees. – **Shoots** creeping, Br. short; **Fl.** 3–4 together, large.

**S.** — v. **grandiflorum** BECK. – **Fl.** 16 to 18 mm ⌀.

**S.** — v. **imbricatum** BECK. – Balkans. – **L.** densely imbricate, broadly ovate.

**S.** — v. **microphyllum** STEFANOFF. – **L.** c. 1 mm long.

**S.** — ssp. **neglectum** (TEN.) MURBECK (*S. n.* TEN., *S. a.* v. *n.* (TEN.) VIS., *S. a.* v. *atlanticum* BALL., *S. a.* v. *morbifugum* CHABERT, *S. a.* v. *majus* MASTERS, *S. mawianum* HORT.). – Central Eur.; As. Minor; Algeria. – **L.** oblong, less fleshy, more projecting.

**S.** — v. **pentagonum** PACKER. – Austria: Carinthia. – Sterile **shoots** with L. in 5 series.

**S.** — v. **robustum** (DOM.) VEL. (*S. r.* DOM.). – Bulgaria. – All parts double the size of the type.

**S.** — v. **sexangulare** (L.) KOCH f. **sexangulare** (*S. s.* L.). – L. in 6 series; **Infl.** more densely leafy.

**S.** — v. — f. **elatum** (PRISZTER) JACOBS. (*S. sexangulare* f. e. PRISZTER). – Taller form from Hungary.

**S.** — v. **umbrosum** DOM. – CSSR. – Variety from shady positions; **shoots** slender, curved, over 20 cm long, laxly leafy.

**S.** — v. **wettsteinii** (FREYN.) HEGI (*S. w.* FREYN.). – Austrian Alps: near Graz. – **L.** broadly ovate, projecting ± at right angles.

**S. acutifolium** LEDEB. (§ 17/11). – USSR: Caucasus. – Resembles **S. album**; **L.** subulate, acute.

**S. adenotrichum** WALL. (§ 17/6) (*S. anoicum* PRAEG., *Cotyledon tenuicaulis* AITCH. et HEMSL., *C. papillosa* AITCH. et HEMSL.). – Himalayas. – ♃; **L.** laxly rosulate, spatulate, obtuse to rounded, glandular-hairy; **Fl.** white.

**S. adolphii** HAMET (§ 11). – Mexico. – Sub-♄ with ascending, thick-fleshy Br.; **L.** lanceolate, obtusely tapering, 3.5 cm long, 1.5 cm across, 6 mm thick, yellowish-green, margins reddish; **Fl.** white.

**S. aetnense** TIN. (§ 20/5). – E. Medit.

**S.** — v. **aetnense** (*S. a.* v. *genuinum* HAMET, *S. skorpilii* VEL., *S. albanicum* BECK., *S. erythrocarpum* PAU). – Small ☉; **L.** oblong, long-spurred, ± flattened, ciliate; **Infl.** simple or dichotomous, Fl. 5-partite.

**S.** — v. **tetramerum** (TRAUTV.) HAMET (*S. t.* TRAUTV., *Macrosepalum turkestanicum* RGL. et SCHMALH.). – **Fl.** 4-partite.

**S. aizoon** L. v. **aizoon** (§ 16) (*S. woodwardii* N. E. BR.). – N.As. – ♃; **St.** erect, up to 80 cm tall; **L.** ovate-lanceolate to linear-lanceolate, obtuse, 5–8 cm long, margins serrate below; **Fl.** many, yellow to orange, in a flat cyme. Hardy.

**S.** — cv. Aurantiacum. – **Fl.** deep orange yellow.

**S.** — v. **latifolium** MAX. – **L.** broader.

**S.** — v. **saxatile** NAKAI. – Lower-growing; **L.** narrower.

**S.** — v. **scabridum** MAX. – All over papillose.

**S. alamosanum** S. WATS. (§ 17/8). – N.W. Mexico. – ♃; sterile **shoots** 1 cm long, densely leafy; **L.** densely appressed, 6 cm long, rather blunt, grey-green, papillose below; **Fl.** shoots ascending, 7–18 cm long, **Fl.** 1 cm ⌀, reddish-white.

**S. alboroseum** BAK. (§ 3) (*S. erythrostictum* MIQ., *S. viridescens* NAKAI). – E.As. – ♃ with thickened **roots**; **shoots** 30–60 cm tall, erect;

L. narrowed teretely towards the base, ovate, bluntly dentate, grey-green, 5–7.5 cm long, 3.5–4 cm across; Fl. greenish-white. Hardy.

**S. — cv. Foliis Variegatis** (*S. a.* f. *f.-v.* RGL., *S. erythrostictum* f. *variegatum* HARA). – L. with a white M.stripe.

**S. — cv. Marginis Variegatis** (*S. a. v. m.-v.* PRAEG.). – L. with mottled margins.

**S. album** L. ssp. **album** (§ 17/11). – Eur.; USSR: Siberia; As. Minor; Medit. – Cushion-forming ♃; L. cylindrical to ovate, the upperside sometimes flattened, 6–12 mm long, dull green, glabrous or with short, papillose hairs, often reddish; Fl. white or red.

**S. — ssp. gypsicolum** (BOISS. et REUT.) MAIRE (*S. g.* BOISS. et REUT.). – Iberian peninsula; Morocco. – L. ± imbricate at the tips of the shoots, ± ovate-rhombic, 6 mm long, minutely papillose-hairy, often reddish; Fl. white.

**S. — ssp. — v. glanduliferum** BALL. (*S. clusianum* GUSS.). – Plant glandular-hairy.

**S. — ssp. — v. — f. purpureum** PAU et FONT Y QUER. – Morocco. – Purple ecotype.

**S. — ssp. teretifolium** (LAMK.) SYME (*S. t.* LAMK.). – Morocco. – L. cylindrical to ovoid-terete, obtuse, glabrous, green; Fl. red to pink.

**S. — ssp. — v. athoum** DC. (*S. a. v. a.* DC.). – L. smaller, thicker than in ssp. **teretifolium**.

**S. — ssp. — v. chloroticum** ROUY et CAM. (*S. a. v. c.* ROUY et CAM.). – Pale green in all parts.

**S. — ssp. — v. micranthum** (BAST.) DC. (*S. m.* BAST., *S. a. v. m.* (BAST.) HEGI, *S. a. v. minutiflorum* PAU). – Fl. many, small.

**S. — ssp. — v. murale** PRAEG. – L. purplish-red; Fl. pink.

**S. — ssp. — v. rhodopaeum** (PODP.) HAYEK (*S. r.* PODP., *S. a. v. r.* (PODP.) HAYEK). – L. still smaller than in v. **athoum**.

**S. — ssp. — v. purpureum** MAIRE (*S. muleyanum* SENNEN et MAURE). – Morocco. – Purple variety.

**S. — ssp. — v. teretifolium** (Pl. 125/1) (*S. a.* ssp. *t.* v. *typicum* FRANCH.). – L. terete.

**S. — ssp. — v. turgidum** DC. – Morocco. – L. turgid.

**S. alfredii** HANCE (§ 17/22). – E.As. – ♃; L. obovate-spatulate or lanceolate-spatulate, narrowed below, often spurred, 1–2.5 cm long, 3–7 mm across; Fl. 15 mm ⌀, yellow.

**S. allantoides** ROSE (Pl. 126/1/c) (§ 11a). – Mexico: Puebla, Oaxaca, ? Hidalgo. – Shoots 10–30 cm tall, 1.5 cm thick below; L. clavate to obovate or deltoid-obovate, rounded above, 2.5 cm long, 6–28 mm across above, 5–15 mm thick, margins broad, rounded, often minutely spotted in maturity; Infl. terminal, the Ped. often zigzagged, bracts rhombic-ovate, Fl. 14–19 mm ⌀, Pet. greenish-white, with a red transverse stripe.

**S. alpestre** VILL. (§ 17/16) (*S. repens* SCHLEICH). – Central Eur.: mts. – Dwarf ♄; L. on sterile shoots ± rosulate, crowded, small, obovate to ± linear, underside convex, glabrous; Fl. greenish-yellow.

**S. alsinefolium** ALL. (§ 17/6). – Italy: Maritime Alps. – Hairy ♃; L. laxly rosulate and crowded on sterile shoots, rhombic-spatulate, abruptly long-petiolate, both surfaces hairy; Infl. 10–20 cm long, Fl. small, white. Hardy.

**S. amecamecanum** PRAEG. (§ 13). – Mexico. – ♄ 15–20 cm tall; shoots smooth; L. ovate-lanceolate, short-spurred, 2 cm long, 6 mm across; Fl. in a flat cyme, small, pale yellow.

**S. anacampseros** L. v. **anacampseros** (§ 3). – Spain to the Tyrol. – ♃; shoots long, prostrate, often rooting; L. sessile, flat, with entire margins, ovate to ± circular, slightly tapered, 12 to 25 mm long, 12–18 mm across, grey, short-spurred; Infl. ± capitate, Fl. 6 mm ⌀, purple. Hardy.

**S. — v. majus** PRAEG. – Larger variety.

**S. andegavense** DC. (§ 20/5) (*Crassula a.* DC.). – W.Eur.; Algeria. – ⊙; L. spherical-ovoid; Fl. white.

**S. anglicum** HUDS. v. **anglicum** (§ 17/11). – W.Eur. – ♃, forming small cushions; L. ovate-elliptical, 4–5 mm long, green, often red; Fl. stellate, white to pink.

**S. — v. melantherum** (DC.) MAIRE (*S. m.* DC.). – N.Afr. – L. with upperside ± furrowed, grey-green; Fl. smaller, pink.

**S. — v. minor** PRAEG. – Smaller in all parts; Fl. deeper red.

**S. angustum** MAX. (§ 3). – W.China; USSR: Caucasus. – ♃; shoots simple, up to 30 cm tall; L. 3–4, verticillate, oblong, short-petiolate, 7 cm long, 2 cm across; Fl. in dense, hemispherical cymes, greenish. Hardy.

**S. anomalum** BRITT. (§ 17/23). – USA: S.W. Calif. – Close to **S. spathulifolium**; L. tapering, margins papillose; Fl. ± sessile.

**S. annuum** L. v. **annuum** (§ 20/6) (*S. saxatile* DC.). – Eur.; As. Minor; Greenland. – ⊙ to ⊙; shoots freely branched, short shoots 25–75 mm long; L. oblong-linear, 6 mm long, pale green, obtuse, slightly spurred; Fl. numerous, small, yellow.

**S. — v. alpinum** BECK. – Austrian Tyrol. – Shoots simple, more rarely branched.

**S. — v. brachysepalum** BECK. – Infl. with more numerous Fl., Sep. small.

**S. — v. epiroticum** BALD. – Greece. – Infl. glandular.

**S. — v. perdurans** MURR. – Side-shoots rooting and therefore ♃.

**S. aoikon** ULRICH (§ 12). – Mexico: Puebla. – Sub-♄ 30 cm ⌀, 10 cm tall; L. spatulate, with a rounded tip, c. 5 cm long, 2 cm across, green, with pink margins; Infl. dense, Fl. c. 1 cm ⌀, yellow. (Acc. JORGE MEYRA identical with **S. purpusii**.)

**S. assyricum** BOISS. (§ 20/6). – Close to **S. annuum**. Insignificant ⊙.

**S. atlanticum** MAIRE v. **atlanticum** (§ 8) (*Monanthes a.* BALL., *M. muralis* HOOK. f., *S. surculosum* COSS.). – N.Afr. – Mat-forming ♃ with fibrous roots; L. 20–30 in small, dense Ros., short-petiolate, spatulate, smooth, underside convex, dark green; Fl. solitary or several together, yellow to rather reddish-brown.

**S.** — v. **fuscum** EMB. (*Monanthes a.* v. *f.* (EMB.) JACOBS.). – Morocco. – **L.** brownish-red.

**S.** — v. **luteum** EMB. (*Monanthes a.* v. *l.* (EMB.) JACOBS.). – **L.** yellowish.

**S. atratum** L. (§ 20/5). – Eur.: mts. – Glabrous, usually deep purple ☉ 3–8 cm tall; **L.** ± cylindrical, narrowly cuneate; **Infl.** of few whitish, greenish or reddish Fl.

**S.** — ssp. **atratum**. – **Shoots** low-growing; **Pet.** cream-coloured with red lines.

**S.** — ssp. **carinthiacum** (HOPPE et PACH.) WEBB (*S. a.* v. *c.* HOPPE et PACH.). – **Shoots** branching from below and higher up; **Fl.** with greenish-yellow, somewhat reddened Pet.

**S.** cv. Autumn Joy. – **S. spectabile** BOREAU × **telephium** L. – Vigorous hybrid to 60 cm with very large corymbs of salmon pink to orange red **Fl.** in autumn.

**S. backebergii** v. POELLN. (§ 17/5). – Central and S.Peru. – Erect, glabrous ♄ up to 7 cm tall; sterile **shoots** short, densely leafy, flowering shoots elongated, roughly hairy; **L.** ± erect or projecting, oblong to ovate-oblong, tip rather rounded to acute, both surfaces somewhat convex, underside little tuberculate, carinate, upperside minutely papillose, c. 3–4 cm long, 2 mm across, bluish-green; **Fl.** yellow.

**S. baileyi** PRAEG. (§ 17/1). – China: Kangsi prov. – Caespitose ♃; **L.** narrowly ovate-spatulate, obtuse to rounded; with margins entire; flowering shoots 5–7 cm tall. Hardy.

**S. barbeyi** HAMET (§ 17/19). – E.As. – Dwarf ♃; **L.** broad-lanceolate, spurred; **Fl.** yellow.

**S. beauverdii** HAMET (§ 17/19). – E.As. – Dwarf ♃; **L.** linear, with a lobed spur, lobes dentate or ciliate.

**S. bellum** ROSE (Pl. 125/2) (§ 17/5) (*S. aleuroides* BITTER, *S. farinosum* ROSE). – Mexico: Durango. – Caespitose ♃; **shoots** short; **L.** ± rosulate, 20–35 mm long, 8–10 mm across, rounded above, light green, pruinose; **Fl.** in a flat, lax cyme, numerous, white.

**S. bergeri** HAMET (§ 17/25). – China: Yunnan. – ♃; **L.** linear-oblong or linear-spatulate, ± acute, 12–35 mm long, spurred; **Infl.** of many yellow Fl.

**S. borissovae** BALKOVSKY (§ 17/18). – USSR: Ukraine. – Caespitose ♃ with a thin rhizome; **shoots** 7–12 cm long, creeping; **L.** oblong-elliptical, obtuse, 3.5–5.5 mm long, c. 2 mm across, greenish-blue; **Infl.** a corymb of many yellow Fl.

**S. bornmuelleri** HAUSSKN. (§ 17/2) (*S. rhodanthum* BORNM.). – Turkey. – Close to **S. listoniae**; ♃; **caudex** with small, tuberous buds; **L.** indistinctly dentate.

**S. botteri** HEMSL. (§ 12). – Mexico: Orizaba. – Sub-♄ with angular shoots up to 12 cm long; **L.** oblanceolate or obovate, petiolate, obtuse, slightly emarginate, green, 14–50 mm long, 6–22 mm across, 2–4 mm thick; **Fl.** in a pendulous Rac., Fl. 14–19 mm ⌀, Pet. with purple or pink spots.

**S. bourgaei** HEMSL. (§ 14) (*S. acre* SESSÉ et MOC.). – Mexico. – Sub-♄ up to 60 cm tall; **shoots** erect, prostrate or pendulous, freely branched, short shoots green or papillose, becoming red; **L.** linear, sessile, sometimes short-spurred, obtuse, green, glabrous or somewhat papillose, often with both surfaces convex, 9–23 mm long, up to c. 2.5 mm across, c. 2 mm thick; **Infl.** of 1–7 Cinc., Fl. up to 17 mm ⌀, white.

**S. bracteatum** DIELS (§ 17/25). – China: Yunnan. – Close to **S. chauveaudii**; cymes tripartite, with large leaflike bracts.

**S.** — v. **emarginatum** FU. – L.tips with entire margins.

**S. brandtianum** v. POELLN. (§ 14). – Mexico. – Caespitose ♃ with horizontal, underground stolons 4–5 mm thick; **shoots** numerous, brownish-grey, 1–2 mm thick; **L.** narrowly oblong or triangular to subulate, tapering, c. 9 mm long, 2 mm across and thick, glossy green, eventually reddish-brown; **Fl.** white.

**S. brevifolium** DC. v. **brevifolium** (Pl. 125/4) (§ 17/11). – Morocco; S.W.Eur. – Caespitose, glabrous ♃; **L.** ovate to spherical, opposite in 4 series, 3–4 mm long, white-pruinose; **Fl.** white. Hardy.

**S.** — v. **induratum** COSSON. – Very rigid and hard.

**S.** — v. **quinquefolium** PRAEG. – **Shoots** twice as long and thick; **L.** in 5 series.

**S. brissemoretii** HAMET (§ 15). – Madeira. – Small ♄ with a woody St.; **L.** crowded at the Br.tips, cylindrical, obtuse; **Fl.** cymose, yellow.

**S. burhamii** (BR.) BGR. (§ 17/5) (*Gormania b.* BR.). – USA: Calif. – ♃; **L.** crowded, rosulate on the sterile shoots, spatulate, obtuse, 1–2 cm long, 1 cm across; **Fl.** whitish-pink.

**S. caducum** R. T. CLAUSEN (§ 17/5). – Mexico: Tamaulipas. – **Shoots** papillose, creeping, branching basally; **L.** ovate-rhombic, tapering, 7–22 mm long, 4–12 mm across, 1–3 mm thick, densely red-spotted, margins minutely crenate; **Fl.** c. 1 cm ⌀, white.

**S. calcaratum** ROSE (§ 17/10). – Mexico: Hidalgo. – Close to **S. potosinum**; ♃, 5–8 cm tall; **Fl.** red.

**S. californicum** BRITT. (§ 17/5). – USA: Calif. – ♃ with small Ros.; **shoots** 15–20 cm tall; **L.** 3 mm long, 8–10 mm across; **Fl.** in a forked cyme, white.

**S. callichroum** BOISS. (§ 20/5). – Iran. – ☉ 2–6 cm tall; **L.** clavate; **Fl.** pinkish-red. Insignificant.

**S. candollei** (DC.) HAMET (§ 20/5) (*Cotyledon sedoides* DC., S. s. (DC.) ROTHM., *Umbilicus s.* DC.). – Portugal. – Insignificant ☉ 3–6 cm tall; **L.** clavate; **Fl.** pink.

**S. cauticolum** PRAEG. (§ 3). – Japan. – ♃ with prostrate, dark purple **shoots**; **L.** ± round to spatulate, 25 mm long, 18 mm across, minutely red-spotted, with about 2 T. on each side; **Fl.** pink. Hardy.

**S. celiae** HAMET (§ 17/19). – China: Yunnan. – ♃ 5 cm tall; **L.** linear, apiculate, 6 mm long, 1 mm across, 0.5 mm thick; **Fl.** greenish-yellow.

**S. cepaea** L. v. **cepaea** (§ 20/1) (*Cepaea caesalpinii* FOURR.). – Central and S.Eur. – ☉ – ⊙, forming a lax Ros.; **shoots** erect, hairy, 30 cm tall; **L.** linear-ovate, often in whorls of 3, red-spotted; **Fl.** in a lax Pan., white.

**S. — v. galioides** (ALL.) DC. (*S. g.* ALL.). – With the lower L. verticillate.

**S. — v. glabrum** VAND. – Glabrous.

**S. — v. tetraphyllum** (SIBTH. et SM.) DC. (*S. t.* SIBTH. et SM.). – L. in whorls of 4.

**S. chauveaudii** HAMET (§ 17/25). – China: Yunnan. – ⚃ with prostrate **shoots**; L. oblong-lanceolate, obtuse, 12–20 mm long, 5 mm across, light green; Fl. yellow.

**S. chloropetalum** R. T. CLAUSEN (§ 13). – Mexico: Oaxaca. – ♄ up to 75 cm tall, Br. papillose, spreading, up to 1 m long; L. short-spurred, 3–29 mm long, 2.5–8 mm across, 1.2 mm thick, yellowish-green, obtuse; Fl. in a divided Infl., 10–12 mm ⌀, Pet. yellowish-green with red stripes.

**S. chontalense** ALEXANDER (§ 17/5). – Mexico: Oaxaca. – ⚃ branching basally; **shoots** fleshy, hairy, at first short with a Ros., eventually elongated up to 15–20 cm long and prostrate, rooting at the nodes; L. broad-cuneate, 10 to 18 mm long, 5–10 mm across, fleshy, thick, ± truncate at the tip, margins rounded and curved upwards, hairy, upperside dark green, underside purplish-red; Fl. c. 1 cm ⌀, whitish-pink.

**S. chrysastrum** HANCE (§ 17/25). – S.China. – ⚃; L. long linear-spatulate, underside convex; Fl. yellow.

**S. churchillianum** ROBYNS. et BOUTIQUE (§ 15). – Trop. Afr.: Uganda. – Erect ♄ up to 40 cm tall; Br. ± 4-angled, bark becoming folded, shoots densely leafy; L. ± sessile, oblong-elliptical to oblong-ovate, 15–30 mm long, 7–14 mm across, fleshy and translucent-membranous, with entire margins and reticulate, reddish veins, red-tipped; Infl. a corymb with yellow Fl. 2.5–3 cm ⌀.

**S. clavifolium** ROSE (§ 17/27). – Mexico. – ⚃ with a branched rhizome 11–33 mm thick, with scale-like L. c. 1.4 mm long; **shoots** with terminal Ros., eventually forming a cushion; L. 5–18 mm long, 1–4 mm across, up to c. 2 mm thick, with a widened petiole; Infl. Cinc. on the elongated shoots, Fl. 7–8 mm ⌀, Sep. clavate-oblanceolate, red-spotted, Pet. with red stripes.

**S. cockerellii** BRITT. (§ 6). – USA: New Mexico. – Dwarf ⚃ with a fleshy taproot; basal L. spatulate, tapering, 12 mm long, 6 mm across, cauline L. linear-lanceolate, acute, 12–15 mm long, 3–6 mm across, green to red; Fl. white.

**S. coeruleum** VAHL (Pl. 125/3) (§ 20/4) (*S. heptapetalum* POIR., *S. azureum* DESF.). – S.Eur.; N.Afr. – ⊙ 5–10 cm tall; **shoots** much branched; L. ovate to ovate-oblong, ± cylindrical, 12 to 18 mm long; Infl. a lax panicle, Fl. pale blue.

**S. compactum** ROSE (§ 17/7). – Mexico: Oaxaca, San Luis Potosí. – ⚃ forming a grey-green mat; **roots** thickened, napiform; L. obovate, obtuse, 3 mm long, densely imbricate in a short or slightly elongated Ros.; Fl. small, ± spherical, white, strongly scented.

**S. compressum** ROSE (§ 12). – Dichotomous sub-♄ 20 cm tall; L. laxly rosulate, 2 cm long, 1 cm across, oblanceolate, obtuse or short-tapering, fleshy, grey-green, margins often cartilaginous; Fl. numerous.

**S. concarpum** FROED. v. **concarpum** (§ 17/25). – China: Yunnan. – ⚃; **shoots** prostrate, 3–5 cm long, flowering shoots erect, 8–9 cm tall; L. on the sterile shoots rosulate, long-petiolate, ± round, tuberculate above, 5–25 mm long, those on the erect flowering shoots conspicuously petiolate, obovate, tuberculate; Fl. in a dense corymb, c. 15 mm ⌀, yellow.

**S. — v. hupehense** FU. – Plant twice as large as the type.

**S. confertiflorum** BOISS. (§ 20/5). – Greece; As. Minor. – Insignificant ⊙ with dense Infl. of many Fl.

**S. confusum** HEMSL. (§ 12). – Mexico. – ♄ 30 cm tall, with ascending Br.; L. oblong-spatulate, obtuse to rounded above, flat, rather thick, 2.5–4 cm long, 12–20 mm across, the petiole with a V-shaped imprint below; Infl. hemispherical, dense, Fl. yellow.

**S. congdonii** EASTW. (§ 21). – USA: Calif. – Insignificant ⊙.

**S. constantinii** HAMET (§ 20/5). – China: Tibet. – Insignificant ⊙.

**S. conzattii** ROSE (§ 15). – Mexico: Oaxaca. – Small ♄ with softly hairy shoots; L. flat, 2–3 cm long, emarginate; Fl. in a short cyme, white with a red centre.

**S. cv. Coral Carpet.** – Garden hybrid up to 8 cm high; L. coral pink; Fl. soft pink.

**S. cormiferum** R. T. CLAUSEN (§ 17/27). – Mexico. – ⊙; in the first year several elliptical or ovate L. develop, also a semi-subterranean Tub. up to 5 mm long; in the dry season these L. die off; in the second rainy season new L. are produced, these being spatulate or oblanceolate, petiolate; **flowering shoot** 6–7 cm tall, papillose, reddish-brown or green with red lines, Fl. pale green or reddish-white. After flowering, the entire plant dies.

**S. craigii** R. T. CLAUSEN (§ 11). – Mexico: Chihuahua. – Small, clump-forming ♄; **shoots** fleshy, creeping; L. sessile, oblong-elliptical, fleshy, rounded above, 2–5 cm long, 9–22 mm across, underside very rounded, reddish-blue; Fl. white.

**S. crassularia** HAMET (§ 20/5) (*Crassula sediforme* SCHWEINF., *S. s.* (SCHWEINF.) HAMET). – Ethiopia; Kenya. – ⊙; L. obovate; Fl. 5-partite.

**S. cremnophila** R. T. CLAUSEN (§ 11) (*S. nutans* ROSE, *Cremnophila n.* ROSE). – Mexico. – Sub-♄; **shoots** erect, prostrate or pendulous, up to 45 cm long, 1–2.8 cm thick; L. obovate or oblanceolate, rounded above, dark green, dull waxy-pruinose, 2–10 cm long, 2–5.5 cm across, 7–17 mm thick; Infl. of many sulphur-yellow Fl. up to 14 mm ⌀.

**S. crenatum** DESF. (§ 17/1). – Turkey. – Close to *S. spurium*; ⚃ with rhombic L., upper ones small; Fl. pink, in a dense cyme. Hardy.

**S. creticum** BOISS. et HELDR. (§ 19). – Crete. – ⊙ or triennial plants; Ros. lax; L. oblong-spatulate, obtuse, 10–12 mm long, 5 mm across; Infl. short, Fl. yellowish.

**S. cupressoides** HEMSL. (§ 17/17). – Mexico: Oaxaca. – ⚃ branched dichotomously from the base; sterile **shoots** forming a cushion, side-shoots cypress-like; **L.** ± ovate, ± imbricate, laxer on flowering shoots, 1.5 mm long; **Fl.** solitary or 2–3 together, light yellow.

**S. cuspidatum** ALEXANDER (§ 13). – Mexico: Chiapas. – **Shoots** glabrous, up to 25 cm long; **L.** narrow-obovate, c. 2 cm long, 1 cm across, 3 mm thick, short-tapering, underside acute-carinate, light yellowish-green to bluish; **Fl.**scape 7 cm long, **Fl.** white, in compact cymes.

**S. cyaneum** RUD. (§ 3). – USSR: E.Siberia. – ♃; **shoots** slender, prostrate, creeping, branched, 5–7.5 cm long; **L.** obovate-oblong, obtuse, with entire margins, grey-green, 1–2 cm long, 5 to 6 mm across; **Fl.** campanulate, up to 13 mm ⌀, pinkish-mauve. Hardy.

**S. daigremontianum** HAMET (§ 17/19). – W.China. – ♃ with **shoots** 8–20 cm tall; **L.** linear, 4–9 mm long, with a 3-lobed spur; **Fl.** yellow.

**S. dasyphyllum** L. (§ 17/11). – N.Afr.; S. and W.Eur.

**S.** – ssp. **dasyphyllum** v. **dasyphyllum** (*S. d.* ssp. *eu-dasyphyllum* MAIRE). – ♃ 2–6 cm tall; **L.** not rosulate on sterile shoots, ovate, upperside ± flat, 5–7 mm long, blue-green, with glandular hairs; **Infl.** of 2–4 white Fl., 5–6-merous. ± hardy.

**S.** – ssp. – v. **alternum** MAIRE (*S. d.* ssp. *eu-d.* v. *a.* MAIRE). – N.Afr. – **L.** spreading, glabrous.

**S.** – ssp. – v. **glanduliferum** (GUSS.) MAIRE (*S. g.* GUSS., *S. d.* ssp. *eu-d.* v. *g.* (GUSS.) MAIRE, *S. corsicum* DUBY). – W.Medit. – Plant densely glandular-hairy.

**S.** – ssp. – v. **mesatlanticum** LIT. et MAIRE (*S. d.* ssp. *eu-d.* v. *m.* LIT. et MAIRE). – Morocco. – **L.** on basal shoots in a dense hemisphere 6 to 7 mm ⌀, ovate-elliptical, pruinose, papillose; **Fl.** with whitish-pink Pet. and yellow anthers.

**S.** – ssp. – v. **suendermannii** PRAEG. (*S. d.* v. *s.* PRAEG.). – Plant and **L.** larger, glandular-hairy; **Fl.** white.

**S.** – ssp. – v. **macrophyllum** ROUY et CAM. – **L.** large, ovate, very thick, 5–10 mm long, glabrous.

**S.** – ssp. **oblongifolium** (BALL.) MAIRE v. **oblongifolium** (*S. d.* v. *o.* BALL.). – **L.** elongated.

**S.** – ssp. – v. **glutinosum** MAIRE. – Morocco. – Plant glandular, glutinous.

**S.** – ssp. – v. **rifanum** MAIRE. – Morocco. – Anthers dark purple.

**S. debile** S. WATS. (§ 17/23) (*Gormania d.* BR., *Echeveria d.* NIELS. et MACBR.). – N.Am. – ♃; **L.** obovate to circular; **Fl.** in cymes, Pet. acute.

**S. dendroideum** Moc. et SESSÉ (§ 12). – Mexico; Guatemala. – Sub-♄; **shoots** erect, up to 60 cm high, pendulous or creeping, Br. spreading; **L.** crowded, ± rosulate on the Br.-tips, sessile or petiolate, elliptical-oblanceolate, spatulate or obovate, or ± circular, obtuse or apiculate, glossy, red or with red blotches along the margins, 4–10 cm long, 14–30 mm across, 2–4 mm thick; **Infl.** cymose with 7–38 cincinni, Fl. yellow.

**S.** – ssp. **dendroideum**. – **L.** spatulate, cylindrical or cuneate below, 41 mm long, 17 mm across, 4 mm thick, margins spotted.

**S.** – ssp. **monticolum** (S. T. BRAND.) R. T. CLAUSEN (*S. m.* S. T. BRAND.). – Mexico. – **L.** oblanceolate, indistinctly petiolate, 6.5–9 cm long, 24–30 mm across.

**S.** – ssp. **parvifolium** R. T. CLAUSEN. – Mexico. – **L.** rather thin, 4–5 cm long, 17 to 18 mm across, oblanceolate, indistinctly petiolate.

**S.** – ssp. **praealtum** (A. DC.) R. T. CLAUSEN (*S. p.* A. DC.). – Mexico. – **L.** oblong-elliptical, sessile, 4.5–6.5 cm long, 16–27 mm across, 2 to 3 mm thick, margins not reddened.

**S. diffusum** S. WATS. (§ 14). – Mexico: Monterrey. – Densely leafy ♄ 15 cm tall, with a fleshy **taproot; L.** subulate, 4–9 mm long; **Fl.** in a lax cyme, white.

**S. divergens** S. WATS. (§ 17/24). – N.Am. – **Shoots** reddish; **L.** obovate or obovate-spatulate, 6 mm long, 3 mm across, 1.5 mm thick; **Infl.** dichotomous, Fl. light yellow.

**S. drymarioides** HANCE (§ 20/2). – E.China. – ⊙ – ⊙, shortly hairy; **shoots** erect, up to 20 cm tall; **L.** rosulate on young plants, ovate to rhombic, distinctly carinate; **Fl.** yellow.

**S. dugueyi** HAMET (§ 17/19). – W.China. – ♃ 6–7.5 cm tall; **L.** long-ovate to triangular, tapering, 2–3 mm long; **Fl.** yellowish.

**S. eastwoodiae** (BRITT.) BGR. (§ 17/5) (*Gormania e.* BRITT.). – USA: Calif. – ♃; **L.** on sterile shoots in a crowded Ros., thick, obtuse, slightly greyish-green; Pet. pointed.

**S. ebracteatum** DC. (§ 4). – Mexico. – ♃ with 1–7 flowering **shoots** 17–25 cm long, with short secondary shoots developing from the base; **Ros.** dense with L. ± round to spatulate-oblong, obtuse, puberulous; **L.** on the flowering shoots ovate to oblong, acute to obtuse, puberulous, often with purple blotches, 1.7–10.5 cm long, 12–17 mm across, 2.4–3.7 mm thick; **Infl.** of 1–8 cincinni, Fl. white.

**S.** – ssp. **ebracteatum** (*Sedastrum e.* (DC.) ROSE, *S. incertum* HEMSL., *Sedastrum i.* (HEMSL.) ROSE, *Sedum chapalense* S. WATS., *Sedastrum c.* (S. WATS.) ROSE, *Sedum cordifolium* SESSÉ et MOC., *Sedastrum rubricaule* ROSE, *Sedum e.* v. *r.* (ROSE) FROED., ? *S. barrancae* M. E. JONES). – Mexico. – **L.** on the flowering shoots 4–5 cm long, ovate to oblong, with erect hairs.

**S.** – ssp. **grandifolium** R. T. CLAUSEN. – **L.** on the flowering shoots 2.5–10.5 cm long, usually elliptical-lanceolate, with prominent cilia, sometimes covered with branched hairs.

**S. eichlamii** L. H. BAILEY (§ 11). – Mexico. – Little branched ♄ 10–20 cm tall; **Br.** 5 to 10 mm ⌀; **L.** crowded on the ends of the Br., subterete, 2–4 cm long, 4–5 mm thick, light green, glossy, with red, rounded tips.

**S. engleri** HAMET (§ 17/22). – China: Yunnan. – ♃; **L.** oblong, obtuse, up to 5 cm long; **Infl.** papillose-tuberculate above, dense, with many Fl.

**S. epidendrum** HOCHST. (§ 15) (*S. schimperi* BRITT.). – Ethiopia. – Small, epiphytic ♄; **L.** linear-oblanceolate, short-spurred, 20–27 mm long, 4–4.5 mm across; **Fl.** few, yellow.

**S. erythraeum** GRISEB. (§ 17/11). – Yugoslavia; Bulgaria; Greece. – ♃, forming low clumps; **L.** 2.5 mm long, 2 mm across, semiterete; **Fl.** pinkish-red, in a short Rac.

**S. esquirolii** LEV. (§ 20/2). – E.China. – Possibly only a variety of **S. daigremontianum**.

**S. ewersii** LODD. v. **ewersii** (§ 3). – W.Himalayas to Mongolia. – ♃; **shoots** simple, ascending, 15–30 cm long; **L.** circular to broad-ovate, sessile, with entire margins, 18 mm long and across; **Infl.** dense, Fl. 12 mm ∅, purplish-pink. Hardy.

**S. — v. homophyllum** PRAEG. – **Shoots** 5 to 7.5 cm long; **L.** mostly ovate, 12–15 mm long, 6–9 mm across, more noticeably grey-green.

**S. — cv.** Turkestanicum. – Smaller cultivar.

**S. farinosum** LOWE (§ 17/11). – Madeira. – ♃ herb or small sub-♄; **Br.** ± erect or creeping, thin, tangled; **L.** dense, in 4–6 R., subulate, obtuse, 8–15 mm long, 2 mm across, both surfaces ± flattened, glabrous, green, slightly pruinose; **Infl.** dichotomous, Fl. white.

**S. feddei** HAMET (§ 17/19). – China. – ♃ 8 to 10 cm tall; **L.** spurred, 6–7 mm long, 2–3 mm across; **Fl.** yellow.

**S. filipes** HEMSL. v. **filipes** (§ 17/4) (*S. trientaloides* PRAEG.). – Burma; China: Hupeh. – ♃ with a thickened taproot; **shoots** 4–8 cm tall; lower **L.** opposite, upper ones larger, in whorls of 2–4; **Infl.** short-dichotomous, Fl. very small.

**S. — v. major** HEMSL. (*S. pseudo-stapfii* PRAEG.). – **L.** larger, in 1–2 whorls.

**S. flaccidum** ROSE (§ 6). – Mexico: Durango. – ♃ with a tuberous **taproot; shoots** smooth, freely branching. 8–10 cm tall; **L.** ovate to narrow-ovate, 5–10 mm long; **Fl.** white.

**S. flexuosum** WETTST. (§ 20/6). – Bulgaria; Albania. – Resembles **S. littoreum; shoots** several, densely leafy, often ☉; **L.** short-lanceolate, short-spurred; **Infl.** dichotomous, with few Fl.

**S. floriferum** PRAEG. (§ 16). – N.E.China. – Bushy ♃; **shoots** 15 cm long, freely branched; **L.** spatulate to ovate-lanceolate, up to 25 mm long, 15 mm across, with margins dentate in the upper third; **Infl.** flat, Fl. 12 mm ∅, yellow. Hardy.

**S. formosanum** N. E. BR. (§ 20/2). – E.China. Taiwan; Canary Is. – ☉; **shoots** prostrate below or erect, 15–25 cm tall, branched; **L.** spatulate, with entire margins, very obtuse, 25 mm long, 12 mm across, light green; **Infl.** large, lax, Fl. light yellow, 12 mm ∅.

**S. forreri** GREENE (§ 20/5). – Mexico: Durango. – Insignificant ☉; **L.** 2–5 mm long.

**S. forsteranum** SM. (§ 17/20) (*S. forsterianum* err. est, *S. rupestre* L. emend. PRAEG., *S. r.* ssp. *f.* (LEJ.) HEGI et SCHMID). – W.Eur.; Morocco. – Mat-forming, glabrous ♃; **shoots** creeping, rooting, branched basally, sterile shoots with dense and ± rosulate L. at the apex, flowering shoots ascending, laxly leafy; **L.** linear, upperside much flattened, short-spurred below, fleshy, green or bluish, more than 1½ times as wide as thick, with the tip curved, apiculate; **Infl.** at first nodding, a cyme of many golden-yellow Fl.

**S. — ssp. elegans** (LEJ.) WARB. (*S. elegans* LEJ., *S. pruinatum* COSSON, LANGE et BALL.). – Morocco. – **L.** pruinose.

**S. — ssp. forsteranum. – L.** green.

**S. frutescens** ROSE (§ 13). – Mexico. – Sub-♄ 80–100 cm tall; **St.** much branched, 6–10 mm thick, the bark eventually peeling and papery; **Br.** minutely papillose; **L.** elliptical-linear, acute or obtuse, underside convex, margins minutely papillose, 14–58 mm long, 3–9 mm across, 1 to 2.4 mm thick; **Fl.** 11–18 mm ∅, white.

**S. fui** ROWL. (§ 20/6) (*S. platyphyllum* FU non ALEXANDER). – China: Yunnan. – Related to **S. leblancae** HAMET. – Glabrous ☉ with opposite branchlets 6.5–8 cm long; upper **L.** oblanceolate, ± petiolate; **Fl.** greenish-yellow.

**S. furfuraceum** MORAN (§ 17/9). – Mexico: San Luis Potosi. – Mat-forming ♃; **shoots** creeping to ascending, sparsely branched throughout, 5–15 cm long, 1.5 mm thick and densely papillose near the tip, papillae orange-brown, eventually black; **L.** crowded at the tips of the shoots, ± triangular-ovate, obtuse, 6 to 11 mm long, 4.5–6 mm across, 4–5 mm thick, dark green to purplish-red, margins and keel bluntly ridged, surface glossy at first, the outer layer eventually ± peeling away; **Fl.** 8 to 10 mm ∅, pink.

**S. fusiforme** LOWE (§ 15). – Madeira. – Small ♄; **Br.** tangled, with spreading sideshoots, 15 cm tall; **L.** rather pointed, narrowing above and below, the upperside somewhat flattened, with a reddish M.stripe, grey-green, up to 2 cm long, 3–4 mm thick; **Fl.** greenish-yellow, with a red centre.

**S. gagei** HAMET (§ 17/19). – China; Sikkim. – Dwarf ♃; **shoots** 4–6 cm tall; **L.** 3.5–5 mm long, up to 1.8 mm across, margins with long papillae; **Fl.** yellowish.

**S. giajai** HAMET (§ 17/22). – China: Szechuan. – ♃ 5–9 cm tall; **L.** obtuse, 5–7.5 mm long; **Infl.** lax.

**S. glabrum** (ROSE) PRAEG. (§ 4) (*Sedastrum g.* ROSE). – Mexico: Saltillo. – ♃; basal **L.** rosulate, 3 cm long, 12 mm across, rather pointed, underside convex; flowering shoots up to 20 cm tall, L. oblong-ovate, thick, 3 cm long, 12 mm across, 6 mm thick; **Fl.** 1 cm ∅, white.

**S. gracile** C. A. MEY. (§ 17/11) (*S. albertii* PRAEG.). – USSR: Caucasus. – Mat-forming ♃; **shoots** slender, leafy, 3–5 cm tall; **L.** linear-oblong, ± sessile, 8 mm long, spurred below; **Infl.** very leafy, Fl. white.

**S. grandipetalum** FROED. (§ 17/17). – Mexico. – ♃ **shoots** creeping, with basal, ± circular, densely imbricate L. with papillose margins; the flowering shoots pendulous, with **L.** lanceolate, elliptical-lanceolate or linear-lanceolate, falcate, with 2 spurs below, obtuse, bluish, sometimes with slightly papillose margins, 9 to 21 mm long, 3.8–8 mm across, 4–6 mm thick; **Fl.** 10–18 mm ∅, yellow.

**S. greggii** HEMSL. (§ 17/17) (*S. diversifolium* ROSE). – Mexico. – ♃ with papillose L. in basal Ros. and thin, glabrous, prostrate or pendulous flowering shoots up to 14 cm long; **Ros.L.** obovate, elliptical or oblanceolate, obtuse; L. on

**Sedum**

the flowering shoots with a rounded tip, sometimes with red blotches, 3–12 mm long, 1.2 to 5 mm across, 1.5–3.2 mm thick; **Infl.** often with vegetative shoots, Fl. c. 1 cm $\varnothing$, sulphur-yellow. ± hardy.

**S. griffithsii** ROSE (§ 17/5). – USA: Arizona. – ♃ with small Ros.; basal **L.** 4–6 mm long, obtuse, both surfaces rough; **Fl.** in a compressed cyme.

**S. grisebachii** HELDR. (§ 20/6). – Yugoslavia; As. Minor. – Insignificant ☉.

**S. griseum** PRAEG. (§ 14). – Mexico. – Sub-♄ 20–30 cm tall; **shoots** erect, sometimes prostrate, freely branched; **Br.** greyish-brown, papillose; **L.** linear or linear-lanceolate, both surfaces convex, sessile, green or yellowish-green, often red, the underside sometimes sparsely hairy, 5–19 mm long, 1.7–2.9 mm across, 1–3 mm thick; **Fl.** 7–13 mm $\varnothing$, Pet. pale yellow or green on the underside.

**S. guadalajarum** S. WATS. (§ 14). – Mexico: Jalisco. – ♄ with a creeping, thickened **taproot**; **shoots** filiform, 20–25 cm long; **L.** flat, rather rounded, 12 mm long, grey-green; **Fl.** greenish-white.

**S. guatemalense** HEMSL. (§ 11) (*S. australe* ROSE). – Guatemala. – Sub-♄; **Br.** terete, often prostrate; **L.** laxly arranged, linear-oblong, rounded above, ± cylindrical, 12–15 mm long, 3 mm thick, green; **Fl.** reddish.

**S. hallii** (BR.) PRAEG. (§ 17/23) (*Gormania h.* BR., *Echeveria h.* NELS. et MACBR.). – USA: Central Calif. – ♃; **L.** on sterile **shoots** often rosulate and crowded, spatulate to obovate; **Fl.** in flat cymes, light yellow.

**S. heckelii** HEMSL. (§ 17/19). – China: E.Tibet. – ♃, close to **S. multicaule**; **L.** broader below, 4–6 mm long, with a blunt spur; **Fl.** yellow. Hardy.

**S. hemsleyanum** ROSE (§ 4) (*S. hemsleanum* ROSE err. est, *Sedastrum h.* ROSE, *Seda. pachucense* C. H. THOMPS., *S. p.* (C. H. THOMPS.) PRAEG., *Seda. painteri* ROSE, *S. p.* (ROSE) BGR.). – Mexico: Puebla. – ♃; flowering **shoots** 1–8, 18–28 cm tall, puberulous, with purple stripes, with secondary shoots from the base, rooting and with L. in a dense Ros.; **Ros.L.** oblong or obovate to linear-lanceolate, puberulous, often tinged red, 13–27 mm long, 4–13 mm across, 3–5 mm thick, L. larger on the flowering shoots; **Fl.** c. 9 mm $\varnothing$, white, with an acrid scent.

**S. hidakanum** TATEWAKI (§ 3). – Japan: Hokkaido. – Mat-forming; **shoots** with a few Br. below, 4–6 cm tall; **L.** thick, sessile, elliptical to ovate, rounded to obtuse at the tip, with entire margins, (2–)4(–5) mm long, 2–4 mm across; **Infl.** of few, white, bisexual Fl. c. 6 mm $\varnothing$.

**S. hintonii** R. T. CLAUSEN (Pl. 125/5) (§ 4). – Mexico. – Mat-forming ♃; **L.** 8–10 in a dense Ros., ovate, much rounded on the underside, 15–16 mm long, 8 mm across, 5–6 mm thick, pale green, densely covered with hairs 3–4 mm long.

**S. hirsutum** L. (§ 17/11). – W.Eur.; Morocco. – Type: ssp. **hirsutum**.

**S.** — ssp. **baeticum** ROUYI v. **baeticum** (*S. h.* v. *b.* ROUYI, *Umbilicus winkleri* WILLK., *S. w.* (WILLK.) DOD., *S. h.* ssp. *w.* FONT Y QUER). – S.Spain. – Like ssp. **hirsutum**, but in general larger and more robust; **Fl.** also larger.

**S.** — ssp. — v. **gattefossei** MAIRE, WEILER et WILCZEK. – Morocco. – **Shoots** elongated; **L.** semiterete, c. 3 mm long, shortly hairy; **Fl.** white.

**S.** — ssp. — v. **jahandiezii** MAIRE. – Morocco. – L. and Fl. smaller, Pet. ± free.

**S.** — ssp. — v. **maroccanum** FONT Y QUER. – Morocco. – Dainty variety; **L.** smaller.

**S.** — ssp. — v. **thermarum** MAIRE, WEILER et WILCZEK. – Morocco. – Forming hemispherical cushions; **L.** flat, compressed, c. 5.5 mm across, covered with rather longer hairs.

**S.** — ssp. — v. **wilczekianum** (FONT Y QUER) JACOBS. (*S. w.* FONT Y QUER). – Morocco. – Fl. with Cor. not rotate, Pet. linear-spatulate, whitish, with a purple M.nerve.

**S.** — ssp. **hirsutum** (*S. h.* ssp. *eu-h.* MAIRE). – France; Iberian peninsula; Italy; Yugoslavia: Dalmatia. – Hairy ♃ 5–10 cm tall; **L.** bluntly ovate-lanceolate, 18 mm long, 3 mm across; **Infl.** of few white or pink Fl.

**S. hispanicum** L. v. **hispanicum** (§ 20/5) (*S. glaucum* WALDST. et KIT., *S. sexfidum* MARSCH. BIEB.). – S. and S.E.Eur.; As. Minor. – Mat-forming, fairly robust ☉, 5–10 cm tall, branching basally, ± grey-green; **L.** linear to oblong-lanceolate, 12–15 mm long, 3 mm thick; **Fl.** 12 mm $\varnothing$, pink to white, (4–)6(–9)-partite.

**S.** — v. **bithynicum** BOISS. – **Shoots** in part ♃.

**S.** — v. **eriocarpum** BOISS. – Carpels glandular-hairy.

**S.** — v. **glabrum** BECK. – Yugoslavia. – Glabrous.

**S.** — v. **leiocarpum** BOISS. – Carpels glabrous.

**S.** — v. **minus** PRAEG. (*S. lydium glaucum* HORT., *S. glaucum* HORT.). – ♃; **L.** 6 mm long; **Infl.** 5 cm tall.

**S.** — v. **polypetalum** BOISS. – Fl. 7–9-partite.

**S. hispidum** DESF. (§ 20/6) (*S. pubescens* VAHL). – Algeria. – Insignificant ☉ herb with long glandular hairs, 5–12 cm tall; **L.** oblong, obtuse; **Fl.** yellow.

**S. horakii** ROHL. (§ 17/16) (*S. alpestre* v. *horakii* ROHL.). – Close to **S. alpestre**; **Fl.** yellowish-red.

**S. hultenii** FROED. (§ 13). – Mexico: Puebla. – Glabrous, caespitose ☉; sterile **shoots** thin, densely leafy, 4–10 cm long, flowering shoots creeping or pendulous, up to 35 cm long; **L.** ± terete below, short-spurred, obovate to circular, with a rounded tip, 6–35 mm long; **Infl.** a lax corymb of few, long-pedicellate, pale yellow Fl. c. 1 cm $\varnothing$.

**S. humifusum** ROSE (§ 17/15). – Mexico: Querétaro. – Mat-forming ♃; **shoots** spherical at first, eventually catkin-like, up to 25 mm long; **L.** flat, obovate, densely imbricate, 2 to 3 mm long, light green, with ciliate margins; **Fl.** solitary, yellow.

**S. hybridum** L. (§ 16). – USSR: Siberia; Mongolia. – ♃; **shoots** evergreen, prostrate, rooting, mat-forming; **L.** spatulate to lanceolate, dentate above, T. often red, 25 mm long, 12 mm across; **Fl.** yellow. Hardy.

**S. involucratum** MARSCH. BIEB. (§ 17/1). – USSR: Caucasus. – Mat-forming ♃ with creeping **shoots**; **L.** ± rosulate, sinuate, crenate, dentate, flat; **Fl.** in a forked cyme, white.

**S. jaccardianum** MAIRE et WILCZEK (§ 8) (*Monanthella j.* (MAIRE et WILCZEK) BGR.). – N.Afr.: Central Atlas. – ♃ with a **taproot** and a dense L.Ros., offsetting; **L.** obovate-spatulate, ± obtuse, tapering abruptly below towards the long petiole, fleshy, margins with soft glandular hairs; **Fl.** 7–10-partite, golden-yellow. (See Monanthella.)

**S. jahandiezii** BATT. (§ 20/5). – Type: ssp. jahandiezii.

**S.** — ssp. **jahandiezii** (*S. j.* ssp. *battandieri* E. et M.). – Morocco. – Small ⊙; **L.** fairly numerous, ovate to elongated; **Infl.** corymbose, Fl. white, 5-partite.

**S.** — ssp. **persicum** MAIRE et SAMUELSSON. – Fl. peach-coloured.

**S. jaliscanum** S. WATS. (§ 20/1) (*S. naviculare* ROSE, *S. synocarpum* FROED.). – Mexico. – ⊙; in the first rainy season developing tiny Tub. and a Ros. of 2–5 petiolate, ovate, round or reniform L., which eventually die off; during the second rainy season red, tuberculate flowering **shoots** up to 26 cm long are produced from the 1.5 cm ⌀ Tub.; seedling **L.** 19–38 mm long, 3–6 mm across, up to 0.5 mm thick, L. on the flowering shoots elliptical-oblanceolate or spatulate, short-spurred, obtuse, sometimes papillose, underside convex, upperside concave, 8–30 mm long, 1–5 mm across, up to 1.5 mm thick; **Infl.** with 1–7 cincinni, Fl. white.

**S. japonicum** SIEB. v. **japonicum** (§ 17/22). – Japan; China. – ♃ 15 cm tall; **shoots** and L. like **S. album**; **L.** linear-oblong, short-spurred, 6 mm long; **Infl.** flat, Fl. yellow.

**S.** — v. **senanense** MAKINO. – Japan. – Shoots more slender; **Fl.** small, often reddish.

**S. kamtschaticum** FISCH. et MEY. (§ 16) (*S. aizoon* v. *k.* HULTEN). – Type: v. **kamtschaticum**.

**S.** — v. **ellacombianum** (PRAEG.) T. B. CLAUSEN (*S. e.* PRAEG.). – Japan: Hokkaido; USSR; Korea; N.E.China. – **Shoots** 15 cm tall; **L.** obovate-spatulate, 3.5–4 cm long, 18 mm across, with 4–6 T. on each side towards the tip; **Fl.** pure yellow.

**S.** — v. **kamtschaticum** (*S. k.* v. *k.* (FISCH. et MEY.) T. B. CLAUSEN). – N.As.; Japan; N. China. – ♃; shoots dwarf, prostrate, at flowering becoming ascending and branching, not exceeding 30 cm tall; **L.** ± ovate to spatulate to oblanceolate, 3–5 cm long, the upper third dentate, margins papillose; **Infl.** lax, of few orange-yellow Fl.

**S.** — v. — cv. Variegatum. – **L.** with white margins.

**S.** — v. **middendorfianum** (MAX.) T. B. CLAUSEN sv. **middendorfianum** (*S. m.* MAX.). – L. linear, little widened above, margins dentate in the upper third, 3.5 cm long, 3 mm across; **Fl.** light yellow.

**S.** — v. — sv. **diffusum** (PRAEG.) JACOBS. (*S. m.* v. *d.* PRAEG.). – Shoots longer, prostrate; L. 25–40 mm long, acute-dentate.

**S. kostovii** STEFANOFF (§ 17/11) v. **kostovii**. – Balkans. – Glabrous ♃; sterile **shoots** short, densely crowded; **L.** linear-elliptical, underside convex, c. 5 mm long, with a crystalline T. at the tip, distinctly spurred at the base, flowering shoots 4–8 cm tall; **Fl.** pale yellow.

**S.** — v. **monocarpum** STEFANOFF. – ⊙ or ⊙; L. narrower and longer.

**S. kotschyanum** BOISS. (§ 20/5). – S.Iran. – Insignificant ⊙; **shoots** branching freely above, glandular-hairy.

**S. krajinae** DON (§ 17/17). – CSSR: S.Slovakia. – Mat-forming ♃, scarcely 10 cm tall; **shoots** erect or ascending, densely leafy; **L.** broadly triangular-ovate to ovate, obtuse, 4.5 to 5 mm long, 3–4 mm across, very thick, fresh green, with a thick spur below; **Fl.** light yellow. Hardy.

**S. kurilense** WOROSCH. (§ 16). – USSR: Kurile Is. – **Caudex** horizontal, thick, fleshy, much branched; **shoots** numerous, ± erect, (6–)10(–15) cm tall, 1–2 mm thick, branched, densely leafy; **L.** thick, obovate to ± spatulate, c. 15 mm long, 7 mm across, obtuse, margins entire except for 7–11 T. towards the tip; **Infl.** a cyme or panicle 2–3 cm ⌀, leafy, Fl. yellow, 8–10 mm ⌀, short-pedicellate.

**S. laconicum** BOISS. et HELDR. ssp. **laconicum** (§ 17/18). – Greece; As. Minor. – Mat-forming ♃; L. cylindrical, glabrous; **Fl.** yellow.

**S.** — ssp. **insulare** (RECH. f.) GREUT. et RECH. f. (*S. idaeum* WEBB). – Crete: high mts. – Smaller subspecies.

**S. lampusae** (KOTSCHY) BOISS. (§ 19) (*Umbilicus l.* KOTSCHY). – Cyprus. – ⊙ or triennial plants with a lax L. Ros.; **L.** obovate-spatulate, short-tapering, up to 5 cm long, 2 cm across; **Infl.** pyramidal, Fl. yellowish-reddish.

**S. lanceolatum** TORREY (§ 17/20). – Type: v. lanceolatum.

**S.** — v. **lanceolatum** (*S. l.* v. *typicum* T. B. CLAUSEN, *S. stenopetalum* v. *s.* T. B. CLAUSEN). – N.Am.: Rocky Mts. – ♃ with Br. from the base; **shoots** 5–6 cm long; **L.** oblong-lanceolate, 10–12 mm long, 2–3 mm across, smooth, margings minutely glandular, serrate (magnifying-glass needed!); **Infl.** 10–15 cm tall, Fl. 15 to 20 mm ⌀. Hardy.

**S.** — v. **nesioticum** (G. B. JONES) T. B. CLAUSEN (*S. n.* G. B. JONES, *S. stenopetalum* v. *n.* (G. B. JONES) T. B. CLAUSEN). – Larger, more robust variety; L. glabrous; **Infl.** ± paniculate.

**S. lancerottense** R. P. MURRAY (§ 15). – Canary Is.: Lanzarote. – Small ♄, becoming slightly woody; **L.** ovate-oblong, 12 mm long, 5 mm across and thick, pale green; **Fl.** in a 2–3-branched cyme, yellow.

**S. laxum** (BR.) BGR. (§ 17/5) (*Gormania l.* BR., *Echeveria gormania* NELS. et MACBR.). – USA: Oregon. – ♃ up to 30 cm tall; **L.** on sterile shoots rosulate, crowded, 2 cm long; **Infl.** paniculate, Fl. red.

**S. leblancae** HAMET (§ 20/6) (*S. dielsii* HAMET). – China: Yunnan, Shensi. – Dainty ⊙ forming at flowering a roundish cushion 10–15 cm ⌀; **shoots** erect, simple, 6 mm thick in the 2nd year, branching, densely tuberculate; **L.** mostly in whorls of 4, or 3 higher up, narrow-linear to spatulate, obtuse, 10 mm long, 1 mm thick, upperside flat; **Fl.** 6 mm ⌀, yellow.

**S. leibergii** BRITT. (§ 17/23). – USA: Oregon, Washington. – Close to **S. spathulifolium**; **L.** oblanceolate to obovate-spatulate; **Fl.** sessile.

**S. lenophylloides** ROSE (§ 6). – Mexico: Monterrey. – Dwarf ⚘ with a woody, ± thickened **caudex; shoots** erect, 5–30 cm tall; **L.** lanceolate, sessile, underside convex, 10 to 15 mm long, 2 mm across, green, becoming pink; **Fl.** greenish-white.

**S. liebmannianum** HEMSL. (§ 17/9). – Mexico: Oaxaca. – ⚘ 5–15 cm tall; **L.** ovate-oblong, red, tapering, imbricate, L. remains giving the shoots a silvery covering; **Fl.** white.

**S. lineare** THUNBG. v. **lineare** (§ 17/25). – China; Japan. – Mat-forming ⚘; **L.** linear-lanceolate, 25 mm long, 3 mm across, light green; **Fl.** in a flat cyme, yellow.

**S.** — v. — f. **variegatum** PRAEG. (Pl. 125/6) (*S. carneum variegatum* HORT.). – **L.** with white margins.

**S.** — v. **robustum** PRAEG. – Robuster, more freely branched variety; **Fl.** larger, paler.

**S. listoniae** VIS. (§ 17/2). – As. Minor. – ⚘; **caudex** with pea-sized, tuberous buds, in autumn with an aerial L.Ros.; **shoots** hairy; **L.** cuneate towards the base, ± spatulate, obovate, ciliate, with entire margins.

**S. littoreum** GUSS. (§ 20/6). – E.Medit.; As. Minor. – ⊙, resembling **S. oreades**; **Infl.** longer, with 2–3 yellow Fl.

**S. longibracteatum** FROED. (§ 20/5). – Syria. – ⊙ with short roots; **shoots** glabrous, much and irregularly branched, 14–17 cm long; **L.** short, obtuse, spurred, narrow-linear, 10–12 mm long; **Infl.** lax, Fl. up to 1 cm ⌀, white, bracts 1–2 cm long, linear, obtuse.

**S. longipes** ROSE (§ 17/5). – Mexico. – Dwarf ⚘; **shoots** creeping, minutely papillose, 12 to 25 cm long; **L.** at first densely rosulate, elliptical to obovate, rounded above, papillose, sometimes red-spotted; **Infl.** lax, Pet. pale green, with red blotches or stripes.

**S.** — ssp. **longipes**. – Ros., where present, with L. compactly arranged; **Fl.** with Pet. 5.5 mm long.

**S.** — ssp. **rosulare** R. T. CLAUSEN. – **Ros.** prominent, with the L. widely spreading; Pet. up to 4.1 mm long.

**S. lucidum** R. T. CLAUSEN (§ 11). – Mexico. – Glabrous sub-♄; **shoots** prostrate, 5–7 mm thick, up to 35 cm long, with erect side-shoots from the base; **L.** long-elliptical or oblong-spatulate, sessile, mucronate, ± flat or flat-convex, underside rounded and unevenly keeled, 7–48 mm long, 6–19 mm across, 3.5–10 mm thick, dark green, glossy; **Fl.** numerous, 9 to 11 mm ⌀, white.

**S. lumholtzii** ROBINS. et FERNALD (§ 17/5). – Mexico: Sonora. – ⚘ with small Ros.; **L.** numerous, obtuse, 8–12 mm long; **Infl.** flat, Fl. white.

**S. lungtsuanense** FU (§ 17/25). – China: Lungtsuan. – ⊙ with fibrous roots; **shoots** erect, hairy, 5–6 cm long, branched; **L.** spatulate, 7–9 mm long, 2.2–3 mm across, hairy, acute; **Fl.** in cymes, c. 1 cm ⌀.

**S. luteoviride** T. B. CLAUSEN (§ 11). – Mexico. – Much-branched sub-♄; **shoots** robust, prostrate at the base, with the tips rigid and erect, green to reddish-green, up to 32 mm long; **L.** in irregular, longitudinal zones, oblong-elliptical to ± spatulate, obtuse, short-spurred, flat, glossy, yellowish-green, with red tips, 5–16 mm long, 3–6 mm across, 1–2 mm thick; **Infl.** of 7–10 yellow Fl.

**S. lydium** BOISS. (§ 17/11). – As. Minor. – Dwarf, densely mat-forming ⚘; **L.** obtuse, narrow-linear, 6 mm long, 1 mm across, grey-green or red, minutely papillose; **Infl.** flat, Fl. white.

**S. madagascariense** PERR. (§ 10). – Madag. – Freely-branched ♄, 50–80 cm tall, with 4-angled Br.; **L.** oblong-spatulate, sessile, rounded above, minutely dentate, 4–5 cm long, 16–22 mm across; **Fl.** 5–10 in a shortly pedunculate cyme, 6-partite, white, suffused red.

**S. madrense** S. WATS. (§ 17/5). – Mexico: Chihuahua. – ⚘ with small Ros.; **shoots** 10 cm tall, densely leafy; **L.** narrow, 6–8 mm long, obtuse.

**S. magellense** TEN. v. **magellense** (§ 17/11) (*S. brutium* N. TERR.). – Italy; Yugoslavia; Greece; As. Minor; Algeria. – Mat-forming ⚘ with lax Ros.; **L.** obovate, 6 mm long, light green; flowering **shoots** 7–10 cm long; **Fl.** white or whitish, in a 25–50 mm long Rac. Hardy.

**S.** — v. **macrostylum** HAL. et BALD. – **L.** spatulate on sterile shoots.

**S. makinoi** MAX. (§ 17/22) (*S. alfredii* v. *makinoi* (MAX.) FROED.). – Japan. – ⊙ or ⚘; sterile **shoots** prostrate, up to 20 cm long, with an annular constriction at the nodes; **L.** spatulate to ± rhombic, 7–18 mm long, 4–9 mm across, up to 1.5 mm thick, obtuse or rounded, short-spurred, with papillose margins; **Fl.** yellow.

**S. margaritae** HAMET (§ 17/22) (*S. chauveaudii* v. *m.* HAMET). – China: Yunnan. – ⚘ with **shoots** 7–15 cm tall; **L.** obtuse, 10–15 mm long, 2.5 mm across; **Fl.** yellow.

**S. maximowiczii** RGL. (§ 16). – Robuster counterpart of **S. aizoon; shoots** prostrate at the base; **Fl.** orange-yellow.

**S. maximum** (L.) SUTER (§ 3). – Eur.; USSR: Caucasus, Siberia. (In 'Fl. Eur.' D. A. WEBB refers this species to **S. telephium** as a ssp.) – ⚘ with a tuberously thickened **caudex; shoots** erect, 50–100 cm tall, branching from near the tip; **L.** opposite or in whorls of 3, cordate and amplexicaul below, broadly ovate, 2–8 cm long, 3–5 cm across, slightly irregularly dentate to entire; **Infl.** dense, Fl. greenish or yellow-white, more rarely white or pale pink. Hardy.

**S.** — ssp. **maximum** (*S. telephium m.* v. *m.* L., *S. m.* ssp. *m.* (L.) KROCKER). – **L.** ± ovate; **Fl.** greenish-white or yellow-white.

**S.** — ssp. — v. **maximum** f. **feminum** HUBER. – **Fl.** pinkish-green, without anthers.

**S. — ssp. — v. — f. atropurpureum** Praeg. – Shoots and L. dark red.

**S. — ssp. — v. glaucopruinosum** (Eklund) Soo (*S. g.* Eklund). – Upper parts of shoots, Infl. and Sep. ± grey-pruinose.

**S. — ssp. — f. versicolor** van Houtte (*S. rodigasii* hort.). – L. with white blotches.

**S. — ssp. ruprechtii** (Jalas) Soo (*S. telephium* ssp. *r.* Jalas, *S. telephium* sensu Borissova in Komarov). – L. as ssp. **maximum**; cauline L. 2–5 cm long, ± circular, dentate or with margins ± entire, bluish-pruinose; Fl. whitish.

**S. mellitulum** Rose (§ 17/10). – Mexico: Chihuahua. – ⚁; shoots prostrate, eventually erect, 7–10 cm tall, often somewhat branched; L. linear-subulate, obtuse, cylindrical, 6–8 mm long, grey-green, densely papillose; Infl. flat, Fl. 12 mm ⌀, white.

**S. mexicanum** Britt. (§ 17/26) (*S. sarmentosum* Masters). – Mexico: near Mexico City. – ⚁; shoots ascending, slender, weak, fresh green, 10–20 cm tall; L. in whorls of 3–5, upper L. alternate, narrow-linear, ± cylindrical, 6 to 15 mm long, light green; Fl. in a flat cyme, golden-yellow.

**S. meyeri-johannis** Engl. v. **meyeri-johannis** (§ 20/2) (*S. volkensii* Engl.). – Tanzania; Kenya. – Glabrous ⊙ – ⊙; shoots slender, 10–40 cm tall, rooting along the entire length; L. oblanceolate, 5–10 mm long, 2–5 mm across; Infl. lax, Fl. yellow.

**S. — v. keniae** Froed. – Kenya. – Epiphytic, with shoots up to 1 m long.

**S. microcarpum** Smith (§ 22). – As. Minor; Cyprus; Syria. – Minute ⊙ herbs; L. cylindrical, obtuse; Fl. in a one-sided spike, white, 3–5-partite.

**S. microstachyum** Boiss. (§ 19). – Cyprus. – ⊙ or triennial, with lax L.Ros.; L. linear-spatulate, obtuse, 2.5 cm long, 6 mm across, with soft, glandular hairs; Fl. reddish.

**S. mingjinianum** Fu (§ 3) (*S. telephium* ssp. *alboroseum* (Bak.) Froed.). – China. – Glabrous ⚁; shoots erect, 20–30 cm tall, simple; L. linear, 2 cm long, 2 mm across, lower L. elliptical-ovate, obtuse, dentate; Infl. a corymb with purple Fl. 6 mm ⌀.

**S. minimum** Rose (§ 6). – Mexico. – ⊙ with small, spherical Tub. 5.5 mm ⌀; shoots mostly underground, leafless, white; Br. creeping, up to 27 cm long; L. clavate at first, up to 15 mm long, up to 4 mm across, up to 3.5 mm thick, those on the flowering shoots oblanceolate-elliptical, semi-cylindrical, short-spurred, up to 15 mm long, 4 mm across, up to 3.3 mm thick; Fl. few together in Cinc., c. 7 mm ⌀, with white, red-striped Pet.

**S. — ssp. delicatulum** (Rose) R. T. Clausen (*S. d.* Rose). – Sep. longer than the Pet. which are united for 4 mm.

**S. — ssp. minimum** (*S. pringlei* v. *minus* Robins. et Seaton). – Sep. usually shorter than the Pet., or as long, Pet. united for 2 mm.

**S. mite** Gilib. v. **mite** (§ 17/19) (*S. sexangulare* auct., *S. boloniense* Lois.). – Eur. – ⚁ closely resembling **S. acre**; L. 3–6 mm long, obtuse, with a small spur at the base, arranged in 6 dense series on sterile shoots; Fl. yellow. Hardy.

**S. — v. montenegrinum** (Horak) Hayek (*S. m.* Horak). – Yugoslavia. – L. much thinner.

**S. modestum** Baill. (§ 20/5). – Morocco. – Small ⊙; shoots glabrous; L. few, flat, broadly spatulate, glabrous; Fl. 7-partite, Sep. and pedicels hairy, Pet. pale red to white.

**S. monregalense** Balb. (§ 17/11) (*S. cruciatum* Desf.). – Italy; S.France, Corsica. – Mat-forming ⚁; shoots 7–15 cm long, hairy; L. in whorls of 4, oblong, obtuse, fleshy, green, papillose above, 7 mm long; Infl. branchlets hairy, Fl. white.

**S. montanum** Perr. et Song (§ 17/20) (*S. rupestre* Willk., *S. r.* L. ssp. *m.* (Perr. et Song) Hegi et Schmid, *S. anopetalum* v. *m.* Rouy et Cam., *S. ochroleucum* ssp. *m.* (Perr. et Song) D. A. Webb). – Pyrenees; W. and S.Alps; Balkans. – Mat-forming ⚁; shoots creeping or pendulous, rooting, branched below, rather thin, flowering shoots ascending, 5–15 cm tall; L. narrow, linear, ± subulate to cylindrical, upperside ± flattened, apiculate, spurred below, fleshy, dark grass-green or reddish-brown, 10 to 15 mm long; Infl. Cinc. of few, golden-yellow Fl.

**S. moranense** H. B. et K. (§ 17/9). – Mexico. – Much-branched, prostrate sub-♄ up to 12 cm tall; shoots glabrous, becoming dark red, putting out aerial roots; L. in 5–6 dense, spiralled R., ovate or lanceolate, short-spurred, obtuse, sometimes papillose, green, often red-striate, 3.6 mm long, almost 3 mm across; Fl. in a Cinc., white.

**S. — ssp. grandiflorum** R. T. Clausen. – Fl. larger than in v. **moranense**.

**S. — ssp. moranense** (*S. arboreum* Masters, *S. m.* v. *a.* (Masters) Praeg., *S. submontanum* Rose). – Fl. 15 mm ⌀.

**S. moranii** R. T. Clausen (§ 17/5) (*Cotyledon glanduliferum* L. F. Henders., *S. g.* (L. F. Henders.) M. E. Peck, *Gormania g.* (L. F. Henders.) Abrams). – USA: S.W.Oregon. – Sterile shoots with Ros. L. spatulate, 4 cm long, 1 cm across, margins glandular, crenate; Infl. glandular-hairy, Fl. c. 3 cm ⌀, yellow.

**S. morganianum** E. Walth. (Pl. 126/2) (§ 11). – Mexico. – Sub-♄; Br. pendulous or creeping, branching basally, rather long, 4 to 6 mm thick; L. numerous, thick, fleshy and swollen, upperside ± flat, ± appressed or spreading, somewhat curved below, oblong-lanceolate, acute, c. 2 cm long, 8 mm thick, light green, bluish-pruinose; Infl. of 6–12 long-pedicellate, light pink to deep scarlet-purple Fl. 2 cm ⌀.

**S. morotii** Hamet (§ 17/22). – China: E.Tibet. – ⚁ 13–15 cm tall; not hairy, otherwise resembling **S. giajai**; L. larger; Infl. dense, Fl. yellow.

**S. multicaule** (§ 17/19). – Himalayas; China; Japan. – Moss-like ⚁ or ⊙, branching freely below, green; L. linear, tapering, 12 mm long, 1.5 mm thick, upperside flattened; Fl. yellow. Hardy.

**S. multiceps** Coss. et Dur. (§ 15). – Algeria. – Small ♄; **L.** very dense, 6–7 mm long, upperside flat, obtuse, very papillose, dying back during the resting period and persisting; **Fl.** yellow, in a 2–3-branched cyme.

**S. muscoideum** Rose (§ 17/17). – Mexico: Oaxaca. – Very close to **S. cupressoides.**

**S. nanum** Boiss. (§ 20/6). – S.Iran. – Insignificant ⊙; **L.** semi-cylindrical, ± acute; **Fl.** yellow.

**S. napiferum** Peyritsch (§ 6). – Mexico: Toluca Prov. – ⚷ with a tuberous taproot; **shoots** smooth, freely branched, 3–4 cm tall; **L.** 3–5 mm long; **Fl.** pink.

**S. nevadense** Coss. (§ 20/5) (*S. jacalambrense* Pau). – Algeria; Spain. – Insignificant ⊙, unbranched or branched only above.

**S. nevii** A. Gray (§ 17/3). – Eastern N.Am. – ⚷ close to **S. ternatum**; **L.** spatulate, with entire margins, bluish-green, 12 mm long, 4 mm across; **Fl.** white. Hardy.

**S. niveum** Davidson (§ 6). – Mexico: Sierra San Pedro. – Roots thickened, **taproot** 10 to 30 cm long, 6 mm thick; old **shoots** up to 20 cm long, 5 mm thick, creeping and rooting, often forming ± diffuse mats; flowering shoots erect to spreading, 2–8.5 cm long, smooth, light green or with some red blotches; **L.** papillose, green, upper L. with red spots, cuneate-obovate to oblanceolate, mostly obtuse, 4–13 mm long, 2–4.5 mm across, 1–2 mm thick, upper L. narrow and longer; **Fl.** 2–3 together on pedicels 1 mm long, Pet. white, each with 2 red blotches at the Cor.-throat.

**S. novakii** Dom. (§ 17/17) (*S. sartorianum* v. *sartorii* Dom.). – Yugoslavia. – ⚷; **shoots** very numerous, slender, weak, with dead L. below; **L.** narrow-oblong or linear, 5 mm long; **Fl.** yellow. Hardy.

**S. nudum** Ait. (§ 15). – Madeira. – Small ♄; **shoots** glabrous, smooth; **L.** obovate-oblong, thick, obtuse, up to 2 cm long, 5 mm across and thick, green and grey-green; **Fl.** greenish-yellow.

**S. nussbaumeranum** Bitter (Pl. 126/1/b) (§ 11). – Mexico: Veracruz. – Resembles **S. adolphii**; sub-♄; **shoots** prostrate, reddish-brown, c. 42 cm long, 7 mm thick; **L.** oblanceolate or elliptical, acute, upperside flat, underside indistinctly carinate, 22–38 mm long, 10–16 mm across, 6–8 mm thick, yellowish-green, yellow or orange; **Infl.** umbel-like, Fl. 14–16 mm ⌀, white.

**S. nuttallianum** Raf. (§ 20/6) (*S. torreyi* Don.). – USA: Missouri, Arkansas, Texas. – Insignificant ⊙ herb.

**S. oaxacanum** Rose (Pl. 126/3) (§ 17/13) (*S. polyrhizum* Praeg.). – Mexico: Oaxaca. – **Shoots** freely branched; **Br.** spreading, rooting, minutely rough-tuberculate; **L.** ovate, thick, flat, 6 mm long, 2 mm thick, grey-green; **Fl.** 1–4, yellow.

**S. obcordatum** R. T. Clausen (§ 17/12). – Mexico: Pico de Orizaba. – Sub-♄ up to 35 cm tall; **shoots** mat-forming, erect or ascending, 4–5 mm thick, Br. readily broken off; **L.** obcordate or ± circular, short-spurred, broadly rounded at the tip, bluish, 7–29 mm long, 6–25 mm across, 1–3.5 mm thick; **Infl.** of 1 to 5 Cinc., Fl. yellow.

**S. obtusatum** A. Gray (§ 17/23) (*Gormania o.* Br., *Echeveria o.* Nels. et Macbr.). – USA: Calif. – ⚷; **L.** on sterile shoots rosulate, crowded, spatulate to obovate-cuneate; **Fl.** yellowish.

**S. obtusifolium** C. A. Mey. (§ 17/2) (*S. properly: proponticum* Aznavour, *S. anatolicum* C. Koch, *S. milii* Bak., *S. gemmiferum* Woron). – As. Minor. – ⚷ with small tuberous buds on the **caudex,** in autumn with aerial Ros.; **shoots** 15–25 cm tall, Infl. glandular-hairy; **L.** ovate, obtuse, 3 cm long, 2 cm across, narrowing below towards the very short petiole.

**S. ochroleucum** Chaix (§ 17/21) (*S. rupestre* Vill., *S. anopetalum* DC., *S. hispanicum* Lam. et DC., *S. r.* L. ssp. *o.* (Chaix) Hegi et Schmid). – S.Eur. to As. Minor. – Mat-forming ⚷; **shoots** creeping, rooting, branched basally, grey-pruinose; **L.** long-persisting, linear to linear-lanceolate, ± cylindrical, apiculate, short-spurred below, grey-pruinose; **Fl.** in a flat cyme, yellowish-white to almost white.

**S. oreades** (Decne.) Hamet (§ 20/6) (*Umbilicus o.* Decne., *Cotyledon o.* C. B. Clarke, *Umb. luteus* Decne., *Umb. spathulatus* Hook. f. et Thoms., *Cot. spathulata* C. B. Clarke, *S. jaeschkei* Kurz). – Himalayas, Tibet. – ⊙ to ⊙; basal **L.** rosulate, narrow-lanceolate, awned; **Infl.** of 1–3 yellow Fl.

**S. oreganum** Nutt. (Pl. 126/4) (§ 17/23) (*Gormania o.* Britt., *Echeveria o.* Nels. et Macbr.). – USA: N.Calif., Alaska. – ⚷; **L.** flat, ± spatulate, those on the sterile, often stolon-like shoots rosulate, crowded, cuneate-spatulate, obtuse, greenish-brown, up to 17 mm long. Hardy.

**S. oxycoccoides** Rose (§ 17/10). – Mexico: Tepic. – Dwarf ⚷ close to **S. potosinum**; **shoots** slender, numerous; **Fl.** red.

**S. oxypetalum** H. B. et K. (§ 13) (*S. arborescens* Sessé et Moc., *S. peregrinum* Sessé et Moc.). – Mexico. – Much-branched ♄ resembling a small ♄, 40–150 cm tall, St. up to 10 cm ⌀, bark papery, peeling, shoots papillose; **L.** oblanceolate to obovate, truncate, minutely papillose, eventually deciduous, 11–57 mm long, 5–19 mm across, 1–1.3 mm thick; **Infl.** forked, Fl. 10–14 mm ⌀, scented, Pet. pink, greenish-white or pale yellow, with pink blotches or stripes.

**S. pachyphyllum** Rose (Pl. 26/1/a) (§ 11). – Mexico: Oaxaca. – Erect sub-♄; **L.** dense, in 5 ± distinct spiralled R., 4 cm long, 6 mm thick, slightly clavate, terete, slightly greyish-green, tip obtuse-rounded, reddish; **Fl.** yellow.

**S. palaestinum** Boiss. (§ 20/6). – S.W.As. – Insignificant ⊙ herb, often glandular-hairy above.

**S. pallescens** Freyn. (§ 3). – S.USSR. – Possibly only a form of **S. telephium** v. **fabaria**; **L.** oblong-elliptical, projecting-dentate in the upper half.

**S. pallidum** Marsch. Bieb. (§ 20/5) (*S. glaucum* v. *p.* Hayek). – W.As. – Insignificant ± hairy or glandular ⊙ herb 5–6 cm tall.

**S. palmeri** S. WATS. (§ 12). − Mexico: Nuevo León. − Sub-♄ 15−22 cm tall; **L.** in lax Ros., spatulate, with a rounded tip, 25 mm long, 15 mm across, bluish-green; **Fl.** in a dichotomous cyme, orange-yellow.

**S. pampaninii** HAMET (§ 17/19). − ♃, close to **S. trullipetalum**; **L.** with smooth margins.

**S. paoshingense** FU (§ 3a). − China: Szechuan. − ♃; flowering **shoots** ascending, simple; **L.** crowded, thick-fleshy, linear-oblanceolate, 3.5 to 4.5 cm long, 5−7.5 mm across, tapering, with entire margins; **Infl.** a corymb with white Fl. 1 cm ⌀.

**S. pedicellatum** BOISS. et REUT. ssp. **pedicellatum** (§ 20/5). − Spain. − Insignificant ☉; **Infl.** freely dichotomously branched.

**S.** − ssp. **lusitanicum** (WILLK. ex MARIZ) LAINZ (*S. p.* v. ? *l.* WILLK. ex MARIZ, *S. willkommianum* R. FDES.). − Portuguese subspecies.

**S. pentastamineum** R. T. CLAUSEN (§ 17/9). − Mexico. − Dwarf ♃; **shoots** creeping or pendulous, minutely papillose, rooting, up to 30 cm long, thin, pink; **L.** elliptical, ± circular, petiolate, short-spurred, with a rounded tip, margins papillose, sometimes red-spotted, 2 to 8 mm long, 2−4.5 mm across, almost 1 mm thick; **Infl.** of 1−5 pale green or yellowish-white Fl. 6−8 mm ⌀, Pet. red-striped.

**S. pilosum** MARSCH. BIEB. (Pl. 126/5) (§ 19) (*Umbilicus pubescens* LEDEB., *S. regeli* HORT.). − As. Minor; Armenian uplands; USSR: Caucasus. − ☉ developing a spherical Ros. during the first year; **L.** linear-spatulate, rather obtuse, ± incurved, 12 mm long, dark green, hairy, ciliate; **Infl.** a flattened sphere, Fl. 6−9 mm ⌀, pinkish-red. Hardy.

**S. pinetorum** BRANDEG. (§ 6) (*Congdonia p.* JEPS.). − USA: Calif. − Dwarf ♃ with small, slender Tub.; **L.** crowded, rosulate, ovate, rather thin; **Fl.** solitary, white.

**S. platyphyllum** ALEXANDER non FU (§ 11a). − Mexico: Oaxaca. − ♄ up to 30 cm tall or shoots creeping, 2 cm thick below; **L.** obovate-spatulate to triangular-obovate, up to 7.5 cm long, 3 cm across, 5 mm thick, the tip rounded to truncate, apiculate or emarginate, lower margins curved slightly inwards, acute or obtuse, both surfaces convex; **Infl.** much-branched, Fl. 13−18 mm ⌀, greenish-white, Pet. with red blotches.

**S. platysepalum** FRANCH. (§ 17/19). − China. − Dwarf ♃; **L.** linear-lanceolate, broader below, acute; **Fl.** yellow.

**S. pluricaule** KUDO (§ 3) (? *S. telephium* v. *p.* MAX., *S. yezoense* MIYADE et TATEWAKI). − USSR: Sakhalin; Japan: Hokkaido. − ♃ with fleshy **roots**; **shoots** erect or prostrate, up to 18 cm long; **L.** ovate to cuneate-spatulate, obtuse above, 7−20 mm long, 4−8 mm across, 1−2 mm thick; **Fl.** in a spherical cyme, 9 to 11 mm ⌀, pink.

**S.** − cv. Rose Carpet. − **L.** dove-grey; **Fl.** soft red.

**S. polystriatum** T. B. CLAUSEN (§ 17/11) (*S. pulchellum* AZNAVOUR). − Turkey. − ♃; **shoots** at first creeping, rooting, eventually ascending, hairy above; **L.** semi-terete, oblong-spatulate and linear, obtuse, ± spurred below, 10−14 mm long, 2.5−4 mm across, upper L. smaller; **Infl.** a corymb, Fl. purplish-violet, Pet. with 11 to 13 stripes.

**S. populifolium** PALLAS (§ 19). − USSR: Siberia. − Sub-♄; **shoots** 20−40 cm tall, simple; **L.** with a petiole c. 2 cm long, ovate-cordate, 2 cm long, flat, green; **Infl.** a flattened sphere, Fl. scented, pale red or white.

**S. porphyreum** KOTSCHY (§ 20/5). − Cyprus. − Insignificant ☉ herb with ± clavate **L.**; **Fl.** red.

**S. potosinum** ROSE (Pl. 126/6) (§ 17/10). − Mexico: San Luis Potosí. − Dwarf ♃; **shoots** creeping, eventually ascending, 7−15 cm tall; **L.** linear, obtuse, terete, short-spurred, 10 to 12 mm long, pale, often pinkish-purple; **Fl.** 15 mm ⌀, white.

**S. pruinatum** BROT. (§ 17/20). − Portugal. − ♃; **L.** linear, tapering, up to 2 cm long, upperside flattened, short-spurred, grey-green; **Fl.** straw-coloured. Hardy.

**S. pruinosum** BRITT. (§ 17/23). − USA: Calif. − Very close to **S. spathulifolium**; densely pruinose plant.

**S. przewalskii** MAX. (§ 20/6). − China: Kansu. − Glabrous ☉; **shoots** erect, often filiform; **L.** appressed, oblong to ovate-oblong, obtuse, up to 2 cm long; **Fl.** yellow.

**S. pseudosubtile** HARA (§ 17/22). − Sikkim. − Glabrous ♃; **shoots** slender, shortly prostrate below, creeping; **L.** in whorls of 3 or 4, or alternate, linear-spatulate, fleshy, flat, 6 to 20 mm long, 1.5−3 mm across, long-tapering below, with a rounded tip, margins rough; flowering shoots 3−12 cm tall; **Infl.** a flattened cyme 2−5 cm ⌀, Fl. short-pedicellate, c. 8 mm ⌀, yellow.

**S. puberulum** S. WATS. (§ 17/5). − Mexico: Chihuahua. − ♃ with small Ros.; **L.** acute, roughly hairy, 7−10 mm long.

**S. pulchellum** MCHX. (§ 17/10). − Eastern N.Am. − Dwarf ♃; **L.** narrow-linear, densely appressed, obtuse, 15 mm long, with a forked spur below; **Infl.** dense, Fl. 4-partite, red.

**S. pulvinatum** R. T. CLAUSEN (§ 13). − Mexico: Oaxaca. − Small ♄; **shoots** at first prostrate, eventually erect, up to 25 cm tall; **L.** oblong-elliptical, 5−11 mm long, 1.5−3 mm across, glossy green, swollen below, underside convex; **Fl.** solitary, 10−12 mm ⌀, white.

**S. pumilum** BENTH. (§ 21). − USA: Calif. − Insignificant ☉.

**S. purdyi** JEPSON (§ 17/5). − USA: S.E.Calif. − ♃; Ros. offsetting, compact, flat; **L.** round to round-spatulate, 2−4 mm long; **Infl.** 7−10 cm tall, Fl. numerous, white.

**S. purpusii** ROSE (§ 12). − Mexico: Veracruz. − Shoots freely branched, rather weak; **L.** oblanceolate, somewhat emarginate, 2−3 cm long; **Infl.** paniculate, dense, Fl. yellow.

**S. pusillum** MICHX. (§ 20/5). − USA: N.Carolina, Georgia. − Insignificant glabrous ☉ branching basally.

**S. quaternum** PRAEG. (§ 17/25). − China: Hupeh, Yunnan. − ♃; **L.** narrow-lanceolate, short-spurred, verticillate on the flowering shoots, up to 1 cm long, 2 mm across; **Fl.** yellow.

**S. quevae** HAMET (§ 13) (*S. falconis* T. S. BRANDEG., *S. arsenii* FROED.). – Mexico. – Sub-♄ up to 35 cm tall with many Br. with a peeling bark, **roots** tuberous; **L.** oblanceolate or more elliptical, spurred, obtuse, minutely papillose, 10–35 mm long, 3–9 mm across, c. 1 mm thick; **Fl.** 10–14 mm ⌀, white.

**S. reflexum** L. v. **reflexum** (§ 17/20) (*S. rupestre* L., *S. r.* ssp. *r.* (L.) HEGI et SCHMID). – Finland; Scandinavia; Central Eur. to USSR: Ukraine. – Mat-forming, glabrous ♃ with fibrous roots; **shoots** creeping, rooting, branched basally, flowering shoots ascending, up to 40 cm tall; **L.** fleshy, linear to ± cylindrical, apiculate, spurred below, grass-green to bluish-pruinose, 10–15 mm long, c. 2 mm across, eventually recurved; **Fl.** many in a cyme, 6–7-partite, golden-yellow.

**S.** — v. **albescens** (HAW.) ROUY et CAM. (*S. a.* HAW.). – Greece: Thessaly. – **Fl.** pale yellow.

**S.** — v. **arrigens** (GRENIER) BRIQUET (*S. a.* GRENIER). – Plant green.

**S.** — v. **aureum** (WIRTGEN) JACOBS. (*S. rupestre* v. *aureum* WIRTGEN). – **L.** truncate above, mostly purplish-red, with a long spur.

**S.** — v. **cristatum** PRAEG. – **Shoots** fasciated.

**S.** — v. **glaucum** DUM. – Plant blue-green.

**S.** — v. **viride** KOCH (*S. rupestre* v. *viride* KOCH). – Plant fresh green.

**S. retusum** HEMSL. (§ 13). – Mexico: San Luis Potosí. – Glabrous ♄ 30 cm tall; **L.** spatulate, deeply emarginate, cuneate-tapering, 2–2.5 cm long, 6–10 mm across, notched at the tip; **Fl.** campanulate, white, with a red centre.

**S. rhodocarpum** ROSE (§ 17/3). – USA: Sierra Nevada. – ♃; **shoots** triangular, alate; **L.** rounded-spatulate to circular, 2 cm long; **Infl.** of few, greenish-red Fl. Hardy.

**S. roborowskii** MAX. (§ 20/2). – China: Kansu. – Glabrous, basally-branched ⊙; **L.** oblong-lanceolate, 15 mm long, 5 mm across, smaller higher up; **Fl.** 4 mm ⌀, campanulate, yellow.

**S. rohleanae** DOM. (§ 17/17) (*S. acre* v. *r.* DOM.). – Yugoslavia. – Close to **S. acre**; **shoots** stiffly erect, little branched; **L.** densely imbricate, ovate-oblong, withering to white; **Fl.** in a dichotomous cyme, yellowish. Hardy.

**S. rosei** HAMET (§ 17/17). – W.China. – Dwarf ♃; **L.** long linear-triangular, acute, 4 to 6 mm long.

**S. rosthornianum** DIELS (§ 3). – W. China: S.Szechuan. – Close to **S. verticillatum**; **shoots** 20–25m tall; **L.** 2–2.5 cm long, 1 cm across; **Fl.** few in an Infl. 7–10 cm long. Hardy.

**S. rosulatum** EDGEW. (§ 17/6) (*Umbilicus radicans* KLOTZSCH). – N.E.Himalayas. – Close to **S. adenotrichum**; basal **L.** narrowed teretely below, spatulate, bluntly dentate, in lax Ros.

**S. rubens** L. v. **rubens** (§ 20/5) (*Crassula r.* L., *Aithales r.* WEBB et BERTH., *Procrassula r.* RAUL.). – W.Eur.; Canary Is.; N.Afr. to the Near East. – ⊙ – ⊙; **shoots** erect, 5–10 cm tall, simple or branched, hairy above; **L.** glabrous, lanceolate to linear, obtuse, 18 mm long; **Infl.** of 2–4 reddish Fl.

**S.** — v. **delicum** (VIERH.) HAYEK (*S. d.* VIERH.). – Greece: Cyclades. – **L.** more spatulate; Pet. yellowish with a red keel.

**S. rubricaule** (ROSE) PRAEG. (§ 4) (*Sedastrum r.* ROSE). – Mexico. – Close to **S. ebracteatum**, possibly only a variety of this; **shoots** more hairy and deeper red.

**S. rubroglaucum** PRAEG. (§ 17/24). – USA: Calif. – ♃; **L.** obovate, rounded and obtuse, narrowed abruptly below into the amplexicaul petiole, the underside much thickened, 2 cm long, 7 mm across, 5 mm thick, blue-green tinged with red; **Fl.** yellow.

**S. rubrotinctum** R. T. CLAUSEN (§ 11) (*S. guatemalense* HORT.). – Mexico. – Sub-♄ 15 to 20 cm tall, branching basally; **shoots** fleshy at first, 2–3 mm thick, at first prostrate, eventually ascending; **L.** ± rosulate, crowded on the tips of the shoots, cylindrical, with a rounded, red tip, 2 cm long, 5 mm across, 4–5 mm thick, light green, suffused reddish-brown all over.

**S.** — cv. Aurora. – **L.** more salmon-coloured, with a silvery sheen. Acc. JOHN G. PUSEY a periclinal chimera, some L. having longitudinal green and white stripes.

**S.** — cv. Ruby Glow. – ♃ to 25 cm high; **L.** ovate, grey-blue; **Fl.** rose-crimson, in corymbs 7–10 cm ⌀.

**S. rubrum** (L.) THELL. (§ 20/5) (*Tillaea r.* L., *Crassula caespitosa* CAV., *S. c.* DC., *Cr. magnolii* DC., *Procrassula m.* GRISEB., *S. m.* BUB., *Aithales caespitosa* WEBB et BERTH., *S. desertihungaricum* SIM.). – Medit. reg.; Hungary; As. Minor. – Insignificant ⊙; **Fl.** white.

**S. ruwenzoriense** BAK. (§ 15) (*S. ducis-apruti* HORT.). – Uganda: Ruwenzori. – Creeping ♄ growing on rocks; **L.** mostly ± round; **Fl.** in dense cymes, brilliant golden-yellow.

**S. sanguineum** BOISS. et HAUSSKN. (§ 20/5). – Iraq. – Insignificant ⊙; **L.** broadly linear; **Infl.** glandular, rough-hairy.

**S. sanhedrianum** BGR. (§ 17/5) (*Gormania retusa* ROSE). – USA: Calif. – ♃ 10–15 cm tall; **L.** flat, emarginate; **Fl.** white.

**S. sarmentosa** BUNGE (§ 17/25). – N.China; Japan. – ♃; **shoots** 30 cm long, pendulous, rooting from the tips; **L.** broad-lanceolate, acute, 2 cm long, light green; **Fl.** yellow.

**S. sartorianum** BOISS. (§ 17/17). – Balkans. – Dwarf ♃ resembling **S. cupressoides**; **L.** not appressed like **Cupressus**. Hardy.

**S.** — ssp. **hillebrandtii** (FENZL) D. A. WEBB (*S. h.* FENZL). – **L.** linear, short-spurred, papillose; **Fl.** golden-yellow.

**S.** — ssp. **ponticum** (VEL.) D. A. WEBB (*S. p.* VEL.). – **Fl.** pale yellow.

**S.** — ssp. **sartorianum**. – **L.** ovate-triangular, obtuse, 4–5 mm long; **Fl.** yellow.

**S.** — ssp. **stribrnyi** (VEL.) D. A. WEBB (*S. s.* VEL.). – **L.** 8–10 mm long, with a larger spur; **Fl.** yellow.

**S. sediforme** (JACQ.) C. PAU v. **sediforme** (Pl. 127/3) (§ 17/20) (*Sempervivum s.* JACQ., *Sed. altissimum* POIR., *S. nicaense* ALL.). – Medit. reg. – ♃; **shoots** 15–60 cm tall; **L.** linear-

lanceolate, apiculate, upperside slightly concave, 12–20 mm long, 3–5 mm across, grey-green; **Infl.** ± spherical, Fl. greenish-white.
**S. — v. montanum** GRISEB. – **Infl.** more crowded.
**S. selskianum** RGL. et MAACK. (§ 16). – N.E. China. – ♃; **taproot** thick; **shoots** 30–45 cm tall, densely hairy; **L.** linear-oblong, 5 cm long, dark green, dentate above midway; **Infl.** large, Fl. light yellow. Hardy.
**S. semiteres** ROSE (§ 17/10). – Mexico: Durango. – Dwarf ♃, close to **S. mellitulum**; **L.** papillose, 1–2 cm long, semi-terete.
**S. sempervivoides** FISCH. (Pl. 127/1) (§ 18) (*S. sempervivum* LEDEB.). – As. Minor. – ☉; **L.** many, in Ros. resembling **Sempervivum**, 25 to 30 mm ⌀; **Infl.** terminal, 25–30 cm tall, with ciliate bracts, cyme broad, paniculate, up to 5 cm ⌀, Fl. crimson. Hardy.
**S. serpentinii** JANCH. (§ 17/11). – Albania. – Dwarf, mat-forming ♃; **L.** ± cylindrical to spherical, 2–4 mm long; **Fl.** purplish-red.
**S. shaeri** L. M. MOORE (§ 17/22). – Central China. – Glabrous ♃; **L.** linear-lanceolate, obtuse, c. 25 mm long; **Infl.** of many Fl.
**S. shastense** BRITT. (§ 17/21). – Mexico: Baja Calif. – Close to **S. stenopetalum**; **L.** 10–15 mm long, 3–5 mm across, papillose, with a long spur; **Fl.** yellow, few in a small cyme.
**S. shigatsense** FROED. (§ 17/25). – China: Tibet. – ? ☉, branching little below, c. 10 cm tall; **L.** ovate, spurred, with an obtuse tip, c. 10 mm long, upper L. smaller, c. 4 mm long; **Infl.** of few, yellow Fl.
**S. shimizuanum** HONDA (§ 3). – Japan: Honshu. – Glabrous ♃; **shoots** clustered, 25 to 30 cm tall, slender, developing adventitious buds above in autumn; **L.** in whorls of 3, oblong, tapering, 2.5–3 cm long, 7–12 mm across, fleshy, margins dentate; **Infl.** of few, small, half-open, green Fl.
**S. sichotense** WOROSCH. (§ 16). – USSR: Primorsk reg. – Glabrous ♃; **shoots** numerous, glabrous, 10–20 cm tall, 1–1.5 mm thick, purplish-brown below, glossy, densely leafy; **L.** thick, oblong-oblanceolate to narrow-lanceolate, 2.5–3.5 cm long, 3–7 mm across, ± obtuse to obtuse-tapering above, margins serrate below; **Infl.** a dense corymb or panicle, 1.5–3.5 cm ⌀, Fl. sessile, 5–8 mm ⌀, greenish-yellow.
**S. sieboldii** SWEET (§ 3). – Japan. – ♃; **shoots** simple, 15–25 cm long, prostrate, red; **L.** in whorls of 3, ± circular, sessile, bluish-grey, with red margins, blunt-dentate above, 17 mm long, 19 mm across; **Fl.** pink.
**S. —** cv. Variegatum. – **L.** with yellowish-white blotches.
**S. smallii** (BRITT.) AHLES (§ 20/?) (*Diamorpha s.* BRITT.). – N.Am.: N.Carolina. – ☉ to ☉ up to 5 cm tall; **shoots** slender, branching below; **L.** alternate, ± terete, oblong, 15 to 20 mm long; **Fl.** small, reddish, in cymes.
**S. somenii** HAMET (§ 20/2) (*S. roborowskii* v. *somenii* HAMET). – China: Yunnan. – ☉ – ☉, dainty, with tangled branches; **shoots** prostrate, 15–20 cm long; **L.** obovate, with entire margins, obtuse, 2.5 cm long, 12 mm across, light green; **Fl.** yellow.

**S. sordidum** MAX. (§ 3). – Japan. – ♃; **shoots** not branched below; **L.** broad-elliptical to roundish-ovate, scarcely dentate, 3–4.5 cm long, 3 cm across; **Fl.** greenish. Hardy.
**S. spathulifolium** HOOK. v. **spathulifolium** (§ 17/23) (*Echeveria s.* L. DE SMET). – USA: Calif.; Brit. Columbia. – **L.** in dense, flat Ros., 2.5 cm long, very obtuse, mucronate, 1 cm across, often suffused red, pruinose; **Fl.** light yellow.
**S. —** cv. Capablanca. – **Ros.** small, white farinose.
**S. — v. major** PRAEG. – Larger variety.
**S. — v. purpureum** PRAEG. – **L.** deep red.
**S. spectabile** BOREAU (Pl. 127/2) (§ 3) (*Anacampseros s.* JORD., *S. pseudospectabile* PRAEG.). – S.Korea; widespread in W.China. – **Roots** fleshy, up to 1 cm thick, from a stout rootstock; **shoots** erect, usually simple, 25 cm tall, 2–8 mm thick, with 9–15 nodes; **L.** 2–4 at each node, mostly elliptical or ovate to circular, blunt or rounded above, lower L. often auriculate, 3 to 9 cm long, 2–2.5 cm across, c. 1–1.5 mm thick, pinnately veined, margins ± entire or irregularly dentate; **Infl.** a corymb, Fl. 5–10 mm ⌀, pale to deep pink. Hardy.
**S. —** cv. Atropurpureum. – **Fl.** deep red.
**S. —** cv. Brilliant. – **Fl.** deep pink.
**S. —** cv. Carmen. – **Fl.** bright carmine-red.
**S. —** cv. Gwendoline Parr. – Sport of cv. Brilliant. Up to 75 cm tall; **L.** on a petiole 1 cm long, ovate-lanceolate, serrate; **Infl.** of up to 300 Fl., Pet. white below, with green stripes.
**S. —** cv. Meteor. – **Fl.** deep carmine-red.
**S. —** cv. Variegatum. – **L.** yellowish-white mottled.
**S. spurium** MARSCH. BIEB. (§ 17/1). – USSR: Caucasus. – Mat-forming ♃; **L.** obovate, 2.5 to 3 cm long, ± rosulate on sterile shoots, green, margins very minutely hairy; **Fl.** white, red or pink. Hardy.
**S. —** cv. Album. – **Fl.** white.
**S. —** cv. Coccineum. – Cultivar with scarlet Fl.
**S. —** cv. Schorbuser Blut. – Trailing, 7 to 10 cm high; **Fl.** deep crimson.
**S. —** cv. Splendens. – Cultivar with crimson Fl.
**S. stahlii** SOLMS (§ 17/12). – Mexico. – Small sub-♄ with spreading Br.; **shoots** spreading or creeping, minutely hairy; **L.** elliptical-oblong to spherical, with a rounded tip, terete, minutely puberulous, usually suffused red, readily detachable, 7–14 mm long, 4–8 mm across, 3–7 mm thick; **Fl.** 13–15 mm ⌀, yellowish-green.
**S. stellarifolium** FRANCH. (§ 20/2). – E. China. – Close to **S. drymarioides**, possibly only a variety of this; laxly hairy.
**S. stellatum** L. (§ 20/3) (*S. deltoideum* TEN.). – S.Eur. – ☉; **shoots** branched, prostrate below, then ascending, 10–15 cm tall; **L.** rounded-spatulate, rounded above, indistinctly dentate, 25 mm long, 15 mm across; **Fl.** purple.
**S. stelliferum** S. WATS. (§ 17/10). – USA: New Mexico, Arizona. – Close to **S. potosinum**; **L.** 4–9 mm long.

**S. stenopetalum** PURSH. (§ 17/20). – N.Am. – Type: v. stenopetalum.

**S. — v. ciliosum** (HOWELL) T. B. CLAUSEN (*S. c.* HOWELL, *S. douglasii* v. *c.* (HOWELL) T. B. CLAUSEN). – USA: Oregon. – **L.** 1–2 cm long, 1.5–3 mm across, ciliate, with short shoots from the axils; Fl. yellow.

**S. — v. radiatum** (S. WATS.) T. B. CLAUSEN (*S. r.* S. WATS., *S. douglasii* v. *r.* (S. WATS.) T. B. CLAUSEN). – USA: Calif., Oregon. – ⊙ – ⊙; **shoots** 7–15 cm tall; **L.** 7–12 mm long, 2–4 mm across, with deciduous buds developing in the axils.

**S. — v. stenopetalum** (*S. s.* v. *typicum* T. B. CLAUSEN, *S. douglasii* HOOK., *S. d.* v. *d.* (PURSH.) T. B. CLAUSEN, *S. subclavatum* HAW.). – Central and western N.Am. – Dwarf ♃; **L.** often in 6 R., linear, tapering, short-spurred, 1 cm long; **Fl.** pale yellow.

**S. steudelii** BOISS. (§ 20/5). – Syria. – Insignificant ⊙; **Br.** with 1–3 whitish-pink Fl.

**S. stevenianum** ROUYI et CAM. (§ 17/1) (*S. roseum* STEV.). – As. Minor; USSR: Caucasus. – Mat-forming ♃; **shoots** filiform, 2–5 cm tall; **L.** obovate, 6 mm long, 1.5 mm across, very obtuse, red-spotted; **Fl.** red. Hardy.

**S. stoloniferum** S. T. GMELIN (§ 17/1). – USSR: Caucasus. – Mat-forming ♃; **shoots** slender, red-striate; **L.** rhombic-spatulate, indistinctly crenate in the upper half, 25 mm long, 12 mm across; **Fl.** stellate, pinkish-red. Hardy.

**S. subtile** MIQ. (§ 17/22). – Japan. – Resembles **S. magellense**; **L.** spatulate, petiolate, 8 mm long, 2–4 mm across; **Fl.** yellow.

**S. susannae** HAMET (§ 17/19). – W.China. – ♃ 5–9 cm tall; **L.** linear-ovate, 3–7 mm long, with an unlobed spur.

**S. taquetii** PRAEG. (§ 3). – Korea. – ♃ with thickened **roots**; **shoots** erect, 31–45 cm tall; branching in the upper third; **L.** sessile, margins upcurved in the lower half, 6.5 cm long, 3 cm across, dark green with red dots; **Fl.** in dense umbels, 15 mm ∅, pale green. Hardy.

**S. tatarinowii** MAX. (§ 3). – W.China. – ♃; **shoots** simple, ± erect, 10–15 cm long; **L.** linear-lanceolate, short-petiolate, margins with several large T.; **Infl.** dense, Fl. 12 mm ∅, pink. Hardy.

**S. techinense** FU (§ 20). – China: Yunnan. – Insignificant ⊙ plants; **L.** linear-lanceolate; **Fl.** red (? yellow).

**S. telephioides** MICHX. (§ 3). – Eastern N.Am. – Close to **S. telephium**; stamens short; nectar scales broad.

**S. telephium** L. (§ 3) (*S. glauco-pruinosum* EKLUND, *S. pseudotelephium* EKLUND). – N. temperate zone. – ♃ with thickened **roots**; **shoots** erect, up to 70 cm tall, simple; **L.** sessile or petiolate, ovate to lanceolate, 5–7.5 cm long, 2–4 cm across, obtuse, irregularly dentate, ovate-oblong; **Fl.** in dense corymbs, pruinose or not. Hardy. Numerous ecotypes.

**S. — ssp. fabaria** (HOCH) SCHINZ et KELLER (*S. f.* KOCH, *S. maximum* ssp. *f.* (KOCH) LÖVE, *Anacampseros vulgaris* HAW., *S. t.* v. *v.* (HAW.) BURNAT). – **L.** narrowed cuneately at the base, usually lanceolate, less coarsely dentate; **Fl.** red.

**S. — ssp. telephium.** – Lower **L.** cuneate, upper ones sessile, with a rounded or truncate base.

**S. — ssp. — v. albiflorum** MAX. (*S. telephium* v. *a.* MAX.). – **Fl.** white.

**S. — ssp. — v. borderi** ROUYI et CAM. (*S. z.* v. *b.* ROUY et CAM.). – **L.** distinctly petiolate, margins irregularly and deeply dentate.

**S. — ssp. — cv.** Munstead Dark Red. – Superior cultivar with deep red **Fl.**

**S. — ssp. — v. purpureum** (L.) SCHINZ et KELLER (*S. telephium* v. *p.* L., *S. t.* ssp. *p.* (L.) SCHINZ et KELLER, *S. purpurascens* KOCH). – **L.** broader, more coarsely dentate, reddish; **Fl.** purplish-red.

**S. — ssp. — f. roseo-variegatum** PRAEG. (*S. t.* f. *r.-v.* PRAEG.). – **L.** light pink at first, eventually green.

**S. — ssp. — cv.** Variegatum.- Up to 30 cm high; **St.** and **L.** streaked white and pink.

**S. tenellum** MARSCH. BIEB. (§ 17/11). – Turkey: Armenian highlands. – Mat-forming ♃ resembling **S. lydium**; **L.** shorter on sterile shoots, obtuse; **Infl.** less flat.

**S. tenuifolium** (SIBTH. et SM.) STROBL. (§ 17/20) (*Sempervivum t.* SIBTH. et SM., *S. amplexicaule* DC., *S. rostratum* TEN.). – Medit. reg. – ♃; **L.** linear, tapering, widened and amplexicaul below, grey-green to whitish, the tip thin, awn-like, curved; **Fl.** large, golden-yellow. Hardy.

**S. ternatum** MICHX. v. **ternatum** (§ 17/3). – Eastern N.Am. – ♃; **shoots** rooting where resting on the soil, 7–15 cm tall; **L.** ovate, rounded or bluntly pointed above, 12–25 mm long, 6–12 mm across, rosulate on the sterile shoots; **Fl.** 4-partite, white. Hardy.

**S. — v. minus** PRAEG. – Smaller variety.

**S. tortuosum** HEMSL. (§ 15) (*S. nelsonii* ROSE, *S. lignicaule* FROED.). – Mexico. – Much-branched, mostly epiphytic sub-♄; **shoots** up to 30 cm long, erect or pendulous, minutely papillose; **L.** mostly deciduous, oblanceolate or elliptical-spatulate, minutely spurred, obtuse and minutely papillose at the tip, green to reddish, 15–40 mm long, 4–11 mm across, up to 2.4 mm thick; **Fl.** up to 19 mm ∅, white.

**S. torulosum** T. B. CLAUSEN (§ 12). – Mexico: Coahuila prov. – Tree-like ♄ up to 1 m tall; **St.** robust, irregularly thickened, 8–10 cm ∅ below, with a ± papery bark; **Br.** dichotomous, erect, ± spirally arranged, torulose, with an apical L.Ros.; **L.** densely crowded, rhombic-spatulate, sessile, acute, 0.5–5.2 cm long, 0.5 to 1.5 cm across, 1–2 mm thick, initially pruinose; **Infl.** terminal, corymbose, Fl. lemon-yellow, on pedicels 12 mm long.

**S. treleasei** ROSE (Pl. 127/4) (§ 11). – Mexico. – Sub-♄; **St.** thick, succulent, eventually woody, little branching; **L.** fairly densely arranged on the tips of the shoots, curved upwards, oblong-ovate, 3 cm long, 15 mm across, 10 mm thick, obtuse, blue-green, grey-pruinose, the underside thickened; **Infl.** ± spherical, Fl. light yellow.

**S. triactina** BGR. (§ 20/2) (*Triactina verticillata* HOOK. f. et THOMS., *S. v.* HAMET). – China; Sikkim. – Branched ⊙; **L.** verticillate, obovate, 12–30 mm long; **Fl.** yellow.

**S. tristriatum** BOISS. (§ 17/11). – Greece, Crete. – Mat-forming ♃; **L.** rosulate on sterile shoots, oblong-spatulate, papillose-pubescent, those on the shoots from the Ros. being obovate, semi-terete; Fl. 6–7 together in a lax cyme.

**S. trullipetalum** HOOK. f. et THOMS. (§ 17/19). – Himalayas; W.China: W.Yunnan. – Dwarf ♃; **shoots** thin, moss-like, freely branched, 6–8 mm tall; **L.** imbricate, tapering, 3–6 mm long, with a 3-lobed spur; Fl. whitish-yellow.

**S. tsugaruense** HARA (§ 3). – Japan: Honshu. – Resembles **S. sieboldii**; **roots** 2–3 mm thick; **shoots** few, mat-forming, 10–40 cm long, prostrate, bluish or brownish-purple; **L.** sessile, often in 3's below, elliptical to ovate, the tip obtuse or rounded, margins irregularly crenate-serrate, 2–4 cm long, 12–35 mm across; Fl. greenish-yellow.

**S. tuberculatum** ROSE (§ 12). – Mexico: Oaxaca. – Small ♄ with reddish tubercles; **L.** 6–12 mm long, obtuse; Fl. white.

**S. tympheum** QUEZEL et CONTANDRIOPOULOS (§ 19). – Crete. – Dwarf ♃; Ros. 2–2.5 cm ⌀; **L.** numerous, spatulate, fleshy, glandular-hairy, yellowish, 10–12 mm long, 3–4 mm across; **Infl.** solitary, 5–6 cm tall, leafy, bipartite above, Fl. red, 12–14 mm ⌀.

**S. variicolor** PRAEG. (§ 17/22). – W.China: Yunnan. – Freely branched, 15 cm tall; **shoots** ♃, green; **L.** deciduous in autumn, oblong-spatulate, 2 cm long, 6 mm across, short-spurred below; Fl. light yellow.

**S. versadense** C. H. THOMPSON (§ 17/5). – Mexico. – ♃; **roots** white; **shoots** hairy, erect or prostrate, 1–7 cm long, flowering shoots glabrous, 3–20 cm long; **L.** obovate or spatulate, truncate above, apiculate, underside convex, 3.2–4.1 cm long, 9–13 mm across, 3–4 mm thick; Fl. delicate pink, 11–14 mm ⌀.

**S. versicolor** (Coss.) MAIRE (§ 20/6) (*S. coeruleum* v. v. HAMET). – N.Afr.: Anti-Atlas mts. – ⊙ 3–15 cm tall; **L.** obtuse, glabrous; **Infl.** glandular-pubescent, Fl. yellow.

**S. verticillatum** L. (§ 3). – Japan; USSR: Kamchatka. – ♃; **shoots** erect, simple or with a few side-shoots, 30–60 cm tall; **L.** in whorls, oblong-lanceolate, 5–7.5 cm long, 2–2.5 cm across, green, with small red spots, petiole 5–6 mm long; Fl. pale green. Hardy.

**S. —** v. **nipponicum** PRAEG. – Japan. – Small variety; **L.** opposite.

**S. victorianum** JANSS. (§ 20). – Burma: Mt. Victoria. – **Shoots** fleshy, erect, 12–16 cm tall; **L.** petiolate, obovate, flat, fleshy, with entire margins but the tip incised, carinate on the underside, 18–28 mm long, 7–15 mm across; Fl. in a flat cyme, 13–20 mm ⌀, deep yellow.

**S. villosum** L. ssp. **villosum** (§ 20/5). – N.Eur. to Algeria; Greenland; Iceland. – Mostly ⊙; **shoots** much branched; **L.** glandular-hairy, linear-oblong, obtuse, 6–12 mm long; Fl. pink, 6 mm ⌀. Hardy.

**S. —** ssp. **aristatum** (EMB. et MAIRE) LAINZ (*S. v.* v. *a.* EMB. et MAIRE, *S. maireanum* SENNEN, *S. paui* SENNEN). – **L.** awned.

**S. —** v. **pentandrum** DC. – Fl. with 5 anthers.

**S. viride** (ROSE) BGR. (§ 11) (*Corynophyllum v.* ROSE). – E.Mexico. – Freely branched ♄ 30 to 40 cm tall; **L.** dense, cylindrical, 3–5 cm long, thick, green; Fl. small, greenish-yellow, sessile.

**S. viscosum** PRAEG. (§ 20/2). – E.China. – Close to **S. drymarioides**, possibly only a variety of this; plant viscous-hairy; Fl. rather larger.

**S. viviparum** MAX. (§ 3). – N.E.China. – ♃; **shoots** erect, branching basally, c. 20 cm tall; **L.** in whorls of 3–4, broadly ovate, obtuse, narrowing into the short petiole, indistinctly and bluntly dentate, grey-green, 5 cm long, 2 cm across; Fl. ± campanulate, white to greenish; during flowering, lateral buds form in the L. axils.

**S. wangii** FU (§ 20). – China: Yunnan. – Inconspicuous ⊙; **L.** linear to lanceolate-ovate; Fl. yellow.

**S. watsonii** (BRITT.) BGR. (§ 17/23) (*Gormania w.* BRITT., *Cotyledon oregonensis* S. WATS., *Echeveria w.* NELS. et MACBR.). – USA: Oregon. – Resembles **S. obtusatum**; **Infl.** dense, Pet. pale yellow.

**S. wenchuanense** FU (§ 17/25). – China: Szechuan, Wenchuan. – ♃; **shoots** 7 cm long, erect, leafy above; **L.** sessile, ovate, flat, 3 to 4 mm long, 1.5–2 mm across, acute to obtuse, spurred below, fleshy, glabrous, with entire margins; Fl. 4 mm ⌀.

**S. woodii** BRITT. (§ 17/23). – USA: Oregon. – Close to **S. spathulifolium**; **L.** very broadly spatulate, obtuse.

**S. wootonii** BRITT. (§ 17/5). – USA: New Mexico, Arizona. – ♃ with small L.Ros.; basal **L.** rough, 8–14 mm long, 2.5–5 mm across, obtuse; those on flowering shoots more acute; Fl. white, in a granular-tomentose cyme.

**S. wrightii** A. GRAY v. **wrightii** (§ 17/5). – USA: Texas; Mexico: Chihuahua. – Dwarf, mat-forming ♃ with a fleshy **taproot**; **L.** crowded, in small Ros., ovate, thick, light green, glossy-papillose; Fl. campanulate, white.

**S. —** v. **beyrichianum** (MASTERS) PRAEG. (*S. b.* MASTERS). – More spreading; **L.** more distant, Ros. laxer.

**S. yabeanum** MAKINO (§ 16). – Japan. – ♃ with a thick **taproot**; **shoots** with flat L. with entire margins; Fl. yellow.

**S. yosemitense** BRITT. (§ 17/23). – USA: Calif. (Yosemite Park). – Close to **S. spathulifolium**; **L.** not pruinose, fresh green, ovate-spatulate; **Infl.** 7–10 cm tall, pedicels short, thick.

**S. yvesii** HAMET (§ 17/25). – China: Szechuan. – ♃ 7–12 cm tall; **L.** sessile, spurred, obovate-linear, 5–9 mm long; Fl. yellow.

**S. zentaro-tashiroi** MAKINO (§ 17/25). – Japan. – ♃; **L.** verticillate on flowering shoots, narrowed teretely below, flat, spatulate-obovate, 12 mm long; Fl. yellow.

**S. zlatiborense** DOM. (§ 17/17). – Yugoslavia. – Close to **S. acre**; **shoots** erect, little branched, 5–8 cm tall; **L.** densely imbricate, broadly ovate or triangular-ovate, 4–4.5 mm long, 2.5 mm across; Fl. yellow. Hardy.

**Sempervivella** STAPF. Crassulaceae. – W.Himalayas. – Dwarf, Ros.-forming ⚈ plants, soon forming small clumps by means of filiform stolons; **L.** flat, thickened on the underside, fleshy; flowering **shoots** lateral, ascending or prostrate, with small bracts; **Fl.** few, in small cymes, white or pink. – Summer: in the open, winter: cold-house. Propagation: seed, cuttings.

**S. acuminata** (DECNE.) BGR. (*Sempervivum a.* DECNE, *S. himalayense* KLOTZSCH, *Sedum moorcroftianum* WALL., *Sed. a.* HAMET). – Himalayas. – **L.** ± lanceolate, awned, glabrous, 3–6 cm long; **Fl.** pinkish-red.

**S. alba** (EDGEW.) STAPF (Pl. 128/1) (*Sempervivum a.* EDGEW., *Sedum sedoides* HAMET). – Himalayas. – **Ros.** 2–3 cm ⌀; **L.** obovate-oblong, rather rounded or rather obtuse above, 1.5 cm long and across, minutely hairy, light green; **Fl.** 12–14 mm ⌀, white or red.

**S. mucronata** (EDGEW.) BGR. (*Sempervivum m.* EDGEW.). – Himalayas. – **L.** lanceolate, 2.5 to 3.5 cm long, with an awned tip; **Fl.** white.

**S. sedoides** (DECNE.) STAPF (*Sempervivum s.* DECNE.). – Kashmir. – Resembles **S. alba**; smaller; **L.** less hairy, narrower, linear-oblong.

**Sempervivum** L. Crassulaceae. – Mts. of Central and S.Eur.; USSR: Caucasus; highlands of Armenia; Turkey; N.W.Afr.: Atlas. – Low ⚈ forming mats or cushions, producing offsets or dividing; **L.** freely spirally arranged in dense Ros., sessile, obovate, lanceolate or acute, thick-fleshy, green, reddish or bluish, glabrous or glandular-hairy, the margins often ciliate with white hairs; **Fl.** in a terminal, erect, dense-flowered, ± many-branched cyme, spreading star-like, 8–20-partite, Pet. entire, white, greenish, yellow, pink, red or purplish; Carp. spreading at the tip; the Infl. with bracts, often villous. Ros. die after flowering in summer. Hardy. – Propagation: seed, division. – The genus was for a time divided into 2 Subgenera: SG. Sempervivum (*Eusempervivum* KOCH) and SG. Jovibarbae DC. The latter SG., acc. to G. HEGI, Fl. Eur. Ed. II vol. IV, part 2, is again isolated from **Sempervivum** as **Diopogon** JORD. et FOURR. q.v. (See Fl. Eur. **I**, 1964; G. HEGI Ill. Fl. M.-Eur., Ed. II, Vol. IV, part II; C. W. MUIRHEAD, Notes Royal Bot. Gard. Edinb., XXVI, No. 3, 1965, p. 279–285; PRAEGER, R. L.: An Account of the Sempervivum Group. London 1932. Reprinted 1967.)

## Classification of the Genus Sempervivum L. by J. A. Huber

**Sect. I. Arachnoidea** LEHM. et SCHNITTSP. – Ros. and St.-L. with a hair-tuft at the apex. – S. arachnoideum. – Pyrenees, Alps, Apennines, Carpathians.

**Sect. II. Sempervivum** (Sect. *Ciliata* LEHM. et SCHNITTSP.). – Ros.-L. without a hair-tuft, surfaces glabrous, glabrescent or short-hairy, margins ciliate.
  a) **SSect. Tectorae** HUBER. – Western branch.
    1. Red-flowering. – S. atlanticum, Afr.: Atlas; S. andreanum, nevadense, vincentei, Spain; S. tectorum, W. and Central Eur. to Hungary.
    2. Yellow-flowering. – S. wulfenii, Alps: Switzerland to Steiermark.
  b) **SSect. Clusianae** HUBER. – Eastern branch.
    1. Red-flowering. – S. borissovae, caucasicum, italicum, marmoreum, Italy: Apennines; Balkans; S. iranicum, W.As.
    2. Yellow-flowering. –
       Filaments red.: S. armenum, W.As.
       Filaments white: S. glabrifolium, ispartae, staintonii, W.As.

**Sect. III. Glandulosa** HUBER (Sect. *Papillosae* and Sect. *Villosae* LEHM. et SCHNITTSP., Sect. *Pubescentia* BAK., 1879. – L. with hairs and glands on both faces.
  a) **SSect. Montanae** HUBER. – Western branch.
    1. Red-flowering. – S. cantabricum, giuseppii, Spain; S. dolomiticum, Alps: Dolomites; S. montanum, Pyrenees, Alps, Carpathians.
    2. Yellow-flowering. – S. grandiflorum, pittonii, Alps.
  b) **SSect. Globiferae** HUBER. – Eastern branch.
    1. Red-flowering. – S. ballsii, erythraeum, kosaninii, macedonicum, Balkans; S. altum, ingwersenii, ossetiense, pumilum, USSR: Caucasus.
    2. Yellow-flowering.
       Filaments reddish: S. kindingeri, octopodes, ruthenicum, thompsonianum, zeleborii, Balkans.
       Filaments white or green: S. ciliosum, leucanthum, Balkans.
       Filaments reddish: S. brevipilum, georgicum, gillianum, transcaucasicum, W.As.
       Filaments white: S. artvinense, davisii, furseorum, minus, W.As.

**S. altum** TURRILL (§ III/b 1). – USSR: Caucasus. – Similar to **S. kosaninii** but much smaller; Fl. with yellow anthers.

**S. andreanum** WALE (§ II/a 1). – Spain. – Ros. 1.5–3 cm ⌀; L. c. 25, inner L. crowded, 19 mm long, 7 mm across, 3 mm thick, obovate, tapering, the tip brownish-red on the underside, back surface glabrous, marginal cilia few, short; Infl. 11 cm tall, hairy, bracts oblong-lanceolate, apiculate, 2 cm long, upperside hairy, cilia short, pale red; Fl. c. 2 cm ⌀, Pet. pale red, dark red below.

**S. arachnoideum** L. (§ I). – Pyrenees, Alps, Apennines, Carpathians. – Ros. 5–50 mm ⌀, mat-forming; L. with white, ± densely tangled, cobwebby hairs at the tips; Fl. 12–23 mm ⌀, pink to crimson or white.

**S. — v. arachnoideum** (Pl. 128/3) (*S. a.* v. *typicum* FIORI). – Alps, Apennines. – Cobwebby hairs ± thick, covering the Ros., which is usually ± closed.

**S. — v.** — cv. Album (*S. a.* v. *album* HORT.). – Fl. white.

**S. — v. bryoides** SCHNITTSP. – Ros. much smaller than v. arachnoideum.

**S. — v. doellianum** (LEHM.) JACCARD (*S. d.* LEHM., *S. d.* v. *d.* JACCARD, *S. a.* v. *glabrescens* WILLK., *S. heterotrichum* SCHOTT, *S. moggridgei* HOOK. f.). – Ros. 2–5 cm ⌀, lax; L. projecting ± ray-like, cobwebby hairs sparse or ± missing; Fl. brilliant red.

**S. — v. oligotrichum** (HAMPE) WETTST. (*S. o.* HAMPE). – Cobweb-hairs very sparse, eventually completely disappearing.

**S. — f. sanguineum** JEANB. et TIMB. – L. very reddened.

**S. — v. tomentosum** (LEHM. et SCHNITTSP.) HAYEK (*S. laggeri* SCHOTT, *S. webbianum* LEHM. et SCHNITTSP., *S. hookeri* VAN HOUTTE). – Ros. flat and compressed spherical; L. densely snow-white, tomentose; Fl. 20–23 mm ⌀.

**S. —** Hybrids: S. × **barbulatum**, × **fauconnettii**, × **fimbriatum**, × **roseum**, × **vaccarii**.

**S. armenum** BOISS. et HUET. v. **armenum** (§ II/b 2) (*S. globiferum* BOISS.). – Armenian highlands. – Ros. large; L. oblong-spatulate, short-tapering, apiculate, both surfaces glabrous, with ciliate margins; Fl. yellow.

**S. — v. insigne** MUIRH. – Turkey. – Ros. 2–3 cm ⌀, slightly open, offsets numerous; L. bluish with purple tips; Fl. with violet filaments.

**S. artvinense** MUIRH. (§ III/b 2). – Turkey. – Ros. 3–4 cm ⌀, densely leafy, side-shoots 2–4 cm long, leafy, densely hairy; L. oblong-lanceolate, c. 1.5 cm long, 7 mm across, thick, green, underside carinate, apiculate, glandular-hairy, with ciliate margins; Infl. 12–13 cm tall, bracts c. 2 cm long, 5–6 mm across, densely hairy, Fl. 2.5 cm ⌀, yellowish-white, with white filaments.

**S. atlanticum** BALL. (§ II/a 1) (*S. tectorum* v. *a.* (BALL.) HOOK.). – Morocco: High Atlas. – Ros. 7–10 cm ⌀; L. red-tipped, minutely glandular at first, eventually glabrous; Fl. 3–4 cm ⌀, 12-partite.

**S. balcanicum** STOY. – Insufficiently known species; acc. Fl. Eur. a variety of **S. erythraeum** or **S. ballsii** or **S. marmoreum**; acc. J. A. HUBER classifiable with **S. marmoreum**.

**S. ballsii** WALE (§ III/b 1). – Albania; Greece. – Ros. 2.5–3.5 cm ⌀, clustering; L. obovate, short-mucronate, c. 18 mm long, ± ciliate, light green to yellowish; Fl. 12-partite, 18–20 mm ⌀, dull pink.

**S. × barbulatum** SCHOTT nm. **barbulatum** (*S. aussendorfferi* HUT., *S. fimbriatum* SCHNITTSP. et LEHM., *S. hausmannii* AUSSEND., *S. hybridum* BRÜGGER). – Hybrid: S. arachnoideum v. a. × S. montanum ssp. montanum. – Alps in an area between the 2 parents. – L. covered with glandular hairs, cobwebby at the tips.

**S. × — nm. delasoieii** (LEHM. et SCHNITTSP.). – ROWL. (*S. d.* LEHM. et SCHNITTSP.). – Hybrid: S. arachnoideum v. doellianum × S. montanum ssp. montanum. – L. with denser cobweb-hairs.

**S. × — nm. noricum** (HAYEK) ROWL. (*S. n.* HAYEK). – Hybrid: S. arachnoideum v. doellianum × S. montanum ssp. stiricaum. – L. with less dense cobweb-hairs.

**S. borissovae** WALE (§ II/b 1). – USSR: Caucasus. – Ros. 3 cm ⌀, half-open, stolons short, slender; L. obovate, stoutly mucronate, 22 mm long, 9 mm across, 3 mm thick, glabrous belowy, green, upperside often brownish-red, robustly ciliate; Pet. purplish red with white margins and red anthers.

**S. brevipilum** MUIRH. (§ III/b 2). – Turkey. – Ros. c. 3 cm ⌀, half-open; L. many, ovate-lanceolate, acute, sea-green, fleshy, with a minute purple mucro, 1–2 cm long, 6–9 mm across, with dense, short, glandular hairs; Infl. 6–9 cm tall, bracts 1–2 cm long, 4–6 mm across, Fl. 1–1.5 cm ⌀, greenish-yellow, with violet filaments.

**S. cantabricum** HUBER et SUENDERM. (§ III/a 1). – Spain. – Ros. half-open, 4–5 cm ⌀, ± spherical, stolons c. 6 cm long, hairy; L. 30 to 40, linear-lanceolate, short-tapering, c. 3.5 cm long, 1 cm across, both surfaces glandular-hairy, margins ciliate, tips dark purple; Fl. dark pink.

**S. caucasicum** RUPR. ex BOISS. (§ II/b 1) (*S. tectorum* v. *c.* RUPR.) – USSR: Caucasus. – Ros. 3.5–5 cm ⌀, with 6–7 stolons 6–8 cm long; L. c. 45, dense, spatulate, abruptly constricted at the brown, distinctly mucronate tip, 2 cm long, 8 mm across, 3 mm thick, both surfaces with a few hairs, green.

**S. × christii** WOLF. – Hybrid: S. grandiflorum × S. montanum ssp. montanum. – Switzerland; N.Italy. – L. green or purple-tipped, ± hairy; Pet. purple with yellowish tips.

**S. ciliosum** CRAIB. f. **ciliosum** (Pl. 128/5) (§ III/b 2) (*S. c.* PANČIC nom. nud., *S. borisii* v. *c.* (PANČIC) HAYEK, *S. wulfenii* v. *skorpioli* VEL.). – Greece. – Ros. spherical, tight; L. oblong-oblanceolate, ash-grey, with 2–4 mm long, glandless hairs along the margins and below the tip, less densely ciliate; Fl. 2.5 cm⌀, greenish-yellow.

**S. — f. borisii** (DEGEN et URUMOV) JACOBS. (*S. b.* DEGEN et URUMOV). – Hairs denser, including along the margins, and clustered brush-like at the tip.

**S. davisii** MUIRH. (§ III/b/2). – Turkey. – Ros. 3–4 cm ⌀, side-shoots 2–3 cm long; L. oblanceolate to obovate, 1.5–2 cm long, 1 cm across, grey-green, with densely woolly, glandular hairs, the tips brownish-red, apiculate; Infl. 10–12 cm tall, bracts numerous, Fl. c. 2 cm ⌀, pale yellow, filaments white.

**S. dolomiticum** FACCH. (§ III/a 1) (*S. tectorum* v. *angustifolium* LEYBOLD). – Italy: Alto Adige. – Ros. spherical; L. with long, dense cilia along the margins, often with brownish-red tips; Infl. red-spotted, shaggy with glandular hairs, Pet. light pink with a brownish-red M. stripe.

**S. erythraeum** VELEN. (§ III/b 1) (*S. montanum* VELEN, *S. cinerascens* PANČIĆ). – Bulgaria. – Ros. spherical, tight, 2–4.5 cm ⌀; L. appearing grey because of the very dense, very short, glandular hairs, margins with distant, short cilia; Fl. pink.

**S. × fauconettii** REUT. nm. **fauconettii**. – Hybrid: **S. arachnoideum** v. **tomentosum** × **S. tectorum** ssp. **alpinum**. – Distribution intermediate to that of the parents. – L. slightly hirsute, ± cobwebby at the tips, at least initially; Fl. intermediate between the parents.

**S. × —** nm. **flavipilum** (HAUSM.) ROWL. (*S. f.* HAUSM., *S. hausmannii* LEHM., *S. mettenianum* HAUSM.). – Hybrid: **S. arachnoideum** v. **arachnoideum** × **S. tectorum** ssp. **glaucum** (TEN.) PRAEG. – With yellowish hairs.

**S. × —** nm. **rubellum** (TIMB.-LAGR.) ROWL. (*S. × r.* TIMB.-LAGR.) – Hybrid: **S. arachnoideum** v. **arachnoideum** × **S. tectorum** ssp. **alpinum**. – With red hairs.

**S. flagelliforme** FISCH. (§ I/1 c). – USSR: Siberia ?. – Ros. offsetting; L. ovate, stoutly mucronate, initially glandular-hairy.

**S. funckii** F. BRAUN. – Certainly of hybrid origin. – Ros. stellate-open, up to 5 cm ⌀; L. obovate, abruptly sinuate-tapering, grey-green, often with a reddish tip, with evanescent glandular hairs especially on the underside, margins white-ciliate, with clustered, curly hairs at the tip; Fl. large, Pet. pinkish-red, with a red M.stripe.

**S. furseorum** MUIRH. (§ III/b 2). – Turkey. – Ros. 3–4 cm ⌀; L. numerous, ovate to oblong-spatulate, tapering, 2.5–3.5 cm long, c. 7 mm across, greenish-grey, with a purple tip, with soft glandular hairs and with densely ciliate margins; Infl. c. 20 cm tall, bracts oblong acute, Fl. c. 2.5 cm ⌀, whitish-green, with white filaments.

**S. georgicum** GURGENIDZE (§ III/b 2) (*S. globiferum* AUCT.). – USSR: Georgia. – Shoots (10–)35(–45) cm tall, densely glandular-hairy, as are also the L. and Ped.; Ros. (3–)6(–7) cm ⌀; L. 2–3 cm long, 1–1.5 cm across, oblong-spatulate, tapering, bracts sessile, alternate, lanceolate, with a red tip; Infl. a corymb or false umbel, Fl. 1.5–2 cm ⌀, greenish-yellow, Pet. hairy, filaments ± violet, pedicels 1–10 mm long.

**S. gillianum** MUIRH. (§ III/b 2). – Turkey. – Ros. 4–6 cm ⌀, stolons short, hairy; L. oblong-spatulate, short-tapering, c. 3 cm long, 1 cm across, 3 mm thick, with a brown, recurved tip, with short, glandular hairs, with ciliate margins; Infl. 12–15 cm tall, bracts numerous, oblong-lanceolate, Fl. 2.5 cm ⌀, greenish-yellow, filaments dark red.

**S. glabrifolium** BORISS. (§ II/b 2). – Turkey. – Close to **S. armenum** but with numerous, short stolons; Ros. L. glandular at first, eventually glabrous, very swollen, olive-green, outer angles of the Ros. tinged deep purple; Fl. with white filaments, Sep.-lobes swollen, ovate, very incurved.

**S. grandiflorum** HAW. (§ III/a 2) (*S. globiferum* GAUD., *S. gaudinii* CHRIST). – Switzerland. – Ros. very large; L. up to 11 cm long, 8–15 mm across, oblong-cuneate, glandular-hairy, ciliate; Infl. with tufted, glandular hairs, Fl. 12–16-partite, Pet. golden-yellow, with violet blotches below.

**S. —** Hybrids: **S.** × **christii**, × **hayekii**, × **vaccarii**.

**S. giuseppii** WALE (§ III/a 1). – Spain. – Ros. 2 cm ⌀, fairly compact, stolons slender; L. obovate, 16 mm long, 7 mm across, 2.5 mm thick, both surfaces covered with dense, short hairs, margins with green, brown-tipped cilia; Infl. compact, hairy, Pet. red, with a narrow, white border.

**S. × hayekii** ROWL. – Hybrid: **S. tectorum** ssp. **glaucum** × **S. grandiflorum**. – Switzerland: Valais. – Vigorous-growing; L. ± glabrous except for cilia; Fl. pale yellowish to purple.

**S. ingwersenii** WALE (§ III/b 1). – USSR: Caucasus. – Ros. 3 cm ⌀, half-open, dense, flat, with numerous offsets; L. c. 35, obovate to ovate, short mucronate, with brownish tips, fleshy, 15 mm long, 7.5 mm across, 3 mm thick, underside convex, very minutely hairy, margins with sparse cilia; Fl. 19 mm ⌀, Pet. pink with a white border, anthers red.

**S. iranicum** BORNM. et GAUBA (§ II/1 b). – Iran. – Close to **S. marmoreum**; expanded Ros. 5–7 cm ⌀, stolons 10 cm long with daughter-Ros.; L. numerous, gradually tapering, 25–30 mm long, 10–15 mm across, blue-green; Fl. 2.5 cm ⌀, Sep. glandular-hairy, Pet. pink with a white border.

**S. ispartae** MUIRH. (§ II/2 b). – Turkey. – Ros. 3–4 cm ⌀, flat; L. few, ovate to oblanceolate, 1.5–2 cm long, 1 cm across, brilliant green to brownish-purple, glabrous, underside carinate, keel and tip with a few, stiff, glandular hairs, margins with dense cilia 1–2 mm long; Infl. 12–15 mm tall, glandular-hairy, bracts oblanceolate, acute, Fl. 2–2.5 cm ⌀, greenish-white with white filaments.

**S. italicum** RICCI (§ II/b 1). – Italy. – Resembles **S. tectorum**; L. densely hairy on both surfaces. Acc. J. A. HUBER this species is referable to the **S. marmoreum** complex.

**S. kindingeri** ADAM. (§ III/b 2). – Macedonia. – Ros. 4–6.5 cm ⌀; L. short-tapering, with fine, long, white hairs; Fl. yellow.

**S. kosaninii** PRAEG. (§ III/b 1). – Macedonia. – Resembles **S. leucanthum**; Ros. 5 cm ⌀; L. ovate-lanceolate; Fl. red.

**S. leucanthum** PANČIC (§ III/b 2). – Bulgaria. – **Ros.** spherical, closed; **L.** spatulate, gradually tapering, with reddened tips; Pet. yellowish-white with a green M.stripe.

**S. macedonicum** PRAEG. (§ III/b 1). – Macedonia. – Resembles **S. montanum;** with long stolons; **Ros.** 3–5 cm ⌀; **L.** light green suffused reddish; Fl. red.

**S. marmoreum** GRISEB. (§ III/b 1) (*S. clusianum* TEN., *S. schlehanii* SCHOTT, *S. assimile* SCHOTT, *S. reginae-amaliae* HELDR. et GUICC. ex HALACHY, *S. banaticum* DOM.). – Balkans; CSSR. – Stolons long; **Ros.** c. 6 cm ⌀, open; **L.** ovate, mucronate, 25 mm long, 12 mm across, usually hairy at first, eventually glabrous, olive-green, often tinged red or brown, margins with stout, recurved cilia; Pet. 1 cm long, red with a white border.

**S. —** ssp. **blandum** (SCHOTT) Soo (*S. b.* SCHOTT. *S. schlehanii* v. *b.* (SCHOTT) HAYEK). – **L.** glabrous.

**S. —** ssp. **marmoreum.** – **L.** hairy at first.

**S. —** ssp. — f. **brunneifolium** (PRAEG.) Soo (*S. schlehanii* f. *b.* PRAEG.). – **L.** brown.

**S. —** ssp. — f. **dinarium** (BECK.) JACOBS. (*S. schlehanii* v. *d.* BECK, *S. m.* v. *d.* (BECK) Soo). – **L.** longer-tapering, much reddened.

**S. —** ssp. — f. **rubicundum** (SCHUR.) Soo (*S. r.* SCHUR., *S. rubrifolium* SCHUR., *S. schlehanii* v. *r.* (SCHUR.) PRAEG.). – **L.** much reddened.

**S. minus** TURRILL v. **minus** (§ III/b 1). – Turkey. – Resembles **S. pumilum; Ros.** 9 to 18 mm ⌀, mat-forming; **L.** oblanceolate, oblong or oblanceolate-elliptical, with a ± shortly pointed tip, c. 8 mm long, 3.5 mm across, 1.5 mm thick, sap-green, both surfaces with short, glandular hairs, purple below, margins glandular-ciliate; Fl. 11–12-partite, 16 mm ⌀, pale yellow.

**S. —** v. **glabrum** WALE f. **glabrum.** – **Ros.** somewhat larger; **L.** eventually glabrous.

**S. —** v. — f. **viridifolium** WALE. – **Ros.** more open; **L.** more crowded, scarcely fleshy, green all over.

**S. montanum** L. (§ III/a 1). – Eur. – Type: ssp. **montanum.**

**S. —** ssp. **burnatii** WETTST. (*S. burnatii* WETTST.). – Piedmont Alps, Maritime Alps, Pyrenees. – **Ros.** open, 15 cm ⌀; **L.** cuneate to oblanceolate, green; Infl. robust.

**S. —** ssp. **carpathicum** WETTST. – Carpathians: Tatra Mts. – **Ros.** large, ± closed, with daughter-Ros. on long stolons; **L.** obovate-oblong, short-tapering; Fl. 6–8 together, crowded.

**S. —** ssp. **montanum** (*S. m.* v. *m.* (L.) WETTST., *S. candollei* ROUY et CAM., ? *S. debile* SCHOTT). – Alps, Pyrenees, Balkans, Caucasus (on igneous rocks). – **Ros.** spherical, closed, 1–2 cm ⌀; **L.** cuneate-oblanceolate, short-tapering, grey-green, densely and uniformly covered with glandular hairs; Pet. 10–12 mm long, light purple to violet, with a distinct M. stripe.

**S. —** ssp. — v. **minimum** (TIMB.-LAGRAVE) HUBER (*S. m.* TIMB.-LAGRAVE, *S. pygmaeum* JEANB. et TIMB.). – Dwarf variety.

**S. —** ssp. — v. **monticolum** LAM. (*S. subalpinum* ROUY). – **L.** narrowing abruptly to a short point, light green; Pet. paler on the underside.

**S. —** ssp. — v. **pallidum** WETTST. (*S. m.* v. *ochroleucum* BEAUVERD). – Fl. yellow or whitish.

**S. —** ssp. **stiriacum** WETTST. (*S. s.* WETTST., *S. funckii* MALY). – Austrian Alps. – **Ros.** larger, opening up in summer; **L.** oblong, rather longer-tapering, with a reddish-brown tip, often with short glandular hairs; Fl. dark violet.

**S. —** ssp. — v. **braunii** (FUNCK) WETTST. (*S. b.* FUNCK). – Fl. yellowish-white.

**S. nevadense** WALE (§ II/a 1). – Spain. – Close to **S. cantabricum; Ros.** compact, flattened, 2.5–3.5 cm ⌀, with short stolons, at first glandular-hairy; **L.** obovate to oblanceolate, apiculate, 12 mm long, 5 mm across, 2.5 mm thick, green, the upperside reddened towards the tip, cilia short; Fl. 2–3 cm ⌀, red.

**S. octopodes** TURRILL (§ III/b 2). – Greece: Macedonia. – **Ros.** ± compact, 19 mm ⌀, stolons slender, glandular-hairy; **L.** oblanceolate to obovate, rather obtuse to acute to short-tapering, c. 7 mm long, 2 mm across, underside convex, ± glandular-hairy, green, margins with glandular cilia, with a cluster of long hairs at the purplish-brown tip; Infl. densely glandular-hairy, Fl. 1 cm ⌀, pale greenish-yellow.

**S. ossetiense** WALE (§ III/b 1). – USSR: Caucasus. – **Ros.** 3 cm ⌀, dense, with robust stolons; **L.** lanceolate to oblong-lanceolate, c. 14 mm long, 6 mm across, 4 mm thick, acute to short-tapering, very convex, green, reddish-brown at the tip, with dense, short, glandular hairs, and rather longer cilia; Fl. 25 mm ⌀, Pet. white, purple below, the underside with fine red lines.

**S. pittonii** SCHOTT, NYM. et KOTSCHY (§ III/a 2). – Austria: Styria. – **Ros.** 3–5 cm ⌀; **L.** 12–20 mm long, 3–8 mm across, scarcely tapered, with ± tufted, glandular hairs and cilia; Fl. 1–3 cm ⌀, Pet. 12–16, pale greenish-yellow.

**S. pumilum** MARSCH. BIEB. (§ III/b 1) (*S. montanum* EICHW.). – USSR: Caucasus. – **Ros.** c. 2 cm ⌀; **L.** short-tapering, long-ciliate; Infl. of 2–7 Fl.

**S.** × **roseum** HUT. nm. **roseum.** – Hybrid: **S. arachnoideum** v. **arachnoideum** × **S. wulfenii.** – Austria, Switzerland, distribution intermediate between the parents. – Resembles **S. wulfenii;** L. tips with cobwebby hairs; Fl. intermediate between the parents in colour.

**S.** × **—** nm. **fimbriatum** (SCHOTT ex HEGI) ROWL. (*S.* × *f.* SCHOTT ex HEGI). – Hybrid: **S. arachnoideum** v. **doellianum** × **S. wulfenii.** – Intermediate; **L.** with long hairs at the tip.

**S.** × **rupicolum** KERN. nm. **rupicolum** (*S. braunii* FACCH., *S. huteri* HAUSM., *S. theobaldii* BRÜGGER). – Hybrid: **S. montanum** × **S. wulfenii.** – Switzerland; Austria: Tyrol. – **Ros.** 4–5 cm ⌀; **L.** oblong-obovate, short-tapering, reddish below, the underside with scattered glandular hairs; Pet. light yellow with red lines.

**S.** × **—** nm. **pernkofferi** (HAYEK) ROWL. (*S. p.* HAYEK). – Hybrid: **S. montanum** ssp. **stiriacum** × **S. wulfenii.** – Austria: Styria, Carinthia. – **L.** oblanceolate, less glandular-hairy; Fl. yellowish-red.

S. ruthenicum SCHNITTSP. et LEHM. (§ III/b 2) (*S. globiferum* L. emend. KOCH, *S. g.* v. *r.* KOCH, *S. r.* KOCH, *S. braunii* LED.). – USSR: Ukraine; Rumania. – **Ros.** large, with stolons 3–5 cm long; **L.** clavate, dark green, short-glandular, margins ciliate; **Pet.** greenish-yellow, eventually purple below.

S. × **schottii** LEHM. et SCHNITTSP. nm. **schottii**. – Hybrid: **S. montanum** ssp. **montanum** × **S. tectorum** ssp. **tectorum**. – French and Swiss Alps, intermediate between the parents in distribution. – Looks like a small **S. tectorum** ssp. **tectorum**; **Ros.** denser; **L.** ± glandular-hairy, with a purple tip; **Fl.** pure purple.

S. × – nm. **rhaeticum** (BRÜGGER) ROWL. (*S. rhaeticum* BRÜGGER). – Hybrid: **S. montanum** ssp. **montanum** × **S. tectorum** ssp. **alpinum**. – **L.** with scattered hairs, ciliate margins and a terminal bristle.

**S. staintonii** MUIRH. (§ II/b 2). – Turkey. – **Ros.** 3–4 cm ∅, flattened, with 3–5 short, very hirsute stolons; **L.** few, 1.5–2 cm long, 7 mm across, 5 mm thick, ovate to obovate, acute, bluish, green or reddish, glabrous apart from glandular cilia, with a red tip; **Infl.** 9–12 cm tall, bracts ovate to lanceolate, acute, red, Fl. c. 2 cm ∅, green with white filaments.

**S. tectorum** L. (§ II/a 1). – Eur. – Type: ssp. **tectorum**. – Acc. C. FAVARGER the classification of this large complex needs further study. – In 'Fl. Eur.' **S. arvernense** LEC. et LAM. is explicitly included with **S. tectorum**.

S. — ssp. **alpinum** (GRISEB. et SCHENK) WETTST, (Pl. 128/2) (*S. a.* GRISEB. et SCHENK, *S. fuscum* SCHNITTSP. et LEHM.). – **Ros.** 2 to 6 cm ∅; **L.** cuneate below, abruptly widened, abruptly tapered, bluish-green, always red below, the tip or the entire L. reddish-brown, margins ciliate; **Fl.** pink.

S. — ssp. **boutignianum** BILLOT et GRENIER (*S. b.* BILL. et GREN.). – This ssp. covers the Pyrenean forms of **S. tectorum**, and is close to ssp. **alpinum** and **S. arvernense**.

S. — ssp. — f. **jordanianum** ROUY et CAM. (*S. pyrenaicum* JORD. et FOURR.). – E. Pyrenees. – Form with **Fl.** 16–22 mm ∅.

S. — ssp. — f. **pallescens** ROUY et CAM. (*S. pyrenaicum* LAM.). – High Pyrenees. – **Fl.** 26 mm ∅.

S. — ssp. **calcareum** (JORD.) CARIOT et ST. LAGER (Pl. 128/4) (*S. s.* JORD.). – Ital. and French Maritime Alps (on limestone). – **Ros.** 4–6 cm ∅; **L.** linear-oblong, short-tapering, 3 cm long, 9–10 mm across, light grey-green with a much-reddened tip, with minute white cilia; **Fl.** pale pink. (In 'Fl. Eur.' given species status as **S. calcareum** JORD., on account of its diploid chromosome complement).

S. — ssp. — cv. Grigg's Surprise (*S. t.* v. c. f. *monstrosum* HORT., *Cotyledon persicus* HORT., *Rosularia persica* HORT. non BGR., *Umbilicus p.* HORT., *U. syriensis* HORT.). – Garden chimaera. – **Ros.** dense; **L.** thicker or ± acutely concave, often semicylindrical, usually less bluish, apiculate, the tip reddened; **Pet.** ± undeveloped, contorted, red, purple or often green.

S. — ssp. **glaucum** (WOHLF.) TEN. (*S. g.* WOHLF., *S. schottii* BAK. nom. illeg. *S. t.* ssp. *schottii* (BAK.) WETTST. ex HEGI et SCHMID, *S. acuminatum* SCHOTT, *S. spectabile* SCHNITTSP. et LEHM.). – S. Alps to Istria (on igneous rocks). – **Ros.** 5–10 cm ∅; **L.** cuneate below, widening abruptly to 2 cm, oblong, short-tapering, grey-green, whitish below, with a reddish-brown and stiffly ciliate tip; **Infl.** with tufted wool, Fl. pink.

S. — ssp. **mettenianum** SCHNITTSP. et LEHM. (*S. m.* SCHNITTSP. et LEHM.). – E. Alps. – Resembles ssp. **tectorum**; **L.** often with recurved tips (acc. H. CORREVON).

S. — ssp. **rupestre** (ROUY et CAM.) BERMANNS (*S. r.* ROUY et CAM.). – Mts. of Central Eur.; USSR: Caucasus; Iran. – **Ros.** 4–9 cm ∅; **L.** oblong or ovate, glabrous, with ciliate margins, the tip with a cluster of white hairs and glandular hairs; **Fl.** red. (Acc. J. A. HUBER this ssp. is the wild state of **S. tectorum** L. Acc. ROUY and CAMUS, 'Fl. de France', there are also the following varieties: **beugsenianum, brachiatum, brevistylum, decoloratum, juratense, praestabilis, violascens**.

S. — ssp. **tectorum** v. **tectorum** (*S. t.* v. *t* (L.) WETTST., *S. t.* L., *S. juratense* JORD.). –. Widespread throughout Eur. – **Ros.** 8–14 cm ∅; **L.** obovate-lanceolate, 4–6 cm long, 1–2 cm across, gradually tapering, whitish below, reddish-brown at the tip; **Infl.** up to 60 cm tall, freely branched. Often found growing on rooftops.

S. — ssp. — cv. Atropurpureum (*S. t.* v. *a.* HORT.). – **L.** dark violet. (Acc. H. CORREVON found in the Vuache in the S. Juras.)

S. — ssp. — cv. Bicolor (*S. t.* v. *b.* HORT.). – **L.** greenish-red.

S. — ssp. — cv. Boissieri (*S. t.* v. *b.* HORT. ex BAKER). – **Ros.** 5–8 cm ∅, with short offsets; **L.** less tapered than in the type.

S. — ssp. — v. **lamottei** (BOREAU) ROUY et FOURR. (*S. l.* BOREAU, *S. t.* v. *l.* (BOREAU) ROUY et FOURR.). – France. – **L.** green; **Fl.** pale. Ecotype of the mts. of med. height.

S. — ssp. — v. **minutum** (Kz.) WILLK. et LANGE (*S. m.* Kz., *S. t.* v. *m.* (Kz.) WILLK. et LANGE). – Spain – Smaller than the type.

S. — ssp. — cv. Ornatum (*S. t.* v. *ornatum* HORT.). – **L.** crimson in the lower 2/3.

S. — ssp. — v. **rhenanum** (HEGI et SCHMID) JACOBS. (*S. t.* v. *r.* HEGI et SCHMID). – Middle Rhine and side-valleys. – Smaller than the type. **Ros.** 6 cm ∅.

S. — ssp. — v. **rubescens** (VOSS) JACOBS. (*S. t.* v. *r.* VOSS). – **L.** pinkish-violet below.

S. — ssp. — cv. Triste (*S. t.* f. *triste* HORT. ex BAKER). – **L.** dark purplish-brown.

S. — ssp. — v. **violascens** (VOSS) JACOBS. (*S. t.* v. *v.* VOSS). – **L.** violet, more bluish-green above, tips brown.

S. —. – Hybrids: S. × **fauconnettii** REUT., S. × **hayekii** ROWL., S. × **schottii** LEHM. et SCHNITTSP., S. × **widderi** LEHM. et SCHNITTSP.

**S. thompsonianum** WALE (§ III/b 2). – Greece: Macedonia. – **Ros.** 1.5–2.5 cm ∅, ± spherical, with basal stolons 6–8 cm long; **L.** ovate-lanceolate, 14 mm long, 4 mm across, 2 mm thick, hairy, mucronate; **Infl.** compact, Fl. 19 mm ⌀, Pet. yellowish to pale purple at the tip, white-bordered midway, greenish-yellow below.

S. **transcaucasicum** MUIRH. (§ III/b 2). - USSR: Caucasus. - **Ros.** 5-7 cm ⌀, dense, with few stolors 2-3 cm long; **L.** obovate to oblanceolate, short-tapering, yellowish-green, with a pink tip, densely and shortly glandular-puberulous, margins ciliate; **Infl.** glandular-puberulous, Pet. greenish-yellow, pale purple below.

S. × **vaccarii** VACC. (*S. v.* WILCZEK nom. nud.). - Hybrid: **S. arachnoideum** v. **arachnoideum** × **S. grandiflorum.** - Switzerland: Valais. - **Ros.** like **S. grandiflorum**, smaller; **L.** tips laxly cobwebby; Pet. pale purple or reddish, often with yellow tips.

S. × **versicolor** VEL. - Hybrid: **S. ? marmoreum** × **S. zeleborii**. - **L.** narrowed towards the base, wider above, stoutly ciliate; **Fl.** pale yellow at first, eventually pale violet.

S. **vincentei** PAU (§ II/a 1). - Spain. - Dubious species, insufficiently described; differentiated from **S. nevadense** by the **L.** of rather different shape and hairy filaments.

S. × **widderi** LEHM. et SCHNITTSP. - Hybrid: **S. tectorum** ssp. **alpinum** × **S. wulfenii.** - Switzerland: above the Bernina Pass. - **Ros.** as **S. wulfenii**; **L.** bluish, with no purple tip; Pet. yellow, red below, or yellow with red stripes.

S. **wulfenii** HOPPE (Pl. 128/6) (§ II/a 2). - Alps. - **Ros.** open, 4-7 cm ⌀; **L.** oblong-spatulate, abruptly tapered, green, red below; **Fl.** 12-18-partite, yellow.

S. —. - Hybrids: **S.** × **roseum,** × **rupicolum,** × **widderi.**

S. **zeleborii** SCHOTT (§ III/b 2) (*S. ruthenicum* KOCH). - Bulgaria, Rumania. - **Ros.** compact, spherical, 4-5 cm ⌀, with daughter-Ros. on short stolons; **L.** pale green or bluish-green, densely puberulous, mostly velvety, mucro small, dark or missing, underside ± distinctly carinate; **Fl.** 2.5 cm ⌀, Pet. pure yellow with a purple base.

S. —. - Hybrid: **S.** × **versicolor-**

**Senecio** L. (*Kleinia* L., *Notonia* DC.). Compositae. - The G. includes c. 1500 species: ♄, ♃, ⚄ and ☉ herbs, with a worldwide distribution. Succulent species occur in S. and N.Afr., Canary Is.; S.W.Eur.; peninsular India; Madag.; Mexico. - The ± succulent spec. include ♃ herbs, ♄ or lianas, often with a tuberous **caudex,** acauline or caulescent; **St.** and **Br.** fleshy, cylindrical and sometimes jointed, sometimes with numerous, dark, parallel lines and a scent of resin; **L.** fleshy, flat, with margins ± entire or lobed, or cylindrical to spherical, sometimes with translucent lines and stripes, or reduced to scales, sometimes deciduous in the resting period; **Fl.** heads terminal, solitary or in corymbs, small, medium or large in size, disc-florets white, yellow or red, ray-florets (where present) yellow. - A number of tree-like but not really succulent species of **Senecio** occur in trop. Afr.: Tanzania, Kenya and the Congo basin, at heights of 3000-4000 m, resembling in appearance the giant **Lobelias** found in the same reg. and the **Espeetias** of S.Am.; HAUMAN has named these **Dendrosenecio.** - Greenhouse, warm; summer: in the open. Propagation: seed, cuttings. Opinions vary as to whether **Senecio** L., **Kleinia** L. and **Notonia** D. C. should be combined in **Senecio** L. or regarded as distinct genera. H. MERXMÜLLER redefines the genera **Senecio** L. and **Kleinia** L. as follows: **Senecio** in its succulent forms always has yellow ray and disc-florets, often with an outer involucre, and truncate stigma-lobes; **Kleinia** always with stem and leaf succulence, without ray-florets, the Cor. of the disc-florets predominantly white and red and stigma-lobes with a hemispherical or short-conical appendage. - See B. K. BOOM in 'Succulenta', **44,** 1965, 114-118: Key to the **Senecio** species found in cultivation. **Kleinia** and **Notonia** are not separated here from **Senecio.**

## Synopsis of the Genus Senecio L. by G. D. Rowley

The following synopsis includes all the **Senecios** described below, and is offered as an approximate guide only in the absence of any modern taxonomic treatment. There is some overlap between certain groups.

Gr. I. - Non-succulent ♄ (trunk undeveloped in **S. brassica**) with palm-like crowns of large petiolate **L. (Dendrosenecio). S. aberdaricus, adnivalis, battiscombei, brassica brassiciformis, ericirosenii, johnstonii, keniodendron.**

Gr. II. - ♄, woody below but without a fleshy, swollen base or rootstock; ultimate branchlets thin, sparingly fleshy, with ± succulent L.
1. Plant thorny .................................................... S. spinescens
1a. Plant unarmed ................................................... 2.
2. L. felted ......................................................... 3.
2a. L. not felted .................................................... 4.
3. L. flat ........................................................ S. medley-woodii
3a. L. ± terete ............................. S. haworthii, pyramidatus, quinquangulatus
4. L. deeply channelled, navicular ............................ S. navicularis
4a. L. ± flat, not navicular:
   a. L. cordate, long-petiolate:
     **S. usambariensis**
   b. L. lanceolate to broad ovate:

S. abyssinicus, antitensis, baronii, barorum, boiteaui, capuronii, crassiusculus, decaryi, eupapposus, melastomifolius, meuselii, neobakeri, paucifolius, polytomus, quartziticolus, rodriguezii, saboureaui, sakalavorus, subsinuatus.

    c. L. linear to subterete
S. hildebrandtii, hirto-crassus, kalambatitrensis, sakamaliensis, subradiatus.

    d. L. terete
S. acutifolius, aloides, barbertonensis, canaliculatus, corymbiferus, spiculosus, talinoides.

    e. L. vertically compressed
S. aizoides, archeri, crassissimus, ficoides.

**Gr. III.** – St.-succulent ♄; St. very fleshy and green or with scale L. only if thinner; L. ± deciduous.

    a. L. flat, 1 cm long or longer:
1. St. strongly tuberculate ............................................. S. papillaris
1a. St. not strongly tuberculate ............................................. 2.
2. L. long petiolate, lobed or incised
S. articulatus, praecox.
2a. L. ± sessile and entire
S. anteuphorbium, cliffordianus, cotyledonis, descoingsii, galpinii, kleinia, lunulatus, neohumbertii.

    b. L. terete:
1. L. felted ............................................. S. scaposus
1a. L. not felted
S. hanburyanus, marnieri, pinguifolius, subulatifolius.

    c. L. scale-like or absent:
S. avasimontanus, deflersii, junceus, longiflorus, mweroensis, pendulus, scottii, stapeliiformis.

**Gr. IV.** – Subshrubs with a swollen St.-base or rootstock.

    a. L. entire:
S. amaniensis, ampliflorus, corymbosus, crassus, leandrii, mesembryanthemoides, sempervivus.

    b. L. notched:
S. fulgens.

**Gr. V.** – Prostate to creeping herbs, some rooting at the nodes.

    a. L. entire, flat or channelled:
S. coccineiflorus, implexus, jacobsenii, nyikensis, serpens.

    b. L. lobed:
S. kleiniiformis.

    c. L. terete:
S. bulbinefolius, chordifolius, cicatricosus, citriformis, crassulifolius, hallianus, herreianus, klinghardtianus, mandraliscae, ovoideus, radicans, rowleyanus, succulentus.

**Gr. VI.** – Caudiciform, with the gnarled, non-jointed caudex half or more above ground; Br. ☉ or undeveloped.
S. ballyi, cephalophorus, grantii, picticaulis.

**Gr. VII.** – Geophytes, with the caudex more than half below ground.

    a. L. flat, cuneate to spatulate:
S. auricula.

    b. L. cordate to peltate, long petiolate:
S. orbicularis, oxyriifolius, tropaeolifolius.

    c. L. terete:
S. acaulis.

**Gr. VIII.** – Climbers with weak, scarcely succulent St.

    a. L. flat, petiolate:
S. subscandens.

    b. L. terete:
S. antandroi.

**Gr. IX.** – Creeping ☉.
S. bakeri.

**S. aberdaricus** R. E. FRIES et TH. FRIES f. (Gr. I.). – Tanzania. – ħ up to 7 m tall; **trunk** sometimes slender, with some dichotomous branching, often covered with dead L.; **L.** in large ± spherical Ros. at the Br.tips, stiff, 40 to 60 cm long, 12–22 cm across, linguiform, the tip mostly rounded, narrowed into the laterally alate petiole c. 15 cm long and c. 3–10 cm across, margins dentate except towards the tip, upperside laxly and sparsely covered with a silky white felt, eventually glabrous, underside densely white felted and woolly, with long hairs on the M.nerve; **Infl.** c. 1 m tall, **Fl.** heads campanulate, c. 14 mm ∅, with linguiform rays.

**S. abyssinicus** (A. RICH.) SCH. BIP. (Gr. II.), (*Notonia a.* A. RICH., *S. a.* (A. RICH.) JACOBS.). – Tanzania; Ethiopia. – ħ up to 1.5 m tall; **L.** thick, elliptical, narrowed towards both ends, sessile, 10 cm long, often with brownish-red markings, densely arranged at the base of the shoots, laxer higher up; **Fl.** heads in Infl. 2.5–5 cm across.

**S. acaulis** (L. f.) SCH. BIP. v. **acaulis** (Gr. VII.) (*Cacalia a.* L. f., *Kleinia a.* (L. f.) DC.). – Cape. – **Caudex** tuberously thickened, offsets short; **shoots** 3–7 cm tall, 8 mm thick, densely leafy; **L.** ± cylindrical, gradually tapering, upperside ± deeply furrowed or grooved, light green, with a short, reddish apiculus, 10–16 cm long, 2–4 mm thick; **Fl.** scape 15–20 cm tall, with a single Infl., Fl. numerous, snow-white, with whitish anthers.

**S. —** v. **burchellii** (DC.) SCH. BIP. (*Kleinia a.* v. *b.* DC.). – **Fl.** Sc. twice as long as the L.

**S. —** v. **ecklonis** (DC.) SCH. BIP. (*Kleinia a.* v. *e.* DC.). – **Fl.** Sc. as long as the L.

**S. acutifolius** DC. (Gr. II.). – Cape: Graaff Reinet D. – Erect, glabrous, branched ħ with cylindrical Br.; **L.** sessile, cylindrical, acute, apiculate.

**S. adnivalis** STAPF (Gr. I.). – Trop. E. Afr. – Resembles **S. johnstonii**; **trunk** short, ± the height of a man, with a cluster of ensiform L. above, or with a thinner trunk dividing above into several, much thicker Br.; **L.** glabrous or with ± long, curly hairs; **Fl.** heads with linguiform ray-florets.

**S. —** v. **adnivalis** (*S. refractisquamatus* DE WILLD., *S. Ianuriensis* DE WILLD.). – Congo Rep.; Uganda. – **L.** ± densely puberulous on the underside, the petiole broadly alate, the blade cuneate below.

**S. —** v. **alticola** (TH. FRIES f.) HEDB. (*S. a.* TH. FRIES f., *S. erici-rosenii* v. *a.* MILDBR., *S. johnstonii* AUCT. non OLIV.). – Congo: Kivu. – **L.** underside ± densely puberulous, the petiole ± broadly alate, densely puberulous with many fine, curly hairs, the lamina cuneate at the base.

**S. —** v. **erioneuron** (COTTON) HEDB.(*S. e.* COTTON, *S. e.* v. *oligochaeta* HAUMAN). – Congo; Uganda. – **L.** glabrous on the underside apart from the M.rib, with a broadly alate petiole.

**S. —** v. **petiolatus** HEDB. – Uganda. – **L.** ± densely hairy on the underside, the petiole not alate, the blade cordate or truncate below.

**S. —** v. **stanleyi** (HAUMAN) HEDB. (*S. s.* HAUMAN). – Congo Rep.; Uganda. – **L.** underside glabrous apart from the M.rib, the petiole not alate.

**S. aizoides** (DC.) SCH. BIP. (Gr. II.) (Pl. 129/3) (*Kleinia a.* DC.). – Cape: Karroo. – Resembles **S. ficoides**; Glabrous ħ with a very short **St.**; **L.** crowded at the tips of the Br., fleshy, laterally compressed, acute, apiculate, 6–7 cm long, 5 mm across; **Fl.** heads 12 mm long.

**S. aloides** DC. (Gr. II.) (*Othonna rhopalophylla* DTR., *S. r.* (DTR.) MERXM.). – S.W.Afr.: S.Lüderitz D. – ħ with spreading Br. up to 40 cm tall, with a smooth, greyish-brown bark; **L.** densely crowded on the short shoots, clavate, up to 25 mm long, 5 mm thick, the tip rounded and mucronate, with 8–10 distinct longitudinal lines, the upperside with a deep longitudinal furrow, L. longer on the long shoots; **Fl.** heads 1–3, disc-florets numerous, ray-florets c. 8.

**S. amaniensis** (ENGL.) JACOBS. (Gr. IV.) (Pl. 129/2) (*Notonia a.* ENGL., *Kleinia a.* (ENGL.) BGR.). – Tanzania. – **Roots** fleshy, up to 40 cm long, 2 cm thick; **St.** several, erect to prostrate, little branched; **L.** oblong to spatulate, narrowed towards the 1 cm long petiole, obtuse, the upperside concave, light green, pruinose, with 3 dark green, longitudinal veins, 9–11 cm long, 3–4 cm across above, fleshy; **Infl.** up to 80 cm long, with 5–8 Br., Fl. many, red to yellow.

**S. ampliflorus** ROWL. (Gr. IV.) (*Notonia grandiflora* DC., *Cacalia g.* WALL., *S. g.* (WALL.) JACOBS., *C. sempervirens* SPR., probably also *N. corymbosa* DC., *S. c.* (DC.) JACOBS., *N. crassissima* WALL., *C. c.* WALL., *S. crassus* (WALL.) JACOBS.). – Peninsular India. – ħ with cylindrical Br.; **L.** in terminal Ros., ovate-lanceolate, with entire margins, ± fleshy; **Fl.** heads large, scarlet.

**S. antandroi** SCOTT ELLIOT (Gr. VIII.) (Pl. 129/1). – S.E. Madag. – Succulent sub-ħ up to 2 m tall, climbing by means of its hooked and curved L.; **Br.** smooth, with L.scars; **L.** spiralled, linear, semicylindrical, hardened and tapering towards the tip which is bent forwards, 4–12 cm long, 3 mm across and thick, 7-angled, the upperside with a narrow groove; **Fl.** heads yellow, 15–20 in a corymb.

**S. anteuphorbium** (L.) SCH. BIP. (Gr. III.) (*Cacalia a.* L., *S. pteroneura* HOOK. f., *Kleinia a.* (L.) DC., *S. a.* (L.) HOOK. f.). – Cape; S. Morocco. – ħ up to 1.5 m tall; **Br.** erect, with long joints, 10–15 mm thick, cylindrical, with slightly prominent L.cushions, grey-pruinose, with darker lines; **L.** lanceolate, 1.5–3.5 cm long, 5–10 mm across, apiculate, grey-green, the upperside grooved; **Fl.**heads 1–3 together, yellowish-white.

**S. —** v. **odorus** (FORSK.) ROWL. (*Cacalia o.* FORSK., *Kleinia o.* (FORSK.) DC., *S. o.* (FORSK.) SCH. BIP.). – S.Arabia: Yemen. – **Br.** 50–70 cm tall, 2–3 cm thick, glaucous; **L.** 6–9 cm long, 2 cm across, 3-veined; **Fl.**heads numerous.

**S. antitensis** BAK. (Gr. II.). – Central Madag. – Branching sub-ħ 20–50 cm tall; **Br.** robust, 5–6 mm thick, eventually becoming woody;

L. numerous, spiralled, sessile, with entire margins, flat, oblanceolate, 4–6 cm long, 13 to 20 mm across, c. 2 mm thick, obtuse to rather acute, apiculate, edges obtuse, whitish; **Fl.** heads with 8 involucral L., Fl. 25, yellow.

**S. archeri** (COMPT.) JACOBS. (Gr. II.) (*Kleinia a.* COMPT.). – Cape: Worcester D. – Glabrous, resinous, aromatic ℏ; **shoots** erect, smooth, cylindrical, 4 mm thick; **L.** variable in size and shape, fleshy, broadly compressed, oblong to ovate, acute, dark green, the upperside with translucent, reticulate, longitudinal lines, 3 to 5 cm long, 4 mm across, 10–12 mm thick; **Fl.** numerous, white. Probably identical with **S. aizoides.**

**S. articulatus** (L. f.) SCH. BIP. v. **articulatus** (Gr. III.) (*Cacalia a.* L. f., *C. laciniata* JACQ., *C. runcinata* LAM., *Kleinia a.* (L. f.) HAW., *K. michelii* HORT.). – Cape: Uitenhage D. – **Shoots** fairly erect, 30–60 cm tall, branching, 15–20 mm thick, cylindrical, easily detachable joints 5–15 cm long, often swollen, with light grey markings; **L.** distant, petiolate, pinnatisect or serrate and incised, light grey; **Fl.** yellowish.

**S. —** v. **globosa** JACOBS. – Joints ± spherical, 10–20 mm long, 10–15 mm ⌀.

**S. auricula** BOURG. (Gr. VII.). – Spain: Prov. Almería. – ♃ with a thick, fleshy, oblique, densely felted-hairy **caudex**; **L.** densely rosulate on 1–3 erect, shortly downy shoots 10–25 cm tall, obovate-cuneate to ± spatulate, narrowing into the short petiole, with 3 T. above, initially somewhat cobweb-haired; **Fl.** heads with 9 to 12 ray-florets.

**S. avasimontanus** DTR. (Gr. III.). – S.W.Afr.: Gr. Namaqualand. – Erect ℏ 60 cm tall; **Br.** 5–7 mm thick, dividing above, densely leafy; **L.** subulate, linear, bluish-green, fleshy, 4 to 9 mm long, 1–4 mm across, shortly tapering; **Fl.** yellowish-white. Acc. H. MERXMÜLLER probably only a form of **S. longiflorus.**

**S. bakeri** SCOTT ELLIOT. (Gr. IX.). – Madag. – Creeping ⊙; **L.** very crowded, fleshy (coastal form), or more distant, thinner (inland form), oblanceolate to spatulate, indistinctly serrate, 2–6 cm long, 10–15 mm across; **Fl.**heads with more than 20 Fl.

**S. ballyi** ROWL. (Gr. VI.) (*Notonia incisifolia* BALLY, *S. i.* (BALLY) JACOBS.). – Kenya. – ♃ with a tuberous **caudex**; **shoots** short, erect, 3–8 cm tall; **L.** 4, fleshy, elliptical, with upturned margins, up to 10 cm long, 4 cm across; the flowering axis elongating to 12–15 cm with scale-like L. above, terminating in 3 **Fl.**heads, Fl. orange.

**S. barbertonensis** KLATT. (Gr. II.). – Transvaal. – Branching ℏ; **Br.** cylindrical, fleshy, eventually woody, 4–6 mm thick; **L.** in dense clusters at the Br.tips, narrowing towards the short petiole, linear, long-tapering, cylindrical, upperside grooved, 4 cm long, 4 mm across, 3 mm thick, the underside with 3 thin, dark longitudinal lines, green with very fine papillose hairs; **Fl.**heads 4 mm ⌀, Fl. golden-yellow.

**S. baronii** H. HUMB. (Gr. II). – Central Madag. – Branching ℏ 50–100 cm tall; **St.** and Br. erect, robust, 5 mm thick, rough with old L. bases; **L.** ± sessile, linear-lanceolate, 10–15 mm long, 2–3 mm across, apiculate, glabrous or minutely hairy, with a distinct M.nerve; **Fl.**heads c. 8 mm ⌀, Fl. yellow.

**S. barorum** H. HUMB. (Gr. II.) (*S. b.* v. *ellipticum* H. HUMB.). – E. and W.Madag. – Branched, glabrous ℏ 50–100 cm tall; **Br.** rough with old L.bases; **L.** spiralled, sessile, with entire margins, oblanceolate to elliptical, obtuse, 1 mm thick, 3-veined, narrowed below, c. 5 cm long, 3 cm across; **Fl.**heads numerous, Fl. yellow.

**S. battiscombei** R. E. FRIES et TH. FRIES f. (Gr. I.). – Tanzania; Kenya: Mt. Kenya. – ℏ; **trunk** up to 6 m tall, branching, the bark splitting, covered with dead L.; **L.** densely rosulate at the Br.tips, 45–60 cm long, 10 to 20 cm across, papery, lanceolate to obovate, acute, narrowing towards the alate petiole, margins entire or minutely dentate, the upperside with a thin, sparse, white felt, the underside yellow-felted, the ribs with a long white beard; **Infl.** ovoid, Fl.heads campanulate-conical, ray-florets linguiform.

**S. boiteaui** H. HUMB. (Gr. II.). – S.Madag. – Branching sub-ℏ 80–100 cm tall; **Br.** erect, 4–6 mm thick, covered with old L.scars; **L.** in spiralled R., sessile, oblanceolate to linear, 5–8 cm long, 6–10 mm across, 2 mm thick, tapering, apiculate, both surfaces with numerous longitudinal veins; **Fl.**heads in a corymb, Fl. numerous, yellow.

**S. brassica** R. H. FRIES et TH. FRIES f. (Gr. I.) (*S. keniensis* BAK. f. p. part., *Lobelia gregorina* BAK. f.). – Tanzania; Kenya: Mt. Kenya. – Dwarf; **St.** short or missing; **L.** in a large, ± spherical, dense Ros., blade c. 20 cm long, 9 cm across, stiff, lanceolate to lanceolate-ovate, acute, narrowing below into the broad petiole, dentate, upperside at first cobwebby, underside densely yellowish-white haired; **Infl.** cylindrical, Fl.heads nodding, ray-florets linguiform.

**S. brassiciformis** R. E. FRIES et TH. FRIES. f. (Gr. I.). – Tanzania. – Dwarf, acauline or caulescent, up to 1 m tall; **L.** in a lax Ros., 30–50 cm long, 5–16 cm across, stiff, oblanceolate to obovate, narrowing below into the broad petiole, margins serrate to dentate, almost glabrous or white-felted; **Infl.** 1 m tall, long-cylindrical, Fl.heads large, ray-florets linguiform.

**S. bulbinefolius** DC. (Gr. V.). – Cape: Lit. Namaqualand; S.W.Afr.: Lüderitz Bay S. – ℏ; **Br.** flexible, cylindrical, often very long, creeping or pendulous; flowering shoots erect, short, up to 10 cm tall; **L.** cylindrical to broad-linear, furrowed midway, 2.5–3 cm long, 3 to 4 mm across, 3–4 mm thick, rather acute, with a few distinct stripes; Fl.heads fairly large, ray-florets 5–8, yellow.

**S. canaliculatus** BOJ. ex DC. (Gr. II.) (*S. petrophilus* KLATT, *S. cicatricosus* BAK., *S. cyclocladus* BAK.). – Central Madag. – Sub-ℏ 15–60 cm tall; **Br.** erect, becoming woody, with narrow L.scars; **L.** sessile, linear-cylindrical, 1.5–7.5 cm long, 1–5 mm across, tapering, apiculate, with fine, parallel veins; **Fl.**heads in corymbs of 3–12, Fl. 30–50, yellow.

**S. capuronii** H. Humb. (Gr. II.). – Central Madag. – Fleshy sub-♄ 40–50 cm tall, branching from the base; **Br.** spreading, ± prostrate, eventually directed upwards, L.scars small; **L.** in spiralled R., sessile, oblanceolate to ± linear, 10–15 cm long, 1.5–2 mm across, tapering, apiculate, 3-veined; **Fl.**heads solitary, Fl. yellow.

**S. cephalophorus** (Compt.) Jacobs. (Gr. VI.) (*Kleinia c.* Compt.). – Cape: Lit. Namaqualand. – Branching ♄ up to 60 cm tall; **shoots** up to 3 cm thick, bluish-green, rough with old L.bases; **L.** ± lanceolate, fleshy, acute, with recurved margins, up to 10 cm long, 13 mm across, 6 mm thick; **Fl.**heads pendulous at first, eventually erect, Fl. light yellow.

**S. chordifolius** Hook. f. (Gr. V.) (*Kleinia c.* Hook. f., *S. c.* (Hook. f.) Jacobs., *S. longifolius* Jacobs. non L.). – Cape: Albert D. – Sparsely-branched sub-♄ c. 30 cm tall; **L.** cylindrical, 20–25 cm long, scarcely 5 mm thick, minutely tapering above, pale green; **Fl.**scape slender, Fl. pale yellow.

**S. cicatricosus** Sch. Bip. (Gr. V.) (*Kleinia breviscapa* DC., *S. b.* Jacobs.). – Cape. – ♄ with short, fleshy, rather hairy **shoots**; **L.** cylindrical, acute, mucronate, 4–5 cm long.

**S. citriformis** Rowl. (Gr. V.) (Pl. 129/4) (*S. pusillus* hort., *S. gracilis* hort.). – Cape: Montagu D. – Forming dense cushions, **rootstock** fleshy; **Br.** ± erect, 5–10 cm long, rarely with adventitious roots; **L.** numerous, usually rosulate, exceedingly succulent, terete, short-fusiform or lemon-shaped, tapering below towards the petiole, obtuse or more rarely apiculate, 15–20 mm long, pale grey, pruinose, with numerous translucent, longitudinal lines; Fl. creamy yellow.

**S. cliffordianus** Hutch. (Gr. III.) (*Kleinia c.* (Hutch.) Adams). – N.Nigeria; Sudan. – Succulent, pruinose ♄ up to 3 m tall; **shoots** thick, fleshy, leafless at flowering, with L.scars; **L.** deciduous, lanceolate, obtuse, thick below, to 10 cm long, glabrous, with **Infl.** terminal, a dense, forked umbel 2–2.5 cm long, bracts linear, c. 3.5 mm long, Fl.heads oblong-cylindrical, 2.5 cm long, involucre pale green, florets white.

**S. coccineiflorus** Rowl. (Gr. V) (*Notonia coccinea* Oliv. et Hiern, *S. c.* (Oliv. et Hiern) Muschl.). – Kenya. – Fleshy, creeping ♄ up to 30 cm long; **L.** crowded, 2.5–7.5 cm long, 1–2 cm across, narrowing towards both ends, bluish; **Fl.**heads of many scarlet Fl.

**S. corymbiferus** DC. (Gr. II) (*S. phonolithicus* Dtr.). – Cape: Lit. Namaqualand; S.W.Afr.: S.Lüderitz D. – ♄ c. 30–100 cm tall with a thick, fleshy, branched St. with a disagreeable smell; **L.** dense at the tips of the 1 cm thick terminal shoots, fleshy, terete, 5–10 cm long, 5–7 mm thick, fusiform, terminating in a slender point, minutely longitudinally striate; **Fl.** in c. 10 cm long Pan. with numerous Fl.heads, ray-florets 5, golden-yellow.

**S. corymbosus** (DC.) Jacobs. non Wall. (Gr. IV.) (*Notonia c.* DC.). – Peninsular India. – ♄ with cylindrical Br.; **L.** broadly elliptical, obtuse. Probably identical with **S. ampliflorus**.

**S. cotyledonis** DC. (Gr. III.). – Cape: Karroo. – ♄; **Br.** dense, with old L.scars; **L.** crowded at the Br.tips, fleshy, linear, carinate, mucronate; **Fl.**heads 10–12, Fl. light yellow.

**S. crassissimus** H. Humb. (Gr. II.). – S.Madag. – Much-branched sub-♄ 50–80 cm tall, St. and Br. rough with old L.scars; **L.** lateral, sessile, short-tapering, apiculate, broadly oval, 4–6 cm long, 2–3 cm across, 3–5 mm thick (the L.-arrangement giving the plant an **Opuntia**-like appearance); **Infl.** 50–100 cm tall, Fl. yellow.

**S. crassiusculus** DC. (Gr. II.). – Cape: Albany D. – Erect ♄; **Br.** cylindrical, striate, corymbose at the tips; **L.** lanceolate, ± incised-dentate, cuneate-tapering below, auriculate, thick.

**S. crassulifolius** (DC.) Sch. Bip. (Gr. V.) (*Kleinia c.* DC., *S. crassulaefolius* (DC.) Jacobs.). – Cape: Uitenhage D. – ♄; **Br.** with old L.scars; **L.** crowded at the Br.tips, semi-terete, acute; **Infl.** dichotomous.

**S. crassus** Jacobs. (Gr. IV.) (*Cacalia crassissima* Wall., *Notonia c.* (Wall.) DC.). – Peninsular India. – Fleshy, bluish ♄; **Br.** cylindrical, **L.** oblong, attenuated towards the base, tapering above. Probably identical with S. **ampliflorus**.

**S. cuneifolius** (L.) Sch. Bip. (*Cacalia c.* L., *Kleinia c.* (L.) DC.). – Insufficiently described by Linnaeus.

**S. decaryi** H. Humb. – (Gr. II.) Central Madag. – Branched ♄ 80–100 cm tall; **Br.** crowded, 4–6 mm thick, with old L.scars; **L.** spiralled, sessile, oblanceolate-linear, 5 to 8 cm long, 6–10 mm across, 2 mm thick, narrowed teretely below, apiculate, upperside concave, both surfaces with longitudinal veins towards the margins, with a reddish tip; **Fl.** yellow.

**S. deflersii** Schwarz (Gr. III.) (*Notonia obesa* Defl.). – S.Arabia: S.Yemen; E. hinterland of Aden, peak of Djebel el 'Ures. – Robust, thick-fleshy, glabrous, with a green, white-spotted bark; **roots** fibrous, thick; **St.** erect, branching, c. 20 cm tall, 6–8 cm thick; **Br.** 10–15 cm long, 3–4 cm ∅, dichotomous, cylindrical, with 3 parallel lines from the L.bases; **L.** scattered, rather thick, linear-subulate, deciduous, 8–10 mm long, 1–1.5 mm thick; **Infl.** erect, Fl. terminal, not known.

**S. descoingsii** (H. Humb.) Jacobs. (Gr. III.) (*Notonia d.* H. Humb.). – Central Madag. – Succulent plant with ± clustered, twisted **roots**; **shoots** long and thin, with few side shoots, 10–30 cm tall, 4–6 mm thick, with internodal furrows, L.scars hardened and prominent; **L.** linear, 1–2 mm long, early deciduous; **Fl.**heads 1–3, Fl. 40–50, pale yellow or whitish.

**S. erici-rosenii** R. E. Fries et Th. Fries f. (Gr. I.) (*S. johnstonii* et *S. adnivalis* auct. p. part. non Oliver nec Stapf). – Tanzania. – ♄ 2–3 m tall; **trunk** with a splitting, rough bark below covered with dead L., dichotomously branched above; **L.** rosulate on the tips of the shoots, broad-lanceolate, thin, 30 to 60 cm long, 12–20 cm across, the underside with tufted hairs, eventually ± glabrous, M.nerve bearded and felty, petiole short and alate, margins dentate; **Infl.** ovoid, c. 50 cm tall, corymbose, Fl.heads numerous, ± campanulate, 18 mm ⌀, rays linguiform.

**S. eupapposus** (CUF.) ROWL. (Gr. II.) (*Kleinia eupapposa* CUF.). – Ethiopia. – Sparingly or squarrosely branched glabrous ♃ 50–70 cm tall; **St.** 30 cm long and 5 mm thick near the top; **L.** fleshy, alternate, ovate-spatulate, narrowing towards the petiole, ± obtuse or minutely apiculate, up to 22 mm long and 5 mm broad; **Fl.** pale rusty red.

**S. ficoides** (L.) SCH. BIP. (Gr. II.) (Pl. 130/1) (*Cacalia f.* L., *Kleinia f.* (L.) HAW.). – Cape. – Intricately branched, prostrate ♄ or sub-♄; **St.** over 1 m long, little branched, finger-thick, with erect tips, light green, white-spotted, with scattered, ± gibbose L.scars each with 5 indistinct lines running downwards, the minute pruinose coating easily removed; **L.** tapering towards both ends, ± compressed laterally, long-tapering, with parallel longitudinal veins, 9–15 cm long, 1.5 cm high, 7 mm wide, sometimes ± laterally grooved; **Infl.** a cyme of many white Fl.heads.

**S. fulgens** NICH. (Gr. IV.) (Pl. 130/5) (*Kleinia f.* HOOK. f., *S. hookerianus* JACOBS.). – Natal. – **Roots** tuberous; **shoots** prostrate, 40–45 cm long, white-spotted and white-pruinose; **L.** spiralled, obspatulate, terete below, 8–9 cm long, 2–3 cm across, upperside furrowed, underside with a prominent M.nerve, light green, violet to grey-pruinose, margins with 2–3 T.; **Fl.** heads 2.5 cm ⌀, **Fl.** red.

**S. galpinii** HOOK. f. (Gr. III.) (*Kleinia g.* HOOK. f., *S. g.* (HOOK. f.) JACOBS.). – Transvaal. – ♄ or sub-♄ 30–60 cm tall; **Br.** finger-thick, Br. and L. light-grey pruinose; **L.** oblanceolate, blunt-tapering, with a distinct M.nerve, 10–15 cm long, 3–3.5 cm across; Fl.heads 3–4, **Fl.** light orange-red.

**S. grantii** (OLIV. et HIERN) SCH. BIP. (Gr. VI.) (*Cacalia g.* OLIV. et HIERN, *Notonia g.* (OLIV. et HIERN) ASCHERS., *Kleinia g.* (OLIV. et HIERN) HOOK. f., *S. longipes* BAK.; acc. HOOKER *S. sempervivus* (FORSK.) SCH. BIP. should also be included). – Tanzania; Ethiopia. – **Caudex** gnarled, tuberous; **shoots** several, 15–20 cm long, prostrate to ascending, slightly reddish, densely leafy; **L.** spatulate-obovate, ± acute, narrowed below into the short petiole, fleshy, with a distinct M.nerve, 5–6 cm long, both surfaces bluish-green.

**S. hallianus** ROWL. (Gr. V.) (Pl. 130/4). – Cape: Prince Albert D. – Cape: Prince Albert D. – **Shoots** short, viscous, developing aerial roots passing over into fusiform or sausage-shaped tubers up to 12 cm long and 12 mm thick; **L.** laxly rosulate, cylindrical-fusiform, short-petiolate, obliquely mucronate, 1.5–2.5 cm long, 2.5–4 mm thick, bluish-pruinose, the upperside with a broad, translucent, longitudinal stripe and 10 narrow stripes; Fl.heads very long, rayless, **Fl.** with a purple style.

**S. hanburyanus** DTR. (Gr. III.) (*Kleinia h.* (DTR.) BGR.). – ? Cape. – ♄ 10 to 15 cm tall; **shoots** finger-thick, branching sparsely from the base, green, with L.scars; **L.** bunched at the Br.tips, long-subulate, minutely acuminate, ± cylindrical, the upperside deeply furrowed towards the base, with translucent, longitudinal veins, c. 22 mm long, 6 mm across, those developed at the end of the growing season short and thick, minutely scaly-felted, 7 mm long, 9 mm thick, ± terete; **Infl.** c. 50 cm tall, **Fl.** agreeably scented.

☞**S. haworthii** (HAW.) SCH. BIP. (Gr. II.) (Pl. 130/2) (*Cacalia tomentosa* HAW., *C. canescens*, *C. h.* SWEET, *Kleinia h.* (HAW.) DC., *K. tomentosa* HAW., *S. h.* HOOK. f.). – Cape: Lit. Namaqualand. – Little branched ♄ up to 30 cm tall; **Br.** erect, sometimes eventually prostrate; **L.** 2–4 cm long, 10–12 mm thick, pointed towards both ends, cylindrical, very short-petiolate, like the Br. densely white-felted; Fl.heads fairly large, Fl. numerous, yellow.

**S.** — cv. Cass's Variety. – Luxuriant cultivar.
**S.** — 'Hans Herre'. – Richtersveld: Karrachab Mts. – **L.** larger, with a flattened tip and up to 4 blunt marginal T.

**S. herreanus** DTR. (Gr. V.) (Pl. 130/3) (*Kleinia h.* DTR. nom. nud., *K. h.* (DTR.) MERXM., *K. gomphophylla* DTR. ex JACOBS.). – S.W.Afr.: Gr. Namaqualand. – Related to **S. radicans**; **shoots** firmer, shorter, up to 5 mm thick; **L.** berry-shaped, with symmetrical, obtuse tips and a small apiculus, up to 15 mm long, 8 mm ⌀, green or tinged reddish, with translucent stripes 2–3 mm across, and numerous translucent, often reddish lines 0.5 mm across.

**S. hildebrandtii** BAK. (Gr. II.). – Central Madag. – Sub-♄ 30–100 cm tall, the main St. simple or a little branched, erect, robust, with L. scars; **L.** spiralled, sessile, semicylindrical, linear, upperside furrowed, tapering, with distinct parallel veins, 1.5–15 cm long, 2 cm across; **Fl.**heads numerous, in a dense corymb, Fl. yellow.

**S. hirto-crassus** H. HUMB. (Gr. II.). – Central Madag. – Sparsely branched ♄ 60–100 cm tall; **Br.** fleshy, minutely rough-haired above, the **L.** similarly; **L.** spiralled, crowded, tapering, 4–5 cm long, 2–3 mm across, veins indistinct; Fl.heads solitary, Fl. yellow.

**S. implexus** BALLY. (Gr. V.). – Kenya: Masai D., coastal prov. – Resembles **S. jacobsenii**; creeping, succulent; **shoots** up to 70 cm long, 6 mm thick, little branched, the tips erect, minutely hairy; **L.** oblanceolate, fleshy, minutely hairy at first, 2.5–6.2 cm long, 9–32 mm across, 2–4 mm thick, acute, apiculate; **Fl.** heads large.

**S. jacobsenii** ROWL. (Gr. V.) (Pl. 131/3) (*S. petraeus* MUSCHLER, *Notonia p.* R. E. FRIES non BOISS. et REUT.). – Kenya; Tanzania. – Creeping ♄, the **shoots** rooting at the nodes, up to 1 cm thick, up to 50 cm long, a little branched; **L.** sessile, very fleshy, oblanceolate, the tip roundish, 5–7.5 cm long, 1–2.5 cm across; Fl.heads 1–3, Fl. orange.

**S. johnstonii** OLIV. (Pl. 131/1) (Gr. I.). – Tanzania: on Mt. Kilimanjaro. – 'Ghost Tree'. – Woody, tree-like, with a thick trunk; **L.** in a

Plate 146. 1. **Astridia ruschii** L. Bol. (Photo: H. Chr. Friedrich); 2. **Aridaria noctiflora** (L.) Schwant. v. **noctiflora** (Photo: as 1); 3. **Astridia dinteri** L. Bol. v. **dinteri**; 4. **Aspazoma amplectens** (L. Bol.) N. E. Br. (Photo: H. Herre).

Plate 147. 1. Bergeranthus multiceps (SALM) SCHWANT.; 2. Bijlia cana N. E. BR.; 3. Berrisfordia khamiesbergensis L. BOL. (Photo: H. HERRE); 4. Braunsia geminata (HAW.) L. BOL. (Photo: G. SCHWANTES).

Plate 148. 1. Psilocaulon marlothii (Pax) Friedr. (Photo: G. Schwantes); 2. Carpanthea pomeridiana (L.) N. E. Br.; 3. Calamophyllum teretiusculum (Haw.) Schwant.

Plate 149. 1. Carpobrotus acinaciformis (L.) L. Bol.; 2. Caryotophora skiatophytoides Leistn.; 3. Carruanthus peersii L. Bol.; 4. C. ringens (L. Bol.) Boom.

Plate 150. 1. Cephalophyllum herrei L. Bol. v. herrei (Photo: H. Herre); 2. C. alstonii Marl. (Photo: as 1); 3. C. subulatoides (Haw.) N. E. Br. (Photo: Kakteen-Haage, Erfurt); 4. Cerochlamys pachyphylla (L. Bol.) L. Bol. v. pachyphylla; 5. Chasmatophyllum musculinum (Haw.) Dtr. et Schwant. (Photo: K. Dinter).

Plate 151. 1. Cheiridopsis borealis L. Bol. (Photo: H. Chr. Friedrich); 2. C. peculiaris N. E. Br.; 3. C. purpurata L. Bol.; 4. C. marlothii N. E. Br. Left: in growth, right: dormant.

Plate 152. 1. **Cheiridopsis roodiae** N. E. Br. (Photo: Kakteen-Haage, Erfurt); 2. **Conicosia pugioniformis** (L.) N. E. Br.

Plate 153. 1. **Conophytum bilobum** (Marl.) N. E. Br. (Photo: G. Schwantes); 2. **C. christiansenianum** L. Bol.; 3. **C. concavum** L. Bol.; 4. **C. compressum** N. E. Br.; 5. **C. circumpunctatum** Schick et Tisch. (Photo: as 1).

dense cluster at the apex of the trunk, lanceolate with a cordate base; **Fl.** in dense thyrses, rays linguiform.

**S. junceus** HARV. (Gr. III.) (*Brachyrhynchos j.* LESS.). – Cape. – ♄ up to 1 m tall; **Br.** slender, erect, 3–5 mm thick, terete, green or grey-green, with lighter stripes; **L.** absent or reduced to small, triangular scales; **Infl.** of 5–7 heads, Fl. yellow.

**S. kalambatitrensis** H. HUMB. (Gr. II.). – Central Madag. – Sub-♄ with few Br., 20 to 40 cm tall; **Br.** erect, rough with old L.bases, densely hairy; **L.** spiralled, sessile, linear, semi-cylindrical, abruptly tapered above, acute, apiculate, 2–3 cm long, 2 mm across, upperside furrowed, the underside with several shallow grooves; **Fl.**heads with 6–9 yellow Fl.

**S. keniodendron** R. E. FRIES et TH. FRIES f. (Gr. I.) (*S. keniensis* BAK. f. p. part.). – Tanzania; Kenya: Mt. Kenya. – ♄ c. 6 m tall; **trunk** simple or rarely dichotomously branched, with a splitting bark, almost completely covered with dead L.; **L.** densely rosulate at the Br.tips, stiff, 35–60 cm long, 8–15 cm across, lanceolate or lanceolate-ovate, tapering towards the broadly alate petiole, margins irregularly dentate except at the tip, upperside initially with lax, white, silky wool; **Infl.** c. 1 m tall, ovoid, Br. white-woolly, Fl.heads dense, large, compressed-conical, without linguiform ray-florets.

**S. kleinia** (L.) LESS. (Gr. III.) (*Cacalia kleinia* L., *K. neriifolia* HAW.). – Canary Is. – ♄ up to 3 m tall, freely verticillately and dichotomously branched, with a trunk 10–20 cm thick with a grey-green bark, fleshy, brittle, with joints 8–40 cm long and 2–4 cm thick, with a light green bark, minutely white-spotted and pruinose, L.scars roundish, with 3 decurrent lines; **L.** linear-lanceolate, short-tapering, narrowed below towards the petiole, 9–15 cm long, 1–2 cm across, grey-green, with a rather thick M.nerve, later deciduous; **Fl.** yellowish-white.

**S.** — cv. Candystick. – **St.** and **L.** variegated in cream and pink. Hort. Rowley.

**S. kleiniiformis** SÜSSENG. (Gr. V.) (Pl. 131/2) (*Kleinia k.* (SÜSSENG.) BOOM, *S. speciosus* HORT., *S. cuneatus* JACOBS. non HOOK., *K. cucullata* BOOM). – S.Afr. – White-pruinose sub-♄; **shoots** ± prostrate but not creeping, Br. erect, up to 60 cm tall; **L.** spatulate, 6 to 10 cm long, 1–3 cm across, mostly ± curved upwards, margins curved inwards, with 1 or more large T., upperside concave to grooved, underside very convex, narrowed teretely below; **Fl.** white or yellow.

**S. klinghardtianus** DTR. (Gr. V.) (Pl. 131/5) (*Othonna pusilla* DTR., *Kleinia p.* (DTR.) MERXM., *S. p.* DTR. ex RANGE nom. nud., *Kleinia p.* DTR. ex RANGE, *S. herreianus* HORT., *S. iosensis* ROWL.). – S.W.Afr.: S.Lüderitz D. to the Orange R.; Cape: Lit. Namaqualand. – Dwarf, herbaceous, caespitose, with a fleshy, thickened root, glabrous apart from small clusters of hairs in the L.axils; **shoots** erect, a little branched, 1–5 cm tall, up to 1 cm thick; **L.** in dense tufts, ± pruinose, lower L. ± spherical, 5–7 mm ⌀, the upper ones navicular, 15 to 35 mm long, c. 10 mm thick, constricted below into a petiole 1–3 mm long, rather tapered above, upperside broadly furrowed, with 5 translucent, longitudinal lines; **Infl.** of a solitary head, Fl. c. 15, white, anthers yellow or light purple.

**S. leandrii** H. HUMB. (Gr. IV.). – Central Madag. – Sub-♄; **caudex** tuberous, ± ovoid, 2–3 cm long, 1–2 cm thick; **shoots** woody, 10–20 cm long, a little branched, with old L. persisting below; **L.** sessile, oblong-lanceolate to oblanceolate, narrowed below, upperside ± concave, margins whitish; **Fl.** heads with yellow Fl.

**S. longiflorus** (DC.) SCH. BIP. v. **longiflorus** (Gr. III.) (Pl. 131/4) (*Kleinia l.* DC., *S. l.* (DC.) OLIV. et HIERN). – Cape: Cradock D.; S.W.Afr.: Gr. Namaqualand; Ethiopia. – Much branched ♄ 45–60 cm tall; **shoots** 6 to 10 mm thick, with round L.scars with 3 decurrent lines; **L.** oblong-linear, tapering, up to 5 cm long, 3–4 mm thick, early deciduous; Fl.whitish, anthers yellow.

**S.** — v. **madagascariensis** (H. HUMB.) ROWL. (*Notonia m.* H. HUMB., *S. m.* (H. HUMB.) JACOBS. non POIRET, *S. humbertii* CHANG.). – E. Madag. – With tangled Br., L.scars short, pointed, mostly with 3 decurrent lines c. 1 mm high, shoots many-angled, green, with white, waxy spots; **Fl.** whitish-green.

**S.** — v. **violaceus** (BGR.) E. A. BRUCE et HUTCHINS. (*Kleinia v.* BGR., *Notonia kleinioides* SCH. BIP., *K. k.* (SCH. BIP.) TAYLOR, *S. k.* (SCH. BIP.) OLIV. et HIERN). – Ethiopia. – **L.** rather more fleshy, mucronate; **Fl.** pale violet.

**S. lunulatus** (CHIOV.) JACOBS. (Gr. III.) (*Monadenium l.* CHIOV., *Notonia l.* (CHIOV.) CHIOV.). – Central Somalia. – **Shoots** fleshy, cylindrical, elongated, variously curved, 15 to 20 mm thick, tapering towards the obtuse tip with 2–3 mm long, lunate-truncate, shortly bifid L.scars; **L.** spiralled, sessile, 5–7 or more at the tip, of the shoots, obovate to oblong-cuneate, bluish, glabrous, brittle, fleshy, 3–7 cm long, 1.5–3.5 cm across, the tip narrow-rounded, apiculate; Fl.heads solitary from the L.axils, conical-campanulate, Fl. golden-yellow.

**S. mandraliscae** (TEN.) JACOBS. (Gr. V.) (*Kleinia m.* TIN.). – ? Cape. – ♄ with fleshy, ± prostrate shoots up to finger-thickness, the bark eventually light grey, with indistinct stripes running down from the L.scars; **L.** crowded towards the tips of the shoots, semi-cylindrical, the upperside rather flat and often grooved, apiculate, with darker stripes and veins, light grey-pruinose, 8–9 cm long, 8 to 11 mm across; **Infl.** a cyme.

**S. marnieri** H. HUMB. (Gr. III.). – Central Madag. – Sub-♄ with tuberous **roots**; **shoots** branched; short shoots densely leafy, ± 10 cm long; L. spiralled, ± sessile, ± linear, cylindrical towards the base, 3.5–4.5 cm long, 3–4 mm across, acute, apiculate, 3-veined; **Fl.** yellow.

**S. medley-woodii** HUTCHINS. (Gr. II.) (Pl. 132/1). – Natal. – ♄ with red, fleshy, woolly **shoots**; **L.** flat, fleshy, obovate, short-tapering, cuneate below, with margins ± undulating, bluntly dentate above, 3.5–5 cm long, 1.5–3 cm across, densely felty at first, becoming glabrous and purple; **Infl.** with 10–13 large yellow ray-florets.

**S. melastomifolius** BAK. v. **melastomifolius** (Gr. II.). – Central Madag. – Sub-♄ 40–60 cm tall; **Br.** becoming woody, with L.scars; **L.** densely crowded, spiralled, rosulate, sessile, oblanceolate, obtuse, 2–6 cm long, 1–1.5 cm across, 3–5-veined, with horny margins; **Fl.**heads of 40–60 yellow Fl.

**S. —** v. **longibracteatus** H. HUMB. – Plant more slender; **L.** more numerous, **Infl.** with long bracts.

**S. mesembryanthemoides** BOJ. ex DC. (Gr. IV.). – Central Madag. – **Roots** tuberous; **shoots** 1–5 cm long; **L.** numerous, rosulate, sessile, fleshy, oblanceolate to ± spatulate, obtuse, thick, 10–15 mm long, 3–5 mm across, upperside concave, glabrous or minutely hairy, veins indistinct; Fl.heads solitary, Fl. yellow.

**S. meuselii** RAUH (Gr. II.). – Central Madag. – Much-branched ♄ up to 1.5 m tall; **Br.** cylindrical, very fleshy; **L.** spiralled, somewhat fleshy, linear-lanceolate, up to 15 cm long, 10 mm across, slightly recurved, tapering to a sharp point, often decurrent into a short petiole, standing vertically, with the upperside narrow and translucent; **Infl.** freely branched, up to 50 cm tall, bracts small, linear, with hairy margins, **Fl.**heads on Ped. 10 mm long, with a few, light yellow Fl.

**S. mweroensis** BAK. (Gr. III.) – Kalongwizi River, Mwero, W. of Lake Tanganyika. – Near **S.anteuphorbium**; **St.** short, cylindrical, fleshy; **L.** minute, linear, fleshy, entire, acute; Ped. 10–15 cm tall; **Fl.** without rays.

**S. navicularis** H. HUMB. (Gr. II.). – Central Madag. – Sub-♄ 60–80 cm tall; **shoots** erect, 4–6 mm thick; **L.** spiralled, broad-linear or oblanceolate-linear, 5–10 cm long, 6–10 mm across, thick, abruptly narrowed above, underside with 2 deep grooves and 11 parallel veins, Y-shaped in cross-section; **Fl.** numerous, yellow.

**S. neobakeri** H. HUMB. (Gr. II.) (*S. vernicosus* BAK. non SCH. BIP.). – Central Madag. – Branched ♄ 1–2 m tall; **Br.** eventually woody, ± angular-alate, with rather prominent L.-cushions; **L.** spiralled, subsessile, rather thick, glabrous, oblanceolate, acute, 3-veined, apiculate, 4–10 cm long, 8–16 mm across; **Fl.** yellow.

**S. neohumbertii** ROWL. (Gr. III.) (*Kleinia humbertii* GUILL., *S.h.* (GUILL.) JACOBS.). – Small, sparsely branched ♄; **shoots** c. 1 cm thick, with the bark becoming brown; **L.** in dense clusters at the tips of the shoots, sessile, linear-oblong, tapering, upperside with a groove 2 mm deep, bluish-green to crimson, pruinose, with numerous, translucent, longitudinal lines. Possibly identical with **S. hildebrandtii.**

**S. nyikensis** BAK. (Gr. V.). – Tanzania. – ♃; **L.** oblanceolate, oblong, narrowed below, 13 to 15 cm long, 25 mm across, fleshy; **Fl.**heads 25 mm ⌀, Fl. orange.

**S. orbicularis** SOND. (Gr. VII.). – Transvaal. – Resembles **S. tropaeolifolius,** and may be identi-

cal. – **L.** peltate, 4–5 cm ⌀, petioles 5–7 cm long.

**S. ovoideus** (COMPT.) JACOBS. (Gr. V.) (*Kleinia o.* COMPT.). – Cape: Ladismith D. – Dwarf, resinous, aromatic ♃; **roots** thick-fleshy, fusiform; **St.** erect, rather succulent, a little branched, cylindrical, 4–6 mm thick; **L.** few, fleshy, with a petiole 2 mm long, narrow, mostly ± ovoid, ± compressed, rather acute above, dark green, with a few longitudinal stripes, 12–24 mm long, 8–10 mm across; **Fl.** white.

**S. oxyriifolius** DC. (Gr. III.). – Cape. – ♄ with creeping **St.** from a gnarled, tuberous rootstock, mat-forming; lower **L.** spatulate, upper ones scutate, irregularly sinuate-dentate, c. 3 cm long, c. 4 cm across, light green, fleshy.

**S. papillaris** (L.) SCH. BIP. (Gr. III.) (*Cacalia p.* L., *Kleinia p.* (L.) HAW., *S. p.* (L.) JACOBS.). – Cape. – ♄ with fleshy **shoots** covered with cylindrical papillae; **L.** lanceolate, flat, bluish, with a distinct M.nerve.

**S. paucifolius** DC. (Gr. II.). – Cape. – ♄; **shoots** thin, laxly leafy; **L.** fleshy, subsessile, oval, 4.5 cm long, 3.5 cm across, light green, waxy-pruinose, with 3 distinct veins, margins often reddish and often with a few small T.

**S. pendulus** (FORSK.) SCH. BIP. (Gr. III.) (*Cacalia p.* FORSK., *Notonia p.* (FORSK.) CHIOV., *N. trachycarpa* KLOTZSCH., *Kleinia p.* (FORSK.) DC., *S. gunninsii* BAK.). – S.Arabia: Yemen; Ethiopia; Somalia. – **Shoots** jointed, curved, prostrate and ascending again, rooting, often growing underground, up to 30 cm long, 1.5 to 2 cm thick, with the dry remains of old L., somewhat laterally compressed, grey-green or brownish, with darker lines; **L.** linear, ± cylindrical, 2 mm long; **Fl.** orange-red or blood-red.

**S. picticaulis** BALLY. (Gr. VI.). – Kenya: central reg.; Tanzania: Tanga; Sudan: Equatoria Prov. – ♄ 15–35 cm tall; **caudex** horizontal, tuberous; **shoots** solitary, erect, sometimes branched, ☉, fleshy, cylindrical or ± clavate, 15–20 cm long, 5 mm thick below and 20 mm above, pale green, with 3 dark green, longitudinal lines, with L. scars; **L.** subulate, up to 5.5 cm long, 3 mm thick, fleshy, green, acute; **Fl.**heads mostly solitary, with 40–50 Fl., tube pale pink, lobes red.

**S. pinguifolius** (DC.) SCH. BIP. (Gr. III.) (*Kleinia p.* DC., *S. p.* (DC.) JACOBS.). – Cape: Lit. Namaqualand; S.W.Afr.: S.Lüderitz D. – Unpleasantly smelling ♃; **shoots** virgate, branching only from the base, 10–20 cm tall and 12–15 mm thick, with a grey bark, densely woolly and felted, with many L.scars, the upper part sap-green, fleshy, leafy; **L.** 5–10 cm long, 3–5 mm thick (larger in cultivation), cylindrical, tapering, upperside with a slight longitudinal furrow; **Fl.** white, with violet anthers.

**S. polytomus** (CHIOV.) JACOBS. emend. BALLY (Gr. II.) (*Kleinia p.* CHIOV.). – Somalia. – ♄ 22–120 cm tall; **St.** cylindrical, fleshy, longitudinally striate, ribbed, with terminal short shoots 3–4 cm long, 5–6 mm thick, branching dichotomously and umbellately from the nodes; **L.** linear-ovate, 5–8 mm long, 2–3 mm ⌀, cuneate and shortly petiolate below, apiculate, petiole remains sometimes persisting, 1 mm high; **Infl.** rather umbel-like, Fl. heads 4–15, Fl. whitish-yellow or purple.

**S. praecox** DC. (Gr. III.) (*Cineraria p.* Cav.). – Mexico. – ♃ up to 2 m tall, **St.** and **Br.** thick; **L.** in crowded clusters at the tips of the shoots, spiralled, 8–9 cm long, 6 cm across below, angular with 3 indentations, thus forming several lobed T., green; **Fl.** inconspicuous.

**S. pyramidatus** DC. (Gr. II.). – Cape: Uitenhage D. – Sub-♃ with erect, thick, fleshy, simple **shoots,** densely leafy below; **L.** subterete, 8–12 cm long, 4–6 mm across, up to 1 cm thick, shortly and obtusely tapered, with a fine grey felty coating, like the rest of the plant; **Infl.** paniculate, Fl.heads numerous, c. 5 cm ⌀, golden-yellow.

**S. quartziticolus** H. Humb. (Gr. II.). – Central Madag. – Erect sub-♃ 20–30 cm tall; **shoots** fleshy, rough with old L.bases; **L.** sessile, elliptical-lanceolate, 12–14 mm long, 2.5–5 mm across, tapering towards both ends, upperside ± concave; **Fl.**heads solitary, Fl. yellow.

**S. quinquangulatus** Sch. Bip. (Gr. II.) (*Kleinia cana* DC.). – Cape. – Like **S. haworthii**; **L.** ovate-acute, 16–19 mm long, 6–8 mm thick.

**S. radicans** (L. f.) Sch. Bip. (Gr. V.) (Pl. 132/4) (*Cacalia r.* L. f., *Kleinia r.* (L. f.) DC., *K. gonoclada* DC., *S. rhopaladenia* Dtr., *S. adenocalyx* Dtr., *K. a.* (Dtr.) Merxm.). – Cape: Karroo; S.W.Afr.: S.Lüderitz, Bethanien D. – ♃; **shoots** creeping, rooting at the nodes, somewhat angular, sparsely branched; **L.** one-sidedly directed upwards, terete, 15–25 mm long, (4–)6(–10) mm thick, narrowed abruptly into a short petiole, tapering or rather obtuse above, apiculate, with indistinct longitudinal stripes, the upperside with a rather broad, longitudinal Wi., dark green, often reddish and ± pruinose; **Fl.**heads 1–4, Fl. white, anthers yellow or light purple.

**S. rodriguezii** Willk. (Gr. II.). – Balearics: Majorca. – ♃; **shoots** 6–12 cm long, branching basally; **L.** sessile, fleshy, oblong or oblong-lanceolate, obtuse, unevenly dentate, 3-veined, upperside grey-green, underside purple; **Infl.** paniculate, Fl.heads with 15–20 pink ray-florets. ± hardy.

**S. rowleyanus** Jacobs. (Gr. V.) (Pl. 132/2). – Southern S.W.Afr. – **Shoots** creeping, thin, with adventitious roots, forming dense mats, weak, green, c. 1 mm thick; **L.** mostly spherical or somewhat compressed-spherical, apiculate, with a translucent, longitudinal stripe 1 mm across; **Fl.** c. 20, cinnamon-scented, without ray-florets, Cor. white, anthers deep brownish-violet.

**S. saboureaui** H. Humb. (Gr. II.) (*S. melastomaefolius* v. *microphyllus* H. Humb.). – Central Madag. – Sub-♃; **shoots** with semicylindrical L.cushions; **L.** spiralled at the end of the shoots, with a petiole up to 1 mm long, ± elliptical-ovate, with indistinct veins; **Fl.** few, yellow.

**S. sakalavorus** H. Humb. (Gr. II.). – W.Madag. – Fleshy to woody, glabrous, freely branched, sub-♃ 20–40 cm tall; **Br.** 20–40 cm tall, cylindrical; **L.** fleshy, oblanceolate, obtuse, 3–5 cm long, 5–7 cm across, much narrowed below, indistinctly 3-veined; **Fl.** whitish, in a corymb.

**S. sakamaliensis** (H. Humb.) H. Humb. (Gr. II.) (*S. antandroi* v. *s.* H. Humb.). – ? Madag. – Sub-♃ 40–100 cm tall; **shoots** erect, robust, 5–6 mm thick, with prominent L.scars; **L.** spiralled, broad-linear, 6–10 mm long, 5–8 mm across, 3–6 mm thick, gradually tapered below, shorter-tapering above, apiculate, upperside deeply furrowed, with many veins below; **Fl.**heads in corymbs, Fl. yellow.

**S. scaposus** DC. v. **scaposus** (Gr. III.). – Cape. – Acauline or very short-stemmed; **L.** rosulate, crowded, linear, ± flatly compressed, especially at the rather spatulately widened tip, 5–8 cm long, younger L. with a cobwebby, white felt, older L. glabrous, green; **Fl.**heads many, Fl. yellow.

**S. —** v. **caulescens** Haw. (*S. calamifolius* Hook., ? *Othonna vestita* DC.). – **St.** rudimentary, thick, c. 4 cm tall; **L.** flat and wider at the tip.

**S. scottii** Balf. f. (Gr. III.). – Socotra. – ♃ 60 cm tall; **shoots** erect, cylindrical, 5 to 8 mm thick; **L.** scale-like, lanceolate, c. 8 mm long, deciduous; **Fl.** terminal.

**S. sempervivus** (Forsk.) Sch. Bip. (Gr. IV.) (*Cacalia s.* Forsk., *Notonia s.* (Forsk.) Aschers., *Kleinia s.* (Forsk.) DC.). – Tanzania; Ethiopia. – ♃; roots tuberous; **shoots** several, erect, c. 10 cm long, 6–8 mm thick, smooth, light grey-pruinose; **L.** numerous, rosulate, crowded, 6.5 cm long, 2.5 cm across, oblanceolate, obtuse, M.nerve projecting and carinate, green, underside reddish; **Fl.**heads solitary or paired, Fl. c. 60, intense red.

**S. serpens** Rowl. (Gr. V.) (*Cacalia repens* L., *Kleinia r.* (L.) Haw., *S. r.* (L.) Muschler, *S. succulentus* Sch. Bip.). – Cape. – ♃ 20 to 30 cm tall and wide; **shoots** 5–7 mm thick, blue-pruinose; **L.** crowded at the tips of the shoots, linear-oblong, rather obtuse, with a short brown mucro, ± cylindrical, upperside compressed or furrowed, 3–4 cm long, 7–8 mm across, minutely light grey or bluish-pruinose; Fl.heads with white Fl.

**S. —** cv. Cristata. – Cultivar with strongly fasciated shoots which occurred in the collection of Albert Baynes, Bradford, England.

**S. spiculosus** (Sheph.) Rowl. (Gr. II.) (Pl. 132/3) (*Kleinia s.* Sheph., *K. cylindrica* Bgr., *S. c.* (Bgr.) Jacobs., *K. spinulosa* Hort.). – S.W.Afr. – **Shoots** erect, up to finger-thick, forming a ♃ 20–60 cm tall; **Br.** green, with a few white spots, the round L.scars tuberculate, with 3 short, dark lines running down from each; **L.** cylindrical, tapered towards both ends, apiculate, light green, with numerous fine, darker nerves, minutely grey-pruinose, 6–8 cm long, 8–9 mm thick; **Fl.**heads in a cyme, Fl. white.

**S. spinescens** Sch. Bip. (Gr. II.) (*Cacalia rigida* Thunbg., *Kleinia r.* (Thunbg.) DC., *S. r.* (Thunbg.) Jacobs.). – Cape. – ♃ with widespread, thorny **Br.**; **L.** ovate to obtuse, flat, with 1–2 T.; **Infl.** terminal.

**S. stapeliiformis** PHILLIPS v. **stapeliiformis** (Gr. III.) (*Kleinia s.* (PHILL.) STAPF, *S. stapeliiformis* STAPF, incorrectly *S. stapelioides*). – Eastern S.Afr. – **St.** erect, branching basally, young shoots at first growing underground, 20 cm or more long, 15–20 mm thick, with 5–7 ± prominent angles along which thorny **L.** 5 mm long are arranged at intervals, the sides between the angles grey, with several dark green, short lines running down from each **L.**base; **Fl.** red, with Ped. 5–6 cm long.

**S.** — v. **minor** ROWL. (*Notonia gregorii* S. MOORE, *S. g.* (S. MOORE) JACOBS.) – Kenya. – In general smaller, with less fleshy **shoots**.

**S. subradiatus** (DC.) SCH. BIP. (Gr. II.) (*Kleinia s.* DC., *S. r.* (DC.) JACOBS.). – Cape: Graaff Reinet D. – Erect, branched ♄ with cylindrical **Br.**; **L.** sessile, linear, fleshy, acute, with entire margins.

**S. subscandens** HOCHST. (Gr. VIII.). – E.Afr. – Climbing ♄; **shoots** striped, slightly pitted; **L.** fleshy, oval, margins incised, with 3–5 dentate lobes, dull bluish-green; **Infl.** terminal. Acc. P. R. O. BALLY this species has been renamed **Crassocephalum bojeri.**

**S. subsinuatus** DC. (Gr. II.). – Cape: Stellenbosch D. – Erect, glabrous sub-♄; **Br.** cylindrical; **L.** rather fleshy, flat, linear-oblong, obtuse, with entire, rather sinuate margins.

**S. subulatifolius** ROWL. (Gr. III.) (*Notonia subulata* BALLY, *S. subulatus* (BALLY) JACOBS.). – Kenya. – **Roots** tuberous; **shoots** simple, erect, 2 cm ⌀ below, pitted, with slightly prominent, tooth-like L.scars, with decurrent, dark green lines; **L.** subulate, up to 4 cm long, fleshy, subpetiolate, deciduous; **Fl.**heads solitary, Fl. c. 50, scarlet.

**S. succulentus** DC. (Gr. V.). – Cape: Karroo. – Glabrous ♄; **shoots** cylindrical, rather fleshy; **L.** linear-oblong, thick, tapering, 3.5 cm long, c. 5 mm across; **Fl.** light yellow.

**S. talinoides** (DC.) SCH. BIP. (Gr. II.) (*Kleinia t.* DC., *S. t.* (DC.) JACOBS.). – Cape. – Erect ♄; **shoots** cylindrical, fleshy; **L.** cylindrical, compressed, fleshy, elongated, tapered, 6 cm long, 3–4 mm thick, bluish-green.

**S. tropaeolifolius** MACOWAN. (Gr. VII.). – S.Afr. – ♄ with tuberous **roots**; **shoots** thin, laxly leafy; **L.** scutate-cordate, irregularly rounded and sinuate, with reddish T., c. 2.5 cm long, fleshy, light green.

**S. usambariensis** MUSCHLER. (Gr. II.). – Tanzania: Usambara D. – ♄ with a few Br.; **shoots** c. 5 mm thick, green, laxly leafy; **L.** with a petiole 2–3 cm long, ± cordate, irregularly lobed, with a few minute, irregularly spaced T., 3–4 cm long and across, fleshy, light green.

**Sesamothamnus** WELW. Pedaliaceae. – Trop. E. Afr.; S.W.Afr.; Angola; Botswana; Ethiopia; Somalia. – Branching, thorny ♄ or small ♄; **St.** smooth, mostly swollen at the base, leafless at flowering; **Br.** ascending, thorny; **L.** deciduous, usually obovate, clustered on short shoots in the axils of the Th. (modified petioles); **Fl.** large, white, pink or yellow, often sweetly scented, few in a Rac., Cal. mostly glandular, Cor.tube long, cylindrical or ± cup-shaped, the mouth widened, spurred or tuberculate below, lobes with margins entire or fringed. – Greenhouse, warm. Propagation: seed.

**S. benguellensis** WELW. – Trop. Afr. – **Caudex** 60 cm or more ⌀, scarcely projecting above the soil, with 4–5 thick **Br.** above, with ascending, thorny shoots 1–1.5 m tall; **L.** oval, 10 to 15 mm long, 8–12 mm across, narrowing towards the petiole; **Fl.** whitish-pink.

**S. busseanus** ENGL. – Kenya; Somalia. – ♄ or small ♄ 2–5 m tall; **St.** swollen, soft-woody, with a papery, peeling, coppery-green bark, Th. numerous, up to 15 mm long, with 2 subsidiary Th. below; **L.** short-petiolate, obovate, 2–5 cm long, 1–2.5 cm across, underside densely glandular; **Fl.** white, or white with a pink tube, all lobes except one distinctly fringed.

**S. guerichii** (ENGL.) BRUCE (*Sigmatosiphon g.* ENGL.). – S.W.Afr.: Outjo. – **St.** succulent, 6 m tall with a papery, peeling bark; **Br.** and short shoots thorny; **Fl.** golden-yellow.

**S. lugardii** N. E. BR. (Pl. 133/1). – N.Transvaal; southern Rhodesia; S.W.Afr. – **Caudex** clavate, often over 2 m ⌀ but rather low, often passing over above into ascending, succulent Br.; **L.** clustered, in the axils of sharp Th.; **Fl.** white, with a long tube.

**S. rivae** ENGL. (*S. erlangeri* ENGL., *S. smithii* (BAK.) STAPF). – Kenya; Ethiopia. – ♄ or small ♄ 2–6 m tall, rather swollen below; **Br.** virgate, rather thorny, Th. 5–9 mm long, swollen below; **L.** short-petiolate, obovate, 2–8 cm long, 1.3–6 cm across, cuneate below, rounded above, both surfaces glandular; **Fl.** white or whitish-brown.

**Sesuvium** L. Aizoaceae.

**S. portulacastrum** L. (Pl. 133/2). – E. Afr. – Succulent ♄ with creeping **shoots**; **L.** oblong, fleshy; **Fl.** short-pedicellate, from the L.axils, purple; the entire plant often red. – Heated greenhouse. Propagation: seed, cuttings.

**Seyrigia** KERAUDREN. Cucurbitaceae. – Madag. – Dioecious, climbing, twining plants; **roots** penetrating deeply, with tuberous swellings; **St.** fleshy, angular, somewhat reddish, ± curved in various directions, grey to white-felted, tendrils simple; **L.** few, often missing, c. 3 mm long, 3-lobed, early deciduous; **Fl.** in Rac., Pet. 5, whitish-yellow to greenish; **Fr.** spherical or obconical, with red flesh. – Greenhouse, warm. Propagation: seed, cuttings.

**S. bosseri** KERAUDREN. – S.Madag. – **St.** thick, 5-angled, longitudinally grooved, with dense, grey, woolly hairs; **Fl.** small, yellowish; **Fr.** a shortly apiculate berry.

**S. gracilis** KERAUDREN. – Madag.: near Tuléar. – **St.** slender, branching, fleshy; side-shoots minutely hairy to ± glabrous; **Fr.** obconical.

**S. humbertii** KERAUDREN (Pl. 133/4). – Madag.: Fiherenana gorge. – **St.** fleshy, furrowed, densely white-felted; **Fr.** ovoid, apiculate.

**S. multiflora** KERAUDREN. – Madag.: near Tsiry. – **St.** fleshy, climbing, minutely grey-felted; **Fr.** ovoid, apiculate.

**Sinocrassula** BGR. Crassulaceae. – Himalayas to W.China. – ☉ or ♃, glabrous or minutely hairy; **L.** rosulate, rather thick, obtuse or tapering, hair-tipped, ± lined, blotched or suffused reddish-brown; flowering shoots erect, ± elongated, bracts lax; **Infl.** paniculate-cymose, more rarely simple, Fl. crowded towards the apex, pedicellate, erect, ± spherical to urn-shaped, Pet. whitish, with vivid red tips. – Summer: in the open, winter: cold-house. Propagation: seed, division.

**S. aliciae** (HAMET) BGR. (*Crassula a.* HAMET). – W.China. – **L.** obtusely obovate, slightly dentate at the tip, T. papillose; **Infl.** erect, simple.

**S. densirosulata** (PRAEG.) BGR. (*Sedum indicum v. d.* PRAEG., *S. i. v. ambigua* HAMET, *Lenophyllum maculatum* HORT.). – W.China: Yunnan. – **L.** numerous, narrowly spatulate, 25 mm long, 7 mm across, very grey, with small red lines; **Infl.** broad to rounded.

**S. indica** (DECNE) BGR. v. **indica** (*Crassula i.* DECNE, *Sedum indicum genuinum* HAMET, *Sed. martinii* LÉV., *Sed. cavalerici* LÉV., *Sed. scallanii* DIELS). – Peninsular India. – **L.** linear-spatulate, long-tapered, 3.5–6 cm long, 10–15 mm across, underside convex, rather greyish-green, spotted and lined; **Infl.** 15–28 cm tall.

**S.** – v. **forrestii** (HAMET) FU (*Sedum indicum v. forrestii* HAMET). – W.China: Yunnan. – **L.** circular or almost so.

**S.** – v. **longistyla** (PRAEG.) JACOBS. (*Sedum l.* PRAEG., *Sed. indicum v. l.* (PRAEG.) FROED., *Sin. l.* (PRAEG.) FU). – W.China. – Carpels with a long style.

**S.** – v. **obtusifolia** (FROED.) FU (*Sedum indicum v. o.* FROED.). – China: N.W.Yunnan. – Ros. very dense and broad; **L.** obtuse.

**S.** – v. **serrata** (HAMET) FU (*Sedum indicum v. s.* HAMET, *Sin. bergeri* JACOBS.). – **L.** obtuse serrate above, hairy.

**S. schoenlandii** (HAMET) FU (*Sedum s.* HAMET). – W.China. – ? ♃; **roots** fibrous; **L.** obovate, cuneate, 4.75–5.5 cm long, 1–1.25 cm across, obtuse, covered above with long, simple hairs; **Infl.** racemose, with lanceolate, acute bracts.

**S. stenostachya** (FROED.) FU (*Sedum s.* FROED.). – China: S.Kansu. – Glabrous, ? ☉; **taproot** rather thick, vertical or creeping; **L.** spur-like below, spatulate-oblong to broadly oblanceolate, somewhat lobed or deeply dentate, tapered, glabrous, 10–15 mm long; **Infl.** of few Fl. in a lax spike.

**S.** – v. **integrifolia** FU. – China: Kansu. – **L.** with entire margins.

**S. yunnanensis** (FRANCH.) BGR. (Pl. 133/3) (*Crassula y.* FRANCH., *Sedum indicum v. y.* HAMET). – W.China: Yunnan. – **L.** 50–70 in dense Ros., lanceolate, tapered, apiculate, upperside roundish, underside very rounded, 12.5–25 mm long, 4.5–6 mm across, dark bluish-green, with minute, white, papillose hairs all over; **Infl.** minutely hairy.

**S.** – cv. Cristata. – Monstrous cultivar.

**Stapelia** L. Asclepiadaceae. – S.Afr., some species in Tanzania, Kenya, E.India. – Succulent ♃; **St.** fleshy, ascending or erect, branched at the base and thus forming clumps, 4-angled, the angles coarsely dentate and often bearing rudimentary and soon deciduous **L.**, glabrous or soft-hairy, green or also reddish; **Fl.** mostly from the base of the St., solitary or several, more rarely from the apices of the St., mostly large and with long pedicels, often with a pungent odour; Cal. 5-partite, the Cal.-lobes tapering, Cor. rotate or broadly campanulate, up to 5-partite halfway or deeper and with spreading or recurved, ± triangular, acute, rather fleshy lobes, variously coloured, ± transversely gibbose and wrinkled, ± hairy or completely glabrous, not rarely with a fleshy annulus (ring) around the corona (§ X), corona (staminal-corona) double. – Greenhouse, warm. Propagation: seeds, cuttings.

## Division of the Genus Stapelia into Sections by A. White and B. L. Sloane

**Sect. I. Stapelluma** WHITE et SLOANE (Woodii-longidens-Group). – St. in habit like Caralluma and Stultitia (formerly Stapelia), spotted, dentate, T. tapering and 1–2 cm long; shape of the Cor. various. – Species: S. discoidea, longidens, melonyae, semota, woodii.

**Sect. II. Podanthes** (HAW.) BGR. (*Podanthes* HAW. as a genus) (Verrucosa-kwebensis-Group). – The name Podanthes means "foot-flower" and refers to the shape of the inner corona-lobes. – Cor. raised somewhat like a disc around the corona, the inner corona-lobes consisting of a single segment, incumbent on the anthers, with or without a dorsal projection; at most thickened knee-like, more rarely somewhat prominent. – Species: S. arenosa, fuscosa, juttae, kwebensis, parvipuncta, parvula, portae-taurinae, rubiginosa, similis, stultitioides, verrucosa.

**Sect. III. Tridentea** (HAW.) BGR. (*Tridentea* HAW. as a genus) (Virescens-gemmiflora-Group). – The name Tridentea refers to the "three-toothed sections" – the only group of the Stapelias characterized primarily by the deeply 3-partite outer corona, the central lobe larger than the lateral ones and either simple, tapering or again dentate and emarginate. – Species: S. auobensis, gemmiflora, marientalensis, pachyrrhiza, peculiaris, umbonata, virescens.

**Sect. IV. Fissirostres** (*Fissirostrum* BGR.) (Rufa-Group). – This title means "cleft-beaks". The erect lobes of the inner corona are split at the tip into two somewhat diverging segments, with a slight dorsal swelling. – Species: S. rufa.

**Sect. V. Gonostemon** (HAW.) WHITE et SLOANE (*Gonostemon* HAW. as a genus) (Deflexa-olivacea – Group). – The name Gonostemon refers to the "knobby corona", the lobes of the inner corona, which pass into a transversely thickened long-acute horn. – Species: S. acuminata, ausana, choanantha, cincta, concinna, deflexa, dinteri, divaricata, dwequensis, erectiflora, flavopurpurea, glandulifera, indocta, jucunda, nouhuysii, olivacea, pearsonii, stricta, surrecta.

**Sect. VI. Stapletonia** (DECNE.) ENDL. (*Stapletonia* DECNE. as a genus) (WHITE and SLOANE wrote "*Stapeltonia*") (Gigantea-hirsuta-Group). – The lobes of the inner corona consist of an inner horn with a ± broad, dorsal wing. – Species: S. ambigua, arnotii, asterias, bergeriana, cylista, desmetiana, flavirostris, forcipes, fuscopurpurea, gariepensis, gettleffii, gigantea, glabricaulis, grandiflora, hirsuta, immelmaniae, leendertziae, maccabeana, macowanii, marlothii, nobilis, nudiflora, peglerae, pillansii, plantii, pulvinata, schinzii, senilis, sororia, tsomoensis, vetula, wilmaniae, youngii.

**Sect. VII. Clavirostres** WHITE et SLOANE (Engleriana-herrei-Group). – The name Clavirostres refers to the 2 "clavate beaks" (horns) of the inner corona, both horns being clavate at the tips. – Species: S. clavicorona, englerana, herrei, neliana.

**Sect. VIII. Caruncularia** (HAW.) DECNE. (*Caruncularia* HAW. as a genus) (Pedunculata-Group). – The name Caruncularia means small fleshy lumps, referring to the inner corona, the lobes of which are 2-horned and one of the two horns or both terminate in a big strongly tuberculate-gibbose knob. – Species: S. longii, longipes, pedunculata, ruschiana.

**Sect. IX. Tromotriche** (HAW.) BGR. (*Tromotriche* HAW. as a genus) (Revoluta-Group). – The name Tromotriche refers to the trembling hairs of the Fl. The lobes of the inner corona are 2-horned and the corona is horizontally constricted and thickened at the centre above the constriction. – Species: S. bella, comparabilis, incomparabilis, prognantha, revoluta, thudichumii.

**Sect. X. Stapelia** (HAW.) BGR. (*Orbea* HAW. as a genus) (Mutabilis-variegata-Group). – This section is difficult to delimit from others. The lobes of the inner corona are 2-horned, with the exception of S. namaquensis; Cor. with a distinct, low or prominent ring (except S. maculosoides). – Species: S. barklyi, × bicolor, × cupularis, × discolor, × hanburyana, lepida, maculosa, maculosoides, mutabilis, namaquensis, pulchella, trifida, variegata.

**S. acuminata** MASS. v. **acuminata** (§ V). – Cape: Lit. Namaqualand. – **Shoots** c. 15 cm tall, 1–2 cm ⌀, with 4 rather compressed, denticulate, minutely hairy angles; **Cor.** rotate, flat, 3.5–4.5 cm ⌀, very deeply cleft, lobes ovate-lanceolate, narrow and long-tapered, glabrous, dark brown with yellowish transverse stripes, lobe-tips dark brown.

**S. — v. brevicuspis** N. E. BR. – Fl. 15–16 mm ⌀, lobes ovate-acute, transverse ridges rather lighter yellow.

**S. ambigua** MASS. v. **ambigua** (§ VI). – Cape: Victoria W., Uitenhage D. – **Shoots** 20–25 cm tall, 2–2.5 cm ⌀, with 4 minutely hairy angles, T. distant; **Cor.** flat, 11–13 cm ⌀, lobes recurved, lanceolate, sparsely rough-haired, glabrous above, margins with projecting cilia, densely puberulous outside.

**S. — v. fulva** SWEET. – Lobes yellowish-green in the lower half, with purplish-brown transverse lines, the upper half purplish-brown.

**S. angulata** TOD. (§ X). – Close to **S. mutabilis** and perhaps a hybrid. – Mat-forming; **shoots** c. 4 cm tall, glabrous, with 4 obtuse angles and with projecting T.; **Fl.** at the base of the shoots, Cor. c. 8 cm ⌀, flat, annulus 18 mm across, pentagonal, lobes broadly ovate, tapered, 25 mm long and across, with irregular, greenish-red, discontinuous transverse lines, and ciliate margins.

**S. arenosa** LUCKH. (§ II). – Cape: Clanwilliam D. – Resembles **S. stultitioides**; **Fl.** with Cor. 3 cm ⌀, thickened inside, annulus not recognizable, lobes 9–10 mm long, 5 mm across below, margins recurved, dark purple to black, with a white zone around the centre.

**S. arnotii** N. E. BR. (§ VI). – Cape: W. Griqualand. – **Shoots** erect, up to 20 cm tall, up to 2.5 cm ⌀, with 4 compressed, green angles, with small T., softly hairy; **Fl.** 2–3 together, Cor. flat, c. 10 cm ⌀, lobes lanceolate, acute, 3.5 cm long, 18–20 mm across, light wine-coloured, the lower half with pale purplish-red hairs, the upper half glabrous, with minute, transverse wrinkles, the tips blackish, the margins revolute, with long white hairs.

**S. asterias** MASS. v. **asterias** (§ VI) (*S. stellaris* HAW., *S. stellata* ST. LAGER). – Cape: Karroo. – **Shoots** crowded, 10–20 cm tall, 2 cm ⌀, with 4 compressed, light green, softly hairy angles, with small T.; **Fl.** 1–5 together, Cor. flat, rotate, 10–11 cm ⌀, lobes projecting star-like, often twisted, narrowly lanceolate, long-tapered, 4–4.5 cm long, 15–18 mm across, dark reddish-brown inside, with minute transverse wrinkles, glossy, with fine white or yellowish, rather undulating, transverse lines, the base sparsely red-hairy, the margins recurved, densely reddish-ciliate.

S. — v. **gibba** N. E. BR. – **Cor.** c. 7.5 cm $\varnothing$, pale to dark wine-red, lobes more acute, 31 mm long, 16 mm across, margins with 5 ochre-coloured or greenish-yellow stripes 3 mm across, directed towards the indentations, without transverse stripes, hairs reddish.

S. — v. **lucida** (DC.) N. E. BR. (*S. lucida* DC., *S. stellaris* JACQ.). – **Cor.** 9 cm $\varnothing$, concolorous purple or purplish-brown inside, very glossy, with transverse wrinkles but no light lines, lobe-tips dull greenish-yellow.

**S. auobensis** NEL (§ III). – S.W.Afr.: Gr. Namaqualand. – **Shoots** 8–11 cm tall, 1 cm $\varnothing$, grey-green, red-spotted, 4-angled, with spine-like T. 10–12 mm long; **Fl.** 5–6 together, Cor. 5 cm $\varnothing$, lobes ovate-linear, 13 mm across, blackish-purple inside, with a purple-spotted, yellow, papillose zone around the tube, the papillae red-tipped.

**S. ausana** DTR. (§ V). – S.W.Afr.: Gr. Namaqualand. – Certainly close to **S. dinteri**; **Fl.** 2–2.5 cm $\varnothing$, flat, yellow, densely brown-spotted, lobes with a very dark red border.

**S. barkleyi** N. E. BR. (§ X). – Cape: Lit. Namaqualand. – **Shoots** 7–10 cm tall, c. 2 cm $\varnothing$, with 4 softly hairy angles, T. stout, projecting; **Fl.** solitary or paired, Cor. 12–16 mm $\varnothing$, flat, underside glabrous, with a stout annulus with 5 indentations, with scattered purple hairs inside and outside, lobes tapered, the upperside with numerous yellow, transverse lines, the tips reddish-brown, the margins ciliate, the ring lighter, with minute, yellow lines.

**S. bella** BGR. (§ IX). – Probably a hybrid: **S. revoluta** × **S. deflexa**. – **Shoots** erect, 15 to 20 cm tall, 15 mm $\varnothing$, with 4 light green angles, somewhat reddish towards the tips, T. projecting, short, pointed; **Fl.** several together, Cor. c. 5 cm $\varnothing$, firm-fleshy, much thickened around the throat, tube narrow, 5-angled, pale inside, sparsely hairy, the limb glossy, dull brownish-red, lobes darker, with flat, transverse wrinkles, triangular-ovate, completely revolute, margins with light brown, ciliate hairs 6–7 mm long. There is an unscented cultivar.

**S. bergerana** DTR. (Pl. 134/2) (§ VI). – S.W. Afr.: Gr. Namaqualand. – **Shoots** 7–10 cm tall, 1.5–2 cm $\varnothing$, with 4 hirsute, alate-compressed angles; **Cor.** flat, rotate, c. 7.5 cm $\varnothing$, lobes ovate, elongated into a long point, underside light reddish-green with 5 darker veins, upperside smooth, slightly glossy, violet to brown, margins with long purple hairs.

**S.** × **bicolor** (DAMM.) BGR. (§ X) (*S. mutabilis* v. *b.* DAMM.). – Garden hybrid of **S. mutabilis**; shoots as **S. variegata**; **Fl.** 6 cm $\varnothing$, lobes broadly triangular-ovate, 18 mm long, 17 mm across, dark brown, with a few yellowish blotches in 5–6 longitudinal R., margins ciliate.

**S. bijliae** PILL. (§ X). – Cape: Mossel Bay D. – **Shoots** 6–8 cm tall, c. 1 cm $\varnothing$, with 4 rather rounded, shortly dentate angles, grey-green, smooth; **Fl.** solitary or paired, Cor. 6.5–7 cm $\varnothing$, flat, annulus inconspicuous, lobes ovate-acute, glabrous outside, slightly wrinkled inside, tube hairy, purplish-brown, with ochre-yellow lines inside along the border.

**S. choanantha** LAVR. et HALL (§ V) (*Stapelianthus c.* (LAVR. et HALL) DYER). – Cape: Calitzdorp D. – **Shoots** prostrate or pendulous, (10–)30(–200) cm long, (6–)10(–12) mm $\varnothing$, bluish-green with brownish blotches, with 4 roundish angles and no perceptible T.; **Fl.** 2–6, Cor. campanulate, tube cup-shaped to 5-angled, pink outside below, spotted above, reddish-purple inside, glabrous all over, 16 mm long, 11 mm $\varnothing$, lobes triangular, acute, 6 mm long, 7 mm across below, margins somewhat revolute, upperside velvety. (W. RAUH and H. P. WERTEL do not accept DYER's transfer to **Stapelianthus**.)

**S. cincta** MARL. (§ V). – Cape: Beaufort West D. – Shoots $\pm$ square, 3–4 cm tall, 10–20 mm $\varnothing$, sides obtuse, almost flat, T. 0.5 mm long; **Fl.** solitary, Cor. 24–36 mm $\varnothing$, the tube hemispherical, lobes rather longer than the tube and 1 cm long, 5–6 mm across, ovate-tapered, rough, not hairy, ochre-coloured, with dense, chestnut-brown blotches and a concolorous border 1.5 to 2 mm wide.

**S. clavicorona** VERDOORN (Pl. 134/4) (§ VII). – Transvaal: Soutpansberg D. – **Shoots** erect, c. 30 cm tall, with 4 compressed angles with velvety, soft hairs and stout T.; Cor. 6 cm $\varnothing$, velvety-haired outside, inside glabrous apart from a few white hairs in the centre, lobes ovate, tapering abruptly to become very thin, 17 mm long, light yellow with purple transverse lines, the margins ciliate in the lower half.

**S. comparabilis** WHITE et SLOANE (§ IX). – Possibly a hybrid; closely resembles **S. incomparabilis**; **Fl.** 9–12 cm $\varnothing$, paler.

**S. concinna** MASS. v. **concinna** (§.V). – Cape: Karroo. – **Shoots** slender, c. 10 cm tall, 6–8 mm $\varnothing$, with 4 green, often reddened, minutely hairy angles with small T.; **Cor.** flat, 3.5 cm $\varnothing$, lobes ovate-tapered, dirty yellow with minute, red, transverse stripes, with reticulate markings at the tips, densely white-hairy all over.

S. — v. **paniculata** (WILLD.) N. E. BR. (*S. p.* WILLD., *Tridentea p.* HAW.). – **Fl.** dull purplish-brown, lobes without yellow transverse lines.

**S. conformis** N. E. BR. v. **conformis** (§ VI). – Cape: Albany, Somerset E., Uitenhage, Bedford D.; ? Natal: Zululand. – **Shoots** erect, 12–25 cm tall, 2–3 cm $\varnothing$, with 4 very compressed, green, often reddish, dentate angles; Cor. 9–10 cm $\varnothing$, with velvety hairs outside, the base plate-shaped, lobes lanceolate, acute, 2.5–3.5 cm long, 16–18 mm across, with minute, transverse wrinkles on the inside apart from the tips, with hairs around the tube and sparsely white-ciliate margins, sulphur-yellow or greenish-yellow below, with irregular, reddish-brown transverse stripes, the margins and the upper half reddish-brown all over, with a yellowish or greenish blotch at the tips.

S. — v. **abrasa** N. E. BR. – **Cor.** not ciliate, colour more intense, lobes for 2/3 of the length with yellowish, transverse stripes, margins with a reddish-brown line 2 mm wide.

**S. × cupularis** N. E. Br. (§ X). – Hybrid: **S. mutabilis × S. variegata.** – Fl. 1–3 together, Cor. spreading, 4.5 cm ⌀, glabrous inside, rough outside, pale lemon-yellow with dense, purple blotches, irregularly confluent into lines, lobes recurved, c. 1 cm long and across, ovate, acute, the margins with clavate hairs.

**S. cylista** Luckh. (§ VI). – Cape: Van Rhynsdorp D. – **Shoots** erect, 14–24 cm tall, 13 mm ⌀, with 4 bluntly rounded, light green, minutely velvety-haired angles and small, indistinct T.; **Fl.** solitary, Cor. 2 cm ⌀, incised 2/3 of the way down, cup-shaped and 6 mm ⌀ below, lobes tapered, 8.5 cm long, 3.5 cm across, flaccid and drooping, light purple outside, indistinctly veined, minutely papillose, ochre-yellow inside, with broken, dull red, transverse lines all over, and light purple hairs towards the tips.

**S. deflexa** Jacq. v. **deflexa** (§ V) (*S. reflexa* Haw., *S. brevirostris* Willd.). – S.Afr.: origin unknown. – **Shoots** erect or ascending, 15–20 cm tall, 10–15 mm ⌀, with 4 minutely hairy angles, with pointed T.; **Fl.** several together, Cor. flat, 6–7 cm ⌀, divided to midway, lobes ovate-lanceolate, long-tapered, very recurved, pale green outside, with reddish longitudinal stripes, brick-red to dirty purplish-brown inside, with a rudimentary annulus below, with minute transverse stripes and scattered hairs, margins ± revolute, with purple cilia.

**S.** – cv. **Atropurpurea.** – **Fl.** darker, with whitish lines between the transverse wrinkles, cilia red and white.

**S.** – v. **brownii** Schinz. – **Shoots** more slender; **Fl.** somewhat smaller, brownish.

**S. desmetiana** N. E. Br. v. **desmetiana** (§ VI). – S.E.Cape: widespread. – **Shoots** ascending, up to 25 cm tall, 3–3.5 cm ⌀, the 4 alate angles distantly sinuate, dentate, puberulous; **Fl.** 2–10 together, Cor. 12–16 cm ⌀, 5-cleft to more than midway, lobes oblong-lanceolate, tapered, 5–7 cm long, 2.5–3 cm across, green or reddened outside, pubescent, pale to brownish red inside, transversely wrinkled, with yellow transverse lines over 2/3 of the length and 5–7 mm long hairs, margins long-ciliate.

**S.** – v. **apicalis** N. E. Br. – Lobes not hairy in the upper part, the yellow transverse lines often extending to the tips.

**S.** – v. **fergusoniae** R. A. Dyer. – Cor. c. 12 cm ⌀, lobes ovate, 4.5 cm long, lemon-yellow white-hairy, not wrinkled all over, rather thick.

**S.** – v. **pallida** N. E. Br. – **Shoots** very robust; lobes purplish-red along the margins and at the tips, otherwise greenish-yellow, without markings, hairs pale grey or whitish.

**S. dinteri** Bgr. v. **dinteri** (§ V). – S.W.Afr.: Gr. Namaqualand. – **Shoots** 8–12 cm tall, 1.5 cm ⌀ below, tapering above, the 4 obtuse angles divided into oblong zones, dull grey-green with brown blotches and spots; **Cor.** 3 cm ⌀, the indentations 5 mm deep, lobes 10 mm across, underside greenish-brown and slightly glossy, with 5 veins, upperside greenish-yellow with dense reddish-brown spots, margins 3 mm across, red, with reddish-brown cilia in the indentations.

**S.** – v. **capensis** Luckh. – Cape: Lit. Namaqualand. – **Cor.** 2.3 cm ⌀, tube flat, with minute purplish-black blotches, lobes with purplish-black borders, not ciliate.

**S.** – v. **pseudocapensis** Luckh. – Cape: Lit. Namaqualand. – **Cor.** 2 cm ⌀, tube deep, covered with scattered, purplish-brown blotches of varying shape, lobes with a narrow, purplish-black border.

**S. discoidea** Oberm. (§ I). – ? Rhodesia. – **Shoots** glabrous, 4-angled, grey-green with red blotches, T. 1 cm long, pointed; **Fl.** solitary, rotate, 4.5 cm ⌀, dark brown in the centre, with a distinct annulus, lobes broadly ovate, 17 mm long, 10 mm across, with distinct transverse folds and dark brown blotches, larger and confluent towards the base, margins with dark red cilia.

**S. × discolor** Tod. – Garden hybrid of **S. mutabilis**; **shoots** 6–12 cm tall, glabrous, with 4 obtuse, projecting, dentate angles; **Cor.** over 5 cm ⌀, annulus pentagonal, 2 cm ⌀, yellowish, slightly red-striate, lobes short-ovate, 18 mm long, 15 mm across below, 18 mm across midway, dull purple, somewhat transversely wrinkled, margins ciliate.

**S. divaricata** Mass. (§ V) (*Gonostemon d.* Haw., *S. pallida* Wendl., *S. pallens* hort.). – **Shoots** slender, 10–18 cm long, 8–12 mm ⌀, tapered above, with 4 blunt, little compressed, fresh green, often reddened angles, T. small, projecting; **Cor.** 3–5 cm ⌀, rotate, deeply 5-cleft, lobes lanceolate, tapered, smooth inside, glossy, pale flesh-coloured, paler around the short, flat tube, the tips green, margins recurved, minutely ciliate.

**S. dwequensis** Luckh. (§ V). – Cape: Ceres D. – **St.** somewhat branched, with 4 acute, green angles with reddish blotches; **Fl.** 2–4 together, Cor. 41 mm ⌀, tube campanulate, 12 mm ⌀ and deep, smooth outside, light green, with broken R. of red spots on the lemon-yellow, spreading, triangular, acute lobes which are 13–15 mm long, 12 mm below, the tube with smaller spots, minutely papillose.

**S. englerana** Schltr. (Pl. 134/3) (§ VII). – S.E.Cape. – **St.** prostrate to ascending, little branched, 15–25 cm tall, 11–13 mm ⌀, with 4 slightly sinuate-dentate angles, grey-green, minutely pubescent; **Fl.** rather fleshy, 5-angled and plate-shaped below, with transverse furrows and wrinkles, chocolate-brown, lobes triangular-ovate, c. 14 mm long, 12 mm across, completely recurved and appressed against the tube, smooth only towards the tips, glabrous apart from small brown hairs along the margins.

**S. erectiflora** N. E. Br. (§ V). – Cape: Clanwilliam D. – **Shoots** slender, 10–15 cm long, c. 1 cm ⌀, with 4 obtuse, pubescent angles, with small blunt T.; **Fl.** in clusters of 2–4, erect, Cor. 5-cleft to midway, lobes revolute and resembling a Turkish fez, 10–13 mm ⌀, purple, with dense white hairs.

**S. flavirostris** N. E. Br. (§ VI) (*S. grandiflora* v. *lineata* N. E. Br.). – Cape: widespread; Lesotho. – **Shoots** erect, up to 17 cm tall, 2–3 cm ⌀, with 4 compressed angles, and small erect T., minutely velvety-pubescent; **Fl.** 1–3 together, Cor. 13–16 cm ⌀, deeply divided, tube short, flat, lobes lanceolate, acute, revolute, 5–7 cm long, 2–3 cm across, softly hairy outside, dull purplish-red, minutely wrinkled all over, the lower half with pale yellow or dull purple, transverse lines, the tips concolorous dull purple, the bottom 1/3 laxly hairy, the margins with red or white cilia.

**S. flavopurpurea** MARL. v. **flavopurpurea** (§ V). – S.E.Cape. – **Shoots** 5–6 cm tall, 12 to 14 mm ⌀, 4-angled, with small, scale-like T.; Cor. 3 cm ⌀, deeply 5-cleft, flat-rotate, lobes linear-lanceolate, 15 mm long, 7 mm across, dull yellow, with irregular, transverse wrinkles, margins revolute, tube flat, short, whitish, with red hairs.

**S. — v. fleckii** (BGR. et SCHLTR.) WHITE et SLOANE (*S. f.* BGR. et SCHLTR.). – S.W.Afr.: Gr. Namaqualand. – **Fl.** 1–4 together, honey-scented, with denser hairs at the tube-mouth, lobes ochre-yellow to greenish-yellow, often pink or reddish.

**S. forcipis** PHILL. et LETTY (§ VI). – Cape: Port Elizabeth D. – **Shoots** c. 14 cm tall, laxly branched, angles very compressed, smooth, light green; **Fl.** solitary, Cor. 7 cm ⌀, green and smooth outside, with whitish-brown hairs inside, lobes ovate, tapered, 28–30 mm long, 18 mm across, with similar hairs below, margins with long, white cilia.

**S. fucosa** N. E. BR. (§ II). – S.E.Cape: Pondoland. – **Shoots** erect, 4–6 cm long, 6 to 9 mm ⌀, very obtusely 4-angled, with pointed T., dark green with purplish-brown blotches; Cor. c. 3 cm ⌀, smooth outside, wrinkled inside, glabrous, flat below, with a 5-angled annulus 8–9 mm across, lobes forming ± an equilateral triangle, 9–10 mm long and wide, with numerous, very dark red blotches 1 mm across, the annulus with blotches denser and ± confluent, margins very deep red.

**S. fuscopurpurea** N. E. BR. (§ VI). – S.Afr.: origin unknown. – **Shoots** erect, 15–20 cm tall, 2–2.5 cm ⌀, with 4 compressed angles and small T., puberulous; Cor. c. 10 cm ⌀, puberulous outside, lobes ovate-lanceolate, acute, dark reddish-brown, minutely transversely wrinkled, with a cushion of long, purple hairs around the tube-mouth, and long cilia along the revolute margins, otherwise glabrous.

**S. gariepensis** PILL. (§ VI). – Cape: Lit. Namaqualand. – **Shoots** prostrate below, erect above, with 4 compressed angles and small T., puberulous; Cor. 8–9 cm ⌀, lobes ovate-lanceolate, tapered, 35–38 mm long, 15–17 mm across, smooth outside, with transverse wrinkles inside, glossy reddish-purple, with distinct, yellow, transverse lines below and dense purple hairs around the tube-mouth, margins with purple cilia.

**S. gemmiflora** MASS. v. **gemmiflora** (Pl. 135/1) (§ III) (*Tridentea g.* HAW., *T. stygia* HAW., *S. s.* SCHULT.). – S.E.Cape, Orange Free State; ? S.W.Afr. – **Shoots** 7–15 cm tall, 15–18 mm ⌀, slightly grey-green, with 4 obtuse angles with erect T.; **Fl.** 1–4 together, Cor. 8 cm ⌀, tube flat, limb rotate, projecting, with dense, minute, transverse wrinkles inside, lobes ovate, acute, 3 cm long, 2 cm across, distinctly 5-veined, blackish-brown or violet-brown, indistinctly marbled, often with several yellow blotches at the lobe-bases, margins with white or brown cilia.

**S. — v. densa** (N. E. BR.) N. E. BR. (*S. hircosa* v. *d.* N. E. BR.). – Cape. – Cor. greenish-yellow, with dense, dark brown blotches except on the margins.

**S. — v. hircosa** (JACQ.) N. E. BR. (*S. hircosa* JACQ., *S. moschata* J. DONN., *Tridentea m.* HAW., *S. hircola* POIRET., *T. h.* SCHULT.). – Cor. with a lighter, yellowish-brown base, with blackish-brown marbling.

**S. gettleffii** POTT. (Pl. 135/2) (§ VI). – Transvaal; ? Botswana. – **St.** erect, branching basally, 20–25 cm tall, 9–13 mm ⌀, velvety-hairy, with 4 dentate angles; **Fl.** 1–3 together, Cor. 14–16 cm ⌀, deeply 5-cleft, greenish-yellow outside, hairy inside, lobes lanceolate, acute, 7 cm long, 2.5 cm across midway, purple inside, with transverse wrinkles, yellowish lines from the base to the tips, and dense, light purple hairs, margins with white or light purple cilia, lobes and tube with yellow blotches.

**S. gigantea** N. E. BR. v. **gigantea** (§ VI). – Natal, Transvaal; Rhodesia. – **Shoots** robust, 15–20 cm tall, over 3 cm ⌀, light green, with alate, compressed angles and small T.; **Fl.** solitary or paired, Cor. 25–35 cm ⌀, flat, deeply divided, tube short, lobes triangular to long tapering, light yellow with numerous minute, thin, rather wavy, red, transverse wrinkles and scattered, reddish hairs, margins ± recurved, with long, white hairs.

**S. — v. pallida** PHILL. – ? Transvaal. – **Fl.** solitary, Cor. 25 cm ⌀, lighter than the type. – ? Hybrid: **S. gigantea** × **S. nobilis**.

**S. glabricaulis** N. E. BR. (§ VI). – Cape: King William's Town, Bathurst D., W.Transkei (Tembuland). – **Shoots** erect, glabrous, 15 to 20 cm tall, dark green, often reddened, with 4 compressed angles, T. small; **Fl.** 3–4 together, Cor. 6–8.5 cm ⌀, 5-cleft to more than midway, tube flat, lobes ovate-lanceolate, tapered, underside glabrous, greenish, with 5 reddish, longitudinal veins, brownish inside with transverse wrinkles, often dull yellow towards the base, or with a broad cushion of long, reddish-brown hairs, lobes glabrous apart from the base and the margins with long red cilia.

**S. glanduliflora** MASS. v. **glanduliflora** (§ V) (*S. glandulifera* WILLD., *S. hispidula* HORNEM., *S. glanduliflora* v. *massonii* BGR., *S. g.* v. *haworthii* BGR.). – Cape: Lit. Namaqualand. – **Shoots** erect, with 4–6 angles, 9–15 cm tall, 12–20 mm ⌀, T. small; **Fl.** solitary or several together, Cor. rotate, 3–3.5 cm ⌀, tube short, lobes ovate-tapered, greenish outside, the base and the lobes greenish with numerous red spots, stripes, transverse wrinkles and white hairs, margins with white cilia.

**S. — v. emarginata** N. E. BR. – Lobes oblong, the tips emarginate or shortly bifid.

**S. grandiflora** Mass. (Pl. 135/4) (§ VI) (*S. spectabilis* Haw., *S. obscura* N. E. Br.). – S.E.Cape, Transvaal. – **Shoots** clavate, 20 to 30 cm tall, 3–4 cm ⌀, densely soft-hairy, with 4 alate, compressed angles, T. distant, erect; **Fl.** 1–3 together, Cor. 15–16 cm ⌀, flat, deeply cleft, tube short, broad, lobes triangular-lanceolate, tapered, transversely wrinkled, tufted, the upper part smooth with reddish or white cilia, underside blue-green, upperside blackish-purple.

**S. × hanburyana** Rüst. et Bgr. (§ X). – Certainly a hybrid of **S. variegata; shoots** up to 10 cm tall, 10–15 mm ⌀, the rounded angles separated by acute furrows, green or grey-green, with stout T.; **Fl.** 1–3 together, with pedicels 4–5 cm long, Cor. 7.5 cm ⌀, lobes triangular-ovate, tapered, soon recurved, upperside transversely wrinkled, with impressed longitudinal veins, whitish-yellow, covered with round, reddish-brown spots merging into lines, the margins darker, with dark cilia, the annulus plate-shaped, 5 mm high, 24 mm across, rather lighter, whitish-yellow.

**S. herrei** Nel (§ VII). – Cape: Lit. Namaqualand. – **Shoots** 7–12 cm tall, twice as thick at the base as at the tip, glabrous, brownish-green, 4-angled; **Fl.** solitary, Cor. 3–4 cm ⌀, campanulate, lobes lanceolate, acute, 1.5 cm long, 10–15 mm across, very wrinkled inside, the wrinkles white, the spaces between light brownish-purple, margins less wrinkled.

**S. hirsuta** L. v. **hirsuta** (§ VI) (*S. pulvinata* J. Donn., *S. sororia* Lodd., *S. laxiflora* Haw., *S. hirsuta* v. *afra* Lindl., *S. lanigera* Loud., *S. villosa* N. E. Br.). – S.E.Cape, Lit. Namaqualand. – **Shoots** slender, up to 20 cm tall, 10 to 15 mm ⌀, sides rather depressed, dirty green, pubescent, T. small; **Fl.** 1–3 together, Cor. deeply divided, lobes ovate-lanceolate, long-tapered, c. 5 cm long, 2.5 cm across, with brownish-red hairs below, up to the middle and along the margins, bluish-green outside, dirty reddish-yellow inside up to midway, with minute, red, ± undulating, transverse stripes, dull red towards the tip.

**S. —** v. **affinis** (N. E. Br.) N. E. Br. (*S. a.* N. E. Br., *S. stellaris* Lodd.). – Cor. 10–11 cm ⌀, hairs in the middle denser, lobes darker brownish-red, with cream coloured transverse stripes, margins with red long cilia.

**S. —** v. **comata** (Jacq.) N. E. Br. (*S. c.* Jacq.). – Cape: Malmesbury, Robertson D. – Cor. as v. **patula;** hairs in the middle denser, woollier and darker purple.

**S. —** v. **depressa** (Jacq.) N. E. Br. (*S. d.* Jacq., *S. sororia* Hook. f., *S. patentirostris* N. E. Br., *S. courcellii* hort. ex N. E. Br., *S. patula* v. *d.* N. E. Br., *Tridentea d.* Schult.). – Segments of the inner corona deeply divided.

**S. —** v. **grata** N. E. Br. – **Shoots** very robust; Cor. 9–10 cm ⌀, lobes broadly ovate, tapering, dark purplish-brown at the tips, the bottom ⅔ lighter, with creamy-yellow, transverse lines, hairs light purple, longer and denser than v. **patula.**

**S. —** v. **longirostris** (N. E. Br.) N. E. Br. (*S. patula* v. *l.* N. E. Br.). – As v. **depressa;** horns of the inner corona very long, patent.

**S. —** v. **lutea** N. E. Br. – **Shoots** pale green; Cor. c. 9 cm ⌀, lobes greenish outside, light yellow inside, without markings, with yellow hairs.

**S. —** v. **patula** (Willd.) N. E. Br. (*S. patula* Willd., *S. sororia* Jacq., *S. elongata* Sweet, *S. variegata* Gouas., *S. rufescens* hort.). – As the type; **Fl.** c. 9 cm ⌀, with cushions of hair only around the tube-mouth, lobes glabrous.

**S. immelmaniae** Pill. (§ VI). – Cape: Piquetberg D. – **Shoots** erect, 12–15 cm tall, 1.5 cm ⌀, with 4 slightly compressed, rounded angles, bluish-green, minutely velvety, T. projecting; **Fl.** 1–6 together, Cor. 4–4.5 cm ⌀, with thin, velvety hairs outside, with purple blotches, tube flat, 11–13 mm ⌀, transversely wrinkled, with purple hairs, olive-green with yellow lines, lobes ovate-lanceolate, 17–19 mm long, 9–10 mm across, purple with yellow transverse lines, with reddish hair, margins with red cilia 1–2 mm long.

**S. incomparabilis** N. E. Br. (§ IX). – S.Afr.: origin unknown. – **Shoots** 8–15 cm tall, 8 to 12 mm thick, 4-angled, minutely pubescent, T. 2–3 mm long; **Fl.** 3–4 together, Cor. 7.5 to 8.5 cm ⌀, broadly cup-shaped and 2.5–3 cm ⌀ below, 12 mm deep, with a 5-angled tube below, lobes ovate, tapered, 3 cm long, glabrous outside, wrinkled inside, purplish-red with the individual wrinkles yellowish or whitish, margins with red clavate hairs.

**S. indocta** Nel (§ V). – Cape: Lit. Namaqualand. – **Shoots** 10–16 cm tall, 16–18 mm ⌀ below, tapering above, the 4 ribs rounded, with short, white hairs, tubercles with subulate T.; **Fl.** mostly in clusters of 4, Cor. rotate, 25 to 30 mm ⌀, tube 5–6 mm long, with slight, greenish-yellow wrinkles interspersed with purplish-red spots and stripes, lobes ovate-lanceolate, sharply tapered, 14 mm long, 8 mm across, wrinkled only below, tipped with white hairs 1 mm long.

**S. jucunda** N. E. Br. v. **jucunda** (§ V). – Cape: W.Griqualand, Herbert D. – **Shoots** 5–8 cm tall, 8–15 mm ⌀, with 4 blunt angles, green or greyish-green, suffused red, T. stout, subulate, sharp, deciduous; **Fl.** 1–3 together, Cor. ± flat, 5-cleft to midway, both surfaces glabrous, lobes triangular-ovate, acute, 6–7 mm long and across, cream-coloured, slightly transversely wrinkled, with reddish-brown blotches, margins ciliate.

**S. —** v. **deficiens** N. E. Br. – **Fl.** greenish to primrose-yellow, with numerous small, purplish-brown blotches, margins with purple hairs up to midway.

**S. juttae** Dtr. (§ II). – S.W.Afr.: Gr. Namaqualand. – **Shoots** 5–12 cm tall, 1–2 cm ⌀, tapering above, minutely hairy; **Fl.** 8–10 together, Cor. 21 mm ⌀, lobes c. 8 mm long, 4 mm across, brown to black, coarsely and transversely wrinkled above, minutely hairy outside, margins recurved.

**S. kwebensis** N. E. BR. v. **kwebensis** (§ II). – Botswana; Transvaal. – **Shoots** c. 13 cm tall, c. 13 mm $\varnothing$, 4-angled, minutely papillose, with short T.; **Fl.** few, Cor. 25–32 mm $\varnothing$, minutely hairy outside, chocolate or chestnut-brown to ochre-coloured inside, the tube small, with a small annulus above, lobes broadly ovate, tapered, 8–11 mm long, 6–8 mm across, transversely wrinkled, minutely hairy above.

**S. —** v. **longipedicellata** BGR. (Pl. 135/3) (*S. l.* N. E. BR.). – **Fl.** long-pedicellate, lobes blackish-brown inside, with numerous dark, $\pm$ undulating transverse wrinkles.

**S. leendertziae** N. E. BR. (§ VI). – Transvaal. – **Shoots** laxly arranged, 8–10 cm tall, 10 to 12 mm $\varnothing$, 4-angled, with slightly furrowed sides, T. small; **Fl.** mostly paired, Cor. 7.5 to 10 cm $\varnothing$, tube cup-shaped, 7–8 cm long, lobes spreading, 6–7 cm long, 3.5 cm across, triangular, terminating in a thin tip, minutely hairy outside, rough and transversely wrinkled inside, with sparse purple hairs around the tube, otherwise dark purple or blackish-purple.

**S. lepida** JACQ. (§ X) (*Podanthes l.* HAW., *Orbea l.* HAW., *S. limosa* S. D.). – S.Afr.: origin unknown. – **Shoots** 4–7 cm long, c. 1 cm $\varnothing$, glabrous; **Fl.** solitary or paired, Cor. c. 3 cm $\varnothing$, the lobes rough inside, the annulus tuberculate, glabrous, sulphur-yellow, with irregular small blotches, otherwise similar to **S. variegata**.

**S. longidens** N. E. BR. (§ I). – Moçambique. – **Shoots** 6–15 cm tall, 10 mm $\varnothing$, with 4 blunt angles, glabrous, green, suffused reddish, T. long-tapered, 12–15 mm long, soft, fleshy; **Fl.** 3 together, Cor. 3.5–4 cm $\varnothing$, tube campanulate, lobes ovate-lanceolate, acute, 14–16 mm long, 9–10 mm across, pale greenish-yellow with reddish-brown blotches, more closely spaced towards the tips.

**S. longii** LUCKH. (§ VIII). – Cape: Hopetown, Jansenville D. – **Shoots** 23 cm tall, 6 mm $\varnothing$, the 4 angles very rounded, glabrous, green, younger shoots curving downwards, T. small; **Fl.** solitary, Cor. 24 mm $\varnothing$, glabrous, green outside, light brown inside, with a few wrinkles and yellowish lines around the mouth of the indistinct tube, lobes 7.5 mm long, margins with purple cilia.

**S. longipes** LUCKH. v. **longipes** (§ VIII). – S.W.Afr.: Witpütz. – Forming compact clumps 40 cm $\varnothing$; **St.** irregularly branched, ascending to prostrate, 5–12 cm tall, 6–16 mm $\varnothing$, with 4 obtuse, scarcely dentate angles, glabrous, greyish-green, often reddened; **Cor.** 6 cm $\varnothing$, densely wrinkled inside, lobes 24 mm long, 8 mm across, widening towards the middle and then tapered, the upper $\frac{2}{3}$ purplish-black, below with white blotches and lines, margins recurved, with red cilia below.

**S. —** v. **namaquensis** LUCKH. – Cape: Lit. Namaqualand. – **Cor.** 7 cm $\varnothing$, lobes rather narrower, less blotched and lined.

**S. maccabeana** WHITE et SLOANE (§ VI). – S.Afr.: origin unknown. – **Shoots** c. 28 cm tall, 3–4 cm $\varnothing$, with 4 very compressed angles, dark green, minutely velvety-hairy, T. with incurved, deciduous, small L.; **Cor.** 15 cm $\varnothing$, the underside minutely hairy and veined, pallid green, reddish above, lobes lanceolate, tapered, 6.5 cm long, 2.2 cm across, greenish-yellow inside, with numerous purple lines, the lower $\frac{2}{3}$ rough, with sparse hairs around the tube-mouth, the margins greenish-yellow above, with 5 distinct, light lines.

**S. macowanii** N. E. BR. (§ VI). – Cape: Albany, Jansenville D. – **Shoots** 15–30 cm tall, 2.5–3 cm $\varnothing$, with 4 alate-compressed angles, distantly dentate, softly hairy; **Fl.** 3–4 together, Cor. 5–7 cm $\varnothing$, pubescent outside, glabrous inside, the tube somewhat 5-angled, with 5 furrows running towards the indentations, lobes ovate, short-tapered, transversely wrinkled, pale greenish-white, with fine, wine-coloured, transverse lines.

**S. maculosa** J. DONN. (§ X) (*S. mixta* J. DONN, *Orbea m.* HAW., *S. maculata* POIRET). – ? Cape. – Possibly a hybrid; **shoots** up to 15 cm tall, with 4 rounded angles, greyish-green, T. 2–3 mm long; **Fl.** solitary or up to 3 together, Cor. flat, c. 7 cm $\varnothing$, lobes triangular-ovate, greenish-yellow, with purplish-red transverse lines and spots, brownish at the tips, the annulus small, rather lighter, margins ciliate.

**S. maculosoides** N. E. BR. (§ X). – Probably a hybrid; **shoots** 8–10 cm tall, 6–8 mm $\varnothing$, the 4 angles rather rounded, dark green, suffused reddish-brown, glabrous, with small T.; **Fl.** 6–8 together, Cor. rotate, 7 cm $\varnothing$, 5-lobed to midway, lobes ovate-lanceolate, glossy outside, pale green, slightly veined, inside with a 5-angled tube 10–12 mm across, its margin somewhat thickened, dark blackish-brown below, with a few brown hairs and yellowish transverse wrinkles, lobes with numerous transverse wrinkles above, the tips and margins concolorous brown.

**S. marientalensis** NEL (§ III). – S.W.Afr.: near Mariental. – **Shoots** 5–7 cm tall, 8 mm $\varnothing$, with 4 prominent angles, tapered above, T. 4 mm long; **Fl.** solitary, Cor. 4.5 cm $\varnothing$, yellowish-green outside, smooth, with 6 dark green ribs, lobes long-triangular, 2.5 cm long, 15 mm across, light yellow below, with numerous yellowish papillae forming an annulus around the tube-mouth, then with purple zones, deep blackish-purple towards the tips, the margins long-ciliate.

**S. marlothii** N. E. BR. (§ VI). – Rhodesia: Matopo Hills. – **Shoots** 10–15 cm tall, 13 to 25 mm $\varnothing$, with compressed angles, minutely hairy; **Fl.** solitary, Cor. 10 cm $\varnothing$, minutely papillose outside, wrinkled and hairy inside, with a short tube 15 mm $\varnothing$, lobes long-lanceolate, very tapered, 4.5 cm long, 13 mm across, hairy.

**S. molonyae** WHITE et SLOANE (§ I). – Kenya: near Nairobi. – **Shoots** 5–10 cm tall, 15 to 20 mm $\varnothing$, with brownish blotches and long T.; **Fl.** 1–3 together, Cor. 4–6 cm $\varnothing$, lobes ovate-triangular, long-tapered, 15–25 mm long, 8 mm across, deep chestnut-brown, with golden-yellow blotches above, with a distinct, flat annulus around the tube, margins purple-haired.

**S. mutabilis** JACQ. (§ X) (*Orbea m.* SWEET, *S. neglecta* TOD., *S. passerinii* TOD., *S. fuscata* HORT., *S. umbiciliata* THURET). – Origin unknown. – **Shoots** up to 15 cm tall, robust, greyish-green, glabrous, with projecting T.; **Fl.** 5 together, Cor. 7 cm $\varnothing$, lobes triangular-ovate, greenish-yellow, with crowded transverse lines and spots, the tips brownish, the annulus small, rather lighter, margins ciliate. Hybrids: **S.** × **bicolor,** × **cupularis,** × **discolor,** etc.

**S. namaquensis** N. E. BR. v. **namaquensis** (§ X) (*S. n.* v. *minor* N. E. BR.). – Cape: Lit. Namaqualand. – **Shoots** creeping or prostrate, 3–8 cm long, 10–15 mm $\varnothing$, with 4 roundish angles, glabrous, green, reddish-striate, T. c. 5 mm long; **Fl.** 1–4 together, Cor. 8–10 cm $\varnothing$, flat, in the centre a 5-angled, thick annulus with its border rolled back to the Fl.-base, the tube with purple hairs inside, lobes broadly ovate, ± long-tapered, 2.5–3 cm long, 2–2.5 cm across, smooth outside, tuberculate-wrinkled inside, pale green to yellow, with purple transverse lines all over, or blotches merging to form an intricate pattern.

**S. —** v. **bidens** N. E. BR. – Lobes abruptly narrowed to a slender tip, minutely ciliate, more blotched.

**S. —** v. **ciliolata** N. E. BR. – Lobe-margins short-ciliate.

**S. neliana** WHITE et SLOANE (§ VII) (*S. tigrina* NEL). – Cape: Lit. Namaqualand. – **Shoots** 10–12 cm tall, 1.5 cm $\varnothing$ below, rather tapered above, glabrous, greyish-green, purple above, with 6 roundish angles, and T. c. 14 mm apart; **Fl.** 2–3 together, Cor. 6 cm $\varnothing$, broadly campanulate, lobes 2 cm long, glabrous outside, greenish-yellow, with greenish-yellow wrinkles inside, the ground-colour between them brownish-purple.

**S. nobilis** N. E. BR. (§ VI). – Transvaal, ? Natal; Moçambique. – **Shoots** branched freely below, 10–15 cm tall, 2 cm $\varnothing$, 4-angled, light green, somewhat hairy, with small T.; **Fl.** solitary or paired, Cor. 20–25 cm $\varnothing$, tube broad-campanulate, lobes ovate-lanceolate, tapered, c. 7 cm long, 3 cm across, dull red outside, minutely hairy, ochre-yellow inside, with fine, blood-red transverse lines, with scattered hairs, margins ciliate.

**S. nouhuysii** PHILL. (Pl. 136/1) (§ V). – Cape: Clanwilliam D. – Forming clumps 30 cm $\varnothing$; **shoots** c. 8 cm tall, with 4 rounded angles, minutely hairy, with small white T.; **Fl.** solitary or paired, Cor. c. 2 cm $\varnothing$, tube cup-shaped, lobes ovate, rather acute, tube and lobes densely red-haired in the lower half, smooth above and outside, the margins hairy above.

**S. nudiflora** PILL. (§ VI). – Cape: Montagu D. – **Shoots** 10–15 cm tall, 10–15 mm $\varnothing$, with 4 slightly compressed, rounded angles, bluish-green, ± glabrous; **Fl.** few in a cluster, Cor. 5 cm $\varnothing$, both surfaces glabrous, tube 6 mm long, plate-shaped, light purplish-brown with several yellow transverse lines, lobes ovate-lanceolate, tapered, transversely wrinkled, with purple lines and pale yellow, transverse lines in the lower half.

**S. olivacea** N. E. BR. (§ V) (*S. cruciformis* HORT.). – Cape: Karroo, Orange Free State. – **Shoots** slender, erect, 7–13 cm tall, c. 1 cm $\varnothing$, with 4 very round angles, minutely soft-haired, with small, appressed T.; **Fl.** 2–6 together, Cor. 3.5–4 cm $\varnothing$, deeply 5-cleft to midway, dirty green outside, with soft hairs, lobes long-ovate, tapered, glabrous inside, light or dark olive-green, or dull red, with dense, brown, transverse wrinkles.

**S. pachyrrhiza** DTR. (§ III). – S.W.Afr.: Gr. Namaqualand. – Cushion-forming, with thick roots; **shoots** 5–25, 4–7 cm tall, 1–2 cm $\varnothing$, with 4 obtuse angles, greyish-green with red blotches, or ± concolorous red; **Fl.** clustered, Cor. 7–7.5 cm $\varnothing$, lobes c. 2 cm long, 23 to 25 mm across, brownish-red outside, inside with red stripes on a yellow ground or concolorous velvety-black, margins with 3 mm long hairs.

**S. parvipuncta** N. E. BR. v. **parvipuncta** (Pl. 136/2) (§ II) (*S. parvipunctata* K. SCHUM.). – Cape: Beaufort W., Laingsburg, Prince Albert D., W.Griqualand. – **Shoots** 5–12 cm tall, 1.5–2 cm $\varnothing$, with 4 rather blunt angles, with short T.; **Fl.** clustered, Cor. rotate, 2.5–3 cm $\varnothing$, lobes triangular-ovate, with an indistinct annulus, pale green outside, with reddish blotches, glabrous inside, minutely wrinkled, pale greenish-yellow, with numerous small or larger red spots, lobes with reddish-brown margins, with reddish cilia.

**S. —** v. **truncata** LUCKH. – Lobe-margins not ciliate, lobe-tips truncate.

**S. parvula** KERS (§ II). – Angola: Moçamedes D. – Mat-forming; **shoots** erect, 1.5 to 4 cm long, 5–7 mm across, with 4 obtuse angles, green, minutely hairy, T. ascending, small, widened; **Fl.** clustered, Ped. c. 1.5 cm long, bracts subulate, 1 mm long, pedicels 1.5–2 cm long, Sep. 2 mm long, Cor. round, 5–6.5 mm $\varnothing$, green outside, minutely hairy, dark brown inside, lobes ovate-triangular, apiculate, slightly rough, 2 mm long and across, the tips minutely hairy, margins with a membranous line and roughly ciliate, outer and inner coronas black.

**S. pearsonii** N. E. BR. (§ V). – S.W.Afr.: Gr. Namaqualand. – **Shoots** 4–8 cm tall, 6–12 mm $\varnothing$, with 4 obtuse angles, minutely hairy, green or greyish-green, with dark brown blotches, slightly dentate; **Cor.** 3.5 cm $\varnothing$, lobes stellate-spreading, lanceolate, acute, 14 mm long, 6–8 mm across, slightly rough inside, glabrous, brownish-purple, minutely hairy outside, margins recurved.

**S. peculiaris** LUCKH. (§ III). – Cape: Lit. Namaqualand, Van Rhynsdorp D. – St. erect, branching from the base, up to 15 cm long, 2 cm $\varnothing$, with 4 tuberculate angles; **L.** deciduous; **Fl.** (1–)2(–3) together at the base of young shoots, on pedicels 12–15 mm long, Sep. 3 mm long, acute, Cor. 2.6 cm $\varnothing$, the centre flat, the corona slightly raised, dull yellowish-green, densely mottled with pale purplish-brown blotches, minutely papillose, lobes 7.5 mm long, acute, with motile, purple, clavate hairs; outer corona-lobes 3.5 mm long, tripartite.

**S. pedunculata** MASS. (§ VIII) (*Caruncularia p.* HAW., *C. jacquinii* SWEET, *C. massonii* SWEET, *C. simsii* MASS., *S. laevis* DECNE). – Cape: Lit.

Namaqualand. – **Shoots** 5–12 cm tall, 6 to 16 mm ⌀, obtusely 4-angled, smooth, glabrous, greyish-green, often reddened, scarcely dentate; **Fl.** 5–6 together, usually solitary or paired, pedicels 10–12 cm long, erect, Cor. 4–4.5 cm ⌀, deeply 5-cleft, lobes lanceolate, with 3 impressed longitudinal veins, glabrous, with a whitish blotch at the base, brownish to pale olive-green or yellowish-green above, margins recurved, with a cluster of dark red, clavate hairs in each indentation, at the base a star of similar hair-clusters.

**S. peglerae** N. E. Br. (§ VI). – Cape: W. Transkei (Tembuland). – **Shoots** c. 15 cm tall, 12 mm ⌀, glabrous, with small T.; **Fl.** 3–4 together. Cor. c. 6 cm ⌀, lobes lanceolate, acute, c. 2.5 cm long, dark brown, transversely wrinkled, glabrous outside and inside, apart from the margins which bear red cilia.

**S. pillansii** N. E. Br. v. **pillansii** (§ VI). – Cape: Laingsburg, Prince Albert D. – **Shoots** 8–13 cm tall, 10–12 mm ⌀, with 4 obtuse, slightly compressed angles, and not very prominent T.; **Fl.** 2–4 together, Cor. 10 to 13 cm ⌀, spreading, star-like, lobes ovate-lanceolate, tapered, underside shortly hairy, dark purplish-brown inside, glabrous, ± smooth, the margins with purple cilia almost up to the tips.

**S. —** v. **attenuata** N. E. Br. – **Cor.** 15–20 cm ⌀, lobes very long-tapered.

**S. plantii** Hook. f. (§ VI). – Cape: Albany D. – **Shoots** 15–20 cm tall, robust, the 4 angles compressed, green, softly hairy, with small T.; **Fl.** 1–3 together, Cor. flat, 10–12 cm ⌀, wide open, 5-cleft to more than midway, lobes ovate-lanceolate, tapered, 4.5 cm long, 2.5 cm across, the upperside sparsely soft-haired at the throat, dull blackish-red, lobes with transverse wrinkles and transverse yellow lines, and broad, dull purple, long-ciliate margins.

**S. portae-taurinae** Dtr. et Bgr. (§ II). – S.W.Afr.: Gr. Namaqualand; Cape: Lit. Namaqualand. – **Shoots** 5–20 cm tall, with 4 rounded angles, minutely hairy, with small T.; **Cor.** c. 2.5 cm ⌀, tube broadly campanulate, with an annular depression at the mouth, lobes ovate-triangular, 9 mm long, 7 mm across, tapered, pale yellow, with brown furrows and tubercles in transverse R., margins recurved.

**S. prognantha** Bally (§ IX). – N.Somalia. – Close to **S. revoluta; shoots** up to 6 cm tall, erect or creeping, 4-angled, with stiff adventitious roots 2 mm thick, with fleshy, sharp, apiculate T.; **Fl.** solitary, Cor. 1.5–3.1 cm ⌀, tube flat, plate-shaped, densely papillose inside, the upper margin constricted into a fleshy disc, lobes triangular with recurved tips, pallid bluish-purple, with hairy margins.

**S. × prometheus** Bgr. (*S. variegata* v. *prometheus* Damman). – Natural hybrid: **S. variegata × S. ?**. – **Cor.** 5.5 cm ⌀, lobes 18 mm long, 15 mm across, lemon-yellow, with brown blotches and margins, annulus 16–17 mm across, 5-angled, ± lemon-yellow, the annular groove light yellow, with spots only on the inner surface, margins with brown cilia.

**S. pulchella** Mass. (§ X) (*Podanthes p.* Haw.). – Cape: Uitenhage, Port Elizabeth, Alexandria, Humansdorp D. – **Shoots** prostrate or ascending, c. 5–8 cm tall, c. 1 cm ⌀, 4-angled, light green, with sharp, projecting T.; **Fl.** several together, Cor. rotate, 5-cleft, 4–5 cm ⌀, lobes triangular-ovate, tapered, minutely transversely wrinkled, with minute brown spots, annulus small, minutely tuberculate, with brownish spots, margins ciliate.

**S. pulvinata** Mass. f. **pulvinata** (§ VI.). – Cape: Lit. Namaqualand. – **Shoots** 10–20 cm tall, 4-angled, dark green to brownish-red, densely soft-hairy, with erect T.; **Fl.** solitary, Cor. 9 cm ⌀, flat-rotate, deeply 5-cleft to midway, lobes triangular-ovate, up to 4 cm long, 4 cm across, short-tapered, dull red with yellowish transverse wrinkles and a dense cushion of reddish hairs in the centre around the subsidiary corona, margins densely ciliate.

**S. —** f. **margarita** (Sloane) Rowl. (*S. margarita* Sloane). – **Cor.** more densely pink haired at the centre, with the apical half of the lobes alone hairless; marginal hairs white instead of pink.

**S. revoluta** Mass. v. **revoluta** (§ IX) (*S. glauca* J. Donn., *S. protensa* Hornem., *Tromotriche r.* Haw., *T. glauca* Haw.). – Cape: Karroo. – **Shoots** up to 30 cm tall, 12–18 mm ⌀, 4-angled, glabrous, light-grey pruinose, with sharp T.; **Fl.** solitary or paired, Cor. fleshy, lobes triangular-ovate, ± recurved, 12–15 mm long and across, glabrous, smooth, rather swollen around the tube, dull wine-coloured, with whitish or greenish-yellow, cordate zones extending outwards to the lobes, margins with brown cilia.

**S. —** v. **fuscata** (Jacq.) N. E. Br. (*S. fuscata* Jacq., *S. glauca* v. *β* Haw.). – Origin unknown. – Lobes longer-tapered, reddish-brown, minutely spotted, marginal hairs longer.

**S. —** v. **tigrida** (Decne.) N. E. Br. (*S. tigrida* Decne., *S. r.* Curtis, *Tromotriche g.* v. *β* Haw.). – **Fl.** rather lighter, with a pale greenish-yellow star in the centre, the bases of the lobes with yellow spots and with shorter transverse lines.

**S. rubiginosa** Nel (§ II). – Cape: Lit. Namaqualand. – **Shoots** 27–50 cm long, 8–10 mm ⌀, with 4 rounded angles, greenish to deep purple, minutely hairy, with sharp, subulate T.; **Cor.** 2 cm ⌀, rotate, lobes ovate, acute, transversely wrinkled below, yellowish-purple, with purple spots, greenish outside with purple stripes, hairy, the margins with red cilia.

**S. rufa** Mass. v. **rufa** (§ IV) (*S. rufescens* S. D.). – Cape: Karroo. – **Shoots** 10–15 cm tall, 1–2 cm ⌀, with obtuse angles, green, often brownish, minutely soft-haired, with small T.; **Fl.** 3–5 together, Cor. 3–4 cm ⌀, tube short, campanulate, lobes triangular, long and thin-tapered above midway, with a prominent, broad, blackish-violet, minutely transversely wrinkled annulus below, both annulus and the lower parts of the lobes with thin, olive-green transverse lines between the wrinkles, minutely hairy at the tips, with ciliate margins.

**S. —** v. **attenuata** N. E. Br. (*S. fissirostris* N. E. Br.). – Lobes longer and narrower.

**S. — v. fissirostris** (JACQ.) WHITE et SLOANE (*S. f.* JACQ.). – Lobes greenish-yellow on the upperside, with red transverse bands.

**S. ruschiana** DTR. (§ VIII). – S.W.Afr.: Gr. Namaqualand. – **Shoots** 10–20 cm tall, with 4 blunt angles, green to reddish-brown; **Fl.** solitary or paired, Cor. broadly campanulate, 36 mm $\varnothing$, lobes 14 mm long, 6–7 mm across, with recurved margins, greenish-red, dark red inside below, with dense, blackish-red, clavate hairs for the first 5.5 mm up from the base, then a white, scarred zone with red spots, the lobes scarred above, reddish-brown, the margins between the lobes with curly, dark reddish-brown, clavate hairs.

**S. schinzii** BGR. et SCHLTR. v. **schinzii** (§ VI). – S.W.Afr.: Gr. Namaqualand. – **Shoots** with alate angles, 8 cm tall, 15–18 mm $\varnothing$, fresh green, minutely soft-haired, with projecting T.; **Fl.** mostly paired, Cor. rotate, flat, 22 cm $\varnothing$, deeply cleft beyond midway, lobes ovate-lanceolate, long-tapered, 10–10.5 cm long, 27 mm across, blackish-brown, glabrous below, minutely transversely wrinkled, the tips smooth, the margins with purple hairs.

**S. — v. angolensis** KERS. – Angola: Moçamedes D. – **Shoots** daintier, thinner, 6–7 cm tall, 5–7 mm thick; young shoots dark reddish-violet; **Fl.** 8–9 cm $\varnothing$.

**S. semota** N. E. BR. (§ I). – Tanzania: near Kondoa. – **Shoots** c. 7.5 cm tall, 12–20 mm $\varnothing$, 4-angled, smooth, spotted, with spreading T. 6–12 mm long; **Fl.** several together, Cor. 4.5 cm $\varnothing$, lobes ovate-lanceolate, tapered, 16 mm long, 8 mm across, upperside rough, chocolate-brown, with light markings, with a dark brown, 5-angled annulus around the tube-mouth, margins with several red cilia.

**S. senilis** N. E. BR. (§ VI). – Cape: Albany D. – **Shoots** c. 30 cm tall, 2–3 cm $\varnothing$, minutely hairy, green, tinged reddish, with 4 very compressed, dentate angles; **Cor.** 11–12.5 cm $\varnothing$, the underside softly hairy, lobes lanceolate, acute, 4.5 to 5 cm long, 2.5–3 cm across, transversely wrinkled inside, the lower ⅔ dull purple or purplish-brown, with a small ochre-coloured or greenish blotch at the tip, densely covered up to midway with soft, pure white hairs 6 mm long, margins white-ciliate.

**S. similis** N. E. BR. (§ II). – Cape: Lit. Namaqualand. – **Shoots** 7–15 cm tall, 6–10 mm $\varnothing$, with 4–6 $\pm$ dentate angles, with fine velvety hairs, deep purple, or ashy-grey and purple-spotted; **Fl.** 3–6 together, Cor. round, 18 mm $\varnothing$, minutely velvety outside, transversely wrinkled inside, glabrous, blackish-purple, lobes 6 mm long and across, ovate-acute.

**S. sororia** MASS. (§ VI) (*S. s.* v. *alia* JACQ., *S. uncinata* JACQ., *S. lunata* SWEET). – S.Afr.: origin unknown. – **Shoots** up to 30 cm tall, over 3 cm $\varnothing$, with 4–5 distantly dentate angles, dark green; **Fl.** usually solitary, Cor. 11–12 cm $\varnothing$, flat, deeply cleft to over midway, lobes ovate-lanceolate, acute, blackish-brown, with yellowish transverse wrinkles, throat and margins with dark, rostrate hairs, margins ciliate.

**S. stricta** SIMS (§ V) (*Gonostemon s.* HAW.). – Cape: origin unknown. – **Shoots** 12 cm tall, 6 to 8 mm $\varnothing$, 4-angled, glabrous (?), with small T.; **Fl.** solitary or paired, Cor. rotate, c. 4.5 cm $\varnothing$, 5-cleft to midway, lobes ovate, short-tapered, flat, dull, pale pinkish-red inside, with rather darker spots, paler below, the margins $\pm$ pale greenish.

**S. stultitioides** LUCKH. (§ II) (*S. beukmanii* LUCKH. in S.Afr. Gard. 1935, 96). – Cape: Clanwilliam D. – **Shoots** c. 9 cm tall, 1 cm $\varnothing$, acute, 4-angled, green, minutely hairy, with sharp T.; **Fl.** clustered, Cor. 4 cm $\varnothing$, minutely hairy outside, green with purple blotches, rough inside, deep purplish-black, with an indistinct annulus in the centre, lobes spreading, ovate, tapering above midway, 15 mm long, 6.5 mm across, with several white hairs below.

**S. surrecta** N. E. BR. v. **surrecta** (§ V). – Cape: Laingsburg, Ceres D. – **Shoots** 7–8 cm tall, c. 1 cm $\varnothing$, with 4 obtuse angles, green, softly hairy; **Fl.** 2–3 together, glabrous inside, rather wrinkled, tube short, bowl-shaped, lobes 9 mm long, 4 mm across, very long-tapered, the underside with 3 distinct veins.

**S. — v. primosii** LUCKH. – Cape: Sutherland D. – Cor. smooth, pale yellow, sometimes uniformly brownish-purple.

**S. thudichumii** PILL. (§ IX). – Cape: Sutherland D. – **Shoots** c. 10 cm tall, up to 1.5 cm $\varnothing$, with 4 obtuse, minutely dentate angles, minutely papillose, with dull purple blotches; **Fl.** solitary, Cor. 1.3 cm $\varnothing$, with a 5-angled depression 8 mm $\varnothing$ in the centre, reticulately wrinkled, purplish-brown, lobes 9 mm long, 10–11 mm across, triangular, acute, recurved, margins pale purplish-brown with motile, clavate hairs.

**S. trifida** TOD. (§ X). – Probably a var. or hybrid of **S. mutabilis; shoots** 5–8 cm tall, 12 to 15 mm $\varnothing$, glabrous, with projecting T.; **Fl.** usually solitary, Cor. 7–8 cm $\varnothing$, lobes ovate-lanceolate, c. 22 mm across the middle, with violet-purple margins and numerous yellow lines, the tips unmarked.

**S. tsomoensis** N. E. BR. (§ VI) (*S. glabricaulis* SCHLTR., *S. t.* HORT. ex WHITE et SLOANE). – Cape: Transkei. – **Shoots** 10–15 cm tall, 10 to 16 mm $\varnothing$, the angles compressed, sinuate, with erect, small T., dull grey, minutely puberulous; **Fl.** 4–9 together, Cor. 6–7.5 cm $\varnothing$, lobes ovate-lanceolate, acute, 2.5–3 cm long, 15 mm across, recurved, dull red, the upper half with slight, often greenish-ochre, transverse wrinkles, the base of the Fl. flat, with soft, red hairs, margins red-ciliate.

**S. umbonata** PILL. (§ III). – Cape: Lit. Namaqualand. – **Shoots** 4–6 cm tall, 10–13 mm $\varnothing$, the angles rounded with roundish T., the sides depressed, bluish-green with red blotches; **Fl.** clustered, Cor. 6.5–7 cm $\varnothing$, glabrous outside, with reddish markings, lobes recurved, triangular-ovate, tapered, 20–23 mm long, glabrous inside, wrinkled, with 9 grooves above, greenish-yellow, with dense purple blotches along the red-ciliate margins.

**S. variegata** L. (§ X). – S.E.Cape, widespread. – Type: v. **variegata.** This species has often hybridised with other species; the **Fl.**-colour is very variable so that the number of variants is exceedingly large. (For a key to the varieties see WHITE et SLOANE, 'The Stapeliae', Vol. II, 700.)

S. — v. **atrata** (TOD.) N. E. BR. (*S. a.* TOD.). – Probably a hybrid; **Cor.** 5.5–7 cm $\varnothing$, lobes ovate, base-colour dark purplish-brown.

S. — v. **atropurpurea** (S. D.) N. E. BR. (*S. a.* S. D., *S. marmorata* HULLE). – Cape: Cape D. – **Cor.** 7–9 cm $\varnothing$, annulus $\pm$ circular, blackish-purple below, with fewer irregular, yellowish, often circular markings.

S. — v. **bufonia** (J. DONN) N. E. BR. (*S. b.* J. DONN, *S. bisulca* J. DONN, *S. orbiculata* J. DONN, *S. ophiuncula* HAW., *Orbea bisulca* HAW., *O. bufonia* HAW., *S. bufonis* LODD., ? *S. bifolia* SCHULT., *S. orbicularis* LODD., *S. bidentata* S. D., *S. monstrosa* STEUD., *S. ciliolulata* TOD. et RÜST., *S. ophioncola* SCHLTR.). – Cape: Cape D. – **Cor.** spreading, annulus flat, lobes dirty brown inside with black blotches and minute, brown, transverse stripes; **Fl.** with a strong, disagreeable smell.

S. — v. **brevicornis** N. E. BR. – **Cor.** 4.5 to 7 cm $\varnothing$, pale yellow or yellowish-green, with very dark blotches and a few yellow, transverse lines.

S. — v. **clypeata** (J. DONN) N. E. BR. (*S. c.* J. DONN, *S. v.* JACQ., *Orbea quinquenervis* HAW., *S. bufonia* SIMS., *O. c.* HAW., *S. q.* SCHULT., *O. q.* LOUD.). – Cape: Cape D. – **Cor.** 4.5 to 7 cm $\varnothing$, flat, with blotches all over, with or without several thin, transverse lines at the base of the minutely ciliate lobes.

S. — v. **conspurcata** (WILLD.) N. E. BR. (*S. c.* WILLD., *S. obliqua* WILLD., *Orbea c.* SCHULT., *Tromotriche obliqua* SWEET, *S. ciliolata* TOD.). – Cape: Cape D. – **Cor.** c. 5 cm $\varnothing$, yellowish inside, $\pm$ covered with blackish-brown blotches, with a large annulus, with minutely ciliate margins.

S. — cv. Cristata. – **Shoots** fasciated.

S. — v. **curtisii** (HAW.) N. E. BR. (Pl. 136/4) (*S. v.* CURTIS, *Orbea c.* HAW., *O. inodora* HAW., *S. c.* SCHULT., *S. inodora* DECNE.). – ? Cape D. – **Cor.** pale greenish-yellow or lemon-yellow, with scattered, small, reddish-brown blotches.

S. — v. **horizontalis** (N. E. BR.) N. E. BR. (*S. h.* N. E. BR.). – ? Cape: Somerset East D. – **Cor.** 6–8 cm $\varnothing$, with the annulus circular to 5-angled, wrinkled all over inside, minutely tuberculate, lobes dull greenish-yellow, with dark brown blotches below around the annulus, dark brown, double lines in the lower half and small brown blotches and a brown M.line towards the tips.

S. — v. **laeta** N. E. BR. (*S. picta* N. E. BR.). – **Cor.** 5–6.5 cm $\varnothing$, lobes not ciliate, with numerous, large, $\pm$ confluent, reddish-brown blotches, those on the annulus blood-red.

S. — v. **marginata** (WILLD.) N. E. BR. (*S. planiflora* v. *m.* WILLD., *S. m.* WILLD., *Orbea m.* SCHULT., *O. p.* v. *m.* G. DON). – **Cor.** fairly large, wrinkled, pale yellow with red margins, the annulus obtusely 5-angled, vivid yellow, wrinkled-tuberculate, with minute violet spots, rather small.

S. — v. **marmorata** (JACQ.) N. E. BR. (*S. m.* JACQ., *Orbea m.* SCHULT.). – **Cor.** 4.5–5 cm $\varnothing$, lobes triangular-ovate, blackish-purple or dark purplish-brown, with several wrinkles and yellowish, longitudinal lines and blotches giving a mottled appearance.

S. — v. **mixta** (MASS.) N. E. BR. (*S. m.* MASS., *Orbea m.* HAW.). – Cape: Robertson D. – **Cor.** 5–7.5 cm $\varnothing$, light, lobes with numerous round blotches, those below being connected by thin lines, with minutely ciliate margins.

S. — v. **pallida** N. E. BR. – Cape: Table Mt. – **Cor.** quite flat, with numerous small to spot-like markings, with scarcely any brown, transverse lines at the lobe-bases, the annulus circular to distinctly 5-angled, lobe-margins not ciliate.

S. — v. **picta** (J. DONN) N. E. BR. (*S. p.* J. DONN, *S. anguinea* JACQ., *Orbea a.* HAW., *O. p.* HAW.). – **Cor.** 5–5.5 cm $\varnothing$, flat, lobes pale yellow or sulphur-yellow, with blackish-red transverse wrinkles and lines and large, confluent blotches, the annulus large, circular, very tuberculate, its margin revolute.

S. — v. **planiflora** (JACQ.) N. E. BR. (*S. v.* JACQ., *S. p.* JACQ., *Orbea p.* HAW., *S. mutabilis* HULLE). – Cape: Table Mt. – **Cor.** flat, 6.5 to 7 cm $\varnothing$, lobes broadly triangular-ovate, acute, dark brown below around the annulus, with yellow, transverse stripes and wrinkles, yellower towards the tips, with large brown blotches, the annulus large, round, with spots and blotches and a revolute margin.

S. — v. **retusa** (HAW.) N. E. BR. (*Orbea r.* HAW., *S. r.* SCHULT.). – **Cor.** very wrinkled inside, yellowish, with blackish-red blotches, especially towards the base, the annulus 5-angled, paler, very wrinkled-tuberculate, uniformly coloured at the base.

S. — v. **rugosa** (J. DONN) N. E. BR. (*S. rugosa* J. DONN, *Tridentea r.* SCHULT., *Orbea r.* SWEET). – ? S.W.Afr. – **Cor.** greenish-yellow, with numerous small, dark reddish-brown blotches and irregular transverse lines on the lobes, the annulus 5-angled, with 5 distinct furrows, lighter yellow with dark brown blotches.

S. — v. **trisulca** (J. DONN) N. E. BR. (*S. trisulca* J. DONN, *Orbea t.* HAW., *S. normalis* LINDL.). – Cape: Cape D. – **Cor.** 7 cm $\varnothing$, flat, yellowish outside, deep yellow inside, transversely wrinkled, with reddish-brown spots and blotches in irregular, longitudinal R., the annulus revolute, rather lighter and with smaller blotches.

S. — v. **variegata** (*S. normalis* JACQ., *Orbea n.* SCHULT., *O. v.* HAW., *S. woodfordiana* SCHULT., *O. w.* HAW., *S. v.* v. *normalis* BGR.). – Mat-forming; **shoots** ascending, 5–10 cm tall, obtusely angled, green to somewhat greyish-green, often reddish, with sharp, projecting T.; **Fl.** 1–5 together, Cor. flat, 5–9 cm $\varnothing$, lobes triangular-ovate, acute, eventually somewhat recurved, smooth outside, glabrous, pale green, transversely wrinkled inside, yellow, with dark brown blotches, scattered or in irregular, longitudinal R., interspersed with transverse and often thin lines of the same colour, the annulus broad, often revolute, circular or very slightly 5-angled, light yellow with small blotches and spots.

S. verrucosa MASS. v. verrucosa (§ II) (*Podanthes v.* HAW., *P. pulchra* v. β HAW.). – S.Afr.: origin unknown. – Shoots 4–8 cm tall, 12 mm ⌀, with 4 obtuse angles, green, glabrous, T. 3 to 6 mm long; Fl. solitary or paired, Cor. rotate, c. 6 cm ⌀, 5-cleft to midway, wrinkled, rough-tuberculate, thickened below into a flat, 5-angled annulus with several hairs, furrowed towards the 5 angles, lobes ovate, sharply tapered, recurved, pale yellow, with blood-red and brown spots.

S. — v. pallescens N. E. BR. – Cape: Somerset East D. – Lobes much tapered, pale yellow, with small, reddish-brown blotches.

S. — v. pulchra (HAW.) N. E. BR. (*S. verrucosa* JACQ., *Podanthes p.* HAW., *S. irrorata* LODD., *S. p.* SCHULT.). – Shoots 6–9 mm ⌀; Fl. somewhat smaller, green outside, with darker lines, sulphur-yellow inside, wrinkled and rough, with numerous red and blackish-red blotches, the centre dark red with a minute yellow border, with a few hairs.

S. — v. punctifera N. E. BR. – Cape: Alexandria D. – Shoots 8–10 mm ⌀; Fl. very pale creamy-yellow, with small, red spots.

S. — v. robusta N. E. BR. (*S. r.* CURTIS). – Cape: Somerset East D. – Robust variety; Fl. as v. pulchra, with a deep pink colour.

S. — v. roriflua (JACQ.) N. E. BR. (*S. r.* JACQ., *S. rugosa* WENDL. *S. wendlandiana* SCHULT., *Orbea w.* SCHULT., *Podanthes r.* SWEET, *Piaranthus r.* DECNE.). – Cape: Somerset East D. – Cor. 5.5–6 cm ⌀, tube broadly campanulate, light yellow or greenish-yellow, with larger and darker blotches than v. pallescens.

S. vetula MASS. v. vetula (§ VI) (*Tridentea v.* HAW.). – Cape: Worcester D. – Shoots 10 to 15 cm tall, c. 13 mm ⌀, with 4 small-dentate angles, puberulous; Fl. 2–3 together, Cor. 6 to 8 cm ⌀, tube campanulate, glabrous, lobes ovate-lanceolate, long-tapered, upperside blackish-red, minutely transversely wrinkled, with 3–5 distinct longitudinal nerves.

S. — v. simsii (HAW.) N. E. BR. (*S. vetula* SIMS, *Tridentea s.* HAW., *S. s.* SCHULT.). – Differentiated only by small characteristics of the inner corona.

S. — v. juvencula (JACQ.) N. E. BR. (*S. j.* JACQ., *Tridentea j.* SWEET). – Differentiated only by small characteristics of the inner corona.

S. virescens N. E. BR. (Pl. 136/3) (§ III). – Cape: Karroo; S.W.Afr.: Gr. Namaqualand. – Shoots 5–8 cm tall, obtusely angled, greyish-green, glabrous, with small T.; Fl. in clusters of 6–9, Cor. 2.5–3 cm ⌀, deeply 5-cleft, lobes curved slightly inwards, ovate-tapered, 10 mm long, 7–8 mm across, glabrous outside, whitish with red blotches, yellowish-green inside, rough-papillose, margins recurved.

S. wilmaniae LUCKH. (§ VI). – Transvaal. – Shoots 9–18 cm tall, 12 mm ⌀, with 4 rounded, indistinctly dentate angles, light green, minutely hairy; Fl. solitary or paired, Cor. cup-shaped, the united part 6.5 cm ⌀, deep chestnut-brown and minutely hairy outside, rough with conspicuous wrinkles inside, deep violet-pink, the wrinkles black, the lower third of the tube with purple hairs, lobes mostly decurved, 3 cm across below, then passing over into the elongated tip, the lower third with dark red cilia.

S. woodii N. E. BR. (§ I). – Natal. – Shoots 4–8 cm tall, 10–15 mm ⌀, green with red blotches, the 4 angles with projecting T. 6 to 12 mm long; Fl. 3 or several together, Cor. 4 cm ⌀, rotate, lobes ovate, tapered, dark brown with a few yellow blotches, very wrinkled, with some cilia in the centre, margins recurved.

S. youngii N. E. BR. (§ VI). – Rhodesia: near Salisbury. – Close to S. gigantea; shoots smaller; Cor. 12.5–15 cm ⌀, lobe-tips shorter, yellow or greenish inside, with purple transverse lines, transversely wrinkled, upper lobe-tips smooth.

**Stapelianthus** CHOUX (*Stapeliopsis* CHOUX). Asclepiadaceae. – S. and S.W.Madag. – Succulent ♃, close to the G. Huernia. This G. was established principally on the form of the outer corona which develops an erect, 5-lobed head from the staminal column; the lobes are ± deeply bifid, united to the lobes of the inner corona, and extending beyond the staminal column. – Greenhouse, warm. Propagation: seed, cuttings. Grafting onto Tub. of Ceropegia woodii is recommended.

S. baylissii LEACH (This species is clearly a Stapelia). – Cape: Steytlerville D. – St. with 4 (rarely 6) sinuate-dentate angles, glabrous, dark green to brownish, branching from below or less often from midway, deeply constricted and jointed, up to 90 cm long, pendulous, 7 to 12 mm thick, with scattered L.scars and the remains of old Ped., with T. up to 5 mm high, 9–13 mm apart; Fl. in Infl. near the tips of the shoots on a pedicel 5–13 mm long, Sep. ovate, tapered, Cor. 5-angled, campanulate, 12–15 mm long, 12–13 mm ⌀ between the lobe-tips, yellowish outside below, deep purple towards the lobe-tips, glabrous or sometimes ± rough, with whitish blotches, reddish-purple inside and completely glabrous, with a soft, silky sheen, the tube 8–10 mm long, c. 7 mm ⌀ below, 9–10 mm ⌀ at the mouth, with a distinct groove towards each lobe-tip, corona dark purple.

S. decaryi CHOUX (Pl. 137/2). – S.Madag.: Fort Dauphin Prov. – St. branching from the base, 10 cm tall, 6–10 mm ⌀, with (5–)6(–8) angles with square or hexagonal L.cushions terminating in a ± spiny, deciduous tip; Fl. usually solitary or few in an Infl., on pedicels 5–8 mm long; Cor. with a tube 10–13 mm long, dark purple inside, lobes 5, broadly triangular, recurved, 7.5–8.5 mm long, 6–7 mm across, intermediate lobes small, lobes and tube yellowish-grey with dark purple blotches, lobes and the upper part of the tube with papillae, each tipped with a minute red hair.

S. hardyi LAVR. – Madag.: near Maromba. – Related to S. montagnacii; Cor. campanulate, very fleshy, c. 13 mm ⌀, glabrous outside but sparsely and minutely tuberculate, very dark purplish-brown near the base which is yellowish-pink, spotted with dark purplish-brown, inside glabrous below, yellowish-pink, with circular, dark purplish-brown spots, elsewhere very dark purplish-brown and densely covered with soft, cylindrical, purplish-brown, clavate hairs 3 mm long; the tube 8–10 mm long, lobes deltoid, acute, c. 9 mm long and broad; intermediate lobes small but distinct; outer corona-lobes divided into 2 sinuate-rounded wings, between these a deltoid, acute T.

**S. insignis** B. Desc. – S.W.Madag. – **Shoots** prostrate, 4-angled, up to 20 cm long, 8 to 10 mm $\varnothing$, reddish-grey with darker blotches, L.cushions oblong, merging into angular ribs, each L.cushion crowned with a very acute L. 1.5–2 mm long; **Fl.** usually solitary, on a pedicel 4–5 mm long, the bud obconical, opening to a small top or mushroom shape, the tube cylindrical-campanulate, very fleshy, c. 7mm $\varnothing$, pale reddish outside, with dark wine-coloured blotches, whitish inside with wine-coloured blotches, widened abruptly to 17–20 mm $\varnothing$, narrowing again towards the tip, greenish outside, with wine-coloured blotches of varying size and darker nerves, black to purple inside, with reticulate, yellow markings, glabrous, the lobes recurved, small, shortly triangular, c. 2 mm long, 3 mm across, blackish-purple with a lighter M.stripe, intermediate lobes still smaller.

**S. madagascariensis** (Choux) Choux (*Stapeliopsis m.* Choux). – S.Madag.: S.E.Tuléar. – **Shoots** prostrate to ascending, rooting from the base, 6–8 mm $\varnothing$, elongated, greyish-green with dark red blotches, with L.cushions in 6 series and tipped with a small L.; **Fl.** few in an Infl., Cor. broadly campanulate, c. 2 cm $\varnothing$, tube 7mm $\varnothing$, pale yellow with wine-coloured blotches, lobes patent, with long recurved tips, broadly triangular, 7–8 mm long, 7 mm across, yellowish, with solitary, wine-coloured blotches, the upperside with red-tipped papillae 2–3 mm long, intermediate lobes small.

**S. montagnacii** (Boit.) Boit. et Bertr. (*Stapelia m.* Boit.). – S.W.Madag. – Possibly only a variety of **S. madagascariensis; shoots** 10mm thick, mostly creeping; **Cor.** flat dish-shaped, whitish, with wine-coloured blotches outside, the markings inside merging into pale red bands, lobes 6 mm long, 6 mm across, with a short tip, the upperside with large, confluent, dark wine-red blotches, the tips whitish, the intermediate lobes very small, scarcely recognizable, papillae as in **S. madagascariensis.**

**S. pilosus** Lavr. et Hardy (Pl. 137/3) (*Trichocaulon decaryi* Choux). – S.Madag. – Forming dense clumps; **shoots** prostrate to ascending, rooting along the creeping parts, 9–12 cm long, 8–12 mm $\varnothing$, vivid green to purplish-brown, L.cushions spherical, small, in densely spiralled R., with a tip 2.5 mm long; **Fl.** paired or several, on pedicels up to 7 mm long, Sep. subulate, Cor. campanulate, fleshy, c. 14 mm $\varnothing$, 14 mm across between the lobes, tube 8–9 mm long and across, lobes spreading, broadly triangular, tapered, 9 mm long, 6–7 mm across, intermediate lobes short, pale yellow outside and inside, with wine-coloured blotches, inside with pale yellow, reddish-tipped papillae ¾ mm long.

**Stapeliopsis** Pillans. Asclepiadaceae. – S. and S.W.Afr. – Succulent ♃; **shoots** 4-angled, thick, dentate, green with purple blotches, minutely papillose-hairy; **Fl.** produced from the base of young shoots, pedicellate, urn-shaped, papillose inside, the papillae tipped with a hair; outer corona erect, cup-shaped or tubular, with 5 obtuse or acute lobes, longer or shorter than the staminal column.

**S. neronis** Pillans. – Cape: Lit. Namaqualand. – **Shoots** crowded, 5–7 cm tall, 3–3.5 cm $\varnothing$, the angles obtuse, very compressed, green, with purple or brown blotches, the T. broadly deltoid, sharp, apiculate; **Fl.** 1–3 together, Sep. acute, 4 mm long, velvety, Cor. urn-shaped, velvety outside, deep purple, fleshy, 17–20 mm long, 1.5–1.75 mm $\varnothing$ at the mouth, tube $\pm$ spherical, with 5 longitudinal grooves below, purple inside, densely papillose, each papilla with a $\pm$ long hair, the throat light purple to white, lobes 4–5 mm long, obtuse or acute, 3.5–4 mm across, ovate-triangular, stiff, white or pale purple.

**S. urniflora** Lavr. (Pl. 137/1). – S.W.Afr.: around Mt. Aruab. – **Shoots** sometimes semi-underground, up to 7 cm long, 2 cm $\varnothing$, greyish-green with brown blotches, minutely papillose, the angles rounded, bluntly dentate, the T. somewhat laterally compressed; **L.** scale-like, minute, early deciduous; **Fl.** 2–4 together, Cor. urn-shaped, 15 mm long, 8 mm $\varnothing$ below, straight, constricted to 5 mm below midway, the tube up to 12 mm long, wine-coloured on both surfaces, glabrous outside, with dense, round papillae inside, each with a long hair, the mouth 2 mm $\varnothing$ inside, lobes triangular, erect, 3 mm long and across.

**Stephania** Lour. Menispermaceae. – Trop. As. – Greenhouse, warm. Propagation: seed.

**S. rotunda** Lour. (Pl. 137/4) (*S. glabra* Miers.). – Himalayas; southern S.Vietnam. – **Caudex** large, $\pm$ spherical, rough, brown, roots fusiform; with several short, thick **Br.** above; each year these produce twining, usually simple, long, terete, glabrous **shoots**, $\pm$ dying back in the resting period; **L.** scutate, triangular-rounded, acute, sinuate, glabrous, alternate, petiolate; **Fl.** lateral, in umbels.

**Streptocarpus** Lindl. Gesneriaceae. – Trop. Afr. – Mostly trop. ♃, with conspicuously large and beautiful Fl. The only $\pm$ succulent species is described below. – Greenhouse, warm, moist. Propagation: seed, division, L.cuttings.

**S. saxorum** Engl. – Tanzania: Usambara. – **Shoots** spreading horizontally, glabrous, fleshy at first, eventually woody, densely leafy; **L.** short-petiolate, soft-fleshy, elliptical, 2.5 cm long, tipped with a thick, translucent Wi. resembling sugar icing; **Infl.** axillary, 7.5 to 15 cm tall, with 1–2 Fl., Cor. 2 cm long, glandular-hairy, white, the margin oblique, pale mauve.

**Stultitia** PHILLIPS (*Stapeliopsis* PHILLIPS non CHOUX). Asclepiadaceae. – Cape. – Taken out of the G. **Stapelia** because the **Fl.** has a distinct and prominent annulus around the tube-mouth, with the outer corona-lobes deeply divided. – Greenhouse, warm. Propagation: seed, cuttings.

**S. conjuncta** WHITE et SLOANE. – N.Transvaal. – **St.** creeping, branched, c. 15 cm long, 8 mm ⌀, 4-angled, glabrous, greyish-green, blotched, with conical T. 1–2 mm long; **Fl.** paired or several together, on pedicels 1 cm long, Cor. spherical-campanulate, creamy-white with pink outside, with a flat tube below 5 mm deep and 14 mm ⌀, with a fleshy annulus 8 mm ⌀ arround its mouth, the rest of the Fl. forming a cup shaped, light chestnut-brown border, 25 mm ⌀ at the mouth, 17 mm ⌀ at the tips, the united part 16 mm deep, the lobes 3 mm long, 10 mm across, creamy-white above, with a groove outside and a raised line inside.

**S. cooperi** (N. E. BR.) PHILLIPS (*Stapelia c.* N. E. BR., *Stapeliopsis c.* (N. E. BR.) PHILLIPS). – S.E.Cape. – **St.** 3.5 cm tall, 8–10 mm ⌀, obtusely 4-angled, glabrous, green or greyish-green, spotted, with T. 5–6 mm long, with 2 lateral T.; **Fl.** up to 10 together, Cor. 3 to 4 cm ⌀, flat, lobes projecting star-like, ovate or lanceolate, tapered, glabrous, green outside with reddish stripes, light purple inside, wrinkled and with small, light yellow tubercles, with minute reddish lines, margins ± recurved, redciliate, the annulus 8–9 mm across, ± round to 5-angled, tuberculate.

**S. hardyi** DYER. – Transvaal: Soutpansberg D. – **St.** forming tangled mounds, spreading, curved, up to 30 cm or more long, rooting, branched, 6–9 mm ⌀, green with purple blotches, with 4 obtuse, tuberculate-dentate angles, the tubercles slightly curved, acute, 1 mm high; **Fl.** solitary or several together, Cor. united for 5–6 mm below the annulus, the lobes in 3 to 4 days attaining 6 mm ⌀, glabrous outside, flat below the annulus, ovate, tapered, 2–2.5 cm long, 13–18 mm across, minutely papillose apart from the tips, sometimes with light yellow markings, otherwise leather-coloured, the annulus cushion-shaped, up to 4 mm high, 18–20 mm ⌀, glabrous, smooth, with an incurved margin.

**S. miscella** (N. E. BR.) LUCKH. (*Stapelia m.* N. E. BR.). – Cape: Jansenville D. – **St.** variable in length, creeping underground and producing at intervals erect shoots 2–3 cm tall, 3–5 mm ⌀, with 4 obtuse angles, glabrous, green, suffused purple, with sharp T. 1–1.5 mm long; **Fl.** solitary, Cor. 11 mm long, circular, without a tube, with a prominent, cushion-like annulus around the centre, glabrous, dark purplish-brown, rather lighter in the centre, lobes 3 mm long and across, acute to tapered, slightly rough inside.

**S. paradoxa** VERD. – Moçambique; Natal. – **St.** indistinctly and obtusely 4-angled, less than 1 cm ⌀, glabrous, greyish-green, with purple spots, T. 1–2 cm long, acute, with 2 minute lateral T. below the tip; **Fl.** c. 5 together, Cor. when open 2–2.4 cm ⌀, tube 8 mm long, urn-shaped below the annulus, cup-shaped above it, dark red, glossy, 4 mm deep, with stiff, red hairs below, the annulus glossy, dark red, with white blotches on the outer margin, c. 2 mm high, 1 mm thick, glabrous, the spreading part with dark red bands on a greenish-white ground, lobes 6 mm long and across, ovate to acute, greenish-white, with dark red blotches in transverse lines, shortly hairy towards the tips, margins with red, motile, clavate hairs.

**S. tapscottii** (VERD.) PHILLIPS (Pl. 137/5) (*Stapelia t.* VERD.). – Botswana: Kalahari; Transvaal; Cape: W.Griqualand. – **St.** 12 cm tall, T. 2 cm long, acute, with 2 smaller T. at the tip; **Fl.** 3–4 together, Cor. 5 cm ⌀, lobes 2 cm long, 1 mm across, ovate-tapered, reddish with white markings and several hairs, annulus 1 cm ⌀.

**S. umbracula** M. D. HENDERSON. – Rhodesia: Bikita D. – **St.** up to 17 cm tall, 1 cm ⌀, greenish-grey with pink blotches, T. 1.75 cm long, 5 mm thick below, tapered, with 2 minute lateral T. 6 mm below the yellow tip; **Fl.** solitary or several together, Cor. recurved from the prominent annulus, thick, fleshy, glabrous outside, yellowish-green, glabrous inside, the annulus 3–5 mm high, 7.5 mm ⌀, it and the lower third of the lobes smooth, leather-coloured, the upper part of the lobes minutely transversely wrinkled, with yellowish-green markings on a leather-coloured ground.

**Suaeda** DUMORT. Chenopodiaceae. – Coasts of Eur., Am., Afr., As. – Plants of sandy beaches, of little importance in cultivation.

**S. fruticosa** (L.) FORSK. (*Chenopodium f.* L., *Salsola f.* L.). – Salt-marshes along the Medit. coasts; central As., southern USSR to India; Austr.; W.Eur.; Madeira; Canary Is.; S.Afr.; N.Am. – Branched ♄ 30–60 cm tall, erect or prostrate and spreading; **L.** numerous, in crowded Ros. at the tips of the shoots, linear to ovate-lanceolate, ± cylindrical, 5–8 mm long, thick, pale green; **Fl.** small, solitary or up to 3 together, sessile, from the L.axils.

**S. maritima** (L.) DUM. (Pl. 138/1) (*Chenopodium m.* L.). – Coasts of Eur. including Great Britain; N.Am. – ☉, semi-succulent, 15 to 30 cm tall; **St.** little branched; **L.** semicylindrical, linear, 15–20 mm long, 2 mm across and thick; **Fl.** mostly 3 together, inconspicuous.

**S. monoica** FORSK. – E.Afr.: Tanzania. – Much-branched, fleshy, evergreen ♄; **L.** sausage-shaped, fleshy, up to 2 cm long; **Fl.** inconspicuous; **Fr.** a red berry.

**Synadenium** BOISS. Euphorbiaceae. – E.Afr. – Succulent, branched ♄ or small ♄ with a milky sap; **L.** alternate, ± fleshy; **Cy.** in cymes at the tip of young shoots or in the axils of forked Br.; cyathophylls paired, at the base of the involucre, not as long as the Cy. or scarcely projecting beyond it; flat cup- or saucer-shaped, with a marginal, annular gland divided into uneven segments by 2–5 notches, but not forming similar, separate glands. – Greenhouse, warm. Propagation: seed, cuttings.

**S. cupulare** (BOISS.) L. C. WHEELER (*Euphorbia c.* BOISS., *E. arborescens* E. MEY., *S. a.* BOISS., *E. synadenia* H. BAILL.). – ('Sheba Valley Death Tree'.) – Natal, Transvaal; Swaziland. – ♄ 1–1.70 m tall; **Br.** and shoots green, eventually woody, covered with L.scars; **L.** ovate-cuneate, 5–10 cm long, 2–4 cm across, acute or short-tapered, the blade slightly compressed, the M.rib alate on the underside, smooth, green; **Infl.** greenish-yellow. Exceedingly poisonous.

**S. grantii** HOOK. f. (Pl. 138/2). – Uganda; Tanzania; Moçambique. – Erect ♄ 2.5–3 m tall; **Br.** 8–15 mm thick; **L.** ± deciduous in winter, ovate-spatulate to obovate, with a rounded to tapered tip, often plicate, margins ± undulating and minutely dentate, leathery-fleshy, 7.5 to 17.5 cm long, 2.5–6 cm across, fresh green, pale green on the underside, M.rib often reddish; **Fl.** small, red.

**Talinopsis** A. GRAY. Portulacaceae. – N.Am. – Summer: in the open, winter: cold-house. Propagation: seed, cuttings.

**T. frutescens** A. GRAY. – Mexico; USA: Texas. – Branched ♄ 18–60 cm tall; **St.** and **Br.** thin; **L.** succulent, opposite, terete, linear; **Infl.** of few sessile, purple Fl.

**Talinum** ADANS. Portulacaceae. – Tropics. – ⚄ herbs or ♄ with **roots** often tuberous, sometimes edible; **St.** very short or elongated, herbaceous or fleshy, sometimes becoming woody, ± branched; **L.** opposite, ovate to ± lanceolate or cylindrical, with entire margins, ± fleshy, smooth, green; **Fl.** lasting only 1 day, terminal, in ± long corymbs, or solitary from the L.axils, pedicellate, Sep. 2, deciduous, Pet. 5 or more, anthers few or numerous. – Greenhouse, light; winter: moderately warm. – Propagation: seed, cuttings.

**T. attenuatum** ROSE et STANDLEY. – Mexico. – Sub-♄; **L.** lanceolate, tapered, cuneate below; **Fl.** 18–20 mm ∅, pink.

**T. aurantiacum** ENGELM. – USA: Texas; Mexico. – ⚄ with fleshy **roots**; **St.** 15–30 cm tall, branched, erect, firm; **L.** linear to linear-elliptical, acute, 15–45 mm long; **Fl.** solitary.

**T. brevicaule** S. WATS. – Mexico. – ⚄ with long, thick roots; **St.** short, scarcely branched; **L.** cylindrical, firm, 12 mm long, subacute; **Fl.** reddish.

**T. caffrum** (THUNBG.) ECKL. et ZEYH. – Kenya. – ⚄ with a tuberous **caudex**; **St.** prostrate; **L.** fleshy; **Fl.** solitary, pale lemon yellow.

**T. carinatum** A. PETER. – Tanzania. – ⚄ with a tuberous **caudex**; **St.** prostrate; **L.** fleshy; **Fl.** solitary, yellow, rarely white.

**T. chrysanthum** ROSE et STANDLEY. – Mexico. – ⚄ with thick, fleshy, tuberous **roots**; **St.** firm, ± 1 m tall, reddish, simple or branched; **L.** numerous, fleshy, obovate to lanceolate, ± 10 cm long, with a rounded tip; **Fl.** light yellow.

**T. confusum** ROSE et STANDLEY. – Mexico. – ⚄ with thick, fleshy **roots**; **St.** erect, branching from the base; **L.** broad-lanceolate, truncate, c. 5 cm long; **Infl.** racemose, Fl. pink.

**T. cuneifolium** (VAHL) WILLD. – E.Afr. – Semi-succulent ⚄ with tuberous roots; **St.** 50–60 cm tall, branched; **L.** fleshy; **Fl.** in a long, lax Rac., c. 2 cm ∅, red to greenish-white.

**T. cymbosepalum** ROSE et STANDLEY. – Mexico. – ⚄ with thick, tuberous **roots**; **St.** branching below, firm, fleshy; **L.** broadly linear, acute, 15–40 mm long; **Fl.** solitary, 30–32 mm ∅, reddish.

**T. diffusum** ROSE et STANDLEY. – Laxly branched sub-♄ 15 cm tall; **St.** firm and fleshy, woody below; **L.** lanceolate-cuneate, very truncate; **Fl.** in Rac., 18 mm ∅, white.

**T. esculentum** DTR. et SCHELLENB. (Pl. 138/3). – S.W.Afr. – ⚄ with a carrot-sized, white, edible Tub.; **St.** erect; **L.** linear-lanceolate, 3–4 cm long, 1 cm across; **Fl.** solitary, c. 12 mm ∅, yellowish.

**T. gracile** ROSE et STANDLEY. – Mexico. – ⚄ with thick, woody **roots**; **St.** thin, much branched, 14 cm tall; **L.** cylindrical, ± 3 cm long, laxly arranged along the shoots; **Fl.** few.

**T. greenmanii** HARSHBG. – Mexico. – ⚄ with thick, fleshy **roots**; **St.** simple, few; **L.** numerous, cylindrical, 4–5 cm long; **Fl.** yellow.

**T. guadalupense** DUDLEY (Pl. 138/4). – Mexico: Is. of Guadalupe. – Firm, fleshy plant; **caudex** thickened, irregularly cylindrical or spherical, gnarled, with a grey, peeling bark; or under favourable conditions forming bushes 30–60 cm tall, **St.** 2–3 cm thick, with the tips rounded; **L.** 10–15, rosulate at the tips of the shoots, ± flat, ovate-spatulate or spatulate, rounded or acute above, 3.5–5.5 cm long, 10 to 14 mm across, fleshy, bluish-green, with a red margin; **Infl.** paniculate, Fl. 25 mm ∅, pink.

**T. lineare** H. B. K. – Mexico. – ⚄ with thick, fleshy **roots**; **St.** branching from the base; **Br.** simple, firm; **L.** linear, thick, fleshy, 12–22 mm long, acute; **Fl.** solitary, 2 cm ∅, yellow.

**T. mexicanum** HEMSL. – Mexico. – ⚄; **St.** thin, much branched; **L.** cylindrical, 5–8 mm long, thin, acute; **Infl.** a cyme with very small Fl.

**T. multiflorum** ROSE et STANDLEY. – Mexico. – ⚄ with tuberous **roots**; **St.** 10 cm tall; **L.** cylindrical, c. 35 mm long, firm, sessile; **Infl.** a cyme with numerous Fl. 12 mm ∅.

**T. napiforme** DC. (*Claytonia tuberosa* MOC. et SESSE). – Mexico. – ⚄ with thick, tuberous **roots**; **St.** rudimentary; **L.** cylindrical, 4–8 cm long, firm, erect; **Infl.** a cyme with white Fl. 14 mm ∅.

**T. okanoganense** ENGLISH (*T. wayae* EASTW.). – N.Am.; Brit. Columbia. – ⚄; **St.** creeping, thin; **L.** minute, cylindrical, fleshy; **Fl.** numerous, translucent, white, with yellow anthers.

**T. oligocarpum** BRANDEG. – Mexico. – ⚄ with a tuberous, spherical **caudex**; **St.** 6–8 cm tall, numerous, much branched, thin, whitish; **L.** linear-oblong, flat, acute, c. 7 cm long; **Infl.** several, cymose, with yellow Fl. 8–10 mm ∅.

**T. palmeri** ROSE et STANDLEY. – Mexico. – ⚄ with thickened, tuberous **roots**; **St.** scarcely discernible; **L.** cylindrical, subacute, 8–9 cm long; **Infl.** 40 cm tall, cymose, with white Fl. 16–20 mm ∅.

**T. paniculatum** (Jacq.) Gaertn. (*Portulaca p.* Jacq., *P. patens* L., *Ruelingia patens* (L.) Ehrh., *T. p.* (L.) Willd.). – Carib. Is.; Mexico. – ♃; **St.** 30–100 cm tall, firm, green, fleshy, simple or branched from the base; **L.** lanceolate to obovate or elliptical, 5–10 cm long, tapered, thick, fleshy; **Infl.** much branched, Fl. pinkish-red.

**T. parvulum** Rose et Standley. – Mexico. – Mat-forming ♃; **roots** thick, tuberous; **St.** only 25 mm tall; **L.** cylindrical, c. 1 cm long, thinly petiolate; **Infl.** cymose, Fl. yellow, c. 6 mm ⌀.

**T. spinescens** Torr. – N.Am. – Sub-♄ up to 30 cm tall; **St.** rather woody, the thorny M.ribs of the Br. persisting on the Br.; **L.** semi-cylindrical, linear, obtuse; **Infl.** of few purple Fl.

**T. taitense** Pax et Vatke. – Kenya. – Climbing ♄; **L.** fleshy, 5 cm long, 2 cm across; **Fl.** yellow.

**T. tenuissimum** Dtr. – S.W.Afr. – Gr. Namaqualand. – ♃ with a fusiform **caudex**, reddish-brown inside; **St.** up to 50 cm tall, sparsely branched, glabrous, cylindrical; **L.** 12–15 mm long, 2–3 mm across, mucronate, with revolute margins; **Fl.** 1 cm ⌀, greenish-yellow.

**T. teretifolium** Pursh. – N.Am. – ♃ with a short fleshy **caudex**; **St.** ± crowded, up to 30 cm tall; **L.** cylindrical; **Infl.** dichotomous, Fl. pinkish-purple.

**T. triangulare** Willd. – W.Indies; S.Am.; Mexico. – ♄ up to 1 m tall; **St.** thick, fleshy, freely branched; **L.** obovate, 4–9 cm long, with a rounded tip, often with a dark border; **Fl.** 25 mm ⌀, white.

**Telfairia** Hook. Cucurbitaceae. – W., E. and S.Afr.; Madag.; Mauritius. – Climbing ♄ with a ± developed **caudex**; **Fl.** unisexual. – Greenhouse, warm. Propagation: seed, cuttings.

**T. occidentalis** Hook. f. – W.Afr.: forests; Sierra Leone; S.E.Nigeria: Calabar; equatorial Guinea: Fernando Po. – **Caudex** thick, fleshy, up to 45 cm ⌀; **shoots** elongated, branched, slender, angular, furrowed; **L.** 3-partite, on a petiole 2–6 cm long, with margins entire to distantly subulate-dentate, slightly rough, the central leaflet 10–15 cm long, 4–7 cm across, the lateral ones much smaller; tendrils robust; **Fl.** broadly campanulate, white, with a purple centre; **Fr.** ovoid, bluntly rostrate, yellowish-green, 40–60 cm long.

**T. pedata** (Smith) Hook. (*Fevillea p.* Smith, *Joliffia africana* Bojer ex Dellie, *Ampelosicyos scandens* Thouras). – E. and S.Afr.: deserts; Zanzibar; Madag.; Mauritius. – **Caudex** fleshy, irregularly jointed, finger-thick; **St.** elongated, corky, 5–10 mm thick, Br. glabrous, angular above, 15–30 cm long; **L.** 5–7-partite, on a petiole 6–10 cm long; tendrils robust; **Fl.** with a purple, thickish Cor., puberulous outside, papillose inside, 4–5 cm across, with long-fringed lobes; **Fr.** oblong, 40–90 cm long, 15 to 25 cm ⌀, with 10–12 ribs.

**Testudinaria** Salisb. Dioscoreaceae. – S.Afr.; Mexico. – 'Elephant's Foot', 'Hottentot Bread'. – **Caudex** large, fleshy, mostly above-ground, with prominent, many-angled, corky tubercles; **shoots** becoming woody, twining, lateral shoots usually projecting; **L.** alternate, ± triangular-cordate, sometimes auriculate; **Fl.** unisexual, small, inconspicuous, greenish-yellow, in Rac.; **seeds** broadly alate. – The synonymy of **Testudinaria** is complicated since "species" have been described on the basis of fragmentary herbarium specimens. In Journ. S.Afr. Bot. XXXIII, 1967, 1–46, E.E.A. Archibald classified the **Testudinaria** species under **Dioscorea** L. Sect. **Testudinaria**. – Greenhouse, warm. Propagation: seed; possibly from cuttings, although this is difficult.

## Key to the Genus Testudinaria by G. D. Rowley after I. H. Burkill & E. E. A. Archibald

| | | |
|---|---|---|
| 1. | Caudex wholly above ground, entire | 2. |
| 1a. | Caudex half or more buried in the ground, lobed below | glauca |
| 2. | Caudex spherical to pyramidal | 3. |
| 2a. | Caudex flat and plate-like | sylvatica 5. |
| 3. | L. blade 7–9-veined, ovate to orbicular, 6–12 cm long and wide (Mexican) | macrostachya |
| 3a. | L. blade 5–7-veined, very broad ovate, up to 26 mm long and 37 mm wide (African) | elephantipes 4. |
| 4. | L. green, not glaucous | f. elephantipes |
| 4a. | L. bluish and glaucous | f. montana |
| 5.[1] | L. firm, with a broad basal sinus and divergent auricles | 6. |
| 5a. | L. thinner, with a narrow basal sinus and auricles not divergent | 8. |
| 6. | L. blade 9-veined, up to 20 mm long and 40 mm broad | v. sylvatica |
| 6a. | L. blade (9–)10(–13)-veined, up to 60 mm long and 80 mm broad | 7. |
| 7. | Pedicel up to 4 mm long | v. paniculata |
| 7a. | Pedicel c. 2 mm long | v. brevipes |
| 8. | Capsule 12–14 mm long along placenta | v. multiflora |
| 8a. | Capsule 20–25 mm long along placenta | v. rehmannii |

[1] The doubtfully distinct var. **lydenbergensis** (Blunden, Hardman and Hind) Rowley differs mainly in having stomata on both leaf surfaces, whereas all the other varieties have stomata on the abaxial surface only.

**T. elephantipes** (L'HER.) LINDL. f. **elephantipes** (Pl. 139/1) (*Tamus e.* L'HER., *Dioscorea e.* (L'HER.) ENGL., *D. elephantopus* SPRENG.). – Cape: near Lit. Kommaggas. – **Caudex** hemispherical or pyramidal above, attaining 1 m $\varnothing$ in age, the bark divided into 6–7-angled tubercles; **St.** with projecting side-shoots; **L.** cordate-triangular, $\pm$ 3-lobed or reniform, reticulate, green; **Fl.** 4 mm $\varnothing$.

**T.** — f. **montana** (BURCH.) ROWL. (*T. m.* BURCH., *Dioscorea m.* v. *glauca* KNUTH). – **L.** bluish-grey and waxy-pruinose.

**T. glauca** MARL. (*Dioscorea hemicrypta* BURCH.). – S.Afr. – **Caudex** semi-underground, the aerial part conical, less deeply furrowed than **T. elephantipes**; **L.** up to 4 cm long, 2.5 cm across, mostly 5-veined.

**T. macrostachya** (BENTH.) ROWL. (*Dioscorea m.* BENTH.). – Mexico. – **Caudex** very similar to that of **T. elephantipes**; **L.** larger, blade 6–12 cm long and broad, with 7–9 veins; petiole 4–6 cm long.

**T. sylvatica** (ECKL.) KNUTH (*Dioscorea s.* ECKL., *D. hederifolia* GRISEB., *D. montana* DUR. et SCHINZ). – S.Afr. – **Caudex** flat and $\pm$ plate-like, 30–60 cm $\varnothing$, aboveground; **St.** long, trailing, branched above; **L.** $\pm$ triangular to cordate, often slightly 3-lobed; **Fl.** 5 mm $\varnothing$.

**T.** — v. **brevipes** (BURTT DAVY) ROWL. (*Dioscorea b.* BURTT DAVY, *D. s.* v. *b.* (BURTT DAVY) BURKILL, *T. paniculata* v. *b.* (BURTT DAVY) ROWL. – **L.** blade large, to 60 mm long by 80 mm broad; **Fl.** subsessile.

**T.** — v. **multiflora** (MARLOTH) ROWL. (*T. m.* MARL., *Dioscorea s.* v. *m.* (MARL.) BURKILL, *D. marlothii* KNUTH). – **Caudex** 10–15 cm thick; **L.** broadly cordate, blade 7.5 cm long, with 9 veins, apiculate, auricles over lapping; **Fr.** small.

**T.** — v. **paniculata** (DUEMM.) ROWL. (*T. p.* DUEMM., *Dioscorea s.* v. *p.* (DUEMM.) BURKILL, *T. montana* f. *p.* (DUEMM.) KUNTZE, *D. m.* v. *p.* (DUEMM.) KNUTH & v. *duemmeri* KNUTH). – **Caudex** domed or flattened, often convex below; **St.** long, trailing, stiff, sparingly branched; **L.** c. 5 cm long and across, broadly reniform or cordate, bluish; **Fl.** numerous, greenish, on pedicels up to 4 mm long.

**T.** — v. **rehmannii** (BAK.) ROWL. (*Dioscorea r.* BAK., *T. r.* (BAK.) ROWL., *D. s.* v. *r.* (BAK.) BURKILL). – **L.** bluntly cordate, with auricles connivent to overlapping, green, glabrous; **Fr.** large, 20–25 mm long.

**T.** — v. **sylvatica.** – **L.** blade small, up to 2 cm long and 4 cm broad, with 9 veins, auricles divergent.

**Thompsonella** BR. et R. (*Echeveria* DC. Sect. *Thompsonella* (BR. et R.) BGR.). Crassulaceae. – Mexico. – Glabrous ♃ plants resembling **Echeveria**, but with Pet. thin and spreading above midway; **Ros.** solitary, sessile or short-stemmed, flat; **Infl.** a slender thyrse, or spicate above or overall, Br. 20–70, in cincinni or dichotomous, **Fl.** $\pm$ sessile, 5-partite, Sep. erect or ascending, $\pm$ equal, clavate, Pet. united below, imbricate in the bud-stage, dark red on the upperside apart from the narrow yellow margins, anthers erect, carpels shortly stalked. Summer: in the open, winter: cold-house. Propagation: seed, division. Text: abbreviated from a contribution by **R. Moran**.

**T. minutiflora** (ROSE) BR. et R. (*Echeveria m.* ROSE, *E. tepeacensis* v. POELLN., *Graptopetalum mexicanum* MATUDA). – Mexico. – **Roots** thickened; **St.** 0.5–1 cm $\varnothing$, rarely 3 cm tall; **L.** 10–20, eventually shrivelling, elliptical to oblanceolate, 2–10 cm long, 1–2.5 cm across, with the upperside grooved, margins straight or curly; flowering shoots 5–35 cm tall; **Infl.** 3–30 cm tall, 1–2.5 cm thick, Cinc. with (1–)3 (–7) Fl., Cor. 6–10 mm $\varnothing$.

**T. platyphylla** ROSE (Pl. 139/2) (*Echeveria planifolia* BGR., *Villadia p.* (ROSE) E. WALTH.). – Mexico. – **Roots** not thickened; **St.** 1–3 cm thick, up to 12 cm tall; **L.** 8–15, oblanceolate, 6 to 12 cm long, 2–4 cm across, flat or grooved on the upperside; flowering shoots 10–40 cm tall; **Infl.** 5–25 cm tall, 2–6 cm across, Cinc. with 1–11 Fl., Cor. 6–8 mm $\varnothing$.

**Tradescantia** L. Commelinaceae. – S.Am. – 30 species are known, but only 3 with $\pm$ succulent L. are mentioned below. – Greenhouse, moderately warm. Propagation: seed, cuttings.

**T. crassifolia** CAV. – Mexico. – ♃ with white cylindrical **roots**; **St.** herbaceous, hard, cylindrical, erect, variously curved, branched, up to 60 cm long, the entire plant with minute, tufted hairs; **L.** alternate, ovate, acute, sessile, sheathing, thick, woolly; **Fl.** in terminal clusters, azure blue.

**T. navicularis** ORTG. (Pl. 139/4) (*Phyodina n.* ORTG.). – N.Peru. – Mat-forming ♃; **St.** creeping, short-jointed, rooting at the nodes, glabrous; **L.** distichous, imbricate, sessile, navicular-plicate, acute, very fleshy, greyish-green, the underside very carinate, with dense violet spots, 1–2 cm long, margins minutely ciliate; **Infl.** cymose, Fl. pink.

**T. sillamontana** MATUDA (*T. pexata* H. E. MOORE). – N.E.Mexico. – ♃ forming small cushions 10–15 cm $\varnothing$; **St.** 5–6 cm long, white-hairy, densely leafy; **L.** distichous, sheathing, amplexicaul, ovate-oblong, obtuse or $\pm$ acute, slightly plicate, c. 2 cm long, 1 cm across below, c. 2 mm thick, green, with dense white hairs; **Fl.** mauve to pink.

**Trematosperma** URB. Icacinaceae. – Somalia. – Greenhouse, warm. Propagation: seed.

**T. cordatum** URB. (Pl. 139/3). – Somalia. – Very succulent; **caudex** above ground, rock-like, acc. P. R. O. BALLY and A. ENGLER up to 3 m $\varnothing$; **St.** with a few glabrous, cylindrical, minutely hairy Br.; **L.** on a petiole 2–3 cm long, cordate, with a rounded tip, 6–8 cm long, 6–7 cm across, the M.nerve standing 1.5–3 mm above the surface, the lateral nerves also very prominent, minutely hairy; **Fl.** 1–3 mm $\varnothing$.

**Trichocaulon** N. E. BR. Asclepiadaceae. – Cape; S.W.Afr. – Very succulent ♃; **St.** soft fleshy, much thickened, spherical or cylindrical, simple or branching from the base, densely tuberculate, the tubercles in part arranged on numerous ribs or spiralled, often tipped with hairs or bristles; **Fl.** small, several together between the tubercles, Cor. flat or dish-shaped, deeply 5-cleft, lobes ovate, acute, variously coloured and blotched. – Greenhouse, warm. Propagation: seed; difficult from cuttings.

**T. alstonii** N. E. BR. – Cape: Lit. Namaqualand. – **St.** c. 15 cm tall, 3–4.5 cm ⌀, bluish, with many, densely tuberculate angles, the tubercles with a stiff Sp. 5–6 mm long; **Fl.** campanulate, 3 mm long and ⌀, lobes 3–4 mm long, sharply tapered.

**T. annulatum** N. E. BR. – Cape: Jansenville, Willowmore D. – **St.** 14–45 cm tall, 3–4.5 cm ⌀, cylindrical, with tubercles 5–6 mm long arranged in 23–30 vertical R. and tipped with a light brown bristle; **Fl.** with rotate Cor. 2 cm ⌀, 5-cleft to less than midway, with a very prominent annulus below, smooth outside, glabrous, dark reddish-brown inside, densely papillose apart from the base, lobes abruptly short-tapered, margins recurved.

**T. cinereum** PILL. – Cape: Lit. Namaqualand. – **St.** 2–5, cylindrical-ovoid, very rounded, glabrous, 4–5.5 cm ⌀, bluish-green, with ± rounded, spineless, tessellate tubercles; **Fl.** with a Cor. 7–9 mm ⌀, glabrous outside, with reddish blotches, lobes triangular, acute, 2–2.5 mm long, pale green, densely whitish-haired, with a distinct, hirsute annulus around the tube-mouth, the tube campanulate, pale green with purple blotches.

**T. clavatum** (WILLD.) H. HUBER (Pl. 140/1) (*Stapelia c.* WILLD., *S. cactiforme* HOOK., *T. c.* (HOOK.) N. E. BR., *T. marlothii* N. E. BR., *T. dinteri* BGR., *T. meloforme* MARL., *T. engleri* DTR., *T. keetmanshoopense* DTR., *T. sinusluederitzii* DTR., *T. sociorum* WHITE et SLOANE). – Cape: Lit. Namaqualand to Bushmanland and Prieska D.; Botswana; S.W.Afr.: Namib. – **St.** solitary or several, obovoid to ± cylindrical, 10–15 cm tall, 4–6 cm ⌀, or ± spherical, 4 to 6 cm ⌀, with dense tubercles in irregular, spiralled R., 4–6-angled to circular, separated by furrows, spineless, grey-green to whitish-grey; **Fl.** borne on the upper parts of the shoots, usually several together, Cor. campanulate-rotate, 6–13 mm ⌀, the tube bowl-shaped, the lobes broadly ovate-triangular, mostly abruptly tapered, whitish-yellow, cream-coloured, light yellow or yellowish-green, with red, reddish-brown or dark brown, often dense spots.

**T. delaetianum** DTR. – S.W.Afr.: Gr. Namaqualand. – **St.** 2–10, 10–20 cm tall, 4–5 cm ⌀, the tubercles in numerous R., with purplish-brown, sharp, thin Sp.; **Fl.** 3–5 together, Cor. broadly campanulate, 12 mm ⌀, brownish-red outside, brownish-yellow inside, with tapered lobes.

**T. flavum** N. E. BR. – Cape: Karroo. – Resembles **T. piliferum**; **St.** c. 15 cm tall, the thin Sp. on the tubercles darker brown; **Cor.** 10–12 mm ⌀, flat, rotate, smooth outside, minutely papillose inside, yellow, the triangular lobes tapered.

**T. grande** N. E. BR. – Cape: Laingsburg D. – **St.** 45–60 cm tall, 5 cm ⌀, branching c. 5 cm above soil-level, grey-green, tubercles in c. 30 vertical R., each tipped with a stiff Sp. 5–6 mm long; **Fl.** 1–3 together, Cor. 14 to 16 mm ⌀, smooth outside, glabrous, lobes and disc densely papillose, greenish-yellow, the tube campanulate, 3 mm long and deep, swollen at the throat, lobes broadly ovate, short-tapered, 5–6 mm long, 5 mm across.

**T. halenbergense** DTR. (Pl. 140/2). – S.W.Afr.: Gr. Namaqualand; Cape: Lit. Namaqualand. – **St.** erect, over 1 m tall, 5–6 cm ⌀, with c. 16 prominent angles with dense, short-conical tubercles, tipped with a brownish-grey, ± curved, hard Sp. 6–7 mm long; **Fl.** c. 15 mm ⌀, lobes triangular, tapered, yellow.

**T. kubusense** NEL (*T. kubusanum* NEL ex JACOBSEN). – Cape: Lit. Namaqualand. – **St.** 16 cm tall, c. 4 cm ⌀, clavate, distinctly 16-ribbed below, the ribs transversely divided, the tubercles in the upper part not dentate; **Fl.** 10–12 together, Cor. 1 cm ⌀, lobes 3 mm long, 2.5 mm across, greenish-white inside, papillose, with brownish-red blotches towards the centre and with a prominent annulus 4 mm ⌀.

**T. officinale** N. E. BR. – S.E.Cape; S.W.Afr. – **St.** 1–8, 20–40 cm tall, 6–7 cm ⌀, bluish-green, with 20–25 irregular ribs and light brown, rigid bristle-Sp. 1 cm long ± directed downwards, with a white tip; **Fl.** solitary, the Cor. without a distinct tube, flat, rotate, 8–9 mm ⌀, reddish-brown, yellow in the centre, minutely hairy inside, lobes triangular-ovate, tapered, 4.5 mm long and across.

**T. pedicellatum** SCHINZ. – S.W.Afr. – **St.** cylindrical, c. 2 cm thick, the tubercles on the angles with a tip 2–3 mm long; **Fl.** 1–4 together, Cor. rotate, 8–10 mm ⌀, dark brownish-red, with a short tube, lobes 4 mm long, 2–2.5 mm across, lanceolate, tapered, glabrous, minutely papillose inside.

**T. perlatum** DTR. – S.W.Afr.: Gr. Namaqualand. – **St.** several, up to 15 cm tall, up to 5 cm ⌀, light greyish-green; **Fl.** in clusters of 2–4, Cor. 3 mm ⌀, the underside yellowish-green with purple blotches, the upperside with dense, white, glossy papillae, the centre with a bowl-like depression.

**T. pictum** N. E. BR. – Cape: Lit. Namaqualand. – **St.** ± spherical or cylindrical-oblong, very rounded, 4–7 cm long, 4–5 cm ⌀, somewhat branched from the base, irregularly tessellate-tuberculate, glabrous, tubercles 10 mm ⌀, ± circular; **Fl.** 2–4 together, Cor. 1 cm ⌀, glabrous, smooth outside, minutely rough inside, pale whitish or yellowish, with brown blotches and short lines, lobes triangular, acute, 2.5 mm long, 3 mm across.

**T. piliferum** (L. f.) N. E. BR. (*Stapelia p.* L. f., *Piaranthus p.* SWEET). – S.E.Cape. – **St.** several, cylindrical, up to 20 cm tall, 4–5 cm ⌀, bluish-green, with tubercles in numerous, vertical R., each with a stiff, brown Sp. 3–5 mm long; **Fl.** foetid, several together, Cor. c. 18 mm ⌀, campanulate to funnel-shaped, light yellowish-red outside, dark reddish-brown inside, minutely tuberculate.

**T. pillansii** N. E. BR. v. **pillansii**. – Cape: Laingsburg D. – **St.** 12–18 cm tall, 3–5 cm ⌀, greyish-green, with many tuberculate angles, the tubercles with a stiff, bristle 4–5 mm long; **Fl.** several together, Cor. 9 mm ⌀, light whitish-yellow, smooth outside, glabrous, lobes triangular-ovate, very acute, up to 3 mm long and across, minutely papillose inside, the tube 3 mm long and across, smooth and glabrous inside.

**T. —** v. **major** N. E. BR. – **St.** and **Fl.** larger.

**T. pubiflorum** DTR. – S.W.Afr.: Gr. Namaqualand. – **Shoots** several, unbranched, c. 15 cm tall, c. 4 cm ⌀, dark bluish-green, the angles regular, with tubercles tipped with a spiny bristle c. 4 mm long; **Fl.** 3–5 together, Cor. flatly saucer-shaped, 11 mm ⌀, minutely hairy in the centre, lobes 4 mm long, 5 mm across, tapered, limb yellowish-green outside, lobes reddish-brown, the inside yellow in the centre, then a yellow zone with red blotches passing into olive-brown to green, margins violet to brown.

**T. rusticum** N. E. BR. – Cape: Kenhardt D. – **St.** 12–13 cm tall, similar to **T. piliferum**, conical, the tubercles in 17 or more vertical R., with a spiny tip 4 mm long; **Cor.** with a tube 2 mm long, both surfaces glabrous, lobes triangular-ovate, acute, 3 mm long, dark reddish-brown, with microscopic pubescence.

**T. simile** N. E. BR. – Cape: Van Rhynsdorp, Prieska, Kenhardt D. – **St.** ± ovoid-spherical, 4.5 cm tall, 4 cm ⌀, with stout tubercles, the youngest ones with a small tip; **Cor.** 9 mm ⌀, glabrous, minutely rough-papillose inside, deep reddish-brown, the tube 2.5 mm long, 5 mm across, bowl-shaped, the lobes projecting, 5 mm long and across, triangular, acute.

**T. somaliense** GUILL. – Somalia. – **St.** 5–8 cm tall, cylindrical, with 12–15 somewhat spirally contorted, tuberculate angles; **Fl.** 7 mm ⌀. (Acc. P. R. O. BALLY perhaps referable to **Echidnopsis**.)

**T. triebneri** NEL. – S.W.Afr.: Swakop near Okandu. – **St.** 30 cm tall, 4 cm ⌀, bluish to greyish green, with 15–16 angles separated by deep furrows 3–4 mm across, the angles with prominent tubercles tipped with a 5 mm long Sp.; **Fl.** in clusters of 8–10, Cor. 1.5 cm ⌀, rotate, the tube 4 mm long, yellow inside, minutely papillose outside, blackish or reddish-purple, lobes triangular, tapered, the margins between the lobes very slightly dentate, lobes and tube with minute conical papillae tipped with a purplish-red Sp.

**T. truncatum** PILL. – Cape: Lit. Namaqualand. – **St.** c. 10 cm tall, cylindrical or ± clavate, 3.5–4 cm ⌀, greyish-green, with irregular, 5-angled, slightly prominent, truncate tubercles; **Cor.** 8 mm ⌀, cream-coloured inside, with purple blotches, tube 1.5 mm deep, with long papillae around the mouth, lobes ovate, acute, 2.5–3 mm long, papillose.

**Tumamoca** ROSE. Cucurbitaceae. – N.Am. – ♃ caudiciform plants with thin, climbing St. – Greenhouse, warm. Propagation: seed.

**T. macdougallii** ROSE. – N.Am.: Arizona to Mexico. – **Caudex** thick, tuberous, in part projecting above-ground; **St.** ⊙, thin, long; **L.** thin, pedate to 3-partite, the lobes deeply cleft or the blade divided into narrow segments; **Fl.** unisexual, the ♂ Fl. in a short Rac. and ♀ Fl. solitary, Cor. pale yellow, lobes narrow; **Fr.** berry-like, spherical, red or yellow, 8–10 mm ⌀.

**Ullucus** CALDAS. Basellaceae. – S.Am. – Greenhouse, cool. Propagation: seed, from Tub.

**U. tuberosus** CALDAS. – Peru; Chile. – **Caudex** tuberous red, edible; **St.** long, twining; **L.** cordate or circular, thick; **Fl.** numerous, small, in slender, lax Rac.

**Umbilicus** DC. Crassulaceae. – Medit. reg. to W.As.; Atlantic central Eur.; Ethiopia. – Dwarf ♃ with tuberous **roots**; **St.** simple; **L.** alternate, petiolate, round-scutate, depressed centrally, green; **Fl.** terminal in a slender Rac., greenish-white. Summer: in the open, winter (dormancy): cold-house. Propagation: seed.

**U. botryoides** HOCHST. (*Cotyledon umbilicus* AUCT., *C. u.* v. *botryoides* (HOCHST. ex A. RICH.) ENGL.). – Ethiopia. – Plant 3–10 cm tall; bracts broad-lanceolate, petiolate; **Fl.** broadly urn-shaped or spherical-campanulate, 5–6 mm long.

**U. chloranthus** HELDR. et SART. (*Cotyledon c.* HAL.). – Yugoslavia: Dalmatia, to Greece: Crete, Cyclades. – Basal **L.** very long-petiolate; **Infl.** paniculately branched from the base, Fl. ± round below, 3 mm long.

**U. citrinus** WOLLEY-DOD (*U. pendulinus* v. *bracteonis* WILLK.). – S.Spain. – Rac. one-sided, **Fl.** projecting or nodding.

**U. erectus** (L.) DC. (*Cotyledon umbilicus-repens* L., *C. u.-veneris* L. acc. HAYEK, *C. lutea* HUDS.). – Yugoslavia: Serbia. – **Caudex** thick, long, creeping; cauline **L.** short-petiolate, dentate; Fl. 12–14 mm long, yellow.

**U. gaditanus** BOISS. – S.Spain; Portugal; Algeria. – Upper cauline **L.** with a small, dentate blade; **Fl.** upright to projecting, dense, in a long Rac.

**U. giganteus** BATT. – Algeria: littoral. – **Fl.** 4 mm ⌀, 10 mm long.

**U. heylandianus** WEBB et BERTH. – Canary Is.: La Palma. – **Fl.** projecting, straw-coloured.

**U. horizontalis** (GUSS.) DC. (*Cotyledon h.* GUSS.). – S.Medit. reg. to Canary Is.; Balkans. – **Fl.** patent, 6 mm long.

**U. intermedius** BOISS. (*Cotyledon i.* STAPF). – Bulgaria to As. Minor; Cyprus; Israel; Egypt. – **Fl.** nodding, 6–8 mm long, yellowish-white, in a dense Rac. 10–16 cm long.

**U. lassithiensis** GAND. (*Cotyledon l.* HAYEK). – Crete. – Basal **L.** large, cordate to circular, upper L. truncate, ovate, sharply serrate; **Fl.** yellowish.

**U. parviflorus** DC. (*Cotyledon p.* SIBTH. et SM., *U. sprunerianus* BOISS.). – Greece: Cyclades, Crete. – Cauline **L.** ± circular or reniform; **Fl.** 4–5 mm long, projecting, in an erect, racemose-paniculate Infl.

**U. patens** POMEL (*Cotyledon umbilicus-veneris* v. *p.* (POMEL) BATT., *U. pendulinus* v. *deflexus* (POMEL) BATT.). – Resembles **U. parviflorus; Fl.** nodding, dense, small.

**U. rupestris** (SALISB.) DANDY v. **rupestris** (Pl. 140/3) (*Cotyledon r.* SALISB., *U. pendulinus* DC., *C. p.* BATT., *C. peltatus* WENDL., *C. tuberosus* HAL., *C. umbilicus tuberosus* L.). – Medit. reg.; As. Minor to Morocco; Portugal; Madeira; Canary Is.; S. and central France; England; Austria: Tyrol; Yugoslavia: Dalmatia; Egypt. – Root-**Tub.** up to 3 cm ⌀; **L.** rosulate at first, long-petiolate, circular, green, L. and the entire plant with reddish lines, becoming brownish-red in full sunshine; **Infl.** 20–40 cm tall, with several reniform, dentate bracts, Fl. nodding, greenish, with red spots.

**U. —** v. **truncatus** (WOLLEY-DOD) ROWL. (*U. pendulinus* v. *t.* WOLLEY-DOD). – S.Spain. – All **L.** laterally petiolate, with a cordate or truncate base.

**U. —** v. **velenovskyi** (ROHL) ROWL. (*U. pendulinus* v. *v.* ROHL). – Yugoslavia: Crna Gora (Black Mts.). – **Infl.** dense, lower bracts spatulate.

**U. schmidtii** BOLLE. – Canary Is.; Cape Verde Is. – **Fl.** projecting, distant, irregularly arranged, dirty yellow, in a lax Rac.

**U. tropaeolifolius** BOISS. – Iraq. – Basal **L.** scutate, long-petiolate; Ped. weak, curved in various directions, with lax, intricate Br., **Fl.** whitish.

**Uncarina** (BAILL.) STAPF (*Harpagophytum* Sect. *Uncarina* BAILL.). Pedaliaceae. – Madag. – Deciduous ♄ or ♄, up to 8 m tall; **St.** up to 30 cm ⌀; **L.** alternate, large, ± emarginate or lobed, probably ± succulent; **Fl.** several, axillary; **Fr.** with barbed hooks. – 9 species are known. Of no significance in cultivation.

**Vauanthes** HAW. Crassulaceae. – S.Afr. – Insignificant ⊙. Propagation: seed.

**V. dichotoma** (L.) KTZE. (Pl. 140/4) (*Crassula d.* L., *V. chloraeflora* HAW., *Grammanthes c.* (HAW.) DC., *G. gentianoides* (LAM.) DC.). – Cape: Cape, Clanwilliam, Van Rhynsdorp D. – Grey, stiff plant; **L.** opposite, oblong or linear, ± fleshy, with entire, pearly margins; **Fl.** in paniculate Infl., yellow or orange.

**Villadia** HAW. Crassulaceae. – Mexico to Peru. – Mostly ♃, more rarely ⊙, often matforming; **roots** often thickened, tuberous; **St.** creeping, often rooting and erect, ± branched; **L.** amplexicaul or sessile, linear or ovate-spatulate, often spurred, small, ± dense; **Fl.** in a simple spike, a panicle or a cyme, the Infl. often dichotomous, sometimes with leaflike bracts, Fl. solitary or 2–3 together, in the axils of the leaflike bracts, small, white, reddish, yellowish, orange or greenish. – Summer: in the open. Propagation: seed. (For the nomenclature see R. T. CLAUSEN, Bull. Torray Bot. Club, LXVII, 1940, 195–198.)

**V. albiflora** (HEMSL.) ROSE (*Cotyledon a.* HEMSL., *Altamiranoa a.* (HEMSL.) E. WALTH.). – Mexico: Oaxaca. – ♃ with smooth **St.; L.** ovate-oblong, obtuse, thick; **Fl.** 5–6 mm long, white.

**V. alpina** (FROED.) JACOBS. (*Altamiranoa a.* FROED.). – Mexico: Ixtacihuatl. – Possibly a dwarf, alpine form of **V. batesii**; cushion-forming ♃; **St.** woody below, freely branched; **L.** broadly spurred, oblong, obtuse, 5–7 mm long; **Fl.** solitary or few together.

**V. andina** (BALL.) BAEHNI et MACBR. (*Sedum a.* BALL.). – Peru. – ♃, mat-forming, freely branched; **L.** hemispherical, very small; **Fl.** blackish-red, c. 4 in a cyme.

**V. batesii** (HEMSL.) BAEHNI et MACBR. v. **batesii** (*Cotyledon b.* HEMSL., *Altamiranoa b.* (HEMSL.) ROSE). – Central Mexico. – ♃; **St.** 10–15 cm tall, reddish, sterile shoots with a L.Ros.; **L.** cylindrical, acute, 1 cm long, minutely rough-tuberculate; **Fl.** reddish, few in a flat cyme.

**V. —** v. **subalpina** (FROED.) ROWL. (*Altamiranoa b.* v. *s.* FROED.). – Mexico. – Glabrous; **L.** smaller, bluntly spurred, truncate.

**V. berillonana** (HAMET) BAEHNI et MACBR. (*Sedum b.* HAMET, *Altamiranoa b.* (HAMET) BGR.). – Peru. – ♃ with **St.** creeping below, eventually erect; **L.** ovate, glabrous, spurred, up to 5 mm long, rather obtuse; **Fl.** few, in a cyme.

**V. calcicola** (ROBINS. et GREENM.) JACOBS. (*Sedum c.* ROBINS. et GREENM., *Altamiranoa c.* ROBINS. et GREENM.) ROSE). – Mexico: San Luis Potosí. – ♃ with prostrate **shoots**; flowering shoots 20–30 cm tall, greyish-green; **L.** numerous, thick, cylindrical, rather obtuse, 10–16 mm long; **Fl.** numerous, greenish-yellow, 5 mm long.

**V. chihuahuensis** (S. WATS.) JACOBS. (*Sedum c.* S. WATS., *Altamiranoa c.* (S. WATS.) ROSE). – Mexico: Chihuahua. – ♃, with tuberous **roots**; **St.** 7–15 cm long, branched above; **L.** 2–3 cm long, obtuse, **Fl.** whitish.

**V. cucullata** ROSE (*Altamiranoa c.* (ROSE) E. WALTH.). – Mexico: Coahuila. – ♃ with very thick **roots**; **St.** 10–30 cm tall; **L.** ± cylindrical, spurred, acute, 2–2.5 cm long; **Fl.** white.

**V. decipiens** (BAK.) JACOBS. (*Cotyledon d.* BAK., *Altamiranoa d.* (BAK.) FROED.). – Peru. – Glabrous ♃, with simple or somewhat branched **St.** c. 20 cm long; **L.** spurred, semicylindrical, obtuse; **Fl.** white, in a dense corymb.

**V. dielsii** BAEHNI et MACBR. (*Cotyledon stricta* DIELS, *Altamiranoa s.* (DIELS) BGR.). – Peru: Ancash. – ♃; **St.** stiffly branched, 10–20 cm tall; **L.** dense, triangular-ovate, rather obtuse, 6–8 mm long, 1–1.5 mm across; **Fl.** 5–8 mm long, whitish.

**V. diffusa** ROSE (*Altamiranoa d.* (ROSE) JACOBS.). – Mexico: Chiapas. – ♃ with freely branched **St.**; **L.** triangular-ovate, 6 mm long and across; **Fl.** 6 mm long, flesh-coloured.

**V. dyvrandae** (HAMET) BAEHNI et MACBR. (*Sedum d.* HAMET, *Altamiranoa d.* (HAMET) BGR.). – Peru: Matucana. – ♃ with glabrous, ascending **St.** 9–14 cm tall; **L.** long-ovate, rather obtuse, 2–6 mm long, spurred; **Infl.** branched, Fl. 4–5 mm long, greenish.

**V. elongata** (ROSE) T. B. CLAUSEN (*Altamiranoa e.* ROSE). – Mexico: Pachuca. – Pubescent ⚃; **St.** slender, soon prostrate and rooting; **L.** linear-ovate, acute, 6 mm long; **Infl.** paniculate, Fl. campanulate, 5 mm long, white or reddish.

**V. fusca** (HEMSL.) JACOBS. (*Sedum f.* HEMSL., *Altamiranoa f.* (HEMSL.) ROSE). – Mexico: San Luis Potosí. – ⚃ with spreading **St.** up to 10 cm tall; **L.** 4–6 mm long, oblong, obtuse; **Fl.** in cymes, whitish.

**V. galeottiana** (HEMSL.) JACOBS. (*Cotyledon g.* HEMSL., *Altamiranoa g.* (HEMSL.) ROSE). – Mexico: Oaxaca. – ⚃; **L.** ± lanceolate-spatulate, 12–18 mm long; **Fl.** white.

**V. goldmanii** (ROSE) BGR. (*Altamiranoa g.* ROSE). – Mexico: Oaxaca. – ⚃ with tuberously thickened **roots**; **St.** ascending, rooting, 5–6 cm tall, densely leafy; **L.** linear, obtuse, 10–12 mm long; **Fl.** few, 6 mm long, yellow and red, in a dense cyme.

**V. grandisepala** (R. T. CLAUSEN) R. T. CLAUSEN (*Sedum g.* R. T. CLAUSEN). – Mexico: Oaxaca. – Erect, branched ♄ c. 15 cm tall; **L.** elliptical-oblong, obtuse to subacute, flat to convex, 10–17 mm long, 3.4–4.4 mm across, 1.2–2.2 mm thick; **Fl.** 3–7, c. 9–12 mm ⌀, Sep. c. 10 mm long, pale green, Pet. green.

**V. grandyi** (HAMET) BAEHNI et MACBR. (*Sedum g.* HAMET, *Altamiranoa g.* (HAMET) BGR.). – Peru: Chachapoyas. – ⚃ with erect, glabrous **St.** up to 10 cm long; **L.** broadly ovate, obtuse, up to 5 mm long, 4.2 mm across, spurred; **Infl.** of few Fl.

**V. guatemalensis** ROSE (Pl. 141/1) (*Altamiranoa g.* (ROSE) E. WALTH.). – Guatemala. – ⚃ with freely branched, prostrate **St.**; **L.** dense, cylindrical, acute, 15–20 mm long; **Fl.** few, lemon-yellow.

**V. hemsleyana** (ROSE) JACOBS. (*Altamiranoa h.* ROSE, *Sedum batesii* HEMSL.). – Mexico: Oaxaca. – ⊙ 3–6 cm tall; **L.** linear-spatulate, 6–10 mm long; **Fl.** white.

**V. imbricata** ROSE (*Altamiranoa i.* (ROSE) E. WALTH., *A. ericoides* JACOBS.). – Mexico: Oaxaca. – Mat-forming ⚃; **St.** a little branched, 2–6 cm long, tapering into a short flowering spike; **L.** densely imbricate and appressed, acute, 5–6 mm long, 2 mm across below, the underside carinate, minutely tuberculate; **Fl.** 4–5 mm long, white.

**V. incarum** (BALL.) BAEHNI et MACBR. (*Cotyledon i.* BALL., *Altamiranoa i.* (BALL.) BGR.). – Peru to Chile. – ⚃; **St.** erect, 15–20 cm tall, branched from the base; **L.** cylindrical; **Fl.** in an erect Rac., yellowish-white to pale reddish.

**V. jurgensenii** (HEMSL.) JACOBS. (*Cotyledon j.* HEMSL., *Altamiranoa j.* (HEMSL.) ROSE). – Mexico. – Hairy ⚃; **St.** erect; **L.** ovate-oblong, obtuse, 3–4 mm long; **Fl.** reddish, 2–3 in a Rac.

**V. levis** ROSE (*Altamiranoa l.* (ROSE) E. WALTH.). – Mexico: Oaxaca. – Glabrous ⚃; **St.** 30–50 cm tall, sparsely branched, with lateral buds below; **Fl.** yellowish-brown.

**V. mexicana** (SCHLTD.) JACOBS. (*Umbilicus m.* SCHLTD., *Cotyledon m.* (SCHLTD.) HEMSL., *Altamiranoa m.* (SCHLTD.) ROSE). – Mexico: Hidalgo. – ⚃ with erect, branched **St.**; **L.** linear; **Fl.** solitary or 2–3 together, white.

**V. minutiflora** ROSE (*Altamiranoa m.* (ROSE) E. WALTH.). – Mexico: Oaxaca. – ⚃ with short, stiff hairs; **St.** 10–20 cm tall; **L.** very numerous, cylindrical, spurred, 6–10 mm long, 1 mm thick; **Fl.** in a lax spike, 3 mm long, whitish.

**V. misera** (LINDL.) R. T. CLAUSEN (*Sedum m.* LINDL., *Cotyledon parviflora* HEMSL., *V. p.* (HEMSL.) ROSE, *Altamiranoa p.* (HEMSL.) JACOBS.). – Mexico. – ⚃; **St.** straight, with a nodding tip; **L.** dense, semicylindrical below; **Fl.** in an interrupted spike, white with red.

**V. nelsonii** ROSE (*Altamiranoa n.* (ROSE) E. WALTH.). – Mexico: Guerrero. – ⚃ with ± rough **St.** 20–30 cm tall; **L.** spatulate, 10 to 15 mm long; **Fl.** 5–6 mm long, white.

**V. nexacana** (FROED.) JACOBS. (*Altamiranoa n.* FROED.). – Mexico: Puebla, Nexaca. – Resembles **V. elongata**, possibly a glabrous variety of this; **St.** very thin, 20–40 cm long, with few Br.; **L.** with a broad, concave spur, semi-ovate, rather obtuse, 4–5 mm long; **Infl.** of few Fl.

**V. painteri** ROSE (*Altamiranoa p.* (ROSE) E. WALTH.). – Mexico: Guadalajara. – ⚃ with woody **St.**; **L.** slender, clavate, pale green; **Fl.**-spike interrupted, 4–10 cm long, Fl. 2 mm long, white.

**V. parva** (HEMSL.) JACOBS. (*Sedum p.* HEMSL., *Altamiranoa p.* (HEMSL.) ROSE). – Mexico: San Luis Potosí. – Freely branched, mat-forming ⚃; **L.** oblong, 3–5 mm long; **Fl.** yellow.

**V. platystyla** (FROED.) R. T. CLAUSEN (*Sedum p.* FROED.). – Mexico: Jalisco. – Glabrous sub-♄; **St.** small; **L.** ± rosulate, spatulate to ± ovate, spurred, tuberculate above midway, acute to apiculate, 9–13 mm long, those on flowering shoots lanceolate, subacute, 10 to 20 mm long, not tuberculate; **Infl.** a dense corymb, Fl. pendulous, ? white.

**V. pringlei** ROSE (*Altamiranoa p.* (ROSE) E. WALTH.). – Mexico: Chihuahua. – ⚃ with **St.** branching freely from the base, 5–15 cm tall; **L.** linear; **Infl.** a spike or panicle, Fl. 6 mm long, whitish.

**V. ramosissima** ROSE (*Altamiranoa r.* (ROSE) E. WALTH.). – Mexico: Puebla, Oaxaca. – ⚃; **St.** woody, freely branching from the base; **L.** projecting; **Fl.** solitary in a lax spike, 4 mm long, pink to flesh-coloured.

**V. ramulosa** (FROED.) JACOBS. (*Altamiranoa r.* FROED.). – Mexico: Rio de Lerma. – **St.** several, thin, elongated, 5–8 cm long; **L.** broadly spurred, linear, tuberculate, obtuse, 6–10 mm long; **Infl.** of few Fl. c. 6 mm long.

**V. reniformis** JACOBS. (*Cotyledon imbricata* DIELS, *Altamiranoa i.* (DIELS) BGR.). – Peru: Catamarca. – ⚃, much branched below; **St.** ascending, 5–10 cm tall; **L.** densely imbricate, broadly triangular or ± reniform, 2–3 mm long, 3–4 mm across; **Fl.** in a cyme, 3–4 mm long, white.

**V. scopulina** (ROSE) JACOBS. (*Altamiranoa s.* ROSE, *V. s.* (ROSE) R. T. CLAUSEN). – Mexico: Puebla. – ⚃; **St.** creeping, much branched, the lower part with scale-like, white remains of old L.; **L.** linear, 4–8 mm long; **Fl.** few on the shoot-tips, white.

**V. squamulosa** (S. WATS.) ROSE (*S. a.* S. WATS., *Cotyledon parviflora* v. *s.* S. WATS., *Altamiranoa s.* (S. WATS.) E. WALTH., *A. s.* (S. WATS.) E. WALTH.). – Mexico: Chihuahua. – ♃ with slender, erect or ascending **St.**; **L.** projecting; **Fl.** pink.

**V. stricta** ROSE (*Altamiranoa s.* (ROSE) E. WALTH., *A. erecta* JACOBS.). – Mexico: Zacatecas. – ♃ with **St.** 10–20 cm tall; **L.** linear; **Infl.** a spike or panicle with white Fl. 3 mm long.

**V. virgata** (DIELS) BAEHNI et MACBR. (*Cotyledon v.* DIELS, *Altamiranoa v.* (DIELS) BGR.). – Peru: Ancash. – ♃ with stiffly erect **St.** and **Br.** 20–25 cm tall; **L.** ovate, 5–8 mm long, 2–3 mm across; **Fl.** in a simple spike, whitish.

**V. weberbaueri** (DIELS) BAEHNI et MACBR. (*Cotyledon w.* DIELS, *Altamiranoa w.* (DIELS) BGR.). – Peru: Amazonas. – ♃; **St.** prostrate, Br. ascending, 10–15 cm tall; **L.** dense, subacute, broadly ovate, 6–9 mm long, 2.5–5 mm across; **Fl.** in a forked cyme, 10 mm long, whitish.

**Welwitschia** HOOK. f. Welwitschiaceae. – Angola; S.W.Afr. – A highly anomalous member of the **Gnetales,** an isolated order intermediate between **Gymnosperms** and **Angiosperms.** Dioecious; not strictly a succulent plant. – Greenhouse, warm. Propagation: seed.
For the hotty disputed nomenclature see: Taxon XXI, iv: 485–489, 1972 and Cact. & Succ. Journ. Am. 42: 200, 1970. Acc. BENSON **Welwitschia mirabilis** is valid.

**W. mirabilis** HOOK. f. (Pl. 141/4) (*Tumboa bainesii* HOOK. f., *W. b.* (HOOK. f.) CARR. – Angola: near Ponta Albina; S.W.Afr.: coastal deserts. – **Caudex** low, woody, obconical in young plants, with a short taproot, the apical surface of the caudex swollen in age, divided by a furrow into 2 halves, often 1–2 m ⌀; **L.** 2 only, long, ligulate, leathery, produced from transverse furrows below the swollen portion, prostrate and lying twisted on the ground, often longitudinally split, continuing to grow from the base, often 2 m or more long, up to 20 cm or more across, green, ± waxy pruinose; flowering **shoots** produced from the apex of the caudex, bearing ♂ or ♀ cones at the tips of forked Br.; **seeds** broadly alate.

**Whitesloanea** CHIOV. Asclepiadaceae. – Somalia. – Close to **Caralluma.** – Greenhouse, warm. Propagation: seed.

**W. crassa** (N.E.BR.) CHIOV. (Pl. 141/3) (*Caralluma c.* N. E. BR., *Drakebrockmania c.*) (N. E. BR.) WHITE et SLOANE). – Somalia: near Odweina. – Dwarf, succulent ♃; **shoots** erect, 10–15 cm tall, 6–7 cm ⌀, with acute, denta teangles, smooth, light green, slightly bluish; **Fl.** borne on young shoots, on pedicels 8 mm long, c. 4 cm ⌀, smooth, whitish-green outside, with purple blotches, light yellow inside, with dark red blotches, the tube with some hairs, campanulate, 12 mm ⌀ and deep, lobes curved to spreading, 18 mm long, 10 mm across, ovate or ovate-lanceolate, with a few hairs between the lobes.

**Xerosicyos** HUMBERT. Cucurbitaceae. – Madag. – Lianas with tendrils; **L.** alternate, elliptical to ± circular, rather fleshy; ♂ Fl. solitary or several, ♀ Fl. in umbellate Infl., Pet. 5, yellowish; **Fr.** spherical to ovoid. – Greenhouse, warm. Propagation: seed, cuttings.

**X. danguyi** HUMBERT (Pl. 141/5). – W. and S.W.Madag. – Dioecious liana up to 5 m tall, branching from the base; **Br.** cylindrical, erect or prostrate, L. and tendrils opposite; **L.** on a petiole 8 mm long and opposite to a tendril, obovate, 4 cm long, c. 3 cm across, ± fleshy; **Fl.** greenish-yellow; **Fr.** small.

**X. decaryi** GUILL. et KERAUDREN (Pl. 141/5). – Madag.: Soalala D. – **Br.** slender, c. 3 mm ⌀; tendrils not very numerous; **L.** on a petiole 2 mm long, elliptical, 2.5 cm long, 1 cm across, 2 mm thick, with a rounded tip, the underside carinate; **Pet.** fleshy, 3 mm long, 2 mm across; **Fr.** 1.5 cm long, 1 cm ⌀.

**X. perrieri** HUMBERT. – W. and S.W.Madag. – Resembles **X. decaryi**; **L.** c. 1 cm long, round to ± ovate, short-tapered, fleshy.

**X. pubescens** KERAUDREN. – Madag. – **L.** petiolate, 3–4 cm long, 2–3 cm across, with dense, velvety hairs. (Probably not succulent.)

**Zehneria** ENDL. Cucurbitaceae. – S.Afr. – Vigorous-growing; the tuberous rhizomes are the only characteristic of succulence. Coldhouse. Propagation: seed.

**Z. hederacea** SOND. – Cape: Krom R., in woodland. – **Taproot** tuberous; **St.** with tendrils, climbing; **L.** cordate, acute, apiculate, c. 5 cm long and across or somewhat broader, margins entire, 3–5-lobed; ♂ **Fl.** few in an elongated, racemose, hairy Infl., ♀ Fl. not known.

**Zygophyllum** L. Zygophyllaceae. – S.W.Afr.; Canary Is. – Small, branched ♄; **L.** mostly sessile, thick-fleshy, sometimes paired or divided, deciduous; **Fl.** numerous, small; **Fr.** berry-like. – Of the many highly succulent species, few are amenable to cultivation. – Greenhouse, warm. Propagation: seed, cuttings.

**Z. clavatum** SCHLTR. et DIELS. – S.W.Afr.: Namib. – Dwarf, branched ♄; **Br.** cylindrical, glabrous, leafy; **L.** simple, thick, obovate-clavate to ± apple-shaped, obtuse, the lower ones 5–7 mm long, 5–6 mm across, the upper ones 4 mm long, 3.5 mm across, on petioles 2.5 mm long; pedicels 2.5 mm long, thickened; **Fl.** scarcely 3 mm ⌀.

**Z. fontanesii** WEBB. et BERTH. (Pl. 141/2). – Canary Is. – Broad ♄ c. 40 cm tall; **Br.** numerous, ascending; **L.** always paired, ± clavate, up to 15 mm long, 5–6 mm across above, bluish-green, fleshy, turning golden-yellow before dropping; **Fl.** numerous, small, stellate, pink; **Fr.** ovoid, golden-yellow to orange-yellow.

**Z. leucocladum** DIELS. – S.W.Afr.: Gr. Namaqualand. – Small ♄; **Br.** cylindrical, stiff, projecting, variously curved, with a whitish bark; **L.** sessile, fleshy, paired, obovate to oblong, 10–15 mm long, 4–5 mm across, with minute Stip.; pedicels 6–8 mm long, thickened above, Fl. c. 1.5 cm ⌀.

**Z. stapfii** SCHINZ. – S.W.Afr.: coastal area around Swakopmund. – ♄ with large, fleshy, orbicular leaflets in pairs on short thick petioles.

# PART II. FAMILY MESEMBRYANTHEMACEAE

Mesembryanthemaceae Fenzl emend. Herre et Volk

**Distribution of the Family:** The dry reg. of southern Afr., with a few G. extending northwards. A few herbaceous species can be found along the Medit. coast of N.Afr., on some Atlantic is., and in the deserts of Arabia, the Middle East and As. Minor. Mes. crystallinum is found on the coasts. of Peru and Calif. The G. Carpobrotus occurs in Austr., New Zealand, Chile, USA: Calif., and Mexico: Baja Calif.; Sarcozona and Lampranthus tegens are found in Austr., and Disphyma australe occurs in Austr. and New Zealand. Mestoklema macrorrhizum grows on Reunion Is., and a species of Delosperma (?) has been discovered at Cape Ste. Marie, Madag.
(Lit.: HERRE, H.: The Genera of the Mesembryanthemaceae, Cape Town, 1971.)

## 1. Systematic Classification of the Mesembryanthemaceae by G. Schwantes[1]

**A. Axillary Placentation** (Pl. 142/1, 2).

  **Sub-Fam. I. Mesembryanthemoideae** IHL., SCHWANT. et STRAKA (*Aptenioideae* SCHWANT.). – Fruits opening when moistened, with numerous seeds in the individual cells.

   **Tribe I: Mesembryanthemeae** (*Aptenieae* SCHWANT.). – Pet. soft, not rigid.

    **Sub-Tribe 1: Apteniinae** SCHWANT. – ♃ herbs; capsule with 4 cells. – Genera: Aptenia, Platythyra.

    **Sub-Tribe 2: Mesembryantheminae** (*Hydrodeinae* SCHWANT.). – ☉ or ⊙ herbs with flat to cylindrical L.; capsule with 5 cells. – Genera: Mesembryanthemum (incl. *Callistigma, Hydrodea, Halenbergia*), Eurystigma, Opophytum, Synaptophyllum.

    **Sub-Tribe 3: Preniinae** SCHWANT. – ♃ herbs, main St. with short Int., flowering shoots elongated; capsule with 4–5 cells. – Genera: Prenia, Sceletium.

    **Sub-Tribe 4: Aridariinae** SCHWANT. (*Nycteranthinae* SCHWANT.). – Erect or creeping ♄, often with tuberous or napiform roots; capsule with 4–5 cells. – Genera: Aridaria, Sphalmanthus.

    **Sub-Tribe 5: Psilocaulinae** JACOBS. (*Brownanthinae* SCHWANT.). – ♄ with L. deciduous or persisting as Th.; capsule with 5 cells. – Genera: Amoebophyllum, Psilocaulon (incl. *Brownanthus*).

   **Tribe 2: Dactylopsideae** SCHWANT. – Fl. in Dactylopsis with stiff Pet.; dwarf or acauline ♄; L. sometimes opposite, with very long Sh., digitate, exceedingly succulent; capsule with 5 cells. – Genera: Aspazoma, Dactylopsis.

  **Sub-Fam. II. Hymenogynoideae** SCHWANT. – ☉ herbs; L. flat; schizocarp with many cells, seeds embedded in the cells (pockets). – Sole G.: Hymenogyne.

  **Sub-Fam. III. Caryotophoroideae** IHL., SCHWANT. et STRAKA. – ♃ herbs with a ♃ taproot; carpels divided into 2, each half with 1 seed (rarely 2), Fr. a schizocarp, the divisions woody. – Sole G.: Caryotophora.

**A. Parietal to Basal Placentation** (Pl. 142/3, 4).

  **Sub-Fam. IV. Ruschioideae** SCHWANT.

   **Tribe 1: Ruschieae** SCHWANT. – Fruit a capsule, opening when moistened, with numerous seeds in each cell.

    **Sub-Tribe 1: Ruschiinae** SCHWANT. – ♄ or acauline; capsule with cells 5 or numerous, with placental tubercles, valve-wings missing or rudimentary, cell-lids often with a closing mechanism. – Genera: Acrodon, Astridia, Bergeranthus, Bijlia, Carruanthus, Eberlanzia, Hereroa, Machairophyllum, Ottosonderia, Rhombophyllum, Ruschia, Stayneria.

---

[1] First published in Sukk. Kd. I, Schweiz. Kakt. Ges. 1947. Revised and amended by H. STRAKA and H.-D. IHLENFELDT in collaboration with G. SCHWANTES. Abbreviated version and further additions: H. JACOBSEN. — No detailed taxonomy can be undertaken in the present work, and the reader is referred to the Bibliography: IHLENFELDT, STRAKA, SCHWANTES.

**Sub-Tribe 2: Leipoldtiinae** Schwant. – ♄ or acauline; capsule with many cells, with cell-lids, sometimes with a closing mechanism. – Genera: Argyroderma, Calamophyllum, Cephalophyllum, Cheiridopsis, Cylindrophyllum, Fenestraria, Leipoldtia, Octopoma, Odontophorus, Polymita, Schlechteranthus, Vanzijlia, Vanheerdea.

**Sub-Tribe 3: Lampranthinae** Schwant. – ♄ or acauline; capsule with 5 cells, with cell-lids and sometimes with placental tubercles. – Genera: Braunsia, Cerochlamys, Dicrocaulon, Disphyma, Ebracteola, Lampranthus, Oscularia.

**Sub-Tribe 4: Jacobseniinae** Schwant. – ♄; L. cylindrical, papillose at first, corpuscular; capsule with 5 cells, with cell-lids and valve-wings (Lampranthus type). – Sole G.: Jacobsenia.

**Sub-Tribe 5: Stoeberiinae** Friedr. – ♄, dwarf or up to 2 m tall; capsules usually 5-celled, valve-wings well developed or rudimentary, placental tubercles cushion-shaped or hooked; in Stoeberia with a well-developed opening mechanism but the capsule not closing on drying, in Ruschianthemum with only a rudimentary opening mechanism; dividing into 5 cells, each usually with 1 seed. – Genera: Ruschianthemum, Stoeberia.

**Sub-Tribe 6: Delospermatinae** Schwant. – Acauline or shrubby, with a ♃ taproot; capsule with 5–6 cells, with or without rudimentary cell-lids. – Genera: Delosperma, Drosanthemum, Ectotropis, Mestoklema, Ruschianthus, Trichodiadema.

**Sub-Tribe 7: Psammophorinae** Schwant. – Acauline or shrubby; L. glutinous, with sand adhering; capsule without cell-lids, with valve-wings (Delosperma type), with 5–8 cells. – Genera: Arenifera, Psammophora.

**Sub-Tribe 8: Erepsiinae** Schwant. – ♄ or acauline, often cushion-forming; Fl. with a ± deep Cal.-tube, some or all stamens and staminodes inclining into the Cal.-tube; capsules with 5 cells, with cell-lids, with or without placental tubercles. – Genera: Erepsia, Kensitia, Nelia, Semnanthe, Smicrostigma.

**Sub-Tribe 9: Nananthinae** Schwant. – Forming Ros. or clumps; L. generally tuberculate; capsule with many cells, with lids, with or without small placental tubercles. – Genera: Aloinopsis, Khadia, Nananthus, Rabiea, Titanopsis.

**Sub-Tribe 10: Pleiospilinae** Schwant. – Acauline; L. with many spots; capsule with many cells, with cell-lids and placental tubercles. – Sole G.: Pleiospilos.

**Sub-Tribe 11: Stomatiinae** Schwant. – Creeping ♄ or acauline; L. tuberculate or papillose; capsule 5-celled, with cell-lids, without placental tubercles. – Genera: Chasmatophyllum, Neohenricia, Neorhine, Rhinephyllum, Stomatium.

**Sub-Tribe 12: Jensenobotryinae** Schwant. – Depressed plants, with long, woody St. as thick as the human arm, creeping over rocks; L. ± spherical; capsule 5-celled. – Sole G.: Jensenobotrya.

**Sub-Tribe 13: Dracophilinae** Schwqnt. – Compact, creeping ♄ with thick L.; capsule with many cells, with well-developed or small, rudimentary cell-lids, or these sometimes missing. – Genera: Dracophilus, Juttadinteria, Namibia.

**Sub-Tribe 14: Lithopinae** Schwant.[1]) – Acauline: seedling-L. forming spherical or conical Bo., later ± separating, apart from Lithops which retains the juvenile form; Fl. without bracts (not Sep.!); capsule with 5 cells, with valve-wings, only Lapidaria having cell-lids. – Genera: Dinteranthus, Lapidaria, Lithops, Schwantesia.

---

[1]) "The above diagnosis suggests that all species with capsules which are not 5-merous should be excluded from the S. Tr. This would at once exclude **Dinteranthus, Lapidaria** and half the **Lithops** species. This point appears to have been overlooked and it is proposed to amend the S. Tr. diagnosis from 5 cells to 5–15 so as to include these plants.
From my own findings, **Dinteranthus puberulus** and many **Lithops** species have a rudimentary cell lid. The genus **Dinteranthus** produces capsules which are up to 15-merous. **Lapidaria** has 6–7 cell parts whilst **Lithops** have capsules with 5–6 cell parts and very occasionally with 7 parts. In two rare instances Dr. de Boer (personal communication) has found a capsule of **Lithops bella** v. **lericheana** with 4 cells and one of **L. steineckeana** with 12 cells.
Dr. N. E. Brown (1931) has reported that **L. fulviceps** can be 4-merous. I have since found single capsules of **L. optica, L. salicola** v. **salicola, L. bella** v. **bella** and another of **L. fulleri** (Gellkop population) all of which were 4-merous. Professor Cole (personal communication) has written saying that he has counted over 7000 seed capsules; 59 of these, representing 28 species and varieties, were 4-merous. He also reports finding a 10-merous capsule of **L. pseudotruncatella** (pulmonuncula) and one 3-merous capsule of **L. ruschiorum.** Acc. to Schwantes (1957) the capsule of the **S. Tr. Lithopinae** is like that of the genus **Delosperma** which is without or with only rudimentary cell lids." **B. Fearn**, May 1973.

**Sub-Tribe 15: Frithiinae** SCHWANT. – Acauline; L. cylindrical, tipped with a Wi.; Pet. united below to form a tube; capsule with 5 cells, without either cell-lids or valve-wings. – Sole G.: Frithia.

**Sub-Tribe 16: Gibbaeinae** SCHWANT. – Acauline or almost so; capsule with 6 to many cells, usually with lids, without placental tubercles. – Genera: Antegibbaeum, Didymaotus, Gibbaeum, Imitaria, Muiria, × Muirio-Gibbaeum.

**Sub-Tribe 17: Conophytinae** SCHWANT. – Acauline or almost so; L. short and thick, or united to form cordate or spherical Bo.; Pet. united to form a tube; capsule with valve-wings, without cell-lids (Delosperma type). – Genera: Berrisfordia, Conophytum, Herreanthus, Oophytum, Ophthalmophyllum.

**Sub-Tribe 18: Faucariinae** SCHWANT. – Acauline: L.-margins usually dentate; capsule with 5 cells; cell-walls opening above and arching over the cells so that the latter have only a small opening as with true cell-lids (Faucaria type). – Genera: Faucaria, Orthopterum.

**Sub-Tribe 19: Malephorinae** SCHWANT. (*Hymenocyclinae* SCHWANT.). – ♄, or ± acauline; capsule with many cells, with cell-lids and valve-wings, in Malephora with adaxial seed-pockets. – Genera: Glottiphyllum, Malephora.

**Sub-Tribe 20: Dorotheanthinae** SCHWANT. – ☉ herbs; L. flat or ± cylindrical; capsule 5-celled, with or without cell-lids. – Genera: Aethephyllum, Dorotheanthus, Micropterum, Pherelobus.

**Sub-Tribe 21: Mitrophyllinae** SCHWANT. – ♄ or acauline; conspicuously heterophyllous; capsule with 5 cells, with cell-lids, with or without rudimentary placental tubercles. – Genera: Diplosoma, Maughaniella, Meyerophytum, Mimetophytum, Mitrophyllum, Monilaria.

**Sub-Tribe 22: Carpantheinae** SCHWANT. – ☉ herbs; L. flat, capsule with many cells, lamellae of the cell-walls erect, pressed over the seed-cells as a waxy connecting arch, thus forming a kind of lid and leaving a narrow slit through which the seeds are released. – Sole G.: Carpanthea.

**Sub-Tribe 23: Scopelogeniae** JACOBS. – Compact, mat-forming ♄; capsule with 5–6 cells, with cell-lids, with or without valve-wings, with or without placental tubercles, the nectary annular, the capsule not closing again once open. – Sole G.: Scopelogena.

**Tribe 2: Apatesieae** SCHWANT. – ☉, ☉ or ♃ herbs; capsule with seed-pockets, i.e. a cavity external to the seed-cell, in which 1–2 seeds develop, the cell-walls erect, without cell-lids, Fr. in part a schizocarp or transitional to this. – Genera: Apatesia, Conicosia, Herrea.

**Tribe 3: Skiatophyteae** (STRAKA) IHL. (*Skiatophytinae* STRAKA as Sub-Tribe). – ☉ to ♃ herbs; L. flat; capsule opening only once and remaining open, a cellular schizocarp, several seeds developing in each cell plus 2 in each seed-pocket. – Sole G.: Skiatophytum.

**Tribe 4: Saphesieae** (SCHWANT.) IHL. (*Saphesiinae* SCHWANT. as Sub-Tribe). – ♄; capsule opening when dry, with 5 cells, each cell with numerous seeds; at first with rudimentary seed-pockets in the cells. – Sole G.: Saphesia.

**Tribe 5: Carpobroteae** SCHWANT. – Creeping ♄; fruit multi-celled, indehiscent, with numerous seeds in each cell; when the Fr. is ripe the Cal. and especially the adjacent Sep.-bases become fleshy; the Fr. is edible and fig-like, hence the original name Ficoides HERMANN. – Genera: Carpobrotus, Sarcozona.

**Genera of uncertain classification:** Amphibolia, Anisocalyx, Enarganthe, Esterhuysenia, Mossia, Namaquanthus, Wooleya, Zeuktophyllum.

**Addendum:** L. BOLUS proposes the division of the Mesembryanthemums into the following main groups: –

**I. Schizocarpi** L. BOL. – Fruit a schizocarp. – Genera: Caryotophora, Hymenogyne.

**II. Carpobroti** (N. E. BR.) L. BOL. (*Carpobrotus* N. E. BR. as G.). – Fr. indehiscent, fleshy, edible. – Genera: Carpobrotus, Sarcozona.

**III. Capsulifera** L. BOL. – Fr. a capsule.
  1. **Central placentation:** to include **Sub-Fam. I. Mesembryanthemoideae.**
  2. **Basal or parietal placentation:** to include **Sub-Fam. IV. Ruschioideae** (without Tribe 5).

## 2. Key to the Genera of Mesembryanthemaceae

by L. Bolus in Notes on Mesembr. III, Apr. 1958, supplemented in H. Jacobsen Handb. Succ. Pl. 1960 & Das Sukkulentenlexikon 1970.

Summary of main groups:

- **A.** Placentation axile (Tab. 142/1, 2).
- **A.** Placentation parietal (Tab. 142/3, 4).
    - **B.** Fr. indehiscent, juicy and edible (Tab. 142/5).
    - **B.** Fr. capsular (Tab. 142/6).
        - **C.** Loculi of capsule without a covering membrane.
        - **C.** Loculi of capsule with a covering membrane.
            - **D.** Loculi without a placental tubercle.
            - **D.** Loculi with a placental tubercle.

**A. Placentation axile** (Tab. 142/1, 2).

| | | |
|---|---|---|
| **1.** | L. flat | **2.** |
| **1a.** | L. terete, subterete or semiterete | **10.** |
| **2.** | Fr. a schizocarp | **3.** |
| **2a.** | Fr. a capsule | **4.** |
| **3.** | ⚄; Pet. white; Fr. consisting of 3–4 nut-like, bilocular, woody mericarps containing a single ovule in each locule | **Caryotophora** |
| **3a.** | ⊙; Pet. yellow; Fr. consisting of 18–21 membranous 1-seeded sections | **Hymenogyne** |
| **4.** | L. persisting as membranous skeletons with margins and veins intact, sometimes until the next season's growth | **Sceletium** |
| **4a.** | L. not as above | **5.** |
| **5.** | Stigmas red | **Mesembryanthemum** (*Callistigma*) |
| **5a.** | Stigmas green or pallid | **6.** |
| **6.** | L. opposite | **7.** |
| **6a.** | L. alternate on the Fl. Br. | **9.** |
| **7.** | ⊙, L. conspicuously united | **Synaptophyllum** |
| **7a.** | ⚄, L. not conspicuously united | **8.** |
| **8.** | L. lanceolate to ovate; Pet. lemon yellow; valves of the expanded capsule winged | **Platythyra** |
| **8a.** | L. cordate or broadly ovate in the upper part; Pet. magenta; valves wingless | **Aptenia** |
| **9.** | Papillae conspicuous; L. of the central tuft (when present) usually forming a Ros.; margins undulate | **Mesembryanthemum** |
| **9a.** | Papillae not, or scarcely, visible; L. of the central tuft erect; margins flat | **Prenia** |
| **10.** | ⚄ plants | **11.** |
| **10a.** | ⊙ or ⊙ plants | **17.** |
| **11.** | L. and Sep. spinescent from an early stage | **Amoebophyllum** |
| **11a.** | L. and Sep. not spinescent from an early stage | **12.** |
| **12.** | Int. entirely enclosed in sheathing L. tissue, or with age becoming evident | **13.** |
| **12a.** | Int. not as above | **14.** |
| **13.** | Shrubby; L. semiterete, up to 3.5 cm long and 8 mm ∅; Sep. very unequal; Cor. fugitive, up to 2.5 cm long | **Aspazoma** |
| **13a.** | Clump-forming; L. terete and finger-like, up to 7 cm long and 2 cm ∅; Sep. nearly equal; Cor. long-lived, up to 1.5 cm long, segments stiff | **Dactylopsis** |
| **14.** | Br. usually constricted at the nodes; L. bases sheathing, persistent; staminodes usually petaloid | **15.** |
| **14a.** | Br. very rarely constricted at the nodes; L. bases not conspicuously persistent; staminodes filamentous, usually absent in Aridaria | **16.** |
| **15.** | L. Sh. conspicuous, with a fringe of deflexed hairs at the base (or at the tip in P. marlothii) | **Psilocaulon** Sect. **Brownanthus** |
| **15a.** | L. Sh. inconspicuous, nude at the base | **Psilocaulon** Sect. **Psilocaulon** |
| **16.** | Papillae scarcely visible; Br. and Ped. woody or wiry; cyme usually dichotomous; Fl. usually nocturnal | **Aridaria** |
| **16a.** | Papillae usually evident; young Br. and Ped. externally herbaceous; Fl. 1–5-nate, usually diurnal | **Sphalmanthus** |
| **17.** | Stigmas broad and somewhat flattened | **Eurystigma** |
| **17a.** | Stigmas subulate or filiform | **18.** |
| **18.** | Expanded capsule with inflexed valve-wings | **Mesembryanthemum** |
| **18a.** | Expanded capsule with spreading or erect valve-wings | **19.** |

19. Cor. not, or slightly, exceeding the Sep., violet-red or citron-yellow; valves of the expanded capsule erect, bifid .................................................. **Mesembryanthemum** (*Hydrodea*)
19a. Cor. well exceeding the Sep., pure white or pale straw-coloured; valves spreading or recurved, not bifid, but wings bifid at the apex ............................................. **20.**
20. Valve wings erect ................................................................. **Opophytum**
20a. Valve-wings spreading ........................... **Mesembryanthemum** (*Halenbergia*)

A. **Placentation parietal** (Tab. 142/3, 4).
   B. Fr. indehiscent, juicy and in part edible (Tab. 142/5) .......................... **21.**
   21. Styles 6–15; stigmas ± plumose; stamens numerous; Fl. large, free of the uppermost L.; seeds smooth or finely tessellate; St. elongated and creeping; L. surface smooth; L. scars short ............................................................................ **Carpobrotus**
   21a. Styles 4–5; stigmas papillate, less plumose; stamens few; Fl. smaller, usually partly included by the enlarged, strongly united bases of the uppermost L.; seeds warty; St. erect or oblique or finally decumbent but not creeping; L. surface warty, with prominent idioblasts; L. scars collar-shaped .................................... **Sarcozona**

   B. Fr. capsular (Tab. 142/6) .................................................................. **C.**
   C. Loculi of the capsule without a covering membrane (or only rudimentary in Micropterum) ............................................................................................. **22.**
      22. Nectary composed of 5 pits or hollows, sometimes surmounted by rudimentary glands ....................................................................................... **23.**
      22a. Nectary not composed of pits or hollows ............................................. **24.**
      23. ☉; herbaceous parts papillose; L. flat, entire or pinnatifid ........ **Micropterum**
      23a. ♃; herbaceous parts not papillose; L. thick, usually acutely keeled and dentate ........................................................................... **Stomatium**
      24. Capsule dehiscing in dry conditions; valves without expanding keels, remaining ± erect (exc. Herrea nelii where the capsule breaks up into separate segments) ... **25.**
      24a. Capsule dehiscing in wet conditions; valves with expanding keels, widely spreading or recurved when fully expanded ........................................................ **27.**
      25. L. flat; stigmas 5; capsule closing again if placed in water .......... **Saphesia**
      25a. L. triquetrous, terete or subterete; stigmas more than 10 ................... **26.**
      26. ☉ growths deciduous; dissepiments reaching to the apex of the valve ... **Herrea**
      26a. ☉ growths not deciduous; dissepiments reaching about halfway up the valve .... ................................................................................................. **Conicosia**
      27. Nectary composed of separate glands, which in Mossia are so close as to form almost an annulus ....................................................................... **28.**
      27a. Nectary annular ............................................................................... **35.**
      28. Pet. united at the base; staminodes and stamens adnate to the Cor. ............. ................................................................................................. **Ruschianthus**
      28a. Pet. not as above ............................................................................. **29.**
      29. L. wholly united to form an ovoid or subglobose body; pubescent with deflexed hairs ................................................................................... **Muiria**
      29a. L. not as above ................................................................................. **30.**
      30. L. fenestrate at the apex ................................................................. **Frithia**
      30a. L. not fenestrate ............................................................................... **31.**
      31. Capsule very small, with thin walls; sometimes breaking up into 5 parts[1]) ...... ................................................................................................. **Ectotropis**
      31a. Capsule not as above ....................................................................... **32.**
      32. Plant compact; Int. concealed ............................................................. **33.**
      32a. Plant ± lax; Int. visible .................................................................. **34.**
      33. L. usually rough (smooth in R. macradenium), not papillose; bracts absent ...... ................................................................................................. **Rhinephyllum**
      33a. L. very smooth; bracts present ......................................................... **Carruanthus**
      34. Creeping; L. dentate ............................................................................ **Mossia**
      34a. Not creeping; L. entire ................................................................... **Delosperma**
      35. Pet. stiff and rather thick, lasting fresh longer than in any other G. exc. Dactylopsis ................................................................................................. **Nelia**
      35a. Pet. not stiff, ± fugitive ..................................................................... **36.**
      36. Sep. and Pet. free to the base ............................................................. **37.**
      36a. Sep. and Pet. united to form a Cal. tube and a Cor. tube ........................... **51.**

---

[1]) Two of the perfect capsules in the type-collection of **Ectotropis** N. E. BR. when expanded were exactly like those of **Delosperma,** exc. that the valve-wings were less ample than is usual in the latter.

| | | |
|---|---|---|
| 37. | L. expanded and flat | 38. |
| 37a. | L. not expanded or flat | 39. |
| 38. | Fl. opening in the afternoon; Pet. yellow; plants ☉, sun-loving | **Apatesia** |
| 38a. | Fl. opening in the morning; Pet. white; plants not ☉, shade-loving ........ **Skiatophytum** | |
| 39. | Plants resembling clusters of grapes on account of the rounded or club-shaped L. ........ **Jensenobotrya** | |
| 39a. | Plants not as above | 40. |
| 40. | Ped. bracteate | 41. |
| 40a. | Ped. bractless | 44. |
| 41. | Herbaceous parts with viscid epidermis to which sand adheres | **Psammophora** |
| 41a. | Herbaceous parts not as above | 42. |
| 42. | L. acutely keeled upwards; Sep. 6 | **Herreanthus** |
| 42a. | L. not acutely keeled; Sep. 5 | 43. |
| 43. | Shrubby; Pet. whitish; staminodes present | **Zeuktophyllum** |
| 43a. | Dwarf, compact plants; Pet. yellow, with or without a central red stripe; staminodes absent | **Nananthus** |
| 44. | Compact plants with Int. enclosed in the L. sheaths | 45. |
| 44a. | Plants elongated, with some Int. visible | 50. |
| 45. | Sep. 4 | **Juttadinteria** |
| 45a. | Sep. 5–10 | 46. |
| 46. | Stigmas 9–25 | **Namibia** |
| 46a. | Stigmas 4–10 | 47. |
| 47. | L. pair forming a united Bo. | 48. |
| 47a. | L. pair not forming a Bo. | **Schwantesia** |
| 48. | Pet. white in the lower half, purple in the upper half | **Oophytum** |
| 48a. | Pet. yellow or white | 49. |
| 49. | Tip of the L. ± flattened; Sep. 4–7 | **Lithops** |
| 49a. | Tip of the L. not flattened, or, if so (as in Dinteranthus pole-evansii & vanzijlii) then Sep. 8–10[1] | **Dinteranthus** |
| 50. | Creeping; L. pairs all alike; Fl. nocturnal | **Neohenricia** |
| 50a. | Erect; L. pairs dimorphic or at Fl. time trimorphic; Fl. diurnal | **Mitrophyllum**[2] |
| 51. | Receptacle usually produced beyond the Ov. to form a tube; Cor.-tube partly adnate to the Cal.-tube | **Ophthalmophyllum** |
| 51a. | Receptacle not produced beyond the Ov.; Cor.-tube free from the Cal.-tube | 52. |
| 52. | ☉ shoots composed of 1 pair of L. | **Conophytum** |
| 52a. | ☉ shoots composed of 2 pairs of L. | **Berrisfordia** |

C. Loculi of the capsule with a covering membrane, which is sometimes incomplete .... **D.**

| | | |
|---|---|---|
| **D.** | Loculi without a placental tubercle (exc. Drosanthemum and Trichodiadema) (Tab. 142/7) | 53. |
| 53. | ☉ plants | 54. |
| 53a. | ♃ plants | 57. |
| 54. | Ov. with 5 processes (lobes) alternating with the stigmas | **Pherelobus** |
| 54a. | Ov. without processes | 55. |
| 55. | L. pinnatifid | **Aethephyllum** |
| 55a. | L. entire | 56. |
| 56. | Stigmas 5, hardening and accrescent with age, persisting throughout the Fr. stage | **Dorotheanthus** |
| 56a. | Stigmas 12–20, withering with the rest of the Fl. before the Fr. matures | **Carpanthea** |
| 57. | L. pairs dimorphic or trimorphic | 58. |
| 57a. | L. pairs all alike (dimorphic in Drosanthemum diversifolium) | 64. |
| 58. | Int. concealed; lower L. pair not always evident | 59. |
| 58a. | Some Int. evident; lower L. always evident | 61. |
| 59. | Upper L. pair spreading or prostrate on the ground and much more united on one side than on the other; Fl. sessile | **Diplosoma** |
| 59a. | Upper L. pair erect, suberect or ascending, symmetrically united below; Fl. usually pedunculate | 60. |
| 60. | Roots fibrous; St. and Br. not moniliform; staminodes present | **Maughaniella** |
| 60a. | Roots ± woody; St. and Br. usually moniliform; staminodes absent | **Monilaria § Monilariae** |

---

[1]) Typical Dinteranthus has 6–7 Sep. and stigmas, but D. pole-evansii & vanzijlii 8–10.
[2]) See the footnote for **Mitrophyllum** SCHWANT., p. 525.

Plate 154. 1. **Conophytum ectypum** N. E. Br. v. **ectypum**; 2. **C. fenestratum** Schwant. (Photo: H. Karstens, G. Schwantes); 3. **C. ficiforme** (Haw.) N. E. Br. (Photo: G. Schwantes); 4. **C. flavum** N. E. Br. (Photo: as 3).

Plate 155. 1. Conophytum frutescens SCHWANT. (Photo: G. SCHWANTES); 2. C. meyerae SCHWANT. v. meyerae; 3. C. minutum N. E. BR. v. minutum (Photo: as 1); 4. C. meyeri N. E. BR. v. meyeri.

Plate 156. 1. **Conophytum mundum** N. E. Br.; 2. **C. pauxillum** (N. E. Br.) N. E. Br. (Photo: G. Schwantes); 3. **C. obcordellum** (Haw.) N. E. Br. v. obcordellum (Photo: as 2); 4. **C. pillansii** Laevis; 5. **C. pearsonii** N. E. Br.; v. pearsonii; 6. **C. praecox** N. E. Br.

Plate 157. 1. Conophytum truncatum (THUNBG.) N. E. BR.; 2. C. ursprungianum TISCH. v. ursprungianum (Photo: W. RAUH); 3. C. velutinum (SCHWANT.) SCHWANT.; 4. C. wittebergense DE BOER (Photo: H. W. DE BOER); 5. C. vagum N. E. BR. (Photo: G. SCHWANTES); 6. C. wettsteinii (HAW.) N. E. BR. v.wettsteinii.

Plate 158. 1. **Cylindrophyllum calamiforme** (L.) Schwant.; 2. **Delosperma pruinosum** (Thunbg.) Ingram; 3. **D. brunnthaleri** (Bgr.) Schwant.; 4. **Dactylopsis digitata** (Ait.) N. E. Br. (Photo: G. Schwantes).

Plate 159. 1. Delosperma hallii L. Bol. (Photo: H. Chr. Friedrich); 2. D. taylorii (N. E. Br.) L. Bol. v. taylorii (Photo: G. Schwantes); 3. & 4. D. sutherlandii (Hook. f.) N. E. Br.

Plate 160. 1. **Dicrocaulon brevifolium** N. E. Br. (Photo: E. Elkan); 2. **Didymaotus lapidiformis** (Marl.) N. E. Br. (Photo: C. A. Luckhoff, G. Schwant.); 3. **Dinteranthus inexpectatus** Jacobs. (Photo: W. Triebner); 4. **D. vanzijlii** (L. Bol.) Schwant. v. **vanzijlii**; 5. **D. microspermus** (Dtr. et Derenb.) Schwant. v. **microspermus** (Photo: G. Schwantes); 6. **D. pole-evansii** (N. E. Br.) Schwant.

Plate 161. 1. **Diplosoma retroversum** (Kensit) Schwant.; 2. **Disphyma crassifolium** (L.) L. Bol.; 3. **Dorotheanthus gramineus** (Haw.) Schwant. f. gramineus; 4. **D. bellidiformis** (Burm. f.) N. E. Br.; 5. **Dracophilus dealbatus** (N. E. Br.) Walgate (Photo: G. Schwantes); 6. **Drosanthemum paxianum** (Schltr. et Diels) Schwant.

| | | |
|---|---|---|
| 61. | L. pairs trimorphic at Fl. time | **Mitrophyllum** (*Conophyllum*)[1] |
| 61a. | L. pairs dimorphic throughout | 62. |
| 62. | Sep. 4; staminodes present | **Dicrocaulon** |
| 62a. | Sep. 5–6; staminodes absent | 63. |
| 63. | Lower L. pairs ovoid in earlier stages; Pet. rather lax, rose or rose-purple | **Meyerophytum** |
| 63a. | Lower L. pairs bead-like; Pet. dense, white or white and rose in the same spec. | **Monilaria** § **Exsertae** |
| 64. | Nectar glands separate or rarely so contiguous as to form almost an annulus | 65. |
| 64a. | Nectary annular or inconspicuous | 74. |
| 65. | Sep. and Pet. connate below; style present | **Imitaria** |
| 65a. | Sep. and Pet. not connate; style absent | 66. |
| 66. | Herbaceous parts papillose | 67. |
| 66a. | Herbaceous parts not papillose | 68. |
| 67. | L. usually crowned with spreading or erect bristles or their vestiges | **Trichodiadema** |
| 67a. | L. without terminal bristles (tubercle in loculi of capsule very rarely present) | **Drosanthemum** |
| 68. | Plants creeping | 69. |
| 68a. | Plants not creeping | 70. |
| 69. | Pet. yellow | **Chasmatophyllum** |
| 69a. | Pet. white | **Mossia** |
| 70. | Fl. in cymes | **Oscularia** |
| 70a. | Fl. solitary | 71. |
| 71. | L. dentate, T. ending in a bristle | **Faucaria** |
| 71a. | L. entire | 72. |
| 72. | Ped. bracteate | 73. |
| 72a. | Ped. bractless | **Gibbaeum** |
| 73. | Pet. yellow | **Rabiea** |
| 73a. | Pet. purplish pink | **Cerochlamys** |
| 74. | Pet. yellow (reddish orange in Titanopsis hugo-schlechteri) | 75. |
| 74a. | Pet. pinkish purple, pink, red, pallid or white | 80. |
| 75. | Stigmas 10 | **Vanheerdea** |
| 75a. | Stigmas 5–7 | 76. |
| 76. | Plants compact; Int. included within the L. sheaths | 77. |
| 76a. | Erect, decumbent or creeping plants with evident Int. | 78. |
| 77. | Glabrous; L. warty or pustulate in the upper part; Sep. and stigmas 6 | **Titanopsis** |
| 77a. | Receptacle and Sep. pubescent; L. without warts or pustules but microscopically granulate; Sep. 7; stigmas 6–7 | **Lapidaria** |
| 78. | L. keel serrate | **Erepsia** |
| 78a. | L. keel entire | 79. |
| 79. | Valves of the expanded capsule winged | **Lampranthus** |
| 79a. | Valves of the expanded capsule not winged | **Scopelogena** |
| 80. | Stigmas 8–14 | 81. |
| 80a. | Stigmas 5–6 | 88. |
| 81. | Pet. clawed, the blade ovate or elliptic, equalling or shorter than the claw | **Kensitia** |
| 81a. | Pet. not as above | 82. |
| 82. | Stigmas minute, completely hidden by staminodes and stamens | 83. |
| 82a. | Stigmas not as above | 84. |
| 83. | L. keels entire | **Smicrostigma** |
| 83a. | L. keels lacerate | **Semnanthe** |
| 84. | Ped. bracteate | 85. |
| 84a. | Ped. bractless | 87. |
| 85. | Plants dwarf, compact; Sep. 5 | **Dracophilus** |
| 85a. | Plants shrubby, 15–30 cm high; Sep. 4 | 86. |
| 86. | Herbaceous parts smooth; stamens not papillate; seeds echinate | **Namaquanthus**)[2] |
| 86a. | Herbaceous parts pubescent; stamens papillate; seeds smooth | **Wooleya** |
| 87. | Staminodes present; valves of the expanded capsule wingless | **Khadia** |
| 87a. | Staminodes absent; valves of the expanded capsule winged | **Anisocalyx** |
| 88. | Herbaceous parts ± papillose | 89. |

---

[1]) See the footnote for **Mitrophyllum** SCHWANT., p. 525
[2]) See footnote under **134a: Namaquanthus**. L. BOL.

88a. Herbaceous parts not papillose .................................... 90.
89. Sep. and stigmas 5 ........................................... **Jacobsenia**
89a. Sep. and stigma 6 ....................... **Drosanthemum** § **Anularia**
90. ☉ growth consisting of 2 pairs of L. with a Fl. on each side of the lower pair terminating the previous year's axillary shoots ............... **Didymaotus**
90a. Habit not as above ............................................. 91.
91. Fl. in copiously branched pedunculate cymes; rootstock massive and sometimes tuberous .............................................. **Mestoklema**
91a. Infl. and rootstock not as above ................................ 92.
92. Herbaceous parts glabrous ..................................... 93.
92a. Herbaceous parts pilose ..................................... **Braunsia**
93. Seeds echinate .............................................. **Braunsia**
93a. Seeds not echinate ............................................ 94.
94. Valves of the expanded capsule winged ................... **Lampranthus**
94a. Valves of the expanded capsule not winged .................. **Khadia**[1])

D. Loculi of the capsule with a placental tubercle (sometimes obscure or absent in Aloinopsis, very small in Antegibbaeum, and uncertain in Calamophyllum). (Tab. 142/8.)

95. Stigmas 5–6 ................................................. 96.
95a. Stigmas more than 6 (in Khadia beswickii 6) .................... 115.
96. Pet. purplish rose, pink, pallid or white ........................ 97.
96a. Pet. yellow, golden or orange-red ............................. 109.
97. Nectary composed of separate glands ..................... **Cerochlamys**
97a. Nectary annular ............................................. 98.
98. Placental tubercle 2-lobed .................................. **Disphyma**
98a. Placental tubercle not 2-lobed ................................ 99.
99. Ped. not spinescent ......................................... 100.
99a. Some or all of the Ped. spinescent ........................ **Eberlanzia**
100. Herbaceous parts with sand adhering to the viscid epidermis .... **Arenifera**
100a. Herbaceous parts not as above ................................ 101.
101. Capsule separating into mericarps in the lower part ...... **Ruschianthemum**
101a. Capsule not as above ......................................... 102.
102. Valves of the expanded capsule winged ........................ 103.
102a. Valves of the expanded capsule not winged ..................... 107.
103. Plants shrubby with evident Int.; cymes often copiously branched ..... 104.
103a. Dwarf, compact plants without evident Int.; Fl. solitary or 1–3-nate ... 105.
104. Capsule not closing again after first opening ................. **Stoeberia**
104a. Capsule not as above ...................................... **Amphibolia**
105. L. cylindrical to subcylindrical; Fl. solitary .................. 106.
105a. L. laterally compressed in the upper part; Fl. finally ternate.... **Ebracteola**
106. Bracts large, sometimes exceeding the Fl.; Sep. 5 .......... **Cylindrophyllum**
106a. Bracts very small; Sep. 6.................................. **Antegibbaeum**
107. Seeds echinate or rough ...................................... **Astridia**
107a. Seeds not echinate or rough .................................. 108.
108. Stigmas plumosely papillate ................................ **Acrodon**
108a. Stigmas not plumosely papillate or, if so, then L. not triquetrous .. **Ruschia**
109. L. pairs dimorphic, or trimorphic at Fl. time; nectary annular ............
       ........ **Mimetophytum**
109a. L. pairs all alike; nectary composed of separate glands ............... 110.
110. Wings of the loculi of the expanded capsule erect ........... **Orthopterum**
110a. Wings not erect ............................................. 111.
111. L. short and thick, polymorphic; valves of expanded capsule erect ... **Bijlia**
111a. L. not as above; valves widely spreading or recurved .............. 112.
112. L. usually triquetrous ....................................... 113.
112a. L. not triquetrous, one of a pair often slightly differing in form from the other ..................................................... 114.
113. Filaments papillate; stigmas 5; keels of valves in expanded capsule diverging from the base; placental tubercle large ..................... **Bergeranthus**
113a. Filaments not papillate; stigmas 5–6; keels parallel below; placental tubercle small ................................................... **Machairophyllum**
114. Placental tubercles with a dividing line in the middle ...... **Rhombophyllum**
114a. Placental tubercle not as above ............................ **Hereroa**
115. L. fenestrate at the apex .................................. **Fenestraria**

---

[1]) **Esterhuysenia** (q.v.) also belongs here.

| | | |
|---|---|---|
| 115 a. | L. not fenestrate at the apex | 116. |
| 116. | Pet. filiform-linear | Polymita |
| 116 a. | Pet. not filiform-linear | 117. |
| 117. | Dwarf, compact plants | 118. |
| 117 a. | Plants with ± elongated Br. with some, at least, of the Int. evident, except in the shrubby Ottosonderia | 129. |
| 118. | Ped. bractless | 119. |
| 118 a. | Ped. bracteate | 120. |
| 119. | Pet. pink or purplish-pink; staminodes present | Khadia |
| 119 a. | Pet. yellow or very rarely white; staminodes absent | Glottiphyllum |
| 120. | L. pairs dimorphic | Cheiridopsis |
| 120 a. | L. pairs all alike | 121. |
| 121. | Receptacle produced above the Ov. into a tube; stamens deflexed; stigmas obscure, forming a viscid circular pulvinus | Argyroderma |
| 121 a. | Receptacle, stamens and stigmas not as above | 122. |
| 122. | Herbaceous parts glabrous | 123. |
| 122 a. | Herbaceous parts pilose or pubescent | 128. |
| 123. | Rootstock (where known) tuberous | Aloinopsis |
| 123 a. | Rootstock not tuberous | 124. |
| 124. | Nectary composed of separate glands | Machairophyllum |
| 124 a. | Nectary annular (unrecorded in Calamophyllum) | 125. |
| 125. | Pet. pallid, rose, rose purple or deep red | 126. |
| 125 a. | Pet. yellow, golden or orange-red | Pleiospilos |
| 126. | Fl. solitary | 127. |
| 126 a. | Fl. not solitary | Cephalophyllum |
| 127. | Sep. 5 | Cylindrophyllum |
| 127 a. | Sep. 4 | Calamophyllum |
| 128. | L. obtusely keeled, dentate or denticulate | Odontophorus |
| 128 a. | L. entire or rarely with the keels only dentate, in which case they are acutely keeled upwards | Cheiridopsis |
| 129. | Infl. persisting for several years and eventually becoming an obconic cyme, the axils of the previous year's bracts normally producing a bracteate Fl. which represents the ⊙ growth | Ottosonderia |
| 129 a. | Infl. not as above | 130. |
| 130. | L. pairs dimorphic | Vanzijlia |
| 130 a. | L. pairs all alike | 131. |
| 131. | Ped. bracteate | 132. |
| 131 a. | Ped. bractless | 142. |
| 132. | Sep. 4 | 133. |
| 132 a. | Sep. 5–6 | 135. |
| 133. | L. entire, 4–6 cm long | 134. |
| 133 a. | L. minutely toothed, less than 1 cm long | Octopoma |
| 134. | Filaments papillate; stigmas 8; seeds smooth | Enarganthe |
| 134 a. | Filaments not papillate; stigmas 8–12; seeds echinate | Namaquanthus[1]) |
| 135. | Sep. 5 | 136. |
| 135 a. | Sep. 6 | Schlechteranthus |
| 136. | L. less than 2 cm long; Fl. sessile or subsessile | Octopoma |
| 136 a. | L. more than 2 cm long; Fl. pedunculate | 137. |
| 137. | Herbaceous parts pubescent | 138. |
| 137 a. | Herbaceous parts glabrous | 139. |
| 138. | L. dentate | Odontophorus |
| 138 a. | L. entire | Cheiridopsis |
| 139. | L. rough, sometimes dentate | Cheiridopsis |
| 139 a. | L. smooth, entire | 140. |
| 140. | Usually decumbent and often creeping; stigmas 10 or more | Cephalophyllum |
| 140 a. | Usually erect or, if decumbent, then not creeping; stigmas 10 or fewer | 141. |
| 141. | L. obtusely keeled; valves of loculi winged | Leipoldtia |
| 141 a. | L. acutely keeled; valves of loculi not winged | Stayneria |
| 142. | L. of a pair dissimilar and unequal in length, the larger more than 5 cm long, sometimes with a marginal tooth or hump, or with a conspicuous white margin | Glottiphyllum |
| 142 a. | L. of a pair similar and about equal in length, less than 5 cm long, entire and without a white margin | Malephora |

---

[1]) **Namaquanthus** is twice represented in the key at **86.** and **134 a.**, because the type plant contains only one capsule which is too old to show whether a placental tubercle is present or not.

## 3. Genera, Intergeneric Hybrids, Species and Lower Taxa, Hybrids, Cultivars, Origins and Descriptions of the Mesembryanthemaceae

The symbol in the heading, after the G.-name – e.g. (§ IV/1/1) – refers to: **The systematic classification of the Mesembryanthemaceae** p.p. – For *Mes.* read *Mesembryanthemum*.

**Acrodon** N. E. Br. (§ IV/1/1). – Dwarf, mat-forming plants; **L.** crowded, rosulate, triangular, with dentate margins; **Fl.** solitary, pedicellate, white, in summer. Greenhouse. Propagation: seed, cuttings.

**A. bellidiflorus** (L.) N. E. Br. v. **bellidiflorus** (Pl. 143/1) (*Mes. b.* L., *Mes. b.* v. *glaucescens* Haw.). – Cape: Mossel Bay D. – **L.** opposite and decussate, united and sheathing below, triangular, laterally compressed above, shortly mucronate, with cartilaginous, dentate angles, greyish-green, 3–5 cm long; **Fl.** 4 cm ∅, Pet. white, with reddish tips.

**A. — v. striatus** (Haw.) N. E. Br. (*Mes. b.* v. *s.* Haw., *Mes. b.* Haw., *Mes. b.* v. *subulatum* Haw.). – Pet. striate.

**A. — v. viridis** (Haw.) N. E. Br. (*Mes. b.* v. *v.* Haw.). – **L.** shorter, green with entire margins.

**A. parvifolia** R. du Plessis. – Cape: Caledon D. – Robust; **Br.** prostrate or creeping, up to 53 cm long, Int. 6–15 mm long, sideshoots up to 12 cm long; **L.** obliquely truncate towards the tip or gradually tapered, margins and keel with 1–4 indistinct T., green to olive-green, reddish-brown at the tip, 12–22 mm long, 2.5 to 4 mm across, 3 mm thick, Sh. 2 mm long; **Fl.** 17–20 mm ∅, Pet. white with a pink border.

**A. subulatus** (Mill.) N. E. Br. (*Mes. s.* Mill., *Mes. bellidiflorum* D.C., *Mes. b.* v. *s.* S.D.). – Cape: Riversdale D. – Resembles **A. bellidiflorus**, but smaller; **L.** with the keel dentate only at the truncate tip; **Fl.** 2.5 cm ∅.

**Aethephyllum** N. E. Br. (§ IV/1/20). – ☉ plants. – In the open. Propagation: seed.

**A. pinnatifidum** (L. f.) N. E. Br. (Pl. 143/2) (*Mes. p.* L. f., *Cleretum p.* (L. f.) L. Bol., *Micropterum p.* (L.f.) Schwant.). – Cape: Cape D. – **Br.** prostrate to ascending, 10 to 20 cm or more long, Int. 1–4 cm long, herbaceous parts green or red, papillose; **L.** 2–4 cm long, 5–20 mm across, pinnately laciniate, vivid green, with glossy papillae; **Fl.** solitary from the L.axils, on a pedicel 5–10 mm long, c. 12 mm ∅, light yellow.

**Aloinopsis** Schwant. (§ IV/1/9). – Plants dwarf, tufted; **roots** tuberous; **St.** mostly with 4–6 L., Int. enclosed in the L.-Sh., herbaceous parts velvety pubescent or glabrous and then variously punctate, the tubercles on the L. similar or differently shaped; **L.** broadly spatulate, ovate, ovate-lanceolate, linear-lanceolate or subclavate, lower surface nearly flat, convex, or obtusely keeled, 15–35 mm long, up to 20 mm broad; **Fl.** stalked, with 2 bracts, 2–4 cm ∅, Pet. 2–3-seriate, yellow, sometimes with a central red stripe, to pink, Fl. open in the afternoon towards evening. – Greenhouse, warm. Propagation: seed, cuttings.

### Key to the Species of the Genus Aloinopsis by L. Bolus

| | | |
|---|---|---|
| 1. | Herbaceous parts pubescent; Pet. without a central red stripe | 2. |
| 1a. | Herbaceous parts glabrous; Pet. with a central red stripe in 4 species | 5. |
| 2. | Mature L. spreading and forming Ros. | 3. |
| 2a. | Mature L. erect or nearly erect | 4. |
| 3. | Pet. 1.5–3 mm broad; stigmas reaching the height of the stamens | **thudichumii** |
| 3a. | Pet. 1–1.5 mm broad; stigmas falling far short of the stamens | **peersii** |
| 4. | L. laterally compressed near the abruptly pointed apex; stigmas 7–9 | **orpenii** |
| 4a. | L. convex on both sides; stigmas 10 | **hilmarii** |
| 5. | L. usually with additional larger white tubercles; Sep. 6; Pet. without a central stripe | 6. |
| 5a. | L. rarely with larger white or pallid tubercles; Sep. 5 | 10. |
| 6. | L. flat on both surfaces, rounded or subtruncate at apex | **malherbei** |
| 6a. | L. ± keeled or convex dorsally | 7. |
| 7. | Some of the tubercles elongated and bristle-like, papillate with minute bristles | **setifera** |
| 7a. | Tubercles not as above | 8. |
| 8. | Margin of upper part of L. often dentate or lobulate; white tubercles often conical | **luckhoffii** |
| 8a. | Margin entire; white tubercles not conical | 9. |
| 9. | Upper surface of L. rounded in the dilated portion; larger white tubercles absent, or unusually inconspicuous if present | **villetii** |
| 9a. | Upper surface of L. reniform or subreniform in the dilated portion; larger white tubercles up to 1 mm ∅ | **lodewykii** |
| 10. | L. flattened, dorsally scarcely convex, larger pallid tubercles concentrated on the apical margin; Pet. rose-purple | **spathulata** |
| 10a. | L. not as before; Pet. yellow or brownish-red | 11. |
| 11. | L. erect | 12. |

**11 a.** L. spreading and forming Ros. .................................................. **14.**
**12.** L. with only the truncate upper part exposed above the ground; Pet. striped with red .....
................ schooneesii v. schooneesii **13.**
**12 a.** L. entirely exposed, laterally compressed, acute; Pet. not striped ................ **orpenii**
**13.** L. keeled across the apex; Pet. nor acure ................ schooneesii v. **willowmorensis**
**13 a.** L. not keeled across apex; Pet. acute ..................... schooneesii v. **acutipetala**
**14.** Pet. striped with red ........................................... **15.**
**14 a.** Pet. not striped ................................................. **17.**
**15.** L. conspicuously broadened in the upper part .......................................... **16.**
**15 a.** L. usually narrowed in the upper part or uniform in breadth ............. **rubrolineata**
**16.** Fl. 1–3-nate; Sep. up to 1.1 cm broad at the base; stamens strongly inflexed ...... **rosulata**
**16 a.** Fl. solitary; Sep. up to 6 mm broad at the base; stamens nearly erect ............ **jamesii**
**17.** L. usually subtruncate or obtuse when flattened by hand; Pet. obtuse, up to 2.5 mm or more broad......................................................................... **loganii**
**17 a.** L. usually acute; Pet. acute, up to 1.7 mm broad ............................. **acuta**

**A. acuta** L. BOL. – Cape: Fraserburg D. – **Caudex** tuberous, up to 17 cm long, 2.5 cm ⌀ above; herbaceous parts glabrous, pale green to reddish-brown, with minutely tuberculate spots; **L.** spatulate, tapered towards the tip, 11–16 mm long, 6–10 mm across, 4 mm thick; **Fl.** 10–15 mm ⌀, lemon-yellow.

**A. hilmarii** (L. BOL.) L. BOL. (*Cheiridopsis h.* L. BOL.). – Cape: Laingsburg D. – Cushions 6.5 cm ⌀; **Br.** crowded, with 2–4 abruptly tapered **L.**, the upperside flat, somewhat widened, the underside rounded, grey, often reddish, velvety, 2–2.5 cm long, 7–9 mm across and thick; **Fl.** 3 cm ⌀, yellow.

**A. jamesii** L. BOL. (*Nananthus j.* (L. BOL.) L. BOL., *N. cradockensis* L. BOL.). – Cape: Cradock D. – **L.** 4–6, up to 18 mm long, 5 mm across below, 8 mm across midway, triangular to long-tapered, 3 mm thick, the underside carinate at the tip, grey-green, rough-tuberculate; **Fl.** 6 mm ⌀, Pet. golden-yellow with a red M.stripe.

**A. lodewykii** L. BOL. (Pl. 144/1) (*Nananthus l.* (L. BOL.) L. BOL.). – Cape: Bushmanland. – **Shoots** with 6–8 L., reniform to semicylindrical on the upperside, 10–15 mm across, 5 mm across below, compressed and widened at the tip, 5–9 mm thick, bluish-green to reddish-brown, tuberculate, the tip-area 17 mm long with 15 to 30 white tubercles 1 mm tall, Sh. 4–5 mm long; **Fl.** 15 mm ⌀, pale pink.

**A. loganii** (L. BOL.) L. BOL. (*Nananthus l.* L. BOL.). – Cape: Laingsburg D. – **Caudex** tuberous; **shoots** 4 mm thick; **L.** 4, spatulate, tapered towards the tip, truncate, the underside convex, brownish-green, with pallid spots, 17–21 mm long, 8–9 mm across at the tip, 3–4 mm across midway, 3 mm thick, Sh. 2 to 3 mm long; **Fl.** 2 cm ⌀, yellow.

**A. luckhoffii** (L. BOL.) L. BOL. (*Titanopsis l.* L. BOL., *Nananthus l.* (L. BOL.) L. BOL.). – Cape: Lit. Namaqualand. – **Ros.** 3–4 cm ⌀; **L.** c. 18 mm long, 4–5 mm across below, 4 to 5 mm thick, the tip-area 12 mm across, triangular, 6 mm thick above, with a slightly chin-like, prominent keel on the underside towards the tip, bluish to grass-green, the underside and the upper ⅓ of the upperside covered with coarse, grey-green tubercles, the angles and the keel each with 5–6 larger tubercles; **Fl.** 25 mm ⌀, light yellow.

**A. malherbei** (L. BOL.) L. BOL. (*Nananthus m.* L. BOL.). – Cape: Calvinia D. – **L.** broadly spatulate to fan-shaped, truncate, 4 mm across below and 12–22 mm across at the tip, the underside somewhat tuberculate, the margins with thick, white tubercles above, the L. otherwise bluish-green; **Fl.** 25 mm ⌀, pale brown to flesh-coloured.

**A. orpenii** (N. E. BR.) L. BOL. (*Mes. o.* N. E. BR., *Prepodesma o.* (N. E. BR.) N. E. BR., *Nananthus o.* (N. E. BR.) L. BOL.). – Cape: W.Griqualand. – Forming thick clumps; **L.** 18–20 mm long, ovate-lanceolate, 7–8 mm across, 5–6 mm thick, the underside with a chin-like keel above, bluish-green, glabrous or puberulous, with numerous prominent spots; **Fl.** 3.5 cm ⌀, yellow, Pet. with reddened tips.

**A. peersii** (L. BOL.) L. BOL. (*Nananthus p.* L. BOL., *Cheiridopsis p.* (L. BOL.) L. BOL., *Deilanthe p.* (L. BOL.) N. E. BR., *Ch. noctiflora* L. BOL., *Mes. canum* (HAW.) BGR. ?, *Nananthus soehlemannii* HGE. jr.) – Cape: Willowmore, Laingsburg, Prince Albert D., Karroo. – **Roots** thick; **L.** 2–4, recurved abruptly at right angles above, 20–22 mm long, 8 mm across below and 15 mm above, obtusely triangular-tapered, the underside carinate towards the tip, smooth, bluish to grey-green, spotted, puberulous (lens needed!); **Fl.** 25 mm ⌀, yellow.

**A. rosulata** (KENSIT) SCHWANT. (Pl. 143/3) (*Mes. r.* KENSIT, *Acaulon r.* (KENSIT) N. E. BR., *Aistocaulon r.* (KENSIT) v. POELLN.). – Cape: Willowmore, Miller D. – **Roots** long, fleshy; **L.** 6–8, rosulate, glossy dark green, 25–30 mm long, 5 mm across below, broadly spatulate and up to 16 mm across above, the underside rounded, slightly carinate above, with obtuse margins, the bluntly rounded tip covered with whitish tubercles; **Fl.** 3–3.5 cm ⌀, yellow.

**A. rubrolineata** (N. E. BR.) SCHWANT. (Pl. 144/2) (*Mes. r.* N. E. BR., *Nananthus r.* (N. E. BR.) N. E. BR., *N. r.* (N. E. BR.) SCHWANT., *N. dyeri* L. BOL., *A. d.* (L. BOL.) L. BOL.). – Cape: Graaff Reinet D. – **Roots** long, fleshy; **L.** 4–6, ± resting on the soil, ovate-acute, c. 25 mm long, 10 mm across below and 20 mm midway, tapered, with the tip broad and rounded, 5 mm thick, grey-green, rough above with white, flat tubercles, the underside carinate towards the tip, the angles rough-tuberculate; **Fl.** yellowish, with a fine red M.line.

**A. schooneesii** L. BOL. v. **schooneesii** (Pl. 144/6) (*Nananthus s.* (L. BOL.) L. BOL.). – Cape: Willowmore D. – **L.** 8–10, broadly spatulate, with the tip rounded-triangular, bluish-green, very small; **Fl.** 10–15 mm ∅, yellowish-red, silky-glossy.

**A.** — v. **acutipetala** L. BOL. – Pet. acute.

**A.** — v. **willowmorensis** L. BOL. – **L.** carinate below the tip.

**A. setifera** (L. BOL.) L. BOL. (*Titanopsis setifera* L. BOL.). – Cape: Lit. Namaqualand. – **Ros.** 2–3 cm ∅, with 2–3 L.pairs.; **L.** 2 cm long, 4 mm thick, 5–6 mm across, with the tip somewhat widened and obliquely rounded-triangular, the underside rough-tuberculate, bluish-green to deep crimson, the angles in the upper part with 5 bristly T. almost 1 mm long, the underside and upperside with 1–3 similar T. at the tip; **Fl.** c. 25 mm ∅, golden-yellow to salmon-pink.

**A. spathulata** (THUNBG.) L. BOL. (*Mes. s.* THUNBG., *Titanopsis s.* (THUNBG.) L. BOL., *T. s.* (THUNBG.) SCHWANT., *Mes. crassipes* MARL., *A. c.* (MARL.) L. BOL., *T. c.* (MARL.) N. E. BR., *Nananthus c.* (MARL.) L. BOL.). – Cape: Sutherland D. – **Roots** fleshy, 3 cm thick, 20 cm long; **Br.** several, short; **L.** 6–8 on one Br., hatchet-shaped, tapered towards the base, apiculate, grey-green, with a tuberculate tip-area and reddened margins; **Fl.** 28–30 mm ∅, deep red.

**A. thudichumii** L. BOL. – Cape: Calvinia, Sutherland D. – **Roots** 12 cm long, 6 cm thick above; herbaceous parts velvety-papillose, reddish-brown; **L.** 2–4 on one Br., oblong, ± tapered towards the tip or oblong-ovate, acute to obtuse, with the underside convex, becoming glabrous, 20–32 mm long, 9–15 mm across, 5 mm thick, Sh. 4–7 mm; **Fl.** c. 3 cm ∅, Pet. yellow with reddish tips.

**A. villetii** (L. BOL.) L. BOL. (Pl. 144/4) (*Nananthus v.* L. BOL.). – Cape: Bushmanland. – **L.** 2 on one Br., 22 mm long, 16 mm across, 7 mm thick, broadly spatulate, ± round above, bluish-green, with dense, white tubercles $1/4$–$1/2$ mm ∅, Sh. 3 mm long; **Fl.** 2 cm ∅, Pet. pale yellow with coppery-red tips.

**Amoebophyllum** N. E. BR. (§ I/1/5). – ♃ plants; **St.** and **Br.** erect, robust; **L.** alternate, subulate, acute, later dying off and persisting as sharp Th.; **Fl.** terminal, in small Infl.

**A. angustum** N. E. BR (*A. roseum* L. BOL.). – Cape: Lit. Namaqualand. – Plant 25–35 cm tall, densely and minutely papillose; **Br.** and short **shoots** virgate, Int. 3–5 mm long; **L.** 12 mm long, 3 mm across below; **Fl.** 2 cm ∅, salmon-pink.

**Amphibolia** L. BOL. (§ IV not hitherto classified more precisely). – ♄; **Br.** prostrate, spreading, or erect; **L.** ± oblong, triangular or ± cylindrical, bluish-green to green, sometimes with a carinate line; **Fl.** in an Infl. or solitary, pedicellate, ± large, pink to purple. – Grows in sand. Summer: in the open, winter: cold-house. Propagation: seed, cuttings.

**A. gydouwensis** (L. BOL.) L. BOL. (*Lampranthus g.* L. BOL.). – Cape: Ceres, ? Clanwilliam D. – ♄ up to 38 cm tall; **St.** up to 1 cm ∅; **Br.** dark, Int. (1–)4(–6) cm long; short **shoots** 6–16 cm long, with 6–10 **L.** which are narrowed above, the keel reduced to a line on the underside, apiculate, bluish-green, minutely spotted, 2.5–3.5 cm long, 3 mm across, 4 mm ∅, Sh. swollen, 3.5 mm long, with impressed lines; **Fl.** in ± umbellate Infl., 18 mm ∅, purplish-pink.

**A. hallii** (L. BOL.) L. BOL. (*Stoeberia h.* L. BOL.). – Cape: Ceres D. – **Br.** prostrate to creeping, up to 30 cm long, Int. 15–25 mm long; **L.** 2–4 on the short shoots, drying and persisting, ± cylindrical, tapered, bluish, 2–3.5 cm long, 4–6 mm across and ∅, Sh. 3–4 mm long; **Fl.** solitary, c.3.5 cm ∅, Pet. pink with darker stripes.

**A. littlewoodii** (L. BOL.) L. BOL. (*Stoeberia l.* L. BOL.). – Cape: Ceres D. – Erect ♄ 26–30 cm tall; **Br.** rigid, virgate, Int. 1.5–2.5 cm long; **L.** tapered towards the obtuse or more rarely hooked tip, green, 10–22 mm long, 3–4 mm across and thick, Sh. 2–3 mm long, with impressed lines; **Fl.** 2–4, laxly arranged, up to 4.3 cm ∅, purplish-pink.

**A. maritima** L. BOL. – Cape: Van Rhynsdorp D. – ♄ covering large areas; **Br.** spreading, 18 cm long, Int. 1.5–3 cm long; **L.** 2–6 on the short shoots, narrowed towards the acute to rather obtuse tip, the keel on the underside reduced to a line, bluish-green to reddish, 1 to 2 cm long, 6–8 mm across, c. 9 mm ∅ above, Sh. 1 mm long; **Fl.** solitary, Pet. 5–6 mm long, pale pink.

**A. stayneri** L. BOL. – Cape: Ceres D. – ♄ c. 7 cm tall; **roots** tuberous, c. 6 cm long, 12 mm ∅ above; younger **Br.** slender, Int. often 1 cm long; **L.** narrowed above midway, acute, apiculate, green, rough, 17 mm long, 1.5–2 mm across and ∅, more rarely ovate to lanceolate-ovate, keel ± indistinct, 6–8 mm long, Sh. 1.5 mm long; **Fl.** solitary, Pet. 6–7 mm long, purplish-pink, staminodes present, paler, purplish-pink above.

**Anisocalyx** L. BOL. (G. of uncertain classification). – Dwarf, papillose ♄ resembling **Drosanthemum**. – Summer: in the open, winter: cold-house. Propagation: seed, cuttings.

**A. vaginatus** (L. BOL.) L. BOL. (*Drosanthemum v.* L. BOL., *A. salarius* L. BOL.). – Cape: Van Rhynsdorp D. – Compact plant 5–8 cm tall; **Br.** covered with dry old L.-Sh., Int. enclosed, rarely up to 1.3 cm long; **L.** 4 on a short shoot, united and sheathing for 7–13 mm, the free part narrowed towards the tip, obtuse to acute, rounded below the tip, 18–30 mm long, 7–11 mm across below, 5–8 mm across midway, bluish-green becoming dirty pink; **Fl.** solitary, pedicellate, 15–20 mm ∅, white, Sep. of different lengths.

**Antegibbaeum** SCHWANT. ex C. WEBER (§ IV/1/16). – Close to the G. **Gibbaeum**. – Greenhouse, warm. Propagation: seed, cuttings.

**A. fissoides** (HAW.) SCHWANT. ex C. WEBER (Pl. 144/5) (*Mes. f.* HAW., *Gibbaeum f.* (HAW.) NEL, *Mes. obtusum* HAW., *G. nelii* SCHWANT. ex JACOBSEN, *Mes. divergens* KENSIT). – Cape: Karroo. – Clump-forming; **St.** short, woody; **Br.** arranged laxly, radially, erect or creeping and rooting, with (1–)4(–5) shoots of uneven length above; each shoot with 2 dissimilar, triangular **L.**, the upperside flat, the underside carinate, ± tuberculate at the obtuse tip, the surface grey-green to reddish, smooth or somewhat wrinkled; **Fl.** sessile, with 2 bracts, 5–6 cm ⌀, light red.

**Apatesia** N. E. BR. (§ IV/2). – Mostly dwarf, ⊙ plants, branched from the base; **L.** opposite, united and sheathing, spatulate to lanceolate, glabrous or minutely papillose; **Fl.** solitary, long-pedicellate, yellow or whitish. – In the open. Propagation: seed.

**A. helianthoides** (AIT.) N. E. BR. (Pl. 144/3) (*Mes. h.* AIT., *Thyrasperma h.* (AIT.) N. E. BR., *T. sabulosa* N. E. BR.). – Cape: Cape D. – Erect, branched, with hairy Br. and short shoots; **L.** opposite, the lower ones spatulate-lanceolate, the upper ones lanceolate, flat, glabrous; **Fl.** terminal or axillary, c. 6 cm ⌀, yellow.

**A. maughanii** N. E. BR. – Cape: Piquetberg D. – 6–10 cm tall; **Br.** short, erect to spreading; **L.** radical, 11–33 mm long, 2–4 mm across, spatulate-obovate to lanceolate, narrowed below towards the petiole and forming an open Sh., glabrous; **Fl.** 3–4 cm ⌀, yellow, with a thin pedicel.

**A. pillansii** N. E. BR. – Cape: Somerset West, beaches. – 10–15 cm tall, glabrous; **St.** short, with 8–10 L.pairs; **L.** narrowed into the long petiole, blade 20–40 mm long, 10 to 15 mm across above, lanceolate-acute; **Fl.** 3–4 cm ⌀, lemon-yellow, whitish inside.

**A. sabulosa** (THUNBG.) L. BOL. (*Mes. s.* THUNBG.). – Cape: Cape D. – Almost acauline, 15–20 cm tall; **Br.** ascending, densely leafy above; **L.** opposite, sessile, lower ones larger, oblong or oblong-spatulate, ± tapered, minutely papillose, c. 2.5 cm long, narrowed towards the 1 cm long petiole; **Fl.** glossy, light yellow, on a pedicel 5–10 cm long.

**Aptenia** N. E. BR. (§ I/1/1). – Dwarf, freely branched sub-♄; **Br.** prostrate, with minute, glossy papillae; **L.** opposite, cordate or lanceolate, ± fleshy, minutely papillose; **Fl.** terminal or lateral, solitary or 3 together, sessile or short-pedicellate, pink to purple. – Summer: in the open, winter: cold-house. Propagation: seed, cuttings.

**A. cordifolia** (L. f.) (SCHWANT. (Pl. 145/4) (*Mes. c.* L. f., *A. c.* (L. f.) N. E. BR., *Litocarpus c.* (L. f.) L. BOL., *Tetracoilanthus c.* (L. f.) RAP. et CAM.). – S.Afr.: E. part of the coastal deserts. – Freely branched; **Br.** prostrate, up to 60 cm long, cylindrical, green, papillose; **L.** cordate-ovate, up to 25 mm long and almost as wide, fresh green, minutely papillose; **Fl.** purplish-red, small.

**A. —** cv. Variegata (*Mes. c. v. v.* HORT.). – **L.** with a cream-coloured border.

**A. lancifolia** L. BOL. – Transvaal: Pietersberg D. – **Br.** prostrate, somewhat constricted at the nodes, Int. 15–45 mm long; **L.** lanceolate to narrow-lanceolate, obtuse or ± acute, upperside grooved, underside bluntly carinate, green, c. 32 mm long, 6 mm across; **Fl.** c. 15 mm ⌀, pink.

**Arenifera** HERRE (§ IV/1/7). – Small ♄; **L.** small, with sand adhering to them; **Fl.** in 3's, pink. – Greenhouse, warm. Propagation: seed, cuttings.

**A. pillansii** (L. BOL.) HERRE (Pl. 145/1) (*Psammophora p.* L. BOL.). – Cape: Lit. Namaqualand; S.W.Afr.: desert. – Erect ♄ 10–25 cm tall, branched densely from the base; **Br.** c. 3 mm thick, Int. 5–15 mm long; **L.** variable, some up to 25 mm long, 4–5 mm across and thick, upperside and sides convex, other L. 6–15 mm long, 7 mm ⌀, indistinctly angled, younger L. green and viscous, older ones bluish-green, rough, ± dry; **Fl.** 22 mm ⌀, pale pink.

**Argyroderma** N. E. BR. (§ IV/1/2). – The text is a shortened version of a contribution by H. Hartmann. – ♃, branched or unbranched plants growing in patches, clusters or solitary, above the ground or sunken into the soil; 2 green or whitish L. per season, bases connate; **L.** finger-, hood-or half-egg-shaped, upperside plane, lower side of L. forming a chin, occasionally keeled; surface of L. smooth, without dots or tubercles, covered with wax; withered L. yellowish or chestnut-coloured, later turning black; of firm or brittle consistence; shape undulate, spreading or appressed (the pair forming a cup); **Fl.** with a distinct Cal.-tube between the top of the Ov. and the base of the stamens and Pet., stamens pendulous; style reduced, stigmata forming a cushion on top of the Ov.; Pet. white, yellow or purple in different shades, also different colours in one Fl.; **Fr.** of the **Leipoldtia** type with big white tubercles, locules with firm complete roofs, valves with narrow wings, 10–24-locular; Cal.-tube persisting; dry bracts yellowish or chestnut-coloured. Cape: Van Rhynsdorp District.

## Division of the Genus Argyroderma in Subgenera

**SG. I. Roodia** (N. E. BR.) H. HARTM. (*Roodia* N. E. BR. as a G.). – Plant highly branched, forming patches with age; **L.** finger-shaped, green, spreading-ascending; old L. yellow, long persisting, spreading; **Fl.** with white inner and purple or (rarely) yellow outer Pet.; **Fr.** ± 12-locular, calyx-lobes longer than the apically constricted calyx-tube; yellow bracts distinctly shorter than the long pedicel. Only species: A. fissum.

**SG. II. Argyroderma.** – Plant unbranched or branched, growing sunken into or above the ground, often in clusters, never in patches; L. whitish, hood-, half-egg-or semiglobose-shaped; old L. long persisting or deciduous, spreading, appressed or undulate, chestnut-coloured or yellow. – Type-species: A. testiculare. – Species: A. congregatum, crateriforme, delaetii, framesii, patens, pearsonii, ringens, subalbum, testiculare.


## Key to the Species and Subspecies of the Genus Argyroderma

1. L. green, finger-shaped (2–4 times as long as broad or thick), spreading; plant highly branched, forming clumps with age; Int. visible; withered L. long persisting, yellow, spreading; Fr. ± 12-locular, pedicel long (2–4 times longer than the ⌀ of the capsule); bracts not reaching the base of the capsule; Cal.-tube convex, apically constricted, shorter than the persisting Cal.-lobes .................................................................................................
**Argyroderma SG. I. Roodia** with a single spec.: **A. fissum**
1a. L. whitish-silvery, hood or half-egg shaped; plant branched or unbranched, no visible Int. ...
........ **Argyroderma SG. II. Argyroderma** 2.

2. Withered L. chestnut-coloured, long persisting; bracts of Fr. entirely chestnut-coloured or spotted, rarely reaching the base of the capsule .................................................. 3.
2a. Withered L. yellowish, long persisting or deciduous, then brittle; bracts of Fr. yellowish, embracing the capsule or at least touching its base .................................. 4.

3. Withered L.-pair forming a cup which holds the assimilating egg-to ball-shaped L.-pair; keel reduced or missing ........................................................................ 5.
3a. Withered L.-pair spreading, assimilating L. hood-shaped, keeled ...................... 6.

4. Bracts embracing the capsule; Cal.-tube of capsule flattened; Fr. less than twice as long as wide; withered L. deciduous or persisting and spreading .......................... 8.
4a. Bracts touching the base of the capsule; Cal.-tube apically narrowed, longer than Cal.-lobes; Fr. twice or more times as long as wide; old yellow L. long persistent, ± appressed; Fl. white, purple or outer Pet. purple and inner Pet. white ............................ **testiculare**

5. Cal.-tube erect ........................................................................ **pearsonii**
5a. Cal.-tube convex ...................................................................... **subalbum**

6. Cal.-tube erect; L. spreading widely, keeled; Fi. wider than the breadth of the L.; L.-pair 20–40 mm long (across the top), 15–20 mm broad; capsule c. 12-locular, bracts chestnut-coloured, just reaching the base of the capsule; Fl. in different shades of purple, white or lemon, different colours in one Fl. common ...................................... **patens**
6a. Cal.-tube convex, narrowed at the top, capsule globose; L.-pair 10–20 mm long (across the top), 5–12 mm broad; Fi. 2–12 mm long; capsule mostly 10–12-locular, bracts entirely chestnut-coloured or spotted, not touching the base of the capsule; Fl. purple, purple and white, or yellow ............................................................. **framesii** 7.

7. L.-pair not longer than 11 mm across the top, not broader than 10 mm; Fi. not wider than 5 mm; bracts with chestnut-coloured dots; Fl. purple, filaments white or purple ........
........ **framesii ssp. framesii**
7a. L.-pair longer than 11 mm (across the top), broader than 10 mm, Fi. wider than 5 mm; bracts of Fr. entirely chestnut-coloured; Fl. purple, purple and white, or yellow .........
........ **framesii ssp. hallii**

8. Withered L. spreading, long persisting; assimilating L. hood-shaped; plant branched, forming clumps; capsule (10–)12(–16)-locular; Fl. mostly yellow, rarely purple ........ **congregatum**
8a. Withered L. undulate, brittle; plant growing sunken into the ground, branched or unbranched
........ 9.

9. Capsule mostly 14–20-locular .................................................... 10.
9a. Capsule (10–)12(–16)-locular; assimilating L. hood-shaped, L.-pair 20–30 mm long (across the top), c. 15 mm broad; Fl. mostly yellow, rarely purple ............... **crateriforme**

10. L. hood-shaped, Fi. as wide as breadth of L.; Fl. purple, or outer Pet. purple, inner Pet. white
........ **ringens**
10a. L. of the shape of half an egg, L.-pair 20–40 mm long (across the top), 15–30 mm broad; Fi. $\frac{1}{5}-\frac{1}{2}$ of breadth; Fl. white, pink, different shades of purple or yellow, frequently several colours in one population ............................................. **delaetii**

**A. congregatum** L. Bol. (SG. II) (*A. angustipetalum* L. Bol., *A. jacobsenianum* Schwant., *A. nortieri* L. Bol., *A. peersii* L. Bol., *A. rooipanense* L. Bol. p.part.). – Plant forming clusters with age; **L.** hood-shaped; old L. yellow, long-persisting, spreading; **Fl.** mostly yellow, in a few populations purple; **Fr.** ± 12-locular, yellow bracts embracing the capsule, Cal.-tube nearly appressed to the surface of the capsule.

**A. crateriforme** (L. Bol.) N. E. Br. (SG. II) (*Mes. crateriforme* L. Bol., *A. pulvinare* L. Bol., *A. subrotundum* L. Bol., *A. testiculare* v. *roseum* (Haw.) N. E. Br.). – Plant branched or unbranched, growing sunken into the ground; **L.** hood-shaped; old L. yellow, undulate, deciduous and brittle; **Fl.** mostly yellow, a few populations purple or white; **Fr.** ± 12-locular, Cal.-tube nearly appressed to the surface of the capsule; yellow bracts enclosing the capsule.

**A. delaetii** Maass (SG. II) (Pl. 145/2) (*A. aureum* L. Bol., *A. australe* L. Bol., *A. blandum* L. Bol., *A. boreale* L. Bol., *A. brevitubum*. L. Bol., *A. carinatum* L. Bol. p.part., *A. citrinum* L. Bol., *A. concinnum* Schwant., *A. cuneatipetalum* L. Bol. p.part., *A. d.* v. *purpureum* Maass, *A. densipetalum* L. Bol., *A. formosum* L. Bol., *A. gregarium* L. Bol., *A. latifolium* L. Bol., *A. lesliei* N. E. Br., *A. leucanthum* L. Bol., *A. longipes* L. Bol., *A. planum* L. Bol., *A. productum* L. Bol., *A. reniforme* L. Bol., *A. rooipanense* L. Bol. p.part., *A. roseum* Schwant. f. *r.*, *A. r.* Schwant. f. *d.* (Maass) Rowl., *A. schuldtii* Schwant., *A. speciosum* L. Bol., *A. splendens* L. Bol., *A. testiculare* v. *luteum* N. E. Br.). – Plant normally not branched, growing sunken into the ground; **L.** half-egg-shaped, variable in shape and size; old L. yellow, undulate, brittle and deciduous; **Fl.** white, purple in different shades or yellow, regularly several colours in one population; **Fr.** mostly 14–18-locular, Cal.-tube nearly appressed to the surface of the capsule, yellow bracts embracing the capsule.

**A. fissum** (Haw.) L. Bol. (SG. I) (Pl. 145/3) (*Mes. f.* Haw., *Cheiridopsis braunsii* Schwant., *Roodia b.* (Schwant.) Schwant., *A. b.* (Schwant.) Schwant., *Mes. brevipes* Schltr., *Roodia b.* (Schltr.) L. Bol., *A. b.* (Schltr.) L. Bol., *R. digitifolia* N. E. Br., *A. d.* (N. E. Br.) Schwant., *A. hutchinsonii* L. Bol., *A. latipetalum* L. Bol. v. *l.*, *A. l.* L. Bol. v. *longitubum* L. Bol., *A. litorale* L. Bol., *A. orientale* L. Bol., ? *Mes. socium* N. E. Br.). – Forming patches with age; **L.** finger-shaped, 5–12 cm long, 8–12 mm wide or thick, green to yellow; outer **Pet.** purple or yellow, inner Pet. white.

**A. framesii** L. Bol. 1934 non 1929 (SG. II). – Plant highly branched, forming clumps with age; **L.** hood-shaped, keeled; **Fl.** purple, purple (outer) and white (inner Pet.) or yellow; **Fr.** globose, Cal.-tube apically constricted, mostly 10–12-locular, bracts not touching the base of the capsule, chestnut-coloured or spotted.

**A.** — ssp. **framesii** (*A. f.* v. *f.*, *A. f.* v. *minus* L. Bol.). – L.-pair not longer across the top than 11 mm, not broader than 10 mm, Fi. not wider than 5 mm; **Fl.** purple or purple and white, bracts with chestnut-coloured spots.

**A.** — ssp. **hallii** (L. Bol.) H. Hartm. (*A. hallii* L. Bol., *A. strictum* L. Bol.). – **L.**-pair 15–20 mm long across the top, 10–15 mm broad, Fi. wider than 5 mm; **Fl.** purple and white, rarely (one population) yellow; bracts entirely chestnut-coloured.

**A.** × **kleijnhansii** L. Bol. (pro spec.). – The type of this spec. resembles very closely hybrids between **A. fissum** and spec. of the **SG. Argyroderma,** which occur freely in habitat and are not rare in cultivation.

**A.** × **necopinum** (N. E. Br.) N. E. Br. (pro spec.) (*Mes. n.* N. E. Br.). – All available material of this "spec." identified by N. E. Brown matches hybrids (see **A.** × **kleijnhansii**). Not rare in cultivation.

**A. patens** L. Bol. (SG. II). – Plant branched, forming clusters with age; **L.** hood-shaped, keeled, Fi. wider than L.-breath; old L. chestnut-coloured, long-persisting, spreading; **Fl.** in different shades of purple to white, rarely lemon; **Fr.** ± 12-locular, Cal.-tube erect, bracts chestnut-coloured, touching the base of the capsule.

**A. pearsonii** (N. E. Br.) Schwant. (SG. II) (*Mes. p.* N. E. Br., *A. amoenum* Schwant., *A. cuneatipetalum* L. Bol. p.part., *A. framesii* L. Bol. 1929 non 1934, *A. luckhoffii* L. Bol., *A. ovale* L. Bol., *A. schlechteri* Schwant., *A. testiculare* v. *p.* N. E. Br.). – Plant unbranched, rarely in clusters, above the ground; **L.** semi-globose, Fi. narrow, L.-pair 20–40 mm long across the top, 15–30 mm broad; old L. chestnut-coloured, long-persisting, appressed to one another, thus forming a column; **Fl.** mostly purple and white, in a few populations orange (outer) and yellow (inner Pet.), very seldom white; **Fr.** ± 12-locular, Cal.-tube erect; bracts chestnut-coloured, touching the base of the capsule.

**A. ringens** L. Bol. (SG. II). – Plants consisting of one or rarely few **Bo.**; **L.** hood-shaped, ascending, Fi. as wide as breadth of L.; old L. yellow, undulate, deciduous and brittle; **Fl.** purple (outer) and white (inner Pet.); **Fr.** mostly 14–18-locular, Cal.-tube nearly appressed to the surface of the capsule; yellow bracts embracing the capsule.

**A. subalbum** (N. E. Br.) N. E. Br. (SG. II) (*Mes. s.* N. E. Br., *A. villetii* L. Bol.). – Plant highly branched and forming clumps with age; **L.** semi-globose, upper surfaces closely pressed together, leaving no Fi.; L.-pair 10 to 20 mm long across the top, c. 10 mm broad; old L. chestnut-coloured, long-persisting, appressed; **Fl.** purple (outer) and white (inner Pet.); **Fr.** globose, Cal.-tube apically constricted, ± 12-locular, chestnut-coloured bracts not reaching the base of the capsule.

**A. testiculare** (Ait.) N. E. Br. (SG. II) (*Mes. t.* Ait., *A. carinatum* L. Bol. p.part.). – Plant not or little branched, growing above the

**Aridaria**

ground; **L.** half-egg-shaped, keeled, Fi. narrow; old L. yellow, long-persisting, ± appressed, thus forming a column; **Fl.** purple, rarely white; **Fr.** 12–18-locular, Cal.-tube gradually narrowed towards the apex, yellow bracts touching the base of the capsule.

**Aridaria** N. E. Br. (§ I/1/4). – Glabrous ♄ with a stout, woody **taproot**; **St.** woody, not herbaceous; **L.** opposite, papillose, usually cylindrical or semicylindrical, very easily detached and thus deciduous; **Fl.** in an Infl. or solitary, pedicellate, white, pink, red, yellow. – Summer: in the open, winter: cold-house. Propagation: seed, cuttings. – Type: **A. noctiflora.**

**A. arcuata** L. Bol. (*Nycteranthus a.* (L. Bol.) Schwant.). – Cape: Lit. Namaqualand. – Laxly branched ♄; **St.** virgate, 50–60 cm long, 4 mm ⌀, Int. 30–35 mm long, with short shoots; **L.** semicylindrical, acute, glossy green to bluish-pruinose, 20–25 mm long, 5 mm thick; **Fl.** 3 cm ⌀, opening at night, scented, Pet. red outside, white inside, with red tips.

**A. barkerae** L. Bol. – Cape: Ceres D. – Small ♄ with **St.** up to 24 cm long, Int. up to 5 cm long; **L.** cylindrical, acute, bluish-green, 20–28 mm long, 5 mm across and ⌀; **Fl.** c. 3 cm ⌀, Pet. white, pale pink outside.

**A. beaufortensis** L. Bol. – Cape: Beaufort West D. – Small ♄; **St.** c. 25 cm long, Int. 15–30 mm long; **L.** narrowed slightly towards the obtuse tip, bluish-green, 15–20 mm long, 2–3 mm ⌀; **Infl.** lax, Fl. 3 cm ⌀, opening at night, scented, Pet. white, lemon-yellow outside.

**A. brevicarpa** L. Bol. (*Nycteranthus b.* (L. Bol.) Schwant.). – Cape: Lit. Namaqualand. – Erect ♄ up to 1 m tall; **St.** virgate, with short shoots in umbels of 4; **L.** semicylindrical, bluish-green, 15–30 mm long, 3 mm thick; **Fl.** 25 mm ⌀, Pet. pale brown on the upperside and white below, yellowish-brown on the underside.

**A. brevifolia** L. Bol. (*Nycteranthus b.* (L. Bol.) Schwant., *Perapentacoilanthus b.* (L. Bol.) Rap. et Cam.). – Cape: Barrydale. – **St.** prostrate, stiff, up to 60 cm long, Int. 1–2 cm long; short **shoots** erect, 4–5 cm long; **L.** semicylindrical, firm, almost 10 mm long, 3 mm thick; **Fl.** solitary, 3 cm ⌀, white.

**A. calycina** L. Bol. (*Nycteranthus c.* (L. Bol.) Schwant.). – Cape: Van Rhynsdorp D. – Stiff, erect ♄ up to 30 cm tall; **St.** robust, up to 12 mm thick, erect; **L.** semicylindrical, acute to rather obtuse, dirty bluish-green, 3–4 cm long, 3–5 mm thick; **Fl.** 4 cm ⌀, red, scented, opening in the evening.

**A. compacta** L. Bol. (*Nycteranthus c.* (L. Bol.) Schwant.). – Cape: Lit. Namaqualand. – Compact, robust ♄ 40 cm tall; **St.** crowded, stiff, 8 mm thick, Int. 1 cm long; short **shoots** 8–10 cm long; **L.** cylindrical, obtuse, glossy bluish-green, 15–22 mm long, 7 mm thick; **Fl.** 5–6 mm ⌀, Pet. pinkish-purple on the upperside, coppery-red on the underside.

**A. debilis** L. Bol. (*Nycteranthus d.* (L. Bol.) Schwant.). – Cape: Van Rhynsdorp D. – Slender ♄; **St.** prostrate, stiff, up to 70 cm long, Int. 10–15 mm long, with erect short shoots 27 cm long; **L.** semicylindrical, obtuse, blue, (15–)20(–25) mm long, 2–4 mm thick; **Fl.** 45 mm ⌀, opening at night, scented, Pet. white on the upperside, pink to brownish-red at the tips on the underside.

**A. dejagerae** L. Bol. (*Nycteranthus d.* (L. Bol.) Schwant.). – Cape: Beaufort West D. – Erect, stiff ♄ 25 cm tall; **St.** 4 mm thick; **L.** spherical or cylindrical-spherical, 1 cm long, 5 mm thick; **Fl.** 25–30 mm ⌀, Pet. white, pink outside at the tips.

**A. elongata** L. Bol. (*Nycteranthus e.* (L. Bol.) Schwant.). – Cape: Lit. Namaqualand. – St. prostrate, up to 64 cm long, up to 6 mm thick, Int. 2–4 cm long, indistinctly papillose; lower L. opposite, the upper ones alternate, narrowing towards the obtuse tip, c. 7 cm long, 1 cm across below, 4 mm thick midway, with a Sh. 1.5 mm long; **Fl.** 30–47 mm ⌀, pale green.

**A. esterhuyseniae** L. Bol. (*Nycteranthus e.* (L. Bol.) Schwant.). – Cape: Bushmanland. – Erect, robust ♄ 16 cm tall; **St.** 25 mm thick; **Br.** grey, Int. 15–30 mm long; **L.** semicylindrical, obtuse, 25–45 mm long, 5 mm across and thick, with a Sh. 1 mm long; **Fl.** in an Infl., 28 mm ⌀, white to pale pink.

**A. floribunda** L. Bol. (*Nycteranthus f.* (L. Bol.) Schwant.). – Cape: Lit. Namaqualand. – Erect, robust ♄; **St.** virgate, Int. 22–50 mm long; **L.** cylindrical, obtuse, bluish, 22 mm long, 4 mm thick; **Fl.** 25–36 mm ⌀, scented, opening at night, white to pale pink.

**A. globosa** L. Bol. (*Nycteranthus g.* (L. Bol.) Schwant.). – Cape: Van Rhynsdorp D. – Robust ♄ c. 30 cm tall; **St.** 6 mm thick, Int. 15–25 mm long; **L.** 9 mm across, 8 mm thick, obtuse; **Fl.** on long side-shoots, almost 5 cm ⌀, Pet. white, pink on the underside.

**A. gracilis** L. Bol. (*Nycteranthus g.* (L. Bol.) Schwant.). – Cape: Van Rhynsdorp D. – Rather weak ♄; **St.** prostrate, 10 cm long, Int. 15–25 mm long; **L.** hemispherical to oblong, obtuse, bluish-green, 4–13 mm long, 3–4 mm thick; **Fl.** 2 cm ⌀, opening at night, Pet. white on the upperside, yellow on the underside.

**A. inaequalis** L. Bol. (*Nycteranthus i.* (L. Bol.) Schwant.). – Cape: Van Rhynsdorp D. – **St.** prostrate, up to 40 cm long, Int. 2.5–3 cm long; **L.** in 5–7 pairs at the lower nodes, opposite, other L. usually alternate, narrowed towards the tip, obtuse to ± acute, 15–45 mm long, 4–6 mm across and thick; **Fl.** 2.5 cm ⌀, outer Pet. salmon-pink, inner ones straw-coloured.

**A. intricata** L. Bol. (*Nycteranthus i.* (L. Bol.) Schwant.). – Cape: Clanwilliam D. – Robust ♄ c. 60 cm tall; **St.** very tangled, 12 mm thick; **shoots** slender; **L.** semicylindrical, obtuse to tapered, 3 cm long, 4 mm thick; **Fl.** 4 cm ⌀, white.

**A. klaverensis** L. Bol. – Cape: Van Rhynsdorp D. – Robust ♄ up to 90 cm tall; **St.** c. 1 cm ⌀; **shoots** c. 6.5 cm long, Int. 11 mm long; **L.** semicylindrical, tapered, green, c. 22 mm long, c. 3 mm across and thick; **Infl.** of 3 white Fl. 15–25 mm ⌀, opening at midday.

**A. leipoldtii** L. Bol. – Cape: Clanwilliam D. – ♃ 17 cm tall; **St.** c. 3 cm thick; **Br.** dense, c. 1.5 cm thick; **shoots** slender, c. 8 cm long, Int. 15–25 mm long; **L.** cylindrical, obtuse to acute, 12–18 mm long, 3.5 mm ⌀; **Infl.** of 3 Fl. 35–40 mm ⌀, Pet. pale below on the upperside, pink above, red on the underside, opening at night.

**A. littlewoodii** L. Bol. – S.W.Afr.: W. of Grünau. – Erect ♃ almost 40 cm tall, c. 90 cm ⌀; **L.** ± cylindrical, somewhat hatchet-shaped midway, 3–25 mm long, 4–5 mm across and thick; **Infl.** 6-branched, Fl. opening at night, up to 4 cm ⌀, Pet. white, the outer ones pink on the underside.

**A. longisepala** L. Bol. (*Nycteranthus l.* (L. Bol.) Schwant.). – Cape: Clanwilliam D. – Erect, slender ♃ c. 30 cm tall; **St.** 3 mm thick, Int. 1–2 cm long; **L.** semicylindrical, obtuse or short-tapered, pale green, 17 mm long, 1.5 mm across, 1.5–3 mm thick; **Fl.** 4 cm ⌀, white.

**A. meridiana** L. Bol. (*Nycteranthus m.* (L. Bol.) Schwant.). – Cape: Clanwilliam D. – Densely branched ♃; **Br.** elongated, bent, Int. 4–15 mm long; **L.** semicylindrical, acute, glossy-papillose, 18 mm long, 2–3 mm thick; **Fl.** 4 cm ⌀, salmon-pink, pale yellowish-pink inside.

**A. meyeri** L. Bol. (*Nycteranthus m.* (L. Bol.) Schwant.). – Cape: Lit. Namaqualand. – ♃ c. 40 cm tall; **St.** 4 mm thick, Int. 10–22 mm long; **L.** ± cylindrical, obtuse to ± acute, pale blue, 7–12 mm long, 3 mm thick; **Fl.** opening at night, Pet. pale pink, coppery-red on the underside.

**A. muirii** N. E. Br. (*Nycteranthus m.* (N. E. Br.) Schwant.). – Cape: Oudtshoorn D. – Erect, much branched ♃ up to 40 cm tall; **St.** opposite, spreading, tangled, Int. 13–26 mm long; **L.** semicylindrical, obtuse, light green to bluish, 13–20 mm long, 2–3 mm thick; **Fl.** opening at night, scented, Pet. 16 mm long, pale yellow with reddish-brown stripes.

**A. multiseriata** L. Bol. (*Nycteranthus m.* (L. Bol.) Schwant. – Cape: Lit. Namaqualand. – ♃ with a thickened **St.**: **Br.** prostrate to ascending, 25 cm long, stiff, Int. 15–25 mm long, papillose at first; **L.** semicylindrical, obtuse, 45 mm long, 5 mm thick; **Fl.** 4 cm ⌀, Pet. in many series, yellow.

**A. mutans** L. Bol. (*Nycteranthus m.* (L. Bol.) Schwant.). – Cape: Cradock D. – Erect, densely branched ♃; **Br.** prostrate to erect, c. 40 cm long, Int. 2–3 cm long; **L.** semicylindrical, obtuse, with the tip curved inwards, pale green, 25–40 mm long, 3–4 mm thick, Sh. 2–3 mm long; **Fl.** 26 mm ⌀, yellow.

**A. nevillei** L. Bol. – S.W.Afr.: Gr. Namaqualand. – Small ♃; **Br.** 16 cm long, Int. 10–18 mm long; **L.** semicylindrical, obtuse, falcate, 15 to 27 mm long, 2–4 mm thick; **Infl.** of 3 Fl. opening in the afternoon, closing in the evening, 3 cm ⌀, white, brownish-red on the underside.

**A. noctiflora** (L.) Schwant.[1]) v. **noctiflora** (Pl. 146/2) (*Mes. n.* L., *Nycteranthus n.* (L.) Necker, *Nyct. n.* (L.) Rothm., *Peratetracoilanthus n.* (L.) Rap. et Cam., *A. n.* (L.) L. Bol.). – S.W.Afr.: widely distributed. – ♃ 80 cm tall, with spreading, cylindrical Br.; **L.** only on the short shoots, crowded, ± cylindrical, obtuse, grey-green, 25–35 mm long, 6–8 mm across; **Fl.** 3–4.5 cm ⌀, white, sweetly scented, opening towards evening.

**A. —** v. **fulva** (Haw.) Herre et Friedr. (*Mes. f.* Haw., *Mes. n.* v. *f.* (Haw.) Salm, *A. f.* (Haw.) Schwant., *Nycteranthus f.* (Haw.) Schwant.). – Smaller in all parts; **L.** slender; **Fl.** smaller, brownish outside, very sweetly scented.

**A. ovalis** L. Bol. (*Nycteranthus o.* (L. Bol.) Schwant.). – Cape: Lit. Namaqualand. – Dwarf ♃; **St.** 1.5 mm thick, Int. 15–20 mm long; **L.** oval, cordate below, dirty brownish-green, 7–13 mm long, 5–8 mm across, 5 mm thick; **Fl.** 3.5 cm ⌀, white, flesh-coloured outside.

**A. paucandra** L. Bol. v. **paucandra** (*Nyct. p.* (L. Bol.) Schwant.). – Cape: Van Rhynsdorp D. – Dwarf ♃; **St.** up to 20 cm long, 1 cm thick; **L.** semicylindrical, curved, obtuse, bluish, brownish below, 5–16 mm long, 3–4 mm thick; **Fl.** solitary, 3 cm ⌀, opening at night, Pet. white and pink-tipped to dirty pink, often coppery outside.

**A. —** v. **gracillima** L. Bol. (*Nycteranthus p.* v. *g.* (L. Bol.) Schwant.). – Slender variety.

**A. pillansii** L. Bol. – Cape: Lit. Namaqualand. – ♃ 24 cm tall; **St.** 16 mm ⌀; **Br.** c. 5 mm ⌀, Int. 1–2 cm long; **L.** cylindrical, c. 12 mm long, 4 mm ⌀; **Fl.** in 3's, opening at night, Pet. almost white, pale pink outside.

**A. serotina** L. Bol. (*Nycteranthus s.* (L. Bol.) Schwant.). – Cape: Lit. Namaqualand. – Laxly branched ♃; **St.** prostrate, stiff, 20–40 cm long; **shoots** 5–7 mm long, with 2 L., or elongated, with 6–10 L.; **L.** ± cylindrical, margins and keel indistinct, obtuse to short-tapered, glossy-papillose at first, bluish-green; **Fl.** 3 cm ⌀, purplish-red.

**A. straminea** (Haw.) Schwant. (*Mes. s.* Haw., *Nycteranthus s.* (Haw.) Schwant.). – Cape: precise origin not known. – **St.** creeping; **L.** cylindrical, bluish; **Fl.** straw-coloured.

**A. subtruncata** L. Bol. (*Nycteranthus s.* (L. Bol.) Schwant.). – Cape: Bushmanland. – Stiff, erect ♃ 25–30 cm tall; **St.** somewhat climbing, 60 cm tall; **shoots** 1–2 mm thick, Int. 10–25 mm long; **L.** spherical to oblong, compressed above, with a rounded tip, 5–14 mm long, 5–7 mm thick; **Infl.** lax, Fl. 3 cm ⌀, upperside white, coppery-red or salmon-pink outside.

---

[1]) See H.-Chr. Friedrich, Der Formenkreis der A. noctiflora (L.) Schwant., Botan. Jahrb. **86**, 1967, 272–279. There are listed as synonyms of **A. noctiflora** (L.) Schwant: *A. arcuata, brevipes, calycina, compacta, debilis, dejagerae, esterhuyseniae, floribunda, fulva, globosa, gracilis, intricata, littlewoodii, longisepala, meridiana, meyeri, paucandra, serotina, straminea, subtruncata, tenuifolia, vespertina.*

**A. tenuifolia** L. Bol. v. **tenuifolia** (*Nycteranthus t.* (L. Bol.) Schwant.). − Cape: Lit. Namaqualand. − **St.** and **Br.** stiff, robust, 1 cm thick; **shoots** slender, densely tangled, somewhat climbing, up to 1 m tall; **L.** thin.

**A.** — v. **speciosa** (L. Bol.) L. Bol. (*Nycteranthus t.* v. *s.* L. Bol.). − **L.** thicker; **Fl.** c. 5 cm ⌀.

**A. vespertina** L. Bol. (*Nycteranthus v.* (L. Bol.) Schwant.). − Cape: Lit. Namaqualand. − Erect ♄ c. 15 cm tall; **St.** woody and thickened below, 15 mm ⌀; **Br.** stiff, contorted, 7–8 mm thick; **shoots** slender, 3.5 cm long, with 6–8 ± spherical **L.** 6–9 mm long, 6 mm thick, pale blue-pruinose; **Fl.** solitary, 2 cm ⌀, opening in the evening, upperside pale pink, underside coppery-red.

**Aspazoma** N. E. Br. (§ I/2). − ♃, succulent, bushy, branched plants; **L.** several on a shoot, alternate and opposite, hemispherical or ± trigonous, amplexicaul, with a tubular **Sh.**; **Fl.** terminal, solitary. − Greenhouse. Propagation: seed, cuttings.

**A. amplectens** (L. Bol.) N. E. Br. (Pl. 146/4) (*Mes. a.* L. Bol.). − Cape: Lit. Namaqualand. − 12–15 cm tall; **L.** 12–30 mm long, **Sh.** 12 mm long; **Fl.** c. 4 cm ⌀, pale straw-coloured.

**Astridia** Dtr. et Schwant. emend. L. Bol. (§ IV/1/1). − Robust, tall, often compact ♄ 25–30 cm tall, often laxly branched; **St.** virgate, Int. up to 4 cm long, main Br. 5–13 mm thick, herbaceous parts ± velvety or glabrous; **L.** erect, patent or widespread, long persistent and drying up, nearly falcate, obtuse or rounded at the tip, mostly narrowed towards the tip, 3–5 times as long as broad, underside indistinctly obtuse or rarely sharply keeled, slightly convex laterally, rarely with the sides very compressed, 3–6 cm long, 1–3 cm thick, Sh. 2–6 mm long; **Fl.** mostly solitary or up to 3 terminally, 4–7.5 cm ⌀, the pedicel very short, with 2 bracts 2 cm long surrounding the receptacle, Pet. white, pink, red, rarely lemon-yellow; **seeds** papillose hairy. − Greenhouse, warm. Propagation: seed, cuttings.

## Key to the Species of the Genus Astridia by L. Bolus

| | | |
|---|---|---|
| 1. | Herbaceous parts glabrous | 2. |
| 1a. | Herbaceous parts ± velvety | 5. |
| 2. | Upper surface of L. lanceolate-acute; staminodes absent | **vanheerdei** |
| 2a. | Upper surface of L. linear; staminodes present | 3. |
| 3. | Upper surface of L. round at the apex | **herrei** |
| 3a. | Upper surface of L. acute or acuminate | 4. |
| 4. | Outer Pet. dense, red, up to 2.2 cm long, 2 mm broad | **longifolia** |
| 4a. | Outer Pet. lax, bright purplish-red, up to 3 cm long, 3 mm broad | **vanbredai** |
| 5. | Pet. white or pale lemon | 6. |
| 5a. | Pet. some shade of red or pink | 10. |
| 6. | Staminodes present | 7. |
| 6a. | Staminodes absent | 9. |
| 7. | Pet. white, stigmas as long as, or shorter than the stamens | 8. |
| 7a. | Pet. pale lemon, stigmas well exceeding the stamens | **citrina** |
| 8. | Upper surface of mature L. ovate-acuminate or lanceolate, apex erect | **dinteri** |
| 8a. | Upper surface of mature L. linear-lanceolate, apex recurved | **alba** |
| 9. | L. strongly compressed laterally, acutely keeled, up to 8 cm or in cultivation 11.7 cm long; filaments white | **hallii** |
| 9a. | L. not compressed, obtusely keeled, sides convex, up to 6 cm long; filaments yellow in the upper part | **ruschii** |
| 10. | Pet. red or reddish orange | 11. |
| 10a. | Pet. pink or salmon-pink | 14. |
| 11. | All the Sep. ± margined, the outer as broad as, or broader than, long | **latisepala** |
| 11a. | Outer Sep. not margined, usually longer than broad | 12. |
| 12. | Lobes of the ovary stellately spreading and reaching the disc | **rubra** |
| 12a. | Lobes of the ovary ± erect, not reaching the disc | 13. |
| 13. | Pet. 6-seriate, reddish-orange; stigmas well exceeding the stamens | **speciosa** |
| 13a. | Pet. 2–3-seriate, red; stigmas shorter than the stamens | **swartportensis** |
| 14. | Pet. pink; stigmas shortly caudate; L. often falcate | 15. |
| 14a. | Pet. salmon-pink; stigmas long-caudate; L. not falcate | **hillii** |
| 15. | Upper surface of L. oblong, usually obtuse or round at the apex | **blanda** |
| 15a. | Upper surface of L. linear or narrowly ovate, acute | **dulcis** |

**A. alba** (L. Bol.) L. Bol. (*A. rubra* v. *a.* L. Bol.). – Cape: Lit. Namaqualand. – **St.** stiff, 14–26 cm long, Int. 13–15 mm long; **L.** narrow-lanceolate, acute, more rarely obtuse, c. 5 cm long, 12 mm across, 13 mm ⌀; bracts obliquely ovate to lanceolate, 1–2 cm long; **Fl.** almost 2 cm ⌀, white.

**A. blanda** L. Bol. v. **blanda.** – S.W.Afr.: origin not known. – **Br.** 23 cm long, Int. 1–3 cm long, herbaceous parts velvety; **L.** lunate, obtuse, the underside bluntly carinate, c. 3.7 cm long, 12 mm across, 15 mm thick; bracts broadly ovate, acute, 7 mm long, 8 mm thick; **Fl.** pale pink, Pet. 6–9 mm long, 1.5 mm across.

**A.** — v. **angusta** L. Bol. – Pet. narrowed at the tip.

**A.** — v. **latipetala** L. Bol. – Pet. 3 mm across.

**A. citrina** (L. Bol.) L. Bol. (*A. rubra* v. *c.* L. Bol.). – S.W.Afr.: southern reg. – **St.** 8–18 cm long, 8 mm ⌀, Int. 2.4 cm long, herbaceous parts rough, white; **L.** linear, acute to obtuse, bluntly carinate, c. 7 cm long, 8 to 10 mm across, c. 19 mm thick; bracts flat, compressed, 11 mm long, 10 mm across, obtuse, with a red border; **Fl.** c. 5 cm ⌀, lemon-yellow.

**A. dinteri** L. Bol. v. **dinteri** (Pl. 146/3) (*Mes. velutinum* Dtr., *A. v.* (Dtr.) Dtr. et Schwant.). – S.W.Afr.: Gr. Namaqualand. – Up to 30 cm tall, with 1 or several **St.**; **roots** thick, woody; **Br.** short; short **shoots** with 2–3 L.pairs; **L.** lunate, with rounded margins, mucronate, with a rounded keel, appearing softly velvety because of the white, acute papillae, 2.5–3.5 cm long, 1.5 cm across, 11–21 mm thick; bracts 16 mm long, 10 mm across, 9 mm thick; **Fl.** 4 cm ⌀, pure white or violet-pink.

**A.** — v. **lutata** L. Bol. (*A. velutina* v. *lutata* L. Bol.). – Cape: Lit. Namaqualand. – **L.** dirty yellow, 2.5–5 cm long, 10–15 mm across; **Fl.** whitish-yellowish.

**A. dulcis** L. Bol. – S.W.Afr.: Gr. Namaqualand. – **St.** stiff, erect, with c. 7 L. pairs, Int. 15–30 mm long, herbaceous parts bluish-green, softly velvety, at first with tufted hairs; **L.** often ± falcate, narrowing towards the tip, acute to obtuse, the keel reduced to a line, 2.5–4.5 cm long, with a Sh. 1.5 mm long, 10 mm across, 8–14 mm thick; bracts obliquely ovate, acute to obtuse, c. 14 mm long, 12 mm across, 8–9 mm thick; **Fl.** c. 5 cm ⌀, pale pink.

**A. hallii** L. Bol. – S.W.Afr.: Gr. Namaqualand. – 30 cm tall; **St.** c. 14 mm thick; **Br.** ascending, Int. 3–5 cm long, herbaceous parts greyish-green, velvety from small, dense hairs; **L.** lanceolate, acute to tapered, acutely carinate, broadest midway in profile, 5–8 cm long, 2–3 mm ⌀, with a Sh. 6 mm long; **Fl.** up to 7.5 cm ⌀, glossy white.

**A. herrei** L. Bol. – Cape: Lit. Namaqualand. – Robust, 30 cm tall; **St.** c. 27 mm ⌀; **Br.** 25 mm ⌀, Int. 1–2 cm long, sideshoots thinner; some **L.** ± thumb-shaped, c. 7.5 cm long, 10 mm across midway, 14 mm thick, other L. ± falcate, 5.7 cm long, 7–9 mm across midway, 16 mm thick, with a Sh. 2–3 mm long, the tip obtuse to rounded, the underside rounded, with the keel reduced to a line, rough, grey, tinged purple, margins and keel initially with red lines; bracts 12–13 mm long, 7 mm across below; **Fl.** red, paler on the underside, Pet. 14–26 mm long.

**A. hillii** L. Bol. – Cape: Lit. Namaqualand. – Compact, c. 18 cm tall; **St.** c. 13 cm long, Int. 5–12 mm long, shoots with 4–6 densely arranged L., Int. enclosed within the L. Sh., herbaceous parts bluish-green, velvety, old L. becoming hard and persisting; **L.** linear, elongate-lanceolate to shortly and broadly obovate, 1.5 cm long, 10 mm or up to 30 mm across midway, slightly narrowed towards the tip, with a Sh. 3 mm long; bracts obliquely acute, 7–10 mm long; **Fl.** salmon-pink, Pet. 14–20 mm long.

**A. latisepala** L. Bol. – Cape: Lit. Namaqualand. – c. 25 cm tall; **Br.** virgate, Int. c. 4 cm long, herbaceous parts bluish-green, minutely velvety; **L.** rarely ± falcate, narrowed towards the acute tip, the underside carinate, 4–6.6 cm long, 5–7 mm across midway, 7–10 mm thick; bracts c. 17 mm long; Pet. 2–3 cm long, the lower half white to pale, the upper half red.

**A. longifolia** (L. Bol.) L. Bol. (*Mes. l.* L. Bol., *Ruschia l.* (L. Bol.) L. Bol., *R. jacobseniana* L. Bol.). – Cape: Lit. Namaqualand. – Erect, up to 24 cm tall; **St.** 7 mm ⌀ below; **Br.** and short **shoots** erect, Int. 5–30 mm long; **L.** ± falcate, with indistinct keel and margins, somewhat compressed laterally, narrowed towards the tip or acute, velvety, bluish-green, 9 cm long, (7–)8(–10) mm across midway, 16 mm thick, Sh. 4 mm long; bracts 4.6 cm long; **Fl.** 5.3 cm ⌀, Pet. scarlet, white below.

**A. rubra** (L. Bol.) L. Bol. (*Mes. r.* L. Bol., *Lampranthus ruber* (L. Bol.) L. Bol., *Ruschia r.* (L. Bol.). – Cape: Lit. Namaqualand. – Erect, 15 cm tall; **St.** stiff, Int. 1–2 cm long; short **shoots** with 2–4 L., Int. short, herbaceous parts velvety; **L.** swollen, scarcely carinate, the tip obliquely acute or ± truncate, 3.5 cm long, 18 mm across; bracts 12 mm long, 15 mm across, 10 mm thick; **Fl.** 4 cm ⌀, Pet. red, white in the centre.

**A. ruschii** L. Bol. (Pl. 146/1). – S.W.Afr.: S. reg. – **St.** 14 cm long, c. 1 cm ⌀, Int. 8–25 mm long, herbaceous parts pale grey, becoming brownish-grey, velvety; **L.** acute to tapered, c. 6 cm long, 8–14 mm across midway, 10–23 mm thick; bracts broadly ovate, c. 2 cm long, with a Sh. 3 mm long, 10 mm ⌀; **Fl.** c. 6 cm ⌀, white.

**A. speciosa** L. Bol. – S.W.Afr.: Gr. Namaqualand. – c. 20 cm tall; herbaceous parts velvety; **L.** oblong-ovate to obliquely ovate to lanceolate, the underside obtusely carinate, the tip obtuse or ± rounded, the L. of a pair unequal in length, up to 6.5 cm long, 16 mm across midway, 20 mm thick; bracts 13 mm long, c. 8 mm ⌀; **Fl.** 6.5 cm ⌀, orange-red.

**A. swartportensis** L. Bol. – Cape: Lit. Namaqualand. – 30 cm tall, herbaceous parts velvety; **L.** lanceolate to linear-lanceolate, tapered to rather obtuse, bluish-green, (3.5–)4(–6) cm long, with a Sh. 2 mm long, 20 mm across midway, 18 mm thick; bracts obliquely ovate, 15 mm long; Fl. 4–5 cm $\varnothing$, red.

**A. vanbredai** L. Bol. – Cape: Lit. Namaqualand. – c. 30 cm tall; **St.** 9 mm $\varnothing$ below; **Br.** each with 6 ± falcate **L.** narrowed towards the acute tip, 4–5 cm long, 6–8 mm across, 8–10 mm thick; bracts flat, compressed, 12 mm long; Pet. in 4 series, 17–30 mm long, 0.5 to 3 mm across, purplish-red, paler below.

**A. vanheerdei** L. Bol. – Cape: Lit. Namaqualand. – 15–20 cm tall, **St.** 1 cm $\varnothing$; **Br.** 8 mm $\varnothing$, Int. 2 cm long, enclosed within the Sh., herbaceous parts bluish-green, glabrous, glossy; one **L.** of a pair ± falcate, acute to rather obtuse, the other more oval to broad-oval, 3 to 6.7 cm long, 12–16 mm across below, 15–22 mm thick midway; bracts 18 mm long, c. 9 mm $\varnothing$ midway; Pet. c. 25 mm long, coppery to purple.

**Bergeranthus** Schwant. (§ IV/1/1). – Acauline, very succulent plants with fleshy **roots**; **L.** densely crowded, opposite and decussate, ± united below, curved somewhat inwards, the upperside flat or ± trough-shaped, the underside semicylindrical, carinate towards the tip, with the keel drawn forward and ± chin-like, with entire margins, smooth, greyish-green, with or without minute, dark spots; **Fl.** solitary or several together, 2–5 cm $\varnothing$, long-pedicellate, with 2 bracts, Pet. yellow, often reddish, Fl. opening in the afternoon, summer. – Greenhouse. – Propagation: seed, cuttings.

**B. addoensis** L. Bol. – Cape: Uitenhage D. – L. lanceolate, tapered, compressed towards the tip, bluish-green, 2–3.5 cm long, 8–10 mm thick, with a Sh. 5 mm long; Fl. c. 25 mm $\varnothing$, yellowish-red.

**B. artus** L. Bol. – Cape: Uitenhage D. – **Shoots** numerous, short, with 6–8 crowded, tapered, yellowish-green **L.**, the underside acutely carinate, 3–4 cm long, 12 mm across, c. 9 mm thick, with a Sh. 8 mm long; Fl. 1–3 together, c. 4 cm $\varnothing$, yellow, opening in the evening, scented.

**B. concavus** L. Bol. – Cape: near E. London. – L. ensiform, the upperside ± oblique, long-tapered, mucronate, the underside acutely keeled and compressed, obliquely round or acute at the tip, green, with crowded dots, 3.5–5 cm long, 1 cm across, 5 mm thick; Fl. 25–30 mm $\varnothing$.

**B. glenensis** N. E. Br. – Orange Free State. – **Shoots** numerous; **L.** crowded on the thick, 4–5 cm long sideshoots, oblong-lanceolate, the underside rounded to ± rounded-carinate, green, glabrous, with dense, dark green spots, 2.5–4 cm long, 5–8 mm across, 4–6 mm thick below; Fl. 3 cm $\varnothing$, glossy yellow.

**B. jamesii** L. Bol. – Cape: Cradock D. – c. 3 cm tall; **shoots** thick; **L.** ensiform, incurved, long-tapered, the underside acutely carinate, laterally compressed, bluish to bluish-green, glabrous, 1–2 cm long, 2.5–4 mm across, 3–4 mm thick, Sh. 3 mm long; Fl. c. 2 cm $\varnothing$.

**B. katbergensis** L. Bol. – Cape: Stokenstrom D. – L. (2–)3(–4.5) cm long, tapered, the underside acutely carinate, 5–8 mm across below, 3–5 mm thick, dirty green; Fl. 3 cm $\varnothing$, yellow.

**B. leightoniae** L. Bol. – Cape: King William's Town D. – L. 2.5–4 cm long, long-tapered, the underside carinate-compressed, 6–10 mm across, 3–4 mm thick, green to dirty green, glossy, densely spotted; Fl. solitary, 3–3.5 cm $\varnothing$, Pet. yellow with red tips.

**B. longisepalus** L. Bol. – Cape: Alexandria D. – L. 5.5–6.5 cm long, with a Sh. 5–8 mm long, long-tapered, carinate on the underside towards the tip, 4 mm across midway, 3 mm thick, bluish-green, purple below, spotted; Fl. c. 3 cm $\varnothing$.

**B. multiceps** (Salm) Schwant. (Pl. 147/1) (*Mes. m.* Salm, *B. firmus* L. Bol.). – Cape: Port Elizabeth, Uitenhage D. – Caespitose, mat-forming; L. 6–8 in a Ros., 2.5–3 cm long, trigonous, 8–10 mm across midway, tapered, minutely awned, smooth, green, the underside obtusely carinate; Fl. 3 cm $\varnothing$, yellow, ± reddish outside.

**B. scapiger** (Haw.) N. E. Br. (*Mes. s.* Haw., *Mes. scapigerum* Eckl. et Zeyh., *B. s.* (Haw.) Schwant.). – Cape: E. London, Uitenhage, Albany, Caledon D. – Mat-forming; L. gradually tapered, 7–12 cm long, one L. of the pair longer than the other, with the angle of the keel pulled forward, 10–18 mm across, dark green, with smooth, pale, cartilaginous margins; Fl. 3–5 together, c. 4–5 cm $\varnothing$, golden-yellow, reddish outside.

**B. vespertinus** (Bgr.) Schwant. (*Mes. v.* Bgr.). – Cape: Uitenhage, Albany, Somerset D. – Mat-forming; **L.** trigonous, tapered, 6 cm long, 5–6 mm across, the underside rounded below, carinate above, grey-green, with darker spots; Fl. 3–5 together, yellow.

**Berrisfordia** L. Bol. (*Berresfordia* L. Bol. sphalm.) (§ IV/1/17). – Very small, clump-forming $\hbar$; **Br.** with 4 L.; **L.** pairs united to form a Bo. resembling **Conophytum**, with numerous small protuberances and T. at the tips. Acc. G. Schwantes intermediate between **Herreanthus** and **Conophytum**. – Greenhouse, warm. Propagation: seed, cuttings.

**B. khamiesbergensis** L. Bol. (Pl. 147/3) (*Conophytum k.* (L. Bol.) L. Bol.). – Cape: Lit. Namaqualand. – **Shoots** thin, with 2–4 obconical Bo. c. 15 mm long, 8 mm across midway, 7 mm across the base of the free parts, with a Sh. 11 mm long, 9 mm $\varnothing$ at the tip, the Fi. acutely incised; Fl. 2 cm $\varnothing$, pink, opening towards evening.

**Bijlia** N. E. Br. (§ IV/1/1). – Mat-forming and very succulent plants. – Greenhouse, warm. Propagation: seed, division.

**B. cana** N. E. Br. (Pl. 147/2). – Cape: Prince Albert D. – **L.** 4–6 in a Ros., opposite and decussate, c. 3 cm long, the upperside 12–15 mm across below, expanding above to 25 mm, then shortly trigonous-tapered, or often rhombic and obliquely tapered, the underside semicylindrical below, carinate and broadly compressed above, the keel pulled ± forward over the 2 cm thick tip, or oblique and pulled over one side, light greyish-green, smooth; **Fl.** short-pedicellate, 35 mm ⌀, yellow.

**Braunsia** Schwant. (*Echinus* L. Bol.) (§ IV/1/3). – Dwarf, creeping ♄ with fibrous **roots**; **Br.** often rooting at the nodes, with 1–2 L.pairs; **L.** short, lunate, triangular, carinate, united to midway, glabrous or velvety, often with scattered spots; **Fl.** terminal on rooting short shoots, ± sessile, pink or white; **seeds** with small appendages, like a hedgehog. – Greenhouse, warm. Propagation: seed, cuttings.

**B. apiculata** (Kensit) L. Bol. (*Mes. a.* Kensit, *Echinus a.* (Kensit) L. Bol., *B. a.* (Kensit) E. Murray, *Mes. binum* N. E. Br., *B. b.* (N. E. Br.) Schwant.). – Cape: Laingsburg D. – c. 20 cm tall, the herbaceous parts velvety; **Br.** stiff, Int. 10–15 mm long, **shoots** short, with 2–4 acutely carinate **L.** 15–30 mm long, 6–9 mm across and thick, the keel acute and entire, it and the margins paler, horny, with small brown hairs, the tip acute to obtuse; **Fl.** 2 cm ⌀, pink.

**B. geminata** (Haw.) L. Bol. (Pl. 147/4) (*Mes. g.* Haw., *Echinus g.* (Haw.) L. Bol., *B. g.* (Haw.) E. Murray, *Mes. mathewsii* L. Bol., *E. m.* (L. Bol.) N. E. Br.). – Cape: Karroo. – **St.** ascending, up to 15 cm long, dichotomous; **L.** triangular, 25 mm long, 14 to 15 mm across and thick, with cartilaginous margins; **Fl.** 4 cm ⌀, white.

**B. nelii** Schwant. (*Echinus n.* (Schwant.) Schwant.). – Cape: origin unknown. – Creeping; **shoots** with 1–2 L.pairs; **L.** lunate, the underside carinate; **Fl.** ± sessile.

**B. stayneri** (L. Bol.) L. Bol. (*Echinus s.* L. Bol.). – Cape: Ceres D., Karroo. – **St.** creeping, 15–30 cm long, Int. alate, compressed, 1–3 cm long; **shoots** erect, 2.5–4 cm long, with 2–4 **L.** c. 1 cm long, with a Sh. 3–5 mm long, 6 mm across midway, 5–7 mm ⌀, obtuse, margins and keel translucent; **Fl.** c. 2.5 cm ⌀, purplish-pink.

**B. varensburgii** (L. Bol.) L. Bol. (*Echinus v.* L. Bol.). – Cape: Bredasdorp D. – c. 8 cm tall; **St.** prostrate; **shoots** erect, up to 25 mm long, with compressed Int.; **L.** 2–4, ± compressed, acutely carinate, obliquely rounded to truncate above, 2 cm long, up to 7 mm across, 8–12 mm ⌀, margins and keel horny, Sh. 3 to 5 mm long; **Fl.** almost 5 cm ⌀, white.

**Calamophyllum** Schwant. (§ IV/1/2). – Short-stemmed to ± acauline, much branched, very succulent plants; **L.** opposite and decussate, rounded below, ± flattened above, the underside with a rounded keel; **Fl.** short-pedicellate, red, summer. – Greenhouse, warm. Propagation: seed, cuttings.

**C. cylindricum** (Haw.) Schwant. (*Mes. c.* Haw.). – S.Afr.: origin not known. – **St.** scarcely 5 cm long, densely and freely branched; **L.** cylindrical to trigonous, apiculate, 7–8 cm long, 8–10 mm thick, greyish-green, somewhat spotted; **Fl.** solitary, c. 3 cm ⌀, red.

**C. teretifolium** (Haw.) Schwant. (*Mes. t.* Haw., *Mes. cylindricum v. t.* Haw., *Mes. c. v. b.* Haw.). – S.Afr.: origin unknown. – **St.** up to 20 cm long; **Br.** prostrate; **L.** upperside flat, underside semicylindrical, with rounded angles, obtuse above, green, spotted, c. 10 cm long; **Fl.** c. 3 cm ⌀, Pet. deep red, white below.

**C. teretiusculum** (Haw.) Schwant. (Pl. 148/3) (*Mes. t.* Haw.). – S.Afr.: origin unknown. – Acauline; **L.** upperside flat, underside round, obtuse above, margins and keel rounded, 5 cm long, 8–10 mm thick.

**Carpanthea** N. E. Br. (§ IV/1/22). – ☉ plants with conspicuously large, yellow **Fl.** – In the open. Propagation: seed.

**C. pilosa** (Haw.) L. Bol. (*Mes. p.* Haw., *Mes. calendulaceum* Haw., *C. c.* (Haw.) L. Bol.). – Cape: Malmesbury D. – 10–15 cm tall with spreading **St.**; Br., pedicel and receptacle hirsute; **L.** opposite, young L. linear-lanceolate, older ones ovate-lanceolate or spatulate, mostly glabrous, 5–11 mm long, 1–2 cm across; **Fl.** solitary, long-pedicellate, 5–6 cm ⌀, Pet. in several series, acute, up to 25 mm long, lemon-yellow.

**C. pomeridiana** (L.) N. E. Br. (Pl. 148/2) (*Mes. p.* L., *C. p.* (L.) Schwant., *Mes. glabrum* Andr., *Mes. p. v. andrewsii* Haw., *Mes. p. v. glabrum* DC., *Mes. candollei* Haw., *Macrocaulon c.* (Haw.) N. E. Br., *Mes. helianthoides* Ait. ex DC.). – S.W.Cape: in sandy reg. – Up to 30 cm tall; shoots, pedicels and Cal. with tufted white hairs; **L.** spatulate or spatulate-lanceolate, narrowing into the broad petiole, 4–10 cm long, 12–25 mm across, margins minutely ciliate; **Fl.** 1–3, terminal, 4–7 cm ⌀, opening in the afternoon, on a pedicel 3–10 cm long, Pet. very numerous, narrow-linear, acute, light golden-yellow.

**Carpobrotus** N. E. Br. (§ IV/5). – Dwarf ♄; **St.** variable in length, creeping, branched, 2-angled, robust, fleshy, becoming woody, rooting from the nodes, Int. ± elongated, flowering shoots ± distinctly spreading, with 2–15 short Int.; **L.** very succulent, opposite, sessile, shortly united, acute-trigonous, ± as thick as broad, glabrous, smooth or almost so, apart from the keel and angles which are sometimes minutely crenate, with small, translucent spots, with a ± hemispherical blister at the base on the upperside; **Fl.** solitary, terminal, opening during the day, usually large, ± sessile to distinctly pedicellate, overtopping the upper L., Pet. numerous, purple or light purple, rarely yellow; flowering period Apr.—Nov.; **Fr.** fig-like, juicy, edible in the following species: **C. deliciosus, edulis, muirii**. (The former generic name *Ficoidea* refers to these species.) – S.Afr.; also Austr., Chile, Mexico and Calif., or introduced and naturalized there. – Summer: in the open, winter: coldhouse. Propagation: seed, cuttings.

**C. acinaciformis** (L.) L. BOL. (Pl. 149/1) (*Mes. a.* L., *C. a.* (L.) SCHWANT., *Abryanthemum a.* (L.) ROTHM., *Mes. rubrocinctum* ECKL. et ZEYH.). – Cape: Cape D., Natal. – **St.** angular, up to 1.5 m long; **Br.** short; **L.** with a blister below, sabre-shaped, compressed, slightly greyish-green, 9 cm long, 1 cm across, 15–20 mm thick, with a much widened keel, the angles cartilaginous, often slightly undulating and rough; **Fl.** 12 cm ⌀, brilliant crimson-purple, opening in the afternoon.

**C. aequilaterus** (HAW.) N. E. BR. (*Mes. a.* HAW., *Mes. exile* HORT. ex HAW., *Mes. chilense* MOL., *C. c.* (MOL.) N. E. BR., *Mes. chiloense* SALM, *Mes. virescens* SALM, *Mes. aequilaterale* WILLD. and HAW., SALM nom. ill., *Mes. a.* v. *chiloense* SALM, *M. a.* v. *decagynum* DC., *C. aequilateralis* (WILLD.) J. M. BLACK, *Mes. nigrescens* HAW.). – Austr.: Queensland, New S.Wales, Victoria, Tasmania; Mexico: Baja Calif.; USA: Calif. – **St.** up to 2 m long, 2–10 mm thick, green, becoming brown, ± compressed, with several angles, usually alate, Int. 0.5–5 cm long; **L.** dull green, ± pruinose, long-tapered, the keel minutely crenate below the tip, 3.5 to 9 cm long, 5–12 mm across and thick; **Fl.** pedicellate, 3.5–8 cm ⌀, Pet. light purple, paler to white at the base.

**C. chilensis** (MOL.) N. E. BR. (*Mes. c.* MOL., *Mes. aequilaterale* v. *chiloense* SALM). – Chile; USA: Calif.; Mexico: Baja Calif. – **St.** creeping; **L.** trigonous, amplexicaul, fleshy; **Fl.** large, violet.

**C. concavus** L. BOL. – Cape: Bredasdorp D. – **St.** 8–10 mm ⌀, Int. 5.5–8 cm long; **L.** mostly narrow, 5–8 mm across, laterally compressed, 11–14 mm thick, bluish-green; **Fl.** 4–5 cm ⌀, purplish-pink.

**C. deliciosus** (L. BOL.) L. BOL. (*Mes. d.* L. BOL.). – Cape: Riversdale D. – **St.** creeping; **Br.** erect; **L.** 8–10, acute-trigonous, tapered, dark greyish-green, 11 cm long, 15 mm thick, the keel often horny, crenate, Sh. 7–10 mm long; **Fl.** 7–8 cm ⌀, pinkish-purple; **Fr.** edible, spherical.

**C. dimidiatus** (HAW.) L. BOL. (*Mes. d.* HAW., *Mes. juritzii* L. BOL., *C. j.* (L. BOL.) L. BOL.). – Natal. – **L.** trigonous, soft, upperside grooved, 8 cm long, 9 mm ⌀ midway; **Fl.** 6 cm ⌀, pinkish-purple.

**C. disparilis** N. E. BR. (*Mes. virescens* S.D.). – Austr.? – **Roots** woody, fibrous; **St.** creeping, thick, somewhat contorted, nodal, angular; **L.** forming a ± equilateral triangle in section, the tip hatchet-shaped, obtuse, apiculate, green to bluish, often with purple blotches, 5.5 cm long, 12 mm across; **Fl.** large, vivid purple.

**C. dulcis** L. BOL. – Cape: Humansdorp D. – Resembles **C. deliciosus**. – **L.** ± sabre-shaped, the upperside and lateral surfaces slightly concave, 15 mm thick; **Fr.** edible, ovoid.

**C. edulis** (L.) BOL. (*Mes. e.* L., *C. e.* (L.) N. E. BR., *C. e.* (L.) SCHWANT., *Mes. acinaciforme β flavum* L., *Mes. e.* v. *f.* (L.) MOSS., *Abryanthemum e.* (L.) ROTHM.). – Cape, Natal; introduced and naturalized in Austr., S.Eur. and Calif. – **St.** up to 2 m long, 8–13 mm thick, with ± emarginate angles, flowering shoots with 2 fleshy Int.; **L.** dull green, curved slightly inwards, tapered above midway, 4–8 cm long, 8–17 mm across and thick; **Fl.** pedicellate, 7 to 8.5 cm ⌀, yellow at first, becoming flesh-coloured to pink.

**C. fourcadei** L. BOL. – Cape: Humansdorp D. – Robust; **St.** creeping, angular, 1 cm ⌀; **L.** arcuate, with the sides carinate and compressed, swollen below, green to bluish-green, 7–9 cm long, 8 mm across, 12 mm thick; **Fl.** 6 cm ⌀, pale pink.

**C. glaucescens** (HAW.) SCHWANT. (*Mes. g.* HAW., *C. g.* (HAW.) N. E. BR.). – Austr.: Queensland, N.S.W., Norfolk Is. – **St.** up to 2 m long, up to 1 cm thick, older parts reddish-brown, flowering shoots spreading to erect, Int. 3 cm and more long; **L.** slightly bluish-pruinose, 3.5–10 cm long, the angles thin, the sides 9–15 mm across; **Fl.** ± sessile, 4–6 cm ⌀, Pet. light purple, lighter to white below.

**C. laevigatus** (L. BOL.) SCHWANT. (*Mes. l.* HAW., *C. l.* (HAW.) N. E. BR.). – S.Afr.: precise origin unknown. – Possibly only a variety of **C. acinaciformis**.

**C. mellei** (L. BOL.) L. BOL. (*Mes. m.* L. BOL.). – Cape: Robertson D. – **L.** ± sabre-shaped, 7–8 cm long, 10 mm across, 15 mm thick; **Fl.** 55 mm ⌀, pink to pale pink.

**C. modestus** S. T. BLAKE. – W. and S.Austr., Victoria. – **St.** up to 35 cm long, 8 mm thick, prostrate, ± creeping, flowering shoots ascending to erect, Int. 0.5–2.5 cm long; **L.** ± bluish-pruinose or tinged reddish-brownish, often arcuate, the angles ± completely smooth, the keel crenate and rough above, 3.5–7 cm long, 4.5–7 mm across at the tip, the sides up to 7 mm across; **Fl.** on a pedicel 7 mm long, c. 3 cm ⌀, Pet. light purple, whitish below.

**C. muirii** (L. BOL.) L. BOL. (*Mes. m.* L. BOL.). – S.W.Cape. – **L.** 5.5–7 cm long, 6 mm across, 7 mm thick, green; **Fl.** 6.5–9 cm ⌀, pinkish-purple; **Fr.** edible (used dried).

**C. pageae** L. BOL. – Cape: Robertson D. – **St.** robust, 12 mm ⌀; **L.** arcuate, pale blue to reddish, 9–11.5 cm long, 2 cm thick, 11 mm across, with the keel minutely serrate towards the tip, with a Sh. 2 mm long; **Fl.** 7 cm ⌀, mauve to pink.

**C. pillansii** L. BOL. – Cape: Caledon D. – **L.** stiff, ± ensiform, 8–10 cm long, 9 mm across, 12 mm thick, green; **Fl.** 6.5–7 cm ⌀, pinkish-purple.

**C. pulleinii** J. M. BLACK. – Austr.: NSW. – **St.** creeping, rooting; **L.** linear-lanceolate, trigonous, with the surface rough-granular, 6–10 cm long, 5–9 mm across; **Fl.** 15–20 mm ⌀, red.

**C. quadrifidus** L. BOL. – Cape: Lit. Namaqualand. – **St.** robust, elongated, 15 mm ⌀; **L.** broader midway, very short-tapered at the tip, with horny margins, bluish, 13.5 cm long, 15 mm across, 18 mm thick, with a Sh. 9 mm long; **Fl.** 13 cm ⌀, white to pale pink, with **C. sauerae** the largest Fl. among the **Mesembryanthemaceae**.

Plate 162. 1. **Enarganthe octonaria** (L. Bol.) Schwant. (Photo: H. Herre); 2. **Eberlanzia spinosa** (L.) Schwant.; 3. **Ebracteola candida** L. Bol.; 4. **Erepsia heteropetala** (Haw.) Schwant. (Photo: H. Chr. Friedrich).

Plate 163. 1. Faucaria bosscheana (BGR.) SCHWANT. v. bosscheana (Photo: G. SCHWANTES); 2. F. paucidens N. E. BR.; 3. F. felina (WESTON) SCHWANT. v. felina (Photo: as 1); 4. F. lupina (HAW.) SCHWANT.; 5. F. tigrina (HAW.) SCHWANT. f. tigrina (Photo: as 1); 6. F. tuberculosa (ROLFE) SCHWANT.

Plate 164. 1. **Frithia pulchra** N. E. Br. v. **pulchra** (Photo: H. Kahl); 2. **Fenestraria aurantiaca** N. E. Br. f. **aurantiaca**; 3. **Gibbaeum album** N. E. Br. f. **album**; 4. **G. angulipes** (L. Bol.) N. E. Br. (Photo: G. C. Nel, G. Schwantes); 5. **G. cryptopodium** (Kensit) L. Bol.; 6. **G. dispar** N. E. Br.

26/B*

Section I/1    Section I/2

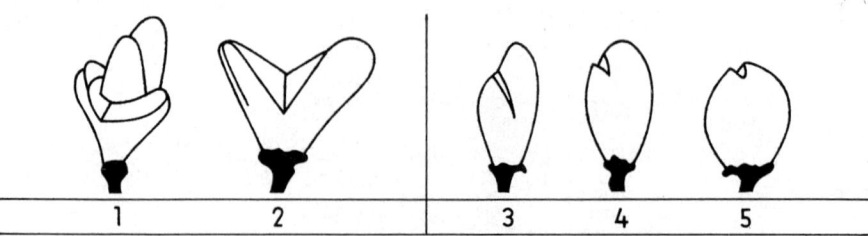

Section II/1    Section II/2

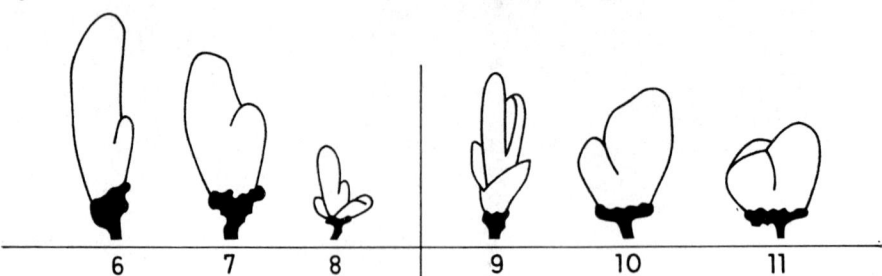

Section II/3

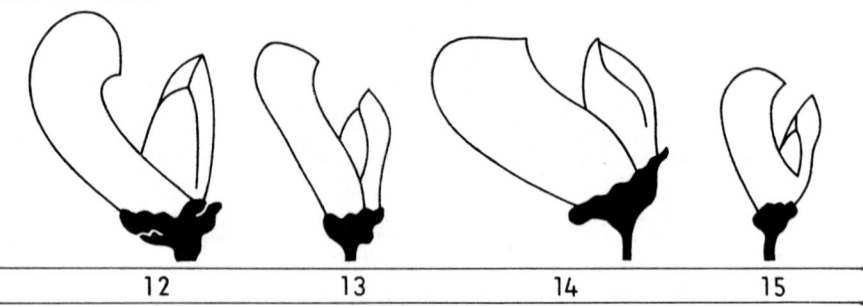

Section III/1    Section III/2

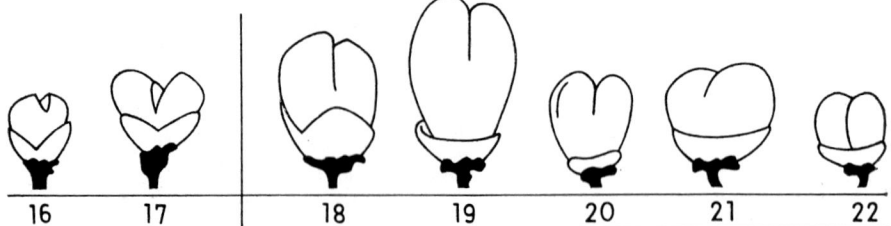

**Plate 165. Division of the Genus Gibbaeum into Sections.**
Gibbaeum
1. luteoviride
2. esterhuyseniae
3. gibbosum
4. cryptopodium
5. pilosulum
6. pubescens
7. shandii
8. geminum
9. angulipes
10. dispar
11. album
12. velutinum
13. haagei
14. schwantesii
15. pachypodium
16. petrense
17. tischleri
18. heathii
19. heathii v. elevatum
20. blackburniae
21. comptonii
22. luckhoffii

Plate 166. 1. Gibbaeum geminum N. E. Br.; 2. G. gibbosum (Haw.) N. E. Br. (Photo: G. Schwantes); 3. G. petrense (N. E. Br.) Tisch.; 4. G. heathii (N. E. Br.) N. E. Br. v. heathii; 5. G. pubescens (Haw.) N. E. Br.; 6. G. velutinum (L. Bol.) Schwant.

Plate 167. 1. Glottiphyllum fragrans (S. D.) Schwant.; 2. G. linguiforme (L.) N. E. Br.

Plate 168. 1. **Mesembryanthemum hypertrophicum** Dtr. (Photo: F. Eberlanz, G. Schwantes); 2. **Hereroa hesperantha** (Dtr.) Dtr. et Schwant. (Photo: H. Chr. Friedrich).

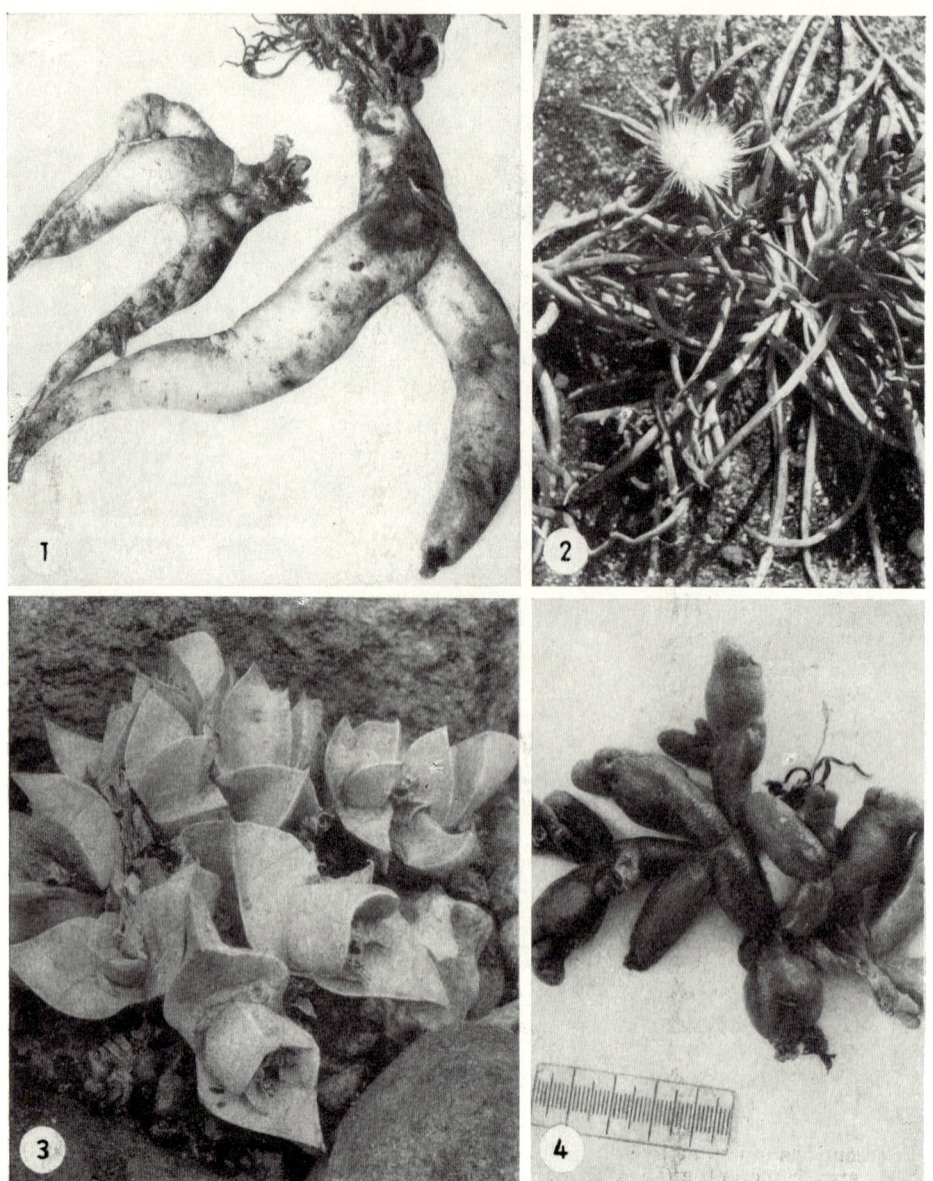

Plate 169. 1. **Herrea nelii** SCHWANT. Thickened roots; 2. **H. nelii** SCHWANT. (Photo: H. STRAKA); 3. **Herreanthus meyeri** SCHWANT.; 4. **Mesembryanthemum cryptanthum** HOOK. f. (Photo: H. HERRE).

**C. rossii** (HAW.) SCHWANT. (*Mes. r.* HAW.). – W. and S.Austr., Victoria, Tasmania. – **St.** up to 1 m long, 3–11 mm thick, flowering shoots ascending to erect, with 2–5 nodes, Int. 0.5 to 3 cm long; **L.** bluish-pruinose, ± curved inwards, tapered, with the keel minutely crenate above, 3.5–10 cm long, (6–)8–(11) mm across; **Fl.** pedicellate, 3.5–5.5 cm ⌀, Pet. light purple, whitish below.

**C. rubrocinctus** (HAW.) SCHWANT. (*Mes. r.* HAW., *C. r.* (HAW.) N. E. BR.). – Close to **C. acinaciformis,** possibly only a variety.

**C. sauerae** SCHWANT. (*C. quadrifidus* v. *roseus* L. BOL., *C. q.* f. *r.* (L. BOL.) ROWL.). – Cape: Darling. – Very robust; **St.** 1 cm thick, 6-angled, Int. 15 mm long; **L.** stiff, up to 12 cm long, linear, up to 1 cm across, the underside sabre-shaped to sharply carinate, up to 2 cm thick, grey to bluish, with the angles often reddened, with Sh. 1 cm long; **Fl.** with **C. quadrifidus** the largest of all Mesembryanthemaceae, 13 cm ⌀, Pet. magenta with a white base.

**C. subulatus** (HAW.) SCHWANT. (*Mes. s.* HAW., *C. s.* (HAW.) N. E. BR.). – S.Afr.: origin unknown. – Close to **C. acinaciformis,** possibly only a variety.

**C. vanzijliae** L. BOL. – Cape: S.W. reg. – **St.** 4-angled, Int. 1.5–6 cm long; short **Br.** 9 cm long, with 6–8 **L.** 5–7 cm long, 6–7 mm across, 11 mm thick, yellowish-green; **Fl.** 8 cm ⌀, a beautiful pink colour.

**C. virescens** (HAW.) SCHWANT. (*Mes. v.* HAW., *Mes. edule* v. *v.* (HAW.) MOSS., *Mes. abbreviatum* HAW., *C. a.* (HAW.) SCHWANT.). – W.Austr. – **St.** up to 1.5 m long, 6–11 mm thick, becoming whitish, flowering shoots with (4–)7(–8) nodes, branched, Int. 0.5–3 cm long; **L.** green, slightly pruinose, curved inwards or straight, tapered, the upperside somewhat concave, the sides convex, with the keel crenate above, 3.5–5 cm long, 7–17 mm across and thick; **Fl.** ± sessile or with a pedicel 5–15 mm long, 4 to 6 cm ⌀, Pet. purple or light purple, white below.

**Carruanthus** (SCHWANT.) SCHWANT. ex N. E. BR. (§ IV/1/1) (*Bergeranthus* SCHWANT. Sub-G. *Carruanthus* SCHWANT.). – Short-stemmed, branched, very succulent, mat-forming plants; **roots** fleshy; **L.** opposite and decussate, crowded at the tips of the shoots, erect to spreading, oblanceolate to clavate, narrowed towards the tip, the upperside flat, the underside carinate, with the keel-angle pulled forward, with ± dentate margins, green to greyish-green; **Fl.** mostly solitary, the pedicel with 2 bracts, Pet. yellow. – Greenhouse. Propagation: seed, cuttings.

**C. peersii** L. BOL. (Pl. 149/3) (*Tischleria p.* SCHWANT.). – Cape: Willowmore D. – **L.** with the upperside expanded, narrowed towards the tip, entire or rarely with 1–2 indistinct T. on each margin, c. 5 cm long, 10 mm across below, elsewhere 15 mm across, 14–16 mm thick; **Fl.** 4 cm ⌀, Pet. yellow with pink tips.

**C. ringens** (L.) BOOM (Pl. 149/4) (*Mes. r.* α *caninum* L., *Mes. caninum* HAW., *Bergeranthus c.* (HAW.) SCHWANT., *C. c.* (HAW.) SCHWANT.). – Cape: Ladismith, Uniondale D. – **Roots** fleshy; **L.** densely crowded, 5–6 cm long, 16–18 mm across, narrower below, oblanceolate to clavate, trigonous, the keel-angle expanded, pulled forward, with the margins ± dentate at the tip, grey-green; **Fl.** 4–5 cm ⌀, Pet. yellow, reddish outside.

**Caryotophora** LEISTN. (§ III). – ♃, glabrous, not very succulent plants; **St.** short; flowering **shoots** ☉, prostrate; **L.** alternate to ± opposite, flat; **Fl.** pedicellate, terminal, large; **Fr.** woody, hard, 3–4-cleft. – Summer: in the open, winter: cold-house. Propagation: seed, root-cuttings.

**C. skiatophytoides** LEISTN. (Pl. 149/2). – Cape: Bredasdorp D.: coastal reg. – Main **root** horizontal, branched; **St.** 3–15 cm long, up to 1 cm ⌀, densely leafy; flowering **Br.** 3–15 cm long, laxly leafy; **L.** amplexicaul, lanceolate-spatulate to narrow-spatulate, with entire margins, obtuse to ± acute, radical **L.** 6 to 16 cm long, 1–3 cm across, those on the Br. 4–6 cm long, 1–2 cm across; **Fl.** 4–6 cm ⌀, on a pedicel 3–6 cm long, Pet. numerous, snow-white, opening during the day, Sept.—Nov.

---

**Cephalophyllum** N. E. BR. (*Cephalophylla* HAW. as a Sect. of the G. *Mesembr.* L.) (§ IV/1/2). – Low, succulent, often caespitose, branched ♄; **Br.** somewhat nodular, often rooting at the nodes, or the plant stemless with stolons (only **C. frutescens** an erect ♄); **L.** partly crowded in tufts (whence the G. name), elongated or cylindrical to trigonous; **Fl.** mostly open at midday in summer, usually 1–3 together, in 3-partitely branched cymes, stalked, the pedicel mostly with 2 bracts, Pet. very large, yellow, golden-yellow, ruby, rose, coppery-red, red, pink to white, filaments dark red or brown, staminodes present only in **C. staminodiosum.** – Beautiful flowering plants. Summer: in the open, winter: in the cold-house. Propagation: seed, cuttings.

## Key to the Groups of the Genus Cephalophyllum by L. Bolus

| | | |
|---|---|---|
| 1. | Heads of L. conspicuous | **2.** |
| 1a. | Heads of L. inconspicuous or absent | **3.** |
| 2. | Primary St. diffuse, elongated, rooting at the nodes – e.g. C. cedrimontanum | **Elongata** |
| 2a. | Primary St. usually more compactly arranged, not rooting at the nodes – e.g. C. pillansii ........ **Decumbentia** | |
| 3. | Primary St. rooting at the nodes – e.g. C. procumbens, C. stayneri | **Reptantia** |
| 3a. | Primary St. not rooting at the nodes | **4.** |

**Cephalophyllum**

4. Primary St. elongated, prostrate normally, or scrambling where support is available – e.g. C. spongiosum .................................................................. **Prostrata**
4a. Primary St. not elongated, ± erect ......................................................... **5.**
5. Plants compact or caespitose, low-growing .................................................. **6.**
5a. Plant frutescent, up to 60 cm high (1 species ± decumbent) – e.g. C. frutescens ...........
.................................................. **Frutescentia**
6. Rootstock woody and thick with age; Int. enclosed in wild plants – e.g. C. subulatoides .....
.................................................. **Abbreviata**
6a. Roots fibrous, where known; Int. ± visible – e.g. C. pulchellum ............... **Caespitosa**

**C. albertinense** (L. Bol.) Schwant. (*Mes. a.* L. Bol., *C. a.* (L. Bol.) L. Bol.). – S.W.Cape. – St. creeping, angular, 30 cm or more long, Int. 10–25 mm long; Br. with 4 L., with short shoots; L. cylindrical or semicylindrical, short-tapered, green, 10–25 mm long, 2–3 mm across, 3–4 mm thick; Fl. solitary, 43 mm ⌀, yellow.

**C. alstonii** Marl. (Pl. 150/2). – Cape: Ceres D. – St. prostrate, over 50 cm long, Int. 5 cm long; L. clustered, ± cylindrical, short-tapered, up to 7 cm long, up to 8 mm across, 7 mm thick, grey-green, with many spots; Fl. 5–8 cm ⌀, dark red, with violet stamens.

**C. anemoniflorum** (L. Bol.) Schwant. (*Mes. a.* L. Bol., *C. a.* (L. Bol.) L. Bol.). – S.W. Cape. – Climbing ♄ with Int. 15–25 mm long; Br. erect, 2–4 cm long, with 2–6 L. 2–2.5 cm long, 5 mm across, semicylindrical below, trigonous towards the tip, mucronate, bluish-green, often reddish; Fl. 52 mm ⌀, salmon-pink.

**C. apiculatum** L. Bol. – Cape: Van Rhynsdorp D. – Robust; St. prostrate, 12 cm long, Int. 1–2 cm long; L. clustered, 8 cm long, 5 mm across and thick, tapered, apiculate, semicylindrical, carinate, cauline L. 2.5–3 cm long, 5 mm thick; Fl. 47 mm ⌀, orange-red, red outside.

**C. artum** L. Bol. – Cape: Ceres D. – Compact, up to 12 cm tall, c. 15 cm ⌀; St. prostrate to creeping, 4–6 cm long, Int. enclosed; L. 4–6 on the short shoots, tapered, convex on the underside and the sides, acute to obtuse above, glossy, 3–5 cm long, 5–9 mm ⌀, with a Sh. 8 mm long; Fl. almost 6 cm ⌀, purplish-pink.

**C. aurantiacum** L. Bol. – Cape: Riversdale D. – Slender; St. prostrate, 13 cm long, Int. 5–10 mm long; L. in the apical tufts cylindrical, 9 cm long, 6 mm thick, long-tapered, apiculate, the cauline L. semicylindrical, green, 4.5–6 cm long, 5 mm thick, with a Sh. 4 mm long; Fl. 33 mm ⌀, golden-yellow, red outside.

**C. aureorubrum** L. Bol. – Cape: Lit. Namaqualand. – St. prostrate, 14–17 cm long; L. in tufts, narrowed towards both ends, bluish-green, 4.5–6 cm long, 6–8 mm across midway, 5–8 mm thick, with a Sh. 6 mm long; Fl. 35 mm ⌀, golden-yellow inside, red outside.

**C. ausense** L. Bol. – S.W.Afr.: Gr. Namaqualand. – St. stiff, creeping, 15 cm long, with 2 acute, narrowly alate angles, Int. 15–25 mm long; L. convex on the upperside, indistinctly carinate on the underside, obtuse, bluish-green to reddish, 2–2.5 cm long, 5 mm thick; Fl. 33–35 mm ⌀, golden-yellow.

**C. baylissii** L. Bol. – Cape: Riversdale D. – Forming laxly branched mats 11 cm ⌀; St. prostrate, up to 4 cm long, Int. 2–10 mm long; L. not recognizably tufted, linear, acute, carinate, up to 3 cm long, up to 4 mm across and thick; Pet. 5–11 mm long, pink, with a dark pink stripe on the underside.

**C. bredasdorpense** L. Bol. – Cape: Bredasdorp D. – Forming mats 20–30 cm ⌀; St. creeping; L. in tufts, 6–8 cm long, 5 mm across midway, 4 mm thick, rounded, with a Sh. 5–6 mm long, cauline L. 2–3 cm long, 4–7 mm across, 4–5 mm thick, short-tapered; Fl. numerous, 3–3.5 cm ⌀, Pet. lemon yellow, striped purplish-red.

**C. brevifolium** L. Bol. – Cape: Van Rhynsdorp. D. – Compact, 5–6 cm tall; St. 5–6 cm long, Int. 4–6 mm long; L. narrowed towards the tip, 2.5–4 cm long, 6–8 mm across and thick midway, with a Sh. 3–4 mm long; Fl. solitary, Pet. 8–13 mm long, yellow to golden-yellow.

**C. caledonicum** L. Bol. – Cape: Caledon D. – L. dark green with numerous small spots, 4 cm long, 2–4 mm across, carinate on the underside; Fl. numerous, 3–3.5 cm ⌀, light yellow.

**C. cauliculatum** (Haw.) N. E. Br. (*Mes. c.* Haw., *Mes. diminutum* v. c. Haw.). – Resembles **C. subulatoides**; St. stolon-like, rooting at the nodes; L. larger, broader at the tips.

**C. cedrimontanum** L. Bol. – Cape: Clanwilliam D. – St. elongated, rooting at the nodes, Int. 2.5–5 cm long; L. in tufts, semicylindrical, long-tapered, bluish-green, up to 9 cm long, 4 mm thick, cauline L. 3.5 cm long, 3–4 mm across and thick; Fl. 3.5–4 cm ⌀, Pet. yellow to lemon-yellow, striped red.

**C. ceresianum** L. Bol. – Cape: Ceres D. – St. 25 cm long, creeping, Int. 12–50 mm long, angular; L. in tufts 8 cm long, cylindrical, long-tapered, 7 mm thick, with a Sh. 5 mm long; Pet. yellow, paler and indistinctly striate on the underside.

**C. clavifolium** (L. Bol.) L. Bol. (*Mes. c.* L. Bol.). – W.Cape. – St. creeping and matforming, elongated, Int. 1–2 cm long; shoots erect, with 4 clavate L., obtuse to short-tapered above, apiculate, with the underside rounded, dirty green, 22 mm long, 3 mm across, 4–5 mm thick, with a Sh. 1.5 mm long; Fl. solitary, 4.5–5 cm ⌀, yellow to coppery-red.

**C. compactum** L. Bol. – Cape: Van Rhynsdorp D. – L. in tufts, slender, narrowed, carinate, 6 mm long, 3 mm across and thick; Fl. yellow, red on the underside.

**C. compressum** L. Bol. – S.W.Afr.: Witpütz. – St. c. 21 cm long, not rooting at the nodes, Int. 3–5 cm long, flat on one side; L. acute, carinate on the underside, 3.5–5 cm long, 6 to

9 mm ⌀ midway, with a Sh. 6 mm long; **Fl.** in an Infl., 3.5–4 cm ⌀, yellow.

**C. confusum** (DTR.) DTR. et SCHWANT. (*Mes. c.* DTR., *C. c.* (DTR.) L. BOL.). – S.W.Afr.: Klinghardt Mts. – Close to **C. rangei**; **St.** not elongated, forming mats 15 cm tall and 20 cm ⌀; **L.** ± trigonous to cylindrical, obtuse, 25–37 mm long, 5 mm thick, with a Sh. 4 mm long; **Fl.** 2.5–3 cm ⌀, golden-yellow.

**C. conicum** L. BOL. – Cape: Calvinia D. – Plant 60 cm ⌀; **St.** creeping, woody, variously curved and 2-angled when young; **Br.** alate, c. 30 cm long; **L.** rounded or obtusely carinate on the underside, 2.5–5.2 cm long, 6 mm across midway and 9 mm ⌀ at the tip, the keel indistinctly serrate, with a Sh. 3–6 mm long; **Fl.** 3–3.5 cm ⌀, Pet. purplish-red, with a narrow, darker stripe.

**C. crassum** L. BOL. – Cape: Clanwilliam D. – **St.** elongated, Int. up to 12 cm long, angular; **L.** densely clustered, 12 cm long, cylindrical, short-tapered, 10 mm thick, cauline L. short and thick, carinate, short-tapered above, 25 mm long, 10 mm thick, bluish-green to green, with a Sh. 4–5 mm long; **Fl.** 6 cm ⌀, Pet. scarlet at the base, otherwise yellow.

**C. cupreum** L. BOL. – Cape: Lit. Namaqualand. – Robust; **St.** creeping, up to 55 cm long, 12 mm thick, Int. 13 mm long; **L.** acute or short-tapered, clavate, with an indistinct keel, pale green, 65 mm long, 13 mm thick; **Fl.** solitary, 78 mm ⌀, Pet. coppery-red, yellow below, pink outside, paler below.

**C. curtophyllum** (L. BOL.) SCHWANT. (*Mes. c.* L. BOL., *C. c.* (L. BOL.) N. E. BR.). – Cape: Ceres Karroo. – **St.** prostrate, 8–13 cm long, Int. 10–25 mm long; **Br.** with 4–6 semicylindrical, acute, apiculate, green **L.** 2–3 cm long, 6 mm across, with a Sh. 3–4 mm long; **Fl.** 37 mm ⌀, pinkish-purple.

**C. decipiens** (HAW.) L. BOL. (*Mes. d.* HAW., *Mes. dubium* L. BOL.). – Cape: Clanwilliam D. – **St.** 30–40 cm long; **Br.** rigid; **L.** semicylindrical, obtusely trigonous above, short-tapered, fresh green, with minute, rough spots, 5 cm long, 6–8 mm across, with a short Sh.; **Fl.** solitary, yellow.

**C. diminutum** (HAW.) L. BOL. (*Mes. d.* HAW., *Mes. d.* v. *pallidum* HAW.). – Cape: Ladismith, Uniondale D. – Plant 3–4 times smaller than **C. subulatoides**; **L.** semicylindrical, trigonous above, with translucent spots; **Fl.** 4 cm ⌀, purplish-red.

**C. dissimile** (N. E. BR.) SCHWANT. (*Mes. d.* N. E. BR., *C. d.* (N. E. BR.) N. E. BR., *Mes. validum* S.D.). – S.Afr.: origin unknown. – **St.** and **Br.** curved, up to 60 cm long; **L.** semicylindrical, rounded-trigonous above, short-tapered, 5–8 cm long, 6–8 mm across, glossy green, often reddened, minutely spotted; **Fl.** yellow.

**C. diversiphyllum** (HAW.) N. E. BR. (*Mes. d.* HAW., *Mes. d.* v. *congestum* SALM, *Mes. diversifolium* HAW., *C. d.* (HAW.) L. BOL., *C. d.* (HAW.) SCHWANT., *Mes. loreum* L. and v. *β* and *γ* L., *C. l.* (HAW.) L. BOL.). – Cape: Clanwilliam D. – Acauline with a large Ros. of opposite and decussate L., producing from their axils stolon-like, rooting side-shoots, each with a terminal Ros.; longest **L.** 8–10 cm long, semicylindrical on the underside, trigonous-compressed, tapered, with minute, rough spots and a short Sh.; **Fl.** mostly solitary, glossy yellow, somewhat reddish on the underside.

**C. dubium** (HAW.) L. BOL. (*Mes. d.* HAW.). – Cape: origin not known. – Very close to **C. tricolorum**; **L.** shorter, semicylindrical to trigonous, rather greyish-green.

**C. ernii** L. BOL. – S.W.Afr.: Kolmans Peak. – Compact, 6 cm tall; **St.** erect; **Br.** with 8 densely arranged **L.** narrowed towards the tip, indistinctly acute, 25 mm long, 7 mm thick, with a Sh. 4 mm long; **Fl.** solitary, 3 cm ⌀, yellow.

**C. framesii** L. BOL. (*C. ramosum* N. E. BR. nom. nud.). – Cape: Van Rhynsdorp D. – **St.** creeping, 17 cm long, Int. 5–10 mm long; **Br.** crowded, with 2–4 dirty green **L.** of unequal length, 2–3 cm long, 5 mm across, 3 mm thick, ± expanded towards the tip, acutely carinate on the underside, with indistinct margins, covered with deciduous scales, with a Sh. 3 mm long; **Fl.** solitary, 4 cm ⌀, pink.

**C. franciscii** L. BOL. – Cape: Clanwilliam D. – Glabrous, c. 9 cm tall, 18 cm ⌀; **L.** erect to projecting, acute, apiculate, c. 8 cm long, 7 mm across and thick, with a Sh. 5 mm long; **Fl.** solitary, on a pedicel 1 cm long, bracts 1.5–2 cm long, Pet. 2–3 cm long, 2–3 mm across, pinkish-purple, with pale pink stamens, deep reddish-brown anthers and pale pink stigma-lobes.

**C. frutescens** L. BOL. v. **frutescens**. – Cape: Lit. Namaqualand. – Erect, stiff ♄ c. 60 cm tall, 90 cm ⌀; **St.** 5–6 mm ⌀, Int. 2–5 cm long; **L.** 4, borne on the short shoots, cylindrical, apiculate, 5–8.5 cm long, 6–9 mm ⌀, with a Sh. 2–8 mm long; **Fl.** c. 6 cm ⌀, yellow to lemon-yellow.

**C. —** v. **decumbens** L. BOL. – **St.** prostrate; Fl. lemon-yellow.

**C. fulleri** L. BOL. – Cape: Kenhardt D. – Compact; **St.** creeping, 7 cm long; **L.** semicylindrical, tapered, 3 cm long, 4 mm thick; **Fl.** 4 cm ⌀, golden-yellow.

**C. goodii** L. BOL. – Cape: Lit. Namaqualand. – **St.** robust, 7 mm thick; **Br.** crowded; **L.** 4, crowded, semicylindrical, truncate above, green, densely and minutely spotted, 25 mm long, 6 mm across, with a Sh. 1–2 mm long; **Fl.** 4.5 cm ⌀, Pet. golden-yellow to coppery-red, striped.

**C. gracile** L. BOL. v. **gracile**. – Cape: Montagu D. – **St.** angular, Int. 1–3 cm long; **Br.** 11 cm long; **L.** cylindrical, concave on the upperside, tapered, yellowish-green, (2–)3.5(–6) cm long, 2–4 mm thick, with a Sh. 3 mm long; **Fl.** 3 cm ⌀, golden-yellow.

**C. —** v. **longisepalum** L. BOL. – **L.** 10.5 cm long, 3.5 mm thick; Sep. 12 mm long, Pet. 14 mm long.

**C. hallii** L. BOL. – Cape: Calvinia D. – Compact, 13 cm tall, 11 cm ⌀; **Br.** covered with the remains of old L.; **L.** narrow-lanceolate, carinate on the underside, obtuse above, olive-green, with horny margins, 3–4 cm long, 6–7 mm across and thick, with a Sh. 4 mm long; Pet. 6–12 mm long, purplish-pink.

**C. herrei** L. Bol. v. **herrei** (Pl. 150/1). – Cape: Lit. Namaqualand. – Erect; **St.** compressed, 17 cm long, Int. 2–4 cm long; **L.** semicylindrical, with keel and margins indistinct, short-tapered, pale bluish-green, spotted, 38 to 50 mm long, 7–9 mm across and thick, with a Sh. 5 mm long; **Fl.** 35 mm $\varnothing$, golden-yellow.

**C.** — v. **decumbens** L. Bol. – **St.** prostrate.

**C. inaequale** L. Bol. – Cape: Lit. Namaqualand. – **Br.** prostrate, nodose, Int. 10–25 mm long; **L.** tapered, carinate, bluish-green, 55 mm long, 8 mm across, 5 mm thick, with a Sh. 8 mm long; **Fl.** 5 cm $\varnothing$, yellow, reddish outside.

**C. insigne** L. Bol. – Cape: Lit. Namaqualand. – Robust; **St.** 2-angled, 28 cm long, Int. 1.5–3 cm long; **Br.** 6–11 cm long, often with 4 **L.** 3–8 cm long, 4–6 mm across and thick, narrowed towards the tip, with a blister on the upperside below, with the keel reduced to a line, and a Sh. 7 mm long; **Fl.** c. 6.2 cm $\varnothing$, red.

**C. kliprandense** L. Bol. – Cape: Van Rhynsdorp D. – **Br.** prostrate to creeping, c. 9 cm long, Int. enclosed; **L.** rounded on the underside, with a short, acute keel, green, 1–3.2 cm long, 5–6 mm across and thick, with a Sh. 3–4 mm long; **Fl.** 3–4 cm $\varnothing$, golden-yellow.

**C. laetulum** L. Bol. – Cape: Lit. Namaqualand. – **St.** prostrate, variously curved, c. 35 cm long, Int. 2–5 cm long; **Br.** 2–6 cm long; **L.** linear, subacute, obtusely carinate on the underside, 3–6 cm long, 3–4 mm across, 4 to 6 mm thick, with a Sh. 3 mm long; **Fl.** 2.5–3.5 cm $\varnothing$, yellow.

**C. latipetalum** L. Bol. – Cape: origin unknown. – Similar to **C. curtophyllum**; **L.** thicker; **Fl.** 4.5 cm $\varnothing$, Pet. 3 mm across, pinkish-purple, white below.

**C. littlewoodii** L. Bol. – Cape: Worcester D. – Slender; **St.** prostrate, c. 18 cm long, Int. 2–6 cm long; terminal **L.** clustered, narrowed teretely below, widened above, acute, bluish-green, c. 6 cm long, 1–2 mm $\varnothing$, cauline L. 2 cm long; **Fl.** 3 cm $\varnothing$, pink.

**C. loreum** (L.) Schwant. (*Mes. l.* L., *C. l.* (L.) N. E. Br., *C. l.* (L.) L. Bol., *Mes. corniculatum* L., *C. c.* (L.) Schwant.). – Cape: origin unknown. – **St.** prostrate, up to 45 cm long; **Br.** curved, nodose; **L.** clustered, trigonous to semicylindrical, rather obtuse, 7–10 cm long, 6–8 mm across, greyish-green, minutely spotted, tipped with a tiny awn; **Fl.** golden-yellow.

**C. maritimum** (L.) Schwant. (*Mes. m.* L. Bol., *C. m.* (L. Bol.) L. Bol.). – Cape: Riversdale D. – **St.** elongated, creeping; **Br.** erect, 7 cm tall; **L.** cylindrical or semicylindrical, tapered, apiculate, reddish, 3.5–6 cm long, 3–6 mm thick, with a Sh. 3–4 mm long; **Fl.** 3.5–4 cm $\varnothing$.

**C. middlemostii** L. Bol. – Cape: Riversdale D. – **St.** elongated, creeping, 2–3-angled, 30 cm long, Int. 1.5–4 cm long; **L.** not clustered, slender, acute to tapered, green, 3.2–4.7 cm long, 1.5 to 3 mm across and thick, with a Sh. 2–3 mm long; **Fl.** 3–4 cm $\varnothing$, purplish-pink.

**C. namaquanum** L. Bol. – Cape: Lit. Namaqualand. – **St.** elongated, angular, Int. 1–2 cm long; **Br.** erect to curved inwards, 5 cm long, with 4–6 shortly tapered, dirty green L. (15–)25 (–40) mm long, carinate on the underside; **Fl.** 35–48 mm $\varnothing$, yellow.

**C. niveum** L. Bol. – Cape: Van Rhynsdorp D. – Dwarf; **St.** with 2–4 swollen, laterally compressed, bluish **L.** 22 mm long, 5 mm across, 7 mm thick, with a serrate keel and a Sh. 4–5 mm long; **Fl.** solitary or in 3's, 4 cm $\varnothing$, snow-white.

**C. pallens** L. Bol. – S.W.Afr.: Gr. Namaqualand. – Compact, c. 13 cm tall, 17 cm $\varnothing$; **St.** prostrate, 9 cm long, Int. enclosed; **L.** narrowed towards the subacute tip, indistinctly carinate on the underside above, pale bluish-green, 5–9.5 cm long, 7 mm across; **Fl.** reddish-yellow.

**C. parviflorum** L. Bol. v. **parviflorum**. – Cape: near Darling. – **St.** c. 3 mm thick, Int. 25 mm long; **L.** carinate on the underside above, acute, 9 cm long, 6 mm across, bluish-green, the upper L. shorter; **Fl.** 3 cm $\varnothing$, Pet. sulphur-yellow with deep yellow tips, pinkish-red on the underside.

**C.** — v. **proliferum** L. Bol. – Cape: Malmesbury D. – **St.** elongated, rooting, with **L.** clustered at the nodes.

**C. paucifolium** L. Bol. – Cape: Ceres Karroo. – **St.** creeping, elongated; **Br.** crowded, erect, with 2–4 cylindrical, tapered, spotted, bluish-green **L.** 6 cm long, 4 mm across, with a Sh. 3–4 mm long; **Fl.** 5 cm $\varnothing$, red.

**C. pillansii** L. Bol. v. **pillansii** (*C. saturatum* L. Bol. nom.nud.). – Cape: Lit. Namaqualand. – Climbing; **L.** cylindrical, acute, dark green, spotted, (5–)6(–16) cm long; **Fl.** very numerous, 4–6 cm $\varnothing$, an attractive yellow, with a red centre.

**C.** — v. **grandiflorum** L. Bol. – **L.** up to 20 cm long; **Fl.** 7.2 cm $\varnothing$.

**C. platycalyx** (L. Bol.) Schwant. (*Mes. p.* L. Bol., *C. p.* (L. Bol.) L. Bol.). – Cape: Karroo. – Prostrate, with Int. 3–5 cm long; **L.** up to 10 cm long, 6 mm across; **Fl.** 1–3, deep pink, shading to white below, with purple anthers.

**C. primulinum** (L. Bol.) Schwant. (*Mes. p.* L. Bol., *C. p.* (L. Bol.) L. Bol.). – Cape: Van Rhynsdorp D. – **St.** climbing, 25 cm and more long, Int. 2–7 cm long; **Br.** with 4–6 long-tapered, pale green **L.** 6 cm long, 7 mm across and thick, rounded on the underside, with a Sh. 3–4 mm long; **Fl.** 6.5 cm $\varnothing$, primrose-yellow, paler inside.

**C. procumbens** (Haw.) L. Bol. (*Mes. p.* Haw.). – Cape: Cape D. – Robust; **St.** elongated, rooting, Int. 10–15 mm long; **L.** clustered, semicylindrical, short-tapered, pinkish to bluish, 3.5 cm long, 5 mm thick, cauline L. shorter; **Fl.** solitary, 5 cm $\varnothing$, golden-yellow.

**C. pulchellum** L. Bol. – Cape: Van Rhynsdorp D. – Shoots with 2 **L.**, the lower ones anceolate, obtuse, 14–16 mm long, 7 mm across, 4–5 mm thick, with a Sh. 5 mm long, or linear, obtuse, 12–20 mm long, 3–4 mm across, 4 mm thick, dirty green, spotted; **Fl.** 26 mm ⌀, pale yellow, sometimes becoming flesh-coloured, anthers purple.

**C. pulchrum** L. BOL. – Cape: Van Rhynsdorp D. – Compact; **St.** 4-angled, with 4 subcylindrical, obtuse, bluish-green **L.** 3.5–4 cm long, 7 mm across, 9 mm thick, with a Sh. 7 mm long; **Fl.** 4.5 cm ⌀, an attractive pink.

**C. punctatum** (HAW.) L. BOL. (*Mes. p.* HAW.). – Cape: origin unknown. – Similar to **C. subulatoides**; **L.** smaller, with large, translucent spots and a white mucro; **Fl.** not known.

**C. purpureo-album** (HAW.) SCHWANT. (*Mes. p.-a.* HAW., *C. p.-a.* (HAW.) L. BOL.) – Cape: probably Riversdale, Albertina D. (acc. L. BOLUS). – **St.** prostrate, 40–50 cm long, slender, curved, with tangled Br., with short Int.; **L.** semicylindrical, fresh green, with numerous small spots and with a reddish mucro, 3–4 cm long, 4 mm across and thick, with a short Sh.; **Fl.** numerous, Pet. white above, red-striate below.

**C. rangei** (ENGL.) L. BOL. (*Mes. r.* ENGL., *Mes. ebracteatum* SCHLTR. et DIELS, *C. e.* (SCHLTR. et DIELS) DTR. et SCHWANT., *C. e.* (SCHLTR. et DIELS) L. BOL.). – Cape: widely scattered over Lit. Namaqualand; occurring frequently in S.W.Afr.: coastal deserts, Lit. Karas Mts., Namib. – Forming small clumps, with stolon-like **St.** up to 70 cm long, with Int. 6 cm long, rooting at the nodes; **L.** subcylindrical, obtuse, sea-green with translucent spots, 3–4 cm long, 5–7 mm thick; **Fl.** 73 mm ⌀, golden-yellow to coppery-red, with purplish-brown anthers.

**C. regale** L. BOL. – Cape: Lit. Namaqualand. – **St.** elongated, Int. 2 cm long; **L.** long-tapered, truncate above, green, densely spotted, 6–9 cm long, 8 mm across, 7 mm thick or sometimes less; **Fl.** 5 cm ⌀, purplish-pink.

**C. rhodandrum** L. BOL. – Cape: Ceres D. – **St.** with 4–6 tapered, bluish to pink, furfuraceous **L.** 35 mm long, 7 mm thick, with a Sh. 7 mm long; **Fl.** 3.5 cm ⌀, coppery-red, paler in the centre.

**C. rigidum** L. BOL. – Cape: Lit. Namaqualand. – **St.** creeping, 30 cm or more long, stiff, angular, with Int. 6 cm long; **L.** semicylindrical, acute, apiculate, bluish-green, 6 cm long, 6 mm thick, with a Sh. 3–5 mm long; **Fl.** 34 mm ⌀, golden-yellow.

**C. roseum** (L. BOL.) L. BOL. (*Leipoldtia r.* L. BOL., *C. calvinianum* L. BOL.). – Cape: Calvinia D. – Compact, c. 12 cm ⌀; **St.** 3–5 cm long, Int. enclosed; **L.** linear, acute, apiculate, rounded on the underside, bluish-green, 5 to 6 cm long, 5 mm across and thick; **Fl.** solitary, Pet. purplish-pink, indistinctly striate.

**C. serrulatum** L. BOL. – Cape: Worcester D. – **St.** prostrate, Int. 15–25 mm long; **L.** rounded or carinate on the underside, narrowed towards the tip, short-tapered, margins and keel serrate, green, 7.6 cm long, 6 mm across and thick, with a Sh. 4 mm long; **Fl.** 5 cm ⌀, purplish-pink.

**C. spongiosum** (L. BOL.) L. BOL. (*Mes. s.* L. BOL.). – Cape: Lit. Namaqualand. – 14 to 28 cm tall; **St.** 35 cm or more long, with nodes 15 mm thick, ± angular, the inside conspicuously spongy, white, Int. 2–5 cm long; **Br.** directed upwards, with 2–6 ± sabre-shaped L., those in each pair being unequal, obtuse or short-tapered, with white scales, 11 cm long, up to 12 mm across, 10–19 mm thick; **Fl.** 6 cm ⌀, scarlet, with golden-yellow anthers.

**C. staminodiosum** L. BOL. – Cape: Lit. Namaqualand. – **St.** elongated, 20 cm and more long, Int. 4 cm long; **L.** semicylindrical, 6 cm long, 8 mm thick, green, those on elongated shoots small, with a Sh. 5 mm long; **Fl.** 47 mm ⌀, white, staminodes present.

**C. stayneri** L. BOL. – Cape: Calvinia D. – Compact, up to 32 cm ⌀; **St.** creeping, up to 22 cm or more long, Int. 2–3 cm long, laterally compressed and narrowly alate; **Br.** erect, with 6 compressed, linear, acute **L.** (3–)4(–5) cm long, 5–6 mm ⌀, with the underside rounded; Pet. up to 24 mm long, glossy and coppery-red above, distinctly striate below, paler or (acc. STAYNER) also yellow, bronze-coloured, pink, mauve and orange.

**C. subulatoides** (HAW.) N. E. BR. (Pl. 150/3) (*Mes. s.* HAW., *Mes. acutum* HAW., *C. a.* (HAW.) N. E. BR., *Mes. subrostratum* WILLD.). – Cape: Caledon, Swellendam, Stormsvley, Barrydale D. – Acauline, densely branched, Ros.-forming; **L.** semicylindrical, short-tapered, 5–7 cm long, 9–10 mm across, ± carinate, with the lateral angles cartilaginous, somewhat greyish-green, with numerous dots; **Fl.** 4 cm ⌀, purplish-red.

**C. tenuifolium** L. BOL. – Cape: near Nieuwoudtville. – **St.** elongated, 30 cm or more long, rooting at the nodes, Int. 3.5 cm long; **L.** in dense clusters, semicylindrical, short-tapered, dirty green, 4–8.5 cm long, 4 mm thick, cauline, L. 4.5 cm long, 3–4 mm thick, with a Sh. 2 mm long; **Fl.** 6–8 cm ⌀, Pet. yellow above, scarlet at the base, with dirty red stripes below.

**C. tricolorum** (HAW.) N. E. BR. (*Mes. t.* HAW., *C. t.* (HAW.) SCHWANT., *Mes. stramineum* WILLD.). – Cape: Clanwilliam, Van Rhynsdorp D. – **St.** prostrate; **Br.** curved, slightly nodose; **L.** cylindrical, long-tapered, light or greyish-green, minutely spotted, 4–8 cm long, 4–6 mm thick, with a short Sh.; **Fl.** 4–5 cm ⌀, Pet. yellow, purple at the base, the underside reddened at the tips, with red filaments and brownish anthers.

**C. truncatum** L. BOL. – Cape: Van Rhynsdorp D. – **Br.** 3 cm long, Int. 12 mm long; **L.** ± expanded, truncate above, 15 mm long, 4–6 mm thick, with a Sh. 2 mm long; **Fl.** solitary, 4.5 cm ⌀, white to pale salmon pink.

**C. uniflorum** L. BOL. – Cape: Lit. Namaqualand. – **St.** elongated, creeping, acutely 2-angled, Int. 15–25 mm long; **Br.** with 4–6 ± clavate, short-tapered, yellowish-green to green **L.** 4 cm long, 5 mm thick, or 6 cm long, 18 mm thick, with a Sh. 4 mm long, with indistinct keel and margins; **Fl.** solitary, 7 cm ⌀, Pet. golden-yellow, pink on the underside.

**C. validum** (Haw.) Schwant. (*Mes. v.* Haw.). – Cape: origin unknown. – **St.** and **Br.** prostrate, curved, up to c. 60 cm long; **L.** semicylindrical, ± rounded to trigonous above, ± obtuse, apiculate, green to reddened, minutely spotted, 5–6 cm long, 6–8 mm across, with a short Sh.; **Fl.** yellow.
**C. vandermerwei** L. Bol. (*Mes. loreum* v. β L., *Mes. l.* v. *congestum* Haw.). – Cape: Swellendam D. – **St.** dense, creeping, up to 20 cm long, Int. 1–3 cm long; **L.** in the clusters long-tapered, in part distinctly carinate, 12 cm long, 5 mm across and thick, cauline L. 3.5 to 6 cm long, 4–6 mm across and thick, with a Sh. 4 mm long; **Fl.** 5 cm ⌀, Pet. yellow to straw-coloured, red-striped outside.
**C. vanheerdei** L. Bol. – Cape: Lit. Namaqualand. – **St.** prostrate, 21 cm long, Int. 1.5–2 cm long, the axillary short shoots giving a compact growth; **L.** semicylindrical, rounded above, bluish-green, 3–8.5 cm long, 5–10 mm ⌀; **Fl.** numerous, almost 10 cm ⌀, yellow.
**C. vanputtenii** L. Bol. (*C. pittenii* L. Bol. sphalm.). – Cape: Piquetberg D. – **St.** elongated, angular, Int. 1–6 cm long; **Br.** with 4 **L.** 5 cm long, 8 mm thick, narrowed towards the base, carinate on the underside, bluish to reddish, with a Sh. 3 mm long; **Fl.** solitary, 8 cm ⌀, yellow, with a white centre.
**C. watermeyeri** L. Bol. – Cape: Van Rhynsdorp D. – **St.** elongated, creeping, Int. 2–3 cm long; **Br.** erect, with 4–6 **L.** 20–35 mm long, 6–7 mm across, 7 mm thick, clavate below, acute above, green to bluish or reddish, with a Sh. 3 mm long; **Fl.** 6–8 cm ⌀, dirty salmon-pink, scented.
**C. worcesterense** L. Bol. – Cape: Worcester D. – Compact or with elongated **St.** 25 cm long, Int. 2–5 cm long; **L.** in the clusters 9.5 cm long, 6–7 mm across and thick, narrowed, indistinctly carinate, long-tapered, cauline L. 2–6 cm long, 5 mm across and thick, with a Sh. 4 mm long; **Fl.** 3 cm ⌀, yellow.

**Cerochlamys** N. E. Br. (§ IV/1/3). – Dwarf, very succulent plants. – Greenhouse, warm. Propagation: seed, cuttings.
**C. pachyphylla** (L. Bol.) L. Bol. v. **pachyphylla** (Pl. 150/4) (*Mes. p.* L. Bol., *C. trigona* N. E. Br.). – Cape: Oudtshoorn, Swellendam D. – Acauline; **L.** c. 2 pairs, opposite and decussate, united below, 5–6 cm long, widened towards the tip and short-tapered, 8–10 mm across below, c. 16 mm across in the uper ⅔, the upperside flat, the underside rounded, carinate above, pulled forward over the tip, smooth, brownish-green, waxy pruinose; **Fl.** on a pedicel 3–4 cm long, purplish-red.
**C. —** v. **albiflora** Jacobs. – **Fl.** white.

**Chasmatophyllum** Dtr. et Schwant. (§ IV/1/11). – Branched, shrubby plants; **St.** eventually prostrate, creeping or elongated, dividing to form a mat, Int. not perceptible or up to 3 mm long; **L.** opposite and decussate, crowded, ± united and sheathing, divaricate, spatulate, the underside semicylindrical or obtusely keeled, the margins and the underside with or without 1–2 obtuse T. towards the tip, vesicular and inflated below, the underside and the upperside above with white tubercles; **Fl.** solitary, terminal, yellow, Aug.-Sept. – Greenhouse, warm. Propagation: seed, cuttings.
**C. braunsii** Schwant. v. **braunsii** – Cape: Laingsburg D. – **Roots** woody; freely and densely branched; **St.** with 1–2 pairs of semiovate, densely white-tuberculate **L.** 7 mm long, 4 mm across; **Fl.** unknown.
**C. —** v. **majus** L. Bol. – **L.** 13 mm long, 4 mm thick; **Fl.** 2 cm ⌀, yellow.
**C. maninum** L. Bol. – Cape: Richmond D. – **St.** with 6–8 pairs of oblong-lanceolate **L.** 10–16 mm long, 2–3 mm across and thick, with minute, pearly tubercles, each margin with one bristly T.; **Fl.** 25 mm ⌀, yellow.
**C. musculinum** (Haw.) Dtr. et Schwant. (Pl. 150/5) (*Mes. m.* Haw., *Stomatium m.* (Haw.) Schwant., *Mes. dinterae* Dtr., *Mes. recumbens* N. E. Br.). – S.W.Afr.: Gr. Namaqualand; Orange Free State. – Forming dense mats; **St.** prostrate; **L.** 1.5–2 cm long, 4–6 mm across, equilaterally trigonous to semicylindrical, obtusely carinate, the keel and angles of the faces with or without 1–2 T., the surface rough-spotted; **Fl.** 15 mm ⌀, Pet. yellow, reddened at the tips.
**C. nelii** Schwant. – Cape: Graaff Reinet D. – **St.** at first erect, eventually prostrate, Int. 3 mm long; **L.** spatulate, trigonous at the tip, 10–12 mm long, 3–4 mm across, 2–3 mm thick, semicylindrical on the underside, obtuse above, tuberculate, without T.; **Fl.** 13 mm ⌀, golden-yellow.
**C. verdoorniae** (N. E. Br.) L. Bol. (*Mossia v.* N. E. Br.). – Orange Free State. – **St.** creeping, rooting at the nodes; **L.** 4–6 mm long, 3–4 mm across, 1–1.5 mm thick, ovate to oblong, with the underside obtusely carinate, bluish-green, rough with prominent spots, the angles with or without 1 T.; **Fl.** 7–8 mm ⌀, light golden-yellow.
**C. willowmorense** L. Bol. (? *Rhinephyllum w.* L. Bol.). – Cape: Willowmore D. – **St.** crowded, woody; **L.** hemispherical, the 2 L. of a pair unequal, 10–14 mm long, 5 mm across and thick, or 8–10 mm long, 3–4 mm ⌀, with the underside rounded-carinate, obtuse, olive-green to brown, with minute, raised spots; **Fl.** 2 cm ⌀, yellow.

**Cheiridopsis** N. E. Br. (§ IV/1/2). – ♃, very succulent, caespitose plants; **shoots** with 1–3 pairs of opposite L.; the L.-pairs that succeed each other are different in form, size and growth, so that one pair is shortly united below, the next pair being united for a quarter to almost all their length, the pairs lying firmly appressed to each other, green to glaucous or whitish, often punctate or immaculate, smooth, more rarely rough, glabrous or fine velvety-hirsute (visible with a lens), drying during the resting period to a sort of Sh. or sleeve (Greek: cheiris) which ± envelops the following pair of L.; **Fl.** solitary, terminal, mostly stalked, the pedicels set with sheathing bracts, Pet. usually yellow, lemon-yellow, rarely golden, white, pink, purplish or red, summer. – Greenhouse, warm. Propagation: seed; cuttings do not grow readily.

## Key to the Groups of the Genus Cheiridopsis by L. Bolus

1.   Br. with ± elongated Int., e.g. C. aspera, C. pilosula ........................ **Elongatae**
1a.  Br. with Int. enclosed in the L.-Sh., or sometimes exserted in cultivation ............... **2.**
2.   Flowering branchlets 4-leaved – e.g. C. bibracteata, C. turbinata ............. **Bibracteatae**
2a.  Flowering branchlets 2-leaved ............................................. **3.**
3.   L. turgidly succulent – e.g. C. pillansii ...................................... **Turgidae**
3a.  L. not turgidly succulent ................................................... **4.**
4.   L.-pairs similar in form, ± hardening with age and not forming "sleeves" .............. **5.**
4a.  L.-pairs dissimilar in that the Sh. forming the "sleeve" is shorter in the lower pair – e.g. C. tuberculata, C. peculiaris ............................................. **Cheiridopsis** (*Eucheiridopsis*)
5.   L. rather soft and succulent, entire, rather short – e.g. C. cuprea, C. speciosa ...... **Cupreae**
5a.  L. rather hard, sometimes dentate, often pallid and among the largest L. in the G. – e.g. C. candidissima, C. denticulata ............................................. **Grandes**

**C. acuminata** L. Bol. – Cape: Lit. Namaqualand. – Robust; **St.** 1 cm thick, with 1–2 pairs of **L.** 5–7 cm long, 15 mm across midway, minutely papillose, with a Sh. 1 cm long; **Fl.** up to 6 cm ⌀, pale yellow.

**C. alata** L. Bol. – Cape: Lit. Namaqualand. – Mat-forming; **St.** with (2–)4(–6) **L.**, the lower ones swollen, rounded or truncate above, 17 mm long, 15 mm ⌀, with a Sh. 5 mm long, the upper ones longer and more slender, pale green, densely spotted; **Fl.** 4–5 cm ⌀, pale yellow.

**C. albiflora** L. Bol. – Cape: Lit. Namaqualand. – c. 7 cm tall, 8 cm ⌀; **St.** 8 mm ⌀; **Br.** 4 mm ⌀; herbaceous parts ± velvety, bluish-green, indistinctly spotted; **L.** 4 on short shoots, narrowed above, acute to obtuse, shortly awned, 4.4 cm long, 12 mm ⌀ above midway, obtusely carinate on the underside, with a Sh. 8 mm across; Pet. in 6 series, 22 mm long, 2 mm across, glossy white.

**C. albirosa** L. Bol. – Cape: Lit. Namaqualand. – Up to 7 cm tall, 12 cm ⌀, herbaceous parts softly puberulous, brownish-green; **L.** lanceolate, acute to rather obtuse, indistinctly dentate, 2–3 cm long, 1 cm across below; **Fl.** c. 5 cm ⌀, Pet. brilliant pink, white below.

**C. altitecta** Schwant. – Cape: Lit. Namaqualand. – Similar to **C. marlothii**; **L.** 2 cm long, 4 mm thick, linear-tapered on the upperside, indistinctly carinate on the underside, bluish to greyish-green, tuberculate, with large spots, with a Sh. 5 mm long, longer than in **C. marlothii**; **Fl.** not known.

**C. ampliata** L. Bol. – Cape: Lit. Namaqualand. – Cushion-forming; **St.** with several pairs of **L.**, the lower ones wider midway, 2–4 cm long, 6 mm across, laterally compressed, 1 cm thick, the upper ones sheathing and united for 10 mm, 22 mm long, 5 mm across midway, pale green, minutely velvety-puberulous; **Fl.** 32 mm ⌀, lemon-yellow.

**C. angustipetala** L. Bol. – Cape: Lit. Namaqualand. – **St.** with 4 **L.**, the lower ones compressed-carinate, 25 mm long, 5 mm across, 6 mm thick, truncate and widened at the tip, ± velvety, densely spotted, with a Sh. 9 mm long; **Fl.** 2.5–3 cm ⌀, Pet. yellow, narrow.

**C. aspera** L. Bol. – Cape: Lit. Namaqualand. – Short-stemmed, with a few **Br.**, each mostly with 2–3 pairs of **L.** 6 cm long, 5–6 mm across and thick, united below to form a round Sh., obtuse, apiculate, with the underside semicylindrical, light green to grey, rough with numerous cartilaginous, whitish dots; **Fl.** not known.

**C. aurea** L. Bol. v. **aurea**. – Cape: Lit. Namaqualand. – **L.** 4.5 cm long, 13 mm across, swollen at the base, often ± oblique, with a rounded tip. grey, velvety, with a Sh. 1 cm long; **Fl.** 3–4 cm ⌀, golden-yellow.

**C. —** v. **lutea** L. Bol. – **L.** 5.5 cm long, 1 cm across, with a Sh. 2.5 cm long, light blue; **Fl.** 8 cm ⌀, yellow.

**C. ausensis** L. Bol. – S.W.Afr.: Aus. – Mat-forming; herbaceous parts of the **Br.** velvety-papillose; **L.** 4, acutely truncate above, with a small blister below, c. 3.5 cm long, 8–10 mm across, juvenile L. with ciliate margins and keel, unequally bifid with lobes 1–6 mm long; **Fl.** yellow.

**C. bibracteata** (Haw.) N. E. Br. (*Mes. b.* Haw., *Mes. rostratum* v. *b.* S.D.). – Cape: Lit. Namaqualand. – Almost acauline, branching from the base; **L.** 6–8 cm long, 8–10 mm across, 6–8 mm thick, united below to form a cylindrical Bo. 13–25 mm long, the free parts narrowed towards the tip, the underside carinate above, one L. of the pair ± pointed, the other more rounded, greyish-green, spotted, with the margins minutely hairy above; **Fl.** 4 cm ⌀, yellow, the pedicel with 2 united and long-sheathing bracts 2–2.5 cm long.

**C. bifida** (Haw.) N. E. Br. (*Mes. b.* Haw., *Mes. multipunctatum* S.D.). – Cape: Lit. Namaqualand. – Mat-forming; **L.** 4–6, soft-fleshy, 5–6 cm long, 6–10 mm across, 5–8 mm thick, semicylindrical on the underside, with a thick Sh., the blades of unequal length, long and acute or shorter and broader, with the keel-angle minutely rough, dull greyish-green, dotted; **Fl.** 4 cm ⌀, yellow.

**C. borealis** L. Bol. (Pl. 151/1). – S.W.Afr.: Gr. Namaqualand. – Cushion-forming; **St.** with 4 compressed **L.**, expanded to 11–13 mm across as seen from the side, 15 mm thick, truncate above, lower L. 3 cm long, with a Sh. 6–8 mm long, 9–11 mm across, the upper L. 2 cm long, 8–9 mm $\varnothing$ at the tip, bluish-green, minutely velvety-puberulous; Fl. c. 5 cm $\varnothing$, yellow.

C. brachystigma L. Bol. – Cape: Lit. Namaqualand. – Cushion-forming; St. with 2 L. 45 mm long, broadly compressed and carinate below, obtuse above, 20 mm thick, 18 mm across, pale green, velvety, with a Sh. 15 mm long; Fl. c. 6 cm $\varnothing$, yellow.

C. breachiae L. Bol. – Cape: Lit. Namaqualand. – 9 cm tall; St. c. 3.5 cm long; lower L. $\pm$ rounded, carinate below the tip, tapered, 6 cm long, with a Sh. 8 mm long, upper L. 3.5 cm long, with a Sh. 2 mm long; Fl. 5.5 cm $\varnothing$, pale greenish-yellow.

C. brevipes L. Bol. – Cape: Lit. Namaqualand. – Cushion-forming; St. with 2–4 L. 4 cm long, 18 mm across, 13 mm thick, with a Sh. 16 mm long, with a large blister below, bluntly tapered, with the underside $\pm$ carinate below the tip, very minutely papillose; Fl. 5 cm $\varnothing$, dirty yellow.

C. brevis L. Bol. – Cape: Calvinia D. – Cushion-forming; St. with 4 bluish-green, minutely velvety L., the lower ones $\pm$ circular, rounded-carinate on the underside, truncate above, 18–20 mm long, 6 mm across and thick, with a Sh. 4 mm long, the upper L. tapered, rounded-carinate, 17 mm long, with a Sh. 7 mm long; Fl. 3–4 cm $\varnothing$, yellow to golden-yellow.

C. brownii Tisch. – Cape: Lit. Namaqualand. – Similar to C. schlechteri but considerably smaller; forming mats 10–15 cm $\varnothing$; St. with 2–4 L. sheathing and united up to midway, with a $\pm$ dentate keel on the underside, 3 to 5 cm long, 1.5–3 cm across, rather less in thickness, greyish-green, smooth, with dots scattered or very dense on the margins and keel; Fl. unknown.

C. candidissima (Haw.) N. E. Br. (Mes. denticulatum v. c. Haw., Mes. c. (Haw.) N. E. Br.). – Cape: Van Rhynsdorp D. – Mat-forming; St. with 1–2 pairs of long-navicular L. united for $\tfrac{2}{5}$ of their length, 8–10 cm long, c. 12 mm across, up to 15 mm thick, with the underside semicylindrical to carinate, with a red apiculus, smooth, whitish-grey, with dark green spots.

C. carinata L. Bol. – Cape: Lit. Namaqualand. – Cushion-forming; St. with 4 minutely velvety L., the lower ones 37 mm long, 8 mm across, 10 mm thick, with a Sh. 7 mm long, the upper ones narrow, laterally compressed, with a roundish keel, obliquely rounded, 5 cm long, with a Sh. 5 mm long; Fl. 6–7 cm $\varnothing$, white.

C. carnea N. E. Br. – Cape: Piquetberg D. – Cushion-forming, up to 5 cm tall; L. in 1–2 pairs, with a Sh. 6–8 mm long, the free part 15–20 mm long, 4–5 mm across and thick, carinate on the underside, obtuse above, with small dots along margins and keel; Fl. 25 mm $\varnothing$, pink.

C. caroli-schmidtii (Dtr. et Bgr.) N. E. Br. (Mes. c.-s. Dtr. et Bgr.). – S.W.Afr.: Gr. Namaqualand. – Mat-forming; St. with several pairs of L. united for $\tfrac{1}{3}$ of their length to form a Bo. 2–4 cm long, the L. of a pair mostly unequally long, 2–2.5 cm across, 12–15 mm thick, with the underside semicylindrical or with a rounded, $\pm$ cartilaginous keel, the tip rounded, apiculate, light grey, spotted; Fl. golden-yellow.

C. cigarettifera (Bgr.) N. E. Br. (Mes. c. Bgr., Mes. vescum N. E. Br.). – Cape: Laingsburg D. – Similar to C. marlothii; L. 15–18 mm long, 2.5–3 mm across, with the underside rounded to $\pm$ carinate, the keel denticulate above, bluish-green, spotted; in the resting period the juvenile L. are completely hidden within the Sh.

C. citrina L. Bol. – Cape: Lit. Namaqualand. – L. pairs very rounded, 45–46 mm long, Sh. 7 mm long, 8–9 mm across; Fl. up to 3.5 cm $\varnothing$, lemon-yellow.

C. compressa L. Bol. – Cape: Lit. Namaqualand. – Cushion-forming; St. with 6 minutely papillose L., the lowest c. 3 cm long, 10 mm thick below the tip, rounded-truncate, with a Sh. 9 mm long, 9 mm across, middle L. 2 cm long, upper ones projecting 10 mm beyond the Sh., compressed, 5.5 mm thick; Fl. 4 cm $\varnothing$, lemon-yellow.

C. cuprea (L. Bol.) N. E. Br. (Mes. c. L. Bol.). – Cape: Van Rhynsdorp D. – Cushion-forming; St. crowded, with 4 densely arranged, trigonous L. 35 mm long, 4 mm across, sheathing and united for 14 mm, truncate above, with serrulate angles; Fl. 3.5–4 cm $\varnothing$, Pet. yellow below, then paler, coppery-red above midway.

C. curta L. Bol. – Cape: Lit. Namaqualand. – Cushion-forming; St. with 4–6 thick, $\pm$ truncate, bluish-green, smooth L. 45 mm long, 9 mm across midway, swollen below, with a Sh. 5–7 mm long, middle L. c. 3 cm long, upper ones with only 15 mm visible; Fl. 3 cm $\varnothing$, yellow.

C. denticulata (Haw.) N. E. Br. v. denticulata (Mes. d. Haw.). – Cape: Lit. Namaqualand. – Acauline; L. subulate-trigonous towards the tip, vesicular and swollen at the base, compressed and expanded, with the keel $\pm$ denticulate above, noticeably grey; Fl. 7–8 cm $\varnothing$, pale straw-coloured.

C. — v. glauca (Haw.) N. E. Br. (Mes. d. v. g. Haw.). – L. bluish.

C. derenbergiana Schwant. – Cape: Lit. Namaqualand. – Mat-forming; L. during the resting period completely hidden in the sleeve-like Sh., 3 cm long, 5–6 mm across and thick, acutely carinate, densely haired, greyish-green; Fl. up to 5.5 cm $\varnothing$, lemon-yellow.

C. difformis (Thunbg.) N. E. Br. (Mes. d. Thunbg., Mes. exiguum N. E. Br.). – Cape: Van Rhynsdorp, Calvinia D. – St. up to 15 cm long, with a few Br., prostrate; L. oblique, with the tip recurved, unequal in length, narrow-linguiform, semicylindrical, 12–14 mm across, 5 and 8 cm long, compressed-carinate, bluish-green, minutely spotted; Fl. yellow.

C. dilatata L. Bol. – Cape: Lit. Namaqualand. – Densely branched, 7 cm tall, 11 cm $\varnothing$; lower L. with a Sh. 1.5 cm long, lanceolate, tapered, c. 5 cm long, 1.5 cm $\varnothing$ midway, $\pm$ truncate, indistinctly dentate, 7 mm $\varnothing$ at the tip, upper L. 4 cm long, 6 mm $\varnothing$; **Fl.** c. 5 cm $\varnothing$, white to rather yellowish.

**C. duplessii** L. Bol. – Cape: Calvinia D. – Compact; **St.** 4 cm long, densely covered with the remains of old L.; **L.** wide below, then narrowed, truncate-rounded at the tip, (4–)5(–7.5) cm long, 7–11 mm across midway, 6–8 mm thick, with a Sh. 3–6 mm long, other L. shorter, bluish-white, minutely papillose; **Fl.** 5–6 cm $\varnothing$, yellow to golden-yellow.

**C. eburnea** L. Bol. – Cape: Lit. Namaqualand. – Cushion-forming; **St.** with 2 L. 28–32 mm long, 7 mm across, 8 mm thick, with a Sh. 12 mm long, truncate and rounded above, pale bluish-grey, minutely spotted; **Fl.** 3.5 cm $\varnothing$, whitish to pale straw-coloured.

**C. excavata** L. Bol. – Cape: Lit. Namaqualand. – Forming mats 12 cm $\varnothing$; **L.** 3–4 cm long, 15–20 mm across below, long-ovate to triangular, with the underside rounded-carinate, with blue dots; **Fl.** 5 cm $\varnothing$, yellow.

**C. framesii** L. Bol. – Cape: Van Rhynsdorp D. – Cushion-forming; **St.** with 4 **L.** 5.5 cm long, 12 mm across, long-tapered, with a short Sh., the underside carinate, the sides and angles with dense, translucent spots, dark green; **Fl.** 4–5 cm $\varnothing$, Pet. yellow with red dots.

**C. gibbosa** Schick et Tisch. – Cape: Lit. Namaqualand. – Short-stemmed and clump-forming; **St.** with 2 pairs of **L.** united for $\frac{1}{2}$ or $\frac{2}{3}$ of their length to form a Bo. 3–4 cm long, 25–30 mm across, 18–20 mm thick, the underside with a roundish keel pulled forward and chin-like, apiculate, bluish to greyish-green, with green spots merging into a line along the margins.

**C. glabra** L. Bol. – Cape: Lit. Namaqualand. – **St.** with 4 tapered **L.** 42 mm long, obtuse to obliquely truncate, with a Sh. 11 mm long, 5–6 mm across, becoming sleeve-like during the resting period; **Fl.** 6–7 cm $\varnothing$, yellow, red on the underside.

**C. grandiflora** L. Bol. – Cape: Lit. Namaqualand. – **St.** with 4 **L.** 7 cm long, 10 mm across midway, 15 mm thick, laterally compressed, with a rounded keel, obtuse to truncate above, pale blue, minutely papillose, with a Sh. 15 mm long; **Fl.** 9 cm $\varnothing$, pale yellow.

**C. hallii** L. Bol. – S.W.Afr.: Lorelei Copper Mine. – Forming mats c. 7 cm $\varnothing$; **St.** with 4 obliquely truncate **L.** 3–4 cm long, with a Sh. 13–18 mm long, the underside obtusely carinate, with a small blister below, bluish-green, spotted, velvety-papillose; **Fl.** 3.5 cm $\varnothing$, Pet. whitish below, yellow above, lemon-yellow on the underside.

**C. herrei** L. Bol. – Cape: Lit. Namaqualand. – Cushion-forming; **St.** with 2–4 **L.** 15–20 mm long, with a Sh. 11 mm long, compressed and truncate at the tip, rounded-carinate, 9–12 mm broad and thick midway, with a large blister at the base, bluish to dirty green, papillose; **Fl.** up to 6 cm $\varnothing$, yellow.

**C. hutchinsonii** L. Bol. – Cape: Van Rhynsdorp D. – Mat-forming; **St.** with 4 L. c. 14 mm long, 6 mm across below, 5 mm thick, with a Sh. 4 mm long, with a rounded keel on the underside, long-tapered, upper L. shorter; **Fl.** 3.5–4.5 cm $\varnothing$, lemon-yellow.

**C. imitans** L. Bol. – Cape: Lit. Namaqualand. – **St.** covered at the base with the remains of old L.; **L.** 2–6, tapered, bluish to reddish, the lowest ones 5–5.5 cm long, 8–9 mm across, 7–9 mm thick, with the underside rounded and keeled and a Sh. 7 mm long, middle **L.** 11–13 mm long, upper ones 5 mm long, 6 mm across, 5 to 7 mm thick; **Fl.** 7–7.5 cm $\varnothing$, yellow.

**C. inaequalis** L. Bol. – Cape: Lit. Namaqualand. – Forming cushions 15 cm $\varnothing$; **St.** with 4–6 **L.**, the lowest 45 mm long, 8 mm across, 5–6 mm thick, rounded to truncate above, with a Sh. 4 mm long, middle L. narrower, 30–35 mm long, upper ones 40 mm long, scarcely projecting from the Sh., bluish-green, velvety, spotted; **Fl.** up to 5 cm $\varnothing$, yellow, Sep. unequal.

**C. inconspicua** N. E. Br. – Cape: Van Rhynsdorp D. – Mat-forming; **St.** 5–6 cm long; **L.** united below to form a Bo. 5–10 mm long, 10 to 12 mm across, 8–10 mm thick, the free parts 48–65 mm long, 8–10 mm across, 7–9 mm thick, with the underside rounded-carinate, bluish-green, with minute dark dots; **Fl.** unknown.

**C. insignis** Schwant. (*C. graessneri* Schick et Tisch., *C. schlechteri* Schwant.). – Cape: Lit. Namaqualand. – **St.** short, thick; shoots with 2 pairs of **L.** 4 cm long, 2 cm across and thick, with a Sh. c. 1.5 cm long, rounded on the underside, apiculate, whitish or bluish-green, with minute velvety hairs, with darker spots.

**C. inspersa** (N. E. Br.) N. E. Br. (*Mes. i.* N. E. Br.). – Cape: origin unknown. – Branching from the base; **St.** 5–18 mm long, 2–3 mm thick; **L.** 5–7 cm long, 7–8 mm across below, 4–5 mm thick, acute, convex-carinate and spotted on the underside, united below for 2–4 cm and forming a Bo., bluish-green to purple; **Fl.** up to 6 cm $\varnothing$, lemon-yellow.

**C. intrusa** L. Bol. – Cape: Calvinia D. – **St.** with 4 **L.** c. 20 mm long, 4 mm across, 5 mm thick, with a Sh. 4 mm long, upper L. longer, $\pm$ tapered, minutely velvety; **Fl.** 4 cm $\varnothing$, lemon-yellow.

**C. latifolia** L. Bol. – Cape: Calvinia D. – **St.** with 6 bluish **L.** with green spots, the lowest 4–6 cm long, 14 mm across, 9 mm thick, with a Sh. 7 mm long, the underside rounded-carinate towards the rounded to truncate tip, the middle L. shorter, the upper ones scarcely projecting beyond the Sh.; **Fl.** 4.5 cm $\varnothing$, yellow.

**C. lecta** (N. E. Br.) N. E. Br. (*Mes. l.* N. E. Br.). – Cape: Van Rhynsdorp D. – Short-stemmed, with a few **Br.**, each usually with 2 pairs of **L.** united to midway into a round Sh., blades 3–4 cm long, 8 mm across, 4–5 mm thick, with the underside semicylindrical to carinate, with the keel pulled forward, mucronate, green, spotted, with the spots merging into a line along the keel and margins; **Fl.** unknown.

**C. leptopetala** L. Bol. – Cape: Lit. Namaqualand. – Robust, 20 cm $\varnothing$; **St.** 5 mm thick, covered with the Sh. of old L.; **L.** 2–4, 5–7 cm long, 7–8 mm across, 7–9 mm thick, with a Sh. 15–17 mm long, with the underside rounded-carinate, with the tip obliquely rounded to acute, yellowish-green to bluish, with indistinct dots; **Fl.** 5 cm $\varnothing$, Pet. pale yellow, pale coppery-red above.

**C. littlewoodii** L. Bol. – Cape: Van Rhynsdorp D. – Robust, 15 cm $\varnothing$; **St.** with 6 L. (4–)5 (–6) cm long, with a Sh. 2 cm long, narrowing above midway, rounded on the underside, or laterally compressed and carinate, with 1–4 T. on the keel above, bluish-green, velvety-papillose, the 3 L.-pairs all differently shaped; **Fl.** c. 6 cm $\varnothing$, coppery-pink, paler in the centre.

**C. longipes** L. Bol. – Cape: Lit. Namaqualand. – **St.** with 6 tapered L. 54–61 mm long, 9 mm across, 8 mm thick, with a Sh. 10 mm long, the underside rounded, with a rounded or acute keel, middle L. shorter, bluish-green, minutely velvety; **Fl.** 46 mm $\varnothing$, yellow.

**C. luckhoffii** L. Bol. – Cape: Lit. Namaqualand. – **St.** with 2–4 variously shaped L., the lower ones slender, concealed within the 1 cm long Sh., the middle ones rounded above, 27 and 20 mm long, 4–6 mm across, 5–6 mm thick, with a Sh. 4 mm long, the upper L. 5–7 mm long, pale blue, minutely velvety; **Fl.** 3 cm $\varnothing$, lemon-yellow.

**C. macrocalyx** L. Bol. – Cape: Lit. Namaqualand. – **St.** with 2–4 green **L.**, the lowest 65 mm long, 9 mm across, 11 mm thick, with a Sh. 12 mm long, widened above, obtusely keeled on the underside, with an obliquely truncate tip, upper L. very rounded below, 5 cm long; **Fl.** 6 cm $\varnothing$, yellow, with a white centre, Cal. very large.

**C. macrophylla** L. Bol. – Cape: Lit. Namaqualand. – Shoots with 2–4 compressed, bluish, velvety **L.**, the lowest 14 cm long, 2 cm across and thick, with a Sh. 3 cm long, rounded-carinate on the underside, tapered, obliquely rounded above, upper L. 8.5 cm long, developing new L. from lateral axillary buds, these being acutely carinate, with the keel serrate-dentate above; **Fl.** 7 cm $\varnothing$, Pet. straw-coloured on the upperside, with the underside coppery-red.

**C. marlothii** N. E. Br. (Pl. 151/4). – Cape: Lit. Namaqualand. – Forming $\pm$ circular mats; **St.** with 2–4 L. 3–6 cm long, one pair sheathing and united for $\frac{1}{3}$ of the length, the other pair shorter, rounded on the underside, acutely carinate towards the tip, with the keel pulled forward chin-like, apiculate, greyish-green, waxy-pruinose, rough with translucent spots especially along the angles; the more united and sheathing L.-pair dries in the resting period into a conical Sh. protecting the immature L.; **Fl.** 12 mm $\varnothing$, lemon-yellow.

**C. meyeri** N. E. Br. v. **meyeri**. – Cape: Lit. Namaqualand. – Very short-stemmed, small, the St. mostly with 2 **L.**-pairs, one pair united into a small, obovoid Bo. 16 mm long, 10 mm across, 6–7 mm thick, not deeply cleft, light greenish-grey, with prominent spots, drying to become papery and membranous, young L. very slightly united, 12–15 mm long, 5–6 mm across, 7–8 mm thick, semicylindrical on the underside.

**C. —** v. **minor** L. Bol. – Smaller variety.

**C. minima** Tisch. – Cape: origin unknown. – Plant 5–6 cm tall; **St.** covered with old L.-Sh., with 1–2 pairs of **L.** 5–7 mm long, 3–5 mm across, 2–3 mm thick, sheathing and united to midway, the free parts obtusely carinate on the underside, truncate, light green, with translucent dots along the margins and keel; **Fl.** on a pedicel 1 cm long.

**C. multiseriata** L. Bol. – Cape: Lit. Namaqualand. – **St.** with 4 narrow, tapered, bluish **L.** 5 cm long, 10 mm across, 12 mm thick, carinate on the underside towards the tip, minutely papillose; **Fl.** 6 cm $\varnothing$, Pet. in several series, golden-yellow.

**C. nelii** Schwant. – Cape: Van Rhynsdorp D. – Mat-forming; **L.** sheathing and united for $\frac{1}{3}$ of the length, 3 cm long, linear, acute above, apiculate, with rounded angles, with the underside rounded-carinate, the keel often tuberculate, greyish-green, dull; **Fl.** not known.

**C. papillata** L. Bol. – Cape: Bushmanland. – **St.** densely covered with dry, persistent L.-Sh., with 2–4 unequal **L.** united to midway, the free parts lanceolate, apiculate, pubescent, eventually glabrous, truncate above, rounded-carinate on the underside, 4–5 cm long, 1.2 cm across and thick; **Fl.** 5 cm $\varnothing$, dark yellow.

**C. parvibracteata** L. Bol. – Cape: Van Rhynsdorp D. – **St.** up to 6 cm long, covered with dry L.-Sh.; **L.** 2–4, the lowest rounded then abruptly tapering, carinate on the underside, below the tip, with convex faces, 12–20 mm long, 6 mm across, 4–5 mm thick, with a Sh. 4 mm long, upper L. slender, 15 mm long; **Fl.** 4 cm $\varnothing$, on a pedicel with bracts 5–7 mm long, Pet. salmon-pink on the upperside, yellow on the underside.

**C. parvula** (Schltr.) N. E. Br. (*Mes. p.* Schltr.). – Cape: Lit. Namaqualand. – $\pm$ mat-forming, c. 5 cm tall; **St.** with 4 linear-oblong, obtuse **L.** 10–17 mm long, 5–15 mm across, with the underside carinate; Pet. pale pink.

**C. paucifolia** L. Bol. – Cape: Lit. Namaqualand. – **St.** with 2–4 L. 13–27 mm long, 5–6 mm across midway, 6–7 mm thick, with a Sh. 9 mm long, 14 mm $\varnothing$ and with a crenate keel below the tip, long-tapered, obliquely truncate, pale green, densely spotted, the upper L. hidden in the Sh.; **Fl.** 36 mm $\varnothing$, yellow.

**C. pearsonii** N. E. Br. – Cape: Lit. Namaqualand. – Cushion-forming, 2.5–4 cm tall; **St.** with 2–4 L. 15–25 mm long, 3–4 mm across and thick, united for $\frac{1}{3}$ of the length, carinate on the underside, acute or obtuse above, the faces spotted; **Fl.** 2.5 cm $\varnothing$, light yellow to slightly reddish.

**C. peculiaris** N. E. Br. (Pl. 151/2). – Cape: Lit. Namaqualand. – Very short-stemmed, usually with a single shoot, more rarely branched; **L.** 2–4, united at the base, $\pm$ resting on the soil, 4–5 cm long and across, 9–10 mm thick, shortly acute above, with the upperside mostly flat, the underside slightly convex, slightly carinate towards the tip, greyish-green, with scattered spots; the L.-pair for the resting period completely united, eventually drying up and becoming papery, with the succeeding, closely appressed L.-pair projecting from the cleft; **Fl.** 3.5 cm ⌀, yellow.

**C. peersii** L. BOL. (*C. victoris* L. BOL.). – Cape: Lit. Namaqualand. – **St.** with 4 bluish-green **L.**, the lowest 6.5–7 cm long, 12 mm thick, rounded on the underside, carinate above, with a Sh. 10 mm long, upper L. 8.5 cm long, acutely carinate; **Fl.** 5 cm ⌀, golden-yellow.

**C. perdecora** N. E. BR. (*C. robusta* L. BOL.). – Origin unknown. – **St.** with 2 **L.** up to 5.5 cm long, up to 17 mm across, 15 mm thick, swollen at the base, with a Sh. 9 mm long; **Fl.** 42 to 58 mm ⌀, Pet. up to 2.5 cm long, with the upperside yellow, then orange, with a red tip, the underside pink to deep red.

**C. pillansii** L. BOL. v. **pillansii** (*C. comptonii* JACOBS.). – Cape: Lit. Namaqualand. – Very short-stemmed, forming large clumps up to 20 cm ⌀; **St.** with 1–2 pairs of **L.** united for ⅓ of the length, forming a compact Bo. 45 mm long and across, 25 mm thick, flat on the upperside, semicylindrical on the underside, rounded-carinate above, whitish-grey with darker spots; **Fl.** 6–8 cm ⌀, light yellow; stigmas 2,6 mm long.

**C. — v. crassa** (L. BOL.) ROWL. (*C. c.* L. BOL.). – More compact in habit; **Fl.** smaller, pale straw yellow, stigmas 17, to 5 mm long.

**C. pilosula** L. BOL. – Cape: Lit. Namaqualand. – Similar to **C. acuminata** but less robust, 7–8 cm tall; **St.** with short Int. and 2–4 **L.** 4–4.5 cm long, c. 11 mm across, 13 mm thick above, narrowed towards the acute tip, carinate on the underside and obliquely truncate above, other L. short-tapered and rounded above, papillose, hairy, with a rather swollen Sh. 8 mm long; **Fl.** 2–3 cm ⌀, Pet. white with yellow tips.

**C. pressa** (N. E. BR.) N. E. BR. (*Mes. p.* N. E. BR.). – Acc. N. E. BROWN possibly a hybrid: **C. rostrata** × **C. tuberculata**. – ± acauline, branching from the base, forming clumps; **St.** with 2–4 **L.**, those of a pair dissimilar, united below to form a Bo. 12–15 mm ⌀, 3.5–6 cm long, 9–14 mm across below, 7–9 mm thick, with the underside convex to obtusely carinate, laterally compressed and widened at the acute to obtuse and apiculate tip, bluish-green with dark green spots; **Fl.** not known.

**C. pulverulenta** L. BOL. – Cape: Bushmanland. – **St.** with 4 narrow **L.** 12 mm long, 7 mm across midway, 8 mm thick, with a Sh. 7 mm long, carinate on the underside, obliquely truncate above, whitish to pale green, spotted, with minute papillae giving the L. a pruinose appearance; **Fl.** 4.5 cm ⌀, yellow.

**C. purpurascens** (S. D.) N. E. BR. v. **purpurascens** (*Mes. p.* S. D.). – Cape: Piquetberg D. – Acauline, branched; **L.** obtusely trigonous, with the keel pulled forward over the tip, grey or ± bluish-grey, eventually red, spotted, with the Sh. very reddened and inflated; **Fl.** yellow.

**C. — v. leipoldtii** L. BOL. – Cape: Malmesbury, Cape D. – **L.** 6 cm long, an attractive red; **Fl.** 4–4.5 cm ⌀, deep yellow.

**C. purpurata** L. BOL. (Pl. 151/3). – Cape: Lit. Namaqualand. – **St.** numerous, with 2 bluish to pale pink **L.** 35 mm long, 11 mm ⌀, with a Sh. 1 cm long, the upperside flat, the underside rounded-carinate; **Fl.** 3.5 cm ⌀, purplish-pink.

**C. quadrifolia** L. BOL. – Cape: Lit. Namaqualand. – **St.** with 4 narrow, spotted, minutely velvety **L.** 34–42 mm long, 7–8 mm across and thick midway, truncate above, with the underside rounded and the keel serrate above, with a Sh. 4–5 mm long; **Fl.** 4.5–5 cm ⌀, lemon-yellow.

**C. quaternifolia** L. BOL. – Cape: Bushmanland. – **St.** with 4–6 narrow **L.**, the lowest 3 cm long, 7 mm across, 5–6 mm thick, rounded to truncate above, with the underside rounded to carinate, with a Sh. 6 mm long, bluish-green, minutely velvety-haired, with a few spots, upper L. acute, shorter; **Fl.** 2.5 cm ⌀, lemon-yellow.

**C. resurgens** L. BOL. – Cape: Lit. Namaqualand. – **Roots** tuberous, 6 cm long; **St.** semi-fleshy, covered with the remains of old L.-Sh., with 2–4 bluish-green **L.** 35 mm long, 3–4 mm across and thick, with a reddish-purple Sh. 11 mm long, upper L. long-tapered; **Fl.** up to 4.5 cm ⌀, yellow.

**C. richardiana** L. BOL. – Cape: Kenhardt D. – St. with 2–4 pale bluish-green **L.** with large spots, the lowest 2–2.5 cm long, 7 mm across and thick, carinate on the underside, truncate above, with a Sh. 6 mm long, upper L. more oblong; **Fl.** 3.5–4 cm ⌀, yellow.

**C. robusta** (HAW.) N. E. BR. (*Mes. r.* HAW.). – Cape: Lit. Namaqualand. – **St.** with 4 unequal **L.** 40 and 35 mm long, 17 mm across, 15 mm thick, with a blister below and a Sh. 9 mm long, with the underside rounded-carinate, minutely velvety, often with shoots developing in the L.axils; **Fl.** 42–58 mm ⌀, Pet. yellow below, orange above.

**C. roodiae** N. E. BR. (Pl. 152/1). – Cape: Van Rhynsdorp D. – Short-stemmed, with a few Br. up to 5 cm long, with 4–5 pairs of **L.** c. 7 cm long, almost 2 cm across below, linear, long-tapered, with a minute, hard tip, acutely carinate on the underside, smooth, greyish-green, with a very firm surface, the margins and keel with a whitish border; **Fl.** 3–4 cm ⌀, yellow.

**C. rostrata** (L.) N. E. BR. (*Mes. r.* L., *Mes. quadrifidum* HAW., *C. q.* (HAW.) SCHWANT.). – Cape: Van Rhynsdorp D. – Acauline, mat-forming; **L.** 5–8 cm long, 12–18 mm across, 8–10 mm thick below and sheathing and united to form a cylindrical Bo. 10–12 mm long, the free part narrowed towards the obtuse, spotted, apiculate tip, greyish-green, rounded-carinate on the underside, keel and margins cartilaginous, rough; **Fl.** large.

**C. rostratoides** (HAW.) N. E. BR. (*Mes. r.* HAW., *Mes. ramulosum* HAW.). – S.Afr.: origin unknown. – Mat-forming; **St.** prostrate in age, 5–7 cm long; **L.** trigonous, obtusely rounded at the tip, inflated below, with a Sh. 2 cm long, with minute, rough spots; Fl. yellow.

**C. rudis** L. Bol. – Cape: Lit. Namaqualand. – St. robust, with Int. 15 mm long, with 2–4 dirty green, rough L. 25–40 mm long, 6 mm across, 7 mm thick, with a Sh. 7 mm long, rounded-carinate on the underside, narrowed on the upperside, rounded to tapered and apiculate above, with serrulate margins; Fl. 3–4 cm ∅, yellow.

**C. scabra** L. Bol. v. **scabra**. – Cape: Ceres D. – Shoots with 6 pale bluish-green L., the lowest 35 mm long, 15 mm across, 10 mm thick, with a Sh. 5 mm long and a large blister below, rounded and rough on the underside with a ± serrate keel, middle L. 20 mm long, upper ones 16 mm long and completely enclosed by the Sh.; Fl. 4.5 cm ∅, Pet. pale yellow on the upperside, pink on the underside.

**C. —** v. **fera** L. Bol. – Cape: Laingsburg D. – L. somewhat shorter, minutely velvety.

**C. schickiana** Tisch. (*C. olivacea* Schwant.). – Cape: Lit. Namaqualand. – Acauline, mat-forming; St. with 1–2 pairs of L. united to midway to form a bifid Bo. 4–6 cm long, 2 to 2.5 cm across, 15–18 mm thick, rounded-carinate on the underside, with 1 T. above, the lower part smooth, otherwise minutely velvet-haired, bluish-green, with translucent spots above.

**C. schlechteri** Tisch. (*C. johannis-winkleri* Schwant.). – Cape: Lit. Namaqualand; S.W. Afr.: S. of Aus. – Mat-forming; L. united into a Bo. 1 cm long, 7 mm across, 5 mm thick, with a Fi. 5 mm deep, smooth, bluish-green, spotted; Fl. small, light violet.

**C. serrulata** L. Bol. – Cape: Lit. Namaqualand. – St. with 4 tapered L., the lowest 45 mm long, 6 mm across, 7 mm thick, rounded or truncate above, with serrate margins and a Sh. 6 mm long, upper L. more acute, 43 mm long, with the margins conspicuously coarsely dentate; Fl. 4.5 cm ∅, lemon-yellow to yellow.

**C. speciosa** L. Bol. – Cape: Lit. Namaqualand. – Cushion-forming; St. with 4 L., the lowest 33 mm long with a blister below and a Sh. 11 mm long, narrow-linear, with the tip expanded, truncate and 11 mm thick, upper L. 22 mm long, all L. minutely papillose; Fl. 5 to 6 cm ∅, coral-red.

**C. splendens** L. Bol. – Cape: Lit. Namaqualand. – Cushion-forming; St. with 4 L., the lowest 4–6 cm long, 15 mm across, expanded on the upperside, truncate above, 17 mm thick midway, acutely carinate on the underside, with a Sh. 12 mm long, upper L. 38 mm long, laterally compressed and thus carinate, grey, minutely papillose; Fl. 5–6 cm ∅, glossy scarlet.

**C. staminodifera** L. Bol. – Cape: Lit. Namaqualand. – L. 27 mm long, 11 mm across, 9 mm thick, with a Sh. 12 mm long, the lower part ± spherical, the free parts 15 mm long, 14 mm across, 12 mm thick, with a truncate-rounded tip, pale green, minutely velvety; Fl. 4 cm ∅, yellow, with filiform staminodes.

**C. subaequalis** L. Bol. – Cape: Lit. Namaqualand. – St. with (2–)4(–6) bluish-green, minutely papillose L. 3.5–4 cm long, 8–9 mm across midway, 10 mm thick, with a Sh. 11 mm long, expanded towards the rounded to truncate, apiculate tip; Fl. 5–6 cm ∅, lemon-yellow.

**C. subalba** L. Bol. – Cape: Lit. Namaqualand. – St. with 4–6 narrow-linear, bluish-green L., the lowest 22 mm long, 5 mm across and thick, with a Sh. 12 mm long, indistinctly carinate on the underside, obtuse at the tip, upper L. 32 mm long, with spots interspersed with tiny scales; Fl. 3.5 cm ∅, straw-coloured.

**C. tenuifolia** L. Bol. – Cape: Van Rhynsdorp D. – St. with 2 L.-pairs of variable shape, truncate above, carinate on the underside, 26 mm long, with a Sh. 6 mm long, 7 mm across, 5 mm thick, or other L. acute and 6 mm long, dirty green to bluish-green; Fl. 4.5 cm ∅, yellow.

**C. truncata** L. Bol. – Cape: Lit. Namaqualand. – St. with 2–4 pale bluish-green, minutely velvet-haired L. 2.5–4 cm long, 9–11 mm across, 14–17 mm thick, with a Sh. 14 mm long, obliquely carinate and truncate at the tip, with the keel indistinctly tuberculate to dentate; Fl. c. 6 cm ∅, yellow.

**C. tuberculata** (Mill.) N. E. Br. (*Mes. t.* Mill., *Mes. rostratum* S. D.). – Cape: Lit. Namaqualand. – Acauline; L. 6–12 cm long, 8–11 mm across, 5–7 mm thick, sheathing and united below into a cylindrical Bo. 15–35 mm long, 10–15 mm thick, the base of the free parts somewhat inflated, round to carinate on the underside, obtuse and apiculate above, bluish to greyish-green, often reddish, with translucent spots, those on the margins confluent, the keel-angle minutely cartilaginous, rough; Fl. 4 cm ∅, yellow.

**C. turbinata** L. Bol. v. **turbinata**. – Cape: Lit. Namaqualand. – St. with 4–6 semicylindrical, bluish-green, minutely velvet-haired L., the lowest 10 cm long, 17 mm across, 12 mm thick, with a Sh. 18 mm long, carinate below the tip, truncate above, the middle L. shorter, the upper ones 8 mm long; Fl. 7 cm ∅, lemon-yellow.

**C. —** v. **minor** L. Bol. – Smaller variety.

**C. turgida** L. Bol. – Cape: Lit. Namaqualand. – St. with 2–4 pale bluish-green, softly velvety L. 2–3 cm long, 11–12 mm thick, 12 mm across, swollen at the obliquely tapered tip, rounded on the underside, with a Sh. 15 mm long; Fl. 4.5 cm ∅, straw-coloured.

**C. umdausensis** L. Bol. – Cape: Lit. Namaqualand. – St. extended, Int. 13 mm long, with 2 L. 6 cm long, 1 cm across, 6 mm thick, with a Sh. 8 mm long, linear on the upperside, acute, the underside rounded-carinate, obliquely truncate, the margins acute, with raised spots, with 1 T. at the tip; Fl. 3.5 cm ∅, lemon-yellow.

**C. vanbredai** L. Bol. – Cape: Lit. Namaqualand. – Forming dense cushions 6 cm high, 13 cm ∅; St. with 2–4 L., the lowest ovate to lanceolate, tapered, apiculate, green, c. 3 cm long, 14 mm across, 8–12 mm thick, with a Sh. 9–13 mm long, obtusely carinate on the underside, the tip with 2 short, awned lobes, upper L.

3.6 cm long; Pet. 7–35 mm long, the upperside very pale pink to white, the underside pale pink.

**C. vanheerdei** L. Bol. – Cape: Lit. Namaqualand. – **St.** with 4–6 linear, bluish-green to bluish **L.** 8.5 cm long, 1 cm across, 1.5 cm thick, swollen at the base, rounded-carinate on the underside, rounded or acute above, with a **Sh.** 2 cm long; **Fl.** 8–11 cm ⌀, Pet. yellow below, salmon-pink above (the largest Fl. of any **Cheiridopsis**-species).

**C. vanzijlii** L. Bol. – Cape: Kenhardt D. – Cushion-forming; **St.** covered with old L.-Sh.; **L.** up to 25 mm long, 14 mm across and thick at the base, with a **Sh.** 9 mm long, flat on the upperside, the underside carinate below the tip, very convex laterally, truncate above, minutely velvet-haired, spotted; **Fl.** 6 cm ⌀, yellow.

**C. velutina** L. Bol. – Cape: Clanwilliam D. – Shoots with 4 pale bluish-green, minutely velvet-haired **L.** 4 cm long, with a **Sh.** 27 mm long, convex-carinate on the underside, the keel serrulate, rounded to truncate at the tip; **Fl.** 6.5 cm ⌀, yellow.

**C. verrucosa** L. Bol. v. **verrucosa** (*C. brevis* Schwant., *C. mirabilis* Schwant., *C. pachyphylla* Schwant.). – Cape: Lit. Namaqualand. – Mat-forming; **St.** with several pairs of greyish-green **L.** united for $\frac{1}{2}$ to $\frac{2}{3}$ of their length, often forming spherical **Bo.** 16–20 mm long and across and 10–15 mm thick, semicylindrical on the underside; **Fl.** 2–2.5 cm ⌀, yellow.

**C. —** v. **minor** L. Bol. – Smaller variety.

**Conicosia** N. E. Br. (§ IV/2). – Succulent plants with erect **St.** up to 30 cm long; **L.** in tufted, dense, tightly spiralled Ros., long-linear, subulate-trigonous, drying up and persisting; **Fl.** borne on lateral shoots, dying off after fruiting and persisting, long-pedicellate, 7 to 8 cm ⌀, yellow, more rarely whitish, unpleasant-smelling, with very long, acute Sep.; commonly self-fertile. Where the Fl. are terminal from the Ros. the entire plant dies off after fruiting. – Summer: in the open, winter: cold-house. Propagation: seed.

**C. alborosea** L. Bol. – Cape: Lit. Namaqualand. – Robust; **St.** c. 20 cm tall, 3 cm ⌀; **Br.** ascending, elongated, c. 1.6 cm ⌀; Ros.**L.** 18 to 27 cm long, cauline L. shorter, up to 1 cm across,

6 mm thick; **Fl.** with numerous white Pet. 1.5–3 cm long, with a pink centre.

**C. australis** L. Bol. – Cape: Bredasdorp D. – Slender, with spreading St.; **roots** napiform; **Br.** 25 cm long, 2–3 mm ⌀, Int. 1.5–3.5 cm long; some **L.** 11 cm long, others 3.5–6 cm long, 1 to 2 mm ⌀; **Fl.** 4–5 cm ⌀, yellow.

**C. bijlii** N. E. Br. – Cape: George D. – **Taproot** fleshy or napiform or cylindrical; Ros.**L.** clustered, 12–18 cm long, 3–5 mm across, 2–4 mm thick, acute-trigonous, green; flowering shoots 10–15 cm long, with smaller L.; **Fl.** 4–5 cm ⌀, Pet. very acute, lemon-yellow, red on the underside.

**C. brevicaulis** (Haw.) Schwant. (*Mes. b.* Haw., *C. b.* (Haw.) N. E. Br.). – Cape: Graff Reinet D. – **St.** 10–15 cm tall; **L.** trigonous, 10–12 cm long, 5–6 mm across, long-tapered; **Fl.** solitary, 5 cm ⌀, sulphur-yellow, glossy.

**C. capensis** (Haw.) N. E. Br. (*Mes. pugioniforme* Haw.). – S.Afr.: origin unknown. – **Roots** thin; **St.** 15 cm or more tall; **L.** up to 40 cm long, compressed-trigonous; **Fl.** 7.5 cm ⌀, straw-coloured.

**C. communis** (Edwards) N. E. Br. (*Mes. c.* Edwards, *Mes. pugioniforme* S. D.). – Cape: Cape D. – **Roots** thin, becoming woody; **St.** up to 12 cm tall; **Br.** elongated, prostrate; flowering Br. creeping; **L.** 10–15 cm long, 4–5 mm across and thick; **Fl.** solitary, 4.5–8 cm ⌀, pure yellow.

**C. coruscans** (Haw.) Schwant. (*Mes. c.* Haw.). – Cape: origin unknown. – Insufficiently known; **St.** ♃; **L.** very long, glistening.

**C. muirii** N. E. Br. – Cape: Riversdale D. – **Taproot** fleshy, cylindrical or radish-like, 20 to 40 cm long; **St.** creeping, 15–30 cm long; **L.** 10–17 cm long, 4–6 mm across, 3–5 mm thick, acute, green, purple below; **Fl.** 4–6 cm ⌀, light lemon-yellow.

**C. pugioniformis** (L.) N. E. Br. (Pl. 152/2) (*Mes. p.* L., *C. p.* (L.) Schwant., *Mes. capitatum* Haw., *C. c.* (Haw.) Schwant.). – **St.** 15–30 cm tall, 1–2 cm thick; **L.** 15–20 cm long, 12 mm across, acute, trigonous, with the upperside grooved and depressed, greyish-green, reddish at the base; **Fl.** 7 cm ⌀, glossy, sulphur-yellow.

**C. pulliloba** N. E. Br. – Cape: Malmesbury D. – Similar to **C. communis**; flowering shoots prostrate, 40–50 cm long; **Fl.** 5–7 cm ⌀, pure yellow.

**Conophytum** N. E. Br. (§ IV/1/17). – Dwarf, highly succulent plants of caespitose habit, with an abbreviated axis and **Br.** which may lengthen with age, but the Int. are always enclosed in the persistent L.-remains, except in a few cases where they are partly visible and elongated to about 1 cm long; **roots** up to 10–15 cm long, the spec. which form St. having a long vertical main root; growth consisting of a small fleshy **Bo.** which may be conical, globose, ovoid, cordate or sub-cylindrical and consists of 2 united L., convex above, flat, impressed, with a small Fi. going ± right across the upper surface or crenate or bilobed, green or brown-red, often dotted, the dots often transparent, often confluent into lines, the upper surface in some spec. is ± fenestrate, epidermis smooth, finely papillose, seldom tuberculate or pilose; **Fl.** solitary, axile, from the Fi. on a ± long pedicel with 2 mostly membranous bracts, the tube mostly whitish, at the mouth in part with staminodes or the Cor.-segments staminodial, the segments white, yellow, ochre-coloured, salmon, pink to violet, Fl. open by day or night at the start of the vegetative season, summer. – Greenhouse, warm, resting time in spring. Propagation: seed, cuttings (single Bo.). (See B. K. Boom: Het geslacht Conophytum N. E. Br. in Succulenta LII, 1973).

## Classification of the Genus Conophytum by A. Tischer

**Genus Conophytum** N. E. Br. (*Conophyton* Haw. nom. prov.; (? Sect.) *Subaphylla* Haw. pro part., Sect. *Minima* Haw. "Horde" *Subacaulis* Haw., Sect. *Sphaeroidea* S.D. (non Haw.) pro part. of the G. Mesembryanthemum L.). – **Type species:** C. truncatum.[1])

**§ I. SG. Fenestrarium** Tisch. (Sect. *Pellucida* Schwant., Sect. *Fenestrata* N. E. Br.). – Plants with single Bo. or caespitose, Int. short, L.-remains papery; **Bo.** soft-fleshy, oblong or cylindrical to oblong-ovoid or obconical, concave above, flat to flat-convex or very shortly bilobed, the upper surface with a zone without chlorophyll ("fenestrate"); ground-colour brownish to earth-coloured, bright or bluish-green to yellowish-green; **Cor.-segments** white or red. Day-flowering. – **Occurrence:** Lit. Namaqualand, Van Rhynsdorp, Kenhardt D. – **Type species:** C. pellucidum.

    **Sect. 1. Pellucida** Schwant. – **Bo.** compact-cylindrical to oblong-obovoid or conical; Wi.-zone ± irregularly distributed over the upper surface of the Bo., partly with opaque islands, upper surface truncate or convex and somewhat roundish, subbilobed, smooth, grooved or tuberculate; ground-colour reddish-brownish to dark olive or grey-green; mouth of the **Cor.-tube** with staminodes or staminodial Cor.-segments, stamens in the lower part of the tube, stigmas and style short. – **Occurrence:** Cape: Lit. Namaqualand, Kenhardt D. – **Type species:** C. pellucidum. – **Species:** C. areolatum, astylum, boreale, cupreatum, fenestratum, koubergense, lilianum, lithopoides, meridianum, pardicolor, pellucidum, rubroniveum, terrestre, terricolor.

    **Sect. 2. Subfenestrata** Tisch. – **Bo.** oblong to compact-cylindrical, obovoid; Wi.-zone on the upper surface distributed rather proportionately, not grooved or tuberculate; ground-colour bright to yellow-green or olive, truncate above or somewhat convex; anthers in the upper part of the **Cor.-tube,** no staminodes, style short, stigmas long. – **Occurrence:** Cape: Lit. Namaqualand, Van Rhynsdorp. – **Type species:** C. pillansii. – **Species:** C. acutum, burgeri, concavum, pillansii, roodiae, smorenskaduense.

    **Sect. 3. Cylindrata** Schwant. – **Bo.** ± cylindrical to oblong-ovoid, to 12 mm long, convex above to subbilobed, smooth; ground-colour yellowish to bright green, not strikingly dotted; **Cor.-segments** white, no staminodes, style long, stigmas short. – **Occurrence:** Cape: Lit. Namaqualand. – **Type species:** C. cylindratum. – **Species:** C. cylindratum, hallii, primosii.

**§ II. SG. Conophytum** (SG. *Euconophytum* Schwant. pro part.; Sect. *Sphaeroidea* N. E. Br. (non Haw.) pro part. of the G. Mesembryanthemum L.; *Derenbergia* Schwant. as G.; SG. *Derenbergia* (Schwant.) Schwant.). – Plants mostly forming cushions, some with longer Int. forming dwarf sub-♄, L.-remains papery, parchment-like or leathery; **Bo.** cylindrical, obconical to ovoid or globose, the upper surface flat to concave or ridged, ± convex or bilobed to subbilobed, not fenestrate above but often with darker, sometimes confluent dots or lines; ground-colour yellowish-green, bright to faint green, bluish-green or olive; **Cor.-segments** white, straw- or wine-coloured, red from bright pink to purplish or flesh-coloured, yellow to orange, salmon to copper-coloured or ochre-yellow. Day- and night-flowering. – **Occurrence:** Cape: Lit. Namaqualand, Kenhardt, Van Rhynsdorp, Calvinia, Piquetberg, Clanwilliam, Malmesbury, Paarl, Worcester, Ceres, Laingsburg, Montagu, Robertson, Swellendam, Riversdale, Oudtshoorn, Ladismith, Prince Albert, Beaufort West, Willowmore, Uniondale, Steytlerville D.; S.W.Afr.: Gr. Namaqualand. – **Type species:** C. truncatum (Thunbg.) N. E. Br.

    **Sect. 1. Cordiformia** (Bgr.) Schwant. emend. Tisch. (Sect. *Cordiformia* Bgr. of the G. Mesembryanthemum L.; (? Sect.) *Biloba* N. E. Br.; *Derenbergia* Schwant. as G.; SG. *Derenbergia* (Schwant.) Schwant.; Ser. *Gracilistyla*, *Velutina* and *Hirta omnia* Schwant. pro part.). – Forming dense or lax cushions, sometimes with elongated Int., forming dwarf sub-♄; **Bo.** cordate, obovoid to cylindrical, rounded above, mostly with ± long, free L.-tips (lobes), epidermis smooth, rough or velvety pilose; ground-colour bright to coppery-green or whitish- to grey-green, the upper surface with or without conspicuous dots, mostly with a darker zone at the ends of the Fi.; **Cor.-segments** white, yellow, orange, salmon to copper-coloured, bright brownish, wine-coloured, pink to purplish. Day- and night-flowering. – **Occurrence:** Cape: Lit. Namaqualand, Kenhardt D.; S.W.Afr.: Gr. Namaqualand. – **Type species:** C. bilobum (Marl.) N. E. Br.

    **SSect. a) Eubiloba** Tisch. (Sect. *Cordiformia* Bgr. of the G. *Mesembryanthemum* L.; Ser. *Cordiformia* (Bgr.) Schwant. of the SG. *Derenbergia* (Schwant.) Schwant.). – Forming dense to lax cushions or dwarf sub-♄, L.-remains stout, papery to leathery; **Bo.** ± truncate, with free L.-tips above or lobes acute or rounded, sometimes drawn out, rarely only a hint of these, Bo. with or without conspicuous dots, with darker zones at the ends of the Fi.;

---

[1]) A type species was not recorded by N. E. Brown. The oldest described species is **C. truncatum** (Thunbg.) N. E. Br. (*Mes. t.* Thunbg. Prod. 88).

**Cor.-segments** white, yellow, orange, salmon to copper-coloured, bright pink to purplish. Day-flowering. – **Occurrence:** Cape: Lit. Namaqualand; S.W.Afr.: Gr. Namaqualand. – **Type species:** C. bilobum.

**Ser. 1. Bo.** generally longer than 20 mm. – **Species:** C. absimile, aequale, albescens, ampliatum, andausanum, angustum, apiatum, bilobum, brevisectum, cauliferum, christiansenianum, citrinum, compressum, conradii, coriaceum, crassum, cupreiflorum, curtum, dilatatum, dissimile, distans, divaricatum, dolomiticum, elishae, excisum, frutescens, gonapense, graciliramosum, grandiflorum, incurvum, insigne, klipbokense, lacteum, largum, lavisianum, laxipetalum, linearilucidum, marnieranum, meyerae, muscosipapillatum, noisabense, notabile, nutaboiense, obtusum, ovatum, piriforme, pluriforme, proximum, regale, simile, simplum, smithersii, sororium, strictum, stylosum, supremum, umdausense.

**Ser. 2. Bo.** generally not more than 20 mm long. – **Species:** C. anjametae, blandum, connatum, cordatum, corniferum, difforme, diversum, extractum, gracile, gracilistylum, helenae, marginatum, parvulum, recisum, retusum, semivestitum, subcylindricum, tantillum, taylorianum, tectum, variabile, violaciflorum.

**SSect. b) Ovigera** TISCH. (Ser. *Ovigera* TISCH. of the SG. *Derenbergia* (SCHWANT.) SCHWANT.). – Forming dense to lax cushions, L.-remains papery to nearly leathery; **Bo.** obovoid to pear-shaped or somewhat cordate, rounded above, the Fi. mostly somewhat sunken and therefore incipiently bilobed, the keel-line rounded; ground-colour greyish or pale green to bluish, not dotted above or with darker dots; **Cor.-segments** white, pink to purplish or yellow. Day-flowering. – **Occurrence:** Cape: Lit. Namaqualand. – **Type species:** C. ovigerum. – **Species:** C. admiralii, altum, anomalum, approximatum, auctum, australe, candidum, corculum, craterulum, ecarinatum, laetum, latum, leopardinum, longibracteatum, luisae, meyeri, nanum, niveum, ovigerum, papillatum, philippii, polyandrum, puberulum, ramosum, semilunulum, tischeri, velutinum.

**SSect. c) Saxetana** SCHWANT. (Ser. *Saxetana* SCHWANT. of the SG. *Derenbergia* (SCHWANT.) SCHWANT.). – Forming dense to lax cushions, L.-remains fine to stout, papery; **Bo.** broad-cordate to oblong-obovoid to nearly fusiform, ± cuneately compressed, bilobed to sub-bilobed, keel-line and edges of the Fi. mostly sharp; ground-colour bright or grass-green to pale green, dotted or not, 1 R. of dots mostly running over the keel-line; **Cor.-segments** short, white, straw-coloured or brownish to wine-coloured, stigmas and style short. Night-flowering. – **Occurrence:** Cape: Lit. Namaqualand; S.W.Afr.: Gr. Namaqualand. – **Type species:** C. saxetanum. – **Species:** C. carpianum, densipunctum, exiguum, graessneri, halenbergense, hians, hirtum, intermedium, loeschianum, misellum, miserum, parvimarinum, quaesitum, rostratum, rubricarinatum, saxetanum, vescum.

**Sect. 2. Truncatum** (Sect. SCHWANT. emend. TISCH., SSer. *Truncatella* SCHWANT. of the Ser. *Carruicola* SCHWANT. of the SG. *Conophytum* pro part.; Ser. *Ficiformia* SCHWANT. of the SG. *Conophytum* pro part.; (? Sect.) *Subaphylla* HAW. pro part., Sect. *Sphaeroidea* SD. of the G. Mesembryanthemum L.). – Forming dense to lax cushions, L.-remains mostly ± stout, papery; **Bo.** obconical to pear-shaped, sometimes somewhat cordate to subbilobed above, epidermis glabrous or somewhat papillose; ground-colour ± bright grey-green, the sides occasionally flushed with reddish, with smaller or larger dots above, occasionally these coalescing into R.; **Cor.-segments** white, straw-coloured, pink to purplish, flesh- or wine-coloured, stigmas and style short. Night-flowering. – **Occurrence:** Cape: Van Rhynsdorp, Clanwilliam, Laingsburg, Worcester, Robertson, Montagu, Swellendam, Prince Albert, Ladismith, Willowmore, Uniondale, Oudtshoorn, Steytlerville D. – **Type species:** C. truncatum. – **Species:** C. calitzdorpense, catervum, ficiforme, koupense, morganii, multipunctatum, novellum, pardivisum, peersii, permaculatum, pisinnum, placitum, prolongatum, rooipanense, stegmanianum, steytlervillense, subglobosum, truncatum, uviforme, viride, viridicatum, wagneriorum, wiggettae.

**Sect. 3. Wettsteiniana** SCHWANT. (Ser. *Wettsteiniana* SCHWANT. of the SG. *Conophytum*). – Forming dense to lax cushions, L.-remains mostly papery, seldom parchment-like to nearly leathery; **Bo.** obconical, turbiniform or pear-shaped to nearly semiglobose, flat or convex above or shallowly concave, epidermis glabrous or papillose; ground-colour yellowish- to grey-green, sea-green or bluish-green, with or without conspicuous dots above; **Cor.-segments** white, pink to purplish or yellow, Fl. with a long style and short stigmas, or with a short style and short stigmas and with staminodes at the mouth of the tube. Day-flowering. – **Occurrence:** Cape: Lit. Namaqualand. – **Type species:** C. wettsteinii.

**SSect. a) Longistyla** TISCH. – Ground-colour bright or pale green to bluish-green; **Fl.** white, pink to purplish or yellow, stigmas short, style long, staminodes absent. – **Occurrence:** Cape: Lit. Namaqualand. – **Type species:** C. wettsteinii. – **Species:** C. avenantii, brevipes, circumpunctatum, doornense, ellipticum, flavum, fragile, fraternum, geyeri, globosum, gratum, inornatum, jacobsenianum, kubusanum, longistylum, luteolum, luteum, marlothii, maximum, middlemostii, nordenstamii, novicium, orbicum, ornatum, praecox, praegratum, rarum, ricardianum, robustum, rubristylum, ruschii, schlechteri, tetracarpum, vanbredai, wettsteinii.

**SSect. b) Minuta** (LITTLEW.) TISCH. (Ser. *Minuta* LITTLEW. of the SG. *Celatostamina* LITTLEW.; Ser. *Wettsteiniana* SCHWANT. of the SG. *Conophytum*). – Ground-colour bright to bluish-green; **Cor.-segments** pink to purplish, stigmas and style short, with staminodes at the mouth of the tube. – **Occurrence:** Cape: Lit. Namaqualand, Van Rhynsdorp D. – **Type species:** C. minutum. – **Species:** C. glabrum, microstoma, minutum, nudum, pearsonii, sellatum, tubatum.

**Sect. 4. Cataphracta** SCHWANT. (Ser. *Cataphracta* SCHWANT. of the SG. *Conophytum*). – Forming dwarf, dense to lax cushions, L.-remains brittle, papery to nearly leathery; **Bo.** obconical to nearly semiglobose, epidermis smooth, glabrous; ground-colour chalky green to whitish or bluish-green, without dots above or with ± distinct dots; **Cor.-segments** yellow, golden-yellow to ochre-coloured; stigmas and style short. Night-flowering. – **Occurrence:** Cape: Lit. Namaqualand, Kenhardt, Van Rhynsdorp, Calvinia D.; S.W.Afr.: Gr. Namaqualand. – **Type species:** C. calculus. – **Species:** C. aequatum, breve, calculus, globuliforme, johannis-winkleri, labiatum, membranaceum, minutiflorum, namiesicum, pageae, paucipunctum, pauperae, productum, pygmaeum, stevens-jonesianum, subrisum, subtile, thudichumii, udabibense, vanzijlii.

**Sect. 5. Colorata** TISCH. (SSer. *Picta* SCHWANT., *Piluliformia* TISCH. and *Tuberculata* SCHWANT. of the Ser. *Carruicola* SCHWANT. of the SG. *Conophytum*). – Forming dense to loose cushions, L.-remains papery; **Bo.** obconical, flat above, shallowly convex or somewhat grooved by the Fi. or crateriform, epidermis smooth; ground-colour pale to dark green, the sides often flushed with red; upper surface with solitary or ± confluent smaller or larger dots, in some species with somewhat prominent brownish-red to dark carmine coloured dots and lines; **Cor.-segments** white, creamy, wine- or flesh-coloured or pink to nearly carmine; stigmas and style short. Night-flowering. – **Occurrence:** Cape: Lit. Namaqualand, Van Rhynsdorp, Clanwilliam, Ceres, Piquetberg, Laingsburg, Montagu, Swellendam D. – **Type species:** C. pictum.

**SSect. a) Picta** SCHWANT. emend. TISCH. (SSer.: *Picta* SCHWANT. and *Piluliformia* TISCH. of the Ser. *Carruicola* SCHWANT. of the SG. *Conophytum*). – Form of **Bo.** mostly obconical, flat above to shallowly convex, dots and R. of dots and lines mostly narrow, dark green or brownish-red to carmine-coloured; **Cor.-segments** white, creamy or flesh- to copper-coloured. – **Occurrence:** Cape: Calvinia, Van Rhynsdorp, Ceres, Laingsburg, Clanwilliam, Montagu, Swellendam D. – **Type species:** C. pictum. – **Species:** C. advenum, albifissum, archeri, assimile, batesii, brevipetalum, complanatum, dispar, edwardii, joubertii, labyrintheum, leviculum, literatum, litorale, longitubum, minimum, muirii, obmetale, occultum, pauxillum, petraeum, pictum, piluliforme, polulum, praecinctum, pusillum, radiatum, rolfii, scitulum, signatum, subconfusum, subincanum, vagum, varians, wittebergense.

**SSect. b) Tuberculata** SCHWANT. (SSer. *Tuberculata* SCHWANT. of the Ser. *Carruicola* SCHWANT. of the SG. *Conophytum*). – **Bo.** obconical, flat above, shallowly convex or grooved or crateriform by the Fi., dots mostly larger than in **SSect. Picta**, often somewhat prominent, dark green to carmine; **Cor.-segments** white, creamy or pink to nearly purplish. – **Occurrence:** Cape: Calvinia, Van Rhynsdorp, Clanwilliam, Ceres, Piquetberg D. – **Type species:** C. obcordellum. – **Species:** C. brevilineatum, ceresianum, clarum, decoratum, divergens, fossulatum, franciscii, germanum, giftbergense, lambertense, longifissum, mundum, notatum, obconellum, obcordellum, parviflorum, praeparvum, rauhii, spectabile, stenandrum, stipitatum, ursprungianum.

**Sect. 6. Minuscula** SCHWANT. emend. TISCH. (Ser. *Minuscula* SCHWANT. pro part. of the SG. *Conophytum*). – Mostly forming dwarf and dense cushions, L.-remains papery; **Bo.** nearly cylindrical to oblong-obovoid, truncate, ± convex above, in some species short, rounded, bilobed, epidermis smooth or papillose; ground-colour olive to grey-green, partly with prominent dark green to brownish-carmine dots or lines above; **Cor.-segments** white, pink to purplish or brownish-reddish, tube long, stigmas and style short, with staminodes at the mouth of the tube. – **Occurrence:** Cape: Van Rhynsdorp, Calvinia, Clanwilliam, Ceres, Paarl D. – **Type species:** C. minusculum. – **Species:** C. bicarinatum, comptonii, edwardsiae, herrei, leipoldtii, luckhoffii, minusculum, rubrolineatum, swanepoelianum, turrigerum.

**Sect. 7. Costata** SCHWANT. emend. TISCH. (Ser. *Costata* SCHWANT., *Verrucosa* SCHWANT. pro part. and *Minuscula* SCHWANT. pro part. of the SG. *Conophytum*). – Mostly forming dense cushions, L.-remains mostly stout, papery to parchmenty; **Bo.** obconical to oblong-ovoid, flat or ± convex above, epidermis smooth; ground-colour bright green, bright grey-green or olive to nearly ochre-coloured, with small prominent dots above or ridged lines, the Fi. bordered with the same lines; **Cor.-segments** yellow, pink or ochre- to nearly flesh-coloured, style long or short, stigmas short. Day- and night-flowering. – **Occurrence:** Cape: Lit. Namaqualand, Kenhardt D.; S.W.Afr.: Gr. Namaqualand. – **Type species:** C. angelicae.

**SSect. a) Costifera** TISCH. (Ser. *Costata* SCHWANT. of the SG. *Conophytum*). – Forming dense cushions, L.-remains papery; **Bo.** obconical, flat to convex above; ground-colour olive to nearly ochre-coloured, with radiating ridged lines above, the Fi. bordered with the same lines; **Cor.-segments** short, ochre to nearly wine-coloured; stigmas and style short. Night-flowering. – **Occurrence:** S.W.Afr.: Gr. Namaqualand. – **Type species:** C. angelicae (sole species).

**SSect. b) Verrucosa** SCHWANT. emend. TISCH. (Ser. *Verrucosa* SCHWANT. pro part. and *Minuscula* SCHWANT. pro part. of the SG. *Conophytum*). – Forming dwarf cushions, L.-remains papery to parchmenty; **Bo.** obconical, ground-colour yellow-green, bright grey-green to olive, with smaller or larger, partly tuberculate dots or ridged lines above, the Fi. mostly bordered with the same lines, epidermis smooth or papillose; **Cor.-segments** yellow or pink; stigmas short, style long. Day-flowering. – **Occurrence:** Cape: Lit. Namaqualand, Kenhardt D. – **Type species:** C. fulleri. – **Species:** C. auriflorum, barbatum, clavatum, ectypum, fulleri, inductum, intrepidum, obscurum, pulchellum, sulcatum, vanheerdei, vitreopapillum.

**Sect. 8. Barbata** SCHWANT. (Ser. *Barbata* SCHWANT. of the SG. *Conophytum*). – Forming dwarf dense cushions; **Bo.** oblong-ovoid or obtuse-obconical, convex above, epidermis pilose; ground-colour bright green to olive, without conspicuous markings; **Cor.-tube** short, segments yellow to ochre-coloured, stigmas and style short. Night-flowering. – **Occurrence:** Cape: Lit. Namaqualand, Van Rhynsdorp D. – **Type species:** C. stephanii. – **Species:** C. depressum, fibuliforme, pubicalyx, stephanii.

**C. absimile** L. BOL. (§ II/1 a/1). – Cape: Lit. Namaqualand.
**C. — v. absimile.** – Up to 4.5 cm tall, laxly branched; **Bo.** 2–2.4 cm long, long-obovoid, up to 12 mm across, lobes mostly unequal, almost acute or rounded, surface minutely velvety-papillose; **Fl.** 21 mm ⌀, light yellow.
**C. — v. majus** L. BOL. – **Bo.** larger and broader.
**C. — v. umbrosum** L. BOL. – **Bo.** up to 5 cm long, lobes 15 mm long, almost equal; **Fl.** up to 3 cm ⌀.
**C. acutum** L. BOL. (§ I/2) (*Ophthalmophyllum a.* (L. BOL.) TISCH.). – Cape: Lit. Namaqualand. – **Bo.** obconical, 11 mm long, with a transparent tip 7–9 mm ⌀, 10 mm ⌀ midway, Fi. 3–4 mm long, surface grass-green; **Fl.** c. 25 mm ⌀, white.
**C. admiralii** L. BOL. (§ II/1 b). – Cape: Lit. Namaqualand. – Forming cushions 4.2 cm tall, 4 cm ⌀, juvenile **Br.** 15–20 mm long; L.-Sh. persisting; **Bo.** long-pyriform, 14 mm long, 5–7 mm thick, minutely papillose, olive-green, broadly elliptical above, very slightly carinate along the 2 mm long Fi. and the margins, with a few dark dots, densely hirsute; **Fl.** ? white. Acc. A. TISCHER a variety of **C. meyeri.**
**C. advenum** N. E. BR. (§ II/5 a) (*C. leightoniae* L. BOL.). – Cape: Montagu D. – **Bo.** obconical-ovoid, 5–7 mm tall, 4–6 mm across, 4–5 mm thick, flat-convex above, crenate, Fi. 1–2 mm long, surface smooth, greyish-green, with a transverse line of dark green or brown spots and solitary spots above, the latter sometimes merging into lines; **Fl.** 4–5 mm ⌀, yellowish-pink; or coppery-red outside, paler and more yellow inside, scented.
**C. aequale** L. BOL. (§ II/1 a/1). – Cape: Lit. Namaqualand. – Compact, 7 cm tall; **Br.** 5 mm ⌀, old L.-Sh. with brownish dots; **Bo.** broadly ovoid, 30–36 mm long, with the same distance between the lobe-tips, 30–34 mm thick, Fi. 7 mm across, lobes acute above, 10–12 mm long, acutely carinate on the underside, pale green, with a few green dots, the transparent section square in shape; **Fl.** 17 mm long, yellow.
**C. aequatum** L. BOL. (§ II/4). – Cape: Lit. Namaqualand. – **Bo.** broadly obovoid, somewhat oblique above, the upperside ± circular, flat or ± convex, 15–18 mm long, 13–17 mm across and thick, Fi. 3–5 mm long; **Fl.** pale yellow. Acc. A. TISCHER probably identical with **C. breve.**
**C. albescens** N. E. BR. (§ II/1 a/1). – Cape: Lit. Namaqualand. – Mat-forming; **Bo.** somewhat compressed laterally, lobes rounded, with an indentation 3–5 mm deep, 25–32 mm tall, 15–18 mm across, rather less in thickness, Fi. 4 mm long; surface light greyish-green, with minute white hairs and large dots; **Fl.** yellow.
**C. albifissum** TISCH. (§ II/5 a). – Origin unknown. – Acauline; **Bo.** obconical, c. 8 mm tall, 4–5 mm across, convex above, pale greyish-green, Fi. 1.5–2 mm long, with white hairs, bordered by a dark line; **Fl.** 5 mm ⌀, white.
**C. altum** L. BOL. (§ II/1 b). – Cape: Lit. Namaqualand.
**C. — v. altum.** – Forming lax cushions up to 9 cm tall; **Bo.** obcordate, somewhat compressed, tapered, 3 cm long, 12 mm thick above, rounded, bluish-green, lobes 5–6 mm high, with a red border; **Fl.** 18 mm ⌀, yellow.
**C. — v. plenum** L. BOL. – Pet. more numerous.
**C. ampliatum** L. BOL. (§ II/1 a/1). – Cape: Lit. Namaqualand. – Old L.-Sh. parchment-like; **Bo.** broadly obovoid to ± spherical or reniform, 18–30 mm tall, 10–13 mm across, lobes rounded, 3–5 mm high, margins and lobe-keels with a purple line, and a square translucent area on the sides; **Fl.** golden-yellow.
**C. andausanum** N. E. BR. (§ II/1 a/1). – Cape: Lit. Namaqualand.
**C. — v. andausanum.** – **Bo.** obovoid, Fi. depressed, forming 2 short lobes 10–15 mm high and across, 9–10 mm thick, dull green, with dark green dots, the lobe-keel and margins with purple or dark green lines; **Fl.** yellow.

**C. — v. immaculatum** L. Bol. – Unspotted variety.

**C. angelicae** (Dtr. et Schwant.) N. E. Br. (§ II/7a) (*Mes. a.* Dtr. et Schwant.). – S.W. Afr.: Gr. Namaqualand. – Forming a dense cushion 3–4 cm ⌀; **Bo.** obconical, **up** to 10 mm tall, circular but rather angular, **upper** ⌀ 6 to 8 mm, the Fi.-margin somewhat prominent, the Fi. 1.5 mm long, with radiating, slightly raised ribs, dirty olive-green to ± ochre-coloured, covered during the resting-period in a white, papery Sh.; **Fl.** 6–8 mm ⌀, light brownish.

**C. angustum** N. E. Br. (§ II/1a/1) (*C. subtenue* L. Bol.). – Cape: Lit. Namaqualand. – Forming lax cushions; **Bo.** ± cylindrical, bilobed and compressed above, 20–25 mm tall, 7–10 mm across, 6–7 mm thick, the lobes 5–8 mm long, 3–4 mm across, slightly carinate below the tip, the surface smooth, glossy, pale green, with distinct dots below and along the indentation and red lines along the lobe-edges; **Fl.** yellow.

**C. anjametae** de Boer (§ II/1a/2). – Cape: Bushmanland. – Cushion-forming; **Bo.** obconical, 2.5–3 cm tall, 1.5 cm ⌀, bluish-green, smooth, glabrous, the rounded, divaricate lobes 4–5 mm high, acutely keeled, the Fi. 4–5 mm deep, c. 5 mm across, with translucent spots along both sides, also with green or reddish dots, with similar dots on the lobes; **Fl.** 2–2.5 cm ⌀, violet to pink.

**C. anomalum** L. Bol. (§ II/1b). – Cape: Lit. Namaqualand. – Mat-forming; **Bo.** pyriform, 1–2 cm tall, 6–9 mm across, 7–10 mm thick, with pale brown lobes 0.5–1 mm high, lobe-tips dirty red, with ± square translucent areas on the sides; **Fl.** yellow.

**C. apiatum** (N. E. Br.) N. E. Br. (§ II/1a/1) (*Mes. a.* N. E. Br., *Derenbergia a.* (N. E. Br.) Schwant.). – Cape: Lit. Namaqualand. – **Bo.** 2–3 together, elongated, up to 4 cm tall, 25 mm across, 15–18 mm thick, the lobes c. 10 mm long, rounded, the surface slightly rough, whitish-green with dark green spots, the margins and tips of the lobes red-lined; **Fl.** 25 mm ⌀, yellow.

**C. approximatum** Lavis (§ II/1b). – Cape: Lit. Namaqualand. – Shrubby, with the **shoots** resting on the ground; **Bo.** up to 50 together, elongate, narrow-ovoid, laterally compressed, 2 cm tall, pale bluish without any recognizable markings, the dry membranes with brown spots, very persistent; **Fl.** 15 mm ⌀, canary-yellow.

**C. archeri** Lavis v. **archeri** (§ II/5a). – Cape: Montagu D. – Mat-forming; **Bo.** 20–25 together, partly concealed in the papery Sh., pyriform, broadly compressed, flat to convex, 19 mm tall, 9 mm across, c. 11 mm thick, with a Fi. 1.5 mm long, greyish-green, reddened on the sides; **Fl.** 5 mm ⌀, wine-coloured.

**C. — v. stayneri** L. Bol. – **Bo.** with a few dots above; **Fl.** white. Acc. A. Tischer a variety of **C. muirii**.

**C. areolatum** Littlew. (§ I/1). – Cape: Lit. Namaqualand. – **Bo.** often 5 together, cordate, oval above, slightly bilobed, the 2 lobes ± unequal, with the Fi. compressed, puberulous, (6–)8(–10) mm high, 6–7 mm thick, with small islands and impressed, pinkish-red dots, the surface otherwise pale brown; **Fl.** white.

**C. assimile** (N. E. Br.) N. E. Br. (§ II/5a) (*Mes. a.* N. E. Br.). – S.Afr.: origin unknown. – Mat-forming; **Bo.** obconical, 10–14 mm tall, 9–12 mm across, 6–9 mm thick, crenate above, ± convex, greyish-green, often purple on the sides, with a line of confluent spots on the lobes, elsewhere with other dark green lines and dots; **Fl.** 15–20 mm ⌀, cream-coloured.

**C. astylum** L. Bol. (§ I/1). – Cape: Van Rhynsdorp D. – **Bo.** elongated to obovoid, broadly elliptical to circular above, 8–10 mm tall, 4 to 5 mm across, 7–8 mm thick, with the Fi. extending over the whole width, brown to dirty green; **Fl.** pink, with no style present.

**C. auctum** N. E. Br. (§ II/2b). – Cape: Lit. Namaqualand. – **Bo.** ± spherical, flat to convex above, sometimes notched, 12–21 mm tall, 9 to 11 mm across, 7–9 mm thick, light glossy green; **Fl.** yellow.

**C. auriflorum** Tisch. (§ II/7b). – Cape: Lit. Namaqualand. – Mat-forming; **Bo.** conical to ± cylindrical, 10–12 mm tall, 4–5 mm across, 3–5 mm thick, circular to elliptical above, flat to convex, with a Fi. 1 mm long and surrounded by a darker zone, the surface dark green, slightly rough, with some dots, reddish on the sides and around the Fi.; **Fl.** 8–10 mm ⌀, golden-yellow.

**C. australe** L. Bol. (§ II/1b). – Cape: Van Rhynsdorp D. – Short shoots with Int. up to 2 mm long; **Bo.** obovoid, with lobes up to 1 mm high, broadly rounded above, 13–17 mm high, 6–8 mm across, with a ciliate Fi. 2–3 mm long; **Fl.** 12–17 mm ⌀, golden-yellow.

**C. avenantii** L. Bol. (§ II/3a). – Cape: Lit. Namaqualand. – Forming a cushion c. 4 cm ⌀, old L.-Sh. persisting and parchment-like, deep brown, with dark dots; **Bo.** broadly obovoid, c. 13 mm tall, 14 mm across above, the upper surface flat to ± convex, with a Fi. 1.5–3 mm long and indistinct, dark green dots; **Fl.** large, an attractive pink.

**C. barbatum** L. Bol. (§ II/7b). – Cape: Lit. Namaqualand. – Plant very small; **Bo.** few together, somewhat clavate, up to 6 mm tall, up to 4.5 mm ⌀ above, the upper surface circular, ± flat, with a fairly deep Fi. 2 mm long, bluish-green, minutely papillose, with the papillae along the Fi. longer, incurved-barbate; **Fl.** c. 2 cm ⌀, pink.

**C. batesii** N. E. Br. (§ II/5a). – Cape: Origin unknown. – Forming small mats; **Bo.** usually 3 on one shoot, the central one larger, obconical, 8–10 mm tall, the upper surface circular, 4 mm ⌀, with a Fi. 0.5–1 mm long, greyish-green, reddish below, with spots above in ± decurrent lines; **Fl.** cream-coloured.

**C. bicarinatum** L. Bol. (§ II/6). – Cape: Ceres D. – Forming cushions 6 cm ⌀; **Bo.** densely crowded, enveloped below in the papery Sh., 12–20 mm tall, 9–13 mm ⌀ above, the lobes acute, 4–5 mm high, compressed, with red borders, often with 2 keels, the surface bluish-green, minutely puberulous; **Fl.** light pink, white outside.

**C. bilobum** (MARL.) N. E. BR. (Pl. 153/1) (§ II/1a/1) (*Mes. b.* MARL., *Derenbergia b.* (MARL.) SCHWANT., *C. exsertum* N. E. BR.). – Cape: Lit. Namaqualand. – Branching in age and forming a mat; **Bo.** somewhat flattened and compressed, cordate, 34–50 mm tall, 20 to 25 mm across, rather less in thickness, the lobe-tips obtuse to rounded, the indentation 7–9 mm deep, the lobe-tips and keels with a red border, the surface greyish-green to whitish-green, slightly papillose; **Fl.** up to 3 cm ⌀, yellow.

**C. blandum** L. BOL. (§ II/1a/2). – Cape: Bushmanland. – Cushion-forming; **Bo.** elongate expanded above, 17–27 mm tall, 6–7 mm across below, 11–14 mm thick above, the lobes compressed-rounded, 4–9 mm high, surface bluish-green, minutely papillose; **Fl.** up to 2 cm ⌀, light pink.

**C. boreale** L. BOL. (§ I/1). – Cape: Bushmanland. – **Bo.** elongate, bilobed, 7–10 mm tall, up to 6 mm across, 9 mm thick at the tip, lobes up to 2 mm high, Wi. scarcely perceptible, Fi. across the entire width; **Fl.** ? white.

**C. breve** N. E. BR. (§ II/4). – Cape: Lit. Namaqualand.

**C. —** v. **breve** (*C. pygmaeum* SCHICK et TISCH.). – **Bo.** obconical, circular, 5–7 mm tall, 5–7 mm ⌀, flat or somewhat convex above, with a Fi. 2–3 mm long, surface smooth, pale bluish-green, with dark dots; **Fl.** 7–8 mm ⌀, yellow.

**C. —** v. **minus** L. BOL. – **Bo.** smaller; **Fl.** up to 10 mm ⌀, ochre-yellow.

**C. brevilineatum** TISCH. (§ II/5b). – Cape: ? Ceres D. – Forming dense cushions; **Bo.** obconical, flat to flat-convex above, with the Fi. scarcely depressed, up to 12 mm tall, up to 10 mm ⌀, surface whitish to greyish-green, the sides reddish, with ± short, dark reddish to brown, radiating lines, branched at the ends, on the upper surface near the margin, interspersed with small dots towards the Fi.; **Fl.** 15 mm ⌀, white to milk-coloured.

**C. brevipes** L. BOL. (§ II/3a). – Cape: Lit. Namaqualand. – Forming hemispherical cushions 10 cm ⌀, up to 7.5 cm high, with the leathery, reddish-brown L.-Sh. persisting; **Bo.** broadly obovoid, circular above, margins and Fi. slightly prominent, 10–13 mm tall, 15 mm ⌀, Fi. 2–3 mm long, surface pale bluish-green with green dots, with the Fi.-zone darker; **Fl.** purplish-pink.

**C. brevipetalum** LAVIS (§ II/5a). – Cape: ? Lit. Karroo. – Dwarf; **Bo.** 7–8 together, depressed-conical, ± flat to slightly convex and 6–7 mm ⌀ above, 11 mm tall, green, conspicuously spotted; **Fl.** reddish.

**C. brevisectum** L. BOL. (§ II/1a/1). – Cape: Lit. Namaqualand. – **Shoots** 6–6.5 cm long, covered with dry parchment-like, rough, pale brown L.-Sh.; **Bo.** obovoid, 18–30 mm tall, 12–17 mm thick at the tip, 6–10 mm across at the Fi. and 8 mm across the Sh. midway, with lobes 1 cm long and 5 mm across below, vesicular with a ± square Wi.; **Fl.** with a Cor. c. 2 cm long, segments golden-yellow, 5–10 mm long.

**C. burgeri** L. BOL. (§ I/2). – Cape: Lit. Namaqualand. – **Bo.** solitary, spherical-conical, 13–18 mm tall, 2 cm ⌀, less above, with a Fi. 3–4 mm long, glossy green, somewhat windowed above; **Fl.**-tube white, with segments purplish-pink above.

**C. calculus** (BGR.) N. E. BR. v. **calculus** (§ II/4) (*Mes. calculus* BGR.). – Cape: Van Rhynsdorp D. – Forming a mat up to 15 cm ⌀; **Bo.** laterally flattened, 16–22 mm tall, rather more across, with a Fi. 3–5 mm long, the surface chalky greyish-green; **Fl.** 12 mm ⌀, Pet. deep yellow with brownish tips.

**C. —** v. **komkansicum** (L. BOL.) RAWÉ (*C. k.* L. BOL.). – Lit. Namaqualand, Komkans. – **Bo.** broadly ovoid, with the upperside flat to slightly convex, lips of the Fi. not thickened.

**C. —** v. **protusum** L. BOL. – **Bo.** smaller than in v. **calculus**, Fi. protuberant, red.

**C. calitzdorpense** L. BOL. (§ II/2) (*C. c.* TISCH.). – Cape: Ladismith D. – **Bo.** numerous, forming clumps, obconical, flat or slightly convex above, 10–15 mm tall, 5–7 mm ⌀ above, the surface greyish-green, with reddish sides and small, dark green dots above, with a Fi. 2 mm long, gaping only slightly; **Fl.** up to 25 mm ⌀, flesh-coloured.

**C. candidum** L. BOL. (§ II/1b). – Cape: Lit. Namaqualand. – Forming cushions 7 cm tall, 10 cm ⌀; **shoots** c. 5 cm long, covered with the remains of dead L.; **Bo.** c. 100 together, pyriform, 12–14 mm tall, 5–7 mm across, 7–9 mm thick, with rounded lobes and a Fi. c. 4 mm long surrounded by a ± square, translucent zone, the surface bluish-green, with minute, papillose hairs; **Fl.** white.

**C. carpianum** L. BOL. (§ II/1c). – Cape: Lit. Namaqualand. – Forming compact mats 1 to 5 cm high, with the old L.-Sh. persisting; **Bo.** ± cylindrical to ovoid to ± square, 6–10 mm tall, 3–4 mm across, 4–7 mm thick, with rounded lobes up to 1 mm high and a Fi. 1.5 mm long, light green, with a few dots; **Fl.** white.

**C. cateryum** (N. E. BR.) N. E. BR. (§ II/2) (*Mes. c.* N. E. BR.). – Cape: Laingsburg D. – Forming ± spherical mats; **Bo.** obconical, ± compressed laterally, almost 2 cm tall, 10 to 12 mm across, 8–10 mm thick, flat to convex above, with a Fi. 4–5 mm long and 1 mm deep, the surface light greyish-green, reddish below, with dots in branched R. above, and scattered dots elsewhere; **Fl.** yellowish-white.

**C. cauliferum** (N. E. BR.) N. E. BR. (§ II/1a/1). – Cape: Lit. Namaqualand.

**C. —** v. **cauliferum** (*Mes. c.* N. E. BR., *Derenbergia c.* (N. E. BR.) SCHWANT.). – Small ℏ with ascending **shoots** 2–3 cm long; **Bo.** 2–3 cm tall, c. 2 cm across, 2 cm thick in the lower ± round part, crenate above, with lobes 3–8 mm long, 7–8 mm thick, with a Fi. 6 mm long, the surface dark green, with indistinct, light dots, the lobe-tips with red spots; **Fl.** yellow to orange.

**C. —** v. **lekkersingense** L. BOL. – Up to 15 cm tall, caulescent in age; **Bo.** 3–3.5 cm tall, 19 to 32 mm ⌀ midway, lobes 1–2 cm high, surface unspotted, sea-green; **Fl.** up to 27 mm ⌀, yellow.

**C. ceresianum** L. Bol. (§ II/5b). – Cape: Ceres D. – **Bo.** densely crowded, narrowly obovoid to obconical, rounded to oval above, 8–10 mm tall, 6–7 mm ⌀, with a Fi. 2–3 mm long, the surface brown to brownish-green, indistinctly puberulous, with 2–3 translucent dots near the Fi. and several spots in transverse lines; Pet. cream-coloured, with brownish-reddish tips.

**C. christiansenianum** L. Bol. (Pl. 153/2) (§ II/1a/1). – Cape: Lit. Namaqualand. – Forming cushions 5–7 cm tall, up to 8 cm ⌀; **Bo.** with lateral shoots from the base, obovoid, 3.5–5 cm tall, 22–28 mm wide above, bilobed above, the lobes c. 5 mm high, acute, with a shortly-haired Fi. 7–12 mm deep with an indistinct blister, the surface rough, bluish-green to pink; **Fl.** yellow, large.

**C. circumpunctatum** Schick et Tisch. (Pl. 153/5) (§ II/3a). – Cape: Lit. Namaqualand. – Mat-forming; **Bo.** conical, circular above, flat, 15 mm tall, 8–12 mm ⌀, with a somewhat depressed Fi. 2–4 mm long surrounded by dots, the surface dark bluish-green, with several dark dots above; **Fl.** 15 mm ⌀, dark red to mauve.

**C. citrinum** L. Bol. (§ II/1a/1). – Cape: Lit. Namaqualand. – Mat-forming; **Bo.** round-ovoid to long-ovoid, 20–24 mm tall, 20 mm across above, lobes 4–5 mm long, with a red-bordered keel, the surface bluish-green, minutely velvety-papillose, with solitary red dots; **Fl.** 15 mm ⌀, lemon-yellow.

**C. clarum** N. E. Br. (§ II/5b). – Cape: Calvinia D. – **Bo.** obconical, broadly rounded or elliptical above, 6–7 mm tall, 4–5 mm across, 4 mm thick, with a Fi. 1–2 mm long, depressed and slightly compressed, the surface light green, with 20–30 dark green dots and a dark green spot and several lines on both sides of the Fi.; **Fl.** ? white.

**C. clavatum** L. Bol. (§ II/7b). – Cape: Lit. Namaqualand. – Slender, with the old L.-Sh. persisting; **Bo.** clavate, rounded above, convex, glossy, deep green, with raised, ± tuberculate dots, some of them forming a border around the Fi.; **Fl.** pink.

**C. complanatum** L. Bol. (§ II/5a). – Cape: Prince Albert D. – **Bo.** obovoid to clavate, cylindrical below, ± circular above, compressed, 7–11 mm tall, 6 mm across above, 7 mm ⌀, with the old, dark brown L.-Sh. persisting, with a slightly compressed Fi. 1.5 mm long and slightly prominent ribs above, spotted; **Fl.** 8 mm ⌀, yellow.

**C. compressum** N. E. Br. (§ II/1a/1) (Pl. 153/4). – Cape: Alexander Bay. – Caulescent in age; **Bo.** cuneate-elongate, bilobed above, up to 5 cm tall, 22 mm across, 7–8 mm thick, the lobes 14–16 mm long, acute-angled, with the underside carinate, laterally compressed, with a R. of red dots along the keel, the surface light greyish-green, with inconspicuous, dense, minute dots and with a dark spot on the side of the indentation; **Fl.** 2 cm ⌀, yellow.

**C. comptonii** N. E. Br. (§ II/6). – Cape: Calvinia D. – Forming small cushions; **Bo.** elongated, often flat-truncate, 6–7 mm tall, 4 mm ⌀ above, bluish-green, tinged brownish, with red, branched lines above and a slightly prominent Fi. 1 mm long; **Fl.** orange-yellow.

**C. concavum** L. Bol. (Pl. 153/3) (§ I/2). – Cape: Lit. Namaqualand. – Mat-forming; **Bo.** 6–8 together, ± obconical, flat-concave above, soft, very succulent, with short, velvety, white hairs, 24–35 mm long, 15 mm across, 19–21 mm thick with a compressed, hirsute Fi. 8 mm long, the surface translucent, light green, the sides purple; **Fl.** 17 mm ⌀, white.

**C. connatum** L. Bol. (§ II/1a/2). – Cape: Lit. Namaqualand; S.W.Afr.: Witpütz. – Mat-forming; **Bo.** on short shoots, ± pyriform, rather truncate above, 14–20 mm tall, 13 mm across above, with lobes scarcely 3 mm long, acute, red-tipped, the surface bluish to yellowish-green, minutely papillose; **Fl.** 12 mm ⌀, yellow.

**C. conradii** L. Bol. (§ II/1a/1). – Cape: Lit. Namaqualand. – Compact; **Bo.** ± elongated, 24–34 mm tall, 9 mm across the Sh. midway, 15 mm thick, with red-bordered lobes up to 12 mm high, the surface very minutely velvety; **Fl.** yellow.

**C. corculum** Schwant. (§ II/1b). – Cape: Lit. Namaqualand. – Mat-forming; **Bo.** cylindrical to cordate, 15–20 mm tall, 6–10 mm across, with lobes 2–4 mm high, with ± reddish angles, the surface yellowish-green, with scarcely perceptible, stiff hairs and with scattered dots; **Fl.** 8 mm ⌀, yellow.

**C. cordatum** Schick et Tisch. (§ II/1a/2). – Cape: Lit. Namaqualand.

**C.** — v. **cordatum** (*C. convexum* L. Bol.). – **Bo.** cordate, 20–25 mm tall, 12–15 mm across, 8–10 mm thick, rounded above, with a Fi. 2–3 mm long, somewhat spotted along both sides, the surface whitish-grey, ± reddish above; **Fl.** 12 mm ⌀, yellow.

**C.** — v. **macrostigma** L. Bol. (*C. m.* (L. Bol.) Schwant.). – **Fl.**-segments and stigmas larger.

**C. coriaceum** L. Bol. (§ II/1a/1). – Cape: Lit. Namaqualand. – Clumps up to 8 cm tall, 11 cm ⌀; **Bo.** obovoid, 20–27 mm tall, 10 to 17 mm across, with lobes 9–15 mm long, rounded to ± tapered above, with purple lobe-margins and keel, the surface very short-papillose; **Fl.** 2.5 cm ⌀, golden-yellow.

**C. corniferum** Schick et Tisch. (§ II/1a/2). – Cape: Lit. Namaqualand. – Mat-forming, caulescent in age; **Bo.** cylindrical below, rather thin, clavate-thickened and cuneately depressed above midway, 15–20 mm tall, 6–8 mm across, with lobes 3–4 mm long, rounded above, with a Fi. 2–3 mm long, the surface dull green, with only 2–3 darker spots below the Fi.; **Fl.** yellow.

**C. crassum** L. Bol. (§ II/1a/1). – Cape: Lit. Namaqualand. – Caulescent, branched; **Bo.** ± long-ovoid, 6–6.5 cm tall, 4.5 cm across midway, bilobed above, the lobes ± truncate or sinuate, the surface sea-green, minutely papillose; **Fl.** up to 18 mm ⌀, yellow.

**C. craterulum** Tisch. (§ II/1b). – Cape: Lit. Namaqualand. – Mat-forming; **Bo.** crowded, ± spherical to apple-shaped, up to 25 mm tall,

20 mm across, 18 mm thick, slightly compressed towards the 3.5 mm long, impressed, rhombic, Fi. which is mostly sunken and crater-like, the surface grey to sea-green, with dense, white hairs; Fl. 12–15 mm ⌀, purplish-crimson.

**C. cupreatum** TISCH. (§ I/1). – Cape: S.Lit. Namaqualand. – Mat-forming; **Bo.** obconical to ± cylindrical, round or slightly oval above, 10–15 mm tall, 6–10 mm ⌀, with a dark green, somewhat depressed Wi. on the upperside and a Fi. 0.5 mm long surrounded by green dots, the surface coppery-brown; **Fl.** 12–15 mm ⌀, white.

**C. cupreiflorum** TISCH. (§ II/1 a/1). – Cape: Bushmanland. – Cushion-forming; **Bo.** laxly arranged, ± cordate, somewhat bilobed above, 22 mm tall, 15 mm across, 12 mm thick, the lobes 4 mm long, with a slightly rounded keel, with several spots in lax R. along the keel-line, the surface dark green with ± large spots, with minute, white dots distributed over the entire Bo.; **Fl.** scarcely opening, Pet. golden-yellow, with deep orange to copper-coloured tips.

**C. curtum** L. BOL. (§ II/1 a/1). – Cape: Lit. Namaqualand. – **Bo.** long-obovoid, 4–5 cm tall, 18–20 mm ⌀ midway, with short-rounded lobes 9–11 mm ⌀ and a red keel-line and broken red lines along the margins, somewhat papillose around the Fi., the surface otherwise smooth; **Fl.** 19 mm ⌀, light yellow.

**C. cylindratum** SCHWANT. – Cape: Lit. Namaqualand. – Mat-forming; **Bo.** long-cylincal, up to 25 mm tall, 10 mm across, hemispherical to convex above, with a Fi. 1.5 mm long, the surface yellowish-green; **Fl.** white.

**C. decoratum** N. E. BR. (§ II/5 b). – Cape: Van Rhynsdorp D. – **Bo.** obconical, 8 mm tall, 7 mm ⌀ above, the upper surface with a branched line perpendicular to the sunken, dark red-bordered Fi., the surface green to reddish, with several ± prominent lines interspersed with solitary, dark green dots; **Fl.** white.

**C. densipunctatum** (TISCH.) L. BOL. (§ II/1 c) (*Derenbergia d.* TISCH.). – S.W.Afr.: Karas Mts. – **Bo.** bilobed above, 17 mm tall, 12 mm ⌀, the lobes laterally compressed and acutely carinate, the surface greyish-green, with dense, small, dark green dots; **Fl.** 10 mm ⌀, white.

**C. depressum** LAVIS (§ II/8). – Cape: Lit. Namaqualand. – **Bo.** several together, 4 mm tall, c. 6 mm across, 5–8.5 mm ⌀, with a Fi. 1–1.5 mm long, the surface earth-coloured, slightly reddish-brown, with white dots and soft hairs; **Fl.** 4–5 mm ⌀, yellowish-pink.

**C. difforme** L. BOL. (§ II/1 a/2). – Cape: Lit. Namaqualand. – Mat-forming, each shoot with several Bo. projecting from the L.-Sh.; **Bo.** ovoid-cordate, bilobed, 18–20 mm tall, c. 13 mm across above, the lobes compressed, 5 mm long, with purple keel and margins, the surface green, with 5 red dots on each side; **Fl.** c. 15 mm ⌀, yellow.

**C. dilatatum** TISCH. (§ II/1 a/1). – Cape: Lit. Namaqualand. – Dwarf, laxly branched ♄ up to 7 cm tall, with the dry L.-Sh. papery and light brown; **Bo.** obovoid, up to 22 mm tall, laterally compressed, bilobed above, 12 mm across, 8 mm thick, the lobes 3 mm long,

acutely angled, mucronate, with a rounded keel, the Fi. in a ± V-shaped notch, with whitish hairs, the Fi.-zone small and indistinct, the Fi.-borders and lobe-tips purple, the surface rough-papillose, whitish or bluish-green to sea-green; **Fl.** c. 2 cm ⌀, light yellow.

**C. dispar** N. E. BR. (§ II/5 a). – S.Afr.: origin unknown. – **Bo.** variously shaped, obconical, ± circular above, flat, with a distinct ridge, 4–12 mm tall, 4–10 mm ⌀, the Fi. compressed, 2–3 mm long, with lateral pits, surrounded by dots arranged in lines, the surface green above, reddened on the sides, with numerous dark green dots; **Fl.** 10–12 mm ⌀, cream-coloured.

**C. dissimile** L. BOL. (§ II/1 a/1). – Cape: Lit. Namaqualand. – Caulescent in age; **Bo.** ± cordate, 25–48 mm tall, up to 26 mm across midway, the lobes 8–23 mm high, 5–12 mm across, compressed, truncate above, the Fi. slightly swollen, with short hairs, the surface rough-papillose, sea-green; **Fl.** 18–24 mm ⌀, light yellow.

**C. distans** L. BOL. (§ II/1 a/1). – Cape: Lit. Namaqualand. – **Bo.** broadly obovoid, compressed, 4–5 cm tall, c. 4 cm ⌀, with lobes 15–20 mm long, the translucent parts of the sides narrowly cuneate to ± square or oblong, the surface rough; **Fl.** yellow. Acc. A. TISCHER possibly identical with **C. simplum**.

**C. divaricatum** N. E. BR. (§ II/1 a/1). – Cape: Lit. Namaqualand. – **Bo.** compressed, cuneate below, bilobed above, 25–30 mm tall, 10 to 25 mm across, 5–10 mm thick, the lobes rounded on the underside, with reddish or green-spotted keels, the surface green to greyish-green, with ± scattered dots; **Fl.** yellow.

**C. divergens** L. BOL. (§ II/5 b). – Cape: Ceres D. – **Bo.** numerous, obconical, circular to oval above, flat or slightly convex, 10–13 mm tall, 7–9 mm across. c. 10 mm thick, the surface olive-green, with irregular, translucent spots, more crowded along the edges; **Fl.** deep crimson.

**C. diversum** N. E. BR. (§ II/1 a/2). – Cape: Lit. Namaqualand. – **Bo.** cordate, very compressed, rounded-carinate above, 20–38 mm tall, up to 18 mm across, 10 mm thick, with lobes 4–8 mm long, the surface greyish-green, with isolated, dark spots; **Fl.** light yellow.

**C. dolomiticum** TISCH. (§ II/1 a/1). – Cape: Lit. Namaqualand. – Forming lax mats up to 7 cm tall, 6 cm ⌀, with dry, leathery, brown L.-Sh.; **Bo.** long-ovoid, much compressed above, cylindrical below, 3–4.5 cm tall, 2.5 cm across midway, 12 mm thick, bilobed with lobes up to 2 cm long, mostly unequal, with the keel-angle rounded, acute or rounded above, sometimes oblique, the angles acute, the Fi.-zone indistinct, the surface bluish to greyish-green, rough to papillose-hairy; **Fl.** 2 cm ⌀, golden-yellow.

**C. doornense** N. E. BR. (§ II/3 a). – Cape: precise source unknown. – Mat-forming; **Bo.** enclosed in the papery, white Sh., 12 mm tall, flat-convex above, 8 mm ⌀, with a Fi. 2 mm long, the surface bluish to greyish-green, with slightly darker dots; **Fl.** 8–10 mm ⌀, pink.

**C. ecarinatum** L. Bol. v. **ecarinatum** (§ II/1b). – Cape: Lit. Namaqualand. – Cushion-forming; **Bo.** obovoid to long-obovoid, bilobed, 22 to 28 mm long, 11 mm across, lobes 4–5 mm high, obtuse to subacute, red-tipped, separated by an indentation 8 mm deep, the Fi. 7 mm long, the lateral, translucent parts ± square, the surface pale bluish-green, with a few dots; **Fl.** white.

**C.** — v. **angustum** L. Bol. – **Bo.** narrower, 15–20 mm tall, c. 9 mm across, 10–18 mm ⌀ above.

**C.** — v. **mutabile** L. Bol. – **Bo.** with obtusely keeled lobes.

**C. ectypum** N. E. Br. (§ II/7b). – Cape: Lit. Namaqualand.

**C.** — v. **brownii** (Tisch.) Tisch. (*C. brownii* Tisch., *C. virens* L. Bol.). – **Bo.** as for v. **ectypum**, reddish to brownish-green, with the ribs more numerous and deep reddish-brown; **Fl.** whitish-pink.

**C.** — v. **ectypum** (Pl. 154/1). – Mat-forming; **Bo.** crowded, obovoid, deeply wrapped in the papery Sh. of the old Bo., 5–10 mm tall, 4 to 6 mm ⌀, circular or elliptical above, with a slightly compressed Fi. 1–2 mm long surrounded by a raised line, with further lines radiating from this, running over the upper surface and down the sides, interspersed with dark green dots, the surface light green to yellowish-green; **Fl.** 8–10 mm ⌀, pink.

**C.** — v. **limbatum** (N. E. Br.) Tisch. (*C. l.* N. E. Br., *C. chloratum* Tisch.). – With an inconspicuous line perpendicular to the Fi. and no other lines, the dots and the line less prominent, the surface light green to yellowish-green; **Fl.** deep pinkish-purple.

**C.** — v. **tischleri** (Schwant.) Tisch. (*C. t.* Schwant.). – As v. **ectypum**; **Fl.** larger, brilliant yellow.

**C. edwardii** Schwant. (§ II/5a) (*C. rubrum* L. Bol.). – Cape: Lit. Namaqualand. – Dwarf, with many shoots, the dry L.-Sh. pale brown; **Bo.** numerous, long-elliptical, 7–10 mm tall, slightly compressed above, 2–4 mm across below, almost circular above, 2–3 mm ⌀, tapered to convex, with the Fi. sunken, 0.5 mm deep, surrounded by hairs and with a few dots, the surface otherwise glabrous, vivid green; **Fl.** 6–8 mm ⌀, red.

**C. edwardsiae** Lavis v. **edwardsiae** (§ II/6). – Cape: Clanwilliam D. – Mat-forming; **Bo.** ± cordate, 12 mm tall, 11 mm ⌀, with a Fi. 1.5 mm long, c. 5 mm deep, the lobes with a line of red dots, the surface bluish, scabrous; **Fl.** 8 mm ⌀, light red.

**C.** — v. **albiflora** Rawé. – Vredenburg D. – Surface smooth, lobes rounded to tapered, acutely carinate above; **Fl.** white.

**C. elishae** (N. E. Br.) N. E. Br. (§ II/1a/1) (*Mes. e.* N. E. Br., *Derenbergia e.* (N. E. Br.) Schwant., *C. dennisii* N. E. Br., *C. nelianum* Schwant., *C. plenum* N. E. Br., *C. springbokense* N. E. Br.). – Cape: Lit. Namaqualand. – Mat-forming; **Bo.** obconical to ± cuneate, with an acute ridge above, up to 25 mm tall, 14 mm across, 10–12 mm thick, the Fi. 5–6 mm deep, thus forming distinct lobes, the sides of the indentation with dark, ± translucent spots, the ridge reddish, the surface green to somewhat bluish with dark dots, minutely rough; **Fl.** 2 cm ⌀, golden-yellow.

**C. ellipticum** Tisch. (§ II/3a). – Cape: Lit. Namaqualand. – Forming small, very dense cushions; **Bo.** obconical, 15 mm tall, 12 mm ⌀ above, somewhat constricted below the circular to elliptical, rather convex upper surface, the Fi. slightly compressed, c. 4 mm long, somewhat papillose, set in a darker zone, the surface dark green to greyish-green, with dark green dots above; **Fl.** c. 25 mm ⌀, yellow.

**C. exiguum** N. E. Br. (§ II/1c). – Cape: Lit. Namaqualand. – **Bo.** cuneate-cordate, compressed, 8–10 mm tall, 6–8 mm across, with the lobes very short and rounded, separated by an indentation 1–1.5 mm deep, the surface light green with some dots; **Fl.** ? white.

**C. excisum** L. Bol. (§ II/1a/1). – Cape: Lit. Namaqualand. – **Bo.** oval to broadly oval, up to 2.3 cm tall, up to 2 cm ⌀ above, with subacute lobes 4–5 mm long and the Fi. set in a ± square zone, the surface short-papillose; **Fl.** 17–19 mm ⌀ golden-yellow.

**C. extractum** Tisch. (§ II/1a/2). – Cape: Lit. Namaqualand. – Mat-forming; **Bo.** cordate, 18 mm tall, 12 mm across, 9 mm thick, compressed, bilobed, the lobes rounded, 2 mm high, chin-like, carinate, with large, dark green spots along the keel-line, the Fi. rhombic, 4 mm long, hirsute, the surface greyish-green, papillose on the upperside; **Fl.** c. 16 mm ⌀, yellow.

**C. fenestratum** Schwant. (Pl. 154/2) (§ I/1). – Cape: Bushmanland. – Close to **C. pellucidum**; the upper surface more flattened and less wrinkled, dark to olive to brownish-green; **Fl.** a magnificent pinkish-purple.

**C. fibuliforme** (Haw.) N. E. Br. (§ II/8) (*Mes. f.* Haw., *Mes. fibulaeforme* Haw. sphalm., *Mes. bolusiae* Schwant. – Cape: Lit. Namaqualand. – Mat-forming; **Bo.** short-obconical, 1 cm tall, 7 mm across, the upper surface circular, slightly convex, the Fi. up to 1.5 mm long, densely haired, bordered by a dark line, the surface greyish-green to mouse-grey, minutely velvety-haired; **Fl.** almost 2 cm ⌀, Pet. white, the upperside pink with a purplish-red stripe.

**C. ficiforme** (Haw.) N. E. Br. (Pl. 154/3) (§ II/2) (*Mes. f.* Haw., *Mes. odoratum* N. E. Br., *C. o.* (N. E. Br.) N. E. Br., *Mes. jugiferum* N. E. Br., *Mes. altile* N. E. Br., *C. a.* (N. E. Br.) N. E. Br., *Mes. pallidum* N. E. Br., *C. p.* (N. E. Br.) N. E. Br.). – Cape: Worcester, Robertson, Montagu D. – Forming compact mats; **Bo.** cordate, obovoid, fig-shaped, often truncate, mostly convex above, with a slight ridge perpendicularly to the Fi., 15–20 mm tall and across, with the variously coloured, dry Sh. persisting, the surface reddish-green on the sides, often with translucent green dots, those on the upper surface arranged in lines and interspersed with red dots, with a translucent zone of confluent dots on both sides of the Fi.; **Fl.** c. 25 mm ⌀, white to pink.

**C. fimbriatum** (Sond.) N. E. Br. (*Mes. f.* Sond.). – Insufficiently known plant; the material collected by Zeyher and Burke under this name is **Ruschia pygmaea** (Haw.) Schwant.

**C. flavum** N. E. Br. (Pl. 154/4) (§ II/3a) (*C. concinnum* Schwant.). – Cape: Lit. Namaqualand. – Mat-forming; **Bo.** obconical, 18–20 mm tall, 6–12 mm ⌀, flat or slightly convex on the upperside, with a short Fi. bordered by a dark green line, the surface green with a few dark green spots; **Fl.** 12–16 mm ⌀, yellow.

**C. fossulatum** Tisch. (§ II/5b). – Cape: origin unknown. – **Bo.** obconical, 15–20 mm tall, circular or elliptical above, 8–10 mm across, 6–9 mm thick, the Fi. hirsute, 2–3 mm long, depressed, so producing 2 lobes, each with a dark green or a brown line, the surface greyish-green, minutely papillose, with several dots on the upper surface; **Fl.** 15 mm ⌀, straw-coloured.

**C. fragile** Tisch. (§ II/3a). – Cape: Lit. Namaqualand. – Mat-forming, with the Bo. densely arranged, the shoots up to 6 cm long in age and easily broken off; **Bo.** obconical, circular and flat above, 18 mm tall, 8 mm ⌀, the Fi. rhombic. 2 mm long, with white hairs, the surface greyish-green to sea-green with dark green spots; **Pet.** translucent pink.

**C. franciscii** L. Bol. (§ II/5b). – Cape: Calvinia D. – Forming mats c. 17 mm high, 2.5 cm ⌀, with the thin, whitish L.Sh. persisting; **Bo.** pyriform, 6–10 mm tall, up to 4 mm ⌀, convex above, the lips of the Fi. ± jaw-like, densely hirsute, the surface bluish-green, with dirty green spots often arranged in 3–5 lines; **Pet.** white below, pale pink above.

**C. fraternum** (N. E. Br.) N. E. Br. (§ II/3a) (*Mes. f.* N. E. Br., *C. leptanthum* L. Bol., *C. f. v. l.* L. Bol.). – Cape: Lit. Namaqualand. – Close to *C.* **marlothii**; forming dense cushions, with thick **shoots** 2–5 cm long, densely covered with old, dry L.-Sh.; **Bo.** numerous, obconical, ± flat on the upperside, mostly circular, with the Fi. not depressed, the surface dull to light green, papillose, with scattered, dark green dots on the upper surface; **Fl.** white to pink.

**C. frutescens** Schwant. (Pl. 155/1) (§ II/1a/1) (*C. salmonicolor* L. Bol., *C. teguliflorum* Tisch.). – Cape: Lit. Namaqualand. – ♄ up to 10 cm tall, with ascending shoots, Int. 10–12 mm long; Bo. 3 cm tall, 15–25 mm across, deeply incised above, with the lobes 8 mm long, with the lobe-margins acutely carinate, the surface dark green with light spots; **Fl.** 25 mm ⌀, deep orange-yellow.

**C. fulleri** L. Bol. (§ II/7b) (*C. wiesemannianum* Schwant.). – Cape: Kenhardt D. – Mat-forming; **Bo.** projecting from the old, whitish L.-Sh., ± cylindrical, 9 mm tall, 6 mm ⌀ above, with round, pale tubercles and a Fi. 3 mm long, smooth below; **Fl.** 15 mm ⌀, pink.

**C. germanum** N. E. Br. (§ II/5b). – Cape: Piquetberg D. – **Bo.** obconical, 8–10 mm tall, 5–8 mm across, 4–6 mm thick, with a 1–2 mm deep indentation above producing 2 small lobes with a Fi. 2–3 mm long, the surface reddish on the sides, grass-green above, with blackish-green or reddish-brown spots, either solitary or in lines; **Fl.** 6 mm ⌀, white.

**C. geyeri** L. Bol. (§ II/3a). – Cape: Lit. Namaqualand. – Mat-forming, with the parchmenty L.-Sh. persisting; **Bo.** obconical, 9 to 10 mm tall, up to 16 mm across and thick, with the upperside flat to ± convex and a Fi. up to 4 mm long, bluish-green; **Fl.** purplish-pink.

**C. giftbergense** Tisch. (§ II/5b). – Cape: Van Rhynsdorp D. – Forming lax cushions 4.5 cm high, 6 cm ⌀; **Bo.** obconical, up to 18 mm tall, somewhat compressed, up to 16 mm across, up to 13 mm thick, flat-convex above, with dark green or crimson spots, saddle-shaped in the centre, with a Fi. up to 5 mm long, the surface light greyish-green, with purple sides.

**C. glabrum** Tisch. (§ II/3b). – Cape: Van Rhynsdorp D. – Mat-forming; **Bo.** obconical, up to 12 mm high, 6–9 mm ⌀, with the upper surface circular, flat or slightly convex, with a Fi. 1.5 mm long, with whitish hairs, the surface green to bluish-green, with indistinct, small dots on the top; **Fl.** 16 mm ⌀, brilliant crimson.

**C. globosum** N. E. Br. (§ II/3a) (*C. obovatum* Lavis., *C. o. v. obtusum* L. Bol.). – Cape: Lit. Namaqualand. – Forming circular mats; **Bo.** 15–20 mm tall and across, the upper surface circular, with a scarcely depressed Fi. almost 5 mm long, the surface green, slightly glossy, unspotted; **Fl.** 12 mm ⌀, delicate pink.

**C. globuliforme** Schick et Tisch. (§ II/4). – Cape: Lit. Namaqualand. – Mat-forming; **Bo.** densely crowded, obconical, with the upper surface convex, 7–10 mm tall, 5–7 mm ⌀ above, with a Fi. 1 mm long, the surface light green, minutely papillose; **Fl.** yellow.

**C. gonapense** L. Bol. (§ II/1a/1). – Cape: Lit. Namaqualand.

**C. — v. gonapense.** – Forming mats up to 5.5 cm high; **shoots** covered with the densely imbricate, dry, old, leathery L.-Sh.; **Bo.** cordate, 2.5–3 cm tall, 11 mm across, 15–36 mm thick, with 2 lobes c. 6 mm long, 6–8 mm across below, rounded-truncate and a narrow purple line along their margins, with a blister at the base, the surface green, with indistinct darker spots, with a cordate, translucent area on the side of the lobes; **Fl.** yellow.

**C. — v. numeesicum** L. Bol. – **Bo.** up to 3.8 cm tall, the transparent areas indistinct.

**C. gracile** N. E. Br. (§ II/1a/2). – Cape: Lit. Namaqualand.

**C. — v. gracile.** – Small, branched ♄; **Bo.** solitary on each shoot, compressed, bilobed, 20–25 mm tall, 8–9 mm across, 6 mm thick, the lobes 3–5 mm long and across, rounded on the underside and at the tip, the surface light green, minutely velvety; **Fl.** yellow.

**C. — v. majusculum** L. Bol. – **Bo.** 3–3.5 cm tall, 1.5 cm ⌀ midway, with lobes 9–10 mm long, subacute; **Fl.** 14–18 mm ⌀.

**C. graciliramosum** L. Bol. (§ II/1a/1). – Cape: Lit. Namaqualand. – ♄ up to 6 cm tall; **shoots** slender, 3 cm long, Int. 4–5 mm long, enclosed in the old L.-Sh.; **Bo.** long-cordate, c. 3 cm tall, 9 mm across, 13–22 mm ⌀ above, with lobes 8–12 mm long, 8 mm wide below, 6–9 mm thick, the surface deep green, purple below; **Fl.** yellow.

**C. gracilistylum** (L. Bol.) N. E. Br. (§ II/1a/2) (*Mes. g.* L. Bol., *Derenbergia g.* (L. Bol.) Schwant., *Mes. augeiforme* Schwant., *D. a.* (Schwant.) Schwant., *Mes. chauviniae* Schwant., *D. c.* (Schwant.) Schwant.). – Cape: Lit. Namaqualand. – **Bo.** oval to cordate, 15–30 mm tall, 10–15 mm across, 9–12 mm thick, the indentation above 2–4 mm deep, the lobe-keels red, the surface bluish-green, with darker green spots and papillose hairs; **Fl.** 15–20 mm $\varnothing$, glossy pink.

**C. graessneri** Tisch. (§ II/1c). – S.W.Afr.: Gr. Namaqualand. – Similar to **C. saxetanum**; **Bo.** obconical, 7–8 mm tall, 3–4 mm across and rather less in thickness, the upper surface hemispherical to convex, with a broad, hirsute indentation surrounded by large dots, somewhat decurrent at the sides, with a Fi. 0.5–1 mm long, the surface greyish-green; **Fl.** creamy-white.

**C. grandiflorum** L. Bol. (§ II/1a/1). – Cape: Lit. Namaqualand. – **Shoots** very short; **Bo.** long-ovoid, 4–5.5 cm tall, 15 mm across midway, 25–30 mm thick, with 2 acute lobes 15–25 mm long, 10–15 mm thick below, 11 mm across, with reddish margins, the surface blue to bluish-green, with large, translucent dots; **Fl.** 4 to 5 cm $\varnothing$, dark yellow.

**C. gratum** (N. E. Br.) N. E. Br. (§ II/3a) (*Mes. g.* N. E. Br., *Mes. jucundum* N. E. Br., *C. j.* (N. E. Br.) N. E. Br., *Mes. nanum* L. Bol.). – Cape: Van Rhynsdorp D. – Forming lax mats, caulescent in age; **Bo.** pyriform, 25 mm tall, 12–15 mm $\varnothing$, the upper surface circular, convex, with scattered dark spots and a Fi. 5–6 mm long, the surface bluish-green with minute dots; **Fl.** up to 22 mm $\varnothing$, glossy red.

**C. hallii** L. Bol. (§ II/3). – Cape: Lit. Namaqualand. – Mat-forming, with the papery, brown to blackish L.-Sh. persisting; **Bo.** long-cylindrical, rounded above, the 2 lobes scarcely perceptible, with the Fi. $\pm$ compressed, 8 to 13 mm tall, 4–6 mm wide, 5–8 mm thick, the surface dirty, light green, softly papillose; **Fl.** white.

**C. halenbergense** (Dtr. et Schwant.) N. E. Br. (§ II/1c) (*Mes. h.* Dtr. et Schwant., *Derenbergia h.* (Dtr. et Schwant.) Schwant., *C. cuneatum* Tisch.). – S.W.Afr.: Gr. Namaqualand. – Mat-forming; **Bo.** cordate above, 8 to 25 mm tall, 7–17 mm across, the lobe-margins somewhat convex, the surface bluish to greyish-green, with numerous dark dots; **Fl.** small, dark straw-coloured.

**C. helenae** Rawé (§ II/1a/2). – Cape: Lit. Namaqualand. – Forming clumps up to 8 cm $\varnothing$; **Bo.** densely crowded, up to 100 together, long-cuneate, up to 15 mm long, 7 mm $\varnothing$, laterally compressed, the surface smooth, green towards the tip, with a few, blackish-green to purple dots, the lobes 2–3 mm long, truncate, 3 mm thick at the base, with a red line along the inner angles and the keel, the Fi. smooth, 2 mm long; **Fl.** 8 mm $\varnothing$, pale purple to purple.

**C. herrei** Schwant. (§ II/6). – Cape: Van Rhynsdorp D. – Mat-forming; **Bo.** obconical, 5 mm tall, 3 mm $\varnothing$, with the upper surface circular, convex, with a Fi. 2–3 mm long with whitish hairs surrounded by a dark zone with an indented margin, the surface green, with a few dark, decurrent lines of dots; **Fl.** c. 2 cm $\varnothing$, deep violet-red.

**C. hians** N. E. Br. v. **hians** (§ II/1c) (*C. elongatum* Schick et Tisch.). – Cape: Lit. Namaqualand. – Clump-forming; **Bo.** ovoid-cordate, c. 12 mm tall, 8 mm across above, laterally compressed, with subacute lobes 3 to 4 mm long, the surface light green, puberulous, with scattered dots merging into lines along the margins; **Fl.** c. 5 mm $\varnothing$, reddish-yellow or white.

**C. —** v. **acuminatum** L. Bol. – Pet. long-tapered, golden-yellow below, with red tips.

**C. hirtum** Schwant. (§ II/1c). – Cape: Lit. Namaqualand. – Forming dense, flat mats; **Bo.** cylindrical, up to 15 mm tall, up to 3 mm thick, the 2 lobes erect, acute, carinate, the Fi. 3 mm long, the surface grey, hirsute above; Pet. yellowish, with coppery tips.

**C. incurvum** N. E. Br. (§ II/1a/1). – Cape: Lit. Namaqualand.

**C. —** v. **incurvum** (*C. subacutum* L. Bol.). – **Bo.** laterally compressed, 32–40 mm tall and 9–13 mm across at the top, 6–10 mm across and 5–6 mm thick below the 2 erect, 6–8 mm long, rounded lobes, with a red line along the lobe-keels and a lateral translucent spot, the surface light green, velvety, with dark green spots below; **Fl.** 15–30 mm $\varnothing$, yellow.

**C. —** v. **leucanthum** (Lavis) Tisch. (*C. leucanthum* Lavis, *C. l.* v. *multipetalum* L. Bol.). – **Fl.** white, Pet. numerous.

**C. indutum** L. Bol. (§ II/7b). – Cape: Lit. Namaqualand. – Plant 5–7 cm tall, with **shoots** 2–3.5 cm long covered with dry L.-Sh.; **Bo.** pyriform, 6–8 mm tall, 4–5 mm $\varnothing$, with a slightly compressed Fi. 1–5 mm long bordered by dots, the surface with short hairs above and several dark green dots above; **Fl.** pinkish-purple.

**C. inornatum** N. E. Br. (§ II/3a). – Cape: Lit. Namaqualand. – **Bo.** obconical, green or greyish-green, often reddened on the sides, 4–6 mm tall, 3–5 mm $\varnothing$, the upper surface convex, circular, with a Fi. 1–2 mm long; **Fl.** pink.

**C. insigne** L. Bol. (§ II/1a/1). – Cape: Lit. Namaqualand. – **Bo.** compressed-cordate, up to 4.5 cm tall, the lobes pustulate, 14 mm long, the surface velvety; **Fl.** on a pedicel with 17 mm long bracts, 4.3 cm $\varnothing$, yellow.

**C. intermedium** L. Bol. (§ II/1c). – Cape: Lit. Namaqualand. – Mat-forming; the old L.-Sh. papery, pale brown; **Bo.** obovoid to cuneate, laterally compressed, 1–1.5 cm tall, 5–7 mm $\varnothing$ above, the upper surface elliptical, the Fi. indistinctly hirsute, the lobes 0.5–1.5 mm long, bordered with a red line, the surface pale green, with a few indistinct dots; **Fl.** coppery-red.

**C. intrepidum** L. Bol. (§ II/7b). – Cape: Augrabis Hills. – Mat-forming, L.-Sh. persisting; **Bo.** obconical, 11 mm tall, 10 mm $\varnothing$ above, the upper surface round, flat, with a small Fi., the surface covered with long, soft papillae; **Fl.** purplish-pink.

**C. jacobsenianum** TISCH. (§ II/3a). – Cape: Lit. Namaqualand. – Forming large clumps; **Bo.** pyriform, 16 mm tall, 10 mm $\varnothing$, the upper surface circular to elliptical, convex, the Fi. sunken, rhombic, bordered by confluent dark green dots, the surface whitish to light greyish-green, with numerous dark green dots in the upper part; **Fl.** 2 cm $\varnothing$, whitish to light pink.

**C. johannis-winkleri** (DTR. et SCHWANT.) N. E. BR. (§ II/4) (*Mes. j.-w.* DTR. et SCHWANT., *C. tabulare* LOESCH et TISCH.). – S.W.Afr.: S.Namib. – Mat-forming; **Bo.** obconical, 18 to 20 mm tall, 20–22 mm $\varnothing$, truncate above, flat, circular, with a Fi. 2–6 mm long, the surface light bluish to greyish-green; **Fl.** yellow.

**C. joubertii** LAVIS (§ II/5a). – Cape: Ladismith D. – Forming mats 23 mm high; **Bo.** long-pyriform to obconical, truncate above, $\pm$ convex, 8.5 mm tall, 2.5–5 mm across, with a Fi. 2 mm long, pale green with a few dots; **Fl.** 8 mm $\varnothing$, pale yellow.

**C. klipbokbergense** L. BOL. (§ II/1a/1). – Cape: Lit. Namaqualand. – **Shoots** up to 3.5 cm long, Int. 4–8 mm long; **Bo.** enclosed in a brown Sh., 47 mm tall, 37 mm $\varnothing$ above, with acute, broadly rounded lobes 15 mm long, 17 mm across, with obtuse, purple margins, the surface bluish-green, minutely papillose; **Fl.** 23 mm $\varnothing$, dark yellow.

**C. koubergense** L. BOL. (§ I/1). – Cape: Bushmanland. – Dry L.-Sh. persisting; **Bo.** compact-pyriform, 17–20 mm tall, 10 mm wide, 14 mm thick above, with an indistinctly hirsute Fi. 4 mm long, the Wi. convex, the surface with a few dots; **Fl.** with white staminodes. (Description incomplete; very close to **C. lithopoides.**)

**C. koupense** TISCH. (§ II/2). – Cape: Laingsburg, Beaufort West D. – Plant forming dense cushions 3.5 cm $\varnothing$ with the L.-Sh. persisting; **Bo.** obconical to ovoid, with the upper surface broadly elliptical, up to 10 mm tall, up to 7 mm across above, 5 mm thick, flat to convex, with a $\pm$ compressed Fi. 2 mm long, the surface light greyish-green to sea-green, with a few spots, more rarely with dots in a R. perpendicularly to the Fi., the sides reddish; **Fl.** c. 1 cm $\varnothing$, light straw-coloured.

**C. kubusanum** N. E. BR. (§ II/3b). – Cape: Lit. Namaqualand. – **Bo.** short-obconical, up to 1 cm tall, 4–6 mm $\varnothing$, the upper surface flat, circular, $\pm$ depressed in the centre, with a darker line surrounding the Fi., the surface dark greyish-green, with numerous dark green dots; **Fl.** c. 2 cm $\varnothing$, light pinkish-mauve.

**C. labiatum** TISCH. (§ II/4). – Cape: Lit. Namaqualand. – Forming lax clumps; **Bo.** obconical, up to 12 mm tall, up to 5 mm $\varnothing$, $\pm$ convex above, with the Fi. scarcely depressed, up to 3 mm long, with the margins rather prominent and labiate, the Fi.-zone dark, the surface whitish-green, with a few dark dots above; Pet. light yellow with red tips.

**C. labyrintheum** (N. E. BR.) N. E. BR. (§ II/5a) (*Mes. l.* N. E. BR.). – S.Afr.: origin unknown. – **Bo.** obconical, 10–14 mm tall, 6 to 8 mm across, 4–6 mm thick, convex and elliptical above, the surface dull greyish-green, with numerous interconnected, brownish-red lines; **Fl.** 9–10 mm $\varnothing$, straw-coloured.

**C. lacteum** L. BOL. (§ II/1a/1). – Cape: Lit. Namaqualand. – Plant up to 6.5 cm tall, with the brownish-blackish L.-Sh. persisting; **Bo.** linear, obtuse to subacute above, with a Sh. 2.8 cm long, lobes 7–18 mm across and $\varnothing$; **Fl.** snow-white.

**C. laetum** L. BOL. (§ II/1b). – Cape: Lit. Namaqualand. – Compact; 3 cm tall, 8 to 9 cm $\varnothing$; **Bo.** cordate, 18 mm tall, 8 mm across, 8 mm thick at the tip, lobes indistinct, 1–2 mm long, broadly rounded, the surface minutely velvety, with indistinct dots and somewhat translucent spots; **Fl.** 15–20 mm $\varnothing$, yellow.

**C. lambertense** SCHICK et TISCH. v. **lambertense** (§ II/5b). – Cape: Clanwilliam D. – Low-growing; **Bo.** c. 10 mm tall, elliptical and 5–7 mm $\varnothing$ above, somewhat depressed towards the 2 mm long Fi., the surface greyish-green, with a darker line and several dots around the Fi., also several short, reddish lines on the upper surface; **Fl.** white.

**C. —** v. **conspicuum** RAWÉ. – Cape: Clanwilliam D. – **Bo.** 12 mm tall, 8–10 mm $\varnothing$, the surface glabrous, olive-green, with dense, prominent lines; **Fl.** 10 mm $\varnothing$.

**C. largum** L. BOL. (§ II/1a/1). – Cape: Lit. Namaqualand. – The persistent L.-Sh. scabrous, brown; **Bo.** broadly obovoid, 6 cm tall, 16 mm across, the lobes 2 cm long, with a blister at the base, 12–20 mm $\varnothing$, the surface rough, with a dirty green line along the lobe-margins and keel; **Fl.** yellow. Acc. A. TISCHER possibly identical with **C. incurvum.**

**C. latum** L. BOL. (§ II/1b). – Cape: Lit. Namaqualand. – **Shoots** 10–15 mm long; **Bo.** long-ovoid, 31 mm tall, 14 mm $\varnothing$ above, convex, with 2 rather obtuse lobes 1–3 mm long, the surface pale bluish-green, papillose; **Fl.** yellow.

**C. lavisianum** L. BOL. (§ II/1a/1) (*C. indefinitum* L. BOL., *C. asperulum* L. BOL., *C. a.* v. *brevistylum* L. BOL., *C. inclusum* L. BOL., *C. barkerae* L. BOL.). – Cape: Lit. Namaqualand. – Cushion-forming; **Bo.** long-cordate, 4 cm long, c. 1.5 cm $\varnothing$, lobes $\pm$ rounded on both sides, without a sharp keel, 10–14 mm long, with the inside of the tips tinged deep pink, the surface smooth to almost imperceptibly papillose or velvety, light sea-green; **Fl.** 20–25 mm long, 20–22 mm $\varnothing$, tube $\pm$ white, Pet. light yellow.

**C. laxipetalum** N. E. BR. (§ II/1a/1). – Cape: Lit. Namaqualand. – **Bo.** obovoid, 25–40 mm tall, 16–18 mm across, 8–10 mm thick, the 2 lobes very compressed, 8–10 mm long, with acute, very reddened angles and keel, the surface bluish-green, scarcely spotted; **Fl.** 15 mm $\varnothing$, golden-yellow.

**C. leipoldtii** N. E. BR. (§ II/6). – Cape: Clanwilliam D. – **Bo.** spherical, 6 mm $\varnothing$, pale brown to reddish, densely puberulous; **Fl.** red.

**C. leopardinum** L. BOL. (§ II/1b). – Cape: Lit. Namaqualand. – Cushions 8–10 cm $\varnothing$, 4 cm tall; **shoots** up to 4 cm long, covered with dry, old L.-Sh.; **Bo.** up to 60 together, broadly obovoid, indistinctly carinate towards the tip, 12 mm tall, 8 mm across and thick, the Fi. 1.5–2 mm long, rhombic, red-spotted, the surface bluish-green, with green dots, minutely papillose above; **Fl.** white to yellow. Acc. A. Tischer probably identical with **C. laetum** or **C. semilunulum.**

**C. leviculum** (N. E. Br.) N. E. Br. (§ II/5a) (*Mes. l.* N. E. Br.). – S.Afr.: origin unknown. – Mat-forming; **Bo.** obconical, 10–15 mm tall, 7–14 mm across, the upper surface very flat to convex, with a Fi. 2–3 mm long surrounded by lines, the surface bluish-green, with darker, often reddish dots and branched lines; **Fl.** 18 mm ⌀, cream-coloured, scented.

**C. lilianum** Littlew. (§ I/1). – Cape: Lit. Namaqualand. – **Bo.** 4–9 together, cordate, 9–11 mm tall, 3–4 mm across, 6–8 mm ⌀, the 2 rounded lobes 0.5–3 mm long, the Wi. convex, ± red to purplish-brown with green spots, the surface otherwise green, with some dots below; **Fl.** with a golden-yellow to brown tube, segments white to pink. (Acc. A. Tischer scarcely differentiated from **C. areolatum.**)

**C. linearilucidum** L. Bol. (§ II/1a/1). – Cape: Lit. Namaqualand. – **Bo.** long-ovoid, c. 45 mm tall, 15 mm wide, 25 mm thick above, with 2 tapered lobes 12 mm long, the Fi.-zone 8 to 12 mm long, the surface pale olive-green, spotted; **Fl.** yellow.

**C. literatum** N. E. Br. (§ II/5a). – Cape: Clanwilliam D. – **Bo.** obconical, somewhat depressed, 12 mm tall, up to 6 mm ⌀, the upper surface flat to flat-convex, greyish-green, with several ± short, brownish-red, radiating, decurrent lines above, the sides brownish-purple; **Fl.** up to 12 mm ⌀, cream-coloured.

**C. lithopoides** L. Bol. (§ I/1). – Cape: Bushmanland. – **Bo.** obovoid, c. 14 mm tall, 7 mm wide, 10 mm thick, the upper surface convex and translucent, the Wi. dirty green, with brown, dendritic markings, the Fi. compressed, 2–3.5 mm long, the surface puberulous; **Fl.** purplish-red.

**C. litorale** L. Bol. (§ II/5a). – Cape: Clanwilliam D. – Forming cushions, with the old, thin, brown, rough L.-Sh. persisting; **Bo.** obovoid to ± pyriform, 10–11 mm tall, 6–8 mm ⌀ the upper surface round to oval, convex, with a Fi. 4 mm long with 4–6 spots, the surface with some spots, those along the margins merging into lines; **Fl.** white to pink.

**C. loeschianum** Tisch. (§ II/1c). – Southern S.W.Afr. – Mat-forming; **Bo.** 5–10 mm tall, 3–6 mm ⌀, the upper surface truncate and distinctly carinate, with the Fi. sharply indented, 1 mm long, the surface yellowish-green, puberulous, with several dots on the keel-line of the short lobes; **Fl.** white.

**C. longibracteatum** L. Bol. (§ II/1b). – Cape: Lit. Namaqualand. – **Bo.** obcordate, 13 mm tall, 8–9 mm ⌀ above, broadly convex, with rounded lobes up to 2 mm long, with green dots along the lobe-angles; **Fl.** dark yellow, with long bracts.

**C. longifissum** Tisch. (§ II/5b). – Cape: Lit. Namaqualand. – **Bo.** obconical, 16–18 mm tall, 10–15 mm across, 6–10 mm thick above, with the Fi. in a wide indentation 6–8 mm long, with dots in a curved line on either side, the surface dark green, with dark green often confluent dots on the lobes, the sides reddish; **Fl.** 2 cm ⌀, white.

**C. longistylum** N. E. Br. (§ II/3a). – Cape: Lit. Namaqualand. – **Bo.** obconical, 7–10 mm tall, 5–8 mm ⌀, the upper surface flat to ± convex, circular to elliptical, with a Fi. 1–2 mm long surrounded by a line of dark green dots, the surface light bluish-green, spotted; **Fl.** 12 mm ⌀, pinkish-red.

**C. longitubum** L. Bol. (§ II/5a). – Cape: Worcester D. – Compact, 3.5 cm ⌀, with the L.-Sh. persisting; **Bo.** obconical to pyriform, 8–11 mm tall, 6–9 mm ⌀, with a ciliate Fi. 2 mm long, the surface dirty green, with spots sometimes arranged in lines; **Fl.** white.

**C. luckhoffii** Lavis (§ II/6). – Cape: Clanwilliam D. – Mat-forming; **Bo.** long-cuneate, broadly compressed, 8 mm tall, up to 5 mm ⌀, the 2 lobes 1.5 mm tall with a reddish-brown line and dots, the Fi. 1.5 mm long, the surface brownish-green; **Fl.** pinkish-purple.

**C. luisae** Schwant. (§ II/1b). – Cape: Lit. Namaqualand. – Forming lax mats; **Bo.** cordate, c. 15 mm tall, 10 mm across, 8 mm thick, the rounded lobes 0.5–2 mm long, puberulous on the inner surface, the lobe-angles dark, the Fi. 1.5–4 mm long, with a darker spot on either side, the surface with a few distinct dots; **Fl.** 18 mm ⌀, yellow.

**C. luteolum** L. Bol. (§ II/3a). – Cape: Lit. Namaqualand.

**C.** – v. **luteolum.** – **Bo.** hidden inside the dry brown Sh., obconical, narrowly oval, the upper surface circular, with dark spots, with a Fi. 1–2 mm long; **Fl.** dark yellow.

**C.** – v. **macrostigma** L. Bol. – The stigma longer than v. **luteolum,** 4–5 mm long.

**C. luteum** N. E. Br. (§ II/3a). – Cape: Lit. Namaqualand. – Cushion-forming; **Bo.** ± pyriform, up to 22 mm tall, up to 15 mm ⌀, the upper surface circular, flat to convex, with the 3 mm long Fi. ± depressed, sea-green to greyish-green, with spots irregularly distributed above, larger along the margin and merging into a zone; **Fl.** c. 25 mm ⌀, yellow to golden-yellow.

**C. marginatum** Lavis v. **marginatum** (§ II/1a/2). – Cape: Bushmanland to southern S.W. Afr. – Forming compact mats; **shoots** 6–8 mm ⌀, old L.-Sh. papery; **Bo.** long-ovoid, smooth, up to 34 mm long, up to 12 mm ⌀ above, lobes up to 10 mm long, carinate or not, the keel-line often red, the surface light green, spotted, the Fi. c. 2 mm long, sometimes translucent; **Fl.** c. 15 mm ⌀, pink.

**C.** – v. **eenkokerense** (L. Bol.) Rawé (*C. e.* L. Bol.). – Eenkoker Mts. – **Bo.** elongated, 25–34 mm long, 9 mm ⌀ below, distinctly expanded above and 15 mm ⌀, lobes indistinctly carinate, truncate above, laterally rounded, 8 to 10 mm long, with dark green spots in a R. along the keel and lobe-margins.

**C.** – v. **karamoepense** (L. Bol.) Rawé (*C. k.* L. Bol., *C. senarium* L. Bol.). – Near Haramoep, Aggeneys, between Namies and Springbok. – **Bo.** 16 mm long, distinctly tapered above and 7 mm ⌀, lobes indistinctly keeled, with all surfaces rounded, 4 mm long, 3–4.5 mm ⌀, distinctly spotted, indistinctly papillose.

**C.** — v. **littlewoodii** (L. Bol.) Rawé (*C. l.* L. Bol.). – Southern S.Afr. – **Bo.** 10–14 mm long, distinctly expanded above, lobes 4.5 mm long, 4.5 mm across below, not carinate, acutely truncate, broadly projecting above.

**C. marlothii** N. E. Br. (§ II/3a). – Cape: Lit. Namaqualand. – In lax cushions; **shoots** short, with numerous Bo., the L.-Sh. persisting; **Bo.** obconical, up to 8 mm tall, up to 5 mm ⌀ above, the upper surface flat to flat-convex, circular, with the ± depressed Fi. surrounded by a darker line, light bluish-green, with a few dots above; **Fl.** c. 1 cm ⌀, Pet. white below, mauve above.

**C. marnieranum** Tisch. et Jacobs. (§ II/1a/1). – Cape: Lit. Namaqualand. – Mat-forming, with short **shoots**; **Bo.** obcordate, up to 2.5 cm tall, up to 15 mm across, up to 10 mm thick, the truncate lobes up to 8 mm long with an acute, often forked keel, the Fi. up to 4 mm long, the surface olive-green, with rather prominent veins and dots on the sides of the Bo., the margins and keels reddish-brown; **Fl.** coppery.

**C. maximum** Tisch. (§ II/3a). – Cape: Lit. Namaqualand; S.W.Afr.: Loreley. – Mat-forming; **Bo.** obconical, 3 cm tall, 25 mm ⌀, the upper surface usually circular, slightly convex, with the 3 mm long Fi. slightly depressed, the surface greyish to bluish-green, with large, distant spots above; **Fl.** 25 mm ⌀, pink.

**C. membranaceum** L. Bol. (§ II/4). – Cape: Lit. Namaqualand. – Plant c. 3 cm tall, 5.5 cm ⌀; **Bo.** semi-ovoid to obovoid, 4–6 mm tall, 4–5 mm ⌀, the upper surface convex, with a slightly compressed Fi. 2.2 mm long, pale green, unspotted; **Fl.** c. 12 mm ⌀, orange-red to coppery, scented.

**C. meridianum** L. Bol. v. **meridianum** (§ I/1). – Cape: Van Rhynsdorp D. – With the papery brown L.-Sh. persisting; **Bo.** long-ovoid, 10 to 13 mm tall, 5–6 mm across, c. 8 mm ⌀, the upper surface convex, with a Fi. c. 1.5 mm long, often indistinctly hirsute, with a small Wi. with brown 'islands', reddish-brown; **Fl.** white, staminodes yellow.

**C.** — v. **pulverulentum** L. Bol. – **Bo.** minutely papillose below, minutely pruinose above, the Fi. c. 2.5 mm long, pubescent.

**C. meyerae** Schwant. (§ II/1a/1). – Cape: Lit. Namaqualand. – **Bo.** few together, long-cordate to long-ovoid, laterally compressed, up to 5 cm tall, 4 cm across, with the 2 lobes variable, usually long, very rounded on the underside, often pubescent, the tips often reddish, the surface dull green to bluish-green, often rough, with a line of dots along the margins and underside of the lobes, otherwise with scarcely perceptible spots all over; **Fl.** yellow.

**C.** — f. **alatum** Tisch. – **Bo.** flatter than f. **meyerae**, lobes flatter and broader.

**C.** — f. **apiculatum** (N. E. Br.) Tisch. (*C. a.* N. E. Br., *C. lekkersingense* L. Bol.). – Lobe-tips curved slightly upwards, with the keel-line broad and pulled towards the outside.

**C.** — f. **asperulum** (L. Bol.) Jacobs. (*C. tumidum* v. *a.* L. Bol.). – **Bo.** rough, the lobes with short papillose hairs.

**C.** — f. **meyerae** (Pl. 155/2) (*C. tumidum* N. E. Br.). – Lobes long, tapered, 2.5 cm long.

**C.** — f. **pole-evansii** (N. E. Br.) Tisch. (*C. p.-e.* N. E. Br.). – Lobes shorter than in f. **meyerae**, broader below, pulled less towards the outside.

**C. meyeri** N. E. Br. (§ II/1b). – Cape: Lit. Namaqualand.

**C.** — v. **meyeri** (Pl. 155/4). – Dwarf ♄; **Br.** mostly prostrate, 6–10 cm long; **Bo.** 20–25 mm tall, 13–15 mm thick, the upper surface with an indentation c. 2–3 mm deep, with a Fi. 4–5 mm long, the lobes very rounded, the surface greyish-green, rough with short hairs; **Fl.** 14 to 16 mm ⌀, yellow, with stigmas 1 mm long; style 8–9 mm long; Sep. 4.

**C.** — v. **quinarium** L. Bol. – Differentiated from v. **meyeri** only by some floral characteristics: Sep. 5, style 1–6 mm long, stigma 5.5 to 6 mm long.

**C. microstigmum** L. Bol. (§ II/3b). – Cape: Lit. Namaqualand. – Plant hemispherical, 5 cm tall, 9 cm ⌀, with the L.-Sh. persisting; **Bo.** pyriform, with the upper surface circular, convex, 10–12 mm tall, 7–10 mm ⌀, with a compressed Fi. 1.5 mm long, minutely papillose, the surface olive-green, spotted; **Fl.** 2 cm ⌀, yellow.

**C. middlemostii** L. Bol. (§ II/3a). – Cape: Lit. Namaqualand. – **Bo.** obconical, 1 cm tall, 9 mm across, 1 cm thick, the upper surface flat, with dense reddish-brown spots and a Fi. 3 mm long; **Fl.** pink. Acc. A. Tischer possibly identical with **C. fragile.**

**C. minimum** (Haw.) N. E. Br. (§ II/5a) (*Mes. m.* Haw.). – Cape: ? Clanwilliam D. – Forming lax mats; **Bo.** obconical, truncate above, up to 1 cm tall, up to 6 mm ⌀, the upper surface flat-convex, circular, sometimes slightly depressed in the centre, with a dark zone surrounding the ciliate Fi., the surface dark sea-green, with darker, brownish-red spots above, merging into short lines, the sides often crimson to reddish; **Fl.** white to cream-coloured.

**C. minusculum** (N. E. Br.) N. E. Br. (§ II/6). – Cape: Clanwilliam D.

**C.** — v. **minusculum** (*Mes. m.* N. E. Br., *C. reticulatum* L. Bol.). – Mat-forming; **Bo.** very numerous, almost completely concealed in the dry L.-Sh., obovoid, 15–16 mm tall, (5–)8(–9) mm ⌀, the upper surface convex, with dark brownish veins and a rib-like line, with a Fi. 1–2 mm long and several tiny tubercles set in a round or square zone, the surface olive-green to purplish-brown, with darker spots and minute dots; **Fl.** dark purple to crimson.

**C.** — v. **reticulatum** (L. Bol.) Rawé (*C. r.* L. Bol., *C. r. v. roseum* Lavis, *C. r. f. r.* (Lavis) Rowl., *C. m. v. roseum* (Lavis) Tisch.). – 6 miles N. of Graafwater. – **Bo.** upperside flat, with distinct, prominent, reticulate markings; **Fl.** brilliant crimson, sometimes pink.

**C. — v. paucilineatum** RAWÉ. – S. of Clanwilliam. – **Bo.** up to 12 mm long, 7 mm across, with markings much less conspicuous or often missing, often with only a line perpendicular to the Fi.

**C. minutiflorum** (SCHWANT.) N. E. BR. (§ II/4) (*Mes. m.* SCHWANT., *Mes. calculus* v. *m.* SCHWANT.). – Cape: Lit. Namaqualand. – **Bo.** obconical, 10–12 mm tall, 10–12 mm across, 8–10 mm thick, the upper surface flat, circular, with a compressed Fi. 4–5 mm long, surface bluish-green; **Fl.** 5–7 mm $\varnothing$, yellow.

**C. minutum** (HAW.) N. E. BR. (§ II/3 b). – Cape: Van Rhynsdorp D.

**C. — v. laxum** L. BOL. – As v. **minutum**; **shoots** elongated; **Fl.** opening wide.

**C. — v. minutum** (Pl. 155/3) (*Mes. m.* HAW., *Mes. thecatum* N. E. BR.). – Forming circular mats; **Bo.** pyriform, 10–12 mm tall, 6–10 mm across, the upper surface circular or elliptical, $\pm$ depressed, sometimes slightly spotted, with a Fi. 2–3 mm long, the surface bluish to greyish-green; **Fl.** 12–15 mm $\varnothing$, pinkish-mauve.

**C. misellum** N. E. BR. (§ II/1 c). – Cape: Lit. Namaqualand. – **Bo.** obovoid, 6–8 mm tall, 4–6 mm $\varnothing$, the upper surface flat, $\pm$ circular, with a slight indentation and a Fi. 1 mm long with a distinct keel along both sides, with a darker line and several dots along the keels and around the Fi., the surface green; **Fl.** light yellow.

**C. miserum** N. E. BR. (§ II/1 c). – Cape: Lit. Namaqualand. – Mat-forming; **Bo.** rounded, broadly ovoid, 15–19 mm tall, 10–12 mm across, with rounded lobes 3–4 mm long, the surface bluish-green, minutely papillose; **Fl.** 5 mm $\varnothing$, light yellow.

**C. morganii** LAVIS (§ II/2). – Cape: Willowmore D. – **Bo.** partially concealed in the old L.-Sh., long-obovoid, 28 mm tall, 13 mm wide, 13 mm thick, with the upperside convex and a compressed Fi. 4.5 mm long, the surface green, with numerous translucent dots; **Fl.** white.

**C. muirii** N. E. BR. (§ II/5 a). – Cape: Ladismith D. – **Bo.** obovoid, 5–6 mm tall, 3–4 mm $\varnothing$, the upper surface flat, $\pm$ circular, slightly indented, with a Fi. 1–2 mm long surrounded by a darker line, the surface light bluish-green or chalky green, with indistinct dots; **Fl.** pale yellow.

**C. multipunctatum** TISCH. (§ II/2). – Cape: Uniondale D. – Mat-forming; **Bo.** pyriform, c. 2 cm tall, 15 mm $\varnothing$, with a compressed Fi. 2–5 mm long, the surface pallid greenish-grey, with numerous dots and lines above; **Fl.** white.

**C. mundum** N. E. BR. (Pl. 156/1) (§ II/5 b). – Cape: Van Rhynsdorp, Clanwilliam D. – Matforming; **Bo.** obconical, 10–12 mm tall and across, truncate above, circular, a little depressed, with a Fi. 3–4 mm long, the surface greyish-green, with raised dots, some of these radiating from the Fi. towards the margins, sometimes merging into lines; **Fl.** whitish to cream-coloured.

**C. muscosipapillatum** LAVIS (§ II/1 a/1) (*C. sitzlerianum* SCHWANT.). – Cape: Lit. Namaqualand. – Branched, c. 7 cm tall, the woody **shoots** 3–5 cm long, covered with dry L.-Sh.; **Bo.** projecting from the L.-Sh., elongate to $\pm$ cordate, laterally compressed, 3.5 cm tall, 2 cm wide, the lobes unequal, truncate, c. 1 cm long, the surface greyish-green to whitish with indistinct green dots, minutely velvety-haired; **Fl.** 3–4 cm $\varnothing$, golden-yellow.

**C. namiesicum** L. BOL. (§ II/4). – Cape: Bushmanland. – **Bo.** broadly obovoid to rounded, slightly convex above, 13–14 mm tall, 16 to 18 mm $\varnothing$; **Fl.** yellow.

**C. nanum** TISCH. (§ II/1 b). – S.Afr.: origin unknown. – **Bo.** up to 1 cm tall, 5–7 mm $\varnothing$, $\pm$ spherical, the upperside very rounded, with a Fi. 1.5 mm long and surrounded by green dots, the surface dark green, slightly velvety-haired; **Fl.** yellow.

**C. niveum** L. BOL. (§ II/1 b). – Cape: Lit. Namaqualand. – Forming clumps up to 10 cm $\varnothing$, with old L.-Sh. persisting; **shoots** up to 2 cm long; **Bo.** pyriform to obovoid, 11–16 mm tall, 8–9 mm wide, 11 mm thick, with papillose-haired lobes 1–1.5 mm high and a Fi. 3–4 mm long, the surface pale bluish-green with indistinct dots; **Fl.** snow-white.

**C. noisabiense** L. BOL. (§ II/1 a/1). – Cape: Lit. Namaqualand. – **Bo.** 40–46 mm tall, 17 mm $\varnothing$ above, the 2 lobes erect, rounded above, 16 mm high, 10 mm across, with reddish, carinate margins, the surface greyish-green, rough to minutely velvety; **Fl.** yellow.

**C. nordenstamii** L. BOL. (§ II/3 a). – Cape: Lit. Namaqualand. – Mat-forming, with the dark brown L.-Sh. persisting; **Bo.** obconical to elongated, the upperside flat to concave, 9 mm tall, 8 mm $\varnothing$ above, the surface pale bluish-green with indistinct darker dots; **Fl.** pinkish-white.

**C. notabile** N. E. BR. (§ II/1 a/1). – Cape: Lit. Namaqualand. – **Bo.** c. 15 mm tall, 10 mm across, the upperside slightly depressed, with a slight ridge, circular to elliptical and $\pm$ rounded, with a Fi. 2–5 mm long with a red spot on either side, the surface light bluish-green; **Fl.** coppery.

**C. notatum** N. E. BR. (§ II/5 b). – S.Afr.: origin unknown. – Forming flat mats; **Bo.** obconical, 20–22 mm tall, 12 mm across, rather less in thickness, the upperside $\pm$ truncate, circular, slightly convex, with a Fi. 2 mm long, the surface dark greyish-green with slightly prominent, scattered, reddish-brown dots, sometimes forming short lines; **Fl.** cream-coloured.

**C. novellum** N. E. BR. (§ II/2). – Cape: Clanwilliam D. – **Bo.** obconical, 6–9 mm tall, 6–8 mm across, 6–7 mm thick, the upperside flat with a distinct ridge, with a Fi. 1–2 mm long and a distinct lateral line, the surface light green, with a few dots; **Fl.** pale yellow.

**C. novicium** N. E. BR. (§ II/3 a). – Cape: Lit. Namaqualand. – Plant forming lax cushions; **Bo.** up to 12 mm tall, up to 9 mm $\varnothing$, slightly constricted below the upper surface which is circular to elliptical to $\pm$ obtusely octagonal, flat or convex, with a slightly depressed Fi. 2 mm long set in a darker zone, the surface light grey to sea-green, with rather darker green dots above, those around the Fi. sometimes merging to form an annulus; **Fl.** up to 2 cm ⌀, yellow to golden-yellow.

**C. nudum** TISCH. (§ II/3b). – Cape: Van Rhynsdorp D. – Forming lax mats; very close to **C. minutum**; **Bo.** cylindrical, up to 1 cm tall, 5 mm ⌀ above, the upperside circular, flat to ± convex, often grooved-depressed, with a Fi. up to 1 mm long with short, white hairs, set in a dark zone, the surface dull green; **Fl.** 12 mm ⌀, mauvish-pink.

**C. nutaboiense** TISCH. (§ II/1a/1) (*C. angustum* L. BOL.). – Cape: Lit. Namaqualand. – **Bo.** compressed, 31 mm tall, 11 mm ⌀ above, with 2 rounded lobes 5 mm long, surface yellowish-green; **Fl.** yellow.

**C. obconellum** (HAW.) SCHWANT. (§ II/5b) (*Mes. o.* HAW.). – Cape: Clanwilliam D. – Cushion-forming; **Bo.** obconical, up to 2 cm tall and across, the upper surface slightly cordate, ridged, slightly convex from the 6–8 mm long, shortly hirsute Fi. towards the outer margin, the surface green to greyish or bluish-green, with numerous prominent, dark green or dark red translucent dots merging into lines or solitary, like tiny tubercles; **Fl.** 8–15 mm ⌀, milky-white to yellowish.

**C. obcordellum** (HAW.) N. E. BR. (§ II/5b) (*Mes. o.* HAW., *Mes. o.* SIMS, *Mes. nevillei* N. E. BR., *C. n.* (N. E. BR.) N. E. BR., *C. klaverense* N. E. BR.). – Cape: Van Rhynsdorp D. – Mat-forming; **Bo.** clavate to obconical, 6–9 mm tall, 4–9 mm ⌀, the upperside circular to reniform, flat to convex or very depressed, with slightly prominent dots often arranged in simple lines, the sides mostly pink to dark red, the surface otherwise light green to greyish or brownish-green, less often pinkish-crimson; **Fl.** 6–15 mm ⌀, white or straw-coloured.

**C. — v. obcordellum f. declinatum** (L. BOL.) TISCH. (*C. d.* L. BOL.). – The upperside depressed saddle-like towards the Fi., with the spots sometimes confluent into short R.; Pet. usually white.

**C. — v. — f. multicolor** (TISCH.) TISCH. (*C. m.* TISCH.). – The dominant colour of the upperside mostly metallic greyish-green, the spots and the sides of the Bo. being dark crimson.

**C. — v. — f. obcordellum** (Pl. 156/3). – **Bo.** clavate, the upperside flat to slightly convex, with rather large, dark green to brownish-red, mostly slightly prominent spots at the tip, these scarcely confluent into lines or R., the dominant colour greyish-green, the sides brownish to deep crimson; Pet. straw-coloured, 9–10 mm long.

**C. — v. parvipetalum** (N. E. BR.) TISCH. (*C. p.* N. E. BR.). – **Bo.** as in the type in shape, colour and markings; Pet. only up to 6 mm long.

**C. — v. — f. picturatum** (N. E. BR.) TISCH. (*C. p.* N. E. BR.). – Spots vivid deep red, mostly confluent into reticulate R. or lines.

**C. obmetale** (N. E. BR.) N. E. BR. (§ II/5a) (*Mes. o.* N. E. BR.). – S.Afr.: origin unknown. – Mat-forming; **Bo.** obconical, 10–12 mm tall, 16–17 mm across, the upperside somewhat truncate and rounded, with a scarcely depressed Fi. 2–3 mm long surrounded by a distinct line from which short, ± branched lines radiate to the margin, the surface greyish-green; **Fl.** 10 to 11 mm ⌀, milky-white.

**C. obscurum** N. E. BR. (§ II/7b). – Cape: Lit. Namaqualand. – Mat-forming; **Bo.** obconical, 12–15 mm tall, the upperside ± circular, 5 to 8 mm ⌀, flat, with a depressed Fi., the surface minutely irregular, dark green or brownish, densely covered with dark green and minute whitish dots; **Fl.** red.

**C. obtusum** N. E. BR. v. **obtusum** (§ II/1a/1) (*C. vlakmynense* L. BOL.). – Cape: Lit. Namaqualand. – Mat-forming; **Bo.** long-ovoid, 25–32 mm tall, 15 mm across, somewhat laterally compressed above, with 2 lobes c. 2 mm long, rounded above, rounded-carinate on the underside, with red keels and tips, the surface minutely rough; **Fl.** 2 cm ⌀, yellow.

**C. — v. amplum** (L. BOL.) RAWÉ (*C. a.* L. BOL.). – **Bo.** 30–50 mm long, lobes 5–10 mm long.

**C. occultum** L. BOL. (§ II/5a). – Cape: Van Rhynsdorp D. – Forming clumps 5 cm ⌀; **Bo.** numerous, 12 mm tall, 5 mm ⌀, the upperside round to elliptical, convex, the surface olive-green, ± earth-coloured, with reddish-brown dots; **Fl.** 8 mm ⌀, pale yellow.

**C. orbicum** N. E. BR. et TISCH. (§ II/3a). – Cape: Lit. Namaqualand. – Cushion-forming; **Bo.** laxly arranged, ± pyriform, 10–15 mm tall, 6–9 mm ⌀ above, the upperside circular, rounded to slightly bilobed, with a purple keel-line over the lobes, the Fi. rhombic, compressed, 2–3 mm long, set in a dark zone, the rest of the surface greyish-green, with some dark green spots; Pet. pink with golden-yellow tips.

**C. ornatum** LAVIS (§ II/3a) (*C. percrassum* SCHICK et TISCH., *C. tinctum* LAVIS). – Cape: Lit. Namaqualand. – **Bo.** short-conical, 25 mm tall, 20 mm ⌀, with the upperside flat to convex, with a slightly depressed Fi. 3 mm long surrounded by a darker line, the surface bluish-green, with few or no spots; **Fl.** 25 mm ⌀, golden-yellow, with red filaments and stigmas.

**C. ovatum** L. BOL. (§ II/1a/1). – Cape: Lit. Namaqualand. – Up to 6.5 cm tall; **Bo.** ± square to broadly ovoid, 15–23 mm tall, 13–22 mm ⌀, with 2 rounded to truncate lobes 4–5 mm high and a Fi. 1.5 mm long, the surface pale green, minutely papillose, with dark green dots; **Fl.** up to 2 cm ⌀, golden-yellow.

**C. ovigerum** SCHWANT. (§ II/1b). – Cape: Lit. Namaqualand. – Mat-forming; **Bo.** tapered-ovoid, up to 1 cm tall, 6–8 mm ⌀, 5–7 mm thick, the upperside very rounded, slightly indented, with a Fi. 2–3 mm long surrounded by dark dots, the surface green; **Fl.** yellow.

**C. pageae** (N. E. BR.) N. E. BR. (§ II/4) (*Mes. p.* N. E. BR., *C. pumilum* N. E. BR., *C. schickianum* TISCH.). – Cape: Lit. Namaqualand. – Forming small mats; **Bo.** obconical, 7–12 mm tall, 4–7 mm ⌀ above, the upperside circular, convex or flat, with a somewhat depressed Fi. 2–4 mm long, the surface whitish-green but reddish on the sides and around the Fi.; **Fl.** c. 12 mm ⌀, Pet. straw-coloured with brownish tips.

**C. papillatum** L. Bol. (§ II/1b). – Cape: Lit. Namaqualand. – Cushions up to 4 cm tall, 8 cm ⌀; **Bo.** slender obovoid, 13–25 mm tall, 10–14 mm across, the lobes up to 3 mm high, with ± rounded tips, the surface velvety-papillose, the Fi. hirsute; **Fl.** 18–20 mm ⌀, light yellow.

**C. pardicolor** Tisch. (§ I/1). – Cape: Bushmanland. – Forming dense cushions with the leathery, dry L.-Sh. persisting; **Bo.** obovoid to conical, somewhat compressed, 8–10 mm tall, 6–10 mm across, 5–7 mm thick, with 2 lobes 1.5–2.5 mm high, the lobe-tips with green spots, the Wi. with a yellowish-brown spot, the surface light brown; **Fl.** c. 2 cm ⌀, white.

**C. pardivisum** Tisch. (§ II/2). – Cape: origin unknown. – **Bo.** obconical, up to 2 cm tall, the upperside 10–13 mm long, 7–10 mm across, broadly ovate, flat, with the 5 mm long Fi. surrounded by darker dots, the surface light greyish-green, velvety-papillose, spotted; **Fl.** white.

**C. parviflorum** N. E. Br. (§ II/5b). – Cape: Clanwilliam D. – **Bo.** clavate to obconical, 5–7 mm tall, 5–6 mm across, 5–7 mm thick, the upperside truncate, circular, with the centre ± depressed and crater-like, with a Fi. 2–3 mm long, the surface greyish-green with blackish-green dots and lines, the sides purple.

**C. — v. impressum** (Tisch.) Tisch. (*C. i.* Tisch.). – **Bo.** like v. **parviflorum** in shape and colour but slightly compressed above; Pet. 8–10 mm long.

**C. — v. parviflorum.** – **Bo.** clavate, the upperside flat to very slightly crater-like and depressed, the spots smaller than **C. obcordellum**, not prominent, scarcely confluent; Pet. up to 6 mm long, straw-coloured.

**C. parvimarinum** L. Bol. (§ II/1c). – Cape: Lit. Namaqualand. – L.-Sh. persisting; **Bo.** broadly obovoid, 6–9 mm tall, c. 8 mm ⌀, the upperside elliptical, convex, with the Fi. indistinctly carinate, 3 mm long, ± jaw-like, with dense hairs inside; **Fl.** white to pink. Acc. A. Tischer probably identical with **C. miserum**.

**C. parvulum** L. Bol. (§ II/1a/2). – Cape: Lit. Namaqualand. – Mat-forming; **Bo.** rounded-elongate, 21 mm tall, 7 mm ⌀, with 2 lobes 3–4 mm high, the surface brownish-purple, with purple dots; **Fl.** 13 mm ⌀, yellow.

**C. paucipunctum** Tisch. (§ II/4). – Cape: Lit. Namaqualand. – Forming dense cushions; **Bo.** obconical, up to 1 cm tall, up to 5 mm ⌀ above, the upper surface circular, ± flat to convex, the Fi. up to 3 mm long, with labiate, protruding margins, the surface whitish-green, with a few dots; **Fl.** c. 1 cm ⌀, brownish-yellow.

**C. pauperae** L. Bol. (§ II/4). – Cape: Lit. Namaqualand. – Forming cushions 6 cm ⌀, with the dry, papery L.-Sh. persisting; **Bo.** obovoid to obconical, 5–6 mm tall, 5.5 mm ⌀, the upperside circular, convex, with a compressed Fi. 2–3 mm long, the surface pale green; **Fl.** pale.

**C. pauxillum** (N. E. Br.) N. E. Br. (Pl. 156/2) (§ II/5a) (*Mes. p.* N. E. Br.). – Cape: Clanwilliam D. – **Bo.** obconical, 10–15 mm tall, 6–9 mm ⌀, the upper surface convex or ± cordate, with a Fi. 1.5–4 mm long, the surface greyish-green, with a few brownish-red or dark green spots and lines; **Fl.** 8–14 mm ⌀, cream-coloured.

**C. pearsonii** N. E. Br. (§ II/3a). – Cape: Van Rhynsdorp D.

**C. — v. latisectum** L. Bol. – Pet. broader than v. **pearsonii**.

**C. — v. pearsonii** (Pl. 156/5) (*Mes. wettsteinii* L. Bol., *C. p. v. minor* N. E. Br., *C. braunsii* Schwant.). – Forming compact cushions; **Bo.** obconical, 8–16 mm tall, 10–18 mm ⌀, the upper surface circular, very slightly flat-convex with slightly projecting margins and a scarcely depressed Fi. 2–4 mm long surrounded by a darker zone with several indistinct dots, dark bluish- to ± yellowish-green; **Fl.** 16–22 mm ⌀, mauvish-pink.

**C. peersii** Lavis (§ II/2). – Cape: Willowmore D. – Close to **C. purpusii**; **Bo.** enclosed in the white, papery Sh., elongated-pyriform, 22 mm high, 11–14 mm ⌀, the upperside ± flat, with a somewhat depressed Fi. 2 mm long, the surface pale yellowish-green with indistinct reddish dots; **Fl.** 25 mm ⌀, creamy-white.

**C. pellucidum** Schwant. (§ I/1) (*C. elegans* N. E. Br., *Lithops. marlothii* N.E.Br., *Ophthalmophyllum m.* (N. E. Br.) Schwant.). – Cape: Lit. Namaqualand. – Forming small mats; **Bo.** 12–14 mm tall, 8–10 mm across, 6–8 mm thick, with 2 rounded lobes above, the surface greenish-brownish, tuberculate, with ochre-coloured markings, often window-like, light-coloured and translucent; **Fl.** 8–10 mm ⌀, white.

**C. permaculatum** Tisch. (§ II/3). – Cape: origin unknown. – Cushion-forming; **Bo.** obconical, 12 mm tall, 5–8 mm ⌀, the upperside flattened, circular, with a ± ciliate Fi. 1–2 mm long, the surface greyish-green, with large, dark green to reddish blotches above and reddish sides; **Fl.** white.

**C. petraeum** N. E. Br. (§ II/5a). – Cape: Laingsburg D. – Mat-forming; **Bo.** obconical, 8–10 mm tall, 3–5 mm ⌀, the upperside elliptical, convex, with a darker line around the Fi. and 1–2 branched lateral lines, the surface bluish-green with very dark red dots arranged in lines; **Fl.** 5 mm ⌀, yellowish-white.

**C. philippii** L. Bol. (§ II/1b). – Cape: Lit. Namaqualand. – Branched, c. 6.5 cm high; **shoots** covered with leathery, dry L.-Sh.; **Bo.** cordate, c. 24 mm tall, 15 mm across, with 2 short, obtuse lobes and a Fi. 2–3 mm long, the surface bluish-green, minutely rough; **Fl.**-tube white, Pet. yellow.

**C. pictum** (N. E. Br.) N. E. Br. (§ II/5a) (*Mes. p.* N. E. Br.). – Cape: Calvinia D. – **Bo.** obconical, 8–15 mm tall, 6–10 mm ⌀ above, the upperside elliptical, depressed, with some spots and vein-like lines, the surface deep green, the sides red; **Fl.** 5–14 mm ⌀, cream-coloured.

**C. pillansii** Lavis (Pl. 156/4) (§ I/2) (*C. edithae* N. E. Br., *Ophthalmophyllum e.* (N. E. Br.) Tisch., *C. lucipunctum* N. E. Br.). – Cape: Van Rhynsdorp D. – **Bo.** 1–3 together, obovoid, 22 mm tall, 22 mm wide, 20 mm thick, the surface light yellowish-green but the sides reddened, the upperside with large, green, translucent dots, sometimes merging to form a Wi., with a depressed Fi. 6 mm long; **Fl.** 25 mm $\varnothing$, purplish-red.

**C. piluliforme** (N. E. Br.) N. E. Br. (§ II/5a) (*Mes. p.* N. E. Br., *C. etaylori* Schwant., *Mes. aggregatum* N. E. Br., *C. a.* N. E. Br.). – Cape: Montagu D. – **Bo.** obconical, the upperside convex, 8–10 mm tall, 5–7 mm $\varnothing$, with a compressed Fi. 1.5–2.5 mm long surrounded by a dark line, the surface grey to greenish, minutely papillose, with rather darker dots; **Fl.** coppery.

**C. piriforme** L. Bol. (§ II/1a/1). – Cape: Lit. Namaqualand. – **Bo.** 2–3 together, with the L.-Sh. persisting, pyriform or obcordate, somewhat compressed, 27–29 mm tall, 10 mm $\varnothing$ above, broader midway, the $\pm$ acute lobes 3 mm long, with carinate, reddish margins, the surface olive-green with reddish spots, rough or minutely velvety; **Fl.** dark yellow.

**C. pisinnum** N. E. Br. (§ II/2). – Cape: Montagu D. – Mat-forming; **Bo.** obconical, 8–10 mm tall, the upperside rounded-truncate and $\pm$ circular, 6–8 mm $\varnothing$, with a somewhat depressed Fi. 2 mm long, the surface grey-green, velvety, indistinctly spotted; **Fl.** 8 mm $\varnothing$, yellowish-white. Acc. A. Tischer a variety of **C. truncatum.**

**C. placitum** (N. E. Br.) N. E. Br. v. **placitum** (§ II/2) (*Mes. p.* N. E. Br.). – Cape: Robertson D. – Mat-forming; **Bo.** obconical, up to 15 mm tall, 10 mm across, slightly bilobed above, with a distinct ridge, the surface green to reddish, with dark dots; **Fl.** 2 cm $\varnothing$, white to pink.

**C. —** v. **pubescens** Littlew. – Cape: Worcester D. – **Bo.** puberulous.

**C. pluriforme** L. Bol. (§ II/1a/1). – Cape: Lit. Namaqualand. – Up to 6 cm tall; **Bo.** oval to long-oval, much tapered at the base, 35–42 mm tall, 22 mm $\varnothing$ above, the 2 lobes rounded, mucronate, up to 1 cm $\varnothing$, the Fi.-zone somewhat clavate, the surface pale sea-green, indistinctly spotted, shortly papillose; **Fl.** 2.5 to 3 cm $\varnothing$, Pet. light yellow with a $\pm$ salmon-pink border.

**C. polulum** N. E. Br. (§ II/5a). – Cape: Laingsburg D. – Forming small mats; **Bo.** obconical, 15 mm tall, 7–10 mm $\varnothing$, somewhat convex above, the surface pale bluish-green, with some dots and spots; **Fl.** 10–14 mm $\varnothing$, whitish.

**C. polyandrum** Lavis (§ II/1b). – Cape: Lit. Namaqualand. – Laxly branched, c. 8.5 cm tall, with the L.-Sh. persisting; **Bo.** elongated-cordate, compressed above, c. 22 mm tall, 12 mm across, with short lobes, the surface olive-green, scarcely spotted; **Fl.** 2 cm $\varnothing$, Pet. whitish with a light pink border.

**C. praecinctum** N. E. Br. (§ II/5a). – Cape: Touwsrivier. – Mat-forming; **Bo.** scarcely projecting from the L.-Sh., obconical, 8–10 mm tall, 6–8 mm $\varnothing$ and flat above, with a red Fi. 3 mm long surrounded by several red lines, the surface light green, with scattered, dark spots; **Fl.** 8 mm $\varnothing$, yellowish-white.

**C. praecox** N. E. Br. (§ II/3a) (Pl. 156/6). – Cape: Lit. Namaqualand. – Mat-forming; **Bo.** obconical, 8–12 mm tall, 8–10 mm $\varnothing$, truncate above, the surface pale grey to bluish-green, with darker dots; **Fl.** 14–20 mm $\varnothing$, white.

**C. praegratum** Tisch. (§ II/3a). – Cape: Lit. Namaqualand. – Forming lax clumps; **Bo.** pyriform, up to 18 mm tall, 8–12 mm $\varnothing$, the top circular, convex, with a slightly depressed, rhombic Fi. surrounded by a dark, spotted zone, the surface light grey to bluish-grey, densely and indistinctly spotted; **Fl.** 11–13 mm $\varnothing$, pink.

**C. praeparvum** N. E. Br. (§ II/5b). – Cape: Lit. Namaqualand.

**C. —** v. **praeparvum.** – Mat-forming; **Bo.** elongated-conical, 10 mm tall, 8 mm wide, 6–7 mm thick, truncate above, with a Fi. 1–2 mm long, the surface bluish-green with a few dark red, raised dots; **Fl.** 5–6 mm $\varnothing$, cream-coloured.

**C. —** v. **roseum** Lavis. – **Fl.** pink.

**C. primosii** Lavis (§ I/3). – Cape: Lit. Namaqualand. – Mat-forming; **Bo.** usually 2 together, projecting from the papery Sh., elongated, 10–17 mm tall, 5 mm $\varnothing$ below, 8 mm $\varnothing$ and rounded above, with 2 very rounded lobes 4 mm long with translucent dots; **Fl.** 11 mm $\varnothing$, white.

**C. productum** L. Bol. (§ II/4). – Cape: Lit. Namaqualand. – L.-Sh. persisting and parchment-like; **Bo.** obovoid, narrowed teretely below, 14–17 mm tall, 10 mm $\varnothing$ above, with a Fi. 2.5–4 mm long; **Fl.** pale pink. Acc. A. Tischer probably identical with **C. breve.**

**C. prolongatum** L. Bol. (§ II/2). – Cape: Van Rhynsdorp D. – Close to **C. framesii; Bo.** rather elongated-pyriform, teretely extended at the base, 15–20 mm tall, c. 7 mm wide, 10 mm thick, the upperside convex, semi-translucent, with the Fi. surrounded by dots; **Fl.** straw-coloured.

**C. proximum** L. Bol. (§ II/1a/1). – Cape: Lit. Namaqualand. – Up to 10 cm tall, caulescent in age; **Bo.** obovoid to long-ovoid, 21–29 mm tall, up to 15 mm $\varnothing$, with a Fi. up to 4 mm long, lobes 5–10 mm long, rounded above, 4–7 mm $\varnothing$, the surface glabrous; **Fl.** c. 22 mm $\varnothing$, golden-yellow to orange.

**C. puberulum** Lavis (§ II/1b). – Cape: Lit. Namaqualand. – ♄ up to 3.5 cm tall; **shoots** 1 cm long; **Bo.** hidden in the old L.-Sh., cordate, slightly compressed, 17 mm tall, 4 mm wide, 8 mm thick, with carinate lobes 4 mm long, with a Fi. 2 mm long, the surface puberulous; **Fl.** yellow.

**C. pubicalyx** Lavis (§ II/8). – Cape: Van Rhynsdorp D. – Dwarf; **Bo.** 15–50 together, projecting only slightly from the L.-Sh., $\pm$ pyriform, very rounded, 5 mm tall, 2 mm $\varnothing$, green to purple, papillose; **Fl.** 2 cm $\varnothing$, reddish.

**C. pulchellum** Tisch. (§ II/7b). – Cape: Lit. Namaqualand. – Mat-forming; **Bo.** obconical, 15 mm tall, 7 mm $\varnothing$, the upper surface flat, circular, with dark green spots and a somewhat compressed, rhombic Fi. 2 mm long surrounded by a dark green line, the surface light to sea-green; **Fl.** 6–7 mm long, white below, with mauve tips.

**C. pusillum** (N. E. Br.) N. E. Br. (§ II/5a) (*Mes. p.* N. E. Br.). – S.Afr.: origin unknown. – Very small plant; **Bo.** obconical, 5–9 mm tall, 3.5–6 mm ⌀, the upperside circular or elliptical, with a Fi. 1–2 mm long, the surface grey-green with brownish-red or dark green, vein-like lines; **Fl.** cream-coloured.

**C. quaesitum** (N. E. Br.) N. E. Br. (§ II/1c) (*Mes. q.* N. E. Br., *Derenbergia q.* (N. E. Br.) Schwant., *C. modestum* L. Bol., *C. q.* Tisch.). – Cape: Lit. Namaqualand; S.W.Afr.: Gr. Namaqualand. – **Bo.** obconical, 10–12 mm tall, 10–15 mm across, 8–12 mm thick, the upper surface carinate, with darker dots in a R. along the keel and a Fi. 2 mm deep, the surface light grey to bluish-green, distantly and indistinctly darker-spotted; **Fl.** 6 mm ⌀, white. L. Bolus maintains *C. modestum* as a distinct species.

**C. radiatum** Tisch. (§ II/5a). – Cape: origin unknown. – **Bo.** 7–10 mm tall, 6 mm ⌀, the upper surface circular, the Fi. 1 mm long, with broad, hirsute lips, with red lines and several minute, broken lines radiating towards the margin, the surface greyish-green; **Fl.** cream-coloured.

**C. ramosum** Lavis (§ II/1b). – Cape: Lit. Namaqualand. – ♄ with **shoots** 3–5 cm long; **Bo.** c. 50 together, pyriform, elongated-compressed, 2–3 cm tall, 1–2 cm across, slightly bilobed, with a red line along the keels, the surface greyish-green, spotted, puberulous; **Fl.** 15 mm ⌀, yellow.

**C. rarum** N. E. Br. (§ II/3a). – Cape: ? Lit. Namaqualand. – Caulescent in age; **Bo.** obconical to ± pyriform, 25 mm tall, 15 mm across, 10–11 mm thick, the upperside ± convex and ± depressed in the centre, with an obtuse ridge and a Fi. 3 mm long surrounded by a dark ring, the surface light greyish-green with a few indistinct dots; **Fl.** pinkish-red.

**C. rauhii** Tisch. (§ II/5b). – Cape: Lit. Namaqualand. – Forming low cushions 5 cm ⌀; **Bo.** obconical, up to 1 cm tall, up to 5 mm across, up to 3.5 mm thick, the upperside flat to slightly convex, circular to oval to ± hexagonal, with radiating lines of slightly raised dots and a short, dark-bordered Fi., the surface dark olive to greyish-brown, rather reddish on the sides; **Pet.** straw-coloured with reddish tips.

**C. recisum** N. E. Br. (§ II/1a/2). – Cape: Lit. Namaqualand. – **Bo.** 14–25 mm tall, 7–10 mm across, 4–5 mm thick below, the 2 tapered, compressed lobes 6–8 mm long with acute margins, margins and lobe-tips purple, the surface dull green, minutely velvety; **Fl.** 15 mm ⌀, light yellow.

**C. regale** Lavis (§ II/1a/1). – Cape: Lit. Namaqualand. – **Bo.** mostly 6–8 together, with transparent, white-haired L.-Sh., long-obcordate, 26–45 mm tall, 10–15 mm ⌀ below and 22–26 mm ⌀ above, the lobes compressed, 6–12 mm long, 5 mm across, with a dark red keel; **Fl.** 15–18 mm ⌀, pink.

**C. retusum** N. E. Br. (§ II/1a/2) (*C. apertum* Tisch.). – Cape: Lit. Namaqualand. – **Bo.** 12 to 24 mm tall, 5–22 mm across above, 4–6 mm thick, laterally compressed, the 2 lobes 2–4 mm long, rounded, carinate, the Fi. surrounded by an indistinct line, the surface light green with solitary, indistinct dots; **Fl.** light yellow.

**C. ricardianum** Loesch et Tisch. (§ II/3a). – Southern S.W.Afr.; Cape: Lit. Namaqualand.

**C. —** ssp. **ricardianum**. – S.W.Afr. – Cushion-forming; **Bo.** obconical, 12–18 mm tall, the circular, mostly flat upper surface 10–15 mm ⌀, the margins often ± projecting, with a hirsute Fi. up to 1 mm long, the surface greyish-green with numerous dark green dots; **Fl.** 7–8 mm ⌀, white.

**C. —** ssp. **rubriflorum** Tisch. – Cape. – Dots larger; **Fl.** purple.

**C. robustum** Tisch. (§ II/3a). – S.W.Afr.: between Witpütz and the Orange Free State. – **Bo.** 25 mm tall, 10–18 mm ⌀, obconical, the upperside convex, the surface bluish-green, with large, dark green dots, especially around the depressed Fi.; **Fl.** 15 mm ⌀, mauve to pink.

**C. rolfii** de Boer (§ II/5a). – Cape: Piquetberg, Elandsberg. – Forming cushions c. 6 cm ⌀; **Bo.** numerous, 10–14 mm tall, obconical, laterally compressed, shortly papillose, the upperside 10 mm long, 6 mm across, slightly convex to ± saddle-shaped, oval, greyish-green, with 3–4 X-, V- or Y-shaped or branched red lines and a very slightly depressed, puberulous Fi. 2–2.5 mm long surrounded by a red line; **Fl.** white to violet-pink, 10–12 mm ⌀.

**C. roodiae** N. E. Br. (§ I/2). – Cape: Van Rhynsdorp D. – Mat-forming; **Bo.** obconical-elongate, 10–12 mm tall, 6–9 mm across, 3 to 5 mm thick, compressed, the 2 lobes equal or not, with translucent spots, the indentation 1.5 mm deep, the surface light green to light yellowish-green; **Fl.** 10–16 mm ⌀, white.

**C. rooipanense** L. Bol. (§ II/2). – Cape: Van Rhynsdorp D. – **Bo.** obovoid, c. 13 mm tall, 10 mm ⌀, the upperside convex, with a slightly compressed Fi. 4 mm long, the surface green, with deep green spots above; **Fl.** pale lemon-yellow. Acc. A. Tischer a form of *C. uviforme*.

**C. rostratum** Tisch. (§ II/1c). – Cape: Lit. Namaqualand. – Forming dense clumps; **Bo.** numerous, cylindrical below, bilobed and somewhat compressed above, up to 20 mm tall, 14 mm across, 10 mm thick, the lobes erect, 12 mm long, rounded outside, the tip acute or chin-like and slightly crenate-incised, with a slightly thickened, ± papillose Fi. 3 mm long set in a dark green zone, the surface whitish-green to greyish-green, indistinctly spotted all over; **Fl.** small, white.

**C. rubricarinatum** Tisch. (§ II/1c). – Cape: Lit. Namaqualand. – **Bo.** obconical, 10 mm tall, 5 mm across, 4 mm thick, the upperside slightly convex, divided by the V-shaped depression of the 2 mm long, hirsute Fi. into 2 small lobes, the surface green, with numerous, minute dots, the keel-line of the lobes crimson, and a reddish border around the Fi.; **Fl.** 6–8 mm ⌀, orange-yellow.

**C. rubristylosum** Tisch. (§ II/3a). – Cape: Lit. Namaqualand. – Clump-forming; **Bo.** obconical, 13 mm tall, c. 6 mm ⌀, the upperside flat, ± elliptical, with a slightly depressed Fi. 1.5 to

Plate 170. 1. **Hymenogyne glabra** (AIT.) HAW. (Photo: G. D. ROWLEY); 2. **Jacobsenia kolbei** (L. BOL.) L. BOL. et SCHWANT.; 3. **Imitaria muirii** N. E. BR.; 4. **Jensenobotrya lossowiana** HERRE.

Plate 171. 1. **Juttadinteria longipetala** L. Bol. (Photo: H. Chr. Friedrich); 2. **Kensitia pillansii** (Kensit) Fedde; 3. **Juttadinteria tetrasepala** L. Bol.; 4. **Khadia beswickii** (L. Bol.) N. E. Br. (Repr.: Fl. Pl. Afr. 1958/5, Drawing: M. M. Page); 5. **Juttadinteria simpsonii** (Dtr.) Schwant. (Photo: as 1).

Plate 172. 1. **Lampranthus roseus** (WILLD.) SCHWANT. (Photo: Kakteen-HAAGE, Erfurt); 2. **Lapidaria margaretae** (SCHWANT.) DTR. et SCHWANT. (Photo: G. SCHWANTES).

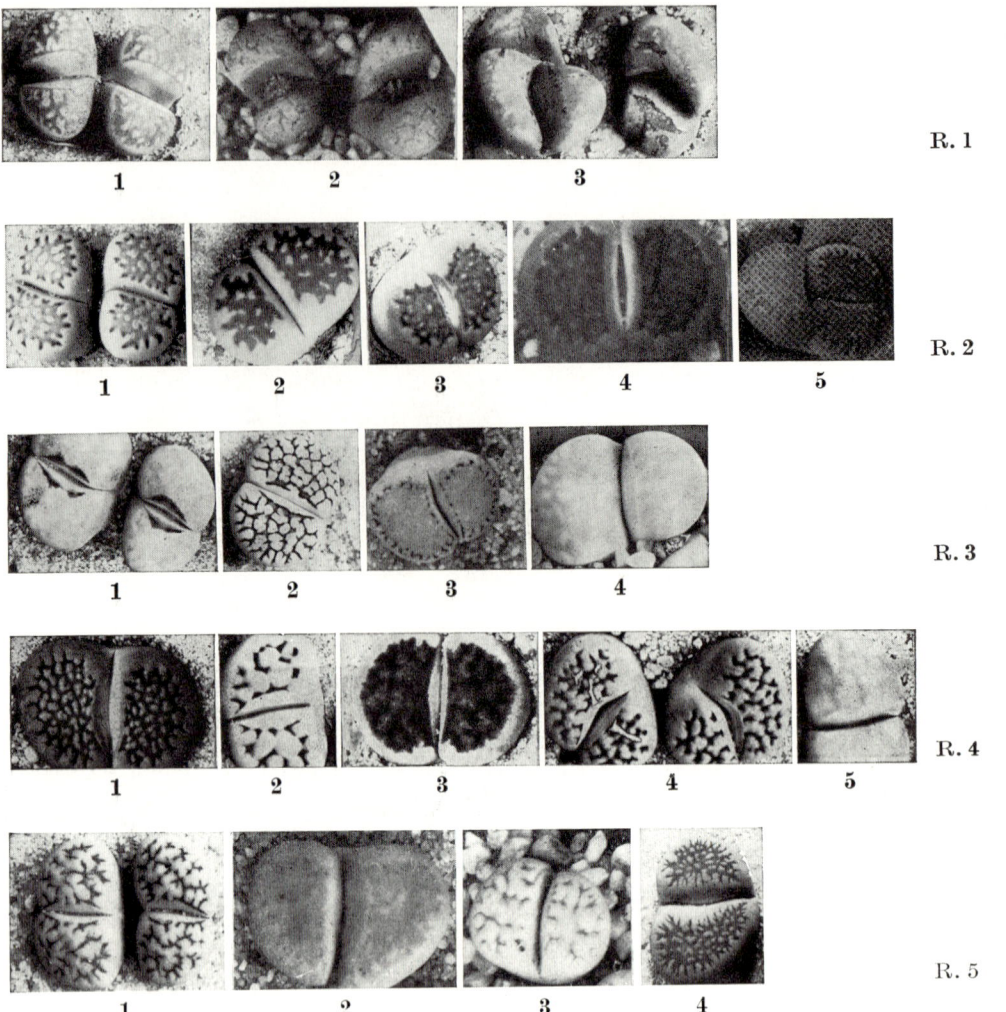

Plate 173. Row 1. 1. Lithops marmorata. 2. L. elisae. 3. L. optica.
Row 2. 1. L. fulleri v. fulleri. 2. L. fulleri v. brunnea. 3. L. fulleri v. ochracea. 4. L. fulleri v. chrysocephala. 5. L. fulleri v. kennedyi.
Row 3. 1. L. julii v. julii. 2. L. julii v. reticulata. 3. L. julii v. rouxii. 4. L. julii v. littlewoodii.
Row 4. 1. L. hallii. 2. L. karasmontana v. karasmontana. 3. L. karasmontana v. summitatum. 4. L. karasmontana v. mickbergensis. 5. L. karasmontana v. opalina.
Row 5. 1, L. erniana v. erniana. 2. L. erniana v. witputzensis. 3. L. erniana v. aiaisensis. 4. L. deboeri.

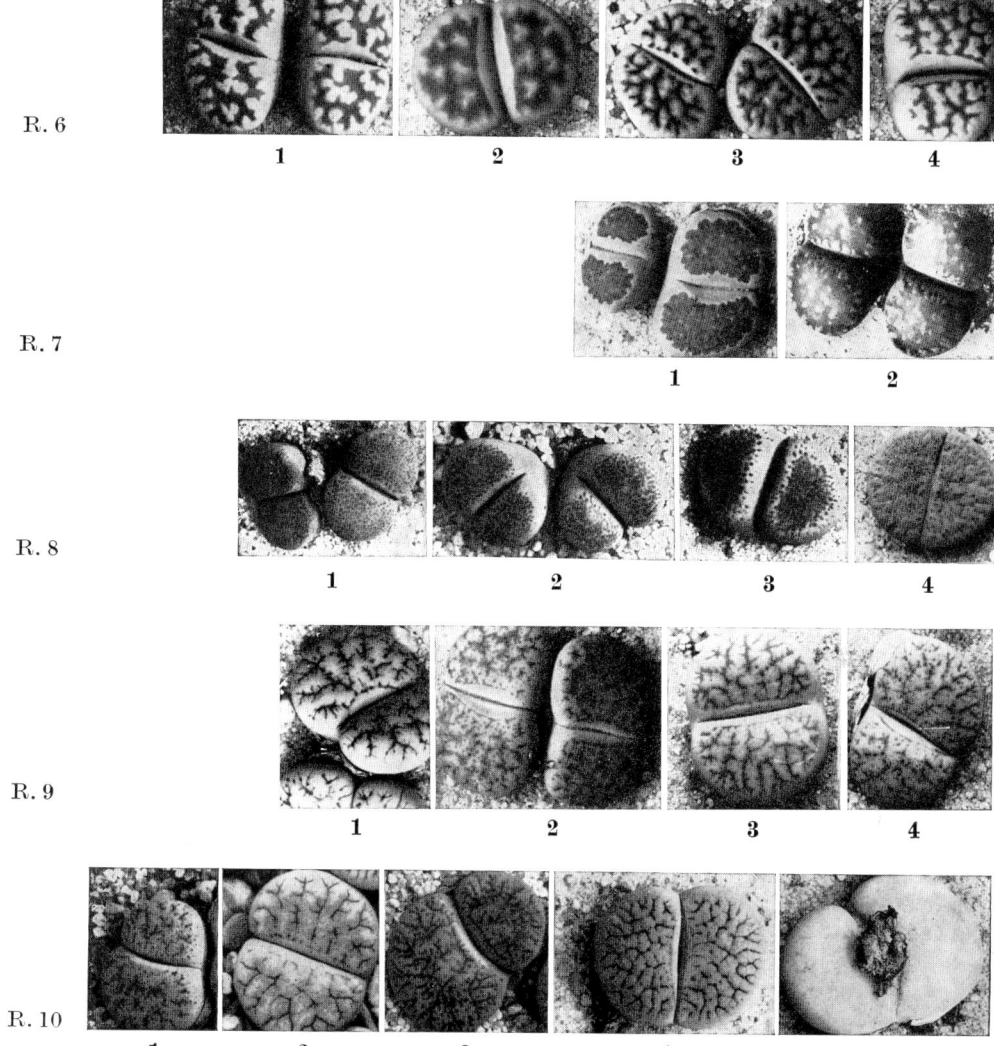

**Plate 174.** Row 6. 1. Lithops bella v. bella. 2. L. bella v. lericheana. 3. & 4. L. bella v. eberlanzii.
Row 7. 1. L. salicola. 2. L. villetii.
Row 8. 1. L. localis v. localis. 2. L. localis v. peersii. 3. L. localis v. terricolor. 4. L. fulviceps.
Row 9. 1. L. werneri. 2. L. pseudotruncatella v. pseudotruncatella. 3. L. pseudotr. v. pulmonuncula. 4. L. pseudotr. v. elisabethae.
Row 10. 1. L. pseudotruncatella v. edithae. 2. L. pseudotr. v. brandbergensis. 3. L. pseudotr. v. mundtii. 4. L. pseudotr. v. dendritica. 5. L. pseudotr. v. volkii.

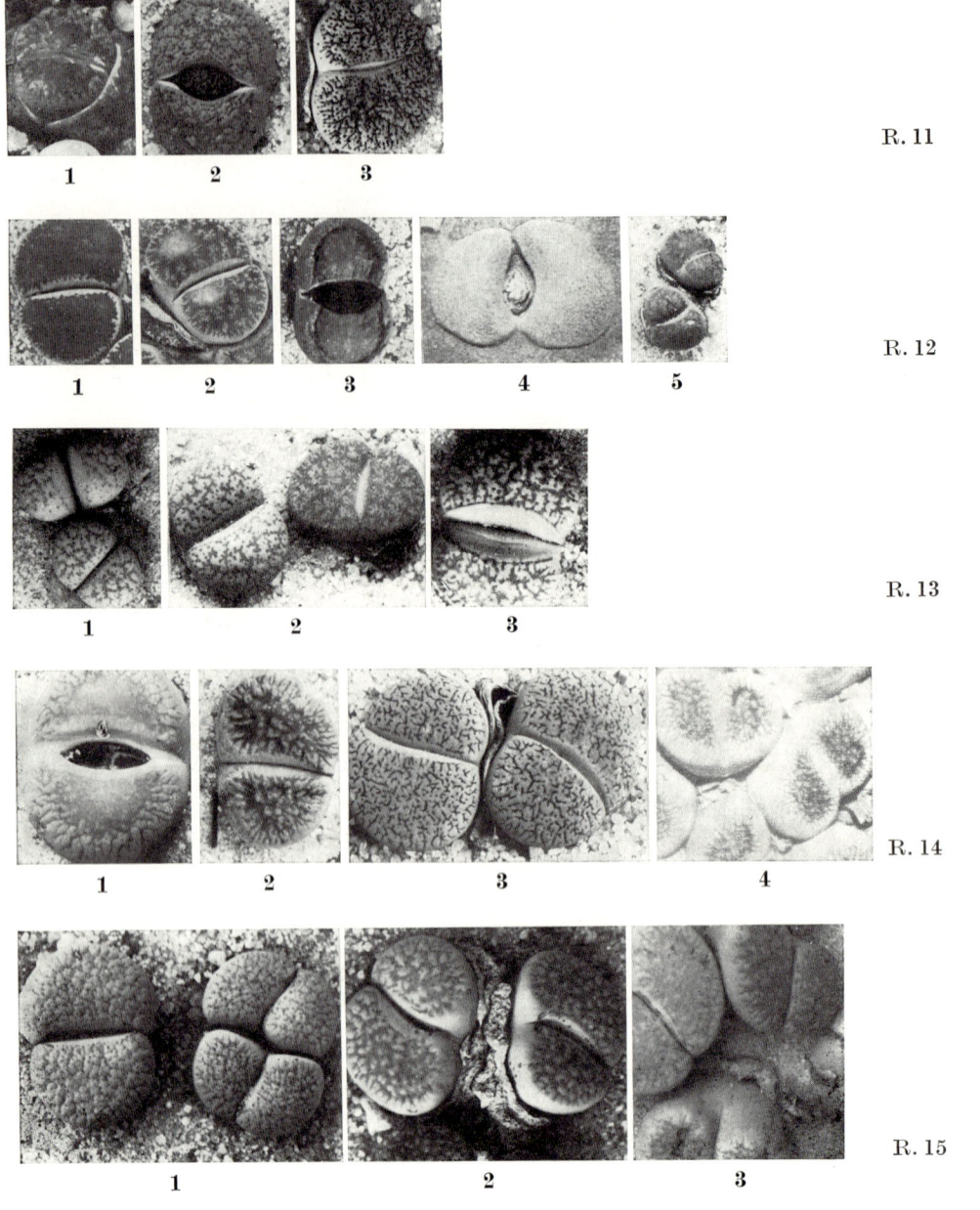

Plate 175. Row 11. 1. Lithops lesliei v. lesliei. 2. L. lesliei v. hornii. 3. L. lesliei v. venteri.
Row 12. 1. L. lesliei v. maraisii. 2. L. lesliei v. luteoviridis. 3. L. lesliei v. rubrobrunnea. 4. L. lesliei v. mariae. 5. L. lesliei v. minor.
Row 13. 1. Lithops franciscii. 2. L. gesinae v. gesinae. 3. L. gesinae v. annae.
Row 14. 1. L. aucampiae v. aucampiae. 2. L. aucampiae v. euniceae. 3. L. aucampiae v. koelemanii. 4. L. aucampiae v. fluminalis.
Row 15. 1. L. verruculosa v. verruculosa. 2. L. verruculosa v. inae. 3. L. verruculosa v. glabra.

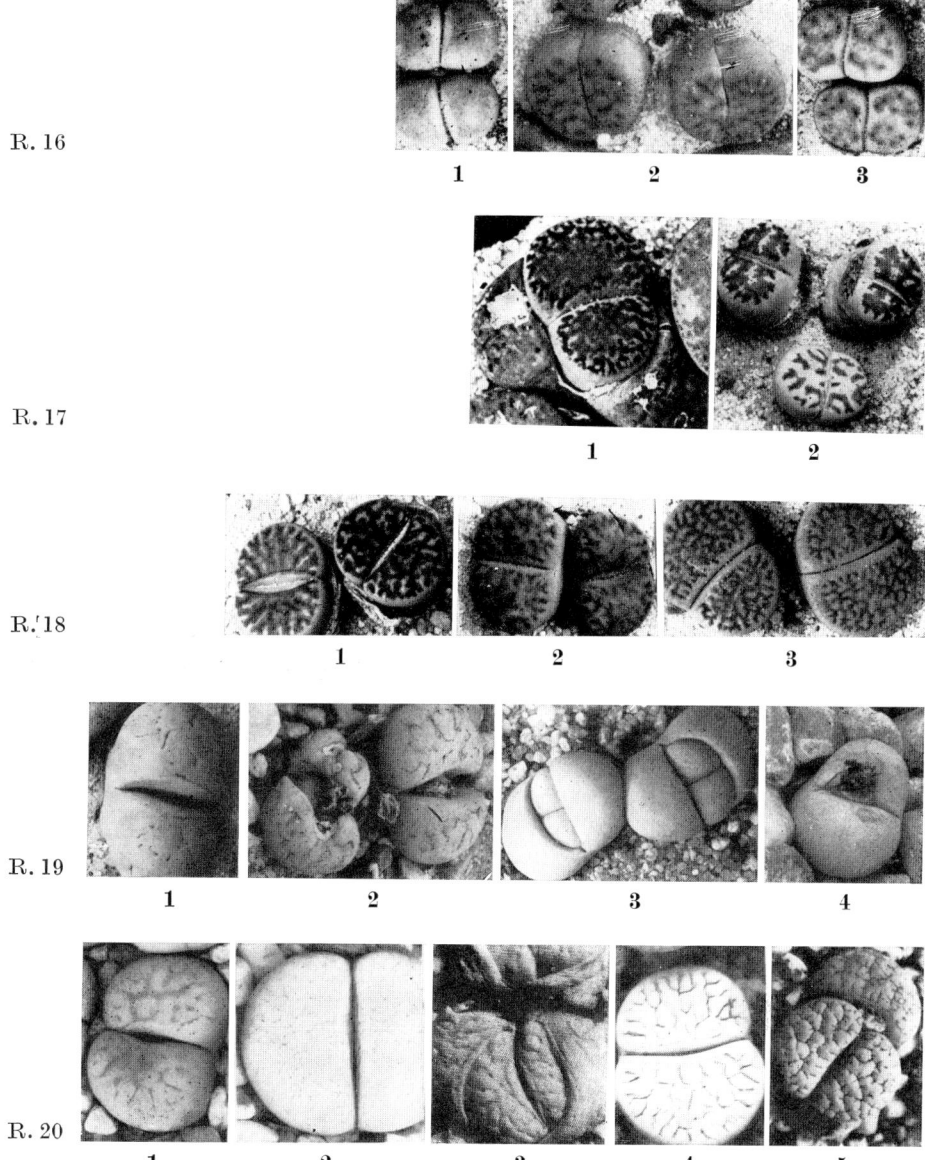

Plate 176. Row 16. **1.** Lithops dinteri v. dinteri. **2.** L. dinteri v. multipunctata. **3.** L. brevis.
Row 17. **1.** L. glaudinae. **2.** L. dorotheae.
Row 18. **1.** L. bromfieldii. **2.** L. insularis. **3.** L. marginata.
Row 19. **1.** L. ruschiorum v. ruschiorum. **2.** L. ruschiorum v. lineata. **3.** L. ruschiorum v. nelii. **4.** L. steineckeana.
Row 20. **1.** L. vallis-mariae v. vallis-mariae. **2.** L. vallis-mariae v. groendraaiensis. **3.** L. vallis-mariae v. margarethae. **4.** L. gracilidelineata v. gracilidelineata. **5.** L. gracilidelineata v. waldronae.

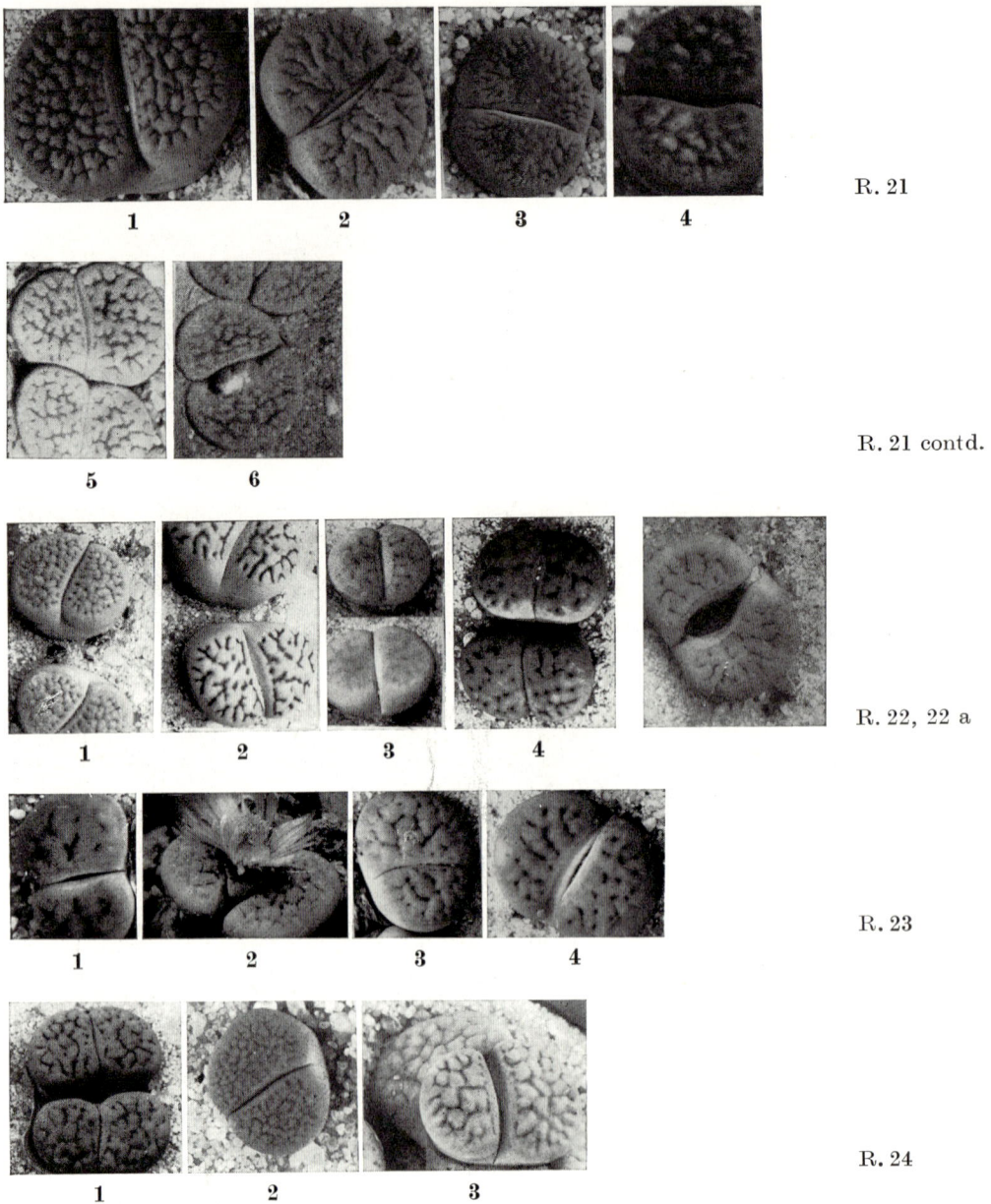

Plate 177. Row 21. 1. Lithops turbiniformis v. turbiniformis. 2. L. turbiniformis v. lutea. 3. L.. turbiniformis v. brunneo-violacea. 4. L. turbiniformis v. subfenestrata. 5. L turbiniformis v. susannae. 6. L. turbiniformis v. elephina.
Row 22. 1. L. christinae. 2. L. mennellii. 3. 4. L. marthae. Row 22a. L. archerae.
Row 23. 1. L. schwantesii v. schwantesii. 2. L. schwantesii v. kunjasensis. 3. L. schwantesii v. rugosa. 4. L. schwantesii v. triebneri.
Row 24. 1. L. schwantesii v. urikosensis. 2. L. schwantesii v. gebseri. 3. L. schwantesii v. nutupdriftensis.

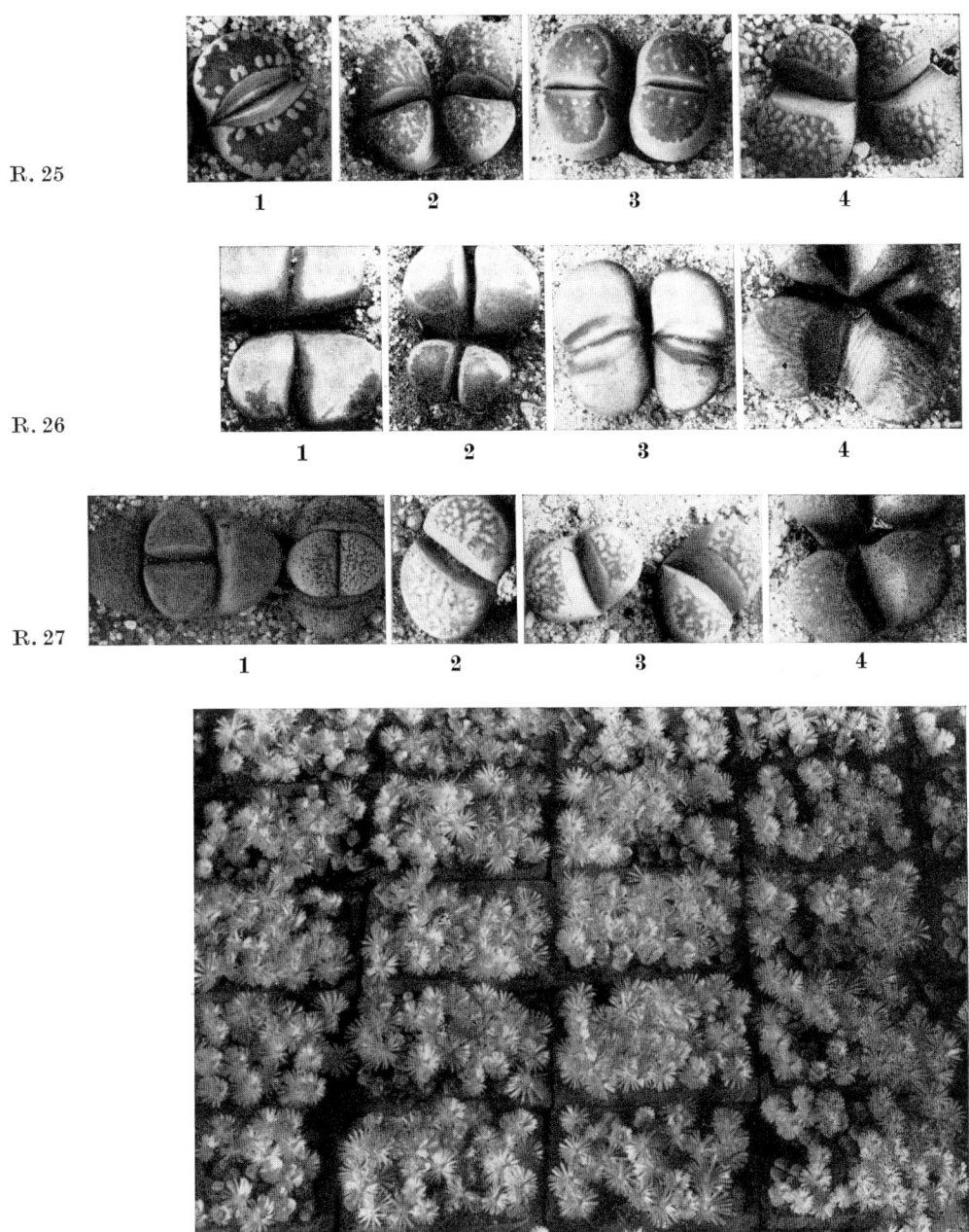

'Flowering stones' in a Japanese nursery.

Plate 178. Row 25. 1. Lithops otzeniana. 2. L. helmutii. 3. L. olivacea. 4. L. weberi.
Row 26. 1. L. meyeri. 2. L. viridis. 3. L. divergens v. divergens. 4. L. divergens v. amethystina.
Row 27. 1. L. herrei v. herrei. 2. L. herrei v. hillii. 3. L. herrei v. geyeri. 4. L. comptonii.

Plate 179. 1. Lithops bella (DTR.) N. E. BR. v. bella; 2. L. fulviceps (N. E. BR.) N. E. BR. (Photo: G. SCHWANTES); 3. L. gracilidelineata DTR. v. gracilidelineata.

Plate 180. 1. Lithops karasmontana v. summitatum (DTR.) DE BOER et BOOM (Photo: G. SCHWANTES; 2. L. julii v. rouxii DE BOER; 3. L. lesliei v. venteri (NEL) DE BOER et BOOM.

Plate 181. 1. Lithops pseudotruncatella (Bgr.) N. E. Br. v. **pseudotruncatella** (Photo: G. Schwantes); 2. **L. turbiniformis** N. E. Br. v. **turbiniformis** (Photo: H. W. de Boer); 3. **L. schwantesii** Dtr. v. **schwantesii**.

Plate 182. 1. **Machairophyllum acuminatum** L. Bol.; 2. **Maughaniella luckhoffii** (L. Bol.) L. Bol. (Photo: G. Schwantes); 3. **Malephora mollis** (Ait.) N. E. Br.; 4. **M. lutea** (Haw.) Schwant.

Plate 183. 1. Mesembryanthemum aitonis JACQ.; 2. M. crystallinum L. (Photo: G. SCHWANTES); 3. M. crystallinum L.

Plate 184. 1. **Mestoklema arboriforme** (Burch.) N. E. Br. (Photo: R. G. Strey); 2. **Meyerophytum meyeri** (Schwant.) Schwant. in the resting period; 3. The same in growth (Photos: 2 & 3: G. Schwantes); 4. **Micropterum herrei** Schwant. (Photo: as 3).

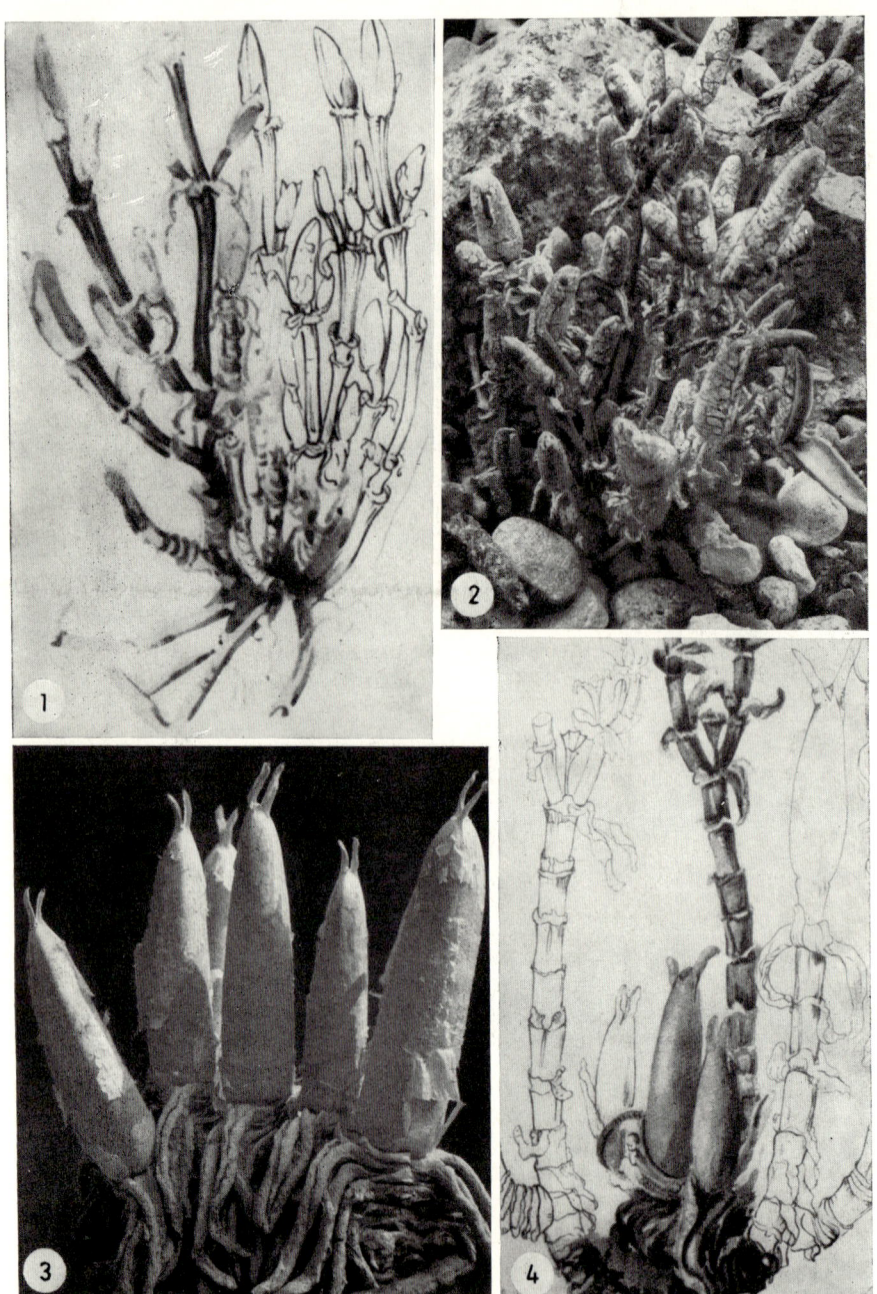

Plate 185. 1. **Mimetophytum parvifolium** L. Bol. (Repr.: L. Bolus, Notes on Mesembr. III, Pl. 87, Drawing: B. O. Carter); 2. **Mitrophyllum gracile** (Schwant.) de Boer; 3. **M. grande** N. E. Br. (Photo: H. W. de Boer); 4. **M. pillansii** N. E. Br. (Repr.: l. c. Pl. 59. Photo: U. C. T. Library).

2.5 mm long bordered by a R. of dots, the surface greyish-green or bluish-green, with rather darker dots above; **Fl.** yellow, style red.

**C. rubrolineatum** RAWÉ (§ II/6). – Cape: Lit. Namaqualand, S.E. of Van Rhynsdorp. – Cushions up to 6 cm ∅; **Bo.** up to 50 together, with short Int., long-conical, up to 14 mm long, up to 7 mm ∅, the upperside flat, circular to oval, with dense, small, glassy papillae, dots and simple or branched, prominent, reddish-purple lines, with a depressed, glabrous Fi. 1 mm long set in a square or oval zone; **Fl.** 2 cm ∅, pinkish-violet.

**C. rubroniveum** L. BOL. (§ I/1). – Cape: Bushmanland. – Mat-forming; **Bo.** hidden in the papery Sh., cylindrical or ± ovoid-cylindrical, 8–15 mm tall, 5–6 mm ∅, the upperside ± circular, with a compressed Fi. 2–3 mm long, the surface green, sometimes with red dots and marks; **Fl.** whitish.

**C. ruschii** SCHWANT. (§ II/3a) (*C. r. v. obtusipetalum* L. BOL.). – Cape: Lit. Namaqualand, Richtersveld. – **Bo.** conical, the upperside ± circular, 22 mm across, almost flat, pale bluish-green, with no spots or these barely perceptible; **Fl.** with obtuse or acute, mauve to pink Pet.

**C. saxetanum** (N. E. BR.) N. E. BR. f. **saxetanum** (§ II/1c) (*Mes. s.* N. E. BR., *Derenbergia s.* (N. E. BR.) SCHWANT., *Mes. fimbriatum* MARL., *Mes. boehmianum* DTR., *C. namibense* N. E. BR.). – S.W.Afr.: Namib, near Lüderitz Bay. – Forming large clumps of 100–200 Bo.; **Bo.** ovoid to ± cylindrical, 6 to 10 mm tall, 2–4 mm ∅, compressed above, with 2 short lobes up to 1.5 mm tall, the Fi. 1 mm long, whitish-haired, open, the surface ± spotted; **Fl.** white.

**C.** – f. **hallianum** ROWL. (*C. s. forma* L. BOL.). – Cape: Lit. Namaqualand. – **Bo.** larger, 7 to 14 mm tall, c. 14 mm ∅, lobes 1.5–2.5 mm long; **Pet.** broad, white to pale yellow.

**C. schlechteri** SCHWANT. (§ II/3a). – Cape: Lit. Namaqualand. – Forming flat mats; **Bo.** obconical to sinuate-cordate, truncate above, with the Fi. 2–3 mm long with white hairs inside, the surface grey to greyish-green with dark green dots; **Fl.** 18 mm ∅, white.

**C. scitulum** (N. E. BR.) N. E. BR. (§ II/5a) (*Mes. s.* N. E. BR.). – Cape: Laingsburg D. – **Bo.** crowded, obconical, 12–14 mm tall, 8 to 14 mm wide, 7–13 mm thick, the upperside slightly convex, broadly elliptical, slightly depressed in the centre, with a Fi. 2 mm long, the surface greyish-green, with several ± connected, reddish-brown, vein-like lines above; **Fl.** 6 to 16 mm ∅, white.

**C. sellatum** TISCH. (§ II/3b). – Cape: Lit. Namaqualand. – Clump-forming; **Bo.** obconical, c. 15 mm tall, the upperside ± rhombic, 4 to 6 mm ∅, flat and with a small saddle, the Fi. 2 mm long, with white hairs, the surface bluish-green, with several bluish-green dots above towards the outer edge; **Fl.** 13 mm ∅, light pink.

**C. semilunulum** TISCH. (§ II/1b). – S.Afr.: origin unknown. – Forming dense mats, with the old L.-Sh. papery and persisting; **Bo.** pyriform or cordate, ± compressed or ± spherical, up to 18 mm tall, 13 mm across above, up to 10 mm thick, with a rounded keel-line, and a scarcely compressed, white-papillose Fi. up to 3 mm long with a dark green zone at each end, and a lunate red mark on either side, the surface glabrous or with flat papillae, whitish-green to dull green, spotted; **Fl.** 20–24 mm ∅, light yellow.

**C. semivestitum** L. BOL. (§ II/1a/2). – Cape: Bushmanland. – **Bo.** 12 together, elongate or somewhat depressed-cylindrical, cuneately compressed above, 22 mm tall, 14 mm wide, 17 mm thick, with acute lobes up to 2 mm long, the surface pale green, ± puberulous above, with green dots along the margins; **Fl.** 12 mm ∅, pink.

**C. signatum** (N. E. BR.) N. E. BR. (§ II/5a) (*Mes. s.* N. E. BR.). – S.Afr.: origin unknown. – Mat-forming; **Bo.** obconical, 8–15 mm tall, 7–10 mm across, 5–8 mm thick, the upperside ± flat, with a Fi. 1.5–2.5 mm long, the surface greyish-green with dark purple or dark green, intricate lines; **Fl.** cream-coloured.

**C. simile** N. E. BR. (§ II/1a/1). – Cape: Lit. Namaqualand. – **Bo.** compressed-cordate, 12 to 20 mm tall, 12–18 mm across, 9–12 mm thick, with a V-shaped indentation above and obtusely keeled lobes, the surface bluish-green or greyish-green, with dark green dots; **Fl.** yellow.

**C. simplum** N. E. BR. (§ II/1a/1). – Cape: Lit. Namaqualand. – **Bo.** ± cylindrical, 4–5 cm tall, 10–12 mm thick, with 2 compressed, rounded lobes 6 mm long and across, the surface green, minutely velvety; **Fl.** yellow.

**C. smithersii** L. BOL. (§ II/1a/1). – Cape: Lit. Namaqualand. – **Bo.** 3.5–4.5 cm tall, 15–17 mm wide above, 17 mm thick, with 2 often unequal, ± acute lobes c. 7 mm long, the surface dark green with indistinct dots; **Fl.** yellow.

**C. smorenskaduense** DE BOER v. **smorenskaduense** (§ I/2). – Cape: Bushmanland. – **Bo.** ovoid to short-cylindrical, 1–2 cm long, 1 to 1.2 cm ∅, with rounded lobes 1–2 mm long and a depressed Fi. 3–6 mm long, the surface green, puberulous, ± glossy, Wi. numerous, small, not translucent, dark green; **Fl.** 23–25 mm ∅, pinkish-mauve to white.

**C.** – f. **rimarium** DE BOER. – **Bo.** lighter, surface less glossy, lobes shorter.

**C. sororium** N. E. BR. (§ II/1a/1). – Cape: Lit. Namaqualand. – **Bo.** cuneate-cylindrical, 25 to 45 mm tall, 12–18 mm across, 8–16 mm thick, with 2 lobes 6–12 mm long and red margins and keel, the surface mostly pale bluish-green, with indistinct green dots; **Fl.** 16–22 mm ∅, yellow.

**C. spectabile** LAVIS (§ II/5b). – Cape: Worcester D. – **Bo.** 12 mm tall, 8 mm wide, 10 mm thick, the upperside flat, circular, green, with prominent reddish-brown and somewhat branched lines, the sides red; **Fl.** 16 mm ∅, pinkish-white.

**C. stegmannianum** L. BOL. (§ II/2). – Cape: Willowmore D. – Forming mats 4–5 cm ∅, with the old L.-Sh. parchment-like, brown, persisting; **Bo.** obconical, 12 mm tall, 10 mm wide, 13 mm thick, the upperside convex, with a hirsute Fi. 3–4 mm long, the surface pale green with indistinct dots above, these sometimes arranged in semicircles; **Fl.** 14 mm ⌀, white to pale pink.

**C. stenandrum** L. Bol. (§ II/5b). – Cape: Van Rhynsdorp D. – Old L.-Sh. parchment-like, grey, persisting; **Bo.** broadly obconical, 8 mm tall, 9 mm ⌀, or obovoid, 11 mm tall, 7–8 mm wide, 11 mm thick, the upperside slightly convex, circular to broadly elliptical, with a Fi. 2–3 mm long, the surface bluish-green, with several brownish-red dots; **Fl.** straw-coloured.

**C. stephanii** Schwant. (§ II/8) (*C. helmutii* Lavis). – Cape: Lit. Namaqualand. – Forming small, slightly convex mats; **Bo.** numerous, obconical, 7 mm tall, 6 mm ⌀, the upperside somewhat convex, with a Fi. up to 1 mm long, the surface greyish-green, somewhat lighter above, entirely covered with dense white hairs up to 1 mm long; **Fl.** 4–5 mm ⌀, whitish.

**C. stevens-jonesianum** L. Bol. (§ II/4). – Cape: Lit. Namaqualand. – Plant 5–8 cm tall, with the reddish-brown L.-Sh. persisting for 1 year; **Bo.** broadly obconical, 1 cm tall and ⌀, the upperside furrowed, with a compressed Fi. 2 mm long, the surface with dark green dots; **Fl.** with a white tube and coppery Pet.

**S. steytlervillense** Tisch. (§ II/2) (*C. orientale* L. Bol.). – Cape: Steytlerville D. – Mat-forming, with the old L.-Sh. papery, grey to brown; **Bo.** obconical, 12 mm tall, 8 mm wide, the upperside ± elliptical, 6 mm thick, flat, with a compressed Fi. 3 mm long, with larger and smaller spots forming a keel-line over the indistinct lobes, the surface light green, the sides reddish; **Fl.** c. 15 mm ⌀, straw-coloured.

**C. stipitatum** L. Bol. (§ II/5b). – Cape: Van Rhynsdorp D. – Forming low cushions; **Bo.** obovoid, much constricted below, 11–14 mm tall, broadly elliptical on the upperside, up to 9 mm across, up to 8 mm thick, with a pubescent Fi., the surface bluish to sea-green, minutely papillose, spotted; **Fl.** 12 mm ⌀, with numerous, acute, straw-coloured Pet. 4–6 mm long with reddish tips.

**C. strictum** L. Bol. (§ II/1a/1). – Cape: Lit. Namaqualand.

**C. —  v. inaequale** L. Bol. – Lobes shorter than in v. **strictum**, unequal in length.

**C. — v. strictum**. – **Bo.** elongate, c. 4 cm tall, 16 mm ⌀ above, with erect lobes 13 mm long, the surface tinged purplish-brown, rough or minutely velvety; **Fl.** golden-yellow.

**C. stylosum** (N. E. Br.) Tisch. (§ II/1a/1) (*Mes. s.* N. E. Br., *C. markoetterae* Schwant.). – Cape: Lit. Namaqualand. – Forming small mats; **Bo.** 10–12 together, up to 5 cm tall, 25 mm across, 5 mm thick, the 2 lobes somewhat compressed, rounded on the outside, separated by an indentation 5–20 mm deep, the surface light green, with the tip and margins of the lobes reddened and scattered spots below the Fi.; **Fl.** yellow.

**C. subconfusum** Tisch. (§ II/5a). – Cape: Ladismith D. – Cushions somewhat lax, with the light brownish L.-Sh. persisting; **Bo.** obconical, truncate above, up to 10 mm tall, the upperside irregularly circular, up to 8 mm ⌀, with the very shortly-papillose Fi. up to 2 mm long, the surface greyish- to ± sea-green, with dark green dots often confluent into short R., ± bullate in appearance; **Fl.** c. 10 mm ⌀, crimson.

**C. subcylindricum** L. Bol. (§ II/1a/2). – Cape: Lit. Namaqualand. – Forming dense mats 5 cm tall, 12 cm ⌀; old L.-Sh. imbricate and persisting for a long time; **Bo.** cylindrical, projecting above the Sh., 15–30 mm tall, 7 mm wide midway, 4 mm wide at the tip and 7–12 mm thick, the 2 lobes 6–7 mm long, rounded to truncate, the surface pale bluish-green, rough; **Fl.** vivid yellow.

**C. subglobosum** Tisch. (§ II/2). – Cape: Uniondale D. – Clump-forming; **Bo.** 2 cm tall, the upperside ± circular, 13 mm ⌀, ± bilobed, the lobes rounded, the minutely puberulous Fi. 3–4 mm long, the surface greyish-green, with large green dots above, and smaller ones below; **Fl.** white.

**C. subincanum** Tisch. (§ II/5a). – Cape: Van Rhynsdorp D. – Forming lax cushions, with the dry L.-Sh. persisting; **Bo.** obconical, up to 10 mm tall, the upperside ± flat to convex, usually broadly elliptical, up to 7 mm across, up to 6 mm thick, with a Fi. 2.5 mm long, the surface light greyish-green, papillose, with dark grey to greyish-brown spots, the Fi. bordered by a R. of dots, mostly with another R. from the Fi. to the margin; **Fl.** white to straw-coloured.

**C. subrisum** (N. E. Br.) N. E. Br. (§ II/4) (*Mes. s.* N. E. Br., *C. poellnitzianum* Schwant., *C. victoris* Lavis, *C. villetii* L. Bol., *C. forrestii* L. Bol., *C. longipetalum* L. Bol., *C. tenuisectum* L. Bol.). – Cape: Lit. Namaqualand. – **Bo.** obconical, up to 15 mm tall and across, the upper surface circular, flat to slightly depressed or convex, the Fi. 3–4 mm long, with slightly protruding lips, with a tiny pit at each end, the surface chalky whitish to bluish-green; Pet. golden-yellow with reddish tips.

**C. subtile** N. E. Br. (§ II/4). – Cape: Lit. Namaqualand. – **Bo.** 4–5 mm long, 3–4 mm ⌀, obconical, the upper surface circular, convex, with a Fi. 1–5 mm long, slightly compressed at the ends, the surface greenish-bluish; **Fl.** yellow.

**C. sulcatum** L. Bol. (§ II/7b). – Cape: Lit. Namaqualand. – Mat-forming; **Bo.** 4–9 together, with the old, thin, longitudinally furrowed, pale brown L.-Sh. persisting, cuneate, up to 16 mm tall, 6 mm across below, 8 mm across above, with 2 acute lobes up to 2.5 mm long, with a distinct line along the margins and raised lines up to 4 mm long on the keel, with c. 7 furrows between the lines, the Fi. 2 mm long; **Fl.** pink to white.

**C. supremum** L. Bol. (§ II/1a/1). – Cape: Lit. Namaqualand. – Forming mats c. 11 cm ⌀, with the L.-Sh. papery, transversely rough outside, persisting; **Bo.** long-ovoid, 34–42 mm tall, 18–28 mm ⌀ above, with 2 lobes 22–26 mm long, 9–12 mm wide midway, up to 27 mm thick, flat to obtuse; **Fl.** with white Cor. and yellow Pet.

**C. swanepoelianum** RAWÉ (§ II/6). – Cape: Calvinia D. – Cushion-forming; **Bo.** long-conical, up to 7 mm long, 4 mm ⌀, the upperside circular, mostly flat, with an indistinct keel, the surface smooth, greyish-green, the Fi. depressed, 1 mm long, sparsely haired, bordered by a line, with prominent, blackish-green, irregular spots and short lines radiating from it; **Fl.** 2 cm ⌀, pinkish-purple.

**C. tantillum** N. E. BR. (§ II/1a/2). – Cape: Lit. Namaqualand. – **Bo.** obconical, 4–10 mm tall, 5–8 mm across, 4–6 mm thick, the upperside with a faint ridge, with 2–9 solitary dots, the Fi. gaping, usually circular and bordered by a red line, with a triangular spot on each side; **Fl.** 12–15 mm ⌀, red.

**C. taylorianum** (DTR. et SCHWANT.) N. E. BR. (§ II/1a/2). – S.W.Afr.: Gr. Namaqualand. – Forming dense, flat cushions 10–15 cm ⌀; **Bo.** somewhat cordate, depressed above, 10–16 mm tall and across, 7–10 mm thick, the indentation on the upperside acute, 2–3 mm deep, the 2 lobes acutely carinate above, the broad sides rather obtuse and drawn upwards chin-like, the Fi. 1.5–3.5 mm long, the keel-line ± cartilaginous, it and the Fi.-border deep crimson, the surface with large and rather prominent, reddish-brown dots above.

**C. – v. ernianum** (LOESCH et TISCH.) DE BOER (*C. e.* LOESCH et TISCH.). – **Bo.**-colour chalky-green, the Fi. deep and gaping, the keel acuter, the spots fewer than in v. **taylorianum** or ± missing; **Fl.** 15–25 mm ⌀, pink to mauve.

**C. – v. taylorianum** (*Mes. t.* DTR. et SCHWANT.). – **Bo.**-colour light green, the sides with a slight pink tinge, the keel distinctly red, the Fi. scarcely gaping, the dots very numerous; **Fl.** 2 cm ⌀, pink. Plants transitional to v. **ernianum** have been observed.

**C. tectum** N. E. BR. (§ II/1a/2). – Cape: Lit. Namaqualand. – **Bo.** cordate, 12–20 mm tall, 9 mm across, 7 mm thick, the acute-angled lobes 1–2 mm high, the surface ± papillose, chalky green, with minute, dark dots above; **Fl.** yellow.

**C. terrestre** TISCH. (§ I/1). – Cape: Lit. Namaqualand. – In lax cushions up to 4 cm ⌀, 3 cm tall, with the leathery, brownish, old L.-Sh. persisting; **Bo.** elongate, oval to ± cylindrical, up to 16 mm tall, 7 mm across above, 6 mm thick, the Fi. up to 3.5 mm long, more rarely across the entire width and up to 5 mm long, making the upperside slightly bilobed, the surface dark olive to dirty brownish-green, murky translucent on the upperside, the Wi. dissolving into solitary spots, often with small islands, the Fi. rimmed with a coloured line or a R. of dots; **Fl.** 2 cm ⌀, pure white.

**C. terricolor** TISCH. (§ I/1). – Cape: Lit. Namaqualand. – Mat-forming; **Bo.** 10–15 mm tall, 6–8 mm across, 3–6 mm thick, with 2 rounded lobes up to 3 mm long, with a Fi. 3 mm long, the surface earth-coloured to greyish-brown, with dark brown, impressed spots above and concolorous dots around them, giving a wrinkled appearance; **Fl.** white.

**C. tetracarpum** LAVIS (§ II/3a). – Cape: Lit. Namaqualand. – ♄ 5–8 cm tall, with **shoots** 25–40 mm long covered with old L.-Sh.; **Bo.** 20–30 together, long-pyriform, up to 18 mm tall, up to 12 mm ⌀, convex above, pale green with dark dots; **Fl.** 13–19 mm ⌀, yellow.

**C. thudichumii** L. BOL. (§ II/4). – Cape: Lit. Namaqualand. – L.-Sh. parchment-like, persisting; **Bo.** broadly obovoid, 11–17 mm tall, 12–15 mm wide and thick, the upperside flat, circular, with a rather swollen, paler Fi. 3–4 mm long; **Fl.** yellow. Acc. A. TISCHER probably identical with **C. subrisum.**

**C. tischeri** SCHICK (§ II/1b). – Cape: Lit. Namaqualand. – Mat-forming, caulescent in age; **Bo.** broadly cordate, somewhat depressed above, 10–13 mm tall, 8–10 mm across, 7–9 mm thick, with the Fi. 2–4 mm long set in an irregularly laciniate, darker zone, the surface greyish-green with dark green dots; **Fl.** 15 mm ⌀, light mauve.

**C. truncatum** (THUNBG.) N. E. BR. (Pl. 157/1) (§ II/2). – Cape: Gamka R. catchment area. – Cushion-forming; **Bo.** obconical, ± convex, the Fi. ± impressed or somewhat gaping, with the lips sometimes slightly protuberant, with an indistinct ridge perpendicular to the Fi., the markings consisting of dense or crowded, small or large, dark green dots, rarely confluent into R.; **Fl.** flesh- to straw-coloured to almost white.

**C. — v. brevitubum** (LAVIS) TISCH. (*C. brevitubum* LAVIS). – Cape: Ladismith D. – **Fl.** somewhat smaller than v. **truncatum f. truncatum,** light flesh-coloured.

**C. — v. truncatum f. truncatum** (*Mes. t.* THUNBG., *C. truncatellum* N. E. BR., *C. cibdelum* N. E. BR., *Mes. albertense* N. E. BR., *C. a.* (N. E. BR.) N. E. BR., *C. spirale* N. E. BR., *Mes. purpusii* SCHWANT., *C. p.* (SCHWANT.) N. E. BR., *Mes. familiare* SCHWANT., *Mes. minusculum* SCHWANT., *Mes. obcordellum* MARL., *Mes. uvaeforme* PURP., *Mes. malleoliforme* SCHWANT.). – Cape: Karroo. – Caulescent in age; **Bo.** 15 mm tall, 13 mm ⌀ above, with the 2–3 mm long Fi. mostly ± depressed and the lips thus somewhat protruding, usually surrounded by a R. of dots, the surface greyish-green to bluish-green, with large and small dots; **Fl.** 14–16 mm ⌀, light straw-coloured to almost white.

**C. — v. — f. parvipunctum** (TISCH.) TISCH. (*C. p.* TISCH.). – Upper surface flatter than f. **truncatum** with the Fi. scarcely depressed, the dots more numerous and smaller.

**C. — v. — f. renniei** (LAVIS) TISCH. (*C. renniei* LAVIS). – With the Fi. rather gaping, bordered by dots.

**C. tubatum** TISCH. (§ II/3b). – Cape: Lit. Namaqualand. – Mat-forming; **Bo.** very crowded, obconical, with the upperside circular, flat or slightly depressed, 12 mm tall, 8 mm ⌀, with the Fi. distinctly pubescent, the surface bluish-green with several darker dots; **Fl.** 15–18 mm ⌀, pinkish-red.

**C. turrigerum** (N. E. BR.) N. E. BR. (§ II/6) (*Mes. t.* N. E. BR., *Derenbergia t.* (N. E. BR.) SCHWANT.). Cape: Malmesbury D. – **Bo.** 5 to 6 together, elongate, 2–3 mm ⌀ below and 8 mm ⌀ above, with 2 very rounded lobes

**Conophytum**

5 mm high, with reddish-brown dots along the lobe-margins, the surface bluish-green; **Fl.** 14 mm ⌀, white.

**C. udabibense** LOESCH et TISCH. (§ II/4). – S.W.Afr. – Udabib Mts. – In compact cushions; **Bo.** obconical, with the upper surface ± circular, rather flat to convex, 5–10 mm ⌀ above, with a slightly depressed Fi. 1–2 mm long with the lips often protruding, the surface chalky whitish-green, unmarked; **Fl.** 8–12 mm ⌀, yellow.

**C. umdausense** L. BOL. (§ II/1 a/1). – Cape: Lit. Namaqualand. – Possibly only a variety of **C. simplum**; **Bo.** shaped like **C. simplum** but the lobes longer, 14 mm long, up to 10 mm ⌀, erect, with a rounded tip, the surface rather velvety-haired; **Fl.** 2 cm ⌀, yellow.

**C. ursprungianum** TISCH. (§ II/5 b). – Cape.

**C.** – v. **stayneri** L. BOL. – Cape: Ceres D. – **Bo.** 6–8 mm tall, 6–7 mm across, 8–12 mm ⌀, with variously arranged dots on the upperside, either scattered or crowded; **Fl.** pink.

**C.** – v. **ursprungianum** (Pl. 157/2). – Cape: Calvinia D. – Cushion-forming; **Bo.** broadly obconical, 22 mm tall, the upperside rhombic to circular, 10–15 mm ⌀, flat to convex, often with a crater-like depression, with a ciliate Fi. 3–4 mm long, the surface greyish-green to chalky green, with dark to blackish-green dots confluent into short lines radiating out towards the margin, markings variable; **Fl.** c. 15 mm ⌀, white.

**C. uviforme** (HAW.) N. E. BR. (§ II/2). – Cape: Van Rhynsdorp, Calvinia, Clanwilliam D., widely distributed. – Caespitose; **Bo.** spherical to pyriform, 10–12 mm tall and across, 6 to 20 mm thick, the Fi. 2 mm long, ± sunken, the surface light yellowish-green with dark to reddish spots, the sides sometimes reddish; **Fl.** 1 cm ⌀, whitish-yellow. Variable species.

**C.** – v. **framesii** (LAVIS) TISCH. (*C. f.* LAVIS). – Van Rhynsdorp, Clanwilliam D. – **Bo.** pyriform, the Fi. V-shaped, slightly sunken, the upperside with distinct spots, not confluent.

**C.** – v. **meleagre** (L. BOL.) TISCH. (*C. m.* L. BOL., *C. colorans* LAVIS, *C. vanrhynsdorpense* SCHWANT., *C. julii* SCHWANT., *C. translucens* N. E. BR., *C. hillii* L. BOL.). – **Bo.** with large, dark spots on the upper surface.

**C.** – v. **uviforme** (*Mes. u.* HAW.). – Van Rhynsdorp, Calvinia D. – **Bo.** spherical to ± pyriform, the upperside ± domed or slightly sunken towards the Fi., marked with ± distinct, small or larger spots which are sometimes confluent into short lines, Bo.-colour green to yellowish or bluish-green, the sides at times tinged with red, when the markings are also reddish.

**C. vagum** N. E. BR. (Pl. 157/5) (§ II/5a). – S.Afr.: origin unknown. – Mat-forming; **Bo.** obovoid, 20 mm tall, the upperside convex, 10 mm wide, 8 mm thick, the surface bluish-green, with red dots and lines above; **Fl.** 1 cm ⌀, cream-coloured.

**C. vanbredai** L. BOL. (§ II/3a). – Cape: Lit. Namaqualand. – Mat-forming, with the parchmenty, brownish L.-Sh. persisting; **Bo.** obconical to pyriform, 17 mm tall, the upperside flat to convex, with a Fi. 3 mm long, the surface bluish-green, the sides reddened; **Pet.** whitish, with purplish-pink tips.

**C. vanheerdei** TISCH. (§ II/7b). – Cape: Bushmanland. – Forming clumps; **Bo.** ovoid to cylindrical, up to 12 mm tall, the upperside almost hemispherical, up to 8 mm ⌀, with a Fi. 1 mm long, the surface yellowish-green with glassy tubercles above; **Fl.** 12 mm ⌀, dark purple.

**C. vanzijlii** LAVIS (§ II/4). – Cape: Kenhardt D. – Mat-forming; **Bo.** obconical, laterally compressed, c. 11 mm tall, 19 mm ⌀, the upperside with a Fi. 4 mm long, the surface whitish-green; **Fl.** 14 mm ⌀, white to pink or yellow.

**C. variabile** L. BOL. (§ II/1 a/2). – Cape: Lit. Namaqualand. – Plant shortly branched; **Bo.** cylindrical or long-obovoid, 18–27 mm tall, 7 to 16 mm across, with 2 lobes 1–5 mm long or scarcely developed, the surface velvety-papillose; **Fl.** 26 mm ⌀, light yellow.

**C. varians** L. BOL. (§ II/5a). – Cape: Van Rhynsdorp D. – Old L.-Sh. persisting; **Bo.** obovoid to cylindrical, 6–15 mm tall, 8 mm wide, c. 10 mm thick, often narrow above, with a pubescent Fi. 3 mm long, the surface variously spotted and lined; **Fl.** whitish.

**C. velutinum** (SCHWANT.) SCHWANT. (Pl. 157/3) (§ II/1 b) (*Derenbergia v.* SCHWANT.). – Cape: Lit. Namaqualand. – **Shoots** branched, covered with old L.-Sh.; **Bo.** long-ovoid, c. 12 mm tall, convex above, up to 8 mm ⌀, with a scarcely depressed Fi., the surface olive-green, minutely velvety-papillose; **Fl.** c. 2 cm ⌀, purplish-pink.

**C. vescum** N. E. BR. (§ II/1c). – Cape: Lit. Namaqualand. – Mat-forming; **Bo.** obconical, up to 12 mm tall, 7–10 mm across, 5–7 mm thick, the upperside with a distinct keel, with the Fi. broadly rhombic, gaping, bordered by spots merging into lines, the surface greyish-green to green, with a few scattered dots; **Fl.** unknown.

**C. violaciflorum** SCHICK et TISCH. (§ II/1 a/2) (*C. geometricum* LAVIS). – Cape: Lit. Namaqualand. – Mat-forming; **Bo.** obcordate, c. 15 mm tall, slightly compressed and 10–11 mm ⌀ above, with scattered dots confluent into lines along the margins of the 2 mm long lobes and the keels, the surface dark green; **Fl.** 15 mm ⌀, violet to pink.

**C. viride** TISCH. (§ II/2). – Cape: Lit. Namaqualand. – Mat-forming; **Bo.** ± cordate, 12 mm tall, 7 mm across above, 5 mm thick, with 2 broadly rounded lobes and a compressed, ± rhombic Fi., the surface green to greyish-green, spotted, with a distinct keel-line; **Fl.** straw- to pale flesh-coloured.

**C. viridicatum** (N. E. BR.) N. E. BR. (§ II/2). – Cape: Laingsburg D.

**C.** – v. **punctatum** N. E. BR. – As for v. **viridicatum,** but with distinct dots, in indistinct lines.

**C.** – v. **viridicatum** (*Mes. v.* N. E. BR.). – **Bo.** obconical, 12–23 mm tall, 10–15 mm across, 9–13 mm thick, the upperside elliptical or circular, with a compressed Fi. 3–6 mm long, the surface green, with scarcely visible dots; **Fl.** white.

**C. vitreopapillum** RAWÉ (§ II/7 b). – Cape: Lit. Namaqualand. – Forming clumps 4 cm $\varnothing$; **Bo.** obconical, 12 mm tall, 7 mm $\varnothing$, the upperside circular, concave to compressed, the surface olive-green, tinged reddish, smooth, densely papillose, the papillae glossy, irregularly branched, translucent, the Fi. compressed, 2 mm long; **Fl.** 7 mm $\varnothing$, pink, with red stigmas.

**C. wagneriorum** SCHWANT. (§ II/2). – Cape: Willowmore D. – Mat-forming; **Bo.** obconical, 8 mm tall, the upperside tuberculate, circular, very convex, 10 mm across, with a Fi. 4 mm long with gaping, labiate margins, the surface whitish to bluish-green, reddish below, with numerous dark green to reddish dots; **Fl.** ? white.

**C. wettsteinii** (BGR.) N. E. BR. (§ II/3 a). – Cape: Lit. Namaqualand, N.E.Richtersveld. – Mat-forming; **Bo.** flattened obconical, 15 mm tall, 22–30 mm across, truncate above, the upperside rather flat or depressed, the Fi. 3 to 4 mm long, the surface dull green or greyish-green with minute whitish dots often interspersed with larger dots, or almost unspotted; **Fl.** violet-purple to light pink, sometimes with a white centre (v. *oculatum*).

**C. — v. speciosum** (TISCH.) TISCH. – Near Gelykswerf, S.E. of Kubus. – **Bo.** bluish or greyish-green, with distinct spots, denser around the Fi.; **Fl.** large, Pet. up to 2 cm long, the underside white, the upperside light pink.

**C. — v. wettsteinii** (*Mes. w.* BGR., *Mes. truncatellum* OTHM., *C. w.* v. *oculatum* L. BOL.). – **Bo.** with the upper surface flat to flattened trough-shaped, scarcely spotted; **Fl.** purplish-violet or light pink with a white centre.

**C. wiggettae** N. E. BR. (§ II/2). – Cape: Oudtshoorn D. – Mat-forming; **Bo.** obconical, 12–18 mm tall, the upper surface circular, flat-convex, up to 18 mm $\varnothing$, with a slightly depressed Fi. 4–6 mm long, the surface greyish-green, with numerous large, often reddish dots, the sides reddish; **Fl.** 15 mm $\varnothing$, whitish.

**C. wittebergense** DE BOER (Pl. 157/4) (§ II/5 a). – Cape: Laingsburg D. – Mat-forming; **Bo.** obconical, 13–18 mm tall, 14–20 mm wide, 10 to 13 mm thick, the upperside circular to oval, convex, with the 3–5 mm long Fi. surrounded by a dark line, the surface sea-green, with thin, reddish-brown, intricately branched lines, and concolorous dots; **Fl.** 14–16 mm $\varnothing$, pale yellow.

**Cylindrophyllum** SCHWANT. (§ IV/1/2). – Dwarf, branched, succulent plants; **St.** eventually dividing, with no visible Int.; **L.** cylindrical, thick, tapered, acute to truncate; **Fl.** on a bracteate pedicel, large, white to pink, summer, opening in the afternoon. – Greenhouse. Propagation: seed, cuttings.

**C. calamiforme** (L.) SCHWANT. (Pl. 158/1) (*Mes. c.* L.). – Cape: Laingsburg D. – Mat-forming; **L.** 6–8, curving, inclined, 5–7 cm long,

8 mm across, $\pm$ cylindrical, obtuse, mucronate, greyish-green with minute dots; **Fl.** 5–7 cm $\varnothing$, yellowish-white.

**C. comptonii** L. BOL. – Cape: Lit. Karroo. – Forming cushions 25 cm $\varnothing$; **L.** erect, cylindrical, tapered, c. 9 cm long, 1 cm $\varnothing$, with one side $\pm$ flattened; **Fl.** 7.5 cm $\varnothing$, silvery-white.

**C. dyeri** L. BOL. – Cape: Jansenville D. – In dense cushions; **St.** 6 cm long; **L.** $\pm$ cylindrical, the upperside flat, c. 7 cm long, 11 mm $\varnothing$, margins rather rough; **Fl.** 3 cm $\varnothing$, pale pink.

**C. hallii** L. BOL. – Cape: Bushmanland. – Cushions up to 20 cm $\varnothing$; **L.** flat on the upperside, indistinctly carinate on the underside, obtuse, 6–7.5 cm long, 5–7 mm wide, 10–12 mm thick; **Fl.** up to 6.5 cm $\varnothing$, creamy-white.

**C. obsubulatum** (HAW.) SCHWANT. (*Mes. o.* HAW.). – S.Afr.: origin unknown. – Resembles **C. calamiforme**; **L.** thickened above, cylindrical, obtuse, greener, spotted.

**C. tugwelliae** L. BOL. – Cape: Prince Albert D. – In dense mats; **L.** 2 on a shoot, $\pm$ cylindrical, tapered, c. 8 cm long, 1 cm wide, 12 mm thick, bluish-green; **Fl.** 5 cm $\varnothing$, light flesh-coloured. Acc. L. BOLUS only a form of **C. comptonii**.

**Dactylopsis** N. E. BR. (§ I/2). – Very dwarf, very succulent plants with a few short, thick **shoots** completely enclosed in the old, papery L.-Sh.; **L.** cylindrical, thick and fleshy, entirely concealed by the L.-Sh. during the long resting-period; **Fl.** solitary, open day and night for several weeks, Pet. very numerous, stiff, acute, staminodial. – Greenhouse, warm. Propagation: seed.

**D. digitata** (AIT.) N. E. BR. (Pl. 158/4) (*Mes. digitatum* AIT., *Mes. digitiforme* THUNBG.). – Cape: Van Rhynsdorp D. – Acauline, forming dense mats; **L.** alternate, 3–4 together on a shoot, c. 8–12 cm long, 2–2.5 cm thick, cylindrical, obtuse, digitate, soft, grey, glabrous; **Fl.** 15–20 mm $\varnothing$, white.

**D. littlewoodii** L. BOL. – Cape: Lit. Namaqualand. – Much smaller than **D. digitata**, with stilt-like **roots** from the main **Br.**; **L.** variable, the first one shortly sheathing, with a short blade, the second and third L. larger, spreading, compressed, long-ovate, c. 28 mm long, 13 mm $\varnothing$, with a Sh. 8 mm long; **Fl.** 14 mm $\varnothing$, white.

**Delosperma** N. E. BR. (§ IV/1/6). – ♃ or rarely ☉ or ⊙, mat-forming, compact or laxly branched plants, erect or with the primary **Br.** ascending, prostrate or more rarely creeping, herbaceous or shrubby, or with ☉ shoots from a tuberous or woody **caudex**; **L.** cylindrical, or semicylindrical or almost so, or flat, shortly united, variously shaped, the herbaceous parts glabrous or with soft hairs or more rarely velvety, apart from **D. pruinosum, longipes** and several similar species which normally have indistinct papillae; **Fl.** solitary or in a lax cyme, opening by day, 2–7 cm $\varnothing$, on a pedicel 1–5 cm long, Pet. numerous, white, or light or darker red, yellow, more rarely coppery to bronze-coloured. – Summer: in the open, winter: coldhouse. Propagation: seed, cuttings.

M. Lavis (S.Afr.) is at present preparing a monograph on this G. See her provisional classification into Sect. and Gr. in Journ. S.Afr. Bot. **XXXII**, 1966, p. 210, l. c. **XXXIII**, 1967, p. 313 and l. c. **XXXV**, 1969, pp. 146–147.

**D. aberdeenense** (L. Bol.) L. Bol. (*Mes. a.* L. Bol.). – Cape: Karroo. – Small, minutely papillose ♄ with dense, often prostrate **Br.**; **L.** semicylindrical, acute, 4–20 mm long, 5 mm across and thick; **Fl.** 15 mm ∅, purplish-red.

**D. abyssinicum** (Rgl.) Schwant. (*Mes. a.* Rgl.). – Ethiopia; Eritrea. – Shrubby, erect, with slender, ascending, reddish-papillose **Br.**; **L.** subcylindrical, obtuse, soft, minutely papillose, 3–4 cm long; **Fl.** pink.

**D. acocksii** L. Bol. v. **acocksii**. – Cape: Sutherland D. – ♄ c. 16 cm tall with thickened, woody **roots**; young **shoots** puberulous; **L.** semicylindrical, obtuse to short-tapered, puberulous, bluish-green, c. 17 mm long, 1.5–2.5 mm ∅; **Fl.** yellow, with Pet. 1.5–2 mm across.

**D.** – v. **luxurians** L. Bol. – Pet. 3 mm across.

**D. acuminatum** L. Bol. – Cape: Albany D. – ♄ 20 cm tall, with a tuberous **caudex** 20 cm long; **Br.** stiff, virgate, 2-angled, with short shoots; **L.** acutely carinate, tapered, 35 mm long, pale bluish-green, 5 mm thick; **Fl.** 2 cm ∅, copper-coloured.

**D. adelaidense** Lavis. – Cape: Fort Beaufort D. – Small ♄ with **Br.** 3–5 cm long; **L.** linear, flat on the upperside, obtuse, with minute, bristle-tipped dots, 16 mm long, 4 to 7 mm across; **Fl.** 3-merous, purplish-pink.

**D. aereum** (L. Bol.) L. Bol. v. **aereum** (*Mes. a.* L. Bol.). – Cape: precise origin unknown. – **Br.** prostrate, elongated, stiff, rough-papillose; **L.** carinate on the underside, laterally compressed, the upperside channelled, minutely papillose, with ciliate hairs below, 20–43 mm long, 4–5 mm across and thick; **Fl.** solitary, 15–24 mm ∅, red.

**D.** – v. **album** (L. Bol.) L. Bol. (*Mes. a.* v. *a.* L. Bol.). – **L.** hirsute; **Fl.** white or pink.

**D. affine** Lavis. – Origin unknown. – **St.** 12–30 cm long, **Br.** 3.5–13.5 cm long; **L.** linear-lanceolate, tapered, 2–5 cm long, 7–11 mm across, c. 4 mm thick; Pet. c. 9 mm long, white, staminodes 7.5 mm long.

**D.** cv. Albert Krejcii. – Garden hybrid. – Parentage unknown.

**D. algoense** L. Bol. – Cape: Uitenhage D. – Resembles **D. lehmannii**; **L.** more obtuse, thick, 10–14 mm across and thick; **Fl.** smaller, white.

**D. aliwalense** L. Bol. – Cape: near Aliwal North. – Compact ♄ with the **St.** eventually projecting and ascending, 6–9 cm long, with slender **Br.**; lower **L.** narrow-ovate to elliptical, 9–13 mm long, 5–6 mm across, the upperside concave, middle **L.** longer, upper ones linear, 2 cm long, 4 mm across, green, indistinctly papillose; **Fl.** solitary, 22 mm ∅, purplish-pink.

**D. alticolum** L. Bol. – Natal: Drakensberg. – ♄ with thick **roots** 18 cm long; **Br.** prostrate, 6–10 cm long; **L.** concave on the upperside, carinate on the underside, acute, 25–45 mm long, 3 mm across and thick; **Fl.** 12 mm ∅, pinkish-purple.

**D. angustifolium** L. Bol. – Cape: Albany D. – Mat-forming; **St.** 2–3 cm ∅; **L.** 4–10 on a shoot, narrow, acute or obtuse to truncate, with the underside carinate, bluish-green, 15–20 mm long, 2.5–4 mm ∅; **Fl.** 2.5–3 cm ∅, Pet. white with pink tips.

**D. angustipetalum** Lavis. – Cape: Albany D. – Mat-forming, 12 cm tall; **L.** obtusely carinate, the upperside slightly concave, 10–15 mm long; **Fl.** c. 2.5 cm ∅, white.

**D. annulare** L. Bol. – Transvaal. – Erect, glossy-papillose ♄ 4–6 cm tall; **Br.** prostrate, 4–7 cm long; **L.** narrowed towards the tip, concave on the upperside, obtuse, 22 mm long, 4 mm across, 2 mm thick; **Infl.** short, compact, with pinkish-purple Fl. 2 cm ∅.

**D. appressum** L. Bol. – Cape: Albany D. – Compact ♄ 6–9 cm tall; **L.** imbricate, appressed, ± ovate below, linear-lanceolate, acute, with the underside sharply carinate, 2.5–3 cm long, 7 mm wide and thick; **Fl.** 3.5 cm ∅, yellow, on a laterally compressed pedicel.

**D. ashtonii** L. Bol. – Transvaal, Orange Free State, Natal. – Dwarf ♄ with a tuberous **caudex**; **Br.** elongated, thick, herbaceous parts often pubescent; **L.** ovate-lanceolate to oblong-linear, obtuse, 45 mm long, 2 mm across, 4 mm thick; **Fl.** 4.5 cm ∅, purplish-pink.

**D. asperulum** (Salm) L. Bol. (*Mes. a.* Salm, *Drosanthemum a.* (Salm) Schwant., *Dros. bredai* L. Bol.). – Cape: Cape, Worcester, Riversdale D. – Bushy ♄ 30–50 cm tall; **St.** stiff, slender, with short shoots; **L.** decurved at the apiculate tip, 12–20 mm long, scarcely 2 mm across; **Fl.** 2.5 cm ∅, pale pink.

**D. ausense** L. Bol. – S.W.Afr.: Gr. Namaqualand. – Fairly erect ♄ 15 cm tall; **Br.** semi-fleshy, 4–5 cm long, with slightly projecting hairs; **L.** 6–8 on a shoot, with the underside rounded, obtuse, ciliate, 15 mm long, 3–4 mm across and thick; **Fl.** 22 mm ∅, pale pink.

**D. basuticum** L. Bol. – Lesotho: Phutta. – Forming mats c. 13 cm ∅; **Br.** minutely papillose; **L.** rounded on the underside, channelled on the upperside, acute, 15–17 mm long, 3.5 mm across and thick, papillose; **Fl.** solitary, 2.8 cm ∅, pink.

**D. brevipetalum** L. Bol. – Cape: Albany D. – Slender ♄ 8 cm tall; **Br.** creeping; **L.** imbricate, concave on the upperside, rounded on the underside, tapered, green, 20–27 mm long, 4 mm across, 3 mm thick; **Fl.** solitary, 8 mm ∅, white.

**D. brevisepalum** L. Bol. v. **brevisepalum**. – Cape: Somerset East D. – Compact ♄ with **Br.** spreading to erect, 9 cm long, rough; **L.** obtuse, narrowed towards the tip, rounded on the underside, green, papillose, 20–25 mm long, 4 mm across and thick; **Fl.** 15–20 mm ∅, white.

**D.** – v. **majus** L. Bol. – Larger in all parts.

**D. britteniae** L. Bol. – Cape: Albany D. – ♄ with elongated **Br.**; **L.** carinate, acute, apiculate, dirty bluish-green, firm-textured, 3 cm long, 7 mm across and thick; **Fl.** solitary, 38 mm ∅, white.

**D. brunnthaleri** (BGR.) SCHWANT. (Pl. 158/3) (*Mes. b.* BGR.). – Origin unknown. – **Br.** 20 to 40 cm tall; **L.** 3.5–4 cm long, flat on the upperside, carinate on the underside above, ciliate below and on the margins, sap-green, papillose; **Infl.** ternate, Fl. 2 cm ⌀, violet-pink.

**D. burtoniae** L. BOL. – Cape: Worcester D. – Densely branched ♄ 15 cm tall; **Br.** ± virgate, rough, minutely hirsute at first; **L.** semicylindrical, tapered, 25 mm long, 3 mm thick, yellowish-green; **Fl.** 2.5–3 cm ⌀, yellow to coppery-red.

**D. caespitosum** L. BOL. f. **caespitosum.** – Cape: Kentani D. – Cushion-forming ♄ with creeping **Br.**; **L.** concave on the upperside, 4 cm long, 4 mm across, 3 mm thick; **Fl.** 22 mm ⌀, white.

**D. — f. roseum** (L. BOL.) L. BOL. (*D. c. v. r.* L. BOL.). – Pet. pink.

**D. calitzdorpense** L. BOL. – Cape: Lit. Karroo. – Densely branched ♄ 10 cm tall; **Br.** 12 cm long, the herbaceous parts with velvety, short hairs; **L.** narrowed towards the tip, 12 to 17 mm long, 5–6 mm across and thick; **Fl.** 24 mm ⌀, white.

**D. calycinum** L. BOL. – Cape: Albany D. – ♄ 10–12 cm tall; **Br.** prostrate, stiff, white; **L.** semicylindrical, obtusely carinate, apiculate, 15–20 mm long, 3 mm thick, yellowish-green; **Fl.** 16 mm ⌀, purple.

**D. carolinense** N. E. BR. v. **carolinense.** – Transvaal: Carolina. – ♄ with a curved **St.** 6–12 cm tall; **L.** linear to linear-lanceolate, (?) papillose; **Fl.** 3 cm ⌀, on a rough, 2-angled pedicel.

**D. — v. compacta** L. BOL. – Orange Free State. – Compact; **caudex** tuberous; **L.** indistinctly papillose, 4–5 cm long, 5–9 mm across; **Fl.** mauvish-pink.

**D. carterae** L. BOL. – Cape: near King William's Town. – Laxly branched ♄; **Br.** erect, virgate, 20 cm long, papillose; **L.** rounded on the underside, acute, c. 3 cm long, 3–4.5 mm thick; **Fl.** in a repeatedly forked Infl., 15 mm ⌀, white.

**D. clavipes** LAVIS. – Orange Free State: Ficksburg. – Mat-forming; **Br.** ascending, 9.5 cm long; **L.** linear, channelled on the upperside, c. 23 mm long, 3–4 mm across, 3 mm thick; **Fl.** solitary, 2.5 cm ⌀, purplish-pink.

**D. cloeteae** LAVIS. – Cape: Fort Beaufort D. – Laxly branched ♄ with curved **Br.** 18 cm long; **L.** linear, tapered, obtuse on the underside, 2–4 cm long, 4 mm across; **Fl.** in an Infl., pinkish-purple.

**D. concavum** L. BOL. – Cape: Graaff Reinet D. – Densely branched ♄ 10–12 cm tall; **Br.** slender; **L.** semicylindrical, concave on the upperside, acute, green, 25 mm long, 3 mm across, 2.5 mm thick; **Fl.** 2 cm ⌀, white.

**D. congestum** L. BOL. – E.Lesotho. – ♄ with **Br.** 13 cm long; **L.** semicylindrical, furrowed on the upperside, acute, 12–18 mm long, 3 mm across and thick; **Fl.** 15 mm ⌀, lemon-yellow.

**D. cooperi** (HOOK. f.) L. BOL. f. **cooperi** (*Mes. c.* HOOK. f., *D. c.* (HOOK. f.) SCHWANT.). – Orange Free State. – Prostrate, freely branched, papillose sub-♄; **L.** ± cylindrical, obtuse, with papillae in longitudinal lines, soft-fleshed, up to 55 mm long, 5 mm thick; **Fl.** 4.5–5 cm ⌀, purplish-red.

**D. — f. bicolor** (L. BOL.) ROWL. (*D. c. v. b.* L. BOL.). – Pet. differently coloured on the two surfaces.

**D. crassuloides** (HAW.) L. BOL. (*Mes. c.* HAW., *Mes. crassulinum* DC.). – Cape: near Grahamstown. – **St.** prostrate, forked; **Br.** filiform, reddish; **L.** narrow-lanceolate, tapered, with the upperside flat-concave, 15–25 mm long, fresh green, papillose; **Fl.** 3 cm ⌀, white.

**D. crassum** L. BOL. (*Drosanthemum robustum* L. BOL.). – Cape: Van Rhynsdorp D. – Robust ♄ 18 cm tall; **St.** erect, 3–7 cm long, with pubescent short shoots; **L.** semicylindrical, obtuse, with soft hairs, green, 15 mm long, 3 mm thick; **Fl.** solitary, 25 mm ⌀, red.

**D. cronemeyeranum** (BGR.) JACOBS. (*Mes. c.* BGR.). – Cape: origin unknown. – Prostrate ♄ with red **Br.** 20–40 cm long; **L.** trigonous, obtuse, apiculate, 15–22 mm long, 5–8 mm across, 5 to 7 mm thick, soft, smooth, green, young L. papillose, fringed; **Fl.** 15–20 mm ⌀, pinkish-violet.

**D. davyi** N. E. BR. – Transvaal. – Resembles **D. herbeum**; **St.** rather weak, thin; **L.** longer; **Fl.** 14 mm ⌀, white.

**D. deleeuwiae** LAVIS. – Orange Free State: Clarens. – Forming mats up to 4 cm ⌀; **Br.** 3–8 mm long; **L.** oblong, flat, with the upperside concave, bristly-papillose, 12–22 mm long, 11 mm across, 5 mm thick; **Fl.** solitary, 4 cm ⌀, pale pink.

**D. dunense** L. BOL. – Cape: East London D. – ♄; **Br.** narrowly 2-winged, up to 30 cm long; **L.** 20–32 mm long, 9 mm across and thick, with the keel decurrent to the Br., with acute margins, concave on the upperside, expanded midway, obtuse, dirty green; **Fl.** 3 cm ⌀, white to pale pink.

**D. dyeri** L. BOL. v. **dyeri.** – Cape: Alice D. – Mat-forming ♄ with elongated **Br.**; **L.** ovate to ovate-lanceolate, 10–15 mm long, 4 mm across, glossy papillose; **Fl.** large, reddish-yellow.

**D. — v. laxum** L. BOL. – Cape: near Tarkastad. – Laxly branched ♄ 5.5 cm tall; **L.** semicylindrical, acute, channelled on the upperside, green, papillose, 25 mm long, 6 mm across; **Fl.** 4 cm ⌀, purplish-red.

**D. ecklonis** (SALM) SCHWANT. v. **ecklonis** (*Mes. e.* SALM., *D. e.* (SALM) L. BOL.). – Cape: on the Swartkops R. – **Br.** prostrate, slender, at first with minute white hairs; **L.** flat-compressed, trigonous, acute, with the upperside grooved, light green, softly hirsute, 25–35 mm long; **Fl.** 16 mm ⌀, white.

**D. — v. latifolia** L. BOL. – **L.** oblong-oval, lanceolate or ovate, 10–27 mm long, 7–14 mm across; Fl. c. 2 cm ⌀, purplish-pink.

**D. edwardsiae** L. BOL. – Cape: Pondoland. – Densely branched, compact, softly haired ♄; **Br.** prostrate, up to 15 cm long; **L.** ovate-lanceolate or linear, 2 or 4 cm long, 10 mm across, 4 mm thick, obtusely keeled, obtuse above; **Fl.** solitary, 1 cm ⌀, yellow with a white centre.

**D. erectum** L. Bol. – Cape: origin unknown. – Erect ♄ 14 cm tall; **Br.** white, papillose; **L.** tapered, rounded on the underside, channelled on the upperside, minutely papillose, 20–25 mm long, 3–4 mm across, 3 mm thick; **Fl.** 12 mm ⌀, pinkish-purple.

**D. esterhuyseniae** L. Bol. – Cape: Uniondale, Humansdorp D. – Compact, mat-forming ♄; **roots** woody, up to 15 mm thick; **Br.** 3–5 cm long; **L.** clavate, obtuse, 15–30 mm long, 4–5 mm across and thick; **Fl.** solitary, 20 to 25 mm ⌀, white.

**D. exspersum** (N. E. Br.) L. Bol. v. **exspersum** (*Mes. e.* N. E. Br., *Drosanthemum e.* (N. E. Br.) Schwant., *Mes. micans* Thunbg.). – Cape: Bokkeveld Mts. – ♄ 15–30 cm tall; **Br.** woody, thin, opposite, with Int. 5–30 mm long, at first ascending, purplish or brownish, eventually variously curved, with the bark becoming brown; **L.** 3–6 mm long, erect or ± spreading, channelled-concave on the upperside, densely papillose, whitish; **Fl.** solitary, terminal, on a pedicel 3–6 mm long, with numerous linear, somewhat reddish Pet.

**D. —** v. **decumbens** L. Bol. – Cape: Worcester D. – Slender ♄ with prostrate **Br.**; **L.** narrowed towards the tip, minutely papillose, 1–2 cm long, 2–3 mm thick; **Fl.** c. 2 cm ⌀, purplish-pink.

**D. ficksburgense** Lavis. – Orange Free State: Ficksburg D. – Papillose ♄ with prostrate **Br.**; **L.** rounded on the underside, flat on the upperside, obtuse, c. 22 mm long, 5 mm or more across, 1–4 mm thick; **Fl.** white, more rarely pink.

**D. floribundum** L. Bol. – Orange Free State: near Smithfield. – Densely branched ♄ 25 cm tall; **L.** semicylindrical, long-tapered, minutely papillose, green, 2–5 cm long, 5 mm thick; **Fl.** 3 cm ⌀, purplish-pink.

**D. framesii** L. Bol. – Transvaal: Johannesburg. – **Caudex** tuberous, 6 cm thick; **Br.** numerous, 5 cm or more long, softly papillose; **L.** tapered, rounded on the underside, concave on the upperside, 2 cm long, 2.5–3 mm across, 2 mm thick; **Fl.** 12 mm ⌀, white.

**D. fredericii** Lavis. – Cape: Uitenhage D. – Densely branched ♄ 27 cm tall; **L.** trigonous, acute, slightly glossy, 10–15 mm long, 2–3 mm across and thick; **Fl.** many in an Infl., 15 mm ⌀, pale.

**D. frutescens** L. Bol. – Cape: Albany D. – Compact ♄ 5 cm tall, with the herbaceous parts velvety; **Br.** variously curved; **L.** ovate to broad-ovate, the upperside flat, pale green, 17 mm long, 9 mm across, very thick; **Fl.** solitary, 14 mm ⌀, white.

**D. galpinii** L. Bol. v. **galpinii**. – Natal. – **St.** thick; **Br.** crowded, 10 cm long; **L.** lanceolate, oblong-lanceolate, ± ovate or spatulate, acute, flat on the upperside, bluntly keeled on the underside, green, with prominent dots and hair-tipped papillae, 4–5 cm long, 21 mm across, 5 mm thick; **Fl.** 4 cm ⌀, pink.

**D. —** v. **minus** L. Bol. – **L.** less tapered, with less distinct cilia; **Fl.** smaller.

**D. gerstneri** L. Bol. – Lesotho. – Erect, densely branched ♄ with the herbaceous parts indistinctly papillose; **L.** narrowed towards the acute to obtuse tip, rounded on the underside, 15–20 mm long, 2–3 mm across, 1–5 mm thick; **Fl.** 18 mm ⌀, purplish-pink.

**D. giffenii** Lavis. – Cape: Albany D. – Stiff ♄ c. 45 cm tall, with woody **roots** 12 mm ⌀; **Br.** 16–27 cm long; **L.** linear, long-tapered, indistinctly carinate on the underside, 22–43 mm long, 3–4 mm across, 2 mm thick, with scattered, minute, bristly papillae; **Fl.** 24 mm ⌀, pinkish-purple.

**D. gracile** L. Bol. (Notes on Mes. III, 319, 1958). – Natal, near Estcourt. – Slender, mat-forming ♄ with prostrate **Br.**; **L.** tapered towards the base, acute, rounded on the underside, 15–25 mm long, 2–3 mm across; **Fl.** solitary, up to 2.5 cm ⌀, pink.

**D. gracillimum** L. Bol. (*D. gracile* L. Bol. Journ. S.Afr. Bot. XXVII, 180, 1961). – Cape: Albany D. – Resembles **D. lehmannii**, but more slender; ♄ up to 10 cm tall; **L.** narrowed towards the tip, acute to tapered, acutely keeled on the underside, 3–6 cm long; **Fl.** c. 3 cm ⌀, white.

**D. gramineum** L. Bol. – Cape: Cradock D. – Erect, branched, with the new season's **Br.** slender, grass-like; **L.** semicylindrical, acute, green, 1–2 cm long, 3 mm across and thick; **Fl.** solitary, 1 cm ⌀, white.

**D. grandiflorum** L. Bol. – Cape: Clanwilliam D. – Erect to semiclimbing ♄ 30–40 cm tall; **Br.** stiff; **L.** narrowed towards the obtuse tip, channelled on the upperside, green, indistinctly papillose, 2–3.5 cm long, 4 mm across, 2–3 mm thick; **Fl.** almost 8 cm ⌀, purplish-pink.

**D. grantiae** L. Bol. – Cape: Alice D. – Resembles **D. vinaceum**; **Fl.** 2 cm ⌀, pinkish-purple, white inside.

**D. gratiae** L. Bol. (*D. longii* L. Bol.). – Cape: Albany, Uitenhage D. – Erect, laxly branched ♄ 25 cm tall; lower **L.** semicylindrical, with the tip hooked and recurved, 3 cm long, upper **L.** 15 mm long, 2 mm across; **Fl.** pink.

**D. guthriei** Lavis. – Cape: Caledon D. – Glabrous, at first forming mats, but the **Br.** eventually up to 62 cm long, with Int. 1–3 cm long, 1–4 mm ⌀; **L.** ascending to erect, trigonous, apiculate, 2 cm long, 4 mm wide and thick, or up to 4.5 cm long, 10 mm across, with a short Sh.; **Fl.** solitary, 2.6 cm ⌀, with Pet. c. 1 cm long, 1.5 mm across, white, on a pedicel 2 cm long.

**D. hallii** L. Bol. (Pl. 159/1). – S.W.Afr.: Gr. Namaqualand. – Resembles **D. pergamentaceum**; compact, dense ♄ 10–15 cm tall; **Br.** 2-angled; **L.** persistent, fleshy, expanded midway, acute to rather obtuse, 4.6 cm long, 15 mm across and thick, silvery-grey to green, minutely rough-tuberculate; **Fl.** solitary or several together, 2.5 cm ⌀, pink.

**D. herbeum** (N. E. Br.) N. E. Br. (*Mes. h.* N. E. Br.). – Transvaal: near Johannesburg, Aliwal North. – ♄; **L.** linear-lanceolate, furrowed on the upperside, 2–4 cm long, 2.5–4 mm across, rather less in thickness; **Fl.** in a compact Infl., 15 mm ⌀, Pet. white with pink tips.

**D. hirtum** (N. E. Br.) Schwant. (*Mes. h.* N. E. Br., *D. h.* (N. E. Br.) L. Bol.). – Origin not known. – **St.** spreading, prostrate, 7.5 to 12 cm long, dull purplish-red, papillose, with rough hairs; **L.** linear, tapered, concave on the upperside, with glistening papillae, the margins and keel hirsute, often purplish-red all over, 4–6 cm long, 3–5 mm across; **Fl.** 4 cm ⌀, light purple.

**D. hollandii** L. Bol. – Cape: Uitenhage D. – ♄; **Br.** creeping, 30 cm long, with 2 narrowly alate angles; **L.** expanded towards the apiculate tip, with acute angles, green, 2 cm long, 5 mm across and thick; **Fl.** up to 23 mm ⌀, white.

**D. imbricatum** L. Bol. – Cape: ? Albany D. – ♄ with creeping **Br.** 20 cm long; **L.** imbricate, acute to tapered, concave on the upperside, 2 cm long, 4 mm across and thick; **Fl.** 2 cm ⌀, pink.

**D. inaequale** L. Bol. – Cape: Bredasdorp D. – Erect, laxly brached ♄; **Br.** virgate; **L.** narrowed towards the acute to rather obtuse tip, 2.7–6 cm long, 3–5 mm across and thick; **Fl.** 3.5–4 cm ⌀, purplish-pink.

**D. incomptum** (Haw.) L. Bol. v. **incomptum** (*Mes. i.* Haw.). – Cape: Uitenhage D. – ♄ up to 60 cm tall; **Br.** often breaking off; **L.** sub-cylindrical, rather tapered, dull greyish-green, minutely papillose, 12–20 mm long, c. 3 mm thick; **Fl.** solitary, 2 cm ⌀, white.

**D.** — v. **ecklonis** (Salm) Jacobs. (*Mes. i.* v. *e.* Salm). – **Br.** more spreading and curved; **L.** flatter on the upperside; Pet. reddish at the base.

**D.** — v. **gracile** L. Bol. – Cape: Prieska D. – **Br.** slender, rough-papillose; **L.** almost glossy-papillose; **Fl.** 1 cm ⌀, glossy white.

**D. inconspicuum** L. Bol. – Cape: Knysna D. – ♄ with a semi-subterranean, robust, creeping **St.**; **Br.** numerous, prostrate to ascending, slender, up to 38 cm long; **L.** tapered, sabre-shaped, recurved at the tip, acutely carinate on the underside, dirty bluish-green, 25–30 mm long, 2–3 mm across, 3 mm thick below; **Fl.** numerous, 6 cm ⌀, with ± translucent Pet.

**D. intonsum** L. Bol. – Cape: Fraserburg D. – Densely branched ♄; **Br.** rough, initially hirsute; **L.** obtusely carinate on the underside, obtuse or short-tapered above, with minute, soft papillae and ciliate margins, 10–15 mm long, 3 mm across and thick; **Fl.** 15 mm ⌀, pale pink.

**D. jansei** N. E. Br. (*D. denticulatum* L. Bol.). – Transvaal: Barberton D. – **St.** creeping, up to 20 cm long; **L.** 15–20 mm long, 3.5–5 mm across, acute; **Fl.** solitary, 2 cm ⌀.

**D. karroicum** L. Bol. – Cape: Graaff Reinet D. – ♄ 9 cm tall; **Br.** spreading to prostrate, 10 to 16 cm long; **L.** narrowed towards both the tip and the base, obtuse or acute, glossy, 15–20 mm long, 3 mm across and thick; **Fl.** 15 mm ⌀, Pet. white with pink tips.

**D. katbergense** L. Bol. v. **katbergense**. – Cape: Stutterheim D. – **Caudex** tuberous, 15 mm thick; **Br.** slender, prostrate, papillose; **L.** lanceolate, acute to tapered, narrowed teretely below, furrowed on the upperside, green, minutely papillose, 2 cm long, 4–6 mm across; **Fl.** 12 mm ⌀, white.

**D.** — v. **amatolense** L. Bol. – Cape: King William's Town D. – **L.** oblong-ovate, 8 mm across.

**D.** — v. **angustifolium** L. Bol. – **L.** more semicylindrical, somewhat furrowed.

**D. klinghardtianum** (Dtr.) Schwant. (*Mes. k.* Dtr.). – S.W.Afr.: Klinghardt reg. – Cushion-forming ♄ 10–15 cm tall, 20 cm ⌀; **Br.** softwoody; **L.** long-ovate, 8–10 mm long, 3–4 mm thick, with a rounded tip, densely papillose; **Fl.** solitary, 17 mm ⌀, white to pink.

**D. knox-daviesii** Lavis. – Transvaal: Johannesburg. – ♄ up to 9 cm tall; **Br.** up to 9 cm long; **L.** semicylindrical, minutely papillose, c. 3 cm long, 5–6 mm across, 4 mm thick; **Fl.** solitary, 33 mm ⌀, purplish-pink.

**D. kofleri** Lavis. – Lesotho: Roma Mission. – Up to 22 cm ⌀; **St.** c. 13 cm long; **Br.** ascending, 2–10 cm long; **L.** ± falcate, slightly narrowed towards the acute tip, green, 4–15 mm long, 2–3 mm across and thick; **Fl.** 1 cm ⌀, white.

**D. lavisiae** L. Bol. v. **lavisiae**. – Cape: Kalahari reg.; Orange Free State. – Mat-forming ♄ with the herbaceous parts olive-green to purple, minutely papillose; **St.** 10 mm ⌀; **Br.** creeping and rooting, 9 cm long; **L.** linear, acute, furrowed on the upperside, 15–20 mm long, 2 mm across; **Fl.** 14–20 mm ⌀, pinkish-purple, with unequal Sep.

**D.** — v. **parisepalum** L. Bol. – Lesotho: Phutta. – Sep. equal, **Fl.** 5–15 mm ⌀.

**D. laxipetalum** L. Bol. – Cape: Albany D. – Slender, densely branched ♄ 20 cm tall; **Br.** spreading to prostrate, 20 cm long; **L.** acutely carinate on the underside, short-tapered or obtuse, 10–15 mm long, 3–4 mm across and thick; **Fl.** 16 mm ⌀, white.

**D. lebombense** (L. Bol.) Lavis (*D. tradescantioides* v. *l.* L. Bol.). – Cape: Zululand, Natal; Swaziland. – Appearance as **D. tradescantioides**; **L.** ovate to circular-ovate, 25–34 mm long, 20–24 mm across, acute or obtuse, cordate at the base, flat on the upperside, with an obtuse keel; **Fl.** 26 mm ⌀, greenish-white or yellow.

**D. leendertziae** N. E. Br. – Transvaal. – Rather weak ♄; **L.** 10–25 mm long, 1–1.5 mm across, flat on the upperside, not very thick; **Fl.** solitary, 15 mm ⌀, golden-yellow.

**D. lehmannii** (Eckl. et Zeyh.) Schwant. (*Mes. l.* Eckl. et Zeyh., *Corpuscularia l.* (Eckl. et Zeyh.) Schwant., *Schoenlandia l.* (Eckl. et Zeyh.) L. Bol., *Mes. sexpartitum* N. E. Br.). – Cape: Uitenhage D., Karroo. – **Br.** 10–25 cm long, 2-angled, reddish; **L.** trigonous with convex sides, apiculate, 12–25 mm long, smaller on the short shoots, 3–4 mm across, grey to greyish-green, smooth; **Fl.** 4 cm ⌀, pale yellow.

**D. leightoniae** Lavis. – Cape: King William's Town D. – Erect, robust ♄ c. 10 cm tall; **St.** woody, short, c. 15 mm ⌀, twice forking, the short **shoots** slender; **L.** tapered, indistinctly carinate on the underside, green to bluish-green, with dry, white papillae; **Fl.** c. 2 cm ⌀, pale pink.

**D. liebenbergii** L. Bol. – Transvaal: Maclear. – Slender ♄ c. 10 cm tall, with outspread **Br.**, the herbaceous parts hirsute; **Br.** prostrate; **L.** semicylindrical, 2–3 cm long, 2–4 mm across and thick; **Infl.** compact, with numerous purplish-red **Fl.** 2 cm ⌀.

**D. lineare** L. Bol. v. **lineare.** – Lesotho. – Slender, erect ♄ 18 cm tall; **Br.** often prostrate, rough, papillose; **L.** linear-semicylindrical, acute, yellowish-green, 20–48 mm long, 3–4 mm across and thick; **Infl.** laxly branched, **Fl.** 2 cm ⌀, white.

**D.** — v. **tenuifolium** L. Bol. – **L.** thinner; **Fl.** 35 mm ⌀, white.

**D. litorale** (Kensit) L. Bol. (*Mes. l.* Kensit). – Cape: Cape D. – Laxly branched herb; **Br.** prostrate, creeping, 35 cm long; **L.** trigonous, the angles with a white border, acute, apiculate, blue, 25–30 mm long, 5–6 mm thick; **Fl.** white.

**D. longipes** L. Bol. – Cape: Clanwilliam D. – Erect ♄; **St.** stiff, reddish-brown, somewhat papillose, up to 31 cm long; **Br.** angular, 6 to 9 cm long; **L.** linear to oblong-lanceolate, obtuse, 14–40 mm long, 2–8 mm across, with round papillae; **Fl.** 2.5–5 cm ⌀, red to yellow.

**D. lootsbergense** Lavis. – Cape: Middelburg D. – Small ♄ with **Br.** 8.5–13.5 cm long; **L.** narrowed towards the obtuse tip, papillose-bristly, 12–22 mm long, 2–3 mm thick; **Fl.** 2.5 cm ⌀, pinkish-purple.

**D. luckhoffii** L. Bol. – Cape: origin unknown. – ♄ 3–5 cm tall; **Br.** slender, prostrate; **L.** acute above, concave on the upperside, 1–2 cm long, 3 mm across and thick; **Fl.** solitary, 22 mm ⌀, pink to purple.

**D. luteum** L. Bol. – Cape: Albany D. – Erect, robust ♄ 20 cm tall; **Br.** woody, stiff; **L.** narrowed towards the tip, obtusely keeled, short-tapered or obtuse above, green, 1–2 cm long, 3 mm across and thick; **Fl.** in a compact Infl., 18 mm ⌀, yellow.

**D. lydenburgense** L. Bol. v. **lydenburgense.** – Transvaal: Lydenburg. – Resembles **D. cooperi**; herbaceous, laxly branched; **Br.** 20 cm long, minutely papillose; **L.** linear, acute, with the upperside flat to furrowed, 3.5–5 cm long, 2 to 5 mm across, 2–3 mm thick; **Infl.** broad, **Fl.** 2–2.5 cm ⌀, purplish-pink.

**D.** — v. **acutipetalum** L. Bol. – Pet. pointed, 3.5 cm long.

**D. macellum** (N. E. Br.) N. E. Br. (*Mes. m.* N. E. Br.). – Transvaal. – **Caudex** cylindrical, fleshy; **St.** several, 4–6 cm long; **L.** narrowed towards the tip, 10–25 mm long, 2–3 mm across, 1.5 mm thick, bluish-green; **Fl.** 18 mm ⌀, light pinkish-purple to red.

**D. macrostigma** L. Bol. – Cape: Swellendam D. – Densely branched, ± mat-forming ♄; **Br.** slender; **L.** tapered, rounded on the underside, 2 mm long, 4–6 mm across, 3 mm thick; **Fl.** solitary, 2 cm ⌀, white or pale pink.

**D. mahonii** (N. E. Br.) N. E. Br. (*Mes. m.* N. E. Br.). – Rhodesia. – ♃ shrubby plant c. 15 cm tall, the young growth covered with crystal-clear papillae; **L.** distant on long shoots, crowded on short shoots, ± cylindrical, acute, furrowed on the upperside, fresh green, 2.5 to 4 cm long, 2–3 mm across and thick; **Fl.** in dichotomous Infl., 24 mm ⌀, light violet-purple.

**D. mariae** L. Bol. – Cape: Bredasdorp D. – ♄ with crowded **Br.** up to 30 cm long; **L.** flat on the upperside, apiculate, tapered, 15 mm long, 4 mm across, 5 mm thick; **Fl.** 1 cm ⌀, purplish-pink.

**D. maxwelliae** L. Bol. – Cape: Willowmore D. – ♄; **Br.** slender, short or elongated; **L.** ± navicular, acutely carinate, 11–20 mm long, 5 mm across, 6 mm thick, expanded midway, short-tapered; **Fl.** 1–3 together, 21 mm ⌀, glossy white.

**D. minimum** Lavis. – Cape: Albany D. – Dwarf, laxly branched ♄ 4–5 cm ⌀; **L.** ovate to ± acute, 4–8 mm long, 4 mm across and thick, to linear and 10–24 mm long, 5 mm wide and thick; **Fl.** almost 3 cm ⌀, whitish to pale pink.

**D. monanthemum** Lavis. – Origin unknown. – Small ♄ with a woody **caudex**; **Br.** slender, 7–18 cm long; **L.** narrowed slightly towards the tip to abruptly acute, the upperside flat to slightly concave, 13–24 mm long, 1.5–4 mm across, c. 3 mm thick; **Fl.** solitary, 24–30 mm ⌀, pinkish-purple.

**D. muirii** L. Bol. – Cape: near E.London. – ± erect ♄; **Br.** spreading, 8–10 cm long, or more prostrate and up to 28 cm long, papillose; **L.** narrowed towards the acute tip, papillose, dirty green, 20–28 mm long, 2–4 mm across, 2–3 mm thick; **Fl.** 25 mm ⌀, pink.

**D. multiflora** L. Bol. – Cape: Humansdorp D. – Minutely velvety-haired ♄; **Br.** prostrate; **L.** semicylindrical to cylindrical, apiculate, 33 mm long, 2 mm across and thick; **Infl.** of many pink **Fl.** 12 mm ⌀.

**D. nakurense** (Engl.) Herre (*Mes. n.* Engl.). – Ethiopia. – ♄ with creeping **Br.**; **L.** linear, obtuse, minutely papillose; **Fl.** solitary, 15 mm ⌀.

**D. neethlingiae** (L. Bol.) Schwant. (*Mes.* L. Bol., *Trichodiadema tenue* L. Bol.). – Cape: Caledon, Riversdale D. – Slender sub-♄ 14 cm tall, with a tuberous **caudex**; **Br.** ± prostrate; **L.** semicylindrical, short-acute, minutely papillose, hirsute at first, 5–15 mm long, 5 mm across, 2.5 mm thick; **Fl.** 15 mm ⌀, purplish-red.

**D. nelii** L. Bol. – Lesotho. – Slender, indistinctly papillose ♄; **Br.** prostrate, creeping, c. 15 cm long; **L.** linear to oblong-linear, acute, 6–15 mm long, 2.5–3.5 mm across; **Fl.** solitary, pink.

**D. nubigenum** (Schltr.) L. Bol. (*Mes. n.* Schltr.). – Orange Free State. – Prostrate, papillose sub-♄; **L.** oblong-elliptical or linear, papillose; **Fl.** 2 cm ⌀, orange-red.

**D. obtusum** L. Bol. – Orange Free State. – Slender ♄ with prostrate, elongated, papillose **Br.**; **L.** 14–17 mm long, 3 mm across and thick, obtuse, green to dirty bluish-green; **Fl.** numerous, 18 mm ⌀, pinkish-purple.

**D. oehleri** (Engl.) Herre (*Mes. o.* Engl.). – Ethiopia. – Creeping and rooting from the **St.**; **Br.** ascending, with numerous short shoots; **L.** linear, semicylindrical, slightly furrowed on the upperside, 16 mm long, 2–3 mm across, 2 mm thick; **Fl.** 1 cm ⌀.

**D. ornatum** N. E. BR. – Cape: Graaff Reinet D. – ♄ with prostrate, papillose, pruinose **Br.** 3.5–7.5 cm long; **L.** linear, tapered, apiculate, 12–25 mm long, 2–3 mm across and thick, green, minutely papillose, often reddish at the tip; **Fl.** 16–20 mm ⌀, light glossy red.

**D. pachyrhizum** L. BOL. v. **pachyrhizum.** – Cape: Zululand. – Erect ♄ 14 cm tall, from a **caudex** 5 cm thick; herbaceous parts velvet-haired; **Br.** 12 cm long, 25 mm thick below; **L.** obtusely keeled, laterally compressed, obtuse to oblique, dirty green, 3 cm long, 3 mm across, 4–5 mm thick; **Fl.** 2 cm ⌀, pinkish-purple.

**D. —** v. **pubescens** L. BOL. – Transvaal. – Herbaceous parts with ± projecting hairs; **Fl.** white.

**D. pageanum** (L. BOL.) L. BOL. (*Mes. p.* L. BOL., *Drosanthemum p.* (L. BOL.) SCHWANT.). – Cape: Montagu D. – Erect, branched ♄ 26 cm tall; **Br.** somewhat hirsute; **L.** cylindrical, 10–15 mm long, 2–3 mm thick, long-tapered, obtuse, with minute papillae tipped with a white hair; **Fl.** 16 mm ⌀, purple.

**D. pallidum** L. BOL. – Cape: origin unknown. – Mat-forming; **Br.** creeping, 20 cm long, densely hirsute; **L.** long ovate to acute, indistinctly carinate, 25 mm long, 13 mm across, more rarely 14 mm long, 17 mm across; **Fl.** solitary, 12 mm ⌀, pale pink.

**D. papillatum** (L. BOL.) L. BOL. (*Drosanthemum p.* L. BOL.). – Cape: Worcester D. – Erect, branched, stiff ♄ 20 cm tall; **Br.** slender, rough-haired; **L.** cylindrical, obtuse, 22 mm long, 4 mm ⌀; **Fl.** 32 mm ⌀, white.

**D. parviflorum** L. BOL. – Cape: Uitenhage D. – ± creeping ♄ 6 cm tall; **L.** convex and acute when viewed from the side, 12 mm long, 3 mm across and thick; **Fl.** 8 mm ⌀, yellow to golden-yellow.

**D. patersoniae** (L. BOL.) L. BOL. (*Mes. p.* L. BOL.). – Cape: Uitenhage D. – Dwarf, branched ♄; **St.** somewhat compressed, hirsute; **L.** ± sabre-shaped, laterally compressed, with ciliate margins, 14 mm long, 5–7 mm across, 6–8 mm thick; **Fl.** 15 mm ⌀, white.

**D. peersii** LAVIS. – Cape: Willowmore D. – Stiff ♄ up to 24 cm tall; **Br.** ascending, c. 17.5 cm long; **L.** semicylindrical, apiculate, 5–14 mm long, 3 mm across, 2–4 mm thick, red; **Fl.** pink.

**D. peglerae** L. BOL. – Cape: Kentani D., Transkei. – ♄ with prostrate **Br.**; **L.** ovate to acute, obtusely keeled, concave on the upperside, green, 15 mm long, 10 mm across midway, 4 mm thick; **Fl.** 12 mm ⌀, pinkish-purple.

**D. pergamentaceum** L. BOL. v. **pergamentaceum.** – Cape: Lit. Namaqualand. – Laxly branched ♄ 30 cm tall; **Br.** crowded; **L.** laterally compressed, rather obtusely carinate, rounded to obliquely truncate at the tip, older L. parchment-like, 7 cm long, 16 mm across, 13–14 mm thick above; **Fl.** solitary, 44 mm ⌀, white.

**D. —** v. **roseum** LAVIS. – **Fl.** 3–5 cm ⌀, pale pink.

**D. pilosulum** L. BOL. – Lesotho. – ♄ with prostrate **Br.**; **L.** furrowed on the upperside, obtuse, laterally compressed, dirty green, with minute, papillose hairs, 15–25 mm long, 2 mm across, 3 mm thick; **Fl.** c. 25 mm ⌀, pink.

**D. platysepalum** L. BOL. – Cape: Albany D. – Robust, densely branched ♄; **Br.** stiff at first, 30 cm long; **L.** like **D. ecklonis,** flat on the upperside, 26 mm long, 10 mm across; **Fl.** 13 mm ⌀, purple.

**D. pondoense** L. BOL. – Cape: Pondoland. – Erect ♄ 17 cm tall; **Br.** indistinctly papillose; **L.** furrowed on the upperside, long-tapered, 2.5–3.3 cm long, compressed midway and 2–3 mm across; **Fl.** 12–15 mm ⌀, pale straw-coloured.

**D. pontii** L. BOL. – Orange Free State. – Erect, laxly branched ♄; **Br.** spreading, often prostrate, 32 cm long, roughly papillose; **L.** long-tapered, 45–70 mm long, 4–5 mm across and thick, with round papillae; **Fl.** 27 mm ⌀, pinkish-purple.

**D. pottsii** (L. BOL.) L. BOL. (*Mes. p.* L. BOL.). – Cape: Kalahari reg.; Orange Free State: Aliwal N. – ♃ with minute, hair-tipped papillae; **St.** short, robust; **Br.** slender, prostrate to ascending; **L.** semicylindrical or cylindrical, compressed above, 10–25 mm long, 3–5 mm thick; **Fl.** 15 mm ⌀, white.

**D. prasinum** L. BOL. – Cape: near Alexandria. – Spreading, sometimes climbing ♄ 12 cm tall; **L.** linear to linear-lanceolate, acute, ± concave on the upperside, 22 mm long, 3 mm across and thick; **Fl.** solitary, purplish-pink.

**D. pruinosum** (THUNBG.) J. INGRAM (Pl. 158/2) (*Mes. p.* THUNBG., *Drosanthemum p.* (THUNBG.) SCHWANT., *Mes. echinatum* LAM., *Mes. e.* AIT., *Del. e.* (AIT.) SCHWANT.). – Cape: Lit. Namaqualand. – Densely bushy ♄ up to 30 cm tall; **Br.** rebranching dichotomously, with white, pointed papillae; **L.** ovate to hemispherical, 10–13 mm long, 6–7 mm across and thick, ± flat on the upperside, sap-green, with large, ± bristle-tipped papillae; **Fl.** solitary, 12–15 mm ⌀, whitish or yellowish.

**D. pubipetalum** L. BOL. – Cape: Robertson D. – Laxly branched ♄ 30 cm tall; **Br.** slender, rough; **L.** narrowed towards the tip, rounded on the underside, papillose, 5–12 mm long, 1 to 2.5 mm thick; **Fl.** 22 mm ⌀, pink.

**D. repens** L. BOL. – Cape: Transkei. – Mat-forming ♄; **Br.** creeping, slender; lower **L.** narrowed teretely below, rounded on the underside, obtuse above, 8–14 mm long, 4 mm across, 3 mm thick, upper **L.** ± spherical, distinctly spotted; **Fl.** solitary, pinkish-red.

**D. reynoldsii** LAVIS. – Orange Free State. – Densely branched ♄ 15–17 cm tall, becoming woody; **Br.** 15 cm long, initially with pink papillae; **L.** ± falcate, slightly narrowed towards the obtuse tip, carinate on the underside, 2–4 cm long, 6 mm across and thick; **Fl.** 4.5 cm ⌀, pinkish-purple.

**D. rileyi** L. BOL. – Transvaal. – Slender, laxly branched ♄ 20 cm tall; **Br.** ± prostrate; **L.** falcate-recurved, linear, obtuse or acute above, laterally convex, glossy green, (1–)2(–3.5) cm long, 2–4 mm across and thick; **Fl.** solitary, c. 2 cm ⌀, Pet. copper-coloured, pink below.

**D. robustum** L. Bol. − Cape: Albany D. − Robust ♄ with pale, woody **Br.; L.** obtusely carinate, linear, acute above, dirty green, 2 to 2.5 cm long, 3−4 mm across and thick; **Fl.** 2 cm ⌀, reddish-yellow.

**D. rogersii** (Schoenl. et Bgr.) L. Bol. v. **rogersii** (*Mes. r.* Schoenl. et Bgr.). − Cape: Albany D. − Laxly branched ♄; **Br.** stiff at first, eventually prostrate, slender, 35 cm long; **L.** long-lanceolate to long-ovate, carinate on the underside, obtuse to short-tapered at the tip, with soft hairs, 3 cm long, 10 mm across, 2 mm thick; **Fl.** 2 cm ⌀, yellow.

**D.** — v. **glabrescens** L. Bol. − **L.** mostly glabrous.

**D. roseopurpureum** Lavis. − Orange Free State; Lesotho. − **St.** c. 17 cm long, with Int. 0.7−4.5 cm long, 3 mm ⌀, herbaceous parts green, papillose; **L.** erect to ascending, semicylindrical to slightly channelled on the upperside, obtuse, 2.5−5 cm long, 4 mm wide and thick; **Fl.** 2−6 in a lax Infl., on a pedicel 2−3 cm long, Pet. 1.5 cm long, 1.5 mm across, purplish-pink.

**D. saturatum** L. Bol. − Cape: Zululand. − ♄ with prostrate, slender **Br.; L.** linear, acute, rounded on the underside, indistinctly papillose, (15−)25(−40) mm long, 3 mm across and thick; **Fl.** 15 mm ⌀, dirty pinkish-purple.

**D. saxicolum** Lavis. − Cape: Humansdorp D. − Forming mats 4 cm tall, c. 11 cm ⌀, herbaceous parts glossy, glabrous; **L.** ascending to projecting, linear, acute, minutely apiculate, 1.3−1.7 cm long, 6−7 mm wide, c. 6 mm thick; **Fl.** solitary, opening in the afternoon, ± sessile or on a pedicel up to 6 mm long, bracts 5−13 mm long, Pet. c. 1 cm long, 1 mm across, pale, with brilliant pink tips.

**D. scabripes** L. Bol. − Cape: Transkei. − Laxly branched ♄ from a **caudex** 1 cm ⌀; **Br.** prostrate; **L.** tapered, rounded on the underside, dirty bluish-green, initially papillose, 2 cm long, 2−3 mm across, 2 mm thick; **Fl.** 27 mm ⌀, pinkish-purple.

**D. smythae** L. Bol. − Natal. − Mat-forming ♄ 3 cm tall; **L.** verticillate, laterally compressed, rounded on the underside, acute above, 10 to 20 mm long, 3 mm across, 3−4 mm thick; **Fl.** 11 mm ⌀, pinkish-purple.

**D. stenandrum** L. Bol. − Cape: Pondoland. − Densely branched ♄; **Br.** slender, initially softly hirsute; **L.** imbricate, obtuse, apiculate, carinate on the underside, green, 10−15 mm long, 4−5 mm across, 3−4 mm thick; **Fl.** 7 mm ⌀, pinkish-purple.

**D. steytlerae** L. Bol. − Rhodesia: near Zimbabwe. − Compact ♄ 9 cm tall; **Br.** rough-papillose; **L.** indistinctly keeled, channelled on the upperside, acute, 35 mm long, 5 mm across and thick; **Fl.** 18−25 mm ⌀, white.

**D. subclavatum** L. Bol. − Cape: Bedford D. − ♄ with a tuberous **caudex**; **Br.** prostrate, slender, 20 cm long; **L.** tapered, flat to ± concave on the upperside, green, 2−3 cm long, 3−4 mm across and thick; **Fl.** solitary, 23 mm ⌀, pale pink to white.

**D. subincanum** (Haw.) Schwant. (*Mes. s.* Haw., *D. s.* (Haw.) L. Bol.). − Cape: Beaufort West, Albany D. − Bushy, becoming prostrate in age; **L.** obtuse, trigonous, short-tapered, apiculate, soft, with minute velvety hairs, 2.5−3 cm long; **Fl.** 16−20 mm ⌀, white.

**D. subpetiolatum** L. Bol. − Cape: origin unknown. − Compact ♄; **Br.** creeping, red, 17 cm or more long; **L.** narrowed teretely below, short-tapered or obtuse at the tip, carinate on the underside, dirty bluish-green or reddish, 15−20 mm long, 5−6 mm across and thick; **Fl.** solitary, 18 mm ⌀, pale pink.

**D. sulcatum** L. Bol. − Cape: near Alexandria. − Laxly branched ♄ c. 10 cm tall; **Br.** spreading, 12 cm long; **L.** narrowed towards the apiculate tip, carinate on the underside, dirty olive-green, 3 cm long, 3 mm across and thick; **Fl.** 1 cm ⌀, purplish-red.

**D. sutherlandii** (Hook. f.) N. E. Br. (Pl. 159/3, 4) (*Mes. s.* Hook. f.). − Transvaal, Natal. − ♃ with a napiform **caudex** 5−8 cm long and up to 1.5 cm thick above; **shoots** ☉, erect or spreading, minutely rough-haired and papillose, 8−15 cm long; **L.** slightly channelled on the upperside, on the underside with a keel-like M.nerve, 5−8 cm long, 2 cm across, fresh green, minutely papillose, with ciliate margins; **Fl.** 6−7 cm ⌀, violet-pink.

**D. suttoniae** Lavis. − Natal: Ladysmith. − ♃ with a tuberous, round **caudex** 3 cm ⌀, with crowded, conical tubercles 7 mm ⌀ towards the tip; **St.** thick; **L.** lanceolate, acute to tapered or ovate and obtuse, dirty green, with minute dots, 3−6 cm long, 10−17 mm across; **Fl.** pink.

**D. taylorii** (N. E. Br., L. Bol. v. **taylorii** (Pl. 159/2) (*Mes. t.* N. E. Br., *Corpuscularia t.* (N. E. Br.) Schwant.). − E.Cape. − Resembles **D. lehmannii**; robuster; **L.** greyish-green to greyish-white.

**D.** — v. **albanense** L. Bol. − Cape: Albany D. − **L.** hemispherical, unequal, 4−7 mm long, 3−5 mm across, or 12 mm long, 5 mm across, bluish-green; **Fl.** 3 cm ⌀, pale pink.

**D. testaceum** (Haw.) Schwant. (*Mes. t.* Haw., *D. t.* (Haw.) L. Bol.). − Cape: Graaff Reinet D. − Erect ♄ 60−90 cm tall; **L.** semicylindrical to trigonous, apiculate, greyish-green, minutely papillose, 12−24 mm long, 2−3 mm across; **Fl.** in an Infl., copper-coloured.

**D. tradescantioides** (Bgr.) L. Bol. (*Mes. t.* Bgr., *Mes. flanaganii* Kensit). − Cape: Komga, Kentani D., Natal: Lydenburg D., Transvaal. − Freely branched, dwarf plant with creeping **St.** and **Br.**, rooting at the nodes; **L.** ovate, tapered, narrowed teretely below, somewhat furrowed on the upperside, with a keel-like M.nerve on the underside, light green, minutely papillose, 25−30 mm long, 12 mm across, 1−2 mm thick; **Fl.** solitary, lateral, 15 mm ⌀, white.

**D. truteri** Lavis. − Cape: Somerset East D. − **Br.** prostrate, c. 17 cm long, often mat-forming; **L.** ovate to oblong, acute, green, slightly glossy, dirty purple, 12−25 mm long, 7−13 mm across, 8 mm thick; **Fl.** up to 3 cm ⌀, glossy white.

**D. uitenhagense** L. Bol. − Cape: Uitenhage D. − Densely branched ♄ with prostrate **St.**

30 cm long; **L.** carinate on the underside, narrowed towards the tip, flat, 25–40 mm long, 3–5 mm across and thick, green; **Fl.** 2 cm ⌀, glossy white.

**D. uncinatum** L. Bol. – Cape: Uitenhage D. – ♄ with **St.** prostrate to creeping, 45 cm long; short **Br.** erect, 7 cm long; **L.** carinate on the underside, tapered, hooked-recurved at the tip, green, 15–30 mm long, 2–4 mm across, 2–3 mm thick; **Fl.** 2 cm ⌀, white.

**D. uniflorum** L. Bol. – Cape: ? Peddie D. – Branched ♄; **St.** prostrate and creeping, 30 cm long, initially compressed and alate; **L.** narrowed towards the tip, with ciliate margins and keel, 2 cm long, 4 mm across, 3 mm thick; **Fl.** solitary, 15 mm ⌀, copper-coloured.

**D. vandermerwei** L. Bol. – Transvaal: Pietersburg D. – **St.** prostrate, 13 cm long, **Br.** erect, 2–5 cm tall, herbaceous parts ± rough with prominent, white papillae; **L.** linear, acute to tapered, 18–28 mm long, 4–5 mm across; **Fl.** 3 cm ⌀, pink.

**D. velutinum** L. Bol. – Natal. – Compact ♄ 5 cm tall, with the herbaceous parts papillose-bristly; **Br.** variously curved; **L.** ovate to broadly ovate, broadly terete below, acutely keeled on the underside, pale green, 17 mm long, 5 mm across; **Fl.** 14 mm ⌀, white.

**D. verecundum** L. Bol. – Cape: Albany D. – Compact, laxly branched ♄ c. 25 cm tall; **St.** robust, **Br.** erect, 5–20 cm long; **L.** ± imbricate, acutely keeled, concave on the upperside, acute, apiculate, green, 15–25 mm long, 8 mm across, 5 mm thick; **Fl.** 14 mm ⌀, purplish-pink.

**D. vernicolor** L. Bol. – Cape: near Alexandria. – Compact, laxly branched ♄; **St.** robust, 25 mm thick; **Br.** stiff, thick; **L.** narrowly ovate, short-tapered to ± lanceolate, apiculate, 6–18 mm long, 3–6 mm across, 3 mm thick; **Fl.** solitary, pink.

**D. versicolor** L. Bol. – Cape: Bedford D. – Branched, spreading ♄ 8 cm tall, with a thick **caudex** 4.5 cm ⌀; **St.** prostrate, narrowly alate, 10 cm long; **L.** narrowed towards the tip, short-tapered, dirty bluish-green, 10–15 mm long, 2–3 mm across and thick; **Fl.** 15 mm ⌀, coppery-red.

**D. vinaceum** (L. Bol.) L. Bol. (Mes. v. L. Bol.). – Cape: origin unknown. – ♄ with creeping, slender **Br.**: **L.** rounded to indistinctly carinate on the underside, acute, minutely papillose, 18–25 mm long, 4 mm across, 2–3 mm thick; **Fl.** 15 mm ⌀, straw-coloured.

**D. virens** L. Bol. – Cape: Riversdale D. – Branched ♄ 20–45 cm tall; **L.** falcate, light green, the indistinct keel on the underside rough at the tip; **Fl.** solitary, 35 mm ⌀, purplish-pink.

**D. vogtsii** L. Bol. – Transvaal: near Pretoria. – Compact ♄, minutely papillose at first, with **Br.** 7–12 cm long; **L.** tapered, concave on the upperside, convex on the underside, 4–6 cm long, 5–7 mm across, 3–4 mm thick; **Fl.** in a cyme, 28 mm ⌀, yellow.

**D. waterbergense** L. Bol. – Transvaal: Waterberg D. – Erect ♄ 10 cm tall; **roots** thickened, herbaceous parts glossy-papillose, green; **L.** narrowed towards the base, acute above, rounded on the underside, 2.5–4 cm long, 3–4 mm across and thick; **Fl.** solitary, 3–4 cm ⌀, purplish-pink.

**D. wethamae** L. Bol. – Orange Free State: near Gumtree. – ♄ 10 cm tall; **Br.** projecting to spreading, 6–12 cm long; **L.** tapered above, rounded on the underside, green, some 20 to 25 mm long, 3 mm thick, other **L.** shorter, 4.5 mm across and thick; **Fl.** 15 mm ⌀, yellow.

**D. wilmaniae** Lavis. – Cape: W.Griqualand. – Densely branched ♄ with **St.** 15 cm long; **L.** linear, narrowed towards the obtuse tip, concave on the upperside, rounded on the underside, 2.5–4 cm long, 3 mm thick; **Fl.** 9–11 mm ⌀, in an Infl.

**D. wiunii** Lavis. – Orange Free State: Langeberg. – Slender, freely branched ♄; **St.** c. 8 cm long; **L.** linear, slightly narrowed towards the obtuse or acute tip, 12–30 mm long, 2–3 mm across and thick; **Fl.** c. 2 cm ⌀, brilliant pink.

**D. zeederbergii** L. Bol. – Transvaal: near Lydenburg. – Erect ♄; **Br.** 17 cm long, eventually prostrate; **L.** narrowed towards the acute or obtuse tip, glossy-papillose, 15 to 50 mm long, 3–4 mm across, 3 mm thick; **Fl.** 25 mm ⌀, purplish-pink.

**D. zoeae** L. Bol. – Cape: Kimberly D. – Erect, scarcely papillose ⊙ to ⊙ plant 15 to 25 cm tall; **St.** virgate, laxly branched; **L.** narrowed above, acute or obtuse, dirty green, 3–3.5 cm long, 3–4 mm across and thick; **Fl.** 15 mm ⌀, whitish.

**D. zoutpansbergense** L. Bol. – Transvaal: Zoutpansberg. – Slender ♄ with the herbaceous parts green, glossy-papillose; **Br.** prostrate, c. 25 cm long; **L.** linear, acute to rather obtuse, convex on the underside, (1–)3(–5) cm long, 2–4 mm across and thick; **Fl.** 17–30 mm ⌀, pinkish-purple.

**Dicrocaulon** N. E. Br. (§ IV/1/3). – Dwarf ♄ with a forked **St.**; **Br.** erect or spreading; **L.** opposite, amplexicaul at the base; **Fl.** with numerous, filiform Pet. Insignificant for cultivation. Propagation: seed, cuttings.

**D. brevifolium** N. E. Br. (Pl. 160/1). – Cape: Van Rhynsdorp D. – Gnarled, dwarf ♄; **Br.** projecting irregularly, with Int. 3–7 mm long, alternating with shorter ones; **L.** 1.5–3 mm long and almost as thick.

**D. humile** N. E. Br. – Cape: Lit. Namaqualand. – **Br.** prostrate, with Int. 1–3 mm long; **L.** 5–7 mm long.

**D. nodosum** N. E. Br. – Cape: Lit. Namaqualand. – Erect ♄; **Br.** with indistinct white dots (remains of papillae), Int. 5–15 mm long; **L.** c. 5–7 mm long; **Fl.** on a long pedicel.

**D. pearsonii** N. E. Br. – Cape: Lit. Namaqualand. – ♄ 6–10 cm tall; **Br.** erect or ascending, with white dots (remains of papillae), with Int. 5–15 mm long; **Fl.** on a pedicel 7–9 mm long.

**D. spissum** N. E. Br. – Cape: Lit. Namaqualand. – **Br.** stiff, erect, with dense side-shoots, with Int. 7 mm long, 2–3 mm thick; **L.** 1.5 to 3 mm long; **Fl.** not known.

**Didymaotus** N. E. BR.' (§ IV/1/16). – Greenhouse, warm, light; dry in early summer for the resting period. Propagation: seed.

**D. lapidiformis** (MARL.) N. E. BR. (Pl. 160/2) (*Mes. d.* MARL.). – Cape: Ceres D., Karroo. – Acauline, highly succulent plant; L. mostly paired, rather shapeless and thick, very fleshy, the upperside triangular, 15–20 mm long, 30 mm across, flat or slightly concave, the underside c. 5 cm long, pulled forward over the tip like a pointed chin, distinctly carinate, the surface whitish grey-green, rough; Fl. borne on both sides of the base of the shoot, pedicellate, white with a pink or red centre.

**Dinteranthus** SCHWANT. (§ IV/1/14). – Acauline, highly succulent plants, branched with age or forming mats; L. very short and thick, upperside flat, nearly as long as broad, underside much rounded, ± semi-ovoid, keeled towards the apex, surface firm, whitish, immaculate or with numerous, not very conspicuous dots, smooth or softly hirsute, rarely finely tuberculate; Fl. solitary, shortly pedicellate, large, yellow, open in the afternoon, summer. – Greenhouse, warm. Resting period: winter. Propagation: seed.

## Key to the Species of the Genus Dinteranthus by L. Bolus

1. Plants entirely glabrous; L. united for about ¾ to nearly the whole of their length ..... **2.**
1a. Plants ± pubescent; L. united for about 1/3 to a little more than 1/2 their length ...... **4.**
2. Sep. 8–10; stigmas 8–15 ..................................................... **3.**
2a. Sep. and stigmas 6 ........................................... **wilmotianus**
3. Plants with age consisting of several growths; L. without a keel, distinctly marked with reddish or purplish dots, some of which coalesce into lines; stigmas 8 ............ **vanzijlii**
3a. Plants usually consisting of a single growth even when old; L. obscurely keeled near the apex, without markings; stigmas 8–15 ........................................... **pole-evansii**
4. Herbaceous parts (i.e. L., pedicels and Sep.) all softly pubescent; L. conspicuously dotted, the dots green; Sep. and stigmas 6 ........................................ **puberulus**
4a. L. glabrous, minutely granulated; pedicels and Sep. softly pubescent; Sep. 6; stigmas 6–9 .... ......... **5.**
5. Upper surface of L. rounded at the apex ................................ **inexpectatus**
5a. Upper surface of L. not rounded at the apex (subacute, acute to very acute) ........... **6.**
6. Pet. obtuse or emarginate; stigmas 6–9 .................... **microspermus** v. **microspermus**
6a. Pet. acute or the inner ones acuminate; stigmas 6 ........... **microspermus** v. **acutipetalus**

**D. inexpectatus** JACOBS. (Pl. 160/3) (*Mes. i.* DTR. nom. nud.). – S.W.Afr.: Gr. Namaqualand. – Very compact; L. smaller and more rounded than **D. microspermus,** with the keel acute and prominent, the surface smooth, grey, with ± translucent greenish dots; Fl. 25 mm ⌀, golden-yellow.

**D. microspermus** (DTR. et DERENB.) SCHWANT. v. **microspermus** (Pl. 160/5) (*Mes. m.* DTR. et DERENB., *Rimaria m.* (DTR. et DERENB.) N. E. BR.). – S.W.Afr.: Gr. Namaqualand. – Solitary or multi-headed; each shoot with 1–2 L.-pairs; L. united up to midway, the free part 25–30 mm long on the upperside, 30 mm across, 20 mm thick, the underside hemispherical, pulled forward chin-like, the surface minutely granular, rough, reddish- to greyish-violet, juvenile L. chalky-white to greyish olive-green; Fl. 4–4.5 cm ⌀, Pet. obtuse, golden-yellow with reddish tips, light yellow on the underside.

**D.** — v. **acutipetalus** L. BOL. – Pet. acute, anthers deep orange-coloured.

**D. pole-evansii** (N. E. BR.) SCHWANT. (Pl. 160/6) (*Mes. p.-e.* N. E. BR., *Rimaria p.-e.* (N. E. BR.) N. E. BR.). – Cape: Prieska D. – Mostly solitary; L. united to midway, c. 4.5 cm long, 22–24 mm thick, very rounded on the underside with the keel scarcely indicated, the surface ± distinctly wrinkled, very minutely granular (lens needed!), dove-grey, often somewhat yellowish or reddish; Fl. 4 cm ⌀, glossy buttercup-yellow.

**D. puberulus** N. E. BR. (*D. punctatus* L. BOL.). – Cape: Bushmanland. – Mat-forming; each shoot with 2 L.-pairs; L. united for ⅓ or ½ of their length, 25–30 mm long, 16 mm across, 13 mm thick, the upperside ± convex, the underside round, scarcely keeled, the surface minutely granular and hirsute (lens needed!), velvety, brownish to greyish-green, with numerous green dots; Fl. 28 mm ⌀, golden-yellow.

**D. vanzijlii** (L. BOL.) SCHWANT. v. **vanzijlii** (Pl. 160/4) (*Lithops v.* L. BOL.). – Cape: Bushmanland. – L. pairs several, almost completely united to form a Bo. 4 cm high, hemispherical on the underside, with the Fi. scarcely depressed, greyish-green, very rarely with faint red dots or lines; Fl. 15 mm ⌀, orange-yellow.

**D.** — v. **lineatus** JACOBS. – Bo. whitish or greyish-green, with distinct dark brown lines and irregular markings.

**D. wilmotianus** L. BOL. – Cape: Gordonia D. – Solitary; L. united for c. ½ of their length, the upperside tapered, the underside rounded, doubly carinate towards the tip, 50–57 mm long, 17–22 mm across above the Sh., 10–15 mm thick, grey, somewhat pink, with dark violet dots ½–¼ mm ⌀; Fl. c. 3 cm ⌀, golden-yellow.

**Diplosoma** SCHWANT. (§ IV/1/21). − Small, ± acauline, ⚄, succulent plants; **roots** numerous, fine; **St.** undeveloped, enclosed within the membranes of dried L. with 2 opposite L. united much further on one side than on the other to form an asymmetrical **Bo.** c. 25 mm or more long, the free parts divaricate, with the tip curved upwards, the upperside flat and ± furrowed, the underside rounded, fleshy, glabrous, green, with translucent dots, drying up during the resting period; **Fl.** terminal, sessile, solitary, pinkish-purple. Greenhouse, warm. Growing season: late summer. Propagation: seed.

**D. leipoldtii** L. BOL. (*Conophyllum nanum* L. BOL., *Mitrophyllum n.* (L. BOL.) N. E. BR., *D. l.* f. *dormiens* L. BOL.). − Cape: Piquetberg, Clanwilliam D. − Resembles **D. retroversum**; **L.** somewhat unequal, 16−18 mm long, c. 6 mm across and thick, slightly twisted, roughly semicircular, with translucent, dark, longitudinal lines.

**D. retroversum** (KENSIT) SCHWANT. (Pl. 161/1) (*Mes. r.* KENSIT). − Cape: Piquetberg D. − **L.** 20−25 mm long, 7−9 mm thick, united by one edge to midway, with a rounded tip and several translucent dots; **Fl.** 15−18 mm ⌀, pinkish-purple.

**Disphyma** N. E. BR. (§ IV/1/3). − Creeping, mat-forming ♄ with the **St.** often rooting at the nodes; **L.** somewhat united at the base, linear, semicylindrical or trigonous, slightly papillose or with ± translucent dots; **Fl.** 1−3 together, pedicellate, whitish, pink or violet, spring or summer, opening in the afternoon. − Summer: in the open, winter: cold-house. Propagation: seed, cuttings.

**D. australe** J. M. BLACK (*Mes. a.* SOL. nom. nud., *D. a.* (SOL.) N. E. BR. comb. ill., *Mes. a.* AIT., *Mes. demissum* WILLD.). − Austr.: coasts of NSW.; Tasmania; New Zealand; Chatham Is. − Freely branched; **St.** c. 30 cm long; **L.** trigonous, 2 cm long, 4 mm across, fresh green, faintly spotted, bluntly tapered; **Fl.** c. 2 cm ⌀, pink.

**D. crassifolium** (L.) L. BOL. (Pl. 161/2) (*Mes. c.* L.). − Cape: Uylenkraal. − **L.** obtusely trigonous, short-tapered, 2.5−3.5 cm long, 5 mm across, dark green, with faint, translucent dots; **Fl.** 4 cm ⌀, pinkish-red.

**D. dunsdonii** L. BOL. − Cape: Caledon D. − **L.** hemispherical, the 2 of a pair unequal, carinate on the underside, 5−9 mm long, 4 mm across and thick, brownish to reddish; **Fl.** 2.5 cm ⌀, whitish.

**Dorotheanthus** SCHWANT. (§ IV/1/20). − Many-branched ⊙ plants; **L.** basal or alternate, narrow or lanceolate, ± glittering-papillose; **Fl.** ± long-pedicellate, very numerous, white, yellow, red or lilac. − Favourite summer-Fl. for the open. − Propagation: seeds.

### Key to the Species of the Genus Dorotheanthus by L. Bolus

1. Pet. few, inconspicuous, much shorter than the Sep., stamens few ............. **apetalus**
1a. Pet. many, showy, equalling or longer or rarely a little shorter than the longest Sep., stamens many ................................................................ **2.**
2. Ov. nearly flat or convex in the middle, lobes obtusely compressed ............... **3.**
2a. Ov. concave above to the very middle, lobes scarcely visible ................. **9.**
3. Pet. usually acute or acuminate ............................................... **4.**
3a. Pet. obtuse .................................................................. **6.**
4. L. obtuse or subacute, up to 1.2 cm broad, Pet. often acute ................... **5.**
4a. L. acute, up to 2.3 cm broad; Pet. usually acuminate, pale yellow ............ **acuminata**
5. Pet. white in the lower half, purple-rose in the upper; filaments and anthers black-purple **tricolor**
5a. Pet. white throughout except for a dark spot at the very base; filaments and anthers red **hallii**
6. Pet. bright rose-pink with darker margins, vittate in the lower part ......... **bellidiformis**
6a. Pet. not as above ............................................................ **7.**
7. Pet. white in the lowest part, deep rose above, up to 3.5 mm broad ......... **bidouwensis**
7a. Pet. entirely yellow or reddish-pink, rarely up to 2.5 mm broad ............... **8.**
8. Pet. and filaments yellow ................................................... **flos-solis**
8a. Pet. reddish-pink; filaments pallid ......................................... **muirii**
9. L. linear or nearly linear; Pet. white in the lower half, rose or bright cerise in the upper **gramineus**
9a. Adult L. oblanceolate, spatulate or obovate; Pet. not as above ............... **10.**
10. Pet. yellowish-hued ........................................................ **oculatus 11.**
10a. Pet. red-hued ............................................................. **12.**
11. Pet. pale buff or usually some shade of yellow, gold or orange, purplish at the very base .... ........ **oculatus v. oculatus**
11a. Pet. copper-flame, margins in the lower part with a darker line, purplish for 2 mm at the base ....... **oculatus v. saldanhensis**
12. Pet. bright rose on the upperside, striped ................................. **littlewoodii**
12a. Pet. purplish-rose, pale to white above .................................... **stayneri**

In the original description the ovary of the following species is not exactly described and therefore these are not included in the key: **D. booysenii, D. rourkei.**

**D. acuminatus** L. BOL. − Cape: Piquetberg D. − Resembles **D. hallii**; **L.** ovate-spatulate, acute, 6−8 cm long, 19−23 mm across; **Fl.** c. 4−4.5 cm ⌀, Pet. narrow, straw-coloured, with red markings at the base.

**D. apetalus** (L. f.) N. E. BR. (*Mes. a.* L. f., *Cleretum a.* (L. f.) N. E. BR., *Mes. copticum* JACQ., *D. c.* (JACQ.) L. BOL., *Stigmatocarpum c.* (JACQ.) L. BOL., *Mes. caducum* AIT., *St. c.* (AIT.) L. BOL., *Psilocaulon c.* (AIT.) N. E. BR.). −

Cape: Cape D. – Glossy, papillose plant; **L.** filiform-semicylindrical to ovate; **Fl.** sessile, with a few insignificant Pet.; ovary conspicuously large, red.

**D. bellidiformis** (BURM. f.) N. E. BR. (Pl. 161/4) (*Mes. b.* BURM. f., *Mes. criniflorum* L. f., *D. c.* (L. f.) SCHWANT., *Cleretum c.* (L. f.) N. E. BR., *Mes. cuneifolium* JACQ., *Cleretum c.* (JACQ.) .N. E. BR., *Micropterum c.* (JACQ.) SCHWANT., *Mes. limpidum* AIT., *Cleretum l.* (AIT.) N. E. BR., *Micropterum l.* (AIT.) SCHWANT., *Mes. spathulifolium* WILLD.). – Cape: Cape D. – **L.** mostly radical, alternate, obovate, narrowed teretely towards the base, 2.5–7 cm long, 6 to 10 mm across, fleshy, rough, papillose; **Fl.** 3–4 cm ∅, Pet. white, pale pink, red, orange or white with red tips.

**D. bidouwensis** L. BOL. – Cape: Clanwilliam D. – **L.** narrow to broadly oblanceolate to long-spatulate, rounded at the tip, 2.5–10 cm long, 6–18 mm across; **Fl.** 2.5–3.5 cm ∅, inner Pet. white, outer ones pink.

**D. booysenii** L. BOL. – Cape: Sutherland D. – Conspicuously papillose; **L.** ± spatulate, rounded above, c. 1.3 cm long; **Fl.** on a pedicel 1.5–3 cm long, Pet. acute to obtuse, white to pale pink, 2.5 cm long, c. 3 mm across, anthers vivid red, stigma-lobes pink.

**D. flos-solis** (BGR.) L. BOL. (*Mes. f.-s.* BGR., *D. luteus* N. E. BR.). – Cape: Calvinia, Van Rhynsdorp D. – **L.** almost rosulate on the shoots, very papillose, linear-lanceolate, obtuse, 0.5–5 cm long, 3–5 mm across; **Fl.** solitary or paired, Pet. glossy sulphur-yellow, stamens purplish-pink.

**D. gramineus** (HAW.) SCHWANT. f. **gramineus** (Pl. 161/3) (*Mes. g.* HAW., *Cleretum g.* (HAW.) N. E. BR., *Mes. clavatum* HAW., *Mes. claviforme* DC., *Mes. pyropaeum* HAW.). – Cape: Cape D. – Branching from the base, up to 10 cm tall; **Br.** papillose, red; **L.** linear, 3–5 cm long, 3 to 5 mm across, rounded on the underside, very papillose, fresh green; **Fl.** 2–2.5 cm ∅, brilliant crimson with a darker centre.

**D. — f. albus** (HAW.) ROWL. (*Mes. pyropaeum* v. *a.* HAW., *D. g.* v. *a.* (HAW.) SCHWANT., *Mes. apetalum* THUNBG., *Mes. lineare* THUNBG.). – Fl. white.

**D. — f. roseus** (HAW.) ROWL. (*Mes. pyropaeum* v. *r.* HAW., *D. g.* v. *r.* (HAW.) SCHWANT.). – Fl. pink.

**D. hallii** L. BOL. – Cape: Malmesbury D. – **L.** narrow-spatulate to oblanceolate, 3.5–6.5 cm long, 5–10 mm across; **Fl.** c. 6 cm ∅, Pet. acute, white to pale straw-coloured, with a red border below.

**D. littlewoodii** L. BOL. – Cape: Worcester D. – **L.** spatulate, c. 5.5 cm long, spatulate, obtuse above, the blade more rarely circular, 9 to 15 mm long; **Fl.** c. 2.5 cm ∅, Pet. brilliant pink above, striped, white below.

**D. muirii** N. E. BR. – Cape: Riversdale D. – **L.** 1–4.5 cm long, 1–4 mm across, linear to spatulate, acute to obtuse; **Fl.** 2–2.5 cm ∅, Pet. pinkish-red with a dark spot below.

**D. oculatus** N. E. BR. v. **oculatus** (*Stigmatocarpus criniflorus* L. BOL., *D. c.* (L. BOL.) BOL.). – Cape: Malmesbury, Piquetberg, Clanwilliam, Calvinia D. – **L.** 1–4.5 cm long, 2 to 10 mm across, spatulate-lanceolate to linear-cuneate, acute above; Pet. mostly light yellow, with a dark red spot at the base, anthers and stigma-lobes red.

**D. —** v. **saldanhensis** L. BOL. – Cape: Malmesbury D. – Pet. coppery-red, lighter below, with dark red stigma-lobes.

**D. rourkei** L. BOL. – Cape: Lit. Namaqualand. – Up to 6 cm tall, 6–12 cm ∅, herbaceous parts distinctly papillose; **L.** spatulate to linear, 2.5–4 cm long, c. 7 mm across; **Fl.** on stalks 1.5 to 6 cm long, Pet. acute to tapered, deep pink, c. 3 cm long, filaments red, anthers yellow.

**D. stayneri** L. BOL. – Cape: Worcester D. – Up to 4 cm tall, 3–5 cm ∅; **L.** narrow-spatulate, c. 2 cm long, c. 6 mm across; **Fl.** c. 5 cm ∅, purplish-pink with a white centre.

**D. tricolor** (WILLD.) L. BOL. (*Mes. t.* WILLD.). – Cape. – **L.** obtuse to ± acute, up to 12 mm across; Pet. white below, purplish-pink above, stigma-lobes and anthers blackish-purple.

**Dracophilus** SCHWANT. (§ 1V/1/13). – Cushion-forming plants; Int. not visible; **L.** numerous, mostly densely crowded, decussate, very fleshy, connate at the base, trigonous, usually of irregular shape, with a few marginal T., bluish-green, minutely rough, without spots; **Fl.** terminal or appearing so, almost hidden in 2 long-united bracts, 2–3 cm ∅, white or pink, summer. – Greenhouse, warm. Propagation: seed, cuttings.

### Key to the Species of the Genus Dracophilus by M. Walgate

| | | |
|---|---|---|
| 1. | L. dentate ................................................................. | **2.** |
| 1a. | L. entire .................................................................. | **3.** |
| 2. | L. 2–3 cm long; Sep. not fleshy; Pet. ± linear; pedicel 18 mm long ........... | **delaetianus** |
| 2a. | L. 3–5 cm long; 2 Sep. fleshy; Pet. attenuate towards the base; pedical 8 mm long ........ ....... | **montis-draconis** |
| 3. | L. 2–3 times as long as broad, thickest near the apex; stigmas usually 10 ....... | **dealbatus** |
| 3a. | L. c. 4 times as long as broad, thickest at the base; stigmas usually 11–12 ........ | **proximus** |

**D. dealbatus** (N. E. Br.) Walgate (Pl. 161/5) (*Mes. d.* N. E. Br., *Juttadinteria d.* (N. E. Br.) L. Bol., *J. rheolens* L. Bol., *Mes. r.* L. Bol., *D. r.* (L. Bol.) Schwant.). – Cape: Lit. Namaqualand; S.W.Afr.: S. of Witpütz. – **St.** very short, mostly with 4 **L.** which are flat on the upperside, ± rounded on the underside, broadly keeled, with entire margins and keel, 23–45 mm long, 15 mm across and thick, light green, smooth; **Fl.** 25–30 mm ⌀, pink, more rarely white.

**D. delaetianus** (Dtr.) Dtr. et Schwant. (*Mes. d.* Dtr., *Juttadinteria d.* (Dtr.) Dtr. et Schwant.). – S.W.Afr.: Gr. Namaqualand. – Cushions 10 cm ⌀; **L.** 4–6, ± rosulate, often oblique, variable in length and thickness, triangular, c. 2–3 cm long, 10–15 mm across and thick, the upperside broadly triangular, the margins with 5–6 T., the keel chin-like, with 2–3 T., the surface scarcely rough, whitish or bluish; **Fl.** 20–22 mm ⌀, violet-pink.

**D. montis-draconis** (Dtr.) Dtr. et Schwant. (*Mes. m.-d.* Dtr., *Juttadinteria m.-d.* (Dtr.) Dtr. et Schwant.). – S.W.Afr.: Gr. Namaqualand. – **Shoots** with 2–3 L.-pairs; **L.** obtusely triangular, curved slightly upwards, 3–5 cm long, 10–15 mm thick, rather oblique, with 1–2 flat tubercles at the tip, the surface bluish-green, minutely rough; **Fl.** solitary, 2.5–3 cm ⌀, white or light pink.

**D. proximus** (L. Bol.) Walgate (*Juttadinteria p.* L. Bol.). – Cape: Lit. Namaqualand; S.W.Afr.: W. of Witpütz. – **St.** 4–6 cm long, with (4–)6(–8) tapered **L.** which are flat or ± convex on the upperside, with rounded sides, margins and keel, 4–5 cm long, 14 mm across and thick, bluish-green, very smooth; **Fl.** 25 mm ⌀, pink.

**Drosanthemum** Schwant. (§ IV/1/16). – Robust, erect, stiff ♄ up to 40 cm high, or **St.** prostrate, slender, pliant, branched, St. and Br. mostly rough with hairlike papillae, or the papillae only sporadic, or puberulous; **rootstock** napiform with age, or not; **L.** decussate, compressed, trigonous to cylindrical, densely covered with bright papillae resembling the Sundew **Drosera**; **Fl.** terminal on lateral short shoots, solitary, in some species in 3's, red or white, up to 4.5 cm ⌀, sometimes with white or black staminodes, summer, open at midday and in the afternoon. – Summer: in the open, winter: in the cold-house. – Propagation: seed, cuttings.

## Key to the Groups and Subgroups of the Genus Drosanthemum by L. Bolus

1. Rootstock a large caudex with age; papillae minute, seen only under a lens. – D. anomalum (sole species) .................................................................. **Gr. I. Anomala**
1a. Rootstock not caudiciform; L. usually conspicuously papillate ........................ 2.
2. Int. and Ped. rough or powdery ............................. **Gr. II. Aspericaulia** 3.
2a. Int. and Ped. hispid with capillary papillae .................. **Gr. III. Hispicaulia** 8.
3. Nectary annular, e.g. D. diversifolium ........................... **SGr. 1. Annularia**
3a. Nectary composed of separate or nearly separate glands ........................... 4.
4. Usually robust or even rigid ♄ 20–40 cm high or, if slender, of lower growth; Fl. meridian . 5.
4a. ♄ with slender St. up to 50 cm high; Fl. pomeridian – e.g. D. lique ........ **SGr. 5. Gracilia**
5. Fl. usually more than 3 cm ⌀ (smaller in D. flavum) .................. **SGr. 2. Speciosa** 6.
5a. Fl. usually 3 cm or less ⌀ (in D. macrocalyx up to 4.5 cm ⌀) ........................ 7.
6. Staminodes absent – e.g. D. flavum
6a. Staminodes white – e.g. D. bicolor  } Subdivisions within SGr. 2.
6b. Staminodes black – e.g. D. micans
7. L. keeled – e.g. D. parvifolium ................................... **SGr. 3. Carinata**
7a. L. dorsally rounded – e.g. D. ambiguum ......................... **SGr. 4. Semiteretia**
8. Staminodes present – e.g. D. stokoei ......................... **SGr. 6. Staminodifera**
8a. Staminodes absent – e.g. D. hispidum ......................... **SGr. 7. Defecta**

**D. acuminatum** L. Bol. – Cape: Montagu D. – Erect, slender, 20 cm tall; **L.** semicylindrical, green, indistinctly papillose, 18 mm long, 2 mm across and thick; **Fl.** golden-yellow, orange inside.

**D. acutifolium** (L. Bol.) L. Bol. (*Mes. capillare v. a.* L. Bol., *D. c. v. a.* (L. Bol.) Schwant.). – Cape: Riversdale D., Lit. Karroo. – Dwarf ♄ with ascending or twisted **St.** 25 cm long; **L.** semicylindrical, short-tapered, green, minutely papillose, 6–8 mm long, 2 mm across, 2–3 mm thick; **Fl.** 12 mm ⌀, purplish-pink.

**D. albens** L. Bol. – Cape: Lit. Namaqualand. – Slender, densely branched; **St.** prostrate, 25 cm long, minutely scaly, bristly-papillose at first; **L.** cylindrical, obtuse, bluish-green, minutely papillose, 10–15 mm long, 4 mm thick; **Fl.** solitary, 25 mm ⌀, white.

**D. albiflorum** (L. Bol.) Schwant. (*Mes. a.* L. Bol.). – Cape: Riversdale D. – Slender, 15 cm tall; **St.** spreading to ascending, roughly papillose; **L.** cylindrical, obtuse, green, minutely papillose, 7–10 mm long, 4 mm thick; **Fl.** solitary, 17 mm ⌀, white.

**D. ambiguum** L. Bol. – Cape: Malmesbury D. – Creeping; **St.** slender, rooting, up to 55 cm long, roughly papillose; **L.** semicylindrical, obtuse, glossy-papillose, 22 mm long, 1.5–2 mm thick; **Fl.** 24 mm ⌀, pinkish-purple.

**D. anomalum** L. Bol. – Cape: Montagu, Robertson D. – **Caudex** tuberous, 7 cm long, 3 cm thick; **St.** tuberously thickened, 13–15 mm thick, short shoots 5–7 cm long, indistinctly papillose, the herbaceous parts green to yellowish-green, with raised dots; **L.** semicylindrical, compressed at the obtuse tip, 16–23 mm long, 2–2.5 mm across, 3–3.5 mm thick; **Fl.** solitary, 15 mm ⌀, golden-yellow.

**D. archeri** L. BOL. – Cape: Laingsburg D. – **St.** prostrate, 20 cm or more long, with papillose hairs; **L.** semicylindrical, rather obtuse, green, with dense, glossy papillae, 15 mm long, 2 to 3.5 mm thick; **Fl.** 24 mm ⌀, purplish-pink.

**D. attenuatum** (HAW.) SCHWANT. (*Mes. a.* HAW., *Mes. striatum* v. *a.* SALM). – Cape: Caledon D. – Resembles **D. striatum** and possibly only a variety; growth weak, more prostrate; Pet. white with red stripes.

**D. aureopurpureum** L. BOL. (*D. micans* v. *aureopurpureum* L. BOL.). – Cape: Swellendam D. – Dwarf; **St.** thin, tangled, up to 22 cm long; Pet. golden-yellow on the underside, a magnificent purple on the upperside.

**D. austricolum** L. BOL. – Cape: Bredasdorp D. – Laxly branched; **St.** ± prostrate, 15 cm long; **Br.** 11 cm long, with projecting hairs; **L.** semicylindrical, obtuse, green, indistinctly papillose, 17 mm long, 3 mm thick; **Fl.** 2 cm ⌀, pinkish-purple.

**D. autumnale** L. BOL. – Cape: Mossel Bay D. – Robust, with tangled Br., 45 cm tall and across; **L.** semicylindrical to broadly ovate, 6 mm long, 2.5 mm thick; **Fl.** 2 cm ⌀, purple.

**D. barkerae** L. BOL. – Cape: Oudtshoorn D. – Erect, with tangled Br., 30–60 cm tall; **St.** stiff, hirsute; **L.** cylindrical, obtuse, glossy-papillose, 5–10 mm long, 3–4 mm thick; **Fl.** 22 mm ⌀, pinkish-purple.

**D. barwickii** L. BOL. – Cape: Swellendam D. – **St.** woody, blackish, herbaceous parts green, with roundish papillae; **L.** laterally compressed, carinate, lanceolate-linear, obtuse, 7–11 mm long, 1.5–2 mm across, 2–3 mm thick, papillose; **Fl.** c. 2 cm ⌀, purplish-pink.

**D. bellum** L. BOL. – Cape: Ceres D. – Laxly branched, up to 30 cm tall; **L.** semicylindrical, glossy-papillose, 15–55 mm long, 4 mm thick; **Fl.** solitary, 5 cm ⌀, yellowish-pink with a white centre.

**D. bicolor** L. BOL. – Cape: Lit. Karroo. – Erect, stiff, branched, c. 30 cm tall; **L.** narrowed towards the tip, short-tapered, rounded on the underside, green to greenish-yellow, indistinctly papillose, 10–18 mm long, 3 mm across and thick; **Fl.** 3 cm ⌀, purplish-red, with a golden-yellow centre.

**D. breve** L. BOL. – Cape: Lit. Namaqualand. – Erect, up to 8 cm tall; **St.** stiff, prostrate, up to 18 cm long, glossy; **L.** semicylindrical, convex on the upperside, obtuse, pale blue, glossy-papillose, 15–21 mm long, 2.5–3.5 mm thick; **Fl.** 22 mm ⌀, whitish-pink.

**D. brevifolium** (AIT.) SCHWANT. (*Mes. b.* AIT.,). – Cape: Uitenhage D. – Over 60 cm tall, with many slender, rough-papillose **St.**; **L.** semicylindrical to trigonous, obtuse, 4–10 mm long, 2–4 mm thick, fresh green, glossy-papillose; **Fl.** 2 cm ⌀, pinkish-red.

**D. calycinum** (HAW.) SCHWANT. (*Mes. c.* HAW.). – Cape: Clanwilliam, Malmesbury D. – Over 30 cm tall; **St.** spreading, hard, thin, densely white-bristly; **L.** cylindrical, rather obtusely tapered, 12–20 mm long, 2 mm thick, bluish-green, minutely glossy-papillose; **Fl.** solitary, 2 cm ⌀, white.

**D. candens** (HAW.) SCHWANT. (*Mes. c.* HAW., *Mes. hispidum* v. *a.* THUNBG.). – Cape: near the coasts. – **St.** widespread, prostrate and rooting, with numerous rough Br. with short hairs; **L.** ± cylindrical, rather obtuse, 8–12 mm long, 2 mm thick, minutely papillose; **Fl.** solitary, 12 mm ⌀, white.

**D. capillare** (L. f.) SCHWANT. (*Mes. c.* L.f.) – Cape: Humansdorp D., Gamtoos River. – **St.** and Br. very slender, filiform, with ± swollen nodes; **L.** 2–4 mm long, 1 mm thick, slightly flattened above, obtuse, papillose; **Fl.** minute, reddish, on pedicels 6–12 mm long.

**D. cereale** L. BOL. – Cape: Ceres D. – **St.** stiff, c. 15 cm long, roughly papillose; **L.** (10–)15(–20) mm long, 2 mm across and thick, with the acute tip slightly recurved; **Fl.** 2.5 cm ⌀, Pet. acute, yellow with red margins.

**D. chrysum** L. BOL. – Cape: Laingsburg D. – Slender, with dense **St.** 10 cm long; **L.** semicylindrical, recurved at the tip, minutely glossy-papillose, 10–12 mm long, 2 mm across and thick; **Fl.** solitary, 4 cm ⌀, golden-yellow.

**D. collinum** (SOND.) SCHWANT. (*Mes. c.* SOND.). – Cape: Bokkeveld D. – 30 cm tall; **St.** erect, straight, filiform; **L.** cylindrical-trigonous, obtuse, minutely papillose, 10 to 16 mm long; **Fl.** 12 mm ⌀, golden-yellow.

**D. comptonii** L. BOL. – Cape: Laingsburg D. – ± erect, stiff, with lax, rough-haired Br. 25 cm long; **L.** semicylindrical, obtuse, indistinctly glossy-papillose, 9 mm long, 3 mm thick; **Fl.** 26 mm ⌀, dull pink.

**D. concavum** L. BOL. – Cape: Calvinia D. – **St.** prostrate to creeping, glabrous; **Br.** 5 cm long, rough-haired; **L.** semicylindrical, obtuse, glossy-papillose, 6–10 mm long, 3 mm thick; **Fl.** 24 mm ⌀, pale pink with a white centre.

**D. crassum** L. BOL. – Cape: Ladismith D. – Stiff, robust, 26 cm tall, with a **St.** 2.5 cm thick; **Br.** variously curved, slender, papillose; **L.** cylindrical to ± spherical, green, indistinctly papillose, 5–8 mm long, up to 3 mm ⌀; **Fl.** solitary, 15–20 mm ⌀, on a pedicel up to 15 mm long, persisting and becoming ± thorny.

**D. croceum** L. BOL. – Cape: Swellendam D. – Dwarf, with dense, slender **Br.**; **L.** semicylindrical, narrowed towards the tip, glossy-papillose, 13 mm long, 2 mm thick; **Fl.** 25 mm ⌀, saffron-yellow.

**D. curtophyllum** L. BOL. – Cape: Lit. Namaqualand. – Compact, 10–30 cm tall; **St.** spreading to erect or creeping, often rooting, glabrous, stiff; **L.** obtuse to ± truncate, minutely glossy-papillose, 5 mm long, 2 mm thick; **Fl.** 15 mm ⌀, pale pink.

**D. cymiferum** L. BOL. – Cape: Calvinia D. – Erect, stiff, glabrous, 10 cm tall; **L.** ± sabre-shaped, obtuse, surface densely papillose, almost granular, 6 mm long, 2.5 mm across and thick; **Infl.** of 3 pale pink, somewhat scented Fl. 20 to 22 mm ⌀.

**D. dejagerae** L. BOL. – Cape: Beaufort West. D. – Erect, stiff, 15 cm tall; **St.** spreading to ascending, initially hirsute and rough; **L.** semicylindrical, laterally somewhat compressed, distinctly papillose, 5–7 mm long, 2.5 mm across, 3.5 mm thick; **Fl.** 2.5 cm ⌀, purplish-pink.

**D. delicatulum** (L. BOL.) SCHWANT. (*Mes. d.* L. BOL.). – Cape: Ladismith, Swellendam D. – Erect, up to 10 cm tall; **St.** stiff; **L.** hemispherical, concave on the upperside, with large papillae, (2–)3(–4) mm long, 2–2.5 mm across and thick; **Fl.** mostly solitary, 8 mm ⌀, pale pink.

**D. diversifolium** L. BOL. (*D. d. forma* L. BOL.). – Cape: Van Rhynsdorp D. – Erect, 12 cm high; **St.** ascending to spreading, papillose-bristly, red; **L.** variously shaped, some hemispherical to oblong, acute to obtuse, papillose, 10–22 cm long, 5 mm thick, other L. semicylindrical, 24–35 mm long, 3–5 mm wide, 4 mm thick; **Fl.** 32–40 mm ⌀, pinkish-purple to white.

**D. duplessiae** L. BOL. – Cape: Oudtshoorn D. – Erect, glabrous, c. 30 cm tall; **St.** projecting to ascending, with slender, rough Br.; **L.** hemispherical to oblong, obtuse, green, glossy-papillose, 6–13 mm long, 3–4 mm thick; **Fl.** almost 3 cm ⌀, pink.

**D. eburneum** L. BOL. – Cape: Karroo. – ± mat-forming; **Br.** prostrate, hirsute; **L.** semicylindrical, narrowed towards the tip, with dense, glossy papillae, 1–2 cm long, 1.5–2 mm thick; **Fl.** 3 cm ⌀, ivory-white.

**D. edwardsiae** L. BOL. – Cape: George D. – Erect, laxly branched, 15 cm tall; **L.** semicylindrical, laterally ± compressed, recurved at the tip, 1–2 cm long, 3 mm across, 4 mm thick; **Fl.** 44 mm ⌀, pinkish-purple.

**D. erigeriflorum** (JACQ.) STEARN (*Mes. e.* JACQ., *Mes. brevifolium* SALM, *Mes. lateriflorum* DC.). – S.Afr.: origin not known. – Much-branched papillose ♄ 20–30 cm tall; **L.** sessile, semi-amplexicaul but not connate in pairs, light green, triquetrous, obtusely truncate at the tip; **Fl.** solitary, terminal on the lateral Br., pale violet.

**D. filiforme** L. BOL. – Cape: Lit. Namaqualand. – Slender; **St.** prostrate, ± variously curved, laxly hirsute, becoming glabrous; **Br.** erect, often tangled; **L.** cylindrical, 10–15 mm long, obtuse, glossy-papillose; **Fl.** 25–30 mm ⌀, pink.

**D. flammeum** L. BOL. – Cape: Robertson D. – Erect, 25 cm tall; **Br.** stiff, indistinctly papillose; **L.** semicylindrical, obtuse, pale bluish-green, indistinctly papillose, 10–17 mm long, 2 mm across and thick; **Fl.** c. 44 mm ⌀, orange to flame-coloured.

**D. flavum** (HAW.) SCHWANT. (*Mes. f.* HAW.). – Cape: on the Swartkops R. – 10–15 cm tall; **St.** numerous, very thin, filiform, curved, papillose at first, later rough-spotted; **L.** ± cylindrical, recurved at the tip, fresh green, crystalline-papillose, 6–8 mm long, 2 mm thick; **Fl.** solitary, 18 mm ⌀, golden-yellow.

**D. floribundum** (HAW.) SCHWANT. (*Mes. f.* HAW., *Mes. hispidum* v. *pallidum* HAW., *Mes. torquatum* HAW.). – Cape: Worcester D.; S.W.Afr.: Gr. Namaqualand. – Cushion-forming; **St.** filiform, prostrate or creeping, freely branched, roughly hirsute; **L.** ± curved, cylindrical, somewhat thickened at the obtuse tip, light green, 12–14 mm long, 2.5 mm thick; **Fl.** very numerous on lateral short shoots, 18 mm ⌀, pale pink.

**D. fourcadei** (L. BOL.) SCHWANT. (*Mes. f.* L. BOL., *D. f.* (L. BOL.) L. BOL.). – Cape: Humansdorp D. – Erect, stiff, 22 cm tall; **St.** and **Br.** covered with the remains of old L.; **L.** laterally compressed, carinate, wider and ± truncate at the tip, minutely papillose, bluish-green, 8 mm long, 3 mm across, 4.5 mm thick above; **Fl.** solitary, 15 mm ⌀, purplish-red.

**D. framesii** L. BOL. – Cape: Ceres, Van Rhynsdorp D. – Erect, compact, 6–7 cm tall; **St.** with white, rough projecting hairs; **L.** cylindrical, obtuse above, 10–13 mm long, 3–4 mm thick; **Fl.** 3 cm ⌀.

**D. fulleri** L. BOL. – Cape: Kenhardt D. – Stiff, erect, partly robust, 20 cm tall; **St.** prostrate, with short hairs; **L.** semicylindrical, rather obtuse, blue, glossy-papillose, 10–15 mm long, 4.5 mm thick; **Fl.** 25–30 mm ⌀, pinkish-purple.

**D. giffenii** (L. BOL.) SCHWANT. v. **giffenii** (*Mes. g.* L. BOL.). – Cape: Montagu D. – Erect; **St.** and **Br.** thick, gnarled, rough; **short shoots** slender, 2 cm long; **L.** semicylindrical, obtuse, pale blue, with prominent, glossy papillae, 4–7 mm long, 2 mm across, 4 mm thick; **Fl.** 14 mm ⌀, pink.

**D.** — v. **intertextum** (L. BOL.) SCHWANT. (*Mes. g.* v. *i.* L. BOL.). – Cape: Riversdale D. – Densely branched; **Br.** numerous, bent, densely tangled; **Fl.** 5–10 together.

**D. glabrescens** L. BOL. – Cape: Calvinia D. – Erect, stiff, 20 cm cr more tall; **St.** brown, with pearl-like papillae; **L.** semicylindrical, ± acute, 7–9 mm long, 3 mm thick; **Fl.** 2 to 3 cm ⌀, purplish-pink.

**D. globosum** L. BOL. – Cape: Montagu D. – Erect, 45 cm tall; **Br.** tangled, stiff, with minute, papillose hairs; **L.** spherical, green, 3 mm thick; **Fl.** 17 mm ⌀, purplish-pink.

**D. godmaniae** L. BOL. – Cape: Lit. Namaquaand. – Erect, 15 cm tall, 20 cm ⌀; **St.** dense, stiff; **Br.** slender, minutely scaly; **L.** cylindrical, obtuse, bluish-green, with round, glossy papillae, 9 mm long, 2.5 mm thick; **Fl.** 14 mm ⌀, pinkish-purple.

**D. gracillimum** L. BOL. – Cape: Uniondale D. – **St.** slender, rather weak, somewhat climbing, variously curved, 20 cm or more long; **L.** semicylindrical, recurved at the tip, 11 to 50 mm long, 1.5 mm thick; **Fl.** 15 mm ⌀, pinkish-purple.

**D. hallii** L. BOL. – Cape: Worcester D. – **St.** erect, 27 cm long; **shoots** 1.5–4 cm long; **L.** linear, acute, rounded on the underside, indistinctly papillose, glossy green, 1–2.6 cm long, c. 2.5 mm across, 1.5–2.6 mm thick; **Fl.** c. 3.5 cm ⌀, yellow.

**D. hirtellum** (HAW.) SCHWANT. (*Mes. h.* HAW.). – Cape: origin not known. – Resembles **D. hispidum** but lower-growing, leafier, younger Br. having recurved, short hairs; **L.** ± cylindrical, minutely rough from the pointed papillae; **Fl.** 3–4 cm ⌀, light red.

**D. hispidum** (L.) SCHWANT. v. **hispidum** (*Mes. h.* L.). – S.W.Afr.: Gr. Namaqualand; Cape: Malmesbury and Montagu D. – Up to 60 cm tall, freely branched, over 1 m across; **St.** erect at first, later drooping and rooting, slender, with rough, projecting, white hairs; **L.** cylindrical, obtuse, light green to reddish, made glossy by the large, transparent papillae, 15–25 mm long, 3–4 mm thick; **Fl.** up to 3 cm ⌀, deep purplish-red, glossy.

**D.** — v. **platypetalum** (HAW.) SCHWANT. (*Mes. h.* v. *p.* HAW.). – Pet. very broad.

**D. inornatum** (L. BOL.) L. BOL. (*Trichodiadema i.* L. BOL., *D. tardum* L. BOL.). – Cape: Lit. Namaqualand. – **St.** lax, slender, densely papillose above, becoming rough; **L.** semicylindrical, obtuse, glossy-papillose, 19 mm long, 3 mm across and thick; **Fl.** solitary, 2 cm ⌀, purplish-pink.

**D. insolitum** L. BOL. – Cape: Caledon D. – Erect, slender, 15 cm tall, laxly branched; **L.** 10–14 mm long, 2 mm thick, with dense but indistinct papillae; **Fl.** on a hirsute pedicel, 3–3.5 cm ⌀, yellow.

**D. intermedium** (L. BOL.) L. BOL. (*Mes. i.* L. BOL.). – Cape: Mossel Bay, Riversdale, Stellenbosch D. – **St.** slender, curved, creeping, thickened at the nodes, shortly hirsute; **L.** concave on the upperside, rounded on the underside, acute, minutely papillose, 7–15 mm long, 1 mm thick; **Fl.** 25 mm ⌀, pinkish-purple.

**D. jamesii** L. BOL. – Cape: Cradock D. – Erect, stiff, c. 25 cm tall; **St.** brown, papillose at first; **L.** semicylindrical, truncate above, bluish, with round, crowded papillae, 6 mm long, 2 mm across, 3 mm thick; **Fl.** 1 cm ⌀, pink.

**D. karrooense** L. BOL. – Cape: Laingsburg D. – Stiff, 20 cm tall, with slender, tangled **Br.**; **L.** ± cylindrical, obtuse, 8–13 mm long, 3 to 4 mm thick; **Fl.** 3 cm ⌀.

**D. latipetalum** L. BOL. – Cape: Van Rhynsdorp D. – **St.** prostrate or creeping, 25 cm or more long, rough with white hairs; **Br.** ascending to erect, with 2–4 L. and an Infl. 3.5 cm across; **L.** ± cylindrical, obtuse, papillose, 13 mm long, 2.5 mm thick; **Fl.** on a densely rough-haired pedicel, 32 mm ⌀, pink, paler within.

**D. lavisiae** L. BOL. – Cape: Bredasdorp D. – Erect, densely branched; **St.** 4 mm thick; **L.** shaped like a long S, recurved at the tip, dirty green, indistinctly papillose, 15–20 mm long, 2–2.5 mm across and thick; **Fl.** red inside, coppery-red outside.

**D. laxum** L. BOL. – Cape: Robertson D. – Erect, stiff, 10 cm tall; **L.** hemispherical to elongate, somewhat compressed laterally, obtuse to truncate above, green, ± glossy-papillose, 3–10 mm long, 3.5 mm across, 5 mm thick, upper L. more slender; **Fl.** solitary, 22 mm ⌀, pink to purplish-pink.

**D. leipoldtii** L. BOL. – Cape: Clanwilliam D.; S.W.Afr.: near Aus. – Erect, stiff, 36 cm tall, bearing many Fl.; **St.** and **Br.** virgate, erect, with translucent papillae; **L.** semicylindrical, obtuse, glossy-papillose, 8 mm long, 2 mm thick; **Fl.** 17 mm ⌀, purplish-pink.

**D. leptum** L. BOL. – Cape: Robertson D. – Up to 25 cm tall; **St.** virgate, up to 34 cm long, with small papillae; **L.** slightly compressed laterally, rounded on the underside, 10–18 mm long, 1–2 mm across and thick; **Fl.** solitary, 2.5–3 cm ⌀, white?

**D. lignosum** L. BOL. – Cape: Ceres D. – Woody, forming cushions 15 cm ⌀; **St.** thick; **L.** semicylindrical, obtuse, glossy-papillose, 8 mm long, 2 mm thick; **Fl.** 24 mm ⌀, pinkish-purple.

**D. lique** (N. E. BR.) SCHWANT. (*Mes. l.* N. E. BR., *Mes. obliquum* HAW., *D. o.* (HAW.) SCHWANT.). – Cape: Swartkops R. – Up to 60 cm tall; **St.** ± erect, filiform, initially papillose, later rough, with tiny white dots; **L.** cylindrical, rather obtuse, 10–16 mm long, scarcely 2 mm thick, fresh green, covered with crystalline papillae; **Fl.** 2 cm ⌀, purplish-red.

**D. littlewoodii** L. BOL. – S.W.Afr.: Namies Kloof. – c. 9 cm tall, 30–35 cm ⌀; **St.** prostrate, c. 16 cm long, with short, rough hairs; **L.** ± cylindrical, rounded at the tip, glossy-papillose, 10–12 mm long, c. 3 mm across, 3–4 mm thick; **Fl.** c. 2 cm ⌀, glossy white.

**D. macrocalyx** L. BOL. – Cape: Lit. Karroo. – Robust, c. 10 cm tall; **St.** 11 mm thick; **Br.** stiff, side **shoots** slender; **L.** obtusely carinate on the underside, narrowed towards the obtuse to acute tip, bluish-green, minutely glossy-papillose, 5–8 mm long, 2 mm across, 4 mm thick; **Fl.** 3.5 cm ⌀, pink.

**D. maculatum** (HAW.) SCHWANT. (*Mes. m.* HAW., *Mes. micans* v. β. HAW.). – Cape: origin unknown. – Over 60 cm tall, freely branched; **St.** opposite and decussate, very thin, hard, ± erect, scarcely rough but covered with spot-like papillae; **L.** standing out horizontally, semicylindrical, with a very obtuse keel, obtuse above, sometimes slightly furrowed on the upperside, green, glossy because of the tiny papillae; insufficiently known species.

**D. marinum** L. BOL. – Cape: Darling, Malmesbury D. – **Br.** elongated, creeping, rooting, slender, hirsute; **L.** laterally compressed and very obtuse, pale green, glossy-papillose, the papillae very small; **Fl.** 33 mm ⌀, pink.

**D. martinii** L. BOL. – Cape: Robertson D. – Small ♄; **L.** often spatulate to linear, obtuse, more rarely ± acute, red, 22–30 mm long, 3 to 7 mm across; Pet. expanded midway to 2.5 mm, yellow.

**D. mathewsii** L. BOL. – Cape: Van Rhynsdorp D. – Erect, slender, 8 cm tall; **St.** shortly hirsute; **L.** ± semicylindrical, obtuse, 12 mm long, 3.5 mm thick; **Fl.** 3 cm ⌀, purplish-pink.

**D. micans** (L.) SCHWANT. (*Mes. m.* L.). – Cape: Swellendam, Worcester D. – Up to 80 cm tall, freely branched; **St.** erect, slender, papillose-glossy at first, becoming rough with white spots; **L.** ± cylindrical, ± flattened on the upperside, fresh green, crystalline-papillose, 12–25 mm long, 2–4 mm thick; **Fl.** 12–15 mm ⌀, purple to somewhat yellowish.

**D. montaguense** L. Bol. – Cape: Montagu D. – Erect, stiff, 17 cm tall; **St.** elongated, slender, papillose; **L.** laterally compressed, flat on the upperside, acutely carinate on the underside, truncate above, pale bluish-green, papillose, margins and keel glossy, 6 mm long, 2 mm across, 3 mm thick; **Fl.** 15 mm ⌀, white.

**D. muirii** L. Bol. – Cape: Riversdale D. – Erect, woody, stiff, 40 cm or more tall; **St.** glossy, dark brown; **Br.** erect, hirsute; **L.** semicylindrical, obtuse, 10 mm long, 2.3 mm thick; **Fl.** 15 mm ⌀, pink.

**D. nordenstamii** L. Bol. – S.W.Afr.: S. of Witpütz. – Mat-forming; **St.** prostrate to creeping, c. 20 cm long, with dense, projecting, long-tapered bristle-hairs; **L.** semicylindrical, obtuse, glossy-papillose, 10–15 mm long, 2.5 mm across, 3 mm thick; **Fl.** c. 2.5 cm ⌀, Pet. white with pink tips.

**D. oculatum** L. Bol. – Cape: Lit. Namaqualand. – Fairly compact, c. 8 cm tall; **St.** prostrate; **L.** semicylindrical, expanded towards the tip, glossy-papillose, 11 mm long, 2.5 mm across, 3.5 mm thick; **Fl.** 28 mm ⌀, Pet. purplish-pink with white stripes on the upperside.

**D. opacum** L. Bol. – Cape: Tulbagh D. – Robust, woody, 10 cm or more high; **St.** prostrate, creeping, 30 cm or more long, papillose, indistinctly hirsute; **L.** laterally compressed, obtuse, minutely and densely papillose, bluish, glossy, 17 mm long, 2 mm across, 2–3 mm thick; **Fl.** 22 mm ⌀, pink.

**D. parvifolium** (Haw.) Schwant. (*Mes. p.* Haw.). – Cape: Lit. Namaqualand, Malmesbury, Bredasdorp, Riversdale D. – 15–20 cm tall; **St.** numerous, filiform, reddish, rough, glossy-papillose at first; **L.** laterally compressed, rather obtuse, fresh green, minutely glossy-papillose, 4 mm long, 2 mm thick; **Fl.** 15 mm ⌀, purplish-red.

**D. pauper** (Dtr.) Dtr. et Schwant. (*Mes. p.* Dtr.). – S.W.Afr.: Gr. Namaqualand. – Stunted, sparsely leafy ♄ up to 30 cm tall; **St.** and **Br.** thin, brown, with solitary papillae; **L.** cylindrical, sap-green, conspicuously papillose, c. 20 mm long, c. 3 mm thick, obtuse; **Fl.** solitary, 16–18 cm ⌀, purplish-red.

**D. paxianum** (Schtlr. et Diels) Schwant. (Pl. 161/6) (*Mes. p.* Schtlr. et Diels, *Mes. paxii* Engl., *Mes. brachyphyllum* Pax, *Mes. luederitzii* Engl., *D. l.* (Engl.) Schwant.). – S.W.Afr.: Lüderitz D. – 10–15 cm tall; **St.** spreading, densely leafy; **L.** projecting ± horizontally, ± cylindrical to lanceolate, subacute, papillose, 1 cm long, 5 mm thick; **Fl.** 15 mm ⌀.

**D. pickhardtii** L. Bol. – Cape: Robertson D. – Slender; **St.** c. 15 cm long, **Br.** 2–5 cm long, indistinctly papillose; **L.** rounded on the underside, subacute, green, 10–15 mm long, c. 2.5 mm across, c. 3 mm thick; **Fl.** c. 4.2 cm ⌀, Pet. 2 mm across, yellow to reddish-orange, inner Pet. paler.

**D. praecultum** (N. E. Br.) Schwant. (*Mes. p.* N. E. Br.). – S.Afr.: near Montagu Baths. – Erect ♄ up to 30 cm tall, densely papillose; **L.** 2–4 mm long, subglobose to short cylindric; **Fl.** very numerous, solitary from the L. axils on pedicels 6–10 mm long; Pet. 6 mm long.

**D. prostratum** L. Bol. – Cape: Clanwilliam D. – **St.** prostrate, stiff, creeping, 37 cm long, hirsute; **Br.** 3 cm long; **L.** semicylindrical, obtuse, pale green, scarcely papillose, glossy, 5–9 mm long, 1.5 mm thick; **Fl.** 3 cm ⌀, an attractive pink.

**D. pulchellum** L. Bol. – Cape: Clanwilliam D. – **St.** dense, elongated, prostrate, rooting, papillose-hirsute initially; **L.** laterally compressed, indistinctly carinate on the underside, pale green, densely papillose, 12 mm long, 1.5 mm across, 2.5 mm thick; **Fl.** 25 mm ⌀, an attractive pink.

**D. pulchrum** L. Bol. – Cape: Worcester D. – Laxly branched, 22 cm tall; **St.** glossy-papillose at first; **L.** semicylindrical, obtuse to short-tapered, papillose, 1–2 cm long, 2.5 mm thick; **Fl.** up to 3 cm ⌀, Pet. orange-red with yellow stripes, inner Pet. pale greenish-yellow.

**D. pulverulentum** (Haw.) Schwant. (*Mes. p.* Haw.). – Cape: Van Rhynsdorp D. – 15 cm tall; **Br.** opposite and decussate, compressed, glossy-papillose; **L.** cylindrical to trigonous, very obtuse above, roughly papillose, slender; **Fl.** deep red.

**D. ramosissimum** (Schltr.) L. Bol. (*Mes. r.* Schltr.) – Cape: Lit. Namaqualand. – Forming dense cushions; **St.** prostrate, rough; **L.** dense, semicylindrical, obtuse, 6–10 mm long, 2 mm thick; **Fl.** 15 mm ⌀, pinkish-red.

**D. roridum** L. Bol. – Cape: Montagu D. – Erect, 25–30 cm tall; **St.** slender, stiff, initially papillose and softly haired, becoming rough eventually; **L.** obtusely keeled, flat on the upperside, rounded or truncate above, densely papillose, pale bluish-green, 12 mm long, 2 mm across, 3–4 mm thick; **Fl.** 2 cm ⌀, pinkish-purple.

**D. roseatum** (N. E. Br.) L. Bol. (*Mes. r.* N. E. Br., *Ruschia comptonii* L. Bol.). – Cape: Van Rhynsdorp D. – Erect, branched, 15 cm tall; **Br.** 2–6 cm long; **L.** navicular, laterally compressed, rounded or ± truncate at the tip, soft, spotted, green to bluish, 10–15 mm long, 4 mm thick above; **Fl.** 1–3 together, 26 mm ⌀, pinkish-purple.

**D. salicolum** L. Bol. – Cape: Piquetberg D. – Densely branched, compact, 25 cm tall; **Br.** slender; **L.** hemispherical, obtuse, with dense, glossy papillae, bluish-green, 3 mm long, 2.5 mm thick, or 6 mm long, 3 mm thick; **Fl.** 14 mm ⌀, pale pink.

**D. schoenlandianum** (Schltr.) L. Bol. (*Mes. s.* Schltr.). – Cape: Van Rhynsdorp D. – **St.** spreading, thin, papillose, reddish-brown; **L.** semicylindrical, obtuse, fresh green, papillose; **Fl.** 25 mm ⌀, creamy-white.

**D. semiglobosum** L. Bol. – Cape: Worcester D. – Erect, with dense, slender **St.** 14–15 cm long; **L.** semicylindrical, slightly compressed, obliquely acuminate, pale bluish-green, 10 to 12 mm long, 1.5 mm across and thick; **Fl.**

25 mm ⌀, yellow, the ovary hemispherical, glossy-papillose.

**D. sessile** (THUNBG.) SCHWANT. (*Mes. s.* THUNBG.). – Cape: between Olifants R. and Bocklandberg. – 30–40 cm tall; **St.** opposite, curved, with numerous **Br.**; **L.** trigonous to spherical, minutely dotted, 2 mm long and across; **Fl.** on a very short pedicel, solitary, red.

**D. speciosum** (HAW.) SCHWANT. (*Mes. s.* HAW.). – Cape: Swellendam, Robertson D. – Up to 60 cm tall; **St.** erect to projecting, papillose at first, later rough, spotted; **L.** semicylindrical, curved, obtuse, fresh green, crystalline-papillose, 12–16 mm long, 4–6 mm thick; **Fl.** solitary, up to 5 cm ⌀, deep orange-red with a greenish centre.

**D. splendens** L. BOL. – Cape: Montagu D. – Erect, c. 20 cm tall; **St.** slender; **L.** concave to flat on the upperside, rounded on the underside, yellowish-green, with glistening papillae, 17 mm long, 4 mm across, 5 mm thick, sometimes 27–30 mm long, 5 mm across and thick; **Fl.** 3 cm ⌀, golden-yellow.

**D. stokoei** L. BOL. – Cape: Cape, Caledon D. – **Br.** dense, projecting, rough-haired; **L.** channelled on the upperside, semicylindrical, obtuse, papillose, 1–2 cm long, 4 mm thick; **Fl.** 17 mm ⌀, pink.

**D. striatum** (HAW.) SCHWANT. v. **striatum** (*Mes. s.* HAW., *Mes. s.* v. *roseum* HAW.). – Cape: Cape D. – Erect, up to 30 cm tall; **St.** erect, rough with projecting bristles; **L.** ± cylindrical, rather obtuse, 20–25 mm long, 3–4 mm thick, green, with large, transparent papillae terminating in minute bristles; **Fl.** 2.5–3 cm ⌀, Pet. pale pink with a red M.-nerve.

**D.** — v. **hispifolium** (HAW.) ROWL. (*Mes. h.* HAW., *Mes. s.* v. *β h.* SALM, *D. h.* (HAW.) SCHWANT., *Mes. tuberculatum* HAW.). – Cape: Stellenbosch, Malmesbury D. – **L.** with bristly papillae directed downwards.

**D.** — v. **pallens** (HAW.) ROWL. (*Mes. s.* v. *p.* HAW., *D. p.* (HAW.) SCHWANT., *Mes. s.* DC., *Mes. s.* v. *pallidum* DC.). – Cape: Caledon D. – Pet. white, with red lines only at the base.

**D. strictifolium** L. BOL. – Cape: Riversdale D. – Erect, mostly stiff, 20–30 cm tall; **St.** thick; **L.** semicylindrical, spreading, acute or rather obtuse, scarcely papillose, 12–23 mm long, 4 mm across and thick; **Fl.** in an Infl., outer Pet. reddish-orange and yellow below, middle ones golden-yellow, inner ones yellow.

**D. subalbum** L. BOL. – Cape: Van Rhynsdorp D. – Erect, 7 cm tall; **St.** erect, 6 cm long, minutely rough; **L.** carinate on the underside, obtuse, with tiny dot-like papillae, ± pruinose, 10 mm long, 4 mm across, 5 mm thick; **Fl.** 2–3 cm ⌀, whitish.

**D. subclausum** L. BOL. – Cape: Bushmanland. – Erect, densely branched, 13–16 cm tall; **St.** slender, stiff, ± glossy, indistinctly papillose; **L.** somewhat compressed, hemispherical, with dense, glossy papillae, 5 mm long, 4 mm thick, or 4 mm long, 2 mm across, 3 mm thick; **Fl.** 12 mm ⌀, dull pink.

**D. subcompressum** (HAW.) SCHWANT. (*Mes. s.* HAW., *D. s.* (HAW.). L. BOL.). – Cape: origin unknown. – Up to 60 cm tall; **St.** erect, slender, ± spreading, with soft hairs, eventually rough, spotted; **L.** semicylindrical, somewhat compressed, with large, glossy papillae only along the angles and at the base, 10–25 mm long, 2.4 mm across; **Fl.** 2 cm ⌀, purplish-red.

**D. subglobosum** (HAW.) SCHWANT. (*Mes. s.* HAW., *Mes. capillare* THUNBG.). – Cape: origin unknown. – **Shoots** filiform; **L.** hemispherical; **Fl.** red.

**D. subplanum** L. BOL. – Cape: Calvinia D. – Dwarf; **St.** spreading, prostrate or creeping, rooting, elongated; **L.** semicylindrical, obtuse, bluish-green, glossy-papillose, 8–14 mm long, 2.5 mm across, 3 mm thick; **Fl.** 26 mm ⌀, pink.

**D. thudichumii** L. BOL. v. **thudichumii** f. **thudichumii.** – Cape: Worcester D. – Erect, woody, almost 1 m tall; flowering **shoots** c. 15 cm long, papillose; **L.** somewhat compressed laterally, indistinctly carinate on the underside, obtuse to subacute, papillose, 12–22 mm long, 2 mm across and thick; **Fl.** 2–3 together, 4 cm ⌀, ivory-white.

**D.** — v. **gracilius** L. BOL. f. **gracilius.** – Slender variety; Pet. more translucent at the base.

**D.** — v. — f. **aurantiaca** L. BOL. – **Fl.** golden-yellow.

**D.** — v. — f. **aurea** L. BOL. – **Fl.** yellow.

**D. torquatum** (HAW.) SCHWANT. (*Mes. t.* HAW.). – Cape: origin unknown. – **St.** intertwined to ± prostrate, slender; **L.** ± cylindrical, obtuse, greyish-green; **Fl.** pale purple, Cal. very hirsute. Possibly identical with **D. floribundum.**

**D. tuberculiferum** L. BOL. – Cape: Worcester D. – Slender; **St.** prostrate, up to 12 cm long, with **Br.** from the L.-axils, all with stout papillae; **L.** cylindrical, obtuse, pale bluish-green, with round papillae, c. 11 mm long, 1.5–2 mm thick; **Fl.** c. 2 cm ⌀, pink.

**D. uniflorum** (L. BOL.) FRIEDR. (*Mes. u.* L. BOL., *Lampranthus u.* (L. BOL.) L. BOL., *Mes. otzenianum* DTR., *Psilocaulon o.* (DTR.) L. BOL., *Lampranthus o.* (L. BOL.) FRIEDR., *Mes. spathulatum* L. BOL., *L. u.* v. *s.* (L. BOL.) L. BOL.). – S.W.Afr.: widely distributed; Cape: Van Rhynsdorp, Beaufort West D. – Forming cushions 40–80 cm across, 30–100 cm tall, with a whitish-grey bark; **St.** and **Br.** bent zigzag fashion, the outermost ones resting on the ground; **L.** very soft-fleshy, cylindrical, obtuse, patent, often curved upwards, 20–25 mm long, 7–9 mm thick; **Fl.** solitary from the upper L.axils, 23–30 mm ⌀, Pet. long-spatulate, purplish-pink.

**D. vandermerwei** L. BOL. – Cape: Swellendam D. – **St.** prostrate, slender, papillose, 40 cm or more long; **Br.** erect, 2–5 cm long; **L.** slender, semicylindrical, obtuse, glossy-papillose, 13 mm long, 1.5 mm thick; **Fl.** 27 mm ⌀, pink.

**D. vespertinum** L. BOL. – Cape: Prince Albert D. – Erect, laxly branched; **St.** minutely rough, slender; **L.** semicylindrical, somewhat glossy-papillose, 3–9 mm long, 2.5–4 mm thick; **Fl.** 14 mm ⌀, white.

**D. — v. suffusum** L. Bol. – Cape: Ceres D. – L. semicylindrical, flat on the upperside, obtuse, 3–7 mm long, 3 mm thick; **Fl.** pale pink, scented.

**D. wittebergense** L. Bol. – Cape: Laingsburg D. – Erect, c. 25 cm tall; **St.** stiff, younger ones slender; **L.** semicylindrical, laterally compressed, obtuse to short-tapered, green, indistinctly papillose, (15–)20(–24) mm long, 2.5–3 mm across and thick; **Fl.** mostly solitary, 25 mm ⌀, purplish-pink.

**D. worcesterense** L. Bol. – Cape: Worcester D. – Erect, branched, 20–30 cm tall, herbaceous parts glossy-papillose; **St.** virgate; **L.** obtuse or short-tapered above, pale green, 10–13 mm long, 2 mm across midway, 2.5 mm thick; **Fl.** 15 mm ⌀, Pet. pale pink, white-striate.

**D. zygophylloides** (L. Bol.) L. Bol. (*Mes. z.* L. Bol., *Lampranthus z.* (L. Bol.) N. E. Br.). – Cape: Clanwilliam D. – Densely branched, 14 cm tall; **St.** prostrate, spreading or ascending, stiff, with short shoots arising from the L. axils; **L.** cylindrical, obtuse, glabrous, green, soft, 10–13 mm long, 2–3 mm thick; **Fl.** numerous, 14 mm ⌀, yellow.

**Eberlanzia** Schwant. (§ IV/1/1). – ♄ with woody **roots**; **St.** erect, rigid; **L.** united somewhat at the base, trigonous to cylindrical, elongated or short and thick, the tip ± obtuse, mucronate, greyish-green or bluish-greenish, with minute dark dots; **Fl.** in a small to medium-large branched Infl., red or pink, May-June, the Infl. or only the Ped. mostly persisting and becoming thorny. – Summer: in the open, winter: cold-house. Propagation: seed, cuttings.

**E. aculeata** (N. E. Br.) Schwant. (*Mes. a.* N. E. Br.). – Cape: Beaufort West D. – Similar to **E. spinosa**; Th. 12–18 mm long; **L.** smaller, shortly connate at the base, flat above, slightly keeled below, obtuse or apiculate; **Fl.** solitary, subsessile, c. 1 cm ⌀.

**E. albertensis** (L. Bol.) L. Bol. (*Ruschia a.* L. Bol.). – Cape: Prince Albert D. – **St.** stiff, 21 cm long, up to 7 mm thick, branched; **Br.** 6–15 cm long, with a terminal Infl.; **L.** flat on the upperside, round on the underside, indistinctly angled and keeled, short-tapered, acute, apiculate, 15–25 mm long, 3–4 mm across, up to 5 mm thick; **Infl.** thorny, Fl. 2.5–3 cm ⌀.

**E. armata** (L. Bol.) L. Bol. (*Ruschia a.* L. Bol., *Mes. a.* (L. Bol.) N. E. Br.). – Cape: Lit. Namaqualand. – Erect, 7–8 cm tall, with very tangled **St.**; **L.** rather variable, flat on the upperside, with convex sides, 9 mm long, 5 mm across and thick, L. of the short shoots longer and narrower; **Infl.** with 2–3 forks, Ped. becoming thorny, Fl. 18 mm ⌀, purplish-pink.

**E. clausa** (Dtr.) Schwant. (*Mes. c.* Dtr.). – S.W.Afr.: Gr. Namaqualand. – Up to 50 cm tall; **St.** numerous, dense, with a white bark; **Br.** ± erect; **L.** flat on the upperside, with a distinct keel, decurved at the apiculate tip, greyish-green, densely and roughly papillose; **Infl.** forked, thorny, Fl. without Pet.

**E. cradockensis** (O. Ktze.) Schwant. (*Mes. c.* O. Ktze.). – Cape: Cradock D. – Robust ♄ c. 40 cm tall; **St.** tangled, new shoots 1–1.5 cm long, spreading, with a brown bark; **L.** glabrous, arranged in 4 superimposed R., 3 mm long, 2 mm across, trigonous, acute; older L. dichotomously divided, with a leafless Th. c. 2 cm long; **Fl.** solitary, ± sessile, 3–4 mm ⌀, red.

**E. disarticulata** (L. Bol.) L. Bol. (*Ruschia d.* L. Bol.). – Cape: Calvinia D. – Erect, stiff, 20 cm tall; **St.** red at first, Br. elongated; **L.** carinate on the underside, with convex sides, acute, apiculate, green, spotted, 13 mm long, 6 mm across and thick, younger L. spherical-ovoid, 4 mm across; **Infl.** becoming thorny, with the pedicels slightly constricted into 2 joints, Fl. 2 cm ⌀, purplish-pink.

**E. divaricata** (L. Bol.) L. Bol. (*Ruschia d.* L. Bol.). – Cape: Kenhardt D. – Stiff, erect, 25 cm or more tall; **St.** rigid, grey, with Br. 3–7 cm long; **L.** keeled on the underside, the upperside flat to convex, short-tapered, apiculate, bluish-green, with dense, small, raised dots, 12 mm long, 3 mm across, 4 mm thick; **Infl.** twice ternate, pedicels 7–15, the intermediate ones forking and becoming thorny, Fl. 15 mm ⌀, pink.

**E. ferox** (L. Bol.) L. Bol. (*Ruschia f.* L. Bol.). – Cape: Cradock D. – Erect, 30–45 cm tall; **St.** stiff; **L.** falcate, flat on the upperside, obtusely carinate on the underside, acute or more rarely obtuse, dirty green, with raised dots, 12–22 mm long, 2.5 mm across, 3 mm thick; **Infl.** with 4–5 dense Br., becoming thorny, Fl. 12 mm ⌀, pinkish-purple.

**E. horrescens** (L. Bol.) L. Bol. v. **horrescens** (*Ruschia h.* L. Bol.). – Cape: Swellendam D. – Erect, c. 30 cm tall; **St.** rigid, 22 cm long; **L.** obtusely carinate, flat on the upperside, obtuse, 6 mm long, 2.5 mm across and thick; **Infl.** irregularly 3-partite, with projecting Th., Fl. 14 mm ⌀, pink.

**E. — v. densa** (L. Bol.) Jacobs. (*Ruschia h. v. d.* L. Bol.). – **Br.** shorter; **L.** hemispherical, 2 mm long; **Th.** numerous.

**E. horrida** (L. Bol.) L. Bol. (*Ruschia h.* L. Bol.). – Cape: Albany D. – Stiff, erect, 15 cm or more tall; **St.** thick; **L.** carinate on the underside, with the upperside and sides convex, short-tapered, green, 7 mm to rarely 1–2 cm long, 2 mm across, 3 mm thick; **Infl.** and Ped. becoming thorny, Fl. 14 mm ⌀, pinkish-purple.

**E. intricata** (N. E. Br.) Schwant. (*Mes. i.* N. E. Br.). – Transvaal: Johannesburg D. – Erect, stiff, thorny, 30 cm or more in height; **St.** opposite or alternate, curved, stout, 3.5 to 5 mm thick, blackish, with short shoots with crowded L. scars, others terminating in projecting, dark grey Th. 10–15 mm long; **L.** flat on the upperside, acutely carinate on the underside, tapered, apiculate, minutely spotted; **Fl.** solitary, 8–9 mm ⌀.

**E. macroura** (L. Bol.) L. Bol. (*Ruschia m.* L. Bol.). – Cape: Prince Albert D. – **St.** stiff, freely branched, 20 cm long, 8 mm thick; **Br.** 7–14 cm long; **L.** flat on the upperside, keeled on the underside, obtuse to acute, apiculate, minutely and densely papillose, bluish, 8 to 11 mm long, 5 mm across and thick; **Infl.** stiff, compact, with terminal Th., Fl. c. 2 cm $\varnothing$, pink.

**E. micrantha** (BGR.) SCHWANT. (*Mes. spinosum* v. m. BGR.). – Cape: near De Aar. – Resembles **E. spinosa**; **Fl.** very numerous, 5 mm $\varnothing$, dark pink.

**E. mucronifera** (HAW.) SCHWANT. (*Mes. m.* HAW., *Mes. pulverulentum* WILLD.). – Cape: Clanwilliam, Van Rhynsdorp D. – Erect, 40–50 cm tall; **St.** straight; **L.** obtusely trigonous, apiculate, grey-green; **Fl.** in 3's, pedicels thorny and persisting. Insufficiently known.

**E. munita** (L. BOL.) SCHWANT. (*Mes. m.* L. BOL.). – Cape: Willowmore D. – Erect, 20 cm tall; **St.** slender, stiff; **L.** falcate, semicylindrical, short-tapered, pale green with dark dots, 18 mm long, 3 mm thick; **Fl.** 14 mm $\varnothing$, purplish-pink.

**E. persistens** (L. BOL.) L. BOL.) (*Ruschia p.* L. BOL.). – Cape: Van Rhynsdorp D. – Stiff, erect, 20 cm tall; **St.** rather tangled; **L.** flat on the upperside, indistinctly carinate on the underside, slightly convex on the sides, short-tapered, green, spotted, 12–17 mm long, 3 mm across, 3–4 mm thick; **Infl.** persisting, becoming thorny, Fl. 2 cm $\varnothing$, pink.

**E. puniens** (L. BOL.) L. BOL. (*Ruschia p.* L. BOL.). – Cape: W.Griqualand, Herbert D. – Erect, stiff; **St.** 26 cm long; **L.** semicylindrical, acute, 10–15 mm long, 3 mm thick; **Infl.** dense, Ped. later thorny. Fl. 1 cm $\varnothing$, purplish-pink.

**E. spinosa** (L.) SCHWANT. (Pl. 162/2) (*Mes. s.* L.). – Cape: Karroo; S.W.Afr.: widely distributed. – Dichotomously branched, 60 cm tall; **St.** tipped with branched Th. arranged perpendicular to one another; **L.** inclined or arcuate and curved upwards, 12–14 mm long, 3–4 mm across, greyish-green with darker dots; **Infl.** branched and re-branched, the intermediate Br. always transformed into a Th. 1–2 cm long, Fl. 2 together on each Th., 15–20 mm $\varnothing$, dark pink. Acc. L. BOLUS, all the thorny species of **Eberlanzia** may belong to **E. spinosa**.

**E. stylosa** (L. BOL.) L. BOL. (*Ruschia s.* L. BOL., *Mes. styliferum* N. E. BR.). – Cape: Lit. Namaqualand. – Erect, stiff, c. 11 cm tall; **St.** 8 mm $\varnothing$ below, densely branched; **L.** flat on the upperside, somewhat convex laterally, obtuse or $\pm$ truncate, bluish-green, 2.5 cm long, 4–6 mm across, 7 mm thick; **Infl.** branched, Ped. becoming thorny, Fl. 21 mm $\varnothing$, pink.

**E. tatasbergensis** L. BOL. – Cape: Lit. Namaqualand. – Erect; **St.** 23 cm long; **L.** flat on the upperside but convex towards the obtuse, apiculate tip, indistinctly carinate on the underside, minutely spotted, bluish-green, 8–14 mm long, 2–3 mm across, 3.5 mm thick; **Infl.** 3-partite, becoming thorny, Fl. 2 cm $\varnothing$, pink.

**E. triticiformis** (L. BOL.) L. BOL. v. **triticiformis** (*Ruschia t.* L. BOL.). – Cape: Van Rhynsdorp, Riversdale D. – Erect, stiff, 30 cm tall; **St.** covered with the remains of old L.; **L.** swollen and obtusely trigonous, apiculate, spotted, 4–8 mm long, 2.5 mm across and thick; **Fl.** solitary, 12 mm $\varnothing$, pink.

**E. —** v. **subglobosa** L. BOL. – Cape: Oudtshoorn to Calitzdorp D. – **L.** hemispherical, 3–4 mm long, obtuse, 3 mm across and thick; **Fl.** 19 mm $\varnothing$.

**E. vanheerdei** L. BOL. – Cape: Lit. Namaqualand. – Erect, compact, 14 cm tall, 9–14 cm $\varnothing$; **St.** 1 cm thick; **Br.** 7 mm thick; **L.** slightly falcate, flat on the upperside, acute to obtuse, laterally compressed, with a rather obtuse keel, 25–30 mm long, 4–5 mm across, 8–10 mm thick; **Infl.** up to 3-partite, with 3–7 Fl., Br. becoming thorny and often only with 1 Fl., Th. up to 2 cm long, Fl. c. 2 cm $\varnothing$, pink.

**E. vulnerans** (L. BOL.) L. BOL. (*Ruschia v.* L. BOL.). – Cape: Hopetown D. – Erect, 20–30 cm tall; **St.** 13 mm $\varnothing$; **Br.** 28 cm long; **L.** obtusely carinate, apiculate, 3–6 mm long, 2 mm across and thick; **Fl.** sessile or in 3's, the middle pedicel then sterile, becoming thorny, Fl. 1 cm $\varnothing$, purplish-pink.

**Ebracteola** DTR. et SCHWANT. (§ IV/1/3). – Low-growing, mat-forming, very succulent plants; **roots** much thickened; Int. not perceptible; **L.** elongated-trigonous, prismatic or cylindrical, sometimes carinate, greyish-green, with dots or minute tubercles; **Fl.** terminal, $\pm$ sessile, with 2 bracts on the short pedicel, Pet. white or pink, summer. – Greenhouse, warm. Propagation: seed, cuttings.

**E. candida** L. BOL. (Pl. 162/3) (? *E. vallispacis* DTR. nom.nud.). – S.W.Afr.: Gr. Namaqualand. – **Roots** napiform, 4.5 cm long, 4 cm $\varnothing$ above; herbaceous parts minutely velvety-papillose, bluish-green; **Br.** densely-crowded, each with 4 **L.** 3–4.5 cm long, 3–5 mm across midway, 7–9 mm thick, linear-tapered on the upperside, laterally compressed and thus acutely carinate, with the sides $\pm$ convex, the tip obliquely rounded, the surface minutely tuberculate; **Fl.** 3–4 cm $\varnothing$, white.

**E. montis-moltkei** (DTR.) DTR. et SCHWANT. (*Mes. m.-m.* DTR., *Mes. renniei* L. BOL., *Ruschia r.* (L. BOL.) SCHWANT. apud JACOBS.). – S.W. Afr.: Damaraland. – **Br.** 6–8 mm thick, 1–2 cm long, with Ros. of alternate and decussate **L.** 2–3 cm long, united for almost 1 cm, the free parts 3–5 mm thick below, 6–7 mm thick above, acutely trigonous, the tip navicular and curved upwards, the upperside lanceolate, the keel $\pm$ distorted, greyish-green, with dense spots; **Fl.** 15 mm $\varnothing$, light violet-pink.

**Ectotropis** N. E. BR. (§ IV/1/6). – ⊙ (?) or small sub-♄, growing among moss, not found in cultivation.

**E. alpina** N. E. BR. – Cape: Fort Beaufort D. – Dwarf; **caudex** tuberous, 6 cm long, 7 mm $\varnothing$, tipped with slender, prostrate, rooting **Br.**, upper parts herbaceous, papillose; **L.** spreading, flat on the upperside, convex on the underside, 6–8 mm long, 1.5 mm thick; **Fl.** solitary, pedicellate, c. 9 mm $\varnothing$, white. – Probably the smallest Mesem.

**Enarganthe** N. E. BR. (G. of dubious classification). – Erect ♄; **L.** succulent. Probably not in cultivation.

**E. octonaria** (L. BOL.) N. E. BR. (Pl. 162/1) *Mes. o.* L. BOL.). – Cape: Lit. Namaqualand. – Erect; **L.** trigonous, with the keel pulled forward towards the tip, 2.5–3 cm long, 5–6 mm thick; **Fl.** solitary, often 5 cm ⌀.

**Erepsia** N. E. BR. (§ IV/1/8). – Glabrous ♄ with erect, 2-angled, compressed **St.**; **L.** somewhat united at the base, compressed-trigonous, the sides often broad, with entire margins or with a serrate, cartilaginous keel; **Fl.** up to 3 together, on a pedicel up to 5 cm long with 2 bracts, Pet. purplish-pink or more rarely yellow, autumn. – Summer: in the open, winter: cold-house. Propagation: seed, cuttings.

**E. anceps** (HAW.) SCHWANT. (*Mes. a.* HAW., *E. a.* (HAW.) L. BOL.). – Cape: Cape D. – 60–80 cm tall; **St.** slender, erect to spreading; **L.** trigonous, tapered and apiculate, green, with translucent dots, 2–3.5 cm long; **Fl.** 3.5 cm ⌀, reddish to red.

**E. aperta** L. BOL. – Cape: Tulbagh, Wellington, Worcester D. – Mat-forming, laxly branched, c. 5 cm tall; **St.** 15 cm long, prostrate; **L.** ± falcate, acutely carinate on the underside, with the sides somewhat convex, with or without an indistinct apiculus, with a few indistinct dots, 5–25 mm long, 2–5 mm across and thick; **Fl.** solitary, 4 cm ⌀, pink. ('aperta' refers to the exposed stigma-lobes.)

**E. aspera** (HAW.) L. BOL. (*Mes. a.* HAW., *E. a.* (HAW.) SCHWANT.). – S.Afr.: origin unknown. – 60–70 cm tall; **St.** stiff, erect, compressed, slender; **L.** equilaterally trigonous, curved and hooked at the tip, slightly greyish-green, rough with tiny, raised dots, 3–4 cm long; **Fl.** 5 cm ⌀, yellow inside.

**E. bracteata** (AIT.) SCHWANT. (*Mes. b.* AIT., *E. b.* (AIT.) L. BOL.). – Cape: Cape D. – 30 to 60 cm tall; **St.** many, erect; **L.** equilaterally trigonous, green, spotted, awn-tipped, 16 to 25 mm long, 2 mm across; **Fl.** red, bracts broadly ovate.

**E. brevipetala** L. BOL. – Cape: Malmesbury D. – Robust, densely branched, c. 20 cm tall, 40 cm ⌀; **L.** compressed-carinate, with the upperside narrow, ± bluish-green, 15–22 mm long, 2 mm across midway, 3 mm thick; **Fl.** 15–20 mm ⌀, purplish-red.

**E. caledonica** L. BOL. – Cape: Caledon D. – Erect, 20 cm tall, branched; **L.** flat on the upperside, the underside with a horny, crenate keel and convex sides, rough with raised dots, bluish-green, 22 mm long, 2–3 mm across midway, 3–4 mm thick; **Fl.** 27 mm ⌀, pale pink.

**E. carterae** L. BOL. v. **carterae**. – Cape: Ceres D. – Erect, branched, c. 26 cm tall; **St.** virgate, narrowly alate; **L.** rounded-carinate on the underside, with a recurved tip, pale green, 12–17 mm long, 2 mm across, 2.5 mm thick; **Fl.** up to 3 cm ⌀, pink.

**E. —** v. **lepta** L. BOL. – Cape: Worcester D. – Slender variety; **Fl.** 2.5 cm ⌀.

**E. compressa** (HAW.) SCHWANT. (*Mes. c.* HAW., *E. c.* (HAW.) L. BOL.). – Cape: Cape D. – Close to **E. aspera**; **L.** shorter, less rough, 2 to 3 cm long, 2 mm across, greyish-green, with translucent, tuberculate dots; **Fl.** red, yellowish inside.

**E. distans** L. BOL. – Cape: Clanwilliam D. – **St.** virgate, compressed, 32 cm long; **L.** falcate, tapered, curved upwards and hooked at the tip, 23 mm long, 2 mm thick; **Fl.** 3 cm ⌀, pinkish-purple.

**E. esterhuyseniae** L. BOL. – Cape: Paarl D. – Slender, 10 cm tall; **Br.** thin; **L.** falcate, tapered, carinate on the underside, 10–15 mm long, 2 mm across and thick; **Fl.** c. 22 mm ⌀, purplish-pink.

**E. gracilis** (HAW.) L. BOL. (*Mes. g.* HAW., *E. g.* (HAW.) SCHWANT., *Mes. stellatum* HORT. HAW.). – Cape: Cape D. – Up to 60 cm tall; **St.** very slender, erect, at first ± compressed, with short shoots; **L.** equilaterally trigonous, tapered, with a tiny, recurved and hooked mucro, fresh green, minutely dotted, 2 cm long; **Fl.** 2–3 together, 3 cm ⌀, purplish-pink.

**E. hallii** L. BOL. – Cape: Malmesbury D. – Erect, compact, 19 cm tall, c. 24 cm ⌀; **St.** tangled, curved to spreading, stiff, rough from old tubercles, c. 18 cm long; **Br.** 4–10 cm long, tuberculate; **L.** acutely carinate on the underside, recurved and hooked at the tip, apiculate, 8–17 mm long, 1.5–3 mm across and thick; **Fl.** c. 15 mm ⌀, white.

**E. heteropetala** (HAW.) SCHWANT. (Pl. 162/4) (*Mes. h.* HAW., *Mes. montanum* SCHLTR., *E. m.* (SCHLTR.) SCHWANT.). – Cape: Stellenbosch D. – Erect; **St.** spreading, 2-angled; **L.** falcate, 2 cm long, the sides 6–8 mm across, compressed, somewhat furrowed on the upperside, the angles cartilaginous and irregularly denticulate, the keel-angle more coarsely dentate, green, spotted, bluish-pruinose; **Fl.** 15 mm ⌀, reddish or whitish.

**E. inclaudens** (HAW.) SCHWANT. (*Mes. i.* HAW.). – Cape D. – **St.** and **Br.** curved-ascending or prostrate, c. 20 cm long with new growth often 2-angled; **L.** sabre-shaped, acute, 15–25 mm long, 6–8 mm across below the tip, glossy green, somewhat reddish, with large, translucent dots, the margins entire but the keel-angle ± denticulate towards the tip; **Fl.** 4 cm ⌀, glossy purplish-violet, Pet. broad.

**E. insignis** (SCHLTR.) SCHWANT. (*Mes. i.* SCHLTR.). – Cape: Ceres D. – Dwarf, branching from the base; **St.** reddish; **L.** trigonous, narrow, with the tip recurved and hooked, with minute, rough dots, 10–15 mm long, up to 3 mm thick; **Fl.** 4–5 cm ⌀, Pet. red with violet stripes and spots.

**E. laxa** L. BOL. – Cape: Caledon, Stellenbosch D. – Slender, 30 cm tall; **St.** spreading to erect; **L.** 12 mm long, 1.5 mm across and thick, with raised dots, pale green; **Fl.** 25 mm ⌀, pinkish-purple.

**E. levis** L. BOL. – Cape: Robertson D. – Erect, c. 30 cm tall; **St.** stiff, virgate; **L.** carinate on the underside, flat on the upperside, linear, tapered, somewhat recurved at the tip, green, 15–25 mm long, 2 mm across, 3.5 mm thick; **Fl.** up to 3 cm ⌀, pink with a white centre.

**E. marlothii** N. E. Br. – Cape: Clanwilliam D. – Erect, slender, 15–22 cm tall; **St.** brownish-yellow, covered with brown dots arranged in lines; **L.** linear-trigonous, acute, spotted, 5 to 10 mm long, 1 mm thick; **Fl.** 2–2.5 cm ⌀, pink to purplish-pink.

**E. mutabilis** (Haw.) Schwant. (*Mes. m.* Haw., *E. m.* (Haw.) L. Bol., *Mes. filamentosum* DC., *Mes. forficatum* Jacq., *Mes. glaucinum* Haw., *Mes. tricolor* Jacq.). – Cape: Cape D. – Erect; **St.** ± curved, initially 2-angled; **L.** trigonous, with the angles straight or undulating, minutely cartilaginous, acute above, with a tiny awn, 15–20 mm long, 4–6 mm across, greyish-green; **Fl.** pink, pale yellow inside.

**? E. nudicaulis** (Bgr.) Jacobs. (*Mes. n.* Bgr.). – W.Cape. – **St.** densely leafy, 13–14 cm long; **L.** trigonous, tapered, apiculate, with rough dots, 15 mm long; **Fl.** 15 mm ⌀, ? red. Acc. L. Bolus probably a form of **E. tuberculata**.

**E. oxysepala** (Schltr. et Bgr.) L. Bol. (*Mes. o.* Schltr. et Bgr., *E. stokoei* L. Bol.). – Cape: Caledon D. – Branched from the base, 13 cm tall; **St.** erect, virgate, compressed; **L.** falcate, laterally compressed, with a serrate keel and the tip short-tapered and recurved, 13–20 mm long, 4 mm across, 5 mm thick; **Fl.** solitary, 2–3 cm ⌀, pink.

**E. pageae** L. Bol. – Cape D. – Stiff, erect, 30 cm tall, with laxly spreading **St.**; **L.** expanded midway, with a recurved tip, with rough dots, 6–15 mm long, 2–3 mm across; **Fl.** 22–25 mm ⌀, violet to pink.

**E. patula** (Haw.) Schwant. (*Mes. p.* Haw., *E. p.* (Haw.) L. Bol.). – Cape: Cape D. – Up to 20 cm tall; **St.** thin, curved, 2-angled; **L.** equilaterally trigonous, tapered, with the tip hooked and recurved, slightly greyish-green, with rough dots, 15–25 mm long; **Fl.** 3 cm ⌀, deep red with a yellow centre.

**E. pentagona** (L. Bol.) L. Bol. (*Mes. p.* L. Bol., *E. p.* (L. Bol.) Schwant.). – Cape: Riversdale D. – Erect, 15–35 cm tall; **St.** virgate, simple or branched, compressed, up to 30 cm long; **L.** linear-lanceolate, tapered, with a recurved, hooked tip, green; **Fl.** up to 4.5 cm ⌀, pinkish-purple.

**E. polita** (L. Bol.) L. Bol. (*Mes. p.* L. Bol., *E. p.* (L. Bol.) Schwant.). – Cape: Ladismith D. – Erect, 18–25 cm tall; **St.** spreading, stiff, up to 12 cm long; **L.** trigonous, ± concave on the upperside, acute to tapered, apiculate, yellowish-green, 15–20 mm long, 3–5 mm across and thick; **Fl.** 2 cm ⌀, pink.

**E. polypetala** (Bgr. et Schltr.) L. Bol. (*Mes. p.* Bgr. et Schltr.). – Cape: Caledon D. – Branched from the base, 10–20 cm tall; **St.** with short shoots; **L.** compressed-trigonous, tapered, with a recurved, apiculate tip; **Pet.** numerous, narrow.

**E. promontorii** L. Bol. – Cape: Cape D. – Slender, rather weak, c. 12 cm tall, with spreading or prostrate **St.** from the base; **L.** carinate on the underside, tapered, dirty green, rough-tuberculate, 1–2 cm long, 2 mm thick; **Fl.** 2.5 cm ⌀, pinkish-purple.

**E. racemosa** (N. E. Br.) Schwant. (*Mes. r.* N. E. Br.). – Cape: origin unknown. – 30–45 cm tall; **St.** mostly alternate, ascending or spreading, 2-angled at first, with reddish or green markings; **L.** obtusely trigonous, acutely keeled, rather obtuse at the tip, with a recurved mucro, green, with tiny, rough dots, 12–22 mm long, 1.5–2 mm across, up to 3 mm thick; **Fl.** in a Rac., 25–28 mm ⌀, deep pinkish-red.

**E. radiata** (Haw.) Schwant. (*Mes. r.* Haw., *E. r.* (Haw.) L. Bol.). – Cape: Cape D. – Resembles **E. bracteata**; **St.** stouter, ± alate at first; **L.** equilaterally trigonous, with a recurved, hooked tip, 15–25 mm long; **Fl.** red.

**E. ramosa** L. Bol. – Cape: Piquetberg, Stellenbosch D. – Erect, 40 cm tall; **St.** laxly arranged, branched; **L.** narrowed towards the acute tip, with a recurved mucro, carinate on the underside, bluish-green, 2 cm long, 1.5–2 mm across and thick; **Fl.** 26–30 mm ⌀, purplish-pink.

**E. roseoalba** L. Bol. – Cape: Malmesbury D. – **St.** prostrate at first, eventually ± erect, 25 cm or more long; **L.** either short-falcate, acute, carinate, 12–16 mm long, 5 mm thick, or narrow-oblong, 2–2.5 cm long, 2–3 mm thick, bluish-green, with rough dots; **Fl.** solitary, 2.5–3 cm ⌀, pink or pale pink, with a white centre.

**E. saturata** L. Bol. – Cape: Piquetberg D. – Erect, 25 cm tall; **St.** virgate, 6–12 cm long; **L.** acutely keeled on the underside, the tip obtuse, recurved and hooked, reddish to dark green, with indistinct dots, 15 mm long, c. 3 mm across and thick; **Fl.** c. 3 cm ⌀, pinkish-purple.

**E. serrata** (L.) L. Bol.) (*Mes. s.* L., *Circandra s.* (L.) N. E. Br.). – Cape: Ceres D. – c. 40 cm tall; **St.** erect; **L.** trigonous-subulate, with prominent dots, with all angles or only the keel serrate-dentate, 7–10 mm long, 6 mm across below; **Fl.** 3 cm ⌀, with the Pet. red-striate above.

**E. steytlerae** L. Bol. – Cape. – Laxly branched, 12–20 cm tall; **St.** compressed, narrowly alate; **L.** laterally compressed, with acute angles, acute to truncate at the ± decurved tip, green, 20–25 mm long, 6 mm across, 10 mm thick; **Fl.** 4 cm ⌀, pinkish-purple.

**? E. tenuicaulis** (Bgr.) Jacobs. (*Mes. t.* Bgr.). – Cape: Tulbagh D. – **St.** slender, erect, elongated, ± contorted, angular; **L.** trigonous, long-tapered, with the tip erect to recurved, with a few dots, c. 3 cm long, 2 mm across; **Infl.** slender, 7–10 cm long.

**E. tuberculata** N. E. Br. (*Mes. nudicaule* Bgr., *E. muirii* L. Bol., *E. restiophila* L. Bol.). – Cape: Cape, Swellendam, Ladismith D. – 50–60 cm tall; **St.** lax, spreading, slender, laterally compressed; **L.** tapered, with a recurved tip, 25 mm long, 2–2.5 mm thick; **Fl.** 22 mm ⌀, pinkish-purple, the Cal. conical, with ± circular tubercles.

**E. urbaniana** (Schltr.) Schwant. (*Mes. u.* Schltr.). – Cape: near Sir Lowry's Pass. – Dwarf; **St.** ascending, 4-angled; **L.** trigonous, obtuse, greyish-green, with minute, rough dots, 5–8 mm long; **Fl.** solitary, 2 cm ⌀, pinkish-red.

**E. villiersii** L. Bol. – Cape: Villiersdorp D. – Densely branched, forming cushions up to 1 m ⌀; **L.** curved into almost an S-shape, narrowed towards the apiculate tip, with the underside rounded to obtusely carinate, 15 to 17 mm long, 1–1.5 mm across and thick; **Fl.** 22–28 mm ⌀, dull pink, with a paler centre.

**Esterhuysenia** L. Bol. (§ IV/1/S.Tr. ?). – Closely related to **Lampranthus** and **Ruschia**, but the capsule differs from **Lampranthus** in having no wings, and from **Ruschia** in having no placental tubercles. – Dwarf plants with very succulent L. – Summer: in the open, winter: cold-house. – Propagation: seed, cuttings.

**E. alpina** L. Bol. – Cape: Worcester D. – Dwarf ♄ up to 33 cm tall and 42 cm ⌀; **roots** creeping, up to 52 cm long, 7 mm thick; **St.** 8 mm thick below, branching and re-branching; **Br. and shoots** up to 4 mm thick, the herbaceous parts minutely tuberculate; **L.** sometimes ± falcate, linear, tapered, apiculate, flat on the upperside, rounded and indistinctly carinate on the underside, with the keel and angles reduced to lines, bluish-green, 8–10 mm long, 1.5–2 mm across and thick, with a Sh. 1.5 mm long; **Fl.** solitary, opening by day, on a short pedicel with 2 acute bracts, Pet. c. 8 mm long, obtuse, pinkish-purple.

**Eurystigma** L. Bol. (§ I/1/2). – Erect, glabrous, ⊙ plants. Rarely found in cultivation. Propagation: seed.

**E. clavatum** (L. Bol.) L. Bol. (*Cryophytum c.* L. Bol.). – Cape: Ceres D. – Up to 15 cm tall; **St.** ascending, gradually thickening towards the clavate tip; **L.** few, drying up during flowering, lower L. opposite, upper ones alternate, ? cylindrical, up to 6.5 cm long; **Fl.** diurnal, lasting up to 25 days, Pet. very narrow, yellow, pedicel and ovary up to 2 cm long, clavate, 14 mm ⌀, the stigma-lobes thick, broadly subulate or mostly cordate.

**Faucaria** Schwant. (§ IV/1/18). – Very succulent plants with a short **St.** becoming elongate in age, branching to form a mat; **roots** fleshy; **L.** 4–6 on one shoot, alternate and decussate, very crowded, united at the base, divaricate, thick-fleshy, semicylindrical at the base, carinate towards the tip and trigonous, shortly rhombic, long-spatulate or lanceolate, the margins usually with stout, often long-awned T., the margins, keel and T. all cartilaginous, the underside often pulled forward chin-like over the tip, the texture firm and the surface glossy or dull, mostly with ± raised, irregular spots; **Fl.** sessile, large, yellow, more rarely pink or white, Aug.–Nov., opening in the afternoon. – Greenhouse, warm. Propagation: seed, cuttings.

**F. acutipetala** L. Bol. – Cape: Somerset D. – **L.** lanceolate, acute to tapered, the margins with 3–5 bristle-tipped T., 35 mm long, 10 mm across, 15 mm thick; **Fl.** golden-yellow.

**F. albidens** N. E. Br. (*F. kendrewensis* N. E. Br.). – Cape: Graaff Reinet D. – **L.** long and triangularly tapering, 25 mm long, 20 mm across, 6–7 mm thick, the angles with 3–5 recurved, whitish, awn-like T., the keel white, horny, the surface fresh green, glossy, with scattered, ± distinct dots; **Fl.** 3–4 cm ⌀, golden-yellow.

**F. bosscheana** (Bgr.) Schwant. v. **bosscheana** (Pl. 163/1) (*Mes. b.* Bgr.). – Cape: Prince Albert, Beaufort West D. – **L.** narrowly lanceolate or acute-rhombic, c. 3 cm long, 10 mm across, glossy green, the angles ± white, cartilaginous, irregularly dentate, with 2–3 T. 2 to 3 mm long on either side; **Fl.** 3–3.5 cm ⌀, golden-yellow.

**F.** – v. **haagei** (Tisch.) Jacobs. (*F. haagei* Tisch.). – Cape: Willowmore D. – **L.** margins and keel white, cartilaginous, sometimes undulating, scarcely dentate.

**F. britteniae** L. Bol. – Cape: Albany D. – **L.** rhombic-ovate, c. 35 mm long, 15 mm across below, 20 mm across midway, with the underside acutely and obliquely carinate above and the tip pulled forward and chin-like, the angles and keel with a horny border 1 mm across, with 3–4 hair-like T. midway.

**F. candida** L. Bol. – Cape: Cradock D. – **L.** tapered on the upperside, with the upper part ± rhombic, obtuse, with a white-lined keel on the underside, 25 mm long, 15 mm across, 10 mm thick, marginal T. 6–7, white, 2 mm long, pink-tipped; **Fl.** c. 4 cm ⌀, white.

**F. coronata** L. Bol. – Cape: Albany D. – **L.** 25–32 mm long, 15–16 mm across, ovate to lanceolate-ovate on the upperside, tapered, the margins white, with 6–8 T. 7 mm long, with a 7 mm long bristle; **Fl.** over 6 cm ⌀, yellow.

**F. cradockensis** L. Bol. – Cape: Cradock D. – **L.** 4.5 cm long, 17 mm across, the upperside acute or ovate-acute, with white, tuberculate dots, the underside rounded and carinate, the keel with 1–2 T., the margins with 5–7 conical T. tipped with a long bristle; **Fl.** c. 6 cm ⌀, yellow.

**F. crassisepala** L. Bol. – Cape: Somerset D. – **L.** either up to 4 cm long, 15 mm across and lanceolate on the upperside, or ovate and 33 to 36 mm long, up to 18 mm across, 10 mm thick, tapered, the surface white-tuberculate, the margins with 6–10 conical-acute T., the underside carinate, the keel sometimes with 1 T.; **Fl.** up to 6.5 cm ⌀, Pet. golden-yellow, purple on the underside.

**F. duncanii** L. Bol. – Cape: Uitenhage D. – **L.** c. 25 mm long, 8 mm across below, navicular, long-tapered, with the underside acutely carinate above, the surface smooth, green, with red dots at the tip, the angles with 6–7 irregularly spaced, pointed T.

**F. felina** (Weston) Schwant. ex Jacobs. v. **felina** (Pl. 163/3) (*Mes. tigrinum* β. f. L., *Mes. f.* Weston, *Mes. f.* Haw.). – Cape: Uitenhage, Graaff Reinet, Albany D. – **L.** oblong-rhombic, ± trigonous and long-tapered, 45 mm long, 15–20 mm across, fresh green to reddish, with indistinct white dots, the margins with 3–5 pointed, fleshy T., the keel slender, white, cartilaginous; **Fl.** c. 5 cm ⌀, golden-yellow.

**F. — v. jamesii** L. Bol. – Cape: Cradock D. – L. rather smaller, the whitish margins having 4–6 T. 1–2 mm long, rarely bristle-tipped.

**F. grandis** L. Bol. – Cape: Albany D. – L. 4–5 cm long, 25 mm across, 8–10 mm thick above, the upperside ± square below, broadly ovate towards the acute tip, the underside broadly compressed and carinate, the bluish-white margins with 9–11 T.; **Fl.** 5–6 cm ⌀, golden-yellow.

**F. gratiae** L. Bol. – Cape: Albany D. – L. in unequal pairs, 18–22 mm long, 13–16 mm across, or 25–35 mm long and 22–27 mm across, the upperside narrowly to obliquely ovate, green with white dots, the underside obtuse, purple, the margins undulating, with indistinct, bristle-tipped T. or lobes; **Fl.** c. 4 cm ⌀, golden-yellow.

**F. hooleae** L. Bol. – Cape: Albany D. – L. 25 mm long, 11 mm across, ovate to lanceolate-ovate on the upperside, tapered, laterally expanded, the surface reddened, with dense, prominent dots, the margins with 5–7 bristle-tipped T.; **Fl.** c. 3 cm ⌀, yellow.

**F. kingiae** L. Bol. – Cape: Albany D. – L. 45–50 mm long, 15 mm across midway, 10 mm across at the tip and 12 mm thick, the upperside lanceolate, tapered, the underside compressed, carinate, the surface bluish-green with dense white dots, the margins with 5–10 T. tipped with a red bristle; **Fl.** 6 cm ⌀, yellow.

**F. latipetala** L. Bol. – Cape: Bedford D. – L. 35 mm long, 10 mm across midway, 10 mm thick, the upperside lanceolate, acute to tapered, the margins with 3–5 small T.; **Fl.** c. 6 cm ⌀, Pet. up to 3 mm across, golden-yellow, reddish on the underside.

**F. laxipetala** L. Bol. – Cape: Albany D. – L. 48 mm long, 13 mm across midway, 10 mm thick, with the apex obtuse in profile and 8 mm thick, square on the upperside below, lanceolate above, tapered, the margins with 3–6 T., the keel white; **Fl.** c. 8 cm ⌀, with the Pet. lax, golden-yellow, reddened on the underside.

**F. longidens** L. Bol. – Cape: Somerset East D. – L. 25–30 mm long, 11–15 mm across midway, with the upperside ovate to lanceolate-ovate, acute or tapered, bluish-green, with raised, rough dots, the underside acutely carinate, the tip 6 mm thick, the margins with 4–6 T., the keel with 1–3 curved T. 2–4 mm long, tipped with an awn 8 mm long; **Fl.** 6.5 cm ⌀, yellow, with a white centre.

**F. longifolia** L. Bol. – Cape: Bedford D. – L. linear-lanceolate, 50–55 mm long, 15 mm across, 7–8 mm thick at the tip, with 1–4 marginal T. 3 mm long; **Fl.** 6–7 cm ⌀, dark yellow.

**F. lupina** (Haw.) Schwant. (Pl. 163/4) (*Mes. l.* Haw.). – Cape: Uitenhage D. – L. lanceolate on the upperside, long triangularly-tapering, with the underside carinate towards the trigonous chin-like tip, the surface fresh green, with tiny rough dots, the margins cartilaginous, with 7–9 recurved T. tapering to fine hairs; **Fl.** 3–3.5 cm ⌀, yellow.

**F. militaris** Tisch. – S.Afr. – Origin unknown. – L. navicular, c. 3 cm long, 8–12 mm thick, with the upperside 15–18 mm across and then tapered, the keel on the underside acute, pulled forward and chin-like, pink, with 2–4 short T., the cartilaginous margins with 4–6 recurved, hair-tipped T. 4–6 mm long and pink above, the surface greyish-green with dots on the underside; **Fl.** 6–7 cm ⌀, glossy golden-yellow with a whitish centre.

**F. montana** L. Bol. – Cape: Cradock D. – L. 4 cm long, 15 mm across, 7 mm thick at the tip, swollen, the upperside ovate to lanceolate and acute above, broadly ovate below, apex blunt with 4 marginal, shortly bristle-tipped T., the surface with green dots and white marks; **Fl.** 6 cm ⌀, golden-yellow.

**F. multidens** L. Bol. v. **multidens.** – Cape: Bedford D. – L. with the upperside lanceolate to linear-lanceolate, acute, c. 4 cm long, 1 cm across, or 5–6 cm long, 13 mm across, others acutely falcate and 38 mm long, 24 mm across, 9 mm thick, rounded on the underside, distinctly spotted, the margins with 3–11 bristle-tipped T.; **Fl.** c. 5 cm ⌀, golden-yellow.

**F. — v. paardeportensis** L. Bol. – Cape: Uitenhage D. – L. lanceolate and tapered on the upperside, compressed and carinate on the underside, obtuse at the tip, 35–55 mm long, 11–16 mm across, 8–13 mm thick, the surface green with translucent, white dots, the keel with white lines, the margins with 5–10 T.; **Fl.** 5.5–6 cm ⌀, dark yellow.

**F. paucidens** N. E. Br. (Pl. 163/2). – Cape: Graaff Reinet D. – L. linear and tapered on the upperside, rounded-carinate on the underside, 3–5 cm long, 8–9 mm across, 5–6 mm thick at the base, with 1–3 marginal T., pale green with dark green dots; **Fl.** 3–4 cm ⌀, yellow.

**F. peersii** L. Bol. – Cape: Jansenville D. – L. rhombic to rounded on the upperside, acute to tapered, square below, carinate on the underside, 2.5–4 cm long, 2–4 cm across above midway, 14 mm thick, the keel and margins distinctly white, the margins with 1–5 large, lobe-like T.; **Fl.** 5.5 cm ⌀, Pet. golden-yellow, pale pink on the underside.

**F. plana** L. Bol. – Cape: origin unknown. – L. linear-lanceolate and tapered on the upperside, 4–4.5 cm long, 8–12 mm across, 4–6 mm thick at the tip, with 4–6 marginal T., the surface bluish-green, with white spots and often with red markings; **Fl.** 4 cm ⌀, dark yellow.

**F. ryneveldiae** L. Bol. – Cape: Bedford D. – L. linear-lanceolate and tapered on the upperside, convex on the underside, compressed-carinate at the tip, 7 cm long, 11 mm across, 8 mm thick, with the margins entire except for 1–2 bristle-tipped T., the surface dirty green with white dots; **Fl.** 6.5 cm ⌀, golden yellow.

**F. smithii** L. Bol. – Cape: Albany D. – L. lanceolate to ovate then tapered above on the upperside, rounded on the underside, 4.5 cm long, 14–17 mm across, 14 mm thick, with 6 bristly marginal T. 1–2 mm long, the surface bluish, spotted; **Fl.** 4 cm ⌀, yellow. Acc. L. Bolus a variety of **F. britteniae.**

**F. speciosa** L. Bol. – Cape: Albany D. – **L.** ovate to broadly ovate on the upperside above, acute to tapered, 25–30 mm long, up to 2.8 cm across, with the underside compressed at the tip and 1 cm thick, with 5–6 large, bristle-tipped marginal T.; **Fl.** up to 8 cm ⌀, Pet. golden-yellow, paler below, with purple tips.

**F. subindurata** L. Bol. – Cape: Peddie D. – **L.** linear to broadly ovate on the upperside above, acute to tapered, square in the lower part, 2–2.5 cm long, 15 mm across, obtuse and 8 mm thick at the tip, with 4–8 marginal T., the surface bluish-green to reddish, the underside mauve with white spots; **Fl.** 2.5 cm ⌀, golden-yellow.

**F. subintegra** L. Bol. – Cape: King William's Town D. – **L.** thick and swollen, ovate on the upperside above, rounded or more rarely sub-acute, square in the lower part, rounded-carinate on the underside, 20–25 mm long, 13 mm across, 10 mm thick, the margins entire or more rarely with 1–3 T.; **Fl.** 4.5 cm ⌀, golden-yellow.

**F. tigrina** (Haw.) Schwant. f. **tigrina** (Pl. 163/5) (*Mes. t.* Haw.). – Cape: Albany D. – **L.** upperside rhombic to ovate, short-tapered, the underside very rounded, with the tip pulled forward and chinlike, 3–5 cm long, 16–25 mm across, the surface greyish-green with numerous white dots arranged in R., the margins with 9–10 stout, recurved, hair-tipped T.; **Fl.** 5 cm ⌀, golden-yellow.

**F. —** f. **splendens** Jacobs. – **L.** conspicuously red in colour.

**F. tuberculosa** (Rolfe) Schwant. (Pl. 163/6) (*Mes. t.* Rolfe). – Cape: Bedford D. – **L.** rhombic to triangular on the upperside, with several T.-like tubercles, 2 cm long, 16 mm across, dark green, with 3 stout marginal T. and several rudimentary ones; **Fl.** 4 cm ⌀, yellow.

**F. uniondalensis** L. Bol. – Cape: Uniondale D. – **L.** lanceolate on the upperside, ± rhombic in the upper part, obtuse to ± carinate on the underside, 4 cm long, 12 mm across, 9–13 mm thick midway, with 4–5 marginal T., the surface bluish with white spots and the keel-angle white; **Fl.** 5 cm ⌀, Pet. yellow, paler below, with reddish tips.

**Fenestraria** N. E. Br. (§ IV/1/2). – Extremely succulent plants with clavate **L.**, translucent at the tips; **Fl.** on a long pedicel with ± reduced bracts, golden-yellow or white. – Greenhouse, warm, in full sunshine. Propagation: seed, leaf-cuttings or by careful division.

**F. aurantiaca** N. E. Br. f. **aurantiaca** (Pl. 164/2). – Cape: Lit. Namaqualand. – Forming cushions up to 10 cm ⌀; **roots** very thin; **L.** inclined, 2–3 cm long, clavate and thickened above, slightly flattened on the upperside, rounded on the underside or sometimes rounded-triangular, convex and circular at the tip which is 6–8 mm ⌀, ± transparent, the surface smooth, whitish, ± reddish below, the upper part with scattered, translucent dots; **Fl.** on a pedicel 4–5 cm long, 3–7 cm ⌀, golden-yellow, Aug.-Sept.

**F. —** f. **rhopalophylla** (Schltr. et Diels) Rowl. (*Mes. r.* Schltr. et Diels, *F. r.* (Schltr. et Diels) N. E. Br.). – Less freely offsetting; **Fl.** smaller, 18–30 mm ⌀, white.

**Frithia** N. E. Br. (§ IV/1/15). – Highly succulent plants, resembling **Fenestraria**. – Greenhouse, warm. Propagation: seed.

**F. pulchra** N. E. Br. v. **pulchra** (Pl. 164/1). – Transvaal: Magalies Mts. – Acauline, offsetting, forming small cushions in age; **L.** 6–8 to a shoot, rosulate, erect, ± clavate to sub-cylindrical, c. 2 cm long, rough, with a translucent Wi.; **Fl.** solitary, sessile or almost so, 9–23 mm ⌀, crimson to purplish with a white centre, or all over pure white.

**F. —** v. **minor** de Boer. – **L.** 10–15 mm long, 3–5 mm ⌀ above, pink to greenish-brown; **Fl.** 1.7 cm ⌀, Pet. light pinkish-purple with crimson tips.

**Gibbaeum** Haw. ex N. E. Br. (§ IV/1/16). – ♃, caespitose, with abbreviated or prostrate **St.**; growth consisting of pairs of **L.** which are united to form an ovate or subglobose Bo. which is slightly notched or distinctly divided into equal or unequal lobes with the cleft at the apex or somewhat below the apex, or the Bo. oblique-ovate or cylindrical with a Fi. on one side, or sometimes the ± connate L. are widely spreading; surface smooth, slightly uneven or ± finely hirsute; **Fl.** autumn to spring, nearly always in the beginning of the growing period, stalked, bractless, Pet. white or violet-red. – Greenhouse, warm. Propagation: seed, cuttings. (Lit.: Nel, G. C.: The Gibbaeum Handbook, edited by Jordaan, P. G., and Shurly, E. W., London, 1953.)

## Classification of the Genus Gibbaeum into Sections by H. D. Wulff

**Sect. I. Archaeogibbaeum** Wulff. – The **cells** of the epidermis normal or characterized by ±distinct projections forming papillae which can finally change to bristle-hairs.

**SSect. 1. Protogibbaeum** Wulff. – **Cells** of the epidermis smooth or somewhat projecting; **L.** of a pair unequally long, united for about ⅓, the surface smooth or slightly papillate. – Species: G. luteoviride, ? esterhuyseniae.

**SSect. 2. Muiriopsis** Wulff. – **Cells** of the epidermis ± papillately projecting or the papillae well formed and often hair-like and elongated; **L.** of a pair united for ⅔ and more, the surface very finely hirsute. – Species: G. cryptopodium, gibbosum, pilosulum.

**Sect. II. Gibbaeum** (*Eugibbaeum* Wulff). – **Cells** of the epidermis with long, simple or stellate hairs like antlers.

**SSect. 1. Gibbaeum** (*Gibbaeotypus* Wulff). – **Cells** of the epidermis with elongated hairs or the hairs stellately branched; **L.** of a pair unequal, the smaller L. almost appressed to the larger one. – Species: G. geminum, pubescens, shandii.

**SSect. 2. Imitariopsis** Wulff. – **Cells** of the epidermis irregular, stellately branched or only terminally forked; **L.** spreading or coalescing to form a ± globose Bo. with a small Fi.; surface with very fine hairs. – Species: G. album, angulipes, dispar.

**SSect. 3. Mentocalyx** (N. E. Br.) Wulff (*Mentocalyx* N. E. Br. as a G.). – **Cells** of the epidermis irregular, stellately branched or antler-like; **L.** large and thick, spreading ± flatly, curved upwards like a bow at the end, keeled; surface pubescent. – Species: G. haagei, pachypodium, schwantesii, velutinum.

**Sect. III. Neogibbaeum** Wulff. – **Cells** of the epidermis conically swollen, equally large, papillae wanting; surface smooth but covered with very fine roundish tubercles visible only by means of a good magnifying-glass.

**SSect. 1. Argeta** (N. E. Br.) Wulff (*Argeta* N. E. Br. as a G.). – **L.** united at the base, spreading, cylindrical to 3-angled. – Species: G. petrense, tischleri.

**SSect. 2. Rimaria** (N. E. Br.) Wulff (*Rimaria* N. E. Br. as a G.). – **L.** connate to form a subglobose Bo., Fi. small. – Species: G. blackburniae, comptonii, heathii, luckhoffii.

**G. album** N. E. Br. f. **album** (Pl. 164/3) (§ II/2). – Cape: Ladismith D. – Clump-forming; **L.** unequal, united to form an obliquely ovoid Bo. 20–25 mm long, 12–14 mm across, rather less in thickness, with the Fi. later ± gaping, the surface densely white-pubescent; **Fl.** 25 mm ⌀, white.

**G.** — f. **roseum** (N. E. Br.) Rowl. (*G. a. v. r.* N. E. Br.). – **Fl.** pink.

**G. angulipes** (L. Bol.) N. E. Br. (Pl. 164/4) (§ II/2) (*Mes. a.* L. Bol.). – Cape: Riversdale D. – **Br.** prostrate; **L.** united for 7–9 mm at the base and 9 mm across, unequal, 23 and 26 mm long, rounded-carinate on the underside, ± tapered above, bluish-green, minutely velvety; **Fl.** 25 mm ⌀, pinkish-purple.

**G. blackburniae** L. Bol. (§ II/2). – Cape: Ladismith D. – Mat-forming; **Bo.** ± spherical, c. 20 mm ⌀, 32 mm long, with a Fi. 6 mm long, greenish-grey to pink; **Fl.** c 3 cm ⌀, pale pink.

**G. comptonii** (L. Bol.) L. Bol. (§ III/2) (*Rimaria c.* L. Bol., *R. c.* f. L. Bol., *G. c.* f. (L. Bol.) Jacobs.). – Cape: Ladismith D. – Mat-forming; **roots** 8 cm long, 2.5 cm thick above; **L.** hemispherical, up to 30 mm long, united for up to 15 mm, mostly unequal, 7–9 mm ⌀ at the tip, perceptibly keeled on the underside, pale bluish-green; **Fl.** c. 3 cm ⌀, pinkish-purple.

**G. cryptopodium** (Kensit) L. Bol. (Pl. 164/5) (§ I/2) (*Mes. c.* Kensit, *Derenbergia c.* (Kensit) Schwant., *Mes. nuciforme* Haw., *D. n.* (Haw.) Schwant., *Conophytum n.* (Haw.) N. E. Br. p. part. ?, *G. molle* N. E. Br., *G. helmiae* L. Bol.). – Cape: Laingsburg, Ladismith Oudtshoorn D. – **Rootstock** branched, firm-fleshy; **shoots** several; **Bo.** ± spherical to ovoid, 16–22 mm tall, 10–16 mm across, 9 to 14 mm thick, with the 2 lobes ± distinctly unequal, ± acute, faintly carinate on the underside, the Fi. 4–5 mm deep, the surface smooth, pale green to reddish; **Fl.** 2.5 cm ⌀, pink.

**G. dispar** N. E. Br. (Pl. 164/6) (§ II/2). – Cape: Ladismith D. – Clump-forming; **Bo.** ovoid, with a very deep Fi., the free parts unequal, sometimes obtusely carinate on the underside, rather thick, the surface greyish-green with a reddish tinge, slightly glossy, minutely velvety; **Fl.** 8–10 mm ⌀, mauvish-red.

**G. esterhuyseniae** L. Bol. (§ I/1 ?) (*G. intermedium* L. Bol. nom. nud.). – Cape: Robertson D. – Compact, 5–7 cm tall, 10 cm ⌀; **St.** 1 cm thick; **L.**-pairs not conspicuously unequal, expanded at the base, shortly narrowed towards the obtuse to subacute tip, in general narrow-lanceolate to broadly ovate, 2.7–5 cm long, 1–2 cm across, up to 16 mm thick; **Fl.** 3.5–5 cm ⌀, pink.

**G. geminum** N. E. Br. (Pl. 166/1) (§ II/1). – Cape: Ladismith D. – Very dwarf ♄; **St.** and **Br.** very short, resting on the soil, forming cushions; **L.**-pairs 2–3 on a shoot, divaricate, unequal, the larger one 15 mm long, 6 mm thick, circular in section, very little compressed, with a distinct ridge over the rounded tip, the smaller L. 4–6 mm long, the surface light greyish-green, with minute whitish hairs (lens needed); **Fl.** 12–15 mm ⌀, red.

**G. gibbosum** (Haw.) N. E. Br. (Pl. 166/2) (§ I/2) (*Mes. g.* Haw., *Mes. perviride* Haw., *G. p.* (Haw.) N. E. Br., *G. marlothii* N. E. Br., *G. muirii* N. E. Br.). – Cape: Ceres, Montagu D. – **Rootstock** woody; **shoots** numerous, forming ± compact clumps 6–15 cm ⌀, 3–6 cm high; **L.** of a pair very unequal, the larger L. curved slightly inwards, semicylindrical, ± flattened above, the underside with 2 keels, rounded at the tip, the small L. only ¼ as long, with its upperside at first lying in the triangular part of the upperside of the other L., with a small Fi., the surface smooth, deep green; **Fl.** 8 to 12 mm ⌀, pinkish-purple.

**G. haagei** Schwant. v. **haagei** (§ II/3). – Cape: Swellendam D. – ± acauline, prostrate, branched; **L.** of a pair unequal, trigonous, acutely keeled, somewhat concave on the upperside, with a Sh. 1.5 cm long, the longer L.

up to 4 cm long, the shorter one with the chin pulled forward, bluish-green, hirsute; **Fl.** up to 3 cm ⌀, red.

**G. — v. parviflorum** L. Bol. – Pet. c. 11 mm long, purplish-pink, distinctly striate below.

**G. heathii** (N. E. Br.) L. Bol. v. **heathii** (Pl. 166/4) (§ III/2) (*Mes. h.* N. E. Br., *Rimaria h.* (N. E. Br.) N. E. Br., *R. dubia* N. E. Br., *G. d.* (N. E. Br.) Jacobs., *Mes. fissum* N. E. Br.). – Cape: Ladismith D. – Mat-forming; **Rootstock** very long, the shoots often rooting as well; **Bo.** ± spherical, 2–3 cm tall, 15–20 mm thick, the 2 hemispherical L. often unequal, united up to midway, with a ± gaping Fi., the surface smooth, grass-green to whitish-green; **Fl.** 3–4 cm ⌀, white to cream-coloured to slightly pink. G. C. Nel includes **G. comptonii, G. blackburniae** and **G. luckhoffii** within this species.

**G. — v. elevatum** (L. Bol.) L. Bol. (*Rimaria h.* v. e. L. Bol.). – **Bo.** taller.

**G. — v. majus** (L. Bol.) L. Bol. (*Rimaria h.* v. m. L. Bol.). – **Bo.** twice as large, whitish-green.

**G. luckhoffii** (L. Bol.) L. Bol. (§ III/2) (*Rimaria l.* L. Bol.). – Cape: Ladismith D. – **Roots** mostly thickend; **Bo.** several together, ± spherical, c. 15 mm tall, 18 mm ⌀, pale bluish-green, with the Fi. only slightly recessed; **Fl.** 2 cm ⌀, pink.

**G. luteoviride** (Haw.) N. E. Br. (§ I/1) (*Mes. l.* Haw., *Mes. p.* v. *b.* Haw., *G. perviride* v. *l.* N. E. Br.). – Cape: Lit. Karroo. – **St.** thin, prostrate, 2–5 cm long; **L.** elongated, semicylindrical, trigonous above, yellowish-green; **Fl.** pale red. Acc. G. C. Nel a variety of **G. gibbosum.**

**G. pachypodium** (Kensit) L. Bol. (§ II/3) (*Mes. p.* Kensit). – Cape: Ladismith D. – **Br.** numerous, radiating to form cushions 25–40 cm ⌀; **roots** c. 1 cm thick, with the remains of old L. persisting; **L.** shortly united below, unequal, trigonous to semicylindrical, slightly carinate on the underside towards the acute or rounded tip, some 6–10 cm long, 5–15 mm ⌀, others 2.5–8 cm long, (4–)5(–10) mm ⌀; **Fl.** pink to reddish.

**G. petrense** (N. E. Br.) Tisch. (Pl. 166/3) (§ III/1) (*Argeta p.* N. E. Br.). – Cape: Riversdale D. – Very dwarf, ± acauline, forming mats or clumps; **roots** fleshy; **shoots** with 1–2 L.-pairs; **L.** united for ⅓ of their length, the underside round at first, the free part acutely carinate, drawn only slightly over the distinctly acute tip, 9–10 mm long, 5–6 mm across, 4–5 mm thick, the surface smooth, firm, whitish-green to greyish-green; **Fl.** 10–15 mm ⌀, reddish.

**G. pilosulum** (N. E. Br.) N. E. Br. (§ I/2) (*Mes. p.* N. E. Br., *Conophytum p.* (N. E. Br.) N. E. Br.). – Cape: Ladismith D. – Mat-forming; **Bo.** obovoid, 25 mm tall, 22 mm across, 16–18 mm thick, the Fi. 10–11 mm long, in a ± eccentric notch 3–4 mm deep, the surface slightly glossy, light green, with very fine, lax, white hairs; during the resting period, the young Bo. are completely hidden in the dry membranes of the old Bo.; **Fl.** 6–7 mm ⌀, mauvish-red.

**G. pubescens** (Haw.) N. E. Br. (Pl. 166/5) (§ II/1) (*Mes. p.* Haw., *G. argenteum* N. E. Br.). – Cape: Ladismith D. – **St.** short, woody, branched, covered with dry old L.; **Br.** with 2–3 L.-pairs; **L.** alternately unequal, the longer L. 3 cm long, cylindrical, somewhat laterally compressed and obliquely carinate above, with the tip mostly hook-like and elevated, up to 15 mm thick, the shorter L. only ⅓ as long, obtuse, the surface whitish-grey, with minute, felty, white hairs; **Fl.** 15 mm ⌀, violet-red.

**G. schwantesii** Tisch. (§ II/3) (*Mentocalyx muirii* N. E. Br., *G. m.* (N. E. Br.) Schwant.). – Cape: Riversdale D. – Resembles **G. velutinum**; L. rather longer, less obliquely keeled, the larger 5–6.5 cm long, 22–33 mm across below, triangular-acute above, 15–28 mm thick on the deeply keeled underside, smaller L. 3.8–5 cm long, 6–8 mm thick on the underside, the L.-pairs closed together at first, the surface soft velvety, dark green, often brownish or grey; **Fl.** pure white.

**G. shandii** (N. E. Br.) N. E. Br. (§ II/1) (*Mes. s.* N. E. Br.). – Cape: Swellendam D. – **St.** 2–3 cm long, prostrate; **L.** in pairs 2–3 together, united at the base, ovoid to semicylindrical, unequal, 3 cm or only 1 cm long, the surface yellowish-green or grey, felty; **Fl.** 12 mm ⌀, reddish.

**G. tischleri** Wulff (§ III/1) (*G. haagei* Schwant. ex Jacobs.). – Cape: origin unknown. – Resembles **G. petrense** but in general larger; **L.** 12–20 mm long, 8 mm across, 6–7 mm thick, bluish or greenish-grey; **Fl.** c. 2 cm ⌀, mauvish-red.

**G. velutinum** (L. Bol.) Schwant. (Pl. 166/6) (§ II/3) (*Mes. v.* L. Bol., *Mentocalyx v.* (L. Bol.) Schwant., *M. v.* (L. Bol.) N. E. Br.). – Cape: Lit. Karroo. – Mat-forming; **Br.** resting on the ground, enclosed in dry old L.; **L.** united at the base, broadly divaricate, ± resting on the soil, unequal, the longer L. 5–6 cm long, 2.5–3 cm across below, tapered to the acute, incurved and ± hooked tip, carinate on the underside, the shorter L. 4 cm long, trigonous, acute, with the acute, oblique keel pulled forward over the side, the surface light grey to greyish-green, with minute whitish hairs (lens needed!); **Fl.** 4–5 cm ⌀, pink.

**Glottiphyllum** Haw. ex N. E. Br. (§ IV/1/19). – Low-growing ♃ very succulent plants; **St.** dichotomously branched; **L.** dense, arranged ± distichously or obliquely alternate and decussate, with 4 or more on a single shoot, the upperside somewhat swollen and blister-like below, mostly 3 or more times longer than broad, semi- to subcylindrical or obliquely linguiform, the L. of a pair ± equal or unequal, tips obtuse, ± curved upwards or acute, thick, soft-fleshy, fresh glossy green or whitish green or ± brownish, some species with translucent dots; **Fl.** solitary, lateral, sessile or on a pedicel 4–6 cm long, very large, glossy yellow, rarely white, Sept.-Jan. – Summer: in the open but under glass, winter: cold-house. Growing-period very short, spring; keep quite dry during the long resting period. –

Propagation: cuttings. Since the plants are self-sterile, true seed is obtained only by cross-pollinating plants from different clones of the species.

**G. 'Album'.** – Cape: Steytlerville, Oudtshoorn D. – A hitherto undescribed plant with white Fl.; possibly a form of **G. herrei**.

**G. angustum** (HAW.) N. E. BR. (*Mes. a.* HAW.). – S.Afr.: origin unknown. – Resembles **G. linguiforme** and acc. A. BERGER only a variety of this; L. linear-linguiform, the 2 L. of a pair ± equal, but one with the tip somewhat pulled forward and chin-like, carinate above; Fl. 6 cm ⌀.

**G. apiculatum** N. E. BR. – Cape: Oudtshoorn D. – L. lanceolate, with a hard, reddish tip, 4–6 cm long, 9–13 mm across below, 6 to 10 mm thick, smooth, grass-green, ± reddish at the base; Fl. 5–6 cm ⌀, yellow.

**G. armoedense** SCHWANT. – Cape: Oudtshoorn D. – St. extremely short; L. recurved, sheathing and united below for 7.5 mm, fresh green, one L. tapered towards the tip, the other carinate above; Fl. 4–5 cm ⌀, yellow.

**G. arrectum** N. E. BR. – Cape: Swellendam D. – L. 4–5 cm long, 8 mm across and thick, acute, with a hard apiculus, green, rounded on the underside and distinctly keeled above; Fl. 4 cm ⌀, golden-yellow.

**G. barrydalense** SCHWANT. – Cape: Barrydale. – ± acauline; L. distichous, c. 2 cm long, 1 cm across, with one L. more concave on the upperside, with the tip carinate on the underside and curved noticeably and gibbosely inwards, apiculate, green, rather rough with numerous ± prominent, translucent protuberances; Fl. yellow.

**G. buffelsvleyense** SCHWANT. – Cape: Ladismith D. – Caulescent; L. broadly linguiform, 4–5 cm long, 2 cm across, distichous, green, with a slight bluish, waxy coating, the tip ± obtuse on one L., ± rounded on the other, rather shorter L.; Fl. 9–10 cm ⌀, yellow, scented.

**G. carnosum** N. E. BR. – Cape: Ladismith D. – Cape: Ladismith D. – **Shoots** crowded; L. distichous, 3.5–6.5 cm long, 15–20 mm across, 9–10 mm thick, ± ligulate, usually falling to one side, grass-green, with one L. incurved and hooked at the tip, the other obliquely carinate; Fl. 6.5 cm ⌀, light yellow.

**G. cilliersiae** SCHWANT. – Cape: Oudtshoorn D. – Acauline or very short-stemmed; L. distichous, oblique and exceedingly broadly linguiform to almost circular, c. 9 cm long, 5 cm across, with a whitish blister at the base, fresh yellowish-green, one L. of the pair thickened and hooked at the tip; Fl. 8–9 cm ⌀, yellow.

**G. compressum** L. BOL. – Cape: Calitzdorp D. – L. erect to ascending, convex on the upperside, obtuse to truncate above, ± unequal with one L. 10 cm long, 2 cm across, 15 mm thick, the other 9 cm long, 23 mm thick; Fl. up to 8 cm ⌀, yellow.

**G. concavum** N. E. BR. – Cape: origin unknown. – **Shoots** numerous, each with 2–3 L.-pairs; L. smooth, grass-green, becoming translucent in age, 2.5–12 cm long, 6–13 mm across, 4–5 mm thick, semicylindrical below, tapered above, one L. concave on the upperside, the other convex; Fl. 4.5–5 cm ⌀, light yellow.

**G. cruciatum** (HAW.) N. E. BR. (*Mes. c.* HAW.). – Cape: origin unknown. – L. obliquely alternate and decussate, falcate, linguiform, 8–10 cm long, 16–20 mm across, semicylindrical, ± thickened on one side, acutely carinate towards the tip, light green; Fl. 6 cm ⌀, yellow.

**G. davisii** L. BOL. – Cape: Ceres D. – L. distichous, eventually prostrate, others ± spreading, oblique, carinate at the obtuse to truncate tip, 3.5 cm long, very compressed, 14 mm across, 8 mm thick, green to yellowish-green; Fl. 5–6 cm ⌀, yellow.

**G. depressum** (HAW.) N. E. BR. (*Mes. d. L., Mes. longum v. declive* HAW., *Mes. l. v. flaccidum* BGR.). – Cape: origin unknown. – L. 3–4 pairs on one shoot, distichous, ± prostrate, elongated, with the tip curved upwards, obliquely carinate, 10 cm long, 25 mm across, 10–12 mm thick, green; Fl. 5.5 cm ⌀, yellow.

**G. difforme** (L.) N. E. BR. (*Mes. d.* HAW.). – Cape: Clanwilliam D. – St. up to 15 cm long, prostrate, with a few Br.; L. curved ± obliquely to one side, narrowly linguiform, glossy, pale green, with minute, translucent dots, unequal, 12–14 mm across, semicylindrical, 5 or 8 cm long, compressed-carinate above, with a T.-like, fleshy protuberance below the tip; Fl. 3.5–4 cm ⌀, yellow.

**G. erectum** N. E. BR. – Cape: Victoria West D. – St. elongated, prostrate, creeping, up to 20 cm long, with a few Br. with 2–3 L.-pairs; L. distichous, inclined, 5.5–10 cm long, 11–22 mm across, 7–8 mm thick, ligulate, obtuse to acute, apiculate, smooth, grass-green; Fl. 4.5–5 cm ⌀, light yellow.

**G. fergusoniae** L. BOL. – Cape: Lit. Karroo. – L. 6–8 on one shoot, ± erect, slightly oblique, one semicylindrical, narrowed towards the tip, the other ± distinctly compressed-carinate, 10–12 cm long, 22 mm across, 12–15 mm thick, soft, dark green; Fl. 5–6 cm ⌀, yellow.

**G. fragrans** (SALM) SCHWANT. (Pl. 167/1) (*Mes. f.* SALM, *Mes. linguiforme v. f.* (SALM) BGR.). – Cape: Ladismith, Swellendam D., Lit. Karroo. – L. obliquely linguiform, 6–8 cm long, 2.5 cm across, 12 mm thick midway, convex on one side with the other extended into a keel, obtuse above; Fl. 8–10 cm ⌀, glossy golden-yellow, perfumed.

**G. framesii** L. BOL. – Cape: Bredasdorp D. – Resembles **G. depressum**, differentiated only by the long-stalked capsule.

**G. grandiflorum** (HAW.) N. E. BR. (*Mes. g.* HAW., *Mes. linguiforme v. g.* (HAW.) BGR.). – Cape: origin unknown. – L. broadly linguiform, 10–15 cm long, 3–4 cm across, up to 15 mm thick, obliquely rounded at the tip, with one margin acute and the other obtuse, slightly convex on the upperside, pale green; Fl. 9–10 cm ⌀, yellow.

**G. haagei** TISCH. – Cape: Oudtshoorn D. – Mat-forming; L. variable in both size and shape, broadest at the base, gradually tapered towards

Plate 186. 1. **Monilaria globosa** (L. Bol.) L. Bol. (Photo: G. C. Crafford, G. Schwantes); 2. **Namibia cinerea** (Marl.) Dtr. et Schwant. (Photo: H. Ströber, G. Schwantes); 3. **Muiria hortenseae** N. E. Br. (Photo: G. Schwantes); 4. **Namaquanthus vanheerdei** L. Bol. (Photo: G. C. Crafford); 5. × **Muiriogibbaeum muirioides** Rowley.

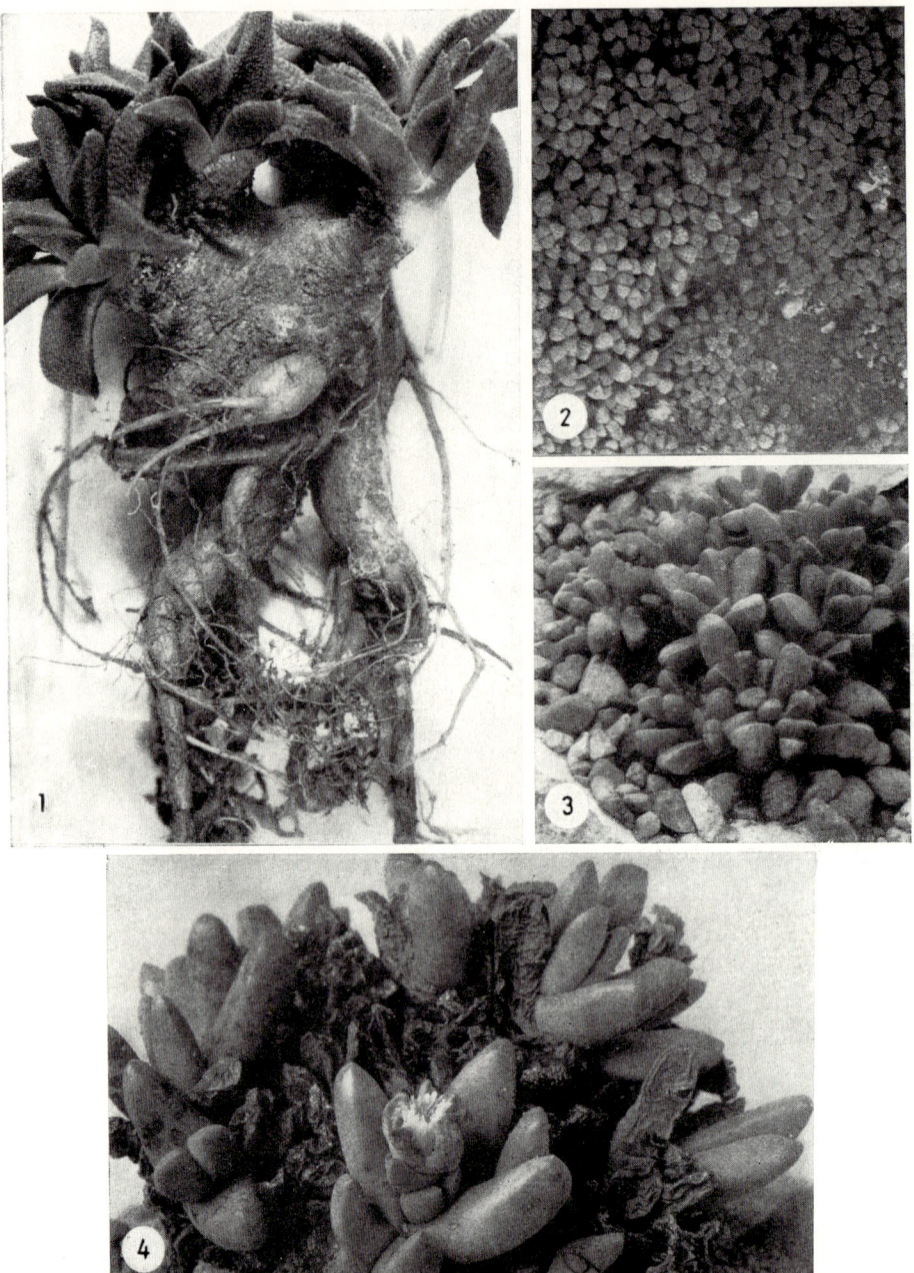

Plate 187. 1. **Nananthus vittatus** (N. E. Br.) Schwant. (Photo: Edrich); 2. **Neohenricia sibbettii** (L. Bol.) L. Bol.; 3. **Neorhine pillansii** (N. E. Br.) Schwant.; 4. **Nelia schlechteri** Schwant. (Photo: G. Schwantes).

Plate 188. 1. **Odontophorus marlothii** SCHWANT.; 2. **Oophytum oviforme** (N. E. BR.) N. E. BR.; 3. **Ophthalmophyllum lydiae** JACOBS.; 4. **O. schuldtii** SCHWANT.; 5. **Orthopterum waltoniae** L. BOL.

30/B*

Plate 189. 1. **Platythyra haeckeliana** (BGR.) N. E. BR.; 2. **Oscularia deltoides** (L.) SCHWANT. v. deltoides (Photo: G. SCHWANTES).

Plate 190. **1. Pleiospilos bolusii** (HOOK. f.) N. E. BR. (Photo: G. SCHWANTES); **2. P. archeri** L. BOL. (Photo: as **1**); **3. P. nelii** SCHWANT.

Plate 191. 1. Pleiospilos simulans (MARL.) N. E. BR. (Photo: G. SCHWANTES); 2. P. prismaticus (MARL.) SCHWANT. (Photo: as 1).

Plate 192. 1. **Psammophora nissenii** (DTR.) DTR. et SCHWANT. (Photo: G. SCHWANTES); 2. **Psilocaulon melanospermum** (BGR.) N. E. BR. (Photo: H. CHR. FRIEDRICH); 3. **P. dinteri** (ENGL. emend. BGR.) SCHWANT. (Photo: as 1); 4. **P. rapaceum** (JACQ.) SCHWANT.

Plate 193. 1. **Rabiea difformis** (L. BOL.) L. BOL. (Photo: Kakteen-HAAGE, Erfurt); 2. **Rhinephyllum schoenlandii** L. BOL. (Photo: G. C. NEL); 3. **Rhombophyllum rhomboideum** (S. D.) SCHWANT.; 4. **R. dolabriforme** (L.) SCHWANT.

the tip, slightly convex or concave, apiculate, greyish-green, tinged opal to mauve, with finely ciliate margins, the larger L. with the underside pulled somewhat forward or obtusely carinate, the other much smaller, thinner, linguiform; **Fl.** 8–10 cm ⌀, glossy golden-yellow.

**G. herrei** L. Bol. – Cape: Worcester D. – **Br.** short; **L.** ± distichous, with one margin resting on the ground, the other compressed above, flat on the upperside, convex on the underside, carinate towards the tip, the keel with minute cilia towards its base, L. of a pair unequal, 6 mm long, 2 mm across, 10 mm thick, bluish-green to reddish; **Fl.** c. 7 cm ⌀, yellow, scented. **G.** 'Album' may be a white-flowered form of this species.

**G. jacobsenianum** Schwant. – Transvaal: S. of Muiskraal. – **L.** unequal, distichous, erect, very thick, somewhat contorted, thick semi-cylindrical, greenish-brown, one L. rounded and curved or ± gibbose at the tip, the other with the tip much pulled forward and very hooked, keeled on the underside; **Fl.** 10 cm ⌀, yellow, slightly scented.

**G. jordaanianum** Schwant. – Cape: Lit. Karroo. – Dwarf, very compact, clump-forming; **L.** glossy, brownish-green to coppery-red, distichous, with the pairs very diverse in shape and size, ± flat or somewhat concave on the upperside, the shorter L. ± curved at the tip, the other with the underside carinate and pulled forward chin-like; **Fl.** yellow.

**G. latifolium** N. E. Br. – Cape: Ladismith D. – **L.** with one margin resting on the soil, the upperside flat, not keeled, the margins ± acute, light green, tinged ± pink, with dark, translucent dots, unequal, 6–8 cm long, 3–4 cm across, 8–15 mm thick; **Fl.** unknown.

**G. latum** (Salm) N. E. Br. v. **latum** (*Mes. linguiforme* v. *l.* Salm, *Mes. l.* (Salm) Haw., *Mes. l.* v. *ascendens* (Salm) Bgr., *M. a.* Salm also Haw., *Mes. obliquum* Willd., *Mes. l.* v. *o.* (Willd.) Bgr., *G. o.* (Willd.) N. E. Br., *Mes. l.* v. *cultratum* sv. *perviride* Salm, *Mes. medium* Haw., *Mes. latum* v. *breve* Haw., *Mes. linguaeforme* Haw., *Mes. l.* v. *subcruciatum* Haw., *Mes. l.* v. *prostratum* Haw., *Mes. l.* v. *assurgens* Haw.). – Cape: Riversdale, Mossel Bay D. – **L.** unequal, linguiform, 7–9 cm long, 2.5 cm across, falcate and curved downwards, thicker and obliquely truncate above, dark green; **Fl.** 5–6 cm ⌀, golden-yellow.

**G. —** v. **cultratum** (Salm) N. E. Br. (*Mes. c.* Salm, *Mes. linguiforme* v. *c.* (Salm) Bgr., *G. c.* (Salm) N. E. Br., *Mes. depressum* Salm, *Mes. l.* v. *d.* (Salm) Bgr., *Mes. linguaeforme* Salm, *Mes. longum* v. *flaccidum* Haw., *Mes. longum* v. *declive* Salm, *Mes. lucidum* Haw., *Mes. linguiforme* v. *declive* Bgr.). – **L.** slightly or not at all curved, 8 cm long, 2.5 cm across, with acute margins, obtuse above.

**G. linguiforme** (L.) N. E. Br. (Pl. 167/2) (*Mes. l.* L., *Mes. scalpratum* Haw., *Mes. l.* v. *s.* (Haw.) Bgr., *Mes. lucidum* Mill.). – Cape: Oudtshoorn D. – **L.** exactly distichous, linguiform, 5–6 cm long, 3–4 cm across, curved slightly upwards above, the lower angle obliquely thickened, the tip bluntly rounded, fresh green, glossy; **Fl.** 7 cm ⌀, golden-yellow. – In cultivation most plants under this name are hybrids.

**G. longipes** (S.D.) N. E. Br.) (*Mes. cruciatum* Salm). – Cape: origin unknown. – **Br.** prostrate or creeping; **L.** green, glossy, with several translucent dots, 6.5–7 cm long, 4–11 mm across, convex on the underside below, laterally compressed and carinate above, one L. obtuse at the tip, the other more acute; **Fl.** 4.5–5 cm ⌀, yellow.

**G. longum** (Haw.) N. E. Br. v. **longum** (*Mes. l.* Haw., *Mes. l.* v. *β* Haw., *Mes. linguiforme* v. *l.* (Haw.) Bgr., *Mes. linguiforme* Salm, *Mes. pustulatum* Haw., *Mes. p.* v. *lividum* Salm, *Mes. linguiforme* v. *p.* (Haw.) Bgr., *G. p.* (Haw.) N. E. Br.). – Cape: Port Elizabeth D. – **L.** ± erect, linguiform, somewhat tapered towards the obtuse tip, flat on the upperside, 7–10 cm long, 2 cm across; **Fl.** 6–8 cm ⌀, golden-yellow.

**G. —** v. **heterophyllum** (Haw.) Rowl. (*Mes. angustum* γ *heterophyllum* Haw., *Mes. h.* Jackson, *G. l.* v. *hamatum* N. E. Br.). – Cape: Uitenhage D. – **L.** curved inwards and hooked at the tip.

**G. marlothii** Schwant. – Cape: Uniondale D. – **L.** distichous, fresh green, with scattered, translucent dots, one L. 6–7 cm long, 12–15 mm across, somewhat concave on the upperside, obliquely carinate, with the tip hooked and incurved, the shorter L. slightly recurved; **Fl.** 7.5 cm ⌀, golden-yellow.

**G. muriii** N. E. Br. – Cape: Riversdale D. – **L.** divaricate, ± prostrate, distichous, 6.5–9 cm long, 15–20 mm across midway, 10–11 mm thick, the sides parallel, the tip obtuse or acute, curved upwards into an obliquely truncate hook, the margin with a protuberance midway, rounded on the underside, smooth, greyish-green; **Fl.** 6.5–7 cm ⌀, light yellow.

**G. neilii** N. E. Br. – Cape: Prince Albert D. – **Shoots** with 2–3 L.-pairs; **L.** distichous, with one angle ± resting on the soil, 11–12 cm long, 3.5–5 cm across, almost 2 cm thick below, linguiform, obliquely truncate with the tip incurved and hooked, slightly glossy, opaline greyish-green, often reddish; **Fl.** 9–10 cm ⌀, yellow.

**G. nelii** Schwant. – Cape: Willowmore D. – Forming rounded clumps; **L.** distichous, ± erect, flat on the upperside, obliquely carinate on the underside, light green, with ± translucent angles, one L. of a pair 4–5 cm long, c. 2 cm across, 12 mm thick, with the rounded tip ± incurved and hooked, the other L. shorter, rounded at the tip; **Fl.** 4 cm ⌀, golden-yellow.

**G. nysiae** Schwant. – Cape: Karroo. – **St.** short; **L.** distichous, very broad, linguiform, deep green, 4 cm long, 2.5 cm across, thick, with one L. slightly tuberculate and carinate at the tip, the other with the keel pulled forward and noticeably hooked; **Fl.** 5 cm ⌀, yellow.

**G. ochraceum** (Bgr.) N. E. Br. (*Mes. o.* Bgr.). – Cape: origin unknown. – Mat-forming;

**Br.** repeatedly forked; **L.** inclined, very soft, 8–10 cm long, trigonous, tapered, obtuse and somewhat thickened above, green, with minute marginal cilia, one L. with a broader tip, with the angle of one side thicker; **Fl.** 3–3.5 cm $\varnothing$, brownish-yellow.

**G. oligocarpum** L. Bol. – Cape: Willowmore D. – **Br.** creeping, prostrate; **L.** distichous, unequal, 4–4.5 cm long, 22 mm across, 10 mm thick at the side, with the tip obtuse or broadly rounded, whitish olive-green, with distinctly prominent dots, indistinctly minutely velvety; **Fl.** 5–6 cm $\varnothing$, yellow.

**G. pallens** L. Bol. – Cape: Jansenville D. – **L.** distichous, arranged obliquely with one margin resting on the ground, 6 cm long, 17 mm across, obtuse to truncate above, with the upper edge concave, pale green; **Fl.** yellow.

**G. parvifolium** L. Bol. – Cape: Ladismith D. – **L.** $\pm$ erect, oblong, somewhat obliquely semicylindrical, 3–4 cm long, 10–12 mm across and thick, acute and apiculate, carinate on the underside towards the tip, green; **Fl.** 8 cm $\varnothing$, glossy golden-yellow.

**G. peersii** L. Bol. – Cape: Prince Albert D. – **Br.** elongated; **L.** with white keel and margins, somewhat glossy, greyish-green, somewhat reddish, $\pm$ erect, unequal, the longer L. subcylindrical, up to 10 cm long, the other one $\pm$ trigonous, up to 11 cm long; **Fl.** 6 cm $\varnothing$, golden-yellow.

**G. platycarpum** L. Bol. – Cape: Riversdale D. – **L.** short, thick, unequal, 3.5 cm long, 18 mm across, with one margin resting on the ground, with the upperside rounded at the tip and the underside carinate, bluish-green; **Fl.** 5 cm $\varnothing$, yellow.

**G. praepingue** (Haw.) N. E. Br. (*Mes. p.* Haw.). – Cape: origin unknown. – **L.** obliquely opposite and decussate, obliquely linguiform, semicylindrical, $\pm$ compressed at the tip, with the margins at first minutely ciliate, 4–7 cm long, 12–15 mm across, smooth, light green; **Fl.** 5 cm $\varnothing$, yellow.

**G. proclive** (Salm) N. E. Br. (*Mes. angustum* v. *heterophyllum* Salm, *Mes. linguiforme* v. *h.* Bgr.). – Cape: Barrydale D. – **Br.** creeping or prostrate; **L.** distichous, $\pm$ decurved, 5–8 cm long, 9–10 mm across, c. 5 mm thick, with upcurved margins, convex on the underside, the larger L. somewhat compressed and carinate; **Fl.** 5 cm $\varnothing$, yellow.

**G. propinquum** N. E. Br. – Cape: Mossel Bay D. – **L.** distichous, linguiform, with a rounded tip, convex on the underside, pale green, 4–7.5 cm long, 15–20 mm across, 6 mm thick; **Fl.** 5 cm $\varnothing$, yellow.

**G. pygmaeum** L. Bol. – Cape: Jansenville D. – **L.** distichous, directed forwards, with the tip resting on the ground, light green, purplish-red in full sun, the upper margins transparent and spotted, 20–35 mm long, 7–10 mm across, broader at the base, $\pm$ semicylindrical, rather concave on the upperside, with the keel on the underside pulled forward over the tip; **Fl.** 4 cm $\varnothing$, yellow. Acc. G. Schwantes this is probably a constant, dwarf form of **G. nelii**.

**G. regium** N. E. Br. – Cape: Oudtshoorn D. – Clump-forming; **shoots** erect, each with 2 L.-pairs; **L.** closed together at first, later $\pm$ divaricate, smooth, light green, one L. 2.5–10 cm long, 10–15 mm across and thick at the base, the other shorter, thinner, rounded on the underside, chinlike above; **Fl.** 4 cm $\varnothing$, yellow.

**G. rosaliae** L. Bol. – Cape: Oudtshoorn D. – Branched, 15 cm tall, 25–30 cm $\varnothing$; **L.** $\pm$ falcate, green, unequal, 7.5 or 9.5 cm long, 10 to 18 mm across and thick, the longer L. acute or more rarely obliquely truncate above, the shorter one tapered, obtuse to $\pm$ acute or rarely rounded, carinate towards the tip; **Fl.** 5–6 cm $\varnothing$, yellow.

**G. rubrostigma** L. Bol. – Cape: Prince Albert D. – **L.** curved, like an inflated blister and semicylindrical below, carinate towards the rounded tip, pale green to yellowish-green, with some dots, $\pm$ unequal, 6–9 cm long, 10 mm across, 12 mm thick below, thicker above, the smaller L. obtuse or acute above; **Fl.** 6 cm $\varnothing$, yellow, with red stigma-lobes.

**G. rufescens** (Haw.) Tisch. (*Mes. r.* Haw., *Mes. depressum* v. *lividum* Haw., *Mes. linguiforme* v. *r.* Haw.). – Cape: origin unknown. – Probably only a very reddened form of **G. latum**.

**G. ryderae** Schwant. – Cape: Calitzdorp D. – **L.** distichous, broadly spatulate to linguiform, 4–6 cm long, 2.5–3 cm across, obtuse, thick, glossy green or often dull, dirty yellowish-green to chalky; **Fl.** 8–9 cm $\varnothing$, golden-yellow.

**G. salmii** (Haw.) N. E. Br. (*Mes. s.* Haw., *Mes. s.* v. *semicruciatum* Haw., *Mes. s.* v. *decussatum* Haw.). – Cape: Lit. Karroo. – **L.** obliquely semicylindrical, curved, narrowed above, fresh green, 7–10 cm long, 16–22 mm across, one L. tapered, the other compressed-carinate, obtusely tapered above; **Fl.** 6 cm $\varnothing$, yellow.

**G. semicylindricum** (Haw.) N. E. Br. (*Mes. s.* Haw., *Mes. bigibberatum* Haw., *Mes. bidentatum* Haw.). – Cape: Jansenville D. – With a short **St.** in age; **Br.** stiffly projecting; **L.** $\pm$ prostrate, slightly incurved, semicylindrical, compressed-carinate at the tip, the margins with small, T.-like prominences, fresh glossy green with faint dots, 4–5 cm long, 5–6 mm across and thick; **Fl.** 4 cm $\varnothing$, golden-yellow.

**G. starkeae** L. Bol. – Cape: Uniondale D. – **L.** distichous, elongated, spreading, with one margin resting on the soil, $\pm$ obtuse or obliquely tapered above, indistinctly bristle-tipped, green, 5.5–7 cm long, (10–)15(–20) mm across, 5–7 mm thick; **Fl.** 6–8 cm $\varnothing$, golden-yellow. Possibly identical with **G. marlothii**.

**G. suave** N. E. Br. – Cape: Ladismith D. – With several shoots, and forming cushions c. 11 cm $\varnothing$; **L.** 3–5 cm long, 15–20 mm across midway, 6–8 mm thick, $\pm$ linguiform, shortly obtuse or obliquely truncate above, grass-green, the L.-pairs lying $\pm$ obliquely across one another, one L. of a pair obliquely carinate on the underside and more rounded above than the other; **Fl.** 6.5 cm $\varnothing$, glossy light yellow, scented.

**G. subditum** N. E. Br. (*Mes. praepingue* Salm). – Cape: origin unknown. – **L.** 3–6 cm long, 6 to 8 mm across, 4 mm thick, laterally curved towards the margin, green, with translucent dots, unequal, one L. concave on the upperside, noticeably rounded on the underside, the L.-pairs lying obliquely across one another; **Fl.** 4–5 cm ⌀.

**G. surrectum** (Haw.) L. Bol. (*Mes. s.* Haw.). – Cape: Calitzdorp D. – **L.** semicylindrical and subulate, like an inflated blister below; **Fl.** not known. – Insufficiently known.

**G. taurinum** (Haw.) N. E. Br. (*Mes. t.* Haw., *Mes. depressum* v. *pallens* Haw., *Mes. angustum* Salm, *Mes. linguiforme* v. *a.* (Salm) Bgr., *Mes. a.* v. *pallidum* Haw.). – Cape: origin unknown. – L.-pairs arranged obliquely across one another; **L.** 4.5–5 cm long, 11–12 mm across, semicylindrical, obliquely rounded on the underside, often ± digitate, yellowish-green, one L. often obliquely compressed-carinate at the tip; **Fl.** yellow.

**G. uncatum** (Salm) N. E. Br. (*Mes. u.* Salm, *Mes. longum* v. *u.* (Salm) Haw., *Mes. linguiforme* v. *u.* (Salm) Bgr., *Mes. longum* v. *atollens* Haw., *Mes. linguiforme* v. *a.* (Haw.) Bgr.). – Cape: origin unknown. – **L.** narrow-linguiform, with a thick, oblique keel and the tip curved upwards and hooked, c. 6 cm long, 1.5 cm across; **Fl.** yellow.

**G. uniondalense** L. Bol. – Cape: Uniondale D. – **Br.** creeping; **L.** distichous, ligulate, with one margin resting on the ground, with the tip compressed and rounded to ± truncate, dirty green, 9.5 cm long, up to 3 cm across, 13 mm thick; **Fl.** yellow.

**Hereroa** (Schwant.) Dtr. et Schwant. (§ IV/1/1) (*Bergeranthus* Schwant. SG. *Hereroa* Schwant.). – Short-stemmed, mat-forming plants or small ♄ with visible Int.; **L.** opposite and decussate, soft-fleshy, ± united and semicylindrical at the base, somewhat laterally compressed and expanded towards the tip, mostly green, with ± large dots; **Fl.** on a pedicel with 2 bracts, several together in a branched Infl. or solitary, medium-large, yellow, often becoming pink, rarely white, summer, opening in the afternoon or at night. – Greenhouse: warm, winter: moderately warm. Propagation: seed, cuttings.

**H. acuminata** L. Bol. – Cape: Ladismith D. – Cushion-forming; **L.** rather unequal, obliquely obtuse above, with serrate margins and keel, rough with raised dots, 5 cm long, 8 mm across, 6 mm thick; **Fl.** c. 3 cm ⌀, yellow, reddish on the underside.

**H. albanensis** L. Bol. (*Ruschia dyeri* L. Bol., *Mes. d.* (L. Bol.) N. E. Br.). – Cape: Albany D. – Compact, stiff ♄ 10 cm tall, 10–16 cm ⌀; **St.** 12 mm thick below; **Br.** spreading to ascending, 9 mm thick, with 6 densely arranged L. on each short shoot; **L.** bluntly carinate, obtuse to truncate above, bluish-green, 15 mm long, 3 mm across and thick, with a parchment-like Sh. 1–4 mm long; **Fl.** solitary, 10–15 mm ⌀, yellow.

**H. aspera** L. Bol. – Cape: Swellendam D. – Small, erect, branched ♄ 8 cm high; **St.** 5 cm long; **L.** semicylindrical, obtuse above, bluish-green, rough with raised dots, 25 mm long, 3 mm across, 3.5 mm thick; **Fl.** solitary, 3.5 cm ⌀, Pet. yellow, with reddened tips.

**H. brevifolia** L. Bol. – Cape: Prince Albert D. – Compact, 20 cm ⌀, 7 cm tall; **St.** 25 mm thick; **Br.** spreading or prostrate; **L.** unequal, expanded midway, truncate or subacute above, with an indistinct keel, the surface somewhat granular, bluish-green, 2–3 cm long, 5–8 mm across, 6–9 mm thick; **Fl.** c. 3 cm ⌀, yellow.

**H. calycina** L. Bol. – Cape: Queenstown D. – Small ♄ with a **St.** 2 cm thick; **Br.** dense with 4–8 bluish-green **L.** 4 cm long, 8 mm across and thick, with distinct dots, the pair dissimilar, one L. carinate on the underside and expanded above, the other laterally keeled, curved, ± falcate; **Fl.** 28 mm ⌀, yellow to rather reddish.

**H. carinans** (Haw.) L. Bol. (*Mes. c.* Haw., *Bergeranthus c.* (Haw.) Schwant., *H. c.* (Haw.) Schwant.). – Cape: W.Griqualand. – Short-stemmed, mat-forming; **L.** semicylindrical below, carinate and expanded above, with the tip somewhat incurved, with obtuse angles, the keel-face 7 mm across, rough, 25–35 mm long, 6 mm across, dull green, with raised dots; **Fl.** 3.5 cm ⌀, yellow.

**H. concava** L. Bol. – Cape: Beaufort West D. – Cushion-forming ♄ 7 cm tall; **L.** slender, acute to tapered, 4–5 or 2–3 mm thick, 3–4 cm long, bluish-green, with ± raised dots; **Fl.** 3 cm ⌀, golden-yellow.

**H. crassa** L. Bol. – Cape: Beaufort West, Laingsburg D. – Compact, 10–12 cm tall; **St.** thick; **L.** thick, swollen like a blister at the base, carinate on the underside, obliquely obtuse to truncate at the tip, with serrate margins and keel, blue, with raised dots, the pair unequal, one L. 25 mm long, 17 mm across, 11 mm thick, the other 2 cm long; **Fl.** 4 cm ⌀, golden-yellow, scented.

**H. dyeri** L. Bol. – Cape: Albany, Willowmore D. – Compact, c. 10 cm ⌀; **rootstock** ± tuberous, 1 cm thick; **St.** short, covered with the remains of old L.; **L.** semicylindrical below, carinate on the underside, tapered towards the expanded, hatchet-shaped and bifid tip, green with raised dots, 5 cm long, 3–10 mm thick; **Fl.** 2.4 cm ⌀, golden-yellow.

**H. fimbriata** L. Bol. – Cape: Prince Albert, Ceres D. – Compact ♄ c. 18 cm tall, 17 cm ⌀; **St.** crowded; **L.** falcate, semicylindrical below, carinate on the underside, expanded towards the obliquely truncate tip, bluish-green, rough, 6–8 mm across midway, 12 mm thick; **Fl.** 4 cm ⌀, golden-yellow, scented.

**H. glenensis** (N. E. Br.) L. Bol. (*Bergeranthus g.* N. E. Br.). – Orange Free State. – Low ♄ c. 10 cm tall; **rootstock** thick, woody; **St.** crowded; **L.** somewhat laterally curved, round on the underside below, rounded-carinate above, green, with dense, raised dots, 2.5–4 cm long, 5–8 mm across, 4–6 mm thick; **Fl.** often several together, 3 cm ⌀, glossy yellow.

**H. gracilis** L. Bol. v. **gracilis** – Cape: Lit. Karroo. – Cushion-forming; **St.** 25 mm long, covered with the remains of old L.; **L.** slender, tapered to ± obtuse, expanded and white at the tip, rounded to indistinctly keeled on the underside, with dense, raised dots, 3.5 cm long, 3–4 mm thick; **Fl.** 25 mm ⌀, yellow.

**H. —** v. **compressa** L. Bol. – Cape: Uniondale D. – **L.** 5 cm long, 6 mm thick, compressed and keel-like.

**H. granulata** (N. E. Br.) Dtr. et Schwant. (*Mes. g.* N. E. Br., *Bergeranthus g.* (N. E. Br.) Schwant., *Mes. carinans* Bgr.). – Cape: Uitenhage D. – **L.** spreading, often prostrate, slightly curved, semicylindrical below, carinate and somewhat expanded above, mucronate, dark green with rough dots, 4–6 cm long, 6 mm across, 2 mm thick; **Fl.** unknown.

**H. herrei** Schwant. – Cape: Graaff Reinet, Griquatown D. – Resembles **H. granulata**; **L.** not expanded above, otherwise only differentiated by the construction of the capsule.

**H. hesperantha** (Dtr.) Dtr. et Schwant. (Pl. 168/2) (*Mes. h.* Dtr., *Mes. bergerianum* Dtr., *Mes. karasbergensis* L. Bol., *H. k.* (L. Bol.) Friedr. nom. prov.). – S.W.Afr.: Gr. Namaqualand. – Erect, stiff, branched ♄ 30 cm tall; **St.** 20 cm long; **Br.** 8–9 cm long, leafy throughout; **L.** swollen-trigonous, oblique to rounded and ± expanded at the tip; **Fl.** solitary, 15–20 mm ⌀, golden-yellow.

**H. incurva** L. Bol. – Cradock. D – Spreading; **L.** acute to tapered, somewhat expanded below, bluish-green to reddish, with crowded dots, 3.5 cm long, 7 mm across, 6 mm thick; **Fl.** solitary, 3–6 cm ⌀, golden-yellow.

**H. joubertii** L. Bol. – Cape: Laingsburg D. – Up to 20 cm ⌀; **St.** densely crowded; **L.** 4–6 on a shoot, bluish to reddish, rounded on the underside, carinate and 1 cm thick towards the obtuse tip, (3–)3.5(–4) cm long, 9 mm across midway; **Fl.** solitary or paired, up to 3.5 cm ⌀, yellow.

**H. latipetala** L. Bol. – Cape: Prince Albert D. – Compact, 7 cm tall, with the **St.** 1 cm thick at the base; **Br.** dense; **L.** falcate, semicylindrical below, obtusely keeled and expanded above, bluish-green, rough, 3.5 cm long, 11 mm across; **Fl.** 3.5 cm ⌀, Pet. broad, golden-yellow.

**H. muirii** L. Bol. – Cape: Swellendam D. – 6–7 cm tall; **L.** 4–6 on a shoot, semicylindrical, rounded-carinate on the underside, with a granular surface, 5 cm long, 8 mm across; **Fl.** 22 mm ⌀, yellow.

**H. nelii** Schwant. – Cape: Ceres D. – **Shoots** with 1–3 L.-pairs; **L.** falcate, indistinctly carinate above on the underside, green, covered with numerous translucent tubercles, 2.5–3 cm long, 5 mm across; **Fl.** 1–3 together, yellow.

**H. odorata** (L. Bol.) L. Bol. (*Mes. o.* L. Bol., *Aridaria o.* (L. Bol.) Schwant., *Nycteranthus o.* (L. Bol.) Schwant.). – Cape: Swellendam, Ladismith, Oudtshoorn D. – Small ♄; **St.** projecting to spreading; **L.** semicylindrical, with indistinct margins and keel, rather tapered, bluish-green, 4 cm long, 6 mm thick; **Infl.** of 3 yellow, scented Fl. 22 mm ⌀, with the Pet. reddish-tipped.

**H. puttkamerana** (Dtr. et Bgr.) Dtr. et Schwant. (*Mes. p.* Dtr. et Bgr., *Bergeranthus p.* (Dtr. et Bgr.) Schwant., *H. pallens* L. Bol., *H. angustifolia* L. Bol.). – S.W.Afr.: Gr. Namaqualand; Cape: Lit. Namaqualand. – Compact, densely leafy plant; **L.** 4–8, inclined, obtusely trigonous, semicylindrical above, slightly glossy, greyish-green, with raised, dark dots especially along the keel, 6–7 cm long, 6–7 mm across; **Fl.** 1–3 together, up to 3 cm ⌀, golden-yellow to orange.

**H. rehneltiana** (Bgr.) Dtr. et Schwant. (*Mes. r.* Bgr., *Bergeranthus r.* (Bgr.) Schwant.). – S.Afr.: origin unknown. – **St.** shortly branched; **L.** somewhat curved inwards or outwards, compressed-carinate in the upper third, the keel c. 1 cm across, the margins rounded, ± obtuse above, with a tiny, cartilaginous mucro, light green with raised, translucent dots, 6–10 cm long, 6 mm thick; **Fl.** 3–7 together, c. 22 mm ⌀, yellow.

**H. stanfordiae** L. Bol. – Cape: Victoria West D. – **St.** densely crowded; **L.** bluish-green, ± distinctly spotted, compressed-carinate below, with the keel densely spotted and serrate, obliquely narrowed at the tip, 25–29 or 27–32 mm long, 5 mm across; **Fl.** 3 cm ⌀, yellow.

**H. stanleyi** (L. Bol.) L. Bol. (*Mes. s.* L. Bol.). – Cape: Willowmore, Prince Albert D. – ♄ 7–9 cm tall, with a **St.** 9 mm ⌀; **L.** 6–8 on a shoot, bluish-green, carinate on the underside, the keel with or without 1–2 T., 10–13 mm long, 3 mm across, 4 mm thick; **Fl.** solitary, 20–24 mm ⌀, scented, Pet. golden-yellow, reddened outside at the tips.

**H. stenophylla** L. Bol. – Cape: Willowmore, Prince Albert, Laingsburg D. – Low-growing; **L.** long-tapered, rounded on the underside, rounded-carinate below the tip, dirty green, 2.5–4 cm long, 3–4 mm across and thick; **Fl.** up to 26 mm ⌀, golden-yellow.

**H. tenuifolia** L. Bol. – Cape: Robertson D. – Erect, laxly branched ♄ 25 cm tall; **St.** stiff, grey; **L.** semicylindrical, obtuse to short-tapered above, carinate on the underside below the tip, bluish-green, rough with minute dots, 4.5 cm long, 5 mm across and thick; **Fl.** c. 3 cm ⌀, yellow, scented.

**H. teretifolia** L. Bol. – Cape: Calvinia, Ceres D. – **St.** 1 cm thick below; **L.** 10 on a shoot, cylindrical to midway, then tapered, obtuse above, green, 5 cm long, 8 mm thick; **Fl.** 3.5 cm ⌀, yellow, scented.

**H. tugwelliae** (L. Bol.) L. Bol. (*Mes. t.* L. Bol., *H. cana* L. Bol., *Bolusanthemum t.* (L. Bol.) Schwant., *Juttadinteria t.* (L. Bol.) Schwant.). – Cape: Karroo. – **St.** thick, woody, short; **Br.** short; **L.** 4–6 on a shoot, ± sabre-shaped, laterally compressed, rounded-carinate, semicylindrical below, ± obliquely rounded at the tip, smooth, bluish-green, c. 6 cm long, 1 cm across, 24 mm thick; **Fl.** 4–5 cm ⌀, yellow, with a whitish centre.

**H. uncipetala** (N. E. Br.) L. Bol. (*Prepodesma u.* N. E. Br.). – Cape: W. Griqualand. – Acauline, 5–7 cm tall; **L.** 4–6 pairs on one shoot, curved, long-tapered, rounded on the underside below, then compressed-carinate, with the keel expanded to 5–8 mm, 3–3.5 cm long, 4 to 5 mm across below, 3–4 mm thick, light greyish-green, with darker dots; **Fl.** 4.5 cm $\varnothing$, light yellow.

**H. willowmorensis** L. Bol. – Cape: Willowmore D. – Branched; **L.** 8 on a shoot, semicylindrical on the underside, indistinctly keeled above, $\pm$ obtuse, green, 22–30 mm or 36 mm long, 6 mm across midway, 6 mm thick; **Fl.** 26 mm $\varnothing$, golden-yellow, paler on the underside.

**H. wilmaniae** L. Bol. v. **wilmaniae.** – Cape: W. Griqualand. – **St.** 1 cm thick below; **Br.** crowded, with 4–6 L.; **L.** long-tapered, semicylindrical below, compressed and 1 cm thick midway, bluish-green with small, raised dots, 4–6 cm long; **Fl.** c. 3 cm $\varnothing$, Pet. in 4 series, golden-yellow.

**H. —** v. **langebergensis** L. Bol. – Pet. in 2 series, unequal.

**Herrea** Schwant. (§ IV/2). – ♃, decumbent, glabrous herbs; **caudex** tuberous, cylindrical, fusiform or irregularly-shaped; **St.** few, later drying up and dropping, often branched, up to 25 cm long, up to c. 7 mm $\varnothing$; **L.** opposite to alternate, terete to semiterete, acute to obtuse at the apex, rarely obtusely keeled on the underside; pedicels straight or curved in age, the old Fl. and Fr. often pendent, **Fl.** up to 10 cm $\varnothing$, open in the afternoon, Pet. very numerous in 4–5 series, bright to golden-yellow in the upper part, rarely white above and dull salmon-pink on the lower surface. – Summer: in the open; winter: the fleshy roots should be kept in dry sand. Propagation: seed.

## Key to the Species of the Genus Herrea by L. Bolus

1. Cor. up to 10 cm $\varnothing$ or very slightly more .................................... **2.**
1a. Cor. up to 12.5 cm $\varnothing$; 2 Sep. much more elongate than the rest, receptacle subcampanulate, stigmas 19 ................................................................. **elongata v. elongata**
2. Sep. shorter, or much shorter, than the Pet. .................................... **3.**
2a. Sep. longer than the Pet., or at least the longest exceeding or nearly equalling them ... **11.**
3. Outer Pet. usually acuminate or long-acuminate .................................... **4.**
3a. Outer Pet. usually acute or emarginate or rarely acuminate ........................... **9.**
4. Receptacle up to 2 cm $\varnothing$; Sep. up to 2 cm long; Cor. 5–8 cm $\varnothing$ ............... **5.**
4a. Receptacle up to 2.5 cm $\varnothing$; Sep. up to 3 cm and 3.2 cm long .................... **6.**
5. Top of Ov. elevated to 0.8 mm in the middle .................................... **brevisepala**
5a. Top of Ov. elevated to 6 mm in the middle .................................... **affinis**
6. Receptacle shortly crateriform, 8 mm long, ribs prominent .................... **laticalyx**
6a. Receptacle globosely obconic or subhemispherical, 1–1.5 cm long .................... **7.**
7. Receptacle grooved downwards and as if stipitate, up to 1.5 cm long ............ **grandis**
7a. Receptacle not as above .................................................... **8.**
8. Receptacle subhemispherical (length unknown) .................... **elongata v. minus**
8a. Receptacle broadly or subglobosely conical, 1 cm long .................... **alboluteä**
9. Receptacle shortly crateriform, subtruncate at the base; Pet. golden in the upper half; stigmas 12 ................................................................. **acocksii**
9a. Receptacle crateriform, obconic-crateriform or obconic, rounded at the base; Pet. not golden; stigmas 14–15 ................................................................. **10.**
10. Ribs on the receptacle prominent; Pet. white in the lower part, pale lemon, pale copper or (in the bud) salmon-pink in the upper part; filaments golden; cone of immature Fr. acute ... ........ **blanda**
10a. Ribs on the receptacle inconspicuous; Pet. white, or pale dull salmon on the underside; cone of immature Fr. obtuse .................................................... **alba**
11. Upper parts of Sep. up to 4 mm $\varnothing$ .................................... **12.**
11a. Upper parts of Sep. usually up to 2.5 mm $\varnothing$ .................................... **13.**
12. L. rounded at the apex; all the Sep. exceeding the Pet.; Pet. up to 1.8 mm, usually 1 mm broad ........ **macrocalyx**
12a. L. acute; 4 Sep. exceeding the Pet.; Pet. up to 3 mm broad .................... **fusiformis**
13. Sep. sometimes conspicuously unequal in length, one of the outer accrescent in mature Fl. and up to 5.3 cm long .................................................... **14.**
13a. Sep. not conspicuously unequal in length .................................... **15.**
14. Receptacle as if stipitate; Sep.-membranes pallid in outer half .................... **plana**
14a. Receptacle (as it appears in dried specimens) rounded at the base; Sep.-membranes entirely black .................................................... **inaequalis**
15. Receptacle shortly crateriform .................................................... **16.**
15a. Receptacle shortly obconic, broadly obconic, globosely obconic or hemispherical ........ **19.**
16. Receptacle rounded at the base .................................................... **17.**
16a. Receptacle truncate at the base .................................................... **18.**
17. Fr. semiglobose in the upper part, or nearly so; Pet. yellow to golden ............ **nelii**
17a. Immature Fr. conical above; Pet. straw or whitish .................... **porcina**

18. Pet. often narrowed downwards, ciliate to the middle or almost to the apex; top of Ov. elevated to 3 mm in the middle; cone of Fr. probably acute ................ **gydouwensis**
18a. Pet. not, or scarcely, narrowed downwards; top of Ov. elevated to 2 mm; cone of immature Fr. obtuse.................................................................. **ronaldii**
19. Receptacle as if somewhat stipitate..................................... 20.
19a. Receptacle not as if stipitate ......................................... 23.
20. Cone of Fr. acute ................................................... 21.
20a. Cone of Fr. obtuse (unknow in **H. caledonica**)........................ 22.
21. Margin of Fr. flat or nearly so ....................................... **stipitata**
21a. Margin of Fr. deeply concave ........................................ **excavata**
22. Receptacle 9 mm long; Pet. up to 0.5 mm broad ......................... **robusta**
22a. Receptacle up to 1.7 cm long; Pet. up to 1.25 mm broad ............... **caledonica**
23. Margin of Fr. deeply lobed; Cor. up to 10 cm ⌀ or Pet. up to 4.3 cm long ...... **albolutea**
23a. Margin of Fr. (where known or described) not lobed; Cor. up to 8 cm ⌀, or Pet. up to 3.3 cm long ............................................................................ 24.
24. Receptacle broadly obconic; Fr. up to 4 cm ⌀, cone obtuse .............. **obtusa**
24a. Receptacle hemispherical or globosely obconic; Fr. (unknown in **H. klaverensis**) up to 5 cm ⌀; cone acute ........................................................................... 25.
25. Caudex white, cylindric or napiform, 1.2 cm ⌀; receptacle 1.2 cm ⌀; top of Ov. concave, elevated in the middle; stigmas 2–4 mm long .................................. **roodiae**
25a. Caudex dark brown, irregular and branched, up to 6–7 cm ⌀; receptacle 1.7 cm ⌀; top of Ov. nearly flat; stigmas 6 mm long ........................................ **klaverensis**

**H. acocksii** L. Bol. – Cape: Bushmanland. – **Caudex** tuberous, c. 5 cm long, 18 mm ⌀; **St.** up to 20 cm long, 6 mm thick; **L.** c. 12 cm long, 3.5 mm across, obtuse above; **Fl.** 5–7 cm ⌀, Pet. yellow, ± white below.

**H. affinis** (N. E. Br.) L. Bol. (*Conicosia a*, N. E. Br., *Mes. elongatum* S.D.). – S.Afr.: origin unknown. – **Caudex** fusiform to ovoid, 7–10 cm long; **L.** 10–15 cm long, 2–3 mm across, channelled to concave on the upperside, rounded on the underside; **Fl.** 7–8 cm ⌀.

**H. alba** L. Bol. – Cape: Lit. Namaqualand. – **Caudex** tuberous, cylindrical, flexuous, over 20 cm long, 3 cm ⌀; **L.** rather bluntly carinate on the underside, obtuse above, c. 15 cm long, 5 mm across and thick; **Pet.** numerous, ± contorted, c. 3 cm long, white, the outer ones dirty pink.

**H. albolutea** L. Bol. – Cape: Calvinia D. – **St.** 22–25 cm long, c. 1 cm thick; **L.** acute to obtuse, up to 18 cm long, 4–6 mm across, 4–7 mm thick; **Fl.** over 10 cm ⌀, Pet. indistinctly ciliate below, white on the upperside, yellow above.

**H. blanda** L. Bol. – Cape: Lit. Namaqualand. – **Caudex** 4.5 cm long, 8 mm thick; **St.** 10–15 cm long; **L.** obtusely semicylindrical, c. 11 cm long, 5 mm thick; **Pet.** numerous, white below, pale copper-coloured to lemon-yellow or salmon-pink above.

**H. brevisepala** L. Bol. – Cape: Van Rhynsdorp D. – **Caudex** tuberous; **St.** semi-fleshy; **L.** obtuse above, 7.5–10 cm long, 7 mm across and thick; **Fl.** 5.5–7 cm ⌀, Pet. yellow with yellowish to white cilia below.

**H. caledonica** L. Bol. – Cape: Caledon D. – **Caudex** tuberous, up to 20 cm long, 1.5–3.5 cm ⌀, brown above; **St.** up to 30 cm long; **L.** semicylindrical below, bluntly carinate on the underside, obtuse to acute above, c. 19 cm long, 4–7 mm across and thick; **Fl.** 9 cm ⌀, Pet. yellow to golden-yellow, ciliate below.

**H. elongata** (Haw.) L. Bol. v. **elongata** (*Mes. e.* Haw., *Conicosia e.* (Haw.) N. E. Br., *C. e.* (Haw.) Schwant., *Mes. e.* v. *β.* Haw., *Mes. pugioniforme* DC.). – S.Afr.: origin unknown. – **Caudex** napiform; **St.** 30 cm or more long, prostrate; **L.** tapered, somewhat channelled on the upperside, rounded on the underside, soft, fresh to greyish-green, 10–15 cm long, 6 mm across; **Fl.** 8 cm ⌀, glossy sulphur-yellow.

**H. — v. minor** (Haw.) L. Bol. (*Mes. e. v. m.* Haw.). – **Fl.** smaller.

**H. excavata** L. Bol. – Cape: Calvinia D. – **Caudex** tuberous, brown, 6.5 cm long, 3.5 cm thick; **St.** 16–30 cm long, 6 mm thick; **L.** obtuse to short-tapered, ± cylindrical, up to 18 cm long, c. 6 mm thick; **Fl.** c. 9 cm ⌀, Pet. numerous, white, yellow to golden-yellow above.

**H. fusiformis** (Haw.) L. Bol. (*Mes. f.* Haw., *Conicosia f.* (Haw.) N. E. Br., *Mes. bicolorum* Klinsmann). – S.Afr.: origin unknown. – **Caudex** fleshy, fusiform or napiform, c. 20 cm long, 1.5 cm thick; **L.** clustered, 8–12 cm long, 4 mm thick, cylindrical, acute, yellowish-green; **Fl.** 6 cm ⌀, sulphur-yellow.

**H. grandis** L. Bol. – Cape: Lit. Namaqualand. – **St.** 8 mm thick; **L.** obtuse above, concave on the upperside, rounded on the underside, c. 16 cm long, 5–7 mm across, 4–5 mm thick; **Fl.** 10 cm ⌀, Pet. yellow below, pale pink above, pink on the underside.

**H. gydouwensis** L. Bol. – Cape: Ceres D. – **St.** 12–14 cm long, 5 mm thick; **L.** obtuse to acute above, bluntly to acutely keeled on the underside, c. 13 cm long, 4–6 mm across and thick; **Fl.** 6–7.5 cm ⌀, Pet. ciliate towards the tips, golden-yellow, with a faint pink border, orange-yellow on the underside.

**H. inaequalis** L. Bol. – Cape: Laingsburg D. – **Caudex** tuberous, either cylindrical, 3 cm long, 8 cm thick, or fusiform, c. 20 cm long, c. 2 cm thick; **St.** c. 25 cm long, 4.5 mm thick; **L.** obtuse to subacute, 14–15 cm long, 3–6 mm thick; **Fl.** c. 6 cm ⌀, yellow.

**H. klaverensis** L. Bol. – Cape: Van Rhynsdorp D. – **Caudex** tuberous, up to 20 cm long,

6–7 cm thick, brown to blackish; **L.** rounded on the underside, up to 20 cm long, 4–5 mm thick; **Fl.** 6–7 cm ⌀, yellow.

**H. laticalyx** L. Bol. – Cape: Lit. Namaqualand. – **St.** fleshy, 17 cm long, up to 9 mm thick; **L.** ± cylindrical, c. 18 cm long, c. 9 mm across and thick; **Fl.** c. 9 cm ⌀, Pet. white below, pale yellow above.

**H. macrocalyx** L. Bol. – Cape: Lit. Namaqualand. – **St.** 16–21 cm long, c. 7 mm thick; **L.** obtuse, ± cylindrical, c. 15 cm long, 7 mm across and thick; **Fl.** 7–8 cm ⌀.

**H. nelii** Schwant. (Pl. 169/1, 2). – Cape: Ceres D. – **Caudex** tuberous, thick, fleshy, white; **L.** 6.5 cm long, 3 mm thick, long-tapered; **Fl.** c. 6 cm ⌀, Pet. very narrow, yellow.

**H. obtusa** L. Bol. – Cape: Robertson D. – **Caudex** tuberous, brown, 12 cm long, divided below into roots 5.5–7 cm long, 1.5 cm thick; **St.** up to c. 40 cm long, 5 mm thick; **L.** obtuse, semicylindrical, c. 23 cm long, 7 mm across; **Fl.** c. 5 cm ⌀, Pet. yellow with golden-yellow tips.

**H. plana** L. Bol. – Cape: Robertson D. – **St.** 17–25 cm long; **L.** semicylindrical, obtuse to subacute, c. 15 cm long, c. 3.5 mm across; **Fl.** almost 8 cm ⌀, yellow.

**H. porcina** L. Bol. – Cape: Lit. Namaqualand. – **Caudex** tuberous, 7.5 cm long, 6.5 cm thick, branched below; **St.** up to 24 cm long, 8 mm thick; **L.** concave on the upperside below, rounded on the underside, obtuse, c. 16 cm long, 1 cm across; **Fl.** almost 6 cm ⌀, yellowish-white.

**H. robusta** (N. E. Br.) L. Bol. (*Conicosia r.* N. E. Br., *H. r.* f. L. Bol., *Mes. pugioniforme* L. Bol.). – Cape: Lit. Namaqualand. – **Caudex** probably tuberous; **L.** 15–22 cm long, 5–7 mm thick, obtuse to ± acute, semicylindrical; **Fl.** 8–8.5 cm ⌀, pale yellow.

**H. ronaldii** L. Bol. – Cape: Laingsburg, Worcester D. – **Caudex** tuberous, 15–18 cm long, c. 6 cm thick; **L.** obtuse, c. 10–18 cm long, 6 mm across; **Fl.** c. 6 cm ⌀, yellow.

**H. roodiae** (N. E. Br.) L. Bol. (*Conicosia r.* N. E. Br.). – Cape: Van Rhynsdorp D. – **Caudex** 15–45 cm long, c. 1.5 cm thick, white; **L.** clustered, long-tapered, 5–12 cm long, 3 to 5 mm across and thick; **Fl.** up to 10 cm ⌀, yellow.

**H. stipitata** L. Bol. – Cape: Robertson D. – **Caudex** tuberous, 7 cm long, c. 4 cm thick, blackish-brown, gnarled, branched; **St.** up to 30 cm long; **L.** 13–23 cm long, up to 7 mm across; **Fl.** not described.

**Herreanthus** Schwant. (§ IV/1/17). – Dwarf, very succulent plants. – Greenhouse, warm. Propagation: seed.

**H. meyeri** Schwant. (Pl. 169/3). – Cape: Lit. Namaqualand. – Mat-forming; **roots** fibrous; **L.** opposite and decussate, thick, trigonous, 4 cm long, 2 cm across, 1.5 cm thick, united for up to 1 cm, triangular and tapered on the upperside, flat, apiculate, slightly carinate on the underside to midway, firm-textured, light bluish-green, smooth with slightly raised dots; **Fl.** ± sessile, 2.5 cm ⌀, white, scented, remaining open night and day, Aug.

**Hymenogyne** (Haw.) N. E. Br. (§ II). – ☉ herbs branching from the base, with large Fl. – Summer: in the open. Propagation: seed.

**H. conica** L. Bol. – Cape: Clanwilliam D., Lit. Namaqualand. – Branched from the base, 11–18 cm tall, up to 30 cm ⌀; **St.** erect, prostrate to creeping, 3 mm thick; **L.** opposite, falcate, sheathing and united below, 4–8 cm long, 4–10 mm across; **Fl.** 6 cm ⌀, pale yellow.

**H. glabra** (Ait.) Haw. (Pl. 170/1) (*Mes. g.* Ait., *Mes. pomeridianum* L. Bol., *H. stephensiae* N. E. Br.). – Cape: Cape, Malmesbury, Tulbagh D. – **St.** weak, prostrate; **L.** on a long petiole, spatulate-lanceolate, green, smooth; **Fl.** glossy, straw-coloured.

**Imitaria** N. E. Br. (§ IV/1/16). – Very succulent plants, **L.** united to form a Bo. – Greenhouse, warm. Propagation: seed, cuttings.

**I. muirii** N. E. Br. (Pl. 170/3) (*Gibbaeum nebrownii* Tisch.). – Cape: Lit. Karroo. – Acauline, clump-forming; **Bo.** 2–8 together, 12–15 mm tall, 10–22 mm across, 8–16 mm thick, with a Fi. running across the entire width, lobes convex, soft, fleshy, the surface smooth, with scarcely perceptible hairs (lens needed!), deep greyish-green or brownish, the lobe-tips ± translucent; **Fl.** solitary, c. 2 cm ⌀, pink, Oct.

**Jacobsenia** L. Bol. et Schwant. (§ IV/1/4). – Compact ♄ with sterile **St.** crowded at the base, flowering shoots erect, ± elongated in age, herbaceous parts minutely papillose, glabrous or velvety; the ☉ shoots bear 2 L.-pairs; **L.** ± cylindrical, ± unequal; **Fl.** pedicellate, up to c. 9 cm ⌀, white or lemon-yellow. – Greenhouse, warm. Propagation: seed, cuttings.

**J. hallii** L. Bol. – Cape: Van Rhynsdorp D. – Robust, branched ♄ up to 15 cm tall, 30 cm ⌀; older **St.** up to 12 mm thick, with short shoots, the Int. enclosed within the L.-Sh.; **L.** flat on the upperside, convex on the sides and the underside, the latter bluntly keeled above, obtuse at the tip, (4–)5(–7) cm long, 1 cm across and thick; **Fl.** up to 8.8 cm ⌀, white to lemon-yellow.

**J. kolbei** (L. Bol.) L. Bol. et Schwant. (Pl. 170/2) (*Mes. k.* L. Bol., *Drosanthemum k.* hort.). – Cape: Van Rhynsdorp D. – Erect ♄; **St.** 24 cm or more long, stiff, Int. 3–5 cm long; **L.** erect, ± cylindrical, somewhat flat on the upperside, obtuse or ± acute at the tip, densely and minutely papillose, green, 2–3.5 cm long, 8–9 mm thick; **Fl.** up to 1.5 cm ⌀, white.

**Jensenobotrya** Herre (§ IV/1/12). – Clump-forming, prostrate ♄ with numerous ± spherical **L.** – Greenhouse, warm. Propagation: seed, cuttings.

**J. lossowiana** Herre (Pl. 170/4). – S.W.Afr.: Lüderitz D. – **St.** very thick; **Br.** short, woody; **shoots** soft, eventually covered with dry old L., Int. 2–3 mm long; **L.** 4–6 on a shoot, opposite, united at the base and 5–7 mm ⌀, swollen towards the tip and 12–15 mm ⌀, ± spherical, mostly reddish, smooth, older L. rather wrinkled; **Fl.** on a pedicel 1 cm long, 2–2.5 cm ⌀, pale pink.

**Juttadinteria** SCHWANT. (§ IV/1/13). – Extremely succulent sub-♄ or caespitose or clump-forming plants; **roots** woody; **L.** decussate, very thick to semi-ovoid or broadly navicular, very short or broad linear or ± triangular or also rhomboidal-spatulate, connate and often vesicularly inflated, trigonous towards the apex, margins and keel rather acute or rounded and often tuberculate or dentate, the lower surface often prolonged forward like a chin, surface rather firm, whitish-grey, light yellow-green, bluish to whitish-green; **Fl.** shortly pedicellate or nearly sessile, medium or large in size, white, lilac-red or violet, Aug. – Greenhouse, warm. Propagation: seeds, cuttings.

### Key to the Species of the Genus Juttadinteria by M. Walgate

1. L. with a fine coat of projecting tubercles, just perceptible to the touch . . . . . . . . . . . . . . . 2.
1a. L. with tubercles convex, flattish or humped, usually smooth-textured . . . . . . . . . . . . . . . 3.
2. Pet. white, c. 11 mm long . . . . . . . . . . . . . . . . . . . . . . . . . . . . . . . . . . . . . . . . . . . . . . . **kovisimontana**
2a. Pet. pink, c. 22 mm long . . . . . . . . . . . . . . . . . . . . . . . . . . . . . . . . . . . . . . . . . . . . . . . **longipetala**
3. L. 2–4 cm long; longer Sep. to 15 mm long . . . . . . . . . . . . . . . . . . . . . . . . . . . . . . . . . . . 4.
3a. L. 5–8 cm long; longer Sep. 15–20 mm long . . . . . . . . . . . . . . . . . . . . . . . . . . . . . . . . . **albata**
4. L.-margin entire . . . . . . . . . . . . . . . . . . . . . . . . . . . . . . . . . . . . . . . . . . . . . . . . . . . . . . . 5.
4a. L.-margin toothed . . . . . . . . . . . . . . . . . . . . . . . . . . . . . . . . . . . . . . . . . . . . . . . . . . . . . 9.
5. L. ± obtuse, ± 2½ times as long as broad . . . . . . . . . . . . . . . . . . . . . . . . . . . . . . . . . . . 6.
5a. L. ± acute, more than 3 times as long as broad . . . . . . . . . . . . . . . . . . . . . . . . . . . . . . **attenuata**
6. L. almost ovoid, upper surface convex . . . . . . . . . . . . . . . . . . . . . . . . . . . . . . . . . . . . . **deserticola**
6a. L. keeled, upper surface flat . . . . . . . . . . . . . . . . . . . . . . . . . . . . . . . . . . . . . . . . . . . . . 7.
7. Pet. 1.8–2.5 mm broad; Fl. subsessile . . . . . . . . . . . . . . . . . . . . . . . . . . . . . . . . . . . . . . 8.
7a. Pet. 0.75–1 mm broad; Ped. 10–20 mm long . . . . . . . . . . . . . . . . . . . . . . . . . . . . . . . . **decumbens**
8. L. dorsally rounded, keel inconspicuous . . . . . . . . . . . . . . . . . . . . . . . . . . . . . . . . . . . . **insolita**
8a. L. laterally compressed, keel very prominent . . . . . . . . . . . . . . . . . . . . . . . . . . . . . . . . **tetrasepala**
9. Stamens pubescent at the base . . . . . . . . . . . . . . . . . . . . . . . . . . . . . . . . . . . . . . . . . . 10.
9a. Stamens glabrous . . . . . . . . . . . . . . . . . . . . . . . . . . . . . . . . . . . . . . . . . . . . . . . . . . . . **ausensis**
10. L. very smooth; Pet. c. 20 mm long . . . . . . . . . . . . . . . . . . . . . . . . . . . . . . . . . . . . . . . **suavissima**
10a. L.-texture rough; Pet. c. 15 mm long . . . . . . . . . . . . . . . . . . . . . . . . . . . . . . . . . . . . . . **simpsonii**

(J. elizae (DTR. et BGR.) L. BOL. is considered a doubtful species and therefore not included in the Key.)

**J. albata** L. BOL. – Cape: Lit. Namaqualand. – St. short, erect; L. 20–25 mm long, scarcely convex on the upperside, 1 cm across below, somewhat expanded and trigonous-tapered above, acutely carinate and broadly compressed on the underside, the surface smooth, whitish to greyish-green with scattered dots, the keel and margins slightly reddish, with the dots merging into a line; Fl. c. 3.5 cm ⌀, white.

**J. attenuata** WALGATE. – S.W.Afr.: Gr. Namaqualand. – Up to 10 cm tall; St. creeping, Br. erect, with 6–8 mostly cylindrical L., with the margins and keel indistinct above, up to 3 cm long, 8 mm across and thick, dark greyish-green, smooth, with red margins and tip; Fl. 3 cm ⌀, white.

**J. ausensis** (L. BOL.) SCHWANT. (*Mes. a.* L. BOL.). – S.W.Afr.: Aus. – L. ovoid-spatulate, the keel on the underside above with 3–6 T., 3 cm long, 13–15 mm across, 10 mm thick, bluish-green; Fl. 43 mm ⌀, white.

**J. decumbens** SCHICK et TISCH. – Cape: Lit. Namaqualand. – Mat-forming; Br. prostrate; L. united for 1/3 of their length, 15–30 mm long, 10–20 mm across, 10–15 mm thick, noticeably convex on the underside and broadly keeled above, with an acute tip, the surface smooth, whitish-green; Fl. 2.5–3 cm ⌀, white.

**J. deserticola** (MARL.) SCHWANT. (*Mes. d.* MARL.). – S.W.Afr.: Namib. – St. short, erect, densely leafy; L. almost round to tapered, with the underside semicylindrical and a scarcely perceptible keel pulled conspicuously forward over the tip, 10–12 mm thick, the surface smooth, whitish to greyish-green, with scattered dots along the angles; Fl. 18 mm ⌀, white.

**J. elizae** (DTR. et BGR.) L. BOL. (*Mes. e.* DTR. et BGR.). – S.W.Afr.: N. of Aus. – St. short, creeping, densely leafy; L. oblong to rhomboid, with dentate margins and keel, whitish, spotted; Fl. white.

**J. insolita** (L. BOL.) L. BOL. (*Mes. i.* L. BOL.). – Cape: Lit. Namaqualand. – St. very short, 7 mm thick; Br. prostrate, c. 9 cm long, 6 mm thick, erect; L. 2–5, dense, ± navicular, with an indistinct, ± lateral keel on the underside, obtuse, pale blue, minutely and densely papillose, 15–27 mm long, 10 mm across and thick; Fl. 32 mm ⌀, whitish.

**J. kovisimontana** (DTR.) SCHWANT. (*Mes. k.* DTR.). – S.W.Afr.: Kovis Mts. – Forming cushions 20 cm ⌀; St. 6–10 cm long; L. 15 to 22 mm long, 10–18 mm thick, 10 mm across below, with a trigonous tip-area 2 cm across at its base, flat tubercles along its margins and a very blunt keel, the surface whitish-grey and densely acute-granulate; Fl. c. 22 mm ⌀, white.

**J. longipetala** L. BOL. (Pl. 171/1) (*Mes. ponderosum* DTR. subnud., *Namibia p.* (DTR.) DTR. et SCHWANT.). – S.W.Afr.: Halenberg. – St. dense, short: L. 4 on a Br., crowded, thick, with a rounded keel, narrowed towards the tip,

4 cm long, 2 cm thick above, 17 mm across, bluish-green to pink; **Fl.** 47 mm $\varnothing$, pale pink.

**J. simpsonii** (DTR.) SCHWANT. (Pl. 171/5) (*Mes. s.* DTR.). – S.W.Afr.: S.Namib. – Sub-♄ with the ascending, densely leafy **St.** forming cushions; **L.** 25–35 mm long, with the upperside 10–12 mm across below, expanded above into a trigonous tip up to 2 cm across and 9 to 15 mm thick, with the keel on the underside pulled laterally forward over the tip and backwardly-directed, acute, short tubercles along the keel and the margins, with reddish **T.** on the upper part of the upperside and the sides, the surface slightly rough, light bluish-green with scarcely perceptible dots; **Fl.** c. 3.5 cm $\varnothing$, white.

**J. suavissima** (DTR.) SCHWANT. (*Mes. s.* DTR.). – S.W.Afr.: S.Namib. – **St.** prostrate or ascending, up to 30 cm tall; **L.** obtusely trigonous, 2–4 cm long, 10–15 mm thick, with the tip ± recurved and several blunt **T.** on the sides above, the surface smooth, light greyish-green, with the scarcely perceptible dots; **Fl.** 4 to 5 cm $\varnothing$, pure white, strongly scented.

**J. tetrasepala** L. BOL. (Pl. 171/3). – Cape: Lit. Namaqualand. – Erect, c. 8 cm tall; **St.** prostrate; **L.** blistered and swollen below, narrowed towards the short-tapered or obtuse tip, keeled on the underside, 15–20 mm long, 9 mm across, 10 mm thick, the surface bluish, with a few dots; **Fl.** 28 mm $\varnothing$, whitish.

**Kensitia** FEDDE (§ IV/1/8). – Shrubby, succulent plants with opposite **L.**; **Fl.** with petaloid staminodes. – Summer: in the open, winter: cold-house. Propagation: seed, cuttings.

**K. pillansii** (KENSIT) FEDDE (Pl. 171/2) (*Mes. p.* KENSIT, *Piquetia p.* (KENSIT) N. E. BR., *P. p.* (KENSIT) SCHWANT.). – Cape: Piquetberg D. – Erect, much-branched ♄ 30–60 cm tall; **St.** forked, reddish; **L.** acutely trigonous, 28 to 33 mm long, 5–7 mm across, 7–10 mm thick, slightly incurved, acute, apiculate, bluish-green; **Fl.** solitary, on a pedicel 8–10 mm long, c. 3.5 cm $\varnothing$, with numerous Pet., the outer ones up to 15 mm long, spatulate, with a white claw and with a purplish-pink oval to ovate blade, the innermost Pet. filiform.

**Khadia** N. E. BR. (§ IV/1/9). – Low ⚃ herbs; **St.** numerous, short, arising from a fleshy rootstock; **L.** opposite, connate at the base, curved, semicylindrical below, trigonous at the apex, smooth, set with transparent dots; **Fl.** solitary, pedicellate, white to pink. – Greenhouse, warm. Propagation: seed, cuttings.

## Key to the Species of the Genus Khadia by L. Bolus

1. Sep. usually 5; placental tubercle absent or not recorded ............................ 2.
1a. Sep. 6; placental tubercles present ................................................ 5.
2. Compact plants with Int. enclosed in the L.-Sh. .................................... 3.
2a. All or some of the St. elongated ................................................... 4.
3. L. in side view abruptly tapering to the apex ........................... **acutipetala**
3a. L. in side view rounded or subtruncate at the apex, or one of a pair acute ........ **nationae**
4. Int. enclosed in L.-Sh. or shortly exserted; stigmas 6 ........................... **nelsoniae**
4a. Int. sometimes 2–3 cm long; stigmas 6–8 ......................................... **borealis**
5. Filaments epapillate; stigmas well exceeding the outer stamens .............. **carolinensis**
5a. Inner filaments papillate; stigmas falling short of the outer stamens ............. **beswickii**

**K. acutipetala** (N. E. BR.) N. E. BR. (*Mes. a.* N. E. BR.). – Transvaal: near Johannesburg. – 4–6 cm tall; **St.** 1–2 cm long with 2–6 **L.** 10 to 22 mm long, 3–8 mm across below, 2–3 mm thick, long-tapered towards the apiculate tip, rounded on the underside, smooth, greyish-green to brownish, with translucent, horny dots along margins and keel; **Fl.** c. 4 cm $\varnothing$, light pinkish-purple.

**K. beswickii** (L. BOL.) N. E. BR. (Pl. 171/4) (*Mes. b.* L. BOL.). – Transvaal: near Johannesburg. – **Rootstock** thickened; **St.** short; **L.** unequal, trigonous, acute above, some 25 mm long, 10 mm across below, others 3.5 cm long, 7–8 mm across; **Fl.** 3.5 cm $\varnothing$, white.

**K. borealis** L. BOL. – Transvaal: Soutpansberg D. – **L.** 4 on a shoot, narrowly tapered, keeled on the underside, bluish-green with translucent dots; **Fl.** c. 2 cm $\varnothing$, pale pink.

**K. carolinensis** (L. BOL.) L. BOL. (*Mes. c.* L. BOL., *Rabiea c.* (L. BOL.) N. E. BR.). – Transvaal: Carolina. – **L.** 4–6 on a shoot, swollen-trigonous, bluish-green with green dots, 6 mm thick midway; **Fl.** c. 4 cm $\varnothing$, white.

**K. nationae** (N. E. BR.) N. E. BR. (*Mes. n.* N. E. BR.). – Transvaal: near Rustenburg. – 5–7 cm tall; **L.** 5–6.5 cm long, 2.5–4.5 mm across, flat on the upperside, with very rounded sides; **Fl.** 2.5 cm $\varnothing$, pink or purplish-pink.

**K. nelsoniae** N. E. BR. – Transvaal: Witte Koppies. – Resembles **K. acutipetala**; **St.** elongated; **L.** somewhat longer, 2.5–4 cm long.

**Lampranthus** N. E. BR. (§ IV/1/3). – Sub-♄; **St.** erect or spreading or prostrate, compressed, glabrous; **L.** shortly connate at the base, numerous, terete or 3-angled, blunt or tapering, ± curved; **Fl.** solitary or several together, terminal or axillary, large or medium-sized, white, pink, red, purple-rose, orange-coloured or yellow, summer. – Summer: in the open; winter: cold-house. Propagation: seed, cuttings.

## Division of the Genus Lampranthus into Sections by G. Schwantes
(More study is needed to arrange all the spec. into the Sect.)

**Sect. 1. Adunci** (SALM) SCHWANT. (*Adunca* SALM as a SSect. of *Mesembryanthemum* L.). – **St.** slender, bent, erect or decumbent; **L.** ± crowded at the end of the Br., semiterete, subulate, somewhat curved inwards, ± bent backwards at the tip; **Fl.** solitary, stalked, small to medium-large, reddish, Pet. arranged in 1 series. – Type species: L. aduncus. – Further species: L. calcaratus, curvifolius, filicaulis, inconspicuus, spiniformis and others.

**Sect. 2. Scabrida** (HAW.) SCHWANT. (*Scabrida* HAW. as a SSect. of *Mesembryanthemum* L.). – Much-branched ♄ with slender, thin **St.** and numerous Br.; **L.** crowded, not large, 2–3 cm long, rather thin, ± spreading, a little united at the base, rough from elevated dots; **Fl.** stalked, 3–5 together, red or violet-pink, staminodes often present. – Type species: L. scaber. – Furtcer species: L. brownii, elegans, emarginatus, glomeratus and others.

**Sect. 3. Lampranthi** (*Tenuifolia* SALM as a SSect. of the G. *Mesembryanthemum* L.). – **St.** numerous, tall, erect or ± spreading, with numerous **Br.**; **L.** linear, trigonous-terete or somewhat depressed, mostly dotted, 2–4 cm long; **Fl.** 3 together, seldom solitary, showy, large, yellow or red, stalks ± long, with bracts. – Type species: L. tenuifolius. – Further species: L. bicoloratus, stenus, variabilis and others.

**Sect. 4. Corallini** (HAW.) SCHWANT. (*Corallina* HAW. as a SSect. of the G. *Mesembryanthemum* L.). – Erect, decussately branched ♄; **L.** trigonous-terete, ± subulate and acuminate, mostly grey, pulverulent and dotted, 4–5 cm long; **Fl.** solitary or rarely 2–3 together (then pale pink to white), large, red, Pet. arranged in several R., the inner Pet. staminodial, stalks long, with bracts. – Type species: L. coralliflorus. – Further species: L. haworthii, productus, stipulaceus, zeyheri and others.

**Sect. 5. Amoeni** (SALM) SCHWANT. (*Amoena* SALM, *Conferta* DC. p.part. as SSect. of the G. *Mesembryanthemum* L.). – ♄ with short, decumbent or elongated erect **St.**; **L.** densely crowded, somewhat united at the base, trigonous, elongated, gradually acuminate and narrowed; **Fl.** long stalked, 1–3 together, large, red. – Type species: L. amoenus. – Further species: L. conspicuus, formosus, spectabilis and others.

**Sect. 6. Blandi** (HAW.) SCHWANT. (*Blanda* HAW., *Conferta* DC. p.part. as SSect. of the G. *Mesembryanthemum* L.). – Much branched ♄ with erect, elongated **St.**; **L.** united at the base, trigonous, acuminate, smooth; **Fl.** mostly 3 together, long-stalked, large, white or reddish, Pet. arranged in several series. – Type species: L. blandus. – Further species: L. turbinatus and others.

**Sect. 7. Aurei** (HAW.) SCHWANT. (*Aurea* HAW. as a SSect. of the G. *Mesembryanthemum* L.). – **St.** erect; **L.** a little united at the base, spreading, trigonous, obtuse; **Fl.** solitary, long-stalked, large, yellow or reddish-yellow. – Type species: L. aureus. – Further species: L. aurantiacus, glaucus and others.

### Survey of the G. Lampranthus according to the Flower-Colour of the Species

Since the shrubby species of Lampranthus are valued as ornamental bedding plants in subtropical gardens, the following list may be of use.

**Pet. white:** L. albus, candidus, densipetalus, dregeanus, dunensis, holensis, macrostigma, mucronatus, praecipitatus, productus, rubroluteus, rupestris, stephanii, watermeyeri.

**Pet. white to pink or bright pink:** L. amabilis, antonii, borealis, calcaratus, dilutus, dulcis, ebracteatus, framesii, galpiniae, intervallaris, lavisii, leipoldtii, lunulatus, middlemostii, monticolus, multiradiatus, polyanthon, productus, roseus, spiniformis, steenbergensis, stenopetalus, stenus, subaequalis, verecundus, watermeyeri.

**Pet. silvery-pink:** L. argenteus, martleyi, simulans.

**Pet. pink to salmon-rose:** L. affinis, arenosus, austricolus, berghiae, blandus, caespitosus, caudatus, cedarbergensis, ceriseus, comptonii v. roseus, convexus, copiosus, creber, edwardsiae, excedens, falcatus, falciformis, franciscii, hallii, hiemalis, immelmaniae, lavisii, leptaleon, lewisiae, liberalis, littlewoodii, longisepalus, macrosepalus, maturus, meleagris, microsepalus, multiseriatus, nardouwensis, nelii, obconicus, occultans, ornatus, paardebergensis, paarlensis, pakhuisensis, parcus, pauciflorus, peacockiae, perreptans, piquetbergensis, primiversus, recurvus, schlechteri, suavissimus, sublaxus, subtruncatus, superans, tegens, thermarus, turbinatus, vernalis, vernicolor, viatorus, virgatus.

**Pet. pink-purple or purplish pink:** L. acutifolius, aestivus, arbuthnotiae, arenicolus, argillosus, brevistamineus, capillaceus, cyathiformis, densifolius, diutinus, egregius, elegans, ernestii, esterhuyseniae, eximius, filicaulis, foliolosus, fugitans, furvus, globosus f. glomeratus, godmaniae, gracilipes, haworthii, henricii, hoerleiniasus, hollandii, laetus, laxifolius, leptosepalus, macrocarpus, mariae, microstigma, montaguensis, neostayneri, paucifolius, peersii, persistens, plautus, pleniflorus, polyanthon, prasinus, productus v. purpureus, prominulus, proximus, purpureus, rabiesbergensis, salicolus, saturatus, sociorum, stayneri, stoloniferus, subrotundus, tenuis, vanheerdei, vredenburgensis, walgatae.

**Pet. violet-pink or purple-violet:** L. emarginatus, zeyheri.

**Pet. lilac:** L. dependens.

**Pet. purple to purplish red:** L. amoenus, conspicuus, globosus, spectabilis, staminodiosus, stipulaceus, villersii, watermeyeri.

**Pet. red:** L. aduncus, coccineus, coralliflorus, inconspicuus, incurvus, tenuifolius.

**Pet. copper-red:** L. magnificus.

**Pet. yellowish salmon-red:** L. antemeridianus.

**Pet. orange:** L. aurantiacus, aureus, brownii, fergusoniae, glaucoides, vanzijliae.

**Pet. golden-yellow:** L. acrosepalus, hurlingii, longistamineus, marcidulus, matutinus, palustris, serpens, sternens.

**Pet. yellow:** L. baylissii, debilis, explanatus, glaucus, inaequalis, promontorii, reptans, variabilis, vereculatus.

**Pet. lemon-yellow:** L. citrinus.

**Pet. bicoloured:** surface pink or red, underside copper-red: L. curvifolius, sauerae; surface yellow, underside red or purple; L. bicolor; surface yellow, underside paler: L. inaequalis; surface purplered, underside reddish yellow; L. salteri; yellow with the tips pink: L. mutans, swartbergensis; white with the tips purplish pink: L. algoensis, formosus, suavissimus v. oculatus, vanputtenii; red with the tips yellow: L. woodburniae.

**Colour of Pet. unknown:** L. altistylus, deflexus, diffusus, leightoniae, rustii, tulbaghensis, vallisgratiae, woodsworthii.

**L. acrosepalus** (L. Bol.) L. Bol. (*Mes. a.* L. Bol.). – Cape: Piquetberg D. – 18–25 cm tall, erect, with virgate **St.; L.** 17–30 mm long, 1–2 mm across, 2 mm thick; **Fl.** solitary, almost 6 cm ⌀, golden-yellow.

**L. acutifolius** (L. Bol.) N. E. Br. (*Mes. a.* L. Bol.). – Cape: Malmesbury D. – Slender, c. 12 cm tall, densely branched; **L.** ± falcate, 10 mm long, 2 mm ⌀, with large dots; **Fl.** 3 cm ⌀, pinkish-purple.

**L. aduncus** (Haw.) N. E. Br. (*Mes. a.* Haw.). – Cape: Cape D. – Low-growing, with curved **St.; L.** ± cylindrical, 15–20 mm long, 2 mm thick, fresh green, with minute, translucent dots; **Fl.** c. 18 mm ⌀, red.

**L. aestivus** (L. Bol.) L. Bol. (*Mes. a.* L. Bol.). – Cape: Caledon D. – Erect, 25 cm tall, with stiff, virgate **St.; L.** ± falcate, 10 to 15 mm long, 3.5 mm thick at the obtuse tip, bluish-green, spotted; **Fl.** 28 mm ⌀, pinkish-purple.

**L. affinis** L. Bol. (*L. a.* forma L. Bol.). – Cape: Prince Albert D. – Erect, c. 20 cm tall or forming cushions; **L.** semicylindrical, 2.5 cm long. 1.5–2 mm ⌀, bluish-green; **Fl.** solitary or in a bipartite Infl., Pet. an attractive pale pink, glossy.

**L. albus** (L. Bol.) L. Bol. (*Mes. a.* L. Bol.). – Cape: Calvinia D. – Erect, 20 cm tall; **L.** falcate, broadest below the tip, 2 cm long, 3 mm across, 6 mm thick, pale blue to whitish; **Fl.** 14 mm ⌀, white.

**L. algoensis** L. Bol. – Cape: near Port Elizabeth. – Erect, stiff, 35 cm tall; **L.** 1–1.4 cm long, 1–2 mm ⌀, apiculate, dark green, with the tip red and recurved; **Fl.** c. 3 cm ⌀, red.

**L. altistylus** N. E. Br. (*Mes. longistylum* L. Bol.). – Cape: Malmesbury D. – Slender **St.** prostrate to creeping; **L.** 4 mm long, 2 mm thick, flat on the upperside, rounded on the sides, green, minutely papillose; **Fl.** solitary, 19 mm ⌀.

**L. amabilis** L. Bol. – Cape: Bredasdorp, Riversdale D. – Laxly branched, erect, up to 30 cm tall; **L.** 1.5–2.3 cm long, 3 mm ⌀, shortly tapered, green; **Fl.** 3.5–5.3 cm ⌀, white or pale pink, more rarely brilliant red.

**L. amoenus** (Salm) N. E. Br. (*Mes. a.* Salm). – Cape: Malmesbury D. – Resembles **L. conspicuus; St.** shorter; **L.** 4 cm long, cylindrical to trigonous, smooth; **Fl.** 3 together, 3–5.4 cm ⌀, a magnificent purple colour.

**L. antemeridianus** (L. Bol.) L. Bol. (*Mes. a.* L. Bol., *Mes. a.* v. *fl. pleno* L. Bol.). – Cape: Riversdale D. – With lax, prostrate **St.** 16 to 20 cm long; **L.** 2.2–5 cm long, 2 mm ⌀, with convex dots on the sides; **Fl.** 4.5 cm ⌀, yellowish to salmon-pink to pink.

**L. antonii** L. Bol. – Cape: Ceres D. – Matforming; **St.** creeping, 5–12 cm long; **L.** linear, acute, apiculate, 2–5 cm long, 2–6 mm ⌀; **Fl.** solitary, 5–6.6 cm ⌀, white to pale pink.

**L. arbuthnotiae** (L. Bol.) L. Bol. (*Mes. a.* L. Bol.). – S.W.Cape. – Slender, with elongated, creeping, rooting **St.; L.** trigonous, 8 mm long, 1 mm ⌀, acute, with raised dots; **Fl.** solitary, 14–16 mm ⌀, pinkish-purple, with a paler centre.

**L. arenicolus** L. Bol. – Cape: Laingsburg D. – Compact, c. 35 cm tall, c. 40 cm ⌀, with the **rootstock** 2.3 cm ⌀ above; **L.** 7–10 mm long, 1 mm across, 1.5 mm thick, hooked and recurved at the tip; **Fl.** in a tripartite Infl., 1.5 cm ⌀, purplish-pink.

**L. arenosus** (L. Bol.) L. Bol. (*Mes. a.* L. Bol.). – Cape: Malmesbury D. – Laxly branched, 25 cm tall, often ± climbing or with creeping **St.; L.** 1.5–2 cm long, 1.5–2.5 mm across, 2–3 mm thick, short-tapered, pale green; **Fl.** 4.5 cm ⌀, pink.

**L. argenteus** (L. Bol.) L. Bol. (*Mes. a.* L. Bol.). – Cape: Clanwilliam D. – c. 23 cm tall, with widespread, stiff **St.; L.** c. 2 cm long, 3–4.5 mm across midway, bluish-green; **Fl.** 3 together, 1.2 cm ⌀, pale pink.

**L. argillosus** L. Bol. (*Mes. tulbaghense* L. Bol., *L. t.* (L. Bol.) N. E. Br.). – Cape: Clanwilliam D. – Robust, 20 cm tall, with stiff **St.; L.** 8–13 mm long, 2 mm across, 2.5 mm thick, obtuse, dirty green; **Fl.** 18 mm $\varnothing$, purplish-pink.

**L. aurantiacus** (DC.) Schwant. (*Mes. a.* DC., *Mes. aurantium* Haw., ? *Mes. glaucoides* Haw., ? *L. g.* (Haw.) N. E. Br.). – Cape: Cape D. – Up to 45 cm tall, sparsely branched; **L.** 2–3 cm long, 4 mm across, ± tapered, apiculate, green, grey-pruinose, minutely rough, spotted; **Fl.** 4–5 cm $\varnothing$, orange-coloured.

**L. aureus** (L.) N. E. Br. (*Mes. a.* L.). – Cape: Cape D. – 30–40 cm tall, with erect **St.; L.** 5 cm and more long, mucronate, fresh green, slightly grey-pruinose, minutely spotted; **Fl.** 6 cm $\varnothing$, glossy, deep orange.

**L. austricolus** (L. Bol.) L. Bol. (*Mes. a.* L. Bol.). – Cape: Cape D. – 15–30 cm tall, erect, with virgate **St.; L.** falcate, 10–12 mm long, 2 mm across, 2.5 mm thick, tapered or obtuse, bluish-green, rough; **Fl.** 28 mm $\varnothing$, pale pink.

**L. baylissii** L. Bol. – Cape: Knysna D. – Spreading, up to 50 cm $\varnothing$, 10–15 cm tall; **L.** ± falcate, c. 2 cm long, 1.5 mm across, 2.5 mm thick below, linear, acute; **Fl.** solitary, up to 2 cm $\varnothing$, vivid yellow.

**L. berghiae** (L. Bol.) L. Bol. (*Mes. b.* L. Bol.). – Cape: Clanwilliam D. – 50 cm tall, with erect **Br.; L.** 2–3 cm long, 2–2.5 mm across midway, 3–4 mm thick, obtuse to short-tapered, bluish-green; **Fl.** 4 cm $\varnothing$, pink to pale pink.

**L. bicolor** (L.) N. E. Br. v. **bicolor** (*Mes. b.* L., *Mes. bicolorus* L., *L. bicolorus* (L.) Jacobs., *Mesembryanthus b.* (L.) Rothm.). – Cape: Cape D. – Up to 30 cm tall; **St.** stiff; **L.** semicylindrical, tapered, trigonous above, 12–25 mm long, 2 mm across, green with translucent dots; **Fl.** c. 3.5 cm $\varnothing$, yellow inside, deep red outside.

**L. — v. inaequalis** (Haw.) Schwant. (*Mes. bicolorum v. i.* Haw.). – Cape: Cape D. – **St.** more prostrate; **Fl.** yellow on both surfaces, with red lines on the Pet.

**L. blandus** (Haw.) Schwant. (*Mes. b.* Haw.). – Cape: Bathurst D. – 30–50 cm tall, with erect **St.**, with a deep red bark; **L.** 3–4 cm long, trigonous, short-tapered, light greyish-green with minute, translucent dots, 3–5 cm long, 3 mm across; **Fl.** 6 cm $\varnothing$, pale pinkish-red.

**L. borealis** L. Bol. – S.W.Afr.: Witpütz. – Erect, glabrous $\hbar$; **St.** 9–18 cm long, Int. up to 2.5 cm long; **L.** variable, lower ones shortly decurrent below, obtuse to obliquely truncate above, lower ones c. 2.4 cm long, 3 mm across, up to 5 mm thick, upper L. thinner; **Fl.** solitary, 4–5 cm $\varnothing$, pale pink.

**L. brevistamineus** (L. Bol.) L. Bol. (*Mes. b.* L. Bol.). – Cape: Ceres D. – Slender, ± spherical, 14 cm tall; **St.** ± contorted; **L.** cuneate to sabre-shaped, short-tapered, 7 mm long, 2 mm $\varnothing$, bluish-green, spotted; **Fl.** 3 cm $\varnothing$, purplish-pink.

**L. brownii** (Hook. f.) N. E. Br. (*Mes. b.* Hook. f.). – Cape. – Very freely branched, 20–30 cm tall; **St.** slender, with numerous short shoots; **L.** 8–10 mm long, semicylindrical, greyish-green, red-mucronate with slightly raised dots; **Fl.** 2 cm $\varnothing$, orange-red, yellow outside, both surfaces becoming light red.

**L. caespitosus** (L. Bol.) N. E. Br. (*Mes. c.* L. Bol.). – Cape: Paarl D. – Mat-forming, herbaceous; **St.** creeping; **L.** acutely trigonous, 1 cm long, 2 mm $\varnothing$, bluish; **Fl.** 10–22 mm $\varnothing$, pink.

**L. — v. luxurians** (L. Bol.) Jacobs. (*Mes. c. v. l.* L. Bol.). – Cape: Worcester D. – **L.** 15 mm long, 4 mm thick, 3 mm across midway; **Fl.** 18 mm $\varnothing$.

**L. calcaratus** (Wolley Dod) N. E. Br. (*Mes. c.* Wolley Dod). – Cape: Cape D. – **St.** ascending; **L.** cylindrical, subulate, 4–8 mm long, faintly spotted, with a short spur below; **Fl.** numerous, 8 mm $\varnothing$, light pink.

**L. candidus** L. Bol. – Cape: Piquetberg D. – Laxly branched, slender, 25–30 cm tall; **L.** ± falcate, 6 cm long, 15 mm across, 2 mm thick, blue, rough; **Fl.** 2 cm $\varnothing$, white.

**L. capillaceus** (L. Bol.) N. E. Br. (*Mes. c.* L. Bol.). – Cape: near Tulbagh. – 14 cm tall; **St.** and pedicels hirsute; **L.** ± falcate, short-tapered, 5 mm long, 1.5 mm across, 2 mm thick, bluish; **Fl.** 2 cm $\varnothing$, purplish-pink, paler on the underside.

**L. caudatus** L. Bol. – Cape: Bredasdorp D. – Erect, robust, stiff, c. 30 cm tall; **L.** 9–14 mm long, 1.5 mm $\varnothing$, shortly and distinctly tapered, bluish-green; **Fl.** 2 cm $\varnothing$, pink to pale pink.

**L. cedarbergensis** (L. Bol.) L. Bol. (*Mes. c.* L. Bol.). – Cape: Clanwilliam D. – Erect, 15 cm tall; **L.** 15–20 mm long, 4 mm across, 5–6 mm thick, acute or obtuse above; **Infl.** of 3–7 pink Fl.

**L. ceriseus** (L. Bol.) L. Bol. (*Mes. c.* L. Bol.). – Cape: Bredasdorp D. – Erect, laxly branched, 11 cm tall; **L.** 2 cm long, 3.5 mm $\varnothing$, acutely keeled on the underside, bluish-green; **Fl.** 3.5 cm $\varnothing$, an attractive pink.

**L. citrinus** (L. Bol.) L. Bol. (*Mes. c.* L. Bol.). – Cape: Malmesbury D. – Laxly branched, 12 cm tall; **L.** tapered, apiculate, 10–15 mm long, 1.5 mm across, 2 mm thick, green, spotted; **Fl.** 48 mm $\varnothing$, lemon-yellow.

**L. coccineus** (Haw.) N. E. Br. (*Mes. c.* Haw., *Mes. bicolor* Curtis). – Cape: Cape D. – 60–90 cm tall, with erect **St.; L.** crowded on the short shoots, trigonous, compressed, rather obtuse, 15–25 mm long, 2 mm across, dull grass-green; **Fl.** up to 4 cm $\varnothing$, intense red.

**L. compressus** L. Bol. – Cape: Clanwilliam D. – Laxly branched, 18 cm tall, 60 cm $\varnothing$; **L.** in unequal pairs, ± lunate, 2.2 cm long, 3–4 mm across midway, 11 mm across at the tip, 7 to 9 mm thick; **Infl.** irregularly tripartite, **Fl.** 2.5 cm $\varnothing$, pale pink.

**L. comptonii** (L. Bol.) N. E. Br. v. **comptonii** f. **comptonii** (*Mes. c.* L. Bol.). – Cape: Van Rhynsdorp D. – Laxly branched; **St.** 25–35 cm long; **L.** falcate, swollen, trigonous, truncate above, with a red apiculus, 4 cm long, 6 mm across, 1 cm thick, green; **Fl.** 27 mm $\varnothing$, white inside, faintly pink outside.

**L. — v. — f. roseus** (L. Bol.) Rowl. (*Mes. c. v. r.* L. Bol., *L. c. v. r.* (L. Bol.) L. Bol.). – Cape: Clanwilliam D. – **Fl.** pink.

**L. — v. angustifolius** (L. Bol.) L. Bol. (*Mes. c. v. a.* L. Bol.). – Cape: Lit. Namaqualand. – **L.** 3 mm across, 5 mm thick.

**L. conspicuus** (Haw.) N. E. Br. (*Mes. c.* Haw.). – Cape: ? Albany D. – Up to 45 cm tall; **St.** almost 2 cm thick; **Br.** curved; **L.** curved inwards, 6–7 cm long, 4–5 mm across, tapered, semicylindrical, green with a red mucro, often spotted; **Fl.** 5 cm ⌀, purplish-red.

**L. convexus** (L. Bol.) L. Bol. (*Mes. c.* L. Bol.). – Cape: Tulbagh D. – Erect, 23 cm tall; **St.** 1 cm thick below, **Br.** stiff; **L.** bluntly keeled on the underside, acute, 15–27 mm long, 2 mm across, 3 mm thick, pale green; **Infl.** much branched, Fl. 17 mm ⌀, pink.

**L. copiosus** (L. Bol.) L. Bol. (*Mes. c.* L. Bol.). – Cape. – Spreading; **St.** prostrate; **L.** laterally compressed, ± falcate, 15–20 mm long, 3 mm across, 6 mm thick; **Fl.** 26 mm ⌀, pink, open day and night.

**L. coralliflorus** (Salm) N. E. Br. (*Mes. c.* Salm, *L. c.* (Salm) Schwant., *Erepsia c.* (Salm) Schwant., *Mes. corallinum* Haw.). – Cape. – 60–90 cm tall, much branched; **Br.** opposite and decussate, rust-coloured; **L.** ± cylindrical, clavately thickened above, rather obtuse, c. 5 cm long, 3–4 mm across, fresh green, grey-pruinose, spotted; **Fl.** solitary, 5 cm ⌀, brilliant red.

**L. creber** (L. Bol.) – Cape: Riversdale D. – 35 cm tall; **St.** 1 cm ⌀; **Br.** reddish-brown; **L.** densely crowded, 6–9 mm long, 1–1.5 mm ⌀, acute, rough; **Fl.** 2.5 cm ⌀, pink.

**L. curviflorus** (Haw.) N. E. Br. (*Mes. c.* Haw., *Mes. blandum* v. *c.* (Haw.) Bgr.). – Cape: Albany D. – Resembles **L. blandus**; **L.** 5–7 cm long, much compressed laterally; **Fl.** 6–7 cm ⌀, white, becoming pink.

**L. curvifolius** (Haw.) N. E. Br. v. **curvifolius** (*Mes. c.* Haw., *L. c.* (Haw.) Schwant., *Mes. ceratophyllum* Willd.). – Cape: Cape D. – Low-growing, with curved, spreading, brown **St.**; **L.** crowded at the tips of the Br., semi-cylindrical, curved, 3 cm long, 4 mm thick, green, spotted; **Fl.** solitary, pink, coppery-red outside.

**L. — v. minor** (Salm) Rowl. (*Mes. curvifolium* v. *m.* Salm, *Mes. flexifolium* Haw., *L. f.* (Haw.) N. E. Br., *Mes. aduncum* Willd., *Ruschia willdenowii* Schwant.). – Cape: Cape D. – **L.** shorter and thinner; **Fl.** paler.

**L. cyathiformis** (L. Bol.) N. E. Br. (*Mes. c.* L. Bol., *L. c.* (L. Bol.) Schwant., *Erepsia c.* (L. Bol.) Schwant.). – Cape: origin unknown. – Up to 22 cm tall; **L.** 25–32 mm long, 4 mm ⌀; **Fl.** 44 mm ⌀, pinkish-purple.

**L. debilis** (Haw.) N. E. Br. (*Mes. d.* Haw.). – Cape. – Close to **L. reptans** and possibly identical.

**L. deflexus** (Ait.) N. E. Br. (*Mes. d.* Ait.). – Cape: Cape D. – Insufficiently known; **L.** trigonous, acute, bluish, rough-spotted; Cal.-lobes membranous.

**L. densifolius** (L. Bol.) L. Bol. (*Mes. d.* L. Bol.). – Cape: Malmesbury D. – Erect, with stiff **St.**; **L.** erect, 17–27 mm long, 2–3 mm ⌀, indistinctly keeled, shortly tapered, bluish-green; **Fl.** 5–6 mm ⌀, pink to purple.

**L. densipetalus** L. Bol. (*Mes. d.* L. Bol.). – Cape: Lit. Namaqualand. – Laxly branched, 26 cm tall; older **St.** prostrate, spreading, 30 cm or more long; **L.** laterally compressed, with a blunt keel, obtuse, 2 cm long, 3 mm thick, bluish; **Fl.** 3.5 cm ⌀, Pet. numerous, white.

**L. dependens** (L. Bol.) L. Bol. (*Mes. d.* L. Bol.). – S.W.Cape. – **St.** elongated, prostrate; **L.** cylindrical above midway, acute, 3.5–5.5 cm long, 3–4 mm thick, minutely glossy-papillose, with indistinct white dots; **Fl.** 6 cm ⌀, mauve.

**L. diffusus** (L. Bol.) N. E. Br. (*Mes. d.* L. Bol.). – Cape: Ceres D. – Spreading, branched; **L.** semicylindrical, 1.5–2 cm long, 2 mm ⌀, bluish, somewhat rough; **Fl.** 27 mm ⌀.

**L. dilutus** N. E. Br. (*Mes. pallidum* L. Bol.). – Cape: Piquetberg D. – With widely spreading **St.** 25 cm long; **L.** ± falcate, with a semi-circular tip, 5–7 mm long, 2 mm across, 1.5 to 3 mm thick, bluish-green, rough; **Fl.** 23 mm ⌀, pale pink to white, almond-scented.

**L. diutinus** (L. Bol.) N. E. Br. (*Mes. d.* L. Bol.). – Cape: Riversdale D. – **St.** elongated, rooting at the nodes, narrowly alate, 3.5–8 cm long; **L.** trigonous, acute, with a red apiculus, 2 cm long, 2.6 mm across, 3.5 mm thick, green; **Fl.** 40–48 mm ⌀, pinkish-purple.

**L. dregeanus** (Sond.) N. E. Br. (*Mes. d.* Sond., *Erepsia d.* (Sond.) Schwant.). – Cape: near Tulbagh. – 60 cm tall; **St.** red, ± angular; **L.** compressed-trigonous, ± denticulate; **Fl.** 1–3 together, white.

**L. dulcis** (L. Bol.) L. Bol. (*Mes. d.* L. Bol.). – Cape: Clanwilliam D. – Erect, 45 cm tall, laxly branched; **L.** incurved, falcate, oblique and acute above, 3.5 cm long, 2.5 mm across, 4 mm thick, bluish-green; **Fl.** 4 cm ⌀, pale pink.

**L. dunensis** (Sond.) L. Bol. (*Mes. d.* Sond., *Mes. macrocalyx* Kensit). – Cape: Cape D. – Compact, with prostrate or creeping **St.**; **L.** acutely carinate, 2.5–4.5 cm long, 3.5 mm ⌀; **Fl.** 3.5 cm ⌀, white.

**L. ebracteatus** L. Bol. – Cape: Van Rhynsdorp D. – Erect, c. 25 cm tall; **L.** ± lunate, rounded below the tip, with a ± acute keel, 1–2 cm long, 3–6 mm across midway, 5–7 mm thick; **Fl.** in a bi- or tripartite Infl., Pet. 7–10 mm long, pale pink.

**L. edwardsiae** (L. Bol.) L. Bol. (*Mes. e.* L. Bol.). – Cape: Clanwilliam D. – 35 cm tall, densely branched; **L.** trigonous, falcate, with a recurved mucro, 2.5 cm long, 2 mm across, 7 mm thick, bluish; **Fl.** 12 mm ⌀, pink.

**L. egregius** (L. Bol.) L. Bol. (*Mes. e.* L. Bol.). – Cape: Montagu D. – C. 40 cm tall; **St.** 32 cm long; **L.** narrowed towards the tip, 2–2.5 cm long, 3–4 mm across, 3–5 mm thick, dirty green; **Fl.** pinkish-purple, paler in the centre.

**L. elegans** (Jacq.) Schwant. (*Mes. e.* Jacq., *Mes. macrocalyx* Kensit, ? *Mes. emarginatoides* Haw., ? *L. e.* (Haw.) N. E. Br., *Mes. retroflexus* Haw., *Mes. deflexum* Haw.). – Cape: Cape, Worcester D. – Up to 30 cm tall; **St.** numerous, spreading, prostrate; **L.** trigonous, thickened above, rather obtuse, with rough, translucent dots, 1–2.5 cm long; **Fl.** 3 together, 2–3 cm ⌀, pink or purplish-pink.

**L. emarginatus** (L.) N. E. Br. v. **emarginatus** (*Mes. e.* L., ? *Mes. flexile* Haw., ? *L. f.* (Haw.) N. E. Br., ? *Mes. imbricans* Haw., ? *L. i.* (Haw.) N. E. Br., ? *Mes. violaceum* DC., ? *L. v.* (DC.) Schwant., *Mes. polyphyllum* Haw.). – Cape: Cape D. – 30–40 cm tall; **St.** erect, compressed, brown; **L.** semicylindrical, mucronate, curved, greyish-green, rough with raised dots, 12–16 mm long, 1–2 mm thick; **Fl.** numerous, 3 cm ⌀, violet to pink.

**L. — v. puniceus** (Jacq.) Schwant. (*Mes. emarginatum* v. *p.* Jacq.). – Larger; **Fl.** deep violet.

**L. ernestii** (L. Bol.) L. Bol. (*Mes. ernestii* L. Bol.). – Cape: Bredasdorp D. – Low-growing; **St.** creeping; **L.** incurved, acute, 14 mm long, 2.5 mm across, 3 mm thick, bluish, rough; **Fl.** 18 mm ⌀, purplish-pink.

**L. esterhuyseniae** L. Bol. – Cape: Lit. Namaqualand. – Erect, up to 25 cm tall, c. 30 cm ⌀; **St.** virgate, 15–20 cm long; **L.** acute, 1–2 cm long, 2 mm ⌀, bluish-green, indistinctly rough; **Fl.** in a simple or bipartite Infl., up to 3 cm ⌀, purplish-pink.

**L. excedens** (L. Bol.) L. Bol. (*Mes. e.* L. Bol.). – Cape: Clanwilliam D. – Erect, c. 16 cm tall; **L.** laterally compressed, ± falcate, obtuse to tapered, apiculate, 2 cm long, 4 mm across, 6 mm thick; **Fl.** 2–3 cm ⌀, an attractive pink, open night and day.

**L. eximius** L. Bol. – Cape: Malmesbury D. – **St.** prostrate to creeping, almost 4.5 cm long; **L.** narrowed towards the shortly tapered tip, 15–22 mm long, 2 mm ⌀; **Fl.** 4.5–5.5 cm ⌀, pinkish-purple, often red at first.

**L. explanatus** (L. Bol.) N. E. Br. (*Mes. e.* L. Bol.). – Cape: Cape D. – **St.** elongated, creeping, 45 cm long; **L.** semicylindrical, acute, 23 mm long, 2 mm ⌀, bluish-green; **Fl.** 44 mm ⌀, yellow.

**L. falcatus** (L.) N. E. Br. v. **falcatus** (*Mes. f.* L.). – Cape. – With many tangled, filiform **St.**; **L.** trigonous, compressed, falcate, 4–6 mm long, greyish-green, spotted; **Fl.** 12–16 mm ⌀, pink, agreeably scented.

**L. — v. galpinii** (L. Bol.) L. Bol. (*Mes. f.* v. *g.* L. Bol.). – Cape: Bredasdorp D. – Erect, 25 cm tall, more rarely mat-forming; **L.** 5 to 7 mm long, 1.5 mm across, 2 mm thick; **Fl.** 17 mm ⌀.

**L. falciformis** (Haw.) N. E. Br. v. **falciformis** (*Mes. f.* Haw.). – Cape: Cape D. – Resembles **L. falcatus** but larger; **L.** 10–15 mm long, 3 mm across, angles acute; **Fl.** 4 cm ⌀, pale pink, numerous.

**L. — v. maritimus** (L. Bol.) L. Bol. (*Mes. m.* L. Bol.). – Cape: Cape D. – c. 15 cm tall, with spreading **St.**, prostrate in age; **L.** short-tapered, with a red tip, otherwise light green, 13–17 mm long, 2.5 mm across, 3.5 mm thick; **Fl.** 18 mm ⌀, pink.

**L. fergusoniae** (L. Bol.) L. Bol. v. **fergusoniae** (*Mes. f.* L. Bol.). – Cape: near Riversdale. – Slender, with elongated, prostrate or creeping **St.**; **L.** compressed, acute, rough, 17 mm long, 1.5 mm thick; **Fl.** 3 cm ⌀, yellow or orange.

**L. — v. crassistigma** L. Bol. – Cape: Bredasdorp D. – Inner Pet. less narrowed, reddish-yellow; stigmas subulate, thick.

**L. filicaulis** (Haw.) N. E. Br. (*Mes. f.* Haw.). – Cape: Cape D. – Low-growing; **St.** filiform, spreading or creeping, often rooting; **L.** semicylindrical, subulate, tapered, c. 2.5 cm long, 3 mm thick, green, faintly spotted; **Fl.** solitary, 2 cm ⌀, an attractive purplish-pink.

**L. foliolosus** L. Bol. – Cape: Caledon D. – Low-growing; **St.** prostrate to creeping and rooting; **L.** dense, acute, 1.5–2 cm long, 1 mm ⌀; **Fl.** 4 cm ⌀, pinkish-purple.

**L. formosus** (Haw.) N. E. Br. (*Mes. f.* Haw.). – Cape. – Low-growing; **St.** short; **L.** upcurved, 5 cm long, 6 mm across, compressed-trigonous, green, with a red mucro; **Fl.** 3 together, 3.5 cm ⌀, a beautiful purplish-red with a white centre.

**L. framesii** (L. Bol.) N. E. Br. (*Mes. f.* L. Bol.). – Cape: Caledon D. – Erect, slender, 20 cm tall; **St.** 2 cm thick; **Br.** spreading, variously curved; **L.** 10 mm long, 1.5 mm across, 2 mm thick, shortly tapered; **Fl.** 26 mm ⌀, pink to white.

**L. franciscii** L. Bol. – Cape: Ceres D. – Erect, robust, 60 cm tall, 90 cm ⌀; **St.** 3 cm ⌀; **Br.** 25–30 cm long; **L.** 2.5–4 cm long, 3–6 mm across, 4–6 mm thick, ± falcate, shortly tapered, rather velvety, pale green; **Infl.** 3–4-partite, with 11–25 pink, scented Fl. 3–4.5 cm ⌀.

**L. fugitans** L. Bol. – Cape: Pondoland. – 5–7 cm tall; older **St.** prostrate, reddish-brown; **L.** mostly falcate, linear, acute, 2–2.8 cm long, c. 3 mm across, c. 4 mm thick, bluish-green; **Fl.** 4.4 cm ⌀, pinkish-purple.

**L. furvus** (L. Bol.) N. E. Br. (*Mes. f.* L. Bol.). – Cape: Caledon D. – Slender, 12 cm tall; **St.** lax, ascending; **L.** bluish, ± falcate, compressed, with an oblique tip, 1 cm long, 2 mm thick; **Fl.** 25 mm ⌀, purplish-pink, with dark purple stigmas.

**L. galpiniae** (L. Bol.) L. Bol. (*Mes. g.* L. Bol.). – Cape: Bredasdorp D. – Erect, 8–10 cm tall; flowering **shoots** slender; **L.** semicylindrical, acute to tapered, green with a reddish tinge, 15–20 mm long, 2 mm thick; **Fl.** solitary, 58 mm ⌀, white, pink inside.

**L. glaucus** (L.) N. E. Br. v. **glaucus** (*Mes. g.* L., *Mesembryanthus g.* (L.) Rothm., *Mes. brachyphyllum* Welw.). – Cape: Cape D. – Up to 30 cm tall, with stiff **St.**; **L.** trigonous, compressed, mucronate, 15–30 mm long, dark green, grey-pruinose, with rough dots; **Fl.** 6–7 mm ⌀, glossy, light yellow.

**L. — v. tortuosus** (Haw.) Schwant. (*Mes. g.* v. *t.* Haw.). – Less robust variety; **St.** curved; **L.** smaller.

**L. globosus** (L. Bol.) L. Bol. (*Mes. g.* L. Bol., *L. g.* forma L. Bol.). – Cape: Van Rhynsdorp D. – Erect, c. 20 cm tall, branched; **L.** falcate, semicylindrical, acute or acuminate, 12 to 25 mm long, 1–2 mm ⌀, pale bluish-green; **Fl.** 3–4.5 cm ⌀, purplish-red.

**L. glomeratus** (L.) N. E. Br. (*Mes. g.* L., *Mes. inflexum* Haw.). – Cape: Cape D. – 20–30 cm tall, erect; **St.** slender, with numerous short shoots; **L.** 12–18 mm long, ± thickened above, compressed-trigonous, green, often reddish, spotted; **Fl.** very numerous, 25 mm ⌀, vivid violet-pink or purple.

**L. godmaniae** (L. Bol.) L. Bol. v. **godmaniae** (*Mes. g.* L. Bol.). – Cape: Lit. Namaqualand. – Erect, stiff, 35 cm or more tall; **L.** compressed and shortly-tapered above, 2–3 cm long, 5 mm ⌀, blue with a red tip; **Fl.** 42 mm ⌀, purplish-pink.

**L.** — v. **grandiflorus** (L. Bol.) L. Bol. (*Mes. g.* v. *g.* L. Bol.). – Cape: Ceres D. – **Fl.** 64 mm ⌀.

**L. gracilipes** (L. Bol.) N. E. Br. f. **gracilipes** (*Mes. g.* L. Bol., *L. g.* (L. Bol.) L. Bol.). – Cape: Piquetberg, Van Rhynsdorp D. – Erect, 20 cm tall; **St.** inclined, 4-angled; **L.** ± falcate, short-tapered, apiculate, 7 mm long, 3 mm across, 4 mm thick; **Fl.** 32 mm ⌀, pinkish-purple.

**L.** — f. **luxurians** L. Bol. – Clanwilliam D. – **Fl.** c. 3.9 cm ⌀, glossy purplish-pink, paler outside.

**L. guthrieae** (L. Bol.) N. E. Br. (*Mes. g.* L. Bol.). – Cape: Ceres D. – Woody, 12 cm tall; **St.** tangled, older ones variously curved; **L.** ± falcate, ± navicular at the shortly tapered tip, 22 mm long, 5–6 mm across, 9 mm thick, minutely scaly, bluish-green; **Fl.** 2 cm ⌀, pinkish-red.

**L. hallii** L. Bol. – Cape: Riversdale D. – 36 cm tall; **St.** eventually prostrate; **L.** laterally compressed, acute, 2–8 cm long, 2.4–5 mm ⌀; **Fl.** 3–3.5 cm ⌀, pink.

**L. haworthii** (Don) N. E. Br. (*Mes. h.* Don, *Erepsia h.* (Don) Schwant.). – Cape: origin not known. – Up to 60 cm tall, freely branched; **L.** 2.5–4 cm long, 4–6 mm across, semicylindrical, tapered, light green, densely light grey-pruinose; **Fl.** up to 7 cm ⌀, light purple.

**L. henricii** (L. Bol.) N. E. Br. (*Mes. h.* L. Bol.). – Cape: Clanwilliam D. – **St.** erect, stiff, 25 cm long; **L.** ± falcate, swollen, trigonous, 9 mm long, 2.5 mm across, 3 mm thick, bluish-green; **Fl.** 12 mm ⌀, pinkish-purple.

**L. hiemalis** (L. Bol.) L. Bol. (*Mes. h.* L. Bol.). – Cape: Clanwilliam D. – Erect, stiff, 40 cm tall; **St.** robust, 6 mm thick; **L.** 35 mm long, 5 mm across, 6 mm thick, with a distinct line along margins and keel, obtuse, dirty green; **Fl.** 15 mm ⌀, pink.

**L. hoerleinianus** (Dtr.) Friedr. (*Mes. h.* Dtr., *Mes. brachyandrum* L. Bol., *L. b.* (L. Bol.) N. E. Br., *L. b.* forma L. Bol.). – S.W.Afr.: Gr. Namaqualand; Cape: Lit. Namaqualand. – ♄ 70 cm tall and across, outer **St.** rooting and ascending, bark light grey to brown, Int. 2.5 to 3.5 cm long; **L.** shortly united, patent, obtusely trigonous, 2.5–3 cm long, 9 mm thick, soft, light bluish-green with a red tip; **Fl.** on a pedicel 6–8 cm long, with a Cal. 1 cm long with 5 densely papillose T., up to 4.5 cm ⌀, violet-pink to purplish-pink, heliotrope-scented.

**L. holensis** L. Bol. – Cape: Van Rhynsdorp D. – Shrubby; **St.** 20 cm long; **L.** acute, rounded on the underside, 4.5–5.5 cm long, 4–5 mm ⌀; **Fl.** solitary, white.

**L. hollandii** (L. Bol.) L. Bol. (*Mes. h.* L. Bol.). – Cape: Alexander D. – Erect, 20 cm tall; **L.** expanded towards the shortly tapered or ± rounded, apiculate tip, 2–2.7 cm long, 2 mm across, 2–3 mm thick, dirty green; **Fl.** 4.5 cm ⌀, an attractive pink to purplish-pink.

**L. hurlingii** (L. Bol.) L. Bol. (*Mes. h.* L. Bol.). – Cape: Robertson D. – Erect, laxly branched, 20–32 cm tall; **L.** bluntly to acutely keeled on the underside, acute above, 2.2–5 cm long, 2.5 mm ⌀, bluish-green, with crowded, raised dots; **Fl.** 24 mm ⌀, golden-yellow.

**L. immelmaniae** (L. Bol.) N. E. Br. (*Mes. i.* L. Bol., *L. i.* (L. Bol.) L. Bol.). – Cape: Malmesbury D. – Erect, 20 cm tall; **St.** ± virgate; **L.** semicylindrical, indistinctly carinate below the acute tip, 22 mm long, 1–2 mm ⌀, bluish-green; **Fl.** 4 cm ⌀, pink.

**L. inaequalis** (Haw.) N. E. Br. (*Mes. i.* Haw.). – Cape. – **Br.** spreading, prostrate; **L.** ± trigonous, green with red spots; Cal.-lobes unequal; **Fl.** yellow, paler outside. Acc. Haworth resembles **L. bicolor.**

**L. inconspicuus** (Haw.) Schwant. (*Mes. i.* Haw.). – Cape. – 40 cm tall; **St.** tangled, curved, spreading, slender; **L.** semicylindrical, tapered, with a reddish, hooked, recurved apiculus, 10–15 mm long, 1 mm across, green, minutely spotted; **Fl.** solitary, 15 mm ⌀, red.

**L. incurvus** (Haw.) Schwant. (*Mes. i.* Haw., *Mes. roseum* v. *confertum* Salm). – Cape: Table Mt. – Resembles **L. roseus**; **Br.** denser, outspread, curved; **L.** grey-pruinose, with rougher dots; **Fl.** deeper red.

**L. intervallaris** L. Bol. – Cape: Clanwilliam D. – Laxly branched; **L.** tapered, obtuse to acute, bluntly keeled on the underside, soft-textured, green, 1.5–2 cm long, 1.5–2 mm across, 2.3 mm thick; **Fl.** 2–2.8 cm ⌀, pale pink.

**L. laetus** (L. Bol.) L. Bol. (*Mes. l.* L. Bol.). – Cape: Montagu D. – Erect, 20 cm or more tall; **St.** robust, woody, tangled, with slender shoots; **L.** acute above, indistinctly keeled, 7–13 mm long, 1.5 mm across, 1 mm thick, dirty green; **Fl.** 2–2.3 cm ⌀, pinkish-purple.

**L. lavisii** (L. Bol.) L. Bol. v. **lavisii** (*Mes. l.* L. Bol.). – Cape: Humansdorp D. – Erect, 25–35 cm tall; **Br.** ± virgate; **L.** keeled on the underside, expanded towards the acute or rather obtuse tip, 2.5–3.5 cm long, 2 mm across, 2–2.5 mm thick, green; **Fl.** 4–6.3 cm ⌀, an attractive pink or pure white, more rarely tinged light pink.

**L.** — v. **concinnus** L. Bol. – Cape: Port Elizabeth D. – Pet. rounded at the tips and spreading.

**L. laxifolius** (L. Bol.) N. E. Br. (*Mes. l.* L. Bol.). – Cape: Riversdale D. – Erect, 15 to 25 cm tall; **St.** laxly leafy; **L.** falcately incurved, keeled on the underside, acute to tapered above, 18–26 mm long, 3 mm across, 2–3 mm thick, yellowish-green; **Fl.** 3–3.5 cm ⌀, pinkish-purple.

**L. leightoniae** (L. Bol.) L. Bol. (*Mes. l.* L. Bol.). – Cape: Clanwilliam D. – Erect, 15 cm tall; **L.** carinate on the underside, shortly tapered, 13 mm long, 2.5 mm across, 3 mm thick, dirty green; **Fl.** solitary, 24 mm $\varnothing$.

**L. leipoldtii** (L. Bol.) L. Bol. (*Mes. l.* L. Bol.). – Cape: Worcester D. – Erect, 20 cm tall; **St.** and **Br.** slender; **L.** bluntly carinate, laterally compressed, 10–15 mm long, 1 mm across, 1.5 mm thick, dirty green with raised dots; **Fl.** (1–)2(–3) together, 34 mm $\varnothing$, pink to pale pink.

**L. leptaleon** (Haw.) N. E. Br. (*Mes. l.* Haw.). – Cape: Somerset West D. – Thin, erect, laxly branched, 12–15 cm tall; **L.** $\pm$ trigonous, somewhat tapered, bluish, with the keel minutely rough, 5–10 mm long, 1–2 mm $\varnothing$; **Fl.** 1–3 together, up to 1.5 cm $\varnothing$, pink; Sep. tapered.

**L. leptosepalus** (L. Bol.) L. Bol. (*Mes. l.* L. Bol.). – Cape: Clanwilliam D. – Erect, 14 cm tall; **St.** slender, virgate, 11–13 cm long; **L.** acute above, keeled on the underside below the tip, 10–15 mm long, 1.5–2 mm $\varnothing$, dirty green, spotted; **Fl.** mostly solitary, 3 cm $\varnothing$, pinkish-purple.

**L. lewisiae** (L. Bol.) L. Bol. (*Mes. l.* L. Bol.). – Cape: Ceres D. – Erect, 7.5 cm tall; **St.** ascending; **L.** narrowed towards the apiculate tip, 12 mm long, 2 mm $\varnothing$, pale green, indistinctly rough; **Fl.** solitary, 3 cm $\varnothing$, pink.

**L. liberalis** (L. Bol.) L. Bol. (*Mes. l.* L. Bol.). – Cape: origin unknown. – Erect, 15–25 cm tall; **L.** narrowed towards the acute, recurved tip, 6–9 mm long, 1.5 mm $\varnothing$, dirty green, rough; **Fl.** solitary, up to 3.5 cm $\varnothing$, pink.

**L. littlewoodii** L. Bol. – Cape: Lit. Namaqualand. – Laxly branched, 23 cm tall, 35 cm $\varnothing$; **St.** up to 16 cm long; **L.** $\pm$ falcate, acute, laterally compressed, 1.25 cm long, 1.5–2 mm across, 2–3 mm thick; **Fl.** 2–3 together, Pet. 5–7 mm long, an attractive pink.

**L. longisepalus** (L. Bol.) L. Bol. (*Mes. l.* L. Bol.). – Cape: Ceres D. – Compact, 7 cm tall, with a whitish bark; **L.** expanded towards the obtuse tip, reddish-brown with large dots, 17 mm long, 6 mm across, 7 mm thick; **Fl.** 25 mm $\varnothing$, Pet. pink, striped on the upperside.

**L. longistamineus** (L. Bol.) N. E. Br. (*Mes. l.* L. Bol.). – Cape: Clanwilliam D. – Stiff, erect, 13 cm tall; **St.** various curved; **L.** bluntly keeled, obtuse above, dirty green, densely spotted, 25 mm long, 6 mm across, 7 mm thick; **Fl.** 3 cm $\varnothing$, golden-yellow.

**L. lunatus** (Willd.) N. E. Br. (*Mes. l.* Willd.). – Cape: Clanwilliam D. – **St.** slender; **L.** crowded on the short shoots, trigonous-compressed, lunate, apiculate, 10–15 mm long, 6 mm across, very grey-pruinose; **Fl.** paniculate, 24 mm $\varnothing$, pale pinkish-red.

**L. lunulatus** (Bgr.). L. Bol. (*Mes. l.* Bgr., *Mes. capornii* f. *fera* and f. *longifolium* L. Bol.). – Cape: Clanwilliam D. – Much branched; **shoots** slender, opposite, brown; **L.** trigonous, obtuse, incurved, spotted; **Fl.** numerous, white, reddish inside.

**L. macrocarpus** (Bgr.) N. E. Br. (*Mes. m.* Bgr.). – Cape: Uitenhage D. – **St.** very tangled, robust, with numerous short shoots; **L.** somewhat expanded above, rather obtuse, with numerous dots, 10–25 mm long, 2–4 mm across; **Fl.** solitary, up to 5 cm $\varnothing$, pinkish-purple.

**L. macrosepalus** (L. Bol.) L. Bol. (*Mes. m.* L. Bol.). – Cape: Humansdorp D. – Low-growing; sterile **St.** short, densely leafy, flowering ones elongated, prostrate; **L.** carinate on the underside, obtuse to short-tapered above, up to 7 cm long, 5 mm across, 7 mm thick; **Fl.** 6 cm $\varnothing$, pink, Sep. 14–21 mm long.

**L. macrostigma** L. Bol. – Cape: Clanwilliam D. – Erect, 22 cm tall; **St.** virgate, brown; **L.** narrowed above, 7–14 mm long, 1.5 mm $\varnothing$, rough; **Fl.** usually solitary, 27 mm $\varnothing$, white.

**L. magnificus** (L. Bol.) N. E. Br. (*Mes. m.* L. Bol.). – Cape: Malmesbury D. – **St.** prostrate to creeping; **L.** semicylindrical, acute, apiculate, 2.5–4.5 cm long, 2–2.5 mm thick, green; **Fl.** solitary, 5 cm $\varnothing$, coppery-red.

**L. marcidulus** N. E. Br. (*Mes. m.* L. Bol.). – Cape: Montagu D. – 20 cm tall with somewhat flaccid **St.**; **L.** green, obliquely trigonous, 2 to 4 cm long, 4–5 mm $\varnothing$; **Fl.** solitary, 3.5 cm $\varnothing$, golden-yellow.

**L. mariae** (L. Bol.) L. Bol. (*Mes. m.* L. Bol.). – Cape: Ceres D. – Erect, 20 cm tall; **St.** stiff, with a red bark; **L.** falcate, rounded on the underside, obtuse to acute above, bluish-green, 7 mm long, 1.5 mm thick; **Fl.** solitary, 14 mm $\varnothing$, pinkish-purple.

**L. martleyi** (L. Bol.) L. Bol. (*Mes. m.* L. Bol.). – Cape: Piquetberg D. – Erect, 30 cm tall; **St.** angular below; **L.** bluntly carinate on the underside, acute above, indistinctly rough, 2.5 cm long, 3 mm $\varnothing$; **Fl.** 4 cm $\varnothing$, silvery-pink.

**L. maturus** N. E. Br. (*Mes. praecox* L. Bol.). – Cape: Paarl D. – Slender, with widespread, prostrate **St.**, occasionally rooting at the nodes; **L.** $\pm$ falcate, laterally compressed, obtuse, 8 mm long, 2 mm thick, bluish-green, rough; **Fl.** solitary, 2 cm $\varnothing$, pink.

**L. matutinus** (L. Bol.) N. E. Br. (*Mes. m.* L. Bol., *L. m.* (L. Bol.) Schwant.). – Cape: Piquetberg D. – **St.** elongated, prostrate, rooting at the nodes; **L.** semicylindrical, acute, yellowish-green, spotted, 15–20 mm long, 2 mm $\varnothing$; **Fl.** solitary, 38 mm $\varnothing$, Pet. golden-yellow with red tips, open only before midday.

**L. maximilianii** (Schltr. et Bgr.) L. Bol. (*Mes. m.* Schltr. et Bgr., *Echinus m.* (Schltr. et Bgr.) N. E. Br., *Braunsia m.* (Schltr. et Bgr.) Schwant., *Mes. apiculatum* v. *muticum* L. Bol., *Mes. phillipsii* L. Bol., *Mes. binum* L. Bol.). – S.W.Cape. – Low-growing, with the herbaceous parts velvety white-haired; **St.** elongated, creeping, up to 17 cm long; **L.** swollen, trigonous, navicular, 15 mm long, 1 cm $\varnothing$ below; **Fl.** solitary, 22 mm $\varnothing$, pink.

**L. meleagris** (L. Bol.) L. Bol. (*Mes. m.* L. Bol.). – Cape: Ceres D. – 4–5 cm tall; **St.** dense, stiff, with rather tangled Br.; **L.** $\pm$ truncate at the tip, indistinctly dentate, bluish-red, with large dots, 12 mm long; **Fl.** 18 mm $\varnothing$, an attractive pink.

**L. microsepalus** L. Bol. – Cape: Tulbagh D. – Small, with **St.** 9–18 cm long; **L.** bluntly keeled, recurved and oblique at the tip, 8–10 mm long,

Plate 194. 1. **Ruschia derenbergiana** (DTR.) C. WEBER (Photo: H. CHR. FRIEDRICH); **2. R. amoena** SCHWANT. (Photo: G. SCHWANTES); **3. R. pygmaea** (HAW.) SCHWANT. (Photo: as 1); **4. R. pusilla** SCHWANT. (Photo: as 1); **5. R. meyeri** SCHWANT. (Photo: as 1).

Plate 195. 1. **Ruschianthus falcatus** L. Bol. (Photo: H. Chr. Friedrich); **2. Ruschia vulvaria** (Dtr.) Schwant. (Photo: K. Dinter); **3. R. robusta** L. Bol. (Photo: G. Schwantes); **4. Saphesia flaccida** (Jacq.) N. E. Br. (Photo: H. Herre acc. Jacquin's original, G. Schwantes).

Plate 196. 1. Sarcozona praecox (F. MUELL.) S. T. BLAKE ex H. J. EICHLER (Repr.: Contr. Qud. Herb. No. 7, 1969, 59, Photo: S. T. BLAKE); 2. Sceletium compactum (HAW.) L. BOL. (Photo: G. SCHWANTES); 3. Schwantesia ruedebuschii DTR. (Photo: as 2).

Plate 197. 1. **Scopelogena vereculata** (L.) L. Bol. (Repr.: A. Berger, Mes. u. Port., 1908 as *Mes. verruculatum* L.); 2. **Semnanthe lacera** (Haw.) N. E. Br.; 3. **Skiatophytum tripolium** (L.) L. Bol.; 4. **Schwantesia succumbens** (Dtr.) Dtr.

Plate 198. **1.** Smicrostigma viride (Haw.) N. E. Br.; **2.** Sphalmanthus splendens (L.) Schwant. (Photo: F. Riviere de Caralt); **3.** Stomatium agninum (Haw.) Schwant. v. agninum; **4.** Sphalmanthus salmoneus (Haw.) N. E. Br. (Photo: H. Straka); **5.** Stomatium meyeri L. Bol.; **6.** S. ermininum (Haw.) Schwant. (Photo: G. Schwantes).

Plate 199. 1. **Stomatium trifarium** L. Bol.; 2. **Synaptophyllum juttae** (Dtr. et Bgr.) N. E. Br. (Photo: H. Herre); 3. **Titanopsis schwantesii** (Dtr.) Schwant. (Photo: G. Schwantes); 4. **T. calcarea** (Marl.) Schwant.; 5. **Trichodiadema barbatum** (L.) Schwant.; 6. **T. densum** (Haw.) Schwant.

Plate 200. **1. Vanzijlia annulata** (BGR.) L. BOL.; **2. Trichodiadema intonsum** (HAW.) SCHWANT. (Photo: G. SCHWANTES); **3. Vanheerdea divergens** L. BOL. (Photo: G. C. NEL); **4. V. roodiae** (N. E. BR.) L. BOL.; **5. Zeuktophyllum suppositum** (L. BOL.) N. E. BR. (Photo: E. ELKAN).

1–1.5 mm $\varnothing$; **Fl.** solitary, c. 2.2 cm $\varnothing$, Sep. 3–4 mm long, Pet. pink.

**L. microstigma** (L. BOL.) N. E. BR. (*Mes. m.* L. BOL.). – Cape: Piquetberg D. – Slender, erect, 40 cm tall, with virgate **St.**; **L.** acute above, bluish-green, indistinctly rough, 15–20 mm long, 1.5 mm $\varnothing$; **Fl.** 3.5–4 cm $\varnothing$, pinkish-purple.

**L. middlemostii** (L. BOL.) L. BOL. (*Mes. m.* L. BOL.). – Cape: Caledon D. – **St.** lax, ± climbing, tangled; **L.** falcate, indistinctly rough, 8 mm long, 1.5 mm across, 2.5 mm thick; **Fl.** 18 mm $\varnothing$, white to pale pink, pink inside, scented.

**L. montaguensis** (L. BOL.) L. BOL. (*Mes. m.* L. BOL.). – Cape: Montagu D. – Stiff, with lax, spreading to prostrate **St.**; **L.** often falcate, slightly swollen, semicylindrical, bluish-green, minutely papillose, 2 cm long, 2–3 mm across, 5 mm thick; **Fl.** in 3's, often solitary, pinkish-purple.

**L. mucronatus** L. BOL. – Cape: Ceres D. – Compact, c. 15 cm tall, 22 cm $\varnothing$; **L.** slightly falcate, acute, apiculate, 2–3.5 cm long, 5 mm across, 4–6 mm thick; **Fl.** 2.5–3 cm $\varnothing$, white.

**L. multiseriatus** (L. BOL.) N. E. BR. (*Mes. m.* L. BOL.). – Cape: Riversdale D. – **St.** elongated, creeping; **L.** bluntly carinate, laterally compressed, bluish-green, 15–18 mm long, 1.5 mm across, 2.5 mm thick; **Fl.** solitary, 4.4 cm $\varnothing$, pink.

**L. mutans** (L. BOL.) N. E. BR. (*Mes. m.* L. BOL.). – Cape: Humansdorp D. – Erect ♄ c. 18 cm tall; **St.** virgate, 1–3 mm thick, Int. 1–2 cm long, shoots rather short; **L.** narrow, ± cylindrical, somewhat hooked at the tip, 15 mm long, 2 mm thick, with a Sh. 1 mm long; **Fl.** on a pedicel 3 cm long, 36 mm $\varnothing$, pink with a yellow centre.

**L. nardouwensis** (L. BOL.) L. BOL. (*Mes. n.* L. BOL.). – Cape: Clanwilliam D. – Erect; **St.** lax, stiff, 23 cm long; **L.** inflated at the base, swollen towards the obtuse tip, 17 mm long, 3 mm across, 4 mm thick; **Fl.** 14 mm $\varnothing$, pink.

**L. nelii** L. BOL. – Cape: Cape D. – Slender, 14 cm tall; **L.** distinctly awned, rounded on the underside, 1–1.7 cm long, 1–1.5 mm $\varnothing$; Pet. lax, in 3 series, 9–10 mm long, dull pink, stigmas thin, long.

**L. neostayneri** L. BOL. – Cape: Piquetberg D. – Erect, 15–30 cm tall, c. 30 cm $\varnothing$; **St.** spreading; **L.** spreading to recurved, narrowed towards the tip, 1–1.5 cm long, 1–1.5 mm $\varnothing$, rough, green; **Fl.** 1–3 together, up to 4.5 cm $\varnothing$, purplish-pink.

**L. obconicus** (L. BOL.) L. BOL. (*Mes. o.* L. BOL.). – Cape: Calvinia D. – Erect, slender, 20–30 cm tall; **L.** carinate on the underside, laterally compressed, indistinctly rough, dirty green, 1 cm long, 2 mm $\varnothing$; **Fl.** 27 mm $\varnothing$, pink.

**L. occultans** L. BOL. – Cape: Worcester D. – Erect, slender, stiff, 30–40 cm tall; **L.** narrow, acute above, bluish-green, 10–17 mm long, 1–1.5 mm thick; **Fl.** 2 cm $\varnothing$, Pet. lax, pink.

**L. ornatus** L. BOL. (*Mes. pageanum* L. BOL.). – Cape: Clanwilliam D. – Robust; **St.** up to 25 cm tall; **L.** falcate, obtuse to ± truncate, apiculate, 3–4 cm long, 12 mm thick below the tip, 7 mm across, bluish-green with reddish angles; **Fl.** 25 mm $\varnothing$, pink.

**L. paardebergensis** (L. BOL.) L. BOL. (*Mes. p.* L. BOL.). – Cape. – Compact, low-growing; **St.** prostrate; **L.** ± falcate, laterally compressed, with the tip truncate and oblique, red, 7–10 mm thick, 2.8 cm long, 5–8 mm across; **Fl.** 2–2.2 cm $\varnothing$, pink.

**L. paarlensis** L. BOL. – Cape: Paarl D. – Up to 46 cm tall, with widespread **St.**; **L.** narrow, acute, green, indistinctly rough, 18–22 mm long, 1.5 mm $\varnothing$ midway; **Fl.** 3–4 cm $\varnothing$, brilliant pink but sometimes white.

**L. pakhuisensis** (L. BOL.) L. BOL. (*Mes. p.* L. BOL.). – Cape: Clanwilliam D. – Erect, 25 cm tall; **St.** spreading to erect; **L.** bluntly keeled on the underside, laterally compressed, obliquely rounded and expanded at the tip, bluish-green, 15 mm long, 2 mm $\varnothing$; **Fl.** solitary, pink.

**L. palustris** (L. BOL.) L. BOL. (*Mes. p.* L. BOL.). – Cape: Stellenbosch D. – Slender, 12–14 cm tall; **L.** keeled on the underside, acute to obliquely acute at the tip, bluish-green, spotted, 6–8 mm long, 2 mm $\varnothing$; **Fl.** 3.8 cm $\varnothing$, golden-yellow.

**L. parcus** N. E. BR. (*Mes. tenue* L. BOL.). – Cape: Caledon D. – Laxly branched, up to 20 cm tall; **L.** indistinctly keeled, almost imperceptibly rough, 4–9 mm long, up to 1.5 mm $\varnothing$; **Fl.** 2 cm $\varnothing$, Pet. pink, with a deeper pink border.

**L. pauciflorus** (L. BOL.) N. E. BR. (*Mes. p.* L. BOL., *L. sparsiflorus* L. BOL.). – Cape: S.W. reg. – 30 cm tall; **St.** lax, prostrate to ascending; **L.** ± falcate-incurved, expanded midway, bluish-green, 18 mm long, 2 mm across, 3 mm thick; **Fl.** solitary, 4 cm $\varnothing$, pink.

**L. paucifolius** (L. BOL.) N. E. BR. (*Mes. p.* L. BOL.). – Cape: Calvinia D. – Erect, 25 cm tall; **St.** virgate, elongated, sparsely leafy; **L.** narrowed towards the acute tip, 17 mm long, 1–2 mm $\varnothing$; **Fl.** 2.6 cm $\varnothing$, purplish-pink.

**L. peacockiae** (L. BOL.) L. BOL. (*Mes. p.* L. BOL.). – Cape: Malmesbury D. – Creeping; **St.** variously curved; **L.** semicylindrical, acute and with a red mucro, bluish-green, spotted, 6–8 mm long, 2 mm across below; **Fl.** 3 cm $\varnothing$, Pet. pink, with a purple border at the tip.

**L. peersii** (L. BOL.) N. E. BR. (*Mes. p.* L. BOL.). – Cape: Clanwilliam D. – Erect, 30 cm tall; **St.** ascending; **L.** trigonous, acute, bluish-green, with short shoots 2–2.5 cm long from the axils; **Fl.** 3.5–4 cm $\varnothing$, Pet. deep orange, changing to red and finally pinkish-purple.

**L. perreptans** L. BOL. (*Mes. humile* L. BOL.). – Cape: Swellendam D. – Low-growing; **St.** prostrate to creeping and rooting, stiff, 20 cm long; **L.** ± falcate, keeled ± laterally on the underside, acute to rounded above, dirty green, 15 mm long, 3 mm across and thick; **Fl.** 43 mm $\varnothing$, pink.

**L. persistens** (L. BOL.) L. BOL. (*Mes. p.* L. BOL.). – Cape: Calvinia D. – Erect, 25 cm tall; **St.** stiff; **L.** recurved at the tip and ±

hooked, with a red apiculus, S-shaped in the lower part, dirty green; **Fl.** 14 mm ∅, pinkish-purple.

**L. piquetbergensis** (L. Bol.) L. Bol. (*Mes. p.* L. Bol.). – Cape: Piquetberg D. – Slender, 10 cm tall, laxly branched; **L.** falcate, truncate and apiculate above, with 2 indistinct marginal T., bluish, 12 mm long, 6 mm ∅; **Fl.** 10 mm ∅, pink.

**L. plautus** N. E. Br. (*Mes. latum* L. Bol.). – Cape: Lit. Namaqualand. – Spreading; **St.** 14 cm long, slender, stiff; **L.** indistinctly keeled, acute, bluish-green, 27 mm long, 5 mm thick; **Fl.** 34 mm ∅, purplish-pink.

**L. pleniflorus** L. Bol. – Cape: Worcester D. – **St.** woody, stiff, brown, c. 35 cm long; **L.** curved hook-like at the tip, 1–2 cm long, 2–3 mm across, 3–5 mm thick; **Fl.** in a bipartite Infl., Pet. numerous, 4–5 mm long, purplish-pink, with a darker border.

**L. plenus** (L. Bol.) L. Bol. (*Mes. p.* L. Bol.). – Cape: Clanwilliam D. – Erect, with spreading, prostrate **St.;** **L.** incurved above, somewhat compressed, short-tapered, green, 10–15 mm long, 2 mm across, 2–3 mm thick; **Fl.** 3.5 cm ∅, pink.

**L. pocockiae** (L. Bol.) N. E. Br. (*Mes. p.* L. Bol.). – Cape: Prince Albert D. – Erect, slender, 9 cm tall; **St.** tangled, stiff, hirsute; **L.** 11 mm long, 2 mm ∅, green with prominent tubercles, with the tip incurved, swollen, acute, channelled on the upperside; **Fl.** 2 cm ∅.

**L. polyanthon** (Haw.) N. E. Br. (*Mes. p.* Haw.). – Cape: origin unknown. – **St.** crowded, spreading, with a red bark; **L.** small, trigonous, obtuse, bluish, rough; **Fl.** numerous, pale pink.

**L. praecipitatus** (L. Bol.) L. Bol. (*Mes. p.* L. Bol.). – Cape: Albany D. – Robust, soon becoming pendulous; **St.** 40 cm long; **L.** rounded on the underside, laterally compressed and obliquely carinate, pale blue to pale green, 17 mm long, 2 mm ∅; **Fl.** 4 cm ∅, white.

**L. prasinus** L. Bol. – Cape: Piquetberg D. – 10–12 cm tall; **St.** slender, spreading; **L.** hemispherically widened in profile, apiculate, leek-green, minutely glossy-papillose, 8 mm long, 4 mm across, 5 mm thick; **Fl.** 12 mm ∅, purplish-pink.

**L. primivernus** (L. Bol.) L. Bol. (*Mes. p.* L. Bol.). – Cape: Piquetberg D. – Erect, 30 cm tall; **St.** stiff; **L.** falcate, laterally compressed, shortly tapered above, 24 mm long, 3 mm across, 9 mm thick; **Infl.** of many Fl. 18 mm ∅, an attractive pink to salmon-pink colour.

**L. productus** (Haw.) N. E. Br. v. **productus** (*Mes. p.* Haw.). – Cape: Uitenhage, Albany D. – 30–80 cm tall, branched; **L.** semicylindrical, ± acute, fresh green, grey-pruinose, minutely spotted, 2.5–4 cm long, 3 mm across; **Fl.** 2.5 cm ∅, pale pink.

**L. —** v. **lepidus** (Haw.) Schwant. (*Mes. l.* Haw., *Mes. p.* v. *l.* (Haw.) S.D.). – Cape: Cape D. – Taller variety; **Fl.** white.

**L. —** v. **purpureus** (L. Bol.) L. Bol. (*Mes. p.* v. *purpureum* L. Bol.). – Cape: Albany D. – **Fl.** 4.4 cm ∅, pinkish-purple.

**L. prominulus** (L. Bol.). L. Bol. (*Mes. p.* L. Bol.). – Cape: George D. – Erect, 35 cm tall; **St.** 30 cm long; **L.** somewhat compressed laterally, with an indistinct keel, expanded and short-tapered above, with the tip recurved, dirty bluish-green with prominent dots, 12 to 20 mm long, 1.5 mm across, 2 mm thick; **Fl.** 2.5 cm ∅, purplish-pink.

**L. promontorii** (L. Bol.) N. E. Br. (*Mes. p.* L. Bol.). – S.W.Cape. – Erect, 15–30 cm tall, with spreading, slender **St.;** **L.** falcate, compressed-trigonous, apiculate, 10–13 mm long, 3 mm across, 3–5 mm thick; **Fl.** 14–23 mm ∅ yellow.

**L. proximus** L. Bol. – Cape: Cape D. – Resembles **L. prominulus,** with the dots on the **L.** much less prominent; Pet. 2 cm long, 2 mm across.

**L. purpureus** L. Bol. – Cape: Clanwilliam D. – 40 cm tall; **St.** lax, virgate; **L.** rounded to bluntly carinate on the underside, narrowed towards the shortly tapered tip, rough, bluish-green, 15–35 mm long, 1–1.5 mm across, 2 mm thick; **Fl.** 28 mm ∅, pinkish-purple.

**L. rabiesbergensis** (L. Bol.) L. Bol. (*Mes. r.* L. Bol.). – Cape: Worcester D. – Spreading; **St.** stiff, flowering shoots slender; **L.** narrow, cylindrical, acute, 15–20 mm long; **Fl.** solitary, 3 cm ∅, purplish-pink.

**L. recurvus** (L. Bol.) Schwant. (*Mes. r.* L. Bol.). – Cape: Tulbagh D. – Spreading, with prostrate **St.;** **L.** ± falcate, compressed and acutely carinate, expanded towards the obliquely rounded tip, 8–10 mm long, 1.5 mm across, 2.5 mm thick; **Fl.** 3 cm ∅, Pet. pink, with deeper pink margins.

**L. reptans** (Ait.) N. E. Br. (*Mes. r.* Ait., *Mes. crassifolium* Thunbg.). – Cape: Cape D. – **St.** prostrate, 10–30 cm long, very slender to filiform; **L.** in dense clusters, narrowed below, apiculate, greyish-green, with raised dots, 15–25 mm long, 4–6 mm across; **Fl.** yellow or white.

**L. roseus** (Willd.) Schwant. (Pl. 172/1) (*Mes. r.* Willd., *Mes. incurvum* v. *r.* DC., ? *Mes. multiradiatum* Jacq., ? *L. m.* (Jacq.) N. E. Br.). – Cape: Cape D. – Erect, spreading, up to 60 cm tall; **L.** compressed-trigonous, apiculate, with ± prominent, translucent dots, 2.5–3 cm long, up to 4 mm across; **Fl.** 4 cm ∅, pale pink.

**L. rubroluteus** (L. Bol.) L. Bol. (*Mes. r.* L. Bol.). – Cape: Piquetberg D. – Erect, 24 to 35 cm tall; **St.** virgate; **L.** laterally compressed, acutely carinate on the underside, expanded at the tip, 1.5–2 cm long, 3 mm ∅ midway; **Fl.** 3.5 cm ∅, glossy white, with orange filaments.

**L. rupestris** (L. Bol.) N. E. Br. (*Mes. r.* L. Bol.). – Cape: Malmesbury D. – 15 cm tall; **St.** spreading, prostrate to creeping, slender; **L.** indistinctly keeled on the underside, short-tapered, pale blue, with raised dots, 12 mm long, 2 mm thick; **Fl.** 2.4 cm ∅, white.

**L. rustii** (Bgr.) N. E. Br. (*Mes. r.* Bgr.). – Cape: Riversdale D. – 30 cm tall; **St.** opposite, erect, slender, brown, with numerous short shoots; **L.** trigonous, acute, with a recurved and hooked apiculus, 14 mm long, 1.5 mm across; Fl. numerous, 2 cm ⌀.

**L. salicolus** (L. Bol.) L. Bol. (*Mes. s.* L. Bol.). – Cape: Piquetberg D. – Erect, 20–30 cm tall; L. rounded on the underside, shortly tapered, 15 mm long, 2 mm across, 3 mm thick; Fl. 4 cm ⌀, purplish-pink.

**L. salteri** (L. Bol.) L. Bol. (*Mes. s.* L. Bol.). – Cape: Bredasdorp D. – 15 cm tall; St. spreading, prostrate, 20 cm long; L. bluntly carinate, ± compressed laterally, acute, green, with raised dots, 15–30 mm long, 1.5–3 mm across, 2–3.5 mm thick; Fl. 3 cm ⌀, Pet. reddish-orange on the upperside, purple on the underside.

**L. saturatus** (L. Bol.) N. E. Br. (*Mes. s.* L. Bol.). – Cape: Clanwilliam, Van Rhynsdorp D. – 25 cm tall, with ascending St. from the base; L. semicylindrical, acute, bluntly keeled on the underside, bluish-green, 4 cm long, 4–5 mm across, 6 mm thick; Fl. 5.6 cm ⌀, pinkish-purple.

**L. sauerae** (L. Bol.) L. Bol. (*L. s.* L. Bol.). – Cape: Darling D. – Erect, 15 cm tall; St. compressed; L. falcate, rounded on the underside, short-tapered, apiculate, dirty bluish-green, with raised dots, 5–10 mm long, 1 mm across, 1.5 mm thick; Fl. solitary, red inside, coppery-red outside.

**L. scaber** (L.) N. E. Br. (*Mes. s.* L.). – Cape: Cape D. – Erect; St. ± angular, 30–45 cm long, stiff; L. linear, compressed-trigonous, short-tapered, green with large, raised dots on the underside, the angles tuberculate-dentate, 2 to 3 cm long; Fl. usually 3 together, c. 3 cm ⌀, violet-pink.

**L. schlechteri** (Zahlbr.) L. Bol. (*Mes. s.* Zahlbr., *Mes. sabulosum* Schltr., *Mes. perspicuum* Bgr., *L. perspicuus* (Bgr.) N. E. Br.). – Cape: Groot Berg Rivier. – St. ± erect, 13 to 20 cm tall; L. ± filiform, acute, indistinctly cylindrical, rough, 1–2 cm long, 1 mm thick; Fl. 5 cm ⌀, salmon-pink.

**L. serpens** (L. Bol.) L. Bol. (*Mes. s.* L. Bol.). – Cape: Caledon D. – St. short, densely leafy; Br. creeping, 25 cm or more long; L. incurved, with an acute, oblique, indistinctly serrate keel on the underside, with a red mucro, 2–3 cm long, 6 mm thick midway, 5 mm across; Fl. solitary, 4.5 cm ⌀, Pet. yellow on the upperside, with golden-yellow stripes, golden-yellow on the underside.

**L. simulans** L. Bol. – Cape: Bredasdorp D. – Erect, slender, laxly branched, 14–26 cm tall; L. narrowed to the acute to obtuse tip, 10 to 17 mm long, 4–6 mm thick; Fl. (1–)2(–3), c. 2 cm ⌀, Pet. silvery-white with pink tips.

**L. sociorum** (L. Bol.) N. E. Br. (*Mes. s.* L. Bol.). – S.W.Cape. – St. elongated, creeping, 25 cm or more long; L. ± falcate, swollen, trigonous to semicylindrical, acute, bluish-green, 10–25 mm long, 3 mm ⌀; Fl. 3–3.5 cm ⌀, pinkish-purple.

**L. spectabilis** (Haw.) N. E. Br. (*Mes. s.* Haw.). – Cape. – St. prostrate, with clustered, upcurved, trigonous, carinate, green L. 5–8 cm long, 6 mm across, with an apiculus and a reddish awn; Fl. 5–7 cm ⌀, purplish-red.

**L. spiniformis** (Haw.) N. E. Br. (*Mes. s.* Haw.). – Cape: Cape D. – 40 cm tall, stiff; St. few, erect, the Br. becoming thorny; L. upcurved, long-subulate, ± cylindrical, fresh green, tuberculate at the tip, up to 5 cm long, 4 mm thick; Fl. solitary, 15 mm ⌀, light red.

**L. staminodiosus** (L. Bol.) Schwant. (*Mes. s.* L. Bol.). – Cape: Clanwilliam D. – Erect, 13 cm tall; L. rounded on the underside, shortly-tapered towards the purplish-red tip, blue, indistinctly rough, 8 mm long, 1.5 mm across and thick; Fl. 12 mm ⌀, purplish-red.

**L. stayneri** (L. Bol.) N. E. Br. (*Mes. s.* L. Bol.). – S.W.Cape. – Compact, robust, c. 20 cm tall, with many spreading St.; L. semicylindrical below, trigonous above, short-tapered with a red mucro, bluish-green, 15 to 25 mm long, 3 mm ⌀; Fl. 3–3.5 cm ⌀, pinkish-purple.

**L. steenbergensis** (L. Bol.) L. Bol. (*Mes. s.* L. Bol.). – Cape: Cape D. – Low-growing, with prostrate St. 10 cm or more long; L. somewhat expanded towards the rounded or truncate tip, acutely keeled on the underside, 23 mm long, 10 mm across midway, 13 mm thick; Fl. 16 mm ⌀, pink with a whitish centre.

**L. stenopetalus** (L. Bol.) N. E. Br. (*Mes. s.* L. Bol.). – S.W.Cape. – Erect, 16–23 cm tall; L. tapered, bluish-green, with raised dots, 1–2 cm long, 2 mm ⌀; Fl. 2 cm ⌀, pale pink.

**L. stenus** (Haw.) N. E. Br. (*Mes. s.* Haw., *Mes. monticolum* L. Bol., *L. m.* (L. Bol.) L. Bol., *L. s.* f. *depauperatissimus* L. Bol.). – Cape: Cape D. – 30–50 cm tall, with many slender, spreading, curved St.; L. subcylindrical, tapered, slightly greyish-green, reddish at the tip, 12–20 mm long, 2 mm across; Fl. 3 cm ⌀, pale pink.

**L. stephanii** (Schwant.) Schwant. (*Mes. s.* Schwant.). – Cape: origin unknown. – Erect, up to 16 cm tall; L. 8–15 mm long, 3 mm across, 4 mm thick, semicylindrical, with light, small, translucent dots; Fl. terminal, 3 cm ⌀, white.

**L. sternens** L. Bol. – Cape: Bredasdorp D. – Low-growing; St. elongated, creeping, slender; L. ± falcate, ± acute, with an indistinct keel and with minute dots, 12–15 mm long, 1.5 mm across, 2 mm thick; Fl. solitary, 3 cm ⌀, golden-yellow.

**L. stipulaceus** (L.) N. E. Br. (*Mes. s.* L., *Erepsia s.* (L.) Schwant., *Mes. disgregum* N. E. Br.). – Cape: Swellendam, Uitenhage D. – 30–40 cm tall; St. with numerous short shoots; L. 4–5 cm long, 3–4 mm across, linear, semicylindrical, trigonous above, apiculate, fresh green, with translucent dots; Fl. 4 cm ⌀ purplish-red.

**L. stoloniferus** L. Bol. – Cape: Sutherland D. – Mat-forming, c. 5 cm tall; St. creeping, up to 10 cm long; L. often falcate, linear, obtuse to acute, with an indistinct and often dentate keel, 1–2 cm long, 3–4 mm ⌀; Fl. solitary, Pet. 8–9 mm long, purplish-pink.

**L. suavissimus** (L. Bol.) L. Bol. (*Mes. s.* L. Bol.). – Cape: Lit. Namaqualand. – Robust, up to 1 m tall; St. 6 mm thick, dark brown; L. ± falcate, bluntly keeled, laterally com-

**Lampranthus**

pressed, short-tapered, bluish-green, 2.5–3.5 cm long, 4 mm ⌀; **Fl.** 4.5–5.5 cm ⌀, an attractive pink, scented.

**L. —** v. **suavissimus** f. **fera** (L. Bol.) L. Bol. (*Mes. s.* f. *f.* L. Bol.). – 40 cm tall; **L.** 18–25 mm long, 2–3 mm ⌀; **Fl.** 53 mm ⌀, purplish-pink, paler inside.

**L. —** v. **oculatus** (L. Bol.) L. Bol. f. **oculatus** (*Mes. s.* v. *o.* L. Bol.). – Pet. white below.

**L. subaequalis** (L. Bol.) L. Bol. (*Mes. s.* L. Bol.). – Cape: Uniondale D. – Erect, 28 cm tall, with the herbaceous parts indistinctly rough; **L.** 8–12 mm long, 1 mm across, 1.5 mm thick, apiculate, very rounded on the underside; **Fl.** 26 mm ⌀, pale pink, Sep. unequal.

**L. sublaxus** (L. Bol.) L. Bol. (*Mes. s.* L. Bol., *Mes. laxum* L. Bol.). – Cape: Calvinia D. – Erect, laxly branched, 16 cm tall; **L.** ± compressed laterally, obtuse or tapered, green, 12 mm long, 1.5 mm across, 2 mm thick; **Fl.** 1–3 together, 2 cm ⌀, pink.

**L. subrotundus** L. Bol. (*Mes. subglobosum* L. Bol.). – Cape: Ceres D. – Erect, c. 20 cm tall; **L.** dirty green, rounded on the underside, with the tip acute to rather obtuse and rough, 9–13 mm long, c. 1 mm across, 1.5 mm thick; **Fl.** 2 cm ⌀, pinkish-purple.

**L. subtruncatus** L. Bol. v. **subtruncatus.** – Cape: Piquetberg D. – 30–45 cm tall; **St.** virgate; **L.** bluish-green, bluntly keeled above, obliquely truncate or obtuse at the tip, 15–20 mm long, 1.5 mm ⌀; **Fl.** 24 mm ⌀, Pet. very numerous, pink.

**L. —** v. **wupperthalensis** L. Bol. – Cape: Clanwilliam D. – **St.** c. 23 cm long, stiff, densely branched; **L.** 2.5 cm long, L.-scars triangular, acute.

**L. superans** (L. Bol.) L. Bol. (*Mes. s.* L. Bol.). – Cape: Clanwilliam, Van Rhynsdorp D. – Erect, laxly branched, 25 cm tall; **St.** stiff; **L.** ± falcate, expanded towards the tip, 12–16 mm long, 4 mm across, 6 mm thick; **Infl.** of many pink Fl. 18 mm ⌀, open day and night.

**L. swartbergensis** (L. Bol.) N. E. Br (*Mes. s.* L. Bol.). – Cape: Prince Albert D. – Erect, 30 cm tall; **St.** virgate; **L.** bluish-green, bluntly keeled on the underside, somewhat rough, 18 mm long, 3 mm ⌀; **Fl.** solitary, 3 cm ⌀, pinkish-purple with a yellow centre.

**L. tegens** (F. Muell.) N. E. Br. (*Mes. t.* F. Muell.). – Austr. – Low-growing, 2.5–7.5 cm tall; **St.** numerous, mat-forming, rather climbing; **L.** trigonous, tapered, pale green, glossy, 1.25–2.5 cm long, 3.4–5 mm across; **Fl.** solitary, c. 10 cm ⌀, pink.

**L. tenuifolius** (L.) N. E. Br. (*Mes. t.* L., *L. t.* (L.) Schwant., *Mesembryanthus t.* (L.) Rothm.). – Cape: Cape D. – Creeping; **St.** spreading, curved; **L.** subulate, tapered, green with translucent dots, 3–4 cm long, 2 mm thick; **Fl.** solitary, c. 4 cm ⌀, vivid red, or orange and purple.

**L. tenuis** L. Bol. – Cape: Cape D. – Glabrous, slender ♄ 26 cm tall; **L.** erect, acute, 6–11 mm long, c. 2 mm ⌀; **Fl.** solitary, 1.6–2 cm ⌀, opening in the afternoon, on a pedicel 5–6 mm long, Pet. c. 9 mm long, 1.5 mm across, pur-

plish-pink, staminodes filiform, deep red, 4 mm long.

**L. thermarum** (L. Bol.) L. Bol. (*Mes. t.* L. Bol.). – Cape: Clanwilliam D. – Erect, 10–20 cm tall; **L.** falcate, laterally compressed, truncate to rounded above, bluish, 16 mm long, 2 mm across, 5 mm thick; **Fl.** 22 mm ⌀, pink, open night and day.

**L. tulbaghensis** (Bgr.) L. Bol. (*Mes. t.* Bgr.). – W.Cape. – 30 cm tall; **St.** robust; **Br.** opposite, with numerous short shoots; **L.** falcate, trigonous, tapered and apiculate above, with rough dots, (5–)7(–10) mm long, 1.5 mm across; **Fl.** 16 mm ⌀.

**L. turbinatus** (Jacq.) N. E. Br. (*Mes. t.* Jacq.). – Cape: Uitenhage D. – **St.** spreading; **L.** crowded, elongated, tapered, trigonous; **Fl.** reddish.

**L. vallis-gratiae** (Schltr. et Bgr.) N. E. Br. (*Mes. v.* Schltr. et Bgr.). – Cape: Caledon D. – Slender, 15–20 cm tall; **St.** opposite; **L.** 4–6 mm long, obtuse, bluish with rough dots; **Fl.** in a small Infl.

**L. vanheerdei** L. Bol. – Cape: Lit. Namaqualand. – Slender; **St.** prostrate, variously curved, c. 35 cm long; **L.** bluish-green, narrowed towards the obtuse tip, 1–1.6 cm long, 1–1.5 mm across, 2 mm thick; **Fl.** c. 2 cm ⌀, pinkish-purple.

**L. vanputtenii** L. Bol. (*Mes. pittenii* L. Bol. sphalm., *L. p.* (L. Bol.) N. E. Br.). – Cape: Clanwilliam D. – Erect, 20–30 cm tall; **St.** thick, stiff; **L.** falcate, thick-fleshy, keeled towards the tip, bluish-green, 2–2.5 cm long, 6 mm across and thick; **Fl.** 7 cm ⌀, purplish-pink with a white centre.

**L. vanzijliae** (L. Bol.) N. E. Br. (*Mes. v.* L. Bol.). – Cape: Worcester D. – **St.** creeping, elongated; **L.** trigonous, tapered, green with translucent dots, 5.5 cm long, 6 mm across in the middle, 10 mm across below; **Fl.** solitary, 6 cm ⌀, golden-yellow or yellowish-orange.

**L. variabilis** (Haw.) N. E. Br. (*Mes. v.* Haw.). – Cape: Saldanha Bay. – Resembles **L. coccineus**; **Br.** more prostrate; **Fl.** yellow at first, later becoming pinkish-red.

**L. verecundus** (L. Bol.) N. E. Br. (*Mes. v.* L. Bol.). – Cape: Riversdale D. – 20 cm tall, with slender **St.**; **L.** ± falcate, bluntly keeled, 1–2 cm long, 1.5 mm across, 3 mm thick; **Fl.** solitary, 24 mm ⌀, pale pink.

**L. vernalis** (L. Bol.) L. Bol. (*Mes. v.* L. Bol.). – Cape: Clanwilliam D. – Laxly branched, older **St.** prostrate, 10–40 cm long; **L.** semicylindrical, short-tapered, fresh green, minutely crystalline-papillose, soft, 10–15 mm long, 2.5 mm thick; **Fl.** 4.4 cm ⌀, an attractive pink.

**L. vernicolor** (L. Bol.) L. Bol. (*Mes. v.* L. Bol.). – Cape: Tulbagh D. – Compact; **St.** lax, spreading to prostrate, with numerous short shoots; **L.** ± falcate, with the tip obliquely rounded, apiculate, 18 mm long, 3–5 mm across, 5–7 mm thick; **Fl.** 18 mm ⌀, an attractive pink, somewhat scented.

**L. viatorum** (L. Bol.) N. E. Br. (*Mes. v.* L. Bol.). – Cape: Lit. Namaqualand. – 15 cm or more tall; **St.** and Br. spreading; **L.** trigonous, obtuse above and awn-tipped, 1–2 cm long, 4 mm thick; **Fl.** solitary, 23 mm $\varnothing$, pink.

**L. villiersii** (L. BOL.) L. BOL. (*Mes. v.* L. BOL.). – Cape: Villiersdorp D. – Erect, c. 12 cm tall; **L.** rough, tapered, 15 mm long, 1.5 mm $\varnothing$; **Fl.** 1–3 together, 24 mm $\varnothing$, purplish-red.

**L. virgatus** L. BOL. – Cape: Clanwilliam D. – Up to 43 cm tall; **St.** virgate, brown, c. 28 cm long; **L.** linear, keeled on the underside, apiculate, green, indistinctly rough, 2–2.5 cm long, 1–1.5 mm across and $\varnothing$; **Fl.** c. 3.4 cm $\varnothing$, brilliant pink.

**L. vredenburgensis** L. BOL. – Cape: Malmesbury D. – **St.** spreading, prostrate, 15 cm long; **L.** in pairs, some lunate, c. 8 mm $\varnothing$, others not lunate, 5–6 mm $\varnothing$, acute; **Fl.** 1–3 together, Pet. in 2–3 series, 8 mm long, sometimes longer, pink to purplish-pink, paler at the base.

**L. walgateae** L. BOL. – Cape: Swellendam D. – Erect, 35 cm tall; **St.** slender; **L.** compressed, expanded above, with a shortly tapered to obtuse tip, dirty bluish-green, 10–17 mm long, 1.5 mm across, 2.5 mm thick; **Fl.** 3 cm $\varnothing$, purplish-pink.

**L. watermeyeri** (L. BOL.) N. E. BR. (*Mes. w.* L. BOL.). – Cape: Clanwilliam D. – Erect, 25–30 cm tall; **L.** semicylindrical, 2–3.5 cm long, up to 6 mm $\varnothing$; **Fl.** c. 5 cm $\varnothing$, pink to purplish-red.

**L. woodburniae** (L. BOL.) N. E. BR. (*Mes. w.* L. BOL.). – Cape: Valley of the Hex R. – Low-growing; **St.** densely crowded, elongated, creeping; **L.** keeled on the underside, bluish with a red mucro, 3–5 cm long, 5–6 mm across, 4–6 mm thick; **Fl.** solitary, 4 cm $\varnothing$, yellow with a red centre.

**L. wordsworthiae** (L. BOL.) N. E. BR. (*Mes. w.* L. BOL.). – Cape: Caledon D. – **St.** 3 cm tall; older **Br.** forming mats; **L.** semicylindrical, acute to tapered, with a red apiculus, 15–28 mm long, 3–5 mm across; **Fl.** 27 mm $\varnothing$.

**L. zeyheri** (SALM) N. E. BR. (*Mes. z.* SALM). – Cape: Uitenhage D. – **Br.** curved, slender, with numerous short shoots; **L.** cylindrical, bluntly tapered, soft, smooth, glossy green, spotted, 4 cm long, 3–4 mm across; **Fl.** 5–6 cm $\varnothing$, purplish-violet.

**Lapidaria** (DTR. et SCHWANT.) SCHWANT. ex N. E. BR. (§ IV/1/14) (*Dinteranthus* SCHWANT. SG. *Lapidaria* DTR. et SCHWANT.). Greenhouse, warm. Propagation: seed.

**L. margaretae** (SCHWANT.) SCHWANT. ex N. E. BR. (Pl. 172/2) (*Mes. m.* SCHWANT., *Dinteranthus m.* (SCHWANT.) SCHWANT., *Argyroderma m.* (SCHWANT.) N. E. BR., *A. roseatum* N. E. BR.) – S.W.Afr.: Gr. Namaqualand. – **St.** short, branching in age to form a mat; **L.** 6–8, united below, divaricate, much thickened, 10–15 mm long, 10 mm across and thick, flat on the upperside, $\pm$ hemispherically convex and acutely carinate on the underside, $\pm$ obtusely trigonous towards the tip, the surface smooth, whitish or reddish-white, with reddish margins; **Fl.** on a broadly compressed pedicel 5–6 cm long, 3–5 cm $\varnothing$, Pet. golden-yellow on the upperside, whitish-yellow on the underside, turning reddish when fading.

**Leipoldtia** L. BOL. (§ IV/1/2). – Glabrous $\hbar$, up to 50 cm tall, mostly erect or with tangled **St.**; in 3 species the Br. are prostrate and elongated, not rooting; **L.** often laterally compressed, obtuse to truncate at the tip, 0.4–4 cm long, with the thickness usually exceeding the width; **Fl.** 1–5 together in an Infl., mostly 2–3 cm $\varnothing$, purplish-pink, more rarely pale pink; pedicels up to 6.5 cm long, with bracts. – Summer: in the open, winter: cold-house. Propagation: seed, cuttings.

**L. amplexicaulis** (L. BOL.) L. BOL. f. **amplexicaulis** (*Mes. a.* L. BOL.). – Cape: Clanwilliam, Calvinia D., Lit. Namaqualand. – **Br.** prostrate, stiff; **L.** $\pm$ sabre-shaped, swollen, trigonous, obtuse above, 12–30 mm long, 6.5 mm across, 5 mm thick; **Fl.** 25 mm $\varnothing$, Pet. purplish-pink, paler on the underside, with an indistinct red stripe.

**L. — f. fera** L. BOL. – Cape: Sutherland D. – Flowering **Br.** 14–19 cm long; **L.** bluntly keeled on the underside, obtuse to rounded at the tip, 15–22 mm long, 4 mm across and thick; **Fl.** 2.5–3 cm $\varnothing$, purplish-pink, with the pedicels persisting.

**L. aprica** (BGR.) L. BOL. (*Mes. a.* BGR.). – Cape: Lit. Namaqualand. – **St.** angular; **Br.** with numerous short shoots; **L.** erect, incurved, compressed-trigonous, expanded above, with the tip shortly and obtusely tapered, apiculate, with acute angles, spotted, 10–13 mm long, c. 3 mm across; **Fl.** on a pedicel 2–4 cm long.

**L. brevifolia** L. BOL. – Cape: Lit. Namaqualand. – 30 cm tall; **St.** rather stiff; **L.** c. 1 cm long, 5 mm thick, with a scarcely discernible keel and angles, with dense, prominent dots; **Fl.** 22 mm $\varnothing$, purplish-pink.

**L. britteniae** (L. BOL.) L. BOL., (*Mes. b.* L. BOL.). – Cape: Robertson D. – **St.** stiff, 20–30 cm long, decurved at the tip; **L.** trigonous, swollen, shortly tapered, bluish-green, spotted, 10–16 mm long, 4 mm $\varnothing$; **Fl.** 28 mm $\varnothing$, pinkish-purple.

**L. calandra** (L. BOL.) L. BOL. (*Mes. c.* L. BOL.). – Cape: origin unknown. – Prostrate; flowering **St.** erect, with 2–4 **L.** with a blister-like swelling at the base, obtuse and apiculate above, green, minutely spotted, 2 cm long, 4 mm across, 5 mm thick; **Fl.** 43 mm $\varnothing$, pale pink.

**L. compacta** L. BOL. – Cape: Lit. Namaqualand. – 7–8 cm tall, with dense **St.** 7 cm long; **L.** bluntly keeled, green, scaly, 15 mm long, 2–4 mm thick; **Fl.** 17–22 mm $\varnothing$, pink.

**L. compressa** L. BOL. v. **compressa** – Cape: Lit. Namaqualand. – **St.** stiff, c. 45 cm long, creeping; **L.** laterally compressed, obtuse above, green to yellowish-green, 2 cm long, 3–4 mm across, 6–7 mm thick; **Fl.** 2.5–3 cm $\varnothing$.

**L. — v. lekkersingensis** L. BOL. – Cape: Lit. Namaqualand. – **St.** c. 19 cm long; **L.** 1–1.8 cm long, 6 mm across, 4–9 mm thick, rounded above; **Fl.** 4.2 cm $\varnothing$, pale pink.

**L. constricta** (L. BOL.) L. BOL. (*Mes. c.* L. BOL.). – Cape: Clanwilliam D., Lit. Namaqualand. – Robust, compact, c. 25 cm tall; **L.** trigonous, acute, bluish, 10–15 mm long, 5 mm

∅; Fl. 1.5–2 cm ∅, Pet. pink with a dirty pink stripe.

**L. framesii** L. Bol. – Cape: Lit. Namaqualand. – Robust, erect, 30 cm tall; **St.** stiff; **L.** obtuse or shortly tapered, 15 mm long, 6 mm ∅; **Fl.** 15 mm ∅, pink.

**L. grandifolia** L. Bol. – Cape: Lit. Namaqualand. – Stiff, sparsely branched, 45 cm tall; **L.** ± compressed, keeled on the underside, with the tip truncate to ± obtuse, bluish-green, often reddened, 2.5–4 cm long, 4–6 mm across, 6–11 mm thick; **Fl.** c. 3 cm ∅, deep pink.

**L. jacobseniana** Schwant. (*Rhopalocyclus nelii* Schwant.). – Cape: Robertson D. – **L.** up to 15 mm long, 4 mm ∅, acutely keeled on the underside, navicular-curved at the tip; **Fl.** 12–15 mm ∅.

**L. klaverensis** L. Bol. – Cape: Van Rhynsdorp D. – Stiff, erect, laxly branched, 20 cm tall; **L.** keeled on the underside, with a short spur below, narrowly tapered, truncate or obliquely obtuse above, dirty green, spotted, 10–13 mm long, 3–4 mm thick; **Fl.** 2 cm ∅, pink.

**L. laxa** L. Bol. – Cape: Lit. Namaqualand. – Laxly branched, c. 25 cm tall; **L.** vesicular-inflated at the base, bluish-green, 3–4 cm long, 4–5 mm across, 5–6 mm thick; **Fl.** 2–3 together, up to 2.6 cm ∅, pale pink.

**L. littlewoodii** L. Bol. – S.W.Afr.: S. of Witpütz. – 25 cm tall, 50–55 cm ∅; older **St.** up to 12 cm long; **L.** obtuse to obliquely truncate at the tip, with the keel reduced to a line, 1–1.3 cm long, 5 mm ∅; **Fl.** in bi- or tripartite Infl., 1.7–2 cm ∅, Pet. purplish-pink, indistinctly striped.

**L. nelii** L. Bol. – Cape: Van Rhynsdorp D. – 16 cm tall; **St.** erect, eventually prostrate; **L.** with a decurrent keel on the underside, bluish-green, 1–2 cm long, 2–3 mm ∅; **Fl.** 2 cm ∅, dirty pink.

**L. pauciflora** L. Bol. – Cape: Calvinia D. – Erect, 35 cm tall; **St.** lax, spreading, slender, stiff, 25–30 cm long; **L.** firm, with rather indistinct keel and margins, shortly tapered above, blue to green or somewhat reddish, slightly spotted, 2–2.5 cm long, 4–5 mm ∅; **Fl.** 26 mm ∅, purplish-pink.

**L. plana** (L. Bol.) L. Bol. (*Mes. planum* L. Bol.). – Cape: Van Rhynsdorp D. – **St.** elongated, prostrate, slender, 10–12 cm long; **L.** rounded-trigonous, 12 mm long, 5–6 mm thick, dirty green, with raised dots; **Fl.** 2.5 cm ∅, pink.

**L. schultzei** (Schltr. et Diels) Friedr. (*Mes. s.* Schltr. et Diels). – S.W.Afr.: Gr. Namaqualand. – Small; **St.** with a peeling bark; **L.** trigonous to thick-semicylindrical, tuberculate, bluish, 15 mm long, 5 mm across, **Fl.** solitary, Cal. with black spots or dots, Pet. 8–10 mm long, 1.5 mm across.

**L. uniflora** L. Bol. – Cape: Lit. Namaqualand. – Erect, 17 cm tall; **St.** virgate; **L.** hemispherical, some 4 mm long, 3 mm thick or some 4–7 mm long, 3–5 mm thick, or sometimes elongated, 11 mm long, 3 mm thick; **Fl.** solitary, 17 mm ∅, pink.

**L. weigangiana** (Dtr.) Dtr. et Schwant. (*Mes. w.* Dtr., *Cephalophyllum w.* (Dtr.) Dtr. et Schwant., *Rhopalocyclus w.* (Dtr.) Dtr. et Schwant.). – S.W.Afr.: Gr. Namaqualand. – Erect, up to 50 cm tall, with 3–20 **St.** 2–3 mm thick with a yellowish-white bark; **L.** patent, 15 mm long, 4 mm thick, navicular, trigonous, bluish-green with dense dots; **Fl.** solitary, 2 cm ∅, violet-pink.

**Lithops** N. E. Br. (§ IV/1/14). – Glabrous, extremely succulent plants, mostly caespitose; **Br·** mostly undeveloped, the ☉ growth consisting of one or more ± obconical Bo. composed of a pair of **L.** which are united for half or almost the whole of their length, the apex of each L. ± semicircular, flat or convex, sometimes almost completely transparent, sometimes with smaller or inconspicuous "Wi.", together with an infinite variety of markings or dots; **Fl.** terminal, solitary, the next season's growth developing from the axil of one, or both, of the L.; pedicels compressed, without bracts, Pet. and stamens yellow or white; nectaries annular, stigmas mostly 5, thin; loculi of the Fr. open. Flowering time July-Nov. – Greenhouse, warm. Propagation: seed, or single Bo. can be rooted. (Lit.: Nel, G. C.: Lithops, Stellenbosch 1947, Cole, D. T. Lithops: a Checklist and Index, Excelsa III: 37–71, 1973).

### Division of the Genus Lithops into SG. according to G. Schwantes[1]
Type species: **L. turbiniformis** (Haw.) N. E. Br.

**§ I. SG. Lithops** (*Xantholithops* Schwant.). – Fi. in young seedlings not extending over the whole width; colour of the Fl. yellow. – Type species: L. turbiniformis. – Further species: L. archerae, aucampiae, brevis, bromfieldii, christinae, comptonii, dinteri, divergens, dorotheae, franciscii,

---

[1] It was Ernst Rusch who first discovered that the two types of seedling in **Lithops** were connected with the two flower colours and that the 'young' forms always had yellow flowers and the 'older' form always had white flowers. The terms 'young' and 'old' refers to the stage of development of the seedlings.
As will be seen later, the correlation does not hold in every case as there are many exceptions. The use of seedling form and flower colour is a useful division of the genus into two groups for the purpose of identification, but I am not convinced that the genus is diphylectic as proposed by Schwantes simply because of these two characters. The division into two S.G. is rather artifical and it is proposed to maintain the two groups as Ser. A. (SG. *Lithops*) and Ser. B. SG *Leucolithops* Schwant.) without subgeneric status. **B. Fearn**, May 1973.

fulviceps, gesinae, glaudinae, gracilidelineata, helmutii, herrei, hookeri, insularis, lesliei, localis, marginata, marthae, mennellii, meyeri, olivacea, otzeniana, pseudotruncatella, ruschiorum, schwantesii, steineckeana, vallis-mariae, verruculosa, viridis, weberi, werneri.

§ II. SG. **Leucolithops** Schwant. – Fi. in young seedlings extending over the whole width; colour of the Fl. white. – Type species: L. karasmontana. – Further species: L. bella, deboeri, elisae, erniana, fulleri, hallii, julii, marmorata, optica, salicola, villetii.

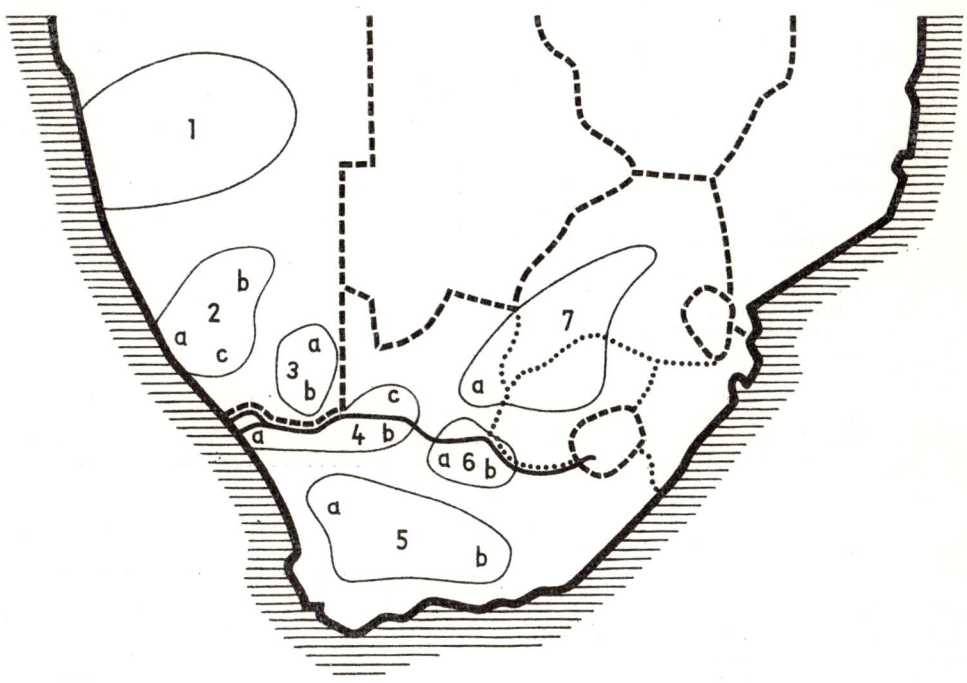

River ▬▬▬▬   International boundaries ▬ ▬ ▬ ▬   Provincial boundaries ••••••••

**Occurrence of the G. Lithops in S. and S.W.Afr.**
Drawn by de Boer and Boom

1. Damaraland

2. Gr. Namaqualand
   a) Luederitz
   b) N. Gr. Namaqualand
   c) S. Gr. Namaqualand

3. S.E. Gr. Namaqualand
   a) Karas Mts.
   b) Warmbad

4. N.W.Cape
   a) Lit. Namaqualand, Lit. Bushmanland
   b) Gr. Bushmanland
   c) Gordonia etc.

5. Central Cape
   a) W. part (Van Rhynsdorp, Ceres D. etc.)
   b) Central (Graaff Reinet, Willowmore D. etc.)

6. N. Cape
   a) Kenhardt D.
   b) Prieska D.

7. Transvaal and adjacent territories
   a) N.W.Cape, W.Griqualand.

## Key to the Species of the Genus Lithops by H. W. De Boer and B. K. Boom[1])

This key applies to adult specimens in the summer; in winter the bodies shrivel too much and the colours are not characteristic. The following terms require some explanation:

WINDOW – A transparent part of the upper surface, sometimes covering the whole top, sometimes branched into well-defined or vague strips which may be very narrow.

ISLAND – The opaque part of the top surface surrounded by the window.

PELLUCID DOTS – Miniature windows of at most 1 mm $\varnothing$, mostly greenish or bluish in colour and scattered over the top; they are not considered here as belonging to the window.

BLUE DOTS – Miniature windows situated under the epidermis, often inconspicuous, never clear-cut, and not to be confused with pellucid dots which belong to the epidermis.

FLOWERS – Where described as, e.g. 5-merous, this implies that there are 5 sepals, 5 styles and 5 fruit-valves.

SEEDS – The size of the seeds is an important characteristic for classification although obviously impracticable for easy identification. The authors recognise 5 groups:

Group 1: 1500–2000 seeds per cubic cm.
Group 2: 4000–5000 seeds per cubic cm.
Group 3: c. 7000 seeds per cubic cm.
Group 4: c. 10000 seeds per cubic cm.
Group 5: c. 15000 seeds per cubic cm.

1. Flowers white; top of the young seedling (3–7 weeks old) ± convex; fissure rather long, separating the two leaves for more than a half ............... **2.**
1*. Flowers yellow or straw-coloured to light orange; top of the young seedlings rather flat, the fissure localised in the form of a rather short aperture in the centre of the top .................................................................... **16.**
2. Fissure wide, in adults mostly deeper than 10 mm; bodies greyish-green, sometimes suffused with pink, or entirely purplish-red; flowers 5- or 6-merous ... **3.**
2*. Fissure not wide, hardly ever deeper than 10 mm: bodies not greyish-green or purplish-red; flowers nearly always 5-merous ........................... **4.**

Pl. 173 Row 1
{
3. Bodies greyish-green to greenish, sometimes suffused with amethyst-colour; window with grey islands; flowers mostly 6-merous; seed-group 5. – Cape: Lt. Namaqualand (4a) .................................................. **marmorata**
3 a. Windows reduced to narrow strips. – Cape: L. Namaqualand (4 a) ... **elisae**
3*. Windows without or with only some islands, or slightly maculate along the margin; flower mostly 5-merous; seed-group 4. – Gt. Namaqualand (2 a) .... **optica**
3*a. Bodies entirely purplish-red ................................. **— f. rubra**
3*b. Bodies pea-sized .......................................... **— f. minor**
}

4. Top with brown or reddish-brown lines (sometimes only within the ultimate branches of the window or along the inner margin) ..................... **5.**
4*. Top without brown lines ................................................. **11.**

5. Top flat with grey margins; window large with some islands, or consisting of broad strips; outer margin with shell-like notches and dark brown spots or lines in the projections between these notches; seed-group 5. – Cape: Kenhardt D. (4b) ........ **fulleri**

Pl. 173 Row 2
{
5 a. Window green, purplish or reddish-brown ................ **— v. fulleri**
5 b. Window brownish, with short ultimate projections; islands smaller, notches broader and shorter, dark brown lines within the projections less distinct ... ........ **— v. brunnea**
5 c. Windows deep ochre-coloured, islands and sides light ochre-coloured ..... ........ **— v. ochracea**
5 d. Window reduced to narrow strips, hence the islands larger; colours as in the type variety; not in cultivation ................. **— v. chrysocephala**
5 e. Window conspicuous only in the juvenile-stage, colour grey-brown ........ ........ **— v. kennedyi**
}

5*. Top slightly convex; outer margin of the window not as in L. fulleri ........ **6.**

---

[1]) "An analytical key for the Genus Lithops" in Nat. Cact. Succ. J. 19: 34–37, 51–55, 1964; revised. NOTE: Designations like 2a, 5b, etc. following the geographical distribution refer to the map on p. 503. Rows (left and column) refer to the illustrations on Pl. 173–178. **L. hookeri** is not included in this key; see explanation under **L. hookeri**.

**6.** Top slightly convex, strikingly light coloured; window often visible; network coloured or not; inner margin nearly always with a brown, sometimes fragmentary line; seed-group 5. – Warmbad (3b) .................................. **julii**

Pl. 173 Row 3
- **6a.** Network only a little darker than the top .................... **— v. julii**
- **6b.** Top whitish, pinkish or bluish-pink, with a conspicuous yellowish-brown impressed network .................................. **— v. reticulata**
- **6c.** Top dark to light grey-pink, network in the centre suffused, marking along the outer margin dark brown, formed as an X or Y, along the inside some dark brown dots or lines................................... **— v. rouxii**
- **6d.** Differs from v. julii by the larger bodies and by the absence of the brown lines along the fissure; hence rather similar to L. karasmontana v. opalina ........ **— v. littlewoodii**

**6\*.** Top not so strikingly light coloured and never with a brown line along the inner margin........................................................... **7.**

**7.** Top with a mostly branched window ................................ **8.**
**7\*.** Window absent ..................................................... **9.**

**8.** Top slightly convex; window not surrounded by a whitish border; brownish markings mostly not very distinct ....................... **bella (see 14a)**
**8\*.** Top flat; window surrounded by a conspicuous whitish border ............ **9.**

**9.** Top flat; ochre-yellow to purplish-brown; grooves and markings continuously running from the outer margin forming a rather fine network; brown lines not broader at the junctions; the grooves sometimes rather broad and confluent, forming windows; seed-group 5. – Prieska (6) ........................ **hallii**
**9\*.** Top slightly but conspicuously convex; grooves mostly not running from the inner to the outer margin, hence the top seldom distinctly reticulate; brown lines often broader at the junctions ........................................ **10.**

Pl. 173 Row 4
**10.** Markings rather wide, not or indistinctly grey-edged; ultimate projections single or bifurcate; seed-group 5. – S.W.Afr.: Karas Mts. (3b) ........ **karasmontana**
- **10a.** Top light greyish-yellow to light reddish-brown, sometimes slightly brownish suffused ................................................. **— v. karasmontana**
- **10b.** Top reddish-brown; network less impressed and without brown lines; perhaps L. gulielmi belongs here ..................... **— v. summitatum**
- **10c.** Top yellowish-red; network finer, more distinct, with darker, sometimes stellate-branched lines........................... **— v. mickbergensis**
- **10d.** Top opaque to light amethyst-coloured, with vague markings .......... ........ **— v. opalina**

**10\*.** Markings narrow, distinctly grey-edged; ultimate projections mostly trifurcate (as chickenlegs); seed-group 3. – S.W.Afr.: Gt. Namaqualand (2c) ..... **erniana**

Pl. 173 Row 5
- **10\*a.** Top brownish-yellow; network fine, dark brown (resembles L. karasmontana v. mickbergensis) ............................ **— v. erniana**.
- **10\*b.** Top light amethyst-coloured with light blue markings, without or with only a few brown lines (resembles L. karasmontana v. opalina) ............... ........ **— v. witputzensis**
- **10\*c.** Top yellowish- to reddish-grey, the marking consisting of narrow, less connected lines and scarcely distinguishable. Perhaps transitional to L. karasmontana v. opalina ....................... **— v. aiaisensis**

**11.** Top slightly convex, bluish-grey; window dark, brownish-grey, mostly with brown to chocolate-coloured islands; base of the lobes in the fissure with a conspicuous pink blister; flowers 6-merous, very late flowering (from the end of October); seed-group 5. – Habitat unknown ..................... **deboeri**
**11\*.** Base of the lobes without a conspicuous pink blister; flowers mostly 5-merous, already finished in October; seed-group 5 or 3 ........................ **12.**

**12.** Markings vague or absent ............................................. **13.**
**12\*.** Markings conspicuous, often reticulate ............................... **14.**

**13.** Top light yellowish-pinkish, with slightly darker, vague markings and hardly ever with brown points or short lines along the inner margin (examine several specimens) ................................. **julii v. littlewoodii (see 6d)**
**13\*.** Top opaque to light amethyst-coloured with vague "karasmontana"-markings, always without a brown line ........... **karasmontana v. opalina (see 10d)**
- **13\*a.** Top as in the preceding variety, but markings finer, often bluish, the ultimate branches bifurcate ........ **erniana v. witputzensis (see 10\*b)**

Lithops 506

14. Top slightly convex; window often impressed, ± transparent, with dark coloured strips of which one is often fragmentarily present along the inner margin and antler-shaped ending in the outer margin; sometimes some brownish-red lines are present; seed-group 3. – S.W.Afr.: Gt. Namaqualand (2c) and Karas Mts. (3a) ........ **bella**

Pl. 174
Row 6

 14a. Strips of the window dark coloured; islands light yellow, pinkish-yellow or light brownish-yellow ................................. — v. **bella**
 14b. Differing from the preceding variety by the darker strips and the darker brownish islands ................................... — v. **lericheana**
 14c. Window less impressed, less transparent, more branched on the outer margin; islands grey to bluish-grey, sometimes brown or purplish; often these characters not simultaneously present; the named L. eberlanzii in our cultures is sometimes L. erniana (see 10) ............... — v. **eberlanzii**
14*. Window large, often with islands ........................................ **15.**

Pl. 174
Row 7

15. Main colour of the bodies lead-grey; window olive-green, mostly with light grey to light greyish-pink coloured islands, mostly with a conspicuous whitish border; flowers 5-merous; seed-group 3. – Cape: S. of Orange River (6b) ...... **salicola**
15*. Main colour of the bodies greenish-yellow to nut-coloured; window more olive-brown, often with greyish to nut-coloured islands; flowers 6-merous; Calvinia (5a) ........ **villetii**

16. Top with blue or greenish dots .......................................... **17.**
16*. Top without pellucid dots ............................................... **23.**

17. Pellucid dots large, rather numerous and regularly scattered; seed-group 3 .. **18.**
17*. Pellucid dots not so large and not regularly scattered ..................... **19.**

18. Top convex, evenly coloured between the large, bluish-green dots and the occasional window; flowers 5-merous; Cap: Cape Prov. (5b) .......... **localis**

Pl. 174
Row 8

 18a. Bodies rather small; top fawn, marked only with pellucid dots, window nearly always absent ................................ — v. **localis**
 18b. Bodies somewhat larger in comparison with the type-variety; top bluish-purple; window mostly absent, but sometimes rather distinct ........ — v. **peersii**
 18c. Bodies as large as in the preceding variety; top greyish- to yellowish-green; window large or rather small, greenish .............. — v. **terricolor**
18*. Top flat or slightly convex, dirty blue to reddish brown, with rather numerous large, dark bluish-green, slightly raised pellucid dots intermixed with tiny red lines; window absent. – S.W.Afr.: Karas Mts. (3a) ............... **fulviceps**

19. Pellucid dots arranged along dendritic markings, which are often red or brown and mostly with blue edges; sometimes the red lines are present only and the pellucid dots only visible by a lens; flowers 6-merous, mostly in July (except L. werneri) ............................................................. **20.**
19*. Pellucid dots not arranged as in 19. or along narrow window-strips (L. lesliei); flowers 5- or 6-merous, from August to October ........................ **21.**

20. Bodies rather small, 1/2 to 2/3 of the size of the next species; top convex, greyish to brown, with 2–3 dendritically branched lines; there are specimens nearly without markings (similar to L. ruschiorum) and specimens with very narrow lines (as in L. gracilidelineata). – S.W.Afr.: N.Damaraland (1) ................. **werneri**
20*. Bodies rather larger; top flat or slightly convex, greyish to brown, with at least 4 dendritically branched lines (these markings sometimes indistinct). – S.W.Afr.: N.Damaraland (1) ................................ **pseudotruncatella**

Pl. 174
Row 9

 A. Edges of the markings usually darker than the rest of the top.
 20*a. Top brownish-grey, dendritic markings brown or brown-red, with dark bluish or grey edges; bodies 20–30 mm ⌀ .......... v. **pseudotruncatella**
 20*a1. Bodies 10–12 mm ⌀ ...................................... v. **alpina**[1])
 20*b. More bluish-grey; top more even brownish grey in the centre, suffused with blue-green; markings and pellucid dots sometimes indistinct ........... — v. **pulmonuncula**
 20*c. Top greyish-blue, mostly coppery suffused; pellucid dots not always distinctly dendritically arranged ................... — v. **elisabethae**

¹) added by H. J.

Pl. 174
Row 10

 20*d. Top light bluish-grey; lines and dots rather distinct; dark bluish edges sometimes indistinct ............................... — v. **edithae**
 20*e. Top evenly yellow-brown with very prominent dark red-brown dendritic markings, rather vague bluish edges and very vague bluish dots ....... ........ — v. **brandbergensis**

B. Edges of markings not or scarcely darker than rest of top.
 20*f. Bodies rather flat; top rather even yellowish brown, with distinct dendritic markings ............................................. — v. **mundtii**
 20*g. Top greyish-brown, with prominent brown dendritic markings; pellucid dots often only visible with a lens or nearly absent; sometimes the top is greyish with very short, brown lines, which, although not connected, are distinctly dendritically arranged (this is in cultivation as L. farinosa, the difference being too small to maintain it as a separate taxon) ....... ........ — v. **dendritica**
 20*h. Bodies greyish-white to almost white; markings indistinct ... — v. **volkii**

21. Top almost flat; fissure very narrow, caused by the closely set lobes; window (at least in typical specimens) consisting of irregular confluent areas; pellucid dots mostly numerous, but never dominating, sometimes dendritically arranged along very narrow window-strips; flowers 5-merous; seed-group 1. – Transvaal, Orange Free State (7) ............................................. **lesliei**

Pl. 175
Row 11

 21a. Bodies dull ferruginous to greenish; window green; islands and margins ferruginous-yellow ................................................ — v. **lesliei**
 21b. Main colour of bodies light ochre; window strongly branched, grey-brown ............................................................ — v. **hornii**
 21c. Main colour of bodies grey; window greenish- to dark grey, mostly divided into narrow dendritic strips; islands and margins greyish-yellow ........ ........ — v. **venteri**

Pl. 175
Row 12

 21d. Sides grey in various shades; window very variable, greenish-brown to brownish-red, strongly branched or covering the whole top; islands and margins ochre-coloured ................................................ — v. **maraisii**
 21e. Main colour of the bodies green; windows light green; islands yellowish-green ............................................................. — v. **luteo-viridis**
 21f. Bodies reddish, sometimes reddish-brown, except the window, which is greenish-brown, less branched caused by the less numerous islands ..... ........ — v. **rubrobrunnea**
 21g. Main colour of bodies dark or lighter grey-orange; window absent; small somewhat impressed dots numerous, often confluent into short transparent lines ............................................................... — v. **mariae**
 21h. Bodies half as large (to ± 1 cm ⌀) and cinnamon coloured except the window .......................................................... — v. **minor**

21*. Top mostly ± convex; fissure not very narrow; window not branched as in the preceding; flowers 6-merous; seed-group 3 ............................ **22.**

Pl. 175
Row 13

 22. Top convex, greyish-yellow, sometimes tinged with green or greenish-brown; window scarcely visible or composed of an irregular network of vague, rather narrow bands; pellucid dots rather small, sometimes inconspicuous; bodies make a clear grey impression. – S.W.Afr.: Gt. Namaqualand (2a) ......... **franciscii**
 22*. Top ± convex, window always distinct and branched, with antler-like branches in the outer margin; pellucid dots conspicuous or absent; general colour of the bodies dirty brownish- to reddish-grey. – S.W.Afr.: Gt. Namaqualand (2b) ... ........ **gesinae**
 22*a. Window dirty green, without vague blue dots, antler-like branches rather few; pellucid dots rather numerous; of the same colour as the window or more bluish ....................................................... — v. **gesinae**
 22*b. Window dark pink to dirty brown, with vague bluish dots; antler-like branches numerous; pellucid dots absent or few ............. — v. **annae**

23. Top strikingly flat and reddish-brown; fissure very narrow. – N.W.Cape (7a) ........ **aucampiae**

Pl. 175
Row 14

 23a. Window dark olive-green to reddish-brown, strongly branched, with numerous antler-like branches in the outer margin and without spots or dots ............................................................. — v. **aucampiae**
 23b. Colour of top dull brown; islands strongly raised, with numerous, impressed, narrow, radially directed grooves on the margin of the top ... ........ — v. **euniceae**
 23c. Window absent; pellucid dots numerous, greenish, usually connected by small, ± impressed red lines ...................... — v. **koelemanii**

Lithops 508

Pl. 175 Row 14
- **23 d.** Window open to half-open, olive-grey to grey-brown, translucent, margins and islands dark greenish grey to brownish green, with grey-green to orange coloured, bifurcate or clavately ending lines (grooves) directed to the margins from the windows .......................... — **v. fluminalis**

**23\*.** Bodies not so reddish-brown coloured ............................. **24.**
**24.** Top with red or brown dots or lines; sometimes only a few are present; flowers 5-, sometimes 6–7-merous ................................................. **25.**
**24\*.** Top never with red or brown dots or lines; flowers always 5-merous ........ **40.**
**25.** Top slightly convex, bluish-green to yellowish-brown, with small raised red (later grey) warts; flowers 5-merous, light brownish-yellow to orange; seed-group 5. – Cape: Kenhard D. (6a) .................................. **verruculosa**

Pl. 175 Row 15
- **25 a.** Window reduced to grooves giving a finely rugose character; top bluish-grey, the fine impressed network of the same colour or a little darker ..... ........ — **v. verruculosa**
- **25 b.** Top with a band-like branched window instead of grooves, bluish-grey to brownish ............................................................. — **v. inae**
- **25 c.** Top light greyish blue; blood-red shining dots (warts) absent; with a big marbled window without islands ......................... — **v. glabra**

**25\*.** Raised red warts absent; flowers yellow ............................... **26.**
**26.** Bodies rather small; window large, sometimes irregularly but seldom distinctly band-like branched (in L. dinteri) ......................................... **27.**
**26\*.** Bodies larger than in the preceding; window absent or band-like branched ... **28.**

**27.** Top flat or slightly convex; window brownish-green, sometimes with yellowish dots and always with 3–15 red dots or little lines, bordered irregularly with brownish-yellow; margin with dark bluish-green dots; seed-group 5. – S.W.Afr.: Warmbad (3 b) ..................................................... **dinteri**

Pl. 176 Row 16
- **27 a.** Top reddish-grey to yellow, with numerous transparent dots (Fenestrella), and 5–12 blood-red dots in the windows, also some blood-red dots which are pulled out in lines ........................................ — **v. dinteri**
- **27 b.** Top ochre-coloured, with a bluish suggestion of windows with 15–25 very striking blood-red dots, pellucid dots and lines absent ................ ........................................................ — **v. multipunctata**

**27\*.** Top flat; window light brownish- or yellowish-green, mostly with 1–3 dull red dots, less marked, usually with some bluish dots, bordered by a sharp, yellowish band; the bluish-green dots are present in the margin or not; seed-group 4. – Cape: Lt. Namaqualand (4a) ............................................. **brevis**

**28.** Top dark coloured, finely papillose; islands reddish-brown; window greenish-grey, large, sometimes branched, with red (sometimes greyish-green) lines and with numerous glossy pellucid points, which are especially distinct if they are seen against the light. – Griqualand West (4c) .............................. **glaudinae**

Pl. 176 Row 17
- **28\*.** Top without the glossy points mentioned above ........................ **29.**
- **29.** Top with a branched window; flowers 5-merous ........................ **30.**
- **29\*.** Top without a window or, if the window is vaguely present, it is never branched; flowers 5- or 7-merous ............................................. **33.**
- **30.** Top slightly convex, smooth to the touch, buff-coloured; window branches dark grey, with red dots and red, sometimes bifurcated lines; colour of the window strongly contrasting against the islands and margin; seed-group 3. – Cape: Lt. Namaqualand (4a) ..................................... **dorotheae**

**30\*.** Top not coloured as in L. dorotheae .................................. **31.**

Pl. 176 Row 18
- **31.** Strips of the window yellowish-brown with bullate, yellowish islands, brownish-red lines and blue points; outer margin with 6–10 bifurcations; plants make at some distance a yellowish red-brown impression; seed-group 3. – Cape: N. of Orange at Upington (4c) ......................................... **bromfieldii**
- **31\*.** Markings finer; outer margin with 10 more bifurcations; look of the plants at some distance not so yellowish-redbrown ................................ **32.**
- **32.** Top slightly bullate but smooth to the touch; window greenish-grey, distinctly transparent; islands yellowish-brown; red lines and dots present (no blue dots); plants make at some distance a rather dark greenish brown impression; seed-group 3. – N. of Orange at Upington (4c) ............................. **insularis**
- **32\*.** Differing from the preceding: top not smooth to the touch, with a fine network; window less transparent; red lines absent or indistinct in the centre of the window and visible only with a lens; plants make a grey-brown impression; seed-group 2. – Hopetown (4b) .................................................. **marginata**

33. Top strongly convex, light coloured; markings absent or consisting of some lines or dots; flowers 6–7-merous, seldom 5-merous; seed-group 5. – S.W.Afr.: Damaraland (1) .................................................... **ruschiorum**

Pl. 176
Row 19

    33 a. Bodies grey, sometimes suffused with amethyst, top with some dark points or short lines; flowers 6–7-merous ................. — v. **ruschiorum**
    33 b. Bodies whitish- to orange-yellow; top with some conspicuous dark, slightly impressed lines; flowers 6–7-merous ................... — v. **lineata**
    33 c. Bodies smaller, rather variable in size, light grey to nearly white; top without markings or with some brown little spots or lines; flowers 5-merous ........ — v. **nelii**
    33 d. Differs from L. ruschiorum by the shorter lobes and the top, which has some scattered greyish-green dots; flowers 7-merous (perhaps a hybrid: L. pseudotruncatella × L. ruschiorum) ................. **steineckeana**

33*. Top less convex, mostly distinctly marked; flowers 5–6-merous ............ **34**.

34. Bodies yellowish to bluish grey; top slightly convex, wrinkly by very fine vermiform and chalky ridges (use a strong lens); markings absent or consisting of very vague little darker coloured (never reddish brown) lines; seed-group 2. – S.W.Afr.: Gt. Namaqualand (2 b) ..................................... **vallis-mariae**

    34 a. Bodies yellowish, somewhat convex, chalky ridges rather conspicuous, markings mostly consisting of diffuse wide lines ..... — v. **vallis-mariae**
    34 b. Bodies bluish-white; top mostly flat, chalky ridges inconspicuous, rugosity consisting of papillae, markings absent, or consisting of narrow interrupted lines ........................................... — v. **groendraaiensis**
    34 c. Bodies grey to flesh-coloured; top tessellate, grooves narrow, brown, interrupted here and there ....................... — v. **margarethae**

34*. Top not wrinkly as in the preceding ................................. **35**.

Pl. 176
Row 20

35. Bodies rather large, strikingly light grey, sometimes yellowish, often slightly suffused with amethyst; top flat or slightly convex; lines very narrow, dark brown, not impressed, sometimes reticulately arranged, sometimes individually running from the inner to the outer margin; a few pellucid dots sometimes present; flowers 6-merous, in August. – S.W.Afr.: Damaraland (1) ........... ...... **gracilidelineata**

    35 a. Top greyish-white, flat or slightly bullate; lines continuously running from the inner to the outer margin, forking branched .. — v. **gracilidelineata**
    35 b. Top light yellowish-grey, strongly bullate; lines very narrow, interrupted and strongly branched ........................... — v. **waldronae**

35*. Top not coloured as in the preceding and lines not so narrow; flowers 5-merous ........ **36**.

36. Grooves distinct, in summer, forming a network with continuous brown to reddish brown lines; blue dots mostly absent; seed-group 2 ...................... **37**.
36*. Grooves indistinct or only present in winter and after dry cultivation; top with irregular markings consisting of interrupted brown or reddish-brown lines; blue dots sometimes suffused, usually present (in L. mennellii absent); seed-group 2 or 3 ............................................................. **38**.

37. Top rather flat, greyish-brown to brownish-red, without dots but with an impressed network of narrow, reddish-brown grooves and a reddish brown line along the fissure; sides brown; there are also forms with broader grooves, ± confluent, forming a window. – Cape (6) ........................... **turbiniformis**

Pl. 177
Row 21

    37 a. Top grey-brown, rusty-brown to red-brown, with a rather dense network of dark coloured grooves ....................... — v. **turbiniformis**
    37 b. Top light grey-yellow to light yellow-brown, the grooves sharply marked, furrows dendroid to tessellately branched to brownish-red ... — v. **lutea**
    37 c. Top fine dark violet-brown, otherwise as var. turbiniformis ............ ........ — v. **brunneo-violacea**
    37 d. Top dark grey-brown, the grooves broad, flat, semipellucid, green-grey, similar to an inconspicuous window ............... — v. **subfenestrata**
    37 e. Top faint greyish yellow or orange-yellow, seldom grey-brown, grooves less than 1 mm broad, grey-green, sometimes grey- to orange-coloured, irregularly branched and bifurcate, in these red dots and short lines visible ............................................. — v. **susannae**
    37 f. Top faint, greyish yellow or orange-yellow, seldom grey-brown, with an irregular network of narrow, pellucid grey-green to greyish-orange coloured grooves and islands in the windows ........... — v. **elephina**

Lithops

|  | | |
|---|---|---|
| | 37*. | Top flat, greyish-yellow, sometimes tinged with blue, with some bluish-green dots; network slightly impressed, consisting of darker grooves and with brownish-red points or little lines; no line along the fissure; sides lead-grey. – S.W.Afr.: Gt. Namaqualand (2b) .................................................. **christinae** |
| | 38. | Top flat, slightly bullate between the grooves, dirty brownish-grey; lines dark brown, interrupted, rather thick and slightly impressed, according to some authors resembling Hebrew script; blue points always absent. – Cape: N. of Orange at Upington (4c) ........................................ **mennellii** |
| Pl. 177 Row 22 | 38*. | Top flat or convex, not coloured and marked as in the preceding, never bullate between the grooves ................................................. **39**. |
| | 39. | Top flat, brownish to greenish-blue, bordered by a light yellow band; grooves absent, markings only consisting of some scarcely visible, brownish-red lines or spots; blue points absent; seed-group 3. – S.W.Afr.: Gt. Namaqualand (2c) .... ....... **marthae** |
| Pl. 177 Row 22a | 39a. | Top slightly convex, bluish to yellowish grey, a yellow band absent, grooves absent or very indistinct, markings very indistinct or consisting of some dendroid darker lines, blue points present, nevertheless rather small and sometimes indistinct. – S.W.Afr.: Gt. Namaqualand, Noukloofberge (2b) ....... **archerae** |
| | 39*. | Top slightly convex, mostly bordered by a yellowish or pinkish band, usually with grooves, markings distinct, consisting of reddish-brown, usually interrupted lines; blue points present (sometimes only a few); seed-group 2. – S.W.Afr.: Gt. Namaqualand (2b) ................................... **schwantesii** |
| Pl. 177 Row 23 | 39*a. | Top smooth, dark greyish yellow-red, with a distinct blue tint in the centre, bordered by a pink band; blue points few to rather numerous .......... — v. **schwantesii** |
| | 39*b. | Similar to preceding but grooves more pronounced ....... — v. **kunjasensis** |
| | 39*c. | Top ± amethyst-coloured, nearly without a pink border; grooves distinct, especially on young leaves; blue points rather numerous .... — v. **rugosa** |
| Pl. 177 Row 24 | 39*d. | Top grey to leather-brown, with a pinkish border; blue points distinct ... — v. **triebneri** |
| | 39*e. | Bodies light grey (as concrete), without a pink border .. — v. **urikosensis** |
| | 39*f. | Top uniformly brown-red without a lighter coloured band, otherwise as v. schwantesii ................................................ — v. **gebseri** |
| | 39*g. | Top peculiarly yellowish-green, brown leather-coloured, otherwise as v. schwantesii ................................ — v. **nutupdriftensis** |
| Pl. 178 Row 25 | 40. | Bodies grey to pink; fissure shallow; window large, greyish-green, sometimes suffused with pink; islands nearly exclusively along the margin, those of the outer margin connected with this, those of the inner margin free or connected with a little line; seed-group 3. – Cape: Brakfontein (5a) .......... **otzeniana** |
| | 40*. | Islands not arranged as in the preceding ............................. **41**. |
| | 41. | Bodies greyish-green; top obliquely declining (one side higher); window large, light green, with light spots and many irregular lighter coloured islands; seed-group 3 (similar to L. marmorata, which has a horizontal top and a white flower). – Cape: Lt. Namaqualand (4a) ................................. **helmutii** |
| | 41*. | Top horizontal or declining to the fissure ............................. **42**. |
| | 42. | Bodies light grey to olive-green, sometimes tinged with pink; window smaller than the top, of the same colour but slightly darker, usually with a few islands; outer margin lighter coloured, crenate; top of the whole body oval (seen from above) and with a narrow fissure; seed-group 5. – Cape: Bushmanland (4b) .... ....... **olivacea** |
| | 42*. | Window as large as the top which is not oval (seen from above) ........... **43**. |
| | 43. | Window mostly branched, always with islands; flowers 5–6-merous ........ **44**. |
| | 43*. | Window entire, without islands or if they are present, the window seldom branched; flowers 5-merous ............................................ **45**. |
| | 44. | Bodies greyish-brown, sometimes suffused with pink; window large, greenish-blue, strikingly branched, with sharply edged, raised islands which make the top ± bullate; also specimens occur with a nearly smooth window with few islands (these are rather similar to L. comptonii, but this never has a greenish-blue window); seed-group 5. – Cape: Calvinia (6a) ..................... **weberi**[1] |

---

[1]) see also L. marginata (**32***); this has red-brown lines in the grooves which are indistinct or absent in some specimens.

Pl. 178
Row 26

- **44\*.** Differing from the preceding: bodies pink- to ochre-coloured; window dark pink to dirty brown; islands less raised; seed-group 3. – **gesinae-** v. **annae** (see **22\*b**)
- **45.** Bodies light bluish-grey (as in Argyroderma) with divergent lobes, which are as broad as high; top usually sickle-shaped, caused by the bent inner margin; window not distinct, sometimes with vague markings; seed-group 3. – Cape: Lt. Namaqualand (4a) .................................................... **meyeri**
- **45\*.** Bodies not light bluish-grey, top not sickle-shaped (or a little); seed-group 2–4 **46.**
- **46.** Bodies greyish-green to pinkish; sides (especially of the seedling) suffused with purple; window green, smooth, without islands; specimens with pinkish shade sometimes ± resemble L. comptonii; seed-group 2. – Cape: Calvinia (6a) ... ....... **viride**
- **46\*.** Window with islands; sometimes only a few present; seed-group 3 or 4 .... **47.**
- **47.** Bodies greyish-green, sometimes very little suffused with purple .......... **48.**
- **47\*.** Bodies purplish .................................................. **49.**
- **48.** Lobes elongated; top flat; ± semilunar (the inner margin bent), on average of unequal size; window with a variable number of whitish islands; seed-group 4. – Cape: Van Rhynsdorp (6a) ........................................ **divergens**
  - **48 a.** Bodies entirely greyish-green; top semilunar ......... **— v. divergens**
  - **48 b.** Bodies amethyst-coloured ......................... **— v. amethystina**
- **48\*.** Differing from L. divergens: lobes not or only a little elongated; top convex, semi-orbicular (the inner margin straight) on average of equal size; seed-group 3. – Cape: Lt. Namaqualand (4a) ...................................... **herrei**

Pl. 178
Row 27

- **48 a.** Top somewhat convex, greyish-green, with indistinct islands ... **— v. herrei**
- **48 b.** As the preceding variety but mostly suffused with a little pink and with rather distinct islands .................................... **— v. hillii**
- **48 c.** Top rather convex, yellowish grey-green, with distinct islands ........ ....... **— v. geyeri**

- **49.** Bodies rather small, reddish, with short slightly convex lobes; window always with small islands; seed-group 4. – Cape: Ceres (6a) ............. **comptonii**
- **49\*.** Differing from the preceding: bodies large, reddish, with elongated, rather flat lobes; window without or with vague islands **divergens** v. **amethystina** (see **48 b**)

The following descriptions of Lithops species should be used in conjunction with the data, including origins, given in the Key. The figure following that for the SG – e.g. (§ I) (3) – refers to the appropriate key-reference. – All species and varieties are shown in the illustrations.

## Key to the Species of the Genus Lithops by B. Fearn

This very simple Key which was published later than the Key by DE BOER and BOOM is divided into sections according to the overall coloration of the top surface. (See Cact. Succ. J. Am. 42: 89–93, 1970 and some alterations.)

- **1.** Top of L. green, greyish-green, whitish-green, olive-green or purplish-green ............. **2.**
- **1 a.** Top of L. not so coloured ................................................... **18.**
- **2.** Pellucid dots present ....................................................... **3.**
- **2 a.** Pellucid dots absent ...................................................... **6.**
- **3.** Top of L. strongly convex, L. not tightly pressed together, Fi. slightly gaping ......... **4.**
- **3 a.** Top of L. not strongly convex, L. tightly pressed together, Fi. not gaping ............. **5.**
- **4.** Pellucid dots regularly scattered over the top surface ....................... **localis**
- **4 a.** Pellucid dots aggregated into indistinct lines ........................... **franciscii**
- **5.** Red or brown lines present, often joined up to form dendritic markings ... **pseudotruncatella**
- **5 a.** Red or brown lines absent .................................................. **lesliei**
- **6.** Top of L. strongly convex ................................................... **7.**
- **6 a.** Top of L. not strongly convex .............................................. **12.**
- **7.** L. tightly pressed together, Fi. not gaping ................................ **olivacea**
- **7 a.** L. not tightly pressed together, Fi. gaping ................................. **8.**
- **8.** Fi. less than 8 mm deep ................................................... **optica**
- **8 a.** Fi. greater than 8 mm deep ................................................. **9.**
- **9.** Fl. white, 6-merous ........................................................ **10.**
- **9 a.** Fl. yellow, 5-merous ...................................................... **11.**
- **10.** Wi. large, islands small, few or absent ................................ **marmorata**
- **10 a.** Wi. absent or obscured by islands and marginal ingrowths ................ **elisae**

| | | |
|---|---|---|
| 11. | Plants always greyish-green, Wi. sometimes reduced by islands and marginal ingrowths | helmutii |
| 11a. | Plants sometimes brownish or purplish-green, Wi. with a few small islands | comptonii |
| 12. | Red dots and/or red or brown lines present | 13. |
| 12a. | Red dots and/or red or brown lines absent | 15. |
| 13. | Top with small raised red dots | verruculosa |
| 13a. | Not as above | 14. |
| 14. | Wi. present; Fl. white | salicola |
| 14a. | Wi. absent; Fl. yellow | archerae |
| 15. | Fl. white, 6-merous | villetii |
| 15a. | Fl. yellow, 5-merous | 16. |
| 16. | Wi. large and distinct, with a few large islands | otzeniana |
| 16a. | Wi. obscured by many small islands and marginal ingrowths | 17. |
| 17. | Wi. not distinct, with vague islands | meyeri |
| 17a. | Wi. with distinct islands | herrei |
| 18. | Top of L. white or pinkish-white | 19. |
| 18a. | Top of L. not so coloured | 25. |
| 19. | Top of L. strongly convex | 20. |
| 19a. | Top of L. not strongly convex | 21. |
| 20. | Wi. absent; Fl. yellow | ruschiorum |
| 20a. | Wi. obscured by islands and marginal ingrowths; Fl. white | elisae |
| 21. | Pellucid dots present | pseudotruncatella |
| 21a. | Pellucid dots absent | 22. |
| 22. | Top of L. wrinkled, with very fine chalky ridges | vallis-mariae |
| 22a. | Not as above | 23. |
| 23. | Wi. present | julii |
| 23a. | Wi. absent | 24. |
| 24. | Top of L. flat | karasmontana |
| 24a. | Top of L. slightly convex | erniana |
| 25. | L. entirely purplish-red | optica v. rubra |
| 25a. | Top of L. not so coloured | 26. |
| 26. | Top of L. brownish | 27. |
| 26a. | Top of L. not so coloured | 48. |
| 27. | Pellucid dots present | 28. |
| 27a. | Pellucid dots absent | 40. |
| 28. | Wi. present | 29. |
| 28a. | Wi. absent | 32. |
| 29. | Top of L. flat | 30. |
| 29a. | Not as above | 31. |
| 30. | Red or brown lines present | bromfieldii |
| 30a. | Red or brown lines absent | lesliei |
| 31. | Pellucid dots numerous, regularly scattered over the top; Wi. small | localis |
| 31a. | Pellucid dots few, Wi. often large | franciscii |
| 32. | Red or brown lines absent | 33. |
| 32a. | Red or brown lines present | 34. |
| 33. | Top of L. flat | aucampiae |
| 33a. | Top of L. convex | localis |
| 34. | Lines joined up to form dendritic markings | 35. |
| 34a. | Not as above | 37. |
| 35. | Top of L. flat | 36. |
| 35a. | Top of L. convex | werneri |
| 36. | Flowering time July | pseudotruncatella |
| 36a. | Flowering time September-October | schwantesii |
| 37. | Top of L. strongly convex | steineckeana |
| 37a. | Top of L. not strongly convex | 38. |
| 38. | Lines and markings impressed so that the top of the L. is bullate and grooved | 39. |
| 38a. | Not as above | glaudinae |
| 39. | Top with numerous large dark bluish-green pellucid dots up to 0.5 mm $\emptyset$ | fulviceps |
| 39a. | Pellucid dots inconspicuous | marginata |
| 40. | Wi. present | 41. |
| 40a. | Wi. absent | 46. |
| 41. | Red dots and lines present | 42. |
| 41a. | Red dots and lines absent | villetii |
| 42. | Lines and markings impressed so that the top is bullate and grooved | 43. |
| 42a. | Lines and markings not impressed, top smooth | 45. |
| 43. | Top with small raised red dots | verruculosa |
| 43a. | Not as above | 44. |

| | |
|---|---|
| 44. Top of L. flat | **julii** |
| 44a. Top of L. slightly convex | **bella** |
| 45. Wi. large, islands few if any | **salicola** |
| 45a. Wi. obscured by islands and marginal ingrowths | **dorotheae** |
| 46. Fl. yellow, 6-merous | **turbiniformis** |
| 46a. Fl. white, 5-merous | 47. |
| 47. Top of L. flat | **karasmontana** |
| 47a. Top of L. slightly convex | **erniana** |
| 48. Pellucid dots absent, Fl. white | **deboeri** |
| 48a. Pellucid dots present, Fl. yellow | 49. |
| 49. Red dots present | 50. |
| 49a. Red dots absent | **werneri** |
| 50. Wi. present | **dinteri** |
| 50a. Wi. absent | **schwantesii** |

The following species accepted by DE BOER and BOOM are dropped in the above treatment by B. FEARN:

L. brevis L. BOL. = **L. dinteri** v. **brevis** (L. BOL.) FEARN
L. christinae DE BOER = **L. schwantesii** v. **christinae** (DE BOER) FEARN
L. divergens L. BOL. = **L. comptonii** v. **divergens** (L. BOL.) FEARN
L. — v. amethystina DE BOER = **L.** — v. — f. **amethystina** (DE BOER) FEARN
L. fulleri N. E. BR. = **L. julii** ssp. **fulleri** (N. E. BR.) FEARN v. **fulleri**
L. — v. brunnea DE BOER = **L.** — ssp. — v. **brunnea** (DE BOER) FEARN
L. — v. kennedyi DE BOER = **L.** — ssp. — v. **kennedyi** (DE BOER) FEARN
L. — v. ochracea DE BOER = **L.** — ssp. — v. **ochracea** (DE BOER) FEARN
L. — v. tapscottii L. BOL. = **L.** — ssp. — v. **fulleri**
L. gesinae DE BOER = **L. franciscii** v. **gesinae** (DE BOER) FEARN
L. — v. annae (DE BOER) DE BOER et BOOM = **L.** — v. **annae** (DE BOER) FEARN (L. annae DE BOER)
L. gracilidelineata DTR. = **L. pseudotruncatella** v. **gracilidelineata** (DTR.) FEARN (L. streyi SCHWANT.)
L. — v. waldronae DE BOER = **L.** — v. — f. **waldronae** (DE BOER) FEARN
L. hallii DE BOER = **L. julii** ssp. **fulleri** v. **hallii** (DE BOER) FEARN (L. salicola v. reticulata DE BOER)
L. inornata DTR. ex JACOBS. = **L. dinteri** v. **marthae** (LOESCH et TISCH.) FEARN
L. insularis L. BOL. = **L. bromfieldii** v. **insularis** (L. BOL.) FEARN
L. marthae LOESCH et TISCH. = **L. dinteri** v. **marthae** (LOESCH et TISCH.) FEARN (L. inornata DTR. ex JACOBS.)
L. mennellii L. BOL. = **L. bromfieldii** v. **mennellii** (L. BOL.) FEARN
L. viridis LUCKH. = **L. comptonii** v. **viridis** (LUCKH.) FEARN
L. weberi NEL = **L. otzeniana** v. **weberi** (NEL) FEARN

Interested readers are referred to the excellent work with photographs by B. K. BOOM: "Een nieuwe tabel voor het geslacht Lithops" in "Succulenta" 50, number 7–12, 1971, also available as a separate edition. – See **Supplement**, Part II, p, 584.

**L. archerae** DE BOER (§ I) **(39a)**. – S.W.Afr.: Noukloof Mts. – **Bo.** c. 3 cm tall, top 2.8 cm long, 2.2 cm across, flat, with 4–5 dendritic lines; **Fl.** yellow. (Classification in the Key dubious: see R. **22a**.)

**L. aucampiae** L. BOL. (§ I) **(23)**. – Transvaal: N.W.Cape. – Large species, clump-forming; **Bo.** obconical, 20–32 mm tall, with the top flat, 20–37 mm wide, 16–30 mm across in the direction of the Fi., sienna to dull brown, ± windowed; **Fl.** 25 mm ∅, yellow, Sept.

**L.** — v. **aucampiae (23a)**. – Transvaal, Cape: W.Griqualand. – Wi. dark olive-green to reddish-brown.

**L.** — v. **euniceae** DE BOER **(23b)**. – Dull brown on top.

**L.** — v. **fluminalis** COLE **(23d)**. – Cape: near Hopetown. – **Bo.** numerous; top surface 20–30 mm long, 16–21 mm across, olive-grey to greyish-brown, Fi. narrow, c. 3 mm deep, margins and islands greenish-grey to brownish-green, with 25–45 radiating, sometimes orange-coloured and doubly-forked lines.

**L.** — v. **koelemanii** (DE BOER) DE BOER et BOOM **(23c)** (*L. k.* DE BOER). – N.W.Cape. – **Bo.** 28–32 mm tall, top surface 34–37 mm long, 27–30 mm across, deep brownish-red, not fenestrate, ± rough.

**L. bella** N. E. BR. (§ II) **(14)**. – S.W.Afr.: Gr. Namaqualand. – Clump-forming; **Bo.** 25–30 mm tall, with the top very convex, 22 mm long, 15 mm across, brownish to yellowish-ochreous, with darker markings, the Fi. 8 mm long, the surface minutely granular; **Fl.** pure white, ± scented, Sept.

**L.** — v. **bella** (Pl. 179/1) **(14a)** (*Mes. b.* (N. E. BR.) DTR.). – S.W.Afr.: near Aus. – Wi. with dark margins and light yellow, pinkish-yellow or light brownish-yellow islands; **Fl.** 25 mm ∅.

**L.** — v. **eberlanzii** (DTR. et SCHWANT.) DE BOER et BOOM **(14c)** (*Mes. e.* DTR. et SCHWANT., *L. e.* (DTR. et SCHWANT.) N. E. BR., *L. halenbergensis* TISCH.). – S.W.Afr.: Kovis Mts. – Wi. less translucent, islands grey to bluish-grey, often ± brownish.

**L. — v. lericheana** (Dtr. et Schwant.) de Boer et Boom **(14b)** (*Mes. l.* Dtr. et Schwant., *L. l.* (Dtr. et Schwant.) Dtr. et Schwant.). – S.W.Afr.: Gr. Karas Mts. – Margins and Wi. darker, islands brownish-yellow; **Fl.** 30 mm ∅, strongly scented.

**L. brevis** L. Bol. (§ I) **(27*).** – Cape: Lit. Namaqualand. – **Bo.** solitary or 2 or more together, conical, flat on top with the Wi. large, light green, with several blood-red lines and spots; **Fl.** yellow. See page 513.

**L. bromfieldii** L. Bol. (§ I) **(31).** – Cape: N. of the Orange R., near Upington. – **Bo.** 4–6 together, obconical, 15 mm tall, the top flat, 15 mm ∅, humped, ochre-brown or dark olive-green, windowed, with brownish-red lines, the Fi. running across the entire surface, the sides sienna-brown; **Fl.** 35–40 mm ∅, yellow.

**L. christinae** de Boer (§ I) **(37*).** – S.W.Afr.: W. of Maltahöhe. – **Bo.** 12–20 mm tall, the top 14–22 mm long, 9–13 mm across, flat-convex, light greyish-yellow, with a network of red lines, the sides light greyish-blue to light greyish-violet; **Fl.** yellow. See page 513.

**L. comptonii** L. Bol. (§ I) **(49).** – Cape: Ceres Karroo, Laingsburg D. – Clump-forming; **Bo.** up to 4 cm ∅, with a deep Fi., the top with a large, dark green, purplish-green or purplish-red Wi., sometimes distinctly divided, with greyish-green margins; **Fl.** 25 mm ∅, yellow.

**L. deboeri** Schwant. (§ II) **(11).** – Origin unknown. – **Bo.** 2–3 cm tall, the sides greyish-blue, the top 17–22 mm long, 13–20 mm across, slightly convex, greyish-blue to brown, the Fi. 7–9 mm deep, gaping, blistered below, the Wi. almost black, irregularly zigzagged, with brownish-red to greyish-blue islands; **Fl.** 2–3 cm ∅, white, Oct.

**L. dinteri** Schwant. (§ I) **(27).** – S.W.Afr.: Gr. Namaqualand. – **Bo.** 2–3 cm tall, the sides milky or pearly-grey, the top somewhat truncate, very convex, 2–3 cm long, 18–20 mm across, reddish or greyish-yellow or ochre-coloured, ± distinctly windowed, with scattered red dots and a Fi. 5–7 mm deep; **Fl.** yellow.

**L. — v. dinteri (27a).** – The top reddish or greyish-yellow, with numerous translucent dots and 5–12 blood-red dots in the Wi.

**L. — v. multipunctata** de Boer **(27b).** – The top ochreous, with 15–25 red dots, translucent dots absent.

**L. divergens** L. Bol. (§ I) **(48).** – Cape: Lit. Namaqualand; S.W.Afr. – **Bo.** solitary or in groups of 2 or more; **L.** unequal, the sides green, the top oblique, ± rounded, the Wi. large, light greyish-green to amethyst, smooth or minutely wrinkled; **Fl.** yellow. See page 513.

**L. — v. amethystina** de Boer **(48b).** – S.W. Afr.: Nuwefontein. – **Bo.** amethyst-coloured.

**L. — v. divergens (48a).** – Cape: Lit. Namaqualand, Van Rhynsdorp D. – Plant greyish-green all over; **L.** rather long, noticeably divaricate.

**L. dorotheae** Nel (§ I) **(30)** (*L. eksteeniae* L. Bol.). – Cape: Bushmanland. – Clustering; **Bo.** 2–3 cm tall, pinkish-yellow; **L.** unequal, the top flat or convex, the large, translucent Wi. having small islands with red dots and lines; **Fl.** yellow.

**L. elisae** de Boer (§ II) **(3a).** – Cape: Lit. Namaqualand. – Often in groups; **Bo.** 15 to 25 mm tall, the L. standing away from one another, equal or unequal, the sides greyish-white, the top 24–35 mm long, 15–22 mm across, convex, with the green-bordered Fi. 10–14 mm long running across the entire surface, with a narrow Wi.; **Fl.** white.

**L. erniana** Loesch et Tisch. (§ II) **(10*).** – S.W.Afr.: Gr. Namaqualand. – Caespitose; **Bo.** compressed, 2 cm tall, the top oval, 25–30 mm long, 18–20 mm across, convex or flattened, reddish-grey, or yellowish-grey to bluish-grey, with branched, reddish-brown to bluish-grey lines and several minute, bluish-green Wi., the Fi. 4–5 mm deep; **Fl.** 3 cm ∅, white.

**L. — v. aiaisensis** de Boer **(10*c).** – S.W.Afr.: Gr. Namaqualand, E. of Aiais. – The top convex, light yellowish to bluish-grey, with dark lines and spots.

**L. — v. erniana (10*a).** – Gr. Namaqualand: Witpütz. – The top yellowish to reddish-grey; markings dark brown, forming a star-shape.

**L. — v. witputzensis** de Boer **(10*b).** – S.W.Afr.: Gr. Namaqualand: Witpütz. – Variable; **Bo.** larger than the type, the top flattened, light yellowish-grey, with dark bluish, broken lines.

**L. franciscii** (Dtr. et Schwant.) N. E. Br. (§ I) **(22)** (*Mes. f.* Dtr. et Schwant.). – S.W.Afr.: S.Namib. – Clump-forming; **Bo.** 15–30 mm tall, the top 12–20 mm ∅, greyish-green tinged brownish or reddish-yellow, with branched, dark lines and dots, merging into a Wi. in the centre, the Fi. 5–10 mm deep; **Fl.** 10–18 mm ∅, yellow.

**L. fulleri** N. E. Br. (§ II) **(5).** – Kenhardt, Upington, Hopetown D. – Clump-forming; **Bo.** 12–14 mm tall, the top 20 mm long, 20 mm across, ± noticeably convex, dove-grey to opal or ochre-coloured, with brownish or greenish to ± violet markings, with islands and reddish-brown dots, the Fi. 5–6 mm deep; **Fl.** 2–3 cm ∅, white. See page 513.

**L. — v. brunnea** de Boer **(5b).** – Cape: near Pofadder. – With dark brown Wi. extending to the margin and reddish-brown markings at the Wi.-tips.

**L. — v. chrysocephala** (Nel) de Boer **(5d)** (*L. c.* Nel). – Cape: Upington D. – **Bo.** usually paired, the top faintly wrinkled, pallid whitish-grey, the Wi. silvery-white, translucent, with dark green, translucent lines and solitary red dots.

**L. — v. fulleri (5a)** (*L. maughanii* N. E. Br.). – Cape: Kenhardt D. – **Bo.** colour dove-grey to opaline, with markings c. 0.5 mm across and the islands ± large.

**L. — v. kennedyi** de Boer **(5e).** – Cape: Bushmanland. – **Bo.** greyish-brown, young plants with distinct Wi. and islands, these eventually indistinct.

**L. — v. ochracea** de Boer **(5c).** – Cape: near Upington. – **Bo.** smaller than the type, with the islands light ochre-coloured and the Wi. dark ochre-coloured.

**L. — v. tapscottii** L. Bol. – Distinctively marked, with reddish-brown spots connected by reddish lines; since the plant is probably not in cultivation it has been excluded from the key.

**L. fulviceps** (N. E. Br.) (Pl. 179/2) (§ I) **(18\*)** (*Mes. f.* N. E. Br., *L. lydiae* Jacobs., *L. l.* L. Bol. nom. err.). – S.W.Afr.: Gr. Namaqualand. – Usually solitary, occasionally 2–4 together; **Bo.** 2.5–3 cm tall, the top flat or slightly convex, 2–2.5 cm long, rather less across, light brown to coffee or rust-coloured, with circular marks interspersed with dark orange lines or spots; **Fl.** c. 3 cm ⌀, yellow, Pet. whitish on the underside.

**L. gesinae** de Boer (§ I) **(22\*).** – S.W.Afr.: Namib. – Grouping in age; **Bo.** 18–22 mm tall, with the sides greyish to ochre-coloured or light violet, the top 22–32 mm long, 14–18 mm across, ± rounded, ochreous or pink to deep ochre-coloured, with a widely or only slightly gaping Fi. 10–12 mm deep, with Wi. consisting or branched dark greyish-green bands, interspersed with minutely windowed islands or with numerous translucent dots; **Fl.** yellow, Sept.-Oct. See page 513.

**L. — v. annae** (de Boer) de Boer et Boom **(22\*b)** (*L. a.* de Boer). – S.W.Afr.: near the Tiras Plateau. – Sides light violet to ochre-coloured, the Fi. only slightly gaping, the top pink to deep ochre-coloured, completely covered with large, deep pink to brown Wi. set with islands and dark bluish-green spots, with their outer borders branched antler-like.

**L. — v. gesinae (22\*a).** – Wi. dirty green, with sparse antler-like Br., with numerous translucent, bluish dots.

**L. glaudinae** de Boer (§ I) **(28).** – Cape: Hay D. – Solitary or double-headed; **Bo.** 1.75 to 2.5 cm tall, with bluish-purple sides, the top 1.5–3 cm long, 1.8–2.4 cm across, slightly convex, reddish-brown, with large, semi-translucent, greenish-grey Wi., branched towards the sides, with red dots and lines, the Fi. slightly gaping, 5 mm deep, running from side to side; **Fl.** yellow.

**L. gracilidelineata** Dtr. (§ I) **(35).** – S.W.Afr.: widely distributed. – Solitary, occasionally double-headed; **Bo.** 15 mm tall, with light yellowish-grey sides, the top flat, circular, 20–22 mm ⌀, with the 3–5 mm deep Fi. running from side to side, the surface coloured almost white or like pale, miky tea to pale yellowish-grey, to yellow and grey to red, divided into c. 30 ± prominent humps by brown or pale reddish-brown lines; **Fl.** 28–30 mm ⌀, yellow, Aug. See page 513.

**L. — v. gracilidelineata** (Pl. 179/3) **(35a)** (*L. streyi* Schwant.)– . S.W.Afr.: Gr. Namaqualand. – **Bo.**-colour varying with locality from almost pure white, through yellow or grey, to pale pink, brick-red or deep red, with dark brown, thin, reticulate lines.

**L. — v. waldronae** de Boer **(35b).** – S.W.Afr.: 45 km S.E. of Walvis Bay to the Vogelsberg. – The top strongly umbonate, pale yellowish-grey, with doubly-branched, pale reddish-brown lines.

**L. hallii** de Boer (§ 2) **(9)** (*L. salicola* v. *reticulata* de Boer). – Cape: S. of Prieska; Orange Free State. – **Bo.** 1.75–2 cm tall, with pale brown sides, the top ± flat, 2–2.5 cm long, 1.4–2 cm across, pale brown with dark brown reticulation and dots, the Fi. slightly gaping; **Fl.** white. – Exceedingly variable species. See page 513.

**L. helmutii** L. Bol. (§ I) **(41).** – Cape: Lit. Namaqualand. – **Bo.** solitary or up to 10–20 together, up to 3 cm ⌀, the top obliquely rounded, with a small Fi. and large, vivid green, often grey-spotted Wi.; **Fl.** 3 cm ⌀, golden-yellow, scented.

**L. herrei** L. Bol. (§ I) **(48\*).** – Cape: Lit. Namaqualand. – Clump-forming; **Bo.** often 10 to 15 together, 2.5 cm tall, with grey sides, the top 1.5 cm ⌀, with the Wi. densely speckled and divided into lines and islands; **Fl.** 1.5–1.8 cm ⌀, yellow, rarely white.

**L. — v. albiflora** Jacobs. – Hort. Bot. Kiel. Probably only an individual aberration; **Bo.** white-striate; **Fl.** white.

**L. — v. geyeri** (Nel) de Boer et Boom **(48\*c)** (*L. g.* Nel). – Cape: Lit. Namaqualand. – **Bo.** smaller than in v. **herrei**, light grey to yellowish, the Wi. with very distinct islands.

**L. — v. herrei (48\*a)** (*L. h.* v. *plena* L. Bol., *L. translucens* L. Bol., *L. elevata* L. Bol.). – Cape: Lit. Namaqualand, Richtersveld; S.W. Afr.: E. of Sendlingsdrift. – The top ± rough, the Wi. greyish-green with faint islands; shrinking greatly in winter and becoming wrinkled.

**L. — v. hillii** (L. Bol.) de Boer **(48\*b)** (*L. h.* L. Bol.). – Cape: Lit. Namaqualand. – **Bo.** several together, with dove-grey sides, the Fi. gaping widely, the top convex, grey with white, scattered dots, or denser and black towards the outer margins, the Wi. with several islands; Pet. white below.

**L. hookeri** (Bgr.) Schwant. (§ I) (*Mes. truncatellum* Hook. f., *Mes. hookeri* Bgr.). – N. E. Brown considered this species identical with L. **turbiniformis** (Gard. Chron. 4. 2. 1922), G. Schwantes considered it identical with **L. terricolor** (Sukk. Kd. **IV**, Schweiz. K. Ges., 1951, 73). A much-disputed species, so not included in the Key. – Cape: Somerset East D. – **Bo.** truncate, grey, the top flat to convex, brownish, tuberculate and uneven; **Fl.** 4 cm ⌀, Pet. straw-coloured with reddened tips.

**L. insularis** L. Bol. (§ I) **(32).** – Cape: Cape: Keimos Is. in the Orange R. near Upington. – **Bo.** solitary or several together, the top flat-convex, the Wi. dark translucent green, smooth or slightly bullate, with numerous blood-red spots or short lines; **Fl.** yellow. See page 513.

**L. julii** (Dtr. et Schwant.) N. E. Br. (§ II) **(6).** – S.W.Afr. – Grouping; **Bo.** 2–3 cm tall, truncate above, the top ± circular, 15–30 mm ⌀, slightly convex, pearly-grey to whitish, faintly opalescent with a reddish tinge, with ± distinct, slightly impressed, branched, somewhat translucent furrows, the deep Fi. running from side to side, with a buff to brown line or dots along each lip; **Fl.** 2–3 cm ⌀, white.

**L. — v. julii (6a)** (*Mes. j.* DTR. nom. subnud., *Mes. j.* DTR. et SCHWANT., *L. j.* v. *pallida* TISCH., *L. lactea* SCHICK et TISCH.). – S.W.Afr.: Gr. Namaqualand. – Markings the same colour as the top surface, with a buff line.

**L. — v. littlewoodii** DE BOER **(6d)**. – S.W.Afr.: Gr. Namaqualand. – **Bo.** almost 3 cm ⌀, with a yellowish-brown line along the lips of the Fi.

**L. — v. reticulata** TISCH. ex DE BOER **(6b)** (*L. j.* v. *r.* TISCH.). – S.W.Afr.: Gr. Namaqualand. – The top whitish, with a ± distinct network of rather broad, deeply impressed, dark yellowish-brown bands, with isolated red dots or short lines.

**L. — v. rouxii** DE BOER (Pl. 180/2) **(6c)**. – S.W.Afr.: Gr. Namaqualand. – Close to v. **reticulata**; the **Bo.**-top dark to light greyish-pink, the reticulation indistinct, at first forming a semi-translucent Wi.

**L. karasmontana** (DTR. et SCHWANT.) N. E. BR. (§ II) **(10)**. – S.W.Afr. – Grouping; **Bo.** 3–4 cm tall, pearly-grey to pale bluish-yellow, the top flat or flattish convex, circular, 15 to 25 mm ⌀, with branched, mostly brownish-buff pits and wrinkles, the deep Fi. running from side to side; **Fl.** 25–35 mm ⌀, glossy, white, Oct.-Nov.

**L. — v. karasmontana (10a)** (*Mes. k.* DTR. et SCHWANT., *Mes. damaranum* N. E. BR., *L. d.* (N. E. BR.) N. E. BR., *L. commoda* DTR. nom. nud.). – S.W.Afr.: Gr. Namaqualand, Lit. Karas Mts. – The top light greyish-green to light rusty brown.

**L. — v. mickbergensis** (DTR.) DE BOER et BOOM **(10c)** (*L. m.* DTR., *L. lateritia* DTR., *L. jacobseniana* SCHWANT.). – S.W.Afr.: Gr. Namaqualand, near Mickberg, Gründoorn. – The top yellowish-red, with ± distinct reticulation and with darker, often stellately branched lines.

**L. — v. opalina** (DTR.) DE BOER et BOOM **(10d)** (*L. o.* DTR., *L. k.* ssp. o. (DTR.) SCHWANT. nom. prov.). – S.W.Afr.: Gr. Namaqualand, Eisterbank. – The top opal to amethyst-coloured, with distinct markings.

**L. — v. summitatum** (DTR.) DE BOER et BOOM (Pl. 180/1) **(10b)** (*L. s.* DTR.) – S.W.Afr.: Gr. Namaqualand, Lit. Karas Mts. – The top reddish-brown, the reticulation slightly impressed, lines absent.

**L. lesliei** (N. E. BR.) N. E. BR. (§ I) **(21)**. – Transvaal, Cape: W.Griqualand, Orange Free State. – Usually solitary or double-headed, rarely several together; **Bo.** 1–4.5 cm tall, the top flat or slightly convex, up to 4 cm long, 1–3 cm across, greyish-yellow, coffee or rust-coloured, sometimes light brown to green, pitted, with a network of dark greenish-brown spots and furrows, the Fi. rather shallow and running from side to side; **Fl.** c. 3 cm ⌀, golden-yellow, very rarely white, Sept.

**L. — v. hornii** DE BOER **(21b)**. – Cape: S. of Kimberley. – The top light ochre-coloured, with the Wi. much branched, greyish-brown.

**L. — v. lesliei (21a)** (*Mes. l.* N. E. BR., *Mes. ferrugineum* SCHWANT.). – Transvaal. – Islands and margins rusty or brownish-yellow; very variable, Wi. often completely smooth.

**L. — v. — f. albiflora** COLE. – Near Cape Warrington. – Pet. white.

**L. — v. — f. albinica** COLE. – Near Cape Warrington. – Wi. distinctly translucent, grass-green, with a yellow sheen and with a pale yellow filigree pattern; Pet. white.

**L. — v. luteoviridis** DE BOER **(21c)**. – Transvaal. – The top yellowish-green, the Wi. light green.

**L. — v. maraisii** DE BOER **(21d)**. – Cape: Barkley West D. – Variable variety; sides grey to purplish-red, tiny islands and margins ochre to brownish-yellow, Wi. green to reddish-brown, variable in size.

**L. — v. mariae** COLE **(21g)**. – Orange Free State: near Boshof. – **Bo.** dark or lighter orange; Wi. absent, small dots numerous.

**L. — v. minor** DE BOER **(21h)**. – Transvaal. – **Bo.** 1.6–1.8 cm long, 1–1.2 cm across, cinnamon-brown, Wi. greenish.

**L. — v. — f. alba** COLE. – HOIT. DESMOND T. COLE, Johannesburg, S.Afr. – Form with white Fl.

**L. — v. rubrobrunnea** DE BOER **(21f)**. – Transvaal. – Sides and small islands reddish-brown, smaller, Wi. greenish-brown, larger.

**L. — v. venteri** (NEL) DE BOER et BOOM (Pl. 180/3) **(21e)** (*L. v.* NEL.). – Cape: W.Griqualand. – Islands and margins greyish-yellow.

**L. localis** (N. E. BR.) SCHWANT. (§ I) **(18)**. – Cape. – **Bo.** solitary or several together and forming a clump, 0.6–3.5 cm tall, truncate above, the top slightly convex, oval, 10 to 30 mm long, 10–22 mm across, ochre-coloured, pink or red to purple or greenish, with small solitary, violet to green, translucent dots, with the Fi. running from side to side, with or without Wi.; **Fl.** 2.5–3 cm ⌀, yellow.

**L. — v. localis (18a)** (*Mes. l.* N. E. BR.). – Cape: Beaufort West D. – **Bo.** 6–12 mm tall, the top 10–15 mm ⌀, with numerous dots, Wi. mostly not present all over.

**L. — v. peersii** (L. BOL.) DE BOER et BOOM **(18b)** (*L. p.* L. BOL.). – Cape: Willowmore, Graaff Reinet, Laingsburg D. – **Bo.** c. 3.5 cm tall, the top 3 cm long, 2 cm across, bluish-purple, often violet, Wi. ± distinct, dots numerous.

**L. — v. terricolor** (N. E. BR.) DE BOER et BOOM **(18c)** (*L. t.* N. E. BR., ? *L. hookeri* (BGR.) SCHWANT. (see **L. h.** (BGR.) SCHWANT.). – Cape: Somerset East, Willowmore D. – The top 25–30 mm long, 20–22 mm across, greyish-green to yellowish-green, with ± large, greenish Wi. and numerous dark green, translucent dots.

**L. marginata** NEL (§ I) **(32\*)** (*L. dabneri* L. BOL.). – Cape: W.Griqualand (acc. NEL: S.W.Afr. err. est). – **Bo.** sometimes in clumps, usually 2 together, the L. unequal, the sides violet, the top slightly convex, very dark olive-green, slightly rugose, the Wi. large, translucent, with numerous blood-red lines and spots; Fl. yellow.

**L. marmorata** (N. E. BR.) N. E. BR. (§ II) **(3)** (*Mes. m.* N. E. BR., *L. diutina* L. BOL.,

*L. framesii* L. Bol., *L. umdausensis* L. Bol.). – Cape: Lit. Namaqualand. – **Bo.** solitary or few together in a clump, up to 30 mm tall, 17 mm across, 20 mm thick, the top slightly convex, greyish-green, with light grey, branched lines, the Fi. c. 10 mm deep; **Fl.** 30 mm ⌀, pure white, scented.

**L. marthae** Loesch et Tisch. (§ I) **(39)** (*L. inornata* Dtr. nom. nud.). – S.W.Afr.: Gr. Namaqualand. – **Bo.** 4–6 together, 2–3 cm tall, the top 1–1.5 cm ⌀, slightly convex, mouse-grey to violet to green or slightly ochreous, with a dull Wi. extending from the Fi., which is 4 mm deep, with several ± scattered, blood-red lines, interspersed with round, red spots; **Fl.** 2.5 cm ⌀, yellow. See page 513.

**L. mennellii** L. Bol. (§ I) **(38)**. – Cape: N. of the Orange R., near Upington. – **Bo.** 17–20 mm tall, 15–18 mm across, 24 mm long, with the Fi. 3 mm deep, extending from side to side, the top slightly convex, pinkish-brown, bullate, with dark brown lines, becoming reticulate towards the margins; **Fl.** almost 3 cm ⌀, golden-yellow. See page 513.

**L. meyeri** L. Bol. (§ I) **(45)**. – Cape: Lit. Namaqualand. – Forming clumps; **Bo.** up to 3 cm tall, 2 cm thick below, with a widely gaping Fi. c. 2 cm deep, the top ± lunate, with the inner face of the L. vesicular, glassy green and translucent, convex; the Bo. dark bluish to greenish-grey, lighter on top; **Fl.** 3–3.5 cm ⌀, yellow.

**L. olivacea** L. Bol. (§ I) **(42)**. – Cape: Lit. Namaqualand, Kenhardt D. – Forming clumps; **Bo.** c. 2 cm tall, 2–2.5 cm across above, c. 15 mm thick, with the Fi. c. 5 mm deep, the top ± circular, dark olive-green to brownish, with a dull Wi.; **Fl.** yellow.

**L. optica** (Marl.) N. E. Br. f. **optica** (§ II) **(3\*)** (*Mes. o.* Marl., *Mes. marginatum* Marl.). – S.W.Afr.: Gr. Namaqualand. – Mat-forming; **Bo.** 20–30 together, up to 3 cm tall, with the Fi. deeply incised to midway, the 2 lobes very convex, divaricate, 10–15 mm often unequal, grey to loam-coloured, rarely purple with an opalescent, ± translucent Wi.; **Fl.** 15–20 mm ⌀, white, Sept.-Oct.

**L. — f. minor** Jacobs. **(3\*b)**. – S.W.Afr.: near Lüderitz. – **Bo.** pea-sized, dark grey in colour, with acuter lobes.

**L. — f. rubra** Tisch. **(3\*a)** (*Mes. o. v. r.* Tisch., *L. o. f. r.* (Tisch.) Rowl., *L. o. v. r.* Tisch., *L. rubra* N. E. Br.). – S.W.Afr.: diamond-mining area S. of Lüderitz Bay; **Bo.** purplish-red; **Fl.**-buds dirty yellowish to reddish, Pet. white, tipped with pink on the upperside.

**L. otzeniana** Nel (§ I) **(40)**. – Cape: Bushmanland. – **Bo.** often 20 together in clumps, 2.5 to 3 cm tall, 1.5–2 cm across, with the Fi. 1–2 cm long, grey-violet, and a large, semi-translucent, greenish to violet Wi. with a light border on each lobe above; **Fl.** 2 cm ⌀, golden-yellow.

**L. pseudotruncatella** (Bgr.) N. E. Br. (§ I) **(20\*)**. – S.W.Afr.: Gr. Namaqualand. – **Bo.** solitary or several together, very compressed, truncate, 1–4 cm tall, with the Fi. ± from side to side, the top convex or ± flat, ± circular, 1–4 cm ⌀, smooth or minutely rough, most commonly pale brownish-grey to pale grey, with dendritic patterning and dots, these markings very variable; **Fl.** 3–3.5 cm ⌀, golden-yellow, very rarely white, July.

**L. — v. alpina** (Dtr.) Jacobs. **(20\*a1)** (*L. a.* Dtr.). – S.W.Afr.: near Lichtenstein. – **Bo.** up to 6 together in a clump, 10–12 mm tall, the top 10–12 mm ⌀, light brown, with minute brown markings and numerous dots.

**L. — v. alta** Tisch. – S.W.Afr.: Damaraland. – Probably an ecotype only; conspicuously tall.

**L. — v. brandbergensis** de Boer **(20\*e)**. – S.W.Afr.: Brandberg. – **Bo.** solitary; the upper surface uniformly yellowish-brown, with very prominent, dark reddish-brown, dendritic lines, the margin greyish-bluish, with indistinct, bluish dots, some of these also along the Fi.

**L. — v. dendritica** (Nel) de Boer et Boom **(20\*g)** (*L. d.* Nel, *L. farinosa* Dtr. nom. nud.). – S.E. of S.W.Afr. – **Bo.** always solitary, extremely rarely in 2's (500 : 1!), the colour variable with the locality, ± white, pink or dark red, the top mostly greyish-brown with distinct dendritic markings.

**L. — v. edithae** (N. E. Br.) de Boer et Boom **(20\*d)** (*L. e.* N. E. Br.). – S.E. of S.W.Afr. – **Bo.** several together, 2.5 cm tall, 2.5 cm ⌀, the top light bluish-grey, with less distinct patterning, with distinctly translucent dots.

**L. — v. elisabethae** (Dtr.) de Boer et Boom **(20\*c)** (*L. e.* Dtr.). – S.W.Afr.: Gr. Waterberg. – **Bo.** in 2's or up to 12 in a clump, 2–3 cm tall, 2–3 cm ⌀, the top greyish-blue, mostly with a coppery tinge, the lines and translucent dots often indistinct.

**L. — v. mundtii** (Tisch.) Tisch. ex Jacobs. **(20\*f)** (*L. m.* Tisch., *L. p.* f. *m.* (Tisch.) Tisch. ex Jacobs., *L. p.* ssp. *m.* (Tisch.) Schwant.). – S.W.Afr.: E. of Windhoek. – Mostly solitary or double-headed; **Bo.** 3–4 cm tall, 3–4 cm ⌀, the top more flattened, rough, uniformly yellowish-brown, with distinct, branched markings.

**L. — v. pseudotruncatella** f. **pseudotruncatella** (Pl. 181/1) **(20\*a)** (*Mes. p.* Bgr., *Mes. truncatellum* Dtr.). – S.W.Afr.: Gr. Namaqualand, widely distributed. – **Bo.** solitary or several together, 2.5–3 cm tall, 2–3 cm ⌀, the top brownish-grey, the distinct, branched brown markings with a dark bluish border.

**L. — v. — f. albiflora** Jacobs. ex Rowley. – Hort. Bot. Kiel. – Probably only a garden mutant; **Fl.** white.

**L. — v. pulmonuncula** (Dtr. ex Jacobs.) Dtr. ex Jacobs. **(20\*e)** (*L. p.* f. *p.* Dtr. ex Jacobs., *L. p.* ssp. *p.* Dtr. ex Jacobs.). – S.W.Afr.: Gr. Namaqualand. – **Bo.** mostly solitary, rarely 2–3 together (100 : 1!), 2–3 cm ⌀, the top uniformly brownish-grey, the Wi. indistinct, brownish-green, the lines and translucent dots often indistinct.

**L. — v. volkii** (Schwant. ex Jacobs.) de Boer et Boom **(20\*h)** (*L. v.* Schwant. ex Jacobs.). – S.E. of S.W.Afr. – **Bo.** solitary or in clumps of

4 together, the individual L. distinctly unequal, up to 4 cm tall, the top flat, up to 3 cm across, pale grey, with indefinite branched lines, with distinct, translucent dots.

**L. ruschiorum** (Dtr. et Schwant.) N. E. Br. (§ I) **(33)**. – S.W.Afr.: coastal deserts. – **Bo.** in clumps of 6–8 (acc. Triebner up to 60!), 7.5–45 mm tall, the top very convex, with a Fi. 5–20 mm deep, milky or pearly-grey to yellowish or brownish-reddish, with the sides often amethyst-coloured, with no markings or with small, acute furrows, with lines running from and often connecting them, furrows and lines translucent, yellowish or red; **Fl.** 2–2.5 cm ⌀, yellow, July-Aug.

**L. — v. lineata** (Nel) de Boer et Boom **(33b)** (*L. l.* Nel). – S.W.Afr.: N. of Walvis Bay. – **Bo.** yellowish-grey to orange-yellow, the top with several distinct, slightly impressed lines.

**L. — v. nelii** (Schwant.) de Boer et Boom **(33c)** (*L. n.* Schwant., *L. r.* ssp. *n.* (Schwant.) Schwant., *L. r.* ssp. *stiepelmanii* Schwant.). – S.W.Afr.: Cape Cross. – **Bo.** 7.5–15 mm tall and across, 5–7.5 mm thick, or 2–4 cm tall and ⌀, pale ochre-coloured or ashy-brown, grey or ± white, the top unmarked or with a few brown spots or short lines.

**L. — v. ruschiorum (33a)** (*Mes. r.* Dtr. et Schwant., *L. r.* ssp. *r.* Schwant., *Mes. ruschii* Dtr. nom. err., *L. pillansii* L. Bol.). – S.W.Afr.: along the W. coast. – **Bo.** 2–4.5 cm tall, 4 cm ⌀, with the Fi. up to 2 cm deep, the top greyish-yellow with several red, ± impressed lines and spots.

**L. salicola** L. Bol. (§ II) **(15)**. – Orange Free State. – **Bo.** solitary, or several in a clump, 2–2.5 cm tall, c. 16 mm across, 20–27 mm thick, with grey sides, the top flat or slightly convex, olive-green, with several large, ± translucent, greenish-grey to dark green to reddish Wi. divided into small islands; **Fl.** c. 2.5 cm ⌀, white.

**L. schwantesii** Dtr. (§ I) **(39)**. – S.W.Afr. – Mat-forming; **Bo.** 1.5–4 cm tall, the top 2–2.5 cm ⌀, ± flat or convex, smooth or ± strongly rugose, slightly glossy, colour as under varieties, with ± distinct blood-red dots or bluish marks and short lines, with or without a light rusty-yellow or pink band 2–5 mm broad; **Fl.** c. 2.5 cm ⌀, yellow.

**L. — v. gebseri** de Boer **(39f)**. – S.W.Afr.: Gr. Namaqualand. – The top brownish-red to dark reddish-brown, tinged somewhat blue, with a fine network of pits, long red lines, red dots and deep-seated, bluish-green dots (lens needed!), older L. rough with lines and dots.

**L. — v. kunjasensis** (Dtr.) de Boer et Boom **(39b)** (*L. k.* Dtr.). – S.W.Afr.: near Kunjas. – **Bo.** 3 cm tall, the top ± flat, 2 cm ⌀, resembling v. **schwantesii** but paler, less noticeably red, more greyish, more conspicuously furrowed.

**L. — v. nutupdriftensis** de Boer **(39g)**. – S.W.Afr.: Gr. Namaqualand. – The top variable in coloration, yellowish-green to brown, older L. greyish-brown, in other plants cement-coloured; markings as for v. **kunjasensis.**

**L. — v. rugosa** (Dtr.) de Boer et Boom **(39c)** *L. r.* Dtr.). – S.W.Afr.: W. of Helmeringhausen. – **Bo.** 1.5–2 cm tall, the top convex, broadly oval, 1.5–2 cm long, 1–1.5 cm across, mostly tuberculate, ± amethyst-coloured, with fairly deep furrows, with numerous distinct, pale green markings.

**L. — v. schwantesii** (Pl. 181/3) **(39a)** (*L. kuibisensis* Dtr. ex Jacobs.). – S.W.Afr.: near Kuibis. – **Bo.** 3–4 cm tall, the top ± flat to slightly convex, 2–2.5 cm ⌀, dark grey to orange with a distinctly blue centre and pink margin, with several or many faint bluish marks (not translucent dots) in the bluish part.

**L. — v. triebneri** (L. Bol.) de Boer et Boom **(39d)** (*L. t.* L. Bol.). – S.W.Afr.: Namib. – **Bo.** 2–3 together, 2 cm tall, the top slightly convex, 2.6 cm long, 1.5 cm across, grey to light leather-coloured or terracotta, with faint, reticulate markings, with scattered red dots or short lines.

**L. — v. urikosensis** (Dtr.) de Boer et Boom **(39e)** (*L. u.* Dtr., *L. gulielmii* L. Bol.). – S.W.Afr.: Urikos, Lit. Karas Mts. – **Bo.** 2 cm tall, the top 2 cm wide, 1.2–1.5 cm ⌀, light grey, with pale brown dots and lines.

**L. steineckeana** Tisch. (§ I) **(33d)**. – S.W.Afr.: Gr. Namaqualand. – **Bo.** solitary or paired, cylindrical to ± ovoid, 22 mm tall, 12–15 mm ⌀, the sides greyish-ochre, the top convex, brownish to ochreous-yellow, with brownish markings, the Fi. compressed, 7 mm long, dividing the surface into 2 convex lobes; **Fl.** c. 3 cm ⌀, golden-yellow. (L. E. Newton in Cact. Succ. J. Gt. Br., **28**: 1966, 28 says this may be a hybrid: **L. ruschiorum** × **L. pseudotruncatella**).

**L. turbiniformis** (Haw.) N. E. Br. (§ I) **(37)**. – Cape: Prieska D. to the Orange R. – **Bo.** solitary or paired, rarely several together, 2–2.5 cm tall, the sides mostly grey to yellowish-brown, the top flattish convex, ± circular, 20–41 mm ⌀, with the Fi. from side to side, colour as under varieties, uneven and tuberculate with ± numerous, reticulate furrows and dark brown, branched lines, in one variety ± windowed; **Fl.** 3.5–4 cm ⌀, yellow.

**L. — v. brunneo-violacea** de Boer **(37c)**. – Cape: W.Griqualand. – The top glossy, dark brownish-violet, with the reticulate furrows still darker.

**L. — v. elephina** Cole **(37f)**. – Cape: near Britstone. – **Bo.** 1–7 together, the upperside slightly rough, 25–41 mm long, 20–27 mm across, with the narrow Fi. 4 mm deep, greyish-yellow to orange-yellow, the Wi. with a network of narrow, translucent channels and islands.

**L. — v. lutea** de Boer **(37b)**. – Cape: Groblerskoop. – **Bo.** light brownish-yellow, with branched, brownish-red furrows.

**L. — v. subfenestrata** de Boer **(37d)**. – Cape: Prieska D. – The top dark greyish-green, the furrows more distant, flatter and broader than in v. **turbiniformis,** with semitranslucent Wi.

**L. — v. susannae** COLE **(37e).** – The top somewhat rough, dull, greyish-yellow or greyish-orange, with very narrow, green to greyish-orange, irregularly branched and doubly forked grooves, with recognizable red dots and short lines.

**L. — v. turbiniformis** (Pl. 181/2) **(37a)** *(Mes. t.* HAW., *L. aurantiaca* L. BOL.). – Cape: Prieska D. – The top greyish-brown, rust-coloured or reddish-brown, with a fairly fine network of deep furrows (see L. **hookeri** (BGR.) SCHWANT.).

**L. vallis-mariae** (DTR. et SCHWANT.) N. E. BR. (§ I) **(34).** – S.W.Afr. – Clump-forming; **Bo.** 2–4 cm tall, the top ± circular, 2–5 cm ⌀, or with the individual L. lunate, slightly convex or flat, rugose or rough with papillae or minute, vermiform crests (lens needed!), yellowish to bluish-white, markings none or consisting of indistinct, blurred, somewhat darker, broad or narrow, broken lines, with a deep Fi.; **Fl.** 2.5–3.5 cm ⌀, yellow.

**L. — v. groendraaiensis** (JACOBS.) DE BOER **(34b)** (*L. pseudotruncatella* v. *g.* JACOBS.). – S.W.Afr.: Groendraai Farm. – **Bo.** bluish-white, the top mostly flat, the vermiform crests indistinct, rough-papillose.

**L. — v. margarethae** DE BOER. – S.W.Afr. – **Bo.** pale grey to flesh-coloured, the top reticulate with narrow, brown furrows, broken intermittently.

**L. — v. vallis-mariae (34a)** *(Mes. v.* DTR. et SCHWANT.). – S.W.Afr.: Alt Mariental. – **Bo.** yellowish, the top ± convex, with distinct vermiform crests and faint lines.

**L. verruculosa** NEL (§ I) **(25).** – N.W.Cape. – **Bo.** solitary or forming clumps, 2–3 cm tall and ⌀, with the Fi. across the entire width, scarcely gaping, surrounded by a distinct line, the top flat or slight convex, with or without recognizable Wi., very rugose, bluish-grey, with or without red or dark grey spots in the wrinkles; **Fl.** orange-yellow.

**L. — v. glabra** DE BOER **(25c).** – Light greyish-blue, the top with a large, mottled Wi., without blood-red dots.

**L. — v. inae** (NEL) DE BOER et BOOM **(25b)** (*L. i.* NEL). – ? Cape: Kenhardt D. – **Bo.** solitary, smaller than in v. **verruculosa,** the top with irregular, darker spots with white dots and the transparent areas confluent into a ribbon-like Wi.

**L. — v. verruculosa (25a).** – Origin not known. – **Bo.** solitary, with the islands on the top mostly not densely crowded and the Wi. reduced to furrows, with numerous, ± prominent, red or grey spots.

**L. villetii** L. BOL. (§ II) **(15\*).** – Cape: Calvinia D. – **Bo.** long-ovoid, 47 mm tall, 30 mm ⌀ above measured over the 16 mm broad lobes which are separated by 7 mm, the Wi. on the upper surface convex, completely clear or with 1–2 spots; **Fl.** white, 3 cm ⌀.

**L. viridis** H. LUCKHOFF (§ I) **(46).** – Cape: Calvinia Karroo. – **Bo.** 2–4 together, or in clumps of up to 10, 1.7–4.8 cm tall, with the Fi. slightly gaping and the free parts (lobes) 1.1–2.3 cm long, 8–16 mm across, greyish-green to olive-green, the upper part and the Wi.-margins with a pinkish tinge, the Wi. large, light green to olive-green, translucent, often with milky islets or reticulate or mottled markings, with several scattered, purple dots; **Fl.** up to 3 cm ⌀, **Pet.** golden-yellow, white below. See page 513.

**L. weberi** NEL (§ I) **(44).** – Cape: Calvinia and Ceres Karroo. – **Bo.** 15–20 mm across above, with the Fi. c. 15 mm across, the top flat or slightly rounded, light green to greyish-green or purplish-green, with several greenish-blue translucent Wi.; **Fl.** yellow. See page 513.

**L. werneri** SCHWANT. ex JACOBS. (§ I) **(20).** – S.W.Afr.: Damaraland. – **Bo.** solitary or several together, in the resting period the size of a pinhead or a pea, during the growing season 10 mm tall, short-conical, 10 mm ⌀, with a Fi. 1–2 mm deep, the top unevenly divided, ± convex, greyish-brownish, with several branched, dark greenish-brown lines and solitary, small, round dots, the sides light greyish-brown; **Fl.** pure yellow.

**Machairophyllum** SCHWANT. (§ IV/1/1). – Compact plants up to 1.20 m ⌀ in age, vegetative parts smooth, shining, pale bluish or whitish-grey, rarely greenish, Int. enclosed; **L.** linear-lanceolate, rarely oblong or ± rhombic, keeled on the underside in the upper part, obliquely truncate in profile, acute, 2.5–10 cm long, 2 cm wide at the base; **Fl.** solitary or in 1–2-partite Infl., 4–6.5 cm ⌀, open in the afternoon or at night, Pet. yellow to orange, often red or pink in the upper part, pedicel with lateral bracts. – Greenhouse, warm. Propagation: seed, cuttings.

## Key to the Species of the Genus Machairophyllum by L. Bolus

1. Fl. 3–7 in an Infl.; Sep. 5–8; stigmas 6–15 .......................................... **2.**
1a. Fl. solitary; Sep. 6–8 ................................................................ **4.**
2. Sep. 5–6, acute or subacuminate; stigmas twice to thrice as many as the Sep............ **3.**
2a. Sep. 6–8, long-acuminate; stigmas the same number as the Sep., or differing probably by only one addition to either ............................................................ **stayneri**
3. L. acutely keeled, whitish-green; Sep. 5–6, stigmas 10–15 ..................... **albidum**
3a. L. obtusely keeled, greyish-green; Sep. 5, stigmas 10 .......................... **cookii**
4. L. less than thrice as long as broad, up to 3 cm long; Fl. vespertine; Pet. up to 2 mm broad; stigmas and Sep. 6 .................................................................. **5.**
4a. L. 3–5 times as long as broad; Fl. nocturnal; Pet. up to 1 mm broad; stigmas and Sep. unequal or rarely equal, 6–8 in number ............................................... **6.**

5. Upper surface of L. nearly oblong, acute or acuminate .................... **brevifolium**
5a. Upper surface of L. rhombic, the lateral angles rounded .................... **latifolium**
6. Pet. usually well exceeding the Sep. in length ............................ **baxteri**
6a. Pet. slightly exceeding the Sep. in length ................................ 7.
7. Sep. up to 1.3 cm long; glands contiguous; top of Ov. flat in the outer half, the lobes acutely compressed ................................................................ **stenopetalum**
7a. Sep. up to 2.2 cm long; glands distant; top of Ov. concave in the outer half, the lobes obtusely compressed ................................................................ **acuminatum**

Omitted from Key, in absence of full data: **M. bijlii, vanbredai.**

**M. acuminatum** L. Bol. (Pl. 182/1). – Cape: near Humansdorp. – **L.** pale green, trigonous, long-tapered, 4–4.5 cm long, 8–11 mm across, 6–10 mm thick; **Fl.** solitary, c. 5 cm ∅, golden-yellow.

**M. albidum** (L.) Schwant. (*Mes. a.* L., *Bergeranthus a.* (L.) Schwant., *Carruanthus a.* (L.) Schwant.). – Cape: Oudtshoorn D. – **L.** incurved, 7–10 cm long, 2 cm across below, carinate-trigonous above, whitish; **Fl.** in 3's, yellow inside, reddish outside.

**M. baxteri** L. Bol. – Cape: near George. – **L.** ± trigonous, 5–7 cm long, 8–14 mm across midway, 8–12 mm thick, bluish-green; **Fl.** solitary, 5–6 cm ∅, red.

**M. bijlii** (N. E. Br.) L. Bol. (*Perissolobus b.* N. E. Br.). – Cape: Prince Albert D. – **L.** trigonous, tapered, the keel and angles acute, 15–20 mm long, 7–10 mm across, 7–9 mm thick, light green, the angles reddish; **Fl.** up to 6 cm ∅, golden-yellow to orange, red outside.

**M. brevifolium** L. Bol. – Cape: Oudtshoorn D. – **L.** 15 mm long, 13 mm across, 12 mm thick, greyish-green with reddish margins, the upperside oblong, acute, the underside oblique, ± rhombic, with a compressed keel; **Fl.** solitary, 35 mm ∅, yellow.

**M. cookii** (L. Bol.) Schwant. (*Mes. c.* L. Bol., *Bergeranthus c.* (L. Bol.) Schwant., *Carruanthus cookii* (L. Bol.) Schwant.). – Cape: Barrydale. – **Br.** with 6 crowded, spreading, trigonous, tapered, bluish-green **L.** 3.5–8 cm long, c. 9 mm across; **Fl.** 4.5 cm ∅.

**M. latifolium** L. Bol. – Cape: Oudtshoorn D. – **L.** 4–6 on a Br., rhombic on the upperside, angles rounded, somewhat horny, carinate on the underside, broadly rounded to truncate at the tip, pale blue, 2.5–3 cm long, 2–2.6 cm across, 1.5 cm thick; **Fl.** 4 cm ∅, yellow.

**M. stayneri** L. Bol. – Cape: Uitenhage D. – Resembles **M. albidum**: flowering **Br.** with 4 bluish-green **L.** 7.5–8 cm long, 1.2–1.6 cm across and thick midway, tapered in the upper half, acute to ± obtuse; **Infl.** branched, mostly with 3 Fl., the terminal one 5–6 cm ∅, inner Pet. pale yellow to whitish-green inside, yellow to golden-yellow outside, outer Pet. golden-yellow to pink.

**M. stenopetalum** L. Bol. – Cape: Willowmore D. – **L.** 2–4 on a Br., narrowly falcate, long-tapered, ± hatchet-shaped and carinate on the underside, 4 cm long, 12 mm across, 8–10 mm thick, pale green; **Fl.** 3.5 cm ∅, yellow to straw-coloured.

**M. vanbredai** L. Bol. – Cape: Somerset East D. – Flowering **Br.** with 6–8 **L.** 3–4.5 cm long sometimes shorter, 1.3–1.7 cm across midway, 1 cm thick, narrowed on the upperside towards the acute tip, rounded on the underside, carinate and obliquely truncate above; **Fl.** solitary, Pet. 8–18 mm long, yellow, red outside above.

**Malephora** N. E. Br. (§ IV/1/19). – Erect or creeping, shrubby, branched plants; **L.** shortly united at the base, trigonous-prismatic, long, semicylindrical, soft-fleshy, slightly bluish-pruinose, unspotted; **Fl.** axillary or terminal, shortly pedicellate, golden-yellow, yellow or pink, up to 5 cm ∅, late summer to winter. – Summer: in the open, winter: cold-house. – Propagation: seed, cuttings.

**M. crassa** (L. Bol.) Jacobs. et Schwant. (*Hymenocyclus c.* L. Bol.). – Cape: Ceres D. – St. prostrate; **L.** erect, swollen, 4 cm long, 13 mm across, 14 mm thick, green, often paler; **Fl.** up to 6 cm ∅, golden-yellow.

**M. crocea** (Jacq.) Schwant. v. **crocea** (*Mes. c.* Jacq., *Hymenocyclus c.* (Jacq.) Schwant., *Crocanthus c.* (Jacq.) L. Bol., *Mes. insititium* Willd.). – Cape: Fraserburg D. – **St.** stout, ± gnarled; **L.** crowded on the short shoots, 2.5 to 4.5 cm long, 6 mm across, indistinctly trigonous, pale green, pruinose; **Fl.** solitary, 3 cm ∅, golden-yellow inside, reddened outside.

**M. —** v. **purpureo-crocea** (Haw.) Jacobs. et Schwant. (*Mes. p.* Haw., *Hymenocyclus p.* (Haw.) Schwant., *Crocanthus p.* (Haw.) L. Bol.). – Cape: Laingsburg D., Lit. Namaqualand. – **Fl.** brilliant red.

**M. englerana** (Dtr. et Bgr.) Schwant. (*Mes. e.* Dtr. et Bgr., *Hymenocyclus e.* (Dtr. et Bgr.) Dtr. et Schwant., *Mes. vernae* Dtr. et Bgr.). – S.W.Afr.: Gr. Namaqualand and S. E. from there. – Dense, very soft ♄, up to 30 cm tall, 50 cm ∅; **L.** 2–4 cm long, obtuse-trigonous, curved, obtuse above, green, slightly waxy, very soft; **Fl.** solitary, 2 cm ∅, glossy orange-yellow inside, orange-red outside.

**M. flavo-crocea** (Haw.) Jacobs. et Schwant. (*Mes. f.-c.* Haw., *Hymenocyclus f.-c.* (Haw.) Schwant., *Mes. purpureo-croceum* v. *f.-c.* Haw., *Mes. croceum* v. *f.-c.* DC.). – Cape: Fraserburg D. – **St.** woody; **L.** ± crowded, cylindrical to semicylindrical, very obtuse above, bluish-pruinose or blue; **Fl.** c. 1.5 cm ∅, yellow.

**M. framesii** (L. Bol.) Jacobs. et Schwant. (*Hymenocyclus f.* L. Bol.). – Cape: Piquetberg D. – **St.** prostrate, 30 cm long, with a whitish bark; **L.** rounded, flat to furrowed on the upperside, bluish-green, 4 cm long, 14 mm thick; **Fl.** 4 cm ∅, glossy yellow.

**M. herrei** (SCHWANT.) SCHWANT. (*Hymenocyclus h.* SCHWANT., *H. smithii* L. BOL.). – Orange Free State. – **St.** mostly prostrate, often rooting; **L.** up to 5 cm long, 5 mm across, trigonous, rounded-carinate, green; **Fl.** axillary, 5 cm ⌀, Pet. golden-yellow, orange on the underside.

**M. latipetala** (L. BOL.) JACOBS. et SCHWANT. (*Hymenocyclus l.* L. BOL.). – Cape: Oudtshoorn D. – Erect ♄; **St.** 10–12 cm long, eventually ± prostrate; **L.** erect, 2.5 cm long, 3 to 5 mm across and thick, blue, tinged red, angles acute, keel indistinct and ciliate; **Fl.** 3–4 cm ⌀, yellow.

**M. lutea** (HAW.) SCHWANT. (Pl. 182/4) (*Mes. l.* HAW., *Hymenocyclus l.* (HAW.) SCHWANT.). – Cape: Lit. Karroo. – Erect ♄; **St.** with numerous short shoots; **L.** spreading, 2.5–4.5 cm long, 4 mm across, compressed-trigonous, yellowish-green, whitish-pruinose; **Fl.** terminal, 2.5 cm ⌀, orange and yellow.

**M. luteola** (HAW.) SCHWANT. (*Mes. l.* HAW., *Hymenocyclus l.* (HAW.) SCHWANT., *Crocanthus l.* (HAW.) L. BOL.). – Cape: Lit. Karroo. – Possibly only a variety of **M. lutea**; **St.** slender, freely branched; **L.** acute; **Fl.** numerous, small, yellow. Insufficiently known.

**M. mollis** (AIT.) N. E. BR. (Pl. 182/3) (*Mes. m.* AIT.). – Orange Free State. – Low-growing; **St.** prostrate, nodose; **L.** trigonous, with distinct angles, obtuse above, greyish-green, minutely dotted, puberulous, 12–15 mm long, 4–6 mm across; **Fl.** solitary, c. 15 mm ⌀, red.

**M. thunbergii** (HAW.) SCHWANT. (*Mes. t.* HAW., *Hymenocyclus t.* (HAW.) L. BOL., *Crocanthus t.* (HAW.) L. BOL., *Corpuscularia t.* (HAW.) SCHWANT., *Mes. laeve* THUNBG.). – Cape: Graaff Reinet, Cradock, Albany, Riversdale, Van Rhynsdorp, Uitenhage, Malmesbury D. – **St.** prostrate, rigid, nodose; **L.** crowded, c. 5 cm or more long, 6–8 mm across below, semicylindrical, obtusely trigonous above, fresh green, spotted; **Fl.** solitary, 3.5–4 cm ⌀, yellow.

**M. uitenhagensis** (L. BOL.) JACOBS. et SCHWANT. (*Hymenocyclus u.* L. BOL., ? *Mes. verruculoides* v. *minus* SOND.). – Cape: Uitenhage D. – Forming dense mats; **St.** creeping, c. 40 cm long, flowering Br. erect, with 2–3 L.-pairs; **L.** 2.5 cm long, 4–5 mm across and thick, ± semicylindrical, green; **Fl.** dirty yellow to coppery-red.

**M. verruculoides** (SOND.) SCHWANT. (*Mes. v.* SOND., *Hymenocyclus v.* SOND., *Hymenocyclus v.* (SOND.) SCHWANT.). – Cape: Cape Peninsula, Table Mt. – Prostrate; **L.** clustered, erect, ± cylindrical below, flat on the upperside, 2.5 to 3.5 cm long, 6 mm across; **Fl.** usually in 3's, rarely, solitary, yellow.

**Maughaniella** L. BOL. (§ IV/1/21) (*Maughania* N. E. BR. non JAUME SAINT-HILAIRE). – Close to the G. **Diplosoma** SCHWANT., differentiated by the L.-Sh. being symmetrically united at the sides, and by the shorter, distinctly papillose stigmas. – Greenhouse.

**M. luckhoffii** (L. BOL.) L. BOL. (Pl. 182/2) (*Monilaria l.* L. BOL., *Maughania l.* (L. BOL.) L. BOL., *Diplosoma l.* (L. BOL.) SCHWANT., *Maughania insignis* N. E. BR.). – Cape: Van Rhynsdorp D. – 1.3–3 cm tall including the Fl.; **caudex** swollen, simple or bipartite, with fibrous roots; **L.** 2, forming a small Bo. c. 16 mm long, sheathing and united for 3 mm, c. 8 mm across and thick, semicylindrical, soft, green, semi-translucent, papillose, the surface eventually reticulate, with the L.-Sh. persisting; **Fl.** pedicellate, 15–20 mm ⌀, snowy-white or pink.

**Mesembryanthemum** L. emend. L. BOL. (§ I/1/2) (*Cryophytum* N. E. BR.). – Type species: **M. nodiflorum** L. (L. BOLUS, Notes on Mes. III, 164–166). – ⊙ to ⊙, ± tall herbs; **St.** usually erect, thick, fleshy, often with a basal Ros. of L., with the **Br.** and branchlets often prostrate, creeping or ascending; **L.** alternate or opposite, shortly united, cylindrical or flat, expanded; **Fl.** solitary or numerous in ± elongated Infl., white, pink, red, more rarely yellow; the entire plant or parts thereof, including the Cal., ± conspicuously glossy-papillose. – Cultivate in the open. – Synonymy: for *Cr.* read *Cryophytum*.

**M. aitonis** JACQ. (Pl. 183/1) (*Cr. a.* (JACQ.) N. E. BR., *Pentacoilanthus a.* (JACQ.) RAP. et CAM., *Perapentacoilanthus a.* (JACQ.) RAP. et CAM., *Gasoul a.* (JACQ.) EICHL., *Mes. angulatum* THUNBG., *Cr. a.* (THUNBG.) SCHWANT., *Mes. volckameri* HAW., *Mes. lanceolatum* HAW. and its varieties, *Mes. crystallophanes* ECKL. et ZEYH.). – Cape: Uitenhage, Albany, Cradock D. – **L.** obovate-spatulate, rather obtuse, c. 5 cm long, 2.5 cm across; **Fl.** red.

**M. alatum** (L. BOL.) L. BOL. (*Cr. a.* L. BOL., *Cr. nanum* N. E. BR.). – Cape: Piquetberg, Malmesbury D. – Robust; **St.** with 6 angles, 2 of these alate; basal **L.** 20 cm long, oblong-ovate with undulating margins, with distinct veins, ± viscous; **Fl.** numerous, 6 cm ⌀, pale green, salmon-pink to flesh-coloured inside.

**M. albatum** L. BOL. – Cape: Lit. Namaqualand. – Erect, up to 30 cm tall; **L.** densely imbricate, 8 cm long, other L. shorter, 21 mm across, with dark dots, distinctly tuberculate; **Fl.** solitary, 35 mm ⌀, white.

**M. alboroseum** L. BOL. – Cape: Lit. Namaqualand. – Erect, glabrous, with robust **St.**; **L.** crowded, 28 mm long, 3–4 mm across, 7–8 mm thick, pale bluish-green; **Fl.** mostly solitary, 3 cm ⌀, white and pink.

**M. annuum** L. BOL. (*Cr. wilmaniae* L. BOL., *Mes. longipapillatum* L. BOL., *Cr. l.* (L. BOL.) L. BOL.). – Cape: Herbert D.; S.W.Afr.: Lüderitz Bay S. – **St.** erect to spreading; **L.** semicylindrical, green, 8–10 mm long, 2 mm across, 1–2 mm thick; **Fl.** in repeatedly forking Infl., 12–15 mm ⌀, with the Cal. very papillose.

**M. barklyi** N. E. BR. (*Cr. b.* (N. E. BR.) N. E. BR., *Platythyra b.* (N. E. BR.) SCHWANT., *Mes. b.* v. *obtusifolium* L. BOL., *Mes. alkalifugum* DTR. nom. nud., *Cr. crassifolium* L. BOL., *Mes. pingue* L. BOL., *Mes. grandifolium* AUCT. non SCHINZ, *Cr. squamulosum* DTR., *Mes. sq.*

(DTR.) DTR.). - Cape: Lit. Namaqualand; S.W.Afr.: Gr. Namaqualand. - Plant up to 60 cm tall, bluish-green, tinged purple; **St.** 4-angled; **L.** expanded-spatulate, lower ones up to 28 cm long, 18 cm across midway, with prominent veins and undulating margins; upper L. smaller; **Fl.** 36 mm $\varnothing$, whitish.

**M. breve** L. BOL. (*Cr. parvum* L. BOL.). - Cape: Lit. Namaqualand. - Up to 8 cm tall; basal **L.** deciduous, sabre-shaped, with undulating margins, 6 cm long, 35 mm across; **Fl.** 2 cm $\varnothing$, white, pink on the underside.

**M. chrysum** L. BOL. (*Cr. aureum* (L. BOL.) L. BOL.). - Cape: Laingsburg D. - **St.** prostrate, 4–20 cm long, with erect Br.; **L.** semicylindrical, furrowed on the upperside, 2 cm long, 2 mm thick; **Fl.** 13 mm $\varnothing$, yellow.

**M. clandestinum** HAW. (*Cr. c.* (HAW.) N. E. BR.). - Origin unknown. - Up to 30 cm tall, with short shoots from the L.-axils; **L.** ovate-lanceolate, 1–2 cm long, 4–10 mm across; **Fl.** small, white, $\pm$ paniculate.

**M. cryocalyx** L. BOL. (*Cr. intermedium* L. BOL.). - Cape: Worcester D. - 10–30 cm tall; **Br.** lax, spreading to prostrate, up to 30 cm long; **L.** linear-oblong to lanceolate to ovate-spatulate, 25–35 cm long, 6–10 mm across; **Fl.** 25 mm $\varnothing$, white.

**M. cryptanthum** HOOK. f. (Pl. 169/4) (*Hydrodea c.* (HOOK. f.) N. E. BR., *Mes. dactylinum* WELW. ex OLIV., *Opophytum d.* (WELW. ex OLIV.) N. E. BR., *Mes. forskahlii* HOCHST., *O. f.* (HOCHST.) N. E. BR., *Mes. sarcocalycanthum* DTR. et BGR., *H. s.* (DTR. et BGR.) DTR., *H. bossiana* DTR., *H. hampdenii* N. E. BR.). - S.W.Afr.: coastal deserts, salt-pans; Red Sea coasts. - **St.** thick, cylindrical, $\pm$ creeping; lower **L.** opposite, flat, oblong, very thick and fleshy, those on the flowering St. $\pm$ cylindrical, $\pm$ elliptical and clavate, 3–4 cm long, 6–8 mm thick, green or red (in the Namib also waxy-yellow); **Fl.** sessile, axillary, 1–2 cm $\varnothing$, white to deep pink or yellowish.

**M. crystallinum** L. (Pl. 183/2, 3) (*Cr. c.* (L.) SCHWANT., *Cr. c.* (L.) N. E. BR., *Gasoul c.* (L.) ROTHM., *Pentacoilanthus c.* (L.) RAP. et CAM., *Perapentacoilanthus c.* (L.) RAP. et CAM., *Mes. glaciale* HAW.). - S.W.Afr.; Cape: Cape D. to Uitenhage D.; introduced to the Medit. Coasts; Canary Is.; USA: Calif. - **St.** spreading, very papillose; **L.** ovate or ovate-spatulate, thick-fleshy, very papillose, with rather undulating margins; **Fl.** 3–5 together, 2–3 cm $\varnothing$, white.

**M. dejagerae** (L. BOL.) L. BOL. (*Cr. d.* L. BOL.). - Cape: Beaufort West D. - Creeping; up to 5 cm tall, round-papillose; **L.** 14 mm long, 3.5 mm across; **Fl.** 2 cm $\varnothing$, white.

**M. excavatum** (L. BOL.) L. BOL. (*Cr. e.* L. BOL.). - Cape: Beaufort West D. - Glossy-papillose; **St.** spreading, 20 cm long; **L.** spatulate, 10–15 mm long, 5 mm across above; **Fl.** solitary, 12 mm $\varnothing$, yellow.

**M. galpinii** (L. BOL.) L. BOL. (*Cr. g.* L. BOL.). - Cape: Calvinia D. - Low-growing; **St.** spreading to prostrate; **L.** linear-spatulate, 2 cm long, 5 mm across above, 2 mm thick; **Fl.** 16 mm $\varnothing$, white, in a small Infl.

**M. guerichianum** PAX (*Cr. g.* (PAX) SCHWANT., *Amoebophyllum g.* (PAX) N. E. BR., *Mes. fenchelii* SCHINZ, *Cr. f.* (SCHINZ) N. E. BR., *Mes. gariusianum* DTR., *Mes. grandifolium* SCHINZ, *Cr. gr.* (SCHINZ) N. E. BR., *Mes. grandiflorum* SCHINZ ex RANGE, *Cr. grandifl.* (SCHINZ) DTR. et SCHWANT. ex RANGE, *Mes. crystallinum* AUCT. non L. - Cape: Alexanderberg; S.W.Afr.: widely distributed. - Very succulent plant; **St.** 5–8 mm thick, cylindrical; basal and lower cauline **L.** in several opposite and decussate pairs, basal L. up to 15 cm long, 8 cm across, ovate to rhombic, narrowed teretely below, upper L. smaller; **Fl.** on a pedicel (1–)2(–4) cm long, 2.5–4 cm $\varnothing$, white or greenish to yellowish-white and pink to light red; Cal., pedicels and L.-undersides small-papillose. Variable species.

**M. hypertrophicum** DTR. (Pl. 168/1) (*Halenbergia h.* (DTR.) DTR., *Perapentacoilanthus h.* (DTR.) RAP. et CAM., *Mes. aquosum* L. BOL., *Opophytum a.* (L. BOL.) N. E. BR.). - S.W.Afr.: Lüderitz Bay South, Warmbad D. in dried-up saltpans; Cape: Lit. Namaqualand. - Forming flat cushions; **St.** spreading, prostrate, forking from the base and above; **L.** digitate to $\pm$ hemispherical, 1–1.3 cm thick, very soft-fleshy and succulent; **Fl.** very numerous on the aging but still fleshy St., 4–6 cm $\varnothing$, white or reddish, with the free Pet.-tips very slender.

**M. inachabense** ENGL. (*Callistigma i.* (ENGL.) DTR. et SCHWANT., *Cr. i.* (ENGL.) N. E. BR., *Cr. arenarium* N. E. BR., *Mes. a.* (N. E. BR.) L. BOL., *Cr. fulleri* L. BOL., *Mes. f.* (L. BOL.) L. BOL., *Mes. caducum* AUCT. non AIT.). - Cape: Kenhardt D. - S.W.Afr.: southern Gr. Namaqualand. - Soft-fleshy, succulent plant; **St.** erect or prostrate or curved and ascending, branching from the base; basal and lower cauline **L.** oblong-spatulate to obovate, 2–5 cm long, 1–2 cm across; **Fl.** mostly very numerous, on a pedicel 5–15 mm long, 1–1.5 cm $\varnothing$, agreeably scented, yellow, rarely white, with red stigmas.

**M. inornatum** L. BOL. (*Cr. cleistum* L. BOL.). - Cape: Laingsburg D. - Up to 20 cm tall; **St.** virgate, purplish-red; **L.** few together, semi-cylindrical, 25 mm long; **Fl.** 30 together in a paniculate cyme, 5–6 mm $\varnothing$, white.

**M. intransparens** L. BOL. v. **intransparens** (*Cr. framesii* L. BOL.). - Cape: Clanwilliam D. - Robust plant; main **St.** erect, up to 50 cm tall, flowering Br. elongated, prostrate, up to 1 m long, 18 mm $\varnothing$, acutely hexagonal; **L.** ovate, acute, 11 cm long, 9 cm across, with undulating margins; **Fl.** c. 30 together, 4 cm $\varnothing$, white.

**M. — v. laxum** (L. BOL.) L. BOL. (*Mes. framesii* v. *l.* L. BOL.). - Cape: Laingsburg D. - **St.** slender, laxly spreading.

**M. karrooense** L. BOL. (*Cr. karroicum* L. BOL.). - Cape: Cradock D. - **St.** spreading to prostrate, c. 30 cm long, 1 cm $\varnothing$; lower **L.** linear, upper ones oblong-lanceolate, with undulating margins, 15 cm long, 5–8 cm across; Pet. filiform, 21 mm long, white.

**M. latisepalum** (L. BOL.) L. BOL. (*Cr. l.* L. BOL.). - Cape: Lit. Namaqualand. - Low-growing plant; basal **L.** thick, c. 14 cm long, 6 cm across, with undulating margins; flowering **St.** up to 45 cm long, 7 mm $\varnothing$; **Fl.** c. 5 cm $\varnothing$, white to pink.

**M. liebendalense** L. Bol. – Cape: Van Rhynsdorp D. – Low-growing, slender plant up to 6.5 cm tall; **L.** linear, flat on the upperside, 6–15 cm long, 1.5–2 mm $\varnothing$; **Fl.** solitary, up to 2 cm $\varnothing$, white to pale pink.

**M. linearifolium** L. Bol. (*Cr. lineare* L. Bol.). – Cape: origin unknown. – Laxly branched, slender plant 40 cm $\varnothing$; **L.** 35 mm long, 2 mm thick; **Fl.** 2 cm $\varnothing$, pale pink.

**M. longipapillosum** Dtr. (*Cr. salteri* L. Bol., *Mes. violense* L. Bol., *Mes. guerichianum* auct. non Pax). – Cape: Lit. Namaqualand; S.W.Afr. – 4–6 cm tall, glossy-papillose; **St.** 4 mm thick; lower **L.** ovate-spatulate, 4 cm long, 18 mm across, upper **L.** $\pm$ rhombic, 2–3 cm long, 18 mm across; **Fl.** 2–2.5 cm $\varnothing$, buttercup-yellow to pale yellow or white, with a greenish style.

**M. louiseae** L. Bol. (*Cr. planum* L. Bol., *Mes. stratum* L. Bol.). – Cape: Caledon D. – Flat-growing plant, resting on the soil, with creeping **St.** and shoots; **L.** lanceolate, acute or obliquely ovate to spatulate, 5–19 mm long, 1.5–6 mm across; **Fl.** 18 mm $\varnothing$, white.

**M. macrophyllum** L. Bol. (*Cr. velutinum* L. Bol.). – Cape: Lit. Namaqualand. – Velvety from papillae with short hairs; **St.** prostrate; **Br.** up to 70 cm long, 4-angled; **L.** c. 32, rosulate, oblong-ovate, the largest in Mesembryanthemaceae, c. 40 cm long, 23 cm across midway, 5 mm thick, with undulating margins, the keel 1 cm high, long-hirsute below; **Infl.** 50 cm tall, 50 cm $\varnothing$, Fl. numerous, 4–5 cm $\varnothing$, mauvish-pink.

**M. macrostigma** L. Bol. (*Cr. maxwellii* (L. Bol.) L. Bol.). – Cape: Ladismith D. – Erect, stiff plant c. 35 cm tall; **St.** prostrate, 50 cm or more long; **L.** crowded at the St. tips, opposite, lower L. oblong, upper ones ovate-oblong, with undulating margins, 11 cm long, 5 cm across; **Fl.** 4.5 cm $\varnothing$, white.

**M. neilsoniae** (L. Bol.) L. Bol. (*Cr. n.* L. Bol.). – Cape: Ceres D. – $\odot$; main **St.** 40 cm tall, 7 mm $\varnothing$, indistinctly hexagonal; flowering Br. elongated, 40 cm long; **L.** linear to oblong-lanceolate, with undulating margins, densely red-crystalline, 15 cm long, 25 mm across; **Infl.** lax, Fl. 6 cm $\varnothing$, red to salmon-pink outside, pale pink inside.

**M. nodiflorum** L. (*Cr. n.* (L.) L. Bol., *Gasoul n.* (L.) Rothm., *Mes. copticum* L., *Cr. gibbosum* N. E. Br.). – S.Eur.; N.Afr.; Arabia; Iran; Baluchistan; Kurdistan; Madeira; Canary Is.; Cape: Lit. Namaqualand and southwards; USA: Calif.; Mexico; Baja Calif. – Up to 20 cm tall, greyish-green, with large papillae; **L.** linear, ciliate below, 1–2.5 cm long, 1–2 mm across; **Fl.** white.

**M. pachypus** L. Bol. (*Cr. crassipes* L. Bol.). – Cape: Van Rhynsdorp D. – Glossy-papillose herb up to 5.5 cm tall; **L.** cylindrical, 3.5 cm long, 6 mm thick; **Fl.** on a pedicel 6–7 mm thick, 20 mm $\varnothing$, Pet. slender, acute, white.

**M. parvipapillatum** L. Bol. (*Cr. acuminatum* L. Bol.). – Cape: Van Rhynsdorp D. – Small-papillose plant; **St.** erect, 18 mm thick below, 30 cm tall, 4–6-angled; **Br.** robust, prostrate, up to 110 cm long; basal **L.** 10 cm long, 3.5 to 5 cm across, distinctly undulating, minutely papillose; **Infl.** terminal, 80 cm across, with c. 100 pale or white Fl. 4 cm $\varnothing$.

**M. paucandrum** L. Bol. (*Cr. rogersii* L. Bol.). – Cape: Calitzdorp, Oudtshoorn D. – Densely branched plant 15–23 cm tall; **St.** prostrate; **L.** linear-tapered, expanded below, 15–20 cm long, 2–4 mm across; **Fl.** 17 mm $\varnothing$, white.

**M. paulum** (N. E. Br.) L. Bol. (*Cr. p.* N. E. Br.). – Cape: Riversdale D. – Low-growing plant scarcely 25 mm tall, with creeping **St.** 10 cm long; **L.** spatulate, flat, 7–10 mm long, 2–3 mm across, minutely papillose; **Fl.** in a lax Infl., 5–6 mm $\varnothing$, white.

**M. pellitum** Friedr. (*M. mollissimum* Dtr. ex Friedr. nom. nud.). – S.W.Afr. – Robust $\odot$, the herbaceous parts with pearl-like papillae 6–8 mm long; **St.** spreading to creeping, 20 to 50 cm long, 1.5 cm thick below and 0.8 cm thick above, Int. 2–10 cm long; cauline **L.** opposite, lyrate, broadly spatulate, $\pm$ amplexicaul below, 7–15 cm long, 4–8 cm across, with $\pm$ undulating margins, upper L. also opposite but narrower; **Fl.** 2–3 together, c. 5 cm $\varnothing$, on a pedicel 1–1.5 cm long, Pet. linear to filiform.

**M. perlatum** Dtr. (*Mes. crystallinum* v. *grandiflorum* Eckl. et Zeyh., *Cr. g.* (Eckl. et Zeyh.) N. E. Br., *Mes. magniflorum* L. Bol., *Derenbergiella luisae* Schwant.). – Cape: Lit. Namaqualand; S.W.Afr.: Gr. Namaqualand. – Resembles **Mes. guerichianum**, but more succulent; **Fl.** on a shorter and thicker pedicel, (4–)5(–6) cm $\varnothing$, papillae more conspicuous, spherical and blister-like, variable in size.

**M. purpureo-roseum** L. Bol. (*Cr. truncatum* L. Bol.). – Cape: Lit. Karroo. – Slender plant 15 cm tall; **L.** clustered at the St.-tips, 6 cm long, 7 mm across, expanded above, cordate at the base, with broadly undulating margins, glossy-papillose; **Fl.** 2.5 cm $\varnothing$, purplish-red.

**M. quinangulatum** L. Bol. (*Cr. pentagonum* L. Bol.). – Cape: Ladismith D. – Erect plant 12 cm tall; **St.** erect to prostrate, indistinctly 3-angled; lower **L.** 14 cm long, upper ones 4–5 cm long, 1–2.5 cm across, with undulating or curly margins; **Fl.** pale pink, white to green in the centre, receptacle 5-angled.

**M. rhodanthum** L. Bol. (*Cr. roseum* L. Bol.). – Cape: Lit. Namaqualand. – **St.** prostrate, up to 15 cm long, with 6–8 L., other St. longer; basal **L.** thick, ovate above, c. 10 cm long, c. 5 cm across, with undulating margins; **Fl.** 28 mm $\varnothing$, pink, green in the centre.

**M. rubroroseum** L. Bol. (*Cr. carinatum* L. Bol.). – Cape: Van Rhynsdorp D. – Up to 30 cm tall; **St.** 4-angled; **L.** thick, 30 cm long, 7 cm across, with undulating margins; **Fl.** 3 cm $\varnothing$, pinkish-red inside, purplish-pink outside.

**M. sedentiflorum** (L. Bol.) L. Bol. (*Cr. s.* L. Bol.). – Cape: Van Rhynsdorp D. – **St.**

ascending, 60 cm long, 22 mm ⌀, with narrowly alate angles; basal **L.** 25 cm long, tapered, with undulating margins, cauline **L.** 18 cm long, 5–9 cm across; **Fl.** many, 4.5 cm ⌀, pale pink, green inside.

**M. setosum** (L. Bol.) L. Bol. (*Cr. s.* L. Bol.). – Cape: Van Rhynsdorp D. – Robust plant, with angular **St.** 60 cm tall; **Br.** 1.20 m long, covered with projecting, papillose hairs 2.2 to 5 mm long; **L.** 10–15 pairs, c. 14 cm long, broadly oval above, with broadly undulating margins; **Fl.** 3.5–5 cm ⌀, pink, pale green inside.

**M. squamulosum** (L. Bol.) L. Bol. (*Cr. s.* L. Bol.). – Cape: Van Rhynsdorp D. – Scaly-papillose plant; **St.** 30 cm tall, 13 mm ⌀ below, furrowed or with rounded angles; **L.** ovate-acute, 3–4 cm long, thick below, 1.5–2 cm wide; **Fl.** 3.6 cm ⌀, alabaster-white.

**M. stenandrum** (L. Bol.) L. Bol. (*Cr. s.* L. Bol.). – Cape: Van Rhynsdorp D. – **St.** prostrate, reddish, 12 cm long; **L.** semicylindrical, 22 mm long, 2 mm thick; **Fl.** 28 mm ⌀, deep pink.

**M. subrigidum** L. Bol. (*Cr. subulatum* L. Bol.). – Cape: Calvinia D. – **St.** 6 cm tall; **Br.** stiff, triangular, red, with indistinct papillae; basal **L.** 10 cm long, 3.5 cm wide, thick, stiff, cauline **L.** 3.5 cm long, ovate above, with undulating margins; **Fl.** 4.5 cm ⌀, pale pink outside, white to green inside.

**M. subtereticaule** L. Bol. (*Cr. calycinum* L. Bol.). – Cape: Van Rhynsdorp D. – Laxly branched plant; **St., Br.** and branchlets ± cylindrical, indistinctly papillose; basal **L.** 8 cm long; **Fl.** 3 cm ⌀, Pet. filiform, white.

**M. subtruncatum** L. Bol. (*Cr. pusillum* L. Bol.). – Cape: Ceres D. – Plant c. 2 cm tall, 6–8 cm across, flat, glossy-papillose; **St.** patent, filiform below, thicker above and directed upwards; basal **L.** spatulate, 12 mm long; **Fl.** sessile, 25 mm ⌀, pink, yellow inside.

**Mestoklema** N. E. Br. (§ IV/1/6). – Small, branched ♄ or ♃, often with a tuberous **caudex**; young **shoots** papillose; **L.** opposite, mostly united at the base, trigonous to cylindrical, minutely papillose or glistening, with the bases persistent like a T. after the L. have fallen; **Fl.** small, in a branched terminal Infl., with the Ped. becoming thorny but not very sharp. – Summer: in the open, winter: cold-house. – Propagation: seed, cuttings.

**M. albanicum** N. E. Br. – Cape: Albany D. – Plant 20–25 cm tall; **St.** prostrate at first, eventually erect; **L.** 5–8 mm long, 1–2 mm thick, minutely papillose; **Fl.** 5 cm ⌀, light red.

**M. arboriforme** (Burch.) N. E. Br. (Pl. 184/1) (*Mes. a.* Burch.). – Cape: W.Griqualand, Philipstown D.; S.W.Afr.: Gr. Namaqualand. – 30–45 cm tall, tree-like, with a distinct **trunk** 2.5 cm thick and freely branched; **shoots** roughly hirsute; **L.** up to 12 mm long, 2 mm thick, minutely hirsute; **Infl.** 5–8 cm across, **Fl.** 5 cm ⌀, yellow to orange.

**M. copiosum** N. E. Br. – Cape: Herbert D. – Small ♄; **L.** 8–12 mm long, 2–3 mm thick, trigonous, minutely papillose; **Fl.** very numerous, 10–12 mm ⌀.

**M. elatum** N. E. Br. – Cape: Albany D. – Tall ♄; **L.** 6–12 mm long, 1 mm thick, compressed-trigonous, minutely papillose; **Infl.** of 15–20 purplish-red Fl. 5–8 mm ⌀.

**M. illepidum** N. E. Br. – Cape: Bedford, Albany D. – ♄ 20–25 cm tall; **L.** 4–5 mm long, 1–1.5 mm thick, compressed-subulate, green, minutely papillose; **Fl.** 4–5 mm ⌀, red.

**M. macrorhizum** (DC.) Schwant. (*Mes. m.* DC., *Mes. napiforme* N. E. Br., *Delosperma n.* (N. E. Br.) Schwant., *D. m.* (Haw.) Schwant.). – Reunion Is. - **Caudex** tuberous; **St.** 2.5–7.5 cm tall; **L.** crowded, obtusely trigonous, resembling those of Salsola kali, both in appearance and in taste; **Fl.** 1–3 together, small, white.

**M. tuberosum** (L.) N. E. Br. v. **tuberosum** (*Mes. t.* L., *Delosperma t.* (L.) Schwant., *D. t.* (L.) L. Bol., *Mes. spinosum* O. Ktze.). – Cape: Uitenhage, Cradock, Middelburg D., Orange Free State; S.W.Afr.: Karas Mts. – Much-branched ♄ 50–70 cm tall, with a hemispherical or tuberous **caudex**; **L.** 10–15 mm long, 2–3 mm across, trigonous, minutely papillose, green; **Fl.** 7–8 mm ⌀, reddish-yellow.

**M. —** v. **macrorrhizum** (Haw.) N. E. Br. (*Mes. m.* Haw., *Delosperma m.* (Haw.) Schwant., *Mes. megarrhizum* Don). – E. Cape: Cape D. – With a very large, tuberous **caudex**; **Fl.** white?

**Meyerophytum** Schwant. (§ IV/1/21). – Rather slender, freely branched, papillose, ♃ plants 7–8 cm high, or 12 cm high including the Fl.; **St.** up to 3 mm ⌀, the lower part often clothed with the crowded remains of previous L., Int. enclosed; ☉ growth consisting of 2–3 pairs of dimorphic **L.**, axillary buds present in the lower L. and in the first pair (when 2 pairs occur) of the upper L.; the lower L. thicker, united for half their length or more, up to 1 cm long, the upper ones united for a third of their length or less, 1.5–2.5 cm long, 2–3 mm across and thick; **Fl.** solitary, up to 3.5 cm ⌀, rose-purple, on a pedicel 3.5 cm long. – Greenhouse, warm. Propagation: seed, cuttings.

## Key to the Species of the Genus Meyerophytum by L. Bolus

1. Sep. usually 5; stigmas 5 .................................................... **2.**
1a. Sep. 6; stigmas 6 ........................................................... **4.**
2. Pet. rose-purple throughout; filaments, anthers and pollen purple ............ **3.**
2a. Pet. white at the base; filaments yellow or white, anthers and pollen yellow ........
........ **meyeri** v. **holgatense**

3. Ov. raised above the level of the disc, lobes conspicuous, obtuse; placental tubercle present ... **meyeri** v. **meyeri**
3a. Ov. deeply concave, lobes inconspicuous (capsule unknown) ................ **tinctum**
4. Ov. slightly elevated; stigmas 1 mm long .................................... **microstigma**
4a. Ov. elevated to 1.5 mm; stigmas 2.5 mm long ............................... **primosii**

**M. meyeri** (SCHWANT.) SCHWANT. v. **meyeri** (Pl. 184/2, 3) (*Mitrophyllum m.* SCHWANT.). − Cape: Lit. Namaqualand. − **St.** 10−12 mm long; **L.** united to form small, plump, round, scarcely cleft **Bo.**, alternating with longer ones 10 mm long and 4 mm thick, dark greenish-yellowish, becoming reddish; **Fl.** flame-coloured.

**M.** — v. **holgatense** L. BOL. − Cape: Lit. Namaqualand. − Pet. white below.

**M. microstigma** (L. BOL.) L. BOL. (*Monilaria m.* L. BOL.). − Cape: Van Rhynsdorp D. − Up to 6 cm tall, glossy-papillose; lower **L.** 3 mm long, with a Sh. 1.5 mm long, upper **L.** 3.5 mm ⌀; **Fl.** 2 cm ⌀, pinkish-purple.

**M. primosii** (L. BOL.) L. BOL. (*Monilaria p.* L. BOL.). − Cape: Van Rhynsdorp D. − **St.** crowded, glossy-papillose, 8 cm tall including the Fl.; **L.** dimorphic, lower ones 3−4 mm long, with a Sh. 3 mm long, carinate on the underside, other **L.** 17−25 mm long, with a Sh. 6 mm long; **Fl.** 2.4 cm ⌀, pinkish-purple.

**M. tinctum** (L. BOL.) SCHWANT. (*Mes. t.* L. BOL., *Depacarpus t.* (L. BOL.) N. E. BR.). − Cape: Lit. Namaqualand. − Resembles **M. meyeri**, differentiated only by the concave, unlobed Ov.

**Micropterum** SCHWANT. (§ IV/1/20.) − ☉, very papillose herbs with fibrous **roots; L.** opposite, soft, lanceolate-spatulate, linear-spatulate or sometimes pinnatifid; **Fl.** solitary, pedicellate, small, yellow. − Summer: in the open. Propagation: seed.

**M. herrei** SCHWANT, (Pl. 184/4). − Cape: Cape D. − Papillose plant with prostrate **St.**; **L.** pinnatifid-lobed, c. 4 cm long, c. 1.5 cm across, terminal lobes rather small; **Fl.** solitary from the L.-axils, 8 mm ⌀, pink.

**M. longipes** (L. BOL.) SCHWANT. (*Cleretum l.* L. BOL.). − Cape: Lit. Namaqualand. − Plant 6−10 cm tall, yellowish-green; **St.** ascending; **L.** spreading, the upper ones erect, spatulate to linear-spatulate, tapered, 35−50 mm long, 5 to 18 mm across; **Fl.** solitary, 42 mm ⌀, yellow.

**M. papulosum** (L. f.) SCHWANT. v. **papulosum** (*Mes. p.* L. f., *Cleretum p.* (L. f.) L. BOL., *Cl. p.* (L. f.) N. E. BR., *Mes. humifusum* AIT., *Mes. oligandrum* KUNZE). − Cape: widely distributed in E. and N. Lit. Namaqualand. − Prostrate, distinctly papillose plant up to 43 cm ⌀; **St.** short, **Br.** and shoots elongated; **L.** lanceolate or linear-lanceolate, 4−7 cm long, 7−12 mm across above midway; **Fl.** solitary, or often alternate along the Br., c. 1 cm ⌀, yellow.

**M.** — v. **multiflorum** SCHWANT. − Free-flowering variety.

**M. puberulum** (HAW.) SCHWANT. (*Mes. p.* HAW., *Cleretum p.* (HAW.) N. E. BR.). − Cape: Sondags R. − Plant 20 cm tall; **St.** spreading, papillose; flowering **Br.** and L.-margins pubescent; **L.** opposite or alternate, obovate-spatulate, 1−2 cm long; **Fl.** solitary, whitish.

**M. schlechteri** SCHWANT. (*Cleretum s.* (SCHWANT.) N. E. BR.). − Cape: Van Rhynsdorp D. − Plant 5−10 cm tall, with spreading **St.**; **L.** spatulate; **Fl.** c. 4 cm ⌀, yellow.

**M. sessiliflorum** (AIT.) SCHWANT. − Cape. − Plant papillose; **Br.** divaricate; **L.** flat, spatulate; **Fl.** sessile.

**M.** — v. **album** (HAW.) JACOBS. (*Mes. s.* v. *a.* HAW., *Cryophytum burchelli* N. E. BR.). − Cape: Graaff Reinet D. − **Fl.** paniculate, white.

**M.** — v. **sessiliflorum** (*Mes. s.* AIT., *Cryophytum s.* (AIT.) N. E. BR., *Cleretum s.* (AIT.) N. E. BR., *Mes. s.* v. *luteum* HAW., *Cl. s.* v. *l.* (HAW.) JACOBS.). − Cape: Van Rhynsdorp D. − **St.** spreading; **Fl.** sessile, yellow.

**Mimetophytum** L. BOL. (§ IV/1/21). − Resembling **Mitrophyllum** in appearance and growth, differentiated by the presence of a tubercle in the Fr.-cell. − Greenhouse, warm. Propagation: seed; difficult from cuttings.

**M. crassifolium** L. BOL. − Cape: Lit. Namaqualand. − Small ♄; **St.** elongated, branched, glossy purple, 21.5 cm long, thickened at the nodes, 12 mm ⌀, Int. 3−7 cm long, 2−4 mm ⌀; **Br.** crowded, 6 mm ⌀; **L.** thick, pale green, obtuse, variously shaped, the first type 32 mm long with a Sh. 17 mm long, the second 7.5 to 8 cm long, 3 cm across and thick, the third type with the pairs united, 53 mm long, with a Sh. 45 mm long; **Fl.** unknown.

**M. parvifolium** L. BOL. (Pl. 185/1). − Cape: Lit. Namaqualand. − ♄ 15 cm tall; **St.** reddish brown, 3−5 mm ⌀, nodes 8 mm ⌀, Int. up to 4.5 cm long; **L.** of the first type ovate or lanceolate, 2−3.5 cm long, the united L.-pairs up to 6 cm long with the free parts 2−3 cm long, L. of the third type 14−18 mm long; **Fl.** yellow.

**Mitrophyllum** SCHWANT.[1]) (§ IV/1/21). − Erect ♄ 12−70 cm high, rarely freely branched; **St.** 5−15 mm ⌀, with the Int. short and enclosed by the persistent L.-bases, or up to 10 cm long; **L.** trimorphic, the third type only present during flowering; sterile Br. and shoots with 2 L.-pairs, the first forming a "mitre" or cone, the second much smaller and united from a quarter or up to the entire length to form a Sh. enclosing the following season's cone-L.; this united L.-pair, borne in the axils of the cone-L., persists until the Sh. is thinner

---

[1]) G. SCHWANTES separated the G. **Mitrophyllum** (capsules without cell-lids) from the G. *Conophyllum* (capsules with cell-lids). DE BOER, BOOM and HERRE subsequently established that the capsules of **Mitrophyllum** also possess cell-lids, so that the two genera have now been reunited.

than tissue-paper, allowing the free part to harden off; L.-pairs of the third type were formerly known as "bracts", and are the second and terminal pair on the flowering Br., which have ± elongated Int.; **Fl.** short-pedicellate, whitish, pink or yellow. Growing season only a few weeks, usually Aug.-Sept. – Greenhouse, warm; keep dry during the long resting period. Propagation: seed, difficult from cuttings.

**M. abbreviatum** L. Bol. (*Conophyllum a.* (L. Bol.) L. Bol.). – Cape: Lit. Namaqualand. – Slender, diffuse, branched, 12 cm tall; **St.** spreading, brown, with Int. 5–25 mm long, 3 mm thick; united **L.** lanceolate to linear-oblong, 12–14 mm long, 6–7 mm across, 3–4 mm thick, cone-**L.** cylindrical, 1.5 cm long, 6 mm ⌀, covered with large papillae; **Fl.** c. 3 cm ⌀, pink to white.

**M. affine** L. Bol. – Cape: Lit. Namaqualand. – **St.** 14 cm long, brownish, 4–5 mm ⌀, Int. 2–4 cm long; united **L.** lanceolate to linear-lanceolate, 29 mm long, 10–11 mm across, 1 cm thick below, cone-**L.** 28 mm long, Sh. 14 mm long; **Fl.** 3 cm ⌀, lemon-yellow.

**M. angustifolium** (L. Bol.) DE BOER (*Conophyllum a.* L. Bol.). – Cape: Lit. Namaqualand. – **St.** numerous, c. 30 cm long, 11 mm ⌀, reddish-brown, with Int. 2.5–4 cm long; united **L.** rounded on the underside, 2.5–9 cm long, with a Sh. 7–10 mm long, 11 mm ⌀ below, cone-**L.** 46–70 mm long, free parts unequal; **Fl.** 10–15 mm ⌀, whitish, becoming lemon-yellow.

**M. articulatum** (L. Bol.) DE BOER (*Conophyllum a.* L. Bol.). – Cape: Lit. Namaqualand. – ♄ 38 cm tall; older **St.** 16 mm ⌀, lower Int. 5–26 mm long, upper ones 35–60 mm long, St. and shoots conspicuously jointed; **L.** 4 cm long; **Fl.** 33 mm ⌀, yellow.

**M. brevisepalum** (L. Bol.) DE BOER (*Conophyllum b.* L. Bol.). – Cape: Lit. Namaqualand. – Flowering **St.** 19 cm long, 9 mm ⌀ below, dirty purple, 2–3 mm ⌀ above, with nodes 6–8 mm ⌀; **L.** 22–33 mm long, with a Sh. 3–17 mm long, L. of the third type mostly smaller, 7 mm long, Sh. 2 mm long; **Fl.** 3 cm ⌀, yellow.

**M. carteranum** (L. Bol.) DE BOER (*Conophyllum c.* L. Bol.). – Cape: Lit. Namaqualand. – Small ♄; **St.** 9.5–10.5 cm long, 14 mm ⌀; **L.** c. 8 cm long, 2.5 cm across below; **Fl.** 4.5 cm ⌀, yellow.

**M. clivorum** (N. E. Br.) SCHWANT. (*Mes. c.* N. E. Br., *Conophyllum c.* (N. E. Br.) SCHWANT., *Schwantesia c.* (N. E. Br.) L. Bol.). – Cape: Lit. Namaqualand. – Plant 25–30 cm tall; **St.** 6–10 mm ⌀, the nodes mostly with an annular thickening, Int. 1–3 cm long, reddish-brown, eventually pale grey; cone-**L.** cylindrical, 2–3 cm long, 7–11 mm thick, the free tips 2.5–5 cm long, 3.5–6 mm across; **L.** of the second type 2–5 cm long, 6–10 mm across, light green, glossy-papillose.

**M. cognatum** (N. E. Br.) SCHWANT. (*Mes. c.* N. E. Br., *Conophyllum c.* (N. E. Br.) SCHWANT., *Schwantesia c.* (N. E. Br.) L. Bol.). – Cape: Lit. Namaqualand. – Up to 15 cm tall; **St.** 6–7 mm thick, viscous, pale grey, with slightly thickened nodes, Int. 5–25 mm long; cone-**L.** cylindrical, 1.5–2.5 cm long, 5–7 mm thick, the free tips spreading, 18–32 mm long, 4–5.5 mm thick; L. of the second type 1.5–3.5 cm long, 6–8 mm across, 3–4 mm thick below, light green, glossy-papillose.

**M. compactum** (L. Bol.) DE BOER v. **compactum** (*Conophyllum c.* L. Bol.). – Cape: Lit. Namaqualand. – Compact bushes 30 cm tall, 35 cm ⌀; **St.** 2 cm ⌀, constricted at the nodes, Int. up to 2 cm long; united **L.** 15–47 mm long, 1 cm ⌀, with a Sh. 8 mm long, cone-L. 5 cm long, united for 2.5 cm, the free part 4 mm thick; **Fl.** 4 cm ⌀, yellow, paler inside.

**M.** — v. **eenrietense** (L. Bol.) DE BOER (*Conophyllum c. v. e.* L. Bol.). – Cape: Lit. Namaqualand. – Cone-**L.** 1.5–3.1 cm long, with a Sh. 2–5 mm long, L. of the second type 3.7 to 7.7 cm long, with a Sh. 1–2 cm long, L. of the third type 2–3.6 cm long, with a Sh. c. 1.5 cm long; **Fl.** 5.3 cm ⌀.

**M. conradii** L. Bol. – Cape: Lit. Namaqualand. – Plant 23 cm tall, branching from the base; **St.** 5–8 mm thick, branched, nodal, lower Int. 10 mm ⌀, upper ones 4–5 cm long, distinctly alate, dark brown, 5 mm thick; **L.** broad-lanceolate to oblong-ovate, pale green, 43–50 mm long, with a Sh. 2 mm long, 17 mm ⌀ below, cone-L. 30–36 mm long, the free part 10–16 mm long; **Fl.** 3 cm ⌀, yellow.

**M. cuspidatum** (L. Bol.) DE BOER (*Conophyllum c.* L. Bol.). – Cape: Lit. Namaqualand. – Plant 12.5 cm tall, **St.** 6 mm ⌀ below; **L.** ± falcate, obtuse, pale green, 10–11 cm long, with a Sh. 12 mm long, 26 mm across below, the free part 5 or 15 mm long; **Fl.** 4 cm ⌀, yellow, paler inside.

**M. dissitum** (N. E. Br.) SCHWANT. (*Mes. d.* N. E. Br., *Conophyllum d.* (N. E. Br.) SCHWANT., *Schwantesia d.* (N. E. Br.) L. Bol.). – Cape: Lit. Namaqualand. – Plant up to 30 cm tall; old **St.** 3–5 mm thick, **Br.** 3 mm thick, thickened at the nodes, viscous, reddish brown, becoming grey, with Int. 2.5–6.5 cm long; cone-**L.** 2.5–4.5 cm long, 7–10 mm thick below, the free parts erect, 10–25 mm long, 4–6 mm across, light green; L. of the second type shortly united, 2–4.5 cm long, 8–11 mm across, light green.

**M. framesii** L. Bol. (*Conophyllum f.* (L. Bol.) L. Bol.). – Cape: Lit. Namaqualand. – Plant 16 cm tall; **St.** 1.3 cm ⌀, brown, with Int. 10–17 mm long, flowering St. 6 mm ⌀, 2–3 cm long; **L.** 7 cm long, with a Sh. 1 cm long, the L. at that point 2.9 cm thick and 1.6 cm across; cone-L. 3.2 cm long, with the free parts slender, tapered; **Fl.** 4.5 cm ⌀, pink.

**M. gracile** (SCHWANT.) DE BOER (Pl. 185/2) (*Conophyllum g.* SCHWANT.). – Cape: Lit. Namaqualand. – Small, very densely branched ♄ c. 15 cm tall; **St.** up to 15 cm long, 2 mm thick, brownish-red, with Int. 5–20 mm long; **L.** 3 cm long, 7 mm ⌀ below, forming an acute triangle; cone-**L.** 5–20 mm long, 5 mm thick, the free part 1–4 mm long, tapered; **Fl.** unknown. ? Identical with **M. abbreviatum**.

**M. grande** N. E. Br. v. **grande** (Pl. 185/3) (*Conophyllum g.* (N. E. Br.) L. Bol., *Mes. proximum* L. Bol.). – Cape: Lit. Namaqualand. – Resembles **M. mitratum**; flowering **Br.** 30–35 cm long, 2-angled, with Int. 2–8 cm long, 4–5 mm ⌀, with reduced L. at the nodes; cone-**L.** c. 20 cm long, 4–5 cm thick below; **Fl.** 4–4.5 cm ⌀, glossy white.

**M.** — v. **compressum** (L. Bol.) de Boer (*Conophyllum g.* v. *c.* L. Bol.). – Cape: Lit. Namaqualand. – **Fl.** with compressed ovary-lobes.

**M. hallii** (L. Bol.) de Boer (*Conophyllum h.* L. Bol.). – Cape: Lit. Namaqualand. – **St.** up to 2 cm ⌀; old **Br.** up to 15 mm ⌀, younger ones 2–3 cm long, 3–5 mm thick; cone-**L.** up to 22 mm long, 6 mm across below, those of the second type up to 12 mm long, with a Sh. 3 mm long, the free parts divergent, L. longer in cultivation; **Fl.** up to 46 mm ⌀, yellow.

**M. herrei** (L. Bol.) de Boer (*Conophyllum h.* L. Bol.). – Cape: Lit. Namaqualand. – Plant up to 70 cm tall, freely branched, with Int. c. 1 cm long, 5 mm thick; cone-**L.** 25 mm long, united ± to midway; L. of the second type c. 3 cm long, 5–6 mm across, cylindrical; **Fl.** 10–12 mm ⌀, yellow.

**M. karrachabense** L. Bol. – Cape: Lit. Namaqualand. – **St.** 12 cm long, 1 cm thick, with Int. 15 mm long; cone-**L.** 5.5 cm long, the free part 4 cm long, 6 mm thick; L. of the second type ± compressed, linear, acute; **Fl.** 35 mm ⌀, light yellow.

**M. kubusanum** L. Bol. (*Conophyllum k.* (L. Bol.) L. Bol.). – Cape: Lit. Namaqualand. – **St.** up to 17 cm long, Int. 2 mm long, 5 mm ⌀; cone-**L.** 2.4 cm long, 11–18 mm ⌀ midway; L. of the second type lanceolate, 6 cm long, 19 mm across midway, with a Sh. 6 mm long; **Fl.** 35 mm ⌀, lemon-yellow.

**M. latibracteatum** (L. Bol.) de Boer f. **latibracteatum** (*Conophyllum l.* L. Bol.). – Cape: Lit. Namaqualand. – Plant up to c. 70 cm tall, 11 cm ⌀ below because of the crowded **St.**, some of them dead; cone-**L.** 7 cm long with a Sh. 56 mm long, L. of the second type narrowed, c. 11 cm long, 4 mm across, pale bluish-green, with a Sh. 17 mm long; "bracts" (third type of L.) ovate, ± rhombic, 5–10 mm long, 5–11 mm across; **Fl.** 5–6 mm ⌀, yellow.

**M.** — f. **fera** (L. Bol.) de Boer (*Conophyllum l. f. f.* L. Bol.). – Sh. of the cone-**L.** 28 mm long; "bracts" (third type of L.) 14 mm long; **Fl.** rather larger.

**M. marlothianum** Schwant. (*Conophyllum m.* (Schwant.) Schwant., *Mes. niveum* L. Bol., *C. niveum* (L. Bol.) Herre). – Cape: Lit. Namaqualand. – Plant 20–40 cm tall; **St.** 18 mm ⌀ below, reddish-brown, lower Int. spherical, upper ones up to 6 cm long; cone-**L.** 2.5–5 cm long, L. of the second type up to 6 cm long, united for 3.5 mm, 13 mm across; **Fl.** up to 4 cm ⌀, glossy, pure white.

**M. mitratum** (Marl.) Schwant. (*Mes. m.* Marl., *Conophyllum m.* (Marl.) Schwant., *M. m.* v. *eburneum* L. Bol.) – Cape: Lit. Namaqualand. – Mat-forming; **St.** thick, soft, becoming woody, with the nodes annularly thickened; cone-**L.** 7–8 cm or more long, 2 cm thick, the free parts obtusely trigonous, 10–12 mm long, light green; L. of the second type 8 to 10 cm long, c. 10 mm thick below, light green, glossy-papillose; **Fl.** 2.5–3 cm ⌀, white, Pet. reddish at the tips.

**M. obtusipetalum** (L. Bol.) de Boer (*Conophyllum o.* L. Bol.). – Cape: Lit. Namaqualand. – Slender plant c. 43 cm tall; **St.** 35–40 cm long, 1 cm ⌀ below, reddish-brown at first, constricted at the nodes, Int. 1–10 cm long; cone-**L.** 4–5 cm long, the free part c. 1.5 cm long, L. of the second type acute or ± obtuse, lanceolate to linear-lanceolate, 3–5 cm long, c. 1 cm across and thick below, with a Sh. 3–5 mm long; "bracts" (third type of L.) 7–11 mm long, glandular-hirsute below; **Fl.** 2 cm ⌀, lemon-yellow to white.

**M. pillansii** N. E. Br. (Pl. 185/4). – Cape: Lit. Namaqualand. – Plant 25–30 cm tall; **St.** 7–9 mm thick, Int. 3–4 mm long, those on flowering shoots 18 mm long; cone-**L.** 4–8 cm long, 2–3 cm thick below, L. of the second type 8–9 cm long, 1.5–2 cm across, 7–15 mm thick below; **Fl.** 3–4 cm ⌀, white.

**M. proximum** (N. E. Br.) Schwant. (*Mes. p.* N. E. Br., *Conophyllum p.* (N. E. Br.) Schwant., *Schwantesia p.* (N. E. Br.) L. Bol.). – Cape: Lit. Namaqualand. – Plant 20–25 cm tall; old **St.** c. 6 mm thick, Br. 4–5 mm thick, viscous, dark violet-grey, Int. thickened, 6 to 25 mm long; cone-**L.** 3–7.5 cm long, 10–18 mm thick below, 6–8 mm thick above, light green, the tips 8–25 mm long, acute; **L.** of the second type widely divergent, 1.5–8 cm long, 8–14 mm across, 6–8 mm thick below, narrowing above, with deciduous T. along the margins.

**M. ripense** (L. Bol.) de Boer (*Conophyllum r.* L. Bol.). – Cape: Lit. Namaqualand. – Plant 15 cm tall; **St.** 8–12 mm ⌀, dirty brown at first, with Int. 2–10 mm long; cone-**L.** 4–7.5 cm long, the free part 2–4.6 cm long, 5 mm ⌀, L. of the second type ± obtuse, 2–4 cm long, 7–13 mm across below, 12 mm thick; "bracts" 3–3.5 cm long; **Fl.** c. 5 cm ⌀, yellow, paler inside.

**M. roseum** L. Bol. (*Conophyllum r.* (L. Bol.) L. Bol.) – Cape: Lit. Namaqualand. – Rootstock c. 1.5 cm ⌀; plant freely branched, the **St.** climbing, thin, with Int. very short or up to 6 cm long; cone-**L.** ovate or ovate-lanceolate, 7–11 cm long or 3.5–6.5 cm long, L. of the second type 1.4–6 cm long, with a Sh. 8–11 mm long, L. of the third type 1.5–4 cm long, with a Sh. 5 mm long; pedicels often sinuous, **Fl.** 5 cm ⌀, deep pink.

**M. tenuifolium** L. Bol. – Cape: Lit. Namaqualand. – Plant 6 cm high; **St.** spreading, with glossy brown Int. 3–9 mm long, 4 mm thick; lower **L.** 5 mm long, 4 mm across, 3 mm thick, with a Sh. 1 mm long, L. of the second type 10–15 mm long, 4 mm thick, with a Sh. 7 mm long, L. of the third type 27 mm long, 3 mm across and thick, with a Sh. 2–4 mm long, 6 mm thick; **Fl.** 15 mm ⌀, primrose-yellow.

**M. vanheerdei** (L. Bol.) DE Boer (*Conophyllum v.* L. Bol.). – Cape: Lit. Namaqualand. – **St.** 15 mm ∅, with Int. 1–9 mm long; flowering **Br.** 18–20 cm long with 5 nodes, the Int. glossy reddish-brown, 5.2 cm long, 3–9 mm ∅; **L.** ovate to oblong-ovate, 12–21 mm long with a Sh. 4mm long, L. on sterile shoots larger c. 4 cm long, 15mm across below, 12 mm thick, with a Sh. 5 mm long; cone-L. 6 cm long, 5 mm ∅, with a Sh. 2 cm long; **Fl.** c. 5 cm ∅, yellow, paler inside.

**Monilaria** Schwant. (§ IV/1/21). – Very low heterophyllous ♄ forming clumps; **St.** short and thick, often jointed like a necklace; **L.** various, those of one type little united, ± semicylindrical with bright papillae, alternating with L. united into subglobose Bo.; **Fl.** long-stalked, white, yellowish or red. – Propagation: seed, cuttings somewhat difficult.

### Division of the Genus Monilaria Schwant. into Sect. by L. Bolus

**Sect. I. Monilaria** (*Eumonilaria* L. Bol.).
Br. ± moniliform, Int. very short, jointed, globose. – Type species: M. moniliformis. – Species: M. brevifolia, chrysoleuca, globosa, moniliformis, nodosa, peersii, pisiformis, polita, salmonea, vestita.

**Sect. II. Exserta** L. Bol. – Br. with extended Int. – Type species: M. ramulosa. – No further species.

**M. brevifolia** L. Bol. (§ I). – Cape: Van Rhynsdorp D. – Flowering **St.** c. 8 cm long, 4 mm thick, Int. 7–9, 0.75–1.5 mm long, younger ones longer; some **L.** 3 mm long, cylindrical, erect, some 23 mm long, 7 mm across, 8 mm thick, with a Sh. 3 mm long, others 27 mm long, minutely glossy-papillose; **Fl.** 4–5 cm ∅, Pet. white with yellow tips or pink.

**M. chrysoleuca** (Schltr.) Schwant. f. **chrysoleuca** (§ I) (*Mes. c.* Schltr., *Mitrophyllum c.* (Schltr.) Schwant., *Conophyllum c.* (Schltr.) Schwant., *Schwantesia c.* (Schltr.) L. Bol., *Mes. scutatum* L. Bol., *C. s.* (L. Bol.) Schwant., *Monilaria s.* (L. Bol.) Schwant.). – Cape: Van Rhynsdorp D. – Plant 6–10 cm tall, with fleshy **St.**; **L.** cylindrical, obtuse, minutely papillose, 3–5 cm long, 5 mm across; **Fl.** c. 3 cm ∅, snow-white.

**M. —** f. **purpurea** (L. Bol.) Rowl. (*M. c. v. p.* L. Bol.). – Cape: Lit. Namaqualand. – **Fl.** pinkish-purple.

**M. globosa** (L. Bol.) L. Bol. (Pl. 186/1) (§ I) (*Conophyllum g.* L. Bol.). – Cape: Lit. Namaqualand. – Plant 6 cm tall, 8 cm ∅; **St.** 15 mm ∅, with **Br.** of one type 1 mm thick, the others approaching to conical, 10–17 mm thick; Bo.-L. c. 2 cm long and almost as thick, others united for 1.5 cm, the free part 2.5 cm long, 8 mm across; **Fl.** 4 cm ∅, white.

**M. moniliformis** (Haw.) Schwant. (§ I) (*Mes. m.* Haw., *Conophyllum m.* (Haw.) Schwant., *Mitrophyllum m.* (Haw.) Schwant., *Schwantesia m.* (Haw.) L. Bol.). – Cape: Van Rhynsdorp D. – Plant 7.5–10 cm tall; **St.** and **Br.** 8–11 mm thick; Bo.-L. oblong to rounded, ± spherical, c. 12 mm long, dark green; secondary L. united for c. 12 mm, this part concealed within the Bo.-L., 10–15 cm long, 4–5 mm across, semicylindrical, soft, papillose at first; **Fl.** white.

**M. nodosa** (Bgr.) Jacobs. (§ I) (*Mes. n.* Bgr.). – W. Cape: Van Rhynsdorp D. – Plant 13–17 cm or more tall; **St.** greyish-brown, branched, Int. 1–2 cm long, 2 mm thick, lateral shoots annular, 1 mm long; Bo.-L. 2 mm long, 3 mm across, secondary L. unequal, 8–12 mm long; **Fl.** solitary.

**M. peersii** L. Bol. (§ I). – Cape: Lit. Namaqualand. – Plant 10–12 cm tall; **St.** and **Br.** 1–2 cm long, Int. 1–10 mm long, 5–11 mm thick, pale brown at first, eventually grey, the bark becoming papery; Bo.-L. short, 7 mm long, 3.5 mm across, 4 mm thick, secondary L. 4–6 cm long, with a Sh. 2–4 mm long, or 2.5 to 4 cm long, 3 mm thick, papillose, bluish-green; **Fl.** 2 cm ∅, white.

**M. pisiformis** (Haw.) Schwant. (§ I) (*Mes. p.* Haw., *Conophyllum p.* (Haw.) Schwant., *Mitrophyllum p.* (Haw.) Schwant., *Schwantesia p.* (Haw.) L. Bol.). – Cape: Van Rhynsdorp D. – **St.** short, 2–3 cm long, freely branched; Bo.-L. very short, pea-sized, secondary L. ± cylindrical, 5–6 cm long, 3–4 mm ∅, glossy papillose; **Fl.** 3 cm ∅, yellow, Pet. reddish with a white border.

**M. polita** L. Bol. (§ I). – Cape: Van Rhynsdorp D. – Plant 13 cm tall; **St.** 7 cm long, jointed, Int. compressed-spherical, 5–7 mm long, 15 mm ∅, the bark parchment-like; Bo.-L. 5 mm long, secondary L. flat on the upperside, rounded on the underside, green, 3–7 cm long, 5–7 mm across; **Fl.** 2–3 cm ∅, white.

**M. ramulosa** (L. Bol.) Schwant. (§ II) (*Schwantesia r.* L. Bol., *M. r.* (L. Bol.) Bol.). – Cape: Lit. Namaqualand. – Plant 10–15 cm tall; **St.** spreading to erect, 7–8 cm long, appearing ± pruinose because of minute papillae, Int. (1–)8(–10) mm long, 4 mm ∅, flowering shoots shorter, with 4 L., sterile shoots with 8 L.; shorter **L.** scarcely 3 mm long, the others 25 mm long, 4 mm across, with a Sh. 1.5 mm long, semicylindrical; **Fl.** 4–5 cm ∅, glossy white.

**M. salmonea** L. Bol. (§ I). – Cape: Van Rhynsdorp D. – Plant 9 cm tall; **St.** with 2–4 L., lower Int. ± compressed, 2–5 mm long, the upper ones up to 16 mm long, 10 mm thick; Bo.-L. 4 mm long, 5 mm ∅, secondary L.

3–3.5 cm long, 5 mm across, with a Sh. 8 mm long, somewhat tuberculate towards the tip, green; **Fl.** 4 cm ⌀, an attractive salmon-pink.

**Mossia** N. E. BR. (G. of dubious classification). – Creeping, succulent, ♃ herb, with elongated **St.** and distinct **Int.**; **L.** opposite, united at the base, trigonous-ovoid; **Fl.** solitary, terminal, shortly pedicellate or ± sessile between the **L.** Probably not in cultivation. – Summer: in the open, winter: cold-house. Propagation: seed, cuttings.

**M. intervallaris** (L. BOL.) N. E. BR. (*Mes. i.* L. BOL.). – Botswana; Transvaal, Orange Free State. – **St.** rooting at the nodes, **Int.** 5–30 mm long, 2 mm thick; **L.** crowded at the St.-tips, 5–8 mm long, 4–5 mm across, 3–4 mm thick, with 1–2 small **T.** on the angles and the keel, conspicuously bluish-green, slightly spotted; **Fl.** c. 12 mm ⌀, pale cream-coloured, opening at night.

**Muiria** N. E. BR. (§ IV/1/16).
**M. hortenseae** N. E. BR. (Pl. 186/3). – Cape: Lit. Karroo, Riversdale D. – Clump-forming, highly succulent plants with short, fibrous **roots**; growth consists of 2 completely united **L.** forming a ± compressed, ovoid to spherical, sometimes somewhat angular **Bo.** 2–2.5 cm high, 15–18 mm thick, with the **Fi.** situated below the apex and scarcely visible until the Fl. forces its way through, very soft-fleshy, light green, densely covered with velvety hairs; **Fl.** solitary, pinkish-white, projecting only slightly beyond the Fi. – Growing-period July-Sept. – Greenhouse, warm, light. – Propagation: seed.

× **Muirio-gibbaeum** JACOBS. (§ IV/1/16). – Natural hybrid: **Muiria hortenseae** × **Gibbaeum album** v. **roseum.**
× **M. muirioides** ROWL. (Pl. 186/5) (*Gibbaeum m.* HERRE n. nud.). – Cape: Lit. Karroo. – Offsetting to form clumps; **roots** woody; **St.** short, thick, densely covered with old, persistent L.-skins; **Bo.** resembling that of **Gibbaeum album,**, 2–2.5 cm long, 2 cm across, 2–2.5 cm thick, with the Fi. 10–12 mm long, 1 mm deep, the surface greenish-brown, puberulous (like **Muiria hortenseae**); **Fl.** white or pink, Dec. – Greenhouse, warm. – Propagation: by careful division!

**Namaquanthus** L. BOL. (G. of dubious classification).
**N. vanheerdei** L. BOL. (Pl. 186/4). – Cape: Lit. Namaqualand. – Erect, stiff, glabrous ♄ 15–20 cm tall; **St.** 10–12 mm thick, covered with the remains of old L.; **L.** erect to ± falcate, ± cylindrical, compressed above, obtuse, dirty bluish-green, with the keel and angles indistinct, rarely ± laterally compressed, 2.5 to 4 cm long, 8–10 mm across, 8–12 mm thick, with a Sh. 2–3 mm long; **Fl.** solitary, up to 6 cm ⌀, Pet. brilliant pinkish-purple, sometimes indistinctly striate, on a pedicel 1 cm long with 2 bracts.

**Namibia** DTR. et SCHWANT. (§ IV/1/13). – Dwarf ♄ forming clumps, with scarcely woody Br.; **L.** in 1–2 pairs at the ends of the Br., very thick and soft, finely rough from white dots; **Fl.** sessile or short stalked, white or violet. – Greenhouse, warm. Propagation: seed, cuttings.

## Key to the Species of the Genus Namibia by M. Walgate

1. Fl. almost sessile; Pet. subacute; stigmas 9–13 .................................... **cinerea**
1a. Ped. 1–1.2 cm long; Pet. obtuse; stigmas 25 ..................................... **pomonae**

**N. cinerea** (MARL.) DTR. et SCHWANT. (Pl. 186/2) (*Mes. c.* MARL., *Juttadinteria c.* (MARL.) SCHWANT.). – S.W.Afr.: S.Namib. – **L.** 13 mm long, 10–12 mm across below, up to 12 mm thick, rather rounded-triangular on the upperside, ± recurved at the tip, navicular and rounded-carinate on the underside, greyish-green, rough; **Fl.** violet.

**N. pomonae** (DTR.) DTR. et SCHWANT. (*Mes. p.* DTR., *Juttadinteria p.* (DTR.) SCHWANT.). – S.W.Afr.: S.Namib. – Forming cushions 20 cm across, 10 cm high; **L.** very crowded, 30 mm long, united for 10–12 mm, 15–18 mm across, 12–14 mm thick, broadly navicular, with distinct angles and an obtuse keel, apiculate, whitish-grey or light grey; **Fl.** 3 cm ⌀, white.

**Nananthus** N. E. BR. (§ IV/1/9). – Dwarf, tufted, glabrous plants 3–5 cm high; **caudex** tuberous, up to 13 cm long, 4–5 cm ⌀; **St.** with 4–6 L., Int. enclosed; **L.** opposite, ascending or spreading, 1.5–5 cm long, widened in the upper part, one side bulging more than the other, widest in the middle, above that square, oblong, linear, ovate or broadly so, subobtuse, acute to acuminate, sometimes conspicuously aristate, the underside obtusely keeled upwards, punctate, the dots often white or whitish; **Fl.** solitary, Pet. in 2–3 series, yellow, with or without a red M.-stripe, on a short pedicel with 2 bracts at the base. – Greenhouse, warm. Propagation: seed, cuttings.

## Key to the Species of the Genus Nananthus by L. Bolus

1. Pet. with a central stripe ............................................... **2.**
1a. Pet. without a central stripe .......................................... **10.**
2. Stigmas exceeding the height of the stamens .......................... **3.**
2a. Stigmas equalling the height of the stamens or falling short of it ..... **4.**

3. L. up to 2.6 cm long and 1.3 cm broad; Pet. acute or subacute; stigmas 10 ..... **wilmaniae**
3a. L. up to 3.2 cm long and 9 mm broad; Pet. obtuse; stigmas 7–10 ............ **pole-evansii**
4. Stamens 7–10 mm long; stigmas 9–10, equalling the stamens, 5–9 mm long ........... 5.
4a. Stamens up to 6 mm long; stigmas usually 6–8, shorter than the stamens, 3–5 mm long ... 9.
5. L. acuminate ................................................................... 6.
5a. L. subacute or obtuse ......................................................... 8.
6. Ov. convex above, the lobes convex ............................................ 7.
6a. Ov. nearly flat above, lobes scarcely convex ................ **transvaalensis v. griquensis**
7. L. 5–6 mm broad in cultivation ............................................. **— v. transvaalensis**
7a. L. when flattened up to 1.1 cm broad ........................................ **— v. latus**
8. L. up to 8 mm broad; Ov. elevated in the centre; stigmas 9 .................. **pallens**
8a. L. up to 12 mm broad; Ov. flat, stigmas 11 ............................ **aloides v. striatus**[1]
9. L. acute or subobtuse; Pet. obtuse .......................................... **vittatus**
9a. L. acuminate; Pet. acute or subacute ........................................ **gerstneri**
10. Stigmas shorter than the stamens ........................................... 11.
10a. Stigmas exceeding the stamens ............................................. 12.
11. L. 1.4–2.6 cm long, up to 5 mm broad ....................................... **broomii**
11a. L. 2.5–3.5 cm long, usually 1 cm broad ................................. **aloides v. aloides**
12. Top of Ov. scarcely elevated ............................................... 13.
12a. Top of Ov. conspicuously elevated; L. to 1.8 cm broad ..................... **cibdelus**
13. L. to 8 mm broad, L. tip more than half covered in pearl-like tubercles .... **margaritiferus**
13a. L. to 2 cm broad, L. tip less than half ................. covered in pearl-like tubercles ........................................................ **aloides v. latus**

**N. aloides** (HAW.) SCHWANT. v. **aloides** (*Mes. a.* HAW., *Aloinopsis a.* (HAW.) SCHWANT., *N. a.* (HAW.) N. E. BR.). – Cape: W.Griqualand; Botswana. – **L.** 6 together, 5 cm long, obliquely lanceolate or narrowly rhombic, often flat-furrowed, carinate-trigonous above, dark green, with numerous prominent, white, tuberculate dots and rough angles; **Fl.** 2.5 cm ⌀, yellow, Aug.-Sept.
**N. —** v. **latus** L. BOL. – **L.** broader.
**N. —** v. **striatus** (L. BOL.) (*Aloinopsis a.* v. *s.* L. BOL.). – Cape: Barkly West D. – **L.** rather shorter; **Fl.** 3–3.5 cm ⌀, Pet. yellow with a darker yellow stripe.
**N. broomii** (L. BOL.) L. BOL. (*Aloinopsis b.* BOL. L.). – Transvaal. – **St.** with up to 10 green, oblique, expanded **L.** 14–16 mm long, 5 mm across midway, 2.5 mm thick, with a truncate, purple tip and the keel laterally compressed; **Fl.** 2 cm ⌀, yellow.
**N. cibdelus** (N. E. BR.) SCHWANT. (*Mes. c.* N. E. BR., *Aloinopsis c.* (N. E. BR.) SCHWANT., *Rabiea c.* (N. E. BR.) N. E. BR., *Mes. aloides* SALM, *Mes. albipunctum* v. *majus* HAW.). – Origin unknown. – Resembles **N. aloides** v. **striatus**; **L.** almost twice as large, to 1.8 cm broad.
**N. gerstneri** (L. BOL.) L. BOL. (*Aloinopsis g.* L. BOL.). – Cape: Aliwal North D. – Older **St.** tuberous, 1–3 cm thick, younger ones with 6 spreading, linear, tapered, apiculate, dirty green **L.** 35 mm long, 3–4 mm thick, expanded midway to 5–6 mm, oblique at the tip, with prominent dots; **Fl.** c. 3 cm ⌀, yellow.
**N. margaritiferus** L. BOL. – ? Botswana: Kalahari. – Plant up to 6 cm ⌀; with a napiform **caudex** 6.5 cm long, up to 1.6 cm ⌀; **L.** 3 cm long, up to 8 mm across, up to 7 mm thick, flat on the upperside, densely covered with pearly tubercles; **Fl.** 4–8 mm long, yellow, with pink tips and somewhat darker yellow lines.

**N. pallens** (L. BOL.) L. BOL. (*Aloinopsis p.* L. BOL.). – Cape: N. of the Orange R., near Upington. – Younger **St.** with up to 10 green **L.** 14–16 mm long, 5 mm across midway, 2.5 mm thick, obliquely expanded above, with the underside obtuse and the keel compressed; **Fl.** 3 cm ⌀, Pet. yellow, indistinctly striped.
**N. pole-evansii** N. E. BR. (*Aloinopsis p.* (N. E. BR.) N. E. BR.). – Cape: Prieska D. – **St.** with 2–3 **L.** pairs; **L.** oblong, long-triangular and apiculate above, navicular on the underside and obliquely keeled, 3 cm long, 8 mm across, 8–9 mm thick, smooth, reddish or bluish-green, waxy pruinose, angles and keel reddened; **Fl.** 26 mm ⌀, glossy golden-yellow, Pet. with a red M.-line.
**N. transvaalensis** (ROLFE) L. BOL. v. **transvaalensis** (*Mes. t.* ROLFE, *Aloinopsis t.* (ROLFE) SCHWANT.). – Transvaal: Boshof D. – Resembles **N. vittatus**; **L.** 3 cm long, 1 cm across, with conspicuously large tubercles crowded along the margins.
**N. —** v. **latus** L. BOL. – **L.** shorter, broader, 2 cm long, 1.3 cm across.
**N. —** v. **griquensis** L. BOL. – Receptacle broader and flatter than in the type.
**N. vittatus** (N. E. BR.) SCHWANT. (Pl. 187/1) (*Mes. v.* N. E. BR., *Aloinopsis v.* (N. E. BR.) SCHWANT., *N. v.* (N. E. BR.) N. E. BR., *Mes. aloides* HORT.). – Orange Free State: Fauresmith D. – **Roots** fleshy; **L.** 6–8 together, the opposite ones unequal, 2–3 cm long, 6–8 mm across, obliquely lanceolate, acute, shortly apiculate, semicylindrical, expanded above, dull green, with tuberculate, raised dots; **Fl.** 2 to 2.5 cm ⌀, light yellow, Pet. with a fine red M.-line.
**N. wilmaniae** (L. BOL.) L. BOL. (*Aloinopsis w.* L. BOL.). – Cape: W.Griqualand. – **Caudex** tuberous; **St.** with 4 dirty olive-green **L.** 22 mm

---

[1] An insufficiently known variety; central stripe of the Pet. not red but yellowish.

long, oblong below and thickened to 6 mm, broadest midway and 13 mm across, ovate above, acute and apiculate, the keel on the underside compressed towards the tip, with pallid dots; **Fl.** 2 cm $\varnothing$, Pet. yellow with red stripes.

**Nelia** SCHWANT. (§ IV/1/8). – Plants ± acauline, forming clumps or domed cushions; **L.** densely crowded, united at the base, 2–4 cm long, linear or rhombic, semicylindrical, ± carinate on the underside, smooth, glossy, green or waxy greyish-green; **Fl.** on a short pedicel with 2 bracts, white or yellowish, May-June, remaining open for a very long time. – Greenhouse, warm. Propagation: seed, cuttings.

**N. meyeri** SCHWANT. v. **meyeri.** – Cape: Lit. Namaqualand. – Low **St.** with 1–2 L.-pairs; **L.** 3 cm long, 1 cm across, 8 mm thick, longtrigonous, acute, sharply keeled up to midway on the underside, bluish-green, glossy; **Fl.** 5 to 6 cm $\varnothing$, whitish-yellow.

**N.** — v. **longipetala** L. BOL. – Pet. very crowded, pale pink.

**N. pillansii** (N. E. BR.) SCHWANT. (*Sterropetalum p.* N. E. BR.). – Cape: Lit. Namaqualand. – **St.** numerous, 5–8 cm tall, each with 2–3 pairs of **L.** 2–3 cm long, 5–8 mm across, 5–6 mm thick, ± digitate below, rounded above, slightly convex on the upperside, round and ± rounded-carinate on the underside, bluish-green or whitish-green; **Fl.** up to 14 mm $\varnothing$, remaining open night and day.

**N. robusta** SCHWANT. – Cape: Lit. Namaqualand. – **L.** 2–4 cm long, 5–12 mm across and thick, linear on the upperside, shortly triangular at the tip, the underside shortly carinate above, semicylindrical below, yellowish-green, waxy, glossy.

**N. schlechteri** SCHWANT. (Pl. 187/4) (*Corpuscularia perdiantha* TISCH.). – Cape: Lit. Namaqualand. – **St.** with 1–2 L.-pairs; **L.** 18 to 25 mm long, 6–8 mm across and thick, with the upperside linear, expanded towards the tip, then shortly acute or rhombic, apiculate, with the underside rounded, acutely keeled above, with the keel pulled forward and chin-like, light bluish-green or whitish to greyish-green, often somewhat reddish; **Fl.** 18 mm $\varnothing$, white.

**Neohenricia** L. BOL. (§ IV/1/11).

**N. sibbettii** (L. BOL.) L. BOL. (Pl. 187/2) (*Mes. s.* L. BOL., *Henricia s.* (L. BOL.) L. BOL.). – Orange Free State: Fauresmith D. – Freely branched, mat-forming, low ♄; **L.** 4 together on a shoot, 10 mm long, 2 mm across below, 5 mm across and convex above, with many tiny white tubercles; **Fl.** on a pedicel 14 mm long, 12 mm $\varnothing$, white, opening at night, strongly scented. – Greenhouse, warm. Propagation: seed, cuttings.

**Neorhine** SCHWANT. (§ IV/1/11).

**N. pillansii** (N. E. BR.) SCHWANT. (Pl. 187/3) *Rhinephyllum p.* N. E. BR.). – Cape: Laingsburg D. – Clump-forming, each **St.** with 2–3 L.-pairs; **L.** suberect to spreading, 10–14 mm long, 5–7 mm across, expanded above and 3–5 mm thick, thinner at the base, somewhat spatulate-clavate, acuminate, with white spots forming a horny margin, the underside rounded, green, with dense white spots; **Fl.** unknown. (Acc. HERRE and VOLK referred back to **Rhinephyllum.**)

**Octopoma** N. E. BR. (§ IV/1/2). – Low-growing, bushily branched ♄ with distinct, short Int.; **St.** succulent at first, eventually woody, often with shoots from the L.-axils; **L.** opposite, ± decurrent on the St., united and sheathing at the base, hemispherical or narrowed towards the apiculate tip, flat on the upperside, with rounded sides, ± distinctly keeled on the underside, often with translucent lines along the keel-angle; **Fl.** 1–3 together on a shoot, sessile or shortly pedicellate, yellow or red. – Summer: in the open, winter: cold-house.

**O. abruptum** (BGR.) N. E. BR. (*Mes. a.* BGR.). – Cape: Clanwilliam D. – Plant 15–20 cm tall, stiff, with erect **St.**; **L.** spreading, 4–8 mm long, 3–6 mm across and thick, trigonous, acute, with denticulate angles, velvety; **Fl.** 25 mm $\varnothing$.

**O. calycinum** (L. BOL.) L. BOL. (*Ruschia c.* L. BOL., *Mes. separatum* N. E. BR.). – Cape: Riversdale D. – Plant 9 cm tall; **St.** crowded, flowering shoots with 4–6 crowded, whitish to blue **L.** 12 mm long, 9 mm $\varnothing$; **Fl.** solitary, 2.5 cm $\varnothing$, pale straw-coloured.

**O. conjunctum** (L. BOL.) L. BOL. (*Ruschia c.* L. BOL.). – Cape: Van Rhynsdorp D. – Plant 17 cm tall, minutely papillose at first; **L.** erect, 8–15 mm long, 3 mm $\varnothing$; **Fl.** solitary, c. 29 mm $\varnothing$, pink.

**O. connatum** (L. BOL.) L. BOL. (*Ruschia c.* L. BOL., *Mes. c.* (L. BOL.) N. E. BR.). – Cape: Lit. Namaqualand. – Compact plant 12 cm tall; **L.** 13–15 mm long, 5 mm across, 4–6 mm thick, sheathing and united for (3–)4(–8) mm, with a translucent keel; **Fl.** solitary, pinkish-purple.

**O. inclusum** (L. BOL.) N. E. BR. (*Mes. i.* L. BOL.). – Cape: Lit. Namaqualand. – Plant 15–20 cm tall; **L.** 2–8 mm long, 4–5 mm across, 2–3 mm thick, triangular-ovate, minutely velvety, with translucent dots; **Fl.** 1–3 together, c. 2 cm $\varnothing$, red.

**O. octojuge** (L. BOL.) N. E. BR. (*Mes. o.* L. BOL., ? *Ruschia o.* (L. BOL.) L. BOL.). – Cape: Riversdale D. – Freely branched plant 5 to 10 cm high; **L.** spreading, 3–7 mm long, 2 to 4 mm $\varnothing$, with translucent dots; **Fl.** solitary, 2.5–3 cm $\varnothing$.

**O. rupigena** (L. BOL.) L. BOL. (*Ruschia r.* L. BOL.). – Cape: Clanwilliam D. – Compact, stiff plant 15–20 cm $\varnothing$, 10–15 cm tall, with tangled, contorted **Br.**; **L.** spreading, 12 mm long, 3–4 mm $\varnothing$; **Fl.** solitary, 24 mm $\varnothing$, brilliant pink.

**O. subglobosum** (L. BOL.) L. BOL. (*Ruschia s.* L. BOL., ? *Mes. reductum* N. E. BR.). – Cape: Lit. Namaqualand. – Stiff plant 12 cm tall; **St.** with 2–6 crowded, hemispherical **L.** 8 mm long, 6–7 mm $\varnothing$, with dark green dots and a serrulate keel; **Fl.** solitary, 15 mm $\varnothing$, purplish-pink.

**Odontophorus** N. E. Br. (§ IV/1/2). – Low ♄ with fleshy **roots**; **St.** ascending or resting on the ground and forming mats; **L.** 1–2 pairs on a shoot, very thick, soft-fleshy, greyish-green, tuberculate, pubescent, dentate; **Fl.** pedicellate, yellow or white. – Growing season: Jan.-June. – Greenhouse.

**O. albus** L. Bol. – Cape: Lit. Namaqualand. – **St.** with 2 erect, olive-green **L.** 43 mm long, 15–17 mm across, Sh. 13 mm long, with soft prominent dots, the margins with rounded T.; **Fl.** 3 cm ⌀, white.

**O. angustifolius** L. Bol. – Cape: Lit. Namaqualand. – **St.** with 4 erect, oblong-linear to linear, tapered **L.** 25–32 mm long, with the tip rhombic and dentate and 3–4 T. along the margins; **Fl.** 6 mm ⌀, lemon-yellow.

**O. herrei** L. Bol. – Cape: Lit. Namaqualand. – Compact, low-growing; **St.** with 2 indistinctly dentate **L.** 27 mm long, 11–14 mm across, 9 to 11 mm thick, acute-oblong on the upperside, expanded on the underside, sheathing and united for 8 mm; **Fl.** 3–4 cm ⌀, dark yellow.

**O. marlothii** N. E. Br. (Pl. 188/1). – Cape: Lit. Namaqualand. – **St.** short, with 2–3 L.-pairs; **L.** 2.5–3.5 cm long, 7–8 mm across and thick, somewhat vesicular and swollen at the base, the side angles with 6–7 rather thick, awned T., grey to dark green, covered with rounded, prominent tubercles tipped with a fine white hair; **Fl.** 3 cm ⌀, yellow.

**O.** (?) **nanus** L. Bol. (*O. areolatus* Marl. nom. nud.). – Cape: Lit. Namaqualand. – Resembles **O. primulinus** but smaller; **L.** crowded, 15 mm long, 10 mm across, 7–9 mm thick, ± ovate on the upperside, with the sides expanded and tuberculate, the margins with stiff T.; **Fl.** white.

**O. primulinus** L. Bol. – Cape: Lit. Namaqualand. – Low-growing, compact plant; **L.** erect to spreading, up to 4 cm long, 18 mm across, 14–16 mm thick, ± swollen at the base, shortly and ± rounded triangular at the tip, the side-angles with 4–5 short, thick T., green, appearing white because of the hair-tipped tubercles; **Fl.** 43 mm ⌀, straw-coloured.

**Oophytum** N. E. Br. (§ IV/1/17). – Resembles **Conophytum**; small ♃ very succulent plants forming mats; growths consist of soft, ovoid **Bo.** concealed during the resting period within the dry, whitish membranes of the old Bo.; **Fl.** mostly white. – Growing-season short, Spring. Greenhouse, warm. – Propagation: seed, cuttings.

**O. nanum** (Schltr.) L. Bol. (*Mes. n.* Schltr.). – Cape: Lit. Namaqualand. – Plant 2 cm high; **Bo.** ± spherical, 5–7 mm ⌀, green, minutely papillose; **Fl.** 1 cm ⌀, Pet. white with reddened tips.

**O. nordenstamii** L. Bol. – Cape: Van Rhynsdorp D. – Resembles the preceding spec.; **St.** thick, with numerous **Bo.** 1.7–2 cm long, 1.3 cm ⌀, with a Fi. 9 mm long, purplish-brown above, puberulous; **Fl.** 3.5 cm ⌀, Pet. white, rarely pink, broader than the preceding spec., with papillose inner filaments.

**O. oviforme** (N. E. Br.) N. E. Br. (Pl. 188/2) (*Mes. o.* N. E. Br., *Conophytum o.* (N. E. Br.) N. E. Br.). – Cape: Van Rhynsdorp D. – **Bo.** 12–20 mm tall, 10–12 mm ⌀, the Fi. small and only slightly gaping, olive-green, often flame-coloured, glossy papillose; **Fl.** 22 mm ⌀, Pet. white below, purplish-pink above.

**Ophthalmophyllum** Dtr. et Schwant. (§ IV/1/17). – Dwarf, stemless plants; **roots** fibrous; **growths** one or few; **Bo.** consisting of 2 L. which are united for most of their length, cylindrical or obconical, very fleshy and juicy, cleft at the end or with a Fi. parted into 2 ± round lobes, surface smooth, mostly shining, often very finely papillose hairy, green, brownish-green or reddish to purple-red, the lobes transparently windowed, the flanks often with transparent dots; **Fl.** solitary out of the Fi., 1–3 cm ⌀, white, pink or lilac-red, pedicel with fleshy bracts; Sept.-Oct. – Greenhouse, warm. Propagation: seed. Acc. R. Rawé O. spec. are included under **Conophytum**.

## Key to the Species of the Genus Ophthalmophyllum by A. Tischer

1. Bo. purple ............................................................................................. 2.
1a. Bo. not purple ...................................................................................... 4.
2. Bo. nearly globose to obtuse conical ........................................................... 3.
2a. Bo. cylindrical to ± elongated ovate, about 1.5 times as long as broad .......... rufescens
3. Bo. to 20 mm tall, to 20 mm ⌀, the Fi.-notch not continuous throughout, lobes to 6 mm tall ........ schuldtii
3a. Bo. to 25 mm tall, to 22 mm ⌀, with the Fi. throughout, lobes to 5 mm long ............................................. latum f. rubrum
4. Bo. reddish-brown ................................................................................. 5.
4a. Bo. not reddish-brown ........................................................................... 9.
5. Bo. cylindrical, somewhat compressed, to 1.5 times as high as broad, somewhat constricted below the lobes which are scarcely convex; Wi. uniform ................................. 6.
5a. Bo. cylindrical to elongated obovate, not constricted below the lobes, 1.5–2 times as high as broad, lobes convex above to somewhat flattened ............................................. 8.
6. Bo. only somewhat longer than broad; Fl. white to pale pink ................... australe
6a. Bo. nearly 1.5 times as long as broad .................................................. 7.
7. Fl. white ............................................................................................. friedrichiae
7a. Fl. pink-purple .................................................................................. dinteri

8. Bo. cylindrical, about twice as high as broad, Wi.-zone convex, Wi. divided into single warty smaller Wi.; Fl. white to pale pink ................................................. **verrucosum**
8a. Bo. somewhat elongated ovate, 1.5 times as high as broad, Wi. somewhat flat, uniform; Fl. white ................................................................................ **carolii**
9. Bo. dark ochreous to copper-coloured, cylindrical, somewhat flattened off, about 1.5 times as high as broad, often somewhat constricted below the lobes, Wi. uniform; Fl. white .... ........ **triebneri**
9a. Bo. not dark ochreous to copper-coloured ............................................... **10.**
10. Bo. light brownish-red, light ochreous to light flesh-coloured to olive ................ **11.**
10a. Bo. green to bluish- or yellowish-green .................................................. **16.**
11. Bo. light brownish-reddish to clay- or light ochreous-coloured; Fl. white to pink ........ ........ **praesectum**
11a. Bo. light flesh-coloured to olive-green .................................................. **12.**
12. Bo. light flesh-coloured to reddish tinted below ........................................ **13.**
12a. Bo. light clay- to olive-coloured ........................................................ **15.**
13. Bo. only a little longer than broad, somewhat constricted below the lobes, surface somewhat velvety-hairy; Fl. pale pink ...................................................... **spathulatum**
13a. Bo. to 1.5 times as long as broad, not constricted below the lobes, surface not or less papillate ........ **14.**
14. Bo. 1–2, somewhat ovoid, the Fi. not throughout, Wi. dull with some smaller Wi. below the Wi.-border; Fl. white ............................................................. **schlechteri**
14a. Bo. 1 to several, more cylindrical and somewhat compressed, Fi. throughout, Wi. somewhat dull, with isolated smaller Wi. only at the border; Pet. white below, pink above .... **lydiae**
15. Bo. somewhat ovoid, to 40 mm tall, narrowed to the apex, somewhat translucent, Wi. clear, convex; Fl. white .................................................................. **maughanii**
15a. Bo. cylindrical, to 20 mm long, Wi. somewhat dull, often somewhat flattened towards the middle; Fl. white to pink ........................................................ **longitubum**
16. Bo. compact-conical, mostly somewhat narrowed above and slightly compressed, to 25 mm tall, scarcely higher than broad ....................................................... **17.**
16a. Bo. nearly cylindrical to elongated ovate ................................................. **19.**
17. Bo. to 22–25 mm tall, yellowish-green to greenish-olive, the Fi. throughout, Wi. rather clear; Fl. diurnal ........................................................................... **18.**
17a. Bo. to 15 mm tall, green to yellowish-green, often somewhat constricted below the lobes, Wi. somewhat dull, often divided into smaller Wi.; Fl. nocturnal ................ **noctiflora**
18. Bo. to 22 mm long, broad obconical, lobes spreading, to 27 mm ⌀; Fl. purple, scented .... ........ **fulleri**
18a. Bo. to 25 mm long, lobes less spreading, Fi. often not throughout, to 22 mm ⌀; Fl. white ........ **latum f. latum**
19. Bo. green, to 15 mm long, to 5 mm ⌀, somewhat constricted below the rounded lobes; Fl. white ................................................................................ **littlewoodii**
19a. Bo. longer than 15 mm ................................................................. **20.**
20. Bo. to 20 mm long, yellowish-green, Wi. often divided into smaller Wi., partly placed on the upper part of the sides; Pet. white below, pink above ................ **subfenestratum**
20a. Bo. longer than 20 mm ................................................................. **21.**
21. Plants consisting of several Bo., ± velvety-haired with the exception of the lobes ....... **22.**
21a. Bo. not velvety-haired below, green to yellowish-green, to 30 mm long, to 15 mm ⌀, somewhat compressed upwards, lobes often ± flattened and with a faint keel in the direction of the Fi.; Fl. pale pink, nearly white ................................................................ **longum**
22. Plants consisting of several cylindrical Bo. to 50 mm long, to 20 mm ⌀, somewhat narrowed above, lobes often somewhat flattened in the direction of the Fi.; Pet. white, often tipped with pink ............................................................................ **pubescens**
22a. Bo. mostly single, to 26 mm long, to 20 mm ⌀, somewhat broadened above, lobes rounded; Pet. white tipped with pink ......................................................... **villetii**

Note: **O. vanheerdei** and **O. haramoepense** are insufficiently known to insert in the Key.

**O. australe** L. Bol. – Cape: Van Rhynsdorp D. – Bo. solitary, ± square, somewhat constricted below the lobes, minutely papillose, green below, dirty purple above, 1.7–2.7 cm long, with fenestrate lobes 3–5 mm long; Pet. 1.2 cm long, pale pink.

**O. carolii** (Lavis) Tisch. (*Conophytum c.* Lavis). – Cape: Lit. Namaqualand. – Bo. solitary, oblong, 22 mm long, 18 mm thick, c. 12 mm across, soft, papillose, with columnar, purplish-brown lobes 6 mm high, 5.5 mm ⌀; Fl. 35 mm ⌀, white. Acc. L. Bolus, Notes on Mes. III, 209, transferred back to **Conophytum**.

**O. dinteri** Schwant. ex Jacobs. – S.W.Afr.: Gr. Namaqualand. – Bo. mostly solitary, ± cylindrical, up to 40 mm tall, 20–22 mm across, 15–16 mm thick, ± oblique above, the Fi. extending from side to side and c. 3 mm deep, smooth, glabrous, the surface ± glossy, dark green to coppery-red, the lobes with a ± translucent Wi.; Fl. up to 30 mm ⌀, mauvish-red.

**O. friedrichiae** (Dtr.) Dtr. et Schwant. (*Mes. f.* Dtr., *Conophytum f.* (Dtr.) Dtr. et Schwant., *Lithops f.* (Dtr.) N. E. Br., *e. Deren-*

**Ophthalmophyllum**

*bergia* f. (DTR.) SCHWANT.). – S.W.Afr.: Gr. Namaqualand; (Cape: Bushmanland, acc. HERRE [?]). – **Bo.** solitary or several together, cylindrical, 25–30 mm long, 14–15 mm across, 10–12 mm thick, the Fi. extending from side to side and 6–7 mm deep, gaping, the lobes rounded, almost transparent, with several light dots below, the surface glabrous, green, eventually coppery-red; **Fl.** 12–20 mm $\varnothing$, white.

**O. fulleri** LAVIS. – Cape: Bushmanland. – **Bo.** mostly solitary, broad-obconical, soft, green to yellowish, 22 mm long, 27 mm across the Fi., with convex, divergent lobes, papillose at the tips; **Fl.** 26 mm $\varnothing$, pinkish-purple, scented.

**O. haramoepense** L. BOL. – Cape: Lit. Namaqualand. – **Bo.** solitary or paired, ovoid to long-obovoid, pale bluish-green, 22 mm long, 11 mm $\varnothing$ in the middle, lobes 2–4 mm long, Wi. convex, 4–5 mm $\varnothing$; **Fl.** purplish-pink. (Not included in the Key.)

**O. latum** TISCH. f. **latum**. – Cape: Lit. Namaqualand. – **Bo.** solitary or paired, very fleshy, mostly conical to $\pm$ cylindrical, c. 25 mm tall, up to 20 mm across and 12 mm thick, thinner above, light green to yellowish-green, the lobes c. 5 mm high, tipped with dark, translucent Wi.; **Fl.** 20 mm $\varnothing$, white.

**O. —** f. **rubrum** (TISCH.) ROWL. (*O. maughanii* v. *rubrum* N. E. BR. nom. nud., *O. l.* v. *rubrum* TISCH.). – **Bo.** deep red.

**O. littlewoodii** L. BOL. – Cape: Lit. Namaqualand. – **Bo.** solitary, cylindrical, 10–16 mm long, $\pm$ tuberculate, the lobes erect, 2–3.4 mm high, tipped with translucent Wi.; Pet. white below, pink above.

**O. longitubum** L. BOL. – Cape: Bushmanland. – **Bo.** solitary or paired, oblong to cylindrical, minutely papillose-hirsute above, 10 to 18 mm long, up to 6 mm across, the lobes erect, 1.5–2 mm high, Wi. convex, translucent, solitary, lateral Wi. also present; Pet. pale pink, rarely white.

**O. longum** (N. E. BR.) TISCH. (*Conophytum l.* N. E. BR., *O. herrei* LAVIS). – Cape: Lit. Namaqualand. – **Bo.** solitary or several together, somewhat compressed-cylindrical, 25–30 mm long, 20 mm across, 14 mm thick, smooth, glabrous, greyish-green, often slightly brownish below, with scattered translucent dots above, the Fi. extending from side to side, the lobes rounded-carinate on the underside; **Fl.** almost white to palest pink.

**O. lydiae** JACOBS. (Pl. 188/3). – Cape: Lit. Namaqualand. – **Bo.** solitary or paired, rarely several together, obconical, 22–25 mm tall, 16 mm $\varnothing$, olive green with minute papillae, the Fi. 5 mm deep, the upperside $\pm$ translucent, the Wi. bordered by light dots 1 mm across; **Fl.** 2 cm $\varnothing$, Pet. white with pink tips.

**O. maughanii** (N. E. BR.) SCHWANT. (*Conophytum m.* N. E. BR., *O. m.* (N. E. BR.) TISCH.). – Cape: Bushmanland, Kenhardt D. – **Bo.** solitary or several together, compressed-cylindrical, the surface translucent, dull, the Fi. 6–8 mm deep, the lobes short, conical, the Wi.-section faintly yellowish-green; **Fl.** 14 mm $\varnothing$, white.

**O. noctiflorum** L. BOL. – Cape: Lit. Namaqualand. – **Bo.** solitary, spherical to almost so, 1.4 cm long, 2 cm $\varnothing$ below the tip, minutely papillose, green, the Fi. compressed, the Wi. large; **Fl.** opening at night, strongly scented, Pet. deep pink, the inner ones white to pale pink.

**O. praesectum** (N. E. BR.) SCHWANT. (*Conophytum p.* N. E. BR., *O. jacobsenianum* SCHWANT.). – Cape: Bushmanland, Kenhardt D. – Clump-forming; **Bo.** 17–30 mm long, 8–18 mm across, 4–12 mm thick, compressed-cylindrical, green, minutely velvety, the lobes only 3–7 mm high, with $\pm$ fenestrate tips; **Fl.** 3 cm $\varnothing$, pinkish-mauve.

**O. pubescens** TISCH. – Cape: Bushmanland. – **Bo.** several together, 3–5 cm tall, 15–20 mm $\varnothing$, slender-cylindrical, light greyish-green, pubescent, the lobes $\pm$ columnar, c. 10 mm high, fenestrate; **Fl.** 2 cm $\varnothing$, Pet. white with pink tips.

**O. rufescens** (N. E. BR.) TISCH. (*Conophytum r.* N. E. BR.). – Cape: Bushmanland. – **Bo.** solitary, 15–24 mm tall, 12–16 mm across, 10–15 mm thick, cylindrical-elliptical, dark purple, spotted, the Fi. 2–5 mm deep, the lobes rounded, fenestrate; **Fl.** 15–30 mm $\varnothing$, creamy-white, opening at night.

**O. schlechteri** SCHWANT. – Cape: Lit. Namaqualand. – **Bo.** solitary or paired, long-ovoid, 3–4 cm tall, 15 mm across, rather less in thickness, the surface dull, green to light flesh-coloured, spotted, the Fi. 5–7 mm across, the lobes conical-tapered, with light Wi.; **Fl.** 25 mm $\varnothing$, white.

**O. schuldtii** SCHWANT. (Pl. 188/4). – S.W.Afr.: Gr. Namaqualand. – **Bo.** solitary, 15–24 mm tall, 12–16 mm across, 10–15 mm thick, cylindrical, elliptical or conical, the surface dark purple, spotted, the Fi. scarcely depressed and 4–6 mm across, the lobes rounded, with transparent Wi.; **Fl.** 2.5–3 cm $\varnothing$, white.

**O. spathulatum** L. BOL. – Cape: Lit. Namaqualand. – **Bo.** solitary, oblong to $\pm$ square, slightly constricted below midway, 2.4–2.7 cm tall, 21–25 mm thick, the surface minutely papillose-hirsute, with brown-windowed lobes 5 mm long; Pet. pale pink.

**O. subfenestratum** (SCHWANT.) TISCH. (*Conophytum s.* SCHWANT., *O. cornutum* SCHWANT., *Con. kennedyi* L. BOL.). – Cape: Bushmanland. – Cushion-forming plant; **Bo.** up to 2.5 cm tall, up to 1.5 cm $\varnothing$, cylindrical to obconical, light green, the Fi. not extending entirely across the surface, the lobes rounded, with conspicuous translucent dots at the tips, the inner faces $\pm$ hirsute towards the Fi.; **Fl.** c. 2 cm $\varnothing$, Pet. white below, light pink above.

**O. triebneri** SCHWANT. – S.W.Afr.: Gr. Namaqualand. – **Bo.** solitary, obconical, 14 to 18 mm tall, 20–22 mm across, 12–14 mm thick, truncate above, the Fi. extending from side to side and 5–6 mm deep, the surface mauvish or brownish-red, more ochre-coloured above, with translucent dots, those on the flattened lobes merging to form Wi.; **Fl.** 26 mm $\varnothing$, strongly scented, Pet. very broad, glossy, white.

**O. vanheerdei** L. Bol. – Cape: Bushmanland. – **Bo.** several together, reddish-brown, broadly obovoid, 3–3.4 cm tall, ± constricted in the middle to 1.3–1.7 cm across, the lobes c. 5 mm high, widely divergent, hirsute on the inside, the Wi. translucent, with small lateral Wi. also present; **Pet.** white to pale pink. (Not included in the Key.)

**O. verrucosum** Lavis. – Cape: Van Rhynsdorp D. – **Bo.** solitary, reddish-brown with numerous translucent dots, cylindrical, 2.7 cm tall, 1.5 cm across, 2.2 cm thick, with a gaping Fi. 3–4.5 mm deep, the lobe-tips distinctly tuberculate; **Fl.** 3 cm ⌀, glossy, white.

**O. villetii** L. Bol. – Cape: Bushmanland. – **Bo.** solitary, green, oblong to ovoid-oblong, 2.2–2.6 cm tall, 1.1 cm across, the divergent lobes 3–7 mm high, distinctly hirsute on the upperside, with convex Wi.; **Pet.** pale, with pink stripes.

**Opophytum** N. E. Br. (§ I/1/2). – Low-growing, erect, fleshy, ☉ herbs with distinct Int.; **L.** ± cylindrical; **Fl.** terminal or in small clusters, whitish. – Not important in cultivation. Grow in the open.

**O. ampliatum** L. Bol. – Cape: Lit. Namaqualand. – Spreading, with **St.** 15–20 cm long; **L.** alternate, cylindrical, obtuse, 3–5 cm long, 5–7 mm ⌀; **Fl.** whitish, with numerous staminodes.

**O. australe** L. Bol. – Cape: Lit. Namaqualand. – **St.** up to 25 cm long, 1 cm ⌀; **L.** 1.2 to 6.6 cm long, 5–8 mm ⌀; **Pet.** very narrow, pink.

**? O. fastigiatum** (Thunbg.) N. E. Br. (*Mes. f.* Thunbg., ? *Perapentacoilanthus f.* (Thunbg.) Rap. et Cam., *Mes. erectum* Haw., *Mes. papuliferum* DC.). – Cape: N. of the Olifants R. – **St.** with opposite **Br.** produced near soil-level and short shoots; **L.** obtuse-ovoid, 10–15 mm long; **Fl.** 1–3 together, white. (Uncertain species, possibly = **Eurystigma**.)

**O. gaussenii** Lereddde. – Algiers. – Resembles **O. theurkauffii** but with golden-yellow stigmas (not white).

**O. theurkauffii** (Jahand. et Maire) Maire (*Mes. t.* Jahand. et Maire, *Aizoon t.* (Jahand. et Maire) Maire). – W. Saharan coasts. – Branching ± radially from the base, **St.** and **Br.** cylindrical; **L.** very succulent, opposite, c. 5 cm long, 1 cm ⌀, light green, often reddish; **Fl.** 15 mm ⌀, yellowish-white, with white stigmas.

**Orthopterum** L. Bol. (§ IV/1/18). – Low-growing succulents resembling **Faucaria**. – Greenhouse, warm. – Propagation: seed, cuttings.

**O. coegana** L. Bol. – Cape: Uitenhage D. – Caulescent, clump-forming; **Br.** with 6–8 opposite, ± erect, linear-lanceolate, green **L.** 3 to 3.5 cm long, 10 mm across, 8 mm thick, with darker dots and a rounded keel on the underside, each margin with one indistinct, awn-tipped T.; **Fl.** pedicellate, 4.5 cm ⌀, golden-yellow.

**O. waltoniae** L. Bol. (Pl. 188/5). – Cape: Albany D. – Similar to the preceding; **L.** ± unequal, 2–3 cm long, 8 mm across, 5 mm thick, apiculate, margins with 1–2 dentate tubercles; **Fl.** sessile, **Pet.** golden-yellow, reddish on the underside.

**Oscularia** Schwant. (§ IV/1/3). – Small sub-♄ with erect or spreading **St.**; **L.** ± united at the base, trigonous, short, expanded above, with a short point, entire or with the margins and keel denticulate, greyish-green; **Fl.** in 3's, shortly pedicellate, pink to red, Feb. – Summer: in the open, winter: cold-house. – Propagation: seed, cuttings.

**O. caulescens** (Mill.) Schwant. (*Mes. c.* Mill., *M. deltoides* v. *simplex* DC.). – Cape: Worcester, Caledon, Stellenbosch D. – **St.** with numerous short shoots; **L.** 2 cm long, 10–11 mm across, triangular, trigonous, entire or with 2–3 T. along the margins towards the tip, light grey-pruinose, margins and T. slightly reddish; **Fl.** 12 mm ⌀, pink, scented.

**O. deltoides** (L.) Schwant. v. **deltoides** (Pl. 189/2) (*Mes. d.* L., *Mes. muricatum* Haw., *Mes. d.* v. *muricatum* (Haw.) Bgr., *O. muricata* (Haw.) Schwant.). – Cape: Tulbagh, Paarl, Caledon D. – Resembles **O. pedunculata** but smaller; **L.** 6–10 mm long, bluish-grey, with sharp T.

**O.** – v. **major** (Weston) Schwant. (*Mes. deltoides* v. *m.* Weston, *O. deltata* v. *m.* (Weston) Schwant., *O. m.* (Weston) Schwant.). – **Fl.** 18 mm ⌀.

**O. pedunculata** (N. E. Br.) Schwant. (*Mes. deltoides* L. v. *p.* N. E. Br., *O. d.* v. *p.* (N. E. Br.) Schwant., *Mes. d.* Mill. ex Bgr., *O. deltata* Schwant., *Mes. deltatum* (Schwant.) Maire et Weill.). – Cape: Tulbagh, Worcester, Caledon D. – **St.** with numerous short shoots; **L.** triangular, trigonous, with 2–4 reddish T. along each angle, light grey, 10–15 mm long, 8 mm across; **Fl.** c. 12 mm ⌀, pale pink, Feb.-Mar.

**Ottosonderia** L. Bol. (§ IV/1/1). – Robust, glabrous ♄; sterile **St.** short, flowering ones elongated; **L.** opposite, tapered to obtuse; **Fl.** pink to pinkish-purple. – Summer: in the open, winter: cold-house. – Propagation: seed, cuttings.

**O. monticola** (Sond.) L. Bol. (*Mes. m.* Sond., *Hymenocyclus m.* (Sond.) Schwant., *Malephora m.* (Sond.) Jacobs. et Schwant., *Mes. bifoliatum* L. Bol.). – Cape: Lit. Namaqualand. – ♄ 22 cm tall, 23 cm ⌀; **St.** 2 cm thick, older **Br.** 14 mm thick, sterile **shoots** 2–4 cm long, flowering shoots 8–15 cm long; **L.** clustered on the short shoots, acute, 4–6 cm long, 4–5 mm across and thick; **Fl.** 2–2.5 cm ⌀, pinkish-purple.

**O. obtusa** L. Bol. – Cape: Lit. Namaqualand. – Similar to the preceding species but lower-growing; **L.** 3.5–4.5 cm long, 1 cm across and thick; **Fl.** c. 2 cm ⌀, pale pink.

**Pherelobus** N. E. Br. (§ IV/1/20). – ☉ plants resembling **Cleretum**, distinguished by the crimson anthers and ovary-lobes, and capsule structure.

**P. maughanii** N. E. Br. v. **maughanii**. – Cape: Calvinia D. – L. spatulately rounded above, or obliquely spatulate; Fl. pale mauve, Pet. reddish-purple at the base.

**P.** — v. **stayneri** L. Bol. – L. broad-spatulate, the blade ± circular, 2 cm and more across, with thick papillae; Fl. up to 5.4 cm $\varnothing$, butter-coloured, turning creamy-white.

**Platythyra** N. E. Br. (§ I/1/1). – $2\!\!\!|$ plants with fleshy **roots**; **St.** herbaceous, prostrate, 3–4-angled; L. opposite, petiolate, ovate or lanceolate; Fl. solitary, pedicellate, yellow. – In the open. – Not important in cultivation. Propagation: seed.

**P. haeckeliana** (Bgr.) N. E. Br. (Pl. 189/1) (*Mes. h.* Bgr., *Peratetracoilanthus h.* (Bgr.) Rap. et Cam., *Mes. ovatum* Thunbg., *Mes. elongatum* Eckl. et Zeyh., *Mes. angulatum* v. *ovatum* Sond.). – Cape: Uitenhage D. – Forming flat cushions 1 m $\varnothing$; **St.** creeping; L. 3–5 cm long, 5–12 mm across, tapering; Fl. on a pedicel 10–15 mm long, c. 8 mm $\varnothing$, pale yellow.

**Pleiospilos** N. E. Br. (§ IV/1/10). – ± acauline, clump-forming, exceedingly succulent plants; L. mostly 1–2 pairs, sometimes 3–4 pairs together, opposite and decussate, very thick, flat on the upperside, very convex on the underside, very obtuse to acute, united at the base and ± rugosely inflated, greyish-green or dark green, with ± translucent dots; Fl. sessile or shortly pedicellate, solitary or several together, large, yellow to orange-coloured, Aug.–Oct., opening in the afternoon, often scented of coconut. – Popularly called 'Living Granite'. – Greenhouse, warm. Propagation: seed. – Plants self-sterile.

**P. archeri** L. Bol. (Pl. 190/2). – Cape: Laingsburg D. – With 1 pair of L. c. 25 mm long, 2 cm across at the base, narrowed towards the rounded tip, keeled on the underside, the tip pulled 8 mm forward on one side, grass-green, often reddish; Fl. sessile, 3.5 cm $\varnothing$, golden-yellow, attractively scented.

**P. beaufortensis** L. Bol. – Cape: Beaufort West D. – L. 2–4 together, ± ovate on the upperside, compressed and carinate below the tip, c. 7 cm long, c. 5 cm across midway, 4 to 4.5 cm thick, pinkish-brown; Fl. 6–7 cm $\varnothing$, Pet. golden-yellow, pale pink below.

**P. bolusii** (Hook. f.) N. E. Br. (Pl. 190/1) (*Mes. b.* Hook. f.). – Cape: Graaff Reinet D. – With one pair of L. 4–7 cm long, the upperside often broader than long, the rounded above clavately thickened above and often drawn chin-like over the upperside for 2–3 cm, 3–3.5 cm thick, reddish or brownish-green, with numerous dots; Fl. 1–4 together, ± sessile, 6–8 cm $\varnothing$, golden-yellow.

**P. borealis** L. Bol. – Orange Free State. – Br. with 2 L.-pairs; L. acute, with a very compressed keel, 19 mm thick, one L. 6 cm long, 16 mm across, the other ± falcate, 7 cm long, 22 mm across; Fl. sessile, solitary, 5 cm $\varnothing$, Pet. golden-yellow on the upperside, reddish on the underside.

**P. brevisepalus** L. Bol. – Cape: Uniondale D. – L. 1–2 pairs, ± dissimilar, abruptly widened midway, up to 32 mm across, 25–30 mm thick, tapered, acute or obtuse, with a lateral keel; Fl. sessile, 38 mm $\varnothing$, dark yellow.

**P. canus** (Haw.) L. Bol. (*Mes. c.* Haw., *Punctillaria c.* (Haw.) L. Bol.). – Cape: Oudtshoorn D. – Insufficiently known. Acc. L. Bolus (Notes on Mes. I, 85) possibly identical with **P. optatus**.

**P. clavatus** L. Bol. – Cape: Laingsburg D. – L. unequal, united for 10 mm, 14 or 18 mm long, 8 mm across, 6–9 mm thick, yellowish-green, scarcely spotted, the upperside ovate-obtuse; Fl. on a clavate pedicel, 25 mm $\varnothing$, golden-yellow, paler inside.

**P. dekenahii** (N. E. Br.) Schwant. (*Punctillaria d.* N. E. Br.). – Cape: Fraserburg D. – With several shoots, each with 2 L.-pairs; L. spreading, c. 6 cm long, 15 mm across at the base and 25 mm midway, then acuminate, acutely keeled and chin-like on the underside, greyish-green, rough with dots, waxy-pruinose, keel and margins reddish; Fl. 3.5 cm $\varnothing$, light yellow.

**P. dimidiatus** L. Bol. – Cape: Laingsburg D. – 2–3 L.-pairs on a shoot; L. 8 cm long, 2 cm across below, expanded midway then acute, acutely keeled on the underside above, greyish-green, keel and margins reddened; Fl. sessile, 6–7 cm $\varnothing$, yellow.

**P. fergusoniae** L. Bol. – Cape: Ladismith D. – St. with 4–6 L. in unequal pairs, 5–7 cm long, 2 cm across and thick, with the underside rounded below and compressed-carinate above, whitish-grey to bluish, rather pruinose to bluish-green; Fl. 4–4.5 cm $\varnothing$, dark yellow.

**P. framesii** L. Bol. – Cape: Willowmore D. – Shoots with 4 L. 3.5–4.5 cm long, 38 mm thick, broadly cordate-ovate above midway, acute, with a compressed keel on the underside, green, with large dots; Fl. 4 cm $\varnothing$, Pet. lax, yellow, white below.

**P. grandiflorus** L. Bol. – Cape: Prince Albert D. – L. in dissimilar pairs, 8–8.5 cm long, 17 mm thick, 18 mm across, acute to rounded, bluish; Fl. almost 10 cm $\varnothing$, Pet. pale yellow, pink on the underside.

**P. hilmarii** L. Bol. – Cape: Ladismith D. – L. paired, 25 mm long, 16 mm across at the base, slightly narrowed towards the tip, the underside semicylindrical and drawn almost 10 mm over the tip, with very rounded angles, reddish-green, with the dark dots ± confluent above into a distinct Wi.; Fl. 24 mm $\varnothing$, golden-yellow.

**P. kingiae** L. Bol. – Cape: Uniondale D. – L. paired, 5 cm long, 10 mm across, 12 mm thick, thickest midway, expanded and rhombic above, acute to tapered, with a rounded tip, keel-like and compressed on the underside; Fl. 4 cm $\varnothing$, yellow.

**P. latifolius** L. Bol. – Cape: Prince Albert D. – Shoots with 2–4 L. 6.5–8 cm long, 44 mm across, up to 21 mm thick, ± rhombic above, thickest midway, narrowed towards the tip, carinate on the underside, concave on the upperside, angles rounded; **Fl.** 6 cm $\varnothing$, yellow with a white centre.

**P. latipetalus** L. Bol. – Cape: Willowmore D. – With 3–4 pairs of **L.** 5–9 cm long, 3.5 cm across, 26 mm thick, rounded on the underside but carinate above, flat to concave on the upperside, obtuse to acute; **Fl.** 2–3 together, 6–8 cm $\varnothing$, yellow with a white centre.

**P. leipoldtii** L. Bol. – Cape: Uniondale D. – **L.** up to 7 cm long, 3 cm across near the apex, 1.5–2 cm across midway, lower **L.** 2 cm thick, others thinner, the upperside flat to convex, abruptly expanded above midway then narrowed and subulate at the tip, dirty green; **Fl.** solitary, 6.5 cm $\varnothing$, yellow with a white centre, scented.

**P. loganii** L. Bol. – Cape: Laingsburg D. – **L.** unequal, long-ovate, obtuse, obtusely keeled below the tip, 18–20 mm long, 11 mm across, 8–10 mm thick, dark purplish-brown; **Fl.** 3 cm $\varnothing$, yellow with a white centre.

**P. longibracteata** L. Bol. – Cape: origin unknown. – Shoots with 2 **L.** 3 cm long, 16 mm across, 8 mm thick, $\pm$ falcate on the upperside, laterally keeled on the underside; **Fl.** on a pedicel with bracts 25 mm long, yellow, white in the centre.

**P. longisepalus** L. Bol. – Cape: Willowmore D. – Shoots with 4 tapered **L.** 6–7 cm long, 17 mm across midway, expanded above that, 14–16 mm thick, with a small recurved tip and a somewhat lateral keel; **Fl.** up to 7 cm $\varnothing$, yellow with a white centre, Sep. very long.

**P. magnipunctatus** (Haw.) Schwant. v. **magnipunctatus** (*Mes. m.* Haw., *Punctillaria m.* (Haw.) N. E. Br.). – Cape: Prince Albert D. – Forming clumps 6–8 cm high; **L.** 2–4 on a shoot, very widely spreading, 3–7 cm long, 9–12 mm across below, 7–11 mm thick, broader and thicker above, convex on the underside but keeled above, with rounded sides, obtuse, green, greyish-green or $\pm$ brownish; **Fl.** solitary, 4.5 to 5 cm $\varnothing$, light yellow.

**P. —** v. **inaequalis** L. Bol. – **L.**-pairs unequal.

**P. minor** L. Bol. – Cape: Willowmore D. – **L.** c. 3 cm long, 10 mm across below, the upper third forming a somewhat rounded triangle 2 cm across, the underside $\pm$ rounded-navicular, greyish-green; **Fl.** 5.5 cm $\varnothing$, yellow, whitish in the centre.

**P. nelii** Schwant. (Pl. 190/3) (*P. tricolor* N. E. Br., *P. pedunculata* L. Bol.). – Cape: Willowmore D. – **L.** quite flat on the upperside, the underside conspicuously rounded and drawn chin-like over the upperside so that the **L.** is hemispherical, the Fi. deep and widely gaping, dark greyish-green with numerous dots; **Fl.** c. 7 cm $\varnothing$, salmon-pink to yellow or $\pm$ orange-coloured.

**P. nobilis** (Haw.) Schwant. (*Mes. n.* Haw., *Punctillaria n.* (Haw.) N. E. Br., *Mes. compactum* Ait., *Punctillaria c.* (Ait.) N. E. Br., *Pl. c.* (Ait.) Schwant.). – Cape: Oudtshoorn D. – Branching to form a mat; **L.**-pairs 3–4 together, 5–6 cm long, broadly linear, expanded above to 1.5–2 cm then short-tapered, the underside with a $\pm$ noticeably sinuate keel towards the trigonous tip, dirty green; **Fl.** solitary, 5 to 6 cm $\varnothing$, yellow.

**P. optatus** (N. E. Br.) Schwant. (*Mes. optatum* N. E. Br., *Punctillaria o.* (N. E. Br.) N. E. Br.). – Cape: Oudtshoorn D. – Forming clumps; **L.** 22–44 mm long, 5–7 mm across, 6–8 mm thick, very rounded on the underside, blunt-carinate below the tip, brownish or reddish-green with a purple tinge to $\pm$ bluish, densely spotted; **Fl.** solitary, 2.5–3 cm $\varnothing$, light yellow.

**P. peersii** L. Bol. – Cape: Willowmore D. – **L.** 3.5–4 cm long, up to 2 cm across, flat on the upperside, indistinctly and very laterally keeled on the underside, green, red or brown; **Fl.** 6–7 cm $\varnothing$, yellow with a white centre.

**P. prismaticus** (Marl.) Schwant. (Pl. 191/2) (*Mes. p.* Marl., *Mes. roodiae* N. E. Br., *Punctillaria r.* (N. E. Br.) N. E. Br., *Pl. r.* (N. E. Br.) Schwant.). – Cape: Ceres D. – **L.** 3.5–4 cm long, 12 mm thick and 3 cm across below, rather broader midway, shortly triangular and obtuse above, the underside flat-convex, rounded-carinate towards the tip, green; **Fl.** 4 cm $\varnothing$, yellow.

**P. purpusii** Schwant. (*Punctillaria p.* (Schwant.) N. E. Br., *Mes. magnipunctatum* Schwant.). – Cape: Willowmore D. – Shoots with 1–2 pairs of **L.** 5–7 cm long, 15 mm across, expanded to 3 cm midway then triangularly tapering and acute, the underside with a rounded keel, $\pm$ sinuate below, dark green; **Fl.** sessile, 9 cm $\varnothing$, yellow.

**P.** cv. Rothii (*Mes.* $\times$ *rothii* Hort.). – Hybrid: **P. bolusii** $\times$ **P. purpusii**; intermediate between the parents.

**P. rouxii** L. Bol. – Cape: Victoria West D. – **L.** distinctly narrowed and $\pm$ falcate on the upperside, tapering, with a lateral keel on the underside, broadest midway, with the sides obliquely truncate, 5–5.5 cm long, 18 mm across, up to 20 mm thick, bluish; **Fl.** 6 cm $\varnothing$, pale yellow.

**P. sesquiuncialis** (N. E. Br.) Schwant. (*Punctillaria s.* N. E. Br., *Punct. magnipunctatus* v. *s.* L. Bol.). – Cape: Prince Albert D. – **L.** short and thick, resembling **P. bolusii**, 3.5 cm long and across, 22–27 mm thick, obtuse on the underside with a robust keel; **Fl.** 3.5 cm $\varnothing$, Pet. lax, yellow.

**P. simulans** (Marl.) N. E. Br. (Pl. 191/1) *Mes. s.* Marl.). – Cape: Graaff Reinet, Aberdeen D. – **L.** usually paired, 6–8 cm long, 5–7 cm across, 1–1.5 cm thick, spreading, ovate-triangular, the upperside flat or trough-like, the underside often thickened towards the tip, never pulled forward or chin-like, reddish, yellowish or brownish-green, conspicuously spotted, slightly undulating and tuberculate; **Fl.** 1 to 4 together, yellow, light yellow or sometimes orange-coloured, perfumed.

**P. sororius** (N. E. Br.) Schwant. (*Mes. s.* N. E. Br., *Punctillaria s.* (N. E. Br.) N. E. Br.). – Cape: Lit. Karroo. – Forming clumps 4.5 to 9 cm high; **L.** in 2–3 pairs, 2.5–5 cm long, 7 to 10 mm across, 5–8 mm thick, curved slightly inwards, the underside keeled above, the upperside expanded above midway and narrowed towards the tip, green with prominent dots; Fl. 4.5–5 cm ⌀, glossy, yellow.

**P. willowmorensis** L. Bol. – Cape: Willowmore D. – Shoots with 2–4 L. 54–65 mm long, those of a pair unequal, lower ones 13–18 mm thick, 16–21 mm across, ± falcate on the upperside, the underside with a broad, lateral keel towards the tip, greenish-purple, conspicuously spotted, upper L. smaller; **Fl.** solitary, up to 7 cm ⌀, Pet. lax, yellow, white below.

**Polymita** L. Bol. (§ IV/1/2). – Small, stiff-branched ℏ; **L.** triangular, with cartilaginous angles; **Fl.** white. – Summer: in the open, winter: cold-house. Propagation: seed, cuttings.

**P. albiflora** (L. Bol.) L. Bol. (*Ruschia a.* L. Bol., *P. pearsonii* N. E. Br.). – Cape: Lit. Namaqualand. – Robust, stiff, 14 cm high; **St.** ascending; **L.** sheathing and united for much of their length, the Sh. with impressed lines, 5 to 12 mm long, the tip projecting, acute, apiculate, 8–10 mm long, 4–6 mm ⌀; **Fl.** 18 mm ⌀, white.

**P. diutina** (L. Bol.) L. Bol. (*Ruschia d.* L. Bol.). – Cape: Lit. Namaqualand. – 10 cm tall; **L.** swollen, keeled on the underside, narrowed towards the obtuse, apiculate tip, the margins denticulate above, blue tinged with pink, 9 mm long, 6 mm across, with a Sh. 5 mm long, lined; **Fl.** 2.4 cm ⌀, white.

**Prenia** N. E. Br. (§ I/1/3). – Low-growing ℏ with a short **St.**; **Br.** elongated, curved, eventually prostrate, papillose; **L.** crowded, ± flat, lanceolate; **Fl.** 1–5 together, large, red or white, July-Oct. – Summer: in the open, winter: cold-house. Propagation: seed, cuttings.

**P. pallens** (Ait.) N. E. Br. v. **pallens** (*Mes. p.* Ait., *Platythyra p.* (Ait.) L. Bol., *Mes. lanceum* Thunbg., *Mes. loratum* Haw., *Mes. expansum* DC.). – Cape: Cape, Clanwilliam D. – **St.** fleshy; **Br.** prostrate, densely leafy at the tips; **L.** 5 cm long, 1 cm across, acute, pale green, grey-pruinose, papillose; **Fl.** 2–3 cm ⌀, white or pink.

**P.** — v. **lutea** L. Bol. – Cape: Malmesbury D. – L. 3–5 cm long, 5–6 mm across; **Fl.** yellow.

**P. relaxata** (Willd.) N. E. Br. (*Mes. r.* Willd., *Platythyra r.* (Willd.) Schwant.). – Cape: Beaufort West D. – **St.** fleshy; **Br.** slender, curved, 30–40 cm long; **L.** linear-lanceolate, up to 5 cm long, 6–8 mm across, green-pruinose, papillose; **Fl.** 3–5 together, over 4 cm ⌀, light red.

**P. sladeniana** (L. Bol.) L. Bol. (*Mes. s.* L. Bol., *Synaptophyllum s.* (L. Bol.) N. E. Br.). – Cape: Lit. Namaqualand. – ☉ with prostrate, elongated **Br.** up to 6 mm ⌀; **L.** broadly elliptical, up to 5.5 cm long, membranous and persistent after withering; **Fl.** 1–3 together, up to 3 cm ⌀, white.

**P. vanrensburgii** L. Bol. (*Perapentacoilanthus v.* (L. Bol.) Rap. et Cam.). – Cape: Bredasdorp D. – Densely branched ℏ with **roots** 15 mm thick; **Br.** up to 1.35 m long, 6–10 mm thick, creeping; **L.** ovate or long-ovate, 5 cm long, 27 mm across, sometimes shorter, with a rounded tip; **Fl.** 4 cm ⌀, white, open at night (in a warm room can remain open night and day).

**Psammophora** Dtr. et Schwant. (§ IV/1/7). – Low-growing, mat-forming, very succulent plants; **St.** ± woody, densely leafy; **shoots** above-ground or hidden in the soil; **L.** short, thick, opposite and decussate, ± trigonous with rounded keel and margins to semicylindrical, acute or expanded at the tip, with the underside pulled forward and chin-like, slightly glossy, bluish to greyish-green, viscous and thus covered with dust and sand; **Fl.** terminal, solitary, pedicellate, violet-pink or white, summer. – Greenhouse, warm. Growing-period: May-Oct. Propagation: seed, cuttings.

**P. herrei** L. Bol. – Cape: Lit. Namaqualand. – L. 2 cm long, 1 cm across below, broader above and short-triangular, convex on the upperside, with the underside rounded but acutely keeled above, the surface rough, brownish-grey; **Fl.** 22–25 mm ⌀, white.

**P. longifolia** L. Bol. – S.W.Afr.: Gr. Namaqualand. – **St.** with 4–6 **L.** 4–4.5 cm long, c. 12 mm across, rather less thick, linear, short-tapered, rounded to rounded-carinate on the underside, the margins very rounded, the surface very rough, light greyish-green to brownish, young L. olive-green; **Fl.** white.

**P. modesta** (Dtr. et Bgr.) Dtr. et Schwant. (*Mes. m.* Dtr. et Bgr.). – S.W.Afr.: near Lüderitz. – Low ℏ c. 5 cm tall, scarcely branched; **L.** 12 mm long, 5–6 mm across, somewhat rounded-triangular, acute, grey-green, slightly reddish, rough; **Fl.** violet.

**P. nissenii** (Dtr.) Dtr. et Schwant. (Pl. 192/1) (*Mes. n.* Dtr.). – S.W.Afr.: Namib. – Low ℏ 5–10 cm across; **Br.** mostly concealed in the soil; **L.** in 2–3 pairs, c. 12–40 mm long, 6 mm across at the base, expanded-triangular at the tip, semicylindrical or with a rounded keel on the underside, greyish-green or whitish or reddish, rough; **Fl.** 12 mm ⌀, white or violet.

**Psilocaulon** N. E. Br. (§ I/1/5). – ☉ or ⊙ plants or erect, ± branched ℏ; **St.** mostly cylindrical, less often 4-angled, Int. often ± spherical, glabrous or puberulous; **L.** early deciduous, small, soft, ± cylindrical, united at the base, Sh. smooth (**Sect. I. Psilocaulon**) or ± laciniate, fringed (**Sect. II. Brownanthus**); **Fl.** small, shortly pedicellate, solitary or several together, white, reddish or yellow. – Summer: in the open, winter: cold-house. Propagation: seed, cuttings. – The G. **Psilocaulon** needs further studies to arrange all species in Sections.

**P. acutisepalum** (Bgr.) N. E. Br. (§ I) (*Mes. a.* Bgr., *P. absimile* N. E. Br.). – Cape: Clanwilliam, Herbert D. – ℏ with erect, minutely papillose **St.** 16–20 cm long, with Int. 1–2 cm long, 2–3 mm ⌀; **L.** early deciduous, leaving a scar; **Fl.** numerous, in 2–3-partite Infl., red.

**P. album** L. Bol. (§ I). – Cape: Van Rhynsdorp D. – Erect ℏ 60 cm tall; **St.** virgate, Int.

10–37 mm long; **L.** ovate, 9 mm long, 3 mm thick; **Fl.** 2.5 cm ⌀, white.

**P. annuum** L. Bol. (§ I). – Cape: Middelburg D. – Blue, soft, ☉ plants 17 cm high, with short papillose hairs; **St.** curved, spreading, 3–5 mm thick, indistinctly alate, with Int. 1–2 cm long; **L.** semicylindrical, 8 mm long, 1.5 mm thick; **Fl.** 16 mm ⌀, pink.

**P. arenosum** (Schinz) L. Bol. (§ I) (*Mes. a.* Schinz, *P. a.* (Schinz) Schwant., *Mes. gymnocladum* Schltr. et Diels, *P. g.* (Schltr. et Diels) Dtr. et Schwant., *Mes. diversipapillosum* Bgr., *P. d.* (Bgr.) N. E. Br.). – S.W.Afr.: Gr. Namaqualand, Namib; Cape: Lit. Namaqualand. – Erect, papillose ♄ 60–90 cm tall; **St.** 4-angled, jointed, with Int. 7–18 mm long, 4 to 7 mm thick; **L.** semicylindrical, 7–10 mm long, densely papillose; **Fl.** numerous, white.

**P. articulatum** (Thunbg.) Schwant. (§ I) (*Mes. a.* Thunbg., *P. a.* (Thunbg.) N. E. Br., *Mes. granulicaule* Haw., *P. g.* (Haw.) Schwant., *P. g.* (Haw.) N. E. Br., *M. g. v. purpurascens* Bgr., *Pentapcoilanthus g.* (Haw.) Rap. et Cam., *Perapentacoilanthus g.* (Haw.) Rap. et Cam., *Mes. puberulum* Dtr. nom. nud., *P. p.* (Dtr.) Dtr. ex Range nom. nud., *M. secundum* Thunb.). – Cape: Beaufort West D., Griqualand W.; S.W.Afr.: Gr. Namaqualand. – Freely branched plant c. 40 cm high; **St.** slender, jointed, with papillose, rough hairs; **L.** linear, cylindrical, pale green, 5–15 mm long, early deciduous; **Fl.** 6–7 mm ⌀, whitish.

**P. baylissii** L. Bol. (§ I). – Cape: Lit. Namaqualand. – ♄ 13–14 cm tall; **St.** up to 23 cm long, 4–5 mm ⌀, constricted at the nodes, Int. 1–24 mm long, herbaceous parts reddish brown, with minute, glossy, crowded dots often arranged in lines, somewhat rough; **L.** small, soon withering; **Fl.** up to 1.5 cm ⌀, pinkish-red.

**P. bicorne** (Sond.) Schwant. (§ I) (*Mes. b.* Sond., *Mes. micranthum* auct. non Haw., *Mes. tenue* auct. non Haw., *P. t.* auct. non (Haw.) Schwant.). – S.W.Afr. – Very variable species; **St.** prostrate, herbaceous, not swollen and jointed or only slightly so, 3–4 mm thick; **L.** subulate, 10–22 mm long, 1–2 mm thick; **Fl.** solitary, rarely 2–3 together, at the tips of very short side-shoots.

**P. bijliae** N. E. Br. (§ ?). – Origin not known. – Erect plants with minutely rough **St.** 3 to 5 mm ⌀. Diagnosis not complete, the further data announced by N. E. Brown (Gard. Chron. 1928, III, **84,** 254) not having been published.

**P. bryantii** L. Bol. (§ I). – Cape: Prieska D. – Plants 9–15 cm tall; **St.** spreading to ascending, 3 mm thick; **L.** semicylindrical, 22 mm long, 1 mm thick; **Fl.** 12 mm ⌀, pink.

**P. calvinianum** L. Bol. (§ I). – Cape: Calvinia D. – Erect plants 20 cm tall, with thick **roots**; herbaceous parts bluish-green with green dots, Int. (4–)10(–15) mm long, 2–3.5 mm thick, constricted-jointed; **L.** cylindrical, 1–2 cm long; **Fl.** 9 mm ⌀, pinkish-purple.

**P. candidum** L. Bol. (§ I). – Cape: Lit. Namaqualand. – Plants c. 33 cm tall, up to 1.30 m ⌀; **St.** slender, 30–40 cm long, (1–)2(–3) mm ⌀, minutely papillose, slightly constricted at the nodes, Int. 1.5–2 cm long; **L.** 1.6 cm long; **Fl.** 1.5 cm ⌀, white to pale pink.

**P. ciliatum** (Ait.) Friedr. (§ II) (*Mes. c.* Ait., *Trichocyclus c.* (Ait.) N. E. Br., *Brownanthus c.* (Ait.) Schwant., *Mes. schenkii* Schinz, *B. s.* (Schinz) Schwant., *T. simplex* N. E. Br. ex Maas, *B. s.* (N. E. Br. ex Maas) Bullock). – Cape: Lit. Namaqualand to Prince Albert D., Karroo; S.W.Afr.: Gr. Namaqualand. – Low-growing, ± branched ♄; **St.** fleshy, green, constricted-jointed, Int. ± spherical, 3–7 mm long, 3–4 mm thick; **L.** with cilia 2–4 mm long at the base, the blade 2–5 mm long, almost cylindrical, minutely papillose; **Fl.** in cymes, small, white.

**P. clavulatum** (Bgr.) N. E. Br. (§ I) (*Mes. c.* Bgr., *P. squamifolium* N. E. Br. nom. nud., *Mes. junceum* Haw. non L. Bol.). – Cape: Lit. Namaqualand. – Papillose ♄; **St.** curving and erect, 10–20 cm long, 1.5–2 mm thick, nodal and jointed, Int. 5–15 mm long; **L.** triangular, obtuse, 5–7 mm long, 1–5 mm across, with ciliate margins; **Fl.** in short Infl.

**P. corallinum** (Thunbg.) Schwant. (§ I) (*Mes. c.* Thunbg., *P. c.* (Thunbg.) N. E. Br.). – Cape: Van Rhynsdorp D., Karroo. – Forked or tripartitely branched, 15–25 cm tall; **St.** 3–5 cm long, green, moniliform, with the Int. elliptical or spherical and swollen, 5–6 mm long, 4–5 mm thick; **L.** subcylindrical; **Fl.** 12–15 mm ⌀, white.

**P. coriarium** (Burch.) N. E. Br. (§ I) (*Mes. c.* Burch., *Ruschia c.* (Burch.) Schwant., *P. stenopetalum* L. Bol.). – Cape: Prieska, Philipstown D. – Suberect ☉ or ♃, 90 cm or more tall; **St.** rather tangled and curved, 4 mm ⌀, Int. 1.5–3 cm long, flowering **Br.** with short shoots; **L.** cylindrical, decurved and indistinctly hooked, 3–10 mm long, 2 mm ⌀; **Fl.** 10 mm ⌀, white.

**P. dejagerae** L. Bol. (§ I). – Cape: Beaufort West D. – Slender, ☉ or ♃ plants; **St.** spreading, 25 cm long, 3 mm thick, Int. 1 cm long, herbaceous parts blue, ± tuberculate; **L.** semicylindrical, with somewhat papillose margins, 10 mm long, 1.5 mm thick; **Fl.** 15 mm ⌀, white inside.

**P. delosepalum** L. Bol. (§ I). – Cape: Van Rhynsdorp D. – Erect plants 30 cm tall; **St.** 3–4 mm thick, with minute dots, jointed, with Int. 13–50 mm long; **L.** 16 mm long, 1.5 mm thick, acute; **Fl.** 16 mm ⌀, white.

**P. dimorphum** (Welw.) N. E. Br. (§ I) (*Mes. d.* Welw.). – Trop. W.Afr.: Angola. – Prostrate plants; **St.** curved, eventually roughly hirsute; **L.** much reduced, 2–2.5 cm long, narrow-linear; **Fl.** small, whitish.

**P. dinteri** (Engl.) Schwant. (§ I) (Pl. 192/3) (*Mes. d.* Engl. emend. Bgr., *P. d.* (Engl.) N. E. Br., *Mes. marinum* Engl. ex Bgr.). – S.W.Afr.: coast near Lüderitz. – Forming large mats; younger **St.** and **Br.** very fleshy, distinctly constricted and jointed, the joints barrel-shaped and swollen, (4–)8(–10) mm ⌀; **L.** cylindrical, up to 1 cm long, 2–3 mm thick; **Fl.** on a short, thick Ped., 3–5 in an Infl.

**P. duthiae** L. Bol. (§ I). – Cape: Cradock D. – Erect plants up to 10 cm tall; **St.** blue, with conical, bristle-tipped papillae, Int. 2–3 mm long; **L.** deciduous; **Fl.** 12 mm ⌀, Pet. white with pink tips.

**P. filipetalum** L. Bol. (§ I). – Cape: Lit. Namaqualand. – **St.** prostrate, 45 cm or more long, 3 mm thick, 5 mm thick at the nodes, Int. 2–3 cm long; **L.** 1–2 cm long, 2–2.5 mm thick; **Fl.** 14 mm ⌀, pale pink with a white centre.

**P. fimbriatum** L. Bol. (§ I) (*P. inachabense* L. Bol.). – S.W.Afr.; Cape: Lit. Namaqualand. – Sub-♄ with the **St.** becoming ± woody, prostrate to curved and ascending, up to 1 m long, with short side-shoots; **L.** cylindrical to slightly trigonous, linear to slightly clavate, up to 15 mm long, greyish-green; **Fl.** c. 1 cm ⌀, very numerous, Pet. white to pale yellowish, often laciniate at the tip.

**P. foliolosum** L. Bol. (§ I). – Cape: Lit. Namaqualand. – Fairly compact, densely branched, 30–50 cm tall and ⌀; **St.** 25 cm long, 5 mm thick, flowering **shoots** 5–7 cm long, Int. 5–10 mm long, indistinctly jointed; **L.** semicylindrical, obtuse, yellowish-green, 13 mm long, 2–3 mm thick; **Fl.** 12 mm ⌀, whitish.

**P. framesii** L. Bol. (§ I). – Cape: Lit. Karroo. – With spreading **St.** 3 mm ⌀, with Int. 15–25 mm long; herbaceous parts with prominent, elliptical to pointed-conical or round papillae; **L.** semicylindrical, short-tapered, densely papillose, 2 cm long, 2 mm thick; **Fl.** 14 mm ⌀, white.

**P. gessertianum** (Dtr. et Bgr.) Schwant. (§ I) (*Mes. g.* Dtr. et Bgr., *P. g.* (Dtr. et Bgr.) N. E. Br., *P. luteum* L. Bol.). – S.W.Afr.: Gr. Namaqualand. – Plant 30 cm tall, with minute papillose hairs; **St.** up to 5 mm thick; **L.** cylindrical, obtuse or acute, 8 mm long, 4 mm ⌀; **Fl.** numerous, 10 mm ⌀, yellow with a white centre, with petaloid staminodes.

**P. glareosum** (Bgr.) Dtr. et Schwant. (§ I) (*Mes. g.* Bgr., *P. g.* (Bgr.) N. E. Br., *P. inconspicuum* L. Bol., *Mes. sinus-redfordiani* Dtr. nom. nud., *P. s.-r.* (Dtr.) Dtr. ex Range nom. nud., *P. planum* L. Bol.). – S.W.Afr.: Gr. Namaqualand. – ⊙ ?, resembling **Salicornia herbacea**. Plant glaucous, spreading; **St.** cylindrical below, young ones jointed, older ones nodose, minutely papillose, joints 1–2 cm long, 3–4 mm ⌀; **L.** triangular, swollen, obtuse, apiculate, 4–10 mm long, early deciduous, leaving a white scar; **Fl.** numerous, 5 mm ⌀, light pink.

**P. godmaniae** L. Bol. v. **godmaniae** (§ I). – Cape: Van Rhynsdorp D. – Low-growing plant; **St.** ascending, 2–3 mm thick, covered with small dots, distinctly jointed, with the Int. 12 mm long, 4 mm thick, those on sterile shoots spherical to oval, 2–5 mm long; **L.** 10–18 mm long, 3 mm thick, acute; **Fl.** 12 mm ⌀, pink.

**P. —** v. **gracile** L. Bol. – Flowering **St.** more slender.

**P. herrei** L. Bol. (§ I). – Cape: Van Rhynsdorp D. – **Br.** spreading, stiff, up to 38 cm long, lower nodes not constricted, Int. 1–2.5 cm long, 1–2 mm ⌀; lateral shoots present; **L.** acute, withering; **Fl.** up to 1 cm ⌀, purplish-pink.

**P. hirtellum** L. Bol. (§ I). – Cape: Malmesbury D. – Erect plant 28 cm tall, herbaceous parts hirsute; **St.** virgate, 5 mm thick, Int. 10–15 mm long, jointed at the nodes; **L.** 23 to 30 mm long; **Fl.** 14 mm ⌀, purplish-pink.

**P. imitans** L. Bol. (§ I). – Cape: Piquetberg D. – As **P. subintegrum**; **Fl.** 16 mm ⌀, pink to pale red.

**P. implexum** N. E. Br. (§ I). – Cape: Riversdale D. – **St.** creeping, tangled, 2 mm thick; **L.** 6–10 mm long, 1.5 mm thick, semicylindrical, acute; **Fl.** 10 mm ⌀, white.

**P. inconstrictum** L. Bol. (§ I). – Cape: Bushmanland. – Plant 25–30 cm tall, herbaceous parts grey to pale blue, with minute dots; **St.** spreading, these and the short shoots constricted at the nodes, Int. 5–50 mm long; **L.** 10–18 mm long, 2–4 mm thick, acute; **Fl.** 10 mm ⌀, white.

**P. junceum** (Haw.) Schwant. (§ I) (*Mes. j.* Haw., *P. j.* (Haw.) N. E. Br., *Peratetracoilanthus j.* (Haw.) Rap. et Cam.). – Cape: Beaufort West, Montagu D. – Plant 70 cm tall; **St.** sparse, jointed, Int. unequal; **L.** 2–3 mm long, 2 mm across; **Fl.** whitish to violet.

**P. kuntzei** (Schinz) Dtr. et Schwant. (§ I) (*Mes. k.* Schinz, *Mes. dinteri* Engl. p. part., *Mes. schlichtianum* auct. non Sond., *P. s.* (auct. non Sond.) Schwant., *P. salicornioides* (auct. non Pax) Schwant.). – S.W.Afr.: coastal reg. – Forming dense cushions near the sea, otherwise a laxer, branched ♄; **St.** distinctly constricted-jointed, spongy-papillose, with a light greyish-green to blackish-grey bark; **L.** cylindrical to semicylindrical, 5–15 mm long; **Fl.** borne at the end of the shoots, solitary or several together, white, shortly pedicellate.

**P. laxiflorum** L. Bol. (§ I). – Cape: Van Rhynsdorp D. – **St.** 25 cm long, green, Int. 2–3 cm long, 3 mm thick, nodes indistinct, jointed; **L.** 2 cm long, membranous below; **Fl.** laxly arranged, 14 mm ⌀, white.

**P. leightoniae** L. Bol. (§ I). – Cape: near Matjiesfontein. – Erect plant with the herbaceous parts papillose; **St.** 24 cm long, 5 mm thick, Int. 10–15 mm long, 2 mm thick; **Fl.** 8 mm ⌀, purplish-pink.

**P. leptarthron** (Bgr.) N. E. Br. (§ I) (*Mes. l.* Bgr.). – W.Cape. – **St.** 3–10 cm long, erect, constricted, joints (4–)6(–20) mm long, (3–)4(–5) mm thick; **L.** semicylindrical, obtuse, minutely papillose, 6–11 mm long, 2 mm across; **Fl.** white.

**P. levynsiae** N. E. Br. (§ I). – ♄ 45 cm tall; **L.** ± trigonous; **Fl.** 1.5 cm ⌀, pink.

**P. lewisiae** L. Bol. (§ I). – Cape: Swellendam D. – Herbaceous parts bluish-green, with minute, prominent papillae; **St.** prostrate, 3 to 4 mm thick, branched, Int. distinctly constricted; **L.** semicylindrical, shortly tapered, 7–10 mm long, 2 mm thick; **Fl.** 14 mm ⌀, pink.

**P. liebenbergii** L. Bol. (§ I). – Cape: Alexandria D. – Erect plant 35 cm tall, freely branched from the base; **St.** 33 cm long, scarcely constricted at the nodes, Int. 2–2.5 cm long, 3 mm ⌀, roughly papillose; **L.** apiculate, 1.9 cm long, 1.5 mm $\varnothing$; Pet. obtuse, 4–8 mm long, white, tinged purple.

**P. lindequistii** (ENGL.) SCHWANT. (§ I) (*Mes. l.* ENGL., *P. l.* (ENGL.) N. E. BR., *Mes. ausanum* DTR. et BGR., *Nycteranthus a.* (DTR. et BGR.) SCHWANT., *Aridaria a.* (DTR. et BGR.) DTR. et SCHWANT., *A. radicans* L. BOL., *N. r.* (L. BOL.) SCHWANT., *Sphalmanthus r.* (L. BOL.) L. BOL.). – S.W.Afr.: Lüderitz Bay S. – ♄ (0.5–)1(–1.5) m tall, but more across; **St.** erect or prostrate, not jointed, ± thickened at the nodes; **L.** subcylindrical, (1.5–)3(–4) cm long, 4–7 mm thick; **Fl.** on a pedicel 1–2 cm long, c. 3 cm $\varnothing$, white to pale yellow. Acc. H.-CHR. FRIEDRICH this species constitutes a transition to **Aridaria.**

**P. littlewoodii** L. BOL. (§ I). – Cape: Van Rhynsdorp D. – Plant up to 16 cm tall, 80 cm $\varnothing$, herbaceous parts minutely glossy-papillose; **St.** c. 35 cm long, with dense, tangled **Br.** constricted at the nodes, Int. 3 mm to 3 cm long, 2–4 mm $\varnothing$; **L.** up to 1 cm long, up to 2.5 mm $\varnothing$, apiculate; **Fl.** 1 cm $\varnothing$, purplish-pink.

**P. longipes** L. BOL. (§ I). – Cape: Lit. Namaqualand. – **St.** thickened; **Br.** prostrate to ascending, 9–18 cm long, 2 mm thick, Int. 15–25 mm long; **L.** obtuse, 10–23 mm long, 2 mm thick; **Fl.** 10 mm $\varnothing$, yellow.

**P. marlothii** (PAX) FRIEDR. (§ II) (Pl. 148/1) (*Mes. m.* PAX, *Trichocyclus m.* (PAX) N. E. BR., *Brownanthus m.* (PAX) SCHWANT., *Mes. solutifolium* BGR., *B. s.* (BGR.) JACOBS.). – S.W.Afr.: Gr. Namaqualand. – Resembles **P. ciliatum;** freely branched plant, Int. 4 mm long; **L.** with a ring of white hairs 5 mm long at the base, blade 6–8 mm long, trigonous; **Fl.** solitary.

**P. melanospermum** (BGR.) N. E. BR. (§ I) (Pl. 192/2) (*Mes. m.* BGR., *P. distinctum* N. E. BR., *Mes. geniculiflorum* AUCT. non L., *P. m.* AUCT. non (L.) N. E. BR.). – S.W.Afr.: Lax sub-♄ up to 1 m tall; **St.** virgate, mostly indistinctly jointed, with blistery-spongy papillae; **L.** semicylindrical, ± channelled on the upperside, oblong-linear to narrowly oblong-lanceolate, 1–2 cm long; **Fl.** on a pedicel 1 cm long, 3–5 together, white. (Acc. H.-CHR. FRIEDRICH possibly referable to **Aridaria.**)

**P. mentiens** (BGR.) N. E. BR. (§ I) (*Mes. m.* BGR., *Mes. fulvum* AUCT. non HAW., *Mes. simile* SOND., *P. s.* (SOND.) N. E. BR., *P. s.* (SOND.) SCHWANT., *Mes. s.* v. *namaquense* SOND., *P. n.* (SOND.) N. E. BR., *P. n.* (SOND.) SCHWANT. emend. L. BOL.). – S.W.Afr.: N.Gr. Namaqualand; Cape: Uitenhage D. – Mostly hemispherical sub-♄ 0.5–1.5 m tall, more across; **St.** becoming woody, once or repeatedly forked, bark white to grey, glabrous; **L.** cylindrical to somewhat trigonous, 1–1.5 cm long, greyish-green, with a straight or ± hooked apiculus; **Fl.** numerous, in ± dense Infl., white to pale pink or red.

**P. mucronulatum** (DTR.) N. E. BR. (§ I) (*Mes. m.* DTR.). – S.W.Afr.: Gr. Namaqualand. – Sub-♄ 10–20 cm tall; **St.** ascending, with shoots only on one side, Int. 2–5 mm long, ± constricted-jointed; **L.** 14 mm long, 2.5 mm thick, mucronate, minutely papillose, with minute white Sp. along the margins (lens needed), bluish-green, very soft; **Fl.** 13 mm $\varnothing$, violet to pink to white.

**P. namibense** (MARL.) FRIEDR. (§ II) (*Mes. n.* MARL., *Trichocyclus n.* (MARL.) N. E. BR. ex MAAS, *Brownanthus n.* (MARL.) BULLOCK). – S.W.Afr.: Namib. – Resembles **P. marlothii,** but without the ring of hairs.

**P. oculatum** L. BOL. (§ I). – Cape: Clanwilliam D. – Plant c. 90 cm tall, the herbaceous parts papillose; **St.** laxly branched, Int. 8 to 20 cm long, 1.5–3 mm $\varnothing$; **L.** furrowed, acute or obtuse, c. 2 cm long, 2–3 mm $\varnothing$; **Fl.** 1.5 to 2 cm $\varnothing$, an attractive pink, with a white centre.

**P. pageae** L. BOL. v. **pageae** (§ I). – S.W. Cape. – **St.** spreading to ascending, the green parts papillose-granular, Int. 7–15 mm long; **L.** 14 mm long, withering; **Fl.** 12 mm $\varnothing$, mauvish-pink.

**P.** — v. **grandiflorum** L. BOL. – Cape: Van Rhynsdorp D. – **Fl.** 18 mm $\varnothing$, pale red.

**P. parviflorum** (JACQ.) SCHWANT. (§ I) (*Mes. p.* JACQ., *P. p.* (JACQ.) L. BOL., *Mes. micranthum* HAW., *P. m.* (HAW.) L. BOL., *Peratetracoilanthus p.* (JACQ.) RAP. et CAM., ? *Mes. tenue* HAW., ? *P. t.* (HAW.) SCHWANT., *P. t.* (HAW.) N. E. BR.). – Cape: Riet Valley and Genadendal. – Plant up to 30 cm tall; **St.** very slender, directed upwards, jointed, soon becoming leafless, rush-like; **L.** 6–15 mm long, linear, pale green; **Fl.** 5–6 mm $\varnothing$, white.

**P. pauciflorum** (SOND.) SCHWANT. (§ I) (*Mes. junceum* v. *p.* SOND.). – S.Afr.: origin unknown. – Close to **P. junceum,** possibly only a variety; **St.** cylindrical, almost leafless, with shoots usually on one side only; **Fl.** 2–4 together.

**P. pauper** L. BOL. (§ II). – Cape: Prince Albert D. – ⊙ ?, erect plant 15 cm tall, herbaceous parts dirty green; **St.** 2 mm thick, Int. 10–25 mm long, jointed at the nodes; **L.** 3 cm long, 1.5 mm thick, with persistent scales 2–4 mm long at the base; **Fl.** few, white.

**P. peersii** L. BOL. (§ I). – Cape: Van Rhynsdorp D. – Sparsely branched plant 12 cm high, constricted at the nodes, Int. ovoid to subovoid, dirty olive-green, 7–10 mm thick; **L.** semicylindrical, shortly tapered, 12 mm long, 2.5 mm $\varnothing$, papillose; **Fl.** 24 mm $\varnothing$, white.

**P. pfeilii** (ENGL.) SCHWANT. (§ I) (*Mes. p.* ENGL., *P. p.* (ENGL.) N. E. BR., *Mes. pseudoausanum* DTR. nom. nud.). – S.W.Afr.: Gr. Namaqualand. – Fairly large sub-♄; **St.** 5 to 6 cm long, Int. up to 2 cm long; **L.** linear-lanceolate, subtrigonous, papillose, 1 cm long, 3 mm across, 2 mm thick; **Fl.** 2.5 cm $\varnothing$.

**P. pillansii** (L. BOL.) FRIEDR. (§ II) (*Trichocyclus p.* L. BOL., *Tr. pubescens* N. E. BR. ex MAAS, *Brownanthus p.* (N. E. BR. ex MAAS) BULLOCK, *Tr. buchubergensis* DTR.). – S.W.Afr.: Lüderitz D.; Cape: Lit. Namaqualand. – c. 12–17 cm tall; **St.** spreading, 4–6 mm thick, Int. 10–15 mm long, the lower ones subclavate, the upper ones spherical-ovoid, 3–6 mm long, papillose; **L.** semicylindrical, rounded to acute, with soft, papillose hairs and cilia 4 mm long at the base, 1–2 cm long, 2–3 mm thick, Sh. 1 mm long; **Fl.** white.

**P. planisepalum** L. Bol. (§ I). – Cape: Bushmanland. – Erect, 20–30 cm tall, with spreading **St.**, herbaceous parts bluish to reddish-brown, glossy, constricted at the nodes, Int. cylindrical, 5–15 mm long, 3–6 mm thick; **L.** semicylindrical, obtuse, 10–18 mm long, 2 mm thick; **Fl.** 14 mm ⌀, pink.

**P. pomeridianum** L. Bol. (§ I). – Cape: Kenhardt D. – ☉; **St.** spreading to prostrate, 20 cm tall, Int. cylindrical, 15–30 mm long, herbaceous parts with round or elongated papillae; **L.** obtuse, membranous below, 24 mm long, 3 mm thick; **Fl.** 2 cm ⌀, pale pink.

**P. rapaceum** (Jacq.) Schwant. (§ I) (Pl. 192/4) (*Mes. r.* Jacq., *P. r.* (Jacq.) L. Bol.). – Cape: Van Rhynsdorp D. – **Caudex** tuberous, thickened; **St.** herbaceous, elongated; **Br.** cylindrical, ± jointed; **L.** 15–20 mm long, cylindrical, spotted; **Fl.** 18–20 mm ⌀, white. – Very dubious species.

**P. rogersiae** L. Bol. (§ I). – Cape: Cradock D. – Erect, bluish, 30–40 cm tall; **St.** spreading, 6 mm thick, Int. 1–2 cm long, 3 mm thick; **L.** short-tapered, 10–25 mm long, 3 mm thick; **Fl.** 17 mm ⌀, white.

**P. roseoalbum** L. Bol. (§ I). – Cape: W. Griqualand. – **St.** spreading, 20–27 cm long, constricted at the nodes, Int. cylindrical, 5 to 20 mm long, herbaceous parts bluish-green, papillose, each papilla tipped with a tiny, pointed hair; **L.** 13 mm long, semicylindrical, with a distinct red tip 3–4 mm long, papillose; **Fl.** 12 mm ⌀, pink with a white centre.

**P. salicornioides** (Pax) Schwant. (§ I) (*Mes. s.* Pax, *P. s.* (Pax) N. E. Br., *Mes. trothai* Engl., *P. t.* (Engl.) N. E. Br., *P. t.* (Engl.) Schwant., *P. woodii* L. Bol.). – S.W.Afr.: coastal reg. – ☉ to ± ♃ herb forming a flat tuft; **St.** erect, branched, ± jointed, papillose; **L.** cylindrical, acute, (5–)10(–25) mm long, 2–4 mm thick, those towards the Infl. smaller, ± scale-like; **Fl.** in lax Infl., whitish.

**P. schlichtianum** (Sond.) Schwant. (§ I) (*Mes. s.* Sond., *P. s.* (Sond.) N. E. Br.). – S.W.Afr.: coast. – Whitish-grey plant; **St.** erect, soft, ± spongy, ± constricted and jointed, Int. 12 mm long; **L.** almost cylindrical; **Fl.** 6 mm ⌀, white.

**P. semilunatum** L. Bol. (§ I). – Cape: Van Rhynsdorp D. (?). – Erect, laxly branched, 40 cm tall; **St.** virgate, 37 cm long, 4 mm thick, flowering shoots alternate, with few Fl.; **L.** acute, 10–17 mm long, Sh. crescent-shaped; **Fl.** 16 mm ⌀.

**P. stayneri** L. Bol. (§ I). – Cape: Calvinia D. – 50 cm tall, 1 m ⌀; **St.** up to 59 cm long, 4 mm ⌀, constricted at the nodes, Int. 5–6 cm long, young shoots 2–4 cm long, herbaceous parts olive-green, minutely papillose; **L.** semicylindrical, apiculate, 4–16 mm long, 2 mm ⌀; **Fl.** 1.6 cm ⌀, pink.

**P. subintegrum** L. Bol. (§ I). – Cape: Malmesbury D. – 25–30 cm tall, 40 cm or more ⌀; **St.** spreading, 7 mm thick, Int. cylindrical, 2 cm long, 5 mm thick; **Br.** tangled, blue, with prominent dots, constricted, jointed; **L.** 2 cm long, 2 mm thick, with a membranous scale at the base, acute, channelled on the upperside; **Fl.** 2 cm ⌀, pink.

**P. subnodosum** (Bgr.) N. E. Br. (§ I) (*Mes. s.* Bgr.). – W.Cape. – Resembles **P. junceum**; **St.** erect, cylindrical, white to grey, Int. 2–3.5 cm long, 4 mm ⌀; immature **Br.** erect to patent, white-papillose; **L.** semicylindrical, 1–2 cm long, early deciduous; **Fl.** in 2–3-partite Infl.

**P. uncinatum** L. Bol. (§ I). – S.W.Afr.: coast. – Freely branched, 35 cm tall; **St.** elongated, curved, Int. 23–50 mm long, herbaceous parts bluish-green, spotted; **L.** cylindrical, hooked and recurved at the tip, 7–12 mm long, 2 mm across; **Fl.** 12 mm ⌀, pale pink with a white centre.

**P. utile** L. Bol. (§ I). – Cape: Laingsburg D. – Erect, 30–45 cm tall; **St.** spreading, 4 mm thick, Int. 5–15 mm long, indistinctly jointed at the nodes; **L.** semicylindrical, acute, 6–12 mm long, pale; **Fl.** an attractive pink.

**P. variabile** L. Bol. (§ I). – Cape: Lit. Namaqualand. – 10–15 cm tall, herbaceous parts bluish-green, with minute dots; **St.** spreading, prostrate, 20 cm or more long, Int. 5–8 mm long, jointed to constricted; **L.** semicylindrical, acute, 15 mm long, 2 mm thick; **Fl.** 11 mm ⌀, pink.

**Rabiea** N. E. Br. (§ IV/1/9). – Dwarf, compact, glabrous plants; **rootstock** fleshy or tuberous; **St.** crowded, with 4–6 L., Int. enclosed in the L.-Sh.; **L.** ascending or spreading, 2 of a pair ± dissimilar, (2–)5(–8) cm long, acute or acuminate, margins parallel or one margin straight and the other curved outwards, often keeled, rarely terete, one L. often dilated and broadened at the tip, often with prominent and conspicuous, white or reddish-brown dots; **Fl.** solitary, nearly sessile, the pedicel enclosed in the bract-Sh., Pet. in 3–4 series, (1.6–)2–2.5(–3.5) cm long, 1.5–2 mm wide, yellow, golden-yellow or orange, the underside often pink or reddish, diurnal. – Greenhouse, warm. Propagation: seed, cuttings.

## Key to the Species of the Genus Rabiea by L. Bolus

1. Upper surface of L. rather suddenly narrowed upwards, usually from above the middle, 1–2.1 cm broad, up to 1.5 cm or, if keel eccentric, 1.8 cm ⌀ ............................. **2.**
1a. Upper surface of L. gradually narrowed upwards from the middle or below the middle, 0.5 to 1.5 cm broad, up to 1.1 cm ⌀ ................................................. **4.**
2. L.-margin entire, apiculus sometimes recurved; Pet. up to about 1.6 cm long ............
 ........ **albinota** v. **albinota**
2a. L.-margins sometimes with a T. or blunt lobe, apiculus erect; Pet. up to 3.5 cm long ..... **3.**

3. Pet. up to 2.7–3.5 cm long; top of Ov. raised in the centre, stigmas up to 5 mm long ....... — v. **longipetala**
3a. Pet. up to 2.5 cm long; top of Ov. not raised; stigmas up to 2.5 mm long .. — v. **microstigm**
4. L. up to 5 mm broad, 3–6 mm ⌀ ............................................................. **albipuncta**
4a. L. up to 1.3 cm broad and 1.1 cm ⌀ ...................................................... **5.**
5. L. dorsally rounded, terete towards the apex; top of Ov. raised in the centre ....... **jamesii**
5a. L. obtusely keeled from near the middle upwards and sometimes acutely keeled and laterally compressed near the apex; top of Ov. not, or scarcely, raised in the centre ............ **6.**
6. Top of Ov. deeply concave ........................................................... **difformis**
6a. Top of Ov. apparently flat or slightly concave .......................................... **7.**
7. One L. of a pair, viewed laterally, slightly incurved towards the smaller L. so as somewhat to resemble a parrot's beak when open, up to 2 cm long; top of Ov. apparently flat ...... **lesliei**
7a. L. not as above, up to 6.5 cm long, top of Ov. slightly concave ................ **comptonii**

**R. albinota** (HAW.) N. E. BR. v. **albinota** (*Mes. a.* HAW., *Aloinopsis a.* (HAW.) SCHWANT., *Nananthus a.* (HAW.) L. BOL.). – Cape: Graaff Reinet D. – L. 6–8 together, up to 10 cm long, 10 mm across, sabre-shaped, trigonous above, apiculate, with spots and prominent dots; Fl. 3.5 cm ⌀, yellow, Sept.
**R.** — v. **longipetala** L. BOL. – Pet. 2.5–3.5 cm long, yellow with red tips.
**R.** — v. **microstigma** L. BOL. – Stigmas shorter than in the v. **albinota**, 2–2.5 mm long.
**R. albipuncta** (HAW.) N. E. BR. v. **albipuncta** (*Mes. a.* HAW., *Aloinopsis a.* (HAW.) SCHWANT., *Nananthus a.* (HAW.) N. E. BR., *N. a.* (HAW.) SCHWANT.). – Orange Free State. – L. 6–8 together, c. 2.5–4 cm long, 5–7 mm across, obliquely short-tapered and apiculate above, carinate on the underside towards the tip, glossy green, rough with numerous, whitish, tuberculate dots; Fl. 3 cm ⌀, Pet. straw or flesh-coloured, with a red M.-line, Nov.
**R.** — v. **major** L. BOL. – Larger variety.
**R. comptonii** (L. BOL.) L. BOL. (*Nananthus c.* L. BOL.). – Orange Free State. – St. 4.5 cm long, 1 cm thick; L. 6 together, projecting-recurved, acute, keeled on the underside, covered with ± prominent dots, 5.5 cm long, 11 mm across and thick; Fl. 3 cm ⌀, Pet. yellow, the tips becoming reddened.

**R. difformis** (L. BOL.) L. BOL. (Pl. 193/1) (*Nananthus d.* L. BOL.). – Cape: Cradock D.; Orange Free State. – L. of a pair unequal and dissimilar, narrowed above and 5 mm thick, or laterally compressed and 8 mm thick, obliquely keeled, obliquely truncate at the tip, dark green, with crowded, prominent, white dots, 4 cm long, up to 9 mm across; Fl. 4 cm ⌀, golden-yellow.
**R. jamesii** (L. BOL.) L. BOL. (*Nananthus j.* L. BOL.). – Cape: Cradock D. – L. 4–6 together, semicylindrical or subcylindrical, with a hardened tip, keeled on the underside above, 2.5–3 cm long, 7–8 mm across and thick, smooth, greyish-green, with distinct tubercles becoming pearl-like in age; Fl. 3.5 cm ⌀, golden-yellow.
**R. lesliei** N. E. BR. – Orange Free State. – Forming clumps 2.5–3 cm tall; L. unequal, 11–12 mm long, 4–5 mm across, 3–4 mm thick, bluntly keeled on the underside, green, ± reddish above, with ± prominent, often blue dots; Fl. 2.5 cm ⌀, orange-coloured.
**? R. tersa** N. E. BR. – Cape: W.Griqualand. – L. in 3 pairs, 22–27 mm long, 2.5–5 mm across, 3–5 mm thick, linear-lanceolate, acute, subacutely keeled on the underside, bluish-green, with minute dots and green spots; Fl. 2.5 to 3 cm ⌀, light yellow, pink on the underside. (Incompletely known and not included in the Key.)


**Rhinephyllum** N. E. BR. (§ IV/1/11). – Short-stemmed, succulent, ♃ plants with enclosed Int., or small branched **shrublets** with visible Int.; **St.** with 2–4 opposite **L.**, clavate or thicker towards the apex, upperside flat, underside round or slightly keeled, rough from the covering of small, hard, whitish tubercles, or smooth and shining; Fl. solitary, terminal, 12–25 mm ⌀, golden-yellow, yellow, yellowish-white or white, nocturnal. – Greenhouse, warm. Propagation: seed, cuttings.

## Key to the Species of the Genus Rhinephyllum by L. Bolus

1. L.-margin entire or with an occasional obscure T. .................................... **2.**
1a. L.-margins variously toothed or lobulate, or sometimes some entire on the same plant .. **10.**
2. Usually erect, compact shrublets up to 15 cm high, Int. visible ..................... **3.**
2a. Caespitose; Int. enclosed ........................................................... **4.**
3. L. in profile semiovate or semicircular, usually 4–7 mm long; Pet. usually 1 mm broad ....... **graniforme**
3a. L. in profile oblong or linear, usually 1–1.4 cm long; Pet. up to 2 mm broad ...... **luteum**
4. L. smooth and polished; Pet. usually less than 1 mm broad ................. **macradenium**
4a. L. ± rough; Pet. 1 mm or more broad .............................................. **5.**
5. L. in profile acute or acuminate ................................................ **frithii**
5a. L. not as above ................................................................. **6.**

| | |
|---|---|
| 6. | L. in profile semicircular, dilated in the upper part, vesicles crowded, small; Pet. white .... **pillansii**[1] |
| 6a. | L. not as above; Pet. yellow .................................................................. 7. |
| 7. | L.-margins sharply defined by a continuous, hard or cartilaginous line; centre of Ov. equalling, or exceeding, the height of the glands; stigmas up to 4 mm long ................ **muirii** |
| 7a. | L.-margins and Ov. not as above; stigmas 5–9 mm long ................................ 8. |
| 8. | Upper surface of L. clavate or spatulate, 6–22 mm long ................................ 9. |
| 8a. | Upper surface of L. linear, narrowed to the tip, to 22 mm long ........... **vanheerdei** |
| 9. | L. mostly 6–9 mm long, the upper surface symmetrical; Fl. 15 mm ⌀, stigmas 5 mm long ... **parvifolium** |
| 9a. | L. 18–22 mm long, the upper surface oblique (i.e. the 2 sides of the midrib unequal in breadth); stigmas 7–9 mm long ......................................................................... **obliquum** |
| 10. | L. in profile usually acute or acuminate ...................................................... **frithii** |
| 10a. | L. in profile usually obtuse or rounded or subtruncate at the apex .................... 11. |
| 11. | L.-keel as well as margins toothed ......................................................... **rouxii** |
| 11a. | L.-keels entire ........................................................................................ 12. |
| 12. | L. of a pair conspicuously unequal in length; nectar-glands distant; stigmas 2–3 mm long ...... **inaequale** |
| 12a. | L. of a pair ± equal in length; nectar-glands almost contiguous; stigmas 5–9 mm long ... 13. |
| 13. | Upper surface of L. semicircular at the apex; stigmas 5 mm long ............. **broomii** |
| 13a. | Upper surface of L. not as above; stigmas 7–9 mm long ................................ 14. |
| 14. | L.-margins finely toothed; Pet. bright yellow; top of Ov. slightly concave ...... **comptonii** |
| 14a. | L.-margins undulate or lobulate; Pet. lemon-yellow, red-tipped; top of Ov. deeply concave ........ **schonlandii** |

**R. broomii** L. Bol. – Cape: Fraserburg D. – Compact plant; **St.** with 4 L., Int. enclosed; lower **L.** spatulate, minutely tuberculate at the tip, with 4–5 marginal T., upper L. hemispherical, densely tuberculate, olive-green, 10 mm long, 7 mm across; **Fl.** 1–2 cm ⌀, yellow.

**R. comptonii** L. Bol. – Cape: Laingsburg D. – **L.** blue-grey or purplish-grey, some toothed; **Pet.** broad, yellow. Incompletely described.

**R. frithii** (L. Bol.) L. Bol. (*Mes. f.* L. Bol., *Peersia f.* (L. Bol.) L. Bol.). – Cape: Grootfontein. – **St.** prostrate, 9 cm long; **L.** of a pair unequal, trigonous, acute, entire or rarely with dentate margins, bluish-green, 2.5–4 cm long, 8 mm across; **Fl.** 1.5 cm ⌀, yellow.

**R. graniforme** (Haw.) L. Bol. (*Mes. g.* Haw., *R. luteum* v. *brevifolium* L. Bol.). – Cape: Prince Albert D. – **St.** 11 mm ⌀; **Br.** numerous, 6 mm thick, Int. 4–7 mm long; **L.** subspherical to grain-shaped, ovoid to oblong-ovoid, acute to obtuse, blue, rough-tuberculate, 5–9 mm long, 3–5 mm ⌀; **Fl.** 15–22 mm ⌀, yellow.

**R. inaequale** L. Bol. v. **inaequale.** – Cape: Graaff Reinet D. – Compact plant; **L.** spatulate to oblique-spatulate, compressed at the tip, those of a pair unequal, 7–12 or 12–22 mm long, 8–10 mm across, 4–7 mm thick, pale bluish-green with crowded dots, margins indistinctly undulating-dentate to entire; **Fl.** 2 cm ⌀, lemon-yellow, Pet. narrow.

**R.** — v. **latipetalum** L. Bol. – **L.** with 4 to 5 lobes towards the tip; Pet. broader.

**R. luteum** (L. Bol.) L. Bol. (*Ruschia l.* L. Bol., *Mes. flavens* (L. Bol.) N. E. Br.). – Cape: Prince Albert D. – Forming mats 5 cm high; **St.** 10 mm thick, **Br.** 6 mm thick, shoots with 4–6 oblong **L.** 10 mm long, 3 mm across and thick, ± obtuse, carinate on the underside, rough, minutely papillose; **Fl.** 12 mm ⌀, yellow.

**R. macradenium** (L. Bol.) K. Bol. (*Mes. m.* L. Bol., *Peersia m.* (L. Bol.) L. Bol.). – Cape: Ceres D. – **St.** with 4–6 subequal, semicylindrical **L.** 2.5–6 cm long, 6–8 mm thick, swollen-trigonous and ± obtuse at the tip; **Fl.** 2.5 cm ⌀, yellowish.

**R. muirii** N. E. Br. – Cape: Lit. Karroo. – **Rootstock** fleshy, developing many **St.** to form clumps; **L.** 10–24 mm long, 5–10 mm across, 4–5 mm thick above, with the underside drawn slightly forward over the tip, green or reddish, the upper part covered with whitish tubercles, margins and keel white, cartilaginous; **Fl.** 12 to 14 mm ⌀, yellowish-white.

**R. obliquum** L. Bol. – Cape: Fraserburg D. – **L.** 4–6 on a shoot, spatulate, oblique, short-tapered and apiculate, rounded on the underside, margins entire or with a distinct lobe, 18–22 mm long, 7–8 mm across, 5 mm thick, dirty green with prominent dots; **Fl.** 2 cm ⌀, yellow, scented.

**R. parvifolium** L. Bol. – Cape: Jansenville D. – **L.** 6–9 mm long, 3–4 mm across, hemispherical on the underside, dark green, minutely tuberculate; **Fl.** 15 mm ⌀, golden-yellow.

**R. rouxii** L. Bol. (*Chasmatophyllum r.* (L. Bol.) L. Bol.). – Cape: Victoria West D. – **St.** slender; flowering **shoots** with 4–6 crowded **L.**, those of a pair unequal, 16 and 20 mm long, 5.5 and 6 mm across, clavate to clavate-spatulate, sides expanded above to 5 mm, mucronate, the underside keeled above, with 2–3 small T.; **Fl.** 18 mm ⌀, yellow.

**R. schonlandii** L. Bol. (Pl. 193/2). – Cape: Fraserburg D. – **St.** 10 mm thick, **Br.** 4 mm thick; **shoots** with 4–6 **L.**, those of a pair unequal, 2 cm long, 12 mm across, broadly spatulate, square below, oblique above, ±

---

[1]) Acc. G. Schwantes = Neorhine pillansii.

rounded and expanded to 6–7 mm, convex on the underside, margins rarely entire, usually with 1–3 indistinct T. or lobes in the upper part; **Fl.** 16 mm ⌀, Pet. yellow with red tips.

**R. vanheerdei** L. BOL. – Cape: Lit. Namaqualand. – Compact, small ♄ 7 cm high, 12 cm ⌀; **St.** 11 mm ⌀, **Br.** 8 mm ⌀; **shoots** with 4 to 6 pale bluish-green, often pink, minutely tuberculate **L.** 3–6.2 cm long, 6–9 mm ⌀ midway, narrowed towards the obtuse to acute tip, rounded or keeled on the underside; **Fl.** 2 cm ⌀, yellow.

**Rhombophyllum** (SCHWANT.) SCHWANT. (§ IV/1/1) (*Bergeranthus* SCHWANT. SG. *Rhombophyllum* SCHWANT.). – Small ♄ or mat-forming, very succulent plants; **roots** often napiform; **L.** densely crowded, opposite and decussate, ± united at the base, semicylindrical, keeled above, with the underside pulled forward and chin-like, the upperside ± linear, expanded towards the middle or ± obliquely rhombic, margins entire or with 1–2 short T., smooth, ± glossy, deep green, with whitish or translucent dots; **Fl.** 3–7 together on a Ped., golden-yellow, June-Sept. – Greenhouse. Propagation: seed, cuttings.

**R. dolabriforme** (L.) SCHWANT. (Pl. 193/4) (*Mes. d. L., Hereroa d.* (L.) L. BOL.). – Cape: E.Karroo. – Forming mats, freely branching, later shrubby and up to 30 cm tall; **L.** 2.5 to 3 cm long, long-tapered on the upperside, the underside semicylindrical below, expanded above to 11–15 mm with a hatched-shaped keel which is pulled forward with a tooth-like tip, grass-green with translucent dots; **Fl.** up to 4 cm ⌀, golden-yellow, June-Aug.

**R. nelii** SCHWANT. – Resembles **R. dolabriforme**; **L.** bifid, 1.5 cm long, 4–6 mm across midway, 6–8 mm across above, pale bluish or greyish-green, with dark dots; **Fl.** almost 4 cm ⌀, yellow.

**R. rhomboideum** (S. D.) SCHWANT. (Pl. 193/3) (*Mes. r.* SALM, *Bergeranthus r.* (SALM) SCHWANT.). – E.Cape. – Acauline, clump-forming; **L.** in 4–5 pairs together, resting on the soil, ± unequal, 2.5–5 cm long, 1–2 cm across, rhombic seen from above, with the underside rounded below, then thickened, keeled towards the tip and pulled forward chin-like, the whitish angles rarely with 1–2 T., dark greyish-green with numerous whitish dots; **Fl.** 3 cm ⌀, golden-yellow, reddish outside, June-Sept.

**R. —  v. groppiorum** W. HEINRICH. – Cultivated plant, perhaps a mutant. – **St.** with 2–3 pairs of trigonous to square **L.** c. 19 mm long, 14 mm across, 9 mm thick, convex on both surfaces, dark green, with a whitish border and dense whitish dots; **Fl.** brilliant yellow.

**Ruschia** SCHWANT. (§ IV/1/1). – Large or small to dwarf ♄; **St.** partly unilaterally branched, prostrate and in some species forming small tufts; **Br.** and twigs often covered with old L. remains or dry persistent Sh.; **L.** amplexicaul with a very long Sh., or L. often only shortly united, the free blades 3-angled, ± long or small and hemispherical, mostly mucronate, the keel entire or armed with blunt T., stiff, texture firm, bluish-green, nearly always set with darker transparent dots, glabrous or ciliate; **Fl.** axillary or terminal, solitary or several together, stalked or sessile, pink, red, violet or white, in summer, open by day or in some species also at night. – Summer: in the open, winter: cold house. Propagation: seed, cuttings.

## Division of the Genus Ruschia into Sections by G. Schwantes

Type species: **R. rupicola.**

**Sect. 1. Rubricauliae** (SALM) SCHWANT. (*Rubricaulia* SALM as a SSect. of *Mesembryanthemum* L.). – Habit as **Lampranthus**, but L. thicker and shorter. – Type species: R. rubricaulis. – Further species: R. filamentosa, serrulata and others.

**Sect. 2. Virgatae** (HAW.) SCHWANT. (*Virgata* HAW. as a SSect. of *Mesembryanthemum* L.). – Shrubby species with fairly short, characteristically compressed L. – Type species: R. virgata. – Further species: R. congesta, staminodiosa and others.

**Sect. 3. Cymbifoliae** SCHWANT. – ♄ with very short, thick L., often with short shoots in the L.-axils so that they resemble **Delosperma lehmannii**; often forming mats or clumps, from which in most species the flowering Br., which have longer Int., grow out. – Type species: R. cymbifolia. – Further species: R. dolomitica, mollis, nobilis, robusta and others.

**Sect. 4. Ruschiellae** SCHWANT. – Stemless species, always mat-forming; flowering shoots much elongated. – Type species: R. ventricosa. – Further species: R. herrei, schlechteri and others.

**Sect. 5. Caespitosae** SCHWANT. – Species forming mats or clumps similar to the foregoing, but the placental tubercle not wide, generally sunken above and stalked like a handle. – Type species: R. longipes. – Further species: R. amoena and others.

**Sect. 6. Tumidulae** (HAW.) SCHWANT. (*Tumidula* HAW. as a SSect. of *Mesembryanthemum* L.). – Shrubby species whose Int. are surrounded by L.-Sh. united to them; L. fairly long like **Lampranthus**. – Type species: R. tumidula. – Further species: R. festiva, umbellata and others.

**Sect. 7. Vaginatae** (SALM) SCHWANT. (*Vaginata* SALM as a SSect. of *Mesembryanthemum* L.). – In general of the same habit as the previous Sect., with a tendency to smaller L. – Type species: R. vaginata. – Further species: R. multiflora and others.

**Sect. 8. Ruschia** (*Uncinata* HAW. as a SSect. of *Mesembryanthemum* L.). – Similar to the species described above with very short, thick L. and L.-Sh. united to the St. – Type species: R. rupicola. – Further species: R. crassa, perfoliata, uncinata and others.

**Sect. 9. Antimimae** (N. E. BR.) SCHWANT. (*Antimima* N. E. BR. as a G.). – Stemless species with particularly short Int.; Fl. sessile and solitary from the pairs of L. – Type species: R. dualis. – Further species not known.

**Sect. 10. Microphyllae** (HAW.) SCHWANT. (*Microphylla* HAW. as a SSect. of *Mesembryanthemum* L.). – Mostly very dwarf, compact shrublets with very small L.-pairs united into long Sh. which, in the dry period, protect the dormant young L.-pair as a dry skin. – Type species: R. microphylla. – Further species: R. androsacea, aristulata, meyeri, pulchella, pumila.

**Sect. 11. Evolutae** SCHWANT. – Dwarf species whose very short L. have a prominent chin. – Type species: R. evoluta. – Further species: R. levynsiae.

**Sect. 12. Marcidae** N. E. BR. – Dimorphic L., normal L. alternate with plant Bo. – Type species: R. pygmaea.

**Note:** In the Sect. **Ruschia** and **Tumidulae** the actual Sh. is usually not as long as it appears to be because the L.-tissue is prolonged beyond the node, and the only way to ascertain where the node is, is by observing the point at which the axillary bud arises. Of course if one dissects the so-called Sh., removing the L.-tissue from the Int., the length of the actual Sh. is clearly revealed.

**R. abbreviata** L. BOL. – Cape: Lit. Namaqualand. – Erect ♄ 8–15 cm tall, herbaceous parts bluish to whitish, velvety-papillose, Int. enclosed; **L.** 2–3 cm long, laterally compressed, keel with 1 indistinct T., Sh. 8 mm long, with impressed lines; **Fl.** 12 mm ⌀, white.

**R. acocksii** L. BOL. – Cape: Sutherland D. – Stiff, laxly branched ♄ 30 cm tall; **St.** 5 mm thick, Int. 1–3 cm long; **L.** 17–32 mm long, narrowed towards the apiculate tip, angles and keel indistinct, surface covered with large dots; Sh. 1 mm long, 4–6 mm thick; **Fl.** 2–2.5 cm ⌀, purplish-pink.

**R. acuminata** L. BOL. (*Mes. exacutum* N. E. BR.). – Cape: Prince Albert D. – ♄ with the herbaceous parts bluish, spotted, becoming minutely rough-papillose; **St.** virgate, 20 cm long; **L.** 15–25 mm long, 7 mm across, 6 mm thick, acute, ± apiculate, the blunt keel rarely indistinctly dentate, the Sh. 4 mm long, 13 mm across above, with impressed lines; **Fl.** 3 cm ⌀, white to pale pink.

**R. acutangula** (HAW.) SCHWANT. (*Mes. a.* HAW., *Mes. vaginatum* v. *a.* BGR.). – Resembles R. vaginata; **L.** rather smaller, not recurved at the tip, or only slightly so.

**R. addita** L. BOL. – Cape: Lit. Namaqualand. – Compact ♄, the herbaceous parts velvety-papillose; **St.** 6.5 cm long, Int. 5 mm long; **L.** narrowed towards the obtuse, apiculate tip, 2.5 cm long, 3 mm across, 4–6 mm thick, Sh. 2 mm long; **Fl.** 3 or more together, 12 mm ⌀, pale pink.

**R. aggregata** L. BOL. – Cape: Lit. Namaqualand. – Erect ♄ with stiff **St.** 35 cm long, Int. 1.5–4 cm long; **L.** 1–2.5 cm long, 3–4 mm across, 5 mm thick, obtuse, distinctly keeled on the underside, glossy green, Sh. 3 mm long, 3 to 4 mm across, 5 mm thick; **Infl.** compact, Fl. 2.5 cm ⌀, white, rather crowded.

**R. alata** L. BOL. – Cape: Van Rhynsdorp D. – ♄ with ascending or prostrate, virgate **St.** 18 cm long, with alate angles, Int. 12 mm long; **L.** 12 mm long, 3 mm ⌀, obtuse, apiculate, green, with prominent dots, margins and keel minutely hirsute, Sh. 1.5 mm long; **Fl.** 1 cm ⌀ pink.

**R. alborubra** L. BOL. – Cape: Lit. Namaqualand. – ♄ 19 cm tall, the herbaceous parts velvety with crowded papillae; **St.** rigid, Int. 10–16 mm long; **L.** 2–2.5 cm long, 8–10 mm across, 9 mm thick, narrowed towards the apiculate tip, Sh. 4–5 mm long; **Fl.** 12 mm ⌀, white, with red staminodes.

**R. altigena** (L. BOL.) L. BOL. (*Mes. a* L. BOL.). – Cape: Laingsburg D. – ♄ with prostrate **St.** 13 cm or more long; **Br.** 5 cm long, densely leafy above, with short shoots from the L.-axils; **L.** 27 mm long, 4 mm ⌀, tapered, keeled, dirty bluish-green; **Fl.** 23 mm ⌀, pinkish-purple.

**R. amicorum** (L. BOL.) SCHWANT. (*Mes. a.* L. BOL.). – Cape: Montagu D. – ♄ 30–40 cm tall; **St.** 5 mm thick, Int. 6–15 mm long; **L.** cylindrical, acute, apiculate, 2.5–4 cm long, 5 mm across, 7 mm thick, bluish-green; **Fl.** 15 mm ⌀, pinkish-purple, scented.

**R. amoena** SCHWANT. (Pl. 194/2). – Cape: Lit. Namaqualand. – Forming hemispherical cushions up to 15 cm ⌀; **L.** 2.5 cm long, united for 8 mm, long-triangular, apiculate, the underside semicylindrical, bluish-green, puberulous, with light dots; **Fl.** 15 mm ⌀, purplish-red.

**R. ampliata** L. BOL. (*Mes. a* (L. BOL.) N. E. BR.). – Cape: Lit. Namaqualand. – Stiff ♄ 25 cm tall; **St.** 22 cm long, 4 mm thick, Int. 10–15 mm long; **L.** 4–6 on a short shoot, crowded, thick, hemispherical, apiculate, 15 mm long, 7 mm ⌀, bluish-green; **Fl.** 18 mm ⌀.

**R. androsacea** MARL. et SCHWANT. (*Mes. a.* MARL. et SCHWANT.). – Cape: Sutherland D. – Low-growing, prostrate; **L.** 4 mm long, 2 mm across, 1.5 mm thick, rounded on the underside, keeled above, slightly tuberculate, whitish-green Fl. small, red.

**R. approximata** (L. BOL.) SCHWANT. (*Mes. a.* L. BOL.). – Cape: Montagu D. – Laxly branched ♄ 9–14 cm high; **St.** 9–14 cm long, Br. 1.5 to 3 cm long; **L.** trigonous, obtuse, apiculate, the keel with 1–3 T. below the tip, 10–14 mm long,

4–5 mm across, Sh. 5–7 mm long; **Fl.** solitary, 2 cm ⌀, pink.

**R. archeri** L. Bol. v. **archeri.** – Cape: Ladismith D. – **St.** virgate, 33 cm long; **L.** 10–12 mm long, 6 mm ⌀, acute, with a Sh. 6–8 mm long, blue, tinged purple; **Fl.** 25 mm ⌀, purplishpink; stigmas 5.

**R.** — v. **sexpartita** L. Bol. – Cape: Swellendam D. – Stigmas 6.

**R. arenosa** L. Bol. – Cape: Calvinia D. – Stiff ♄ 20–30 cm tall; **St.** erect, Int. 1–4 cm long; **L.** flat on the upperside, keeled on the underside, obtuse to acute, bluish-green, 23 mm long, 3 mm across, 4 mm thick, Sh. 2 mm long; **Fl.** 15 mm ⌀, pink.

**R. aristata** L. Bol. – Cape: Uitenhage D. – Erect, stiff ♄ 60 cm tall; **St.** 4 mm thick, Int. 10–15 mm long; **L.** of a pair dissimilar, one L. tapered, the other falcate, ± truncate, laterally compressed, acutely keeled, bluish green, with crowded dots, 4 cm long, 4 mm across, Sh. 4 mm long; **Fl.** 3 cm ⌀, Pet. numerous, sometimes dentate, bristle-tipped, red to yellow.

**R. aspera** L. Bol. – Cape: Bushmanland. – Stiff, erect ♄ 22 cm tall; **St.** robust, Int. 5 to 13 mm long; **L.** ± spherical to broad-ovoid, 3–4 mm long, 2–2.5 mm thick, dirty green, rough with prominent dots, Sh. 1.5 mm long; **Fl.** 1 cm ⌀, purplish-red.

**R. atrata** L. Bol. (*Mes. a.* (L. Bol.) N. E. Br.). – Cape: Lit. Namaqualand. – Erect ♄ 28 cm tall; **St.** and **Br.** stiff, dark, young Br. with 2–6 L. at the tip; **L.** 15 mm long, 4 mm ⌀, rather obtuse, with minute velvety hairs, bluishgreen, Sh. 2 mm long; **Fl.** 1 cm ⌀, pink.

**R. axthelmiana** (Dtr.) Schwant. (*Mes. a.* Dtr., *Mes. uncinellum* auct.). – S.W.Afr.: Rehoboth, Keetmanshoop D.; Cape: Upington D. – Much-branched ♄ up to 50 cm tall, Int. 2–3 cm long; **L.** sheathing and united for c. 2 cm, the free part c. 2 cm long, trigonous, apiculate, 2–3 mm ⌀, light bluish-green, with green dots, the keel with (2–)4(–6) T., the upperside often with 1–2 marginal T.; **Fl.** solitary, 18 mm ⌀, an attractive violet-pink.

**R. beaufortensis** L. Bol. – Cape: Beaufort West D. – Erect ♄ 12 cm tall; **St.** 5 mm thick, Int. enclosed; **L.** 4–6 on young shoots, (10–)15(–20) mm long, acutely keeled on the underside, margins and keel with large, red-apiculate T., bluish-green with a reddish tinge, Sh. 15 mm long, with impressed lines; **Fl.** 16 mm ⌀, pink.

**R. bicolorata** L. Bol. (*Mes. bicoloratum* (L. Bol.) N. E. Br. – Cape: Lit. Namaqualand. – Erect, stiff ♄ 40 cm tall; **St.** 5 mm thick, Int. 2–3 cm long; **L.** bluish-green, minutely scaly, dimorphic, some 15 mm long, 7 mm ⌀, expanded midway, obtuse or acute, keel reduced, other L. twice as large, Sh. 2 mm long, with impressed lines; **Fl.** 2 cm ⌀, white.

**R. biformis** (N. E. Br.) Schwant. (*Mes. b.* N. E. Br.). – Cape: Swellendam D. – Forming a mat 2.5 cm high; **L.** dimorphic, either 2–7 mm long, 2–3 mm across, 2–2.5 mm thick, trigonous, acute, or united to form a Bo. 2–5 mm long, greyish-green, tinged reddish, spotted; **Fl.** unknown, on a pedicel 3–4 mm long.

**R. bijliae** L. Bol. (*Mes. b.* (L. Bol.) N. E. Br.). – Cape: Prince Albert D. – ♄; **St.** pink at first, becoming bluish to whitish, with green dots, 6–9 cm tall and erect or prostrate and up to 27 cm long; **L.** acutely trigonous, obtuse to truncate, with 1–3 marginal T., up to 11 mm long, 5 mm ⌀, Sh. 2 mm long; **Fl.** 18 mm ⌀, pink.

**R. bina** L. Bol. – Cape: Clanwilliam D. – Erect ♄ 9 cm high; **St.** brown, eventually white, with the remains of old L. at the nodes; **L.** dimorphic, some semi-ovoid, 5 mm long, 3 mm ⌀, others ± compressed, short-tapered or obtuse, 8 mm long, 2 mm ⌀, bluish-green, margins and keel somewhat translucent; **Fl.** 16 mm ⌀, pink.

**R. bipapillata** L. Bol. – Cape: Van Rhynsdorp D. – **St.** stiff, 3 mm thick, flowering Br. erect, 22 cm or more long, Int. 2–3 cm long; **L.** either slender, 15–20 mm long, with indistinct keel and margins, or 3 cm long, 2.5 cm across, 3 mm thick, Sh. 3–5 mm long, with impressed lines; **Infl.** pyramidal, 10 cm long, 7 cm ⌀, Fl. numerous, with pinkish-purple, indistinctly striate Pet. 5–7 mm long, filaments pink, papillose-ciliate below, barbate-papillose midway.

**R. bolusiae** Schwant. (*R. rigida* L. Bol., *Mes. albicuta* N. E. Br.). – Cape: Clanwilliam D. – Stiff, robust ♄ 12 cm high; **St.** spreading, 7 mm thick, **Br.** 15 mm long, with a white bark, with 4–6 crowded L., Int. 14 mm long; **L.** semicylindrical, narrowed towards the tip, 2–3 mm long, 4 mm across, 5 mm thick, Sh. 2 mm long, with impressed lines; **Fl.** 2 cm ⌀, pink.

**R. bracteata** L. Bol. – Cape: Lit. Namaqualand. – **St.** lax, prostrate, up to 25 cm long, 3 mm thick, Int. 2.5–3 cm long; **L.** bluish-green, spotted, dissimilar, some 9 mm long, short and thickened, others 14 mm long, 4–5 mm ⌀, keeled on the underside, short-tapered, Sh. 2 mm long, with impressed lines; **Fl.** 15 mm ⌀, Pet. white, with an indistinct pink stripe, on a pedicel with 2 large bracts.

**R. brakdamensis** (L. Bol.) L. Bol. (*Mes. b.* L. Bol.). – Cape: Brakdam. – Stiff ♄; **St.** and Br. prostrate, 23 cm long, 6 mm thick; **L.** longunited, semicylindrical, trigonous and apiculate above, 7 mm long, 4 mm across, 5 mm thick, green with darker dots; **Fl.** 13 mm ⌀, pinkishpurple.

**R. breekpoortensis** L. Bol. – Cape: Lit. Namaqualand. – Stiff ♄ 13 cm tall, herbaceous parts velvety-papillose, bluish, often red; **St.** stiff, Int. 10–15 mm long; **L.** 7–12 mm long, 3–4 mm ⌀, apiculate, Sh. 2 mm long; **Fl.** 15 to 20 mm ⌀, pink.

**R. brevibracteata** L. Bol. – Cape: Lit. Namaqualand. – ± erect ♄; **St.** prostrate or ± climbing, 40 cm long, 7 mm thick; **L.** semicylindrical, acute, 3–3.5 cm long, 4 mm ⌀, Sh. 3–5 mm long, with impressed lines; **Infl.** 4–5 cm long, Fl. 13 mm ⌀, Pet. purplish-pink, indistinctly striate.

**R. brevicarpa** L. Bol. – Cape: Clanwilliam D. – Forming mats 2 cm high; **St.** up to 19 cm long, Int. 5–10 mm long; **L.** 4–6 on a young shoot, circular to ± ovoid to long-ovoid, semicircular to circular in profile, obtuse, margins and keel horny, bluish-green, velvety with papillose hairs, margins ciliate, 4—12 mm long, with a Sh. 2—3 mm long; **Fl.** 15 mm $\varnothing$, pink.

**R. brevicollis** (N. E. Br.) Schwant. (*Mes. b.* N. E. Br.). — Cape: Ladismith D. — Mat-forming plant; **L.** ± free, 4—6 mm long, 1.5 to 2 mm $\varnothing$, acutely keeled on the underside above, pale bluish-green, with indistinct dots; **Fl.** 12 mm $\varnothing$, pink.

**R. brevicyma** L. Bol. — Cape: Worcester D. — Erect ♄ 45 cm tall; **St.** 4—6 mm thick, Int. short; **L.** narrowed towards the tip, with a serrulate keel on the underside, bluish-green with raised dots, 2 cm long, 2 mm across, 3 mm thick, Sh. 6 mm long, with impressed lines; **Pet.** 4—6 mm long, pink, with a darker stripe.

**R. brevifolia** L. Bol. — Cape: Lit. Namaqualand. — ♄; **St.** stiff, virgate, black to dark brown, c. 31 cm long, Int. 1—1.5 cm long, Br. 25 cm long, with short shoots; **L.** variable, either 6—7 mm long, 4 mm across and early deciduous, or oblong-ovate, obtuse, keeled on the underside, 8—10 mm long, with a Sh. 2 mm long; **Fl.** in 1—3-partite Infl., 2 cm $\varnothing$, purplish-pink.

**R. brevipes** L. Bol. v. **brevipes** (*Mes. firmum* N. E. Br.). — Cape: Prince Albert D. — Erect, robust, 30—45 cm tall; **St.** stiff, rather tangled, virgate, 20 cm long, 5 mm thick, Int. 5—10 mm long; **L.** ± navicular, obtuse, 8—11 mm long, 6 mm thick, with a Sh. 2 mm long; **Fl.** 18 mm $\varnothing$, flesh-coloured, on a pedicel 1—3 mm long.

**R.** — v. **gracilis** L. Bol. — **St.** more slender; **L.** small; **Fl.** on a pedicel 4 mm long.

**R. britteniae** L. Bol. (*Mes. iteratum* N. E. Br.). — Cape: Albany D. — Resembles **R. parvifolia**: **L.** with a dentate keel, without any impressed lines at the base, 10 mm long, 4 mm $\varnothing$.

**R. burtoniae** L. Bol. — Cape: Calvinia D. — Erect, stiff, 28 cm tall; old **St.** 8 mm thick, with a grey to silvery bark, Int. 1—2 cm long; **L.** semicylindrical, rounded on the underside, indistinctly keeled below the acute tip, bluish-green, 3.5—5 cm long, the Sh. swollen, with impressed lines, 2.5 mm long, 3—4 mm thick; **Fl.** 19 mm $\varnothing$, pink.

**R. calcarea** L. Bol. — Cape: Hopetown D. — **St.** spreading to prostrate, stiff, 5 cm long, 3 mm thick; **L.** narrowed towards the obliquely truncate, apiculate tip, the keel on the underside oblique, with several T., blue with a pinkish tinge, 10—13 mm long, 2.5—3 mm thick, 2 mm across, Sh. 2 mm long, with impressed lines; **Fl.** 1 cm $\varnothing$, purplish-pink.

**R. calcicola** (L. Bol.) L. Bol. (*Mes. c.* L. Bol.). — Cape: Riversdale D. — **St.** elongated, prostrate, Int. 15—20 mm long; **L.** acute to tapered, with a compressed keel, bluish-green, 2.5—3 cm long, 3 mm $\varnothing$; **Fl.** 18 mm $\varnothing$, pinkish-purple.

**R. callifera** L. Bol. — Cape: Prince Albert D. — ♄ 13 cm tall; **St.** stiff, prostrate, 4 mm thick; **L.** 10—15 mm long, 5 mm $\varnothing$, the keel on the underside with 2—3 T. near the apiculate tip, each margin with 1 T., blue with dirty green dots, velvety, Sh. 2 mm long, with impressed lines; **Fl.** 1 cm $\varnothing$, pink.

**R. campestris** (Burch.) Schwant. (*Mes. c.* Burch.). — Cape: Sutherland D. — Erect plant c. 20 cm tall, herbaceous parts dark green or reddish; **St.** virgate, up to 17 cm long, Int. up to 2.5 cm long, 2—3 mm $\varnothing$, flowering Br. short, with a solitary Fl.; **L.** narrowed towards the recurved tip, 2 cm long, Sh. 4 mm long; **Fl.** up to 2.5 cm $\varnothing$, cherry-red.

**R. canonotata** (L. Bol.) Schwant. (*Mes. c.* L. Bol.). — Orange Free State, Cape: Kalahari, W.Griqualand, Kimberley D. — Slender plant 17 cm tall; **St.** spreading, **Br.** ascending, 15 cm long, Int. 2 cm long; **L.** trigonous, acute, with (2—)3(—5) T. along the keel, blue with raised dots, Sh. trigonous, white; **Fl.** 15 mm $\varnothing$, pink.

**R. capornii** (L. Bol.) Schwant. (*Mes. c.* L. Bol., *Lampranthus c.* (L. Bol.) L. Bol.). — Cape: Lit. Namaqualand, Tulbagh, Clanwilliam D. — Erect ♄ up to 29 cm tall; **St.** slender, semicylindrical, 2 mm $\varnothing$; **L.** 1 cm long, 4 mm $\varnothing$, trigonous, obtuse; **Fl.** 8 mm $\varnothing$, pinkish-purple, with purple stigmas.

**R. carolii** (L. Bol.) Schwant. (*Mes. c.* L. Bol.). — Cape: Clanwilliam, Robertson, Montagu D. — **St.** elongated, prostrate, 5 mm thick, Int. 1—2 cm long; **L.** obtuse-trigonous, recurved at the tip, bluish-green with green dots, 10 cm long, 4—5 mm across, 5—6 mm thick, with a furrowed Sh. 1 cm long; **Fl.** 20—23 mm $\varnothing$, Pet. pink with a pinkish-purple stripe.

**R. caudata** L. Bol. — Cape: Malmesbury D. — Erect ♄ 25—40 cm tall; **St.** 2—5 mm thick, Int. 2—5 cm long; **L.** narrowed towards the tip, 15—45 mm long, 2.5 mm thick, dirty green, Sh. 2—3 mm long; **Fl.** 2 cm $\varnothing$, pinkish-purple.

**R. cedarbergensis** L. Bol. — Cape: Clanwilliam D. — Erect ♄; **St.** 25 cm long, 5 mm thick, Int. 1—3 cm long; **L.** narrowed towards the recurved tip and keeled on the underside, 15—25 mm long, 2—4 mm $\varnothing$, Sh. 4—6 mm long, with impressed lines; **Pet.** 6.5 mm long, pinkish-purple, indistinctly striped.

**R. ceresiana** Schwant. (*R. c.* L. Bol., *R. congesta* L. Bol.). — Cape: Ceres D. — Erect ♄; **St.** 14—25 cm long, 5 mm thick, Int. 2—2.5 cm long; **L.** shortly and obliquely tapered, bluish, compressed and bluntly keeled on the underside, the sides narrowed towards the tip, 36 mm long, the Sh. swollen, 5 mm long, with impressed lines; **Pet.** 8 mm long, purplish-pink with an indistinct stripe.

**R. cincta** (L. Bol.) L. Bol. (*Mes. c.* L. Bol.). — Cape: Riversdale D. — **St.** prostrate, 20 cm or more long, Int. 1—2 cm long; **L.** recurved and hooked at the tip, trigonous, margins serrulate, bluish-green, 2 cm long, 5 mm across, 6 mm, thick; **Fl.** 22 mm $\varnothing$, Pet. pink with a red border.

**R. clavata** L. Bol. — Cape: Van Rhynsdorp D. — Slender ♄ 20 cm tall; **St.** lax, prostrate, 1.5 mm thick, Int. 5—15 mm long; **L.** hemispherical, 3 mm long and thick, ± navicular; **Fl.** 11 mm $\varnothing$, pinkish-purple; receptacle clavate.

**? R. clavellata** (Haw.) Jacobs. (*Mes. c.* Haw.). — Austr. — **St.** and **Br.** slender, rooting, with spreading, very brittle shoots; **L.** thickened and clavate above, trigonous, obtuse, green with translucent dots; **Fl.** vivid purplish-red.

**R. cleista** L. Bol. (*Mes. c.* (L. Bol.) N. E. Br.). – Cape: Lit. Namaqualand. – Slender, stiff ♄ 10 cm tall; **St.** lax, branched, Int. 10 to 15 mm long; **L.** obliquely spherical to clavate, apiculate, dirty green, spotted, 1 cm long, 6 mm ⌀; **Fl.** never opening.

**R. compacta** L. Bol. – Cape: Lit. Namaqualand. – Compact ♄ 8 cm tall, herbaceous parts pale blue, velvety-papillose; **St.** slender, Int. 5–10 mm long; **L.** 1 cm long, obtuse to short-tapered, Sh. 4 mm long, 3 mm ⌀; **Fl.** 17 mm ⌀, pink.

**R. complanata** L. Bol. – Cape: Cradock D. – Erect, glabrous, 20 cm tall; **St.** 5 mm thick, Int. 5–10 mm long; upper **L.** 8 mm long, 2 mm thick, with a Sh. 4–6 mm long, with impressed lines, lower **L.** 5–6 mm long, 2.5 mm ⌀, obtusely keeled on the underside, with a hook-like tip, dirty green, spotted; **Fl.** 8–11 mm ⌀, white.

**R. compressa** L. Bol. – Cape: Lit. Namaqualand. – **St.** 7–8 cm long, Int. 4–20 mm long, herbaceous parts minutely velvety-papillose; **L.** linear, carinate on the underside, with a Sh. 3–4 mm long, 6–8 mm thick; **Fl.** 2.1 cm ⌀, pink.

**R. concava** L. Bol. (*Mes. ciliolatum* N. E. Br.). – Cape: Ceres D. – ♄ 12 cm tall; **St.** spreading to prostrate, 16 cm long, the bark red, eventually brown, Int. 1–2 cm long, herbaceous parts minutely velvety; lower **L.** 3 mm long, 2.5 mm thick, with a ciliate keel on the underside, Sh. 1.5 mm long, upper **L.** longer and narrower; **Fl.** 15 mm ⌀, purplish-pink.

**R. concinna** L. Bol. (*Mes. comptum* N. E. Br.). – Cape: Malmesbury D. – Erect ♄ 7 cm tall; **St.** and **Br.** 2 mm thick, Int. 4 mm long; **L.** of a pair unequal, green, bristle-tipped, with ± ciliate angles, lower **L.** semi-ovate, 2–5 mm long, upper ones 1.5 mm thick, 4 mm long, with a Sh. 1 mm long; **Fl.** 15 mm ⌀, pale pink, pinkish-purple outside.

**R. condensa** (N. E. Br.) Schwant. (*Mes. c.* N. E. Br.). – Cape: Montagu D. – Resembles **R. propinqua**; **L.** 6–8 mm long, not hirsute; **Fl.** 1 cm ⌀, pink.

**R. congesta** (Salm) L. Bol. (*Mes. c.* Salm). – Cape: Uitenhage, Albany D. – **St.** hard, woody, thin, erect, 2-angled, soon becoming prostrate; **L.** 2–3 cm long, as long as the Int., compressed-trigonous, narrowed towards the hooked, apiculate tip, greyish-green, stout, spotted, with minutely rough angles; **Fl.** 3 cm ⌀, pink.

**R. constricta** L. Bol. – Cape: Bredasdorp D. – Densely branched ♄; **L.** yellowish-green, narrowed towards the tip, the keel with 1 to 3 indistinct T., 27 mm long, 7 mm ⌀; Pet. 8–11 mm long, pale pink on the upperside below, paler above, coppery or salmon-pink on the underside.

**R. copiosa** L. Bol. – Cape: Clanwilliam D. – Densely branched, cushion-forming ♄ 30 to 35 cm tall; **St.** 5 mm thick, Int. 2–5 cm long; **L.** narrowed above, short-tapered at the tip, ± keeled on the underside, 3–4 cm long, 3–4 mm thick, with a Sh. 2 mm long; Pet. 6 mm long, pinkish-purple, indistinctly striate below.

**R. costata** L. Bol. – Cape: Montagu D. – **St.** prostrate, 2–5 mm thick, elongated, Int. 2.5 to 3.5 cm long; **L.** bluish-green, lower ones narrowed towards the tip, with an indistinct keel, 2.5–3.5 cm long, upper ones 6 cm long, Sh. 5–6 mm long, with impressed lines; **Fl.** 2 cm ⌀, purplish-pink.

**R. crassa** (L. Bol.) Schwant. (*Mes. c.* L Bol.). – Cape: Laingsburg, Prince Albert D. – Robust ♄ with the pale bluish-green herbaceous parts covered in soft, short hairs; **St.** ascending to prostrate, Int. enclosed; **L.** swollen, obtuse, apiculate, the keel on the underside with 1 indistinct T., 1–2 cm long, almost as wide; **Fl.** 22 mm ⌀, white.

**R. crassifolia** L. Bol. – Cape: Lit. Namaqualand. – Resembles **R. valida**; plant 7 cm tall, 15 cm ⌀; sterile **St.** and **Br.** densely crowded, Int. enclosed, herbaceous parts velvety with small papillose hairs, flowering Br. alate-compressed, with Int. 1–2 cm long; **L.** c. 3.7 cm long, 1.2 cm across, shortly tapered, apiculate, obtusely keeled, the margins translucent, red-lined; **Fl.** 2.5 cm ⌀, white.

**R. crassisepala** L. Bol. v. **crassisepala**. – Cape: Van Rhynsdorp D. – Erect, stiff, 20 cm tall; **St.** 4 mm thick, Int. 2–3 cm long; **L.** narrowed towards the tip, rounded on the underside, green, 15–30 mm long, 4 mm ⌀; **Fl.** 17 mm ⌀, purplish-pink, Sep. very thick.

**R.** – v. **major** L. Bol. – More robust; **Fl.** 33 mm ⌀, purplish-pink.

**R. crassuloides** L. Bol. (*Mes. crassulifolium* N. E. Br.). – Cape: Lit. Namaqualand. – Slender, mat-forming, 10 cm tall; older **St.** creeping, elongated, nodose, 4 mm thick, Int. 15 mm long; **L.** hemispherical to oblique-clavate, obtuse, minutely velvety, 5–8 mm long, 4–5 mm ⌀, Sh. 1.5 mm long; **Fl.** 8 mm ⌀, white.

**R. cupulata** (L. Bol.) Schwant. (*Mes. c.* L. Bol.). – Cape: Malmesbury D. – Erect, robust, 20 cm or more tall; **St.** ascending, 6 mm thick, Int. 15–30 mm long; **L.** acute, apiculate, with a horny, serrulate keel on the underside, 15–20 mm long, 3–4 mm ⌀, Sh. slightly swollen, 3–5 mm long, with an indistinct longitudinal line; **Fl.** 28 mm ⌀.

**R. curta** (Haw.) Schwant. (*Mes. c.* Haw., *Mes. vaginatum* v. c. (Haw.) Bgr.). – Resembles **R. vaginata**; **L.** usually longer than the Int.; **Fl.** solitary on short shoots.

**R. cyathiformis** L. Bol. – Cape: Lit. Namaqualand. – ♄ 40 cm tall; **St.** lax, stiff, 4 mm thick, Int. 15–30 mm long; **L.** indistinctly angled and keeled on the underside, expanded and rather obtuse above, swollen below, 2–2.5 cm long, 5 mm ⌀, yellowish-green to bluish, minutely spotted, Sh. 2.5 mm long, with impressed lines; **Fl.** 15 mm ⌀, pink.

**R. cymbifolia** (Haw.) L. Bol. (*Mes. c.* Haw., *Mes. deceptum* N. E. Br.). – Cape: Riversdale D. – **St.** filiform, directed upwards; **L.** navicular, obtuse, greyish-green, spotted. – Insufficiently known.

**R. dasyphylla** (Schltr.) Schwant. (*Mes. d.* Schltr.). – Cape: Clanwilliam D. – ♄ 20 cm tall; **St.** 2-angled; **L.** ovoid-trigonous, 3–5 mm long, the keel on the underside and the margins minutely ciliate-denticulate; **Fl.** 12 mm ⌀.

**R. decumbens** L. Bol. (*Mes. deflectum* N. E. Br.). – Cape: Ladismith D. – Robust ♄; **St.** prostrate; **Br.** with 4–6 crowded, unequal **L.** 10–14 mm or 18 mm long, 6 mm ⌀, bluntly keeled, rounded towards the tip, bluish-purple, rough, with green dots, Sh. 3–4 mm long; **Fl.** solitary, 3.5 cm ⌀, pinkish-purple.

**R. decurrens** L. Bol. (*Mes. d.* (L. Bol.) N. E. Br.). – Cape: Van Rhynsdorp, Clanwilliam D. – Erect ♄; **St.** stiff, 13 cm long, 3 mm ⌀, Int. 1–2 cm long; **L.** variable, some 8 mm long, 5 mm ⌀, hemispherical, others 18 mm long, 4 mm across, 8 mm thick, obliquely tapered, apiculate, minutely velvety; **Fl.** 16 mm ⌀.

**R. decurvans** L. Bol. – Cape: Van Rhynsdorp, Clanwilliam D. – Erect, robust, 45 cm tall; **St.** stiff, Int. 3–4 cm long; **L.** narrowed above, acute to obtuse, bluish-green, 46–50 mm long, 5 mm ⌀, Sh. swollen, 3–7 mm long, with impressed lines; **Infl.** of many Fl. 20–26 mm ⌀, Pet. pinkish-purple, indistinctly striate.

**R. deflecta** L. Bol. – Cape: Lit. Namaqualand. – Plant 13 cm high, 15 cm ⌀; sterile **St.** 8 cm long, with c. 14 L., Int. 5–15 mm long; **L.** variously shaped, lanceolate to linear, obtuse to acute, apiculate, green, with hirsute margins and keel, some L. c. 11 mm long, 2–3 mm across, 2 mm thick, Sh. 1–5 mm long, other L. broadly ovate, obtuse, 3 mm long, or broadly ovate, acute, 5 mm long, with a parchmenty Sh. 3 mm long; Pet. 5–7 mm long.

**R. dejagerae** L. Bol. (*Mes. d.* (L. Bol.) N. E. Br.). – Cape: Beaufort West D. – Compact ♄ 7 cm tall; **St.** stiff; **L.** semicylindrical, apiculate, keeled on the underside, bluish-green, minutely velvety, 11 mm long, 3 mm ⌀, Sh. swollen; **Fl.** 17 mm ⌀, pink.

**R. dekanahii** (N. E. Br.) Schwant. (*Mes. d.* N. E. Br.). – Cape: Fraserburg D. – Dwarf plant 2–2.5 cm high; **L.** 4–10 mm long, 1.5–3.5 mm ⌀, trigonous, acute, keeled on the underside, blue with dark green dots; **Fl.** 12 mm ⌀, pinkish-purple.

**R. deminuta** L. Bol. – S.W.Afr.: near Aus. – Compact, hemispherical, branched ♄; **L.** densely crowded, imbricate, semi-ovoid, apiculate, with a horny keel, pale bluish-green, ± prickly, 3–5 mm long, 3–4 mm ⌀, Sh. 1 mm long; **Fl.** c. 12 mm ⌀, pink.

**R. densiflora** L. Bol. – Cape: Clanwilliam D. – ♄ 45 cm tall, up to 1.20 m ⌀; **St.** c. 40 cm long, Int. often 4–5 cm long, 4.5 mm ⌀; **L.** acute, apiculate, rounded on the underside, 2.5 to 4.2 cm long, Sh. 4–5 mm long, with impressed lines; **Infl.** tripartite, Fl. c. 12 mm ⌀, an attractive mauve.

**R. depressa** L. Bol. – Cape: origin unknown (? Uitenhage D.). – **St.** stiff, prostrate to creeping, 13 mm long, 3 mm thick; **L.** narrowed towards the truncate, apiculate tip, the keel with 1–2 T., the margins with 1 T., bluish, 8 mm long, 3–4 mm ⌀, Sh. 4 mm long, with impressed lines; **Fl.** 15 mm ⌀, pink.

**R. derenbergiana** (Dtr.) C. Weber. (Pl. 194/1) (*Mes. d.* Dtr., *Ebracteola d.* (Dtr.) Dtr. et Schwant., *Bergeranthus d.* (Dtr.) Schwant.). – S.W.Afr.: Damaraland. – Forming cushions 20 cm ⌀; **rootstock** thick, 20 cm long; **St.** with 2–3 pairs of **L.** 3–4 cm long, obtuse-trigonous, obtuse, with the sides hatchet-shaped above and expanded to 10 mm across, light bluish-green, densely spotted, Sh. 2–3 mm long; **Fl.** 2–2.5 cm ⌀, light pink.

**R. dichotoma** L. Bol. – Cape: Lit. Namaqualand. – ♄ c. 16 cm tall; **St.** lax, older ones prostrate, Int. 2.5–3 mm long; **L.** cylindrical, obtuse, bluish with green dots, soft, 10–18 mm long, 3 mm thick, with a ± swollen Sh.; **Fl.** numerous, 8 mm ⌀, purplish-pink.

**R. dichroa** (Rolfe) L. Bol. v. **dichroa** (*Mes. d.* Rolfe, *Cylindrophyllum d.* (Rolfe) Schwant.). – Cape: Clanwilliam D. – **St.** robust, often 20 cm long, prostrate, with short shoots; **L.** 4–6 together, linear-oblong, trigonous, thick, bluntly keeled on the underside, with denticulate margins, blue with crowded dots, 4 to 6 cm long, 6–9 mm across; **Fl.** 3.5–4 cm ⌀, white to pink.

**R. —** v. **alba** L. Bol. – **St.** creeping; **Fl.** 5.3 cm ⌀, Pet. unequal, red, white below.

**R. dielsiana** (Bgr.) Jacobs. (*Mes. d.* Bgr.). – W.Austr.: Albany. – **St.** spreading, 3 mm thick, grey, Int. 3–5 cm long; **L.** equilaterally trigonous, tapered, with a hooked apiculus, the angles minutely horny and acute, with minute, rough dots, 3–4 cm long, 3 mm across; **Fl.** 15 mm ⌀, purple.

**R. dilatata** L. Bol. – Cape: Van Rhynsdorp D. – ♄ 30 cm tall; **St.** lax, flowering shoots virgate, 10–17 cm long, Int. 1–2 cm long; **L.** narrowed towards the obliquely truncate, apiculate tip, yellowish-green, with large, green dots, 1 cm long, 3 mm ⌀, Sh. 1 mm long; **Fl.** 2 cm ⌀, pink.

**R. distans** (L. Bol.) L. Bol. (*Mes. d.* L. Bol.). – Cape: Clanwilliam D. – Erect plant 10–12 cm tall; **St.** ascending, with narrowly alate angles, the nodes distant, Int. 15–30 mm long; **L.** variable, some 7–10 mm long, united to midway, cap-shaped, indistinctly keeled, acute, 6 mm thick, others with the free part trigonous, 4 to 8 mm long, 1–2 mm thick; **Fl.** 15 mm ⌀, purplish-pink.

**R. diversifolia** L. Bol. (*Mes. variifolia* N. E. Br.). – Cape: Worcester D. – Plant 14 cm tall; **St.** prostrate, stiff, 25 cm long, red, eventually grey, Int. 3–4 cm long; basal **L.** 7 cm long, 12 mm thick, acutely keeled on the underside, truncate above, Sh. 5 mm long, with impressed lines, L. on the Br. falcate, 34–50 mm long, 6 mm across and thick, with horny angles; Fl. 28 mm ⌀, Pet. pinkish-purple, striate.

**R. dolomitica** (Dtr.) Dtr. et Schwant. (*Mes. d.* Dtr., *Corpuscularia d.* (Dtr.) Schwant., *Mes. buchubergense* Dtr., *R. b.* Dtr.). – S.W. Afr.: Lüderitz D. – Forming cushions c. 20 cm ⌀, 10 cm high, with 1–10 stiff, erect **St.** up to 25 cm high in age, densely covered with broadly obovate **Bo.** borne on the terminal shoots, consisting of 2 stiffly erect **L.** up to 25 mm long, united almost to midway, 6–10 mm thick, broadly navicular, apiculate, the rounded keel and acute margins long-papillose, the surface otherwise minutely papillose; **Fl.** 10 mm ⌀, pink.

**R. drepanophylla** (SCHLTR. et BGR.) L. BOL. v. **drepanophylla** (*Mes. d.* SCHLTR. et BGR., *Lampranthus d.* (SCHLTR. et BGR.) N. E. BR., *Mes. mallesoniae* L. BOL., *R. m.* (L. BOL.) L. BOL.). – Cape: Clanwilliam D. – Plant up to 15 cm tall; **St.** and **Br.** ascending, 4-angled, with Int. 1.7 cm long; **L.** falcate, swollen-trigonous, apiculate, with the apiculus, margins and keel all deep purple, 1–1.5 cm long, 5 mm across, 6 mm ⌀; **Fl.** solitary, 2 cm ⌀, pinkish-purple.

**R. —** v. **sneeubergensis** L. BOL. – **Fl.** in tripartite Infl., Pet. pale pink.

**R. dualis** (N. E. BR.) L. BOL. (*Mes. d.* N. E. BR., *Argyroderma d.* (N. E. BR.) N. E. BR., *Antimima d.* (N. E. BR.) N. E. BR.). – Cape: Van Rhynsdorp D. – Forming clumps 3–5 cm tall; **L.** paired, united for 1/3 of their length, 2 cm long, 5 mm ⌀, flat on the upperside, the underside semicylindrical below, keeled above, whitish-grey, with cartilaginous margins; **Fl.** 15 mm ⌀, deep pink.

**R. dubitans** (L. BOL.) L. BOL. (*Mes. d.* L. BOL., *Lampranthus d.* (L. BOL.) L. BOL., *Mes. uncum* v. *gydouwense* L. BOL., *Lampr. u.* v. *g.* (L. BOL.) SCHWANT., *R. leightoniae* L. BOL., *R. marginata* L. BOL.). – Cape: Ceres, Clanwilliam D. – Stiff plant c. 16 cm tall; **St.** and **Br.** erect, Int. enclosed, 2–15 mm long; **L.** semicylindrical, with a hooked apiculus, indistinctly keeled on the underside, dirty green, spotted, 12–50 mm long, 3–4 mm ⌀; **Fl.** 12 mm ⌀, pink.

**R. duplessiae** L. BOL. – Cape: Oudtshoorn D. – Robust ♄; **St.** crowded; **L.** bluish-green, unequal, c. 6 cm long, 9 mm across, 8 mm thick, obliquely obtuse to acute, with 4–7 T. along the acute keel on the underside, the margins with up to 15 T.; Pet. 16–19 mm long, pale fleshcoloured, with a red stripe.

**R. duthiae** (L. BOL.) SCHWANT. (*Mes. d.* L. BOL.). – S.W.Cape. – **St.** creeping, Int. 2–3.5 cm long. flowering Br. erect, 7–10 cm long, with 6 crowded L. above, with short shoots; **L.** trigonous, acute, 2–8 cm long, 7 mm across, 8 mm thick, green; Pet. pinkish-purple, with a darker stripe.

**R. edentula** (HAW.) L. BOL. (*Mes. e* HAW., *Echinus e.* (HAW.) N. E. BR., *Braunsia e.* (HAW.) N. E. BR., *Mes. perfoliatum* MILL.). – Cape. – ♄; **L.** trigonous, thick, almost without T.; **Fl.** unrecorded. Insufficiently known.

**R. ebracteata** L. BOL. – Cape: Lit. Namaqualand. – Erect, stiff plant 8 cm tall; **St.** 18 cm long, Int. 5–15 mm long; **L.** 4 together, crowded on the young shoots, expanded above, rounded with an indistinct apiculus, rounded on the underside, bluish-green, ± velvety, 7–9 mm long, 3–4 mm across, 5 mm ⌀, Sh. 1 mm long; **Fl.** 1 cm ⌀, pale pink.

**R. elevata** L. BOL. – Cape: Lit. Namaqualand. – 3 cm tall; **St.** crowded, 4 cm long; flowering **Br.** with 2 **L.** 7 mm long, 4 mm ⌀, obtuse, apiculate, with small tubercles above, indistinctly keeled, Sh. 1 mm long; **Fl.** solitary, 12 mm ⌀, white; capsule elevated above.

**R. elineata** L. BOL. (*Mes. e.* (L. BOL.) N. E. BR.). – Cape: Lit. Namaqualand. – Slender ♄; **St.** lax, elongated to climbing, 40 cm or more long, Int. 25–40 mm long, herbaceous parts densely papillose; **L.** narrowing towards the shortly tapered, hooked tip, 15–20 mm long, 4 mm ⌀, Sh. swollen, 1 mm long; **Fl.** numerous, 12 mm ⌀, pink.

**R. emarcidens** L. BOL. – Cape: Sutherland D. – **St.** erect, stiff, 21–23 cm long, Int. 5–10 mm long, flowering **Br.** very soft, with short shoots from the L.-axils; **L.** 7–10 mm long, 2 mm ⌀, acute, long-tapered, with an acute, horny, crenate keel on the underside; Fl. 2 cm ⌀, purplish-pink.

**R. erecta** (L. BOL.) SCHWANT. (*Mes. e.* L. BOL.). – Cape: Lit. Namaqualand. – Erect ♄ 60 cm tall; **St.** stiff, Int. 2.5–3 cm long; **L.** 2 to 2.5 cm long, 3–4 mm thick, cylindrical, shorttapered, apiculate, with a slightly swollen Sh. 2 mm long; **Infl.** lax, Fl. 12 mm ⌀, Pet. purplishpink with a purple stripe.

**R. erosa** L. BOL. – Cape: Prince Albert D. – Stiff, compact, 11 cm tall, 18 cm ⌀; young **St.** 3–5 cm long, Int. 5–10 mm long; **L.** oblong, obtuse, indistinctly angled and keeled, bluishgreen, tuberculate at first, 8–14 mm long, 3 mm across, Sh. 2–3 mm long; Fl. 20–23 mm ⌀, pink.

**R. esterhuyseniae** L. BOL. – Cape: Uniondale D. – Erect plant 12–17 cm tall; **St.** virgate, Int. 10–15 mm long; **L.** stiff, narrowing towards the ± recurved tip, with an acute, decurrent keel on the underside, 7–11 mm long, 2 mm thick; Fl. solitary, 3.6 cm ⌀, pinkish-purple.

**R. evoluta** (N. E. BR.) L. BOL. (*Mes. e.* N. E. BR.). – Cape: Van Rhynsdorp D. – Forming a dense, small mat; **St.** numerous, with 2–3 pairs of small, hemispherical, bluish-green **L.** with papery, white Sh.; **Fl.** solitary, 15 to 17 mm ⌀, brilliant pink.

**R. excedens** L. BOL. – Cape: Van Rhynsdorp D. – Compact, freely branched, 12–24 cm tall; **St.** woody, Int. 5–15 mm long; **L.** oval, rounded on the underside, hemispherical to ovate in profile, 4–8 mm long, 3–4 mm ⌀, Sh. 1.5 mm long; **Fl.** 10–12 mm ⌀, pinkish-purple.

**R. exigua** L. BOL. – Cape: Montagu D. – Compact ♄; **St.** 6 cm or more long; **L.** 4 on a shoot, lower ones hemispherical, upper ones elongated, bluntly keeled, narrowing towards the tip, 10 mm long, 3 mm across, 3.5 mm thick, Sh. 2 mm long; **Fl.** solitary, 15–20 mm ⌀, pink.

**R. exsurgens** L. BOL. – Cape: Calvinia D. – ♄ 7 cm high; **St.** prostrate or creeping, elongated, Int. 5–13 mm long, 2–3 mm thick; **L.** dissimilar, lower ones lunate, 6 mm long, 4 mm ⌀, Sh. white, 1.5 mm long, upper L. shortly tapering, 8 mm long, 3 mm thick, Sh. 1 mm long; **Fl.** 15 mm ⌀, pinkish-purple.

**R. extensa** L. BOL. – Cape: Lit. Namaqualand. – Stiff, laxly branched ♄; **St.** erect to prostrate, 85 cm long, Int. 4–4.5 cm long; **L.** bluish-green, narrowing above, indistinctly angled and keeled, 3–3.5 cm long, 3 mm ⌀, Sh. 1.5 mm long, with impressed lines; **Fl.** 12 mm ⌀, purplish-pink.

**R. fenestrata** L. Bol. – Cape: Van Rhynsdorp D. – Forming mats 4 cm high; **L.** 6–8 on the sterile shoots, lower L. hemispherical, with a translucent, fenestrate tip, apiculate, the angles minutely papillose, 4–6 mm long, 2 mm thick, Sh. 2 mm long, upper L. longer, narrower, the keel and angles bristly, 7 mm long, Sh. 3 mm long; **Fl.** 15 mm ⌀, pink.

**R. fergusoniae** L. Bol. (*Mes. f.* (L. Bol.) N. E. Br.). – Cape: Riversdale D. – Mat-forming; **St.** 15 mm long, Int. enclosed; **L.** 4 together on young shoots, 2–4 mm long, 2 mm ⌀, acute to rounded, keeled on the underside, convex on the sides; **Fl.** solitary, 1 cm ⌀, purplish-pink, with a white centre.

**R. festiva** (N. E. Br.) Schwant. (*Mes. f.* N. E. Br., *Mes. tumidulum* Sond.). – Cape: Van Rhynsdorp, Malmesbury D. – 30 cm tall; **St.** branched above, young growth papillose; **L.** semicylindrical, 1.5 mm thick, acute, indistinctly keeled on the underside, bluish-green, St.-L. deciduous; **Fl.** 15 mm ⌀.

**R. filamentosa** (L.) L. Bol. (*Mes. f. L.*). – Cape: Cape D. – **St.** with angular, prostrate **Br.** bearing short shoots; **L.** 4–5 cm long, 14 mm across, trigonous, short-tapered, apiculate, the keel-angle curved, the angles cartilaginous and serrate; **Fl.** solitary, 5 cm ⌀, Pet. an attractive purplish-pink, the margins and underside paler.

**R. filipetala** L. Bol. (*Mes. f.* (L. Bol.) N. E. Br.). – Cape: Clanwilliam D. – Erect ♄; **St.** 30 cm long, Int. 2.5–3.5 cm long; **L.** 23 mm long, 3–4 mm ⌀, with a hooked tip, keeled on the underside, Sh. 2 mm long; **Fl.** numerous, 15 mm ⌀, Pet. filiform.

**R. firma** L. Bol. – Cape: Van Rhynsdorp D. – Erect, stiff ♄ 40 cm or more tall; **St.** erect, 30 cm long, Int. 3.4–5 cm long, with a grey bark; **L.** ± falcate, with the tip recurved and apiculate and thus S-shaped, keeled on the underside, yellowish-green, firm-textured, 7 cm long, 4 mm across, 5 mm thick, the Sh. swollen, with impressed lines; **Fl.** numerous, 18 mm ⌀, Pet. purplish-pink, indistinctly striped.

**R. floribunda** L. Bol. – Cape: Van Rhynsdorp D. – Erect, stiff ♄; **St.** 30 cm long, Int. 3–5 cm long; old **L.** 3.2–3.5 cm long, others 6–7.5 cm long, indistinctly keeled on the underside, Sh. swollen, with impressed lines; **Infl.** of many pink Fl., Pet. 9–10 mm long, with one faint stripe.

**R. forficata** (L.) L. Bol. (*Mes. f. L.*, *Erepsia f.* (L.) Schwant., *Mes. purpureostylum* L. Bol., *Erepsia p.* (L. Bol.) Schwant., *Mes. filamentosa* v. *anceps* DC.). – Cape: Swellendam D. – **St.** elongated, creeping, compressed-alate, 5–7 mm thick, thickened at the nodes, the bark purple, eventually turning brown; **Br.** erect; **L.** trigonous, laterally compressed, truncate, the margins entire or indistinctly dentate, olive-green, some 6.5 cm long, others 3–4 cm long, 9 mm across, 16 mm thick; **Fl.** 3 cm ⌀, Pet. pinkish-pink, striped purple.

**R. fourcadei** L. Bol. (*Mes. f.* (L. Bol.) N. E. Br., *R. f.* f. *luxurians* L. Bol. n. prov., *R. heteropetala* L. Bol., *R. globularis* L. Bol., *Eberlanzia g.* (L. Bol.) L. Bol.). – Cape: Uniondale, Ladismith, Prince Albert D. – Erect ♄ 12 cm high; **St.** stiff; **L.** green with large dots, rounded and apiculate above, acutely keeled on the underside, 12–15 mm long, 4–5 mm thick, 4 mm across; **Fl.** 26 mm ⌀, pink.

**R. framesii** L. Bol. (*Mes. f.* (L. Bol.) N.E. Br.). – Cape: Jansenville D. – Erect, compact plant flattened above, 7 cm high; **St.** slender, with 4–10 L. crowded at the tips and encl sed Int.; **L.** 3–4 mm long, 3 mm across, 2 mm thick hemispherical, obtuse, with a translucent line along keel and margins, pale green to ± whitish, Sh. 2 mm long; **Fl.** solitary, 18 mm ⌀, pink.

**R. fredericii** (L.Bol.) L. Bol. (*Mes. f. L.Bol.*). – S.W.Cape. – **St.** spreading; **Br.** ascending to erect, Int. 3–3.5 cm long; **L.** 2 cm long, 3 mm ⌀, semicylindrical, short-tapered, bluish; **Fl.** 1 cm ⌀, purple.

**R. frutescens** (L. Bol.) L. Bol. (*Mes. f.* L. Bol.). – Cape: Van Rhynsdorp D. – Erect ♄ up to 2.5 m high; **St.** stiff, 6 mm ⌀; **L.** spreading to incurved, semicylindrical or swollen-trigonous, up to 5.5 cm long, 1.2 cm ⌀; **Infl.** 4–5-partite, Pet. 7–8 mm long, whitish.

**R. fugitans** L. Bol. – Cape: Lit. Namaqualand. – **St.** elongated, prostrate, 7 mm thick, Int. 15–20 mm long, flowering **shoots** erect, 15 cm tall; **L.** semicylindrical, rounded or acute, bluish-green, 4–6.5 cm long, 3–5 mm thick, Sh. 3 mm long, with impressed lines; Pet. 7 mm long, pinkish-purple, with a narrow stripe.

**R. fulleri** L. Bol. – Cape: Kenhardt D. – Compact ♄; **St.** 13 mm thick above; **Br.** 3 cm long; **L.** 4–6 on the shoots, those of a pair unequal, 18–25 mm long, 3–4 mm across, 6 mm thick, pale blue, ± keeled on the underside, laterally compressed, acute, bristle-tipped; **Fl.** 4 cm ⌀, pink.

**R. gemina** L. Bol. (*Mes. g.* (L. Bol.) N. E. Br.). – Cape: Laingsburg D. – Plant 6 cm tall; **St.** crowded, 5 mm thick; **Br.** with 2 L., with short shoots from the axils; **L.** 23–35 mm long, 5–6 mm across and thick, pale bluish-green, tapered, apiculate, with red margins and keel; **Fl.** solitary, 3 cm ⌀, pink.

**R. geminiflora** (Haw.) Schwant. (*Mes. g.* Haw., *Mes. geminatum* Jacq.). – Cape: Cape D. – **St.** slender, 60–90 cm long, prostrate, often rooting, Int. 2–3 cm long; **L.** 5 cm long, 4 mm across, obliquely united at the base, equilaterally trigonous, tapered, with the margins and keel denticulate, greyish-green with translucent dots; Pet. 6 mm long, pale pink with a dark M.-line.

**R. gibbosa** L. Bol. – Cape: Lit. Namaqualand. – Forming mats 3 cm high; **St.** erect to ascending, Int. 2 mm long, bark whitish, flowering **shoots** 1.5 cm long, with 2–4 bluish-green L. 5–10 mm long, 3 mm thick, narrowing towards the shortly tapered tip, indistinctly keeled, Sh. 1.5 mm long; **Fl.** solitary, 2 cm ⌀, pink.

**R. glauca** L. Bol. (*Mes. valens* N. E. Br.). – Cape: Lit. Namaqualand. – Robust, erect ♄ 18 cm tall; **St.** rigid, Int. 12–30 mm long; **L.** 2 cm long, 5 mm across, 8 mm thick, bluish, velvety, expanded towards the obtuse or shortly tapered tip, recurved-apiculate, keeled on the underside, 2 cm long, 5 mm across, 8 mm thick, Sh. 2 mm long; **Fl.** 18 mm $\varnothing$, pink.

**R. goodiae** L. BOL. (*Mes. g.* (L. BOL.) N. E. BR.). Cape: Lit. Namaqualand. – Robust ♄ 16 cm tall; **St.** 12 mm thick; **Br.** prostrate, tangled, 18 cm long, Int. 2–9 mm or rarely 3–4 cm long, flowering **shoots** erect, 6 cm long, with 2–4 bluish-green, tapered, carinate **L.** 2 to 2.5 cm long, 3 mm thick, Sh. 2–3 mm long; **Fl.** 3 cm $\varnothing$, an attractive pink.

**R. gracilipes** L. BOL. – Cape: Lit. Namaqualand. – **St.** stiff, up to 38 cm long, Int. (1–)1.5(–3.5) cm long; **L.** 16–21 mm long, 3–4 mm $\varnothing$, green, narrowed towards the obtuse tip, rounded to keeled on the underside; **Infl.** with 1–3 Br., Fl. c. 1.5 cm $\varnothing$, pinkish-purple.

**R. gracilis** L. BOL. (*Mes. exile* N. E. BR.). – Cape: Riversdale D. – ♄ up to 45 cm tall; **St.** prostrate, up to 60 cm long, Int. 1–3 cm long, **shoots** erect, 5–8 cm long; **L.** ± slender, hamate above, bluntly keeled, green, 40 mm or 15 to 25 mm long, 2–3 mm thick, Sh. 2–3 mm long; **Fl.** 1 cm $\varnothing$, Pet. pink to pale pink, with an indistinct stripe.

**R. gracillima** L. BOL. – Cape: Calvinia D. – Slender ♄; **St.** elongated, prostrate, 20 cm long, 1.5 mm thick, Int. 1–2 cm long, flowering shoots 2.5–5 cm long; **L.** slender, with a recurved bristle-tip, with hirsute margins and keel, green, 13 mm long, 2.5 mm thick, Sh. 2 mm long; **Fl.** 1 cm $\varnothing$, purplish-pink.

**R. granitica** (L. BOL.) L. BOL. (*Mes. g.* L. BOL.). – Cape: Lit. Namaqualand. – **St.** slender, prostrate to creeping, 11 cm long, flowering **shoots** erect, with 2–4 oblong, obtuse, apiculate, bluish-green **L.** 5–11 mm long, 4 to 5 mm across, keeled on the underside, with red angles; **Fl.** solitary, 12 mm $\varnothing$, pink.

**R. gravida** L. BOL. – Cape: Lit. Namaqualand. – Erect, stiff ♄ 35 cm tall; **St.** ascending, bluish-green, ± tinged pink, herbaceous parts velvety with small papillae; **L.** 23–25 mm long, 10 mm across, 12 mm thick below the tip, expanded above, obtuse to truncate, keeled on the underside, Sh. 5 mm long; **Fl.** 28 mm $\varnothing$, brilliant or pale pink.

**R. griquensis** (L. BOL.) SCHWANT. (*Mes. g.* L. BOL.). – Cape: Kimberley D. – ♄ 20 cm high; **St.** lax, 12–18 cm long, Int. 3 cm long; **L.** 3 cm long, 2–3 mm $\varnothing$, trigonous, acute, apiculate, the keel with (1–)2(–3) distinct T., Sh. 1 cm long; **Fl.** 22 mm $\varnothing$, pink.

**R. grisea** (L. BOL.) SCHWANT. (*Mes. g.* L. BOL.). – Cape: ? Montagu D. – ♄ 30 cm high; **St.** lax, stiff, 30 cm long, Int. 10–15 mm long; **Br.** erect, 15–40 mm long; **L.** 2–4 mm long, with 2 T. at the tip, Sh. 2 mm long, with an impressed line; **Fl.** 14 mm $\varnothing$, pale pink.

**R. hallii** L. BOL. – Cape: Ceres Karroo. – Slender plant 2.5–8 cm high; **St.** elongated, prostrate to creeping and rooting, 14 cm long, Int. 2.5 cm long, sterile shoots 1.5 cm long, with the L. densely imbricate; **L.** dimorphic, lower pair hemispherical, 3–4 mm long, 3 mm across, upper pair oblong, 5 mm long, 3 mm across, margins and keel hirsute; **Fl.** solitary, 15 to 18 mm $\varnothing$, purplish-pink.

**R. hamata** (L. BOL.) SCHWANT. (*Mes. h.* L. BOL.). – Cape: Kalahari reg.; Orange Free State; Fauresmith D. – Robust ♄ 30 cm high; **St.** 25 mm thick at the base; **Br.** spreading, stiff, 12 mm thick; **L.** 4–7 cm long, 2 mm thick, with a recurved apiculus, faintly keeled, with prominent dots; **Fl.** solitary, 12 mm $\varnothing$.

**R. hamatilis** L. BOL. (*Mes. h.* (L. BOL.) N. E. BR.). – Cape: Worcester D. – **St.** prostrate, 8–12 cm long, Int. 1–3 cm long; **L.** velvety, narrowed and curved towards the 1 mm long bristly tip, dimorphic with one pair short, broad, faintly keeled, 10–15 mm long, 6 mm across, 4 mm thick, Sh. 4 mm long, the second pair 2 cm long, 2 mm $\varnothing$, acutely keeled on the underside; **Fl.** 14 mm $\varnothing$, Pet. pinkish-purple, striped on the upperside.

**R. haworthii** JACOBS. et ROWL. (*Mes. dilatatum* HAW.). – Cape. – ♄ 60–90 cm tall; **Br.** erect to spreading, somewhat angled; **L.** united at the base, with short shoots from the axils, erect to spreading, trigonous, broadened to the tip, shortly and bluntly pointed, arched and recurved, 3–4 cm long, 6 mm wide, soft fleshy, green, grey puberulous and with numerous translucent dots, margins somewhat crenate; **Fl.** terminal or lateral on pedicels 2.5 cm long, c. 2 cm $\varnothing$, pale pink.

**R. herrei** SCHWANT. (*Mes. h.* (SCHWANT.) N. E. BR.). – Cape: Ceres D., Bushmanland. – Forming dense, hemispherical mats; **St.** short, with 1–2 pairs of **L.** 22 mm long, 4 mm $\varnothing$, sheathing and united for up to 5 mm, rounded to keeled on the underside, margins and keel translucent, bluish-green, with large translucent dots.

**R. hexamera** L. BOL. v. **hexamera** (*Mes. h.* (L. BOL.) N. E. BR.). – Cape: Lit. Namaqualand. – Erect, robust ♄; **St.** 8 mm $\varnothing$ at the base; sterile **Br.** short, with 4–6 L., Int. 25 mm long; **L.** ± navicular, acutely keeled on the underside, minutely velvety-haired, 4 cm long, 24 mm across, 14 mm thick, Sh. 3 mm long; **Fl.** solitary, 16 mm $\varnothing$, white to pale pink.

**R. —** v. **longipetala** L. BOL. – Cape: Lit. Namaqualand. – Pet. longer, pink.

**R. holensis** L. BOL. – Cape: Van Rhynsdorp D. – Erect, stiff, 18–21 cm tall, c. 25 cm $\varnothing$; **St.** 8–10 mm $\varnothing$; **Br.** c. 5 mm $\varnothing$, Int. 1–2.5 cm long, sterile Br. with 2–4 **L.** 1–1.8 cm long, 4–5 mm across, 5–6 mm thick, with the keel reduced to a line; **Fl.** 1–3 together, Pet. numerous, 1.5 cm long, purplish-pink.

**R. hutchinsonii** L. BOL. (*Mes. h.* (L. BOL.) N. E. BR.). – Cape: Malmesbury D. – Plant 10 cm tall; **St.** lax, ascending to prostrate, 20 cm long, with a white bark, Int. 10–15 mm long, flowering **shoots** erect, tangled, 5–9 cm long; **L.** 15 mm long, 6 mm thick, obtuse, with a rounded keel, pale green to blue, tinged pink; **Fl.** solitary, 19 mm $\varnothing$, pink.

**R. imbricata** (HAW.) SCHWANT. (*Mes. i.* HAW.). – Cape: origin unknown. – Resembles **R. multiflora** and possibly only a variety; **L.** greener; **Fl.** smaller, less numerous, pedicel covered by 4–8 bracts up to the Cal.

**R. impressa** L. Bol. (*Mes. i.* (L. Bol.) J. Ingram). – Cape: Lit. Karroo. – Compact plant 6 cm high; **St.** spreading; **L.** 15 mm long, 5 mm thick, pale bluish-green, obtuse, bluntly keeled, with somewhat translucent margins, Sh. 4 mm long, with an impressed longitudinal line; **Fl.** 1 cm ⌀, pink.

**R. inclaudens** L. Bol. – Cape: Robertson D. – Compact, dense ♄; **St.** stiff, up to 21 cm long, 4 mm ⌀, Int. 1–2 cm long; **L.** linear to oblong, acute, sharply tapered, apiculate, rounded on the underside, 1.3–2 cm long, 5–7 mm ⌀, Sh. 4 mm long, decurrent for 7 mm, with an impressed line; **Fl.** solitary, 2.5–3 cm ⌀, pinkish-purple.

**R. inclusa** L. Bol. (*Mes. amplexum* (L. Bol.) N. E. Br.). – Cape: Laingsburg D. – Compact plant 6–8 cm high; **St.** ascending to prostrate, nodes 2–4 mm thick, Int. 5–7 mm thick; **L.** 4–6 on the flowering shoots, with the upper **L.** enclosing the receptacle, 4 mm long, 5 mm across, 3 mm thick, with impressed lines 3 mm long at the base, the acute keel on the underside with 2 T., the margins with 1–2 T.; **Fl.** 15 mm ⌀, purplish-pink.

**R. inconspicua** L. Bol. (*Mes. latens* N. E. Br.). – Cape: Lit. Namaqualand. – Erect, 15 cm tall, 18 cm ⌀; **St.** spreading; **Br.** erect; **L.** 7–10 mm long, 4–6 mm across, 5 mm thick, bluish, with pale green and white dots, rounded or obtuse, keeled on the underside, Sh. 5–7 mm long; **Fl.** 12 mm ⌀, white.

**R. incumbens** L. Bol. – Cape: Clanwilliam D. – Erect, robust, 45 cm tall; **St.** stiff, Int. 3 to 5 cm long; **L.** 2.5–4 cm long, 4 mm ⌀, bluish-green, narrowing to the shortly tapered tip, faintly keeled on the underside, Sh. 4 mm long; **Infl.** 7 cm long, Fl. 15 mm ⌀, pink.

**R. incurvata** L. Bol. (*Mes. i.* (L. Bol.) N. E. Br.). – Cape: Clanwilliam D. – Erect, robust, 50 cm tall; **St.** 25 mm thick; **Br.** stiff, 7 mm ⌀, Int. 4 cm long; **L.** 4 cm long, 5 mm ⌀, shortly tapered, apiculate, faintly keeled on the underside, Sh. 6 mm long; **Fl.** 2 cm ⌀, Pet. pale pink, striped.

**R. indecora** (L. Bol.) Schwant. (*Mes. i.* L. Bol. – Cape: Cape D. – Erect, robust, 30 cm tall; **St.** spreading to ascending, stiff, **Br.** short, Int. 7–25 mm long; **L.** 21 mm long, 3 mm ⌀, narrowed towards the hamate, recurved tip, keeled on the underside, bluish, the swollen Sh. with an impressed line; **Fl.** 8–10 mm ⌀, white.

**R. indurata** (L. Bol.) Schwant. (*Mes. i.* L. Bol.). – Cape: Karroo. – **St.** creeping, stiff, 5 mm thick, young shoots with 4 L., Int. short, enclosed; **L.** united for 3 mm, the free parts trigonous, 8–10 mm long, 3 mm across, apiculate, bluish-green, with darker dots; **Fl.** solitary, 2 cm ⌀, pink.

**R. insidens** L. Bol. (*Mes. i.* (L. Bol.) N. E. Br.). – Cape: Calvinia D. – Erect plant 9 to 10 cm tall, ± dichotomously branched; **St.** stiff, flowering shoots 15–30 mm long, Int. 5 mm long, nodes 8–10; **L.** 5–6 mm long, 1 mm thick, acute, apiculate, minutely velvety, ciliate, blue, tinged pink; **Fl.** 18 mm ⌀, pink.

**R. intermedia** L. Bol. – Cape: Piquetberg D. – Erect ♄ 22–24 cm tall; **St.** virgate, stiff, Int. 15–35 mm long; **L.** 2–3 cm long, 3 mm across, 3–4 mm thick, green, ± reddened, apiculate, faintly keeled on the underside, the angles often serrulate, Sh. 4–6 mm long; **Infl.** compact, Pet. 10–12 mm long, pinkish-purple striped.

**R. intervallaris** L. Bol. – Cape: Van Rhynsdorp D. – Stiff, robust ♄; **St.** spreading to prostrate, 28 cm long, Int. 3–4 cm long, flowering **shoots** erect, with 4 L., Int. 5–30 mm long; **L.** bluish-green, papillose, dimorphic, lower L. lanceolate, acute, 3.5 cm long, 11 mm across, 9 mm thick, with a Sh. 3–4 mm long, upper L. linear, acutely keeled, apiculate, 15 mm long, 3 mm ⌀; **Fl.** solitary, 24 mm ⌀, pinkish-purple.

**R. intrusa** (Kensit) L. Bol. (*Mes. i.* Kensit). – Cape: Swellendam, Robertson D. – ♄ 6–7 cm tall; **St.** robust; **Br.** with 2–4 L. 55 mm long, 9 mm thick, swollen-trigonous, obtuse, entire, green with white transverse lines; **Fl.** solitary, 4 cm ⌀, pinkish-purple.

**R. ivorii** (N. E. Br.) Schwant. (*Mes. i.* N. E. Br.). – Cape: Fraserburg D. – Mat-forming ♄ 2–2.5 cm high, with the older L. withering and persisting; **L.** dimorphic, one pair 4 mm long, hemispherical, the other pair united into a spherical Bo. and enclosing a further pair, 3 mm thick, with green dots; **Fl.** solitary, 10 to 12 mm ⌀, an attractive purple.

**R. kakamasensis** L. Bol. – Cape: Bushmanland. – Erect ♄ 30 cm tall, the herbaceous parts bluish, rough with stiff, crowded papillae; **St.** stiff, Int. 2–3.5 cm long, 5–7 mm thick; **L.** 2 cm long, 4–6 mm thick, tapered, the tip oblique, with one T., apiculate, with indistinct keel and margins, Sh. short, with impressed lines 1 to 2 mm long; **Fl.** 26 mm ⌀, pink.

**R. karrachabensis** L. Bol. – Cape: Lit. Namaqualand. – **St.** 9 cm long, 3 mm thick, Int. 12–20 mm long; **L.** 2–4 on the flowering shoots, 10–15 mm long, 3.5 mm across, 5 mm thick, keeled on the underside, apiculate, pale green, Sh. 3 mm long, with short shoots from the L.-axils bearing semi-ovate, shortly tapered L.; **Fl.** 1–3 together, 16 mm ⌀, pink.

**R. karroidea** L. Bol. (*Mes. mutabile* Sond.). – Cape: Laingsburg, Beaufort West D. – Laxly branched, with the herbaceous parts velvety-papillose, pale blue; **St.** spreading to prostrate, rooting, elongated, Int. 15–20 mm long; **shoots** 4-angled, with 4 **L.** in dissimilar pairs, the lower ovate-oblong, rounded to short-tapered and apiculate, keeled on the underside, the keel and the angles translucent, 11 mm long, 5 mm across, 3 mm thick, Sh. 1.5 mm long, the upper pair smaller, compressed, 7 mm long, 2 mm thick, the keel and angles with pectinate, horny hairs; **Fl.** 15 mm ⌀, pink.

**R. karrooica** (L. Bol.) L. Bol. (*Mes. k.* L. Bol.). – Cape: Karroo. – Robust, erect ♄ 30 cm tall; **St.** and **Br.** erect; **L.** 8–21 mm long, 3–5 mm ⌀, tapered, apiculate, with the clavate Sh. 8–11 mm thick, enclosing the Int.; **Fl.** solitary, 5 cm ⌀, Pet. purplish-pink with a purple stripe.

**R. kenhardtensis** L. BOL. – Cape: Kenhardt D. – Erect ♄; **St.** 26 cm long, 6 mm thick, Int. 10–15 mm long; **Br.** 2–3 cm long, with 4–6 **L.** 5 mm long, 3 mm thick, obliquely truncate above, with the keel on the underside denticulate at the tip, Sh. 3 mm long, 2.5 mm ⌀, with impressed lines 1.5 mm long; **Fl.** 10 mm ⌀, pale pink.

**R. klaverensis** (L. BOL.) SCHWANT. (*Mes. k.* L. BOL.). – Cape: Van Rhynsdorp D. – Dwarf herb with a fishy smell; **St.** thick; **Br.** short and crowded or elongated and prostrate, Int. 2 cm long, flowering **shoots** 2–4 cm long; **L.** 10 to 15 mm long, 4 mm across, 6 mm thick, obtuse, keeled on the underside above, bluish-green; **Fl.** 16 mm ⌀, purplish-pink.

**R. klipbergensis** L. BOL. – Cape: Darling D. – Erect, laxly branched ♄ 33 cm tall; **St.** often incurved above, 30 cm long, Int. 1–5 cm long, 1.5–4 mm ⌀; **L.** linear, acute, keeled, ± S-shaped, hamate-recurved above, 2.5–3.5 cm long, the Sh. with impressed lines 1–2 mm long; **Fl.** in 3-partite Infl. 1.5 cm ⌀, pink.

**R. knysnana** (L. BOL.) L. BOL. v. **knysnana** (*Mes. k.* L. BOL.). – Cape: Knysna D. – Erect ♄ 30–40 cm tall; **St.** virgate, 17–34 cm long, Int. 1–6 cm long; **Br.** 3–4 cm long; **L.** 3.3 cm long, 5 mm across, 6 mm thick, green, acute to obtuse, apiculate, acutely keeled on the underside; **Fl.** 2.5 cm ⌀, pale pink to purplish-pink.

**R. —** v. **angustifolia** L. BOL. – **L.** narrower, 4 mm ⌀, Sh. shorter.

**R. koekenaapensis** L. BOL. – Cape: Van Rhynsdorp D. – Resembles **R. obtusifolia**; **L.** rarely ± recurved, c. 5 mm ⌀, with a thin line along the angles and keel; Pet. pink.

**R. komkansica** L. BOL. – Cape: Lit. Namaqualand. – Compact ♄ c. 60 cm tall; **St.** stiff, 17 to 23 cm long, Int. 1–2 cm long, herbaceous parts with papillose, velvety hairs; **L.** 2 on flowering shoots, linear, obtuse, apiculate, indistinctly keeled, 2.5 cm long, 4 mm across, 5 mm thick, Sh. 1.5 mm long; **Fl.** in 1–3-partite Infl., Pet. 8 mm long, purplish-pink.

**R. kuboosana** L. BOL. – Cape: Lit. Namaqualand. – **St.** limp, 7–14 cm long, Int. 1.5–3.5 cm long; **L.** linear, rounded above, indistinctly keeled, green, rough with dots, 3–4.5 cm long, 5–7 mm ⌀, Sh. c. 4 mm long; **Fl.** 2–2.5 cm ⌀, purplish-pink.

**R. langebaanensis** L. BOL. – Cape: Malmesbury D. – Robust ♄; **St.** prostrate, stiff, Int. 2–3 cm long; **L.** 3.5 cm long, 4 mm across, 5 mm thick, green to light blue, keeled on the underside, tipped with a curved mucro, Sh. 7 mm long, with an impressed line; Pet. 5 to 10 mm long, pinkish-purple, faintly striped.

**R. lapidicola** L. BOL. – Cape: Clanwilliam D. – Freely branched ♄ 30 cm tall; **St.** 1–4 mm ⌀, Int. 4–25 mm long; **L.** ± falcate, acute, with a blunt keel, green, 9–15 mm long, 3–6 mm across, 3–7 mm thick; **Fl.** 15 mm ⌀, purplish-pink.

**R. lavisii** L. BOL. – Cape: Stellenbosch D. – Erect ♄ 40 cm tall; **St.** inclined; **L.** 7 cm long, 5 mm across, 6 mm thick, pale bluish-green, keeled on the underside, laterally compressed, rounded above; **Fl.** 22 mm ⌀.

**R. lawsonii** (L. BOL.) L. BOL. (*Mes. l.* L. BOL.). – Cape: Hay D. – Dense, robust, matforming ♄; **St.** 3 mm thick, flowering **shoots** slender, with 4–6 **L.** 4–6 mm long, 3 mm ⌀, acute, apiculate, keeled on the underside, Sh. 1 mm long; **Fl.** 1 cm ⌀, pink.

**R. laxa** (WILLD.) SCHWANT. (*Mes. l.* WILLD.). – Origin unknown. – **St.** very tangled, very slender, creeping; **L.** shorter than the Int., compressed-trigonous, the keel and angles denticulate, fresh green with minute, tuberculate dots; **Fl.** reddish. Insufficiently known.

**R. laxiflora** L. BOL. – Cape: Lit. Namaqualand. – Erect, laxly branched ♄ 50 cm tall and ⌀; **St.** ascending, Int. 2–5 cm long; **L.** semicylindrical, obtuse, green to bluish with green dots, 2–3 cm long, 4 mm thick, the Sh. 2 mm long, swollen, with impressed lines; **Infl.** lax, Fl. 2 cm ⌀, pink.

**R. laxipetala** L. BOL. (*Mes. l.* (L. BOL.) N. E. BR.). – Cape: Laingsburg D., Ceres Karroo. – ♄ 7 cm high; **St.** prostrate, nodes green, 3 mm thick, Int. 5 mm long; **L.** 3–4 mm long, 2 mm thick, bluish-green, oblique and with 2 T. at the tip, ± acutely keeled on the underside; **Fl.** 18 mm ⌀, Pet. lax, pinkish-purple.

**R. leipoldtii** L. BOL. (*Mes. l.* (L. BOL.) N. E. BR.). – Cape: Worcester D. – Resembles **R. peersii** but more slender, laxly branched, Int. 1 cm long; **L.** apiculate, broader, Sh. 2–3 mm long; **Fl.** 16 mm ⌀.

**R. leptocalyx** L. BOL. – Cape: Riversdale D. – ♄ 8 cm high; **St.** prostrate, narrowly alate, 14 cm long, Int. 3–9 mm long; **L.** 18 mm long, 5 mm across, 7 mm thick, laterally compressed, acutely keeled, apiculate; Pet. 8 mm long, pinkish-red, faintly striped.

**R. leptophylla** L. BOL. – Cape: Caledon D. – Dense ♄ with older **St.** tuberously thickened, 12 mm ⌀; **Br.** with 2 L. and short shoots from the L.-axils; **L.** green to bluish, 5 cm long, 3 mm ⌀, the keel with 1–2 indistinct, bristle-tipped T. on the underside, acute, apiculate, Sh. 8 mm long; Pet. 9 mm long, pale salmon-pink with a narrow stripe, red on the underside.

**R. lerouxii** (L. BOL.) L. BOL. (*Mes. l.* L. BOL., *Lampranthus l.* (L. BOL.) N. E. BR.). – Cape: Lit. Namaqualand. – **St.** prostrate, 3 mm thick, Int. 15–20 mm long; **Br.** narrowly alate, with 2–4 trigonous to swollen, acute, apiculate, bluish-green **L.** 6 cm long, 4–8 mm across, 6–8 mm thick, keeled on the underside, Sh. 3–5 mm long; Fl. 5.6 cm ⌀, Pet. purplish-pink with a faint stripe.

**R. leucanthera** (L. BOL.) L. BOL. (*Mes. l.* L. BOL.). – Cape: Prince Albert D., Karroo. – Low ♄ with short **St.** and short shoots from the L.-axils; **L.** swollen, trigonous, ± acute, blue, spotted, velvety-puberulous, 10 mm long, 4 mm ⌀, Sh. 3 mm long; **Fl.** solitary, 14–20 mm ⌀, pinkish-purple.

**R. leucosperma** L. BOL. – Cape: Lit. Namaqualand. – Erect, slender ♄ 10–20 cm tall; **St.** and **Br.** stiff, at first softly hirsute, Int. 10 mm long; **L.** 4–7 mm long, 4 mm thick, hemispherical, obtuse, bluish-green, velvety, margins and keel indistinctly translucent, Sh. 2 mm long; **Fl.** 12 mm ⌀, pink to purplish-pink.

**R. levynsiae** (L. Bol.) Schwant. (*Mes. l.* L. Bol.). – Cape: between Ceres and Calvinia. – Low ♄ up to 2.5 cm high; **St.** 3 mm thick; **Br.** covered with dry old L., with short shoots; **L.** of the lower pair spherical, divergent at 45°, with indistinct margins, bluish with green dots, rough, united during the resting period into a Bo. and enclosed in the papery Sh.

**R. limbata** (N. E. Br.) Schwant. (*Mes. l.* N. E. Br.). – Cape: Saldanha Bay. – Low, mat-forming ♄ 2–3 cm high; **L.** 5–10 mm long, 5–7 mm across, 4–5 mm thick, trigonous-ovoid, greyish-green, with brown margins and keel; **Fl.** 14–16 mm ⌀, purplish-red.

**R. lineolata** (Haw.) Schwant. (*Mes. l.* Haw.). – Cape: Caledon, Riversdale, Humansdorp D. – Low, freely branched ♄; **St.** 2–3 mm thick, Int. 2–2.5 cm long; **L.** mostly enclosed in the previous year's Sh., 2–4 mm long, up to 2 mm across, acutely keeled on the underside, angles set with indistinct horny hairs, Sh. 1.5 mm long; **Fl.** 2 cm ⌀, purple.

**R. lisabeliae** L. Bol. – Cape: Van Rhynsdorp D. – Forming mats up to 90 cm ⌀; **St.** prostrate, stiff, up to 60 cm long, Int. 2–3.5 cm long, 2–4 mm ⌀, with short shoots from the L.-axils; **L.** 2–3.2 cm long, 3–5 mm ⌀, green, narrowed towards the indistinctly apiculate tip, the keel reduced to a line; **Fl.** in a cyme, 2 cm ⌀, purplish-pink.

**R. littlewoodii** L. Bol. – Cape: Ceres D. – Low ♄ 3 cm high; **roots** napiform, 9 cm long, 1 cm ⌀ above; herbaceous parts minutely tuberculate; **St.** c. 2.5 mm ⌀; **Br.** with 6 crowded L., Int. enclosed; **L.** lanceolate, acute to obtuse, apiculate, bluntly keeled on the underside, 1–1.8 cm long, 2–3 mm across, 2–4 mm thick, Sh. 2–4 mm long; **Fl.** 1–1.5 cm ⌀, purplish-pink.

**R. lodewykii** L. Bol. – Cape: Bushmanland. – ♄ c. 3 cm high, 13 cm ⌀; **St.** robust, 7 mm thick; **Br.** 5 mm thick, **shoots** 1.5 mm thick, Int. densely covered with the imbricate L.-remains; **L.** projecting from the old Sh., laterally compressed, with horny angles, 2–4 mm long, 1.5 to 2 mm across; **Fl.** 15 mm ⌀.

**R. loganii** L. Bol. – Cape: Laingsburg D. – Forming mats 5 cm high; **St.** 3 mm thick, Int. 2 cm long; **L.** 2 together on a shoot, broadly ovate, rounded on the underside, minutely papillose, bristle-tipped, bluish-green, 9 mm long, 5 mm across, 3 mm thick, Sh. 3 mm long, L. on sterile shoots narrower, ciliate; **Fl.** 1 cm ⌀, purplish-pink.

**R. lokenbergensis** L. Bol. – Cape: Calvinia D. – Erect ♄ 13–16 cm high; **St.** c. 1.5 cm ⌀; **Br.** c. 5 mm ⌀, Int. c. 2.5 cm long, herbaceous parts velvety-papillose; **L.** oblong-ovate to linear-lanceolate, keeled on the underside, bluish-green, with translucent margins, 5–15 mm long, 4–5 mm across, 4–6 mm thick; **Fl.** 1.5–2 cm ⌀, brilliant pink.

**R. longifolia** L. Bol. (*R. graminea* Jacobs.). – Cape: Ladismith D. – Grass-like in appearance, forming large cushions; **St.** elongated; **L.** narrowed towards the tip, laterally compressed, with 1–4 T. on the keel, dirty green, up to 7.5 cm long, 6 mm across, 4–5 mm thick, Sh. 9 mm long; **Fl.** long-pedicellate, Pet. up to 19 mm long, pinkish-red, faintly striped.

**R. longipes** L. Bol. (*Mes. l.* (L. Bol.) N. E. Br.). – Cape: Lit. Namaqualand. – ♄ 7 cm high, 15 cm ⌀; **St.** crowded, with 4 L. and short shoots from the L.-axils, or elongated, with 6 L., Int. 2 cm long; **L.** narrowing towards the ± oblique tip, 4.5 cm long, 9 mm across at the base and sheathing and united for 5 mm; **Fl.** 16 mm ⌀, pink.

**R. luckhoffii** L. Bol. – Cape: Ceres D. – Creeping ♄ up to 5 cm high; **St.** 7 mm thick; older **Br.** 5 mm thick, Int. 1–4 mm long, flowering **shoots** 3 cm long, with 4 dissimilar L., lower ones hemispherical, with translucent keel and margins, minutely hirsute, 2–3 mm long, 2 mm ⌀, Sh. 1 mm long, upper L. 5–6 mm long, 3 mm across, 2 mm thick; **Fl.** 15 mm ⌀, pink.

**R. macowanii** (L. Bol.) Schwant. (*Mes. m.* L. Bol.). – Cape: Cape Peninsula. – Laxly branched ♄ 15–20 cm high; **St.** prostrate, Int. 2–3 cm long; **L.** ± swollen, slightly oblique, keeled on the underside, 21–35 mm long, 4 mm ⌀, Sh. 5 mm long; **Fl.** 22 mm ⌀, Pet. pink, with a darker stripe.

**R. macrophylla** L. Bol. – Cape: Robertson D. – Robust ♄ with **St.** 12 mm thick and densely crowded **Br.**; **L.** acute, compressed above into a dentate keel, 6–9.5 cm long, 12 mm ⌀, Sh. 7–9 mm long; **Fl.** 31 mm ⌀, Pet. pale pink, with a muddy stripe, pink on the underside.

**R. maleolens** L. Bol. (*Mes. m.* (L. Bol.) N. E. Br.). – Cape: Lit. Namaqualand. – Robust ♄ with a fishy smell; **St.** stiff, 21 cm long, Int. 15 mm long, **Br.** with 4–8 L., with short shoots from the L.-axils; **L.** shortly tapered or acute, bluntly keeled, minutely velvety, covered with prominent papillae, 12 to 15 mm long, 3–4 mm thick, Sh. 4 mm long; **Fl.** 2.5 cm ⌀, pink.

**R. mariae** L. Bol. – Cape: Steytlerville D. – Forming mats only 3 cm high including the Fl.; **St.** 4–5 mm thick, flowering **shoots** with 2–4 L., Int. enclosed; **L.** shortly apiculate, margins entire or with 1–2 T., acutely keeled, blue, tinged pink, 6–9 mm long, 2.5 mm ⌀, Sh. 3 mm long; **Fl.** 16 mm ⌀, purplish-pink.

**R. marianae** (L. Bol.) Schwant. (*Mes. m.* L. Bol.). – W.Cape. – ♄ 31 cm high; **L.** swollen-trigonous, acute, apiculate, bluish-green, 5 to 10 cm long, 4–7 mm across, sheathing and united for 7–15 mm; **Fl.** 4 cm ⌀, an attractive scarlet.

**R. mathewsii** L. Bol. – Cape: Malmesbury D. – Erect ♄; **St.** stiff, rigid, 2–4-angled, 18 cm long, 4 mm thick, Int. 10–15 mm long, older L. withered and persisting, with short shoots from the axils; **L.** laterally compressed, with a translucent keel on the underside, hirsute above, bristle-tipped, pale bluish-green, finely velvety-haired, 15 mm long, 3–4 mm thick; **Fl.** solitary, 16 mm ⌀, pink.

**R. maxima** (Haw.) L. Bol. (*Mes. m.* Haw., *Astridia m.* (Haw.) Schwant.). – Cape: Clanwilliam D. – ♄ 30 cm or more high; **L.** crowded, lunate, trigonous, much compressed laterally, with a curved keel, grey to whitish-grey, with numerous translucent dots, 45 mm long, sides 20 mm across; **Fl.** up to 2 cm $\varnothing$, pink.

**R. maxwellii** L. Bol. (*Mes. m.* (L. Bol.) N. E. Br.). – Cape: Willowmore D. – Robust ♄ with the **St.** 9 mm thick at the base, 6 cm tall; **Br.** 6 mm thick, Int. 1 cm long, flowering **shoots** with 2–4 L., Int. enclosed, herbaceous parts velvety; **L.** carinate on the underside, narrowing towards the shortly tapered, apiculate tip, 17 mm long, 3 mm $\varnothing$, Sh. 3 mm long; **Fl.** solitary, 2 cm $\varnothing$, pink.

**R. menniei** L. Bol. – Cape: Malmesbury D. – Small ♄; **St.** 4 mm $\varnothing$, **shoots** short, with c. 10 L., Int. 1–1.7 cm long; **L.** linear, acute, apiculate, with microscopic horny cilia along the margins and keel, c. 1.5 cm long, 2 mm across, 1.5 mm thick; Pet. 4–6 mm long, pink above, sometimes purple on the underside.

**R. mesklipensis** L. Bol. – Cape: Lit. Namaqualand. – ♄ 6 cm high; **St.** alate-angular, Int. 5–7 mm long; **L.** withering and persisting, apiculate, bluish, rough-papillose, dissimilar, the lower ones hemispherical, 4 mm across and long, upper ones smaller, keel and margins slightly translucent, Sh. 1 mm long; **Fl.** 17 mm $\varnothing$, purplish-pink.

**R. meyerae** Schwant. – Cape: Lit. Namaqualand. – ♄ up to 5 cm high; **St.** covered by closely appressed, yellowish-brown L.-Sh.; **L.** semi-ovate, dark green or bluish to whitish-green, with translucent margins and keel, 4 mm long, 2.5 mm across; **Fl.** up to 1.5 cm $\varnothing$, violet-pink.

**R. meyeri** Schwant. (Pl. 194/5) (*Mes. m.* (Schwant.) N. E. Br.). – Cape: Lit. Namaqualand. – Erect ♄ 20–40 cm high; **St.** with numerous Br.; **L.** 10 mm long, sheathing and united for 3 mm, 4 mm across, 3 mm thick, greyish-green, with the acute keel and the angles minutely hirsute and translucent; during the resting period the old L.-Sh. completely conceal the new L.; **Fl.** red.

**R. microphylla** (Haw.) Schwant. (*Mes. m.* Haw.). – Cape: Paarl D. – **St.** numerous, erect, 7–10 cm tall, thin, Int. short, with numerous short shoots; **L.** trigonous, swollen at the base, apiculate, with the minutely cartilaginous, often minutely ciliate keel pulled somewhat forward, 4–5 mm long, 2 mm across, green, with large, translucent dots; **Fl.** solitary, 2 cm $\varnothing$, Pet. white with pink tips.

**R. middlemostii** L. Bol. – Cape: Lit. Namaqualand. – Robust, erect ♄ 30–35 cm tall; **St.** 1 cm $\varnothing$ at the base; **Br.** c. 29 cm long, 8 mm $\varnothing$, Int. 1.5–2.5 cm long, **shoots** with 2–4 L., herbaceous parts velvety-papillose; **L.** acute, indistinctly keeled, dissimilar, 1–1.5 cm long, 7–10 mm across, 10 mm thick, if growing in the shade $\pm$ compressed, with a $\pm$ acute keel, longer and broader; Pet. 6–11 mm long, pink.

**R. milleflora** L. Bol. – Cape: Lit. Namaqualand. – Flowering **St.** 17 cm long, erect, much branched, Int. 5–10 mm long; **L.** $\pm$ falcate, rounded on the underside, 7–14 mm long, 2–4 mm across, c. 5 mm thick; **Infl.** much branched, Fl. numerous, c. 1 cm $\varnothing$, pinkish-purple.

**R. minutiflora** L. Bol. – Cape: Clanwilliam D. – Compact ♄; **St.** c. 7 mm $\varnothing$; **Br.** in part creeping, 7–8 cm tall, c. 13 mm $\varnothing$, short **shoots** with 4 L. and 2 bracts; **L.** $\pm$ apiculate, with ciliate angles, red, dissimilar, lower L. ovate, c. 3 mm long, 2.5 mm across, upper ones lanceolate, c. 4 mm long, 1.5 mm across; **Fl.** 12–15 mm $\varnothing$, purplish-pink.

**R. misera** (L. Bol.) L. Bol. (*Mes. m.* L. Bol.). – S.W.Cape. – Laxly branched ♄ c. 37 cm tall; **St.** spreading, slender, Int. 1.5–3 cm long; **L.** falcate, $\pm$ swollen-trigonous, shortly tapered, apiculate, 1.7 cm long, 2.5 mm across, Sh. 2 mm long; Fl. numerous, Pet. missing.

**R. modesta** L. Bol. f. **modesta**. – S.W.Afr.: Lüderitz D. – Erect, woody ♄ 45 cm tall; **St.** 8 mm thick, Int. 10–25 mm long, **Br.** elongated, velvety-papillose; **L.** 2 on a shoot, $\pm$ navicular, mucronate, bluish-green, with red keel and margins, 13–20 mm long, 8 mm across midway, 9 mm thick; **Fl.** opening only slightly, Pet. 4 mm long, purplish-pink.

**R.** — f. **glabrescens** L. Bol. – Glabrous form.

**R. mollis** (Bgr.) Schwant. (*Mes. m.* Bgr., *Corpuscularia m.* (Bgr.) Schwant., *Mes. leve* N. E. Br.). – Cape: origin unknown. – Low ♄ with prostrate **St.**; **L.** trigonous with rounded sides and a distinct line along the angles, obtuse, minutely hirsute, greyish-green, 12–15 mm long, 4–6 mm across; Fl. solitary, 15 mm $\varnothing$, red.

**R. montaguensis** L. Bol. – Cape: Montagu D. – Erect, stiff ♄; **St.** 30 cm long, 3–4 mm thick, Int. 2–4 cm long, young **shoots** with 4–6 bluish-green **L.** 13–20 mm long, 4 mm $\varnothing$, narrowing towards the tip, with indistinct keel and angles, Sh. with impressed lines; Pet. 7 mm long, purplish-pink, faintly striped.

**R. mucronata** (Haw.) Schwant. (*Mes. m.* Haw.). – Cape: Malmesbury D. – Very small, freely branched ♄; **St.** spreading, prostrate, c. 5 cm long; **L.** semicylindrical, trigonous above, awn-tipped, greyish-green, with large prominent dots, minutely rough, 6 mm long; **Fl.** solitary, 2 cm $\varnothing$, purplish-pink.

**R. muelleri** (L. Bol.) Schwant. (*Mes. m.* L. Bol., *Mes. psammophilum* Dtr., *R. p.* (Dtr.) Dtr. et Schwant.), *Mes. ruschianum* Dtr., *R.r.* (Dtr.) Dtr. et Schwant.). – S.W.Afr.: Lüderitz D. – ♄ up to 1.5 m or more tall; **St.** stiff, 13 mm thick, Int. 25–30 mm long, **Br.** ascending, Int. 1–6 cm long; **L.** narrowing towards the tip, bluish-green, 3.5–5 cm long, 5 mm across, 6 mm thick, sheathing and united for 4–5 mm; **Fl.** 10 mm $\varnothing$, Pet. pink, with a dark pink stripe.

**R. muiriana** (L. Bol.) Schwant. (*Mes. m.* L. Bol.). – Cape: Ladismith D. – Erect ♄ 15–20 cm tall; **St.** 1 cm thick at the base; **Br.** 2–6 mm thick, Int. 5–20 mm long; **L.** acute, entire, 25 mm long, 3 mm $\varnothing$, Sh. 6–7 mm long; **Fl.** 13 mm $\varnothing$, purplish-pink.

**R. multiflora** (Haw.) Schwant. (*Mes. m.* Haw.). – Cape: Swellendam, Robertson D. – ♄ up to 1 m tall, 2 m $\varnothing$; **St.** erect, freely forking, $\pm$ hexagonal at first; **L.** equilaterally trigonous, apiculate, sheathing and united, slightly greyish-green, with translucent dots, c. 3 cm long, 3–4 mm across; **Fl.** numerous, up to 3 cm ⌀, white.

**R. muricata** L. BOL. (*Mes. unidens* AUCT., *Mes. pellanum* N. E. BR.). – S.W.Afr.: Lüderitz D.; Cape: Bushmanland, Kenhardt and Upington D. – Compact ♄; **St.** prostrate, sometimes rooting, 30 cm long, **Br.** erect, Int. enclosed, with short shoots from the L.-axils; **L.** obtuse, keeled on the underside with 4–6 T. in the upper part, blue, tinged pink, 4–7 mm long, Sh. 3 mm long, with an impressed line; **Fl.** solitary, 12 mm ⌀.

**R. mutica** L. BOL. (*Mes. m.* (L. BOL.) N. E. BR.). – Cape: Ceres D. – Resembles **R. triquetra**; **L.** shorter, thicker, one pair projecting 2–3 mm beyond the Sh., the other 4–5 mm long, not apiculate.

**R. namaquana** L. BOL. v. **namaquana** (*Mes. n.* (L. BOL.) N. E. BR.). – Cape: Lit. Namaqualand. – Erect, branched ♄ 10–15 cm tall; **St.** stiff, Int. 10 mm long; **L.** obtuse, apiculate, broadest above midway, keeled on the underside, with hair-tipped dots, 1 cm long, 4 mm ⌀; **Fl.** 10 mm ⌀, pale pink.

**R. —** v. **quinqueflora** L. BOL. – Cape: Lit. Namaqualand. – ♄ 25 cm tall; **St.** 2–3 mm thick; **L.** obliquely spherical-clavate, velvety-papillose, spotted, 7–10 mm long, 4–5 mm ⌀, Sh. 1 mm long, with an impressed line; **Fl.** 8 mm ⌀, opening in the evening.

**R. namusmontana** FRIEDR. (*R. foliolosa* L. BOL. non (HAW.) SCHWANT.). – S.W.Afr.: Lorelei Copper Mine, Namus Mts. – Erect, densely leafy ♄ 30 cm tall; **St.** c. 21 cm long, 5 mm thick, Int. 5–20 mm long, sterile shoots with 4–6 crowded, linear, acute **L.** 5–10 mm across, compressed towards the tip; **Fl.** solitary or in 2–3-partite Infl., Pet. 7–9 mm long, pink.

**R. nana** L. BOL. (*Mes. reductum* N. E. BR. p. part.). – Cape: Laingsburg D. – Mat-forming ♄ c. 4 cm high; **St.** 2 mm thick, densely leafy; **L.** ± ovate, short-tapered, 3–5 mm long, 3 mm thick, Sh. 2 mm long; **Fl.** 26 mm ⌀, white.

**R. nelii** SCHWANT. – Origin unknown. – Very small ♄, Int. very short; **L.** ovate, ± hemispherical on the underside, bluish or greyish-green, with large dots, with translucent margins and keel, up to 5 mm long, 5 mm across, 2 mm thick; **Fl.** not known.

**R. neovirens** SCHWANT. (*Mes. virens* HAW.). – Cape: Uitenhage D. – **St.** slender, 40–60 cm tall, **Br.** compressed; **L.** compressed-trigonous, short-tapered, with the angles smooth and the keel ± rough, green, with numerous dots, 2.5–3.5 cm long, 6–7 mm across; **Fl.** 4 cm ⌀, red.

**R. nieuwerustensis** L. BOL. – Cape: Van Rhynsdorp D. – **St.** ascending to prostrate, 10–15 cm long, Int. 15–20 mm long; **L.** semicylindrical, obtuse, green, 2.5 cm long, 6 mm thick, Sh. 2.5 mm long, with an impressed line; Fl. pinkish-purple.

**R. nobilis** SCHWANT. – Cape: Lit. Namaqualand. – Clump-forming plant; **St.** short, **Br.** rudimentary; **L.** navicular, trigonous, acutely keeled on the underside above, greyish-green, grey-puberulous, united for 5–7 mm at the base, 3 cm long, 1 cm ⌀; **Fl.** 17 mm ⌀, mauvish-pink.

**R. nonimpressa** L. BOL. – Cape: Cradock D. – **St.** prostrate to creeping, 27 mm long, Int. 15 mm long; **L.** blunt or obliquely truncate, the keel on the underside with 1–3 indistinct T., spotted, 10–12 mm long, 3–4 mm ⌀, with a short Sh.; **Fl.** 15 mm ⌀, purplish-pink.

**R. nordenstamii** L. BOL. – Cape: Van Rhynsdorp D. – Small ♄ 3.5 cm high, creeping from the base, with a **St.** 1 cm long; **Br.** up to 1.5 cm long, covered below with old L.-Sh., with 2–4 L. above, Int. enclosed, herbaceous parts ± velvety-hirsute; **L.** long-ovate, obliquely rounded, carinate, with the angles ± translucent, bluish-green with faint dots, 1–2 cm long, 5 mm ⌀, L. smaller on flowering shoots; **Fl.** solitary, 8 to 10 mm ⌀, purplish-pink.

**R. obtusa** L. BOL. (*Mes. hebes* N. E. BR.). – Cape: Lit. Namaqualand. – Stiff ♄ with rigid **St.** 40 cm long, Int. 35–40 mm long; **Br.** ascending, with 4–6 obtuse to rounded, apiculate, bluntly keeled, bluish-green **L.** 20–25 mm long, 3–8 mm thick, Sh. 3–5 mm long, with an impressed line; **Fl.** 18 mm ⌀, purplish-pink.

**R. obtusifolia** L. BOL. – Cape: Van Rhynsdorp D. – Compact ♄ 7 cm tall, 10 cm ⌀, with a **St.** 8 mm ⌀; **St.** c. 6 cm long, Int. 5–15 mm long; the herbaceous parts velvety; flowering **shoots** with 2–4 **L.** 1–1.3 cm long, 3–4 mm ⌀, rounded above and on the underside, Sh. 1–5 mm long; **Fl.** solitary, c. 1.6 cm ⌀, purplish-pink, open at noon.

**R. odontocalyx** (SCHLTR. et DIELS) SCHWANT. (*Mes. o.* SCHLTR. et DIELS, *Mes. rupicolum* ENGL., *R. r.* (ENGL.) SCHWANT., *Mes. steingroeveri* PAX, *R. st.* (PAX) SCHWANT., *Mes. sabulicolum* DTR., *R. s.* DTR., *Mes. pseudorupicolum* DTR., *R. p.* (DTR.) DTR. et SCHWANT. nom. nud., *R. banardii* L. BOL., *Mes. b.* (L. BOL.) N. E. BR., *R. nivea* L. BOL., *Mes. n.* (L. BOL.) N. E. BR.). – S.W.Afr.: Malta Heights, Lüderitz, Bethanien, Keetmanshoop, Warmbad D.; Cape: Lit. Namaqualand. – Laxly branched ♄ 30 cm tall, smelling of fish; **St.** ascending, herbaceous parts softly velvety with white papillae, nodes 1 cm ⌀, Int. 2.5–3 cm long; **L.** thick, indistinctly keeled, oblique and with a blunt T. at the tip, the margins with 2–3 indistinct T., bluish-green; **Fl.** 2–3 cm ⌀, white.

**R. orientalis** L. BOL. – Cape: Albany D. – Erect, laxly branched ♄ 20 cm tall; **St.** 3 mm thick, Int. 1–3 cm long, young **shoots** 1–1.5 cm long, with 4–6 crowded, dissimilar, apiculate, dirty green **L.** 2 cm long, 4 mm across, 4 mm thick at the base, with the acute keel on the underside ± decurrent, rarely horny and serrulate; **Fl.** solitary, 22 mm ⌀, pinkish-purple.

**R. orsmondiae** L. BOL. – Cape: Kenhardt D. – Laxly branched ♄ 8 cm high; **St.** prostrate, Int. 5 mm long, covered with whitish L.-Sh., herbaceous parts pruinose; **L.** acute, the keel on the underside entire or with 3–5 slender T., faintly spotted, 12–20 mm long, 2.5 mm across, Sh.

3–4 mm long, with an impressed line; **Fl.** 2 cm ∅, an attractive pink.

**R. oviformis** L. Bol. – Cape: Lit. Namaqualand. – ♄ 6 cm high; **St.** slender, Int. 7 mm long, Sh. persisting, herbaceous parts minutely papillose; **L.** paired, ovate, 5 mm long, 3 mm across, 2 mm thick, with somewhat translucent margins, or 4 L. together, when the upper pair is laterally compressed; **Fl.** 11 mm ∅, pale to darker pink.

**R. pakhuisensis** L. Bol. – Cape: Clanwilliam D. – Stiff ♄ 30 cm tall; **St.** 4 mm thick, Int. 15–20 mm long; **L.** swollen or slightly compressed, obtuse, 8–10 mm long, 3 mm across, 4 mm thick, Sh. 1 mm long; **Fl.** 12 mm ∅, red.

**R. pallens** L. Bol. – Cape: Lit. Namaqualand. – Erect ♄ 60 cm tall; **St.** 5 mm thick, Int. 25–35 mm long, **Br.** ascending, Int. 5 cm long; **L.** narrowed towards the tip, indistinctly keeled on the underside, 3.5–5 cm long, 4–5 mm ∅, Sh. 3–4 mm long, with an impressed line; **Fl.** 22 mm ∅, pale pink.

**R. papillata** L. Bol. (*Mes. p.* (L. Bol.) N. E. Br.). – Cape: Brakfontein. – ♄ 17 cm tall; **St.** elongated, Int. 2–2.5 cm long; **L.** 2–4 on the shoots, acute above, with an obtuse, ± translucent keel on the underside, rough-papillose, with longer papillae along the margins; **Fl.** 15 mm ∅, pink.

**R. paripetala** (L. Bol.) L. Bol. v. **paripetala** (*Mes. p.* L. Bol.). – Cape: Lit. Namaqualand. – Robust ♄ 30–60 cm tall; **St.** stiff, 20 cm long, 8 mm thick, Int. 10–25 mm long, **Br.** with 2–6 apiculate, minutely velvety **L.** 20–32 mm long, 12 mm ∅, narrowed above, convex on the underside and keeled above, Sh. 7 mm long; **Fl.** 27 mm ∅.

**R. —** v. **occultans** L. Bol. – **L.** 15 mm long, 5 mm across, 4 mm thick; **Fl.** 12 mm ∅.

**R. parvibracteata** L. Bol. (*Mes. p.* (L. Bol.) N. E. Br.). – Cape: Lit. Namaqualand. – Erect ♄ with stiff, virgate **St.** 28 cm long, Int. 10–15 mm long, 4-angled, with a scarred bark; **L.** obtuse above, expanded as seen in profile, 7–9 mm long, 2–4 mm thick, Sh. 2–3 mm long; **Fl.** 1 cm ∅, pinkish-purple.

**R. parvifolia** L. Bol. – Cape: Karroo. – Compact, stiff ♄ 8 cm high; **St.** inclined, **Br.** with 4–6 L. crowded at the tip, with short shoots from the L.-axils; **L.** with a spinescent apiculus, the margins translucent, with one T. above, bluish with green dots, 6–8 mm long, 3 mm ∅, Sh. 1 mm long; **Fl.** solitary, 2 cm ∅, pink.

**R. patens** L. Bol. (*Mes. diplosum* N. E. Br.). – Cape: ? Piquetberg D. – ♄ 25 cm high; **St.** virgate, red, Int. 15–35 mm long; **L.** eventually withering and persisting, shortly tapering and apiculate, with a serrate keel on the underside, 15–20 mm long, 3 mm across and thick, Sh. 2 mm long; **Fl.** solitary, Pet. purplish-pink, with a darker stripe.

**R. patulifolia** L. Bol. – Cape: Van Rhynsdorp D. – ♄ 30 cm high; **St.** virgate, stiff, Int. 1 cm long, **Br.** short, with 2 green, densely spotted **L.** 7 mm long, 3–4 mm ∅, expanded towards the tip, bluntly keeled on the underside, Sh. scarcely 1 mm long, with an impressed line; Fl. solitary, 12 mm ∅, pinkish-purple.

**R. pauciflora** L. Bol. – Cape: Piquetberg **D.** – **St.** elongated, creeping, 40 cm long, compressed, flat, Int. 1.5–5 cm long; **Br.** erect, with 4 compressed, apiculate, green **L.** 2.5 cm long, 4 mm across, 6 mm thick, the keel on the underside serrate above, Sh. 3 mm long, with an impressed line; **Fl.** solitary, Pet. 12–14 mm long, pinkish-purple, indistinctly striped.

**R. paucifolia** L. Bol. – Cape: Calvinia D. – Erect ♄ c. 21 cm high, with a **St.** 5 mm ∅; **Br.** covered with old L.-Sh.; short shoots with 2 ovate **L.** 4–5 mm long, 3 mm across, convex on the underside, with a translucent line along the margins; **Fl.** 1 cm ∅, purplish-pink.

**R. paucipetala** L. Bol. – Cape: Lit. Namaqualand. – Densely branched ♄ c. 24 cm tall; **St.** up to 20 cm long, Int. enclosed; **L.** 5–10 mm long, 4 mm across, 5 mm thick, obtuse, the keel on the underside reduced to a line, green with a minutely tuberculate surface, Sh. 2 mm long; **Fl.** 10–12 mm ∅, Pet. few, purplish-pink.

**R. pauper** L. Bol. – Cape: Calvinia D. – Forming mats up to 22 cm high, 11 cm ∅; **St.** 11 mm ∅, older **Br.** covered with dry L.-Sh., Int. enclosed or 1–2.5 cm long, young Br. with 4–8 lanceolate to linear, acute, apiculate **L.** 5–15 mm long, 1.5–3 mm across, Sh. 2–3 mm long; **Fl.** 2–2.5 cm ∅, purplish-pink.

**R. peersii** L. Bol. (*Mes. p.* (L. Bol.) N. E. Br.). – Cape: Riversdale D. – Compact ♄ up to 6 cm high; **St.** 4 cm long; **Br.** with 4 bluish-green **L.** 2 cm long, 4 mm across, 3 mm thick, acutely keeled on the underside, immature L. sheathing and united for 8 mm, older ones persisting; **Fl.** solitary, 22 mm ∅, Pet. purplish-pink, striped on the upperside.

**R. perfoliata** (Mill.) Schwant. (*Mes. p.* Mill.). – Cape: Prince Albert D., Karroo. – **St.** with a few Br.; **L.** trigonous, compressed above, with a reddish apiculus, the keel with 1–2 T., 1–1.5 cm long, firm-textured, light grey, slightly reddish, Sh. 2–2.5 mm long; **Fl.** solitary, 2.5 cm ∅, pinkish-red.

**R. persistens** L. Bol. (J. S. Afr. Bot. **XXIX**, 16; non Notes on Mes. **II**, 334). – Cape: Montagu, Ladismith D. – Compact ♄ 7 cm high, 15 to 16 cm ∅; **St.** thick; Br. with many **shoots** which are covered below with dead L.; **L.** 5–6 mm long, 2.5 mm ∅, narrowed towards the acute tip, green with translucent dots which eventually become pearly, Sh. 0.5 mm long; **Fl.** 12 to 15 mm ∅, outer Pet. paler, inner ones vivid pink, with an indistinct stripe.

**R. phylicoides** L. Bol. (*Mes. p.* (L. Bol.) N. E. Br.). – Cape: Lit. Namaqualand. – Robust ♄ with stiff **St.**; **Br.** rigid, Int. 5–15 mm long; **L.** acute to tapered, apiculate, 5–10 mm long, 1.5 mm thick, Sh. 1 mm long; **Fl.** 8 to 20 mm ∅, white.

**R. pilosula** L. Bol. – Cape: Lit. Namaqualand. – Compact, robust ♄ 24 cm high; **St.** 16 mm thick at the base, **Br.** spreading, 12 to 20 cm long, Int. 3–20 mm long, herbaceous parts velvety-papillose; **L.** subfalcate, obtuse or obliquely rounded above, acutely or bluntly keeled, blue with indistinct green dots, 3 to 4.5 cm long, 6–8 mm across, 10–15 mm thick, Sh. 15 mm long; **Fl.** in a compact Infl., c. 4 cm ⌀, nocturnal.

**R. pinguis** L. Bol. – Cape: Lit. Namaqualand. – ♄ 10 cm high; **St.** erect, Int. 1 cm long; **L.** thick, plump, indistinctly angled, obtuse above, with the keel ± decurrent, bluish-green, velvety with hair-like papillae, 25–30 mm long, 13 mm thick, 8 mm across, Sh. 3 mm long, with an impressed line; **Fl.** paired, 22 mm ⌀, white.

**R. piscodora** L. Bol. (*Mes. p.* (L. Bol.) N. E. Br.). – Cape: Prince Albert D. – ♄ 20 cm high, smelling of fish; **St.** 15 mm thick at the base, **Br.** elongated, younger **shoots** with 4 L., Int. 1–3 cm long; **L.** obtuse to acute, bluntly keeled, soft, pale blue, 2 cm long, 4 mm across, 5 mm thick; **Fl.** 24 mm ⌀, pink.

**R. polita** L. Bol. – Cape: Laingsburg D. – Forming cushions 7 cm high; **St.** and **Br.** densely leafy; **L.** narrowing towards the shortly tapered tip, carinate on the underside, keel and margins ± horny, pale blue, 11 mm long, 5 mm ⌀, Sh. 6 mm long; **Fl.** 22 mm ⌀, pink.

**R. pollardii** Friedr. – S.W.Afr.: Lüderitz D. – Erect, much branched ♄ c. 30 cm high; **St.** ± 2-angled, c. 10 cm long, Int. 7–15 mm long, young **Br.** short, with 2–4 oblong to ± clavate, apiculate, pale green, minutely pruinose **L.** 1–3 cm long, 4–7 mm across, 4–10 mm thick, swollen at the base, the margins reduced to a translucent line, Sh. 1–2 mm long, with an impressed line; **Fl.** solitary, Pet. 12–15 mm long, white.

**R. primosii** L. Bol. – Cape: Bushmanland. – Forming mats 3 cm high; flowering **shoots** with 2–4 tapered, obtuse, bluish-green, rough-papillose **L.** 11 mm long, 3 mm thick, Sh. 4 mm long; **Fl.** 12 mm ⌀, pink.

**R. profunda** L. Bol. – Cape: Piquetberg D. – Erect, laxly branched ♄ 36 cm high; **St.** cylindrical, 4 mm thick, young Br. compressed-alate, Int. 1–3 cm long; **L.** falcate and curved, or semicylindrical to obliquely acute, with a horny, serrate keel on the underside, yellowish-green, 10–17 mm long, 3–4 mm across, 5 mm thick, Sh. 2–3 mm long; **Fl.** 17 mm ⌀, pink to purplish-pink.

**R. prolongata** L. Bol. – Cape: Sutherland D. – Mat-forming plant; **St.** up to 9 cm long; **Br.** up to 40 cm long, creeping, Int. 1–3 cm long, herbaceous parts velvety-papillose; **L.** 4, crowded on the shoots, lanceolate to ovate, acute, apiculate, keeled on the underside, with ciliate angles and keel, 5–8 mm long, Sh. 1–4 mm long; **Fl.** 15–20 mm ⌀, purplish-pink.

**R. promontorii** L. Bol. (*Mes. pansifolium* N. E. Br.). – Cape: Cape D. – Erect ♄ 12 cm high; **St.** stiff, Int. 2 cm long; **L.** acute, with a decurrent keel on the underside, green to reddish, 25 mm long, 12 mm across, 10 mm thick, Sh. 1 mm long; **Fl.** 3 cm ⌀, Pet. pink with a purple stripe.

**R. propinqua** (N. E. Br.) Schwant. (*Mes. p.* N. E. Br.). – Cape: Montagu D. – Low ♄; some **L.** ± free, 5–6 mm long, 2.5–3 mm across, 2 mm thick, others united, acutely keeled on the underside, with puberulous margins and keel, bluish-green; **Fl.** 8–16 mm ⌀, Pet. pale pink, with a dirty pink stripe.

**R. prostrata** L. Bol. (*Mes. p.* (L. Bol.) N. E. Br.). – Cape: Clanwilliam D. – Robust ♄ forming dense mats c. 1 cm high; **St.** stiff, 30 cm long, rooting, Int. 7 mm long; **L.** 4 on the short shoots, 3 mm long, 2.5 mm across and thick, puberulous, with a blunt, distinctly spotted keel on the underside, Sh. 2 mm long; **Fl.** sessile.

**R. pulchella** (Haw.) Schwant. v. **pulchella** (*Mes. p.* Haw., *R. p.* (Haw.) L. Bol., *Mes. canescens* Haw., *Mes. aristulatum* Sond., *R. a.* (Sond.) Schwant.). – Cape: Cape D. – Very freely branched ♄; **St.** prostrate, curved, with short Br. from the nodes; **L.** equilaterally trigonous, ± obtuse, with an awned tip, the keel-angle minutely ciliate, greyish-green, spotted, 7–10 mm long, 3 mm across; **Fl.** 3.5 cm ⌀, pale pinkish-red.

**R. —** v. **caespitosa** L. Bol. – Cape: Stellenbosch D. – **St.** creeping, rooting, 18 cm or more long; **L.** rather longer than the type; **Fl.** 2 cm ⌀, pink.

**R. pulvinaris** L. Bol. – Cape: Steynsburg D. – Forming spherical cushions 7 cm high, 18 cm ⌀; **St.** 8 mm ⌀, **Br.** 3 cm long including the Fl., with 4–6 apiculate, bluish-green **L.** 13 mm long, 2 mm thick, narrowing towards the tip, Sh. 2 mm long; **Fl.** pinkish-purple.

**R. pumila** L. Bol. (*Mes. p.* (L. Bol.) N. E. Br.). – Cape: Calvinia D. – Very low ♄, the herbaceous parts with small, densely crowded, pearly, tubercles; **L.** concealed by the dry Sh. – Insufficiently known.

**R. punctulata** (L. Bol.) L. Bol. (*Mes. p.* L. Bol., *Lampranthus p.* (L. Bol.) L. Bol.). – Cape: Calvinia D. – Laxly branched ♄ 10 cm high, 20–30 cm ⌀, the herbaceous parts with large, prominent papillae; **St.** stiff, Int. 5–8 mm long, with the old L.-Sh. persisting; **L.** semi-cylindrical, apiculate, green to reddish, 14 mm long, 3 mm thick, Sh. 3 mm long; **Fl.** 14 mm ⌀, purplish-pink.

**R. pungens** (Bgr.) Jacobs. (*Mes. p.* Bgr., *Mes. cymosa* L. Bol., *R. c.* (L. Bol.) L. Bol.). – Cape: Bushmanland. – Erect ♄; **St.** robust, Int. 2–2.5 cm long, ± swollen; **L.** cylindrical to trigonous, obtuse, apiculate, 2–3 cm long; **Infl.** branched, Fl. 15 mm ⌀.

**R. pusilla** Schwant. (Pl. 194/4) (*Mes. parvum* N. E. Br.). – Cape: Lit. Namaqualand. – Forming flat mats up to 7 cm ⌀; **St.** enclosed in the L.-Sh.; **L.** 4 mm long, 3 mm across, 2 mm thick, obtuse, green, with tiny, acute tubercles and dots confluent into a line along the keel on the underside.

**R. putterilii** (L. Bol.) L. Bol. (*Mes. p.* L. Bol.). – Orange Free State. – **St.** stiff, ascending to prostrate, 18 cm long, alate when immature, Int. 5–25 mm long; **L.** swollen at the base, obtuse and apiculate above, the keel and margins with cartilaginous hairs, 8–10 mm long, 3.5 mm ⌀, Sh. 1.5 mm long; **Fl.** 15 mm ⌀, pinkish-purple.

**R. pygmaea** (HAW.) SCHWANT. (Pl. 194/3) (*Mes. p.* HAW., *Mes. fimbriatum* SOND., *Conophytum f.* (SOND.) N. E. BR.). – Cape: Worcester, Laingsburg D. – Forming small, flat mats; **St.** very short, with 1–2 dissimilar **L.**-pairs, the first pair 4–5 mm long, 2–3 mm thick, united ± to the tip, green, the skin shrivelling, becoming parchment-like and enclosing the second pair, the L. of which are free, ovate or lanceolate, with a roundish keel on the underside; **Fl.** 18 mm ⌀.

**R. quadrisepala** L. BOL. – Cape: Lit. Karroo. – Forming cushions 25 cm ⌀, 10 cm high; **St.** 14 mm ⌀; **Br.** 1 cm ⌀; flowering **shoots** 4 to 6 cm long, with 4–6 hemispherical **L.** 3–5 mm long, 3–4 mm thick, the keel and angles minutely scaly and covered with faint dots, Sh. 3–4 mm long, with an impressed line; **Fl.** solitary, 3 cm ⌀, pale pink.

**R. quartzitica** (DTR.) DTR. et SCHWANT. (*Mes. q.* DTR., *Corpuscularia q.* (DTR.) SCHWANT., *R. argentea* L. BOL.). – S.W.Afr.: S.Namib. – Many-stemmed ♄ 15 cm high; **L.** 10–11 mm long, 5 mm across, 4 mm thick, obtusely trigonous, with indistinct keel and angles, with a red mucro, minutely white-papillose; **Fl.** 6 to 8 mm ⌀, purplish-red.

**R. radicans** L. BOL. (*Mes. pronum* N. E. BR.). – Cape: Clanwilliam D. – **St.** elongated, rooting at the nodes, 33 cm long, Int. 3.5 cm long; **Br.** with 2–4 L., with short shoots; **L.** swollen at the base, keeled on the underside above, minutely velvety, 15 mm long, 6 mm across, Sh. 2 mm long; **Fl.** solitary, 18 mm ⌀, pink.

**R. rariflora** L. BOL. – Cape: Clanwilliam D. – **St.** stiff, spreading to prostrate, Int. 2–5 cm long; **L.** semicylindrical, acute, bluish-green, 4–8 cm long, shorter on the shoots, 5 mm thick, Sh. 8 mm long, with a faint, impressed line; **Fl.** few, 24 mm ⌀, pink.

**R. rigens** L. BOL. (*Mes. dressianum* INGRAM). – Orange Free State; Cape: Hofmeyr. – Cushion-forming ♄; **St.** rigid, stiff, creeping; **shoots** densely leafy; **L.** tapered, the keel on the underside with a sharp T. under the tip, 25–30 mm long, 5 mm thick, Sh. 3 mm long; **Fl.** 22 mm ⌀, pink.

**R. rigida** (HAW.) SCHWANT. (*Mes. r.* HAW., *Mes. parviflorum* HAW., *R. p.* (HAW.) SCHWANT.). – Cape: Caledon, Swellendam D. – Freely branched, bushy ♄ 40 cm high; **St.** thin, filiform; **L.** trigonous, shorter than the Sh., light green, with tiny, translucent dots, rough along the keel, Sh. 5–10 mm long; **Fl.** 1 cm ⌀, white.

**R. rigidicaulis** (HAW.) SCHWANT. (*Mes. r.* HAW.). – Cape: near Langevalley. – Sparingly branched ♄; **St.** compressed; **L.** equilaterally trigonous, ± obtuse with a red mucro, green, freely spotted; **Fl.** 3 cm ⌀, Pet. pink, with a red M.-nerve.

**R. robusta** L. BOL. (Pl. 195/3). – Cape: Lit. Namaqualand. – Robust ♄ 50 cm high; **St.** virgate, black, 20–40 cm long; flowering **shoots** dense, 1–2.5 cm long; **L.** obtuse, faintly keeled, dirty bluish-green with crowded dots, 6–9 mm long, 3.5 mm ⌀, Sh. 2.5 mm long; **Fl.** solitary, 18 mm ⌀, pinkish-purple.

**R. roseola** (N. E. BR.) SCHWANT. (*Mes. r.* N. E. BR.). – Cape: Montagu D. – Low-growing ♄; **L.** tapered, obtuse, ± keeled on the underside, bluish-green, with translucent dots, 6 to 8 mm long, 3 mm across, 2.5 mm thick, some L. ± free, others united to midway; **Fl.** 15 mm ⌀, pale pink.

**R. rostella** (HAW.) SCHWANT. (*Mes. r.* HAW., *Mes. rostellatum* DC.). – Cape: Ladismith D. – **St.** prostrate, ± nodose, Int. short, with numerous short shoots; **L.** semicylindrical, sheathing and united at the base, greyish-green, with minute, prominent dots, 12–18 mm long, 4 mm across; **Fl.** 2–2.5 cm ⌀, white or purplish-pink.

**R. rubricaulis** (HAW.) L. BOL. (*Mes. r.* HAW.). – Cape: Cape D. – Small ♄; **St.** angular, reddish; **L.** trigonous, short-tapered, ± swollen below, with cartilaginous margins and a denticulate keel, slightly greyish-green, 2.5–4 cm long, 4–8 cm across; **Fl.** solitary, 3 cm ⌀, light purple.

**R. rupis-arcuatae** (DTR.) FRIEDR. (*Mes. r.-a.* DTR., *Stoeberia r.-a.* (DTR.) DTR. et SCHWANT., *R. perforata* L. BOL., *Mes. p.* (L. BOL.) N. E. BR., ? *R. saginata* L. BOL.). – S.W.Afr.: Lüderitz D. – ♄ up to 50 cm high; **St.** 10 cm long; **L.** patent, scarcely united, shortly navicular, obtuse, bluish-green with a few dots; **Fl.** 15 mm ⌀, ± white.

**R. ruralis** (N. E. BR.) SCHWANT. (*Mes. r.* N. E. BR.). – Transvaal. – Low ♄; **St.** 2.5 to 5 cm high, thick, freely branching from the base; **Br.** creeping, (2–)4(–15) cm high; **L.** 10–15 mm long, 2–4 mm across, 2–2.5 mm thick, acute, laterally compressed, the acute keel on the underside with or without 1–3 T. above, with translucent dots; **Fl.** 14 mm ⌀.

**R. salteri** L. BOL. – Cape: Ceres D. – Laxly branched ♄; **St.** prostrate, 10–15 cm long, Int. 1–3.5 cm long; **L.** 4 on the shoots, compressed, the keel on the underside dentate below the tip, older L. swollen, 15 mm long, 5 mm ⌀, Sh. 2–3 mm long; Pet. 7–13 mm long, pink, striped on the upperside.

**R. sandbergensis** L. BOL. (*Mes. herrei* L. BOL., *Lampranthus h.* (L. BOL.) L. BOL.). – Cape: Lit. Namaqualand. – Erect ♄; **St.** crowded; **L.** swollen at the base, expanded towards the obtuse tip, keeled on the underside, 3 cm long, 6 mm across, 11 mm thick, rarely longer and broader, the Sh. 9 mm long, with an impressed line; **Fl.** solitary, 56 mm ⌀, pale salmon-pink, with a yellowish centre.

**R. sarmentosa** (HAW.) SCHWANT. v. **sarmentosa** (*Mes. s.* HAW.). – Cape: Cape D. – **St.** 60 cm or more long, prostrate, rooting from the nodes which are 3 cm apart; **L.** equilaterally trigonous, ± obtuse, apiculate, with minutely rough margins, light green, spotted, 2.5–5 cm long, 6 mm across, Fl. in 3's, Pet. 8 mm long, reddish, with a dark M.-nerve.

**R. —** v. **rigida** (SALM) SCHWANT. (*Mes. s. v. r.* SALM). – **St.** robuster, Int. shorter; **L.** rather narrower; **Fl.** scarcely 2 cm ⌀.

**R. saturata** L. BOL. (*Mes. atrocinctum* N. E. BR.). – Cape: Tulbagh D. – Erect ♄ 9 cm high;

St. and Br. erect to spreading, stiff, shoots 5 cm long, covered with the remains of old L., Int. 1 cm long; L. 4 on a shoot, with short shoots from the axils, short-tapered, the angles ciliate and with a translucent line and ± confluent dots, 1 cm long, 1–1.5 mm ⌀; Fl. 18 mm ⌀, dirty pink, deep purple outside.

**R. saxicola** L. Bol. (*Mes. s.* (L. Bol.) N. E. Br.). – Orange Free State: Fauresmith. – Erect ♄ 30 cm high; St. sometimes elongated, Int. 10–25 mm long; L. ± narrowing towards the tip, with acute, horny, serrulate angles and a decurrent keel, bluish-green, 15–25 mm long, 4–5 mm ⌀, Sh. 1.5 mm long; Fl. 2 cm ⌀, pinkish-purple.

**R. schlechteri** Schwant. (*Mes. s.* (Schwant.) N. E. Br.). – Cape: Lit. Namaqualand. – Forming dense, convex mats, with taproots; St. short, with 1–2 pairs of L. 2–9 mm long, 4 mm across, 3 mm thick, sheathing and united for up to 5 mm, scarcely keeled on the underside, bluish-green, rough.

**R. schneiderana** (Bgr.) L. Bol. (*Mes. s.* Bgr., *R. hollowayana* L. Bol., *Mes. rugulosum* Bgr. ex Range, *R. pillansii* L. Bol., *R. spathulata* L. Bol.). – S.W.Afr.: Lüderitz D. – Erect ♄ 30 cm high; St. 5 mm thick, Int. 7–15 mm long; L. broadest midway, long-tapering or obtuse, apiculate, bluntly or indistinctly keeled on the underside, velvety with small, hair-tipped papillae, spotted, older L. pruinose or rough, bluish, tinged pink, 13–17 mm long, 4–6 mm across, 8–11 mm thick, Sh. 1.5 mm long; Fl. 15 mm ⌀, purplish-pink.

**R. schollii** (S.D.) Schwant. v. **schollii** (*Mes. s.* S.D., *Mes. recurvum* Haw., *Mes. aduncum* Jacq.). – Cape: Stellenbosch D. – ♄ 10–20 cm high; St. erect, curved, slender, with short Int.; L. subcylindrical, tapered, with a recurved tip, fresh green, spotted, 15–20 mm long, 2 mm across; Fl. solitary, 18 mm ⌀, red.

**R. — v. caledonica** (L. Bol.) Schwant. (*Mes. s. v. c.* L. Bol.). – Cape: Caledon D. – St. creeping; L. trigonous, acute; Fl. 15–32 mm ⌀, pinkish-purple.

**R. sedoides** (Dtr. et Bgr.) Friedr. (*Mes. s.* Dtr. et Bgr., *Eberlanzia s.* (Dtr. et Bgr.) Schwant.). – S.W.Afr.: Gr. Namaqualand. – ♄ up to 40 cm high, with many erect St.; Int. 15–20 mm long, 2–3 mm thick; L. rounded-trigonous, obtuse, somewhat united below, green, spotted, 18 mm long, 9–11 mm across, 5–8 mm thick; Infl. patently branched, Fl. ± spherical, seldom opening, Pet. missing.

**R. semidentata** (S.D.) Schwant. (*Mes. s.* Salm). – Cape: on the Gamka R. – St. erect, stiff, forking; L. equilaterally trigonous, tapered, the keel-angle expanded and with 2–4 T., whitish-grey, 2–3 cm long; Fl. almost 4 cm ⌀, red.

**R. semiglobosa** L. Bol. – Cape: Van Rhynsdorp D. – Robust, laxly branched ♄ 20 cm high; St. stiff, Int. 2–5 cm long; L. narrowing shortly towards the obtuse tip, rounded on the underside, green, 18–31 mm long, 8 mm across, 7 mm thick, Sh. 4 mm long; Fl. solitary or in 3's, 2 cm ⌀, purplish-pink.

**R. senaria** L. Bol. (*Mes. s.* (L. Bol.) N. E. Br.). – Cape: Lit. Namaqualand. – Laxly branched ♄; St. 30 cm or more long, Int. 25–30 mm long, herbaceous parts minutely velvety, spotted; L. oblique-clavate, expanded towards the tip, rarely short-tapering, 35 mm long, some L. shorter, 5–8 mm ⌀, Sh. 2 mm long, with an impressed line; Fl. 26–30 mm ⌀, purplish-pink.

**R. serrulata** (Haw.) Schwant. (*Mes. s.* Haw.). – Cape: origin unknown. – St. and Br. at first erect, eventually ± prostrate; L. 4 cm long, trigonous, thick, with cartilaginous, serrulate margins; Fl. c. 3 cm ⌀, red.

**R. simulans** L. Bol. – Cape: Van Rhynsdorp D. – Erect, slender ♄ 5 cm high; St. with 8 to 10 L., Int. 4–7 mm long; L. 9 mm long, 2 mm thick, apiculate, faintly keeled, green with indistinct dots, Sh. 5 mm long; Fl. 22 mm ⌀, Pet. pinkish-purple, faintly striped.

**R. singula** L. Bol. – Cape: W.Griqualand. – Compact ♄; St. with 4 L., with short shoots from the axils, Int. enclosed, 2–4 mm thick; L. narrowing above, tapered, apiculate, laterally compressed, keeled on the underside, with horny margins, bluish, up to 5 cm long, 7–9 mm thick, Sh. 1 cm long; Fl. 35–40 mm ⌀, pale pink.

**R. sobrina** (N. E. Br.) Schwant. (*Mes. s.* N. E. Br.). – Cape: Riversdale D. – Low-growing, mat-forming ♄; L. sometimes united, keeled on the underside, bluish-green, 3–5 mm long, 2 mm across, 1.5 mm thick; Pet. 8 mm long, pinkish-purple, with a dirty red stripe.

**R. socium** (N. E. Br.) Schwant. (*Mes. s.* N. E. Br.). – Cape: Van Rhynsdorp D. – ♄ 2.5 cm high, branching from the base; L. 2–4 on a shoot, semicylindrical, faintly keeled below the tip on the underside, bluish-green, 8–20 mm long, 7–9 mm across, 5–6 mm thick, sheathing and united for 8–10 mm; Fl. unknown.

**R. solida** (L. Bol.) L. Bol. v. **solida** (*Mes. s.* L. Bol.). – Cape: Van Rhynsdorp D. – St. 2 cm high; Br. prostrate, (15–)20(–40) cm long, Int. 15–20 mm long, covered with dry L.-Sh., shoots 3–5 cm long, with 2 cylindrical, obtuse, bluish-green L. 25 mm long, 7 mm across, with darker dots; Fl. 2 cm ⌀, pinkish-purple.

**R. — v. stigmatosa** L. Bol. – Shoots with 4 L., flowering shoots with only 2; L. 5 cm long, 8 mm ⌀, swollen at the base; Fl. with subulate, bristly-acuminate stigmas.

**R. solitaria** L. Bol. – Cape: Van Rhynsdorp D. – Erect, laxly branched ♄ 25 cm high; Int. 2–4 cm long; L. 2(–4) on flowering shoots, narrowing towards the obtuse, apiculate tip, rounded on the underside, minutely rough-tuberculate, dirty green, some 3.2 cm, others 1.5–2 cm long, 3–4 mm ⌀, Sh. 1.5 mm long; Fl. solitary, Pet. 5–6 mm long, purplish-pink, indistinctly striped.

**R. spinescens** L. Bol. – Cape: Laingsburg D. – Erect, robust ♄ 20 cm high; St. thick; Br. short, with 4–6 crowded, obtuse, apiculate L. 10(–15) mm long, 3 mm ⌀, expanded above, bluntly keeled or with the keel reduced to a line, Sh. 2 mm long; pedicels spinescent, Fl. 16 mm ⌀, purplish-pink.

**R. staminodiosa** L. Bol. (*Mes. s.* (L. Bol.) N. E. Br.). – Cape: Riversdale, Montagu D. – Erect ♄ 10–15 cm high; **St.** stiff; **L.** acute, the keel on the underside acute, decurrent and spurlike, bluish-green, 15 mm long, 5 mm ⌀, Sh. 2 mm long; **Fl.** 25 mm ⌀, purplish-pink.

**R. stayneri** L. Bol. – Cape: Calvinia D. – Forming mats up to 4.5 cm high, 6–7.5 cm ⌀; **St.** up to 13 mm ⌀; **Br.** woody, stiff, covered with old L.-Sh., Int. enclosed or 1–2.5 cm long, herbaceous parts velvety; **L.** lanceolate to linear, acute, apiculate, acutely keeled on the underside, 5–15 mm long; **Fl.** 1–1.5 cm ⌀, purplish-pink.

**R. stellata** L. Bol. (*Mes. stellans* N. E. Br.). – Cape: Laingsburg D. – ♄ 10 cm high; **St.** 5 cm high; **L.** 2 on the flowering shoots, obtuse, apiculate, sometimes ± velvety, with a ± decurrent keel on the underside, 10–13 mm long, 3.5–4 mm ⌀; pedicels persisting and spinescent, tipped with the star-like, spreading, hardened Cal.-lobes, **Fl.** 14 mm ⌀, Pet. stellate, purplish-pink.

**R. stenopetala** L. Bol. – Cape: Van Rhynsdorp D. – ♄ 9 cm high, 10 cm ⌀; **St.** 6–8 cm long; **Br.** 3–5 cm long, with 2 linear, obtuse, bluntly keeled, dirty green **L.** 1.5–2 cm long, Sh. 2–3 mm long; **Fl.** solitary, 1–4 cm ⌀, Pet. pink, paler on the upperside.

**R. stenophylla** (L. Bol.) L. Bol. (*Mes. s.* L. Bol., *Marlothistella uniondalensis* Schwant.). – Cape: Uniondale D. – Mat-forming, with fleshy, napiform **roots**; Int. not evident; **L.** up to 4.5 cm long, narrow, acute, ± semicylindrical, spotted; **Fl.** 35 mm ⌀, pinkish-purple.

**R. stokoei** L. Bol. – Cape: Caledon D. – Compact ♄ 5 cm high; **St.** stiff, herbaceous parts minutely papillose, velvety, with prominent dots; upper **L.** united for 11 mm, 3 mm thick, 4 mm across at the base, the free part 9 mm long, keeled below the tip, apiculate, lower Sh. 5 mm long, 3 mm thick, united for 3 mm; **Fl.** 22 mm ⌀, Pet. pinkish-purple, faintly striped.

**R. stricta** L. Bol. v. **stricta** (*Mes. sublunulatum* N. E. Br.). – Cape: Lit. Namaqualand. – Erect, sparingly branched ♄ 30–90 cm high; **St.** and **Br.** stiff, rigid, Int. 25–30 mm long; **L.** 2–6, crowded on the Br., laterally compressed, semicylindrical, with a horny line along the decurrent keel on the underside, pale bluish-green, lower L. 25 mm long, 12 mm thick, upper pairs unequal, 15 mm long, 7 mm across, 9 mm thick, Sh. 1.5 mm long; **Fl.** unknown.

**R. —** v. **turgida** L. Bol. – **L.** swollen, with a red line along the keel and margins; **Fl.** 24 mm ⌀, Pet. pink, with a red stripe.

**R. strubeniae** (L. Bol.) Schwant. (*Mes. s.* L. Bol.). – Cape: Piquetberg D. – Erect ♄ 75 cm high; **St.** compressed, red; **L.** shortly united, navicular to trigonous, laterally compressed, with serrate margins, green, 4–6 cm long, 5–6 mm across, 12 mm thick; **Fl.** 32 mm ⌀, Pet. pink, with a purple M.-nerve.

**R. suaveolens** L. Bol. – Cape: Van Rhynsdorp D. – Erect, stiff ♄ c. 50 cm high; **St.** 5 mm thick, Int. 15–45 mm long; **L.** narrowing towards the recurved tip, acutely keeled on the underside, blue, velvety, 3 cm long, 4 mm across, Sh. 5 mm long, with an impressed line; **Fl.** numerous, 15 mm ⌀, pink, scented.

**R. subaphylla** Friedr. – S.W.Afr. – Hemispherical ♄ c. 30 cm high, 50 cm ⌀, herbaceous parts bluish to bluish-green, minutely papillose; **St.** spreading to divergent, covered with old L.-Sh., the current season's **Br.** erect, constricted-jointed, Int. (1–)1.5(–2) cm long, 3–7 mm thick; **L.** dissimilar, those on young shoots with the blade free, clavate to cylindrical, 5–10 mm long, 3–4 mm ⌀, the L. on older shoots completely enclosed in the Sh.; pedicel much thickened, 5 mm long, **Pet.** 3–4 mm long, linear, white.

**R. subpaniculata** L. Bol. – Cape: Clanwilliam D. – Erect ♄; **St.** 15–29 cm long, Int. 2–6 cm long; **L.** narrowing towards the recurved tip, keeled on the underside above, 3 cm long, 3 mm across, with an impressed line at the base; **Infl.** ± paniculate, Pet. 6 mm long, pink, faintly striped.

**R. subsphaerica** L. Bol. – Cape: Lit. Namaqualand. – Erect ♄ 25 cm high; **St.** virgate, 22 cm long, Int. 5–8 mm long; **L.** 2 on a shoot, hemispherical, 4–5 mm long, 3–4 mm thick, the keel and angles lined, green with minute dots, Sh. 2.5 mm long, with an impressed line; **Fl.** solitary, 13 mm ⌀, purplish-pink.

**R. subteres** L. Bol. – Cape: Worcester D. – Erect ♄; **St.** 21 cm long, Int. 4 cm long, **Br.** erect; **L.** semicylindrical, swollen, obtuse, bluish-green, 2.5–3 cm long, 5 mm thick; Pet. 6 mm long, purplish-pink.

**R. subtruncata** L. Bol. v. **subtruncata**. – Cape: Calvinia. D – **St.** creeping to climbing, 25 cm long, Int. 2 cm long, flowering **shoots** 2–4 cm long; **L.** ± dissimilar, some 6 mm long, 3 mm thick, Sh. 2 mm long, other L. 6–10 mm long, with the margins indistinctly hirsute-dentate, bluish-green, almost velvety; **Fl.** 25 mm ⌀, pink.

**R. —** v. **minor** L. Bol. – Cape: Sutherland D. – Mat-forming ♄; lower **L.** 3–5 mm long, 1.5 to 2 mm ⌀, upper L. obliquely ovate.

**R. succulenta** L. Bol. (*Mes. s.* (L. Bol.) N. E. Br.). – Cape: Lit. Namaqualand. – ♄ 11 cm high; **St.** prostrate, 22 cm long, Int. 15–35 mm long; **Br.** with 4–8 laxly arranged, fleshy, bluntly keeled, obtuse **L.** 3 cm long, 4 mm ⌀, with green dots; Fl. solitary, 25 mm ⌀, Pet. pink, with a red stripe on the upperside.

**R. tardissima** L. Bol. – Cape: Calvinia D. – Compact ♄; **St.** prostrate, stiff, compressed, 40 cm long, Int. 5–6 cm long; **L.** acute, keeled on the underside, 5–7 cm long, 5 mm across, 6 mm thick; **Fl.** slow in developing, Pet. 14 to 17 mm long, pink with a purple stripe.

**R. tecta** L. Bol. – Cape: Malmesbury D. – Erect, stiff ♄ 57 cm tall; **St.** 6 mm thick, Int. 25–80 mm long; **L.** S-shaped, narrowing towards the tip, obtusely keeled on the underside, dirty green, 9.5 cm long, 4 mm ⌀, Sh. 15 mm long, with an indistinct, impressed line; **Fl.** many, Pet., 17 mm long, purplish-pink, white below.

**R. tenella** (HAW.) SCHWANT. (*Mes. t.* HAW., *Mes. rigidum* v. *t.* HAW.). – Cape: Karroo. – Close to **R. rigida,** and probably only a variety; **St.** more slender, Int. and **L.** longer.

**R. testacea** L. BOL. – Cape: Van Rhynsdorp D. – **St.** 24 cm long, Int. 4–4.5 cm long; **L.** falcate, cylindrical, obtuse or shortly tapered, bluish, 3.5–5.5 cm long, 4.5 mm across, 6 mm thick; **Infl.** of many brick-red Fl. 1 cm ∅.

**R. tetrasepala** L. BOL. – Cape: Van Rhynsdorp D. – Laxly branched ♄ 13 cm high; **St.** robust, Int. 4 mm long; **L.** 4 on a shoot, dissimilar, lower ones oblique-ovoid, 4 mm long and across, 5 mm thick, Sh. 3 mm long, upper **L.** ± compressed and obliquely truncate above, 9 mm long, 2.5 mm across and thick; **Fl.** solitary, 2 cm ∅, pink, Sep. 4.

**R. thomae** L. BOL. (*Mes. stokoei* L. BOL., *Lampranthus s.* (L. BOL.) L. BOL.). – Cape: Caledon D. – Low, compact ♄; **St.** elongated, prostrate, Int. 1–2 mm long; **L.** sabre-shaped, acute, faintly keeled below the tip, bluish-green, minutely papillose, 14 mm long, 8 mm across and thick; **Fl.** solitary, 23 mm ∅, purplish-pink.

**R. —** v. **microstigma** L. BOL. – Cape: Worcester D. – Fl. 3–4.5 cm ∅, with short stigmas.

**R. translucens** L. BOL. (*Mes. t.* (L. BOL.) N. E. BR.). – Cape: Lit. Namaqualand. – Laxly branched ♄; **St.** prostrate, stiff, 50 cm long; **L.** cylindrical, swollen at the base, obtuse above, pale green, almost translucent, with darker dots, 25–32 mm long, 8–11 mm thick, Sh. 2 mm long; **Fl.** 4 mm ∅, Pet. transparent.

**R. tribracteata** L. BOL. – Cape: Lit. Namaqualand. – Laxly branched ♄ 30 cm high; **St.** stiff, Int. 2–5 cm long; **L.** mostly semicylindrical, obtuse, green, 5 mm ∅, Sh. 4 mm long; pedicels with 3 bracts 6–14 mm long, **Fl.** 18 mm ∅.

**R. triflora** L. BOL. (*Mes. t.* (L. BOL.) N. E. BR.). – Cape: Clanwilliam D. – Compact ♄ 10 cm high; **St.** 4 mm thick, Int. 2–5 mm long, enclosed in the dry L.-Sh.; **L.** cylindrical, obtuse, bluish-green, 3 cm long, 6 mm thick, Sh. swollen, 2 mm long, with an impressed line; **Fl.** in 3's, 16 mm ∅, pink.

**R. triquetra** L. BOL. (*Mes. t.* (L. BOL.) N. E. BR.). – Cape: Clanwilliam D. – Erect ♄ 15 cm high; **St.** stiff, rather tangled, Int. 10 mm long, **Br.** 25 mm long, Int. 6 mm long; **L.** 10 on a shoot, exactly triquetrous, with a decurrent keel, keel and margins with minute papillose hairs, 8 mm long, 2 mm ∅; **Fl.** solitary, 17 mm ∅, Pet. pink, white-striped.

**R. truteri** L. BOL. – Cape: Somerset East D. – **St.** virgate, compressed, ± alate, 11–19 cm long, Int. 3–4 cm long; **L.** narrowing towards the tip, the keel on the underside reduced to a line, rarely faintly dentate, or immature **L.** with 3–5 T., 2 cm long, 3–4 mm ∅; **Fl.** solitary, Pet. 13 mm long, purplish-pink.

**R. tuberculosa** L. BOL. (*Mes. enormis*, N. E. BR.). – Cape: Clanwilliam D. – Robust, laxly branched ♄ c. 8 cm high; **St.** prostrate, stiff, branched, Int. 5–15 mm long, flowering **shoots** 3 cm long; **L.** 2 on a shoot, narrowing to the short-tapered tip, swollen at the base, keeled on the underside, with a faint line along the angles, bluish-green, 15 mm long, 4 mm ∅; **Fl.** solitary, 19 mm ∅, pink.

**R. tumidula** (HAW.) SCHWANT. (*Mes. t.* HAW.[1]), *Mes. foliolosum* HAW., *R. f.* (HAW.) SCHWANT., *Mes. imbricatum* v. *rubrum* HAW., *Mes. multiflorum* v. *rubrum* HAW.). – Cape: Swellendam, Cape, Malmesbury D.; S.W.Afr.: Lüderitz D. – Freely branched ♄ up to 60 cm high; young **St.** ± compressed; **L.** swollen at the base, linear-trigonous, ± obtuse, apiculate, with smooth angles, dull greyish-green, spotted, 2.5 cm or more long, 4 mm across; **Fl.** numerous, 2 cm ∅, pink.

**R. turnerana** L. BOL. – Cape: Van Rhynsdorp D. – Forming mats 2–4 cm high; **St.** 5 to 10 mm ∅; **Br.** covered with dry old L.; **L.** oblong to ± oval, acute to rounded above, with ciliate margins and keel, bluish-green, 1.5 cm long, Sh. 5 mm long; **Fl.** 15 mm ∅, pink.

**R. uitenhagensis** (L. BOL.) SCHWANT. (*Mes. u.* L. BOL.). – Cape: Uitenhage D. – Erect, densely branched ♄ 25–30 cm high; **St.** cylindrical; **L.** ± semicylindrical, with a hamate, recurved, apiculate tip, bluish-green, spotted, 5–9 mm long, 2–3 mm ∅; **Fl.** solitary, 14 mm ∅, pink.

**R. umbellata** (L.) SCHWANT. (*Mes. u.* L.). – Cape: Karroo. – ♄ 60–80 cm high; **St.** inclined; **L.** longer than the Int., 5–7 cm long, 4–6 mm across, obtusely trigonous, recurved at the tip, fresh green, spotted, Sh. thickened and amplexicaul; **Fl.** numerous, 3 cm ∅, white.

**R. unca** (L. BOL.) L. BOL. (*Mes. u.* L. BOL., *Lampranthus u.* (L. BOL.) SCHWANT.). – Cape: Calvinia D. – Erect ♄; **St.** spreading, 25 cm long; **Br.** short, with numerous bluish-green **L.** 10–25 mm long, 3 mm ∅, united and sheathing for 3–4 mm, tipped with a hooked bristle; **Fl.** solitary, 18 mm ∅, pale pink.

**R. uncinata** (L.) SCHWANT. (*Mes. u.* L., *Mes. uncinellum* HAW., *R. uncinella* (HAW.) SCHWANT., ? *Mes. perfoliatum* v. *integrifolium* L. BOL. nom. nud.). – Cape: Karroo, Van Rhynsdorp D.; S.W.Afr.: Karas Mts. – **St.** elongated, curved, prostrate, with Br. and shoots only on one side; **Br.** 4–5 mm thick; **L.** with a long Sh. completely enclosing the shoot, 4–8 mm long, trigonous to cylindrical, apiculate, with 1–2 short T. on the keel, greyish-green, spotted; **Fl.** 2 cm ∅, pinkish-red. Hardy outdoors.

**R. unidens** (HAW.) SCHWANT. (*Mes. u.* HAW.). – Cape: Albany D. – ♄ 30–40 cm high; **St.** spreading; **L.** compressed-trigonous, obliquely truncate above, with 1 T. directed downwards, apiculate, whitish-grey, spotted, 2–5 cm long, Sh. 1.5 mm long; **Fl.** solitary, 2 cm ∅, pink.

**R. utilis** (L. BOL.) L. BOL. v. **utilis** (*Mes. u.* L. BOL., *Lampranthus u.* (L. BOL.) SCHWANT.). –

---

[1] N. E. BROWN and L. BOLUS regard *Mes. tumidulum* HAW. as a synonym of *Mes. umbellatum* L., but G. SCHWANTES's interpretation is followed here.

Cape: Lit. Namaqualand, Piquetberg D.– Erect, densely branched ♄ up to 2 m high; **St.** 5 mm thick; **Br.** slender, Int. 1–4 mm long; **L.** ± expanded towards the obtuse tip, 10 to 15 mm long, 2–5 mm thick, Sh. 1 mm long; Fl. 1 cm ⌀, white.

**R. — v. giftbergensis** L. BOL. – ♄ 1.2–1.5 m tall; **L.** falcate, laterally compressed, 2 cm long, 3–4 mm across, 5–7 mm thick; Fl. pale pink.

**R. vaginata** (HAW.) SCHWANT. (*Mes. v.* HAW.). – Cape. – Rounded ♄ 60–90 cm high; St. inclined, freely forking; **L.** sheathing and united for 1–2 cm, enclosing the Int., trigonous, with a recurved mucro, the keel-angle minutely rough; Fl. numerous, 2–2.5 cm ⌀, white.

**R. valida** SCHWANT. (*R. robusta* SCHWANT.). – Cape: Lit. Namaqualand. – **L.** in crowded pairs, thick, navicular, bluntly tapering, ± distinctly keeled, velvety-papillose, 25 mm long, 5 mm across, 10 mm thick, Sh. 5 mm long; Fl. solitary, 1.8 cm ⌀.

**R. vanbredai** L. BOL. – Cape: Uniondale D. – ♄ c. 12 cm high, c. 35 cm ⌀; St. slender, tangled, 14–17 cm long, Int. 2.5–3.5 cm long; **Br.** c. 15 cm long; **L.** falcate, ± rounded at the tip, upperside linear; Fl. 1–1.5 cm ⌀, purplish-pink.

**R. vanderbergii** L. BOL. – Cape: Graaff Reinet D. – Erect, laxly branched ♄ 20 cm high; **St.** 15–20 cm long, Int. 15–20 mm long, young **shoots** 2–3.5 cm long, with 4–6 compressed-carinate, apiculate, bluish-green, spotted **L.** 5–9 mm long, 2–3 mm ⌀, with 1–2 slender T. on the margins and keel above, Sh. 4 mm long; Fl. 15 mm ⌀, purplish-pink.

**R. vanheerdei** L. BOL. – Cape: Lit. Namaqualand. – Prostrate, tangledh; **St.** slender, stiff, up to 30 cm long, Int. 3–5 cm long; **shoots** 11 to 14 cm long; **L.** narrowing towards the obtuse to acute tip, rounded on the underside, 3 to 4 cm long, 6–8 mm ⌀, Sh. 3.5 mm long; Fl. 2–4 cm ⌀, white.

**R. vanniekerkiae** L. BOL. – Cape: Ladismith D. – **St.** up to 4 mm thick; flowering **shoots** with 2 swollen, bluish-green to pink-tinged **L.** 4–7 mm long, 2–3 mm ⌀, obliquely keeled and with one indistinct T. at the tip; Fl. 18 mm ⌀, pink.

**R. vanzijlii** L. BOL. – Cape: Kenhardt D. – Stiff, compact ♄ 6 cm high; **St.** 9 mm thick, Int. 2–10 mm long, **Br.** short, with 2–4 **L.** 1 cm long, 3 mm across, 2.5 mm thick, keeled on the underside, narrowing towards the shortly tapered and apiculate tip, bluish-green with crowded dots, Sh. 2 mm long; Fl. solitary, 13 mm ⌀, purplish-pink.

**R. varians** L. BOL. – Cape: Lit. Namaqualand. – **St.** spreading to ascending, 18 cm long, Int. 12–50 mm long; **Br.** 2.5–3 cm long; **L.** acute, bluntly keeled, blue to reddish, 20 to 27 mm long, 4 mm ⌀; Fl. 1–3 together.

**R. velutina** L. BOL. (*Mes. lacuniatum* N. E. BR.). – Cape: Lit. Namaqualand. – Erect ♄ 30 cm high, herbaceous parts velvety with small papillae; **St.** cylindrical, Int. 15–20 mm long; **L.** ± clavate, obtuse, apiculate, ± keeled towards the tip, bluish-green, 20–25 mm long, 6 mm across, 7 mm thick, Sh. 1.5 mm long, with an impressed line; Fl. 8 mm ⌀, pink.

**R. ventricosa** (L. BOL.) SCHWANT. (*Mes. v.* L. BOL., *Cheiridopsis v.* (L. BOL.) N. E. BR.). – Cape: Van Rhynsdorp D. – **St.** crowded, 4–5 cm long, covered with old L.-Sh.; **Br.** with 2–4 L., with short shoots and a basal Sh.; **L.** (4–)8(–9) cm long, 13–17 mm across, 10–13 mm thick, swollentrigonous, obtuse or acute, bluish-green to grey, Sh. 18 mm long; Fl. solitary, with ventricose bracts on the pedicels, Pet. purplish-pink.

**R. verruculosa** L. BOL. (*Mes. v.* (L. BOL.) N. E. BR.). – Cape: Worcester D. – Compact, erect ♄ 7 cm high; **St.** 2 mm ⌀, Int. 5 mm long, herbaceous parts tuberculate with small papillae; lower **L.** semi-ovate, 4 mm long, 2 mm ⌀, upper L. 1 mm ⌀, bristle-tipped, Sh. 1 mm long; Fl. 12 mm ⌀, purplish-pink.

**R. versicolor** L. BOL. (*Mes. coloratum* N. E. BR.). – Cape: Lit. Namaqualand. – Robust ♄; **St.** ascending to spreading, 40 cm long, Int. 22–25 mm long; **L.** cylindrical, tapered above, bluish-green, 95 mm long, 8 mm thick, Sh. swollen, 4 mm long; Fl. 2 cm ⌀, pink.

**R. victoris** (L. BOL.) L. BOL. (*Mes. v.* L. BOL.). – Cape: Clanwilliam D. – Densely branched ♄; **St.** spreading to prostrate, 6 mm or more ⌀, up to 8 cm long; **L.** acutely trigonous above, obtuse, the keel rather oblique, velvety, 2–4 –(6.5) cm long, 6 mm across and thick; Fl. 15 mm ⌀, Pet. pinkish with a purple stripe.

**R. villetii** L. BOL. – Cape: Van Rhynsdorp D. – **St.** and **Br.** crowded, covered below with the remains of old L.; **L.** 2 on young shoots, ovate, subacute to obtuse, with the keel on the underside and the margins ± horny, pale blue, tinged pink, 16 mm long, 10 mm across, 7 mm thick, Sh. 7 mm long; Fl. 2 cm ⌀, pinkish-purple.

**R. virens** L. BOL. (*Mes. dumosum* N. E. BR.). – Cape: Mossel Bay, Riversdale D. – Erect ♄ 25–28 cm high; **St.** 12 mm thick; **Br.** ascending to spreading, Int. 15–20 mm long; **L.** laterally compressed, with acute angles, rounded and ± mucronate above, with crowded, dark green dots, 20–28 mm long, 6 mm across, 10–22 mm thick below the tip; Fl. 18 mm ⌀, pink.

**R. virgata** (HAW.) L. BOL. (*Mes. v.* HAW.). – Cape: George, Uniondale D. – ♄ 20–50 cm high; **St.** thin, erect, eventually prostrate, 2-angled at first; **L.** trigonous, narrowing towards the apiculate, hooked tip, somewhat greyish-green, minutely spotted, 15–20 mm long; Fl. solitary, 2–2.5 cm ⌀, red.

**R. viridifolia** L. BOL. – Cape: Lit. Namaqualand. – Robust ♄; **St.** elongated to prostrate, Int. 15–25 mm long, flowering **Br.** densely leafy, 6–12 cm long; **L.** semicylindrical with indistinct angles, acute, apiculate, green with dense, translucent dots, 2–4 cm long, 7 mm thick Sh. 3 mm long, with an impressed line; Fl. 3 cm ⌀, pinkish-purple.

**R. vulvaria** (DTR.) SCHWANT. (Pl. 195/2) (*Mes. v.* DTR.). – S.W.Afr.: Gr. Namaqualand. – Densely branched, hemispherical ♄ 30–40 cm high; **L.** sheathing and united for

10–13 mm, free parts 15–20 mm long, apiculate, with slender, acute papillae, scarcely spotted, with 1 T. on the keel below the tip; **Fl.** solitary, 2 cm ⌀, violet-pink.

**R. watermeyeri** L. Bol. – Cape: Van Rhynsdorp D. – Erect ♃ 9 cm high; **St.** 2 cm ⌀ at the base; **Br.** robust, 9 mm thick; **shoots** 8 cm long, flowering shoots with 2–4 obtuse, bluntly keeled **L.** 16 mm long, 5 mm ⌀, Sh. 1–2 mm long; **Fl.** 15 mm ⌀, pink.

**R. wilmaniae** (L. Bol.) L. Bol. v. **wilmaniae** (*Mes. w.* L. Bol.). – Cape: W.Griqualand, Hay D. – ♃ with a tuberous **caudex** 15 mm thick; **St.** with 6–10 crowded, trigonous, acute, laterally keeled **L.** 2–3 cm long, 8–10 mm across, 7–9 mm thick, bluish-green, tinged pink, densely green-spotted; **Fl.** 2–2.5 cm ⌀, pink.

**R.** — v. **angustifolia** L. Bol. – Transvaal. – L. more slender, 4 cm long, 3–4 mm across, 4 mm thick.

**R.** — v. **vermeuleniae** (L. Bol.) L. Bol. (*Mes. v.* L. Bol.). – Cape: Karroo. – **St.** very short; **L.** 10 mm across below, 6 mm across midway and 3 mm at the tip, 4–7 mm thick; **Fl.** white.

**R. wittebergensis** (L. Bol.) Schwant. (*Mes. w.* L. Bol.). – Cape: Laingsburg D., Karroo. – Low, cushion-forming ♃ 2.5 cm high; **St.** bearing short shoots with 4 L.; **L.** navicular, white-tuberculate especially on the keel on the underside and the margins, 4 mm long, 4 mm across, 2 mm thick, Sh. 1 mm long; **Fl.** 14 mm ⌀, pinkish-purple.

**Ruschianthemum** Friedr. (§ IV/1/5). – ♃ resembling **Stoeberia**; **L.** fleshy, opposite, ± united at the base, spreading to ascending, flat on the upperside, the margins ± angular, ± keeled below the tip; **Infl.** terminal, dichotomous; **Fl.** numerous, on pedicels with 2 bracts, small, with short, filiform Pet.; **Fr.** a dehiscent capsule. – Summer: in the open, winter: coldhouse. Propagation: seed, cuttings.

**R. gigas** (Dtr.) Friedr. (*Mes. g.* Dtr., *Stoeberia g.* (Dtr.) Dtr. et Schwant., *Ruschia micropetala* L. Bol., *Mes. m.* (L. Bol.) L. Bol.). – S.W.Afr.: Gr. Namaqualand; Cape: Lit. Namaqualand. – Dichotomously branched ♃ up to 80 cm high; **L.** 35–40 mm long, 7–9 mm across, 12–16 mm thick, ± hatchet-shaped, obtusely trigonous, obtuse with a red mucro, dark green to brownish-red, with sparse dots, soft-fleshy; **Pet.** very narrow, white.

**Ruschianthus** L. Bol. (§ IV/1/6). – Small ♃ related to the G. **Delosperma**; Pet. united at the base, staminodes coherent with the stamens, the stigmas conspicuously shorter. – Summer: in the open, winter: cold-house. Propagation: seed, cuttings.

**R. falcatus** L. Bol. (Pl. 195/1). – S.W.Afr.: Numeis. – ♃ 5–7 cm high, 9–13 cm ⌀; **roots** woody to ± tuberous; herbaceous parts minutely rough, pale bluish-green; **St.** densely covered with the remains of old L.; **L.** 4 on the flowering shoots, falcate, acute, convex on the sides below and ± compressed above, with an indistinct, somewhat pinkish keel, expanded below midway and rounded below the tip, the L. of a pair unequal, one L. 3–4.5 cm long, 5–6 mm across midway, c. 1.8 cm ⌀, the other one smaller; **Fl.** on a pedicel 1–2 cm long, with 2 leaf-like bracts, Pet. 1.4 mm long, pale lemon-yellow.

**Saphesia** N. E. Br. (§ IV/4). – ♃ succulents with a long **caudex; St.** branched, woody, with distinct Int.; **L.** opposite, not united at the base, flat, simple; **Fl.** solitary, terminal, long-pedicellate. – Greenhouse: requires a deep pot. – Propagation: difficult from seed, woody portions of the **shoots** can be used as cuttings.

**S. flaccida** (Jacq.) N. E. Br. (Pl. 195/4) (*Mes. f.* Jacq.). – Cape: Cape, Malmesbury D., rare. – **Caudex** up to 1.50 m long, black, brittle, producing woody **St.** 10–20 cm long; **L.** at first in a basal Ros. and deciduous at flowering, limp, sessile, spreading, linear-lanceolate, tapered, rounded-keeled on the underside, green, 4–5 cm long, 3–5 mm across; Fl. c. 2 cm ⌀, Pet. very numerous, very acute, snow-white (yellow acc. Jacquin).

**Sarcozona** J. M. Black (§ IV/5). – Small, glabrous, succulent ♃, branching repeatedly from almost every node; lower **Br.** elongated, prostrate, rarely rooting at the nodes, 2-angled at first, eventually woody and polygonal; **L.** opposite, sessile, narrow, ± incurving, united at the base, acutely trigonous, as thick as broad, mostly asymmetrical, thickened below on the upperside, gradually narrowing towards the tip where the keel is curved abruptly upwards, the keel and angles ± tuberculate to dentate, with translucent dots, leaving a collar-shaped scar after withering; flowering **shoots** with 1(–3) nodes; **Fl.** variable in size, light purple or very rarely white, diurnal, solitary, sessile or almost so, enclosed in the bases of the upper L. which unite to form a bract-cup; **Fr.** juicy, purple or pale. – Cultivation as for **Carpobrotus**.

**S. bicarinata** S. T. Blake. – S.Austr. – C. 10 cm high; **St.** (3–)6(–7) mm thick, light brown, eventually pale; **L.** 2.5–5 cm long, 5 to 9 mm across the upperside, with the sides 4–9 mm across, narrowing towards the tip, dull green, tinged ± deep red or purple, the angles and keel red at first, ± dentate at the tip, with conspicuous, translucent dots, specially on the upperside below; **Fl.** 1–3 together, c. 3 cm ⌀; Cal.-tube and Fr. acutely double-keeled; **Fr.** purple.

**S. praecox** (F. Muell.) S. T. Blake ex H. J. Eichler (Pl. 196/1) (*Mes. p.* F. Muell., *Carpobrotus pulleinei* J. M. Black, *S. p.* (J. M. Black) J. M. Black). – W. and S.Austr., NSW, Victoria. – Plant ± pruinose; **St.** 3.5–8 mm thick, mostly cylindrical, pale green below, Int. expanded above, channelled, angular and light grey-pruinose at the tip; flowering Br. usually with only 1 node; **L.** almost knife-shaped to lanceolate, narrowing towards the tip, with or without a mucro, the angles and keel irregularly dentate to tuberculate-dentate, covered with prominent dots and so appearing blistered, 8–10 cm long, 4.5 mm across, 5–6 mm thick; bract-cup 3-ribbed with lobes 0.5–3 cm long; Fl. 2.5–6 cm ⌀, Pet. white at the base or uniformly so.

**Sceletium** N. E. Br. (§ I/1/3). – Small sub-♄ with a short **St.**; **Br.** long or short, spreading or creeping, papillose when immature; **L.** united at the base, ovate-lanceolate, acute or tapered, the M.-rib with 2 reticulate, lateral, principal veins, persisting as skeletons after the leaf-fabric has withered, covered by the dry, transparent epidermis; **Fl.** 1–3 together on thick pedicels or elongated flowering Br., large, white, pale yellow or pink. – Greenhouse, warm, with complete rest in winter. Propagation: seed, cuttings.

**S. albanense** L. Bol. – Cape: Albany D. – Flowering **Br.** elongated, 23 cm long, Int. 2–3.5 cm long, the bark flaccid and uneven; **L.** 6–10 on a shoot, linear to lanceolate, acute to tapered, glossy-papillose, 2–3 cm long, 5–9 mm across, Sh. 2–3 mm long; **Fl.** 4–5 cm ⌀, white.

**S. anatomicum** (Haw.) L. Bol. (*Mes. a.* Haw., *Tetracoilanthus a.* (Haw.) Rap. et Cam.). – Cape: Cradock D. – **Br.** slender, gnarled in age; **L.** oblong-lanceolate, shortly tapered, glossy with transparent papillae, 2–2.5 cm long, 8–10 mm across; **Fl.** 4–5 cm ⌀, white.

**S. archeri** L. Bol. – Cape: Matjiesfontein D. – **Br.** elongated, slender, Int. 12–30 mm long; **L.** ovate-lanceolate to lanceolate, tapered, dirty green, indistinctly papillose, 15 mm long, 5–7 mm across, Sh. 1–2 mm long; **Fl.** 28 mm ⌀, very pale straw-coloured.

**S. boreale** L. Bol. – Cape: Bushmanland. – **Br.** 5 mm thick at the base; **L.** oblong-lanceolate, 2–2.5 cm long, 10 mm across, acute to tapered, sheathing and united for 4 mm; **Fl.** 3 cm ⌀, Pet. white, salmon-pink above.

**S. compactum** L. Bol. (Pl. 196/2). – Cape: Laingsburg D. – Compact ♄ 7 cm high; **L.** imbricate, narrow-ovate, rarely to ± lanceolate, tapered, 12–15 mm long, 6–10 mm across, veins of the dry L. slender, reticulate; **Fl.** 3 cm ⌀, ivory-coloured.

**S. concavum** (Haw.) Schwant. (*Mes. c.* Haw., *Tetracoilanthus c.* (Haw.) Rap. et Cam.). – Cape: origin unknown. – **Br.** slender, curved, slightly gnarled; **L.** oblong-lanceolate, tapered, 2–2.5 cm long, up to 1 cm across, fresh green, minutely papillose; **Fl.** c. 3.5 cm ⌀, white.

**S. crassicaule** (Haw.) L. Bol. (*Mes. c.* Haw., *Pentacoilanthus c.* (Haw.) Rap. et Cam.). – Cape: origin unknown. – **Br.** densely leafy; **L.** linear-lanceolate, furrowed on the upperside, green, glossy-papillose, 5 cm long, 6 to 10 mm across; **Fl.** c. 4 cm ⌀, pale straw-coloured.

**S. dejagerae** L. Bol. – Cape: Beaufort West D. – Plant 10 cm high; **St.** 8 mm thick at the base; **Br.** spreading to prostrate, 29 cm long, Int. 15–35 mm long, with a white, membranous bark, **shoots** 4 mm thick, Int. 5.5 cm long, herbaceous parts papillose; **L.** lanceolate-ovate, tapered, 36 mm long, 14 mm across, rarely ovate and shorter; **Fl.** 35 mm ⌀, white, rarely pale pink.

**S. emarcidum** (Thunbg.) L. Bol. (*Mes. e.* Thunbg.). – Cape: Calvinia, Fraserburg D., Lit. Namaqualand. – Laxly branched ♄ up to 30 cm high; **Br.** spreading, up to 45 cm long, Int. 2.3 cm long; **L.** ovate to ovate-lanceolate, acute, 2–4 cm long, 1–2 cm across, with the veins persisting as a conspicuous, branched, stiff skeleton after the L. has withered; **Pet.** snow-white, up to 17 mm long, 1 mm across.

**S. expansum** (L.) L. Bol. (*Mes. e. L., Pentacoilanthus e.* (L.) Rap. et Cam., *Mes. tortuosum* DC.). – Cape: Uniondale, Willowmore D. – Up to 30 cm high, with a fleshy **St.**; **Br.** slender, curved, prostrate; **L.** lanceolate, tapered, the M.-nerve keeled on the underside, fresh green, glossy-papillose, c. 4 cm long, 15 mm across; Fl. 4–5 cm ⌀, dirty yellow.

**S. framesii** L. Bol. – Cape: Calvinia D. – Compact plant; **Br.** 5–10 mm thick, with 6–8 crowded, imbricate, ovate-lanceolate to oblong-lanceolate, acute, thick, yellowish-green **L.** 10–15 mm long, 8 mm across, Sh. 3 mm long; **Fl.** 45 mm ⌀, white.

**S. gracile** L. Bol. – Cape: Willowmore D. – **Br.** prostrate, 5–6 cm long; **L.** ovate-lanceolate, acute, 2.5 cm long, 1 cm across, Sh. 2 mm long; Fl. 34 mm ⌀, pale yellow with a white centre.

**S. joubertii** L. Bol. – Cape: Ladismith D. – Compact plant; **Br.** prostrate, 7 cm long; **L.** imbricate, ovate, tapered, bluntly keeled, either 3.5 cm long, 2 cm across or 4 cm long, 18 mm across; **Fl.** 3.8 cm ⌀, straw-coloured.

**S. namaquense** L. Bol. v. **namaquense** – Cape: Lit. Namaqualand. – **Br.** prostrate, 26 cm long; **L.** imbricate, lanceolate-linear to oblong or ovate-lanceolate, green, slender-veined, 25–32 mm long, 6–12 mm across, Sh. 4 mm long; Fl. almost 4 cm ⌀, pale pink.

**S. —** v. **subglobosum** L. Bol. – Variety with subglobose receptacle and narrower Sep.

**S. ovatum** L. Bol. – Cape: Laingsburg, Montagu D. – **Br.** up to 21.5 cm long, Int. 15–35 mm long; **L.** usually thick, ovate, ± cordate at the base, obtuse, nerves and keel imperceptible, 17–22 mm long, 10–13 mm across, Sh. 3 mm long; Fl. 25–30 mm ⌀, straw-coloured.

**S. regium** L. Bol. – Cape: Malmesbury, Piquetberg, Clanwilliam D. – Robust plant; **caudex** tuberous, 8.5 cm or more long, up to 2.5 cm ⌀; **St.** 1 cm long, **Br.** 4–10 cm long; **L.** variable, either lanceolate, 6.5 cm long, 2.2 cm across, or ovate-lanceolate to ovate, acute, the Sh. indistinct or up to 3 mm long; Fl. 5.5 cm ⌀, yellow to golden-yellow.

**S. rigidum** L. Bol. – Cape: Prince Albert, Laingsburg D. – **Br.** ± erect, stiff, ± 4-angled, with a white, scaly bark, Int. 12–20 mm long; **L.** lanceolate to ovate-lanceolate, acute, 2 cm long or less, 7 mm across, Sh. 2 mm long, L. of the short shoots oblong-lanceolate, 10 mm long, 5 mm across, with large papillae; **Fl.** 22 mm ⌀, white.

**S. strictum** L. BOL. – Cape: Ladismith, Willowmore D., Karroo. – **Br.** crowded, short or elongated, stiff, erect, 15 cm long, bark membranous, Int. 5 cm long; **L.** ovate-lanceolate, acute, covered with round papillae, basal L. 4 cm long, upper ones 2 cm long, 9 mm across, 3 mm thick; **Fl.** white.

**S. subvelutinum** L. BOL. – Cape: Worcester D. – **Br.** prostrate, 23 cm long including the Fl., Int. 15–25 mm long; **L.** oblong-lanceolate, 16 mm long, 6 mm across at the base, Sh. 1 mm long; **Fl.** 34 mm ⌀, white to pale straw-coloured.

**S.** — v. **luxurians** L. BOL. – More luxuriant variety; **caudex** tuberous; **L.** 11 mm across; **Fl.** c. 5 cm ⌀.

**S. tortuosum** (L.) N. E. BR. (*Mes. t.* L., *Pentacoilanthus t.* (L.) RAP. et CAM., ? *S. varians* (HAW.) L. BOL., ? *Mes. v.* HAW.). – Cape: Karroo. – Resembles **S. concavum**; **L.** ovate-lanceolate, obtusely tapered, channelled on the upperside, keeled on the underside, fresh green, minutely papillose, c. 2.5 cm long, 1 cm across; **Fl.** 4–5 cm ⌀, yellowish-white.

**S. tugwelliae** L. BOL. – Cape: Prince Albert D. – Compact plant; **Br.** 5–7 cm long, Int. short; **L.** 26 mm long, 14 mm across, oblong-ovate, acute, bluntly keeled on the underside; **Fl.** 4 cm ⌀, white.

**Schlechteranthus** SCHWANT. (§ IV/1/2). – Low-growing ♄ with conspicuously united, succulent **L.**; **Fl.** solitary, sessile or shortly pedicellate, the pedicel with 2 bracts, Pet. purplish-pink or pale pink. – Summer: in the open, winter: cold-house. Propagation: seed, cuttings.

**S. hallii** L. BOL. – Cape: Lit. Namaqualand. – Erect, compact, stiff ♄ 25 cm high and ⌀; St. c. 13 mm ⌀, Int. 7–15 mm long; **L.** 2–4 on a shoot, acute, rarely obtuse, keeled or only indistinctly so, glossy, pale bluish-green, blistered at the base, 15–25 mm long, 8–10 mm ⌀, Sh. 6 mm long and decurrent for 1 cm, with an impressed line; **Fl.** solitary, sessile, the bracts 7–10 mm long, with membranous margins, Pet. 7–10 mm long, white, ± pink above.

**S. maximilianii** SCHWANT. – Cape: Lit. Namaqualand. – Erect ♄ 5–6 cm high; **roots** stout, woody; St. very crowded, without visible Int.; **L.** united to midway into a Bo. 5 mm long and across, apiculate above, keel with 1–2 T., smooth, glossy, bluish to reddish; **Fl.** solitary, 11 mm ⌀, with bracts clasping the pedicel which is 5 mm long, Pet. purplish-pink.

**Schwantesia** DTR. (non L. BOLUS) (§ IV/1/14). – Many-headed, cushion-forming, highly succulent plants, Int. enclosed in the L.-Sh.; **St.** up to 1 cm ⌀; **Br.** with age densely clothed with the hardened remains of previous years' L., up to 7 mm ⌀, flowering branchlets with 2 erect or ascending, acute, obtuse or somewhat rounded, glabrous or velvety, bluish or pale bluish-green, often marbled **L.** 2.5–5 cm long, 5–17 mm across, those of a pair unequal, entire or dentate or lobulate, acutely and eccentrically keeled on the underside, margins and keel often red, the upperside flat and there ± oblique, often rounded; **Fl.** solitary, on a pedicel 1–2.5 cm long, 2.8–5.7 cm ⌀, yellow or golden-yellow, rarely orange. – Greenhouse, warm. Propagation: seed, cuttings.

## Key to the Species of the Genus Schwantesia by L. Bolus

| | | |
|---|---|---|
| 1. | Herbaceous parts, especially when young, velvety on account of minute hairs terminating the papillae ("pearls") | 2. |
| 1a. | Herbaceous parts glabrous | 3. |
| 2. | L. usually dentate or lobed near the apex | **marlothii** |
| 2a. | L. entire | **triebneri** |
| 3. | L. usually dentate or lobed near the apex | 4. |
| 3a. | L. entire | 5. |
| 4. | L. marbled with white; Ov.-lobes abruptly elevated from the disc, obtusely compressed .... **ruedebuschii** | |
| 4a. | L. not marbled; Ov.-lobes gradually elevated from the disc | **australis** |
| 5. | L., viewed laterally, acute or acuminate | **pillansii** |
| 5a. | L., viewed laterally, usually obtuse, rounded or somewhat truncate | 6. |
| 6. | L. marbled with white; Ov.-lobes abruptly elevated from the disc | **borcherdsii** |
| 6a. | L. not marbled; Ov.-lobes elevated from well within the disc | 7. |
| 7. | L. up to 3.5 cm long | 8. |
| 7a. | L. up to 5.5 cm long | 10. |
| 8. | L. up to 3.5 cm long; Pet. emarginate | 9. |
| 8a. | L. up to 2 cm long; Pet. entire | **herrei** v. **minor** |
| 9. | Fl. 32–42 mm ⌀ | **— v. herrei f. herrei** |
| 9a. | Fl. up to 55 m ⌀ | **— f. major** |
| 10. | Pet. usually acute or acuminate | **acutipetala** |
| 10a. | Pet. (where known) obtuse | 11. |
| 11. | Fl. 3–3.5 cm ⌀, Pet. yellow | **succumbens** |
| 11a. | Fl. 5.5 cm ⌀; Pet. golden or orange | **speciosa** |

**S. acutipetala** L. Bol. − Cape: Lit. Namaqualand. − **L.** 42 mm long, 15 mm across, 5 mm thick, sharply tapered, flat on the upperside, semicircular on the underside, acutely angled, bluish-grey; **Fl.** 4 cm ⌀, yellow.

**S. australis** L. Bol. − Cape: Bushmanland. − **St.** often with 2 pairs of **L.** up to 5 cm long, 8−10 mm across, flat on the upperside, angles and keel entire or with 1−3 small T., often with 1−4 short lobes at the tip, whitish or greyish-blue; **Fl.** 3.5−4 cm ⌀, yellow.

**S. borcherdsii** L. Bol. − Cape: Upington D. − **St.** with (2−)4(−6) variously shaped **L.** 3−4 cm long, up to 12 mm across, truncate to rounded at the tip, flat on the upperside, keeled on the underside; **Fl.** 3.5−4 cm ⌀, yellow.

**S. herrei** L. Bol. v. **herrei** f. **herrei**. − Cape: Lit. Namaqualand; S.W.Afr.: Witpütz. − Forming compact cushions; **St.** with 2−3 pairs of **L.** 2.5−3.5 cm long, 16 mm across, with an entire acute keel pulled forward chinlike towards the obtuse tip, often with several T., pale bluish green to chalky green, smooth; **Fl.** 32−42 mm ⌀, yellow; Pet. emarginate.

**S.** — v. — f. **major** Rowl. (*S. loeschiana* Tisch.) − S.W.Afr.: Gr. Namaqualand. − **Fl.** up to 5.5 cm ⌀.

**S.** — v. **minor** L. Bol. − S.W.Afr.: Lüderitz D. − **L.** 9−20 mm long, 10 mm ⌀; **Fl.** 2−2.5 cm ⌀, Pet. entire.

**S. marlothii** L. Bol. − Cape: Lit. Namaqualand. − **L.** almost 4 cm long, c. 9 mm across, 5 mm thick, expanded towards the tip and with 8 red-bordered T. or lobes, rarely entire, semicircular on the underside, whitish to bluish-green, often with red dots and marks; **Fl.** 3−4 cm ⌀, yellow.

**S. pillansii** L. Bol. − Cape: Kenhardt D. − **St.** with 2 bluish-green, obliquely acute to tapered **L.** 5.3 cm long, 5 mm across, 8 mm thick, obliquely keeled on the underside, with a red line along the keel; **Fl.** 3.2 cm ⌀, yellow.

**S. ruedebuschii** Dtr. (Pl. 196/3). − S.W.Afr.: Gr. Namaqualand. − Forming round mats; **L.** 3−5 cm long, 10−12 mm across, 10 mm thick at the base, navicular, the margins ± rounded, suffused bluish-green with white mottling, the tips expanded and obtuse, with 3−7 thick, broad, blue, brown-tipped T. up to 4 mm long; **Fl.** 3.5−4 cm ⌀, light yellow.

**S. speciosa** L. Bol. − Cape: Bushmanland. − **L.** 4.5 cm long, sometimes longer, 12−17 mm across, 18−20 mm thick, entire, acute to tapered, laterally keeled, Sh. 1 cm long; **Fl.** 5.5 cm ⌀, golden-yellow.

**S. succumbens** (Dtr.) Dtr. (Pl. 197/4) (*Mes. s.* Dtr.). − S.W.Afr.: S. of Warmbad. − **L.** 5 to 6 cm long, 15 mm across at the base, somewhat expanded above, apiculate, 1 cm thick below, obliquely keeled on the underside above, the keel and margins acute, whitish to bluish-green; **Fl.** 3−3.5 cm ⌀, yellow.

**S. triebneri** L. Bol. − Cape: Bushmanland. − **St.** with 3 pairs of **L.** c. 4.6 cm long, 1 cm across, 5 mm thick, very expanded and mucronate above, with the underside round, keeled towards the tip, whitish to bluish-green or yellowish, with red dots and red angles; **Fl.** 4−5 cm ⌀, yellow.

**Scopelogena** L. Bol. (§ IV/1/23). − Robust, compact to mat-forming ♄ 30 cm or more high; **St.** up to 7 mm ⌀; **L.** fleshy, ± falcate, flat to slightly convex on the upperside, rounded to bluntly keeled on the underside, somewhat compressed laterally, rounded to obtuse to shortly acute above, 3−5 cm long, 3−8 mm ⌀; **Fl.** 7−15 together in a laxly branched Infl., on a pedicel 4−15 mm long with 2 bracts at the base, up to 2 cm ⌀, yellow. − Summer: in the open, winter: cold-house. Propagation: seed, cuttings.

**S. gracilis** L. Bol. − Cape: Swellendam, Riversdale D. − Slender ♄ up to 120 cm ⌀; **St.** up to 2.5 cm ⌀; **Br.** erect, spreading to prostrate; **L.** slender, trigonous to cylindrical, ± obtuse; **Fl.** small, yellow, scented.

**S. vereculata** (L.) L. Bol. (Pl. 197/1) (*Mes. verruculatum* L., the spelling with 2 r's being incorrect, *Ruschia v.* (L.) Rowl., *Lampranthus v.* (L.) L. Bol.). − Cape: Cape D. − ♄ up to 30 cm high; **St.** stout, erect, Int. short; **L.** crowded, trigonous to cylindrical, ± obtuse, apiculate, 2.5−3.5 cm long, 6−8 mm across, soft, fresh green, grey-pruinose; **Fl.** c. 1.5 cm ⌀, yellow.

**Semnanthe** N. E. Br. (§ IV/1/8). − Small ♄. − Summer: in the open, winter: cold-house. Propagation: seed, cuttings.

**S. lacera** (Haw.) N. E. Br. v. **lacera** (Pl. 197/2) (*Mes. l.* Haw,. *Mes. carinatum* Vent., *Mes. gladiatum* Jacq., *Mes. falcatum* Thunbg., *Mes. dentatum* Kern.). − Cape: Paarl D. − ♄ 60 to 80 cm high; **St.** spreading, robust, 2-angled; **L.** ± sabre-shaped, acute, apiculate, united at the base, strongly laterally compressed, trigonous, margins cartilaginous, denticulate, the keel-angle laciniate or cartilaginous-dentate, light green, grey-pruinose with translucent dots, 3−5 cm long, 8−11 mm across; **Fl.** solitary or paired, shortly pedicellate, 4−5 cm ⌀, glossy pinkish-red.

**S.** — v. **densipetala** L. Bol. − Cape: Malmesbury D. − Pet. more crowded, ± spatulate, deep pinkish-purple.

**Skiatophytum** L. Bol. (§ IV/3). − ⊙ to ♃ herbs; **L.** flat, oblanceolate-spatulate; **Fl.** pedicellate, terminal, white. − Summer: in the open, winter: cold-house. Shade-loving. − Propagation: seed.

**S. tripolium** (L.) L. Bol. (Pl. 197/3) (*Mes. t.* L., *Gymnopoma t.* (L.) N. E. Br., *Mes. expansum* Thunbg.). − Cape: Cape, Clanwilliam D. − **St.** short, thick; **Br.** up to 20 cm long, simple; **L.** densely rosulate at the base of the St., more distant on the Br., oblanceolate-spatulate, expanded at the sheathing base, scarcely fleshy, with a thick M.-nerve, smooth, glossy, the margins undulating at first, 5−8 cm long, 15−20 mm across; **Fl.** c. 2 cm ⌀, snow-white.

**Smicrostigma** N. E. Br. (§ IV/1/8). – Shrubby succulents. – Summer: in the open, winter: cold-house. Propagation: seed, cuttings.

**S. viride** (Haw.) N. E. Br. (Pl. 198/1) (*Mes. v.* Haw., *Erepsia v.* (Haw.) L. Bol., *Mes. integrum* L. Bol., *Ruschia i.* (L. Bol.) Schwant.). – Cape: Montagu, Robertson, Riversdale D. – ♄ up to 40 cm high; St. erect, slender, becoming woody, Int. indistinct; L. opposite, sheathing and united for 20–25 mm, the blade shorter, slightly trigonous, with a tiny red apiculus, smooth, green, with minute, translucent dots; Fl. solitary, terminal, ± sessile, 3 cm ⌀, pink.

**Sphalmanthus** N. E. Br. (§ I/1/4) (see *Nycteranthus*, paragraph 2). – Low, bushy sub-♄; **root-stock** partly large, tuberous; **Br.** sometimes partly creeping, rooting at the nodes, papillose, very soft fleshy; **L.** opposite, often crowded and little united at the base, linear, mostly terete or semiterete; **Fl.** solitary or 3 together in a 3-partite cyme, small to M. size, whitish, pink to red, summer. – Summer: in the open; winter: cold-house. Propagation: seed, cuttings.
Type species: **S. canaliculatus.**

## Division of the Genus Sphalmanthus into SG. by L. Bolus

**SG. I. Trichotoma** (Haw.) L. Bol. (*Trichotoma* Haw., *Geniculiflora* DC. and *Noctiflora* Haw. p. part as Sect. of *Mesembryanthemum* L.). – Bushy ♄; **L.** trigonous or nearly terete, often dorsally recurved at the apex, mostly minutely papillate; **Fl.** on 3-nate branched cymes, small, white or red, Pet. in one R., staminodes present. – Type species: S. trichotomus. – Further species: S. brevipalus, decurvatus, decussatus, defoliatus, delus, englishiae, framesii, geniculiflorus, hallii, prasinus, pumilus, quarziticus, radicans, rejuvenalis, rhodandrus, stayneri, strictus, suffusus, tetragonus, watermeyeri.

**SG. II. Digitiflora** (Haw.) L. Bol. (*Digitiflora* Haw. and *Splendentia* Salm as Sect. of *Mesembryanthemum* L.; acc. G. Schwantes: *Nycteranthus* Necker emend. Schwantes, SG. II *Neoaridaria* Schwant. p. part.). – Glabrous ♄ with copious opposite Br., if creeping without a tuberous rootstock, very watery and juicy; **L.** opposite on numerous branchlets, little connate at the base, nearly terete and only slightly trigonous, smooth, green, not dotted, very soft fleshy; **Fl.** solitary or 3 together, M.-sized, scented in some spec. which open at night, whitish, yellowish to red, Pet. in several R., staminodes absent. – Type species: S. splendens. – Further species: S. acuminatus, albicaulis, bijliae, blandus, celans, constrictus, flexuosus, fourcadei, leptopetalus, nothus, pentagonus, plenifolius, primulinus, rabiesbergensis, reflexus, roseus, spinuliferus, striatus, subaequans, subpatens, sulcatus, umbelliflorus, vernalis.

**SG. III. Sphalmanthus** (*Eusphalmanthus* L. Bol., *Spinulifera* Haw. and *Crassulina* Salm p. part as Sect. of *Mesembryanthemum* L.; acc. G. Schwantes: *Nycteranthus* Necker emend. Schwant., SG. *Sphalmanthus* (N. E. Br.) Schwant.). – Low sub-♄; **roots** partly tuberous; **Br.** long, creeping, rooting at the nodes and there forming Tub.; **L.** opposite, in the flowering reg. alternate, nearly terete or semiterete, upper surface slightly grooved, papillate, the papillae often prickly; **Fl.** solitary or several together, greenish to yellowish or reddish. – Type species: S. canaliculatus. – Further species: S. abbreviatus, acocksii, albertensis, anguineus, arenicolus, auratus, baylissii, carneus, caudatus, commutatus, congestus, crassus, deciduus, dinteri, dyeri, glanduliferus, godmanii, gratiae, grossus, gydouwensis, herbertii, humilis, latipetalus, laxipetalus, laxus, leipoldtii, lignescens, ligneus, littlewoodii, longipapillatus, longispinulus, longitubus, macrosiphon, nanus, nitidus, obtusus, oculatus, olivaceus, oubergensis, platysepalus, pomonae, praecox, rabiei, recurvus, salmoneus, saturatus, scintillans, sinuosus, straminicolor, subpetiolatus, tenuiflorus, vanheerdei, vigilans, viridiflorus, willowmorensis.

**SG. IV. Phyllobolus** (N. E. Br.) L. Bol. (*Phyllobolus* N. E. Br. as a G.). – Plants with a tuberous rootstock or caudex, with numerous papillate shoots above; **L.** all or only those of the flowering reg. alternate and papillate; **Fl.** nearly sessile or short stalked, dirty reddish to yellowish to green-yellow. – Type species: S. resurgens. – Further species: S. micans.

In the following synonyms, for *Nyct.* read *Nycteranthus*.

**S. abbreviatus** (L. Bol.) L. Bol. (§ III) (*Aridaria a.* L. Bol., *Nyct. a.* (L. Bol.) Schwant.). – Cape: Lit. Namaqualand. – Caudex bipartite and tuberous, 3 cm long, 15 mm thick; herbaceous parts papillose; St. prostrate, 8 cm long, Int. 5–12 mm long; L. yellowish-green, 12 mm long, 9 mm across, 4 mm thick, Sh. 1.5 mm long; Fl. 2 cm ⌀, dirty yellow.

**S. acocksii** L. Bol. (§ III). – Cape: Calvinia D. – Caudex tuberous, woody, 7 cm long, up to 1 cm ⌀; herbaceous parts round-papillose; St. woody, stiff, 26 cm long, Int. 1–4.5 cm long, shoots spreading; L. opposite, linear to linear-lanceolate, bluish-green, 1.5–3 cm long, 2–3 mm across; Fl. c. 2 cm ⌀, pink.

**S. acuminatus** (Haw.) L. Bol. (§ II) (*Mes. a.* Haw., *Aridaria a.* (Haw.) Schwant., *Nyct. a.* (Haw.) Schwant., *Perapentacoilanthus a.* (Haw.) Rap. et Cam., *Mes. flexuosum* S.D., *Mes. sulcatum* S.D.). – S.Afr.: origin unknown. – Erect ♄ up to 80 cm high; St. numerous, erect; L. fairly numerous, trigonous to semicylindrical, shortly tapering, green; **Fl.** solitary, terminal, glossy white, becoming faintly pink.

**S. albertensis** (L. BOL.) L. BOL. (§ III) (*Aridaria a.* L. BOL., *Nyct. a.* (L. BOL.) SCHWANT.). – Cape: Prince Albert D. – Compact, papillose ♄; **caudex** tuberous, round to oval, 3 cm high, 7 mm ⌀; **Br.** fleshy, creeping, 2.5–3 cm long; **L.** opposite, cylindrical, obtuse, 13–22 mm long, 3–4 mm across, 5 mm thick; **Fl.** 2.5 cm ⌀, dirty salmon-pink.

**S. albicaulis** (HAW.). L. BOL. (§ II) (*Mes. a.* HAW., *Aridaria a.* (HAW.) N. E. BR., *Nyct. a.* (HAW.) SCHWANT.). – Cape: origin unknown. – **St.** slender, white; **L.** semicylindrical, subulate, apiculate; **Fl.** white.

**S. anguineus** (L. BOL.) L. BOL. (§ III) (*Aridaria a.* L. BOL., *Nyct. a.* (L. BOL.) SCHWANT.). – Cape: Lit. Namaqualand. – Erect ♄ 26 cm high; **caudex** tuberous below, 8 mm thick; herbaceous parts very papillose; **St.** suberect, nodose; **L.** rather obtuse, pale blue tinged pink, 3.5 cm long, 5–8 mm thick; **Fl.** 22 mm ⌀, yellowish-pink.

**S. arenicolus** (L. BOL.) L. BOL. (§ III) (*Aridaria a.* L. BOL., *Nyct. a.* (L. BOL.) SCHWANT.). – Cape: Lit. Namaqualand. – **Caudex** thickened to tuberous; **St.** elongated, creeping, stiff, papillose, Int. 15–30 mm long; **L.** alternate, cylindrical, obtuse, blue tinged pink, glossy papillose, 2–3 cm long, 4 mm ⌀; **Fl.** 25 mm ⌀, pink.

**S. auratus** (SOND.) L. BOL. (§ III) (*Mes. a.* SOND., *Mes. aureum* THUNBG., *Aridaria aurea* (THUNBG.) L. BOL., *Nyct. aureus* (THUNBG.) SCHWANT.). – Cape: Van Rhynsdorp D. – Compact ♄; **St.** elongated, nodose, Int. 2 to 2.5 cm long; **L.** cylindrical, thick, obtuse, papillose, 2–2.5 cm long, 6 mm thick; **Fl.** numerous, pale pink, yellow inside.

**S. baylissii** L. BOL. (§ III). – Cape: Lit. Namaqualand. – Very tangled ♄; **caudex** thickened, 2.3 cm ⌀ above; **St.** stiff, woody, Int. 1–3.5 cm long; **shoots** with the herbaceous parts distinctly papillose; **L.** 10, opposite, acute to tapered or obtuse, bluish-green, 17 mm long, 2.5 mm ⌀, sometimes persisting and spinescent; **Fl.** salmon-pink.

**S. bijliae** (N. E. BR.) L. BOL. (§ II) (*Aridaria b.* L. BOL,. *Nyct. b.* (N. E. BR.) SCHWANT.). – Cape: Prince Albert D. – Small, erect ♄; **St.** opposite, divergent; **L.** 14–20 mm long, 4 mm ⌀, semicylindrical or indistinctly trigonous, obtuse; **Fl.** in a lax Infl.

**S. blandus** (L. BOL.) L. BOL. (§ II) (*Aridaria b.* L. BOL., *Nyct. b.* (L. BOL.) SCHWANT.). – Cape: Laingsburg D. – ± climbing ♄ up to 20 cm high; **St.** tangled, Int. 25 mm long; **L.** channelled on the upperside, rounded on the underside, tapered, 25–32 mm long, 4–6 mm across; **Fl.** 4 cm ⌀, pink to pinkish-red.

**S. brevisepalus** (L. BOL.) L. BOL. v. **brevisepalus** (§ I) (*Aridaria b.* L. BOL., *Nyct. b.* (L. BOL.) SCHWANT.). – Cape: Lit. Namaqualand. – Erect ♄; **St.** thick-fleshy, appearing ± spiny because of the remains of old L., 11 mm thick, Int. 1–2 cm long; **L.** opposite, channelled on the upperside, narrowing towards the obtuse tip, green, 3 cm long, 5 mm across; **Fl.** 5–7 cm ⌀, white to straw-coloured.

**S. — v. ferus** (L. BOL.) L. BOL. (*Aridaria b.* v. *f.* L. BOL., *Nyct. b.* v. *f.* (L. BOL.) SCHWANT., *A. parvisepala* L. BOL. = nom. err. for *brevisepala,* A. *gibbosa* L. BOL.). – Plant 30–60 cm high; **St.** ± tuberculate; **L.** semicylindrical; **Fl.** 22 mm ⌀, pale yellow.

**S. canaliculatus** (HAW.) N. E. BR. (§ III) (*Mes. c.* HAW., *Aridaria c.* (HAW.) FRIEDR., *Nyct. c.* (HAW.) SCHWANT., *Mes. calycinum* ECKL. et ZEYH., *S. c.* (ECKL. et ZEYH.) L. BOL., *Mes. salmoneum* SALM, *Mes. longispinulum* SOND., *Mes. reflexum* WILLD.). – Cape: Cape, Stellenbosch D. – **St.** freely branched, prostrate, **shoots** ascending; **L.** linear, ± obtuse, very convex on the underside, channelled on the upperside, green, papillose, of medium size; **Fl.** solitary, pale red or flesh-coloured.

**S. carneus** (HAW.) N. E. BR. (§ III) (*Mes. c.* HAW., *Nyct. c.* (HAW.) SCHWANT., *Mes. grossum* HAW.). – S.Afr.: origin unknown. – Close to S. **grossus,** possibly identical.

**S. caudatus** (L. BOL.) N. E. BR. (§ III) (*Mes. c.* L. BOL., *Aridaria c.* (L. BOL.) L. BOL., *Nyct c..* (L. BOL.) SCHWANT.). – Cape: Worcester D. – **Caudex** tuberous; **St.** prostrate, 9 cm long, Int. 5–10 mm long; **L.** alternate, semicylindrical, acute, channelled on the upperside, 5–8 mm long, 3 mm thick; **Fl.** 5–6 cm ⌀, straw-coloured.

**S. celans** (L. BOL.) L. BOL. (§ II) (*Aridaria c.* L. BOL., *Nyct. c.* (L. BOL.) SCHWANT.). – Cape: Montagu D. – **St.** 16–18 cm long, Int. 1–3 cm long; **L.** cylindrical, obtuse, 18–23 mm long, 2.5–4 mm ⌀; **Fl.** pink inside, paler in the centre.

**S. commutatus** (BGR.) N. E. BR. (§ III) (*Mes. c.* BGR., *Nyct. c.* (BGR.) SCHWANT., *Mes. longispinulum* S.D.). – S.Afr.: origin unknown. – **St.** and **Br.** fleshy, older Br. nodose, covered with dry, ± spinescent L.; **L.** semicylindrical, channelled on the upperside at the base, tapered, grey-papillose, 2.5–3 cm long, 4 mm across; Fl. 3.5 cm ⌀, creamy-white.

**S. congestus** (L. BOL.) SCHWANT. (§ III) (*Aridaria c.* L. BOL., *Nyct. c.* (L. BOL.) SCHWANT.). – Cape: Van Rhynsdorp D. – Compact plant, forming mats 4–8 cm high; **St.** and **Br.** thick, **shoots** erect, 2 cm long, with the Int. enclosed and 6–8 crowded, obtuse, bluish-green **L.** 2–3 cm long, 5 mm across, 1 mm thick, narrowed towards the tip, with prominent papillae and a Sh. 3 mm long; **Fl.** 28 mm ⌀, lemon-yellow.

**S. constrictus** (L. BOL.) L. BOL. (§ II) (*Aridaria c.* L. BOL., *Nyct. c.* (L. BOL.) SCHWANT.). – Cape: origin unknown. – Weak, slender ♄ with prostrate **St.** 25 cm or more long; **shoots** constricted above the nodes, Int. 2–3 mm long; **L.** cylindrical, constricted at the base, with a recurved tip, 10–17 mm long, 2 mm thick; **Fl.** 20–26 mm ⌀, white.

**S. crassus** L. BOL. (§ III). – Cape: Lit. Namaqualand. – **Caudex** thick, tuberous, up to 7 cm ⌀; **St.** up to 1 cm ⌀, Int. 2–2.5 cm long; herbaceous parts with round papillae; **L.** linear, obtuse, convex on the underside, 3.5 to 5.5 cm long; **Fl.** whitish-green, turning pale yellow and finally pale pink.

**S. deciduus** (L. Bol.) L. Bol. (§ III) (*Aridaria d.* L. Bol., *Nyct. d.* (L. Bol.) Schwant.). – Cape: Lit. Namaqualand. – Robust ♄ c. 40 cm high; **St.** 3.5 cm thick; **Br.** virgate; **L.** deciduous, obtuse, sometimes spiralled, dissimilar, lower L. 6–12 mm long, larger ones 15–17 mm long, 3–4 mm ⌀, rounded, channelled on the upper side, membranous at the base; **Fl.** 25 mm ⌀, salmon-pink to straw-coloured.

**S. decurvatus** (L. Bol.) L. Bol. (§ I) (*Aridaria d.* L. Bol., *Nyct. d.* (L. Bol.) Schwant.). – Cape: Lit. Namaqualand. – Much branched ♄; **St.** very curved; **L.** opposite; **Fl.** 2–2.5 cm ⌀, salmon-pink.

**S. decussatus** (Thunbg.) L. Bol. (§ I) (*Mes. d.* Thunbg.). – Cape: Bockland. – ♄ 30–60 cm high; **St.** opposite, papillose; **L.** semicylindrical, obtuse, ± furrowed on the upperside, green, papillose, 25 mm long, 2 mm across; **Fl.** white.

**S. defoliatus** (Haw.) L. Bol. (§ I) (*Mes. d.* Haw., *Aridaria d.* (Haw.) Schwant., *Nyct. d.* (Haw.) Schwant., *Peratetracoilanthus d.* (Haw.) Rap. et Cam., *Mes. clavatum* Jacq., ? *Mes. horizontale* Haw., ? *A. h.* (Haw.) Schwant., ? *Nyct. h.* (Haw.) Schwant.). – Cape: origin unknown. – ♄ 40 cm high; **St.** few, erect; **L.** semicylindrical, obtuse, slightly bluish-green, 2.5–3.5 cm long, 6 mm across; **Fl.** white to straw-coloured.

**S. delus** (L. Bol.) L. Bol. (§ I) (*Mes. d.* L. Bol., *Aridaria d.* (L. Bol.) L. Bol., *Perapentacoilanthus d.* (L. Bol.) Rap. et Cam., *Nyct. d.* (L. Bol.) Schwant.). – Cape: highlands. – Erect ♄ c. 11 cm high; **St.** tuberously thickened at the base; **Br.** ascending, 7–9 cm long, covered with the spiny remains of dead L.; **L.** cylindrical, obtuse, 15–20 mm long, 4 mm thick; **Fl.** 2 cm ⌀, purplish-pink.

**S. dinteri** (L. Bol.) L. Bol. (§ III) (*Aridaria d.* L. Bol., *Nyct. d.* (L. Bol.) Schwant., *Mes. melanospermum* Dtr., *S. m.* (Dtr.) Schwant., sphalm.). – S.W.Afr.: Namib. – ♄ up to 25 cm high, 30 cm ⌀; **St.** lax, Int. 12–20 mm long; **L.** very soft, succulent, slightly pruinose, with papillae in 30–35 longitudinal R., very obtuse, trigonous, c. 3.5 cm long, 7 mm thick; **Fl.** numerous, 2 cm ⌀, sand-coloured.

**S. dyeri** (L. Bol.) L. Bol. (§ III) (*Aridaria d.* L. Bol., *Nyct. d.* (L. Bol.) Schwant.). – Cape: Albany D. – **Roots** tuberous; **St.** sometimes prostrate, 11 cm long; **L.** furrowed on the upperside, cylindrical towards the tip, long tapering, green, faintly glossy-papillose, 4 cm long, 5 mm thick, 7 mm across at the base; **Fl.** solitary, 4 cm ⌀, yellow.

**S. englishiae** (L. Bol.) L. Bol. (§ I) (*Mes. e.* L. Bol., *Aridaria e.* (L. Bol.) N. E. Br., *Nyct. e.* (L. Bol.) Schwant.). – Cape: Robertson D. – **St.** creeping, up to 50 cm long; **shoots** erect; **L.** crowded, semicylindrical, obtuse, glossy to pruinose, up to 3.5 cm long, 5–7 mm across; **Fl.** c. 4 cm ⌀, white. Possibly only a variety of **S. defoliatus.**

**S. flexuosus** (Haw.) L. Bol. (§ II) (*Mes. f.* Haw., *Aridaria f.* (Haw.) Schwant., *Nyct. f.* (Haw.) Schwant.). – Cape: Lit. Karroo. – **St.** slender, spreading, curved; **L.** with short shoots from the axils, compressed-cylindrical, apiculate, glossy green, 2–3 cm long; **Fl.** solitary, 4 cm ⌀, white.

**S. fourcadei** (L. Bol.) L. Bol. (§ II) (*Aridaria f.* L. Bol., *Nyct. f.* (L. Bol.) Schwant.). – Cape: Uniondale D. – Erect ♄; **St.** 10 cm long, Int. 2–4 cm long; **L.** semicylindrical, acute, dirty green, 3 cm long, 3 mm thick, Sh. short; **Fl.** 4 cm ⌀, white.

**S. framesii** (L. Bol.) L. Bol. (§ I) (*Aridaria f.* L. Bol., *Nyct. f.* (L. Bol.) Schwant.). – Cape: Van Rhynsdorp D. – Erect ♄ 9–10 cm high, herbaceous parts yellowish-green, papillose; **St.** thick; shoots 7 cm long, with Int. 5–9 mm long; **L.** 8, soft, semicylindrical, tapering, apiculate, 2–4 cm long, 4 mm thick, Sh. short; **Fl.** 24 mm ⌀, straw-coloured.

**S. geniculiflorus** (L.) L. Bol. (§ I) (*Mes. g.* L., *Aridaria g.* (L.) N. E. Br., *Nyct. g.* (L.) Schwant., *Peratetracoilanthus g.* (L.) Rap. et Cam., *Mes. brachiatum* Ait.). – Cape: Cradock D. – Freely branched ♄; **St.** spreading or pendulous, 50–150 cm long, papillose, forking or dividing tripartitely, jointed; **L.** semicylindrical, often furrowed on the upperside, rather obtusely tapered, green, minutely papillose, 2 to 2.5 cm long, 3 mm across; **Fl.** 22 mm ⌀, greenish-yellow.

**S. glanduliferus** (L. Bol.) L. Bol. (§ III) (*Aridaria g.* L. Bol., *Nyct. g.* (L. Bol.) Schwant.). – Cape: Lit. Namaqualand. – **Caudex** tuberous; herbaceous parts glossy-papillose; **St.** prostrate at first, 25 cm long, covered with the spiny remains of dead L., Int. 12–50 mm long; **L.** narrowing towards the subacute tip, rounded on the underside, 3–5 cm long, 5 mm thick; **Fl.** 35 mm ⌀, pink tinged yellow, coppery-red in the centre.

**S. godmaniae** (L. Bol.) L. Bol. (§ III) (*Aridaria g.* L. Bol., *Nyct. g.* (L. Bol.) Schwant.). – Cape: Van Rhynsdorp D. – **St.** prostrate to creeping, 50 cm or more long, Int. 10–15 mm long; **L.** with short shoots from the axils, narrowing towards the acute tip, bluish-green, glossy-papillose, minutely scaly, 25 mm long, 3 mm ⌀; **Fl.** numerous, 3 cm ⌀, straw-coloured.

**S. gratiae** (L. Bol.) L. Bol. (§ III) (*Aridaria g.* L. Bol., *Nyct. g.* (L. Bol.) Schwant.). – Cape: Albany D. – **Caudex** obconical, 2–4 cm long, 3 cm thick above; **St.** 4–10 cm long, Int. 3–7 cm long; **L.** concave on the upperside, flat above, rounded on the underside, laterally compressed, bluish-green, ± glossy-papillose, 22 mm long, 3 mm thick, Sh. 1 mm long; **Fl.** 3.5 cm ⌀, light yellow.

**S. grossus** (Ait.) N. E. Br. (§ III) (*Mes. g.* Ait., *Aridaria g.* (Ait.) Friedr., *Nyct. g.* (Ait.) Schwant., *Perapentacoilanthus g.* (Ait.) Rap. et Cam.). – Cape: Swartkops R. – **Shoots** with the spiny remains of old L.; **L.** semicylindrical, 2.5–4 cm long, 4 mm across; **Fl.** 2 cm ⌀, pale flesh-coloured, becoming yellowish.

**S. gydouwensis** L. Bol. (§ III). – Cape: Ceres D. – **Caudex** robust, ± tuberous, with old **St.** persisting above; **Br.** prostrate, slender, Int. 1–3 cm long; herbaceous parts papillose; **L.**

linear-lanceolate, flat, obtuse to acute, 2–4 cm long, 3–7 mm across, Sh. 1 mm long; **Fl.** with Pet. 10–13 mm long, lemon-yellow, white-tipped.

**S. hallii** L. Bol. (§ I). – Cape: Lit. Namaqualand. – **St.** spreading, ± stiff, prostrate, c. 22 cm long, Int. 1.5–3 cm long, herbaceous parts densely hirsute, green; **L.** tapering towards the acute to ± obtuse tip, rounded on the underside, 1–2.5 cm long, 2 mm across, 3 mm thick, Sh. 1 mm long; **Fl.** c. 15 mm ⌀, purplish-red.

**S. herbertii** N. E. Br. (§ III) (*Aridaria h.* (N. E. Br.) Friedr., *Nyct. h.* (N. E. Br.) Schwant.). – Cape: Van Rhynsdorp D. – **Caudex** ovoid-elliptical, tuberous, fleshy; **St.** 1–2, ± desiccating in the resting period, branched above, 35–55 mm long, longer in cultivation; **L.** channelled on the upperside, convex on the underside, 7–12 mm long, 2–4 mm across, 1.5–2 mm thick, papillose, green to reddish; **Fl.** 16 mm ⌀, lemon-yellow.

**S. herrei** L. Bol. (§ III). – Cape: Lit. Namaqualand. – **Caudex** fleshy, very thick; **St.** numerous, up to 50 cm long, Int. 2.5–5 cm long, herbaceous parts indistinctly papillose; **L.** narrowed above, slightly concave on the upperside, rounded on the underside, c. 5 cm long, 5 mm ⌀; **Fl.** c. 3 cm ⌀, pale green, diurnal.

**S. humilis** L. Bol. (§ III). – Cape: Van Rhynsdorp D. – **Caudex** tuberous, 1–2 cm long, 5–12 mm ⌀; **St.** short, 3–5 cm long; **L.** narrowed above, furrowed on the upperside, 1–2 cm long, 4–6 mm across, 4–5 mm ⌀; **Fl.** 2–3 cm ⌀, white, outer Pet. 1.75 mm across, the inner staminodial ones 0.5 mm across.

**S. latipetalus** (L. Bol.) L. Bol. (§ III) (*Aridaria l.* L. Bol., *Nyct. l.* (L. Bol.) Schwant.). – Cape: Lit. Namaqualand. – Papillose ♄; **St.** prostrate, 17 cm or more long, with the thorny remains of old L.; **L.** rounded on the underside, flat to convex on the upperside, acute, green, 25–37 mm long, 4 mm ⌀; **Fl.** 16 mm ⌀, fleshy-pink.

**S. laxipetalus** (L. Bol.) L. Bol. (§ III) (*Aridaria l.* L. Bol., *Nyct. l.* (L. Bol.) Schwant.). – Cape: Worcester D. – **Caudex** tuberous, 22 mm thick; **St.** 5 cm long, 1 cm thick; **Br.** slender, up to 10 cm long, Int. 5–15 mm long, herbaceous parts glossy-papillose; **L.** tapered, rounded on the underside, 2 cm long, 3 mm thick, Sh. 0.5 mm long; **Fl.** 33 mm ⌀, pale salmon-pink.

**S. laxus** (L. Bol.) N. E. Br. (§ III) (*Aridaria l.* L. Bol., *Nyct. l.* (L. Bol.) Schwant.). – Cape: Lit. Namaqualand. – **Caudex** thickened, 2 cm ⌀; **St.** prostrate, elongated, stiff; **L.** semicylindrical, 4 cm long, 4–7 mm thick, spinescent; **Fl.** 2–4 cm ⌀.

**S. leipoldtii** L. Bol. (§ III). – Cape: Calvinia D. – **Caudex** thick, 2 cm ⌀; **St.** prostrate, 18 cm long, Int. 7–20 mm long; **L.** linear-lanceolate, c. 4 cm long, 1–1.5 cm wide, 5 mm thick; **Fl.** c. 3 cm ⌀, pink.

**S. leptopetalus** (L. Bol.) L. Bol. (§ II) *Aridaria l.* L. Bol., *Nyct. l.* (L. Bol.) Schwant., *Nyct. l.* (L. Bol.) Friedr.). – Cape: Ladismith D. – ♄ 25 cm high; **St.** dense, prostrate, elongated; **L.** semicylindrical, acute, bluish-green, 1.5–3 cm long, 2 mm thick; **Fl.** 38 mm ⌀, pale pink.

**S. ligneus** (L. Bol.) L. Bol. (§ III) (*Aridaria l.* L. Bol., *Nyct. l.* (L. Bol.) Schwant.). – Cape: Bushmanland. – **Caudex** thickened to tuberous, 8 mm ⌀; **St.** spreading to prostrate, woody, ± angled; **Br.** with short shoots, herbaceous parts bluish-green, papillose; **L.** semicylindrical, obtuse or shortly tapered, 17 mm long, 2–3 mm thick; **Fl.** 25 mm ⌀, yellow.

**S. littlewoodii** L. Bol. (§ III). – Cape: Worcester D. – **Roots** woody; **St.** 6 mm ⌀; **Br.** prostrate to creeping, up to 37 cm long, Int. 2.5–4.5 cm long; **shoots** 7–15 cm long; **L.** flat on the upperside, rounded on the underside, obtuse, 2–3.5 cm long, 4–6 mm ⌀; **Fl.** 2.5 cm ⌀, pale pink.

**S. longipapillatus** L. Bol. (§ III). – Cape: Lit. Namaqualand. – **St.** dense, prostrate, Int. 1.5–4 cm long, herbaceous parts with projecting papillae; **L.** linear, convex on the underside, 2–5 cm long, 3–5 mm across; **Fl.** solitary, rarely in 3's, 2.5 cm ⌀, pale straw-coloured.

**S. longispinulus** (Haw.) N. E. Br. (§ III) (*Mes. l.* Haw., *Aridaria l.* (Haw.) L. Bol., *Nyct. l.* (Haw.) Schwant., *Perapentacoilanthus l.* (Haw.) Rap et Cam.). – Cape: near Rietvley. – **Caudex** tuberous, partly above-ground, 12 to 24 mm thick; **St.** 9–13 cm long, prostrate; **L.** semicylindrical, channelled on the upperside, acute or pungent-acute, papillose, green, dead L. persisting as Th. 4–5 mm long; **Fl.** 10–15 mm ⌀, light yellow.

**S. longitubus** (L. Bol.) L. Bol. (§ III) (*Aridaria l.* L. Bol., *Nyct. l.* (L. Bol.) Schwant.). – Cape: Van Rhynsdorp D. – Erect ♄ 20 cm high; **St.** with thickened nodes 10–15 mm thick, herbaceous parts softly hirsute, flowering **shoots** 10 cm long, with the pungent remains of 6–8 dead L., Int. 10–15 mm long; **L.** semicylindrical to cylindrical, tapered, green, stout, 3.5 cm long, 2 mm across; **Fl.** 34 mm ⌀, pale yellow.

**S. macrosiphon** (L. Bol.) L. Bol. (§ III) (*Aridaria m.* L. Bol., *Nyct. m,* (L. Bol.) Schwant.). – Cape: Van Rhynsdorp D. – **Caudex** tuberous; **St.** 3 cm thick; **Br.** elongated, 15 cm long; **L.** narrowed towards the obtuse tip, dirty green, papillose; **Fl.** 35 mm ⌀, light yellow.

**S. micans** L. Bol. (§ IV). – Cape: Calvinia D. – Close to **S. resurgens**; plant up to 20 cm ⌀; **caudex** large, fleshy, long, tipped with c. 20 crowded, unbranched **shoots** resting on the soil, covered with glossy, acute papillae; **L.** tapered, green, 4.5–7 cm long, 1.5–3 mm ⌀; **Fl.** greenish-yellow. – Needs a long resting period.

**S. nanus** L. Bol. (§ III). – Cape: Lit. Namaqualand. – Appearance similar to **S. humilis**; **caudex** oblong to ± spherical, 2–2.5 cm long, up to 18 mm ⌀, tipped with 1–5 prostrate **Br.** 3.5–7.5 cm long; **L.** 2.5 cm long; **Fl.** c. 33 mm ⌀, pink, nocturnal.

**S. nitidus** (Haw.) L. Bol. (§ III) (*Mes. n.* Haw., *Drosanthemum n.* (Haw.) Schwant.,

*Mes. brachiatum* DC.). – Cape: Swartkops R. – ♄ up to 45 cm high; **St.** spreading, nodose; **L.** semicylindrical, ± channelled on the upperside, ± obtuse, light green, glossy-papillose, 2–2.5 cm long, 4 mm across; **Fl.** 15 mm ⌀, yellow.

**S. nothus** (N. E. Br.) Schwant. (§ II) (*Mes. n.* N. E. Br., *Aridaria n.* (N. E. Br.) N. E. Br., *Nyct. n.* (N. E. Br.) Schwant., *Mes. acuminatum* Salm). – Cape: Robertson D. – Erect ♄ up to 80 cm high; **St.** spreading, glossy green at first, becoming greyish-brown; **L.** 1–2 cm long, c. 2 mm across, projecting, semicylindrical, with a tiny, slender awn, smooth, unspotted; **Fl.** solitary, terminal, remaining open, 3 cm ⌀, white suffused with rather dirty pink.

**S. obtusus** (L. Bol.) L. Bol. (§ II) (*Aridaria o.* L. Bol., *Nyct. o.* (L. Bol.) Schwant.). – Cape: Lit. Namaqualand (?). – **St.** prostrate, Int. 15–25 mm long, flowering **shoots** ascending, 15 cm long, Int. 4 cm long, herbaceous parts pale green to pink or yellowish, very papillose; **L.** narrowed towards the obtuse to truncate tip, variable in length, up to 2 cm long, 4 mm thick, with a short Sh.; **Fl.** 17 mm ⌀, pale salmon-pink.

**S. oculatus** (N. E. Br.) N. E. Br. (§ III) (*Mes. o.* N. E. Br., *Aridaria o.* (N. E. Br.) L. Bol., *Nyct. o.* (N. E. Br.) Schwant.). – Cape: Lit. Namaqualand. – Habit of **S. viridiflorus**, but **L.** eventually drying and persisting, similar to those of **Sceletium anatomicum**, with large, flat papillae in longitudinal R.

**S. olivaceus** (Schltr. et Bgr.) L. Bol. (§ III) (*Mes. o.* Schltr. et Bgr., ? *Prenia o.* (Schltr. et Bgr.) Jacobs.). – S.W.Cape: Karree Mts. – **Caudex** tuberous, 4 cm long, 2 cm ⌀; **St.** densely leafy, 9–10 cm long, Int. 5 cm long; **L.** cylindrical, 25 mm long, obtuse; **Fl.** numerous, 10–12 mm ⌀, Pet. numerous.

**S. oubergensis** (L. Bol.) L. Bol. (§ III) (*Aridaria o.* L. Bol., *Nyct. o.* (L. Bol.) Schwant.). – Cape: Montagu D. – **Caudex** tuberous, 2–4 cm thick; **Br.** spreading, Int. (5–)15(–25) mm long, papillose; **L.** narrowed towards the obtuse tip, bluish-green, 23 to 50 mm long, 3–4 mm across, Sh. 2 mm long; **Fl.** 2–3 cm ⌀, salmon-pink or a little paler.

**S. pentagonus** (L. Bol.) L. Bol. v. **pentagonus** (§ II) (*Aridaria p.* L. Bol., *Nyct. p.* (L. Bol.) Schwant.). – Cape: Albany D. – ± climbing ♄ with spreading **St.** 15 cm long, Int. 1–2 cm long; **L.** semicylindrical, obtuse to shortly tapered, green to pale bluish-green, 22 mm long, 3 mm ⌀, with a short Sh.; **Fl.** c. 3 cm ⌀, orange to golden-brown.

**S. — v. occidentalis** (L. Bol.) L. Bol. (*Aridaria p.* v. *o.* L. Bol., *Nyct. p.* v. *o.* (L. Bol.) Schwant.). – Cape: Kenhardt D. – **L.** 10–15 mm long, 4 mm thick; **Fl.** 4 cm ⌀, pale salmon-pink.

**S. platysepalus** (L. Bol.) L. Bol. (§ III) (*Aridaria p.* L. Bol., *Nyct. p.* (L. Bol.) Schwant.). – Cape: Montagu D. – **Caudex** spherical, tuberous, 14 mm thick; **St.** 6 cm long, Int. 5–13 mm long; **L.** lanceolate to oblong-ovate, papillose, 10–15 mm long, 4 to 7 mm across, Sh. 1.5 mm long; **Fl.** 2 cm ⌀, pale straw-coloured.

**S. plenifolius** (N. E. Br.) L. Bol. (§ II) (*Mes. p.* N. E. Br., *Aridaria p.* (N. E. Br.) Stearn, *Nyct. p.* (N. E. Br.) Schwant., *Mes. fastigiatum* Haw., *Aridaria f.* (Haw.) Schwant.). – S.Afr.: origin unknown. – ♄ up to 1 m high; **St.** erect, with shoots from the L.-axils; **L.** ± cylindrical, ± obtuse, soft-fleshy, light green to slightly greyish-green, 10–16 mm long, 2 mm across; **Fl.** 3.5 cm ⌀, white, yellowish-white outside.

**S. pomonae** (L. Bol.) L. Bol. (§ III) (*Aridaria p.* L. Bol., *Nyct. p.* (L. Bol.) Schwant.). – S.W.Afr.: Pomona. – **Caudex** tuberous; **Br.** elongated, 20 cm long, Int. 5–20 mm long; **L.** narrowed towards the tip, convex on the underside, bluish-green, glossy and rough-papillose, 3 cm long, 6 mm across, eventually drying and persisting; **Fl.** 3 cm ⌀, white to light yellow.

**S. praecox** L. Bol. (§ III). – Cape: Lit. Namaqualand. – **Caudex** woody, 5 cm long, 11 mm ⌀; **St.** c. 80 cm long, papillose, with short shoots from the L.-axils; **L.** truncate to rounded above, 2.5–3.5 cm long, 1.5–3 mm ⌀; **Fl.** pale salmon-pink to pale brown, filaments red, anthers yellow.

**S. prasinus** (L. Bol.) L. Bol. (§ I) (*Aridaria p.* L. Bol., *Nyct. p.* (L. Bol.) Schwant.). – Cape: Walle Kraal. – Erect ♄ 30 cm high; **caudex** thickened below, ± tuberous, 2.5 cm ⌀; **St.** thick, ± spiny above from the old L.-remains, with short shoots from the L.-axils; **Br.** 15 cm long, Int. 1–3 cm long; **L.** tapered, semicylindrical, obtuse, green, papillose, 2–7 cm long, 6–7 mm across, 6 mm thick; **Fl.** 25 mm ⌀, greenish.

**S. primulinus** (L. Bol.) L. Bol. (§ II) (*Aridaria p.* L. Bol., *Nyct. p.* (L. Bol.) Schwant.). – Cape: Uitenhage, Willowmore D. – ♄ up to 30 cm high, 80 cm ⌀; **St.** crowded, 3.6 mm thick, Int. 10–25 mm long; **L.** semicylindrical, furrowed on the upperside, obtuse to shortly tapered, 15–23 mm long, 3 mm thick; **Fl.** 5 cm ⌀, primrose-yellow.

**S. pumilus** (L. Bol.) L. Bol. (§ I) (*Aridaria p.* L. Bol., *Nyct. p.* (L. Bol.) Schwant.). – Cape: Laingsburg D. – Low-growing ♄ with a tuberous **caudex**; **St.** 2 cm long, papillose, with 6 crowded, cylindrical **L.** 12 mm long, 4 mm thick, spherically expanded at the tip, Sh. 1 mm long; **Fl.** 11 mm ⌀, pink.

**S. quartziticus** (L. Bol.) L. Bol. (§ I) (*Aridaria q.* L. Bol., *Nyct. q.* (L. Bol.) Schwant.). – Cape: Lit. Namaqualand. – Erect ♄ c. 30 cm high; **St.** stiff, 3–3.5 cm long; **L.** semicylindrical, obtuse, minutely papillose, pale blue to white, 15–22 mm long, 6 mm thick, with a scarcely recognizable Sh., eventually persistent and thorny; **Fl.** 2 cm ⌀, salmon-pink.

**S. quaternus** (L. Bol.) L. Bol. (§ I) (*Aridaria q.* L. Bol., *Nyct. q.* (L. Bol.) Schwant.). – Cape: Van Rhynsdorp D. – Erect ♄ 23 cm high; **St.** 27 mm thick, **Br.** 5 mm thick, covered with the remains of old L.; **L.** 15 mm long, 4 mm ⌀, rounded on the underside, expanded towards the obtuse tip, yellowish-green, indistinctly papillose, ± soft; **Fl.** 2 cm ⌀, yellow to red, with a white centre.

**S. rabiei** (L. Bol.) N. E. Br. (§ III) (*Aridaria r.* L. Bol., *Nyct. r.* (L. Bol.) Schwant.). – Orange Free State. – Compact, dwarf, papillose; **caudex** tuberous, cylindrical; **St.** fleshy, with 6–8 crowded, cylindrical, tapered **L.** 3 cm long, 4 mm thick; **Fl.** solitary, 35 mm ⌀, pale straw-coloured.

**S. rabiesbergensis** (L. Bol.) L. Bol. (§ II) (*Aridaria r.* L. Bol., *Nyct. r.* (L. Bol.) Schwant.). – Cape: Worcester D. – **St.** crowded, 3 mm thick, Int. 1–3 cm long; **L.** cylindrical, tapered, 2 cm long, 2–3 mm thick; **Fl.** solitary or in 3's, 3.5 cm ⌀, pale pink.

**S. recurvus** (L. Bol.) L. Bol. (§ III) (*Aridaria r.* L. Bol., *Nyct. r.* (L. Bol.) Schwant.). – Cape: Bushmanland. – **Caudex** tuberous; **St.** erect, indistinctly papillose, 4–6 cm long, Int. 1 cm long; **L.** ± amplexicaul, tapered, green, papillose, 16–27 mm long, 2–4 mm across, 2–3 mm thick; **Fl.** 3 cm ⌀, pink.

**S. reflexus** (Haw.) L. Bol. (§ II) (*Mes. r.* Haw., *Aridaria r.* (Haw.) N. E. Br., *Mes. fastigiatum* v. *r.* Haw., *Mes. pallens* Jacq., *Mes. f.* v. *p.* Haw., *Mes. longistylum* DC., *Aridaria l.* (DC.) Schwant., *Nyct. l.* (DC.) Schwant., *Mes. l.* v. *purpurascens* DC., *Mes. pallescens* v. β and v. α Haw.). – S.Afr.: origin unknown. – **St.** erect, freely branched; **L.** linear, subulate, acute, ± keeled, fresh green, minutely papillose; **Fl.** 2 cm ⌀, pale pink.

**S. rejuvenalis** L. Bol. (§ II). – Cape: Van Rhynsdorp D. – Low ♄ c. 25 cm high and ⌀; **St.** woody, 18 mm ⌀; **Br.** numerous, crowded, slender, 7.5–13 cm long, Int. 2–3 cm long; **L.** slightly tapered towards the tip, rounded on the underside, c. 3 cm long, c. 5 mm ⌀; **Fl.** opening at midday, Pet. red, tinged yellow outside, salmon-pink inside.

**S. resurgens** (Kensit) L. Bol. (§ IV) (*Mes. r.* Kensit, *Aridaria r.* (Kensit) L. Bol., *Phyllobolus r.* (Kensit) Schwant., *Ph. r.* (Kensit) L. Bol.). – Cape: Matjiesfontein, Calvinia, Clanwilliam, Willowmore D. – **Caudex** tuberous, bearing many shoots; **L.** ± prostrate, rosulate, flat on the upperside, rounded on the underside, with translucent papillae, 4–5 cm long, c. 4 mm thick; **Fl.** solitary, c. 2 cm ⌀, pale yellow, diurnal. – Long resting-period, summer.

**S. rhodandrus** (L. Bol.) L. Bol. (§ I) (*Aridaria r.* L. Bol., *Nyct. r.* (L. Bol.) Schwant.). – Cape: Calvinia D. – Erect ♄ 25 cm high, with **St.** 40 cm long; **Br.** 25 mm long; **L.** cylindrical, acute, glossy-papillose, dirty green to reddish, 17 mm long, 2.5 mm thick, Sh. 1 mm long; **Fl.** 35 mm ⌀, pale pink.

**S. roseus** (L. Bol.) L. Bol. (§ II) (*Aridaria r.* L. Bol., *Nyct. r.* (L. Bol.) Schwant.). – Cape: Oudtshoorn D. – Stiff, erect ♄ 20 cm or more high, the herbaceous parts yellowish-green, papillose; **L.** semicylindrical, shortly tapering, older ones 13 mm long, 6 mm thick, others 3 cm long, 4 mm thick; **Fl.** 4 cm ⌀, an attractive pink, paler inside.

**S. salmoneus** (Haw.) N. E. Br. (Pl. 198/4) (§ III) (*Mes. s.* Haw., *Nyct. s.* (Haw.) Schwant., *Mes. canaliculatum* Salm). – S.Afr.: origin unknown. – **St.** prostrate, up to 20 cm long; **L.** linear-lanceolate, tapered towards both ends, ± channelled on the upperside, very convex on the underside, fresh green, papillose, 12–20 mm long; **Fl.** c. 15 mm ⌀, yellowish to reddish.

**S. saturatus** (L. Bol.) L. Bol. (§ III) (*Aridaria s.* L. Bol., *Nyct. s.* (L. Bol.) Schwant.). – Cape: Clanwilliam D. – **Caudex** tuberous, 12 mm thick; **St.** creeping to prostrate, 20 cm or more long, Int. 12 mm long; **L.** obtuse, rounded on the underside, concave on the upperside, bluish, papillose, 35 mm long, 5 mm across; **Fl.** 2–3 cm ⌀, golden-yellow.

**S. scintillans** (Dtr.) Dtr. et Schwant. (§ III) (*Mes. s.* Dtr., *Aridaria s.* (Dtr.) Friedr., *Nyct. s.* (Dtr.) Schwant. ex Jacobs., *Perapentacoilanthus s.* (Dtr.) Rap. et Cam., *S. fragilis* N. E. Br., *N. f.* (N. E. Br.) Schwant. ex Jacobs., *A. f.* (N. E. Br.) Friedr., *Amoebophyllum rangei* N. E. Br., *Arid. r.* (N. E. Br.) Friedr., *Arid. varians* L. Bol., *N. v.* (L. Bol.) Schwant. ex Jacobs., *S. v.* (L. Bol.) L. Bol.). – S.W.Afr.: S.Namib; Cape: Lit. Namaqualand, Richtersveld. – Forming mats over 50 cm ⌀, very papillose; **L.** semicylindrical, rounded above, flat on the upperside, 2–5 cm long, 2–4 mm thick; **Fl.** in lax spikes, up to 2 cm ⌀, salmon-pink to brownish-red.

**S. sinuosus** (L. Bol.) L. Bol. (§ III) (*Mes. s.* L. Bol., *Nyct. s.* (L. Bol.) Schwant.). – Cape: Lit. Namaqualand. – **St.** creeping, much and variously curved, Int. short, flowering **shoots** short, erect; **L.** semicylindrical, ± channelled on the upperside, obtuse, 1–2.5 cm long, 5 mm ⌀; **Fl.** 2.5 cm ⌀, pale straw-coloured.

**S. spinuliferus** (Haw.) L. Bol. (§ III) (*Mes. s.* Haw., *Aridaria s.* (Haw.) N. E. Br., *Nyct. s.* (Haw.) Schwant., *Peratetracoilanthus s.* (Haw.) Rap. et Cam.). – S.Afr.: origin unknown. – **St.** and **Br.** c. 30 cm high, fleshy and thickened, swollen at the nodes but not thorny, with spreading shoots, very papillose at first; **L.** linear, very convex on the underside, compressed at the tip, soft-fleshy, fresh green, glossy-papillose, 2.5–4 cm long, 4–6 mm across; **Fl.** 3 cm ⌀, pale yellowish. The name 'spinuliferus' is misleading!

**S. splendens** (L.) L. Bol. (Pl. 198/2) (§ II) (*Mes. s.* L., *Aridaria s.* (L.) Schwant., *Nyct. s.* (L.) Schwant., *Perapentacoilanthus s.* (L.) Rap. et Cam., *Pentacoilanthus s.* (L.) Rap. et Cam.). – Cape: Steytlerville. – Freely branched ♄; **St.** prostrate and cushion-forming, with a watery, sappy appearance; **L.** cylindrical, not united, recurved at the tip, light green, smooth, with several dark green tubercles, slightly pruinose, 1.5–2 cm long; **Fl.** 3.5–4 cm ⌀, yellowish-white.

**S. stayneri** L. Bol. (§ I). – Cape: Calvinia D. – Spreading, laxly branched ♄ 16 cm high, 58 cm ⌀; **St.** ± prostrate, slender, Int. 2–3 cm long, herbaceous parts minutely hirsute; **L.** flat on the upperside, rounded on the underside and towards the tip, c. 1.5 cm long, Sh.

2 mm long, persistent; **Infl.** lax, of many Fl. c. 15 mm ⌀, Pet. red, inner ones golden-yellow to yellow.

**S. straminicolor** (L. BOL.) L. BOL. (§ III) (*Aridaria s.* L. BOL., *Nyct. s.* (L. BOL.) SCHWANT., *Aridaria straminea* L. BOL.). – Cape: Lit. Namaqualand. – Erect ♄ from a tuberous **caudex; St.** prostrate, papillose, 35 cm long; **L.** tapered, concave on the upperside, rounded on the underside, papillose, 3 cm long, 4 mm across, 3 mm thick; **Fl.** 35 mm ⌀, straw-coloured.

**S. striatus** (L. BOL.) L. BOL. (§ II) (*Aridaria s.* L. BOL., *Nyct. s.* (L. BOL.) SCHWANT.). – Cape: origin unknown. – **St.** 3 mm thick, Int. 1.5–3 cm long; **L.** narrowed towards the acute tip, furrowed on the upperside, bluntly keeled, green, 4 cm long, 4 mm across, 3–4 mm thick; **Fl.** 45 mm ⌀, Pet. pale pink, with a dark stripe on the upperside.

**S. strictus** (L. BOL.) L. BOL. (§ I) (*Aridaria s.* L. BOL., *Nyct. s.* (L. BOL.) SCHWANT.). – Cape: Lit. Namaqualand. – Erect, stiff ♄ 32 cm high; **caudex** tuberously thickened at the base, c. 18 mm ⌀; **L.** cylindrical, tapered, 3.5 cm long, 3–4 mm thick; **Fl.** 25 mm ⌀, salmon-pink.

**S. suaveolens** (L. BOL.) JACOBS. (§ III) (*Mes. s.* L. BOL., *Cryophytum s.* (L. BOL.) J. INGRAM, *S. lignescens* L. BOL.). – S.W.Afr.: Warmbad D. – Spreading, papillose ♄ up to 50 cm high, 70 cm ⌀; **St.** 2 cm ⌀ at the base; **Br.** 25–30 cm long, 5–15 mm thick, Int. 1–2 cm long; **L.** ± amplexicaul, narrowed towards the tip, rounded on the underside, obtuse to acute, 1–1.5 cm long, 2–3 mm across; **Fl.** numerous, 2–3 cm ⌀, white, scented.

**S. subaequans** (L. BOL.) L. BOL. (§ II) (*Aridaria s.* L. BOL., *Nyct. s.* (L. BOL.) SCHWANT.). – S.Afr.: origin unknown. – Erect, densely branched ♄; **L.** laterally compressed, flat on the upperside, rounded on the underside, 2–2.5 cm long, 3–4 mm across, 3–3.5 mm thick; **Fl.** 4 cm ⌀, pale brown to flesh-coloured, paler inside.

**S. subpatens** (L. BOL.) L. BOL. (§ II) (*Aridaria s.* L. BOL., *Nyct. s.* (L. BOL.) SCHWANT.). – Cape: Graaff Reinet D. – Erect ♄; **St.** 1.5–2 mm thick, Int. 7–25 mm long; **L.** narrowed towards the shortly tapered tip, rounded on the underside, 12–20 mm long, 2.5 mm ⌀; **Fl.** 3 cm ⌀, white inside, pale salmon-pink outside.

**S. subpetiolatus** (L. BOL.) L. BOL. (§ III) (*Aridaria s.* L. BOL.). – Cape: Clanwilliam D. – **Caudex** tuberous, 4.5 cm ⌀ above; **St.** semi-underground, c. 9 mm ⌀; **Br.** prostrate, slender; **L.** cylindrical or flat on the upperside, narrowing below for 2–8 mm and towards the tip, green to bluish-green, ± papillose, 2–3.6 cm long, 2–3 mm ⌀; Pet. 21–26 mm long, straw-coloured to pale pink. Acc. L. BOLUS this species constitutes a transition between § III and § IV.

**S. suffusus** (L. BOL.) L. BOL. (§ I) (*Aridaria s.* L. BOL., *Nyct. s.* (L. BOL.) SCHWANT.). – Cape: Lit. Namaqualand. – Erect ♄; **St.** virgate, indistinctly 4-angled, pink, pruinose, Int. 4–5 cm long; lower **L.** cylindrical, obtuse, 3 cm long, upper L. 1.5–2 cm long, 8 mm thick, Sh. 1–1.5 mm long, persistent; **Fl.** 2 cm ⌀, white to pink.

**S. sulcatus** (HAW.) L. BOL. (§ II) (*Mes. s.* HAW., *Aridaria s.* (HAW.) SCHWANT., *Nyct. s.* (HAW.) SCHWANT., *Perapentacoilanthus s.* (HAW.) RAP. et CAM.). – Cape: Willowmore D. – ♄ over 60 cm high, with erect, straight **St.**; **L.** linear-subulate, incurving, ± furrowed on the upperside, 1.5–2.5 cm long, 2–3 mm across; **Fl.** mostly solitary, 4.5 cm ⌀, white.

**S. tenuiflorus** (JACQ.) N. E. BR. (§ III) (*Mes. t.* JACQ., *Nyct. t.* (JACQ.) SCHWANT.). – S.Afr.: origin unknown. – **L.** 25–45 mm long, spreading or recurved; **Fl.** purplish-red. Treated by SONDER and BERGER as *Mes. viridiflorum* AIT. = **S. viridiflorus** (AIT.) N. E. BR.

**S. tetragonus** (THUNBG.) L. BOL. (§ I) (*Mes. t.* THUNBG., *Aridaria t.* (THUNBG.) L. BOL., *A. t.* (THUNBG.) SCHWANT., *Nyct. t.* (THUNBG.) SCHWANT., *Peratetracoilanthus t.* (THUNBG.) RAP. et CAM., ? *Aridaria luteoalba* L. BOL., ? *Nyct. l.* (L. BOL.) SCHWANT.). – Cape: Lit. Namaqualand, Willowmore D., Orange Free State; S.W.Afr.: Gr. Namaqualand, Gibeon D. – Freely branched ♄ c. 40 cm high; **St.** erect, somewhat 4-angled; **L.** incurving, flat on the upperside, obtuse, deciduous, 12–24 mm long; **Fl.** 2 cm ⌀, yellowish-white.

**S. tetramerus** (L. BOL.) L. BOL. v. **tetramerus** (§ I) (*Aridaria t.* L. BOL., *Nyct. t.* (L. BOL.) SCHWANT.). – Cape: Lit. Namaqualand. – Erect ♄ with a woody, thick and ± tuberous **caudex; St.** 15 mm thick at the base; **Br.** ascending to ± prostrate, stiff, somewhat thickened at the nodes, Int. 15–25 mm long; **L.** obtuse to subacute and red at the tip, convex on the upperside, rounded on the underside, with ± distinct margins, minutely papillose, bluish-green, 15–20 mm long, 4 mm ⌀; **Fl.** 2 cm ⌀, straw-coloured.

**S. —** v. **parviflorus** (L. BOL.) L. BOL. (*Aridaria t.* v. *p.* L. BOL., *Nyct. t.* v. *p.* (L. BOL.) SCHWANT.). – **Fl.** nocturnal, c. 10 mm ⌀, flesh-coloured inside, red outside.

**S. trichotomus** (THUNBG.) L. BOL. (§ I) (*Mes. t.* THUNBG., *Aridaria t.* (THUNBG.) L. BOL., *Dicrocaulon t.* (THUNBG.) N. E. BR., *Nyct. t.* (THUNBG.) SCHWANT., *Mes. ebracteatum* L. BOL., *Aridaria e.* (L. BOL.) L. BOL., *Nyct. e.* (L. BOL.) SCHWANT., *A. e.* v. *brevipetala* L. BOL., *Nyct. e.* v. *b.* (L. BOL.) SCHWANT.). – Cape: Clanwilliam, Van Rhynsdorp D. – Erect ♄; **St.** stiff, tangled and curved, with numerous short shoots; **L.** rounded-trigonous, obtuse, c. 12 mm long; **Fl.** solitary, purplish-pink, inner Pet. white.

**S. umbelliflorus** (JACQ.) L. BOL. (§ II) (*Mes. u.* JACQ., *Aridaria u.* (JACQ.) SCHWANT., *Nyct. u.* (JACQ.) SCHWANT.). – Cape: origin unknown. – ♄ up to 80 cm high; **St.** numerous, spreading, with numerous short shoots; **L.** compressed-cylindrical, ± clavately thickened above, ± obtuse, somewhat greyish-green, 15–25 mm long, 2–3 mm across; **Fl.** 3.5 cm ⌀, Pet. whitish, with yellowish-tips.

**S. vanheerdei** L. BOL. (§ III). – Cape: Lit. Namaqualand. – Stiff ♄ c. 30 cm high; **roots**

woody, 8 mm $\varnothing$; **St.** erect, up to 27 cm long, herbaceous parts papillose, scarcely glossy; **L.** flaccid, semicylindrical, obtuse, 27–38 mm long, 2–5 mm $\varnothing$, with buds in the axils; **Pet.** 10 to 16 mm long, white below, an attractive pink above; very free-flowering.

**S. vernalis** (L. Bol.) L. Bol. (§ II) (*Aridaria v.* L. Bol., *Nyct. v.* (L. Bol.) Schwant.). – Cape: Ceres D. – Stiff, densely branched ♄ c. 25 cm high; **St.** 3 mm thick, Int. 10–15 mm long; **Br.** 7–13 cm long, Int. 5–10 mm long; **L.** semicylindrical, obtuse to shortly tapered, 10–17 mm long, 3 mm thick; **Fl.** 4 cm $\varnothing$, pale pink.

**S. vigilans** (L. Bol.) L. Bol. (§ III) (*Mes. v.* L. Bol., *Nyct. v.* (L. Bol.) Schwant., *Mes. hesperanthum* L. Bol.). – Cape: Riversdale, Montagu, Worcester, Laingsburg, Malmesbury D., Karroo. – Papillose ♄ up to 20 cm high; **St.** ascending to spreading, Int. 1–3 cm long; **L.** linear, semicylindrical, subacute, minutely papillose, 2–3 cm long, 2–3 mm across; **Fl.** nocturnal, sweetly scented, 2.5 cm $\varnothing$, Pet. white below, flesh-coloured above.

**S. viridiflorus** (Ait.) N. E. Br. (§ III) (*Mes. v.* Ait., *Aridaria v.* (Ait.) L. Bol., *Nyct. v.* (Ait.) Schwant., *Perapentacoilanthus v.* (Ait.) Rap. et Cam., *Aridaria peersii* L. Bol., *Nyct. p.* (L. Bol.) Schwant., *Mes. trichosanthum* Bgr.). – Cape: Lit. Namaqualand. – **St.** and **Br.** $\pm$ fleshy, nodose, c. 40 cm long, prostrate; **L.** semicylindrical, tapered, with minute papillose hairs, 2.5–4 cm long, 4–6 mm across, persisting as Th.; **Fl.** over 2 cm $\varnothing$, Pet. filiform, pale green or dirty reddish.

**S. watermeyeri** (L. Bol.) L. Bol. (§ I) (*Aridaria w.* L. Bol., *Nyct. w.* (L. Bol.) Schwant.). – Cape: Van Rhynsdorp D. – Erect ♄ with the **St.** thickened at the base; **Br.** up to 25 cm long, **L.** semicylindrical, concave on the upperside, obtuse, minutely papillose, dirty green, 2 to 2.5 cm long, 7 mm thick; persisting after withering; **Fl.** 1 cm $\varnothing$, straw-coloured.

**S. willowmorensis** (L. Bol.) L. Bol. (§ III) (*Aridaria w.* L. Bol., *Nyct. w.* (L. Bol.) Schwant.). – Cape: Willowmore D. – Papillose ♄; **St.** prostrate, 10–20 cm long, Int. 1–4 cm long; **L.** acute, convex on the underside, concave or flat on the upperside, 2 cm long, 2 mm $\varnothing$; **Fl.** 22 mm $\varnothing$, white to straw-coloured.

**Stayneria** L. Bol. (§ IV/1/1). – Close to the G. **Ruschia**; erect, robust ♄; **roots** woody; **L.**

$\pm$ falcate; **Fl.** in 3's, large; stigma-lobes 7–9, very rarely 6; **Fr.** a capsule which, once open, does not again close completely. – Summer: in the open, winter: cold-house. Propagation: seed, cuttings.

**S. neilii** (L. Bol.) L. Bol. (*Ruschia n.* L. Bol., *S. littlewoodii* L. Bol.). – Cape: Robertson D. – Up to 1 m high; **St.** 7 mm thick, Int. 3–6 mm long; **L.** curved, flat on the upperside, with an acute, serrate keel on the underside, acute to obliquely truncate above, green, 5.5–7 cm long, 4–6 mm across, 8–9 mm thick, Sh. with impressed lines 7 mm long; **Infl.** compact, Fl. on a pedicel 5–20 mm long with bracts 12 mm long, Pet. 17 mm long, purplish-pink to white.

**Stoeberia** Dtr. et Schwant. emend. Friedr. (§ IV/1/5). – ♄ with spreading **St.** and **Br.**; **L.** opposite, somewhat united at the base, fleshy, navicular to $\pm$ clavate, flat on the upperside, emarginate, not distinctly angled, somewhat keeled on the underside below the tip; **Infl.** terminal, dichasially branched, Fl. on a pedicel with 2 bracts, Pet. in only one series, white, red or absent. – Summer: in the open, winter: cold-house. Propagation: seed, cuttings.

**S. apetala** L. Bol. – S.W.Afr.: mouth of the Orange R. – Plant 17 cm high, the sterile **Br.** densely leafy; **shoots** 12 cm long, Int. 3 cm long; **L.** apiculate, with the keel often reduced to a line, larger L. $\pm$ clavate, less often $\pm$ spatulate, c. 2 cm long, 3–4 mm across, the tip up to 8 mm $\varnothing$, Sh. 2–3 mm long; **Pet.** missing.

**S. beetzii** (Dtr.) Dtr. et Schwant. v. **beetzii** (*Mes. b.* Dtr.). – S.W.Afr.: E. Namib. – Spreading, up to 50 cm high, with Int. 2.2–5 cm long; **L.** $\pm$ clavate, with a recognizable keel-line, greyish-green with small dots, 3–3.5 cm long, 6–7 mm across, 7–8 mm thick; **Fl.** 10–12 mm $\varnothing$, white.

**S.** — v. **arborescens** Friedr. (*Mes. (Pentaschista) arborescens* Dtr. nom. nud.). – S.W.Afr.: Gr. Namaqualand. – **L.** larger; **Infl.** with the Fl. more numerous, somewhat larger, white to red.

**S. carpii** Friedr. (*Ruschianthemum merenskyanum* Dtr. nom. nud., *Jensenobotrya vanheerdei* L. Bol. sphalm.). – S.W.Afr.: Gr. Namaqualand; Cape: Lit. Namaqualand. – ♄ up to 1 m high, with spreading, somewhat angular **St.**, Int. 1–9 cm long; **L.** $\pm$ clavate, 2–5 cm long, 1–10 mm across, 8–20 mm thick, soft, whitish-grey; **Fl.** c. 2 cm $\varnothing$, white.

**Stomatium** Schwant. (§ IV/1/11). – Caespitosely branched, very short-stemmed, glabrous plants; **L.** of a pair often unequal, 4–6 to one growth, decussate, connate and $\pm$ vesicularly inflated below, thick fleshy, short 3-angled or broadly spatulate or elongated lanceolate, semicylindrical in transverse section at the base, lower surface keeled towards the apex and often drawn forwards like a chin, the margins set with $\pm$ short and broad T., epidermis dull, soft, $\pm$ transparently tuberculate; **Fl.** sessile or short stalked, of M. size, yellow, in only 1–2 spec. whitish or pink, open at night, fragrant, summer. – Greenhouse: warm. Propagation: seed, cuttings.

## Division of the Genus Stomatium by O. H. Volk

### Sect. I Stomatium

**Fl.** with a short or long tube formed from the united Sep. and Pet., sessile or short stalked, M.-sized, without bracts; surface of L. dull, epidermis soft. – Type species: S. suaveolens. – Species: S. acutifolium, alboroseum, beaufortense, bolusiae, braunsii, bryantii, deficiens, difforme, duthiae, erminium, fulleri, geoffreyi, gerstneri, grandidens, integrum, jamesii, latifolium, leve, loganii, meyeri, middelburgense, murinum, mustellinum, niveum, patulum, paucidens, peersii, pluridens, pyrodorum, resedolens, ronaldii, rouxii, ryderae, suaveolens, suricatum, trifarium, villetii, viride.

**Sect. II Agnirictus** (SCHWANT.) VOLK (*Agnirictus* SCHWANT. as a G.). – **Fl.** without such a tube; epidermis of L. rough from fine elevations. – Type species: S. lesliei. – Species: S. agninum, lesliei.

**S. acutifolium** L. BOL. (§ I). – Cape: Beaufort West D. – **L.** 8 on a **Br.** , ± linear, acute to almost truncate at the tip, 20–24 mm long, 5 mm across, 4–6 mm thick, with 2–3 bristle-tipped T. 1.5 mm long on each margin, keel with 1–3 T.; **Fl.** 24 mm ⌀.

**S. agninum** (HAW.) SCHWANT. v. **agninum** (Pl. 198/3) (§ II) (*Mes. a.* HAW., *Agnirictus a.* (HAW.) SCHWANT.). – Cape: origin unknown. – **L.** soft, oblong, rather obtuse, very convex on the underside and keeled-trigonous, the margins often with 3–5 short T., dull greyish-green, rough with green papillae, 4–5 cm long, 10 to 15 mm across; **Fl.** 2.5 cm ⌀, light yellow.

**S. —** v. **integrifolium** (SALM) VOLK (*Mes. a.* v. *i.* SALM, *Agnirictus a.* v. *i.* (SALM) SCHWANT.). – **L.** entire.

**S. alboroseum** L. BOL. (§ I). – Cape: Bushmanland. – **Shoots** very numerous, with 6–8 entire, obtuse and rounded-truncate, greyish-green, densely spotted, somewhat tuberculate **L.** up to 25 mm long, 7 mm across, 7 mm thick at the tip, with an indistinct, lateral keel; **Fl.** c. 2 cm ⌀, white or pink.

**S. angustifolium** L. BOL. (§ I). – Cape: Somerset East D. – **L.** acute on the upperside, scarcely narrowed midway, finely tapered, with 3–4 marginal T. and a spotted and dentate keel, 18–26 mm long, 6–7 mm across, 5–7 mm thick; **Fl.** 24 mm ⌀, yellow.

**S. beaufortense** L. BOL. (§ I). – Cape: Beaufort West D. – **L.** oblong, acute in profile with a rounded tip, blistered and inflated below, bluish-green, 32–34 mm long, 14 mm across, 8–9 mm thick, the keel entire, each margin with 3–4 T., Sh. 12 mm long; **Fl.** 3 cm ⌀, yellow.

**S. braunsii** L. BOL. (§ I). – Cape: Clanwilliam D. – **St.** with 6 densely spotted unequal **L.** 22–24 mm or 24–28 mm long, 11 mm across, 4–7 mm thick, with 3–4 bristle-tipped T. along each margin towards the tip; **Fl.** 24 mm ⌀, yellow, nocturnal.

**S. bryantii** L. BOL. (§ I). – Cape: Prieska D. – **L.** 18–20 mm long, 10–11 mm across, 5–8.5 mm thick, Sh. 5–6 mm long, acute to obtuse, bluish, with prominent dots, with 2–4 marginal T. and 2–5 small T. on the keel; **Fl.** 2.5 mm ⌀, yellow.

**S. deficiens** L. BOL. (§ I). – Cape: Cradock D. – **L.** narrowing above midway, acute, indistinctly keeled, green, papillose, margins and keel entire or serrate-papillose, 12–22 mm long, 5–8 mm across, Sh. 5–7 mm long; **Fl.** not known.

**S. difforme** L. BOL. (§ I). – Cape: Laingsburg D. – **St.** with 2–5 dissimilar **L.**, one being fan-shaped on the upperside, the other shorter, 26 mm long, 16 mm across, 9 mm thick, with white dots and 13 marginal T. 1.5 mm long, the keel with 1–3 indistinct T.; **Fl.** 22 mm ⌀, scented, nocturnal.

**S. duthiae** L. BOL. (§ I). – Cape: Middelburg D. – **St.** with 4–6 bluish, pink-tinged **L.** in unequal pairs, with 3 marginal T. 1 mm long and crowded, prominent, pearl-like dots (lens needed), one L. 13 mm long, 11 mm across, 8 mm thick, Sh. 5 mm long, the other shorter, narrower; **Fl.** 18 mm ⌀, yellow.

**S. ermininum** (HAW.) SCHWANT. (Pl. 198/6) (§ I) (*Mes. e.* HAW.). – Cape: Cradock D. – **L.** 16–20 mm long, 8–12 mm across, rather obtuse, the underside convex below and keeled above, light greyish-green, minutely rough with numerous prominent dots, margins with 3–4 short T. towards the tip; **Fl.** 2–2.5 cm ⌀, yellow, perfumed.

**S. fulleri** L. BOL. (§ I). – Cape: Kenhardt D. – **St.** with 6–8 **L.** c. 30 mm long, c. 12 mm across at the base, expanded above, obtuse, very convex on the underside, with 3–4 marginal T. above, grass-green, rough with light grey dots, expecially on the underside; **Fl.** c. 2 cm ⌀, yellowish-white.

**S. geoffreyi** L. BOL. (§ I). – Cape: Cradock D. – **St.** with 6–8 **L.** 2 cm long, 9–12 mm across, 4–5 mm thick, the keel indistinctly dentate at the tip, margins with 4–6 white, bristle-tipped T.; **Fl.** 15 mm ⌀, Pet. pale straw-coloured, the tips reddish outside.

**S. gerstneri** L. BOL. (§ I). – Cape: Herschel D. – **L.** 20–24 mm long, 8–9 mm across, 6 mm thick, Sh. 8–10 mm long, acute, with 1–2 indistinct T. and parallel veins on the upperside, laterally keeled, green, with prominent dots; **Fl.** 2 cm ⌀, yellow.

**S. grandidens** L. BOL. (§ I). – Cape: origin unknown. – **L.** tapered, obliquely obtuse or rounded at the tip, with 2–4 T. on each margin up to 3 mm long, laterally keeled on the underside, indistinctly dentate at the tip, bluish, with prominent dots, 3–4 cm long, 14 mm across, Sh. 1 mm long; **Fl.** 25 mm ⌀, yellow to lemon-yellow.

**S. integrum** L. BOL. (§ I). – Fort Beaufort D. – **St.** creeping, 5–7 cm long; **L.** oblong to ovate-oblong, acute, rounded on the underside, keeled towards the tip, margins entire or with 1–3 very indistinct T., greenish-blue, spotted, 15–25 mm long, 6–11 mm across, 4–6 mm thick, Sh. 5 to 7 mm long; **Fl.** 2 cm ⌀, yellow.

**S. jamesii** L. BOL. (§ I). – Cape: Cradock D. – **St.** 12 cm long, 5 mm thick, rooting; **Br.** with 6 **L.**, with short shoots; **L.** 19 mm long, 9 mm across, 6 mm thick, Sh. 5 mm long, ± spatulate, with 3 bristle-tipped T. on the underside along the keel, blue, rough with conical, pearly, white-tipped dots; **Fl.** 2 cm ⌀, yellow, nocturnal, scented.

**S. latifolium** L. Bol. (§ I). – Cape: Fraserburg D. – **St.** with 4 bluish-purple **L.** 18 mm long, expanded above to 15 mm across, upper L. smaller, with 2–4 marginal T. and 1–2 indistinct T. along the keel, with minute pearly dots and 2–3 tubercles on the underside; **Fl.** 2 cm ⌀, with red-tipped Pet.

**S. lesliei** (Schwant.) Volk (§ II) (*Agnirictus l.* Schwant.). – Cape: Cradock D. – **L.** up to 4 cm long, 15 mm across, 8 mm thick, trigonous, tapered, with minute marginal T. 2–3 mm apart, scarcely keeled on the underside, greyish-green, minutely rough; **Fl.** 25 mm ⌀, yellow.

**S. leve** L. Bol. (§ I). – Cape: ? Cradock D. – **L.** oblong to linear-oblong on the upperside, with 2–4 bristle-tipped T., obliquely obtuse, dissimilar, 25–30 mm long, 9 mm across midway and 12 mm across above, one L. 7 mm thick at the tip, the shorter one less thick and convex on the underside; **Fl.** 2 cm ⌀, yellow.

**S. loganii** L. Bol. (§ I). – Cape: Laingsburg D. – **St.** with 8, rarely 12, crowded, oblong **L.** 12–20 mm long, 7–8 mm across broadening to 11 mm midway, c. 6 mm thick below the rounded tip, entire or with 2–3 indistinct, bristle-tipped, marginal T., Sh. 6 mm long; **Fl.** 16 mm ⌀, yellow, open evenings and nights, scented.

**S. meyeri** L. Bol. (Pl. 198/5) (§ I). – Cape: Lit. Namaqualand. – **L.** ± linear, 18–20 mm long, c. 6 mm across, only slightly expanded towards the tip, semicylindrical on the underside, slightly keeled above, with the keel pulled forward, light greyish-green with coarse, rough dots, the angles ± conspicuously tuberculate at the tip; **Fl.** 24 mm ⌀, Pet. white, pale pink outside.

**S. middelburgense** L. Bol. (§ I). – Cape: Middelburg D. – Resembles **S. duthiae**; **Fl.** 2 cm ⌀, pale lemon-yellow to straw-coloured.

**S. murinum** (Haw.) Schwant. (§ I) (*Mes. m.* Haw.). – Cape: Fraserburg D. – Freely branched, bushy; **L.** projecting and incurved, thickened above, the keel pulled forward, the margins and keel each with 3 T., greyish-green with white, transparent, prominent dots; **Fl.** 20–25 mm ⌀, yellow.

**S. mustellinum** (S.D.) Schwant. (§ I) (*Mes. m.* S.D.). – Orange Free State; Botswana. – **St.** prostrate; **L.** rosulate, up to 2 cm long, 1 cm across, rhombic to broadly spatulate, obtuse, semicylindrical on the underside, keeled towards the tip, with 5–7 small, acute T. on the angles, light greyish-green, rough with minute dots; **Fl.** 2 cm ⌀, yellow.

**S. niveum** L. Bol. (§ I). – Cape: Bushmanland. – **St.** with 4 bluish-green, dissimilar **L.** 2 cm long, 1 cm across, spatulate, 6 mm thick at the tip, with prominent, white dots and 3 indistinct marginal T.; **Fl.** 18 mm ⌀, white, open evenings and nights.

**S. patulum** Jacobs. (§ I). – Cape: origin unknown. – **L.** 18–20 mm long, 5 mm across below, forming a triangle 8 mm across above, semicylindrical on the underside below, obliquely keeled above, light greyish-green with rough dots and 2–9 pointed tubercles on the upper part of the angles, the keel less tuberculate, pulled somewhat forward.

**S. paucidens** L. Bol. (§ I). – Cape: Britstown. – **L.** unequal, 22–32 or 18–22 mm long, 7–9 mm across, acute, apiculate, with a lateral keel, compressed above and 8 mm thick, margins with 1–3 indistinct T., Sh. 7 mm long; **Fl.** 25 mm ⌀, yellow.

**S. peersii** L. Bol. (§ I). – Cape: Victoria West D. – Compact plant; **St.** with 4–6 **L.**, the lowest oblong, obtuse, mucronate, 15 mm long, 8 mm across, upper L. narrower, 5–6 mm thick, Sh. 6 mm long, margins with 2–3 T., less often entire; **Fl.** 2 cm ⌀, yellow.

**S. pluridens** L. Bol. (§ I). – Cape: Lit. Namaqualand, Bushmanland. – **L.** blistered and swollen at the base, narrowing and rounded to truncate or oblique and ± expanded at the tip, blue with prominent dots, 32–43 mm long, 13 mm across midway, 7–9 mm across above, 6–9 mm thick, margins with 4–6 T., keel with 4–5 T., all T. bristle-tipped; **Fl.** 15–20 mm ⌀.

**S. pyrodorum** (Diels) L. Bol. (§ I) (*Mes. p.* Diels). – Cape: Calvinia D. – **L.** fleshy to leathery, papillose, broadly ovate to obovate, with short, cartilaginous, marginal T., 15–20 mm long, 9–12 mm across; **Fl.** light yellow, smelling of apples.

**S. resedolens** L. Bol. (§ I). – Cape: Sutherland D. – **Shoots** with 4–6 rounded to obtuse, mucronate **L.** 20–25 mm long, up to 12 mm across, 6 mm thick, blistered and swollen at the base, with 3 and 5 marginal T. 1–2.5 mm long and 1–3 T. on the keel, Sh. 8–11 mm long; **Fl.** 24 mm ⌀ yellow, mignonette-scented.

**S. ronaldii** L. Bol. (§ I). – Cape: Cradock D. – **L.** scarcely expanded midway, with 1–2 indistinctly bristle-tipped T. on the margins and the keel at the tip, bluish-green with crowded, prominent, pearly dots, 7 mm long, 5 mm across, larger on cultivated plants; **Pet.** almost filiform.

**S. rouxii** L. Bol. (§ I). – Cape: Carnarvon D. – **L.** ± constricted midway, obliquely tapered, mucronate, with 3 marginal T., the keel expanded at the tip to 5–7 mm, with 1–3 T. interspersed with large, prominent dots, 20 to 25 mm long, 9 mm across midway, 12 mm across above, Sh. 5–6 mm long; **Fl.** 18 mm ⌀, yellow.

**S. ryderae** L. Bol. (§ I). – Cape: Cradock D. – **L.** ± constricted on the upperside, acute or obliquely acute, mucronate, with 2–3 minute marginal T., bluish-green with crowded dots, unequal, 27 or 24 mm long, 7 and 10 mm across, Sh. 8 mm long; **Fl.** 25 mm ⌀, yellow.

**S. suaveolens** Schwant. (§ I) (*Mes. s.* Schwant. nom. nud.). – Cape: Sutherland D. – **L.** very soft-fleshy, thickened and expanded above, very rounded on the underside and pulled conspicuously forward over the tip, margins entire or with 1–5 T., light greyish-green to copper-coloured with light, rough dots, 1–2 cm long, 9–15 mm across; **Fl.** yellow.

**S. suricatum** L. Bol. (§ I). – Cape: Laingsburg D. – **Shoots** with 4 light green, indistinctly green-spotted **L.** 24–30 mm long, 7 mm across,

6 mm thick, rounded truncate, indistinctly keeled below the tip, with 4–6 indistinct marginal T. and a Sh. 8 mm long; Fl. 2 cm ⌀, lemon-yellow.

**S. trifarium** L. Bol. (Pl. 199/1) (§ I). – Cape: Fraserburg D. – L. laterally compressed, carinate on the underside above, with 3–4 marginal and 2–4 keel T., bluish-green, densely spotted, 15–25 mm long, 6–9 mm across, 8 mm thick; Fl. 24 mm ⌀, yellow, open evenings and nights.

**S. villetii** L. Bol. (§ I). – Cape: Beaufort West. – L. linear on the upperside, with a blistery swelling at the base, obliquely acute above, indistinctly keeled, with 1–3 marginal T., bluish, tinged pink, ± hirsute, unequal, the larger L. of the pair in profile scarcely expanded and truncate, 21 and 25 mm long, 5–7 mm across, 5–6 mm thick; Fl. 18 mm ⌀, Pet. yellow, red-tipped.

**S. viride** L. Bol. (§ I). – Cape: Beaufort West D. – L. oblong to linear-oblong, acute or obtuse, constricted midway, keeled on the underside above, entire or with 2–4 marginal and 1–3 keel T., green, rough-spotted, 25–35 mm long, 11 to 14 mm across, Sh. 13 mm long; Fl. 26 mm ⌀, light yellow.

**Synaptophyllum** N. E. Br. (§ I/1/2). – ○ plants. – Cultivate in the open. – Propagation: seed.

**S. juttae** (Dtr. et Bgr.) N. E. Br. (Pl. 199/2) (*Mes. j.* Dtr. et Bgr.). – S.W.Afr.: Lüderitz D. – Branching from the base, glabrous, glossy-papillose, fresh purplish-red or salmon-pink; St. repeatedly forked, 2–4 mm thick, Int. 3 cm long, ± alate; L. opposite, perfoliate below, rounded-ovate to elliptical, obtuse, 2 cm long, 13–15 mm across, 2 mm thick; Fl. in small Infl., shortly pedicellate, 6–7 mm ⌀, white.

**Titanopsis** Schwant. (§ IV/1/9). – Short-stemmed plants forming dense mats or clumps; roots fleshy; St. with Ros. of 6–8 or more opposite and decussate, densely crowded, narrowly to broadly spatulate L., expanded and triangular above, with a bluish-green to reddish or yellowish-white surface, the tip-area covered with rounded or ± square tubercles; Fl. solitary, ± sessile or shortly pedicellate, medium-sized, yellow or orange. – Growing season: summer: Greenhouse, warm. Propagation: seed or cuttings.

**T. calcarea** (Marl.) Schwant. (Pl. 199/4) (*Mes. c.* Marl.). – Cape: W. Griqualand, Prieska D. – Ros. 6–8 cm ⌀; L. spatulate, triangularly truncate at the tip, rounded in the lower part, somewhat whitish or greenish-bluish, with reddish to greyish-white tubercles, up to 25 mm long, 8 mm across below, 12 mm across above; Fl. 2 cm ⌀, golden-yellow to almost orange.

**T. fulleri** Tisch. – Cape: W.Griqualand, Prieska D. – St. numerous, Ros. with 5–6 pairs of L. 20–22 mm long, 4 mm across at the base, expanded above to 10 mm across, then tapered to a rounded triangle, with a rounded keel on the underside, the surface attractively bluish-green to reddish with faint, dark, prominent dots and greyish-brown tubercles on a purple ground along the margins; Fl. c. 16 mm ⌀, dark yellow.

**T. hugo-schlechteri** (Tisch.) Dtr. et Schwant. v. **hugo-schlechteri** (*Mes. h.-s.* Tisch., *Verrucifera h.-s.* (Tisch.) N. E. Br., *Mes. astridae* Dtr.). – S.W.Afr.: Gr. Namaqualand. – Plant up to 5 cm ⌀; L. 12–15 mm long, 3–4 mm across and thick at the base, 4–6 mm across below the equilaterally triangular tip, rounded to carinate on the underside, slightly glossy, reddish to greyish-green or brownish, the tip-area ± perpendicular to the blade and covered with numerous large and small, light grey or brown to reddish tubercles; Fl. yellow.

**T. —** v. **alboviridis** Dtr. – L. whitish-green.

**T. luederitzii** Tisch. (*Verrucifera l.* (Tisch.) N. E. Br.). – S.W.Afr.: Gr. Namaqualand. – Resembles **T. schwantesii**; L. rather narrower, less expanded below the rounded tip which is covered with large tubercles 1 mm across, ochre-yellow, reddish at the base; Fl. yellow.

**T.** cv. Primosii. – Resembles **T. schwantesii**; Fl. 2.5 cm ⌀, Pet. canary-yellow, with flesh-coloured tips.

**T. schwantesii** (Dtr.) Schwant. (Pl. 199/3) (*Mes. s.* Dtr., *Mes. concinnum* N. E. Br., *Verrucifera s.* (Dtr.) N. E. Br.). – S.W.Afr.: Karas Mts. – L. up to 3 cm long, 3–7 mm across at the base, expanded somewhat above and then abruptly to 12 mm at the rounded-triangular tip, rounded on the underside, surface light grey to bluish-green excepting the tip-area and both surfaces in the upper part which are covered with vesicular, rounded, yellowish-brown tubercles; Fl. 15–18 mm ⌀, light yellow,

**Trichodiadema** Schwant. (§ IV/1/6). – ♄ with woody or tuberous roots; St. long, slender and arched, or short and then of ± tufted habit; L. little connate at the base, semiterete to ± terete, glistening with papillae, the tip usually with a cluster of ± radiate, spreading bristles; Fl. solitary, shortly pedicellate, large to small, white, yellow or red, spring to late autumn. – Summer: in the open, winter: cold-house, resting period. – Propagation: seed, cuttings.

### Key to the Groups and SGroups of the Genus Trichodiadema by L. Bolus

1. Papillae tipped with a soft hair – e.g. T. strumosum, fergusoniae .............. **STRUMOSA**
1a. Papillae not as above ................................................. **ESTRUMOSA** 2.
2. L. ± conspicuously tipped with bristles................................ **SETIFERA** 3.
2a. L. not, or very obscurely, tipped with bristles – e.g. T. attonsum ............... **ATTONSA**

3. Bristles radiating to form a diadem .................................... **Radiantia 4.**
3a. Bristles erect, forming a tuft – e.g. T. intonsum ............................... **Stricta**
4. Plants low-growing, densely compact or loosely branched – e.g. T. densum, T. barbatum ....
................. **Barbata**
4a. Plants with elongated virgate Br. – e.g. T. setuliferum, T. pomeridianum ........... **Virgata**

**T. attonsum** (L. BOL.) SCHWANT. (*Mes. a.* L. BOL.). – S.W.Cape: Karroo. – ♄ 10–15 cm high, with spreading **St.; Br.** hirsute, with Int. 7–10 mm long; **L.** 10–13 mm long, 3–4 mm ⌀, tipped with 1–3 erect bristle-hairs, the papillae rounded-ovoid, with one short hair; **Fl.** 16 to 20 mm ⌀, white.

**T. aureum** L. BOL. – Cape: Willowmore D. – Plant 12 cm high; **roots** c. 13 cm long, tuberous below, the taproot 9 cm long, 3.5 cm ⌀; **St.** woody, 8 mm ⌀; **Br.** up to 8 cm long, young Int. papillose-hairy, 5–10 mm long; **L.** 5–6 mm long, 2 mm across, obtuse, green, tipped with white bristles 3 mm long; **Fl.** with Pet. 6–8 mm long, yellow to golden-yellow.

**T. barbatum** (L.) SCHWANT. (Pl. 199/5) (*Mes. b.* L.). – Cape: Karroo; S.W.Afr.: Karas Mts. – **Roots** napiform; **St.** prostrate; **L.** 8–12 mm long, 3–4 mm across, with 8–10 black bristles at the tip, appearing greyish-green because of the pointed papillae; **Fl.** 3 cm ⌀, deep red.

**T. bulbosum** (HAW.) SCHWANT. (*Mes. b.* HAW.). – Cape: near Port Elizabeth. – **Caudex** tuberous; **St.** 20 cm high; **L.** 5–8 mm long, 2.5 to 3 mm thick, grey-papillose, with 8–11 white bristles at the tip; **Fl.** 2 cm ⌀, deep red.

**T. burgeri** L. BOL. – Cape: Oudtshoorn D. – Mat-forming; **caudex** tuberous, up to 3.3 cm ⌀; **St.** thick, 8–12 cm high; **Br.** 1–2 cm long, with old L.-Sh. below enclosing the Int.; **L.** densely imbricate, 5–12 mm long, Sh. 4 mm long, with 14 white, radiating bristles at the tip and narrowly ovoid, ± bristle-tipped papillae; **Fl.** 2.4 cm ⌀, reddish, the Sep. covered with tiny bristles up to 3 mm long.

**T. calvatum** L. BOL. – Cape: Robertson D. – **Caudex** tuberous, 6 mm thick at the base; **St.** 13 cm long, 12 mm thick; **Br.** 7 cm long, Int. 1 cm long, papillose; **L.** 1–2 cm long, 3 mm across, obliquely tapered, not bristle-tipped, bluish-green, papillose; **Fl.** 24 mm ⌀, pink, pedicel and Cal. hirsute.

**T. concinnum** L. BOL. – Cape: Alexandria D. – Plant 3.5–10 cm high; **roots** woody, napiform, up to 6.5 cm long; **St.** 7 mm ⌀; **Br.** erect to ascending, Int. 1 cm long; **shoots** variously curved; **L.** 5–9 mm long, 1–2 mm ⌀, whith small bristles at the base, tipped with 8 bristles c. 2 mm long; pedicel softly bristly, Pet. 11 mm long, pinkish-purple.

**T. decorum** (N. E. BR.) STEARN (*Mes. d.* N. E. BR.). – Origin not known. – **St.** erect or ascending, white-papillose, terminating in a pedicel; **L.** bluish-green, curved at the base, also at the tip which has a cluster of 7–9 radiating white hairs or is covered with pearly papillae; **Pet.** mauvish-red, Cal.-lobes each with a tuft of hairs.

**T. densum** (HAW.) SCHWANT. (Pl. 199/6) (*Mes. d.* HAW., *Mes. barbatum* v. *d.* WILLD.,

*Mes. b.* v. *c.* L.). – Cape: Karroo. – **Caudex** fleshy, thickened; **St.** short, mat-forming; **L.** 15–20 mm long, 4–5 mm thick, green, covered with pointed papillae, tipped with a cluster of 20–25 fairly long, white bristles; **Fl.** 4–5 cm ⌀, crimson.

**T. emarginatum** L. BOL. – Cape: Lit. Karroo. – Mat-forming ♄; **St.** thick, 14 mm ⌀ at the base; **L.** 15 mm long, 6 mm across, 4 mm thick, papillose, tipped with 4–10 white hairs 2 mm long; **Fl.** 3 cm ⌀, pink.

**T. fergusoniae** L. BOL. – Cape: Riversdale D. – **Caudex** tuberous, fusiform, 12 cm long, 2.5 cm ⌀ above; **St.** 2–3, 12 cm long; **Br.** crowded, with 4–6 crowded, acute, soft, papillose-haired **L.** 9 mm long, 2.5 mm across, 2 mm thick, the papillae mostly subulate, with a white hair; **Fl.** colour not known.

**T. fourcadei** L. BOL. – Cape: Humansdorp D. – **Caudex** tuberous, semi-underground; **St.** ± curved, spreading, 8–15 cm long; **Br.** with white hairs, Int. 6–9 mm long; **L.** 15–25 mm long, 2–3 mm ⌀, acute, pubescent, tipped with brown bristle-hairs 2 mm long; **Fl.** 2 cm ⌀, white to yellowish.

**T. gracile** L. BOL. v. **gracile.** – Cape: Caledon D. – **Caudex** ± fusiform, 6 cm long, 15 mm thick; **St.** slender, 5–7 cm long, spreading to prostrate, Int. 7–9 mm long; **L.** 14 mm long, 3 mm ⌀, indistinctly hair-tipped, with rounded papillae; **Fl.** 18 mm ⌀, pink.

**T. — v. piliferum** L. BOL. – Cape: Swellendam D. – Pedicel and receptacle densely hirsute.

**T. — v. setiferum** L. BOL. – **L.** with bristle-tipped papillae; **Fl.** 28 mm ⌀, pedicel and receptacle papillose.

**T. hallii** L. BOL. – Cape: Lit. Karroo. – Compact plant 6 cm high; **caudex** tuberous; **St.** 4 mm thick, herbaceous parts bristly-papillose; **L.** 13–18 mm long, 4 mm ⌀, tipped with 7–10 brown hairs, Sh. 3 mm long; **Fl.** 46 mm ⌀, pink, with a white centre.

**T. hirsutum** (HAW.) STEARN (*Mes. h.* HAW., *Mes. barbatum* v. *b.* L., *Mes. stellatum* DC., *Mes. barbatum* v. *h.* WILLD. ?). – Origin unknown. – Close to **T. stellatum**; compact or laxly branched plant; pedicel hirsute.

**T. imitans** L. BOL. – Origin unknown. – **St.** spreading, virgate, angular, hirsute, 35 cm long, Int. 10–35 mm long, herbaceous parts with bristle-tipped papillae; **L.** 15–25 mm long, 2–3 mm ⌀, bluish-green, the bristles at the tip forming a distinct diadem, Sh. 1–2 mm long; **Fl.** 25–30 mm ⌀, white.

**T. intonsum** (HAW.) SCHWANT. (Pl. 200/2) (*Mes. i.* HAW.). – Cape: Swartkops R. – **St.** prostrate; **L.** 12–13 mm long, 4 mm thick, semicylindrical, covered with acute papillae, tipped with 8–10 brownish bristles, the margins ± ciliate; **Fl.** 2 cm ⌀, reddish to whitish, the Cal.-tube with white hairs.

**T. littlewoodii** L. Bol. f. **littlewoodii**. – S.W. Afr.: Gr. Namaqualand. – ♄ 20 cm high; **caudex** tuberous, up to 6 cm long, 2.5 cm ⌀, with c. 16 secondary Tub. c. 9 cm long below; **St.** densely crowded, c. 19 cm long, virgate, Int. 1–3 cm long; **L.** 10–27 mm long, 3–5 mm ⌀, sparsely barbate at the tip, with bristle-tipped papillae along the margins, Sh. 1 mm long; **Fl.** c. 16 mm ⌀, pink.

**T. —** f. **alba** L. Bol. – Pet. white.

**T. marlothii** L. Bol. – Cape: Barrydale, Riversdale D. – **St.** 12 mm thick at the base; **Br.** spreading to prostrate, up to 1 cm thick, 12 cm long; **shoots** 3–5 cm long, with 8–10 crowded, bluish-green, papillose **L.** 17 mm long, 6 mm across, 5 mm thick, tipped with brown bristles 4 mm long; **Fl.** 32 mm ⌀, pink.

**T. mirabile** (N. E. Br.) Schwant. v. **mirabile** (*Mes. m.* N. E. Br.). – Cape: Laingsburg, Uitenhage D., Karroo. – Bushy plant 7–8 cm high; **St.** covered with white, bristly hairs with a spur-like protuberance at the base; **L.** 12 to 26 mm long, 4–6 mm thick, obtuse, green, covered with rhombic, acute papillae, tipped with (8–)12(–14) erect, dark brown, stiff bristles 2 mm long; **Fl.** 4 cm ⌀, white.

**T. —** v. **leptum** L. Bol. – Cape: Karroo. – ♄ 9–12 cm high; **caudex** ± tuberous; **St.** virgate; **L.** 10–18 mm long; **Fl.** 25–52 mm ⌀.

**T. obliquum** L. Bol. – Cape: Beaufort West D. – **St.** 11 cm long, branched, rough with dense, bristle-tipped papillae; **L.** 15–25 mm long, 3 mm ⌀, obliquely obtuse, with c. 6 papillae at the tip; Pet. 5–6 mm long.

**T. occidentalis** L. Bol. – Cape: Bredasdorp D. – Mat-forming plant; **St.** glabrous; **L.** 11 mm long, 3 mm ⌀, bluish-green, papillose, tipped with 2–3 short, white bristles; **Fl.** 15 mm ⌀, pink.

**T. olivaceum** L. Bol. – Cape: Somerset East D. – **St.** prostrate, 10 cm long, Int. 5–10 mm long, with papillose hairs; **L.** 7–9 mm long, 3 mm thick, covered with oval, stiffly bristle-tipped papillae and tipped with 6–12 white bristle-hairs 3 mm long; **Fl.** 24 mm ⌀, white.

**T. orientale** L. Bol. – Cape: Uitenhage D. – **St.** woody, 9 mm thick at the base; **Br.** spreading, prostrate, 10–19 cm long, Int. 4–9 mm long; **shoots** papillose-haired, 3 cm long; **L.** 10–14 mm long, 5 mm across, tipped with mostly black bristles up to 2 mm long, densely covered with pointed or bristle-tipped papillae; **Fl.** 22 mm ⌀, white, tinged pink.

**T. peersii** L. Bol. – Cape: Willowmore D. – Erect ♄ 7–9 cm high; **St.** spreading, glabrous; **shoots** 7 cm long, Int. 2–5 mm long; **L.** 5–8 mm long, 4 mm thick, ± spherical, papillose, tipped with 8–9 pale brown bristles 3 mm long; **Fl.** 38 mm ⌀, white.

**T. pomeridianum** L. Bol. – Orange Free State. – Erect, stiff ♄ 30 cm high from a tuberous **caudex** 5 cm long; **St.** lax, ascending, virgate, rough with white papillae; **L.** semi-cylindrical, 15 mm long, 2 mm thick, pale green, densely papillose, tipped with 3–8 white, radiating hairs 1.5 mm long; **Fl.** 13 mm ⌀, purplish-pink.

**T. pygmaeum** L. Bol. – Cape: Swellendam D. – Forming mats 1–1.5 cm high, c. 5 cm ⌀; **roots** slender; **St.** c. 2 cm long, Int. enclosed; **shoots** 1–2 cm long; **L.** c. 6 mm long, 3 mm across, often with indistinct, hairlike bristles; **Fl.** 2 cm ⌀, pink.

**T. rogersiae** L. Bol. – Cape: Cradock D. – Erect plant c. 10 cm high; **St.** ascending, papillose-haired; **Br.** 4–5 cm long, Int. 2–7 mm long, with 6–10 **L.** 16 mm long, 5 mm thick, covered with large, round papillae, Sh. 1.5 mm long; **Fl.** 32 mm ⌀, pinkish-purple.

**T. rupicolum** L. Bol. – Cape: Uitenhage D. – Mat-forming plant; **St.** glabrous; **shoots** short, densely leafy; **L.** 3–6 mm long, 2–3 mm thick, felty-papillose, tipped with 1–6 radiating, brownish-white hairs; **Fl.** 13 mm ⌀, whitish to pale pink.

**T. ryderae** L. Bol. – Cape: Bedford D. – Erect ♄ c. 8 cm high; **St.** stiff, with dense, minute, pearly dots, Int. 3–8 mm long; **L.** green, purple above, covered with large, round, hair-tipped papillae, with indistinct bristle-hairs at the tip; **Fl.** 14 mm ⌀, purplish-pink.

**T. schimperi** (Engl.) Herre (*Mes. s.* Engl.). – Ethiopia; S.W.Afr.: near Nauchas. – Creeping ♄; **St.** ascending, rooting at the nodes, Int. 10 to 12 mm long; **Br.** 2–3 cm long; **L.** obtuse, papillose, tipped with 6–8 tiny bristles; **Fl.** 12 mm ⌀.

**T. setuliferum** (N. E. Br.) Schwant. v. **setuliferum** (*Mes. s.* N. E. Br.). – Cape: Somerset East D. – **Roots** woody, fibrous; **St.** prostrate, 30 cm or more long, green, roughly papillose; **L.** 25–35 mm long, 3–5 mm thick, obtuse, covered with glossy, bristle-tipped papillae, tipped with a cluster of 7–16 curved, white, radiating hairs; **Fl.** violet.

**T. —** v. **niveum** L. Bol. – Cape: Victoria West D. – **Fl.** 22 mm ⌀, white.

**T. stayneri** L. Bol. – Cape: Uitenhage D. – ♄ 4 cm high, 7–10 cm ⌀; **caudex** tuberous, up to 7 mm ⌀; **St.** spreading, minutely rough-papillose, 6 cm long, Int. 1–4 mm long; **shoots** with 4 crowded, obtuse **L.** 7–11 mm long, 2.5 mm ⌀, covered with minute papillae tipped with fragile bristles 1–2 mm long; **Pet.** 6–9 mm long, purplish-pink.

**T. stellatum** (Mill.) Schwant. (*Mes. s.* Mill.). – Cape: Lit. Karroo. – Forming mats 5–10 cm ⌀; **roots** fleshy; **L.** c. 10 mm long, 3–4 mm across, greyish-green, rough-papillose, tipped with 12–15 bristles c. 3–4 mm long; **Fl.** 3 cm ⌀, light violet to red.

**T. stelligerum** (Haw.) Schwant. (*Mes. s.* Haw., *Mes. barbatum* v. β Haw., *Mes. b.* Curtis,. – Cape: Uitenhage, Mossel Bay D. – **St.** tangled, curved, glabrous; **L.** 10–15 mm long, 5–6 mm thick, light green, covered with glossy, flat papillae, obtuse at the tip, with 5–10 white or brownish, radiating bristles; **Fl.** large, Pet. light pruple.

**T. strumosum** (Haw.) L. Bol. (*Mes. s.* Haw.). – Cape: origin not known. – **St.** scarcely 5 cm high, erect or ± prostrate, with a yellowish-brown bark; **L.** 12–18 mm long, tapered, fresh green, glossy; **Fl.** up to 2 cm ⌀, pale yellow to white.

**Vanheerdea** L. Bol. (*Vanheerdia* L. Bol.) (§ IV/1/2). – Highly succulent, cushion-forming plants resembling the extremely succulent **Gibbaeum** spec.; L. always symmetrical, united to form nearly globose to elongated Bo., the ± indistinct keels and edges mostly denticulate, only one spec. with indistinct Wi.; Fl. up to 3 together, long-pedicellate, yellow, open in the afternoon. Growing-period: spring. Greenhouse: warm. Propagation: seed, cuttings.

## Key to the Species of the Genus Vanheerdea by L. Bolus

1. L. glabrous, depressed and flattened or slightly convex at the apex and having a translucent area or Wi. .................................................................... **primosii**
1a. L. not as above ............................................................. **2.**
2. Mature L. united for 2/3 of their length or more, not laterally compressed at the apex .... **3.**
2a. Mature L. united for 1/2 their length or more, laterally compressed at the apex .. **divergens**
3. L. 3–4 times longer than broad; the greatest ∅ of the Bo. less than 1/2 its length .......... ........ **angusta**
3a. L. less than twice as long as broad; the greatest ∅ of the Bo. nearly equal to its length ...... ........ **roodiae**

**V. angusta** (L. Bol.) L. Bol. (*Rimaria a.* L. Bol.). – Cape: Lit. Namaqualand. – Bo. slender, up to 7 cm long, 15–20 mm across the Sh., 5–10 mm thick and 9–14 mm across at the base, the free parts 8–14 mm long, obtuse to ± acute above, the keel on the underside and the angles with 1–3 T., reddish-brownish, minutely velvety; Fl. 35 mm ∅, yellow to golden-yellow.

**V. divergens** (L. Bol.) L. Bol. (Pl. 200/3) (*Rimaria d.* L. Bol.). – Cape: Bushmanland. – L. pairs 4–6 cm long, united for 20–25 mm, 25–30 mm across, 16–25 mm thick, broadly compressed above, the keel and angles denticulate, greyish-green, very reddened during the resting period; Fl. 2–4 cm ∅, golden-yellow.

**V. primosii** L. Bol. (*Rimaria p.* (L. Bol.) L. Bol.). – Cape: Bushmanland. – Bo. 35 mm long, 23 mm ∅, with the Fi. gaping and 5 mm deep, the free parts 37 mm long, convex to ± truncate at the tip which has a pale green, marbled Wi., the surface otherwise pale violet; Fl. 25 mm ∅, golden-yellow.

**V. roodiae** (N. E. Br.) L. Bol. (Pl. 200/4) (*Rimaria r.* N. E. Br.). – Cape: Lit. Namaqualand. – Bo. 22–25 mm high, the L. united for 18–22 mm, the free parts 12–14 mm ∅, oval or hemispherical, broadly rounded on the underside, ± distinctly keeled, the angles and keel ± distinctly dentate, smooth, green, puberulous, at first with ± translucent dots; Fl. 25 mm∅, orange-yellow.

**Vanzijlia** L. Bol. (§ IV/1/2). – Low ♄ with wiry, prostrate or creeping St.; L. in dissimilar pairs; Fl. solitary, white to purple. – Summer: in the open, winter: cold-house. Propagation: seed, cuttings.

**V. angustipetala** (L. Bol.) N. E. Br. (*Mes. a.* L. Bol., *V. annulata* L. Bol.). – Cape: Clanwilliam D. – ♄ 12.5 cm high; St. ascending; shoots with 4 L., those of the first pair united for 1/3 of their length, the second pair longer-united, trigonous, rounded at the tip, ± keeled, pale green, 14 mm long, 4 mm ∅; Fl. 4–6 cm ∅, pale pink or white.

**V. annulata** (Bgr.) L. Bol. (Pl. 200/1) (*Mes. a.* Bgr., *V. a.* (Bgr.) N. E. Br.). – Cape: Van Rhynsdorp, Piquetberg D. – Resembles **V. angustipetala**; Int. 4–32 mm long, each bearing 1 L. pair and ± constricted at the base of the old, persistent L.-Sh. which is 10 mm long, the free part of the L. up to 25 mm long, acute, apiculate; Fl. up to 5 cm ∅.

**V. rostella** L. Bol. – Cape: Malmesbury D. – St. creeping, c. 15 cm long, pink to whitish; L. persisting and papery after withering, the pairs united for 4 mm, the upper pair 22 mm long, 3 mm ∅, compressed and rostrate, acute, purple to green, spotted; Fl. 3 cm ∅, purplish-pink.

**Wooleya** L. Bol. (G. of uncertain status). – Small ♄ resembling **Namaquanthus**, the herbaceous parts very pruinose. – Summer: in the open, winter: cold-house. Propagation: seed, cuttings.

**W. farinosa** (L. Bol.) L. Bol. (? *Namaquanthus f.* L. Bol.). – Cape: Lit. Namaqualand. – Compact, robust, erect ♄ 30 cm high; St. 5–10 mm ∅, with the Int. enclosed in the L.-Sh. or more rarely 1 cm long; L. expanded at the base, flat to convex on the upperside, rounded on the underside, with the keel reduced to a line and the sides expanded at the tip, unequal, 34 mm long or 15–25 mm long, 6–9 mm across, with the dry Sh. persisting; Fl. solitary, with pale pink, purple-striped Pet. 13–17 mm long, on a 4-angled pedicel 6–7 mm long with bracts 1.5 mm long at its base.

**Zeuktophyllum** N. E. Br. (G. of uncertain status). – Small ♄, rare in cultivation.

**Z. suppositum** (L. Bol.) N. E. Br. (Pl. 200/5) (*Mes. s.* L. Bol.). – Cape:Riversdale D., Lit. Namaqualand. – Robust plant c. 9 cm high; St. ascending, with the old dry L. persisting; L. 4–6 on a St., crowded, 3–4 cm long, 5–10 mm across, 12 mm thick, the pairs united for 2–5 mm, trigonous, swollen, somewhat obliquely keeled, bluish-purple; Fl. solitary, 3 cm ∅, yellowish-pink.

# SUPPLEMENT

**Part I** (see also p. 585)

**Aloe abyssicola** Lavr. et A. S. Bilaidi (§ IV/C/16) in Cact. y Succ. Journ. Am. **43**, 206–207, 1971.
**A. bella** Rowl. (*A. pulchra* Lavr. in Cact. y Succ. Journ. Am. **45**, 118–119, 1973 non Jacq.).
**Caralluma caudata** N. E. Br. ssp. **caudata** in Bothalia **XI**, 1–2, 134, 1973.
— — ssp. **rhodesiaca** Leach in l. c.
× **Carapelia** Rowl. Asclepiadaceae. – Intergeneric hybrid: **Caralluma** R. Br. × **Stapelia** L. – For an analysis of naturally-occurring hybrids see L. C. Leach in Bothalia, **XI**, 133–137, 1973.
× **C. tarantuloides** (R. A. Dyer) Rowl. (*Stapelia t.* R. A. Dyer). – Hybr.: **Caralluma caudata** × **Stapelia gigantea.** – Rhodesia: Gwanda D.; N.Transvaal. – Intermediate between the parents; **Fl.** 15 cm ⌀, with attenuated Cor.-lobes covered with soft purple hairs; occurs both wild and where the parents are grown together.
× **C. 1.** – Hybr.: **Caralluma caudata** × **Stapelia kwebensis** (?). – Rhodesia: Sabi Valley. – Intermediate between the supposed parents.
**Crassula compacta** Schoenl. – Supplement to the description: Stemless ♃; L. in Ros., broad obovate; **Infl.** up to 12 cm high, thyrsoid, Fl.white.
**C. streyi** Toelken (II/2). – S.Natal. – Resembles **C. multicava** but the **Fl.** are always 5-merous.
**Echeveria coccinea** (Cav.) DC. (§ I) (*Cotyledon c.* Cav., *E. longifolia* Hort. fide Herb. Kew, *C. pubescens* Bak., *E. p.* Schltd.) Supplement of synonyms in Succulenta, **53**, 16–18, 1974.
**E. —** cv. Tortuosa (*E. pubescens* cv. *T., E. p.* cv. *Recurvata*). – Hybr.: **E. coccinea** × **E. bifida** in l. c.
**E. cv. Pulvicox.** – Hybr.: **E. pulverulenta** × **E. coccinea** in l. c.
**Endadenium** Leach, Euphorbiaceae, in Garcia de Orta, Sér. Bot. Lisbon, **1**, 31, 1973.
**E. gossweileri** (N. E. Br.) Leach (*Monadenium g.* N. E. Br.) in l. c. p. 32.
**Euphorbia tirucalli** L. and its Typification, Synonymy and Relationship, by L. C. Leach in Kirkia, **9**, 69–86, 1973.
**Haworthia pulchella** Bayer (§ ?) in Journ. S.Afr. Bot. **39**, 252, 1973.
**H. serrata** Bayer (§ XIV) in l. c. 249.
**Huernia verekeri** v. **angolensis** Leach in l. c. **40**, 19, 1974.
**Lewisia megarhiza** (Hemsley) Macbride in Journ. Scott. Rock Gard. Club, **13**, 4, 294–6, 1973.
**Monadenium cancellii** Leach (§ II c/1) in Garcia de Orta, Sér. Bot. Lisbon, **1**, 35, 1973.
**Pectinaria mirkinii** J. Pillans in Journ. S.Afr. Bot. **V**, 64, 1954.
**Sempervivum ciliosum** Craib. – Hybr.: **S.** × **praegeri** Rowl.
**S.** × **praegeri** Rowl. – Hybr.: **S. ciliosum** × **S. erythraeum** in Repert. Plant. Succ. XXIII, 1974.
**Stapelia villetae** Luckh. in S.Afr. Gard. **28**, 228, 1938.
**Stultitia araysiana** Lavr. et A. S. Bilaidi in Cact. y Succ. Journ. Am. **43**, 207–7, 1971.
**Trichocaulon mossamedense** Leach in Journ. S.Afr. Bot. **40**, 15, 1974.

**Part II.**

The following comb. and var. nov. are noted in **LITHOPS:** A Checklist and Index by D. T. Cole in Excelsa No. 3, 1973, 37–71:

**Lithops aucampiae** v. **koelemanii** (de Boer) Cole (*L. k.* de Boer, *L. a.* v. *k.* (de Boer) de Boer nom. inv.).
**L. bromfieldii** v. **glaudinae** (de Boer) Cole (*L. g.* de Boer).
**L. comptonii** v. **weberi** (Nel) Cole (*L. w.* Nel).
**L. dinteri** v. **frederici** Cole (as var. nov. in Madoqua, Ser. I. No. 7, 34, 1973).
**L. fulleri** v. **rouxii** (de Boer) Cole (*L. julii* v. *r.* de Boer).
**L. fulviceps** v. **lactinea** Cole.
**L. hallii** v. **ochracea** (de Boer) Cole (*L. fulleri* v. *o.* de Boer).
**L. karasmontana** v. **lericheana** (Dtr. et Schwant.) Cole (*Mes. l.* Dtr. et Schwant., *L. l.* (Dtr. et Schwant.) N. E. Br., *L. bella* v. *l.* (Dtr. et Schwant.) de Boer et Boom).
**L. —** v. **tischeri** Cole (*L. fossulifera* Tisch. nom. nud.).
**L. ruschiorum** v. **lineata** (Nel) Cole (*L. r.* v. *l.* (Nel) de Boer et Boom nom. inv.).
**L. schwantesii** v. **marthae** (Loesch et Tisch.) Cole (*L. m.* Loesch et Tisch., *L. inornata* Dtr., *L. dinteri* v. *m.* (Loesch et Tisch.) Fearn (as var. nov. in Madoqua, Ser. I. No. 7, 35, 1973).
**L. turbiniformis** v. **dabneri** (L. Bol.) Cole (*L. d.* L. Bol.).
**L. —** v. **marginata** (Nel) Cole (*L. m.* Nel).
**L. villetii** v. **kennedyi** (de Boer) Cole (*L. fulleri* v. *k.* de Boer).

## Part I. Continuation

**Agave americana** L. v. **americana** in H. S. GENTRY, The Agave Family in Sonora, p. 84, 1972 and the following new spec. and var. in the same book: **A. americana** v. **expansa** (JACOBI) GENTRY *(A. expansa* JACOBI) p. 80. – **A. aktites** GENTRY (§ III/14 ?) p. 148–150. – **A. felgeri.** GENTRY (§ II/2) p. 60–62. – **A. fortiflora** GENTRY (§ III/5) p. 122–126. – *A. hartmannii* S. WATS. nom. confusum, p. 54. – **A. jaibolii** GENTRY (§ III/7/3) p. 89–94. – **A. multifilifera** GENTRY (§ II/2) p. 46–50. – **A. ocahui** GENTRY (§ II/7/b) p. 72–76. – **A. parviflora** TORR. ssp. **parviflora** (§ II/2) p. 54. – **A. parviflora** ssp. **flexifolia** GENTRY p. 56–57. – **A. pelona** GENTRY (§ II/5 ?) p. 76–80. – **A. polianthiflora** GENTRY (§ II/2) p. 51–54. – **A. shrevei** GENTRY ssp. **shrevei** (§ III/6) p. 111. – **A. shrevei** ssp. **matapensis** GENTRY p. 115–117. – **A. zebra** GENTRY (§ III/17 ?) p. 126–130.
**Aloe andongensis** BAK. v. **repens** LEACH in Journ. S. Afr. Bot. **40,** 115, 1974.
**A. erinacea** HARDY (§ ?) in Bothalia **10,** 366, March 1971.
**A. lepida** LEACH (§ IV/C/18a) in Journ. S. Afr. Bot. **40,** 102, 1974.
**A. procera** LEACH (§ IV/D/23) in l. c. **40,** 117, 1974.
**A. scorpioides** LEACH (§ IV/C/19c) in l. c. **40,** 106, 1974.
**A. vallaris** LEACH (§ IV/C/19c) in l. c. **40,** 111, 1974.
**Duvalia galgallensis** LAVR. in Cact. Succ. Journ. Am. **XLVI,** 184, 1974.
**Echeveria agavoides** LEM. v. **multifida** E. WALTH. in E. WALTHER, Echeveria. 85. 1972. and the following new. spec. and var. in the same book: **E. agavoides** v. **prolifera.**
E. WALTH. p. 85. – **E. angustifolia** E. WALTH. (§ 14) p. 211. – **E. bella** ALEXAND. v. **major** E. WALTH. p. 358. – **E. colorata** E. WALTH. (§ 8) p. 91. – **E. cornuta** E. WALTH. (§ 11) p. 133. – **E. dactylifera** E. WALTH. (§ 18) p. 179. – **E. elegans** ROSE v. **hernandonis** E. WALTH. p. 99. – **E. elegans** v. **tuxpanensis** E. WALTH. p. 101. – **E. erubescens** E. WALTH. (§ 14) p. 263. – **E. lindsayana** E. WALTH. (§ 8) p. 90. – **E. lutea** ROSE v. **fuscata** E. WALTH. p. 259. – **E. moranii** E. WALTH. (§ 4) p. 347. – **E. nodulosa** (BAK.) OTTO v. **minor** E. WALTH. p. 317. – **E. proxima** E. WALTH. (§ 4) p. 350. – **E. pulidonis** E. WALTH. (§ 8) p. 122. – **E. reglensis** E. WALTH. (§ 11) p. 133. – **E. sanchez-mejoradae** E. WALTH. (§ 8) p. 108. – **E. sayulensis** E. WALTH. (§ 18) p. 151. – **E. shaviana** E. WALTH. (§ 13) (new descr.) p. 270. – **E. skinneri** E. WALTH. (§ 2) p. 307. – **E. tenuifolia** E. WALTH. (§ 14) p. 264. – **E. westii** E. WALTH. (§ 4) p. 361.
**Echeveria recurvata** CARRUTHERS (§ ?) in Bull. A. S. P. S. **9.** No. 2, 56, 1974.
**Echidnopsis bavazzani** LAVR. in Cact. Succ. Journ. Am, **XLVI,** 181, 1974.
**E. chrysantha** LAVR. v. **filipes** LAVR. in l. c. 184.
**E. mijerteina** LAVR, v. **marchandii** LAVR. in l. c. 184.
**E. stellata** LAVR. in l. c. 182.
**Portulaca pachyrrhiza** GAGN. in Bull. Soc. Bot. France, **56,** 41, 1909.
**P. pilosa** L. ssp. **decipiens** (v. POELLN.) GEES. (*P. d.* v. POELLN) in Blumea, **17,** 297, 1969.
**P.** — ssp. **grandiflora** (HOOK.) GEES. (*P. g.* HOOK.) in l. c. 297.
**P.** — ssp. **lakhonensis** (GAGN.) GEES. (*P. l.* GAGN.) in l. c. 299.
**P.** — ssp. **okinawensis** (WALK. et TAW.) GEES. (*P. o.* WALK. et TAW.) in l. c. 298.
**P.** — ssp. **papulosa** (SCHLECHTD.) GEES. (*P. p.* SCHLECHTD. ex v. POELLN.) in l. c. 298.
**P.** — ssp. **sundaensis** (v. POELLN.) GEES. (*P. s.* v. POELLN.) in l. c. 298.
**P.** — ssp. **villosa** (CHAM.) GEES. (*P. v.* CHAM.) in l. c. 297–298.
**Rosularia davisii** MUIRHEAD in Notes Royal Bot. Gard. Edinburgh, **31,** 393, 1972.
**Sedum hewittii** CHAMBERLAIN (§ 20/5) in l. c. 325.
**S. hispanicum** v. **planifolium** CHAMBERLAIN in l. c. 325.
**Sempervivum heuffelii** SCHOTT v. **bulgaricum** CHESHM. in Bot. Zurn (Leningrad) **54,** 474, 1969.
**Testudinaria sylvatica** (ECKL.) KNUTH v. **lydenbergensis** (BL., HD. et HIND) ROWL. in Bot. J. Linn. Soc., **64,** 445, 1974 *(Diosc.).*
**Sarcostemma vanlessenii** LAVR. in Nat. Cact. Succ. Journ. **29,** 35, 1974.

Notice: Plate $\frac{8}{5}$ in this Lexicon shows acc. H. S. GENTRY in „The Agave Family in Sonora', Washington 1972, not **Agave parviflora** TORR. but the new **Agave polianthiflora** GENTRY.

## Bibliography of Literature

GENTRY, H. S. The Agave Family in Sonora. Agriculture Handbook No. 399. Agric. Res. Serv. U. S. Dept. of Agricult., Washington, D. C. March 1972
WALTHER, E. Echeveria. California Academy of Sciences. San Francisco 1972

# BIBLIOGRAPHY
of Literature used by the author

## A. Periodical publications

**Austria:** Annals K. u. K. Naturw. Hofmuseum, Vienna, vol. 15, 1900.
**France:** Bulletin de la Société Histoire Naturelle de l'Afrique du nord, Algeria, from vol. 12/14, 1921/23.
Cactus, Revue périodique de l'Association Française des amateurs de Cactées et Plantes Grasses, Paris, from vol. 1, 1946.
**Germany – Federal Republic:** Monatsschrift für Kakteenkunde (now Kakteen und andere Sukkulenten), Deutsche Kakteen-Gesellschaft, from 1891.
ENGLER, A., Botanische Jahrbücher, Berlin.
Gartenflora, Deutsche Gartenbaugesellschaft, Berlin, from vol. 76, 1927.
Gartenwelt, Paul Parey, Berlin/Hamburg, from vol. 32, 1928.
MÖLLERS Deutsche Gärtner-Zeitung, from vol. 41, 1926.
Mitteilungen aus der Botanischen Staatssammlung München, from vol. 3, 1951.
**Germany – Democratic Republic:** FEDDES Repertorium specierum novarum regni vegetabilis, Berlin, vol. 1, 1905 to vol. 49, 1940.
**Great Britain:** Bulletin of the African Succulent Plant Society, London, from vol. 1, 1966.
CURTIS' Botanical Magazine, London, from vol. 1, 1787.
Gardener's Chronicle, London, from 1920.
Journal of Botany, British and Foreign, by RENDLE, London, from vol. 65, 1927.
Quarterly Bulletin of the Alpine Garden Society, London, vols. 9–10, 1941–42.
The Cactus and Succulent Journal of Great Britain (formerly Cactus Journal), The Cactus and Succulent Society of Great Britain, from vol. 1, 1932.
The National Cactus and Succulent Journal, The National Cactus and Succulent Society, from vol. 1, 1946.
**International:** I.O.S. Bulletin, International Organisation for Succulent Plant Study, Wageningen, Netherlands, from vol. 1, 1963.
Repertorium Plantarum Succulentarum, British Section of I.O.S., (now publ. by Abbey Garden Press, Santa Barbara, California, U.S.A.), from no. 1, 1951.
Succulentarum Japonia, Japanese Section of I.O.S., Tokyo, from vol. 1, 1958.
Taxon, Journal of the International Association for Plant Taxonomy, Utrecht, Netherlands, from vol. 1, 1952.
**Japan:** Shaboten, Shaboten-Sha, Kanagawa, from no. 16, 1957.
The Journal of the Cactus and Succulent Society of Japan, from no. 36, 1952.
The Study of Cacti, Journal of the Desert Plant Society of Japan, from no. 1, 1958.
**Mexico:** Cactaceas y Suculentas Mexicanas, Organo de la Sociedad Mexicana de Cactologia, Mexico, from vol. 1, 1955/56.
**Netherlands:** Cactussen en Vetplanten, Maanblad van de Nederlandsche Vereeniging van Cactussen en Vetplanten, Amsterdam, from vol. 1, 1935.
Succulenta, Maanblad van de Nederlands-Belgische Vereeniging van Liefhebbers van Cactussen en andere Vetplanten, Hamersveld, from vol. 9, 1927.
**Rhodesia:** Kirkia, Government Herbarium, Salisbury, vols. 1–5, 1961–65.
**South Africa (Republic of):** Bothalia, Pretoria, from vol. 1, 1921.
Bulletin of Miscellaneous Information, Annals of the University of Stellenbosch.
Journal of South African Botany, Kirstenbosch, from vol. 18, 1952.
South African Gardening and Country Life, Wynberg, C.P., from vol. 17, 1927.
The Flowering Plants of (South) Africa, Pretoria, vols. 2–36, 1922–64.
Transactions of the South African Philosophical Society, Cape Town, from vol. 1, 1909/10.
**Switzerland:** Mitteilungen der Schweizerischen Kakteen-Gesellschaft, Zurich, 1942–56.
Sukkulentenkunde, Jahrbuch der Schweizerischen Kakteen-Gesellschaft, Zurich, from vol. 1, 1947.
**U.S.A.:** Desert Plant Life, Desert Magazine Publishing Company, Pasadena, California, 1929–52.
Euphorbia Review, The International Euphorbia Society, Los Angeles, California, 1935–37.
Journal of the Cactus and Succulent Society of America (now Cactus and Succulent Journal), Santa Barbara, California, from vol. 1, 1929.

## B. Books and individual important papers

BAKER, J. G., Contribution to the Flora of Madagascar. In Journ. Linn. Soc., Bot., London, vol. **20,** 1882/83.
BALLY, P. R. O., East African Succulents. In Journ. E.Afr. and Uganda Nat. Hist. Soc., Nairobi, vols. **15–18,** 1940–46.
- Three Euphorbias as Timber Trees. In Journ. E.Afr. Hist. Soc. **22:** 105–106, 1954.
- The Genus Monadenium. Benteli, Berne, 1961.
- Miscellaneous notes on the flora of Tropical East Africa, including descriptions of new taxa. In Candollea, several papers from vol. **17,** 1959, onwards.
BERGER, A., Sukkulente Euphorbien. Eugen Ulmer, Stuttgart, 1907.
- Mesembrianthemen und Portulacaceen. Eugen Ulmer, Stuttgart, 1908.
- Aloineae. In A. ENGLER, Das Pflanzenreich, vol. **33,** Engelmann, Leipzig, 1908.
- Stapelieen und Kleinien. Eugen Ulmer, Stuttgart, 1910.
- Die Agaven. Gustav Fischer, Jena, 1914.
- Crassulaceae. In A. ENGLER & K. PRANTL, Die Nat. Pflanzenfamilien, Edn. 2, **18a:** 352–483, Engelmann, Leipzig, 1930.
BLAKE, S. T., A Revision of Carpobrotus and Sarcozona in Australia, genera allied to Mesembryanthemum (Aizoaceae). In Contr. Queensland Herbarium No. 7, Brisbane, 1969.
BOLUS, H. M. L., Novitates Africanae. In Annals of the Bolus Herbarium, C.U.P., London, vols. **1–4,** 1915–1928.
- Notes on Mesembryanthemum and Some Allied Genera. Univ. Cape Town. Vols. 1–3, 1928–1958. Continued in Journ. S.Afr. Bot., vols. **25–35,** 1959–1969.
BORISSOVA, A., Genus Rhodiola. In Flora U.S.S.R. **9:** 24, 1939.
BOSS, G., Die Pflanzenwelt bei Lüderitzbucht, 1935.
BRADLEY, R., Collected Writings on Succulent Plants, with an Introduction by G. D. ROWLEY. Gregg Press, London, 1964.
BROWN, N. E., New and old species of Mesembryanthemum, with critical notes. In Journ. Linn. Soc., Bot., London, **45:** 53–140, 1920.
- New species of and critical notes on Mesembryanthemum and allied genera. In Bull. Misc. Information, Kew, 1929.
- Many papers in Kew Bulletin, 1878–1929: see bibliography in BROWN, TISCHER & KARSTEN, 1931.
BROWN, N. E., TISCHER, A., & KARSTEN, M. C., Mesembryanthema. Edited by E. J. LABARRE. L. Reeve, Ashford, 1931.
BULLOCK, A., Notes on Mesembryanthemaceae. In Bull. Misc. Information, Kew, 1937: pp. 492–7, 1938: pp. 153–161.
BURCHARD, O., Beiträge zur Ökologie und Biologie der Kanarienpflanzen. In Bibliotheca Botanica **98:** 1–262. Stuttgart, 1929.
BURTT-DAVY, J., A Manual of the Flowering Plants and Ferns of the Transvaal, with Swaziland, S.Africa. London 1926.
CHEVALIER, A., Les Euphorbes crassulascentes de l'Ouest et du Centre Africain et leurs usages. In Rev. Bot. Appl. **13:** 529–570, Paris 1933.
CHOUX, P., Études biologiques sur les Asclépiadacées de Madagascar. In Ann. Mus. Col. Marseille, ser. 3, **2:** 211–464, Marseilles 1914.
- Les Didiéréacées xérophytes de Madagascar. In Mém. Acad. Malgache, vol. **17,** 1934.
CHRISTIAN, H. B., Notes on African Aloes. In Rhodesia Agric. Journ. Bull. 876, Salisbury 1933.
CLAUSEN, R. T., Sedums of the Trans-Mexican Volcanic Belt: An Exposition of Taxonomic Methods. Comstock Publ. Assn., Ithaca 1959.
CROIZAT, L., De Euphorbio Antiquorum Atque Officinarum. New York, 1934.
CUATRECASAS, J., Las especies del género Espeletia. In Boll. Inst. Catalana Hist. Nat., vol. **37,** Barcelona 1937.
CUFODONTIS, G., The species of Kalanchoe occurring in Ethiopia and Somalia Republic. In Webbia **19:** 711–744, 1965.
- Kalanchoe crenata (Andrews) Haw. In Bull. Jard. Bot. Bruxelles **27:** 713–8, 1957 (1958).
- Drei neue Arten von Kalanchoe aus Kenya und Tanzania. In Österr. Bot. Zeits. **114:** 149–155, Vienna 1967.
- Über Kalanchoe integra (Med.) O. Kuntze und ihre Beziehung zu K. crenata (Andr.) Haw. In l. c. **116:** 312–320, 1969.
DENIS, M., Les Euphorbes des Iles Australes d'Afrique. Impr. Nemourienne, Nemours, 1921.
DESCOINGS, B., Notes taxonomiques et descriptives sur quelques Asclépiadées Cynanchées (Asclépiadacées) aphylles de Madagascar. In Adansonia **1:** 299–342, 1961.
- Une vigne sauvage, Cissus laza. In Nat. Malgache **11:** 189–190, 1959.
- Un genre méconnu de Vitacées: Compréhension et distinction des genres Cissus L. et Cyphostemma (Planch.) Alston. In Notulae Systematicae **16:** 113–125, 1960.
- Cyphostemma (Vitacées) nouveaux de Madagascar. In Bull. Soc. Bot. France **109:** 266–276, **110:** 149–155, 392–397, Paris 1962–3.

- Cissus (Vitacées) nouveaux de Madagascar. In Bull. Soc. Bot. France **112**: 467–480, **113**: 88–100, Paris 1965–6.
- DINTER, K., Neue und wenig bekannte Pflanzen Deutsch-Südwest-Afrikas. Okahandja 1914.
- Botanische Reisen in Deutsch-Südwest-Afrika. In Feddes Repertorium, Suppl. vol. **3,** Posen 1918.
- Sukkulentenforschung in Südwestafrika. Parts I and II. In Feddes Repertorium Suppl. vol. **23** & Suppl. vol. **53.** 1923, 1928.
- Beiträge zur Flora von Südwest-Afrika. In Feddes Repertorium **19**: 122–160. 1923.
- DRAKE DEL CASTILLO, M. E., Didierea. In Bull. Soc. Linn. Paris, vol. **1.**
- DRESSLER, R. L., The Genus Pedilanthus (Euphorbiaceae). In Contr. Gray Herbarium, no. **182.** Cambridge, Mass., 1957.
- ELLIOTT, R., The Genus Lewisia. In Bull. Alpine Gard. Soc. **34**: 1–76. London 1966.
- ENGLER, A., Botanische Jahrbücher. From vol. 27.
- Das Pflanzenreich. W. Engelmann, Leipzig, 1908.
- Die Pflanzenwelt Afrikas. Leipzig 1910.
- Syllabus der Pflanzenfamilien. Edn. 12. Vols. **1** and **2**, Bornträger, Berlin 1954/1964.
- ENGLER, A., & PRANTL, K., Die Natürlichen Pflanzenfamilien. Edn. 2. Engelmann, Leipzig 1930.
- FRANÇOIS, E., Plantes de Madagascar. In Mem. Acad. Malgache, vol. **34,** 1937.
- FRIES, R. E., and TH. C. E., Die Riesen-Lobelien Afrikas. In Svensk Bot. Tidskrift **16**: 384–416, Uppsala 1922.
- Über die Riesen-Senecionen der Afrikanischen Hochgebirge. In l. c. 321–340.
- FU SHU-HSIA, Species et Combinationes Novae Crassulacearum Sinicarum. In Acta Phytotaxonomica Sinica 1965: 12.
- GOULD, S. W., Family Names of the Plant Kingdom. International Plant Index vol. **1,** New Haven & New York 1962.
- GROENEWALD, B. H., Die Alwyne van Suid-Afrika, Suidwes-Afrika, Portugees Oos-Afrika, Swaziland, Basoetoland en 'n speziale Ondersoek van die Klassifikasie, Chromosome en Areale van die Aloe maculatae. Pieterburg, Transvaal 1941.
- GUILLAUMIN, A., & HUMBERT, H., Plantes Nouvelles, rares ou critiques des serres du Museum. In Bull. Mus. Nat. Hist. Nat., Paris 1940 onwards.
- HANSEN, A., Die Pflanzendecke der Erde. Leipzig 1920.
- HARVEY, W. H., SONDER, O. W., & THISELTON-DYER, W. T., Flora Capensis. Vols. **1–6,** Dublin 1859–1920.
- HARTMANN, H., Untersuchungen zur Morphologie und Systematik der Gattung Argyroderma N. E. BR. (Mesembryanthemaceae FENZL). I.O.S. Bull. vol. III, No. 3, 1974.
- HAWORTH, A. H., Complete Works on Succulent Plants. Gregg Press, London 1965. Vol. **1**: Biographical and bibliographical Introduction by W. T. STEARN, Observations on the genus Mesembryanthemum (1794–5); Vol. **2**: Miscellanea Naturalia (1803), A new arrangement of the genus Aloe (1804); Vol. **3**: Synopsis Plantarum Succulentarum (1812); Vol. **4**: Supplementum Plantarum Succulentarum (1819), Saxifragearum Enumeratio, accedunt Revisiones Plantarum Succulentarum (1821); Vol. **5**: Series of essays in The Philosophical Magazine (1823–1831).
- HEDBERG, O., Afroalpine Vascular Plants. Uppsala 1957.
- HERRE, H., The Genera of the Mesembryanthemaceae. Tafelberg-Uitgewerks Beperk. Cape Town, 1971.
- see JACOBSEN, H., HERRE, H., & VOLK, O. H.
- HIGGINS, V., Crassulas in Cultivation. Blandford Press, London 1964.
- HOCHREUTHER, B. P. C., Un nouveau Baobab, Revision du genre Adansonia. In Ann. Cons. Jard. Bot. Geneva 1907–1908: 136–143.
- HUBER, H., Revision der Gattung Ceropegia. In Mem. Soc. Brot. **12**: 1–203, Coimbra 1957.
- HUBER, J. A., Zur Systematik der Gattung Sedum L. Landshut 1926.
- HUMBERT, H., Flore de Madagascar et des Comores: Famille Composées, Famille Didieréacées. Mus. Nat. Hist. Nat., Paris 1963.
- HUTCHINSON, J., A Botanist in Southern Africa. Gawthorn, London 1946.
- IHLENFELDT, H.D., Entwicklungsgeschichtliche, morphologische und systematische Untersuchungen an Mesembryanthemen. In Feddes Repertorium **63**: 1–104, 1960.
- –, SCHWANTES, G., and STRAKA, H., Die höheren Taxa der Mesembryanthemaceae. In Taxon **11**: 52–56, 1962.
- and STRAKA, H., Über die systematische Stellung und Gliederung der Mesembryanthemen. In Ber. Deutsch. Bot. Ges. **74**: 485–492, 1962.
- JACOBSEN, H., Die Sukkulenten. Paul Parey, Berlin 1933.
- Succulent Plants. Williams & Norgate, London 1935.
- Verzeichnis der Arten der Gattung Mesembryanthemum L. nebst deren abgetrennten Gattungen. In Feddes Repertorium **106**: 1–198, Berlin 1938; Supplements 1 & 2, in l. c. **106** Suppl.: 1–34, 1939.
- Handbuch der sukkulenten Pflanzen. V.E.B. Gustav Fischer Verlag, Jena 1954–5.
- A Handbook of Succulent Plants. Blandford Press Ltd., London 1960.
- Das Sukkulentenlexikon, V.E.B. Gustav Fischer Verlag, Jena 1970.
- –, HERRE, H., and VOLK, O. H., Mesembryanthemaceae. Eugen Ulmer, Stuttgart 1950.

Jumelle, H., and Perrier de la Bathie, H., Les Asclépiadacées aphylles de l'Ouest de Madagascar. In Rev. Gen. Bot. **11**: 257, 1911.
Karsten, M. C., see Brown, N. E., Tischer, A., & Karsten, M. C.
Keraudren, M., Une Cucurbitacée aphylle de Madagascar: Seyrigia gen. nov. In Bull. Soc. Bot. France **107** no. 7–8, 1960.
- Un Nouveau Xerosicyos Malgache. In Not. Syst., Mus. Hist. Nat. **16** fasc. 1–2, Paris 1960.
- Le genre Kedrostis Medik. (Cucurbitaceae) à Madagascar. In Bull. Soc. Bot. France **108**: 241–242, Paris 1961.
- Seyrigia bosseri, espèce nouvelle de Cucurbitacée Malgache. In l. c. **109**: 101–102, Paris 1962.
- Pachypodes et Baobabs à Madagascar. In Rev. Sci. Nat. **55**, 1963.
- Revision du genre Zehneria à Madagascar. In Adansonia **4**: 331–7, Paris 1964.
- Le Genre Moringa en Afrique et à Madagascar. In Webbia **19**: 815–824, Florence 1964.
- Famille (185e) Cucurbitacées. In Flore de Madagascar et des Comores. Mus. Nat. Hist. Nat., Paris 1966.
Labarre, E. J., see Brown, N. E., Tischer, A., & Karsten, M. C.
Leach, L. C., Stapeliae from South Tropical Africa, Part V. In Bothalia **10**: 45–54, Pretoria 1969.
- Euphorbiae Succulentae Angolenses: II. In Bol. Soc. Brot. **43**: 163–182, Coimbra 1969.
Leandri, J., Un Pachypodium (Apocynacées) nouveau de l'Ouest de Madagascar. In Bull. Soc. Bot. France **81**: 141, Paris 1934.
- Contribution à l'etude des Euphorbiacées de Madagascar. IX. Groupe de Euphorbia pyrifolia et observations sur la section Goniostema et seq. In Not. Syst. **11**, Paris 1945.
- Contribution à l'etude des Euphorbiacées de Madagascar. X. Euphorbes du groupe Diacanthium. In Not. Syst. **12**: 156–164, 1946.
- see also Ursch, E., & Leandri, J.
Legrand, C. D., Las Especies Americanas de Portulaca. In An. Mus. Hist. Nat. Montevideo, ser. 2, **7**: 1–147, 1962.
Liebenberg, L. C. C., A revision of the South African species of Adenia. In Bothalia **3**: 513–570, Pretoria 1939.
Lowe, R. T., A Manual of the Flora of Madeira. London 1868.
Maatsch, R., Eine neue Stapelie und ihre Benennung (bella). In Gartenwelt **64**: 223, Hamburg 1964.
Merxmüller, H., Die sukkulenten Senecionen Südwestafrikas. In Bot. Not. **119**: 121–135, Lund 1966.
Muirhead, C. W., Sempervivum globiferum. In Notes Roy. Bot. Gard. Edinburgh **26**: 279–285, 1965.
Nel, G. C., Lithops. Univ. Press, Stellenbosch 1946.
- The Gibbaeum Handbook. Edited by P. G. Jordaan & E. W. Shurly. Blandford Press Ltd., London 1953.
Pareys Blumengärtnerei. Publ. Paul Parey, Berlin-Hamburg. Edn. 1: Edited by C. Bonstedt, 1931; Edn. 2: Edited by F. Encke, 1960.
Pax, F., and Hoffman, K., Aizoaceae. In A. Engler & K. Prantl, Die nat. Pflanzenfamilien, Edn. 2, **16c**: 179–233, Leipzig & Berlin 1934.
Perrier de la Bathie, P., Les Pachypodium de Madagascar. In Bull. Soc. Bot. France **81**: 297–318, Paris 1934.
- Les Adansonia de Madagascar et leur utilisation. In Rev. Bot. Appliq. **33**, 1963.
- and Hochreuter, B. P. G., Bombacacées. In H. Humbert, Flore de Madagascar et des Comores, Paris 1955.
Phillips, E. P., The Genera of South African Flowering Plants. Cape Town 1926; Edn. 2: Govt. Printer, Pretoria 1951.
Pichon, M., Révision des Apocynacées des Mascareignes et des Seychelles. XXI, Genre Pachypodium. In Mém. Inst. Sci. Madagascar **2**: 98–125, 1949. (Includes Classification des Apocynacées, pp. 108–9.)
Poisson, H., Notes sur un Pachypodium nouveau du nord de Madagascar. In Bull. Acad. Malgache, n.ser. 3, 1919: 237–239.
- Notes sur un Pachypodium nouveau de la région Diego-Suarez. In l. c.: 235–236.
- Contribution à l'étude des Pachypodium malgaches. In l. c.: 235–246, 1918.
- Nouvelle contribution à l'étude des Pachypodium malgaches. In l. c., n. ser. vii: 1–10, 1924.
- and Decary, R., Nouvelles observations sur les Pachypodium malgaches. In l. c., n. ser. iii, 1917.
Range, P., Zur Flora SW.Afrikas (Conspectus Florae namaensis). In Sitzungsber. Ges. Nat.Freunde, Lübeck 1939.
Rappa, F., L'evoluzione della Capsula dei Mesembriantemi. In Malpighia **26**: 73–94, Catania 1913.
- N. E. Brown: la sua opera e la mia priorità. In Lav. Inst. Bot. Giard. Col. Palermo, vol. **14**, 1952.
-, and Camarrone, V., Mesembrianthemum e Mesembryanthemum, una rivendicazione. In l. c., v. **14**, 1952.
- - La classificazione naturale delle Mesembrianthemaceae. In l. c. **18**: 3–24, 1960.

– – Mesembrianthemum: la classificazione naturale (Primi approcci di collaborazione). In l. c. **18**: 3–48, 1962.
RAUH, W., Morphologische, entwicklungsgeschichtliche, histogenetische und anatomische Untersuchungen an den Sprossen der Didiereaceen. In Akad. Wiss. Lit., 1956: No. 6: 1–104.
– Weitere Untersuchungen an Didieraceen. 1. Teil: Beitrag zur Kenntnis der Wuchsformen der Didiereaceen, unter besonderer Berücksichtigung neuer Arten. In Sitzungsber. Heidelb. Akad. Wiss. 1961: 183–300.
– Didieréacées. In H. HUMBERT, Flore de Madagascar et des Comores, Paris 1963.
– Die großartige Welt der Sukkulenten. Paul Parey, Hamburg und Berlin 1967.
RAWÉ, R., Succulents in the Veld. Howard Timmins, Cape Town 1968.
RESENDE, F., Über zwei neue Haworthia. In Feddes Repert. **48**: 113, 1940.
– Estudos Caryologicos nas Aloineae. In Bol. Soc. Brot., vol. **14,** Lisbon 1940.
– Suculentas Africanas. In l. c. **15**: 159–162, 1941.
– Suculentas Africanas III. In Mem. Soc. Brot. **2**: 1–119, 1943.
– Hibridos intergenericos e interspecificos em Kalanchoideae. I: In Bol. Soc. Port. Cienc. Nat. **6**: 241–244, 1956; II: In Rev. Biol. **1**: 157, 1957.
– Über die Brutknospen-Entwicklung bei Bryophyllum und ihre Vererbung. In Ber. Deut. Bot. Ges. **72**: 3–10, 1959.
RESENDE, F., and PINTO-LOPES, J., Suculentas Africanas IV. In Port. Acta Biol. (B) **2**: 177–196, 1946.
– and POELLNITZ, K. VON, Suculentas Africanas II. In Rev. Broteria Ser. Cienc. Nat. **11**: 49–55, 1942.
– and VIANA, M. J., Photo- and Thermoperiodism, their interaction for the phenogenesis. I. On some morphogenetic aspects in the vegetative and floral stage (Kalanchoideae). In Port. Acta Biol. (A) **9**: 159–210, 1965.
REYNOLDS, G. W., The Aloes of South Africa. Trustees of the Aloes of S.Africa Book Fund, Johannesburg 1950.
– The Aloes of Nyasaland. Nyasaland Society, Blantyre 1954.
– Les Aloes de Madagascar. In Nat. Malgache **10** (Supplement): 1–156, Tananarive 1958.
– The Aloes of Tropical Africa and Madagascar. Trustees of the Aloes Book Fund, Mbabane 1966.
ROSE, J. N., and STANDLEY, P. C., The Genus Talinum in Mexico. In Contr. U.S. Nat. Herb. **13**: 281–288, Washington 1911.
ROEDER, W. VON, Die Sukkulenten. Frankh'sche, Verlagshandlung, Stuttgart 1931.
SCHWANTES, G., The Cultivation of the Mesembryanthemaceae. Blandford Press, London 1954. (This work includes an Introduction by E. SHURLY in which all of the author's previous publications on the Mesembryanthemaceae are listed; see also Monatsschrift D. K .G. **6**: 130 (1955), **12**: 19 (1961), **13**: 78 (1962).)
– Flowering Stones and Mid-Day Flowers. Ernest Benn, London 1957.
– see also: IHLENFELDT, H. D., SCHWANTES, G., & STRAKA, H.
STRAKA, H., Anatomische und Entwicklungsgeschichtliche Untersuchungen an Früchten paraspermer Mesembryanthemen. In Nova Acta Leopoldina **17**: 127–190, Leipzig 1955.
– see also: IHLENFELDT, H. D., SCHWANTES, G., & STRAKA, H.
SVENTENIUS, E. R., Contribución al Conocimiento de la Flora Canaria. In Bull. Inst. Nac. Invest. Agron. **79**: 175–194, Madrid 1946.
– Additamentum ad Floram Canariensem. Vol. **1.** Madrid 1960.
TISCHER, A., see: BROWN, N. E., TISCHER, A., & KARSTEN, M. C.
TRELEASE, W., and YUNCKER, T. G., Piperaceae of Northern South America. Univ. Illinois Press, Urbana 1950.
TUYAMA, T., Plantae Boninense Novae vel Criticae, XII. In Bot. Mag. **53**, 1933.
URSCH, E., and LEANDRI, J., Les Euphorbes Malgaches Epineuses et Charnues du Jardin Botanique de Tsimbazaza. In Mem. Inst. Sci. Madagascar **5**: 109–185, 1954.
VATKE, G. C. W., Reliquiae Rutenbergianae VI. In Abh. Naturw. Vereins Bremen, vol. **9**: 125, 1887.
VOLK, O. H., see: JACOBSEN, H., HERRE, H., and VOLK, O. H.
WALTHER, E., Echeveria: Conspectus Serierum. In Leafl. West. Bot. **9**: 1–4, San Francisco 1959.
WARMING, E., Ökologische Pflanzengeographie. Berlin 1914.
WHITE, A., and SLOANE, B. L., The Stapelieae. Edn. 1: Pasadena 1933; Edn. 2: Pasadena 1937.
WHITE, A., DYER, R. A., and SLOANE, B. L., The Succulent Euphorbieae (Southern Africa). Abbey Garden Press, Pasadena 1941.
WILDE, W. J. J. O. DE, A survey of the species of the genus Adenia (Passifloraceae) in Madagascar, with some combinations, the description of some taxa, and a key to the species. Adansonia **10,** 1970: 111–126.
WULFF, H. D., Die Polysomatie des Wurzelperiblems der Aizoaceen. In Ber. Deutsch. Bot. Ges. **58**: 400–410, Berlin 1940.
– Untersuchungen zur Zytologie und Systematik der Aizoaceae, Subtribus Gibbaeinae SCHWANT. In Bot. Archiv, vol. **45**, Leipzig 1944.
YUNCKER, T. G., The Piperaceae of **Argentina**, Bolivia and Chile. In Lilloa, vol. **37,** Tucuman 1953.

# INVALID DESIGNATIONS

Basionyms, Homonyms, Synonyms, nomina nuda, insufficiently known species in alphabetic order of the Genera in Parts I and II.
The valid designations are shortened. For the abbreviation Mesembr. read Mesembryanthemaceae.
Genera marked with ● belong to **Part II.** Family Mesembryanthemaceae.


**Abromeitiella** MEZ. Bromeliaceae.
A. pulvinata MEZ    A. chlor.
**Abryanthemum** NECKER. Mesembr. ●
A. acinaciforme (L.) ROTHM.    Carpobr. a.
— edule (L.) ROTHM.    — e.
**Acaulon** N. E. BR. Mesembr. ●
A. rosulatum (KENSIT)
   N. E. BR.    Al. r.
**Adansonia** L. Bombacaceae.
A. bozy JUM. et PERR.    A. za v. b.
— rubrostigma JUM. et PERR.    — fon. v. r.
— za JUM. et PERR.    — za v. b.
**Adenia** FORSK. Passifloraceae.
A. angustisecta BURTT-DAVY    A. digit.
— buchananii HARMS et ENGL.    — —
— multiflora POTT.    — —
— pseudoglobosa VERDC.    — glob. ssp. p.
— — ssp. curvata VERDC.    — — v. c.
**Adenium** ROEM. et SCHULT. Apocynaceae.
A. arabicum BALF. f.    A. obes. ssp. o.
— coetanum STAPF    — — ssp. —
— honghel A. D. C.    — — ssp. o.
— lugardii N. E. BR.    — ol.
— micranthum STAPF    — ob. ssp. o.
— multiflorum BALF. f.    — — ssp. soc.
— — KLOTZSCH    — — ssp. mult.
— namaquanum (WYLEY ex
   HARV.) WELW.    Pachyp. n.
— socotranum VIERH.    A. obes. ssp. s.
— speciosum FENZL    — obes. ssp. o.
**Adromischus** LEM. Crassulaceae.
A. anticordatum v. POELLN.    A. antid.
— bolusii v. karrooensis
   (SCHOENL.) JACOBS.    — bol.
— casmithianus v. POELLN.
   nom. ill.    — nan.
— clavifolius HORT. ex
   v. ROED.    — coop.
— — v. POELLN. p. min. part.    — pach.
— — v. ROED.    — fest.
— cooperi v. POELLN.
— cristatus BGR. p. min. part.    — zeyh.
— cuneatus LEM.    Cot. pap. v. p.
— — v. POELLN.    A. hal.
— filicaulis (ECKL. et ZEYH.)
   C. A. SMITH    — mam. v. fil.
— fusiformis ROLFE    — — f. fus.
— hemisphaericus LEM. ex
   JACOBS.    — rot.
— hoerleinianus (DTR.)
   v. POELLN.    Cot. schaef.
— humilis (MARL.) v. POELLN.    A. hum.
— jasminiflorus (SALM) LEM.    — car.
— keilhackii WERDERM.    Cot. schaef.
— kleinioides C. A. SMITH    A. mam. v. rub.
— leucothrix C. A. SMITH    Cot. l.

A. maculatus BGR. ex ENGL.
   et PRANTL.    A. rhomb. v. b.
— mamillaris v. rubra
   v. POELLN.    — mam. v. ruber
— marlothii (SCHOENL.) BGR.    — — v.
— montium-klinghardtii
   (DTR.) BGR.    — hem.
— mucronatus LAM.    Cot. und. v. m.
— pauciflorus P. C. HUTCHIS.    A. nan.
— procurvus (N. E. BR.)
   v. POELLN.    — pr.
— rhombifolius BGR.    — rh. v. sp.
— robustus LEM.    — trifl.
— rotundifolius (HAW.)
   C. A. SMITH    A. rot.
— rupicolus C. A. SMITH    — trig.
— schaeferianus (DTR.) BGR.    Cot. schaef.
— — v. keilhackii
   (WERDERM.) v. POELLN.    — —
— sphenophyllus C. A. SMITH    A. rhomb. v. sph.
— trigynus v. POELLN. p. part.    — alst.; — umb.
— turgidus nom. nud.    — lieb.
— vanderheydeni HORT. ex
   BGR.    — clav.
**Aeonium** WEBB et BERTH. Crassulaceae.
A. barbatum WEBB et BERTH.    A. × hybr.
— bentejui WEBB    — spath.
— berthelotianum BOLLE    — tab.
— bethencourtianum (WEBB)
   WEBB    Aichr. b.
— caespitosum (C. SM.)
   WEBB et BERTH.    A. sims.
— chrysanthum (HOCHST)
   BGR.    — leucobl. v. l.
— cruentum WEBB et BERTH.    — spath. v. c.
— doramae WEBB    — manr.
— exsul BORNM.    — can. v. c.
— × floribundum BGR.    — × hybr.
— macrolepum WEBB    — tab.
— meyerheimii BOLLE    — gland.
— strepsicladum WEBB et
   BERTH.    — spath. v. s.
— tortuosum PIT. et PROUST    — lindl.
Acc. PRAEGER spontaneous hybrids:
A. canariensis × A. cuneatum    A. × bramw.
— castello-paivae × A. sub-
   planum    — × castello-plan.
— — × A. viscatum    — × brav.
— ciliatum × A. haworthii    — × ten.
— — × A. holochrysum    — × kunk.
— — × A. nobile    — × splend.
— — × A. palmense    — × jun.
— — × A. rubicum    — × ten.
— glandulosum × A. gluti-
   nosum    — × low.

**INVALID DESIGNATIONS** Aeonium–Agave 592

A. goochiae × A. palmense
— haworthii × A. urbicum
— hierrense × A. palmense
— — × A. valverdense
— lindleyi × A. tabulaeforme
— palmense × A. valverdense
— percarneum × A. undulatum
— — × A. virgineum
— saundersii × A. subplanum
— simsii × A. manriqueorum
— — × A. undulatum
— smithii × A. spathulatum
— subplanum × A. viscatum
— tabulaeforme × A. urbicum

**Aethephyllum** N. E. Br. Mesembr. ●
A. (N. E. Br.) Schwant. is an invalid SG. of Micropterum Schwant.

**Agave** L. Agavaceae
A. affinis Trel.
— aloinea C. Koch
— altissima Zumagl.
— americana Auct.
— — Dill.
— — Griseb. p. part.

— — Humboldt
— — L. v. latifolia Torr. sensu Bigelow
— — v. — Torr. sensu Emory
— — v. picta Trel.
— — Lam.
— — Urban
— — sobolifera Hermann
— amoena Hort. ex Lem. ex Jacobi
— amurensis Jacobi
— applanata Trel.
— — v. huachucencis (Bak.) Mulf.
— — v. parryi Mulf.
— — v. — Purp.
— aquariensis Trel. nom. nud.
— argyrophylla Hort.
— asperrima J. Mulf.
— atrovirens Bak.
— — Karw. v. stigmatophylla Bgr.
— — E. Neub.
— attenuata Auct.
— — Salm v. compacta Jacobi
— — v. serrulata Terr.
— avellanidens Trel.
— bakeri Rose
— beaucarnei Lem.

— beaulueriana Jacobi
— besseriana Hort. ex Bak.
— — Jacobi
— — candida Hort.
— hystryx Hort. ex Hook.
— longifolia glauca Jacobi
— — — viridis Jacobi
— brachystachys v. pubescens Terr.

A. × santos.
— × hawbic.
— × jacobs.
— × ombr.
— × anag.
— × lamb.

— × boll.
— × lems.
— × sanctis.
— × svent.
— × praeg.
— × rowl.
— × vegam.
— × tabulic.

A. sobr. v. s.
— sart.
— am.
— ing.
— barb.
— duss.; egg.; med.; obd.
— coc.

— grac.

— parr. v. p.
— pict.
— sob.
— ant. v. a.
— sob.

— cren.
— xyl. v. x.
— grac.

— parr. v. h.
— — v. p.
— seb. ?
— mckelv.
Furcr. roez.
A. asp.
— atr. v. salm.

— — v. a.
— — v. cochl.
— cern. v. c.

— — v. —
— — v. s.
— seb.
— karw.
— kerch. v. brev.

— franz.
— macr. v. m.
— — v. —
— — v. —
— — v. —
— — v. —
— — v. —

— pub.

A. brandegei Trel.
— brevicornuta Gentry
— bulbifera Salm; sensu Baker
— caeciliana Bgr.
— caerulescens Salm
— caespitosa Tod.
— californica Bak.
— calodonta Bgr.
— campaniflora Trel.
— candelabrum Tod.
— cantala Jacobi
— cantula Roxb.
— carchariodonta Pampanini
— caribaea Bak.
— carminis Trel.
— celsiana Hort. ex Bgr.
— — Jacobi
— cerulata Trel.
— chisosensis Müller
— chloracantha Salm
— chrysantha Peeb.
— coarctata Jacobi
— coccinea Hort.
— — Roezl.
— coelum Hort. ex Besauc.
— compacta Hort. ex Besauc.
— concinna Bak.
— conduplicata Jacobi et Bouché. Insufficiently known spec.
— connochaetodon Trel.
— consideranti Duch.
— consociata Trel.
— corderoyi Bak.
— cornuta Hort.
— couesii Englm.
— cubensis Jacq.
— — v. striata Hort.
— dealbata Lem. ex Jacobi
— dentiens Trel.
— desertii Orc.
— desmetiana Hort.
— difformis Bgr.
— disceptata J. R. Drumm.
— dyckii Hort. ex Besauc.
— eborispina Hester
— elegans Hort. ex Salm
— elizae Bgr.
— elliptica Hort. ex Besauc.
— elongata Bgr.
— ensifera Jacobi
— erubescens Ellemeet
— excelsa Bak.
— — Jacobi
— filamentosa Salm
— filifera v. adornata Scheidw.
— — v. major Hort.
— — v. pannosa Scheidw.
— — v. schidigera Terr.
— flaccida Jacobi
— — Salm
— flavescens Salm
— — v. macroacantha Jacobi
— flavovirens Jacobi
— foetida L.

A. prom.
— mesc.

— cant.
— asp.
— univ. v. c.
— sart.
— falc. v. f.
— fenzl.
— aur.
— cant.
— — —
— — —
— niv. v. c.
— car.
— sobr. v. s.
— am. v. marg.
— cels.
— sobr. v. s.
— glom.
— micr.
— palm. v. c.
— mitr.
— lat.
— fer.
— cern. v. c.
— macr. v. m.

— sobr. v. ros.
— vict. reg.
— des.
— karw.
— xyl. v. x.
— parr. v. p.
Furcr. c.
— sell. v. marg.
A. das. v. d.
— des.
— — —
— horr. v. h.
— tri. v. sub.
— cv. Leop.
— atr. v. salm.
— uth. v. e.
— pot. v. p.
— desm.
— cern. v. c.
— fourc.
— niv. v. het.
— rub. ?
— ang. v. a.
— — v. —
— filif. v. f.
— schid. v. s.
— filif. v. f.
— schid. v. s.
— — v. s.
— fenzl.
— rub.
— macr. v. m.
— — v. —
— ang. v. a.
Furcr. gig. v. g.

## Agave  INVALID DESIGNATIONS

A. fourcroydes JACOBI
— franceschiana TREL.
— franzosinii NISSEN et RICASOLA
— frederici ('friderici') BGR.
— funifera LEM.
— geminiflora v. filamentosa HOOK.
— — v. knightiana DRUMM.
— — v. ? sonorae TORR.
— ghiesbreghtii v. peacockii (CROUCHER) TERR.
— gilbeyi BGR.
— — HORT. HAAGE et SCHMIDT
— glaucescens HOOK.
— grandidentata JACOBI
— guadalajara TREL.
— guedemeyeri HOULLET. Insufficiently known spec.
— gustaviana HORT. ex BAK.
— hartmannii S. WATS.
— haynaldii v. utahensis TREL.
— henriquesii BAK.
— heteracantha BAK.
— — BGR.
— — v. glomeruliflora ENGELM.
— heterodon HORT.
— hookeri HORT. ex BESAUC.
— — JACOBI
— horrida HORT.
— — LEM. v. micracantha BAK.
— — v. triangularis BAK.
— houlletiana CELS. ex JACOBI
— houlettii JACOBI
— huachucensis BAK.
— humilis ROEM.
— hybrida HORT.
— haynaldii TOD.
— hystrix HORT. CELS.
— inaequidens C. KOCH
— ingens v. picta (S. D.) BGR.
— intrepida GREENM.
— ixtli C. KOCH
— — v. elongata BAK.
— — v. excelsa TERR.
— — HORT.
— ixtlioides HOOK. ?
— — LEM. ex JACOBI
— jacobiana S. D.
— jaquiniana SCHULT.
— kaibabensis MCKELVEY
— katharinae BGR.
— keratto HAW.
— kerchovei HORT. ex JACOBI
— — LEM. v. inermis BAK.
— kerratto BAK.
— killisckii HORT.
— knightiana DRUMM.
— kochii JACOBI
— laevis HORT. ex BAK.
— latifolia HORT. ex SALM
— laurentiana BGR.
— laxa KARW. et OTTO
— laxifolia BAK.

A. marm. ?
— desm.
— franz.
— —
Hesp. f.
A. gem. v. g.
— — v. atr.
— schott. v. s.
— peac.
— horr. v. h.
— v. g.
— att.
— obsc.
— megal.
— max.
— parv.
— uth. v. u.
— peac.
— univ. v. lat.
— — v. h.
— glom.
— cren.
— dens.
— mesc.
— nol. tang.
— obsc.
— triang. v. t.
— sis. v. s.
— v. —
— parr. v. h.
— brach.
— xyl. v. vitt.
— vitt. ?
— strict.
— mesc.
— pict.
— das. v. dealb.
— marm. ?
— fourcr.
— ang. v. a.
— fourcr.
— ang. v. a.
— fourcr.
— atr. v. a.
— ang. v. a.
— uth. v. k.
— mesc.
— kar.
— triang. v. t.
— — v. i.
— pol.
— horr. v. h.
— gem. v. atr.
— xylon. v. x.
— sis. v. s.
— pot. v. p.
— legr. v. l.
— cant.
— dec.

A. lechuguilla v. glomeriflora (ENGELM.) TREL.
— lehmannii JACOBI
— linearis JACOBI
— longifolia HORT.
— — v. picta RGL.
— lophantha SCHIEDE ex KUNTH
— — v. angustifolia BGR.
— — v. brevifolia JACOBI
— — v. coerulescens (SALM) JACOBI
— — v. gracilior JACOBI
— — v. latifolia BGR.
— — v. pallida BGR.
— — v. poselgeri (S. D.) BGR.
— — v. subcanescens JACOBI
— — v. taumalipasana BGR.
— lurida AIT.
— — β. AIT.
— — v. jacquiniana SALM
— — JACQ.
— macracantha HERB.

— — v. macrantha TERR.
— macroculmis TREL.
— maculata ENGELM.

— — FORSK. - Acc. REYNOLDS an insufficiently known species.

— marcusae HORT. ex TREL.
— marcusii L. DE SMET
— marensii HORT. ex TREL.
— maximiliana BAK.
— — HORT. ex BESAUC.
— mayoensis GENTRY
— mexicana LAM. p. part.
— — v. picta CELS.
— micracantha SALM v. albidior SALM
— micrantha BAK.
— milleri SALM v. picta VAN HOUTTE
— miradorensis JACOBI
— mitis SALM
— — v. albicans TERR.
— morrisii AUCT.
— mulfordiana TREL.
— myriacantha HORT. ex BESAUC.
— nelsonii TREL.
— neoamericana WOOTON et STANDLEY
— nevadensis (ENGELM. ex GREENM. et ROUSH) HESTER
— newberryi ENGELM.
— nickelsii Hort. ROLLINS. Goss.
— nigromarginata HORT.

— noackii JACOBI
— noah TREL.
— oblongata JACOBI
— oligophylla BAK.
— orcuttiana TREL.

A. glom.
— atr. v. a.
— macr. v. m. ?
— fourcr.
— pict.
— univ. v. u.
— — v. a.
— — v. b.
— — v. c.
— — v. g.
— — v. l.
— lech.
— —
— univ. v. s.
— — v. t.
— ver. cr.
— ang. v. a.
— — v. —
— — v. —
— macroac. v. m.
— macr.
— lat.
— maculosa v. m.

— parr. v. p.
— — v. —
— — v. —
— fenzl.
— xyl. v. lat.
— vilm.
— ver. cr.; lur.
— pict.
— albic.
— flacc.
— pict.
— desm.
— micr.
— albic.
— miss.
— schott. v. s.
— unc.
— desert.
— parr. v. p.

— utah. v. n.
— — v. discr.
— ferd. reg.
— fil. v. fil. f. ort.
— sart.
— scabr.
— micr.
— macr. v. m.
— shaw.

## INVALID DESIGNATIONS  Agave

| | | | |
|---|---|---|---|
| A. ortgiesiana Tod. | A. fil. v. fil. f. ort. | A. salm. v. mitraeformis Cels. | A. mitr. |
| — — brevifolia Hort. | — v. — f. — | — — v. recurvata Jacobi | — atr. v. salm. |
| — ottonis Jacobi | — atr. v. a. | — saponaria Lindl. | — brach. |
| — owenii J. M. Johnston | — pacif. | — sartori v. caespitosa Terr. | — sart. |
| — pachyacantha Trel. | — shaw. | — scapoides Greenm. et Roush | — utah. v. ut. |
| — parayi Hort. | — parr. v. p. | — schnittspahnii Jacobi | — appl. |
| — parrayi Hort. ex Trel. | — v. — | — scolymus Bgr. | — fenzl. |
| — parryi Engelm. v. couesii (Engelm. ex Trel.) Kearn. et Peeb. | | — — Karw. | — pot. v. p. |
| | | — — Knuth. Insufficiently known spec. | |
| — — Haage et Schmidt | — v. — | — seemanniana Besauc. | — seem. |
| — paryi Hort. ex Trel. | — v. — | — serrulata Karw. ? | — rub. |
| — paucifolia Bak. | — macr. v. m. | — shawii Brandeg. | — shaw. |
| — — Tod. | — falc. v. f. | — — Engelm. v. sebastiana (Greenm.) Gentry | — seb. |
| — paupera Bgr. | — desm. | — silvestris Hort. | — conj. ? |
| — pendula Schnittsp. | — sart. | — simonis Hort. | — pum. |
| — perbella Hort. | — xyl. v. vitt. | — sobolifera v. serrulata Terr. | — rub. |
| — perplexans Trel. | — fil. v. comp. | — spectabilis Hort. ex Besauc. | — cern. v. c. |
| — picta Bgr. | — ing. | | |
| — polyacantha Haw. – Unclarified spec. possibly identical with A. polyacantha Jacobi | | — — Tod. | — elong. |
| | | — spicata DC. | — brach. |
| | | — spiralis Brandeg. | — decip. |
| | | — striata Hook. | — stri. v. rec. |
| — — Jacobi v. densiflora Terr. | — dens. | — — Zucc. v. californica Terr. | — falc. v. f. |
| — polyanthoides Cham. et Schltr. | — brach. | — — v. echinoides Bak. | — ech. |
| | | — — v. stricta Bak. | — strict. |
| — poselgeri Salm | — lech. | — subfalcata Jacobi ? | — macr. v. m. |
| — — v. kerchovei Terr. | — kerch. v. k. | — subsimplex Trel. | — sobr. v. s. |
| — potatorum C. Koch | — mitr. | — sudburyensis Bak. | — macr. v. m. |
| — — Hort. | — — | — sullivani Trel. | — fourcr. |
| — prainiana Bgr. | — ver. cr. | — taylori Besauc. | — schid. v. t. |
| — pringlei Engelm. | — des. | — tehuacanensis Karw. | — pot. v. versch. |
| — protuberans Engelm. | — gutt. | | |
| — pruinosa Lem. | — ellem. | — — Kunth. | — atr. v. a. |
| — pseudofilifera Ross. et Lanza | — fil. v. fil. ? | — theometel Zucc. | — vivip. ? |
| | | — todarei Bak. | — marm. |
| — pulchra Hort. ex Salm | — pot. v. p. | — toneliana Bak. | — vitt. |
| — pumila Simon ex Besauc. | — — ? | — treleasei Toumey | — schott. v. t. |
| — punctata Salm | — rub. | — troubestzkoyana Hort. ex Baker | — marm. |
| — recurva Zucc. | — striat. v. r. | | |
| — regeli Hort. ex Besauc. | — horr. v. h. | — tuberosa Miller | Furcr. tub. |
| — regeliana Hort. | — — v. — | — univittata v. subscandens Jacobi sphalm. | |
| — — Jacobi | — desm. | | |
| — — Hort. ex Jacobi | — triang. v. t. | — utahensis v. scaphoides (Greenm. et Roush) M. E. Jones | |
| — reginae Hort. ex Bgr. | — mel. | | |
| — rigida de Spin. | — ang. v. a. | | A. utah. v. ut. |
| — — Hort. | — fourcr. | — vera-crucis Haw. | — ver. cr. |
| — — v. longifolia Engelm. | — — | — vera-cruz Drum. et Prain | — am. |
| — — v. sisalana Engelm. | — sis. v. s. | — — v. jaquiniana Ashers. | — ang. v. a. |
| — rigidissima Jacobi | — tri. v. a. | — vernae Bgr. | — ver. cr. |
| — roezliana v. peacockii (Croucher) Trel. | — peac. | — verschaffeltii Lem. | — pot. v. versch. |
| — roezlii Hort. | — purp. | — vestita S. Wats. | — schid. v. s. |
| — rohanii Hort. | — peac. | — vexans Trel. | — dact. v. vex. |
| — roseana Trel. | — sobr. v. r. | — victoriae-reginae v. laxior Bgr. | — ferd. reg. |
| — rubrocincta Jacobi | — sart. | | |
| — rumphii Jacobi | — cant. | — — f. nickelsii (Hort. Rol.-Goss.) Trel. | — — — |
| — rupicola Rgl. | — micr. ? | | |
| — salm-dyckii Bak. | — kerr. | — villae-pirottii Hort. | — villar. |
| — salmiana Otto ex Salm | — atr. v. salm. | — virens Besauc. | — cern. v. c. |
| — — v. angustifolia Bgr. | — — | — vivipara Dalz. et Gibs. | — cant. |
| — — v. glauca Becker | — — | — — Lam. p. part. | — seb.; ant. v. a. |
| — — v. gracilispina Rol.-Goss. | — grac. | — — Oldendorp | — miss. |

A. vivipara SALM
— — v. woodrowii HORT. ex WATSON
— — WIGHT
— weingartii BGR.
— whitackeri HORT.
— wightii DRUMM. et PRAIN
— willdingeri TOD. ex BAKER
— wislizenii ENGELM.
— xylonacantha v. mediopicta TREL.
— yaquiana TREL.
**Agnirictus** SCHWANT. Mesembr. ●
A. agninus (HAW.) SCHWANT.
— — v. integrifolius (SALM) SCHWANT.
— lesliei SCHWANT.
Agnirictus (SCHWANT.) VOLK
**Aichryson** WEBB et BERTH. Crassulaceae
A. dichotomum (DC.) WEBB et BERTH.
— domesticum PRAEG.
— immaculatum WEBB

— mollii PIT.
— pachycaulon BOLLE

— parlatorei BOLLE
— pulchellum C. A. MEY.

— pulvinatum BURCH.
— radicescens WEBB
— sedifolium WEBB
— tortuosum PRAEG.
— — (DC.) WEBB et BERTH.
Acc. PRAEGER spontaneous hybrids:
A. laxum × A. porphyrogenetos
— — × A. punctatum
— porphyrogenetos × A. punctatum

**Aistocaulon** v. POELLN. Mesembr. ●
A. rosulatum (KENSIT) v. POELLN.
**Aizoanthemum** DTR. ex FRIEDR. Aizoaceae
A. bossii DTR.
— sphingis DTR.
— stellatum DTR.
**Aizoon** L. Aizoaceae
A. dinteri SCHINZ
— galenioides FENZL ex SOND.
— mossamedense WELW.
**Aldasorea** HORT. ex HAAGE et SCHMIDT Crassulaceae
A. goochiae, guttata, percarneum, saundersii, strepsicladum HORT. ex HAAGE et SCHMIDT are ancient horticultural names for Aeonium with the same species-epithets.
**Aloe** L. Liliaceae
A. abyssinica BGR.
— — HOOK. f.

A. barb.
— ang. cv. Marg. w.
— — v. a.
— samal.
— atr. v. salm.
— ang. v. a.
— willd.
— scabr.

— xyl. v. vitt.
— pacif.

Stom. a. v. a.

— — v. int.
— l.
Stom. Sect. II

A. lax.
Aeon. d.
Aichr. punct. v. pach.
Mon. ict.
Aichr. punct. v. p.
— — v. punct.
Aeon. spath. v. cr.
— tort.
— —
— sed.
— tort.
— dom.


Aichr. × praeg.
— × interm.

— × bramw.


Aloin. r.

A. dint.
— —
— —

Aizoanth. d.
— —
— —










A. eleg.
— adrig.

A. abyss. LAM. v. peacockii BAK.
— — v. percrassa BAK.
— — SALM
— acinacifolia JACQ.
— — v. angustifolia SALM
— — v. laetevirens SALM
— — v. venusta SALM
— acuminata HAW.
— — v. major S. D.
— — LAM.
— aethiopica (SCHWEINF.) BGR.
— affinis POLE EVANS ?
— africana MILL. v. angustior HAW.
— — v. latifolia HAW.
— — SALM
— agrophila REYN.
— albicans HAW.

— albispina HAW.
— albocincta HAW.
— albopicta HORT. Lig.
— altilinea ROEM. et SCHULT.
— — v. mucronata (HAW.) v. POELLN.
— amanensis BGR.
— ambigens CHIOV. Insufficiently known spec.
— amoena PILL.
— angiens DE WILLD.
— — kitaliensis REYN.
— angulata WILLD.
— — v. striata WILLD.
— — v. truncata WILLD.
— angustifolia GROENW.
— — HAW.
— — SALM
— — v. laevis SALM
— anomala HAW.
— arabica LAM.
— — SALM
— arachnoides LAM.
— — v. klugii SALM
— — v. pellucens SALM
— — MILL.
— arachnoides AIT.
— — v. reticulata KER.
— — v. translucens HAW.
— arborea MEDIC.
— arborescens v. milleri BGR.
— aristata ROEM. et SCHULT.
— aspera HAW.
— asperiuscula ROEM. et SCHULT.
— atherstonei BAK.
— atrivirens DC.
— attenuata HAW.
— — v. clariperla SALM ex ROEM. et SCHULT.
— aurantiaca BAK.
— bainesii v. barberae (DYER) BAK.
— bamangwatensis SCHOENL.
— barberae DYER

A. eleg.
— percr.
— camp.
Gast. a. v. a.
— trig. v. t.
— v. —
— ac. v. ven.
A. hum. v. a.
— — v. sub.
Gast. verr. v. v.
A. eleg.
— parvibr. v. p.
— afr.
— —
— princ.
— boyl.
Haw. margin. v. m.
A. mitr. v. a.
— striat.
— camp.
Haw. alt.

— mucr. v. m.
A. laterit.


— fram.
— laterit.
— —
Gast. ang. v. a.
— — v. str.
— — v. tr.
A. vand.
— afr.
Gast. ang.
— — v. l.
Haw. rec.
A. barb.
— micr.
Haw. ar.
— —
— —
— —
— —
— ret. v. ret.
— tr.
A. arb.
— — v. a.
Haw. ar.
Astr. a. v. a.

Haw. a. v. a.
A. plur.
Haw. herb.
— a.

— — v. c.
A. striatula

— bain.
— zebr.
— bain.

38*

## INVALID DESIGNATIONS Aloe

A. barteri Bak. p. part.
—— v. dahomensis A. Chev.
—— v. lutea A. Chev.
—— v. sudanica A. Chev.
— barteri Schnell
— baumii Engl.
— bayfieldii S. D.
— × bedinghausii Radl
— × beguinii Radl
— × — v. perfectior Radl
— bicolor Roem. et Schult.
— boastii Letty
— bolusii Bak.
— bourea Roem. et Schult.
— bowieana Roem. et Schult.
—— v. formosa Salm
—— S. D.
— brachyphylla Salm
— brevifolia Haw.
—— v. serra Bgr.
—— Roem. et Schult.
— brevis Roem. et Schult.
— brunneo-punctata Engl. et Gilg
— brunnthaleri Bgr.
— bulbicaulis Christian
— bullulata Jacq.
— candelabrum Engl. et Drude
— candicans Roem. et Schult.
—— v. linita Salm
—— v. nitens Salm
— capitata v. trachyticola Perr.
— carinata Ker.
—— v. laevior Salm
—— Mill.
—— v. acinaciformis DC.
— carowii Reyn.
— cascadensis Ktze.
— cernua Tod.
— charlesii nom. nud.
— chimanimaniensis Christ.
— chinensis Bak.
— chlorantha (Haw.) Roem. et Schult.
— × chludowii Beguin
— ciliaris f. mutans gigas Res.
— clariperla Roem. et Schult.
— coarctata Salm
— commelinii Willd.
— commutata Engl.
— comphylosiphon Bgr.
— concinna Bak.
—— Roem. et Schult.
—— Spreng.
— congesta Salm
— conspurcata Salm
— cooperi Hort. de Smet
— corbisieri de Willd.

A. buettn.; schweinf. v. s.
——
— schweinf. v. s.
— buettn.
— macroc. v. m.
— zebr.
× Gast.'haw. b.
× Gastr. b.
× — b.
× —— nm. p.
Gast. b.
A. chort. v. b.
— afr.
Cham. afr.
Gast. pict.
—— v. f.
Cham. afr.
Gast. brev.
A. dist.
— brev. v. s.
Gast. brev.
Haw. margar. v. min.

A. nutt.
— micr.
— buettn.
Astr. b.

A. thrask.

Gast. cand.
——
— ac. v. n.

A. trach.
Gast. glab.
— car. v. strig.
—— v. car.
—— v. —
A. slad.
— striat.
— capit. v. c.

— swynn.
— barb.

Haw. chl. v. c.
× Gastr. beg. nm. ch.
A. gig.
Haw. att. v. c.
— c. v. c.
A. mitr. v. c.
— macr. v. maj.
— lat.
— squarr.
Haw. visc. v. c.
——
— c.
Gast. c.
A. lept. v. l.
— nutt.

A. cordifolia Roem. et Schult.
— crassifolia Roem. et Schult.
— croucheri Hook. f.
— curta Haw.
— cylindrica v. rigida Lam.
— cymbifolia Schrad.
— cymbiformis Haw.
— debrana Christian
— decipiens Roem. et Schult.
— decora Schoenl.
— defalcata Chiov. – Somal. – Acc. Reynolds a mixture of two species, so the name must be disregarded.
— deltoidea Hook. f.
— deltoideodonta v. contigua Perr.
—— f. latifolia Perr.
—— sf. variegata Boit.
—— f. longifolia Perr.
—— v. intermedia Perr.
—— v. typica Perr.
— dentata Pers.
— denticulata Roem. et Schult.
— dependens Steud.
— depressa Haw.
— × derbetzii Hort. Bgr.
— dictyoides Roem. et Schult.
— disticha L.
—— v. plicatilis L.
—— N. L. Burm.
—— Mill.
—— Roem. et Schult.
—— v. latifolia Kunth.
—— Thunbg.
— dorsalla Haw.
— echinata Willd.
— edulis A. Chev.

— elizae Bgr. Insufficiently known spec.
— ellenbergeri Guill.
— elongata Murray
—— Salm
— engleri Bgr.
— ensifolia Roem. et Schult.
— erecta Salm
—— v. laetevirens Salm
—— v. minor Salm
— eru Bgr.
—— v. cornuta Bgr.
—— v. hookeri Bgr.
—— f. erecta, f. glauca, f. maculata et f. parvipunctata Hort.
— everettiae nom. nud.
— excavata Willd.
— expansa Haw.
—— v. paulo major Haw.
— eylesii Christian
— fasciata Salm
—— v. major Salm

Haw. c.
Gast. nigr. v. c.
— cr.
Haw. tort. v. c.
— rig. v. r.
— cymb.
——
A. percr.
Gast. dec.
A. clav.

Astr. d. v. d.

A. cont.
——

——
——
— subacut.
— delt. v. d.
Lomat. purp.

Gast. dist. v. d.
A. pend.
— brev. v. d.
× Gastr. d.

Gast. ret.
— dist. v. d.
A. plic.
——
— sap. v. s.
Gast. dist.
—— v. l.
— verr.
A. lin. v. l.
— hum. v. e.
— macroc. v. m.

— arist. v. a.
— barb.
Gast. trig. v. t.
A. sec.
Gast. ac. v. e.
Haw. margar. v. min.
—— v. l.
—— v. m.
A. camp.
——
— adrig.

— camp.

Gast. exc.
Haw. rig. v. e.
—— v. rig.
A. rhod.
Haw. f. v. f.
— subf. v. s.

## Aloe INVALID DESIGNATIONS

A. fasc. v. major SALM ex ROEM. et SCHULT. — A. f. v. f.
— ferox v. erythrocarpa BGR. — fer.
— — v. galpinii (BAK.) BGR. — —
— — v. hanburyi BAK. — —
— — v. incurva BAK. — —
— — v. subferox (SPRENG.) BAK.
— — v. xanthocarpa BGR. — spect.
— flava PERSOON — barb.
— flavispina HAW. — mitr. v. f.
— floramaculata CHRISTIAN — sec.
— forbesii BALF. f. — parr.
— formosa ROEM. et SCHULT. — pict. v. f.
— frutescens SALM — arb. v. f.
— fruticosa LAM. — —
— fulgens TOD. — princ.
— galpinii BAK. — fer.
— gariusana DTR. — gariep.
— glabra SALM and v. minor et v. major SALM — Gast. gl.
— glabrata SALM — Haw. gl. v. g.
— glauca v. elatior S. D. and v. humilior S. D., v. major HAW., v. minor HAW. — A. glauc. v. g.
— — v. spinosior HAW. — — v. mur.
— gloriosa RADL. — × Lomat. g.
— gracilis BAK. — A. comm.
— granata ROEM. et SCHULT. — Haw. margar. v. min.
— — v. laetevirens SALM — — v. — sv. l.
— — v. major SALM — — v. —
— — v. minima SALM — — v. — sv. pol.
— — v. polyphylla HAW. — — v. — sv. —
— grandidentata HORT. PAN. — A. comm. v. c.
— guttata SALM — Gast. pseud. v. gl.
— hanburyana NAUD. — A. striat.
— hanburyi A. BORZI — × ant.
— herbacea DC. — Haw. ret. v. ret.
— — MILL. — herb.
— — × holtzei RADL — × Gast'haw. h.
— horrida HAW. — A. fer.
— — × hoyeri RADL — × Lomat. h.
— humilis KER. — A. hum. v. ac.
— — (L.) MILL. v. candollei BAK.
— — v. echinata sv. minor S. D. — — v. h.
— — v. suberecta sv. semi-guttata HAW. — — v. e.
— hybrida SALM — — v. semig.
— — v. asperior SALM — Haw. hybr.
— imalotensis REYN. — — —
— × imbricata BGR. — A. cant.
— imbricata HAW. — × Gastr. i.
— indica ROYLE — Haw. i.
— indurata ROEM. et SCHULT. — A. barb.
— intermedia HAW. — Haw. visc. v. i.
— — v. asperrima SALM — Gast. verr. v. i.
— — REYN. — — v. a.
— jex-blakeae CHRISTIAN — A. subacut.
— johnstonii BAK. — rusp. v. r.
— juttae DTR. — micrac.
— krausii BAK. v. minor BAK. — microst.
— albid.

A. krausii SCHOENL.
— laetepunctata ROEM. et SCHULT.
— laetevirens SALM
— laevis SALM
— lanzae TOD.
— × lapaixii RADL
— × lauchii RADL
— laxiflora N. E. BR.
— laxissima REYN.
— leptocaulon BOJ. nom. nud.
— limpida SCHULT.
— lingua KER.
— — LINK
— — v. angulata HAW.
— — SALM
— — v. angustifolia SALM
— — v. — AIT.
— — v. brevifolia SALM
— — v. crassifolia AIT.
— — v. latifolia SALM
— — v. — WILLD.
— — v. longifolia HAW.
— — v. major SALM
— — THUNBG.
— linguaeformis L. f.
— linguiformis DC.
— liniata ROEM. et SCHULT.
— longiaristata ROEM. et SCHULT.
— lugardiana BAK.
— luntii BAK.
— lusitanica GROENW.
— × lynchii BAK.
— macowanii BAK.
— maculata KER.
— — THUNBG.
— — v. angustior S. D.
— — v. obliqua AIT.
— — v. pulchra AIT.
— maculosa LAM.
— magnidentata VERD. et CHRISTIAN
— mamillaris v. major SALM
— margaritifera DC.
— — MILL.
— — SPRENG.
— — v. β v. min. part.
— — v. β v. max. part.
— — v. major AIT.
— — v. — DC.
— — v. — WILLD.
— — v. maxima HAW.
— — v. media AIT.
— — v. minima AIT.
— — v. minor WILLD.
— marginalis DC.
— marginata LAM.
— — WILLD.
— marmorata HAW.
— marshallii WOOD. et EVANS
— melsetterensis CHRISTIAN
— micrantha LK. et OTTO

A. albid.
Gast. l.
Haw. l.
Gast. ang. v. l.
A. barb.
× Gastr. l.
Gast. × l.
A. grac. v. g.
— transv. v. t.
— ant.
Haw. mucr. v. l.
Gast. mac. v. mac.
— ang. v. laev.
— sulc.
— dist. v. d.
— — v. —
— ang.
— obt.
— nigr. v. c.
— dist. v. l.
— ang. v. ang.
— v. —
— dist. v. lat.
A. plic.
— —
Gast. sulc.
— cand.

A. ar. v. a.
— zebr.
— inerm.
— komat.
× Gastr. l.
— striatul.
Gast. pulch.
— mac. v. mac.
— — v. fall.
— — v. mac.
— pulch.
A. obsc.

— megal.
Haw. margar. v. minor
— — v. minim. f. m.
— — v. margar.
— — v. minim.
— — v. —
— — v. minor
— — v. margar.
— — v. —
— — v. —
— — v. max.
— — v. minor
— — v. m.
— — v. m.
Lom. purp.
Haw. m. v. m.
Lom. purp.
Gast. pict. v. form.
A. kniph.
— swynn.
— microc.

# INVALID DESIGNATIONS  Aloe

| | | | |
|---|---|---|---|
| A. minor ROEM. et SCHULT. | Haw. margar. v. m. | A. parvula PERR. | A. perr. |
| — mirabilis SALM | — m. | — peacockii BGR. | — eleg. |
| — mitraeformis S. D. | A. mitr. v. xanth. | — pentagona HAW. | Astr. p. v. p. |
| — — v. angustior LAM. | — dist. | — — v. spiralis SALM | — — v. sp. |
| — — v. brevifolia SIMS | — — | — percrassa BGR. v. saganeitiana BGR. | A. eleg. |
| — — v. humilior HAW. | — mitr. v. comm. | — SCHWEINF. | — trichos. v. t. |
| — mitriformis v. elatior HAW. | — v. m. | — — v. menachensis SCHWEINF. | — men. |
| — — v. humilior WILLD. | — dist. | — × perfectior BGR. | × Gastr. beg. nm. p. |
| — — v. spinosior HAW. | — nob. | — perfoliata L. v. barbadensis L. | A. barb. |
| — mketiensis CHRISTIAN | — nutt. | | |
| — mollis ROEM. et SCHULT. | Gast. m. | — — v. vera WILLD. | — — |
| — montana SCHINZ | A. dich v. m. | — —THUNBG. | — fer. |
| — × mortolensis BGR. | × Gastr. m. | — — v. africana AIT. | — afr. |
| — mucronata SCHULT. | Haw. m. v. m. | — — v. arborescens SOL. | — arb. |
| — muiri MARL. | A. lin. v. m. | — — v. brevifolia AIT. | — dist. |
| — multifaria ROEM. et SCHULT. | Haw. mir. | — — v. ferox AIT. | — fer. |
| — muricata HAW. | A. fer. | — — v. humilis L. | — hum. v. h. |
| — — SCHULT. | — glauc. v. m. | — — v. lineata AIT. | — lin. v. l. |
| — mutica ROEM. et SCHULT. | Haw. ret. v. m. | — — v. mitraeformis AIT. | — mitr. v. m. |
| — mwanzana CHRISTIAN | A. macros. | — — v. obscura AIT. | — obsc. |
| — myriacantha v. minor BGR. | — alb. | — — v. purpurascens AIT. | — purp. |
| — natalensis WOOD. et EVANS | — arb. v. n. | — — v. saponaria AIT. | — sap. v. s. |
| — ngongensis CHRISTIAN | — rab. | — — v. serrulata AIT. | — serr. |
| — nigra SCHULT. | Haw. n. v. n. | — — v. succotrina AIT. | — succ. |
| — nigricans HAW. | Gast. n. v. n. | — — v. vera L. | — barb. |
| — — v. crassifolia SALM | — — v. c. | — × pethamensis BAK. | × Gastr. p. |
| — — v. fasciata SALM | fasc. v. f. | — picta THUNBG. | A. obsc. |
| — — v. marmorata SALM | — nigr. v. m. | — pienaarii POLE EVANS | — crypt. |
| — nitens BAK. | A. rup. | — planifolia BAK. | Gast. p. |
| — — ROEM. et SCHULT. | Gast. ac. v. n. | — — ROEM. et SCHULT. | Haw. p. |
| — nitida KER. | — nit v. parv. | — platyphylla BAK. | A. zebr. |
| — — SALM | — — v. n. | — plicatilis v. major S. D. | — plic. |
| — — v. brevifolia SALM | — obt. | — pluridens v. beckeri SCHOENL. | — plur. |
| — — v. grandipunctata SALM | — nit. v. g. | | |
| — — v. obtusa SALM | — obt. | — pluripuncta ROEM. et SCHULT. | Gast. ac. v. a. |
| — — v. parvipunctata SALM | — nit. v. p. | | |
| — nobilis BAK. | A. stans | — pongolensis REYN. | A. parvibr. v. p. |
| — × nowotnyi RADL | × Gastr. n. | — — v. zuluensis REYN. | — — v. z. |
| — obliqua DC. | Gast. pulch. | — postgenita ROEM. et SCHULT. | — brev. v. p. |
| — — HAW. | — mac. v. mac. | | |
| — — v. fallax ROEM. et SCHULT. | — — v. f. | — princae BGR. Insufficiently known spec. | |
| — — JACQ. | — nigric. v. n. | — prolifera HAW. | — — v. post. |
| — obscura BGR. et SCHOENL. | A. runc. | — — v. major S. D. | — — v. — |
| — — WILLD. | Gast. exc. | — × prorumpens BGR. | × Gastr. p. |
| — obtusa ROEM. et SCHULT. | — o. | — pseudoafricana S. D. | A. afr. |
| — obtusifolia SALM | — o. | — pseudoangulata SALM | Gast. subcar. v. s. |
| — oligophylla BAK. | Lom. o. | | |
| — oligopila BAK. | A. percr. | — — v. striata SALM | — — v. str. |
| — orpeniae SCHOENL. | — her. v. orp. | — pseudoferox SALM | A. fer. |
| — paedegona BGR. | — buettn. | — pseudonigricans SALM | Gast. p. v. p. |
| — pallescens HAW. ? | — serr. | — pseudotortuosa SALM | Haw. visc. v. p. |
| — pallida ROEM. et SCHULT. | Haw. p. v. p. | — — v. elongata SALM | — — v. torq. |
| — pallidiflora BGR. | A. greath. | — pulchra JACQ. | Gast. p. |
| — paludicola A. CHEV. | — buettn. | — pumilio JACQ. | Haw. ret. v. ret. |
| — paniculata JACQ. | — striat. | — — v. arachnoidea L. | — ar. |
| — papillosa SALM | Haw. p. v. p. | — — v. margaritifera δ L. | — margar. |
| — — v. maxima SALM | — — v. — | — — v. — β L. | — — v. minor |
| — — v. minor SALM | — — v. semip. | — — v. — γ L. | — — v. minim. |
| — — v. semipapillosa ROEM. et SCHULT. | — — v. — | — — v. — THUNBG. p. part. | — — v. —; sv. minor; v. marg. |
| — paradoxa HORT. Paris | A. het. | | |
| — parva ROEM. et SCHULT. | Haw. tess. v. p. | — punctata HAW. | A. var. v. v. |
| — parvispina SCHOENL. | A. mitr. v. m. | — purpurea LAM. | Lom. p. |

598

A. pygnantha MacOwan
— × quehli Radl
— racemosa Lam.
— radula Jacq.
— — v. major Salm
— — v. margaritifera Salm ex Haw.
— — v. minor Salm
— — Kerr.
— ramosa Haw.
— × rebutii Bgr.
— recurva Haw.
— recurvifolia Groenw.
— reinwardtii Salm
— repens Roem. et Schult.
— retusa L.
— — v. mutica Salm
— rhodocantha DC.
— rhodocincta Hort.
— rigida DC.
— — Jacq.
— — Ker.
— — Salm et v. expansa & v. minor Salm
— rossii Tod.
— rubescens DC.
— rubrolutea Schinz
— rugosa Salm ex Roem. et Schult.
— — v. perviridis Salm
— sahundra Boj.
— salmdyckiana Roem. et Schult.
— saponaria v. latifolia Haw.
— — v. minor Haw.
— — v. obscura Haw.
— scaberrima Salm ex Roem. et Schult.

— scabra Roem. et Schult.
— schimperi Schweinf.
— schinzii Bak.
— schlechteri Schoenl.
— schmidtiana Rgl.
— schweinfurthii Bak. ex Bgr.
— — Hook.
— semiglabrata Roem. et Schult.
— semimargaritifera Salm

— — v. glabrata Salm
— — v. major Salm
— — — v. maxima L.
— — v. minor Salm
— — v. multiperlata Roem. et Schult.

— — v. multipapillosa Salm

— sempervivoides Perr.
— serra DC.
— setosa Roem. et Schult.
— — v. major Roem. et Schult.

A. rup.
× Gastr. q.
Gast. verr. v. v.
Haw. rad.
— sub.

— subatt.
— rug. v. rug.
— att. v. att.
A. purp.
× Gastr. r.
Haw. r.
A. al.
Haw. r.
Gast. r.
Haw. r. v. r.
— — v. m.
A. glauc.
— striat.
Haw. r. v. r.
— tort. v. ps.
— — v. tort.

— rig. v. exp.
A. delt.
— barb.
— lit.

Haw. r. v. r.
— — v. p.
A. div. v. d.

— princ.
— lut.
— sap.
— obsc.

Gast. verr. v. asp.
Haw. s.
A. percr.
— lit.
— clav.
— coop.

— eleg.
— —

Haw. s.
— margar. v. max.
— — v. subalb.
— — v. max. sv. maj.
— — v. —
— subatt.

— margar. v. max. sv. m.
— — v. — sv. —

A. parv.
— brev. v. depr.
Haw. setat. v. s.
— — v. m.

A. setosa v. media Roem. et Schult.
— — v. nigricans Roem. et Schult.
— × simoniana Del.
— sinuata Thunbg.
— × smaragdina Hort. Bgr.
— socotrina Mass.
— soleiana Christian
— somaliensis C. Wright. Incorrect author citation for A. s. Watson
— spicata Bak.
— spinulosa S. D.
— spiralis Haw.

— — L.
— spirella Haw.
— — v. quinquangularis Salm
— stephaninii Chiov.
— stenophylla Roem. et Schult.
— — Schultes
— straussii Bgr. cult. Insufficiently known.
— striatula v. caesia f. haworthii Res.
— strigata Roem. et Schult.
— subalbicans Salm

— — v. acuminata Salm
— — v. laevior Salm
— subattenuata Salm
— subcarinata Salm
— suberecta Haw.
— — v. Haw.
— subfasciata Salm ex Roem. et Schult.
— subferox Spreng.
— subnigricans Haw.

— — v. canaliculata Salm
— — Kunth
— subrigida Roem. et Schult.
— subtortuosa Roem. et Schult.
— — Spreng.
— subtuberculata Haw.
— subulata Salm
— subverrucosa Salm
— — v. grandipunctata Salm
— — v. parvipunctata Salm
— succotrina Garsault
— — Lam. v. purpurascens Ker.
— — v. saxigena Bgr.
— sulcata Salm
— supralaevis Haw.
— — v. β hanburyi Bak.
— tenuior v. glaucescens A. Zahlbr.
— termetophila de Wild
— tessellata Salm
— thraskii de Wild
— tidmarshii (Schoenl.) Mueller

Haw. setat. v. m.
— — v. n.
× Gastr. s.
A. purp.
× Gastr. s.
A. fer.
— lat.

— camp.
— mitr. v. s.
Astr. pent. v. sp.
— sp.
— pent. v. sp.
— — v. tor.
A. rusp.

Haw. ang. v. a.
— montic.

A. striat. v. c.
Gast. car. v. st.
Haw. margar. v. s.
— — v. — sv. a.
— — v. — sv. l.
— s.
Gast. s. v. s.
A. hum. v. s.
— — v. ac.

Haw. s. v. s.
A. fer.
Gast. pseud. v. ps.
— — v. c.
— — v. ps.

Haw. tort. v. ps.

— visc. v. ps.
— tort. v. t.
A. hum. v. s.
Haw. s.
Gast. s. v. s.
— — v. —
— — v. p.
A. s.

— purp.
— succ.
Gast. s.
A. fer.
— marl. v. m.

— ten. v. t.
— greath.
Haw. t.
A. × lut.

— cil. f. t.

A. torquata ROEM. et SCHULT. Haw. visc. v. t.
— — v. laevior SALM — — v. —
— torrei v. wildii REYN. A. wild.
— tortuosa HAW. Haw. t. v. t.
— translucens HAW. — t.
— triangularis MEDIC. — visc.
— trichosantha v. menachensis
    SCHWEINF. A. men.
— tricolor BAK. — comm. v. t.
— — HAW. Haw. ven.
— trigona ROEM. et SCHULT. Gast. t. v. t.
— — v. elongata SALM — — v. —
— — v. minor ROEM. et
    SCHULT. — — v. —
— tripetala MED. A. plic.
— trivialis A. CHEV. — schweinf.
    v. s.
— — v. lutea and v. simplex
    A. CHEV. — —
— tuberculata HAW. — hum. v. ech.
— turgida BAK. Astr. delt. v. t.
— — ROEM. et SCHULT. Haw. t. v. t.
— ucriae TERR. f. A. arb. v. u.
— umbellata SALM — sap.
— vahontsohy PERR. — div. v. d.
— vaotsohy DECORSE et
    POISS. — — v. —
— — v. rosea R. DECARY — — v. r.
— variegata DTR. — aus.
— venosa LAM. Haw. v.
— venusta ROEM. et SCHULT. Gast. ac. v. ven.
— vera 'L.' AUCT. A. barb.
— — v. aethiopica
    (SCHWEINF.) BGR. — eleg.
— — v. angustifolia
    SCHWEINF. — off. v. a.
— — v. chinensis BGR. — barb.
— — v. lanzae BGR. — —
— — v. littoralis KOENIG ex
    BAKER — —
— — v. officinalis (FORSK.)
    BAK. — off. v. o.
— — v. puberula SCHWEINF. — × pub.
— — v. wratislaviensis HORT. — barb.
— — MILL. — succ.
— verrucosa MED. Gast. v. v. v.
— — MILL. — — v. —
— — v. striata SALM — — v. str.
— viscosa L. Haw. v.
— — v. indurata SALM — — v. i.
— — v. major SALM — — v. conc.
— vittata ROEM. et SCHULT. Gast. fasc. v. f.
— — v. latifolia SALM — — v. —
— — v. laxa ROEM. et
    SCHULT. — — v. l.
— vulgaris LAM. A. barb.
— wooliana POLE EVANS — woolley.
— xanthacantha WILLD. — mitr. v. x.
— — SALM — — v. mitr.
— yuccaefolia GRAY Hesp. parv.
— zanzibarica MILNE-
    REDHEAD A. squarr.
— zeyheri GRAY Gast. z.
**Aloinella** (BGR.) LEMÉE
    (A. BGR. as Sect. I of the
    gen. Aloe L.). Liliaceae.

A. haworthioides (BAK.)
    LEMÉE Aloe haw. v. h.
**Aloinopsis** SCHWANT. Mesembr. ●
A. albinota (HAW.) SCHWANT. Rab. a. v. a.
— albipuncta (HAW.)
    SCHWANT. — a. v. a.
— aloides (HAW.) SCHWANT. Nan. a. v. a.
— — v. striata L. BOL. — — v. str.
— — broomii L. BOL. — b.
— cibdela (N. E. BR.)
    SCHWANT. — c.
— crassipes (MARL.) L. BOL. A. spath.
— dyeri L. BOL. — rubr.
— gerstneri L. BOL. Nan. g.
— pallens L. BOL. — p.
— pole-evansii (N. E. BR.)
    N. E. BR. — p. e.
— transvaalensis (ROLFE)
    SCHWANT. — t. v. t.
— vittata (N. E. BR.)
    SCHWANT. — v.
— wilmaniae L. BOL. — w.
**Altamiranoa** ROSE. Crassulaceae.
A. albiflora (HEMSL.)
    E. WALTH. Vill. a.
— alpina FROED. — a.
— batesii (HEMSL.) ROSE — b. v. b.
— — v. subalpina FROED. — — v. s.
— berillonana (HAMET) BGR. — b.
— calcicola ROBINS. et
    GREENM. — c.
— chihuahuensis (S. WATS.)
    ROSE — ch.
— cucullata (ROSE) E. WALTH. — c.
— decipiens (BAK.) FROED. — d.
— diffusa (ROSE) JACOBS. — d.
— dyvrandae (HAMET) BGR. — d.
— elongata ROSE — e.
— erecta JACOBS. — strict.
— ericoides JACOBS. — imbr.
— fusca (HEMSL.) ROSE — f.
— galeottiana (HEMSL.) ROSE — g.
— goldmannii ROSE — g.
— grandyi (HAMET) BGR. — g.
— guatemalensis (ROSE)
    E. WALTH. — g.
— hemsleyana ROSE — h.
— imbricata (DIELS) BGR. — renif.
— — (ROSE) E. WALTH. — i.
— incarum (BALL.) BGR. — i.
— jurgensii HEMSL. — j.
— levis (ROSE) E. WALTH. — l.
— mexicana (SCHLTD.) ROSE — m.
— minutiflora (ROSE)
    E. WALTH. — m.
— nelsonii (ROSE) E. WALTH. — n.
— nexacana FROED. — n.
— painteri (ROSE) E. WALTH. — p.
— parva (HEMSL.) ROSE — p.
— parviflora (HEMSL.)
    JACOBS. — p.
— pringlei (ROSE) E. WALTH. — p.
— ramosissima (ROSE)
    E. WALTH. — r.
— ramulosa FROED. — r.
— scopulina ROSE — s.

A. squamulosa (S. Watts.)
 E. Walth.                             Vill. s.
— stricta (Diels) Bgr.                 — diels.
— — (Rose) E. Walth.                   — s.
— virgata (Diels) Bgr.                 — v.
— weberbaueri (Diels) Bgr.             — w.
**Amoebophyllum** N. E. Br. Mesembr. ●
A. guerichianum (Pax)
 N. E. Br.                             Mes. g.
— rangei N. E. Br.                     Sphalm. scint.
— roseum L. Bol.                       A. ang.
**Anacampseros** L. Portulacaceae.
A. arachnoides hort.                   A. ruf.
— — (Haw.) Sims v. rubens
 Sond.                                 — a. v. a.
— asperula Bgr.                        — — v. hisp.
— avasmontana Dtr.                     — alb. v. a.
— — v. caespitosa v. Poelln.           — — v. a.
— baeseckei v. crinita Dtr.            — crin.
— filamentosa de Wild                  — ruf.
— — (Haw.) Sims v. depau-
 perata Bgr.                           — dep.
— hispidula Bgr.                       — ar. v. h.
— intermedia Haw.                      — tel.
— — Nich.                              — fil.
— linguaefolia Lem.                    Echev. l.
— margaretae Dtr.                      A. tom. v. m.
— omburensis Dtr.                      — ombon.
— quinaria E. Mey. v.
 schmidtiana Bgr.                      — schm.
— — Kraus                              — —
— ramosa E. Mey. MS. ex
 Sonder                                Tal. caffr.
— rotundifolia Sweet                   A. tel.
— rubens hort.                         — ruf.
— spectabilis Jord.                    — ar. v. r.
— varians Haw.                         — tel.
**Anisocalyx** L. Bol. Mesembr. ●
A. salarius L. Bol.                    A. vag.
**Antimima** N. E. Br. Mesembr. ●
A. dualis (N. E. Br.) N. E. Br. Rusch. d.
**Apicra** Haw. Liliaceae.
A. aspera (Haw.) Willd.                Astr. a. v. a.
— — v. major Haw.                      — — v. m.
— bicarinata Haw.                      × Astrow. bic.
— — Resende                            Haw. res.
— bullulata (Jacq.) Willd.             Astr. b.
— congesta (Salm) Bak.                 — c.
— deltoidea (Hook. f.) Bak.            — d. v. d.
— — v. intermedia Bgr.                 — — v. i.
— — v. turgida (Bak.) Bgr.             — — v. t.
— egregia v. Poelln.                   — e. v. e.
— foliolosa (Willd.) Haw.              — f.
— jacobseniana v. Poelln.              Poelln.rubr.v.j.
— pentagona (Haw.) Haw.                Astr. p. v. p.
— — v. bullulata Bak.                  — bull.
— — v. spirella (Haw.) Bak.            — p. v. sp.
— — v. torulosa Haw.                   — — v. t.
— — v. willdenowii Bak.                — — v. spiral.
— rubriflora L. Bol.                   Poelln. r.
— skinneri Bgr.                        × Astrow. bic.
                                        nm. s.
— spiralis (L.) Bak.                   Astr. sp.
— — Willd.                             — pent. v. sp.
— turgida Bak.                         — t.
**Apicra** Willd. Liliaceae.
A. anomala Willd.                      Haw. rec.

A. arachnoides Willd.                  Haw. a.
— attenuata Willd.                     — a.
— cymbaefolia Willd.                   — c.
— expansa Willd.                       — rig. v. r.
— fasciata Willd.                      — f. v. f.
— granata Willd.                       — margar. v.
                                         minim.
— margaritifera Willd.                 — — v. m.
— — v. major Willd.                    — — v.
— — v. maxima Willd.                   — pap. v. pap.
— — v. minor Willd.                    — margar. v. m.
— mirabilis Willd.                     — m.
— nigra Haw.                           — n. v. n.
— patula Willd.                        — rig. v. exp.
— pseudorigida Salm                    — tort. v. p.
— pumilio Willd.                       — ret. v. ret.
— radula Willd.                        — r.
— recurva Willd.                       — r.
— reticulata Willd.                    — r. v. r.
— retusa Willd.                        — r. v. r.
— tortuosa Willd. p. part.             — t. v. t.; visc.
                                         v. ps.
— translucens Willd.                   — t.
— tricolor Willd.                      — ven.
**Aptenia** N. E. Br. Mesembr. ●
A. cordifolia (L. f.) N. E. Br.        A. c.
**Apteranthes** Mik. Asclepiadaceae.
? A. cylindrica Decne.                 Echidn. cer. v.
                                         obsc.
A. gussoneana Boiss.                   Car. eur. v. conf.
— — Mik.                               — — ssp. g.
— tessellata Decne.                    Echidn. cer. v. c.
× **Apworthia** v. Poelln. Liliaceae.
A. fardeniana v. Poelln. nom.
 nud.                                  × Astrow.
                                        'Fard.'
**Argeta** N. E. Br. Mesembr. ●
A. petrensis N. E. Br.                 Gibb. p.
Argeta (N. E. Br.) Wulff               — Sect. I, Sub-
                                         Sect. III
**Argyroderma** N. E. Br. Mesembr. ●
A. amoenum Schwant.                    A. pears.
— angustipetalum L. Bol.               — congreg.
— aureum L. Bol.                       — delaet.
— australe L. Bol.                     — —
— blandum L. Bol.                      — —
— boreale L. Bol.                      — —
— braunsii (Schwant.)
 Schwant.                              — fiss.
— brevipes (Schltr.)L. Bol.            — —
— brevitubum L. Bol.                   — delaet.
— carinatum L. Bol. p. part.           — —, testic.
— citrinum L. Bol.                     — —
— concinnum Schwant.                   — —
— cuneatipetalum L. Bol.
 p. part.                              — —, pears.
— delaetii Maass v. pur-
 pureum Maass                          — d.
— densipetalum L. Bol.                 — —
— digitifolium (N. E. Br.)
 Schwant.                              — fiss.
— duale N. E. Br.                      Rusch. d.
— formosum L. Bol.                     A. delaet.
— framesii L. Bol. 1929 non
 1934 nom. ill.                        — pears.
— — v. f.                              — f. ssp. f.
— — v. minus L. Bol.                   — — ssp. —

**INVALID DESIGNATIONS** Argyroderma-Aridaria

| | | | |
|---|---|---|---|
| A. gregarium L. Bol. | A. delaet. | A. caudata (L. Bol.) L. Bol. | Sphalm. c. |
| — hallii L. Bol. | — fram. ssp. h. | — celans L. Bol. | — c. |
| — hutchinsonii L. Bol. | — — | — congesta L. Bol. | — c. |
| — jacobsenianum Schwant. | — congr. | — constricta L. Bol. | — c. |
| — latifolium L. Bol. | — delaet. | — decidua L. Bol. | — d. |
| — latipetalum L. Bol. | — fiss. | — decurvata L. Bol. | — d. |
| — — v. longitubum L. Bol. | — — | — defoliata (Haw.) Schwant. | — d. |
| — lesliei N. E. Br. | — delaet. | — dela (L. Bol.) L. Bol. | — d. |
| — leucanthum L. Bol. | — — | — dinteri L. Bol. | — d. |
| — litorale L. Bol. | — fiss. | — dyeri L. Bol. | — d. |
| — longipes L. Bol. | — delaet. | — ebracteata (L. Bol.) N. E. Br. | — trich. |
| — luckhoffii L. Bol. | — pears. | — — v. brevipetala L. Bol. | — — |
| — margaretae (N. E. Br.) N. E. Br. | Lapid. m. | — englishiae (L. Bol.) N. E. Br. | — e. |
| — nortieri L. Bol. | A. congreg. | — fastigiata (Haw.) Schwant. | — plen. |
| — octophyllum (Haw.) Schwant. (Mes. o. Haw.). - Haworth himself cited plants of A. fissum and A. delaetii by this name, and it has hence been used for very different plants. The name is rejected and should not be used any longer. | | — flexuosa (Haw.) Schwant. | — f. |
| | | — fourcadei L. Bol. | — f. |
| | | — fragilis (N. E. Br.) Friedr. | — scint. |
| | | — framesii L. Bol. | — f. |
| | | — fulva (Haw.) Schwant. | A. noct. v. f. |
| | | — geniculiflora (L.) N. E. Br. | Sphalm. g. |
| | | — gibbosa L. Bol. | — brev. f. fer. |
| | | — glandulifera L. Bol. | — g. |
| — orientale L. Bol. | — fiss. | — godmaniae L. Bol. | — g. |
| — ovale L. Bol. | — pears. | — gratiae L. Bol. | — g. |
| — peersii L. Bol. | — congreg. | — grossa (Ait.) Friedr. | — g. |
| — planum L. Bol. | — delaet. | — herbertii (N. E. Br.) Friedr. | — g. |
| — productum L. Bol. | — — | | |
| — pulvinare L. Bol. | — craterif. | — horizontalis (Haw.) Schwant. | — defol. ? |
| — reniforme L. Bol. | — delaet. | | |
| — roseatum N. E. Br. | Lap. marg. | — latipetala L. Bol. | — l. |
| — roseum f. delaetii (Maas) Rowl. | A. delaet. | — laxa L. Bol. | — l. |
| | | — laxipetala L. Bol. | — l. |
| — — f. roseum | — — | — leptopetala L. Bol. | — l. |
| — rooipanense L. Bol. p. part. | — congreg., delaet. | — lignea L. Bol. | — l. |
| | | — longispinula (Haw.) L. Bol. | — l. |
| — schlechteri Schwant. | — pears. | — longistyla (DC.) Schwant. | — refl. |
| — schuldtii Schwant. | — delaet. | — longituba L. Bol. | — l. |
| — speciosum L. Bol. | — — | — luteoalba L. Bol. | — tetr. |
| — splendens L. Bol. | — — | — macrosiphon L. Bol. | — m. |
| — subrotundum L. Bol. | — craterif. | — noctiflora (L.) L. Bol. | A. n. v. n. |
| — testiculare v. luteum N. E. Br. | — delaet. | — notha (N. E. Br.) N. E. Br. | Sphalm. n. |
| | | — obtusa L. Bol. | — o. |
| — — v. pearsonii N. E. Br. | — pears. | — oculata (N. E. Br.) L. Bol. | — o. |
| — — v. roseum (Haw.) N. E. Br. | — craterif. | — odorata (L. Bol.) Schwant. | Herer. o. |
| | | — oubergensis L. Bol. | Sphalm. o. |
| — — v. roseum N. E. Br. | — — | — 'parvisepala' f. fera L. Bol. | A. brev. f. f. |
| — villetii L. Bol. | — subalb. | — peersii L. Bol. | Sphalm. virid. |
| | | — pentagona L. Bol. | — p. v. p. |
| **Aridaria** N. E. Br. Mesembr. ● | | — — v. occidentalis L. Bol. | — — v. o. |
| A. abbreviata L. Bol. | Sphalm. a. | — platysepala L. Bol. | — p. |
| — acuminata (Haw.) Schwant. | — a. | — plenifolia (N. E. Br.) Stearn | — p. |
| — albertensis L. Bol. | — a. | — pomonae L. Bol. | — p. |
| — albicaulis L. Bol. | — a. | — prasina L. Bol. | — p. |
| — anguinea L. Bol. | — a. | — primulina L. Bol. | — p. |
| — arenicola L. Bol. | — a. | — pumila L. Bol. | — p. |
| — aurea (Thunbg.) L. Bol. | — aurat. | — quarzitica L. Bol. | — qu. |
| — ausana (Dtr. et Bgr.) Dtr. et Schwant. | Psil. lind. | — quaterna L. Bol. | — qu. |
| — bijliae N. E. Br. | Sphalm. b. | — rabiei L. Bol. | — r. |
| — blanda L. Bol. | — b. | — rabiesbergensis L. Bol. | — r. |
| — brevisepala L. Bol. | — b. v. b. | — radicans L. Bol. | — r. |
| — — f. fera L. Bol. | — — v. f. | — rangei (N. E. Br.) Friedr. | Psil. lind. |
| — canaliculata (Haw.) Friedr. | — c. | — recurva L. Bol. | Sphalm. r. |
| | | — 'recurvata' L. Bol. ex Jacobsen | — decurv. |

A. reflexa (HAW.) N. E. BR.   Sphalm. r.
— resurgens (KENSIT) L. BOL.  — r.
— rhodandra L. BOL.           — r.
— rosea L. BOL.               — r.
— saturata L. BOL.            — s.
— scintillans (DTR.) FRIEDR.  — s.
— spinulifera (HAW.) N. E. BR.— s.
— splendens (L.) SCHWANT.     — s.
— straminea L. BOL.           — s.
— straminicolor L. BOL.       — s.
— striata L. BOL.             — s.
— stricta L. BOL.             — s.
— subaequans L. BOL.          — s.
— subpatens L. BOL.           — s.
— subpetiolata L. BOL.        — s.
— suffusa L. BOL.             — s.
— sulcata (HAW.) SCHWANT.     — s.
— tetragona (THUNBG.)
    SCHWANT.                  — t.
— tetramera L. BOL.           — t. v. t.
— — v. parviflora L. BOL.     — — v. p.
— trichotoma (THUNBG.)
    L. BOL.                   — t.
— umbelliflora (JACQ.)
    SCHWANT.                  — u.
— varians L. BOL.             — scint.
— vernalis L. BOL.            — v.
— viridiflora (AIT.) L. BOL.  — v.
— watermeyeri L. BOL.         — w.
— willowmorensis L. BOL.      — w.
**Aristanthus** SCHWANT. nom.
    nud. Mesembr. ●           — Lampr.
**Arthrocnemum** MOQ. Chenopodiaceae.
A. coralloides Loscos et PARDO Microc. c.
— fruticosum TORR.            Allenr. occ.
— macrostachys TORR.          — —
**Asclepias** L. Asclepiadaceae.
A. aphylla BOJ.               Cyn. dec.
— — THUNBG.                   — a.
**Astridia** DTR. et SCHWANT. Mesembr. ●
A. maxima (HAW.) SCHWANT.     Rusch. m.
— rubra v. alba L. BOL.       A. alb.
— — v. citrina L. BOL.        — citr.
— velutina (DTR.) DTR. et
    SCHWANT.                  — dint. v. d.
— — v. lutata L. BOL.         — — v. l.
**Astroloba** UITEW. Liliaceae.
A. bicarinata (HAW.) UITEW.   × Astrow. b.
— skinneri (BGR.) UITEW.      × — nm. s.

**Avonia** E. MEY. Portulacaceae. Anac. Sect. I
**Beaucarnea** LAM. Agavaceae.
B. bigelovii BAK.             Nol. b.
— glauca ROEZL.               B. strict.
— oedipus ROSE                — grac.
— purpusii ROSE               — strict.
— recurvata LEM.              Nol. r.
— — v. stricta BAK.           B. s.
— tuberculata ROEZL.          Nol. rec.
**Bergeranthus** SCHWANT. Mesembr. ●
B. albidus (L.) SCHWANT.      Mach. a.
— caninus (HAW.) SCHWANT.     Carr. c.
— carinans (HAW.) SCHWANT.    Herer. c.
— cookii (L. BOL.) SCHWANT.   Mach. c.
— derenbergianus (DTR.)
    SCHWANT.                  Rusch. d.

B. ignavus SCHWANT. nom.
    nud.
— puttkamerianus (DTR. et
    BGR.) SCHWANT.            Herer. puttk.
— rehneltianus (BGR.)
    SCHWANT.                  — r.
— rhomboideus (S. D.)
    SCHWANT.                  Rhomb. r.
— scapigerus (HAW.)
    SCHWANT.                  Berg. s.
B. SG. Carruanthus SCHWANT.   Carr.
B. SG. Hereroa SCHWANT.       Herer.
B. SG. Rhombophyllum
    SCHWANT.                  Rhomb.
**Berrisfordia** L. BOL. Mesembr. ●
B. (L. BOL.) SCHWANT.; B.
    (L. BOL.) TISCH. – Deleted
    SG. of Conophytum N. E.
    BR.
**Beschorneria** KUNTH. Agavaceae.
B. argyrophylla HORT.         B. dek.
— floribunda HORT.            Furcr. roezl.
— multiflora HORT.            — bed.
**Blephanthera** SCHLTR. Asclepiadaceae.
B. dinteri SCHLTR.            Brachyst.bleph.
— edulis SCHLTR.              — —
**Bolusanthemum** SCHWANT. Mesembr. ●
B. tugwelliae (L. BOL.)
    SCHWANT.                  Herer. t.
**Boucerosia** WIGHT et ARN. Asclepiadaceae.
B. aaronis HART.              Car. a.
— acutangula DECNE.           — retr. v. a.
— adenensis DEFL.             — a.
— aucheri DECNE.              — tub.
— aucheriana CHEV.            — a.
— — HOOK. f.                  — tub.
— awdeliana DEFL.             — a.
— campanulata WIGHT           — umb.
— cicatricosa DEFL.           — c.
— crenulata WIGHT et ARN.     — c.
— cylindrica BROGN.           Echidn. damm.
— decaisneana LEM.            Car. d.
— dentata (FORSK.) DEFL.      — —
— diffusa WIGHT               — d.
— edulis EDGEW.               — e.
— europaea (GUSS.) HOOK. f.   — e.
— forskahlii DECNE. p. part.  — ciatr.; quadr.
— gussoneana (MIK.) HOOK. f.  — eur. ssp. g.
— hispanica DE COINCY         — munb. v. h.
— incarnata N. E. BR.         — i. v. i.
— indica DECNE.               — i.
— lasiantha WIGHT             — umb.
— mamillaris N. E. BR.        — m.
— maroccana HOOK. f.          — eur. v. m.
— munbyana DECNE.             — m. v. m.
— — v. hispanica DE COINCY    — munb. v. h.
— — v. — JIM. et IBAN.        — eur. v. conf.
— pauciflora WIGHT            — p.
— penicillata DEFL.           — p. v. p.
— quadrangula DECNE.          — qu.
— russeliana COURBON          — retr. v. r.
— simonis HORT. ex BGR.       — eur. v. s.
— sinaica DECNE.              — s.
— socotrana BALF. f.          — s.
— stocksiana BOISS.           — edul.
— tombuctuensis CHEV.         — retr. v. r.

INVALID DESIGNATIONS   Boucerosia-Cacalia                                                                    604

B. truncato-coronata
  SEDGWICK                       B. t.-c.
— umbellata WIGHT et ARN.        — u.
**Bowiea** HAW. Liliaceae.
B. africana HAW.                 Cham. a.
— myriacantha HAW.               Al. m.
**Brachystelma** R. BR. Asclepiadaceae.
B. barbertonensis HORT.          B. barberiae
— brevipedicellatum TURILL       — dint.
— caudatum N. E. BR.             — crisp.
— cinereum (SCHLTR.) N. E. BR.   — circ.
— filiformis HARV.               — —
— galpinii (SCHLTR.) N. E. BR.   — —
— grossarthii DTR.               — arn.
— hirsutum E. MEY.               Raph. h.
— ovatum OLIV.                   B. circ.
— pallidum (SCHLTR.) N. E. BR.   — —
— ringens E. A. BRUCE            — dint.
— spathulatum DECNE.             — crisp.
— — LINDL.                       
— undulatum (SCHLTR.) N. E.
  BR.                            — circ.
— zeyheri (SCHLTR.) N. E. BR.    — —
**Braunsia** SCHWANT. Mesembr. ●
B. apiculata (KENSIT)
  E. MURRAY                      B. ap.
— bina (N. E. BR.) SCHWANT.      — —
— edentula (HAW.) N. E. BR.      Rusch. e.
— geminata (HAW.) E. MURRAY      B. g.
— maximiliani (SCHLTR. et
  BGR.) SCHWANT.                 Lampr. m.
**Brownanthus** SCHWANT. Mesembr. ●
B. ciliatus (AIT.) SCHWANT.      Psil. c.
— marlothii (PAX) SCHWANT.       — m.
— namibensis (MARL.)
  BULLOCK                        — n.
— pubescens (N. E. BR. ex
  MAAS) BULLOCK                  — pill.
— schenkii (SCHINZ)
  SCHWANT.                       — cil.
— simplex (N. E. BR. ex
  MAAS) BULLOCK
— solutifolius (BGR.) JACOBS.    — marl.
**Bryophyllum** SALISB. Crassulaceae.
B. adelae (HAMET) BGR.           Kal. a.
— aliciae (HAMET) BGR.           — pub. v. p.
— ambrense (PERR.) BGR.          — unifl. v. u.
— beauverdii (HAMET) BGR.        — b.
— bouvetii (HAMET et PERR.)
  BGR.                           — b.
— calycinum SALISB.              — pinn. v. p.
— costantinii (HAMET) BGR.       — beauv.
— crenatum BAK.                  — lax.
— × crenodaigremontianum
  RES. et VIANI                  — 'Crenodaigr.'
— daigremontianum (HAMET
  et PERR.) BGR.                 — d.
— delagonense (ECKL. et
  ZEYH.) W. SCHINZ               — tub.
— jueli (HAMET et PERR.)
  BGR.                           — beauv.
— macrochlamys (PERR.) BGR.      — m.
— manginii (HAMET et PERR.)
  NOTHDURFT                      — m.
— — v. triploidea (BOIT. et
  MANN.) NOTHDURFT               — — v. t.

  miniatum (HILS. et BOJ.)
    BGR.                        Kal. m. v. m.
— pinnatum (LAM.) KURZ           — p. v. p.
— — v. calcicola (BAK.) BGR.     — — v. c.
— porphyrocalyx (BAK.) BGR.      — p. v. p.
— rosei (HAMET et PERR.) BGR.    — r. v. r.
— rubellum BAK.                  — r.
— scandens BGR.                  — beauv.
— schizophyllum (BAK.) BGR.      — sch.
— streptanthum (BAK.) BGR.       — s.
— suarezense (PERR.) BGR.        — s.
— sulphureum (BAK.) BGR.         — porph. v. p.
— tsaratananense (PERR.)
  BGR.                           — rol. bon.
— tubiflorum HARVEY              — t.
— uniflorum (STAPF) BGR.         — u. v. u.
— verticillatum (SCOTT
  ELLIOT) BGR.                   — tub.
B. (SALISB.) BOIT. et MANN.      Kal. Sect. II
**Bulbine** L. Liliaceae.
B. bulbosa (R. BR.) HAW.         Bulbinops. b.
— caulescens L. ex STEUD.        B. frut. v. f.
— — v. incurva BAK.              — — v. i.
— — v. rostrata (JACQ.)
  v. POELLN.                     — — v. r.
— concinna BAK.                  — min.
— dielsii v. POELLN.             — croc. ?
— inops N. E. BR.                — min.
— latifolia (L. f.) SPRENG.      — lat.
— longiscapa (JACQ.) WILLD.      — alt.
— orchioides DRÉGE. - A very
  imperfectly known herbarium
  species.
— praemorsa (JACQ.) ROEM. et
  SCHULT.                        — praem.
— pugioniformis (JACQ.) ROEM.
  et SCHULT.                     — pug.
— semibarbata (R. BR.) HAW.      Bulbinops. s.
**Bulliardia** DC. Crassulaceae.
Generic synonym for **Crassula**
  L. Sect. I/1, 2, 6, 8. - Disregarded species since they are not succulent plants.
**Bursera** 'JACQ.' L. emend.
  TRIANA et PLANCH. Burseraceae.
B. morelensis RAMIREZ            B. micr.
— pubescens S. WATS.             Pachyc. disc.
                                 v. p.
**Byrnesia** ROSE. Crassulaceae.
B. weinbergii ROSE               Graptop. parag.
**Cacalia** CASS. Compositae.
C. acaulis L. f.                 Sen. a. v. a.
— anteuphorbium L.               — a.
— articulata HAW.                — a. v. a.
— canescens WILLD.               — haw.
— crassissima WALL.              — c.
— cuneifolia L.                  — c.
— cylindrica LAM.                Oth. c.
— ficoides L.                    Sen. f.
— grandiflora WALL.              — ampl.
— granti OLIV. et HIERN          — g.
— haworthii SWEET                — h.
— kleinia L.                     — k.
— laciniata JACQ.                — art. v. a.
— odora FORSK.                   — ant. v. o.

## Cacalia-Carpobrotus INVALID DESIGNATIONS

C. papillaris L. — Sen. p.
— pendula FORSK. — p.
— radicans — r.
— repens L. — serp.
— rigida THUNBG. — r.
— runcinata LAM. — art. v. a.
— sempervirens SPR. — ampl.
— semperviva FORSK. — s.
— tomentosa HAW. — haw.
**Calandrinia** H. B. et K. Portulacaceae.
C. caulescens H. B. et K. — C. cil.
— elegans HORT. — disc.
— lindleyana HORT. — spec.
— speciosa HORT. — menz.
— — LEHM. — disc.
**Callistigma** DTR. et SCHWANT. Mesembr. ●
C. inachabensis (ENGL.) DTR.
  et SCHWANT. — Mes. i.
**Caralluma** R. BR. Asclepiadaceae.
C. acutangula DECNE. — C. retr. v. a.
— affinis DE WILD — eur. v. a.
— arachnoidea BALLY nom.
  nud.
— atrosanguinea N. E. BR. — Huerniops. a.
— attenuata WIGHT — C. adsc. v. a.
— ausana DTR. — hott. v. maj.
— australis NEL — mel. v. m.
— brownii DTR. et BGR. — lut. ssp. vag.
— campanulata N. E. BR. — umb.
— carnosa sensu SCHWEICK. — keith.
— chibensis LUCKH. — cand. v. ch.
— codonoides K. SCHUM. — spec.
— commutata ssp. eu-commutata JEHAND. et MAIRE — dec.
— — ssp. hesperidum JEHAND.
  et MAIRE — —
— comptum SCHLTR. — Piar. c.
— confusa FONT Y QUER — C. eur. v. c.
— corrugata N. E. BR. — socotr.
— crassa N. E. BR. — Whitesl. c.
— decora SCHLTR. — Piar. d.
— dentata (FORSK.) BGR. — s. C. dec.
— edulis CHEV. — C. mour.
— elata CHIOV. — priog.
— europaea ssp. gussoniana
  (MIK.) MAIRE (Apteranthus
  g. MIK., Boucerosia g. (MIK.)
  HOOK. f., Stapelia g. (MIK.)
  JACQ. f. ex LINDL.). – Unclarified ssp.
— — ssp. — v. decipiens
  MAIRE nom. nud. in Cat. Pl.
  Maroc. IV (1941) 1097
— — ssp. maroccana v. barrueliana MAIRE nom. nud. in
  l. c.
— fimbriata HOOK. f. — adsc. v. att.
— — WALLICH — v. f.
— fosteri PILL. — keith.
— geminata SCHLTR. — Piar. g.
— grandidens VERD. — C. mac. v. m.
— grivana SCHLTR. — Piar. g.
— hahnii NEL — C. lut. ssp. vag.
— hirtiflora N. E. BR. — retr. v. h.
— kalaharica NEL — knob.
— — v. langii (WHITE et
  SLOANE) WHITE et SLOANE — —

C. langii WHITE et SLOANE — C. knob.
— lasiantha N. E. BR. — umb.
— lateritia N. E. BR. — lut.
— — v. stevensonii WHITE et
  SLOANE — —
— leendertziae N. E. BR. — mel. v. m.
— lutea v. lateritia (N. E. BR.)
  NEL — l. v. l.
— — v. vansonii (BREMEK. et
  OBERM.) LUCKH. — — v. —
— maroccana BGR. — eur. v. aff.
— — N. E. BR. — — v. m.
— — v. confusa SEEM. et
  MAIRE — — v. —
— meintjesiana LAVR. — baldr.
— montana R. A. DYER et
  E. A. BRUCE — Echidn. m.
— namaquana WELW. ex
  JACOBS. — C. winkl.
— nebrownii BGR. — lut. ssp. vag.
— — v. discolor NEL — — ssp. —
— — v. pseudonebrownii (DTR.)
  WHITE et SLOANE — — ssp. —
— parviflora SCHLTR. — dep.
— piaranthoides OBERM. — schweinf.
— plurifasciculata nom. nud. — cong.
— pseudo-nebrownii DTR. — lut. ssp. vag.
— punctata SCHLTR. — Piar. p.
— quadrangula DI CAPUA — C. dicap.
— rangeana DTR. et BGR. — mac. v. brev.
— rangei DTR. et BGR. — — v. —
— respiciens K. SCHUM. — retr. v. r.
— rivae CHIOV. — socotr.
— robusta N. E. BR. — pen. v. r.
— rosengreenii VIERH. — socotr.
— rubiginosa WERD. — mel. v. m.
— russeliana (COURBON) CUF. — retr. v. r.
— schweickerdtii OBERM. — keith.
— scutellata DEFL. — Echidn. s.
— serpentina NEL — C. s.
— serrulata SCHLTR. — Piar. dec.
— simonis HORT. ex BGR. — C. eur. ssp.
  guss. v. s.
— tessellata PILL. — Echidn. fram.
— tombuctuensis N. E. BR. — C. retr.
— turneri E. A. BRUCE — disc. ssp. t.
— vaga (N. E. BR.) WHITE et
  SLOANE — lut. ssp. v.
— vansonii BREMEK. et
  OBERM. — — ssp. l.
**Carpanthea** N. E. BR. Mesembr. ●
C. calendulacea (HAW.) L. BOL. — C. pil.
— pomeridiana (L.) SCHWANT. — pom.
**Carpobrotus** N. E. BR. Mesembr. ●
C. abbreviatus (HAW.)
  SCHWANT. — C. vir.
— acinaciformis (L.)
  SCHWANT. — a.
— aequilateralis (WILLD. nom.
  ill.) J. M. BLACK — a.
— chilensis (MOL.) N. E. BR. — —
— disparilis N. E. BR. — —
— edulis (L.) N. E. BR. — e.
— — (L.) SCHWANT. — —
— glaucescens (HAW.) N. E.
  BR. — g.
— juritzii (L. BOL.) L. BOL. — dim.

C. laevigatus (Haw.) N. E. Br. C. l.
— peersii L. Bol. ex Jacobs. sphalm.
— pulleinei J. M. Black — Sarcoz. praec.
— quadrifidus v. roseus L. Bol. C. sauer.
— — f. roseus (L. Bol.) Rowl.
— rossii (Haw.) N. E. Br. — r.
— rubrocinctus (Haw.) N. E. Br. — r.
— subalatus (Haw.) N. E. Br. — s.

**Carruanthus** Schwant. Mesembr.
C. albidus (L.) Schwant. — Mach. a.
— caninus (Haw.) Schwant. — C. ring.
— cookii (L. Bol.) Schwant. — Mach. c.

**Caruncularia** Haw. Asclepiadaceae.
C. aperta Sweet — Carall. a.
— jacquinii Sweet — Stap. ped.
— massonii Sweet — —
— pedunculata Haw. — —
— simii Sweet — —
C. (Haw.) DC. — Stap. Sect. VIII

**Catevala** Medic. Liliaceae.
C. arachnoidea Med. — Haw. a.
— retusa Med. — r. v. r.

**Cepaea** Fourr. Crassulaceae.
C. caesalpinii Fourr. — Sed. c.

**Cephalophyllum** (Haw.) N.E.Br. Mesembr.
C. acutum (Haw.) L. Bol. — C. sub.
— albertinense (L.Bol.) L. Bol. — a.
— anemoniflorum (L. Bol.) L. Bol. — a.
— calvinianum L. Bol. — ros.
— comptonii (L. Bol.) N. E. Br. Acc. L. Bolus an invalid renaming.
— confusum (Dtr.) L. Bol. — c.
— corniculatum (L.) L. Bol. — lor.
— — (L.) Schwant. — —
— curtophyllum (L. Bol.) N. E. Br. — c.
— densum N. E. Br. Acc. L. Bolus an invalid designation.
— dissimile (N. E. Br.) N. E. Br. — d.
— diversifolium (Haw.) L. Bol. — d.
— — (Haw.) Schwant. — —
— ebracteatum (Schltr. et Diels) Dtr. et Schw. — rang.
— — (Schltr. et Diels) L. Bol. — —
— loreum (Haw.) L. Bol. — div.
— — (L.) N. E. Br. — l.
— maritimum (L. Bol.) L. Bol. — m.
— pittenii L. Bol. sphalm. — vanp.
— platycalyx (L. Bol.) L. Bol. — p.
— primulinum (L. Bol.) L. Bol. — p.
— purpureo-album (Haw.) L. Bol. — p. a.
— ramosum (L. Bol.) N. E. Br. — fram.
— saturatum L. Bol. nom. nud. — pill. v. p.

C. tricolorum (Haw.) Schwant. — C. t.
— weigangianum (Dtr.) Dtr. et Schwant. — Leip. w.

**Cerochlamys** N. E. Br. Mesembr.
C. trigona N. E. Br. — C. pach. v. p.

**Ceropegia** L. Asclepiadaceae.
C. acacietorum Schltr. et Dtr. — C. pach.
— acuminata Roxb. — bulb.
— albertina S. Moore — arist. v. a.
— albisepta v. viridis (Choux) H. Huber — vir. v. v.
— apiculata Schltr. — dist. ssp. lug.
— assimilis N. E. Br. — canc.
— barbertonensis N. E. Br. — wood. ssp. w.
— barklyi v. tugelensis N. E. Br. — b.
— botrys K. Schum. — subaph.
— boussingaultifolia Dtr. — nit.
— brosima E. A. Bruce et Bally — vign.
— brownii Ledger — dent. v. b.
— conrathii Schltr. — flor.
— constricta N. E. Br. — nil.
— crispata N. E. Br. — crass.
— cynanchoides Schltr. — rac. ssp. sec.
— cyrtoidea Werderm. — dist. ssp. lug.
— debilis N. E. Br. — wood. ssp. d.
— decaryi Choux — s. C. vir.
— decidua E. A. Bruce — C. eury.
— decumbens Bally — s. C. nil.
— elegans × C. sandersonii — C. hybr.
— estelleana R. A. Dyer — fimbr.
— evelynae E. A. Bruce — s. C. succ.
— fortuita R. A. Dyer — afr. ssp. f.
— galpinii Schltr. — rend.
— gardneri Thwaites ex Hook. — el. v. g.
— gossweileri S. Moore — s. C. dent.
— grandis E. A. Bruce — s. — nil.
— hastata N. E. Br. — C. wood. ssp. w.
— haygarthii Schltr. — dist. v. h.
— helicoides Choux — s. C. virid.
— infundibuliformis E. Mey. — C. fil.
— ledgeri N. E. Br. — eleg.
— linearis ssp. debilis (N. E. Br.) H. Huber — wood. ssp. d.
— — ssp. woodii (Schltr.) H. Huber — — ssp. w.
— lugardae N. E. Br. — dist. ssp. l.
— lushii Gran. — bulb.
— meyeri-arthuri Herter — hybr.
— monteiroae Hook. f. — sand.
— mozambicensis Schltr. — nil.
— nuda Hutch. — subaph.
— patersoniae N. E. Br. — zeyh.
— phalangium Dtr. — Kinep. sch.
— plicata E. A. Bruce — C. nil.
— rothii Gürke — rad. × sand.
— schoenlandii N. E. Br. — wood. ssp. w.
— secamonoides S. Moore — rac. v. s.
— serpentina E. A. Bruce — stap. v. s.
— setifera Schltr. — rac. ssp. s.
— — v. natalensis N. E. Br. — — ssp. —
— similis N. E. Br. — eleg.
— smithii Henders. — rad. v. s.

Ceropegia-Conophyllum  **INVALID DESIGNATIONS**

C. squamulata DECNE. — Echidn. s.
— tentaculata N. E. BR. — C. mult. v. t.
— tenuis N. E. BR. — wood. ssp. w.
— thorncroftii N. E. BR. — crass.
— tristis HUTCH. — dist. ssp. lug.
— tuberculata DTR. — crass.
— tubulifera DEFL. — var.
— undulata N. E. BR. — pach. v. u.
— variegata v. cornigera H. HUBER — de-vecc. v. d.-v.
— volubilis v. crassicaulis H. HUBER — set.
— wightii GRAN. et WIGHT — afr. v. a.

**Chamaerhodiola** (FISCH. et MEY.) NAKAI. Crassulaceae.
C. (FISCH. et MEY.) NAKAI — Rhod. Sect. 2

**Chasmatophyllum** DTR. et SCHWANT. Mesembr. ●
C. rouxii L. BOL. — Rhin. r.

**Cheiridopsis** N. E. BR. Mesembr. ●
C. braunsii SCHWANT. — Arg. fiss.
— brevis SCHWANT. — C. verr. v. v.
— comptonii L. BOL. — pill. v. p.
— crassa L. BOL. — — v. c.
— graessneri SCHICK et TISCH. — ins.
— hilmari L. BOL. — Al. h.
— johannis-winkleri SCHWANT. — C. schlecht.
— mirabilis SCHWANT. — verr. v. v.
— noctiflora L. BOL. — Al. peers.
— olivacea SCHWANT. — C. schick.
— pachyphylla SCHWANT. — verr. v. v.
— peersii (L. BOL.) L. BOL. — Al. p.
— quadrifida (HAW.) SCHWANT. — C. rostrata
— robusta L. BOL. — perd.
— schlechteri SCHWANT. — ins.
— ventricosa (L. BOL.) N. E. BR. — Rusch. v.
— victoris L. BOL. — C. peers.

**Circandra** N. E. BR. Mesembr. ●
C. serrata (L.) N. E. BR. — Ereps. s.

**Cissus** DC. Vitaceae.
C. amboensis SCHINZ — Cyph. fleck.
— bainesii (HOOK. f.) GILG et BRANDT — b.
— betaeformis CHIOV. — b.
— cactiformis GILG — C. quad.
— cirrhosa (THUNBG.) WILLD. — Cyph. c.
— crameriana SCHINZ — c.
— crassifolia (BAK.) PLANCH — C. rot.
— currori HOOK. f. — Cyph. c.
— dinteri SCHINZ — C. nymph.
— fleckii SCHINZ — Cyph. f.
— hereroensis SCHINZ — h.
— juttae DTR. et GILG — j. v. j.
— — v. ternatus JACOBS. — — v. t.
— macropus WELW. — m.
— marlothii DTR. — fleck.
— migiurtinorum CHIOV. — m.
— quinata DREGE — C. hyp.
— rotundifolia VAHL — quadr.
— rupicola GILG et BRANDT — Cyph. r.
— seitziana GILG et BRANDT — s.
— succulenta BURTT DAVY — C. quadr.
— uter EXELL et MEDONCA — Cyph. u.
C. L. Sect. Cyphostemma PLANCH. — Cyph.

**Cleretum** N. E. BR. Mesembr. ●
C. apetalum (L. f.) N. E. BR. — Dor. a.
— criniflorum (L. f.) N. E. BR. — bell.
— cuneifolium (JACQ.) N. E. BR. — —
— gramineum (HAW.) N. E. BR. — g. f. g.
— limpidum (AIT) N. E. BR. — bell.
— longipes L. BOL. — Micr. l.
— papulosum (L.) L. BOL. — p. v. p.
— — (L. f.) N. E. BR. — —v. —
— pinnatifidum (L. f.) L. BOL. — Aeth. p.
— puberulum (HAW.) N. E. BR. — Micr. p.
— schlechteri (SCHWANT.) N. E. BR. — s.
— sessiliflorum (AIT.) N. E. BR. — s.

**Congdonia** JEPSON. Crassulaceae.
C. pinetorum JEPSON — Sed. p.

**Conicosia** N. E. BR. Mesembr. ●
C. affinis N. E. BR. — Herr. a.
— brevicaulis (HAW.) N. E. BR. — C. b.
— capitata (HAW.) SCHWANT. — pug.
— elongata N. E. BR. — Herr. e. v. e.
— — (HAW.) SCHWANT. — — v. —
— fusiformis (HAW.) N. E. BR. — f.
— pugioniformis (L.) SCHWANT. — C. p.
— robusta N. E. BR. — Herr. r.
— roodiae N. E. BR. — r.

**Conophyllum** SCHWANT. Mesembr. ●
C. abbreviatum (L. BOL.) L. BOL. — Mitr. a. ?
— angustifolium L. BOL. — a.
— articulatum L. BOL. — a.
— brevisepalum L. BOL. — b.
— carterianum L. BOL. — c.
— chrysoleucum (SCHLTR.) SCHWANT. — Mon. c. v. c.
— clivorum (N. E. BR.) SCHWANT. — Mitr. c.
— compactum L. BOL. — c. v. c.
— — v. eenrietense L. BOL. — — v. e.
— cuspidatum L. BOL. — c.
— dissitum (N. E. BR.) SCHWANT. — d.
— framesii (L. BOL.) L. BOL. — f. ?
— globosum L. BOL. — Mon. g.
— gracile SCHWANT. — Mitr. g.
— grande (N. E. BR.) L. BOL. — g. v. g.
— — v. compressum L. BOL. — — v. c.
— hallii L. BOL. — h.
— herrei L. BOL. — h.
— kubusanum (L. BOL.) L. BOL. — k. ?
— latibracteatum L. BOL. — l. v. l.
— — f. fera L. BOL. — l. v. f. ?
— moniliforme (HAW.) SCHWANT. — Mon. m.
— nanum L. BOL. — Dipl. leip.
— niveum (L. BOL.) HERRE — Mitr. n. ?
— obtusipetalum L. BOL. — o.
— pisiforme (HAW.) SCHWANT. — Mon. p.
— proximum (N. E. BR.) SCHWANT. — Mitr. p.

**INVALID DESIGNATIONS** Conophyllum-Conophytum

C. ripense L. Bol.
— roseum (L. Bol.) L. Bol.
— scutatum (L. Bol.)
  Schwant.
— vanheerdei L. Bol.
**Conophytum** N. E. Br. Mesembr.
C. aggregatum N. E. Br.
— albertense (N. E. Br.)N. E.
  Br.
— altile (N. E. Br.) N. E. Br.
— amplum L. Bol.
— angustum L. Bol.
— apertum Tisch.
— apiculatum N. E. Br.
— asperulum L. Bol.
— — v. brevistylum L. Bol.
— barkerae L. Bol.
— bolusiae Schwant.
— braunsii Schwant.
— brevitubum Lavis
— brownii Tisch.
— calitzdorpense Tisch.
— caroli Lavis
— chloratum Tisch.
— cibdelum N. E. Br.

— colorans Lavis
— concinnum Schwant.
— convexum L. Bol.
— cornutum (Schwant.)
  Schwant.
— cuneatum Tisch.
— declinatum L. Bol.

— dennisii N. E. Br.
— edithae N. E. Br.
— eenkokerense L. Bol.
— elegans N. E. Br.
— elongatum Schick et Tisch.
— ernianum Lösch et Tisch.
— etaylorianum Schwant.
— exsertum N. E. Br.
— fissum Lavis nom. nud.
— forresteri L. Bol.
— framesii Lavis
— fraternum v. leptanthum
  L. Bol.
— friedrichiae (Dtr.)
  Schwant.
— geometricum Lavis
— helmutii Lavis
— hildegardii nom. nud.
— hillii L. Bol.
— impressum Tisch.
— inclusum L. Bol.
— indefinitum L. Bol.
— jucundum (N. E. Br.) N. E.
  Br.
— julii Schwant.
— karamoepense L. Bol.
— kennedyi L. Bol.
— khamiesbergense (L. Bol.)
  L. Bol.
— klaverense N. E. Br.
— komkasicum L. Bol.
— leightoniae L. Bol.

Mitr. r.
— r. ?

Mon. chrys.
Mitr. v.

C. pil.

— trunc. v. t.
  f. t.
— fic.
— obt. v. a.
— nut.
— ret.
— mey. v. a.
— lav.
— —
— —
— fib.
— pears.
— trunc. v. b.
— ect. v. b.
— c.
Ophth. c.
C. ect. v. limb.
— trunc. v. t.
  f. t.
— uvaef. v. mel.
— flav.
— cord. v. cord.

Ophth. subf.
C. hal.
— obcord. v. o.
  f. d.
— elish.
— pill.
— marg. v. e.
— pell.
— hians v. h.
— tayl. v. e.
— pilul.
— bil.

— subris.
— uvaef. v. f.

— f.

Ophth. f.
C. viol.
— steph.

— uvaef. v. mel.
— parv. v. i.
— lavis.
— —

— grat.
— uvaef. v. mel.
— marg. v. k.
Ophth. subf.

Berrisf. k.
C. obcord.
— calc. v. k.
— adv.

C. lekkersingense L. Bol.
— leptanthum L. Bol.
— leucanthum Lavis
— — v. multipetalum L. Bol.
— limbatum N. E. Br.
— littlewoodii L. Bol.
— longipetalum L. Bol.
— longum N. E. Br.
— lucipunctum N. E. Br.
— macrostigma (L. Bol.)
  Schwant.
— markoetterae Schwant.
— maughanii N. E. Br.
— meleagre L. Bol.
— minusculum v. roseum
  (Lavis) Tisch.
— modestum L. Bol.
— multicolor Tisch.

— namibense N. E. Br.
— nelianum Schwant.
— nevillei (N. E. Br.) N. E.
  Br.
— nuciforme (Haw.) N. E. Br.
— obovatum Lavis
— — v. obtusum L. Bol.
— odoratum N. E. Br.
— orientale L. Bol.
— ovatum Lavis sphalm.
— oviforme (N. E. Br.) N. E.
  Br.
— pallidum (N. E. Br.) N. E.
  Br.
— parvipetalum (N. E. Br.)
  N. E. Br.
— parvipunctum Tisch.

— pearsonii v. minor N. E.
  Br.
— percrassum Schick et
  Tisch.
— perpusillum (Haw.) N. E.
  Br. (Mes. p. Haw.). – Acc.
  A. Tischer this spec. cannot
  now be identified.
— picturatum N. E. Br.
— pilosulum (N. E. Br.) N. E.
  Br.
— plenum N. E. Br.
— poellnitzianum Schwant.
— pole-evansii N. E. Br.
— praesectum N. E. Br.
— pumilum N. E. Br.
— purpusii (Schwant.) N. E.
  Br.

— pygmaeum Schick et Tisch.
— quarziticum Tisch.
— renniei Lavis

— reticulatum L. Bol.
— — v. roseum Lavis
— — f. — (Lavis) Rowl.
— rubrum L. Bol.
— rufescens L. Bol.
— ruschii v. obtusipetalum
  L. Bol.

C. mey. v. ap.
— frat.
— inc. v. l.
— — v. —
— ect. v. l.
— marg. v. l.
— subr.
Ophth. l.
C. pill.

— cord. v. m.
— styl.
Ophth. m.
C. uvaef. v. m.

— m. v. ret.
— quaes.
— obcord. v. o.
  f. m.
— saxet. f. s.
— elish.

— obcord.
Gibb. crypt.
C. glob.
— —
— ficif.
— steytl.
— obovat.

Ooph. o.

C. ficif.

— obcord. v. p.
— trunc. v. t.
  f. p.

— p. v. p.

— orn.

— obc. v. o. f. p.

Gibb. p.
C. elish.
— subr.
— mey. v. p. e.
Ophth. p.
C. pag.

— trunc. v. t.
  f. t.
— brev.
— quaes.
— trunc. v. t.
  f. r.
— min. v. r.
— — v. —
— — v. —
— edw.
Ophth. r.

C. r.

608

## Conophytum-Cotyledon INVALID DESIGNATIONS

C. salmonicolor L. Bol.
— saxetanum (N. E. Br.) N. E. Br. forma L. Bol.
— schickianum Tisch.
— senarium L. Bol.
— sitzlerianum Schwant.
— speciosum Tisch.
— spirale N. E. Br.
— springbokense N. E. Br.
— subacutum L. Bol.
— subfenestratum Schwant.
— subtenue L. Bol.
— tabulare Lösch et Tisch.
— teguliflorum Tisch.
— tenuisectum L. Bol.
— tinctum Lavis
— tischleri Schwant.
— translucens N. E. Br.
— triebneri Schwant.
— truncatellum (Haw.) N. E. Br.
— tumidum N. E. Br.
— — v. asperulum L. Bol.
— vanrhynsdorpense Schwant.
— victoris Lavis
— villetii L. Bol.
— virens L. Bol.
— vlakmynense L. Bol.
— wettsteinii (Bgr.) N. E. Br. v. oculatum L. Bol.
— — N. E. Br.
— wiesemannianum Schwant.

**Corallocarpus** (Schweinf.) Hook. f. Cucurbitaceae.
C. fenzlii Hook. f.

**Corpuscularia** Schwant. Mesembr. ●
C. cymbiformis (Haw.) Schwant.
— dolomitica (Dtr.) Schwant.
— lehmannii (Eckl. et Zeyh.) Schwant.
— mollis (Bgr.) Schwant.
— perdiantha Tisch.
— quarzitica (Dtr.) Schwant.
— taylori (N. E. Br.) Schwant.
— thunbergii (Haw.) Schwant.

**Corynephyllum** Rose. Crassulaceae.
— viride Rose

**Cotyledon** L. Crassulaceae.
C. abramsii (Rose) Fedde
— acuminata (Rose) Fedde
— acutifolia Bak.
— adunca Bak.
— aegyptiaca Lam.
— agavoides Bak.
— aggregata Makino
— aizoon (Fenzl) Schoenl.
— albida (Rose) Fedde
— albiflora (Rose) Fedde
— — Hemsl.
— aloides (Rose) Fedde
— alstonii Schoenl. et Bak. f.

C. frut.
— s. f. hall.
— pag.
— marg. v. kar.
— musc.
— wettst. v. s.
— trunc. v. t. f. t.
— elish.
— inc.
Ophth. s.
C. ang.
— joh. winkl.
— frut.
— subr.
— orn.
— ect. v. t.
— uvaef. v. mel.
— marg.
— truncat.
— meyr. v. m.
— — v. a.

— uvaef. v. mel.
— subris.
— —
— ect. v. br.
— obtus.

— w. v. w.
— pears.
— full.

C. c.

s. Mes. c.

Rusch. d.

Del. l.
Rusch. m.
Nel. schl.
Rusch. qu.

Del. t.
Mal. t.

Sed. v.

Dudl. a.
— a.
Echev. a.
Pachyph. hook.
Kal. integr.
Ech. a. v. a.
Orost. mal.
Ros. a.
Dudl. a.
— a.
Vill. a.
Dudl. sax.
Adr. a.

C. alternans Haw.
— angulata f. fol. minoribus E. Mey.
— angustiflora (Rose) Fedde
— arborescens Mill.
— atropurpurea Bak.
— attenuata S. Wats.
— ausana Dtr.
— barnardiana (Britt.) Fedde
— batesii Hemsl.
— bicolor Hort. Herb. Kew
— bifida Hemsl.
— boehmeri (Makino) Makino
— bolusii Schoenl. and v. karrooensis Schoenl.
— bracteolata (Lk., Klotzsch et Otto) Bak.
— brandegei (Rose) Fedde
— brasiliensis Vell.
— brauntonii (Rose) Fedde
— breviceps Fedde sphalm.
— breviflora (Boiss.) Maire
— — ssp. intermedia (Boiss. et Reut.) Maire
— — ssp. — v. rubella Batt.

— — ssp. salzmannii (Boiss.) Maire
— — ssp. — v. flaviflora Batt.

— — ssp. rhodantha Maire

— bryceae (Britt.) Fedde
— cacalioides Eckl. et Zeyh.
— caespitosa Haw.
— — v. paniculata Jeps.
— californica Bak.
— campanulata Harv.
— canaliculata Bak.
— canalifolia Haw.
— candelabrum (Rose) Fedde
— candida (Britt.) Fedde
— carnicolor Bak.
— caryophyllacea Burm. f.
— chiclensis Ball.
— clavifolia Bak.
— — Bgr.
— coccinea Cav.
— compacta (Rose) Fedde
— congesta (Britt.) Fedde
— cooperi Bak.
— — v. immaculata Schoenl. et Bak.
— corderoyi Bak.
— crassifolia Salisb.
— crenata Auct. non Vent.
— — ('Andr.') Vent.
— cristata Harv. p. min. part.
— — Haw.
— cucullata Schoenl.

Adr. mac.

C. pap. v. p.
Dudl. cym.
Crass. a.
Adr. crist.
Echev. a.
C. orb. v. a.

Dud. cym. ssp. min.
Vill. b. v. b.
Echev. nod.
— b.

Orost. b.

Adr. b.

Echev. b.
Dudl. ac.
Kal. cren. v. ver.
Dudl. lanc.
— brevip.
Pist. b.

— — ssp. i.
— — ssp. — v. r.

— — ssp. s.

— — ssp. — v. f.

— — ssp. — v. r.
Dudl. cand.
C. eckl.
Dudl. c.
— cym. ssp. set.
— caesp.
C. teret.
Echev. c.
C. cor.

Dudl. c.
— c.
Echev. c.
Adr. c.
Echev. c.
Adr. c.
× Pachyv. c. v. c.
Echev. c.
Dudl. far.
— lanc.
Adr. c.

— pachyl.
Echev. ag. v. c.
Adr. hem.
Kal. cren. v. c.
— int. v. ver.

Adr. clav.
— c.
— tric.

39 Lexicon of Succulent Plants

## INVALID DESIGNATIONS Cotyledon

| | | | |
|---|---|---|---|
| C. cultrata (Rose) Fedde | Dudl. c. | C. hispida Lam. | Muciz. h. |
| — cuneata E. Mey. | C. pap. v. p. | — hoerleiniana Dtr. | C. schaef. |
| — — Harv. | — pill. | — — v. schaeferi Dtr. | — — |
| — — Thunbg. | — pap. v. p. | — horizontalis Guss. | Umb. h. |
| — curviflora Sims | — grand. | — humilis (Rose) Fedde | Dudl. cym. v. setch. |
| — cymosa Bak. | Dudl. cym. ssp. c. | — — Marl. | Adr. h. |
| — deficiens Hochst. et Steudel | Kal. int. | — imbricata Diels | Vill. renif. |
| | | — incarum Ball. | — i. |
| — — Forsk. | — schimp. | — insularis (Rose) Fedde | Dudl. vir. |
| — delicata (Rose) Fedde | Dudl. sax. v. al. | — insignis N. E. Br. | Kal. eliz. |
| — densiflora (Rose) Fedde | — d. | — integra Med. | — i. |
| — desoides DC. | Sed. cand. | — interjecta Haw. | C. grand. |
| — devensis N. E. Br. | Echev. acut. | — iwarenge Makino | Orost. i. |
| — dichotoma Haw. | C. ret. | — japonica Max. | — er. v. j. |
| — dielsii Schltr. | — dec. v. d. | — jasminiflora Salm | Adr. car. |
| — eastwoodiae (Rose) Fedde | Dudl. far. | — jurgensenii Hemsl. | Vill. j. |
| — edulis (Nutt.) Brewer et Wats. | — e. | — juttae Dtr. | C. ret. |
| | | — laciniata L. | Kal. l. |
| — — v. attenuata Jeps. | — att. | — lanceolata (Nutt.) Brewer et Wats. | Dudl. l. |
| — elata Haw. | C. orb. v. a. | | |
| — elegans N. E. Br. | Echev. harms. | — — v. saxosa (M. E. Jones) Jeps. | — — v. s. |
| — engleri Dtr. et Bgr. | C. orb. v. e. | | |
| — erubescens (Max.) Franch. et Savat. | Orost. e. v. e. | — — Forsk. | Kal. l. |
| | | — laxa Br. et R. | Dudl. caesp. |
| — eurychlamys Diels | Echev. e. | — — v. cymosa (Lem.) Jeps. | — cym. ssp. c. |
| — excelsa Diels | — e. | — — v. nevadensis (S. Wats.) Jeps. | — — ssp. — |
| — farinosa Bak. | Dudl. f. | | |
| — fascicularis Ait. | C. pan. | — — v. paniculata Jeps. | — — ssp. setch. |
| — filicaulis Eckl. et Zeyh. | Adr. mam. v. f. | — — v. setchellii Jeps. | — — ssp. — |
| — flanaganii Schoenl. et Bak. f. | C. dec. v. d. | — libanotica (L.) Labill. | Ros. l. v. l. |
| | | — linearis Greene | Dudl. l. |
| — — v. karrooensis Schoenl. et Bak. f. | — — v. — | — linguaefolia Lem. | Echev. l. |
| | | — linguiformis R. Br. | Dudl. caesp. |
| — flavida Fourc. | — — v. f. | — lingula S. Wats. | — far. |
| — fourcadei Schoenl. | — — v. f. | — lurida Haw. | Echev. rac. v. r. |
| — fulgens Bak. | Echev. f. | — — (Rose) Fedde | Dudl. lanc. |
| — fusiformis Rolfe | Adr. mam. v. f. | — maculata Bak. | Adr. rhomb. v. bak. |
| — galeottiana Hemsl. | Vill. g. | | |
| — gibbiflora Moc. et Sesse ex DC. | Echev. g. v. g. | — — Eckl. et Zeyh. | — hem. |
| | | — — Salm | — m. |
| — gigantea Rose | Dudl. cym. v. g. | — malacophylla Pallas | Orost. m. |
| — glandulifera L. F. Henders. | Sed. mor. | — — v. boehmeri Makino | — b. |
| — glauca Bak. | Echev. g. | — mamillaris Bak. | Adr. m. v. m. |
| — glutinosa Schoenl. | C. pap. v. g. | — — L. f. | — — |
| — gracilis Bgr. | — jacobs. | — marianae Marl. | — m. v. m. |
| — — Harv. | — pap. v. rob. | — marlothii Schoenl. | — mam. v. m. |
| — — Semen et Maurico nom. nud. | | — maximiliana Schltr. | C. eckl. |
| | | — meyeri Harv. | — pap. v. p. |
| — grandiflora (Rose) Fedde | Dudl. sax. ssp. al. | — mexicana (Schltd.) Hamet | Vill. m. |
| | | — minor (Rose) Fedde | Dudl. cym. ssp. m. |
| — grandis Morren | Echev. gibb. | | |
| — grayi Bak. | — pan. | — minuta Komorov | Orost. m. |
| — greenei (Rose) Fedde | Dudl. g. | — modesta Bornm. | Ros. m. |
| — hallii (Rose) Fedde | — lanc. | — montium-klinghardtii Dtr. | Adr. hem. |
| — hassei (Rose) Fedde | — h. | — mucizonia Ortega | Muciz. hisp. |
| — helleri (Rose) Fedde | — caesp. | — — ssp. ortega Maire v. hispida (Lamk.) Perez Lara sv. euhispida Maire | — — |
| — hemisphaerica Harv. p. min. part. | Adr. rhomb. v. r. | | |
| | | — mucronata Bak. | Echev. m. |
| — — L. | — h. | — — Lam. | C. und. v. m. |
| — herrei Barker | — h. | — nana Marl. | Adr. hum. |
| — heterophylla Schoenl. | C. ladism. | — — N. E. Br. | — n. |
| — hispanica L. | Pist. h. | — nevadensis S. Wats. | Dudl. cym. ssp. c. |
| — — v. flaviflora Maire | — — v. f. | | |
| — — v. maculata Maire | — — v. m. | — nodulosa Bak. | Echev. n. |
| — — v. purpurea Maire | — — v. p. | — nubigena Brand. | Dudl. n. |

- C. nuda BAK.
- — nudicaulis ABRAMS
- — — MURR.
- — nussbaumeriana v. POELLN.
- — oblonga HAW.
- — orbiculata v. viridis DTR.
- — orcuttii (ROSE) FEDDE
- — oreades C. B. CLAUSEN
- — oregonensis S. WATS.
- — ovata HAW.
- — — MILL.
- — ovatifolia (BRITT.) FEDDE
- — pachyphytum BAK.
- — palmeri S. WATS.
- — paniculata (JEPS.) FEDDE
- — pannosa BAK.
- — papillaris L. f. v. tricuspidata SALM ex DC.
- — — SCHOENL.
- — papillosa AITCH. et HEMSL.
- — paraguayensis N. E. BR.
- — parishii (BRITT.) FEDDE
- — — (ROSE) FEDDE
- — parviflora HEMSL.
- — — v. squamulosa S. WATS.
- — parvula BURCH.
- — peacockii BAK.
- — peltata WENDL.
- — pendulina BATT.
- — persica HORT.
- — peruviana BAK.
- — phillipsiae MARL.
- — pinnata LAM.
- — pistorinia ORT.
- — plattiana JEPS.
- — polycephala MAKINO
- — pringleyi S. WATS.
- — procurva N. E. BR.
- — pseudogracilis v. POELLN.
- — pubescens BAK.
- — pulverulenta (NUTT.) BAK.
- — pulvinata HOOK. f.
- — pumila BAK.

- — — (ROSE) FEDDE

- — purpurea HAW.
- — purpusii K. SCHUM.

- — quitensis BAK.
- — ramosa HAW.
- — ramosissima S. D. ex HAW.
- — — v. woodii SCHOENL. et BAK.
- — reflexa WILLD.
- — retusa BAK.
- — rhombifolia BAK.
- — — ECKL. et ZEYH.
- — — HAW.
- — — v. spathulata N. E. BR.
- — rigida (ROSE) FEDDE
- — rigidiflora (ROSE) FEDDE
- — robusta (BRITT.) FEDDE

- Echev. n.
- Dudl. dens.
- Kal. integr.
- Adr. n.
- C. orb. v. o.
- — — v. engl.
- Dudl. att. v. o.
- Sed. o.
- — wats.
- C. orb. v. o.
- Crass. argent.
- Dudl. cym. v. o.
- Pachyph. bract.
- Dudl. p.
- — cym. v. setch.
- Kal. erioph.

- C. tric.
- — p. v. glut.
- Sed. aden.
- Graptop. p.
- Dudl. att. v. orc.
- — lanc.
- Vill. mis.
- — squam.
- Adr. hum.
- Echev. p.
- Umbil. rup. v. r.
- — —
- Semp. tect. ssp. calc. cv.
- Echev. p.
- Adr. p.
- Kal. p. v. p.
- Pist. hisp.
- Dudl. cym. ssp. c.
- Orost. erub. v. b
- Echev. p.
- Adr. p.
- C. pap. v. rob.
- Echev. p.
- Dudl. rig.
- — p. ssp. p.
- Echev. glauc. v. p.
- Dudl. cym. ssp. min.
- C. grand.
- Dudl. cym. ssp. c.
- Echev. qu.
- C. orb.
- — salm. v. s.
- — — v. w.
- Dudl. caesp.
- Echev. r.
- Adr. r. v. sphen.
- — hemisph.
- — r. v. r.
- — trig.
- Dudl. r.
- — r.
- — cym. ssp. min.

- C. roseata BAK.
- — rotundifolia HAW.
- — rubens BRANDEG.
- — rugens FEDDE sphalm.
- — rupestris SALISB.
- — rusbyi GREENE
- — salzmannii SCHOENL.

- — samium URW.
- — saxosa M. E. JONES
- — schaeferi DTR.
- — schaffneri S. WATS.

- — scheeri BAK.
- — schoenlandii PHILLIPS
- — schuldtiana v. POELLN.
- — secunda BAK.
- — semenowii O. et R. FEDTSCH.
- — semiteres (ROSE) FEDDE
- — sempervivum MARSCH. BIEB.
- — septentrionalis (ROSE) FEDDE
- — setchellii JEPS.
- — sheldonii (ROSE) FEDDE
- — sikokiana MAKINO
- — simensis BR.
- — spathulata C. B. CLARKE
- — speciosa HORT. ex v. ROEDER
- — spinosa L.
- — sprucei BAK.
- — spuria L.
- — stolonifera BAK.
- — stricta DIELS
- — strictiflora BAK.
- — subrigida ROBINS. et SEAT.
- — subspicata BAK.
- — subulifera BAK.
- — tardiflora BONPL.
- — tenuicaulis (ROSE) FEDDE
- — tenuis (ROSE) FEDDE
- — teretifolia v. subglaber HARV.
- — thraskiae (ROSE) FEDDE
- — triflora SALM
- — — THUNBG.
- — trigyna BURCH.
- — — BURTT-DAVY p. max. part.
- — — SCHOENL.
- — tuberculosa LAM.
- — tuberosa HAL.
- — — v. botryoides (HOCHST. ex A. RICH.) ENGL.
- — — v. patens (POMEL) BATT.
- — — — repens L.
- — — — tuberosa L.
- — — veneris L.
- — ungulata LAM.
- — virens (ROSE) FEDDE
- — virescens SCHOENL. et BAK. f.
- — virgata DIELS

- Echev. r.
- Adr. r.
- Dudl. r.
- — ing.
- Umbil. r. v. r.
- Graptop. r.
- Pist. brev. ssp. s.
- Ros. serr.
- Dudl. s.
- C. s.
- Echev. teret. v. s.
- — s.
- Adr. s.
- — s.
- Echev. s.

- Rhod. s.
- Dudl. × s.

- Ros. pest.

- Dudl. far.
- — cym. ssp. s.
- — — ssp. c.
- Met. s.
- Afrov. s.
- Sed. or.

- Adr. fest.
- Orost. s.
- Echev. s.
- C. pan.
- Echev. s.
- Vill. diels.
- Echev. s.
- — s.
- — bic. v. b.
- — teret. v. t.
- C. pan.
- Sed. aden.
- Dudl. abr. ssp. a.

- C. t.
- Adr. rhomb. v. sph.
- — t.
- — t.

- — umbr.
- — alst.
- — grand.
- Umb. rup. v. r.

- — —
- — pat.
- — er.
- — rupr.
- — er.
- C. dec. v. d.
- Dudl. v.

- C. macr. v. v.
- Vill. v.

## INVALID DESIGNATIONS Cotyledon-Crassula

C. viridis Haw.
— viscida S. Wats.
— — v. insularis Jepson
— weberbaueri Diels
— whitei Schoenl. et Bak. f.
— woodii Schoenl. et Bak.
— zeyheri Harv.
C. S.G. Echeveria (DC.) Benth. et Hook.
C. Sect. Echeveria (DC.) Schoenl.
C. Sect. Gibbiflorae Bak.
C. Sect. Paniculatae Schoenl.
C. Sect. Racemosae Bak.
C. Sect. Secundae Bak.
C. Sect. Spicatae Bak.
**Courantia** Lem. Crassulaceae.
C. echeverioides Lem.
— rosea Rose
— roseata Bak.
C. Lem.
C. (Lem.) Bgr.
**Crassula** L. Crassulaceae.
C. abyssinica A. Rich.
— aitonii Br. et R.
— albiflora Sims
— aliciae Hamet
— aloides N. E. Br.
— alpestre auct. non Thunbg.
— alstonii auct. non Marl.
— × andegavensis Boom
— andegavensis DC.
— anguinea Harv.
— argentea L. f.
— argyrophylla v. ramosa auct. non Schoenl.
— arta hort. et auct. plur. non Schoenl.
— aurosensis Dtr. nom. nud.
— avasmontana Dtr.
— barklyi N. E. Br.
— bibracteata Haw.
— biconvexa (Eckl. et Zeyh.) Harv.
— bolusii Hook. f.
— brevifolia auct. non Harv. p. part.
— — hort.
— — (Eckl. et Zeyh.) Schoenl. = C. brachyphylla Adams. in the disregarded Ser. I.
— caespitosa Cav.
— caffra L.
— canescens Haw.
— capitata hort.
— centauroides Harv.
— cephalophora v. dubia Schoenl.
— — v. tayloriae (Schoenl.) Schoenl.
— cernua N. E. Br.
— coccinea L.
— cooperi v. major hort.
— — v. robusta Schoenl.

C. macr. v. v.
Dudl. v.
— vir.
Vill. w.
C. cor.
— salm. v. w.
Adr. z.

Echev.
—
— § 18
s. Adr. § I; Cot. § I/2
Echev. § 4
— § 11
— § 3

Echev. ros.
— —
— —
— § 3
— § 3

C. alb.
— cord.
— deject.
Sinocr. a.
C. acinac.

— depend.
— namib.
— × pulv.
Sed. a.
C. lyc. v. l.
— port. ?

— ser.

— pleg.

— nod.
— ter.
— acutif.

— dew.
— coop. v. c. ?

— comm.; rupr.
— rupr.

Sed. rubr.
C. tetr.
— ceph.
Roch. od.
C. marg.

— cotyl.

— —
— rac.
Roch. c.
C. pict.
— —

C. 'corallina' hort.
— cordata Lodd.
— cornuta auct. non Schoenl. et Bak. f.
— corpusculariopsis Boom
— corymbulose auct. non Link
— cotyledon Jacq.
— cymosa Bergius
— — sensu Schoenl.
— deceptrix Schoenl.
— decipiens N. E. Br.
— decussata Salisb. nom. nud.
— deltoidea auct. non Harv. nec Thunbg.
— — Harv.
— deminuta Diels
— dewinteri Friedr. p. part.
— dichotoma L.
— divaricata Eckl. et Zeyh.
— dubia Schoenl.
— eendornensis Dtr. nom. nud.
— elata N. E. Br.
— enantiophylla Bak. f.
— ericoides hort.
— falx Ldgr. nom. inval.
— flavovirens Pill.
— fragilis Schoenl.
— fragillima Dtr.
— fruticulosa L.
— galpinii Schoenl.
— gracilis hort.
— guachabensis Merxm.
— harveyi Britt. et Bak. f.
— — v. dependens (Bol.) Schoenl.
— hispida Schoenl. et Bak. f.
— hofmeyeriana Dtr.
— hookeri hort.
— hybrida hort. Graeser
— imbricata Ait.
— impressa N. E. Br.
— indica Decne.
— interrupta v. glabrifolia (Harv.) Schoenl.
— involucrata Schoenl.
— jacobseniana v. Poelln.
— jasminea Sims
— johannis-winkleri Ldgr.
— laxa Schoenl.
— liquiritiodora Dtr. nom. nud.
— lucens Gram.
— lucida Lam.
— lycopodioides v. acuminata Jacobs.
— — v. variegata H. Hall
— maculata hort. ex v. Roeder
— magnolii DC.
— malladrae Chiov.
— mannii Hook. f.
— margaritifera (Eckl. et Zeyh.) Harv.
— marginata Thunbg.
— mariae Thunbg.
— maritima Schoenl.
— massonii Britt. et Bak. f.

C. das.
— spat.

— arta
— hum.
— thyrsifl.
— arb.
— subul.
— arenic.
— decept.
— tect.
— rob.

— plegm.
— art.
— setul.
— seric.
Vau. d.
C. muric.
— hum.
— glabrif.
— nod.
— —
— lyc. v. purp.
— falcat.
— brev.
— pub.
— mont. dr.
— tetr.
— peploid.
— schmidt. v. s.
— nod.
— depend.

— —
— puberul.
— aus.
— schmidt. v. s.
— × graes.
— lyc. v. l.
— schmidt. v. s.
Sinocr. i. v. i.

— glabrif.
C. lineolat.
— ericoid.
Roch. j.
C. falcat.
— depend.

— hum.
— port.
— spat.

— lyc. v. l.
— — v. v.

Adr. m.
Sed. rubr.
Hypag. ab.
C. alb.

— smuts.
— pelluc.
— nod.
— filic.
— alp.

| | | | |
|---|---|---|---|
| C. massonioides DIELS | C. comp. | C. sedioides MILL. | C. orbic. |
| — merxmuelleri FRIEDR. | — hott. | — selago DTR. | — transv. |
| — mesembrianthoides AUCT. non SCHOENL. et BAK. f. | — namib. | — septas THUNBG. — simiana HORT. | — capens. — cor. |
| — — SCHOENL. et BAK. f. | — hum. | — — SCHOENL. | — das. |
| — mesembryanthemoides DTR. et BGR. | — namib. | — sphaeritis HARV. — stachygera ECKL. et ZEYH. | — subul. — revers. |
| — milleriana BURTT-DAVY | — rubic. | — subulata HOOK. | — trans. |
| — monticola N. E. BR. | — rup. | — swaziensis SCHOENL. | — arg. v. s. |
| — montis-moltkei DTR. | — depend. | — tayloriae SCHOENL. | — cotyl. |
| — muscosa THUNBG. | — lyc. v. l. | — telephioides HAW. | — crenul. |
| — nitida SCHOENL. | — port. | — torquata BAK. | — — |
| — nivalis HARV. | — nem. | — trachysantha (ECKL. et ZEYH.) HARV. | — mesembr. |
| — obliqua AIT. | — argent. | | |
| — — ANDR. | — falcat. | — triebneri 'SCHOENL.' ex JACOBS. | — lued. |
| — obvallata THUNBG. | — deject. | | |
| — odoratissima ANDR. | Roch. od. | — undata HAW. | — dej. |
| — ovata (MILL.) DRUCE | C. portul. | — uniflora SCHOENL. | — filic. |
| — pachyphylla SCHOENL. | — cong. | — variabilis N. E. BR. | — alp. |
| — parvifolia E. A. BRUCE | — phytur. | — versicolor BURCH. | Roch. v. |
| — pearsonii SCHOENL. | — brevifol. | — verticillata L. nom. confus. | Sed. rubr. ? |
| — pectinata CONRATH | — nod. | — weissii N. E. BR. | C. flabell. |
| — pellucida L. | — marg. | — whyteana SCHOENL. | — nyik. |
| — pentandra v. phytura (MILDBR.) HEDB. | — phytur. | — yunnanensis FRANCH. **Cremophila** L. Crassulaceae. | Sinocr. y. |
| — perfoliata L. v. albiflora HARV. | — perfol. v. p. | C. nutans ROSE **Crocanthus** L. BOL. Mesembr.● | Sed. n. |
| — — v. miniata TOELKEN | — falcat. | C. croceus (JACQ.) L. BOL. | Mal. c. |
| — — SCOP. | — perfor. | — luteolus (HAW.) L. BOL. | — l. |
| — — SCHOENL. p. part. | — falc. | — purpureo-croceus (HAW.) L. BOL. | — croc. v. p. c. |
| — perfossa DC. | — perfor. | | |
| — — v. minor HORT. | — neal. | — thunbergii (HAW.) L. BOL. | — t. |
| — — LAM. | — perfor. | **Cryophytum** N. E. BR. Mesembr. ● | |
| — portulacaria L. | Port'caria a. | C. acuminatum L. BOL. | Mes. parv. |
| — profusa HOOK. f. | C. marg. | — aitonis (JACQ.) N. E. BR. | — a. |
| — pruinosa sensu HARV. | — pust. | — alatum (JACQ.) N. E. BR. | — a. |
| — pseudocolumnaris DTR. | — phlegm. | — angulatum (THUNBG.) SCHWANT. | — ait. |
| — pseudolycopodioides DTR. et SCHINZ | — lyc. v. p. | — arenarium N. E. BR. | — inach. |
| — punctata HORT. | — multic. | — aureum L. BOL. | — chrys. |
| — — L. | — arb. | — barklyi (N. E. BR.) N. E. BR. | — b. |
| — — AUCT. non LAM. | — rupr. | | |
| — purpusii HORT. | — lyc. v. p. | — bijliae N. E. BR. | — ait. forma ? |
| — quadrifida BAK. | — multic. | — burchellii N. E. BR. | Micr. sess. v. alb. |
| — radicans HARV. | — r. | | |
| — ramosa HARV. | — macow. v. m. | — calycinum L. BOL. | Mes. subteret. |
| — — THUNBG. | — subul. | — carinatum L. BOL. | — rubr. |
| — ramuliflora L. v. pulchella (HAW.) SCHOENL. | — revers. | — clandestinum (HAW.) N. E. BR. | — c. |
| — rapacea SCHOENL. nom. nud. | — mesembr. | — — L. BOL. — clavatum L. BOL. | s. C. conj. Euryst. c. |
| — rehmannii BAK. f. | — cotyled. | — cleistum L. BOL. | Mes. inorn. |
| — retropilosa BITTER sphalm. | — revers. | — conjectum (L. BOL.) N. E. BR. (C. clandestinum L. BOL.). - Insufficiently described species. | |
| — rhomboidea N. E. BR. | — delt. | | |
| — rubens L. | Sed. r. | | |
| — rubicunda HORT. | × C. pulv. | | |
| — rupestris minor HORT. | C. neal. | — crassifolium L. BOL. | — barkl. |
| — scabra v. minor SCHOENL. | — pruin. | — crassipes (L. BOL.) L. BOL. | — pachyp. |
| — scabrella HARV. | — — | — crystallinum (L.) N. E. BR. | — c. |
| — scheppigiana DIELS | — setul. | — — (L.) SCHWANT. | — — |
| — schlechteri SCHOENL. | — — | — dejagerae L. BOL. | — d. |
| — schoenlandii AUCT. JACOBS. Handb. I, 329 | — namib. | — excavatum L. BOL. — fenchelii (SCHINZ) N. E. BR. | — e. — guer. |
| — — JACOBS. Nat. C. S. Journ. **10,** 80 | — humil. | — framesii L. BOL. — — v. laxum L. BOL. | — intr. v. i. — — v. l. |
| — schweinfurthii DE WILD | — alba | — fulleri L. BOL. | — inach. |
| — sediformis SCHWEINF. | Sed. crass. | — galpinii L. BOL. | — g. |

**INVALID DESIGNATIONS** Cryophytum-Derenbergia

C. gibbosum N. E. Br.    Mes. nod.
— glaucum Dtr.    — barkl.
— grandiflorum (Eckl. et Zeyh.) N. E. Br.    — perl.
— — (Schinz) Dtr. et Schwant. ex Range    — guer.
— grandifolium (Schinz) Dtr. et Schwant.    — —
— — (Schinz) N. E. Br.    — —
— guerichianum (Pax) Schwant.    — —
— inachabense (Engl.) N. E. Br.    — i.
— intermedium L. Bol.    — cryoc.
— karroicum L. Bol.    — karr.
— latisepalum L. Bol.    — l.
— lineare L. Bol.    — l.
— longipapillatum (L. Bol.) L. Bol.    — ann.
— maxwellii L. Bol.    — macr.
— nanum N. E. Br.    — alat.
— neglectum N. E. Br.    probably Aridaria
— nelsoniae L. Bol.    Mes. n.
— nodiflorum (L.) L. Bol.    — n.
— parvum L. Bol.    — brev.
— paulum N. E. Br.    — p.
— pentagonum L. Bol.    — quin.
— planum L. Bol.    — louis.
— pusillum L. Bol.    — subtr.
— rogersii L. Bol.    — pauc.
— roseum L. Bol.    — rhod.
— salteri L. Bol.    — longip.
— sessiliflorum L. Bol.    — sedent.
— — (Ait.) N. E. Br.    Micr. s.
— setosum L. Bol.    Mes. s.
— squamulosum Dtr.    — barkl.
— — L. Bol.    — s.
— stenandrum L. Bol.    — s.
— suaveolens (L. Bol.) J. Ingram    Sphalm. s.
— subulatum L. Bol.    Mes. subrig.
— suffruticosum L. Bol.    Aridaria ?
— truncatum L. Bol.    Mes. purp.
— velutinum L. Bol.    — macr.
— wilmaniae L. Bol.    — ann.
**Curtogyne** Haw. Crassulaceae.
C. dejecta Jacq.    Crass. d.
— undulata DC.    — —
**Cylindrophyllum** Schwant. Mesembr. ●
C. dichroum (Rolfe) Schwant.    Rusch. d. v. d.
**Cynanchum** L. Asclepiadaceae.
C. aphyllum L.    Sarcost. vim.
— bojerianum (Decne.) Choux    C. dec.
— crispum Jacq.    Fock. c.
— humbertii Choux    C. amp.
— sarcostemmatoides K. Schum.    — aph.
**Dactylanthes** Haw. Euphorbiaceae.
D. anacantha Haw.    Euph. trid.
— globosa Haw.    — g.
— hamata Haw.    — h.
— patula Haw.    — ornit.
— tuberculata Haw.    — t.
**Dasylirion** Zucc. Agavaceae.
D. acrotrichum Bak.    D. a.

D. bigelovii Torr.    Nol. b.
— glaucum Corr.    D. glauc.
— gracile Planch.    — acr.
— juncifolium Hort.    — long.
— laxiflorum Bak.    — serr.
— quadrangulatum S. Wats.    — long.
— strictum MacBride    Beauc. s.
**Decabelone** Decne. Asclepiadaceae.
D. angolensis N. E. Br.    D. grand.
— sieberi Pfersd.    — eleg.
**Decanema** Decne. Asclepiadaceae.
D. bojeriana Decne.    Cyn. dec.
— grandiflorum Jum. et Perr.    Folots. g.
— luteifluens Jum. et Perr.    Cyn l.
**Decanemopsis** Cost. et Gall. Asclepiadaceae.
D. aphylla Cost. et Gall.    Sarcost. mad.
**Decodontia** Haw. Asclepiadaceae.
Synonym for Huernia R. Br.

**Deilanthe** N. E. Br. Mesembr. ●
D. peersii (L. Bol.) N. E. Br.    Al. p.
**Delosperma** N. E. Br. Mesembr. ●
D. bredai L. Bol. sphalm.    Dros. b.
— cooperi (Hook. f.) L. Bol. v. bicolor L. Bol.    D. c. f. b.
— — (Hook. f.) Schwant.    — — v. c.
— denticulatum L. Bol.    — jans.
— echinatum (Ait.) Schwant.    — pruin.
— ecklonis (Salm) L. Bol.    — e. v. e.
— gracile L. Bol. ex Journ. S.Afr. Bot. **XXVII**, 180 (non Notes on Mes. **III**, 319)    — gracill.
— hirtum (N. E. Br.) L. Bol.    — h.
— linneae L. Bol. ex Jacobsen sphalm.
— longii L. Bol.    — grat.
— macrorrhizum (Haw.) Schwant.    Mest. tub. v. m.
— napiforme (N. E. Br.) Schwant.    — macr.
— subincanum (Haw.) L. Bol.    D. s.
— testaceum (Haw.) L. Bol.    — t.
— tradescantioides v. lebomboense L. Bol.    — leb.
— tuberosum (L.) L. Bol.    Mest. t. v. t.
— — (L.) Schwant.    — — v. —
**Depacarpus** N. E. Br. Mesembr. ●
D. tinctus (L. Bol.) N. E. Br.    Meyer. t.
**Derenbergia** Schwant. Mesembr. ●
D. apiata (N. E. Br.) Schwant.    Con. a.
— augeiformis Schwant.    — grac.
— biloba (Marl.) Schwant.    — b.
— caulifera (N. E. Br.) Schwant.    — c.
— chauviniae (Schwant.) Schwant.    — grac.
— cryptopoda (Kensit) Schwant.    Gibb. c.
— densipuncta Tisch.    Con. d.
— elishae (N. E. Br.) Schwant.    — e.
— friedrichiae (Dtr.) Schwant.    Ophth. f.
— gracilistyla (L. Bol.) Schwant.    Con. g.

D. halenbergensis DTR. et
  SCHWANT. Con. h.
— nuciformis (HAW.)
  SCHWANT. Gibb. crypt.
— quaesita (N. E. BR.)
  SCHWANT. Con. q.
— saxetana (N. E. BR.)
  SCHWANT. — s. f. s.
— turrigera (N. E. BR.)
  SCHWANT. — t.
— velutina SCHWANT. — v.
D. (SCHWANT.) SCHWANT. Con. SG. II,
  Sect. 1
**Derenbergiella** SCHWANT. Mesembr. ●
D. luisae SCHWANT. Mes. perl.
**Desmidorchis** EHRENB. Asclepiadaceae.
D. acutangula DECNE. Car. retr.
— crenulata DECNE. — c.
— forskahlii DECNE. — cicatr.
— pauciflora DECNE. — p.
— quadrangula DECNE. — q.
— retrospiciens EHRENB. — r. v. r.
— umbellata DECNE. — u.
**Diamorpha** NUTT. Crassulaceae.
D. pusilla NUTT. D. cym.
— smallii BRITT. Sed. s.
**Dichaelia** SCHLTR. Asclepiadaceae.
D. cinerea SCHLTR. Brach. circ.
— filiformis SCHLTR. — —
— forcipata SCHLTR. — —
— galpinii SCHLTR. — —
— macra SCHLTR. — —
— microphylla S. MOORE — —
— natalensis SCHLTR. — sand.
— pallida SCHLTR. — circ.
— pygmaea SCHLTR. — p.
— undulata SCHLTR. — u.
— zeyheri SCHLTR. — z.
**Dicrocaulon** N. E. BR. Mesembr. ●
D. trichotomum (THUNBG.)
  N. E. BR. Sphalm. t.
**Didierea** H. BAILL. Didiereaceae.
D. ascendens DRAKE All. a.
— mirabilis H. BAILL. D. mad.
— procera DRAKE All. p.
**Dinteranthus** SCHWANT. Mesembr. ●
D. margaretae (SCHWANT.)
  SCHWANT. Lap. m.
— punctatus L. BOL. D. pub.
D. SG. Lapidaria (SCHWANT.)
  DTR. et SCHWANT. Lap.
**Diopogon** JORD. et FOURR. Crassulaceae.
D. allionii JORD. et FOURR. D. hirt. ssp. a.
— arenarius (KOCH) LEUTE — — ssp. a.
— — ssp. allioni (JORD. et
  FOURR.) LEUTE — — ssp. a.
— — ssp. pseudohirtus LEUTE — — ssp. p.
— austriacus JORD. et FOURR. — — ssp.
  h. f. a.
— globiferus (L.) LEUTE — — ssp. bor.
— stramineus JORD. et FOURR. — heuff. v. s.
**Dioscorea** L. Dioscoreaceae.
D. brevipes BURTT DAVY Test. sylv. v. b.
— elephantipes (L'HER.) ENGL. — el.
— elephatopus SPRENG. — —
— hederifolia GRISEB. — sylv.
— hemicrypta BURCH. — glauc.

D. macrostachys BENTH. Test. m.
— marlothii KNUTH — sylv. v. mult.
— montana DUR. et SCHINZ — —
— — v. duemmeri KNUTH — — v. pan.
— — v. glauca KNUTH — el. v. mont.
— — v. paniculata (DUEMM.)
  KNUTH — syl. v. p.
— rehmannii BAK. — — v. r.
— sylvatica ECKL. — —
— — v. brevipes (BURTT DAVY)
  BURKHILL — — v. b.
— — v. multiflora (MARL.)
  BURKHILL — — v. m.
— — v. paniculata (DUEMM.)
  BURKHILL — — v. p.
— — v. rehmannii (BAK.)
  BURKHILL — — v. r.
**Diotostemon** S. D. Crassulaceae.
D. clavifolia BGR. × Pachyv.
  cv. C.
— glauca HORT. × — cv. —
— hookeri S. D. Pachyph. h.
D. (S. D.) BGR. — Sect. III
**Diplosoma** SCHWANT. Mesembr. ●
D. leipoldtii f. dormiens L. BOL. D. l. dormant
  form
— luckhoffii (L. BOL.)
  SCHWANT. Maugh. l.
**Disphyma** N. E. BR. Mesembr. ●
D. australe (SOL.) N. E. BR. D. a.
**Dorotheanthus** SCHWANT. Mesembr. ●
D. copticus (JACQ.) L. BOL. D. apet.
— criniflorus (L. f.) SCHWANT. — bell.
— gramineus v. albus (HAW.)
  SCHWANT. — g. f. a.
— — v. roseus (HAW.)
  SCHWANT. — — f. r.
— luteus N. E. BR. — fl. sol.
**Dorstenia** PLUM. ex L. Moraceae.
D. braunii 'BALLY' ex JACOB-
  SEN sphalm.
— ophioglossoides BUREAU D. born. v. o.
— peltata ENGL. — — v. trop.
— telekii SCHWEINF. — — v. t.
— tropaeolifolia BUREAU — — v. t.
— unyika ENGL. et WARB. — wall. v. u.
**Dracophilus** SCHWANT. Mesembr. ●
D. rheolens (L. BOL.)
  SCHWANT. D. dealb.
**Drakebrockmania** WHITE et SLOANE. Asclepia-
  daceae.
D. crassa (N. E. BR.) WHITE
  et SLOANE Whitesl. c.
**Drepanostemma** JUM. et PERR. Asclepiadaceae.
D. luteum JUM. et PERR. Sarcost. dec.
**Drosanthemum** SCHWANT. Mesembr. ●
D. asperulum (SALM)
  SCHWANT. D. a.
— bredai L. BOL. — asp.
— capillare v. acutifolium
  (L. BOL.) SCHWANT. — acut.
— expersum (N. E. BR.)
  SCHWANT. — e. v. e.
— fourcadei (L. BOL.) L. BOL. — f.
— hermannii (PAX) SCHWANT.
  (Mes. h. PAX nom. nud.)
  Known only from a picture.

D. hispifolium (Haw.)
    Schwant. — D. striat. v. h.
— kolbei nom. nud. — Jacobs. k.
— luederitzii (Engl.)
    Schwant. — D. par.
— micans v. aureopurpureum
    L. Bol. — aureop.
— — v. atropurpureum L. Bol.
    ex Jacobsen sphalm.
— nitidum (Haw.) Schwant. — Sphalm. n.
— obliquum (Haw.) Schwant. — D. liqu.
— pageanum (L. Bol.)
    Schwant. — Del. p.
— pallens (Haw.) Schwant. — D. striat. v. p.
— pruinosum (Thunbg.)
    Schwant. — p.
— robustum L. Bol. — crass.
— subcompressum (Haw.)
    L. Bol. — s.
— tardum L. Bol. — inorn.
— vaginatum L. Bol. — Anis. v.

**Dudleya** Britt. et Rose. Crassulaceae.
D. aloides Rose — D. sax. v. a.
— angustiflora Rose — cym. v. c.
— arizonica Rose — pulv. v. a.
— attenuata ssp. typica
    Moran — att. v. a.
— bernardiana Britt. — cym. ssp. min.
— brandegei Rose — acum.
— brauntonii Rose — lanc.
— bryceae Britt. — cand.
— collomiae Rose — sax. ssp. c.
— compacta Rose — far.
— congesta Britt. — lanc.
— cotyledon (Jacq.) Br. et R. — caesp.
— delicata Rose — sax. v. al.
— eastwoodiae Rose — far.
— echeverioides D. A. Johansen — green.
— eximia D. A. Johansen — ing.
— gigantea Rose — cym. ssp. g.
— goldmanii Rose — — ssp. min.
— grandiflora Rose — sax. v. al.
— hallii Rose — lanc.
— helleri Rose — caesp.
— hoffmannii D. A. Johansen — green.
— humilis Rose — cym. ssp. setch.
— lagunensis (Munz) E.
    Walth. — pulv. v. ariz.
— laxa (Lindl.) Br. et R. — caesp.
— — auct. non (Lindl.) Br.
    et R.
— lingula (S. Wats.) Br. et R. — far.
— lurida Rose — lanc.
— minor Rose — cym. v. m.
— moranii D. A. Johansen — albifl. ?
— murina Eastw. — abr. ssp. m.
— nevadensis (S. Wats.) Br.
    et R. — cym. ssp. c.
— nudicaulis (Abrams)
    Moran — dens.
— ovatifolia Britt. — cym. ssp. o.
— paniculata (Jeps.) Br. et R. — cym. ssp. setch.

D. parishii Rose — D. lanc.
— plattiana (Jeps.) Br. et R. — cym. ssp. p.
— pulverulenta ssp. typica
    Moran — pulv. ssp. p.
— pumila Rose — cym. ssp. m.
— purpusii (K. Schum.) Br.
    et R. — — ssp. c.
— reflexa Britt. — lanc.
— regalis D. A. Johansen — green.
— robusta Britt. — lanc.
— rusbyi (Greene) Br. et R. — Graptop. r.
— septentrionalis Rose — D. far.
— setchellii (Jeps.) Br. et R. — cym. ssp. s.
— sheldonii Rose — — ssp. c.
— tenuis Rose — abr. ssp. a.
— viridicata D. A. Johansen — ing. ?
— xantii Rose — nubig.

**Duvalia** Haw. Asclepiadaceae.
D. andreaeana Rauh — Huern. a.
— dentata N. E. Br. — D. pol. v. p.
— hirtella Sweet — rad. v. h.
— — v. minor N. E. Br. — — v. m.
— — v. obscura N. E. Br. — — v. o.
— jacquiniana Sweet — el. v. e.
— mastodes Sweet — comp.
— procumbens R. A. Dyer — Huern. p.
— propinqua Bgr. — D. recl.
— replicata Sweet — rad. v. r.
— tanganyikensis Bruce et
    Bally — Huern. t.
— transvaalensis Schltr. — D. pol. v. t.
— — v. parviflora L. Bol. — — v. t.

**Dyckia** Schult. Bromeliaceae.
D. gigantea C. Koch — D. alt.
— laxiflora Mast.
— princeps Hort. — brev.
— — Lem. — alt.
— ramosa Hort.

**Eberlanzia** Schwant. Mesembr. ●
E. globularis (L. Bol.) L. Bol. — Rusch. fourc.
— hospitalis (Dtr.) Schwant.
    nom. nud.
— sedoides (Dtr. et Bgr.)
    Schwant. — s.

**Ebracteola** Dtr. et Schwant. Mesembr. ●
E. derenbergiana (Dtr.) Dtr.
    et Schwant. — Rusch. d.
— vallis-pacis Dtr. — E. cand.

× **Echephytum** Gossot. Crassulaceae.
× E. albo-mucronatum
    (Gossot) Jacobs. — × Pachyv. cv. A. M.
× — paradoxum (Gossot)
    Jacobs. — × — cv. P.
× — sempervivoides Gossot — × — cv. S.

**Echeveria** DC. Crassulaceae.
E. abramsii (Rose) Bgr. — Dudl. a.
— × acaulis Gossot — × Graptov. cv. A.
— acuminata Bgr. — Dudl. a.
— adunca (Bak.) Otto — Pachyph. hook.
— akontiophylla Werd. — E. alp.
— alba Hort. — albic.
— albida (Rose) Bgr. — Dudl. vir.
— albiflora Bgr. — — a.

# Echeveria INVALID DESIGNATIONS

E. albo-carinata HORT.
— aloides (ROSE) BGR.
— amadorana BGR.
— amethystina HORT. ex JACOBS.
— angusta v. POELLN.
— angustiflora (ROSE) BGR.
— anthonyi BGR.
— argentea LEM.
— arizonica HORT.
— — (ROSE) KEARN. et PEEB.
— attenuata PURP. nom. nud.
— — (WATS.) BGR.
— backebergii v. POELLN.
— bartramii (ROSE) KEARN. et PEEB.
— × bergeriana HORT.
— bernardiana (BRITT.) BGR.
— bernhardiana FOERST. et LEM.
— bicolor v. subspicata (BAK.) E. WALTH.
— bifurcata ROSE
— brachyantha SPRAG.
— bracteosa (LK., KLOTZSCH et OTTO) LINDL. et PAXT.
— brandegei (ROSE) BGR.
— brauntonii (ROSE) BGR.
— brevipes (ROSE) BGR.
— bryceae BGR.
— caerulescens GOSSOT
— caespitosa (HAW.) DC.
— californica BAK.
— calophana HORT. angl. ex LEM.
— × calva GOSSOT
— campanulata KUNZE
— candelabrum (ROSE) BGR.
— candida BGR.
— clavifolia BGR.
— — HORT. ex BERGER
— × clavifolia DEL. ex MORREN
— clavifolia HORT.
— — v. cristata HORT.
— collomiae (ROSE) KEARN. et PEEB.
— compacta BGR.
— congesta BGR.
— cooperi OTTO
— corderoyi MORR.
— coruscans HORT. ex v. ROEDER
— cotyledon (JACQ.) NELS. et MACBR.
— cultrata BGR.

× Pachyv. scheid. cv. Alb.
Dudl. sax. ssp. a.
— cym. ssp. gig.
Graptop. a.
E. subrig.
Dudl. cym. ssp. c.
— a.
— pulv. ssp. p.
Graptop. parag.
Dudl. pulv. v. a.
— a.
— —
E. chicl.
Graptop. b.
× Pachyv. cv. Sod.
Dudl. cym. ssp. min.
E. gibb. v. g.
— b. v. b.
— teret. v. b.
× Pachyv. cv. Cl.
Pachyph. b.
Dudl. ac.
— lanc.
— b.
— cand.
× Graptov. cv. C.
Dudl. c.
— —
E. acut.
× Graptov. cv. C.
E. grandifl.
Dudl. c.
— c.
× Pachyv. cv. C.
× — cv. —
× — cv. —
× — cv. Clavat.
× — cv. Cr.
Dudl. sax.
— far.
— lanc.
Adr. c.
E. ag. v. c.
Cot. c.
Dudl. caesp.
— c.

E. cuspidata J. A. PURP.
— cymosa LEM.
— × dasyphylla Hort. Bot. Copenhagen
— debilis NELS. et MACBR.
— delicata (ROSE) BGR.
— densiflora (ROSE) BGR.
— derenbergii major HORT.
— desmetiana L. DE SMET
— devensis (N. E. BR.) M. L. GREENE
— diaboli BGR.
— discolor L. DE SMET
— eastwoodiae BGR.
— edulis (NUTT.) BGR.
— — v. attenuata JEPS.
— elegans BGR.
— — ROSE v. kesselringiana v. POELLN.
— — v. simulans (ROSE) v. POELLN.
— elongata (ROSE) BGR.
— farinosa LINDL.
— farinulenta LEM.
— × fruticosa ROLL.
— × fucifera ROLL.
— × fusifera HORT.
— gibbiflora v. metallica HORT.
— — v. typica BGR.
— glauca v. tolucensis (ROSE) v. POELLN.
— × glauco-metallica HORT.
— globosa HORT. ex MORREN
— gloriosa ROSE
— goldmanii (ROSE) BGR.
— gormannii NELS. et MACBR.
— × gossotii ROWL.
— grandiflora (ROSE) BGR.
— grandis MORREN
— grayi BAK.
— greenei BGR.
— × haageana HORT. ex WALTH. nom. ill.
— hallii NELS. et MACBR.
— — (ROSE) NELS. et MACBR.
— harmsii v. multiflora E. WALTH.
— hassei (ROSE) BGR.
— × haworthioides GOSSOT
— helleri BGR.
— holwayi ROSE
— hookeri (S. D.) LEM.
— hoveyi ROSE
— ingens ROSE
— — (ROSE) BGR.
— insularis BGR.
— jepsonii NELS. et MACBR.
— lagunensis MUNZ

E. parr.
Dudl. c.

E. × deros.
Sed. d.
Dudl. sax. ssp. al.
— d.
— × deros.
E. peac.
— acut.
Dudl. cym. ssp. setch.
E. nod.
Dudl. far.
— e.
— — ssp. a.
E. harms.
— albic.
— sim.
Dudl. e.
— f.
— —
× Pachyv. cv. Gl.
× — cv. F.
× — cv. Sobr.
E. viol.
— gibb. v. g.
— tol.
— × imbr.
— glauc.
— rubr.
Dudl. cym. ssp. min.
Sed. lax.
E. × pseudo-lancifol.
Dudl. sax. ssp. al.
E. gibb. v. g.
— pan.
Dudl. g.
E. × graessn.
Sed. h.
Dudl. lanc.
E. long.
Dudl. h.
× Graptov. cv. H.
Dudl. caesp.
E. acut.
Pachyph. h.
E. 'Hoveyi'
— cuenc.
Dudl. i.
— vir.
— cym. ssp. setch.
— pulv. ssp. ar.

**INVALID DESIGNATIONS  Echeveria**

E. × lanceolata Gossot — E. × pseudo-lancif. — E. parishii (Rose) Bgr. — Dudl. lanc.
— lanceolata Nutt. — Dudl. l. — parva Bgr. — cym. ssp. min.
— — v. aloides (Rose) Munz — sax. ssp. a. — pauciflora (Rose) Bgr. — p.
— — v. compacta Jeps. — — ssp. — — perelegans Bgr. — E. eleg.
— — v. incerta Jeps. — — ssp. gig. — pfersdorfii Hort. ex Morren — stolon.
— — v. lurida (Rose) Munz — lanc. — planifolia Bgr. — Thomps. plat.
— — v. saxosa (M. E. Jones) Jeps. — sax. ssp. s. — plattiana (Jeps.) Nels. et Macbr. — Dudl. cym. ssp. c.
— laxa auct. non Lindl. — cym.
— — Lindl. — caesp. — pulverulenta Nutt. ex Torr. et Gray — pulv. ssp. p.
— — v. cymosa (Lem.) Purp. — cym. ssp. c.
— — v. minor (Rose) Jeps. — — ssp. m. — — v. arizonica (Rose) Clokey — — ssp. a.
— — v. nevadensis (S. Wats.) Jeps. — pumila Schltd. — E. glauc. v. p.
— — v. paniculata Jeps. — — ssp. c. — purpusii Britt. — micr.
— — v. setchellii Jeps. — — ssp. setch. — — (K. Schum.) Wittm. — Dudl. cym. ssp. c.
— linearis (Greene) Bgr. — — ssp. —
— lingula (S. Wats.) Nels. et Macbr. — l. — pusilla Bgr. — E. amoen.
— longifolia Hort. Herb. Kew — far. — quitensis v. gracilior Sod. — qu.
— lucida Steud. — E. cocc. — — v. sprucei (Bak.) v. Poelln. — spruc.
— lurida Haw. — rac. v. l. — reflexa (Britt.) Bgr. — Dudl. lanc.
— magnifica Hort. — — v. r. — retusa Lindl. — E. fulg.
— × Pachyv. cv. Cl. — rhombifolia Otto — Adr. r. v. r.
— × metallica Hort. — E. cv. Met. — rigida (Rose) Bgr. — Dudl. r.
— metallica Lem. — gibb. v. m. — rigidiflora (Rose) Bgr. — r.
— minor (Rose) Bgr. — Dudl. cym. v. m — robusta (Britt.) Bgr. — lanc.
— minutiflora (Br. et R.) Rose — Thomps. m. — rosei Nels. et Macbr. — E. subrig.
— minutifoliolata v. Poelln. — Graptop. pach. — rubens (Brandeg.) Bgr. — Dudl. r.
— × mirabile Del. — × Pachyv. cv. M. — rusbyi (Greene) Nels. et Macbr. — Graptop. r.
— misteca L. — E. nod. — sanguinea Morr. — E. atr.
— monicae Bgr. — Dudl. lanc. — saxosa (M. E. Jones) Nels. et Macbr. — Dudl. s.
— × morreniana Del. — × Pachyv. m. — × scaphophylla Bgr. — E. × scaph.
— navicularis L. — E. nud. — schaffneri Rose — panic.
— neglecta v. Poelln. — chicl. — — (S. Wats.) E. Walth. — teret. v. s.
— nevadensis (S. Wats.) Nels. et Macbr. — Dudl. cym. — × scheideckeri de Smet — × Pachyv. scheid. cv. Sch.
— nobilis Hort. — × Pachyv. cv. Clav. — × striata Haage jr. — × — cv. Alboc.
— nuda v. montana (Rose) v. Poelln. — E. mont. — scopulorum Rose — E. obt.
— nudicaulis (Abrams) Munz — Dudl. dens. — secunda v. byrnesii Rose — byrn.
— obscura (Rose) Bgr. — E. ag. v. a. — — v. glauca Otto — glauc.
— obtusata Nels. et Macbr. — Sed. o. — — v. pumila Morr. — — v. p.
— obtusifolia v. scopulorum (Rose) v. Poelln. — E. o. — — v. — Otto — — v. —
— orcuttii Rose — Dudl. att. v. o. — semiteres (Rose) Bgr. — Dudl. × s.
— oregana Nels. et Macbr. — Sed. o. — septentrionalis (Rose) Bgr. — far.
— ovatifolia (Britt.) Bgr. — Dudl. cym. ssp. c. — sessiliflora v. pinetorum (Rose) v. Poelln. — E. pin.
— pachyphytoides L. de Smet ex Morren — × Pachyv. cv. P. — setchellii (Jeps.) Nels. et Macbr. — Dudl. cym. ssp. s.
— pachyphytum (Bak.) Morr. — Pachyph. bract. — sheldonii Bgr. — cym. ssp. c.
— palensis Bgr. — Dudl. att. v. orc. — × sobrina Bgr. — × Pachyv. cv. S.
— palmeri (S. Wats.) Nels. et Macbr. — — × sodalis Bgr. — × — cv. S.
— — Rose — p. — × spathulata Del. ex Morren — × — cv. S.
— × paradoxa Gossot — E. subrig. — spathulifolia L. de Smet — Sed. s.
— × Pachyv. cv. P. — spilota Kunze — E. sec.
— paraguayensis Hort. ex v. Poelln. — — × spiralis Del. — × Dudlev. s.
— × splendens Haage jr. — s. E. fulg.
— Graptop. p. — sprucei (Bak.) Bgr. — E. s.

## Echeveria-Euphorbia INVALID DESIGNATIONS

E. sturmiana v. POELLN.
— subsessilis ROSE
— subspicata (BAK.) BGR.
— subulifolia MORR.
— tenuis (ROSE) BGR.

— tepeacensis v. POELLN.
— × titubans GOSSOT

— traskiae (ROSE) BGR.
— trianthina ROSE
— × undulata HORT.
— venezuelensis ROSE
— virens (ROSE) BGR.
— viridiflora ROSE
— viscida (S. WATS.) BGR.
— — v. insularis JEPS.
— watsonii NELS. et MACBR.
— weinbergii HORT. ex ROSE
— whitei ROSE
— xantii (ROSE) BGR.
— yuccoides MORR.
— zahnii cv. Hoveyi
E. Ser. Amoenae E. WALTH.
E. Ser. Australes E. WALTH.
E. Ser. Bracteolatae E. WALTH.
E. § Clavatae E. WALTH.
E. Ser. Elatae E. WALTH.
E. Ser. Grandes E. WALTH.
E. Ser. Retusae E. WALTH.
E. Ser. Urbiniae E. WALTH.
E. Ser. Vestitae E. WALTH.
E. 1. Sect. Oliveranthus (ROSE) BGR.
E. 2. Sect. Euecheveria BGR. p. part.
E. 3. Sect. Urbinia (BR. et R.) BGR.
E. 4. Sect. Dudleya (BR. et R.) BGR.
E. 5. Sect. Stylophyllum (BR. et R.) BGR.
E. 6. Sect. Courantia (LEM.) BGR.
E. 7. Sect. Thompsonella (BR. et R.) BGR.
E. (DC.) BENTH. et HOOK. as SG. of the Genus Cotyledon L.
E. (DC.) SCHOENL. as Sect. of the Genus Cotyledon L.
**Echidnopsis** HOOK. Asclepiadaceae.
E. cylindrica K. SCHUM.
— dammanniana SCHWEINF.
— — SPRENG. v. brunnea DAMM.
— golathii SCHWEINF.
— quadrangula DEFL.
— somalensis N. E. BR.
— tessellata K. SCHUM.
**Echinothamnus** ENGL. Passifloraceae.
E. pechuelii ENGL.
**Echinus** L. BOL. non LOUR. Mesembr. ●
E. apiculatus (KENSIT) L. BOL. Brauns. a.
— edentulus (HAW.) N. E. BR. Rusch. e.
— geminatus (HAW.) L. BOL. Brauns. g.
— mathewsii N. E. BR.

E. nod.
— peac.
— bicol. v. b.
— teret. v. t.
Dudl. abr. ssp. a.
Thomps. min.
× Graptov. cv. T.

Dudl. t.
E. bifid.
— cv. Car.
— bic. v. b.
Dudl. v.
E. heter.
Dudl. v.
— vir.
Sed. w.
Graptov. par.
E. chil.
Dudl. nub.
E. ag. v. a.
— cv. H.
E. § 7
E. § 2
E. § 2
E. § 15
E. § 2
E. § 18
E. § 18
E. § 9
E. § 1

E. § 1

E. § 3, 4, 7, 11

E. § 8

Dudl. SG. 2

Dudl. SG. 1

E. § 3

Thomps.

E.

E.

E. cer. v. obsc.
— nub.

E. cer. v. b.
Car. pen. v. p.
— qu.
E. damm.
— cer. v. c.

Adenia p.

— —

E. maximiliani (SCHLTR. et BGR.) N. E. BR.
— nelii (SCHWANT.) SCHWANT.
— stayneri L. BOL.
— varensburgii L. BOL.
**Edithcolea** N. E. BR. Asclepiadaceae.
E. sordida N. E. BR.
**Elaphrium** JACQ. Burseraceae.
E. hindsianum BENTH.
— microphyllum ROSE
**Erepsia** N. E. BR. Mesembr. ●
E. anceps (HAW.) L. BOL.
— aspera (HAW.) SCHWANT.
— bracteata (AIT.) L. BOL.
— compressa (HAW.) L. BOL.
— coralliflora S. D.
— cyathiformis (L. BOL.) SCHWANT.
— dregeana (SOND.) SCHWANT.
— forficata (L.) SCHWANT.
— gracilis (HAW.) SCHWANT.
— haworthii (DON) SCHWANT.
— montana (SCHLTR.) SCHWANT.
— muiri L. BOL.
— mutabilis (HAW.) L. BOL.
— patula (HAW.) L. BOL.
— pentagona (L. BOL.) SCHWANT.
— polita (L. BOL.) SCHWANT.
— purpureostyla (L. BOL.) SCHWANT.
— radiata (HAW.) L. BOL.
— restiophila L. BOL.
— stipulacea (L.) SCHWANT.
— stokoei L. BOL.
— viridis (HAW.) L. BOL.
**Eupedilanthus** BENTH. Euphorbiaceae.
E. macrocarpus BENTH.
**Euphorbia** L. Euphorbiaceae.
E. abyssinica v. mozambiquensis BOISS.
— — v. tetragona SCHWEINF.
— adenensis DEFL.
— alcicornis HORT. non BAK.
— alluaudii DRAKE
— ammak SCHWEINF.
— anacantha HAW.
— anacampseroides LAM.

— angularis sensu BOUGHEY
— — sensu ENGLER
— — sensu EYLES
— — sensu JACOBSEN p. part.
— — sensu N. E. BR. p. part.
— — sensu PAX p. part.
— — sensu WHITE, DYER et SLOANE p. part.
— antankara J. LEANDRI
— × anticaffra LOTSY et GODD.
— antiquorum E. MEY.
— arborescens E. MEY.
— — S. D.
— armata THUNBG.
— atrispina N. E. BR.

Lampr. m.
Brauns. n.
— st.
— v.

E. grand. v. g.

Burs. h.
— m.

E. a.
— a.
— b.
— c.
Lampr. c.

— c.
— d.
Rusch. f.
E. g.
Lampr. h.

E. het.
— tub.
— m.
— p.

— p.
— p.

Rusch. forf.
E. r.
— r.
Lampr. st.
E. st.
Smicr. v.

Pedil. m.

E. ang.
— acr.
— bals. ssp. a.
— ramipr.
— leucod.
— candel. v. c.
— trid.
Ped. tith. ssp. pad.
E. coop. v. cal.
— — v. c.
— v. —
— lem.
— —
— coop. v. uss.

— — v. c.; lem.
— pach.

— × both. v. a.
— hamat.
Synad. cup.
E. grandid.
— loric.
— stellaesp. v. a.

**INVALID DESIGNATIONS Euphorbia**

E. — v. viridis White, Dyer et Sloane
— barteri N. E. Br.
— basutica Marl.
— baumii Pax
— beaumeriana Hook. f. et Coss.
— beillei Chev.
— bellica Hiern.
— bengulensis Pax
— bergeriana Dtr.
— bevilaniensis L. Croiz.
— biglandulosa Willd.
— bojeri Hook.
— brachiata Boiss. p. part.
— — E. Mey. p. part.
— bracteata Jacq.
— breoni L. Nois.
— brunellii Chiov.
— bupleurifolia E. Mey.
— burmannii v. karrooensis Boiss.
— bussei Pax
— cactus sensu Schweinf. p. part.
— canaliculata Lam.
— — Lodd.

— canariensis Forsk.
— — Thunbg. ?
— candelabrum Welw.
— captiosa N. E. Br.
— caput-medusae E. Mey.
— — Lam.
— — L. p. part.
— — v. delta L.
— carinata Lodd.

— cereiformis Hort.
— — L. v. submammillaris Bgr.
— — K. Schum.
— cervicornis Boiss.
— ciliolata Pax
— cirsioides Cost. et Gall.
— clava E. Mey.
— clavata Salisb.
— commelinii DC.
— commiphora Dtr.
— cooperi Bgr.
— — sensu Cannon
— — Hutchins.
— coronata Thunbg.
— crassipes Marl. p. part.
— crispata Lem.
— cupularis Boiss.
— cynanchoides Drake
— decaryana Croiz.
— decussata Boiss. p. part.
— × dentonii Hort.
— dinteri Bgr.
— discreta N. E. Br.
— divaricata Jacq.
— drupifera Tronn. ex Schum.
— — v. elastica (Poiss.) Chev.
— echinata S. D.

E. stellaesp. v. v.
— kamer.
— clavar. v. c.
— mont.

— off. v. b.
Elaeoph. b.
E. viros. ssp. v.
— trich.
— gariep.
— mil. v. b.
— burm.
— mil. v. m.
— muric.
— perp.
Pedil. b.
E. mil. v. b.
— rub. v. b.
— oxyst.

— karr.
— coop. v. uss.

— ang.
— clava
Pedil. tith. ssp. t.
E. parc.
— led. v. dreg.
— consp.
— fer.
— dec.
— berg.
— clav.
— pugn.
Pedil. tith. ssp. t.
E. subm.

— —
— fimbr.
— ham.
— transv.
— stenocl.
— bub.
— clava
— cap.-m.
— guer.
— candel. v. c.
— —
— conf.
— clav.
— albert.
— lem.
Synad. c.
E. orth.
— hed.
— mund.
— 'W. D.'
— viros. ssp. v.
— wood.
— dendr.

Elaeoph. d. v. d.
— — v. e.
E. cereif.

E. eendornensis Dtr.
— elastica Marl.
— — Poiss.

— ellenbeckii Chiov.
— elliptica Thunbg.
— — v. undulata Boiss.
— engleriana Dtr. nom. nud.
— enneagona Bgr.
— — Haw.
— enopla Bgr.
— — v. dentata Bgr.
— enterophora Drake
— ephredroides Boiss.
— erosa Bgr.
— — Willd.
— falsa N. E. Br.
— fasciculata N. E. Br.
— fimbriata Hort. ex Lem.
— — N. E. Br.
— fleckii Pax nom. nud. ?
— fournieri Reb. hort. Lyon
— fraterna N. E. Br. p. part.
— frickiana N. E. Br.
— fructuspina Sweet
— fructus-pini Mill.
— — Sweet p. part.
— fusca Phil.
— galpinii Pax
— gardenifolia Hort.
— gilbertii Bgr.
— glomerata Hort. ex Bgr.
— goetzei Pax
— gossweileri Pax
— grandicornis sensu Berger p. part.
— — sensu N. E. Br. p. part.
— grandidens Adlam
— — sensu Burtt Davy
— — sensu Goebel
— — Sim.
— grandifolia Haw.

— grandis Lem.
— griseola v. robusta Pax ex Engl. nom. nud.
— gynophora Pax
— habanensis Hort.
— hastisquamata N. E. Br.
— heliocothela Hort. gall.
— — Lem.
— heptagona Bgr.
— hermentiana Lem. sphalm.
— hernandez-pachecoi Caballero
— heteracantha Pax
— heteropoda Pax
— hislopii N. E. Br.
— huttonae N. E. Br.
— hystrix Jacq.
— — Marl.
— impervia Bgr.
— indecora N. E. Br.
— inermis v. laniglans N. E. Br.

E. fusc.
— dreg.
Elaeoph. drup. v. e.
Monad. stell.
E. silen.
— crisp.

— aggr.
— fimbr.
— hept. v. h.
— — v. d.
— xyloph.
— rect.
— fimbr.
— cereif.
— mel.
— schoenl.
— lem.
— nesem.

— leucon.
— dek.; strang.
— pseudogl.
— berg.
— cap.-m.
— berg.
— dus.
— transv.
— und.
— micr.
— glob.
— transv.
— trichad.

— lem.
— brev.
— candel. v. c.
— coop. v. c.
— grandic.
— triang.
Elaeoph. drup. v. d.
E. abyss.

— gris.
— esp.
s. E. lactea
— caterv.
— und.
— niv.
— pent.
— trig.

— ech. v. h. p.
— subs. v. s.
Monad. h.
E. mil. v. h.
— inerm. v. h.
— lor.
— eust.
— heter.
— decuss.

— esc.

Euphorbia  INVALID DESIGNATIONS

- E. infausta N. E. Br.
- — infesta Pax
- — ingens E. Mey. ex Boiss.
- — insulae-europae Pax
- — isalensis Leandri
- — kalaharica Marl.
- — karasmontana Dtr. nom. nud.
- — latimammillaris L. Croiz. p. part.
- — leandriana P. Boit.
- — ledinii N. E. Br.
- — leonensis N. E. Br.
- — lohanensis H. Bn.
- — lombardensis Nel
- — longibracteata Pax
- — madagascariensis Comm.
- — magnidens Haw. ex S.D.
- — mainiana H. Poiss.
- — maintyi Decorse
- — mammillaris L. v. spinosior Bgr.
- — — Haw. p. part.
- — mammillosa Lem.
- — marientali Dtr. ex Range
- — marlothii N. E. Br.
- — — Pax
- — medusae Thunbg.
- — meloniformis Link
- — meyeri Nel
- — mogadorensis hort. germ.
- — morinii Bgr.
- — mundii R. A. Dyer
- — napoides Pax
- — natalensis hort. ex Bgr.
- — neumannii hort.
- — nykiae sensu Pax
- — — sensu Reynolds
- — — Werth.
- — oblongicaulis Bak.
- — odontophylla Willd.
- — officinarum v. arboreum Forsk.
- — oxystegia Bak.
- — parvimamma Bgr.
- — passa N. E. Br.
- — patula Sweet
- — pentagona Royle
- — pfersdorfii hort.
- — pimeleodendron Pax
- — pistaefolia Boiss.
- — platymammillaris L. Croiz.
- — polyacantha sensu Hiern. p. part.
- — polygona Mars. et Scop.
- — pomiformis Thunbg.
- — procumbens Meerb.
- — — Mill.
- — — N. E. Br.
- — — Sweet
- — pugniformis Bak.
- — pyrifolia Lam. p. part.
- — racemosa E. Mey.
- — radiata Boiss.
- — — E. Mey.
- — — Thunbg.

- E. mel.
- — triac. v. t.
- — candel.
- — sten.
- — isal.
- — avasm. v. a.

- — avasmontana v. a.
- — fimbr.; nes.
- — bup.
- — led. v. dreg.
- Elaeoph. l.
- E. orth.
- — micr.
- — mont.
- — leuc.
- — grandid.
- — isal.
- — int. v. m.

- — fimbr.
- — —
- — squarr.
- — m.
- — m.
- — mont.
- — cap. m.
- — mel.
- — nel.
- — resin. v. r.
- — hept. v. h.
- — caterv.
- — hadr.
- — candel. v. c.
- — mil. v. br.
- — coop. v. uss.
- — halip.
- — lem.
- — hadr.
- — cereif.

- — candel.
- — bub.
- — berg.
- — wood.
- — ornith.
- — royl.
- — subm.
- — robecc.
- — eckl.
- — fimbr.

- — dek.; strang.
- — fimbr.
- — melof.
- — stellat.
- — pugn. ?
- — wood.
- — pugn.
- — wood.
- — comm.
- — rhomb.
- — clava
- — rest.
- — stellata

- E. regis-jubae Webb et Berth.
- — rhipsaloides Glaz. nom. nud.
- — — Welw.
- — rhombifolia v. cymosa, v. laxa et v. triceps N. E. Br.
- — rubella v. exigua Bally sphalm.
- — ruspolii Chiov.
- — sagittaria Marl.
- — san salvador hort.
- — schizoclada H. Baill. is a Didierea (acc. Denis)
- — schubei Pax
- — scolopendria Donn.
- — — Haw.
- — scopoliana Steudel
- — serpentina hort.
- — similis Bgr.
- — spec. 1 White
- — splendens Boj. ex Hook.
- — — v. betsileana Leandri
- — — v. bevilanensis (L. Croiz.) Leandri
- — — v. — f. rubro-striata Drake et Castillo
- — — v. bojeri Cost. et Gall.
- — — ssp. — M. Denis
- — — v. breoni (L. Nois.)
- — — v. hislopii (N. E. Br.) Leandri
- — — v. imperatae Leandri
- — — v. mainiana (H. Poiss.) Leandri
- — — v. typica Leandri
- — — v. — f. platyacantha Leandri
- — — v. tananarivae Leandri
- — — v. vulcanii Leandri
- — stapelioides Herre
- — stapfii Bgr.
- — stegmatica Nel
- — strangulata sensu Hutchinson
- — stuhlmannii Schweinf. ex Volkens
- — subapoda H. Baill.
- — subfalcata Hiern.
- — synadenia H. Baill.
- — taitensis Pax non Boiss.
- — tessellata Sweet
- — tetracantha Pax
- — tetragona Bak.
- — — Sim. p. part.
- — tirucallii Thunbg. p. part.

- — tithymaloides L.
- — — v. myrtifolia L.
- — — β padifolia L.
- — triacantha Ehrenb.
- — trigona Roxb.
- — uncinata DC.
- — ussanguensis N. E. Br.
- — varians Haw.

- E. obt. v. r. j.

- — tir. v. r.
- — —

- — rhomb.

- — r. v. brun.
- — rob.
- — sag. v. s.
- — res. v. r.

- Monad. s.
- E. stellata
- — —
- — fimbr.
- — inerm. v. i.
- — candel. v. c.
- — coop. v. cal.
- — mil. v. s.
- — — v. b.

- — — v. b.
- — — v. — f. r. st.
- — — v. b.
- — — v. —

- — — v. b.

- — — v. h.
- — — v. i.

- — isal.
- — mil. v. spl.

- — — v. — f. p.

- — — v. t.
- — — v. v.
- — herr.
- — heter.
- — oxyst.

- — coop. v. uss.

- — heter.
- — prim.
- — trichad.
- Synad. cup.
- E. ndur.
- — cap. m.
- — nyass.
- — gilb.
- — pent.
- — maur.; arc.; burm.; decuss.
- Pedil. t. ssp. t.
- — — ssp. —
- — — ssp. —
- E. t. v. t.
- — barnh.
- — stellata
- — coop. v. u.
- — nivul.

## INVALID DESIGNATIONS  Euphorbia-Gibbaeum                                                    622

E. vepretorum DRAKE
— viminalis BURM. f.
— — L.
— — MILL.
— viperina BGR.
— virosa v. coerulescens BGR.
— winkleri PAX
— xanthadenia DENIS
**Faucaria** SCHWANT. Mesembr. ●
F. armstrongii nom. nud.
— haagei TISCH.
— kendrewensis L. BOL.
— tigrina superba HORT.
**Fenestraria** N. E. BR. Mesembr. ●
F. rhopalophylla (SCHLTR. et
  DIELS) N. E. BR.
**Ficoidaceae** JUSS., F. JUSS. emend. HUTCHINSON, F. JUSS. emend. ROWL., Ficoidea MIRB., Ficoides HERMANN have all been superseded. Mesembr. ●
**Fockea** ENDL. Asclepiadaceae.
F. capensis ENDL.
— glabra DECNE.
**Fouquieria** H. B. et K. Fouquieriaceae.
F. columnaris KELLOG
— gigantea ORCUTT
— peninsularis NASH
— spinosa H.B. et K.
— — TORREY
**Fourcroya** SCHULT. Agavaceae. Furcraea VENT.
**Frerea** DALZIEL. Asclepiadaceae.
F. indica DALZIEL
F. (DALZIEL) ROWL.
**Furcraea** VENT. Agavaceae.
F. commelinii SALM
— cubensis v. lindenii HORT.
— flavoviridis HOOK.
— foetida HAW.
— gigantea BOLDINGER
— hexapetala URBAN
— interrupta HORT.
— lindenii JACOBI
— madagascariensis HAW.
— pubescens BAK.
— selloa v. edentata TREL.
— spinosa TARZ.–TOZZ.
— variegata HORT.
— watsoniana HORT.
**Gasoul** ADANS. Mesembr. ●
G. aitonis (JACQ.) EICHL.
— crystallinum (L.) ROTHM.
— nodiflorum (L.) ROTHM.
× **Gasteraloe** GUILL. Liliaceae.
× — pfrimmeri GUILL.
**Gasteria** DUVAL. Liliaceae.
G. acinacifolia v. patula HAW.
— — v. pluripunctata BAK.
— angulata HAW.
— antandroi R. DECARY
— bayfieldii (S.D.) BAK.
— bijliae v. POELLN.
— carinata ROWL.
— crassifolia HAW.
— denticulata HAW.
— disticha v. angulata BAK.
— — v. angustifolia BAK.

E. orth. ssp. v.
— burm.
Sarcost. v.
E. tir. v. t.
— inerm. v. i.
— coer.
— obov.
— mah. v. x.



F. bossch. v. h.
— alb.
— tigr.



F. aur. f. r.









F. crisp.
— ed.




Idr. c.
— —
F. dig.
— fasc.
— spl.



Car. fr.
Car. Sect. I


F. gig. v. will.
— sell. v. marg.
— — v. s.
— gig. v. g.
Ag. v. grol.
F. cub.
— tub.
— sell. v. marg.
— gig. v. g.
— und.
— s. v. marg.
— tub.
— gig. v. m.-p.
— — v. — —


Mes. a.
— c.
— n.

× Gastrol. p.


G. a. v. nit.
— — v. acin.
— sulc.
Al. a.
× Gast haw. b.
G. parv.
— bates.
— nigr. v. c.
— dist. v. dist.
— angul. v. a.
— angust.

G. — v. conspurcata BAK.
— — v. major HAW.
— ensifolia HAW.
— excelsa BAK.
— formosa HAW.
— × holtzei (RADL) BGR.
— intermedia HAW.
— — v. asperrima HAW.
— — v. repens SALM
— laetepunctata BAK.
— laevis HAW.
— latifolia BAK.
— — HAW.
— lingua BGR.
— — v. angulata HAW.
— linita HAW.
— longifolia HAW.
— minima HORT.
— natalensis BAK.
— nigricans v. guttata BAK.
— — v. platyphylla BAK.
— — v. polyspila BAK.
— — v. subnigricans BAK.
— nitens HAW.
— obliqua HAW.
— — JACQ.
— parva HAW.
— × peacockii BAK.
— × pethamensis BAK.
— pluripunctata HAW.
— pulchra PHILLIPS
— striata HAW.
— strigata HAW.
— subnigricans HAW.
— — v. canaliculata (SALM)
  BGR.
— — v. torta BAK.
— subverrucosa v. grandi-
  punctata HAW.
— — v. marginata HORT.
— trigona v. elongata HAW.
— — v. minor HAW.
— venusta HAW.
— verrucosa v. scaberrima
  (S.D.) BAK.
— vroomii HORT. nom. nud.
× **Gastrolea** E. WALTH. Liliaceae.
× G. chludowii (RADL)
  E. WALTH.

× — perfectior (BGR.)
  E. WALTH.
**Geranium** L. Geraniaceae.
G. aculeatum PAT.
— echinatum THUNBG.
— tetragonum L. f.
**Gerrardanthus** HARV. Cucurbitaceae.
G. grandiflorus COGN. v. lobatus
  COGN.
— portentosus NAUD. ex
  DURIEU
**Gibbaeum** HAW. Mesembr. ●
G. argenteum N. E. BR.
— comptonii f. L. BOL.
— digitiforme (THUNBG.)
  N. E. BR.

G. consp.
— d. v. lat.
— ac. v. e.
— fuscop.
— pict. v. f.
× Gast'haw. h.
G. verr. v. i.
— — v. a.
— rep.
— laetip.
— ang. v. l.
— mac. v. mac.
— excav.
— dist. v. a.
— sulc.
— cand.
— ang. v. a.
— lilip.
— crouch.
— pseud. v. gl.
— mac. v. mac.
— fasc. v. p.
— pseud. v. ps.
— ac. v. n.
— mac. v. mac.
— nigr.
— parvif.
× Gastrol. p.
× — p.
G. ac. v. ac.
— lilip.
— subc. v. s.
— car. v. s.
— pseudon.

— — v. c.
— dict.


— subv. v. s.
— bates.
— trig.
— —
— ac. v. v.

— verr. v. asp.
— lutz.



× G. beg. nm.
  ch.

× — — nm. p.

Pelarg. ech. v. e.
— — v. —
— t.


G. lob.

— macr.

G. pub.
— c.

Dact. d.

G. dubium (N. E. Br.)
  Jacobs.                                   G. heath.
— fissoides (Haw.) Nel                    Anteg. f.
— haagei Schwant. ex
  Jacobsen                                 G. tischl.
— helmiae L. Bol.                         — crypt.
— intermedium L. Bol. nom.
  nud.                                    — esterh.?
— marlothii N. E. Br.                     — gibb.
— molle N. E. Br.                         — crypt.
— muiri N. E. Br.                         — gibb.
— — (N. E. Br.) Schwant.                  — schwant.
— muirioides Herre MS.                    Muir. Gibb. m.
— nebrownii Tisch.                        Imit. m.
— nelii Schwant.                          Anteg. fiss.
— perviride (Haw.) N. E. Br.              G. gibb.
— — v. luteoviride N. E. Br.              — lut.
**Globulea** Haw. Crassulaceae.
G. canescens Haw.                         Crass. ceph.
— clavifolia E. Mey.                      — c.
— hispida Haw.                            — h.
— mesembryanthoides Haw.                  — m.
**Glottiphyllum** N. E. Br. Mesembr. ●
G. cultratum (S.D.) N. E. Br.             G. lat. v. c.
— longum v. hamatum N. E.
  Br.                                     — l. v. het.
— obliquum (Willd.) N. E.
  Br.                                     — — v. l.
— pustulatum (Haw.) N. E.
  Br.                                     — long. v. l.
**Gonostemon** Haw. Asclepiadaceae.
G. divaricata Haw.                        Stap. d.
— gordonii Sweet                          Hood. g.
— strictum Haw.                           Stap. s.
G. (Haw.) White et Sloane                 Stap. Sect. V.
**Gormania** Britt. Crassulaceae.
G. burhamii Britt.                        Sed. b.
— debilis Britt.                          — d.
— eastwoodiae Britt.                      — e.
— glandulifera (L. F. Henders.)
  Abrams                                  — mor.
— hallii Britt.                           — h.
— laxa Britt.                             — l.
— obtusata Britt.                         — o.
— oregona Britt.                          — o.
— retusa Rose                             — sandh.
— watsonii Britt.                         — w.
**Grammanthes** DC. Crassulaceae.
G. chloraefolia (Haw.) DC.                Vau. dich.
— gentianoides (Haw.) DC.                 — —
**Graptopetalum** Rose. Crassulaceae.
G. byrnesia E. Walth.                     G. parag.
— goldii Matuda                           Sed. allant.
— mexicanum Matuda                        Thomps. min.
— orpetii E. Walth.                       G. rusb.
— weinbergii (Rose)
  E. Walth.                               — parag.
G. (Rose) Bgr.                            Sed. Sect. 7.
**Greenovia** Webb et Berth. Crassulaceae.
G. gracilis Bolle                         G. dodr.
— quadrantalis Webb                       — aiz.
**Guillauminia** A. Bertrand. Liliaceae.
G. albiflora (Guill.) Bertr.              Aloe a.
**Gymnopoma** N. E. Br. Mesembr. ●
G. tripolium (L.) N. E. Br.               Skiat. t.

**Gynicidium** Necker. Mesembr. ●
A genus-name which acc.
  G. Schwantes includes the
  complex of the SF. Mesembry-
  anthemoideae.
**Halenbergia** Dtr. Mesembr. ●
H. hypertrophica (Dtr.) Dtr.              Mes. h.
**Harpagophytum** DC. ex Meissn. Pedaliaceae.
H. pinnatifidum Engl.                     Pterod. spec.
**Hasseanthus** Rose. Crassulaceae.
H. blochmaniae (Eastw.)
  Rose                                    Dudl. b. ssp. b.
— — v. brevifolia Moran                   — — ssp. br.
— — v. insularis Moran                    — — ssp. i.
— elongatus Rose                          — mult.
— kessleri Davids.                        — blochm. ssp.
                                            b.
— multicaulis Rose                        — m.
— nesioticus Moran                        — n.
— variegatus (S. Wats.) Rose              — v.
— — v. blochmaniae Jeps.                  — bl. ssp. b.
— — v. elongatus (Rose) Jeps.             — multic.
H. (Rose) Moran                           Dudl. S. G. 3.
**Haworthia** Duval. Liliaceae.
H. affinis Bak.                           H. bilin. v. a.
— agavoides Zant. et
  v. Poelln.                              — sord. v. a.
— albanensis Schoenl.                     — ang. v. a.
— albicans Haw.                           — margin. v. m.
— — v. laevis Haw.                        — — v. l.
— — v. ramifera Haw.                      — — v. r.
— — v. virescens Haw.                     — — v. v.
— — v. Zantner                            —· uitew.
— altilinea Bgr. p. min. part.            — mucr. v. m.
— — Haw. v. bicarinata
  Triebn.                                 — — v. b.
— — v. brevisetata v. Poelln.             — — v. mucr.
— — v. inermis v. Poelln.                 — — v. limp.
                                            f. i.
— — v. limpida (Haw.)
  v. Poelln.                              — — v. —
— — v. — f. acuminata
  v. Poelln.                              — — v. — f. a.
— — v. — f. inconfluens
  v. Poelln.                              — — v. — f. i.
— — v. — f. inermis
  v. Poelln.                              — — v. — f. i.
— — v. — f. typica
  v. Poelln.                              — — v. —
— — v. morrisiae v. Poelln.               — — v. m.
— — v. — f. subglauca
  v. Poelln.                              — — v. — f. s.
— — v. mucronata (Haw.)
  v. Poelln.                              — —
— — v. polyphylla (Bak.)
  v. Poelln.                              — — v. p.
— — v. — f. minor Triebn.                 — — v. — f. m.
— — v. setulifera Triebn. et
  v. Poelln.                              — — v. s.
— — v. typica v. Poelln.                  — alt. v. a.
— angolensis Bak.                         Chort. a.
— angustifolia Bak.                       H. mont.
— arachnoidea (L.) Duv. v.
  minor Haw.                              — set. v. maj.
— arachnoides (Ait.) Haw.                 — ar.
— aspera Haw.                             Astrol. a. v. a.
— — v. major (Haw.) Parr                  — — v. m.

# INVALID DESIGNATIONS  Haworthia 624

H. atrovirens (DC.) Haw.
— attenuata v. argyrostigma
  f. typica (Bak.) Bgr.
— — v. britteniana f. typica
  v. Poelln.
— — v. minor Salm
— — v. typica Haw.
— bicarinata (Haw.) Parr
— bicarinata (Haw.) Parr
— bijliae v. Poelln.
— — v. joubertii v. Poelln.
— blackburniae v. Poelln.
— brevis Haw.

— britteniana v. Poelln.
— broteriana Res.
— bullulata (Jacq.) Parr.
— chalwinii Marl. et Bgr.
— clariperla Haw.
— coarctata v. haworthii Res.
— — v. — f. major Res.
— — v. pseudocoarctata
  (v. Poelln.) Res.
— — v. sampaiana Res.
— columnaris Bak.
— concava Haw.
— concinna Haw.
— confusa v. Poelln.
— congesta (Salm) Parr
— curta Haw.
— cuspidata v. Poelln.
— cymbiformis Bgr.
— — (Haw.) Duv. v. obtusa
  (Haw.) Bak.
— — v. typica Triebn. et
  v. Poelln.
— deltoidea (Hook. f.) Parr
— — v. intermedia (Uitew.)
  Parr
— — v. turgida (Bak.) Parr
— denticulata Haw.
— dielsiana v. Poelln.
— distincta N. E. Br.
— diversifolia v. Poelln.
— dodsoniana (Uitew.) Parr
— egregia (v. Poelln.) Parr
— — v. fardeniana (Uitew.)
  Parr
— eilyae v. poellnitziana Res.
— engleri Dtr.
— erecta Haw.

— expansa Salm.
— — v. major Haw.
— fallax Haw.
— fasciata v. caespitosa Bgr.
— — v. major Haw.
— — v. — (Salm) Bgr.
— — v. subconfluens
  v. Poelln.
— fergusoniae v. Poelln.
— foliolosa Willd.
— gigas v. Poelln.
— gordoniana v. Poelln.
— gracilidelineata v. Poelln.
— granata Haw.

H. herb.

— att. v. a. f. a.

— — v. b. f. b.
— — v. clarip.
— — v. a.
— — v. a.
× Astrow. b.
H. set. v. b.
— — v. j.
— corr.
— margar. v.
  min.
— att. v. b.
— samp. f. b.
Astrol. b.
H. reinw. v. c.
— att. v. c.
— c. v. c.
— — v. —

— green. v. p.
— samp. v. s.
— obt. v. c.
— cymb.
— visc. v. c.
— ten. v. c.
Astrol. c.
H. tort. v. c.
— turg. v. sub.
— cusp.

— obt. v. o.

— c. v. c.
Astrol. d. v. d.

— — v. i.
— — v. t.
H. alt. v. d.
— obt. v. d.
— ven.
— nigr. v. d.
Astrol. d.
— e. v. e.

— — v. f.
H. e. v. e.
— tess. v. e.
— margar. v.
  min.
— rig. v. e.
— — v. rig.
— reinw. v. f.
— att. v. c.
— subfasc. v. s.
— f. v. f. f. m.

— fasc. v. f. f. s.
— set. v. bijl.
Astrol. f.
H. set. v. g.
— obt. v. g.
— bil. v. g.
— margar. v.

H. granata v. polyphylla
  Haw.
— greeni f. bakeri Res.
— gweneana Parr
— harlandiana Parr
— herrei v. poellnitzii Res.
— hilliana v. Poelln.
— imbricata Parr
— indurata Haw.
— inermis v. Poelln.

— kingiana v. Poelln.
— limifolia v. marlothiana
  (Marl.) Res.
— — v. schultzeana Res.
  sphalm.

— limpida Haw.
— longiaristata v. Poelln.
— longifolia v. Poelln. sphalm.
— major Duv.
— Salm

— margaritifera v. erecta
  (Haw.) Bak.
— — v. granata (Willd.)
  Bak.

— — v. — sv. laetevirens
  (Salm) Bgr.
— — sv. minor Bgr.

— — v. — sv. polyphylla
  (Haw.) v. Poelln.

— — v. maxima sv. major
  Haw.

— — v. — sv. multipapillosa
  Salm

— — v. minor Salm

— — v. semimargaritifera
  Salm

— — v. — (Salm) Bak.
— — v. — sv. major (Salm)
  Bgr.

— — v. — sv. multiperlata
  (Haw.) v. Poelln.

— — v. typica Bgr.
— marginata Hort.
— maxima Duv.

— minima Bak.
— — v. confusa v. Poelln.
— — v. major v. Poelln.

— — Haw.
— minor Duv.
— minutissima v. Poelln.
— multifaria Haw.
— mutica Haw.
— nigricans Haw.

H. marg. v. min.
  sv. p.
— g. f. g.
Astrol. spir.
— herr.
H. h. v. h.
— umbr. v. h.
Astrol. spir.
H. visc. v. i.
— mucr. v.
  limp. f. i.
— subf. v. k.

— l. v. l.

— — v.
  schuldt.
— mucr. v. l.
— set. v. xyph.

— margar.
— fasc. v. f. f.
  m.

— m. v. min.

— — v. — sv.
  min.

— — v. — sv. l.
— — v. — sv.
  pol.

— — v. — sv.
  —

— — v. m. sv.
  m.

— — v. — sv.
  multiperl.
— — v. min.
  sv. pol.

— — v. max.
  sv. m.
— — v. —

— — v. — sv.
  m.

— — v. —
  sv. m.
— — v. marg.
— nigr. v. schm.
— margar. v.
  m. sv. m.
— ten. v. t.
— — v. c.
— margar. v.
  min.
— — v. min.
— — v. m.
— tess. v. m.
— mir.
— ret. v. m.
Gast. dec.

# Haworthia-Huernia INVALID DESIGNATIONS

H. olivetteana Parr
— parva Haw.
— paynii v. Poelln.
— peacockii Hort. Kew
— pellucens Haw.
— — v. delicatula Bgr.
— pentagona Haw.
— — v. spiralis (Salm) Parr
— — v. spirella (Haw.) Parr
— — v. torulosa (Haw.) Parr
— pilifera Bak.
— — v. columnaris (Bak.) v. Poelln.
— — v. dielsiana v. Poelln.
— — v. — f. acuminata v. Poelln.
— — v. gordoniana v. Poelln.
— — v. salina v. Poelln.
— — v. stayneri v. Poelln.
— polyphylla Bak.
— pseudotessellata v. Poelln.
— pseudotortuosa Salm
— pumila Haw.
— radula v. asperior Haw.
— — v. laevior Salm ex Haw.
— — v. magniperlata Haw.
— — v. minor Salm
— — v. pluriperlata Haw.
— reinwardtii v. pseudocoarctata v. Poelln.
— — v. pulchra v. Poelln.
— — v. typica v. Poelln.
— reticulata Troll
— revendettii Hort.
— rubriflora (L. Bol.) Parr
— — v. jacobseniana (v. Poelln.) Parr
— ryneveldiae v. Poelln.
— schmidtiana v. Poelln.
— — v. angusta v. Poelln.
— — v. diversifolia v. Poelln.
— — v. — f. nana v. Poelln.
— — v. elongata v. Poelln.
— — v. pusilla v. Poelln.
— — v. suberecta v. Poelln.
— semimargaritifera Haw.
— — v. major Haw.
— — v. maxima Haw.
— — v. minor Haw.
— — v. multiperlata Haw.
— setata v. subinermis v. Poelln.
— shieldsiana Parr
— spiralis Haw.
— stayneri v. Poelln.
— — v. salina v. Poelln.
— stenophylla Bak.
— subfasciata v. argyrostigma Bak.
— subrigida Bak.

× Astrow. bic. nm. skinn.
H. tess. v. p.
— pall. v. p.
— kew.
— transl.
— — v. d.
Astrol. p. v. p.
— — v. s.
— — v. s.
— — v. t.
H. obt. v. p.

— — v. c.
— — v. d.

— — v. d. f. a.
— v. g.
— — v. s.
— — v. s.
— mucr. v. p.
— tess. v. infl.
— visc. v. p.
— herb.
— rug. v. r.
— subul.
— subatt.
— rug.
— rad.

— green. v. p.
— r. v. maj.
— r. v. r.
— obt. v. diels.
— r.
Poelln. r.

— — v. j.
H. nigr. v. n.
— — v. s.
— — v. a.
— — v. d.
— — v. — f. n.
— — v. e.
— — v. p.
— — v. s.
— margar. v. max.
— — v. — sv. m.
— — v. — sv. m.
— subatt.
— margar. v. max. sv. m.

— ar.
Astrol. turg.
— pent.
H. obt. v. s.
— — v. s.
Chort. s.

H. att. v. a. f. a.
— tort. v. pseud.

H. subspicata Bak.
— taylori Barner MS.
— tenuifolia Engl.
— tessellata v. haworthii Res.
— — v. typica Haw.
— torquata Haw.
— tortella Haw.
— truncata v. maughanii (v. Poelln.) Fearn
— — f. normalis v. Poelln.
— ubomboensis Verd.
— unicolor v. Poelln.
— virescens Haw.
— viscosa v. typica Bgr.
— whitesloaneana v. Poelln.

— xyphiophylla Bak.
× **Haworthiogasteria** Kondo et Megata nom. nud.
**Henricia** L. Bol. Mesembr. ●
H. sibbettii (L. Bol.) L. Bol.
**Herrea** Schwant. Mesembr. ●
H. robusta (N. E. Br.) L. Bol. forma L. Bol.
**Hereroa** Dtr. et Schwant. Mesembr. ●
H. angustifolia L. Bol.
— cana L. Bol.
— dolabriformis (L.) L. Bol.
— karasbergensis (L. Bol.) Friedr. nom. prov.
— pallens L. Bol.
**Hertia** Less. Compositae.
H. crassifolia Less.
**Hesperaloe** Engelm. Agavaceae.
H. yuccaefolia Engelm.
**Hoodia** Sweet. Asclepiadaceae.
H. dinteri Schltr. nom. nud.
— juttae Dtr.
— senilis Jacobs.
— similis Dtr. nom. nud.
**Hoya** R. Br. Asclepiadaceae.
H. bicarinata A. Gray
— coriacea Ldl.
— dalrympleana F. v. Muell.
— paxtonii Hort.
**Huernia** R. Brown. Asclepiadaceae.
H. appendiculata Bgr.
— arabica N. E. Br.
— barbata v. crispa Loud.
— bicampanulata sensu Bremek. et Oberm.
— — Verd.
— blackbeardae R. A. Dyer ex Jacobs.
— crispa Haw.
— flava N. E. Br.
— lentiginosa Haw.
— macrocarpa Bgr.
— — Taubert
— multangula Schweinf.
— nigeriana Lavr.
— penzigii N. E. Br.
— — v. arabica Bgr.
— — v. schimperi Bgr.
— — v. schweinfurthii Bgr.
— repens Lavr.

Chort s.
H. stark.
Chort. t.
H. t. v. t.
— — v. —
— visc. v. t.
— tortuos. v. t.

— maugh.
— t. v. t.
— lim. v. u.
— ar.
— margin. v. v.
— v. v. v.
— schmidt. v. w.
— set. v. x.

× Gasterhaw.

Neohenr. s.

H. r.

H. puttk.
— tugw.
Rhomb. d.

H. hesp.
— puttk.

H. cheir.

H. parv.

H. bain. v. j.
Decab. gr.

H. austr.
— mult.
— austr.
— bell.

H. hystr. v. a.
— macr. v. a.
— b.

— long. v. l.
— kirk.

— zebr. v. magn.
— barb.
— prim. v. rug.
— gutt.
— m. v. penz.
— conc.
s. Car. dec.
H. volk. v. n.
— macr. v. p.
— — v. a
— — v. macr.
— — v. s.
— volk. v. r.

40 Lexicon of Succulent Plants

# INVALID DESIGNATIONS  Huernia-Kalanchoe

| | | | |
|---|---|---|---|
| H. rogersii R. A. Dyer | H. ocul. | J. hirta v. adenophora (Borb.) Löve | Diop. h. |
| — scabra N. E. Br. | — brev. v. s. | — — ssp. glabrescens (Sabr.) Soo et Jav. | — — ssp. h. v. g. |
| — — v. cornuta N. E. Br. | — — v. c. | | |
| — — v. immaculata N. E. Br. | — — v. i. | | |
| — — v. longituba N. E. Br. | — — v. l. | | |
| — — v. quinta Phill. | — quint. | — — ssp. hirta | — — ssp. h. |
| — sprengeri Hort. | Car. schw. | — — ssp. tatrensis (Dom.) Löve | — — |
| — Schweinf. | — s. | | |
| — tavaresii Welw. | Decab. el. | — sobolifera (Sims) Opiz | — — ssp. bor. |
| — verekeri Stent. v. stevensoniae White et Sloane | — v. v. v. | **Juttadinteria** Schwant. Mesembr. ● J. cinerea (Marl.) Schwant. | Namib. c. |
| — volkartii Gossw. | — v. | — dealbata (N. E. Br.) L. Bol. | Drac. d. |
| **Hutchinia** Wight et Arn. Asclepiadaceae. H. indica Wight et Arn. | Car. i. | — delaetiana (Dtr.) Dtr. et Schwant. | — d. |
| **Hydrodea** N. E. Br. Mesembr. ● | | — montis-draconis (Dtr.) Dtr. et Schwant. | — m. d. |
| H. bossiana Dtr. | Mes. crypt. | | |
| — cryptantha (Hook. f.) N. E. Br. | — — | — pomonae (Dtr.) Schwant. | Namib. p. |
| | | — proxima L. Bol. | Drac. p. |
| — hampdenii N. E. Br. | — — | — rheolens L. Bol. | — dealb. |
| — sarcocalycantha (Dtr. et Bgr.) Dtr. | — — | — tugwelliae (L. Bol.) Schwant. | Herer. t. |
| **Hymenocyclus** Dtr. et Schwant. Mesembr. ● | | **Kalanchoe** Adans. Crassulaceae. | |
| H. crassa L. Bol. | Mal. c. | K. acutifolia Haw. | K. lac. |
| — croceus (Jacq.) Schwant. | — c. | — aegyptiaca DC. | — — |
| — englerianus (Dtr. et Bgr.) Dtr. et Schwant. | — e. | — — (Lam.) DC. | — integr. |
| | | — afzeliana Britt. | — — v. cren. |
| — flavo-croceus (Haw.) Schwant. | — f. c. | — aliciae Hamet | — pub. v. p. |
| | | — alternae Defl. | Cot. barb. |
| — framesii L. Bol. | — f. | — — Persoon | K. lac. |
| — herrei Schwant. | — h. | — ambrensis Perr. | — unifl. v. u. |
| — latipetalus L. Bol. | — l. | — amplexicaulis Baill. | — camp. |
| — luteolus (Haw.) Schwant. | — l. | — angolensis N. E. Br. | — vel. |
| — luteus (Haw.) Schwant. | — l. | — angustifolia A. Rich. | — lac. |
| — monticolus (Sond.) Schwant. | Ott. m. | — antonasyana Drake et Castillo | — org. |
| | | — beniensis de Wild | s. K. glauc. |
| — purpureo-croceus (Haw.) L. Bol. | Mal. cr. v. p. c. | — bequaertii de Wild | K. dens. v. d. |
| | | — bergeri v. typica Mann et Boit. | — b. v. b. |
| — smithii L. Bol. | — herr. | | |
| — thunbergii (Haw.) L. Bol. | — t. | — brachycalyx Bak. | — trich. |
| — uitenhagensis L. Bol. | — u. | — — Engl. | — dens. v. subp. |
| — verruculoides (Sond.) L. Bol. | — v. | — — A. Rich. | — lanc. v. l. |
| **Hymenogyne** (Haw.) N. E. Br. Mesembr. ● | | — brasilica (Vell.) Stellenf. | — integr. v. ver. |
| H. stephensiae N. E. Br. | H. gl. | | |
| **Jatropha** L. Euphorbiaceae. | | | |
| J. cathartica Teran et Berland | J. berl. | — brasiliensis Camb. | — — v. — |
| | | — brevicaulis Bak. | — plum. |
| — lacerti Silva Manso | — ell. | — carnea N. E. Br. | — lac. |
| — officinalis Mart. | — — | — ceratophylla Haw. | — — |
| — opifera Mart. | — — | — citrina v. erythraea Schweinf. | — citr. |
| — woodii v. kuntzei Pax | — — v. w. | | |
| **Jensenobotrya** Herre. Mesembr. ● | | — coccinea Auct. non Welw. ex Britt. | — integr. v. ver. |
| J. vanheerdei L. Bol. | Stoeb. carp. | | |
| **Jovibarba** Opiz. Crassulaceae. | | | |
| J. allionii (Jord. et Fourr.) D. A. Webb | Diop. hirt. ssp. a. | — — Welw. ex Britt. | — — v. i. |
| | | — — v. subsessilis Britt. | — — v. s. |
| | | — constantinii Hamet | — beauv. |
| — arenaria (Koch) Opiz | — — ssp. a. | — crenata (Andr.) Haw. | — integr. v. c. |
| — — ssp. pseudohirta (Leute) Holub | — — ssp. p. | — — v. coccinea Cuf. | — — v. i. |
| | | — — v. collina Engl. | — lanc. v. l. |
| — heuffelii (Schott) A. et D. Löve | — h. | — — v. crenata | — integr. v. c. |
| | | — — v. verea (Jacq.) Cuf. | — — v. v. |
| — — ssp. glabra (Beck.) et Szysz) Holub | — — v. g. | — — Broun. et Massey | s. K. glauc. |
| | | — — Hamet | K. laxifl. |
| — hirta (Juslen) Opiz | — h. | — crusinii Duv. | — vel. |

K. deficiens (Forsk.) Aschers.
et Schweinf.
— deflersii Gagnepain
— delagonensis Eckl. et Zeyh.
— diversa N. E. Br.
— dyeri N. E. Br.
— ellacombei de Wild
— fedtschenkoi v. typica
  Boit. et Mann.
— floribunda Wight et Arn.
— glaberrima Volkens
— glandulosa Hochst. ex
  Rich.
— — v. bengulensis Engl.
— — v. rhodesica Bak. f. ex
  Rendle
— glaucescens v. deficiens
  (Forsk.) Senn.
— globulifera v. coccinea Perr.
— goetzei Engl.
— grandiflora A. Rich.
— — v. angustipetala Engl.
— gregaria Dtr.
— guignardii Hamet et Perr.
— heterophylla Wight
— holstii Engl.
— homblei de Wild
— — v. reducta de Wild
— humbertii Mann. et Boit.
— integra auct. non Kubtze
— jongmansii Humbert
— jueli Hamet et Perr.
— kelleriana Schinz
— kirkii N. E. Br.

— laciniata auct. non (L.) DC.
— — (L.) DC. v. brachycalyx
  (Rich.) Chiov.
— — Wight
— lateritia Engl.
— lentiginosa Cuf.

— macrantha Bak.
— macrodon nom. nud.
— magnidens N. E. Br.
— mandrakensis Perr.
— marinellii Pamp.
— marmorata v. maculata
  Terr.
— — f. somaliensis (Bak.)
  Pamp.
— miniata v. decaryana Perr.
— — v. glandulosa Perr.
— — v. pubescens Perr.
— — v. tsinjoarivensis Perr.
— — v. typica Boit. et Mann.
— modesta Kotschy et Arn.
— multiceps Bak.
— multiflora Schinz
— ndorensis Schweinf.
— neumannii Engl.
— nyikae Engl.
— — auct. non Engl.
— panduriformis (Bak.)
  Baill.
— parviflora (Bak.) Baill.

K. integr.
— bent.
— tub.
— lac.
— quart.
— lanc. v. l.

— f. v. f.
— lanc. v. l.
— densifl. v. d.

— lanc. v. g.
— — v. l.

— v. —

— integr.
— blossf.
— lanc. v. l.
— marm.
— —
— lanc. v. l.
— beauv. v. g.
— lanc.
— glauc.
— lanc. v. l.
— — v. l.
— marm.
— i. v. cren.
— j.
— beauv.
— marm.
— integr. v.
  subs.
— — v. cren.


— lanc. v. l.
s. K. schweinf.
K. vel.
— dens. v.
  subsp.
— marm.
— pinn.
— lac.
— pelt. v. m.
s. K. glauc.

K. marm.

— —
— pseudoc.
— pub. v. p.
— — v. —
— — v. —
— min. v. m.
— lanc. v. l.
— pum.
— brachyl.
— lac.
— pet. v. p.
— grandifl.
— hemsl.

— camp. v. c.
— — v. —

K. petitiaesii A. Rich. ex
  Jacq.
— petitiana Andrews
— — auct. non Richard
— — Hort.

— — v. salmonea Hort.
— pilosa Bak.
— — Hort.
— porphyrocalyx v. typica
  Mann. et Boit.
— prasina N. E. Br.
— pruinosa Dtr.
— pubescens v. typica Mann.
  et Boit.
— — v. — f. reducta Humb.
— — R. Br.
— quartiana v. micrantha
  Pamp.
— ritchiana Dalziel

— robynsiana Hamet
— rohlfsii Engl.
— rosea Clarke
— rosei v. typica Humb.
— rutshurensis Lebr. et
  Touss.
— scandens (Perr.) Bak.
— schweinfurthii Penzig
— somalensis Bak.
— spathulata DC.
— stapfii Hamet et Perr.
— stenosiphon Britt.
— subamplectens nom. nud.
— subpeltata Bak.
— sulphurea (Bak.) Bak.
— teretifolia Defl.
— tieghemii Hamet
— tsaratananensis Perr.
— uniflora v. typica Mann.
  et Boit.
— vantieghemii Hamet
— verea Andrews
— verticillata Scott-Elliot
— welwitschii Britt.
× **Kalorochea** Veitch ex Rowl. Crassulaceae.
× K. langleyensis Veitch
× — cv. Capitola
**Kalosanthes** Haw. Crassulaceae.
K. coccinea Haw.
— jasminea Sims
— odoratissima Haw.
**Kedrostis** Medic. Cucurbitaceae.
K. grossulariaefolia Presl.
**Kitchingia** Bak. Crassulaceae.
K. amplexicaulis Bak.
— campanulata Bak.
— gracilipes Bak.
— laxiflora Bal.
— mandrakensis (Perr.) Bak.
— miniata Bak.
— panduriformis Bak.
— parviflora Bak.
— peltata Bak.
— — v. stapfii (Hamet et
  Perr.) Bgr.

K. petit.
— dens. v. d.
— cren. v. c.
— longifl. v.
  cocc.
— — v. l.
— lanc.
— toment.

— p. v. p.
— baum.
— brachyl.

— p. v. p.
— — v. — f. r.
— lanc. v. l.

— petit. v. p.
— lanc. v.
  gland.
— lug.
— lac.
— —
— r. v. r.

— marm.
— beauv.
— lac.
— marm.
— tom.
— pelt. v. s.
— lac.
— var.
— min. v. m.
— porph.
— bent.
— laxifl.
— rol. bon.

— u. v. u.
— behar.
— cren. v. c.
— tubifl.
— lac.
× Rocheass. l.
× — —



Roch. c.
— j.
— o.

K. glauc.

Kal. camp. v. c.
— — v. —
— g.
— l.
— pelt. v. m.
— m. v. m.
— camp. v. c.
— — v. —
— p.

— — v. s.

| | | | |
|---|---|---|---|
| K. porphyrocalyx BAK. | Kal. p. v. p. | K. violacea BGR. | Sen. long. v. v. |
| — schizophylla BAK. | — s. | **Kosaria** FORSK. Moraceae. | |
| — streptantha BAK. | — s. | K. foetida FORSK. | Dorst. f. |
| — sulphurea BAK. | — porph. v. p. | — palmata SCHWEINF. MS. | — p. |
| — uniflora (BAK.) BOIT. et MANN. | — u. v. u. | — tropaeolifolia (BUREAU) SCHWEINF. | — barn. v. t. |
| K. (BAK.) BOIT. et MANN. | Kal. Sect. I. | **Lampranthus** N. E. BR. Mesembr. ● | |
| **Kleinia** L. Compositae. | | L. affinis L. BOL. forma L. BOL. | L. a. v. a. |
| K. acaulis DC. | Sen. a. v. a. | — bicolorus (L.) JACOBS. | — bicolor v. b. |
| — adenocalyx (DTR.) MERXM. | — rad. | — brachyandrus (L. BOL.) N. E. BR. | — hoerl. |
| — aizoides DC. | — a. | | |
| — amaniensis (ENGL.) BGR. | — a. | — — forma L. BOL. | — — |
| — anteuphorbium DC. | — a. | — capornii (L. BOL.) L. BOL. | Rusch. c. |
| — archeri COMPT. | — a. | — drepanophyllus (SCHLTR. et BGR.) N. E. BR. | — d. v. d. |
| — articulata HAW. | — a. v. a. | | |
| — breviscapa DC. | — ciatr. | — dubitans (L. BOL.) L. BOL. | — d. |
| — cana DC. | — quinqu. | — dyckii (BGR.) N. E. BR. (Mes. d. BGR. nom. nud.). | — d. |
| — cephalophora COMPT. | — c. | | |
| — chordifolia HOOK. f. | — c. | — emarginatoides (HAW.) N. E. BR. | L. eleg. |
| — cliffordiana (HUTCH.) ADAMS. | — c. | — erratus (S.D.) N. E. BR. (Mes. virens S.D.). – An insufficiently known species. | |
| — crassulaefolia BAK. | Oth. carn. | | |
| — crassulifolia DC. | Sen. c. | | |
| — cucullata BOOM | — klein. | | |
| — cuneifolia (L.) DC. | — c. | — flexifolius (HAW.) N. E. BR. | — curv. v. min. |
| — cylindrica BGR. | — spic. | — flexilis (HAW.) N. E. BR. | — emarg. v. e. ? |
| — eupapposa CUF. | — e. | — glaucoides (HAW.) N. E. BR. | — aur. |
| — ficoides HAW. | — f. | — globosus (L. BOL.) L. BOL. forma L. BOL. | — g. |
| — fulgens HOOK. f. | — f. | | |
| — galpinii HOOK. f. | — g. | — gracilipes (L. BOL.) L. BOL. | — g. v. g. |
| — gomphophylla DTR. | — herr. | — gydouwensis L. BOL. | Amoeb. g. |
| — gonoclada DC. | — rad. | — herrei (L. BOL.) L. BOL. | Rusch. sand. |
| — grantii (OLIV. et HIERN) HOOK. f. | — g. | — imbricans (HAW.) N. E. BR. | L. emarg. v. e. ? |
| | | — immelmaniae (L. BOL.) L. BOL. | — i. |
| — hanburyana (DTR.) BGR. | — h. | — lerouxii (L. BOL.) N. E. BR. | Rusch. l. |
| — haworthii DC. | — h. | — matutinus (L. BOL.) SCHWANT. | L. m. |
| — herreiana (DTR.) MERXM. | — h. | | |
| — humbertii GUILL. | — neoh. | — monticolus (L. BOL.) L. BOL. | — sten. |
| — kleiniaeformis (SUESSENG.) BOOM | — k. | — multiradiatus (JACQ.) N. E. BR. | — ros. ? |
| — kleinioides (SCH. BIP.) M. T. TAYLOR | — long. v. viol. | — otzenianus (DTR.) FRIEDR. | Dros. un. |
| | | — perspicuus (BGR.) N. E. BR. | L. schl. |
| — longiflora DC. | — — v. l. | — pittenii (L. BOL.) N. E. BR. | — vanp. |
| — mandraliscae TIN. | — m. | — punctatus (L. BOL.) L. BOL. | Rusch. p. |
| — michelii HORT. | — art. v. a. | — ruber (L. BOL.) L. BOL. | Astr. r. |
| — neriifolia HAW. | — kleinia | — sparsiflorus L. BOL. | L. pauc. |
| — odora (FORSK.) DC. | — ant. v. o. | — stenus v. depauperatissimus L. BOL. | — st. |
| — ovoidea COMPT. | — o. | | |
| — pendula DC. | — p. | — stokei (L. BOL.) L. BOL. | Rusch. thom. |
| — pinguifolia DC. | — p. | — tenuifolius (L.) SCHWANT. | L. t. |
| — polycotoma CHIOV. | — p. | — tulbaghensis (L. BOL.) N. E. BR. | — arg. |
| — (?) pugioniformis DC. | Cot. ? | | |
| — pteroneura DC. | Sen. ant. | — uncus (L. BOL.) SCHWANT. | Rusch. u. |
| — pusilla (DTR.) MERXM. | — klingh. | — — v. gydouwensis (L. BOL.) SCHWANT. | — dubit. |
| — — DTR. ex RANGE | — — | | |
| — radicans L. f. | — r. | — uniflorus (L. BOL.) L. BOL. | Dros. u. |
| — repens (L.) HAW. | — serp. | — — v. spathulatus (L. BOL.) L. BOL. | — — |
| — rigida (THUNBG.) DC. | — spin. | | |
| — semperviva (FORSK.) DC. | — s. | — utilis (L. BOL.) SCHWANT. | Rusch. u. v. u. |
| — spicata HORT. | — spicul. | — vereculatus (L.) L. BOL. | Scopol. v. |
| — spiculosa SHEPH. | — — | — violaceus (DC.) SCHWANT. | L. emarg. ? |
| — spinulosa HORT. | — — | — zygophylloides (L. BOL.) L. BOL. | Dros. z. |
| — stapeliiformis (PHILL.) STAPF | — s. | | |
| — subradiata DC. | — s. | **Leipoldtia** L. BOL. Mesembr. ● | |
| — talinoides DC. | — t. | L. herrei (SCHWANT.) SCHWANT. (Rhopalocyclus | |
| — tomentosa HAW. | — haw. | | |

h. Schwant.). – Cape: Robertson D. – Insufficiently known.
L. rosea L. Bol. — Ceph. r.
**Lenophyllum** Rose. Crassulaceae.
L. maculatum Hort. — Sinocr. dens.
**Leptaloe** Stapf. Liliaceae.
L. albida Stapf — Aloe a.
— blyderivierensis Groenw. — min. v. b.
— minima (Bak.) Stapf — — v. m.
— myriacantha (Haw.) Stapf — m.
— parviflora (Bak.) Stapf — p.
— saundersiae Reyn. — s.
L. Bgr. — Aloe Sect. 3.
**Leptosyne** DC. Compositae.
L. gigantea Kellog — Coreop. g.
**Lewisia** Pursh. Portulacaceae.
L. alba Kellog — L. red. v. r.
— heckneri Morton — cot. v. h.
— minor Rydberg — red. v. m.
— tweedei A. Gray — t.
**Lithocaulon** Bally. Asclepiadaceae.
L. cubiforme Bally — Pseudol. c.
— sphaericum Bally — mig.
**Lithops** N. E. Br. Mesembr. ●
L. alpina Dtr.
— annae de Boer — L. pseud. v. a.
— aurantiaca L. Bol. — ges. v. a.
— chrysocephala Nel — turb. v. t.
— commoda Dtr. nom. nud. — full. v. c.
— dabneri L. Bol. — kar. v. k.
— damarana (N. E. Br.) N. E. Br. — marg.
— dendritica Nel — kar. v. k.
— diutina L. Bol. — pseud. v. d.
— eberlanzii (Dtr. et Schwant.) N. E. Br. — marm.
— edithae N. E. Br. — bell. v. e.
— eksteeniae L. Bol. — pseud. v. e.
— elevata L. Bol. — dor.
— elisabethae Dtr. — herr. v. h.
— farinosa Dtr. nom. nud. — pseud. v. e.
— fossulifera Tisch. nom. nud. — — v. dendr.
— framesii L. Bol. — marm.
— friedrichiae (Dtr.) N. E. Br. — Ophth. f.
— geyeri Nel — L. herr. v. g.
— gulielmi L. Bol. — schwant. v. ur. (acc. D. T. Cole v. schwant.)
— halenbergensis Tisch. — bell. v. eb.
— herrei v. plena L. Bol. — h. v. h.
— hillii L. Bol. — — v. h.
— inae Nel — verr. v. i.
— inornata Dtr. nom. nud. — marth.
— jacobseniana Schwant. — kar. v. mick.
— julii v. pallida Tisch. — j. v. j.
— karasmontana ssp. opalina (Dtr.) Schwant.
— koelemanii de Boer — k. v. o.
— kuibisensis Dtr. ex Jacobs. — auc. v. k.
— kunjasensis Dtr. — schwant. v. s
— lactea Schick et Tisch. — — v. k.
— lateritia Dtr. — jul. v. j.
— lericheana (Dtr. et Schwant.) Dtr. et Schwant. — kar. v. mick.
— bell. v. l.

L. lineata Nel
— lydiae Jacobs.
— marlothii N. E. Br.
— maughanii N. E. Br.
— mickbergensis Dtr.
— mundtii Tisch.
— nelii Schwant.
— opalina Dtr.
— optica v. rubra Tisch.
— peersii L. Bol.
— pillansii L. Bol.
— pseudotruncatella v. groendraaiensis Jacobs.
— — f. mundtii Tisch. ex Jacobs.
— — ssp. — (Tisch.) Schwant.
— — f. pulmonuncula Dtr. ex Jacobs.
— ssp. — Dtr. ex Jacobs.
— rubra N. E. Br.
— rugosa Dtr.
— ruschiana N. E. Br. sphalm.
— ruschiorum ssp. nelii (Schwant.) Schwant.
— — ssp. ruschiorum Schwant.
— — ssp. stipelmanii Schwant.
— salicola v. reticulata de Boer
— schickiana nom. nud.
— streyi Schwant.
— summitatum Dtr.
— terricolor N. E. Br.
— translucens L. Bol.
— triebneri L. Bol.
— umdausensis L. Bol.
— urikosensis Dtr.
— ursulae nom. nud.
— vanzijlii L. Bol.
— venteri Nel
— volkii Schwant. ex Jacobs.
**Lithocarpus** L. Bol. Mesembr. ●
L. cordifolius (L. f.) L. Bol.
**Lobelia** L. Campanulaceae.
L. gregoriana Bak. f.
**Lomatophyllum** Willd. Liliaceae.
L. aloiflorum Nich.
— borbonicum Willd.
— marginatum Hoffmgg.
**Machoda** Welw. Passifloraceae.
M. huillensis Welw.
**Macrocaulon** N. E. Br. Mesembr. ●
M. candollei (Haw.) N. E. Br.
**Macrosepalum** Regel et Schmalh. Crassulaceae.
M. turkestanicum Regel et Schmalh.
**Mahafalia** Jum. et Perr. Asclepiadaceae.
M. nodosa Jum et Perr.
**Malephora** N. E. Br. Mesembr. ●

L. rusch. v. l.
— fulv.
Con. pell.
L. full.
— kar. v. m.
— pseud. v. m.
— rusch. v. n.
— kar. v. o.
— o. f. r.
— loc. v. p.
— rusch. v. r.
— vall. m. v. g.
— pseud. v. m.
— — v. —
— — v. p.
— — v. p.
— opt. f. r.
— schwant. v. r.
— r. v. n.
— r. v. r.
— r. v. nel.
— hall.
— grac. v. g.
— kar. v. s.
— loc. v. t.
— herr. v. h.
— schwant. v. t.
— marm.
— schwant. v. u.
— kar. v. mickb.
Dint. v.
L. lesl. v. v.
— pseud. v. v.
Apt. c.
Sen. brass.
L. purp.
— — —
— — —
Adenia h.
Carpanth. pom.
Sed. aet. v. tetr.
Cyn. n.

**INVALID DESIGNATIONS** Malephora-Mesembryanthemum

M. monticola (SOND.) JACOBS.
et SCHWANT. Ott. m.
**Manettia** ADANS. – Homonym of Manettia MUTIS (Rubiaceae), generic synonyms of Aridaria N. E. BR. – Mesembr. ●
**Manfreda** SALISB. Agavaceae.
M. brachystachys (CAV.) ROSE Ag. b.
— brunnea (S. WATS.) ROSE — b.
— debilis (S. WATS.) ROSE — d.
— elongata ROSE — grac.
— guttata (JACOBI et BOUCHÉ)
ROSE — g.
— jaliscana ROSE — j.
— maculata ROSE — pub.
— maculosa ROSE — m.
— oliverana ROSE — o.
— planifolia (S. WATS.) ROSE — p.
— pratensis ROSE — p.
— revoluta ROSE — r.
— singuliflora (S. WATS.) ROSE — s.
— undulata (KLOTZSCH) ROSE — u.
— variegata (JACOBI) ROSE — v.
— virginica (L.) SALISB. — v. v. v.
M. (SALISB.) BAK. Ag. SG. I.
**Marah** KELLOG. Cucurbitaceae.
M. gilensis GREENE Echinoc. g.

**Marlothistella** SCHWANT. Mesembr. ●
M. uniondalensis SCHWANT. Rusch. sten.
**Maughania** N. E. BR. (non JAUME SAINT HILAIRE). Mesembr. ●
M. insignis N. E. BR. Maugh'ella luckh.
— luckhoffii (L. BOL.) L. BOL. — —
**Maximowiczia** COGN. (non RUPR.). Cucurbitaceae.
M. insularis BRANDEG. Iberv. i.
— lindheimeri v. tenuisecta
(A. GRAY) COGN. — ten.
— sonorae S. WATS. — s.
— — v. brevicaulis I. M.
JOHNSTON — ins.
— — v. peninsularis I. M.
JOHNSTON — son. v. p.
— — v. sonorae — v. s.
**Mentocalyx** N. E. BR. Mesembr. ●
M. muiri N. E. BR. Gibb. schwant.
— velutinus (L. BOL.)
N. E. BR. — v.
— — (L. BOL.) SCHWANT. — —
M. (N. E. BR.) WULFF Gibb. Sect. II.
SSect. 3.

**Mesembrianthemum** BREYNE. Former spelling of **Mesembryanthemum,** derived from "mesembria" (Greek) = noon, and "anthemion (anthemon)"[1] (Greek) = flower, i.e. "midday flowers", as they were called by BREYNE (1669), although this name only makes sense if all species flower by day. DILLENIUS (DILLEN) changed the name to Mesembryanthemum, so as to give it a different sense and thus include the night-flowering species as well. He derived it from "mesos" (Greek) = in the centre, and "embryon" = ovaries, and "anthemon"[1] = flower. The word "embryon", now used in Botany to denote only the embryo, was formerly also used by botanists to denote the ovary. DILLENIUS used the phrase: 'Flos nempe qui embryo est in medio', i.e. Fl. with an ovary in the centre. Unfortunately LINNAEUS in 1753 continued to use the spelling with 'y' which makes little sense since innumerable other Fl. also have a central ovary. According to the Rules of Nomenclature, the spelling with 'y' is obligatory (see Kew Bull. 1928, 113, 287).

*Mesembrianthemaceae* LOWE = **Mesembryanthemaceae** FENZL emend. HERRE et VOLK.
*Mesembrianthemaceae* LOWE emend. HERRE et VOLK = as the previous.
*Mesembrinae* LINK. is invalid.
*Mesembryaceae* LINDL. is invalid.
*Mesembryanthemaceae* BAILL. is invalid.
*Mesembryanthemaceae* J. TORREY is invalid.
*Mesembryanthemeae* ENDLICHER is invalid.
*Mesembryanthemeae* FENZL is invalid.
*Mesembryanthemum* DILL.: see Mesembrianthemum BREYNE.
*Mesembryanthemum* L.: see Synonyms.

*Mesembryanthemum* L. emend. N. E. BR. – Type species: *Mes. umbellatum* L., hence **Ruschia** and **Lampranthus** = *Mesembryanthemum* (Gard. Chron. 1925 I/212 and l.c. 1930 I/32) = invalid (see Synonyms).
*Mesembryanthemum* L. emend. SCHWANT. – Type species: **Mes. tenuifolium** L., hence only **Lampranthus** = *Mesembryanthemum* (Journ. of the D.K.G. 1927/28) = invalid (see Synonyms).
*Mesembryum* ADANS. – Earlier, invalid name.
Mesembryanthemums, Mesems. = colloquial names for species of the family **Mesembryanthemaceae.**

**Mesembryanthemum** L.; L. emend. N. E. BR.;
L. emend. SCHWANT. Mesembr. ●
M. abbreviatum HAW. Carpobr. vir.
— aberdeenense L. BOL. Del. a.
— abruptum BGR. Oct. a.
— abyssinicum RGL. Del. a.
— acinaciforme DC. Semn. lac.

M. acinaciforme L. Carpobr. a.
— — β flavum L. — ed.
— acrosepalum L. BOL. Lampr. a.
— aculeatum N. E. BR. Eberl. a.
— acuminatum HAW. Sphalm. a.
— — S.D. — noth.
— acutangulum HAW. Rusch. a.

[1]) anthemon (Greek) = anthemum (Lat.), used in **Mesembryanthemum.**

## Mesembryanthemum  INVALID DESIGNATIONS

M. acutifolium L. Bol. — Lampr. a.
— acutipetalum N. E. Br. — Khad. a.
— acutisepalum Bgr. — Psil. a.
— acutum Haw. — Ceph. a.
— aduncum Haw. — Lampr. a.
— — Jacq. — Rusch. scholl. v. s.
— — Willd. — Lampr. curv. v. min.
— aequilaterale Haw. — Carpobr. a.
— — v. chiloense S.D. — chil.
— — v. decagynum DC. — aequ.
— — Willd. also Haw. — — —
— aequilaterum Haw.
— aerum L. Bol. — Del. a. v. a.
— — v. album L. Bol. — — — v. a.
— aestivum L. Bol. — Lampr. a.
— aggregatum Haw. nom. nud. — — a.
— — N. E. Br. — Con. pil.
— agninum Haw. — Stom. a. v. a.
— — v. integrifolium Salm — — v. i.
— albertense N. E. Br. — Con. trunc. v. t. f. t.
— albertinense L. Bol. — Ceph. a.
— albicaule Haw. — Sphalm. a.
— albicuta N. E. Br. — Rusch. bol.
— albidum L. — Mach. a.
— albiflorum L. Bol. — Dros. a.
— albinotum Haw. — Rab. a. v. a.
— albipunctum Haw. — — a. v. a.
— — v. major Haw. — Nan. cibd.
— album L. Bol. — Lampr. a.
— alkalifugum Dtr. nom. nud. — M. barkl.
— aloides Haw. — Nan. a. v. a.
— — S.D. — — cibd.
— alsinifolium Don — M. cland.
— altile N. E. Br. — Con. fic.
— amicorum L. Bol. — Rusch. a.
— amoenum S.D. — Lampr. a.
— amplectens L. Bol. — Aspaz. a.
— amplexicaule L. Bol. — Leip. a. f. a.
— amplexum N. E. Br. — Rusch. incl.
— ampliatum (L. Bol.) N. E. Br. — — a.
— anatomicum Haw. — Scel. a.
— anceps Haw. — Ereps. a.
— androsaceum (Marl. et Schwant.) N. E. Br. — Rusch. a.
— anemoniflorum L. Bol. — Ceph. a.
— angelicae Dtr. et Schwant. — Con. a.
— angulatum Thunbg. — M. ait.
— — v. ovatum Sond. — Plat. haeck.
— angulipes L. Bol. — Gibb. a.
— angustipetalum L. Bol. — Vanz. a.
— angustum Haw. — Glott. a.
— — v. pallidum Haw. — — taur.
— — S.D. — — —
— — v. heterophyllum γ Haw. — — long. v. het.
— — v. — Salm — — procl.
— annulatum Bgr. — Vanz. a.
— antemeridianum L. Bol. — Lampr. a.
— — f. flore pleno L. Bol. — — —
— apetalum L. f. — Dor. a.
— — Thunbg. — — gram. f. alb.
— apiatum N. E. Br. — Con. a.

M. apiculatum Kensit — Brauns. a.
— — v. muticum L. Bol. — Lampr. max.
— apodanthum Schltr. et Diels. - Fedde Repert. VIII, 148. – An insufficiently described species and therefore not assigned to any other genus of the family.
— approximatum L. Bol. — Rusch. a.
— apricum Bgr. — Leip. a.
— aquosum L. Bol. — M. hyp.
— arborescens Dtr. — Stoeb. beetz. v. a.
— arboriforme Burch. — Mest. a.
— arbuthnotiae (L. Bol.) L. Bol. — Lampr. a.
— arenarium (N. E. Br.) L. Bol. — M. inach.
— arenosum L. Bol. — Lampr. a.
— — Schinz — Psil. a.
— argenteum L. Bol. — Lampr. a.
— aristulatum Sond. — Rusch. pulch. v. p.
— armatum (L. Bol.) N. E. Br. — Eberl. a.
— articulatum Thunbg. — Psil. gran.
— ascendens Haw. — Glott. lat. v. l.
— — S.D. — — — v. —
— asperulum Haw. — Dros. pulch. ?
— — Salm — Del. a.
— asperum Haw. — Ereps. a.
— assimile N. E. Br. — Con. a.
— astridiae Dtr. — Tit. h. schl.
— atratum (L. Bol.) N. E. Br. — Rusch. a.
— atrocinctum N. E. Br. — — sat.
— attenuatum Haw. — Dros. a.
— attonsum L. Bol. — Trich. a.
— augeiforme Schwant. — Con. grac.
— aurantiacum DC. — Lampr. a.
— — Haw. — — —
— auratum Sond. — Sphalm. a.
— aureum L. — Lampr. a.
— — Thunbg. — Sphalm. aurat.
— ausanum Dtr. et Bgr. — Psil. lind.
— ausense L. Bol. — Jutt. a.
— australe Ait. — Disph. a.
— — Sol. nom. nud. — — —
— austricolum L. Bol. — Lampr. a.
— axthelmianum Dtr. — Rusch. a.
— barbatum L. — Trich. b.
— — v. β Haw. — — stell.
— — v. β L. — — hirs.
— — v. γ L. — — dens.
— — v. densum Willd. — — —
— — Curtis — — stell.
— — v. hirsutum Willd. — — hirs.
— barklyi v. obtusifolium L. Bol. — M. b.
— barnardii (L. Bol.) N. E. Br. — Rusch. od.
— beetzii Dtr. — Stoeb. b. v. b.
— bellidiflorum DC. — Acrod. sub.
— — Haw. — — b. f. str.
— — L. — — — v. b.
— — v. glaucescens Haw. — — v. —
— — v. striatum Haw. — — v. str.

**INVALID DESIGNATIONS  Mesembryanthemum**

M. bellidifl. v. subulatum HAW.  Acr. s. v. striat.
— — v. — SALM            — sub.
— — v. viride HAW.        — b. v. v.
— bellidiforme BURM. f.   Dor. b.
— bellum (N. E. BR.) DTR. Lith. b. v. b.
— bergerianum DTR.        Herer. hesp.
— berghiae (L. BOL.) L. BOL.  Lampr. b.
— beswickii L. BOL.       Khad. b.
— bibracteatum HAW.       Cheirid. b.
— bicoloratum (L. BOL.) N. E. BR.  Rusch. b.
— bicolor CURTIS          Lampr. cocc.
— — L.                    — b. v. b.
— bicolorum KLINSMANN     Herer. fug.
— — L.                    Lampr. bicolor v. b.
— — v. inaequale HAW.     — — v. i.
— bicorne SOND.           Psil. b.
— bidentatum HAW.         Glott. sem.
— bifidum HAW.            Cheir. b.
— bifoliatum L. BOL.      Ott. mont.
— biforme N. E. BR.       Rusch. b.
— bigibberatum HAW.       Glott. sem.
— bijliae (L. BOL.) L. BOL.  Rusch. b.
— bilobum MARL.           Con. b.
— binum L. BOL.           Lampr. max.
— — N. E. BR.             Brauns. apic.
— blandum HAW.            Lampr. b.
— — v. curviflorum (HAW.) BGR.  — curv.
— boehmianum DTR.         Con. sax. f. s.
— bolusiae (L. BOL.) SCHWANT. ex JACOBS. nom. ill.
— — HOOK. f.              Pleiosp. b.
— bosscheanum BGR.        Fauc. b. v. b.
— brachiatum AIT.         Sphalm. gen.
— — DC.                   — nit.
— brachyandrum L. BOL.    Lampr. hoerl.
— brachyphyllum PAX       Dros. pax.
— — WELW.                 Lampr. glauc.
— bracteatum AIT.         Ereps. b.
— brakdamense L. BOL.     Rusch. b.
— buchubergense DTR.      — dol.
— brevicaule HAW.         Conic. b.
— brevicolle HAW.         Rusch. b.
— brevifolium AIT.        Dros. b.
— — S.D.                  — erig.
— brevipes SCHLTR.        Argyr. fiss.
— brevistamineum L. BOL.  Lampr. b.
— britteniae L. BOL.      Leip. b.
— brownii HOOK. f.        Lampr. b.
— brunthaleri BGR.        Del. b.
— bulbosum HAW.           Trich. b.
— caducum AIT.            Dor. apet.
— — AUCT. non. AIT.       M. inach.
— caespitosum L. BOL.     Lampr. c.
— — f. luxurians L. BOL.  — — v. l.
— calamiforme L.          Cyl. c.
— calandrum L. BOL.       Leip. c.
— calcaratum WOLLEY-DOD   Lampr. c.
— calcareum MARL.         Tit. c.
— calcicolum L. BOL.      Rusch. c.
— calculus BGR.           Con. c.
— — v. minutiflorum (SCHWANT.) SCHWANT.  — — v. m.
— calendulaceum HAW.      Carpanth. pil.

M. calycinum ECKL. et ZEYH.  Sphalm. canal.
— — HAW.                  Dros. c.
— campestre BURCH.        Rusch. c.
— canaliculatum HAW.      Sphalm. c.
— — S.D.                  — salm.
— candens HAW.            Dros. c.
— candidissimum N. E. BR. Cheir. c.
— candollei HAW.          Carpanth. pom.
— canescens HAW.          Rusch. pulch. v. p.
— caninum HAW.            Carr. ring.
— — v. paucidentatum SALM
  sv. vulpium HAW.        s. next line
— — v. vulpinum HAW. (Mes. vulpinum HAW., Mes. caninum v. paucidentatum SALM). Acc. SCHWANTES
— a synonym               Carr. ring.
— α caninum L.            — — —
— caninum L. ex L. BOL. Notes on Mes. I, 90
— canonotatum L. BOL.     Rusch. c.
— canum HAW.              Pleiosp. c.
— — HAW. ex A. BERGER, Mes. u. Port. 279. Insufficiently descr. acc. L. BOLUS  Bijl. c. ?
— capillaceum L. BOL.     Lampr. c.
— capillare L.            Dros. c.
— — v. acutifolium L. BOL. — subgl.
— — capitatum EDWARDS     Conic. comm.
— — HAW.                  — pugn.
— capornii L. BOL.        Rusch. c.
— — f. fera L. BOL.       Lampr. lun.
— — v. longifolium L. BOL. — — —
— carinans BGR.           Herer. gran.
— — HAW.                  — c.
— carinatum VENT.         Semn. lac. v. l.
— carneum HAW.            Sphalm. c.
— caroli L. BOL.          Rusch. c.
— carolinense L. BOL.     Khad. c.
— caroli-schmidtii DTR. et BGR.  Cheir. c.-s.
— catervum N. E. BR.      Con. c.
— caudatum L. BOL.        Sphalm. c.
— caulescens MILL.        Osc. c.
— cauliculatum HAW.       Ceph. c.
— cauliferum N. E. BR.    Con. c.
— cedarbergense L. BOL.   Lampr. c.
— ceratophyllum WILLD.    — curvifol. v. c.
— ceriseum L. BOL.        — c.
— chauvinae SCHWANT.      Con. grac.
— chilense MOLINA         Carpobr. aequ.
— chiloense SALM          — — —
— chloratum TISCH.        Con. ect. v. limb.
— chrysoleucum SCHLTR.    Mon. c. v. c.
— cibdelum N. E. BR.      Nan. c.
— cigarettiferum BGR.     Cheir. c.
— ciliatum AIT. also THUNBG.  Psil. c.
— ciliolatum N. E. BR.    Rusch. conc.
— cinctum L. BOL.         — c.
— cinereum MARL.          Namib. c.
— citrinum L. BOL.        Lampr. c.
— clausum DTR.            Eberl. c.
— clavatum HAW.           Dor. gram. f. g.
— — JACQ.                 Sphalm. def.

| | | | |
|---|---|---|---|
| M. clavellatum HAW. | Rusch. c. ? | M. cryptopodium KENSIT | Gibb. c. |
| — clavifolium L. BOL. | Ceph. c. | — crystallinum AUCT. non L. | M. guer. |
| — claviforme DC. | Dor. gram. f. g. | — — L. v. grandiflorum | |
| — clavulatum BGR. | Psil. c. | ECKL. et ZEYH. | — perl. |
| — cleistum (L. BOL.) N. E. BR. | Rusch. c. | — crystallophanes ECKL. et | |
| — clivorum N. E. BR. | Mitr. c. | ZEYH. | — ait. |
| — coccineum HAW. | Lampr. c. | — cultrata S.D. | Glott. lat. v. c. |
| — cognatum N. E. BR. | Mitr. c. | — cuneifolium JACQ. | Dor. bell. |
| — collinum SOND. | Dros. c. | — cupreum L. BOL. | Cheir. c. |
| — coloratum N. E. BR. | Rusch. vers. | — cupulatum L. BOL. | Rusch. c. |
| — communis EDWARDS | Conic. c. | — curtophyllum L. BOL. | Ceph. c. |
| — commutatum BGR. | Sphalm. c. | — curtum HAW. | Rusch. c. |
| — compactum AIT. | Pleiosp. nob. | — curviflorum HAW. | Lampr. c. |
| — compressum HAW. | Ereps. c. | — curvifolium HAW. | — c. v. c. |
| — comptonii L. BOL. | Lampr. c. v. c. f. c. | — — v. minum SALM | — flex. |
| | | — cyathiforme L. BOL. | — c. |
| — — v. angustifolium L. BOL. | — — v. a. | — cylindricum HAW. | Cal. c. |
| — — v. roseum L. BOL. | — — v. c. f. r. | — — v. α HAW. | — teret. |
| — comptum N. E. BR. | Rusch. conc. | — — v. teretifolium HAW. | — — |
| — concavum HAW. | Scel. c. | — cymbifolium HAW. | Rusch. c. |
| — concinnum N. E. BR. | Tit. schwant. | — cymbiforme HAW. (Corpusc. | |
| — condensum N. E. BR. | Rusch. c. | c. (HAW.) SCHWANT.). – | |
| — confusum DTR. | Ceph. c. | Insufficiently known | |
| — congestum SALM | Rusch. c. | species. | |
| — connatum (L. BOL.) N. E. BR. | Octopom. c. | — cymosa L. BOL. | Rusch. pung. |
| — conspicuum HAW. | Lampr. c. | — dactylinum WELW. ex OLIV. | M. crypt. |
| — — S.D. | s. Lampr. dyck. | — damaranum N. E. BR. | Lith. kar. v. k. |
| — constrictum L. BOL. | Leip. c. | — dasyphyllum SCHLTR. | Rusch. d. |
| — convexum L. BOL. | Lampr. c. | — dealbatum N. E. BR. | Drac. d. |
| — cookii L. BOL. | Mach. c. | — debile HAW. | Lampr. d. |
| — cooperi HOOK. f. | Del. c. v. c. | — deceptum N. E. BR. | Rusch. cymb. |
| — copiosum L. BOL. | Lampr. c. | — decipiens HAW. | Ceph. d. |
| — copticum JACQ. | Dor. apet. | — decorum N. E. BR. | Trich. d. |
| — — L. | M. nod. | — decurrens (L. BOL.) N. E. BR. | Rusch. d. |
| — coralliflorum S.D. | Lampr. c. | — decussatum THUNBG. | Sphalm. d. |
| — coralliforme VEST. ex A. BERGER, Mes. u. Port. 292 is an unclarified species. | | — deflexum AIT. | Lampr. d. |
| | | — — HAW. | — el. |
| | | — defoliatum HAW. | Sphalm. d. |
| | | — dejagerae (L. BOL.) N. E. BR. | Rusch. d. |
| — corallinum HAW. | Lampr. cor'fl. | | |
| — — THUNBG. | Psil. c. | — dekenahii N. E. BR. | — d. |
| — cordifolium L. f. | Apt. c. | — delaetianum DTR. | Drac. d. |
| — coriarium BURCH. | Psil. c. | — delicatulum L. BOL. | Dros. d. |
| — corniculatum L. | Ceph. lor. | — deliciosum L. BOL. | Carpobr. d. |
| — coruscans HAW. | Conic. c. | — deltatum (SCHWANT.) MAIRE et WEILL | Osc. ped. |
| — cradockense O. KTZE. | Eberl. c. | | |
| — crassicaule HAW. | Scel. c. | — deltoides L. | — d. |
| — crassifolium L. | Disph. c. | — — v. muricatum (HAW.) BGR. | — — |
| — — THUNBG. | Lampr. rept. | | |
| — crassipes MARL. | Al. spath. | — — v. pedunculatum N. E. BR. | — ped. |
| — crassulifolium N. E. BR. | Rusch. crass'oid. | | |
| | | — deltoides MILL. ex BGR. | — — |
| — crassulina DC. | Del. c. | — — v. majus WESTON | — delt. v. m. |
| — crassuloides HAW. | — c. | — delum L. BOL. | Sphalm. d. |
| — crassum L. BOL. | Rusch. c. | — demissum WILLD. | Disph. austr. |
| — crateriforme L. BOL. | Arg. c. | — densifolium L. BOL. | Lampr. d. |
| — criniflorum L. f. | Dor. bell. | — densipetalum L. BOL. | — d. |
| — crispum HAW. (M. crispatum HAW. is a plant of the family Compositae.) | | — densum HAW. | Trich. d. |
| | | — dentatum KERN. | Semn. lac. v. l. |
| | | — denticulatum HAW. | Cheir. d. v. d. |
| — croceum JACQ. | Mal. c. | — — v. candidissimum HAW. | — cand. |
| — — v. flavicroceum DC. | — flav. | — — v. canum HAW. is perhaps a ssp. of | |
| — — v. - HAW. | — — | | — dent. |
| — cronemeyerianum BGR. | Del. c. | — — v. glaucum HAW. | — — v. g. |
| — cruciatum HAW. | Glott. c. | — dependens L. BOL. | Lampr. d. |
| — — SALM | — longip. | — depressum HAW. | Glott. d. |

## INVALID DESIGNATIONS  Mesembryanthemum

M. depress. v. lividum Haw.
— — v. pallens Haw.
— — S.D.
— derenbergianum Dtr.
— deserticolum Marl.
— dichroum Rolfe
— dielsianum Bgr.
— difforme L. (Haw. sphalm.)
— — Thunbg.
— diffusum L. Bol.
— digitatum Ait.
— digitiforme Thunbg.
— dilatatum Haw.
— dimidiatum Haw.
— diminutum Haw.
— — v. cauliculatum Haw.
— — v. pallidum Haw.
— dimorphum Welw.
— dinterae Dtr.
— dinteri Engl. emend. Bgr.
— — Engl. p. part.
— diplosum N. E. Br.
— disgregum N. E. Br.
— dissimile N. E. Br.
— dissitum N. E. Br.
— distans L. Bol.
— diutinum L. Bol.
— divergens Kensit
— diversifolium Haw.
— — v. congestum Salm
— — L.
— diversipapillosum Bgr.
— diversiphyllum Haw.
— dolabriforme L.
— dolomiticum Dtr.
— dregeanum Sond.
— drepanophyllum Schltr. et Bgr.
— dressianum J. Ingram
— duale N. E. Br.
— dubitans L. Bol.
— dubium Haw.
— — L. Bol.
— dulce L. Bol.
— dumosum N. E. Br.
— dunense Sond.
— duthiae L. Bol.
— dyckii Bgr.
— dyeri (L. Bol.) N. E. Br.
— eberlanzii Dtr. et Schwant.
— ebracteatum L. Bol.
— — Schltr. et Diels
— — Pax nom. nud.
— echinatum Ait.
— — Lam.
— ecklonis Salm
— edentulum Haw.
— edule L.
— — v. flavum (L.) Moss.
— — v. virescens Moss.
— edwardsiae L. Bol.
— egregium L. Bol.
— elegans Jacq.
— elineatum (L. Bol.) N. E. Br.
— elishae N. E. Br.

Glott. ruf.
— taur.
— lat. v. cultr.
Rusch. d.
Jutt. d.
Rusch. d. v. d.
— d.
Glott. d.
Cheir. d.
Lampr. d.
Dact. d.
— —
Rusch. haw.
Carpobr. d.
Ceph. d.
— caul.
— dim.
Psil. d.
Chasm. musc.
Psil. kuntz.
— dint.
Rusch. pat.
Lampr. stip.
Ceph. d.
Mitr. d.
Rusch. d.
Lampr. d.
Anteg. fiss.
Ceph. d.
— —
s. Ceph. dens.
Psil. ar.
Ceph. d.
Rhomb. d.
Rusch. d.
Lampr. d.

Rusch. d. v. d.
— trig.
— d.
— d.
Ceph. d.
— dec.
Lampr. d.
Rusch. vir.
Lampr. d.
Rusch. d.
s. Lampr. d.
Herer. alb.

Lith. bell. v. e.
Sphalm. trich.
Ceph. rang.

Del. pruin.
— —
— e. v. e.
Rusch. e.
Carpobr. e.
— —
— —
Lampr. e.
— e.
— e.

Rusch. e.
Con. e.

M. elizae Dtr. et Bgr.
— elongatum Eckl. et Zeyh.
— — Haw.
— — v. β Haw.
— — v. minimum Haw.
— — Salm
— — Thunbg.
— emarcidum Thunbg.
— emarginatoides Haw.
— emarginatum L.
— — v. puniceum Jacq.
— englerianum Dtr. et Bgr.
— englishiae L. Bol.
— enormis N. E. Br.
— erectum Haw.
— — hort. ex A. Berger, Mes. u. Port. 292 is a doubtful species.
— — L. Bol.
— erigeriflorum Jacq.
— ernestii L. Bol.
— evolutum N. E. Br.
— exacutum (L. Bol.) N. E. Br.
— excedens L. Bol.
— exiguum N. E. Br.
— exile hort. ex Haw.
— — N. E. Br.
— expansum DC.
— — L.
— — Thunbg.
— explanatum L. Bol.
— expersum N. E. Br.
— falcatum L.
— — v. galpinii L. Bol.
— — Thunbg.
— falciforme Haw.
— familiare Schwant.
— fastigiatum Haw.
— — v. pallens Haw.
— — v. reflexum Haw.
— — Thunbg.
— felinum Haw.
— — Weston
— fenchelii Schinz
— fergusoniae L. Bol.
— — (L. Bol.) N. E. Br.
— ferrugineum Schwant.
— festivum N. E. Br.
— fibulaeforme Haw.
— ficiforme Haw.
— filamentosa DC.
— — v. anceps DC.
— — L.
— filicaule Haw.
— filiforme Thunbg. is not a Mesembr.
— filipetalum (L. Bol.) N. E. Br.
— fimbriatum Marl.
— — Sond.[1]
— firmum N. E. Br.
— fissoides Haw.
— fissum Haw.

Jutt. e.
Platyth. e.
Herrea e. v. e.
— — v. —
— — v. m.
— aff.
Platyth. haeck.
Scel. e.
Lampr. e. ?
— — v. e.
— — v. p.
Mal. e.
Sphalm. e.
Rusch. e.
Opoph. fast. ?

Rusch. e.
Dros. e.
Lampr. e.
Rusch. e.

— e.
Lampr. e.
Cheir. diff.
Carpobr. aeq.
Rusch. grac.
Pren. pall. v. p.
Scel. e.
Skiat. trip.
Lampr. e.
Del. e. v. e.
Lampr. f. v. f.
— — v. g.
Semn. lac. v. l.
Lampr. f. v. f.
Con. trunc. v. t. f. t.
Sphalm. plen.
— refl.
— —
Opoph. f.
Fauc. f. v. f.
— — v. —
M. guer.
Lampr. f. v. f.
Rusch. f.
Lith. lesl. v. l.
Rusch. f.
Con. f.
— f.
Ereps. mut.
Rusch. forf.
— fil.
Lampr. f.

Rusch. f.
Con. sax. v. s.
Rusch. pygm.
— brevip. v. b.
Anteg. f.
Arg. f.

[1] See also note under *Conophytum fimbriatum*

## Mesembryanthemum INVALID DESIGNATIONS

M. fissum N. E. Br.
— flaccidum Jacq.
— — L. Bol.
— flanaganii Kensit
— flavens (L. Bol.) N. E. Br.
— flavocroceum Haw.
— flavum Haw.
— flexifolium Haw.

— flexile Haw.

— flexuosum Haw.
— — S.D.
— floribundum Haw.
— floriferum N. E. Br.
— flos-solis (Bgr.) L. Bol.
— foliolosum Haw.
— forficatum Jacq.
— — L.
— formosum Haw.
— forskahlii Hochst.
— fourcadei L. Bol.
— — (L. Bol.) N. E. Br.
— fragrans S.D.
— framesii L. Bol.
— — (L. Bol.) N. E. Br.
— fraternum N. E. Br.
— fredericii L. Bol.
— friedrichiae Dtr.
— frithii L. Bol.
— frutescens L. Bol.
— fulleri L. Bol.
— fulviceps N. E. Br.
— fulvum Haw.
— — Auct. non Haw.
— furvum L. Bol.
— fusiforme Haw.
— galpiniae L. Bol.
— gariusianum Dtr.
— geminatum Haw.
— — Jacq.

— geminiflorum Haw.
— geminum (L. Bol.) N. E. Br.
— — N. E. Br. sphalm.
— geniculiflorum L.
— — Auct. non L.
— gessertianum Dtr. et Bgr.
— giffenii L. Bol.
— — v. intertextum L. Bol.
— gigas Dtr.
— glabrum Ait.
— — Andr.
— glaciale Haw.
— gladiatum Jacq.
— glareosum Bgr.
— glaucescens Haw.
— glaucinum Haw.
— glaucoides Haw.
— glaucum L.
— — v. tortuosum Haw.
— glebula Schwant.
— globosum L.

Gibb. heath. v. h.
Saph. f.
Lampr. marc.
Del. trad.
Rhin. lut.
Mal. f.
Dros. f.
Lampr. curv. v. min.
— emarg. v. e. ?
Sphalm. f.
— acum.
Dros. f.
Rusch. pill.
Dor. f. s.
Rusch. tum.
Ereps. mut.
Rusch. f.
Lampr. f.
M. crypt.
Dros. f.
Rusch. f.
Glott. f.
Lampr. f.
Rusch. f.
Con. f.
Rusch. f.
Ophth. f.
Rhin. f.
Rusch. f.
M. inach.
Lith. f.
Arid. noct. v. f.
Psil. ment.
Lampr. f.
Herrea f.
Lampr. g.
M. guer.
Brauns. g.
Rusch. gem'fl.
— g.
— g.

Sphalm. g.
Psil. mel.
— g.
Dros. g. v. g.
— — v. i.
Ruschianth. g.
Hymen. g.
Carpanth. pom.
M. cryst.
Semn. lac.
Psil. g.
Carpobr. g.
Ereps. mut.
Lampr. aur.
— g.
— — v. t.
Con. lev.
Lampr. g.

M. glomeratum L.
— godmaniae L. Bol.
— — v. grandiflorum L. Bol.
— goodiae (L. Bol.) N. E. Br.
— gracile Haw.
— gracilipes L. Bol.
— gracilistylum L. Bol.
— gramineum Haw.
— grandiflorum Haw.
— — Schinz ex Range
— — Auct. non Schinz
— — Schinz
— graniforme Haw.
— graniticum L. Bol.
— granulatum N. E. Br.
— granulicaule Haw.
— — v. purpurascens Bgr.
— gratum N. E. Br.
— griquense L. Bol.
— griseum L. Bol.
— grossum Ait.
— — Haw.
— guerichianum Auct. non Pax
— guinanense Klotzsch ex A. Berger, Mes. u. Port. 283 is a doubtful species.
— guthriae L. Bol.
— gymnocladum Schltr. et Diels
— haeckelianum Bgr.
— halenbergense Dtr. et Schwant.
— hamatile (L. Bol.) N. E. Br.
— hamatum L. Bol.
— harazianum Defl. ex A. Berger, Mes. u. Port. 293 is a doubtful species.
— haworthii Don
— heathii N. E. Br.
— hebes N. E. Br.
— helianthoides Ait.
— — v. glaucum Sond.
— — Ait. ex de Candolle, Pl. Gr. t. 135 (1799)
— henricii L. Bol.
— herbeum N. E. Br.
— hermannii Pax nom. nud.
— herrei L. Bol.
— — (Schwant.) N. E. Br.
— hesperanthum Dtr.
— — L. Bol.
— heteropetalum Haw.
— heterophyllum Haw. acc. N. E. Br.

— — Jacks.
— hexamerum (L. Bol.) N. E. Br.
— hiemale L. Bol.
— hirsutum Haw.
— hirtellum Haw.
— hirtum N. E. Br.
— hispidum L.
— — v. α Thunbg.

Lampr. g.
— g. v. g.
— — v. g.
Rusch. g.

Ereps. g.
Lampr. g. v. g.
Con. g.
Dor. g.
Glott. g.
M. guer.
— barkl.
— guer.
Rhin. g.
Rusch. g.
Herer. g.
Psil. g.
— art.
Con. g.
Rusch. g.
— e.
Sphalm. g.
— carn.

— long.


Lampr. g.

Psil. ar.
Platyth. h.

Con. h.

Rusch. h.
— h.


Lampr. h.
Gibb. h. v. h.
Rusch. obt.
Apat. h.
— — ?

Carpanth. pom.
Lampr. h.
Del. h.
Dros. h.
Rusch. sandh.
— h.
Herer. h.
Sphalm. vig.
Ereps. h.

Glottiphyllum spec.?
— long. v. het.

Rusch. h. v. h.
Lampr. h.
Trich. h.
Dros. h.
Del. h.
Dros. h. v. h.
— cand.

**INVALID DESIGNATIONS Mesembryanthemum**

| | | | |
|---|---|---|---|
| M. hispid. v. pallidum Haw. | Dros. florib. | M. julii Dtr. | Lith. j. v. j. |
| — — v. platypetalum Haw. | — hisp .v. p. | — — Dtr. et Schwant. | — — v. - |
| — hispifolium Haw. | — str. v. h. | — junceum Haw. | Psil. j. |
| — hoerleinianum Dtr. | Lampr. h. | — — v. pauciflorum Sond. | — pauc. |
| — hollandii L. Bol. | — h. | — — L. Bol. | — clavul. |
| — hookeri Bgr. | Lith. h. | — juritzii L. Bol. | Carpobr. dim. |
| — horizontale Haw. | Sphalm. def. ? | — juttae Dtr. et Bgr. | Syn. j. |
| — hospitale Dtr. nom. nud. | Eberl. h. | — karasbergense L. Bol. | Herer. hesp. |
| — hugo-schlechteri Tisch. | Tit. h.-s. v. h.-s. | — karasmontanum Dtr. et | |
| — humifusum Ait. | Micr. pap. v. p. | Schwant. | Lith. kar. v. k. |
| — humile Haw. ex A. Berger, Mes. u. Port. 209, 292 is a doubtful species. | | — karroicum L. Bol. | Rusch. k. |
| | | — klaverense L. Bol. | — k. |
| | | — klinghardtianum Dtr. | Del. k. |
| — — L. Bol. | Lampr. perr. | — knysnanum L. Bol. | Rusch. k. v. k. |
| — hurlingii L. Bol. | — h. | — kolbei L. Bol. | Jacobs. k. |
| — hutchinsonii (L. Bol.) N. E. Br. | Rusch. h. | — kovisimontanum Dtr. | Jutt. k. |
| | | — kuntzei Schinz | Psil. k. |
| — hybridum Haw. ex A. Berger, Mes. u. Port. is a lost hybrid. | | — labyrintheum N. E. Br. | Con. l. |
| | | — lacerum Haw. | Semn. l. v. l. |
| | | — lacuniatum N. E. Br. | Rusch. vel. |
| — hypertrophicum Dtr. | Hal. h. | — laetum L. Bol. | Lampr. l. |
| — ignavum Schwant. nom. nud. | | — laeve Ait. ex N. E. Brown, New and old spec. of Mesembr. 103. Present status unclarified. | |
| — imbricans Haw. | Lampr. emarg. v. e. | | |
| — imbricatum Haw. | Rusch. i. | — — L. Bol. nom. nud. | |
| — — v. rubrum Haw. | — tumid. | — — Thunbg. | Mal. thunb. |
| — immelmaniae L. Bol. | Lampr. i. | — laevigatum Haw. | Carpobr. l. |
| — impressum (L. Bol.) J. Ingram | Rusch. i. | — lanceolatum Haw. incl. vars. | M. ait. |
| — inachabense Engl. | Call. i. | — lanceum Thunbg. | Pren. pall. v. p. |
| — inaequale Haw. | Lampr. i. | — lapidiforme Marl. | Didym. l. |
| — inclaudens Haw. | Ereps. i. | — latens N. E. Br. | Rusch. inconsp. |
| — inclusum L. Bol. | Oct. i. | — lateriflorum DC. | Dros. erig. |
| — incomptum Haw. | Del. i. v. i. | — latum (S.D.) Haw. | Glott. l. v. l. |
| — — v. ecklonis Salm | — — v. e. | — — v. breve Haw. | — — v. — |
| — inconspicuum Haw. | Lampr. i. | — — L. Bol. | Lampr. plaut. |
| — incurvatum (L. Bol.) N. E. Br. | Rusch. i. | — lavisii L. Bol. | — l. v. l. |
| | | — lawsonii L. Bol. | Rusch. l. |
| — incurvum Haw. | Lampr. i. | — laxifolium L. Bol. | Lampr. l. |
| — — v. roseum DC. | — ros. | — laxipetalum (L. Bol.) N. E. Br. | Rusch. l. |
| — indecorum L. Bol. | Rusch. i. | | |
| — induratum L. Bol. | — i. | — laxum L. Bol. | Lampr. subl. |
| — inexpectatum Dtr. nom. nud. | Dint. i. | — — Willd. | Rusch. l. |
| | | — lectum N. E. Br. | Cheir. l. |
| — inflexum Haw. | Lampr. glom. | — lehmannii Eckl. et Zeyh. | Del. l. |
| — insidens (L. Bol.) N. E. Br. | Rusch. i. | — leightonii L. Bol. | Lampr. l. |
| — insigne Schltr. | Ereps. i. | — leipoldtii L. Bol. | — l. |
| — insititum Willd. | Mal. croc. | — — (L. Bol.) N. E. Br. | Rusch. l. |
| — insolitum L. Bol. | Jutt. i. | — lene N. E. Br. | — moll. |
| — inspersum N. E. Br. | Cheir. i. | — lepidum Haw. | Lampr. prod. v. l. |
| — integrum L. Bol. | Smicr. vir. | | |
| — intermedium L. Bol. | Dros. i. | — leptaleon Haw. | — l. |
| — intervallare L. Bol. | Moss. i. | — leptarthron Bgr. | Psil. l. |
| — intonsum Haw. | Trich. i. | — leptosepalum L. Bol. | Lampr. l. |
| — intricatum N. E. Br. | Eberl. i. | — lericheanum Dtr. et Schwant. | Lith. bell. v. l. |
| — introrsum Haw. sec. Hook. f. sphalm. | Trich. intons. | | |
| | | — lerouxiae L. Bol. | Rusch. l. |
| — intrusum Kensit | Rusch. i. | — lesliei N. E. Br. | Lith. l. v. l. |
| — invalidum N. E. Br. (insufficiently descr.) | Del. spec.? | — leucantherum L. Bol. | Rusch. l. |
| | | — leviculum N. E. Br. | Con. l. |
| — iteratum N. E. Br. | Rusch. britt. | — levynsiae L. Bol. | Rusch. l. |
| — ivori N. E. Br. | — i. | — lewisiae L. Bol. | Lampr. l. |
| — johannis-winkleri Dtr. et Schwant. | Con. j. w. | — liberale L. Bol. | — l. |
| | | — limbatum N. E. Br. | Rusch. l. |
| — jucundum N. E. Br. | — grat. | — limpidum Ait. | Dor. bell. |
| — jugiferum N. E. Br. | — ficif. | — lindequistii Engl. | Psil. l. |

## Mesembryanthemum INVALID DESIGNATIONS

| | | | |
|---|---|---|---|
| M. lineare Thunbg. | Dor. gram f. alb. | M. longum v. declive Haw. | Glott. depr. |
| — lineolatum Haw. | Rusch. l. | — — v. — S.D. | — lat. v. cultr. |
| — linguaeforme Haw. | Glott. lat. v. l. | — — v. flaccidum Haw. | — — v. - |
| — — v. assurgens Haw. | — — v. - | — — v. purpurascens Haw. is an unclarified name. | |
| — — v. prostratum Haw. | — — v. - | — — v. uncatum (S.D.) Haw. | — unc. |
| — v. rufescens Haw. | — ruf. | — loratum Haw. | Pren. pall. v. p. |
| — — v. subcruciatum Haw. | — lat. v. l. | — loreum L. | Ceph. l. |
| — — S.D. | — — v. cultr. | — — red. fl. | — dim. |
| — linguiforme L. | — l. | — — yellow fl. | — divers. |
| — — v. angustum (Haw.) Bgr. | — ang. | — — v. congestum Haw. | — vand. |
| | | — — v. L. Bol. nom. nud. | |
| — — v. - (S.D.) Bgr. | — taur. | — — v. δ L. acc. to N. E. Br. | Ceph. divers. |
| — — v. ascendens (S.D.) Bgr. | — lat. v. l. | — — v. δ L. acc. to L. Bolus | — vand. |
| — — v. attollens (Haw.) Bgr. | — unc. | — — v. β & γ L. | — divers. |
| — — v. cultratum (S.D.) Bgr. | — lat. v. c. | — lucidum Haw. | Glott. lat. v. cultr. |
| — — v. - sv. perviride Salm | — — v. l. | | |
| — — v. declive Bgr. | — — v. cult. | — — Mill. | — ling. |
| — — v. depressum Bgr. | — — v. - | — luederitzii Engl. | Dros. pax. |
| — — v. flaccidum Bgr. | — depr. | — lunatum Willd. | Lampr. l. |
| — — v. fragrans Bgr. | — fragr. | — lunulatum Bgr. | — l. |
| — — v. grandiflorum (Haw.) Bgr. | — grand. | — lupinum Haw. | Fauc. l. |
| | | — luteolum Haw. | Mal. l. |
| — — v. heterophyllum Bgr. | — procl. | — luteum Haw. | — l. |
| — — v. latum S.D. | — lat. v. l. | — macellum N. E. Br. | Del. m. |
| — — v. longum (Haw.) Bgr. | — long. v. l. | — macowanii L. Bol. | Rusch. m. o. |
| — — v. - sv. attollens Haw., sv. declive Haw., sv. purpurascens Haw. are synonyms. | — long. | — macradenium L. Bol. | Rhin. m. |
| | | — macrocalyx Kensit | Lampr. dun. |
| | | — macrocarpum Bgr. | — m. |
| | | — macrorrhizum DC. | Mestokl. m. |
| | | — — Haw. | — tub. |
| — — v. obliquum (Willd.) Bgr. | — lat. v. l. | — macrosepalum L. Bol. | Lampr. m. |
| | | — maculatum Haw. | Dros. m. |
| — — v. — sv. assurgens Haw., sv. prostratum Haw., sv. subcruciatum Haw. are synonyms. | — lat. | — magnificum L. Bol. | Lampr. m. |
| | | — magniflorum L. Bol. | M. perl. |
| | | — magnipunctatum Haw. | Pleiosp. m. v. m. |
| | | — — Schwant. | — purp. |
| — — v. purpurascens nom. nud. | | — magnipunctum Burch. is not a Mesembr. | |
| — — v. pustulatum (Haw.) Bgr. | — long. | — mahonii N. E. Br. | Del. m. |
| — — v. — sv. lividum Salm | — long. | — maleolens (L. Bol.) N. E. Br. | Rusch. m. |
| — — v. rufescens Haw. | — ruf. | — malleoliforme Schwant. | Con. trunc. v. t. f. t. |
| — — v. scalpratum (Haw.) Bgr. | — ling. | — mallesoniae L. Bol. | Rusch. drep. v. d. |
| — — v. uncatum (Haw.) Bgr. | — unc. | — margaretae Schwant. | Lap. m. |
| — — S.D. | — long. v. l. | — marginatum Haw. ex A. Berger, Mes. u. Port. 117 is not a spec. of Mesembr. | |
| — lique N. E. Br. | Dros. l. | | |
| — litorale Kensit | Del. l. | | |
| — locale N. E. Br. | Lith. l. v. l. | | |
| — longifolium L. Bol. | Astr. l. | | |
| — longipapillatum( L. Bol.) L. Bol. | M. ann. | — — Ktze. ex A. Berger, l. c. 118 is an unclarified spec. | |
| — longipes (L. Bol.) N. E. Br. | Rusch. l. | | |
| — longisepalum L. Bol. | Lampr. l. | — — Marl. | Lith. opt. f. o. |
| — longispinulosum Sond. | Sphalm. can. | — mariae L. Bol. | Lampr. m. |
| — longispinulum Haw. | — l. | — marianae L. Bol. | Rusch. m. |
| — S.D. | — comm. | — marinum Bgr. ex Engl. | Psil. dint. |
| — longistamineum L. Bol. | Lampr. l. | — maritimum L. Bol. Ann. Bol. Herb. IV, 2 | Ceph. m. |
| — longistylum DC. | Sphalm. refl. | | |
| — — v. purpurascens DC. | — — | — L. Bol. in Notes on Mes. III, 55 | Lampr. falc. v. m. |
| — — L. Bol. | Lampr. alt. | | |
| — longum Haw. | Glott. l. v. l. | | |
| — — v. δ Haw. | — — v. l. | — marlothii Pax | Psil. m. |
| — — v. angustum Haw. is an unclarified name. | | — marmoratum N. E. Br. | Lith. m. |
| | | — marthleyi L. Bol. | Lampr. m. |
| — — v. attollens Haw. | — unc. | — mathewsii L. Bol. | Brauns. gem. |

**INVALID DESIGNATIONS** Mesembryanthemum

| | | | |
|---|---|---|---|
| M. matutinum L. Bol. | Lampr. m. | M. muticum (L. Bol.) | |
| — maximum Haw. | Rusch. m. | N. E. Br. | Rusch. m. |
| — maximiliani Schltr. | Echin. m. | — nakurense Engl. | Del. n. |
| — — Schltr. et Bgr. | Lampr. m. | — namaquanum (L. Bol.) | |
| — maxwellii (L. Bol.) | | N. E. Br. | Rusch. n. v. n. |
| N. E. Br. | Rusch. m. | — namaquense Sond. Insufficiently known spec. | |
| — megarhizum Don | Mestokl. tub. v. m. | — namibense Marl. | Psil. n. |
| — melanospermum Bgr. | Psil. m. | — nanum L. Bol. | Con. grat. |
| — — Dtr. | Sphalm. dint. | — — Schltr. | Ooph. n. |
| — meleagre L. Bol. | Lampr. m. | — napiforme N. E. Br. | Mestokl. macr. |
| — mellei L. Bol. | Carpobr. m. | — nardouwense L. Bol. | Lampr. n. |
| — mentiens Bgr. | Psil. m. | — nationae N. E. Br. | Khad. n. |
| — meyeri (Schwant.) N. E. Br. | Rusch. m. | — necopinum N. E. Br. | Arg. × n. |
| | | — neethlingiae L. Bol. | Del. n. |
| — micans L. | Dros. m. | — nevillei N. E. Br. | Con. obcord. |
| — — Thunbg. | Del. exp. v. e. | — nigrescens Haw. | Carpobr. aequ. |
| — — v. δ Haw. | Dros. mac. | — nissenii Dtr. | Psamm. n. |
| — micranthum Haw. | Psil. parv. | — nitidum Haw. | Sphalm. n. |
| — — auct. non Haw. | — bic. | — niveum (L. Bol.) N. E. Br. | Rusch. edent. |
| — micropetalum (L. Bol.) N. E. Br. | Ruschianth. g. | — nobile Haw. | Pleiosp. n. |
| | | — noctiflorum L. | Arid. n. v. n. |
| — microphyllum Haw. | Rusch. m. | — — v. fulvum S.D. | — — v. f. |
| — microspermum Dtr. et Derenbg. | Dint. m. v. m. | — nodosum Bgr. | Mon. o. |
| | | — nothum N. E. Br. | Sphalm. n. |
| — microstigmum L. Bol. | Lampr. m. | — nubigenum Schltr. | Del. n. |
| — middlemostii L. Bol. | — m. | — nuciforme Haw. | Gibb. crypt. ? |
| — minimum Haw. | Con. m. | — nudicaule Bgr. | Ereps. n. ? |
| — minusculum N. E. Br. | — m. | — obconellum Haw. | Con. o. |
| — — Schwant. | — trunc. v. t. f. t. | — obconicum L. Bol. | Lampr. o. |
| | | — obcordellum Haw. | Con. o. |
| — minutiflorum Schwant. | — m. | — — Marl. | — trunc. v. t. f. t. |
| — minutum Haw. | — m. | | |
| — mirabile N. E. Br. | Trich. m. v. m. | — — Sims | — o. |
| — miserum L. Bol. | Rusch. m. | — obliquum Haw. | Dros. liqu. |
| — mitratum Marl. | Mitr. m. | — — Willd. | Glott. lat. v. l. |
| — modestum Dtr. et Bgr | Psamm. m. | — obmetale N. E. Br. | Con. o. |
| — molle Ait. | Mal. m. | — obsubulatum Haw. | Cyl. o. |
| — — Bgr. | Rusch. m. | — obtusum Haw. | Anteg. fiss. |
| — mollissimum Dtr. ex Friedr. | M. pell. | — ochraceum Bgr. | Glott. o. |
| | | — octojuge L. Bol. | Octop. o. |
| — moniliforme Haw. | Mon. m. | — octonarium L. Bol. | Enarg. o. |
| — montaguense L. Bol. | Lampr. m. | — octophyllum Haw. | Arg. o. |
| — montanum Schltr. | Ereps. het. | — — v. δ Haw. | — nec. |
| — monticolum L. Bol. | Lampr. sten. | — — v. γ roseum Haw. | — ros. |
| — — Sond. | Ottos. m. | — oculatum N. E. Br. | Sphalm. o. |
| — montis-draconis Dtr. | Drac. m. d. | — odontocalyx Schltr. et Diels | Rusch. o. |
| — montis-moltkei Dtr. | Ebract. m. m. | | |
| — mucronatum Haw. | Rusch. m. | — odoratum L. Bol. | Herer. o. |
| — mucroniferum Haw. | Eberl. m. | — — N. E. Br. | Con. ficif. |
| — mucronulatum Dtr. | Psil. m. | — oehleri Engl. | Del. o. |
| — muelleri L. Bol. | Rusch. m. | — oligandrum Kunze | Micr. pap. v. p. |
| — muirianum L. Bol. | — m. | — olivaceum Schltr. et Bgr. | Sphalm. o. |
| — muiri L. Bol. | Carpobr. m. | — optatum N. E. Br. | Pleiosp. o. |
| — multiceps Salm. | Berg. m. | — opticum Marl. | Lith. o. f. o. |
| — multiflorum Haw. | Rusch. m. | — — v. rubrum Tisch. | — — f. r. |
| — — v. rubrum Haw. | — tum. | — opulentum L. Bol. probably a ssp. of | Rusch. virg. |
| — multipunctatum S.D. | Cheir. bif. | | |
| — multiradiatum Jacq. | Lampr. ros. ? | — orpenii N. E. Br. | Al. o. |
| — multiseriatum L. Bol. | — m. | — otzenianum Dtr. | Dros. unifl. |
| — munitum L. Bol. | Eberl. m. | — ovatum Thunbg. | Platyth. haeck. |
| — muricatum Haw. | Osc. delt. | — oviforme N. E. Br. | Ooph. o. |
| — musculinum Haw. | Chasm. m. | — oxysepalum Schltr. et Bgr. | Ereps. o. |
| — mustellinum S.D. | Stom. m. | | |
| — mutabile Haw. | Ereps. m. | — paardebergense L. Bol. | Lampr. p. |
| — — Sond. | Rusch. karr. | — pachyphyllum L. Bol. | Cerochl. p. v. p. |
| — mutans L. Bol. | Lampr. m. | — pachypodium Kensit | Gibb. p. |

M. pageae N. E. Br. — Con. p.
— pageanum L. Bol. Ann. Bol. Herb. IV, 99. — Del. p.
— — L. Bolus, Notes on Mes. II, 87. — Lampr. orn.
— pakhuisense L. Bol. — — p.
— pallens Ait. — Pren. p. v. p.
— — Jacq. — Sphalm. refl.
— pallescens v. α et v. β — — —
— pallidum L. Bol. — Lampr. dil.
— — N. E. Br. — Con. ficif.
— palustre L. Bol. — Lampr. p.
— pansifolium N. E. Br. — Rusch. prom.
— papillatum N. E. Br. — — p.
— papuliferum DC. — Opophyt. fast. ?
— papulosum L. f. — Micr. p. v. p.
— paripetalum L. Bol. — Rusch. p. v. p.
— parvibracteatum (L. Bol.) N. E. Br. — — p.
— parviflorum Haw. — — rig.
— — Jacq. — Psil. p.
— parvifolium Haw. — Dros. p.
— parvipetalum N. E. Br. — Con. obcord. v. p.
— parvulum Schltr. — Cheirip.
— parvum N. E. Br. — Rusch. pus.
— patersoniae L. Bol. — Del. p.
— patulum Haw. — Ereps. p.
— pauciflorum L. Bol. — Lampr. p.
— paucifolium L. Bol. — — p.
— pauper Dtr. — Dros. p.
— pauxillum N. E. Br. — Con. p.
— paxianum Schltr. et Diels — Dros. p.
— paxii Engl. — — p.
— peacockii L. Bol. — Lampr. p.
— pearsonii N. E. Br. — Arg. p.
— peersii L. Bol. — Lampr. p.
— — (L. Bol.) N. E. Br. — Rusch. p.
— pellanum N. E. Br. — — muric.
— pentagonum L. Bol. — Ereps. p.
— perfoliatum Mill. — Rusch. p.
— — Mill. v. — — edent.
— — v. integrifolium L. Bol. — — unc. ?
— perforatum (L. Bol.) N. E. Br.
— perpusillum Haw. — Con. p.
— persistens L. Bol. — Lampr. p.
— perspicuum Bgr. — — schlecht.
— perviride Haw. — Gibb. gibb.
— pfeilii Engl. — Psil. p.
— phillipsii L. Bol. — Lampr. max.
— phylicoides (L. Bol.) N. E. Br. — Rusch. p.
— pictum N. E. Br. — Con. p.
— pillansii Kensit — Kensit. p.
— pilosulum N. E. Br. — Gibb. p.
— pilosum Haw. — Carpanth. p.
— piluliforme N. E. Br. — Con. p.
— pingue L. Bol. — M. barkl.
— pinnatifidum L. f. — Aeth. p.
— pinnatum Thunbg. — — —
— piquetbergense L. Bol. — Lampr. p.
— piscodorum (L. Bol.) N. E. Br. — Rusch. p.

M. pisiforme Haw. — Mon. p.
— pittenii L. Bol. sphalm. — Lampr. vanp.
— placitum N. E. Br. — Con. p. v. p.
— planum L. Bol. — Leip. p.
— platycalyx L. Bol. — Ceph. p.
— plenifolium N. E. Br. — Sphalm. p.
— plenum L. Bol. — Lampr. p.
— pocockiae L. Bol. — — p.
— pole-evansii N. E. Br. — Dint. p.-e. v. p.-e.
— politum L. Bol. — Ereps. p.
— polyanthon Haw. — Lampr. p.
— polypetalum Bgr. et Schltr. — Ereps. p.
— polyphyllum Haw. — Lampr. emarg. v. e.
— pomeridianum L. — Carpanth. p.
— — v. andrewsii Haw. — — —
— — v. glabrum DC.
— — L. Bol. — Hymen. glabr.
— pomonae Dtr. — Namib. p.
— ponderosum Dtr. — Jutt. long.
— pottsii L. Bol. — Del. p.
— praecipitatum L. Bol. — Lampr. p.
— praecox L. Bol. — — mat.
— — F. Muell. — Scorz. p.
— praecultum N. E. Br. — Dros. p.
— praepingue Haw. — Glott. p.
— — S.D. — — subdit.
— pressum N. E. Br. — Cheir. p.
— primivernus L. Bol. — Lampr. p.
— primulinum L. Bol. — Ceph. p.
— prismaticum Marl. — Pleiosp. p.
— procumbens L. Bol. — Ceph. p.
— productum Haw. — Lampr. p. v. p.
— — v. lepidum (Haw.) S.D. — — v. l.
— — v. purpureum L. Bol. — — v. p.
— prominulum L. Bol. — — p.
— promontori L. Bol. — — p.
— pronum N. E. Br. — Rusch. rad.
— proquinquum N. E. Br. — — p.
— prostratum (L. Bol.) N. E. Br. — — p.
— proximum L. Bol. — Mitr. grand. v. g.
— — N. E. Br. — — p.
— pruinosum Thunbg. — Del. p.
— psammophilum Dtr. — Rusch. muell.
— pseudoausanum Dtr. nom. nud. — Psil. pfeil.
— pseudorupicolum Dtr. — Rusch. odont.
— pseudotruncatella Bgr. — Lith. p. v. p. f. p.
— puberulum Haw. — Micr. p.
— — Dtr. nom. nud. — Psil. art.
— pugioniforme DC. — Herrea elong. v. e.
— — Haw. — Conic. cap.
— — L. — — p.
— — L. Bol. — Herrea rob.
— — Salm — Conic. comm.
— — v. purpureum Haw. and v. carneum Haw. are species of Chrysanthemum.
— pulchellum Haw. — Rusch. p. v. p.
— pulverulentum Haw. — Dros. p.
— — Willd. — Eberl. mucr.

**INVALID DESIGNATIONS Mesembryanthemum** 640

M. pulvinatum O. Ktze. ex
A. Berger, Mes. u. Port.
74. Insufficiently described
spec.
— pumilum (L. Bol.)
N. E. Br. — Rusch. p.
— punctatum Haw. — Ceph. p.
— punctulatum L. Bol. — Rusch. p.
— pungens Bgr. — p. ?
— purpurascens S.D. — Cheir. p. v. p.
— purpureo-album Haw. — Ceph. p. a.
— purpureo-croceum Haw. — Mal. p. c.
— — v. flavo-croceum Haw. — fl. c.
— purpureostylum L. Bol. — Rusch. forf.
— purpusii Schwant. — Con. trunc.
— pusillum N. E. Br. — p.
— pustulatum Haw. — Glott. long. v. l.
— — v. lividum S.D. — — v. —
— putterilii L. Bol. — Rusch. p.
— puttkamerianum Dtr. et
Bgr. — Herer. p.
— pygmaeum Haw. — Rusch. p.
— pyrodorum Diels — Stom. p.
— pyropaeum Haw. — Dor. gram. f. g.
— — v. album Haw. — — f. a.
— — v. roseum Haw. — — f. r.
— quadrifidum Haw. — Cheir. rostrata
— quaesitum N. E. Br. — Con. qu.
— quarziticum Dtr. — Rusch. qu.
— rabiesbergense L. Bol. — Lampr. r.
— racemosum N. E. Br. — Ereps. r.
— radiatum Haw. — — r.
— ramosissimum Schltr. — Dros. r.
— ramulosum Haw. — Cheir. rostr.
— rangei Engl. — Ceph. r.
— rapaceum Jacq. — Psil. r.
— rectum Haw. sphalm. =
M. erectum Haw. — Opoph. fast. ?
— recumbens N. E. Br. — Chasm. musc.
— recurvum Haw. — Rusch. scholl.
v. s.
— — L. Bol. — Lampr. r.
— — Moench. ex N. E. Brown,
New and old spec. of Mes.
123. An insufficiently described species.
— reductum N. E. Br., Gard.
Chron. 1930, I/23 is an
unclarified combination for Rusch.subgl.,R. nana
— reflexum Haw. — Sphalm. r.
— — Willd. — — canal.
— rehmannii Schinz, Fedde,
Repert. XIX, 192. Insufficiently known species.
— rehneltianum Bgr. — Herer. r.
— relaxatum Willd. — Pren. r.
— renniei L. Bol. — Ebract. m. m.
— reptans Ait. — Lampr. r.
— resurgens Kensit — Sphalm. r.
— retroflexum Haw. — Lampr. el.
— retroversum Kensit — Dipl. r.
— rheolens L. Bol. — Drac. dealb.
— rhopalophyllum Schltr. et
Diels — Fen. aur. f. r.
— rigidicaule Haw. — Rusch. r.
— rigidum Haw. — — r.

M. rigid. v. tenellum Haw. — Rusch. ten.
— ringens v. caninum L. — Carr. r.
— — v. felinum L. — Fauc. f.
— robustum Haw. — Cheir. r.
— rogersii Schoenl. et Bgr. — Del. r. v. r.
— roodiae N. E. Br. — Pleiosp. prism.
— roseatum N. E. Br. — Dros r.
— roseolum N. E. Br. — Rusch. r.
— roseum Willd. — Lampr. r.
— — v. confertum Salm — inc.
— rossii Haw. — Carpobr. r.
— rostellum DC. — Rusch. r.
— — Haw. — — r.
— rostratoides Haw. — Cheir. r.
— rostratum L. — — r.
— — v. bibracteatum Salm — bibr.
— — Salm — tuberc.
— rosulatum Kensit — Al. r.
— × rothii Hort. — Pleiosp. × r.
— rubricaule Haw. — Rusch. r.
— rubrocinctum Eckl. et
Zeyh. — Carpobr. acin.
— — Haw. — — r.
— rubrolineatum N. E. Br. — Al. r.
— rubroluteum L. Bol. — Lampr. r.
— rubrum L. Bol. — Astr. r.
— rufescens Haw. — Glott. r.
— rugulosum Bgr. ex Range — Rusch. schneid.
— rupestre L. Bol. — Lampr. r.
— rupicolum Engl. — Rusch. odont.
— rupis-arcuatae Dtr. — — r. a.
— rurale N. E. Br. — — r.
— ruschianum Dtr. — — muell.
— ruschii Dtr. et Schwant.
sphalm.
— rustii Bgr. — Lampr. r.
— sabulosum Schltr. — — schlecht.
— — Thunbg. — Apat. s.
— sabulicola Dtr. — Rusch. odont.
— salicolum L. Bol. — Lampr. s.
— salicornioides Pax — Psil. s.
— salmii Haw. — Glott. s.
— — v. decussatum Haw. — — —
— — v. semicruciatum Haw. — — —
— salmoneum Haw. — Sphalm. s.
— — Salm — — can.
— salteri L. Bol. — Lampr. s.
— sarcocalycanthum Dtr. et
Bgr. — M. crypt.
— sarmentosum Haw. — Rusch. s. v. s.
— — v. rigidum Salm — — v. r.
— saturatum L. Bol. — Lampr. s.
— sauerae L. Bol. — — s.
— saxetanum N. E. Br. — Con. s. f. s.
— saxicolum (L. Bol.) N. E.
Br. — Rusch. s.
— scabrum L. — Lampr. s.
— scalpratum Haw. — Glott. ling.
— scapiger Haw. — Berg. s.
— scapigerum Eckl. et Zeyh. — — s.
— schenkii Schinz — Psil. cil.
— × schickii Pointdexter.
Hybr.: Glott. ling. × Del.
pruin. Evidently not still in
cultivation.
— schimperi Engl. — Trich. s.
— schlechteri N. E. Br. — Rusch. s.

# Mesembryanthemum INVALID DESIGNATIONS

M. schlechteri ZAHLBR. — Lampr. s.
— schlichtianum SOND. — Psil. s.
— — AUCT. non SONDER — — kuntz.
— schneiderianum BGR. — Rusch. s.
— schoenlandianum SCHLTR. — Dros. s.
— schollii S.D. — Rusch. s. v. s.
— — v. caledonicum L. BOL. — — — v. c.
— schultzei SCHLTR. et DIELS — Leip. s.
— schwantesii DTR. — Tit. s.
— scintillans DTR. — Sphalm. s.
— scitulum N. E. BR. — Con. s.
— scutatum L. BOL. — Mon. chrys. v. c.
— secundum THUNBG. — Psil. art.
— sedoides DTR. et BGR. — Rusch. s.
— semicylindricum HAW. — Glott. s.
— semidentatum S.D. — Rusch. s.
— senarium (L. BOL.) N. E. BR. — — s.
— separatum N. E. BR. — Octop. cal.
— serpens L. BOL. — Lampr. s.
— serratum L. — Ereps. s.
— serratulum HAW. — Rusch. s.
— sessile THUNBG. — Dros. s.
— sessiliflorum AIT. — Micr. s. v. s.
— — v. album HAW. — — — v. a.
— — v. luteum HAW. — — — v. l.
— setuliferum N. E. BR. — Trich. s. v. s.
— sexpartitum N. E. BR. — Del. lehm.
— shandii N. E. BR. — Gibb. s.
— sibbettii L. BOL. — Neohenr. s.
— signatum N. E. BR. — Con. s.
— simile SOND. — Psil. ment.
— — v. namaquense SOND. — —
— simpsonii DTR. — Jutt. s.
— simulans MARL. — Pleisp. s.
— sinus-redfordiani DTR. nom. nud. — Psil. glar.
— sladenianum L. BOL. — Pren. s.
— sobrinum N. E. BR. — Rusch. s.
— sociorum L. BOL. — Lampr. s.
— socium L. BOL. — Rusch. s.
— — N. E. BR. — Rusch. s. (acc. N. E. BROWN = Arg. fiss.)
— solidum L. BOL.
— solutifolium BGR. — Rusch. s. v. s.
— sororium N. E. BR. — Psil. marl.
— spathulatum L. BOL. — Pleiosp. s.
— — THUNBG. — Dros. s.
— spathulifolium WILLD. — Al. s.
— speciosum HAW. — Dor. bell.
— spectabile HAW. — Dros. s.
— spinescens PAX in FEDDE, Repert. XIX, 236 is an insufficiently known spec. — Lampr. s.
— spiniforme HAW.
— spinosum L. — — s.
— — v. micranthum PAX — Eberl. s.
— — O. KTZE. — —
  — Mestokl. tub. v. t.
— spinuliforme HAW. — Sphalm. s.
— splendens L. — — s.
— spongiosum L. BOL. — Ceph. s.
— squamulosum (DTR.) DTR. — M. barkl.
— staminodiosum L. BOL. — Lampr. s.
— — (L. BOL.) N. E. BR. — Rusch. s.

M. stanleyi L. BOL. — Herer. s.
— stayneri L. BOL. — Lampr. s.
— steenbergense L. BOL. — — s.
— steingroeveri PAX — Rusch. od.
— stellans N. E. BR. — — stellat.
— stellatum DC. — Trich. hirs.
— — HORT. HAW. — Ereps. grac.
— — MILL. — Trich. s.
— stelligerum HAW. — — s.
— stenopetalum L. BOL. — Lampr. s.
— stenophyllum L. BOL. — Rusch. s.
— stenum HAW. — Lampr. s.
— stephanii SCHWANT. — — s.
— stipulaceum L. — — s.
— stokoei L. BOL. — Rusch. thom.
— stramineum HAW. — Arid. s.
— — WILLD. — Ceph. tric.
— stratum L. BOL. — M. louis.
— striatum DC. — Dros. s. v. pall.
— — HAW. — — s. v. s.
— — v. attenuatum SALM — — att.
— — v. hispifolium SALM — — s. v. h.
— — v. pallidum DC. — — — v. p.
— — v. roseum HAW. — — — v. striat.
— strictum HAW. ex A. BERGER, Mes. u. Port. 131 is an insufficiently known spec.
— strictum L. BOL. — Arg. fram. ssp. hall.
— strubeniae L. BOL. — Rusch. s.
— strumosum HAW. — Trich. s.
— styliferum N. E. BR. — Eberl. s.
— stylosum N. E. BR. — Con. s.
— suaveolens L. BOL. — Sphalm. s.
— — SCHWANT. nom. nud. — Stom. s.
— suavissimum DTR. — Jutt. s.
— — L. BOL. — Lampr. s.
— — f. fera L. BOL. — — — v. s. f. f.
— — oculatum L. BOL. — — — v. o. f. o.
— subaequale L. BOL. — — s.
— subalatum HAW. — Carpobr. s.
— subalbum N. E. BR. — Arg. s.
— subcompressum HAW. — Dros. s.
— subglobosum HAW. — — s.
— — L. BOL. — Lampr. subrot.
— subincanum HAW. — Del. s.
— sublunulatum N. E. BR. — Rusch. strict. v. s.
— subnodosum BGR. — Psil. s.
— subrisum N. E. BR. — Con. s.
— subrostratum HAW. — Ceph. subul.
— subulatoides HAW. — — s.
— subulatum MILL. — Acr. s.
— succulentum (L. BOL.) N. E. BR. — Rusch. s.
— succumbens DTR. — Schwant. s.
— sulcatum HAW. — Sphalm. s.
— — S.D. — — acum.
— superans L. BOL. — Lampr. s.
— suppositum L. BOL. — Zeukt. s.
— surrectum HAW. — Glott. s.
— sutherlandii HOOK. f. — Del. s.
— swartbergense L. BOL. — Lampr. s.
— taurinum HAW. — Glott. t.
— taylorianum DTR. et SCHWANT. — Con. t. v. t.

41 Lexicon of Succulent Plants

## INVALID DESIGNATIONS  Mesembryanthemum

| | | | |
|---|---|---|---|
| M. tegens F. MUELL. | Lampr. t. | M. tumidulum HAW. | Rusch. t. |
| — tenellum HAW. | Rusch. t. | — — SOND. | — fest. |
| — tenue HAW. | Psil. parvifl. | — turbinatum JACQ. | Lampr. t. |
| — — AUCT. non HAW. | — bic. | — turbiniforme HAW. | Lith. t. v. t. |
| — — L. BOL. | Lampr. parc. | — turrigerum N. E. BR. | Con. t. |
| — tenuicaule BGR. | Ereps. t. ? | — uitenhagense L. BOL. | Rusch. u. |
| — tenuiflorum JACQ. | Sphalm. t. | — umbellatum L. | — u. |
| — tenuifolium L. | Lampr. t. | — umbelliflorum JACQ. | Sphalm. u. |
| — teretifolium HAW. | Cal. t. | — uncatum S.D. | Glott. u. |
| — teretiusculum HAW. | — t. | — uncinatum L. | Rusch. u. |
| — testaceum HAW. | Del. t. | — uncinellum AUCT. | — axth. |
| — testiculare AIT. | Arg. t. | — — HAW. | — uncinat. |
| — — v. γ HAW. (Misc. 1803) | — delaet. | — uncum L. BOL. | — u. |
| — testiculatum JACQ. sphalm. | | — — v. gydouwense L. BOL. | — dub. |
| — tetragonum THUNBG. | Sphalm. t. | — unidens HAW. | — mur. |
| — tetrasepalum RGL. ex A. BERGER, Mes. u. Port. 294 is an insufficiently known spec. | | — uniflorum L. BOL. | Dros. u. |
| | | — urbanianum SCHLTR. | Ereps. u. |
| | | — utile L. BOL. | Rusch. u. v. u. |
| | | — uviforme HAW. | Con. u. v. u. |
| | | — — PURP. | — trunc. v. t. f. t. |
| — thecatum B. E. NR. | Con. min. | — vaginatum HAW. | Rusch. v. |
| — theris PERR. ex A. BERGER, Mes. u. Port. 294 is an insufficiently known spec. | | — — v. acutangulum (HAW.) BGR. | — acut. |
| — thermarum L. BOL. | Lampr. t. | — — v. curtum (HAW.) BGR. | — curt. |
| — theurkauffii JAHAND. et MAIRE | Opoph. t. | — valens N. E. BR. | — glauc. |
| | | — validum HAW. | Ceph. v. |
| — thunbergii HAW. | Mal. t. | — — S.D. | — diss. |
| — tigrinum HAW. | Fauc. t. f. t. | — vallis gratiae SCHLTR. et BGR. | Lampr. v. g. |
| — — β felinum L. | — fel. v. f. | | |
| — tinctum L. BOL. | Meyer. t. | — vallis mariae DTR. et SCHWANT. | Lith. v. m. v. v. m. |
| — torquatum HAW. | Dros. t. | | |
| — tortuosum DC. | Scel. exp. | — vanzijliae L. BOL. | Lampr. v. |
| — — L. | — t. | — variabile HAW. | — v. |
| — tradescantioides BGR. | Del. t. | — varians HGW. | Scel. tort. |
| — translucens (L. BOL.) N. E. BR. | Rusch. t. | — variifolium N. E. BR. | Rusch. div. |
| | | — velutinum DTR. | Astr. dint. v. d. |
| — transvaalense ROLFE | Nan. t. v. t. | — — L. BOL. | Gibb. v. |
| — trichosanthum BGR. | Sphalm. virid. | — ventricosum L. BOL. | Rusch. v. |
| — trichotomum THUNBG. | — t. | — verecundum L. BOL. | Lampr. v. |
| — tricolor JACQ. | Ereps. mut. | — vermeuleniae L. BOL. | Rusch. wilm. v. v. |
| — — WILLD. | Dor. t. | | |
| — — v. album HAW. | — gram. v. a. | — vernae DTR. et BGR. | Mal. engl. |
| — — v. roseum HAW. | — — v. r. | — vernale L. BOL. | Lampr. v. |
| — tricolorum HAW. | Ceph. t. | — vernicolor L. BOL. | — v. |
| — triflorum (L. BOL.) N. E. BR. | Rusch. t. | — verruculatum L. | Scopol. v. |
| | | — verruculoides SOND. | Mal. v. |
| — tripolium L. | Skiat. t. | — — v. minus SOND. | — uit. |
| — triquetrum (L. BOL.) N. E. BR. | Rusch. t. | — verruculosum (L. BOL.) N. E. BR. | Rusch. v. |
| — triticiforme L. BOL. | Eberl. t. | | |
| — trothai ENGL. | Psil. sal. | — versicolor HAW. ex A. BERGER, Mes. u. Port. 149 is an insufficiently known spec. | |
| — truncatellum DTR. | Lith. pseud. v. p. f. p. | | |
| | | — vescum N. E. BR. | Cheir. cig. |
| — — HAW. can no longer be typified. | | — vespertinum BGR. | Berg. v. |
| | | — viatorum L. BOL. | Lampr. v. |
| — — HOOK. f. | — hook. | — victoris L. BOL. | Rusch. v. |
| — — OTHMER | Con. wettst. | — vigilans L. BOL. | Sphalm. v. |
| — truncatum THUNBG. | — t. v. t. f. t. | — villersii L. BOL. | Lampr. v. |
| — tuberculatum DC. | Dros. strict. v. hisp. | — villosum L. belongs to the genus Aizoon. | |
| — — MILL. | Cheir. t. | | |
| — tuberculosum ROLFE | Fauc. t. | — vinaceum L. BOL. | Del. v. |
| — tuberosum L. | Mestokl. t. v. t. | — violaceum DC. | Lampr. emarg. v. e. ? |
| — tugwelliae L. BOL. | Herer. t. | | |
| — tulbaghense BGR. | Lampr. t. | — — HAW. | — emarg. |
| — — L. BOL. | — arg. | — virens HAW. | Rusch. neov. |

Mesembryanthemum-Nananthus INVALID DESIGNATIONS

M. virens SALM — Lampr. err.
— virescens HAW. — Carpobr. v.
— — S.D. — aequ.
— virgatum HAW. — Rusch. v.
— viride HAW. — Smicr. v.
— viridicatum N. E. BR. — Con. v. v. v.
— viridiflorum AIT. — Spahlm. v.
— vittatum N. E. BR. — Nan. v.
— volckameri HAW. — M. ait.
— vulpinum HAW. s. M. caninum v. vulp. HAW.
— vulvarium DTR. — Rusch. v.
— watermeyeri L. BOL. — Lampr. w.
— weigangianum DTR. — Leip. w.
— wettsteinii BGR. — Con. w.
— — L. BOL. — pears.
— wilmaniae L. BOL. — Rusch. w. v. w.
— wittebergense L. BOL. — w.
— woodburniae L. BOL. — Lampr. w.
— wordsworthii L. BOL. — w.
— zeyheri S.D. — z.
— zygophylloides L. BOL. — Dros. z.
**Mesembryanthus** NECKER. Mesembr. ●
M. bicolor (L.) ROTHM. — Lampr. b. v. b.
— glaucus (L.) ROTHM. — g.
— tenuifolius (L.) ROTHM. — t.
**Micropterum** SCHWANT. Mesembr. ●
M. cuneifolium (JACQ.) SCHWANT. — Dor. bell.
— limpidum (AIT.) SCHWANT. — —
— pinnatifidum (L. f.) SCHWANT. — Aeth. p.
— sessiliflorum v. luteum (HAW.) JACOBS. — M. s. v. s.
**Mitrophyllum** SCHWANT. Mesembr. ●
M. chrysoleucum (SCHLTR.) SCHWANT. — Mon. c. v. c.
— meyeri SCHWANT. — Meyer. m. v. m.
— mitratum v. eburneum L. BOL. — M. m.
— moniliforme (HAW.) SCHWANT. — Mon. m.
— nanum (L. BOL.) N. E. BR. — Dipl. leip.
— niveum L. BOL. — M. marl.
— pisiforme (HAW.) SCHWANT. — Mon. p.
— schickianum TISCH. — Dipl. leip. ?
— scutatum SCHWANT. — Mon. chrys.
**Modecca** LAM. Passifloraceae.
M. apiculata MAST. — Adenia a.
— digitata HARV. — —
— gummifera (HARV.) HARV et SOND. — g.
— lobata JACQ. — l.
— senensis MAST. — digit.
**Monadenium** PAX. Euphorbiaceae.
M. aculeolatum PAX — M. ech. v. e.
— asperrimum PAX — — v. -
— guentheri CHIOV. — stell.
— lunulatum CHIOV. — Sen. l.
— pedunculatum MANSF. — M. ech. v. glabr.
— pubibundum BALLY — simpl. v. p.
— succulentum SCHWEICK. — stap. v. st.
— zavatteri CHIOV. — ell. v. caul.
**Monanthes** HAW. Crassulaceae.
M. agriostaphys CHRIST — M. lax.
— atlantica HOOK. f. — Sed. a. v. a.

M. atl. v. fusca (EMB.) JACOBS. — Sedum a. v. f.
— — v. lutea (EMB.) JACOBS. — — v. l.
— chlorotica BORNM. — M. lax. v. c.
— muralis HOOK. f. — Sed. atl. v. a.
— pallens v. silensis PRAEG. — M. sil.
Acc. PRAEGER spontaneous hybrids:
M. anagensis × M. laxiflora — M. × anagifl.
— brachycaulon × M. laxiflora — — × titoph.
— — × M. pallens — — × interm.
— — × M. silensis — — × pum.
— laxiflora × M. pallens — — × svent.
— — × M. silensis — — × burch.
— polyphylla × M. subcrassicaulis — — × polycaul.
— silensis × M. polyphylla — — × siloph.
**Monilaria** SCHWANT. Mesembr. ●
M. chrysoleuca v. purpurea L. BOL. — M. v. f. p.
— luckhoffii L. BOL. — Maug'ell. l.
— microstigma L. BOL. — Meyer. m.
— primosii L. BOL. — p.
— ramulosa (L. BOL.) L. BOL. — Mon. r.
— scutata (L. BOL.) SCHWANT. — chrys. v. c.
— vestita SCHWANT. nom. nud.
— watermeyeri (L. BOL.) L. BOL. (Schwantesia w. L. BOL. nom. nud.). Invalid combination.
**Monothylaceum** D. DON Asclepiadaceae.
M. gordonii G. DON — Hood. g.
**Monsonia** L. Geraniaceae.
M. burmannii DC. — Sarcoc. b.
— cotyledonis L. — Pelarg. c.
— l'heritieri DC. — Sarcoc. l' h. v. l. 'h.
— mossamedensis WELW. — m.
— multifida E. MEY. — m.
— patersonii DC. — p.
— spinosa BURM. — s.
**Moringa** BURM. Moringaceae.
M. ovalifoliolata DTR. et BGR. — M. o.
**Mossia** N. E. BR. Mesembr. ●
M. intervallis (L. BOL.) N. E. BR. Gard. Chron. 1930/I, 151 is a misprint for M. intervallaris.
— verdoorniae N. E. BR. — Chasm. v.
**Myrmecodia** JACK. Rubiaceae.
M. antonii BECC. — M. ech. ?
— inermis DC. — — —
— tuberosa JACK. — — —
**Namaquanthus** L. BOL. Mesembr. ●
N. farinosus L. BOL. — Wool. f.
**Namibia** DTR. et SCHWANT. Mesembr. ●
N. ponderosa (DTR.) DTR. et SCHWANT. — Jutt. longip.
**Nananthus** N. E. BR. Mesembr. ●
N. albinotus (HAW.) L. BOL. — Rab. a. v. a.
— albipunctus (HAW.) N. E. BR. — — a. v. a.
— — (HAW.) SCHWANT. — — v. a.
— aloides (HAW.) N. E. BR. — N. a. v. a.

**INVALID DESIGNATIONS** Nananthus-Nycteranthus

N. comptonii L. Bol. — Rab. c.
— cradockensis (L. Bol.) L. Bol. — Al. jam.
— crassipes (Marl. )L. Bol. — spath.
— difformis L. Bol. — Rab. d.
— dyeri L. Bol. — Al. rubr.
— jamesii L. Bol. — Rab. j.
— — (L. Bol.) L. Bol. — Al. j.
— lodewykii (L. Bol.) L. Bol. — l.
— loganii (L. Bol.) L. Bol. — l.
— luckhoffii (L. Bol.) L. Bol. — l.
— malherbei L. Bol. — m.
— orpenii (N. E. Br.) L. Bol. — o.
— peersii L. Bol. — p.
— rubrolineatus (N. E. Br.) N. E. Br. — r.
— — (N. E. Br.) Schwant. — —
— schooneesii (L. Bol.) L. Bol. — s. v. s.
— soehlemannii Hge. jr. — peers.
— villetii L. Bol. — v.
— vittatus (N. E. Br.) N. E. Br. N. v.
**Neoaridaria** Schwant. Mesembr. ●
N. is the SG. of the invalid Genus II Nycteranthus Necker.
**Nolina** Mchx. Agavaceae.
**Nycteranthus** Necker. Mesembr. ●

N. hystrix Hort. — Beauc. grac.
— stricta nom. nud. — s.
**Notonia** DC. Compositae.
N. abyssinica A. Rich. — Sen. a.
— amaniensis Engl. — a.
— coccinea Oliv. et Hiern — c.
— corymbosa DC. — c.
— crassissima DC. — c.
— descoingsii H. Humb. — d.
— grandiflora (Wall.) DC. — ampl.
— grantii (Oliv. et Hiern) Aschers. — g.
— gregori Moore — stap. v. min.
— incisifolia Bally — ball.
— kleinioides Sch. Bip. — long. v. viol.
— lunulata (Chiov.) Chiov. — l.
— madagascariensis Humb. — long. v. m.
— obesa Defl. — defl.
— pendula (Forsk.) Chiov. — p.
— petraea R. E. Fries — jacobs.
— semperviva (Forsk.) Aschers. — s.
— subulata Bally — sub'fol.
— trachycarpa Klotzsch — pend.
Not. (DC.) Hook. f. is a Sect. of the invalid Gen. Kleinia L.


Nycteranthus Necker is an old generic name (1790), derived from the Greek $\gamma\nu\xi$ = night, and $\check{\alpha}\nu\tau\sigma\varsigma$ = flower, proposed by Necker to cover the night-flowering Mesembryanthemum species such as Mes. geniculiflorum L. and Mes. noctiflorum L.

In 1950 G. Schwantes emended this early generic name and included Aridaria N. E. Br., later also Sphalmanthus N. E. Br. This was intended to include Haworth's former **Mes**. Sections: § Spinulifera, § Digitiflora, § Trichotoma, § Noctiflora and § Geniculiflora p. part. Schwantes divided Nycteranthus Necker emend. Schwant. as follows: – SG. I. Nycteranthus; SG. II. Neoaridaria Schwant.; SG. III. Sphalmanthus (N. E. Br.) Schwant. (Sphalmanthus N. E. Br. as G.)

The validity of Nycteranthus Necker has been much disputed. The Congress on Nomenclature, Montreal 1960, did not accept the name as valid because Necker's genera could not be harmonized with the rules of the "Code of Nomenclature". See Rowley and Stearn in Nat. Cact. & Succ. J. **VIII**, 1953, 45; "Taxon", **VII**, IX, 1957, 261–262; Journ. S.Afr. Bot. **XXVII**, 1962, 228–229 and **XXX**, 1964, 43.

Acc. L. Bolus, l. c. **XXVIII**, 228–229, the G. Aridaria N. E. Br. is restricted to Mesembr. Sect. Noctiflora Haw., while Mesembr. Sect. Digitiflora Haw., Trichotoma Haw. and Spinulifera Haw. p. part. constitute the G. Sphalmanthus N. E. Br.

The G. Phyllobolus N. E. Br. is also referred to the G. Sphalmanthus N. E. Br.

All Nycteranthus names are thus reduced to synonymy.


N. abbreviatus (L. Bol.) Schwant. — Sphalm. a.
— acuminatus (Haw.) Schwant. — a.
— albertensis (L. Bol.) Schwant. — a.
— albicaulis (Haw.) Schwant. — a.
— anguineus (L. Bol.) Schwant. — a.
— arcuatus (L. Bol.) Schwant. — Arid. a.
— arenicolus (L. Bol.) Schwant. — Sphalm. a.

N. aureus (Thunbg.) Schwant. — a.
— ausanus (Dtr. et Bgr.) Schwant. — Psil. lineol.
— bijliae (N. E. Br.) Schwant. — Sphalm. b.
— blandus (L. Bol.) Schwant. — b.
— brevicarpus (L. Bol.) Schwant. — Arid. b.
— brevifolius (L. Bol.) Schwant. — b.
— brevisepalus (L. Bol.) Schwant. — Sphalm. b. v. b.

Nycteranthus INVALID DESIGNATIONS

N. calycinus (L. Bol.)
Schwant. — Arid. c.
— canaliculatus (Haw.)
Schwant. — Sphalm. c.
— carneus (Haw.) Schwant. — c.
— caudatus (L. Bol.)
Schwant. — c.
— celans (L. Bol.) Schwant. — c.
— commutatus (Bgr.)
Schwant. — c.
— compactus (L. Bol.)
Schwant. — Arid. c.
— congestus (L. Bol.)
Schwant. — Sphalm. c.
— constrictus (L. Bol.)
Schwant. — c.
— debilis (L. Bol.) Schwant. — Arid. d.
— deciduus (L. Bol.)
Schwant. — Sphalm. d.
— decurvatus (L. Bol.)
Schwant. — d.
— defoliatus (Haw.)
Schwant. — d.
— dejagerae (L. Bol.)
Schwant. — Arid. d.
— delus (L. Bol.) Schwant. — Sphalm. d.
— dinteri (L. Bol.) Schwant. — d.
— dyeri (L. Bol.) Schwant. — d.
— ebracteatus (L. Bol.)
Schwant. — trich.
— — v. brevipetalus (L. Bol.)
Schwant. — —
— elongatus (L. Bol.)
Schwant. — Arid. e.
— englishiae (L. Bol.)
Schwant. — Sphalm. e.
— esterhuyseniae (L. Bol.)
Schwant. — Arid. e.
— flexuosus (Haw.) Schwant. — Sphalm. f.
— floribundus (L. Bol.)
Schwant. — Arid. f.
— fragilis (N. E. Br.)
Schwant. — Sphalm. scint.
— framesii (L. Bol.)
Schwant. — f.
— fulvus (Haw.) Schwant. — Arid. noct. v. f.
— geniculiflorus (L.) Necker — Sphalm. g.
— — (L.) Schwant. — —
— glanduliferus (L. Bol.)
Schwant. — g.
— globosus (L. Bol.)
Schwant. — Arid. g.
— godmaniae (L. Bol.)
Schwant. — Sphalm. g.
— gracilis (L. Bol.) Schwant. — Arid. g.
— grossus (Ait.) Schwant. — Sphalm. g.
— herbertii (N. E. Br.)
Schwant. — h.
— horizontalis (Haw.)
Schwant. — defol. ?
— inaequalis (L. Bol.)
Schwant. — Arid. i.
— intricatus (L. Bol.)
Schwant. — i.
— latipetalus (L. Bol.)
Schwant. — Sphalm. l.

N. laxipetalus (L. Bol.)
Schwant. — Sphalm. l.
— laxus (L. Bol.) Schwant. — l.
— leptopetalus (L. Bol.)
Friedr. — l.
— — (L. Bol.) Schwant. — —
— ligneus (L. Bol.) Schwant. — l.
— longisepalus (L. Bol.)
Schwant. — Arid. l.
— longispinulus (Haw.)
Schwant. — Sphalm. l.
— longistylus (DC.) Schwant. — refl.
— longitubus (L. Bol.)
Schwant. — l.
— luteoalbus (L. Bol.)
Schwant. — l.
— macrosiphon (L. Bol.)
Schwant. — m.
— meridianus (L. Bol.)
Schwant. — Arid. m.
— meyeri (L. Bol.) Schwant. — m.
— muiri (N. E. Br.) Schwant. — m.
— multiseriatus (L. Bol.)
Schwant. — m.
— mutans (L. Bol.) Schwant. — m.
— noctiflorus (L.) Necker — n. v. n.
— — (L.) Rothm. — — v. —
— nothus (N. E. Br.)
Schwant. — Sphalm. n.
— obtusus (L. Bol.) Schwant. — o.
— oculatus (N. E. Br.)
Schwant. — o.
— oubergensis (L. Bol.)
Schwant. — o.
— ovalis (L. Bol.) Schwant. — Arid. o.
— parvisepalus (L. Bol.)
Schwant. — Sphalm. brev.
— — f. ferus (L. Bol.)
Schwant. — —
— paucandrus (L. Bol.)
Schwant. — Arid. p. v. p.
— — v. gracillimus (L. Bol.)
Schwant. — — v. g.
— peersii (L. Bol.) Schwant. — Sphalm. virid.
— pentagonus (L. Bol.)
Schwant. — p. v. p.
— — v. occidentalis (L. Bol.)
Schwant. — — v. o.
— platysepalus (L. Bol.)
Schwant. — p.
— plenifolius (N. E. Br.)
Schwant. — p.
— pomonae (L. Bol.)
Schwant. — p.
— prasinus (L. Bol.)
Schwant. — p.
— primulinus (L. Bol.)
Schwant. — p.
— pumilus (L. Bol.)
Schwant. — p.
— quarziticus (L. Bol.)
Schwant. — qu.
— quaternus (L. Bol.)
Schwant. — qu.
— rabiei (L. Bol.) Schwant. — r.
— rabiesbergensis (L. Bol.)
Schwant. — r.

N. radicans (L. Bol.)
  Schwant. — Psil. lind.
— recurvus (L. Bol.)
  Schwant. — Sphalm. r.
— rhodandrus (L. Bol.)
  Schwant. — — r.
— salmoneus (Haw.)
  Schwant. — — s.
— saturatus (L. Bol.)
  Schwant. — — s.
— scintillans (Dtr.) Schwant. — s.
— serotinus (L. Bol.)
  Schwant. — Arid. s.
— sinosus (L. Bol.) Schwant. Sphalm. s.
— spinuliferus (Haw.)
  Schwant. — — s.
— splendens (L.) Schwant. — s.
— stramineus (Haw.)
  Schwant. — Arid. s.
— straminicolor (L. Bol.)
  Schwant. — Sphalm. s.
— striatus (L. Bol.) Schwant. — s.
— strictus (L. Bol.) Schwant. — s.
— subaequans (L. Bol.)
  Schwant. — — s.
— subpatens (L. Bol.)
  Schwant. — — s.
— subtruncatus (L. Bol.)
  Schwant. — Arid. s.
— suffusus (L. Bol.)
  Schwant. — — s.
— sulcatus (Haw.) Schwant. — s.
— tenuiflorus (Jacq.)
  Schwant. — Sphalm. t.
— tenuifolius (L. Bol.)
  Schwant. — Arid. t. v. t.
— — v. speciosus L. Bol. — — v. s.
— tetragonus (Thunbg.)
  Schwant. — Sphalm. t. v. t.
— tetramerus (L. Bol.)
  Schwant. — — t.
— — v. parviflorus (L. Bol.)
  Schwant. — — v. p.
— trichotomus (Thunbg.)
  Schwant. — — t.
— umbelliflorus (Jacq.)
  Schwant. — — u.
— varians (L. Bol.) Schwant.
  ex Jacobs. — scint.
— vernalis (L. Bol.)
  Schwant. — — v.
— vespertinus (L. Bol.)
  Schwant. — Arid. v.
— vigilans (L. Bol.) Schwant. Sphalm. v.
— viridiflorus (Ait.) Schwant. — v.
— watermeyeri (L. Bol.)
  Schwant. — — w.
— willowmorensis (L. Bol.)
  Schwant. — — w.

**Oliveranthus** Rose. Crassulaceae.
O. elegans Rose — Ech. harms.
Oliv. Rose — Ech. Ser. 1

**Oliverella** Rose. Crassulaceae.
O. elegans Rose — Ech. harms.
Oliv. Rose — Ech. Ser. 1

**Ophiocaulon** Hook. f. Passifloraceae.
O. firingalavense Drake — Adenia f.

Oph. (Hook.) Harms. is a Sect.
  of the Gen. Adenia Drake
**Ophthalmophyllum** Dtr. et Schwant.
  Mesembr. ●
O. acutum (L. Bol.) Tisch. — Con. a.
— cornutum Schwant. — O. subf.
— edithae N. E. Br. — Con. pill.
— herrei Lavis — O. long.
— jacobsenianum Schwant. — praes.
— marlothii (N. E. Br.)
  Schwant. — Con. pell.
— maughanii (N. E. Br.)
  Schwant. v. rubrum N. E.
  Br. — O. lat. f. r.
— — (N. E. Br.) Tisch. — m.
— pillansii L. Bol. nom. nud.? Con. p.

**Opophytum** N. E. Br. Mesembr. ●
O. aquosum (L. Bol.) N. E.
  Br. — Mes. hyp.
— dactylinum (Welw.) ex
  Oliv. — — crypt.
— forskahlii (Hochst.) N. E.
  Br. — — —
— speciosum N. E. Br. nom.
  nud.

**Orbea** Haw. Asclepiadaceae.
O. anguinea Haw. — Stap. var. v.
  pict.
— bisulca Haw. — — — v. buf.
— bufonia Haw. — — v. —
— clypeata Haw. — — — v. c.
— conspurcata Schlt. — — — v. c.
— curtisii Haw. — — — v. c.
— inodora Haw. — — — v. —
— lepida Haw. — — l.
— maculosa Haw. — — 'm.'
— marginata Schult. — — var. v. m.
— marmorata Schult. — — — v. m.
— normalis Schult. — — — v. v.
— picta Haw. — — — v. p.
— planiflora Haw. — — — v. p.
— — v. marginata G. Don — — — v. m.
— quinquenervia Haw. — — — v. clyp.
— quinquenervis Loud. — — — v. —
— retusa Haw. — — — v. r.
— rugosa Haw. — — — v. r.
— trisulca Haw. — — — v. t.
— variegata Haw. — — — v. v.
— wendlandiana Schult. — verr. v. verr.
— woodfordiana Haw. — — v. —
Orb. (Haw.) Bgr. — Stap. Sect. X.

**Orostachys** Fisch. Crassulaceae.
O. aggregatus (Makino) Hara O. mal.
— — v. boehmeri (Makino)
  Ohwi — boehm.
— — v. roseus Sugaya — mal.
— filifera (Nakai) Nakai — — —
— furusei Ohwi — — —
— genkaiensis Ohwi — — —
— japonicus (Max.) Bgr. — er. v. j.
— kanboensis Ohwi — min.
— malacophyllus v. japonicus
  (Franch. et Savat) Bgr. — mal.
— polycephalus (Makino)
  Hara — er. v. p.
— sikokianus (Makino) Ohwi Meter. s.

P. tetragonus (THUNBG.) RAP.
et CAM. Sphalm. t.
**Perissolobus** N. E. BR. Mesembr. ●
P. bijlii N. E. BR. Mach. b.
**Petrophyes** WEBB et BERTH. Crassulaceae.
P. agriostaphys WEBB et
  BERTH. Monanth. lax.
    v. l.
— brachycaulon WEBB et
  BERTH. — b. v. b.
— icteria WEBB — i.
— microbotrys BOLLE et WEBB — m.
— muralis WEBB — m.
— pallens WEBB — p. v. p.
— polyphylla WEBB — p.
— purpurascens BOLLE et
  WEBB — p.
— × titophila BOLLE — × t.
**Phyllobolus** N. E. BR. Mesembr. ●
P. lesliei N. E. BR. In-
  sufficiently known spec.
— pearsonii N. E. BR. sphalm.
— pubicalyx N. E. BR. In-
  sufficiently known spec.
— resurgens (KENSIT) N. E.
  BR. Sphalm. r.
— — (KENSIT) SCHWANT. — —
**Phylloma** KER.-GAWL. Liliaceae.
P. aloiflorum KER. Lomat. purp.
**Piaranthus** R. BR. Asclepiadaceae.
P. aridus G. DON Car. a.
— fascicularis HORT. Echidn. cer.
    v. c.
— incarnatum G. DON Car. i. v. i.
— — v. albus G. DON — — v. a.
— nebrownii DTR. Piar. pulchr.
    v. n.
— parviflorus SWEET Car. p.
— piliferus SWEET Trichoc. p.
— pullus R. BR. Car. mam.
— ramosus SWEET — r.
— rorifluus DEC. Stap. verr. v. v.
— serrulatus N. E. BR. P. dec.
— streyianus NEL Car. mac. v.
    brev.
**Pilea** LINDL. Urticaceae.
P. globosa WEDD. P. serp.
— subcrenata WEDD. — —
— thymifolia BLUME — —
**Pistorinia** DC. Crassulaceae.
P. breviflora COSS. et BAL. P. b.
— salzmannii BOISS. — — v. s.
**Piquetia** N. E. BR. Mesembr. ●
P. pillansii (KENSIT) N. E. BR. Kensit. p.
— — (KENSIT) SCHWANT. — —
**Platykeleba** N. E. BR. Asclepiadaceae.
P. insignis N. E. BR. Sarcost. i.
**Platythyra** N. E. BR. Mesembr. ●
P. barklyi (N. E. BR.)
  SCHWANT. Mes. b.
— pallens (AIT.) SCHWANT. Pren. p. v. p.
— relaxata (WILLD.)
  SCHWANT. — r.
**Pleiospilos** N. E. BR. Mesembr. ●
P. compacta (AIT.) SCHWANT. P. nob.
— herrei L. BOL. nom. nud.

P. pedunculata L. BOL. P. nel.
— roodiae (N. E. BR.)
  SCHWANT. — prism.
— tricolor N. E. BR. — nel.
**Plumeria** L. Apocynaceae.
P. acutifolia POIR. P. acum.
**Podanthes** HAW. Asclepiadaceae.
P. ciliata DON. Diploc. c.
— geminata NICH. Piar. g.
— incarnata SWEET Car. i. v. i.
— lepida HAW. Stap. l.
— pulchella HAW. — p.
— pulchra HAW. — verr. v. p.
— — HAW. v. β — —
— roriflua DON. — — v. v.
— verrucosa HAW. — — v. v.
Pod. (HAW.) WHITE et SLOANE Stap. Sect. II
**Portulaca** L. Portulacaceae.
P. anacampseros L. Anac. tel.
— arachnoides HAW. — a. v. a.
— filamentosa HAW. — f.
— fruticosa THUNBG. Port'aria afr.
— lanceolata HAW. Anac. l.
— meridiana L. f. P. quadr.
— microphylla A. RICH. — —
— paniculata JACQ. Talin. p.
— patens L. MANT. — —
— poellnitziana WERDERM. et
  JACOBS. P. jacobs.
— prostrata DOMB. Calandr. umb.
— rubens HAW. Anac. arachn.
    v. a.
**Portulacaria** JACQ. Portulacaceae.
P. namaquensis SOND. Cerar. n.
— pygmaea PILL. — p.
**Procrassula** GRISEB. Crassulaceae.
P. magnoli GRISEB. Sed. rubr.
— rubens JORD. et RAUL. — r.
**Prosopostelma** BAILL. Asclepiadaceae.
P. aculeata B. DESC. Folots. a.
— grandiflora CHOUX — flor.
— madagascariensis JUM. et
  PERR. — m.
**Polymita** L. BOL. Mesembr. ●
P. pearsonii N. E. BR. P. alb.
**Prenia** N. E. BR. Mesembr. ●
P. olivacea (SCHLTR. et BGR.)
  JACOBS. Sphalm. o.
**Prepodesma** N. E. BR. Mesembr. ●
P. orpenii (N. E. BR.) N. E.
  BR. Al. o.
— uncipetala N. E. BR. Herer. u.
**Psammophora** DTR. et SCHWANT. Mesembr. ●
P. pillansii L. BOL. Arenif. p.
**Pseudolithos** BALLY. Asclepiadaceae.
P. sphaericus (BALLY) BALLY P. mig.
**Psilocaulon** N. E. BR. Mesembr. ●
P. absimile N. E. BR. P. acut.
— arenosum (SCHINZ)
  SCHWANT. — a.
— articulatum (THUNBG.)
  N. E. BR. — a.
— asperulum N. E. BR. nom.
  nud.
— caducum (AIT.) N. E. BR. Dor. ap.
— corallinum (THUNBG.)
  N. E. BR. P. c.

# INVALID DESIGNATIONS Psilocaulon-Rimaria

P. densum N. E. Br. nom. nud.
— dinteri (Engl.) N. E. Br.    P. d.
— distinctum N. E. Br.    — mel.
— diversipapillosum (Bgr.)
   N. E. Br.    — aren.
— fasciculatum N. E. Br.
   nom. nud.
— geniculiflorum Auct. non
   (L.) N. E. Br.    — mel.
— gessertianum (Dtr. et Bgr.)
   N. E. Br.    — g.
— glareosum (Bgr.) N. E. Br.    — g.
— granulicaule (Haw.)
   Schwant.    — art.
— gymnocladum (Schltr. et
   Diels) Dtr. et Schwant.    — g.
— inachabense L. Bol.    — fimbr.
— inconspicuum L. Bol.    — glar.
— junceum (Haw.) N. E. Br.    — j.
— lindequistii (Engl.) N. E.
   Br.    — l.
— luteum L. Bol.    — gess.
— micranthum (Haw.) L. Bol. — parv.
— namaquense (Sond.) N. E.
   Br.    — ment.
— — (Sond.) Schwant.    — —
— otzenianum (Dtr.) L. Bol.    Dros. unifl.
— parviflorum (Jacq.) L. Bol.    P. p.
— pfeilii (Engl.) N. E. Br.    — p.
— planum L. Bol.    — glar.
— pubescens N. E. Br. nom.
   nud.
— puberulum (Dtr.) ex Range
   nom. nud.    — art.
— rapaceum (Jacq.) L. Bol.    — r.
— salicornioides (Pax) N. E.
   Br.    — s.
— — Auct. non (Pax)
   Schwant.    — kuntz.
— schlichtianum (Sond.)
   N. E. Br.    — s.
— — Auct. non (Sond.)
   Schwant.    — kuntz.
— simile (Sond.) N. E. Br.    — ment.
— — (Sond.) Schwant.    — —
— sinus-redfordiani Dtr. ex
   Range nom. nud.    — glar.
— squamifolium N. E. Br.    — clavul.
— stenopetalum L. Bol.    — cor.
— tenue Auct. non (Haw.)
   Schwant.    — bic.
— — (Haw.) N. E. Br.    ? — parvifl.
— — (Haw.) Schwant.    ? — —
— trothai (Engl.) N. E. Br.    — salic.
— — (Engl.) Schwant.    — —
— woodii L. Bol.    — —
**Pterodiscus** Hook. Pedaliaceae.
P. heterophyllus Stapf    P. kell.
— somaliensis (Bak.) Stapf    — rusp.
— welbyi Stapf    — —
**Pteronia** L. Compositae.
P. carnosa Muschler    P. succ.
**Pteropentacoilanthus** Rappa et Camarrone.
   Mesembr. ●
P. fastigiatus (Thunbg.) Rap.
   et Cam.    Opoph. f.

P. hypertrophicus (Dtr.) Rap.
   et Cam.    Mes. h.
**Pterotetracoilanthus** Rappa et Camarrone.
   Mesembr. ●
Designations of species are not published.
**Punctillaria** N. E. Br. Mesembr. ●
P. cana (Haw.) L. Bol.    Pleiosp. c.
— compacta (Ait.) N. E. Br.    — nob.
— dekenahi N. E. Br.    — d.
— magnipunctata (Haw.)
   N. E. Br.    — m. v. m.
— — v. sesquiuncialis L. Bol.    — sesq.
— nobilis (Haw.) N. E. Br.    — n.
— optata (N. E. Br.) N. E. Br.    — o.
— purpusii (Schwant.) N. E.
   Br.    — p.
— roodiae (N. E. Br.) N. E.
   Br.    — prism.
— sesquiuncialis N. E. Br.    — s.
— sororia (N. E. Br.) N. E.
   Br.    — s.
**Puya** Mol. Bromeliaceae.
P. rubricaulis Miers.    P. coer.
— suberosa Mol.    — chil.
— wytei Hook. f.    — alp.
**Quaqua** N. E. Br. Asclepiadaceae.
Q. hottentotorum N. E. Br.    Car. h. v. h.
**Rabiea** N. E. Br. Mesembr. ●
R. carolinensis (L. Bol.)
   N. E. Br.    Khad. c.
— cibdela (N. E. Br.) N. E. Br. Nan. c.
**Raphionacme** Schltr. Asclepiadaceae.
R. divaricata Harv.    R. hirs.
**Rhinephyllum** N. E. Br. Mesembr. ●
R. luteum v. brevifolium
   L. Bol.    R. gran.
— pillansii N. E. Br.    Neorh. p.
— willowmorense L. Bol.    Chasm. w.
**Rhodiola** L. Crassulaceae.
R. alaskana Rose    R. ros. ssp. a.
— arctica Boriss.    — — ssp. a.
— asiatica D. Don.    — quadr.
— atuntsuensis (Praeg.) Fu    — —
— forrestii (Hamet) Fu    — yunn. v. f.
— henryi (Diels) Fu    — — v. h.
— integrifolia Raf.    — ros. ssp. i.
— kirilowii v. altum Froed.    — longic.
— neo-mexicana Br.    — ros. ssp. n. m.
— nobilis (Franch.) Fu    — quadr.
— polygama (Rydb.) Br. et R. — ros. ssp. p.
— roanensis Br.    — — ssp. r.
— scabrida (Franch.) Fu    — quadr.
— tachiroi Franch. et Sav.    — ros. ssp. t.
— wallichiana (Hook. f.) Fu    — crass.
— — v. cholaensis (Praeg.)
   Fu    — — v. c.
Rhod. (L.) Scop.    Sed. Sect. 1
     (delete)
**Rhopalocyclus** Schwant. Mesembr. ●
R. herrei Schwant.    Leip. h.
— nelii Schwant.    — jacobs.
— weigangianus (Dtr.) Dtr.
   et Schwant.    — w.
**Rimaria** N. E. Br. Mesembr. ●
R. angusta L. Bol.    Vanh. a.
— comptonii L. Bol.    Gibb. c.
— — forma L. Bol.    — —

O. spinosus v. erubescens
  (MAX.) BGR.    O. er. v. e.
**Oscularia** SCHWANT. Mesembr. ●
O. deltata SCHWANT.    O. ped.
— — v. major (WESTON)
  SCHWANT.    — — v. m.
— deltoides v. pedunculata
  (N. E. BR.) SCHWANT.    — ped.
— major (WESTON) SCHWANT.    — delt. v. m.
— muricata (HAW.) SCHWANT.    — —
**Othonna** L. Compositae.
O. aeonioides DTR.    O. furc.
— barkerae COMPT.    — osteosp.
— bulbosa α L.    — tub.
— — β L.    — dent.
— — WILLD.    — tub.
— calthoides MILL.    Hert. cheirif.
— cheirifolia L.    — —
— coronopifolia THUNBG.    O. arb.
— crassicaulis COMPT.    — protect.
— crassifolia HARV.    — cap.
— — L. ?    Hert. cheirif.
— dinteri MUSCHLER ex DTR.    — ciliat.
— filicaulis ECKL.    O. cap.
— floribunda SCHLTR. ex
  RANGE    — sparsifl.
— frutescens L.    — othonnit.
— lamulosa SCHINZ    — retrofr.
— lasiocarpa (DC.) SCH. BIP.    — —
— linifolia L. f.    — stenoph.
— litoralis DTR.    — retrofr.
— macrocephala MUSCHLER
  nom. nud.    Hert. ciliat.
— minima DC.    O. cacal.
— papillosa DTR.    — sedif.
— papulosa DTR. nom. nud.    — —
— pillansii HUTCHINS.    — cacal.
— pusilla DTR.    Sen. klingh.
— pygmaea COMPT.    O. cacal.
— rhopalophylla DTR.    Sen. aloid.
— schaeferi MUSCHLER ex DTR.
  nom. nud.    O. retrofr.
— schlechteriana DTR. non.
  succ. ?    Euryops aspar.
— surculosa MUSCHLER ex DTR.
  nom. nud.    O. retrofr.
— tenuissima HAW.    — cyl.
— torulosa MUSCHLER ex DTR.
  nom. nud.    — —
— vestita DC.    Sen. scap. v.
     caul. ?
**Othonnopsis** L. Compositae.
O. cheirifolia (L.) BENTH. et
  HOOK.    Hert. c.
**Pachydendron** HAW. Liliaceae.
P. africanum HAW.    Al. a.
— — v. angustum HAW.    — —
— — v. latum HAW.    — —
— angustifolium HAW.    — —
— ferox HAW.    — f.
— principis HAW.    — p.
— pseudoferox HAW.    — fer.
— — S. D.    — —
— supralaeve HAW.    — —
**Pachyphytum** LINK, KLOTZSCH et OTTO. Crassulaceae.
P. aduncum (BAK.) ROSE    P. hook.

P. amethystinum ROSE    Grapt. a.
— chloranthum E. WALTH.    Ech. het.
— heterosepalum (ROSE)
  E. WALTH.    — —
— lingua HORT. ex BAK.    — ling'fol.
— linguaefolium HORT. ex
  GOSSOT    s. × Pachyv.
     cv. Semp.
— × pachyphytoides
  (L. DE SMET) BGR.    × — cv. Pach.
— roseum HORT. ex BAK.    P. hook.
— sodale (BGR.) ROSE    × Pachyv. cv.
     Sod.
— uniflorum ROSE    P. hook.
Pach. LINK, KLOTZSCH et OTTO
  Sect. Echeveriopsis
  E. WALTH.    Ech. Ser. 12
**Pachypodium** LDL. Apocynaceae.
P. baronii v. erythraeum
  H. POISS.    P. b. v. b.
— — v. typicum M. PICHON    — — v. —
— brevicalyx (PERR.)
  M. PICHON    — dens. v. b.
— cactipes K. SCHUM.    — ros. v. c.
— champenoisianum BOISS.    — lam. v. ram.
— drakei COST. et BOIS.    — ros. v. r.
— giganteum ENGL.    — leal. v. l.
— glabrum G. DON nom. ill.    — bisp.
— griquense L. BOL.    — succ.
— horombense H. POISS.    — vos. v. h.
— jasminiflorum L. BOL.    — —
— lamerei v. typicum DRAKE    — l. v. l.
— menabeum LEANDR.    — — v. ram.
— meridionale M. PICHON    — rut.
— ramosum COST. et BOIS.    — lam. v. r.
— rosulatum v. delphinense
  PERR.    — ris. v. r.
— — v. typicum COST. et
  BOIS.    — — v. —
— rutenbergianum v. perrieri
  H. POISS.    — r. v. p.
— — v. sofiense H. POISS.    — sof.
— — v. typicum PERR.    — rut. v. r.
— saundersii N. E. BR.    — leal. v. s.
— tomentosum G. DON nom.
  ill.    — succ.
— tuberosum LINDL.    — —
— — v. loddigesii A. DC.    — bisp.
— — sensu LODD.    — —
— windsori H. POISS.    — bar. v. w.
× **Pachyrantia** E. WALTH. Crassulaceae.
× P. clavifolia (BGR.)
  E. WALTH.    × Pachyv.
     cv. C.
× — echeverioides (BGR.)
  E. WALTH.    × — cv. —
× **Pachyveria** HAAGE et SCHMIDT. Crassulaceae.
× P. haagei HORT.    × P. cv. Glauc.
**Passchanthus** BURCH. Passifloraceae.
P. pechuelii ENGL.    Adenia p.
**Pectinaria** HAW. Asclepiadaceae.
P. mammillaris SWEET    Car. m.
**Pedalium** ROYEN. Pedaliaceae.
P. busseanum (ENGL.) STAPF    Pterod. ang.
**Pedilanthus** NECKER. Euphorbiaceae.

**INVALID DESIGNATIONS** Pedilanthus-Peratetracoilanthus

P. anacampseroides (LAM.)
  KL. et GK.
— angustifolius POIT.
— aphyllus BOISS.
— articulatus (KL. et GK.)
  BOISS.
— bahamensis MILLSP.
— campester BRANDEG.
— camporum STANDL. et
  STEYERM.
— carinatus SPRENG.
— deamii MILLSP.
— fendleri MOISS.
— ghiesbreghtii BAILL.
— greggii MILLSP
— grisebachii MILLSP. et
  BRITT.
— gritensis ZAHLBR.
— houlletii BAILL.
— involucratus (KL. et GK.)
  BOISS.
— irensis BRITT.
— itzaeus MILLSP.
— jamaicensis MILLSP. et
  BRITT.
— latifolius MILLSP. et BRITT.
— macradenius DONNELL
— olson-sefferi MILLSP.
— parasiticus KL. et GK.
— pavonis (KL. et GK.) BOISS.
— peritropoides MILLSP.
— personatus CROIZ.
— petraeus BRANDEG.
— pringlei ROBINS.
— purpusii BRANDEG.
— ramosissimus BOISS.
— retusus BENTH.
— rubescens BRANDEG.
— smallii MILLSP.
— spectabilis ROBINS.
**Peersia** L. BOL. Mesembr. ●
P. frithii L. BOL.
— macradenium (L. BOL.)
  L. BOL.
**Pelargonium** L'HER. Geraniaceae.
P. alternans AUCT. non
  WENDL.
— amabile DTR.
— chelidonifolium SALISB.
— crythmifolium AUCT. non
  J. S. SMITH
— damarense R. KNUTH
— eberlanzii DTR.
— ferulaceum AUCT. non
  (BURM. f.) WILLD.
— — (BURM. f.) WILLD. v.
  polycephalum E. MEY.
— graniticum R. KNUTH
— hamatum JACQ.
— jacobi R. A. DYER
— millefoliolatum SWEET
— mirabile DTR.
— monsoniaefolium DTR.
— paradoxum DTR.
— sisonifolium BAK.

P. tith. ssp.
  pad.
— — ssp. a.
— cymb.
— bract.
— tith. ssp. b.
— — ssp. t.
— — ssp. t.
— — ssp. t.
— ssp. —
— ssp. —
— ssp. calc.
— brach.
— tith. ssp.
  jam.
— — ssp. t.
— — ssp. —
— bract.
— tith. ssp. t.
— — ssp. par.
— — ssp. j.
— — ssp. par.
— calc.
— bract.
— tith. ssp. p.
— bract.
— palm.
— nod.
— tith. ssp. t.
— — ssp. —
— calc.
— tith. ssp. par.
— — ssp. r.
— bract.
— tith. ssp. s.
— bract.

Rhin. f.
— m.

P. crythm.
— sibt.
— fulg.
— cerat.
— roess.
— sibt.
— cerat.
— carn.
— sibt.
— ech. v. e.
— klingh.
— trist. v. dauc.
— crass.
— cort.
— klingh.
— carn.

**Pentacoilanthus** RAPPA et CAMARRONE.
  Mesembr. ●
P. aitonis (JACQ.) RAP. et
  CAM.                                  Mes. a.
— crassicaulis (HAW.) RAP. et
  CAM.                                  Scel. c. —
— crystallinus (L.) RAP. et
  CAM.                                  Mes. c.
— expansus (L.) RAP. et CAM.  Scel. e.
— granulicaulis (HAW.) RAP.
  et CAM.                               Psil. g.
— splendens (L.) RAP. et CAM. Sphalm. s.
— tortuosus (L.) RAP. et CAM. Scel. t.
**Pentaschista** DTR. nom. nud. (Pentaschisma).
  Mesembr. ●
P. arborescens DTR. nom. nud. Stoeb. beetz.
                                        v. a.
**Peperomia** RUIZ. et PAV. Piperaceae.
P. nummulariaefolia H. B.
  et K.                                 P. rot.
— — v. pubescens C. DC.       — — v. pil.
— umbiliciata KUNTH.          — camp.
— — RUIZ. et PAV.             — —
— v. macrophylla C. DC.       — —
**Perapentacoilanthus** RAPPA et CAMERRONE.
  Mesembr. ●
— acuminatus (HAW.) RAP. et
  CAM.                                  Sphalm. a.
— aitonis (JACQ.) RAP. et CAM. Mes. a.
— brevifolius (L. BOL.) RAP.
  et CAM.                               Arid. b.
— crystallinum (L.) RAP. et
  CAM.                                  Mes. c.
— delus (L. BOL.) RAP. et
  CAM.                                  Sphalm. d.
— fastigiatus (THUNBG.) RAP.
  et CAM.                               Opoph. f.
— granulicaulis (HAW.) RAP.
  et CAM.                               Psil. g.
— grossus (AIT.) RAP. et CAM. Sphalm. g.
— longispinulus (HAW.) RAP.
  et CAM.                               — l.
— scintillans (DTR.) RAP. et
  CAM.                                  — s.
— splendens (L.) RAP. et CAM. — s.
— sulcatus (HAW.) RAP. et
  CAM.                                  — s.
— varensburgii (L. BOL.) RAP.
  et CAM.                               Pren. v.
— viridiflorus (AIT.) RAP. et
  CAM.                                  Sphalm. v.
**Peratetracoilanthus** RAPPA et CAMARRONE.
  Mesembr. ●
P. defoliata (HAW.) RAP. et
  CAM.                                  Sphalm. d.
— geniculiflorus (L.) RAP. et
  CAM.                                  — g.
— haeckelianus (BGR.) RAP.
  et CAM.                               Platyth. h.
— junceus (HAW.) RAP. et
  CAM.                                  Psil. j.
— noctiflorus (L.) RAP. et CAM. Arid. n. v. n.
— parviflorus (JACQ.) RAP. et
  CAM.                                  Psil. p.
— spinuliferus (HAW.) RAP. et
  CAM.                                  Sphalm. s.

R. divergens L. Bol.
— dubia N. E. Br.
— heathii (N. E. Br.) N. E. Br.
— — v. elevata L. Bol.
— — v. major L. Bol.
— luckhoffii L. Bol.
— microspermus (Dtr. et Derenb.) N. E. Br.
— pole-evansii (N. E. Br.) N. E. Br.
— primosii L. Bol.
— roodiae N. E. Br.
Rim. (N. E. Br. p. gen.) Wulff

**Rochea** DC. Crassulaceae.
R. albiflora DC.
— falcata DC.
— flava L.
— microphylla E. Mey.
— perfoliata DC.

**Roodia** N. E. Br. Mesembr. ●
R. braunsii (Schwant.) Schwant.
— brevipes (Schltr.) L. Bol.
— digitifolia N. E. Br.

**Rosularia** (DC.) Stapf. Crassulaceae.
R. chrysantha (Boiss.) Takht.
— parvifolia Froed. et Sam.
— persica Hort. non (Boiss.) Bgr.
— sempervivum (Marsch. Bieb.) Bgr.

**Ruelingia** Ehrh. Portulacaceae.
R. anacampseros Ehrh.
— angustifolia Haw.
— arachnoides Haw.
— filamentosa Haw.
— lanceolata Haw.
— polyphylla Haw.
— rubens Haw.
— rufescens Haw.

**Ruschia** Schwant. Mesembr. ●
R. albertensis L. Bol.
— albiflora L. Bol.
— argentea L. Bol.
— aristulata (Sond.) Schwant.
— armata L. Bol.
— barnardii L. Bol.
— buchubergensis Dtr.
— calycina L. Bol.
— capulata (L. Bol.) Schwant. sphalm.
— ceresiana L. Bol.
— comptonii L. Bol.
— congesta L. Bol.
— conjuncta L. Bol.
— connata L. Bol.
— coriacea (Burch.) Schwant.
— cymosa (L. Bol.) Schwant.

Vanh. d.
Gibb. heath. v. h.
— — v. —
— — v. e.
— — v. m.
— l.
Dint. m. v. m.

— p.-e.
Vanh. p.
— r.

Gibb. Sect. III Ssect. 2.

Crass. dej.
— falc.
— f.
R. jasm.
Crass. falc.

Arg. fiss.
— —

R. pall.
— lib. v. pub.

Semp. tect. ssp. calc. cv. G. S.

R. pest.

Anac. tel.
— a.
— a. v. a.
— f.
— l. v. l.
— p.
— arachn. v. a.
— r.

Eberl. a.
Polym. a.
R. quarz.

— pulch. v. p.
Eberl. a.
R. odont.
— dolom.
Octop. c.

R. cupul.
— c.
Dros. ros.
R. ceres.
Octop. c.
— c.

Psil. stenop.
R. pung.

R. derenbergiana (Dtr.) L. Bol. nom. ill.
— disarticulata L. Bol.
— diutina L. Bol.
— divaricata L. Bol.
— dyeri L. Bol.
— ferox L. Bol.
— foliolosa L. Bol.
— — (Haw.) Schwant.
— fourcadei f. luxurians L. Bol. nom. prov.
— globularis L. Bol.
— graminea L. Bol.
— heteropetala L. Bol.
— hollowayana L. Bol.
— horrescens L. Bol.
— — v. densa L. Bol.
— horrida L. Bol.
— integra (L. Bol.) Schwant.
— jacobseniana L. Bol.
— leightoniae L. Bol.
— longifolia (L. Bol.) L. Bol.
— macroura L. Bol.
— mallesoniae (L. Bol.) L. Bol.
— marginata L. Bol.
— micropetala L. Bol.
— neilii L. Bol.
— nivea L. Bol.
— octojugis (L. Bol.) L. Bol.
— parviflora (Haw.) Schwant.
— perforata L. Bol.
— persistens L. Bol. ex Notes on Mes. II, 334; non Journ. S.Afr. Bot. XXIX, 16.
— pillansii L. Bol.
— psammophila (Dtr.) Dtr. et Schwant.
— pseudorupicola Dtr. et Schwant.
— pulchella (Haw.) L. Bol.
— puniens L. Bol.
— renniei (L. Bol.) Schwant.
— rigida L. Bol.
— robusta Schwant.
— rubra (L. Bol.) L. Bol.
— rupigena L. Bol.
— ruschiana (Dtr.) Dtr. et Schwant.
— sabulicola Dtr.
— saginata L. Bol.
— spathulata L. Bol.
— steingroeveri (Pax) Schwant.
— stylosa L. Bol.
— subglobosa L. Bol.
— uncinella (Haw.) Schwant.
— vereculata (L.) Rowl.
— vulnerans L. Bol.
— willdenowii (Willd.) Schwant.

Eberl. d.
Polymit. d.
Eberl. d.
Herer. alb.
Eberl. f.
R. nanusm.
— tumid.

— fourc.
— —
— longif.
— fourc.
— schneid.
Eberl. h.
— — v. d.
— h.
Smicr. vir.
Astr. long.
R. dub.
Astr. l.
Eberl. m.

R. drep. v. d.
— dub.
Ruschianth. gig.

Stayn. n.
R. odont.
Octop. o.

R. rig.
— rup. arc.?

Eberl. p.
R. schneid.

— muell.

— odont.
— p.
Eberl. p.
Ebract. m. m.
R. bol.
— val.
Astr. r.
Octop. r.

R. muell.
— odont.
— rup. arc.?
— schneid.

— odont.
Eberl. s.
Octop. s.
R. uncinata
Scopol. v.
Eberl. v.

Lampr. curv. v. min.

**Ruschianthemum** FRIEDR. Mesembr. ●
R. merenskyanum DTR. nom.
nud. Stoeb. carp.
**Salicornia** L. Chenopodiaceae.
S. brachystachya MEYER S. eur. f. b.
— herbacea L. — —
— — f. brachystachya
(MEYER) JACOBS. — — f. b.
— — f. stricta (DUMORT.)
JACOBS. — — f. s.
— occidentalis GREENE Allenr. o.
— patula DUVAL-JOUWE S. eur. f. br.
— stricta DUMORT. — — f. s.
— tenuis BENTH. Pachycorn. t.
**Sansevieria** THUNBG. Agavaceae.
S. chinensis GENTIL S. lib.
— cornui GÉR. et LABR. — seneg.
— gentili MATTEI — lib.
— guineensis BAK. — met.
— — GÉR. et LABR. — trif.
— — WILLD. — thyrs.
— hahnii HORT. — trif. cv. H.
— laurentii DE WILD — — v. l.
— thunbergii MATTAI — aeth.
— zeylanica REDOUTÉ
— — ROXB. — roxb.
**Sarcocaulon** (DC.) SWEET. Geraniaceae.
S. ernii DTR. nom. nud.
— lorrei nom. nud. ?
— marlothii ENGL. S. moss.
**Sarcocodon** N. E. BR. Asclepiadaceae.
S. speciosus N. E. BR. Car. s.
**Sarcophagophilus** DTR. Asclepiadaceae.
S. winkleri DTR. Car. w.
— winklerianus DTR.
**Sarcostemma** R. BR. Asclepiadaceae.
S. implicatum JUM. et PERR. Cyn. i.
— tetrapterum TURCZ. — aph.
**Sarcozona** J. M. BLACK. Mesembr. ●
S. pulleinei (J. M. BLACK)
J. M. BLACK S. praec.
**Sceletium** N. E. BR. Mesembr. ●
S. varians (HAW.) L. BOL. ? S. tort.
**Schizobasopsis** FR. MACBR. Liliaceae.
S. kilimandscharica (MILDBR.)
BARSCHUSS Bow. k.
— volubilis (HARV. et HOOK. f.)
BARSCHUSS — v.
**Schoenlandia** L. BOL. Mesembr. ●
S. lehmannii (ECKL. et ZEYH.)
L. BOL. Del. l.
**Schwantesia** DTR. Mesembr. ●
S. loeschiana TISCH. S. herr. v. h. f. maj.
**Schwantesia** L. BOL. (Monilaria SCHWANT. p. part.; Mitrophyllum SCHWANT. p. part.). Mesembr. ●
S. chrysoleuca (SCHLTR.)
L. BOL. Mon. c.
— clivora (N. E. BR.) L. BOL. Mitr. c.
— cognata (N. E. BR.) L. BOL. — c.
— dissita (N. E. BR.) L. BOL. — d.
— moniliformis (HAW.) L. BOL. Mon. m.
— pisiformis (HAW.) L. BOL. — p.
— proxima (N. E. BR.) L. BOL. Mitr. p.
— ramulosa (L. BOL.) L. BOL. Mon. r.
— scutata L. BOL. — chrys. v. c.

S. watermeyeri L. BOL. nom.
nud. s. Mon. w.
**Scytanthus** HOOK. Asclepiadaceae.
S. burkei HOOK. Hood. curr.
— currori HOOK. — —
— gordonii HOOK. — burk.
**Sedastrum** ROSE. Crassulaceae.
S. chalapense (S. WATS.) ROSE Sedum ebr.
ssp. e.
— ebracteatum (DC.) ROSE — — ssp. —
— glabrum ROSE — g.
— hemsleyanum ROSE — h.
— incertum (HEMSL.) ROSE — ebr. ssp. e.
— pachuense C. H. THOMPS. — hemsl.
— painteri ROSE — —
— rubricaule ROSE — ebr. ssp. e.
Sed. (ROSE) BGR. Sed. Sect. 4
**Sedella** BR. et ROSE. Crassulaceae.
Sed. (BR. et R.) BGR. s. Sed. Sect. 21
**Sedum** L. Crassulaceae.
S. abyssinicum (HOCHST.)
HAMET Hypag. a.
— acre v. atlanticum BALL. S. a. ssp. negl.
— — v. majus MASTERS — — ssp. —
— — v. morbifugum CHABERT — — ssp. —
— — v. neglectum (TEN.) VIS. — — ssp. —
— — v. rohleanae DOM. — rohl.
— — SESSÉ et MOC. S. bourg.
— acuminatum HAMET Semp'ell. a.
— aetnense v. genuinum
HAMET S. a. —
— affine (SCHRANK.) MAX. Pseudos. a.
— aizoides DC. Aeon. a.
— aizoon v. kamtschaticum
HULTER S. kamt.
— alaskanum ROSE Rhod. ros.
ssp. a.
— albanicum BECK. S. aet. v. a.
— alberti PRAEG. — grac.
— — RGL. Pseudos. aff.
— albescens HAW. S. refl. v. a.
— alboroseum f. fol. varieg.
RGL. — a. cv. Fol. Var.
— — v. marginae-varieg.
PRAEG. — — cv. Marg. Var.
— album v. athoum DC. — a. ssp. ter. v. a.
— — v. chloroticum ROUYI et
CAM. — — ssp. —
v. c.
— — v. micranthum (BAST.)
HEGI — — ssp. —
v. m.
— — v. minutiflorum PAU — — ssp. —
v. —
— — v. murale PRAEG. — — ssp. —
v. m.
— — v. rhodopaeum (PODB.)
HAYEK — — ssp. —
v. r.
— — ssp. teretifolium v.
typicum FRANCH. — a. ssp. t. v. t.
— aleuroides BITTER — bell.
— alfredi v. makinoi (MAX.)
FROED. — mak.

## Sedum INVALID DESIGNATIONS

S. algidum LEDEB. — Rhod. a.
— — v. altaicum MAX. — — v. a.
— — v. jeniseense MAX. — — v. j.
— — v. tanguticum MAX. — — v. t.
— alpestre v. horaki ROHL. — S. hor.
— alsium FROED. — Rhod. a.
— altissimum POIR. — S. sedif.
— amplexicaule DC. — ten.
— anatolicum C. KOCH — obtus.
— andinum BALL. — Villad. a.
— anoicum PRAEG. — S. aden.
— anopetalum DC. — ochr.
— — v. montanum ROUY et CAM. — mont.
— arborescens SESSÉ et MOC. — oxyp.
— arboreum MASTERS — moran. v. m.
— arcticum (BORISS.) RONNING — Rhod. ros. ssp. a.
— arrigens GRENIER — S. refl. v. a.
— arsenii FROED. — quer.
— asiaticum DC. — Rhod. quadr.
— — WALL. et C. B. CLARKE p. part. — crass.; quadr.
— atropurpureum TURCZ. — ros. ssp. a.
— atuntsuense PRAEG. — quadr.
— atypicum BGR. — Graptop. pach.
— australe ROSE — S. guat.
— azureum DESF. — coer.
— balanense LIMPR. — Rhod. balf.
— balfouri HAMET — —
— barnesianum PRAEG. — hum.
— barrancae M. E. JONES — S. ebr. v. e.
— batesii HEMSL. — Vill. hemsl.
— berillonanum HAMET — b.
— beyrichianum MASTERS — S. wright. v. b.
— bhutanense PRAEG. — Rhod. b.
— bicolor H. B. et K. — Ech. b. v. b.
— blochmaniae EASTW. — Dudl. b. v. b.
— boehmeri (MAKINO) MAKINO — Orost. b.
— boloniense LOIS. — S. mit.
— bouvieri HAMET — Rhod. him. v. b.
— brachystylum FROED. — quadr.
— brevipetiolatum FROED. — b.
— brutium N. TERR. — mag.
— bupleuroides WALL. ex HOOK. f. et THOMS. — b.
— caerulesens HAW. – Unclarified species.
— caeruleum VATK. — S. coer.
— caespitosum DC. — rubr.
— calcicolum ROBINS. et GREENM. — Vill. c.
— carneum variegatum HORT. — S. lin. v. l. f. v.
— cavalirici LEV. — Sinocr. ind. v. i.
— chapalense S. WATS. — S. abr. ssp. e.
— chanetii LEV. — Orost. c.
— chauveaudii v. margaritae HAMET — S. marg.
— chihuahuense S. WATS. — Vill. c.
— chilonense O. KTZE. — Ech. c.
— chrysanthemifolium LÉVL. — Rhod. c.
— chrysanthum HAMET — pall.
— — v. aizoon HAMET — aiz.
— ciliosum HOWELL — S. sten. v. c.
— clavifolium BGR. — × Pachyv. 'C.'
— clusianum GUSS. — S. alb. v. gland.

S. coccineum HORT.
— — ROYLE
— coeruleum v. versicolor HAMET
— concinnum PRAEG.
— cooperi PRAEG.
— cordifolium SESSÉ et MOC.
— corsicum DUBY
— cotyledon JACQ.
— crassipes AUCT.
— — HOOK. f. et THOMS.
— — v. cholanense PRAEG.
— — v. cretinii (HAMET) FROED.
— — v. stephanii (CHAM.) FROED.
— — WALL. nom. nud.
— cretinii HAMET
— cruciatum DESF.
— dasyphyllum ssp. eu-dasyphyllum MAIRE
— — ssp. – v. alternum MAIRE
— — ssp. — v. glanduliferum (GUSS.) MAIRE
— — ssp. macrophyllum v. dyris MAIRE.
— — ssp. — v. mesatlanticum LIT. et MAIRE
— — v. oblongifolium BALL.
— — v. suendermannii PRAEG.
— delicatulum ROSE
— delicum VIERH.
— deltoideum TEN.
— desertihungaricum SIM.
— dielsianum LIMPR. f.
— dielsii HAMET
— discolor FRANCH.
— divaricatum AIT.
— diversifolium ROSE
— doratocarpum FROED.
— douglasii HOOK.
— — v. ciliosum (HOWELL) T. B. CLAUSEN
— — v. douglasii (PURSH.) T. B. CLAUSEN
— — v. radiatum (S. WATS.) T. B. CLAUSEN
— drucei GRAEBNER
— ducis-apruti ART.
— dumulosum FRANCH.
— durisii HAMET
— dyvrandae HAMET
— ebracteatum v. rubricaule (ROSE) FROED.
— edule NUTT.
— elegans LEJ.
— ellacombianum PRAEG.
— elongatum (ROSE) FEDDE
— — LEDEB.

S. spur. scarlet fl.
Rhod. quadr.
S. vers.
Rhod. c.
— bhut.
S. ebr. ssp. c.
— dasyph.
Dudl. caesp.
Rhod. c. v. chol.
— —
— — v. chol.
— — v. c.
— — v. s.
— —
— — v. c.
S. monr.
— d. ssp. d.
— — ssp. — v. a.
— — ssp. — v. g.
— — ssp. m. v. rifan.
— — ssp. m. v. m.
— — ssp. o.
— — ssp. s.
— min. ssp. d.
— rub. v. d.
— stell.
— rubr.
Rhod. d.
S. lebl.
Rhod. d.
Aichr. d.
S. gregg.
Rhod. fast.
S. sten. v. s.
— — v. c.
— — v. sten.
— — v. r.
— acre
— ruw.
Rhod. d. f. d.
— d.
Vill. d.
S. e. ssp. e.
Dudl. e.
S. forst.
— kamt. v. e.
Dudl. mult.
Rhod. ros. v. e.

**INVALID DESIGNATIONS  Sedum**  654

S. elong. WALL. nom. nud.
— elymeiticum (BOISS. et HELDR.) HAMET
— erubescens (MAX.) OHWI
— — v. japonicum (MAX.) OHWI
— — v. polycephalum (MAKINO) OHWI
— erythrocarpum PAU
— erythrostictum MIQ.
— — f. variegatum HARA

— eurycarpum FROED.
— euryphyllum FROED.
— fabaria KOCH
— falconis T. S. BRANDEG.
— farinosum ROSE
— farreri W. W. SMITH
— fastigiatum HOOK. f. et THOMS.
— filiferum S. WATS.
— fimbriatum (TURCZ.) FRANCH.
— fuscum HEMSL.
— galeottianum (HEMSL.) HAMET
— galioides ALL.
— gattefossei BATT. et JAHAND.
— gelidum LEDEB.
— geminiferum WORON
— gertrudianum EASTW.

— glaciale CLARION
— glanduliferum GUSS.

— — (L. F. HENDERS.) M. E. PECK
— glaucopruinosum EKLUND

— glaucum HORT.
— — WALDST. et KIT.
— v. pallidum HAYEK
— globulariaefolium (FENZL) HAMET
— grandisepalum R. T. CLAUSEN
— grandyi HAMET
— graptopetalum BGR.
— guatemalense HORT.
— guttatum ROSE
— gypsicolum BOISS. et HELDR.
— henryi DIELS

— heptopetalum POIR.
— heterodontum HOOK. f. et THOMS.
— hillebrandtii FENZL
— himalense PRAEG.
— hirsutum ssp. eu-hirsutum MAIRE
— — ssp. winkleri FONT Y QUER
— — v. baeticum ROUYI
— hirtum LOUDON
— hispanicum HAMET

Rhod. hook.

Ros. e.
Orost. e. v. e.

— — v. j.

— — v. p.
S. aet. v. gen.
— alboros.
— cv. Fol. Var.
Rhod. e.
— e.
S. tel. ssp. f.
— quev.
— bell.
Rhod. dum. f. f.

— f.
Graptop. f.

Orost. f.
Vill. f.

— g.
S. cep. v. g.

Aichr. g.
Rhod. fast. v. g.
S. obt.
Dudl. bloch. ssp. b.
S. acr. v. g.
— das. ssp. d. v. g.

— mor.
— max. ssp. m. v. g.
— hisp. v. min.
— —
— pall.

Ros. g.

Vill. g.
— g.
Graptop. pus.
S. rubrot.
Lenoph. g.

S. alb. ssp. g.
Rhod. yunn. v. h.
S. coer.

Rhod. h.
S. sart. ssp. h.
Rhod. h.

S. h. ssp. h.

— — ssp. baet.
— ssp. —
Diop. h.
Pist. h.

S. hispanicum LAG.
— — LAM. et DC.
— hobsonii HAMET
— horridum PRAEG.
— humile HOOK. f. et THOMS.
— idaeum WEBB
— incertum HEMSL.
— indicum v. ambigua HAMET
— — v. forrestii HAMET
— — v. genuineum HAMET
— — v. longistylum (PRAEG.) FROED.
— — v. obtusifolium FROED.
— — v. serratum HAMET
— — v. yunnanense HAMET
— ishidae MAKINO
— iwarenge (MAKINO) MAKINO
— jaeschkei KURZ
— jahandiezii ssp. battandieri M. et E.
— japonicolum MAKINO
— jacalambrense PAU
— juparense FROED.
— karpeleseae HAMET
— kirilowii RGL.
— — v. altum FROED.
— — v. aurantiacum RGL.
— — v. rubrum PRAEG.
— komarovii (BORISS.) CHU
— lagascae PAU
— lanceolatum v. typicum T. B. CLAUSEN
— laxiflorum DC.
— leveilleanum HAMET
— levii HAMET
— libanoticum L.
— liciae HAMET
— livenii (LEDEB.) HAMET
— lignicaule FROED.
— likiangense FROED.
— limuloides PRAEG.

— linearifolium ROYLE
— — v. balfouri HAMET
— — v. dielsianum (LIMPR. f.) HAMET
— — v. forrestii HAMET
— — v. ovatisepalum HAMET
— — v. sacrum (PRAIN ex HAMET) HAMET
— — v. sinuatum (ROYLE ex EDGEW.) HAMET
— — v. tieghemi HAMET
— longicaule PRAEG.
— longistylum PRAEG.
— lydium v. glaucum HORT.
— macrocarpum PRAEG.
— macrolepis FRANCH.
— madagascaricum CLUS.
— magnolii BUB., also DC.
— maireanum SENNEN
— malacophyllum (PALLAS) STEUD.
— — v. iwarenge (MAKINO) FROED.
— malladrae CHIOV.

Muciz. lagasc.
S. ochr.
Rhod. h.
— quadr.
— h.
S. lac. ssp. ins.
— ebr. ssp. e.

Sinocr. dens.
— — v. f.
— — v. i.

— — v. l.
— — v. o.
— — v. s.
— yunn.
Rhod. him. v. i.
Orost. i.
S. oread.

— j. ssp. j.
Orost. er. v. jap.
S. nev.
Rhod. j.
— k.
— k.
— longic.
— kir. v. a.
— — v. r.
— k.
Muciz. l.

S. l. v. l.
Mon. l. v. l.
S. sikok.
Rhod. hum.
Ros. l. v. l.
Rhod. l.
Pseudos. l.
S. tort.
Rhod. l.
Orost. fimbr. v. ram. sv. l.
Rhod. l.
— chrys.

— diels.
— chrys.
— ovat. v. o.

— sacr.

— sin.
— tiegh.
— l.
Sinocr. ind. v. l.
S. hisp. v. min.
Rhod. m.
— m.
Kal. pinn. v. p.
S. rubr.
— vill. ssp. ar.

Orost. m.

— iw.
Hypag. ab.

S. martinii Sev. — Sinocr. ind. v. i.
— maweanum Hort. — S. acr. ssp. negl.
— maximum ssp. fabaria (Koch) Löve — tel. ssp. fab.
— megalacanthum Froed. — Rhod. rot.
— megalophyllum Froed. — m.
— melanantherum DC. — S. angl. ssp. m.
— mengtsianum Ulbr. — Rhod. yunn. v. y.
— micranthum Bast. — S. alb. ssp. ter. v. m.
— middendorfianum Max. — kamtsch. v. m.
— — v. diffusum Praeg. — — v. — sv. d.
— milii Bak. — obt.
— miserum Lindl. — Vill. m.
— montenegrinum Horak — S. mit. v. m.
— monticolum S. T. Brandeg. — dendr. ssp. m.
— moorcroftianum Wall. — Semp'ell. acum.
— moranense v. arboreum (Mast.) Praeg. — S. m. ssp. m.
— mossii Hamet — Rhod. balf.
— mucizonia (Ort.) Hamet — Muciz. hisp.
— — (Ort.) Rothm. — — —
— mucronatum Edgew. — Rhod. lin.
— muleyanum Sennen et Maure — S. alb. ssp. ter. v. purp.
— multicaule (Rose) Fedde — Dudl. m.
— naviculare Rose — S. jalisc.
— neglectum Ten. — acr. ssp. n.
— nelsonii Rose — tort.
— nesioticum G. N. Jones — lanc. v. n.
— nicaeense All. — diff.
— nobile Franch. — Rhod. quadr.
— nuristanicum Kitamaru — n.
— nutans Rose — S. cremn.
— oblongirhizum Bgr. — Dudl. mult.
— ochroleucum Sims — S. spur. white fl.
— — ssp. montanum (Perr. et Song.) D. A. Webb — mont.
— oppositifolium (Led.) Hamet — Chiast. o.
— orichalcum W. W. Sm. — Rhod. balf.
— orientoasiaticum Makino — Meter. sik.
— pachucense Praeg. — S. hemsl.
— pachyclades Aitch. et Hemsl. — Rhod. prim. v. p.
— painteri (Rose) Bgr. — S. hemsl.
— paraguayense (N. E. Br.) Bullock — Graptop. p.
— parvum Hemsl. — Vill. p.
— paucifolium Edgew. — Rhod. lin.
— paui Sennen — S. vill. ssp. arist.
— peregrinum Sessé et Moc. — oxyp.
— petiolatum Froed. — Rhod. p.
— platyphyllum Hamet sphalm.
— platyphyllum Froed. — Vill. p.
— — Fu — Sed. fui
— pleurogynanthum Hand.-Maz. — Rhod. p.

S. polycephalum (Makino) Makino — Orost. er v. p.
— polygamum Rydb. — Rhod. ros. ssp. p.
— polytrichoides Kom. — komar.
— ponticum Vel. — S. sart. v. p.
— praegerianum W. W. Smith — Rhod. hobs.
— prainii Hamet — hum.
— prealtum A. DC. — S. dendr. ssp. p.
— primuloides Franch. — Rhod. p.
— — v. pleurogynanthum (Hand.-Maz.) Froed. — pleur.
— pringlei v. ? minus
— Robins. et Seatl. — S. min. ssp. m.
— progressum Diels — Rhod. euryc.
— proponticum Aznavour — S. obt.
— pruinatum Cosson, Lange et Ball. — forst.
— pseudospectabile Praeg. — spect.
— pseudo-stapfii Praeg. — filip. v. maj.
— pseudo-telephium Eklund — tel. Immature f.
— pubescens Vahl — hisp.
— pulchellum Aznavour — polystr.
— purpurascens Koch — tel. v. purp.
— purpureo-viride Praeg. — Rhod. p.
— purpusii (K. Schum.) Kuntze — Dudl. cym. ssp. c.
— pyramidale Praeg. — Orost. chan.
— quadrifidum Pall. — Rhod. qu.
— quitense H. B. et K. — Ech. qu.
— radiatum S. Wats. — S. sten. v. r.
— radicosum Boiss. — Ros. pan.
— rariflorum N. E. Br. — Rhod. dum. f. d.
— recticaule (Boriss.) Wendelbo
— regeli Hort. — r.
— repens Schleich. — S. pil.
— rhodanthum A. Gray — alp.
— rhodiola DC. — Rhod. r.
— rhodocarpum Podp. — ros.
— roanense Br. — S. alb. ssp. ter. v. r.
— roborowskii v. someni Hamet — Rhod. ros. ssp. r.
— robustum Domin — S. som.
— — Praeg. — acr. v. r.
— rodigasii Hort. — Rhod. long. v. r.
— roseum (L.) Scop. — S. max. ssp. m. v. vers.
— — v. alaskanum (Rose) Max. — Rhod. r.
— — v. atropurpureum (Turcz.) Praeg. — — ssp. a.
— — v. elongatum (Led.) Max. — — ssp. a.
— — v. integrifolium (Raf.) Max. — — ssp. e.
— — v. microphyllum Froed. — — ssp. i.
— — v. neo-mexicanum (Br.) Max. — — ssp. m.
— — v. polygamum (Rydb.) Max. — — ssp. n. m.
— — v. roanense (Br.) Max. — — ssp. p.
— — v. roseum — — ssp. r.
— — v. sino-alpinum Froed. — — ssp. r.
— — ssp. s. a.

**INVALID DESIGNATIONS Sedum**

S. roseum v. tachiroi (FRANCH. et SAV.) MAX.
— — v. vulgare MAX.
— — v. — sv. continentale MAX.
— — v. — sv. maritimum MAX.
— — STEV.
— rostratum TEN.
— rotundatum HEMSL.
— — v. oblongum MARQ. et SHAW.
— rupestre L.
— — v. aureum WIRTGEN
— — ssp. elegans (LEJ.) HEGI et SCHMID
— — ssp. montanum (PERR. et SONG.) HEGI
— — ssp. ochroleucum (CHAIX) HEGI
— — reflexum (L.) HEGI et SCHMID
— — v. viride KOCH
— — L. emend. PRAEG.
— — VILL.
— — WILLK.
— sacrum PRAIN et HAMET
— sangpo-tibetanum FROED.
— sanctae monicae BGR.
— sarmentosa MASTERS
— sartorianum v. novaki DOM.
— saxatile DC.
— scabridum FRANCH.
— scallanii DIELS
— schimperi BRITT.
— schoenlandii HAMET
— sediforme HAMET
— sedoides HAMET
— — (DC.) ROTHM.
— semenowii (RGL. et HERD.) MAST.
— — v. kansuense FROED.
— sempervivum (MARSCH. BIEB.) HAMET
— — v. glabrum HAMET
— — LEDEB.
— sexangulare AUCT. non L.
— — L.
— — f. elatum PRISZTER
— sexfidum MARSCH. BIEB.
— sikokianum (MAKINO) HAMET
— sinicum DIELS
— sinuatum ROYLE ex EDGEW.
— skorpilii VELEN
— smithii HAMET
— soboliferum BREHM.
— spinosum THUNBG.
— splendens HORT.
— squamulosum S. WATS.
— stapfii HAMET

Rhod. r. ssp. t.
— — ssp. r.
— — ssp. — v. c.
— — ssp. — v. m.
S. stev.
— ten.
Rhod. r.
— —
S. refl.
— — v. a.
— forst.
— mont.
— ochr.
— refl.
— — v. v.
— forst.
— glauc.
— mont.
Rhod. s.
— s. t.
Dudl. mult.
S. max.
— nov.
— ann. v. a.
Rhod. quadr.
Sinocr. i. v. i.
S. epid.
Sinocr. s.
S. crassul.
Semp'ell. alb.
S. cand.
Rhod. s.
— kans.
Ros. pest.
— pers.
S. semp.
— mit.
— acr. v. s. f. s.
— — v. — f. e.
— hispan.
Meter. s.
Rhod. s.
— s.
S. aetn. v. a.
Rhod. s.
Diop. hirt. v. bor.
Orost. er. v. jap.
S. spur. v. fl. rubr.
Vill. sq.
Rhod. s.

S. stenopetalum v. nesioticum (G. N. JONES) T. B. CLAUSEN
— — v. stenopetalum T. B. CLAUSEN
— — v. typicum T. B. CLAUSEN
— stenostachyum FROED.
— stephanii CHAM.
— stribrnyi VEL.
— subclavatum HAW.
— submontanum ROSE
— suboppositum MAX.
— — v. telephioides MAX.
— surculosum COSS.
— synocarpum FROED.
— talisiense FROED.
— telephium ssp. alboroseum (BAK.) FROED.
— — ssp. maximum (L.) KROCKER.
— — ssp. purpureum (L.) SCHINZ et KELLER
— — ssp. ruprechtii JALAS
— — v. albiflorum MAX.
— — v. borderi ROUYI et CAM.
— — v. maximum L.
— — v. pluricaule MAX.
— — v. purpureum L.
— — f. roseo-variegatum PRAEG.
— — v. vulgare (HAW.) BURMAT
— — sensu BORISSOV ex KOMAROV
— teretifolium LAMK.
— — MOC. et SESSÉ
— tetramerum TRAUTV.
— tetraphyllum SIBTH. et SM.
— texanum J. G. SMITH
— tibeticum HOOK. f. et THOMS.
— tieghemi HAMET
— torreyi DON
— trientaloides PRAEG.
— trifidum HOOK. f., CLARKE et PRAEG. p. part.
— — v. balfouri HAMET
— — v. forrestii HAMET
— — WALL.
— tuberosum COSS. et LETOURN.
— valerianoides DIELS
— variegatum S. WATS.
— venustum PRAEG.
— verticillatum HAMET
— villosum L. v. aristatum EMB. et MAIRE

S. lanc. v. n.
— — v. l.
— — v. —
Sinocr. s.
Rhod. crass. v. s.
S. sart. ssp. str.
— sten. v. sten.
— mor. v. m.
Rhod. s.
— — v. t.
S. atl. v. a.
— jalisc.
Rhod. fast.
S. ming.
— max. ssp. m.
— tel. ssp. t. v. p.
— max. ssp. r.
— tel. ssp. t. v. a.
— — ssp. — v. b.
— max. ssp. m.
— plur.
— tel. ssp. t. v. p.
— — ssp. — f. r. v.
— — ssp. fab.
— max. ssp. rupr.
— alb. ssp. t.
Ech. t. v. t.
S. aetn. v. t.
— cap. v. t.
Lenoph. t.
Rhod. t.
— t.
S. nutt.
— fil.
Rhod. sin.; R. ovat. v. o.
— chrys.
— —
— trif.
— t.
— yunn. v. v.
Dudl. v.
Rhod. v.
S. triac.
— vill. ssp. a.

S. viridescens Nakai
— volkensii Engl.
— wallichianum Hook.
— weinbergii (Rose) Bgr.
— wettsteinii Freyn.
— wilczekianum Font y Quer
— willkommianum R. Fdes.
— winkleri (Willk.) W. Dod
— woodwardii N. E. Br.
— yezoense Miyade et Tatewaki
— yunnenense Franch.
— — v. forrestii Hamet
— — v. henryi (Diels) Hamet
— — v. muliense Froed.
— — v. oxyphyllum Froed.
— — v. papillocarpum Froed.
— — v. rotundifolium Froed.
— — v. strictum Froed.
— — v. valerianoides (Diels) Hamet

**Sempervivum** L. Crassulaceae.
— abyssinicum Hochst.
— acuminatum Decne.
— — Schott
— adenophorum Borb.
— admontense hort. Sündermann nom. nud. ?
— agriostaphys O. Ktze.
— aizoides DC.
— aizoon Christ
— album Edgew.
— allionii (Jord. et Fourr.) Nyam
— alpinum Griseb. et Schenk
— annuum Chr. Sm.
— arachnoideum v. glabrescens Willk.
— — v. typicum Fiori
— arboreum L.
— — v. albovariegatum West.
— — v. atropurpureum Nich.
— — v. luteovariegatum West.
— — v. variegatum G. Don
— arenarium Koch
— arvernense Lec. et Lamot.
— assimile Schott
— aureum C. Sm.
— aussendorfferi Hut.
— balsamiferum Webb et Berth.
— banaticum Domokos
— barbatum C. Sm.
— barretii Menez. ex Praeg.
— blandum Schott
— bollei Christ

S. alb. ros.
— meyr. joh.
Rhod. crass.
Graptop. par.
S. acr. v. w.
— hirs. ssp. baet. v. w.
— pedic. ssp. lus.
— hirs. ssp. baet.
— aiz. v. a.

— pluric.
Rhod. y. v. y.
— — v. f.
— — v. h.
— — v. forr.
— — v. yunn.
— pap.
— rot.
— yunn. v. forr.

— — v. v.

Hypag. a.
Semp'ell. a.
S. tect. ssp. glauc.
Diop. hirt.

— — ssp. h. f. austr.
Mon. lax. v. l.
Aeon. a.
Green. a.
Semp'ell. a.

Diop. hirt. ssp. a.
S. tect. ssp. a.
Aichr. lax.

S. a. v. doell.
— — v. a.
Aeon. a.
— — cv. A.
— — cv. A.

— — cv. L.
— — cv. A. v.
Diop. hirt. ssp. a.
s. S. tect.
S. marm.
Green. a.
S. × barb. nm. b.

Aeon. b.
S. marm.
Aeon. spath. v. s.
Aichr. vill.
S. marm. ssp. b.
Aichr. b.

S. borisii Degen et Urumov
— — v. ciliosum (Pančic) Hayek
— boutingnianum Bill. et Green.
— brassenii Schur.
— braunii Facch.

— — Funck
— — Ledeb.
— burchardii Praeg.
— burnatii Wettst.
— caespitosum C. Sm.
— calcareum Jord.
— calciforme Haw.
— canariense L.
— — v. christii Burch.
— — ssp. longithyrsum Burch. nom. nud.
— — v. virgineum Burch.
— candollei Rouy et Cam.
— castello-paivae (Bolle) Christ
— christii Praeg.
— chrysanthum Hochst.
— — v. glandulosum Chiov.
— ciliare Haw.
— — β hybridum Haw.
— ciliatum C. Sm.
— — Willd.
— ciliosum Pančic nom. nud.
— clusianum Ten.
— compactum A. DC.
— cuneatum (Webb et Berth.) Christ
— debile Schott
— decorum (Webb et Bolle) Christ
— delasoieii Lehm. et Schnittsp.

— dichotomum DC.
— diplocyclum Burch.
— divaricatum Ait.
— — v. politum Lowe
— dodrantale Willd.
— doellianum Schnittsp. et Lehm.
— domesticum Praeg.
— dumosum Lowe
— × fimbriatum Schott ex Hegi
— fimbriatum Schnittsp. et Lehm.
— × flavipilum Hausm.
— funckii Maly

— fuscum Schnittsp. et Lehm.

— gaudinii Christ
— glabrescens Sabr.
— glabrum Beck. et Szysz

S. cil. f. b.

— — f. c.

— tect. ssp. b.
Diop. heuff.
S. × rupic. nm. r.
— mont. v. stir. sv. b.
— ruth.
Aeon. b.
S. mont. ssp. b.
Aeon. sims.
S. tect. ssp. c.
Green. aur.
Aeon. c. v. c.
— palm.

— long.
— virg.
S. mont. ssp. m.

Aeon. c. p.
— palm.
— leucobl. v. l.
— — v. g.
— sims.
— × hybr.
— sims.
— c.
S. c. f. c.
— marm.
Aeon. tab.

— c.
S. mont. ssp. m.

Aeon. d.

S. × barb. nm. d.
Aichr. lax.
Green. d.
Aichr. d. v. d.
— — v. —
Green. d.

S. arachn.
Aeon. d.
Aichr. d.

S. × ros. nm. f.

S. × barb. nm. b.
— × fauc. nm. f.
— mont. ssp. stir.
— tect. ssp. alp.
— grand.
Diop. hirt. ssp. h. f. g.
— heuff. v. g.

**INVALID DESIGNATIONS Sempervivum**

S. glandulosum AIT.
— glaucum WOHLF.
— globiferum AUCT.
— — BOISS.
— — GAUD.
— — JACQ.

— — L. emend. KOCH
— — v. ruthenicum KOCH
— — REICHENB.

— glutinosum AIT.
— goochiae (WEBB et BERTH.) CHRIST
— gracile CHRIST
— hausmannii AUSSEND.

— — LEHM.

— haworthii S. D.
— heterotrichum SCHOTT

— heuffelii SCHOTT
— — v. glabrum BECK et SZYSZ
— — v. kapaonikense (PANČIC) J. A. HUBER
— — v. patens (GRISEB. et SCHENK) J. A. HUBER
— — v. — f. glaucum HORT.
— — v. reginae-amaliae (HELDR. et SART. et BAKER) BAK.
— — v. stramineum (JORD. et FOURR.) BAK.
— hierrense MURR.
— hillebrandti SCHOTT, NYM. et KOTSCHY

— himalayense KLOTZSCH
— hirsutum POLLINI

— hirtellum SCHOTT
— hirtum ALL.
— — JUSLEN
— — v. austriacum JORD. et FOURR.
— — v. — f. glaucum HORT.
— — v. — f. major HORT.
— — v. glabriusculum PARL.
— — v. hillebrandtii (SCHOTT, NYM. et KOTSCHY) HAYEK
— — — ssp. neilreichii (SCHOTT, NYM. et KOTSCHY) O. SCHWARZ

— — v. — (SCHOTT, NYM. et KOTSCHY) HAYEK

— — ssp. patens (GRISEB. et SCHENK) STOJANOFF et STEFANOFF
— — v. pumilum BERT.
— — v. raripilum HORT.

— — SIBTH. et SMITH

Aeon. g.
S. tect. ssp. g.
— georg.
— arm. v. a.
— grand.
Diop. hirt. ssp. h.
S. ruth.
— —
Diop. hirt. ssp. bor.
Aeon. g.

— g.
Green. dodr.
S. × barb. nm. b.
— × fauc. nm. flav.
Aeon. h.
S. arachn. v. doell.
Diop. h.

— — v. g.

— — v. k.

— — v. p.
— — v. — cv. G.

— — v. r. a.

— — v. s.
Aeon. h.
Diop. hirt. ssp. h. f. h.
Semp'ell. ac.
Diop. h. ssp. all.
— — ssp. ar.
— — ssp. all.
— —

— — ssp. a.
— — ssp. — f. a.
— — ssp. — f.—
— — ssp. ar.
— — ssp. h. f. h.

— — ssp. —
v. n.

— — ssp. —
v. n.

— heuff. v. p.
— hirt. ssp. ar.
— — ssp. h. cv. R.
— heuff.

S. hirtum STERNBG.
— — WIMM. et GRAB.
— holochrysum (WEBB et BERTH.) CHRIST
— hookeri HORT. VAN HOUTTE
— huteri HAUSM.

— hybridum BRÜGGER

— — SWEET
— juratense JORD.
— kochii FACCH.

— kapaonikense PANČIC
— laggeri SCHOTT

— lamottei BOREAU

— laxum HAW.
— leucoblepharum (WEBB) HUTCHINSON et E. A. BRUCE
— lindleyi WEBB et BERTH.
— lineare HAW.
— manriqueorum (BOLLE) CHRIST
— marmoreum v. dinaricum (BECK.) SOO
— masferreri HILLBR.
— mettianum HAUSM.

— — SCHNITTSP. et LEHM.
— minimum TIMB.-LAGRAVE

— minutum KZ.

— moggridgei HOOK. f.

— monanthes AIT.
— — v. subcrassicaule O. KTZE.
— montanum EICHW.
— — L. v. m. (L.) WETTST.
— — v. ochroleucum BEAUVERD.
— — VELEN.
— mucronatum EDGEW.
— mutabile SCHLTR. et BREITER

— neilreichii SCHOTT

— nobile PRAEG.
— × noricum HAYEK

— oligotrichum HAMPE
— pachycaulon (BOLLE) CHRIST
— palmense (WEBB et CHRIST) CHRIST
— parviflorum CHRIST
— patens GRISEB. et SCHENK

— percarneum MURR.

Diop. hirt. ssp. ar.
— — ssp. bor.

Aeon. h.
S. ar. v. tom.
— × rup. nm. r.
— × barb. nm. b.
Aeon. × hybr.
S. tect. ssp. t.
Diop. hirt. ssp. ar.
— heuff. v. k.
S. arachn. v. tom.
— tect. ssp. t. v. l.
Aichr. l.

Aeon. l. v. l.
— l.
— spath. v. s.

— m.

S. m. ssp. m. v. d.
Aeon. sed.
S. fauv. nm. flacc.
— tect. ssp. m.
— mont. ssp. m. v. m.
— tect. ssp. t. v. m.
— arachn. v. doell.
Mon. pol.

— subcr.
S. pum.
— m. ssp. m.

— — ssp. — v. pall.
— erythr.
Semp'ell. m.

Aeon. arb. cv. 'A'.
Diop. hirt. ssp. h. v. n.
Aeon. n.
S. × barb. nm. n.
— arachn. v. o.

Aichr. punct. v. p.
Aeon. p.
Aichr. p.
Diop. heuff. v. p.
Aeon. p.

S. × pernkofferi Hayek
— porphyrogenetos Christ
— punctatum C. Sm.
— pygmaeum C. Sm.
— — Jeanb. et Timb.
— pyrenaicum Jord. et Fourr.
— — Lamotte
— radicescens Lowe
— reginae-amaliae Heldr. et Sart. ex Baker
— — Heldr. et Guicc. ex Halachy
— rhaeticum Brügger
— rhenanum (Hegi et Schmid) Law.
— × rubellum Timb.-Lagr.
— rubicundum Schur.
— rubrifolium Schur.
— rupestre Rouy et Cam.
— rhutenicum Koch
— saundersii (Bolle) Christ
— schlehanii Schott
— — v. blandum (Schott) Hayek
— — f. brunneifolium Praeg.
— — v. dinaricum Beck.
— — v. rubrifolium (Schub.) Praeg.
— schottii Bak. nom. ill.
— sedifolium Christ
— sediforme Jacq.
— sedoides Decne.
— simense Hochst.
— simonskaianum Degen.
— simsii Sweet
— soboliferum Fleisch. et Lindem.
— — Heldr. et Sart.
— — Sims
— — f. glaucum Hort.
— — f. major Murr.
— spathulatum Hornem.
— spectabile Schnittsp. et Lehm.
— stiriacum Wettst.
— stramineum (Jord. et Fourr.) Bak.
— strepsicladum Bornem.
— subalpinum Rouy
— tabulaeforme Haw.

S. × rup. nm. p.
Aichr. p.
— p. v. p.
Aeon. tort.
S. mont. ssp. m. v. min.
— tect. ssp. bout. f. jord.
— — ssp. — f. pall.
Aeon. tort.
Diop. heuff. v. r. a.
S. marm.
— × schott. nm. r.
s. S. tect. ssp. t. v. r.
S. × fauc. v. r.
— marm. ssp. m. f. r.
— — ssp. — f. r.
— tect. ssp. r.
— r.; S. zel.
Aeon. s.
S. marm.
— — ssp. b.
— — ssp. — f. b.
— — ssp. — f. d.
— — ssp. — f. r.
— tect. ssp. gl.
Aeon. s.
Sed. s.
Semp'ell. s.
Afrov. s.
Diop. hirt.
Aeon. s.
Diop. hirt.
— — ssp. ar. ?
— — ssp. bor.
— — ssp. — cv. G.
— — ssp. — f. m.
Aeon. s. v. s.
S. tect. ssp. gl.
— mont. ssp. s.
Diop. heuff. v. s.
Aeon. spath. v. s.
S. mont. ssp. m. v. mont.
Aeon. t.

S. tatrense Dom.
— tectorum ssp. schottii (Bak.) Wettst., Hegi et Schmid
— — v. angustifolium Leybold
— — v. atlanticum (Ball.) Hook.
— — v. atropurpureum Hort.
— — v. bicolor Hort.
— — v. boissieri Hort. ex Baker
— — v. calcareum (Jord.) Cariot et St. Lager
— — v. — f. monstrosum Hort.
— — v. caucasicum Rupr.
— — v. lamottei (Boreau) Rouy et Fourr.
— — v. minutum (Kz.) Willk. et Lange
— — v. ornatum Hort.
— — v. rhenanum Hegi et Schmid
— — v. rubens Voss
— — v. tectorum (L.) Wettst.
— — v. triste Hort. ex Baker
— — v. violascens Voss
— tenuifolium Sibth. et Sm.
— theobaldii Brügger
— tomentosum Lehm. et Schnittsp.
— tortuosum Ait.
— — DC.
— — Link
— undulatum Webb et Berth.
— urbicum C. Sm.
— valverdense Praeg.
— velutinum N. E. Br.
— villosum Ait.
— — Haw.
— — Lindl.
— virgineum (Webb et Christ) Christ
— viscatum (Bolle) Christ
— viscosum Webb
— webbianum Lehm. et Schnittsp.
— wulfenii v. skorpioli Vel.
— youngianum Bourg.
Senecio L. Compositae.
S. abyssinicus (A. Rich.) Jacobs.
— adenocalyx Dtr.
— adnivalis Auct. non Stapf

Diop. hirt.
S. tect. ssp. gl.
— dolom.
— atl.
— tect. ssp. t. cv. A.
— — ssp. t. cv. B.
— — ssp. — cv. B.
— — ssp. calc.
— — ssp. — cv. G. S.
— couc.
— tect. ssp. t. v. l.
— — ssp. - v. m.
— — ssp. — v. o.
— — ssp. — v. r.
— — ssp. — v. r.
— — ssp. —
— — ssp. — cv. T.
— — ssp. — v. v.
Sed. t.
S. × rup. nm. r.
— arachn. v. t.
Aeon. t.
— dom.
— lindl.
— u.
— u.
— v.
— × v.
Aichr. v.
Aeon. tort.
— spath.
— v.
— v.
— lindl.
S. arachn. v. tom.
— cil. f. c.
Aeon. und.
S. a.
— rad.
— erici-ros.

**INVALID DESIGNATIONS** Senecio-Stapelia

S. alticola TH. FRIES f.
— antandroi v. sakamaliensis H. HUMB.
— anteuphorbium (L.) HOOK. f.
— barorum v. ellipticus H. HUMB.
— breviscapus (DC.) JACOBS.
— cactaeformis KLATT
— calamifolius HOOK.
— chordifolius (HOOK. f.) JACOBS.
— cicatricosus BAK.
— coccineus (OLIV. et HIERN) MUSCHL.
— crassulaefolius (DC.) JACOBS.
— cuneatus JACOBS. non HOOK. f.
— — SCH. BIP. ex. JACOBSEN sphalm.
— cyclocladus BAK.
— cylindricus (BGR.) JACOBS.
— erici-rosenii v. alticola MILDBR.
— erioneuron COTTON
— — v. oligochaeta HAUMAN
— galpinii (HOOK. f.) JACOBS.
— gracilis HORT.
— grandiflorus (WALL.) JACOBS.
— gregori S. MOORE
— gunninsi BAK.
— haworthii HOOK. f.
— herreianus HORT.
— hookerianus JACOBS.
— humbertii CHANG.
— — (GUILL.) JACOBS.
— incisifolius BALLY
— iosensis ROWL.
— johnstonii AUCT. non OLIV.
— keniensis BAK. p. part.
— kleinioides (SCH. BIP.) OLIV. et HIERN
— lanurinensis DE WILD.
— longiflorus (DC.) OLIV. et HIERN
— longifolius JACOBS. non L.
— longipes BAK.
— madagascariensis (H. HUMB.) JACOBS. non POIRET
— melastomaefolius v. microphyllus H. HUMB.
— odorus (FORSK.) DC.
— papillaris (L.) JACOBS.
— petraeus MUSCHLER non BOISS. et HOOK.
— petrophilus KLATT
— phonolithicus DC.
— pinguifolius (DC.) JACOBS.
— pteroneura HOOK. f.
— pusillus DTR. ex RANGE
— — HORT.
— refractisquamatus DE WILD.
— repens (L.) MUSCHLER
— rhopaladenia DTR.

S. adn. v. alt.

— sak.
— a.

— b.
— ciatr.
Othonn. grav.
S. scap. v. caul.

— ch.
— can.

— cocc'fl.

— c.

— klein.

— —
— canal.
— spic.

— adn. v. alt.
— — v. e.
— — v. —
— g.
— citr.

— ampl.
— stap. ssp. min.
— pend.
— h.
— klingh.
— fulg.
— long. v. mad.
— neohumb.
— ball.
— klingh.
— eric. ros.
— brass.;
— ken.

— long. v. viol.
— adn. v. a.

— l. v. l.
— chord.
— grant.

— long. v. m.

— sabour.
— art. v. o.
— p.

— jacobs.
— canan.
— corymb.
— p.
— ant.
— kling.
— citr.
— adn. v. a.
— serp.
— rad.

S. rhopalophyllus (DTR.) MERXM.
— rigidus (THUNBG.) JACOBS.
— speciosus HORT. non AUCT.
— stanleyi HAUMAN
— stapeliformis STAPF
— stapelioides sphalm.
— subscandens HOCHST. ex JACOBS. sphalm.
— subradiatus (DC.) JACOBS.
— subulatus (BALLY) JACOBS.
— succulentus FORSK.
— — SCH. BIP.
— talinoides (DC.) JACOBS.
— variegatum HORT.
— vernicosus HORT. non SCH. BIP.
— vestitus BGR. sphalm.
**Sesamothamnus** WELW. Pedaliaceae.
S. erlangeri ENGL.
— smithii (BAK.) STAPF
**Sigmatosiphon** SCHLTR. Pedaliaceae.
S. guerichii ENGL.
**Sinocrassula** BGR. Crassulaceae.
S. bergeri JACOBS.
— longistyla (PRAEG.) FU
**Siphonostelma** SCHLTR. Asclepiadaceae.
S. stenophyllum SCHLTR.
**Spathulopetalum** CHIOV. Asclepiadaceae.
S. dicapuae CHIOV.
**Sphaeritis** ECKL. et ZEYH. Crassulaceae.
S. biconvexa ECKL. et ZEYH.
— incana ECKL. et ZEYH.
— subaphylla ECKL. et ZEYH.
— trachysantha ECKL. et ZEYH.
**Sphalmanthus** N. E. BR. Mesembr. ●
S. calycinus (ECKL. et ZEYH.) L. BOL.
— fragilis N. E. BR.
— lignescens L. BOL.
— melanospermus (DTR.) DTR. et SCHWANT.
**Stapelia** L. Asclepiadaceae.
S. adscendens ROXB.
— affinis N. E. BR.
— albo-castanea MARL.
— anemoniflora DEFL.
— ango A. RICH.
— anguinea JACQ.
— aperta MASS.
— arida MASS.
— articulata AIT.
— atrata TOD.
— atropurpurea S. D.
— atrosanguinea N. E. BR.
— barbata MASS.
— — SCHLTR.
— beukmanii LUCKH.

— bidentata S. D.
— bifolia SCHULTES
— bisulca J. DONN
— brevirostris WILLD.
— bufonia J. DONN
— — SIMS
— bufonis LODD.
— cactiformis HOOK.

S. aloid.
— spin.
— klein.
— adn. v. s.
— st.
— —

— subcanesc.
— s.
— subul'fol.
Inul. crithm.
S. serp.
— t.
— stap.

— neobak.
— scap. v. caul.

S. riv.

Sesam. g.

S. ind. v. serr.
— — v. l.

Brachyst. s.

Car. d.

Crass. dew.
— i.
— s.
— mesembr.

S. can.
— scint.
— suav.

— dint.

Car. a. v. a.
S. hirs. v. a.
Car. a. c.
— a.
— a.
S. virg. v. pict.
Car. a.
— a.
Pect. a.
S. var. v. a.
— — v. a.
Huerniops. a.
Huern. b.
— retic.
Luckh. b.;
S. stult.
S. var. v. buf.
— — v. —
— — v. —
— defl. v. d.
— var. v. b.
— — v. clyp.
— — v. b.
Trichoc. clav.

## Stapelia INVALID DESIGNATIONS

S. caespitosa DC.
— — MASS.
— campanulata MASS.
— — SIMS
— caroli-schmidtii DTR. et BGR.
— caudata THUNBG.
— chrysostephana DEFL.
— ciliata THUNBG.
— ciliolata TOD.
— cililutata TOD. et RÜST.
— clavata DECNE.
— — WILLD.
— clavigera JACQ.
— clypeata J. DONN
— connata JACQ.
— compacta SCHULT.
— conspurcata WILLD.
— cooperi N. E. BR.
— corderoyi HOOK. f.
— courcellii HORT. ex N. E. BR.
— crassa J. DONN
— cruciformis HORT.
— curtisii SCHULT.
— cylindrica HORT.

— decora MASS.
— dentata FORSK.
— depressa JACQ.
— digitaliflora PFERSD.
— dummeri N. E. BR.
— echinata HORT. ex N. E. BR.
— elegans MASS.
— elongata SWEET
— europaea GUSS.
— — WEBB
— fimbriata WALL.
— fissirostris JACQ.
— — N. E. BR.
— fleckii BGR. et SCHLTR.
— furcata N. E. BR.
— fuscata HORT.
— geminata MASS.
— glabricaulis SCHLTR.
— glandulifera WILLD.
— glanduliflora v. haworthii BGR.
— — v. massonii BGR.
— glauca J. DONN
— gordonii HOOK.
— — MASS.
— grandiflora v. lineata N. E. BR.
— gussoneana (MIK.) JACQ. f. ex LINDL.
— guttata MASS.
— hircola POIRET.
— hircosa JACQ.
— hirsuta v. afra LINDL.
— hirtella JACQ.
— hispidula BORNEM.
— horizontalis N. E. BR.
— humilis MASS.
— hystrix HOOK. f.
— incarnata L. f.
— inodora DECNE.

Duv. rad. v. hirt.
— — c.
Huern. c.
— clavig.

Car. alb. cast.
Brachyst. crisp.
Car. c.
Dipl. c.
S. var. v. consp.
— — v. buf.
Huern. clavig.
Trichoc. c.
Huern. c.
S. var. v. c.
— hirs. v. c.
Duv. c.
S. var. v. c.
Stult. c.
Duc. c.
S. hirs. v. depr.
Huern. ret.
S. oliv.
— var. v. c.
Echidn. cer. v. c.
Piar. d.
s. Car. dec.
S. hirs. v. d.
Decab. el.
Car. d.
Duv. pol. v. p.
— e. v. e.
S. hirs. v. pat.
Car. e.
— — v. conf.
— adc. v. f.
S. ruf. v. flavir.
— — v. att.
— flavop. v. f.
Car. mel. v. m.
S. mutab.
Piar. g.
S. tsom.
— gland'fl. v. g.

— — v. —
— — v. —
— revol. v. r.
Hood. g.
— —

S. flavir.

Car. eur. ssp. g.
Huern. g.
S. gemm. v. h.
— — v. —
— hirs. v. h.
Duv. rad. v. h.
S. gland. v. g.
— var. v. h.
Huern. h.
— h. v. h.
Car. i. v. i.
S. var. v. curt.

S. intermedia N. E. BR.
— irrorata LODD.

— jacquiniana SCHULT.
— juvencula SWEET
— knobeli PHILL.
— laevis DECNE.
— lanigera LOUD.
— laxiflora HAW.
— lentiginosa CURTIS
— limosa S. D.
— longipedicillata N. E. BR.
— lucida DC.
— lunata SWEET
— macrocarpa A. RICH.
— maculata POIRET
— mammillaris L.
— margarita SLOANE
— marmorata HULLE
— — JACQ.
— mastodes JACQ.
— melanantha SCHLTR.
— miscella N. E. BR.
— mixta J. DONN
— — MASS.
— monstrosa STEUD.
— montagnaci BOTT.
— moschata J. DONN

— mutabilis HULLE
— — JACQ. v. bicolor DAMM.
— namaquensis v. minor N. E. BR.
— neglecta TOD.
— normalis JACQ.
— — LINDL.
— obliqua WILLD.
— obscura N. E. BR.
— ocellata JACQ.
— ophioncula SCHLTR.
— ophiuncula HAW.
— orbicularis LODD.
— orbiculata J. DONN
— pallens HORT.
— pallida WENDL.
— paniculata WILLD.
— parviflora MASS.
— parvipunctata K. SCHUM.
— passerinii TOD.
— patentirostris N. E. BR.
— patula WILLD.
— — v. depressa N. E. BR.
— — v. longirostris N. E. BR.
— picta J. DONN
— — N. E. BR.
— pilifera L. f.
— planiflora JACQ.
— — v. marginata WILLD.
— polita HORT. angl. ex N. E. BR.
— protensa HORNEM.
— pruinosa MASS.
— pulchra SCHULT.
— pulla AIT.
— pulvinata J. DONN
— punctata MASS.
— quadrangula FORSK.

Car. i.
S. verr. v. pulchr.
Duv. eleg. v. e.
S. var. v. j.
Car. k.
S. ped.
— hirs. v. h.
— — v. —
Huern. gutt.
S. lep.
— kweb. v. l.
— ast. v. l.
— sor.
Huern. m. v. m.
S. maculos.
Car. m.
S. pulv. v. m.
— var. v. atr.
— — v. m.
Duv. comp.
Car. m. v. m.
Stult. m.
S. × macul.
— var. v. m.
— — v. buf.
Stap'anth. m.
S. gemm. v. hirc.
— var. v. plan.
— bic.

— nam. v. n.
— mut.
— var. v. v.
— — v. tris.
— — v. consp.
— grand.
Huern. o.
S. var. v. buf.
— — v. —
— — v. —
— — v. —
— divar.
— —
— conc. v. p.
Car. p.
S. p. v. p.
— mutab.
— hirs. v. depr.
— — v. p.
— — v. d.
— — v. l.
— var. v. p.
— — v. laet.
Trichoc. p.
S. var. v. p.
— — v. m.

Duval. p. v. p.
S. revol. v. r.
Car. p. v. p.
S. verr. v. p.
Car. mam.
S. hirs.
Piar. p.
Car. qu.

## INVALID DESIGNATIONS Stapelia-Talinum

S. quinquenervis SCHULT.
— radiata JACQ.
— — LINK
— — SIMS
— ramosa MASS.
— reclinata MASS.
— — SIMS
— reflexa HAW.
— replicata JACQ.
— reticulata MASS.
— retusa SCHULT.
— revoluta CURTIS
— rogersii L. BOL.
— roriflua JACQ.
— rufescens HORT.
— — S. D.
— rugosa J. DONN
— — WENDL.
— serrulata JACQ.
— simsii SCHULT.
— sororia HOOK. f.
— — JACQ.
— — LODD.
— — MASS. v. alia JACQ.
— spectabilis HAW.
— sprengeri (LEM.) A. CHEV.
— stellaris HAW.
— — JACQ.
— — LODD.
— stellata ST. LAGR.
— stygia SCHULT.
— subulata FORSK.
— tapscottii WERD.
— thuretii CROUCHER
— tigrida DECNE.
— tigrina BGR. nom. nud.
— — NEL
— tomentosa HORT. ex WHITE et SLOANE
— trisulca J. DONN
— tuberosa MEERBURG
— umbellata ROXB.
— umbilicata THURET
— uncinata JACQ. f.
— vaga sensu HUBER p. part.
— — N. E. BR.
— variegata CURTIS
— — GOUAS
— — JACQ. (1778)
— — JACQ. (1806–19)
— — L. v. normalis BGR.
— — v. prometheus DAMMANN
— venusta JACQ.
— — MASS.
— verrucosa CURTIS
— — JACQ.
— vetula SIMS
— villosa N. E. BR.
— wendlandiana SCHULT.
— woodfordiana SCHULT.
**Stapelianthus** CHOUX. Asclepiadaceae.
S. chonanthus (LAVR. et HALL) DYER
**Stapeliopsis** CHOUX. Asclepiadaceae.
S. ballyi J. MARN.-LAP.
— madagascariensis CHOUX
**Stapeliopsis** PHILLIPS. Asclepiadaceae.

S. var. v. clyp.
Duv. el. v. e.
— recl. v. r.
— r. v. r.
Car. r.
Duv. r. v. r.
— rad. v. hirt.
S. defl. v. d.
Duv. rad. v. r.
Huern. r.
S. var. v. r.
— r. v. tigr.
Car. r.
S. verr. v. r.
— hirs. v. pat.
— r. v. r.
— var. v. r.
— — v. rorifl.
Piar. dec.
S. vet. v. e.
— hirs. v. depr.
— — v. pat.
— — v. h.
— sor.
— grand.
Car. dec.
S. ast. v. a.
— — v. luc.
— hirs. v. aff.
— ast. v. a.
— gemm.
Car. s.
Stult. t.
Huern. t.
S. revol. v. t.

— nel.

— tsom.
— var. v. t.
Brachyst. crisp.
Car. u.
S. mut.
— sor.
Car. lut. ssp. l.
— — ssp. —
S. verr. v. curt.
— hirs. v. pat.
— — v. plan.
— — v. clyp.
— var. v. v.
— × prom.
Huern. gutt.
— ven.
S. v. v. rob.
— — v. pulchr.
— vet. v. sims.
— hirs. v. h.
— verr. v. rorifl.
— var. v. v.


Echidn. b.
Stap'anth. m.

S. cooperi (N. E. BR.) PHILLIPS
**Stayneria** L. BOL. Mesembr. ●
S. littlewoodii L. BOL.
**Stenadenium** PAX. Euphorbiaceae.
St. spinescens PAX
**Stephania** LOUR. Menispermaceae.
S. glabra MIERS.
**Sterigmanthe** KL. et GARCKE. Euphorbiaceae.
S. bojeri KL. et GARCKE
— splendens KL. et GARCKE
**Sterropetalum** L. BOL. Mesembr. ●
S. pillansii N. E. BR.
**Stigmatocarpum** L. BOL. Mesembr. ●
S. caducum (AIT.) L. BOL.
— copticum (JACQ.) L. BOL.
— criniflorum L. BOL.
**Stoeberia** DTR. et SCHWANT. emend. FRIEDR. Mesembr. ●
S. gigas (DTR.) DTR. et SCHWANT.
— hallii L. BOL.
— littlewoodii L. BOL.
— micropetala (L. BOL.) L. BOL.
— rupis-arcuatae (DTR.) DTR. et SCHWANT.
**Stomatium** SCHWANT. Mesembr. ●
S. bolusiae SCHWANT. nom. nud.
— ermininum SCHWANT. ex L. BOL.
— musculinum SCHWANT. ex L. BOL.
**Streptanthes** NUTT. Cruciferae.
S. crassicaulis TORR.
— inflatus GREENE
**Stylophyllum** BR. et R. Crassulaceae.
S. albidum ROSE
— anomalum DAVIDS.
— attenuatum (WATS.) BR. et R.
— coronatum FROED.
— densiflorum ROSE
— edule (NUTT.) BR. et R.
— hassei ROSE
— insulare FROED.
— — ROSE
— nudicaule ABRAMS.
— orcuttii ROSE
— parishii BRITT.
— × semiteres ROSE
— traskiae ROSE
— virens ROSE
— viscidum (S. WATS.) BR. et R.
Styl. (BR. et R.) MORAN
**Synadenium** BOISS. Euphorbiaceae.
S. arborescens BOISS.
**Synaptophyllum** N. E. BR. Mesembr. ●
S. sladenium (L. BOL.) N. E. BR.
**Talinum** ADANS. Portulacaceae.
T. anacampseros WILLD. also MNCH.
— arachnoides AIT.
— filamentosa AIT.
— lanceolatum LINK.

Stult. c.

S. neil.

Monad. s.

S. rot.

Euph. mil. v. m.
— — v. spl.

Nel. p.

Dor. apet.

— ocul. v. o.


Ruschianth. g.
Amphib. h.
— l.


Ruschianth. gig.

Rusch. r. a.


S. bolus.

Chasm. m.

Caul. c.
— i.

Dudl. vir.
— a.


— a.
— anomal.
— d.
— e.
— h.
— anomal.
— vir.
— dens.
— att. ssp. o.
— — ssp. —
— × s.
— t.
— v.

— v.
Dudl. SG. I

S. cup.


Pren. s.


Anac. tel.
— a. v. a.
— f.
— l. v. l.

Stapelia c.

T. linguaeforme LINK. Echev. l.
— patens (L. MART.) WILLD. T. pan.
— polyandrum HOOK. Calandr. p.
— polyphyllum LINK. Anac. p.
— retusum WILLD. — arachn. v. a.
— rubens HORT. ex STEUD. — — v. —
— umbellatum RUIZ. et PAV. Calandr. u.
— wayae EASTW. T. okan.
**Tavaresia** WELW. Asclepiadaceae.
T. angolensis WELW. Decab. el.
— barklyi (T. DYER) N. E. BR. — b.
— grandiflora (K. SCHUM.)
 BGR. — g.
— meintjesii R. A. DYER — m.
**Terebinthus** TOURN. Anacardiaceae.
T. multifolia ROSE Burs. micr.
**Testudinaria** SALISB. Dioscoreaceae.
T. montana BURCH. T. el. v. m.
— — v. paniculata KUNTZE — sylv. v. p.
— multiflora MARL. — — v. m.
— paniculata DUEMM. — — v. p.
— — v. brevipes (BURTT DAVY)
 ROWL. — — v. b.
— rehmannii (BAK.) ROWL. — — v. r.
**Tetracoilanthus** RAPPA et CAMERRONE.
 Mesembr. ●
T. anatomicus (HAW.) RAP. et
 CAM. Scel. a.
— concavus (HAW.) RAP. et
 CAM. — c.
— cordifolius (L. f.) RAP. et
 CAM. Apt. c.
Thyracosperma N. E. BR.
 sphalm. Thyrasperma
**Thyrasperma** N. E. BR. Mesembr. ●
T. helianthoides N. E. BR. Apat. h.
— sabulosa (AIT.) N. E. BR. — —
**Tillaea** L. Crassulaceae.
T. alsinoides HOOK. f. Crass. a.
**Tischleria** SCHWANT. Mesembr. ●
T. peersii SCHWANT. Carr. p.
**Titanopsis** SCHWANT. Mesembr. ●
T. crassipes (MARL.) N. E. BR. Al. spath.
— luckhoffii L. BOL. — l.
— setifera L. BOL. — s.
— spathulata (THUNBG.)
 L. BOL. — s.
— — (THUNBG.) SCHWANT. — s.
**Tithymalus** BOISS. Euphorbiaceae.
An invalid G. the species of
which have been referred to
**Pedilanthus** NECKER. The
numerous synonyms cannot
be mentioned here.
**Triactina** HOOK. f. et THOMS. Crassulaceae.
T. verticillata HOOK. f. et
 THOMS. Sed. v.
**Trichocaulon** N. E. BR. Asclepiadaceae.
T. cactiforme (HOOK. f.)
 N. E. BR. T. clav.
— columnare NEL Echidn. c.
— decaryi CHOUX Stap'anth. pil.
— dinteri BGR. T. clav.
— engleri DTR. — —
— karasmontana DTR. nom.
 nud.

T. keetmanshoopense DTR. T. clav.
— kubusanum NEL ex
 JACOBSEN — kubus.
— marlothii N. E. BR. — clav.
— meloforme MARL. — —
— sinus-luederitzii DTR. — —
— sociorum WHITE et SLOANE — —
**Trichocyclus** N. E. BR. Mesembr. ●
T. buchubergensis DTR. Psil. pill.
— ciliatus (AIT.) N. E. BR. — c.
— marlothii (PAX) N. E. BR. — m.
— namibensis (MARL.) N. E.
 BR. — n.
— pillansii L. BOL. — p.
— pubescens N. E. BR. ex
 MASS. — —
— simplex N. E. BR. ex MASS. — cil.
**Trichodiadema** SCHWANT. Mesembr. ●
T. echinatum (AIT.) L. BOL. Del. e.
— inornatum L. BOL. Dros. i.
— tenue L. BOL. Del. neethl.
**Tridentea** HAW. Asclepiadaceae.
T. depressa SCHULT. Stap. hirs. v. d.
— gemmiflora HAW. — g.
— hircosa SCHULT. — — v. h.
— juvencula SWEET — vet. v. j.
— moschata HAW. — gemm. v.
 hirc.
— paniculata HAW. — conc. v. p.
— rugosa SCHULT. — var. v. r.
— simsii HAW. — vet. v. s.
— stygia HAW. — gemm.
— vetula HAW. — v. v. v.
Trid. (HAW.) BGR. Stap. Sect. II
**Tromotriche** HAW. Asclepiadaceae.
T. ciliata SWEET Diploc. c.
— glauca HAW. Stap. rev. v. r.
— — v. β HAW. — — v. fusc.
— obliqua SWEET — var. v. consp.
— pruinosa HAW. Car. p. v. p.
— revoluta HAW. Stap. r. v. r.
Trom. (HAW.) BGR. Stap. Sect. IX
**Tumboa** WELW. Welwitschiaceae.
T. bainesii HOOK. f. Welw. mir.
**Umbilicus** DC. Crassulaceae.
U. affinis SCHRANK. Pseudos. a.
— aizoon FENZL Ros. a.
— chrysanthus BOISS. — pall.
— elymaiticus BOISS. et
 HELDR. — e.
— erubescens MAX. Orost. e. v. e.
— fimbriatus TURCZ. — f.
— glaber RGL. et WINKLER Ros. g.
— globulariaefolius FENZL — g.
— haussknechtii BOISS. et
 REUT. — h.
— hispida (LAM.) DC. Muciz. h.
— inermis LEDEB. Orost. mal.
— lampusae KOTSCHY Sed. l.
— leucanthus LEDEB. Orost. thyrs.
— libanoticus (L.) DC. Ros. l. v. l.
— — v. glaber BOISS. — pers.
— lievenii LEDEB. Pseudos. l.
— linearifolius A. FRANCH. Rhod. sem.
— lineatus BOISS. Ros. l.
— linifolius OST.-SACK. Rhod. sem.
— luteus DECNE. Sed. orad.

U. malacophyllus (PALLAS)
DC. Orost. m.
— mexicanus SCHLTD. Vill. m.
— oppositifolius LEDEB. Chiast. o.
— oreades DECNE. Sed. o.
— pallidus SCHOTT et
KOTSCHY Ros. p.
— paniculatus RGL. et
SCHMALH. — p.
— pendulinus DC. U. rup. v. r.
— — v. bracteonis WILLK. — citr.
— — v. deflexus (POMEL)
BATT. — pat.
— — v. truncatus WOLLEY-
DOD — rup. v. t.
— — v. velenovskyi ROHL. — — v. v.
— persicus BOISS. Ros. p.
— — HORT. Semp. tect. ssp.
calc. cv. G. S.
— pestalozzae BOISS. Ros. semp.
— platyphyllus BOISS. — pan.
— — SCHENK — p.
— pubescens LEDEB. Sed. pil.
— radicans KLOTZSCH — ros.
— ramosissimus MAX. Orost. fimbr.
v. r.
— samium (URV.) DC. Ros. serr.
— sedoides DC. Sed. caud.
— semenowii RGL. et HERD. Rhod. s.
— semiensis G. GAY. Afrov. sim.
— serratus (L.) DC. Ros. s.
— spathulatus HOOK. f. et
THOMAS Sed. oread.
— spinosus DC. Orost. s.
— sprunerianus BOISS. U. parv.
— stamineus LEDEB. Orost. mal.
— syriensis HORT. Semp. tect. ssp.
calc. cv. G. S.
— thyrsiflorus (FISCH.) DC. Orost. t.
— turkestanicus RGL. et
WINKLER
— winkleri WILLK. Ros. t.
Sed. hirs. ssp.
baet.
**Uncaria** BURCH. Pedaliaceae. s. Harpagophytum
**Urbinia** BR. et R. Crassulaceae.
U. agavoides ROSE Ech. ag. v. a.
— corderoyi ROSE — — v. c.
— lurida ROSE — tobar.
— obscura ROSE — agav.
— purpusii ROSE — purp.
Urb. BR. et R. Ech. Ser. 8
× **Urbino-Echeveria** v. LAREN. Crassulaceae.
Hybr.: Urbinia × Echeveria.

× U.-E. angustata v. LAREN Ech. × setor.
cv. V.
× **Urbiphytum** GOSSOT. Crassulaceae. — Hybr.
Urbinia × Pachyphytum s. × Pachyveria
× U. albo-mucronatum
GOSSOT × Pachyv.
cv. A.
**Vanzijlia** L. BOL. Mesembr. ●
V. annulata (BGR.) N. E. BR. V. a.
— — L. BOL. — ang.
**Vauanthes** HAW. Crassulaceae.
V. chloraeflora HAW. V. dich.
**Verrucifera** N. E. BR. Mesembr. ●
V. hugo-schlechteri (TISCH.)
N. E. BR. Tit. h.-s. v. h.-s.
— luederitzii (TISCH.) N. E.
BR. — l.
— schwantesii (DTR.) N. E. BR. — s.
**Villadia** HAW. Crassulaceae.
V. parviflora (HEMSL.) ROSE V. mis.
— platyphylla (BR. et R.)
E. WALTH. Thomps. p.
— scopulina (ROSE) R. T.
CLAUSEN V. s.
— texana ROSE Lenoph. t.
**Virchowia** VATKE. Asclepiadaceae.
V. africana VATKE Echidn. v.
**Vitis** L. Vitaceae.
V. bainesii HOOK. f. Cyphost. b.
— cirrhosa THUNBG. — c.
— crassifolia BAK. Ciss. rot.
— gastropus WELW. et
PLANCH. Cyphost. macr.
— nymphaefolia WELW. ex
BAK. Ciss. n.
— quadrangularis WALL. — qu.
— rotundifolia DEFLERS — rot.
— succulenta GALPIN — cact.
**Voharanga** COST. et GALL. Asclepiadaceae.
V. madagascariensis COST. et
BOIS. Cynanch. m.
**Vohemaria** BUCH. Asclepiadaceae.
V. implicatum (JUM. et PERR.)
JUM. et PERR. Cynanch. i.
— messeri BURCH. — m.
**Vossia** ADANS. Mesembr. ●
This old generic name is ill-
defined and hence has been
dropped (See Kakt. u. a.
Sukk. **3**, 1953, No. 4).
**Welwitschia** HOOK. f. Welwitschiaceae.
W. bainesii (HOOK. f.) CARR. W. mirab.
**Whitesloanea** CHIOV. Asclepiadaceae.
W. migiurtinorum CHIOV. Pseudol. m.

The **Reference of Literature** for all valid genera, species and lower taxa, hybrids, as well as all basionyms, homonyms and synonyms is preserved in a register in the property of the author or later in the property of the German Cactus Society (Deutsche Kakteengesellschaft, Frankfurt/M.).